Quantum Field Theory III: Gauge Theory

Eberhard Zeidler

Quantum Field Theory III: Gauge Theory

A Bridge between Mathematicians and Physicists

Springer

Eberhard Zeidler
Max Planck Institute
for Mathematics in the Sciences
Inselstr. 22-26
04103 Leipzig
Germany

ISBN 978-3-642-22420-1 e-ISBN 978-3-642-22421-8
DOI 10.1007/978-3-642-22421-8
Springer Heidelberg Dordrecht London New York

Library of Congress Control Number: 2006929535

Mathematics Subject Classification (2010): 35-XX, 47-XX, 49-XX, 51-XX, 55-XX, 81-XX, 82-XX

Cover design: deblik, Berlin

Printed on acid-free paper

Springer is part of Springer Science+Business Media (www.springer.com)

TO KRZYSZTOF MAURIN
IN GRATITUDE

Preface

Sein Geist drang in die tiefsten Geheimnisse der Zahl, des Raumes und der Natur; er maß den Lauf der Gestirne, die Gestalt und die Kräfte der Erde; die Entwicklung der mathematischen Wissenschaft eines kommenden Jahrhunderts trug er in sich.[1]

Lines under the portrait of Carl Friedrich Gauss (1777–1855)
in the German Museum in Munich

Force equals curvature.
The basic principle of modern physics

A theory is the more impressive, the simpler are its premises, the more distinct are the things it connects, and the broader is the range of applicability.

Albert Einstein (1879–1955)

Textbooks should be attractive by showing the beauty of the subject.
Johann Wolfgang von Goethe (1749–1832)

The present book is the third volume of a comprehensive introduction to the mathematical and physical aspects of modern quantum field theory which comprises the following six volumes:

Volume I: Basics in Mathematics and Physics
Volume II: Quantum Electrodynamics
Volume III: Gauge Theory
Volume IV: Quantum Mathematics
Volume V: The Physics of the Standard Model
Volume VI: Quantum Gravitation and String Theory.

It is our goal to build a bridge between mathematicians and physicists based on challenging questions concerning the fundamental forces in

- the macrocosmos (the universe) and
- the microcosmos (the world of elementary particles).

[1] His mind pierced the deepest secrets of numbers, space, and nature; he measured the orbits of the planets, the form and the forces of the earth; in his mind he carried the mathematical science of a coming century.

The six volumes address a broad audience of readers, including both undergraduate and graduate students, as well as experienced scientists who want to become familiar with quantum field theory, which is a fascinating topic in modern mathematics and physics, full of many crucial open questions.

For students of mathematics, detailed knowledge of the physical background helps to enliven mathematical subjects and to discover interesting interrelationships between quite different mathematical topics. For students of physics, fairly advanced mathematical subjects are presented that go beyond the usual curriculum in physics. The strategies and the structure of the six volumes are thoroughly discussed in the Prologue to Volume I. In particular, we will try to help the reader to understand the basic ideas behind the technicalities. In this connection, the famous ancient story of Ariadne's thread is discussed in the Preface to Volume I:

> *In terms of this story, we want to put the beginning of Ariadne's thread in quantum field theory into the hands of the reader.*

There are four fundamental forces in the universe, namely,

- gravitation,
- electromagnetic interaction (e.g., light),
- strong interaction (e.g., the binding force of the proton),
- weak interaction (e.g., radioactive decay).

In modern physics, these four fundamental forces are described by

- Einstein's theory of general relativity (gravitation), and
- the Standard Model in elementary particle physics (electromagnetic, strong, and weak interaction).

The basic mathematical framework is provided by gauge theory:

> *The main idea is to describe the four fundamental forces by the curvature of appropriate fiber bundles.*

In this way, the universal principle *force equals curvature* is implemented. There are many open questions:

- A mathematically rigorous quantum field theory for the quantized version of the Standard Model in elementary particles has yet to be found.
- We do not know how to combine gravitation with the Standard Model in elementary particle physics (the challenge of quantum gravitation).
- Astrophysical observations show that 96 percent of the universe consists of both dark matter and dark energy. However, both the physical structure and the mathematical description of dark matter and dark energy are unknown.

> *One of the greatest challenges of the human intellect is the discovery of a unified theory for the four fundamental forces in nature based on first principles in physics and rigorous mathematics.*

In the present volume, we concentrate on the *classical aspects* of gauge theory related to curvature. These have to be supplemented by the crucial, but elusive quantization procedure. The quantization of the Maxwell–Dirac system leads to quantum electrodynamics (see Vol. II). The quantization of both the full Standard Model in elementary particle physics and the quantization of gravitation will be studied in the volumes to come.

> *One cannot grasp modern physics without understanding gauge theory, which tells us that the fundamental interactions in nature are based on parallel transport, and in which forces are described by curvature, which measures the path-dependence of the parallel transport.*

Gauge theory is the result of a fascinating long-term development in both mathematics and physics. Gauge transformations correspond to a change of potentials, and physical quantities measured in experiments are invariants under gauge transformations. Let us briefly discuss this.

Gauss discovered that the curvature of a two-dimensional surface is an intrinsic property of the surface. This means that the Gaussian curvature of the surface can be determined by using measurements on the surface (e.g., on the earth) without using the surrounding three-dimensional space. The precise formulation is provided by Gauss' theorema egregium (the egregious theorem). Bernhard Riemann (1826–1866) and Élie Cartan (1859–1951) formulated far-reaching generalizations of the theorema egregium which lie at the heart of

- modern differential geometry (the curvature of general fiber bundles), and
- modern physics (gauge theories).

Interestingly enough, in this way,

- Einstein's theory of general relativity (the curvature of the four-dimensional space-time manifold), and
- the Standard Model in elementary particle physics (the curvature of a specific fiber bundle with the symmetry group $U(1) \times SU(2) \times SU(3)$)

can be traced back to Gauss' theorema egregium.

In classical mechanics, a large class of forces can be described by the differentiation of potentials. This simplifies the solution of Newton's equation of motion and leads to the concept of potential energy together with energy conservation (for the sum of kinetic and potential energy). In the 1860s, Maxwell determined that the computation of electromagnetic fields can be substantially simplified by introducing potentials for both the electric and the magnetic field (the electromagnetic four-potential).

Gauge theory generalizes this by describing forces (interactions) by the differentiation of generalized potentials (also called connections).

The point is that gauge transformations change the generalized potentials, but not the essential physical effects.

Physical quantities, which can be measured in experiments, have to be invariant under gauge transformations.

Parallel to this physical situation, in mathematics the *Riemann curvature tensor* can be described by the differentiation of the Christoffel symbols (also called connection coefficients or geometric potentials). The notion of the Riemann curvature tensor was introduced by Riemann in order to generalize Gauss' theorema egregium to higher dimensions. In 1915, Einstein discovered that the Riemann curvature tensor of a four-dimensional space-time manifold can be used to describe gravitation in the framework of the theory of general relativity.

The basic idea of gauge theory is the transport of physical information along curves (also called parallel transport).

This generalizes the parallel transport of vectors in the three-dimensional Euclidean space of our intuition.

In 1917, it was discovered by Levi-Civita that the study of curved manifolds in differential geometry can be based on the notion of parallel transport of tangent vectors (velocity vectors).

In particular, curvature can be measured intrinsically by transporting a tangent vector along a closed path. This idea was further developed by Élie Cartan in the 1920s (the method of moving frames) and by Ehresmann in the 1950s (the connection of both principal fiber bundles and their associated vector bundles). The very close relation between

- gauge theory in modern physics (the transport of local $SU(2)$-phase factors investigated by Yang and Mills in 1954), and
- the formulation of differential geometry in terms of fiber bundles in modern mathematics

was only noticed by physicists in 1975 (see T. Wu and C. Yang, Concept of non-integrable phase factors and global formulation of gauge fields, Phys. Rev. **D12** (1975), 3845–3857).

The present Volume III on gauge theory and the following Volume IV on quantum mathematics form a unified whole. The two volumes cover the following topics:

Volume III: Gauge Theory

Part I: The Euclidean Manifold as a Paradigm

Chapter 1: The Euclidean Space E_3 (Hilbert Space and Lie Algebra Structure)
Chapter 2: Algebras and Duality (Tensor Algebra, Grassmann Algebra, Clifford Algebra, Lie Algebra)
Chapter 3: Representations of Symmetries in Mathematics and Physics
Chapter 4: The Euclidean Manifold $\mathbb{E}^3$
Chapter 5: The Lie Group $U(1)$ as a Paradigm in Harmonic Analysis and Geometry
Chapter 6: Infinitesimal Rotations and Constraints in Physics
Chapter 7: Rotations, Quaternions, the Universal Covering Group, and the Electron Spin
Chapter 8: Changing Observers – A Glance at Invariant Theory Based on the Principle of the Correct Index Picture
Chapter 9: Applications of Invariant Theory to the Rotation Group
Chapter 10: Temperature Fields on the Euclidean Manifold $\mathbb{E}^3$
Chapter 11: Velocity Vector Fields on the Euclidean Manifold $\mathbb{E}^3$
Chapter 12: Covector Fields on the Euclidean Manifold $\mathbb{E}^3$ and Cartan's Exterior Differential – the Beauty of Differential Forms

Part II: Ariadne's Thread in Gauge Theory

Chapter 13: The Commutative Weyl $U(1)$-Gauge Theory and the Electromagnetic Field
Chapter 14: Symmetry Breaking
Chapter 15: The Noncommutative Yang–Mills $SU(N)$-Gauge Theory
Chapter 16: Cocycles and Observers
Chapter 17: The Axiomatic Geometric Approach to Vector Bundles and Principal Bundles

Part III: Einstein's Theory of Special Relativity

Chapter 18: Inertial Systems and Einstein's Principle of Special Relativity
Chapter 19: The Relativistic Invariance of the Maxwell Equations
Chapter 20: The Relativistic Invariance of the Dirac Equations and the Electron Spin

Part IV: Ariadne's Thread in Cohomology

Chapter 21: Exact Sequences
Chapter 22: Electrical Circuits as a Paradigm in Homology and Cohomology
Chapter 23: The Electromagnetic Field and the de Rham Cohomology.

Volume IV: Quantum Mathematics

Part I: The Hydrogen Atom as a Paradigm

Chapter 1: The Non-Relativistic Hydrogen Atom via Lie Algebra, Gauss's Hypergeometric Functions, von Neuman's Functional Analytic Approach, the Weyl–Kodaira Theory, Gelfand's Generalized Eigenfunctions, and Supersymmetry
Chapter 2: The Dirac Equation and the Relativistic Hydrogen Atom via the Clifford Algebra of the Minkowski Space

Part II: The Four Fundamental Forces in the Universe

Chapter 3: Relativistic Invariance and the Energy–Momentum Tensor in Classical Field Theories
Chapter 4: The Standard Model for Electroweak and Strong Interaction in Particle Physics
Chapter 5: Gravitation, Einstein's Theory of General Relativity, and the Standard Model in Cosmology

Part III: Lowest-Order Radiative Corrections in Quantum Electrodynamics (QED)

Chapter 6: Dimensional Regularization for the Feynman Propagators in QED (Quantum Electrodynamics)
Chapter 7: The Electron in an External Electromagnetic Field (Renormalization of Electron Mass and Electron Charge)
Chapter 8: The Lamb Shift

Part IV: Conformal Symmetry

Chapter 9: Conformal Transformations According to Gauss, Riemann, and Lichtenstein
Chapter 10: Compact Riemann Surfaces
Chapter 11: Minimal Surfaces
Chapter 12: Strings and the Graviton
Chapter 13: Complex Function Theory and Conformal Quantum Field Theory

Part V: Models in Quantum Field Theory

Part VI: Distributions and the Epstein–Glaser Approach to Perturbative Quantum Field Theory

Part VII: Nets of Operator Algebras and the Haag–Kastler Approach to Quantum Field Theory

Part VIII: Symmetry and Quantization – the BRST Approach to Quantum Field Theory

Part IX: Topology, Quantization, and the Global Structure of Physical Fields

Part X: Quantum Information.

Readers who want to understand modern differential geometry and modern physics as quickly as possible should glance at the Prologue of the present volume and at Chaps. 13 through 17 on Ariadne's thread in gauge theory.

Cohomology plays a fundamental role in modern mathematics and physics.

> *It turns out that cohomology and homology have their roots in the rules for electrical circuits formulated by Kirchhoff in 1847.*

This helps to explain why the Maxwell equations in electrodynamics are closely related to cohomology, namely, de Rham cohomology based on Cartan's calculus for differential forms and the corresponding Hodge duality on the Minkowski space. Since the Standard Model in particle physics is obtained from the Maxwell equations by replacing the commutative gauge group $U(1)$ with the noncommutative gauge group $U(1) \times SU(2) \times SU(3)$, it should come as no great surprise that de Rham cohomology also plays a key role in the Standard Model in particle physics via the theory of characteristic classes (e.g., Chern classes which were invented by Shing-Shen Chern in 1945 in order to generalize the Gauss–Bonnet theorem for two-dimensional manifolds to higher dimensions).

It is our goal to show that the gauge-theoretical formulation of modern physics is closely related to important long-term developments in mathematics pioneered by Gauss, Riemann, Poincaré and Hilbert, as well as Grassmann, Lie, Klein, Cayley, Élie Cartan and Weyl. The prototype of a gauge theory in physics is Maxwell's theory of electromagnetism. The Standard Model in particle physics is based on the principle of local symmetry. In contrast to Maxwell's theory of electromagnetism, the gauge group of the Standard Model in particle physics is a noncommutative Lie group. This generates additional interaction forces which are mathematically described by Lie brackets.

We also emphasize the methods of invariant theory. In terms of physics, different observers measure different values in their experiments. However, physics does not depend on the choice of observers. Therefore, one needs both an invariant approach and the passage to coordinate systems which correspond to the observers, as emphasized by Einstein in the theory of general relativity and by Dirac in quantum mechanics. The appropriate mathematical tool is provided by invariant theory.

Acknowledgments. In 2003, Jürgen Tolksdorf initiated a series of four International Workshops on the state of the art in quantum field theory and the search for a unified theory concerning the four fundamental interactions in nature. I am very grateful to Felix Finster, Olaf Müller, Marc Nardmann, and Jürgen Tolksdorf for organizing the workshop *Quantum Field Theory and Gravity*, Regensburg, 2010. The following three volumes contain survey articles written by leading experts:

F. Finster, O. Müller, M. Nardmann, J. Tolksdorf, and E. Zeidler (Eds.), Quantum Field Theory and Gravity: Conceptual and Mathematical Advances in the Search for a Unified Framework, Birkhäuser, Basel (to appear).

B. Fauser, J. Tolksdorf, and E. Zeidler (Eds.), Quantum Field Theory – Competitive Methods, Birkhäuser, Basel, 2008.

B. Fauser, J. Tolksdorf, and E. Zeidler (Eds.), Quantum Gravitation: Mathematical Models and Experimental Bounds, Birkhäuser, Basel, 2006.

These three volumes are recommended as supplements to the material contained in the present monograph. For stimulating discussions and guidance, I would like to thank Sergio Albeverio, Christian Bär, Helga Baum, Christian Brouder, Romeo Brunetti, Detlef Buchholz, Christopher Deninger, Michael Dütsch, Claudia Eberlein, Kurusch Ebrahimi-Fard, William Farris, Bertfried Fauser, Joel Feldman, Chris Fewster, Felix Finster, Christian Fleischhack, Hans Föllmer, Alessandra Frabetti,

Klaus Fredenhagen, Harald Fritzsch, Jörg Fröhlich, Peter Gilkey, José Gracia-Bondía, Ivor Grattan-Guiness, Harald Grosse, Stefan Hollands, Arthur Jaffe, Jürgen Jost, Hans Kastrup, Jerzy Kijowski, Klaus Kirsten, Dirk Kreimer, Elliott Lieb, Jan Louis, Dieter Lüst, Kishore Marathe, Matilde Marcolli, Michael Oberguggenberger, Robert Oeckl, Norbert Ortner, Frédéric Patras, Frank Pfäffle, Klaus Rehren, Vincent Rivasseau, Hartmann Römer, Gerd Rudolph, Manfred Salmhofer, Erhard Scholz, Klaus Sibold, Harold Steinacker, Jürgen Tolksdorf, Armin Uhlmann, Rainer Verch, Peter Wagner, Raimar Wulkenhaar, Hans Wußing, and Jacob Yngvason.

I am very grateful to the staff of the Max-Planck Institute for Mathematics in the Sciences, Leipzig, for kindly supporting me in various aspects. In particular, I would like to cordially thank my secretary Regine Lübke (for her invaluable support), Kerstin Fölting (for her meticulous production of the graphics), Katarzyna Baier, Ingo Brüggemann, and the library team (for their perfect cooperation), Oliver Heller, Rainer Kleinrensing, and the computer team (for their continued support), as well as, Michaela Krieger–Hauwede (for patiently answering my LaTeX questions). Thanks also to the staff of the Springer Publishing House in Heidelberg, Ruth Allewelt, Joachim Heinze, and Martin Peters, for their harmonious collaboration. The LaTeX file was written by myself; therefore, any errors should be laid at my door.

This volume is gratefully dedicated to Professor Krzysztof Maurin in Warsaw. As a young man, I learned from him that mathematics, physics, and philosophy form a unity; they represent marvellous tools for the human intellect in order to approximate step by step the better understanding of the real world, and they have to serve the well-being of human society.

My hometown, Leipzig, is full of the music composed by Johann Sebastian Bach, who worked in Leipzig's Saint Thomas church from 1723 until his death in 1750. In the Preface of his book *Electroweak and Strong Interaction: An Introduction to Theoretical Particle Physics,* Springer, Berlin, 1996, my colleague Florian Scheck from Mainz University adapted Bach's dedication to his "Well-Tempered Clavier" from 1722:

> Written and composed for the benefit and use of young physicists and for the particular diversion of those already advanced in this study.

I would like to use the same quotation, replacing 'physicists' with 'mathematicians and physicists.'

I hope that readers will get a feel for the unity of mathematics and the unity of science. In 1915, John Dewey wrote in his book *The School and Society,* The University of Chicago Press, Chicago, Illinois: "We do not have a series of stratified earths, one of which is mathematical, another physical, another historical, and so on. We should not be able to live very long in any one taken by itself. We live in a world where all sides are bound together; all studies grow out of relations in the one great common world."

Leipzig, Spring 2011 Eberhard Zeidler

Contents

Part IV. Ariadne's Thread in Cohomology

Prologue

Geometry is the knowledge of what eternally exists.

Plato of Athen (428–348 B.C.)

He who understands geometry may understand anything in this world.

Galileo Galilei (1564–1642)

The way of people to the laws of nature are not less admirable than the laws themselves.

Johannes Kepler (1571–1630)

In humbleness, we have to admit that if 'number' is a product of our imagination, 'space' has a reality outside of our imagination, to which *a priori* we cannot assign its laws.

Gauss (1777–1855) in a letter to Bessel, 1840

This prologue should help the reader to understand the sophisticated historical development of gauge theory in mathematics and physics. We will not follow a strict logical route. This will be done later on. At this point, we are going to emphasize the basic ideas. It is our goal to show the reader how the methods of modern differential geometry work in the case of Einstein's theory of general relativity, which describes the gravitational force in nature. In particular, we want to show how

- the language of physicists created by Einstein and used in most physics textbooks (based on the use of local space-time coordinates) and
- the language of mathematicians used in modern textbooks on differential geometry (based on the invariant – i.e., coordinate-free – formulation)

are related to each other. This should help physicists to enter modern differential geometry. One cannot grasp modern physics without understanding gauge field theory which tells us the following crucial facts:

- interactions in nature are based on the *parallel transport* of physical information;
- forces are described by *curvature* which measures the path-dependence of the parallel transport.

Here, we will discuss the following points:

- an interview with the Nobel prize laureate Chen Ning Yang (born 1922) on the history of modern gauge theory,
- Einstein's theory of general relativity on gravitation,
- changing observers in the universe and tensor calculus,
- the Riemann curvature tensor and the beauty of Gauss' theorema egregium,
- two fundamental variational principles in general relativity,

E. Zeidler, *Quantum Field Theory III: Gauge Theory*,
DOI 10.1007/978-3-642-22421-8_1,

- symmetry and Felix Klein's invariance principle in geometry (a glance at the history of invariant theory in the 19th century),[1]
- Einstein's principle of general relativity and invariants – the geometrization of physics (the paradigm of higher-dimensional cartography),
- gauge transformations:
  - Einstein's gauge transformation in the theory of both special relativity and general relativity (change of the observer),
  - Dirac's unitary gauge transformations in the Hilbert space approach to quantum mechanics (change of the observer by changing the measurement device),
  - Yang's gauge transformation by changing the local phase factor of the wave function,
  - the $U(1)$-gauge transformation in classical electrodynamics and quantum electrodynamics,
  - the $U(1) \times SU(2)$ gauge transformations in electroweak interaction,
  - the $SU(3)$ gauge transformations in strong interaction (quantum chromodynamics),
  - the $U(1) \times SU(2) \times SU(3)$ gauge transformations in the Standard Model in particle physics,
  - the conformal gauge transformations in string theory,
  - Élie Cartan's gauge transformations in his method of moving frames (change of the frame),
- construction of invariants by the universal index killing principle,
- Lie's intrinsic tangent vectors,
- Élie Cartan's algebraization of calculus and infinitesimals,
- Riemann's invariant sectional curvature and the geometric meaning of Riemann's curvature tensor,
- Levi-Civita's parallel transport and the geometric meaning of the Riemann curvature tensor,
- two fundamental approaches in differential geometry:
  - Gauss' method of symmetric tensors, and
  - Cartan's method of antisymmetric tensors,
- Yang's matrix trick (the relation between the Einstein equations in general relativity and the Maxwell–Yang–Mills equations), and Cartan's calculus for matrices with differential forms as entries,
- Cartan's structural equations:
  - local structural equations,
  - global structural equations,
- partial covariant derivative and the classical Ricci calculus,
- the Lie structure behind curvature,
- the generalized Riemann curvature tensor in modern mathematics and physics,
- parallel transport of physical information and curvature,
- the modern language of fiber bundles in mathematics and physics,
- summary of typical applications,
- perspectives (instantons and gauge theory, conformal symmetry and twistors, the Seiberg–Witten equations and the quark confinement, the Donaldson theory for 4-dimensional manifolds, Morse theory and Floer homology, quantum cohomology, J-holomorphic curves, Frobenius manifolds, Ricci flow and the Poincaré conjecture).

[1] One has to distinguish between the German mathematician Felix Klein (1849–1925) and the Swedish physicist Oskar Klein (1894–1977) (one of the authors of the Klein–Fock–Gordon equation in quantum field theory).

The classical formulas (0.13) and (0.14) on page 11 for defining the Riemann curvature tensor via Christoffel symbols for the metric tensor are clumsy. The development of modern differential geometry was essentially influenced by the desire of mathematicians to get insight into the true structure of curvature. This led to a better understanding of curvature and to far-reaching generalizations which proved to be useful in modern physics. The basic paper in mathematics is due to:

C. Ehresmann, Les connexions infinitésimales dans un espace fibré differentiable (in French) (The infinitesimal connections in a differentiable fiber bundle), Colloque de Topologie, Bruxelles, 1950, pp. 29–55.

Charles Ehresmann (1905–1979) based his theory on Élie Cartan's work created in the 1920s.[2] The first textbook on modern differential geometry was written by:

S. Kobayashi and K. Nomizu, Foundations of Differential Geometry, Vols. 1, 2, Wiley, New York, 1963.

We also recommend:

Y. Choquet-Bruhat, C. DeWitt-Morette, and M. Dillard-Bleick, Analysis, Manifolds, and Physics. Vol. 1: Basics; Vol. 2: 92 Applications, Elsevier, Amsterdam, 1996.

T. Frankel, The Geometry of Physics, Cambridge University Press, Cambridge, 2004.

S. Novikov and T. Taimanov, Geometric Structures and Fields, Amer. Math. Soc., Providence, Rhode Island, 2006.

As an introduction to the theory of general relativity based on the use of local coordinates, we recommend the classical Lecture Notes by

P. Dirac, General Theory of Relativity, Princeton University Press, 1996 (70 pages)

together with

Ø. Grøn and S. Hervik, Einstein's Theory of General Relativity: with Modern Applications in Cosmology, Springer, New York, 2007.

Both the invariant formulation and the formulation in terms of local coordinates is discussed in great detail in the classic textbook by

C. Misner, K. Thorne, and A. Wheeler, Gravitation, Freeman, San Francisco, California, 1973.

For the sophisticated mathematical problem of solving the initial-value problem for the Einstein equations on the gravitational field, we recommend:

P. Crusciel and H. Friedrich, The Einstein Equations and the Large Scale Behavior of Gravitational Fields: 50 Years of the Cauchy Problem in General Relativity, Birkhäuser, Boston, 2004.

Y. Choquet–Bruhat, General Relativity and the Einstein Equations, Oxford University Press, 2008.

As a comprehensive modern textbook, we recommend:

[2] One has to distinguish between the great French geometer Élie Cartan (1869–1951), who strongly influenced the development of modern differential geometry, and his famous son Henri Cartan (1904–2008), who made important contributions to algebra, analysis, and topology (e.g., homological algebra, the theory of analytic functions of several complex variables, and the cohomology of sheaves).

T. Padmanabhan, Gravitation: Foundations and Frontiers, Cambridge University Press, 2010.

The nature of dark matter is one of the great open problems in physics. We refer to:

G. Bertone (Ed.), Particle Dark Matter, Cambridge University Press, 2010.

An Interview with Chen Ning Yang on the History of Modern Gauge Theory

To begin with, let us quote some parts of an interview given by the physicist Chen Ning Yang answering the questions of Dianzhou Zhang:[3]

Zhang: Chen Ning Yang (born 1922 in Hefei, China), one of the twentieth century's great theoretical physicists, shared the Nobel prize in physics with Tsung-Dao Lee in 1957 for their joint contribution to parity non-conservation in weak interaction. Mathematicians, however, know Yang best for the Yang–Mills gauge field theory and the Yang–Baxter equation. After Einstein and Dirac, Yang is perhaps the twentieth-century physicist who has had the greatest impact on the development of mathematics ...
While a student in Kunming (China) and Chicago, Yang was impressed with the fact that gauge invariance determined all electromagnetic interactions. This was known from the works in the years 1918–1929 of Weyl, Fock, and London, and through later review papers by Pauli. But by the 1940s and the early 1950s, it played only a minor and technical role in physics. In Chicago, Yang tried to generalize the concept of gauge invariance to non-Abelian groups (the gauge group for electromagnetism being the Abelian group $U(1)$). In analogy with Maxwell's equations he tried

$$\mathcal{F}_{\alpha\beta} = \partial_\alpha \mathcal{A}_\beta - \partial_\beta \mathcal{A}_\alpha,$$

where A_α are matrices ($\alpha, \beta = 0, 1, 2, 3$). As Yang pointed out later on, "This led to a mess, and I had to give up."
In 1954, as a visiting physicist at Brookhaven National Laboratory on Long Island, New York, Yang returned once again to the idea of generalizing gauge invariance. His officemate was Robert Mills, who was about to finish his Ph.D. degree at Columbia University, New York City. Yang introduced the idea of non-Abelian gauge field to Mills, and they decided to add a quadratic term:[4]

$$\boxed{\mathcal{F}_{\alpha\beta} = \partial_\alpha \mathcal{A}_\beta - \partial_\beta \mathcal{A}_\alpha + \mathcal{A}_\alpha \mathcal{A}_\beta - \mathcal{A}_\beta \mathcal{A}_\alpha.} \tag{0.1}$$

That cleared up the "mess" and led to a beautiful new field theory.[5]
Zhang: *Did you study gauge field theory continuously after 1954?*

---

[3] D. Zhang, N. C. Yang and contemporary mathematics, Mathematical Intelligencer **15** (1993), Springer, New York, pp. 13–21. Reprinted with permission.

[4] Here, the point x of the space-time manifold has the (local) coordinates x^0, x^1, x^2, x^3 with $x^0 = ct$ (t time, c velocity of light in a vacuum). Furthermore, the symbol ∂_α denotes the partial derivative $\frac{\partial}{\partial x^\alpha}$.

[5] C. Yang and R. Mills, Conservation of isotopic spin and isotopic spin invariance, Phys. Rev. **96** (1954), 191–195.

Yang: Yes, I did ... In the late 1960s, I began a new formulation of gauge field theory through the approach of non-integrable phase factors. It happened that one semester I was teaching general relativity, and I noticed that the formula (0.1) in gauge field theory and the formula

$$R^{\delta}_{\alpha\beta\gamma} = \partial_{\alpha}\Gamma^{\delta}_{\beta\gamma} - \partial_{\beta}\Gamma^{\delta}_{\alpha\gamma} + \Gamma^{\delta}_{\alpha\mu}\Gamma^{\mu}_{\beta\gamma} - \Gamma^{\delta}_{\beta\mu}\Gamma^{\mu}_{\alpha\gamma} \tag{0.2}$$

with $\alpha, \beta, \gamma, \delta = 0, 1, 2, 3$ for the Riemann curvature tensor in Riemannian geometry are not just similar – they are, in fact, the same if one makes the right identification of symbols.[6] It is hard to describe the thrill I felt at understanding the point.

Zhang: *Is that the first time that you realized the relation between gauge theory and differential geometry?*

Yang: I had noticed the similarity between Levi-Civita's parallel displacement and non-integrable phase factors in gauge fields. But the exact relationship was appreciated by me only when I realized that the formula (0.1) in gauge field theory and the Riemann formula (0.2) are the same. With an appreciation of the geometrical meaning of gauge theory, I consulted Jim Simons, a distinguished geometer, who was then the chairman of the Mathematics Department at Stony Brooke (Long Island, New York). He said gauge theory must be related to connections on fiber bundles. I then tried to understand fiber-bundle theory from such books as Steenrod's "The Topology of Fiber Bundles," Princeton University Press, 1951, but I learned nothing. The language of modern mathematics is too cold and abstract for a physicist.

Zhang: *I suppose only mathematicians appreciate the mathematical language of today.*

Yang: I can tell you a relevant story. About ten years ago, I gave a talk on physics in Seoul, South Korea. I joked "There exist only two kinds of modern mathematics books: one which you cannot read beyond the first page and one which you cannot read beyond the first sentence. The *Mathematical Intelligencer* later reprinted this joke of mine. But I suspect many mathematicians themselves agree with me.

Zhang: *When did you understand bundle theory?*

Yang: In early 1975, I invited Jim Simons to give us a series of luncheon lectures on differential forms and bundle theory. He kindly accepted the invitation, and we learned about de Rham's theorem, differential forms, patching and so on ...

Zhang: Simon's lecture helped Wu and Yang to write a famous paper in 1975.[7] In this paper, they analyzed the intrinsic meaning of electromagnetism, emphasizing especially its global topological aspects. They discussed the mathematical meaning of the Aharonov–Bohm experiment and of the Dirac magnetic monopole. They exhibited a dictionary on the translation of terminologies used in mathematics and physics. Half a year later, Isadore Singer of the Massachusetts Institute of Technology (MIT, Cambridge, Massachusetts) visited Stony Brooke and discussed these matters with Yang at length. Singer had been an undergraduate student in physics and a graduate student in mathematics in the 1940s. He wrote in 1985: "Thirty years later I found myself lecturing on gauge theories, beginning with the Wu and Yang dictionary and ending with instantons,

[6] This will be shown on page 35 under the heading "Yang's matrix trick."

[7] T. Wu and C. Yang, Concept of non-integrable phase factors and global formulation of gauge fields, Phys. Rev. **D12** (1975), 3845–3857.

that is, self-dual connections. I would be inaccurate to say after studying mathematics for thirty years, I felt prepared to return to physics."
Yang: In 1975, impressed with the fact that gauge fields are connections on fiber bundles, I drove to the house of Shing-Shen Chern (1911–2004) in El Cerrito near Berkeley (California) ... I said I found it amazing that gauge theory are exactly connections on fiber bundles, which the mathematicians developed without reference to the physical world. I added "This is both thrilling and puzzling, since you mathematicians dreamed up these concepts out of nowhere." Chern immediately protested "No, no. These concepts were not dreamed up. They were natural and real."
Zhang: *The Yang–Baxter equation*

$$A(u)B(u+v)A(v) = B(v)A(u+v)B(u)$$

appearing in statistical mechanics is just a simple equation for matrix functions. Why does it have such great importance?
Yang: In the simplest situation, the Yang–Baxter equation has the form

$$ABA = BAB.$$

This is the fundamental equation of Artin (1898–1962) for the braid group. The braid group is, of course, a record of the history of permutations. It is not difficult to understand that the history of permutations is relevant to many problems in mathematics and physics. Looking at the developments of the last six or seven years, I got the feeling that the Yang–Baxter equation is the next pervasive algebraic equation after the Jacobi identity

$$[A,[B,C]] + [B,[C,A]] + [C,[A,B]] = 0.$$

The study of the Jacobi identity has, of course, led to the whole of Lie algebra and its relationship to Lie groups that govern symmetry in nature.

Zhang: Yang–Mills theory and the Yang–Baxter equation both figure prominently in today's score mathematics. One can see this by the Fields medals awarded in 1986 and 1990. Simon Donaldson was awarded a Fields medal at the International Congress of Mathematicians held in Berkeley in 1986. Sir Michael Atiyah spoke on Simon Donaldson's work: "Together with the important work of Michael Freedman (another Fields medal winner in 1986), Donaldson's result implied that there exist 'exotic' four-dimensional spaces which are topologically but not differentially equivalent to the standard Euclidean four-dimensional space $\mathbb{R}^4$ ... Donaldson's results are derived from the Yang–Mills equations of theoretical physics which are nonlinear generalizations of Maxwell's equations. In the Euclidean case the solution to the Yang–Mills equations giving the absolute minimum are of special interest and called instantons."
There were four Fields medalists in 1990: Vladimir Drinfeld, Vaughan Jones, Shigefumi Mori, and Edward Witten. The work of three of them was related to the Yang–Mills equations

$$\boxed{-D * \mathcal{F} = *\mathcal{J}, \qquad D\mathcal{F} = 0}$$

and/or the Yang–Baxter equation (see Sect. 15.4).

(i) We should mention Drinfeld's pioneering work with Yuri Manin on the construction of instantons. These are solutions to the Yang–Mills equations which can be thought of as having particle-like properties of localization and size. Drinfeld's interest in physics continued with his investigation of the Yang–Baxter equation.
(ii) Jones opened a whole new direction upon realizing that under certain conditions solutions of the Yang–Baxter equation could be used for constructing invariants of links ... The theory of quantum groups (i.e., deformations of classical Lie groups based on non-commutative Hopf algebras) was devised by Jimbo and Drinfeld to produce solutions of Yang–Baxter equations.
(iii) Witten described in these terms the invariants of Donaldson and Floer (extending the earlier ideas of Atiyah) and generalized the Jones polynomials to the case of an arbitrary ambient three-dimensional manifold.

We note with amusement that there were complaints that the plenary lectures at the International Congress of Mathematicians in Kyoto, 1990, were heavily slanted toward the topics of mathematical physics: "Everywhere we heard quantum group, quantum group, quantum group!" ...

Yang: Many theoretical physicists are, in some ways, antagonistic to mathematics, or at least have a tendency to downplay the value of mathematics. I do not agree with these attitudes. I have written:[8] "Perhaps of my father's influence, I appreciate mathematics. I appreciate the value judgement of the mathematician, and I admire the beauty and power of mathematics: there are ingenuity and intricacy in tactical maneuvers, and breathtaking sweeps in strategic campaigns. And, of course, miracle of miracles, some concepts in mathematics turn out to provide the fundamental structures that govern the physical universe!"

In the present volume, we will show that the Yang–Mills equations generalize the Maxwell equations in electromagnetism.

Einstein's Theory of General Relativity on Gravitation

We set

$$R_{\alpha\beta} = \kappa_G (T_{\alpha\beta} - \tfrac{1}{2} g_{\alpha\beta} T), \qquad \alpha, \beta = 0, 1, 2, 3.$$

This completes the general theory of relativity as a logical structure. The postulate of relativity in its most general form, which makes the space-time coordinates meaningless parameters, leads necessarily to a certain form of gravitational theory which explains the motion of the Perihelion of the planet Mercury.

Anyone who has really grasped the general theory of relativity, will be captured by its beauty. It is a triumph of the general differential calculus, which was created by Gauss (1777–1855), Riemann (1826–1866), Christoffel (1829–1900), Ricci-Curbastro (1853–1925), Bianchi (1856–1928), and Levi-Civita (1873–1941).[9]

Albert Einstein, 1915

[8] C. Yang, Selected Papers, Freeman, San Francisco, 1983.

[9] A. Einstein, On general relativity. The field equations of gravitation. Reports on the meetings of the Prussian Academy of Sciences (Berlin) on November 11 and December 2, 1915 (in German).

The two fundamental Einstein equations. In 1915, motivated by the study of classical differential geometry, Einstein based his theory of general relativity on the Riemann curvature tensor of the four-dimensional space-time manifold $\mathcal{M}^4$. The points P of $\mathcal{M}^4$ are called space-time points or events. Einstein's fundamental equations read as follows:[10]

(i) The equation of motion for the gravitational field:

$$\boxed{R_{\alpha\beta} = \kappa_G(T_{\alpha\beta} - \tfrac{1}{2}g_{\alpha\beta}T), \qquad \alpha,\beta = 0,1,2,3.} \tag{0.3}$$

Here, the universal constant $\kappa_G := \frac{8\pi G}{c^4}$ depends on Newton's gravitational constant G and the velocity c of light in a vacuum.

(ii) The equation of motion for the trajectories $x = x(\sigma), \sigma_0 \le \sigma \le \sigma_1$, of celestial bodies (e.g., planets, the sun, stars, or galaxies) and light rays:

$$\boxed{\ddot{x}^\gamma = -\dot{x}^\alpha \Gamma^\gamma_{\alpha\beta}\,\dot{x}^\beta, \qquad \gamma = 0,1,2,3.} \tag{0.4}$$

This equation generalizes Newton's classical equation of motion.[11]

Let us discuss (i) and (ii). We choose an arbitrary observer which uses the local space-time coordinates x^0, x^1, x^2, x^3 in order to describe events. The local coordinates are obtained by measurements of space positions and time. By convention, we write x instead of (x^0, x^1, x^2, x^3). Different observers may use completely different methods for measuring space positions and time. The locality means that the real numbers x^0, x^1, x^2, x^3 do not represent coordinates for the global universe, but only for a sufficiently small spatial region and a sufficiently small time interval. The crucial change of local coordinates will be considered below. We set

- $\partial_\alpha := \frac{\partial}{\partial x^\alpha}$ (partial derivative), and
- $\dot{x}^\alpha(\sigma) := \frac{d}{d\sigma}x^\alpha(\sigma)$ (derivative with respect to the real parameter σ).

Arc length and proper time. The curve

$$C : x = x(\sigma), \qquad \sigma_0 \le \sigma \le \sigma_1,$$

with the real parameter σ, describes a family of events, for example, the motion of a planet or the motion of a light ray. The length of the curve C is given by the integral

$$\boxed{l(C) := \int_{\sigma_0}^{\sigma_1} \sqrt{g_{\alpha\beta}(x(\sigma))\,\dot{x}^\alpha(\sigma)\,\dot{x}^\beta(\sigma)}\,d\sigma.} \tag{0.5}$$

Here, the functions $x \mapsto g_{\alpha\beta}(x)$ are called the components of the metric tensor with respect to the local coordinates x^0, x^1, x^2, x^3. For the motion of a planet (resp. light ray), we get $l(C) > 0$ (resp. $l(C) = 0$). The length of the curve $l(C)$ does not depend on the choice of the local coordinate system. If the trajectory $x = x(\sigma), \sigma_0 \le \sigma \le \sigma_1$, describes the motion of a spaceship, then $l(C)/c$ is the proper time of the flight, which is measured by the crew in the spaceship during the flight.

We want to show that Einstein's equation (0.3) represents an equation for computing the components $g_{\alpha\beta}$ of the metric tensor which govern the measurement of spatial distances and proper times. We postulate that

---

[10] We will use the Einstein summation convention, that is, we sum over equal upper and lower Greek indices from 0 to 3.

[11] Explicitly, this reads as $\ddot{x}^\gamma(\sigma) = -\dot{x}^\alpha(\sigma)\Gamma^\gamma_{\alpha\beta}(x(\sigma))\,\dot{x}^\beta(\sigma)$.

$$g_{\alpha\beta} = g_{\beta\alpha} \qquad \text{for all} \quad \alpha, \beta = 0, 1, 2, 3.$$

In order to distinguish between the time-like coordinate x^0 and the space-like coordinates x^1, x^2, x^3, we assume that the following definiteness conditions are always satisfied:

$$g_{00} > 0, \quad \begin{vmatrix} g_{00} & g_{01} \\ g_{10} & g_{11} \end{vmatrix} < 0, \quad \begin{vmatrix} g_{00} & g_{01} & g_{02} \\ g_{10} & g_{11} & g_{12} \\ g_{20} & g_{21} & g_{22} \end{vmatrix} > 0, \quad g := \begin{vmatrix} g_{00} & g_{01} & g_{02} & g_{03} \\ g_{10} & g_{11} & g_{12} & g_{13} \\ g_{20} & g_{21} & g_{22} & g_{23} \\ g_{30} & g_{31} & g_{32} & g_{33} \end{vmatrix} < 0. \tag{0.6}$$

In particular, these conditions are satisfied if $g_{\alpha\beta}(x) \equiv \eta_{\alpha\beta}$ for all indices $\alpha, \beta = 0, 1, 2, 3$. Here, we introduce the so-called Minkowski symbol $\eta_{\alpha\beta}$ given by

$$(\eta_{\alpha\beta}) := \begin{pmatrix} 1 & 0 & 0 & 0 \\ 0 & -1 & 0 & 0 \\ 0 & 0 & -1 & 0 \\ 0 & 0 & 0 & -1 \end{pmatrix}. \tag{0.7}$$

In this special case, the rescaled arc length $l(C)/c$ from (0.5) is the proper time of a spaceship which moves freely, that is, the gravitational force vanishes.

The structure of the Einstein equations. Because of the symmetry properties of $g_{\alpha\beta}$, $R_{\alpha\beta}$, and $T_{\alpha\beta}$, we obtain the following:

- The Einstein equations (0.3) for the gravitational field represent a nonlinear system of 10 second-order partial differential equations for the 10 unknown functions

$$g_{00}; \quad g_{10}, g_{11}; \quad g_{20}, g_{21}, g_{22}; \quad g_{30}, g_{31}, g_{32}, g_{33}$$

  which depend on the space-time variables x^0, x^1, x^2, x^3.
- The Einstein equations (0.4) for the motion of planets and light rays represent a nonlinear system of 4 ordinary differential equations for the 4 unknown functions

$$x^\alpha = x^\alpha(\sigma), \qquad \alpha = 0, 1, 2, 3,$$

  which depend on the real parameter σ living in the interval $[\sigma_0, \sigma_1]$.

Changing the observer and Einstein's principle of general relativity. The following considerations are crucial for understanding the philosophy of Einstein's theory of general relativity. The Einstein equations (0.3) and (0.4) are formulated in terms of local coordinates x^0, x^1, x^2, x^3. In terms of physics, the local coordinates describe the measurements of space positions and time positions carried out by an observer. The theory only makes sense if the following hold:

- the Einstein equations (0.3) and (0.4) are valid for all observers (i.e., for all choices of local coordinates), and
- we know the transformation laws for all the quantities under changing the local coordinates of observers:

$$\boxed{(x^0, x^1, x^2, x^3) \mapsto (x^{0'}, x^{1'}, x^{2'}, x^{3'}).} \tag{0.8}$$

To simplify notation, we briefly write $x \mapsto x'$ or $x' = x'(x)$.

(P) In terms of physics, Einstein postulated that: Physics does not depend on the choice of observers. This is Einstein's principle of general relativity.

(M) In terms of mathematics, Einstein's principle of general relativity is realized by the use of tensor calculus introduced in the second half of the 19th century.

Let us discuss this. The main points are

- the key transformation laws (0.11) and (0.12) in tensor calculus, and
- the mnemonic principle of index killing for constructing invariants.

To begin with, let us consider a typical example.

Invariance of the arc length and the proper time as a paradigm. Naturally enough, we postulate that

The rescaled arc length (i.e., the proper time) $l(C)/c$ possesses an invariant meaning.

This means that, under a change of local coordinates (0.8), the arc length $l(C)$ remains unchanged, that is, it does not depend on the choice of the observer. Explicitly, we have

$$l(C) = \int_{\sigma_0}^{\sigma_1} \sqrt{g_{\alpha\beta}(x(\sigma))\,\dot{x}^\alpha(\sigma)\dot{x}^\beta(\sigma)}\,d\sigma$$
$$= \int_{\sigma_0}^{\sigma_1} \sqrt{g_{\alpha'\beta'}(x'(\sigma))\,\dot{x}^{\alpha'}(\sigma)\dot{x}^{\beta'}(\sigma)}\,d\sigma. \tag{0.9}$$

Here, for the indices $\alpha = 0, 1, 2, 3$, the equation

$$x^\alpha = x^\alpha(\sigma), \qquad \sigma_0 \le \sigma \le \sigma_1,$$

of the curve C is transformed into the equation $x^{\alpha'} = x^{\alpha'}(\sigma), \sigma_0 \le \sigma \le \sigma_1$. Explicitly, we set $x'(\sigma) := x'(x(\sigma))$. We want to show that the transformation law

$$\boxed{g_{\alpha'\beta'}(x') = g_{\alpha\beta}(x)\cdot\frac{\partial x^\alpha(x')}{\partial x^{\alpha'}}\frac{\partial x^\beta(x')}{\partial x^{\beta'}}} \tag{0.10}$$

implies the invariance relation (0.9). Here, we sum over equal upper and lower indices from 0 to 3. In order to prove (0.9), observe that the chain rule yields

$$g_{\alpha\beta}\frac{dx^\alpha}{d\sigma}\cdot\frac{dx^\beta}{d\sigma} = g_{\alpha\beta}\frac{\partial x^\alpha}{\partial x^{\alpha'}}\frac{dx^{\alpha'}}{d\sigma}\cdot\frac{\partial x^\beta}{\partial x^{\beta'}}\frac{dx^{\beta'}}{d\sigma} = g_{\alpha'\beta'}\frac{dx^{\alpha'}}{d\sigma}\cdot\frac{dx^{\beta'}}{d\sigma}$$

along the curve C. Using the square root and integrating this over the parameter interval $[\sigma_0, \sigma_1]$, we get the claim (0.9).

Tensorial transformation laws – Ariadne's thread in tensor calculus. The argument above is a special case of the tensor calculus which allows us to construct invariant expressions under a change of local coordinates, in a general setting. This will be thoroughly studied in Chap. 8. At this point, we would like to discuss the basic ideas. By the chain rule of calculus, we get the following two key transformation laws of tensor calculus:

- $\frac{dx^{\alpha'}}{d\sigma} = \frac{\partial x^{\alpha'}}{\partial x^\alpha}\cdot\frac{dx^\alpha}{d\sigma}$ (derivative with respect to the real parameter σ), and
- $\frac{\partial\Theta}{\partial x^{\alpha'}} = \frac{\partial\Theta}{\partial x^\alpha}\cdot\frac{\partial x^\alpha}{\partial x^{\alpha'}}$ (partial derivative)

where we sum over $\alpha = 0, 1, 2, 3$. More precisely, taking the arguments explicitly into account, this reads as

$$\frac{dx^{\alpha'}(\sigma)}{d\sigma} = \frac{\partial x^{\alpha'}(x(\sigma))}{\partial x^\alpha}\cdot\frac{dx^\alpha(\sigma)}{d\sigma} \tag{0.11}$$

and

$$\frac{\partial \Theta(x')}{\partial x^{\alpha'}} = \frac{\partial \Theta(x)}{\partial x^{\alpha}} \cdot \frac{\partial x^{\alpha}(x')}{\partial x^{\alpha'}}. \tag{0.12}$$

Here, the local coordinate x' corresponds to x, that is, $x' = x'(x)$. Moreover, we assume that the real-valued function $P \mapsto \Theta(P)$ is an invariant function on the four-dimensional space-time manifold $\mathcal{M}^4$. This means that the value of Θ at the point P only depends on the event P, but not on the choice of the local coordinates which describe the event. Explicitly, $\Theta(x') = \Theta(x)$ where x' and x are related to each other by $x' = x'(x)$.[12] Let us introduce the following terminology:

- the velocity components $\dot{x}^\alpha$ form a *contravariant* tensorial family, and
- the partial derivatives $\partial_\alpha \Theta$ of the invariant function Θ form a *covariant* tensorial family,

In the general case, the family of functions,

$$x \mapsto T^{\alpha_1 \ldots \alpha_m}_{\beta_1 \ldots \beta_n}(x), \qquad \alpha_1, \ldots, \alpha_m, \beta_1, \ldots, \beta_n = 0, 1, 2, 3,$$

is called a tensorial family of type (m, n) iff the functions are transformed like the product

$$\dot{x}^{\alpha_1} \cdots \dot{x}^{\alpha_m} \partial_{\beta_1} \Theta \cdots \partial_{\beta_n} \Theta$$

under a change of local coordinates. Such a tensorial family is also called m-fold contravariant and n-fold covariant. For example, by (0.10), $g_{\alpha\beta}$ transforms like the product

$$\boxed{\partial_\alpha \Theta\, \partial_\beta \Theta.}$$

Therefore, the components $g_{\alpha\beta}$ of the metric tensor form a two-fold covariant tensorial family.

The components of the Riemann curvature tensor. As in classical differential geometry, let us introduce the following Christoffel symbols:[13]

$$\Gamma^{\gamma}_{\alpha\beta} := \tfrac{1}{2}(\partial_\alpha g_{\beta\sigma} + \partial_\beta g_{\alpha\sigma} - \partial_\sigma g_{\alpha\beta}) g^{\sigma\gamma}, \qquad \alpha, \beta, \gamma = 0, 1, 2, 3. \tag{0.13}$$

Following Riemann, we define the components of the Riemann curvature tensor by setting[14]

$$\boxed{R^{\delta}_{\alpha\beta\gamma} := \partial_\alpha \Gamma^{\delta}_{\beta\gamma} - \partial_\beta \Gamma^{\delta}_{\alpha\gamma} + \Gamma^{\delta}_{\alpha\mu} \Gamma^{\mu}_{\beta\gamma} - \Gamma^{\delta}_{\beta\mu} \Gamma^{\mu}_{\alpha\gamma}} \tag{0.14}$$

where $\alpha, \beta, \kappa, \gamma, \delta = 0, 1, 2, 3$. This yields the following quantities:

- $R_{\alpha\beta\gamma\delta} := R^{\sigma}_{\alpha\beta\gamma} g_{\sigma\delta}$ (components of the metric Riemann curvature tensor),
- $R_{\alpha\delta} := R_{\alpha\beta\gamma\delta} g^{\beta\gamma}$ (components of the Ricci curvature tensor),
- $R := R_{\alpha\delta} g^{\alpha\delta}$ (scalar curvature – trace of the Ricci tensor),
- $G_{\alpha\beta} := R_{\alpha\beta} - \frac{1}{2} R g_{\alpha\beta}$ (components of the Einstein tensor),
- $T := T_{\alpha\beta} g^{\alpha\beta}$.

---

[12] For example, it turns out that the trace T of the energy-momentum tensor, and the scalar curvature R are invariant functions on the space-time manifold $\mathcal{M}^4$.

[13] By definition, the symbol $(g^{\alpha\beta})$ represents the inverse matrix to $(g_{\alpha\beta})$.

[14] Mnemonically, the position of the indices κ and λ of $\Gamma^{\kappa}_{\alpha\lambda}$ and $R^{\kappa}_{\alpha\beta\lambda}$ is dictated by the symmetric formulation of the Yang matrix trick which will be discussed on page 35.

The functions $x \mapsto T_{\alpha\beta}(x)$ represent the components of the energy-momentum tensor. Furthermore, we set $T := T_{\alpha\beta}g^{\alpha\beta}$ (trace of the energy momentum tensor). In terms of physics, the energy-momentum tensor describes the distribution of mass and energy in the universe.

The gravitational force corresponds to the curvature of the four-dimensional space-time manifold. Observe the following:

- The first Einstein equation (0.3) on page 8 describes the crucial fact that the mass and energy distributions in the universe influence the curvature of the four-dimensional space-time manifold $\mathcal{M}^4$. Equation (0.3) is equivalent to

$$G_{\alpha\beta} = \kappa_G T_{\alpha\beta}, \qquad \alpha, \beta = 0, 1, 2, 3. \tag{0.15}$$

- As we will discuss below, the second Einstein equation (0.4) on page 8 tells us that the motion of a celestial body or a light ray corresponds to a geodesic line of the curved 4-dimensional space-time manifold $\mathcal{M}^4$.

The vanishing of the gravitational force. The local vanishing of the Riemann curvature tensor corresponds to the vanishing of the gravitational force. In fact, suppose that for all indices $\alpha, \beta, \gamma, \delta = 0, 1, 2, 3$ we have

$$R^{\delta}_{\alpha\beta\gamma}(x) \equiv 0 \tag{0.16}$$

on some neighborhood of the point P_0 of the space-time manifold $\mathcal{M}^4$. Using Riemann's classical argument, one can show that (0.16) implies

$$g_{\alpha\beta}(x) \equiv \eta_{\alpha\beta}, \qquad \alpha, \beta = 0, 1, 2, 3 \tag{0.17}$$

on a sufficiently small neighborhood of the point P_0. More precisely, relation (0.17) is valid after a change of local coordinates if necessary. It follows from equation (0.17) that the Christoffel symbols vanish on a sufficiently small neighborhood of P_0. Then the Einstein equations (0.4) of motion read locally as

$$\ddot{x}^{\gamma}(\sigma) \equiv 0, \qquad \gamma = 0, 1, 2, 3.$$

This equation has straight lines as solutions. For example, this corresponds to a trivial motion of space ships without any acceleration. In terms of physics, this means that the observer does not measure any gravitational force. Observe that:

If equation (0.16) *is valid with respect to a fixed local coordinate system, then it is valid in every local coordinate system.*

This follows from the crucial (highly nontrivial) fact that the components

$$R^{\delta}_{\alpha\beta\gamma}$$

of the Riemann curvature tensor form a tensorial family. That is, they are transformed like the product

$$\dot{x}^{\delta} \cdot \partial_{\alpha}\Theta \cdot \partial_{\beta}\Theta \cdot \partial_{\gamma}\Theta$$

under a change of local coordinates. This will be proved later on.

Einstein's local equivalence principle. Observe the following special feature. In contrast to the components $R^{\delta}_{\alpha\beta\gamma}$ of the Riemann curvature tensor, the Christoffel symbols $\Gamma^{\delta}_{\alpha\beta}$ do *not* form a tensorial family. For a given point P_0 of the space-time manifold $\mathcal{M}^4$, it is always possible to choose a specific local coordinate system such that the Christoffel symbols vanish at the point P_0, that is,

$$\Gamma^{\delta}_{\alpha\beta}(P_0) = 0, \qquad \alpha, \beta, \delta = 0, 1, 2, 3. \tag{0.18}$$

However, as a rule, this condition is not valid in all local coordinate systems. To illustrate this by a simple example, consider an elevator which goes down with the acceleration a. If a is equal to the gravitational acceleration (i.e., $a = 9.81\mathrm{m/s}^2$), then an observer inside the elevator does not feel anymore the gravitational field of earth. Einstein called this the local equivalence principle. This principle tells us that the gravitational force can be locally compensated by passing to an accelerated reference system. Mathematically, the local equivalence principle corresponds to (0.18). Finally, set

$$T^{\kappa}_{\alpha\beta} := \Gamma^{\kappa}_{\alpha\beta} - \Gamma^{\kappa}_{\beta\alpha}.$$

In contrast to the Christoffel symbols themselves, the so-called torsion functions $T^{\kappa}_{\alpha\beta}$ form a tensorial family. In fact, the torsion functions vanish identically, that is,

$$T^{\kappa}_{\alpha\beta} \equiv 0.$$

In other words, the Christoffel symbols are symmetric with respect to the lower indices in every local coordinate system: $\Gamma^{\kappa}_{\alpha\beta} \equiv \Gamma^{\kappa}_{\beta\alpha}$ for all indices $\alpha, \beta, \kappa = 0, 1, 2, 3$.

Dark energy. The components $T_{\alpha\beta}$ of the energy-momentum tensor allow the following decomposition:

$$T_{\alpha\beta} := T^{\mathrm{class}}_{\alpha\beta} + T^{\mathrm{CDM}}_{\alpha\beta} + T^{\mathrm{DE}}_{\alpha\beta}$$

where we us the following terminology:

- $T^{\mathrm{class}}_{\alpha\beta}$ (classical mass and energy),
- $T^{\mathrm{CDM}}_{\alpha\beta}$ (cold dark matter),
- $T^{\mathrm{DE}}_{\alpha\beta} = -\eta_{\mathrm{DE}} \cdot g_{\alpha\beta}$ (dark energy),
- η_{DE} (density of dark energy),
- κ_G (universal coupling constant for gravitation),
- $\Lambda = \kappa_G \cdot \eta_{DE}$ (cosmological constant).

The quantities under consideration possess the following physical dimensions:

- $g_{\alpha\beta}$ (dimensionless),
- $R_{\alpha\beta}$ (1/length^2),
- $T_{\alpha\beta}$ (energy density = energy/length^3),
- κ_G (length/energy),[15]
- Λ (1/length^2).

This will be studied in Volume IV.

> *Surprisingly enough, only 4 percent of the total mass and energy of our universe are of classical type.*

Moreover, 70 percent of the total amount of energy of the universe consist of dark energy. The remaining 26 percent consist of cold dark matter.[16]

In Volume IV, we will use explicit solutions of the two Einstein equations (0.3) and (0.4) in order to study the following physical problems:

- the motion of the semi-axis of the planet Mercury (i.e., the slow rotation of the Perihelion of Mercury),
- the deflection of light in the gravitational field of the sun,
- the red shift in the spectrum of light caused by the gravitational field of the earth,

---

[15] In the SI system, $\kappa_G = 2.07 \cdot 10^{-43}\mathrm{m/J}$. This shows that the gravitational interaction is very weak compared to the scale used in daily life.

[16] See G. Börner, The Early Universe, Springer, 2003, and S. Weinberg, Cosmology, Oxford University Press, 2008.

- the Big Bang, the red shift in the spectrum of distant galaxies (Hubble effect), and the accelerated expansion of the universe,
- black holes,
- the low-energy background radiation as a relict of the Big Bang.

The symmetry properties of the components of the Riemann curvature tensor. Let $\alpha, \beta, \gamma, \delta = 0, 1, 2, 3$. Then:

- $g_{\alpha\beta} = g_{\beta\alpha}$ (symmetry of the metric tensor),
- $\Gamma^{\delta}_{\alpha\beta} = \Gamma^{\delta}_{\beta\alpha}$ (symmetry of the Christoffel symbols),
- $R_{\alpha\beta} = R_{\beta\alpha}$ (symmetry of the Ricci tensor),
- $T_{\alpha\beta} = T_{\beta\alpha}$ (symmetry of the energy-momentum tensor).

Furthermore, the following hold:

(A1) $R^{\delta}_{\alpha\beta\gamma} = -R^{\delta}_{\beta\alpha\gamma}$ (interchanging α with β).

(A2) $R^{\delta}_{\alpha\beta\gamma} + R^{\delta}_{\beta\gamma\alpha} + R^{\delta}_{\gamma\alpha\beta} = 0$ (Ricci identity – cyclic permutation of the indices α, β, γ). This is equivalent to $R^{\delta}_{[\alpha\beta\gamma]} = 0$ (antisymmetrization with respect to α, β, γ).

(A3) $\partial_\mu R^{\delta}_{\alpha\beta\gamma} + \partial_\alpha R^{\delta}_{\beta\mu\gamma} + \partial_\gamma R^{\delta}_{\mu\beta\gamma} = 0$ (Bianchi identity – cyclic permutation of the indices μ, α, β). This is equivalent to $\partial_{[\mu} R^{\delta}_{\alpha\beta]\gamma} = 0$ (antisymmetrization with respect to μ, α, β).

In order to get further symmetry properties, we consider the components of the so-called metric Riemann curvature tensor

$$\boxed{R_{\alpha\beta\gamma\delta} := R^{\sigma}_{\alpha\beta\gamma} g_{\sigma\delta}}$$

by lowering the upper index δ of $R^{\delta}_{\alpha\beta\gamma}$. Then:

(B1) $R_{\alpha\beta\gamma\delta} = -R_{\beta\alpha\gamma\delta}$ (interchanging α with β),

(B2) $R_{\alpha\beta\gamma\delta} = -R_{\alpha\beta\delta\gamma}$ (interchanging γ with δ),

(B3) $R_{\alpha\beta\gamma\delta} = R_{\gamma\delta\alpha\beta}$ (interchanging α, β with γ, δ),

(B4) $R_{\alpha\beta\gamma\delta} + R_{\beta\gamma\alpha\delta} + R_{\gamma\alpha\beta\delta} = 0$ (metric Ricci identity – cyclic permutation of α, β, γ). This is equivalent to $R_{[\alpha\beta\gamma]\delta} = 0$ (antisymmetrization with respect to α, β, γ).

(B5) $\partial_\mu R_{\alpha\beta\gamma\delta} + \partial_\alpha R_{\beta\mu\gamma\delta} + \partial_\beta R_{\mu\alpha\gamma\delta} = 0$ (metric Bianchi identity – cyclic permutation of α, β, γ). This is equivalent to $\partial_{[\mu} R_{\alpha\beta]\gamma\delta} = 0$ (antisymmetrization with respect to α, β, γ).

The relation (B4) (resp. (B5)) is obtained from (A2) (resp. (A3)) by lowering the upper index κ.

The metric Riemann curvature tensor has $4^4 = 256$ components $R_{\alpha\beta\gamma\delta}$. However, by (B1) through (B3), this large number of components is reduced to 20 independent components. Mnemonically, the symmetry properties of $R_{\alpha\beta\gamma\delta}$ motivate the definition of the components of the Ricci tensor by setting

$$R_{\alpha\delta} := R_{\alpha\beta\gamma\delta} g^{\beta\gamma}.$$

In fact, up to sign, this is the only possibility to get a nontrivial expression. In fact,

$$R_{\alpha\beta\gamma\delta} g^{\alpha\beta} = 0, \quad R_{\alpha\beta\gamma\delta} g^{\gamma\delta} = 0, \quad R_{\alpha\beta\gamma\delta} g^{\alpha\delta} = R_{\beta\alpha\delta\gamma} g^{\alpha\delta} = R_{\beta\gamma},$$

and $R_{\alpha\beta\gamma\delta} g^{\alpha\gamma} = -R_{\beta\alpha\gamma\delta} g^{\alpha\gamma} = -R_{\beta\delta}$.

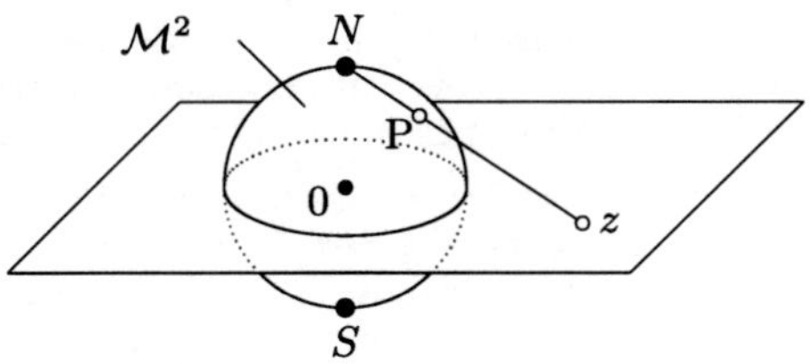

Fig. 0.1. Stereographic projection

The Riemann Curvature Tensor and the Beauty of the Gauss Theorema Egregium

For the next remarks, let us pass to the special case of a smooth 2-dimensional surface $\mathcal{M}^2$ which is embedded into the 3-dimensional Euclidean manifold (e.g., the surface of earth). We will use the same formulas for the components $R^{\alpha}_{\beta\gamma\delta}$ of the Riemann curvature tensor as introduced above, but now the indices $\alpha, \beta, \gamma, \delta$ only run from 1 to 2.

The surface theory of Gauss (1777–1855) was strongly influenced by Gauss' work as a surveyor. Under great physical pains, Gauss worked from 1821 to 1825 as a land surveyor in the kingdom of Hannover in the northern part of Germany. It almost led to his physical exhaustion. In 1822, he submitted his prize memoir "General solution of the problem of mapping parts of a given surface onto another given surface in such a way that image and pre-image become similar in their smallest parts" to the Royal Society of Sciences in Copenhagen (Denmark) for which he received the official prize. What was the importance of his work?

The mapping of surfaces onto one another, which satisfy certain given properties, is a basic problem of cartography, in particular the reproduction of parts of surfaces of the earth in plane geographic charts. Intuitively, it is impossible, for example, to map parts of the surface of the earth onto the plane and preserve the length. Therefore, one has to look for other mappings. Of great practical use are the conformal maps, that is, the angle-preserving maps. Angle preservation of geographical charts is important in navigation, that is, in determining routes of ships in charts. It turns out that conformal maps are also similar in the small. Special cases of conformal maps from the surface of the earth onto the plane are stereographic projections (see Fig. 0.1), which were already known to the Greeks, and the projection of Mercator (1512–1594) is still being used in the cartography of today. Gauss succeeded in finding a procedure to determine all conformal maps in the small for analytic surfaces.

The study of conformal maps in the large began with the Ph.D. thesis of Bernhard Riemann (1826–1866), which was written in 1851. Riemann's Ph.D. thesis contains the development of complex function theory including the famous Riemann mapping theorem. When writing his prize memoir, Gauss had apparently already worked on a more general surface theory, because he added the following Latin saying to his title page:

> Ab his via sterniture ad maiora.[17]

The development of the general surface theory, however, was difficult, though the basic ideas were known to Gauss since 1816. On February 19, 1826, he wrote to Olbers:

---

[17] From here the path to something more important is prepared.

> I hardly know any period in my life, where I earned so little real gain for truly exhausting work, as during this winter. I found many, many beautiful things, but my work on other things has been unsuccessful for months.

Finally, on October 8, 1827, Gauss presented the general surface theory. The title of his paper was "Disquisitiones generales circa superficies curvas" (Investigations about curved surfaces). The most important result of this masterpiece in the mathematical literature is the *theorema egregium* – the egregious theorem. As a crucial quantity, Gauss introduced the *Gaussian curvature* $K(P)$ of a 2-dimensional surface $\mathcal{M}^2$ at the point P. For a sphere of radius r, Gauss defined

$$K(P) := \frac{1}{r^2}.$$

This tells us that the larger the radius is, the smaller is the Gaussian curvature of the sphere. By an approximation argument, Gauss generalized the curvature definition for the sphere to general surfaces $\mathcal{M}^2$. In particular, the Gaussian curvature of a hyperboloid is negative (see Sect. 9.6.3). Gauss' definition used the surrounding 3-dimensional Euclidean space. This is called an extrinsic definition. Motivated by his practical work as land surveyor, Gauss posed the following fundamental question:

> *Is it possible to compute the Gaussian curvature K of a 2-dimensional surface by only using measurements on the surface?*

After a long fight, Gauss found that the answer is "yes"! He discovered the following sophisticated formula:

$$\boxed{K(P) = \frac{R_{1221}(P)}{g(P)}} \tag{0.19}$$

where $g := g_{11}g_{22} - (g_{12})^2$. This is the famous theorema egregium. Let us discuss this. By (0.13) and (0.14) on page 11, the following hold:

> *The Gaussian curvature K is an intrinsic property of the 2-dimensional surface; it depends on the components $g_{\alpha\beta}$ of the metric tensor and their first and second partial derivatives with respect to the local coordinates.*

In fact, Gauss did not explicitly use the Riemann curvature tensor, but in terms of the modern terminology, his key formula can be written as (0.19).[18] Concerning cartography, the theorema egregium tells us in rigorous terms that it is impossible to introduce geographic charts which are length preserving after rescaling. Indeed, one can show that length preserving maps preserve the components of the metric tensor. In turn, such maps preserve the Gaussian curvature. Finally, note that the Gaussian curvature of the sphere is positive, but the Gaussian curvature of the Euclidean plane vanishes.

> *Gauss' theorema egregium had an enormous impact on the development of modern differential geometry and modern physics culminating in the principle "force equals curvature." This principle is basic for both Einstein's theory of general relativity on gravitation and the Standard Model in elementary particle physics.*

[18] C. Gauß, Disquisitiones generales circa superficies curvas, Göttinger Nachr. **6**, 99–146 (1827) (in Latin). English translation: General Investigations of Curved Surfaces, Raven Press, New York.
See also P. Dombrowski, 150 years after Gauss' 'Disquisitiones generales circa superficies curvas', Astérisque **62** (1979).

In order to understand the intuitive meaning of both the components $R_{\alpha\beta}$ of the Ricci tensor and the scalar curvature R on the 2-dimensional surface $\mathcal{M}^2$, observe that the components of the Riemann curvature tensor read as

$$R_{\alpha\beta\gamma\delta} = K(g_{\alpha\delta}g_{\beta\gamma} - g_{\alpha\gamma}g_{\beta\delta}), \qquad \alpha,\beta,\gamma,\delta = 1,2.$$

Since $g_{\alpha\beta} = g_{\beta\alpha}$, we get the following symmetry properties

$$R_{\alpha\beta\gamma\delta} = -R_{\beta\alpha\gamma\delta} = -R_{\alpha\beta\delta\gamma} = R_{\gamma\delta\alpha\beta}$$

for all indices $\alpha,\beta,\gamma,\delta = 1,2$. Therefore, the $2^4 = 16$ components of the Riemann curvature tensor reduce to one essential component, namely,

$$R_{1221} = K(g_{11}g_{22} - g_{12}^2).$$

In fact, we have $R_{1221} = -R_{2121} = -R_{1212} = -R_{2112}$. The remaining 12 components vanish identically. For example, it follows from $R_{\alpha\beta\gamma\delta} = -R_{\beta\alpha\gamma\delta}$ that $R_{1112} = 0$. In order to simplify notation, let us introduce an orthogonal local coordinate system, that is, we have the special case where $g_{12} = g_{21} = 0$. Hence $g = g_{11}g_{22}$, and

$$g^{11} = (g_{11})^{-1}, \quad g^{22} = (g_{22})^{-1}, \quad g^{12} = g^{21} = 0.$$

This implies that:

- $R_{11} = g_{11}K, R_{22} = g_{22}K$, and $R_{12} = R_{21} = 0$ (Ricci tensor),
- $R = 2K$ (Ricci (or scalar) curvature).

Thus, the scalar curvature R is twice the Gaussian curvature K.

Heat conduction and the Riemann curvature tensor. Let x denote the tuple (x^1, x^2, x^3) of Cartesian coordinates. In the late 1850s, the Paris Academy posed the following problem: Find conditions such that the inhomogeneous heat conduction equation

$$\frac{\partial\Theta(x)}{\partial t} = \sum_{j,k=1}^{3} g^{jk}(x)\,\partial_j\partial_k\Theta(x) \tag{0.20}$$

for the temperature Θ can be locally transformed into the standard heat conduction equation

$$\frac{\partial\Theta(y)}{\partial t} = \sum_{j=1}^{3} \partial_j^2\Theta(y) \tag{0.21}$$

by a change $x \mapsto y$ of local coordinates[19] near the given point x_*. In terms of physics, equation (0.20) describes the heat conduction in an inhomogeneous material. If there exists such a coordinate transformation, the solution of the original complicated equation (0.20) can be reduced to the well-known solutions of the simpler equation (0.21). In 1861, Riemann solved this problem. He proved that:

> *The transformation is possible iff the Riemann curvature tensor R^l_{ijk} vanishes in a small neighborhood of the point x_*, that is, $R^l_{ijk}(x) \equiv 0$ for all $i,j,k,l = 1,2,3$.*

[19] More precisely, we make the assumption that all the eigenvalues of the real symmetric (3×3)-matrix $(g^{jk}(x))$ are positive for all points $x \in \mathbb{R}^3$, and the local coordinate change $x \mapsto y$ is a local diffeomorphism on some open neighborhood of the point x_*.

In his famous 1854 lecture on the foundations of geometry, Riemann described the Riemann curvature tensor only in intuitive terms. In his 1861 paper, Riemann published the precise analytic formula of the Riemann curvature tensor for the first time.[20] In the textbook by M. Spivak, A Comprehensive Introduction to Differential Geometry, Vol. 2, Publish or Perish, Boston, one finds seven variants of the proof of Riemann's solution of the Paris Academy problem.

Riemann died in 1866 at the age of 40. His collected works fill only one volume. But his ideas, revealing deep connections between analysis, topology, and geometry, profoundly influenced the mathematics and physics of the 20th century. This is described in the beautiful book by K. Maurin, The Riemann Legacy: Riemannian Ideas in Mathematics and Physics of the 20th Century, Kluwer, Dordrecht.

The importance of conformal maps. Conformal mappings are essential for both classifying Riemann surfaces and proving the existence of minimal surfaces with prescribed boundary curves (the problem of Plateau (1801–1883) on soap bubbles spanned by a metallic frame).[21]

> *Conformal mappings play also a fundamental role in modern physics, namely, in string theory and conformal quantum field theory.*

The point is that the principle of critical action in string theory is invariant under conformal mappings (which represent the gauge transformations in string theory). In 2-dimensional conformal quantum field theory, the conformal symmetry strongly restricts the structure of possible correlation functions (i.e., Green's functions). Two Riemann surfaces $\mathcal{M}$ and $\mathcal{N}$ are called conformally equivalent iff there exists a conformal diffeomorphism

$$\chi : \mathcal{M} \to \mathcal{N}.$$

Let $\dim_{\mathbb{R}} \mathcal{M}_g$ denote the real dimension of the space of all compact Riemann surfaces of genus g modulo conformal equivalence. By considering the description of $\mathcal{M}_g$ by real parameters called moduli, Riemann suggested that

$$\dim_{\mathbb{R}} \mathcal{M}_g = 6g - 6 \ \text{ if } g = 2, 3, \ldots, \quad \dim_{\mathbb{R}} \mathcal{M}_1 = \infty, \quad \dim_{\mathbb{R}} \mathcal{M}_0 = 0. \tag{0.22}$$

This was the beginning of the sophisticated theory of moduli spaces which describe the set of given geometric (or algebraic) structures up to equivalence via symmetry groups. The rigorous proof of theorem (0.22) can be given in the setting of Teichmüller spaces.[22]

In what follows, we will pass back to the 4-dimensional space-time manifold $\mathcal{M}^4$ used in Einstein's theory of general relativity.

Two Fundamental Variational Principles

Einstein formulated the final form of the field equations (0.3) for the gravitational field in a meeting of the Prussian Academy of Sciences (Berlin) on November 25, 1915. Five days before, on November 20, 1915 in Göttingen, Hilbert lectured on an axiomatic approach based on a variational principle. A changed version of this

[20] B. Riemann, Mathematical remarks answering a question asked by the famous Paris Academy, pp. 391–404. In: B. Riemann, Collected Mathematical Works, Teubner, Leipzig, and Springer, New York, 1990.

[21] For the solution of the Plateau problem, Ahlfors (1907–1996) was awarded the Fields medal in 1936.

[22] See J. Jost, Compact Riemann Surfaces, Springer, 2006. We also refer to M. Schlichenmaier, An Introduction to Riemann Surfaces, Algebraic Curves, and Moduli Spaces, Springer, New York, 2008.

lecture was published by Hilbert in March 1916. For example, in the special case of a universe without any matter, Hilbert's variational problem reads as

$$\int_{\mathcal{C}} R\sqrt{|g|}\, d^4x = \text{critical!} \tag{0.23}$$

Using the volume form $v_{\mathcal{M}^4} := \sqrt{|g|}\, dx^0 \wedge dx^1 \wedge dx^2 \wedge dx^3$ of the space-time manifold $\mathcal{M}^4$, the variational problem (0.23) corresponds to

$$\boxed{\int_{\mathcal{C}} R \cdot v_{\mathcal{M}^4} = \text{critical!}} \tag{0.24}$$

Here, $\mathcal{C}$ is a nonempty open subset of $\mathcal{M}^4$ with compact closure. The variational problem concerns all smooth metric tensors which are fixed on the boundary $\mathcal{C}$.

Surprisingly enough, the variational problem (0.24) *is the simplest invariant variational problem related to the Riemann curvature tensor.*

We will show in Volume IV that every solution of (0.24) satisfies the Euler–Lagrange equation

$$R_{\alpha\beta} = 0, \qquad \alpha, \beta = 0, 1, 2, 3. \tag{0.25}$$

This coincides with the first fundamental Einstein equation

$$R_{\alpha\beta} = \kappa_G(T_{\alpha\beta} - \tfrac{1}{2} g_{\alpha\beta} T), \qquad \alpha, \beta = 0, 1, 2, 3 \tag{0.26}$$

for vanishing energy-momentum tensor, $T_{\alpha\beta} \equiv 0$. The general case is obtained from (0.24) by adding the source term $\kappa_G(T_{\alpha\beta} - \frac{1}{2} g_{\alpha\beta} T)$.

The principle of critical arc length and geodesic lines. Consider the variational principle

$$\boxed{\int_{\sigma_0}^{\sigma_1} g_{\alpha\beta}(x(\sigma))\, \dot{x}^\alpha(\sigma) \dot{x}^\beta(\sigma)\, d\sigma = \text{critical!}} \tag{0.27}$$

Here, we vary over all smooth curves $C : x = x(\sigma)$ with fixed initial points and fixed endpoints. The solutions of (0.27) satisfy the Euler-Lagrange equations

$$\ddot{x}^\gamma(\sigma) = -\dot{x}^\alpha(\sigma) \Gamma^\gamma_{\alpha\beta}(x(\sigma))\, \dot{x}^\beta(\sigma), \qquad \gamma = 0, 1, 2, 3, \tag{0.28}$$

which represent the equations of motion (0.4) for celestial bodies and light rays in general relativity. For the motion of particles which travel with a velocity smaller than that of light (e.g., the motion of planets), we can also use the variational principle of critical arc length:

$$l(C) = \int_{\sigma_0}^{\sigma_1} \sqrt{g_{\alpha\beta}(x(\sigma))\, \dot{x}^\alpha(\sigma) \dot{x}^\beta(\sigma)}\, d\sigma = \text{critical!} \tag{0.29}$$

The solutions satisfy the equations of motion (0.28) if the parameter σ is the proper time. Einstein wrote in a letter of October 1912:

> At the moment I am only concerned with the gravitational problem and I hope to overcome all the difficulties with the help of a local friend and mathematician, Marcel Grossmann (1878–1936). But it is true that, never in my life, I have worked so hard, and I am filled with a great respect for mathematics. In its subtle parts, I have regarded it, in my simplicity, as pure luxury.

The complete story of the competition between Einstein and Hilbert can be found in the newspaper article by J. Renn, Einstein, Hilbert, and the magic scrap of paper, Frankfurter Allgemeine Zeitung, November 20, 2005 (in German). Renn emphasizes the priority of Einstein's contributions to the creation of the theory of general relativity.

Symmetry and Klein's Invariance Principle in Geometry

> Felix Klein (1849–1925) emphasized the importance of invariants in geometry.
> Sophus Lie (1842–1899) discovered the importance of the linearization principle due to Newton (1643–1727) and Leibniz (1646–1716) for constructing invariants in differential geometry via Lie algebras and Lie groups.
> Élie Cartan (1859–1951) combined the methods of Gauss (1777–1855) and Riemann (1826–1866) in order to describe curvature based on the ideas due to Klein and Lie.
>
> Folklore

Klein's Erlangen program and gauge theory in physics. In the 19th century, numerous new geometries emerged in mathematics (e.g., non-Euclidean geometry and projective geometry). Missing was a general principle for classifying geometries. In 1869, the young German mathematician Felix Klein (1849–1925) and the young Norwegian mathematician Sophus Lie (1842–1899) met each other in Berlin and became close friends. Klein and Lie extensively discussed the classification problem for geometry. They agreed that symmetry groups play a distinguished role. In his 1872 Erlangen program, Felix Klein formulated the following general principle:

Geometry is the invariant theory of transformation groups.

In physics, gauge theory corresponds to a special case of this principle:

Gauge theory studies the invariants of physical fields under both space-time transformations and gauge transformations.

The main goal of gauge theory is the formulation of

- variational principles (principle of critical action) and
- partial differential equations (Euler–Lagrange equations)

which are invariant under both space-time transformations and gauge transformations. Such invariant variational principles and differential equations appear in:

(a) electrodynamics (the Maxwell equations),
(b) the Standard Model in elementary particle physics,
(c) the theory of general relativity (e.g., the Standard Model in cosmology).

In this connection, our main goal is

to create a differential calculus which respects both space-time transformations and gauge transformations.

It was the beautiful idea of Élie Cartan to combine curvature in differential geometry with local symmetry. Nowadays we know that precisely this idea is basic for modern physics, too.

A glance at the history of invariant theory. Invariant theory was created in the 19th century by George Boole (1815–1864), James Sylvester (1814–1897), and Arthur Cayley (1821–1895). Hermann Weyl wrote:[23]

[23] H. Weyl, Invariants, Duke Math. J. **5** (1939), 489–502.

> The theory of invariants came into existence about the middle of the nineteenth century somewhat like Minerva:[24] a grown-up virgin, mailed in the shining armor of algebra, she sprang forth from Cayley's Jovian head. Her Athens over which she ruled and which she served as a tutelary and beneficent goddess was projective geometry.

Cayley was a master in doing long computations and in inventing algorithms. A brief history of invariant theory can be found in the introduction of Peter Olver's book: Classical Invariant Theory, Cambridge University Press, 1999. We also refer to Felix Klein's famous book: Development of Mathematics in the 19th Century, Math. Sci. Press, New York, 1979.

The goal of invariant theory. We are given a mathematical object $\mathcal{O}$ and a symmetry group $\mathcal{G}$ which transforms the object $\mathcal{O}$. The final goal is to construct $\mathcal{G}$-invariants of $\mathcal{O}$. That is, we are looking for quantities which are assigned to $\mathcal{O}$ and which are invariant under the action of the symmetry group $\mathcal{G}$. Moreover, we are interested in determining a complete system of invariants. By definition, a system of $\mathcal{G}$ invariants of $\mathcal{O}$ is called complete iff it uniquely determines the object $\mathcal{O}$ up to symmetry operations contained in the group $\mathcal{G}$.

A typical example. Consider the quadratic equation

$$\boxed{ax^2 + 2bxy + dy^2 = 1, \qquad (x, y) \in \mathbb{R}^2} \tag{0.30}$$

with the real coefficients a, b, d. The theorem of principal axes tells us that there exists a rotation $(x, y) \mapsto (\xi, \eta)$ such that equation (0.30) is transformed into

$$\boxed{\alpha\xi^2 + \beta\eta^2 = 1, \qquad (\xi, \eta) \in \mathbb{R}^2} \tag{0.31}$$

with the real coefficients α and β. Moreover, we have the invariants

$$ad - b^2 = \alpha\beta \quad \text{and} \quad a + d = \alpha + \beta.$$

This immediately implies the following two statements:

(i) Ellipse: If $ad - b^2 > 0$ and $a + d > 0$, then $\alpha > 0$ and $\beta > 0$. Thus, equation (0.30) represents an ellipse.

(ii) Hyperbola: If $ad - b^2 < 0$, then α and β have different signs. For example, $\alpha > 0$ and $\beta < 0$. Thus, equation (0.30) represents a hyperbola.

The point is that the matrix

$$A := \begin{pmatrix} a & b \\ b & d \end{pmatrix}$$

has the eigenvalues α, β. This means that the equation

$$\boxed{\det(A - \lambda I) = \lambda - \text{tr}(A)\lambda + \det(A) = 0} \tag{0.32}$$

has the zeros α and β. This will be studied in Sect. 3.8.1 on page 200 (theorem of principal axes). Equation (0.32) was used by Lagrange (1736–1813) in order to compute the long-time (secular) perturbations of the orbit of a planet under the influence of the other planets. Therefore, this equation is called the secular equation. Gauss used the two invariants $\det(A)$ and $\text{tr}(A)$ in order to define the Gaussian curvature and the mean curvature of a 2-dimensional surface, respectively (see Sect. 9.6.3 on page 628). James Sylvester (1814–1897) said in 1864:[25]

[24] Minerva was the ancient Roman goddess of wisdom and the art, identified with the Greek goddess Athena.

[25] J. Sylvester, Collected Mathematical Papers, Vol. 2, p. 380, Cambridge University Press, 1864.

> As all roads lead to Rome so I find in my own case at least that all algebraic inquiries, sooner or later, end at the Capitol of modern algebra over whose shining portal is inscribed the Theory of Invariants.

Invariant theory is essential for modern physics. In the present volume we will encounter invariant theory again and again.

Einstein's Principle of General Relativity and Invariants – the Geometrization of Physics

> Einstein emphasized the importance of invariants in physics.
>
> Folklore

The components $R^{\alpha}_{\beta\gamma\delta}$ of the Riemann curvature tensor depend on the choice of local space-time coordinates x^0, x^1, x^2, x^3, that is, they depend on the choice of the observer. Recall that Einstein's principle of general relativity tells us that:

Physics is independent of the choice of the observer.

This means that proper physical quantities have to be independent of the choice of local coordinates. As an example, choose two events P_0 and P_1 (e.g., the depart P_0 and the return P_1 of a space ship to earth). Consider the difference

$$\Delta t := \frac{x^0(P_1)}{c} - \frac{x^0(P_0)}{c}$$

where $x^\alpha(P), \alpha = 0, 1, 2, 3$, denotes the local coordinates of the event P. The quantity Δt has the physical dimension of time. But, as a rule, Δt has not an immediate physical meaning because it depends on the choice of the local coordinates for describing the measurements. In contrast to this, the proper time $l(C)/c$ (i.e., the rescaled arc length) introduced by (0.5) on page 8 does not depend on the choice of local coordinates, and hence it possesses an invariant meaning called the proper time interval which can be measured by physical experiments. We refer to the twin paradox considered in Sect. 18.4.3 on page 926.

In the theory of general relativity, transformations of local space-time coordinates are called gauge transformations. Using this term, one can say that

Einstein wanted to construct his theory of general relativity in such a way that it is gauge invariant.

In other words, starting with his philosophical principle of general relativity, Einstein was looking for a mathematical approach which describes invariants in terms of local coordinates. The prototype of such an approach is given by cartography.

Cartography as a paradigm. In cartography, parts of the surface of earth are described by local geographic charts collected in a geographic atlas. The Euclidean coordinates of each chart are called local coordinates of earth. Obviously, geometric properties of the surface of earth do not depend on the choice of the geographic charts, for example, the distance of two points on the surface of earth does not depend on the choice of local coordinates. Geometric properties are invariants with respect to the possible choices of local coordinates.

Intuitively spoken, Einstein looked for higher-dimensional cartography.

His friend – the mathematician Marcel Grossmann (1878–1936) – told him that Riemann generalized Gauss' theory of cartography to higher dimensions and that there exists a well-developed calculus for higher-dimensional manifolds, namely, the Ricci calculus due to Gregorio Ricci-Curbastro (1853–1925). By the help of

Grossmann, Einstein studied the Ricci calculus and he applied it to his theory of gravitation.

The geometrization of physics. Geometry is a mathematical model for describing both invariant geometric properties and their representation by local coordinates. In ancient times, one only considered invariant geometric properties. The description of geometric properties by coordinates dates back to René Descartes (1596–1650). In 1667 Descartes published his "Discours de la méthode" which contains, among a detailed philosophical investigation and its application to the sciences, the foundation of analytic geometry (e.g., the use of Cartesian coordinates).[26]

Einstein geometrized gravitation in his 1915 theory of general relativity. Quantum mechanics was geometrized by Dirac, as a unitary geometry of Hilbert spaces. In the introduction to his book "The Principles of Quantum Mechanics," Clarendon Press, Oxford, 1930, the young Dirac (1902–1984) wrote:

> The important things in the world appear as invariants ... The things we are immediately aware of are the relations of these invariants to a certain frame of reference ... The growth of the use of transformation theory, as applied first to relativity and later to the quantum theory, is the essence of the new method in theoretical physics.

Finally, note that the Standard Model in particle physics starts from a classical field theory which is closely related to the geometry of specific fiber bundles.

Invariant formulation of the fundamental Einstein equations. To begin with, let us introduce the following notation:

- $\mathbf{v} = v^\alpha \partial_\alpha$ (intrinsic tangent vector),
- $\mathbf{g} := g_{\alpha\beta} dx^\alpha \otimes dx^\beta$ (metric tensor field),
- $\langle \mathbf{u}|\mathbf{v}\rangle := \mathbf{g}(\mathbf{u}, \mathbf{v})$ (indefinite Hilbert inner product), $\langle \mathbf{u}|\mathbf{v}\rangle = g_{\alpha\beta} u^\alpha v^\beta$,
- $\mathrm{Ric}(\mathbf{g}) = R_{\alpha\beta} dx^\alpha \otimes dx^\beta$ (Ricci tensor field), $R_{\alpha\beta} = R_{\alpha\kappa\lambda\beta} g^{\kappa\lambda}$,
- $R = R_{\alpha\beta} g^{\alpha\beta}$ (scalar curvature or Ricci curvature – trace of the Ricci tensor),
- $\mathbf{T} = T_{\alpha\beta} dx^\alpha \otimes dx^\beta$ (energy-momentum tensor field),
- $\mathrm{tr}(\mathbf{T}) = T_{\alpha\beta} g^{\alpha\beta}$ (trace of the energy-momentum tensor field), $\mathrm{tr}(\mathbf{T}) \equiv T^\alpha_\alpha$,
- $\mathbf{G} = \mathrm{Ric}(\mathbf{g}) - \frac{1}{2} R\mathbf{g}$ (the Einstein tensor field).

Then, in general relativity, the two fundamental Einstein equations (0.3) and (0.4) on page 8 read as follows in an invariant way:

(i) The equation of motion for the gravitational field:

$$\boxed{\mathrm{Ric}(\mathbf{g}) = \kappa_G(\mathbf{T} - \tfrac{1}{2}\,\mathrm{tr}(\mathbf{T}) \cdot \mathbf{g}).} \tag{0.33}$$

This is equivalent to: $\mathbf{G} = \kappa_G \mathbf{T}$.

(ii) The equation of motion for the trajectories $x = x(\sigma), \sigma_0 \le \sigma \le \sigma_1$, of celestial bodies (e.g., planets, the sun, stars, or galaxies) and light rays:

$$\boxed{\frac{D\dot{x}(\sigma)}{d\sigma} = 0, \qquad \sigma_0 \le \sigma \le \sigma_1.} \tag{0.34}$$

Let us discuss (i).[27] To begin with, fix the event P. Choose the local coordinates $x = (x^0, x^1, x^2, x^3)$. Assume that the event P has the local coordinate x_P. Let us

---

[26] Descartes' Latin name was Cartesius.

[27] The covariant derivative $\frac{D\dot{x}(\sigma)}{d\sigma}$ will be discussed on page 43. The invariance of the two fundamental equations (0.33) and (0.34) follows from the principle of killing indices to be discussed on page 29.

start with the trajectory $x = x(\sigma)$ which passes through the point P, that is, $x(0) = x_P$. Set $v^\alpha := \dot{x}^\alpha(0)$ if $\alpha = 0, 1, 2, 3$. Moreover, we define the differential operator

$$\mathbf{v} := v^\alpha \partial_\alpha.$$

We want to show that this is an invariant notion, that is,

$$\boxed{\mathbf{v} = v^\alpha \partial_\alpha = v^{\alpha'} \partial_{\alpha'}} \tag{0.35}$$

with the tensorial transformation laws $v^{\alpha'} = \frac{\partial x^{\alpha'}(x)}{\partial x^\alpha} v^\alpha$ and $\partial_{\alpha'} = \frac{\partial x^\alpha(x')}{\partial x^{\alpha'}} \partial_\alpha$.

To this end, let us consider different local coordinates $x^{0'}, x^{1'}, x^{2'}, x^{3'}$ given by the transformation $x' = x'(x)$ together with the inverse transformation $x = x(x')$. Then the curve $x = x(\sigma)$ corresponds to $x' = x'(x(\sigma))$. Let $\Theta = \Theta(P)$ be a given invariant real-valued function, that is,

$$\Theta(x(\sigma)) = \Theta(x(x'(\sigma))).$$

Differentiating this with respect to the real parameter σ at the value $\sigma = 0$, we get

$$\frac{\partial \Theta}{\partial x^\alpha} \frac{dx^\alpha(0)}{d\sigma} = \frac{\partial \Theta}{\partial x^\beta} \frac{\partial x^\beta}{\partial x^{\alpha'}} \frac{dx^{\alpha'}(0)}{d\sigma} = \frac{\partial \Theta}{\partial x^{\alpha'}} \frac{dx^{\alpha'}(0)}{d\sigma},$$

by the chain rule. Hence

$$\partial_\alpha \Theta \cdot v^\alpha = \partial_\alpha \Theta \cdot \dot{x}^\alpha(0) = \partial_{\alpha'} \Theta \cdot \dot{x}^{\alpha'}(0) = \partial_{\alpha'} \Theta \cdot v^{\alpha'}.$$

This yields $(v^\alpha \partial_\alpha)\Theta = (v^{\alpha'} \partial_{\alpha'})\Theta$ for all smooth functions Θ. This is the claim (0.35).

Lie's intrinsic tangent vectors on the space-time manifold $\mathcal{M}^4$. Note the following:

> *By definition, tangent vectors of the 4-dimensional space-time manifold $\mathcal{M}^4$ are linear differential operators of first order with constant coefficients.*

Moreover, smooth tangent vector fields on $\mathcal{M}$ are linear differential operators of first order with smooth coefficient functions. This definition dates back to the work of Lie in the second half of the 19th century. For a moment, the definition sounds strange. Let us discuss this. Following Gauss, we have to distinguish between the extrinsic and the intrinsic approach to differential geometry. To illustrate this, consider a 2-dimensional sphere $\mathcal{M}^2$ embedded in the 3-dimensional Euclidean manifold (e.g., the surface of earth).

- Extrinsic tangent vectors: Intuitively, the tangent plane $T_P\mathcal{M}^2$ of the sphere $\mathcal{M}^2$ at the point P is a 2-dimensional plane in the 3-dimensional Euclidean manifold. This plane is orthogonal to the normal vector of the sphere $\mathcal{M}^2$ at the point P. In this setting, the definition of $T_P\mathcal{M}^2$ is based on the surrounding Euclidean manifold $\mathbb{E}^3$. Such a definition is called an extrinsic one. Extrinsic tangent vectors of the sphere are precisely the position vectors of the Euclidean manifold $\mathbb{E}^3$ at the point P which are orthogonal to the external normal unit vector of the sphere at the point P.
- Intrinsic tangent vectors: We will show on page 529 that there exists a natural linear isomorphism

$$T_P\mathcal{M}^2 \simeq D_P\mathcal{M}^2 \tag{0.36}$$

between the linear space $T_P\mathcal{M}^2$ of extrinsic tangent vectors and the linear space $D_P\mathcal{M}^2$ of linear differential operators (of first order with constant coefficients) on the sphere $\mathcal{M}^2$ at the point P. Intuitively, these differential operators act on temperature fields $\Theta : \mathcal{M}^2 \to \mathbb{R}$ on the sphere. The notion of a linear differential operator (of first order) on the sphere does not use the surrounding Euclidean manifold. Therefore, motivated by (0.36), linear differential operators contained in $D_P\mathcal{M}^2$ are called intrinsic tangent vectors.

In the case of a sphere in the 3-dimensional Euclidean manifold, the close relation between extrinsic tangent vectors, the directional derivative, intrinsic tangent vectors, and derivations is discussed in Sect. 8.15 on page 529. Motivated by velocity vectors of fluids on earth (e.g., rivers and oceans), tangent vectors are also called velocity vectors.

In the case of the 4-dimensional space-time manifold $\mathcal{M}^4$, we do not want to use any surrounding space. Therefore, we intrinsically describe tangent vectors by linear differential operators. The symbol $T_P\mathcal{M}^4$ denotes the set of all tangent vectors of the space-time manifold $\mathcal{M}^4$ at the point P. This is called the tangent space of $\mathcal{M}^4$ at the point P. Naturally enough, suppose that $\Theta : \mathcal{M}^4 \to \mathbb{R}$ is a real-valued function on the space-time manifold $\mathcal{M}^4$. Then we define

$$\mathbf{v}_P(\Theta) := v^\alpha(x)\,\partial_\alpha\Theta(x)$$

where x denotes the local coordinate of P. This definition does not depend on the choice of local coordinates. The linear differential operator

$$\mathbf{v}_P := v^\alpha(x)\,\partial_\alpha$$

is called a tangent vector of the space-time manifold $\mathcal{M}^4$ at the point P.

Élie Cartan's algebraization of calculus and infinitesimals. In a heuristic manner, Newton (1643–1727) and Leibniz (1646–1716) used "infinitesimally small" quantities possessing the typical properties $dx > 0$ and "$dx^2 = 0$." In fact, there are *no* real numbers which possess such strange properties.[28] Observe that:

In modern differential geometry, differentials like dx^α are well-defined mathematical objects, namely, linear functionals on the tangent space.

That is, $dx^\alpha \in T_P^*(\mathcal{M}^4)$. Let us discuss this. If $\mathbf{v} = v^\alpha\partial_\alpha$ is an element of the tangent space $T_P\mathcal{M}^4$, then we define[29]

$$dx^\alpha(\mathbf{v}) := v^\alpha.$$

Thus, the linear functional dx^α assigns to the (abstract) tangent vector $\mathbf{v}$ the real value v^α measured by physical experiment. The linear functionals

$$\omega : T_P\mathcal{M}^4 \to \mathbb{R}$$

are precisely given by the linear combinations $\omega = a_\alpha dx^\alpha$ where the symbols a_0, a_1, a_2, a_3 are fixed, but otherwise arbitrary real numbers. These linear functionals form the dual space to the tangent space $T_P\mathcal{M}^4$ which is called the cotangent space $T_P^*\mathcal{M}^4$ of the space-time manifold $\mathcal{M}^4$ at the point P. Tensor products will

[28] In non-standard analysis, one rigorously introduces infinitesimally small numbers ι which are contained in a field extension $^*\mathbb{R}$ of the classical field $\mathbb{R}$ of real numbers, and which have the property that $0 < \iota < \varepsilon$ for all positive real numbers ε. In addition, $\iota^2 > 0$ (see Sect. 4.6 of Vol. II).

[29] Note that $dx^\alpha, \mathbf{v}$, and v^α depend on the choice of the point P.

be thoroughly studied in Sect. 2.1.2. For example, the symbol $dx^\alpha \otimes dx^\beta$ denotes a bilinear map from $T_P\mathcal{M}^4 \times T_P\mathcal{M}^4$ to $\mathbb{R}$ given by

$$(dx^\alpha \otimes dx^\beta)(\mathbf{u},\mathbf{v}) := dx^\alpha(\mathbf{u}) \cdot dx^\beta(\mathbf{v}) = u^\alpha v^\beta$$

for all tangent vectors $\mathbf{u},\mathbf{v} \in T_P(\mathcal{M}^4)$. Élie Cartan introduced the antisymmetric wedge product of differentials by setting

$$\boxed{dx^\alpha \wedge dx^\beta := dx^\alpha \otimes dx^\beta - dx^\beta \otimes dx^\alpha.}$$

Hence

$$(dx^\alpha \wedge dx^\beta)(\mathbf{u},\mathbf{v}) = dx^\alpha(\mathbf{u})\, dx^\beta(\mathbf{v}) - dx^\alpha(\mathbf{v})\, dx^\alpha(\mathbf{u}) = u^\alpha v^\beta - v^\alpha u^\beta.$$

In particular, we obtain

$$\boxed{dx^\alpha \wedge dx^\alpha = 0, \qquad \alpha = 0,1,2,3,} \tag{0.37}$$

which replaces the heuristic relation "$(dx^\alpha)^2 = 0$" used by Newton and Leibniz.

Gauge Transformations

Some important gauge groups. For the convenience of the reader, let us start with summarizing some matrix notation which will be used again and again in this volume. Let $N = 1, 2, \ldots$

- The group $GL(N,\mathbb{C})$ (resp. $GL(N,\mathbb{R})$) consists of all complex (resp. real) invertible $(N \times N)$-matrices. Furthermore, we have the following chain of subgroups: $SU(N) \subseteq U(N) \subseteq GL(N,\mathbb{C})$.
- $G \in U(N)$ iff $G \in GL(N,\mathbb{C})$ and $G^{-1} = G^\dagger$.
- $G \in SU(N)$ iff $G \in U(N)$ and $\det G = 1$.

In particular, the group $U(1)$ consists of all complex numbers z with $|z| = 1$.

Einstein's gauge transformations in the theory of general relativity. Suppose that the event P corresponds to the local coordinates $x = (x^0, x^1, x^2, x^3)$ and $x' = (x^{0'}, x^{1'}, x^{2'}, x^{3'})$. As a preparation, let us introduce the following matrices:

$$v = \begin{pmatrix} v^0 \\ v^1 \\ v^2 \\ v^3 \end{pmatrix}, \quad v' = \begin{pmatrix} v^{0'} \\ v^{1'} \\ v^{2'} \\ v^{3'} \end{pmatrix}, \quad \partial\Theta = \begin{pmatrix} \partial_0\Theta \\ \partial_1\Theta \\ \partial_2\Theta \\ \partial_3\Theta \end{pmatrix}, \quad \partial'\Theta = \begin{pmatrix} \partial_{0'}\Theta \\ \partial_{1'}\Theta \\ \partial_{2'}\Theta \\ \partial_{3'}\Theta \end{pmatrix}, \tag{0.38}$$

$$G(P) = \begin{pmatrix} G_0^{0'} & G_1^{0'} & G_2^{0'} & G_3^{0'} \\ G_0^{1'} & G_1^{1'} & G_2^{1'} & G_3^{1'} \\ G_0^{2'} & G_1^{2'} & G_2^{2'} & G_3^{2'} \\ G_0^{3'} & G_1^{3'} & G_2^{3'} & G_3^{3'} \end{pmatrix}, \quad G(P)^{-1} = \begin{pmatrix} G_{0'}^0 & G_{1'}^0 & G_{2'}^0 & G_{3'}^0 \\ G_{0'}^1 & G_{1'}^1 & G_{2'}^1 & G_{3'}^1 \\ G_{0'}^2 & G_{1'}^2 & G_{2'}^2 & G_{3'}^2 \\ G_{0'}^3 & G_{1'}^3 & G_{2'}^3 & G_{3'}^3 \end{pmatrix}. \tag{0.39}$$

Here, we set $G_\alpha^{\alpha'} := \frac{\partial x^{\alpha'}(x)}{\partial x^\alpha}$ and $G_{\alpha'}^\alpha := \frac{\partial x^\alpha(x')}{\partial x^{\alpha'}}$. By the chain rule,

$$G_\alpha^{\alpha'} G_{\beta'}^\alpha = \frac{\partial x^{\alpha'}}{\partial x^\alpha}\frac{\partial x^\alpha}{\partial x^{\beta'}} = \frac{\partial x^{\alpha'}}{\partial x^{\beta'}} = \delta_{\beta'}^{\alpha'}.$$

Therefore, the matrix $(G_{\alpha'}^\alpha)$ is the inverse matrix to $G(P) := (G_\alpha^{\alpha'})$. By (0.35), we get $v^{\alpha'} = G_\alpha^{\alpha'} v^\alpha$. Hence

$$\boxed{v' = G(P)v.} \tag{0.40}$$

This is called the gauge transformation of the components of the tangent vector $\mathbf{v} = v^\alpha \partial_\alpha = v^{\alpha'} \partial_{\alpha'}$. The matrix $G(P)$ is contained in the Lie group $GL(4, \mathbb{R})$. The transformation formula $\partial_{\alpha'}\Theta = \frac{\partial x^\alpha}{\partial x^{\alpha'}} \partial_\alpha \Theta = G_{\alpha'}^\alpha \partial_\alpha \Theta$ reads as

$$\boxed{(\partial'\Theta)^d = (\partial\Theta)^d \, G(P)^{-1}.} \tag{0.41}$$

For example, this implies the invariance relation (0.35). In fact, we get

$$(\partial'\Theta)^d v' = (\partial\Theta)^d G(P)^{-1} G(P) v = (\partial\Theta)^d v = \mathbf{v}(\Theta).$$

Einstein's gauge transformations in the special theory of relativity. In the theory of special relativity, gauge transformations correspond to a change of inertial systems. Then the matrix $G(P)$ from (0.40) represents a Lorentz transformation (see Chap. 18).

Dirac's gauge transformations in quantum mechanics. We want to discuss Dirac's quotation mentioned on page 23. To this end, let X be a complex n-dimensional Hilbert space. The unit vectors ψ of X are called physical states. Choose a complete orthonormal system $e_1, \ldots, e_n$ of X. Then the Fourier expansion reads as

$$\psi = \sum_{j=1}^n \langle e_j | \psi \rangle e_j.$$

In this setting, we have to distinguish between

- the (invariant) physical state ψ, and
- the local coordinates $\langle e_j|\psi\rangle, j = 1, \ldots, n$, of ψ (also called the Feynman probability amplitudes of ψ).

In terms of physics, the choice of the orthonormal basis $e_1, \ldots, e_n$ corresponds to a measurement device. If the given quantum particle is in the state ψ, then the real number

$$|\langle \psi | e_j \rangle|^2$$

is the probability for measuring the particle in the state e_j of the measurement device.

In Dirac's setting of quantum mechanics, a gauge transformation corresponds to a change of the measurement device.

That is, in terms of mathematics, we pass from the orthonormal basis $e_1, \ldots, e_n$ to the orthonormal basis $e_{1'}, \ldots, e_{n'}$. This basis change can be described by a unitary transformation[30]

$$G : X \to X$$

---

[30] Recall that the linear operator $G : X \to X$ is called unitary iff it is bijective and it preserves inner products. The group of all unitary transformations of the Hilbert space X is denoted by $U(X)$ (unitary group of X).

defined by $Ge_j := e_{j'}$ for $j = 1, \ldots, n$. In this sense, quantum mechanics corresponds to the unitary geometry of the Hilbert space X. By definition, this is the theory of invariants under the unitary transformation group $U(X)$. For example, the transition amplitude $\langle\psi|\varphi\rangle$ is a unitary invariant. We also define

$$dx^j(\psi) := \langle e_j|\psi\rangle, \qquad j = 1, \ldots, n.$$

Here, the linear functional $dx^j : X \to \mathbb{C}$ assigns to the physical state ψ the local coordinate $\langle e_j|\psi\rangle$ (probability amplitude) which depends on the choice of the measurement device.

This argument can be immediately generalized to complex infinite-dimensional separable Hilbert spaces. Such spaces possess a countable orthonormal basis $e_1, e_2, \ldots$ (also called complete orthonormal system).

In the theory of special relativity, gauge transformations correspond to a change of inertial systems, which corresponds to a change of a pseudo-orthonormal basis of the Minkowski space which is an indefinite Hilbert space (see Chap. 18).

Yang's gauge transformations via local phase factors. We are given the function $\psi : \mathbb{R}^4 \to \mathbb{C}$. Consider the transformation formula

$$\psi_+(x) = G(x)\psi(x), \qquad x \in \mathbb{R}^4 \tag{0.42}$$

where $G(x)$ is an element of the Lie group $U(1)$. Explicitly,

$$G(x) = e^{i\chi(x)}, \qquad x \in \mathbb{R}^4.$$

Here, the so-called phase $\chi(x)$ is a real number which depends on the choice of the space-time point x in $\mathbb{R}^4$. Therefore, $G(x)$ is called a local phase factor according to Yang. In terms of physics, the function ψ is the wave function of an electron, and the map $\psi(x) \mapsto \psi_+(x)$ given by (0.42) is called a gauge transformation. In the Standard Model in particle physics, this situation is generalized in the following way:

- The function $\psi : \mathbb{R}^4 \to \mathbb{C}^N$ describes the basic particles (i.e., quarks and leptons),
- and the local phase factor $G(x)$ is an element of the Lie group $GL(N, \mathbb{C})$.

More precisely, we have $G(x) \in \mathcal{G}$ for all $x \in \mathbb{R}^4$. Here, $\mathcal{G}$ is a subgroup of $GL(N, \mathbb{C})$ which is isomorphic to the Lie group $U(1) \times SU(2) \times SU(3)$. The latter group is called the gauge group of the Standard Model in particle physics. Summarizing, we will encounter the following groups as gauge groups:

- $SO(1,3)$ – Lorentz group (Einstein's theory of special relativity),
- $U(1)$ (Maxwell's theory of electromagnetism),
- $SU(2)$ (the original Yang–Mills theory for the local isospin phase factor),
- $U(1) \times SU(2)$ (electroweak interaction),
- $SU(3)$ (strong interaction – quantum chromodynamics),
- $U(1) \times SU(2) \times SU(3)$ (Standard Model in particle physics – combining electroweak interaction with strong interaction).

Observe that the gauge groups $U(1), SU(2)$, and $SU(3)$, as well as their direct product $U(1) \times SU(2) \times SU(3)$ are compact Lie groups, whereas the gauge group $SO(1,3)$ is not compact, but only locally compact. The representations of compact groups are much easier to handle than the representations of non-compact, locally compact Lie groups.

Furthermore, note the following. Let $\mathrm{Diff}(\mathcal{M}^4)$ denote the group of all diffeomorphisms $\chi : \mathcal{M}^4 \to \mathcal{M}^4$ of the space-time manifold $\mathcal{M}^4$ onto itself.

- According to Einstein's principle of general relativity, physical quantities have to be invariant under the group $\mathrm{Diff}(\mathcal{M}^4)$. In contrast to the finite-dimensional Lie groups $U(1), SU(2), SU(3), SO(1,3)$, the group $\mathrm{Diff}(\mathcal{M}^4)$ is an infinite-dimensional generalized Lie group.
- String theory is based on conformal symmetry (like the theory of minimal surfaces, the theory of Riemann surfaces, and the conformal quantum field theory). On an infinitesimal level, this is described by an infinite-dimensional Lie algebra called the Virasoro algebra.

Therefore, the theory of finite-dimensional and infinite-dimensional groups (resp. Lie algebras) and their invariants play a fundamental role in modern physics. Important contributions to this topic were made by Élie Cartan, Weyl (compact Lie groups), Wigner, Bargmann, Gelfand, and Harrish–Chandra (noncompact groups), and Victor Kac (infinite-dimensional Lie algebras).

Construction of Invariants by the Principle of Killing Indices

> Mnemonically, the principle of killing indices works on its own.
>
> Folklore

Fix $n, m = 1, 2, \ldots$ In what follows we will sum over equal upper and lower Greek indices from 0 to 3. Let

$$T^{\alpha_1 \ldots \alpha_m}_{\beta_1 \ldots \beta_n}, \qquad S^{\beta_1 \ldots \beta_n}_{\alpha_1 \ldots \alpha_m},$$

and $U^{\gamma_1 \ldots \gamma_r; \alpha_1 \ldots \alpha_m}_{\delta_1 \ldots \delta_s; \beta_1 \ldots \beta_n}$ be tensorial families on the 4-dimensional space-time manifold $\mathcal{M}^4$. Set

$$V^{\gamma_1 \ldots \gamma_r}_{\delta_1 \ldots \delta_s} := U^{\gamma_1 \ldots \gamma_r; \alpha_1 \ldots \alpha_m}_{\delta_1 \ldots \delta_s; \beta_1 \ldots \beta_n} S^{\beta_1 \ldots \beta_n}_{\alpha_1 \ldots \alpha_m}.$$

Then, we have the following three very useful principles for constructing invariants:

(K1) $T^{\alpha_1 \ldots \alpha_m}_{\beta_1 \ldots \beta_n} S^{\beta_1 \ldots \beta_n}_{\alpha_1 \ldots \alpha_m}$ is an invariant function on $\mathcal{M}^4$.

(K2) $V^{\gamma_1 \ldots \gamma_r}_{\delta_1 \ldots \delta_s}$ is a tensorial family (r-fold contravariant and s-fold covariant).

(K3) $T^{\beta_1 \ldots \beta_m}_{\alpha_1 \ldots \alpha_n} \, dx^{\alpha_1} \otimes \cdots \otimes dx^{\alpha_n} \otimes \partial_{\beta_1} \otimes \cdots \otimes \partial_{\beta_m}$ is an invariant mathematical object denoted by $\mathbf{T}$.

Let us explain the meaning of the mathematical object $\mathbf{T}$.

(i) Consider first the special case where

$$\mathbf{T} := T^{\alpha\beta}_{\gamma} dx^{\gamma} \otimes \partial_{\alpha} \otimes \partial_{\beta}.$$

Fixing the point P of the manifold $\mathcal{M}^4$, we get

$$\mathbf{T}_P := T^{\alpha\beta}_{\gamma}(P)\, dx^{\gamma} \otimes \partial_{\alpha} \otimes \partial_{\beta}.$$

As usual, the tensor product $\partial_\alpha \otimes \partial_\beta$ of two differential operators acts on smooth functions $(x, y) \mapsto f(x, y)$ by setting

$$(\partial_\alpha \otimes \partial_\beta) f(x, y) = \frac{\partial}{\partial x^\alpha} \frac{\partial}{\partial y^\beta} f(x, y) = \frac{\partial^2 f(x, y)}{\partial x^\alpha \partial y^\beta}.$$

Here, $x = (x^0, x^1, x^2, x^3)$ and $y = (y^0, y^1, y^2, y^3)$. Now choose a tangent vector $\mathbf{u} := u^\sigma(P)\partial_\sigma$ of $\mathcal{M}^4$ at the point P. Then

$$\mathbf{T}_P(\mathbf{u}) = T^{\alpha\beta}_{\gamma}(P)\, dx^{\gamma}(\mathbf{u}) \otimes \partial_\alpha \otimes \partial_\beta = T^{\alpha\beta}_{\gamma}(P) u^{\gamma}(P)\, \partial_\alpha \otimes \partial_\beta.$$

Finally, we obtain

$$\boxed{\mathbf{T}_P(\mathbf{u}) = T_\gamma^{\alpha\beta}(P)u^\gamma(P)\,\frac{\partial^2}{\partial x^\alpha \partial y^\beta}.}$$

This is a linear differential operator of second order with coefficients which depend on the point P.

(ii) Consider now the general case of (K3). For all tangent vectors $\mathbf{u}_1, \ldots, \mathbf{u}_n$ of the manifold $\mathcal{M}^4$ at the fixed point P, we get the following linear differential operator of mth-order:

$$\mathbf{T}_P(\mathbf{u}_1, \ldots, \mathbf{u}_n) = T^{\alpha_1 \ldots \alpha_m}_{\beta_1 \ldots \beta_n}(P)\, u_1^{\beta_1}(P) \cdots u_n^{\beta_n}(P)\, \frac{\partial^m}{\partial x_1^{\alpha_1} \cdots \partial x_m^{\alpha_m}}.$$

Here, $\mathbf{u}_j = u_j^\alpha(P)\partial_\alpha$ and $x_k := (x_k^0, x_k^1, x_k^2, x_k^3)$ where $j = 1, \ldots, n$. Moreover, $k = 1, \ldots, m$. The map

$$P \mapsto \mathbf{T}_P$$

is called a tensor field of type (m, n) on the space-time manifold $\mathcal{M}^4$. The elementary proof of statements (K1) through (K3) based on the chain rule will be given in Chap. 8.

Observe the following. By the Einstein convention, we sum over equal upper and lower Greek indices from 0 to 3.

The essential feature is that both the expressions from (K1) *and* (K3) *have no free indices anymore.*

The summation kills the indices. Therefore, we summarize (K1) through (K3) under the slogan "principle of killing indices."

Examples. Taking for granted that $g_{\alpha\beta}, g^{\alpha\beta}$, $T_{\alpha\beta}$, and $R^\delta_{\alpha\beta\gamma}$ are tensorial families, which will be shown in Chap. 8, it follows from (K2) that the following expressions are tensorial families:

- $R_{\alpha\beta\gamma\delta} = R^\sigma_{\alpha\beta\gamma} g_{\sigma\delta}$,
- $R_{\alpha\delta} = R_{\alpha\beta\gamma\delta} g^{\beta\gamma}$,
- $R = R_{\alpha\beta} g^{\alpha\beta} = R^\alpha_\alpha$ (trace).

In particular, R is an invariant function on the space-time manifold $\mathcal{M}^4$, by (K1). In addition, it follows from (K3) that the following expressions are invariantly defined tensor fields on the space-time manifold $\mathcal{M}^4$:

- $\mathbf{g} := g_{\alpha\beta}\, dx^\alpha \otimes dx^\beta$ (metric tensor field),
- $\mathrm{Ric}(\mathbf{g}) := R_{\alpha\beta}\, dx^\alpha \otimes dx^\beta$ (Ricci tensor field),
- $\mathbf{R} := R^\delta_{\alpha\beta\gamma}\, dx^\alpha \otimes dx^\beta \otimes dx^\gamma \otimes \partial_\delta$ (Riemann curvature tensor field),
- $\mathcal{R} := R_{\alpha\beta\gamma\delta}\, dx^\alpha \otimes dx^\beta \otimes dx^\gamma \otimes dx^\delta$ (metric Riemann curvature tensor field).

In what follows, we want to discuss the geometric meaning of both the Riemann curvature tensor field and the metric Riemann curvature tensor field.

Levi-Civita's parallel transport of velocity vectors. In 1917, Levi-Civita (1873–1941) published a fundamental paper.[31]

[31] T. Levi-Civita, The notion of parallel transport in manifolds and its geometric consequences for the Riemann curvature, Rend. Palermo **42** (1917), 73–205 (in Italian).

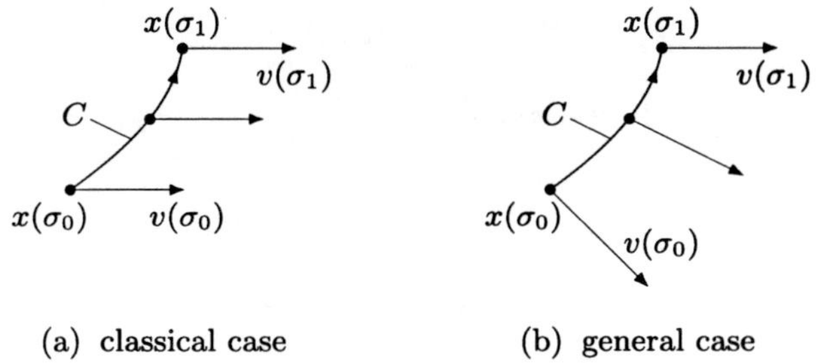

Fig. 0.2. Parallel transport of the vector v along the curve C

He proved that the Riemann curvature tensor can be obtained by parallel transport of velocity vectors along a sufficiently small closed curve.[32]

This observation was crucial for the development of modern differential geometry and modern physics. Let us discuss this. Consider a curve C on the space-time manifold $\mathcal{M}^4$. With respect to a given local coordinate system, the curve C is given by the equation

$$x = x(\sigma), \qquad \sigma_0 \leq \sigma \leq \sigma_1.$$

In what follows we will write

$$v = \begin{pmatrix} v^0 \\ v^1 \\ v^2 \\ v^3 \end{pmatrix} \qquad \text{and} \qquad \mathbf{v} = v^\alpha \partial_\alpha.$$

That is, the components of the velocity vector $\mathbf{v}$ (which lives in the tangent space $T_P\mathcal{M}^4$) are the entries of the column matrix v (which represents a vector in $\mathbb{R}^4$). If the curve C is a geodesic line, then it satisfies the differential equation

$$\ddot{x}^\gamma(\sigma) = -\dot{x}^\alpha(\sigma)\Gamma^\gamma_{\alpha\beta}(x(\sigma))\,\dot{x}^\beta(\sigma), \qquad \sigma_0 \leq \sigma \leq \sigma_1, \ \ \gamma = 0, 1, 2, 3.$$

Setting $v^\gamma := \dot{x}^\gamma(\sigma)$, we get the differential equation

$$\dot{v}^\gamma(\sigma) = -\dot{x}^\alpha(\sigma)\Gamma^\gamma_{\alpha\beta}(x(\sigma)) \cdot v^\beta(\sigma), \quad \sigma_0 \leq \sigma \leq \sigma_1, \ \ \gamma = 0, 1, 2, 3. \tag{0.43}$$

Introducing the real (4×4)-matrix $\mathcal{A}_\alpha := (\Gamma^\gamma_{\alpha\beta})$, equation (0.43) can be written as

$$\boxed{\dot{v}(\sigma) = -\dot{x}^\alpha(\sigma)\mathcal{A}_\alpha(x(\sigma)) \cdot v(\sigma), \qquad \sigma_0 \leq \sigma \leq \sigma_1.} \tag{0.44}$$

This is called the equation of parallel transport. Now to the point. We replace the geodesic line by a general curve C, and we consider general solutions $v = v(\sigma)$ of (0.44). In this setting, we say that the velocity vector $v(\sigma)$ is parallel transported

[32] Sophus Lie motivated his approach to differential geometry by the physical picture of the flow of fluid particles on, say, a sphere (e.g., an ocean on earth). In this setting, tangent vectors of the sphere represent velocity vectors of fluid particles. Therefore, in this monograph, tangent vectors of manifolds will be synonymously called velocity vectors.

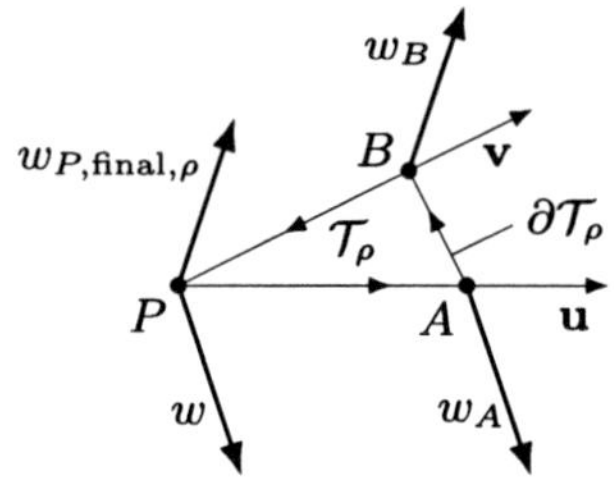

Fig. 0.3. Parallel transport of the vector w along the loop $\partial\mathcal{T}_\rho$

along the curve C. Note that this is a generalization of the classical parallel transport of vectors in the Euclidean manifold (Fig. 0.2). Furthermore, our definition of parallel transport is chosen in such a way that, as a special case, the tangent vectors of a geodesic line C are parallel along this line. The simple geometric meaning of the parallel transport of velocity vectors on a sphere (based on orthogonal projection) will be discussed in Sect. 9.5 on page 593.

Now let us use the notion of parallel transport in order to define the linear operator $\Pi_C : T_P\mathcal{M}^4 \to T_P\mathcal{M}^4$ given by

$$\boxed{\Pi_C \mathbf{v}(\sigma_0) := \mathbf{v}(\sigma_1).}$$

More precisely, we proceed as follows (Fig. 0.2(b)):

- We are given the velocity vector $\mathbf{v}(\sigma_0) \in T_P\mathcal{M}^4$. Choosing a fixed local coordinate system, the vector $\mathbf{v}(\sigma_0)$ corresponds to the coordinate matrix $v(\sigma_0)$.
- Solving the differential equation (0.44) of parallel transport, we get the function $v = v(\sigma)$ along the curve $C : x = x(\sigma)$.
- Finally, by definition, the velocity vector $\mathbf{v}(\sigma_1)$ corresponds to $v(\sigma_1)$.

The following fact is crucial. We will prove in Chap. 8 that the transformation law for the Christoffel symbols implies that:

The definition of the operator Π_C does not depend on the choice of the local coordinate system.

In other words, the parallel transport of a velocity vector along a curve C is a geometric property of the space-time manifold $\mathcal{M}^4$. In terms of physics, this corresponds to the transport of physical information.

The geometric meaning of the Riemann curvature tensor via parallel transport. Fix the point P of the space-time manifold $\mathcal{M}^4$. We are given the two tangent vectors $\mathbf{u}, \mathbf{v} \in T_P\mathcal{M}^4$ at the point P. It is our goal to construct the operator

$$\mathbf{F}_P(\mathbf{u}, \mathbf{v}) : T_P\mathcal{M}^4 \to T_P\mathcal{M}^4$$

which measures the curvature of $\mathcal{M}^4$ at the point P with respect to the plane spanned by $\mathbf{u}$ and $\mathbf{v}$. To this end, choose a fixed, but otherwise arbitrary chart (i.e., a local coordinate system), and consider the situation pictured in Fig. 0.3.

- Fix the scaling factor $\varrho > 0$. Let $\mathcal{T}_\varrho$ denote the triangle spanned by the vectors ϱu and ϱv, and let $\partial\mathcal{T}_\varrho$ denote the positively oriented boundary curve of $\mathcal{T}_\varrho$.
- Parallel transport of the given vector $\mathbf{w}$ at the initial point P along the closed curve $PABP = \partial\mathcal{T}_\varrho$ yields the vector $w_{P,\mathrm{final};\varrho} = \Pi_{\partial\mathcal{T}_\varrho} w$ at the final point P.

The following hold.

- Let $\mathrm{meas}(\mathcal{T}_\varrho)$ denote the Euclidean measure of the triangle $\mathcal{T}_\varrho$. There exists the limit
$$\lim_{\varrho \to 0} \frac{\Pi_{\partial \mathcal{T}_\varrho} w}{\mathrm{meas}(\mathcal{T}_\varrho)} = w_{P,\mathrm{final}}.$$
- The tangent vector $\mathbf{w}_{P,\mathrm{final}}$ corresponding to the coordinate matrix $w_{P,\mathrm{final}}$ does not depend on the choice of the local coordinates.

We define
$$\boxed{\mathbf{F}_P(\mathbf{u},\mathbf{v})\mathbf{w} := \mathbf{w}_{P,\mathrm{final}}.}$$
The operator $\mathbf{w} \mapsto \mathbf{F}(\mathbf{u},\mathbf{v})\mathbf{w}$ is called the Riemann curvature operator.

In physics, we want to use real numbers which can be measured in physical experiments. In order to pass from the Riemann curvature operator to real numbers, it is quite natural to use the (indefinite) inner product on the tangent space $T_P\mathcal{M}^4$ of the space-time manifold at the point P. Explicitly, fix a tangent vector $\mathbf{z} \in T_P\mathcal{M}^4$, and consider the inner product
$$\boxed{\mathcal{R}_P(\mathbf{u},\mathbf{v};\mathbf{w},\mathbf{z}) := \langle \mathbf{F}_P(\mathbf{u},\mathbf{v})\mathbf{w}|\mathbf{z}\rangle.} \tag{0.45}$$
This is equal to $\mathbf{g}_P(\mathbf{F}_P(\mathbf{u},\mathbf{v})\mathbf{w},\mathbf{z})$. The map
$$(\mathbf{u},\mathbf{v},\mathbf{w},\mathbf{z}) \mapsto \mathcal{R}_P(\mathbf{u},\mathbf{v},\mathbf{w},\mathbf{z})$$
is a 4-linear map of the form
$$\mathcal{R}_P : T_P\mathcal{M}^4 \times T_P\mathcal{M}^4 \times T_P\mathcal{M}^4 \times T_P\mathcal{M}^4 \to \mathbb{R}.$$
This map is called the metric Riemann curvature tensor.

Let $\mathbf{u} = u^\alpha \partial_\alpha$ (together with similar representations of $\mathbf{v}, \mathbf{w}$, and $\mathbf{z}$). With respect to local coordinates, we get the following symmetric formulas:

- $\mathbf{F}(\mathbf{u},\mathbf{v})\mathbf{w} = (R^\kappa_{\alpha\beta\gamma}\, u^\alpha v^\beta w^\gamma)\, \partial_\kappa$,
- $\mathcal{R}_P(\mathbf{u},\mathbf{v},\mathbf{w},\mathbf{z}) = R_{\alpha\beta\gamma\delta}\, u^\alpha v^\beta w^\gamma z^\delta$,
- $\mathrm{Ric}(\mathbf{u},\mathbf{z}) := \langle \mathbf{F}(\mathbf{u},\partial_\beta)\partial_\gamma|\mathbf{z}\rangle\, g^{\beta\gamma} = R_{\alpha\beta}u^\alpha v^\beta$ (averaging).

Here, the real coefficients are given by

- $R^\delta_{\alpha\beta\gamma} := dx^\delta(\mathbf{F}(\partial_\alpha,\partial_\beta)\partial_\gamma)$,
- $R_{\alpha\beta\gamma\delta} := \mathcal{R}_P(\partial_\alpha,\partial_\beta,\partial_\gamma,\partial_\delta)$,
- $R_{\alpha\delta} := R_{\alpha\beta\gamma\delta}g^{\beta\gamma}$.

Symmetries of the Riemann curvature tensor. We have:

- $\mathbf{F}(\mathbf{u},\mathbf{v})\mathbf{w} = -\mathbf{F}(\mathbf{v},\mathbf{u})\mathbf{w}$,
- $\mathbf{F}(\mathbf{u},\mathbf{v})\mathbf{w} + \mathbf{F}(\mathbf{v},\mathbf{w})\mathbf{u} + \mathbf{F}(\mathbf{w},\mathbf{u})\mathbf{v} = 0$ (cyclic permutation),
- $\mathcal{R}_P(\mathbf{u},\mathbf{v};\mathbf{w},\mathbf{z}) = -\mathcal{R}_P(\mathbf{v},\mathbf{u};\mathbf{w},\mathbf{z})$,
- $\mathcal{R}_P(\mathbf{u},\mathbf{v};\mathbf{w},\mathbf{z}\rangle = -\mathcal{R}_P(\mathbf{u},\mathbf{v};\mathbf{z},\mathbf{w}\rangle$,
- $\mathcal{R}_P(\mathbf{u},\mathbf{v};\mathbf{w},\mathbf{z}) = \mathcal{R}_P(\mathbf{w},\mathbf{z};\mathbf{u},\mathbf{v})$,
- $\mathrm{Ric}(\mathbf{g})(\mathbf{u},\mathbf{v}) = \mathrm{Ric}(\mathbf{g})(\mathbf{v},\mathbf{u})$.

For the coefficients $R_{\alpha\beta\gamma\delta}$ and $R_{\alpha\beta}$, this yields the symmetry relations summarized on page 14.

Riemann's sectional curvature and the geometric meaning of the Riemann curvature tensor. Let $\mathbf{u} = u^\alpha\partial_\alpha$ and $\mathbf{v} = v^\beta\partial_\beta$ be linearly independent tangent vectors of the space-time manifold $\mathcal{M}^4$ at the point P. The sectional curvature of $\mathcal{M}^4$ at the point P is defined by [33]

[33] Naturally enough, we assume that $v_P(\mathbf{u},\mathbf{v}) \neq 0$.

$$\boxed{K_P(\mathbf{u},\mathbf{v}) := \frac{\mathcal{R}_P(\mathbf{u},\mathbf{v},\mathbf{v},\mathbf{u})}{\upsilon_P(\mathbf{u},\mathbf{v})^2}} \tag{0.46}$$

where $\upsilon_P(\mathbf{u},\mathbf{v})^2 := \mathbf{g}_P(\mathbf{u},\mathbf{u})\mathbf{g}_P(\mathbf{v},\mathbf{v}) - \mathbf{g}_P(\mathbf{u},\mathbf{v})^2$. If $\mathbf{u}$ and $\mathbf{v}$ are space-like vectors, then $\upsilon_P(\mathbf{u},\mathbf{v})$ is the surface area of the parallelogram spanned by the two vectors $\mathbf{u}$ and $\mathbf{v}$ at the point P. It turns out that this sectional curvature only depends on the 2-dimensional plane $\mathcal{P}$ spanned by the tangent vectors $\mathbf{u}$ and $\mathbf{v}$.

- If two 2-dimensional submanifolds $\mathcal{S}$ and $\mathcal{S}'$ of the space-time manifold $\mathcal{M}^4$ have the same tangent plane at the point P, then they possess the same sectional curvature at the point P.
- If the plane $\mathcal{P}$ is space-like, then the sectional curvature coincides with the Gaussian curvature of $\mathcal{S}$ and $\mathcal{S}'$ at the point P.

This sectional curvature was introduced by Riemann in his seminal lecture "On the hypotheses which lie at the foundation of geometry" in 1854.[34]

> *It was the idea of Riemann to describe the curvature of a higher-dimensional manifold $\mathcal{M}$ at the point P by studying the Gaussian curvature of all possible two-dimensional submanifolds $\mathcal{M}^2$ of $\mathcal{M}$ at the point P.*

Let $\mathcal{K}(\mathbf{u},\mathbf{v})_P := \upsilon_P(\mathbf{u},\mathbf{v})^2 K_P(\mathbf{u},\mathbf{v})$. Then

$$\begin{aligned}\mathcal{R}(\mathbf{u},\mathbf{v},\mathbf{w},\mathbf{z}) = {} & \mathcal{K}(\mathbf{u}+\mathbf{z},\mathbf{v}+\mathbf{w}) - \mathcal{K}(\mathbf{u}+\mathbf{z},\mathbf{v}) - \mathcal{K}(\mathbf{u}+\mathbf{z},\mathbf{w}) \\ & -\mathcal{K}(\mathbf{u},\mathbf{v}+\mathbf{w}) - \mathcal{K}(\mathbf{z},\mathbf{v}+\mathbf{w}) + \mathcal{K}(\mathbf{u},\mathbf{w}) + \mathcal{K}(\mathbf{z},\mathbf{v}) \\ & -\mathcal{K}(\mathbf{v}+\mathbf{z},\mathbf{u}+\mathbf{w}) + \mathcal{K}(\mathbf{v}+\mathbf{z},\mathbf{u}) + \mathcal{K}(\mathbf{v}+\mathbf{z},\mathbf{w}) \\ & -\mathcal{K}(\mathbf{v},\mathbf{u}+\mathbf{w}) + \mathcal{K}(\mathbf{z},\mathbf{u}+\mathbf{w}) - \mathcal{K}(\mathbf{v},\mathbf{w}) - \mathcal{K}(\mathbf{z},\mathbf{u}).\end{aligned}$$

This key formula tells us that the sectional curvature determines the Riemann curvature tensor which describes the whole curvature.

Two Fundamental Approaches to Differential Geometry

There exist the following two different approaches to differential geometry, namely,

(I) Gauss' approach by means of symmetric tensors, and

(II) Élie Cartan's approach by means of antisymmetric tensors (also called differential forms).

The Einstein equation $\mathrm{Ric}(\mathbf{g}) = \kappa_G(\mathbf{T} - \frac{1}{2}\,\mathrm{tr}(\mathbf{T})\cdot\mathbf{g})$ (for the motion of the gravitational field by means of the symmetric Ricci tensor $\mathrm{Ric}(\mathbf{g})$) is formulated in the spirit of Gauss (see page 23). In what follows, we will study Cartan's approach based on the structural equation and its integrability condition (the Bianchi identity). We will sketch the following ideas:

- Yang's matrix trick,
- Cartan's local structural equation, and
- Cartan's global structural equation.

Modern differential geometry is based on Cartan's approach. The decisive advantage of Cartan's approach is that it allows the use of symmetry groups (also called gauge groups) in a very flexible way.

---

[34] B. Riemann, Über die Hypothesen, welche der Geometrie zugrundeliegen, Göttinger Abhandlungen **13** (1854), 272–287 (in German). An English translation can be found in M. Spivak, A Comprehensive Introduction to Differential Geometry, Vol. 2, Publish or Perish, Boston.

Yang's Matrix Trick

While preparing a lecture on Einstein's general relativity theory in the 1960s, Yang discovered that the fundamental equations for the components of the Riemann curvature tensor,

$$\boxed{R^{\delta}_{\alpha\beta\gamma} := \partial_\alpha \Gamma^{\delta}_{\beta\gamma} - \partial_\beta \Gamma^{\delta}_{\alpha\gamma} + \Gamma^{\delta}_{\alpha\mu}\Gamma^{\mu}_{\beta\gamma} - \Gamma^{\delta}_{\beta\mu}\Gamma^{\mu}_{\alpha\gamma},} \tag{0.47}$$

coincide with the Yang–Mills field equations

$$\boxed{\mathcal{F}_{\alpha\beta} := \partial_\alpha \mathcal{A}_\beta - \partial_\beta \mathcal{A}_\alpha + [\mathcal{A}_\alpha, \mathcal{A}_\beta]} \tag{0.48}$$

with the Lie product $[\mathcal{A}_\alpha, \mathcal{A}_\beta] := \mathcal{A}_\alpha\mathcal{A}_\beta - \mathcal{A}_\beta\mathcal{A}_\alpha$. Here, the indices $\alpha, \beta, \gamma, \delta$ run from 0 to 3. Observe that formula (0.47) can be regarded as a generalization of Gauss' theorema egregium to higher dimensions. In order to obtain Yang's result, let us introduce the following matrices:

$$\mathcal{A}_\alpha := (\Gamma^{\delta}_{\alpha\gamma}), \qquad \mathcal{F}_{\alpha\beta} := (R^{\delta}_{\alpha\beta\gamma})$$

where the upper index δ numbers the rows, and the lower index γ numbers the columns. Explicitly, this means the following:

(i) Christoffel matrices (connection matrices):

$$\mathcal{A}_\alpha := \begin{pmatrix} \Gamma^0_{\alpha 0} & \Gamma^0_{\alpha 1} & \Gamma^0_{\alpha 2} & \Gamma^0_{\alpha 3} \\ \Gamma^1_{\alpha 0} & \Gamma^1_{\alpha 1} & \Gamma^1_{\alpha 2} & \Gamma^1_{\alpha 3} \\ \Gamma^2_{\alpha 0} & \Gamma^2_{\alpha 1} & \Gamma^2_{\alpha 2} & \Gamma^2_{\alpha 3} \\ \Gamma^3_{\alpha 0} & \Gamma^3_{\alpha 1} & \Gamma^3_{\alpha 2} & \Gamma^3_{\alpha 3} \end{pmatrix}.$$

(ii) Riemann curvature matrices:

$$\mathcal{F}_{\alpha\beta} := \begin{pmatrix} R^0_{\alpha\beta 0} & R^0_{\alpha\beta 1} & R^0_{\alpha\beta 2} & R^0_{\alpha\beta 3} \\ R^1_{\alpha\beta 0} & R^1_{\alpha\beta 1} & R^1_{\alpha\beta 2} & R^1_{\alpha\beta 3} \\ R^2_{\alpha\beta 0} & R^2_{\alpha\beta 1} & R^2_{\alpha\beta 2} & R^2_{\alpha\beta 3} \\ R^3_{\alpha\beta 0} & R^3_{\alpha\beta 1} & R^3_{\alpha\beta 2} & R^3_{\alpha\beta 3} \end{pmatrix}.$$

Using the multiplication of matrices, it follows immediately that the Yang–Mills field equation (0.48) corresponds to the Riemann equation (0.47).

Cartan's Local Structural Equation

> In order to get insight in differential geometry, use differential forms and employ their invariance properties.
>
> Folklore

In order to kill indices, let us define the following differential forms:

- $\mathcal{A} := \mathcal{A}_\alpha dx^\alpha$ (local connection form on the space-time manifold $\mathcal{M}^4$),
- $\mathcal{F} := \frac{1}{2}\mathcal{F}_{\alpha\beta}\, dx^\alpha \wedge dx^\beta$ (local curvature form on $\mathcal{M}^4$).

Here, $\mathcal{F}_{\alpha\beta} = -\mathcal{F}_{\beta\alpha}$ for all indices $\alpha, \beta = 0, 1, 2, 3$. Note that $\mathcal{A}$ and $\mathcal{F}$ are differential forms with the (4×4)-matrices $\mathcal{A}_\alpha$ and $\mathcal{F}_{\alpha\beta}$ as coefficients. This can also be written as

$$\mathcal{A} = (\omega^{\delta}_{\gamma}), \qquad \mathcal{F} = (\Omega^{\delta}_{\gamma})$$

with the differential forms

- $\omega^\delta_\gamma := \Gamma^\delta_{\alpha\gamma}\, dx^\alpha$ (Cartan's local connection forms), and
- $\Omega^\delta_\gamma := \frac{1}{2} R^\delta_{\alpha\beta\gamma}\, dx^\alpha \wedge dx^\beta$ (Cartan's local curvature forms).

Calculus for matrices with differential forms as entries. If a, b, c, d and e, f, g, h are complex numbers, then we have the following classical matrix product:

$$\begin{pmatrix} a & b \\ c & d \end{pmatrix} \begin{pmatrix} e & f \\ g & h \end{pmatrix} = \begin{pmatrix} ae + bg & af + bh \\ ce + dg & cf + dh \end{pmatrix}.$$

Now suppose that all the entries $a, b, \ldots$ are differential forms. Then the wedge products $a \wedge e, \ldots$ of entries are well defined. This motivates the following definition of the wedge product of matrices with differential forms as entries:

$$\begin{pmatrix} a & b \\ c & d \end{pmatrix} \wedge \begin{pmatrix} e & f \\ g & h \end{pmatrix} := \begin{pmatrix} a \wedge e + b \wedge g & a \wedge f + b \wedge h \\ c \wedge e + d \wedge g & c \wedge f + d \wedge h \end{pmatrix}. \tag{0.49}$$

That is, we merely replace the classical product of entries by the wedge product of entries.

Cartan's local structural equation. We claim that the differential forms $\mathcal{A}$ and $\mathcal{F}$ satisfy the following two elegant equations:

(C) Cartan's local structural equation:

$$\boxed{\mathcal{F} = d\mathcal{A} + \mathcal{A} \wedge \mathcal{A}.} \tag{0.50}$$

(B) Bianchi's local identity (integrability condition to (C)):

$$\boxed{d\mathcal{F} = \mathcal{F} \wedge \mathcal{A} - \mathcal{A} \wedge \mathcal{F}.} \tag{0.51}$$

Proof. Ad (C). This is nothing else than a clever reformulation of the classical curvature relation

$$R^\delta_{\alpha\beta\gamma} := \partial_\alpha \Gamma^\delta_{\beta\gamma} - \partial_\beta \Gamma^\delta_{\alpha\gamma} + \Gamma^\delta_{\alpha\mu}\Gamma^\mu_{\beta\gamma} - \Gamma^\delta_{\beta\mu}\Gamma^\mu_{\alpha\gamma}$$

in terms of Cartan's calculus of differential forms invented by Cartan in 1899.[35] In fact, it follows from $\omega^\delta_\gamma = \Gamma^\delta_{\beta\gamma} dx^\beta$ and the antisymmetry of the Grassmann product (also called the wedge product) $dx^\alpha \wedge dx^\beta = -dx^\beta \wedge dx^\alpha$ that

$$\begin{aligned} d\omega^\delta_\gamma &= d\Gamma^\delta_{\beta\gamma} \wedge dx^\beta = \partial_\alpha \Gamma^\delta_{\beta\gamma}\, dx^\alpha \wedge dx^\beta \\ &= \tfrac{1}{2}(\partial_\alpha \Gamma^\delta_{\beta\gamma} - \partial_\beta \Gamma^\delta_{\alpha\gamma})\, dx^\alpha \wedge dx^\beta \end{aligned}$$

and

$$\begin{aligned} \omega^\delta_\mu \wedge \omega^\mu_\gamma &= \Gamma^\delta_{\alpha\mu}\Gamma^\mu_{\beta\gamma}\, dx^\alpha \wedge dx^\beta \\ &= \tfrac{1}{2}(\Gamma^\delta_{\alpha\mu}\Gamma^\mu_{\beta\gamma} - \Gamma^\delta_{\beta\mu}\Gamma^\mu_{\alpha\gamma})\, dx^\alpha \wedge dx^\beta. \end{aligned}$$

Using $\Omega^\delta_\gamma = \frac{1}{2} R^\delta_{\alpha\beta\gamma}\, dx^\alpha \wedge dx^\beta$, we obtain Cartan's system of structural equations

$$\boxed{\Omega^\delta_\gamma = d\omega^\delta_\gamma + \omega^\delta_\mu \wedge \omega^\mu_\gamma, \qquad \gamma, \delta = 0, 1, 2, 3.} \tag{0.52}$$

---

[35] Grassmann (1809–1877), Élie Cartan (1869–1951).

In order to kill the indices, we introduce the matrices $\mathcal{A} = (\omega^\delta_\gamma)$ and $\mathcal{F} = (\Omega^\delta_\gamma)$. Then Cartan's system (0.52) can be elegantly written as

$$d\mathcal{F} = d\mathcal{A} + \mathcal{A} \wedge \mathcal{A}.$$

Ad (B). We will use the following properties of the calculus for differential forms (with real coefficients). Let ω, ϱ and τ be differential forms of degree p, r, and s, respectively. Then, the wedge product has the following properties:

- $(\omega \wedge \varrho) \wedge \tau = \omega \wedge (\varrho \wedge \tau)$ (associativity),
- $\omega \wedge \varrho = (-1)^{pr} \varrho \wedge \omega$ (graded anticommutativity),
- $d\omega = 0$ (the Poincaré cohomology rule),
- $d(\omega \wedge \varrho) = d\omega \wedge \varrho + (-1)^p \omega \wedge d\varrho$ (the graded Leibniz product rule).

These properties induce the corresponding rules for the wedge product of matrices. In particular, applying the Cartan differential "d" to (C), we get

$$d\mathcal{F} = d(d\mathcal{A}) + d\mathcal{A} \wedge \mathcal{A} - \mathcal{A} \wedge d\mathcal{A}.$$

By the Poincaré cohomology rule, $d(d\mathcal{A}) = 0$. Since $\mathcal{F} = d\mathcal{A} + \mathcal{A} \wedge \mathcal{A}$, we get

$$d\mathcal{F} = (\mathcal{F} - \mathcal{A} \wedge \mathcal{A}) \wedge \mathcal{A} - \mathcal{A} \wedge (\mathcal{F} - \mathcal{A} \wedge \mathcal{A}) = \mathcal{F} \wedge \mathcal{A} - \mathcal{A} \wedge \mathcal{F}.$$

□

Gauge transformations. Let us consider the change

$$x^{\alpha'} = x^{\alpha'}(x^0, x^1, x^2, x^3), \qquad \alpha = 0, 1, 2, 3$$

of local coordinates. We want to determine the transformation laws for the family of Christoffel symbols $\Gamma^\kappa_{\alpha\lambda}$ and the family of Riemann symbols $R^\kappa_{\lambda\alpha\beta}$. We will use the notation introduced on page 26. Let us first describe an elementary brute-force method based on completely elementary, but lengthy computations based on the chain rule in classical differential calculus.

- We start with the transformation law[36]

  $$g_{\alpha'\beta'} = G^\alpha_{\alpha'} G^\beta_{\beta'} g_{\alpha\beta}$$

  for the components $g_{\alpha\beta}$ of the metric tensor. Using matrices, this means that

  $$(g_{\alpha'\beta'}) = G^d (g_{\alpha\beta}) G.$$

- For the inverse matrix, we get

  $$(g^{\alpha'\beta'}) = (g_{\alpha'\beta'})^{-1} = G^{-1}(g_{\alpha\beta})^{-1}(G^d)^{-1} = G^{-1}(g^{\alpha\beta})(G^{-1})^d.$$

  This implies

  $$g^{\alpha'\beta'} = G^{\alpha'}_\alpha G^{\beta'}_\beta \cdot g^{\alpha\beta}.$$

  Thus, $g^{\alpha\beta}$ is a tensorial family of type $(2,0)$.
- We use $\Gamma^\delta_{\alpha\beta} := \frac{1}{2}(\partial_\alpha g_{\beta\sigma} + \partial_\beta g_{\alpha\sigma} - \partial_\sigma g_{\alpha\beta}) g^{\sigma\kappa}$ in order to get the transformation formula for the Christoffel symbols:

  $$\Gamma^{\delta'}_{\alpha'\beta'} = G^{\delta'}_\delta G^\alpha_{\alpha'} G^\beta_{\beta'} \cdot \Gamma^\delta_{\alpha\beta} - G^\alpha_{\alpha'} G^\beta_{\beta'} (\partial_\alpha G^{\delta'}_\beta). \qquad (0.53)$$

[36] More precisely, $g_{\alpha'\beta'}(x') = G^\alpha_{\alpha'}(x) G^\beta_{\beta'}(x) g_{\alpha\beta}(x)$.

- From (0.47), we obtain the following transformation formula for the Riemann symbols:

$$R^{\delta'}_{\alpha'\beta'\gamma'} = G^{\delta'}_{\delta} G^{\alpha}_{\alpha'} G^{\beta}_{\beta'} G^{\gamma}_{\gamma'} \cdot R^{\delta}_{\alpha\beta\gamma}. \tag{0.54}$$

Relation (0.47) tells us that $R^{\delta}_{\alpha\beta\gamma}$ forms a tensorial family of type $(1,3)$. Moreover, it follows from (0.53) that the Christoffel family $\Gamma^{\delta}_{\alpha\beta}$ is not a tensorial family.[37] In contrast to this brute-force method, a much simpler proof of (0.53) and (0.47) will be given in Chap. 8 based on the inverse index principle(see page 505). More elegantly, using the language of matrices, the transformation formulas (0.53) and (0.47) can be written in the following way:

(i) $\dot{x}' = G\dot{x}$ (transformation law for the velocity components $\dot{x}^{\alpha}(\sigma)$),
(ii) $\mathcal{A}' = G\mathcal{A}G^{-1} - (dG)G^{-1}$ (transformation law for the connection form), and
(iii) $\mathcal{F}' = G\mathcal{F}G^{-1}$ (transformation law for the curvature form).[38]

Here, we set $\mathcal{A}(x) := \mathcal{A}_{\alpha}(x)dx^{\alpha} = (\Gamma^{\delta}_{\alpha\gamma}(x)dx^{\alpha})$ and

$$\mathcal{A}'(x') := \left(\Gamma^{\delta'}_{\alpha'\gamma'}(x')\, dx^{\alpha'}\right), \qquad x' = x'(x).$$

Note that $\mathcal{A}(x)$ is a matrix with the differential forms $\Gamma^{\delta}_{\alpha\gamma}(x)dx^{\alpha}$ as entries; the upper index δ numbers the rows, and the lower index γ numbers the columns. Furthermore,

$$\mathcal{F} := \tfrac{1}{2}\mathcal{F}_{\alpha\beta}\, dx^{\alpha} \wedge dx^{\beta} = \left(\tfrac{1}{2}R^{\delta}_{\alpha\beta\gamma}\, dx^{\alpha} \wedge dx^{\beta}\right),$$

and

$$\mathcal{F}'(x') := \left(\tfrac{1}{2}R^{\delta'}_{\alpha'\beta'\gamma'}(x')\, dx^{\alpha'} \wedge dx^{\beta'}\right).$$

Finally, recall that $G = (G^{\delta'}_{\delta})$ and $G^{-1} = (G^{\delta}_{\delta'})$. The proof of (ii), (iii) above will be given in Chap. 8.

Cartan's Global Structural Equation

Extend the space-time manifold $\mathcal{M}^4$ to its frame bundle.
Folklore

In order to arrive at a global approach, we pass from the local structural equation $\mathcal{F} = d\mathcal{A} + \mathcal{A} \wedge \mathcal{A}$ considered in (0.50) to the global structural equation

$$\boxed{\mathsf{F} = \mathsf{DA}.} \tag{0.55}$$

Moreover, we add the global integrability condition

$$\boxed{\mathsf{DF} = 0} \tag{0.56}$$

which is called the global Bianchi identity. Let us discuss this.

Extension of the space-time manifold $\mathcal{M}^4$ to the frame bundle $F(\mathcal{M}^4)$. The two differential forms

[37] The Christoffel family is also called the connection family.
[38] More precisely, $x'(\sigma) = G(x(\sigma))x(\sigma)$,

$$\mathcal{A}'(x') = (G\mathcal{A}G^{-1} - dG \cdot G^{-1})(x),$$

and $\mathcal{F}'(x') = (G\mathcal{F}G^{-1})(x)$. Note that the prime refers to the transformed coordinates $x' = (x^{1'}, \ldots, x^{n'})$, but it does not refer to any derivative.

- A (global connection form), and
- F (global curvature form)

are not defined on the space-time manifold, but on the frame bundle. By definition, the frame bundle $F(\mathcal{M}^4)$ of the space-time manifold $\mathcal{M}^4$ consists of all the tuples

$$(P, \mathbf{b}_0, \mathbf{b}_1, \mathbf{b}_2, \mathbf{b}_3) \tag{0.57}$$

where P is an arbitrary point of $\mathcal{M}^4$, and $\mathbf{b}_0, \mathbf{b}_1, \mathbf{b}_2, \mathbf{b}_3$ is an arbitrary basis of the tangent space $T_P\mathcal{M}^4$ of $\mathcal{M}^4$ at the point P. Choose local coordinates (x^0, x^1, x^2, x^3) which live in the open set U of $\mathbb{R}^4$. Since $\partial_0, \partial_1, \partial_2, \partial_3$ form a basis of the tangent space $T_P\mathcal{M}^4$, there exist real numbers $G_\alpha^\beta(x)$ depending on x such that

$$\mathbf{b}_\alpha = G_\alpha^\beta(x)\partial_\beta, \qquad \alpha = 0, 1, 2, 3.$$

Introducing the matrix $G(x) := (G_\alpha^\beta(x))$, we get the matrix equation

$$(\mathbf{b}_0, \mathbf{b}_1, \mathbf{b}_2, \mathbf{b}_3) = (\partial_0, \partial_1, \partial_2, \partial_3)\; G(x)$$

where $G(x)$ is an invertible real (4×4)-matrix, that is, $G(x) \in GL(4, \mathbb{R})$. The tuple $(x, G(x))$ is called the local coordinate of the point (0.57) of the frame bundle $F(\mathcal{M}^4)$. Obviously,

$$(x, G(x)) \in U \times GL(4, \mathbb{R}).$$

Parallel transport on the space-time manifold in terms of the frame bundle. We are given the curve

$$C : P = P(\sigma), \qquad \sigma_0 \le \sigma \le \sigma_1,$$

on the space-time manifold $\mathcal{M}^4$. With respect to local coordinates, this curve corresponds to the map $\sigma \mapsto x(\sigma)$. Consider the differential equation

$$\dot{G}(\sigma) = -\dot{x}^\alpha(\sigma)\mathcal{A}_\alpha(x(\sigma)) \cdot G(\sigma), \qquad \sigma_0 \le \sigma \le \sigma_1, \tag{0.58}$$

with the initial condition $G(\sigma_0) = I$. Let $G = G(\sigma)$ be the unique solution of the initial-value problem for (0.58). We are given the tangent vector $\mathbf{v}_0 = v_0^\alpha \partial_\alpha$ at the point $P(\sigma_0)$. We set $v(\sigma) := G(\sigma)v_0$. Then the differential equation of parallel transport

$$\dot{v}(\sigma) = -\dot{x}^\alpha(\sigma)\mathcal{A}_\alpha(x(\sigma)) \cdot v(\sigma), \qquad \sigma_0 \le \sigma \le \sigma_1,$$

is satisfied. Consequently, setting

$$\mathbf{v}(\sigma) := v^\alpha(\sigma)\partial_\alpha,$$

we obtain the tangent vector $\mathbf{v}(\sigma)$ at the curve point $P(\sigma)$, and this family represents a parallel transport of velocity vectors along the curve C. We will show later on that this parallel transport can be used to define quite naturally the differential form A, the covariant differential DA and hence the curvature form F on the frame bundle.

Restriction of the global curvature form F to the local curvature form $\mathcal{F}$ and gauge transformations. We want to discuss briefly the relation between the global differential forms A and F on the frame bundle $F(\mathcal{M}^4)$ and the (local) differential forms $\mathcal{A}$ and $\mathcal{F}$ on the space time-manifold $\mathcal{M}^4$. Setting

$$s(x) := (x, I),$$

we get the map $s : U \to U \times GL(4, \mathbb{R})$. The corresponding pull-backs of the differential forms yield

- $s^*\mathsf{A} = \mathcal{A}$, and
- $s^*\mathsf{F} = \mathcal{F}$.

Moreover, for given function $x \mapsto G(x)$, we set

$$s(x) := (x, G(x)).$$

This corresponds to the choice of frames depending on the point P related to the local coordinate x. Then we get:

- $s^*\mathsf{A}(x) = G(x)\mathcal{A}(x)G(x)^{-1} - dG(x) \cdot G(x)^{-1}$, and
- $s^*\mathsf{F}(x) = G(x)\mathcal{F}(x)G(x)^{-1}$.

This yields precisely the gauge transformation formulas for $\mathcal{A}$ and $\mathcal{F}$ summarized on page 38, namely,

- $\mathcal{A}'(x) = G(x)\mathcal{A}(x)G(x)^{-1} - dG(x) \cdot G(x)^{-1}$, and
- $\mathcal{F}'(x) = G(x)\mathcal{F}(x)G(x)^{-1}$.

Élie Cartan's method of moving frames. The procedure sketched above is called Cartan's method of moving frames. In terms of mathematics, Cartan's global structural equation (0.55) represents a generalization of the basic formula (0.47) which relates the Riemann curvature tensor to the Christoffel symbols and their first partial derivatives.

In terms of physics, the global curvature form F corresponds to the gravitational force. The structural equation (0.55) relates the gravitational force F (i.e., the global curvature form) to the so-called potential A (i.e., the global connection form). In physics, the use of the frame bundle in general relativity is called the tetrad formalism.

The general form of Cartan's approach (based on frame bundles) allows us to transform the local coordinates of the space-time manifold $\mathcal{M}^4$ and the local coordinates of the frames in a separate way. This yields optimal flexibility.

Covariant Partial Derivative and the Classical Ricci Calculus

> Replace the classical derivatives $\ddot{x}^\alpha(\sigma)$ and $\partial_\alpha v^\beta$ by the covariant derivatives $\frac{D\dot{x}(\sigma)}{d\sigma}$ and $\nabla_\alpha v^\beta$, respectively.
> In contrast to the classical partial derivative ∂_α, the covariant partial derivative ∇_α has the very useful property that it sends tensorial families again to tensorial families.
>
> Folklore

Classical identities for partial derivatives. Let $\Theta : U \to \mathbb{R}$ and

$$u^\kappa, v^\kappa, w^\kappa : U \to \mathbb{R}^4, \qquad \kappa = 0, 1, 2, 3$$

be smooth functions where U is a nonempty open subset of $\mathbb{R}^4$ (e.g., $U = \mathbb{R}^4$). Choose the coordinates x^0, x^1, x^2, x^3 on $\mathbb{R}^4$, and recall the notation $\partial_\alpha := \frac{\partial}{\partial x^\alpha}$ for the partial derivative with respect to the variable x^α. Set

$$\boxed{d_{v(x)}\Theta(x) := v^\alpha(x)\partial_\alpha\Theta(x).}$$

This is called the directional derivative of the function Θ at the point x with respect to the direction $v(x)$. Let $x = x(\sigma)$ be a smooth curve with $x^\alpha(0) = x_0^\alpha$ and $\dot{x}^\alpha(0) = v^\alpha$ for all $\alpha = 0, 1, 2, 3$. Then, by the chain rule,

$$d_v\Theta(x_0) = \frac{d\Theta(x(\sigma))}{d\sigma}_{|\sigma=0}.$$

This motivates the designation "directional derivative." Furthermore, set

$$v = \begin{pmatrix} v^0 \\ v^1 \\ v^2 \\ v^3 \end{pmatrix}, \quad w = \begin{pmatrix} w^0 \\ w^1 \\ w^2 \\ w^3 \end{pmatrix}, \quad d_{v(x)}w(x) := \begin{pmatrix} v^\alpha(x)\partial_\alpha w^0(x) \\ v^\alpha(x)\partial_\alpha w^1(x) \\ v^\alpha(x)\partial_\alpha w^2(x) \\ v^\alpha(x)\partial_\alpha w^3(x) \end{pmatrix}.$$

Here, $d_{v(x)}w(x)$ is called the directional derivative of the function $w : U \to \mathbb{R}^4$ at the point x with respect to the direction $v(x)$. Finally, introduce the Lie product

$$\boxed{[v, w] := d_v w - d_w v.} \tag{0.59}$$

In addition, we introduce the symbol $[\partial_\alpha, \partial_\beta]_-$ in the sense of linear operators. That is, we set $[\partial_\alpha, \partial_\beta]_- v := \partial_\alpha\partial_\beta v - \partial_\beta\partial_\alpha v = 0$. Then we have the following trivial identities, which will be generalized to nontrivial identities later on:

(I) Trivial Lie product: $[\partial_\alpha, \partial_\beta]_- = 0$.
(II) Trivial Jacobi identity:

$$([\partial_\alpha, [\partial_\beta, \partial_\gamma]_-]_- + ([\partial_\beta, [\partial_\gamma, \partial_\alpha]_-]_- + ([\partial_\gamma, [\partial_\alpha, \partial_\beta]_-]_-)v = 0. \tag{0.60}$$

(III) Trivial Bianchi identity:

$$(\partial_\alpha[\partial_\beta, \partial_\gamma]_- + \partial_\beta[\partial_\gamma, \partial_\alpha]_- + \partial_\gamma[\partial_\alpha, \partial_\beta]_-)v = 0. \tag{0.61}$$

(IV) The key Lie relation:[39]

$$\boxed{d_u(d_v w) - d_v(d_u w) = d_{[u,v]}w.} \tag{0.62}$$

(V) The Lie algebra $C^\infty(U, \mathbb{R}^4)$: The real linear space $C^\infty(U, \mathbb{R}^4)$ of all smooth functions

$$v : U \to \mathbb{R}^4$$

forms a real Lie algebra with respect to the Lie product (0.59). Explicitly, this means that $C^\infty(U, \mathbb{R}^4)$ is a real linear space. Furthermore, for all functions $u, v, w \in C^\infty(U, \mathbb{R}^4)$ and all real numbers λ, μ, the following hold:

- $[u, v] \in C^\infty(U, \mathbb{R}^4)$ (consistency),
- $[\lambda u + \mu v, w] = \lambda[u, w] + \mu[v, w]$ (distributivity),
- $[v, w] = -[w, v]$ (antisymmetry),
- $[u, [v, w]] + [v, [w, u]] + [w, [u, v]] = 0$ (Jacobi identity).

(VI) The Lie algebra $D(C^\infty(U, \mathbb{R}^4))$: By definition, the symbol $D(C^\infty(U, \mathbb{R}^4))$ denotes the set of of all linear differential operators

$$v^\alpha(x)\partial_\alpha, \qquad x \in U$$

with smooth coefficient functions $v^\alpha : U \to \mathbb{R}$. This coincides with the set of all differential operators

[39] Leibniz (1646–1717), Jacobi (1804–1851), Lie (1842–1899), Bianchi (1856–1928).

$$d_v : C^\infty(U, \mathbb{R}^4) \to C^\infty(U, \mathbb{R}^4)$$

where $v \in C^\infty(U, \mathbb{R}^4)$. The real linear space $D(C^\infty(U, \mathbb{R}^4))$ becomes a real Lie algebra with respect to the Lie product $[d_u, d_v]_-$ in the sense of operators. That is,

$$[d_u, d_v]_- w = d_u(d_v w) - d_v(d_u w) \tag{0.63}$$

for all $w \in C^\infty(U, \mathbb{R}^4)$. Observe the following crucial fact for the Lie theory of partial differential equations. By (IV),

$$[d_u, d_v]_- w = d_{[u,v]} w.$$

Consequently, $[d_u, d_v]_-$ is not a second-order differential operator, but only a first-order partial differential operator because of the cancellation of partial derivatives of second order. Therefore, $[d_u, d_v]_- w$ is an element of the linear space $D(C^\infty(U, \mathbb{R}^4))$. The Jacobi identity

$$[A, [B, C]_-\ ]_- w + [B, [C, A]_-\ ]_- w + [C, [A, B]_-\ ]_- w = 0$$

on $D(C^\infty(U, \mathbb{R}^4))$ follows from (0.63). This is a special case of the general fact that the Jacobi identity is always satisfied for linear operators on linear spaces (see (0.66)).

(VII) Leibniz rule: For all $v \in C^\infty(U, \mathbb{R}^4)$ and all smooth functions $\Theta : U \to \mathbb{R}$, we have the product rule:

$$d_v(\Theta w) = (d_v \Theta) w + \Theta d_v w. \tag{0.64}$$

(VIII) Differential: Set $dw := \partial_\alpha w \, dx^\alpha$. Then

$$d_v w = (dw)(v).$$

In fact, $dx^\alpha(v) = v^\alpha$. Hence $dw(v) = v^\alpha \partial_\alpha w$.

Proof. All the statements follow by using elementary computations based on the key relation

$$\boxed{\partial_\alpha \partial_\beta = \partial_\beta \partial_\alpha, \qquad \alpha, \beta = 0, 1, 2, 3.} \tag{0.65}$$

This is the mnemonic formulation of the commutativity property

$$\frac{\partial}{\partial x^\alpha}\left(\frac{\partial \Theta}{\partial x^\beta}\right) = \frac{\partial}{\partial x^\beta}\left(\frac{\partial \Theta}{\partial x^\alpha}\right)$$

for the partial derivatives of smooth real-valued functions Θ.

For example, let us prove the key Lie relation (IV). Observe first that

$$d_u(d_v w) = u^\alpha \partial_\alpha (v^\beta \partial_\beta w) = u^\alpha v^\beta \partial_\alpha \partial_\beta w + u^\alpha \partial_\alpha v^\beta \partial_\beta w.$$

Similarly, $d_v(d_u w) = v^\alpha u^\beta \partial_\alpha \partial_\beta w + v^\alpha \partial_\alpha u^\beta \partial_\beta w$. The crucial point is that the second partial derivatives cancel each other, by (0.65).[40] Finally, we get

$$d_u(d_v w) - d_v(d_u w) = (u^\alpha \partial_\alpha v^\beta - v^\alpha \partial_\alpha u^\beta) \partial_\beta w = d_{[u,v]} w.$$

---

[40] It turns out that this is the main trick of Lie's approach to analysis and differential geometry. In addition, the tensorial property of the Riemann curvature tensor is based on a similar cancellation of second order partial derivatives.

This is (IV). Furthermore, the Leibniz rule (VII) follows from

$$d_v(\Theta w) = v^\alpha \partial_\alpha(\Theta w) = (v^\alpha \partial_\alpha \Theta)w + \Theta \cdot v^\alpha \partial_\alpha w = (d_v\Theta)w + \Theta d_v w.$$

□

In the present case, the Jacobi identity and the Bianchi identity are trivial consequences of the classical commutativity property (0.65) for partial derivatives. For more general situations, note the following. If $A, B, C : X \to X$ are linear operators on the linear space X, then cyclic permutation yields

$$\begin{aligned}&[A,[B,C]_-]_- + [B,[C,A]_-]_- + [C,[A,B]_-]_- \\ &\quad = A(BC - CB) - (BC - CB)A + B(CA - AC) - (CA - AC)B \\ &\quad +C(AB - BA) - (AB - BA)C = 0. \end{aligned} \tag{0.66}$$

This means that the Jacobi identity is always satisfied for linear operators. Moreover, antisymmetrization yields

$$\begin{aligned}\mathrm{Alt}(ABC) &:= \tfrac{1}{6}(ABC - ACB + BCA - BAC + CAB - CBA) \\ &= \tfrac{1}{6}(A[B,C]_- + B[C,A]_- + C[A,B]_-).\end{aligned}$$

The trouble with classical partial derivatives in the theory of general relativity. As a rule, the relations above are not invariant under general nonlinear coordinate transformations (i.e., by using local diffeomorphisms for local coordinates). In terms of physics, this means that the relations above do not possess any physical meaning.

As we will show, this lack of invariance can be cured by replacing partial derivatives by covariant partial derivatives.

Then the trivial Lie product does not vanish anymore, but the corresponding generalization determines the Riemann curvature tensor. The following sketched material will thoroughly be studied in Chap. 8.

The trouble with acceleration. Let $C : P = P(\sigma), \sigma_0 \le \sigma \le \sigma_1$, be a smooth curve on the space-time manifold $\mathcal{M}^4$ given by the equation $x = x(\sigma)$ with respect to a local coordinate system. As a rule, the equation

$$\ddot{x}^\gamma(\sigma) = 0, \qquad \sigma_0 \le \sigma \le \sigma_1, \qquad \gamma = 0,1,2,3 \tag{0.67}$$

is not invariant under a change of local space-time coordinates. Set

$$\frac{D\dot{x}^\gamma(\sigma)}{d\sigma} := \ddot{x}^\gamma(\sigma) + \dot{x}^\alpha(\sigma)\Gamma^\gamma_{\alpha\beta}(x(\sigma))\,\dot{x}^\beta(\sigma).$$

This is called the covariant derivative of $\sigma \mapsto \dot{x}(\sigma)$ with respect to the real parameter σ. It turns out that $\frac{D\dot{x}^\gamma(\sigma)}{d\sigma}$ is a tensorial family, in contrast to $\ddot{x}^\gamma$. The equation

$$\frac{D\dot{x}^\gamma(\sigma)}{d\sigma} = 0$$

describes geodesic lines. By the principle of index killing, the following vector functions possess an invariant meaning:

- $\mathbf{v}(\sigma) := \dot{x}^\gamma(\sigma)\partial_\gamma$ (velocity vector of the curve C),
- $\mathbf{a}(\sigma) := \frac{D\dot{x}^\gamma(\sigma)}{d\sigma}\,\partial_\gamma$ (acceleration vector of the curve C).

The partial covariant derivative as a key tool. We set

- $\nabla_\alpha v^\beta := \partial_\alpha v^\beta + \Gamma^\beta_{\alpha\lambda} v^\lambda$, and
- $\nabla_\alpha v_\beta = \partial_\alpha v_\beta - \Gamma^\lambda_{\alpha\beta} v_\lambda$.

More generally, we define

$$\nabla_\alpha T^{\alpha_1 \ldots \alpha_m}_{\beta_1 \ldots \beta_n} := \partial_\alpha T^{\alpha_1 \ldots \alpha_m}_{\beta_1 \ldots \beta_n} + \sum_{r=1}^{m} \Gamma^{\alpha_r}_{\alpha\lambda} T^{\alpha_1 \ldots \lambda \ldots \alpha_m}_{\beta_1 \ldots \beta_n} - \sum_{s=1}^{n} \Gamma^{\mu}_{\alpha\beta_s} T^{\alpha_1 \ldots \alpha_m}_{\beta_1 \ldots \mu \ldots \beta_n}.$$

Here, we replace the index α_r (resp. β_s) by λ (resp. μ). In addition, for an invariant function Θ, we define $\nabla_\alpha \Theta := \partial_\alpha \Theta$.

The covariant partial derivative ∇_α has the crucial property that it preserves tensorial families:

- If v^β is a tensorial family of type $(1,0)$, then $\nabla_\alpha v^\beta$ is a tensorial family of type $(1,1)$.
- If w_β is a tensorial family of type $(0,1)$, then $\nabla_\alpha v_\beta$ is a tensorial family of type $(0,2)$.
- If $T^{\alpha_1 \ldots \alpha_m}_{\beta_1 \ldots \beta_n}$ is a tensorial family of type (m,n), then $\nabla_\alpha T^{\alpha_1 \ldots \alpha_m}_{\beta_1 \ldots \beta_n}$ is a tensorial family of type $(m, n+1)$.

Let us mention two typical examples. Choose $\alpha, \beta, \gamma, \delta, \mu = 0,1,2,3.$[41] It follows from the key relation

$$\boxed{\nabla_\alpha(\nabla_\beta v^\delta) - \nabla_\beta(\nabla_\alpha v^\delta) = R^\delta_{\alpha\beta\gamma} v^\gamma} \tag{0.68}$$

that the Riemann curvature tensor measures the noncommutativity of the covariant partial derivative. The relation

$$\nabla_\mu R^\delta_{\alpha\beta\gamma} + \nabla_\alpha R^\delta_{\beta\mu\gamma} + \nabla_\beta R^\delta_{\mu\alpha\gamma} = 0 \tag{0.69}$$

represents the Bianchi identity in covariant formulation. This is based on cyclic permutation of the indices μ, α, β. The relation (0.69) is equivalent to

$$\boxed{\nabla_{[\mu} R^\delta_{\alpha\beta]\gamma} = 0} \tag{0.70}$$

which represents an antisymmetrization with respect to the indices μ, α, β. The Ricci identity reads as

$$R^\delta_{\alpha\beta\gamma} + R^\delta_{\beta\gamma\alpha} + R^\delta_{\gamma\alpha\beta} = 0.$$

This is equivalent to

$$\boxed{R^\delta_{[\alpha\beta\gamma]} = 0} \tag{0.71}$$

which represents an antisymmetrization with respect to the lower indices α, β, γ. Moreover, we have

$$\boxed{\nabla_\alpha g_{\beta\gamma} = 0.} \tag{0.72}$$

This is called the Ricci lemma. Consider the Bianchi identity (0.69). We will show later on that Ricci's lemma allows us to lower the index δ. This yields

[41] By the Einstein convention, we sum over equal upper and lower indices from 0 to 3.

$$\nabla_\mu R_{\alpha\beta\gamma\delta} + \nabla_\alpha R_{\beta\mu\gamma\delta} + \nabla_\beta R_{\mu\alpha\gamma\delta} = 0 \tag{0.73}$$

based on cyclic permutation of the indices μ, α, β. Hence

$$\nabla_{[\mu} R_{\alpha\beta]\gamma\delta} = 0.$$

The very useful Ricci calculus principle of replacing partial derivatives by covariant partial derivatives. Consider the partial differential equation

$$\Box\Theta = g^{\alpha\beta}\partial_\alpha\partial_\beta\Theta$$

in a fixed local coordinate system. For example, if $g_{\alpha\beta} = \eta_{\alpha\beta}$ (for all indices), and if we write ct, x, y, z instead of x^0, x^1, x^2, x^3, respectively, then we obtain the wave equation

$$\Box\Theta = \frac{1}{c^2}\frac{\partial^2\Theta}{\partial t^2} - \frac{\partial^2\Theta}{\partial x^2} - \frac{\partial^2\Theta}{\partial y^2} - \frac{\partial^2\Theta}{\partial z^2}. \tag{0.74}$$

It is our goal to rewrite (0.74) in such a way that it is valid in arbitrary local coordinate systems. To this end, we need an invariant expression which coincides with (0.74) in the special (x^0, x^1, x^2, x^3)-coordinate system chosen above with $g_{\alpha\beta} = \eta_{\alpha\beta}$. Since the Christoffel symbols vanish identically in this special coordinate system, we can replace the partial derivatives by the corresponding covariant partial derivatives. This yields the invariant formulation

$$\boxed{\Box\Theta = g^{\alpha\beta}\nabla_\alpha\nabla_\beta\Theta.}$$

By the index killing principle, this expression is valid in each local coordinate system provided we use the tensorial family $g^{\alpha\beta}$. This simple trick can be used in order to write all the equations appearing in theoretical physics in such a way that they are valid in each coordinate system. For example, in Sect. 19.3.1 we will use this trick in order to formulate the Maxwell equations in electrodynamics with respect to an arbitrary space-time coordinate system. Observe that, for the Dirac equation of the relativistic electron, one has to replace tensorial families by spinorial families (see Vol. IV).

The Lie Structure behind Curvature

Let the symbol $\mathrm{Vect}(\mathcal{M}^4)$ denote the space of all smooth velocity vector fields $\mathbf{v}$ on the space-time manifold $\mathcal{M}^4$. With respect to local coordinates, we write $\mathbf{v} = v^\alpha\partial_\alpha$. In what follows, let $\mathbf{u}, \mathbf{v}, \mathbf{w} \in \mathrm{Vect}(\mathcal{M}^4)$. Sophus Lie frequently used the fact that

$$u^\alpha\partial_\alpha v^\beta - v^\alpha\partial_\alpha u^\beta$$

forms a tensorial family on $\mathcal{M}^4$. By index killing, we get the invariant expression

$$\boxed{\mathcal{L}_\mathbf{u}\mathbf{v} = (u^\alpha\partial_\alpha v^\beta - v^\alpha\partial_\alpha u^\beta)\,\partial_\beta.}$$

This is called the Lie derivative of the velocity field $\mathbf{v}$ with respect to the velocity field $\mathbf{u}$.

The Lie algebra $\mathrm{Vect}(\mathcal{M}^4)$. Using the Lie product

$$[\mathbf{u}, \mathbf{v}] := \mathcal{L}_\mathbf{u}\mathbf{v},$$

the linear space $\mathrm{Vect}(\mathcal{M}^4)$ of smooth velocity vector fields on $\mathcal{M}^4$ becomes a real Lie algebra. Explicitly, this means that for all $\mathbf{u}, \mathbf{v}, \mathbf{w} \in \mathrm{Vec}(\mathcal{M}^4)$ and all real numbers λ, μ, we have:

- $[\lambda\mathbf{u} + \mu\mathbf{v}, \mathbf{w}] = \lambda[\mathbf{u}, \mathbf{w}] + \mu[\mathbf{v}, \mathbf{w}]$ (distributivity),
- $[\mathbf{v}, \mathbf{w}] = -[\mathbf{w}, \mathbf{v}]$ (antisymmetry),
- $[\mathbf{u}, [\mathbf{v}, \mathbf{w}]] + [\mathbf{v}, [\mathbf{w}, \mathbf{u}]] + [\mathbf{w}, [\mathbf{u}, \mathbf{v}]] = 0$ (Jacobi identity).

The physical meaning of the Lie derivative in terms of the flow of fluid particles will be considered in Sect. 11.2.

The covariant differential of a velocity field. For $\mathbf{u}, \mathbf{v} \in \mathrm{Vect}(\mathcal{M}^4)$, we have the tensorial family $\nabla_\alpha v^\beta$. Using the index killing principle, we get the invariant expression

$$\boxed{D\mathbf{v} := (\nabla_\alpha v^\beta)\, dx^\alpha \otimes \partial_\beta.}$$

This tensor field of type $(1,1)$ on the space-time manifold $\mathcal{M}^4$ is called the covariant differential of the velocity vector field $\mathbf{v}$. Furthermore, we set

$$D_\mathbf{u}\mathbf{v} := (D\mathbf{v})(\mathbf{u}).$$

This is called the covariant directional derivative of the velocity vector field $\mathbf{v}$ on $\mathcal{M}^4$ with respect to the velocity vector field $\mathbf{u}$. Explicitly,

$$D_\mathbf{u}\mathbf{v} = (u^\alpha \nabla_\alpha v^\beta)\, \partial_\beta.$$

Two key relations for velocity vector fields. For all smooth velocity vector fields $\mathbf{u}, \mathbf{v}, \mathbf{w} \in \mathrm{Vect}(\mathcal{M}^4)$ on the space-time manifold $\mathcal{M}^4$, we have

$$[\mathbf{u}, \mathbf{v}] = D_\mathbf{u}\mathbf{v} - D_\mathbf{v}\mathbf{u}, \tag{0.75}$$

and

$$\mathbf{F}(\mathbf{u}, \mathbf{v})\mathbf{w} = (D_\mathbf{u}D_\mathbf{v} - D_\mathbf{v}D_\mathbf{u} - D_{[\mathbf{u},\mathbf{v}]})\mathbf{w}. \tag{0.76}$$

In terms of geometry, the Riemann curvature operator $\mathbf{F}(\mathbf{u}, \mathbf{v})$ was introduced on page 33 by using parallel transport of a velocity vector along a sufficiently small closed path. In terms of analysis, relation (0.76) connects the Riemann curvature operator with the covariant directional derivative. Equivalently, relation (0.75) reads as

$$\mathcal{L}_\mathbf{u}\mathbf{v} = D_\mathbf{u}\mathbf{v} - D_\mathbf{v}\mathbf{u}.$$

This connects the Lie derivative with the covariant directional derivative. In terms of the covariant directional derivative $D_\mathbf{u}\mathbf{v}$, we get the following two key formulas for fixed smooth velocity vector fields $\mathbf{u}, \mathbf{v}$ on the 4-dimensional space-time manifold $\mathcal{M}^4$ equipped with the metric tensor $\mathbf{g}$:

$$\boxed{D_\mathbf{u}\mathbf{v} - D_\mathbf{v}\mathbf{u} - [\mathbf{u}, \mathbf{v}] = 0} \tag{0.77}$$

and

$$\boxed{\mathbf{F}(\mathbf{u}, \mathbf{v}) = D_\mathbf{u}D_\mathbf{v} - D_\mathbf{v}D_\mathbf{u} - D_{[\mathbf{u},\mathbf{v}]},} \tag{0.78}$$

where $\mathbf{F}(\mathbf{u}, \mathbf{v})$ represents the Riemann curvature operator.[42] Introducing Weyl's torsion operator

$$\mathbf{T}(\mathbf{u}, \mathbf{v}) := D_\mathbf{u}\mathbf{v} - D_\mathbf{v}\mathbf{u} - [\mathbf{u}, \mathbf{v}],$$

[42] Gauss (1777–1855), Riemann (1826–1866), Lie (1842–1899), Ricci-Curbastro (1853–1925), Élie Cartan (1869–1951), Levi-Civita (1873–1941), Einstein (1879–1955), Weyl (1885–1955).

the first key relation (0.77) reads as

$$\mathbf{T}(\mathbf{u},\mathbf{v}) = 0.$$

That is, the torsion vanishes in Riemannian and pseudo-Riemannian geometry. But note that there are more general geometries where the torsion does not vanish. It turns out that the two key relations (0.77) and (0.78) govern Riemannian and pseudo-Riemannian geometry. If the metric tensor $\mathbf{g}$ is constant on $\mathcal{M}^4$, then we have the situation of Einstein's theory of special relativity where the gravitational force vanishes.[43] Then the two key relations (0.77) and (0.78) pass over to the two classical relations

- $d_{\mathbf{u}}\mathbf{v} - d_{\mathbf{v}}\mathbf{u} - [\mathbf{u},\mathbf{v}] = 0$, and
- $d_{\mathbf{u}}d_{\mathbf{v}} - d_{\mathbf{v}}d_{\mathbf{u}} - d_{[\mathbf{u},\mathbf{v}]} = 0$,

respectively. Here, $d_{\mathbf{u}}\mathbf{v}$ denotes the classical directional derivative. Relation (0.76) allows far-reaching generalizations which represent the fundamental principle

force = curvature

in modern physics. This will be studied in the present volume.

Parallel Transport and the Covariant Directional Derivative

We are given the smooth curve

$$C : P = P(\sigma), \quad \sigma_0 < \sigma < \sigma_0. \tag{0.79}$$

Let $\sigma_0 < 0 < \sigma_1$. Set $P_0 := P(0)$. We want to characterize both the covariant directional derivative $D_{\mathbf{u}}\mathbf{v}(P_0)$ and the covariant derivative $\frac{D\mathbf{v}(0)}{d\sigma}$ with respect to the real parameter σ by a limiting process based on parallel transport.

Covariant directional derivative. For a smooth classical real-valued function $f : \mathbb{R} \to \mathbb{R}$, the derivative $\dot{f}(0)$ at the point $\sigma = 0$ is given by

$$\dot{f}(0) = \lim_{\sigma\to 0} \frac{f(\sigma) - f(0)}{\sigma}.$$

As we will show later on, the generalization to the covariant directional derivative reads as follows:

$$\boxed{D_{\mathbf{u}}\mathbf{v}(P_0) = \lim_{\sigma\to 0} \frac{\Pi_\sigma^{-1}\mathbf{v}(P(\sigma)) - \mathbf{v}(P_0)}{\sigma}.} \tag{0.80}$$

Let us discuss this. We are given the smooth velocity vector field $\mathbf{v} = \mathbf{v}(P)$ on the space-time manifold $\mathcal{M}^4$. Fix the point $P_0 \in \mathcal{M}^4$, and fix the tangent vector $\mathbf{u} \in T_{P_0}\mathcal{M}^4$ at the point P_0. We choose a smooth curve C as given by (0.79) which passes through the point P_0 at the parameter value $\sigma = 0$ and has the tangent vector $\mathbf{u}$ at P_0. With respect to a local coordinate system, the curve C is given by $x = x(\sigma)$ where $x(0)$ corresponds to the point P_0 and $\mathbf{u} = \dot{x}^\alpha(0)\partial_\alpha$. Let us now consider the parallel transport along the curve C. To this end, we introduce the operator

$$\Pi_\sigma : T_{P_0}\mathcal{M}^4 \to T_{P(\sigma)}\mathcal{M}^4.$$

---

[43] In this special case, the Christoffel symbols vanish, and the covariant partial derivative coincides with the classical partial derivative, $\nabla_\alpha = \partial_\alpha$.

By definition, the tangent vector $\Pi_\sigma \mathbf{w}$ (at the point $P(\sigma)$) is obtained from the tangent vector $\mathbf{w}$ (at the point P_0) by using parallel transport from the point P_0 to the point $P(\sigma)$ along the curve C. This parallel transport is reversed by the inverse operator

$$\Pi_\sigma^{-1} : T_{P(\sigma)}\mathcal{M}^4 \to T_{P_0}\mathcal{M}^4.$$

This completes the explanation of the notation used in (0.80).

The geometric intuition behind (0.80) will be explained in Chap. 9.5 by considering the situation on a sphere. Naively, one would use the limit

$$\lim_{\sigma\to 0} \frac{\mathbf{v}(P(\sigma)) - \mathbf{v}(P_0)}{\sigma}.$$

This expression can be computed by using local coordinates. However, it turns out that, as a rule, the result depends on the choice of the local coordinates, that is, the expression does not possess any invariant geometric (or physical) meaning. Roughly speaking, the reason for this is the fact that the tangent vectors $\mathbf{v}(P_0)$ and $\mathbf{v}(P(\sigma))$ live in different tangent spaces. In order to be able to compute the vector difference in the same tangent space $T_{P_0}\mathcal{M}^4$, we replace the tangent vector $\mathbf{v}(P(\sigma))$ at the point $P(\sigma)$ by the parallel transported tangent vector $\Pi_\sigma^{-1}\mathbf{v}(P(\sigma))$ at the point P_0.

> *From the physical point of view, it is impossible to compare physical quantities (e.g., fields) at different space-time points without using the transport of physical information.*

This observation is crucial for gauge theory in modern physics (the theory of general relativity and the Standard Model in particle physics).

Covariant derivatives with respect to the real parameter σ. Let the family $\mathbf{v} = \mathbf{v}(\sigma)$ of velocity vectors be given along the curve C from (0.79), that is, the vector $\mathbf{v}(\sigma)$ lives in the tangent space $T_{P(\sigma)}\mathcal{M}^4$ for all $\sigma \in]\sigma_0, \sigma_1[$. We define

$$\boxed{\frac{D\mathbf{v}(0)}{d\sigma} := \lim_{\sigma\to 0} \frac{\Pi_\sigma^{-1}(\mathbf{v}(\sigma)) - \mathbf{v}(P_0)}{\sigma}.} \tag{0.81}$$

Similarly, we define the covariant derivative $\frac{D\mathbf{v}(\sigma)}{d\sigma}$ at the parameter $\sigma \in]\sigma_0, \sigma_1[$. With respect to local coordinates, we have

$$\mathbf{v}(\sigma) = v^\gamma(\sigma)\partial_\gamma, \qquad \frac{D\mathbf{v}(\sigma)}{d\sigma} = \frac{Dv^\gamma(\sigma)}{d\sigma}\,\partial_\gamma, \qquad \sigma \in]\sigma_0, \sigma_1[$$

together with

$$\frac{Dv^\gamma(\sigma)}{d\sigma} := \dot{v}^\gamma(\sigma) + \dot{x}^\alpha(\sigma)\Gamma^\gamma_{\alpha\beta}(x(\sigma)) \cdot v^\beta(\sigma), \qquad \gamma = 0, 1, 2, 3.$$

Here, the curve C corresponds to $x^\gamma = x^\gamma(\sigma)$ with $\sigma_0 < \sigma < \sigma_1$. By definition, the family $\mathbf{v} = \mathbf{v}(\sigma)$ of velocity vectors is parallel along the curve C iff

$$\boxed{\frac{D\mathbf{v}(\sigma)}{d\sigma} = 0 \qquad \text{for all} \quad \sigma \in]\sigma_0, \sigma_1[.}$$

As we will show later on, this ordinary differential equation describes the transport of physical information in gauge theory (theory of general relativity and the Standard Model in physics).

Parallel transport respects the inner product. Suppose that the two smooth velocity vector fields $\mathbf{v} = \mathbf{v}(\sigma)$ and $\mathbf{w} = \mathbf{w}(\sigma)$ are given along the curve

$C : P = P(\sigma)$, $\sigma_0 < \sigma < \sigma_1$. Then, for all $\sigma \in]\sigma_0, \sigma_1[$, we get the following generalized Leibniz product rule:

$$\frac{d}{d\sigma}\langle \mathbf{v}(\sigma)|\mathbf{w}(\sigma)\rangle = \left\langle \frac{D\mathbf{v}(\sigma)}{d\sigma}|\mathbf{w}(\sigma)\right\rangle + \left\langle \mathbf{v}(\sigma)|\frac{D\mathbf{w}(\sigma)}{d\sigma}\right\rangle. \tag{0.82}$$

This implies that, for all smooth velocity vector fields $\mathbf{u}, \mathbf{v}, \mathbf{w}$ on the space-time manifold $\mathcal{M}^4$, we have

$$d_{\mathbf{u}}\langle \mathbf{v}|\mathbf{w}\rangle = \langle D_{\mathbf{u}}\mathbf{v}|\mathbf{w}\rangle + \langle \mathbf{v}|D_{\mathbf{u}}\mathbf{w}\rangle.$$

Mnemonically, we write

$$d\langle \mathbf{v}|\mathbf{w}\rangle = \langle D\mathbf{v}|\mathbf{w}\rangle + \langle \mathbf{v}|D\mathbf{w}\rangle.$$

In particular, if $\mathbf{v} = \mathbf{v}(\sigma)$ and $\mathbf{w} = \mathbf{w}(\sigma)$ are parallel along the curve C, then $\frac{D\mathbf{v}(\sigma)}{d\sigma} = \frac{D\mathbf{w}(\sigma)}{d\sigma} \equiv 0$. By (0.82),

$$\langle \mathbf{v}(\sigma)|\mathbf{w}(\sigma)\rangle = \text{const} \qquad \text{for all} \quad \sigma \in]\sigma_0, \sigma_1[.$$

This means that the parallel transport of tangent vectors (i.e., velocity vectors) on the 4-dimensional space-time manifold preserves the (indefinite) inner product.

The covariant differential for general tensor fields. Later on, starting from the covariant differential $D\mathbf{v}$ of the velocity vector field $\mathbf{v}$, we will introduce the covariant differential $D\mathbf{T}$ for general smooth tensor fields $\mathbf{T}$ by using

- $(D\omega)(\mathbf{v}) = \omega(D\mathbf{v})$ (duality between vector fields $\mathbf{v}$ and covector fields ω), and
- $D(\mathbf{T} \otimes \mathbf{S}) = D\mathbf{T} \otimes \mathbf{S} + \mathbf{T} \otimes D\mathbf{S}$ (the Leibniz product rule for tensor fields $\mathbf{T}$ and $\mathbf{S}$).

Explicitly, for the given tensor field

$$\mathbf{T} = T^{\beta_1 \ldots \beta_m}_{\alpha_1 \ldots \alpha_n}\, dx^{\alpha_1} \otimes \cdots \otimes dx^{\alpha_n} \otimes \partial_{\beta_1} \otimes \cdots \otimes \partial_{\beta_m},$$

we obtain

$$D\mathbf{T} = \nabla_\alpha T^{\beta_1 \ldots \beta_m}_{\alpha_1 \ldots \alpha_n}\, dx^\alpha \otimes dx^{\alpha_1} \otimes \cdots \otimes dx^{\alpha_n} \otimes \partial_{\beta_1} \otimes \cdots \otimes \partial_{\beta_m}.$$

The Generalized Riemann Curvature Tensor in Modern Mathematics and Physics

Let $GL(N, \mathbb{C})$ denote the Lie group of all invertible complex $(N \times N)$-matrices. This is a real manifold of dimension $2N^2$. Moreover, let $gl(N, \mathbb{C})$ denote the Lie algebra to the Lie group $GL(N, \mathbb{C})$. Explicitly, the real Lie algebra $gl(N, \mathbb{C})$ consists of all complex $(N \times N)$-matrices. The following generalization of gauge theory is crucial for the Standard Model in particle physics. Note the following:

The structural equation

$$\mathcal{F}_{\alpha\beta} = \partial_\alpha \mathcal{A}_\beta - \partial_\beta \mathcal{A}_\alpha + \mathcal{A}_\alpha \mathcal{A}_\beta - \mathcal{A}_\beta \mathcal{A}_\alpha \tag{0.83}$$

with $\alpha, \beta = 0, 1, 2, 3$ *makes sense if* $\mathcal{A}_0, \mathcal{A}_1, \mathcal{A}_2, \mathcal{A}_3$ *are complex* $(N \times N)$*-matrices with* $N = 1, 2, \ldots$

This means that the matrices $\mathcal{A}_0, \mathcal{A}_1, \mathcal{A}_2, \mathcal{A}_3$ are contained in the real Lie algebra $gl(N, \mathbb{C})$. More generally, let the symbol $\mathcal{G}$ denote a closed subgroup of the Lie group $GL(N, \mathbb{C})$.

Then $\mathcal{G}$ is a Lie subgroup of $GL(N,\mathbb{C})$.

Furthermore, let $\mathcal{LG}$ denote the real Lie algebra to the Lie group $\mathcal{G}$. Choose

$$\mathcal{A}_\alpha \in \mathcal{LG}, \qquad \alpha = 0,1,2,3.$$

Then, the Lie product $[\mathcal{A}_\alpha, \mathcal{A}_\beta] := \mathcal{A}_\alpha\mathcal{A}_\beta - \mathcal{A}_\beta\mathcal{A}_\alpha$ is also contained in the Lie algebra $\mathcal{LG}$, and hence $\mathcal{F}_{\alpha\beta}$ is contained in the Lie algebra $\mathcal{LG}$. That is, both

- the connection form $\mathcal{A} = \mathcal{A}_\alpha\, dx^\alpha$ and
- the curvature form $\mathcal{F} = \frac{1}{2}\mathcal{F}_{\alpha\beta}\, dx^\alpha \wedge dx^\beta$

are differential forms with values in the Lie algebra $\mathcal{LG}$. Here, $\mathcal{G}$ is called the gauge group, and $\mathcal{LG}$ is called the gauge Lie algebra. For example, choose

$$\mathcal{G} = SU(N) \qquad \text{and} \qquad \mathcal{LG} = su(N), \qquad N = 2,3,\ldots$$

Here, the Lie group $SU(N)$ consists of all complex $(N\times N)$-matrices A with $AA^\dagger = I$ and the determinant $\det(A) = 1$. This is called the special unitary group (on the complex Hilbert space $\mathbb{C}^N$). The real Lie algebra $su(N)$ to the Lie group $SU(N)$ consists of all complex $(N \times N)$-matrices with $A + A^\dagger = 0$ and the trace $\mathrm{tr}(A) = 0$. Passing to components, we get

- $\mathcal{A}_\alpha = (\gamma^K_{\alpha L})$, and
- $\mathcal{F}_{\alpha\beta} = (r^K_{L\alpha\beta})$.

For the indices, we have $\alpha, \beta = 0,1,2,3$, and $K, L = 1,\ldots,N$. Here, the functions $\gamma^K_{\alpha L}$ (resp. $r^K_{L\alpha\beta}$) are called the generalized Christoffel symbols (resp. the components of the generalized Riemann curvature tensor).

Cartan's global approach. As we will show later on, it turns out that, as a rule, the connection form $\mathcal{A}$ and the curvature form $\mathcal{F}$ are not invariant under local gauge transformations. Therefore, Cartan introduced

- the global connection form A with values in the Lie algebra $\mathcal{LG}$, and
- the global curvature form F with values in the Lie algebra $\mathcal{LG}$

on an appropriate principal fiber bundle $\mathcal{P}(\mathcal{M}^4)$ over the space-time manifold $\mathcal{M}^4$. Then Cartan's local structural equation (0.83) is generalized to the global structural equation

$$\boxed{\mathsf{F} = \mathsf{DA}.} \qquad (0.84)$$

In terms of physics, this global structural equation represents the most elegant mathematical formulation of the principle "force equals curvature." In terms of mathematics, equation (0.84) represents a far-reaching generalization of Gauss' theorema egregium. Every differential form F represented by (0.84) satisfies the integrability condition

$$\boxed{\mathsf{DF} = 0} \qquad (0.85)$$

which is called the (global) Bianchi identity (see Chap. 17).

Parallel Transport of Physical Information and the Local Phase Factor

Let us sketch the basic ideas. We are given the curve $x = x(\sigma)$, $\sigma_0 \le \sigma \le \sigma_1$, on the space-time manifold $\mathcal{M}^4$. The ordinary differential equation

$$\boxed{\dot{G}(\sigma) = -\dot{x}^\alpha(\sigma)\mathcal{A}_\alpha(x(\sigma)) \cdot G(\sigma)} \tag{0.86}$$

with the initial condition $G(\sigma_0) = I$ (unit matrix) is called the differential equation of parallel transport (for the phase factor G). Since $\mathcal{A}_\alpha(P)$ is contained in the Lie algebra $\mathcal{LG}$ for all indices α and all points P of the space-time manifold, the solution $\sigma \mapsto G(\sigma)$ has the property that $G(\sigma)$ is an element of the Lie group $\mathcal{G}$ for all values of the parameter σ. As we will show later on, the differential equation (0.86) can be used in order to define the covariant differential DA by means of the classical differential $d\mathsf{A}$ and a suitable projection defined on the tangent spaces of the appropriate principal fiber bundle (horizontal tangent vectors; see Chap. 17.

The local phase factor of a physical field ψ. Let

$$\psi(x) = \begin{pmatrix} \psi^1(x) \\ \vdots \\ \psi^N(x) \end{pmatrix}$$

be a complex column matrix with N rows which depends on the space-time point x. In terms of physics, the function $x \mapsto \psi(x)$ describes a physical field (e.g., the wave function of an electron in the Standard Model in particle physics). Let $\sigma \mapsto G(\sigma)$ be the unique solution of the differential equation (0.86) of parallel transport. By definition,

$$\psi(\sigma) := G(\sigma)\psi(x(\sigma_0)), \qquad \sigma_1 \le \sigma \le \sigma_1. \tag{0.87}$$

This equation describes the parallel transport of the physical field ψ along the curve $x = x(\sigma)$, $\sigma_0 \le \sigma \le \sigma_1$. Here, $G(\sigma)$ is called the phase factor of the physical field ψ at the space-time point $x(\sigma)$. Using this terminology, the equation (0.86) describes the parallel transport of the local phase function $\sigma \mapsto G(\sigma)$. Differentiating equation (0.87) with respect to the parameter σ, we get the following differential equation of parallel transport for the physical field ψ :

$$\boxed{\dot{\psi}(\sigma) = -\dot{x}^\alpha(\sigma)\mathcal{A}_\alpha(x(\sigma))\,\psi(\sigma), \qquad \sigma_0 \le \sigma \le \sigma_1.}$$

In terms of physics, this differential equation describes the transport of physical information via phase factor.

The Modern Language of Fiber Bundles in Mathematics and Physics

> Physical fields are sections of fiber bundles. The qualitative (i.e., topological) structure of physical fields is determined by the topological structure of the corresponding fiber bundles. The prototype of a fiber bundle is the tangent bundle of a manifold.
>
> Folklore

The tangent bundle and tangent vector fields. Consider a tangent vector field $\mathbf{v} = \mathbf{v}(P)$ on the space-time manifold $\mathcal{M}^4$. Such a field assigns to each point P of the manifold $\mathcal{M}^4$ the tangent vector $\mathbf{v}(P)$ at the point P, that is, $\mathbf{v}(P) \in T_P\mathcal{M}^4$. In order to handle conveniently tangent vector fields as mathematical objects, we will describe them as maps of the form

$$\boxed{s : \mathcal{M}^4 \to T\mathcal{M}^4,} \tag{0.88}$$

which are called sections s of the tangent bundle $T\mathcal{M}^4$. Let us discuss this basic notion in modern mathematics. By definition, the tangent bundle $T\mathcal{M}^4$ of the manifold $\mathcal{M}^4$ consists of all the ordered pairs

$$(P, \mathbf{v})$$

where P is an arbitrary point of $\mathcal{M}^4$, and $\mathbf{v}$ is an arbitrary tangent vector of $\mathcal{M}^4$ at the point P. Briefly,

$$T\mathcal{M}^4 := \{(P, \mathbf{v}) : \ \ P \in \mathcal{M}^4, \ \mathbf{v} \in T_P\mathcal{M}^4\}.$$

The map $s : \mathcal{M}^4 \to T\mathcal{M}^4$ is called a section of the tangent bundle $T\mathcal{M}^4$ iff the image $s(P)$ has the form $(P, \mathbf{v}(P))$ with $\mathbf{v}(P) \in T_P\mathcal{M}^4$ for all points $P \in \mathcal{M}^4$.

Let us add some more terminology used in modern mathematics. Setting $\pi(P, \mathbf{v}) := P$, we get the so-called projection map

$$\pi : T\mathcal{M}^4 \to \mathcal{M}^4.$$

The pre-image $\mathcal{F}_P := \pi^{-1}(P)$ is called the fiber of the tangent bundle $T\mathcal{M}^4$ over the base point P. Explicitly,

$$\mathcal{F}_P := \{(P, \mathbf{v}) : \ \mathbf{v} \in T_P\mathcal{M}^4\}.$$

There exists a one-to-one correspondence $T_P\mathcal{M}^4 \leftrightarrow \mathcal{F}_P$. Therefore, the tangent spaces of the manifold $\mathcal{M}^4$ can be identified with the fibers of the tangent bundle $T\mathcal{M}^4$. If $P \neq Q$, then $\mathcal{F}_P \cap \mathcal{F}_Q = \emptyset$. We have

$$T\mathcal{M}^4 = \bigcup_{P \in \mathcal{M}^4} \mathcal{F}_P.$$

That is, the tangent bundle $T\mathcal{M}^4$ is the union of the pairwise disjoint fibers $\mathcal{F}_P$. Observe that the map $s : \mathcal{M}^4 \to T\mathcal{M}^4$ is a section of the tangent bundle iff the following diagram is commutative:

$$\begin{array}{ccc} \mathcal{M}^4 & \xrightarrow{\ s\ } & T\mathcal{M}^4 \\ & {}_{\text{id}}\searrow & \downarrow \pi \\ & & \mathcal{M}^4. \end{array} \tag{0.89}$$

That is, $\pi(s(P)) = P$ for all $P \in \mathcal{M}^4$. In other words, the section s respects fibers, that is, $s(P)$ lives in the fiber $\mathcal{F}_P$ for all $P \in \mathcal{M}^4$. Synonymously, we will use the notions

- tangent vector and
- velocity vector.

In particular, tangent vector fields are also called velocity vector fields. In a quite natural way, we are able to introduce local coordinates on the tangent bundle. To this end, we choose an arbitrary local coordinate system of the basis manifold $\mathcal{M}^4$. We assign to the point P of $\mathcal{M}^4$ the local coordinate $x = (x^0, x^1, x^2, x^3)$. Moreover, we have $\mathbf{v} = v^\alpha \partial_\alpha$. Finally, we assign to the point $(P, \mathbf{v})$ of the tangent bundle $T\mathcal{M}^4$ the local coordinates

$$(x^0, x^1, x^2, x^3, v^0, v^1, v^2, v^3).$$

This way, the tangent bundle $T\mathcal{M}^4$ becomes a real 8-dimensional manifold. By definition, the velocity vector field $\mathbf{v} = \mathbf{v}(P)$ is smooth on $\mathcal{M}^4$ iff the corresponding section $s : \mathcal{M}^4 \to T\mathcal{M}^4$ is a smooth map between the two manifolds $\mathcal{M}^4$ and $T\mathcal{M}^4$. The symbol $\mathrm{Vect}(\mathcal{M}^4)$ denotes the set of all smooth tangent vector fields on $\mathcal{M}^4$.

The cotangent bundle and cotangent vector fields. Now let us pass from tangent vectors to the dual objects called cotangent vectors. Precisely the linear functionals

$$\omega : T_P\mathcal{M}^4 \to \mathbb{R}$$

on the tangent space $T_P\mathcal{M}^4$ are called cotangent vectors of the manifold $\mathcal{M}^4$ at the point P. By definition, the set of all these linear functionals forms the dual space $T_P^*\mathcal{M}^4$ to the tangent space $T_P\mathcal{M}^4$. This dual space $T_P^*\mathcal{M}^4$ is called the cotangent space of the manifold $\mathcal{M}^4$ at the point P. We have $\omega \in T_P^*\mathcal{M}^4$ iff there exist real numbers $\omega_0, \omega_1, \omega_2, \omega_3$ such that

$$\omega = \omega_\alpha dx^\alpha.$$

Recall that $dx^\alpha(\mathbf{v}) := v^\alpha$ for all tangent vectors $\mathbf{v} = v^\alpha \partial_j$ at the point P. A cotangent vector field

$$\omega = \omega(P) \qquad \text{on } \mathcal{M}^4$$

assigns to each point P of $\mathcal{M}^4$ the cotangent vector $\omega(P)$. By definition, the cotangent bundle $T^*\mathcal{M}^4$ consists of all ordered pairs

$$(P, \omega)$$

where P is an arbitrary point of $\mathcal{M}^4$, and ω is an arbitrary cotangent vector of $\mathcal{M}^4$ at the point P. Briefly,

$$T^*\mathcal{M}^4 := \{(P, \omega) : \ P \in \mathcal{M}^4, \omega \in T_P^*(\mathcal{M}^4)\}.$$

The map

$$s : \mathcal{M}^4 \to T^*\mathcal{M}^4$$

is called a section of the cotangent bundle $T^*\mathcal{M}^4$ iff the image $s(P)$ has the form $(P, \omega(P))$ with $\omega(P) \in T_P^*\mathcal{M}^4$ for all points $P \in \mathcal{M}^4$. Setting $\pi(P, \omega) := P$, we get the so-called projection map

$$\pi : T^*\mathcal{M}^4 \to \mathcal{M}^4.$$

For the pre-image, we obtain $\pi^{-1}(P) = \{P\} \times T_P^*\mathcal{M}^4$. This is called the fiber $\mathcal{F}_P$ over the base point P. Explicitly,

$$\mathcal{F}_P = \{(P, \omega) : P \in \mathcal{M}^4, \ \omega \in T_P^*\mathcal{M}^4\}.$$

Therefore, the cotangent spaces of the manifold $\mathcal{M}^4$ can be identified with the fibers of the cotangent bundle $T^*\mathcal{M}^4$. The map

$$s : \mathcal{M}^4 \to T^*\mathcal{M}^4$$

is called a section of the cotangent bundle iff the following diagram is commutative:

$$\begin{array}{ccc} \mathcal{M}^4 & \xrightarrow{s} & T^*\mathcal{M}^4 \\ & {\scriptstyle \mathrm{id}} \searrow & \downarrow {\scriptstyle \pi} \\ & & \mathcal{M}^4. \end{array} \tag{0.90}$$

That is, $\pi(s(P)) = P$ for all $P \in \mathcal{M}^4$. Sections of the cotangent bundle $T^*\mathcal{M}^4$ correspond to cotangent vector fields $\omega = \omega(P)$ on $\mathcal{M}^4$. Synonymously, we use the following notions:

- cotangent vector,
- velocity covector (or briefly covector),
- differential 1-form.

In particular, cotangent vector fields are also called velocity covector fields (or fields of differential 1-forms). Using $\omega = \omega_\alpha dx^\alpha$, we assign to the point (P, ω) of the cotangent bundle $T^*\mathcal{M}^4$ the local coordinates

$$(x^0, x^1, x^2, x^3, \omega_0, \omega_1, \omega_2, \omega_3).$$

This way, the cotangent bundle $T^*\mathcal{M}^4$ becomes a real 8-dimensional manifold. By definition, the cotangent vector field $\omega = \omega(P)$ is smooth on $\mathcal{M}^4$ iff the corresponding section $s : \mathcal{M}^4 \to T^*\mathcal{M}^4$ is a smooth map. The symbol $\mathrm{Covect}(\mathcal{M}^4)$ denotes the set of all smooth cotangent vector fields on $\mathcal{M}^4$.

Tensor bundles and tensor fields. By definition, a tensor of type $(0, 2)$ of the manifold $\mathcal{M}^4$ at the point P is a bilinear map of the form

$$\mathbf{g} : T_P^*\mathcal{M}^4 \times T_P^*\mathcal{M}^4 \to \mathbb{R}.$$

This means that $\mathbf{g} = g_{\alpha\beta}\, dx^\alpha \otimes dx^\beta$ where $g_{\alpha\beta}$ are fixed, but otherwise arbitrary real numbers for all indices $\alpha, \beta = 0, 1, 2, 3$. The tensor bundle $T_2^0(\mathcal{M}^4)$ consists of all the ordered pairs

$$(P, \mathbf{g})$$

where P is an arbitrary point of $\mathcal{M}^4$, and $\mathbf{g}$ is an arbitrary tensor of type $(0, 2)$. The map $s : \mathcal{M}^4 \to T_2^0\mathcal{M}^4$ is called a section of the tensor bundle $T_2^0\mathcal{M}^4$ iff the following diagram is commutative:

$$\begin{array}{ccc} \mathcal{M}^4 & \xrightarrow{s} & T_2^0\mathcal{M}^4 \\ & {\scriptstyle \mathrm{id}} \searrow & \downarrow {\scriptstyle \pi} \\ & & \mathcal{M}^4. \end{array} \tag{0.91}$$

Here, $\pi(P, \mathbf{g}) := P$. Sections of the tensor bundle $T_2^0\mathcal{M}^4$ correspond to tensor fields $\mathbf{g} = \mathbf{g}(P)$ of type $(0, 2)$ on $\mathcal{M}^4$. Using $\mathbf{g} = g_{\alpha\beta}\, dx^\alpha \otimes dx^\beta$, we assign to the point $(P, \mathbf{g})$ of the tensor bundle $T_2^0\mathcal{M}^4$ the local coordinates

$$(x^\gamma, g_{\alpha\beta})_{\alpha,\beta,\gamma=0,1,2,3}.$$

This way, the tensor bundle $T_2^0\mathcal{M}^4$ becomes a real 20-dimensional manifold. The symbol $\bigotimes_2^0(\mathcal{M}^4)$ denotes the set of all smooth sections $s : \mathcal{M}^4 \to T_2^0\mathcal{M}^4$ of the tensor bundle $T_2^0\mathcal{M}^4$ (i.e., the set of all smooth tensor fields of type $(0, 2)$ on the space-time manifold $\mathcal{M}^4$).

Analogously, we will introduce the tensor bundle $T_n^m \mathcal{M}^4$, and the set $\bigotimes_n^m(\mathcal{M}^4)$ of all smooth tensor fields of type (m,n) on $\mathcal{M}^4$.

The symbol $\bigwedge^m(\mathcal{M}^4)$ denotes the set of all smooth antisymmetric tensor fields of type $(m,0)$ on $\mathcal{M}^4$. As we will show later on, such tensor fields coincide with fields of differential m-forms on $\mathcal{M}^4$.

The topological structure of velocity vector fields and characteristic classes. It turns out that the qualitative behavior of physical fields depends on the topological properties of the corresponding fiber bundles. In this connection, characteristic classes of fiber bundles are the most important topological invariants which govern the topological structure of physical fields. To illustrate this, let us consider the case of velocity vector fields $\mathbf{v} = \mathbf{v}(P)$ on the n-dimensional unit sphere $\mathbb{S}^n$. We want to study the following three basic questions:

- the existence of stagnation points,
- the maximal number of linearly independent continuous velocity vector fields (in the global sense), and
- the existence of a global parallel transport.

By definition, the velocity vector field $\mathbf{v} = \mathbf{v}(P)$ has the critical (or stagnation) point P_0 iff $\mathbf{v}(P_0) = 0$.

Stagnation points. The Euler characteristic of the sphere $\mathbb{S}^n$ is given by

$$\chi(\mathbb{S}^n) = 1 + (-1)^n, \qquad n = 1, 2, \ldots$$

Let n be even. There exists a smooth field $\omega = \omega(P)$ of differential n-forms on the sphere $\mathbb{S}^n$ such that we have the integral representation

$$\boxed{\chi(\mathbb{S}^n) = \int_{\mathbb{S}^n} \omega, \qquad n = 2, 4, 6, \ldots}$$

Physicists call the Euler characteristic $\chi(\mathbb{S}^n)$ a topological charge. By definition, the Euler class $[\omega]$ of the tangent bundle $T\mathbb{S}^n$ is the set

$$[\omega] := \{\omega + d\nu\}$$

where ν is an arbitrary smooth differential $(n-1)$-form on $\mathbb{S}^n$. In other words,

$$[\omega] \in H^n(\mathbb{S}^n), \qquad n = 1, 2, \ldots$$

That is, the Euler class $[\omega]$ is an element of $H^n(\mathbb{S}^n)$ (the nth de Rham cohomology group of the sphere $\mathbb{S}^n$). Note that the generalized Stokes integral theorem tells us that

$$\int_{\mathbb{S}^n} d\nu = \int_{\partial \mathbb{S}^n} \nu = 0,$$

since the boundary $\partial \mathbb{S}^n$ of the sphere $\mathbb{S}^n$ is empty (see page 729). Therefore,

$$\chi(\mathbb{S}^n) = \int_{\mathbb{S}^n} \omega + d\nu,$$

that is, the Euler characteristic of the sphere only depends on the Euler class of the sphere. Let $n = 1, 2, 3, \ldots$ A special case of the Poincaré–Hopf theorem tells us that:

A continuous velocity vector field without stagnation points exists on the n-dimensional sphere iff $\chi(\mathbb{S}^n) = 0$.

Explicitly, this means the following:

- If n is even, then every continuous velocity vector field on $\mathbb{S}^n$ has a stagnation point.
- If n is odd, then there exists a continuous velocity vector field on $\mathbb{S}^n$ which has no stagnation point.

The theorem of Adams. By definition, an m-field on the sphere $\mathbb{S}^n$ is a family of m continuous velocity vector fields $\mathbf{v}_1, \ldots, \mathbf{v}_m$ on $\mathbb{S}^n$ with the property that the tangent vectors $\mathbf{v}_1(P), \ldots, \mathbf{v}_m(P)$ are linearly independent at each point P in $\mathbb{S}^n$. The number

$$\mathrm{Span}(T\mathbb{S}^n)$$

is the maximal number m such that an m-field exists on $\mathbb{S}^n$. We have:

- $\mathrm{Span}(T\mathbb{S}^n) = 0$ if n is even (i.e., $n = 2, 4, 6, \ldots$),
- $\mathrm{Span}(T\mathbb{S}^1) = 1$, $\mathrm{Span}(T\mathbb{S}^3) = 3, \mathrm{Span}(T\mathbb{S}^5) = 1$, $\mathrm{Span}(T\mathbb{S}^7) = 7$.

The deep theorem of Adams tells us the precise result:[44]

$$\mathrm{Span}(T\mathbb{S}^n) = 8a + 2^b - 1, \qquad n = 1, 2, \ldots$$

Here, the nonnegative integers a and b with $b \leq 3$ are uniquely determined by the prime number factorization $n + 1 = 2^{4a+b} \cdot c$, where c is a positive odd integer.

For example, if $n = 7$, then $8 = 2^3 \cdot 1$. Hence $a = 0, b = 3, c = 1$. This implies $\mathrm{Span}(T\mathbb{S}^7) = 2^3 - 1 = 7$.

Global parallel transport. The sphere $\mathbb{S}^n$ is called parallelizable iff the condition $\mathrm{Span}(T\mathbb{S}^n) = n$ is satisfied.

The sphere $\mathbb{S}^n$ is parallelizable iff $n = 1, 3, 7$.

Let us discuss this. Suppose that $n = 1, 3, 7$. We want to show that there exists a global parallel transport. In fact, there exist continuous velocity fields $\mathbf{v}_1, \ldots, \mathbf{v}_n$ on $\mathbb{S}^n$ such that the vectors

$$\mathbf{v}_1(P), \ldots, \mathbf{v}_n(P)$$

form a basis of the tangent space $T_P\mathbb{S}^n$ for all points $P \in \mathbb{S}^n$. Fix the point P_0 and choose a fixed tangent vector $\mathbf{v}_0 \in T_{P_0}$. Then there exist uniquely determined real numbers $v^1, \ldots, v^n$ such that

$$\mathbf{v}_0 = \sum_{j=1}^{n} v^j \mathbf{v}_j(P_0).$$

Naturally enough, the global parallel transport of the vector $\mathbf{v}_0$ to the arbitrary point P of $\mathbb{S}^n$ is defined by

$$\mathbf{v}(P) := \sum_{j=1}^{n} v^j \mathbf{v}_j(P).$$

Triviality of the tangent bundle. The tangent bundle $T\mathbb{S}^n$ is called trivial iff $\mathbb{S}^n$ is parallelizable, that is, $n = 1, 3, 7$. Then every tangent vector $\mathbf{v}$ at the point P can uniquely be represented by the formula

$$\mathbf{v} = \sum_{j=1}^{n} v^j \mathbf{v}_j(P)$$

[44] J. Adams, Vector fields on spheres, Ann. Math. **75** (1962), 603–632.

where $v^1, \ldots, v^n$ are arbitrary real numbers. This means that we have a global coordinate system for the tangent bundle $T\mathbb{S}^n$. There exists a bijection between the points $(P, \mathbf{v})$ of the tangent bundle and the points

$$(P, v^1, \ldots, v^n)$$

of the product bundle $\mathbb{S}^n \times \mathbb{R}^n$. Summarizing:

> *The topology of fiber bundles allows us to distinguish between trivial and nontrivial physical fields.*

We expect that essential physical effects are related to nontrivial topological structures.

Typical examples. In the present Volume III, we will study the following applications of gauge theory:

- gauge theory on the 3-dimensional Euclidean manifold (Sect. 9.4),
- gauge theory on the sphere (e.g., the surface of earth) as a paradigm (Sect. 9.5),
- the relation between Gauss' surface theory and Levi-Civita's parallel transport (Sect. 9.5),
- the relation between Gauss' surface theory and Cartan's method of moving frames (Sect. 9.5),
- the main theorem on velocity vector fields on the Euclidean manifold – the classic predecessor of gauge theory in physics (Sect. 12.10.3),
- Maxwell's theory of electromagnetism as a commutative $U(1)$-gauge theory on the Minkowski manifold (Chap. 13 and Chaps. 18–23),
- the noncommutative $SU(N)$-gauge theory due to Yang and Mills (Chap. 15).

The general axiomatic approach to gauge theory will be discussed in Chap. 17. In Volume IV, we will study

- the Standard Model in particle physics with a representation of the product group $U(1) \times SU(2) \times SU(3)$ as gauge group, and
- the general theory of relativity for gravitation (connection on the tangent bundle of the pseudo-Riemannian space-time manifold),
- minimal surfaces and the conformal gauge symmetry,
- string theory and the conformal gauge symmetry.

Perspectives

The discussion above displays fruitful relations between mathematics and physics. Let us add some further quotations.

Instantons and gauge theory:

> From 1977 onward my interest moved in the direction of gauge theories and the interactions between geometry and physics. I had for many years a mild interest in theoretical physics, stimulated on many occasions by lengthy discussions with George Mackey from Harvard University. However, the stimulus in 1977 came from two other sources. On the one hand, Singer told me about the Yang–Mills equations, which through the influence of Yang were just beginning to percolate into mathematical circles. During his stay in Oxford in early 1977, Singer, Hitchin, and I took a serious look at the self-duality equations. We found that a simple application of the Atiyah–Singer index theorem gave the formula for the number of instanton parameters ... The other stimulus came from the presence in Oxford of

Roger Penrose and his group working on relativistic spinor calculus and twistor theory.[45]

Sir Michael Atiyah, 1988

Conformal symmetry and twistors:

A new type of algebra for Minkowski space-time is described, in terms of which it is possible to express any conformally or Poincaré covariant operation. The elements of the algebra (twistors) are combined according to tensor-type rules, but they differ from tensors or spinors in that they describe locational properties in addition to directional ones.

Twistor algebra will have the same type of universality, in relation to the conformal group, that the well-known and highly effective two-component spinor algebra of van der Waerden has, in relation to the Lorentz group. Twistors are, in fact, the "spinors" which are relevant to the six-dimensional space whose (pseudo)-rotation group is isomorphic to the conformal group of ordinary space-time.[46]

Roger Penrose, 1967

The Seiberg–Witten equations and the quark confinement:

Riemannian, symplectic and complex geometry are often studied by means of solutions to systems of nonlinear differential equations, such as the equations of geodesics, minimal surfaces, Einstein's curved universe, pseudo-holomorphic curves and Yang–Mills connections. For studying such equations, a unified technology has been developed, involving analysis on infinite-dimensional manifolds.

A striking application of the new technology is Donaldson's theory of "anti-self-dual" connections on $SU(2)$-bundles over four-manifolds, which applies the Yang–Mills equations from mathematical physics to shed light on the relationship between the classification of topological and smooth four-manifolds. This reverses the expected direction of application from topology to differential equations to mathematical physics. Even though the Yang–Mills equations are only mildly nonlinear, a prodigious amount of nonlinear analysis is necessary to fully understand the properties of the space of solutions.

At our present state of knowledge, understanding smooth structures on topological four-manifolds seems to require nonlinear as opposed to linear partial differential equations. It is therefore quite surprising that there is a set of partial differential equations which are even less nonlinear than the Yang–Mills equation, but can yield many of the most important results from Donaldson's theory. These are the Seiberg–Witten equations ...

During the 1980's, Simon Donaldson used the Yang–Mills equations, which had originated in mathematical physics, to study the differential topology

---

[45] See M. Atiyah, Collected Works, vol. V: Gauge Theories, Clarendon Press, Oxford, 1988. Reprinted by permission of Oxford University Press.

[46] Reprinted with permission from R. Penrose, Twistor algebra, J. Math. Phys. **8**(2) (1967), 345–366. Copyright 1967, American Institute of Physics.
N. Hitchin, G. Segal, and R. Ward, Integrable Systems, Twistors, Loop Groups, and Riemann Surfaces, Oxford University Press, 1999.
J. Frauendiener and R. Penrose, Twistors and general relativity, pp. 479–505. In: B. Engquist and W. Schmid (Eds.), Mathematics Unlimited—2001 and Beyond, Springer, New York, 2001.

of four-manifolds. Using work of Michael Freedman, he was able to prove theorems of the following type:

- There exist many compact four-manifolds which have no smooth structure.
- There exist many pairs of compact four-manifolds which are homeomorphic but not diffeomorphic.

The nonlinearity of the Yang-Mills equations presented difficulties, so many techniques within the theory of nonlinear partial differential equations had to be developed. Donaldson's theory was elegant and beautiful, but the details were difficult for beginning students to master.
In the fall of 1994, the physicist Edward Witten proposed a set of equations which give the main results of Donaldson's theory in a far simpler way than had been thought possible... The Seiberg–Witten equations give rise to new invariants of four-dimensional smooth manifolds called the Seiberg–Witten invariants. The key point is that homeomorphic smooth four-manifolds may have quite different Seiberg–Witten invariants ... Shortly after the Seiberg–Witten invariants were discovered, several striking applications were found concerning
(i) the proof of the Thom conjecture on the smooth embedding of compact Riemann surfaces into two-dimensional complex projective spaces $\mathbb{P}^2_{\mathbb{C}}$,
(ii) obstructions to the existence of a Riemannian geometry with positive curvature on manifolds, and
(iii) the existence of pseudo-holomorphic curves on symplectic manifolds.[47]

John Moore, 1996

This quotation refers to the following two fundamental papers on the quark confinement:

N. Seiberg and E. Witten, Electric-magnetic duality, monopole condensation, and confinement in $N = 2$ supersymmetric Yang–Mills theory, Nuclear Phys. **B426** (1994), 19–52.

N. Seiberg and E. Witten, Monopoles, duality and chiral symmetry breaking in $N = 2$ supersymmetric QCD, Nucl. Physics **B431** (1994), 485–550.

These two papers concern the computation of models which describe electrically and magnetically charged supersymmetric particles at low energies in the setting of gauge theory. The Seiberg–Witten equations use the spin structure of manifolds called spin manifolds, and they generalize the Landau–Ginzburg equation in superconductivity. This can be found in:

C. Nash, Topology and Physics – a Historical Essay, pp. 359–415. In: I. James (Ed.), History of Topology, Oxford University Press, 1999.

M. Atiyah, The Dirac equation and geometry, pp. 108–124. In: P. Goddard (Ed.), Paul Dirac – the Man and his Work, Cambridge University Press, 1998.

E. Witten, Physical law and the quest for mathematical understanding, Bull. Amer. Math. Soc. **40** (2003), 21–30.

---

[47] J. Moore, Lectures on Seiberg–Witten Invariants, Springer, Berlin, 1996 (reprinted with permission). We recommend these Lecture Notes as an introduction to the study of the Seiberg–Witten equations.

E. Witten, From superconductors and four-manifolds to weak interaction, Bull. Math. Soc. **44** (2007), 361–391.

J. Jost, Riemannian Geometry and Geometric Analysis, Springer, Berlin, 2008.

H. Lawson and M. Michelsohn, Spin Geometry, Princeton University Press, 1994.

Furthermore, we refer to:

R. Stern, Instantons and the topology of four-manifolds, Mathem. Intelligencer **5**(3) (1983), 39–44.

K. Marathe, Topics in Physical Mathematics, Springer, London, 2010.

D. Freed and K. Uhlenbeck, Instantons and Four-Manifolds, Springer, New York, 1984.

S. Donaldson and P. Kronheimer, The Geometry of Four-Manifolds, Oxford University Press, 1990.

S. Donaldson, The Seiberg–Witten equations and 4-manifold topology, Bull. Amer. Math. Soc. **33** (1996), 45–70.

J. Morgan, The Seiberg–Witten Equations and Applications to the Topology of Four-Manifolds, Princeton University Press, 1996.

T. Friedrich, Dirac Operators in Riemannian Geometry, Amer. Math. Soc., Providence, Rhode Island, 2000.

P. Kronheimer and T. Mirowka, Monopoles and Three-Manifolds, Cambridge University Press, 2007.

Concerning Morse theory, Floer homology, and quantum cohomology, we refer to the following monographs:

J. Jost, Riemannian Geometry and Geometric Analysis, Springer, Berlin, 2008.

M. Schwarz, Morse Homology, Birkhäuser, Basel, 1993.

D. McDuff and D. Salamon, J-holomorphic Curves and Quantum Cohomology, Amer. Math. Soc., Providence, Rhode Island, 1994.

Yu. Manin, Frobenius manifolds, quantum cohomology and moduli spaces, Amer. Math. Soc., Providence, Rhode Island, 1999.

S. Donaldson, Floer Homology Groups, Cambridge University Press, 2002.

J. Kock and I. Vainsencher, An Invitation to Quantum Cohomology: Kontsevich's Formula for Plane Curves, Birkhäuser, Basel, 2006.

S. Novikov, Topological Library, Vol. 1: Cobordisms and their Applications, Vol. 2: Characteristic Classes and Smooth Structures, World Scientific, Singapore, 2007/09.

The spectrum of elliptic Dirac operators on manifolds and noncommutative geometry. The original Dirac equation for the relativistic electron is a first-order system of partial differential equations of hyperbolic type. In geometry, one uses an elliptic variant of the Dirac operator by passing to imaginary time. For example, the Seiberg–Witten equations in geometry are nonlinear equations related to the elliptic Dirac differential operator. In 1985, Alain Connes created noncommutative geometry. The decisive analytic information comes from the spectrum of an elliptic Dirac operator on a compact manifold. References can be found on page 346.

The Millennium Prize Problem in quantum field theory. One of the seven Millennium Prize Problems concerns the Yang–Mills gauge theory. This is described in:

A. Jaffe and E. Witten, Quantum Yang–Mills theory, pp. 129–152. In: J. Carlson, A. Jaffe, and A. Wiles (Eds.), The Millennium Prize Problems, Amer. Math. Soc., Providence, Rhode Island, 2006.

The problem is to show that there exists a gap between the ground state energy and the first excitation energy of a Yang–Mills quantum field. The award for solving this problem will be one million dollars. Nowadays it is completely open how to attack this problem.

The Seiberg–Witten equations and the Weinstein conjecture. Let us mention a recent beautiful and deep application of the Seiberg–Witten equations to dynamical systems on 3-dimensional contact manifolds. Recall first the classical Poincaré theorem saying that every continuous velocity vector field on a two-dimensional sphere has a zero. In terms of physics, this corresponds to a stationary point of the velocity vector field. Such a point represents a (trivial) closed orbit of the corresponding flow of fluid particles. In other words, the Poincaré theorem tells us that:

Every continuous velocity vector field on a 2-dimensional sphere has a closed orbit.

We want to generalize this to the 3-dimensional sphere. It turns out that this is a hard problem. First of all note that there exist continuous velocity vector fields on the 3-dimensional sphere which have no closed orbits. Recently, Taubes proved the following theorem:

Every smooth Reeb velocity vector field on a real compact oriented 3-dimensional manifold (without boundary) has a closed orbit.

This tells us that the Weinstein conjecture is true in three dimensions. Let us explain the notation. A Reeb velocity vector field $\mathbf{v}$ on the 3-dimensional manifold $\mathcal{M}$ is given by the equations

$$d\omega(\mathbf{v}) = 0, \qquad \omega(\mathbf{v}) = 1 \qquad \text{on } \mathcal{M}$$

where ω is a contact form on $\mathcal{M}$, that is, ω is a smooth 1-form on $\mathcal{M}$ with the property $(d\omega \wedge \omega)(P) \neq 0$ for all points P of $\mathcal{M}$.[48]

The point of departure is the Taubes theorem on the relation between the Seiberg–Witten theory and the Gromov theory on pseudo-holomorphic curves. This theorem relates the Seiberg–Witten invariants of real symplectic 4-dimensional manifolds to counts of holomorphic curves. In the present case, we have to deal with 3-dimensional manifolds. To this end, Taubes uses a 3-dimensional variant of the Seiberg–Witten theory combined with Floer homology. We refer to:

M. Hutchings, Taubes's proof of the Weinstein conjecture in dimension three, Bull. Amer. Math. Soc. **47**(1) (2010), 73–126.

C. Taubes, Seiberg–Witten and Gromov Invariants for Symplectic 4-Manifolds, International Press, Boston, 2000.

C. Taubes, The Seiberg–Witten equations and the Weinstein conjecture I, II: Geom. Topol. **11** (2007), 2117–2002, **13** (2009), 1337–1417.

Taubes discovered in the 1990s that

[48] An introduction to Lie's contact geometry can be found in Sect. 5.7 of Vol. II.

The Seiberg–Witten theory and Gromov's theory of pseudo-holomorphic curves are equivalent in some sense.

This relates apparently completely different physical and mathematical topics with each other.

- In terms of physics, the Seiberg–Witten equation is closely related to the Landau–Ginzburg equation for describing phase transitions in condensed matter (e.g., superconductors).
- The Gromov theory of pseudo-holomorphic curves generalizes the Cauchy–Riemann differential equations for holomorphic functions.

We refer to:

I. Vekua, Generalized Analytic Functions, Pergamon Press, London, 1962.

M. Gromov, Pseudo-holomorphic curves in symplectic manifolds, Invent. Math. **82** (1985), 307–347.

H. Hofer and E. Zehnder, Symplectic Invariants and Hamiltonian Dynamics, Birkhäuser, Basel, 1994.

D. McDuff and D. Salamon, J-holomorphic Curves and Quantum Cohomology, Amer. Math. Soc., Providence, Rhode Island, 1994.

D. McDuff and D. Salamon, Introduction to Symplectic Topology, Clarendon Press, Oxford, 1998.

We will show in Sect. 2.3.2 on page 129 that the Cauchy–Riemann equations are closely related to the Clifford algebra $\bigvee(E_2)$ of the Euclidean plane E_2 (i.e., the algebra of quaternions). It turns out that

The Dirac equation for the relativistic electron is based on the Clifford algebra $\bigvee(M_4)$ of the 4-dimensional Minkowski space M_4 (flat space-time manifold). Based on the theory of Clifford algebras, the Dirac equation represents a generalization of the classical Cauchy–Riemann equations.

In his 1851 Ph.D. thesis, Riemann used the Cauchy–Riemann differential equations in order to create the geometric theory of holomorphic functions based on the notion of conformal map and the idea of the Riemann surface for describing analytic continuation in a global setting. Riemann was strongly motivated by ideas coming from physics (e.g., electricity). He used physical intuition in order to motivate the existence of global analytic functions on compact Riemann surfaces. Riemann's successors filled the gaps in Riemann's arguments step by step. The final form of the theory was published by

H. Weyl, The Concept of a Riemann Surface (in German), Teubner, Leipzig, 1913. New edition with commentaries supervised by R. Remmert, Teubner, Leipzig, 1997. English edition: Addison Wesley, Reading, Massachusetts, 1955.

In terms of mathematics, the following topics lurk behind the equivalence of the Seiberg–Witten theory and the Gromov theory of pseudo-holomorphic curves:

- symplectic geometry, fixed-point theorems for symplectic maps (i.e., higher-dimensional versions of Poincaré's last theorem), geometrical optics, and Hamiltonian mechanics (e.g., celestial mechanics) (see Vol. II),
- Lie's contact geometry and the Legendre transformation for thermodynamical potentials (see Vol. II),
- Clifford algebras, spin geometry, the Dirac equation for the relativistic electron in flat and curved space-times, symmetry breaking and the Higgs particle in weak interaction, and the quark confinement.

The creation of Floer homology was essentially motivated by the following fundamental paper:

E. Witten, Supersymmetry and Morse theory, J. Diff. Geom. **17** (1982), 661–692.

We refer to the basic paper by

A. Floer, Witten's complex and infinite-dimensional Morse theory, J. Diff. Geometry **30** (1989), 207–221.

Morse theory studies the relation between the critical points of energy functionals $E : \mathcal{M} \to \mathbb{R}$ and the topology of the manifold $\mathcal{M}$. As a nice introduction to modern Morse theory based on Floer homology, we recommend:

J. Jost, Riemannian Geometry and Geometric Analysis, fifth edition, Chapter 6, Springer, Berlin, 2008.

We also recommend the monographs on Morse homology and its applications by Schwarz (1993) and Donaldson (2002) quoted on page 60.

The language of bundles. The following equations of physical theories possess a similar structure:

- the basic equations of quantum electrodynamics (see Chap. 11 of Vol. II),
- the Landau–Ginzburg equation,
- the Dirac equation for the relativistic electron,
- the Yang–Mills equation,
- the Seiberg–Witten equation,
- the Standard model in particle physics (see Vol. IV).

In order to display the similarities, one has to use variational problems based on the principle of critical action. Then the Lagrangians possess a similar structure which depends on the choice of both

- the symmetry group $\mathcal{G}$ (the curvature of the principal fiber bundle $\mathcal{P}$ with the structure group $\mathcal{G}$), and
- the spaces of physical fields (sections of vector bundles which are associated to $\mathcal{P}$ via representations of the symmetry group $\mathcal{G}$).

The symmetry group of quantum electrodynamics (resp. of the Yang–Mills equations) is the group $U(1)$ (resp. $SU(2)$). The Dirac equation and the Seiberg–Witten equations are based on so-called spin groups (universal covering groups of the Lorentz group $O(1,3)$ and the rotation groups $SO(N)$) which are closely related to the spin of the particles.

The interactions correspond to the curvature of the principal fiber bundle.

The basic ideas will be studied in Chap. 15 (Ariadne's thread in gauge theory).

Gauge Potentials, moduli spaces, correlation functions of quantum fields, and Feynman path integrals. The basic formula

$$\mathsf{F} = D\mathsf{A}$$

describes the relation between the interaction forces F and the gauge potential A by means of the first-order differential operator D. Note the following peculiarity. By a local gauge transformation, we understand a change of the local bundle coordinates of the corresponding principal bundle with the symmetry group $\mathcal{G}$. By a global gauge transformation, roughly speaking, we understand a diffeomorphism $f : \mathcal{P} \to \mathcal{P}$ of the bundle space $\mathcal{P}$ of the principal bundle which is generated by the action of the symmetry group $\mathcal{G}$ (also called gauge group) on the principal fiber bundle. There arises the following question:

Which is the qualitative (topological) and quantitative structure of the space of all gauge potentials A *up to global gauge transformations?*

In other words, one has to investigate the moduli space $\mathrm{Mod}(\mathsf{A})$ of all gauge potentials A (called connections in mathematics). More precisely, $\mathrm{Mod}(\mathsf{A})$ is the space of all equivalence classes of connections modulo global gauge transformations. The space $\mathrm{Mod}(\mathsf{A})$ is called the moduli space of gauge potentials (connections). The investigation of moduli spaces needs sophisticated topological tools. The point is that, as a rule, moduli spaces are not smooth structures; they are not manifolds, but they possess singularities. In a natural way, such geometric objects arise in algebraic geometry (e.g., the curve $x^2 - y^2 = 0$ has a singularity at the point $(0, 0)$). We refer to:

S. Abhyankar, Algebraic Geometry for Scientists and Engineers, Amer. Math. Soc., Providence, Rhode Island, 1990.

In terms of quantum field theory, the knowledge of the moduli space is fundamental for computing correlation functions via Feynman path integrals:

$$\int_{[\mathsf{A}]\in \mathrm{Mod}(\mathsf{A})} \mathrm{e}^{\mathrm{i}S([\mathsf{A}])/\hbar}\, \mathcal{D}([\mathsf{A}]). \tag{0.92}$$

Here, one has to sum (i.e., to integrate) over all the possible physical states $[\mathsf{A}]$ (i.e., over all the elements of the moduli space). The statistical weight $\mathrm{e}^{\mathrm{i}S([\mathsf{A}])/\hbar}$ depends on the action $S([\mathsf{A}])$ corresponding to the physical state $[\mathsf{A}]$ (equivalence class of connections modulo global gauge transformations). As an introduction to moduli spaces in gauge theory, we recommend:

K. Marathe and G. Martucci, The Mathematical Foundations of Gauge Theories, North-Holland, Amsterdam, 1992.

K. Marathe, Topics in Physical Mathematics, Springer, London, 2010.

G. Naber, Topology, Geometry, and Gauge Fields, Springer, New York, 1997.

Fundamental papers on this subject can be found in:

M. Atiyah, Collected Works, Vol. 5: Gauge Theories, Cambridge University Press, 2004.

Topological quantum field theory. The basic idea of topological quantum field theory is to choose special functionals $\mathsf{A} \mapsto S([\mathsf{A}])$ in order to obtain topological invariants by using Feynman path integrals of the type (0.92). These integrals are computed by means of the method of stationary phase. This is an approximative method. However, there exists a rigorous result which shows that, in a special model, the method of stationary phase in lowest order yields the precise value of the integral:

J. Duistermaat and G. Heckmann, On the variation in the cohomology in the symplectic form of the reduced phase space, Invent. Math. **69** (1982), 259–268; **72** (1983), 153.

We refer to:

E. Witten, Topological quantum field theory, Commun. Math. Phys. **117** (1988), 353–386.

E. Witten, Witten's Lectures on Three-Dimensional Topological Quantum Field Theory. Edited by Sen Hu, World Scientific, Singapore 1999.

In Sect. 23.8, we will sketch how the Jones polynomials (i.e., topological invariants in knot theory) can be obtained by the method of topological quantum field theory due to Witten. This approach is based on the Chern–Simons gauge theory on the 3-dimensional sphere.

Historical remarks on moduli spaces and modular forms. The moduli space $\mathrm{Mod}_g(\mathsf{R})$ of compact Riemann surfaces R of genus g consists of all equivalence classes of compact Riemann surfaces of genus g modulo conformal equivalence. This space was first studied by Riemann who determined the finite dimension of this space:

- If $g \geq 2$, then the real dimension of $\mathrm{Mod}(\mathsf{R})_g$ is equal to $6g - 6$. This corresponds to the conformal classification of algebraic curves parametrized by sophisticated automorphic functions which were investigated by Poincaré and Klein at the end of the 19th century.
- If $g = 0$, then the moduli space $\mathrm{Mod}(\mathsf{R})_0$ consists of precisely one point which corresponds to the Riemann sphere; this sphere is conformally equivalent to the one-dimensional complex projective space $\mathbb{P}^1_{\mathbb{C}}$.
- If $g = 1$, then the dimension of $\mathrm{Mod}(\mathsf{R})_1$ is equal to two. This corresponds to the conformal classification of elliptic curves parametrized by elliptic functions. The theory of elliptic functions was created by Legendre, Gauss, Jacobi, and Weierstrass at the end of the 18th century and in the 19th century.

In the setting of Teichmüller theory, the moduli space $\mathrm{Mod}(\mathsf{R})_g$ is studied in:

J. Jost, Compact Riemann Surfaces: An Introduction to Contemporary Mathematics, Springer, Berlin, 2006.

Furthermore, we refer to:

F. Klein, Development of Mathematics in the 19th Century, Math. Sci. Press, New York, 1979.

K. Maurin, The Riemann Legacy: Riemannian Ideas in Mathematics and Physics of the 20th Century, Kluwer, Dordrecht, 1997.

A. Hurewitz and R. Courant, Lectures on Complex Function Theory and Elliptic Integrals (in German), Springer, Berlin, 1964.

D. Lawden, Elliptic Functions and Applications, Springer, New York, 1989.

L. Ford, Automorphic Functions, Chelsea, New York, 1972.

T. Apostol, Introduction to Analytic Number Theory, Springer, New York, 1986.

T. Apostol, Modular Functions and Dirichlet Series in Number Theory, Springer, New York, 1990.

M. Waldschmidt, P. Moussa, J. Luck, and C. Itzykson (Eds.), From Number Theory to Physics, Springer, New York, 1995 (survey articles).

J. Bruinier, G. van der Geer, G. Harder, and D. Zagier, The 1-2-3 of Modular Forms, Lectures at a Summer School in Nordfjordeid, Norway, Springer, Berlin, 2008 (survey articles).

The theory of elliptic curves possesses a very rich structure. The spectacular proof of Fermat's last theorem by Andrew Wiles in 1995 was based on recent progress for elliptic curves (see the discussion in the Prologue to Vol. I on page 17):

S. Singh, Fermat's Last Theorem: The Story of a Riddle that Confounded the World's Greatest Minds for 358 Years, Fourth Estate, London, 1997.

F. Diamond and J. Shurman, A First Course in Modular Forms, Springer, Berlin, 2005 (the modularity theorem).

Fermat's last theorem is a consequence of the so-called modularity theorem:

All rational elliptic curves arise from modular forms.

This fundamental result was conjectured by Taniyama, Shimura, and Weil in the 1950s and 1960s. For a special class of elliptic curves, the theorem was proven by Wiles in order to get the proof of Fermat's last theorem. The correctness of the general modularity theorem was proven by Breuil, Conrad, Diamond, and Taylor in 2001:

A. Wiles, Modular elliptic curves and Fermat's last theorem, Ann. Math. **142** (1995), 443–551.

C. Breuil, B. Conrad, F. Diamond, and R. Taylor, On the modularity of elliptic curves over the field $\mathbb{Q}$ of rational numbers: wild 3-adic exercises, J. Amer. Math. Soc. **14**(4) (2001), 843–939.

An introduction to the field Q_p of p-adic numbers and the adelic ring $\mathsf{A}_{\mathbb{Q}}$ will be given in Sect. 4.6.7 on page 332. We will also briefly discuss the relation of p-adic numbers and the adelic ring to mathematical models motivated by physics (chaos and turbulence or dark matter in cosmology). This is called adelic physics.

The moduli space of compact Riemann surfaces is used in string theory in order to compute the Feynman path integral for the free bosonic string. This was first done by:

A. Polyakov, Quantum geometry of bosonic strings, Phys. Lett. **103B** (1981), 207.

A. Polyakov, Quantum geometry of fermionic strings, Phys. Lett. **103B** (1981), 213.

A detailed computation can be found in:

B. Hatfield Quantum Field Theory of Point Particles and Strings, Addison-Wesley, Redwood City, California.

For a rigorous approach based on the mathematical theory of Riemann surfaces, we refer to:

J. Jost, The Bosonic String: A Mathematical Treatment, International Press, Boston, 2001.

The Ricci flow and the Poincaré conjecture:

The system of ordinary differential equations

$$\boxed{\frac{\partial \mathbf{g}}{\partial \sigma} = -\mathrm{Ric}(\mathbf{g})} \tag{0.93}$$

defines the Ricci flow $\mathbf{g} = \mathbf{g}(P, \sigma)$, $P \in \mathcal{M}$, on the n-dimensional (compact) Riemannian manifold $\mathcal{M}$. This is a family of metric tensors on $\mathcal{M}$ which depends on the real parameter $\sigma \in [\sigma_0, \sigma_1]$. In terms of local coordinates, equation (0.93) reads as

$$\frac{\partial g_{ij}(x,\sigma)}{\partial \sigma} = -R_{ij}(g(x,\sigma)), \qquad x \in \mathcal{M},\ \sigma \in [\sigma_0, \sigma_1]$$

where $i, j = 1\ldots, n$. This equation generalizes the heat equation

$$\frac{\partial \Theta(P,\sigma)}{\partial \sigma} = -\Delta\Theta(P,\sigma), \qquad P \in \mathcal{M},\ \sigma \in [\sigma_0, \sigma_1]$$

for the temperature field $\Theta = \Theta(P, \sigma)$ on the manifold $\mathcal{M}$. Here, the real parameter σ denotes time. It was the ingenious idea of Perelman to solve the Poincaré conjecture by deforming appropriate 3-dimensional manifolds to a 3-dimensional sphere by means of the Ricci flow. We refer to:

B. Chow and D. Knopf, The Ricci Flow: An Introduction, Amer. Math. Soc., Providence, Rhode Island, 2004.

B. Chow, P. Lu, and L. Ni, Hamilton's Ricci Flow, Amer. Math. Soc., Providence, Rhode Island, 2006.

H. Cao, S. Yau, and X. Zhu, Structure of Three-dimensional Space: The Poincaré and Geometrization Conjectures, International Press, Boston, 2006.

J. Morgan and G. Tian, Ricci Flow and the Poincaré Conjecture, Amer. Math. Soc., Providence, Rhode Island/Clay Mathematics Institute, Cambridge, Massachusetts, 2007.

D. O'Shea, The Poincaré Conjecture: In Search of the Shape of the Universe, Walker, New York, 2007.

Modern differential geometry in physics. A lot of material on modern differential geometry and its applications to physics can be found in:

V. Ivancevic and T. Invancevic, Differential Geometry: A Modern Introduction, World Scientific, Singapore, 2007.

A lot of historical material including the history of gauge field theory is contained in the survey article by:

H. Kastrup, On the advancement of conformal transformations and their associated symmetries in geometry and theoretical physics, Annalen der Physik **17** (2008), 631–690.

For the relations between mathematics and physics in the history of quantum field theory, we recommend:

W. Nahm, Conformal field theory: a bridge over troubled waters, pp. 571–604. In: A. Mitra (Ed.), Quantum Field Theory: A 20th Century Profile, Indian National Science Academy and Hindustan Book Agency, India, 2000.

For recent progress in quantum field theory based on close relations between mathematics and physics, we refer to:

C. Bär and K. Fredenhagen, Quantum Field Theory in Curved Space-Times, Springer 2009.

1. The Euclidean Space E_3 (Hilbert Space and Lie Algebra Structure)

In the occupation with mathematical problems, a more important role than generalization is played – I believe – by specialization.

David Hilbert, Paris Lecture, 1900

1.1 A Glance at History

We need an analysis which is of geometric nature and describes physical situations as directly as algebra describes quantities.

Gottfried Wilhelm Leibniz (1646–1716)

One has to distinguish between

- the Euclidean space E_3 (a set of vectors), and
- the Euclidean manifold $\mathbb{E}^3$ (a set of points).

The Euclidean space E_3 is a real 3-dimensional Hilbert space equipped with the inner product

$$\langle \mathbf{x}|\mathbf{y}\rangle := \mathbf{x}\mathbf{y}$$

of vectors $\mathbf{x}, \mathbf{y}$. Additionally, the Euclidean space E_3 is a Lie algebra equipped with the vector product

$$[\mathbf{x}, \mathbf{y}] := \mathbf{x} \times \mathbf{y}.$$

The Euclidean manifold $\mathbb{E}^3$ is a real 3-dimensional Riemannian manifold whose tangent spaces (consisting of velocity vectors) are isomorphic to the Hilbert space E_3. The theory of Euclidean temperature fields, velocity fields, and velocity covector fields (differential forms) refers to the Euclidean manifold $\mathbb{E}^3$.

Nowadays, the notion of 'vector' is one of the basic notions in mathematics and physics. However, the history of the theory of vectors and dual vectors (called covectors) is very strange and involved. In an implicit manner, the notion of vector slowly emerged in ancient static equilibrium physics as

- force $\mathbf{F}$,
- torque $\mathbf{x} \times \mathbf{F}$ (vector product) of a force, and
- work $\mathbf{F}\Delta\mathbf{x}$ (inner product) of a force.

Newton (1646–1727) and his successors noticed implicitly that in the dynamics of planets the following vectors play a crucial role:

- $\dot{\mathbf{x}}(t)$ (velocity vector/tangent vector),
- $\ddot{\mathbf{x}}(t)$ (acceleration vector),
- $m\dot{\mathbf{x}}(t)$ (momentum vector), and
- $\mathbf{x}(t) \times m\dot{\mathbf{x}}(t)$ (angular momentum vector).

E. Zeidler, *Quantum Field Theory III: Gauge Theory*,
DOI 10.1007/978-3-642-22421-8_2,

Here, t denotes time. In 1845, the notion of 'vector' was explicitly introduced by Hamilton (1805–1865), as a special 3-dimensional case of his 4-dimensional quaternions.

In the nineteenth century, mathematicians and physicists discovered that it is very convenient to leave the three-dimensional Euclidean space E_3. In 1844, Grassmann (1809–1877) introduced the exterior (or alternating) product

$$\mathbf{x} \wedge \mathbf{y}$$

which lives outside the Euclidean space E_3. This exterior product generates the exterior algebra (or Grassmann algebra) $\bigwedge(E_3)$ which is an extension of E_3. Explicitly, $\mathbf{x} \wedge \mathbf{y}$ is an antisymmetric bilinear form on E_3 given by

$$(\mathbf{x} \wedge \mathbf{y})(\mathbf{a}, \mathbf{b}) := (\mathbf{xa})(\mathbf{yb}) - (\mathbf{xb})(\mathbf{ya}) \qquad \text{for all} \quad \mathbf{a}, \mathbf{b} \in E_3.$$

The exterior product is related to the tensor product $\mathbf{x} \otimes \mathbf{y}$ by the relation

$$\mathbf{x} \wedge \mathbf{y} = \mathbf{x} \otimes \mathbf{y} - \mathbf{y} \otimes \mathbf{x}.$$

Here, $\mathbf{x} \otimes \mathbf{y}$ is a bilinear form on E_3 defined by

$$(\mathbf{x} \otimes \mathbf{y})(\mathbf{a}, \mathbf{b}) := (\mathbf{xa})(\mathbf{yb}) \qquad \text{for all} \quad \mathbf{a}, \mathbf{b} \in E_3.$$

The dual space E_3^d to the Euclidean space E_3 generates the Grassmann algebra $\bigwedge(E_3^d)$, and hence we get the algebra of alternating differential forms $\Lambda(\mathbb{E}^3)$ on the Euclidean manifold $\mathbb{E}^3$. Élie Cartan (1869–1951) based his beautiful calculus of alternating differential forms on the algebra $\Lambda(\mathbb{E}^3)$ by adding the differential operator[1]

$$d : \Lambda(\mathbb{E}^3) \to \Lambda(\mathbb{E}^3).$$

We will show that:

> *Élie Cartan's calculus for alternating differential forms represents the most effective generalization of the classic calculus due to Newton and Leibniz to functions of several variables, and hence to physical fields.*

Hodge (1903–1975) introduced the star operator $* : \bigwedge(X) \to \bigwedge(X)$ where we set $X := E_3$ or $X := E_3^d$. This way, Hodge obtained the dual differential operator

$$d^* : \Lambda(\mathbb{E}^3) \to \Lambda(\mathbb{E}^3)$$

given by $d^*\omega := (-1)^p *^{-1} d(*\omega)$ for p-differential forms ω. Setting $\Delta := dd^* + d^*d$, this yields the Laplacian

$$\Delta : \Lambda(\mathbb{E}^3) \to \Lambda(\mathbb{E}^3),$$

which plays a crucial role in modern differential geometry together with the Dirac operator, the Yang–Mills operator, and the Seiberg–Witten operator (generalized Landau–Ginzburg operator). The Laplacian is basic for differential topology (the de Rham cohomology theory on Riemannian manifolds).

In 1843, Hamilton (1805–1865) introduced 4-dimensional objects $\mathbf{x} + a$ (vector plus real number) which he called quaternions. These quaternions form a skew-field which extends the field of complex numbers. Hamilton and Cayley (1821–1895) showed independently how the Euler formula for the rotations of the 3-dimensional Euclidean space E_3 can be very elegantly formulated in terms of 4-dimensional quaternions (see Sect. 7.1 on page 425). In a quite natural way, this leads to the universal covering group $SU(2)$ of the Lie group $SO(3)$ (which is isomorphic to the rotation group $SU(E_3)$ of the Euclidean space E_3). The point is that:

[1] The symbol $\Lambda(\mathbb{E}^3)$ denotes the real linear space of all smooth differential forms on the Euclidean manifold $\mathbb{E}^3$ (see page 701).

The groups $SU(2)$ and $SO(3)$ have isomorphic Lie algebras.

That is, the groups are locally isomorphic near the unit element. However, the groups $SU(2)$ and $SO(3)$ are not globally isomorphic. In fact, the group $SU(2)$ is simply connected, whereas the group $SO(3)$ does not have this topological property.

The local theory of Lie groups was created by Lie (1842–1899) in his seminal work. The global theory of Lie groups was strongly influenced by Chevalley (1909–1984). We will show later on that the quaternions and the more general Clifford algebra of the Minkowski space are crucial for understanding the spin of electrons and other elementary particles. This was discovered by Pauli (1900–1958) and Dirac (1902–1984). The general theory of Clifford algebras, based on a generalization of the product

$$\mathbf{x} \vee \mathbf{y} := \mathbf{x} \wedge \mathbf{y} - \mathbf{x}\mathbf{y}, \qquad \mathbf{x}, \mathbf{y} \in E_3,$$

lies at the heart of modern spin geometry.

In 1905, Einstein (1879–1955) published his theory of special relativity. In 1908, Minkowski (1864–1909) emphasized in a famous lecture that the theory of special relativity can be understood best by extending the Hilbert space E_3 to a four-dimensional indefinite Hilbert space M_4 called the Minkowski space. In this setting, the classic equations in electrodynamics due to Maxwell (1831–1879) become the equations

$$d^*F = -\mu_0 \mathcal{J}, \qquad dF = 0$$

for the differential 2-form F of the electromagnetic field and the differential 1-form $\mathcal{J}$ for the electric 4-current on the 4-dimensional Minkowski manifold $\mathbb{M}^4$.

Curvature. The Euclidean manifold $\mathbb{E}^3$ allows a global parallel transport of vectors. This reflects the flatness of $\mathbb{E}^3$. Using a general transport of frames, it is possible to assign nontrivial curvature and torsion to the Euclidean manifold $\mathbb{E}^3$. This method of moving frames due to Élie Cartan is the prototype for modern gauge theory which is basic for both modern differential geometry and the Standard Model in particle physics. Intuitively, a moving frame on the Euclidean manifold represents the motion of a rigid body under the influence of an external force.

Duality. In what follows, we will encounter the following important dualities:

- Riesz duality and Lie–Cartan duality between velocity fields and velocity covector fields (differential forms),
- Levi-Civita duality,
- Weyl duality, and
- Hodge duality.

1.2 Algebraic Basic Ideas

The intuitive model. Consider the three-dimensional space of our intuition. Fix a point O called the origin. By definition, position vectors $\mathbf{x}$ have the origin as initial point and the point P as final point (Fig 1.1(a)). We also write

- $\mathbf{x} = \overrightarrow{OP}$ and $P = O + \mathbf{x}$.

By definition, the Euclidean space $E_3(O)$ is the space of all position vectors $\mathbf{x}$ starting at the point O. This is a real 3-dimensional linear space. To simplify notation, we write E_3 instead of $E_3(O)$. By definition, the Euclidean manifold $\mathbb{E}^3$ consists of all the points P. The motion of a particle can be described either by the equation

$$P = P(t), \qquad t_0 \leq t \leq t_1$$

or by the equivalent equation $\mathbf{x} = \mathbf{x}(t)$ (Fig. 1.1(b)). In order to introduce Cartesian coordinates, choose a system $\mathbf{i}, \mathbf{j}, \mathbf{k}$ which has the following properties:

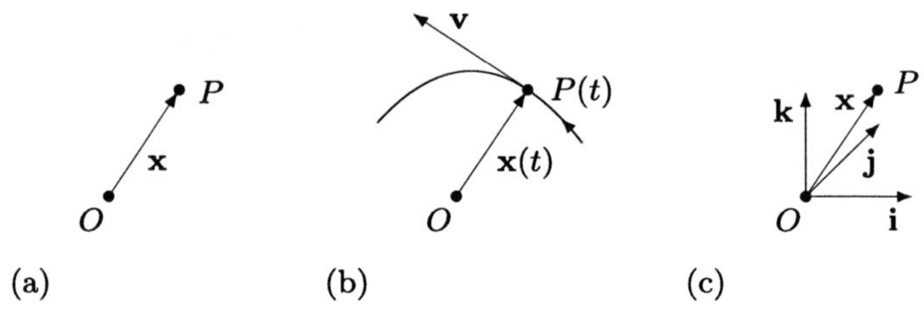

Fig. 1.1. Motion of a particle

- $\mathbf{i}, \mathbf{j}, \mathbf{k}$ are unit vectors which are pairwise orthogonal to each other;
- $\mathbf{i}, \mathbf{j}, \mathbf{k}$ are positively oriented (Fig. 1.1(c)).

The position vector $\mathbf{x}$ allows the unique representation

$$\mathbf{x} = x\mathbf{i} + y\mathbf{j} + z\mathbf{k}.$$

Here, the real numbers x, y, z are called the Cartesian coordinates of the point P. As we will see below, for the inner product, we have

$$\mathbf{x}\mathbf{x}' = xx' + yy' + zz'.$$

Thus, the map $\mathbf{x} \mapsto (x, y, z)$ yields the Hilbert space isomorphism $E_3 \simeq \mathbb{R}^3$, and the map $P \mapsto (x, y, z)$ yields the bijection $\mathbb{E}^3 \simeq \mathbb{R}^3$.[2]

Einstein's summation convention. In this chapter, we sum over equal upper and lower Latin indices from 1 to 3. For example,

$$\mathbf{e}_i \times \mathbf{e}_j = c_{ij}^k \mathbf{e}_k = \sum_{k=1}^{3} c_{ij}^k \mathbf{e}_k, \qquad i, j = 1, 2, 3.$$

1.2.1 Symmetrization and Antisymmetrization

> Symmetrization and antisymmetrization play a crucial role in mathematics and physics (e.g., for constructing invariants). For example, bosons (e.g., photons) are based on symmetrization, whereas fermions (e.g., electrons) are based on antisymmetrization.
>
> Folklore

The sign of a permutation and its generalization. Let $i_k, j_k = 1, \ldots, n$ if $k = 1, 2, \ldots, n$. There exists precisely one symbol $\varepsilon^{i_1 i_2 \ldots i_n}_{j_1 j_2 \ldots j_n}$ which has the following two properties:

- $\varepsilon^{12\ldots n}_{12\ldots n} = 1$ (normalization condition), and
- $\varepsilon^{i_1 i_2 \ldots i_n}_{j_1 j_2 \ldots j_n}$ changes sign if two upper (resp. lower) indices are transposed.

[2] Note that in this monograph, Cartesian coordinate systems are always positively oriented. Otherwise, we will use the term 'reflected' Cartesian coordinate system.

For example, $\varepsilon_{12}^{12} = -\varepsilon_{21}^{12} = 1$ and $\varepsilon_{11}^{12} = \varepsilon_{22}^{12} = 0$. We also write

$$\varepsilon_{j_1 j_2 \ldots j_n} := \varepsilon^{1\,2\,\ldots\,n}_{j_1 j_2 \ldots j_n}, \qquad \varepsilon^{i_1 i_2 \ldots i_n} := \varepsilon^{i_1 i_2 \ldots i_n}_{1\,2\,\ldots\,n}.$$

Let $\mathcal{N} := \{1, 2, \ldots, n\}$. By definition, precisely the bijective maps

$$\pi : \mathcal{N} \to \mathcal{N}$$

are called permutations. The value of $\varepsilon_{\pi(1)\pi(2)\ldots\pi(n)}$ is called the sign of the permutation π denoted by $\operatorname{sgn}(\pi)$. For example, in the special case where $n = 2$ with $\pi(1) := 2$ and $\pi(2) := 1$, we have $\operatorname{sgn}(\pi) = \varepsilon_{21} = -1$.

Symmetrization. Let $\{a_{ij}\}$ be a family of complex numbers with the indices $i, j = 1, 2, \ldots, n$. For $n = 2$, we define

$$\boxed{(\text{Sym } a)_{12} := \frac{1}{2}(a_{12} + a_{21}).}$$

In the general case, we set

$$(\text{Sym } a)_{12\ldots n} = \frac{1}{n!} \sum_{\pi} a_{\pi(1)\pi(2)\ldots\pi(n)}. \tag{1.1}$$

Here, we sum over all permutations π of the elements $1, 2, \ldots, n$.

Antisymmetrization. Taking the sign of permutations into account, we define the antisymmetrized quantities

$$\boxed{(\text{Alt } a)_{12} := \frac{1}{2}(a_{12} - a_{21}),}$$

and

$$(\text{Alt } a)_{12\ldots n} := \frac{1}{n!} \sum_{\pi} \operatorname{sgn}(\pi) \cdot a_{\pi(1)\pi(2)\ldots\pi(n)}. \tag{1.2}$$

Equivalently, $(\text{Alt } a)_{12\ldots n} = \frac{1}{n!} \sum_{i_1, i_2, \ldots, i_n = 1}^{n} \varepsilon^{i_1 i_2 \ldots i_n} a_{i_1 i_2 \ldots i_n}$.

1.2.2 Cramer's Rule for Systems of Linear Equations

> The starting point for linear and multilinear algebra is the problem of solving linear systems of equations.
>
> Folklore

The prototype of Cramer's rule. Consider the equation

$$Ax = b \tag{1.3}$$

where A and b are given complex numbers. If $A \neq 0$, then the unique solution of (1.3) is given by

$$\boxed{x = A^{-1} b.} \tag{1.4}$$

The goal of Cramer's rule is to generalize this formula to systems of n equations with n unknowns. To begin with, consider the special case where $n = 2$:

$$\begin{aligned} a_{11}x_1 + a_{12}x_2 &= b_1, \\ a_{21}x_1 + a_{22}x_2 &= b_2. \end{aligned} \tag{1.5}$$

Here, we are given the complex numbers a_{ij} and b_j with $i, j = 1, 2$. We are looking for the complex numbers x_1, x_2. If $a_{11}a_{22} - a_{12}a_{21} \neq 0$, then the unique solution of (1.5) reads as

$$x_1 = \frac{b_1 a_{22} - b_2 a_{12}}{a_{11}a_{22} - a_{12}a_{21}}, \qquad x_2 = \frac{a_{11}b_2 - a_{21}b_1}{a_{11}a_{22} - a_{12}a_{21}}. \tag{1.6}$$

In fact, multiplying the first (resp. second) equation of (1.5) by a_{22} (resp. $-a_{12}$) and adding up, we get

$$(a_{11}a_{22} - a_{12}a_{21})x_1 = b_1 a_{22} - b_2 a_{12}.$$

This yields (1.6). Defining the determinant

$$\begin{vmatrix} a_{11} & a_{12} \\ a_{21} & a_{22} \end{vmatrix} := a_{11}a_{22} - a_{12}a_{21},$$

Cramer's rule (1.6) reads as

$$x_1 = \frac{\begin{vmatrix} b_1 & a_{12} \\ b_2 & a_{22} \end{vmatrix}}{\begin{vmatrix} a_{11} & a_{12} \\ a_{21} & a_{22} \end{vmatrix}}, \qquad x_2 = \frac{\begin{vmatrix} a_{11} & b_1 \\ a_{21} & b_2 \end{vmatrix}}{\begin{vmatrix} a_{11} & a_{12} \\ a_{21} & a_{22} \end{vmatrix}}.$$

In 1693, Leibniz (1646–1716) considered a system of two linear equations for three unknowns. He used a symbolic method. This was the forerunner of determinant theory. Determinants were first systematically used by Maclaurin (1698–1746) and Cramer (1704–1752) in about 1750 in order to solve linear systems of n linear equations with n unknowns. The theory of matrices was created by Cayley (1821–1897) in 1855. He introduced the matrix symbol

$$A = \begin{pmatrix} a_{11} & a_{12} \\ a_{21} & a_{22} \end{pmatrix}, \tag{1.7}$$

and he wrote the system (1.5) in the following form:

$$\begin{pmatrix} a_{11} & a_{12} \\ a_{21} & a_{22} \end{pmatrix} \begin{pmatrix} x_1 \\ x_2 \end{pmatrix} = \begin{pmatrix} b_1 \\ b_2 \end{pmatrix}, \tag{1.8}$$

or briefly, $Ax = b$. Motivated by this special case, Cayley defined the matrix product of two square matrices by setting

$$\begin{pmatrix} a_{11} & a_{12} \\ a_{21} & a_{22} \end{pmatrix} \begin{pmatrix} b_{11} & b_{12} \\ b_{21} & b_{22} \end{pmatrix} := \begin{pmatrix} c_{11} & c_{12} \\ c_{21} & c_{22} \end{pmatrix} \tag{1.9}$$

with $c_{ij} := \sum_{s=1}^{2} a_{is}b_{sj}$. Mnemonically, c_{ij} is obtained by multiplying the ith row of the first factor with the jth column of the second factor:

$$c_{ij} = \begin{pmatrix} a_{i1} & a_{i2} \end{pmatrix} \begin{pmatrix} b_{1j} \\ b_{2j} \end{pmatrix} = a_{i1}b_{1j} + a_{i2}b_{2j}.$$

For the matrix A from (1.7), the symbol $\det(A)$ denotes the determinant of A:

$$\det(A) = \det \begin{pmatrix} a_{11} & a_{12} \\ a_{21} & a_{22} \end{pmatrix} := \begin{vmatrix} a_{11} & a_{12} \\ a_{21} & a_{22} \end{vmatrix} = a_{11}a_{22} - a_{12}a_{21}.$$

Writing equation (1.9) as $AB = C$, an explicit computation yields

$$\det(AB) = \det(A)\det(B). \tag{1.10}$$

This is the crucial product property of determinants (see Sect. 2.11.3 for the general case). If $\det(A) \neq 0$, let us define the matrix

$$A^{-1} := \frac{1}{\det(A)} \begin{pmatrix} a_{22} & -a_{12} \\ -a_{21} & a_{11} \end{pmatrix}. \tag{1.11}$$

Setting $I := \begin{pmatrix} 1 & 0 \\ 0 & 1 \end{pmatrix}$ (unit matrix), we get the matrix product $A^{-1}A = I$ which justifies the notation A^{-1}. One also checks easily that $AA^{-1} = I$.

Theorem 1.1 *If* $\det(A) \neq 0$, *then the equation* (1.4), *that is,* $Ax = b$, *has the unique solution* $x = A^{-1}b$.

Proof. (I) Uniqueness: If $Ax = b$, then $x = Ix = A^{-1}Ax = A^{-1}b$. Thus, if the solution x exists, it has necessarily the form $x = A^{-1}b$.

(II) Existence: $A(A^{-1}b) = Ib = b$. □

Theorem 1.1 represents Cramer's rule (1.6) in the language of matrices. The point is that the results above can be generalized to systems of n equations with n unknowns by introducing the general notion of determinant. We will do this next.

1.2.3 Determinants and the Inverse Matrix

> The general Laplace expansion formula for determinants from 1772 is equivalent to the associativity of the Grassmann alternating product from 1844.
>
> Folklore

Let $A = (a_{jk})$ be a complex $(n \times n)$-matrix with $n = 1, 2, \ldots$ We define the determinant $\det(A)$ of the matrix A by setting

$$\det(A) := \sum_{\pi} \operatorname{sgn}(\pi)\, a_{1\pi(1)}a_{2\pi(2)} \cdots a_{n\pi(n)}. \tag{1.12}$$

Here, we sum over all permutations π of the indices $1, 2, \ldots, n$. Equivalently,

$$\det(A) = \sum_{i_1,\ldots,i_n=1}^{n} \varepsilon^{i_1 i_2 \ldots i_n} a_{1i_1}a_{2i_2} \cdots a_{ni_n}.$$

Multilinear functionals. The determinant is the prototype of an antisymmetric multilinear functional. Let us discuss this. Choose $\mathbb{K} = \mathbb{R}$ (field of real numbers) or $\mathbb{K} = \mathbb{C}$ (field of complex numbers). Let X be a linear space over $\mathbb{K}$. The function

$$f : X \times X \to \mathbb{K}$$

is called bilinear iff it is linear with respect to each argument. That is, we have

$$f(\alpha x + \beta y, z) = \alpha f(x, z) + \beta f(y, z)$$

and $f(z, \alpha x + \beta y) = \alpha f(z, x) + \beta f(z, y)$ for all $x, y, z \in X$ and all $\alpha, \beta \in \mathbb{K}$. In addition, f is called symmetric iff it remains invariant under a transposition of arguments, that is,

$$f(x, y) = f(y, x) \qquad \text{for all} \quad x, y \in X.$$

Moreover, f is called antisymmetric iff it changes sign under a transposition of arguments, that is,

$$f(x, y) = -f(y, x) \qquad \text{for all} \quad x, y \in X.$$

Similarly, the functional $f : X \times \cdots \times X \to \mathbb{K}$ with n factors X is called n-linear iff it is linear with respect to each argument. In addition, f is called symmetric (resp. antisymmetric) iff it remains unchanged under an arbitrary transposition of two arguments (resp. it changes sign under a transposition of two arguments).

Let $n = 1, 2, \ldots$ The symbol $gl(n, \mathbb{C})$ denotes the space of all complex $(n \times n)$-matrices A. Set

$$f(A) := \det(A).$$

Using elementary properties of permutations, one shows that the determinant $\det(A)$ has the following properties:

(i) The function $f : gl(n, \mathbb{C}) \to \mathbb{C}$ is n-linear and antisymmetric with respect to the columns of the matrix A.
(ii) $f(I) = 1$ (normalization condition).
(iii) $\det(A^d) = \det(A)$.

Here, A^d denotes the dual matrix to $A = (a_{ij})$ with the elements $(A^d)_{ij} = a_{ji}$ for all indices i, j. In other words, relation (iii) tells us that the determinant does not change if we interchange rows and columns of the matrix A. For example,

$$\begin{vmatrix} a_{11} & a_{12} \\ a_{21} & a_{22} \end{vmatrix} = \begin{vmatrix} a_{11} & a_{21} \\ a_{12} & a_{22} \end{vmatrix}, \qquad \begin{vmatrix} a_{11} & a_{12} \\ a_{21} & a_{22} \end{vmatrix} = - \begin{vmatrix} a_{12} & a_{11} \\ a_{22} & a_{21} \end{vmatrix},$$

and

$$\begin{vmatrix} \lambda a_{11} + \mu b_{11} & a_{12} \\ \lambda a_{21} + \mu b_{21} & a_{22} \end{vmatrix} = \lambda \begin{vmatrix} a_{11} & a_{12} \\ a_{21} & a_{22} \end{vmatrix} + \mu \begin{vmatrix} b_{11} & a_{12} \\ b_{21} & a_{22} \end{vmatrix}.$$

The special Laplace expansion formula. By an elementary computation, it follows from the definition of the determinant that

$$\begin{vmatrix} a_{11} & a_{12} & a_{13} \\ a_{21} & a_{22} & a_{23} \\ a_{31} & a_{32} & a_{33} \end{vmatrix} := a_{11} \begin{vmatrix} a_{22} & a_{23} \\ a_{32} & a_{33} \end{vmatrix} - a_{12} \begin{vmatrix} a_{21} & a_{23} \\ a_{31} & a_{33} \end{vmatrix} + a_{13} \begin{vmatrix} a_{21} & a_{22} \\ a_{31} & a_{32} \end{vmatrix}.$$

This is a special case of the Laplace expansion theorem. For $n = 2, 3, \ldots$ and fixed $i = 1, \ldots, n$, the general formula reads as

$$\det(A) = \sum_{j=1}^{n} a_{ij}\mathcal{A}_{ij}. \tag{1.13}$$

Here, the determinant $\mathcal{A}_{ij}$ is obtained from the matrix A after cancelling the ith row and the jth column and multiplying the corresponding determinant by $(-1)^{i+j}$. Here, the subdeterminant $\mathcal{A}_{ij}$ of the determinant $\det(A)$ is called the adjunct to the element a_{ij}.

The inverse matrix. The Laplace expansion theorem can be used in order to construct the inverse matrix. To this end, suppose that we are given the complex $(n \times n)$-matrix A with $\det(A) \neq 0$. Define the $(n \times n)$-matrix

$$A^{-1} := \tfrac{1}{\det(A)}\,(\mathcal{A}_{ij})^d. \tag{1.14}$$

Recall that the $(n \times n)$-matrix $I = (\delta_{ij})$ is called the unit matrix.

Proposition 1.2 $A^{-1}A = AA^{-1} = I.$

Proof. For all indices $i, k = 1, \ldots, n$, we have the orthogonality relation

$$\sum_{j=1}^{n} a_{kj}\mathcal{A}_{ij} = \delta_{ik}\det(A). \tag{1.15}$$

If $i = k$, then this is the Laplace expansion of $\det(A)$. If $i \neq k$, this is the Laplace expansion of some matrix B which is obtained from A by replacing the ith row of A by the kth row of A. Therefore, $\det(B) = 0$. Relation (1.13) implies the claim. □

Cramer's rule. Consider the system

$$\sum_{j=1}^{n} a_{ij}x_j = b_j, \qquad i = 1, \ldots, n \tag{1.16}$$

with given complex numbers a_{ij} and b_j where $i, j = 1, \ldots, n$. In the language of matrices, this system reads as

$$Ax = b$$

with the complex $(n \times n)$-matrix $A = (a_{ij})$. If $\det(A) \neq 0$, then the unique solution of (1.16) is given by $x = A^{-1}b$. The proof proceeds as the proof of Theorem 1.1.

The general Laplace expansion theorem. As an example, let us consider the complex (4×4)-matrix

$$A := \begin{pmatrix} a_{11} & a_{12} & a_{13} & a_{14} \\ a_{21} & a_{22} & a_{23} & a_{24} \\ a_{31} & a_{32} & a_{33} & a_{34} \\ a_{41} & a_{42} & a_{43} & a_{44} \end{pmatrix}.$$

In contrast to the special Laplace expansion formula, it is also possible to represent the (4×4)-determinant $\det(A)$ with the aid of (2×2)-subdeterminants. Explicitly,

$$\det(A) = \mathcal{B}_{12}\mathcal{C}_{34} - \mathcal{B}_{13}\mathcal{C}_{24} + \mathcal{B}_{14}\mathcal{C}_{23} + \mathcal{B}_{23}\mathcal{C}_{14} - \mathcal{B}_{24}\mathcal{C}_{13} + \mathcal{B}_{34}\mathcal{C}_{12}.$$

Here, we set

$$\mathcal{B}_{ij} := \begin{vmatrix} a_{1i} & a_{1j} \\ a_{2i} & a_{2j} \end{vmatrix}, \qquad \mathcal{C}_{kl} := \begin{vmatrix} a_{3k} & a_{3l} \\ a_{4k} & a_{4l} \end{vmatrix}.$$

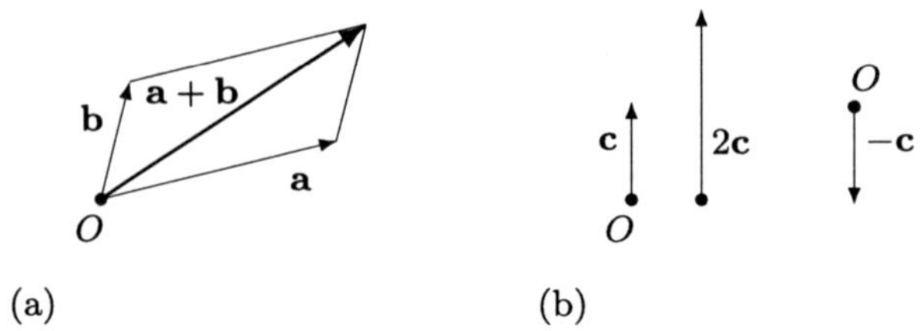

Fig. 1.2. Linear structure

This is the prototype of the general Laplace expansion formula. For a complex $(n \times n)$-matrix $A = (a_{ij})$ with $n = 2, 3, \ldots$, the general formula reads as follows:

$$\boxed{\det(A) = \sum_C (-1)^{\sigma(R)+\sigma(C)} \det(A_C^R) \det(A_{C^*}^{R^*}).} \tag{1.17}$$

More precisely, we fix the ordered subset R of the index set $\mathcal{N} := \{1, 2, \ldots, n\}$, and we sum over all ordered subsets C of the index set $\mathcal{N}$. The symbol R^* (resp. C^*) denotes the ordered subset of $\mathcal{N}$ which is complementary to R (resp. C). The symbol

$$A_C^R$$

stands for that submatrix of A which has precisely the elements of R (resp. C) as row (resp. column) indices. Finally $\sigma(R)$ (resp $\sigma(C)$) is the sum of the elements of the set R (resp. C). For example, if $\mathcal{N} := \{1, 2, 3, 4\}$ and $R := \{1, 2\}$, $C := \{2, 4\}$, then $R^* = \{3, 4\}$ and $C^* = \{1, 3\}$. Furthermore,

$$\mathcal{B}_{24}\mathcal{C}_{13} = \det(A_C^R) \det(A_{C^*}^{R^*}).$$

For the proof, we refer to Problem 3.2 on page 312.

1.2.4 The Hilbert Space Structure

The linear structure of vectors. Let $\mathbf{a}$ and $\mathbf{b}$ be two position vectors at the origin O. The vector sum

$$\mathbf{a} + \mathbf{b}$$

is given by the parallelogram construction pictured in Fig 1.2(a). From the physical point of view, this corresponds to the superposition of the forces $\mathbf{a}$ and $\mathbf{b}$.

Furthermore, let $\mathbf{c}$ be a position vector at the origin O. The length of $\mathbf{c}$ is denoted by $|\mathbf{c}|$. If $\lambda > 0$, then the vector $\lambda\mathbf{c}$ has the same direction as the vector $\mathbf{c}$ and the length $\lambda|\mathbf{c}|$. The vector $-\mathbf{c}$ is a position vector at the point O of length $|c|$ which points in the opposite direction of $\mathbf{c}$ (Fig. 1.2(b)). If $\lambda < 0$, then the vector $\lambda\mathbf{c}$ has the same direction as $-\mathbf{c}$ and the length $|\lambda| \cdot |\mathbf{c}|$ The zero vector $\mathbf{0}$ has the length zero.

Using the sum $\mathbf{a}+\mathbf{b}$ *and the product* $\lambda\mathbf{c}$*, the Euclidean space* $E_3(O)$ *becomes a real linear space.*

The definition of a linear space can be found in Sect. 7.3 of Vol. I. The family $\mathbf{a}_1, \ldots \mathbf{a}_m$ of position vectors at the origin O is called linearly dependent iff real numbers $\alpha_1, \ldots, \alpha_m$ exist with $\alpha_1^2 + \ldots + \alpha_m^2 \neq 0$ such that

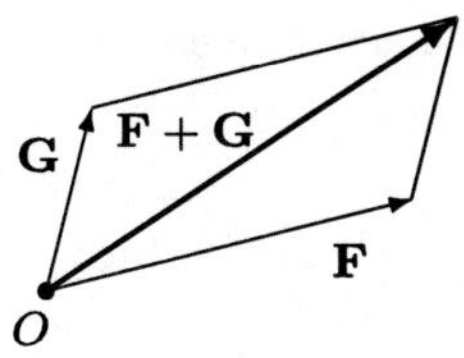

Fig. 1.3. Superposition of the forces **F** and **G**

$$\alpha_1 \mathbf{a}_1 + \ldots + \alpha_m \mathbf{a}_m = 0.$$

Otherwise, the family $\mathbf{a}_1, \ldots \mathbf{a}_m$ is called linearly independent. To illustrate the geometric meaning, let $\mathbf{a}, \mathbf{b}, \mathbf{c}$ be vectors at the origin O. Then the following hold:

- The vectors $\mathbf{a}, \mathbf{b}$ are linearly independent iff they span a parallelogram of nonvanishing area (Fig. 1.5 on page 82).
- The vectors $\mathbf{a}, \mathbf{b}, \mathbf{c}$ are linearly independent iff they span a parallelepiped of nonvanishing volume (Fig. 1.6 on page 83).

In the Euclidean space E_3, the maximal number of linearly independent vectors is equal to three. For example, the three vectors $\mathbf{i}, \mathbf{j}, \mathbf{k}$ pictured in Fig. 1.1(c) on page 72 are linearly independent.

In terms of physics, the addition of vectors corresponds to the superposition of forces (Fig. 1.3).

The inner product. For two position vectors $\mathbf{a}$ and $\mathbf{b}$ at the origin O, we define the inner product by setting

$$\mathbf{ab} := |\mathbf{a}| \cdot |\mathbf{b}| \cdot \cos \gamma.$$

The angle γ is to be chosen in such a way that $0 \leq \gamma \leq \pi$ (Fig. 1.4). In particular, $|\mathbf{a}| = \sqrt{\mathbf{a}^2}$. If $\mathbf{a}$ and $\mathbf{b}$ are nonzero vectors, then they are orthogonal to each other iff

$$\mathbf{ab} = 0.$$

The inner product fits best the orthogonality properties in Euclidean geometry. This is the intuitive root for the Hilbert space geometry in quantum physics. Choose the vectors $\mathbf{i}, \mathbf{j}, \mathbf{k}$ as pictured in Fig. 1.1 on page 72. Then

$$\mathbf{ij} = \mathbf{jk} = \mathbf{ki} = 0, \qquad \mathbf{i}^2 = \mathbf{j}^2 = \mathbf{k}^2 = 1.$$

We set $\langle \mathbf{a}|\mathbf{b} \rangle := \mathbf{ab}$. Then $||\mathbf{a}|| := \sqrt{\langle \mathbf{a}|\mathbf{a} \rangle} = \sqrt{\mathbf{a}^2} = |\mathbf{a}|$.

Theorem 1.3 *The Euclidean space E_3 becomes a 3-dimensional real Hilbert space with respect to the inner product $\langle \mathbf{a}|\mathbf{b} \rangle$.*

The definition of Hilbert spaces can be found in Sect. 7.4 of Vol. I. We have to prove the following for all $\mathbf{a}, \mathbf{b}, \mathbf{c} \in E_3$ and all real numbers α:

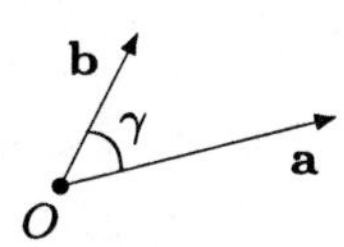

Fig. 1.4. Inner product **ab**

- The space E_3 is a real 3-dimensional linear space.
- $\langle \mathbf{a}|\mathbf{b}\rangle \in \mathbb{R}$ and $\langle \mathbf{a}|\mathbf{a}\rangle \geq 0$.
- $\langle \mathbf{a}|\mathbf{a}\rangle = 0$ iff $\mathbf{a} = \mathbf{0}$.
- $\langle \mathbf{a}|\mathbf{b}\rangle = \langle \mathbf{b}|\mathbf{a}\rangle$.
- $\langle \alpha\mathbf{a}|\mathbf{b}\rangle = \alpha\langle \mathbf{a}|\mathbf{b}\rangle$ and $\langle \mathbf{a}+\mathbf{b}|\mathbf{c}\rangle = \langle \mathbf{a}|\mathbf{c}\rangle + \langle \mathbf{b}|\mathbf{c}\rangle$.

For the elementary proof, see Problem 3.5.

The Fourier series. At the beginning of the 20th century, Hilbert (1862–1943) generalized the space $\mathbb{R}^n$ to infinite dimensions in order to create a general theory of integral equations. In this context, he noticed that the classical Fourier series is closely related to the notion of orthonormal basis in Hilbert space. In the setting of the Hilbert space $L_2(-\pi,\pi)$ of square-integrable functions $f : [-\pi,\pi] \to \mathbb{R}$, the precise theorem was proven by E. Fischer and F. Riesz in 1907 (see F. Riesz and B. Nagy, Functional Analysis, Frederyck Ungar, New York, 1978.) At this point, we will discuss this for the Euclidean space E_3, and we will explain the relation to the Dirac calculus used in quantum mechanics. Let $\mathbf{e}_1, \mathbf{e}_2, \mathbf{e}_3$ be an orthonormal basis of the Euclidean space E_3, that is,[3]

$$\langle \mathbf{e}_j|\mathbf{e}_k\rangle = \delta_{jk}, \qquad j,k = 1,2,3.$$

In order to be able to use the convenient Einstein summation convention (see page 72), let us set $\mathbf{e}^k := \mathbf{e}_k$. Then $\langle \mathbf{e}^k|\mathbf{e}_l\rangle = \delta^k_l$ for $k,l = 1,2,3$. The real numbers

$$a^k := \langle \mathbf{e}^k|\mathbf{a}\rangle, \qquad k = 1,2,3$$

are called the Fourier coefficients of the vector $\mathbf{a}$ (with respect to $\mathbf{e}_1, \mathbf{e}_2, \mathbf{e}_3$). We also write $a_k := \langle \mathbf{e}_k|\mathbf{a}\rangle$, that is, $a_k = a^k$ if $k = 1,2,3$. Finally, for fixed k, we define the linear functional $dx^k : E_3 \to \mathbb{R}$ by setting

$$dx^k(\mathbf{a}) := \langle \mathbf{e}^k|\mathbf{a}\rangle, \qquad k = 1,2,3. \tag{1.18}$$

In other words, dx^k assigns to every vector $\mathbf{a}$ the Fourier coefficient a^k. For all vectors $\mathbf{a}, \mathbf{b} \in E_3$, the following hold:[4]

Theorem 1.4 (i) $\mathbf{a} = \langle \mathbf{e}^k|\mathbf{a}\rangle\, \mathbf{e}_k$ *(Fourier series).*

(ii) $\langle \mathbf{a}|\mathbf{b}\rangle = \langle \mathbf{a}|\mathbf{e}^k\rangle\langle \mathbf{e}_k|\mathbf{b}\rangle$ *(Parseval equation).*

Equivalently, $\mathbf{a} = a^k\mathbf{e}_k$ *and* $\langle \mathbf{a}|\mathbf{b}\rangle = a^k b_k = a^k\delta_{kl}b^l$.

Proof. Ad (i). Since $\mathbf{e}_1, \mathbf{e}_2, \mathbf{e}_3$ is a basis of E_3, there exist real numbers a^1, a^2, a^3 such that $\mathbf{a} = a^s\mathbf{e}_s$. Hence $\mathbf{e}^k\mathbf{a} = a^s\mathbf{e}^k\mathbf{e}_s = a^s\delta^k_s = a^k$.

Ad (ii). $\mathbf{ab} = (a^k\mathbf{e}_k)(b^l\mathbf{e}_l) = a^kb^l\delta_{kl} = a^kb_k$. □

Matrix elements of a linear operator. Let $\mathbf{A}, \mathbf{B}, \mathbf{C} : E_3 \to E_3$ be linear operators,

$$\mathbf{A}(\alpha\mathbf{a} + \beta\mathbf{b}) = \alpha\mathbf{A}\mathbf{a} + \beta\mathbf{A}\mathbf{b} \qquad \text{for all} \quad \mathbf{a},\mathbf{b} \in E_3, \quad \alpha,\beta \in \mathbb{R}.$$

The Fourier coefficients a^1, a^2, a^3 of a vector $\mathbf{a}$ can be regarded as the coordinates of $\mathbf{a}$. We want to introduce coordinates A^k_l of the operator $\mathbf{A}$. To this end, we set

$$A^k_l := \langle \mathbf{e}^k|A\mathbf{e}_l\rangle, \qquad k,l = 1,2,3.$$

The real (3×3)-matrix (A^k_l) is called the coordinate matrix of the operator $\mathbf{A}$. Here, k (resp. l) is the row (resp. column) index.

---

[3] Recall that $\delta_{jk} := 0$ if $j \neq k$, and $\delta_{jj} := 1$ Moreover, we set $\delta^{jk} = \delta^k_j := \delta_{jk}$.

[4] Parseval des Chénes (1755–1836), Fourier (1768–1830).

Proposition 1.5 (i) $\mathbf{A}\mathbf{e}_l = A^k_l \mathbf{e}_k$, $l = 1,2,3$.
(ii) *The operator equation* $\mathbf{a} = \mathbf{A}\mathbf{b}$ *corresponds to the matrix equation* $a^k = A^k_l b^l$, $k = 1,2,3$. *Explicitly,*

$$\begin{pmatrix} a^1 \\ a^2 \\ a^3 \end{pmatrix} = \begin{pmatrix} A^1_1 & A^1_2 & A^1_3 \\ A^2_1 & A^2_2 & A^2_3 \\ A^3_1 & A^3_2 & A^3_3 \end{pmatrix} \begin{pmatrix} b^1 \\ b^2 \\ b^3 \end{pmatrix}.$$

(iii) $\langle \mathbf{a}|\mathbf{A}\mathbf{b}\rangle = a_k A^k_l b^l$ *for all* $\mathbf{a}, \mathbf{b} \in E_3$.
(iv) *The operator product equation* $\mathbf{C} = \mathbf{A}\mathbf{B}$ *corresponds to the matrix product equation* $C^k_l = A^k_s B^s_l$.

Proof. Ad (i). Since $\mathbf{e}_1, \mathbf{e}_2, \mathbf{e}_3$ form a basis of E_3, there exist real numbers α^s_l such that $\mathbf{A}\mathbf{e}_l = \alpha^s_l \mathbf{e}_s$. Hence $\alpha^k_l = \alpha^s_l \delta^k_s = \alpha^s_l \mathbf{e}^k \mathbf{e}_s = \mathbf{e}^k(\mathbf{A}\mathbf{e}_l) = A^k_l$.

Ad (ii). $\mathbf{A}\mathbf{b} = \mathbf{A}(b^l \mathbf{e}_l) = (A^k_l b^l)\mathbf{e}_k$. This is equal to $\mathbf{a} = a^k \mathbf{e}_k$.

Ad (iii). $\langle \mathbf{a}|\mathbf{A}\mathbf{b}\rangle = \langle a_s \mathbf{e}^s | A^k_l b^l \mathbf{e}_k\rangle = a_s A^k_l b^l \delta^s_k = a_k A^k_l b^l$.

Ad (iv). $\mathbf{A}(\mathbf{B}\mathbf{b}) = \mathbf{A}(B^s_l b^l \mathbf{e}_s) = (A^k_s B^s_l b^l)\mathbf{e}_k$. This is equal to $\mathbf{C}\mathbf{b} = C^k_l b^l \mathbf{e}_k$. Hence $C^k_l = A^k_s B^s_l$. □

The adjoint operator. Let $\mathbf{A} : E_3 \to E_3$ be a linear operator. We are looking for a linear operator $A^\dagger : E_3 \to E_3$ with the characteristic symmetry property

$$\langle \mathbf{A}^\dagger \mathbf{a}|\mathbf{b}\rangle = \langle \mathbf{a}|\mathbf{A}\mathbf{b}\rangle \qquad \text{for all} \quad \mathbf{a}, \mathbf{b} \in E_3. \tag{1.19}$$

Proposition 1.6 (i) *There exists precisely one linear operator on* E_3 *with the property* (1.19). *This operator* $\mathbf{A}^\dagger$ *is called the adjoint operator to* $\mathbf{A}$.

(ii) *For the matrix elements, we get* $(A^\dagger)^k_l = A^l_k$ *if* $k, l = 1,2,3$.

(iii) *The linear operator* $\mathbf{A} : E_3 \to E_3$ *is called self-adjoint iff* $\mathbf{A}^\dagger = \mathbf{A}$. *This is the case iff the matrix* (A^k_l) *is symmetric, that is,* $A^k_l = A^l_k$ *for all* $k, l = 1,2,3$.

Proof. (I) Uniqueness. Suppose that the operators $\mathbf{B}$ and $\mathbf{C}$ have the property (1.19). Then $\langle(\mathbf{B}-\mathbf{C})\mathbf{a}|\mathbf{b}\rangle = 0$ for all $\mathbf{a}, \mathbf{b} \in E_3$. This implies $(\mathbf{B}-\mathbf{C})\mathbf{a} = 0$ for all $\mathbf{a} \in E_3$. Hence $\mathbf{B} = \mathbf{C}$.

(II) Existence. Parallel to $\mathbf{A}\mathbf{b} = (A^k_l b^l)\mathbf{e_k}$, define $\mathbf{B}\mathbf{a} := (a_k A^k_l)\mathbf{e}^l$. Then we obtain $B^r_s = \langle \mathbf{e}^r|\mathbf{B}\mathbf{e}_s\rangle = \delta_{ks} A^k_l \mathbf{e}^r \mathbf{e}^l = A^s_l \delta^{rl} = A^s_r$. Moreover, set $\mathbf{A}^\dagger := \mathbf{B}$. Then the desired relation (1.19) is satisfied. □

1.2.5 Orthogonality and the Dirac Calculus

In the setting of Dirac's approach to quantum mechanics, the vector $\mathbf{a}$ (resp. the self-adjoint operator $\mathbf{A}$) represents a quantum state (resp. an observable). The choice of an orthonormal basis $\mathbf{e}_1, \mathbf{e}_2, \mathbf{e}_3$ corresponds to the choice of an observer. The goal is to compute the real coordinates of $\mathbf{a}$ and $\mathbf{A}$ which are related to measurement processes. The key formula is given by the following decomposition of the identity operator $I : E_3 \to E_3$:

$$\boxed{I = |\mathbf{e}_k\rangle\langle\mathbf{e}^k|\,.} \tag{1.20}$$

This leads immediately to the following formulas:

- $|\mathbf{a}\rangle = I|\mathbf{a}\rangle = |\mathbf{e}_k\rangle\langle\mathbf{e}^k|\mathbf{a}\rangle$ (Fourier series).

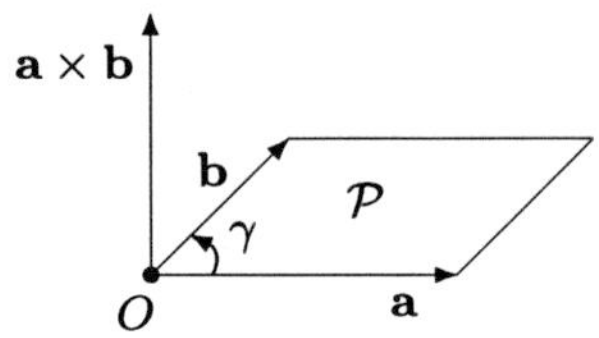

Fig. 1.5. Vector product $\mathbf{a} \times \mathbf{b}$

- $\langle \mathbf{a}|\mathbf{b}\rangle = \langle \mathbf{a}| \cdot I|\mathbf{b}\rangle = \langle \mathbf{a}|\mathbf{e}_k\rangle\langle \mathbf{e}^k|\mathbf{b}\rangle$ (Parseval equation).
- $\langle \mathbf{e}^k|\mathbf{A}\mathbf{B}\mathbf{e}_l\rangle = \langle \mathbf{e}^k|\mathbf{A}I\mathbf{B}\mathbf{e}_l\rangle = \langle \mathbf{e}^k|\mathbf{A}\mathbf{e}_s\rangle\langle \mathbf{e}^s|\mathbf{B}\mathbf{e}_l\rangle$ (matrix product formula).

The simple trick is to insert the identity operator I and to use the so-called completeness relation (1.20). These formulas coincide with the formulas considered in Sect. 1.2.4. A more general variant of the Dirac calculus based on covectors and duality, will be considered in Sect. 2.11.7 on page 171. The Dirac calculus was introduced by Dirac in his monograph *The Principles of Quantum Mechanics*, Clarendon Press, Oxford, 1930.

1.2.6 The Lie Algebra Structure

Definition of the vector product. Consider two position vectors $\mathbf{a}$ and $\mathbf{b}$ at the origin O which are not linearly dependent (Fig. 1.5). Then the vector product $\mathbf{a}\times\mathbf{b}$ is uniquely determined by the following properties:

- $\mathbf{a} \times \mathbf{b}$ is a position vector at the point O which is perpendicular to the plane spanned by the vectors $\mathbf{a}$ and $\mathbf{b}$.
- The length of $\mathbf{a} \times \mathbf{b}$ is equal to the area $A(\mathcal{P})$ of the parallelogram $\mathcal{P}$ spanned by $\mathbf{a}$ and $\mathbf{b}$. Explicitly,
$$A(\mathcal{P}) = |\mathbf{a}| \cdot |\mathbf{b}| \cdot \sin\gamma.$$
The counter-clockwise oriented angle γ points from $\mathbf{a}$ to $\mathbf{b}$ with $0 < \gamma < \pi$.
- The three vectors $\mathbf{a}, \mathbf{b}, \mathbf{a} \times \mathbf{b}$ form a right-handed system.

Otherwise, we set $\mathbf{a} \times \mathbf{b} := \mathbf{0}$. Furthermore, let us introduce the so-called volume product
$$\boxed{(\mathbf{abc}) := (\mathbf{a} \times \mathbf{b})\mathbf{c}.}$$

By elementary geometry, the nonnegative number $|(\mathbf{abc})|$ is the volume of the parallelepiped spanned by the three vectors $\mathbf{a}, \mathbf{b}, \mathbf{c}$ (Fig. 1.6). The sign of $(\mathbf{abc})$ is positive (resp. negative) iff $\mathbf{a}, \mathbf{b}, \mathbf{c}$ form a right-handed (resp. left-handed) system. Obviously, the following hold:

- The vectors $\mathbf{a}, \mathbf{b}$ are linearly dependent iff $\mathbf{a} \times \mathbf{b} = 0$.
- The vectors $\mathbf{a}, \mathbf{b}, \mathbf{c}$ are linearly dependent iff $(\mathbf{abc}) = 0$.

For example, let $\mathbf{i}, \mathbf{j}, \mathbf{k}$ be a right-handed orthonormal system (see Fig. 1.1(c) on page 72). Then:

$$\mathbf{i} \times \mathbf{j} = \mathbf{k}, \quad \mathbf{j} \times \mathbf{i} = -\mathbf{k}, \quad \mathbf{i} \times \mathbf{i} = 0. \tag{1.21}$$

Using the cyclic permutation $\mathbf{i} \Rightarrow \mathbf{j} \Rightarrow \mathbf{k} \Rightarrow \mathbf{i}$, we also get:

- $\mathbf{j} \times \mathbf{k} = \mathbf{i}$, $\mathbf{k} \times \mathbf{j} = -\mathbf{i}$, and $\mathbf{j} \times \mathbf{j} = 0$,
- $\mathbf{k} \times \mathbf{i} = \mathbf{j}$, $\mathbf{i} \times \mathbf{k} = -\mathbf{j}$, and $\mathbf{k} \times \mathbf{k} = 0$.

Let us introduce the Lie product $[\mathbf{a}, \mathbf{b}] := \mathbf{a} \times \mathbf{b}$.

(a) (**abc**) > 0 (b) (**abc**) < 0

Fig. 1.6. Volume product

Theorem 1.7 *The Euclidean space* E_3 *becomes a real 3-dimensional Lie algebra with respect to the Lie product* $[\mathbf{a}, \mathbf{b}]$. *This Lie algebra is denoted by* $(E_3)_{\text{Lie}}$.

The proof will be given below. Explicitly, this theorem means that, for all vectors $\mathbf{a}, \mathbf{b}, \mathbf{c}$ and all real numbers α, β, the following hold:

- E_3 is a real 3-dimensional linear space.
- $[\mathbf{a}, \mathbf{b}] = -[\mathbf{b}, \mathbf{a}]$ (anticommutativity),
- $[\alpha\mathbf{a} + \beta\mathbf{b}, \mathbf{c}] = \alpha[\mathbf{a}, \mathbf{c}] + \beta[\mathbf{b}, \mathbf{c}]$ (distributivity),
- $[\,[\mathbf{a}, \mathbf{b}],\ \mathbf{c}] + [\,[\mathbf{b}, \mathbf{c}],\ \mathbf{a}] + [\,[\mathbf{c}, \mathbf{a}],\ \mathbf{b}] = 0$ (Jacobi identity).

Note that the associative law is not valid for the Lie product $[\mathbf{a}, \mathbf{b}]$. In some sense, the Jacobi identity can be regarded as a substitute for the missing associative law. Suppose that we are given the right-handed (resp. left-handed) orthonormal system $\mathbf{e}_1, \mathbf{e}_2, \mathbf{e}_3$ (with orientation number $\iota = 1$ (resp. $\iota = -1$)). Then

$$[\mathbf{e}_i, \mathbf{e}_j] = c_{ij}^k \mathbf{e}_k, \qquad i, j = 1, 2, 3 \tag{1.22}$$

with the so-called structure constants $c_{ij}^k := \iota\varepsilon^{ijk}$ for $i, j, k = 1, 2, 3$ of the Lie algebra $(E_3)_{\text{Lie}}$. By the way, $(E_3)_{\text{Lie}}$ is the simplest nontrivial (i.e., noncommutative) real Lie algebra. We will show in Chap. 7 that:

> *The Lie algebra* $(E_3)_{\text{Lie}}$ *is isomorphic to the Lie algebra* $su(E_3)$ *which consists of all the infinitesimal rotations of the Euclidean space* E_3.

In other words, the Lie algebra $(E_3)_{\text{Lie}}$ is obtained from the group of rotations $SU(E_3)$ of the Euclidean space E_3 by linearization.

Properties of the vector product. For all vectors $\mathbf{a}, \mathbf{b}, \mathbf{c}, \mathbf{d}, \mathbf{x}, \mathbf{y}, \mathbf{z}$ and all real numbers α, β, the following hold:

Proposition 1.8 (i) *The map* $(\mathbf{a}, \mathbf{b}, \mathbf{c}) \mapsto (\mathbf{abc})$ *is antisymmetric and 3-linear.*

(ii) $(\alpha\mathbf{a} + \beta\mathbf{b}) \times \mathbf{c} = \alpha(\mathbf{a} \times \mathbf{c}) + \beta(\mathbf{b} \times \mathbf{c})$ (distributive law).

(iii) *Let* $\mathbf{i}, \mathbf{j}, \mathbf{k}$ *be a right-handed orthonormal system. Suppose that*

$$\mathbf{a} = a^1\mathbf{i} + a^2\mathbf{j} + a^3\mathbf{k}, \qquad \mathbf{b} = b^1\mathbf{i} + b^2\mathbf{j} + b^3\mathbf{k},$$

and $\mathbf{c} = c^1\mathbf{i} + c^2\mathbf{j} + c^3\mathbf{k}$. *Then*

$$\mathbf{a} \times \mathbf{b} = \begin{vmatrix} \mathbf{i} & \mathbf{j} & \mathbf{k} \\ a^1 & a^2 & a^3 \\ b^1 & b^2 & b^3 \end{vmatrix}. \tag{1.23}$$

Explicitly, $\mathbf{a} \times \mathbf{b} = (a^2b^3 - a^3b^2)\mathbf{i} + (a^3b^1 - a^1b^3)\mathbf{j} + (a^1b^2 - a^2b^1)\mathbf{k}$. *Moreover,*

$$(\mathbf{abc}) = \begin{vmatrix} a^1 & a^2 & a^3 \\ b^1 & b^2 & b^3 \\ c^1 & c^2 & c^3 \end{vmatrix}. \tag{1.24}$$

(iv) *Let* $\mathbf{e}_1, \mathbf{e}_2, \mathbf{e}_3$ *be a right-handed (resp. left-handed) orthonormal system (with the orientation number* $\iota = 1$ *(resp.* $\iota = -1$*)). Suppose that* $\mathbf{a} = a^i\mathbf{e}_i$, $\mathbf{b} = b^i\mathbf{e}_i$, *and* $\mathbf{c} = c^i\mathbf{e}_i$. *Set* $\mathbf{e}^k := \mathbf{e}_k$. *Then*

$$\mathbf{a} \times \mathbf{b} = \iota\varepsilon_{ijk}a^ib^j\mathbf{e}^k, \qquad (\mathbf{abc}) = \iota\varepsilon_{ijk}a^ib^jc^k. \tag{1.25}$$

Note that these formulas depend on the orientation ι *of the basis.*

(v) *The Bunyakovski–Cauchy–Schwarz equation:*

$$(\mathbf{ab})^2 + (\mathbf{a} \times \mathbf{b})^2 = |\mathbf{a}|^2|\mathbf{b}|^2. \tag{1.26}$$

This implies the Bunyakovski–Cauchy–Schwarz inequality

$$|\mathbf{ab}| \le |\mathbf{a}| \cdot |\mathbf{b}|, \tag{1.27}$$

and $|\mathbf{ab}| = |\mathbf{a}| \cdot |\mathbf{b}|$ *iff* $\mathbf{a} \times \mathbf{b} = 0$.

(vi) *The Grassmann expansion formula:*

$$\mathbf{a} \times (\mathbf{b} \times \mathbf{c}) = (\mathbf{ac})\mathbf{b} - (\mathbf{ab})\mathbf{c}. \tag{1.28}$$

(vii) *The Jacobi identity:*

$$(\mathbf{a} \times \mathbf{b}) \times \mathbf{c} + (\mathbf{b} \times \mathbf{c}) \times \mathbf{a} + (\mathbf{c} \times \mathbf{a}) \times \mathbf{b} = \mathbf{0}. \tag{1.29}$$

(viii) *The Lagrange identity:*

$$(\mathbf{a} \times \mathbf{b})(\mathbf{c} \times \mathbf{d}) = \begin{vmatrix} \mathbf{ac} & \mathbf{ad} \\ \mathbf{bc} & \mathbf{bd} \end{vmatrix} = (\mathbf{ac})(\mathbf{bd}) - (\mathbf{ad})(\mathbf{bc}). \tag{1.30}$$

(ix) *The volume identity:*

$$(\mathbf{a} \times \mathbf{b}) \times (\mathbf{c} \times \mathbf{d}) = (\mathbf{abd})\mathbf{c} - (\mathbf{abc})\mathbf{d} = (\mathbf{acd})\mathbf{b} - (\mathbf{bcd})\mathbf{a}.$$

(x) *The Gram determinant:*[5]

$$(\mathbf{abc})(\mathbf{xyz}) = \begin{vmatrix} \mathbf{ax} & \mathbf{ay} & \mathbf{az} \\ \mathbf{bx} & \mathbf{by} & \mathbf{bz} \\ \mathbf{cx} & \mathbf{cy} & \mathbf{cz} \end{vmatrix}. \tag{1.31}$$

Therefore, the three vectors $\mathbf{a}, \mathbf{b}, \mathbf{c}$ *are linearly dependent iff* $(\mathbf{abc})^2 = 0$, *that is,*

$$\begin{vmatrix} \mathbf{a}^2 & \mathbf{ab} & \mathbf{ac} \\ \mathbf{ba} & \mathbf{b}^2 & \mathbf{bc} \\ \mathbf{ca} & \mathbf{cb} & \mathbf{c}^2 \end{vmatrix} = 0.$$

[5] Gram (1850–1916)

Proof. Ad (i). Since $(\mathbf{abc}) = (\mathbf{a} \times \mathbf{b})\mathbf{c}$, this expression is linear with respect to the third argument $\mathbf{c}$. If two vectors are transposed, then the orientation changes and hence $(\mathbf{abc})$ changes sign (Fig. 1.6 on page 83). Thus, $(\mathbf{abc})$ is linear with respect to each argument.

Ad (ii). By (i), $\{(\alpha\mathbf{a} + \beta\mathbf{b}) \times \mathbf{c}\}\mathbf{x}$ is equal to

$$((\alpha\mathbf{a} + \beta\mathbf{b})\mathbf{cx}) = \alpha(\mathbf{acx}) + \beta(\mathbf{bcx}).$$

Moreover, $\{\alpha\mathbf{a} \times \mathbf{c} + \beta\mathbf{b} \times \mathbf{c})\}\mathbf{x}$ is equal to

$$((\alpha\mathbf{a})\mathbf{cx}) + ((\beta\mathbf{b})\mathbf{cx}) = \alpha(\mathbf{acx}) + \beta(\mathbf{bcx}).$$

Thus, for all vectors $\mathbf{x} \in E_3$,

$$\{(\alpha\mathbf{a} + \beta\mathbf{b}) \times \mathbf{c}\}\mathbf{x} = \{\alpha\mathbf{a} \times \mathbf{c} + \beta\mathbf{b} \times \mathbf{c}\}\mathbf{x}.$$

Therefore, $(\alpha\mathbf{a} + \beta\mathbf{b}) \times \mathbf{c} = \alpha\mathbf{a} \times \mathbf{c} + \beta\mathbf{b} \times \mathbf{c}$.

Ad (iii). By the distributive law (ii) and $\mathbf{i} \times \mathbf{j} = -(\mathbf{j} \times \mathbf{i}) = \mathbf{k}, \ldots$, we get

$$(a^1\mathbf{i} + a^2\mathbf{j} + a^3\mathbf{k})(b^1\mathbf{i} + b^2\mathbf{j} + b^3\mathbf{k}) = (a^1b^2 - a^2b^1)\mathbf{k} \ldots$$

Ad (iv)–(ix). See Problem 3.4 on page 313. □

1.2.7 The Metric Tensor

For all vectors $\mathbf{a}, \mathbf{b} \in E_3$, we define

$$\boxed{\mathbf{g}(\mathbf{a}, \mathbf{b}) := \mathbf{ab}.} \tag{1.32}$$

The bilinear symmetric functional $\mathbf{g} : E_3 \times E_3 \to \mathbb{R}$ is called the metric tensor of the Euclidean space E_3.

1.2.8 The Volume Form

For all vectors $\mathbf{a}, \mathbf{b}, \mathbf{c} \in E_3$, we define

$$\boxed{v(\mathbf{a}, \mathbf{b}, \mathbf{c}) := (\mathbf{abc}).} \tag{1.33}$$

Since $v(\mathbf{a}, \mathbf{b}, \mathbf{c})$ is equal to the oriented volume spanned by the three vectors $\mathbf{a}, \mathbf{b}, \mathbf{c}$, the 3-linear antisymmetric functional $v : E_3 \times E_3 \times E_3 \to \mathbb{R}$ is called the volume form of the Euclidean space E_3 (see Fig. 1.6 on page 83).

Both the metric tensor $\mathbf{g}$ and the volume form v play a crucial role in the theory of n-dimensional Riemannian manifolds. The generalization to pseudo-Riemannian manifolds is the key to Einstein's theory of general relativity.

1.2.9 Grassmann's Alternating Product

The alternating (or exterior) product was introduced by Grassmann in his "Ausdehnungslehre" (Theory of extensions) from 1844. In the Euclidean space E_3, the exterior product is closely related to the vector product. However, in contrast to the vector product in E_3, Grassmann's alternating product makes sense in linear spaces of arbitrary finite dimension. Unfortunately, the contemporaries of Grassmann (1809–1877) did not understand this ingenious approach. The crucial point is that the linear dependence of vectors in n-dimensional linear spaces can be characterized by the vanishing of appropriate antisymmetric multilinear functionals which are closely related to determinants.

Folklore

The alternating product for vectors. Let $\mathbf{a}, \mathbf{b}, \mathbf{c} \in E_3$. By definition, the alternating product $\mathbf{a} \wedge \mathbf{b}$ is an antisymmetric bilinear functional $\mathbf{a} \wedge \mathbf{b} : E_3 \times E_3 \to \mathbb{R}$ given by

$$(\mathbf{a} \wedge \mathbf{b})(\mathbf{x}, \mathbf{y}) := \begin{vmatrix} \mathbf{ax} & \mathbf{ay} \\ \mathbf{bx} & \mathbf{by} \end{vmatrix} = (\mathbf{ax})(\mathbf{by}) - (\mathbf{ay})(\mathbf{bx}).$$

Similarly, the alternating product $\mathbf{a} \wedge \mathbf{b} \wedge \mathbf{c}$ is an antisymmetric 3-linear functional $\mathbf{a} \wedge \mathbf{b} \wedge \mathbf{c} : E_3 \times E_3 \times E_3 \to \mathbb{R}$ given by

$$(\mathbf{a} \wedge \mathbf{b} \wedge \mathbf{c})(\mathbf{x}, \mathbf{y}, \mathbf{z}) := \begin{vmatrix} \mathbf{ax} & \mathbf{ay} & \mathbf{az} \\ \mathbf{bx} & \mathbf{by} & \mathbf{bz} \\ \mathbf{cx} & \mathbf{cy} & \mathbf{cz} \end{vmatrix}.$$

This product is called alternating, since it has the following typical property:

$$\boxed{\mathbf{a} \wedge \mathbf{b} = -\mathbf{b} \wedge \mathbf{a}.}$$

The relation to the vector product is very close. In fact, we have
- $(\mathbf{a} \wedge \mathbf{b})(\mathbf{x}, \mathbf{y}) = (\mathbf{a} \times \mathbf{b})(\mathbf{x} \times \mathbf{y})$, and
- $(\mathbf{a} \wedge \mathbf{b} \wedge \mathbf{c})(\mathbf{x}, \mathbf{y}, \mathbf{z}) = (\mathbf{abc})(\mathbf{xyz})$.

Therefore, we get the following result.

Theorem 1.9 (i) *The vectors* $\mathbf{a}, \mathbf{b}$ *are linearly dependent iff* $\mathbf{a} \wedge \mathbf{b} = 0$.
(ii) *The vectors* $\mathbf{a}, \mathbf{b}, \mathbf{c}$ *are linearly dependent iff* $\mathbf{a} \wedge \mathbf{b} \wedge \mathbf{c} = 0$.

Moreover, we get the key relation

$$\boxed{\mathbf{a} \wedge \mathbf{b} \wedge \mathbf{c} = v(\mathbf{a}, \mathbf{b}, \mathbf{c}) \cdot (\mathbf{i} \wedge \mathbf{j} \wedge \mathbf{k}).} \tag{1.34}$$

This means that, for all $\mathbf{x}, \mathbf{y}, \mathbf{z} \in E_3$,

$$(\mathbf{a} \wedge \mathbf{b} \wedge \mathbf{c})(\mathbf{x}, \mathbf{y}, \mathbf{z}) = v(\mathbf{a}, \mathbf{b}, \mathbf{c}) \cdot (\mathbf{i} \wedge \mathbf{j} \wedge \mathbf{k})(\mathbf{x}, \mathbf{y}, \mathbf{z}).$$

In fact, $(\mathbf{a} \wedge \mathbf{b} \wedge \mathbf{c})(\mathbf{x}, \mathbf{y}, \mathbf{c}) = (\mathbf{abc})(\mathbf{xyz})$ and $(\mathbf{ijk}) = 1$.

Corollary 1.10 *Suppose that the pairs* $\mathbf{a}, \mathbf{b}$ *and* $\mathbf{c}, \mathbf{d}$ *are linearly independent position vectors at the origin O, that is,* $\mathbf{a} \wedge \mathbf{b} \neq 0$ *and* $\mathbf{c} \wedge \mathbf{d} \neq 0$. *Then:*
(i) $\mathbf{a} \wedge \mathbf{b} = \mathbf{c} \wedge \mathbf{d}$ *iff* $\mathbf{a} \times \mathbf{b} = \mathbf{c} \times \mathbf{d}$.
(ii) *The pairs of vectors* $\mathbf{a}, \mathbf{b}$ *and* $\mathbf{c}, \mathbf{d}$ *span the same plane iff* $(\mathbf{abc}) = (\mathbf{abd}) = 0$.

Note that condition (ii) means that the vectors $\mathbf{c}$ and $\mathbf{d}$ are orthogonal to the nonzero vector $\mathbf{a} \times \mathbf{b}$.

1.2.10 Perspectives

Choose a fixed (x, y, z)-Cartesian coordinate system with the (right-handed) orthonormal system $\mathbf{i}, \mathbf{j}, \mathbf{k}$. Let $\mathbf{a} := a^1\mathbf{i} + a^2\mathbf{j} + a^3\mathbf{k}$. We introduce the special linear functionals $dx, dy, dz : E_3 \to \mathbb{R}$ by setting

$$dx(\mathbf{a}) := a^1, \quad dy(\mathbf{a}) := a^2, \quad dz(\mathbf{a}) := a^3.$$

(i) The alternating product for linear functionals. Let $F, G, H : E_3 \to \mathbb{R}$ be linear functionals on the Euclidean space E_3. For all vectors $\mathbf{a}, \mathbf{b}, \mathbf{c} \in E_3$, we set:

- $(F \wedge G)(\mathbf{a}, \mathbf{b}) := F(\mathbf{a})G(\mathbf{b}) - F(\mathbf{b})G(\mathbf{a})$.
- $(F \wedge G \wedge H)(\mathbf{a}, \mathbf{b}, \mathbf{c})$ is equal to the determinant

$$\begin{vmatrix} F(\mathbf{a}) & G(\mathbf{a}) & H(\mathbf{a}) \\ F(\mathbf{b}) & G(\mathbf{b}) & H(\mathbf{b}) \\ F(\mathbf{c}) & G(\mathbf{c}) & H(\mathbf{c}) \end{vmatrix}.$$

Obviously, $F \wedge G = -G \wedge F$. In particular, the symbol $dx \wedge dy$ denotes the bilinear functional $dx \wedge dy : E_3 \times E_3 \to \mathbb{R}$ given by

$$(dx \wedge dy)(\mathbf{a}, \mathbf{b}) = dx(\mathbf{a})dy(\mathbf{b}) - dx(\mathbf{b})dy(\mathbf{a}) \qquad \text{for all} \quad \mathbf{a}, \mathbf{b} \in E_3.$$

Furthermore, $dx \wedge dy = -dy \wedge dx$. The product $dx \wedge dy \wedge dz : E_3 \times E_3 \times E_3 \to \mathbb{R}$ is an antisymmetric 3-linear functional. Explicitly, $(dx \wedge dy \wedge dz)(\mathbf{a}, \mathbf{b}, \mathbf{z})$ is equal to

$$\begin{vmatrix} dx(\mathbf{a}) & dy(\mathbf{a}) & dz(\mathbf{a}) \\ dx(\mathbf{b}) & dy(\mathbf{b}) & dz(\mathbf{b}) \\ dx(\mathbf{c}) & dy(\mathbf{c}) & dz(\mathbf{c}) \end{vmatrix} = \begin{vmatrix} a^1 & a^2 & a^3 \\ b^1 & b^2 & b^3 \\ c^1 & c^2 & c^3 \end{vmatrix} = (\mathbf{abc}). \tag{1.35}$$

Consequently, $dx \wedge dy \wedge dz$ is equal to the volume form:

$$\boxed{\upsilon = dx \wedge dy \wedge dz.} \tag{1.36}$$

This is true for any Cartesian coordinate system.[6]

In modern mathematics, the Cartan calculus of alternating differential forms plays a crucial role. This will be studied later on. At this point, let us only mention that the alternating product $dx \wedge dy$ is the starting point of the Cartan calculus. Note that:

> *In the Cartan calculus, the differentials dx, dy, dz are well-defined mathematical objects, namely, linear functionals on the Euclidean space E_3.*

This avoids the trouble with the heuristic concept of Leibniz's 'infinitesimals' in the history of mathematics and physics.

(ii) Hodge duality on the Euclidean space E_3. In Hodge theory, one defines the Hodge $*$-operation by setting:

- $*\mathbf{i} := \mathbf{j} \wedge \mathbf{k}$, and $*(\mathbf{j} \wedge \mathbf{k}) := \mathbf{i}$,
- $*(\mathbf{i} \wedge \mathbf{j} \wedge \mathbf{k}) = 1$, and $*1 = \mathbf{i} \wedge \mathbf{j} \wedge \mathbf{k}$.

---

[6] For a reflected Cartesian coordinate system, we get $\upsilon = -dx \wedge dy \wedge dz$.

Obviously, $**(.) = (.)$. Further definitions of the $*$-operation are obtained by using the cyclic permutation

$$\mathbf{i} \Rightarrow \mathbf{j} \Rightarrow \mathbf{k} \Rightarrow \mathbf{i}.$$

For example, $*\mathbf{j} = \mathbf{k} \wedge \mathbf{i}$. Finally, the $*$-operation is extended to real linear combinations by using the linearity principle. For example,

$$\begin{aligned} *\mathbf{a} &= *(a^1\mathbf{i} + a^2\mathbf{j} + a^3\mathbf{k}) = a^1 * \mathbf{i} + a^2 * \mathbf{j} + a^3 * \mathbf{k} \\ &= a^1\mathbf{j} \wedge \mathbf{k} + a^2\mathbf{k} \wedge \mathbf{i} + a^3\mathbf{i} \wedge \mathbf{j}. \end{aligned}$$

(iii) Hodge duality for linear functionals. Replacing $\mathbf{i}, \mathbf{j}, \mathbf{k}$ by dx, dy, dz, respectively, we define:

- $*dx = dy \wedge dz$ and $*(dy \wedge dz) = dx$,
- $*(dx \wedge dy \wedge dz) = 1$, and $*1 = dx \wedge dy \wedge dz$,

and so on. In particular, for the volume form we get

$$\boxed{v = *1.} \tag{1.37}$$

The following remark is crucial. The definition of the Hodge $*$-operation given above depends on the choice of the Cartesian coordinate system. However, in Sect. 2.7 we will show the following:

The Hodge $*$-operation can be introduced in a geometric setting which is independent of the choice of a Cartesian coordinate system.

This is the reason for the importance of Hodge duality in modern mathematics.

(v) The tensor product of linear functionals. Let $F, G, H : E_3 \to \mathbb{R}$ be linear functionals. For all vectors $\mathbf{a}, \mathbf{b}, \mathbf{c} \in E_3$, we set:

- $(F \otimes G)(\mathbf{a}, \mathbf{b}) := F(\mathbf{a})G(\mathbf{b})$.
- $F \otimes G \otimes H := F(\mathbf{a})G(\mathbf{b})H(\mathbf{c})$.

For example, the metric tensor of the Euclidean space E_3 allows the representation

$$\boxed{\mathbf{g} = dx \otimes dx + dy \otimes dy + dz \otimes dz.} \tag{1.38}$$

This is true for any Cartesian coordinate system (resp. reflected Cartesian coordinate system). In fact, $(dx \otimes dx)(\mathbf{a}, \mathbf{b}) = dx(\mathbf{a})dx(\mathbf{b}) = a^1b^1$. Similarly,

$$(dx \otimes dx + dy \otimes dy + dz \otimes dz)(\mathbf{a}, \mathbf{b}) = a^1b^1 + a^2b^2 + a^3b^3 = \mathbf{ab}.$$

(vi) The tensor product of vectors. Let $\mathbf{a}, \mathbf{b}, \mathbf{c}$ be vectors in E_3. for all linear functionals $F, G, H : E_3 \to \mathbb{R}$, we set:

- $(\mathbf{a} \otimes \mathbf{b})(F, G) := F(\mathbf{a})G(\mathbf{b})$.
- $\mathbf{a} \otimes \mathbf{b} \otimes \mathbf{c} := F(\mathbf{a})G(\mathbf{b})H(\mathbf{c})$.

Note that this definition respects duality. In fact,

- $(F \otimes G)(\mathbf{a}, \mathbf{b}) = (\mathbf{a} \otimes \mathbf{b})(F, G)$, and
- $(F \otimes G \otimes H)(\mathbf{a}, \mathbf{b}, \mathbf{c}) = (\mathbf{a} \otimes \mathbf{b} \otimes \mathbf{c})(F, G, H)$.

(vii) The Riesz duality generated by the inner product on the Euclidean space E_3. Let E_3^d denote the dual Euclidean space consisting of all the linear functionals $F : E_3 \to \mathbb{R}$. There exists a crucial one-to-one correspondence

$$F \Longleftrightarrow \mathbf{a}$$

between the linear functionals $F \in E_3^d$ and the vectors $\mathbf{a} \in E_3$ given by

$$F(\mathbf{x}) := \mathbf{ax} \qquad \text{for all} \quad \mathbf{x} \in E_3$$

(see Sect. 1.4). We write $F := \aleph(\mathbf{a})$. This way,

- the bilinear functional $\mathbf{a} \otimes \mathbf{b} : E_3^d \times E_3^d \to \mathbb{R}$ is transformed into
- the bilinear functional $\mathbf{a} \otimes \mathbf{b} : E_3 \otimes E_3 \to \mathbb{R}$ given by[7]

$$(\mathbf{a} \otimes \mathbf{b})(\mathbf{x}, \mathbf{y}) := (\mathbf{ax})(\mathbf{by}) - (\mathbf{ay})(\mathbf{bx}) \qquad \text{for all} \quad \mathbf{x}, \mathbf{y} \in E_3.$$

Similarly, the 3-linear functional $\mathbf{a} \otimes \mathbf{b} \otimes \mathbf{c} : E_3^d \times E_3^d \times E_3^d \to \mathbb{R}$ is transformed into a 3-linear functional of the form

$$\mathbf{a} \otimes \mathbf{b} \otimes \mathbf{c} : E_3 \times E_3 \times E_3 \to \mathbb{R}.$$

Tensor products in elementary particle physics. Note that in elementary particle physics, the tensor product

$$\mathbf{a} \otimes \mathbf{b}$$

is used in order to describe mathematically the quantum state of a composite particle (e.g., a meson) which consists of the quantum states $\mathbf{a}$ and $\mathbf{b}$ of two single particles (e.g., a quark-antiquark pair). Similarly,

$$\mathbf{a} \otimes \mathbf{b} \otimes \mathbf{c}$$

describes a composite particle (e.g., a proton) which consists of three single particles (e.g., three quarks). More precisely, one has to use linear combinations of tensor products. Furthermore, tensor products are systematically used in order to attribute additional properties to given objects (e.g., this way, one attributes the color to quarks, as additional degrees of freedom).

1.3 The Skew-Field $\mathbb{H}$ of Quaternions

> Both the two-dimensional field $\mathbb{C}$ of complex numbers and the three-dimensional Euclidean space E_3 can be extended to the four-dimensional skew-field $\mathbb{H}$ of quaternions.
>
> Folklore

In this section, we will use the terminology on algebras introduced in Vol. II and summarized on page 116 of the present volume.

[7] To simplify notation, we use the same symbol $\mathbf{a} \otimes \mathbf{b}$ for the two different bilinear functionals on $E_3^d \times E_3^d$ and on $E_3 \times E_3$.

1.3.1 The Field $\mathbb{C}$ of Complex Numbers

> It is almost impossible for anyone today who already hears at school about $\mathrm{i} = \sqrt{-1}$ being a solution of the equation $x^2 + 1 = 0$ to understand what difficulties the complex (that is, imaginary) numbers presented to mathematicians and physicists in former times ...
> Imaginary quantities make their first appearance during the Renaissance. In 1539, Giralmo Cardano (1501–1576), a mathematician and renowned physician in Milan (Italy), learned from Nicolò Tartaglia (1506–1559) a process for solving cubic equations; in 1545 he broke his promise never to divulge the secret to anyone... It is not clear whether Cardano was led to complex numbers through cubic or quadratic equations.[8]
>
> Reinhold Remmert, 1995

Cardano's solution formula. Consider the cubic equation

$$x^3 = px + q$$

where p and q are real numbers. In his *Ars magna* from 1545, Cardano published the following solution formula

$$x = \sqrt[3]{\frac{q}{2} + \sqrt{d}} + \sqrt[3]{\frac{q}{2} - \sqrt{d}} \tag{1.39}$$

where $d := \frac{q^2}{4} - \frac{p^3}{27}$ is the so-called discriminant.

In his book *L'Algebra* published in Bologna in 1572, Bombelli (1526–1572) introduced fundamental computational rules for complex numbers including[9]

$$\boxed{\sqrt{-1}\sqrt{-1} = -1.}$$

For example, he used $(2 \pm \sqrt{-1})^3 = 2 \pm 11\sqrt{-1}$ in order to get

$$\sqrt[3]{2 \pm \sqrt{-121}} = 2 \pm \sqrt{-1}.$$

Applying the Cardano solution formula (1.39) to the equation

$$x^3 = 15x + 4, \tag{1.40}$$

Bombelli obtained the formal solution

$$x = \sqrt[3]{2 + \sqrt{-121}} + \sqrt[3]{2 - \sqrt{-121}}$$

which yields $x = (2 + \sqrt{-1}) + (2 - \sqrt{-1}) = 4$. This is a classical solution of (1.40). As a big surprise for Bombielli and his contemporaries, the formal approach based on 'mystical' complex numbers yields a solution which is a classical (non-mystical) number.

---

[8] R. Remmert, Complex numbers. In: H. Ebbinghaus et al. (Eds.), Numbers, Springer, New York, pp. 55–96 (reprinted with permission). We recommend reading this beautiful article.

[9] The symbol i for $\sqrt{-1}$ was used by Euler (1707–1783) since 1777.

Gauss' proof of the fundamental theorem of algebra. In his dissertation in 1799, Gauss gave a proof of the fundamental theorem of algebra.[10] The fundamental theorem of algebra says that, for each polynomial

$$p(x) = a_0 + a_1 x + \ldots + a_{n-1} x^{n-1} + x^n$$

with real coefficients $a_0, a_1, \ldots, a_{n-1}$, there exist complex numbers $x_1, \ldots, x_n$ such that we have the factorization

$$p(x) = (x - x_1)(x - x_2) \cdots (x - x_n).$$

This implies that the equation $p(x) = 0$ has precisely the solutions $x_1, \ldots, x_n$. The big surprise is that we only have to add the solution i of the very special equation $x^2 + 1 = 0$ in order to get the field $\mathbb{C}$ of complex numbers which contains all the solutions of all possible polynomial equations with real coefficients.[11] Gauss' proof was of topological nature. In 1920, Alexander Ostrowski (1893–1986) pointed out in a comment on Gauss' dissertation[12] that Gauss used geometrical properties of real algebraic curves which are neither proved in the dissertation itself nor had been proved in the pre-Gaussian literature. Ostrowski showed that Gauss' proof can be completed by using nontrivial tools from modern algebraic geometry.

Let us emphasize that the most elegant proof of the fundamental theorem can be based on Liouville's theorem saying that a bounded holomorphic function on the complex plane $\mathbb{C}$ is a constant.[13] The proof goes like this. We are given a polynomial $p = p(z)$ with complex coefficients of degree ≥ 1. It is sufficient to show that p has at least one complex zero. If this is not true, then the quotient $\frac{1}{p}$ is a constant function by Liouville's theorem; this is a contradiction.[14]

1.3.2 The Galois Group Gal($\mathbb{C}|\mathbb{R}$) and Galois Theory

Consider the extension $\mathbb{R} \subseteq \mathbb{C}$ of the field $\mathbb{R}$ of real numbers to the field $\mathbb{C}$ of complex numbers. By definition, the map

$$S : \mathbb{R} \to \mathbb{C}$$

is called a symmetry map of the field extension iff it is a field automorphism which has all the points of the basic field $\mathbb{R}$ as fixed points. This is a subgroup of the group of all automorphisms of the field $\mathbb{C}$. This subgroup is called the Galois group Gal($\mathbb{C}|\mathbb{R}$) of the field extension $\mathbb{R} \subseteq \mathbb{C}$.

Proposition 1.11 *The field extension $\mathbb{R} \subseteq \mathbb{C}$ has precisely the two symmetry transformations $x + y\mathrm{i} \mapsto x + y\mathrm{i}$ (identical transformation) and $x + y\mathrm{i} \mapsto (x + y\mathrm{i})^\dagger$ (conjugation).*

The Galois group Gal($\mathbb{C}|\mathbb{R}$) *is isomorphic to the group $\mathbb{Z}_2 := \{1, -1\}$.*

A complex number $x + y\mathrm{i}$ is real iff it is a fixed point of all the elements of the Galois group, that is, $(x + y\mathrm{i})^\dagger = x + y\mathrm{i}$.

---

[10] The sophisticated history of this theorem is thoroughly discussed in R. Remmert, The fundamental theorem of algebra. In: H. Ebbinghaus et al. (Eds.), Numbers, Springer, New York, 1995, pp. 77–122.

[11] In fact, this remains true for polynomials with complex coefficients.

[12] See C. Gauss, *Collected Works*, Vol. 10.2.

[13] This is the special case of the following topological theorem: A holomorphic function $h : \mathcal{R} \to \mathbb{C}$ on a compact Riemann surface $\mathcal{R}$ is a constant.

[14] Liouville (1809–1882)

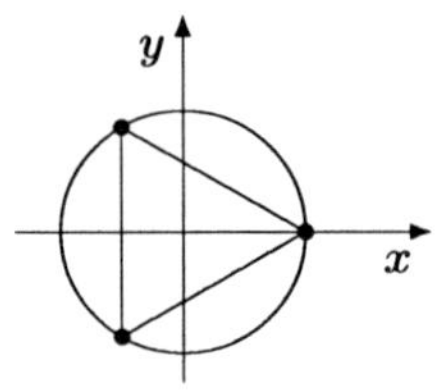

Fig. 1.7. The cyclotomic equation $x^3 = 1$

Proof. Since $\mathrm{i}^2 + 1 = 0$, and since S is an automorphism, we get $S(\mathrm{i})^2 + 1 = 0$. Hence either $S(\mathrm{i}) = \mathrm{i}$ or $S(\mathrm{i}) = -\mathrm{i}$. Consequently, S is either the identical map or we have $S(x + y\mathrm{i}) = x - y\mathrm{i}$ for all $x, y \in \mathbb{R}$. □

As we will discuss below, the main theorem of Galois theory is a far-reaching generalization of Prop. 1.11.

Gauss' investigation of the cyclotomic equation and cyclotomic fields. In 1796, the young Gauss proved that it is possible to construct a regular 17-gon with a ruler and a compass. This was an open problem since ancient times. More general, Gauss studied the cyclotomic equation

$$x^n = 1 \tag{1.41}$$

where $n = 2, 3, \ldots$ By definition, the corresponding cyclotomic field is the smallest subfield of the field of complex numbers $\mathbb{C}$ which contains the field $\mathbb{Q}$ of rational numbers and the n solutions

$$\mathrm{e}^{2\pi k\mathrm{i}/n}, \qquad k = 0, 1, \ldots, n-1.$$

These solutions lie on the unit circle, and they form a regular n-gon (Fig. 1.7). The cyclotomic field of (1.41) is denoted $\mathbb{Q}(\mathrm{e}^{2\pi\mathrm{i}/n})$. Gauss studied the structure of cyclotomic fields and reduced the solution of (1.41) to the solution of simpler equations. For example, if $p = 17$, then the solution of (1.41) can be reduced to the successive solution of quadratic equations. This was a breakthrough in algebra.

The explicit solution of polynomial equations and Abel's theorem. The fundamental theorem of algebra only proves the existence of solutions. But the main goal of mathematicians was to solve polynomial equations by explicit formulas. To illustrate this, consider the equation

$$a_0 + a_1 x + \ldots + a_{n-1} x^{n-1} + x^n = 0 \tag{1.42}$$

with arbitrary real coefficients $a_0, \ldots, a_{n-1}$. This equation is called solvable iff the solutions can be obtained from the coefficients by performing rational operations (addition, subtraction, multiplication, division) and by extracting roots. In the 16th century, it was known that the equation (1.42) is solvable if $n = 2, 3, 4$. In 1826, Abel (1802–1829) proved the following crucial result:

Theorem 1.12 *The equation* (1.42) *is not solvable if* $n \geq 5$.

The sophisticated proof was based on Gauss' theory of cyclotomic fields. Galois (1811–1832) studied the papers of Lagrange, Gauss, and Abel. He discovered a completely new approach to the investigation of polynomial equations called Galois theory. The tragical life of Galois, who died in a dual at the age of 21 and wrote down the most important results of his theory on the eve of the dual in a letter to a friend, is described in the marvellous book of the student of Einstein, Leopold Infeld, *Whom the Gods Love*, Whittlesly, New York, 1984. Let us discuss the basic ideas of Galois theory.

Field extensions which are Galois. Consider a field extension

$$F \subseteq E. \tag{1.43}$$

Then the extended field E is a linear space over F. The dimension of this space is called the degree of the extension and denoted $\dim_F(E)$. The extension $F \subseteq E$ is called finite iff $\dim_F(E)$ is finite. For example, the extension $\mathbb{Q} \subseteq \mathbb{R}$ from the field of rational numbers to the field of real numbers is infinite, but the extension $\mathbb{R} \subseteq \mathbb{C}$ from the field $\mathbb{R}$ of real numbers to the field $\mathbb{C}$ of complex numbers is finite with $\dim_{\mathbb{R}} \mathbb{C} = 2$.

The Galois group of the extension (1.43), denoted $\mathrm{Gal}(E|F)$, is by definition the group of all automorphisms $S : F \to E$ which fix the elements of E. The symbol $|\mathrm{Gal}(E|F)|$ denotes the number of elements of $\mathrm{Gal}(E|F)$. A finite field extension (1.43) is called *Galois* iff the number of elements of the Galois group is equal to the degree of the field extension,

$$\boxed{|\mathrm{Gal}(E|F)| = \deg_{\mathbb{F}}(E).}$$

Such an extension has the crucial property that the basic field F can be described by symmetry. More precisely, for a finite Galois extension (1.43), the following hold:

$$x \in F \quad \text{iff} \quad S(x) = x \ \text{ for all } S \in \mathrm{Gal}(E|F).$$

For example, the extension $\mathbb{R} \subseteq \mathbb{C}$ is Galois of degree 2, by Prop. 1.11.

The main theorem of Galois theory. Let $F \subseteq E$ be a finite field extension which is Galois. The subfield $\mathcal{I}$ of E is called an intermediate field iff $F \subseteq \mathcal{I} \subseteq E$.

Theorem 1.13 *There exists a one-to-one correspondence*

$$\mathcal{I} \mapsto \mathcal{G}$$

between the intermediate fields $\mathcal{I}$ of the field extension $F \subseteq E$ and the subgroups $\mathcal{G}$ of the Galois group $\mathrm{Gal}(E|F)$*. Here, the subgroup $\mathcal{G}$ consists of all the elements of* $\mathrm{Gal}(E|F)$ *which fix the elements of the field $\mathcal{I}$.*

The extension $F \subseteq \mathcal{I}$ is Galois iff $\mathcal{G}$ is a normal subgroup of $\mathrm{Gal}(E|F)$*. In this case, the Galois group* $\mathrm{Gal}(\mathcal{I}|F)$ *of the extension $F \subseteq \mathcal{I}$ is equal to the quotient group* $\mathrm{Gal}(E|F)/\mathcal{G}$.

Corollary 1.14 *A finite Galois field extension $F \subseteq E$ has no proper intermediate Galois field extension $F \subseteq \mathcal{I}$ iff the Galois group* $\mathrm{Gal}(E|F)$ *is simple.*[15]

[15] Recall that a group is called simple iff it has no nontrivial normal subgroups. Nowadays all the finite simple groups are known. See D. Gorenstein, Classifying the finite simple groups, Bull. Amer. Math. Soc. **14** (1986), 1–98. Mathematicians needed about 150 years in order to get the full classification. The final breakthrough was initiated by ideas coming from quantum field theory (see Sect. 17.5 of Vol. I about the monster group, vertex algebras, and physics). Already in the 19th century, Cayley proved that each finite group is isomorphic to the subgroup of some permutation group.

For example, Abel's Theorem 1.12 above is a consequence of the fact that if $n \geq 5$, then the permutation group $\mathcal{S}_n$ of n elements has the simple noncommutative normal subgroup $\mathcal{A}_n$ of even permutations.[16]

The prototype of a quadratic number field. Consider the equation

$$x^2 = 2. \tag{1.44}$$

Let $\mathbb{Q}(\sqrt{2})$ denote the smallest subfield of $\mathbb{C}$ which contains the field $\mathbb{Q}$ of rational numbers and the solutions $\sqrt{2}, -\sqrt{2}$ of (1.44). The elements of $\mathbb{Q}(\sqrt{2})$ have precisely the form

$$r + s\sqrt{2}$$

where r and s are arbitrary rational numbers. This is a field, since we have the relation $(r + s\sqrt{2})(r - s\sqrt{2}) = r^2 + 2s^2$. Thus, $r + s\sqrt{2}$ is invertible iff $r^2 + s^2 \neq 0$. The Galois group of the field extension $\mathbb{Q} \subseteq \mathbb{Q}(\sqrt{2})$ consists of the identical map $r+s\sqrt{2} \mapsto r+s\sqrt{2}$ and the conjugation map $r+s\sqrt{2} \mapsto r-s\sqrt{2}$. This Galois group is isomorphic to the simple group $\mathbb{Z}_2 = \{1, -1\}$. If $\mathbb{Q} \subseteq \mathcal{I} \subseteq \mathbb{Q}(\sqrt{2})$ is an intermediate field extension, then we have either $\mathcal{I} = \mathbb{Q}$ or $\mathcal{I} = \mathbb{Q}(\sqrt{2})$, for dimensional reasons.

The prototype of a transcendental field extension. Note that the number π is transcendental, that is, π is not the zero of a polynomial equation with integer coefficients. The set of all the quotients

$$\frac{a_0 + a_1\pi + \ldots + a_m\pi^m}{b_0 + b_1\pi + \ldots + b_n\pi^n}, \qquad m, n = 0, 1, \ldots$$

forms a field denoted by the symbol $\mathbb{Q}(\pi)$ which is an infinite extension of the field $\mathbb{Q}$ of rational numbers. More precisely, $a_0, \ldots, a_m, b_0, \ldots, b_n$ are rational numbers with $b_0^2 + \ldots + b_n^2 \neq 0$.

Perspectives. Galois theory is a powerful tool of modern mathematics. For example, Wiles' seminal proof of Fermat's Last Theorem used methods from Galois theory.[17] One of the famous open problems in mathematics is the classification of all possible field extensions. For commutative Galois groups, this is related to Hilbert's class field theory. For noncommutative Galois groups, this is related to the Langlands program. We refer to V. Varadarajan, Euler through Time: A New Look at Old Themes, Amer. Math. Soc., Providence, Rhode Island, 2006, together with J. Bernstein and S. Gelbart (Eds.), An Introduction to the Langlands Program, Birkhäuer, Boston, 2003.

1.3.3 A Glance at the History of Hamilton's Quaternions

The geometric Gauss model of complex numbers. For more than 250 years, complex numbers were considered to be useful, but mystic objects. Leibniz (1646–1716) wrote:

> From the irrationals are born the impossible or imaginary quantities whose nature is very strange but whose usefulness is not to be despised.

[16] As an introduction to Galois theory and its applications to classical problems, we recommend E. Zeidler, Oxford Users' Guide to Mathematics, Oxford University Press, 2004. Furthermore, we refer to B. van der Waerden, Algebra, Vol. 1, Chap. 6, Frederyck Ungar, New York, 1975, and K. Spindler, Abstract Algebra and Applications, Vol. II, Marcel Dekker, New York, 1994.

[17] A. Wiles, Modular elliptic curves and Fermat's Last Theorem, Ann. Math. **142** (1995), 443–551.

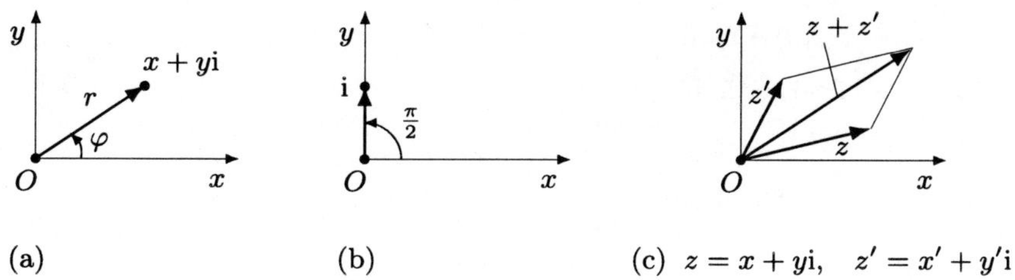

Fig. 1.8. Complex numbers

Euler (1707–1783) wrote:

> It is clear therefore that the square roots of negative numbers cannot be reckoned among the possible numbers; consequently we have to say that the square roots of negative numbers are impossible. This circumstance leads us to the concept of numbers, which by their very nature are impossible, and which are commonly called imaginary numbers or fancied numbers because they exist only in our fancy or imagination.

In 1831, Gauss (1777–1855) published his geometric model which visualized complex numbers. This was the breakthrough in understanding complex numbers as reasonable mathematical objects. Gauss used a Cartesian (x, y)-coordinate system. In modern terminology, the following hold:

- The complex number $x + y\mathrm{i}$ corresponds to a vector in the plane that points from the origin $(0, 0)$ to the point (x, y) (see Fig. 1.8).
- In polar coordinates, we have

$$x + y\mathrm{i} = r(\cos\varphi + \mathrm{i}\sin\varphi), \qquad -\pi < \varphi \leq \pi$$

  where $r := \sqrt{x^2 + y^2}$ is the length, and φ is the angle of the vector.
- The addition of complex numbers corresponds to the addition of vectors.
- The product $(x + y\mathrm{i})(x' + \mathrm{i}y')$ of complex numbers is a complex number of length rr' and angle $\varphi + \varphi'$.

Reinhold Remmert writes:[18]

> "You have made possible the impossible" is a phrase used in a congratulatory address made to Gauss in 1849 by the Carolinum in Brunswick (now the Technical University) on the occasion of the 50-year jubilee of his doctorate.

The algebraic Hamilton model of complex numbers. Hamilton showed that there exists a purely algebraic model. He considered the set $\mathbb{C}$ of all the tuples (x, y) where x and y are real numbers x, y, and he defined the following operations:

- $(u, v) + (x, y) := (u + x, v + y)$ (sum),
- $(u, v)(x, y) := (ux - vy, uy + vx)$ (product).

By explicit computation, one checks easily that the multiplication is distributive, associative, and commutative, and so on. This way, $\mathbb{C}$ becomes a field called the field of complex numbers.[19] Let $\mathcal{R}$ denote the set of all the pairs $(x, 0)$ in $\mathbb{C}$. It follows from

[18] R. Remmert, Complex Numbers, pp. 55–96. In: H. Ebbinghaus et al. (Eds.), Numbers, Springer, New York, 1991 (reprinted with permission).

[19] The definition of a field can be found on page 179 of Vol. II.

$$(x,0)+(y,0)=(x+y,0), \qquad (x,0)(y,0)=(xy,0)$$

that $\mathcal{R}$ is a subfield of $\mathbb{C}$. The map $\chi : \mathbb{R} \to \mathbb{C}$ given by

$$\chi(x) := (x,0)$$

is an injective field morphism. Consequently, the field $\mathbb{R}$ of real numbers is isomorphic to the subfield $\mathcal{R}$. Let us write x instead of $(x,0)$. Furthermore, set $\mathrm{i} := (0,1)$. Since $(0,1)(0,1) = (-1,0)$, we get the key relation

$$\boxed{\mathrm{i}^2 = -1.}$$

Moreover, it follows from $(x,0)+(y,0)(0,1) = (x,y)$ that every complex number (x,y) can be written as $x+y\mathrm{i}$. This justifies the usual computation rules for complex numbers. Hamilton (1805–1865) published this model in 1835.

The question of Hamilton's boys. Once a day, Hamilton told his two young sons that he was able to multiply doublets

$$(x,y),$$

namely, complex numbers $x+yi$. His sons asked him whether he could multiply triplets

$$(x,y,z).$$

For a long time, Hamilton tried to invent such a multiplication. But he was not successful. He considered generalized complex numbers

$$x+yi+zj$$

with real coefficients x,y,z and $i^2 = -1$. The problem is to define the products j^2, ij, ji in a suitable way. After many trials, Hamilton had the idea to pass to quadruplets

$$x+yi+zj+wk.$$

But still he had trouble to find the right products of the basis elements $1,i,j,k$. In a letter to his son a few months before his death in 1865, Hamilton wrote the following:[20]

> Every morning on my coming down to breakfast, your brother and yourself used to asked me: "Well, Papa can you multiply triplets?" Whereto I was always obliged to reply, with a sad shake of my head: "No, I can only add and subtract them" ...
>
> But on the 16th of October, 1843, which happened to be a Council day of the Royal Irish Academy – I was walking in to attend and to preside, and your mother was talking with me, along the Royal Canal ... and although she talked with me now and then, yet an undercurrent of thought was going on in my mind, which gave at last a result, whereof ... I felt at once the importance. An electric circuit seemed to close, and a spark flashed forth, the herald of many long years to come of definitely directed thought and work ... I pulled out on the spot a pocket-book, which still exists, and made an entry there and then. Nor could I resist the impulse - unphilosophical as it may have been – to cut with a knife on a stone of Brougham Bridge, as we passed it, the fundamental formula for quaternions:
>
> $$i^2 = j^2 = k^2 = ijk = -1.$$

[20] B. van der Waerden, History of Algebra, Springer, Berlin, 1985 (reprinted with permission).

The following proposition shows that it is not possible to multiply triplets by preserving the associative law.

Proposition 1.15 *It is impossible to extend the real 2-dimensional associative algebra $\mathbb{C}$ to a real 3-dimensional associative algebra $\mathcal{A}$.*

Proof. Assume that there exists a real algebra $\mathcal{A}$ with the three basis elements $1, i, j$ where 1 is the unit element, and $i^2 = -1$. Thus, the elements Q of $\mathcal{A}$ have the form

$$Q = x + yi + zj \tag{1.45}$$

where x, y, z are real numbers which are uniquely determined by Q. In particular, choose $Q := ji$. By the associative law, $Qi = (ji)i = j(ii) = -j$. Hence

$$\begin{aligned} -j &= (x + yi + zj)i = xi - y + zji \\ &= xi - y + z(x + yi + zj) = zx - y + (x + zy)i + z^2 j. \end{aligned}$$

From $-j = \ldots + z^2 j$ we get $-1 = z^2$ where z is a real number. This is impossible. □

Theorem 1.16 *It is possible to extend the real 2-dimensional associative algebra $\mathbb{C}$ to a real 4-dimensional associative algebra $\mathbb{H}$ with the unit element* 1.

This algebra is uniquely determined by the existence of a basis $1, i, j, k$ for which the following Hamilton product formulas are valid: $i^2 = j^2 = k^2 = ijk = -1$.

Corollary 1.17 *The elements of $\mathbb{H}$ called quaternions have the form*

$$Q = x + yi + zj + wk \tag{1.46}$$

with real coefficients x, y, z, w, and we have the following product formulas: $i^2 = j^2 = k^2 = -1$ and

$$ij = -ji = k, \quad jk = -kj = i, \quad ki = -ik = j. \tag{1.47}$$

These rules allow us to compute arbitrary products. For example,

$$(1 + 3j)(i + 2k) = i + 2k + 3ji + 6jk = i + 2k - 3k + 6i = 7i - k.$$

Introducing the conjugate quaternion $Q^\dagger := x - yi - zj - wk$, we get the crucial formula

$$QQ^\dagger = Q^\dagger Q = x^2 + y^2 + z^2 + w^2. \tag{1.48}$$

Mnemonically, it is sufficient to know the product rules $i^2 = -1$ and $ij = -ji = k$. The remaining product rules are obtained by the aid of the cyclic permutation $i \Rightarrow j \Rightarrow k \Rightarrow i$.

Proof. (I) Uniqueness. If there exists such an algebra $\mathbb{H}$, then the elements of $\mathbb{H}$ have the form (1.46). It follows from $i^2 = j^2 = k^2 = ijk = -1$ together with the associative law that

$$-k = (ijk)k = (ij)(kk) = -ij.$$

Moreover, $ijk = -1$ implies $(ii)jk = -i$. Hence $jk = i$. In turn, $ji = (jj)k = -k$. Similarly, we get all the other product formulas mentioned in Corollary 1.17.

(II) Existence. Since the multiplication of quaternions is based on the multiplication rules for the basis elements $1, i, j, k$, the validity of the distributive law is obvious. We have to show that the multiplication is associative. For example, we have

$$(ij)k = i(jk).$$

In fact, $(ij)k = k^2 = -1$ and $i(jk) = ii = -1$. Similarly, we obtain that the associative law is satisfied for all the products of basis elements. □

Corollary 1.18 *The algebra* $\mathbb{H}$ *is a skew-field.*

Proof. For each nonzero quaternion Q, set

$$P := \frac{Q^\dagger}{QQ^\dagger} = \frac{x - yi - zj - wk}{x^2 + y^2 + z^2 + w^2}.$$

Then, $QP = PQ = 1$. Thus, P is the inverse element to Q. We will write Q^{-1} instead of P. □

Hamilton (1805–1865) introduced his quaternions as four-dimensional objects in 1843. Two years later, the special quaternions

$$yi + zj + wk$$

were called 'vectors' by Hamilton. This way, the following 3-dimensional vector operations emerged: **ab** (inner product), $\mathbf{a} \times \mathbf{b}$ (vector product), $\mathbf{grad}\,\Theta$ (gradient of the temperature field Θ), $\operatorname{div}\mathbf{v}$ (divergence of the velocity vector field $\mathbf{v}$), and $\mathbf{curl}\,\mathbf{v}$ (curl of the velocity vector field $\mathbf{v}$.) This 3-dimensional vector calculus was popularized by the physicist Gibbs (1839–1903) who worked at Yale University (New Haven, Connecticut) and who made fundamental contributions to statistical physics. For example, the Maxwell equations in electrodynamics are essentially based on $\operatorname{div}\mathbf{E}, \mathbf{curl}\,\mathbf{E}, \operatorname{div}\mathbf{B}, \mathbf{curl}\,\mathbf{B}$ for the electromagnetic field $\mathbf{E}, \mathbf{B}$. In the 1850s, Cayley (1821–1895) developed the matrix calculus. In particular, he showed that quaternions can be realized by complex (2×2)-matrices. This elegant approach to quaternions will be considered in Sect. 1.3.5. In 1878, Clifford (1845–1879) generalized the algebra of quaternions to Clifford algebras.

Quaternions in modern physics. Hamilton was convinced that quaternions would be important for physics. However, deep relations of quaternions to physics were only discovered in the twentieth century:

- In 1908, Minkowski (1864–1909) formulated the geometrization of Einstein's theory of special relativity. Minkowski emphasized that the theory of special relativity can be understood best in terms of a special 4-dimensional geometry (the Minkowski geometry of the Minkowski space). We will show in Sect. 1.3.9 that the indefinite metric of the Minkowski space M_4 is intimately related to quaternions. Einstein's theory of special relativity is based on the Lorentz group which describes the change of space and time coordinates of inertial systems. The infinitesimal Lorentz group is the 6-dimensional real Lie algebra $sl(2, \mathbb{C})$ which is related to the 3-dimensional Lie group $su(2)$ by the direct sum formula

$$\boxed{sl(2, \mathbb{C}) = su(2) \oplus su(2).}$$

  Here, the Lie algebra $su(2)$ of infinitesimal rotations is isomorphic to the Lie algebra $(E_3)_{\mathrm{Lie}}$, which represents the simplest nontrivial Lie algebra.
- In 1927, Pauli (1900–1958) created the theory of the non-relativistic spinning electron. For describing the electron spin, he introduced the so-called Pauli matrices which are intimately related to the Lie algebra $su(2)$. We will show in Sect. 1.3.4 that the Pauli spin matrices and the Lie algebra $su(2)$ can be understood

best by using Cayley's approach to quaternions via the Cayley quaternionic matrices

$$q_1, q_2, q_3.$$

For the electron spin, it is important that the Lie algebra $su(2)$ is not only the Lie algebra of the rotation group $SO(3)$, but $su(2)$ is also the Lie algebra of the group $U(1, \mathbb{H})$ which consists of all the quaternions of length one. In other words, $U(1, \mathbb{H})$ corresponds to the unit sphere in the 4-dimensional real Hilbert space $\mathbb{H}$ of quaternions. We have the group isomorphism

$$U(1, \mathbb{H}) \simeq SU(2).$$

This simply connected group $SU(2)$ is the universal covering group of the not simply connected rotation group $SO(3)$. This means that there exists a surjective group morphism

$$\mu : SU(2) \to SO(3) \tag{1.49}$$

with the kernel $\{I, -I\}$. Hence we get the group isomorphism

$$SO(3) \simeq SU(2)/\{I, -I\}.$$

The map μ from (1.49) generalizes the map

$$\nu : \mathbb{R} \to U(1)$$

given by $\nu(\varphi) := e^{i\varphi}$ for all $\varphi \in \mathbb{R}$.

- In 1928, Dirac (1902–1984) used the Clifford algebra of the Minkowski space in order to combine the theory of special relativity with quantum mechanics. This way, Dirac created the theory of the relativistic electron. This theory shows that the electron spin is a relativistic effect.
- The Standard Model in particle physics was created in the 1960s and early 1970s. This model is based on the gauge group

  $$U(1) \times SU(2) \times SU(3).$$

  Here, $U(1) \times SU(2)$ corresponds to the electroweak interaction, and $SU(3)$ corresponds to the strong interaction.
- In about 1900, Élie Cartan (1869–1951) showed that the complexification $sl_{\mathbb{C}}(2, \mathbb{C})$ of the Lie algebra $su(2)$ is the building block for all complex semi-simple Lie algebras. Graphically this corresponds to Dynkin diagrams and root diagrams which are widely used in elementary particle physics (see Chap. 3).
- The group $SU(2)$ and the theory of the spinning electron are also closely related to the Hopf fibration of the 3-dimensional unit sphere $\mathbb{S}^3$ (see Sect. 5.7.2 of Vol. I). This construction proves the existence of nontrivial continuous mappings $f : \mathbb{S}^3 \to \mathbb{S}^2$. In turn, this is related to the Chern class of a special $U(1)$-principal fiber bundle. The theory of $U(1)$-fiber bundles was created by Weyl (1885–1955) in 1929 in order to reformulate Maxwell's theory of the electromagnetic field (see Chap. 13).

1.3.4 Pauli's Spin Matrices and the Lie Algebras $su(2)$ and $sl(2, \mathbb{C})$

We want to study the relation between special Lie matrix algebras and quaternions. The following matrices

$$\sigma^0 := \begin{pmatrix} 1 & 0 \\ 0 & 1 \end{pmatrix}, \quad \sigma^1 := \begin{pmatrix} 0 & 1 \\ 1 & 0 \end{pmatrix}, \quad \sigma^2 := \begin{pmatrix} 0 & -\mathrm{i} \\ \mathrm{i} & 0 \end{pmatrix}, \quad \sigma^3 := \begin{pmatrix} 1 & 0 \\ 0 & -1 \end{pmatrix} \tag{1.50}$$

are called the Pauli matrices. These matrices are self-adjoint, and they possess the following properties:

- $\sigma^1\sigma^2 = -\sigma^2\sigma^1 = \mathrm{i}\sigma^3$ and $\sigma^1\sigma^1 = \sigma^0$.
- $\sigma^1\sigma^2 - \sigma^2\sigma^1 = 2\mathrm{i}\sigma^3$ (Lie relation),
- $\sigma^1\sigma^2 + \sigma^2\sigma^1 = 0$ (Clifford relation).

The matrices

$$S_k := \tfrac{\hbar}{2}\sigma^k, \qquad k = 1, 2, 3$$

are called the Pauli spin matrices. They satisfy the following commutation relations:

$$S_1S_2 - S_2S_1 = \hbar\mathrm{i}S_3.$$

These are the commutation relations for angular momentum in quantum mechanics. Further relations are obtained by using the cyclic permutation $1 \Rightarrow 2 \Rightarrow 3 \Rightarrow 1$. For example, $\sigma^2\sigma^3 = -\sigma^3\sigma^2 = \mathrm{i}\sigma^1$ and $\sigma^2\sigma^2 = \sigma^0$.

The complex matrix algebra $\mathbb{M}_{\mathbb{C}}(2,2;\mathbb{C})$. Let $\mathbb{M}(2,2;\mathbb{C})$ denote the set of all complex (2×2)-matrices. With respect to the usual matrix operations

$$\alpha A + \beta B, \quad AB, \quad A^\dagger$$

for all $A, B \in \mathbb{M}(2,2;\mathbb{C})$ and all $\alpha, \beta \in \mathbb{C}$, the set $\mathbb{M}(2,2;\mathbb{C})$ becomes a complex 4-dimensional associative noncommutative $*$-algebra with the unit matrix σ^0 as unit element. This complex algebra is denoted by $\mathbb{M}_{\mathbb{C}}(2,2;\mathbb{C})$ (see page 116).

The real matrix algebra $\mathbb{M}_{\mathbb{R}}(2,2;\mathbb{C})$. With respect to the matrix operations

$$\alpha A + \beta B, \quad AB, \quad A^\dagger$$

for all $A, B \in \mathbb{M}(2,2;\mathbb{C})$ and all $\alpha, \beta \in \mathbb{R}$, the set $\mathbb{M}(2,2;\mathbb{C})$ becomes a real 8-dimensional associative noncommutative $*$-algebra with the unit matrix σ^0 as unit element (see page 116). This real algebra is denoted by $\mathbb{M}_{\mathbb{R}}(2,2;\mathbb{C})$. Let us introduce the Cayley matrices

$$\boxed{q_0 := \sigma^0, \quad q_k := -\mathrm{i}\sigma^k, \qquad k = 1, 2, 3.} \tag{1.51}$$

Explicitly,

$$q_0 := \begin{pmatrix} 1 & 0 \\ 0 & 1 \end{pmatrix}, \quad q_1 := \begin{pmatrix} 0 & -\mathrm{i} \\ -\mathrm{i} & 0 \end{pmatrix}, \quad q_2 := \begin{pmatrix} 0 & -1 \\ 1 & 0 \end{pmatrix}, \quad q_3 := \begin{pmatrix} -\mathrm{i} & 0 \\ 0 & \mathrm{i} \end{pmatrix}. \tag{1.52}$$

The basis theorem. Set $\mathcal{A} := \mathbb{M}_{\mathbb{R}}(2,2;\mathbb{C})$. An elementary argument shows the following.

Proposition 1.19 (i) *The matrices $\sigma^0, \sigma^1, \sigma^2, \sigma^3$ form a basis of the real 4-dimensional linear space $\{A \in \mathcal{A} : A^\dagger = A\}$.*

(ii) *The matrices $\sigma^1, \sigma^2, \sigma^3$ form a basis of the real 3-dimensional linear space $\{A \in \mathcal{A} : A^\dagger = A, \ \mathrm{tr}(A) = 0\}$.*

(iii) *The matrices $\mathrm{i}\sigma^0, q_1, q_2, q_3$ form a basis of the real 4-dimensional linear space $\{A \in \mathcal{A} : A^\dagger = -A\}$.*

(iv) *The matrices q_1, q_2, q_3 form a basis of the real 3-dimensional linear space $\{A \in \mathcal{A} : A^\dagger = -A, \ \mathrm{tr}(A) = 0\}$.*

(iv) *The matrices $\sigma^0, \sigma^1, \sigma^2, \sigma^3, \mathrm{i}\sigma^0, q_1, q_2, q_3$ form a basis of the real 8-dimensional linear space $\mathcal{A}$.*

The Lie algebra $su(2)$. Recall that $\mathcal{A} := M_{\mathbb{R}}(2,2;\mathbb{C})$. Define

$$[A,B]_- := AB - BA, \qquad A,B \in \mathcal{A}.$$

With respect to the product $[A,B]_-$, the linear space $\mathcal{A}$ becomes a real 8-dimensional Lie algebra. Set

$$su(2) := \{A \in \mathcal{A} : A^\dagger = -A, \ \ \mathrm{tr}(A) = 0\}.$$

The elements of $su(2)$ have the form $\alpha q_1 + \beta q_2 + \gamma q_3$ where α, β, γ are real numbers.

Proposition 1.20 *$su(2)$ is a real 3-dimensional Lie subalgebra of $\mathcal{A}$ with*

$$[q_1,q_2]_- = 2q_3, \qquad [q_2,q_3]_- = 2q_1, \quad [q_3,q_1]_- = 2q_2. \tag{1.53}$$

The Lie algebra $sl(2,\mathbb{C})$. Define $sl(2,\mathbb{C}) := \{A \in \mathcal{A} : \mathrm{tr}(A) = 0\}$.

Proposition 1.21 *$sl(2,\mathbb{C})$ is a real 6-dimensional Lie subalgebra of $\mathcal{A}$ with the basis $q_1, q_2, q_3, \mathrm{i}q_1, \mathrm{i}q_2, \mathrm{i}q_3$.*

Proof. The elements of $sl(2,\mathbb{C})$ have the form

$$\alpha q_1 + \beta q_2 + \gamma q_3 + \varrho(\mathrm{i}q_1) + \sigma(\mathrm{i}q_2) + \tau(\mathrm{i}q_3)$$

with real coefficients $\alpha, \beta, \gamma, \varrho, \sigma, \tau$. It follows from (1.53) that all the products $[A,B]_-$ of the six basis elements $q_1, q_2, q_3, \mathrm{i}q_1, \mathrm{i}q_2, \mathrm{i}q_3$ lie in $sl(2,\mathbb{C})$. □

1.3.5 Cayley's Matrix Approach to Quaternions

The following definition is basic for quaternions. Let $\mathcal{H}$ denote the set of all the matrices

$$\boxed{A = \alpha q_0 + \beta q_1 + \gamma q_2 + \delta q_3} \tag{1.54}$$

with real numbers $\alpha, \beta, \gamma, \delta$. Explicitly,

$$A = \begin{pmatrix} \alpha - \delta\mathrm{i} & -\gamma - \beta\mathrm{i} \\ \gamma - \beta\mathrm{i} & \alpha + \delta\mathrm{i} \end{pmatrix} \tag{1.55}$$

with $\det(A) = \alpha^2 + \beta^2 + \gamma^2 + \delta^2$. The matrix q_0 is the unit matrix, and the matrices q_1, q_2, q_3 have the following product properties:

- $(q_1)^2 = -q_0$ and $q_1 q_2 = -q_2 q_1 = q_3$,
- $(q_2)^2 = -q_0$ and $q_2 q_3 = -q_3 q_2 = q_1$,
- $(q_3)^2 = -q_0$ and $q_3 q_1 = -q_1 q_3 = q_2$,
- $q_0^\dagger = q_0, q_1^\dagger = -q_1, q_2^\dagger = -q_2, q_3^\dagger = -q_3$.

Hence $A^\dagger = \alpha q_0 - \beta q_1 - \gamma q_2 - \delta q_3$. Finally, for all A, B let us introduce the inner product $\langle A|B\rangle := \frac{1}{2}\mathrm{tr}(AB^\dagger)$ together with the corresponding norm $||A|| := \sqrt{\langle A|B\rangle}$. Then[21]

$$\langle q_j|q_k\rangle = \delta_{jk}, \qquad j,k = 0,1,2,3.$$

This means that q_0, q_1, q_2, q_3 represents an orthonormal basis of $\mathcal{H}$. This implies

$$||A|| = \sqrt{\alpha^2 + \beta^2 + \gamma^2 + \delta^2}.$$

Hence $||A|| = \sqrt{\det(A)}$. Moreover, $A^\dagger A = AA^\dagger = ||A||^2 q_0$.

[21] For example, $\langle q_1|q_1\rangle = \frac{1}{2}\mathrm{tr}(q_1 q_1^\dagger) = \frac{1}{2}\mathrm{tr}(-q_1^2) = \frac{1}{2}\mathrm{tr}(q_0) = 1.$

Theorem 1.22 (i) *$\mathcal{H}$ is a 4-dimensional real associative algebra with the unit matrix q_0 as unit element; the algebra $\mathcal{H}$ is isomorphic to the algebra $\mathbb{H}$.*

(ii) *$\mathcal{H}$ is a skew-field.*

(iii) *$\mathcal{H}$ is a real Hilbert space.*

(iv) *$||AB|| = ||A|| \cdot ||B||$ for all $A, B \in \mathcal{H}$ (product rule).*

(iv) *$\mathcal{H}$ is a real C^*-algebra. For all $A, B \in \mathcal{H}$ and all real numbers α, β, we have $||A^\dagger|| = ||A||$ and*

$$(\alpha A + \beta B)^\dagger = \alpha A^\dagger + \beta B^\dagger, \quad (AB)^\dagger = B^\dagger A^\dagger, \quad (A^\dagger)^\dagger = A.$$

Proof. Ad (i). Note that $\mathcal{A}$ is a real associative algebra. Since we have the relations $q_1 q_2 = -q_2 q_1 = q_3, q_1^\dagger = -q_1$, and so on, $\mathcal{H}$ is a subalgebra of $\mathcal{A}$. The map

$$1 \mapsto q_0, \quad i \mapsto q_1, \quad j \mapsto q_2 \quad k \mapsto q_3$$

yields the algebra isomorphism $\mathbb{H} \simeq \mathcal{H}$.

Ad (ii). If $A \neq 0$, then $\det(A) \neq 0$. Hence the inverse matrix A^{-1} exists.

Ad (iii). $\det(AB) = \det(A)\det(B)$, and $||A|| = \sqrt{\det(A)}$.

□

Euler's "four squares theorem." Choose the quaternions $Q = \alpha + \beta i + \gamma j + \delta k$ and $P = a + bi + cj + dk$. Then $||Q||^2 = \alpha^2 + \beta^2 + \gamma^2 + \delta^2$. It follows from the product rule $||Q||^2 \cdot ||P||^2 = ||QP||^2$ that the product

$$(\alpha^2 + \beta^2 + \gamma^2 + \delta^2)(a^2 + b^2 + c^2 + d^2)$$

of two four-squares sums is equal to the following sum of four-squares sums

$$(\alpha a - \beta b - \gamma c - \delta d)^2 + (\alpha b + ba + \gamma d - \delta c)^2$$
$$+(\alpha c + \gamma a + \delta b - \beta d)^2 + (\alpha d + \delta a + \beta c - \gamma b)^2.$$

Using Cayley's octonions, this formula can be extended to the sum of 8 squares. But Hurwitz (1859–1919) proved in 1898 that it is impossible to extend this to n sums of squares with $n > 8$. Such formulas are used in additive number theory.[22]

1.3.6 The Unit Sphere $U(1, \mathbb{H})$ and the Electroweak Gauge Group $SU(2)$

Let $N = 1, 2, \ldots$ By definition, the group $U(N)$ consists of all complex $N \times N$ matrices A with $\det(A) \neq 0$ and $A^{-1} = A^\dagger$. Moreover, let us define the group $SU(N) := \{A \in U(N) : \det(A) = 1.\}$.

Proposition 1.23 $SU(2) = \{A \in \mathcal{H} : ||A|| = 1\}$.

Proof. For the given complex (2×2)-matrix $A = \begin{pmatrix} a & b \\ c & d \end{pmatrix}$, we have $\det(A) = ad - bc$. Suppose that $\det(A) \neq 0$. Then

$$A^\dagger = \begin{pmatrix} a^\dagger & c^\dagger \\ b^\dagger & d^\dagger \end{pmatrix}, \qquad A^{-1} = \frac{1}{\det(A)} \begin{pmatrix} d & -b \\ -c & a \end{pmatrix}.$$

---

22 The proof can be found in H. Ebbinghaus et al. (Eds.), Numbers, Springer, New York, 1995, Chap. 10.

If $A \in SU(2)$, then $A^\dagger = A^{-1}$ implies $a^\dagger = d$ and $b^\dagger = -c$. Hence $A \in \mathcal{H}$, by (1.55), and $||A|| = \sqrt{\det(A)} = 1$.

Conversely, if $A \in \mathcal{H}$ and $||A|| = 1$, then $\det(A) = 1$, and $A^\dagger = A^{-1}$. Hence $A \in SU(2)$. □

Proposition 1.23 tells us that the group $SU(2)$ is isomorphic to the group $U(1,\mathbb{H})$ of all the quaternions of length one. This is the unit sphere in the 4-dimensional real Hilbert space $\mathbb{H}$.

1.3.7 The Four-Dimensional Extension of the Euclidean Space E_3

We want to study the relation between quaternions and vectors which played a crucial role in the history of vector calculus. To this end, we consider the direct sum $\mathbb{R} \oplus E_3$ which consists of all the sums

$$\alpha + \mathbf{a}$$

where α is a real number and $\mathbf{a}$ is a vector in E_3. We define the following operations:

- $(\alpha + \mathbf{a}) + (\beta + \mathbf{b}) := (\alpha + \beta) + (\mathbf{a} + \mathbf{b})$ (sum),
- $(\alpha + \mathbf{a}) \vee (\beta + \mathbf{b}) := \alpha\beta + \alpha\mathbf{b} + \beta\mathbf{a} - \mathbf{ab} + \mathbf{a} \times \mathbf{b}$ (product),
- $(\alpha + \mathbf{a})^\dagger := \alpha - \mathbf{a}$ (conjugation),
- $\langle \alpha + \mathbf{a} | \beta + \mathbf{b} \rangle := \alpha\beta + \mathbf{ab}$ (inner product),
- $||\alpha + \mathbf{a}|| := \sqrt{\alpha^2 + \mathbf{a}^2}$,
- $\Re(\alpha + \mathbf{a}) := \alpha$ (real part),
- $\Im(\alpha + \mathbf{a}) = \mathbf{a}$ (imaginary part).

Choose a right-handed orthonormal basis $\mathbf{i}, \mathbf{j}, \mathbf{k}$ of the Euclidean space E_3. Then:

- $\mathbf{i} \vee \mathbf{i} = -1, \quad \mathbf{i} \vee \mathbf{j} = -\mathbf{j} \vee \mathbf{i} = \mathbf{k}$,
- $\mathbf{j} \vee \mathbf{j} = -1, \quad \mathbf{j} \vee \mathbf{k} = -\mathbf{k} \vee \mathbf{j} = \mathbf{i}$,
- $\mathbf{k} \vee \mathbf{k} = -1, \quad \mathbf{k} \vee \mathbf{i} = -\mathbf{i} \vee \mathbf{k} = \mathbf{j}$.

For example, it follows from $\mathbf{ij} = 0$ and $\mathbf{i} \times \mathbf{j} = \mathbf{k}$, as well as $\mathbf{j} \times \mathbf{i} = -\mathbf{i} \times \mathbf{j}$ that $\mathbf{i} \vee \mathbf{j} = \mathbf{k}$ and $\mathbf{j} \vee \mathbf{i} = -\mathbf{i} \vee \mathbf{j}$.

Proposition 1.24 *The algebra $\mathbb{R} \oplus E_3$ is isomorphic to the algebra $\mathbb{H}$ of quaternions.*

The isomorphism is given by the map $1 \mapsto 1, \quad \mathbf{i} \mapsto i, \quad \mathbf{j} \mapsto j, \quad \mathbf{k} \mapsto k$.

The product $\mathbf{a} \vee \mathbf{b} = -\mathbf{ab} + \mathbf{a} \times \mathbf{b}$ is called the Clifford product of the vectors $\mathbf{a}$ and $\mathbf{b}$. We have the so-called Clifford relation

$$\mathbf{a} \vee \mathbf{b} + \mathbf{b} \vee \mathbf{a} = -2\mathbf{ab} \qquad \text{for all} \quad \mathbf{a}, \mathbf{b} \in E_3. \tag{1.56}$$

In particular, this implies the so-called square-root relation

$$\mathbf{a} \vee \mathbf{a} = -\mathbf{a}^2 \qquad \text{for all} \quad \mathbf{a} \in E_3. \tag{1.57}$$

Moreover,

$$\mathbf{a} \vee \mathbf{b} - \mathbf{b} \vee \mathbf{a} = 2\mathbf{a} \times \mathbf{b} \qquad \text{for all} \quad \mathbf{a}, \mathbf{b} \in E_3. \tag{1.58}$$

1.3.8 Hamilton's Nabla Operator

Choose a right-handed Cartesian (x, y, z)-coordinate system with the right-handed orthonormal basis $\mathbf{i}, \mathbf{j}, \mathbf{k}$. By definition, the symbol

$$\nabla := \mathbf{i}\frac{\partial}{\partial x} + \mathbf{j}\frac{\partial}{\partial y} + \mathbf{k}\frac{\partial}{\partial z}$$

is called Hamilton's nabla operator.[23] Let $\mathbf{v} = a\mathbf{i} + b\mathbf{j} + c\mathbf{k}$. Using the quaternionic product, we get

$$\nabla \vee (\Theta + \mathbf{v}) = \mathbf{grad}\,\Theta - \operatorname{div}\mathbf{v} + \mathbf{curl}\,\mathbf{v}$$

with

- $\mathbf{grad}\,\Theta := \nabla\Theta = \Theta_x\mathbf{i} + \Theta_y\mathbf{j} + \Theta_z\mathbf{k}$ (gradient of the temperature field Θ),
- $\operatorname{div}\mathbf{v} := \nabla\mathbf{v} = a_x\mathbf{i} + b_y\mathbf{j} + c_z\mathbf{k}$ (divergence of the velocity vector field $\mathbf{v}$),
- $\mathbf{curl}\,\mathbf{v} := \nabla \times \mathbf{v} = (c_y - b_z)\mathbf{i} + (a_z - c_x)\mathbf{j} + (b_x - a_y)\mathbf{k}$ (curl of the velocity vector field $\mathbf{v}$).

In Sect. 9.1.3 on page 560, we will show that $\mathbf{grad}\,\Theta$, $\operatorname{div}\mathbf{v}$, and $\mathbf{curl}\,\mathbf{v}$ possess an invariant meaning which does not depend on the choice of the Cartesian coordinate system.

1.3.9 The Indefinite Hilbert Space $\mathbb{H}$ and the Minkowski Space

We equip the 4-dimensional linear space $\mathbb{R} \oplus E_3$ with the indefinite inner product

$$\langle \alpha + \mathbf{a} |\, \beta + \mathbf{b}\rangle_- := \Re\big((\alpha + a) \vee (\beta + \mathbf{b})\big) = \alpha\beta - \mathbf{ab}.$$

This way, we get a 4-dimensional real indefinite Hilbert space which is isomorphic to the Minkowski space M_4. This space will be studied in Chap. 18 in connection with Einstein's theory of special relativity. By the way, the Hilbert space structure on $\mathbb{R} \oplus E_3$ is obtained by the inner product

$$\langle \alpha + \mathbf{a} |\, \beta + \mathbf{b}\rangle = \Re\big((\alpha + \mathbf{a}) \vee (\beta + \mathbf{b})^\dagger\big) = \alpha\beta + \mathbf{ab}.$$

1.4 Riesz Duality between Vectors and Covectors

> The Hilbert space E_3 possesses a natural duality which is based on the Riesz operator $\aleph : E_3 \to E_3^d$. This is the special case of a general theorem in functional analysis.
>
> Folklore

This section will be based on the following key formula

$$\boxed{F(\mathbf{x}) := \langle \mathbf{a}|\mathbf{x}\rangle \qquad \text{for all} \quad \mathbf{x} \in E_3} \tag{1.59}$$

where $\langle \mathbf{a}|\mathbf{x}\rangle$ denotes the inner product on the Hilbert space E_3. Let $\mathbf{e}_1, \mathbf{e}_2, \mathbf{e}_3$ be an orthonormal basis of the Euclidean space E_3. Recall the definition of the special linear functionals $dx^j \in E_3^d$ given by $dx^j(\mathbf{a}) := a^j$ where $j = 1, 2, 3$.

The dual linear space E_3^d. Recall that the symbol E_3^d denotes the set of all linear functionals $F : E_3 \to \mathbb{R}$.

[23] Nabla was a Phoenician string instrument (800 B.C.).

Proposition 1.25 *The dual space E_3^d is a real 3-dimensional linear space with the basis dx^1, dx^2, dx^3.*

Proof. (I) Linear structure. If $F, G \in E_3^d$ and $\alpha, \beta \in \mathbb{R}$, then we define

$$(\alpha F + \beta G)(\mathbf{x}) := \alpha F(\mathbf{x}) + \beta G(\mathbf{x}) \qquad \text{for all} \quad \mathbf{x} \in E_3.$$

Then, $\alpha F + \beta G \in E_3^d$. This way, the set E_3^d becomes a real linear space.

(II) Linear combination. Let b_1, b_2, b_3 be given real numbers. Then the linear combination

$$F := b_k dx^k \tag{1.60}$$

is a linear functional, that is, $F \in E_3^d$. We want to show that every element of E_3^d can be represented this way. In fact, let $F \in E_3^d$. Introduce the so-called coordinates b_1, b_2, b_3 of F by setting $b_k := F(\mathbf{e}_k)$, $k = 1, 2, 3$. Then

$$F(\mathbf{a}) = F(a^k \mathbf{e}_k) = a^k F(\mathbf{e}_k) = b_k dx^k(\mathbf{a}).$$

Hence we get (1.60).

(III) Basis. It remains to show that the elements dx^1, dx^2, dx^3 are linearly independent in E_3^d. To this end, suppose that

$$a_k dx^k = 0.$$

This implies $0 = a_k dx^k(\mathbf{e}_j) = a_k \delta_j^k = a_j$. Hence $a_1 = a_2 = a_3 = 0$. □

In mathematics, duality is systematically designated by the prefix 'co'. For example, the elements of the dual space E_3^d are called covectors, and the system of the three covectors dx^1, dx^2, dx^3 is called the cobasis related to $\mathbf{e}_1, \mathbf{e}_2, \mathbf{e}_3$. The linear functional $a_k dx^k$ is also called a differential 1-form (with constant coefficients).

The Riesz duality operator $\aleph$. For given vector $\mathbf{a} \in E_3$, the key formula (1.59) yields the linear functional $F \in E_3^d$. We set $\aleph(\mathbf{a}) := F$.

Theorem 1.26 *The Riesz operator $\aleph : E_3 \to E_3^d$ is a linear isomorphism.*

Proof. (I) Injectivity. If $\aleph(\mathbf{a}) = 0$, then $\langle \mathbf{a}|\mathbf{a}\rangle = 0$. Hence $\mathbf{a} = 0$.

(II) Surjectivity. Explicitly, we have $\aleph(\mathbf{a}) = a_k dx^k$ where $a_k := a^k$. By Prop. 1.25, each functional $F \in E_3^d$ can be represented as $\aleph(\mathbf{a})$ for some vector $\mathbf{a}$. □

The dual Hilbert space E_3^d. The Riesz operator $\aleph$ can be used in order to equip the dual space E_3^d with an inner product in a quite natural manner. To this end, we set

$$\langle F|G\rangle_{E_3^d} := \langle \mathbf{a}|\mathbf{b}\rangle \qquad \text{for all} \quad F, G \in E_3^d$$

where $\mathbf{a} := \aleph^{-1}(F)$ and $\mathbf{b} := \aleph^{-1}(G)$.

The bidual space E_3^{dd}. By definition, the space E_3^{dd} is the dual space to E_3^d. For given vector $\mathbf{a}$, define

$$\Phi(F) := F(\mathbf{a}) \qquad \text{for all} \quad F \in E_3^d.$$

Then, $\Phi : E_3^d \to \mathbb{R}$ is a linear functional on the dual space E_3^d. Set $\chi(\mathbf{a}) := \Phi$.

Proposition 1.27 *The map $\chi : E_3 \to E_3^{dd}$ is a linear isomorphism.*

The proof proceeds similarly as the proof of Theorem 1.26.

The inner product between vectors and covectors. For the vector $\mathbf{v}$ and the covector $F \in E_3^d$, we define the inner product

$$\mathbf{v}\rfloor F := F(\mathbf{v}).$$

This is a real number. Similarly, for the n-linear functional $F : E_3 \times \cdots \times E_3 \to \mathbb{R}$ with $n = 2, 3, \ldots$, we define the inner product $\mathbf{v}\rfloor F$ by setting

$$(\mathbf{v}\rfloor F)(\mathbf{x}_2, \ldots, \mathbf{x}_n) := F(\mathbf{v}, \mathbf{x}_2, \ldots, \mathbf{x}_n) \qquad \text{for all} \quad \mathbf{x}_2, \ldots, \mathbf{x}_n \in E_3.$$

Obviously, $\mathbf{v}\rfloor F$ is an $(n-1)$-linear functional on E_3. Instead of $\mathbf{v}\rfloor F$, we also write $i_{\mathbf{v}}F$. Mnemonically, the symbol $i_{\mathbf{v}}$ stands for 'insert $\mathbf{v}$'.

The general theorem in Hilbert spaces. Choose $\mathbb{K} = \mathbb{R}$ or $\mathbb{K} = \mathbb{C}$. Let X be a Hilbert space over $\mathbb{K}$. For given $a \in X$, define

$$F(x) := \langle a|x\rangle \qquad \text{for all} \quad x \in X.$$

Then, the map $F : X \to \mathbb{K}$ is a linear continuous functional on the Hilbert space X with the norm $||F|| = ||a||$.

Theorem 1.28 *Every linear continuous functional on X can be obtained this way.*

This famous theorem was independently proven by Fryges Riesz and Maurice Fréchet in 1907 in the special case of the Hilbert space $X = L_2(\mathbb{R})$. Note that one has to distinguish between the Hungarian mathematician Fryges (Frédéric) Riesz (1880–1956), who worked in Budapest (Hungary), and his brother Marcel Riesz (1886–1969), who worked in Lund (Sweden). Fryges Riesz is one of the founders of functional analysis at the beginning of the twentieth century. Analytic renormalization dates back to the papers of Marcel Riesz on Riemann–Liouville integrals and their applications to fundamental solutions of linear hyperbolic partial differential equations (written in the 1940s).

An elementary proof (together with many important applications) can be found in E. Zeidler, Applied Functional Analysis, Vol. 1: Applications to Mathematical Physics, Springer, New York, 1995. It turns out that Theorem 1.28 is equivalent to the unique solvability of the variational problem

$$||x - x_0|| = \min!, \qquad x \in L, \tag{1.61}$$

where the element $x_0 \in X$ is given, and L is a given closed linear subspace of the Hilbert space X. In geometric terms, the solution x of (1.61) corresponds to the foot of the perpendicular from the point x_0 to the plane L. In analytic terms, the minimum problem (1.61) allows the justification of the famous Dirichlet principle for the Laplacian via generalized derivatives and Sobolev spaces (see Sect. 10.4.9 of Vol. II).

1.5 The Heisenberg Group, the Heisenberg Algebra, and Quantum Physics

The deformation of mathematical structures is fundamental for modern physics. This concerns both Heisenberg's quantum physics and Einstein's theory of relativity.

> Quantum physics is closely related to a deformation of the classical additive group structure of the Euclidean space E_3 via the Heisenberg group.[24]
>
> Folklore

We want to show that the Heisenberg group and the corresponding Heisenberg algebra can be realized isomorphically in completely different ways. Behind these isomorphisms (also called faithful representations), there lurks the passage from classical mechanics to quantum mechanics discovered by Heisenberg (1901–1976) in 1925. The terms 'Heisenberg group' and 'Heisenberg algebra' were coined in the 1970s. However, the Heisenberg algebra was already implicitly used by Jacobi in 1843 (Poisson brackets in celestial mechanics).

The Heisenberg group $\mathcal{G}_{\text{Heis}}(\mathbf{k})$. We want to equip the Euclidean space E_3 with a product. To this end, we choose a unit vector $\mathbf{k} \in E_3$. For every vector $\mathbf{x} \in E_3$, we have the orthogonal splitting

$$\mathbf{x} = (\mathbf{x}\mathbf{k})\mathbf{k} + \mathbf{x}^\perp.$$

Here, the vector $\mathbf{x}^\perp$ is the orthogonal projection of the vector $\mathbf{x}$ onto the linear subspace $L^\perp(\mathbf{k})$ of E_3 which is orthogonal to $\mathbf{k}$. We define the product

$$\boxed{\mathbf{x} \cdot \mathbf{y} := \mathbf{x} + \mathbf{y} + \tfrac{1}{2}(\mathbf{x}^\perp \times \mathbf{y}^\perp).} \tag{1.62}$$

For all $\mathbf{x}, \mathbf{y}, \mathbf{z} \in E_3$, we have the associative law[25]

$$\mathbf{x} \cdot (\mathbf{y} \cdot \mathbf{z}) = (\mathbf{x} \cdot \mathbf{y}) \cdot \mathbf{z}.$$

This way, the Euclidean space E_3 becomes a Lie group with the unit element $\mathbf{x} = 0$, and the inverse element to $\mathbf{x}$ equals $-\mathbf{x}$. This group is called the Heisenberg group $\mathcal{G}_{\text{Heis}}(\mathbf{k})$.

> *The Heisenberg group $\mathcal{G}_{\text{Heis}}(\mathbf{k})$ is a deformation of the additive group of the Euclidean space E_3 equipped with the group operation $\mathbf{x} + \mathbf{y}$.*

To see this, fix the real parameter ε, and define the product

$$\mathbf{x} \cdot \mathbf{y} := \mathbf{x} + \mathbf{y} + \tfrac{1}{2}\varepsilon(\mathbf{x}^\perp \times \mathbf{y}^\perp).$$

This way, for any ε, the Euclidean space E_3 becomes a group. The Heisenberg group $\mathcal{G}_{\text{Heis}}(\mathbf{k})$ corresponds to $\varepsilon = 1$, and the classical additive group E_3 is obtained by setting $\varepsilon = 0$.

The Heisenberg algebra $\mathcal{A}_{\text{Heis}}(\mathbf{k})$. The Euclidean space E_3 becomes a Lie algebra by introducing the Lie product

$$\boxed{[\mathbf{x}, \mathbf{y}] := \mathbf{x}^\perp \times \mathbf{y}^\perp.} \tag{1.63}$$

This Lie algebra is called the Heisenberg algebra $\mathcal{A}_{\text{Heis}}(\mathbf{k})$.[26]

---

[24] Note that the passage from classical mechanics to Einstein's theory of special relativity is based on a deformation of the Galilean group to the Lorentz group. Moreover, Einstein's theory of general relativity is a deformation of Newton's theory on gravitation.

[25] Note that $(\mathbf{x}^\perp \times \mathbf{y}^\perp)^\perp = 0$.

[26] Note that the Jacobi identity $[[\mathbf{x}, \mathbf{y}], \mathbf{z}] + [[\mathbf{y}, \mathbf{z}], \mathbf{x}] + [[\mathbf{z}, \mathbf{x}], \mathbf{y}] = 0$ is trivially satisfied, since $[[\mathbf{x}, \mathbf{y}], \mathbf{z}] = (\mathbf{x}^\perp \times \mathbf{y}^\perp)^\perp \times \mathbf{z}^\perp = 0$.

Proposition 1.29 *In the sense of the general theory of Lie groups, the Lie algebra of the Heisenberg group $\mathcal{G}_{\rm Heis}(\mathbf{k})$ is the Heisenberg algebra $\mathcal{A}_{\rm Heis}(\mathbf{k})$.*

Proof. Fix $\mathbf{x}, \mathbf{y} \in E_3$. Consider the trajectories

$$G(t) := t\mathbf{x}, \qquad H(t) := t\mathbf{y}, \qquad t \in \mathbb{R}$$

in the group $\mathcal{G}_{\rm Heis}(\mathbf{k})$. The tangent vectors at the unit element $\mathbf{x} = 0$ are $\dot{G}(0) = \mathbf{x}$ and $\dot{H}(0) = \mathbf{y}$. The commutator reads as

$$C(t,s) := G(t)H(s)G(t)^{-1}H(s)^{-1}, \qquad t, s \in \mathbb{R}.$$

According to the general theory of Lie groups, the Lie product is given by the partial derivative

$$[\mathbf{x}, \mathbf{y}] := C_{st}(0,0).$$

It follows from

$$C(t,s) = \left(t\mathbf{x} + s\mathbf{y} + \tfrac{1}{2}ts(\mathbf{x}^\perp \times \mathbf{y}^\perp)\right) \cdot \left(-t\mathbf{x} - s\mathbf{y} + \tfrac{1}{2}ts(\mathbf{x}^\perp \times \mathbf{y}^\perp)\right)$$

that $C(t,s) = ts(\mathbf{x}^\perp \times \mathbf{y}^\perp) + \ldots$ Hence $[\mathbf{x}, \mathbf{y}] = \mathbf{x}^\perp \times \mathbf{y}^\perp$. □

The Lie group $\mathcal{G}_{\rm Heis}(\mathbf{k})$ has the same topology as the Euclidean space E_3. Thus, $\mathcal{G}_{\rm Heis}(\mathbf{k})$ is arcwise connected and simply connected. Consequently, the Heisenberg group $\mathcal{G}_{\rm Heis}(\mathbf{k})$ is the universal covering group of the Lie algebra $\mathcal{A}_{\rm Heis}(\mathbf{k})$.

The relation to quantum mechanics. Let $\mathbf{i}, \mathbf{j}, \mathbf{k}$ be a right-handed orthonormal basis of the Euclidean space E_3. Then

$$[\mathbf{i}, \mathbf{j}] = \mathbf{k}, \quad [\mathbf{j}, \mathbf{k}] = 0, \quad [\mathbf{k}, \mathbf{i}] = 0.$$

Let $\mathcal{D}(\mathbb{R})$ be the complex linear space of all the smooth complex-valued functions $\psi : \mathbb{R} \to \mathbb{C}$ which have compact support. Motivated by quantum mechanics, define the linear operators $Q, P, K : \mathcal{D}(\mathbb{R}) \to \mathcal{D}(\mathbb{R})$ by setting

$$(Q\psi)(x) := x\psi(x), \quad (P\psi)(x) := -\mathrm{i}\hbar\frac{d\psi(x)}{dx}, \quad K\psi(x) := \mathrm{i}\hbar\psi(x), \quad x \in \mathbb{R}.$$

Then

$$\boxed{[Q,P]_- = K, \quad [P,K]_- = 0, \quad [Q,K]_- = 0} \tag{1.64}$$

where $[Q,P]_- := QP - PQ$, and so on.[27] The real linear combinations $\alpha Q + \beta P + \gamma K$ with $\alpha, \beta, \gamma \in \mathbb{R}$ form a Lie algebra $\mathcal{L}$. The map

$$\alpha\mathbf{i} + \beta\mathbf{j} + \gamma\mathbf{k} \mapsto \alpha Q + \beta P + \gamma K$$

is a Lie algebra isomorphism from $\mathcal{A}_{\rm Heis}$ onto $\mathcal{L}$. In terms of physics, the operator P (resp. Q) is the momentum (resp. position) operator of a quantum particle on the real line (see Sect. 7.4.3 of Vol. II). For the history of the Born–Heisenberg–Jordan commutation relation $[Q,P]_- = K = \mathrm{i}\hbar I$, see Sect. 1.3 of Vol. I. These commutation relations did not explicitly appear in Heisenberg's fundamental paper in 1925.[28]

---

[27] In fact, $PQ\psi = -\mathrm{i}\hbar(x\psi(x))' = -\mathrm{i}\hbar(\psi + x\psi') = -K\psi + QP\psi$.

[28] W. Heisenberg Quantum-theoretical re-interpretation of kinematics and mechanical relations (in German), Z. Physik **33**, 879–893 (English translation in: B. van der Waerden (Ed.), Sources of Quantum Mechanics (1917–1926), Dover, New York, 1968, pp. 261–276).

The Poisson–Lie algebra and quantization. Let $\mathcal{D}(\mathbb{R}^2)$ denote the real linear space of real-valued smooth functions $f : \mathbb{R}^2 \to \mathbb{R}$ which have compact support. Define the Poisson bracket

$$\{f,g\}(q,p) := \frac{\partial f(q,p)}{\partial q}\frac{\partial g(q,p)}{\partial p} - \frac{\partial f(q,p)}{\partial p}\frac{\partial g(q,p)}{\partial q}.$$

The linear space $\mathcal{D}(\mathbb{R}^2)$ becomes an infinite-dimensional Lie algebra equipped with the Lie product $\{f,g\}$. In particular, let us consider the following three functions: $f(q,p) := q, g(q,p) := p, \mathbf{1}(q,p) := 1$. Then

$$\{q,p\} = \mathbf{1}, \quad \{p,\mathbf{1}\} = 0, \qquad \{\mathbf{1},q\} = 0. \tag{1.65}$$

Thus, the linear functions $\alpha q + \beta p + \gamma\mathbf{1}$ with real coefficients α, β, γ form a real 3-dimensional Lie algebra which is isomorphic to the Heisenberg algebra $\mathcal{A}_{\mathrm{Heis}}(\mathbf{k})$. The isomorphism is given by

$$\alpha q + \beta p + \gamma\mathbf{1} \mapsto \alpha Q + \beta P + \gamma K.$$

Historically, the relations (1.65) were used by Jacobi (1804–1851) in his 1843 lectures on celestial mechanics.[29] In 1925 Dirac (1902–1984) emphasized that the quantization of classical mechanics can be based on the passage from Poisson brackets (1.65) to Lie brackets (1.64).[30]

Components. Let $\mathbf{x} = x\mathbf{i} + y\mathbf{j} + z\mathbf{k}$, and $\mathbf{y} = \xi\mathbf{i} + \eta\mathbf{j} + \zeta\mathbf{k}$. Then the product $\mathbf{x}\cdot\mathbf{y}$ corresponds to

$$(x,y,z)\cdot(\xi,\eta,\zeta) := (x+\xi, y+\eta, z+\zeta+\tfrac{1}{2}(x\eta - y\xi)). \tag{1.66}$$

The set of all the tuples $(x,y,z) \in \mathbb{R}^3$ equipped with the product (1.66) forms a Lie group $\mathcal{G}_{\mathrm{Heis}}(\mathbb{R}^3)$ which is isomorphic to the Heisenberg group $\mathcal{G}_{\mathrm{Heis}}(\mathbf{k})$. The isomorphism is given by the map $\mathbf{x} \mapsto (x,y,z)$.

Similarly, the Lie product $[\mathbf{x},\mathbf{y}]$ corresponds to

$$[(x,y,z),(\xi,\eta,\zeta)] = (0,0,x\eta - y\xi). \tag{1.67}$$

The set of all the tuples $(x,y,z) \in \mathbb{R}^3$ equipped with the Lie product (1.67) forms a Lie algebra $\mathcal{A}_{\mathrm{Heis}}(\mathbb{R}^3)$ which is isomorphic to the Heisenberg algebra $\mathcal{A}_{\mathrm{Heis}}(\mathbf{k})$. The isomorphism is given by the map $\mathbf{x} \mapsto (x,y,z)$.

Lie matrix groups. Let $n = 1,2,\ldots$ The symbol $GL(n,\mathbb{C})$ (resp. $GL(n,\mathbb{R})$) denotes the Lie group of all complex (resp. real) invertible $(n\times n)$-matrices equipped with the matrix product.[31]

The symbol $gl(n,\mathbb{C})$ (resp. $gl(n,\mathbb{R})$) denotes the real Lie algebra of all real (resp. complex) $(n\times n)$-matrices equipped with the Lie product

$$[A,B]_- := AB - BA.$$

In the sense of the general theory of Lie groups, the Lie algebra to $GL(n,\mathbb{C})$ (resp. $GL(n,\mathbb{R})$ is $gl(n,\mathbb{C})$ (resp. $gl(n,\mathbb{R})$).

---

[29] C. Jacobi, Jacobi Lectures on Dynamics, Hindustan Books Agency, India, 2009.

[30] P. Dirac, The fundamental equations of quantum mechanics, Proc. Roy. Soc. A **109** (1925), 642–653.

[31] If we do not expressively state the opposite, Lie groups are to be understood as real manifolds (i.e., real Lie groups). Basic properties of Lie algebras and Lie groups can be found in Chap. 7 of Vol. I.

By definition, a Lie matrix group is a closed subgroup of $GL(n, \mathbb{C})$. Important examples of Lie matrix groups can be found in Sect. 7.5ff of Vol. I. General properties of Lie matrix groups related to universal covering groups and general representation theory are summarized on page 1084.

The matrix representation of the Heisenberg group and the Heisenberg algebra. It turns out that the Heisenberg group is isomorphic to a well-known classical Lie group studied by Lie (1842–1899) in the 1870s. The set of all the real matrices

$$G = \begin{pmatrix} 1 & x & z \\ 0 & 1 & y \\ 0 & 0 & 1 \end{pmatrix}, \qquad x, y, z \in \mathbb{R}$$

equipped with the matrix product forms the Lie group $SUT(3, \mathbb{R})$ (special upper triangular matrices). The set of all the real matrices

$$A = \begin{pmatrix} 0 & x & z \\ 0 & 0 & y \\ 0 & 0 & 0 \end{pmatrix}, \qquad x, y, z \in \mathbb{R}$$

equipped with the Lie product $[A, B]_- := AB - BA$ forms a real 3-dimensional Lie algebra $sut(3, \mathbb{R})$.

Proposition 1.30 (i) *The map $G \mapsto (x, y, z - \frac{1}{2}xy)$ is a Lie group isomorphism from the Lie group $SUT(3, \mathbb{R})$ onto the Heisenberg group $\mathcal{G}_{\mathrm{Heis}}(\mathbb{R}^3)$.*

(ii) *The map $A \mapsto (x, y, z)$ is a Lie algebra isomorphism from the Lie algebra $sut(3, \mathbb{R})$ onto the Heisenberg algebra $\mathcal{A}_{\mathrm{Heis}}(\mathbb{R}^3)$.*

Proof. Ad (i). The matrix product

$$\begin{pmatrix} 1 & x & z \\ 0 & 1 & y \\ 0 & 0 & 1 \end{pmatrix} \begin{pmatrix} 1 & \xi & \zeta \\ 0 & 1 & \eta \\ 0 & 0 & 1 \end{pmatrix} = \begin{pmatrix} 1 & x+\xi & z+\zeta+x\eta \\ 0 & 1 & y+\eta \\ 0 & 0 & 1 \end{pmatrix}$$

corresponds to

$$(x+\xi, y+\eta, z+\zeta+x\eta - \tfrac{1}{2}(x+\xi)(y+\eta)).$$

This is equal to

$$(x, y, z - \tfrac{1}{2}xy) \cdot (\xi, \eta, \zeta - \tfrac{1}{2}\xi\eta) = (x+\xi, y+\eta, z+\zeta - \tfrac{1}{2}xy - \tfrac{1}{2}\xi\eta + \tfrac{1}{2}(x\eta - y\xi)).$$

Ad (ii). The Lie product

$$\begin{pmatrix} 0 & x & z \\ 0 & 0 & y \\ 0 & 0 & 0 \end{pmatrix} \begin{pmatrix} 0 & \xi & \zeta \\ 0 & 0 & \eta \\ 0 & 0 & 0 \end{pmatrix} - \begin{pmatrix} 0 & \xi & \zeta \\ 0 & 0 & \eta \\ 0 & 0 & 0 \end{pmatrix} \begin{pmatrix} 0 & x & z \\ 0 & 0 & y \\ 0 & 0 & 0 \end{pmatrix} = \begin{pmatrix} 0 & 0 & x\eta - y\xi \\ 0 & 0 & 0 \\ 0 & 0 & 0 \end{pmatrix}$$

corresponds to $(0, 0, x\eta - y\xi)$ which is equal to $[(x, y, z), (\xi, \eta, \zeta)]$. □

Solvability of the Heisenberg algebra. One has to distinguish between solvable and semisimple Lie algebras. This will be studied in Chap. 3. For example, the quark model in strong interaction is based on the representations of semismple

Lie algebras (see Sect. 3.14). As we will show in Sect. 3.17.3, the Heisenberg algebra is a solvable Lie algebra.

The Birkhoff–Heisenberg quotient group. Let $U(1)$ denote the multiplicative group of all complex numbers z with $|z| = 1$. Consider again the Lie group $SUT(3, \mathbb{R})$. Let $\mathcal{G}$ be the set of all the tuples

$$(x, y, z), \qquad x, y \in \mathbb{R}, \ z \in U(1)$$

equipped with the product

$$(x, y, z)(\xi, \eta, \zeta) := (x + \xi, y + \eta, \ z\zeta \cdot \mathrm{e}^{\mathrm{i}x\eta}).$$

This way, the set $\mathcal{G}$ becomes a Lie group with the unit element $\mathbf{1} := (0, 0, 1)$. The map $\varrho : SUT(3, \mathbb{R}) \to \mathcal{G}$ given by

$$\begin{pmatrix} 1 & x & \varphi \\ 0 & 1 & y \\ 0 & 0 & 1 \end{pmatrix} \mapsto (x, y, \mathrm{e}^{\mathrm{i}\varphi})$$

is a Lie group epimorphism. The kernel $\ker(\varrho) = \varrho^{-1}(\mathbf{1})$ consists precisely of all the matrices

$$\begin{pmatrix} 1 & 0 & 2\pi n \\ 0 & 1 & 0 \\ 0 & 0 & 1 \end{pmatrix}, \qquad n \in \mathbb{Z}.$$

This implies that we have the Lie group isomorphism

$$\mathcal{G} \simeq SUT(3, \mathbb{R}) / \ker(\varrho).$$

The Lie group $\mathcal{G}$ is called the Birkhoff–Heisenberg quotient group.[32] Since $\ker(\varrho)$ is a discrete subgroup of the Lie group $SUT(3, \mathbb{R})$ which is isomorphic to the Heisenberg group $\mathcal{G}_{\mathrm{Heis}}(\mathbb{R}^3)$, the Lie group $\mathcal{G}$ has the same Lie algebra as $\mathcal{G}_{\mathrm{Heis}}(\mathbb{R}^3)$, namely, the Heisenberg algebra $\mathcal{A}_{\mathrm{Heis}}(\mathbb{R}^3)$. Whereas $\mathcal{G}_{\mathrm{Heis}}(\mathbb{R}^3)$ is arcwise connected and simply connected, the Lie group $\mathcal{G}$ is not simply connected, since the group manifold of $U(1)$ (i.e., the unit circle) is not simply connected.

In 1936, Garrett Birkhoff proved that the group $\mathcal{G}$ is a Lie group which is not a Lie matrix group.[33] This was the first example which showed that one has to distinguish between Lie matrix groups and the abstract notion of a Lie group. This will be discussed in the Appendix on page 1085.[34] For the proof, we refer to B. Hall, Lie Groups, Lie Algebras, and Representations: An Elementary Introduction, Springer, New York, 2003, p. 314.

The Ado theorem. In contrast to the distinction between Lie matrix groups and abstract Lie groups, such a distinction drops out for Lie algebras. The Ado theorem tells us that:[35]

---

[32] The notion of quotient group is introduced in Sect. 4.1.3 of Vol. II.

[33] Garrett Birkhoff (1911–1996) was the son of George Birkhoff (1884–1944). Both mathematicians were professors at Harvard University, Cambridge, Massachusetts.

[34] G. Birkhoff, Lie groups simply isomorphic to no linear group, Bull. Amer. Math. Soc. **42** (1936), 883–888.

[35] D. Ado, The representation of Lie algebras by matrices, Usphehi Mat. Nauk **2** (1947), 159–173 (in Russian); Am. Math. Soc. Transl. No. **2** (1949).

Every real (or complex) finite-dimensional Lie algebra is isomorphic to a Lie algebra of square matrices.

Let $gl(n,\mathbb{K})$ be the Lie algebra of all $(n\times n)$-matrices with entries in $\mathbb{K}=\mathbb{R},\mathbb{C}$ equipped with the Lie product $[A,B]_-:=AB-BA$. Then every Lie algebra over $\mathbb{K}$ is isomorphic to a Lie subalgebra of $gl(n,\mathbb{K})$ for some $n=1,2,\ldots$. The sophisticated proof of the Ado theorem can be found in N. Jacobson, Lie Algebras, Dover, New York, 1979 (standard textbook on Lie algebras).

1.6 The Heisenberg Group Bundle and Gauge Transformations

The Heisenberg group $\mathcal{G}_{\mathrm{Heis}}(\mathbf{k})$ depends on the choice of the unit vector $\mathbf{k}$. Our goal is to introduce a global geometric object which describes this dependence. This motivates the introduction of the mathematical concept of bundles. Note that bundles are basic for modern geometry and modern physics. The set

$$S(E_3):=\{\mathbf{k}\in E_3:\ \ |\mathbf{k}|=1\}$$

is called the unit sphere of the Euclidean space E_3. The family

$$\{\mathcal{G}_{\mathrm{Heis}}(\mathbf{k})\}_{\mathbf{k}\in S(E_3)}$$

of Heisenberg groups indexed by the unit vector $\mathbf{k}$ is called the abstract Heisenberg group bundle. In order to equip this with a manifold structure, consider the set

$$\mathsf{B}:=\{(\mathbf{k},\mathbf{x}):\ \ \mathbf{k}\in S(E_3),\ \mathbf{x}\in\mathcal{G}_{\mathrm{Heis}}(\mathbf{k})\},$$

and define $\pi(\mathbf{k},\mathbf{x}):=\mathbf{k}$. The surjective smooth map

$$\pi:\mathsf{B}\to S(E_3)$$

is called the Heisenberg group bundle over the unit sphere $S(E_3)$ with the bundle space B. Moreover, the preimage

$$\pi^{-1}(\mathbf{k})=\{\mathbf{k}\}\times\mathcal{G}_{\mathrm{Heis}}(\mathbf{k})$$

is called the fiber over $\mathbf{k}$.

Local coordinates. In order to introduce local coordinates on the bundle space B, choose an open set $\mathcal{U}$ of the sphere $S(E_3)$ which is different from $S(E_3)$ (e.g., $\mathcal{U}=S(E_3)\setminus\{\mathbf{k}_0\}$ where $\mathbf{k}_0$ is a fixed unit vector). For every unit vector $\mathbf{k}\in\mathcal{U}$, choose a right handed orthonormal system

$$\mathbf{i}(\mathbf{k}),\ \ \mathbf{j}(\mathbf{k}),\ \ \mathbf{k} \tag{1.68}$$

such that the maps $\mathbf{k}\mapsto\mathbf{i}(\mathbf{k})$ and $\mathbf{k}\mapsto\mathbf{j}(\mathbf{k})$ are smooth. Because of $\mathcal{U}\neq S(E_3)$ such a family of frames exists. If $\mathbf{x}\in\mathcal{G}_{\mathrm{Heis}}(\mathbf{k})$, then

$$\mathbf{x}=x\mathbf{i}(\mathbf{k})+y\mathbf{j}(\mathbf{k})+z\mathbf{k}.$$

Let $(\mathbf{k},\mathbf{x})\in\mathsf{B}$ with $\mathbf{k}\in\mathcal{U}$. The map $\chi:\pi^{-1}(\mathcal{U})\to\mathcal{U}\times\mathcal{G}_{\mathrm{Heis}}(\mathbb{R}^3)$ given by

$$\boxed{\chi(\mathbf{k},\mathbf{x}):=(\mathbf{k},x,y,z)}$$

assigns to the bundle point $(\mathbf{k}, \mathbf{x})$ the bundle coordinate $(\mathbf{k}, x, y, z)$ where we have $(x, y, z) \in \mathcal{G}_{\text{Heis}}(\mathbb{R}^3)$.

Gauge transformation. The local coordinates $(\mathbf{k}, x, y, z)$ of the bundle point $(\mathbf{k}, \mathbf{x})$ depend on the choice of the frame family (1.68). Let us pass to a different frame family

$$\mathbf{i}^+(\mathbf{k}), \;\; \mathbf{j}^+(\mathbf{k}), \;\; \mathbf{k}.$$

Then we have the matrix equation

$$\begin{pmatrix} \mathbf{i}^+(\mathbf{k}) \\ \mathbf{j}^+(\mathbf{k}) \\ \mathbf{k} \end{pmatrix} = G(\mathbf{k}) \begin{pmatrix} \mathbf{i}(\mathbf{k}) \\ \mathbf{j}(\mathbf{k}) \\ \mathbf{k} \end{pmatrix} \tag{1.69}$$

where the matrix $G(\mathbf{k}) \in SO(3)$ describes a rotation about the axis $\mathbf{k}$ (i.e., the real (3×3)-matrix $G(\mathbf{k})$ satisfies $G(\mathbf{k})G(\mathbf{k})^d = I$ and $\det G(\mathbf{k}) = 1$). Then the change of local coordinates of the bundle point $(\mathbf{k}, \mathbf{x})$ is given by

$$\begin{pmatrix} x^+ \\ y^+ \\ z \end{pmatrix} = G(\mathbf{k}) \begin{pmatrix} x \\ y \\ z \end{pmatrix}. \tag{1.70}$$

This follows from

$$\mathbf{x} = x^+\mathbf{i}^+(\mathbf{k}) + y^+\mathbf{j}^+(\mathbf{k}) + z\mathbf{k} = x\mathbf{i}(\mathbf{k}) + y\mathbf{j}(\mathbf{k}) + z\mathbf{k}$$

and $(\mathbf{i}^+(\mathbf{k}), \mathbf{j}^+(\mathbf{k}), \mathbf{k})(x^+, y^+, z)^d = (\mathbf{i}(\mathbf{k}), \mathbf{j}(\mathbf{k}), \mathbf{k})G^d(x^+, y^+, z)^d$. Hence

$$(x^+, y^+, z)^d = (G^d)^{-1}(x, y, z)^d, \qquad (G^d)^{-1} = G.$$

The transformation (1.70) of local coordinates of bundle points is called a gauge transformation. In terms of physics, the change of the orthonormal frame family (1.68) describes the change of observers. Using local bundle coordinates, the Heisenberg group bundle space B becomes a real 6-dimensional manifold.

Poincaré's topological obstruction. In the 1890s, Poincaré (1854–1912) discovered the crucial topological fact that it is impossible to construct a continuous tangent vector field on a 2-dimensional sphere which is different from zero at all the points of the sphere. Therefore, it is impossible to introduce globally a family of orthonormal frames (1.68) for all vectors $\mathbf{k} \in S(E_3)$ which depends smoothly on $\mathbf{k}$. In other words, it is not possible to introduce global coordinates on the Heisenberg group bundle space B. Locally, the bundle space B looks like the product set

$$\mathcal{U} \times \mathcal{G}_{\text{Heis}}(\mathbb{R}^3)$$

where $\mathcal{U}$ is an open neighborhood of a point on the unit sphere $S(E_3)$. However, globally, the bundle space B is different from the product set $S(E_3) \times \mathcal{G}_{\text{Heis}}(\mathbb{R}^3)$. Intuitively, we say that

The Heisenberg group bundle is a nontrivial (also called twisted) bundle over the unit sphere of the Euclidean space with the Lie group $\mathcal{G}_{\text{Heis}}(\mathbb{R}^3)$ as typical fiber.

As we will show later on, the theory of bundles is the decisive mathematical tool of gauge theory. Nontrivial topological properties of bundles are responsible for crucial properties of physical fields in gauge theory. Mathematically, nontrivial topological properties are described by nontrivial characteristic classes (i.e., nonvanishing elements of cohomology groups).

Further reading. A detailed study of the Heisenberg group and its representations along with applications to quantum physics can be found in G. Folland, Harmonic Analysis in Phase Space, Princeton University Press, 1989. Heisenberg group bundles are studied in E. Binz and S. Pods, The Geometry of Heisenberg Groups: With Applications in Signal Theory, Optics, Quantization, and Field Quantization, Amer. Math. Soc., Providence, Rhode Island, 2008.

2. Algebras and Duality (Tensor Algebra, Grassmann Algebra, Clifford Algebra, Lie Algebra)

> Both the two-dimensional and the three-dimensional Euclidean space E_2 and E_3, respectively, possess an extraordinarily rich algebraic structure whose generalization plays a fundamental role in modern mathematics and physics.
> There is a beautiful interplay between symmetry and antisymmetry, Hilbert spaces, Lie algebras, Grassmann algebras, and Clifford algebras, enriched by Riesz duality (vectors and covectors), as well as Hodge duality (the Hodge $*$-operator).
>
> Folklore

Operator algebras play a fundamental role in algebraic quantum field theory. In order to understand this, one has first to understand the crucial algebraic structures of the Euclidean space. The point is that relevant products possess an invariant meaning, that is, they are independent of the choice of a basis of the Euclidean space.

2.1 Multilinear Functionals

> In terms of mathematics, tensor products $\mathbf{x} \otimes \mathbf{y}$ and alternating (Grassmann) products $\mathbf{x} \wedge \mathbf{y}$ generalize the product of polynomials. In terms of elementary particle physics, such products describe composite particles.
>
> Folklore

2.1.1 The Graded Algebra of Polynomials

Algebras play a fundamental role in quantum physics. Typical properties of algebras can be understood best by using polynomial algebras as a paradigm. Let $\mathbb{K} = \mathbb{R}$ (real numbers) or $\mathbb{K} = \mathbb{C}$ (complex numbers). Let the symbol $\mathbb{K}[x]$ denote the set of all polynomials

$$P(x) := \alpha_0 + \alpha_1 x + \alpha_2 x^2 + \ldots + \alpha_n x^n, \qquad n = 0, 1, 2, \ldots$$

with the coefficients $\alpha_0, \alpha_1, \ldots \in \mathbb{K}$. If $\alpha_n \neq 0$, then the integer n is called the degree of the polynomial P. Precisely the polynomials of the form $\alpha_k x^k$ with fixed $k = 0, 1, 2, \ldots$ are called monomials. For all polynomials $P, Q, R \in \mathbb{K}[x]$ and all numbers $\alpha, \beta \in \mathbb{K}$, we have two operations, namely,

- $\alpha P + \beta Q$ (linear combination), and
- PQ (product).

The product is distributive, that is,

E. Zeidler, *Quantum Field Theory III: Gauge Theory*,
DOI 10.1007/978-3-642-22421-8_3,

- $(\alpha P + \beta Q)R = \alpha PR + \beta QR$, and
- $R(\alpha P + \beta Q) = \alpha RP + \beta RQ$.

Setting $B(P,Q) := PQ$, the distributive law is equivalent to the bilinearity of the map B.

Algebra. Setting $\mathcal{A} := \mathbb{K}[x]$, the following hold:

(A1) The set $\mathcal{A}$ is a linear space over $\mathbb{K}$.
(A2) There exists a bilinear map $B : \mathcal{A} \times \mathcal{A} \to \mathcal{A}$.
(A3) The product is associative, that is, $(PQ)R = P(QR)$ for all $P, Q, R \in \mathcal{A}$.
(A4) The product is commutative, that is, $PQ = QP$ for all $P, Q \in \mathcal{A}$.
(A5) There exists a unit element, denoted $\mathbf{1}$, such that $\mathbf{1}P = P\mathbf{1} = P$ for all $P \in \mathcal{A}$.[1]

The following definition is crucial. The set $\mathcal{A}$ is called an algebra over $\mathbb{K}$ iff conditions (A1) and (A2) are satisfied. Moreover, the algebra $\mathcal{A}$ is called associative (resp. commutative) iff the condition (A3) (resp. (A4)) is satisfied.[2] The algebra is called unital iff $\mathcal{A} \neq \{0\}$ and condition (A5) is satisfied.

Observe that an algebra is not necessarily associative or commutative. For example, as a rule, a Lie algebra is neither associative nor commutative. If the algebra $\mathcal{A}$ is unital, then the field $\mathbb{K}$ can be regarded as a subset of $\mathcal{A}$, and the unit element $\mathbf{1}$ can be identified with the unit element 1 of the field $\mathbb{K}$. This will be shown in Problem 3.7.

Subalgebra. A subset $\mathcal{S}$ of the algebra $\mathcal{A}$ is called a subalgebra iff $\mathcal{S}$ is a linear subspace of $\mathcal{A}$, and $\mathcal{S}$ is invariant under multiplication, that is, $P, Q \in \mathcal{S}$ and $\alpha, \beta \in \mathbb{K}$ imply $\alpha P + \beta Q \in \mathcal{S}$ and $PQ \in \mathcal{S}$. For example, the set $\mathbb{K}$ is a subalgebra of $\mathbb{K}[x]$.

The subset $\mathcal{J}$ of the algebra $\mathcal{A}$ is called an ideal (or an invariant subalgebra) iff it is a subalgebra of $\mathcal{A}$ with the additional property that $ab, bc \in \mathcal{J}$ for all $b \in \mathcal{J}$ and all $a, c \in \mathcal{A}$.

For given subset $\mathcal{S}$ of the algebra $\mathcal{A}$, the symbol $\mathcal{A}(\mathcal{S})$ denotes the smallest subalgebra of $\mathcal{A}$ which contains the set $\mathcal{S}$. Explicitly, the subalgebra $\mathcal{A}$ consists of all the finite sums of products

$$P_1 P_2 \cdots P_n$$

with $P_1, \ldots, P_n \in \mathcal{S}$ and $n = 1, 2, \ldots$

Algebra morphism. Let $\mathcal{A}$ and $\mathcal{B}$ be algebras over $\mathbb{K}$. The map

$$\mu : \mathcal{A} \to \mathcal{B}$$

is called an algebra morphism iff it is linear and it respects products, that is $\mu(PQ) = \mu(P)\mu(Q)$ for all $P, Q \in \mathcal{A}$.

Bijective algebra morphisms are called algebra isomorphisms.

In the special case where $\mathcal{B} = \mathcal{A}$, the isomorphism (resp. morphism) $\mu : \mathcal{A} \to \mathcal{A}$ is called an algebra automorphism (resp. algebra endomorphism).

Direct sum. Let $\mathcal{A}$ be a linear space over $\mathbb{K}$. We write

$$\mathcal{A} = \bigoplus_{n=0}^{\infty} \mathcal{A}_n \tag{2.1}$$

iff $\mathcal{A}_0, \mathcal{A}_1, \ldots$ are linear subspaces of $\mathcal{A}$, and each element P of $\mathcal{A}$ can be uniquely written as a sum of the form

---

[1] This unit element is always uniquely determined. In fact, if there is another unit element $\mathbf{1}'$, then $\mathbf{1}' = \mathbf{1}'\mathbf{1} = \mathbf{1}$.

[2] The algebra is called real (resp. complex) iff $\mathbb{K} = \mathbb{R}$ (resp. $\mathbb{K} = \mathbb{C}$).

$$P = P_0 + P_1 + P_2 + \ldots$$

where $P_n \in \mathcal{A}$ for all indices n, and at most a finite number of summands $P_0, P_1, \ldots$ is different from zero.

Graded algebra. The algebra $\mathcal{A}$ is called graded iff the linear space $\mathcal{A}$ allows the direct sum decomposition (2.1), and the product has the property that

$$P \in \mathcal{A}_m, \; Q \in \mathcal{A}_n \qquad \text{always implies} \qquad PQ \in \mathcal{A}_{m+n}.$$

We say that the elements of $\mathcal{A}_m$ have the degree m, and the degree is additive with respect to multiplication. For example, the polynomial algebra $\mathbb{K}[x]$ is graded. Here, $\mathcal{A}_n$ consists of the monomials of degree n.

Polynomials of two variables. By definition, the set $\mathbb{K}[x, y]$ consists of all the polynomials

$$P(x, y) = \alpha_0 + \alpha_{10}x + a_{01}y + \alpha_{20}x^2 + \alpha_{11}xy + \alpha_{02}y^2 + \ldots$$

where $\alpha_0, \alpha_{10}, \ldots \in \mathbb{K}$, and at most a finite number of coefficients does not vanish. The polynomials of the form $\alpha_{kl}x^k y^l$ are called monomials of degree $k + l$. The set $\mathbb{K}[x, y]$ equipped with the usual multiplication of polynomials represents a commutative and associative algebra. This algebra is also graded with respect to the degree of monomials.

Similarly, one constructs the polynomial algebra $\mathbb{K}[x_1, x_2, \ldots, x_N]$ of N variables $x_1, \ldots, x_N$. This algebra over $\mathbb{K}$ is commutative, associative, and graded.

Division algebra. By definition, a division algebra is a nontrivial algebra (i.e., $\mathcal{A} \neq \{0\}$) with the additional property that, for given $a, b \in \mathcal{A}$ with $a \neq 0$, the two equations

- $ax = b, \; x \in \mathcal{A}$, and
- $ya = b, \; y \in \mathcal{A}$

have unique solutions.

Fields and skew-fields. The definition of fields and skew-fields can be found on page 179 of Vol. II. Roughly speaking, a skew-field is equipped with the following operations: addition, associative and distributive multiplication (including a unit element), and division. If the multiplication is commutative, then the skew-field is called a field.

- Every associative division algebra with unit element is a skew-field.
- Every commutative associative division algebra with unit element is a field.

The prototype of a field (resp. skew-field) is the set $\mathbb{R}$ of real numbers (resp. the set $\mathbb{H}$ of quaternions).

$*$-Algebra. An algebra $\mathcal{A}$ over $\mathbb{K}$ is called a $*$-algebra (star algebra) iff there exists a map $* : \mathcal{A} \to \mathcal{A}$ such that for all A, B and all α, β the following hold:

- $*(\alpha A + \beta B) = \alpha^\dagger * A + \beta^\dagger * B$,
- $*(AB) = *B * A$ and $*(*A) = A$.

For example, let $n = 2, 3, \ldots$ The set $\mathbb{M}(n, n; \mathbb{C})$ of complex $(n \times n)$-matrices forms an associative noncommutative $*$-algebra over $\mathbb{K}$ with unit element. In this connection, we use the following matrix operations:

$$\alpha A + \beta B, \quad AB, \quad *A := A^\dagger$$

for all $A, B \in \mathbb{M}(n, n; \mathbb{C})$ and all $\alpha, \beta \in \mathbb{K}$. The unit matrix I represents the unit element of the algebra.

2.1.2 Products of Multilinear Functionals

In terms of algebra, quantum processes are described by noncommutative mathematical structures. In what follows, we want to construct associative algebras which are not commutative. These algebras were introduced in the second half of the 19th century.

Ariadne's thread in tensor algebra. Let X be a linear space over $\mathbb{K}$, where $\mathbb{K} = \mathbb{R}$ or $\mathbb{K} = \mathbb{C}$. Let $F, G : X \to \mathbb{K}$ be linear functionals. Recall that the symbol X^d denotes the dual space to X; this is the space of all the linear functionals $F : X \to \mathbb{K}$. This is a linear space over $\mathbb{K}$ (see Sect. 7.9.1 of Vol. I). The basic idea of tensor algebra is to introduce the tensor product $F \otimes G$ by setting

$$(F \otimes G)(x, y) := F(x)G(y) \qquad \text{for all} \quad x, y \in X.$$

Obviously, $F \otimes G$ is a bilinear functional on X. Symmetrization and antisymmetrization yield the symmetrized tensor product

$$F \odot G := F \otimes G + G \otimes F$$

and the antisymmetrized (or alternating) tensor product

$$F \wedge G := F \otimes G - G \otimes F,$$

respectively. Explicitly, for all $x, y \in X$, we get:

- $(F \odot G)(x, y) = F(x)G(y) + F(y)G(x)$, and
- $(F \wedge G)(x, y) = F(x)G(y) - F(y)G(x)$.

We want to generalize this to multilinear functionals. For $n = 1, 2, \ldots$, let $M_n(X)$ denote the set of all n-linear functionals

$$F : X \times \cdots \times X \to \mathbb{K}.$$

Moreover, we set $M_0(X) := \mathbb{K}$. We also introduce the direct sum

$$M(X) := \bigoplus_{n=0}^{\infty} M_n(X).$$

That is, the elements of $M(X)$ are tuples of the form

$$(F_0, F_1, F_2, \ldots) \tag{2.2}$$

where $F_k \in M_k(X)$ for all indices $k = 0, 1, 2, \ldots$, and at most a finite number of the entries $F_0, F_1, \ldots$ is different from zero. The set $M(X)$ becomes a linear space over $\mathbb{K}$ in a natural way. To simplify notation, we write $F_0 + F_1 + F_2 + \ldots$ instead of (2.2).

The tensor product $\otimes$ for multilinear functionals and the algebra $M(X)$. Fix $m, n = 1, 2, \ldots$ For multilinear functionals $F \in M_m(X)$ and $G \in M_n(X)$, we define

$$\boxed{(F \otimes G)(x_1, \ldots, x_{m+n}) := F(x_1, \ldots, x_m)G(x_{m+1}, \ldots, x_{m+n})} \tag{2.3}$$

for all arguments $x_1, \ldots, x_{m+n} \in X$. Obviously, $F \otimes G \in M_{m+n}(X)$. If $m = 0$ or $n = 0$, then we define $F \otimes G := FG$.

Proposition 2.1 *The $\otimes$-product is distributive and associative.*

This is an easy consequence of the definition. In a quite natural way, the $\otimes$-product can be extended to the direct sum $M(X)$ by setting

$$(F_0 + F_1 + \ldots) \otimes (G_0 + G_1 + \ldots) = F_0 \otimes G_0 + (F_0 \otimes G_1 + F_1 \otimes G_0) + \ldots$$

This way, the linear space $M(X)$ becomes an algebra over $\mathbb{K}$ which contains the dual space X^d of the original linear space X:

$$X^d \subseteq M(X).$$

The alternating product $\wedge$ for antisymmetric multilinear functionals and the algebra $M_{\text{anti}}(X)$. For antisymmetric multilinear functionals $F \in M_m(X)$ and $G \in M_n(X)$, we define $(F \wedge G)(x_1, \ldots, x_{m+n})$ by

$$\frac{1}{m!n!} \sum_{\pi} \operatorname{sgn}(\pi) F(x_{\pi(1)}, \ldots, x_{\pi(m)}) G(x_{\pi(m+1)}, \ldots, x_{\pi(m+n)}) \tag{2.4}$$

for all arguments $x_1, \ldots, x_{m+n} \in X$. Here, we sum over all the permutations π of the indices $1, \ldots, m+n$. Obviously, $F \wedge G \in M_{m+n}(X)$, and $F \wedge G$ is antisymmetric. If $m = 0$ or $n = 0$, then we define $F \wedge G := FG$. It follows immediately from the definition that

$$\boxed{F \wedge G = (-1)^{mn} G \wedge F.} \tag{2.5}$$

We say that the $\wedge$-product is graded anticommutative.

Theorem 2.2 *The alternating product* (2.4) *is distributive and associative.*

Proof. Consider first the special case where $F, G, H : X \to \mathbb{K}$ are linear functionals. Then $(F \wedge (G \wedge H))(x, y, z)$ is equal to

$$F(x) \begin{vmatrix} G(y) & G(z) \\ H(y) & H(z) \end{vmatrix} - F(y) \begin{vmatrix} G(x) & G(z) \\ H(x) & H(z) \end{vmatrix} + F(z) \begin{vmatrix} G(x) & G(y) \\ H(x) & H(y) \end{vmatrix}.$$

Moreover, $((F \wedge G) \wedge H)(x, y, z)$ is equal to

$$\begin{vmatrix} F(x) & F(y) \\ G(x) & G(y) \end{vmatrix} H(z) - \begin{vmatrix} F(x) & F(z) \\ G(x) & G(z) \end{vmatrix} H(y) + \begin{vmatrix} F(y) & F(z) \\ G(y) & G(z) \end{vmatrix} H(x).$$

By the Laplace expansion theorem, the two expressions are equal to the determinant

$$\begin{vmatrix} F(x) & F(y) & F(z) \\ G(x) & G(y) & G(z) \\ H(x) & H(y) & H(z) \end{vmatrix}.$$

The reader should convince herself/himself that the general statement is equivalent to the general Laplace expansion theorem (1.17) on page 78. □

In order to get an algebra, introduce the following direct sum

$$M_{\text{anti}}(X) := \bigoplus_{n=0}^{\infty} M_{n,\text{anti}}(X),$$

where $M_{n,\text{anti}}(X)$ denotes the linear space of all antisymmetric n-linear functionals on X. In particular, $M_{0,\text{anti}}(X) := \mathbb{K}$. In a quite natural way, the $\wedge$-product can be extended to the direct sum $M_{\text{anti}}(X)$ by setting

$$(F_0 + F_1 + \ldots) \wedge (G_0 + G_1 + \ldots) = F_0 \wedge G_0 + (F_0 \wedge G_1 + F_1 \wedge G_0) + \ldots$$

This way, the linear space $M_{\text{anti}}(X)$ becomes an algebra over $\mathbb{K}$ which contains the dual space X^d of the original linear space X:

$$X^d \subseteq M_{\text{anti}}(X).$$

The symmetrized tensor product $\odot$ for symmetric multilinear functionals and the algebra $M_{\text{sym}}(X)$. For given symmetric multilinear functionals $F \in M_m(X)$ and $G \in M_n(X)$, we define $(F \wedge G)(x_1, \ldots, x_{m+n})$ by

$$\frac{1}{m!n!} \sum_{\pi} F(x_{\pi(1)}, \ldots, x_{\pi(m)}) G(x_{\pi(m+1)}, \ldots, x_{\pi(m+n)}) \tag{2.6}$$

for all arguments $x_1, \ldots, x_{m+n} \in X$. Here, we sum over all the permutations π of the indices $1, \ldots, m+n$. Obviously, $F \odot G \in M_{m+n}(X)$, and $F \odot G$ is symmetric. If $m = 0$ or $n = 0$, then we define $F \odot G := FG$. It follows immediately from the definition that

$$\boxed{F \odot G = G \odot F.} \tag{2.7}$$

Proposition 2.3 *The symmetrized tensor product* (2.6) *is distributive, associative, and commutative.*

The proof proceeds analogously to the proof of Theorem 2.2. Introduce the direct sum

$$M_{\text{sym}}(X) := \bigoplus_{n=0}^{\infty} M_{n,\text{sym}}(X),$$

where $M_{n,\text{sym}}(X)$ denotes the linear space of all symmetric n-linear functionals on X. In particular, $M_{0,\text{sym}}(X) := \mathbb{K}$. In a quite natural way, the $\odot$-product can be extended to the direct sum $M_{\text{sym}}(X)$ by setting

$$(F_0 + F_1 + \ldots) \odot (G_0 + G_1 + \ldots) = F_0 \odot G_0 + (F_0 \odot G_1 + F_1 \odot G_0) + \ldots$$

This way, the linear space $M_{\text{sym}}(X)$ becomes an algebra over $\mathbb{K}$ which contains the dual space X^d of the original linear space X:

$$X^d \subseteq M_{\text{sym}}(X).$$

2.1.3 Tensor Algebra

We want to construct the tensor algebra $\bigotimes(X)$ of the linear space X over $\mathbb{K}$. To begin with, note that there exists an injective linear map $j : X \to (X^d)^d$ given by

$$j(x)(F) := F(x) \qquad \text{for all} \quad F \in X^d.$$

Therefore, we can identify the set X with a subset of $(X^d)^d$. In this sense, $X \subseteq M(X^d)$. By definition, the tensor algebra $\bigotimes(X)$ is the smallest subalgebra of $M(X^d)$ which contains the set $X \cup \mathbb{K}$. Explicitly, the elements of $\bigotimes(X)$ are finite sums where the summands are elements of $\mathbb{K}$, or elements of X, or products of the form

$$x_1 \otimes x_2 \otimes \cdots \otimes x_n$$

with $x_1, \ldots, x_n \in X$ and $n = 2, 3, \ldots$. Explicitly,

$$(x_1 \otimes x_2 \otimes \cdots \otimes x_n)(F_1, F_2, \ldots, F_n) = F_1(x_1)F_2(x_2)\cdots F_n(x_n)$$

for all linear functionals $F_1, F_2, \ldots, F_n \in X^d$. That is, the map

$$x_1 \otimes x_2 \otimes \cdots \otimes x_n : X^d \times X^d \times \cdots \times X^d \to \mathbb{K}$$

is n-linear.

2.1.4 Grassmann Algebra (Alternating Algebra)

By definition, the Grassmann algebra (also called the alternating algebra) $\bigwedge(X)$ of the linear space X over $\mathbb{K}$ is the smallest subalgebra of $M_{\text{anti}}(X^d)$ which contains the set $X \cup \mathbb{K}$. Explicitly, the elements of $\bigwedge(X)$ are finite sums where the summands are elements of $\mathbb{K}$, or elements of X, or products of the form

$$x_1 \wedge x_2 \wedge \cdots \wedge x_n$$

with $x_1, \ldots, x_n \in X$ and $n = 2, 3, \ldots$. Explicitly,

$$x_1 \wedge x_2 = x_1 \otimes x_2 - x_2 \otimes x_1 \qquad \text{for all} \quad x_1, x_2 \in X.$$

Moreover, for all $x_1, \ldots, x_n \in X$,

$$x_1 \wedge x_2 \wedge \cdots \wedge x_n = \sum_\pi \operatorname{sgn}(\pi) \cdot x_{\pi(1)} \otimes x_{\pi(2)} \otimes \cdots \otimes x_{\pi(n)}.$$

2.1.5 Symmetric Tensor Algebra

By definition, the symmetric tensor algebra $\bigotimes_{\text{sym}}(X)$ of the linear space X over $\mathbb{K}$ is the smallest subalgebra of $M_{\text{sym}}(X^d)$ which contains the set $X \cup \mathbb{K}$. Explicitly, the elements of $\bigotimes_{\text{sym}}(X)$ are finite sums where the summands are elements of $\mathbb{K}$, or elements of X, or finite products of the form

$$x_1 \odot x_2 \odot \cdots \odot x_n$$

with $x_1, \ldots, x_n \in X$ and $n = 2, 3, \ldots$. Explicitly,

$$x_1 \odot x_2 = x_1 \otimes x_2 + x_2 \otimes x_1 \qquad \text{for all} \quad x_1, x_2 \in X.$$

Moreover, for all $x_1, \ldots, x_n \in X$,

$$x_1 \odot x_2 \odot \cdots \odot x_n = \sum_\pi x_{\pi(1)} \otimes x_{\pi(2)} \otimes \cdots \otimes x_{\pi(n)}.$$

2.1.6 The Universal Property of the Tensor Product

> Products on real (resp. complex) linear spaces can be reduced to linear operators on tensor products. This way, multilinear algebra can be reduced to linear algebra.
>
> Folklore

The tensor product $X \otimes Y$. Let $K = \mathbb{R}$ or $\mathbb{K} = \mathbb{C}$, and let X and Y be linear spaces over $\mathbb{K}$. We want to construct the linear space $X \otimes Y$ which is called the tensor product of X and Y. In addition, we want to show that the tensor product $X \otimes Y$ is distinguished by a universal property. To begin with, let $M_2(X^d, Y^d)$ denote the space of all bilinear functionals

$$F : X \times Y \to \mathbb{K}.$$

This is a linear space over $\mathbb{K}$. For given $x \in X$ and $y \in Y$, we set

$$(x \otimes y)(F, G) := F(x)G(y) \qquad \text{for all} \quad F \in X^d, G \in Y^d.$$

Then, $x \otimes y \in M_2(X^d, Y^d)$. In particular, we have the distributive laws

- $(\alpha w + \beta x) \otimes y = \alpha(w \otimes y) + \beta(x \otimes y)$, and
- $x \otimes (\alpha y + \beta z) = \alpha(x \otimes y) + \beta(x \otimes z)$

for all $w, x \in X, y, z \in Y$, and $\alpha, \beta \in \mathbb{K}$. By definition, the symbol $X \otimes Y$ denotes the smallest linear subspace of $M_2(X^d, Y^d)$ which contains all the special bilinear functionals $x \otimes y$. Explicitly, the elements of $X \otimes Y$ are finite sums of the form

$$x_1 \otimes y_1 + x_2 \otimes y_2 + \ldots + x_n \otimes y_n, \qquad n = 1, 2, \ldots \tag{2.8}$$

with $x_1, \ldots, x_n \in X$ and $y_1, \ldots y_n \in Y$. Two sums of this form represent the same element in $X \otimes Y$ iff they represent the same bilinear functional $B : X^d \times Y^d \to \mathbb{K}$.

Proposition 2.4 *Let $b_1, \ldots, b_m$ and $c_1, \ldots, c_n$ be linearly independent elements of X and Y, respectively. Then the mn products $b_i \otimes c_j$ with $i = 1, \ldots, m$ and $j = 1, \ldots, n$ are linearly independent elements of the tensor product $X \otimes Y$.*

Corollary 2.5 *If the linear spaces X and Y have finite dimension, then we have the product rule* $\dim(X \otimes Y) = \dim(X)\dim(Y)$ *for the dimensions.*

Proof. (I) Proof of the proposition. Suppose that

$$\sum_{i=1}^{m} \sum_{j=1}^{n} a^{ij} b_i \otimes c_j = 0. \tag{2.9}$$

The key idea is to choose linear functionals $F^1, \ldots, F^m \in X^d$ such that[3]

$$F^k(b_i) = \delta_i^k, \qquad i, k = 1, \ldots, m.$$

Then, $F^1, \ldots, F^m$ is a system of linearly independent elements of X^d called the dual system to $b_1, \ldots, b_m$. Similarly, let $G^1, \ldots, G^n \in Y^d$ be the dual system to $c_1, \ldots, c_n$. This implies the key relation

$$(b_i \otimes c_j)(F^k, G^l) = F^k(b_i)G^l(c_j) = \delta_i^k \delta_j^l$$

[3] Such a system always exists, by Problem 4.13 of Vol. II.

for all possible indices. Applying this to (2.9), we get $\sum_{k,l} a^{kl}\delta_i^k\delta_j^l = 0$. Hence $a^{ij} = 0$ for all possible indices.

(II) Proof of the corollary. Let $b_1, \ldots, b_m$ and $c_1, \ldots, c_n$ be a basis of X and Y, respectively. Let $F \in X^d$ and $G \in Y^d$. Then $F^1, \ldots, F^m$ (resp. $G^1, \ldots, G^n$) is a basis of X^d (resp. Y^d). Hence $F = \sum_{k=1}^m \alpha_k F^k$ and $G = \sum_{l=1}^n \beta_l G^l$. For an arbitrary bilinear functional $B : X^d \times Y^d \to \mathbb{K}$, we get

$$B(F,G) = \sum_{k,l} \alpha_k \beta_l B(F^k, G^l) = \sum_{k,l} B(F^k, G^l)(b_k \otimes c_l)(F,G).$$

Thus, each element B of $M_2(X^d, Y^d)$ allows the representation

$$B = \sum_{k=1}^{m}\sum_{l=1}^{n} B(F^k, G^l)\, b_k \otimes c_l. \tag{2.10}$$

By Prop. 2.4, the mn elements $b_k \otimes c_j$ with $k = 1, \ldots, m, j = 1, \ldots, n$ are linearly independent. By (2.10), they form a basis of $X \otimes Y$. □

Comparing elements of $X \otimes Y$. We want to decide wether two elements of the form (2.8) represent the same element of $X \otimes Y$. For example, consider the equation

$$x \otimes y = u \otimes w + v \otimes z. \tag{2.11}$$

Choose a basis $b_1, \ldots, b_n$ of the linear space $\operatorname{span}\{x, u, v\}$ (resp. a basis $c_1, \ldots, c_m$ of $\operatorname{span}\{y, w, z\}$). Then, x is a linear combination of $b_1, \ldots, b_n$, and so on. By the distributive law,

$$x \otimes y = \sum_i x^i b_i \otimes \sum_j y^j c_j = \sum_{i,j} \alpha^{ij}\, b_i \otimes c_j.$$

Similarly, $u \otimes w + v \otimes z = \sum_{i,j} \beta^{ij}\, b_i \otimes c_j$. By Prop. 2.4, we have (2.11) iff $\alpha^{ij} = \beta^{ij}$ for all indices $i = 1, \ldots, m$ and $j = 1, \ldots, n$. The general case proceeds analogously.

The universality property. The following commutative diagram contains the basic idea:

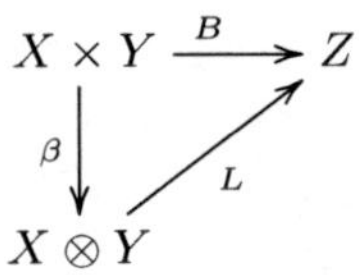

(2.12)

Let $\mathbb{K} = \mathbb{R}$ or $\mathbb{K} = \mathbb{C}$, and let X, Y, Z be arbitrary linear spaces over $\mathbb{K}$. Define

$$\beta(x,y) := x \otimes y.$$

The bilinear map $\beta : X \times Y \to X \otimes Y$ is called the canonical bilinear map. If the operator $L : X \otimes Y \to Z$ is linear, and the diagram (2.12) is commutative, then the map $B = L \circ \beta$ is bilinear because of the distributive law for $x \otimes y$. The following proposition shows that also the converse is true.

Proposition 2.6 *For each bilinear map $B : X \times Y \to Z$, there exists a unique linear map $L : X \otimes Y \to Z$ such that the diagram (2.12) is commutative.*

Before proving this, let us discuss the result. Each bilinear map $B : X \times Y \to Z$ represents a product on the spaces X and Y with values in the space Z. We want to describe all possible products. The result above solves this problem. By the aid of the factorization

$$B = L \circ \beta,$$

each product can be represented by the canonical product β and a linear map $L : X \otimes Y \to Z$. Therefore, the tensor product $X \otimes Y$ is called a universal product of linear spaces. In the next section, we will show that there exists precisely one such universal product of linear spaces. This underlines the importance of the tensor product.

Proof. Step 1: Assume that the dimension of the linear spaces X and Y is finite.

(I) Uniqueness. Suppose that there exists a linear operator L with $B = L \circ \beta$. Then $L(b_i \otimes c_j) = B(b_i, c_j)$ for all possible indices. This yields the uniqueness of L.

(II) Existence. Let $b_1, \ldots, b_m$ and $c_1, \ldots, c_n$ be a basis of X and Y, respectively. Define

$$L\left(\sum_{i,j} \alpha^{ij} b_i \otimes c_j\right) := \sum_{i,j} \alpha^{ij} B(b_i, c_j).$$

Then

$$(L \circ \beta)(x, y) = L(x \otimes y) = L\left(\sum_{i,j} x^i y^j b_i \otimes c_j\right) = \sum_{i,j} x^i y^j B(b_i, c_j) = B(x, y).$$

Step 2: Assume that the linear spaces X and Y have arbitrary dimension. Replace the finite basis above by a general basis as defined in Problem 3.9. Use the following fact: If S_X (resp. S_Y) is a basis of X (resp Y), then the set

$$\{b \otimes c : \ b \in S_X, c \in S_Y\}$$

is a basis of $X \otimes Y$, by Prop. 2.4. Replace the argument from Step 1 by defining $L(b \otimes c) := B(b, c)$ for all $b \in S_X, c \in S_Y$. □

2.1.7 Diagram Chasing

As a paradigm, we want to show that the tensor product $X \otimes Y$ is uniquely determined by the universal property discussed in Prop. 2.6 above. The key is the following commutative diagram:

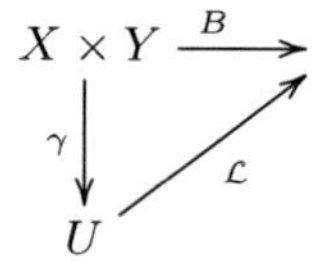

(2.13)

Let $\mathbb{K} = \mathbb{R}$ or $\mathbb{K} = \mathbb{C}$, and let X, Y, Z be arbitrary linear spaces over $\mathbb{K}$. Define

$$\beta(x, y) := x \otimes y.$$

The linear space U over $\mathbb{K}$ is called a universal object for the family of bilinear maps $B : X \times Y \to Z$ iff there exists a bilinear map $\gamma : X \times Y \to U$ such that, for every bilinear map $B : X \times Y \to Z$, there exists a unique linear operator $\mathcal{L}$ such that the diagram (2.13) is commutative.

Theorem 2.7 *There exists a unique universal object for the family of bilinear maps* $B : X \times Y \to Z$. *This object coincides with the tensor product* $X \otimes Y$.[4]

Proof. (I) Existence. By Prop. 2.6 above, the tensor product $X \otimes Y$ is a universal object.

(II) Uniqueness. Let U be a universal object. We want to show that U is linearly isomorphic to $X \otimes Y$. In fact, choosing $Z := U$ and $B := \gamma$, it follows from (2.12) that we have the commutative diagram:

$$\begin{array}{ccc} X \times Y & \xrightarrow{\;\gamma\;} & U \\ {\scriptstyle\beta}\big\downarrow & \nearrow{\scriptstyle L} & \\ X \otimes Y & & \end{array} \tag{2.14}$$

Similarly, choosing $Z = X \otimes Y$ and $B := \beta$, it follows from (2.13) that we have the commutative diagram:

$$\begin{array}{ccc} X \times Y & \xrightarrow{\;\beta\;} & X \otimes Y \\ {\scriptstyle\gamma}\big\downarrow & \nearrow{\scriptstyle \mathcal{L}} & \\ U & & \end{array} \tag{2.15}$$

Hence $\gamma = L \circ \beta$ and $\beta = \mathcal{L} \circ \gamma$. We want to show that the linear map $L : X \otimes Y \to U$ is a linear isomorphism. To this end, we will study the linear map $\mathcal{L} \circ L$. It follows from

$$(\mathcal{L} \circ L) \circ \beta = \mathcal{L} \circ (L \circ \beta) = \mathcal{L} \circ \gamma = \beta$$

that we have the commutative diagram:

$$\begin{array}{ccc} X \times Y & \xrightarrow{\;\beta\;} & X \otimes Y \\ {\scriptstyle\beta}\big\downarrow & \nearrow{\scriptstyle \mathcal{L} \circ L} & \\ X \otimes Y & & \end{array} \tag{2.16}$$

Let us add the trivial commutative diagram:

$$\begin{array}{ccc} X \times Y & \xrightarrow{\;\beta\;} & X \otimes Y \\ {\scriptstyle\beta}\big\downarrow & \nearrow{\scriptstyle \mathrm{id}} & \\ X \otimes Y & & \end{array} \tag{2.17}$$

Comparing (2.16) with (2.17), the universal property of $X \otimes Y$ yields $\mathcal{L} \circ L = \mathrm{id}$.

Analogously, we get $L \circ \mathcal{L} = \mathrm{id}$. Consequently, the map $L : X \otimes Y \to U$ is a linear isomorphism. □

The point is that mathematical objects can be defined in completely different ways. For example, in algebraic topology it is frequently a highly nontrivial task to prove that different objects are indeed isomorphic (e.g., homology groups, cohomology groups, or homotopy groups). Here, diagram chasing is very useful and effective.

---

[4] Naturally enough, uniqueness of the universal object means that the linear space U is uniquely determined up to linear isomorphism.

2.2 The Clifford Algebra $\bigvee(E_1)$ of the One-Dimensional Euclidean Space E_1

The idea of Clifford algebra is basic for Dirac's theory of the relativistic electron, and hence it is crucial for the fundamental fermions in the Standard Model in particle physics.

Folklore

We want to generalize the notion of complex numbers to real finite-dimensional linear spaces X equipped with a symmetric bilinear form B. The goal is to extend the linear space X to a real associative unital algebra, denoted $\bigvee_B(X)$, such that the product satisfies the relation

$$\boxed{\mathbf{a} \vee \mathbf{a} = B(\mathbf{a}, \mathbf{a}) \qquad \text{for all} \quad \mathbf{a} \in X.} \tag{2.18}$$

Here, the algebra $\bigvee_B(X)$ (with the Clifford product $\vee$) is called the Clifford algebra of the linear space X with respect to the bilinear form B. By Problem 3.7, we may assume that the field $\mathbb{R}$ is a subset of the Clifford algebra, and the real number 1 is the unit element of the Clifford algebra. Moreover,

$$\alpha \vee \mathbf{a} = \mathbf{a} \vee \alpha = \alpha\mathbf{a} \qquad \text{for all} \quad \mathbf{a} \in X,\ \alpha \in \mathbb{R}.$$

The Clifford relation (2.18) implies[5]

$$\mathbf{a} \vee \mathbf{b} + \mathbf{b} \vee \mathbf{a} = 2B(\mathbf{a}, \mathbf{b}) \qquad \text{for all} \quad \mathbf{a}, \mathbf{b} \in X. \tag{2.19}$$

Note that in the special case where $B \equiv 0$, we have $\mathbf{a} \vee \mathbf{b} = -\mathbf{b} \vee \mathbf{a}$. In this case, the Clifford product $\mathbf{a} \vee \mathbf{b}$ is identical with the alternating (or Grassmann) product $\mathbf{a} \wedge \mathbf{b}$, and the Clifford algebra becomes the alternating (or Grassmann) algebra $\bigwedge(X)$ of the linear space X.

Before studying the general case in Sect. 2.12, let us consider the special cases where $X = E_1, E_2, E_3$, and $B(\mathbf{a}, \mathbf{b}) := -\mathbf{ab}$ is the negative inner product. We will show that

$$\bigvee(E_1) = \mathbb{C}, \qquad \bigvee(E_2) = \mathbb{H}, \qquad \bigvee(E_3) = \mathbb{H} \times \mathbb{H},$$

up to isomorphism. From the physical point of view, the Clifford algebra $\bigvee(M_4)$ of the Minkowski space M_4 is crucial for Dirac's theory of the relativistic electron.

Convention. In what follows, isomorphic algebras will be identified with each other. If we claim that an algebra is uniquely determined by certain conditions, then this is to be understood in the sense of "up to isomorphism."

Let E_1 denote the one-dimensional Euclidean space E_1. Using the unit vector $\mathbf{i}$, we have $E_1 = \{x\mathbf{i} : \ x \in \mathbb{R}\}$ (Fig. 2.1).

Proposition 2.8 *There exists precisely one real associative unital algebra, denoted $\bigvee(E_1)$, which contains the one-dimensional Euclidean space E_1, and the algebra product $\vee$ satisfies the Clifford relation*

$$\mathbf{a} \vee \mathbf{a} = -\mathbf{a}^2 \qquad \textit{for all} \quad \mathbf{a} \in E_1.$$

The algebra $\bigvee(E_1)$ is called the Clifford algebra of the one-dimensional Euclidean space E_1. We have $\bigvee(E_1) = \mathbb{C}$. Thus, the dimension of $\bigvee(E_1)$ is equal to two.

[5] In fact, $(\mathbf{a} + \mathbf{b}) \vee (\mathbf{a} + \mathbf{b}) = B(\mathbf{a} + \mathbf{b}, \mathbf{a} + \mathbf{b})$ yields

$$\mathbf{a} \vee \mathbf{a} + \mathbf{a} \vee \mathbf{b} + \mathbf{b} \vee \mathbf{a} + \mathbf{b} \vee \mathbf{b} = B(\mathbf{a}, \mathbf{a}) + B(\mathbf{a}, \mathbf{b}) + B(\mathbf{b}, \mathbf{a}) + B(\mathbf{b}, \mathbf{b}).$$

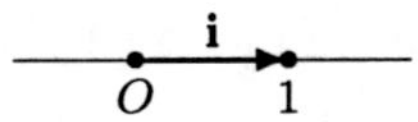

Fig. 2.1. One-dimensional Euclidean space E_1

Proof. (I) Uniqueness. Suppose that there exists such an algebra denoted $\mathcal{A}$. Then the elements of $\mathcal{A}$ are finite sums of the form

$$\alpha + \beta\mathbf{i} + \gamma\mathbf{i} \vee \mathbf{i} + \kappa\mathbf{i} \vee \mathbf{i} \vee \mathbf{i} + \ldots$$

with real coefficients $\alpha, \beta, \gamma, \kappa$. By the Clifford relation, $\mathbf{i} \vee \mathbf{i} = -1$. Moreover, $\mathbf{i} \vee \mathbf{i} \vee \mathbf{i} = -\mathbf{i}$, and so on. Therefore, the elements of $\mathcal{A}$ are of the form

$$\alpha + \beta\mathbf{i}, \qquad \alpha, \beta \in \mathbb{R}$$

with $\mathbf{i} \vee \mathbf{i} = -1$. This determines uniquely the algebra.

(II) Existence. The algebra $\mathbb{C}$ of complex numbers has the desired properties, and the map $\alpha + \beta\mathbf{i} \mapsto \alpha + \beta i$ is an algebra isomorphism from $\mathcal{A}$ onto $\mathbb{C}$. □

2.3 Algebras of the Two-Dimensional Euclidean Space E_2

Consider a Cartesian (x, y, z)-coordinate system of the three-dimensional Euclidean space E_3 with the orthonormal basis $\mathbf{i}, \mathbf{j}, \mathbf{k}$. Let E_2 denote the linear hull spanned by the vectors $\mathbf{i}$ and $\mathbf{j}$. The space E_2 is a real 2-dimensional Hilbert space with the orthonormal basis $\mathbf{i}, \mathbf{j}$. The elements of E_2 have the form $\mathbf{x} = x\mathbf{i} + y\mathbf{j}$ where x, y are real numbers (Fig. 2.2). We define

$$dx(\mathbf{x}) := x, \qquad dy(\mathbf{x}) := y \qquad \text{for all} \quad \mathbf{x} \in E_2.$$

Then, $dx : E_2 \to \mathbb{R}$ and $dy : E_2 \to \mathbb{R}$ are linear functionals on E_2. This means that $dx, dy \in E_2^d$. All the linear functionals $F : E_2 \to \mathbb{R}$, that is, all the elements of the dual space E_2^d are given by

$$F = \alpha dx + \beta dy, \qquad \alpha, \beta \in \mathbb{R}.$$

If we choose another Cartesian (x', y', z')-coordinate system of the 3-dimensional Euclidean space E_3, then we get a different space E_3'. But the two Hilbert spaces E_3 and E_3' are isomorphic. Therefore, the choice of the Cartesian coordinate system does not matter.

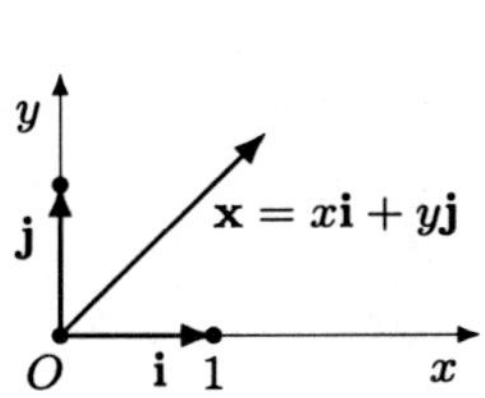

Fig. 2.2. Two-dimensional Euclidean space

2.3.1 The Clifford Algebra $\bigvee(E_2)$ and Quaternions

Proposition 2.9 *There exists precisely one real associative unital algebra of maximal dimension, denoted $\bigvee(E_2)$, which contains the two-dimensional Euclidean space E_2, and the algebra product $\vee$ satisfies the Clifford relation*

$$\mathbf{a} \vee \mathbf{a} = -\mathbf{a}^2 \qquad \textit{for all} \quad \mathbf{a} \in E_2. \tag{2.20}$$

The algebra $\bigvee(E_2)$ is called the Clifford algebra of the two-dimensional Euclidean space E_2. We have $\bigvee(E_2) = \mathbb{H}$. Thus, the dimension of $\bigvee(E_2)$ is equal to four.

Proof. (I) Uniqueness. Suppose that there exists such an algebra denoted $\mathcal{A}$. The elements of $\mathcal{A}$ are finite sums of $\vee$-products with an arbitrary number of factors, that is,

$$\alpha + \mathbf{a} + \mathbf{b} \vee \mathbf{c} + \ldots + \mathbf{e} \vee \mathbf{f} \vee \mathbf{g} + \ldots$$

Here, $\alpha \in \mathbb{R}$, and $\mathbf{a}, \mathbf{b} \ldots \in E_2$. Representing the vectors in E_2 by the basis vectors $\mathbf{i}, \mathbf{j}$, we get

$$\varrho + \sigma\mathbf{i} + \tau\mathbf{j} + \mu\mathbf{i} \vee \mathbf{j} + \nu\mathbf{j} \vee \mathbf{i}$$

where $\varrho, \ldots, \nu$ are real numbers. Note that products possessing more than two factors can be reduced to products with two factors or less than two factors, by the Clifford relations,

$$\mathbf{i} \vee \mathbf{i} = \mathbf{j} \vee \mathbf{j} = -1, \qquad \mathbf{j} \vee \mathbf{i} = -\mathbf{i} \vee \mathbf{j}.$$

For example, $\mathbf{i} \vee \mathbf{j} \vee \mathbf{i} = -\mathbf{i} \vee \mathbf{i} \vee \mathbf{j} = \mathbf{j}$. Finally, replacing $\mathbf{j} \vee \mathbf{i}$ by $-\mathbf{i} \vee \mathbf{j}$, the elements of $\mathcal{A}$ can be represented by

$$\alpha + \beta\mathbf{i} + \gamma\mathbf{j} + \lambda\mathbf{i} \vee \mathbf{j} \tag{2.21}$$

where $\alpha, \beta, \gamma, \lambda$ are real numbers. Thus, for the dimension of $\mathcal{A}$ we get $\dim(\mathcal{A}) \leq 4$.

If $\dim(\mathcal{A}) = 4$, then $1, \mathbf{i}, \mathbf{j}, \mathbf{i} \vee \mathbf{j}$ are linearly independent. Consequently, the coefficients $\alpha, \beta, \gamma, \lambda$ from (2.21) are uniquely determined. Thus, all the algebras $\mathcal{A}$ with $\dim(\mathcal{A}) = 4$ are isomorphic. It remains to show that such an algebra exists.

(II) Existence. Recall that the algebra $\mathbb{H}$ of quaternions consists of all the sums $\alpha + \mathbf{a}$ with $\alpha \in \mathbb{R}$ and $\mathbf{a} \in E_3$ (see Sect. 1.3.7). Let $\mathbf{a} \in E_2$. Then

$$\mathbf{a} \vee \mathbf{a} = -\mathbf{a}^2 + \mathbf{a} \times \mathbf{a} = -\mathbf{a}^2.$$

Therefore, the real 4-dimensional associative unital algebra $\mathbb{H}$ satisfies the Clifford relation. □

If we do not demand that the dimension of the algebra is maximal, then there exists another real associative unital algebra which satisfies the Clifford relation (2.20). This algebra consists of all the elements

$$\alpha + \beta\mathbf{i} + \gamma\mathbf{j}, \qquad \alpha, \beta, \gamma \in \mathbb{R}$$

equipped with the products $\mathbf{i} \vee \mathbf{i} := \mathbf{j} \vee \mathbf{j} = -1$ and $\mathbf{i} \vee \mathbf{j} = \mathbf{j} \vee \mathbf{i} := 0$. This yields a real 3-dimensional associative unital commutative Clifford algebra which contains the space E_2.

2.3.2 The Cauchy–Riemann Differential Equations in Complex Function Theory

Basic differential equations. The Cauchy–Riemann differential equations

$$u_x = v_y, \quad u_y = -v_x \tag{2.22}$$

are basic for the study of holomorphic functions. Let $\mathcal{U}$ be an open subset of the complex plane $\mathbb{C}$ (e.g., an open disc). The smooth function $f : \mathcal{U} \to \mathbb{C}$ with

$$f(z) = u(x,y) + v(x,y)\mathrm{i}$$

and $z = x + y\mathrm{i}$ is holomorphic iff it satisfies the Cauchy–Riemann differential equations (2.22) on $\mathcal{U}$. The smooth function $g : \mathcal{U} :\to \mathbb{C}$ with $g(z) = a(x,y) + \mathrm{i}b(x,y)$ is called antiholomorphic iff the function $z \mapsto g(z)^\dagger$ is holomorphic. This is equivalent to the anti-Cauchy–Riemann differential equations

$$a_x = -b_y, \qquad a_y = b_x. \tag{2.23}$$

It is well-known that holomorphic and antiholomorphic functions are harmonic. We want to show that this property is related to the square-root property of Clifford algebras.

Introducing the Poincaré differential operators

$$\frac{\partial}{\partial \overline{z}} := \frac{1}{2}\left(\frac{\partial}{\partial x} + \mathrm{i}\frac{\partial}{\partial y}\right), \qquad \frac{\partial}{\partial z} := \frac{1}{2}\left(\frac{\partial}{\partial x} - \mathrm{i}\frac{\partial}{\partial y}\right),$$

the Cauchy–Riemann differential equations (2.22) are equivalent to

$$\frac{\partial f}{\partial \overline{z}} = 0,$$

whereas the anti-Cauchy–Riemann differential equations (2.23) are equivalent to

$$\frac{\partial g}{\partial z} = 0.$$

Reformulation in terms of the Clifford algebra $\mathbb{H}$ of quaternions. Recall the definition of the Cayley matrices:

$$q_0 := \begin{pmatrix} 1 & 0 \\ 0 & 1 \end{pmatrix}, \quad q_1 := \begin{pmatrix} 0 & -\mathrm{i} \\ -\mathrm{i} & 0 \end{pmatrix}, \quad q_2 := \begin{pmatrix} 0 & -1 \\ 1 & 0 \end{pmatrix}, \quad q_3 := \begin{pmatrix} -\mathrm{i} & 0 \\ 0 & \mathrm{i} \end{pmatrix}.$$

The key idea is to introduce the first-order differential operator

$$P\psi := \left(q_1\frac{\partial}{\partial x} + q_2\frac{\partial}{\partial y}\right)\psi, \qquad \psi := \begin{pmatrix} f \\ g \end{pmatrix}.$$

Explicitly,

$$P\psi = -2\mathrm{i}\begin{pmatrix} \frac{\partial g}{\partial z} \\ \frac{\partial f}{\partial \overline{z}} \end{pmatrix}.$$

Moreover, recall the definition of the Laplacian $\Delta\psi := -\frac{\partial^2}{\partial x^2} - \frac{\partial^2}{\partial y^2}$.

Proposition 2.10 $P^2\psi = \Delta\psi$.

This tells us that the operator P is the square root of the Laplacian Δ.
Proof. We will use the Clifford relations

$$(q_1)^2 = (q_2)^2 = -q_0, \qquad q_1q_2 = -q_2q_1. \tag{2.24}$$

From $P^2 = \left(q_1\frac{\partial}{\partial x} + q_2\frac{\partial}{\partial y}\right)^2$, we get

$$P^2 = (q_1)^2\frac{\partial^2}{\partial x^2} + (q_1q_2 + q_2q_1)\frac{\partial^2}{\partial x\partial y} + (q_2)^2\frac{\partial^2}{\partial x^2} = \Delta.$$

□

Summarizing, we obtain the following.

Theorem 2.11 *The differential equation*

$$P\psi = 0 \qquad \text{on } \mathcal{U} \tag{2.25}$$

is equivalent to the Cauchy–Riemann equation $\frac{\partial f}{\partial \bar{z}} = 0$ *on* $\mathcal{U}$ *and the anti-Cauchy–Riemann equation* $\frac{\partial g}{\partial z} = 0$ *on* $\mathcal{U}$. *If* ψ *is a solution of* (2.25), *then* ψ *is harmonic, that is,* $\Delta\psi = 0$ *on* $\mathcal{U}$. *Hence* $\Delta f = \Delta g = 0$ *on* $\mathcal{U}$.

Note that the Clifford relations (2.24) generate a Clifford algebra which consists of all the matrices

$$\alpha q_0 + \beta q_1 + \gamma q_2 + \kappa q_3, \qquad \alpha, \beta, \gamma, \kappa \in \mathbb{R}.$$

This Clifford algebra is isomorphic to the algebra $\mathbb{H}$ of quaternions.

Perspectives. The investigation of the Laplacian is crucial for modern mathematics and physics. There are two important approaches for reducing the second-order differential operator Δ to first-order systems:

(i) Square root P of the Laplacian and Clifford algebras: $P^2 = \Delta$. Then the equation $P\psi = 0$ implies $\Delta\psi = 0$ (spin geometry).
(ii) Hodge duality for the alternating (Grassmann) algebra: $\Delta = d\delta + \delta d$. The system $d\omega = 0, \delta\omega = 0$ implies $\Delta\omega = 0$.

If the Riemannian metric is replaced by a pseudo-Riemannian metric (e.g., the Minkowski metric), then the Laplacian Δ is replaced by the wave operator $\Box$.

The approach (i) concerns the theory of complex analytic functions, the Dirac equation of the relativistic electron, fermions in the Standard Model in particle physics, the Seiberg–Witten equations (generalized Landau–Ginzburg equations), Kähler geometry, and string theory.

The approach (ii) concerns the main theorem in vector analysis, (computation of vector fields by given divergence (sources) and curl), the Maxwell equations, the Yang–Mills equations, and the Standard Model in elementary particle physics (gauge theory).

> *It turns out that all of these equations are powerful tools in order to investigate the structure of manifolds and to describe fundamental processes in physics.*

In fact, there exist interesting relations between (i) and (ii). For example, in the Standard Model in particle physics, the fermions (i.e., the basic particles – quarks and leptons) are described by (i), whereas the bosons (i.e., the interaction particles – photon, vector bosons, and gluons) are described by (ii). In the 1950s, Kähler developed a differential calculus which combines (i) with (ii) (the exterior–interior calculus).

We will study this later on. As a general reference, we recommend J. Jost, Riemannian Geometry and Geometric Analysis, 5th edn., Springer, New York, 2008, and E. Kähler, Mathematical Works, de Gruyter, Berlin, 2004.

2.3.3 The Grassmann Algebra $\bigwedge(E_2)$

Proposition 2.12 *The elements of the alternating algebra $\bigwedge(E_2)$ of the two-dimensional Euclidean space E_2 have the form*

$$\alpha + \beta\mathbf{i} + \gamma\mathbf{j} + \varrho\mathbf{i} \wedge \mathbf{j}$$

with uniquely determined real coefficients $\alpha, \beta, \gamma, \varrho$.

This means that $1, \mathbf{i}, \mathbf{j}, \mathbf{i} \wedge \mathbf{j}$ form a basis of $\bigwedge(E_2)$, that is, $\dim \bigwedge(E_2) = 4$.
Proof. Let $\mathbf{a}, \mathbf{b}, \mathbf{c} \in E_2$. By Sect. 2.1.2, $\mathbf{a} \wedge \mathbf{b}$ is an antisymmetric bilinear form given by

$$(\mathbf{a} \wedge \mathbf{b})(F, G) = F(\mathbf{a})G(\mathbf{b}) - F(\mathbf{b})G(\mathbf{a}) \qquad \text{for all} \quad F, G \in E_2^d.$$

Hence $\mathbf{a} \wedge \mathbf{b} = -\mathbf{b} \wedge \mathbf{a}$, and $\mathbf{a} \wedge \mathbf{a} = 0$. Moreover, by definition of dx and dy, it follows from $dx(\mathbf{i}) = dy(\mathbf{j}) = 1$ and $dx(\mathbf{j}) = dy(\mathbf{i}) = 0$ that

$$(\mathbf{i} \wedge \mathbf{j})(dx, dy) = dx(\mathbf{i})dy(\mathbf{j}) - dx(\mathbf{j})dy(\mathbf{i}) = 1.$$

Hence $\mathbf{i} \wedge \mathbf{j} \neq 0$. Furthermore,

$$\mathbf{a} \wedge \mathbf{b} \wedge \mathbf{c} = 0.$$

In fact, using the basis $\mathbf{i}, \mathbf{j}$, the product $\mathbf{a} \wedge \mathbf{b} \wedge \mathbf{c}$ can be reduced to the sum of products of $\mathbf{i}, \mathbf{j}$ with three factors. Such a product has at least two equal factors, for example, $\mathbf{i} \wedge \mathbf{j} \wedge \mathbf{i}$. Finally, note that alternating products with two equal factors always vanish.

Every element of $\bigwedge(E_2)$ is a finite sum of $\wedge$-products of vectors. Using the basis $\mathbf{i}, \mathbf{j}$ and the distributive law, each element of $\bigwedge(E_2)$ can be represented by

$$\alpha + \beta\mathbf{i} + \gamma\mathbf{j} + \varrho\mathbf{i} \wedge \mathbf{j}, \quad \alpha, \beta, \gamma, \varrho \in \mathbb{R}.$$

In order to show that the coefficients $\alpha, \beta, \gamma, \varrho$ are uniquely determined, we have to show that $1, \mathbf{i}, \mathbf{j}, \mathbf{i} \wedge \mathbf{j}$ are linearly independent. To prove this, assume that

$$\alpha + \beta\mathbf{i} + \gamma\mathbf{j} + \varrho\mathbf{i} \wedge \mathbf{j} = 0.$$

Since $\bigwedge(E_2)$ is a direct sum, we get $\alpha = 0$, $\beta\mathbf{i} + \gamma\mathbf{j} = 0$, and $\varrho\mathbf{i} \wedge \mathbf{j} = 0$. Hence $\beta = \gamma = 0$, and $\varrho = 0$. □

2.3.4 The Grassmann Algebra $\bigwedge(E_2^d)$

Let $F, G \in E_2^d$. By Sect. 2.1.2,

$$(F \wedge G)(\mathbf{a}, \mathbf{b}) = F(\mathbf{a})G(\mathbf{b}) - F(\mathbf{b})G(\mathbf{a}) \qquad \text{for all} \quad \mathbf{a}, \mathbf{b} \in E_2.$$

Proposition 2.13 *The elements of the alternating algebra $\bigwedge(E_2)$ of the dual space E_2^d have the form*

$$\alpha + \beta dx + \gamma dy + \varrho dx \wedge dy$$

with uniquely determined real coefficients $\alpha, \beta, \gamma, \varrho$.

This means that $1, dx, dy, dx \wedge dy$ form a basis of $\bigwedge(E_2)$, that is, $\dim \bigwedge(E_2) = 4$. The proof proceeds similarly to the proof of Prop. 2.12.

2.3.5 The Symplectic Structure of E_2

Let X be a real finite-dimensional linear space of even dimension. The bilinear form $B : X \times X \to \mathbb{R}$ is called nondegenerate iff

- it follows from $B(x_0, x) = 0$ for all $x \in X$ that $x_0 = 0$, and
- it follows from $B(x, x_0) = 0$ for all $x \in X$ that $x_0 = 0$.

The bilinear form B is called symplectic iff it is antisymmetric and nondegenerate.

The volume form $dx \wedge dy$ of E_2. If $\mathbf{a}, \mathbf{b} \in E_2$, then

$$\mathbf{a} \times \mathbf{b} = (a^1\mathbf{i} + a^2\mathbf{j}) \times (b^1\mathbf{i} + b^2\mathbf{j}) = (a^1b^2 - a^2b^1)\mathbf{k}.$$

Here, $a^2b^1 - b^2a^1$ is the (oriented) area of the parallelogram spanned by the vectors $\mathbf{a}$ and $\mathbf{b}$ (Fig. 1.5 on page 82). Furthermore,

$$(dx \wedge dy)(\mathbf{a}, \mathbf{b}) = dx(\mathbf{a})dy(\mathbf{b}) - dx(\mathbf{b})dy(\mathbf{a}) = a^1b^2 - b^1a^2.$$

Therefore, $dx \wedge dy$ is called the volume form of the linear space E_2. This is a symplectic form. The linear space E_2 equipped with the symplectic form $dx \wedge dy$ is called a symplectic space.

Symplectic transformations. The linear operator $A : E_2 \to E_2$ is called symplectic iff $(dx \wedge dy)(A\mathbf{a}, A\mathbf{b}) = (dx \wedge dy)(\mathbf{a}, \mathbf{b})$ for all vectors $\mathbf{a}, \mathbf{b} \in E_2$, that is, the volume form is preserved.

Proposition 2.14 *The linear operator $A : E_2 \to E_2$ is symplectic iff* $\det(A) = 1$.

Proof. Using $A\mathbf{a} = (A_1^1a^1 + A_2^1a^2)\mathbf{i} + (A_1^2a^1 + A_2^2a^2)\mathbf{j}$ and $\det(A) := \begin{vmatrix} A_1^1 & A_2^1 \\ A_1^2 & A_2^2 \end{vmatrix}$, we get

$$(dx \wedge dy)(A\mathbf{a}, A\mathbf{b}) = (A_1^1A_2^2 - A_2^1A_1^2)(a^1b^2 - b^1a^2) = \det(A) \cdot (dx \wedge dy)(\mathbf{a}, \mathbf{b}).$$

We will show in Sect. 2.11.3 on page 167 that the determinant $\det(A)$ is independent of the choice of the basis of the space E_2. □

2.3.6 The Tensor Algebra $\bigotimes(E_2)$

The elements of the tensor algebra $\bigotimes(E_2)$ of the two-dimensional Euclidean space E_2 are finite sums of the form

$$\alpha + \beta \mathbf{i} + \gamma \mathbf{j} + \lambda \mathbf{i} \otimes \mathbf{i} + \mu \mathbf{i} \otimes \mathbf{j} + \nu \mathbf{j} \otimes \mathbf{i} + \varrho \mathbf{j} \otimes \mathbf{j} + \sigma \mathbf{i} \otimes \mathbf{i} \otimes \mathbf{i} + \ldots$$

with uniquely determined real coefficients $\alpha, \beta, \ldots$ This means that the linearly independent elements

$$1, \ \mathbf{i}, \mathbf{j}, \quad \mathbf{i} \otimes \mathbf{i}, \ \mathbf{i} \otimes \mathbf{j}, \ \mathbf{j} \otimes \mathbf{i}, \ \mathbf{j} \otimes \mathbf{j}, \quad \mathbf{i} \otimes \mathbf{i} \otimes \mathbf{i}, \ldots$$

form a basis of the infinite-dimensional tensor algebra $\bigotimes(E_2)$.

2.3.7 The Tensor Algebra $\bigotimes(E_2^d)$

The elements of the tensor algebra $\bigotimes(E_2^d)$ of the dual space E_2^d are finite sums of the form

$$\alpha + \beta dx + \gamma dy + \lambda dx \otimes dx + \mu dx \otimes dy + \nu dy \otimes dx + \varrho dy \otimes dy + \sigma dx \otimes dx \otimes dx + \ldots$$

with uniquely determined real coefficients $\alpha, \beta, \ldots$ Therefore, $\dim \bigotimes(E_2^d) = \infty$.

2.4 Algebras of the Three-Dimensional Euclidean Space E_3

2.4.1 Lie Algebra

The space E_3 is a real 3-dimensional Lie algebra equipped with the vector product $\mathbf{a} \times \mathbf{b}$ (see Sect. 1.2.6 on page 82).

2.4.2 Tensor Algebra

The elements of the tensor algebra $\bigotimes(E_3)$ of the 3-dimensional Euclidean space E_3 are finite sums of the form

$$\alpha + \beta \mathbf{i} + \gamma \mathbf{j} + \varrho \mathbf{k} + \lambda \mathbf{i} \otimes \mathbf{i} + \mu \mathbf{i} \otimes \mathbf{j} + \ldots + \sigma \mathbf{i} \otimes \mathbf{i} \otimes \mathbf{i} + \ldots$$

with uniquely determined real coefficients $\alpha, \beta, \ldots$ This means that the linearly independent elements

- $1, \ \mathbf{i}, \mathbf{j}, \mathbf{k}$,
- $\mathbf{i} \otimes \mathbf{i}, \ \mathbf{j} \otimes \mathbf{j}, \ \mathbf{k} \otimes \mathbf{k}, \ \mathbf{i} \otimes \mathbf{j}, \ \mathbf{j} \otimes \mathbf{i}, \ \mathbf{i} \otimes \mathbf{k}, \ \mathbf{k} \otimes \mathbf{i}, \ \mathbf{j} \otimes \mathbf{k}, \ \mathbf{k} \otimes \mathbf{j}$,
- $\mathbf{i} \otimes \mathbf{i} \otimes \mathbf{i}, \ldots$

form a basis of the infinite-dimensional tensor algebra $\bigotimes(E_3)$. Recall that, for given vectors $\mathbf{a}, \mathbf{b} \in E_3$, we have

$$(\mathbf{a} \otimes \mathbf{b})(F, G) = F(\mathbf{a})G(\mathbf{b}) \qquad \text{for all} \quad F, G \in E_3^d.$$

This means that $\mathbf{a} \otimes \mathbf{b}$ is a real bilinear functional on the product space $E_3^d \times E_3^d$. If we define

$$(\mathbf{a} \otimes \mathbf{b})(\mathbf{x}, \mathbf{y}) := (\mathbf{a}\mathbf{x})(\mathbf{b}\mathbf{y}) \qquad \text{for all} \quad \mathbf{x}, \mathbf{y} \in E_3,$$

then $\mathbf{a} \otimes \mathbf{b}$ can be identified with a real bilinear functional on the product space $E_3 \times E_3$.

2.4.3 Grassmann Algebra

Proposition 2.15 *The elements of the alternating algebra $\bigwedge(E_3)$ of the three-dimensional Euclidean space E_3 have the form*

$$\alpha + \beta\mathbf{i} + \gamma\mathbf{j} + \kappa\mathbf{k} + \lambda\mathbf{i}\wedge\mathbf{j} + \mu\mathbf{j}\wedge\mathbf{k} + \nu\mathbf{k}\wedge\mathbf{i} + \varrho\mathbf{i}\wedge\mathbf{j}\wedge\mathbf{k}$$

with uniquely determined real coefficients $\alpha, \beta, \ldots, \varrho$.

This means that $1, \mathbf{i}, \mathbf{j}, \mathbf{k}, \mathbf{i}\wedge\mathbf{j}, \mathbf{j}\wedge\mathbf{k}, \mathbf{k}\wedge\mathbf{i}, \mathbf{i}\wedge\mathbf{j}\wedge\mathbf{k}$ form a basis of $\bigwedge(E_3)$, that is, $\dim\bigwedge(E_3) = 8$. The computation of the $\wedge$-product is governed by the relation

$$\mathbf{a}\wedge\mathbf{b} = -\mathbf{b}\wedge\mathbf{a} \qquad \text{for all} \quad \mathbf{a}, \mathbf{b} \in E_3$$

and by the associativity law. In particular, $\mathbf{a}\wedge\mathbf{a} = 0$. For example,

$$(\mathbf{i}\wedge\mathbf{k})\wedge\mathbf{j} = \mathbf{i}\wedge(\mathbf{k}\wedge\mathbf{j}) = -\mathbf{i}\wedge(\mathbf{j}\wedge\mathbf{k}) = -\mathbf{i}\wedge\mathbf{j}\wedge\mathbf{k},$$

and $\mathbf{i}\wedge\mathbf{j}\wedge\mathbf{i} = \mathbf{i}\wedge(\mathbf{j}\wedge\mathbf{i}) = -\mathbf{i}\wedge(\mathbf{i}\wedge\mathbf{j}) = -(\mathbf{i}\wedge\mathbf{i})\wedge\mathbf{j} = 0$. The proof of Prop. 2.15 proceeds as the proof of Prop. 2.12 on page 131. In terms of the tensor product, we have

$$\mathbf{a}\wedge\mathbf{b} = \mathbf{a}\otimes\mathbf{b} - \mathbf{b}\otimes\mathbf{a}.$$

Explicitly, this means that

$$(\mathbf{a}\wedge\mathbf{b})(F, G) = F(\mathbf{a})G(\mathbf{b}) - F\mathbf{b})G(\mathbf{a}) \qquad \text{for all} \quad F, G \in E_3^d.$$

Thus, $\mathbf{a}\wedge\mathbf{b}$ is a real bilinear antisymmetric functional on the product space $E_3^d \times E_3^d$. If we define

$$(\mathbf{a}\wedge\mathbf{b})(\mathbf{x}, \mathbf{y}) := (\mathbf{ax})(\mathbf{by}) - (\mathbf{ay})(\mathbf{bx}) \qquad \text{for all} \quad \mathbf{x}, \mathbf{y} \in E_3,$$

then $\mathbf{a}\wedge\mathbf{b}$ can be identified with a real bilinear antisymmetric functional on the product space $E_3 \times E_3$.

2.4.4 Clifford Algebra

Proposition 2.16 *There exists precisely one real associative unital algebra of maximal dimension, denoted $\bigvee(E_3)$, which contains the three-dimensional Euclidean space E_3, and the algebra product $\vee$ satisfies the Clifford relation*

$$\mathbf{a}\vee\mathbf{a} = -\mathbf{a}^2 \qquad \textit{for all} \quad \mathbf{a} \in E_3.$$

The algebra $\bigvee(E_3)$ is called the Clifford algebra of the three-dimensional Euclidean space E_3. We have $\bigvee(E_3) = \mathbb{H}\times\mathbb{H}$. Thus, the dimension of $\bigvee(E_3)$ is equal to eight.

Proof. (I) Uniqueness. Suppose that there exists such an algebra denoted $\mathcal{A}$. Then we have the Clifford relation

$$\mathbf{a}\vee\mathbf{b} + \mathbf{b}\vee\mathbf{a} = -2(\mathbf{ab}) \qquad \text{for all} \quad \mathbf{a}, \mathbf{b} \in E_3.$$

It follows as in the proof of Prop. 2.9 that the elements of $\mathcal{A}$ are sums of the form

$$\alpha + \beta\mathbf{i} + \gamma\mathbf{j} + \kappa\mathbf{k} + \lambda\mathbf{i}\vee\mathbf{j} + \mu\mathbf{j}\vee\mathbf{k} + \nu\mathbf{k}\vee\mathbf{i} + \varrho\mathbf{i}\vee\mathbf{j}\vee\mathbf{k} \tag{2.26}$$

with real coefficients $\alpha, \beta, \ldots, \varrho$. Thus, for the dimension we get $\dim(\mathcal{A}) \leq 8$. If the elements $1, \mathbf{i}, \mathbf{j}, \mathbf{k}, \mathbf{i}\vee\mathbf{j}, \mathbf{j}\vee\mathbf{k}, \mathbf{k}\vee\mathbf{i}, \mathbf{i}\vee\mathbf{j}\vee\mathbf{k}$ are linearly independent, then $\dim(\mathcal{A}) = 8$, and the coefficients $\alpha, \beta, \ldots, \varrho$ are uniquely determined. Now we show that $\mathbb{H} \times \mathbb{H}$ is such a model.

(II) Existence. We will use the space of quaternions $\mathbb{H}$ as introduced in Sect. 1.3.7. The linear space $\mathbb{H} \times \mathbb{H}$ consists of all the ordered pairs (q, p) with the quaternions $q, p \in \mathbb{H}$. By definition,

$$\alpha(q,p) + \beta(r,s) = (\alpha q + \beta r, \alpha p + \beta s) \qquad \text{for all} \quad q,p,r,s \in \mathbb{H},\ \alpha, \beta \in \mathbb{R}.$$

The real linear space $\mathbb{H} \times \mathbb{H}$ has the basis

$$(1,0),\ (\mathbf{i},0),\ (\mathbf{j},0),\ (\mathbf{k},0),\ (0,1),\ (0,\mathbf{i}),\ (0,\mathbf{j}),\ (0,\mathbf{k}).$$

Thus, $\dim(\mathbb{H} \times \mathbb{H}) = 8$. Furthermore, we define the product

$$(q,p) \vee (r,s) := (q \vee r, p \vee s) \qquad \text{for all} \quad q,p,r,s \in \mathbb{H}$$

based on the $\vee$-product for quaternions. This way, the direct product $\mathbb{H} \times \mathbb{H}$ becomes a real 8-dimensional associative algebra with the unit element $(1,1)$. For $\mathbf{a} \in E_3$, we get

$$(\mathbf{a}, -\mathbf{a}) \vee (\mathbf{a}, -\mathbf{a}) = (\mathbf{a} \vee \mathbf{a}, \mathbf{a} \vee \mathbf{a}) = (-\mathbf{a}^2, -\mathbf{a}^2) = -\mathbf{a}^2(1,1).$$

Thus, identifying $\mathbf{a}$ with $(\mathbf{a}, -\mathbf{a})$, the algebra $\mathbb{H} \times \mathbb{H}$ contains the space E_3, and the Clifford relation is satisfied. □

2.5 Algebras of the Dual Euclidean Space E_3^d

2.5.1 Tensor Algebra

The elements of the tensor algebra $\bigotimes(E_3^d)$ of the dual space E_2^d are finite sums of the form

$$\alpha + \beta dx + \gamma dy + \lambda dx \otimes dx + \mu dx \otimes dy + \nu dy \otimes dx + \ldots + \sigma dx \otimes dx \otimes dx + \ldots$$

with uniquely determined real coefficients $\alpha, \beta, \ldots$ The $\otimes$-product is governed by the distributive law and the associative law.

2.5.2 Grassmann Algebra

The elements of the Grassmann (or alternating) algebra $\bigwedge(E_3^d)$ of the dual space E_3^d have the form

$$\alpha + \beta dx + \gamma dy + \kappa dz + \lambda dx \wedge dy + \mu dy \wedge dz + \nu dz \wedge dx + \varrho dx \wedge dy \wedge dz$$

with uniquely determined real coefficients $\alpha, \beta, \ldots, \varrho$. This means that

$$1, dx, dy, dz, \quad dx \wedge dy,\ dy \wedge dz,\ dz \wedge dx, \quad dx \wedge dy \wedge dz$$

form a basis of $\bigwedge(E_3)$, that is, $\dim \bigwedge(E_3) = 8$. The $\wedge$-product is governed by the distributive law, the associative law, and by the relations

$$dx \wedge dy = -dy \wedge dx, \quad dy \wedge dz = -dz \wedge dy, \quad dz \wedge dx = -dx \wedge dz$$

together with $dx \wedge dx = dy \wedge dy = dz \wedge dz = 0$. The Grassmann algebra $\bigwedge(E_3^d)$ is graded, that is, we have the direct sum decomposition

$$\bigwedge(E_d^3) = \bigoplus_{p=0}^{3} \bigwedge^p(E_3^d)$$

where $\bigwedge^p(E_3^d)$ is the real linear space of p-differential forms. Explicitly, the elements of $\bigwedge^p(E_3^d)$, $p = 0, 1, 2, 3$, have the form

- α,
- $\beta dx + \gamma dy + \kappa dz$,
- $\lambda dx \wedge dy + \mu dy \wedge dz + \nu dz \wedge dx$,
- $\varrho dx \wedge dy \wedge dz$,

respectively. Here, the coefficients $\alpha, \beta, \ldots, \varrho$ are arbitrary real numbers.

2.6 The Mixed Tensor Algebra

It was discovered in the nineteenth century that one has to distinguish between contravariant and covariant tensors. Let us discuss this.

The tensor product $E_3 \otimes E_3^d$. Let $\mathbf{b}_1, \mathbf{b}_2, \mathbf{b}_3$ be a basis of the Euclidean space E_3. In what follows, we will sum over equal upper and lower indices from 1 to 3. The cobasis, that is, the basis dx^1, dx^2, dx^3 of the dual space E_3^d is defined by

$$dx^i(v^j \mathbf{b}_j) := v^i, \qquad i = 1, 2, 3.$$

The elements of the tensor product $E_3 \otimes E_3^d$ have the form

$$A^i_j \mathbf{b}_i \otimes dx^j \tag{2.27}$$

where the real coefficients $A^1_1, A^1_2, A^1_3, \ldots$ are uniquely determined. The products $\mathbf{b}_i \otimes dx^j$, $i, j = 1, 2, 3$, form a basis of the real 9-dimensional linear space $E_3 \otimes E_3^d$. Recall that

$$A^i_j \mathbf{b}_i \otimes dx^j : E_3^d \times E_3 \to \mathbb{R}$$

is a bilinear functional. Explicitly, for all linear functionals $F \in E_3^d$ and all vectors $\mathbf{a} \in E_3$, we have

$$(A^i_j \mathbf{b}_i \otimes dx^j)(F, \mathbf{a}) = A^i_j \cdot F(\mathbf{b}_i) dx^j(\mathbf{a}).$$

In the sense of the general definition given below, the tensor (2.27) is called of type (1,1) (i.e., 1-fold contravariant and 1-fold covariant).

The space $\mathrm{End}(E_3)$. The symbol $\mathrm{End}(E_3)$ denotes the space of linear operators $A : E_3 \to E_3$. Since $\mathbf{b}_1, \mathbf{b}_2, \mathbf{b}_3$ is a basis of E_3, there exist uniquely determined real numbers A^i_j such that

$$A\mathbf{b}_j = A^i_j \mathbf{b}_i, \qquad j = 1, 2, 3.$$

Hence $A\mathbf{v} = A(v^j \mathbf{b}_j) = (A^i_j v^j)\mathbf{b}_i$ for all $\mathbf{v} \in E_3$. We define

$$(A^i_j \mathbf{b}_i \otimes dx^j)(\mathbf{v}) := A^i_j dx^j(\mathbf{v}) \mathbf{b}_j = A^i_j v^j \mathbf{b}_i = A\mathbf{v} \qquad \text{for all} \quad \mathbf{v} \in E_3.$$

Consequently, the map $A \mapsto A^i_j \mathbf{b}_i \otimes dx^j$ is a linear isomorphism from the real linear space $\mathrm{End}(E_3)$ onto the tensor product $E_3 \otimes E_3^d$. We briefly write

$$\boxed{A = A^i_j \mathbf{b}_i \otimes dx^j.}$$

In particular, the identical operator $I : E_3 \to E_3$ is given by

$$I = \mathbf{b}_j \otimes dx^j. \tag{2.28}$$

This is the key relation of the Dirac calculus (see Sect. 2.11.7 on page 171).

Multilinear functionals of mixed type. Fix $\mathbb{K} = \mathbb{R}$ or $\mathbb{K} = \mathbb{C}$. Choose linear spaces X and Y over $\mathbb{K}$. For $m, n = 1, 2, \ldots$, let $M_{m,n}(X, Y)$ denote the set of all $(m+n)$-linear functionals

$$F : (X \times \cdots \times X) \times (Y \times \cdots \times Y) \to \mathbb{K}$$

with m factors X and n factors Y. Explicitly, this is a map of the form

$$(x_1, \ldots, x_m; y_1, \ldots, y_n) \mapsto F(x_1, \ldots, x_m; y_1, \ldots, y_n).$$

Moreover, we set $M_{0,0}(X) := \mathbb{K}$ together with $M_{1,0} := X$ and $M_{0,1} := Y$. Naturally enough, $M_{m,0}(X, Y) := M_m(X)$ and $M_{0,n}(X, Y) := M_n(Y)$. We also introduce the direct sum

$$M(X, Y) := \bigoplus_{m,n=0}^{\infty} M_{m,n}(X, Y).$$

That is, the elements of $M(X, Y)$ are tuples of the form

$$(F_{0,0}, F_{1,0}, F_{0,1}, F_{1,1}, F_{2,0}, \ldots) \tag{2.29}$$

where $F_{k,l} \in M_{k,l}(X, Y)$ for all indices $k, l = 0, 1, 2, \ldots$, and at most a finite number of the entries $F_{0,0}, F_{1,0}, \ldots$ is different from zero. The set $M(X, Y)$ becomes a linear space over $\mathbb{K}$ in a natural way. To simplify notation, we replace (2.29) by

$$F_{0,0} + F_{1,0} + F_{0,1} + F_{1,1} + F_{2,0} + \ldots$$

The product of multilinear functionals of mixed type. In addition, it is possible to introduce a product in a quite natural way. To this end, suppose that $F \in M_{m,n}(X, Y)$ and $G \in M_{k,l}(X, Y)$. We introduce the product $F \otimes G$ by defining

$$(F \otimes G)(x_1, \ldots, x_m, x_{m+1}, \ldots, x_{m+k}; y_1, \ldots, y_n, y_{n+1}, \ldots, y_{n+l})$$

by the product

$$F(x_1, \ldots, x_m; y_1, \ldots, y_n) G(x_{m+1}, \ldots, x_{m+k}; y_{n+1}, \ldots, y_{n+l})$$

for all $x_1, \ldots, x_{m+k} \in X$ and all $y_1, \ldots, y_{n+l} \in Y$. Thus, $F \otimes G \in M_{m+k,n+l}(X, Y)$. Equipped with the product $F \otimes G$, the linear space $M(X, Y)$ becomes an associative algebra over $\mathbb{K}$. The number 1 is the unit element of the algebra.

Tensors of type (m, n). Let X be a finite-dimensional linear space over $\mathbb{K}$. Set

$$\bigotimes\nolimits_n^m (X) := M_{m,n}(X^d, X).$$

Let $b_1, \ldots, b_r$ be a basis of the linear space X, and let $dx^1, \ldots dx^r$ be the corresponding cobasis.[6] In what follows, we will sum over equal upper and lower indices from 1 to r. Then the elements of $\bigotimes_n^m(X)$ have the form

$$\boxed{t^{i_1 \ldots i_m}_{j_1 \cdots j_n} \cdot b_{i_1} \otimes \cdots \otimes b_{i_m} \otimes dx^{j_1} \otimes \cdots \otimes dx^{j_n}.} \tag{2.30}$$

All the uniquely determined numbers

[6] Recall that $dx^i(b_j) = \delta^i_j$ if $i, j = 1, \ldots, r$.

$$t^{i_1 \dots i_m}_{j_1 \dots j_n} \in \mathbb{K}$$

are called the tensor components. Recall that

$$(b_{i_1} \otimes \cdots \otimes b_{i_m} \otimes dx^{j_1} \otimes \cdots \otimes dx^{j_n})(F_1, \dots, F_m; a_1, \dots, a_n)$$

is equal to the product

$$F_1(b_{i_1}) \cdots F_m(b_{i_m}) \cdot dx^{j_1}(a_1) \cdots dx^{j_n}(a_n)$$

for all $F_1, \dots, F_m \in X^d$ and all $a_1, \dots, a_n \in X$. The tensor (2.30) is called m-fold contravariant and n-fold covariant.

In terms of physics, the tensor components $t^{i_1 \dots i_m}_{j_1 \dots j_n}$ are numbers measured in physical experiments (e.g., the components of the electromagnetic field). The change of tensor components corresponds to the change of the observer.

Transformation of the tensor components. Consider the change

$$b_{i'} = A^i_{i'} b_i, \qquad i' = 1', \dots r'$$

of the basis vectors. The inverse transformation is given by

$$b_i = A^{i'}_i b_{i'}, \qquad i = 1, \dots, r$$

with $A^{i'}_i A^j_{i'} = \delta^j_i$ and $A^i_{i'} A^{j'}_i = \delta^{j'}_{i'}$. For the cobasis, this implies the transformation law

$$dx^{i'} = A^{i'}_i dx^i, \qquad i' = 1', \dots, r'.$$

In fact, $dx^{i'}(b_{j'}) = A^{i'}_i dx^i(A^j_{j'} b_j) = A^{i'}_i A^j_{j'} \delta^i_j = A^{i'}_i A^i_{j'} = \delta^{i'}_{j'}$. The tensor (2.30) is transformed into the expression

$$t^{i'_1 \dots i'_m}_{j'_1 \dots j'_n} b_{i'_1} \otimes \cdots \otimes b_{i'_m} \otimes dx^{j'_1} \otimes \cdots \otimes dx^{j'_n}$$

with the new tensor components

$$t^{i'_1 \dots i'_m}_{j'_1 \dots j'_n} = A^{i'_1}_{i_1} \cdots A^{i'_m}_{i_m} A^{j_1}_{j'_1} \cdots A^{j_n}_{j'_n} \cdot t^{i_1 \dots i_m}_{j_1 \dots j_n}.$$

The tensor algebra $\bigotimes(X, X^d)$. We define

$$\bigotimes(X, X^d) := \bigoplus_{m,n=0}^{\infty} \bigotimes\nolimits^m_n (X).$$

Explicitly, the elements of the associative unital algebra $\bigotimes(X, X^d)$ are finite sums of the form (2.30).

2.7 The Hilbert Space Structure of the Grassmann Algebra (Hodge Duality)

It is crucial that the Grassmann algebra can be equipped with two additional structures, namely,

- the Hilbert space structure (Hodge theory), and
- the Clifford algebra structure (Kähler theory).

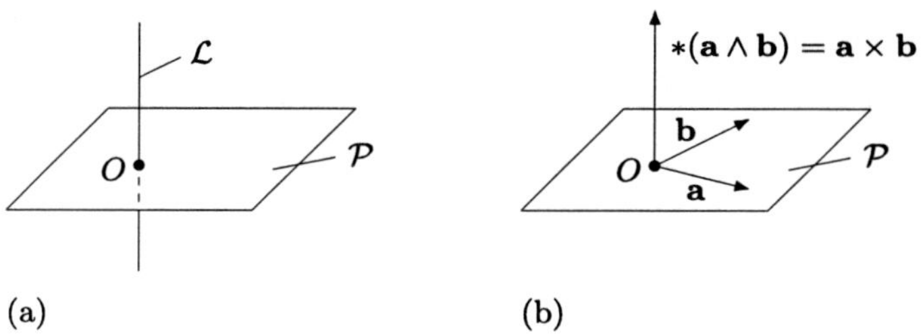

Fig. 2.3. Hodge duality

In terms of physics, Hodge theory is closely related to the Laplacian and spectral geometry, whereas Kähler theory is closely related to the Dirac equation for the relativistic electron and spin geometry.

Motivation. The simple geometric meaning of Hodge duality is pictured in Fig. 2.3. We are given the pair $(\mathbf{a}, \mathbf{b})$ of vectors $\mathbf{a}$ and $\mathbf{b}$ in the plane $\mathcal{P}$ through the origin O, and we assign to this pair the line $\mathcal{L}$ which is perpendicular to the plane $\mathcal{P}$ and passes through the origin (orthogonal complement of the plane). Finally the dual vector $*(\mathbf{a}, \mathbf{b})$ is equal to the vector product $\mathbf{a} \times \mathbf{b}$. More precisely, we will define $*(\mathbf{a} \wedge \mathbf{b}) := \mathbf{a} \times \mathbf{b}$ (see (2.34)).

Let us study this in terms of the Grassmann algebras $\bigwedge(E_3)$ and $\bigwedge(E_3^d)$. In Sect. 2.7.3, we will discuss the relation to multivectors. The approach concerning $\bigwedge(E_3^d)$ is the prototype of the crucial Hodge duality for alternating differential forms to be studied later on.[7] First we will introduce Hodge duality in terms of a fixed Cartesian (x, y, z)-coordinate system with the right-handed orthonormal basis $\mathbf{i}, \mathbf{j}, \mathbf{k}$. Then we will show that the definition does not depend on the choice of the Cartesian coordinate system. Note the crucial fact that:

Hodge duality critically depends on orientation.

Later on, we will see that Hodge duality only makes sense on oriented manifolds.

2.7.1 The Hilbert Space $\bigwedge(E_3)$

The elements ω of $\bigwedge(E_3)$ have the form

$$\omega = \alpha + a\mathbf{i} + b\mathbf{j} + c\mathbf{k} + A\,\mathbf{j} \wedge \mathbf{k} + B\,\mathbf{k} \wedge \mathbf{i} + C\,\mathbf{i} \wedge \mathbf{j} + \gamma\,\mathbf{i} \wedge \mathbf{j} \wedge \mathbf{k}$$

with unique real coefficients $\alpha, a, b, c, A, B, C, \gamma$. Equipped with the inner product

$$\langle \omega | \omega' \rangle := \alpha\alpha' + aa' + bb' + cc' + AA' + BB' + CC' + \gamma\gamma',$$

the alternating algebra $\bigwedge(E_3)$ becomes a real 8-dimensional Hilbert space. Furthermore, we define the linear Hodge star operator $* : \bigwedge(E_3) \to \bigwedge(E_3)$ by setting

$$*\omega := \alpha\,\mathbf{i} \wedge \mathbf{j} \wedge \mathbf{k} + a\,\mathbf{j} \wedge \mathbf{k} + b\,\mathbf{k} \wedge \mathbf{i} + c\,\mathbf{i} \wedge \mathbf{j} + A\mathbf{i} + B\mathbf{j} + C\mathbf{k} + \gamma.$$

In particular, $*1 := \mathbf{i} \wedge \mathbf{j} \wedge \mathbf{k}$, $*\mathbf{i} := \mathbf{j} \wedge \mathbf{k}$, and

[7] Hodge (1903–1975) introduced Hodge duality in order to get a differential calculus which allows to generalize Riemann's theory of algebraic functions to higher dimensions. See W. Hodge, The Theory and Applications of Harmonic Integrals, Cambridge University Press, 1941 (second revised edition 1951).

$$*(\mathbf{j} \wedge \mathbf{k}) := \mathbf{i}, \quad *(\mathbf{i} \wedge \mathbf{j} \wedge \mathbf{k}) := 1.$$

Obviously, for all $\boldsymbol{\omega}, \boldsymbol{\omega}' \in \bigwedge(E_3)$, we have

$$\langle *\boldsymbol{\omega} | *\boldsymbol{\omega}' \rangle = \langle \boldsymbol{\omega} | \boldsymbol{\omega}' \rangle, \tag{2.31}$$

and

$$\langle *\boldsymbol{\omega} | \boldsymbol{\omega}' \rangle = \langle \boldsymbol{\omega} | *\boldsymbol{\omega}' \rangle. \tag{2.32}$$

In other words, the Hodge star operator is unitary and self-adjoint. Furthermore,

$$*\boldsymbol{\omega} \wedge \boldsymbol{\omega}' = \boldsymbol{\omega} \wedge *\boldsymbol{\omega}' = \langle \boldsymbol{\omega} | \boldsymbol{\omega}' \rangle \, (*1). \tag{2.33}$$

Finally, $**\boldsymbol{\omega} = \boldsymbol{\omega}$.

2.7.2 The Hilbert Space $\bigwedge(E_3^d)$

The considerations above can be immediately translated to the dual situation of $\bigwedge(E_3^d)$. The elements ω of $\bigwedge(E_3^d)$ have the form

$$\begin{aligned} \omega = \alpha + a\, dx + b\, dy + c\, dz + A\, dy \wedge dz + B\, dz \wedge dx \\ + C\, dx \wedge dy + \gamma\, dx \wedge dy \wedge dz \end{aligned}$$

with unique real coefficients $\alpha, a, b, c, A, B, C, \gamma$. Recall that $v = dx \wedge dy \wedge dz$ is the volume form of E_3. Equipped with the inner product

$$\langle \omega | \omega' \rangle := \alpha\alpha' + aa' + bb' + cc' + AA' + BB' + CC' + \gamma\gamma',$$

the alternating algebra $\bigwedge(E_3^d)$ becomes a real 8-dimensional Hilbert space. Furthermore, we define the linear Hodge star operator $* : \bigwedge(E_3^d) \to \bigwedge(E_3^d)$ by setting

$$\begin{aligned} *\omega := \alpha dx \wedge dy \wedge dz + a\, dy \wedge dz \\ + b\, dz \wedge dx + c\, dx \wedge dy + A\, dx + B\, dy + C\, dz + \gamma. \end{aligned}$$

In particular, $*1 := v = dx \wedge dy \wedge dz$, $*dx := dy \wedge dz$ and

$$*(dy \wedge dz) := dx, \quad *(dx \wedge dy \wedge dz) := 1.$$

For all $\omega, \omega' \in \bigwedge(E_3^d)$, we have

- $\langle *\omega | *\omega' \rangle = \langle \omega | \omega' \rangle$,
- $\langle *\omega | \omega' \rangle = \langle \omega | *\omega' \rangle$,
- $*\omega \wedge \omega' = \omega \wedge *\omega' = \langle \omega | \omega' \rangle \, (*1)$, and
- $**\omega = \omega$.

In particular, the Hodge star operator $* : \bigwedge(E_3^d) \to \bigwedge(E_3^d)$ is unitary and self-adjoint. Setting $x^1 := x, x^2 := y, x^3 = z$, we get:

- $*\omega := \frac{1}{3!}\, \varepsilon_{ijk}\omega\, dx^i \wedge dx^j \wedge dx^k$ if ω is a real number (0-form),
- $*\omega = \frac{1}{2!}\varepsilon_{ijk}\omega^i dx^j \wedge dx^k$ if $\omega = \omega_i dx^i$ (1-form),
- $*\omega = \frac{1}{2!}\varepsilon_{ijk}\omega^{ij} dx^k$ if $\omega = \frac{1}{2!}\omega_{ij} dx^i \wedge dx^j$ (2-form),
- $*\omega = \frac{1}{3!}\varepsilon_{ijk}\omega^{ijk}$ if $\omega = \frac{1}{3!}\omega_{ijk}\, dx^i \wedge dx^j \wedge dx^k$ (3-form).

Here, we sum over equal upper and lower indices from 1 to 3, and we assume that ω_{ij} and ω_{ijk} are antisymmetric with respect to the indices (e.g., $\omega_{ij} = -\omega_{ji}$). Moreover, we set $\omega^i := \omega_i, \omega^{ij} := \omega_{ij}$, and $\omega^{ijk} := \omega_{ijk}$. These formulas are special cases of general formulas used in classical tensor calculus (see (8.66) on page 470).

The definitions above depend on the choice of the Cartesian coordinate system. The point is that the definitions are indeed independent of the choice of the Cartesian coordinate system. To show this, there exist two different approaches:

(i) Use the index killing principle of the classical tensor calculus, or
(ii) use an invariant definition.

In what follows, we will discuss (ii). We postpone (i) to Sect. 9.1.

The invariant approach for $\bigwedge(E_3^d)$. We proceed as follows:

- If $F : E_3 \to \mathbb{R}$ is a linear functional, then there exists a unique vector in E_3, denoted $\mathbf{a}_F$, such that[8]
$$F(\mathbf{x}) = \mathbf{a}_F \mathbf{x} \qquad \text{for all} \quad \mathbf{x} \in E_3.$$
The bilinear antisymmetric functional $*F : E_3 \times E_3 \to \mathbb{R}$ is defined by
$$(*F)(\mathbf{x}, \mathbf{y}) := \mathbf{a}_F(\mathbf{x} \times \mathbf{y}) \qquad \text{for all} \quad \mathbf{x}, \mathbf{y} \in E_3.$$
For example, if $F = dx$, then $\mathbf{a}_F = \mathbf{i}$, and $*F = dy \wedge dz$.
- If the bilinear functional $B : E_3 \times E_3 \to \mathbb{R}$ is antisymmetric, then there exists a unique vector in E_3, denoted $\mathbf{b}_B$, such that
$$B(\mathbf{x}, \mathbf{y}) = \mathbf{b}_B(\mathbf{x} \times \mathbf{y}) \qquad \text{for all} \quad \mathbf{x}, \mathbf{y} \in E_3.$$
The linear functional $*B : E_3 \to E_3$ is defined by
$$(*B)(\mathbf{x}) = \mathbf{b}_B \mathbf{x} \qquad \text{for all} \quad \mathbf{x} \in E_3.$$
For example, if $B := dx \wedge dy$, then $\mathbf{b}_B = \mathbf{k}$, and $*B = dz$.
- If the trilinear functional $T : E_3 \times E_3 \times E_3 \to \mathbb{R}$ is antisymmetric, then there exists a unique real number, denoted γ_T, such that $T = \gamma_T v$ (multiple of the volume form v). We define $*T := \gamma_T$.
- If α is a real number, then we define $*\alpha := \alpha v$.

Thus, the Hodge $*$-operator reflects the fact that linear and multilinear functionals on the Euclidean space E_3 can be represented in a simple way by the aid of vectors, inner products of vectors, and vector products.

Every element ω of $\bigwedge(E_3^d)$ can be uniquely represented as
$$\omega = \alpha + F + B + T,$$
where α is a real number, F is a linear functional on E_3, B (resp. T) is a bilinear (resp. trilinear) antisymmetric functional on E_3. We define:

- $*\omega := *\alpha + *F + *B + *T$, and
- $\langle \omega | \omega' \rangle := \alpha\alpha' + \mathbf{a}_F \mathbf{a}_{F'} + \mathbf{b}_B \mathbf{b}_{B'} + \gamma_T \gamma_{T'}$.

This coincides with the definitions given above in terms of a Cartesian coordinate system.

The invariant approach for $\bigwedge(E_3)$. Replacing E_3^d by E_3, we obtain the following:

[8] In terms of the Riesz duality operator, $\mathbf{a}_F = \aleph^{-1}(F)$.

- Let $\mathbf{a} \in E_3$. This vector generates the linear functional $\mathbf{a} : E_3^d \to \mathbb{R}$ given by
$$\mathbf{a}(\mathcal{F}) := \mathbf{a}\mathbf{a}_{\mathcal{F}} \qquad \text{for all} \quad \mathcal{F} \in E_3^d.$$
The bilinear antisymmetric functional $*\mathbf{a} : E_3^d \times E_3^d \to \mathbb{R}$ is defined by
$$*\mathbf{a}(\mathcal{F}, \mathcal{G}) := \mathbf{a}(\mathbf{a}_{\mathcal{F}} \times \mathbf{a}_{\mathcal{G}}) \qquad \text{for all} \quad \mathcal{F}, \mathcal{G} \in E_3^d.$$
For example, $*\mathbf{i} = \mathbf{j} \wedge \mathbf{k}$.
- If the bilinear functional $\mathcal{B} : E_3^d \times E_3^d \to \mathbb{R}$ is antisymmetric, then there exists a unique vector in E_3, denoted $\mathbf{b}_{\mathcal{B}}$, such that
$$\mathcal{B}(\mathcal{F}, \mathcal{G}) = \mathbf{b}_{\mathcal{B}}(\mathbf{a}_{\mathcal{F}} \times \mathbf{a}_{\mathcal{G}}) \qquad \text{for all} \quad \mathcal{F}, \mathcal{G} \in E_3^d.$$
The linear functional $*\mathcal{B} : E_3^d \to E_3^d$ is defined by
$$(*\mathcal{B})(\mathcal{F}) := \mathbf{b}_{\mathcal{B}}\mathbf{a}_{\mathcal{F}} \qquad \text{for all} \quad \mathcal{F} \in E_3^d.$$
For example, if $\mathcal{B} := \mathbf{i} \wedge \mathbf{j}$, then $*\mathcal{B} = \mathbf{k}$.
- We define the special trilinear functional $\mathcal{T}_0 : E_3^d \times E_3^d \times E_3^d \to \mathbb{R}$ by setting
$$\mathcal{T}_0(\mathcal{F}, \mathcal{G}, \mathcal{H}) := (\mathbf{a}_{\mathcal{F}}\mathbf{a}_{\mathcal{G}}\mathbf{a}_{\mathcal{H}}) \qquad \text{for all} \quad \mathcal{F}, \mathcal{G}, \mathcal{H} \in E_3^d.$$
If the trilinear functional $\mathcal{T} : E_3^d \times E_3^d \times E_3^d \to \mathbb{R}$ is antisymmetric, then there exists a unique real number, denoted $\gamma_{\mathcal{T}}$, such that $\mathcal{T} = \gamma_{\mathcal{T}}\mathcal{T}_0$.[9] In this case, we define $*\mathcal{T} := \gamma_{\mathcal{T}}$.
- If α is a real number, then we define $*\alpha := \alpha\mathcal{T}_0$.

Every element $\boldsymbol{\omega}$ of $\bigwedge(E_3)$ can be uniquely represented as
$$\boldsymbol{\omega} = \alpha + \mathbf{a} + \mathcal{B} + \mathcal{T},$$
where α is a real number, $\mathbf{a}$ is an element of E_3 regarded as a linear functional on E_3^d, $\mathcal{B}$ (resp. $\mathcal{T}$) is a bilinear (resp. trilinear) antisymmetric functional on E_3^d. We define:

- $*\boldsymbol{\omega} := *\alpha + *\mathbf{a} + *\mathcal{B} + *\mathcal{T}$, and
- $\langle \boldsymbol{\omega} | \boldsymbol{\omega}' \rangle := \alpha\alpha' + \mathbf{a}\mathbf{a}' + \mathbf{b}_{\mathcal{B}}\mathbf{b}_{\mathcal{B}'} + \gamma_{\mathcal{T}}\gamma_{\mathcal{T}'}$.

This coincides with the definitions given above in terms of a Cartesian coordinate system.

2.7.3 Multivectors

> Multivectors possess an immediate geometric meaning. In 1844, it was the ingenious idea of Grassmann (1809–1877) to introduce real linear combinations of multivectors. Such real chains of multivectors are nice algebraic objects which can be regarded as families of weighted multivectors.
> In particular, such chains are the fundamental objects in modern algebraic topology and differential topology for constructing homology and cohomology groups.[10]
>
> Folklore

[9] Note that trilinear functionals on a three-dimensional linear space are unique up to a multiplicative real constant (see Sect. 2.11.2).

[10] This will be studied in Vol. IV on quantum mathematics.

Bivectors. If $\mathbf{a}, \mathbf{b} \in E_3$, then the Grassmann product $\mathbf{a} \wedge \mathbf{b}$ is called a bivector. In other words, bivectors are special real bilinear antisymmetric functionals on E_3.

Proposition 2.17 $\mathbf{a} \wedge \mathbf{b} = \mathbf{a}' \wedge \mathbf{b}'$ *iff* $\mathbf{a} \times \mathbf{b} = \mathbf{a}' \times \mathbf{b}'$.

Proof. For all $\mathbf{x}, \mathbf{y} \in E_3$, we have $(\mathbf{a} \wedge \mathbf{b})(\mathbf{x}, \mathbf{y}) = (\mathbf{a} \times \mathbf{b})(\mathbf{x} \times \mathbf{y})$. □

In terms of geometry, we have $\mathbf{a} \wedge \mathbf{b} = \mathbf{a}' \wedge \mathbf{b}'$ iff the following hold:

- The vectors $\mathbf{a}, \mathbf{b}$ and $\mathbf{a}', \mathbf{b}'$ span the same plane through the origin O (Fig. 2.3 on page 139).
- The parallelograms spanned by the vectors $\mathbf{a}, \mathbf{b}$ and $\mathbf{a}', \mathbf{b}'$ have the same area, denoted $||\mathbf{a} \wedge \mathbf{b}||$, and the same orientation.

The Hodge $*$-operator sends the bivector $\mathbf{a} \wedge \mathbf{b}$ to the vector

$$* \, (\mathbf{a} \wedge \mathbf{b}) := \mathbf{a} \times \mathbf{b}. \tag{2.34}$$

In terms of geometry, this means that the vector $*(\mathbf{a} \wedge \mathbf{b})$ is perpendicular to the plane spanned by the vectors $\mathbf{a}, \mathbf{b}$, its length is equal to the area of the parallelogram spanned by $\mathbf{a}, \mathbf{b}$, and the three vectors $\mathbf{a}, \mathbf{b}, *(\mathbf{a} \wedge \mathbf{b})$ are positively oriented (in the nondegenerate case).

Real chains of bivectors. Let $\alpha_1, \ldots, \alpha_n \in \mathbb{R}$. The real linear combination

$$\alpha_1(\mathbf{a}_1 \wedge \mathbf{b}_1) + \cdots + \alpha_k(\mathbf{a}_k \wedge \mathbf{b}_k)$$

can be regarded as a family of bivectors where $\mathbf{a}_j \wedge \mathbf{b}_j$ is weighted with the real number α_j, $j = 1, 2, \ldots, k$. Two such chains represent the same object iff they represent the same real antisymmetric bilinear functional on the Euclidean space E_3.

Trivectors. By definition, the products $\mathbf{a} \wedge \mathbf{b} \wedge \mathbf{c}$ with $\mathbf{a}, \mathbf{b}, \mathbf{c} \in E_3$ are called trivectors.

Proposition 2.18 $\mathbf{a} \wedge \mathbf{b} \wedge \mathbf{c} = \mathbf{a}' \wedge \mathbf{b}' \wedge \mathbf{c}'$ *iff the vectors* $\mathbf{a}, \mathbf{b}, \mathbf{c}$ *and* $\mathbf{a}', \mathbf{b}', \mathbf{c}'$ *span the same volume, and they have the same orientation (in the nondegenerate case).*

Proof. For all $\mathbf{x}, \mathbf{y}, \mathbf{z} \in E_3$, we have $(\mathbf{a} \wedge \mathbf{b} \wedge \mathbf{c})(\mathbf{x}, \mathbf{y}, \mathbf{z}) = (\mathbf{abc})(\mathbf{xyz})$. Thus, $\mathbf{a} \wedge \mathbf{b} \wedge \mathbf{c} = (\mathbf{abc})\upsilon$ where υ is the volume form of E_3. □

The Hodge $*$-operator sends the trivector $\mathbf{a} \wedge \mathbf{b} \wedge \mathbf{c}$ to the real number

$$*(\mathbf{a} \wedge \mathbf{b} \wedge \mathbf{c}) = (\mathbf{abc})$$

which equals the oriented volume spanned by the three vectors $\mathbf{a}, \mathbf{b}, \mathbf{c}$ (Fig. 1.6 on page 83).

Historical Remarks

In 1844, Grassmann (1809–1877) wrote his book *Calculus of Extensions* (in German). This book was full of new ideas, but Grassmann's style of writing made hard reading. Grassmann did not formulate precise definitions, and he gave only a few genuine proofs. Typically, Grassmann started from the very beginning with higher-dimensional spaces, and he developed tools for this general situation. For example, the vector product $\mathbf{a} \times \mathbf{b}$ cannot be generalized to higher dimensions in a straightforward manner. But the Grassmann product $\mathbf{a} \wedge \mathbf{b}$ is well-defined for arbitrary dimensions. It was Grassmann's tragedy that his contemporaries did not understand the power of his new ideas. Nowadays Grassmann algebras lie at the heart of modern mathematics. We refer to J. Dieudonné, The tragedy of Grassmann, Linear and Multilinear Algebra **8**(1) (1979), 1–14.

2.8 The Clifford Structure of the Grassmann Algebra (Exterior–Interior Kähler Algebra)

2.8.1 The Kähler Algebra $\bigwedge(E_3)_\vee$

Again let us consider the Grassmann algebra $\bigwedge(E_3)$ equipped with the $\wedge$-product (Grassmann product or exterior product). It is our goal to introduce an additional $\vee$-product (Clifford product or interior product) by setting

$$\boxed{\mathbf{a} \vee \mathbf{b} := \mathbf{a} \wedge \mathbf{b} - \mathbf{ab}} \tag{2.35}$$

and $\alpha \vee \mathbf{a} = \mathbf{a} \vee \alpha := \alpha\mathbf{a}$ for all $\mathbf{a}, \mathbf{b} \in E_3$ and all real numbers α. This implies the Clifford relation

$$\mathbf{a} \wedge \mathbf{b} + \mathbf{b} \wedge \mathbf{a} = -2(\mathbf{ab}) \qquad \text{for all} \quad \mathbf{a}, \mathbf{b} \in E_3.$$

The key observation is that

$$\mathbf{a} \vee \mathbf{b} = \mathbf{a} \wedge \mathbf{b} \qquad \text{if } \mathbf{ab} = 0.$$

In other words, the inner product $\mathbf{a} \vee \mathbf{b}$ coincides with the exterior product $\mathbf{a} \wedge \mathbf{b}$ if the vector $\mathbf{a}$ is orthogonal to the vector $\mathbf{b}$. However, in contrast to $\mathbf{a} \wedge \mathbf{a} = 0$, we get $\mathbf{a} \vee \mathbf{a} = -\mathbf{a}^2$. This observation can be used in order to extend the interior product quite naturally to more than two factors.

Proposition 2.19 *Using the key relation* (2.35), *the Grassmann algebra* $\bigwedge(E_3)$ *can be additionally equipped with the structure of a Clifford algebra. This way, we get the Kähler algebra* $\bigwedge(E_3)_\vee$.

Proof. Choose an orthonormal basis $\mathbf{i}, \mathbf{j}, \mathbf{k}$ of E_3. By (2.35), we get

$$\mathbf{i} \vee \mathbf{i} = \mathbf{j} \vee \mathbf{j} = \mathbf{k} \vee \mathbf{k} = -1.$$

Again by (2.35), for two different factors of basis vectors, the $\vee$-product coincides with the $\wedge$-product. For example, $\mathbf{i} \vee \mathbf{j} = \mathbf{i} \wedge \mathbf{j}$ and $\mathbf{j} \vee \mathbf{i} = \mathbf{j} \wedge \mathbf{i}$. By definition, for three different factors of basis vectors, the $\vee$-product coincides with the $\wedge$-product. For example,

$$\mathbf{i} \vee \mathbf{j} \vee \mathbf{k} := \mathbf{i} \wedge \mathbf{j} \wedge \mathbf{k}. \tag{2.36}$$

This allows us to compute all the $\vee$-products. For example,

$$\mathbf{i} \vee \mathbf{j} \vee \mathbf{i} \vee \mathbf{k} = -\mathbf{i} \vee \mathbf{i} \vee \mathbf{j} \vee \mathbf{k} = \mathbf{j} \vee \mathbf{k} = \mathbf{i} \wedge \mathbf{k}.$$

This way, the real 8-dimensional associative algebra $\bigwedge(E_3)$ with respect to the exterior $\wedge$-product (Grassmann algebra) becomes an associative algebra with respect to the interior $\vee$-product (Clifford algebra), too. The real number 1 is the unit element for both the $\wedge$-product and the $\vee$-product.

Note that the $\vee$-product is independent of the choice of the orthonormal basis. This follows from the fact that the key relation (2.35) possesses an invariant meaning. Moreover, under a change $\mathbf{i}, \mathbf{j}, \mathbf{k} \Rightarrow \mathbf{i}', \mathbf{j}', \mathbf{k}'$ of the orthonormal basis, the $\wedge$-product and the $\vee$-product are transformed the same way, by the distributive law. Thus, for example, the relation (2.36) implies $\mathbf{i}' \vee \mathbf{j}' \vee \mathbf{k}' = \mathbf{i}' \wedge \mathbf{j}' \wedge \mathbf{k}'$. □

2.8.2 The Kähler Algebra $\bigwedge(E_3^d)_\vee$

The Riesz duality operator $\aleph : E_3 \to E_3^d$ represents a linear isomorphism from the Euclidean space E_3 to its dual space E_3^d. This isomorphism can be used in order to equip the Grassmann algebra $\bigwedge(E_3^d)$ with an additional $\vee$-product such that we obtain the Kähler algebra $\bigwedge(E_3^d)_\vee$ which is isomorphic to the Kähler algebra $\bigwedge(E_3)_\vee$ with respect to both the $\wedge$-product and the $\vee$-product. To this end, we replace the fixed orthonormal basis $\mathbf{i}, \mathbf{j}, \mathbf{k}$ (and their $\vee$-products) by the corresponding linear functionals dx, dy, dz (and their $\vee$-products). For example,

$$dx \vee dx = dy \vee dy = dz \vee dz = -1,$$

and $dx \vee dy = dx \wedge dy$, as well as $dx \vee dy \vee dz = dx \wedge dy \wedge dz$. Since the $\vee$-product on $\bigwedge(E_3)_\vee$ is independent of the choice of the orthonormal basis, the $\vee$-product on $\bigwedge(E_3^d)_\vee$ is also independent of the choice of the basis dx, dy, dz.

The Kähler algebra $\bigwedge(E_3^d)_\vee$ is a real 8-dimensional associative algebra with respect to both the $\wedge$-product and the $\vee$-product. The real number 1 is the unit element with respect to both the products.

2.9 The C^*-Algebra End(E_3) of the Euclidean Space

Let End(E_3) denote the space of all linear operators $A : E_3 \to E_3$. Define the operator norm

$$||A|| := \max_{|\mathbf{x}|\leq 1} |A\mathbf{x}|.$$

Equipped with this norm, the real linear space End(E_3) becomes a Banach space and, more general, a real unital C^*-algebra. By Sect. 7.16.3 of Vol. II, for all real numbers α, β and all linear operators $A, B \in \text{End}(E_3)$, this means the following:

- $(\alpha A + \beta B)^\dagger = \alpha A^\dagger + \beta B^\dagger$ (linearity),
- $(AB)^\dagger = B^\dagger A^\dagger$, and $(A^\dagger)^\dagger = A$,
- $||AB|| \leq ||A|| \, ||B||$,
- $||A^\dagger A|| = ||A||^2$ and $||AA^\dagger|| = ||A||^2$,
- $||I|| = 1$ (unitality).

This is the special case of the following general result:

> *If X is a real (resp. complex) Hilbert space, then the space* End(X) *of all linear continuous operators $A : X \to X$ is both a real (resp. complex) Banach space and a C^*-algebra.*

Naturally enough, in the case of a complex space X, we have the linearity condition above to replace by the following antilinearity condition:

$$(\alpha A + \beta B)^\dagger = \alpha^\dagger A^\dagger + \beta^\dagger B^\dagger \qquad \text{for all} \quad \alpha, \beta \in \mathbb{C}, \; A, B \in \text{End}(X).$$

The proof can be found in Zeidler, Applied Functional Analysis: Applications to Mathematical Physics, Springer, Berlin, 1995, Sect. 5.18.

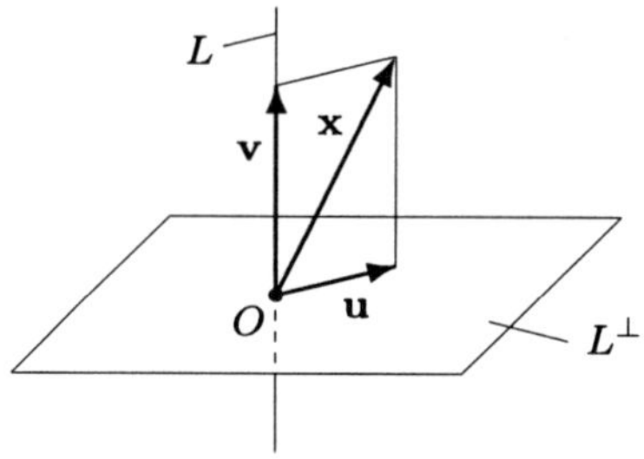

Fig. 2.4. Orthogonal complement

2.10 Linear Operator Equations

The question of which of several proofs is the simplest and most natural ordinarily cannot be decided in and of itself, but only a consideration of whether the basic principles are capable of generalization and useful for further research, will give us a sure reply.

David Hilbert

2.10.1 The Prototype

For the linear operator $A : E_3 \to E_3$, let us introduce the following notions:

- $\ker(A) := \{\mathbf{x} \in E_3 : \ A\mathbf{x} = 0\}$ (kernel of A),
- $\mathrm{im}(A) := \{A\mathbf{x} : \ \mathbf{x} \in E_3\}$ (image or range of A),
- $\mathrm{rank}(A) := \dim(\mathrm{im}(A))$ (rank of A).

Orthogonal decomposition. Let L be a linear subspace of the Euclidean space E_3 (Fig. 2.4). By definition, the orthogonal complement $L^\perp$ to L consists of all the vectors in E_3 which are orthogonal to the vectors of L, that is,

$$L^\perp := \{\mathbf{u} \in E_3 : \ \mathbf{uv} = 0 \ \text{ for all } \ \mathbf{v} \in L\}.$$

For every vector $\mathbf{x} \in E_3$, we have the unique decomposition

$$\boxed{\mathbf{x} = \mathbf{u} + \mathbf{v}, \quad \mathbf{u} \in L, \mathbf{v} \in L^\perp.}$$

That is, $E_3 = L \oplus L^\perp$. Obviously, $(L^\perp)^\perp = L$.

The Fredholm alternative by means of the adjoint operator. For given linear operator $A : E_3 \to E_3$, we want to solve the linear operator equation

$$\boxed{A\mathbf{x} = \mathbf{y}, \qquad \mathbf{x} \in E_3.} \tag{2.37}$$

We are given the vector $\mathbf{y}$ in E_3. We are looking for the vector $\mathbf{x}$ in E_3. The trick is to add the homogeneous adjoint equation

$$A^\dagger \mathbf{z} = 0, \qquad \mathbf{z} \in E_3. \tag{2.38}$$

Recall that the adjoint operator $A^\dagger$ to the linear operator $A : E_3 \to E_3$ is the uniquely determined linear operator $A^\dagger : E_3 \to E_3$ which satisfies the relation

$$\langle \mathbf{x} | A^\dagger \mathbf{z} \rangle = \langle A\mathbf{x} | \mathbf{z} \rangle \qquad \text{for all} \quad \mathbf{x}, \mathbf{z} \in E_3.$$

This is equivalent to (2.39).

Proposition 2.20 *Equation* (2.37) *has a solution iff* $\mathbf{yz} = 0$ *for all solutions* $\mathbf{z}$ *of* (2.38).

Corollary 2.21 *If* $\mathbf{x}_0$ *is a solution of* (2.37), *then* $\mathbf{x}_0 + \ker(A)$ *is the solution set of* (2.37) *with the dimension*

$$\dim(\ker(A)) = \dim(\ker(A^\dagger)) = 3 - \operatorname{rank}(A).$$

Corollary 2.22 *If* $A\mathbf{x} = 0$ *implies* $\mathbf{x} = 0$, *then equation* (2.37) *has a unique solution for every given* $\mathbf{y} \in E_3$.

This Corollary tells us that uniqueness implies existence.
Proof. The idea of proof is to use the orthogonal decomposition

$$E_3 = \ker(A) \oplus \ker(A)^\perp = \operatorname{im}(A) \oplus \operatorname{im}(A)^\perp.$$

(I) For every $\mathbf{y} \in \operatorname{im}(A)$, the modified equation

$$A\mathbf{x} = \mathbf{y}, \qquad \mathbf{x} \in \ker(A)^\perp$$

has a unique solution. In fact, if $A\mathbf{x} = \mathbf{y}$ and $A\mathbf{x}' = \mathbf{y}$, then $A(\mathbf{x} - \mathbf{x}') = 0$. Hence $\mathbf{x} - \mathbf{x}' \in \ker(A)$. Moreover, $\mathbf{x} - \mathbf{x}' \in \ker(A)^\perp$. Therefore, $\mathbf{x} - \mathbf{x}' = 0$.
(II) By (I), the restricted operator $A : \ker(A)^\perp \to \operatorname{im}(A)$ is bijective. Hence

$$\operatorname{rank}(A) = \dim \operatorname{im}(A) = \dim \ker(A)^\perp = 3 - \dim \ker(A).$$

(III) We show that $\operatorname{im}(A)^\perp = \ker(A^\dagger)$. To this end, we will use the relation

$$\mathbf{x}(A^\dagger \mathbf{z}) = (A\mathbf{x})\mathbf{z} \qquad \text{for all} \quad \mathbf{x}, \mathbf{z} \in E_3 \tag{2.39}$$

which defines the adjoint operator. If $\mathbf{z} \in \operatorname{im}(A)^\perp$, then $(A\mathbf{x})\mathbf{z} = 0$. This yields $\mathbf{x}(A^\dagger \mathbf{z}) = 0$ for all $\mathbf{x} \in E_3$. Hence $A^\dagger \mathbf{z} = 0$. Conversely, if $\mathbf{z} \in \ker(A^\dagger)$, then $0 = \mathbf{x}(A^\dagger \mathbf{z}) = (A\mathbf{x})\mathbf{z}$ for all $\mathbf{x} \in E_3$. Thus, $\mathbf{z} \in \operatorname{im}(A)^\perp$. □

The corollaries are immediate consequences of the proof. In particular, the proof shows that if $\ker(A) = \{0\}$, then the operator $A : E_3 \to \operatorname{im}(A)$ is bijective. Hence $\dim \operatorname{im}(A) = 3$. This implies $\operatorname{im}(A) = E_3$.

The Fredholm alternative by means of the dual operator. Let us replace the homogeneous adjoint equation (2.38) by the homogeneous dual equation

$$A^d F = 0, \qquad F \in E_3^d. \tag{2.40}$$

Proposition 2.23 *Equation* (2.37) *has a solution iff* $F(\mathbf{y}) = 0$ *for all solutions* F *of* (2.40).

Recall that the dual operator $A^d : E_3^d \to E_3^d$ to the linear operator $A : E_3 \to E_3$ is defined by

$$(A^d F)(\mathbf{x}) = F(A\mathbf{x}) \qquad \text{for all} \quad F \in E_3^d,\ \mathbf{x} \in E_3.$$

In particular, the operator A^d is linear.
Proof. Using the bijective Riesz duality operator $\aleph : E_3 \to E_3^d$, the claim is a consequence of Prop. 2.20. In fact, it follows from

$$(A^\dagger \mathbf{z})\mathbf{x} = \mathbf{z}(A\mathbf{x}) = \aleph(\mathbf{z})(A\mathbf{x}) = (A^d \aleph(\mathbf{z}))(\mathbf{x}) \qquad \text{for all} \quad \mathbf{x} \in X$$

that $\aleph(A^\dagger \mathbf{z}) = A^d \aleph(\mathbf{z})$. Consequently, the diagram

$$\begin{array}{ccc} E_3 & \xrightarrow{A^\dagger} & E_3 \\ \aleph \downarrow & & \downarrow \aleph \\ E_3^d & \xrightarrow{A^d} & E_3^d \end{array}$$

is commutative. Therefore, $A^\dagger \mathbf{z} = 0$ iff $A^d \aleph(\mathbf{z}) = 0$. Moreover, $\aleph(\mathbf{z})\mathbf{y} = \mathbf{zy}$. □

2.10.2 The Grassmann Theorem

> The Grassmann $\wedge$-product allows us to study elegantly both linear independence of vectors and the solution of linear equations.
>
> Folklore

Choose $\mathbb{K} = \mathbb{R}$ or $\mathbb{K} = \mathbb{C}$, and let $n = 1, 2, \ldots$ Consider the n-dimensional linear space X over $\mathbb{K}$. Fix $k = 1, 2, \ldots, n$.

Theorem 2.24 *The vectors $a_1, \ldots, a_k$ are linearly independent iff $a_1 \wedge \cdots \wedge a_k \neq 0$.*

Proof. (I) If the vectors $a_1, \ldots, a_k \in X$ are linearly independent, then there exist linear functionals $F^i, \ldots, F^k \in X^d$ such that $F^i(a_j) = \delta^i_j$ if $i, j = 1, \ldots, k$ (see Problem 3.8). Hence

$$(a_1 \wedge \cdots \wedge a_k)(F^1, \ldots, F^k) = F^1(a_1)F^2(a_2) \cdots F^k(a_k) = 1.$$

This implies $a_1 \wedge \cdots \wedge a_k \neq 0$.

(II) If $a_1, \ldots, a_k$ are linearly dependent, then there exists some vector, say a_1, such that

$$a_1 = \alpha_2 a_2 + \ldots + \alpha_k a_k, \qquad \alpha_2, \ldots, \alpha_k \in \mathbb{K}.$$

Then

$$a_1 \wedge a_2 \wedge \cdots \wedge a_k = \alpha_2 a_2 \wedge a_2 \wedge \cdots \wedge a_k + \ldots + \alpha_k a_k \wedge a_2 \wedge \cdots \wedge a_k = 0,$$

since always two factors coincide. □

Proposition 2.25 *Let $\alpha, \beta \in \mathbb{K}$ with $\alpha \neq 0$. If $a_1, a_2, \ldots, a_k$ are linearly independent, then so is $\alpha a_1 + \beta a_2, a_2, \ldots, a_k$.*

Proof. $(\alpha a_1 + \beta a_2) \wedge a_2 \wedge \cdots a_k = \alpha(a_1 \wedge a_2 \wedge \cdots \wedge a_k) \neq 0$. □

Theorem 2.26 *The linearly independent vector families $a_1, \ldots, a_k$ and $b_1, \ldots, b_k$ span the same linear subspace of X iff there exists a nonzero number γ such that*

$$a_1 \wedge \cdots \wedge a_k = \gamma(b_1 \wedge \cdots \wedge b_k). \tag{2.41}$$

Proof. If $a_1, \ldots, a_k$ are linear combinations of the vectors $b_1, \ldots, b_k$, then the distributive law yields (2.41). For example,

$$a_1 \wedge a_2 = (\alpha b_1 + \beta b_2) \wedge (\lambda b_1 + \mu b_2) = (\alpha\mu - \beta\lambda)\, b_1 \wedge b_2.$$

Since $a_1 \wedge a_2 \neq 0$, we get $\alpha\mu - \beta\lambda \neq 0$.

Conversely, assume (2.41). Suppose that the vector b_1 is not an element of $\text{span}\{a_1, \ldots, a_k\}$. By Theorem 2.24,

$$a_1 \wedge \cdots \wedge a_k \wedge b_1 \neq 0.$$

By (2.41), $a_1 \wedge \cdots \wedge a_k \wedge b_1 = \gamma(b_1 \wedge \cdots \wedge b_k) \wedge b_1 = 0$, since two factors coincide. This is the desired contradiction. □

The dual space X^d. Fix $n = 1, 2, \ldots$ Let X be an n-dimensional linear space over $\mathbb{K} = \mathbb{R}, \mathbb{C}$. Recall that, by definition, the dual space X^d consists of all the linear functionals $F : X \to \mathbb{K}$.

Proposition 2.27 *There exists a linear isomorphism $\aleph : X \to X^d$.*

Proof. (I) Choose a basis $b_1, \ldots, b_n$ of X, and define $dx^i \left(\sum_{j=1}^n x^j b_j\right) := x^i$. Then, we have the following biorthogonality relations:

$$dx^i(b_j) = \delta^i_j, \qquad i, j = 1, \ldots, n. \tag{2.42}$$

It follows from $F\left(\sum_{j=1}^n x^j b_j\right) = \sum_{j=1}^n x^j F(b_j)$ that

$$F = \sum_{j=1}^n F(b_j)\, dx^j.$$

Conversely, let $F := \sum_{j=1}^n \beta_j dx^j$ with real coefficients $\beta_1, \ldots, \beta_n$. Then, $F \in X^d$ and $\beta_j = F(b_j)$ for $j = 1, \ldots, n$.

(II) Define $\aleph\left(\sum_{j=1}^n x^j b_j\right) := \sum_{j=1}^n x^j dx^j$. Obviously, the operator $\aleph : X \to X^d$ is surjective and injective. □

The bidual space X^{dd}. Let X^{dd} denote the dual space of the dual space X^d. Briefly, $X^{dd} := (X^d)^d$.

Proposition 2.28 *There exists a linear isomorphism $\nu : X \to X^{dd}$.*

Proof. Let $x \in X$. Define $\nu(x)(F) := F(x)$ for all $F \in X^d$.

(I) The map $\nu : X \to X^d$ is injective. In fact, if $F(x) = F(y)$ for all $F \in X^d$, then $dx^j(x) = dx^j(y)$. Hence $x^j = y^j$ for all indices $j = 1, \ldots, n$. This implies $x = y$.

(II) The elements $\nu(b_1), \ldots, \nu(b_n)$ are linearly independent. In fact, suppose that $\sum_{j=1}^n \beta^j \nu(b_j) = 0$. Then

$$0 = \sum_{j=1}^n \beta^j \nu(b_j)(dx^i) = \sum_{j=1}^n \beta^j dx^i(b_j) = \beta^i, \quad i = 1, \ldots, n.$$

(III) By Prop. 2.27, $\dim(X^{dd}) = \dim X^d = \dim X = n$. Thus, $\nu(b_1), \ldots, \nu(b_n)$ is a basis of X^{dd}. Consequently, the map $\nu : X \to X^{dd}$ is surjective. □

Note that Prop. 2.28 is true for all finite-dimensional linear spaces, but not for all infinite-dimensional linear spaces. For example, the proposition is valid for reflexive Banach spaces, but not for general Banach spaces. See "further reading" on page 157.

The direct sum and the codimension of a linear subspace (Fig. 2.5). Let X be an n-dimensional linear space over $\mathbb{K} = \mathbb{R}, \mathbb{C}$ with $n = 1, 2, \ldots$. We are given

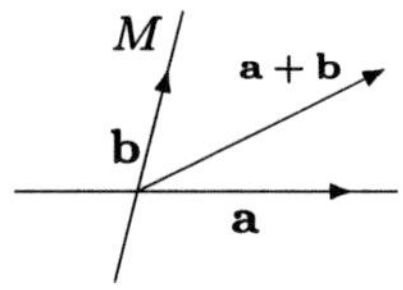

Fig. 2.5. Direct sum

the m-dimensional linear subspace L of X. Then there exists a linear subspace M such that

$$X = L \oplus M.$$

This means that, for every vector $\mathbf{x}$ in X, there exists the unique sum representation

$$\mathbf{x} = \mathbf{a} + \mathbf{b}, \qquad \mathbf{a} \in L,\ \mathbf{b} \in M.$$

Figure 2.5 shows that the complementary linear subspace M is not uniquely determined by the original linear subspace L. However, we have the linear isomorphism

$$M \simeq X/L$$

where X/L is the linear quotient space of X with respect to L (see Sect. 4.1.4 of Vol. II). Recall that the codimension of L is defined by

$$\mathrm{codim}(L) =: \dim(X/L).$$

In particular, $\dim(M) = \mathrm{codim}(L) = \dim X - \dim L$. Hence

$$\dim(X) = \dim(L) + \mathrm{codim}(L).$$

Recall that X^d denotes the space of all the linear functionals $F : X \to \mathbb{K}$. This linear space X^d is called the dual space to X.

Theorem 2.29 *The codimension of L is equal to the dimension of the linear subspace*

$$\{F \in X^d : \ F(x) = 0 \ \text{ on } \ L\} \tag{2.43}$$

of the dual space X^d.

Proof. Let $b_1, \ldots, b_r$ be a basis of L. We extend this to a basis $b_1, \ldots, b_n$ of X, and we construct the linear functionals $F^{r+1}, \ldots, F^n \in X^d$ by setting

$$F^i\left(\sum_{j=1}^n x^j b_j\right) = x^i, \qquad i = r+1, \ldots, n.$$

Hence $F^i(x) = 0$ on L if $i = r+1, \ldots, n$. The functionals $F^{r+1}, \ldots, F^n$ are linearly independent, and every functional F of the set (2.43) allows the representation $F = \alpha_{r+1}F^{r+1} + \ldots + \alpha_n F^n$ with real coefficients $\alpha_{r+1}, \ldots, \alpha_n$. □

Linear manifold. The subset $\mathcal{L}$ of the linear space X is called a linear manifold iff there exist both an element $x_0 \in X$ and a linear subspace L of X such that

$$\mathcal{L} = x_0 + L.$$

(a) straight line ($\dim \mathcal{L} = 1$) (b) plane ($\dim \mathcal{L} = 2$)

Fig. 2.6. Linear manifolds

Explicitly, $\mathcal{L} = \{x_0 + x : x \in L\}$. Intuitively, the linear subspace L passes through the origin O, whereas $x_0 + L$ is obtained from L by the translation x_0 (Fig. 2.6). By definition, the dimension of $\mathcal{L}$ is equal to the dimension of L. This dimension is well-defined. In fact, suppose that

$$\mathcal{L} = x_0 + L = x_1 + M$$

where $x_1 \in X$, and M is a linear subspace of X. Then $x_1 = x_0 + x$ for some $x \in L$. This implies $M \subseteq L$. Similarly, $L \subseteq M$. hence $L = M$. Linear manifolds are also called affine manifolds.

Next we want to show that Theorems 2.26 and 2.29 imply almost immediately the fundamental results for linear operator equations on finite-dimensional linear spaces without using matrices and determinants. This is important for generalizations to Fredholm operator equations in infinite-dimensional Banach spaces which comprehend large classes of linear integral equations and linear partial differential equations.[11]

2.10.3 The Superposition Principle

> Nonlinear problems describe interactions in nature. In contrast to nonlinear problems, linear problems enjoy the superposition principle. The advantage of perturbation theory is, that it reduces nonlinear problems to linear problems. Therefore, perturbation theory is widely used in physics.
>
> Folklore

We want to study the linear operator equation

$$\boxed{Ax = y, \qquad x \in X.} \tag{2.44}$$

We assume the following:

(H) We are given the linear operator $A : X \to Y$, where X and Y are finite-dimensional linear spaces over $\mathbb{K} = \mathbb{R}, \mathbb{C}$.

Concerning our problem (2.44), we are given $y \in Y$, and we are looking for $x \in X$. There arise the following two questions:

- What is the dimension d of the solution set?
- What is the number s of solvability conditions for the given right-hand side y.

Our goal is to compute d and s. The final answer will be:

---

[11] See "further reading" on page 157.

(i) $d = \dim X - \text{rank } A$,
(ii) $s = \dim Y - \text{rank } A$.

Here, we use the following terminology:

- $\ker A := \{x \in X : Ax = 0\}$ (kernel of the operator A). This is the solution set of the homogenous equation $Ax = 0$.
- $\text{im } A = A(X) := \{Ax \in Y : x \in X\}$ (image or range of the operator A). This is the set of all elements y in Y for which the equation $Ax = y$ has a solution.
- $\text{rank } A := \dim(\text{im } A)$ (rank of the operator A).
- $\text{ind } A := \dim(\ker A) - \text{codim}(\text{im } A)$ (index of the operator A).

The importance of the index. The classical theory for finite-dimensional linear operator equations uses the rank A of the operator A which can be explicitly computed by either the Gauss elimination method or the computation of determinants, as we will show below. However, in the infinite-dimensional case, the rank is infinite. Therefore, one has to pass to the index $\text{ind } A$ which is finite for Fredholm operators. In the finite-dimensional case, we always have

$$\text{ind } A = \dim X - \dim Y. \tag{2.45}$$

Thus, the index is completely stable. If $\dim X = \infty$ and $\dim Y = \infty$, then formula (2.45) becomes meaningless.

However, there are many important cases concerning integral- and differential equations where the **index** *remains a well-defined* **finite** *quantity; the index has the crucial property that it remains unchanged under fairly large perturbations (e.g., under compact perturbations).*

This allows us to compute the index by deforming the operator A into a simpler operator. One of the most important mathematical results obtained in the 20th century is the Atiyah–Singer index theorem which expresses the index of elliptic differential operators (and certain classes of integral operators) on compact manifolds $\mathcal{M}$ in terms of topological invariants of $\mathcal{M}$ (see Sect. 5.6.9 of Vol. I). In terms of the index, the formulas above for d and s read as:

(i) $d = \dim(\ker A)$,
(ii) $s = \dim(\ker A^d) = d - \text{ind } A$.

Equivalently,

$$\boxed{\text{ind } A = d - s.} \tag{2.46}$$

That is, concerning the original equation (2.44), the index measures the difference between the dimension d of the linear solution manifold and the number s of linearly independent solvability conditions.

Uniqueness implies existence. For example, suppose we know that $\text{ind } A = 0$ which happens frequently. Then the index formula (2.46) above tells us that $s = d$. If $d = 0$ (uniqueness), then $s = 0$ (no solvability condition). Therefore, we get the following crucial result:

If $\text{ind } A = 0$*, then uniqueness implies existence.*

As we will show below, this is true in the finite-dimensional case iff $\dim X = \dim Y$ (e.g., $Y = X$). But it is also true for large classes of differential- and integral equations (see "further reading" on page 157). Note that, as a rule, it is much simpler to prove the uniqueness of a solution, than the existence. Therefore, the principle "uniqueness implies existence" for Fredholm operator equations of index zero is extremely useful.

The linear solution manifold. Assume (H) above.

Theorem 2.30 *If x_{special} is a solution of the linear operator equation* (2.44)*, then the complete solution set of* (2.44) *is the linear manifold* $x_{\text{special}} + \ker A$.

The dimension d of the solution manifold is given by $d = \dim X - \operatorname{rank} A$.

Proof. Choose a linear subspace M of X such that

$$X = \ker A \oplus M.$$

Then the restriction $A : M \to \operatorname{im} A$ is a linear isomorphism. In fact, if $Ax = Ax'$ with $x, x' \in M$, then $A(x - x') = 0$. Hence $x - x' \in M$ and $x, x' \in \ker A$. This implies $x = x'$. Therefore, the map $A : M \to \operatorname{im} A$ is injective and obviously surjective. The linear isomorphism $M \simeq \operatorname{im} A$ yields

$$\dim M = \operatorname{rank} A.$$

Moreover, $\dim(\ker A) + \dim M = \dim X$. Consequently,

$$\dim(\ker A) + \operatorname{rank} A = \dim X.$$

□

If $b_1, \ldots, b_d$ is a basis of $\ker A$, then the solution set of (2.44) has the form

$$\boxed{x = x_{\text{special}} + \sum_{j=1}^{d} \gamma_j b_j, \qquad \gamma_1, \ldots, \gamma_d \in \mathbb{K}.} \tag{2.47}$$

That is, if $\mathbb{K} = \mathbb{R}$ (resp. $\mathbb{K} = \mathbb{C}$), then the solution set of the equation $Ax = y$ depends on d real (resp. complex) parameters. In particular, if $y = 0$, then we may choose $x_{\text{special}} = 0$. Mnemonically, we briefly say:

General solution of the inhomogeneous equation $Ax = y$
$=$ *special solution of the inhomogeneous equation*
$+$ *general solution of the homogeneous equation* $Ax = 0$.

2.10.4 Duality and the Fredholm Alternative

> In the winter semester 1900/01 Holmgren, who had come from Uppsala (Sweden) to study under Hilbert in Göttingen, held a lecture in Hilbert's seminar on Fredholm's work on linear integral equations which had been published the previous year.[12] This was a decisive day in Hilbert's life. He took up Fredholm's new discovering with great zeal, and combined it with his variational methods (concerning the Dirichlet principle for the Laplace equation). In this way, he succeeded in creating a uniform theory which solved outstanding open problems of this time: boundary-value problems and boundary-eigenvalue problems for partial differential equations.[13]
>
> Otto Blumenthal, 1932

The dual operator. Consider the linear operator

$$A : X \to Y$$

[12] Fredholm (1866–1927)

[13] O. Blumenthal, Hilbert's biography (in German). In: D. Hilbert, Collected Works, Vol. 3, pp. 388–429, Springer, Berlin (12th edn. 1977).

where X and Y are linear spaces over $\mathbb{K} = \mathbb{R}, \mathbb{C}$. We want to construct a linear operator of the form

$$A^d : Y^d \to X^d.$$

The operator A^d sends the given linear functional $F \in Y^d$ to the linear functional $A^d F \in X^d$. The precise definition reads as

$$\boxed{(A^d F)(x) := F(Ax) \qquad \text{for all} \quad x \in X.}$$

The Fredholm alternative. We assume that X and Y are finite-dimensional linear spaces over $\mathbb{K}$ where $\mathbb{K} = \mathbb{R}$ or $\mathbb{K} = \mathbb{C}$.

Theorem 2.31 *For given $y \in Y$, the equation*

$$Ax = y, \qquad x \in X \tag{2.48}$$

has a solution iff $F(y) = 0$ for all solutions F of the homogeneous dual equation

$$A^d F = 0, \quad F \in Y^d.$$

The number s of linearly independent solvability conditions for (2.48) *is given by*

$$s = \dim Y - \operatorname{rank} A.$$

Moreover, we have duality invariance of the rank:

$$\operatorname{rank} A = \operatorname{rank} A^d. \tag{2.49}$$

The intuitive meaning of the fundamental rank relation (2.49) will be discussed on page 162 under the heading "the philosophy of linear systems."

Corollary 2.32 *For the index,* $\operatorname{ind} A^d = -\operatorname{ind} A$. *Furthermore,*

$$s = \dim(\ker A^d) = \dim(\ker A) - \operatorname{ind} A. \tag{2.50}$$

In addition, $s = \operatorname{codim}(\operatorname{im} A)$.

In particular, this corollary implies the following for the linear operator $A : X \to Y$:

- A is injective iff $\dim(\ker A) = 0$ (i.e., the equation $Ax = y$ has at most one solution $x \in X$),
- A is surjective iff $\dim(\ker A) = \operatorname{ind} A$ (i.e., the equation $Ax = y$ has at least one solution $x \in X$ for every given $y \in Y$).[14]
- A is bijective iff $\dim(\ker A) = \operatorname{ind} A = 0$ (i.e., the equation $Ax = y$ has a unique solution $x \in X$ for every given $y \in Y$).
- The equation $Ax = y$ is well posed iff $\dim(\ker A) = \operatorname{ind} A = 0$ (i.e., for every given $y \in Y$, the equation $Ax = y$ has a unique solution $x \in X$, and the solution depends continuously on y).
- If $\dim Y = \dim X$ and the homogeneous equation $Ax = 0, x \in X$ has only the trivial solution $x = 0$, then the inhomogeneous equation $Ax = y$ is well posed (uniqueness implies existence).[15]

---

[14] This shows that a linear operator $A : X \to Y$ with negative index, that is, $\dim X < \dim Y$, can never be surjective.

[15] Observe that $\operatorname{ind} A = 0$ and $\dim(\ker A) = 0$ by assumption.

Proof. Let us prove Theorem 2.31 and Corollary 2.32.

(I) By definition of the dual operator,

$$\ker A^d = \{F \in Y^d : \; F(y) = 0 \; \text{ for all } \; y \in \text{im } A\}. \tag{2.51}$$

In fact, if $F \in \ker A^d$, then $A^d F = 0$. Hence $F(Ax) = 0$ for all $x \in X$. Therefore, $F(y) = 0$ for all $y \in \text{im } A$. The converse is also true.

(II) By Theorem 2.29, it follows from (2.51) that $\dim(\ker A^d) = \text{codim}(\text{im } A)$. Therefore,

$$\dim(\ker A^d) = \dim Y - \text{rank } A. \tag{2.52}$$

(III) By Theorem 2.30,

$$\dim(\ker A) + \text{rank } A = \dim X.$$

Replacing the operator A by the dual operator A^d, we get

$$\dim(\ker A^d) + \text{rank } A^d = \dim Y^d.$$

Since $\dim Y^d = \dim Y$, it follows from (2.52) that $\text{rank } A^d = \text{rank } A$.

(IV) By (2.51), the number s of linearly independent solvability conditions is equal to $\dim(\ker A^d)$. Hence $s = \dim Y - \text{rank } A$.

Finally, $\text{ind} A = \dim X - \dim Y$. Analogously, $\text{ind } A^d = \dim Y^d - \dim X^d$. Therefore, $\text{ind } A^d = \dim Y - \dim X = -\text{ind } A$. □

In what follows let us sketch two prototypes of infinite-dimensional problems.

The Fredholm alternative for integral equations. Consider the integral equation

$$\psi(x) - \int_0^1 \mathcal{K}(x,y)\psi(y)dy = f(x), \qquad 0 \le x \le 1 \tag{2.53}$$

together with the homogeneous equation

$$\psi(x) - \int_0^1 \mathcal{K}(x,y)\psi(y)dy = 0, \qquad 0 \le x \le 1, \tag{2.54}$$

and the dual homogeneous equation

$$\varphi(x) - \int_0^1 \mathcal{K}(y,x)\varphi(y)dy = 0, \qquad 0 \le x \le 1 \tag{2.55}$$

with the transposed kernel function. The kernel function $\mathcal{K} : [0,1] \times [0,1] \to \mathbb{R}$ is assumed to be continuous. Let $C[0,1])$ denote the space of all continuous functions

$$\psi : [0,1] \to \mathbb{R}.$$

This is a real Banach space equipped with the norm $||\psi|| := \max_{0 \le x \le 1} |\psi(x)|$. We are given the function $f \in C[0,1]$.

Proposition 2.33 (i) *Equation* (2.53) *has a solution* $\psi \in C[0,1]$ *iff*

$$\int_0^1 f(x)\varphi(x)dx = 0$$

for all solutions $\varphi \in C[0,1]$ *of* (2.55).

(ii) *The homogeneous equation* (2.54) *and the homogeneous dual equation* (2.55) *have the same finite number d of solutions in the space* $C[0,1]$). *Let* $\psi_1, \ldots, \psi_d$ *and* $\varphi_1, \ldots, \varphi_d$ *be a basis of the solution spaces of* (2.54) *and* (2.55), *respectively.*

(iii) *If* ψ_{special} *is a solution of* (2.53), *then the general solution of* (2.53) *reads as*

$$\psi(x) = \psi_{\text{special}}(x) + \sum_{j=1}^{d} \gamma_j \psi_j(x), \qquad 0 \le x \le 1$$

where $\gamma_1, \ldots, \gamma_d$ *are arbitrary real coefficients. If* $d = 0$, *then the solution is unique, that is,* $\psi = \psi_{\text{special}}$.

Obviously, the solvability condition (i) can be replaced by the following condition: Equation (2.53) has a solution $\psi \in C[0,1]$ iff the d solvability conditions

$$\int_0^1 f(x)\varphi_j(x)dx = 0, \qquad j = 1, \ldots, d$$

are satisfied. If the kernel is symmetric, that is, $\mathcal{K}(x,y) = \mathcal{K}(y,x)$ for all $x, y \in [0,1]$, then the two equations (2.54) and (2.55) coincide. From the abstract point of view, the integral equation (2.53) corresponds to the operator equation

$$(I - K)\psi = f, \qquad \psi \in C[0,1].$$

Here, the linear operator $K : C[0,1] \to C[0,1]$ is defined by

$$(K\psi)(x) := \int_0^1 \mathcal{K}(x,y)\psi(y)dy, \qquad 0 \le x \le 1.$$

The point is that the linear operator $K : C[0,1] \to C[0,1]$ is compact. Therefore, the operator $I - K : C[0,1] \to C[0,1]$ is Fredholm with the index

$$\operatorname{ind}(I - K) = \operatorname{ind} I = 0.$$

Here, we use the general theorem that the identity operator $I : C[0,1] \to C[0,1]$ on a Banach space is Fredholm of index zero. Moreover, the compact perturbation of a Fredholm operator is again a Fredholm operator, and the index remains unchanged.

The Fredholm alternative for boundary-value problems. Consider the boundary value problem

$$\psi''(x) + \mu\psi(x) = f(x), \qquad -\pi \le x \le \pi, \quad \psi(-\pi) = \psi(\pi) = 0 \tag{2.56}$$

together with the homogeneous problem

$$\psi''(x) + \mu\psi(x) = 0, \qquad -\pi \le x \le \pi, \quad \psi(-\pi) = \psi(\pi) = 0. \tag{2.57}$$

Here, μ is a fixed real number. Let $C[-\pi, \pi]$ denote the Banach space of all continuous functions

$$\psi : [-\pi, \pi] \to \mathbb{R} \tag{2.58}$$

equipped with the norm $||\psi||_C := \max_{-\pi \le x \le \pi} |\psi(x)|$. Moreover, let $C^2[-\pi, \pi]$ denote the Banach space of all twice continuously differentiable functions (2.58), that is, $\psi, \psi', \psi'' \in C[-\pi, \pi]$. This is a real infinite-dimensional Banach space equipped with the norm $||\psi|| := ||\psi||_C + ||\psi'||_C + ||\psi''||_C$. Set

$$\psi_n(x) := \sin nx, \quad \mu_n := n^2, \qquad n = 1, 2, \ldots$$

We are given the function $f \in C[-\pi, \pi]$.

Proposition 2.34 (i) *Let $\mu \neq \mu_n$ for all $n = 1, 2, \ldots$ Then the homogenous equation* (2.57) *has only the trivial solution $\psi \equiv 0$. For every $f \in C[-\pi, \pi]$, the inhomogeneous equation* (2.56) *has a unique solution $\psi \in C^2[-\pi, \pi]$.*

(ii) *Let $\mu = \mu_n$ for fixed $n = 1, 2, \ldots$. Then the homogenous equation* (2.57) *has the solution $\psi = \gamma\psi_n$ where γ is an arbitrary real number. The inhomogeneous problem* (2.56) *has a solution iff*

$$\int_{-\pi}^{\pi} f(x)\psi_n(x)dx = 0.$$

If ψ_{special} is a solution of (2.56), *then the general solution of* (2.57) *reads as*

$$\psi(x) = \psi_{\text{special}}(x) + \gamma\psi_n(x), \qquad -\pi \leq x \leq \pi, \qquad \gamma \in \mathbb{R}.$$

This is a problem of index zero. In fact, let d (resp. s) denote the dimension of the solution space (resp. the number of linearly independent solvability conditions). In case (i), we have $d = 0$ and $s = 0$. Hence index $=d - s = 0$. In case (ii), we have $d = 1$ and $s = 1$. Hence index $= d - s = 0$.

Further reading. For a detailed study of linear operator equations (including generalized duality via dual pairs) and their applications to integral equations and differential equations, we recommend:

E. Zeidler, Applied Functional Analysis, Vol. 1: Applications to Mathematical Physics, Vol 2: Main Principles and Their Applications, Springer, New York, 1995.

2.10.5 The Language of Matrices

In finite-dimensional linear spaces, the theory of linear operators is equivalent to matrix theory.
However, this equivalence is destroyed in infinite-dimensional spaces. Here, linear operators have priority, since the passage from linear operators to infinite matrices may cause a loss of crucial information about the domain of definition of the operator. This was emphasized by John von Neumann in about 1930 when creating the mathematical foundations of quantum mechanics based on the spectral theory of self-adjoint operators in Hilbert spaces.[16]

Folklore

Let us fix either $\mathbb{K} := \mathbb{R}$ (real matrix elements) or $\mathbb{K} = \mathbb{C}$ (complex matrix elements). We are given $\mathbf{y} \in Y$. Our goal is to transform the linear operator equation

$$\boxed{A\mathbf{x} = \mathbf{y}, \qquad \mathbf{x} \in X} \tag{2.59}$$

into the matrix equation

$$\boxed{\mathcal{A}x = y, \qquad x \in \mathbb{K}^n} \tag{2.60}$$

where $y \in \mathbb{K}^m$ is given. Explicitly, the matrix equation (2.60) reads as[17]

[16] J. von Neumann, Mathematical Foundations of Quantum Mechanics (in German), Springer, Berlin, 1932 (reprint 1996). English edition: Princeton University Press, 1955. See also the discussion on page 430 of Vol. II.

[17] We will use the convention that the upper index of a matrix entry is always the row index.

$$\begin{pmatrix} A_1^1 & A_2^1 & \dots & A_n^1 \\ A_1^2 & A_2^2 & \dots & A_n^2 \\ \vdots & \vdots & \dots & \vdots \\ A_1^m & A_2^m & \dots & A_n^m \end{pmatrix} \begin{pmatrix} x^1 \\ x^2 \\ \vdots \\ x^n \end{pmatrix} = \begin{pmatrix} y^1 \\ y^2 \\ \vdots \\ y^m \end{pmatrix}. \tag{2.61}$$

Equivalently,

$$\sum_{j=1}^{n} A_j^i x^j = y^i, \qquad i = 1, \dots, m. \tag{2.62}$$

Here, we choose $x^j, y^i, A_j^i \in \mathbb{K}$ for $i = 1, \dots, m$ and $j = 1, \dots, n$. This corresponds to

$$x \in \mathbb{K}^n, \quad y \in \mathbb{K}^m, \qquad n, m = 1, 2, \dots$$

Our goal is to obtain the following commutative diagram:

$$\begin{array}{ccc} X & \xrightarrow{A} & Y \\ {\scriptstyle \tau(X)}\downarrow & & \downarrow{\scriptstyle \tau(Y)} \\ \mathbb{K}^n & \xrightarrow{\mathcal{A}} & \mathbb{K}^m. \end{array} \tag{2.63}$$

In order to get this, let $n := \dim X$ and $m := \dim Y$.

- We choose a basis $\mathbf{b}_1(X), \dots, \mathbf{b}_n(X)$ of the space X. Every element $\mathbf{x}$ of the linear space X can be uniquely represented as

$$\mathbf{x} = \sum_{j=1}^{n} x^j \mathbf{b}_j(X), \qquad x^1, \dots, x^n \in \mathbb{K}.$$

  We assign to $\mathbf{x}$ the column matrix

$$x = \begin{pmatrix} x^1 \\ \vdots \\ x^n \end{pmatrix}.$$

  Set $\tau(\mathbf{x}) := x$. Then the map $\tau : X \to \mathbb{K}^n$ is a linear isomorphism. In what follows, we will write $\tau(X)$ instead of τ.
- Analogously, we construct the linear isomorphism $\tau(Y) : Y \to \mathbb{K}^m$.

The matrix elements A_j^i of the linear operator $A : X \to Y$ are defined by

$$\boxed{A\mathbf{b}_j(X) = \sum_{i=1}^{m} A_j^i \mathbf{b}_i(Y), \qquad j = 1, \dots, n} \tag{2.64}$$

where $\mathbf{b}_1(Y), \dots, \mathbf{b}_m(Y)$ is a basis of Y. Therefore, the numbers $A_j^i \in \mathbb{K}$ are uniquely determined by the operator A. In particular, we get

$$A\mathbf{x} = A\left(\sum_{j=1}^{n} x^j \mathbf{b}_j(X)\right) = \sum_{i=1}^{m}\sum_{j=1}^{n} A_j^i x^j \mathbf{b}_i(Y),$$

and $\mathbf{y} = \sum_{i=1}^{m} y^i \mathbf{b}_i(Y)$. This yields (2.62).

The matrix product. Let $\mathcal{A} = (A^i_j)$ be an $(m \times n)$-matrix with entries in $\mathbb{K}$, and let $\mathcal{B} = (B^k_i)$ be an $(l \times m)$-matrix with entries in $\mathbb{K}$. By definition, the entries of the product matrix $\mathcal{C} := \mathcal{B}\mathcal{A}$ are given by

$$\boxed{C^k_j := \sum_{i=1}^{m} B^k_i A^i_j, \qquad k = 1, \ldots, l, j = 1, \ldots n.} \tag{2.65}$$

For a simple example of matrix multiplication, see (1.9) on page 74. This definition is justified by the following proposition.

Proposition 2.35 *Let X, Y, Z be linear spaces of dimension $n, m, l = 1, 2, \ldots$ We are given the linear operators $A : X \to Y$ and $B : Y \to Z$ with the corresponding matrices $\mathcal{A}$ and $\mathcal{B}$, respectively. Then the following diagram is commutative:*

$$\begin{array}{ccccc} X & \xrightarrow{A} & Y & \xrightarrow{B} & Z \\ \downarrow & & \downarrow & & \downarrow \\ \mathbb{K}^n & \xrightarrow{\mathcal{A}} & \mathbb{K}^m & \xrightarrow{\mathcal{B}} & \mathbb{K}^l. \end{array} \tag{2.66}$$

Here, the vertical maps are linear isomorphisms.

This proposition tells us that the matrix product $\mathcal{B}\mathcal{A}$ corresponds to the operator product BA.

Proof. Let $Ax = y$ and $By = z$. It follows from

$$z^k = \sum_{i=1}^{m} B^k_i y^i, \qquad y^i = \sum_{j=1}^{n} A^i_j x^j$$

that $z^k = \sum_{j=1}^{n} C^k_j x^j$. □

The dual matrix equation. Next let be given the linear functional $G \in X^d$. Our goal is to transform the dual linear operator equation

$$\boxed{A^d F = G, \qquad F \in Y^d} \tag{2.67}$$

into the dual matrix equation

$$\boxed{\mathcal{A}^d \mathcal{F} = \mathcal{G}, \qquad \mathcal{F} \in \mathbb{K}^m} \tag{2.68}$$

where $\mathcal{G} \in \mathbb{K}^n$. Here, by definition, the dual matrix $\mathcal{A}^d$ is obtained from the original matrix $\mathcal{A}$ by transposing rows and columns.[18] Explicitly, equation (2.68) reads as

$$\begin{pmatrix} A^1_1 & A^2_1 & \ldots & A^m_1 \\ A^1_2 & A^2_2 & \ldots & A^m_2 \\ \vdots & \vdots & \ldots & \vdots \\ A^1_n & A^2_n & \ldots & A^m_n \end{pmatrix} \begin{pmatrix} F_1 \\ F_2 \\ \vdots \\ F_m \end{pmatrix} = \begin{pmatrix} G_1 \\ G_2 \\ \vdots \\ G_n \end{pmatrix}. \tag{2.69}$$

[18] Explicitly, if $A = (A^i_j)$, then $(A^d)^i_j := A^j_i$. Moreover, we set $(A^c)^i_j := (A^i_j)^\dagger$, and $A^\dagger = (A^c)^d$. The matrix $A^d, A^c, A^\dagger$ is called the dual (or transposed), conjugate-complex, adjoint matrix to A, respectively. In particular, the entries of the matrix A^c are the conjugate-complex entries of the matrix A.

This means that

$$\sum_{i=1}^{m} A^i_j F_i = G_j, \qquad j = 1, \ldots, n. \tag{2.70}$$

Proposition 2.36 *The operator equation* (2.67) *is equivalent to* (2.70), *that is, from* (2.63) *we get the following commutative diagram:*

$$\begin{array}{ccc} X^d & \xleftarrow{A^d} & Y^d \\ {\scriptstyle \tau(X^d)}\downarrow & & \downarrow{\scriptstyle \tau(Y^d)} \\ \mathbb{K}^n & \xleftarrow{\mathcal{A}^d} & \mathbb{K}^m. \end{array} \tag{2.71}$$

Here, the maps $\tau(X^d)$ and $\tau(Y^d)$ are linear isomorphisms.

Recall that $(A^dF)(x) = F(Ax)$ for all $x \in X$. Thus, the equation (2.67) means that $F(Ax) = G(x)$ for all $x \in X$.

Proof. We will use the linear functionals $dy^1, \ldots, dy^m$ which form a basis of the dual space Y^d, and which are dual to the basis $\mathbf{b}_1(Y), \ldots \mathbf{b}_m(Y)$ of the space Y, that is,

$$dy^i(\mathbf{b}_j(Y)) = \delta^i_j, \qquad i, j = 1, \ldots, m.$$

Every element F of Y^d can be uniquely represented as

$$F = \sum_{i=1}^{m} F_i dy^i, \qquad F_1, \ldots, F_m \in \mathbb{K}.$$

We assign to F the column matrix

$$\mathcal{F} = \begin{pmatrix} F_1 \\ \vdots \\ F_m \end{pmatrix}.$$

Setting $\tau(F) := \mathcal{F}$, the map $\tau : Y^d \to \mathbb{K}^m$ is a linear isomorphism. In diagram (2.71), we write $\tau(Y^d)$ instead of τ. Similarly,

$$G = \sum_{j=1}^{n} G_j dx^j, \qquad G_1, \ldots, G_n \in \mathbb{K}.$$

It follows from $A^dF = G$ that $(A^dF)(\mathbf{b}_j(X)) = G(\mathbf{b}_j(X)) = G_j$. Hence

$$G_j = F(A\mathbf{b}_j(X)) = F\left(\sum_{i=1}^{m} A^i_j \mathbf{b}_i(Y)\right) = \sum_{i=1}^{m} A^i_j F_i.$$

This is the desired equation (2.70). □

The product rule. Interchanging indices, from the product formula (2.65) we get

$$C^j_k := \sum_{i=1}^{m} A^j_i B^i_k, \qquad k = 1, \ldots, l, j = 1, \ldots n. \tag{2.72}$$

This means that

$$\boxed{(\mathcal{B}\mathcal{A})^d = \mathcal{A}^d\mathcal{B}^d.}$$

Furthermore, from (2.66) we obtain the following commutative diagram:

$$\begin{array}{ccccc} X^d & \xleftarrow{\mathcal{A}^d} & Y^d & \xleftarrow{\mathcal{B}^d} & Z^d \\ \downarrow & & \downarrow & & \downarrow \\ \mathbb{K}^n & \xleftarrow{\mathcal{A}^d} & \mathbb{K}^m & \xleftarrow{\mathcal{B}^d} & \mathbb{K}^l. \end{array} \tag{2.73}$$

Here, the vertical maps are linear isomorphisms.

The rank of a matrix. Consider the $(m \times n)$-matrix

$$\mathcal{A} = \begin{pmatrix} A_1^1 & A_2^1 & \ldots & A_n^1 \\ \vdots & \vdots & \ldots & \vdots \\ A_1^m & A_2^m & \ldots & A_n^m \end{pmatrix} \tag{2.74}$$

where all the entries A_j^i are elements of $\mathbb{K}$. This represents a linear operator

$$\mathcal{A} : \mathbb{K}^n \to \mathbb{K}^m \tag{2.75}$$

given by (2.61). By definition, the rank of the matrix $\mathcal{A}$ is the rank of the linear operator (2.75). The columns (resp. rows) of the matrix $\mathcal{A}$ are denoted by $A_1, \ldots, A_n$ (resp. $A^1, \ldots, A^m$). Thus,

$$\mathcal{A} = (A_1, \ldots, A_n) = \begin{pmatrix} A^1 \\ \vdots \\ A^m \end{pmatrix}.$$

Proposition 2.37 *The rank of the matrix* $\mathcal{A}$, $\mathrm{rank}(\mathcal{A})$, *is equal to the maximal number of linearly independent columns of* $\mathcal{A}$.

Proof. We have $b \in \mathrm{im}(\mathcal{A})$ iff $b = x^1 A_1 + \ldots + x^n A_n$ with coefficients $x_1, \ldots, x_n \in \mathbb{K}$. □

Let us introduce the following terminology:

- The column rank of the matrix $\mathcal{A}$ is the maximal number of linearly independent columns of $\mathcal{A}$.
- The row rank of $\mathcal{A}$ is the maximal number of linearly independent rows of $\mathcal{A}$.
- The determinant rank of $\mathcal{A}$ is the maximal order of nonvanishing subdeterminants of the matrix $\mathcal{A}$.

For example, the determinant rank of the matrix

$$\mathcal{A} := \begin{pmatrix} 0 & 1 & 1 \\ 0 & 0 & a \end{pmatrix}$$

is equal to 2 (resp. 1) if $a \neq 0$ (resp. $a = 0$). Note that $\begin{vmatrix} 1 & 1 \\ 0 & a \end{vmatrix} = a.$

The Grassmann criterion. Let $\mathbf{b}^1, \ldots, \mathbf{b}^n$ be a basis of the linear space X over $\mathbb{K}$. We are given $\mathbf{a}^i := \sum_{j=1}^n A_j^i \mathbf{b}^j$ where $i = 1, \ldots, m$.

Proposition 2.38 *The vectors* $\mathbf{a}^1, \ldots, \mathbf{a}^m$ *are linearly independent iff the determinant rank of the matrix* $\mathcal{A}$ *from* (2.74) *is equal to* m.

The vectors $\mathbf{a}^1, \ldots, \mathbf{a}^m$ *are linearly dependent iff the determinant rank of the matrix* $\mathcal{A}$ *from* (2.74) *is less than* m.

This proposition is an immediate consequence of Theorem 2.24 on page 148. In order to illustrate this, consider the special case where

$$\mathcal{A} := \begin{pmatrix} A_1^1 & A_2^1 & A_3^1 \\ A_1^2 & A_2^2 & A_3^2 \end{pmatrix}.$$

Then

$$\begin{aligned} \mathbf{a}^1 \wedge \mathbf{a}^2 &= (A_1^1\mathbf{b}^1 + A_2^1\mathbf{b}^2 + A_3^1\mathbf{b}^3) \wedge (A_1^2\mathbf{b}^1 + A_2^2\mathbf{b}^2 + A_3^2\mathbf{b}^3) \\ &= \begin{vmatrix} A_1^1 & A_2^1 \\ A_1^2 & A_2^2 \end{vmatrix} \mathbf{b}^1 \wedge \mathbf{b}^2 + \begin{vmatrix} A_1^1 & A_3^1 \\ A_1^2 & A_3^2 \end{vmatrix} \mathbf{b}^1 \wedge \mathbf{b}^3 + \begin{vmatrix} A_2^1 & A_3^1 \\ A_2^2 & A_3^2 \end{vmatrix} \mathbf{b}^2 \wedge \mathbf{b}^3. \end{aligned}$$

Consequently, $\mathbf{a}^1 \wedge \mathbf{a}^2 = 0$ iff all the determinants are equal to zero. Moreover, $\mathbf{a}^1 \wedge \mathbf{a}^2 \neq 0$ iff at least one determinant is different from zero.

Proposition 2.39 *The rank of the matrix* $\mathcal{A}$ *from* (2.74) *is equal to to the determinant rank of* $\mathcal{A}$.

This is an immediate consequence of Props. 2.37 and 2.38. Transposing rows and columns of the matrix $\mathcal{A}$ and noting that subdeterminants remain unchanged, we get the following result.

Corollary 2.40 $\mathrm{rank}(\mathcal{A}) = \mathrm{rank}(\mathcal{A}^d)$.

Summarizing this, we obtain that the following quantities of the matrix $\mathcal{A}$ coincide: the rank, the column rank, the determinant rank, and the row rank.

The rank theorem for linear systems. Consider the linear system

$$\sum_{j=1}^{n} A_j^i x^j = b^i, \qquad i = 1, \ldots, m. \tag{2.76}$$

Theorem 2.41 *The system* (2.76) *has a solution iff the rank of the coefficient* $(m \times n)$*-matrix* $\mathcal{A} = (A_j^i)$ *equals the rank of the extended* $(m \times (n+1))$*-matrix* $(\mathcal{A}, b)$.

Proof. The system (2.76) has a solution iff b depends linearly on the columns of the matrix $\mathcal{A}$. Now use the Grassmann criterion (Prop. 2.38). □

The philosophy of linear systems. By the Fredholm alternative (Theorem 2.31 on page 154), the linear system

$$\begin{aligned} \alpha x + \beta y + \gamma z &= b, \\ \lambda x + \mu y + \nu z &= c, \qquad x, y, z \in \mathbb{K} \end{aligned} \tag{2.77}$$

has a solution iff

$$(u, v)\begin{pmatrix} \alpha & \beta & \gamma \\ \lambda & \mu & \nu \end{pmatrix} = 0 \quad \text{implies} \quad (u, v)\begin{pmatrix} b \\ c \end{pmatrix} = 0.$$

This means that the relations between the rows of the coefficient matrix of (2.77) are the same as the relations between the terms b, c of the right-hand side of (2.77). This quite natural principle is valid for all finite-dimensional linear systems. The rank theorem (Theorem 2.41) is nothing else than a reformulation of this principle.

2.10.6 The Gaussian Elimination Method

> The Gaussian elimination method is a universal method for solving finite-dimensional linear matrix equations on computers.
>
> Folklore

In order to explain the basic idea of the Gaussian elimination method, let us consider the linear system (2.77).
Case 1: Suppose that not all the coefficients $\alpha, \beta, \gamma, \lambda, \mu, \nu$ vanish. After transposing rows and using a permutation of the unknowns x, y, z, if necessary, we may assume that $\alpha \neq 0$.[19] Multiplying the first row of (2.77) by $-\lambda$ and the second row by α and using subtraction, we get

$$\begin{aligned} \alpha x + \beta y + \gamma z &= b, \\ \mu' y + \nu' z &= c'. \end{aligned} \tag{2.78}$$

After division, we may assume that $\alpha = 1$.

Case 1.1: Suppose that one of the coefficients μ' and ν' does not vanish. After a permutation of the unknowns y, z and after division, if necessary, we may assume that $\mu' = 1$. Hence

$$\begin{aligned} x + \beta y + \gamma z &= b, \\ y + \nu' z &= c'. \end{aligned}$$

This yields $y = c' - \nu' z$ and $x = b - \beta y - \gamma z$. In this case, the general solution depends on the free parameter z. That is, the dimension of the solution space is equal to 1. The elimination method reduces the coefficient matrix of (2.77) to the following triangular matrix

$$\begin{pmatrix} 1 & \beta & \gamma \\ 0 & 1 & \nu' \end{pmatrix}.$$

Case 1.2: Suppose that $\mu' = \nu' = 0$. There exists a solution of (2.78) iff $c' = 0$. In this case, the general solution of (2.78) depends on the two parameters y and z. That is, the dimension of the solution space is equal to 2. The coefficient matrix is reduced to the following degenerate triangular matrix

$$\begin{pmatrix} 1 & \beta & \gamma \\ 0 & 0 & 0 \end{pmatrix}.$$

Case 2: Suppose that all the coefficients $\alpha, \beta, \gamma, \lambda, \mu, \nu$ vanish. Then the system (2.77) has a solution iff $b = c = 0$. In the latter case, all the tuples $(x, y, z) \in \mathbb{K}^3$ represent solutions of (2.77). That is, the dimension of the solution space is equal to 3.

For general linear systems, we proceed analogously. This elimination method is used in order to solve huge linear systems by the aid of computers. Note that this Gaussian elimination method does not change the rank of the coefficient matrix and the rank of the extended coefficient matrix of the original linear system (2.76).

---

[19] In order to minimize the round-off errors in the computer program, we proceed in such a way that $|\alpha|$ is maximal compared with $|\beta|, |\gamma|, \ldots$

2.11 Changing the Basis and the Cobasis

Both the covariant and the contravariant transformation law play a key role in classical tensor calculus.

Folklore

Einstein's summation convention. In this section, we sum over equal upper and lower indices from 1 to n. For example,

$$\mathbf{v} = v^i \mathbf{b}_i = v^{i'} \mathbf{b}_{i'} \tag{2.79}$$

means $\mathbf{v} = \sum_{i=1}^n v^i \mathbf{b}_i = \sum_{i=1}^n v^{i'} \mathbf{b}_{i'}$. Here, i and i' are regarded as different indices. An index is called free iff we do not sum over it. For example, the index i' from (2.80) below is free.

The covariant transformation law for the basis vectors. Fix $\mathbb{K} = \mathbb{R}$ or $\mathbb{K} = \mathbb{C}$. Consider the n-dimensional linear space X over $\mathbb{K}$. Let us pass from the basis $\mathbf{b}_1, \ldots, \mathbf{b}_n$ of X to the new basis $\mathbf{b}_{1'}, \ldots, \mathbf{b}_{n'}$ by means of the so-called *covariant* transformation law

$$\boxed{\mathbf{b}_{i'} = T^i_{i'} \mathbf{b}_i, \qquad i' = 1', \ldots, n'} \tag{2.80}$$

together with the inverse transformation law

$$\mathbf{b}_i = T^{i'}_i \mathbf{b}_{i'}, \qquad i = 1, \ldots, n. \tag{2.81}$$

Here, all the transformation coefficients $T^i_{i'}$ and $T^{i'}_i$ are elements of $\mathbb{K}$. Substituting (2.81) into (2.80) and vice versa, we get

$$T^i_{i'} T^{i'}_j = \delta^i_j, \quad T^{i'}_i T^i_{j'} = \delta^{i'}_{j'} \tag{2.82}$$

for all free indices.

The contravariant transformation law for the velocity components. For the components of the velocity vector $\mathbf{v}$ from (2.79) we obtain the so-called *contravariant* transformation law

$$\boxed{v^{i'} = T^{i'}_i v^i, \qquad i' = 1', \ldots, n'.} \tag{2.83}$$

In fact, $v^i \mathbf{b}_i = v^i T^{i'}_i \mathbf{b}_{i'} = v^{i'} \mathbf{b}_{i'}$.

Note that the lower (resp. upper) index of $\mathbf{b}_i$ *(resp* v^i*) indicates the covariant (resp. contravariant) transformation law.*

Let us summarize. Using the language of matrices, the two fundamental transformation laws can be formulated as

$$\begin{pmatrix} v^{1'} \\ \vdots \\ v^{n'} \end{pmatrix} = \mathcal{T} \begin{pmatrix} v^1 \\ \vdots \\ v^n \end{pmatrix}, \qquad \begin{pmatrix} \mathbf{b}_{1'} \\ \vdots \\ \mathbf{b}_{n'} \end{pmatrix} = (\mathcal{T}^d)^{-1} \begin{pmatrix} \mathbf{b}_1 \\ \vdots \\ \mathbf{b}_n \end{pmatrix} \tag{2.84}$$

with the transformation matrices

$$\mathcal{T} := \begin{pmatrix} T^{1'}_1 & T^{1'}_2 & \ldots & T^{1'}_n \\ T^{2'}_1 & T^{2'}_2 & \ldots & T^{2'}_n \\ \vdots & \vdots & \ldots & \vdots \\ T^{n'}_1 & T^{n'}_2 & \ldots & T^{n'}_n \end{pmatrix}, \qquad (\mathcal{T}^d)^{-1} = \begin{pmatrix} T^1_{1'} & T^2_{1'} & \ldots & T^n_{1'} \\ T^1_{2'} & T^2_{2'} & \ldots & T^n_{2'} \\ \vdots & \vdots & \ldots & \vdots \\ T^1_{n'} & T^2_{n'} & \ldots & T^n_{n'} \end{pmatrix}. \tag{2.85}$$

Thus, the contravariant transformation law for the velocity components $v^1, \ldots, v^n$ and the covariant transformation law for the basis vectors $\mathbf{b}_1, \ldots, \mathbf{b}_n$ corresponds to the matrix $\mathcal{T}$ and the so-called contragredient matrix $(\mathcal{T}^d)^{-1}$ (i.e., the inverse of the transposed matrix), respectively.

The contravariant transformation law for the cobasis functionals. It follows from

$$dx^i(\mathbf{v}) = dx^i(v^j\mathbf{b}_j) = v^i$$

and $dx^{i'}(\mathbf{v}) = dx^{i'}(v^{j'}\mathbf{b}_{j'}) = v^{i'}$ that the cobasis $dx^1, \ldots, dx^n$ transforms like the velocity components $v^1, \ldots, v^n$. Hence

$$\boxed{dx^{i'} = T_i^{i'} dx^i, \qquad i' = 1', \ldots, n'.} \tag{2.86}$$

The covariant transformation law for the covector components. Let $F \in X^d$. It follows from

$$F = F_i dx^i = F_{i'} dx^{i'}$$

and $F_i dx^i = F_i T_{i'}^i dx^{i'}$ that

$$\boxed{F_{i'} = T_{i'}^i F_i, \qquad i' = 1', \ldots, n'.} \tag{2.87}$$

That is, the covector components $F_1, \ldots, F_n$ transform like the basis vectors $\mathbf{b}_1, \ldots, \mathbf{b}_n$ which is indicated by the choice of the lower index.

2.11.1 Similarity of Matrices

For a linear operator, the change of the basis of the underlying linear space corresponds to a similarity transformation of the matrices.

Folklore

The space End(X). Let X be an n-dimensional linear space over $\mathbb{K} = \mathbb{R}, \mathbb{C}$. Let $\mathrm{End}(X)$ denote the space of all the linear operators $A : X \to X$ (i.e., endomorphisms of X). We want to show that, quite naturally, there exists a linear isomorphism of the form

$$\boxed{\mathrm{End}(X) \simeq X \otimes X^d.} \tag{2.88}$$

To this end, choose the basis $\mathbf{b}_1, \ldots, \mathbf{b}_n$ and $dx^1, \ldots, dx^n$ of X and X^d, respectively. Then the n^2 tensor products $\mathbf{b}_i \otimes dx^j$ with $i, j = 1, \ldots, n$ form a basis of the linear space $X \otimes X^d$ over $\mathbb{K}$. Define

$$(\mathbf{b}_i \otimes dx^j)(\mathbf{v}) := dx^j(\mathbf{v})\mathbf{b}_i \qquad \text{for all} \quad \mathbf{v} \in X.$$

For every $A \in \mathrm{End}(X)$, this implies

$$\boxed{A = A_j^i \mathbf{b}_i \otimes dx^j,} \tag{2.89}$$

which yields the desired linear isomorphism (2.88). In fact, we have

$$(A_j^i \mathbf{b}_i \otimes dx^j)(\mathbf{v}) = A_j^i dx^j(\mathbf{v})\mathbf{b}_i = A_j^i v^j \mathbf{b}_i = A\mathbf{v}.$$

The transformation law of matrices. Let us pass from the basis $\mathbf{b}_1, \ldots, \mathbf{b}_n$ to the new basis $\mathbf{b}_{1'}, \ldots, \mathbf{b}_{n'}$. Then it follows from

$$A = A^i_j \mathbf{b}_i \otimes dx^j = A^{i'}_{j'} \mathbf{b}_{i'} \otimes dx^{j'}$$

and $A^i_j \mathbf{b}_i \otimes dx^j = (A^i_j T^{i'}_i T^j_{j'}) \mathbf{b}_{i'} \otimes dx^{j'}$ that

$$\boxed{A^{i'}_{j'} = T^{i'}_i T^j_{j'} A^i_j.} \tag{2.90}$$

This means that the matrix elements A^i_j of the linear operator A transform like the product $v^i F_j$. For the matrices $\mathcal{A} = (A^i_j)$ and $\mathcal{A}' = (A^{i'}_{j'})$, we get

$$\boxed{\mathcal{A}' = \mathcal{T}\mathcal{A}\mathcal{T}^{-1}.} \tag{2.91}$$

Alternatively, this follows from the fact that the operator equation $\mathbf{w} = A\mathbf{v}$ corresponds to the matrix equation

$$\begin{pmatrix} w^1 \\ \vdots \\ w^n \end{pmatrix} = \mathcal{A} \begin{pmatrix} v^1 \\ \vdots \\ v^n \end{pmatrix},$$

and hence

$$\begin{pmatrix} w^{1'} \\ \vdots \\ w^{n'} \end{pmatrix} = \mathcal{T} \begin{pmatrix} w^1 \\ \vdots \\ w^n \end{pmatrix} = \mathcal{T}\mathcal{A} \begin{pmatrix} v^1 \\ \vdots \\ v^n \end{pmatrix} = \mathcal{T}\mathcal{A}\mathcal{T}^{-1} \begin{pmatrix} v^{1'} \\ \vdots \\ v^{n'} \end{pmatrix}.$$

Similarity invariants of square matrices. By definition, two real (resp. complex) $(n \times n)$-matrices $\mathcal{A}$ and $\mathcal{A}'$ are called similar iff there exists an invertible real (resp. complex) $(n \times n)$-matrix $\mathcal{T}$ such that the relation (2.91) above is satisfied. This is an equivalence relation for the elements of the linear matrix space $\mathbb{M}(n, n; \mathbb{K})$. Similar matrices describe the same linear operator. Invariants of similar matrices are invariants of the corresponding linear operator. For example, this concerns the determinant and the trace of a linear operator, as we will show below.

2.11.2 Volume Functions

The theory of determinants is equivalent to the theory of volume functions.

Folklore

We want to show that the theory of determinants becomes extremely simple if one uses the notion of volume function. Let X be an n-dimensional linear space over $\mathbb{K} = \mathbb{R}, \mathbb{C}$, where $n = 1, 2, \ldots$ By definition, a volume function on X is an n-linear functional $V : X \times \cdots \times X \to \mathbb{K}$ with the additional property that V vanishes if two arguments coincide (e.g., $V(x_1, x_1, x_3, \ldots, x_n) = 0$). For example, if $\mathbf{a}, \mathbf{b}, \mathbf{c}$ are vectors in E_3, then

$$V_0(\mathbf{a}, \mathbf{b}, \mathbf{c}) := (\mathbf{abc})$$

is a volume function. This is the volume of the parallelepiped spanned by the vectors $\mathbf{a}, \mathbf{b}, \mathbf{c}$ (see Fig. 1.6 on page 83). The following proposition shows that, as a special case, all the possible volume functions on E_3 are of the form $V = \gamma V_0$ where γ is a real number.

Proposition 2.42 *The space of volume functions on X is one-dimensional.*

Explicitly, the following hold. Let $b_1, \ldots, b_n$ be a basis of X. There exists precisely one volume function on X, denoted V_0, such that $V_0(b_1, \ldots, b_n) = 1$. All the volume functions on X have the form $V = \gamma V_0$ where $\gamma = V(b_1, \ldots, b_n)$.
Proof. (I) Every volume function is antisymmetric. For example, let $n = 3$. It follows from $V(x, x, z) = 0$ that

$$0 = V(x + y, x + y, z) = V(x, y, z) + V(y, x, z),$$

and hence $V(x, y, z) = -V(y, x, z)$.

Conversely, every n-linear antisymmetric function $V : X \times \cdots \times X \to \mathbb{K}$ is a volume function.

(II) Let $n = 2$. If V is a volume function, then[20]

$$V(\alpha^i b_i, \beta^j b_j) = \alpha^i \beta^j V(b_i, b_j) = \alpha^i \beta^j \varepsilon_{ij} V(b_1, b_2).$$

In particular, $V_0(\alpha^i b_i, \beta^j b_j) = \alpha^i \beta^j \varepsilon_{ij} V_0(b_1, b_2) = \alpha^i \beta^j \varepsilon_{ij}$. This tells us that we have $V = V(b_1, b_2)V_0$. The general case proceeds similarly. □

2.11.3 The Determinant of a Linear Operator

Theorem 2.43 *Let $A : X \to X$ be a linear operator on the n-dimensional linear space X over $\mathbb{K} = \mathbb{R}, \mathbb{C}$ where $n = 1, 2, \ldots$ There exists precisely one number in $\mathbb{K}$, denoted* $\det(A)$, *such that*

$$V(Ax_1, \ldots, Ax_n) = \det(A) \cdot V(x_1, \ldots, x_n) \quad \text{for all } x_1, \ldots, x_n \in X \tag{2.92}$$

and all volume functions V on X.

The number $\det(A)$ is called the determinant of the linear operator A. The point is that $\det(A)$ is independent of the choice of a basis.
Proof. Set $W(x_1, \ldots, x_n) := V_0(Ax_1, \ldots, Ax_n)$. Since W is a volume function, there exists a number γ_0 such that $W = \gamma_0 V_0$. Set $\det(A) := \gamma_0$. Finally, observe that, for a general volume function, we have $V = \gamma V_0$. □
The determinant possesses the following three properties:

(i) $\det(I) = 1$ if $I : X \to X$ is the identity operator.
(ii) $\det(AB) = \det(A)\det(B)$ if $A, B : X \to X$ are linear operators.
(iii) $\det(A) \neq 0$ iff the linear operator $A : X \to X$ is invertible. In addition, we have $\det(A^{-1}) = (\det(A))^{-1}$.

Proof. Ad (i). This follows immediately from (2.92).

Ad (ii). Note that $V(ABx_1, \ldots, ABx_n)$ is equal to

$$\det(A) \cdot V(Bx_1, \ldots, Bx_n) = \det(A)\det(B) \cdot V(x_1, \ldots, x_n).$$

Ad (iii). If A is invertible, then $AA^{-1} = I$ implies $\det(A)\det(A^{-1}) = 1$.

Conversely, assume that $\det(A) \neq 0$, and A is not invertible. Then there exists an element $b_1 \neq 0$ such that $Ab_1 = 0$. Extend the vector b_1 to a basis $b_1, \ldots, b_n$ of X. Then

$$V_0(Ab_1, \ldots, Ab_n) = \det(A) \cdot V_0(b_1, \ldots, b_n).$$

Choosing the volume function V_0 such that $V_0(b_1, \ldots, b_n) = 1$, the relation $V_0(Ab_1, \ldots, Ab_n) = V_0(0, \ldots) = 0$ yields a contradiction. □

---

[20] We sum over $i, j = 1, 2$.

Let $\mathcal{A} = (A^i_j)$ be the matrix of the linear operator $A : X \to X$ with respect to the basis $b_1, \ldots, b_n$. Then

$$\boxed{\det(A) = \det(\mathcal{A}).}$$

To prove this, choose $n = 2$. Then $V(Ab_1, Ab_2)$ is equal to

$$V(A^i_1 b_i, A^j_2 b_j) = A^i_1 A^j_2 V(b_i, b_j) = A^i_1 A^j_2 \varepsilon_{ij} V(b_1, b_2) = \det(\mathcal{A}) V(b_1, b_2).$$

The general case proceeds analogously.

In particular, this implies the Grassmann relation

$$Ax_1 \wedge \cdots \wedge Ax_n = \det(A) \cdot x_1 \wedge \cdots \wedge x_n \quad \text{for all } x_1, \ldots, x_n \in X. \tag{2.93}$$

In fact, choosing a basis and the matrix elements of the linear operator A, the antisymmetry of the Grassmann product yields (2.93) with $\det(A)$ replaced by $\det(\mathcal{A})$.

2.11.4 The Reciprocal Basis in Crystallography

> The reciprocal basis is crucial in crystallography. For the lattice of a given crystal structure, spanned by the vectors $\mathbf{b}_1, \mathbf{b}_2, \mathbf{b}_3$, the reciprocal basis $\mathbf{b}^1, \mathbf{b}^2, \mathbf{b}^3$ corresponds to the reciprocal lattice of a reciprocal crystal structure.
>
> Folklore

The prototype in the Euclidean space E_3. Let $\mathbf{b}_1, \mathbf{b}_2, \mathbf{b}_3$ be a right-handed basis of the Euclidean space E_3. We are looking for vectors $\mathbf{b}^1, \mathbf{b}^2, \mathbf{b}^3$ with the property

$$\mathbf{b}^i \mathbf{b}_j = \delta^i_j, \qquad i, j = 1, 2, 3. \tag{2.94}$$

Proposition 2.44 *The system* (2.94) *has a unique solution which is called the reciprocal basis* $\mathbf{b}^1, \mathbf{b}^2, \mathbf{b}^3$ *to* $\mathbf{b}_1, \mathbf{b}_2, \mathbf{b}_3$.

Proof. Set $g_{ij} := \mathbf{b}_i \mathbf{b}_j$. By the Gram determinant (1.31) on page 84,

$$\det(g_{ij}) = (\mathbf{b}_1 \mathbf{b}_2 \mathbf{b}_3)^2.$$

Thus, $\det(g_{ij}) \neq 0$. Consequently, the symmetric matrix $\mathcal{G} = (g_{ij})$ has an inverse matrix $\mathcal{G}^{-1} = (g^{ij})$ which is also symmetric, that is,

$$g^{is} g_{sj} = \delta^i_j, \qquad i, j = 1, 2, 3.$$

(I) Uniqueness. Setting $\mathbf{b}^i := b^{is} \mathbf{b}_s$, we get $b^{is} g_{sj} = \delta^i_j$. Hence $b^{is} = g^{is}$.

(II) Existence. The vectors $\mathbf{b}^i := g^{is} \mathbf{b}_s$, $i = 1, 2, 3$, satisfy equation (2.94). □

Obviously, the three vectors

$$\mathbf{b}^1 = \frac{\mathbf{b}_2 \times \mathbf{b}_3}{(\mathbf{b}_1 \mathbf{b}_2 \mathbf{b}_3)}, \quad \mathbf{b}^2 = \frac{\mathbf{b}_3 \times \mathbf{b}_1}{(\mathbf{b}_1 \mathbf{b}_2 \mathbf{b}_3)}, \quad \mathbf{b}^3 = \frac{\mathbf{b}_1 \times \mathbf{b}_2}{(\mathbf{b}_1 \mathbf{b}_2 \mathbf{b}_3)}$$

satisfy condition (2.94). Therefore, they represent the reciprocal basis. In terms of the Riesz duality operator $\aleph : E_3 \to E_3^d$, we get

$$dx^i = \aleph(\mathbf{b}^i), \qquad i = 1, 2, 3.$$

We want to generalize this to higher dimensions.

The metric tensor. Let X be a real n-dimensional Hilbert space with the inner product $\langle .|.\rangle$. The metric tensor $\mathbf{g}$ is defined by

$$\mathbf{g}(\mathbf{v},\mathbf{w}) := \langle \mathbf{v}|\mathbf{w}\rangle \qquad \text{for all} \quad \mathbf{v},\mathbf{w}\in X.$$

Let $\mathbf{b}_1,\ldots,\mathbf{b}_n$ be a basis of X. Setting $g_{ij} := \langle \mathbf{b}_i|\mathbf{b}_j\rangle$, we get

$$\boxed{\mathbf{g} = g_{ij}\,dx^i \otimes dx^j.}$$

In fact, it follows from $\mathbf{v} = v^i\mathbf{b}_i$ and $\mathbf{w} = w^j\mathbf{b}_j$ that

$$(g_{ij}dx^i \otimes dx^j)(\mathbf{v},\mathbf{w}) = g_{ij}dx^i(\mathbf{v})dx^j(\mathbf{w}) = g_{ij}v^iw^j = \langle \mathbf{v}|\mathbf{w}\rangle = \mathbf{g}(\mathbf{v},\mathbf{w}).$$

Changing the basis from $\mathbf{b}_1,\ldots,\mathbf{b}_n$ to $\mathbf{b}_{1'},\ldots,\mathbf{b}_{n'}$, we obtain

$$\mathbf{g} = g_{ij}dx^i \otimes dx^j = g_{i'j'}dx^{i'} \otimes dx^{j'}.$$

It follows from $g_{ij}dx^i \otimes dx^j = g_{ij}T^i_{i'}T^j_{j'}dx^{i'} \otimes dx^{j'}$ that

$$\boxed{g_{i'j'} = T^i_{i'}T^j_{j'}g_{ij}, \qquad i',j' = 1',\ldots,n'.} \tag{2.95}$$

Thus, the components g_{ij} of the metric tensor transform like the product F_iF_j of covector components. Introducing the symmetric matrix $\mathcal{G} = (g_{ij})$, the transformation law (2.95) reads as

$$\mathcal{G}' = (\mathcal{T}^{-1})^d\mathcal{G}\mathcal{T}^{-1}. \tag{2.96}$$

Alternatively, this follows from the fact that $\mathbf{g}(\mathbf{v},\mathbf{w})$ is equal to

$$\begin{aligned}(v^1\ldots v^n)\ \mathcal{G}\begin{pmatrix} w^1\\ \vdots \\ w^n\end{pmatrix} &= (v^{1'}\ldots v^{n'})\ (\mathcal{T}^{-1})^d\mathcal{G}\mathcal{T}^{-1}\begin{pmatrix} w^{1'}\\ \vdots \\ w^{n'}\end{pmatrix}\\ &= (v^{1'}\ldots v^{n'})\ \mathcal{G}'\begin{pmatrix} w^{1'}\\ \vdots \\ w^{n'}\end{pmatrix}.\end{aligned}$$

The dual metric tensor. In the special case where $\mathbf{b}_{1'},\ldots,\mathbf{b}_{n'}$ is an orthonormal basis of the real Hilbert space X, we get $g_{i'j'} = \delta_{i'j'}$, and $\det\mathcal{G}' = 1$. By (2.96), $\det\mathcal{G}' = (\det\mathcal{T}^{-1})^2\det\mathcal{G}$. Hence $\det(\mathcal{G}) \neq 0$. Therefore the inverse matrix $\mathcal{G}^{-1} = (g^{ij})$ exists. The dual metric tensor $\mathbf{g}^d$ is defined by

$$\mathbf{g}^d = g^{ij}\mathbf{b}_i \otimes \mathbf{b}_j.$$

Moreover, we define the reciprocal basis to $\mathbf{b}_1,\ldots,\mathbf{b}_n$ by setting

$$\boxed{\mathbf{b}^i := g^{ij}\,\mathbf{b}_j, \qquad i = 1,\ldots,n.} \tag{2.97}$$

Obviously, we have the following biorthogonal relation:

$$\langle \mathbf{b}^i | \mathbf{b}_j \rangle = \delta^i_j, \qquad i,j = 1,\ldots,n.$$

The latter equation determines the reciprocal basis in a unique way. This follows as in the proof of Prop. 2.44 on page 168.

Matrix elements of linear operators. Since $A\mathbf{b}_i = A^j_i \mathbf{b}_j$, the matrix element can be obtained by

$$A^i_j = \langle \mathbf{b}^i | A\mathbf{b}_j \rangle, \qquad i,j = 1,\ldots,n. \tag{2.98}$$

Alternatively,

$$A^i_j = dx^i(A\mathbf{b}_j), \qquad i,j = 1,\ldots,n. \tag{2.99}$$

2.11.5 Dual Pairing

The Riesz duality operator. Let X be a linear n-dimensional Hilbert space over $\mathbb{K} = \mathbb{R}, \mathbb{C}$, where $n = 1, 2, \ldots$ For every $\mathbf{v} \in X$, we define

$$\aleph(\mathbf{v})(\mathbf{w}) := \langle \mathbf{v} | \mathbf{w} \rangle \qquad \text{for all} \quad \mathbf{w} \in X.$$

The Riesz duality operator $\aleph : X \to X^d$ is an antilinear bijective operator, that is,

$$\aleph(\alpha\mathbf{u} + \beta\mathbf{v}) = \alpha^\dagger \aleph(\mathbf{u}) + \beta^\dagger \aleph(\mathbf{v}) \qquad \text{for all} \quad \alpha, \beta \in \mathbb{K}, \ \mathbf{u}, \mathbf{v} \in X.$$

Let $\mathbf{b}_1, \ldots, \mathbf{b}_n$ be a basis of X. Recall that $dx^i(\sum_{j=1}^n v^j \mathbf{b}_j) := v^i$, $i = 1, \ldots, n$. Hence $dx^i(\mathbf{b}_j) = \delta^i_j$ if $i, j = 1, \ldots, n$. Setting $\mathbf{b}^i := \aleph^{-1}(dx^i)$, we get

$$\langle \mathbf{b}^i | \mathbf{b}_j \rangle = \delta^i_j, \qquad i,j = 1,\ldots,n.$$

This way, we obtain the reciprocal basis $\mathbf{b}^1, \ldots, \mathbf{b}^n$ to $\mathbf{b}_1, \ldots, \mathbf{b}_n$.

The dual pairing. For the linear functional $F \in X^d$ and the vector $\mathbf{w} \in X$, we define the dual pairing

$$\langle F | \mathbf{w} \rangle_d := F(\mathbf{w}).$$

Let $\mathbf{v} \in X$. If $F = \aleph(\mathbf{v})$, then

$$\langle F | \mathbf{w} \rangle_d = \langle \mathbf{v} | \mathbf{w} \rangle. \tag{2.100}$$

To simplify notation, we will write $\langle F | \mathbf{w} \rangle$ instead of $\langle F | \mathbf{w} \rangle_d$. This coincides with the convention used in physics, as we will show in Sect. 2.11.7 on the Dirac calculus.

2.11.6 The Trace of a Linear Operator

> The trace of linear operators is critically used in statistical physics.
>
> Folklore

The trace of a matrix. Let $\mathcal{A}, \mathcal{B}, \mathcal{C}, \mathcal{A}_1, \ldots, \mathcal{A}_m$ be complex $(n \times n)$-matrices. The trace of the matrix $\mathcal{A}$ is defined by

$$\mathrm{tr}(\mathcal{A}) := A^1_1 + A^2_2 + \ldots + A^n_n. \tag{2.101}$$

This is the sum of the main-diagonal entries. Using the Einstein sum convention, $\mathrm{tr}(\mathcal{A}) = A^i_i$. It follows from the definition (2.65) of the matrix product that

$$\boxed{\mathrm{tr}(\mathcal{AB}) = \mathrm{tr}(\mathcal{BA}).} \tag{2.102}$$

Hence $\mathrm{tr}(\mathcal{A}(\mathcal{BC})) = \mathrm{tr}((\mathcal{BC})\mathcal{A})$. More generally,

$$\mathrm{tr}(\mathcal{A}_1 \cdots \mathcal{A}_{m-1}\mathcal{A}_m) = \mathrm{tr}(\mathcal{A}_m\mathcal{A}_1 \cdots \mathcal{A}_{m-1}), \qquad m = 2, 3, \ldots \tag{2.103}$$

This tells us that the trace of a product of square matrices remains unchanged under a cyclic permutation of the factors. Moreover, we have

$$\mathrm{tr}(\mathcal{A}^d) = \mathrm{tr}(\mathcal{A}), \qquad \mathrm{tr}(\mathcal{A}^\dagger) := (\mathrm{tr}(\mathcal{A}))^\dagger. \tag{2.104}$$

Recall that $(\mathcal{A}^d)^i_j := \mathcal{A}^j_i$ and $(\mathcal{A}^\dagger)^i_j := (\mathcal{A}^j_i)^\dagger$.

The intrinsic trace as a similarity invariant. Let $A : X \to X$ be a linear operator. Let $\mathcal{A}$ be the matrix assigned to A with respect to a fixed basis $\mathbf{b}_1, \ldots, \mathbf{b}_n$ of X. We define the trace of the linear operator by setting

$$\mathrm{tr}(A) := \mathrm{tr}(\mathcal{A}).$$

The point is that this definition does not depend on the choice of the basis. In fact, if we pass to another basis, we assign to A the similar matrix $\mathcal{A}' = \mathcal{TAT}^{-1}$. However,

$$\mathrm{tr}(\mathcal{TAT}^{-1}) = \mathrm{tr}(\mathcal{T}^{-1}\mathcal{TA}) = \mathrm{tr}(\mathcal{A}).$$

Using the dual pairing and the reciprocal basis, we get

$$\mathrm{tr}(A) = dx^i(A\mathbf{b}_i) = \langle dx^i | A\mathbf{b}_i\rangle = \langle \mathbf{b}^i | A\mathbf{b}_i\rangle.$$

2.11.7 The Dirac Calculus

Mnemonically, the Dirac calculus works perfectly.
Folklore

We want to provide a bridge between

- the language of physicists (the Dirac calculus based on bra-vectors, ket-vectors, and their products), and
- the language of mathematicians (based on covectors, vectors, and dual pairings, as well as tensor products).

The modified Einstein sum convention. Since the Dirac calculus does not distinguish between upper and lower indices, we modify the Einstein sum convention by postulating that we sum over two equal indices from 1 to n. For example, the crucial Dirac completeness relation for the identity operator,

$$\boxed{I = |i\rangle\langle i|,} \tag{2.105}$$

explicitly means $I = \sum_{i=1}^n |i\rangle\langle i|$. In terms of mathematics, this corresponds to the following sum of tensor products: $I = \mathbf{b}_i \otimes dx^i = \sum_{i=1}^n \mathbf{b}_i \otimes dx^i$.

Terminology. Let X be an n-dimensional linear Hilbert space over $\mathbb{K} = \mathbb{R}, \mathbb{C}$. We choose:

- $\mathbf{b}_1, \ldots, \mathbf{b}_n$ (basis of the original space X),
- $dx^1, \ldots, dx^n$ (cobasis of the dual space X^d),
- $\mathbf{b}^1, \ldots, \mathbf{b}^n$ (reciprocal basis of X),
- $\mathbf{e}_1, \ldots, \mathbf{e}_n$ (orthonormal basis of X).

Note that if $\mathbf{b}_i = \mathbf{e}_i$ for all i, then $\mathbf{b}^i = \mathbf{e}_i$ for all i. Physicists use the following symbols:

- $|\mathbf{w}\rangle$ (replacing the vector $\mathbf{w} \in X$),
- $\langle F|$ (replacing the covector $F \in X^d$),
- $\langle F|A$ (replacing the dual operator A^d applied to F, i.e., A^dF),
- $|i\rangle$ (replacing the basis vector $\mathbf{b}_i$),
- $\langle i|$ (replacing the cobasis vector dx^i),
- $\langle \mathbf{v}|$ (replacing the special covector $F = \aleph(\mathbf{v})$ where $\mathbf{v} \in X$),
- $\langle \mathbf{v}|A|\mathbf{w}\rangle$ (replacing the inner product $\langle \mathbf{v}|A\mathbf{w}\rangle$),
- $\langle F|A|\mathbf{w}\rangle$ (replacing $F(A\mathbf{w})$).

Moreover, we use the following products:

- $\langle F| \cdot |\mathbf{w}\rangle := \langle F|\mathbf{w}\rangle = F(\mathbf{w})$ (dual pairing),
- $\langle \mathbf{v}| \cdot |\mathbf{w}\rangle := \langle \mathbf{v}|\mathbf{w}\rangle$ (inner product on the Hilbert space X),
- $\langle F|A \cdot |\mathbf{w}\rangle = \langle F| \cdot A|\mathbf{w}\rangle := \langle F|A|\mathbf{w}\rangle$.

In particular, we have the biorthogonality relation

$$\boxed{\langle i|j\rangle = \delta^i_j, \qquad i,j = 1,\ldots,n.} \tag{2.106}$$

Moreover, $\langle i| = \langle \mathbf{b}^i|$ if $i = 1,\ldots,n$. Relation (2.106) summarizes both $dx^i(\mathbf{b}_j) = \delta^i_j$ and $\langle \mathbf{b}^i|\mathbf{b}_j\rangle = \delta^i_j$. Motivated by the word 'bracket' for the symbol $\langle .|.\rangle$,

- the symbol $\langle F|$ is called bra-vector (or costate), and
- the symbol $|\mathbf{w}\rangle$ is called ket-vector (or state).

In the special case, where the basis $\mathbf{b}_1,\ldots,\mathbf{b}_n$ equals the orthonormal basis $\mathbf{e}_1,\ldots,\mathbf{e}_n$, we get

- $|i\rangle = |\mathbf{e}_i\rangle$ and $\langle i| = \langle \mathbf{e}_i|$.
- The biorthogonality relation (2.106) passes over to the orthogonality relation $\langle \mathbf{e}_i|\mathbf{e}_j\rangle = \delta_{ij}$, $i,j = 1,\ldots,n$.

Finally, we postulate that there exists a †-operation (called dagger operation) which has the following properties:

- $\alpha^\dagger$ (conjugate-complex number assigned to $\alpha \in \mathbb{K}$),
- $\alpha^{\dagger\dagger} = \alpha$,
- $|\mathbf{v}\rangle^\dagger := \langle \mathbf{v}|$ (replacing the Riesz duality $\mathbf{v} \mapsto \aleph(\mathbf{v})$ where $\mathbf{v} \in X$),
- $\langle \mathbf{v}|^\dagger = |\mathbf{v}\rangle^{\dagger\dagger} = |\mathbf{v}\rangle$,
- $A^\dagger$ (linear operator $A^\dagger : X \to X$ adjoint to the linear operator $A : X \to X$).

For all possible objects a, b (i.e., numbers α, β, vectors $|\mathbf{v}\rangle, |\mathbf{w}\rangle$, covectors $\langle \mathbf{v}|, \langle \mathbf{w}|$, and linear operators $A, B : X \to X$), we have the following rules:

- $(\alpha a + \beta b)^\dagger = \alpha^\dagger a^\dagger + \beta^\dagger b^\dagger$ (antilinearity),
- $a^{\dagger\dagger} = a$ (doubling rule),
- $(ab)^\dagger = b^\dagger a^\dagger$ (anti-product rule).

For example, in the language of the Dirac calculus,

$$\langle \mathbf{v}|\mathbf{w}\rangle^\dagger = (\langle \mathbf{v}| \cdot |\mathbf{w}\rangle)^\dagger = |\mathbf{w}\rangle^\dagger \cdot \langle \mathbf{v}|^\dagger = \langle \mathbf{w}| \cdot |\mathbf{v}\rangle = \langle \mathbf{w}|\mathbf{v}\rangle.$$

This shows that the Dirac calculus fits the basic property $\langle \mathbf{v}|\mathbf{w}\rangle^\dagger = \langle \mathbf{w}|\mathbf{v}\rangle$ of the inner product on the Hilbert space X. Furthermore,

$$\begin{aligned}\langle \alpha\mathbf{v}| + \langle \beta\mathbf{w}| &= |\alpha\mathbf{v}\rangle^\dagger + |\beta\mathbf{w}\rangle^\dagger = (\alpha|\mathbf{v}\rangle)^\dagger + (\beta|\mathbf{w}\rangle)^\dagger \\ &= \alpha^\dagger|\mathbf{v}\rangle^\dagger + \beta^\dagger|\mathbf{w}\rangle^\dagger = \alpha^\dagger\langle \mathbf{v}| + \beta^\dagger\langle \mathbf{w}|.\end{aligned}$$

This corresponds to the antilinearity of the Riesz duality operator $\aleph : X \to X^d$ on page 170.

Applications of Dirac's completeness relation. The mnemonic elegance of the Dirac calculus relies on the completeness relation (2.105) above. The point is that the Dirac calculus allows us to pass quickly from vectors $\mathbf{v}$, covectors F, and linear operators A to components v^i, co-components F_i, and matrix elements A^i_j, respectively.

- For all vectors $\mathbf{v} \in X$, we have $|\mathbf{v}\rangle = I|\mathbf{v}\rangle$. Hence, by (2.105),
$$|\mathbf{v}\rangle = |i\rangle\langle i|\mathbf{v}\rangle.$$
This corresponds to $\mathbf{v} = v^i\mathbf{b}_i$ with $v^i := \langle i|\mathbf{v}\rangle = \langle \mathbf{b}^i|\mathbf{v}\rangle$. Observe that this encodes the two relations $v^i = dx^i(\mathbf{v})$ and $v^i = \langle \mathbf{b}^i|\mathbf{v}\rangle$.
- For all covectors $F \in X^d$, we have $\langle F| = \langle F|I$. Hence
$$\langle F| = \langle F|i\rangle\langle i|.$$
This corresponds to $F = F_i dx^i$ with $F_i := \langle F|i\rangle = F(\mathbf{b}_i)$.
- For all vectors $\mathbf{v} \in X$, we have $\langle \mathbf{v}| = \langle \mathbf{v}|I$. Hence
$$\langle \mathbf{v}| = \langle \mathbf{v}|i\rangle\langle i|. \tag{2.107}$$
Explicitly, $\langle \mathbf{v}| = \langle \mathbf{v}|\mathbf{b}_i\rangle\langle \mathbf{b}^i|$. Applying the †-operation, we get $\langle \mathbf{v}|^\dagger = \langle \mathbf{b}^i|^\dagger\langle \mathbf{v}|\mathbf{b}_i\rangle^\dagger$. Hence
$$|\mathbf{v}\rangle = |\mathbf{b}^i\rangle\langle \mathbf{b}_i|\mathbf{v}\rangle.$$
This corresponds to $\mathbf{v} = v_i\mathbf{b}^i$ with $v_i = \langle \mathbf{b}_i|\mathbf{v}\rangle$, and $v_i^\dagger = \langle \mathbf{v}|\mathbf{b}_i\rangle = \langle \mathbf{v}|i\rangle$. In terms of mathematics, this implies $\aleph(\mathbf{v}) = v_i^\dagger\aleph(\mathbf{b}^i) = v_i^\dagger dx^i$ which coincides with (2.107). This tells us that the identity operator $I = |\mathbf{b}_i\rangle\langle \mathbf{b}^i|$ can also be dually represented as
$$I = |\mathbf{b}^i\rangle\langle \mathbf{b}_i|. \tag{2.108}$$
In terms of mathematics, this reflects the fact that the reciprocal basis to the reciprocal basis $\mathbf{b}^1, \ldots, \mathbf{b}^n$ is the original basis $\mathbf{b}_1, \ldots, \mathbf{b}_n$.
- For all vectors $\mathbf{v}, \mathbf{w} \in X$, we have $\langle \mathbf{v}| \cdot |\mathbf{w}\rangle = \langle \mathbf{v}| \cdot I|\mathbf{w}\rangle$. Hence
$$\boxed{\langle \mathbf{v}|\mathbf{w}\rangle = \langle \mathbf{v}|i\rangle\langle i|\mathbf{w}\rangle.}$$
This corresponds to the generalized Parseval equation $\langle \mathbf{v}|\mathbf{w}\rangle = v_i^\dagger w^i$. The classical Parseval equation is obtained in the special case where $\mathbf{b}_1, \ldots, \mathbf{b}_n$ equals an orthonormal basis $\mathbf{e}_1, \ldots, \mathbf{e}_n$. Then $\langle \mathbf{v}|\mathbf{w}\rangle = \langle \mathbf{v}|\mathbf{e}_i\rangle\langle \mathbf{e}_i|\mathbf{w}\rangle$.
- For all covectors $F \in X^d$ and all vectors $\mathbf{w} \in X$, we have $\langle F| \cdot |\mathbf{w}\rangle = \langle F| \cdot I|\mathbf{w}\rangle$. Hence
$$\langle F|\mathbf{w}\rangle = \langle F|i\rangle\langle i|\mathbf{w}\rangle.$$
This corresponds to $F(\mathbf{w}) = F_i w^i$.
- Let $A : X \to X$ be a linear operator. The linear operator equation $A\mathbf{w} = \mathbf{v}$ implies
$$\boxed{\langle i|A|j\rangle\langle j|\mathbf{w}\rangle = \langle i|\mathbf{v}\rangle, \qquad j = 1, \ldots, n.}$$
This corresponds to the matrix equation $A^i_j w^j = v^i$ with $A^i_j := \langle i|A|j\rangle$. In other words, $A^i_j = dx^i(A\mathbf{b}_j) = \langle \mathbf{b}^i|A\mathbf{b}_j\rangle$.

- For linear operators $A, B : X \to X$, we get

$$\boxed{\langle i|AB|j\rangle = \langle i|A|s\rangle\langle s|B|j\rangle, \qquad i,j = 1,\ldots,n.}$$

This corresponds to the product formula $(AB)^i_j = A^i_s B^s_j$ for matrix elements.
- For the linear operator $A : X \to X$, the dual operator equation $A^d F = G$ reads as $\langle F|A = \langle G|$. Hence

$$\langle F|i\rangle\langle i|A|j\rangle = \langle G|j\rangle, \qquad j = 1,\ldots,n.$$

This corresponds to the matrix equation $A^i_j F_i = G_j$.
- For a linear operator $A : X \to X$ and all vectors $\mathbf{v}, \mathbf{w} \in X$, it follows from

$$\langle A\mathbf{w}|\mathbf{v}\rangle = \langle \mathbf{v}|A\mathbf{w}\rangle^\dagger = (\langle \mathbf{v}| \cdot A|\mathbf{w}\rangle)^\dagger = |\mathbf{w}\rangle^\dagger A^\dagger \cdot \langle \mathbf{v}|^\dagger = \langle \mathbf{w}|A^\dagger \cdot |\mathbf{v}\rangle$$

that $\langle A\mathbf{w}|\mathbf{v}\rangle = \langle \mathbf{w}|A^\dagger \mathbf{v}\rangle$.
- The operator equation $A^\dagger \mathbf{v} = \mathbf{w}$ implies

$$\langle i|A^\dagger|j\rangle\langle j|\mathbf{v}\rangle = \langle i|\mathbf{w}\rangle, \qquad i = 1,\ldots,n.$$

This corresponds to the matrix equation

$$(A^\dagger)^i_j v^j = w^i, \qquad i = 1,\ldots,n$$

with $(A^\dagger)^i_j := \langle i|A^\dagger|j\rangle = \langle \mathbf{b}^i|A^\dagger \mathbf{b}_j\rangle$. If $\mathbf{b}_1,\ldots,\mathbf{b}_n$ represents an orthonormal basis $\mathbf{e}_1,\ldots,\mathbf{e}_n$, then

$$(A^\dagger)^i_j = \langle \mathbf{e}_i|A^\dagger \mathbf{e}_j\rangle = \langle A\mathbf{e}_i|\mathbf{e}_j\rangle = \langle \mathbf{e}_j|A\mathbf{e}_i\rangle^\dagger = (A^j_i)^\dagger.$$

2.12 The Strategy of Quotient Algebras and Universal Properties

> Equivalence classes and the corresponding quotient structures (e.g., quotient groups, quotient algebras, quotient fields) appear everywhere in mathematics.
>
> Folklore

We will use the method of equivalence classes studied in Chap. 4 of Vol. II.

Cauchy's approach to complex numbers. In 1847, motivated by the Gauss method of residue classes modulo a prime number (see Sect 4.1.1 of Vol. II), Cauchy introduced complex numbers in the following way. In modern terminology, Cauchy starts with the ring $\mathbb{R}[x]$ of all the polynomials

$$a_0 + a_1 x + \ldots + a_n x^n$$

with real coefficients $a_0,\ldots,a_n$. Let $\mathcal{J}$ denote the ideal generated by the polynomial $1 + x^2$. Explicitly, this ideal consists of all the products $(x^2+1)p(x)$ with $p \in \mathbb{R}[x]$. For polynomials $p, q \in \mathbb{R}[x]$, we write

$$p(x) \sim q(x) \quad \text{iff} \quad p(x) - q(x) \in \mathcal{J}.$$

This is an equivalence relation which respects the addition and multiplication of polynomials. That is, $p \sim r$ and $q \sim s$ imply $p + q \sim r + s$ and $pq \sim rs$. The equivalence classes $[p(x)]$ form a ring, namely, the quotient ring $\mathbb{R}[x]/\mathcal{J}$ equipped with the operations

- $[p(x)] + [q(x)] := [p(x) + q(x)]$ (sum), and
- $[p(x)][q(x)] := [p(x)q(x)]$ (product).

In particular, $x^2 + 1 \sim 0$ implies $[x^2 + 1] = [0]$. The class $[x]$ satisfies the equation

$$[x]^2 = -[1].$$

Moreover, if $n = 1, 2, \ldots$, then we get

- $[x]^{2n} = ([x]^2)^n = (-1)^n$, and
- $[x]^{2n+1} = [x]^{2n}[x] = (-1)^n[x]$.

Thus,

$$[p(x)] = [a_0 + a_1 x + \ldots a_n x^n] = [a_0] + [a_1][x] + \ldots + [a_n][x]^n.$$

Consequently, $[p(x)] = [a] + [b][x]$ where a and b are real numbers. This corresponds to

$$a + bi$$

in the usual notation for complex numbers. Frequently, it is convenient not to use the equivalence classes, but the original objects together with additional relations. In the present case, one works with polynomials $p(x)$ and adds the relation $x^2 + 1 = 0$. For example,

$$x^2 x^2 = (x^2 + 1 - 1)(x^2 + 1 - 1) = (-1)(-1) = 1.$$

The construction of general Clifford algebras. Fix $\mathbb{K} = \mathbb{R}$ or $\mathbb{K} = \mathbb{C}$. Let $B : X \times X \to \mathbb{K}$ be a bilinear symmetric functional on the n-dimensional linear space X over $\mathbb{K}$, $n = 1, 2, \ldots$ Consider the tensor algebra $\bigotimes(X)$, and let J denote the smallest ideal of $\bigotimes(X)$ which contains all the elements

$$x \otimes x - B(x, x), \qquad x \in X. \tag{2.109}$$

For $t, s \in \bigotimes(X)$, we write $t \simeq s$ iff $t - s \in J$. This is an equivalence relation. The corresponding equivalence classes $[t]$ form the quotient algebra $\bigotimes(X)/J$ which is called the Clifford algebra, denoted $\bigvee(X)$, of the linear space X with respect to the bilinear form B. The product on $\bigvee(X)$ is given by

$$[t] \vee [s] := [t \otimes s].$$

If $x \in X$ (resp. $\alpha \in \mathbb{K}$), then we identify $[x]$ with x (resp. $[\alpha]$ with α). By (2.109), $[x] \vee [x] = [x \otimes x] = B(x, x)$. Hence we get the so-called Clifford relation

$$\boxed{x \vee x = B(x, x) \qquad \text{for all} \quad x \in X.}$$

Replacing x by $x + y$, we get

$$(x + y) \vee (x + y) = B(x + y, x + y) = B(x, x) + B(x, y) + B(y, x) + B(y, y).$$

Hence

$$\boxed{x \vee y + y \vee x = 2B(x, y) \qquad \text{for all} \quad x, y \in X.}$$

The Clifford algebra $\bigvee(X)$ has the following properties:

(i) $\bigvee(X)$ is an associative unital algebra over $\mathbb{K}$ which contains both X and $\mathbb{K}$.

(ii) $\bigvee(X)$ has the following universal property. If $L : X \to \mathcal{A}$ is a linear map into the algebra $\mathcal{A}$ over $\mathbb{K}$ with the property $L(x)^2 = B(x,x)$ for all $x \in X$, then there exists an algebra morphism $\mu : \bigvee(X) \to \mathcal{A}$ such that the following diagram is commutative:

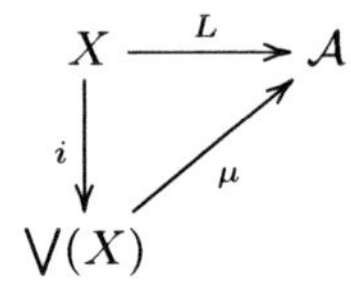

(2.110)

Here, $i(x) := x$ for all $x \in X$.

(iii) By Lagrange's principal axis theorem, there exists a basis $b_1, \ldots, b_n$ of X such that $B(b_k, b_l) = 0$ for all $k \neq l$. Hence

$$b_k \vee b_l + b_l \vee b_k = 0, \qquad k \neq l.$$

The elements $1, b_1, \ldots, b_n$ and all the products

$$b_{i_1} \vee \cdots \vee b_{i_r}$$

with $i_1 < \ldots < i_r$ and $r = 2, \ldots n$ form a basis of the Clifford algebra $\bigvee(X)$. Thus, $\dim \bigvee(X) = 2^n$. For the proof, see Problem 3.15.

The construction of Grassmann algebras. In the special case where $B \equiv 0$, the Clifford algebra $\bigvee(X)$ coincides with the Grassmann algebra $\bigwedge(X)$. Here, we have only to replace the $\vee$-product symbol by $\wedge$.

2.13 A Glance at Division Algebras

Only low-dimensional algebras possess a nice structure.
Folklore

2.13.1 From Real Numbers to Cayley's Octonions

We want to show that complex numbers, quaternions, and octonions can be constructed in a similar way by the inductive process

$$\mathbb{R} \Rightarrow \mathbb{C} \Rightarrow \mathbb{H} \Rightarrow \mathbb{O}.$$

To this end, we start with some set $\mathcal{R}$, and we equip the product set

$$\mathcal{C} := \{(u,v) : u, v \in \mathcal{R}\}$$

with the following operations:

- $(u,v) + (x,y) := (u+x, v+y)$ (sum),
- $(u,v)(x,y) := (ux - y^\dagger v, vx^\dagger + yu)$ (product),
- $(u,v)^\dagger := (u^\dagger, -v)$ (conjugation),
- $\langle (u,v) | (x,y) \rangle := \langle u|x\rangle + \langle v|y\rangle$ (inner product),
- $||(u,v)|| := \sqrt{\langle u|u\rangle + \langle v|v\rangle}$ (norm).

Using conjugation, we further define:

- $\Re(u,v) := \frac{1}{2}((u,v) + (u,v)^\dagger))$ (real part),

- $\Im(u,v) := \frac{1}{2}\big((u,v) - (u,v)^\dagger\big)$ (imaginary part),
- $\Re(\mathcal{C}) := \{(u,v) \in \mathcal{C} : \ \Im(u,v) = 0\}$,
- $\Im(\mathcal{C}) := \{(u,v) \in \mathcal{C} : \ \Re(u,v) = 0\}$.

Proposition 2.45 (i) *If $\mathcal{R} = \mathbb{R}$, then $\mathcal{C}$ is isomorphic to the field $\mathbb{C}$ of complex numbers.*

(ii) *If $\mathcal{R} = \mathbb{C}$, then $\mathcal{C}$ is isomorphic to the skew-field $\mathbb{H}$ of quaternions.*

(iii) *If $\mathcal{R} = \mathbb{H}$, then $\mathcal{C}$ is a real 8-dimensional non-associative algebra called the algebra $\mathbb{O}$ of octonions with the unit element* $(1,0)$.

Proof. Ad (i). This is Hamilton's construction of $\mathbb{C}$.

Ad (ii). One checks easily that the map $\chi : \mathbb{H} \to \mathcal{C}$ given by

$$\chi(\alpha + \beta i + \gamma j + \delta k) := (\alpha + \beta \mathrm{i}, \gamma + \delta \mathrm{i})$$

is an algebra isomorphism. For example,

$$\chi(i) = (\mathrm{i}, 0), \quad \chi(j) = (0,1), \quad \chi(k) = (0, \mathrm{i}),$$

and $\chi(i)\chi(j) = (\mathrm{i},0)(0,1) = (0,\mathrm{i}) = \chi(k)$.

Ad (iii). By construction, the product is distributive. The following example shows that the associative law for the multiplication of octonions is violated. In fact,

$$(0,1) \cdot (0,i)(0,j) = -(0,1)(k,0) = (0,k).$$

However, $(0,1)(0,i) \cdot (0,j) = (i,0)(0,j) = -(0,k)$. □

Corollary 2.46 *For all octonions, the following hold:*

(i) $\big((u,v)(x,y)\big)^\dagger = (x,y)^\dagger (u,v)^\dagger$ *and* $||(u,v)^\dagger|| = ||(u,v)||$,

(ii) $||(u,v)(x,y)|| = ||(u,v)|| \cdot ||(x,y)||$ *(product rule).*

This follows from tedious, but elementary computations. Statement (ii) generalizes Euler's "four squares theorem" to eight squares. Recall that the statements (i) and (ii) of Corollary 2.46 are also valid for $\mathbb{C}$ and $\mathbb{H}$.

Let $\mathcal{C} = \mathbb{C}, \mathbb{H}, \mathbb{O}$. On the set $\Im(\mathcal{C})$, we define the following product:

$$(u,v) \times (x,y) := \tfrac{1}{2}\big((u,v)(x,y) - (x,y)(u,v)\big).$$

If $\mathcal{C} = \mathbb{H}$, then the product corresponds to the vector product on the 3-dimensional Euclidean space E_3, by (1.58). In the case where $\mathcal{C} = \mathbb{O}$, we call this product the octionic vector product.[21]

2.13.2 Uniqueness Theorems

We want to show that, among all possible algebras, the algebras $\mathbb{R}, \mathbb{C}, \mathbb{H}$, and $\mathbb{O}$ are distinguished by special properties. We have the following theorems:[22]

- The algebras $\mathbb{R}$, $\mathbb{C}$, and $\mathbb{H}$ are the only real finite-dimensional associative division algebras (Frobenius 1877).

---

[21] If $\mathcal{C} = \mathbb{C}$, then the product is trivial, that is, $(u,v) \times (x,y) = 0$. This corresponds to $\alpha \mathrm{i} \times \beta \mathrm{i} = \frac{1}{2}(\alpha \mathrm{i} \beta \mathrm{i} - \beta \mathrm{i} \alpha \mathrm{i}) = 0$.

[22] We will not distinguish between isomorphic algebras. The proofs can be found in H. Ebbinghaus et al., Numbers, Springer, New York, 1995.

- The algebras $\mathbb{R}$ and $\mathbb{C}$ are the only real finite-dimensional commutative division algebras (Heinz Hopf 1940).
- The algebras $\mathbb{R}$ and $\mathbb{C}$ are the only real commutative associative normed division algebras (Mazur 1938, Gelfand 1941).
- The algebra $\mathbb{C}$ is the only complex associative normed division algebra (Mazur 1938, Gelfand 1941).[23]
- The algebra $\mathbb{O}$ of octonions is the only real non-associative unital division algebra which has the following properties:
  - There holds $x(xy) = x^2y$ and $(xy)y = xy^2$ for all $x, y \in \mathcal{A}$ (weak associativity).
  - Every element x of $\mathcal{A}$ satisfies a quadratic equation $x^2 + \alpha x + \beta = 0$ with real coefficients α and β (Zorn 1933).
- Let $\mathcal{A}$ be a real finite-dimensional unital algebra. In addition, suppose that $\mathcal{A}$ is a Hilbert space, and we have the product rule

$$||xy|| = ||x|| \cdot ||y|| \qquad \text{for all} \quad x, y \in \mathcal{A}.$$

  Then, $\mathcal{A}$ is one of the algebras $\mathbb{R}, \mathbb{C}, \mathbb{H}, \mathbb{O}$ (Hurwitz 1898).

Finally, let us study possible generalizations of the classical vector product. Let $\mathcal{A}$ be a real algebra which is a Hilbert space of dimension greater than one. Suppose that there exists a distributive product $x \times y$ for all $x, y \in \mathcal{A}$, which has the following properties for all $x, y, z \in \mathcal{A}$:

- $x \times y = -y \times x$ and $\langle x \times y|z\rangle = \langle x|y \times z\rangle$.
- If $||x|| = ||y|| = 1$ and $\langle x|y\rangle = 0$, then $||x \times y|| = 1$.

Then this product is either the classical vector product on the Euclidean space E_3 or the octionic vector product on the 7-dimensional real linear space $\Im(\mathbb{O})$ (Eckmann 1942).

2.13.3 The Fundamental Dimension Theorem

The algebras $\mathbb{R}, \mathbb{C}, \mathbb{H}, \mathbb{O}$ are real division algebras of dimension $1, 2, 4, 8$, respectively.

Theorem 2.47 *Every real finite-dimensional division algebra has the dimension* $1, 2, 4,$ *or* 8.

This famous theorem was independently proven by Milnor and Kervaire in 1958. The proof uses a deep result from topology, namely, the periodicity theorem for the homotopy groups of Lie groups discovered by Bott in 1958. For a topological proof of Theorem 2.47 based on Bott's periodicity theorem and K-theory, we refer to the beautiful paper by F. Hirzebruch, Division algebras and topology. In: H. Ebbinghaus et al., Numbers, Springer, New York, 1995, pp. 281–302.

Historical remarks. In ancient times, mathematicians used rational (and partly irrational) numbers and their operations, namely, addition, subtraction, multiplication, and division. An important prerequisite for the development of algebraic thinking was the transition from the calculation with numbers to the use of letters representing indefinite quantities. This revolution in mathematics was carried out by François Viète (Vieta) (1540–1603).

In order to solve hard mathematical problems, mathematicians invented new mathematical objects together with operations which can be regarded as generalized addition, subtraction, multiplication, or division. The modern structural theory of

[23] The real (resp. complex) algebra $\mathcal{A}$ is called normed iff it is a real (resp. complex) normed space, and we have the inequality $||xy|| \leq ||x|| \cdot ||y||$ for all $x, y \in \mathcal{A}$.

algebra has its origin in lectures of Emmy Noether (1882–1935) in Göttingen and Emil Artin (1898–1962) in Hamburg in the twenties of the twentieth century and was presented in a monograph for the first time in the book *Modern Algebra* by Bartel Leendert van der Waerden (1903–1998) which appeared in 1930 (in German). It is typical of modern algebra that even the quality of the symbols used can be left indeterminate, leading to a genuine theory of the operations.

However, the basis for this work was laid in the nineteenth century. Important impulses were given by Gauss (theory of quadratic forms in number theory, cyclotomic fields), Abel (algebraic functions and algebraic equations), Galois (group theory and algebraic equations), Riemann (genus of Riemann surfaces and divisors of algebraic functions), Weierstrass (algebraic numbers), Cayley (invariant theory and matrix calculus), Kummer and Dedekind (ideal theory), Felix Klein (group theory and geometry), Lie (Lie groups and Lie algebras), Picard (algebraic surfaces), Poincaré (algebraic surfaces, topological manifolds, and automorphic functions), Kronecker (number fields), Hensel (p-adic numbers and p-adic number fields), Camille Jordan (general group theory), Frobenius and Schur (representation theory of groups), Killing and Élie Cartan (structure of Lie algebras), Minkowski (lattices and the geometry of numbers), and Hilbert (number fields and invariant theory).

Theorem 2.47 tells us that, for algebras (also called hypercomplex number systems), the existence of the operation of division is extremely restrictive. The reason for that is of topological nature, as was first discovered by Heinz Hopf in 1940.

3. Representations of Symmetries in Mathematics and Physics, and Elementary Particles

> Representations of symmetries with the aid of linear operators (e.g., matrices) play a crucial role in modern physics. In particular, this concerns the linear representations of groups, Lie algebras, and quantum groups (Hopf algebras).
>
> Folklore

3.1 The Symmetric Group as a Prototype

Fix $n = 1, 2, \ldots$ Consider a set $\mathcal{S}$ of n elements

$$e_1, e_2, \ldots, e_n.$$

The bijective maps $\pi : \mathcal{S} \to \mathcal{S}$ are called permutations. For example, if $n = 3$, then the symbols

$$\pi := \begin{pmatrix} 1\,2\,3 \\ 3\,2\,1 \end{pmatrix}, \qquad \sigma := \begin{pmatrix} 1\,2\,3 \\ 3\,1\,2 \end{pmatrix}$$

correspond to the maps

$$\pi(1) := 3, \;\; \pi(2) := 2, \;\; \pi(3) := 1, \qquad \sigma(1) := 3, \;\; \sigma(2) := 1, \;\; \sigma(3) := 2,$$

respectively. The composition $\pi \circ \sigma$ corresponds to the product of permutations:

$$\begin{pmatrix} 1\,2\,3 \\ 3\,2\,1 \end{pmatrix} \begin{pmatrix} 1\,2\,3 \\ 3\,1\,2 \end{pmatrix} = \begin{pmatrix} 3\,1\,2 \\ 1\,3\,2 \end{pmatrix} \begin{pmatrix} 1\,2\,3 \\ 3\,1\,2 \end{pmatrix} = \begin{pmatrix} 1\,2\,3 \\ 1\,3\,2 \end{pmatrix}.$$

For example, $(\pi \circ \sigma)(1) = \pi(\sigma(1)) = \pi(3) = 1$. Note that the product of permutations is carried out from right to left. This convention corresponds to the notation $(BA)x = B(Ax)$ for linear operators A and B.

The group of all the permutations of the numbers $1, 2, \ldots, n$ is called the symmetric group $Sym(n)$. The numbers of elements of $Sym(n)$ is equal to $n!$ We write $|Sym(n)| = n!$ For example, $|Sym(2)| = 2$, and

$$|Sym(3)| = 1 \cdot 2 \cdot 3 = 6, \quad |Sym(4)| = 1 \cdot 2 \cdot 3 \cdot 4 = 24.$$

The formula due to Stirling (1692–1770) tells us that

$$n! \simeq \sqrt{2\pi n} \left(\frac{n}{e}\right)^n, \qquad n \to \infty.$$

In general, the symbol $|\mathcal{G}|$ denotes the order of the finite group $\mathcal{G}$ (i.e., the number of group elements).

Cycles. The language of cycles is very useful for describing permutations in an effective way.

E. Zeidler, *Quantum Field Theory III: Gauge Theory*,
DOI 10.1007/978-3-642-22421-8_4,

Cycles are the atoms of permutations.

Let $\{i_1, i_2, \ldots, i_k\}$ be a subset of $\{1, 2, \ldots, n\}$. The cycle symbol

$$(i_1 i_2 \ldots i_k)$$

describes a permutation of the elements $1, 2, \ldots, n$ by $i_1 \mapsto i_2 \mapsto \ldots \mapsto i_k \mapsto i_1$. The other numbers remain fixed. This is called a cycle of length k (or briefly a k-cycle). For example, if $n = 4$, then the 3-cycle

$$(123)$$

corresponds to the permutation $1 \mapsto 2 \mapsto 3 \mapsto 1$ and $4 \mapsto 4$. The symbol (1) describes the identical permutation. Obviously, disjoint cycles commute with each other. For example,

$$(12)(534) = (534)(12).$$

The main result reads as follows:

Every permutation of the elements $1, 2, \ldots, n$ can be written as a product of disjoint cycles. This product is unique up to the order of the cycles.

For example,

$$\pi = \begin{pmatrix} 1\,2\,3\,4\,5\,6\,7\,8\,9 \\ 2\,3\,1\,4\,6\,5\,8\,7\,9 \end{pmatrix} = (4)(9)(56)(78)(123). \tag{3.1}$$

Let $\pi \in Sym(n)$. We assign to the permutation π the symbol

$$\boxed{1^{m(1)} 2^{m(2)} \cdots n^{m(n)}} \tag{3.2}$$

iff the disjoint cycle product of π contains exactly $m(r)$ r-cycles where the index r runs from 1 to n. The symbol (3.2) is called the cycle symbol of the permutation π. The number $m(r)$ is called the multiplicity of r-cycles in the disjoint cycle product of π. For example, the permutation (3.1) has the cycle symbol $1^2 2^2 3^1$. Obviously, for the cycle symbol (3.2), we have the following partition of the group order n:

$$\boxed{n = 1 \cdot m(1) + 2 \cdot m(2) + \ldots + n \cdot m(n).}$$

We will show below that the cycle symbols and the corresponding Young diagrams and Young tableaux play a crucial role in the representation theory of the symmetric group $Sym(n)$.

The parity of a permutation. The 2-cycles (ij) are called transpositions; they interchange i and j whereas the other numbers remain unchanged. For example,

$$(13) = \begin{pmatrix} 1\,2\,3\,4 \\ 3\,2\,1\,4 \end{pmatrix}.$$

Every permutation can be written as a product of transpositions. For example,

$$(123) = (31)(12) = (12)(23). \tag{3.3}$$

Obviously, this product representation is not unique, but the number of factors is either even or odd. Then the permutation is called either even or odd. The function

$$\prod_{i<j} (x_i - x_j)$$

of the n real variables $x_1, \ldots, x_n$ remains unchanged (resp. changes sign) under an even (resp. odd) permutation of the variables. Let $\pi \in Sym(n)$. We define

$$\operatorname{sgn}(\pi) := \begin{cases} 1 & \text{if } \pi \text{ is even,} \\ -1 & \text{if } \pi \text{ is odd.} \end{cases}$$

This is called the sign (or the parity) of the permutation π. The map

$$\operatorname{sgn} : Sym(n) \to \{1, -1\}$$

is a group morphism. This means that

$$\operatorname{sgn}(\pi\sigma) = \operatorname{sgn}(\pi)\operatorname{sgn}(\sigma), \qquad \operatorname{sgn}(\pi^{-1}) = (\operatorname{sgn}(\pi))^{-1} = \operatorname{sgn}(\pi).$$

If π is a k-cycle, then $\operatorname{sgn}(\pi) = (-1)^{k+1}$.

Conjugacy classes of the symmetric group $Sym(n)$**.** For two permutations $\pi, \sigma \in Sym(n)$, we write

$$\pi \sim \sigma$$

iff there exists a permutation $\tau \in Sym(n)$ such that $\pi = \tau\sigma\tau^{-1}$. This is an equivalence relation. We say that π is equivalent to σ. The equivalence classes $[\pi]$ are called conjugacy classes. If $\tau \in Sym(n)$, then

$$\tau(i_1 i_2 \ldots i_k)\tau^{-1} = (\tau(i_1)\tau(i_2)\ldots\tau(i_k)). \tag{3.4}$$

For example, $(12)(321)(12)^{-1} = (12)(132)(21) = (312)$.

Two elements $\pi, \sigma \in Sym(n)$ are conjugate iff they have the same cycle symbols.

Examples. The symmetric group $Sym(2)$ contains the two elements (1) and (12) with the cycle symbols $1^1 2^0$ and $1^0 2^1$, respectively. There are the two conjugacy classes $[(1)]$ and $[(12)]$. We have the signs $\operatorname{sgn}(1) = 1$ and $\operatorname{sgn}(12) = -1$.

The symmetric group $Sym(3)$ consists of the six permutations

$$(1), \quad (12), \quad (23), \quad (31), \quad (123), \quad (132)$$

with the signs $\operatorname{sgn}(1) = \operatorname{sgn}(123) = \operatorname{sgn}(132) = 1$ (even permutations), and $\operatorname{sgn}(12) = \operatorname{sgn}(23) = \operatorname{sgn}(31) = -1$ (odd permutations). Elements are conjugate iff they possess the same cycle symbol. Thus, there exist the following three conjugacy classes

$$[(1)], \quad [(12), (23), (31)], \quad [(123), (132)].$$

We will show below that the number of equivalence classes of irreducible representations of $Sym(n)$ is equal to the number of conjugacy classes of $Sym(n)$. In turn, this is equal to the number of partitions of the group order n; this coincides with the number of Young diagrams for n (see (3.36) on page 219.)

The universality of the symmetric groups. The following theorem is due to Cayley (1821–1895):

Every finite group is a subgroup of $Sym(n)$ for some positive integer n.

The proofs about permutations can be found in K. Spindler, Abstract Algebra and Applications, Vol. 1, Sects. 22, 23, Marcel Dekker, New York, 1994.

3.2 Incredible Cancellations

Iterative methods. One of the main tasks in elementary particle physics is to compute the cross sections of scattering processes observed in particle accelerators. To this end, physicists use Feynman diagrams and the procedure of renormalization. This leads to extremely complicated computations based on computer algebra. Fortunately enough, there occur incredible cancellations. The prototype is the following well-known formula

$$\boxed{(1-a)(1+a+a^2+\ldots+a^n) = 1-a^{n+1}, \qquad n=1,2,\ldots} \tag{3.5}$$

for all complex numbers a. This implies

$$\frac{1-a^{n+1}}{1-a} = 1+a+a^2+\ldots+a^n \tag{3.6}$$

for all complex numbers a with $a \neq 1$. Formally, we write

$$(1-a)(1+a^2+a^3+\ldots) = 1 \tag{3.7}$$

where a is regarded as a variable. Note the following:

- If a is a complex number with $|a| < 1$, then the formula (3.7) together with
$$\frac{1}{1-a} = 1+a+a^2+a^3+\ldots \tag{3.8}$$
makes sense in terms of a convergent power series expansion.
- The equations (3.7) and (3.8) make sense in terms of formal power series expansions with respect to the variable a.

However, observe that the relation (3.5) also makes sense for all the mathematical quantities which are elements of an associative unital algebra (e.g., a is a square matrix, or a is a linear operator on a linear space). The relation (3.5) is crucial because of its connection with the iterative method

$$\boxed{x_{n+1} = ax_n + y, \qquad n=0,1,2,\ldots} \tag{3.9}$$

Here, the object y is given, and we choose the starting point $x_0 := 0$. Successively, we obtain $x_1 = y$, $x_2 = y + ay$, $x_3 = y + ay + a^2y, \ldots$ Formally, the expression

$$x = (1+a+a^2+a^3+\ldots)y \tag{3.10}$$

is a solution of the equation

$$\boxed{x = ax + y,} \tag{3.11}$$

by (3.5). Rigorously, the equation (3.11) has a unique solution if X is a Banach space (e.g., a Hilbert space), and a is a linear operator $a : X \to X$ with the norm property $||a|| < 1$. Then, for given $y \in X$, the series (3.10) converges in the Banach space X, and it represents the unique solution $x \in X$ of the equation (3.11).[1] As a generalization of (3.6) let us mention the identity

$$(a-b)(a^n + a^{n-1}b + a^{n-2}b^2 + \ldots + ab^{n-1} + b^n) = a^{n+1} - b^{n+1}, \qquad n=1,2,\ldots$$

Table 3.1. Symmetric Schur polynomials in two variables

$\lvert\lambda\rvert = 1$	$\mathcal{S}_{(1,0)}(a,b) = a + b$
$\lvert\lambda\rvert = 2$	$\mathcal{S}_{(1,1)}(a,b) = ab, \qquad \mathcal{S}_{(2,0)}(a,b) = a^2 + ab + b^2$
$\lvert\lambda\rvert = 3$	$\mathcal{S}_{(3,0)}(a,b) = a^3 + a^2b + ab^2 + b^3$

which is valid for complex numbers a and b (or, more generally, for elements a, b of an associative commutative unital algebra).

Symmetric Schur polynomials. We want to generalize (3.7). Our goal is the identity

$$(1-ac)(1-ad)(1-bc)(1-bd) \cdot P(a,b,c,d) = 1 \tag{3.12}$$

due to Cauchy (1789–1857). Here,

$$P(a,b,c,d) = 1 + \sum_{k=1}^{\infty} \sum_{|\lambda|=k} \mathcal{S}_\lambda(a,b)\, \mathcal{S}_\lambda(c,d)$$

where we define

$$\mathcal{S}_\lambda(a,b) := \frac{\begin{vmatrix} a^{1+\lambda_1} & b^{1+\lambda_1} \\ a^{\lambda_2} & b^{\lambda_2} \end{vmatrix}}{a-b}. \tag{3.13}$$

Alternatively, we write

$$\frac{1}{(1-ac)(1-ad)(1-bc)(1-bd)} = P(a,b,c,d). \tag{3.14}$$

Here, we set $\lambda := (\lambda_1, \lambda_2)$ where λ_1 and λ_2 are integers with $\lambda_1 \ge \lambda_2 \ge 0$. Furthermore, we set $|\lambda| := \lambda_1 + \lambda_2$. The functions $\mathcal{S}_\lambda$ are special cases of the symmetric Schur polynomials. In fact, if we set $a = b$, then the determinant defining $\mathcal{S}_\lambda(a,b)$ vanishes. Thus, it can be divided by $a - b$. Note that $\mathcal{S}_\lambda$ is a homogeneous polynomial of degree $|\lambda|$. For example, $\mathcal{S}_{(1,0)}(a,b) = \frac{a^2-b^2}{a-b} = a + b$. Using Table 3.1, we get the following beautiful cancellation formulas (3.12) and (3.14) with

$$\begin{aligned} P(a,b,c,d) &= 1 + (a+b)(c+d) + (ab)(cd) + (a^2+ab+b^2)(c^2+cd+d^2) \\ &\quad + \mathcal{S}_{(3,0)}(a,b)\, \mathcal{S}_{(3,0)}(c,d) + \sum_{k=4}^{\infty} \sum_{|\lambda|=k} \mathcal{S}_\lambda(a,b)\, \mathcal{S}_\lambda(c,d) \end{aligned}$$

where $\mathcal{S}_{(3,0)}(a,b) = a^3 + a^2b + ab^2 + b^3$ is the completely symmetric polynomial of degree 3 with respect to the variables a and b.

---

[1] This is a special case of the famous fixed-point theorem due to Banach (1892–1945) which is also valid for nonlinear operator equations. Many applications of Banach's fixed-point theorem (also called the contraction principle) to quite different problems can be found in Zeidler (1986), Vol. I, quoted on page 1089.

Proposition 3.1 *The relations* (3.12) *and* (3.14) *are valid in terms of formal power series expansions with respect to the variables* a, b, c, d.

For the proof, we refer to Problem 3.21 on page 319.

The Kepler equation, Lagrange's inversion formula, Hopf algebras, and quantum field theory. In order to solve the Kepler equation for the time-dependence of the orbits of planets, Lagrange (1736–1813) invented his famous inversion formula. This can be formulated in terms of Hopf algebras. We refer to the discussion of this fascinating piece of mathematics in Sect. 3.4 of Vol. II. These formulas are the prototype for relations used in renormalization theory. We refer to:

> M. Gracia-Bondia, The Epstein–Glaser Approach to Quantum Field Theory, Lecture Notes, AIP Conference Proceedings, 809, pp. 24-43, Mexiko City, 2005. Internet: `http://arxiv.org/hep-th/0408145`

3.3 The Symmetry Strategy in Mathematics and Physics

> One of the main tasks of mathematics and physics is to simplify extremely long computations by getting insight into the symmetry structure of the expressions.
>
> Folklore

It happens quite often in mathematics and physics that there appears a nice final result after many cancellations during the process of computation. The experience shows that the cancellations are the result of hidden symmetry properties. Therefore, it is wise to look for a simpler approach based on symmetry. Let us explain this by considering an example. Suppose that we have to compute the quotient

$$f(a,b,c) := \frac{b^4c^2 - c^4b^2 + c^4a^2 - a^4c^2 + a^4b^2 - b^4a^2}{(a-b)(a-c)(b-c)}.$$

After an elementary, but lengthy computation, we get the very symmetric result

$$f(a,b,c) = (a+b)(a+c)(b+c).$$

Therefore, we are looking for a method which exploits symmetry. In fact, the function f can be written as

$$f(a,b,c) = \frac{D(a,b,c)}{(a-b)(a-c)(b-c)}, \qquad D := \begin{vmatrix} a^4 & b^4 & c^4 \\ a^2 & b^2 & c^2 \\ 1 & 1 & 1 \end{vmatrix}.$$

For the determinant, we get

$$D = \begin{vmatrix} 0 & b^4 - a^2b^2 & c^4 - a^2c^2 \\ a^2 & b^2 & c^2 \\ 1 & 1 & 1 \end{vmatrix} = \begin{vmatrix} 0 & b^2(b^2-a^2) & c^2(c^2-a^2) \\ 0 & b^2-a^2 & c^2-a^2 \\ 1 & 1 & 1 \end{vmatrix},$$

by adding a multiple of the second (resp. third) row to the first (resp. second) row. Hence

$$D = (b^2 - a^2)(c^2 - a^2) \begin{vmatrix} b^2 & c^2 \\ 1 & 1 \end{vmatrix} = (a-b)(a-c)(b-c)(a+b)(a+c)(b+c).$$

This is the classic trick for computing determinants of the Vandermonde type. Hence $f(a,b,c) = (a+b)(a+c)(b+c)$. Note that the function f coincides with the symmetric Schur polynomial $\mathcal{S}_{(2,1,0)}(a,b,c)$ (see Table 3.14 on page 273).

3.4 Lie Groups and Lie Algebras

> Lie groups describe finite symmetries or symmetries which smoothly depend on a finite number of real parameters. Lie algebras are the linearization of Lie groups at the unit element. The passage from Lie groups to Lie algebras simplifies considerably the approach. Lie algebras are frequently called infinitesimal symmetries.
>
> Folklore

The Standard Model in elementary particle physics is based on the following Lie groups:

- the unitary groups $U(1), SU(2), SU(3)$, and
- the Poincaré group $P(1,3)$ (the symmetry group of Einstein's theory of special relativity).

The basic definitions in the theory of Lie groups and Lie algebras can be found in Chapter 7 of Volume I.

If we do not explicitly state the contrary, Lie groups are real finite-dimensional manifolds, and their Lie algebras are real Lie algebras.

The Lie group $U(1)$ as a paradigm. The set of all complex numbers z with $|z| = 1$ forms a group with respect to multiplication. This group is called the Lie group $U(1)$. In fact, if $w, z \in U(1)$, then $wz \in U(1)$. The group $U(1)$ will be studied in Chap. 5. The set

$$u(1) := \{\mathrm{i}\varphi : \ \varphi \in \mathbb{R}\}$$

of purely imaginary numbers is a real linear space. Equipped with the Lie product $[a,b]_- := ab - ba = 0$ for all $a, b \in u(1)$, the set $u(1)$ becomes a real commutative Lie algebra. The linearization

$$\mathrm{e}^{\mathrm{i}\varphi} = 1 + \mathrm{i}\varphi + O(\varphi^2), \qquad \varphi \to 0, \quad \varphi \in \mathbb{R}$$

relates the elements $\mathrm{e}^{\mathrm{i}\varphi}$ of the Lie group $U(1)$ to the elements $\mathrm{i}\varphi$ of the Lie algebra $u(1)$ corresponding to $U(1)$. This is the main idea behind the theory of Lie groups and Lie algebras (see Fig. 5.1 on page 356).

The unitary group $U(n)$. Fix $n = 1, 2, \ldots$ The Lie group $U(n)$ consists of all the complex invertible $(n \times n)$-matrices G with $G^{-1} = G^\dagger$.[2] The real Lie algebra $u(n)$ consists of all the complex $(n \times n)$-matrices A with $A^\dagger = -A$ (skew-adjoint matrices). As for all matrix Lie algebras, the Lie product on $u(n)$ is given by

$$\boxed{[A,B]_- := AB - BA \qquad \text{for all} \quad A, B \in u(n).}$$

The map

[2] This is equivalent to $GG^\dagger = I$.

$$A \mapsto e^A \tag{3.15}$$

is a surjective map from $u(n)$ onto $U(n)$. Conversely, the map

$$G \mapsto \ln G \tag{3.16}$$

is a diffeomorphism from the open neighborhood $\{G \in U(n) : ||G - I|| < 1\}$ of the unit element I of the group $U(n)$ onto an open neighborhood of the zero element in $u(n)$. The real Lie algebra $u(n)$ is called the Lie algebra to the Lie group $U(n)$.

In other words, a sufficiently small neighborhood of the unit element of the Lie group $U(n)$ can be parametrized by the matrices of a sufficiently small neighborhood of the origin of the Lie algebra $u(n)$.

This is typical for the relation between Lie groups and their Lie algebras. Much mathematical material with applications to physics can be found in:

J. Schwinger, On angular momentum, U.S. Atomic Energy Commission, Report NYO-3071, 1952. Reprinted in L. Biedenharn and H. van Dam, Quantum Theory of Angular Momentum, Academic Press, New York, 1965.

J. Louck, Unitary Symmetry and Combinatorics, World Scientific, Singapore, 2008.

The special unitary group $SU(n)$. The Lie group $SU(n)$ consists of all the matrices $G \in U(n)$ with $\det G = 1$. The real Lie algebra $su(n)$ consists of all the matrices $A \in u(n)$ with $\mathrm{tr}(A) = 0$ (traceless). The map (3.15) is a surjective map from $su(n)$ onto $SU(n)$. Conversely, the map (3.16) is a diffeomorphism from the open neighborhood $\{G \in SU(n) : ||G - I|| < 1\}$ of the unit element I of the Lie group $SU(n)$ onto an open neighborhood of the zero element in $su(n)$. The real Lie algebra $su(n)$ is called the Lie algebra to the Lie group $SU(n)$.

The general linear group $GL(n, \mathbb{C})$. All the complex (resp. real) invertible $(n \times n)$-matrices form the Lie group $GL(n, \mathbb{C})$ (resp. $GL(n, \mathbb{R})$). The real Lie algebra $gl(n, \mathbb{C})$ (resp. $gl(n, \mathbb{R})$) consists of all the complex (resp. real) $(n \times n)$-matrices equipped with the Lie product $[A, B]_-$. Here, $gl(n, \mathbb{C})$ (resp. $gl(n, \mathbb{R})$) is called the Lie algebra to the Lie group $GL(n, \mathbb{C})$ (resp. $GL(n, \mathbb{R})$). Note that, in contrast to $gl(n, \mathbb{C})$, the symbol

$$gl_{\mathbb{C}}(n, \mathbb{C})$$

denotes the complex Lie algebra of all the complex $(n \times n)$-matrices equipped with the Lie product $[A, B]_-$.

The special linear group $SL(n, \mathbb{C})$. Let $\mathbb{K} = \mathbb{R}, \mathbb{C}$. The Lie group $SL(n, \mathbb{K})$ consists of all the matrices $G \in GL(n, \mathbb{K})$ with $\det G = 1$. The real Lie algebra $sl(n, \mathbb{K})$ consists of all the matrices $A \in gl(n, \mathbb{K})$ with $\mathrm{tr}(A) = 0$. The real Lie algebra $sl(n, \mathbb{K})$ is the Lie algebra to the Lie group $GL(n, \mathbb{K})$. The complex Lie algebra $sl_{\mathbb{C}}(n, \mathbb{C})$ consists of all the matrices $A \in gl_{\mathbb{C}}(n, \mathbb{C})$ with $\mathrm{tr}(A) = 0$. In particular, the Lie group $GL(1, \mathbb{R})$ consists of all the nonzero real numbers, whereas we have $SL(1, \mathbb{R}) = \{1, -1\}$.

Classification of Lie groups. We have to distinguish between

- compact Lie groups, and
- locally compact (but not compact) Lie groups.

A Lie group is called compact iff it is a compact topological space. In particular, a subset of a finite-dimensional real or complex linear space is compact iff it is bounded and closed. For example, finite groups and the groups $U(n), SU(n)$ are compact Lie groups. The Poincaré group $P(1,3)$ is locally compact, but not compact. For example, if the Lie group $\mathcal{G}$ is compact and arcwise connected, then the map $A \mapsto e^A$ from the Lie algebra $\mathcal{LG}$ to the Lie group $\mathcal{G}$ is surjective. This is not always true for locally compact Lie groups (e.g., this is not true for the Lie group $SL(2,\mathbb{C})$; see Problem 3.20).

The theory of locally compact Lie groups is much more difficult than the theory of compact Lie groups.

In particular, one needs sophisticated tools from functional analysis (see Vol. IV on quantum mathematics). The Lie groups $GL(n,\mathbb{C}), GL(n,\mathbb{R}), SL(n,\mathbb{C}), SL(n,\mathbb{R})$ are locally compact, but not compact. See the historical remarks in Sect. 3.21 on 279.

Classification of Lie algebras. One has to distinguish between solvable and semisimple Lie algebras (see Sect. 3.17.1). For example, the Lie algebra $su(n)$ is semisimple. At the end of the 19th century, Killing and Élie Cartan classified all the complex simple Lie algebras. This yields a classification of all complex semisimple Lie algebras (see Sect. 3.17.2). The Heisenberg algebra reflects the commutation relations in quantum mechanics; it is solvable and not semisimple (see Sect. 3.17.3).

3.5 Basic Notions of Representation Theory

The reader should study this section parallel to Sect. 3.6 on the representations of the group $Sym(2)$ and its applications in biology, chemistry, and physics.

3.5.1 Linear Representations of Groups

Irreducible representations are the atoms of representations.
Folklore

Let X be a real or complex linear space, and let the symbol $GL(X)$ denote the group of all bijective linear operators $A : X \to X$. This group is also called the automorphism group of the linear space X.

We want to realize a given group $\mathcal{G}$ as transformation group on the linear space X.

By definition, a linear representation of the group $\mathcal{G}$ on the linear space X is a group morphism

$$\varrho : \mathcal{G} \to GL(X).$$

Explicitly, we assign to the group element $G \in \mathcal{G}$ the invertible linear operator $\varrho(G) : X \to X$ such that

$$\boxed{\varrho(GH) = \varrho(G)\varrho(H) \qquad \text{for all} \quad G, H \in \mathcal{G}.}$$

This way, the group element G acts on the linear space X by the transformation $x \mapsto y$ where

$$y := \varrho(G)x \qquad \text{for all} \quad x \in X.$$

Mnemonically, one also briefly writes $y = Gx$. In particular, if $\mathbf{1}$ is the unit element of the group $\mathcal{G}$, then $\varrho(\mathbf{1})$ is the identity operator I on X. For the inverse group element G^{-1}, we have $\varrho(G^{-1}) = \varrho(G)^{-1}$. The representation ϱ is called finite-dimensional iff the dimension of the linear space X is finite. The dimension of the linear space X is called the degree $\deg(\varrho)$ of the representation ϱ. The following definitions are basic:

- Invariant subspace of the representation: The linear subspace Y of X is called invariant under the representation ϱ iff all the operators $\varrho(G)$ map Y into Y, that is, we have
$$\varrho(G) : Y \to Y \qquad \text{for all} \quad G \in \mathcal{G}.$$
- Faithful representation: ϱ is called faithful iff it is injective.
- Irreducible representation: ϱ is called irreducible iff the invariant subspaces of ϱ are trivial (i.e., only X and $\{0\}$ are invariant subspaces of ϱ, and $X \neq \{0\}$).
- Completely reducible representation: ϱ is called completely reducible iff there exists a direct sum decomposition
$$X = X_1 \oplus X_2 \oplus \ldots \oplus X_m$$
where $X_1, \ldots, X_m$ are linear subspaces of X which are irreducible with respect to the representation ϱ.
- Unitary representation: If X is a Hilbert space, then the representation ϱ is called unitary iff all the operators $\varrho(G)$ are unitary, that is, for every group element G in $\mathcal{G}$, we have
$$\langle \varrho(G)x | \varrho(G)y \rangle = \langle x|y \rangle \qquad \text{for all} \quad x, y \in X.$$
Let $U(X)$ denote the group of all unitary operators $A : X \to X$. Then, a unitary representation ϱ is a group morphism of the form
$$\varrho : \mathcal{G} \to U(X).$$
Unitary representations of compact Lie groups play a fundamental role in quantum physics. The symbol $SU(X)$ denotes a specific subgroup of $U(X)$; by definition, $SU(X)$ is the component of the unit element I of $U(X)$. If the complex Hilbert space X has the finite dimension $n = 1, 2, \ldots$, then the compact Lie group $U(X)$ is isomorphic to the group $U(n)$ of complex unitary $(n \times n)$-matrices. Moreover, $SU(X)$ is isomorphic to the matrix group $SU(n)$.
- Continuous representation: The linear representation $\varrho : \mathcal{G} \to GL(X)$ of the Lie group $\mathcal{G}$ on the Hilbert space X is called continuous iff, for each element x of X, the map
$$G \mapsto \varrho(G)x$$
is a continuous map from $\mathcal{G}$ to X.
- The direct sum $\varrho \oplus \sigma$ and the tensor product $\varrho \otimes \sigma$ of representations: Let $\varrho : \mathcal{G} \to GL(X)$ and $\sigma : \mathcal{G} \to GL(Y)$ be representations. Then we define[3]
$$(\varrho \oplus \sigma)(G) := \varrho(G) \oplus \sigma(G) \qquad \text{and} \qquad (\varrho \otimes \sigma)(G) := \varrho(G) \otimes \sigma(G)$$
for all $G \in \mathcal{G}$. The tensor product $\varrho \otimes \sigma$ is also called the Kronecker product.
- Character functions due to Frobenius: If $\varrho : \mathcal{G} \to GL(X)$ is a linear representation on the finite-dimensional linear space X, then we set
$$\chi_\varrho(G) := \operatorname{tr} \varrho(G).$$

The function $\chi_\varrho : \mathcal{G} \to \mathbb{C}$ is called the character of the representation ϱ. For characters, we have the following rules:

---

[3] Recall that if $A : X \to X$ and $B : Y \to Y$ are linear operators, then we define
$$(A \oplus B)(x + y) := Ax + By \qquad \text{and} \qquad (A \otimes B)(x \otimes y) := Ax \otimes By$$
for all $x \in X, y \in Y$. This way, we get linear operators $A \oplus B : X \oplus Y \to X \oplus Y$ and $A \otimes B : X \otimes Y \to X \otimes Y$, by linear extension.

- $\chi_{\varrho\oplus\sigma} = \chi_\varrho + \chi_\sigma$ (sum rule),
- $\chi_{\varrho\otimes\sigma} = \chi_\varrho \chi_\sigma$ (product rule), and
- $\chi_\varrho(\mathbf{1}) = \dim(X)$.

Example (rotations about an axis). Consider the group $\mathcal{G}$ of the rotations of the 3-dimensional Euclidean space E_3 about a fixed axis through the origin. Choose a right-handed Cartesian (x, y, z)-coordinate system with the orthonormal basis $\mathbf{i}, \mathbf{j}, \mathbf{k}$. Assume that we rotate about the z-axis. Set

- $Z := \operatorname{span}\{\mathbf{k}\}$ (z-axis),
- $Z^\perp := \operatorname{span}\{\mathbf{i}, \mathbf{j}\}$ ((x, y)-plane).

Then we have the decomposition:

$$E_3 = Z \oplus Z^\perp.$$

The 1-dimensional (resp. 2-dimensional) linear subspace Z (resp. $Z^\perp$) of E_3 is an invariant subspace under the action of the group $\mathcal{G}$ on E_3. Moreover, Z and $Z^\perp$ have no proper linear subspaces which are invariant under the action of the rotation group $\mathcal{G}$. In other words, the linear representation ϱ of the rotation group $\mathcal{G}$ on the 3-dimensional Euclidean space E_3 has the following properties:

- ϱ is irreducible on both the z-axis Z and the orthogonal (x, y)-plane $Z^\perp$;
- ϱ is completely reducible;
- ϱ is unitary, since rotations do not change the inner product on E_3.

Irreducible (resp. completely reducible) representations are also called simple (resp. semisimple).

Equivalent representations. Fix $\mathbb{K} = \mathbb{R}, \mathbb{C}$. Let X and Y be linear spaces over $\mathbb{K}$. The linear representations $\varrho : \mathcal{G} \to GL(X)$ and $\mu : \mathcal{G} \to GL(Y)$ of the group $\mathcal{G}$ are called equivalent iff there exists a linear isomorphism $J : X \to Y$ such that the diagram

$$\begin{CD} X @>{\varrho(G)}>> X \\ @V{J}VV @VV{J}V \\ Y @>{\mu(G)}>> Y \end{CD} \tag{3.17}$$

is commutative for all elements G of the group $\mathcal{G}$. Explicitly, $\varrho(G) = J^{-1}\mu(G)J$. In terms of matrices, this means that there exists a basis in X and Y such that the corresponding matrix elements of the linear operators $\varrho(G)$ and $\mu(G)$ are the same for all $G \in \mathcal{G}$.

Theorem 3.2 *Every continuous linear representation of a compact Lie group $\mathcal{G}$ on a complex finite-dimensional Hilbert space X is completely reducible and equivalent to a unitary representation.*

If, in addition, the group $\mathcal{G}$ is commutative, then the linear irreducible representations of $\mathcal{G}$ act on one-dimensional linear spaces.

Typical examples of compact Lie groups are finite groups, the rotation group $SO(3)$ of the 3-dimensional Euclidean space, and the gauge groups $U(1), SU(2)$, and $SU(3)$ of the Standard Model in particle physics. For finite groups, every linear representation is continuous. Theorem 3.2 tells us that it is sufficient to study the irreducible representations. The proof of Theorem 3.2 can be found in B. Simon, Representations of Finite and Compact Groups, page 23 and page 156, Amer. Math. Soc., Providence, Rhode Island, 1996.

We will classify irreducible representations up to equivalence.

Dual representation. In what follows let $n = 1, 2, \ldots$ The so-called dual representation

$$G \mapsto (G^{-1})^d$$

of the group $GL(n, \mathbb{C})$ assigns to each matrix $G \in GL(n, \mathbb{C})$ the so-called contragredient matrix $(G^{-1})^d$. This is indeed a representation because of

$$\left((GH)^{-1}\right)^d = \left(H^{-1}G^{-1}\right)^d = (G^{-1})^d (H^{-1})^d \qquad \text{for all} \quad G, H \in GL(n, \mathbb{C}).$$

The passage from the matrix G to $(G^{-1})^d$ plays a crucial role in invariant theory, as we will show in Chap. 8. In fact, this is the basis of the fundamental principle of the correct index picture in tensor analysis. In Sect. 3.14.2 on page 236, we will show that the passage from quarks to antiquarks corresponds to a passage from the group $SU(3)$ to its dual representation.

This can be generalized. Let $\varrho : \mathcal{G} \to GL(X)$ be a representation of the group G on the linear space X. Define

$$\varrho^d(G) := \left(\varrho(G)^{-1}\right)^d \qquad \text{for all} \quad G \in X.$$

Then the map $G \mapsto \varrho^d(G)$ is a representation of the group $\mathcal{G}$ on the dual linear space X^d. This representation

$$\varrho^d : \mathcal{G} \to GL(X^d)$$

is called the dual representation to ϱ.

Complex-conjugate representation. The so-called complex-conjugate representation

$$G \to G^c$$

of the group $GL(n, \mathbb{C})$ assigns to each matrix $G \in GL(n, \mathbb{C})$ the complex-conjugate matrix G^c. This is a representation because of

$$(GH)^c = G^c H^c \qquad \text{for all} \quad G, H \in GL(n, \mathbb{C}).$$

Complex-dual representation. The so-called complex-dual representation

$$G \to (G^{-1})^\dagger$$

of the group $GL(n, \mathbb{C})$ assigns to each matrix $G \in GL(n, \mathbb{C})$ the matrix $(G^{-1})^\dagger$. This is a representation because of

$$\left((GH)^{-1}\right)^\dagger = \left(H^{-1}G^{-1}\right)^\dagger = (G^{-1})^\dagger (H^{-1})^\dagger \qquad \text{for all} \quad G, H \in GL(n, \mathbb{C}).$$

As a typical application, we will study later on the spinor calculus which is based on the dual, complex-conjugate, and complex-dual representations of the symplectic group $SL(2, \mathbb{C})$ (the universal covering group of the proper Lorentz group in Einstein's theory of special relativity). The complex-conjugate and complex-dual representations play also a crucial role in complex differential geometry (e.g., Kähler manifolds in string theory).

Let us generalize this. Suppose that $\varrho : \mathcal{G} \to GL(X)$ is a representation of the group $\mathcal{G}$ on the finite-dimensional Hilbert space X. Define

$$\sigma(G) := \left(\varrho(G)^{-1}\right)^\dagger \qquad \text{for all} \quad G \in \mathcal{G}.$$

This yields the complex-dual representation $\sigma : \mathcal{G} \to GL(X)$ of the group $\mathcal{G}$ on X. Setting

$$\mu(G) := (\varrho(G)^\dagger)^d \qquad \text{for all} \quad G \in \mathcal{G},$$

we get the complex-conjugate representation $\mu : \mathcal{G} \to GL(X^d)$ of the group $\mathcal{G}$ on the dual space X^d.[4]

3.5.2 Linear Representations of Lie Algebras

Let $\mathcal{L}$ be a real (resp. complex) Lie algebra. A linear representation of $\mathcal{L}$ on the real (resp. complex) linear space X is a Lie algebra morphism

$$\varrho : \mathcal{L} \to gl(X)$$

where $gl(X)$ denotes the set of all linear operators $\mathsf{A} : X \to X$ equipped with the Lie product

$$[\mathsf{A}, \mathsf{B}]_- := \mathsf{AB} - \mathsf{BA} \qquad \text{for all} \quad \mathsf{A}, \mathsf{B} \in gl(X).$$

Explicitly, we assign to every element A of $\mathcal{L}$ a linear operator

$$\varrho(A) : X \to X$$

such that the Lie product is respected, that is,

$$\varrho(\,[A, B]) = [\varrho(A), \varrho(B)]_- \qquad \text{for all} \quad A, B \in \mathcal{L}.$$

The notions 'invariant linear subspace', 'irreducible representation', 'completely irreducible representation', and 'equivalent representation' are defined analogously as for groups above. Let X be a finite-dimensional Hilbert space. The representation ϱ of the Lie algebra $\mathcal{L}$ is called skew-adjoint iff all the operators $\varrho(A)$ are skew-adjoint, that is,

$$\langle \varrho(A)x|y\rangle = -\langle x|\varrho(A)y\rangle \qquad \text{for all} \quad A \in \mathcal{L}, \;\; x, y \in X.$$

Recall that a linear operator $C : X \to X$ on the finite-dimensional Hilbert space X is skew-adjoint iff $C^\dagger = -C$.

Adjoint representation of the Lie $\mathcal{L}$ algebra on itself. For fixed $A \in \mathcal{L}$, define

$$\mathrm{ad}(A)B := [A, B] \qquad \text{for all} \quad B \in \mathcal{L}.$$

This yields the linear operator $\mathrm{ad}(A) : \mathcal{L} \to \mathcal{L}$. The map $A \mapsto \mathrm{ad}(A)$ is a representation

$$\mathrm{ad} : \mathcal{L} \to gl(\mathcal{L}).$$

This is a consequence of the Jacobi identity (see Problem 3.24). The adjoint representation reveals the intrinsic symmetry of the Lie algebra $\mathcal{L}$.

Dual representation of a Lie algebra. The dual representation of the real Lie algebra $gl(n, \mathbb{C})$ assigns to each matrix $A \in gl(n, \mathbb{C})$ the matrix $-A^d$. This is indeed a representation because of

$$-[A, B]_-^d = (BA - AB)^d = A^d B^d - B^d A^d = [-A^d, -B^d]_-$$

for all $A, B \in gl(n, \mathbb{C})$. This can be generalized. Let $\varrho : \mathcal{L} \to gl(X)$ be a representation of the real (resp. complex) Lie algebra $\mathcal{L}$ on the real (resp. complex) Hilbert space X. Set

---

[4] Note that, for a matrix $G \in GL(n, \mathbb{C})$, we have $G^c = (G^\dagger)^d$.

$$\varrho^d(A) := -\varrho(A)^d \qquad \text{for all} \quad A \in \mathcal{L}.$$

The map $A \mapsto -\varrho(A)^d$ yields the so-called dual representation $\varrho^d : \mathcal{L} \to gl(X^d)$ of the Lie algebra $\mathcal{L}$ on the dual space X^d.

The adjoint representation of a Lie group on its Lie algebra. As a prototype, consider the Lie group $GL(n, \mathbb{C})$. Fix $G_0 \in GL(n, \mathbb{C})$. Set

$$\varrho(G_0)A := G_0 A G_0^{-1} \qquad \text{for all} \quad A \in gl(n, \mathbb{C}).$$

The map $G_0 \to \varrho(G_0)$ is a linear representation of the group $GL(n, \mathbb{C})$ on the real linear space $gl(n, \mathbb{C})$. This is indeed a representation because of

$$(G_0 H_0)A(G_0 H_0)^{-1} = G_0(H_0 A H_0^{-1})G_0^{-1}.$$

The trouble with infinite-dimensional representations. In this chapter, we restrict ourselves to representations on finite-dimensional linear spaces. Every finite-dimensional linear space can be equipped with the structure of a Hilbert space. Unfortunately, if we pass to infinite-dimensional Hilbert spaces, then there occur technical difficulties. This is related to the fact that the generators of unitary groups on infinite-dimensional complex Hilbert spaces are self-adjoint operators which, as a rule, are not defined on the total Hilbert space, as noted by von Neumann and Stone in the late 1920s. Observe that the use of infinite-dimensional Hilbert spaces is unavoidable in quantum field theory, as was first shown by Wigner in 1939 (see page 286). This is part of the mathematical trouble encountered in quantum field theory.

3.6 The Reflection Group $\mathcal{Z}_2$ as a Prototype

> Representations of the symmetry group $\mathcal{Z}_2 := \{1, -1\}$ play a crucial role in physics, chemistry, and biology.
>
> Folklore

3.6.1 Representations of $\mathcal{Z}_2$

The simplest nontrivial group is the (multiplicative) group $\mathcal{Z}_2 := \{1, -1\}$ which consists of the real numbers 1 and -1 with $(-1)(-1) = 1$. The group $\mathcal{Z}_2$ is finite, and hence it is a compact Lie group with the trivial Lie algebra $\mathcal{L}\mathcal{Z}_2 = \{0\}$. The group is isomorphic to the following (multiplicative) groups:

- The reflection group $\{I, -I\}$ of the Euclidean space E_3. This group consists of the identity map $\mathbf{x} \mapsto \mathbf{x}$ and the reflection map $\mathbf{x} \mapsto -\mathbf{x}$.[5]
- The permutation group $Sym(2) = \{(1), (12)\}$. This group consists of the identical permutation (1) (i.e., $1 \mapsto 1$ and $2 \mapsto 2$) and the cyclic permutation (12) (i.e., $1 \mapsto 2$ and $2 \mapsto 1$). The group $Sym(2)$ is isomorphic to the group $\mathcal{Z}_2$. The isomorphism is given by $(1) \mapsto 1, (12) \mapsto -1$.

Moreover, the multiplicative group $\mathcal{Z}_2$ is also isomorphic to the additive group $\mathbb{Z}_2 := \{0, 1\}$ with $1 + 1 = 0$, and $0 + 1 = 1 + 0 = 1, 0 + 0 = 0$. The group isomorphism $\mu : \mathcal{Z}_2 \to \mathbb{Z}_2$ is given by $\mu(1) := 0$ and $\mu(-1) := 1$.

Isomorphic groups possess the same linear representations. Let us study the linear representations of the group $Sym(2)$ on the complex Hilbert space X of finite dimension $n = 1, 2, \ldots$

[5] The isomorphism $\chi : \mathcal{Z}_2 \to \{I, -I\}$ is given by $\chi(1) := I$ and $\chi(-1) := -I$.

Theorem 3.3 *The representation ϱ of the group $Sym(2)$ on X is unitary iff there exists an orthonormal basis $e_1, \ldots, e_n$ of the Hilbert space X such that*

$$\varrho(\pi)e_j = \lambda_j e_j, \qquad j = 1, \ldots, n \tag{3.18}$$

where $\lambda_j = \pm 1$ for all j.

We say that the basis vector e_j has positive (resp. negative) parity iff $\lambda_j = 1$ (resp. $\lambda_j = -1$).

Proof. Suppose that ϱ is a unitary representation. Since the operator $\varrho(\pi)$ is unitary, it possesses an orthonormal basis of eigenvectors $e_1, \ldots, e_n$, by a general result in linear algebra. If $\varrho(\pi)x = \lambda x$, $x \neq 0$, then $\varrho(\pi)\varrho(\pi) = \varrho(\pi^2) = \varrho(\mathrm{id}) = I$. Hence

$$x = \varrho(\pi)^2 x = \varrho(\pi)(\varrho(\pi)x) = \lambda^2 x.$$

This implies $\lambda^2 = 1$, that is, $\lambda = \pm 1$. Conversely, equation (3.18) yields a representation of $Sym(2)$. □

Now let X be a real or complex linear space of dimension $n = 1, 2, \ldots$. Then it follows from Theorem 3.2 that:

> *Every linear representation ϱ of the group $Sym(2)$ on X is given by (3.18) where $e_1, \ldots, e_n$ is a basis of the linear space X. This space can be equipped with a Hilbert space structure such that ϱ becomes a unitary representation.*

In other words, every representation of the commutative group $Sym(2)$ on X is completely reducible, and it acts irreducibly on one-dimensional linear subspaces of X. In addition, ϱ is equivalent to a unitary representation. This result allows a far-reaching generalization to compact Lie groups. This Peter–Weyl theory will be studied in Vol. IV on quantum mathematics.

Consider a fixed right-handed (x, y, z)-Cartesian coordinate system on the Euclidean manifold $\mathbb{E}^3$. The transformation

$$(x, y, z) \mapsto (-x, -y, -z)$$

is called a reflection at the origin $(0, 0, 0)$. In terms of matrices, we have

$$\begin{pmatrix} x' \\ y' \\ z' \end{pmatrix} = \begin{pmatrix} -1 & 0 & 0 \\ 0 & -1 & 0 \\ 0 & 0 & -1 \end{pmatrix} \begin{pmatrix} x \\ y \\ z \end{pmatrix}.$$

The real (3×3)-matrices I (unit matrix) and $-I$ form a multiplicative group which is called the reflection group. The map $\pm I \mapsto \pm 1$ is an isomorphism of the reflection group onto the multiplicative group $\mathcal{Z}_2 = \{1, -1\}$.

3.6.2 Parity of Elementary Particles

The parity of an elementary particle describes the transformation of the particle state under space reflections. Suppose that we describe the given quantum system by a complex finite-dimensional or infinite-dimensional Hilbert space X. In addition, we assume that there exists a linear unitary representation

$$\varrho : \{I, -I\} \to GL(X).$$

Suppose that x is an eigenvector of $\varrho(-I)$, that is,

$$\varrho(-I)x = \lambda x, \qquad x \in X \setminus \{0\}.$$

Then it follows from $(-I)(-I) = I$ that $\varrho(-I)\varrho(-I) = \varrho(I) = I$. Hence $x = \lambda^2 x$. Thus, $\lambda^2 = 1$. By definition, the state x has positive (resp. negative) parity iff $\lambda = 1$ (resp. $\lambda = -1$).

This idea can be used in order to construct representations of the group $\{I, -I\}$. Suppose that the infinite-dimensional Hilbert space X is separable. Then there exists an orthonormal basis $e_1, e_2, \ldots$ of X. We choose numbers $\lambda_j = 1$ or $\lambda_j := -1$. Define

$$\varrho(-I)e_j := \lambda_j e_j, \qquad \varrho(I)e_j := e_j, \qquad j = 1, 2, \ldots$$

This way, we obtain a continuous unitary representation $\varrho : \{I, -I\} \to GL(X)$.

3.6.3 Reflections and Chirality in Nature

There are molecules in nature that exist in right-handed and left-handed versions like two gloves. Such molecules are said to be chiral; molecules lacking such handedness are called achiral. In biological systems on earth, the essential molecules are single-handed. Experimentally, the handedness of molecules can be detected by the fact that linearly polarized light is rotated either clockwise or counterclockwise by chiral molecules. For example, natural amino acids are left-handed. Right-handed amino acids only exist outside earth in cosmic clouds. Fix the numbers $\omega > 0$ and $v > 0$. Choose $\chi = 1$ or $\chi = -1$. The equation

$$x = \cos \omega t, \quad y = \sin \omega t, \quad z = \chi v t, \quad t \in \mathbb{R} \tag{3.19}$$

describes a screw line with positive (resp. negative) chirality if $\chi = 1$ (resp. $\chi = -1$). The reflection $x \mapsto -x, y \mapsto -y, z \mapsto -z$ yields

$$x = -\cos \omega t = \cos \omega(t + t_0), \quad y = -\sin \omega t = \sin \omega(t + t_0), \quad z = -\chi v t, \quad t \in \mathbb{R}$$

with $t_0 := \pi/\omega$. This changes the chirality of the screw line.

3.6.4 Parity Violation in Weak Interaction

Consider a fixed physical process. Apply a space reflection to this process. If the reflected process is not realized in nature, then we say that parity is violated in this process. The classic example of parity violation is the β-decay of cobalt ${}^{60}_{27}\mathrm{Co}$. This was experimentally discovered by Mrs. Wu in 1957. See the discussion in Sect. 2.7 of Vol. I.

The Wu experiment showed that parity can be violated in weak interaction.

Theoretically, parity violation in weak interaction was investigated by Lee (born 1926) and Yang (born 1922); they were awarded the Nobel prize in physics in 1957.

3.6.5 Helicity

The earth is the prototype for the motion of a rotating body. Such a body is characterized by the momentum vector $\mathbf{p}$ (direction of the motion) and the rotation axis $\mathbf{n}$ (unit vector). The vector $\mathbf{S} := s\mathbf{n}$ is called the spin vector where s is the angular momentum of the body. The number

$$\zeta := \frac{\mathbf{Sp}}{|\mathbf{p}|}$$

is called the helicity of the body. In classical mechanics, the helicity may depend on time. In elementary particle physics, as a rule, one uses particle states with time-independent helicity. Here, the vector **S** describes the intrinsic angular momentum of the elementary particle called spin. We refer to Sect. 2.6 of Vol. I (general discussion) and Chap. 15 of Vol. II (application to quantum electrodynamics). In 1957, Goldhaber and his collaborators established experimentally that the helicity of the neutrino is negative. Physicists say that the neutrino is a left-handed particle. For a detailed history of neutrino physics including parity violation and the weak interaction force, we recommend C. Sutton, Spaceship Neutrino, Cambridge University Press, 1992.

3.7 Permutation of Elementary Particles

3.7.1 The Principle of Indistinguishability of Quantum Particles

In contrast to classical particles (e.g., planets), quantum particles (e.g., molecules of a gas or elementary particles) cannot be distinguished by labels. In terms of psychology, quantum particles are not individuals. In terms of mathematics, this means that

Quantum states of several quantum particles have to be invariant under permutations of the particles.

This underlines the importance of the symmetric group $Sym(n)$ for quantum physics. In statistical physics, one has to count the number of different states. Because of the principle of indistinguishability for quantum particles, classical statistics and quantum statistics produce different results. For low temperatures, one has to use quantum statistics.

3.7.2 The Pauli Exclusion Principle

Concerning elementary particles, we have to distinguish between

- fermions (half-integer spin, e.g., electrons, neutrinos, quarks), and
- bosons (integer spin, e.g., photons, gluons, the vector bosons $W^{\pm}, Z$).

The crucial Pauli exclusion principle says that

Two fermions are never in the same quantum state.

Suppose that the function $\psi : \mathbb{R}^2 \to \mathbb{C}$ is antisymmetric, that is,

$$\psi(x, y) = -\psi(y, x) \qquad \text{for all} \quad x, y \in \mathbb{R}.$$

This implies $\psi(x, x) = 0$ for all $x \in \mathbb{R}$. The same idea motivates the fact that we describe fermions (resp. bosons) by quantum states which are antisymmetric (resp. symmetric) with respect to permutations of the quantum particles (see the states b and f on page 198). Many physical phenomena in nature can be explained by using the Pauli exclusion principle. For example, the periodic table of chemical elements is a consequence of the Pauli exclusion principle combined with quantum mechanics (see van der Waerden (1932) quoted on page 282). We recommend:

W. Pauli, Exclusion principle, Lorentz group and reflection of space-time and charge, pp. 30–51. In: W. Pauli (Ed.), Niels Bohr and the Development of Physics, Pergamon Press, New York, 1955.

R. Streater and A. Wightman, PCT, Spin, Statistics, and All That, Benjamin, New York, 1968.

I. Duck and E. Sudarshan, Pauli and the Spin-Statistics Theorem, World Scientific, Singapore, 1997.

3.7.3 Entangled Quantum States

Let x, y be an orthonormal basis of the complex 2-dimensional Hilbert space X. Then the four tensor products

$$x \otimes x, \quad x \otimes y, \quad y \otimes x, \quad y \otimes y$$

form an orthonormal basis of the 4-dimensional complex Hilbert space $X \otimes X$. In terms of physics, we consider x and y as the states of two elementary particles. Interchanging x with y, we obtain a linear representation of the group $Sym(2)$ on the product space $X \otimes X$. Explicitly, for $\pi = (12)$ we get

$$\varrho(\pi)\,(\alpha x \otimes x + \beta x \otimes y + \gamma y \otimes x + \delta y \otimes y) = \alpha x \otimes x + \beta y \otimes x + \gamma x \otimes y + \delta y \otimes y$$

for all complex numbers $\alpha, \beta, \gamma, \delta$. The four states

- $b := \frac{1}{\sqrt{2}}\,(x \otimes y + y \otimes x)$ (bosonic state),
- $f := \frac{1}{\sqrt{2}}\,(x \otimes y - y \otimes x)$ (fermionic state),
- $u := \frac{1}{\sqrt{2}}(x \otimes x + y \otimes y)$,
- $v := \frac{1}{\sqrt{2}}\,(x \otimes x - y \otimes y)$

are invariant under the operator $\varrho(\pi)$, up to sign. Consequently, the splitting

$$\boxed{X \otimes X = \mathrm{span}\{b\} \oplus \mathrm{span}\{f\} \oplus \mathrm{span}\{u\} \oplus \mathrm{span}\{v\}}$$

of the product space $X \otimes X$ into one-dimensional, pairwise orthogonal, linear subspaces corresponds to the decomposition of the representation ϱ into irreducible representations.

From the physical point of view, the state b (resp. f) describes a boson (resp. fermion) which is obtained by composing the particle x with the particle y. The symmetry of b (resp. antisymmetry of f) with respect to a permutation of the two particles x and y reflects the indistinguishability of the two particles. The state $x \otimes y$ violates the principle of indistinguishability of the two particles. Hence the state $x \otimes y$ does not possess any physical meaning. This is the prototype of group-theoretical arguments widely used in elementary particle physics. We will encounter this quite often in the volumes of this monograph. The states b and f are called entangled states. Such states are related to the Einstein–Podolski–Rosen (EPR) paradox.[6] Entangled states play a key role in quantum information. We will study this in Vol. IV. At this point, we refer to M. Nielsen and I. Chuang, Quantum Computation and Quantum Information, Cambridge University Press, 2001. See also M. Freedman et al., Topological quantum computation, Bull. Amer. Math. Soc. **40**(1) (2003), 31–39.

[6] A. Einstein, B. Podolski, and N. Rosen, Can quantum-mechanical description of physical reality be considered complete? Phys. Rev. **47** (1935), 777–780.

3.8 The Diagonalization of Linear Operators

Representation theory is based on the study of invariant linear subspaces. In this connection, one has to understand first the structure of the invariant linear subspaces of a single linear operator. Let us discuss this. In what follows, we assume that

X is an n-dimensional complex Hilbert space with $n = 1, 2, \ldots$

By definition, the linear operator $D : X \to X$ is called diagonalizable iff there exist both a basis $b_1, \ldots, b_n$ of X and complex numbers $\lambda_1, \ldots, \lambda_n$ such that

$$\boxed{Db_j = \lambda_j b_j \qquad \text{for all} \quad j = 1, \ldots, n.}$$

By linearity,

$$D(x^1 b_1 + x^2 b_2 + \ldots + x^n b_n) = \lambda_1 x^1 b_1 + \lambda_2 x^2 b_2 + \ldots + \lambda_n x^n b_n$$

for all $x \in X$ where $x = x^1 b_1 + x^2 b_2 + \ldots + x^n b_n$. This means that the basis vectors $b_1, \ldots, b_n$ are eigenvectors of the operator D, and the complex numbers $\lambda_1, \ldots, \lambda_n$ are the eigenvalues of D. The main result reads as follows.

Theorem 3.4 *The linear operator $D : X \to X$ is diagonalizable iff the linear hull of the eigenvectors of D is equal to X. In particular, self-adjoint, skew-adjoint, and unitary operators are diagonalizable.*

This is a special case of the Jordan normal form to be considered below. Diagonalizable operators are also called completely reducible by eigenspaces.

The language of matrices. Let $A : X \to X$ be a linear operator. Fix a basis $b_1, \ldots, b_n$ of the space X. Let

$$\mathcal{A} = (A^i_j)$$

be the matrix corresponding to the operator A with respect to $b_1, \ldots, b_n$ (see (2.64) on page 158). Then the operator A is diagonalizable iff there exists a complex invertible $(n \times n)$-matrix $\mathcal{T}$ such that

$$\mathcal{T}\mathcal{A}\mathcal{T}^{-1}$$

is a diagonal matrix $\mathrm{diag}(\lambda_1, \ldots, \lambda_n)$.[7] Here, the complex numbers $\lambda_1, \ldots, \lambda_n$ are the eigenvalues of A. The transformation $\mathcal{A} \mapsto \mathcal{T}\mathcal{A}\mathcal{T}^{-1}$ is called a similarity transformation of the matrix $\mathcal{A}$. For an arbitrary complex $(n \times n)$-matrix, the eigenvalues $\lambda_1, \ldots, \lambda_n$ are defined as the zeros of Lagrange's secular equation

$$\det(\mathcal{A} - \lambda I) = 0, \qquad \lambda \in \mathbb{C}.$$

Observe that the eigenvalues of the matrix $\mathcal{A}$ are invariants under similarity transformations. This follows from

$$\det(\mathcal{T}\mathcal{A}\mathcal{T}^{-1} - \lambda I) = \det(\mathcal{A} - \lambda I)$$

by noting that $\det(\mathcal{T}^{-1}) = (\det \mathcal{T})^{-1}$ together with

$$\det(\mathcal{T}\mathcal{A}\mathcal{T}^{-1} - \lambda I) = \det(\mathcal{T}(\mathcal{A} - \lambda I)\mathcal{T}^{-1}) = \det(\mathcal{T})\det(\mathcal{A} - \lambda I)\det(\mathcal{T}^{-1}).$$

---

[7] If $n = 2$, then $\mathcal{T}\mathcal{A}\mathcal{T}^{-1} = \begin{pmatrix} \lambda_1 & 0 \\ 0 & \lambda_2 \end{pmatrix}$.

Simultaneous diagonalization. Consider a family $A_\gamma : X \to X$ of linear operators on X where the index γ lives in some index set Γ. By definition, the operator family $\{A_\gamma\}$ is called simultaneously diagonalizable iff there exist a basis $b_1, \ldots, b_n$ of the space X and complex numbers $\lambda_{1,\gamma}, \ldots, \lambda_{n,\gamma}$ such that

$$A_\gamma b_{j,\gamma} = \lambda_{j,\gamma} b_{j,\gamma} \qquad \text{for all} \quad j = 1, \ldots, n, \ \gamma \in \Gamma.$$

In this case, we get

$$A_\gamma A_\varrho = A_\varrho A_\gamma \qquad \text{for all} \quad \gamma, \varrho \in \Gamma. \tag{3.20}$$

Theorem 3.5 *Every family $A_\gamma : X \to X$ of self-adjoint (resp. unitary) operators with the commutativity property* (3.20) *is simultaneously diagonalizable.*

The same result is true if we replace the self-adjointness of the operators by skew-adjointness. The eigenvalues of a self-adjoint (resp. skew-adjoint) operator are real (resp. purely imaginary). Moreover, the eigenvalues of a unitary operator lie on the unit circle.

In terms of square matrices, the following hold. Fix $n = 2, 3, \ldots$ Let $\{\mathcal{A}_\gamma\}_{\gamma\in\Gamma}$ be a family of complex self-adjoint (resp. unitary) $(n \times n)$-matrices such that

$$\mathcal{A}_\gamma \mathcal{A}_\varrho = \mathcal{A}_\varrho \mathcal{A}_\gamma \qquad \text{for all} \quad \gamma, \varrho \in \Gamma.$$

Then there exists a complex invertible $(n \times n)$-matrix $\mathcal{T}$ such that

$$\mathcal{T} \mathcal{A}_\gamma \mathcal{T}^{-1}$$

is a diagonal matrix for all indices $\gamma \in \Gamma$. The same is true for skew-adjoint matrices.

Further reading. The proofs of the theorems above and below can be found in K. Spindler, Abstract Algebra and Applications, Vol. 1, Chap. 10, Marcel Dekker, New York, 1994. For the Jordan normal form and its generalization to infinite-dimensional Banach spaces, we also refer to F. Riesz and B. Nagy, Functional Analysis, Chap. IV, Frederyck Ungar, New York, 1978.

3.8.1 The Theorem of Principal Axes in Geometry and in Quantum Theory

> In 1904, Hilbert (1862–1941) generalized the Cauchy–Hermite theorem of principal axes for finite-dimensional self-adjoint matrices to a certain class of infinite-dimensional self-adjoint matrices which correspond to compact operators. In 1929, von Neumann (1903–1957) proved the theorem of principal axes for linear self-adjoint (bounded or unbounded) operators in infinite-dimensional Hilbert spaces. The final form of John von Neumann's spectral theorem on the unitary equivalence of self-adjoint operators to diagonal operators lies at the heart of quantum mechanics.[8] There is a fascinating historical development from conic sections in ancient times to John von Neumann's spectral theorem in functional analysis and its applications to harmonic analysis (i.e., the Fourier transform and its generalizations) and to quantum physics.
>
> Folklore

[8] We will study this in Vol. IV on quantum mathematics.

Conic sections as a prototype of the theorem of principal axes. Consider the equation

$$\boxed{ax^2 + 2bxy + dy^2 = 1, \qquad x, y \in \mathbb{R}} \tag{3.21}$$

where the real numbers a, b, d are given. Let us introduce the two matrices

$$\mathcal{A} := \begin{pmatrix} a & b \\ b & d \end{pmatrix}, \qquad u = \begin{pmatrix} x \\ y \end{pmatrix}.$$

Note that the real matrix $\mathcal{A}$ is self-adjoint, that is, $\mathcal{A}^\dagger = \mathcal{A}$. The eigenvalue problem

$$\mathcal{A}u = \lambda u, \qquad u \neq 0$$

has the solutions λ_1, λ_2 given by Lagrange's secular equation $\det(\mathcal{A} - \lambda I) = 0$. Explicitly,

$$\begin{vmatrix} a - \lambda & b \\ b & d - \lambda \end{vmatrix} = \lambda^2 - (a + d)\lambda + ad - b^2 = 0.$$

We have

$$\mathrm{tr}(\mathcal{A}) = a + d = \lambda_1 + \lambda_2, \quad \det(\mathcal{A}) = ad - b^2 = \lambda_1 \lambda_2.$$

The theorem of principal axes tells us that the eigenvalues λ_1, λ_2 are real, and that there exists a real invertible (2×2)-matrix $\mathcal{T}$ such that

$$\mathcal{T}\mathcal{A}\mathcal{T}^{-1} = \begin{pmatrix} \lambda_1 & 0 \\ 0 & \lambda_2 \end{pmatrix}.$$

In addition, the matrix $\mathcal{T}$ is an element of the Lie group $SO(2)$, that is, $\det \mathcal{T} = 1$ and $\mathcal{T}^{-1} = \mathcal{T}^d$. The original equation (3.21) reads as $u^d \mathcal{A}u = 1$. Using the rotation $u' = \mathcal{T}u$, we get the transformation

$$u^d \mathcal{A}u = u'^d \mathcal{T}\mathcal{A}\mathcal{T}^{-1} u'.$$

Consequently, equation (3.21) passes over to the rotated equation

$$\lambda_1 x'^2 + \lambda_2 y'^2 = 1.$$

This implies the following:

- If $ad - b^2 > 0$ and $a + d > 0$, then $\lambda_1 > 0$ and $\lambda_2 > 0$. Hence equation (3.21) describes an ellipse.
- If $ad - b^2 < 0$, then $\lambda_1 \lambda_2 < 0$, and hence equation (3.21) describes a hyperbola.

This is a typical argument of invariant theory. It is not necessary to know explicitly the eigenvalues λ_1 and λ_2. It is sufficient to know how the eigenvalues are related to the coefficients of the matrix $\mathcal{A}$. In ancient times, Appolonius of Perga (ca. 260–190 B.C.) wrote eight books about conic sections based on purely geometric arguments.

Linear self-adjoint operators on a finite-dimensional Hilbert space. Let $A : X \to X$ be a linear self- adjoint operator on the complex finite-dimensional Hilbert space X of positive dimension n.

Theorem 3.6 *There exist both an orthonormal basis $b_1, \ldots, b_n$ of the Hilbert space X and real numbers $\lambda_1, \ldots, \lambda_n$ such that*

$$Ab_j = \lambda_j b_j \qquad \textit{for all} \quad j = 1, \ldots, n.$$

The proof can be found in E. Zeidler, Applied Functional Analysis. Vol. 1: Applications in Mathematical Physics, Sect. 4.2, Springer, New York, 1997.

The Morse index. Suppose that all the eigenvalues of the operator A are different from zero. By definition, the Morse index $\mu(A)$ of the operator A is equal to the number of the negative eigenvalues of A. For example, if all the eigenvalues of A are positive, then $\mu(A) = 0$. In terms of quantum physics, the operator A describes an observable. If we measure the observable A in the state b_j, then we get the mean value $\langle b_j|Ab_j\rangle = \lambda_j$ with the mean fluctuation

$$(\Delta A)^2 = \langle b_j|(A - \lambda_j I)^2 b_j\rangle = 0.$$

Therefore, the eigenvalue λ_j is called a sharp value of the observable A.

3.8.2 The Schur Lemma in Linear Representation Theory

Fix $\mathbb{K} = \mathbb{R}, \mathbb{C}$, and let X and Y be finite-dimensional linear spaces over $\mathbb{K}$. Let $\varrho : \mathcal{G} \to GL(X)$ and $\mu : \mathcal{G} \to GL(Y)$ be irreducible representations of the group $\mathcal{G}$ on X and Y, respectively.

Lemma 3.7 *Suppose that there exists a linear morphism $J : X \to Y$ such that the diagram* (3.17) *on page 191 is commutative. Then:*

(i) *ϱ and μ are equivalent (i.e., J is a linear isomorphism) or we have the trivial situation $J = 0$.*

(ii) *If $X = Y$ and $\mathbb{K} = \mathbb{C}$, then there exists a complex number λ such that $J = \lambda I$.*

Proof. Ad (i). It follows from $\varrho(G)J = J\varrho(G)$ that $Jx = 0$ implies $J\varrho(G)x = 0$. Thus, $\ker(J)$ is an invariant linear subspace of the irreducible representation ϱ. Hence $\ker(J) = X$ or $\ker(J) = \{0\}$. Moreover, $J(X)$ is an invariant linear subspace of the irreducible representation μ. Hence $J(X) = Y$ or $J(X) = \{0\}$.

Ad (ii). Choose a complex number λ such that $\det(J - \lambda I) = 0$, and apply (i) to $J - \lambda I$. □

3.8.3 The Jordan Normal Form of Linear Operators

> The name of Camille Jordan is well known to all mathematicians of my generation because of his excellent "Cours d'analyse", a considerably enlarged elaboration of his lectures given at the École Polytechnique in Paris...[9]
>
> Jordan's monumental work of 667 pages "Traité des substitutions et des équations algébrique" on group theory and Galois theory, published in 1870 by Gauthier-Villars, Paris, is a masterpiece of mathematical architecture.
>
> Bartel Leendert van der Waerden, 1984

[9] One has to distinguish between the mathematician Camille Jordan (1838–1922) and the physicist Pascal Jordan (1902–1980) – one of the founders of quantum mechanics.

This quotation is taken from B. van der Waerden, A History of Algebra: From al-Khwarizmi to Emmy Noether, Springer, New York, 1984 (reprinted with permission). On the history of the Jordan normal form, see T. Hawkins, Weierstrass and the theory of matrices, Archive for History of Exact Sciences **17** (1977), 119–163. The Jordan normal form of matrices was published by Weierstrass (1815–1897) and Jordan in 1868 and 1870, respectively.

Prototype of the Jordan normal form. Consider the complex (2×2)-matrix

$$\mathcal{A} = \begin{pmatrix} a & b \\ c & d \end{pmatrix}.$$

Case 1: Suppose that the matrix $\mathcal{A}$ has two different eigenvalues λ_1 and λ_2. Then there exists a complex invertible (2×2)-matrix $\mathcal{T}$ such that

$$\mathcal{T}\mathcal{A}\mathcal{T}^{-1} = \begin{pmatrix} \lambda_1 & 0 \\ 0 & \lambda_2 \end{pmatrix}. \tag{3.22}$$

Case 2: Suppose that $\mathcal{A}$ has precisely one eigenvalue λ_1, and the eigenvector equation $\mathcal{A}u = \lambda_1 u, u \neq 0$, has precisely one linearly independent solution u. Then there exists a complex invertible (2×2)-matrix $\mathcal{T}$ such that

$$\mathcal{T}\mathcal{A}\mathcal{T}^{-1} = \begin{pmatrix} \lambda_1 & 1 \\ 0 & \lambda_1 \end{pmatrix}. \tag{3.23}$$

In this case, the matrix $\mathcal{A}$ is not diagonalizable.
Case 3: Suppose that $\mathcal{A}$ has precisely one eigenvalue λ_1, and the eigenvector equation $\mathcal{A}u = \lambda_1 u, u \neq 0$, has two linearly independent solutions u. Then

$$\mathcal{A} = \begin{pmatrix} \lambda_1 & 0 \\ 0 & \lambda_1 \end{pmatrix}.$$

The general theorem. Let $A : X \to X$ be a linear operator on the complex finite-dimensional linear space X of positive dimension. Then the operator A is completely reducible. That is, there exists a direct sum decomposition

$$X = X_1 \oplus X_2 \oplus \ldots \oplus X_m$$

such that every X_j is an invariant irreducible linear subspace of the operator A. More precisely, for any index j, the reduced operator

$$A : X_j \to X_j$$

has precisely one eigenvalue λ_j and precisely one linearly independent eigenvector b_j on X_j. In addition, there exists a basis $c_1 = b_j, c_2, \ldots, c_k$ of X_j (depending on j with $k \geq 1$) such that

$$\boxed{Ab_j = \lambda_j b_j, \qquad Ac_r = \lambda_j c_r + c_{r-1}, \qquad r = 2, \ldots, k.}$$

Two linear operators $A, B : X \to X$ are called similar iff there exists a linear bijective operator $T : X \to X$ with

$$A = TBT^{-1}.$$

This is precisely the case iff the operators A and B have the same eigenvalues and the invariant spaces $X_j(A)$ and $X_j(B)$, $j = 1, \ldots, m$, have the same dimensions, after reordering if necessary.

Summarizing, the eigenvalues and the dimension of the invariant irreducible linear subspaces form a complete invariant system for linear operators on X, up to similarity.

If X is a Hilbert space, and the operator A is self-adjoint, skew-adjoint, or unitary, then all the spaces X_j are one-dimensional eigenspaces of A. In other words, the operator A is diagonalizable.

3.8.4 The Standard Maximal Torus of the Lie Group $SU(n)$ and the Standard Cartan Subalgebra of the Lie Algebra $su(n)$

> Maximal commutative subgroups (resp. maximal commutative Lie subalgebras) of $SU(n)$ (resp. $su(n)$) know much about the structure and the representations of $SU(n)$ (resp. $(su(n))$. In terms of physics, this determines the quantum numbers in the $SU(3)$-model of quarks, baryons, and mesons (strong interaction in the Standard Model of particle physics).
>
> Folklore

Fix $n = 1, 2, \ldots$ The set of all the $(n \times n)$-diagonal matrices

$$\mathrm{diag}(\mathrm{e}^{\mathrm{i}\alpha_1}, \mathrm{e}^{\mathrm{i}\alpha_2}, \ldots, \mathrm{e}^{\mathrm{i}\alpha_n}), \qquad \alpha_1, \alpha_2, \ldots, \alpha_n \in \mathbb{R}$$

forms a commutative subgroup of the group $SU(n)$ denoted by $\mathcal{C}(SU(n))$. Since this subgroup is diffeomorphic to the direct product $\mathbb{S}^1 \times \cdots \times \mathbb{S}^1$ of n unit circles, the subgroup $\mathcal{C}(SU(n))$ is called the standard maximal torus of the Lie group $SU(n)$. Note that

Every element of $SU(n)$ is similar to one element of the standard maximal torus of $SU(n)$.

In other words, the conjugacy class of every element of $SU(n)$ contains an element of the standard maximal torus.

The set of all the $(n \times n)$-diagonal matrices

$$\mathrm{diag}(\mathrm{i}\alpha_1, \mathrm{i}\alpha_2, \ldots, \mathrm{i}\alpha_n), \qquad \alpha_1, \alpha_2, \ldots, \alpha_n \in \mathbb{R}$$

forms a commutative Lie subalgebra of $su(n)$. This is called the standard Cartan subalgebra $\mathcal{C}(su(n))$ of $su(n)$.

3.8.5 Eigenvalues and the Operator Strategy for Lie Algebras (Adjoint Representation)

In 1888, Killing (1847–1923) wrote a fundamental paper on the classification of semisimple Lie algebras. From his academic teacher Weierstrass (1815–1897) in Berlin, Killing learned the relation between eigenvalues and normal forms of matrices. He wanted to apply this technique to the structure theory of semisimple Lie algebras by studying the eigenvalues and normal forms of the matrices corresponding to the adjoint representation of the Lie algebra on itself. Recall that the adjoint representation $\mathrm{ad} : \mathcal{L} \to gl(\mathcal{L})$ of the real (resp. complex) Lie algebra $\mathcal{L}$ is given by the operator

$$\mathrm{ad}(A)B := [A, B] \qquad \text{for all} \quad B \in \mathcal{L} \tag{3.24}$$

where $A \in \mathcal{L}$. Killing's operator strategy was also critically used by Élie Cartan (1868–1951) in his 1894 thesis.

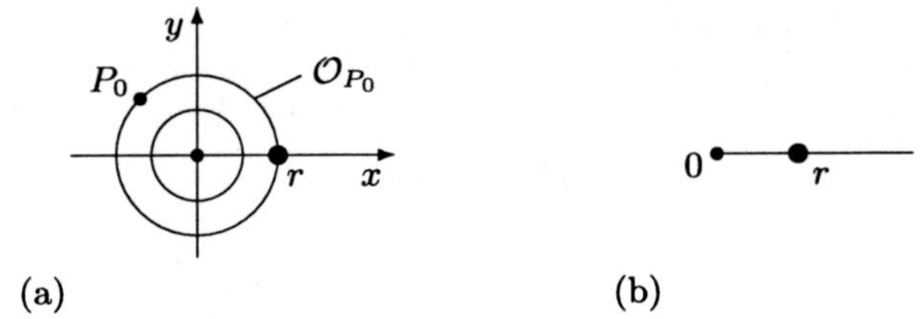

Fig. 3.1. Orbits and orbit space

3.9 The Action of a Group on a Physical State Space, Orbits, and Gauge Theory

> Physical states are equivalent iff they lie on the same orbit generated by the action of the gauge group. Therefore, gauge theory has to be based on orbit spaces (also called moduli spaces in mathematics). This complicates substantially the mathematical theory. As a rule, moduli spaces are not manifolds; they possess singularities. Surfaces or more general varieties with singularities are studied in algebraic geometry.
>
> Folklore

Let S be a set. We regard the elements of this set as physical states. We want to describe the action of a symmetry group $\mathcal{G}$ on the space S. We call the group $\mathcal{G}$ a gauge group. Consider first all the bijective maps

$$\sigma : S \to S. \tag{3.25}$$

With respect to the composition of maps, all the maps (3.25) form a group denoted by $Sym(S)$. This group is called the symmetry group of the set S. Explicitly, if $\sigma, \sigma' \in Sym(S)$, then

$$(\sigma\sigma')(P) = \sigma(\sigma'(P)) \qquad \text{for all} \quad P \in S.$$

For example, if the set S has n elements, then the group $Sym(S)$ is isomorphic to the symmetric group $Sym(n)$ of all the permutations of n elements. Next suppose that we are given a group $\mathcal{G}$. We say that the group $\mathcal{G}$ acts on the set S iff there exists a group morphism

$$\mu : \mathcal{G} \to Sym(S).$$

Explicitly, we assign to every element G of the group $\mathcal{G}$ a bijective map $\mu_G : S \to S$, and the group multiplication corresponds to the composition of maps, that is,

$$\mu_{GH}(P) = \mu_G(\mu_H(P)) \qquad \text{for all} \quad P \in S,$$

and all $G, H \in \mathcal{G}$. For fixed group element G, the map μ_G sends the point P_0 in S to the point P in S. Naturally enough, the set

$$\mathcal{O}_{P_0} := \{\mu_G(P_0) : \ G \in \mathcal{G}\}$$

is called the orbit of the point P_0 under the action of the group $\mathcal{G}$. Two points P and Q are called equivalent,

$$P \sim Q,$$

iff they lie on the same orbit. This is an equivalence relation. The corresponding equivalence classes $[P]$ (i.e., the orbits) form a group denoted by $S/\mathcal{G}$.

Example. Suppose that the set S is the Euclidean plane equipped with a Cartesian (x, y)-coordinate system. Let $\mathcal{G}$ denote the group of all the rotations of the plane about the origin. Then the orbits are circles about the origin,

$$x^2 + y^2 = r^2,$$

parametrized by the nonnegative number r (Fig. 3.1 on page 205). There exists a one-to-one relation between the orbits and the interval $[0, \infty[$ given by the map $O_{P_0} \mapsto r$. The orbit space $S/\mathcal{G} = [0, \infty[$ is not a manifold, but only a manifold with boundary. The origin $(0, 0)$ corresponds to a degenerate orbit; this is the boundary point of the orbit space.

Orbit spaces play a fundamental role in the quantization of gauge field theories. The basic ideas are discussed in Sect. 16.5ff of Vol. I (path integral and the Faddeev–Popov ghost, BRST-symmetry and quantization, cohomology).

3.10 The Intrinsic Symmetry of a Group

> The intrinsic symmetry of a group $\mathcal{G}$ is governed by the action of the group on itself via the adjoint representation. The orbits of this action are the so-called conjugacy classes of $\mathcal{G}$ which are of fundamental importance for the linear representations of $\mathcal{G}$. The corresponding symmetry group is the group $\mathrm{Aut}_{inner}(\mathcal{G})$ of inner automorphisms of the original group $\mathcal{G}$.
>
> Folklore

Conjugacy classes of the group $\mathcal{G}$. Let $\mathcal{G}$ be an arbitrary group. For group elements G and H of $\mathcal{G}$, we write

$$G \sim H$$

iff there exists a group element G_0 of $\mathcal{G}$ such that $G = G_0 H G_0^{-1}$. We say that G is conjugate to H. This is an equivalence relation. The corresponding equivalence classes $[H]$ are called conjugacy classes of $\mathcal{G}$. If the group $\mathcal{G}$ is commutative, then every conjugacy class of $\mathcal{G}$ contains precisely one group element.

We will show below that the number of conjugacy classes of the finite group $\mathcal{G}$ is precisely the number of essential irreducible representations of $\mathcal{G}$.

Therefore, in order to get geometric inside, let us show that the conjugacy classes of the group $\mathcal{G}$ are precisely the orbits of the action of the group $\mathcal{G}$ on itself by inner automorphisms. As we will show, the inner automorphisms of a group describe the intrinsic symmetries of a group.

Inner automorphisms and the intrinsic symmetry of a group. The map

$$\mathsf{A} : \mathcal{G} \to \mathcal{G} \tag{3.26}$$

is called a group automorphism iff it is a bijective group morphism. Using the composition of maps, the set of all the automorphisms (3.26) form a group $\mathrm{Aut}(\mathcal{G})$ which is called the automorphism group (or the symmetry group) of the originally group $\mathcal{G}$. For understanding the intrinsic geometry of the group $\mathcal{G}$, we have to use special automorphisms which are called inner automorphisms. For a given group element $G_0 \in \mathcal{G}$, we set

$$\mathsf{A}_{G_0}(G) := G_0 G G_0^{-1} \qquad \text{for all} \quad G \in \mathcal{G}.$$

Then the map $\mathsf{A}_{G_0} : \mathcal{G} \to \mathcal{G}$ is a special automorphism called an inner automorphism generated by the group element G_0. The set $\{\mathsf{A}_{G_0}\}_{G_0 \in \mathcal{G}}$ forms a subgroup of $\mathrm{Aut}(\mathcal{G})$ which is called the group of inner automorphisms of $\mathcal{G}$ (or the intrinsic symmetry group of $\mathcal{G}$). This group is denoted by $\mathrm{Aut}_{\mathrm{inner}}(\mathcal{G})$.

A subgroup $\mathcal{H}$ of $\mathcal{G}$ is called a normal subgroup iff it is invariant under inner automorphisms, that is, $G_0 H G_0^{-1} \in \mathcal{H}$ for all $H \in \mathcal{H}$ and all $G_0 \in \mathcal{G}$. Recall that the set of normal subgroups knows all about the group morphisms from $\mathcal{G}$ onto another group. In fact, the morphism theorem for groups tells us that if the map

$$\mu : \mathcal{G} \to \mathcal{G}^+ \tag{3.27}$$

is a surjective group morphism, then the kernel $\mu^{-1}(\mathbf{1})$ is a normal subgroup of $\mathcal{G}$, and we have the group isomorphism

$$\boxed{\mathcal{G}^+ \simeq \mathcal{G}/\mu^{-1}(\mathbf{1})}$$

(see Sect. 4.1.3 of Vol. II). The conjugacy class $[H]$ of the group element $H \in \mathcal{G}$ is equal to

$$[H] := \{\mathsf{A}_{G_0}(H) : \quad G_0 \in \mathcal{G}\}.$$

Therefore, $[H]$ consists of all the group elements which are obtained from H by transport via all possible inner automorphisms. In other words, $[H]$ is an orbit of the action of $\mathcal{G}$ on itself, and this orbit contains the element H.

A special invariant subgroup of the group $\mathcal{G}$ is the center of $\mathcal{G}$ denoted by center($\mathcal{G}$). By definition, the group element G_0 of $\mathcal{G}$ is an element of center($\mathcal{G}$) iff

$$G_0 G = G G_0 \qquad \text{for all} \quad G \in \mathcal{G}.$$

Equivalently, $G_0 G G_0^{-1} = G$ for all $G \in \mathcal{G}$. Thus, G is an element of the center of $\mathcal{G}$ iff it is a fixed point of all the inner automorphisms of $\mathcal{G}$. The group $\mathcal{G}$ is commutative iff center$(\mathcal{G}) = \mathcal{G}$.

The center of a group measures its deviations from a commutative group.

Set $\varrho(G_0) := \mathsf{A}_{G_0}$. Then $\varrho(G_0 G_1) = \varrho(G_0)\varrho(G_1)$.[10] Therefore, the map

$$\varrho : \mathcal{G} \to \mathrm{Aut}_{\mathrm{inner}}(\mathcal{G}) \tag{3.28}$$

is a group endomorphism with the kernel $\varrho^{-1}(\mathrm{id}) = \mathrm{center}(\mathcal{G})$. Hence we have the group isomorphism

$$\mathrm{Aut}_{\mathrm{inner}}(\mathcal{G}) = \mathcal{G}/\mathrm{center}(\mathcal{G}).$$

The map ϱ from (3.28) is a representation of the group $\mathcal{G}$ on itself. More precisely, ϱ is called the adjoint representation of $\mathcal{G}$ on itself.

3.11 Linear Representations of Finite Groups and the Hilbert Space of Functions on the Group

> Whoever understands the representation theory of finite groups can understand everything in the representation theory of compact and locally compact Lie groups and their applications in physics.
> The main trick is to study the Hilbert space $L_2(\mathcal{G})$ of complex-valued functions $\psi : \mathcal{G} \to \mathbb{C}$ on the finite group $\mathcal{G}$. This is closely related to special functions called characters and to the group algebra $\mathbb{C}[\mathcal{G}]$).
>
> Folklore

---

[10] In fact, $G_0(G_1 G G_1^{-1})G_0^{-1} = (G_0 G_1)G(G_0 G_1)^{-1}$.

In what follows we will summarize important results. Some of the proofs are highly nontrivial (e.g., the proof of the Frobenius character formula for symmetric groups; see page 276). Important consequences of the Frobenius character formula are the Frobenius reciprocity theorem for the characters of representations induced by subgroups and the branching rule for symmetric groups. For the proofs, we refer to:

B. Simon, Representations of Finite and Compact Groups, Amer. Math. Soc., Providence, Rhode Island, 1996.

This book is based on lectures given at Princeton University and the California Institute of Technology in Pasadena. The advantage of this textbook is that it emphasizes the close relations between finite groups and compact Lie groups, and it contains many examples which are useful in physics. As an elementary introduction to representation theory emphasizing the applications to spectroscopy in quantum mechanics, we recommend the classic monograph by B. van der Waerden, Group Theory and Quantum Mechanics, Springer, New York, 1974 (German edition, 1932). Concerning symmetric functions, we refer to:

C. Procesi, Lie Groups: An Approach Through Invariants and Representations, Springer, New York, 2007.

I. Macdonald, Symmetric Functions and Hall Polynomials, Oxford University Press, 1995.

Further references can be found on page 537. Combinatorial methods are fundamental for the renormalization of quantum fields. Here, Hopf algebras play a key role. The algebraic point of view is emphasized in the BFFO-approach to quantum field theory. We refer to:

C. Brouder, B. Fauser, A. Frabetti, and R. Oeckl, Quantum field theory and Hopf algebra cohomology. J. Phys. A: Math. Gen. 37 (2004), 5895–5927. Internet: `http://www.arXiv:hep-th/0311253`

R. Caroll, Fluctuations, Information, Gravity and the Quantum Potential, Chap. 8, Kluwer, Dordrecht, 2005 (summary of the BFFO-approach).

Consider a finite group $\mathcal{G}$. The number of group elements of $\mathcal{G}$ is called the order of $\mathcal{G}$; this is denoted by $|\mathcal{G}|$. The Lagrange theorem tells us the following:

If $\mathcal{H}$ is a subgroup of $\mathcal{G}$, then $|\mathcal{H}|$ is a divisor of $|\mathcal{G}|$.

For example, if $\mathcal{H}$ is a subgroup of $Sym(3)$ with $|Sym(3)| = 6$, then the number of elements of $\mathcal{H}$ is never 4 or 5.

For a given finite group $\mathcal{G}$, the main goal is to get a complete system of irreducible representations.

By definition, a system $\varrho_1, \varrho_2, \ldots, \varrho_m$ of irreducible representations of $\mathcal{G}$ is called complete iff the following hold:

- Every $\varrho_j : \mathcal{G} \to GL(X_j)$ is an irreducible representation of $\mathcal{G}$ on a finite-dimensional complex linear space X_j. We set $d_j := \dim X_j$, and this dimension of the representation space X_j is called the degree of the representation ϱ_j.
- Every irreducible representation $\varrho : \mathcal{G} \to GL(X)$ on a finite-dimensional complex linear space X is equivalent to one of the representations $\varrho_1, \ldots, \varrho_m$.
- If $j \neq k$, then ϱ_j is not equivalent to ϱ_k.

Theorem 3.8 *Let $\mathcal{G}$ be a finite group. Then:*

(i) *Every linear representation $\varrho : \mathcal{G} \to GL(X)$ on a finite-dimensional complex linear space X is completely reducible and unitarily equivalent (theorem of Maschke (1853–1908)).*

(ii) *If m is the number of conjugacy classes of the group $\mathcal{G}$, then there exists a complete system $\varrho_1, \ldots, \varrho_m$ of irreducible representations.*

(iii) *The degrees $d_1, \ldots, d_m$ of $\varrho_1, \ldots, \varrho_m$, respectively, are divisors of the group order $|\mathcal{G}|$. More precisely, they are divisors of the quotient $|\mathcal{G}|/|\text{center}(\mathcal{G})|$.*

(iv) $d_1^2 + d_2^2 + \ldots + d_m^2 = |\mathcal{G}|$ *(theorem of Burnside (1852–1927)).*

For the permutation group $Sym(2)$ with $|Sym(2)| = 2$ it follows from the unique decomposition $2 = 1^2 + 1^2$ and (iv) that $m = 2$ and $d_1 = d_2 = 1$. Similarly, for $Sym(3)$ with $|Sym(3)| = 6$, it follows from the unique decomposition $6 = 1^2+1^2+2^2$ and (iv) that $m = 3$ with $d_1 = d_2 = 1$ and $d_3 = 2$. The explicit form of complete systems of irreducible representations of $Sym(2)$ and $Sym(3)$ will be studied on page 214.

The group algebra $\mathbb{C}[\mathcal{G}]$ of $\mathcal{G}$ and the regular representation of $\mathcal{G}$. Let $\mathcal{G}$ be a finite group consisting of the elements $G_1, \ldots, G_k$ where $G_1 := \mathbf{1}$. By definition, the set $\mathbb{C}[\mathcal{G}]$ consists of all the symbols

$$\alpha_1 G_1 + \ldots + \alpha_k G_k, \qquad \alpha_1, \ldots, \alpha_k \in \mathbb{C}.$$

Naturally enough, this is a k-dimensional complex linear space which becomes a complex unital algebra by introducing the product

$$(\alpha G + \beta H)(\mu R + \nu S) := \alpha\mu \cdot GR + \alpha\nu \cdot GS + \beta\mu \cdot HR + \beta\nu \cdot HS.$$

This algebra is called the group algebra of $\mathcal{G}$. For all $G \in \mathcal{G}$ and all $A \in \mathbb{C}[\mathcal{G}]$, we define

$$\varrho(G)A := GA.$$

This way, we obtain the so-called regular representation ϱ of the finite group $\mathcal{G}$ on its group algebra $\mathbb{C}[\mathcal{G}]$. The regular representation ϱ knows all about the irreducible representations of $\mathcal{G}$, up to equivalence.

> *More precisely, every irreducible representation of the group $\mathcal{G}$ on a finite-dimensional complex linear space is equivalent to an irreducible component of the regular representation ϱ.*

The Hilbert space $L_2(\mathcal{G})$ of complex-valued functions on the group $\mathcal{G}$. Let $L_2(\mathcal{G})$ denote the space of all complex-valued functions $\psi : \mathcal{G} \to \mathbb{C}$ on the group $\mathcal{G}$. This set becomes a complex Hilbert space equipped with the inner product

$$\boxed{\langle\psi|\varphi\rangle := \frac{1}{|\mathcal{G}|} \sum_{G \in \mathcal{G}} \psi(G)^\dagger \varphi(G).}$$

The function $f \in L_2(\mathcal{G})$ is called a class function iff $f(G) = f(AGA^{-1})$ for all elements G, A of the group $\mathcal{G}$. For example, characters are class functions. All the class functions form a linear subspace $L_2^{cl}(\mathcal{G})$ of the Hilbert space $L_2(\mathcal{G})$. The following theorem tells us that the characters know all about the irreducible representations of the group $\mathcal{G}$. This was discovered by Frobenius in about 1900. Let X be a finite-dimensional complex linear space of positive dimension.

- Two representations ϱ and σ of the finite group $\mathcal{G}$ on X are equivalent iff they have the same character.
- The representation ϱ of $\mathcal{G}$ on X is irreducible iff $\langle\chi|\chi\rangle = 1$.
- If ϱ and ϱ' are inequivalent irreducible representations of $\mathcal{G}$ on X, then $\langle\chi|\chi'\rangle = 0$.

Here, χ and χ' denotes the character of ϱ and ϱ', respectively.

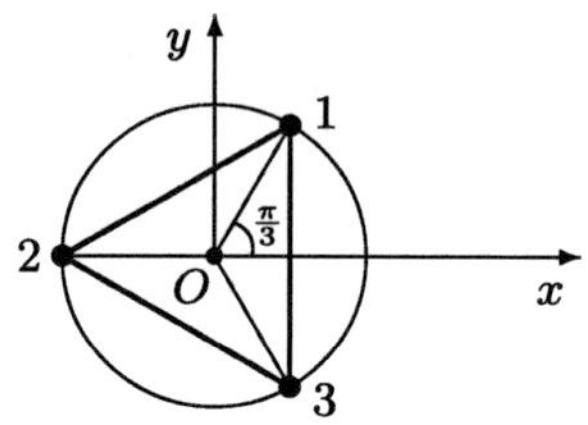

Fig. 3.2. The symmetry group D_3 of an equilateral triangle

Theorem 3.9 *Let $\varrho_1, \ldots, \varrho_k$ be a system of representations of the finite group $\mathcal{G}$ on finite-dimensional complex linear spaces. Then this is a complete system of irreducible representations of $\mathcal{G}$ iff the characters are an orthonormal basis of the Hilbert space $L_2^{cl}(\mathcal{G})$ of class functions.*

The crucial orthogonality relations for the matrix elements of irreducible representations. Let $\varrho : \mathcal{G} \to GL(X)$ be an irreducible unitary representation on the complex Hilbert space X of finite dimension $m = 1, 2, \ldots$ Let $e_1, \ldots, e_m$ be an orthonormal basis of X. Define the matrix elements

$$\varrho_{ij}(G) := \langle e_i | \varrho(G) e_j \rangle, \qquad i, j = 1, \ldots, m$$

of the linear operator $\varrho(G) : X \to X$ with respect to the basis $e_1, \ldots, e_m$. Then

$$\boxed{\frac{1}{|\mathcal{G}|} \sum_{G \in \mathcal{G}} \varrho_{ij}(G)^\dagger \varrho_{kl}(G) = \frac{1}{m}\, \delta_{ik}\delta_{jl}, \qquad i, j, k, l = 1, \ldots, m.} \tag{3.29}$$

The symmetry group D_3 of an equilateral triangle. We want to show that:

> *The group D_3 is a finite subgroup of the orthogonal group $O(2)$, and D_3 is a faithful and irreducible linear representation of the permutation group $Sym(3)$ on the Euclidean plane.*

Consider Fig. 3.2. The transformation

$$\begin{pmatrix} x' \\ y' \end{pmatrix} = A \begin{pmatrix} x \\ y \end{pmatrix}, \qquad A = \begin{pmatrix} a_{11} & a_{12} \\ a_{21} & a_{22} \end{pmatrix}$$

is called an orthogonal transformation of the Euclidean plane $\mathbb{E}^2$ iff the real matrix A is orthogonal, that is, $AA^d = I$. By definition, the orthogonal group $O(2)$ consists of all the real (2×2)-matrices A which are orthogonal. Choose the angle $\alpha := \frac{2\pi}{3}$. Set

$$R_\alpha := \begin{pmatrix} \cos\alpha & -\sin\alpha \\ \sin\alpha & \cos\alpha \end{pmatrix}, \quad R_{2\alpha} := \begin{pmatrix} \cos 2\alpha & -\sin 2\alpha \\ \sin 2\alpha & \cos 2\alpha \end{pmatrix}, \quad I_\pm := \begin{pmatrix} 1 & 0 \\ 0 & \pm 1 \end{pmatrix},$$

and

$$R_\alpha I_- = \begin{pmatrix} \cos\alpha & \sin\alpha \\ \sin\alpha & -\cos\alpha \end{pmatrix}, \qquad R_{2\alpha} I_- = \begin{pmatrix} \cos 2\alpha & \sin 2\alpha \\ \sin 2\alpha & -\cos 2\alpha \end{pmatrix}.$$

Geometrically, the matrix R_α (resp. $R_{2\alpha}$) describes a counter-clockwise rotation of the Euclidean plane with the angle α (resp. 2α) about the origin (Fig. 3.2), whereas the matrix I_- corresponds to the reflection $(x, y) \mapsto (x, -y)$ with respect to the x-axis. The six matrices

$$\boxed{I,\ R_\alpha,\ R_{2\alpha},\ I_-,\ R_\alpha I_-,\ R_{2\alpha} I_-} \tag{3.30}$$

form a finite subgroup of the orthogonal group $O(2)$ denoted by D_3. This group is also called the symmetry group of the equilateral triangle pictured in Fig. 3.2. Alternatively, the group D_3 can be described by the permutation group $Sym(3)$ of the vertices $1, 2, 3$ of the triangle. More precisely, we have the group isomorphism

$$\varrho : Sym(3) \to D_3$$

given by the following maps:

- $(1) \mapsto I, \quad (12) \mapsto R_{2\alpha} I_-, \quad (23) \mapsto R_\alpha I_-, \quad (31) \mapsto I_-,$
- $(123) \mapsto R_\alpha, \quad (132) \mapsto R_{2\alpha}.$

This way, we obtain a representation of the group $Sym(3)$ on the Euclidean plane. This faithful representation is irreducible. In fact, there is no straight line through the origin which is left invariant by the transformation group D_3 of the Euclidean plane.

The group D_3 as a subgroup of the matrix group $GL(2, \mathbb{C})$. The six real (2×2)-matrices $I,\ R_\alpha,\ R_{2\alpha},\ I_-,\ R_\alpha I_-,\ R_{2\alpha} I_-$ form a subgroup of the group $GL(2, \mathbb{C})$ of complex (2×2)-matrices. This subgroup coincides with D_3. Since $Sym(3)$ is isomorphic to D_3, we get the injective group morphism

$$\varrho_3 : Sym(3) \to GL(2, \mathbb{C}) \tag{3.31}$$

with the group isomorphism $\varrho_3(Sym(3)) \simeq D_3$. Finally, let us consider the character χ of the faithful representation ϱ_3. Obviously, for the trace of the matrices of the group D_3 we get

$$\mathrm{tr}(I) = 2, \ \ \mathrm{tr}(R_\alpha) = \mathrm{tr}(R_{2\alpha}) = 1, \ \ \mathrm{tr}(I_-) = \mathrm{tr}(R_\alpha I_-) = \mathrm{tr}(R_{2\alpha} I_-) = 0.$$

Note that $\cos\alpha = \cos 2\alpha = \frac{1}{2}$ if $\alpha = 2\pi/3$. This implies

$$\chi_{(1)} = 2, \ \ \chi_{(123)} = \chi_{(132)} = 1, \ \ \chi_{(31)} = \chi_{(23)} = \chi_{(12)} = 0. \tag{3.32}$$

The character $\chi : Sym(3) \to \mathbb{C}$ is an element of the Hilbert space $L_2(Sym(3))$. Explicitly,

$$\langle \chi | \chi \rangle = \frac{1}{6} \sum_{G \in Sym(3)} \chi_G^\dagger \chi_G = \frac{1}{6}(2^2 + 1^2 + 1^2) = 1.$$

Thus, the representation (3.31) of the symmetric group $Sym(3)$ on the complex linear space $\mathbb{C}^2$ is irreducible.

3.12 The Tensor Product of Representations and Characters

The decomposition of tensor products of representations into irreducible components plays a crucial role in physics. This concerns the symmetry classification of composed particles (e.g., baryons and mesons as composed particles consisting of quarks and antiquarks, the spectra of molecules, the periodic table of chemical elements, the physical properties of crystals or atomic nuclei).

The main tool is the Fourier analysis of the characters in terms of the complete orthonormal system of irreducible characters.

The classic textbooks on applications of group theory to physics contain large tables used by physicists. The sketch words are Clebsch–Gordan coefficients, Racah-coefficients, and the Littlewood–Richardson rules. Nowadays the computations are done via computer algebra. Let us explain the basic ideas.

Fourier coefficients and multiplicities of irreducible representations. Let $\varrho_1, \ldots, \varrho_m$ be a complete system of irreducible linear representations of the finite group $\mathcal{G}$. Let

$$\varrho : \mathcal{G} \to GL(X)$$

be an arbitrary linear representation of the finite group $\mathcal{G}$ on the complex finite-dimensional Hilbert space X of positive dimension with the character function $\chi_\varrho : \mathcal{G} \to \mathbb{C}$. The representation ϱ is completely reducible. This means that there exists a direct sum decomposition

$$X = X^{(1)} \oplus X^{(2)} \oplus \ldots \oplus X^{(r)}$$

such that the restriction of the operators $\varrho(G) : X \to X$, $G \in \mathcal{G}$, to the linear subspace $X^{(k)}$ yields an irreducible representation

$$\varrho : \mathcal{G} \to GL(X^{(k)}), \qquad k = 1, \ldots, r$$

of the group $\mathcal{G}$ on the space $X^{(k)}$, and this irreducible representation is equivalent to some ϱ_j. We want to know which irreducible representations appear. To this end, we compute the Fourier expansion

$$\chi_\varrho = \sum_{j=1}^{m} c_j \chi_j$$

with $c_j = \langle \chi_j | \chi_\varrho \rangle = \frac{1}{|\mathcal{G}|} \sum_{G \in \mathcal{G}} \chi_j(G)^\dagger \chi_\varrho(G)$. Then we have

$$\boxed{\varrho = c_1\varrho_1 \oplus c_2\varrho_2 \oplus \ldots \oplus c_m\varrho_m.} \tag{3.33}$$

This mnemonic formula means that the decomposition of X contains the irreducible representation ϱ_j iff $c_j \neq 0$, and the value c_j tells us how many times the irreducible representation ϱ_j appears. For example, $\varrho = 2\varrho_1 \oplus \varrho_2$ means that $\varrho = \varrho_1 \oplus \varrho_1 \oplus \varrho_2$. This implies

$$\dim X = \sum_{j=1}^{m} c_j \deg(\varrho_j).$$

Recall that $\deg(\varrho_j) = \dim X^{(k)}$ if ϱ_j acts on $X^{(k)}$.

The tensor product of representations. Let

$$\sigma : \mathcal{G} \to GL(X), \qquad \mu : \mathcal{G} \to GL(Y)$$

be representations of the finite group $\mathcal{G}$ on the complex finite-dimensional Hilbert spaces X and Y, respectively. This generates the tensor product of representations $\sigma \otimes \mu$ given by

$$\sigma \otimes \mu : \mathcal{G} \to GL(X \otimes Y).$$

Explicitly, if $x_1, \ldots, x_k$ and $y_1, \ldots, y_l$ is a basis of X and Y, respectively, then

$$(\sigma \otimes \mu)(G) \cdot (x_r \otimes y_s) := \sigma(G)x_r \otimes \mu(G)y_s, \qquad r = 1, \ldots, k, \; s = 1, \ldots, l.$$

Observe that the representation does not depend on the choice of the basis vectors on X and Y. The trick is to use the following product formula for the character $\chi_{\sigma\otimes\mu}$ of the representation $\sigma \otimes \mu$:

$$\boxed{\chi_{\sigma\otimes\mu} = \chi_\sigma \chi_\mu.}$$

Then the decomposition of $\sigma \otimes \mu$ follows from the Fourier method (3.33) above.

Example. Choose a right-handed (x, y)-Cartesian coordinate system in the Euclidean plane with the right-handed orthonormal basis $\mathbf{i}, \mathbf{j}$. Set $X := \text{span}(\mathbf{i}, \mathbf{j})$. Consider the reflection group $\mathcal{Z}_2 = \{1, -1\}$. Define

$$\sigma(\pm 1)(x\mathbf{i} + y\mathbf{j}) := \pm\,(x\mathbf{i} + y\mathbf{j}), \qquad x, y \in \mathbb{R}.$$

This yields the representation $\sigma : \mathcal{Z}_2 \to GL(X)$. Explicitly,

$$\sigma(\pm 1) \cdot \begin{pmatrix} \mathbf{i} \\ \mathbf{j} \end{pmatrix} = \begin{pmatrix} \pm 1 & 0 \\ 0 & \pm 1 \end{pmatrix} \begin{pmatrix} \mathbf{i} \\ \mathbf{j} \end{pmatrix}.$$

Thus, the linear operator $\sigma(\pm 1)$ corresponds to the (2×2)-matrix $\pm I$ with the trace $\text{tr}(\pm I) = \pm 2$. This yields the character function

$$\chi_\sigma(\pm 1) = \pm 2.$$

Applied to the tensor product $X \otimes X$, we get the representation

$$\sigma \otimes \sigma : \mathcal{Z}_2 \to GL(X \otimes X).$$

Explicitly, $\sigma(\pm 1) \cdot (\mathbf{i} \otimes \mathbf{j}) = (\pm \mathbf{i}) \otimes (\pm \mathbf{j}) = \mathbf{i} \otimes \mathbf{j}$, and so on. Hence

$$(\sigma(\pm 1) \otimes \sigma(\pm 1)) \cdot \begin{pmatrix} \mathbf{i} \otimes \mathbf{i} \\ \mathbf{i} \otimes \mathbf{j} \\ \mathbf{j} \otimes \mathbf{i} \\ \mathbf{j} \otimes \mathbf{j} \end{pmatrix} = \begin{pmatrix} 1\,0\,0\,0 \\ 0\,1\,0\,0 \\ 0\,0\,1\,0 \\ 0\,0\,0\,1 \end{pmatrix} \begin{pmatrix} \mathbf{i} \otimes \mathbf{i} \\ \mathbf{i} \otimes \mathbf{j} \\ \mathbf{j} \otimes \mathbf{i} \\ \mathbf{j} \otimes \mathbf{j} \end{pmatrix}.$$

Thus, the linear operator $\sigma(\pm 1) \otimes \sigma(\pm 1)$ corresponds to the (4×4)-unit matrix I with the trace $\text{tr}(I) = 4$. This yields the character function

$$\chi_{\sigma\otimes\sigma}(\pm 1) = \chi_\sigma(\pm 1)\chi_\sigma(\pm 1) = 4.$$

The group $\mathcal{Z}_2$ is isomorphic to the permutation group $Sym(2)$. By Table 3.3 on page 215, the irreducible character functions χ_1 and χ_2 of $\mathcal{Z}_2$ read as $\chi_1(\pm 1) = 1$ and $\chi_2(\pm 1) = \pm 1$. This implies

$$c_j = \langle \chi_{\sigma\otimes\sigma} | \chi_j \rangle = \tfrac{1}{2}(\chi_\sigma(1)^2 \chi_j(1) + \chi_\sigma(-1)^2 \chi_j(-1)), \quad j = 1, 2.$$

Hence $c_1 = 4$ and $c_2 = 0$. Thus,

$$\sigma \otimes \sigma = 4\varrho_1 = \varrho_1 \oplus \varrho_1 \oplus \varrho_1 \oplus \varrho_1.$$

This corresponds to the decomposition

$$X \otimes X = X_1 \oplus X_2 \oplus X_3 \oplus X_4$$

with $X_1 := \text{span}(\mathbf{i} \otimes \mathbf{i})$, $X_2 := \text{span}(\mathbf{i} \otimes \mathbf{j})$, $X_3 := \text{span}(\mathbf{j} \otimes \mathbf{i})$, and $X_4 := \text{span}(\mathbf{j} \otimes \mathbf{j})$. The one-dimensional linear spaces X_1, X_2, X_3, X_4 are invariant under the action of the reflection group $\mathcal{Z}_2$ on $X \otimes X$.

Table 3.2. Irreducible representations of the group $Sym(2)$

irreducible representation	character
$\varrho_1 : Sym(2) \to \{1\}$	χ_1
$\varrho_2 : Sym(2) \to \{1, -1\} \qquad (\varrho_2(G) := \mathrm{sgn}(G))$	χ_2

3.13 Applications to the Symmetric Group $Sym(n)$

Physically, partitions play a crucial role in the theory of multi-particle systems. The basic idea is to describe the physics of the total system by the physics of all possible subsystems.
Mathematically, partitions govern the representation theory of permutation groups. The number of essential irreducible linear representations of a permutation group is equal to the number of orbits of the permutation group with respect to the inner automorphisms; this is equal to the number of conjugacy classes. In turn, this is equal to the number of partitions of the group order. Graphically, this equals the number of Young diagrams.
Folklore

3.13.1 The Characters of the Symmetric Group $Sym(2)$

Recall that the group $Sym(2)$ consists of the identical permutation $\mathbf{1}$ and the transposition $\pi := (12)$ with $\pi(1) = 2$ and $\pi(2) := 1$. Here, $\pi^2 = \mathbf{1}$. There are two obvious group morphisms, namely,

- $\sigma : Sym(2) \to \{1\}$ (trivial morphism), and
- $\mathrm{sgn} : Sym(2) \to \{1, -1\}$ (sign of the permutations; $\mathrm{sgn}(\mathbf{1}) = 1$ and $\mathrm{sgn}(\pi) = -1$).

We want to show how this fits the general theory of linear representations of finite groups. The group $Sym(2)$ is commutative. Therefore, it has the two conjugacy classes $[\mathbf{1}]$ and $[\pi]$. By the general theory, a complete system of irreducible linear representations of $Sym(2)$ consists of two elements. These two irreducible representations are given on the real line $\mathbb{R}$ by setting:

- $\varrho(\mathbf{1})x := x$ and $\varrho(\pi)x := x$ for all $x \in \mathbb{R}$ (this trivial representation corresponds to $\sigma : Sym(2) \to \{1\}$), and
- $\varrho(\mathbf{1})x := x$ and $\varrho(\pi)x := -x$ for all $x \in \mathbb{R}$ (this reflection corresponds to the signature map $\mathrm{sgn} : Sym(2) \to \{1, -1\}$).

Next we want to discuss the relation to the group algebra $\mathbb{C}[Sym(2)]$ which consists of all the symbols

$$\alpha \cdot \mathbf{1} + \beta \cdot \pi, \qquad \alpha, \beta \in \mathbb{C}.$$

The regular representation $\varrho : Sym(2) \to \mathbb{C}[Sym(2)]$ is given by the identity map $\varrho(1) = \mathrm{id}$ on $\mathbb{C}[Sym(2)]$ and by the map

$$\varrho(\pi)(\alpha\mathbf{1} + \beta\pi) := \pi(\alpha\mathbf{1} + \beta\pi) = \alpha\pi + \beta\mathbf{1} \qquad \text{for all} \quad \alpha, \beta \in \mathbb{C}.$$

Table 3.3. Irreducible characters of the group $Sym(2)$

value of the character χ_j	group element G	
	(1)	(12)
$\chi_1(G)$	1	1
$\chi_2(G)$	1	−1

In order to decompose the representation ϱ into irreducible components, let us introduce the new basis

$$\pi_+ := \mathbf{1} + \pi, \qquad \pi_- := \mathbf{1} - \pi$$

of $\mathbb{C}[Sym(2)]$. Because of $\pi^2 = \mathbf{1}$ we get $\pi\pi_+ = \pi_+$ and $\pi\pi_- = -\pi_-$. Therefore,

$$\varrho(\pi)(\alpha_+\pi_+ + \alpha_-\pi_-) = \alpha_+\pi_+ - \alpha_-\pi_- \qquad \text{for all} \quad \alpha_\pm \in \mathbb{C}. \tag{3.34}$$

Set $X := \mathbb{C}[Sym(2)]$. Let us introduce the two projection operators $P_\pm : X \to X$ defined by

$$P_+(\alpha_+\pi_+ + \alpha_-\pi_-) := \alpha\pi_+, \quad P_-(\alpha_+\pi_+ + \alpha_-\pi_-) := \alpha_-\pi_-, \qquad \alpha_\pm \in \mathbb{C}.$$

This yields the direct sum decomposition

$$\boxed{X = P_+(X) \oplus P_-(X).}$$

By (3.34), the one-dimensional linear subspaces $P_+(X)$ and $P_-(X)$ are invariant under the regular representation ϱ. Thus, we get the decomposition

$$\varrho = \varrho_+ \oplus \varrho_-$$

where ϱ_+ and ϱ_- are irreducible representations of $Sym(2)$ on $P_+(X)$ and $P_-(X)$, respectively. Hence

$$(\dim P_+(X))^2 + (\dim P_+(X))^2 = 1^2 + 1^2 = |Sym(2)|$$

where $|Sym(2)| = 2$ is the number of group elements. This is a special case of the Burnside theorem on page 209. Finally, let us discuss the characters of ϱ_+ and ϱ_-. The Hilbert space $L_2(Sym(2))$ consists of all the functions $\psi, \varphi : Sym(2) \to \mathbb{C}$ with the inner product

$$\langle\psi|\varphi\rangle := \tfrac{1}{2}\left(\psi(\mathbf{1})^\dagger\varphi(\mathbf{1}) + \psi(\pi)^\dagger\varphi(\pi)\right).$$

Since the conjugacy classes of the group $Sym(2)$ contain precisely one element, the Hilbert space $L_2^{cl}(Sym(2))$ of class functions coincides with $L_2(Sym(2))$. We have

- $\chi_+(\mathbf{1}) = \chi_+(\pi) = 1$ (character χ_+ of ϱ_+), and
- $\chi_-(\mathbf{1}) = 1, \chi_-(\pi) = -1$ (character of ϱ_-).

This follows from $\varrho_-(\mathbf{1})\pi_- = \pi_-$ and $\varrho_-(\pi)\pi_- = -\pi_-$. Thus, the operators $\varrho_-(\mathbf{1})$ and $\varrho_-(\pi)$ defined on the one-dimensional linear space span(π_-) correspond to the multiplication with the real numbers 1 and -1, respectively. Therefore,

$$\langle\chi_+|\chi_+\rangle = 1, \quad \langle\chi_-|\chi_-\rangle = 1, \quad \langle\chi_+|\chi_-\rangle = 0.$$

Thus, the characters χ_+ and χ_- form an orthonormal basis of the Hilbert space $L_2^{cl}(Sym(2))$ as predicted by the general theory. The relation to the method of Young tableaux will be explained on page 222.

Table 3.4. Irreducible representations of the group $Sym(3)$

irreducible representation		character
$\varrho_1 : Sym(3) \to \{1\}$		χ_1
$\varrho_2 : Sym(3) \to \{1, -1\}$	$(\varrho_2(G) := \mathrm{sgn}(G))$	χ_2
$\varrho_3 : Sym(3) \to GL(2, \mathbb{C})$	$(\varrho_3(Sym(3)) \simeq D_3)$	χ_3

3.13.2 The Characters of the Symmetric Group $Sym(3)$

Recall that the symmetric group $Sym(3)$ consists of the following three conjugacy classes:

$$[(1)],\ [(12),\ (23),\ (31)],\ [(123),\ (132)].$$

By the general theory, a complete system of irreducible representations of $Sym(3)$ on complex Hilbert spaces contains three elements. Note that we already know three inequivalent irreducible representations of $Sym(3)$, namely,

- $\varrho_1 : Sym(3) \to \{1\}$ (trivial symmetric representation on the Gaussian plane $\mathbb{C}$: $\varrho_1(G)z := z$ for all $z \in \mathbb{C}$ and all $G \in Sym(3)$),
- $\varrho_2 : Sym(3) \to \{1, -1\}$ (nontrivial antisymmetric representation on the Gaussian plane: $\varrho_2(G)z = \mathrm{sgn}(G)z$ for all $z \in \mathbb{C}$ and all $G \in Sym(3)$),
- $\varrho_3 : Sym(3) \to GL(2, \mathbb{C})$ (see (3.31) on page 211).

For the dimensions of the complex representation spaces $\mathbb{C}, \mathbb{C}, \mathbb{C}^2$, we get

$$(\dim \mathbb{C})^2 + (\dim \mathbb{C})^2 + (\dim \mathbb{C}^2)^2 = 1^2 + 1^2 + 2^2 = 6 = |Sym(3)|.$$

This is a special case of the Burnside theorem on page 209. Note that the characters $\chi_k(G) := \mathrm{tr}(\varrho_k(G))$ read as follows:

- $\chi_1(G) = 1$ for all $G \in Sym(3)$,
- $\chi_2(G) = \mathrm{sgn}(G)$ for all $G \in Sym(3)$,
- $\chi_3(\mathbf{1}) = 2, \chi_3(G) = 0$ if $G = (12), (23), (31)$, and $\chi_3(G) = 1$ if $G = (123), (132)$, by (3.32).

Table 3.5. Irreducible characters of the group $Sym(3)$

value of the character χ_j	group element G		
	(1)	(12), (23), (31)	(123), (132)
$\chi_1(G)$	1	1	1
$\chi_2(G)$	1	-1	1
$\chi_3(G)$	2	0	1

The complex Hilbert space $L_2(Sym(3))$ consists of all the complex-valued functions $\psi, \varphi : Sym(3) \to \mathbb{C}$ equipped with the inner product

$$\langle\psi|\varphi\rangle := \tfrac{1}{6} \sum_{G \in Sym(3)} \psi(G)^\dagger \varphi(G).$$

The function $\psi \in L_2(Sym(3))$ is a class function iff it is constant on the conjugacy classes, that is,

$$\psi(G) = \text{const if } G = (12), (23), (31) \text{ and } \psi(G) = \text{const if } G = (123), (132).$$

The class functions form a linear subspace $L_2^{cl}(Sym(3))$ of the Hilbert space $L_2(Sym(3))$. By the general theory, the characters χ_1, χ_2, χ_3 represent an orthonormal basis of $L_2^{cl}(Sym(3))$. In fact,

$$\langle\chi_k|\chi_k\rangle = 1, \qquad \langle\chi_k|\chi_l\rangle = 0, \qquad k = 1, 2, 3,\ l \neq k.$$

In addition, note that the order of the subgroups of $Sym(3)$ is a divisor of the group order 6, by the Lagrange theorem on page 208. Thus, the order of a subgroup is equal to $1, 2, 3, 6$. Explicitly, the proper subgroups of $Sym(3)$ are given by

$$\{(1)\},\ \{(1), (12)\},\ \{(1), (23)\},\ \{(1), (31)\},\ A_3 := \{(1), (123), (132)\}.$$

The center of $Sym(3)$ is trivial (i.e., it is equal to (1)). Thus, we have the group isomorphism

$$\mathrm{Aut}_{\mathrm{inner}}(Sym(3)) \simeq Sym(3).$$

All the automorphisms of the group $Sym(3)$ are inner automorphisms. The only normal subgroups of $Sym(3)$ are the trivial subgroups $\{(1)\}$, $Sym(3)$ and the non-trivial subgroup A_3 (i.e., the subgroup of even permutations). If

$$\mu : Sym(3) \to \mathcal{G}$$

is a surjective group morphism, then μ is either trivial (i.e., μ is an isomorphism or $\mathcal{G}$ contains only the unit element) or $\mu(G) = \mathrm{sgn}(G)$ for all $G \in Sym(3)$ and $\mathcal{G} = \{1, -1\}$, up to isomorphisms of $\mathcal{G}$.

Theorem 3.10 *If $\varrho : Sym(3) \to GL(X)$ is a representation of the permutation group $Sym(3)$ on the finite-dimensional complex linear space X of positive dimension, then there exists a direct sum decomposition*

$$X = X_1 \oplus X_2 \oplus \ldots \oplus X_m$$

into one-dimensional or two-dimensional linear subspaces $X_1, \ldots, X_m$ of X such that, for any index j, the restriction $\varrho(G) : X_j \to X_j$ of $\varrho(G)$ to the linear subspace X_j yields an irreducible representation of $Sym(3)$ on X_j which is equivalent to ϱ_1, ϱ_2 or ϱ_3.

3.13.3 Partitions and Young Frames

Step 1: Partitions. Let $n = 1, 2, \ldots$ By definition, a partition of the number n is a tuple $(n_1, n_2, \ldots, n_k)$ of positive integers with

$$n = n_1 + n_2 + \ldots + n_k, \qquad n_1 \geq n_2 \geq \ldots \geq n_k \geq 1.$$

Table 3.6. Young frames of the group $Sym(2)$

partition	Young frame	disjoint cycle product	
$2 = 2$	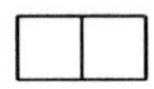	(12)	2^1
$2 = 1 + 1$	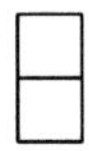	$(1)(2)$	1^2

Graphically, this is represented by the Young frame which consists of k rows where the first row has n_1 boxes, the second row has n_2 boxes, and so on. We assign to every Young frame the symbol

$$1^{m(1)}2^{m(2)}\cdots n^{m(n-1)}n^{m(n)}.$$

This tells us that the Young frame has $m(1)$ rows with 1 box, $m(2)$ rows with 2 boxes, and so on. Note that the length of the rows is increasing from bottom to top. For example, we assign to the partition

$$12 = 4 + 4 + 2 + 1 + 1$$

the following Young frame

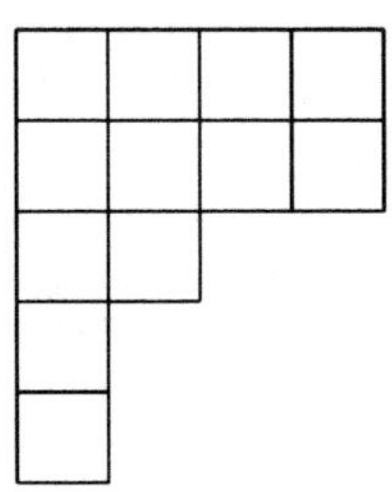

$$\eta_1 = 4, \quad \eta_2 = 4, \quad \eta_3 = 2, \quad \eta_4 = 1, \quad \eta_5 = 1 \tag{3.35}$$

$1^2 2^1 3^0 4^2$

The symbol $1^2 2^1 3^0 4^2$ tells us that there are

- 2 rows with 1 box,
- 1 row with 2 boxes,
- no row with 3 boxes, and
- 2 rows with 4 boxes.

Instead of $1^2 2^1 3^0 4^2$ we briefly write $1^2 2^1 4^2$.

- The two partitions $2 = 1 + 1$ and $2 = 2$ correspond to the Young frames depicted in Table 3.6.
- The three partitions $3 = 1 + 1 + 1, 3 = 2 + 1, 3 = 3$ of the number $n = 3$ are represented by the following three Young frames:

Table 3.7. Young tableaux of the group $Sym(2)$

Young tableau	$\mathcal{H}$	$\mathcal{V}$	Young symmetrizer
[1 | 2]	$Sym(2)$	(1)	$\mathsf{S}_+ = \frac{1}{2}\big((1) + (12)\big)$
[1] over [2]	(1)	$Sym(2)$	$\mathsf{S}_- = \frac{1}{2}\big((1) - (12)\big)$

(i) $3 = 1 + 1 + 1$ (ii) $3 = 2 + 1$ (iii) $3 = 3$

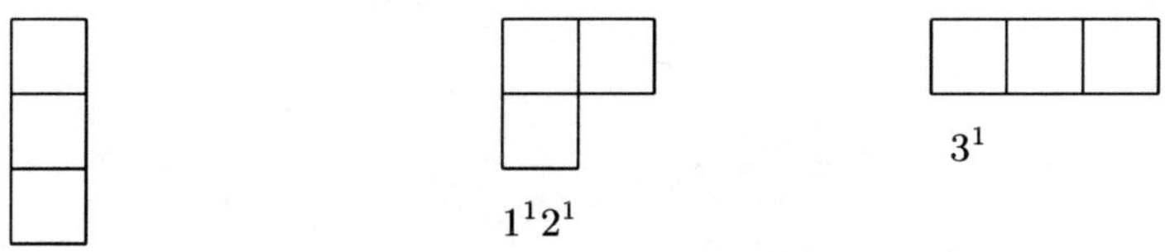

1^3 $1^1 2^1$ 3^1 (3.36)

Step 2: Dual Young frame. By definition, the dual Young frame $\mathcal{Y}^d$ is obtained from the original Young frame $\mathcal{Y}$ by interchanging the rows with the columns. For example, the frame (iii) from (3.36) is dual to the frame (i), and vice versa. The frame (ii) is self-dual. Note the following. As we will discuss below, for every positive integer n, the corresponding Young frames $\mathcal{Y}$ are in one-to-one correspondence to the irreducible representations of the group $Sym(n)$, up to equivalence. Therefore, the characters $\chi_{\mathcal{Y}}$ of the irreducible representations of $Sym(n)$ can be labelled by the Young frames. For the dual Young frame, we get the duality formula

$$\chi_{\mathcal{Y}^d}(\pi) = \operatorname{sgn}(\pi)\, \chi_{\mathcal{Y}}(\pi) \qquad \text{for all} \quad \pi \in Sym(n).$$

This symmetry property of characters saves time when computing the characters.

Step 3: Young standard tableaux. From Young frames we pass over to Young standard tableaux by inserting the numbers $1, 2, \ldots, n$ in the following way:

- Every number $1, 2, \ldots, n$ appears precisely once.
- The numbers of the rows are increasing from left to right.
- The numbers of the columns are increasing from top to bottom.

For example, if $n = 3$, then the standard tableaux read as follows:

(i) $w = 1$ (ii) $w = 2$ (iii) $w = 1$

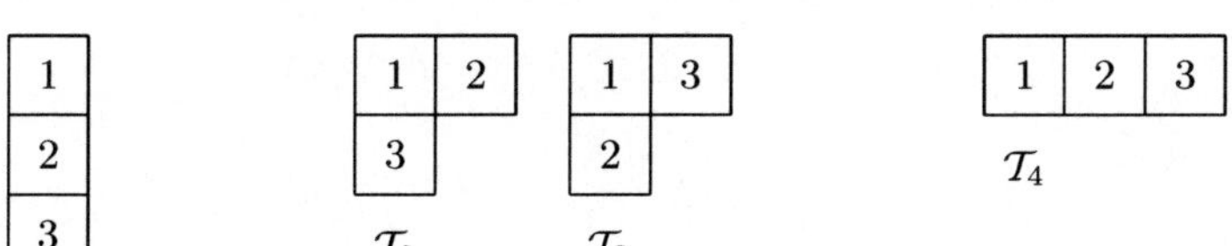

$\mathcal{T}_1$ $\mathcal{T}_2$ $\mathcal{T}_3$ $\mathcal{T}_4$ (3.37)

By definition, the weight w of a Young frame is the number of standard tableaux corresponding to the Young frame.

Step 4: The crucial Young symmetrizers. We assign to every standard tableau $\mathcal{T}$ of weight w the element S of the group algebra $\mathbb{C}[Sym(n)]$ given by

$$\boxed{\mathsf{S} := \frac{w}{n!}\left(\sum_{\pi\in\mathcal{V}} \operatorname{sgn}(\pi)\cdot\pi\right)\left(\sum_{\pi\in\mathcal{H}}\pi\right).} \tag{3.38}$$

Explicitly, the following hold:

- $\mathcal{H}$ is the maximal subgroup of $Sym(n)$ which leaves invariant the rows of the standard tableau (horizontal invariance).
- $\mathcal{V}$ is the maximal subgroup of $Sym(n)$ which leaves invariant the columns of the standard tableau (vertical invariance).

The Young symmetrizer has the crucial property that

$$\mathsf{S}^2 = \mathsf{S}.$$

This means that S is an idempotent element of the group algebra. The Young symmetrizers for the groups $Sym(2)$ and $Sym(3)$ can be found in Table 3.7 and 3.9 on pages 219 and 222, respectively.

The main result. For the symmetric group $Sym(n)$ with $n = 1, 2, \ldots,$ we have the following main theorem.

Theorem 3.11 *Using Young symmetrizers, it is possible to construct explicitly a complete system*

$$\varrho_j : Sym(n) \to GL(n_j, \mathbb{C}), \quad j = 1, \ldots, m$$

of irreducible representations of the symmetric group $Sym(n)$. These irreducible representations $\varrho_1, \ldots, \varrho_m$ are in one-to-one correspondence to the Young frames for the partitions of the group order n.

The dimension n_j is equal to the weight w_j of the Young frame $\mathcal{Y}_j$ corresponding to ϱ_j. We have $\sum_{j=1}^m w_j^2 = n!$ by the Burnside theorem.

For example, consider (3.37) on page 219: the group $Sym(3)$ has 3 Young frames with weights $w = 1, 2, 1$. This tells us that there is a complete system $\varrho_1, \varrho_2, \varrho_3$ of irreducible representations of $Sym(3)$ which act on complex linear spaces X_1, X_2, X_3 of dimension $1, 2, 1$, respectively. Here,

$$1^2 + 2^2 + 1^2 = 3!$$

The main result above shows that it is important to know the number of Young frames and their weights. In this connection, the following formulas are useful:

- The partition formula (number of Young frames): The number $p(n)$ of partitions of the positive integer n is given by the generating function

$$\begin{aligned}\sum_{k=0}^{\infty} p(d)z^k &= \prod_{m=1}^{\infty}\frac{1}{1-z^m}\\ &= (1+z+z^2+\ldots)(1+z^2+z^4+\ldots)(1+z^3+\ldots).\end{aligned} \tag{3.39}$$

Table 3.8. Young frames of the group $Sym(3)$

partition	Young frame	disjoint cycle product	
$3 = 3$	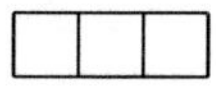	$(123),\ (132)$	3^1
$3 = 2 + 1$	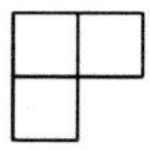	$(12)(3),\ (23)(1),\ (31)(2)$	$2^1 1^1$
$3 = 1 + 1 + 1$		$(1)(2)(3)$	$1^1 1^1 1^1$

This formula is valid for all complex numbers z with $|z| < 1$. Euler (1707–1783) discovered the following recursion formula:

$$p(n) = \sum_{k=1}^{n} (-1)^{k+1} \big(p(n - \omega(k)) + p(n - \omega(-k))\big), \qquad n = 0, 2, \ldots$$

where $\omega(k) = \frac{1}{2}(3k^2 - k)$. Furthermore, $p(0) = p(1) := 1$, and $p(n) := 0$ if $n < 0$. For example,

$$p(2) = p(1) + p(0) = 2, \;\; p(3) = p(2) + p(0) = 3, \;\; p(4) = p(3) + p(2),$$

and $p(200) = 3\ 972\ 999\ 029\ 388$. In 1918, Hardy (1877–1947) and Ramanujan (1887–1920) discovered the following asymptotic formula:

$$\ln p(n) \simeq \pi \sqrt{\frac{2n}{3}}, \qquad n \to \infty.$$

- The weight formula. Set $l_j := n_j + r - j$, $j = 1, \ldots, r$. If the Young frame has rows of length $n_1 \geq n_2 \geq \ldots \geq n_r \geq 1$, then it has the weight

$$w := n! \, \frac{\prod_{i<k}(l_i - l_k)}{l_1! l_2! \cdots l_r}.$$

Lexicographic order of partitions (Young frames). For two partitions of the positive integer n, we write

$$(n_1, n_2, \ldots, n_k) > (m_1, m_2, \ldots, m_l)$$

iff the first nonzero difference is positive. For example, if $n = 3$, then

$$(3) > (2, 1) > (1, 1, 1).$$

Concerning (3.36), this corresponds to the ordering (iii) > (ii) > (i) of the Young frames.

Lexicographic order of Young tableaux. For a fixed tableau, we read the numbers of the first row from left to right, then there follow the numbers of the

Table 3.9. Young tableaux of the group $Sym(3)$

Young tableau	weight w	$\mathcal{H}$	$\mathcal{V}$	Young symmetrizer
$\begin{array}{\|c\|c\|c\|}\hline 1 & 2 & 3 \\ \hline\end{array}$	1	$Sym(3)$	(1)	$\mathsf{S}_1 = \frac{1}{6} \sum_{\pi \in Sym(3)} \pi$
$\begin{array}{\|c\|c\|}\hline 1 & 2 \\ \hline 3 \\ \cline{1-1}\end{array}$	2	$(1), (12)$	$(1), (13)$	$\mathsf{S}_2 = \frac{1}{3}\Big((1) - (13)\Big)\Big((1) + (12)\Big) = \frac{1}{3}\Big((1) + (12) - (13) - (123)\Big)$
$\begin{array}{\|c\|c\|}\hline 1 & 3 \\ \hline 2 \\ \cline{1-1}\end{array}$	2	$(1), (13)$	$(1), (12)$	$\mathsf{S}_3 = \frac{1}{3}\Big((1) - (12)\Big)\Big((1) + (13)\Big) = \frac{1}{3}\Big((1) + (13) - (12) - (132)\Big)$
$\begin{array}{\|c\|}\hline 1 \\ \hline 2 \\ \hline 3 \\ \hline\end{array}$	1	(1)	$Sym(3)$	$\mathsf{S}_4 = \frac{1}{6} \sum_{\pi \in \mathcal{V}} \mathrm{sgn}(\pi) \cdot \pi$
$\mathsf{S} = \frac{w}{3!}\left(\sum_{\pi \in \mathcal{V}} \mathrm{sgn}(\pi) \cdot \pi\right)\left(\sum_{\pi \in \mathcal{H}} \pi\right)$				

second row from left to right, and so on. If $\mathcal{T}_i$ and $\mathcal{T}_j$ are tableaux corresponding to the same Young frame, then we write

$$\mathcal{T}_i < \mathcal{T}_j$$

iff the first non-zero difference of the tableau-numbers is positive. For example, we have $\mathcal{T}_2 < \mathcal{T}_3$ in (3.37).

3.13.4 Young Tableaux and the Construction of a Complete System of Irreducible Representations

Application to the symmetric group Sym(2). The elements of the group algebra $\mathbb{C}[Sym(2)]$ are $\alpha(1) + \beta(12)$ where α, β are complex numbers. By Table 3.7 on page 219, we have the two Young symmetrizers $\mathsf{S}_\pm$ given by

$$\mathsf{S}_+ = \tfrac{1}{2}(\,(1) + (12)), \quad \mathsf{S}_- = \tfrac{1}{2}(\,(1) - (12))$$

with the properties

$$\mathsf{S}_+ + \mathsf{S}_- = I, \qquad \mathsf{S}_\pm^2 = \mathsf{S}_\pm, \qquad \mathsf{S}_+\mathsf{S}_- = 0$$

where I denotes the identity operator. In fact,

$$((1) + (12))((1) - (12)) = (1) - (12)(12) = (1) - (1) = 0. \tag{3.40}$$

Hence $\mathsf{S}_+\mathsf{S}_-\pi = 0$ for all $\pi \in Sym(2)$. The Young symmetrizers $\mathsf{S}_\pm$ coincide with the projection operators $P_\pm$ studied on page 214.

Application to the symmetric group Sym(3). We will use the S_j-tableaux of Table 3.9. Let us discuss the role of the Young symmetrizers

$$\mathsf{S}_j, \quad j = 1, 2, 3, 4$$

defined by (3.38).

Ad (i). The Young symmetrizer S_1. The group $Sym(3)$ leaves invariant the row of the S_1-tableau. Hence $\mathcal{H} = Sym(3)$. Moreover, only the trivial permutation (1) leaves invariant the columns. Hence $\mathcal{V} = \{(1)\}$. This implies

$$\mathsf{S}_1 = \frac{1}{3!} \sum_{\pi \in Sym(3)} \pi.$$

Ad (ii). The Young symmetrizer S_2. The group $\mathcal{H} = \{(1), (12)\}$ leaves invariant the rows of the S_2-tableau, and the group $\mathcal{V} = \{(1), (13)\}$ leaves invariant the column. Since $w = 2$, we get

$$\mathsf{S}_2 = \frac{2}{3!}\big((1) - (13)\big)\big((1) + (12)\big) = \tfrac{1}{2}((1) + (12) - (13) - (123)).$$

Ad (iii). The Young symmetrizer S_3. By using duality, $\mathcal{H} = \{(1), (13)\}$ and $\mathcal{V} = \{(1), (12)\}$. Hence

$$\mathsf{S}_3 = \frac{2}{3!}\big((1) - (12)\big)\big((1) + (13)\big) = \tfrac{1}{3}((1) + (13) - (12) - (132)).$$

Ad (iv). The Young symmetrizer S_4. Since $\mathcal{V} = Sym(3)$ and $\mathcal{H} = \{(1)\}$, we obtain

$$\mathsf{S}_4 = \frac{1}{3!} \sum_{\pi \in Sym(3)} \mathrm{sgn}(\pi) \cdot \pi.$$

The main trick is to set $X_j := \mathbb{C}[(Sym(3)]\mathsf{S}_j$ and to use the direct sum decomposition

$$\boxed{\mathbb{C}[Sym(3)] = X_1 \oplus X_2 \oplus X_3 \oplus X_4} \tag{3.41}$$

of the group algebra $\mathbb{C}[Sym(3)]$. For the dimensions, we get $\dim X_1 = \dim X_4 = 1$ and $\dim X_2 = \dim X_3 = 2$. Explicitly, the space X_j consists of all the products $\sigma\mathsf{S}_j$ where $\sigma \in Sym(n)$. The linear subspace X_j has the simple, but crucial property that, for every element π of $Sym(3)$, we get

$$\pi\tau \in X_j \qquad \text{for all} \quad \tau \in X_j.$$

In terms of algebra, X_j is a left ideal of the group algebra $\mathbb{C}[(Sym(n)]$. In terms of representation theory, the linear subspace X_j, $j = 1, 2, 3, 4$, is invariant under the regular representation ϱ. In fact, if $\pi \in Sym(n)$, then

$$\varrho(\pi)\tau = \pi\tau \in X_j \qquad \text{for all} \quad \tau \in X_j.$$

Let

$$\varrho^{(j)}(\pi) : X_j \to X_j, \qquad \pi \in Sym(3)$$

denote the restriction of $\varrho(\pi)$ to X_j (i.e., $\varrho^{(j)}(\pi)\sigma = \pi\sigma$ for all $\sigma \in X_j$). Then the map

$$\varrho^{(j)} : Sym(3) \to GL(X_j)$$

represents an irreducible representation of the group $Sym(3)$ on X_j. Moreover $\varrho^{(1)}, \varrho^{(2)}, \varrho^{(3)}, \varrho^{(4)}$ is equivalent to the irreducible representation $\varrho_1, \varrho_3, \varrho_3, \varrho_2$ of the group $Sym(3)$, respectively (see Table 3.4 on page 216).

This is Theorem 3.11 on page 220 for the special case where $n = 3$. Let us sketch the proof. Set

$$\sigma_\pm := \tfrac{1}{3!}\big(\,(1) \pm (12) \pm (13) \pm (23) + (123) + (132)\big),$$

that is, $\sigma_+ = \mathsf{S}_1$ and $\sigma_- = \mathsf{S}_4$.

Step 1: The Young symmetrizers. We first show that

$$\boxed{\sum_{k=1}^{4} \mathsf{S}_k = I, \quad \mathsf{S}_j^2 = \mathsf{S}_j, \quad \mathsf{S}_i\mathsf{S}_j = 0, \qquad i,j = 1,2,3,4,\ i \neq j} \tag{3.42}$$

where I denotes the unit element (1) of the group algebra. Note that

$$(12)\sigma_\pm = (12) \pm (1) \pm (132) \pm (123) + (23) + (13) = \pm\sigma_\pm.$$

Similarly, we get

$$\pi\sigma_\pm = \operatorname{sgn}(\pi)\cdot\sigma_\pm \qquad \text{for all} \quad \pi \in Sym(3). \tag{3.43}$$

It is not necessary to explicitly compute this. Observe that the map $\pi \mapsto \pi\sigma$ is a permutation of the group elements of $Sym(3)$. Moreover, note the signature rule $\operatorname{sgn}(\pi\sigma) = \operatorname{sgn}(\pi)\operatorname{sgn}(\sigma)$. This implies that

$$\pi\mathsf{S}_1 = a\sigma_+ \qquad \text{for all} \quad \pi \in Sym(3)$$

where the complex number a depends on π. Thus, $X_1 = \{b\sigma_+ : b \in \mathbb{C}\}$. Similarly, we get

$$X_4 = \{b\sigma_- : b \in \mathbb{C}\}.$$

Consequently, $\dim X_j = 1$ if $j = 1, 4$.

By (3.43), $\sigma_\pm^2 = \sigma_\pm$. Hence $\mathsf{S}_j^2 = \mathsf{S}_j$ if $j = 1, 4$. Moreover, it follows from $((1) + (12))((1) - (12)) = 0$ that

$$\mathsf{S}_2\mathsf{S}_3 = \tfrac{1}{9}\big(\,(1) - (13)\big)\cdot\big(\,(1) + (12)\big)\big(1 - (12)\big)\cdot\big(\,1 + (13)\big) = 0.$$

The other claims are obtained similarly.

Step 2: The direct sum decomposition (3.41). Let $\pi \in Sym(3)$. Then we get the sum representation

$$\pi = \pi\cdot(1) = \pi\mathsf{S}_1 + \pi\mathsf{S}_2 + \pi\mathsf{S}_3 + \pi\mathsf{S}_4$$

where $\mathsf{S}_j \in X_j$ for all j. This representation is unique. In fact, let

$$\pi = \alpha_1 + \alpha_2 + \alpha_3 + \alpha_4, \qquad \alpha_j \in X_j, \quad j = 1,2,3,4.$$

Then $\alpha_1 = \beta\mathsf{S}_1$ for some $\beta \in Sym(3)$. By (3.42), $\mathsf{S}_j\mathsf{S}_1 = 0$ if $j \neq 1$. Hence

$$\pi\mathsf{S}_1 = \alpha_1\mathsf{S}_1 = \beta\mathsf{S}_1^2 = \beta\mathsf{S}_1 = \alpha_1.$$

Similarly, $\alpha_j = \pi\mathsf{S}_j$ if $j = 1,2,3,4$. We recommend the reader to complete the proof by checking all the claims via explicit computation.

Application to the symmetric group Sym(n). Fix $n = 2, 3, \ldots$

- Setting $X_j := [\mathbb{C}(Sym(n)]\mathsf{S}_j$, we get the direct sum decomposition
$$\mathbb{C}[Sym(n)] = X_1 \oplus X_2 \oplus \ldots \oplus X_m$$
of the group algebra $\mathbb{C}[Sym(n)]$ where m equals the number of Young frames to $Sym(n)$ (i.e., m equals the number of partitions of n).
- Every linear subspace X_j of the group algebra is invariant under the regular representation $\varrho : Sym(n) \to \mathbb{C}(Sym(n))$.
- The restriction $\varrho^{(j)}(\pi) : X_j \to X_j$ of ϱ to X_j yields an irreducible representation of the symmetric group $Sym(n)$ on X_j.
- Every irreducible representation of $Sym(n)$ on a finite-dimensional complex Hilbert space is equivalent to one of the representations $\varrho^{(j)}$.

If we choose precisely one standard tableaux to every Young frame, then the construction described above yields a complete system of irreducible representations of the symmetric group $Sym(n)$. Therefore, the irreducible representations of $Sym(n)$ are in one-to-one correspondence to the Young frames of $Sym(n)$. This yields Theorem 3.11.

In the late 1920s, John von Neumann simplified the original proof of Frobenius given in 1903. This polished proof of Theorem 3.11 can be found in B. van der Waerden, Modern Algebra, Vol. 2, Sect. 110, Frederyck Ungar, New York, 1975. We also refer to H. Weyl, The Theory of Groups and Quantum Mechanics, Sect. V.C, Dover, New York 1931.

3.14 Application to the Standard Model in Elementary Particle Physics

3.14.1 Quarks and Baryons

> Composed n-quark states correspond to irreducible representations of the quark symmetry group $SU(3)$ on tensor products of the quark Hilbert space. In terms of physics, the basis vectors of the $SU(3)$-invariant linear subspaces of the tensor product describe multiplets of elementary particles. The invariant subspaces are obtained by using the Young symmetrizers corresponding to the Young tableau of the symmetric group $Sym(n)$. The crucial basis vectors which describe particle states are obtained by regular fillings of the Young frames. This is a special case of the tensor method in representation theory due to Weyl.[11]
>
> Folklore

There are precisely six quarks in the Standard Model in particle physics, namely, u (up), d (down), c (charm), s (strange), t (top), b (bottom) (for more details, see Sect. 2.4 of Vol. I). These quarks are ordered by three so-called generations, namely,
$$\begin{pmatrix} u \\ d \end{pmatrix}, \quad \begin{pmatrix} c \\ s \end{pmatrix}, \quad \begin{pmatrix} t \\ b \end{pmatrix}.$$

[11] The tensor method corresponds to the construction of tensors with certain symmetry properties by symmetrization and antisymmetrization of $T^{i_1 \ldots i_r}_{j_1 \ldots j_s}$ with respect to selected indices. The prototype of this procedure is the passage from T_{ij} to $\frac{1}{2}(T_{ij} - T_{ji})$.

One has to distinguish between the flavors of quarks and the inner degrees of freedom of a single quark called the color of the quark (red, green, or blue). The flavors concern the names u, d, c, s, t, b of the quarks. In what follows, we will only consider the three quarks u, d, s. These three quarks were considered in the early history of the quark model. That is, we restrict ourselves to the flavor u, d, s. The colors will be discussed later on. For the quark states, we write

$$e_1 = |u\rangle, \quad e_2 = |d\rangle, \quad e_3 = |s\rangle.$$

We want to construct composed states of the quarks. Our goal is to motivate the proton state

$$|p\rangle = \frac{1}{\sqrt{2}}\left(|u\rangle|u\rangle|d\rangle - |d\rangle|d\rangle|u\rangle\right)$$

with $\langle p|p\rangle = 1$. The proton consists of u-quarks and d-quarks. In the language of mathematics, we write

$$\boxed{|p\rangle = \frac{1}{\sqrt{2}}(e_1 \otimes e_1 \otimes e_2 - e_2 \otimes e_2 \otimes e_1).} \tag{3.44}$$

Note that this is an entangled state of u-quarks and d-quarks. We want to show that this entanglement is based on irreducible representations of the quark flavor symmetry group $SU(3)$ on linear subspaces of the tensor product $X \otimes X \otimes X$ where $X = \text{span}(e_1, e_2, e_3)$.

The Hilbert space X of the quarks. Let X be a 3-dimensional complex Hilbert space with the orthonormal basis e_1, e_2, e_3. The elements of X are given by

$$\psi = \psi_1 e_1 + \psi_2 e_2 + \psi_3 e_3, \qquad \psi_1, \psi_2, \psi_3 \in \mathbb{C}$$

with the inner product $\langle \psi|\varphi\rangle := \sum_{k=1}^{3} \psi_k^\dagger \varphi_k$.

The compact Lie group $SU(3)$ as the flavor symmetry group of the quarks u, d, s. Let $SU(X)$ be the group of all the linear unitary operators

$$\mathsf{U} : X \to X$$

with $\det \mathsf{U} = 1$. This means that the corresponding matrix $(\langle e_i|\mathsf{U}e_j\rangle)_{i,j=1,2,3}$ is an element of the group $SU(3)$. There exists a natural isomorphism between the groups $SU(X)$ and $SU(3)$. We postulate:

> *The physics of the quarks u, d, s is invariant under the symmetry group $SU(3)$.*

The Hilbert space $X \otimes X \otimes X$ of composed quarks. Set $Z := X \otimes X \otimes X$. For the elements

$$\Psi = \sum_{i,j,k=1}^{3} \Psi_{ijk} e_i \otimes e_j \otimes e_k, \qquad \Psi_{ijk} \in \mathbb{C}$$

of the 27-dimensional complex Hilbert space Z, we introduce the inner product

$$\langle \Psi|\Phi\rangle := \sum_{i,j,k=1}^{3} \Psi_{ijk}^\dagger \Phi_{ijk}.$$

We will frequently use the following fact. If $\mathsf{U} : X \to X$ is a linear operator on the Hilbert space X, then the operator can be uniquely extended to an operator $\mathsf{S} : Z \to Z$ by setting[12]

$$\mathsf{U}\left(\sum_{i,j,k=1}^{3} \Psi_{ijk} e_i \otimes e_j \otimes e_k\right) := \sum_{i,j,k=1}^{3} \Psi_{ijk} \mathsf{U}(e_i \otimes e_j \otimes e_k) \qquad (3.45)$$

where

$$\mathsf{U}(e_i \otimes e_j \otimes e_k) := \mathsf{U}e_i \otimes e_j \otimes e_k + e_i \otimes \mathsf{U}e_j \otimes e_k + e_i \otimes e_j \otimes \mathsf{U}e_k.$$

In particular, if $\mathsf{U} \in SU(X)$, then the corresponding operator $\mathsf{U} : Z \to Z$ acts on the tensor product $Z = X \otimes X \otimes X$. This way, we get a representation

$$\varrho : SU(X) \to GL(Z)$$

of the group $SU(X)$ on the Hilbert space Z. Since the group $SU(3)$ is isomorphic to the group $SU(X)$, the group $SU(3)$ acts on Z, too.

For the Standard Model in particle physics, it is important that the compact Lie groups $SU(3)$ and $Sym(3)$ act on the Hilbert space Z of three composed quarks.

This will allow us to apply Weyl's tensor method in the representation theory of the classical compact Lie groups. Let us now discuss the action of $Sym(3)$ on Z.

Permutations of the quarks and the action of the symmetric group Sym(3) on Z. Let $\pi \in Sym(3)$. We define

$$\pi\left(\sum_{i,j,k=1}^{3} \Psi_{ijk} e_i \otimes e_j \otimes e_k\right)$$

by the aid of the corresponding permutation of the quarks e_i, e_j, e_k. For example,

$$(12)(e_i \otimes e_j \otimes e_k) = e_j \otimes e_i \otimes e_k, \qquad (123)(e_i \otimes e_j \otimes e_k) = e_j \otimes e_k \otimes e_i.$$

The main trick is to use the direct sum decomposition

$$\boxed{Z = \mathsf{S}_1(Z) \oplus \mathsf{S}_2(Z) \oplus \mathsf{S}_3(Z) \oplus \mathsf{S}_4(Z)}$$

where S_j, $j = 1, 2, 3, 4$, are the Young symmetrizers of the symmetric group $Sym(3)$ to be found in Table 3.9 on page 222. The point is that the general theory due to Weyl tells us the following.

Theorem 3.12 *The linear subspaces $\mathsf{S}_j(Z)$ of Z, $j = 1, 2, 3, 4$, are invariant irreducible linear subspaces of the Hilbert space Z under the action of the group $SU(X)$.*

We will show below that the dimensions are

$$\dim Z = \dim 27 = 10 + 8 + 8 + 1,$$

that is, $\dim \mathsf{S}_j(Z) = 10, 8, 8, 1$ if $j = 1, 2, 3, 4$, respectively. Physicists briefly write

$$\mathbf{3} \otimes \mathbf{3} \otimes \mathbf{3} = \mathbf{10} \oplus \mathbf{8} \oplus \mathbf{8} \oplus \mathbf{1}. \qquad (3.46)$$

---

[12] To simplify notation, we write U instead of $\mathsf{U} \otimes \mathsf{U} \otimes \mathsf{U}$.

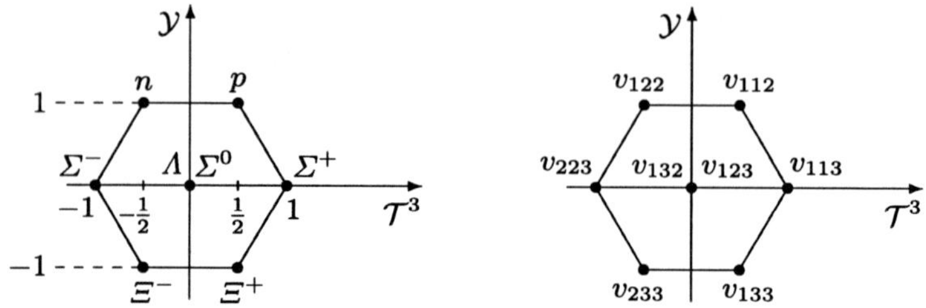

Fig. 3.3. The baryon octet of the proton p and the neutron n: $R(1,1)$

In terms of physics, the space $\mathsf{S}_2(Z)$ contains 8 basis vectors which correspond to an 8-multiplet (octet) of 8 elementary particles which possess similar masses and similar physical properties (see Fig. 3.3). These particles are the following baryons:

$$p\,(\text{proton}),\;\; n\,(\text{neutron}), \Sigma^+, \Sigma^-, \Sigma^0, \Xi^+, \Xi^-, \Lambda.$$

Note that the linear subspace $\mathsf{S}_j(Z)$ of Z is invariant under the action of the Young symmetrizer S_j if $j = 1,2,3,4$. This follows from the fact that $\mathsf{S}_j^2 = \mathsf{S}_j$, that is, Young symmetrizers are projection operators.

Irreducible representations of the group $SU(X)$ **on the complex Hilbert space Z of three-quark states.** Let us discuss how to get a basis of the linear space $\mathsf{S}_j(Z)$. We will use the explicit form of Young symmetrizers S_j from Table 3.9 on page 222.

(i) The linear space $\mathsf{S}_1(Z)$ (see Fig. 3.4). Set $e_{ijk} := e_i \otimes e_j \otimes e_k$. We have

$$\mathsf{S}_1(e_{ijk}) = \frac{1}{3!} \sum_{\pi \in Sym(3)} \pi(e_{ijk}) = u_{ijk}.$$

This is the symmetrization of e_{ijk}. Explicitly,

$$u_{ijk} := \tfrac{1}{6}(e_{ijk} + e_{ikj} + e_{jki} + e_{jik} + e_{kij} + e_{kji}).$$

A basis of the space $\mathsf{S}_1(Z)$ is given by

$$u_{111}, u_{112}, u_{113}, u_{122}, u_{123}, u_{133}, u_{222}, u_{223}, u_{233}, u_{333}.$$

Thus, the dimension of $\mathsf{S}_1(Z)$ is equal to 10. Graphically, the indices of $u_{111}, u_{112}, \ldots$ are given by

$$\begin{array}{cccc}
\boxed{1\,|\,1\,|\,1} & \boxed{1\,|\,1\,|\,2} & \boxed{1\,|\,1\,|\,3} & \\[1em]
\boxed{1\,|\,2\,|\,2} & \boxed{1\,|\,2\,|\,3} & \boxed{1\,|\,3\,|\,3} & \\[1em]
\boxed{2\,|\,2\,|\,2} & \boxed{2\,|\,2\,|\,3} & \boxed{2\,|\,3\,|\,3} & \boxed{3\,|\,3\,|\,3}
\end{array} \qquad (3.47)$$

Note that the diagrams (3.47), (3.49), and (3.50) are the Young frames of the Young symmetrizers S_j filled in with the numbers $1, 2, 3$. Here, the rows are not decreasing from left to right, and the columns are increasing from top

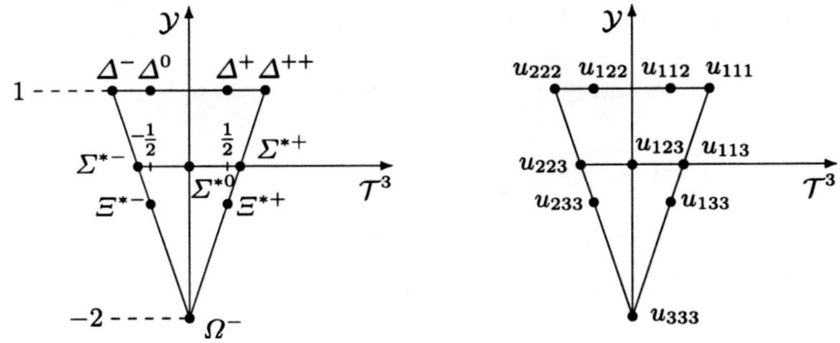

Fig. 3.4. Baryon decuplet: $R(3,0)$

to bottom. Such fillings are called regular fillings of Young frames. From the physical point of view, the 10 basis vectors $u_{111}, u_{112}, u_{113}\ldots$ correspond to a 10-multiplet (decuplet) of 10 baryons depicted in Fig. 3.4. For example, the state

$$u_{112} = \tfrac{1}{3}(e_1 \otimes e_1 \otimes e_2 + e_1 \otimes e_2 \otimes e_1 + e_2 \otimes e_1 \otimes e_1)$$

consists of two u-quarks e_1 and one d-quark e_2. The additive quantum numbers $\mathcal{T}^3$ and $\mathcal{Y}$ will be computed below (see Table 3.10 on page 232).

(ii) The linear space $\mathsf{S}_2(Z)$ (see Fig. 3.3). We have

$$\mathsf{S}_2 e_{ijk} = \tfrac{1}{3}\big(\,(1) + (12) - (13) - (123)\big)e_{ijk} = v_{ijk} \tag{3.48}$$

where

$$v_{ijk} := \tfrac{1}{3}(e_{ijk} + e_{jik} - e_{kji} - e_{jki}).$$

These vectors can be used in order to construct a basis of the linear subspace $\mathsf{S}_2(Z)$. In fact, the following eight vectors

$$v_{112}, v_{122}, v_{132}, v_{113}, v_{123}, v_{133}, v_{223}, v_{233}$$

form a basis of $\mathsf{S}_2(Z)$. Thus, the dimension of $\mathsf{S}_2(Z)$ is 8. Graphically, the indices are obtained by the regular filling of the corresponding Young frame:

$$\begin{array}{cccc}
\begin{array}{|c|c|}\hline 1 & 1 \\ \hline 2 \\ \cline{1-1}\end{array} &
\begin{array}{|c|c|}\hline 1 & 2 \\ \hline 2 \\ \cline{1-1}\end{array} &
\begin{array}{|c|c|}\hline 1 & 3 \\ \hline 2 \\ \cline{1-1}\end{array} &
\begin{array}{|c|c|}\hline 1 & 1 \\ \hline 3 \\ \cline{1-1}\end{array} \\[2em]
\begin{array}{|c|c|}\hline 1 & 2 \\ \hline 3 \\ \cline{1-1}\end{array} &
\begin{array}{|c|c|}\hline 1 & 3 \\ \hline 3 \\ \cline{1-1}\end{array} &
\begin{array}{|c|c|}\hline 2 & 2 \\ \hline 3 \\ \cline{1-1}\end{array} &
\begin{array}{|c|c|}\hline 2 & 3 \\ \hline 3 \\ \cline{1-1}\end{array}
\end{array} \tag{3.49}$$

(iii) The linear space $\mathsf{S}_3(Z)$. We have

$$\mathsf{S}_3 e_{ijk} = \tfrac{1}{3}\big(\,(1) - (12) + (13) - (132)\big)e_{ijk} = w_{ijk}$$

where $w_{ijk} := \frac{1}{3}(e_{ijk} - e_{jik} + e_{kji} - e_{kij})$. The eight vectors

$$w_{112}, w_{122}, w_{132}, w_{113}, w_{123}, w_{133}, w_{223}, w_{233}$$

form a basis of $\mathsf{S}_3(Z)$. Thus, the dimension of $\mathsf{S}_3(Z)$ is equal to 8. Since the Young symmetrizers S_2 and S_3 belong to the same Young frame, the indices of the basis vectors $w_{112}, \ldots$ are the same as for the basis vectors $v_{112}, \ldots$ in (ii).

(iv) The linear space $\mathsf{S}_4(Z)$. We have

$$\mathsf{S}_4(e_{ijk}) = \frac{1}{3!} \sum_{\pi \in Sym(3)} \operatorname{sgn}(\pi) \cdot \pi(e_{ijk}) = z_{ijk}.$$

This is the antisymmetrization of e_{ijk}. Explicitly,

$$z_{ijk} := \tfrac{1}{6}(e_{ijk} - e_{ikj} + e_{jki} - e_{jik} + e_{kij} - e_{kji}).$$

The one-dimensional linear space $\mathsf{S}_4(Z)$ has the basis vector z_{123}. Graphically, this corresponds to the diagram

$$\begin{array}{|c|}\hline 1 \\ \hline 2 \\ \hline 3 \\ \hline \end{array} \tag{3.50}$$

The Clebsch–Gordan coefficients. Consider the baryon octet (Fig. 3.3 on page 228). Then

$$v_{112} = \sum_{i,j,k=1}^{3} C_{112}^{ijk}\, e_i \otimes e_j \otimes e_k,$$

by (3.48). The real numbers C_{112}^{ijk} are called the Clebsch–Gordan coefficients of v_{112}. For example, $C_{112}^{112} = \frac{2}{3}$. Similarly, the Clebsch–Gordan coefficients are defined for $u_{ijk}, v_{ijk}, w_{ijk}$. In terms of physics, the Clebsch–Gordan coefficients tell us how to obtain the particle states of a multiplet from the product states $e_i \otimes e_j \otimes e_k$. As a rule, one uses normalized particle states. The advantage is that the Clebsch–Gordan coefficients only depend on the symmetry group, but not on the concrete physical situation.

The Clebsch–Gordan coefficients provide physicists with very useful symmetry information which can be computed via group theory in a universal way, without using details of the specific physical situation.

The Lie algebra $su(3)$. For the physical interpretation of the preceding approach, the Lie algebra $su(3)$ of the Lie group $SU(3)$ plays a fundamental role in order to construct observables (i.e., self-adjoint operators on the quark Hilbert space X whose eigenvalues are quantum numbers). Set

$$e_1 := \begin{pmatrix} 1 \\ 0 \\ 0 \end{pmatrix}, \quad e_2 := \begin{pmatrix} 0 \\ 1 \\ 0 \end{pmatrix}, \quad e_3 := \begin{pmatrix} 0 \\ 0 \\ 1 \end{pmatrix}.$$

Then the Hilbert space X (i.e., the quark space) is equal to $\mathbb{C}^3$. The group $SU(3)$ acts on X by matrix multiplication. The elements G of $SU(3)$ have the form

$$G := \mathrm{e}^A, \qquad A \in su(3).$$

Here, $A = \ln G$ for all $G \in SU(3)$ with $||I - G|| < 1$. Recall that the complex (3×3)-matrix A is an element of $su(3)$ iff $A^\dagger = -A$ and $\operatorname{tr}(A) = 0$. The map $G \mapsto \ln G$ is a diffeomorphism from an open neighborhood of the unit element I of $SU(3)$ onto an open neighborhood of the zero element of the Lie algebra $su(3)$ (see Sect. 7.7 of Vol. I). Let us introduce the following eight Gell-Mann matrices:

$$\lambda_1 = \begin{pmatrix} 0 & 1 & 0 \\ 1 & 0 & 0 \\ 0 & 0 & 0 \end{pmatrix}, \quad \lambda_2 = \begin{pmatrix} 0 & -\mathrm{i} & 0 \\ \mathrm{i} & 0 & 0 \\ 0 & 0 & 0 \end{pmatrix}, \quad \lambda_3 = \begin{pmatrix} 1 & 0 & 0 \\ 0 & -1 & 0 \\ 0 & 0 & 0 \end{pmatrix}, \quad \lambda_4 := \begin{pmatrix} 0 & 0 & 1 \\ 0 & 0 & 0 \\ 1 & 0 & 0 \end{pmatrix} \qquad (3.51)$$

and

$$\lambda_5 := \begin{pmatrix} 0 & 0 & -\mathrm{i} \\ 0 & 0 & 0 \\ \mathrm{i} & 0 & 0 \end{pmatrix}, \ \lambda_6 := \begin{pmatrix} 0 & 0 & 0 \\ 0 & 0 & 1 \\ 0 & 1 & 0 \end{pmatrix}, \ \lambda_7 := \begin{pmatrix} 0 & 0 & 0 \\ 0 & 0 & -\mathrm{i} \\ 0 & \mathrm{i} & 0 \end{pmatrix}, \ \lambda_8 := \frac{1}{\sqrt{3}} \begin{pmatrix} 1 & 0 & 0 \\ 0 & 1 & 0 \\ 0 & 0 & -2 \end{pmatrix}.$$

These matrices are self-adjoint and traceless. Set $A_k := -\mathrm{i}\lambda_k$. Then the eight matrices $A_k, k = 1, \ldots, 8$ form a basis of the real Lie algebra $su(3)$ with the commutation rules

$$[A_j, A_k]_- = \sum_{l=1}^{3} c_{jkl}\, A_l, \qquad j, k = 1, \ldots, 8.$$

The numbers c_{jkl} are called the structure constants of the Lie algebra $su(3)$. Explicitly, c_{jkl} is antisymmetric with respect to the indices j, k, l. Moreover,

$$c_{123} = 2, \quad c_{147} = c_{246} = c_{257} = c_{345} = -c_{156} = -c_{367} = 1, \quad c_{458} = c_{678} = \sqrt{3}.$$

Remark. Generally, physicists do not use the commutation rules for the real Lie algebra $su(3)$. They like to use self-adjoint matrices which correspond to physical observables in quantum mechanics. Therefore, physicists use the commutation rules

$$[\lambda_j, \lambda_k]_- = 2\mathrm{i} \sum_{l=1}^{3} f_{jkl} \lambda_l, \qquad j, k = 1, \ldots, 8$$

for the self-adjoint Gell-Mann matrices λ_k which are not elements of the real Lie algebra $su(3)$. However, the matrices $-\mathrm{i}\lambda_k$ are skew-adjoint and traceless, and hence they are elements of $su(3)$. The distinction between the different versions vanishes if we pass to the complexfication of the real Lie algebra $su(3)$. This is the complex Lie algebra $sl_{\mathbb{C}}(3, \mathbb{C})$ of all complex traceless (3×3)-matrices. This will be discussed below.

Isospin component $\mathcal{T}^3$, hypercharge $\mathcal{Y}$, and weight diagrams. We introduce the self-adjoint operators $\mathsf{T}^3, \mathsf{Y} : X \to X$ defined by

$$\boxed{\mathsf{T}^3 := \frac{\lambda_3}{2}, \qquad \mathsf{Y} := \frac{\lambda_8}{\sqrt{3}}.}$$

Hence

$$\mathsf{T}^3 = \begin{pmatrix} \frac{1}{2} & 0 & 0 \\ 0 & -\frac{1}{2} & 0 \\ 0 & 0 & 0 \end{pmatrix}, \qquad \mathsf{Y} = \begin{pmatrix} \frac{1}{3} & 0 & 0 \\ 0 & \frac{1}{3} & 0 \\ 0 & 0 & -\frac{2}{3} \end{pmatrix}. \qquad (3.52)$$

Then, the three quark states e_1, e_2, e_3 are common eigenvectors of the operators T^3 and Y with

$$\mathsf{T}^3 e_j = \mathcal{T}^3(e_j)\, e_j, \qquad \mathsf{Y} e_j = \mathcal{Y}(e_j)\, e_j, \qquad j = 1, 2, 3.$$

The eigenvalue $\mathcal{T}^3(e_j)$ (resp. $\mathcal{Y}(e_j)$) is called the third component of the isospin (resp. the hypercharge) of the quark e_j. The values can be found in Table 3.10.

Table 3.10. The quantum numbers of quarks, antiquarks, and of composed quark states

quantum numbers (weight tuple $(\mathcal{T}^3, \mathcal{Y})$)	$\mathcal{T}^3$	$\mathcal{Y}$	Q	
u-quark e_1	$\frac{1}{2}$	$\frac{1}{3}$	$\frac{2}{3}e$	
d-quark e_2	$-\frac{1}{2}$	$\frac{1}{3}$	$-\frac{1}{3}e$	
s-quark e_3	0	$-\frac{2}{3}$	$-\frac{1}{3}e$	
$\bar{u}$-antiquark $\bar{e}^1$	$-\frac{1}{2}$	$-\frac{1}{3}$	$-\frac{2}{3}e$	
$\bar{d}$-antiquark $\bar{e}^2$	$\frac{1}{2}$	$-\frac{1}{3}$	$\frac{1}{3}e$	
$\bar{s}$-antiquark $\bar{e}^3$	0	$\frac{2}{3}$	$\frac{1}{3}e$	
proton p	$\frac{1}{2}$	1	e	uud
neutron n	$-\frac{1}{2}$	1	0	udd
meson π_+	1	0	e	$u\bar{d}$

The tuple $(\mathcal{T}^3(e_j), \mathcal{Y}(e_j))$ is called the weight tuple of e_j. The values are depicted in Fig. 3.5. For example, the u-quark e_1 has the weight tuple

$$(\mathcal{T}^3(u), \mathcal{Y}(u)) = (\tfrac{1}{2}, \tfrac{1}{3}).$$

For two weight tuples, we write

$$\boxed{(\mathcal{T}^3, \mathcal{Y}) < (\mathcal{T}^3_+, \mathcal{Y}_+)} \tag{3.53}$$

iff either $\mathcal{T}^3 < \mathcal{T}^3_+$ or $\mathcal{T}^3 = \mathcal{T}^3_+, \mathcal{Y} < \mathcal{Y}_+$. For example, the u-quark e_1 has the highest weight tuple in Fig. 3.5. In a natural way, the operators T^3 and Y can be extended to the Hilbert space Z of three composed quarks. Explicitly, by (3.45), we get

$$\mathsf{T}^3(e_i \otimes e_j \otimes e_k) = (\mathcal{T}^3(e_i) + \mathcal{T}^3(e_j) + \mathcal{T}^3(e_k))(e_i \otimes e_j \otimes e_k).$$

Thus, $\mathcal{T}^3$ is an additive quantum number. The same is true for the hypercharge $\mathcal{Y}$. In particular, by (3.44), for the proton we get

$$\boxed{\mathsf{T}^3|p\rangle = \tfrac{1}{2}|p\rangle, \qquad \mathsf{Y}|p\rangle = |p\rangle.}$$

Hence the proton has the weight tuple $(\mathcal{T}^3, \mathcal{Y}) = (\frac{1}{2}, 1)$. This coincides with the value indicated in Fig. 3.3.

The Cartan subalgebra of the Lie algebra $su(3)$**.** The diagonal matrices $\mathrm{i}\mathsf{T}^3 = \mathrm{i}\lambda_3/2$ and $\mathrm{i}Y = \mathrm{i}\lambda_8/\sqrt{3}$ are elements of the Lie algebra $su(3)$. The linear hull $\mathrm{span}(\mathsf{T}^3, \mathsf{Y})$ forms a subalgebra of the real Lie algebra $su(3)$ which is called the Cartan subalgebra $\mathcal{C}(su(3))$ of $su(3)$.

The Hilbert space structure of the real Lie algebra $su(3)$**.** For the Gell-Mann matrices, we have the following trace formulas:

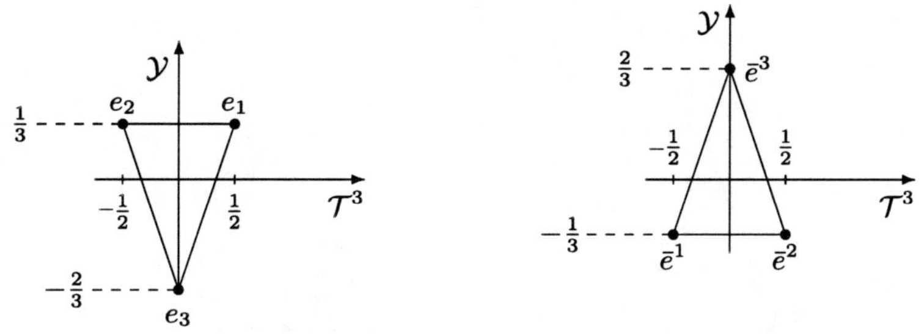

(a) **quarks** $e_1 = u$, $e_2 = d$, $e_3 = s$ (b) **antiquarks** $\bar{e}^1 = \bar{u}$, $\bar{e}^2 = \bar{d}$, $\bar{e}^3 = \bar{s}$

Fig. 3.5. The quark triplet $R(1,0)$ and the antiquark triplet $R(0,1)$

$$\mathrm{tr}(\lambda_i \lambda_j) = 2\delta_{ij}, \qquad i,j = 1,\ldots,8.$$

We set

$$\langle A|B\rangle := -6\,\mathrm{tr}(AB) \qquad \text{for all} \quad A, B \in su(3).$$

We introduce the normalization factor 6 in order to get $\langle A|B\rangle = -K(A,B)$ for all $A, B \in su(3)$ where K is the Killing form of $su(3)$ to be introduced below. The real linear space $su(3)$ becomes a Hilbert with respect to the inner product $\langle A|B\rangle$. For $B_j := -\frac{\mathrm{i}}{\sqrt{12}}\lambda_j$, we get

$$\langle B_i|B_j\rangle = \delta_{ij}, \qquad i,j = 1,\ldots,8.$$

Thus, the matrices $B_1, B_2, \ldots, B_8$ form an orthonormal basis of $su(3)$.

The irreducible representations of the Lie group $SU(3)$**.** The complete system of irreducible representations of the group $SU(3)$ can be labelled by $R(q,p)$ where $q, p = 0, 1, 2, \ldots$ Here, the irreducible representation $R(q,p)$ of $SU(3)$ acts on a complex Hilbert space of dimension $\deg R(q,p)$ where

$$\boxed{\deg R(q,p) = \tfrac{1}{2}(q+1)(p+1)(q+p+2).} \tag{3.54}$$

This is equal to the number of particles of the corresponding multiplet.[13] For example,

- $\deg R(0,0) = 1, \deg R(1,0) = \deg(R(0,1) = 3, \deg R(1,1) = 8,$
  $\deg R(2,0) = \deg R(2,0) = 6, \deg R(3,0) = \deg R(0,3) = 10.$

Physicists use the following notation:

- $R(0,0)) = \mathbf{1},\ R(1,0) = \mathbf{3},\ R(0,1) = \bar{\mathbf{3}},\ R(1,1) = \mathbf{8},$
  $R(2,0) = \mathbf{6},\ R(0,2) = \bar{\mathbf{6}},\ R(3,0) = \mathbf{10},\ \ R(0,3) = \overline{\mathbf{10}}.$

The corresponding multiplets with the weight tuples $(\mathcal{T}^3, \mathcal{Y})$ can be found in Figs. 3.4 through 3.6. Table 3.11 on page 240 contains the weight tuples of $R(1,0)$ through $R(3,0)$. In particular, $R(0,0)$ denotes the trivial representation $\varrho : Y \to Y$ on the one-dimensional complex linear space Y where $\varrho(G)$ is the identity operator for all $G \in SU(3)$.

Note that the diagram of $R(q,p)$ *is obtained from the diagram of* $R(p,q)$ *(dual diagram) by a reflection at the origin.*

[13] In terms of mathematics, the relation (3.54) is a special case of Weyl's dimension formula for the degrees of irreducible representations of compact Lie groups (see Vol. IV).

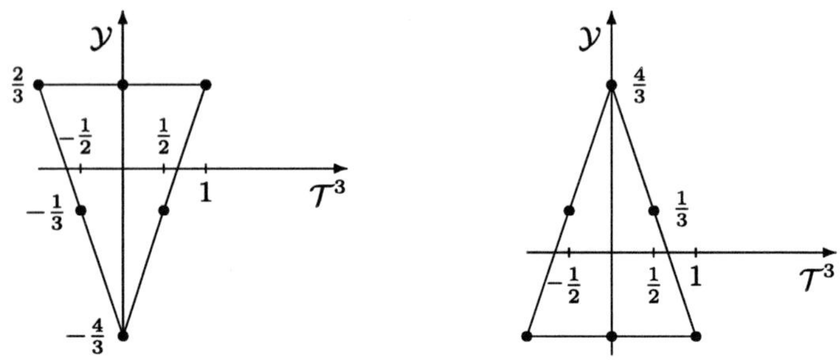

(a) $R(2,0)$ (b) $R(0,2)$

Fig. 3.6. Duality of weight diagrams

This means that the weight tuples $(\mathcal{T}^3, \mathcal{Y})$ are replaced by $(-\mathcal{T}^3, -\mathcal{Y})$ (see Fig. 3.6). The highest weight of $R(q,p)$ is equal to

$$(\mathcal{T}^3, \mathcal{Y}) = (\tfrac{1}{2}(q+p), \tfrac{1}{3}(q-p)).$$

Élie Cartan's adjoint representation of the real Lie algebra $su(3)$ **on itself, and the Killing form.** Fix $A \in su(3)$. Define

$$\boxed{\mathrm{ad}(A)B := [A,B]_- \qquad \text{for all} \quad B \in su(3).}$$

This yields the linear operator $\mathrm{ad}(A) : su(3) \to su(3)$. The map $A \mapsto \mathrm{ad}(A)$ yields a representation of the Lie algebra $su(3)$ on the real linear space $su(3)$. Define

$$K(A,B) := \mathrm{tr}(\mathrm{ad}(A)\,\mathrm{ad}(B)) \qquad \text{for all} \quad A, B \in su(3).$$

Explicitly,

$$K(A,B) = 6\,\mathrm{tr}(AB) \qquad \text{for all} \quad A, B \in su(3).$$

Élie Cartan's adjoint representation of the group $SU(3)$ **on its Lie algebra** $su(3)$**.** Fix $G \in SU(3)$. Define

$$\boxed{\mathrm{Ad}(G)B := GBG^{-1} \qquad \text{for all} \quad B \in su(3).}$$

The map $G \mapsto \mathrm{Ad}(G)$ is a representation of the group $SU(3)$ on the real linear space $su(3)$. This follows from $(GH)B(GH)^{-1} = G(HBH^{-1})G^{-1}$. This representation is called the adjoint representation

$$\mathrm{Ad} : SU(3) \to GL(su(3))$$

of $SU(3)$ on $su(3)$. Fix $A \in su(3)$. Set $G(t) := \mathrm{e}^{tA}$ for all $t \in \mathbb{R}$. Then $G(t) \in SU(3)$ for all $t \in \mathbb{R}$. Differentiating $\mathrm{Ad}(G(t))B$ with respect to t at the point $t = 0$, we get

$$\frac{d}{dt}\mathrm{Ad}(G(t))B_{|t=0} = AB - BA = [A,B]_- = \mathrm{ad}(A)B.$$

This way, we obtain the adjoint representation ad of $su(3)$ on itself. One can show that the adjoint representation of $SU(3)$ on $su(3)$ is irreducible, and it corresponds to $R(1,1)$. Graphically, it is described by Fig. 3.8, which corresponds to the baryon octet of the proton. According to the general theory of semisimple Lie algebras, the quantum numbers $(\mathcal{T}^3, \mathcal{Y})$ of the adjoint representation of $SU(3)$ on $su(3)$ are called the root tuples of the Lie algebra $su(3)$; they play a fundamental role in the structure theory of semisimple Lie algebras (see Sect. 3.15.2).

Isospin and the Gell-Mann–Okubo mass formula. Finally, let us introduce the isospin operator

$$|\mathsf{T}|^2 := \sum_{j=1}^{3} (\mathsf{T}^j)^2$$

on the quark space X. Explicitly, we obtain the diagonal operator

$$|\mathsf{T}|^2 = \frac{3}{4}\begin{pmatrix} 1 & 0 & 0 \\ 0 & 1 & 0 \\ 0 & 0 & 0 \end{pmatrix}.$$

Thus, for $j = 1, 2,$, we get

$$|\mathsf{T}|^2 e_j = \mathcal{T}(\mathcal{T}+1)e_j, \ \mathcal{T} = \tfrac{1}{2} \quad \text{and} \quad |\mathsf{T}|^2 e_3 = \mathcal{T}(\mathcal{T}+1)e_3, \ \mathcal{T} = 0.$$

Therefore, physicists assign to the u-quark e_1 and d-quark e_2 (resp. to the s-quark e_3) the isospin $\mathcal{T} = \frac{1}{2}$ (resp. $\mathcal{T} = 0$). The isospin is not an additive quantum number. For the proton and the neutron, we get

$$\boxed{|\mathsf{T}|^2 \, |p\rangle = \mathcal{T}(\mathcal{T}+1)\, |p\rangle, \quad |\mathsf{T}|^2 \, |n\rangle = \mathcal{T}(\mathcal{T}+1)\, |n\rangle, \qquad \mathcal{T} = \tfrac{1}{2}.}$$

Thus, the proton and the neutron have the same isospin $\mathcal{T} = \frac{1}{2}$, but the different isospin components $\mathcal{T}^3 = \frac{1}{2}$ and $\mathcal{T}^3 = -\frac{1}{2}$, respectively. In 1962 Okubo proposed the following mass formula

$$m = a + b\mathcal{Y} + c\big(\mathcal{T}(\mathcal{T}+1) - \tfrac{1}{4}\mathcal{Y}^2\big). \tag{3.55}$$

This is the mass m of a particle of isospin $\mathcal{T}$ and hypercharge $\mathcal{Y}$ if the particle is contained in a baryon multiplet.[14] The free coefficients a, b, c have to be chosen in such a way that they fit best the experimental values of the baryon multiplet under consideration. For the baryon octet of the neutron n (Fig. 3.3), we get

$$m_n + m_{\Xi^0} = \tfrac{3}{2} m_\Lambda + \tfrac{1}{2} m_{\Sigma^0}$$

which fits fairly well the experimental values of 2255 MeV/c^2 (left-hand side) and 2270 MeV/c^2 (right-hand side). Gell-Mann used the Okubo mass formula in order to predict the mass $m = 1672\text{MeV}/c^2$ of the unknown Ω^- particle (see Fig. 3.4). This particle was discovered in 1964. Note that the mass formula (3.55) does not take the electromagnetic interaction into account. Physicists assume that the electromagnetic interaction causes additional mass differences.

Generalization to the group $SU(n)$. Fix $n = 2, 3, \ldots$ The tensor method also works if we replace the group $SU(3)$ by the group $SU(n)$. Then we start with the n-dimensional complex Hilbert space X with the orthonormal basis $e_1, \ldots, e_n$, and the $Sym(n)$-Young symmetrizers yield the decomposition of the tensor products into linear subspaces which have the following two crucial properties: they are invariant under the action of the group $SU(n)$, and the corresponding representation of $SU(n)$ is irreducible.

[14] S. Okubo, Note on unitary symmetry in strong interaction, Progr. Theor. Phys. **27** (1962), 949–966.

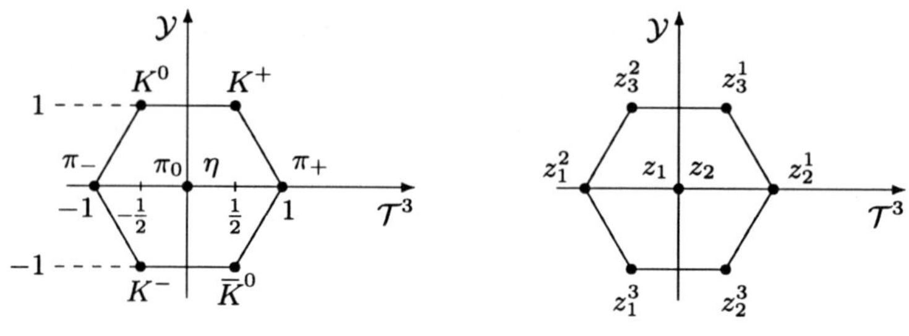

Fig. 3.7. Meson octet of π_+: $R(1,1)$

3.14.2 Antiquarks and Mesons

> Particles and antiparticles in nature are mathematically described by duality. The dual weight diagram to the weight diagram of the quarks u, d, s corresponds to the antiquarks $\bar{u}, \bar{d}, \bar{s}$.
>
> Folklore

The Einstein summation convention. In this section, we will sum over equal upper and lower indices from 1 to 3. For example

$$\boxed{t_m^m = t_1^1 + t_2^2 + t_3^3.}$$

This expression is called a trace. As we will show below, the distinction between upper and lower indices is crucial for quickly detecting invariant expressions via the principle of the correct index picture (see Chap. 8). In particular, traces will play the decisive role. Our goal is to motivate Fig. 3.5 on page 233. To begin with, we set

$$\bar{e}^1 := \begin{pmatrix} 1 \\ 0 \\ 0 \end{pmatrix}, \quad \bar{e}^2 := \begin{pmatrix} 0 \\ 1 \\ 0 \end{pmatrix}, \quad \bar{e}^3 := \begin{pmatrix} 0 \\ 0 \\ 1 \end{pmatrix}.$$

The dual Hilbert space X^d of antiquarks. Let X^d denote the dual space to the 3-dimensional complex Hilbert space X of quarks. The two complex linear spaces X and X^d are isomorphic to $\mathbb{C}^3$. Let us describe the space X^d as a 3-dimensional complex Hilbert space with the orthonormal basis $\bar{e}^1, \bar{e}^2, \bar{e}^3$ which corresponds to the antiquarks $\bar{u}, \bar{d}, \bar{s}$, respectively. In addition, we will describe the flavor symmetry of the antiquarks $\bar{u}, \bar{d}, \bar{s}$ by means of the dual representation

$$G \mapsto (G^{-1})^d$$

of the quark flavor group $SU(3)$. Explicitly, the representation $\varrho : SU(3) \to GL(X^d)$ is given by

$$\boxed{\varrho(G)\,(c_j\bar{e}^j) := (G^{-1})^d(c_j\bar{e}^j), \qquad c_1, c_2, c_3 \in \mathbb{C}.} \tag{3.56}$$

Since we have $G = \mathrm{e}^A$ with $A \in su(3)$, we get $(G^{-1})^d = \mathrm{e}^{-A^d}$. This induces the dual representation

$$A \mapsto -A^d$$

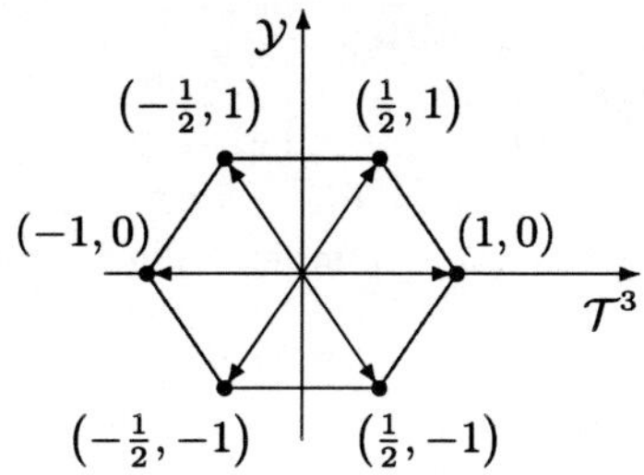

Fig. 3.8. Root tuples of the Lie algebra $su(3)$

of the Lie algebra $su(3)$ on the antiquark space X^d. Explicitly, we get the map $\sigma : su(3) \to gl(X^d)$ with

$$\sigma(A)\,(c_j\bar{e}^j) := -A^d(c_j\bar{e}^j), \qquad c_1, c_2, c_3 \in \mathbb{C}.$$

Since both the isospin operator T^3 and the hypercharge operator Y from (3.52) are multiples of elements in the Lie algebra $su(3)$, we define

$$\sigma(\mathsf{T}^3) := -(\mathsf{T}^3)^d = -\mathsf{T}^3, \qquad \sigma(\mathsf{Y}) := -(\mathsf{Y}^3)^d = -\mathsf{Y}^3.$$

To simplify notation, following the convention used by physicists, we replace $\sigma(\mathsf{T}^3)$ and $\sigma(\mathsf{Y})$ by T^3 and Y, respectively. This way, we get

$$\mathsf{T}^3\bar{e}^1 := -\mathcal{T}^3(e_1)\,\bar{e}^1, \qquad \mathsf{Y}\bar{e}^1 = -\mathcal{Y}(e_1)\,\bar{e}^1.$$

The same is true if we replace e_1 by e_2 or e_3 (resp. $\bar{e}^1$ by $\bar{e}^2$ or $\bar{e}^1$). This yields the quantum numbers $(\mathcal{T}^3, \mathcal{Y})$ indicated in Fig. 3.5(b) on page 233. Observe that the antiquark diagram (b) is obtained from the quark diagram (a) by reflection at the origin. The irreducible representation of the group $SU(3)$ on the antiquark space X^d is denoted by $R(0,1)$ or $\bar{\mathbf{3}}$ (dual representation to the quark representation $R(1,0)$).

The reduction of the tensor product $X \otimes X^d$. We want to show that

$$R(1,0) \otimes R(0,1) = R(1,1) \oplus R(0,0). \tag{3.57}$$

Physicists briefly write

$$\boxed{\mathbf{3} \otimes \bar{\mathbf{3}} = \mathbf{8} \oplus \mathbf{1}.} \tag{3.58}$$

The representation of the group $SU(3)$ on the tensor product $X \otimes X^d$ is given by

$$\varrho(G)\,(e_j \otimes \bar{e}^k) := Ge_j \otimes \bar{e}^k + e_j \otimes (G^{-1})^d\,\bar{e}^k, \qquad j,k = 1,2,3$$

for all $G \in SU(3)$. Furthermore, for all $A \in su(3)$, the map

$$\sigma(A)(e_j \otimes \bar{e}^k) := Ae_j \otimes \bar{e}^k + e_j \otimes (-A^d\bar{e}^k), \qquad j,k = 1,2,3$$

describes the representation of the real Lie algebra $su(3)$ on $X \otimes X^d$. Set

$$z_j^k := e_j \otimes \bar{e}^k.$$

Every element of $X \otimes X^d$ can be written as

$$t_k^j z_j^k = \left(t_k^j - \tfrac{1}{3} t_m^m \delta_k^j\right) z_j^k + \tfrac{1}{3} t_m^m \delta_k^j z_j^k$$

where the coefficients t_k^j are complex numbers. This corresponds to the decomposition

$$X \otimes X^d = W \oplus Y$$

with $W := \{s_k^j z_j^k : s_j^j = 0\}$ and $Y := \{a z_j^j : a \in \mathbb{C}\}$. Thus, $\dim Y = 1$. Since $\dim(X \otimes X^d) = 9$, we get $\dim W = 8$.

Proposition 3.13 *The linear subspace Y is invariant under the action of the group $SU(3)$.*

Proof. We will use a simple argument which is crucial for the invariant theory. Let $G \in SU(3)$. The basic transformation formulas

$$\begin{pmatrix} e_{1'} \\ e_{2'} \\ e_{3'} \end{pmatrix} = G \begin{pmatrix} e_1 \\ e_2 \\ e_3 \end{pmatrix}, \qquad \begin{pmatrix} \bar{e}^{1'} \\ \bar{e}^{2'} \\ \bar{e}^{3'} \end{pmatrix} = (G^{-1})^d \begin{pmatrix} \bar{e}^1 \\ \bar{e}^2 \\ \bar{e}^3 \end{pmatrix}$$

can be written as

$$e_{j'} = G_{j'}^j e_j, \quad \bar{e}^{j'} = G_j^{j'} \bar{e}^j, \quad G_{j'}^j G_j^{k'} = \delta_{j'}^{k'},$$

since $GG^{-1} = I$. The same is true if we interchange the index j with j'. It follows from $t_k^j (e_j \otimes \bar{e}^k) = t_k^j G_j^{j'} G_{k'}^k (e_{j'} \otimes \bar{e}^{k'}) = t_{k'}^{j'} (e_{j'} \otimes \bar{e}^{k'})$ that

$$t_{k'}^{j'} = G_j^{j'} G_{k'}^k t_k^j.$$

Hence $t_{j'}^{j'} = \delta_j^k t_k^j$. This implies the key relation

$$\boxed{t_{j'}^{j'} = t_j^j}$$

called the trace invariance. The same argument yields $e_{j'} \otimes \bar{e}^{j'} = e_j \otimes \bar{e}^j$, that is, $z_{j'}^{j'} = z_j^j$. □

Similarly, one shows that the linear subspace W is invariant under the action of $SU(3)$. Furthermore, one can show that W and Y are irreducible linear subspaces of the linear space $X \otimes X^d$ under the action of the group $SU(3)$. The following eight vectors

- $z_1^2, z_2^1, z_1^3, z_3^1, z_2^3, z_3^2$,
- $z_1 := 2^{-1/2}(z_1^1 - z_2^2), \quad z_2 := 6^{-1/2}(z_1^1 + z_2^2 - 2z_3^3)$

form an orthonormal basis of W. By Fig. 3.5 on page 233,

$$\mathsf{T}^3 z_1^2 = \left(T^3(e_1) + T^3(\bar{e}^2)\right) z_1^2 = \left(\tfrac{1}{2} + \tfrac{1}{2}\right) z_1^2 = z_1^2, \quad \mathsf{Y} z_1^2 = \left(\mathcal{Y}(e_1) + \mathcal{Y}(\bar{e}^2)\right) z_1^2 = 0.$$

Similarly, we get all the quantum numbers depicted in Fig. 3.7 on page 236. It can be shown that the action of $SU(3)$ on W (resp. Y) corresponds to $R(1,1)$ (resp. $R(0,0)$).

3.14.3 The Method of Highest Weight for Composed Particles

The method of highest weight is used by physicists in elementary particle physics because of its simplicity and elegance.

Folklore

In elementary particle physics, one wants to construct all the possible multiplets of hadrons (baryons and mesons). To this end, one considers all the tensor products $X_1 \otimes X_2 \otimes \cdots \otimes X_n$ of quark spaces ($X_j = X$) and antiquark spaces ($X_k = \bar{X}$), and one decomposes them into a direct sum

$$\boxed{X_1 \otimes X_2 \otimes \cdots \otimes X_n = Y_1 \oplus Y_2 \oplus \cdots \oplus Y_m}$$

of linear subspaces in such a way that there exist irreducible representations of the group $SU(3)$ on $Y_1, \ldots, Y_m$ (complete reducibility). This can be explicitly done by using Young symmetrizers, as we explained above. However, if one only wants to know the size of the multiplets (i.e., the dimensions of the linear invariant subspaces $Y_1, \ldots, Y_m$,) then one can quickly obtain this by using the method of highest weight. Using Table 3.11, this method proceeds as follows:

Step 1: Compute the quantum numbers $\mathcal{T}^3$ and $\mathcal{Y}$ of the tensor products of quarks and antiquarks. Note that $\mathcal{T}^3$ and $\mathcal{Y}$ are additive quantum numbers.

Step 2: Choose the quantum tuple $(\mathcal{T}_0^3, \mathcal{Y}_0)$ of highest weight (see (3.53) on page 232). Concerning this highest weight, there exists precisely one irreducible representation $R(q_0, p_0)$ of $SU(3)$. Explicitly,

$$\boxed{q_0 = \mathcal{T}_0^3 + \tfrac{3}{2}\mathcal{Y}_0, \quad p_0 = \mathcal{T}_0^3 - \tfrac{3}{2}\mathcal{Y}_0.} \tag{3.59}$$

Cancel the weight tuples corresponding to $R(q_0, p_0)$ (see Table 3.11).

Step 3: Determine the highest weight $(\mathcal{T}_1^3, \mathcal{Y}_1)$ of the remaining pairs of quantum numbers. This yields the irreducible representation $R(q_1, p_1)$, and so on.

Example 1. We want to show that

$$R(1,0) \otimes R(0,1) = R(1,1) \oplus R(0,0). \tag{3.60}$$

Physicists briefly write

$$\boxed{\mathbf{3} \otimes \bar{\mathbf{3}} = \mathbf{8} \oplus \mathbf{1}.}$$

This corresponds to (3.58) above (meson octet). In terms of mathematics, this means that

$$X \otimes \bar{X} = Y_8 \oplus Y_1$$

where $X = \text{span}(e_1, e_2, e_3)$ (quark space) and $\bar{X} = \text{span}(\bar{e}^1, \bar{e}^2, \bar{e}^3)$ (antiquark space). The group $SU(3)$ acts on the complex Hilbert spaces X and $\bar{X}$, and hence it acts on the 9-dimensional complex Hilbert space $X \otimes \bar{X}$. Moreover, the group $SU(3)$ acts on the 8-dimensional linear subspace Y_8 via the irreducible representation $R(1,1)$, and $SU(3)$ acts on the one-dimensional linear subspace Y_1 via the trivial representation $R(0,0)$. In fact, we have

$$\mathsf{T}^3(e_j \otimes \bar{e}^k) = (\mathcal{T}^3(e_j) + \mathcal{T}^3(\bar{e}^k))(e_j \otimes \bar{e}^k),$$
$$\mathsf{Y}(e_j \otimes \bar{e}^k) = (\mathcal{Y}(e_j) + \mathcal{Y}(\bar{e}^k))(e_j \otimes \bar{e}^k).$$

Choose $j = 1$ and $k = 2$. By Fig. 3.5 on page 233, we get

Table 3.11. The group $SU(3)$

irreducible representation of the group $SU(3)$	ordered weight tuples (quantum number $(\mathcal{T}^3, \mathcal{Y})$)
$R(0,0) = \mathbf{1}$	$(0,0)$
$R(1,0) = \mathbf{3}$	$(\frac{1}{2}, \frac{1}{3})$, $(0, -\frac{2}{3})$, $(-\frac{1}{2}, \frac{1}{3})$, (quarks; Fig. 3.5)
$R(0,1) = \bar{\mathbf{3}}$	$(\frac{1}{2}, -\frac{1}{3})$, $(0, \frac{2}{3})$, $(-\frac{1}{2}, -\frac{1}{3})$, (antiquarks; Fig. 3.5)
$R(2,0) = \mathbf{6}$	$(1, \frac{2}{3})$, $(\frac{1}{2}, -\frac{1}{3})$, $(0, \frac{2}{3})$, $(0, -\frac{4}{3})$, $(-\frac{1}{2}, -\frac{1}{3})$, $(-1, \frac{2}{3})$
$R(0,2) = \bar{\mathbf{6}}$	$(1, -\frac{2}{3})$, $(\frac{1}{2}, \frac{1}{3})$, $(0, \frac{4}{3})$, $(0, -\frac{2}{3})$, $(-\frac{1}{2}, \frac{1}{3})$, $(-1, -\frac{2}{3})$
$R(1,1) = \mathbf{8}$	$(1,0)$, $(\frac{1}{2}, 1)$, $(\frac{1}{2}, -1)$, $(0,0)$, $(0,0)$, $(-\frac{1}{2}, 1)$, $(-\frac{1}{2}, -1)$, $(-1,0)$ (8-multiplet of the proton; Fig. 3.3)
$R(3,0) = \mathbf{10}$	$(\frac{3}{2}, 1)$, $(1,0)$, $(\frac{1}{2}, 1)$, $(\frac{1}{2}, -1)$, $(0,0)$, $(0,-2)$, $(-\frac{1}{2}, 1)$, $(-\frac{1}{2}, -1)$, $(-1,0)$, $(-\frac{3}{2}, 1)$ (10-multiplet of the Ω^-particle; Fig. 3.4)

$$\mathcal{T}^3(e_1 \otimes \bar{e}^2) = \mathcal{T}^3(e_1) + \mathcal{T}^3(\bar{e}^2) = \tfrac{1}{2} + \tfrac{1}{2} = 1, \ \ \mathcal{Y}(e_1 \otimes \bar{e}^2) = \tfrac{1}{3} - \tfrac{1}{3} = 0.$$

For the nine vectors $e_j \otimes \bar{e}^k$, $j, k = 1, 2, 3$, we get the following 9 tuples $(\mathcal{T}^3, \mathcal{Y})$ ordered by weight:

$$(1,0), (\tfrac{1}{2}, 1), (\tfrac{1}{2}, -1), (0,0), (0,0), (-\tfrac{1}{2}, 1), (-\tfrac{1}{2}, -1), (-1,0), (0,0).$$

The tuple $(1,0)$ has the highest weight, in the sense of (3.53). This yields the irreducible representation $R(1,1)$ because of (3.59). By Table 3.11, $R(1,1)$ corresponds to the 8 tuples

$$(1,0), (\tfrac{1}{2}, 1), (\tfrac{1}{2}, -1), (0,0), (0,0), (-\tfrac{1}{2}, 1), (-\tfrac{1}{2}, -1), (-1,0).$$

It remains the tuple $(0,0)$ which corresponds to $R(0,0)$. This way, we get (3.60).

Example 2. We want to show that

$$R(1,0) \otimes R(1,0) \otimes R(1,0) = R(3,0) \oplus R(1,1) \oplus R(1,1) \oplus R(0,0). \quad (3.61)$$

Physicists briefly write

$$\boxed{\mathbf{3} \otimes \mathbf{3} \otimes \mathbf{3} = \mathbf{10} \oplus \mathbf{8} \oplus \mathbf{8} \oplus \mathbf{1}.}$$

This is (3.46). To get (3.61) by means of the method of highest weight, note that

$$\begin{aligned} \mathsf{T}^3(e_i \otimes e_j \otimes e_k) &= (\mathcal{T}^3(e_i) + \mathcal{T}^3(e_j) + \mathcal{T}^3(e_k))\,(e_i \otimes e_j \otimes e_k), \\ \mathsf{Y}^3(e_i \otimes e_j \otimes e_k) &= (\mathcal{Y}(e_i) + \mathcal{Y}(e_j) + \mathcal{Y}(e_k))\,(e_i \otimes e_j \otimes e_k). \end{aligned}$$

For the 27 states $e_i \otimes e_j \otimes e_k$, $i, j, k = 1, 2, 3$, we get 27 tuples of quantum numbers $(\mathcal{T}^3, \mathcal{Y})$. The vector $e_1 \otimes e_1 \otimes e_1$ yields the tuple $(\frac{3}{2}, 1)$ which has the highest weight. This corresponds to $R(3,0)$ because of (3.59). Taking away the 10 tuples $(\mathcal{T}^3, \mathcal{Y})$ belonging to $R(3,0)$ by Table 3.11, the tuple $(1,0)$ has the highest weight of all the remaining tuples. This corresponds to $R(1,1)$. Taking away the tuples belonging to $R(1,1)$, the tuple $(1,0)$ has the highest weight of the remaining tuples. Again taking away the tuples belonging to $R(1,1)$, the tuple $(0,0)$ remains; this corresponds to $R(0,0)$. This way, we get (3.61) (see Problem 3.25 on page 320).

From the physical point of view, many examples in elementary particle physics can be found in W. Greiner and B. Müller, Quantum Mechanics: Symmetries, Springer, New York, 1996. From the mathematical point of view, many examples about specific Lie algebras and their representations are studied in B. Simon (1996) quoted on page 208, B. Hall, Lie Groups, Lie Algebras, and Representations: An Elementary Introduction, Springer, New York, 2003, W. Fulton and J. Harris, Representation Theory: A First Course, Springer, Berlin, 1991. We also recommend S. Sternberg, Group Theory and Physics, Cambridge University Press, 1995; this textbook combines mathematics with applications to physics.

3.14.4 The Pauli Exclusion Principle and the Color of Quarks

In the Standard model of particle physics, there are six quarks: u (up), d (down), c (charme), s (strange), b (bottom), t (top). Quarks possess the spin $\frac{1}{2}\hbar$, that is, they are fermions. By the Pauli exclusion principle, two quarks of a particle system cannot be in the same state. This can be described by states which are antisymmetric under the permutation of quarks and antiquarks.

Hidden internal degrees of freedom of quarks and antiquarks. Historically, violation of the Pauli exclusion principle was caused by the naive description of the Δ^{++} baryon (see Fig. 3.4 on page 229). To solve this problem, the particle was described by the state $|\Delta^{++}, S_z = \frac{3}{2}\rangle$ given by

$$\frac{1}{\sqrt{6}} \sum_{i,j,k=1}^{3} \varepsilon_{ijk} \big(|u\rangle \otimes |i\rangle \otimes |\tfrac{1}{2}\rangle \big) \otimes \big(|u\rangle \otimes |j\rangle \otimes |\tfrac{1}{2}\rangle \big) \otimes \big(|u\rangle \otimes |k\rangle \otimes |\tfrac{1}{2}\rangle \big). \tag{3.62}$$

Here, the symbol $|\frac{1}{2}\rangle$ (resp. $|-\frac{1}{2}\rangle$) describes a spin state of spin $\frac{1}{2}\hbar$ (resp. $-\frac{1}{2}\hbar$) with respect to the z-axis of a fixed Cartesian coordinate system. The crucial point is that

The three symbols $|1\rangle$, $|2\rangle$, $|3\rangle$ describe three additional internal degrees of freedom of quarks.

These degrees of freedom have fancy names. They are called the colors of the quarks, namely, $|1\rangle$ (red), $|2\rangle$ (green), $|3\rangle$ (blue). In mathematics, additional degrees of freedom are described by tensor products. Using the Dirac calculus, physicists briefly write

$$|\Delta^{++}, S_z = \tfrac{3}{2}\rangle = \frac{1}{\sqrt{6}} \sum_{i,j,k=1}^{3} \varepsilon_{ijk} |u_i^{\uparrow}, u_j^{\uparrow}, u_k^{\uparrow}\rangle.$$

Note that the state is antisymmetric with respect to the color indices i, j, k because of the antisymmetric coefficient ε_{ijk}. In this setting, the 3-dimensional complex Hilbert space X with the basis vectors e_1, e_2, e_3 (quark space) has to be replaced by the 18-dimensional complex linear space $\mathcal{X}$ with the basis vectors

$$e_i \otimes |j\rangle \otimes |S_z\rangle, \qquad i, j = 1, 2, 3, \;\; S_z = \pm\tfrac{1}{2}.$$

Recall that $e_1 = |u\rangle, e_2 = |d\rangle$, and $e_3 = |s\rangle$. For example, the tensor product $e_1 \otimes |2\rangle \otimes |\frac{1}{2}\rangle$ describes the state of a green u-quark which has the spin $\frac{1}{2}\hbar$ in direction of the z-axis. We postulate that

The group $SU(3)$ is the symmetry group of the quark colors.

Whereas electromagnetic interaction is based on photons which see the electric charge of the particles, strong interaction is based on eight gluons which see the colors of quarks. Based on the $SU(3)$ color symmetry, we introduce the color charge operator Y_c defined by

$$\mathsf{Y}_c|j\rangle = \eta_j|j\rangle, \qquad \eta_1 = \eta_2 := \tfrac{1}{3}, \ \eta_3 := -\tfrac{2}{3}.$$

From the mathematical point of view, the operator Y_c has the same properties as the hypercharge operator Y. Physically, this operator concerns the completely different phenomenon of quark colors. For example,

$$\boxed{\mathsf{Y}_c(\ |1\rangle \otimes |2\rangle \otimes |3\rangle\ ) = (\tfrac{1}{3} + \tfrac{1}{3} - \tfrac{2}{3})\,(\ |1\rangle \otimes |2\rangle \otimes |3\rangle\ ) = 0.}$$

A state is called colorless (or white) iff it is an eigenstate of the color charge operator Y_c with eigenvalue zero. For example,

$$\mathsf{Y}_c|\Delta^{++}, S_z = \tfrac{3}{2}\rangle = 0.$$

Therefore, the particle Δ^{++} is white. In quantum chromodynamics, one postulates that

Baryons and mesons are white.

This crucial property restricts seriously the possible states of elementary particles including color. Physicists say that baryons and mesons are color singlets. Observe that the state (3.62) above does not violate the Pauli exclusion principle, since it is antisymmetric with respect to a permutation of the colors. Baryons and mesons are called hadrons.[15]

Physical evidence for colors. In 1971, the color of quarks was introduced by Fritzsch (born 1943) and Gell-Mann (born 1929). Experimentally, the color hypothesis is established by the following facts:

- the lifetime of the neutral pion π_0, and
- the rate of hadron production in electron-positron annihilation.

Let us briefly discuss the lifetime problem. In the early 1960s, Gell-Mann created a new approach to quantum field theory called current algebras. To sketch the basic idea, consider first the classical equation

$$\varrho_t + \operatorname{div} \mathbf{J} = 0$$

which describes the conservation of charge. Here, ϱ is the electric charge density, and $\mathbf{J}$ is the electric current density vector. In the heuristic current algebra approach, the operator-valued quantum field function

$$\psi = \psi(\mathbf{x}, t)$$

is used in order to construct operators $\boldsymbol{\varrho}$ and $\mathbf{J}$ which satisfy appropriate commutation relations governed by the presupposed symmetries. In terms of physics, the

[15] The Greek word *hadrós* means 'strong'. Hadrons are elementary particles which are governed by strong interaction.

operators $\boldsymbol{\varrho}$ and $\mathbf{J}$ describe the currents of elementary particles in particle accelerators. The point is that this method allows us to compute the lifetime of particles. For example, the measured lifetime of the neutral pion π_0 is $0.83 \cdot 10^{-16}$ seconds. Applying the technique of current algebras to the early quark model, the computed lifetime was $7.5 \cdot 10^{-16}$ second; this value is wrong by the factor 9. However, including the additional three color degrees of freedom of quarks, one obtains the correct value of the lifetime. Roughly speaking, the three colors of the quarks decrease the lifetime of the π_0 by the factor 9. More details can be found in H. Fritzsch, Quarks: The Stuff of Matter, Allen Lane, Penguin Books, London, 1983. For current algebras, we refer to S. Treiman, R. Jackiw, and D. Gross (Eds.), Lectures on Current Algebras and its Applications, Princeton University Press, 1972.

Quark dynamics and gauge theory. So far, we have only considered quantum states which do not depend on space and time variables. In 1971/72, quantum chromodynamics was created by Fritzsch (born 1943), Gell-Mann (born 1929), and Leutwyler (born 1938), as a quantum field theory for describing strong interaction in nature. This quantum field depends on space and time. Nowadays, quantum chromodynamics is part of the Standard Model in elementary particle physics. The basic papers on quantum chromodynamics are:

H. Fritzsch and M. Gell-Mann, Quarks and what else? Proceedings of the XVIth International Conference on High Energy Physics, Chicago **2** (1972), 135–165 (based on current algebras).

H. Fritzsch, M. Gell-Mann, and H. Leutwyler, Advantages of the color octet gluon picture, Phys. Lett. **47B** (1973), 365–368 (Lagrangian of a gauge field theory).

One has to distinguish between

- quantum chromodynamics (QCD) as a classical field theory, and
- the quantization of the classical field theory.

The classical approach is an $SU(3)$-gauge theory which is well established from the mathematical point of view. This will be studied in Chap. 15. For the quantized version, a rigorous approach is missing. Physicists use the universal method of path integrals combined with the method of perturbation theory. We recommend:

O. Nachtmann, Elementary Particle Physics: Concepts and Phenomena, Springer, Berlin, 1990.

M. Böhm, A. Denner, and H. Joos, Gauge Theories of the Strong and Electroweak Interaction, Teubner, Stuttgart, 2001.

S. Narison, Quantum Chromodynamics as a Theory of Hadrons: From Partons to Confinement, Cambridge University Press, 2004.

P. Langacker, The Standard Model and Beyond, CRC Press, Boca Raton, Florida, 2010.

I. Ioffe, V. Fadin, and L. Lipatov, Quantum Chromodynamics: Perturbative and Non-Perturbative Aspects, Cambridge University Press, 2010.

Quark confinement. Quarks possess two important properties: the quark confinement and the asymptotic freedom of quarks. In Vol. II, we started quantum electrodynamics by considering free electrons, positrons, and photons. Then we computed physical processes by switching on the weak electromagnetic interaction. The situation changes completely in strong interaction.

There are no free quarks.

This is called the quark confinement. A complete theoretical understanding of this phenomenon is still missing. The following two fundamental papers are devoted to the confinement problem:

N. Seiberg and E. Witten, Electric-magnetic duality, monopole condensation, and confinement in $N = 2$ supersymmetric Yang–Mills theory, Nuclear Phys. **B426** (1994), 19–52.

N. Seiberg and E. Witten, Monopoles, duality and chiral symmetry breaking in $N = 2$ supersymmetric QCD, Nucl. Physics **B431**, 485–550.

From the mathematical point of view, these two papers opened a new door to the Donaldson theory on four-dimensional manifolds by using the Seiberg–Witten equation. Donaldson used the Yang–Mills equations in order to prove that the classic space $\mathbb{R}^4$ can be equipped with exotic manifold structures which are not diffeomorphic to the classical manifold structure of $\mathbb{R}^4$. It turns out that the Donaldson theory can be based on the Seiberg–Witten equation which has nice compactness properties, in contrast to the Yang–Mills equation. We refer to:

J. Moore, Lectures on Seiberg–Witten Invariants, Springer, Berlin, 1996 (nice introduction).

J. Morgan, The Seiberg–Witten Equations and Applications to the Topology of Four-Manifolds, Princeton University Press, 1996.

Asymptotic freedom. This phenomenon means that we have the following strange physical situation:

If the energy goes to infinity, quarks behave like free particles.

This means that the method of perturbation theory in strong interaction works best if the energy is extremely high. For low energies, perturbation theory does not work very well for strong interaction, in contrast to electroweak interaction. Therefore, physicists are interested in inventing non-perturbative methods. The asymptotic freedom in strong interaction was discovered by Gross (born 1941), Politzer (born 1949), and Wilczek (born 1951) in 1973. For this discovery, the three physicists were awarded the 2004 Nobel prize in physics. The basic papers are:

D. Gross and F. Wilczek, Asymptotically free gauge theories, Phys. Rev. **D9** (4) (1973), 980–993.

D. Politzer, Reliable perturbative results for strong interactions? Phys. Rev. **30** (2) (1973), 1346–1349.

3.15 The Complexification of Lie Algebras

In mathematics and physics, the theory is frequently simplified by passing from real Lie algebras to complex Lie algebras. The complexification $\mathcal{L}_{\mathbb{C}}$ of a real Lie algebra $\mathcal{L}$ has the same dimension as the original Lie algebra $\mathcal{L}$, $\dim \mathcal{L}_{\mathbb{C}} = \dim \mathcal{L}$. In contrast to this, for the realification $\mathcal{L}_{\mathbb{R}}$ of the complex Lie algebra $\mathcal{L}_{\mathbb{C}}$, we get $\dim \mathcal{L}_{\mathbb{R}} = 2 \dim \mathcal{L}_{\mathbb{C}}$.

Folklore

Complexification and realification. As a prototype, consider the set $\mathbb{C}$ of complex numbers with respect to the usual multiplication ab. This is a 1-dimensional complex algebra with the basis element 1. However, $\mathbb{C}$ is also a 2-dimensional real algebra with the basis elements 1 and i. In order to emphasize the difference, we denote the two algebras by $\mathbb{C}$ and $\mathbb{C}_{\mathbb{R}}$. Observe that the dimensions are different. We have

- $\dim_{\mathbb{R}} \mathbb{C}_{\mathbb{R}} = 2$ (real dimension),
- $\dim_{\mathbb{C}} \mathbb{C} = 1$ (complex dimension).

For the real linear spaces $\mathbb{C}_{\mathbb{R}}$ and $\mathbb{R}^2$, we have the linear isomorphism

$$\mathbb{C}_{\mathbb{R}} \simeq \mathbb{R}^2.$$

The complex 1-dimensional linear space $\mathbb{C}$ is called the complexification of the real 1-dimensional linear space $\mathbb{R}$. Moreover, the real 2-dimensional linear space $\mathbb{R}^2$ is called the realification of the complex 1-dimensional space $\mathbb{C}$. Similarly, the real linear space $\mathbb{R}^2$ consists of all the tuples

$$(a, b), \qquad a, b \in \mathbb{R}.$$

The complexification of $\mathbb{R}^2$ is the space $\mathbb{C}^2$ which consists of all the tuples

$$(a, b), \qquad a, b \in \mathbb{C}.$$

Let $a = \alpha + \mathrm{i}\beta$ and $b = \gamma + \mathrm{i}\delta$ where $\alpha, \beta, \gamma, \delta \in \mathbb{R}$. If we regard the complex linear space $\mathbb{C}^2$ as a real linear space, then we obtain the realification of $\mathbb{C}^2$ which consists of all the linear combinations

$$\alpha(1,0) + \beta(\mathrm{i},0) + \gamma(0,1) + \delta(0,\mathrm{i}), \qquad \alpha, \beta, \gamma, \delta \in \mathbb{R}.$$

Note that $(1,0)$ and $(\mathrm{i},0)$ are linearly dependent on $\mathbb{C}^2$ because of $(\mathrm{i},0) = \mathrm{i}(1,0)$. But $(1,0)$ and $(\mathrm{i},0)$ are linearly independent if we only allow real linear combinations. In fact, if

$$\alpha(1,0) + \beta(\mathrm{i},0) = 0, \qquad \alpha, \beta \in \mathbb{R},$$

then $(\alpha + \beta\mathrm{i}, 0) = 0$, and hence $\alpha = \beta = 0$. Therefore, the realification of the complex 2-dimensional linear space $\mathbb{C}^2$ is a real 4-dimensional linear space which is isomorphic to $\mathbb{R}^4$. Similarly, let $n = 1, 2, \ldots$

- The complexification of the real n-dimensional linear space $\mathbb{R}^n$ is the complex n-dimensional linear space $\mathbb{C}^n$.
- The realification of the complex n-dimensional space $\mathbb{C}^n$ is a real $2n$-dimensional space which is isomorphic to $\mathbb{R}^{2n}$.

Complexification $\mathcal{L}_{\mathbb{C}}$ of a real Lie algebra $\mathcal{L}$. Let $\mathcal{L}$ be a real Lie algebra. Then the tensor product $\mathbb{C} \otimes \mathcal{L}$ is a complex linear space by setting

$$\alpha(w \otimes A) + \beta(z \otimes B) := (\alpha w) \otimes A + (\beta z) \otimes B, \qquad \alpha, \beta \in \mathbb{C}.$$

In addition, $\mathbb{C}\otimes\mathcal{L}$ becomes a complex Lie algebra $\mathcal{L}_{\mathbb{C}}$ by introducing the Lie product

$$[w \otimes A, z \otimes B] := (wz) \otimes [A, B], \qquad w, z \in \mathbb{C}, \quad A, B \in \mathcal{L}.$$

One can show that these definitions do not depend on the choice of the representatives.

Example. The real Lie algebra $sl(2,\mathbb{C})$ of complex traceless (2×2)-matrices has the basis

$$B_+ := \begin{pmatrix} 0 & 1 \\ 0 & 0 \end{pmatrix}, \quad B_- := \begin{pmatrix} 0 & 0 \\ 1 & 0 \end{pmatrix}, \quad C := \begin{pmatrix} 1 & 0 \\ 0 & -1 \end{pmatrix}$$

with the commutation relations

$$\boxed{[C, B_+]_- = 2B_+, \quad [C, B_-]_- = -2B_-, \quad [B_+, B_-]_- = C.} \tag{3.63}$$

The complexification of $sl(2,\mathbb{C})$ has the basis

$$1 \otimes B_+, \quad 1 \otimes B_-, \quad 1 \otimes C.$$

Set $\chi(z_+ \otimes B_+ + z_- \otimes B_- + z \otimes C) := z_+ B_+ + z_- B_- + zC$. It follows from

$$z_+ B_+ + z_- B_- + zC = 0, \qquad z_+, z_-, z \in \mathbb{C}$$

that $z_+ = z_- = z = 0$. Thus, the complex Lie algebra $\mathcal{L}_{\mathbb{C}}$ is isomorphic to the complex Lie algebra spanned by the matrices B_+, B_-, C. This coincides with the complex Lie algebra $sl_{\mathbb{C}}(2, \mathbb{C})$ of all complex traceless (2×2)-matrices.

Realification $\mathcal{L}_{\mathbb{R}}$ of a complex Lie algebra $\mathcal{L}$. Let $\mathcal{L}$ be a complex Lie algebra. Restricting to real linear combinations, we obtain a real Lie algebra denoted by $\mathcal{L}_{\mathbb{R}}$. If $B_1, \ldots, B_m$ is a basis of $\mathcal{L}$, then $B_1, \ldots, B_m, \mathrm{i}B_1, \ldots, \mathrm{i}B_m$ is a basis of $\mathcal{L}_{\mathbb{R}}$.

Examples. Observe that we have to distinguish between the following notions:

- The 3-dimensional real Lie algebra $sl(2, \mathbb{R})$ consists of all the real traceless (2×2)-matrices. The matrices B_+, B_-, C form a basis of $sl(2, \mathbb{R})$.
- The 6-dimensional real Lie algebra $sl(2, \mathbb{C})$ consists of all the complex traceless (2×2)-matrices. The six matrices $B_+, B_-, C, \ \mathrm{i}B_+, \ \mathrm{i}B-, \ \mathrm{i}C$ form a basis of $sl(2, \mathbb{C})$.
- The 3-dimensional complex Lie algebra $sl_{\mathbb{C}}(2, \mathbb{C})$ consists of all the complex traceless (2×2)-matrices. The three matrices B_+, B_-, C form a basis of $sl_{\mathbb{C}}(2, \mathbb{C})$.
- The 3-dimensional real Lie algebra $su(2)$ consists of all the complex skew-adjoint traceless (2×2)-matrices. The three matrices

$$\mathrm{i}\sigma^1 = \mathrm{i}(B_+ + B_-), \quad \mathrm{i}\sigma^2 = B_+ - B_-, \quad \mathrm{i}\sigma^3 = \mathrm{i}C$$

form a basis of $su(2)$. Here, the three self-adjoint matrices

$$\sigma^1 := \begin{pmatrix} 0 & 1 \\ 1 & 0 \end{pmatrix}, \quad \sigma^2 := \begin{pmatrix} 0 & -\mathrm{i} \\ \mathrm{i} & 0 \end{pmatrix}, \quad \sigma^3 := \begin{pmatrix} 1 & 0 \\ 0 & -1 \end{pmatrix} \tag{3.64}$$

are called the Pauli matrices.
- The Pauli matrices $\sigma^1, \sigma^2, \sigma^3$ form a basis of the 3-dimensional real linear space of complex self-adjoint traceless (2×2)-matrices.
- The six matrices $\sigma^1, \sigma^2, \sigma^3, \ \mathrm{i}\sigma^1, \ \mathrm{i}\sigma^2, \ \mathrm{i}\sigma^3$ form a basis of the real Lie algebra $sl(2, \mathbb{C})$.
- The complexification $su(2)_{\mathbb{C}}$ of the 3-dimensional real Lie algebra $su(2)$ is the 3-dimensional complex Lie algebra $sl_{\mathbb{C}}(2, \mathbb{C})$.
- The complexification $sl(2, \mathbb{C})_{\mathbb{C}}$ of the 3-dimensional real Lie algebra $sl(2, \mathbb{C})$ is the 3-dimensional complex Lie algebra $sl_{\mathbb{C}}(2, \mathbb{C})$.
- The realification $(sl_{\mathbb{C}}(2, \mathbb{C}))_{\mathbb{R}}$ of the 3-dimensional complex Lie algebra $sl_{\mathbb{C}}(2, \mathbb{C})$ is the 6-dimensional real Lie algebra $sl(2, \mathbb{C})$.

3.15.1 Basic Ideas

Élie Cartan's theory of semisimple Lie algebras and their representations is based on the notions 'root' and 'weight'. We want to explain the origin of these notions. To this end, we will consider diagonal matrices.

The group $\mathcal{G}$. Consider all the diagonal matrices

$$G := \begin{pmatrix} \mathrm{e}^{\lambda_1} & 0 & 0 \\ 0 & \mathrm{e}^{\lambda_2} & 0 \\ 0 & 0 & \mathrm{e}^{\lambda_3} \end{pmatrix}$$

where $\lambda_1, \lambda_2, \lambda_3$ are complex numbers with $\lambda_1 + \lambda_2 + \lambda_3 = 0$. Thus, $\det G = 1$, and hence $\mathcal{G}$ is a commutative subgroup of $SL(3, \mathbb{C})$.

The Lie algebra $\mathcal{LG}$. Let $\mathcal{LG}$ denote the set of all the matrices

$$A := \begin{pmatrix} \lambda_1 & 0 & 0 \\ 0 & \lambda_2 & 0 \\ 0 & 0 & \lambda_3 \end{pmatrix}$$

where $\lambda_1, \lambda_2, \lambda_3$ are given as above. Because of $\lambda_1 + \lambda_2 + \lambda_3 = 0$, every matrix A is traceless. Thus, the commutative complex Lie algebra $\mathcal{LG}$ is a Lie subalgebra of $sl_{\mathbb{C}}(3, \mathbb{C})$. Using linearization, every matrix $G \in \mathcal{G}$ can be written as

$$G = I + A + r,$$

where the remainder r goes to zero if $|\lambda_1| + |\lambda_2| + |\lambda_3| \to 0$. Consequently, the complex Lie algebra $\mathcal{LG}$ is the complexification of the Lie algebra to the Lie group $\mathcal{G}$. Choose the matrix E_{ij} defined in (3.65) below. Then

$$\boxed{[A, E_{ij}]_- = (\lambda_i - \lambda_j)E_{ij}, \qquad i, j = 1, 2, 3}$$

for all $A \in \mathcal{LG}$. Define

$$\alpha_{ij}(A) := \lambda_i - \lambda_j, \qquad A \in \mathcal{LG}.$$

This way, we get a linear functional

$$\boxed{\alpha_{ij} : \mathcal{LG} \to \mathbb{C}, \qquad i, j = 1, 2, 3, \;\; i \neq j}$$

which is called a root functional (or briefly a root).

The representation of the group $\mathcal{G}$. Fix $N = 1, 2, \ldots$ Let G denote the group of all the complex $(N \times N)$-matrices

$$\varrho(G) := \begin{pmatrix} e^{\Lambda_1} & 0 & 0 & \ldots & 0 \\ 0 & e^{\Lambda_2} & 0 & \ldots & 0 \\ \vdots & & \ddots & & \vdots \\ 0 & 0 & 0 & \ldots & e^{\Lambda_N} \end{pmatrix}$$

where

$$\Lambda_i := m_{i1}\lambda_1 + m_{i2}\lambda_2 + m_{i3}\lambda_3, \qquad i = 1, 2, 3,$$

and all the coefficients m_{ij} are integers. The map

$$\varrho : \mathcal{G} \to \mathsf{G}$$

is a group morphism. Thus, ϱ is a representation of the group $\mathcal{G}$ on $\mathbb{C}^N$.[16] The numbers $\Lambda_1, \ldots, \Lambda_N$ are called the weights of the matrix $\varrho(G)$.

The Lie algebra $\mathcal{L}\mathsf{G}$ and weights. The linearization of $\varrho(G)$ yields $I + \mathsf{A}$ with

[16] Here, we assume that the matrix $\varrho(G)$ acts on column matrices $c = (c_1, \ldots, c_N)^d$ by means of the matrix multiplication $\varrho(G)c$.

$$\mathsf{A} := \begin{pmatrix} \Lambda_1 & 0 & 0 & \ldots & 0 \\ 0 & \Lambda_2 & 0 & \ldots & 0 \\ \vdots & & \ddots & & \vdots \\ 0 & 0 & 0 & \ldots & \Lambda_N \end{pmatrix}.$$

The eigenvalues of this matrix are precisely the weights of $\varrho(G)$. The matrix A is an element of the complexification of the Lie algebra to the group G.

This approach is trivial. However, Cartan related successfully the investigation of the group $SL(3,\mathbb{C})$ and the Lie algebra $sl_{\mathbb{C}}(3,\mathbb{C})$ to a subgroup and a Lie subalgebra of diagonal matrices, respectively. Let us sketch this.

3.15.2 The Complex Lie Algebra $sl_{\mathbb{C}}(3,\mathbb{C})$ and Root Functionals

> Root functionals allow a complete classification of semisimple complex Lie algebras. The basic idea is to investigate the common eigenvalues of the operator family $\{\mathrm{ad(C)}\}$ where C runs through the Cartan subalgebra $\mathcal{CL}$ of the given Lie algebra $\mathcal{L}$. In terms of physics, the evaluation of root functionals with respect to a basis of the Cartan subalgebra yields tuples of quantum numbers which classify particles.
>
> Folklore

The complex Lie algebra $sl_{\mathbb{C}}(3,\mathbb{C})$. Set

$$\boxed{\mathcal{L} := sl_{\mathbb{C}}(3,\mathbb{C}).}$$

The complex Lie algebra $\mathcal{L}$ consists of all the complex traceless (3×3)-matrices. We want to discuss the main ideas of Cartan's general theory for complex semisimple Lie algebras by considering the special case $\mathcal{L}$. In particular, we want to help the reader to understand the relation between the quark model and the approach to semisimple Lie algebras used in mathematics. The general theory will be considered in Volume IV on quantum mathematics.[17] For the Gell-Mann matrices $\lambda_1,\ldots,\lambda_8$, it follows from

$$a_1\lambda_1 + a_2\lambda_2 + \ldots + a_8\lambda_8 = 0, \qquad a_1, a_2, \ldots, a_8 \in \mathbb{C}$$

that $a_j = 0$ for all $j = 1,\ldots,8$. Thus, the matrices $\lambda_1,\ldots,\lambda_8$ form a basis of the complex Lie algebra $\mathcal{L}$. Since the matrices $\mathrm{i}\lambda_1,\ldots,\mathrm{i}\lambda_8$ form a basis of the real Lie algebra $su(3)$, the complex Lie algebra $\mathcal{L}$ is the complexification of the real Lie algebra $su(3)$. The Killing form of the Lie algebra $\mathcal{L}$ reads as[18]

$$\boxed{K(A,B) := 6\,\mathrm{tr}(AB) \qquad \text{for all} \quad A,B\in\mathcal{L}.}$$

For the Gell-Mann matrices, we get

$$K(\lambda_i,\lambda_j) = 12\delta_{ij}, \qquad i,j = 1,\ldots,8.$$

In order to get a simpler basis of the Lie algebra $\mathcal{L}$, set

[17] We recommend B. Simon (1996) quoted on page 208 and A. Kirillov, jr., An Introduction to Lie Groups and Lie Algebras, Cambridge University Press, 2008.

[18] The Killing form of $sl_{\mathbb{C}}(n,\mathbb{C})$, $n = 2,3,\ldots$, is equal to $K(A,B) = 2n\,\mathrm{tr}(AB)$ for all $A,B\in sl_{\mathbb{C}}(n,\mathbb{C})$.

$$E_{11} := \begin{pmatrix} 1 & 0 & 0 \\ 0 & 0 & 0 \\ 0 & 0 & 0 \end{pmatrix}, \quad E_{12} := \begin{pmatrix} 0 & 1 & 0 \\ 0 & 0 & 0 \\ 0 & 0 & 0 \end{pmatrix}, \quad \ldots \quad , E_{33} := \begin{pmatrix} 0 & 0 & 0 \\ 0 & 0 & 0 \\ 0 & 0 & 1 \end{pmatrix}. \tag{3.65}$$

That is, the entries of the (3×3)-matrix E_{ij} are equal to zero, except for the number 1 at the (ij)-place. The eight matrices

$$E_{12}, E_{13}, E_{21}, E_{23}, E_{31}, E_{32}, \quad E_{11} - E_{22}, \quad E_{22} - E_{33}$$

form a basis of the complex Lie algebra $\mathcal{L} = sl_{\mathbb{C}}(3, \mathbb{C})$.

The standard Cartan subalgebra. By definition, the standard Cartan subalgebra $\mathcal{CL}$ of the Lie algebra $\mathcal{L} = sl_{\mathbb{C}}(3, \mathbb{C})$ consists of all the diagonal matrices

$$C = \begin{pmatrix} d_1 & 0 & 0 \\ 0 & d_2 & 0 \\ 0 & 0 & d_3 \end{pmatrix}, \qquad d_1, d_2, d_3 \in \mathbb{C}$$

with $\mathrm{tr}(C) = d_1 + d_2 + d_3 = 0$. Define

$$\boxed{\delta_j(C) := d_j, \qquad j = 1, 2, 3.} \tag{3.66}$$

Below we will use the linear functional $\delta_j : \mathcal{CL} \to \mathbb{C}$ as a potential for the root functional. The two matrices

$$\mathsf{T}^3, \mathsf{Y} \in \mathcal{CL}$$

form a basis of the standard Cartan subalgebra $\mathcal{CL}$. The matrix T^3 (resp. Y) is called the operator of the third component of the isospin (resp. of hypercharge).[19] Alternatively, the two matrices $E_{11} - E_{22}, E_{22} - E_{33}$ form a basis of the standard Cartan subalgebra $\mathcal{CL}$, too.

The main goal. It is our main goal to simplify the commutation relations of the complex Lie algebra $\mathcal{L} = sl_{\mathbb{C}}(3, \mathbb{C})$ by using

$$\boxed{[C_j, B_j]_- = 2B_j, \quad [C_j, B_{-j}]_- = -2B_{-j}, \quad [B_j, B_{-j}]_- = C_j} \tag{3.67}$$

if $j = \pm 1, \pm 2, \pm 3$. In addition, we want to get the direct sum decomposition

$$\mathcal{L} = \mathcal{CL} \oplus (\mathcal{CL})^{\perp}$$

where

- the six linearly independent elements $B_1, B_2, B_3, B_{-1}, B_{-2}, B_{-3}$ of the Lie algebra $\mathcal{L}$ span the linear subspace $(\mathcal{CL})^{\perp}$ of $\mathcal{L}$, and
- the six elements $C_1, C_2, C_3, C_{-1}, C_{-2}, C_{-3}$ of $\mathcal{L}$ span the 2-dimensional standard Cartan subalgebra $\mathcal{CL}$.
- Consequently, the eight elements $\mathsf{T}^3, \mathsf{Y}, B_1, B_2, B_3.B_{-1}, B_{-2}, B_{-3}$ form a basis of the complex Lie algebra $\mathcal{L}$.

Explicitly, we choose

- $B_1 := E_{12}, B_{-1} = E_{21}, B_2 := E_{23}, B_{-2} := E_{32}, B_3 := E_{13}, B_{-3} := E_{31}$,

[19] The explicit form can be found in (3.52) on page 231. Note that the self-adjoint matrices T^3 and Y are not elements of the Lie algebra $su(3)$. However, the passage to the complexification $sl_{\mathbb{C}}(3, \mathbb{C})$ allows us to include the crucial physical quantities T^3 and Y into the Lie algebra setting.

- $C_1 := E_{11} - E_{22}, C_2 := E_{22} - E_{33}, C_3 := E_{11} - E_{33}$,
- $C_{-1} = -C_1, C_{-2} = -C_2, C_{-3} := -C_3$.

In fact, this yields (3.67).

Fix the index $j = 1, 2, 3$. It follows from the commutation relations (3.67) that, for fixed index $j = 1, 2, 3, -1, -2, -3$, the map

$$(B_+, B_-, C) \mapsto (B_j, B_{-j}, C_j)$$

yields a Lie algebra isomorphism from $sl_{\mathbb{C}}(2, \mathbb{C})$ onto the subalgebra of $\mathcal{L}$ spanned by the matrices B_j, B_{-j}, C_j. In this sense, the complex Lie algebra $sl_{\mathbb{C}}(2, \mathbb{C})$ is the building block of the Lie algebra $\mathcal{L} = sl(3, \mathbb{C})_{\mathbb{C}}$. In order to obtain the nice commutation relations (3.67) in a general setting, Élie Cartan used root functionals. Let us discuss this.

Root functionals. First let us translate the nice commutation relations (3.67) into the language of eigenvalues. To this end, fix $A \in \mathcal{L}$, and set

$$\text{ad}(A)(B) := [A, B]_- \qquad \text{for all} \quad B \in \mathcal{L}.$$

This yields the linear operator $\text{ad}(A) : \mathcal{L} \to \mathcal{L}$. The complex number λ is an eigenvalue of the linear operator $\text{ad}(A)$ iff there exists a non-zero element B of $\mathcal{L}$ such that $\text{ad}(B) = \lambda B$. Equivalently,

$$[A, B]_- = \lambda B.$$

By definition, a root functional α of the complex Lie algebra $\mathcal{L}$ is a linear non-zero functional

$$\alpha : \mathcal{CL} \to \mathbb{C}$$

on the standard Cartan subalgebra $\mathcal{CL}$ of $\mathcal{L}$ such that there exists a non-zero element $B \in \mathcal{L}$ with

$$\text{ad}(C)(B) = \alpha(C)B \qquad \text{for all} \quad C \in \mathcal{CL}.$$

In other words, the values $\alpha(C)$ of the root functional α are the eigenvalues of all the operators $\text{ad}(C)$, $C \in \mathcal{CL}$, with respect to a common eigenvector B. Root functionals are briefly called roots of the Lie algebra $\mathcal{L}$.

Proposition 3.14 *The complex Lie algebra $sl_{\mathbb{C}}(3, \mathbb{C})$ has precisely six roots α_{ij} with $i, j = 1, 2, 3$ and $i \neq j$. Explicitly,*

$$\alpha_{ij}(C) = d_i - d_j \qquad \textit{for all} \quad C \in \mathcal{CL}$$

where d_1, d_2, d_3 are the diagonal elements of the matrix C.

Proof. Fix $i, j = 1, 2, 3$ with $i \neq j$. Then

$$[C, E_{ij}]_- = (d_i - d_j)E_{ij} \qquad \text{for all} \quad C \in \mathcal{CL},$$

and $[C, C']_- = 0$ for all $C, C' \in \mathcal{CL}$. This yields the claim. □

Let us change the notation by setting

$$\alpha_1 := \alpha_{12},\ \alpha_2 := \alpha_{23},\ \alpha_3 := \alpha_{13},\ \alpha_{-1} := -\alpha_1,\ \alpha_{-2} := -\alpha_2,\ \alpha_3 := -\alpha_3.$$

The root potential and cohomology. Using the functional δ_i from (3.66), we get

$$\boxed{\alpha_{ij} := \delta_i - \delta_j, \qquad i, j = 1, 2, 3,\ i \neq j.} \tag{3.68}$$

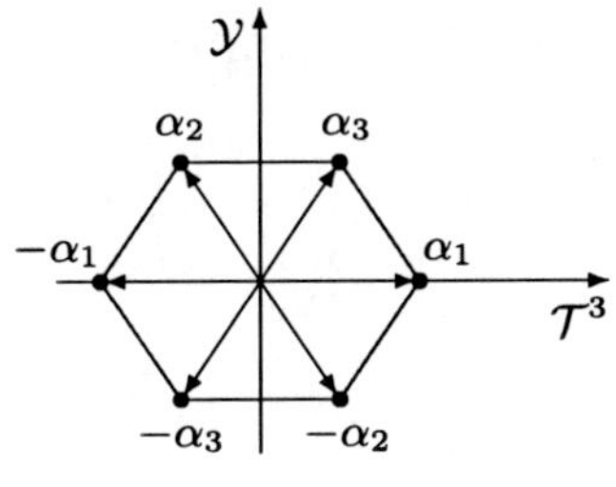

Fig. 3.9. Roots of the Lie algebra $sl_{\mathbb{C}}(3, \mathbb{C})$

The linear functionals $\delta_i : \mathcal{CL} \to \mathbb{C}$ are called root potentials. In Sect. 16.8.4 of Vol. I, we have discussed the Heisenberg relation

$$\hbar\omega_{nm} = E_n - E_m, \qquad n, m = 0, 1, \ldots, n > m$$

between the energy levels E_n of the electron in a molecule and the angular frequencies ω_{nm} of the emitted light. This represents a simple form of cohomology. Similarly, there lurks cohomology behind (3.68). In Vol. IV on quantum mathematics, we will show that cohomology plays a crucial role not only in topology, but also in the theory of Lie algebras. We refer to V. Varadarajan, Lie Groups, Lie Algebras, and Their Representations, Springer, New York, 1984.

The passage from root functionals to physics (quantum numbers). Since the diagonal matrices T^3 and Y from (3.52) on page 231 form a basis of the standard Cartan subalgebra $\mathcal{CL}$, we have

$$\alpha_j(a\mathsf{T}^3 + b\mathsf{Y}) = a\mathcal{T}^3 + b\mathcal{Y} \qquad \text{for all} \quad a, b \in \mathbb{C}$$

with the so-called root tuples

$$(\alpha_j(\mathsf{T}^3), \alpha_j(\mathsf{Y})) = (\mathcal{T}^3, \mathcal{Y}), \qquad j = \pm 1, \pm 2, \pm 3. \tag{3.69}$$

Obviously, there is a one-to-one correspondence between the root functionals α_j and the root tuples. Therefore, the root tuples are also called roots. Using (3.52), we get

$$\alpha_1(\mathsf{T}^3) = \alpha_{12}(\mathsf{T}^3) = \delta_1(\mathsf{T}^3) - \delta_2(\mathsf{T}^3) = \tfrac{1}{2} + \tfrac{1}{2} = 1,$$

and $\alpha_1(\mathsf{Y}) = \alpha_{12}(\mathsf{Y}) = \delta_1(\mathsf{Y}) - \delta_2(\mathsf{Y}) = \frac{1}{3} - \frac{1}{3} = 0$. The same way, we obtain the root tuples of the complex Lie algebra $sl_{\mathbb{C}}(3, \mathbb{C})$:

$$(1, 0), (-\tfrac{1}{2}, 1), (\tfrac{1}{2}, 1), (-1, 0), (\tfrac{1}{2}, -1), (-\tfrac{1}{2}, -1)$$

corresponding to the roots $\alpha_1, \alpha_2, \alpha_3, -\alpha_1, -\alpha_2, -\alpha_3$, respectively.

This is depicted in Fig. 3.9. Observe that this diagram coincides with the baryon octet of the proton (Fig. 3.3 on page 228). The roots α_1, α_2 form a basis of the six roots, with respect to integer coefficients. In fact,

$$\alpha_3 = \alpha_1 + \alpha_2, \qquad \alpha_{-1} = -\alpha_1, \ \alpha_{-2} = -\alpha_2, \ \alpha_{-3} = -\alpha_3.$$

Moreover, the roots $\alpha_1, \alpha_2, \alpha_3$ (resp. $-\alpha_1, -\alpha_2, -\alpha_3$) are called positive (resp. negative). Finally, the roots α_1 and α_2 are called simple; they are positive, but they cannot be represented as the sum of two positive roots, in contrast to α_3.

The Cartan matrix and the Dynkin diagram. Using the root basis α_1, α_2, the matrix

$$\mathrm{Cartan}(sl_{\mathbb{C}}(3,\mathbb{C})) := (\alpha_i(C_j))_{i,j=1,2} = \begin{pmatrix} 2 & -1 \\ -1 & 2 \end{pmatrix}$$

is called the Cartan matrix of the Lie algebra $sl_{\mathbb{C}}(3,\mathbb{C})$. Note that $\alpha_1(C_2) = \alpha_2(C_1)$. By convention, this is depicted by the Dynkin diagram from Fig. 3.10.

Geometric root system and its Weyl symmetry group. Fix $n = 2, 3, \ldots$ Let $\mathcal{R}$ be a finite subset of $\mathbb{R}^n$ which has the following properties:

(R1) The linear hull of $\mathcal{R}$ is equal to $\mathbb{R}^n$, and the origin 0 is not an element of the set $\mathcal{R}$.

(R2) Symmetry: For each $\alpha \in \mathcal{R}$, there is a linear bijective map

$$s_\alpha : \mathbb{R}^n \to \mathbb{R}^n$$

which leaves the set $\mathcal{R}$ invariant. Moreover, s_α sends α to -α, and the fixed points of the map s_α form an $(n-1)$-dimensional linear subspace of $\mathbb{R}^n$.

(R3) $s_\alpha(\beta) - \beta$ is an integer multiple of α for all $\alpha, \beta \in \mathcal{R}$.

Then the set $\mathcal{R}$ is called a root system of $\mathbb{R}^n$. The group of all invertible real $(n \times n)$-matrices generated by all of the symmetry maps $s_\alpha, \alpha \in \mathcal{R}$, is called the Weyl group $W(\mathcal{R})$ of the root system $\mathcal{R}$.

Example. The six roots depicted in Fig. 3.9 form a root system. The map s_{α_1} is the reflection at the $\mathcal{Y}$-axis. Analogously, every map s_{α_j} is a reflection at a straight line passing through the origin and being orthogonal to the root α_j where $j = \pm 1, \pm 2, \pm 3$. Thus, the Weyl group has six elements, and it is isomorphic to the symmetry group of an equilateral triangle (Fig. 3.2 on page 210). In turn, this is isomorphic to the symmetric group $Sym(3)$.

The general theory of the classification of complex semisimple Lie algebras can be based on analyzing the geometry of root systems. This is equivalent to the classification of Cartan matrices. In geometric terms, this corresponds to the classification of Dynkin diagrams. We will discuss this in Vol. IV. We refer to J. Serre, Complex Semisimple Lie Algebras, Springer, Berlin, 2001.

3.15.3 Representations of the Complex Lie Algebra $sl_{\mathbb{C}}(3,\mathbb{C})$ and Weight Functionals

Representation theory is governed by the highest weight.
Folklore

Again set $\mathcal{L} := sl_{\mathbb{C}}(3,\mathbb{C})$. Let $\varrho : \mathcal{L} \to gl(X)$ be a representation of the complex Lie algebra $sl_{\mathbb{C}}(3,\mathbb{C})$ on the finite-dimensional complex linear space X. By definition, a weight functional w of $\mathcal{L}$ is a linear functional

$$w : \mathcal{CL} \to \mathbb{C}$$

on the standard Cartan subalgebra $\mathcal{CL}$ of $\mathcal{L}$ such that there exists a non-zero vector $x \in X$ with

$$\varrho(C)x = w(C)x \qquad \text{for all} \quad C \in \mathcal{CL}.$$

In other words, the values $w(C)$ of the weight functional w are the eigenvalues of all the operators $\varrho(C)$ with respect to a common eigenvector x. Weight functionals are briefly called weights of the Lie algebra $\mathcal{L}$.

The importance of the highest weight. The weight w of a representation of the complex Lie algebra $\mathcal{L}$ is called a highest weight iff $w + \alpha$ is not a weight for all positive roots $\alpha = \alpha_1, \alpha_2$. The main result reads as follows.

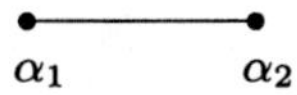

Fig. 3.10. Dynkin diagram of the complex Lie algebra $sl_{\mathbb{C}}(3,\mathbb{C})$

Theorem 3.15 *There is a one-to-one correspondence between the irreducible representations of the complex Lie algebra $sl_{\mathbb{C}}(3,\mathbb{C})$ on finite-dimensional complex linear spaces and the tuples*

$$(q,p),$$

where q and p are non-negative integers.

More precisely, for a given tuple (q,p), there exists an irreducible representation ϱ of $\mathcal{L}$ with the highest weight w such that

$$w(\mathsf{T}^3) = \tfrac{1}{2}(q+p), \quad w(\mathsf{Y}) = \tfrac{1}{3}(q-p).$$

This way, we get all the possible irreducible representations of $\mathcal{L}$, up to equivalence. The tuple (q,p) determines uniquely the irreducible representation up to equivalence. In terms of geometry, the irreducible representations are labelled by the subset

$$\{(q,p): \ q,p \in \mathbb{Z}, \quad q,p \geq 0\}$$

of the lattice $\mathbb{Z} \times \mathbb{Z}$. In terms of physics, the irreducible representation of $sl_{\mathbb{C}}(3,\mathbb{C})$ corresponds to the $(\mathcal{T}^3,\mathcal{Y})$-diagram of the irreducible representation $R(q,p)$ of the group $SU(3)$ with the highest weight $(\mathcal{T}^3,\mathcal{Y}) = (\frac{1}{2}(q+p), \frac{1}{3}(q-p))$. For $R(1,0)$ (quarks), $R(0,1)$ (antiquarks), $R(1,1)$ (baryon octet of the proton), see Figs. 3.3 and 3.5 on page 228. The sophisticated proof of the theorem can be found in B. Hall, Lie Groups, Lie Algebras, and Representations: An Elementary Introduction, Sect. 5.3, Springer, New York, 2003. There exists a perfect theory of weights for the representation of complex semisimple Lie algebras.

Weyl's unitarian trick. Observe that the diagrams for the weight tuples $(\mathcal{T}^3,\mathcal{Y})$ of the irreducible representations $R(q,p)$ of the Lie group $SU(3)$ coincide with the diagrams for the weight tuples of the irreducible representations of the Lie algebra $sl_{\mathbb{C}}(3,\mathbb{C})$. The reason for this is the fact that the Lie algebra $su(3)$ of $SU(3)$ has the complexification $sl_{\mathbb{C}}(3,\mathbb{C})$. This way, the study of the Lie algebra $sl_{\mathbb{C}}(3,\mathbb{C})$ can be reduced to the study of the compact Lie group $SU(3)$ whose irreducible representations are equivalent to unitary representations. This simplifies substantially the approach. Weyl used this argument quite often; he coined this the 'unitarian trick.'

In particular, it follows from the unitarian trick that the representations of the complex Lie algebra $sl_{\mathbb{C}}(3,\mathbb{C})$ on finite-dimensional complex linear spaces are completely reducible.

3.16 Classification of Groups

3.16.1 Simplicity

The morphisms of a simple group are trivial. Simple groups can be regarded as the atoms in group theory.

Folklore

A group $\mathcal{G}$ is called trivial iff it consists only of the unit element (i.e., $\mathcal{G} = \{\mathbf{1}\}$). Consider the group morphism

$$\boxed{\mu : \mathcal{G} \to \mathcal{H}.} \tag{3.70}$$

This is a map which respects the group structure (i.e., $\mu(GH) = \mu(G)\mu(H)$ for all $G, H \in \mathcal{G}$). The image $\mu(\mathcal{G})$ of a group morphism is a subgroup of $\mathcal{H}$. The morphism μ is called trivial iff the image $\mu(\mathcal{G})$ is trivial or it is isomorphic to the group $\mathcal{G}$.

A group is called simple iff it is not trivial and there are only trivial morphisms defined on the group.

A normal subgroup $\mathcal{N}$ of the group $\mathcal{G}$ is called trivial iff $\mathcal{N} = \{\mathbf{1}\}$ or $\mathcal{N} = \mathcal{G}$.

A nontrivial group is simple iff it has only trivial normal subgroups.

More precisely, the kernel $\mu^{-1}(\mathbf{1})$ of the group morphism (3.70) is a normal subgroup of $\mathcal{G}$ and we have the group isomorphism

$$\mathcal{G}/\mu^{-1}(\mathbf{1}) \simeq \mu(\mathcal{G}).$$

For the proofs, see Sect. 4.1.3 of Vol. II.[20]

Example. Choose the integer $m \geq 2$. Let

$$\mathcal{Z}_m := \{\mathrm{e}^{2\pi k \mathrm{i}/m} \; : \; k = 0, 1, \ldots, m-1\}.$$

This is a cyclic group of order m. Note that $G^m = 1$ if $G := \mathrm{e}^{2\pi\mathrm{i}/m}$. The group $\mathcal{Z}_m$ is simple iff m is a prime number. Similarly, we define

$$\mathbb{Z}_m = \{0, 1, \ldots, m-1\}.$$

This is an additive group with respect to the usual addition of integers by adding the relation $m = 0$. For example, if $m = 5$, then $3+4 = 2$, since $3+4 = 7 = 5+2 = 2$. The group $\mathbb{Z}_m$ is also denoted by $\mathbb{Z}/m$ or $\mathbb{Z}/\mathrm{mod}\ m$.

A finite group is simple if its order is a prime number. A commutative group is simple iff it is cyclic of prime order (i.e., it is isomorphic to $\mathcal{Z}_m$ where m is a prime number). The group $Sym(2)$ is simple. The groups $Sym(n)$ are not simple if $n = 3, 4, \ldots$ The subgroup A_n of all the even permutations of n elements is a normal subgroup of $Sym(n)$. If $n \geq 5$, then A_n is simple.

The additive group $\mathbb{Z}$ of integers is not simple. For example, the even integers form a proper normal subgroup.

The classification of finite simple groups was only completed in about 1980. The largest finite simple group is the Monster group which is closely related to conformal quantum field theory (see Sect. 17.5 of Vol. I). Mathematicians needed more than 100 years in order to get the final classification of finite simple groups. Note that the axioms for a group are extremely simple.

---

[20] Recall that a normal subgroup $\mathcal{N}$ of $\mathcal{G}$ is a subgroup with $GNG^{-1} \in \mathcal{N}$ for all $N \in \mathcal{N}$ and $G \in \mathcal{G}$. Moreover, recall the following classification of group morphisms: surjective (resp. injective) group morphisms are called epimorphisms (resp. monomorphisms). Bijective group morphisms are called isomorphisms. Consider now the special case where the image group $\mathcal{H}$ coincides with the original group $\mathcal{G}$. In this special case, group morphisms (resp. group isomorphisms) are called endomorphisms (resp. automorphisms). The same classification will be used for linear spaces, Hilbert spaces, algebras, Lie algebras, rings, and modules. The general setting in terms of category theory will be studied in Vol. IV on quantum mathematics.

It turns out that the innocently looking group axioms describe a huge variety of mathematical models.

This phenomenon is typical for mathematics. A complete mathematical classification of all groups (i.e., the classification of all possible symmetries) is hopeless at present.

3.16.2 Direct Product and Semisimplicity

If $\mathcal{G}$ and $\mathcal{H}$ are groups, then the product set $\mathcal{G} \times \mathcal{H} := \{(G, H) : G \in \mathcal{G}, H \in \mathcal{H}\}$ becomes a group by introducing the product

$$(G, H)(G', H') := (GG', HH'), \qquad G, G' \in \mathcal{G}, \; H, H' \in \mathcal{H}.$$

This is called the direct product group $\mathcal{G} \times \mathcal{H}$. Analogously, the direct product group of a finite number of groups is defined.

A group is called semisimple (or completely reducible) iff it is isomorphic to the direct product of a finite number of simple groups.

For example, the symmetric $Sym(3)$ is not simple, but semisimple, since we have the group isomorphism

$$Sym(3) \simeq \mathcal{Z}_2 \times \mathcal{Z}_3.$$

The direct product $\mathcal{G}_1 \times \mathcal{G}_2$ is called trivial iff one of the factors is trivial. A group $\mathcal{G}$ is called indecomposable iff it admits no non-trivial decomposition $\mathcal{G} = \mathcal{G}_1 \times \mathcal{G}_2$.

- A finite commutative group is indecomposable iff it is cyclic of prime power order (i.e., the group is isomorphic to $\mathcal{Z}_m$ where $m = p^k$, p is a prime number, and k is a positive integer).
- Every non-trivial finite commutative group $\mathcal{G}$ is a direct product of indecomposable finite commutative groups $\mathcal{G} = \mathcal{G}_1 \times \cdots \times \mathcal{G}_m$.

The proof can be found in C. Curtis and I. Reiner, Representation Theory of Finite Groups and Associative Algebras, p. 12, Interscience, New York, 1962.

3.16.3 Solvablity

Solvable groups are closely related to commutative groups.
Folklore

By definition, the commutant $\mathcal{G}'$ of a group $\mathcal{G}$ is the smallest subgroup of $\mathcal{G}$ which contains all the so-called commutators

$$GHG^{-1}H^{-1}, \qquad G, H \in \mathcal{G}.$$

The symbol $\mathcal{G}''$ denotes the commutant of $\mathcal{G}'$, and so on. By definition, a group is called solvable iff some iterated commutant is trivial. This means that there is some positive integer n such that

$$\boxed{\mathcal{G}^{(n)} = \{\mathbf{1}\}.}$$

If the group $\mathcal{G}$ is commutative, then $\mathcal{G}' = \{\mathbf{1}\}$. Thus, solvability generalizes the commutativity of groups. A group $\mathcal{G}$ is solvable iff there exists a finite sequence

$$\mathcal{G}_0 = \{\mathbf{1}\} \subseteq \mathcal{G}_1 \subseteq \mathcal{G}_2 \subseteq \ldots \subseteq \mathcal{G}_m = \mathcal{G}$$

of subgroups $\mathcal{G}_k$ of $\mathcal{G}$ such that $\mathcal{G}_k$ is a normal subgroup of $\mathcal{G}_{k+1}$ and the quotient group $\mathcal{G}_{k+1}/\mathcal{G}_k$ is commutative if $k = 0, 1, \ldots, m-1$.

Examples. Every finite group of prime power order is solvable.

- The symmetric group $Sym(n)$ is solvable if $n = 2, 3, 4$;
- $Sym(n)$ is not solvable if $n = 5, 6, \ldots$; this is a consequence of the simplicity of the group A_n if $n \geq 5$.

This fact is important for classic Galois theory. Galois (1811–1832) showed that the non-solvability of the symmetric group $Sym(n)$ for $n \geq 5$ is responsible for the non-solvability of algebraic equations of order $n \geq 5$ by radicals.

A deep result in group theory is the 1963 Feit–Thompson theorem which says that

Every finite group of odd order is solvable.

This theorem was conjectured by Burnside in 1911.[21]

3.16.4 Semidirect Product

In mathematics and physics, one encounters much more semidirect products of groups than direct products.

Folklore

The group $E_+(3)$ of proper Euclidean motions as a paradigm. Consider the transformation

$$\boxed{y = a + Gx \qquad \text{for all} \quad x \in \mathbb{R}^3} \tag{3.71}$$

with fixed matrices $G \in SO(3)$ and $a \in \mathbb{R}^3$. Here, $SO(3)$ denotes the Lie group of all the special orthogonal matrices $G \in SO(3)$, that is, G is a real (3×3)-matrix with $GG^d = I$ and $\det G = 1$. In terms of geometry, let us use a right-handed Cartesian (x^1, x^2, x^3)-coordinate system of the Euclidean manifold $\mathbb{E}^3$ and set

$$x := \begin{pmatrix} x^1 \\ x^2 \\ x^3 \end{pmatrix}, \quad y := \begin{pmatrix} y^1 \\ y^2 \\ y^3 \end{pmatrix}, \quad a := \begin{pmatrix} a^1 \\ a^2 \\ a^3 \end{pmatrix}.$$

Then the transformation (3.71) describes the combination of a rotation $x \mapsto Gx$ with a translation $x \mapsto a + x$. All the transformations (3.71) form a group called the group of proper Euclidean motions. In fact, if $y = a + Gx$ and $x = b + Hu$, then

$$y = (a + Gb) + GHu.$$

If we use the symbol (a, G) for the transformation (3.71), then the composition of transformations corresponds to the product rule

$$\boxed{(a, G)(b, H) = (a + Gb, GH).}$$

This group is denoted by the symbol

$$E_+(3) = \mathbb{R}^3 \rtimes SO(3),$$

and it is called the semidirect product of the translation group $\mathbb{R}^3$ with the rotation group $SO(3)$.

[21] Walter Feit (born 1930). For his contributions to group theory, John Thompson (born 1932) was awarded the Fields medal in 1970, the Wolf prize in 1992, and the Abel prize in 2008.

- The unit element of $\mathbb{R}^3 \rtimes SO(3)$ is given by $(0, I)$. In fact, $(a, G)(0, I)$ is equal to $(a + G0, GI) = (a, G)$.
- For the inverse element, we get $(a, G)^{-1} = (-G^{-1}a, G^{-1})$. In fact,

$$(a, G)(-G^{-1}a, G^{-1}) = (a - GG^{-1}a, GG^{-1}) = (0, I).$$

As a preparation for the generalization considered below, set

$$\mathcal{N} := \{(a, I) : a \in \mathbb{R}^3\}, \qquad \mathcal{G} := \{(0, G) : G \in SO(3)\}.$$

Then, $\mathcal{N}$ is a normal subgroup of $\mathbb{R}^3 \rtimes SO(3)$ which is isomorphic to the first factor $\mathbb{R}^3$, and $\mathcal{G}$ acts on $\mathbb{R}^3$. Mnemonically, the asymmetry of the product symbol $\mathbb{R}^3 \rtimes SO(3)$ distinguishes between the different properties of the two factors.

Roughly speaking, the second factor acts on the first factor which is isomorphic to a normal subgroup.

In order to translate the semidirect product $\mathbb{R}^3 \rtimes SO(3)$ into a matrix group, consider all the real (4×4)-matrices

$$\{a, G\} := \begin{pmatrix} G & a \\ 0 & 1 \end{pmatrix}, \quad a \in \mathbb{R}^3,\ G \in SO(3). \tag{3.72}$$

For the matrix product, we get

$$\begin{pmatrix} G & a \\ 0 & 1 \end{pmatrix} \begin{pmatrix} H & b \\ 0 & 1 \end{pmatrix} = \begin{pmatrix} GH & a + Gb \\ 0 & 1 \end{pmatrix}.$$

Thus, $\{a, G\}\{b, H\} = \{a + Gb, GH\}$. Consequently, all the matrices (3.72) form a subgroup of the group $GL(4, \mathbb{R})$ of real invertible (4×4)-matrices which is isomorphic to $\mathbb{R}^3 \rtimes SO(3)$. The isomorphism is given by the map $\{a, G\} \mapsto (a, G)$. If we set $G = I$ (resp. $a = 0$), then the matrices (3.72) are restricted to a group which is isomorphic to the translation group $\mathbb{R}^3$ (resp. the rotation group $SO(3)$). Note that the direct product $\mathbb{R}^3 \times SO(3)$ consists of all the symbols (a, G) with $a \in \mathbb{R}^3$ and $G \in SO(3)$. The multiplication law reads as

$$(a, G)(b, H) = (a + b, GH).$$

Thus, the direct product $\mathbb{R}^3 \times SO(3)$ differs from the semidirect product $\mathbb{R}^3 \rtimes SO(3)$.

The general definition. Let $\mathcal{N}$ and $\mathcal{G}$ be (multiplicative) groups. Recall that the direct product $\mathcal{N} \times \mathcal{G}$ consists of all the symbols (a, G) with $a \in \mathcal{N}$ and $G \in \mathcal{G}$ equipped with the multiplication

$$(a, G)(b, H) = (ab, GH).$$

Now let us consider a slight modification. By definition, the semidirect product $\mathcal{N} \rtimes_\varrho \mathcal{G}$ consists of all the symbols (a, G) with $a \in \mathcal{N}$ and $a \in \mathcal{T}$ together with the multiplication rule

$$\boxed{(a, G)(b, H) = (a\varrho(G)b, GH).} \tag{3.73}$$

In order to explain the meaning of ϱ, let

$$\mathsf{A} : \mathcal{N} \to \mathcal{N}$$

be a bijective group morphism (i.e., A is a group automorphism). With respect to the composition of maps, all the automorphisms of $\mathcal{N}$ form a group $\mathrm{Aut}(\mathcal{N})$ called the automorphisms group $\mathrm{Aut}(\mathcal{N})$ of $\mathcal{N}$. Now to the point. We choose a group morphism

$$\boxed{\varrho : \mathcal{G} \to \mathrm{Aut}(\mathcal{N}).}$$

This means that the group $\mathcal{G}$ acts on the group $\mathcal{N}$. Note that if $a, b \in \mathcal{N}$ and $G \in \mathcal{G}$, then $\varrho(G)b \in \mathcal{N}$, and hence

$$a\varrho(G)b \in \mathcal{N}.$$

By a straightforward argument, one can show that the group axioms are satisfied for the multiplication law (3.73) (see Sect. 7.5 of Vol. I). In particular, we have

- $(\mathbf{1}, \mathbf{1})$ (unit element), and
- $(a, G)^{-1} = (\varrho(G)^{-1}a^{-1}, G^{-1})$ (inverse element).

This way, we get the group $\mathcal{N} \rtimes_\varrho \mathcal{G}$ as claimed above.

Obviously, the direct product $\mathcal{N} \times \mathcal{G}$ is a special case of the semidirect product $\mathcal{N} \rtimes_\varrho \mathcal{G}$ if we choose the trivial representation $\varrho(G) = I$ for all $G \in \mathcal{G}$.

Products of subgroups. Many applications concern the direct or semidirect product of subgroups of a given group. Let us discuss this. Suppose that $\mathcal{N}$ and $\mathcal{G}$ are subgroups of the group G with trivial intersection (i.e., $\mathcal{N} \cap \mathcal{G} = \{\mathbf{1}\}$). Consider the set

$$\mathcal{N} \cdot \mathcal{G} := \{aG : \ a \in \mathcal{N}, \ G \in \mathcal{G}\}.$$

Proposition 3.16 (i) Direct product: *If $\mathcal{N} \cdot \mathcal{G} = \mathcal{G} \cdot \mathcal{N}$, then $\mathcal{N} \cdot \mathcal{G}$ is a subgroup of G which is isomorphic to the direct product of $\mathcal{N}$ with $\mathcal{G}$, that is,*

$$\mathcal{N} \cdot \mathcal{G} \simeq \mathcal{N} \times \mathcal{G}. \tag{3.74}$$

For all $P \in \mathcal{N} \cdot \mathcal{G}$, the decomposition

$$P = aG, \qquad a \in \mathcal{N}, \ G \in \mathcal{G} \tag{3.75}$$

is unique, and the isomorphism (3.74) *is given by the map $P \mapsto (a, G)$. Moreover, $\mathcal{N}$ and $\mathcal{G}$ are normal subgroups of $\mathcal{N} \cdot \mathcal{G}$. For the quotient groups, we have the following group isomorphisms:*

$$(\mathcal{N} \cdot \mathcal{G})/\mathcal{N} \simeq \mathcal{G}, \qquad (\mathcal{N} \cdot \mathcal{G})/\mathcal{G} \simeq \mathcal{N}.$$

These isomorphisms are given by $aG \mapsto (\mathbf{1}, G)$ and $aG \mapsto (a, \mathbf{1})$, respectively.[22]

(ii) Semidirect product: *If $\mathcal{N}$ is a normal subgroup of G, then $\mathcal{N} \cdot \mathcal{G}$ is a subgroup of G which is isomorphic to the semidirect product of $\mathcal{N}$ with $\mathcal{G}$, that is,*

$$\mathcal{N} \cdot \mathcal{G} \simeq \mathcal{N} \rtimes_\varrho \mathcal{G}. \tag{3.76}$$

For all $P \in \mathcal{N} \cdot \mathcal{G}$, the decomposition (3.75) *is unique, and the isomorphism* (3.76) *is given by the map $P \mapsto (a, G)$. In addition, the map ϱ is defined by $\varrho(G)b := GbG^{-1}$ for all $b \in \mathcal{N}$, $G \in \mathcal{G}$. We have the group isomorphism*

$$(\mathcal{N} \cdot \mathcal{G})/\mathcal{N} \simeq \mathcal{G}.$$

This isomorphism is given by the map $aG \mapsto G$.

[22] The notion of quotient group, quotient ring, quotient algebra can be found in Sect. 4.1.3 of Vol. II.

Proof. Ad (i). If $aG = bH$ with $a, b \in \mathcal{N}$, and $G, H \in \mathcal{G}$, then $b^{-1}a = HG^{-1}$. Since $b^{-1}a \in \mathcal{N}$ and $HG^{-1} \in \mathcal{G}$, we get $b^{-1}a = \mathbf{1}$ and $HG^{-1} = \mathbf{1}$. Hence $a = b$ and $G = H$. Thus, the decomposition (3.75) is unique. Set $\chi(aG) := (a, G)$. Since $bG = Gb$, we get

$$\chi(aG \cdot bH) = \chi(ab \cdot GH) = (ab, GH).$$

Therefore, the map $\chi : \mathcal{N} \cdot \mathcal{G} \to \mathcal{N} \times \mathcal{G}$ is an isomorphism.

Ad (ii). The trick is to use the decomposition

$$aG \cdot bH = a(GbG^{-1}) \cdot GH = a\varrho(G)b \cdot GH.$$

Hence $\chi(aG \cdot bH) = (a\varrho(G)b, GH) = (a, G)(b, H)$. Thus, $\chi : \mathcal{N} \cdot \mathcal{G} \to \mathcal{N} \rtimes_\varrho \mathcal{G}$ is an isomorphism. □

Examples. (i) $O(3) = O(1) \times SO(3)$. In fact, every real (3×3)-matrix G with $GG^d = I$ (i.e., $G \in O(3)$) can be written as $G = (\pm I)H$ with $\det H = 1$. Hence $H \in SO(3)$. Moreover, $O(1) = \{1, -1\}$. The claim follows from Prop. 3.16(i).

(ii) $Sym(n) = A_n \rtimes Sym(2)$, $n = 3, 4, \ldots$ The subgroup A_n of $Sym(n)$ consists of all the even permutations. This is a normal subgroup. Every permutation π in the symmetric group $Sym(n)$ can be written as $\pi_{\mathrm{even}}\pi_0$ where $\pi_0 := (12)$ if π is odd, and $\pi_0 := (1)$ if π is even. The claim follows from Prop. 3.16(ii).

(iii) The Euclidean group of motions $E(3) = \mathbb{R}^3 \rtimes O(3)$ (or the Euclidean isometry group) consist of all the transformations

$$y = a + Gx, \qquad x \in \mathbb{R}^3, \tag{3.77}$$

where $a \in \mathbb{R}^3$ and $G \in O(3)$. By (i), the group $O(3)$ is generated by rotations and the reflection $x \mapsto -x$ at the origin.

(iv) The Poincaré group $P(1,3) = \mathbb{R}^4 \rtimes O(1,3)$ consists of all the transformations $y = a + Gx, x \in \mathbb{R}^4$, where $a \in \mathbb{R}^4$ and $G \in O(1,3)$ (Lorentz group). The Poincaré group is the symmetry group of both Einstein's theory of special relativity and quantum field theory (see Sect. 18.3.2).

(v) The affine group $A(3) = \mathbb{R}^3 \rtimes GL(3, \mathbb{R})$. This group consists of all the transformations $y = a + Gx, x \in \mathbb{R}^3$, where $a \in \mathbb{R}^3$ and $G \in GL(3, \mathbb{R})$.

The language of exact sequences. Semidirect products can be defined for many structures in mathematics. To this end, one uses exact sequences which split. The very effective language of exact sequences will be studied in Vol. IV on quantum mathematics. Coming from algebraic topology and algebraic geometry, this language is fundamental for modern mathematics.

3.17 Classification of Lie Algebras

3.17.1 The Classification of Complex Simple Lie Algebras

The Lie algebra morphism

$$\mu : \mathcal{L} \to \mathcal{M}$$

is called trivial iff the image $\mu(\mathcal{L})$ is equal to zero or isomorphic to $\mathcal{L}$.

A real or complex Lie algebra $\mathcal{L}$ is called simple iff it is not commutative and there are only trivial morphisms defined on the Lie algebra.

Equivalently, $\mathcal{L}$ is simple iff it is not commutative and it has only the trivial ideals $\{0\}$ and $\mathcal{L}$. In his famous 1894 thesis, by completing earlier seminal work of Killing (1847–1923), Élie Cartan (1869–1951) used the simple Lie algebra $sl_{\mathbb{C}}(2,\mathbb{C})$ as a building block in order to prove that there are precisely the following *simple* finite-dimensional complex Lie algebras:[23]

- $sl_{\mathbb{C}}(n,\mathbb{C})$, $n \geq 2$;
- $sp_{\mathbb{C}}(2n,\mathbb{C})$, $n \geq 2$;
- $so_{\mathbb{C}}(n,\mathbb{C})$, $n \geq 7$;
- E_6, E_7, E_8, F_4, G_2 (exceptional Lie algebras).

These Lie algebras are pairwise different, that is, they are pairwise not isomorphic. In the list above, the following Lie algebras are missing: the 'symplectic' Lie algebra $sp_{\mathbb{C}}(2,\mathbb{C})$, and the 'orthogonal' Lie algebras $so_{\mathbb{C}}(n,\mathbb{C})$, $n = 2,3,4,5,6$. In this connection, note that the Lie algebra $so_{\mathbb{C}}(2,\mathbb{C})$ is commutative, and hence not simple. Furthermore, we have the Lie algebra isomorphism

$$so_{\mathbb{C}}(4,\mathbb{C}) \simeq sl_{\mathbb{C}}(2,\mathbb{C}) \times sl_{\mathbb{C}}(2,\mathbb{C}).$$

Therefore, the Lie algebra $so_{\mathbb{C}}(4,\mathbb{C})$ is not simple, but semisimple. In addition, we have the following Lie algebra isomorphisms:

- $so_{\mathbb{C}}(3,\mathbb{C}) \simeq sl_{\mathbb{C}}(2,\mathbb{C}) \simeq sp_{\mathbb{C}}(2,\mathbb{C})$,
- $so_{\mathbb{C}}(5,\mathbb{C}) \simeq sp_{\mathbb{C}}(4,\mathbb{C})$,
- $so_{\mathbb{C}}(6,\mathbb{C}) \simeq sl_{\mathbb{C}}(4,\mathbb{C})$.

Explicitly, we use the following notation:

- $gl_{\mathbb{C}}(n,\mathbb{C})$ consists of all complex $(n \times n)$-matrices equipped with the Lie product $[A,B]_- := AB - BA$. The following Lie algebras refer to this Lie product.
- $sl_{\mathbb{C}}(n,\mathbb{C})$ consists of all traceless matrices in $gl_{\mathbb{C}}(n,\mathbb{C})$.
- $A \in so_{\mathbb{C}}(n,\mathbb{C})$ iff $A \in sl_{\mathbb{C}}(n,\mathbb{C})$ and $A = -A^d$.
- $A \in sp_{\mathbb{C}}(2n,\mathbb{C})$ iff $A \in sl_{\mathbb{C}}(2n,\mathbb{C})$ and $AJ = -JA^d$ where $J := \begin{pmatrix} 0 & I_n \\ -I_n & 0 \end{pmatrix}$.[24]

The complex Lie algebras mentioned above have the following complex dimensions for $n = 1,2,\ldots$:

- $\dim gl_{\mathbb{C}}(n,\mathbb{C}) = n^2$, $\dim sl_{\mathbb{C}}(n,\mathbb{C}) = n^2 - 1$,
- $\dim sp_{\mathbb{C}}(2n,\mathbb{C}) = n(2n+1)$, $\dim so_{\mathbb{C}}(n,\mathbb{C}) = \frac{1}{2}n(n-1)$,
- $\dim G_2 = 14$, $\dim F_4 = 52$, $\dim E_6 = 78$, $\dim E_7 = 133$, $\dim E_8 = 248$.

In the late 1880s, Killing summarized important properties of the exceptional Lie algebras, but he did not prove their existence. This was done by Élie Cartan in his 1894 thesis. To this end, Cartan had to check numerous Jacobi identities. But he left this to the reader. A complete a priori proof was given by Harrish-Chandra (1923–1983) and Chevalley (1909–1984) around 1948.

The exceptional Lie algebras in physics. Some physicists believe that the exceptional Lie algebras play a crucial role in describing nature. For example, the exceptional Lie algebra E_8 appears in string theory. Moreover, the exceptional Lie group G_2 is fundamental in a variant of the Standard Model in particle physics

[23] É. Cartan, Sur la structure des groupes de transformation fini et continus, Thèse, Paris, 1894. Cartan studied at the École Normale Supérieur in Paris. In 1912 he became professor at the Sorbonne in Paris.

[24] Here, I_n denotes the $(n \times n)$-unit matrix.

based on Cayley's octonions. Finally, it is possible to introduce the algebra E_{10} as a generalized exceptional Lie algebra, and E_{10} is related to some models in quantum gravity. We refer to:

A. Kleinschmidt and H. Nicolai, E10 Cosmology, J. High Energy Physics **137** (2006), 0601.

Tables for the exceptional Lie algebras can be found in J. Tits, Tables for Simple Lie Groups and Their Representations, Springer, Berlin, 1967 (in German). See also page 547. For the sophisticated proof of the classification theorem for complex simple Lie algebras, we refer to N. Jacobson, Lie Algebras, Dover, New York, 1979, and to Bourbaki (2001) quoted on page 281.

3.17.2 Semisimple Lie Algebras

Historical remarks. A major problem in representation theory is to prove the complete reducibility of all the finite-dimensional representations. In his 1894 thesis, Cartan proved the complete reducibility of the representations of the Lie algebra $sl(2,\mathbb{C})$... In 1913, Cartan constructed the irreducible representations of all the complex simple Lie algebras. Weyl noticed that there was a gap in Cartan's argument who implicitly used the unproved complete reducibility of the representations. In 1924, Weyl combined the approach due to Cartan and to Hurewitz (1859–1919) and Schur (1875–1941) in order to prove the complete reducibility of all the finite-dimensional representations of complex semisimple Lie algebras...[25]

Ideals and quotient Lie algebras. Set $\mathbb{K}=\mathbb{R}$ (real numbers) or $\mathbb{K}=\mathbb{C}$ (complex numbers). Let $\mathcal{L}$ be a finite-dimensional Lie algebra over $\mathbb{K}$. By definition, an ideal $\mathcal{J}$ of $\mathcal{L}$ is a Lie subalgebra which has the property that

$$[A,B]\in\mathcal{J} \qquad \text{for all} \quad A\in\mathcal{J}, B\in\mathcal{L}.$$

Ideals of a Lie algebra play the same crucial role as normal subgroups of a group (see Sect. 4.1.3 of Vol. II). For example, let $C,D\in\mathcal{L}$. Define

$$C\sim D \quad \text{iff} \quad C-D\in\mathcal{J}.$$

This is an equivalence relation The equivalence classes $[C]$ form a Lie algebra over $\mathbb{K}$ denoted by $\mathcal{L}/\mathcal{J}$. Explicitly, we set

$$\alpha[C]+\beta[D]:=[\alpha C+\beta D], \qquad [\,[C],[D]\,]:=[\,[C,D]\,]$$

for all $C,D\in\mathcal{L}$, and all $\alpha,\beta\in\mathbb{K}$. The point is that these definitions do not depend on the choice of the representatives. Equivalently, the quotient Lie algebra $\mathcal{L}/\mathcal{J}$ can be obtained by using the original Lie algebra $\mathcal{L}$, and by setting $A=0$ for all $A\in\mathcal{J}$. For example, if $A\in\mathcal{J}$, and $B,C\in\mathcal{L},\alpha\in\mathbb{K}$, then

$$A+B=B, \quad [A+B,C]=[B,C], \qquad \alpha A=0.$$

The direct product of Lie algebras. Let us consider the simplest method for constructing new Lie algebras. Fix $\mathbb{K}=\mathbb{R}$ or $\mathbb{K}=\mathbb{C}$. Let $\mathcal{A}$ and $\mathcal{B}$ be Lie algebras over $\mathbb{K}$. Naturally enough, the product set

$$\mathcal{A}\times\mathcal{B}:=\{(A,B):\ A\in\mathcal{A}, B\in\mathcal{B}\}$$

becomes a Lie algebra over $\mathbb{K}$ by setting

[25] Weyl used invariant measures of Lie groups due to Hurewitz; these measures were forerunners of the Haar measure introduced by Haar (1885–1933) in 1933.

- $\alpha(A, B) + \gamma(C, D) := (\alpha A + \gamma C, \alpha B + \gamma D)$ (linear combination),
- $[\,(A, B),\ (C, D)\,] := (\ [A, C],\ [B, D]\ )$ (Lie product)

for all $A, C \in \mathcal{A}$ and $B, D \in \mathcal{B}$, as well as for all $\alpha, \gamma \in \mathbb{K}$. This Lie algebra is called the direct product $\mathcal{A} \times \mathcal{B}$ of the Lie algebras $\mathcal{A}$ and $\mathcal{B}$. Similarly, we define the direct product $\mathcal{A}_1 \times \cdots \times \mathcal{A}_m$ of a finite number of Lie algebras $\mathcal{A}_1, \ldots, \mathcal{A}_m$ over $\mathbb{K}$.

Semisimplicity. Let $\mathcal{L}$ be a real (resp. complex) finite-dimensional Lie algebra different from $\{0\}$.

$\mathcal{L}$ is called semisimple iff it is isomorphic to the direct product of a finite number of simple real (resp. complex) Lie algebras.

The trivial Lie group $\{0\}$ is not simple. By usual convention, $\{0\}$ is called semisimple. Observe that the complex Lie algebras $gl_{\mathbb{C}}(N, \mathbb{C})$, $N = 1, 2, \ldots,$ are neither simple nor semisimple. In fact, we have the direct sum[26]

$$gl_{\mathbb{C}}(N, \mathbb{C}) = sl_{\mathbb{C}}(N, \mathbb{C}) \oplus \{zI : z \in \mathbb{C}\}, \qquad N = 2, 3, \ldots$$

of two ideals. Hence we have the Lie algebra isomorphism

$$gl_{\mathbb{C}}(N, \mathbb{C}) \simeq sl_{\mathbb{C}}(N, \mathbb{C}) \times \mathbb{C}$$

(see statement (ii) on page 264). The first factor is simple, but the second factor – the complex one-dimensional Lie algebra $\mathbb{C}$ – is commutative, and hence not simple.

Weyl's fundamental theorem. In 1925, Weyl proved the following key result for the representation theory of semisimple Lie algebras.[27]

Theorem 3.17 *The finite-dimensional representations of a complex finite-dimensional semisimple Lie algebra are completely reducible.*

Since semisimple Lie algebras are direct products of simple Lie algebras, Weyl's theorem tells us that all the finite-dimensional representations of a complex semisimple Lie algebra are known. This is a consequence of Cartan's complete classification of all the simple complex Lie algebras and their irreducible representations.

Algebraic characterization of semisimplicity in terms of ideals. Let $\mathcal{L}$ be a real (resp. complex) Lie algebra different form $\{0\}$. The Lie algebra $\mathcal{L}$ is semisimple iff one of the following equivalent properties holds:

- $\mathcal{L}$ is the direct product of simple real (resp. complex) Lie algebras.
- $\mathcal{L}$ is the direct sum of simple ideals of $\mathcal{L}$.
- $\mathcal{L}$ has no commutative ideal different from $\{0\}$.
- Every representation of $\mathcal{L}$ over a finite-dimensional real (resp complex) linear space is completely reducible.
- The radical of $\mathcal{L}$ is trivial (i.e., equal to $\{0\}$).

The notion of radical will be introduced in the next section.

---

[26] This corresponds to the decomposition $A = (A - N^{-1}\operatorname{tr}(A)I) + N^{-1}\operatorname{tr}(A)I$ for all $A \in gl_{\mathbb{C}}(N, \mathbb{C})$.

[27] H. Weyl, Representation theory for continuous semisimple groups by linear transformations I, II, III, Math. Zeitschrift **23** (1925), 271–309; **24**, 328–376, 377–395, 789–791 (in German). See H. Weyl, Collected Works, Vol. II, pp. 543–647, Springer, Berlin, 1968.

3.17.3 Solvability and the Heisenberg Algebra in Quantum Mechanics

> Solvable Lie algebras ar close to both upper triangular matrices and commutative Lie algebras. In contrast to this, semisimple Lie algebras are as far as possible from being commutative. By Levi's decomposition theorem, any Lie algebra is built out of a solvable and a semisimple one. The nontrivial prototype of a solvable Lie algebra is the Heisenberg algebra.
>
> Folklore

Let $\mathcal{L}$ be a real or complex Lie algebra. By definition, the commutant $\mathcal{L}'$ of $\mathcal{L}$ is the smallest Lie subalgebra of $\mathcal{L}$ which contains all the Lie products

$$[A, B], \qquad A, B \in \mathcal{L}.$$

Moreover, $\mathcal{L}'' = (\mathcal{L}')'$, and so on. The Lie algebra $\mathcal{L}$ is said to be solvable iff some iterated commutant is trivial, that is, there exists a positive integer n such that

$$\mathcal{L}^{(n)} = \{0\}.$$

Every commutative Lie algebra is solvable.

Example. The real Lie algebra $sut(2, \mathbb{R})$ consisting of all the matrices

$$\begin{pmatrix} 0 & x \\ 0 & 0 \end{pmatrix}, \qquad x \in \mathbb{R}$$

is solvable. In fact, it follows from

$$\begin{pmatrix} 0 & x \\ 0 & 0 \end{pmatrix}\begin{pmatrix} 0 & \xi \\ 0 & 0 \end{pmatrix} - \begin{pmatrix} 0 & \xi \\ 0 & 0 \end{pmatrix}\begin{pmatrix} 0 & x \\ 0 & 0 \end{pmatrix} = \begin{pmatrix} 0 & 0 \\ 0 & 0 \end{pmatrix}$$

that $sut(2, \mathbb{R})' = 0$. Similarly, we get $sut(3, \mathbb{R})'' = 0$. Thus, the Lie algebra $sut(3, \mathbb{R})$ is solvable.

> *Since the Heisenberg algebras $\mathcal{A}_{\mathrm{Heis}}(\mathbf{k})$ and $\mathcal{A}_{\mathrm{Heis}}(\mathbb{R}^3)$ are isomorphic to $sut(3, \mathbb{R})$, they are solvable.*[28]

In contrast to this, the real Lie algebra $sl(2, \mathbb{R})$ consisting of all the matrices

$$\begin{pmatrix} z & x \\ y & -z \end{pmatrix}, \qquad x, y, z \in \mathbb{R}$$

is not solvable, since $sl(2, \mathbb{R})' = sl(2, \mathbb{R})$ because of the commutation relations (3.63). The Lie algebra $sl(2, \mathbb{R})$ is semisimple.

A finite-dimensional real or complex Lie algebra $\mathcal{L}$ is solvable iff there exists a finite sequence

$$\mathcal{L}_0 = \{0\} \subseteq \mathcal{L}_1 \subseteq \mathcal{L}_2 \subseteq \ldots \subseteq \mathcal{L}_m = \mathcal{L}$$

of Lie subalgebras $\mathcal{L}_k$ of $\mathcal{L}$ such that $\mathcal{L}_k$ is an ideal of $\mathcal{L}_{k+1}$ and the quotient Lie algebra $\mathcal{L}_{k+1}/\mathcal{L}_k$ is commutative if $k = 0, 1, \ldots m-1$.

Lie's theorem. The following hold:[29]

---

[28] See Sect. 1.5 on page 106.

[29] For the proof, see A. Kirillov, jr., An Introduction to Lie Groups and Lie Algebras, Cambridge University Press, 2008, p. 94.

- Let $\varrho : \mathcal{L} \to gl(X)$ be a representation of the finite-dimensional complex solvable Lie algebra $\mathcal{L}$ on the finite-dimensional complex linear space X different from $\{0\}$. Then there exists a basis of X such that all the linear operators $\varrho(\mathcal{A})$ correspond to upper triangular matrices.
- If ϱ is irreducible, then the dimension of X is equal to one.

Thus, in contrast to semisimple Lie algebras, the irreducible representations of a solvable Lie algebra are trivial.

Élie Cartan's theorem. Fix $n = 1, 2, \ldots$ Let $\mathcal{L}$ be a real (resp. complex) Lie algebra consisting of $(n \times n)$-matrices. Then, the following hold:

- $\mathcal{L}$ is solvable iff $\mathrm{tr}(A[B,C]_-) = 0$ for all $A, B, C \in \mathcal{L}$.
- $\mathcal{L}$ is semisimple iff the Killing form
$$K(A,B) := \mathrm{tr}(\mathrm{ad}(A)\,\mathrm{ad}B), \qquad A, B \in \mathcal{L}$$
is non-degenerate, that is, it follows from $K(A,B) = 0$ for all $B \in \mathcal{L}$ that $A = 0$.

The radical. If $\mathcal{L}$ is a real or complex Lie algebra, then there is a unique solvable ideal of $\mathcal{L}$ which contains any other solvable ideal of $\mathcal{L}$. This solvable ideal is called the radical $\mathrm{rad}(\mathcal{L})$ of $\mathcal{L}$.

$\mathcal{L}$ is semisimple iff the radical of $\mathcal{L}$ is trivial, $\mathrm{rad}(\mathcal{L}) = \{0\}$.

The quotient Lie algebra $\mathcal{L}/\mathrm{rad}(\mathcal{L})$ is always semisimple. In other words, the radical measures the deviation of a Lie algebra from being semisimple.

3.17.4 Semidirect Product and the Levi Decomposition

> The fundamental Levi decomposition of a Lie algebra is the prototype of a semidirect product of Lie algebras.
>
> Folklore

We are given the real or complex Lie algebra $\mathcal{L}$. Suppose that $\mathcal{J}$ and $\mathcal{S}$ are Lie subalgebras of $\mathcal{L}$. We write

$$\boxed{\mathcal{L} = \mathcal{J} \oplus \mathcal{S}} \tag{3.78}$$

iff this is true in the sense of linear spaces, that is, every element A of $\mathcal{L}$ can be uniquely written as

$$A = B + C, \qquad B \in \mathcal{J},\ C \in \mathcal{S}.$$

Let $B, B' \in \mathcal{J}$ and $C, C' \in \mathcal{S}$. For the Lie product, we get

$$[B + C, B' + C'] = [B, B'] + [B, C'] + [C, B'] + [C, C'].$$

(i) Semidirect product. If $\mathcal{J}$ is an ideal of $\mathcal{L}$, then (3.78) is called a semidirect product, and we write $\mathcal{L} = \mathcal{J} \rtimes \mathcal{S}$. Then $[B, C'], [C, B'] \in \mathcal{J}$. Hence

$$[B + B', C + C'] = ([B, B'] + [B, C'] + [C, B']) \oplus [C, C'].$$

Thus, the quotient Lie algebra $\mathcal{L}/\mathcal{J}$ is isomorphic to the Lie algebra $\mathcal{S}$.

(ii) Direct product. If $\mathcal{J}$ and $\mathcal{S}$ are ideals of $\mathcal{L}$, then

$$[B, C] = 0 \qquad \text{for all} \quad B \in \mathcal{J}, C \in \mathcal{S}.$$

In fact, $[B, C] \in \mathcal{J}$ and $[B, C] \in \mathcal{S}$. Hence $[B, C] = 0$. Thus

$$[B + C, B' + C'] = [B, B'] + [C, C'].$$

Therefore, we have the Lie algebra isomorphism $\mathcal{L} \simeq \mathcal{J} \times \mathcal{S}$. This isomorphism is given by the map $B + C \mapsto (B, C)$.

Levi's decomposition theorem. Let $\mathcal{L}$ be a real or complex Lie algebra. Then there exists a semisimple Lie subalgebra $\mathcal{S}$ of $\mathcal{L}$ such that we have the direct sum

$$\boxed{\mathcal{L} = \mathrm{rad}(\mathcal{L}) \oplus \mathcal{S}.}$$

The semisimple quotient Lie algebra $\mathcal{L}/\mathrm{rad}(\mathcal{L})$ is isomorphic to $\mathcal{S}$. Consequently, for any real or complex Lie algebra $\mathcal{L}$, we have the semidirect product

$$\boxed{\mathcal{L} = \mathrm{rad}(\mathcal{L}) \rtimes \mathcal{S}}$$

where $\mathrm{rad}(\mathcal{L})$ is a solvable Lie algebra, and $\mathcal{S}$ is a semisimple Lie algebra. In addition $\mathrm{rad}(\mathcal{L})$ is an ideal of $\mathcal{L}$.

A variant of the fundamental Levi theorem was conjectured by Wilhelm Killing in the late 1880s, and it was proved by Eugenio Levi (1883–1917) in 1905. Levi also formulated the famous Levi problem on the characterization of the domains of holomorphy in the theory of analytic functions of several complex variables. In 1911, Levi introduced the parametrix as an approximate fundamental solution of partial differential equations (i.e., an approximate Green's function); parametrices play a key role in the modern theory of pseudo-differential operators.[30] Eugenio Levi was the younger brother of Beppo Levi (1875–1961) who worked in algebraic geometry, analysis, number theory, and set theory. In First World War (1914–1918), Eugenio was killed as a soldier.

Infinitesimal Lie groups. We are given the Lie group $\mathcal{G}$. Let the symbol $\mathcal{LG}$ denote the Lie algebra to $\mathcal{G}$. For historical reasons, we call $\mathcal{LG}$ the infinitesimal Lie group to $\mathcal{G}$.

> *The Lie group $\mathcal{G}$ is said to be infinitesimally solvable (resp. infinitesimally semisimple) iff its Lie algebra is solvable (resp. semisimple).*

For example, the infinitesimal Lie group corresponding to the Lie group $SU(3)$ is nothing else than the real Lie algebra $su(3)$ which is semisimple. Therefore, the Lie group $SU(3)$ is said to be infinitesimally semisimple.

Let us make some historical remarks. Sophus Lie (1842–1899) used the term *continuous transformation group* for local Lie groups in the neighborhood of the unit element. He only studied the local behavior of Lie groups by means of infinitesimal transformation groups called Lie algebras in modern terminology. The term *Lie group* was introduced by Élie Cartan around 1930, whereas the term *Lie algebra* was introduced by Hermann Weyl in his celebrated lectures given at the Institute for Advanced Study (IAS) in Princeton during the academic year 1933/34. The classical theory can be found in

> S. Lie and F. Engel, Theory of Transformation Groups (in German), Vols. 1–3, Teubner, Leipzig, 1888. Reprint: Chelsea Publ. Company, 1970.

The main collaborator of Lie, Friedrich Engel (1861–1941), discovered that the exponential map

$$\exp : \mathcal{L} \to \mathcal{G}, \qquad A \mapsto \mathrm{e}^A$$

with $\mathcal{L} := sl(2, \mathbb{C})$ is surjective for the quotient group $\mathcal{G} := SL(2, \mathbb{C})/\{1, -1\}$, but it is not surjective for the group $\mathcal{G} := SL(2, \mathbb{C})$. Note that both the groups have the same Lie algebra $sl(2, \mathbb{C})$. This example shows that the Lie algebra does not determine the global behavior of the corresponding Lie group.

---

[30] See Yu. Egorov, A. Komech, and M. Shubin, Elements of the Modern Theory of Partial Differential Equations, Springer, New York, 1999.

3.17.5 The Casimir Operators

The flow of ideas from physics to mathematics and vice versa is crucial.
Folklore

In 1931, the young Dutch physicist Casimir introduced the so-called Casimir operator for the rotation group $SO(3)$ in order to study the orbital angular momentum and the spin in quantum mechanics. In 1935, Casimir (1909–2000) and van der Waerden (1903–1998) used the idea of Casimir operators in order to give the first purely algebraic proof of the complete reducibility of finite-dimensional representations of complex semisimple Lie algebras.[31] For the Lie group $SU(2)$, the Casimir operator possesses an immediate physical interpretation in terms of the electron spin (see Sect. 7.3 on page 427). At this point, let us sketch the introduction of the two Casimir operators of the group $SU(3)$ used frequently by physicists. In this connection, the Schur lemma will play a crucial role. Following the notation used in physics, we start with the commutation relations

$$[\lambda_j, \lambda_k]_- = \sum_{l=1}^{8} f_{jkl}\lambda_l, \qquad j,k = 1,\ldots,8$$

and the anticommutation relations

$$[\lambda_j, \lambda_k]_+ = \tfrac{4}{3}\delta_{jk}I + 2\sum_{l=1}^{8} d_{jkl}\lambda_l, \qquad j,k = 1,\ldots,8$$

for the Gell-Mann matrices $\lambda_1, \ldots, \lambda_8$. Set $B_j := -\frac{\mathrm{i}}{2}\lambda_j$. Then, the matrices $B_1, \ldots, B_8$ form a basis of the real Lie algebra $su(3)$ with the commutation relations

$$[B_j, B_k]_- = \sum_{l=1}^{8} f_{jkl}B_l.$$

Suppose now that we have an irreducible unitary representation

$$\varrho : SU(3) \to GL(X)$$

of the group $SU(3)$ on the finite-dimensional complex Hilbert space X. In particular, for fixed B_j, the one-parameter subgroup

$$U(t) := \mathrm{e}^{tB_j}, \qquad t \in \mathbb{R}$$

of $SU(3)$ is transformed into the one-parameter subgroup $\{\varrho(U(t)) : t \in \mathbb{R}\}$. Differentiation with respect to time t at the point $t = 0$ yields the operator

$$\hat{B}_j = \frac{d\varrho(U(t))}{dt}_{|t=0}.$$

[31] H. Casimir and B. van der Waerden, Algebraic proof of the complete reducibility of the representations of semisimple Lie groups, Math. Annalen **111** (1935), 1–11 (in German). The following volume is dedicated to the memory of Casimir on the occasion of his 100th birthday and the 150th birthday of the Riemann hypothesis: G. van Dijk and and M. Wakayama (Eds.), Casimir Force, Casimir Operator, and the Riemann Hypothesis, de Gruyter, Berlin, 2010. For the relation between the Casimir force (i.e., the Casimir effect) in quantum field theory and the Riemann–Epstein zeta function, see Sect. 6.6 of Vol. I.

Hence $\varrho(U(t)) = e^{t\hat{B}_j}$ for all $t \in \mathbb{R}$. The map $B_j \to \hat{B}_j, j = 1, \ldots 8$, generates the representation

$$\sigma : su(3) \to gl(X)$$

of the Lie algebra $su(3)$. Therefore, all the operators $\hat{B}_j$ satisfy the same commutation relations as the matrices B_j. Thus,

$$[\hat{B}_j, \hat{B}_k]_- = \sum_{l=1}^{8} f_{jkl} \hat{B}_l. \tag{3.79}$$

Finally, setting $F_j := \mathrm{i}\hat{B}_j$, we define the two Casimir operators

$$C_1 := \sum_{j=1}^{8} F_j^2, \qquad C_2 := \sum_{j,k,l=1}^{8} d_{jkl} F_j F_k F_l.$$

Now to the point. First it can be shown that

$$[C_1, F_j]_- = 0, \qquad [C_2, F_j]_- = 0, \qquad j = 1, \ldots, 8. \tag{3.80}$$

This is a consequence of (3.79). Since the representation ϱ is irreducible, the Schur lemma on page 202 tells us that there exist numbers c_1 and c_2 such that

$$C_r = c_r I, \qquad r = 1, 2.$$

Since the representation ϱ is unitary, all the operators $\hat{B}_j$ are skew-adjoint. Hence all the operators F_j are self-adjoint. Thus, C_1 is self-adjoint. Since the coefficients d_{jkl} are symmetric with respect to the indices j, k, l, it follows from

$$(F_j F_k F_l)^\dagger = F_l^\dagger F_k^\dagger F_j^\dagger = F_l F_k F_j$$

that the Casimir operator C_2 is also self-adjoint. Consequently, c_1 and c_2 are real numbers.

The crucial nontrivial property of the Casimir operators is that the so-called quantum numbers c_1 and c_2 characterize the irreducible representation ϱ.

That is, two irreducible representations of $SU(3)$ are equivalent iff they possess the same quantum numbers c_1 and c_2. For example, an explicit computation shows that the adjoint representation $R(1,1)$ of $SU(3)$ has the quantum numbers $c_1 = 3$ and $c_2 = 0$ (see Problem 3.26).

3.18 Symmetric and Antisymmetric Functions

Symmetric and antisymmetric functions play a fundamental role in mathematics and physics, for example, in representation theory and topology (e.g. the construction of topological invariants like characteristic classes).

Folklore

For the material of this section, we refer to I. Macdonald, Symmetric Functions and Hall Polynomials, Oxford University Press, 1995, and to C. Procesi, Lie Groups. An Approach Through Invariants and Representations, Springer, New York, 2007.

The prototype. The theory of symmetric polynomials has a long tradition in mathematics dating back to the problem of solving polynomial equations in the 16th century. There are the following two key formulas:

- $(x - x_1)(x - x_2) = x^2 - ax + b$ with $a := x_1 + x_2$ and $b := x_1 x_2$, and
- the determinants due to Vandermonde (1735–1796):

$$\begin{vmatrix} x_1 & x_2 \\ 1 & 1 \end{vmatrix} = x_1 - x_2, \qquad \begin{vmatrix} x_1^2 & x_2^2 & x_3^2 \\ x_1 & x_2 & x_3 \\ 1 & 1 & 1 \end{vmatrix} = (x_1 - x_2)(x_1 - x_3)(x_2 - x_3).$$

The polynomial $P(x_1, x_2) = a_0 + a_1 x_1 + a_2 x_2 + \ldots$ with complex coefficients $a_0, a_1, \ldots$ is called symmetric (resp. antisymmetric) iff

$$P(x_1, x_2) = P(x_2, x_1)$$

(resp. $P(x_1, x_2) = -P(x_2, x_1)$). The following hold:

- Every polynomial $P(x_1, x_2)$ can be uniquely written as the sum of a symmetric polynomial and an antisymmetric polynomial:

$$P(x_1, x_2) = \tfrac{1}{2}\big(P(x_1, x_2) + P(x_2, x_1)\big) + \tfrac{1}{2}\big(P(x_1, x_2) - P(x_2, x_1)\big). \quad (3.81)$$

- Every symmetric polynomial $P(x_1, x_2)$ can be written as a polynomial of the elementary symmetric functions $a(x_1, x_2) := x_1 + x_2$ and $b(x_1, x_2) := x_1 x_2$.
- Every antisymmetric polynomial $P_{\mathrm{asym}}(x_1, x_2)$ can be written as a product of the form $(x_1 - x_2)P_0(x_1, x_2)$ where $P_0(x_1, x_2)$ is a symmetric polynomial.

For example, choose $D(x_1, x_2) := (x_1 - x_2)^2$. Then

$$D(x_1 - x_2) = (x_1 + x_2)^2 - 4x_1 x_2 = a^2 - 4b.$$

This implies that the equation

$$x^2 - ax + b = 0, \qquad a, b \in \mathbb{C}$$

has two different zeros iff $a^2 - 4b \neq 0$.

Our goal is to generalize this to polynomials of n variables. In this connection, partitions will play a crucial role.

3.18.1 Symmetrization and Antisymmetrization

Fix $n = 1, 2, \ldots$ Let $\mathbb{C}[x_1, x_2, \ldots, x_n]$ denote the space of all polynomials

$$a_0 + a_1 x_1 + \ldots + a_n x_n + a_{12} x_1 x_2 + \ldots$$

with respect to the variables $x_1, \ldots, x_n$ and complex coefficients $a_0, a_1, \ldots$ Moreover, let

$$\mathbb{C}[x_1, x_2, \ldots] = \mathbb{C} \oplus \mathbb{C}[x_1] \oplus \mathbb{C}[x_1, x_2] \oplus \ldots$$

This is the space of all the polynomials of arbitrary order with complex coefficients. This is an infinite-dimensional complex linear space and a commutative complex algebra with respect to the usual addition and multiplication of polynomials. If $P \in \mathbb{C}[x_1, \ldots, n]$, then we define

$$(\mathsf{Sym}(P))(x_1, \ldots, x_n) := \frac{1}{n!} \sum_{\pi \in Sym(n)} P(x_{\pi(1)}, \ldots, x_{\pi(n)}),$$

$$(\mathsf{Asym}(P))(x_1, \ldots, x_n) := \frac{1}{n!} \sum_{\pi \in Sym(n)} \operatorname{sgn} \pi(n) \cdot P(x_{\pi(1)}, \ldots, x_{\pi(n)}).$$

The polynomial $P(x_1, \ldots, n)$ is called symmetric (resp. antisymmetric) iff it does not change (resp. it changes sign) under a transposition of two arguments. This is equivalent to $\mathsf{Sym}(P) = P$ (resp. $\mathsf{Asym}(P) = P$). The prototype of an antisymmetric polynomial is the Vandermonde polynomial

$$\mathcal{V}(x_1, \ldots, x_n) := \prod_{1 \leq i < j \leq n} (x_i - x_j).$$

If $P(x_1, \ldots, x_n)$ is a symmetric polynomial, then the product

$$P_{\text{asym}}(x_1, \ldots, x_n) = \mathcal{V}(x_1, x_2, \ldots, x_n) P(x_1, x_2, \ldots, x_n) \qquad (3.82)$$

is an antisymmetric polynomial. The point is that every antisymmetric polynomial can be obtained this way.

The importance of Young symmetrizers. If $P \in \mathbb{C}[x_1, x_2]$, then we have the unique decomposition

$$P = \mathsf{Sym}(P) + \mathsf{Asym}(P)$$

which corresponds to (3.81). The point is that the polynomial space $\mathbb{C}[x_1, x_2, \ldots, x_n]$ is not spanned by the symmetric and antisymmetric polynomials if $n \geq 3$. In this case, we need further 'elementary' symmetries which are provided by the Young symmetrizers. To explain this, let us consider the case where $n = 3$. We choose the Young symmetrizers $\mathsf{S}_j, j = 1, 2, 3, 4$, from Table 3.9 on page 222. Then, for every polynomial $P(x_1, x_2, x_3)$, we get

- $\mathsf{S}_1 P = \mathsf{Sym}(P)$ and $\mathsf{S}_4 P = \mathsf{Asym}(P)$,
- $(\mathsf{S}_2 P)(x_1, x_2, x_3) = \frac{1}{3}(P(x_1, x_2, x_3) + P(x_2, x_1, x_3) - P(x_3, x_2, x_1) - P(x_2, x_3, x_1))$,
- $\mathsf{S}_3 P(x_1, x_2, x_3) = \frac{1}{3}(P(x_1, x_2, x_3) + P(x_3, x_2, x_1) - P(x_2, x_1, x_3) - P(x_3, x_1, x_2))$.

For example, this follows from $\mathsf{S}_2 = \frac{1}{3}((1) + (12) - (13) - (123))$ by applying the permutations to the arguments of P. One checks explicitly that

$$\boxed{P = \mathsf{S}_1 P + \mathsf{S}_2 P + \mathsf{S}_3 P + \mathsf{S}_4 P,}$$

and $\mathsf{S}_j^2 = \mathsf{S}_j$ for all j. This way, we get the direct sum decomposition

$$\mathbb{C}[x_1, x_2, x_3] = \bigoplus_{j=1}^{4} \mathsf{S}_j(\mathbb{C}[x_1, x_2, x_3]).$$

By definition, the polynomial P is called S_j-symmetric iff $\mathsf{S}_j P = P$. Note that the polynomial $\mathsf{S}_j P$ is always S_j-symmetric. In fact, $\mathsf{S}_j(\mathsf{S}_j P) = \mathsf{S}_j P$. Similarly, we get the decomposition

$$\mathbb{C}[x_1, \ldots, x_n] = \bigoplus_j (\mathsf{S}_j \mathbb{C})[x_1, \ldots x_n], \qquad n = 2, 3, \ldots$$

where the Young symmetrizers S_j are obtained as for $Sym(3)$ by using the Young tableaux. Note that the Young symmetrizers are constructed by the superposition of symmetrization and antisymmetrization procedures with respect to certain variables. This is governed by the symmetry groups $\mathcal{H}$ and $\mathcal{V}$ of the corresponding Young tableau (see Table 3.9 on page 222). Finally, we add a multiplicative constant which guarantees that $\mathsf{S}_j^2 = \mathsf{S}_j$.

Table 3.12. Elementary symmetric polynomials

$x = (a, b)$
$\mathcal{E}_0(x) = 1, \quad \mathcal{E}_1(x) = a + b, \quad \mathcal{E}_2(x) = ab$
$n = 3, \quad x = (a, b, c)$
$\mathcal{E}_0(x) = 1, \quad \mathcal{E}_1(x) = a + b + c,$ $\mathcal{E}_2(x) = ab + ac + bc, \quad \mathcal{E}_3(x) = abc$

3.18.2 Elementary Symmetric Polynomials

The elementary symmetric polynomials are homogeneous polynomials which can be regarded as the atoms of symmetric polynomials.

Folklore

Fix $n = 2, 3, \ldots$ Set $x := (x_1, \ldots, x_n)$. The elementary symmetric functions $\mathcal{E}_k(x_1, \ldots, x_n)$, $k = 0, 1, \ldots, n$, are defined by the formula

$$\prod_{i=1}^{n}(z - x_i) = \mathcal{E}_0(x)z^n - \mathcal{E}_1(x)z^{n-1} + \mathcal{E}_2(x)z^{n-2} - \ldots + (-1)^n\mathcal{E}_n(x).$$

For example, $\mathcal{E}_0(x) := 1, \mathcal{E}_1(x) = x_1 + x_2 + \ldots + x_n$,

$$\mathcal{E}_2(x) = \sum_{1 \le i < j \le n} x_i x_j, \quad \mathcal{E}_3(x) = \sum_{1 \le i < j < k \le n} x_i x_j x_k, \quad \ldots, \quad \mathcal{E}_n(x) = x_1 x_2 \cdots x_n.$$

Special cases of these functions were used by Viète (Vieta)(1540–1603). The general case was studied by Newton (1643–1727). The main theorem on symmetric polynomials tells us the following:

The elementary symmetric functions $\mathcal{E}_k(x_1, \ldots, x_n), k = 0, 1, \ldots, n$, are homogeneous polynomials of degree k which generate the complex algebra $\mathbb{C}_{\mathrm{sym}}[x_1, \ldots, x_n]$ of symmetric polynomials with n variables.

This means that every symmetric polynomial $P(x_1, \ldots, x_n)$ with complex coefficients can be uniquely represented as a polynomial of elementary symmetric polynomials with complex coefficients.

Example. Consider the equation

$$z^3 - az^2 + bz - c = 0$$

with complex coefficients a, b, c. This equation has 3 different zeros x_1, x_2, x_3 iff the so-called discriminant

$$D = -4a^3c + a^2b^2 + 18abc - 4b^3 - 27c^2 \tag{3.83}$$

does not vanish. To prove this, set $D := \big((x_1 - x_2)(x_1 - x_3)(x_1 - x_3)\big)^2$. This is a symmetric polynomial with respect to x_1, x_2, x_3. Therefore, D can be written as a polynomial with respect to $1, a, b, c$. Explicitly, we get (3.83).

3.18.3 Power Sums

The prototype of a power sum is the polynomial $\mathcal{P}_2(a,b) = a^2 + b^2$. In the general case, we set

$$\mathcal{P}_k(x_1,\ldots,x_n) = x_1^k + x_2^k + \ldots + x_n^k$$

where $k = 1, 2, \ldots$ For $k = 0$, we set $\mathcal{P}_0(x) := 1$. The following hold:

The power sums $\mathcal{E}_k(x_1,\ldots,x_n), k = 0, 1, \ldots, n$, are homogeneous polynomials of degree k which generate the complex algebra $\mathbb{C}_{\text{sym}}[x_1,\ldots,x_n]$ of symmetric polynomials with n variables.

Newton discovered general recursion formulas for representing power sums as polynomials of elementary symmetric polynomials, and vice versa. For example, $\mathcal{P}_0(x) = \mathcal{E}_0(x)$, and

$$\mathcal{P}_1 = \mathcal{E}_1, \quad \mathcal{P}_2 = \mathcal{E}_1^2 - 2\mathcal{E}_2, \quad \mathcal{P}_3 = \mathcal{E}_1^3 - 3\mathcal{E}_1\mathcal{E}_2 + 3\mathcal{E}_3.$$

Conversely,

$$\mathcal{E}_1 = \mathcal{P}_1, \quad \mathcal{E}_2 = \tfrac{1}{2}\mathcal{P}_1^2 - \tfrac{1}{2}\mathcal{P}_2, \quad \mathcal{E}_3 = \tfrac{1}{6}\mathcal{P}_1^3 - \tfrac{1}{2}\mathcal{P}_1\mathcal{P}_2 + \tfrac{1}{3}\mathcal{P}_3.$$

3.18.4 Completely Symmetric Polynomials

Elementary symmetric polynomials can be replaced by completely symmetric polynomials as generating polynomials for symmetric polynomials.

Folklore

The prototype of a completely symmetric polynomial of degree 2 is the polynomial $\mathcal{C}_2(a,b) = a^2 + ab + b^2$. This is a homogenous polynomial of degree 2 which contains all the possible terms of degree 2 equipped with the coefficient 1. Let $x = (x_1,\ldots,x_n)$. It is convenient to introduce the monomial symbol

$$\boxed{x^\alpha := x_1^{\alpha_1} x_2^{\alpha_2} \cdots x_n^{\alpha_n}}$$

where $\alpha = (\alpha_1,\ldots,\alpha_n)$ is a tuple of non-negative integers. We set $|\alpha| = \alpha_1 + \ldots + \alpha_n$. A polynomial $P(x_1,\ldots,x_n)$ of degree k can be elegantly written as

$$P(x) = \sum_{|\alpha|\le k} a_\alpha x^\alpha$$

where all the coefficients a_α are complex numbers. By definition, the polynomial

$$\mathcal{C}_k(x) = \sum_{|\alpha|=k} a_\alpha x^\alpha$$

is called a completely symmetric polynomial of degree k. In particular, we get $\mathcal{C}_0(x) = 1$ and $\mathcal{C}_1(x) = x_1 + \ldots + x_n$. Further examples can be found in Table 3.13.

The completely symmetric polynomials $\mathcal{C}_k(x_1,\ldots,x_n), k = 0, 1, \ldots, n$, are homogeneous polynomials of degree k which generate the complex algebra $\mathbb{C}_{\text{sym}}[x_1,\ldots\ldots,x_n]$ of symmetric polynomials with n variables.

Table 3.13. Completely symmetric polynomials

$x = (a, b)$
$\mathcal{C}_0(x) = 1, \quad \mathcal{C}_1(x) = a + b, \quad \mathcal{C}_2(x) = a^2 + b^2 + ab$
$x = (a, b, c)$
$\mathcal{C}_0(x) = 1, \quad \mathcal{C}_1(x) = a + b + c$
$\mathcal{C}_2(x) = a^2 + b^2 + c^2 + ab + ac + bc$
$\mathcal{C}_3(x) = a^3 + b^3 + c^3 + a^2b + a^2c + b^2c + ab^2 + ac^2 + bc^2 + abc$

3.18.5 Symmetric Schur Polynomials

The symmetric Schur polynomials are homogeneous polynomials which have the fundamental property that they form a basis of the complex linear space of symmetric polynomials. Elementary symmetric polynomials generate the Schur polynomials by computing products and complex linear combinations. The same is true for completely symmetric polynomials. The explicit formulas were found by Jacobi and his student Trudi in 1841.

Folklore

The antisymmetric basis polynomials. Let $\mu = (\mu_1, \ldots, \mu_n)$ be a tuple of integers such that $\mu_1 > \mu_2 > \ldots > \mu_n \geq 0$. We define

$$\mathcal{V}_\mu(x) := \begin{vmatrix} x_1^{\mu_1} & x_2^{\mu_1} & \ldots & x_n^{\mu_1} \\ x_1^{\mu_2} & x_2^{\mu_2} & \ldots & x_n^{\mu_2} \\ \vdots & \vdots & \ldots & \vdots \\ x_1^{\mu_n} & x_2^{\mu_n} & \ldots & x_n^{\mu_n} \end{vmatrix} \tag{3.84}$$

where $x = (x_1, \ldots, x_n)$, and $n = 1, 2, \ldots$. Here, $\mathcal{V}_\mu(x)$ is called an antisymmetric Vandermonde polynomial. The following hold:

All the possible antisymmetric Vandermonde polynomials form a basis of the complex linear space $\mathbb{C}_{\mathrm{asym}}[x_1, x_2, \ldots]$ of antisymmetric polynomials with complex coefficients.

The symmetric basis polynomials. Fix $n = 1, 2, \ldots$ Let $\lambda = (\lambda_1, \ldots, \lambda_n)$ be a tuple of non-negative integers with $\lambda_1 \geq \lambda_2 \geq \ldots \geq \lambda_n \geq 0$. Furthermore, set $\nu = (n-1, n-2, \ldots, 1, 0)$. Define

$$\mathcal{S}_\lambda(x) := \frac{\mathcal{V}_{\nu+\lambda}}{\prod_{1 \leq i < j \leq n}(x_i - x_j)} \tag{3.85}$$

where

$$\mathcal{V}_{\nu+\lambda} := \begin{vmatrix} x_1^{n-1+\lambda_1} & x_2^{n-1+\lambda_1} & \ldots & x_n^{n-1+\lambda_1} \\ x_1^{n-2+\lambda_2} & x_2^{n-2+\lambda_2} & \ldots & x_n^{n-2+\lambda_2} \\ \vdots & \vdots & \ldots & \vdots \\ x_1^{\lambda_n} & x_2^{\lambda_n} & \ldots & x_n^{\lambda_n} \end{vmatrix} \tag{3.86}$$

Table 3.14. Symmetric Schur polynomials

$n = 2, \quad x = (a, b)$	partition
$\mathcal{S}_{(1,1)}(x) = \mathcal{E}_2(x) = ab$	$2 = 1 + 1$
$\mathcal{S}_{(2,0)}(x) = \mathcal{C}_2(x) = a^2 + b^2 + ab$	$2 = 2$
$n = 3, \quad x = (a, b, c)$	
$\mathcal{S}_{(1,1,1)}(x) = \mathcal{E}_3(x) = abc$	$3 = 1 + 1 + 1$
$\mathcal{S}_{(2,1,0)}(x) = (a + b)(a + c)(b + c)$	$3 = 2 + 1$
$\mathcal{S}_{(3,0,0)}(x) = \mathcal{C}_3(x) \qquad$ (see Table 3.13)	$3 = 3$

Hence $\mathcal{V}_{\nu+\lambda} = \mathcal{V}_\nu \mathcal{S}_\lambda$.

All the possible symmetric Schur polynomials form a basis of the complex linear space $\mathcal{C}_{\text{sym}}[x_1, x_2, \ldots]$ *of symmetric polynomials with complex coefficients.*

The label λ of a Schur polynomial is a uniquely defined partition of the number $|\lambda|$. For example, $\lambda = (4, 3, 2, 2, 0, 0)$ corresponds to the partition $11 = 4 + 3 + 2 + 2$. Therefore, the symmetric Schur polynomials are labelled by partitions; they play the key role in the famous Frobenius character formula for the symmetric groups.

Fundamental properties of symmetric Schur polynomials. The following hold:

(i) The 1841 Jacobi–Trudi determinant formula:

$$\boxed{\mathcal{S}_\lambda(x) = \det(\mathcal{C}_{\lambda_i - i + j}(x))_{1 \le i,j \le n}.} \tag{3.87}$$

In addition, we have the dual formula

$$\mathcal{S}_\lambda(x) = \det(\mathcal{E}_{\lambda'_i - i + j})_{1 \le i,j \le m} \tag{3.88}$$

where the Young frame to the n-tuple $\lambda' = (\lambda'_1, \ldots, \lambda'_m)$ is dual to the Young frame to $\lambda = (\lambda_1, \ldots, \lambda_n)$. For example, let $x = (a, b)$. Then

$$\mathcal{S}_{(1,1)}(a, b) = \begin{vmatrix} \mathcal{C}_1(x) & \mathcal{C}_2(x) \\ \mathcal{C}_0(x) & \mathcal{C}_1(x) \end{vmatrix} = \begin{vmatrix} a + b & a^2 + b^2 + ab \\ 1 & a + b \end{vmatrix} = ab.$$

(ii) The 1934 Littlewood–Richardson product formulas: One can show that

$$\mathcal{S}_{(3)}(x)\mathcal{S}_{(2)}(x) = \mathcal{S}_{(3,2)}(x) + \mathcal{S}_{(4,1)}(x) + \mathcal{S}_{(5)}(x).$$

This is a special case of the general formula

$$\boxed{\mathcal{S}_\lambda(x)\mathcal{S}_\mu(x) = \sum_\nu c^\nu_{\lambda\mu} \mathcal{S}_\nu(x).} \tag{3.89}$$

Since the symmetric Schur polynomials form a basis of the polynomial algebra, the product $\mathcal{S}_\lambda \mathcal{S}_\mu$ is a linear combination of the form (3.89). The main problem is to compute explicitly the coefficients $c^{\nu}_{\lambda\mu}$ which are nonnegative integers. In 1934, D. Littlewood and Richardson invented an elegant combinatorial rule for computing $c^{\nu}_{\lambda\mu}$ by using generalized Young tableaux. This Littlewood–Richardson rule is a highlight in combinatorial mathematics. Complete proofs of this rule only appeared in the 1970s.[32] For details, we refer to MacDonald (1995) and Procesi (2007) quoted on page 208. See also B. Sagan, The Symmetric Group: Representations, Combinatorial Algorithms, and Symmetric Functions, Springer, Berlin, 2001, and J. Louck, Unitary Symmetry and Combinatorics, World Scientific, Singapore, 2008.

(iii) The Cauchy generating function: see (3.93) on page 276.

3.18.6 Raising Operators and the Creation and Annihilation of Particles

The raising operators in combinatorial mathematics can be regarded as simplified models for particle creation and particle annihilation in quantum field theory.

Folklore

We want to show that Schur functions can be generated by a special combinatorics which is an appropriate superposition of simulated particle creation and particle annihilation processes.

Prototype. Let $\lambda = (\lambda_1, \lambda_2)$ where λ_1 and λ_2 are integers. We define the raising operator R_{12} by setting

$$R_{12}\lambda := (\lambda_1 + 1, \lambda_2 - 1).$$

Consider a partition (λ_1, λ_2) with $\lambda_2 \geq \lambda_1 \geq 0$ and integers λ_1, λ_2. As a special case of Theorem 3.18 below, the Schur function $\mathcal{S}_\lambda(x)$ is given by the formula

$$\mathcal{S}_\lambda(x) = (1 - R_{12})\mathcal{C}_\lambda(x), \qquad x \in \mathbb{R}^2.$$

This is to be understood as

$$\mathcal{S}_\lambda(x) = \mathcal{C}_\lambda(x) - \mathcal{C}_{R_{12}\lambda}(x).$$

Here, we set

$$\mathcal{C}_{R_{12}\lambda}(x) := \mathcal{C}_{\lambda_1+1}(x)\mathcal{C}_{\lambda_2-1}(x),$$

and we use the convention that $\mathcal{C}_{\lambda_2-1} = 0$ if $\lambda_2 - 1 < 0$.

Examples. (i) Let $\lambda = (1,0)$. Then $R_{12}\lambda = (2,-1)$. Hence

$$\mathcal{S}_{(1,0)}(x) = \mathcal{C}_{(1,0)}(x) - \mathcal{C}_{(2,-1)}(x) = \mathcal{C}_{(1,0)}(x) = x_1 + x_2.$$

(ii) Let $\lambda = (1,1)$. Then $R_{12}\lambda = (2,0)$. Hence

$$\begin{aligned}\mathcal{S}_{(1,1)}(x) &= \mathcal{C}_{(1,1)}(x) - \mathcal{C}_{(2,0)}(x)\\ &= (x_1 + x_2)(x_1 + x_2) - (x_1^2 + x_1x_2 + x_2^2) = x_1x_2.\end{aligned}$$

[32] One has to distinguish between John Littlewood (1885–1977) and Dudley Littlewood (1903–1979). Famous books are: G. Hardy, J. Littlewood, and G. Pòlya, Inequalities, Cambridge University Press, 1988 (first edition 1934).
D. Littlewood, The Theory of Group Characters and Matrix Representations of Groups, Amer. Math. Society, Providence, Rhode Island, 2006 (originally published by Clarendon Press, Oxford, 1940).

The general case. Fix $n = 2, 3, \ldots$ Let $\lambda = (\lambda_1, \ldots, \lambda_n)$ be an n-tuple of integers. For $i < j$, we define the raising operator R_{ij} by setting

$$R_{ij}\lambda := (\lambda_1, \ldots, \lambda_i + 1, \ldots, \lambda_j - 1, \ldots, \lambda_n).$$

In terms of physics, the operator R_{ij} creates one particle in the ith state, and it annihilates one particle in the jth state. We will also consider products of raising operators. For example,

$$R_{23}R_{12}(\lambda_1, \lambda_2, \lambda_3) = R_{23}(\lambda_1 + 1, \lambda_2 - 1, \lambda_3) = (\lambda_1 + 1, \lambda_2, \lambda_3 - 1).$$

In order to organize the products of raising operators in a convenient way, let us also use the symbol

$$\prod_{1 \le i < j \le n} (1 - R_{ij}) = 1 + \mathsf{R}$$

where the remainder R is the sum of the corresponding products of raising operators. For example,

$$\prod_{1 \le i < j \le 3} (1 - R_{ij}) = (1 - R_{12})(1 - R_{13})(1 - R_{23}) = 1 + \mathsf{R},$$

where

$$\mathsf{R} := -R_{12} - R_{13} - R_{23} + R_{12}R_{13} + R_{12}R_{23} + R_{13}R_{23} - R_{12}R_{13}R_{23}.$$

Let $\lambda = (\lambda_1, \ldots, \lambda_n)$ be a partition, that is, $\lambda_n \ge \ldots \ge \lambda_1 \ge 0$.

Theorem 3.18 *The Schur function $\mathcal{S}_\lambda$ is given by the mnemonic formula*

$$\mathcal{S}_\lambda(x) = \prod_{1 \le i < j \le n} (1 - R_{ij})\mathcal{C}_\lambda(x), \qquad x \in \mathbb{R}^n.$$

This is to be understood as $\mathcal{S}_\lambda(x) = \mathcal{C}_\lambda(x) - \mathcal{C}_{\mathsf{R}\lambda}(x)$.

Here, we set $\mathcal{C}_{\mathsf{R}\lambda} := 0$ if the tuple $\mathsf{R}\lambda$ contains at least one negative integer. The proof can be found in MacDonald (1995), page 43, quoted on page 208.

3.19 Formal Power Series Expansions and Generating Functions

> Generating functions are used in physics in order to encode the properties of multi-particle systems in statistical physics and quantum field theory. The Feynman path integral encodes the correlations (i.e., the Green functions) of a quantum field. The main task is to decode the information.
>
> Folklore

Let us summarize the key formulas which are to be understood in the sense of a formal power series expansion.

(i) Elementary symmetric functions:

$$E(t) = \sum_{k=0}^{\infty} \mathcal{E}_k t^k = \prod_{i=1}^{\infty}(1 + x_i t). \tag{3.90}$$

For example, this yields $\mathcal{E}_0 = 1$, $\mathcal{E}_1 = x_1 + x_2 + \ldots$, and $\mathcal{E}_2 = x_1x_2 + x_1x_3 + \ldots$ To get $\mathcal{E}_2(x_1, x_2)$, we have to set $x_3 = x_4 = \ldots = 0$. Then, $\mathcal{E}_2(x_1, x_2) = x_1x_2$, and so on.

(ii) Completely symmetric functions:

$$C(t) = \sum_{k=0}^{\infty} \mathcal{C}_k t^k = \prod_{i=1}^{\infty} \frac{1}{1 - x_i t}. \tag{3.91}$$

(iii) Symmetric power functions:

$$P(t) = \sum_{r=1}^{\infty} \mathcal{P}_r t^{r-1} = \frac{d}{dt} \ln C(t). \tag{3.92}$$

(iv) Symmetric Schur polynomials (Cauchy's formula):

$$\prod_{i,j=1}^{n} \frac{1}{1 - x_i x_j} = 1 + \sum_{k=1}^{\infty} \sum_{|\lambda|=k} \mathcal{S}_\lambda(x)\, \mathcal{S}_\lambda(y). \tag{3.93}$$

Concerning (3.93), we fix $n = 1, 2, \ldots$, and we set $x := (x_1, \ldots, x_n)$, as well as $y := (y_1, \ldots, y_n)$. Furthermore, we choose the n-tuple $\lambda = (\lambda_1, \ldots, \lambda_n)$ of integers with $\lambda_1 \geq \lambda_2 \geq \ldots \geq \lambda_1 \geq 0$, and $|\lambda| := \lambda_1 + \ldots + \lambda_n$.

(v) The Bell polynomials $B_{n,k}$: The function $\exp\left(z \sum_{m=1}^{\infty} x_m \frac{t^m}{m!}\right)$ is equal to the sum

$$1 + \sum_{n=1}^{\infty} \sum_{k=1}^{n} z^k B_{n,k}(x_1, x_2, \ldots, x_{n+1-k}). \tag{3.94}$$

This is discussed in Sect. 3.4.3 of Vol. IV together with applications. We also refer to L. Comtet, Advanced Combinatorics, Reidel, Dordrecht, 1974.

3.19.1 The Fundamental Frobenius Character Formula

Using symmetric Schur polynomials as a basis, products of power sums are the generating functions for the characters of the symmetric group *Sym*(n).

Folklore

The Frobenius character formula tells us that the product

$$\prod_{r=1}^{n} (x_1^r + x_2^r + \ldots + x_n^r)^{m(r)}$$

of symmetric power polynomials is equal to the sum

$$\sum_{\mathcal{Y}} \chi_{\lambda_{\mathcal{Y}}} (1^{m(1)} 2^{m(2)} \cdots n^{m(n)}) \cdot \mathcal{S}_{\lambda_{\mathcal{Y}}}(x_1, \ldots, x_n) \tag{3.95}$$

$2 = 1 + 1$ $\qquad$ $2 = 2$

(a) $\lambda_{\mathcal{F}} = (1, 1)$ $\qquad$ (b) $\lambda_{\mathcal{F}} = (2, 0)$

Fig. 3.11. The frame tuples $\lambda_{\mathcal{F}}$ of the symmetric group $Sym(2)$

of symmetric Schur polynomials where the uniquely determined coefficients $\chi_{\lambda_{\mathcal{Y}}}(\ldots)$ are the characters of a complete system of irreducible representations of the symmetric group $Sym(n)$. Let us discuss this.

(i) Partitions: Fix $n = 2, 3, \ldots$ Choose a partition $n = n_1 + \ldots + n_k$ with

$$n_1 \geq n_2 \geq \ldots \geq n_k \geq 1.$$

Construct the corresponding Young frame $\mathcal{Y}$ and the corresponding frame n-tuple

$$\lambda_{\mathcal{Y}} := (n_1, n_2, \ldots, n_k, 0, \ldots, 0).$$

Here, in order to get an n-tuple, we fill in zeros if needed (see Figs. 3.11 and 3.12). In (3.95), we sum over all partitions (i.e., Young frames) $\mathcal{Y}$ of n. Recall that the partitions $\mathcal{Y}$ of n are in one-to-one correspondence to a complete system of irreducible representations of the symmetric group $Sym(n)$. Let $\chi_{\lambda_{\mathcal{Y}}}$ denote the character of the irreducible representation corresponding to the Young frame $\mathcal{Y}$.

(ii) Group elements and cycles: We want to compute the value $\chi_{\lambda_{\mathcal{Y}}}(\pi)$ where $\pi \in Sym(n)$. To this end, we determine the cycle symbol $1^{m(1)}2^{m(2)} \cdots n^{m(n)}$ of π (see (3.2) on page 182).

(iii) Characters: Compute $\chi_{\lambda_{\mathcal{Y}}}(1^{m(1)}2^{m(2)} \cdots n^{m(n)})$ by the Frobenius formula (3.95). Then

$$\chi_{\lambda_{\mathcal{Y}}}(\pi) = \chi_{\lambda_{\mathcal{Y}}}(1^{m(1)}2^{m(2)} \cdots n^{m(n)}).$$

In particular, $\chi_{\lambda_{\mathcal{Y}}}(\pi)$ only depends on the cycle symbol of π. This reflects the fact that the character is a class function, that is, $\chi_{\lambda_{\mathcal{Y}}}(\pi) = \chi_{\lambda_{\mathcal{Y}}}(\sigma)$ if π and σ are elements of the same conjugacy class of $Sym(n)$.

Example ($Sym(2)$). The group $Sym(2)$ has the elements (1) and (12) with the disjoint cycle products

$3 = 1 + 1 + 1$ $\qquad$ $3 = 2 + 1$ $\qquad$ $3 = 3$

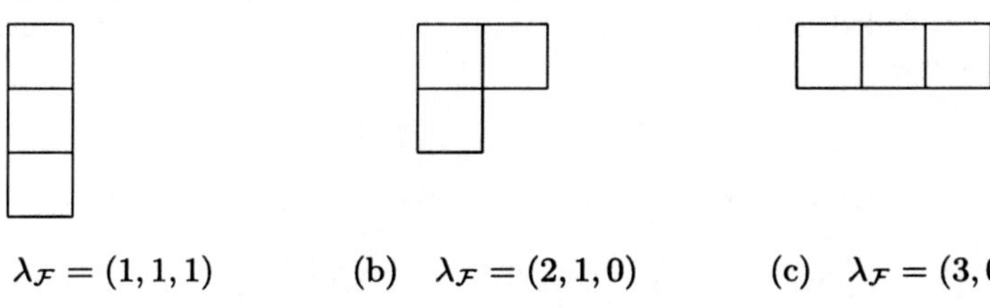

(a) $\lambda_{\mathcal{F}} = (1, 1, 1)$ $\qquad$ (b) $\lambda_{\mathcal{F}} = (2, 1, 0)$ $\qquad$ (c) $\lambda_{\mathcal{F}} = (3, 0, 0)$

Fig. 3.12. The frame tuples $\lambda_{\mathcal{F}}$ of the symmetric group $Sym(3)$

$$(1) = (1)(2), \qquad (12) = (12).$$

This yields the cycle symbols $(1) = 1^2 2^0$ and $(12) = 1^0 2^1$. By Fig. 3.11, there are two Young frames which we label by the tuples $(1,1)$ and $(2,0)$. By the Frobenius formula (3.95), we get

- $(x_1 + x_2)^2 = \chi_{(1,1)}(1^2 2^0)\, \mathcal{S}_{(1,1)} + \chi_{(2,0)}(1^2 2^0)\, \mathcal{S}_{(2,0)}$,
- $x_1^2 + x_2^2 = \chi_{(1,1)}(1^0 2^1)\, \mathcal{S}_{(1,1)} + \chi_{(2,0)}(1^0 2^1)\, \mathcal{S}_{(2,0)}$.

Explicitly, we have $\mathcal{S}_{(1,1)} = x_1 x_2$ and $\mathcal{S}_{(2,0)} = x_1^2 + x_1 x_2 + x_2^2$, by Table 3.1 on page 185. Hence

$$\chi_{(1,1)}(1^2 2^0) = \chi_{(2,0)}(1^2 2^0) = 1, \qquad \chi_{(1,1)}(1^0 2^1) = -1,\ \chi_{(2,0)}(1^0 2^1) = 1.$$

This implies

$$\chi_{(2,0)}(\,(1)\,) = \chi_{(2,0)}(\,(12)\,) = 1, \qquad \chi_{(1,1)}(\,(1)\,) = 1,\ \chi_{(1,1)}(\,(12)\,) = -1,$$

which coincides with Table 3.3 on page 215.

3.19.2 The Pfaffian

For all complex numbers a, we have $\det \begin{pmatrix} 0 & a \\ -a & 0 \end{pmatrix} = a^2$. The point is that the determinant is a square. This result can be generalized. Let $n = 2, 4, 6, \ldots$ If A is a complex skew-symmetric $(n \times n)$-matrix, that is, $A^d = -A$, then

$$\det A = (\mathrm{Pf}(A))^2$$

where $\mathrm{Pf}(A)$ is a polynomial in the entries of the matrix A.[33] If B is an arbitrary complex $(n \times n)$-matrix, then

$$\mathrm{Pf}(BAB^d) = (\det B) \cdot \mathrm{Pf}(A).$$

The proof can be found in K. Spindler, Abstract Algebra and Applications, Vol. 1, p. 342, Marcel Dekker, New York, 1994.[34] The Pfaffian plays a crucial role in the formulation of the famous Gauss–Bonnet–Chern theorem (see Vol. IV).

3.20 Frobenius Algebras and Frobenius Manifolds

Frobenius algebra. The theory of Frobenius algebras was started by Brauer (1901–1977) in the early 1930s. A finite-dimensional complex associative unital algebra $\mathcal{A}$ is called a Frobenius algebra iff there exists a non-degenerate bilinear map $B : \mathcal{A} \times \mathcal{A} \to \mathbb{C}$ with

$$B(ab, c) = B(a, bc) \qquad \text{for all} \quad a, b, c \in \mathcal{A}.$$

Examples. (i) The set of complex numbers $\mathbb{C}$ is a Frobenius algebra by setting

$$B(a, b) := ab, \qquad a, b \in \mathbb{C}.$$

[33] If n is odd, then $\det A = 0$.

[34] Pfaff (1765–1825) discovered the genius of the young Gauss (1777–1855).

In fact, we have the associative law $(ab)c = a(bc)$.

(ii) The group algebra $\mathbb{C}[\mathcal{G}]$ of a finite group $\mathcal{G}$ is a Frobenius algebra. To this end, we define

$$B\left(\sum_{j=1}^{m} \alpha_j G_j\right) := \alpha_1 \qquad \text{for all} \quad \alpha_1, \ldots, \alpha_m \in \mathbb{C}$$

where G_1 is the unit element of $\mathcal{G}$.

Duality. In terms of representation theory, Frobenius algebras are distinguished by the following duality property. Let $\mathcal{A}$ be a finite-dimensional complex associative unital algebra. Then $\mathcal{A}$ is a Frobenius algebra iff the following two representations of $\mathcal{A}$ are equivalent:

(i) The regular representation on $\mathcal{A}$: $\varrho(a)b := ab$ for all $b \in \mathcal{A}$ (and all $a \in \mathcal{A}$).

(ii) The dual regular representation: $(\varrho_*(a)f)(b) := f(ba)$ for all $f \in \mathcal{A}^d$ (and all $a, b \in \mathcal{A}$). [35]

The proof can be found in Curtis and Reiner, Representation Theory of Finite Groups and Associative Algebras, Wiley, 1962, Chapter IX.

Frobenius manifold. By definition, a Frobenius manifold is an n-dimensional complex manifold $\mathcal{M}$ such that, for all points $P \in \mathcal{M}$, the tangent space T_P is equipped with both the structure of a complex Hilbert space and the structure of a Frobenius algebra such that the inner product on $T_P\mathcal{M}$ has the additional product property

$$\langle \mathbf{u} \cdot \mathbf{v} | \mathbf{w} \rangle_P = \langle \mathbf{u} | \mathbf{v} \cdot \mathbf{w} \rangle_P$$

for all tangent vectors $\mathbf{u}, \mathbf{v}, \mathbf{w} \in T_P\mathcal{M}$. Here, $\mathbf{u} \cdot \mathbf{v}$ denotes the product of the Frobenius algebra $T_P\mathcal{M}$. In order to get strong mathematical results, one postulates additional properties of the covariant derivative with respect to the Levi-Civita connection (e.g., if $\mathbf{1}_P$ denotes the unit element of the Frobenius algebra $T_P\mathcal{M}$, then the covariant derivative of the tangent vector field $P \mapsto \mathbf{1}_P$ vanishes on $\mathcal{M}$, $\nabla \mathbf{1} = 0$.)

Further reading. We recommend the basic paper by

B. Dubrovin, Geometry of Two-Dimensional Field Theories, pp. 120–348. In: Donagi et al. (Eds.), Integrable Systems and Quantum Groups, Springer, Berlin, 1993

together with the following monographs:

Yu. Manin, Frobenius Manifolds, Quantum Cohomology, and Moduli Spaces, Amer. Math. Soc., Providence, Rhode Island, 1999.

C. Hertling, Frobenius Manifolds and Moduli Spaces for Singularities, Cambridge University Press, 2002.

J. Kock, Frobenius Algebras and Two-Dimensional Topological Quantum Field Theories, Cambridge University Press, 2003.

3.21 Historical Remarks

Symmetry and hence the theory of Lie groups and Lie algebras lie at the heart of modern mathematics and physics.

[35] Recall that the dual algebra $\mathcal{A}^d$ consists of all the linear functions $f : \mathcal{A} \to \mathbb{C}$ which are also multiplicative, that is, $f(ab) = f(a)f(b)$ for all $a, b \in \mathcal{A}$.

Celestial mechanics and quantum mechanics. Symmetry played a fundamental role in the history of celestial mechanics. The rotational symmetry of the gravitational field of the sun (i.e., the $SO(3)$-symmetry) implies conservation of angular momentum, and hence the planets move in a plane. This was known to Newton (1643–1727). Lagrange (1736–1813) used symmetry properties in order to find special solutions of the 3-body problem.[36] For the motion of planets in terms of classical mechanics, there exists an additional conservation law which says that the so-called Laplace vector is fixed. This is responsible for the crucial fact that, in classical mechanics, the great semi-axis of a planet does not rotate.

> *This corresponds to a hidden $SO(4)$-symmetry in the classical motion of planets.*

In contrast to the classical approach, the 1916 Schwarzschild solution in Einstein's theory of general relativity predicts mathematically that the great semi-axis of the planet Mercury rotates with the angle of 43 seconds per 100 years; this is indeed observed by astronomers (see Vol. IV).

In 1924, Lenz (1888–1957) used the Laplace vector in order to study the spectrum of the hydrogen atom in the setting of the semiclassical Bohr–Sommerfeld quantization from 1913/1916.[37] In 1925, Pauli (1900–1958) was assistant of Lenz (1988–1957) in Hamburg. Pauli used the $SO(4)$-symmetry in order to compute, in a purely algebraic way, the discrete spectrum of the hydrogen atom based on the 1925 matrix quantum mechanics due to Heisenberg, Born, and Jordan. As an essential ingredient, Pauli replaced the Poisson brackets of classical mechanics by the Lie brackets of the Lie algebra $so(4)$ of the Lie group $SO(4)$. Shortly after Pauli, Schrödinger (1887–1961) published his partial differential equation (Schrödinger equation) in 1926; following Weyl's advice, Schrödinger used the theory of singular differential operators in order to compute both the discrete and the continuous spectrum of the hydrogen atom in quantum mechanics. We will study the approach to the spectrum of the hydrogen atom due to Pauli, Schrödinger, and Weyl in Vol. IV on quantum mathematics. There we will also discuss the relation to the work of von Neumann (spectral theory of self-adjoint operators in Hilbert spaces), Kodaira (the Weyl–Kodaira theory), and Gelfand (generalized eigenfunctions and the Gelfand triplet).

Crystals and Lie algebras. In 1830, the mineralogist Hessel (1796–1872) classified the crystals. He found that there are 32 crystallographic classes. All of them are realized by crystals in nature. In 1890, based on a correspondence by letter, Fedorov (1853–1919) and Schoenflies (1853–1928) proved that there are 230 crystal groups, up to equivalence. Quasicrystals were mathematically predicted by Penrose (born 1931) in 1974; they were experimentally established in 1984. Quasicrystals contain an ordered structure, but the patterns are subtle.[38]

---

[36] See R. Abraham and J. Marsden, Foundations of Mechanics, Addison-Wesley, Reading, Massachusetts, 1978.
W. Neutsch and K. Scherer, Celestial Mechanics: An Introduction to Classical and Contemporary Methods, Wissenschaftsverlag, Mannheim, 1992.
D. Boccaletti and G. Pucacco, Theory of Orbits, Vols. 1, 2, Springer, Berlin, 1996.

[37] In 1922, Runge (1856–1927) used the Laplace vector in the numerical computation of planetary orbits. Therefore, the Laplace vector is also called the Laplace–Runge–Lenz vector. Note that one has to distinguish between the physicists Heinrich Lenz (1804–1865) (the Lenz rule in electromagnetism) and Wilhelm Lenz (1888–1957).

[38] See S. Novikov and T. Taimanov, Geometric Structures and Fields, Chap. 6, Amer. Math. Soc., Providence, Rhode Island, 2006.

The theory of Lie groups and Lie algebras was created by Sophus Lie (1842–1899). He studied symmetries which depend smoothly on a finite number of real or complex parameters. Lie discovered the fundamental fact that

> *The local behavior of a Lie group is completely determined by the linearization of the group at the unit element (the Lie algebra).*

The semisimple Lie algebras were classified by Killing (1847–1923) in 1888 and by Élie Cartan (1869–1951) in his famous 1894 thesis in Paris. Interestingly enough, the theory of semisimple Lie algebras is closely related to the theory of crystallic groups. The sketch words are: abstract root system, Weyl group, Coxeter group, Dynkin diagram. In 1941, Witt proved that there is a one-to-one correspondence between appropriate geometric root systems and semisimple Lie algebras. In 1944, Dynkin introduced Dynkin diagrams for classifying geometric root systems.[39] This will be studied in Vol. IV on quantum mathematics. The point is that

> *There exists a perfect correspondence between completely different mathematical structures, namely, semisimple Lie algebras and a discrete geometric structure (root system, Coexter group, Dynkin diagram).*

Such surprising correspondences also appear in

- the 'equivalent' description of 4-manifolds by the Yang–Mills equations (Donaldson theory) and the Seiberg–Witten equations (Seiberg–Witten theory), and
- in modern string theory (mirror symmetry, duality between strong and weak interaction).

It is a sophisticated task for the future of mathematics to better understand the mathematical core behind such correspondences.

Seminal contributions to the theory of Lie groups and Lie algebras were made by Weyl (1885–1955) in the 1920s and 1930s. This culminated in his monograph

> H. Weyl, The Classical Groups: Their Invariants and Representations, Princeton University Press, 1938 (8th edition, 1973).

Weyl created modern harmonic analysis as a generalization of the classic Fourier analysis (based on the translation group) to more general symmetries. The global theory of Lie groups based on topology and the theory of algebraic groups – Lie groups and Lie algebras over general fields and skew-fields (e.g., finite fields or quaternions) – was created in the 20th century. One of the heroes was Chevalley (1909–1984) who wrote the first monograph on the global theory of Lie groups:

> C. Chevalley, Theory of Lie Groups, Princeton University Press, 1946 (15th edition, 1999).

The modern theory of Lie groups and Lie algebras can be found in the monumental treatise:

> N. Bourbaki, Lie Groups and Lie Algebras, Vols. 1, 2, Springer, New York 1989/2002.

The electron spin. In 1927, Pauli used implicitly the Lie algebra $su(2)$ in order to describe mathematically the electron spin (see Sect. 7.3):

> W. Pauli, On the spinning electron in a magnetic field, Z. Phys. **43** (1927), 603–623 (in German).

---

[39] Coexter (1907–2003), Witt (1911–1991), Dynkin (born 1924).

The crucial point is that the rotation group $SO(3)$ and its universal covering group $SU(2)$ have isomorphic Lie algebras; but only the $SU(2)$-symmetry is responsible for the electron spin (see Sect. 7.3). Dirac (1902–1984) discovered in 1928 that the existence of the electron spin is a consequence of Einstein's theory of special relativity combined with quantum mechanics (see Sect. 20.3):

P. Dirac, The quantum theory of the electron, Proc. Royal Soc. London **A117** (1928), 610–624; **A118**, 351–361.

The spectra of atoms and molecules. In the late 1920s, group theory was used in order to understand the structure of the spectra of atoms and molecules in terms of quantum mechanics. Here, a crucial role was played by the electron spin and Pauli's exclusion principle.[40] As an introduction, we recommend:

B. van der Waerden, Group Theory and Quantum Mechanics, Springer, New York 1974 (German edition, 1932).

Furthermore, we refer to:

H. Weyl, The Theory of Groups and Quantum Mechanics, Dover, New York, 1931 (German edition, 1929).

E. Wigner, Group Theory and its Applications to the Quantum Mechanics of Atomic Spectra, Academic Press, New York, 1959 (German edition, 1931).

G. Drake (Ed.), Springer Handbook of Atomic, Molecular, and Optical Physics, Springer, Berlin, 2005.

Heisenberg's isospin. In 1932, motivated by Pauli's 1927 spin theory, Heisenberg (1901–1976) considered the proton and the neutron as two so-called isospin states of one particle called nucleon, and he assigned the isospin $\frac{1}{2}$ (resp. $-\frac{1}{2}$) to the proton (resp. neutron). The proton p and the neutron n have similar masses. Explicitly,

$$m_p = 938 \text{ MeV}/c^2, \qquad m_n = 940 \text{ MeV}/c^2.$$

Heisenberg assumed that this mass difference is caused by electromagnetic interaction; the neutron has no electric charge, whereas the proton has the positive electric charge e. In terms of mathematics, the isospin is based on the Lie group $SU(2)$ and its Lie algebra $su(2)$.

The quark model. In the 1950s, physicists noticed that the idea of the isospin can be refined. Based on scattering experiments, physicists assigned further quantum numbers to elementary particles like strangeness or hypercharge (see Sect. 2.4 of Vol. I). This way, elementary particles can be grouped into multiplets. Fig 3.3 on page 228 shows the octet of the proton. Based on the inspection of particle multiplets, in 1961 Gell–Mann (born 1929) and Ne'eman (1925–2006) independently emphasized the importance of the symmetry group $SU(3)$ in strong interaction.[41] As hypothetical particles, quarks were introduced independently by Gell-Mann and Zweig in 1964. Gell-Mann and Zweig (born 1937) postulated that

Baryons consist of quarks, whereas mesons are quark-antiquark pairs.

---

[40] In 1945, Pauli (1900–1958) was awarded the Nobel prize in physics for the 'exclusion principle', also called the Pauli principle.

[41] M. Gell-Mann and Y. Neéman, The Eightfold Way, Benjamin, New York, 1964. The name 'Eightfold Way' was suggested by analogy with the Eightfold Path of Buddhism because of the frequent occurrence of 8-multiplets (e.g., the octet of the proton).

They used the fact that the additive quantum numbers $\mathcal{T}^3$ and $\mathcal{Y}$ from Fig. 3.5 on page 233 explain the quantum numbers $\mathcal{T}^3$ and $\mathcal{Y}$ from Fig. 3.3 on page 228 if one assumes that the particles consist of three quarks. For example, if we assume that the proton consists of two u-quarks and one d quark, then we get $\mathcal{Y} = \frac{2}{3} + \frac{1}{3} = 1$ for the proton. The electric charge Q of the particles is given by the key formula

$$\boxed{Q = e\left(\mathcal{T}^3 + \frac{\mathcal{Y}}{2}\right), \qquad e > 0.}$$

For example, the three quarks u, d, s have the electric charges $\frac{2}{3}e, -\frac{1}{3}e, -\frac{1}{3}e$, respectively. Further details can be found in Sect. 2.6.2 of Vol. I.

The term 'quark' was coined by Gell-Mann. He used the name of ghostly beings from the novel *Finnegan's Wake* by James Joyce.[42] The great success of the 'Eightfold Way' was the prediction of the particle Ω^- (by complementing the symmetry of the baryon octet from Fig. 3.4 on page 229) and the experimental discovery of the particle Ω^- at the Brookhaven National Laboratory, New York, in 1964. In the late 1960s, scattering experiments between electrons and protons were carried out at the linear accelerator SLAC of the Stanford University in California. These experiments revealed an internal structure of the proton. To explain this, Feynman (1918–1988) created his parton model of the proton. In fact, Feynman's partons coincide with Gell–Mann's ghostly quarks. This way the existence of quarks was experimentally established. In 1969 Murray Gell-Mann (born 1929) was awarded the Nobel prize in physics for his contributions and discoveries concerning the classification of elementary particles and their interactions. The Russian chemist Mendeleev (1834–1907) discovered periodicity properties of the chemical elements presented in his famous periodic table (see Table 2.5 of Vol. I). In the late 1920s, the periodic table was justified by quantum mechanics (Pauli's exclusion principle and the $SU(2)$-symmetry of the electron spin).

Gell-Mann is called the Mendeleev of the 20th century.

Many years ago, Murray Gell-Mann told the author in a personal conversation that he invented his quark approach without knowing the representation theory of Lie algebras. He discovered the necessary mathematics by himself, motivated by physical intuition. Later on in Paris, Murray Gell-Mann told Jean-Piere Serre his approach, and he learned from Serre that there exists a beautiful mathematical theory strongly influenced by the work of Élie Cartan and Hermann Weyl.[43] Yuval Ne'eman, a graduate student of Abdus Salam (1926–1996) in the late 1950s, writes:[44]

> The mathematical formulation of the $SU(3)$-model drew on a (then) little-known branch of the theory of groups, a mathematical theory whose previous applications in physics had dealt with the symmetries of crystals. This special branch had been developed in the nineteenth century by the Norwegian mathematician Sophus Lie (1842–1899). The 'Lie groups' which

[42] The Irish novelist James Joyce (1882–1941) is noted for his experimental use of language and exploration of new literally methods in such large works of fiction as *Ulysses* (1922) and *Finnegans Wake* (1939). See Encyclopedia Britannica. In German the word 'Quark' means liquid cheese. In German colloquial language, the word 'Quark' is also used for 'nonsense'.

[43] Jean-Pierre Serre (born 1923) was awarded the Fields medal in 1954 and the Abel prize in 2003 for his seminal contributions to algebra and topology.

[44] Y. Ne'eman and Y. Kirsh, The Particle Hunters, Cambridge University Press, 1996 (reprinted with permission).

for about 100 years had found relatively few practical applications, became the cornerstone of the new physical theory.

A key role in the classification of complex semisimple Lie algebras is played by Dynkin diagrams (see Sect. 3.15.2). Eugene Dynkin (born 1924; Cornell University, Ithaca, New York) writes:[45]

> I was eleven when my family was exiled from Leningrad (Saint Petersburg) to Kasakhstan and I was thirteen when my father, one of millions of Stalin's victims, disappeared in the Gulag. It was almost a miracle that I was admitted (at the age of sixteen) to Moscow University. Every step in my professional career was difficult because of the fate of my father, in combination with my Jewish origin, made me permanently undesirable for the party authorities at the university. Only special efforts by A. N. Kolmogorov,[46] who put, more than once, his influence at stake, made it possible for me to progress through the graduate school to a teaching position at Moscow University ...
>
> I worked at Gelfand's seminar on Lie groups and at Kolmogorov's seminar on Markov chains. Both were important for me as a research mathematician. Gelfand requested that I review the Weyl–van der Waerden papers on semisimple Lie groups. I found them very difficult to read, and I tried to find my own ways. It came to my mind that there is a natural way to select a set of generators for a semisimple Lie algebra by using simple roots (i.e., roots which cannot be represented as a sum of positive roots). Since the angle between any two simple roots can be equal only to $90^\circ, 120^\circ, 135^\circ, 150^\circ$, a system of simple roots can be represented by a simple diagram. An article was submitted to *Matematicheskii Sbornik* in October 1944. Only a few years later, when recent literature from the West reached Moscow, I discovered that similar diagrams have been used by Coxeter for describing crystallographic groups...[47] I was flattered when Yuval Ne'eman (1925–2006) told me that his work on elementary particle physics was based on my dissertation, which he had read in one of the London libraries.[48]

Many results in the theory of Lie groups and Lie algebras can be traced back to the forgotten ingenious work by Killing (1847–1923) in the late 1880s. This is described in the following article: A. Coleman, The greatest mathematical paper of all time, Math. Intelligencer **11**(3) (1989), 29–38. For the history of quarks and quantum chromodynamics, we refer to:

H. Fritzsch, Quarks, Penguin, London, 1983.

Y. Ne'eman and Y. Kirsh, The Particle Hunters, Cambridge University Press, 1996.

M. Veltman, Facts and Mysteries in Elementary Particle Physics, World Scientific, Singapore, 2003.

Tian Yu Cao, From Current Algebra to Quantum Chromodynamics, Cambridge University Press, 2010.

---

[45] 1993 Steele Prizes, Career Award, Notices of the Amer. Math. Soc. **40** (1993), 975–977 (reprinted with permission).

[46] Kolmogorov (1903–1987), Wolf prize in mathematics 1980, Gelfand (1913–2009), Wolf prize in mathematics 1978, Coexter (1907–2003).

[47] See H. Coxeter, Generators and Relations for Discrete Groups, Springer, Berlin, 1965, and J. Humphreys, Reflection Groups and Coxeter Groups, Cambridge University Press, 1990.

[48] E. Dynkin, The structure of semisimple Lie algebras, Uspehi Mat. Nauk **2** (1947), pp. 59–127. Am. Math. Transl. **17** (1950).

Finite groups. The representation theory for finite groups was independently created by Fjodor Molin (1861–1941), Georg Frobenius (1849–1917), and William Burnside (1852–1927) in about 1900. Important contributions were made by Alfred Young (1873–1940) and Issai Schur (1875–1941). The representation theory of finite groups and algebras was completed by a fundamental paper of Emmy Noether based on her lectures given in Göttingen in 1927/28:

E. Noether, Hypercomplex quantities and representation theory, Math. Ann. **30** (1929), 641–692 (in German).

This paper had a profound influence on the development of modern algebra. Bartel Leendert van der Waerden (1903–1998) attended the lectures given by Emmy Noether (1882–1935) and Emil Artin (1998–1962) in Göttingen in the 1920s, and he used this material for writing his seminal book *Modern Algebra*, Vols 1, 2, Springer, Berlin, 1930 (in German). (English edition: Frederyck Ungar, New York, 1975).

Finite groups are special cases of compact Lie groups. For example, the rotation group $SO(3)$ of the 3-dimensional Euclidean space or the gauge groups

$$U(1), \quad SU(2), \quad SU(3)$$

of the Standard Model in elementary particle physics are compact Lie groups. In his famous 1894 thesis on Lie groups and Lie algebras, Élie Cartan (1869–1951) completed the classification of semi-simple Lie algebras initiated by Killing (1847–1923) in the late 1880s. The representation theory for compact Lie groups and its relation to functional analysis was created by Weyl (1885–1955) in the 1920s. A highlight is the Peter–Weyl theorem which will be investigated in Vol. IV on quantum mathematics:

F. Peter and H. Weyl, On the completeness of the irreducible representations of compact continuous groups, Math. Ann. **97** (1927), 737–755 (in German).

In the 1930s, Weyl worked together with the algebraist Brauer (1901–1977) who took over a professorship at Harvard University (Cambridge, Massachusetts) in 1951. After the establishment of the Nazi regime in Germany in 1933, Weyl – the successor of Hilbert in Göttingen – left Germany, and he became a member of the Institute for Advanced Study in Princeton (New Jersey), where also Einstein (1979–1955), Gödel (1906–1978), and von Neumann (1903–1957) worked. Motivated by Dirac's theory of the relativistic electron and Élie Cartan's geometric spinors, Brauer and Weyl wrote the fundamental paper:

R. Brauer and H. Weyl, Spinors in n dimensions, Amer. J. Math. **57** (1935), 425–449.

Here, they used Clifford algebras in order to construct the universal covering group $Spin(n)$ of the n-dimensional rotation group $SO(n)$, $n = 3, 4, \ldots$ This paper represents the algebraic core of modern spin geometry. We refer to:

H. Lawson and M. Michelsohn, Spin Geometry, Princeton University Press, 1994.

The spinor calculus on the Minkowski space (based on the representations of the symplectic group $SL(2, \mathbb{C})$) was invented by

B. van der Waerden, Spinor analysis, Ges. Wiss. Göttingen **1929**, pp. 100–109 (in German).

In 1928 Dirac proved the relativistic invariance of the Dirac equation by using the commutation relations for his γ-matrices (Clifford algebra). Motivated by Dirac's 1928 paper, van der Waerden (1903–1998) invented his spinor calculus in order to prove the relativistic invariance of the Dirac equation in the spirit of classic tensor analysis. In the paper

B. van der Waerden and L. Infeld, The wave equation of the electron in general relativity, Akad. Wiss. Berlin, Math.-Phys. Klasse **9** (1933), pp. 308–401 (in German),

the covariant differentiation was invented in spinor calculus (via appropriate Christoffel symbols). In modern terminology, a $SL(2,\mathbb{C})$-connection (also called spin connection) was introduced on curved space-time manifolds.

Wigner's unitary representations of the Poincaré group and locally compact Lie groups. The prototype of a locally compact, but not compact Lie group, is the Poincaré group $P(1,3)$ which is the symmetry group of Einstein's theory of special relativity. In contrast to compact Lie groups, the Poincaré group $P(1,3)$ has no finite-dimensional unitary representations.

This mathematical fact indicates that infinitely many degrees of freedom are crucial for quantum fields.

The mathematical trouble in quantum field theory is mainly caused by the infinite number of degrees of freedom. In 1939, the physicist Wigner classified the infinite-dimensional unitary representations of the Poincaré group which play the key role in the classification of elementary particles:

E. Wigner, On unitary representations of the inhomogeneous Lorentz group, Ann. Math. **40** (1939), 149–204.

This marked the beginning of the representation theory for locally compact Lie groups created in the 1940s and 1950s by Gelfand (1913–2009), Naimark (1909–1978), Mackey (1916–2006), and Harish-Chandra (1923–1983). We refer to:

I. Gelfand, R. Minlos, and Ya. Shapiro, Representations of the Rotation and Lorentz Groups and Their Applications, Pergamon Press, New York, 1963.

M. Naimark, Linear Representations of the Lorentz Group, Macmillan, New York, 1964.

M. Naimark, Normed Rings, Noordhoff, Groningen, 1964.

K. Maurin, Generalized Eigenfunction Expansions and Unitary Representations of Topological Groups, Polish Scientific Publishers, Warsaw, 1968.

A. Knapp, Representation Theory of Semisimple Groups: An Overview Based on Examples, Princeton University Press, 1986.

V. Varadarajan, Geometry of Quantum Theory, Springer, New York, 2007.

We also refer to:

E. Wigner, Philosophical Reflections and Syntheses. Annotated by G. Emch, Springer, New York, 1995.

In 1963, Eugene Wigner (1902–1995) was awarded the Nobel prize in physics for his contributions to the theory of atomic nucleus and the elementary particle, particularly through the discovery and application of fundamental symmetry principles. For the history of group theory and representation theory, we recommend:

B. van der Waerden, A History of Algebra, Springer, Berlin, 1985.

A. Borel, Essays in the History of Lie Groups and Algebraic Groups, Amer. Math. Soc., Providence, Rhode Island, 2001.

H. Wußing, The Genesis of the Abstract Group Concept, Dover, New York, 2007.

For a survey on modern developments in representation theory (e.g., the Langlands program in number theory), we refer to:

V. Varadarajan, Euler through Time: A New Look at Old Themes, Amer. Math. Soc., Providence, Rhode Island, 2006.

3.22 Supersymmetry

> In mathematics one frequently distinguishes between even and odd objects (e.g. even and odd integers). This is the prototype of a $\mathbb{Z}_2$-grading. In modern physics, $\mathbb{Z}_2$-graded mathematical structures are called super structures.
>
> Folklore

3.22.1 Graduation in Nature

In elementary particle physics, we distinguish between

- fermions (particles with half-integer spin quantum number), and
- bosons (particles with integer spin quantum number).

In the Standard Model in particle physics, the basic particles (quarks, the electron, neutrinos) are fermions, whereas the messenger particles (the photon, the vector bosons W^+, W^-, Z, and the eight gluons) are bosons.[49] The hypothetical Higgs particle ($s = 0$) and the hypothetical graviton ($s = 2$) are bosons.

> *In terms of physics, supersymmetry postulates a perfect symmetry between fermions and bosons in the very early universe. In terms of mathematics, supersymmetry is based on graded Lie algebras called Lie super algebras.*

In the present universe, supersymmetry is not observed. But it is possible, that there exist relicts of supersymmetry. At the LHC (Large Hadron Collider/CERN, Geneva, Switzerland), physicists plan to perform experiments which establish the existence of supersymmetric particles.

3.22.2 General Strategy in Mathematics

> The important point about super objects is that whenever an operation (e.g., a product) changes the order of two odd elements, a minus sign is introduced.[50]
>
> Jürgen Jost, 2009

Linear super space. Let X be a real or complex linear space. We call this a linear super space iff there exist linear subspaces X_0 and X_1 such that

$$\boxed{X = X_0 \oplus X_1.}$$

The elements of X_0 are called even, and the non-zero elements of X_1 are called odd. In physics, X_0 (resp. X_1) is called the bosonic (resp. fermionic) part of X. Precisely, the elements of X_0 and X_1 are called graded. For graded elements x, the so-called additive parity is defined by setting

---

[49] Note that if s is the spin quantum number, then the following hold: For any given oriented axis, there exists a quantum state of the particle which possesses the spin $s\hbar$ in direction of the axis.

[50] J. Jost, Geometry and Physics, Springer, 2009 (reprinted with permission). We recommend this book as an introduction to supersymmetry. Furthermore, we recommend D. Freed, Five Lectures on Supersymmetry, Amer. Math. Soc., Providence, Rhode Island, 1999. Further references can be found on page 543.

$$|x| := \begin{cases} 0 & \text{if} \quad x \text{ is even,} \\ 1 & \text{if} \quad x \text{ is odd.} \end{cases}$$

The number $(-1)^{|x|}$ is called the (multiplicative) parity of x. If X and Y are real (resp. complex) linear super spaces, then a morphism

$$\mu : X \to Y$$

is a linear map which preserves the parity of graded elements.

Supercommutative super algebra. By definition, a real (resp. complex) supercommutative super algebra $\mathcal{A}$ is a real (resp. complex) algebra with the following additional properties: $\mathcal{A}$ is a linear super space, and the multiplication of graded elements respects parity. Explicitly, this means the following for all graded elements a and b:

- If a and b have the same parity, then the products ab and ba are even.
- If a and b have different parities, then the products ab and ba are odd.
- The unit element is even.
- $ab = (-1)^{|a|\cdot|b|}ba$ (supercommutativity).

Explicitly, this means that:

- $ab = ba$ if either a and b are even or a and b have different parity.
- $ab = -ba$ if a and b are odd.

Standard example (Grassmann algebra). The Grassmann algebra $\Lambda(E_3^d)$. Let α, β, γ be real numbers. Define:

- α and $\alpha dx \wedge dy + \beta dx \wedge dz + \gamma dy \wedge dz$ are even.
- The non-zero elements $\alpha dx + \beta dy + \gamma dz$ and $\alpha dx \wedge dy \wedge dz$ are odd.

Then the Grassmann algebra $\Lambda(E_3^d)$ of differential forms on the Euclidean space E_3 is a super algebra which is supercommutative.

Super Lie algebra. Let $\mathcal{A}$ be a real or complex linear super space. We are given a bilinear map $(a, b) \mapsto [a, b]$ from $\mathcal{A} \times \mathcal{A} \to \mathcal{A}$ which has the following properties for all graded elements a, b, and c:

- $[a, b] = -(-1)^{|a|\cdot|b|}[b, a]$ (superantisymmetry),
- $\alpha[a, [b, c]] + \beta[b, [c, a]] + \gamma[c, [a, b]] = 0$ (super Jacobi identity).

Here, $\alpha := (-1)^{|a|\cdot|c|}$, $\beta := (-1)^{|b|\cdot|a|}$, and $\gamma := (-1)^{|c|\cdot|b|}$.

3.22.3 The Super Lie Algebra of the Euclidean Space

It is our goal to extend the real Lie algebra $(E_3)_{\text{Lie}}$ introduced on page 83 to a real super Lie algebra $\text{Sup}(E_3)$ consisting of all the sums

$$\boxed{\mathbf{a} + Q,}$$

where $\mathbf{a} \in E_3$. The elements Q are called *supercharges*. They live in a finite-dimensional real linear space $\mathcal{Q}$ called supercharge space. By definition, an element $\mathbf{a} + Q$ of $\text{Sup}(E_3)$ is called graded iff either $Q = 0$ or $\mathbf{a} = 0$. For graded elements, we introduce a sign defined by

$$\text{sgn}\,\mathbf{a} := 0 \quad \text{and} \quad \text{sgn}\,Q := 1.$$

By defining supercharges in an appropriate way, we want to introduce a so-called super vector product

$$\boxed{(\mathbf{a}+Q)\times(\mathbf{b}+P) := \mathbf{a}\times P + Q\times\mathbf{b} + \mathbf{a}\times\mathbf{b} + Q\times P,}$$

which has the following properties for all super vectors $u, v, w \in \mathrm{Sup}(E_3)$, all vectors $\mathbf{a}, \mathbf{b}, \mathbf{c} \in E_3$, and all real numbers α, β :

(i) *Distributive laws*: $(\alpha u + \beta v) \times w = \alpha(u \times v) + \beta(v \times w)$. Similarly,

$$w \times (\alpha u + \beta v) = \alpha(w \times u) + \beta(w \times v).$$

(ii) *Graduation*: $\mathbf{a} \times \mathbf{b}$ and $Q \times P$ live in E_3, whereas $\mathbf{a} \times Q$ and $Q \times \mathbf{a}$ live in the supercharge space $\mathcal{Q}$.

(iii) *Superantisymmetry*:

$$\mathbf{a}\times\mathbf{b} = -\mathbf{b}\times\mathbf{a}, \quad \mathbf{a}\times Q = -Q\times\mathbf{a}, \quad Q\times P = P\times Q.$$

(iv) *Super Jacobi identities*:

$$\mathbf{a}\times(\mathbf{b}\times\mathbf{c}) + \mathbf{b}\times(\mathbf{c}\times\mathbf{a}) + \mathbf{c}\times(\mathbf{a}\times\mathbf{b}) = 0,$$
$$-\mathbf{a}\times(\mathbf{b}\times Q) + \mathbf{b}\times(Q\times\mathbf{a}) - Q\times(\mathbf{a}\times\mathbf{b}) = 0,$$
$$-\mathbf{a}\times(Q\times P) - Q\times(P\times\mathbf{a}) - P\times(\mathbf{a}\times Q) = 0.$$

This means that, for graded elements in $\mathrm{Sup}(E_3)$, the following hold true.[51]

(a) *Graduation* : $\mathrm{sgn}(uv) \equiv \mathrm{sgn}\, u + \mathrm{sgn}\, v \mod 2$.

(b) *Superantisymmetry*: $u \times w = (-1)^\alpha w \times u$.

(c) *Super Jacobi identity*:

$$(-1)^\alpha u\times(v\times w) + (-1)^\beta v\times(w\times u) + (-1)^\gamma w\times(u\times v) = 0.$$

Here, $\alpha := \mathrm{sgn}\ u\ \mathrm{sgn}\ w$, and $\beta := \mathrm{sgn}\ v\ \mathrm{sgn}\ u$, as well as $\gamma := \mathrm{sgn}\ w\ \mathrm{sgn}\ v$. By setting, $[u, w]_- := u \times w$ for all $u, w \in \mathrm{Sup}(E_3)$, we get a real super Lie algebra.

To construct such an algebra in a nontrivial way, let $\dim \mathcal{Q} = 2$. Choose an orthonormal basis $\mathbf{e}_1, \mathbf{e}_2, \mathbf{e}_3$ in E_3 and a basis Q_1, Q_2 in $\mathcal{Q}$. Define the products of the basis elements by

$$\begin{pmatrix} Q_1\times Q_1 & Q_1\times Q_2 \\ Q_2\times Q_1 & Q_2\times Q_2 \end{pmatrix} := \sum_{j=1}^3 \sigma^j\sigma^2\mathbf{e}_j$$

and

$$\begin{pmatrix} \mathbf{e}_j\times Q_1 \\ \mathbf{e}_j\times Q_2 \end{pmatrix} := -\frac{1}{2}\sigma^j\begin{pmatrix} Q_1 \\ Q_2 \end{pmatrix}, \quad j = 1, 2, 3.$$

Here, we use the Pauli matrices

$$\sigma^1 := \begin{pmatrix} 0 & 1 \\ 1 & 0 \end{pmatrix}, \quad \sigma^2 := \begin{pmatrix} 0 & -\mathrm{i} \\ \mathrm{i} & 0 \end{pmatrix}, \quad \sigma^3 := \begin{pmatrix} 1 & 0 \\ 0 & -1 \end{pmatrix}. \tag{3.96}$$

Hence

$$\begin{pmatrix} Q_1\times Q_1 & Q_1\times Q_2 \\ Q_2\times Q_1 & Q_2\times Q_2 \end{pmatrix} = \begin{pmatrix} \mathbf{e}_2 + \mathrm{i}\mathbf{e}_1 & -\mathrm{i}\mathbf{e}_3 \\ -\mathrm{i}\mathbf{e}_3 & \mathbf{e}_2 - \mathrm{i}\mathbf{e}_1 \end{pmatrix}.$$

An explicit computation shows that our product definition fulfills (i) through (iv) above.

[51] Recall that $a \equiv b \mod 2$ iff the difference $a - b$ is even.

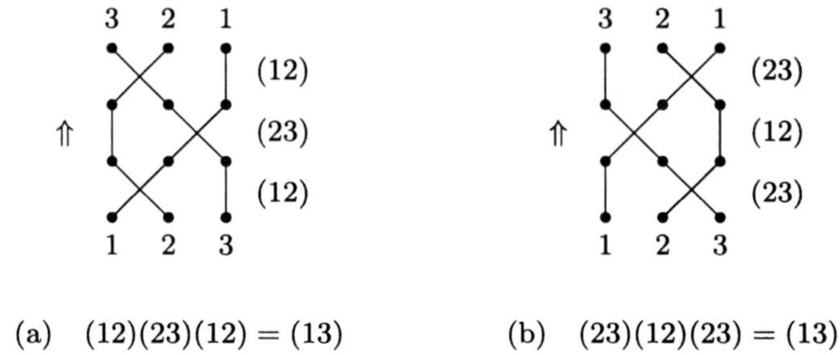

Fig. 3.13. Braid relation

3.23 Artin's Braid Group

In terms of physics, Artin's braid groups describe interactions in nature based on a special type of twisting.

Folklore

3.23.1 The Braid Relation

Consider the symmetric group $Sym(3)$ of the permutations of the numbers $1, 2, 3$. Choosing the cyclic permutations $a := (12)$ and $b := (23)$, we have the so-called braid relation

$$aba = bab. \tag{3.97}$$

This terminology is motivated by Fig. 3.13.

The free group $FG(2)$ generated by two letters. We are given two letters a, b. We add the symbols $a^{-1}, b^{-1}, \mathbf{1}$. Let $FG(2)$ denote the set of all words with the five letters $a, b, a^{-1}, b^{-1}, \mathbf{1}$. The product of words corresponds to the composition of words. Finally, we add the relations $aa^{-1} = a^{-1}a = \mathbf{1}$ and $c\mathbf{1} = \mathbf{1}c = \mathbf{1}$ if $c = a, b, \mathbf{1}$. This way, the set $FG(2)$ becomes a group of infinite order. For example,

$$aba^{-1} \cdot ab\mathbf{1} = aba^{-1}ab\mathbf{1} = ab\mathbf{1}b\mathbf{1} = abb.$$

The braid group $Braid(3)$ with three strands. If we add the braid relation

$$aba = bab$$

and the corresponding relation $a^{-1}b^{-1}a^{-1} = b^{-1}a^{-1}b^{-1}$, then the group $FG(2)$ passes over to the so-called braid group $Braid(3)$. For example,

$$abab^{-1} = babb^{-1} = ba\mathbf{1} = ba.$$

This group was introduced by Artin in 1925.[52] The maps $a \mapsto (12)$ and $b \mapsto (23)$ induce a surjective group morphism $\mu : Braid(3) \to Sym(3)$, which is not an isomorphism.

[52] E. Artin, Theory of braids, Hamburger Abh. **4** (1925), 47–72 (in German).

3.23.2 The Yang–Baxter Equation

> The Yang–Baxter equation first came up in a paper by Yang[53] as a factorization condition for the scattering S-matrix in the many-body problem in one dimension and in a work of Baxter on exactly solvable models in statistical mechanics. The Yang–Baxter equation also played an important rôle in the quantum inverse scattering method created around 1978–79 by Faddeev, Sklyanin, and Takhtadjian for the construction of quantum integrable systems. Attempts to find R-matrices (i.e., solutions of the Yang–Baxter equation) in a systematic way have led to the theory of quantum groups created by Drinfeld, Jimbo, and Woronowicz in the late 1980s ... The derivation of knot invariants from quantum groups and, more generally, from ribbon categories first appeared in papers by Reshitikin and Turaev in the early 1990s.[54]
>
> Christian Kassel, Marc Rosso, and Vladimir Turaev, 1997

> The Yang–Baxter equation $ABA = BAB$ models a controlled deviation from the commutativity relation $AB = BA$.
>
> Folklore

Let X be a linear space over $\mathbb{K} = \mathbb{R}, \mathbb{C}$. The Yang–Baxter equation reads as

$$\boxed{ABA = BAB, \qquad A, B \in GL(X).} \tag{3.98}$$

We are looking for two bijective linear operators $A, B : X \to X$ which satisfy the relation (3.98).

The flip operator F. Fix $N = 1, 2, \ldots$ Let $b_1, \ldots, b_N$ be a basis of the complex linear space Y. Define

$$F(b_i \otimes b_j) := b_j \otimes b_i, \qquad i, j = 1, \ldots, N.$$

This can be uniquely extended to the linear operator $F : Y \otimes Y \to Y \otimes Y$. Finally, set

$$A := F \otimes \mathrm{id}_Y \qquad \text{and} \qquad B := \mathrm{id}_Y \otimes F.$$

This yields the bijective linear operators $A, B : X \to X$ where $X := Y \otimes Y \otimes Y$.

Proposition 3.19 *The operators A and B are solutions of the Yang–Baxter equation* (3.98).

Proof. This is a consequence of the braid relation (3.97). Indeed,

$$ABA(b_i \otimes b_j \otimes b_k) = AB(b_j \otimes b_i \otimes b_k) = A(b_j \otimes b_k \otimes b_i) = b_k \otimes b_j \otimes b_i.$$

This coincides with $BAB(b_i \otimes b_j \otimes b_k)$. □

[53] C. Yang, Some exact results for the many-body problem in one dimension with repulsive delta-function interaction, Phys. Rev. Lett. **19** (1967), 1312–1315.
R. Baxter, Exactly Solved Models in Statistical Mechanics, Academic Press, New York, 1982.

[54] C. Kassel, M. Rosso, and V. Turaev, Quantum Groups and Knot Invariants, Société Mathématique de France, 1997 (reprinted with permission).
V. Turaev, Quantum Invariants of Knots and 3-Manifolds, de Gruyter, Berlin, 1994.

The deformed flip operator F_q. Let $q \in \mathbb{C} \setminus \{0\}$. We want to construct a one-parameter family of solutions of the Yang–Baxter equation by deforming the flip operator. To this end, we define

$$F_q(b_i \otimes b_j) := \begin{cases} qb_i \otimes b_i & \text{if} \quad i = j \\ b_j \otimes b_i & \text{if} \quad i < j \\ b_j \otimes b_i + (q - q^{-1})b_i \otimes b_j & \text{if} \quad i > j. \end{cases}$$

In the special case where $q = 1$, we obtain the flip operator F above.

Constructing the operators $A_q, B_q : X \to X$ as above by replacing the flip operator F by the deformed flip operator F_q, we obtain solutions A_q, B_q of the Yang–Baxter equation (3.98) if the deformation parameter q is a nonzero complex number.

Linear representations of the braid group. The Yang–Baxter equation can be used in order to construct linear representations of the braid group $Braid(3)$. In fact, let A, B be solutions of the Yang–Baxter equation (3.98). In a natural way, the maps $a \mapsto A$ and $b \mapsto B$ induce a linear representation

$$\varrho : Braid(3) \to GL(X)$$

of the braid group $Braid(3)$ on the linear space X. For example,

$$\varrho(aba^{-1}b) := ABA^{-1}B.$$

3.23.3 The Geometric Meaning of the Braid Group

Isotopic embeddings. The mathematical theory of knots, links, and braids is based on the notion of isotopic embedding. By definition, an embedding of the topological space X into the topological space Y is an injective continuous map

$$\chi : X \to Y$$

with the additional property that the map χ is a homeomorphism from X onto the image $\chi(X)$.[55] By definition, the two embeddings $\chi : X \to Y$ and $\mu : X \to Y$ are called isotopic iff there exists a continuous map

$$H : X \times [0,1] \to Y$$

such that $H(x, 0) = \chi(x)$ and $H(x, 1) = \mu(x)$ for all points $x \in X$, and the map

$$H(., t) : X \to Y$$

is an embedding for all times $t \in [0,1]$. Intuitively, the embedding χ is regularly deformed into the embedding μ during the time interval $[0,1]$.

The braid group $Braid(n)$ with n strands. Fix $n = 3, 4, \ldots$ Similarly, as for the group $FG(2)$ on page 290, we define the group $FG(n-1)$ which is freely generated by $n-1$ letters $a_1, \ldots, a_{n-1}$. We obtain the braid group $Braid(n)$ from $FG(n-1)$ by adding the following relations:

- $a_i a_j = a_j a_i$ for all $i, j = 1, \ldots n-1$ with $|i - j| \geq 2$.
- $a_i a_{i+1} a_i = a_{i+1} a_i a_{i+1}$ if $i = 1, \ldots, n-2$.

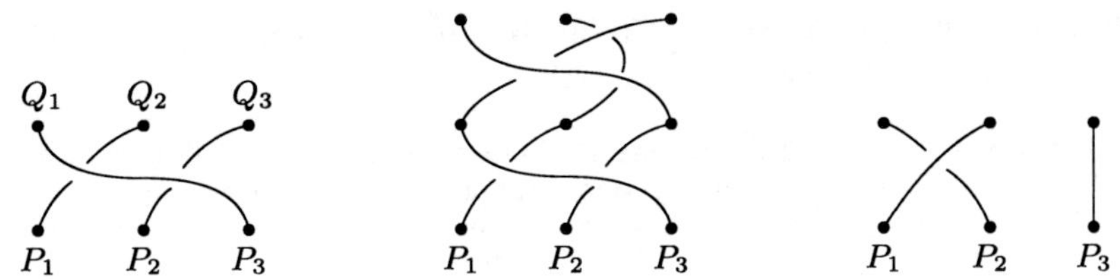

(a) (b) (c) (P_1P_2)

Fig. 3.14. Braids

In addition, let $Braid(2)$ denote the free group generated by one letter a, that is, $Braid(2) := \{\mathbf{1}, a, a^2, a^3, \ldots\}$. Finally, let us set $Braid(1) = Braid(0) := \{\mathbf{1}\}$.

Geometric interpretation of the braid group. We want to discuss Fig. 3.14. Recall that a continuous (resp. smooth) embedding

$$\chi : [0,1] \to \mathbb{E}^3$$

is called a Jordan curve (resp. smooth Jordan curve).[56] Moreover, the continuous (resp. smooth) embedding

$$\chi : \mathbb{S}^1 \to \mathbb{E}^3$$

is called a closed Jordan curve (resp. smooth closed Jordan curve). Two Jordan curves (resp. smooth Jordan curves) are called isotopic (resp. smoothly isotopic) iff the corresponding embeddings are isotopic. The same terminology is used for closed Jordan curves.

Fix the points $P_1, \ldots, P_n$ where $n = 3, 4, \ldots$ A braid is a collection of n smooth disjoint Jordan curves with starting points $P_1, \ldots, P_n$ and end points $Q_1, \ldots, Q_n$, but possibly in a different order (Fig. 3.14(a)). Multiplication of braids corresponds to composition of braids as depicted in Fig. 3.14(b). We allow deformations. More precisely, we consider the braids up to isotopy. In addition, we pass to closed Jordan curves through the given points $P_1, \ldots, P_n$, by identifying the starting points and the end points of the Jordan curves.

The group of isotopy classes of smooth closed Jordan curves is isomorphic to the braid group $Braid(n)$.

The $n-1$ generators are those braids where only *two threads* are interchanged (Fig. 3.14(c)). These $n-1$ isotopy classes of braids are denoted by $(P_1P_2), (P_2P_3), \ldots,$ and $(P_{n-1}P_n)$.

[55] If we replace the terms 'topological space', 'continuous map', and 'homeomorphism' by 'manifold', 'smooth map', and 'diffeomorphism', respectively, then we obtain the notion of a 'smooth embedding'. In the same sense, isotopic smooth embeddings are to be understood.

[56] One has to distinguish between the mathematician Camille Jordan (1838–1922) (who made important contributions to group theory and analysis, and who studied 'Jordan curves') and the physicist Pascal Jordan (1902–1980) (who made important contributions to quantum mechanics and studied 'Jordan algebras'.)

3.23.4 The Topology of the State Space of n Indistinguishable Particles in the Plane

Recall that particles in quantum physics loose their individuality; they are indistinguishable. This fact is crucial for quantum statistics (Pauli principle). We want to study the topology of the corresponding state space. To this end, we define the set

$$\mathbb{C}_*^n := \{(z_1, \ldots, z_n) : z_i \neq z_j \text{ if } i \neq j\}, \qquad n = 3, 4, \ldots$$

This is called the state space of n distinguishable particles. The symmetric group $Sym(n)$ acts on the topological space $\mathbb{C}_*^n$. We write

$$(z_1, \ldots, z_n) \sim (z_{i_1}, \ldots z_{i_n})$$

iff $i_1, \ldots i_n$ is a permutation of $1, \ldots, n$. The set $\mathbb{C}_*^n/Sym(n)$ of all the corresponding equivalence classes becomes a topological space equipped with the quotient topology. Recall that Poincaré's fundamental group $\pi_1(X)$ of a topological space X measures the connectivity of X (see Sect. 4.4.5 of Vol. II). For example, the arcwise connected topological space X is simply connected iff the fundamental group is trivial, that is, $\pi_1(X) = \{0\}$. In 1925 Artin proved the following.[57]

Theorem 3.20 *The fundamental group of the state space $\mathbb{C}_*^n/Sym(n)$ is isomorphic to the braid group with n strands: $\pi_1(\mathbb{C}_*^n/Sym(n)) \simeq Braid(n)$.*

Physical interpretation. We consider n particles in the Gaussian plane $\mathbb{C}$ described by the coordinates $z_1, \ldots, z_n \in \mathbb{C}$. We assume that there are no collisions, and the particles are *indistinguishable.* Then the equivalence class $[(z_1, \ldots, z_n)]$ describes a state of the n particles. Artin's theorem shows that the state space has a complicated topological structure.

Generalization. In 1994, Fulton and McPherson proved a far-reaching generalization of Artin's theorem concerning the homotopy of the state space $\mathbb{C}_*^n$ of n *distinguishable* particles and the generalization to algebraic varieties.

> W. Fulton and R. MacPherson, A compactification of configuration spaces, Ann. of Math. **139** (1994), 183–225.

The introduction of this paper begins as follows:[58]

> The aim of this article is to describe and study a natural compactification of the configuration space of n distinct labeled points in a nonsingular algebraic variety X. We give an explicit description of the degenerate configurations added in the compactification, and we give presentations of the cohomology ring of the compactifications and all strata at infinity. As an application, when X is compact, we determine the rational homotopy type of the configuration spaces in terms of invariants of X, a problem with a long history in topology.

This new result from algebraic geometry was used by Hollands in order to master the singularities of operator product expansions in curved space-time:

> S. Hollands, The operator product expansion for perturbative quantum field theory in curved space-time, Commun. Math. Phys. **273** (2007), pp. 1–36.

---

[57] See the footnote on page 290.

[58] Reprinted with permission.

3.24 The HOMFLY Polynomials in Knot Theory

> Given a projection of a knot, it is possible to decide in finitely many steps if it is equivalent to an unknot. This question was answered affirmatively by Wolfgang Haken in 1961.[59] He proposed an algorithm ... However, because of its complexity it has not implemented on a computer even after 40 years. We would like to add that in 1974 Haken and Appel solved the famous Four-Color problem for planar maps (posed by Francis Guthrie in 1852) by making essential use of a computer program to study the thousands of cases that needed to be checked.[60]
>
> Kishore Marathe, 2001

We want to discuss the HOMFLY polynomials which contain the classical Alexander polynomials from 1928 and the Jones polynomials from 1985 as special cases. The name HOMFLY refers to the names of the authors of the 1985 basic paper about these polynomials.[61]

Intuitively, a knot is a closed curve in the 3-dimensional Euclidean manifold without self-intersections; two knots are called equivalent iff they can be deformed into each other in the 3-dimensional Euclidean manifold $\mathbb{E}^3$, by avoiding self-intersections during the deformation process. More precisely, by definition, a knot is a smooth embedding

$$\boxed{f : \mathbb{S}^1 \to \mathbb{E}^3} \tag{3.99}$$

of the unit circle $\mathbb{S}^1$ into $\mathbb{E}^3$, that is, the image $f(\mathbb{S}^1)$ of the smooth map f is a submanifold of $\mathbb{E}^3$, and the induced map

$$f : \mathbb{S}^1 \to f(\mathbb{S}^1)$$

is a diffeomorphism. Two knots $f, g : \mathbb{S}^1 \to \mathbb{E}^3$ are called equivalent (or ambient isotopic) iff there exists an orientation-preserving diffeomorphism

$$F : \mathbb{E}^3 \to \mathbb{E}^3$$

which maps the set $f(\mathbb{S}^1)$ onto the set $g(\mathbb{S}^1)$. The knot is called an unknot iff it is equivalent to the unit circle.

A link is a finite collection of pairwise disjoint knots. Note that knots are special links. Graphically, links are represented by projections onto a fixed plane with crossings marked as over and under (Fig. 3.15).

Compactification of the three-dimensional Euclidean manifold. To simplify the mathematical situation, we will compactify the Euclidean manifold $\mathbb{E}^3$. In complex function theory, one compactifies the Gaussian plane $\mathbb{C}$ by the Riemann sphere $\mathbb{S}^2$ based on stereographic projection (Fig. 0.1 on page 15). Similarly, we replace the 3-dimensional Euclidean manifold $\mathbb{E}^3$ by the compact 3-dimensional

---

[59] W. Haken, Theory of normal surfaces, Acta math. **105** (1961), 245–375.
K. Appel and W. Haken, The solution of the four-color-map problem, Scientific American, September 1977, 108–121.

[60] K. Marathe, A chapter in physical mathematics: theory of knots in the sciences, pp. 873–888. In: B. Enquist and W. Schmid (Eds.), Mathematics Unlimited – 2001 and Beyond, Springer, Berlin 2001 (reprinted with permission).

[61] HOMFLY: D. Hoste, A. Ocneanu, W. Millett, P. Freyd, W. Lickorish, D. Yetter, A new polynomial invariant of knots and links, Bull. Amer. Math. Soc. **12** (1985), 239–246.

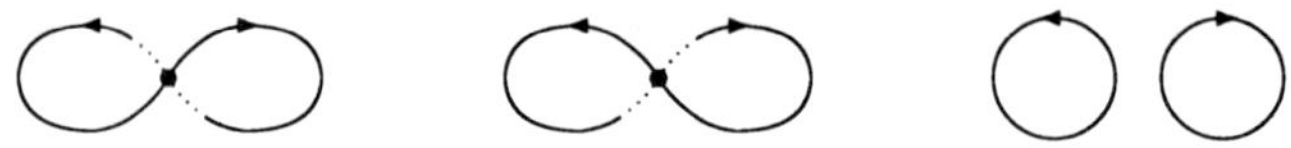

(a) L_+ (b) L_- (c) L_0

Fig. 3.15. Knots and links

unit sphere $\mathbb{S}^3$. This implies that we replace the knot map (3.99) by the smooth embedding

$$\boxed{f : \mathbb{S}^1 \to \mathbb{S}^3.}$$

We will write $K := f(\mathbb{S}^1)$. Using some motion, if necessary, we can assume that the compact knot set $f(\mathbb{S}^1)$ lies outside some open neighborhood of the North Pole N (also denoted by ∞). This means that, after stereographic projection, the knot set lies in some ball of $\mathbb{E}^3$.

Knot invariants. The main problem of knot theory is to decide whether a knot is nontrivial, that is, it is not equivalent to the unknot. To solve this problem, knot invariants are introduced. By definition, a knot invariant does not change if we pass to an equivalent knot. There exist the following knot invariants:

(i) the crossing number: this is defined to be the minimal number of crossings in any projection of the knot onto a plane;
(ii) the fundamental group $\pi_1(\mathbb{S}^3 \setminus K)$ of the knot complement $\mathbb{S}^3 \setminus K$;
(iii) the HOMFLY polynomials which yield the Alexander polynomials and the Jones polynomials.

In particular, if the HOMFLY polynomial of a knot is different from 1, then the knot is not trivial.

The recursive construction of the HOMFLY polynomials. These polynomials depend on the two variables x and y. They are defined recursively by the so-called skein relations:

$$\boxed{y^{-1}H_+(x,y) - yH_-(x,y) = xH_0(x,y).} \qquad (3.100)$$

For the unknot, we define $H := 1$. The skein relation refers to three diagrams which differ only at one crossing point as depicted in Fig. 3.16(a), (b), (c). For a given link, the idea of computing the HOMFLY polynomial is to successively change the link diagram at crossing points according to Fig. 3.16 in order to finally obtain the unknot. The corresponding skein relations then yield recursive formulas for the desired HOMFLY polynomial. For an oriented link L, we get

- the Alexander polynomial $A_L(t, t^{-1}) := H(x, 1)$ with $x := t^{1/2} - t^{-1/2}$, and
- the Jones polynomial $J_L(y) := H_L(x, y)$ with $x := y^{1/2} - y^{-1/2}$.

In particular, for the links L_+, L_-, L_0 depicted in Fig. 3.15, the skein relation with $H_+ = H_- = 1$ yields

$$y^{-1} - y = xH_0(x,y).$$

Thus, for the link L_0, we get the following polynomials:

$$H_0(x,y) = \frac{y^{-1} - y}{x}, \qquad A_0 = 0, \qquad J_0(y) = -y^{1/2} - y^{-1/2}.$$

Considering Fig. 3.15, the link L_+ (resp. L_-) is not equivalent to the link L_0, since the corresponding Jones polynomial $J_+ = 1$ (resp. $J_- = 1$) is different from J_0.

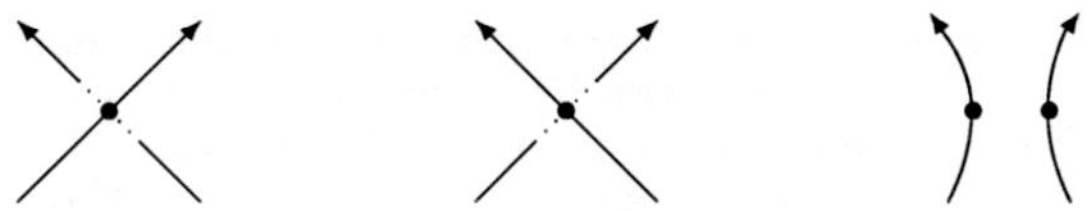

(a) L_+ (b) L_- (c) L_0

Fig. 3.16. Crossing points

Chirality of knots. A knot is called chiral iff it is not equivalent to its mirror image. If K_{mirror} is the mirror image of the knot, then the Jones polynomial of the knot has the property

$$J_{K_{\text{mirror}}}(y) = J_K(y^{-1}).$$

Observe that the Jones polynomial is not always symmetric in y and y^{-1}. Therefore, in contrast to the Alexander polynomials, the Jones polynomials frequently allow us to discover the chirality of knots. For example, the link L_0 from Fig. 3.15 is chiral. If we change the orientation of the two curves, then the Jones polynomial changes sign.

The details can be found in the modern textbook by V. Manturov, Knot Theory, Chapman & Hall, CRC, Boca Raton, Florida, 2004.

3.25 Quantum Groups

> Algebraic deformation represents a method for the quantization of Lie groups and Lie algebras. Quantum groups are not groups, but Hopf algebras. They are obtained by deforming the coordinate Hopf algebra of a group.
>
> Folklore

3.25.1 Quantum Mechanics as a Deformation

The crucial commutation relation. The mechanics of a quantum particle on the real line is governed by the Heisenberg–Born–Jordan commutation relation[62]

$$\boxed{QP - PQ = \mathrm{i}\hbar I.} \tag{3.101}$$

In the classical case, Q and P are real numbers which describe the position and the momentum of the particle, respectively. The passage from classical mechanics to quantum mechanics corresponds to the passage from $QP - PQ = 0$ to the relation (3.101) which shows that Q and P do not commute. This is an algebraic deformation which depends on the parameter $\hbar := h/2\pi$ where h is the Planck quantum of action.

Quantum physics is governed by noncommutative mathematical structures.

The deformation parameter q. In terms of physics, we set

$$q := \mathrm{e}^{h/S_0}$$

[62] For the interesting history of this relation, see page 64 of Vol. I.

where $S_0 := 1\text{Js}$ (Joule second) and $h = 6.626 \cdot 10^{-34}\text{Js}$. The quantity S_0 describes the typical value of the action (energy times time) for processes running in daily life, whereas the Planck quantum of action h refers to the typical action of quantum processes. We have the tiny dimensionless quantity $h/S_0 = 6.626 \cdot 10^{-34}$. The limit

$$q \to 1$$

corresponds to the passage from quantum mechanics to classical mechanics. In order to model quantization in algebraic terms, let us study the passage from the commutative relation $xy = yx$ to the noncommutative relation $xy = qyx$ with $q \neq 1$.

3.25.2 Manin's Quantum Planes $\mathbb{R}_q^2$ and $\mathbb{C}_q^2$

In algebraic geometry, geometric objects are described by algebraic objects. In this connection, the coordinate algebra of a geometric object plays a crucial role.

- In classical algebraic geometry, the coordinate algebras are quotient algebras of *commutative* polynomial algebras.
- In noncommutative algebraic geometry, the coordinate algebras are quotient algebras of *noncommutative* polynomial algebras (also called generalized polynomial algebras).

We want to deform the classical Euclidean plane $\mathbb{E}^2$ by deforming

- the coordinate algebra $\mathbb{K}[x,y]/xy - yx$ of $\mathbb{E}^2$
- to the coordinate algebra $\mathbb{K}[x,y]/xy - qyx$ which is also called the quantum plane to the Euclidean plane $\mathbb{E}^2$ over the field $\mathbb{K}$. Naturally enough, this quantum plane is denoted by $\mathbb{K}_q^2$.

Let us discuss this. To begin with, we choose a fixed field $\mathbb{K}$. For example, let $\mathbb{K} = \mathbb{R}, \mathbb{C}, \mathbb{Q}$ or $\mathbb{K} = \mathbb{Z}/\text{mod}2$ (the field of real numbers, complex numbers, rational numbers, or the finite field $\{0, 1\}$ with $1 + 1 = 0$, respectively). Let $\mathbb{K}[x,y]$ denote the algebra over $\mathbb{K}$ of all the generalized polynomials

$$a + bx + cy + dxx + exy + fyx + gyy + hxxx + kxxy + lxyx + \ldots$$

with coefficients $a, b, \ldots$ in the field $\mathbb{K}$. Note that the order of the factors x and y is crucial (i.e., we have to distinguish between xy and yx). Therefore, we speak of generalized polynomials.[63] If J is a two-sided ideal in $\mathbb{K}[x,y]$, then we write

$$p(x,y) \sim q(x,y)$$

iff $p(x,y) - q(x,y) \in J$. The equivalence classes $[p]$ form an algebra over $\mathbb{K}$ denoted by $\mathbb{K}[x,y]/J$ (see Sect. 4.1.3 of Vol. II). The use of coordinate algebras in algebraic geometry is based on Hilbert's fundamental results in commutative algebra – the basis theorem, the 'Nullstellensatz' (zero theorem), and the theory of constraints (syzygies) – published in about 1890.[64]

Example 1 (Euclidean plane $\mathbb{E}^2$). By definition, the coordinate algebra $\mathcal{C}(\mathbb{E}^2, \mathbb{K})$ of $\mathbb{E}^2$ over the field $\mathbb{K}$ consists of all the generalized polynomials in $\mathbb{K}[x,y]$ together with the relation

---

[63] Polynomials are obtained from this by adding the relation $xy = yx$.

[64] See K. Spindler, Abstract Algebra with Applications, Vol. II, Marcel Dekker, 1994, and D. Eisenbud, Commutative Algebra with a View to Algebraic Geometry, Springer, New York, 1994. See also the references quoted on page 418.

$$xy - xy = 0. \tag{3.102}$$

For example, $exy + fyx = (e + f)xy$. In other words, $\mathcal{C}(\mathbb{E}^2, \mathbb{K})$ is the polynomial algebra over the field $\mathbb{K}$ with the generators x and y.

Example 2 (unit circle $\mathbb{S}^1$). The classical equation of $\mathbb{S}^1$ reads as

$$x^2 + y^2 - 1 = 0. \tag{3.103}$$

By definition, the coordinate algebra $\mathcal{C}(\mathbb{S}^1, \mathbb{K})$ of $\mathbb{S}^1$ over the field $\mathbb{K}$ consists of all the generalized polynomials in $\mathbb{K}[x, y]$ together with the relations (3.102) and (3.103). For example,

$$x + y + x^2 + xy + yx + y^2 = x + y + 2xy + 1.$$

Example 3 (Manin's quantum plane $\mathcal{C}_q(\mathbb{E}^2, \mathbb{K})$). Fix the nonzero element q of $\mathbb{K}$. By definition, the deformed algebra $\mathcal{C}_q(\mathbb{E}^2, \mathbb{K})$ over $\mathbb{K}$ consists of all the generalized polynomials in $\mathbb{K}[x, y]$ together with the relation[65]

$$\boxed{xy - qyx = 0.}$$

For example, $exy + fyx = (eq + f)yx$. If $q \neq 1$, then the 'coordinates' x and y are noncommutative quantities. This yields a noncommutative geometry by using the following general strategy:

Reformulate geometric properties in terms of the coordinate algebra

$$\mathcal{C}(\mathbb{E}^2, \mathbb{K}) = \mathbb{K}[x, y]/xy - yx$$

and replace this by the deformed algebra $\mathcal{C}_q(\mathbb{E}^2, \mathbb{K}) = \mathbb{K}[x, y]/xy - qyx$.

We briefly write $\mathbb{K}_q^2$ instead of $\mathcal{C}_q(\mathbb{E}^2, \mathbb{K})$. Mnemonically, let us use the following terminology:

- The real algebra $\mathbb{R}_q^2$ is called the real quantum plane.
- The complex algebra $\mathbb{C}_q^2$ is called the complex quantum plane.

This algebraic approach to noncommutative geometry was proposed by Manin in 1988.[66] This approach also allows us to incorporate supersymmetry in a natural manner.

Example 3 (quantum super plane $\mathrm{Sup}(\mathbb{K}_q^2)$). This is an algebra over the field $\mathbb{K}$ which consists of all the generalized polynomials in $\mathbb{K}[x, y, \xi, \eta]$ together with the following relations:

- $xy = qyx$ (bosonic relation),
- $\xi\eta = -q\eta\xi, \quad \xi^2 = 0, \quad \eta^2 = 0$ (fermionic relations), and
- $x\xi = -\xi x, x\eta = -\eta x, y\xi = -\xi y, y\eta = -\eta y$ (mixed products).[67]

---

[65] This coincides with the quotient algebra $\mathbb{K}[x, y]/J$ where J is the smallest two-sided ideal of $\mathbb{K}[x, y]$ which contains the element $xy - qyx$. This ideal is well-defined as the intersection of the ideal $\mathbb{K}[x, y]$ with all the ideals which contain the polynomial $xy - qyx$.

[66] See Yu. Manin, Quantum Groups and Non-Commutative Geometry, Centre des Recherches Mathématiques, Université de Montréal, 1988, and
Yu. Manin, Topics in Noncommutative Geometry, Princeton University Press, 1991.

[67] This means that $\mathrm{Sup}_q(\mathbb{E}^2, \mathbb{K})$ is the quotient algebra $\mathbb{K}[x, y, \xi, \eta]/J$ where J is the smallest two-sided ideal of $\mathbb{K}[x, y, \xi, \eta]$ which contains $xy - qyx$, $\xi\eta + q\eta\xi$, ξ^2, η^2, as well as $x\xi + \xi x, \quad x\eta + \eta x, \quad y\xi + \xi y, \quad y\eta + \eta y$.

The special case where $q = 1$ is called super plane. For example,

$$(\xi + \eta)^2 = (\xi + \eta)(\xi + \eta) = \xi^2 + \xi\eta + \eta\xi + \eta^2 = (1 - q)\eta\xi.$$

In what follows, we will use the definition of Hopf algebras and the typical properties of such algebras (see the detailed discussion in Chap. 3 of Vol. II).

3.25.3 The Coordinate Algebra of the Lie Group $SL(2, \mathbb{C})$

The Lie group $SL(2, \mathbb{C})$. By definition, $SL(2, \mathbb{C})$ consists of all complex matrices

$$A = \begin{pmatrix} A^1_1 & A^1_2 \\ A^2_1 & A^2_2 \end{pmatrix}$$

with $\det A = 1$. As we will show later on, this is the universal covering group of the orthochronous Lorentz group which plays a fundamental role in relativistic quantum physics. The inverse matrix reads as

$$A^{-1} = \begin{pmatrix} A^2_2 & -A^1_2 \\ -A^2_1 & A^1_1 \end{pmatrix}.$$

The coordinate algebra $\mathcal{C}(SL(2, \mathbb{C}))$ of the Lie group $SL(2, \mathbb{C})$. We want to describe the group $SL(2, \mathbb{C})$ in a dual way by some Hopf algebra. This will allow us to perform algebraic deformations. To this end, we fix the indices $i, j = 1, 2$, and we define

$$\boxed{c^i_j(A) := A^i_j \qquad \text{for all} \quad A \in SL(2, \mathbb{C}).}$$

This is called the ij-coordinate map $c^i_j : SL(2, \mathbb{C}) \to \mathbb{C}$ on the Lie group $SL(2, \mathbb{C})$. This map assigns to the operator $A \in SL(2, \mathbb{C})$ the complex number A^i_j.

> *In terms of physics, the coordinate map c^i_j assigns to the observable A the complex number A^i_j which can be measured in a physical experiment.*[68]

Fix c^i_j. For all $A, B \in SL(2, \mathbb{C})$, we define:

- $(\Delta c^i_j)(A, B) := c^i_j(AB)$ (coproduct Δ);
- $(Sc^i_j)(A) := c^i_j(A^{-1})$ (coinverse S);
- $\varepsilon(c^i_j) = c^i_j(I)$ (counit).

The coinverse is also called the antipode. Noting that the matrix product is given by

$$(AB)^i_j = A^i_1 B^1_j + A^i_2 B^2_j,$$

and noting that we have $(c^i_k \otimes c^k_j)(A, B) = c^i_k(A) c^k_j(B) = A^i_k B^k_j$, we get the coproduct

$$\Delta c^i_j = c^i_1 \otimes c^1_j + c^i_2 \otimes c^2_j, \qquad i, j = 1, 2,$$

the coinverse

$$\begin{pmatrix} Sc^1_1 & Sc^1_2 \\ Sc^2_1 & Sc^2_2 \end{pmatrix} = \begin{pmatrix} c^2_2 & -c^1_2 \\ -c^2_1 & c^1_1 \end{pmatrix},$$

[68] In the sense of Plato's famous cave parable from the seventh book of his "Politea", coordinate maps project abstract quantities from the 'world of ideas' into the 'real world'.

and the counit

$$\begin{pmatrix} \varepsilon(c_1^1) & \varepsilon(c_2^1) \\ \varepsilon(c_1^2) & \varepsilon(c_2^2) \end{pmatrix} = \begin{pmatrix} 1 & 0 \\ 0 & 1 \end{pmatrix}.$$

In a natural way, we extend the functions Δ, S, and ε to complex linear combinations and to products of coordinate maps. Explicitly, we define

- $\Delta(\alpha c_j^i + \beta c_l^k) := \alpha \Delta c_j^i + \beta \Delta c_l^k$, and
- $\Delta(c_j^i c_l^k) := (\Delta c_j^i)(\Delta c_l^k)$

for all complex numbers α, β and all coordinate maps. Further definitions are obtained by replacing Δ by either S or ε. By Sect. 3.5.3 of Vol. II, we get the following.

Proposition 3.21 *The polynomial algebra $\mathbb{C}[c_1^1, c_2^1, c_1^2, c_2^2]$ becomes a complex Hopf algebra with respect to Δ, S, ε introduced above.*

To simplify notation in the following, we set

$$\begin{pmatrix} a & b \\ c & d \end{pmatrix} := \begin{pmatrix} c_1^1 & c_2^1 \\ c_1^2 & c_2^2 \end{pmatrix},$$

that is, $a := c_1^1$, and so on. This yields (3.104) and (3.105) below with $q = 1$.

3.25.4 The Quantum Group $SL_q(2, \mathbb{C})$

Fix the nonzero complex number q which is called the deformation parameter. Our goal is to construct a Hopf algebra $\mathcal{C}_q(SL(2, \mathbb{C}))$ which is isomorphic to $\mathcal{C}(SL(2, \mathbb{C}))$ if $q = 1$. Traditionally, $\mathcal{C}_q(SL(2, \mathbb{C}))$ is called the q-quantum group to the Lie group $SL(2, \mathbb{C})$. Let us start with the generalized polynomial algebra $\mathbb{C}[a, b, c, d]$. The elements of this algebra are finite sums of words of the form

$$\alpha a_1 a_2 \ldots a_m, \qquad m = 1, 2, \ldots$$

where α is a complex number, and every a_j is equal to a, b, c, or d.[69] For the letters a, b, c, d, we add the following relations:

- $ab = qba, ac = qca, ad - dc = (q - q^{-1})bc,$
- $bd = qdb, bc = cb, cd = qdc.$

Note that this reduces to commutativity relations if $q = 1$. This way, we pass from the polynomial algebra $\mathbb{C}[a, b, c, d]$ to a complex algebra denoted by $\mathcal{C}_q[a, b, c, d]$. In addition, we introduce the maps Δ, S, ε by setting:

$$\begin{pmatrix} \Delta(a) & \Delta(b) \\ \Delta(c) & \Delta(d) \end{pmatrix} := \begin{pmatrix} a \otimes a + b \otimes c & a \otimes b + b \otimes d \\ c \otimes a + d \otimes c & c \otimes b + d \otimes d \end{pmatrix}, \tag{3.104}$$

$$\begin{pmatrix} S(a) & S(b) \\ S(c) & S(d) \end{pmatrix} := \begin{pmatrix} d & -q^{-1}b \\ -qc & a \end{pmatrix}, \qquad \begin{pmatrix} \varepsilon(a) & \varepsilon(b) \\ \varepsilon(c) & \varepsilon(d) \end{pmatrix} := \begin{pmatrix} 1 & 0 \\ 0 & 1 \end{pmatrix} \tag{3.105}$$

and $\det_q \begin{pmatrix} a & b \\ c & d \end{pmatrix} := ad - qbc$ (quantum determinant). Mnemonically,

[69] By convention, in the special case where $m = 0$, this word reads as α.

$$\begin{pmatrix} \Delta(a) & \Delta(b) \\ \Delta(c) & \Delta(d) \end{pmatrix} = \begin{pmatrix} a & b \\ c & d \end{pmatrix} \otimes \begin{pmatrix} a & b \\ c & d \end{pmatrix} = \begin{pmatrix} a\otimes a + b\otimes c & a\otimes b + b\otimes d \\ c\otimes a + d\otimes c & c\otimes b + d\otimes d \end{pmatrix}.$$

Theorem 3.22 *For any nonzero complex number q, the algebra $\mathcal{C}_q(SL(2,\mathbb{C}))$ becomes a complex Hopf algebra which is isomorphic to the Hopf algebra $\mathcal{C}(SL(2,\mathbb{C}))$ if $q = 1$.*

To prove this, one has to check that all the axioms for a Hopf algebra are satisfied, by elementary, but lengthy computations (see Sect. 3.3.2 of Vol. II). For example, we define

$$\Delta(ab) := \Delta(a)\Delta(b),$$

and, naturally enough, the product $\Delta(a)\Delta(b)$ is computed by using the relation $(a \otimes b)(c \otimes d) := ac \otimes bd$, and so on. We refer to Problem 3.17.

3.25.5 The Quantum Algebra $sl_q(2,\mathbb{C})$

The complex algebra $U_q sl_{\mathbb{C}}(2,\mathbb{C})$. Fix the deformation parameter q as a complex number with $q \neq 0, \pm 1$. By definition, the complex algebra $U_q sl_{\mathbb{C}}(2,\mathbb{C})$ is obtained from the generalized polynomial algebra $\mathbb{C}[A_+, A_-, C, C^{-1}]$ by adding the following relations:

$$CC^{-1} = C^{-1}C = 1, \qquad A_+A_- - A_-A_+ = \frac{C - C^{-1}}{q - q^{-1}}, \tag{3.106}$$
$$CA_+C^{-1} = q^2 A_+, \qquad CA_-C^{-1} = q^{-2}A_-.$$

The Hopf algebra structure of $U_q sl_{\mathbb{C}}(2,\mathbb{C})$. Interestingly enough, the algebra $U_q sl_{\mathbb{C}}(2,\mathbb{C})$ can be equipped with the additional structure of a Hopf algebra. It is sufficient to define the maps Δ, S, ε for the generators. We define the coproduct

- $\Delta(C) := C \otimes C, \quad \Delta(A_+) := A_+ \otimes C + 1 \otimes A_+,$
- $\Delta(A_-) := A_- \otimes 1 + C^{-1} \otimes A_-,$

the coinverse

- $S(C) := C^{-1},\ S(A_+) := -A_+C^{-1},\ S(A_-) = -CA_-,$

and the counit

- $\varepsilon(C) := 1,\ \varepsilon(A_+) = \varepsilon(A_-) := 0.$

Theorem 3.23 *This way, the algebra $U_q sl_{\mathbb{C}}(2,\mathbb{C})$ becomes a Hopf algebra.*

This complex Hopf algebra is denoted by $sl_q(2,\mathbb{C})$. Recall that this is well-defined for complex numbers q with $q \neq 0, \pm 1$. The proof of the theorem is based on elementary, but clumsy computations (see Problem 3.18).

Next we want to study how the quantum algebra $sl_q(2,\mathbb{C})$ is related to the original algebra $U sl_{\mathbb{C}}(2,\mathbb{C})$ (universal enveloping algebra of the complex Lie algebra $sl_{\mathbb{C}}(2,\mathbb{C})$). We want to show that

$$\boxed{\lim_{q\to 1} U_q sl_{\mathbb{C}}(2,\mathbb{C}) = U sl_{\mathbb{C}}(2,\mathbb{C}),} \tag{3.107}$$

in the sense of an appropriate notion of convergence. To this end, let us rigorously introduce operations in the algebra $\mathcal{A}_\infty[h]$. Setting $q = e^h$, we will study the limit $h \to 0$.

The algebra $\mathcal{A}_\infty[h]$. Let $\mathcal{A}$ be an algebra over the field $\mathbb{K}$ (e.g., a real or complex algebra). The infinite tuple

$$(a_0, a_1, a_2, \ldots)$$

with $a_j \in \mathcal{A}$ for all indices j is called a formal power series with coefficients in the algebra $\mathcal{A}$. We define linear combinations, products, differentiation, and limits in the following way:

- $\alpha(a_0, a_1, \ldots) + \beta(b_0, b_1, \ldots) := (\alpha a_0 + \beta b_0, \alpha a_1 + \beta b_1, \ldots) \quad (\alpha, \beta \in \mathbb{K})$;
- $(a_0, a_1, \ldots)(b_0, b_1, \ldots) := (a_0 b_0, \;\; a_0 b_1 + a_1 b_0, \ldots)$;
- $(a_0, a_1, a_2, \ldots)' := (a_1, 2a_2, \ldots)$;
- $\lim_{h \to 0}(a_0, a_1, \ldots) = a_0$.

This way, the set $\mathcal{A}_\infty[h]$ of all formal power series with coefficients in $\mathcal{A}$ becomes an algebra over $\mathbb{K}$.[70]

The limit $h \to 0$. We want to show that the relations on the algebra $sl_q(2, \mathbb{C})$ pass over to the relations on $Usl_{\mathbb{C}}(2, \mathbb{C})$, in the sense of the calculus for formal power series. Since the calculus respects the algebra operations, it is sufficient to investigate the limits for the relations (3.106) of the generators of $sl_q(2, \mathbb{C})$. We have to show that, as $h \to 0$, (3.106) passes over to

$$BA_\pm - A_\pm B = \pm 2A_\pm, \quad A_+A_- - A_-A_+ = B. \tag{3.108}$$

Let us start with $CA_+C^{-1} = \mathrm{e}^{2h}A_+$. We replace C by e^{hD}, and we assume that the relation

$$\mathrm{e}^{hD}A_+\mathrm{e}^{-hD} = \mathrm{e}^{2h}A_+$$

holds in the sense of the algebra $\mathcal{A}_\infty[h]$ where $\mathcal{A} := Usl_{\mathbb{C}}(2, \mathbb{C})$, and $A_+, A_-, D \in \mathcal{A}$. Note that e^{hD} stands for $1 + Dh + \frac{1}{2!}D^2h^2 + \ldots$ Differentiation of the formal power series (with respect to h) yields

$$\mathrm{e}^{hD}(DA_+ - A_+D)\mathrm{e}^{-hD} = 2\mathrm{e}^{2h}A_+.$$

Letting $h \to 0$ and setting $D = B$, we obtain $BA_+ - A_+B = 2A_+$. Similarly, from (3.106) we get the remaining relations of (3.108).

3.25.6 The Coaction of the Quantum Group $SL_q(2, \mathbb{C})$ on the Quantum Plane $\mathbb{C}_q^2$

The classical action of the Lie group $SL(2, \mathbb{C})$ on the complex plane $\mathbb{C}^2$ is given by the matrix formula

$$\begin{pmatrix} a & b \\ c & d \end{pmatrix}\begin{pmatrix} x \\ y \end{pmatrix} = \begin{pmatrix} ax + by \\ cx + dy \end{pmatrix}.$$

We want to generalize this to

- a coaction of the quantum group $SL_q(2, \mathbb{C})$ on
- the quantum plane $\mathbb{C}_q^2$.

[70] Mnemonically, we write $f(h) = a_0 + a_1 h + a_2 h^2 + \ldots$ Then the definitions above correspond to the usual operations for power series expansions. For example, $f'(h) = a_1 + 2a_2 h + \ldots$ motivates the definition of $(a_0, a_1, \ldots)' := (a_1, 2a_2, \ldots)$.

Recall that $SL_q(2,\mathbb{C})$ and $\mathbb{C}_q^2$ are complex Hopf algebras with the generators a, b, c, d and x, y, respectively. The key definition reads as

$$\begin{pmatrix} a & b \\ c & d \end{pmatrix} \otimes \begin{pmatrix} x \\ y \end{pmatrix} := \begin{pmatrix} a \otimes x + b \otimes y \\ c \otimes x + d \otimes y \end{pmatrix}.$$

This motivates to introduce the map φ given by

$$\varphi(x) := a \otimes x + b \otimes y, \qquad \varphi(y) := c \otimes x + d \otimes y.$$

The following statement can be obtained by an explicit computation.

Proposition 3.24 $\varphi(x)\varphi(y) := q\varphi(y)\varphi(x)$.

This means that the map φ respects the relation $xy = qyx$ on the quantum plane. This motivates the extension

$$\varphi(\alpha x + \beta y) := \alpha\varphi(x) + \beta\varphi(y), \qquad \varphi(xy) := \varphi(x)\varphi(y).$$

3.25.7 Noncommutative Euclidean Geometry and Quantum Symmetry

> Quantum physics enforces the study of noncommutative mathematical structures.
>
> Folklore

The passage from classical mechanics to quantum mechanics corresponds to a passage from commutative function algebras to noncommutative operator algebras for the observables (e.g., energy, momentum, angular momentum). The basic idea of noncommutative geometry is

- to replace classical geometric objects by appropriate commutative algebras and
- to deform these into noncommutative algebras which are called objects of noncommutative geometry.

Our considerations above concern the noncommutative geometry of the Euclidean plane. Let us summarize the main points:

(i) Real quantum plane $\mathbb{R}_q^2$

- The Euclidean space E_2 is a 2-dimensional real Hilbert space which is isomorphic to the 2-dimensional real Hilbert space $\mathbb{R}^2$.
- The Euclidean plane $\mathbb{E}^2$ is a 2-dimensional real manifold which is in one-to-one correspondence to $\mathbb{R}^2$.
- The real quantum plane $\mathbb{R}_q^2$ with $q \in \mathbb{R} \setminus \{0\}$ is an infinite-dimensional real Hopf algebra.
- The algebra $\mathbb{R}_1^2$ is isomorphic to the infinite-dimensional coordinate algebra of the Euclidean plane $\mathbb{E}^2$ over the field $\mathbb{R}$ of real numbers.

(ii) Complex quantum plane $\mathbb{C}_q^2$

- $\mathbb{C}^2$ is a 2-dimensional complex Hilbert space.
- The complex quantum plane $\mathbb{C}_q^2$ with $q \in \mathbb{C} \setminus \{0\}$ is an infinite-dimensional complex Hopf algebra.
- The algebra $\mathbb{C}_1^2$ is isomorphic to the infinite-dimensional complex coordinate algebra of the Euclidean plane $\mathbb{E}^2$ over the field $\mathbb{C}$ of complex numbers.

(iii) Quantum group $SL_q(2,\mathbb{C})$

- $SL(2,\mathbb{C})$ is a 3-dimensional complex Lie group and a 6-dimensional real Lie group.
- The quantum group $SL_q(2,\mathbb{C})$ with $q \in \mathbb{C}\setminus\{0\}$ is an infinite-dimensional complex Hopf algebra which coacts on the quantum plane $\mathbb{C}^2_q$.
- The algebra $SL_1(2,\mathbb{C})$ is isomorphic to the infinite-dimensional complex coordinate Hopf algebra of $SL(2,\mathbb{C})$.

(iv) Quantum algebra $sl_q(2,\mathbb{C})$

- $sl_{\mathbb{C}}(2,\mathbb{C})$ is a 3-dimensional complex Lie algebra whose realification $sl(2,\mathbb{C})$ is a 6-dimensional real Lie algebra.
- The quantum algebra $sl_q(2,\mathbb{C})$ with $q \in \mathbb{C} \setminus \{0, 1, -1\}$ is an infinite-dimensional complex Hopf algebra.
- By definition, $sl_1(2,\mathbb{C})$ is the universal enveloping algebra $Usl_{\mathbb{C}}(2,\mathbb{C})$ of the complex Lie algebra $sl_{\mathbb{C}}(2,\mathbb{C})$; this universal enveloping algebra is an infinite-dimensional complex Hopf algebra.
- $\lim_{q\to 1} sl_q(2,\mathbb{C}) = sl_1(2,\mathbb{C})$, in the sense of formal power series expansions.

(v) Quantum symmetry

- The classical $SL(2,\mathbb{C})$-symmetry of the complex plane $\mathbb{C}^2$ is described by the action of the Lie group $SL(2,\mathbb{C})$ on $\mathbb{C}^2$ via the map $u \mapsto Au$ from $\mathbb{C}^2 \to \mathbb{C}^2$. This is the symplectic geometry on $\mathbb{C}^2$. In fact, introducing the matrix
$$J := \begin{pmatrix} 0 & 1 \\ -1 & 0 \end{pmatrix},$$
we get the following: *The complex* (2×2)*-matrix* A *is an element of* $SL(2,\mathbb{C})$ *iff it is symplectic, that is, the symplectic form*
$$u^d Ju, \qquad u \in \mathbb{C}^2$$
is invariant under the transformation $u \mapsto Au$. To prove this, note that the invariance condition
$$(Au)^d J(Au) = u^d Ju \qquad \text{for all} \quad u \in \mathbb{C}^2$$
is equivalent to $u^d(A^d JA)u = u^d Ju$ for all $u \in \mathbb{C}^2$. In turn, this is equivalent to $A^d JA = J$. This means that
$$\begin{pmatrix} a & c \\ b & d \end{pmatrix}\begin{pmatrix} 0 & 1 \\ -1 & 0 \end{pmatrix}\begin{pmatrix} a & b \\ c & d \end{pmatrix} = \begin{pmatrix} 0 & ad-bc \\ bc-ad & 0 \end{pmatrix} = \begin{pmatrix} 0 & 1 \\ -1 & 0 \end{pmatrix}.$$
Hence $\det A = ad - bc = 1$. Therefore, the Lie group $SL(2,\mathbb{C})$ is also denoted by $Sp(2,\mathbb{C})$ (complex symplectic group).[71]
This symmetry is behind the spinor calculus in relativistic quantum physics.
- The Lie algebra $sl(2,\mathbb{C})$ is obtained from the Lie group $SL(2,\mathbb{C})$ by linearization at the unit element (see Sect. 7.7 of Vol. I). Intuitively, the Lie algebra $sl(2,\mathbb{C})$ describes the symplectic geometry of the space $\mathbb{C}^2$ on an infinitesimal level.
- The quantum group $SL_q(2,\mathbb{C})$ coacts on the complex quantum plane $\mathbb{C}^2_q$. This can be viewed as a generalized symplectic geometry.

[71] By definition, the real symplectic group $Sp(2,\mathbb{R})$ consists of all the real (2×2)-matrices A with $A^d JA = J$. The same argument as above shows that $Sp(2,\mathbb{R}) = SL(2,\mathbb{R})$.

- Finally, the quantum algebra $sl_q(2, \mathbb{C})$ can be considered as an infinitesimal variant of the symplectic geometry of the quantum plane $\mathbb{C}_q^2$.

Further reading on Lie groups, Lie algebras, and quantum groups. See page 534.

3.26 Additive Groups, Betti Numbers, Torsion Coefficients, and Homological Products

> Additive groups play a crucial role in algebraic topology. For example, homology and cohomology groups are used in order to introduce Betti numbers and torsion coefficients as fundamental topological invariants.
>
> Folklore

We are going to study additive groups and more general modules which allow many applications in mathematics and physics. The definition of an additive group is extremely simple. Nevertheless, additive groups possess a rich algebraic structure which is the subject of homological algebra. This structure was discovered in the 1940s by studying homology and cohomology groups with arbitrary coefficents (see Vol. IV).

The nontriviality of additive groups is closely related to torsion which corresponds to the existence of periodic elements. In topology, the torsion of homology and cohomology groups describes geometric twists.

At this point, we only summarize some elementary results which allow far-reaching generalizations in homological algebra.

Cyclic groups. The prototype of an additive group is the additive group $\mathbb{Z}$ of integers with respect to addition $m+n$ for all $m, n \in \mathbb{Z}$. In general, a commutative group is called an additive group iff the group operation is denoted by the symbol $+$, and the unit element is denoted by 0. An additive group $\mathcal{G}$ is called cyclic iff it is generated by precisely one element g_1, that is, every element g of $\mathcal{G}$ can be represented as

$$g = ng_1, \qquad n \in \mathbb{Z}.$$

For example, the group element $g_1 = 1$ generates the additive group of integers $\mathbb{Z}$.

A cyclic group is isomorphic either to the infinite cyclic group $\mathbb{Z}$ or to the finite cyclic group $\mathbb{Z}_m$ of order $m = 1, 2, \ldots$

By definition, the group $\mathbb{Z}_m$ consists of the m elements $0, 1, 2, \ldots, m-1$ where we add the relation $m = 0$. For example, $1 + (m-1) = m = 0$. Note that $\mathbb{Z}_1 = \{0\}$. In what follows we will briefly write $\mathbb{Z}_1 = 0$.

Direct sums. The direct sum $\mathcal{G} \oplus \mathcal{H}$ of the additive groups $\mathcal{G}$ and $\mathcal{H}$ is defined as for linear spaces. Explicitly, $\mathcal{G} \oplus \mathcal{H}$ consists of all the pairs

$$(g, h), \qquad g \in \mathcal{G}, h \in \mathcal{H}$$

with $(g, h) + (g', h') = (g + g', h + h')$.

Generators and free additive groups. The system $\{g_j\}_{j \in J}$ of elements g_j of the additive group $\mathcal{G}$ is called a generating system of $\mathcal{G}$ iff every element g of $\mathcal{G}$ can be represented as

$$g = \sum_{j \in J} \alpha_j g_j$$

where all the coefficients α_j are integers, and at most a finite number of coefficients α_j is different from zero. In particular, $\mathcal{G}$ is called finitely generated iff it possesses a finite generating system. The additive group $\mathcal{G}$ is called free iff every group element g determines uniquely all the coefficients α_j. Then the generating system $\{g_j\}_{j\in J}$ is called a basis of $\mathcal{G}$. For example, the additive group $\mathbb{Z}$ is free, whereas all the groups $\mathbb{Z}_m$ are not free. For example, in order to see that the group $\mathbb{Z}_2$ is not free, note that $2g = 4g = 0$ for all $g \in \mathbb{Z}_2$.

Betti numbers and torsion coefficients. Consider the finite direct sum

$$\boxed{\mathcal{G} = \mathbb{Z} \oplus \mathbb{Z} \oplus \ldots \oplus \mathbb{Z} \oplus \mathbb{Z}_{n_1} \oplus \mathbb{Z}_{n_2} \oplus \ldots \oplus \mathbb{Z}_{n_r}} \tag{3.109}$$

where $n_1, \ldots, n_r$ are integers with $1 < n_1 \le n_2 \le \ldots \le n_r$, and n_k is a divisor of n_{k+1} for all $k = 1, \ldots, r-1$. The direct sum $\mathcal{G}$ is a finitely generated additive group.

- The number of summands $\mathbb{Z}$ is called the Betti number β (or the rank) of the additive group $\mathcal{G}$.
- The numbers $n_1, \ldots, n_r$ are called the torsion coefficients of the group $\mathcal{G}$.

For example, the additive group

$$\mathbb{Z} \oplus \mathbb{Z}_2 \oplus \mathbb{Z}_4$$

has the Betti number (or the rank) $\beta = 1$ and the torsion coefficients $2, 4$. The additive group $\mathbb{Z} \oplus \mathbb{Z}$ has the Betti number $\beta = 2$, and the torsion coefficients drop out. To simplify notation, we write[72]

$$\boxed{\mathcal{G} = \mathbb{Z}^\beta \oplus \mathbb{Z}_{n_1} \oplus \mathbb{Z}_{n_2} \oplus \ldots \oplus \mathbb{Z}_{n_r}.} \tag{3.110}$$

The main theorem on finitely generated additive groups. The following theorem tells us that the direct sum (3.109) describes the normal form of finitely generated additive groups.

Theorem 3.25 *Every finitely generated additive group is isomorphic to a finite direct sum of cyclic groups.*

Two finitely generated additive groups are isomorphic iff they have the same Betti numbers and the same torsion coefficients.

The proof can be found in K. Spindler, Abstract Algebra and Applications, Sect. 25, Vol. 1, Marcel Dekker, New York, 1994.

Periodic elements and torsion. The element $g = 1$ of $\mathbb{Z}_2$ has the property that

$$2g = 0.$$

We say that the element g has the period 2. In general, the element g of the additive group $\mathcal{G}$ is called periodic iff $g \neq 0$ and there exists a positive integer m such that

$$mg = 0.$$

The element $g = 1$ in $\mathbb{Z}_m$ with $m = 2, 3, \ldots$ has the minimal period m. An additive group $\mathcal{G}$ is called torsion free iff it has no periodic elements. For example, $\mathbb{Z}$ is torsion free. Using the decomposition (3.110), we set

- $F(\mathcal{G}) := \mathbb{Z}^\beta$ (free part of $\mathcal{G}$), and

[72] By definition, $\mathbb{Z}^0 := 0, \mathbb{Z}^1 := \mathbb{Z}, \mathbb{Z}^2 = \mathbb{Z} \oplus \mathbb{Z}, \ldots$

- $T(\mathcal{G}) := \mathbb{Z}_{n_1} \oplus \ldots \oplus \mathbb{Z}_{n_r}$ (torsion part of $\mathcal{G}$).

Then we get the group decomposition

$$\boxed{\mathcal{G} = F(\mathcal{G}) \oplus T(\mathcal{G}).} \tag{3.111}$$

A finitely generated additive group $\mathcal{G}$ is torsion free iff it is free. In turn, $\mathcal{G}$ is free iff $\mathcal{G} = F(\mathcal{G})$ (i.e., $T(\mathcal{G}) = 0$), and the torsion coefficients drop out.

The dual additive group. Let $\mathcal{G}$ be an additive group. The set of all the additive group morphisms

$$f : \mathcal{G} \to \mathbb{Z}$$

forms an additive group $\mathcal{G}^d$ which is called the dual group to $\mathcal{G}$. We also write $\mathrm{Hom}(\mathcal{G}, \mathbb{Z})$ instead of $\mathcal{G}^d$. For example,

$$\mathrm{Hom}(\mathcal{G}, \mathbb{Z}) = \mathbb{Z} \quad \text{and} \quad \mathrm{Hom}(\mathbb{Z}_m, \mathbb{Z}) = 0, \qquad m = 1, 2, \ldots$$

Proof. For example, let $m = 2$. If $f : \mathbb{Z}_2 \to \mathbb{Z}$ is a morphism, then we obtain $2f(1) = f(1) + f(1) = f(1+1) = f(0) = 0$. Hence $f = 0$. □

The homomorphism product Hom. Let $\mathcal{G}$ and $\mathcal{H}$ be additive groups. Let $\mathrm{Hom}(\mathcal{G}, \mathcal{H})$ denote the set of all the additive group morphisms[73]

$$f : \mathcal{G} \to \mathcal{H},$$

then $\mathrm{Hom}(\mathcal{G}, \mathcal{H})$ is an additive group which is called the homomorphism product of $\mathcal{G}$ and $\mathcal{H}$. This product respects direct sums, that is, we have the following distributive laws:

$$\mathrm{Hom}(\mathcal{G} \oplus \mathcal{G}', \mathcal{H}) = \mathrm{Hom}(\mathcal{G}, \mathcal{H}) \oplus \mathrm{Hom}(\mathcal{G}', \mathcal{H}), \tag{3.112}$$

and $\mathrm{Hom}(\mathcal{G}, \mathcal{H} \oplus \mathcal{H}') = \mathrm{Hom}(\mathcal{G}, \mathcal{H}) \oplus \mathrm{Hom}(\mathcal{G}, \mathcal{H}')$.

Examples. (i) $\mathrm{Hom}(\mathbb{Z}, \mathcal{H}) \simeq \mathcal{H}$ for all additive groups $\mathcal{H}$.

(ii) $\mathrm{Hom}(\mathbb{Z} \oplus \mathbb{Z}, \mathcal{H}) = \mathrm{Hom}(\mathbb{Z}, \mathcal{H}) \oplus \mathrm{Hom}(\mathbb{Z}, \mathcal{H}) = \mathcal{H} \oplus \mathcal{H}$.

(iii) Hom $(\mathbb{Z}, \mathbb{Z}_m) = \mathbb{Z}_m$ for all $m = 1, 2, \ldots$

(iv) Hom $(\mathbb{Z}_m, \mathbb{Z}) = 0$ for all $m = 1, 2, \ldots$

(v) Hom $(\mathbb{Z}_m, \mathbb{Z}_n) = \mathbb{Z}_{(m,n)}$ for all $m, n = 1, 2, \ldots$ (where (m, n) is the greatest common divisor of m and n).[74]

(vi) Hom $(\mathbb{Z}, \mathbb{F}) = \mathbb{F}$ and Hom $(\mathbb{Z}_m, \mathbb{F}) = 0$, $m = 1, 2, \ldots,$ if $\mathbb{F}$ is a field or skew-field (e.g., $\mathbb{F} = \mathbb{Z}_2, \mathbb{Z}_p$ (p prime number), $\mathbb{Q}$ (rational numbers), $\mathbb{R}$ (real numbers), $\mathbb{H}$ (quaternions)).

(vii) If $\mathcal{G}$ is finitely generated, then Hom $(\mathcal{G}, \mathbb{Z}) = F(\mathcal{G})$ (free part of the additive group $\mathcal{G}$).

Proof. Ad (i). Let $f : \mathbb{Z} \to \mathcal{H}$ be a group morphism. Set $\chi(f) := f(1)$. The map

$$\chi : \mathrm{Hom}(\mathbb{Z}, \mathcal{H}) \to \mathcal{H}$$

[73] In modern mathematics, the term 'morphism' is used for all mathematical structures in the setting of category theory (see Vol. IV). In the classical textbook by B. van der Waerden, Modern Algebra (in German), Springer, Berlin, 1930 (English edition: Frederyk Ungar, New York, 1975), the term 'homomorphism' is used for morphisms of groups, rings, modules, and fields. Therefore, 'morphisms' and 'homomorphisms' are synonymous terms nowadays. For historical reasons, in homological algebra one always uses 'Hom' instead of 'Mor' in the theory of categories.

[74] For example, Hom $(\mathbb{Z}_m, \mathbb{Z}_m) = \mathbb{Z}_m$, Hom $(\mathbb{Z}_4, \mathbb{Z}_6) = \mathbb{Z}_2$, and Hom $(\mathbb{Z}_3, \mathbb{Z}_5) = 0$.

is a group morphism. If $f(1) = 0$, then $f(1+1) = 0$, and so on. Hence $f = 0$. Thus, $\ker(\chi) = 0$. Consequently, the map χ is an isomorphism (see Sect. 4.1.3 of Vol. II). The other claims are proved similarly. □

The computation of homology and cohomology groups with general coefficients. In homological algebra, one introduces two other products Tor $(\mathcal{G}, \mathcal{H})$ and Ext $(\mathcal{G}, \mathcal{H})$ for computing the homology groups $H_k(X, \mathcal{H})$ and cohomology groups $H^k(X, \mathcal{H})$ of topological spaces X with coefficients in the additive group $\mathcal{H}$ (e.g., $\mathcal{H} = \mathbb{Z}$, $\mathbb{Z}_m$, $\mathbb{Q}$ (rational numbers), $\mathbb{R}$ (real numbers), $\mathbb{H}$ (quaternions)). It is sufficient to know the homology groups $H_k(X, \mathbb{Z})$ with integer coefficients in order to get all the other homology and cohomology groups. In this connection, one uses the universal coefficient formulas in homology and cohomology (see Vol. IV). At this point, we only summarize Tor and Ext for cyclic groups. The groups Tor $(\mathcal{G}, \mathcal{H})$ and Ext $(\mathcal{G}, \mathcal{H})$ measure how the torsion of the group $\mathcal{G}$ fits the torsion of the group $\mathcal{H}$.

The torsion product Tor of cyclic groups. Let $m, n = 1, 2, \ldots$

- Tor $(\mathbb{Z}_m, \mathbb{Z}_n) = \mathbb{Z}_{(m,n)}$ (e.g., $\text{Tor}(\mathbb{Z}_m, \mathbb{Z}_m) = \mathbb{Z}_m$).
- $\text{Tor}(\mathbb{Z}, \mathbb{Z}) = \text{Tor}(\mathbb{Z}, \mathbb{Z}_m) = \text{Tor}(\mathbb{Z}_m, \mathbb{Z}) = 0$.
- Tor $(\mathbb{Z}, \mathbb{F})$ = Tor $(\mathbb{Z}_m, \mathbb{F}) = 0, m = 1, 2, \ldots$, if $\mathbb{F}$ is a field or skew-field (e.g., $\mathbb{F} = \mathbb{Z}_2, \mathbb{Z}_p$ (p prime number), $\mathbb{Q}$ (rational numbers), $\mathbb{R}$ (real numbers), $\mathbb{H}$ (quaternions)).
- If $\mathcal{G}$ is a finitely generated additive group of the form (3.109), then

$$\text{Tor}(\mathcal{G}, \mathbb{Z}_m) = \mathbb{Z}_{(n_1,m)} \oplus \ldots \oplus \mathbb{Z}_{(n_r,m)}$$

where (n_j, m) is the greatest common divisor of n_j and m. The product Tor is commutative.

The extension product Ext of cyclic groups. Let $m, n = 1, 2, \ldots$

- $\text{Ext}(\mathbb{Z}_m, \mathbb{Z}_n) = \mathbb{Z}_{(m,n)}$ (e.g., $\text{Ext}(\mathbb{Z}_m, \mathbb{Z}_m) = \mathbb{Z}_m$).
- $\text{Ext}(\mathbb{Z}_m, \mathbb{Z}) = \mathbb{Z}_m$.
- $\text{Ext}(\mathbb{Z}, \mathbb{Z}) = \text{Ext}(\mathbb{Z}, \mathbb{Z}_m) = 0$.
- $\text{Ext}(\mathbb{Z}, \mathbb{F}) = \text{Ext}(\mathbb{Z}_m, \mathbb{F}) = 0$, $m = 1, 2, \ldots$, if $\mathbb{F}$ is a field or skew-field (e.g., $\mathbb{F} = \mathbb{Z}_2, \mathbb{Z}_p$ (p prime number), $\mathbb{Q}$ (rational numbers), $\mathbb{R}$ (real numbers), $\mathbb{H}$ (quaternions)).
- If $\mathcal{G}$ is a finitely generated additive group, then $\text{Ext}(\mathcal{G}, \mathbb{Z}) = T(\mathcal{G})$ (torsion part of $\mathcal{G}$). Hence

$$\boxed{\mathcal{G} = \text{Hom}(\mathcal{G}, \mathbb{Z}) \oplus \text{Ext}(\mathcal{G}, \mathbb{Z}).}$$

Moreover, $\text{Ext}(\mathcal{G}, \mathbb{F}) = 0$ if $\mathbb{F}$ is a field or skew-field.

3.27 Lattices and Modules

Linear spaces, additive groups, rings, fields, and algebras have the common feature that they are modules over some ring. Modules of chains (e.g., electric circuits) play a crucial role in algebraic topology. Moreover, modules are critically used in the representation theory for groups and algebras (e.g. Lie algebras). The prototype of a left module over the ring $\mathbb{Z}$ of integers is a lattice.

Folklore

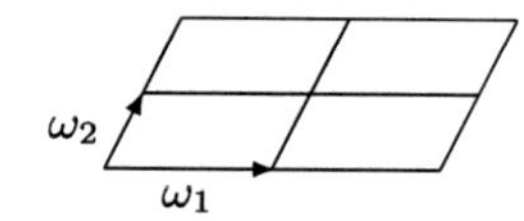

Fig. 3.17. Lattice

Lattice. Consider the complex numbers ω_1, ω_2 as depicted in Fig. 3.17. The set of all the complex numbers

$$\alpha_1\omega_1 + \alpha_2\omega_2,$$

where α_1 and α_2 are arbitrary integers, is called a lattice generated by the complex numbers ω_1 and ω_2. Such a lattice is the prototype of a left module with integers as coefficients, in the sense of the definition given below.

A chain as the prototype of a module. Fix $n = 1, 2, \ldots$ Choose the symbols $s_1, \ldots, s_n$. By a chain with integer coefficients, we understand the set of all the expressions

$$\boxed{\alpha_1 s_1 + \alpha_2 s_2 + \ldots + \alpha_n s_n} \tag{3.113}$$

with integers $\alpha_1, \alpha_2, \ldots, \alpha_n$. The addition of chains and the multiplication of chains with integers are defined in a natural way:

- $(\alpha_1 s_1 + \ldots + \alpha_n s_n) + (\beta_1 s_1 + \ldots + \beta_n s_n) := (\alpha_1 + \beta_1)s_1 + \ldots + (\alpha_n + \beta_n)s_n$,
- $\gamma(\alpha_1 s_1 + \ldots + \alpha_n s_n) := (\gamma\alpha_1)s_1 + \ldots + (\gamma\alpha_n)s_n$

for all integers $\gamma, \alpha_1, \ldots, \alpha_n, \beta_1, \ldots, \beta_n$. In the sense of the definition given below, the chains form a left module with integer coefficients (i.e., a left $\mathbb{Z}$-module). In topology, one frequently adds relations to chains, for example,

$$s_1 + s_1 = 0.$$

In other words, $2s_1 = 0$. This way, it is possible to describe non-orientable manifolds (e.g., the Möbius strip or the projective space $\mathbb{P}^2$).

In algebraic topology, it is very useful to introduce chains (3.113) where the coefficients $\alpha_1, \ldots, \alpha_n$ are not integers, but elements of a ring R (e.g., $\mathbb{Z}_2$, $\mathbb{Q}$ (rational numbers), $\mathbb{R}$ (real numbers), $\mathbb{C}$ (complex numbers), $\mathbb{H}$ (quaternions)). This flexibility is used in order to obtain new topological invariants (topological charges). This motivates the following general definition.

Definition of left R-modules. Let R be a ring. The set $\mathcal{G}$ is called a left R-module iff it is an additive group equipped with a product αg such that, for all $\alpha, \beta \in R$ and all $g, h \in \mathcal{G}$, the following hold:

(C) αg is a uniquely determined element of $\mathcal{G}$ (consistency);
(D) $\alpha(g + h) = \alpha g + \alpha h$ and $(\alpha + \beta)g = \alpha g + \beta g$ (distributivity);
(A) $\alpha(\beta g) = (\alpha\beta)g$ (associativity).

If the ring R has a unit element 1, then we postulate that $1g = g$ for all elements g of $\mathcal{G}$.[75]

Examples. Let $\mathbb{K} = \mathbb{R}, \mathbb{C}$. Every linear space $\mathcal{G}$ over $\mathbb{K}$ is a left $\mathbb{K}$-module. Mnemonically, a left R-module can be regarded as a generalized linear space over the ring R.

---

[75] Similarly, if we replace the product αg by the product $g\alpha$, then we obtain the definition of a right R-module. As a rule, we will use left modules.

Every ring R is automatically a left R-module. Here, the product αg with $\alpha, g \in R$ is nothing other than the ring product.

The simplest left $\mathbb{Z}$-module is the additive cyclic group $\mathbb{Z}_2 = \{0, 1\}$ with the operations

$$1 + 1 = 0, \quad 0 + 1 = 1 + 0 = 1, \quad 0 + 0 = 0.$$

Moreover, let $g \in \mathbb{Z}_2$, and let m be an integer. Then we define the product $mg := 0$ if m is even (resp. $mg := g$ if m is odd). Analogously, $\mathbb{Z}_m$ is a left $\mathbb{Z}$-module.

Additive groups as left $\mathbb{Z}$-modules. If $\mathcal{G}$ is an additive group, then $\mathcal{G}$ is a left $\mathbb{Z}$-module by introducing the product mg with $g \in \mathcal{G}, m \in \mathbb{Z}$ in a natural way: we set

$$2g := g + g, \quad (-1)g := -g, \quad (-2)g := -g - g, \quad \text{and so on.} \tag{3.114}$$

Conversely, every left R-module is an additive group. This tells us that left $\mathbb{Z}$-modules can be identified with additive groups.

Morphisms. Suppose that $\mathcal{G}$ and $\mathcal{H}$ are left R-modules. By definition, a left R-module morphism is a map $\chi : \mathcal{G} \to \mathcal{H}$ which respects the sum and the product. Explicitly,

$$\chi(\alpha g + \beta h) = \alpha\chi(g) + \beta\chi(h) \qquad \text{for all} \quad g, h \in \mathcal{G},\ \alpha, \beta \in R.$$

The symbol $\mathrm{Hom}_R(\mathcal{G}, \mathcal{H})$ denotes the set of all left R-module morphisms

$$\chi : \mathcal{G} \to \mathcal{H}.$$

Naturally enough, the set $\mathrm{Hom}_R(\mathcal{G}, \mathcal{H})$ is a left R-module equipped with the linear combination $\alpha\chi + \beta\mu$. Here, we set

$$(\alpha\chi + \beta\mu)(g) := \alpha\chi(g) + \beta\mu(g) \qquad \text{for all} \quad g \in \mathcal{G},$$

and all $\chi, \mu \in \mathrm{Hom}_R(\mathcal{G}, \mathcal{H}), \alpha, \beta \in R$.

Precisely the bijective left R-module morphisms are called left R-module isomorphisms. We have the following isomorphisms of left R-modules:

- $\mathrm{Hom}_{\mathbb{Z}}(\mathbb{Z}, \mathbb{Z}) \simeq \mathbb{Z}$ and $\mathrm{Hom}_{\mathbb{Z}}(\mathbb{Z}_2, \mathbb{Z}) \simeq 0$,
- $\mathrm{Hom}_{\mathbb{Z}}(\mathbb{Z}_m, \mathbb{Z}_n) \simeq \mathbb{Z}_{(m,n)}$ for all $m, n = 1, 2, \ldots$ where (m, n) denotes the greatest common divisor of m and n.

These isomorphisms are given by the map $\chi \mapsto \chi(1)$. For example, if the map $\chi : \mathbb{Z}_2 \to \mathbb{Z}$ is a $\mathbb{Z}$-morphism, then $1+1 = 0$ implies $\chi(1+1) = 0$. Thus, $\chi(1)+\chi(1) = 2\chi(1) = 0$, and hence $\chi = 0$.

Submodules. The subset $\mathcal{S}$ of a left R-module $\mathcal{G}$ is called a submodule of $\mathcal{G}$ iff it is a left R-module with respect to the operations in $\mathcal{G}$. This is the case iff $\mathcal{S}$ is a subgroup of $\mathcal{G}$ and $\alpha g \in \mathcal{S}$ for all $g \in \mathcal{G}, \alpha \in R$.

Free modules. The system $\{g_j\}_{j \in J}$ of elements g_j of the left R-module $\mathcal{G}$ is called a generating system of $\mathcal{G}$ iff every element g of $\mathcal{G}$ can be represented as

$$g = \sum_{j \in J} \alpha_j g_j$$

where $\alpha_j \in R$ for all $j \in J$, and at most a finite number of coefficients α_j is different from zero. If, in addition, the coefficients α_j are always uniquely determined, then the system $\{g_j\}_{j \in J}$ is called a basis of $\mathcal{G}$. The left R-module $\mathcal{G}$ is called free iff it has a basis.

Direct sum of modules. If $\mathcal{G}$ and $\mathcal{H}$ are left R-modules, then the direct sum $\mathcal{G} \oplus \mathcal{H}$ is defined in the same ways as for linear spaces. Explicitly, we set

$$(g,h)+(k,l) := (g+k, h+l), \qquad \alpha(g,h) := (\alpha g, \alpha h)$$

for all $g,k \in \mathcal{G}, h,l \in \mathcal{H}$, and all $\alpha \in R$. This way, the set of all ordered pairs

$$(g,h) \qquad \text{with } g \in \mathcal{G}, h \in \mathcal{H}$$

becomes a left R-module denoted by $\mathcal{G} \oplus \mathcal{H}$.

The operator module $\mathrm{Op}(X)$ of a linear space X. Let X be a real or complex linear space. Then the set $\mathrm{End}(X)$ of all linear operators $A : X \to X$ is a ring denoted by R. The operator A sends the element x of X to the element Ax of X. Using the product

$$Ax \qquad \text{with } x \in X, A \in R,$$

the linear space X becomes a left R-module which is denoted by $\mathrm{Op}(X)$.

The representation module X_ϱ. Let ϱ be a representation of the ring R on the linear space X. That is, the map

$$\varrho : R \to \mathrm{End}(X)$$

is a ring morphism. This means that the ring R is realized by linear operators on the linear space X. Then $\varrho(\alpha) : X \to X$ is a linear operator for every element α of the ring R. This yields the product

$$\alpha x := \varrho(\alpha)x, \qquad x \in X, \;\; \alpha \in R.$$

This way, the linear space X becomes a left R-module denoted by X_ϱ; this is called the representation module of ϱ. Obviously, X_ϱ is a submodule of $\mathrm{Op}(X)$.

Problems

3.1 *The determinant of the dual matrix.* Let A be a real or complex $(n \times n)$-matrix. Show that $\det(A^d) = \det(A)$. Solution: Use the definition

$$\det(A) = \sum_\pi \mathrm{sgn}(\pi)\; a_{1\pi(1)} \cdots a_{n\pi(n)}. \tag{3.115}$$

In order to pass from a_{ij} to a_{ji}, replace the permutation π by the inverse permutation π^{-1} and observe that $\mathrm{sgn}(\pi^{-1}) = \mathrm{sgn}(\pi)$.

3.2 *The general Laplace expansion theorem.* Prove (1.17) on page 78. Hint: Write the sum (3.115) as

$$\sum_{i_1<i_2<\ldots<i_p} \; \sum_{\pi(i_1,\ldots i_p)} \; \sum_{\pi(i_{p+1},\ldots,i_n)} \mathrm{sgn}(i_1 \ldots i_n) \cdot a_{1i_1} \ldots a_{ni_n}.$$

Here, $i_{p+1} < \ldots < i_n$. The symbol $\pi(i_1,\ldots,i_p)$ means that we sum over all permutations of the numbers $i_1,\ldots,i_p$. Moreover, $\mathrm{sgn}(i_1,\ldots,i_n)$ is the signum of the permutation $(1,\ldots,n) \mapsto (i_1 \ldots i_n)$.

3.3 *Associativity of the alternating product.* Prove Theorem 2.2 on page 119. Hint: Use the general Laplace expansion theorem.
Remark. If one introduces the Grassmann algebra in an abstract manner by using a quotient algebra, then the validity of the associative law of the $\wedge$-product is obvious, and this implies immediately the general Laplace expansion formula. For the approach via quotient algebras, see Sect. 2.12 on page 174.

3.4 *The vector product.* Prove Prop. 1.8 (iv)–(x) on page 83.
Hints: Ad (iv). Use $\mathbf{e}_1 \times \mathbf{e}_2 = \iota \mathbf{e}_3$.
Ad (v). Use $(\mathbf{ab})^2 = |\mathbf{a}|^2|\mathbf{b}|^2 \cos^2 \varphi$ and

$$(\mathbf{a} \times \mathbf{b})^2 = |\mathbf{a}|^2 \cdot |\mathbf{b}|^2 \cdot \sin^2 \varphi$$

together with $\cos^2 \varphi + \sin^2 \varphi = 1$.
Ad (vi). Without any loss of generality, choose $\mathbf{b} := \mathbf{k}$, and use (iii).
Ad (vii). Use (vi).
Ad (viii), (ix). Use an explicit computation based on (iii).
Ad (x). This is an immediate consequence of the product formula for determinants.

3.5 *Proof of Theorem 1.3* on page 79. Hint: In order to prove the distributive law, use the elementary geometric fact that the orthogonal projection $P(\mathbf{a}+\mathbf{b})$ of a vector sum $\mathbf{a}+\mathbf{b}$ is equal to the sum $P\mathbf{a}+P\mathbf{b}$ of the orthogonal projections of the summands.

3.6 *Symmetric and antisymmetric products.* Let $\mathcal{A}$ be an associative algebra. Let $n = 2, 3, \ldots$ and $a_1, \ldots, a_n \in \mathcal{A}$. Show the following:
(i) Symmetry. If $ab = ba$ for all $a, b \in \mathcal{A}$, then the product of an arbitrary number of factors is symmetric under a permutation of the factors.
(ii) Antisymmetry. Suppose that $a_1 a_2 \cdots a_n = 0$ if two factors coincide. Then the product of an arbitrary number of factors is antisymmetric with respect to a permutation of the factors.
Hint: Ad (a). $abc = (ab)c = (ba)c = bac = bca$.
Ad (b). $(a+b)(a+b) = 0$ implies $a^2 + ab + ba + b^2 = 0$. Since $a^2 = b^2 = 0$, we get $ab = -ba$.

3.7 *Unital algebras.* Let $\mathcal{A}$ be a unital algebra over $\mathbb{K} = \mathbb{R}$ or $\mathbb{K} = \mathbb{C}$ with $\mathcal{A} \neq \{0\}$. Let $\mathbf{1}$ (resp. $\mathbf{0}$) denote the unit element (resp. zero element) of $\mathcal{A}$. Let $a \vee b$ denote the product on $\mathcal{A}$. Show that the following hold:
(i) There exists an injective field morphism $\chi : \mathbb{K} \to \mathcal{A}$ given by $\chi(\lambda) := \lambda \mathbf{1}$. Therefore, the field $\mathbb{K}$ can be regarded as a subset of $\mathcal{A}$ by identifying λ with the element $\lambda \mathbf{1}$.
(ii) For all $\lambda, \mu \in \mathbb{K}$ and all $a \in \mathcal{A}$, we have

$$\lambda \vee a = a \vee \lambda = \lambda a, \qquad \lambda \vee \mu = \lambda \mu.$$

In particular, the unit element $\mathbf{1}$ can be identified with the unit element 1 of the field $\mathbb{K}$.
Solution: Ad (i). Let $\chi(\lambda) = \mathbf{0}$. Assume that $\lambda \neq 0$. Then $\mathbf{1} = \mathbf{0}$. Choose a nonzero element a of $\mathcal{A}$. Then

$$a = \mathbf{1} \vee a = (\lambda \mathbf{1}) \vee a = \lambda(\mathbf{1} \vee a) = \lambda a$$

for all $\lambda \in \mathbb{K}$. Hence $a = 0$, a contradiction.
Ad (ii). Note that $(\lambda \mathbf{1}) \vee a = \lambda(\mathbf{1} \vee a) = \lambda a$, and $a \vee (\lambda \mathbf{1}) = \lambda(a \vee \mathbf{1}) = \lambda a$. Moreover, $\lambda \mathbf{1} \vee \mu \mathbf{1} = \lambda\mu(\mathbf{1} \vee \mathbf{1}) = \lambda\mu\mathbf{1}$.

3.8 *Basis extension and biorthogonal system.* Let X be a finite-dimensional linear space over $\mathbb{K} = \mathbb{R}, \mathbb{C}$ of dimension $n = 1, 2, \ldots$ We are given linearly independent vectors $a_1, \ldots, a_k$. Show that:
(i) The system $a_1, \ldots, a_k$ can be extended to a basis $a_1, \ldots, a_n$ of X.
(ii) There exist linear functionals $F^1, \ldots, F^n \in X^d$ such that $F^i(a_j) = \delta^i_j$ for all $i, j = 1, \ldots, n$.
Solution: Ad (i). If $a_1, \ldots, a_k$ is not a basis, then there exists a vector b which is not linearly dependent on $a_1, \ldots, a_k$. Set $a_{k+1} := b$. After a finite number of steps, we get the desired basis.
Ad (ii). Use (i) and set $F^i(\sum_{j=1}^n x^j a_j) := x^i$.

3.9 *Hamel basis.* Choose $\mathbb{K} = \mathbb{R}, \mathbb{C}, \mathbb{Q}$ (field of real, complex, rational numbers). Let X be a linear space over $\mathbb{K}$. A nonempty subset S of X is called a basis of X iff every element x of X can be uniquely represented as

$$x = \sum_{s \in S} \alpha_s s$$

where we have $\alpha_s \in \mathbb{K}$ for all $s \in S$, and at most a finite number of the coefficients α_s is different of zero. This implies that every finite subset of elements of S is linearly independent. Show the following:
(i) For given nonempty set S, there exists a linear space over $\mathbb{K}$ which has the set S as a basis. This set is denoted by span(S).[76]
(ii) Every linear space X over $\mathbb{K}$ has a basis.
Solution: Ad (i). By definition, the set span(S) consists of all the maps

$$x : S \to \mathbb{K}$$

which are different from zero at most on a finite subset of S. Naturally enough, this set of maps becomes a linear space over $\mathbb{K}$. For fixed $s_0 \in S$, we define

$$b_{s_0}(t) := \begin{cases} 1 & \text{if} \quad t = s_0 \\ 0 & \text{otherwise.} \end{cases}$$

The family $\{b_s\}_{s \in S}$ is a basis of span(S). This basis is in one-to-one correspondence to S.
Ad (ii). Hint: Apply Zorn's lemma (see page 248 of Vol. II). Use the ordered family $\mathcal{F} := \{Y\}$ of all the linear subspaces Y of X which have a basis. Show that the maximal element of $\mathcal{F}$ coincides with X. See S. Lang, Algebra, Springer, New York, 2002, page 139.
Historical remark. The set $\mathbb{R}$ of real numbers is a linear space over $\mathbb{R}$. The unit element 1 represents a basis of the real linear space $\mathbb{R}$. Thus, $\dim(\mathbb{R}) = 1$. The situation changes dramatically if we consider $\mathbb{R}$ as a linear space over the field $\mathbb{Q}$ of rational numbers. This infinite-dimensional linear space is denoted by $\mathbb{R}_{\mathbb{Q}}$. Hamel (1877–1954) proved that $\mathbb{R}_{\mathbb{Q}}$ has a basis. However, an explicit basis is still unknown. Each basis of $\mathbb{R}_{\mathbb{Q}}$ is called a Hamel basis.

3.10 *The tensor product as a linear quotient space.* For pedagogical reasons, in Chap. 2 we have used concrete mathematical objects (namely, bilinear functionals) in order to introduce the tensor product $X \otimes Y$ (see page 122). There exists an equivalent abstract approach to the tensor product based on equivalence classes (more precisely, on linear quotient spaces).[77] In what follows, let us discuss this abstract approach and its equivalence to the concrete approach used in Sect. 2.1.6, including the universality of the tensor product.
Choose $\mathbb{K} = \mathbb{R}$ or $\mathbb{K} = \mathbb{C}$. Let X, Y, and Z be linear spaces over $\mathbb{K}$. By definition, the elements of the set span$(X \times Y)$ have the form

$$a = \sum_{i=1}^{n} \alpha_i (x_i, y_i)$$

where $\alpha_1, \ldots, \alpha_n \in \mathbb{K}$, and $n = 1, 2, \ldots$ In a natural way, the set span$(X \times Y)$ can be equipped with the structure of a linear space over $\mathbb{K}$. Let $\mathcal{J}$ denote the smallest linear subspace of the linear space span$(X \times Y)$ over $\mathbb{K}$ which contains all the elements

[76] For the empty set S, we define span$(\emptyset) := \{0\}$.

[77] The importance of equivalence classes in mathematics including linear quotient spaces is thoroughly discussed in Chap. 4 of Vol. II.

- $(\alpha u + \beta x, y) - \alpha(u, y) - \beta(x, y)$, and
- $(x, \alpha y + \beta z) - \alpha(x, y) - \beta(x, z)$,

where $u, x \in X$, $y, z \in Y$, and $\alpha, \beta \in \mathbb{K}$. For $a, b \in \text{span}(X \times Y)$, let us introduce the equivalence relation

$$a \sim b \quad \text{iff} \quad a - b \in \mathcal{J}.$$

The equivalence classes $[a]$ form the quotient space

$$\text{span}(X \times Y)/\mathcal{J}.$$

This linear space over $\mathbb{K}$ will be denoted by the symbol $X \otimes_{\mathbb{K}} Y$.[78] We define

$$x \otimes y := [(x, y)].$$

Let $a \in \text{span}(X \times Y)$, that is, $a = \sum_{i=1}^{n} \alpha_i (x_i, y_i)$. Then

$$[a] = \sum_{i=1}^{n} \alpha_i [(x_i, y_i)] = \sum_{i=1}^{n} \alpha_i x_i \otimes y_i.$$

We also define the bilinear functional $j(a) : X^d \times Y^d \to \mathbb{K}$ by setting

$$j(a)(F, G) := \sum_{i=1}^{n} \alpha_i F(x_i) G(y_i), \qquad F \in X^d, G \in Y^d.$$

Finally, let $B : X \times Y \to Z$ be a bilinear map. Define

$$L(a) := \sum_{i=1}^{n} \alpha_i B(x_i, y_i).$$

Prove the following:
(i) $(\alpha u + \beta x) \otimes y = \alpha u \otimes y + \beta x \otimes y$, and $x \otimes (\alpha y + \beta z) = \alpha x \otimes y + \beta x \otimes z$ (distributive laws).
(ii) If $a \sim b$, then $j(a) = j(b)$ and $L(a) = L(b)$.
(iii) The linear operators $j : X \otimes Y \to M_2(X^d, Y^d)$ and $L : X \otimes Y \to Z$ are well defined.
(iv) There exists a unique universal object for bilinear maps $B : X \times Y \to Z$, and this object coincides with the tensor product $X \otimes Y$ introduced above.
(v) If $b_1, \ldots, b_m$ (resp. $c_1, \ldots c_n$) are linearly independent elements of X (resp. Y), then the mn tensor products $b_j \otimes c_k$ with $j = 1, \ldots, m$ and $k = 1, \ldots, n$ are linearly independent elements of $X \otimes Y$.
(vi) If S_X (resp. S_Y) is a basis of X (resp. Y), then the product set $S_X \otimes S_Y$ defined by $\{b \otimes c : \ b \in S_X, c \in S_Y\}$ is a basis of $X \otimes Y$.
(vii) The map $j : X \otimes Y \to M_2(X^d, Y^d)$ is injective, and this map represents a linear isomorphism between the tensor product $X \otimes Y$ defined above and the tensor product defined in Sect. 2.1.6 on page 122.

Solution: Ad (i). Use the fact that $a \in \mathcal{J}$ implies $[a] = 0$.
Ad (ii). If $a \in \mathcal{J}$, then $j(a) = L(a) = 0$.
Ad (iii). Set $j([a]) := j(a)$ and $L([a]) := L(a)$. This is well defined by (ii).

[78] To simplify notation, we briefly write $X \otimes Y$ instead of $X \otimes_{\mathbb{K}} Y$, if a misinterpretation is excluded.

Ad (iv). Set $\beta(x,y) := x \otimes y$. Then $B = L \circ \beta$. This shows that $X \otimes Y$ is a universal object. The uniqueness of the universal object follows as in the proof of Theorem 2.7 on page 125 by diagram chasing.
Ad (v), (vi). Use the same argument as in the proof of Prop. 2.4 on page 122.
Ad (vii). Let the symbol $(X \otimes Y)_{\mathrm{bil}}$ denote the tensor product defined in Sect 2.1.6 via bilinear functionals. Then the map $j : X \otimes Y \to (X \otimes Y)_{\mathrm{bil}}$ is surjective. In order to prove the injectivity, let $j([a]) = 0$. Using a basis and the distributive law, we get

$$\sum_{(b,c)\in S_X \times S_Y} a(b,c)j(b \otimes c) = 0.$$

Hence

$$\sum_{(b,c)\in S_X \times S_Y} a(b,c)j(b \otimes c)(F,G) = 0 \qquad \text{for all} \quad F \in X^d, G \in Y^d.$$

Note that, by the definition of a basis, at most a finite number of real coefficients $a(b,c)$ is different from zero. Therefore, we can use the same argument as in the proof of Prop. 2.4 in order to show that all the coefficients $a(b,c)$ vanish. Hence $[a] = 0$.

3.11 *The product $L \otimes M$ of linear operators.* Let $\mathbb{K} = \mathbb{R}$ or $\mathbb{K} = \mathbb{C}$. We are given the linear operators $L : X \to X$ and $M : Y \to Y$ where X and Y are linear spaces over $\mathbb{K}$. Show that there exists precisely one linear operator $N : X \otimes Y \to X \otimes Y$ with

$$N(x \otimes y) = Lx \otimes My \qquad \text{for all} \quad x \in X, y \in Y.$$

This operator is denoted by $L \otimes M$. Hence $(L \otimes M)(x \otimes y) = Lx \otimes My$.
Solution: (I) Uniqueness. Suppose that such an operator exists. Then, by linearity,

$$N\left(\sum_{j=1}^{m} x_j \otimes y_j\right) = \sum_{j=1}^{m} Lx_j \otimes My_j.$$

(II) Existence: We define

$$N\left(\sum_{j=1}^{m} \alpha_j(x_j, y_j)\right) := \sum_{j=1}^{m} \alpha_j(Lx_j, My_j).$$

One checks easily that the linearity of N implies that $Na \in \mathcal{J}$ if $a \in \mathcal{J}$. Consequently, if $a \sim b$, then $Na \sim Nb$. That is, the definition of N respects the equivalence relation. Therefore, passing to equivalence classes, the definition $N[a] := [Na]$ makes sense. This is the desired operator.

Alternatively, one can argue as follows. We define

$$N\left(\sum_{j=1}^{m} x_j \otimes y_j\right) := \sum_{j=1}^{m} Lx_j \otimes My_j.$$

We have to guarantee that different finite sums, which represent the same element, yield the same operator value. To this end, note that by linearity we get

$$L(u+x) \otimes My = (Lu + Lx) \otimes My = Lu \otimes My + Lx \otimes My,$$

and $L(\alpha x) \otimes y = (\alpha Lx) \otimes y = \alpha(Lx \otimes y)$. Hence

$$N\big((u+x)\otimes y - u\otimes y - x\otimes y\big) = 0.$$

This tells us that our definition of the operator N respects the defining relation

$$(u+x)\otimes y - u\otimes y - x\otimes y = 0.$$

Analogously, we get that the defining relation

$$x\otimes(\alpha v + \beta y) - \alpha(x\otimes v) - \beta(x\otimes y) = 0$$

is respected. Thus, we are done.

3.12 *The derived operator L_{der}.* Let $L : X \to X$ be a linear operator on the linear space over $\mathbb{K} = \mathbb{R}, \mathbb{C}$. Show that there exists precisely one linear operator $L_{\text{der}} : X\otimes X \to X\otimes X$ such that we have the Leibniz rule

$$L_{\text{der}}(x\otimes y) = (L_{\text{der}}x)\otimes y + x\otimes(L_{\text{der}}y) \qquad \text{for all} \quad x, y \in X.$$

Hint: Argue as in Problem 3.11.

3.13 *The contraction operator.* Let X be a linear space over $\mathbb{K} = \mathbb{R}, \mathbb{C}$. Show that there exists precisely one linear operator $\mathsf{C} : X^d \otimes X \to \mathbb{K}$ which satisfies the relation

$$\mathsf{C}(F\otimes x) = F(x) \qquad \text{for all} \quad F \in X^d,\ x \in X.$$

Hint: Argue as in Problem 3.11.

3.14 *Tensor products and the complexification of a real linear space.* Let X be a real n-dimensional linear space with the basis $b_1, \ldots, b_n$. The elements x of X are given by

$$x = \alpha_1 b_1 + \ldots + \alpha_n b_n$$

where the real coefficients $\alpha_1, \ldots, \alpha_n$ are uniquely determined by x. By definition, the complexification $\mathcal{X}_{\mathbb{C}}$ of X consists of all the expressions

$$\beta_1 b_1 + \ldots + \beta_n b_n$$

where $\beta_1, \ldots, \beta_n$ are complex numbers. Then $\mathcal{X}_{\mathbb{C}}$ is a complex linear space which contains the space X as a subset. One shows easily that this construction does not depend on the choice of the basis $b_1, \ldots, b_n$. In a basis-independent approach, the complexification can be obtained in the following way. We define the tensor product

$$\boxed{X_{\mathbb{C}} := \mathbb{C}\otimes_{\mathbb{R}} X.} \tag{3.116}$$

Here, the space $\mathbb{C}$ of complex numbers is regarded as a real two-dimensional linear space.[79] The elements of $\mathbb{C}\otimes_{\mathbb{R}} X$ are finite sums of the form

$$a := \beta_1\otimes x_1 + \ldots + \beta_n\otimes x_n, \qquad x_k \in X,\ \beta_k \in \mathbb{C},\ k = 1, 2, \ldots$$

For $\alpha \in \mathbb{C}$, we define $\alpha(\beta\otimes x) := \alpha\beta\otimes x$. More general,

$$\alpha a := \alpha\beta_1\otimes x_1 + \ldots + \alpha\beta_n\otimes x_n.$$

Show that the following hold:

---

[79] The method of extending algebraic structures to new domains of coefficients based on tensor products like in (3.116) is frequently used in modern mathematics (e.g., the construction of homology groups and cohomology groups for different coefficient rings in algebraic topology).

(i) The definition of the product αa does not depend on the representation of the element a.
(ii) The space $X_{\mathbb{C}}$ is a complex linear space.
(iii) The map $j : X \to X_{\mathbb{C}}$ given by $j(x) := 1 \otimes x$ is an injective linear morphism if we regard $X_{\mathbb{C}}$ as a real linear space. Therefore, the set X can be identified with a subset of $X_{\mathbb{C}}$.
(iv) If $b_1, \ldots, b_n$ are linearly independent elements of the real linear space X, then $1 \otimes b_1, \ldots, 1 \otimes b_n$ are linearly independent elements of $X_{\mathbb{C}}$. In particular, if $\dim_{\mathbb{R}} X = n$ with $n = 0, 1, 2, \ldots$, then $\dim_{\mathbb{C}} X_{\mathbb{C}} = n$.
(v) The complex linear space $X_{\mathbb{C}}$ is linearly isomorphic to the complex linear space $\mathcal{X}_{\mathbb{C}}$ constructed above.
(vi) If $X = \mathbb{R}$, then $X_{\mathbb{C}} = \mathbb{C}$.
Hint: Ad (i). For example, let $\beta \otimes x = \beta_1 \otimes x_1 + \beta_2 \otimes x_2$. We have to show that

$$\alpha\beta \otimes x = \alpha\beta_1 \otimes x_1 + \alpha\beta_2 \otimes x_2.$$

To this end, we choose a basis of the subset $\mathrm{span}\{x, x_1, x_2\}$ of X, and we choose the basis $1, \mathrm{i}$ of $\mathbb{C}$. Reducing the tensor products to the products of basis elements, the equality is obvious by the distributive law for tensor products.

3.15 *The general Clifford algebra.* Prove the properties of $\bigvee(X)$ summarized on page 175. Hint: See B. van der Waerden, Algebra, Vol. 2, Frederyck Ungar, New York, 1975, and T. Friedrich, Dirac Operators in Riemannian Geometry, Chap. 1, Amer. Math. Soc., Providence, Rhode Island, 2000.

3.16 *The Hilbert space* $\mathrm{End}(X)$. Let X be an n-dimensional Hilbert space over $\mathbb{K} = \mathbb{R}, \mathbb{C}$. Let $\mathrm{End}(X)$ denote the space of all linear operators $A : X \to X$. Define both the inner product

$$\langle A|B\rangle := \mathrm{tr}(AB^\dagger) \qquad \text{for all} \quad A, B \in \mathrm{End}(X)$$

and the corresponding norm $||A|| := \sqrt{\langle A|A\rangle}$. Show that $\mathrm{End}(X)$ is both a Hilbert space over $\mathbb{K}$ and a C^*-algebra. For the definition of C^*-algebras, see Sect. 7.16.3 of Vol. II. In particular, we have $||AB|| \le ||A||\,||B||$, as well as $||A^\dagger|| = ||A||$ and $||AA^\dagger|| = ||A^\dagger A|| = ||A||^2$ for all $A, B \in \mathrm{End}(X)$.
Hint: Show that $||A||^2 = \sum_{i,j=1}^n |A^i_j|^2$ for the matrix elements A^i_j of A. This implies $||AB||^2 \le ||A||^2||B||^2$ by using the Schwarz inequality. In order to understand the simple idea of the proof, consider first (2×2)-matrices.

3.17 *The Hopf algebra* $SL_q(2, \mathbb{C})$. Prove Theorem 3.22 on page 302.
Hint: See A. Klimyk and K. Schmüdgen, Quantum Groups and Their Representations, Sect. 4.1, Springer Berlin, 1997.

3.18 *The Hopf algebra* $sl_q(2, \mathbb{C})$. Prove Theorem 3.23 on page 302.
Hint: See A. Klimyk and K. Schmüdgen, Quantum Groups, Sect. 3.1, Springer, Berlin, 1997. This monograph contains a lot of material on both the general theory and concrete examples. In particular, one finds the Drinfeld–Jimbo quantum algebra $U_q(\mathcal{L})$ for arbitrary finite-dimensional complex semi-simple Lie algebras $\mathcal{L}$.

3.19 *Simple Lie algebras.* Prove that the complex Lie algebra $sl_{\mathbb{C}}(2, \mathbb{C})$ and the real Lie algebra $sl(2, \mathbb{R})$ are simple.

3.20 *The exponential map is not always surjective.* Prove that the map

$$\exp : sl(2, \mathbb{C}) \to SL(2, \mathbb{C})$$

given by $A \mapsto \mathrm{e}^A$ is not surjective.

Hint: Show that there is no matrix $A \in sl(2, \mathbb{C})$ with $\mathrm{e}^A = \begin{pmatrix} -1 & 1 \\ 0 & -1 \end{pmatrix}$.

See Hein (1990), p. 116 (quoted in Problem 3.19).

3.21 *Symmetric Schur polynomials.* Prove Prop. 3.1 on page 186.
Hint: Noting that $\frac{1}{1-ac} = \sum_{k=0}^{\infty}(ac)^k$, we get

$$D := \begin{vmatrix} \sum_{n=0}^{\infty}(ac)^n & \sum_{n=0}^{\infty}(ad)^n \\ \sum_{n=0}^{\infty}(bc)^n & \sum_{n=0}^{\infty}(bd)^n \end{vmatrix} = \begin{vmatrix} \frac{1}{1-ac} & \frac{1}{1-ad} \\ \frac{1}{1-bc} & \frac{1}{1-bd} \end{vmatrix}.$$

This yields

$$D = \frac{(a-b)(c-d)}{(1-ac)(1-ad)(1-bc)(1-bd)}.$$

Alternatively, since the determinant D is additive in the rows, we obtain that D is an infinite sum of determinants of the form

$$\begin{vmatrix} (ac)^k & (ad)^k \\ (bc)^l & (bd)^l \end{vmatrix} = a^k b^l \begin{vmatrix} c^k & d^k \\ c^l & d^l \end{vmatrix}.$$

Summing up this, we get an infinite sum of products of determinants which appear in the definition (3.13) of Schur polynomials. Finally, dividing this by $(a-b)(c-d)$, we get the claim. See C. Procesi, Lie Groups: An Approach Through Invariants and Representations, Springer, New York, 2007, p. 33.

3.22 *The irreducible representations of the group Sym(3) via Young tableaux.* Complete the proof sketched on page 224.

3.23 *The Gell-Mann matrices.* Show that the multiples $i\lambda_1, \ldots, i\lambda_8$ of the Gell-Mann matrices $\lambda_1, \ldots \lambda_8$ from (3.51) on page 231 form a basis of the real Lie algebra $su(3)$.
Hint: One has to show that every complex (3×3)-matrix A with $A = -A^\dagger$ and $\mathrm{tr}(A) = 0$ can be uniquely represented as $A = \sum_{j=1}^{8} a_j \cdot i\lambda_j$ with $a_j \in \mathbb{R}$ for all j. This yields a system of 8 equations. An explicit computation shows that the coefficient determinant of this system is different from zero. See W. Greiner and B. Müller, Quantum Mechanics: Symmetries, Sect. 7.1, Springer, New York, 1996.
The vanishing of the determinant also shows that the Gell-Mann matrices $\lambda_1, \ldots, \lambda_8$ are linearly independent over the field $\mathbb{C}$ of complex numbers. Explicitly, it follows from $\sum_{k=1}^{8} a_k = 0$ with $a_k \in \mathbb{C}$ for all k that $a_k = 0$ for all indices $k = 1, \ldots, 8$.

3.24 *The adjoint representation of a Lie algebra.* Let $\mathcal{L}$ be a real Lie algebra. Fix $A \in \mathcal{L}$. Define

$$\mathrm{ad}(A)(B) := [A, B] \qquad \text{for all} \quad B \in \mathcal{L}.$$

This way, we get the linear operator $\mathrm{ad}(A) : \mathcal{L} \to \mathcal{L}$. Use the Jacobi identity in order to show that the map $A \mapsto \mathrm{ad}(A)$ is a representation of the Lie algebra $\mathcal{L}$ on the linear space $\mathcal{L}$. In other words, the map

$$\mathrm{ad} : \mathcal{L} \to gl(\mathcal{L})$$

is a Lie algebra morphism.
Solution: The Jacobi identity reads as

$$[\,A, [B, C]\,] + [\,B, [C, A]\,] + [\,C, [A, B]\,] = 0, \qquad A, B, C \in \mathcal{L}.$$

We have to show that

$$[\mathrm{ad}(A), \mathrm{ad}(B)]C = \mathrm{ad}([A, B])C, \qquad A, B, C \in \mathcal{L}.$$

This follows from

$$\begin{aligned}\mathrm{ad}(A)\big(\mathrm{ad}(B)C\big) - \mathrm{ad}(B)\big(\mathrm{ad}(A)C\big) &= \mathrm{ad}(A)[B,C] - \mathrm{ad}(B)[A,C] \\ &= [\,A,[B,C]\,] - [\,B,[A,C]\,] = [\,[A,B],C] = \mathrm{ad}([A,B])C.\end{aligned}$$

3.25 *Method of highest weight.* Show that $\mathbf{3}\otimes\mathbf{3} = \mathbf{6} + \bar{\mathbf{3}}$. Moreover, check (3.60).

3.26 *The Casimir operators of the group SU(3).* Prove the commutation relations (3.80) for the Casimir operators C_1 and C_2. Moreover, show that the adjoint representation $R(1,1)$ of the group $SU(3)$ has the quantum numbers $c_1 = 3$ and $c_2 = 0$.
Hint: The explicit computations can be found in W. Greiner and W. Müller, Quantum Mechanics: Symmetries, 7.7/7.8, Springer, New York, 1995, together with many interesting examples in elementary particle physics (e.g., the computation of decay rates for particles based on the $SU(3)$ symmetry).

3.27 *Tensor representation.* Let $\varrho_1, \varrho_2, \varrho_3$ be a complete system of irreducible representations of the permutation group $Sym(3)$ (see Table 3.5 on page 216). Show that

$$\varrho_3 \otimes \varrho_3 = \varrho_1 \oplus \varrho_2 \oplus \varrho_3.$$

Hint: Use the Fourier method (3.33) on page 212.

4. The Euclidean Manifold $\mathbb{E}^3$

4.1 Velocity Vectors and the Tangent Space

Tangent spaces are spaces of velocity vectors.
Folklore

Let us use the notation introduced at the beginning of Sect. 1.2 on page 71. Consider the motion

$$P = P(t), \qquad t \in \mathbb{R}$$

of a particle (Fig. 4.1). Equivalently, we write

$$\mathbf{x} = \mathbf{x}(t), \qquad t \in \mathbb{R}. \tag{4.1}$$

Here, $\mathbf{x}(t)$ denotes the position vector starting at the origin O at time t with the terminal point $P(t) = O + \mathbf{x}(t)$. Let $E_3(P)$ denote the space of all the position vectors starting at the point P. This is a real 3-dimensional Hilbert space equipped with the inner product $\langle \mathbf{u}|\mathbf{w}\rangle_P := \mathbf{uw}$ and the norm $|\mathbf{u}|_P := \sqrt{\langle \mathbf{u}|\mathbf{u}\rangle_P}$ for all $\mathbf{u}, \mathbf{w} \in E_3(P)$.

Velocity vector. By definition, the velocity vector $\mathbf{v} = \dot{\mathbf{x}}(t)$ of the motion (4.1) at the point $P(t)$ is given by

$$\boxed{\dot{\mathbf{x}}(t) := \lim_{\Delta t \to 0} \frac{\mathbf{x}(t+\Delta t) - \mathbf{x}(t)}{\Delta t}.} \tag{4.2}$$

This limit is to be understood in the sense of the Hilbert space $E_3(P(t))$, that is,

$$\lim_{\Delta t \to 0} \left| \dot{\mathbf{x}}(t) - \frac{\mathbf{x}(t+\Delta t) - \mathbf{x}(t)}{\Delta t} \right|_{P(t)} = 0.$$

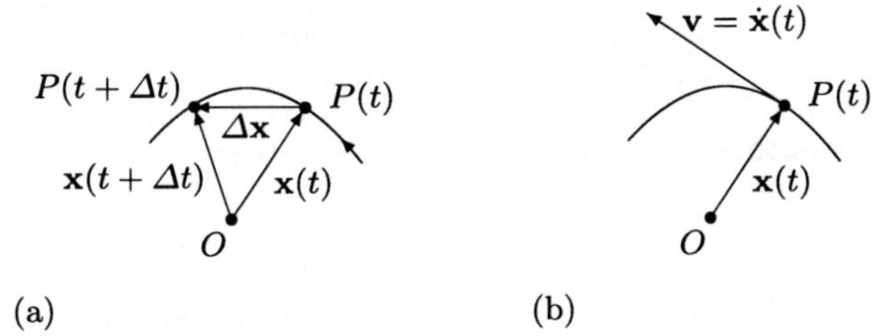

Fig. 4.1. Velocity vector

E. Zeidler, *Quantum Field Theory III: Gauge Theory*,
DOI 10.1007/978-3-642-22421-8_5,

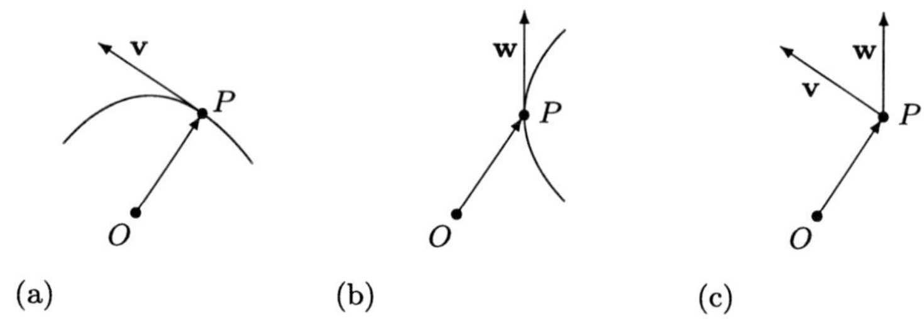

Fig. 4.2. The tangent space $T_P\mathbb{E}^3$

Here, after parallel transport, the vector $\Delta\mathbf{x} := \mathbf{x}(t+\Delta t) - \mathbf{x}(t)$ is regarded as a position vector starting at the point $P(t)$ (Fig. 4.1). Therefore, $\dot{\mathbf{x}}(t) \in E_3(P(t))$. Explicitly, if $\mathbf{i}, \mathbf{j}, \mathbf{k}$ is the right-handed orthonormal basis of a Cartesian coordinate system (Fig. 1.1(c) on page 72), then the position vector $\mathbf{x}(t)$ starting at the origin O can be written as

$$\mathbf{x}(t) = x(t)\,\mathbf{i} + y(t)\,\mathbf{j} + z(t)\,\mathbf{k}.$$

To avoid technicalities, we assume that the functions $t \mapsto x(t), y(t), z(t)$ are smooth for all times $t \in \mathbb{R}$. Note that this assumption does not depend on the choice of the Cartesian coordinate system. Then

$$\dot{\mathbf{x}}(t) = \dot{x}(t)\,\mathbf{i}_{P(t)} + \dot{y}(t)\,\mathbf{j}_{P(t)} + \dot{z}(t)\,\mathbf{k}_{P(t)}, \tag{4.3}$$

where $\mathbf{i}_{P(t)}, \mathbf{j}_{P(t)}, \mathbf{k}_{P(t)}$ are position vectors starting at the point $P(t)$, which are obtained from the position vectors $\mathbf{i}, \mathbf{j}, \mathbf{k}$ starting at the origin O, respectively, by parallel transport (Fig. 4.3).

Tangent vector. Recall that, by definition, the 3-dimensional Euclidean manifold $\mathbb{E}^3$ consists of all the points P which can be represented as $P = O + \mathbf{x}$ with $\mathbf{x} \in E_3(O)$. Intuitively, $\mathbb{E}^3$ consists of all the points of the 3-dimensional space in the naive sense. Moreover, by definition, the tangent space $T_P\mathbb{E}^3$ of the Euclidean manifold $\mathbb{E}^3$ is the space of all possible velocity vectors $\mathbf{v}$ at the point P. This means that we consider all the possible smooth motions of particles which pass through the point P, and we take all the possible velocity vectors of these trajectories at the point P.

Velocity vectors are also called tangent vectors.

In particular, choosing the motion along straight lines, we get

$$T_P\mathbb{E}^3 = E_3(P).$$

That is, the tangent space $T_P\mathbb{E}^3$ coincides with the Hilbert space $E_3(P)$.

The reader should note the following. In order to understand the theory of general manifolds later on, it is wise to distinguish between the tangent space $T_P\mathbb{E}^3$ at the point P and the tangent space $T_Q\mathbb{E}^3$ at the point Q if $P \neq Q$. Of course, in the special case of the Euclidean manifold $\mathbb{E}^3$, the velocity vectors

$$\mathbf{v} \in T_P\mathbb{E}^3$$

can be identified with the velocity vectors $\mathbf{w} \in T_Q\mathbb{E}^3$ by a global parallel transport. However, such a global parallel transport is not possible anymore on general manifolds (e.g., in Einstein's theory of general relativity). Therefore, we will use a language which remains valid when passing to the general case. For example, we will use the operator $\Pi_{\Delta t}$ of parallel transport in (4.5) below.

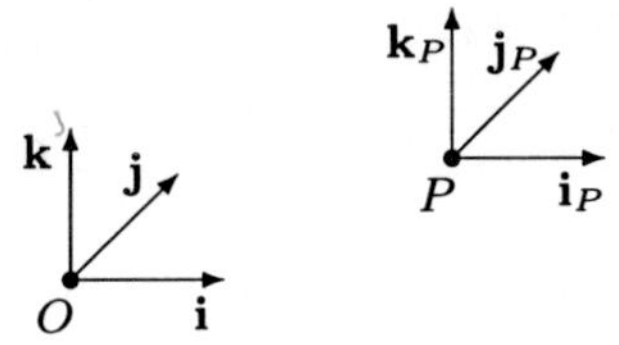

Fig. 4.3. Right-handed orthonormal basis of the tangent space $T_P\mathbb{E}^3$

4.2 Duality and Cotangent Spaces

Cotangent spaces are the dual spaces to tangent spaces.
Folklore

The dual space $T_P^*\mathbb{E}^3$ to the tangent space $T_P\mathbb{E}^3$ is called the cotangent space of the Euclidean manifold $\mathbb{E}^3$ at the point P. The elements ω of the cotangent space $T_P\mathbb{E}_3$ are linear functionals

$$\omega : T_P\mathbb{E}^3 \to \mathbb{R}.$$

Here, ω is also called a covector. In a Cartesian coordinate system, we have

$$\mathbf{v} = u\,\mathbf{i}_P + v\,\mathbf{j}_P + w\,\mathbf{k}_P$$

for all tangent vectors (velocity vectors) $\mathbf{v} \in T_P\mathbb{E}^3$ at the point P. Here, $u, v, w \in \mathbb{R}$. Define

$$dx(\mathbf{v}) := u, \quad dy(\mathbf{v}) := v, \quad dz(\mathbf{v}) := w.$$

Then, dx, dy, dz is a basis of the cotangent space $T_P^*\mathbb{E}^3$. That is, for every $\omega \in T_P^*\mathbb{E}^3$ we get

$$\omega = \alpha dx + \beta dy + \gamma dz$$

where the coefficients α, β, γ are uniquely determined real numbers. Explicitly, we have $\omega(\mathbf{v}) = \alpha u + \beta v + \gamma w$. The numbers α, β, γ are called the coordinates of the covector ω.

4.3 Parallel Transport and Acceleration

The acceleration vector $\ddot{\mathbf{x}}(t)$ of the motion (4.1) at time t is defined by

$$\ddot{\mathbf{x}}(t) := \lim_{\Delta t \to 0} \frac{\dot{\mathbf{x}}(t + \Delta t) - \dot{\mathbf{x}}(t)}{\Delta t}. \tag{4.4}$$

This limit is to be understood in the sense of the Hilbert space (tangent space) $T_{P(t)}\mathbb{E}^3$. This means the following:

- The velocity vector $\dot{\mathbf{x}}(t)$ lies in the Hilbert space $T_{P(t)}\mathbb{E}^3$, whereas
- the velocity vector $\dot{\mathbf{x}}(t + \Delta t)$ lies in the Hilbert space $T_{P(t+\Delta t)}\mathbb{E}^3$.

In order to get the limit (4.4) in the Hilbert space $T_{P(t)}\mathbb{E}^3$, we move

- the position vector $\dot{\mathbf{x}}(t + \Delta t)$ starting at the point $P(t + \Delta t)$ to
- the corresponding position vector $\Pi_{-\Delta t}\dot{\mathbf{x}}(t + \Delta t)$ starting at the point $P(t)$, by using parallel transport in the classical sense.

Then we get

$$\ddot{\mathbf{x}}(t) = \lim_{\Delta t \to 0} \frac{\Pi_{-\Delta t}\dot{\mathbf{x}}(t+\Delta t) - \dot{\mathbf{x}}(t)}{\Delta t}. \tag{4.5}$$

Hence $\ddot{\mathbf{x}}(t) \in T_{P(t)}\mathbb{E}^3$. Explicitly, similarly as in (4.3), we obtain

$$\ddot{\mathbf{x}}(t) = \ddot{x}(t)\,\mathbf{i}_{P(t)} + \ddot{y}(t)\,\mathbf{j}_{P(t)} + \ddot{z}(t)\,\mathbf{k}_{P(t)}.$$

4.4 Newton's Law of Motion

Suppose that we are given a force field

$$\mathbf{F} = \mathbf{F}(P), \qquad P \in \mathbb{E}^3$$

on the Euclidean manifold $\mathbb{E}^3$. Here, the force $\mathbf{F}(P)$ acting at the point P is a position vector starting at the point P. That is, the force $\mathbf{F}(P)$ is an element of the tangent space $T_P\mathbb{E}^3$. For the trajectory $P = P(t), t \in \mathbb{R}$, of a particle of positive mass m, Newton's law of motion in classical mechanics reads as

$$\boxed{m\ddot{\mathbf{x}}(t) = \mathbf{F}(P(t)), \qquad t \in \mathbb{R}.} \tag{4.6}$$

Using a Cartesian coordinate system, we obtain that

$$m(\ddot{x}(t)\,\mathbf{i}_{P(t)} + \ddot{y}(t)\,\mathbf{j}_{P(t)} + \ddot{z}(t)\,\mathbf{k}_{P(t)})$$

is equal to

$$F(P(t))\,\mathbf{i}_{P(t)} + G(P(t))\,\mathbf{j}_{P(t)} + H(P(t))\,\mathbf{k}_{P(t)}.$$

This is equivalent to the system

$$\begin{aligned} m\ddot{x}(t) = F(x(t),y(t),z(t)), \quad m\ddot{y}(t) = G(x(t),y(t),z(t)), \\ m\ddot{z}(t) = H(x(t),y(t),z(t)), \qquad t \in \mathbb{R}. \end{aligned} \tag{4.7}$$

4.5 Bundles Over the Euclidean Manifold

> In modern differential geometry, velocity vector fields, covector fields, and tensor fields on manifolds are described by sections of the tangent bundle, the cotangent bundle, and the tensor bundle, respectively. Later on, this will allow us to study the global (i.e., topological) properties of physical fields.
>
> Folklore

4.5.1 The Tangent Bundle and Velocity Vector Fields

By definition, the tangent bundle $T\mathbb{E}^3$ of the Euclidean manifold $\mathbb{E}^3$ consists of all the pairs

$$(P, \mathbf{v})$$

where $P \in \mathbb{E}^3$ and $\mathbf{v} \in T_P\mathbb{E}^3$. The map

$$\boxed{s : \mathbb{E}^3 \to T\mathbb{E}^3} \tag{4.8}$$

is called a section of the tangent bundle iff $s(P) = (P, \mathbf{v}(P))$ and $\mathbf{v}(P) \in T_P\mathbb{E}^3$ for all $P \in \mathbb{E}^3$. In other words, this is a velocity vector field (also called tangent vector field) which assigns to each point P of the Euclidean manifold $\mathbb{E}^3$ the velocity vector $\mathbf{v}(P)$ at the point P.

The tangent bundle $T\mathbb{E}^3$ is a real 6-dimensional manifold.

To see this, consider a right-handed Cartesian coordinate system. Then the point P corresponds to the position vector

$$\mathbf{x} = x\mathbf{i} + y\mathbf{j} + z\mathbf{k}$$

starting at the origin O, and the velocity vector $\mathbf{v}$ at the point P is given by

$$\mathbf{v} = u\,\mathbf{i}_P + v\,\mathbf{j}_P + w\,\mathbf{k}_P.$$

Naturally enough, we assign to the point $(P, \mathbf{v})$ of the tangent bundle $T\mathbb{E}^3$ the coordinates $(x, y, z; u, v, w)$. The section (4.8) is called smooth iff it is a smooth map from the Euclidean manifold $\mathbb{E}^3$ into the manifold $T\mathbb{E}^3$. Intuitively, this means that the velocity vector field $P \mapsto \mathbf{v}(P)$ depends smoothly on the point P. Equivalently, the map

$$(x, y, z) \mapsto (u(x, y, z), v(x, y, z), w(x, y, z))$$

is smooth on $\mathbb{R}^3$. Note that this smoothness property does not depend on the choice of the Cartesian coordinate system.

4.5.2 The Cotangent Bundle and Covector Fields

By definition, the cotangent bundle $T^*\mathbb{E}^3$ of the Euclidean manifold $\mathbb{E}^3$ consists of all the pairs

$$(P, \omega)$$

where $P \in \mathbb{E}^3$ and $\omega \in T_P^*\mathbb{E}^3$. The map

$$s : \mathbb{E}^3 \to T^*\mathbb{E}^3 \tag{4.9}$$

is called a section of the cotangent bundle iff $s(P) = (P, \omega(P))$ and $\omega(P) \in T_P^*\mathbb{E}^3$ for all $P \in \mathbb{E}^3$. In other words, this is a covector field (also called cotangent vector field) which assigns to each point P of the Euclidean manifold $\mathbb{E}^3$ the covector $\omega(P) \in T^*\mathbb{E}^3$ at the point P.

The cotangent bundle $T^*\mathbb{E}^3$ is a real 6-dimensional manifold.

To see this, consider a Cartesian coordinate system. Then the point P corresponds to the position vector

$$\mathbf{x} = x\mathbf{i} + y\mathbf{j} + z\mathbf{k}$$

starting at the origin O, and the covector ω at the point P is given by

$$\omega = \alpha\, dx + \beta\, dy + \gamma\, dz.$$

Naturally enough, we assign to the point (P, ω) of the cotangent bundle $T^*\mathbb{E}^3$ the coordinates $(x, y, z; \alpha, \beta, \gamma)$. The section (4.9) is called smooth iff it is a smooth map from the Euclidean manifold $\mathbb{E}^3$ into the manifold $T^*\mathbb{E}^3$. Intuitively, this means that the covector field $P \mapsto \omega(P)$ depends smoothly on the point P. Equivalently, the map

$$(x, y, z) \mapsto (\alpha(x, y, z), \beta(x, y, z), \gamma(x, y, z))$$

is smooth on $\mathbb{R}^3$. Note that this smoothness property does not depend on the choice of the Cartesian coordinate system.

4.5.3 Tensor Bundles and Tensor Fields

By definition, the tensor bundle $T^m_n(\mathbb{E}^3)$ of type (m, n) of the Euclidean manifold $\mathbb{E}^3$ consists of all the pairs

$$(P, \mathcal{T})$$

where $P \in \mathbb{E}^3$ and $\mathcal{T} \in \bigotimes^m_n(T_P\mathbb{E}^3)$. Here, the symbol $\bigotimes^m_n(T_P\mathbb{E}^3)$ denotes the linear space of the tensors of type (m, n) on the tangent space $T_P\mathbb{E}^3$. The map

$$s : \mathbb{E}^3 \to T^m_n(\mathbb{E}^3) \tag{4.10}$$

is called a section of the tensor bundle of type (m, n) iff

$$s(P) = (P, \mathcal{T}(P))$$

and $\mathcal{T}(P) \in \bigotimes^m_n(T_P\mathbb{E}^3)$ for all $P \in \mathbb{E}^3$. In other words, this is a tensor field which assigns to each point P of the Euclidean manifold $\mathbb{E}^3$ the tensor $\mathcal{T}(P) \in \bigotimes^m_n(T_P\mathbb{E}^3)$ at the point P.

The tensor bundle $T^m_n(\mathbb{E}^3)$ is a real manifold of dimension $3 + 3(m + n)$.

To see this, consider a Cartesian coordinate system. Then the point P corresponds to the position vector $\mathbf{x} = x\mathbf{i} + y\mathbf{j} + z\mathbf{k}$ starting at the origin O, and the tensor $\mathcal{T}$ at the point P is given by

$$\mathcal{T} = \sum_{i_1,\ldots,i_m,j_1,\ldots,j_n=1}^{3} \mathcal{T}^{i_1\ldots i_m}_{j_1\ldots j_n} \cdot \mathbf{b}_{i_1}(P) \otimes \cdots \otimes \mathbf{b}_{i_m}(P) \otimes dx^{j_1} \otimes \cdots \otimes dx^{j_n}.$$

Here, $\mathbf{b}_1 := \mathbf{i}_P, \mathbf{b}_2 := \mathbf{j}_P, \mathbf{b}_3 := \mathbf{k}_P$, and $dx^1 = dx, dx^2 := dy, dx^3 := dz$. Naturally enough, we assign to the point $(P, \mathcal{T})$ of the tensor bundle $T^m_n(\mathbb{E}^3)$ the $3+3(m+n)$ coordinates

$$(x, y, z; \mathcal{T}^{i_1\ldots i_m}_{j_1\ldots j_n}), \qquad i_1, \ldots, i_m, j_1, \ldots, j_n = 1, 2, 3.$$

The section (4.10) is called smooth iff it is a smooth map from the Euclidean manifold $\mathbb{E}^3$ into the manifold $T^m_n(\mathbb{E}^3)$. Intuitively, this means that the tensor field $P \mapsto \mathcal{T}(P)$ depends smoothly on the point P. Equivalently, all the maps

$$(x, y, z) \mapsto \mathcal{T}^{i_1\ldots i_m}_{j_1\ldots j_n}(x, y, z), \qquad i_1, \ldots, i_m, j_1, \ldots, j_n = 1, 2, 3$$

are smooth on $\mathbb{R}^3$. Note that this smoothness property does not depend on the choice of the Cartesian coordinate system.

4.5.4 The Frame Bundle

By definition, the frame bundle $F\mathbb{E}^3$ of the Euclidean manifold $\mathbb{E}^3$ consists of all the quadruples

$$(P, \mathbf{b}_1, \mathbf{b}_2, \mathbf{b}_3)$$

where $P \in \mathbb{E}^3$, and the vectors $\mathbf{b}_1, \mathbf{b}_2, \mathbf{b}_3$ form a basis of the tangent space $T_P\mathbb{E}^3$ at the point P. The map

$$s : \mathbb{E}^3 \to F\mathbb{E}^3 \tag{4.11}$$

is called a section of the frame bundle iff

$$s(P) = (P, \mathbf{b}_1(P), \mathbf{b}_2(P), \mathbf{b}_3(P))$$

and the vectors $\mathbf{b}_1(P), \mathbf{b}_2(P), \mathbf{b}_3(P))$ form a basis of $T_P\mathbb{E}^3$ for all $P \in \mathbb{E}^3$. In other words, the section (4.11) is a map which fixes a basis at each point P of the Euclidean manifold.

The frame bundle $F\mathbb{E}^3$ is a real 12-dimensional manifold.

To see this, consider a Cartesian coordinate system with the right-handed orthonormal basis $\mathbf{i}, \mathbf{j}, \mathbf{k}$ at the origin O. Then the point P corresponds to the position vector

$$\mathbf{x} = x\mathbf{i} + y\mathbf{j} + z\mathbf{k}.$$

Moreover, the basis $\mathbf{b}_1, \mathbf{b}_2, \mathbf{b}_3$ at the point P is given by the matrix equation

$$\begin{pmatrix} \mathbf{b}_1 \\ \mathbf{b}_2 \\ \mathbf{b}_3 \end{pmatrix} = \begin{pmatrix} G_1^1 & G_2^1 & G_3^1 \\ G_1^2 & G_2^2 & G_3^2 \\ G_1^3 & G_2^3 & G_3^3 \end{pmatrix} \begin{pmatrix} \mathbf{i}_P \\ \mathbf{j}_P \\ \mathbf{k}_P \end{pmatrix}.$$

Here, the real (3×3)-matrix $\mathcal{G} = (G_j^i)$ is invertible. Naturally enough, we assign to the point $(P, \mathbf{b}_1, \mathbf{b}_2, \mathbf{b}_3)$ of the frame bundle $F\mathbb{E}^3$ the 12 coordinates $(x, y, z; G_1^1, G_2^1, \ldots, G_3^3)$. The section (4.11) is called smooth iff it is a smooth map from the Euclidean manifold $\mathbb{E}^3$ into the manifold $F(\mathbb{E}^3)$. Equivalently, the map

$$(x, y, z) \mapsto (G_1^1(x, y, z), G_2^1(x, y, z), \ldots, G_3^3(x, y, z))$$

is smooth on $\mathbb{R}^3$. Note that this smoothness property does not depend on the choice of the Cartesian coordinate system.

4.6 Historical Remarks

4.6.1 Newton and Leibniz

Newton (1643–1727) motivated his differential calculus by the kinematics of bodies. He used the terms

- 'fluent' x (a magnitude that flows in time) for 'variable depending on time', and
- 'fluxion' $\dot{x}$ for 'time derivative'.

He also introduced the symbol $\ddot{x}$ for the second time derivative. Moreover, he used the symbols $o, xo, \dot{x}o$ which correspond to the differentials $dt, dx, d\dot{x}$ in modern terminology, respectively. Newton's differential calculus can be found in I. Newton, *Methodus fluxiorum et serierum infinitorum.* The Latin manuscript was ready in 1671, but the book only appeared ten years after Newton's death as an English translation under the title, *The Method of Fluxions and Infinite Series with its Application to the Geometry of Curve-Lines,* London, 1736. Newton's calculus was first published in his treatise

Philosophiae Naturalis Principia Mathematica, 1687 (in Latin).[1]

This was one of the most important single works in the history of modern science.

In 1672, Leibniz (1646–1716) visited Paris. Huygens (1629–1695) introduced the young Leibniz into his theory of curves. In the next years, Leibniz was fascinated by mathematics. In 1684, he published an article in the newly founded Leipzig Acta Eruditorum entitled *A new method for maxima and minima as well for tangents* (in Latin). In this article, Leibniz introduced the symbol dx satisfying the sum rule $d(x+y) = dx + dy$ and the product rule

$$\boxed{d(xy) = (dx)y + x(dy)} \tag{4.12}$$

called the Leibniz rule today. In 1686, Leibniz published his article *On a deeply hidden geometry* where he used the symbol $\int$ for integration. This symbol was designed by resembling the letter 'S' for 'summation'.

Élie Cartan's calculus of alternating differential forms. This beautiful and extremely elegant calculus was invented by Cartan (1869–1951) in 1899; Cartan used the wedge product $a \wedge b$ introduced by Grassmann (1809–1877) in 1844. We will extensively study this in Chap. 12. The key relations are the graded Leibniz rule

$$\boxed{d(\omega \wedge \mu) = d\omega \wedge \mu + (-1)^p \omega \wedge d\mu} \tag{4.13}$$

and the Poincaré cohomology rule

$$\boxed{d(d\omega) = 0.} \tag{4.14}$$

Here, ω and μ are differential forms of degree p and q, respectively.

The main theorem of calculus. The fundamental relation

$$\int_0^1 f'(x)dx = f(1) - f(0)$$

and its far-reaching generalization to the Stokes integral theorem

$$\boxed{\int_{\mathcal{M}} d\omega = \int_{\partial\mathcal{M}} \omega}$$

for differential forms ω on a manifold $\mathcal{M}$ will be discussed in Sect. 12.1.1.

The Fréchet derivative and the calculus on infinite-dimensional function spaces. This theory was initiated by Fréchet (1878–1973) at the beginning of the 20th century. We refer to Zeidler (1986), Vol. I, quoted on page 1089.

Further reading. Concerning the history of calculus, we recommend:

[1] The Principles of Natural Philosophy.

S. Chandrasekhar, Newton's Principia for the Common Reader, Oxford University Press, 1997.

N. Rescher, Leibniz: An Introduction to his Philosophy, Rowman and Littlefield, Totowa, New Jersey, 1979.

M. Kline, Mathematical Thought from Ancient to Modern Times, Vols. 1–3, Oxford University Press, 1990.

N. Bourbaki, Elements of the History of Mathematics, Springer, New York, 1994.

J. Dieudonné, Abrégé d'histoire des mathématiques 1700–1900, I, II, Hermann, Paris, 1978. Extended German edition in one volume: Vieweg, Wiesbaden, 1985.

J. Dieudonné, History of Algebraic Geometry, 400 B.C.–1985 A.D., Chapman, New York, 1985.

J. Dieudonnè, A History of Algebraic and Differential Topology, 1900–1960, Birkhäuser, Boston, 1989.

J. Dieudonné, History of Functional Analysis, 1900–1975, North-Holland, Amsterdam, 1981.

F. Klein, Development of Mathematics in the 19th Century, Math. Sci. Press, New York, 1979.

V. Varadarajan, Euler Through Time: A New Look at Old Themes, Amer. Math. Soc., Providence, Rhode Island, 2006.

I. James (Ed.), History of Topology, Elsevier, Amsterdam, 1999.

A. Pietsch, History of Banach Spaces and Linear Operators, Birkhäuser, Boston, 2007.

H. Wußing, 6000 Years of Mathematics: a Cultural Journey through Time, Springer, Heidelberg, 2009 (in German).

E. Zeidler, Some Reflections on the Future of Mathematics. In: H. Wußing (2009), Vol. II, Chap. 12 (in German).

4.6.2 The Lebesgue Integral

The Lebesgue integral was the key to modern analysis based on functional analysis in the twentieth century.

Folklore

In his 1851 Ph.D. thesis, Riemann (1826–1866) founded complex analysis based on the so-called Cauchy–Riemann differential equations. In his 1854 habilitation thesis, Riemann investigated the convergence of Fourier series. In order to define the Fourier coefficients by integrals, he introduced the so-called Riemann integral

$$\int_0^1 f(x)dx = \lim_{n\to\infty} \sum_{j=1}^{n} f(j\Delta x)\Delta x$$

where $\Delta x = \frac{1}{n}$. In his seminal 1902 thesis, Lebesgue (1875–1941) founded modern measure and integration theory.[2] In contrast to the Riemann integral, the Lebesgue integral has the crucial property that the limit relation

[2] H. Lebesgue, Integral, length, area, Annali Mat. Pura Appl. **7** (1902), 231–359 (in French).

$$\boxed{\lim_{n\to\infty}\int_0^1 f_n(x)dx = \int_0^1 \lim_{n\to\infty} f_n(x)dx}$$

holds under quite natural assumptions. In particular, this implies that the function space $L_2(0,1)$ becomes not only a complex pre-Hilbert space with respect to the inner product

$$\langle f|g\rangle := \int_0^1 f(x)^\dagger g(x)dx,$$

but a Hilbert space. The Lebesgue integral is basic for the modern functional-analytic approach to linear and nonlinear partial differential equations and integral equations. In particular, von Neumann's rigorous approach to the Schrödinger equation in quantum mechanics via Hilbert space theory is based on the use of the Lebesgue integral (see Chapter 7 of Volume II). For further reading, we recommend:

V. Zorich, Analysis I, II, Springer, New York, 2003.

E. Stein and R. Shakarchi, Princeton Lectures in Analysis. I: Fourier Analysis, II: Complex Analysis, III: Measure Theory, Princeton University Press, 2003.

H. Triebel, Higher Analysis, Barth, Leipzig, 1989.

G. Gustafson and I. Sigal, Mathematical Concepts of Quantum Mechanics, Springer, New York, 2003.

4.6.3 The Dirac Delta Function and Laurent Schwartz's Distributions

Generalizing Dirac's idea of the 'Dirac delta function', Laurent Schwartz (1915–2002) created the theory of distributions (generalized functions) in 1945. In this setting, the Dirac delta distribution is given by the functional

$$\delta(\varphi) := \varphi(0) \qquad \text{for all} \quad \varphi \in \mathcal{D}(\mathbb{R}).$$

Mnemonically, this is written as $\delta(\varphi) = \int_{\mathbb{R}} \delta(x) f(x)dx$. The distributions $T \in \mathcal{D}'(\mathbb{R})$ have the extremely nice property that they possess derivatives of any order, in contrast to classical functions. For a detailed study of distributions and their relation to the Dirac delta function, we refer to Chapters 11 and 12 of Volume I.

4.6.4 The Algebraization of the Calculus

In the history of mathematics, there emerged the following three possibilities for the algebraization of the classical differential calculus:

- derivations: Leibniz (1646–1716), Élie Cartan (1861–1951), Kähler, (1906–2000), Weil (1906–1996),
- formal power series expansions: Lagrange (1736–1813), Weierstrass (1815–1897), Hensel (1861–1941) (p-adic numbers), and
- nonstandard analysis: Robinson (1918–1974), Nelson (born 1932) (see Section 4.6 of Volume II).

In what follows we want to discuss some basic ideas.

4.6.5 Formal Power Series Expansions and the Ritt Theorem

By definition, a formal power series expansion over the field $\mathbb{C}$ of complex numbers is a symbol of the form

$$a_0 + a_1 x + a_2 x^2 + \ldots, \qquad a_j \in \mathbb{C},\ j = 0, 1, 2, \ldots$$

Quite naturally, we define the sum and the product by setting

- $(a_0 + a_1 x + \ldots) + (b_0 + b_1 x + \ldots) := a_0 + b_0 + (a_1 + b_1)x + \ldots,$
- $(a_0 + a_1 x + \ldots)(b_0 + b_1 x + \ldots) := a_0 b_0 + (a_0 b_1 + a_1 b_0)x + \ldots$

This way, we obtain the ring $\mathcal{P}_{\mathbb{C}}[x]$ of formal power series expansions.[3] In addition, we define the derivative

$$(a_0 + a_1 x + a_2 x^2 + a_3 x^3 + \ldots)' = a_1 + 2a_2 x + 3a_3 x^2 + \ldots$$

in a purely algebraic way. Such formal power series expansions were used by Lagrange and Weierstrass.

> *In quantum field theory, perturbation theory (including renormalization) is based on formal power series expansions with respect to small physical constants.*

A crucial theorem proved by Ritt in 1916 tells us that a formal power series expansion is always the asymptotic expansion of some analytic function defined in some conic domain (see Section 15.5.2 of Volume I).

From the analytic point of view, the machinery of formal power series expansions with applications to the solution of nonlinear systems of equations (bifurcation theory) can be found in:

> M. Vainberg and V. Trenogin, Theory of Branching of Solutions of Nonlinear Equations, Noordhoof, Leyden, The Netherlands, 1974 (the Weierstrass preparation theorem, resultants, Kronecker's elimination method).

The point is that a perfect theory for nonlinear equations only works if one gives up the search for convergent solutions. A summary of important results (solution of branching equations) can be found in Zeidler (1996), Vol. I, pp. 430–437 (quoted on page 1089). From the algorithmic point of view, we refer to:

> D. Cox, J. Little, and D. O'Shea, Using Algebraic Geometry, Springer, New York, 1998.

4.6.6 Differential Rings and Derivations

Erich Kähler (1906–2000) emphasized in 1958 that it is promising to establish the differential calculus on a purely algebraic basis.[4] Let R be a commutative ring with unit element. The ring R' is called a differential ring to R iff $\mathsf{R} \subseteq \mathsf{R}'$, and there exists a map $d : \mathsf{R} \to \mathsf{R}'$ such for all $a, b \in \mathsf{R}$ the following hold:

[3] If we replace the complex coefficients $a_0, a_1, \ldots$ by elements of a commutative ring R with unit element (e.g., the ring $\mathbb{Z}$ of integers), then we obtain the ring $\mathcal{P}_{\mathsf{R}}[x]$.

[4] E. Kähler, Algebra and Differential Calculus, Akademie-Verlag, Berlin, 1958 (in German). See also E. Kähler, Collected Works, pp. 282–387, de Gruyter, Berlin, 2003, and E. Kähler, Geometria Aritmetica, Annali di Matematica Pura et Applicata, Serie IV, Tomo XLV (1958) (in Italian).

- $d(a+b) = da + db$ (sum rule),
- $d(ab) = (da)b + a(db)$ (Leibniz rule), and
- $(da)(db) = 0$ (algebraic infinitesimals).

Define $\sigma(a) := a + da$ for all $a \in \mathsf{R}$. Then, for all $a, b \in \mathsf{R}$, we have

- $\sigma(a+b) = \sigma(a) + \sigma(b)$, and
- $\sigma(ab) = \sigma(a)\sigma(b)$.

That is, σ is a ring morphism from R to R', which was coined an 'infinitesimal motion' of the ring R by Kähler. In fact, it follows from $(da)(db) = 0$ that

$$\sigma(a)\sigma(b) = (a+da)(b+db) = ab + (da)b + a(db) = ab + d(ab) = \sigma(ab).$$

The prototype of a differential ring. By definition, the ring $\mathcal{P}_{\mathbb{C}}[x]'$ consists of all the symbols

$$p + qdx, \qquad p, q \in \mathcal{P}_{\mathbb{C}}[x]$$

together with the relation $(dx)^2 = 0$. Moreover, we define

$$d(p + q\, dx) := p'(x)\, dx.$$

In particular, this implies $d(dx) = 0$. The ring $\mathcal{P}_{\mathbb{C}}[x]'$ is a differential ring to the ring $\mathcal{P}_{\mathbb{C}}[x]$ of formal power series expansions.

Derivations. Let M be a left module over the commutative ring R with unit element (see page 309). The map $d : \mathsf{R} \to \mathsf{M}$ is called a derivation iff the following hold for all $a, b \in \mathsf{R}$:

- $d(a+b) = da + db$ (sum rule), and
- $d(ab) = (da)b + a(db)$ (Leibniz rule).

There exists a universal derivation on R which generates all the possible derivations from R to M. The proof based on tensor products can be found in S. Lang, Algebra, Sect. XIX.3, Springer, New York, 2002.

4.6.7 The p-adic Numbers

> God made the integers, all the rest is the work of Man.
>
> Leopold Kronecker (1823–1891)

> The p-adic numbers were invented at the beginning of the twentieth century by Kurt Hensel (1861–1941). The aim was to make the methods of power series expansions, which play a dominant role in the theory of complex-valued functions, available to the theory of numbers as well.[5]
>
> Jürgen Neukirsch, 1991

Many stuctural properties of classical numbers allow far-reaching generalizations. Let us discuss a few points.

The extension of the semi-ring $\mathbb{N}$ of natural numbers to the ring $\mathbb{Z}$ of integers and K-theory. The set $\mathbb{N}$ of natural numbers $0, 1, 2, 3, \ldots$ forms a semiring with respect to addition and multiplication. This semiring can be extended to the ring $\mathbb{Z}$ of integers. Interestingly enough, the corresponding construction can be generalized to more general objects (e.g., linear spaces, vector bundles) by means of

[5] J. Neukirsch, The p-adic numbers, pp. 155–178. In: H. Ebbinghaus et al. (Eds.), Numbers, Springer, New York, 1991 (reprinted with permission).

K-theory which was invented by Grothendieck (born 1928) in the 1950s; algebraic and topological K-theory represent a powerful tool of modern mathematics (see Section 4.4.9 of Volume II).

The extension of the ring $\mathbb{Z}$ of integers to the field $\mathbb{Q}$ of rational numbers is the prototype of the extension of a commutative ring without zero divisors to a quotient field (see Sect. 4.1.3 of Volume II).

Cantor's extension of the field $\mathbb{Q}$ of rational numbers to the field $\mathbb{R}$ of real numbers. In 1872, Cantor (1845– 1918) invented a method for completing the field of rational numbers $\mathbb{Q}$ to the field $\mathbb{R}$ of real numbers. Cantor considered sequences (a_n) of rational numbers, and he introduced the following equivalence relation:

$$(a_n) \sim (b_n)$$

iff for every $\varepsilon > 0$ there exists a natural number $n_0(\varepsilon)$ such that $|a_n - b_n| < \varepsilon$ for all $n \geq n_0(\varepsilon)$. In other words, the sequence $(a_n - b_n)$ is a Cauchy sequence. The corresponding equivalence classes $[(a_n)]$ form the field $\mathbb{R}$ of real numbers, and the set $\mathbb{Q}$ is a dense subfield of $\mathbb{R}$. Cantor's method is used in functional analysis in order to extend a metric space X to a complete metric space $\overline{X}$. Here, the original space X is a dense subset of the extended space $\overline{X}$, which is uniquely determined up to isometries.[6]

The extension of the field of rational numbers $\mathbb{Q}$ to the field Q_p of p-adic numbers. The symbol P denotes the set of prime numbers $p = 2, 3, \ldots$. Fix $p \in \mathsf{P}$. We want to apply Cantor's method to a different metric $|.|_p$ for the field of rational numbers. To begin with, let us consider an example. Using the unique prime number decomposition of natural numbers, we get

$$\frac{18}{40} = \frac{2 \cdot 3^2}{2^3 \cdot 5} = 2^{-2} \cdot \frac{3^2}{5}.$$

We describe this decomposition by writing

$$\left|\frac{18}{40}\right|_2 := 2^2.$$

Similarly, if r is a nonzero rational number, then we get

$$r = p^m \cdot \frac{a}{b}$$

where the exponent m is an integer, and the integers a and b are not divisible by the prime number p. Then we set

$$\boxed{|r|_p := p^{-m}.}$$

Moreover, let $|r|_p := 0$ if $r = 0$. Obviously, $|n|_p \leq 1$ if n is an integer. This is a rather strange result compared with the classical absolute value.

Applying Cantor's method to the metric $|.|_p$ on the field $\mathbb{Q}$, we get the field Q_p of p-adic numbers. This is a complete metric space which contains the set $\mathbb{Q}$ of rational numbers as a dense subfield. In 1918, Ostrowski (1893–1986) proved the following theorem.[7]

---

6 The definition of a complete metric space can be found in Section 5.10 of Volume II. The proof of the completion theorem for metric spaces is given in the textbook by V. Zorich, Analysis II, Sect. 9.5.2, Springer, Berlin, 2003.

7 A. Ostrowski, On some solutions of the functional equation $\varphi(xy) = \varphi(x)\varphi(y)$, Acta math. **41** (1918), 271–284. We refer to N. Koblitz, p-adic Numbers, p-adic Analysis, and Zeta Functions, p. 3, Springer, New York, 1984.

Theorem 4.1 *The absolute value $|.|$ and all the p-adic metrics $|.|_p$ with $p \in \mathsf{P}$ are the only valuations of the field of rational numbers $\mathbb{Q}$ (up to equivalent valuations).*

Let us discuss this. By definition, the function $\varphi : \mathbb{Q} \to [0, \infty[$ is called a valuation of the field $\mathbb{Q}$ of rational numbers iff the following hold for all $r, s \in \mathbb{Q}$:

- $\varphi(r) \geq 0$ and $\varphi(r) = 0$ iff $r = 0$ (definiteness),
- $\varphi(r+s) \leq \varphi(r) + \varphi(s)$ (triangle inequality),
- $\varphi(rs) = \varphi(r)\varphi(s)$ (product rule).

Two valuations φ and ψ of $\mathbb{Q}$ are called equivalent iff there exists a positive real number σ such that $\varphi(r) = \psi(r)^\sigma$ for all $r \in \mathbb{Q}$.

From the philosophical point of view, Ostrowski's theorem tells us that the field $\mathbb{R}$ of real numbers and the fields Q_p of p-adic numbers are the only reasonable field extensions of the field $\mathbb{Q}$ of rational numbers.

The p-adic valuation $|.|_p$ possesses the crucial additional property that

$$|r+s|_p \leq \max\{|r|_p, |s|_p\} \qquad \text{for all} \quad r, s \in \mathbb{Q}.$$

This property is typical for a so-called non-Archimedean valuation. This implies that if $a, b \in \mathsf{Q}_p$ with $|a|_p \leq 1$ and $|b|_p \leq 1$, then $|a+b|_p \leq 1$ and $|ab|_p \leq 1$. This is basic for the definition of the adelic ring on page 337.

There arises the question whether the p-adic numbers play a role in nature.

We will come back to this in Sect. 4.6.9.

Infinite series in p-adic analysis. The following proposition is not true for real numbers.

Proposition 4.2 *The infinite series $\sum_{k=0}^{\infty} b_k$ with $b_k \in \mathsf{Q}_p, k = 0, 1, 2, \ldots,$ is convergent in Q_p iff $\lim_{k\to\infty} b_k = 0$ in Q_p.*

Proof. Set $s_n := \sum_{k=0}^{n} b_k$. If $\lim_{n\to\infty} s_n = s$, then

$$\lim_{n\to\infty} b_n = \lim_{n\to\infty} (s_n - s_{n-1}) = 0.$$

Conversely, if $\lim_{k\to\infty} b_k = 0$, then

$$|s_{n+m} - s_n|_p = \left| \sum_{k=n+1}^{n+m} b_k \right|_p \leq \max_{n+1\leq k\leq n+m}\{|b_k|_p\} < \varepsilon$$

if $n \geq n_0(\varepsilon)$. Thus, (s_n) is a Cauchy sequence. Since the metric space Q_p is complete, (s_n) is convergent. □

As a crucial example, consider the special infinite series

$$\boxed{x = \frac{a_{-n}}{p^n} + \frac{a_{-n+1}}{p^{n-1}} + \ldots + \frac{a_{-1}}{p} + a_0 + a_1 p + a_2 p^2 + \ldots} \tag{4.15}$$

where $n = 1, 2, \ldots$, and $a_j = 0, 1, \ldots, p-1$ for all indices j. By Prop. 4.2, this series is convergent, and it represents a p-adic number a.[8]

Conversely, every p-adic number x can be uniquely represented by (4.15).

The proof can be found in N. Koblitz (1984), Chap. 1, quoted on page 333. For example, choose $p = 2$:

[8] Note that $|a_k p^k|_p = |a_k|_p |p^k|_p = |a_k|_p p^{-k} \leq p^{-k}$.

- $21 = 1 + 4 + 16 = 1 + p^2 + p^4$,
- $\frac{5}{8} = \frac{1}{8} + \frac{4}{8} = \frac{1}{p^3} + \frac{1}{p}$.

Classification of p-adic numbers. Fix the prime number p. Let $x \in \mathsf{Q}_p$ be a p-adic number.

- x is called an integer p-adic number iff we have the representation (4.15) with $a_{-n} = \ldots = a_{-1} = 0$. This is equivalent to $|x|_p \leq 1$.
- The set Z_p of integer p-adic numbers forms a ring without zero divisors; the corresponding quotient field is isomorphic to the field Q_p of p-adic numbers. Every $x \in \mathsf{Q}_p$ can be uniquely represented as the quotient

$$x = \frac{y}{p^n}, \qquad y \in \mathsf{Z}_p, \;\; n = 0, 1, 2, \ldots$$

- The p-adic number x is a rational number iff the sequence (a_k) of digits in (4.15) is periodic from some point onwards.

The ring Z_p of integer p-adic numbers as the projective limit of the Gaussian residue class rings $\mathbb{Z}_{p^k}$ as $k \to \infty$. Let us start with an example.[9] Fix the prime number $p = 2$. The decomposition

$$39 = 1 + 2 + 4 + 32 = 1 + p + p^2 + p^5$$

can be written as a sequence of congruences:

$$39 \equiv 1 \mod p, \quad 39 \equiv 1 + p \mod p^2, \quad 39 \equiv 1 + p + p^2 \mod p^3, \ldots$$

Thus, he sequence $([1]_p, [1+p]_{p^2}, [1+p+p^2]_{p^3}, \ldots)$ can be uniquely assigned to the integer 39. Obviously, we have the chain

$$\mathbb{Z}_p \supseteq \mathbb{Z}_{p^2} \supseteq \mathbb{Z}_{p^3} \supseteq \ldots$$

for all prime numbers p, with the embedding maps $\iota_{k+1} : \mathbb{Z}_{p^{k+1}} \to \mathbb{Z}_{p^k}$ for all $k = 1, 2, \ldots$. By definition, the projective limit

$$Z_p := \lim_{k \to \infty} \text{ proj } \mathbb{Z}_{p^k}$$

consists of all the sequences

$$([x_1]_p, [x_2]_{p^2}, [x_3]_{p^3}, \ldots)$$

with $\iota_{k+1}([x_{k+1}]_{p^{k+1}}) = [x_k]_{p^k}$ for all $k = 1, 2, \ldots$[10] Using componentwise addition and multiplication, the limit set Z_p becomes a ring. The map

$$\sum_{k=0}^{\infty} a_k p^k \mapsto ([a_0]_p, [a_0 + a_1 p]_{p^2}, \ldots)$$

yields the ring isomorphism $\mathsf{Z}_p \simeq Z_p$.

---

[9] We will use the classical terminology for Gaussian congruences introduced in Sect. 4.1.1 of Volume II. Let $a, b \in \mathbb{Z}$ be integers. Recall that we write

$$a \equiv b \mod p^k, \qquad k = 1, 2, \ldots$$

iff the difference $a - b$ is divisible by p^k. This is an equivalence relation. The corresponding equivalence classes, denoted by $[a]_{p^k}$, form the ring $\mathbb{Z}_{p^k}$.
For example, the Gaussian ring $\mathbb{Z}_2$ consists of the two elements $0, 1$ with $1+1 = 0$. Observe that the Gaussian ring $\mathbb{Z}_m$ is also denoted by $\mathbb{Z}/m\mathbb{Z}$ or $\mathbb{Z}/\text{mod } m$ in the literature.

[10] The general definition of inductive (direct) and projective (inverse) limits of mathematical structures can be found in Sect. 4.5.5 of Vol. II.

4.6.8 The Local–Global Principle in Mathematics

Many mathematical results describe either the local behavior (e.g., the derivative of a function) or the global behavior (e.g., topological properties) of mathematical objects. In the history of mathematics, there emerged the following two crucial situations where the local behavior completely determines the global behavior, namely:

(i) holomorphic and analytic functions, and
(ii) the Minkowski–Hasse theorem on quadratic Diophantine equations.

Here, item (i) concerns the analytic continuation. For example, in quantum field theory, the S-matrix describes scattering processes for elementary particles. Global properties of the S-matrix are determined by the local behavior via analytic continuation. We refer to:

A. Barut, The Theory of the Scattering Matrix, MacMillan, New York, 1967.

Analytic continuation is closely related to the theory of Riemann surfaces based on the theory of vector bundles and sheafs:

M. Schlichenmaier, An Introduction to Riemann Surfaces, Algebraic Curves, and Moduli Spaces, Springer, Berlin, 2008.

O. Forster, Lectures on Riemann Surfaces, Springer, Berlin, 1981.

J. Jost, Compact Riemann Surfaces: An Introduction to Contemporary Mathematics, Springer, Berlin, 1997.

The Minkowski–Hasse theorem concerns the following quadratic Diophantine equation:[11]

$$\boxed{a_1x_1^2 + a_2x_2^2 + \ldots + a_nx_n^2 = 0.} \tag{4.16}$$

Here $a_1, a_2, \ldots, a_n$ are rational numbers, all non-vanishing. We are looking for rational solutions $x_1, \ldots, x_n \in \mathbb{Q}$.

Theorem 4.3 *The equation* (4.16) *has a non-trivial solution in* $\mathbb{Q}$ *iff it has non-trivial solutions in* $\mathbb{R}$ *and in* Q_p *for all prime numbers* p.

The solutions in $\mathbb{Q}$ (resp. in $\mathbb{R}$ and Q_p for all prime numbers p) are called global (resp. local) solutions of the original problem (4.16). Therefore, the Minkowski–Hasse theorem tells us that the existence of local solutions implies the existence of a global solution. This is a variant of the local-global principle in number theory. The proof of Theorem 4.3 can be found in Z. Borevič and I. Šafarevič, Number Theory, Sect. 7, Academic Press, New York, 1967.

The classical version of Theorem 4.3 in terms of congruences was proven by Minkowski. It reads as follows: *The quadratic Diophantine equation* (4.16) *has a nontrivial rational solution* $x_1, \ldots, x_n \in \mathbb{Q}$ *if it has a nontrivial real solution* $x_1, \ldots, x_n \in \mathbb{R}$, *and all the congruences*

$$a_1x_1^2 + a_2x_2^2 + \ldots + a_nx^n \equiv 0 \mod p^m$$

with arbitrary prime numbers p *and arbitrary exponents* $m = 1, 2, \ldots$ *have nontrivial integer solutions* $x_1, \ldots, x_n \in \mathbb{Z}$.

Mnemonically, the prime numbers are the atoms of the integers, and the integer p-adic numbers are the atoms of the rational numbers.

[11] Diophantus of Alexandria (ca. 250 A.D.), Minkowski (1864–1909), Hasse (1898–1979).

4.6.9 The Global Adelic Ring

> On the fundamental level our world is neither real, nor p-adic, it is adelic. For some reasons, reflecting the physical nature of our kind of living matter (e.g., the fact that we are built of massive particles), we tend to project the adelic picture onto its real side (by using real numbers). We can equally well spiritually project it upon the non-Archimedean side and calculate most important things arithmetically (by using p-adic numbers).
> The relation between "real" and "arithmetical" pictures of the world is that of complementarity, like the relation between conjugate variables (in Hamiltonian physics).
> Of course, one is not obliged to take this metaphysics seriously. A skeptical reader can still use it as a guiding principle in a mathematical study of string theory.[12]
>
> Yuri Manin, 1987

Motivated by the Minkowski–Hasse local-global principle above, we define the product

$$\mathcal{A} := \mathbb{R} \times \prod_{p \in \mathsf{P}} \mathbb{Q}_p.$$

Recall that the symbol P denotes the set of prime numbers. Furthermore, by definition, the set $\mathbb{A}_{\mathbb{Q}}$ consists of all the sequences $(x, x_2, x_3, x_5, \ldots)$ in $\mathcal{A}$ with

$$|x_p|_p \le 1 \qquad \text{for almost all } p \in \mathsf{P}. \tag{4.17}$$

Here 'almost all' means that the inequality (4.17) is violated for at most a finite number of prime numbers. The set $\mathbb{A}_{\mathbb{Q}}$ of sequences becomes a ring (called the adelic ring or the ring of adeles) if we define the sum and the product componentwise. Adeles were introduced by Chevalley (1909–1984) in the 1930s. The multiplicative invertible elements in $\mathbb{A}_{\mathbb{Q}}$ are called ideles. For all nonzero rational numbers r, we have the universal product formula

$$|r| \prod_{p \in \mathsf{P}} |r|_p = 1$$

which shows that the absolute value $|r|$ can be computed by means of the p-adic valuations $|r|_p$. The famous product formula

$$\frac{\pi^2}{6} \prod_{p \in \mathsf{P}} \left(1 - \frac{1}{p^2}\right) = 1$$

relates the transcendental number π to p-adic numbers. This Euler formula can also be written as

---

[12] Yu. Manin, Reflections on arithmetical physics. In: Conformal Invariance and String Theory, Poiana Brasov, 1987, Academic Press, 1989, pp. 293–303 (reprinted with permission). See also Yu. Manin, Selected Papers, pp. 518–528, World Scientific, Singapore, 1996.
Adelic methods were used by Yu. Manin, The partition function of the Polyakov string can be expressed in terms of theta functions, Phys. Lett. **B172** (1986), 184–186. We also recommend: Yu. Manin, Strings, Math. Intelligencer **11**(2) (1989), 59–65.

$$\frac{\pi^2}{6} = \zeta(2) = \sum_{n=1}^{\infty} \frac{1}{n^2} = \prod_{p \in \mathsf{P}} \left(1 - \frac{1}{p^2}\right)^{-1}.$$

Further reading. As an introduction written for physicists, we recommend:

M. Schlichenmaier, An Introduction to Riemann Surfaces, Algebraic Curves, and Moduli Spaces, Springer, Berlin, 2008.

The appendix contains an introduction to p-adic numbers. Additionally, we recommend:

N. Koblitz, p-adic Numbers, p-adic Analysis, and Zeta Functions, Springer, New York, 1984.

Z. Borevič and I. Šafarevič, Number Theory, Academic Press, New York, 1967.

W. Stein, A Brief Introduction to Classical and Adelic Algebraic Number Theory, Lecture Notes, Harvard University, Cambridge, Massachusetts, 2004. Internet: http://sage.math.washington.edu/Spring2004/129/ant.pdf

Furthermore, we refer to:

J. Serre, A Course in Arithmetic, Springer, New York, 1973.

J. Serre, Local Fields, Springer, New York, 1979.

A. Weil, Basic Number Theory, Springer, Berlin, 1974.

S. Lang, Algebraic Number Theory, Springer, New York, 1986.

S. Lang, Introduction to Algebraic and Abelian Functions, Springer, New York, 1995.

S. Lang, Introduction to Diophantine Approximation, Springer, Berlin, 1995.

V. Vladimirov, V. Volovich, and E. Zelenov, p-Adic Analysis and Mathematical Physics, World Scientific, Singapore, 1994.

A. Khrennikov, p-Adic Valued Distributions in Mathematical Physics, Kluwer, Dordrecht, 1994.

A. Kedlaya, p-Adic Differential Equations, Cambridge University Press, Cambridge, 2010.

The study of physical models based on adelic methods can be found in:

B. Dragovich, P. Frampton, and C. Urosevic, Classical p-adic space time, Mod. Phys. Lett, **AA5** (1990), 1521–1528.

B. Dragovich, Adelic harmonic oscillator, Int. J. Mod. Phys. **A10** (1995), 2349–2365.

G. Djordevic and B. Dragovich, p-adic and adelic path integrals, Proceedings of the XIth Yugoslavian Conference on Nuclear Physics in Studenica, Yugoslavia, 1998, pp. 312–315.

G. Djordevic, B. Dragovich, and L. Nesic, p-adic and adelic free relativistic particle, Mod. Physics Lett. **7** (1999), 150–154.

B. Dragovich, Non-archimedean geometry and physics on adelic spaces, 2003. Internet: http://www.arxiv:math-phys/0306023

R. Schmidt, Arithmetic gravity and Yang-Mills theory: An approach to adelic physics via algebraic spaces. Ph.D. thesis, University of Münster (Germany), 2008. Internet: http://www.arxiv:hep-th/0809.3579

The following paper concerns dark matter:

B. Dragovich, p-adic and adelic cosmology, p-adic origin of dark energy and dark matter, 2006. Internet: http://www.arxiv:hep-th/0602044

4.6.10 Solenoids, Foliations, and Chaotic Dynamical Systems

The topology of the p-adic field $\mathbb{Q}_p$ is weird.[13]
William Stein, 2004

The way to chaos via period doubling is universal in nature.
Mitchell Feigenbaum, 1978

Totally disconnected sets. A manifold looks locally like a Euclidean space. Sullivan introduced a generalization of this notion called solenoid. A solenoid looks locally like the product

$$\mathcal{U} \times \mathcal{C}$$

where $\mathcal{U}$ is an open set in $\mathbb{R}^n$, and $\mathcal{C}$ is a totally disconnected set.[14] A topological space is called totally disconnected iff it has only trivial connected subsets, that is, the nonempty components are points. The prototype of such a pathological set was introduced by Cantor in 1883. The Cantor set $\mathcal{C}$ consists precisely of all the real numbers which allow the representation

$$\frac{a_1}{3} + \frac{a_2}{3^2} + \frac{a_3}{3^3} + \frac{a_4}{3^4} + \ldots$$

where either $a_k = 0$ or $a_k = 2$ for every $k = 1, 2, \ldots$ Intuitively, the Cantor set $\mathcal{C}$ is obtained from the unit interval $[0, 1]$ by taking away the open subinterval $]\frac{1}{3}, \frac{2}{3}[$ of length $\frac{1}{3}$. From the remaining intervals $[0, \frac{1}{3}]$ and $[\frac{2}{3}, 1]$, we are taking away the open subintervals

$$\left]\frac{1}{3^2}, \frac{2}{3^2}\right[\quad \text{and} \quad \left]\frac{2}{3} + \frac{1}{3^2}, \frac{2}{3} + \frac{2}{3^2}\right[$$

of length $\frac{1}{3^2}$, respectively, and so on. The connected subsets of the real line are intervals. Obviously, the Cantor set $\mathcal{C}$ does not contain any subinterval of finite length. Therefore, the compact Cantor set $\mathcal{C}$ is totally disconnected.

The field $\mathbb{Q}_p$ of p-adic numbers is a totally disconnected metric space.

The proof can be found in Stein (2004) quoted on page 338.

Renormalization and chaotic dynamical systems. Historically, it was discovered in the 1970s that very simple discrete dynamical systems may possess a highly chaotic structure. Let us mention the following three fundamental papers:

M. Feigenbaum, Quantitative universality for a class of nonlinear transformations, J. Stat. Physics **19** (1978), 25–52.

O. Lanford, A computer-assisted proof of the Feigenbaum conjectures, Bull. Amer. Math. Soc. **6** (1982), 427–434.

J. Milnor and W. Thurston, On iterated maps of the interval, Lecture Notes in Mathematics **1342** (1988), 465–563, Springer, Berlin.

We also recommend the following monographs:

---

13 See W. Stein (2004) quoted on page 338.

14 The author would like to thank Christopher Deninger (Münster) for drawing his attention to the notion of 'solenoid'.

J. Jost, Dynamical Systems: Examples of Complex Behavior, Springer, Berlin, 2005.

W. de Melo and S. van Strien, One-Dimensional Dynamics, Springer, Berlin, 1993.

J. Milnor, Dynamics in One Complex Variable: Introductory Lectures, Vieweg, Wiesbaden, 2000.

P. Schuster, Deterministic Chaos: An Introduction, Weinheim (Germany), Physik-Verlag, 1994.

J. Cardy, Scaling and Renormalization in Statistical Physics, Cambridge University Press, 1997.

W. McComb, Renormalization Methods: A Guide for Beginners, Oxford University Press, 2007.

Finally, we recommend the following two survey articles:

P. Cvitanović, Circle maps: irrationally winding, pp. 631–658. In: M. Waldschmidt et al. (Eds.), From Number Theory to Physics, Springer, New York, 1995.

J. Yoccoz, An introduction to small divisor problems, pp. 659–679. In: M. Waldschmidt et al. (Eds) (see above).

Small divisors are related to the appearance of resonances (e.g., in celestial mechanics) which cause trouble in perturbation theory. For solenoids, we refer to:

D. Sullivan, Linking the universalities of Milnor–Thurston, Feigenbaum, and Ahlfors–Bers, pp. 543–564. In: R. Goldberg and A. Phillips (Eds.), Topological Methods in Modern Analysis, Publish or Perish, Houston, Texas, 1993.

D. Sullivan, Bounds, quadratic differentials, and renormalization conjectures, pp. 417–466. In: F. Browder (Ed.), Mathematics into the Twenty-First Century, Amer. Math. Soc. Providence, Rhode Island, 1992.

D. Sullivan, On the foundation of geometry, analysis, and the differentiable structure for manifolds, pp. 89–92. In: A. Banyaga et al. (Eds.), Topics in Low-Dimensional Topology, World Scientific, Singapore, 1999.

Foliations, Hopf algebras, and renormalization in quantum field theory. Solenoids are also related to foliations. A deep result is the generalization of the Atiyah–Singer index theorem to foliations:

A. Connes and H. Moscovici, Hopf algebras, cyclic cohomology and the transverse index theorem, Commun. Math. Phys. **198** (1998), 199–246.

This is also closely related to renormalization in quantum field theory (see Sect. 19.3 of Vol. II).

The main theorem on discrete dynamical systems. Let us consider the following iterative method

$$\boxed{x_{n+1} = f(x_n), \quad n = 0, 1, 2, \ldots} \tag{4.18}$$

where the real number x_0 is given. Suppose that $f : \mathbb{R} \to \mathbb{R}$ is a given real-valued function. We assume that there exists a real number x_* such that

$$x_* = f(x_*). \tag{4.19}$$

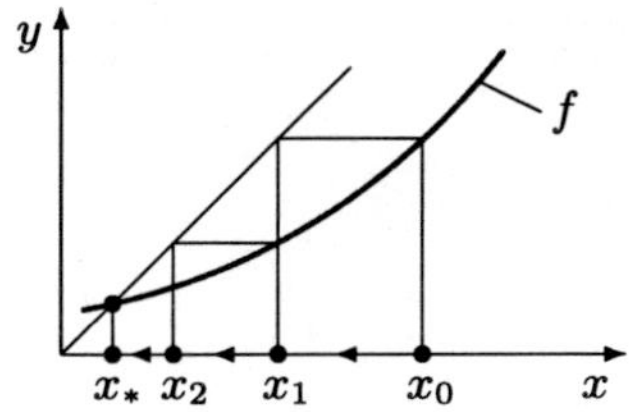

Fig. 4.4. Asymptotically stable fixed point x_*

In addition, we assume that the function f is smooth on some open interval containing the point x_*. In order to get a physical interpretation, regard x_n as the position of a particle on the real line at time $n\Delta t$ for fixed $\Delta t > 0$. Then the sequence (x_n) describes the dynamics of the particle. We call (x_n) a discrete dynamical system. The point x_* is called a fixed point of the function f. In terms of physics, x_* represents an equilibrium state of the particle. In fact, if $x_0 = x_*$, then $x_n = x_*$ for all $n = 0, 1, 2, \ldots$ The fixed point x_* is called asymptotically stable iff

$$\lim_{n\to\infty} x_n = x_*$$

for all initial points x_0 in a sufficiently small open neighborhood of the point x_*. In terms of numerical mathematics, this means that the iterative method (x_n) is convergent if the initial point x_0 is sufficiently near the solution x_* of the equation (4.19). In terms of physics, the asymptotically stable equilibrium state x_* is an attractor. In Fig. 4.4, the fixed point x_* corresponds to the intersection point between the curves $y = x$ (diagonal) and $y = f(x)$. The iterative method for getting $x_0, x_1, x_2, \ldots$ represents graphically a simple geometric procedure. Motivated by Fig. 4.4, we expect the following result.

Theorem 4.4 *If $|f'(x_*)| < 1$, then the fixed point x_* is asymptotically stable.*

This is a special case of the Banach fixed-point theorem formulated by Banach (1892–1945) in 1922. The proof can be found in Zeidler (1986) (Sect. 1.4 of Vol. I) quoted on page 1089.

Prototype. Fix the real parameter, and consider the discrete dynamical system

$$x_{n+1} = ax_n, \qquad n = 0, 1, 2, \ldots, \qquad x_0 \in \mathbb{R}$$

with the equilibrium point $x_* = 0$. Let $x_0 \neq 0$. Explicitly,

$$x_{n+1} = a^{n+1} x_0, \qquad n = 0, 1, 2, \ldots$$

Then the following hold (Fig. 4.5):

- If $|a| < 1$, then $\lim_{n\to\infty} x_n = 0$ ($x_* = 0$ is asymptotically stable/attractor).
- If $|a| > 1$, then $\lim_{n\to\infty} |x_n| = \infty$ ($x_* = 0$ is unstable/repeller).
- If $|a| = 1$, then the equilibrium point $x_* = 0$ has a so-called critical behavior. Arbitrarily small perturbations of the parameter a may change dramatically the behavior of the system (attractor or repeller).

Feigenbaum's numerical turbulence model. In the late 1970s, Feigenbaum studied the iterative method

$$\boxed{x_{n+1} = 4\lambda x_n(1 - x_n), \qquad n = 0, 1, 2, \ldots, \qquad x_0 \in [0, 1]} \tag{4.20}$$

with the parameter $\lambda \in]0, 1]$ by using a pocket calculator.

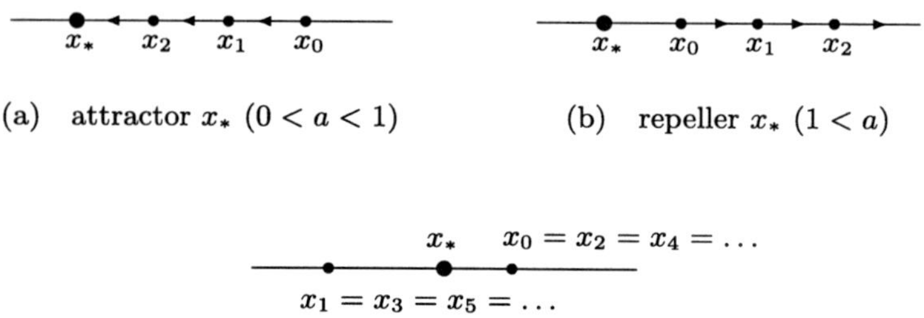

(a) attractor x_* $(0 < a < 1)$ (b) repeller x_* $(1 < a)$

(c) critical behavior $(a = -1)$

Fig. 4.5. Discrete dynamical system: $x_{n+1} = ax_n, \ n = 0, 1, \ldots; \ x_* = 0$

Computer experiments show that the behavior of the dynamical system becomes more and more complex if the parameter λ increases, that is, the system approaches turbulence (chaos).

There exist critical parameters λ called bifurcation parameters where the system changes its qualitative behavior. These critical parameters model the Reynolds numbers in real turbulence experiments. Specifically, the Feigenbaum model approaches chaos by producing a cascade of new periodic solutions via permanent period doubling.

Fixed points of iterated maps and stable periodic motions. Let us discuss the Feigenbaum approach in greater detail. To this end, for fixed parameter $\lambda \in]0, 1]$, we introduce the operator

$$A_\lambda(x) := 4\lambda x(1 - x), \qquad 0 \le x \le 1,$$

and we will study the iterated operators A_λ^k with the exponents $k = 2, 3, \ldots$ Note that the discrete dynamical system (x_n) is given by the equation

$$x_n := A_\lambda^n(x_0), \qquad n = 0, 1, 2, \ldots, \qquad x_0 \in [0, 1].$$

For any fixed parameter $\lambda \in [0, 1]$, the operator

$$\boxed{A_\lambda : [0, 1] \to [0, 1]}$$

maps the unit interval $[0, 1]$ into itself; we have $A_\lambda(0) = 0$ together with the maximal value $A_\lambda(\frac{1}{2}) = \lambda$. In what follows, we will always assume that $x_0, x_* \in [0, 1]$.

(i) Stable equilibrium states. Suppose that x_* is a fixed point of the operator A_λ, that is,

$$A_\lambda(x_*) = x_*.$$

This means that if we choose the initial position $x_0 := x_*$, then $x_n = x_*$ for all $n = 0, 1, , 2 \ldots$ In terms of physics, x_* is an equilibrium state of the dynamical system. The fixed point x_* of A_λ is called asymptotically stable iff

$$\lim_{n \to \infty} A_\lambda^n(x_0) = x_*$$

for all initial positions x_0 in a sufficiently small neighborhood of x_*.

(ii) Stable periodic solutions. Suppose that x_* is a fixed point of the iterated operator A_λ^2, that is,

$$A_\lambda^2(x_*) = x_*.$$

If we choose the initial position x_*, then

$$x_{2m} = x_*, \qquad m = 0, 1, 2, \ldots,$$

that is, the motion of the dynamical system has the period 2. Such a solution is called a 2-cycle. This periodic motion is called asymptotically stable iff x_* is a stable fixed point of the iterated operator A_λ^2. Similarly, the fixed points of A_λ^k with $k = 3, 4, \ldots$ are called k-cycles.

The Feigenbaum scenario. We want to consider the iterative method (4.20) for increasing parameters λ.

- Step 1: For $0 < \lambda < \frac{1}{4}$, the operator A_λ has exactly the fixed point $x = 0$. This point is asymptotically stable and becomes critical for $\lambda = \frac{1}{4}$.
- Step 2: For $\frac{1}{4} < \lambda < \frac{1}{2}$, the fixed point $x = 0$ is unstable. There appears a new asymptotically stable fixed point $x_\lambda := 1 - \frac{1}{4\lambda}$.
- Step 3: If λ crosses the value $\frac{1}{2}$, then the equilibrium state x_λ becomes unstable, and two asymptotically stable 2-cycles arise from x_λ. It is possible to find a parameter value $\lambda_1 > 0$ such that $x = \frac{1}{2}$ becomes a fixed point of $A_{\lambda_1}^2$. This corresponds to a 2-cycle departing at $x = \frac{1}{2}$.
- Step 4: Continuing the growth of the parameter λ, we permanently get new cycles at critical λ-parameter values with period doubling.
- Step 5: It is of special significance that there is a sequence $0 < \lambda_1 < \lambda_2 < \ldots \leq 1$ such that $x = \frac{1}{2}$ is a fixed point of A_λ^m with $\lambda = \lambda_k$ and $m = 2^k$, $k = 1, 2, \ldots$ For fixed parameter λ_k, let us consider the corresponding dynamical system

$$x_{n+1} = 4\lambda_k x_n(x_n - 1), \qquad n = 0, 1, 2, \ldots, \qquad x_0 := \tfrac{1}{2}.$$

This system has the period 2^k, that is, $x_{2^k} = x_0$, and we set

$$d_k := x_0 - x_{2^{k-1}} = \tfrac{1}{2} - A_{\lambda_k}^{2^{k-1}}\left(\tfrac{1}{2}\right), \qquad k = 1, 2, \ldots$$

Note that 2^{k-1} is half of the period 2^k.

Feigenbaum computed the limits

$$\lim_{k \to \infty} \frac{\lambda_{k+1} - \lambda_k}{\lambda_{k+2} - \lambda_{k+1}} = \delta, \tag{4.21}$$

and

$$\lim_{k \to \infty} \frac{d_k}{d_{k+1}} = -\alpha. \tag{4.22}$$

He obtained the values $\delta = 4.6692016$ and $\alpha = 2.5029079$ called the Feigenbaum numbers.

Lanford's computer-assisted proof of Feigenbaum's universality conjecture. Feigenbaum conjectured that the numbers δ and α possess a universal meaning in physics. This means that there exists a broad class of dynamical systems with period doubling at critical parameter values such that the limits (4.21) and (4.22) hold. Experiments establish this conjecture.[15] For a class of iterative methods, Feigenbaum's conjecture was proven by Lanford in the paper quoted on page 339 above. The universality hypothesis of physicists claims the following:

Critical phenomena in nature are governed by a few universal laws.

This is a surprising fact. Universality laws in thermodynamics were first formulated by Landau in 1937.[16] For example, let M denote the magnetization of ferromagnetic material. There exists a critical temperature T_{crit} called the Curie temperature where the spontaneous magnetization of the material is lost if the temperature is greater than T_{crit}. If the temperature T is near the critical temperature T_{crit}, then the following hold:

- If $T < T_{\text{crit}}$, then $M = M_0 \cdot \left(1 - \frac{T}{T_{\text{crit}}}\right)^{\alpha}$ (spontaneous magnetization).
- If $T > T_{\text{crit}}$, then $\chi_{\text{magn}} = \chi_0 \cdot \left(\frac{T}{T_{\text{crit}}} - 1\right)^{\beta}$ (Curie–Weiss law).

The exponents $\alpha := \frac{1}{2}$ and $\beta := -1$ are called critical exponents. For the magnetic field strength vector $\mathbf{B}$, we have

$$\mathbf{B} = \mu_0 \mathbf{H} + \mu_0 \mathbf{M} = \mu_0 (1 + \chi_{\text{magn}})\, \mathbf{H}.$$

Here, we use the following notation: μ_0 magnetic constant of the vacuum, χ_{magn} magnetic susceptibility (dimensionless quantity), $\mathbf{H}$ derived magnetic field vector, $\mathbf{M}$ magnetization (magnetic moment vector per volume), $M = |\mathbf{M}|$. In particular, $\mathbf{M} = \chi_{\text{magn}}\, \mathbf{H}$. The Curie–Weiss law was discovered by Pierre Curie (1859–1906) at the beginning of the 20th century. For iron, we have $T_{\text{crit}} = 1017$ Kelvin. One of the main tasks of statistical physics is to compute critical exponents. As an introduction to the theory of critical phenomena in physics, we recommend:

C. Domb, The Critical Point: A Historical Introduction to the Modern Theory of Critical Phenomena, Taylor & Francis, London, 1996.

This topic is closely related to the theory of the renormalization group. In 1982, Kenneth Wilson (born 1935) was awarded the Nobel prize in physics for his theory of critical phenomena in connection with phase transitions (see also page 981).

Feigenbaum's renormalization trick. Feigenbaum used the following heuristic rescaling method called the renormalization trick. Set

$$g_r(y) := \lim_{k \to \infty} (-1)^k \alpha^k A^{2^k}_{\lambda_k + r} \left(\frac{y}{(-1)^k \alpha^k} \right), \qquad r = 1, 2, \ldots$$

---

[15] In 1981, Libchaber and Maurer carried out the following experiment. A small box, containing liquid ^{4}He helium, is heated from the bottom with the temperature $T = \lambda$, where the temperature is constant at the top. If we measure the temperature at a fixed internal point of the box, then we get periodic curves $T = T(t)$ (t time) with period doubling for the critical temperatures $\lambda_1 < \lambda_2 < \lambda_3$. The experimenters measured $\lambda_1, \lambda_2, \lambda_3$, and they found $\delta = 3.5 \pm 1.5$. Universality theory predicts $\delta = 4.7$. Note that it is difficult to measure arbitrarily high critical temperatures λ_k because there is a lot of noise in the temperature curve, since the system is already near chaos. See A. Libchaber and J. Maurer, Experimental study of hydrodynamic instabilities in the Rayleigh–Bénard experiment: helium in a small box. In: T. Riste (Ed.), NATO Advanced Study Institute on Nonlinear Phenomena at Phase Transitions and Instabilities, Plenum, New York, 1981.

[16] In 1962, Lev Landau (1908–1968) was awarded the Nobel prize in physics for his pioneering theories for condensed matter, especially liquid helium.

where $y := x - \frac{1}{2}$. This yields the recursion formula

$$g_{r-1}(y) = -\alpha g_r \left(g_r \left(\frac{y}{\alpha} \right) \right), \qquad r = 1, 2, \ldots$$

For $r \to +\infty$, we formally get the key functional equation

$$\boxed{g(y) = -\alpha g \left(g \left(\frac{y}{\alpha} \right) \right).}$$

Feigenbaum determined numerically the unique solution g and α of this functional equation. Since this equation does not depend on the details of the original iterative method (4.21), Feigenbaum conjectured that there must exist a universal law.

More about the relation of the Feigenbaum approach to the fixed-point theory for iterated maps can be found in Zeidler (1986) (Chap. 17 of Vol. I) quoted on page 1089.

4.6.11 Period Three Implies Chaos

> The Li–York paper is one of the immortal gems in the literature of mathematics.[17]
>
> Freeman Dyson, 2010

Let us consider the iterative method

$$\boxed{x_{n+1} = F(x_n), \qquad n = 0, 1, \ldots, \quad x_0 \in \mathbb{R}} \tag{4.23}$$

where the function $F : \mathbb{R} \to \mathbb{R}$ is continuous. We regard the sequence (x_n) as a dynamical system on the real line $\mathbb{R}$ with discrete time. Choosing the time scale $\Delta t > 0$, the 'system' is at the point x_n on the real line $\mathbb{R}$ at time $t_n := n\Delta t$. We are given the initial position x_0 of the 'system' at the initial time $t_0 = 0$. We will write

$$x_1 = F^1(x_0), \quad x_2 = F^2(x_0), \quad x_3 = F^3(x_0), \ldots$$

Let $m = 1, 2, 3, \ldots$ Recall that the point $p \in \mathbb{R}$ is called periodic of period m iff

$$p = F^m(p)$$

and $k = 1, \ldots, m-1$ implies $p \neq F^k(p)$. Moreover, the point $a \in \mathbb{R}$ is called asymptotically periodic iff there is a periodic point p such that

$$\lim_{n \to \infty} (F^n(a) - F^n(p)) = 0. \tag{4.24}$$

Theorem 4.5 *Suppose that there exists a point of period three. Then for each integer $m = 1, 2, \ldots,$ there is a point of period m. Furthermore, there is an uncountable subset of points x in $\mathbb{R}$ which are not even asymptotically periodic.*

The proof of this famous theorem in chaos theory can be found in T. Li and J. Yorke, Period three implies chaos, American Mathematical Monthly **82** (1975), 985–992.

---

[17] F. Dyson, Birds and frogs in mathematics and physics, Einstein lecture 2008, Notices Amer. Math. Soc. **56** (2) (2009), 212–223. We recommend reading this article about two different philosophical approaches to mathematics and physics.

4.6.12 Noncommutative Geometry and the Standard Model in Particle Physics

It was discovered by Connes and Lott in 1990 that differential calculi in noncommutative geometry can be used in order to get the Standard Model in particle physics. This is studied in:

A. Connes and J. Lott, Particle models and noncommutative geometry, Nucl. Phys. **B** (Proc. Suppl.) **18** (1990), 29–47.

As an introduction to noncommutative geometry, we recommend:

J. Várilly, Lectures on Noncommutative Geometry, European Mathematical Society, Zurich, 2006.

J. Mignaco, C. Sigaud, A. da Silva, and F. Vanhecke, The Connes–Lott program on the sphere, Rev. Math. Phys. **9** (1997), 689–718.

M. Gracia-Bondia, M., J. Várilly, and H. Figueroa, Elements of Noncommutative Geometry, Birkhäuser, Boston, 2001.

F. Scheck, W. Wend, and H. Upmeier (Eds.), Noncommutative Geometry and the Standard Model of Elementary Particle Physics, Springer, Berlin, 2003 (collection of survey articles).

Furthermore, we refer to the following monographs:

A. Connes and M. Marcolli, Noncommutative Geometry, Quantum Fields, and Motives, Amer. Math. Soc., Providence, Rhode Island, 2008,

M. Marcolli, Feynman Motives: Renormalization, Algebraic Varieties, and Galois Symmetries, World Scientific, Singapore, 2009,

and to the following papers:

A. Chamseddine and A. Connes, Universal formula for noncommutative geometry actions: Unification of gravity and the Standard Model, Phys. Rev. Lett. **77** 48680484871 (1996).

A. Chamseddine and A. Connes, Why the Standard Model, J. Geom. Phys. **58**:38 (2008). Internet: `http://www.arXiv:0706.3690` [hep-th]

A. Chamseddine and A. Connes, Conceptual explanation for the algebra in the noncommutative approach to the Standard Model, Phys. Rev. Lett. **99**:191601 (2007). Internet: `http://www.arXiv:0706.3690` [hep-th]

A. Chamseddine and A. Connes (1997), The spectral action principle, Commun. Math. Phys. **186** (1997), 731–750.

A. Chamseddine, A. Connes, and M. Marcolli, Gravity and the Standard Model with neutrino mixing, Adv. Theor. Math. Phys. **11**:991 (2007). Internet: `http://www.hep-th/0610241`

A. Chamseddine and A. Connes, Scale invariance in the spectral action, J. Math. Phys. **47**:063504. Internet: `http://www.hep-th/0512169`

F. Hanisch, F. Pfäffle, and C. Stephan, The spectral action for Dirac operators with skew-symmetric torsion (applied to the Standard Model), 2010. Internet: `arXiv:0911.5074` [hep-th]

The physics of the Standard Model is investigated in:

P. Langacker, The Standard Model and Beyond, CRC Press, Boca Raton, Florida, 2010.
Supplementary material: `http://www.sns.ias.edu/ pgl/SMB/`

In order to describe physical models in noncommutative space-times, deformations of the classical calculus can be used. We refer to:

S. Woronowicz, Twisted $SU(2)$ group, an example of noncommutative differential calculus, Publ. RIMS, Kyoto Univ. **23** (1987), 117–181.

S. Woronowicz, Differential calculus on compact matrix pseudogroups (quantum groups), Commun. Math. Phys. **122** (1989), 125–170.

J. Wess and B. Zumino, Covariant differential calculus on the quantum hyperplane, Nucl. Phys. **B** (Proc. Suppl.) **18** (1991), 302–312.

J. Wess, Gauge theories on noncommutative space-time treated by the Seiberg–Witten method, pp. 179–192. In: U. Carow-Watamura et al. (Eds.), Quantum Field Theory and Noncommutative Geometry, Springer, Berlin, 2005.

O. Kovras (Ed.), Current Topics in Quantum Field Research, Nova Science Publisher, New York, 2005.

H. Wachter, Towards a q-deformed quantum field theory, pp. 261–281. In: B. Fauser, J. Tolksdorf, and E. Zeidler (Eds.) (2008), Quantum Field Theory – Competitive Methods, Birkhäuser, Basel.

A. Schmidt, Towards a q-deformed supersymmetric field theory, pp. 283–300. In: B. Fauser, J. Tolksdorf, and E. Zeidler (Eds.) (2008).

4.6.13 BRST-Symmetry, Cohomology, and the Quantization of Gauge Theories

The quantization of gauge theories is a nontrivial procedure which is complicated by constraints for the potentials. The prototype is given by quantum electrodynamics where one uses the Gupta–Bleuler method invented in 1950 in order to eliminate unphysical states which correspond to longitudinal (also called virtual) photons (see Sect. 12.4.4 of Volume II). It was discovered by Becchi, Rouet, Stora and independently by Tyutin that the quantization process of gauge theories is governed by a hidden symmetry called the BRST-symmetry:

C. Becchi, A. Rouet, and R. Stora, Renormalization of the Abelian Higgs–Kible model, Commun. Math. Phys. **52** (1975), 127–162.

C. Becchi, A. Rouet, and R. Stora, Renormalization of gauge theories, Annals of Physics **98** (1976), 287–321.

The BRST-symmetry generates an operator Q with the crucial cohomology property

$$Q^2 = 0.$$

The operator Q can be used in order to distinguish between physical and unphysical quantum states. The basic idea is discussed in Sect. 16.7 of Volume I. We will come back to this in Volume IV. At this point, we only refer to the following references:

A. Das, Lectures on Quantum Field Theory, World Scientific, Singapore, 2008.

O. Piguet and S. Sorella, Algebraic Renormalization, Springer, Berlin, 1995.

M. Henneaux and C. Teitelboim, Quantization of Gauge Systems, Princeton University Press, 1993.

4.6.14 Itô's Stochastic Calculus

The pure/applied division of probability theory (as of mathematics in general) is nonsense.

David Williams, 1978[18]

In 1987 Kiyosi Itô (1915–2008) received the Wolf prize in mathematics. The laudatio states that "he has given us a full understanding of the infinitesimal development of Markov sample paths. This may be viewed as Newton's law in the stochastic realm, providing a direct translation between the governing partial differential equation (the diffusion equation) and the underlying probabilistic mechanism (for the motion of particles). Its main ingredient is the differential and integral calculus of functions of Brownian motion. The resulting theory is a cornerstone of modern probability, both pure and applied." The reference to Newton (1643–1727) stresses the fundamental character of Itô's contribution to Markov processes.[19] Let us also mention Leibniz (1646–1716) in order to emphasize the fundamental importance of Itô's work from another point of view. In fact Itô's approach can be seen as a natural extension of Leibniz's algorithmic formulation of the differential calculus. In a manuscript written in 1675 Leibniz argues that the whole differential calculus can be developed out of the basic product rule[20]

$$d(XY) = XdY + YdX.$$

...It was Itô who discovered how these rules can be modified in such a way that they generate a highly efficient calculus for the non-differentiable trajectories of the particles of a diffusion process ...
Already in the 1960s engineers discovered that Itô's calculus provides the right concepts and tools for analyzing the stability of dynamical systems perturbed by noise and to deal with problems of filtering and control. When I was an instructor at MIT (Massachusetts Institute of Technology) in 1969/70, stochastic analysis did not appear in any course offered in the Department of Mathematics. But I counted four courses in electrical engineering and two in aeronautics and astronautics in which stochastic differential equations played a role (e.g., the motion of satellites and spaceships under random perturbations) ...
In the seventies the relevance of Itô's work was also recognized in physics and in particular in quantum field theory. When I came to ETH (Federal Institute of Technology) Zurich in 1977, Barry Simon (Princeton University) gave a series of lectures for Swiss physicists on path integral techniques which included the construction of Itô's integral for Brown-

[18] From the Preface of the first edition of the textbook by L. Rogers and D. Williams, Diffusions, Markov Processes, and Martingales, Vol. 1: Foundations, Vol 2: Itô Calculus, Cambridge University Press, 2000.

[19] Gauss (1777–1855) (normal Gaussian distribution), Markov (1856–1922) (Markov processes), Einstein (1879–1955) (Brownian motion), Wiener (1894–1964) (Brownian motion and the Wiener path integral), Kolmogorov (1903–1987) (foundation of the modern theory of probability including the theory of stochastic processes), Feynman (1918–1988) (path integral in quantum physics). See Sect. 7.5 of Vol. II. Itô's first paper appeared in 1942 (in Japanese).

[20] We will show in Chap. 8 that the Leibniz rule is also basic for gauge theory.

ian motion, an introduction to stochastic calculus, and applications to Schrödinger operators with magnetic fields.[21]

Hans Föllmer, 2006

The classic Leibniz differential. Let $x = x(t), t \geq 0$, be a smooth function $x : [0, \infty[\to \mathbb{R}$. In terms of physics, this function describes the classic motion of a particle on the real line. Here, $x(t)$ is the position of the particle at time t. By the chain rule,

$$\frac{d}{dt}\left(\tfrac{1}{2}x(t)^2\right) = x(t)\,\frac{dx(t)}{dt}.$$

Then

$$\int_0^t \frac{d}{dt}\left(\tfrac{1}{2}x(t)^2\right) dt = \tfrac{1}{2}x(t)^2 - \tfrac{1}{2}x(0)^2.$$

According to Leibniz, we write

$$\boxed{d\left(\tfrac{1}{2}x(t)^2\right) = x(t)\,dx(t), \qquad t \geq 0}$$

and

$$\int_0^t d(\tfrac{1}{2}x(t)^2) = \int_0^t x(t)\,dx(t) = \tfrac{1}{2}x(t)^2 - \tfrac{1}{2}x(0)^2, \qquad t \geq 0.$$

Brownian motion. In contrast to the classical particle motion, let us now consider the random motion $X = X(t), t \geq 0$, of a particle on the real line, that is, $X(t)$ is a real-valued random variable parametrized by the parameter $t \in [0, \infty[$. In terms of physics, $X(t)$ describes the random position of the particle on the real line at time t. Let $0 \leq \tau \leq t$. Fix the real number $\sigma > 0$ called the diffusion coefficient, and fix the position x_*. We assume the following:

(i) $X(0) = x_*$ (i.e., the particle is located at the point x_* at time $t = 0$).

(ii) The random variable $X(t) - X(\tau)$ possesses a normal Gaussian distribution with the mean value $\overline{X(t) - X(\tau)} = 0$, and the dispersion

$$\overline{(X(t) - X(\tau))^2} = \sigma^2 \cdot (t - \tau).$$

This means the following. Suppose that the particle is located at the point x_0 at time $\tau \geq 0$. Then the real number

$$p(X(t) \in J) := \frac{1}{\sigma\sqrt{2\pi(t-\tau)}} \int_J \mathrm{e}^{-(x-x_0)^2/2\sigma^2(t-\tau)} dx$$

is the probability for finding the particle in the interval J at time t. In physics, this is called the transition probability of the particle for passing from the position x_0 at time τ to the interval J at time t. In mathematics, this is called a conditional probability.

(iii) If $0 \leq t_0 < t_1 < t_2, \ldots, t_n$ with $n = 1, 2, \ldots$, then the random variables

$$X(t_1) - X(t_0), \quad X(t_2) - X(t_1), \quad \ldots, \quad X(t_n) - X(t_{n-1})$$

are independent.

[21] H. Föllmer, On Kiyosi Itô's work and its impact, pp. 109–123, Proceedings of the International Congress of Mathematicians, Madrid 2006, European Mathematical Society, Zurich, 2007 (reprinted with permission). In 2006, Itô was awarded the newly founded Gauss prize for applications of mathematics.

This stochastic process is called Brownian motion (or Wiener process).

Stochastic causality and the Markov property of the Brownian motion. Let us consider a system of many particles on the real line (e.g., a fluid). Let $\varrho(x,t)$ denote the particle density at the point x at time t. Fix the initial time $t_0 \geq 0$. We assume that the given density function $x_0 \mapsto \varrho(x_0, t_0)$ (at the initial time t_0) is smooth, and it has compact support. Then the density function $x \mapsto \varrho(x,t)$ at time t is given by

$$\varrho(x,t) = \int_{-\infty}^{\infty} \mathcal{P}(x,t;x_0,t_0)\varrho(x_0,t_0)\, dx_0, \qquad x \in \mathbb{R}, \ t > t_0$$

with the Feynman propagator kernel

$$\mathcal{P}(x,t;x_0,t_0) := \frac{\mathrm{e}^{-(x-x_0)^2/2\sigma^2(t-t_0)}}{\sigma\sqrt{2\pi(t-t_0)}}, \qquad t > t_0.$$

In mathematics, the Feynman propagator kernel $\mathcal{P}$ is called the diffusion kernel (or the heat kernel). The density function ϱ is a solution of the diffusion equation

$$\boxed{\frac{\partial \varrho(x,t)}{\partial t} = \frac{\sigma^2}{2}\frac{\partial^2 \varrho(x,t)}{\partial x^2}, \qquad x \in \mathbb{R},\ t > t_0.}$$

The Feynman propagator kernel $(x,t) \mapsto \mathcal{P}(x,t;x_0,t_0)$ is a solution of the Einstein–Fokker–Planck equation

$$\frac{\partial \mathcal{P}(x,t;x_0,t_0)}{\partial t} = \frac{\sigma^2}{2}\frac{\partial^2 \mathcal{P}(x,t;x_0,t_0)}{\partial x^2}, \qquad x, x_0 \in \mathbb{R},\ t > t_0 \tag{4.25}$$

which is also called the Kolmogorov forward equation. The Kolmogorov backward equation reads as

$$\frac{\partial \mathcal{P}(x,t;x_0,t_0)}{\partial t_0} = -\frac{\sigma^2}{2}\frac{\partial^2 \mathcal{P}(x,t;x_0,t_0)}{\partial x_0^2}, \qquad x, x_0 \in \mathbb{R},\ t > t_0.$$

For the propagator kernel, we also have the Chapman–Kolmogorov equation

$$\boxed{\mathcal{P}(x,t;x_0,t_0) = \int_{-\infty}^{\infty} \mathcal{P}(x,t;y,s)\ \mathcal{P}(y,s;x_0,t_0)\, dy, \qquad t \geq s \geq t_0.}$$

In physics, this is called the Feynman propagator kernel equation. Let us introduce the operator $P(t,t_0)$ by setting

$$\varrho(t) := P(t,t_0)\varrho(t_0), \qquad t \geq t_0. \tag{4.26}$$

Here, the symbol $\varrho(t)$ denotes the density function $x \mapsto \varrho(x,t)$ at time t. Then we get the so-called propagator operator equation

$$\boxed{P(t,t_0) = P(t,s)P(s,t_0), \qquad t \geq s \geq t_0.} \tag{4.27}$$

This equation reflects causality. In fact, set $\varrho(t) := P(t,t_0)\varrho(t_0)$ together with $\varrho(s) := P(s,t_0)\varrho(t_0)$. By (4.27),

$$\varrho(t) = P(t,s)\varrho(s) = P(t,t_0)\varrho(t_0).$$

This tells us that we have the causal chain $\varrho(t_0) \Rightarrow \varrho(s) \Rightarrow \varrho(t)$. In the theory of probability, the equation (4.27) reflects the so-called Markov property of the Brownian motion. Intuitively, the stochastic dynamics of a physical system has the Markov property if the state $\psi(t_0)$ of the system at time t_0 determines the possible states $\psi(t)$ at a later time $t > t_0$ by transition probabilities $p(t, t_0)$ which only depend on the initial time t_0 and the final time t, but they do not depend on the history of the physical system before time t_0. Such stochastic processes are called Markov processes. In other words, Markov processes forget their history.

The Brownian motion is homogeneous in space and time. Explicitly, setting $\mathcal{P}_0(x,t) := \mathcal{P}(x, t; 0, 0)$ we get

$$\mathcal{P}(x, t; x_0, t_0) = \mathcal{P}_0(x - x_0, t - t_0).$$

Similarly, setting $P_0(t) := P(t, 0)$, we obtain

- $P(t + s) = P(t)P(s)$ if $t, s \geq 0$, and
- $P(0) = I$ (identity operator).

In terms of functional analysis, the family $\{P(t)\}_{t \geq 0}$ of operators forms a semi-group.[22]

The stochastic Itô differential. Consider the Brownian motion $t \mapsto X(t)$ with the diffusion coefficient $\sigma := 1$. As a prototype of the Itô calculus, let us discuss the Itô formula

$$\tfrac{1}{2}X(t)^2 - \tfrac{1}{2}X(0)^2 = \int_0^t X(\tau)\, dX(\tau) + \int_0^t d\tau. \tag{4.28}$$

Mnemonically, we write

$$\boxed{d\left(\tfrac{1}{2}X(t)^2\right) = X(t)\, dX(t) + dt.} \tag{4.29}$$

This corresponds to the classic Leibniz formula up to the additional term dt which is caused by the non-vanishing dispersion, $\overline{(X(t) - X(0))^2} = t$. Observe that the integral

$$\mathcal{I}(t) := \int_0^t X(\tau)\, dX(\tau)$$

is not a classic integral, since $X(\tau)$ is a random variable. In order to define the integral $\mathcal{I}(t)$, Itô used a decomposition $t_0 := 0 < t_1 < \ldots < t_n := t$ with $t_k := kt/n$, and he introduced the finite sum

$$\mathcal{I}_n := \sum_{k=0}^{n-1} X(t_k) \cdot (X(t_{k+1}) - X(t_k))$$

of random variables. Then the Itô integral $\mathcal{I}(t)$ is defined by the following stochastic limit

$$\lim_{n\to\infty} \overline{(\mathcal{I}(t) - \mathcal{I}_n)^2} = 0.$$

In particular, the Itô integral $\mathcal{I}(t)$ is a random variable at time t.

[22] As an introduction to semigroup theory in functional analysis and its applications to partial differential equations, we recommend Zeidler (1986), Vol. IIA, Chap. 19, quoted on page 1089. A detailed study of Markov processes based on the functional analytic Hille–Yosida semigroup theory can be found in K. Itô (2004), (2006) quoted on page 353.

Let us generalize this. Suppose that we are given the classic smooth function $(x,t) \mapsto F(x,t)$ from $\mathbb{R}^2$ to $\mathbb{R}$. Let $F_x(x,t)$ denote the classic partial derivative $\frac{\partial F(x,t)}{\partial x}$. Set

$$\boxed{Y(t) := F(X(t),t), \qquad t \geq 0.}$$

Then the random variable $Y(t)$ at time t is given by the following Itô integral

$$Y(t) = Y(0) + \int_0^t F_x(X(\tau),\tau)\, dX(\tau) + \int_0^t \left(F_t(X(\tau),\tau) + \tfrac{1}{2}F_{xx}(X(\tau),\tau)\right) d\tau.$$

Mnemonically, we write

$$\boxed{dY(t) = F_x(X(t),t))\, dX(t) + F_t(X(t),t))\, dt + \tfrac{1}{2}F_{xx}(X(t),t)\, dt.}$$

This is the Itô formula which differs from the classic Leibniz formula by the additional term $\frac{1}{2}F_{xx}(X(t),t)\, dt$.

The Wiener measure, the Wiener path integral, and Brownian motion. Modern theory of probability was founded by Kolmogorov in 1933.[23] Kolmogorov used measure theory as the basic tool. In this setting, Brownian motion is described by the Wiener measure μ on the space $C_0[0,T]$ of all continuous functions

$$x : [0,T] \to \mathbb{R}$$

with $x(0) = 0$. Here, we fix the time interval $[0,T]$ and the initial position $x(0) = 0$ of the particle. The elements $t \mapsto x(t)$ of $C_0[0,T]$ are the possible trajectories of the Brownian motion of a particle. If $\mathcal{S}$ is a subset of $C_0[0,T]$, then the measure

$$\mu(\mathcal{S})$$

is the probability for finding the trajectory $t \mapsto x(t)$ in the set $\mathcal{S}$. Fix time $t \in [0,T]$. The random variable $X(t)$ is defined by

$$X(t)(x) := x(t) \qquad \text{for all} \quad x \in C_0[0,T].$$

The mean value is given by the Wiener path integral

$$\overline{X(t)} := \int_{x \in C_0[0,T]} X(t)(x)\, d\mu(x) = 0.$$

Here, we integrate over all trajectories $t \mapsto x(t)$ which are elements of the space $C_0[0,T]$. The classic construction of the Wiener measure μ invented by Wiener in 1923 can be found in Sect. 7.11.4 of Vol. II. The integral

$$\int_{x \in C_0[0,T]} F(x)\, d\mu(x)$$

for real valued-functions $F : C_0[0,T] \to \mathbb{R}$ on the space $C_0[0,T]$ of trajectories with respect to the Wiener measure μ is called the Wiener path integral.

[23] A. Kolmogorov, Foundations of the Theory of Probability, Springer, Berlin 1933 (in German). English edition: Chelsea, New York, 1950. An introduction to the modern theory of probability based on Kolmogorov's axioms can be found in E. Zeidler, Oxford Users' Guide to Mathematics, Chap. 6, Oxford University Press, 2004.

Concerning the Itô calculus, let us mention that almost all trajectories of the Brownian motion are not differentiable (i.e., the probability for finding a differentiable trajectory is equal to zero, in the sense of the Wiener measure μ). Therefore, Itô could not use the classic methods for introducing stochastic differentiation and integration.

> *In the 1940s, Feynman (1918–1988) emphasized that propagators are of fundamental importance for quantum physics (quantum mechanics, quantum statistics, and quantum field theory).*

Feynman's basic idea was to describe the motion of quantum particles by a statistics over classical trajectories. To this end, he used the Feynman path integral (see Chap. 7 of Vol. II). The Schrödinger equation is a diffusion equation in imaginary time. Therefore, the Feynman approach to quantum physics has to be based on diffusion processes in imaginary time. In particular, the Feynman path integral is a 'Wiener path integral in imaginary time'. This causes serious mathematical difficulties. In fact, this is the reason why the Feynman path integral approach frequently lacks mathematical rigor. However, there are situations which can be handled rigorously. We refer to:

S. Albeverio and S. Mazucchi, A survey on mathematical Feynman path integrals: construction, asymptotics, applications. In: B. Fauser, J. Tolksdorf, and E. Zeidler (Eds.), Quantum Field Theory: Competitive Models, Birkhäuser, Basel, 2009, pp. 49–66.

S. Albeverio, Yu. Kondratiev, Yu. Kositzsky, and M. Röckner, The Statistical Mechanics of Quantum Lattice Systems: a Path Integral Approach, European Mathematical Society, Zurich, 2009.

Further reading. Classic material can be found in:

K. Itô, On stochastic differential equations, Mem. Amer. Math. Soc. **4** (1951), 1–51.

K. Itô, Stochastic Processes, Lectures given at Aarhus University, Denmark, Springer, Berlin, 2004.

K. Itô, Essentials of Stochastic Processes, Amer. Math. Soc. Providence, Rhode Island, 2006.

K. Itô, Selected Papers. Edited by D. Stroock and S. Varadhan, Springer, New York, 1986.

As an introduction, we recommend:

Yu. Rozanov, Introductory Probability Theory, Prentice-Hall, Englewood Cliffs, 1969,

D. Williams, Probability with Martingales, Cambridge University Press, 1991,

L. Rogers and D. Williams, Diffusions, Markov Processes, and Martingales, Vol. 1: Foundations, Vol. 2: Itô Calculus, Cambridge University Press, 1978

together with

B. Simon, Functional Integration and Quantum Physics, Academic Press, New York, 1979.

D. Stroock, Markov Processes from K. Itô's Perspective, Princeton University Press, 2003.

L. Evans, An Introduction to Stochastic Differential Equations, Lectures held at the University of California at Berkeley, 2005.
Internet: `http://math.berkeley.edu/~evans/SDE.course.pdf`

B. Øksendal, Stochastic Differential Equations, Springer, New York, 1998.

Furthermore, we refer to:

A. Dynkin, Markov Processes, Vols. 1, 2, Springer, Berlin, 1965.

H. McKean, Stochastic Integrals, Academic Press, New York, 1979.

M. Freidlin, Functional Integration and Partial Differential Equations, Princeton University Press, 1985.

I. Karatzas and S. Shreve, Brownian Motion and Stochastic Calculus, Springer, New York, 1988.

N. Ikeda and S. Watanabe, Stochastic Differential Equations and Diffusion Processes, North-Holland, Amsterdam 1989.

H. Kunita, Stochastic Flows and Stochastic Differential Equations, Cambridge University Press, 1990.

O. Kallenberg, Foundations of Modern Probability, Springer, New York, 1997.

Summary of measure theory:

E. Zeidler, Nonlinear Functional Analysis and its Applications, Vol. IIB, Appendix, Springer, New York, 1997.

5. The Lie Group $U(1)$ as a Paradigm in Harmonic Analysis and Geometry

For understanding the gauge groups of the Standard Model in elementary particle physics, one has to understand the low-dimensional Lie groups. Maxwell's theory of the electromagnetic field and quantum states are intimately related to the commutative Lie group $U(1)$. The passage to electroweak (resp. strong) interaction in the Standard Model of particle physics is obtained by replacing the commutative gauge group $U(1)$ by the non-commutative gauge groups $SU(2)$ (resp. $U(3)$).[1]

Folklore

The theory of Lie groups and Lie algebras is nothing else than a far-reaching generalization of Euler's exponential function. The simplest case is the Lie group $U(1)$ defined by

$$U(1) := \{z \in \mathbb{C} : |z| = 1\}$$

equipped with the usual multiplication of complex numbers. Equivalently,

$$U(1) = \{e^{i\varphi} : \varphi \in \mathbb{R}\}.$$

The set $U(1)$ is a real one-dimensional manifold, namely, the unit circle. This manifold is called the group manifold of the Lie group $U(1)$.[2] In particular, a Lie group $\mathcal{G}$ is called compact iff $\mathcal{G}$ is a compact manifold. For example, the Lie group $U(1)$ is compact. In fact, the unit circle is a compact manifold.

A Lie group $\mathcal{G}$ is called locally compact iff it is a locally compact manifold, that is, every point has a compact neighborhood. The prototype for a locally compact Lie group is the additive Lie group $\mathbb{R}$ of real numbers (real line).

5.1 Linearization and the Lie Algebra $u(1)$

The linearization of $U(1)$ at the unit element 1 reads as

$$e^{i\varphi} = 1 + i\varphi + o(\varphi), \qquad \varphi \to 0.$$

This yields the Lie algebra

$$u(1) := \{i\varphi : \varphi \in \mathbb{R}\}$$

of the Lie group $U(1)$. Here, $u(1)$ is a real one-dimensional linear space equipped with the (trivial) Lie product $[V, W] := 0$ for all $V, W \in u(1)$. In terms of geometry,

[1] See Chap. 15 on page 843.

[2] The definition of a Lie group (resp. Lie algebra) can be found in Sect. 7.8 (resp. Sect. 7.6) of Vol. I.

E. Zeidler, *Quantum Field Theory III: Gauge Theory*,
DOI 10.1007/978-3-642-22421-8_6,

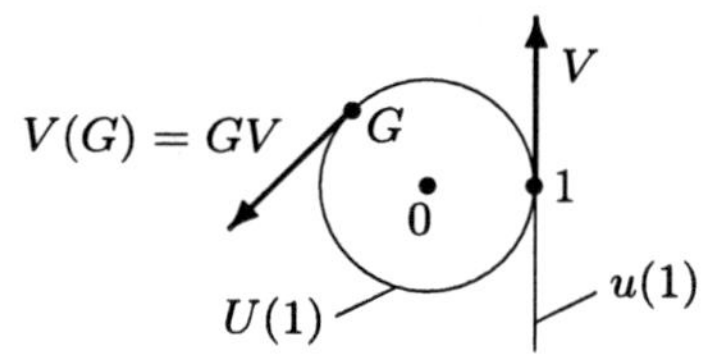

Fig. 5.1. The Lie group $U(1)$

the Lie algebra $u(1)$ can be identified with the tangent space of the Lie group $U(1)$ at the unit element 1 (see Fig. 5.1). The Lie algebra $u(1)$ becomes a real one-dimensional Hilbert space with respect to the inner product

$$\langle V|W\rangle := -VW \qquad \text{for all} \quad V, W \in u(1).$$

Note that $V = \mathrm{i}\varphi$ and $W = \mathrm{i}\psi$ are purely imaginary numbers (i.e., $\varphi, \psi \in \mathbb{R}$). Hence $\langle V|W\rangle = \varphi\psi$.

5.2 The Universal Covering Group of $U(1)$

Define $\chi(\varphi) := \mathrm{e}^{\mathrm{i}\varphi}$ for all real numbers φ. Euler's addition theorem for the exponential function, $\mathrm{e}^{\mathrm{i}(\varphi+\psi)} = \mathrm{e}^{\mathrm{i}\varphi}\mathrm{e}^{\mathrm{i}\psi}$, tells us that

$$\chi(\varphi + \psi) = \chi(\varphi)\chi(\psi) \qquad \text{for all} \quad \varphi, \psi \in \mathbb{R}.$$

Consequently, the map

$$\boxed{\chi : \mathbb{R} \to U(1)} \tag{5.1}$$

is a surjective group morphism from the (additive) group $\mathbb{R}$ onto the group $U(1)$. The group $\mathbb{R}$ (real line) is called the universal covering group of the Lie group $U(1)$. Let

$$f : U(1) \to \mathbb{C}$$

be a function on the Lie group $U(1)$. Setting $F(\varphi) := f(\chi(\varphi))$, we obtain the 2π-periodic function

$$F : \mathbb{R} \to \mathbb{C}.$$

This way, the functions on the Lie group $U(1)$ can be described by 2π-periodic functions on the real line and vice versa. This simple trick is frequently used in mathematics and physics.

5.3 Left-Invariant Velocity Vector Fields on $U(1)$

Left translations of the Lie group $U(1)$**.** Fix the group element $G \in U(1)$. We define

$$L_G H := GH \qquad \text{for all} \quad H \in U(1).$$

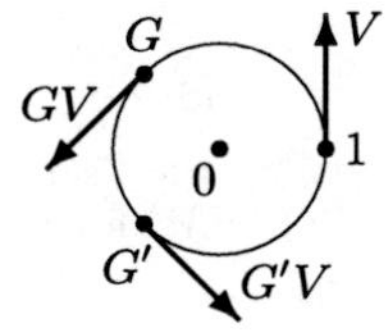

Fig. 5.2. Left-invariant velocity vector field

The transformation $L_G : U(1) \to U(1)$ is called a left translation of the Lie group $U(1)$.[3]

The internal symmetry properties of the Lie group $U(1)$ are described by left translations.

Velocity vector fields. Sophus Lie used velocity vector fields in order to study the structure of Lie groups. In terms of physics, the tangent vectors of the unit circle $U(1)$ are called velocity vectors. Consider a fixed velocity vector V at the unit element 1, that is, $V \in u(1)$. Define

$$V(G) := GV \qquad \text{for all} \quad G \in U(1).$$

This velocity vector field is left invariant, that is,

$$V(GH) = G \cdot V(H) \qquad \text{for all} \quad G, H \in U(1).$$

Obviously, this construction yields a one-to-one relation between the elements V of the Lie algebra $u(1)$ and the left-invariant velocity vector fields on the Lie group $U(1)$ (Fig. 5.2).

5.3.1 The Maurer–Cartan Form of $U(1)$

Fix the point G of the Lie group $U(1)$. Consider the smooth function

$$z(t) := \mathrm{e}^{\mathrm{i}\varphi(t)}, \qquad t_0 \le t \le t_1, \quad z(0) = G \tag{5.2}$$

with the angle $\varphi(t) \in \mathbb{R}$ for all times t. In terms of physics, the function $t \mapsto z(t)$ describes the motion of a particle on the unit circle which passes through the point G at time $t = 0$.[4] Differentiation with respect to time t yields the velocity vector $W = \dot{z}(0)$ at the point G. Explicitly,

$$\dot{z}(0) = z(0)\mathrm{i}\dot{\varphi}(0).$$

Hence $\dot{z}(0) = GV$ with $V := \mathrm{i}\dot{\varphi}(0)$. Consequently, the vectors of the tangent space $T_G U(1)$ of the Lie group $U(1)$ at the point G have the form

$$W = GV, \qquad V \in u(1).$$

[3] Similarly, the transformation $R_G : U(1) \to U(1)$ given by $R_G H := HG$ for all $H \in U(1)$ is called a right translation of the Lie group $U(1)$. Because of the commutativity of the group $U(1)$, left translations and right translations coincide. The situation changes in the case of noncommutative groups.

[4] We assume that $t_0 < 0 < t_1$.

We define

$$\mathsf{M}_G(W) := V \qquad \text{for all} \quad W \in T_G U(1).$$

The map $\mathsf{M}_G : T_G U(1) \to u(1)$ is called the Maurer–Cartan form of the Lie group $U(1)$ at the point G.[5] Setting $dG(W) := W$ for all $W \in T_G U(1)$, we get

$$\boxed{\mathsf{M}_G = G^{-1} dG.}$$

We say that M is a differential 1-form on the Lie group $U(1)$ with values in the Lie algebra $u(1)$. We will show later on that

Gauge field theory is based on differential forms with values in some Lie algebra.

5.3.2 The Maurer–Cartan Structural Equation

Applying the Cartan differential, it follows from $d\mathsf{M}_G = dG^{-1} \wedge dG = -G^{-2} dG \wedge dG$ and $dG \wedge dG = 0$ that

$$\boxed{d\mathsf{M} = 0 \qquad \text{on } U(1).}$$

This is a (trivial) special case of the famous Maurer–Cartan structural equation in the theory of Lie groups which governs the (local) structure of Lie groups. In particular, the general Maurer–Cartan structural equation is used in order to prove that a real n-dimensional Lie group $\mathcal{G}$ is locally trivial (i.e, it is locally isomorph to $\mathbb{R}^n$) iff the Lie algebra $\mathcal{LG}$ of $\mathcal{G}$ is commutative (i.e., $[V, W] = 0$ for all $V, W \in \mathcal{LG}$).[6] In particular, the Lie group $U(1)$ has the same local topological structure as the real line $\mathbb{R}$, but the global topological structure of $U(1)$ and $\mathbb{R}$ is different. In fact, the real line $\mathbb{R}$ is simply connected, whereas the unit circle $U(1)$ is not simply connected.

5.4 The Riemannian Manifold $U(1)$ and the Haar Measure

We assign to the trajectory (5.2) the arc length

$$s(t) := \int_{t_0}^{t} \frac{1}{2\pi} \frac{d\varphi(\tau)}{d\tau}\, d\tau$$

between the points $z(t_0)$ and $z(t)$. This is the (normalized) arc length on the unit circle. Hence $\frac{ds}{dt} = \frac{1}{2\pi}\frac{d\varphi}{dt}$. Mnemonically, in terms of classical analysis, we write

$$ds = \frac{1}{2\pi}\, d\varphi.$$

---

[5] Joseph Fourier (1768–1830), Sophus Lie (1842–1899), Élie Cartan (1859–1951), Ludwig Maurer (1859–1927), Alfred Haar (1885–1933), Hermann Weyl (1885–1955).

[6] For the proof, see the textbook by Y. Choquet-Bruhat, C. DeWitt-Morette, and M. Dillard-Bleick, Analysis, Manifolds, and Physics. Vol. 1, Elsevier, Amsterdam, 1996, page 209. The proof uses the Poincaré cohomology rule for differential forms.

In terms of modern mathematics, the differential form

$$\upsilon := \frac{1}{2\pi}\, d\varphi$$

is called the volume form of the Lie group $U(1)$. In particular, $\int_{U(1)} \upsilon = 1$.

Consider now the normalized measure μ on the unit circle induced by arc length, that is,

$$\int_{U(1)} d\mu = \frac{1}{2\pi} \int_{-\pi}^{\pi} d\varphi = 1.$$

This measure is called the Haar measure of the Lie group $U(1)$; it has the characteristic property that it is invariant under left translations. This means that

$$\int_{L_G \Omega} d\mu = \int_{\Omega} d\mu$$

for all measurable subsets Ω (with respect to the Haar measure) of the Lie group $U(1)$ and all group elements $G \in U(1)$.

5.5 The Discrete Fourier Transform

5.5.1 The Hilbert Space $L_2(U(1))$

Using the Haar measure, let us introduce the inner product

$$\langle f|g\rangle := \int_{U(1)} f^\dagger g\, d\mu.$$

That is, $\langle f|g\rangle = \frac{1}{2\pi}\int_{-\pi}^{\pi} f(\varphi)^\dagger g(\varphi) d\varphi$. This yields the complex Hilbert space $L_2(U(1))$. The elements of $L_2(U(1))$ are measurable functions $f, g : U(1) \to \mathbb{C}$ (with respect to the Haar measure, e.g., continuous functions) with $\langle f|f\rangle < \infty$.[7] Two functions f and g represent the same element of $L_2(U(1))$ iff they differ at most on a subset of $U(1)$ which has the Haar measure zero (e.g., the exceptional set is a finite or countable subset of $U(1)$). Setting

$$e_k(\varphi) := \mathrm{e}^{ik\varphi}, \qquad -\pi \le \varphi \le \pi, \quad k = 0, \pm 1, \pm 2, \ldots,$$

we get

$$\langle e_k|e_l\rangle = \delta_{kl}, \qquad k, l = 0, \pm 1, \pm 2, \ldots$$

The functions $e_0, e_1, e_{-1}, \ldots$ form a complete orthonormal system on the Hilbert space $L_2(U(1))$. For every function $f \in L_2(U(1))$, we define the Fourier transform $\hat{f} : \mathbb{Z} \to \mathbb{C}$ by setting

$$\boxed{\hat{f}(k) := \langle e_k|f\rangle = \int_{U(1)} e_{-k} f\, d\mu, \qquad k = 0, \pm 1, \pm 2, \ldots}$$

[7] Note that a function $f : U(1) \to \mathbb{C}$ is measurable with respect to the Haar measure iff the 2π-periodic function $f : \mathbb{R} \to \mathbb{C}$ ($\varphi \mapsto f(\varphi)$) is measurable with respect to the Lebesgue measure on the real line $\mathbb{R}$.

Proposition 5.1 (i) *The function $f : U(1) \to \mathbb{C}$ is smooth iff, for every positive integer n, there exists a constant $C(n)$ such that we have the growth condition*

$$|\hat{f}(k)| \leq \frac{C(n)}{1+|k|^n} \qquad \textit{for all} \quad k = 0, \pm 1, \pm 2, \ldots \tag{5.3}$$

(ii) *The measurable function $f : U(1) \to \mathbb{C}$ (with respect to the Haar measure on $U(1)$) is an element of the Hilbert space $L_2(U(1))$ iff $\sum_{k\in\mathbb{Z}} |\hat{f}(k)|^2 < \infty$.*

For the proof, we refer to Problem 5.1. If the growth condition (5.3) is satisfied, then the inverse Fourier transform $\hat{f} \mapsto f$ is given by the absolutely convergent series

$$f(\varphi) = \sum_{k\in\mathbb{Z}} \hat{f}(k) e_k(\varphi) \qquad \text{for all} \quad \varphi \in [-\pi, \pi]$$

with the Fourier coefficients $\hat{f}(k)$. In addition, for $n = 1, 2, \ldots$, we get the absolutely convergent series:[8]

$$\boxed{\frac{d^n f(\varphi)}{d\varphi^n} = \sum_{k\in\mathbb{Z}} (\mathrm{i}k)^n \hat{f}(k) \cdot e_k(\varphi) \qquad \text{for all} \quad \varphi \in [-\pi, \pi].} \tag{5.4}$$

5.5.2 Pseudo–Differential Operators

Motivated by (5.4), we define

$$\boxed{(Af)(\varphi) := \sum_{k\in\mathbb{Z}} \sigma(k)\hat{f}(k) \cdot e_k(\varphi), \qquad \varphi \in [-\pi, \pi].} \tag{5.5}$$

We assume that the function $\sigma : \mathbb{Z} \to \mathbb{C}$ satisfies the following growth condition

$$|\sigma(k)| = O(|k|^n), \qquad |k| \to \infty$$

where n is a fixed positive integer.

Proposition 5.2 *By* (5.5)*, we get the linear operator $A : C^\infty(U(1)) \to C^\infty(U(1))$.*

This follows from 5.3. The operator A is called a pseudo-differential operator with the symbol σ. Note that the operator A corresponds to the multiplication operator

$$k \mapsto \sigma(k)\hat{f}(k)$$

in the Fourier space (i.e., in the space of Fourier coefficients).
Example. Fix $m = \pm 1, \pm 2, \ldots$. Choose the symbol

$$\sigma(k) := (\mathrm{i}k)^m, \qquad k = \pm 1, \pm 2, \ldots, \qquad \sigma(0) := 0.$$

If $m = 1, 2 \ldots$, then

$$Af = \left(\frac{d}{d\varphi}\right)^m f \qquad \text{for all} \quad f \in C^\infty(U(1)),$$

[8] Recall that $\mathbb{Z} := \{0, \pm 1, \pm 2, \ldots\}$ denotes the set of integers.

in the classical sense.[9] If $m = -1, -2, \ldots$, then the operator A is denoted by $\left(\frac{d}{d\varphi}\right)^{-1}, \left(\frac{d}{d\varphi}\right)^{-2}, \ldots$, respectively. We get

$$\left(\frac{d}{d\varphi}\right)^{-n}\left(\frac{d}{d\varphi}\right)^{n} f = f - \hat{f}(0) \qquad \text{for all} \quad f \in C^\infty(U(1)), \quad n = 1, 2, \ldots$$

The general theory of pseudo-differential operators was created by J. Kohn and L. Nirenberg, An algebra of pseudo-differential operators, Commun. Pure Appl. Math. **18** (1965), 269–305. The basic idea is to replace the discrete Fourier transform by the Fourier integral transform. We recommend:

Yu. Egorov, A. Komech, and M. Shubin, Elements of the Modern Theory of Partial Differential Equations, Springer, New York, 1999 (survey).

S. Alinhac and P. Gérard, Pseudo-Differential Operators and the Nash–Moser Theorem, Amer. Math. Soc. Providence, Rhode Island, 2007.

G. Hsiao and W. Wendland, Boundary Integral Equations, Springer, New York, 2008.

M. Shubin, Pseudo-Differential Operators and Spectral Theory, Springer, Berlin, 2001.

5.5.3 The Sobolev Space $W_2^m(U(1))$

Let m be a real number. By definition, the space $W_2^m(U(1))$ consists of all the measurable functions $f : U(1) \to \mathbb{C}$ (with respect to the Haar measure) such that

$$\sum_{k\in\mathbb{Z}, k\neq 0} |k^m \hat{f}(k)|^2 < \infty.$$

5.6 The Group of Motions on the Gaussian Plane

Proper motions. Fix the complex numbers $g \in U(1)$ and $a \in \mathbb{C}$. Define the transformation

$$\boxed{T(g,a)z := gz + a \qquad \text{for all} \quad z \in \mathbb{C}.} \tag{5.6}$$

This is a so-called proper Euclidean motion on the Gaussian plane $\mathbb{C}$ of complex numbers. Explicitly, this is the superposition of a translation $z \mapsto z + a$ and a rotation $z \mapsto gz$. In particular, if $g = \mathrm{e}^{\mathrm{i}\varphi}$ and $a = 0$, then we get the rotation

$$\begin{aligned} z' &= x' + y'\mathrm{i} = \mathrm{e}^{\mathrm{i}\varphi} z = (\cos\varphi + \mathrm{i}\sin\varphi)(x + y\mathrm{i}) \\ &= (x\cos\varphi - y\sin\varphi) + \mathrm{i}(x\sin\varphi + y\cos\varphi). \end{aligned}$$

Hence

$$\begin{pmatrix} x' \\ y' \end{pmatrix} = \begin{pmatrix} \cos\varphi & -\sin\varphi \\ \sin\varphi & \cos\varphi \end{pmatrix} \begin{pmatrix} x \\ y \end{pmatrix}. \tag{5.7}$$

[9] The space $C^\infty(U(1))$ coincides with the space of smooth 2π-periodic functions $f : \mathbb{R} \to \mathbb{C}$.

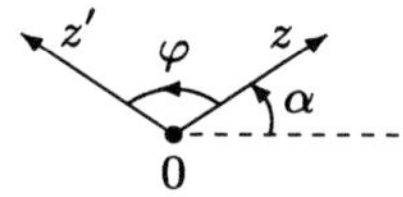

Fig. 5.3. Rotation of the Gaussian plane

The transformation matrix reads as

$$A = \begin{pmatrix} \cos\varphi & -\sin\varphi \\ \sin\varphi & \cos\varphi \end{pmatrix}. \tag{5.8}$$

Formula (5.7) describes a rotation of the Gaussian plane by the angle φ (see Fig. 5.3). If $\varphi > 0$, then we get a counter-clockwise rotation.

Improper motions. In contrast to (5.6), the transformation

$$S(g,a)z := gz^\dagger + a \qquad \text{for all} \quad z \in \mathbb{C} \tag{5.9}$$

is called an improper Euclidean motion on the Gaussian plane $\mathbb{C}$ if $g \in U(1)$ and $a \in \mathbb{C}$ are fixed complex numbers. In particular, the transformation $z \mapsto z^\dagger$ is a reflection at the x-axis, $(x,y) \mapsto (x,-y)$. Whereas proper motions (5.6) preserve the orientation, improper motions (5.9) change the orientation. Proper and improper motions form a Lie group called the group of Euclidean motions on the Gaussian plane $\mathbb{C}$.

Let us translate this into the language of operators on the Euclidean Hilbert space E_2.

5.7 Rotations of the Euclidean Plane

In terms of operator theory, rotations in planar Euclidean geometry correspond to orientation-preserving unitary operators of the real two-dimensional Hilbert space E_2.

Folklore

The unitary group $U(E_2)$ of the Hilbert space E_2. Let E_2 be a fixed two-dimensional linear subspace of the Hilbert space E_3. We call E_2 the Euclidean plane. The operator

$$U : E_2 \to E_2$$

is called unitary iff it is linear and it preserves the inner product on E_2, that is,

$$\langle U\mathbf{x}|U\mathbf{y}\rangle = \langle \mathbf{x}|\mathbf{y}\rangle \qquad \text{for all} \quad \mathbf{x},\mathbf{y} \in E_2.$$

Here, $\langle \mathbf{x}|\mathbf{y}\rangle := \mathbf{xy}$. If $U, V \in U(E_2)$, then $UV \in U(E_2)$. In fact,

$$\langle UV\mathbf{x}|UV\mathbf{y}\rangle = \langle V\mathbf{x}|V\mathbf{y}\rangle = \langle \mathbf{x}|\mathbf{y}\rangle.$$

Consequently, $U(E_2)$ is a group called the unitary group of E_2.

The special unitary group $SU(E_2)$. By definition, the set $SU(E_2)$ consists of all operators $U \in U(E_2)$ which preserve the orientation. This is a subgroup of $U(E_2)$. Choose a right-handed orthonormal system $\mathbf{i},\mathbf{j},\mathbf{k}$ of E_3 such that the vectors $\mathbf{i},\mathbf{j}$ span E_2. We want to show that the group $SU(E_2)$ coincides with the group of rotations on E_2, and the operators B of $U(E_2)$ have precisely the form

$$B = RA$$

where $A \in SU(E_2)$, and R is either the identity operator or a reflection operator with respect to a straight line passing through the origin.

Proposition 5.3 *We have $U \in U(E_2)$ iff there exists a real number φ such that the linear operator $U : E_2 \to E_2$ is given by*

$$\begin{aligned} U\mathbf{i} &= \cos\varphi\, \mathbf{i} + \sin\varphi\, \mathbf{j}, \\ U\mathbf{j} &= \gamma(-\sin\varphi\, \mathbf{i} + \cos\varphi\, \mathbf{j}) \end{aligned} \tag{5.10}$$

where either $\gamma = 1$ or $\gamma = -1$. In addition, $U \in SU(E_2)$ iff $\gamma = 1$.

Proof. Let $U \in U(E_2)$. Since the operator U preserves the inner product, the vectors $U\mathbf{i}, U\mathbf{j}$ form an orthonormal system. Thus, $U\mathbf{i}$ is a unit vector. This yields the first line of (5.10). Since the unit vector $U\mathbf{j}$ is orthogonal to $U\mathbf{i}$, we get the second line of (5.10). Conversely, one checks easily that every linear operator with the property (5.10) is unitary. Finally note that the pair $U\mathbf{i}, U\mathbf{j}$ has the same orientation as $\mathbf{i}, \mathbf{j}$ iff $\gamma = 1$. □

Rotations. The linear operator $U : E_2 \to E_2$ given by

$$\boxed{U\mathbf{x} := \cos\varphi \cdot \mathbf{x} + \sin\varphi \cdot (\mathbf{k} \times \mathbf{x}), \qquad \mathbf{x} \in E_2} \tag{5.11}$$

coincides with (5.10) iff $\gamma = 1$. This is a rotation by the angle φ (see Fig. 6.1(a) on page 373). Setting $\mathbf{x}' = U\mathbf{x}$, we get

$$\begin{aligned} x'\mathbf{i} + y'\mathbf{j} &= U(x\mathbf{i} + y\mathbf{j}) = x \cdot U\mathbf{i} + y \cdot U\mathbf{j} \\ &= (x\cos\varphi - y\sin\varphi)\mathbf{i} + (x\sin\varphi + y\cos\varphi)\mathbf{j}. \end{aligned}$$

This coincides with (5.7).

Infinitesimal rotations and the Lie algebra $su(E_2)$. Linearization of the operator U from (5.11) with respect to the small rotation angle φ yields

$$\boxed{U\mathbf{x} = \mathbf{x} + \varphi(\mathbf{k} \times \mathbf{x}), \qquad \varphi \to 0.}$$

The operator $T_{\varphi\mathbf{k}}$ defined by

$$T_{\varphi\mathbf{k}}\mathbf{x} := \varphi(\mathbf{k} \times \mathbf{x}), \qquad \mathbf{x} \in E_2$$

is called an infinitesimal rotation of the Euclidean plane E_2 with the rotation angle φ. The set of all those infinitesimal rotations is denoted by $su(E_2)$. Equipped with the Lie product

$$[T_{\varphi\mathbf{k}}, T_{\psi\mathbf{k}}]_- := T_{\varphi\mathbf{k}}T_{\psi\mathbf{k}} - T_{\psi\mathbf{k}}T_{\varphi\mathbf{k}},$$

the real one-dimensional linear space $su(2)$ becomes a commutative (i. e. trivial) Lie algebra. In fact, it follows from $\mathbf{k} \times \mathbf{k} = 0$ that

$$T_{\varphi\mathbf{k}}T_{\psi\mathbf{k}}\mathbf{x} = \varphi\mathbf{k} \times (\psi\mathbf{k} \times \mathbf{x}) = 0 \qquad \text{for all} \quad \mathbf{x} \in E_2,\ \varphi, \psi \in \mathbb{R}.$$

Hence $[T_{\varphi\mathbf{k}}, T_{\psi\mathbf{k}}]_- = 0$ for all elements of $su(2)$. Set $\varrho(\mathrm{i}\varphi) := T_{\varphi\mathbf{k}}$. The map

$$\varrho : u(1) \to su(E_2)$$

is a Lie algebra isomorphism (i.e., a linear bijective map which respects the Lie product).[10] We write

[10] See Sect. 7.6 of Vol. I.

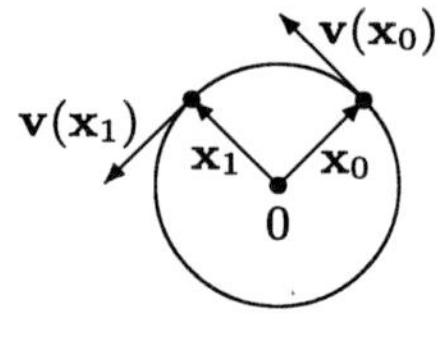

Fig. 5.4. Killing velocity vector field

$$u(1) \simeq su(E_2).$$

The map $\varphi \mapsto \cos\varphi \cdot \mathbf{x} + \sin\varphi\ (\mathbf{k} \times \mathbf{x})$ sends the elements of the Lie group $U(1)$ to the corresponding rotations of the Euclidean plane E_2. This yields the group isomorphism $U(1) \simeq SU(E_2)$.

In order to get an interpretation in terms of physics, replace the angle φ by time t. Let $L(E_2, E_2)$ denote the space of all the linear operators $A : E_2 \to E_2$. This is a real Banach space.[11]

The exponential map. Motivated by (5.11), we define

$$\boxed{U(t)\mathbf{x} := \cos t \cdot \mathbf{x} + \sin t \cdot (\mathbf{k} \times \mathbf{x}) \qquad \text{for all} \quad \mathbf{x} \in E_2,}$$

and all times $t \in \mathbb{R}$. Setting $A := \dot{U}(0)$, we get

$$\boxed{A\mathbf{x} = \mathbf{k} \times \mathbf{x}.}$$

Proposition 5.4 $U(t) = \mathrm{e}^{tA}$ *for all* $t \in \mathbb{R}$.

Proof. It follows from $\dot{U}(t)\mathbf{x} = -\sin t \cdot \mathbf{x} + \cos t \cdot (\mathbf{k} \times \mathbf{x})$ and the vector formula $\mathbf{k} \times (\mathbf{k} \times \mathbf{x}) = \mathbf{k}(\mathbf{kx}) - \mathbf{x}(\mathbf{kk}) = -\mathbf{x}$ for all $\mathbf{x} \in E_2$ that

$$\dot{U}(t) = AU(t), \quad t \in \mathbb{R}, \quad U(0) = I. \tag{5.12}$$

This is a differential equation in the Banach space $L(E_2, E_2)$ which has the unique solution $U(t) = \mathrm{e}^{tA}$ for all $t \in \mathbb{R}$ (see Sect. 7.7 of Vol. I). □

Physical interpretation. For fixed position vector $\mathbf{x}_0 \in E_2$, we set

$$\mathbf{x}(t) := \mathrm{e}^{tA}\mathbf{x}_0 \qquad \text{for all} \quad t \in \mathbb{R},$$

and $\mathbf{v}(\mathbf{x}_0) := \dot{\mathbf{x}}(0)$. Hence

$$\mathbf{v}(\mathbf{x}_0) = A\mathbf{x}_0.$$

Explicitly, $\mathbf{v}(\mathbf{x}_0) := \mathbf{k} \times \mathbf{x}_0$. That is, the vector $\mathbf{v}(\mathbf{x}_0)$ is perpendicular to $\mathbf{x}_0$, and it has the same length as the vector $\mathbf{x}_0$ (Fig. 5.4).

In terms of physics, the trajectory $t \mapsto \mathbf{x}(t)$ describes the counter-clockwise rotation of a fluid particle with the angular velocity $\omega = 1$. By (5.12), we have the equation of motion

$$\dot{\mathbf{x}}(t) = \mathbf{v}(\mathbf{x}(t)), \qquad t \in \mathbb{R}, \quad \mathbf{x}(0) = \mathbf{x}_0. \tag{5.13}$$

The velocity vector field $\mathbf{x} \mapsto \mathbf{v}(\mathbf{x})$ of the fluid is called the Killing velocity vector field of the rotation group $SU(E_2)$ on the Euclidean plane E_2.[12] Observe that

[11] The norm on $L(E_2, E_2)$ is the operator norm given by $||B|| := \max_{||\mathbf{x}|| \le 1} ||B\mathbf{x}||$ (see Sect. 7.13 of Vol. I)

[12] Killing (1847–1923)

$\mathbf{v}(\mathbf{x}) = A\mathbf{x}$ and $A \in su(E_2)$. Therefore, the Lie algebra $su(2)$ of the rotation group $SU(E_2)$ describes the Killing velocity vector field.

Flow. Define $F_t := e^{tA}$. Then, $F_0 := I$, and

$$F_{t+s} = F_t F_s \qquad \text{for all} \quad t, s \in \mathbb{R}, \tag{5.14}$$

by the addition theorem $e^{(t+s)A} = e^{tA}e^{sA}$ of the exponential function. The family $\{F_t\}$ is called a one-parameter group (or a flow). In particular,

$$(F_t)^{-1} = F_{-t} \qquad \text{for all} \quad t \in \mathbb{R}.$$

The language of matrices. If $U : E_2 \to E_2$ is a linear operator, then there exists a uniquely determined real (2×2)-matrix $\mathcal{U}$ such that we have the following matrix product

$$(U\mathbf{i}, U\mathbf{j}) = (\mathbf{i}, \mathbf{j})\,\mathcal{U}.$$

Here, $\mathcal{U}$ is called the matrix of the operator U with respect to the orthonormal basis $\mathbf{i}, \mathbf{j}$ of E_2. Explicitly,

$$\mathcal{U} = \begin{pmatrix} \langle \mathbf{i}|U\mathbf{i}\rangle & \langle \mathbf{i}|U\mathbf{j}\rangle \\ \langle \mathbf{j}|U\mathbf{i}\rangle & \langle \mathbf{j}|U\mathbf{j}\rangle \end{pmatrix} = \begin{pmatrix} U^1_1 & U^1_2 \\ U^2_1 & U^2_2 \end{pmatrix}.$$

Setting $\mathbf{y} = U\mathbf{x}$, as well as $\mathbf{x} = x^1\mathbf{i} + x^2\mathbf{j}$, and $\mathbf{y} = y^1\mathbf{i} + y^2\mathbf{j}$, we obtain the transformation formula

$$\begin{pmatrix} y^1 \\ y^2 \end{pmatrix} = \mathcal{U} \begin{pmatrix} x^1 \\ x^2 \end{pmatrix}.$$

In particular, the matrix $\mathcal{U}$ to the operator U from (5.10) reads as

$$\mathcal{U} = \begin{pmatrix} \cos\varphi & -\gamma\sin\varphi \\ \sin\varphi & \gamma\cos\varphi \end{pmatrix}.$$

We have $\mathcal{U}^d\mathcal{U} = I$, that is, the matrix $\mathcal{U}$ is orthogonal.[13] Moreover, $\det\mathcal{U} = \gamma = \pm 1$.

Proposition 5.5 *The linear operator $U : E_2 \to E_2$ is unitary iff the corresponding matrix $\mathcal{U}$ is orthogonal.*

Proof. The operator U is unitary iff

$$U^\dagger U = I. \tag{5.15}$$

This follows from $\langle \mathbf{x}|\mathbf{y}\rangle = \langle U\mathbf{x}|U\mathbf{y}\rangle = \langle U^\dagger U\mathbf{x}|\mathbf{y}\rangle$ for all $\mathbf{x}, \mathbf{y} \in E_2$, which is equivalent to $U^\dagger U\mathbf{x} = \mathbf{x}$ for all $\mathbf{x} \in E_2$. It follows from (5.15) that

---

[13] Recall that the $(n \times n)$-matrix $\mathcal{A}$ is called unitary iff $\mathcal{A}^\dagger\mathcal{A} = I$. This implies

$$1 = \det I = \det\mathcal{A}^\dagger \cdot \det\mathcal{A} = (\det\mathcal{A})^\dagger \cdot \det\mathcal{A}.$$

Thus, for unitary matrices, we have $|\det\mathcal{A}| = 1$.
If the entries of the matrix $\mathcal{A}$ are real numbers, unitary matrices are called orthogonal. Then $\mathcal{A}^\dagger = \mathcal{A}^d$, and hence $\det\mathcal{A} = \pm 1$.

$$1 = \det I = \det U^\dagger \det U = (\det U)^\dagger \det U = (\det U)^2.$$

Hence $\det U = \pm 1$. In the language of matrices, equation (5.15) is equivalent to

$$\mathcal{U}^\dagger \mathcal{U} = I.$$

Since the matrix $\mathcal{U}$ is real, we have $\mathcal{U}^\dagger = \mathcal{U}^d$. □

The orthogonal group $O(2)$. The symbol $O(2)$ denotes the set of all the real (2×2)-matrices which are orthogonal. This is a multiplicative group. In fact, if $\mathcal{U}, \mathcal{V} \in O(2)$, then $(\mathcal{U}\mathcal{V})^d(\mathcal{U}\mathcal{V}) = \mathcal{V}^d(\mathcal{U}^d\mathcal{U})\mathcal{V} = \mathcal{V}^d\mathcal{V} = I$. By Prop. 5.5, we have $\mathcal{U} \in O(2)$ iff there exist a real number φ and a number $\gamma = \pm 1$ such that

$$\mathcal{U} = \begin{pmatrix} \cos\varphi & -\gamma \sin\varphi \\ \sin\varphi & \gamma\cos\varphi \end{pmatrix}. \tag{5.16}$$

The map $U \mapsto \mathcal{U}$ yields a group isomorphism from $U(E_2)$ onto $O(2)$. We write

$$\boxed{U(E_2) \simeq O(2).}$$

The special orthogonal group $SO(2)$. By definition, $\mathcal{U} \in SO(2)$ iff $\mathcal{U} \in O(2)$ and $\det \mathcal{U} = 1$. It follows from $\mathcal{U}, \mathcal{V} \in SO(2)$ that $\mathcal{U}\mathcal{V} \in SO(2)$.[14] Therefore, $SO(2)$ is a subgroup of $O(2)$. We have $\mathcal{U} \in SO(2)$ iff there exists a real number φ such that $\mathcal{U} = \mathcal{U}(\varphi)$ with

$$\mathcal{U}(\varphi) := \begin{pmatrix} \cos\varphi & -\sin\varphi \\ \sin\varphi & \cos\varphi \end{pmatrix}. \tag{5.17}$$

The map $e^{i\varphi} \mapsto \mathcal{U}(\varphi)$ yields the group isomorphism

$$U(1) \simeq SO(2).$$

In addition, it follows from $\det U = \det \mathcal{U}$ that the map $U \mapsto \mathcal{U}$ yields the group isomorphism

$$\boxed{SU(E_2) \simeq SO(2).}$$

The Lie algebra $so(2)$. Linearization of the matrix $\mathcal{U}(\varphi)$ from (5.17) for a small rotation angle φ yields

$$\mathcal{U}(\varphi) = I + \varphi\mathcal{U}'(0) + o(\varphi), \qquad \varphi \to 0$$

with

$$\mathcal{U}'(0) = \begin{pmatrix} 0 & -1 \\ 1 & 0 \end{pmatrix}.$$

Let $so(2)$ denote the set of all real (2×2)-matrices which are antisymmetric and trace-free. That is, $\mathcal{A}^d = -\mathcal{A}$ and $\mathrm{tr}(\mathcal{A}) = 0$. We have $\mathcal{A} \in so(2)$ iff there exists a real number φ such that $\mathcal{A} = \varphi\mathcal{U}'(0)$. Setting

$$[\mathcal{A}, \mathcal{B}]_- := \mathcal{A}\mathcal{B} - \mathcal{B}\mathcal{A} \qquad \text{for all} \quad \mathcal{A}, \mathcal{B} \in so(2),$$

[14] Note that $\det(\mathcal{U}\mathcal{V}) = \det\mathcal{U} \det\mathcal{V} = 1$ and $\det\mathcal{U}^{-1} = (\det\mathcal{U})^{-1} = 1$.

the real one-dimensional linear space $so(2)$ is a commutative Lie algebra. In fact, $[\mathcal{A},\mathcal{B}]_- = 0$ for all $\mathcal{A},\mathcal{B} \in so(2)$. The map $T_{\varphi\mathbf{k}} \mapsto \varphi\mathcal{U}'(0)$ yields the Lie algebra isomorphism

$$\boxed{su(E_2) \simeq so(2).}$$

The exponential map. Let $\mathcal{U}(\varphi) \in SO(2)$. Differentiation with respect to the rotation angle φ yields

$$\frac{d\mathcal{U}(\varphi)}{d\varphi} = \begin{pmatrix} -\sin\varphi & -\cos\varphi \\ \cos\varphi & -\sin\varphi \end{pmatrix}.$$

Setting $\varphi = 0$, we get $\mathcal{A} := \begin{pmatrix} 0 & -1 \\ 1 & 0 \end{pmatrix}$. Thus, we obtain the key differential equation

$$\boxed{\frac{d\mathcal{U}(\varphi)}{d\varphi} = \mathcal{A}\mathcal{U}(\varphi), \quad \varphi \in \mathbb{R}, \quad \mathcal{U}(0) = I} \tag{5.18}$$

which has the unique solution

$$\mathcal{U}(\varphi) = \mathrm{e}^{\varphi\mathcal{A}} \qquad \text{for all} \quad \varphi \in \mathbb{R}. \tag{5.19}$$

The addition theorem for the matrix exponential function yields

$$\mathcal{U}(\varphi + \psi) = \mathcal{U}(\varphi)\mathcal{U}(\psi) \qquad \text{for all} \quad \varphi, \psi \in \mathbb{R}.$$

Explicitly,

$$\begin{pmatrix} \cos(\varphi+\psi) & -\sin(\varphi+\psi) \\ \sin(\varphi+\psi) & \cos(\varphi+\psi) \end{pmatrix} = \begin{pmatrix} \cos\varphi & -\sin\varphi \\ \sin\varphi & \cos\varphi \end{pmatrix} \begin{pmatrix} \cos\psi & -\sin\psi \\ \sin\psi & \cos\psi \end{pmatrix}.$$

This is equivalent to the addition theorem for trigonometric functions. For example,

$$\sin(\varphi + \psi) = \sin\varphi \, \cos\psi + \cos\varphi \, \sin\psi \qquad \text{for all} \quad \varphi, \psi \in \mathbb{R}.$$

This shows that the well-known classical addition theorems for the trigonometric functions are rooted in the rotation group of the Euclidean plane. From (5.19) we get the following result.

Proposition 5.6 *We have* $\mathcal{U} \in SO(2)$ *iff there exists a matrix* $\mathcal{A} \in so(2)$ *such that* $\mathcal{U} = \mathrm{e}^{\mathcal{A}}$.

The reflection group. The set $\mathcal{Z}_2 := \{1, -1\}$ is a multiplicative group called the multiplicative cyclic group of order two. Let

$$I := \begin{pmatrix} 1 & 0 \\ 0 & 1 \end{pmatrix}.$$

Since $\det(\pm I) = 1$, the set $\{I, -I\}$ is a subgroup of $SO(2)$. The map $\pm I \mapsto \pm 1$ yields the group isomorphism

$$\{I, -I\} \simeq \mathcal{Z}_2.$$

Define $\mathcal{R}_- := \begin{pmatrix} 1 & 0 \\ 0 & -1 \end{pmatrix}$. The set $R_{\text{refl}} := \{I, \mathcal{R}_-\}$ is a subgroup of $O(2)$ called the special reflection group. The map $\mathcal{R}_- \mapsto -1, I \mapsto 1$ yields the group isomorphism

$$R_{\text{refl}} \simeq \mathbb{Z}_2.$$

Set $\chi(\mathcal{U}) := \det \mathcal{U}$. The map $\chi : O(2) \to \mathbb{Z}_2$ is a surjective group morphism with the kernel

$$\ker(\chi) = \{\mathcal{U} \in O(2) : \det \mathcal{U} = 1\} = SO(2).$$

By the morphism theorem for groups,[15] the kernel $SO(2)$ is a normal subgroup of $O(2)$, and we have the group isomorphism

$$\boxed{O(2)/SO(2) \simeq \mathbb{Z}_2.}$$

Local and global behavior. We want to show that the Lie groups $O(2)$ and $SO(2)$ have the same local structure near the unit element, but they possess different global topological structures. It follows from (5.16) that there exists a neighborhood $\mathcal{N}(I)$ of the unit matrix I such that

$$O(2) = SO(2) \qquad \text{on } \mathcal{N}(I).$$

Consequently, the linearization of the orthogonal group $O(2)$ at the unit element I is the same as the corresponding linearization of $SO(2)$ at I. This implies that

$$o(2) = so(2)$$

where $o(2)$ (resp, $so(2)$) is the Lie algebra of $O(2)$ (resp. $SO(2)$).

Proposition 5.7 *The Lie group $SO(2)$ is arcwise connected, but the Lie group $O(2)$ is not arcwise connected. The group $O(2)$ has two components.*

Proof. The map $\varphi \mapsto \mathcal{U}(\varphi)$ is continuous on $\mathbb{R}$ (i.e., the matrix elements depend continuously on φ). Consequently, two elements of $SO(2)$ can always be connected by a continuous curve.

The elements I and $\mathcal{R}_-$ cannot be connected by a continuous curve in $O(2)$. Otherwise, the determinant of the matrices is continuous along the curve. Since this determinant only attains the values ± 1, the determinant is constant along the curve. Hence $\det I = \det \mathcal{R}_-$. However, $\det \mathcal{R}_- = -1$. This is the desired contradiction.

The two sets $SO(2) = \{\mathcal{U}(\varphi) : \varphi \in \mathbb{R}\}$ and

$$O_-(2) := \{\mathcal{U}(\varphi)\mathcal{R}_- : \varphi \in \mathbb{R}\}$$

are arcwise connected. This way, we get the two components of $O(2)$ which contain the elements I and $\mathcal{R}_-$, respectively. Therefore the two components are characterized by the sign of the determinant of the matrices. This sign is equal to 1 (resp. -1) on $SO(2)$ (resp. $O_-(2)$). □

---

[15] See Sect. 4.1.3 of Vol. II.

5.8 Pontryagin Duality for $U(1)$ and Quantum Groups

Duality plays a crucial role in the theory of topological groups; characters generalize the exponential function.

Folklore

Characters. Let $\mathcal{G}$ be an arbitrary group. A character of $\mathcal{G}$ is a group morphism

$$\chi : \mathcal{G} \to U(1),$$

that is, $\chi(gh) = \chi(g)\chi(h)$ for all $g, h \in \mathcal{G}$. This generalizes the functional equation of the exponential function. In a natural way, the set $\mathcal{G}'$ of all the characters of the group $\mathcal{G}$ can be equipped with a group structure by using the usual product of functions. The unit element $\mathbf{1}$ of $\mathcal{G}'$ is the constant character: $\mathbf{1}(g) := 1$ for all $g \in \mathcal{G}$. The commutative group $\mathcal{G}'$ is called the dual group to $\mathcal{G}$.

Examples. We have the following group isomorphisms:

- $\mathbb{R}' \simeq \mathbb{R}$;
- $\mathbb{Z}' \simeq U(1)$ and $U(1)' \simeq \mathbb{Z}$.

Here, $\mathbb{R}$ and $\mathbb{Z}$ denotes the additive group of real numbers and integers, respectively. Note the biduality relations

$$\boxed{U(1)'' \simeq U(1), \qquad \mathbb{Z}'' \simeq \mathbb{Z}}$$

which are the prototypes of far-reaching duality theorems in harmonic analysis. Let us discuss this.[16]

Ad $\mathbb{R}'$. Fix $r \in \mathbb{R}$. Define

$$\chi_r(x) := \mathrm{e}^{\mathrm{i}rx} \qquad \text{for all} \quad x \in \mathbb{R}.$$

Then, $\chi_r(x+y) = \chi_r(x)\chi_r(y)$ for all $x, y \in \mathbb{R}$. Hence χ_r is a character of $\mathbb{R}$. In fact, there are no other characters, that is,

$$\mathbb{R}' = \{\chi_r : \ r \in \mathbb{R}\},$$

and the map $r \mapsto \chi_r$ yields the group isomorphism $\mathbb{R} \simeq \mathbb{R}'$.

Ad $\mathbb{Z}'$. Fix $z \in U(1)$. Define

$$\chi_z(n) := z^n \qquad \text{for all} \quad n \in \mathbb{Z}.$$

Since $\chi_z(m+n) = \chi_z(m)\chi_n(z)$ for all $m, n \in \mathbb{Z}$, the function χ_z is a character on $\mathbb{Z}$. More precisely,

$$\mathbb{Z}' = \{\chi_z : \ z \in U(1)\},$$

and the map $z \mapsto \chi_z$ yields the group isomorphism $U(1) \simeq \mathbb{Z}'$.

Ad $U(1)'$. Fix $n \in \mathbb{Z}$. Define

$$\chi_n(z) := z^n \qquad \text{for all} \quad z \in U(1).$$

Then $U(1)' = \{\chi_n : \ n \in \mathbb{Z}\}$, and the map $n \mapsto \chi_n$ yields the group isomorphism $\mathbb{Z} \simeq U(1)'$.

Pontryagin duality. In 1934, Pontryagin proved the following general theorems:[17]

[16] The proofs of the following statements can be found in L. Pontryagin, Topological Groups, Gordon and Breach, 1966.

[17] L. Pontryagin, The theory of topological commutative groups, Ann. of Math. **35** (1934), 361–388.

(i) If $\mathcal{G}$ is a locally compact, commutative topological group, then so is the dual group $\mathcal{G}'$.
(ii) If $\mathcal{G}$ is a compact (resp. discrete) commutative topological group, then the dual group $\mathcal{G}'$ is discrete (resp. compact), and $\mathcal{G}'' \simeq \mathcal{G}$ (group isomorphism).

Recall that a topological space X is called compact iff every open covering of X contains a finite subcover (i.e., finitely many of these open sets already cover the set X). Moreover, X is called locally compact iff every point of X has a compact neighborhood. For example, $\mathbb{R}, \mathbb{C}, \mathbb{Z}, \mathbb{R}^n, \mathbb{C}^n, n = 1, 2, \ldots$, are locally compact, and $U(1)$ is compact. Moreover, $\mathbb{Z}$ is locally compact and discrete, but not compact. A group is called a topological group iff

- it is a topological space (see Sect. 5.5 of Vol. I),
- the product map $(g, h) \mapsto gh$ is a continuous map from $X \times X \to X$, and
- the map $g \mapsto g^{-1}$ is continuous from X to X.

Every Lie group is a topological group.

Tannaka–Krein duality. The Krein–Tannaka theory generalizes the Pontryagin theorem to *noncommutative* compact Lie groups. The point is that one has to replace the dual group by a real Hopf algebra. For details, see Sect. 3.5.4 of Vol. II.

Quantum groups. It is possible to generalize sophistically the Tannaka–Krein duality to quantum groups. See Woronowicz (1987) and Timmermann (2007) quoted on page 546.

Problems

5.1 *Proof of Proposition 5.1.* Solution: Ad (i). Let $k \in \mathbb{Z}$, $k \neq 0$. Integration by parts, yields

$$2\pi \hat{f}(k) = \int_{-\pi}^{\pi} \mathrm{e}^{imk\varphi} f(\varphi)\, d\varphi = \frac{\mathrm{i}}{\mathrm{i}k} = \int_{-\pi}^{\pi} \mathrm{e}^{imk\varphi} f(\varphi)\, d\varphi.$$

Hence $|\hat{f}(k)| \leq \text{const} \cdot k^{-1}$.
Ad (ii). Study the proof in E. Zeidler, Applied Functional Analysis: Application to Mathematical Physics, Sect. 3.2, Springer, New York, 1995.

6. Infinitesimal Rotations and Constraints in Physics

Constraints play a crucial role in the history of physics. This concerns the lever principle due to Archimedes of Syracus (287–212 B.C.), the principle of virtual work due to d'Alembert (1717–1783), the multiplier rule due to Lagrange (1736–1813), the principle of least squares and the principle of least constraint due to Gauss (1777–1855), the motion of rigid bodies and the spinning top, the spinning electron due to Pauli (1900–1958), and the BRST-quantization of gauge fields in quantum field theory.[1] The main idea is to simplify the constraints by passing to infinitesimal constraints which are closely related to the linearization of Lie groups via Lie algebras.

Folklore

6.1 The Group $U(E_3)$ of Unitary Transformations

Operators. The operator $A : E_3 \to E_3$ is called unitary iff it is linear and it respects the inner product, that is,

$$\langle A\mathbf{x}|A\mathbf{y}\rangle = \langle \mathbf{x}|\mathbf{y}\rangle \qquad \text{for all} \quad \mathbf{x}, \mathbf{y} \in E_3. \tag{6.1}$$

The symbol $U(E_3)$ denotes the set of all unitary operators $A : E_3 \to E_3$. We have

$$A \in U(E_3) \qquad \text{iff} \qquad A^\dagger A = I. \tag{6.2}$$

In fact, it follows from (6.1) that

$$\langle \mathbf{x}|A^\dagger A\mathbf{y}\rangle = \langle A\mathbf{x}|A\mathbf{y}\rangle = \langle \mathbf{x}|\mathbf{y}\rangle \qquad \text{for all} \quad \mathbf{x}, \mathbf{y} \in E_3.$$

Hence $A^\dagger A = I$. Conversely, $A^\dagger A = I$ implies (6.1).

If $A \in U(E_3)$, *then* $\det(A) = \pm 1$.

In fact, $I = A^\dagger A$ implies $1 = \det I = \det A^\dagger \det A = (\det A)^\dagger \det A = |\det A|^2$.

Consequently, if $A \in U(E_3)$, then A is invertible. This tells us that

$$A \in U(E_3) \qquad \text{iff} \qquad A^{-1} = A^\dagger. \tag{6.3}$$

Hence $A \in U(E_3)$ iff $AA^\dagger = I$.

[1] The letters 'BRST' stand for Becchi, Rouet, Stora, and Tyutin. See C. Becchi, A. Rouet, and R. Stora, Renormalization of the Abelian Higgs–Kible model, Commun. Math. Phys. **52** (1975), 127–162; Renormalization of gauge theories, Annals of Physics **98** (1976), 287–321.

E. Zeidler, *Quantum Field Theory III: Gauge Theory*,
DOI 10.1007/978-3-642-22421-8_7,

Theorem 6.1 *The set $U(E_3)$ forms a group.*

Proof. If $A, B \in U(E_3)$, then we get $(AB)^\dagger AB = B^\dagger(A^\dagger A)B = B^\dagger IB = I$. Hence $AB \in U(E_3)$. Moreover, $(A^{-1})^\dagger A^{-1} = AA^\dagger = I$. Thus, $A^{-1} \in U(E_3)$. □

By definition, the set $SU(E_3)$ contains all the elements A of $U(E_3)$ with $\det(A) = 1$. If $\det A = \det B = 1$, then $\det(AB) = \det A \det B = 1$. Therefore, $SU(3)$ is a subgroup of $U(E_3)$.

Matrices. Let $\mathbf{e}_1, \mathbf{e}_2, \mathbf{e}_3$ be a right-handed orthonormal basis of the Euclidean Hilbert space E_3. We are given the unitary operator $A \in U(E_3)$. Set

$$\mathcal{A}^i_j := \langle \mathbf{e}_i | A\mathbf{e}_j \rangle, \qquad i,j = 1,2,3.$$

For the real matrix $\mathcal{A} = (A^i_j)$ it follows from $A^\dagger A = I$ that

$$\mathcal{A}^\dagger \mathcal{A} = I.$$

Consequently, the linear operator $A : E_3 \to E_3$ is unitary iff the matrix $\mathcal{A}$ is orthogonal, that is, $\mathcal{A} \in O(3)$. Moreover, we have $A \in SU(E_3)$ iff $\mathcal{A} \in SO(3)$, that is, $\mathcal{A} \in O(3)$ and $\det \mathcal{A} = 1$. The map

$$A \mapsto \mathcal{A}$$

yields the group isomorphisms $U(E_3) \simeq O(3)$ and $SU(E_3) \simeq SO(3)$. Since the matrix groups $O(3)$ and $SO(3)$ are Lie groups, the isomorphic groups $U(E_3)$ and $SU(E_3)$ are also Lie groups.

Theorem 6.2 *Precisely the rotations are the elements of the Lie group $SU(E_3)$.*

Proof. Let $A \in SU(3)$. The eigenvalue equation

$$A\mathbf{x} = \lambda\mathbf{x}, \qquad \mathbf{x} \in E_3,\ \mathbf{x} \neq 0 \tag{6.4}$$

is equivalent to the matrix equation

$$\mathcal{A}x = \lambda x$$

with respect to the orthonormal basis $\mathbf{e}_1, \mathbf{e}_2, \mathbf{e}_3$. This implies

$$\det(\mathcal{A} - \lambda I) = 0.$$

This equation has the form

$$\lambda^3 + a\lambda^2 + b\lambda + c = 0 \tag{6.5}$$

with real coefficients a, b, c. If $\lambda_1, \lambda_2, \lambda_3$ are the solutions of (6.5), then

$$c = \lambda_1\lambda_2\lambda_3 \quad \text{and} \quad c = \det \mathcal{A} = \det A = 1.$$

Since the operator A is unitary, $|\lambda_j| = 1$ for all j.[2] Moreover, since the coefficients a, b, c are real, the numbers $\lambda_j^\dagger$ are also eigenvalues. Thus, there exists at least one real eigenvalue, say λ_1, and $\lambda_1 = 1$. By (6.4), there exists a unit vector $\mathbf{k}$ such that $A\mathbf{k} = \lambda_1\mathbf{k} = \mathbf{k}$. Supplementing the vector $\mathbf{k}$ to a right-handed orthonormal system $\mathbf{i}, \mathbf{j}, \mathbf{k}$, the operator A is unitary on the linear subspace E_2 spanned by the vectors $\mathbf{i}$ and $\mathbf{j}$. Consequently, the corresponding matrix has the form

[2] Note that $\langle \mathbf{x}|\mathbf{x}\rangle = \langle A\mathbf{x}|A\mathbf{x}\rangle = \langle \lambda\mathbf{x}|\lambda\mathbf{x}\rangle = \lambda^\dagger\lambda\langle \mathbf{x}|\mathbf{x}\rangle$.

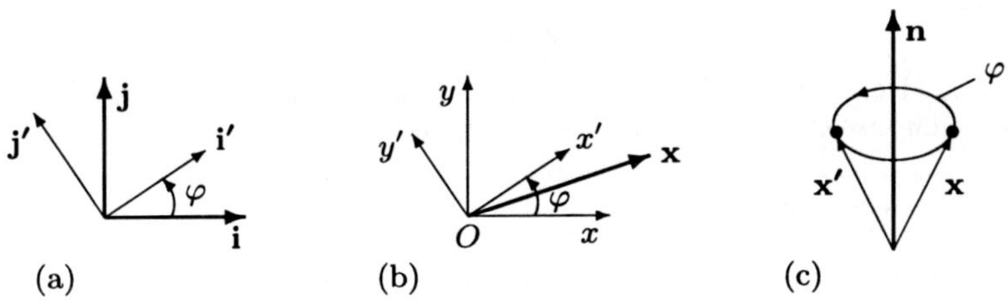

Fig. 6.1. Rotation

$$\mathcal{A} = \begin{pmatrix} B & 0 \\ 0 & \lambda_1 \end{pmatrix}$$

where $B \in SO(2)$. Explicitly, there exists a real parameter φ such that

$$\mathcal{A} = \begin{pmatrix} \cos\varphi & -\sin\varphi & 0 \\ \sin\varphi & \cos\varphi & 0 \\ 0 & 0 & 1 \end{pmatrix}.$$

Consequently, the operator A describes a rotation about the axis vector $\mathbf{k}$ with the rotation angle φ.

Conversely, a similar argument shows that all the rotations are unitary operators with $A \in SU(E_3)$. □

The linearization of the matrix $\mathcal{A}$ with respect to the small rotation angle φ reads as

$$\mathcal{A} = \begin{pmatrix} 1 & 0 & 0 \\ 0 & 1 & 0 \\ 0 & 0 & 1 \end{pmatrix} + \begin{pmatrix} 0 & -\varphi & 0 \\ \varphi & 0 & 0 \\ 0 & 0 & 0 \end{pmatrix} + O(\varphi^2), \qquad \varphi \to 0.$$

We refer to (7.10).

6.2 Euler's Rotation Formula

Let $\mathbf{n}$ be a unit vector. The Euler formula

$$\boxed{\mathbf{x}' = \cos\varphi \cdot \mathbf{x} + \sin\varphi \cdot (\mathbf{n} \times \mathbf{x}) + (1 - \cos\varphi)(\mathbf{xn})\mathbf{n}} \tag{6.6}$$

describes the counter-clockwise rotation of the vector $\mathbf{x}$ about the axis $\mathbf{n}$ with rotation angle φ (Fig.6.1(c)). To check the Euler rotation formula (6.6), consider a right-handed orthonormal basis $\mathbf{i}, \mathbf{j}, \mathbf{k}$ with $\mathbf{k} = \mathbf{n}$. Then $\mathbf{k}' = \mathbf{k}$, and the Euler formula passes over to the well-known rotation formula in the Euclidean plane depicted in Fig. 6.1(a):

$$\mathbf{i}' = \cos\varphi\,\mathbf{i} + \sin\varphi\,\mathbf{j}, \qquad \mathbf{j}' = -\sin\varphi\,\mathbf{i} + \cos\varphi\,\mathbf{j}. \tag{6.7}$$

The decomposition of the vector $\mathbf{x}$ with respect to the Cartesian basis $\mathbf{i}, \mathbf{j}, \mathbf{k}$ and the rotated Cartesian basis $\mathbf{i}', \mathbf{j}', \mathbf{k}'$ reads as

$$\mathbf{x} = x\mathbf{i} + y\mathbf{j} + z\mathbf{k} = x'\mathbf{i}' + y'\mathbf{j}' + z'\mathbf{k}'$$

with $\mathbf{k} = \mathbf{k}'$. Hence $x' = \mathbf{x}\mathbf{i}'$ and $y' = \mathbf{x}\mathbf{j}'$, implying the transformation formula for the corresponding Cartesian coordinate systems,

$$\boxed{x' = \cos\varphi \cdot x + \sin\varphi \cdot y, \quad y' = -\sin\varphi \cdot x + \cos\varphi \cdot y, \quad z' = z.} \tag{6.8}$$

If x^+, y^+, z^+ are the components of the rotated vector $\mathbf{x}'$ with respect to the original orthonormal basis $\mathbf{i}, \mathbf{j}, \mathbf{k}$, that is, $\mathbf{x}' = x^+\mathbf{i} + y^+\mathbf{j} + z^+\mathbf{k}$, then

$$x^+ = \cos\varphi \cdot x - \sin\varphi \cdot y, \quad y^+ = \sin\varphi \cdot x + \cos\varphi \cdot y, \quad z^+ = z. \tag{6.9}$$

The transformation formula (6.9) (resp. (6.8)) is called an active (resp. passive) rotation by physicists. The passage from (6.9) to (6.8) corresponds to $\varphi \Rightarrow -\varphi$.

6.3 The Lie Algebra of Infinitesimal Rotations

In the 1870's, Sophus Lie discovered that the investigation of symmetries in geometry can be essentially simplified by studying the linearization of transformation groups which are called infinitesimal transformations. This is the basic idea behind studying Lie groups with the help of Lie algebras. Let us apply the idea of linearization to Euler's rotation formula. Taylor expansion yields $\sin\varphi = \varphi + O(\varphi^2)$ and $\cos\varphi = 1 + O(\varphi^2)$ as $\varphi \to 0$. Thus, for small rotation angles φ and all vectors $\mathbf{x}$, the Euler rotation formula (6.7) reads approximately as

$$\boxed{\mathbf{x}' = \mathbf{x} + \varphi T_{\mathbf{n}}\mathbf{x} + O(\varphi^2), \qquad \varphi \to 0.} \tag{6.10}$$

Here, we set $T_{\mathbf{n}}\mathbf{x} := \mathbf{n} \times \mathbf{x}$ for all $\mathbf{x} \in E_3$. All the transformations $T_{\mathbf{n}}$ with $\mathbf{n} \in E_3$ are called infinitesimal rotations. If $\mathbf{n}^2 = 1$, then the transformation $T_{\varphi\mathbf{n}} = \varphi T_{\mathbf{n}}$ is called an infinitesimal rotation with the rotation axis $\mathbf{n}$ and the rotation angle φ. The Jacobi identity tells us that

$$\mathbf{m} \times (\mathbf{n} \times \mathbf{x}) - \mathbf{n} \times (\mathbf{m} \times \mathbf{x}) = (\mathbf{m} \times \mathbf{n}) \times \mathbf{x}.$$

This implies the key relation for infinitesimal rotations,

$$\boxed{T_{\mathbf{m}\times\mathbf{n}} = T_{\mathbf{m}}T_{\mathbf{n}} - T_{\mathbf{n}}T_{\mathbf{m}},} \tag{6.11}$$

which is valid for all vectors $\mathbf{m}, \mathbf{n} \in E_3$. Therefore, all the infinitesimal rotations $T_{\mathbf{n}}$ form a real Lie algebra denoted by $su(E_3)$. The map

$$\mathbf{n} \mapsto T_{\mathbf{n}}$$

is an isomorphism from the Lie algebra $(E_3)_{\text{Lie}}$ of vectors equipped with the vector product onto the Lie algebra $su(E_3)$.

Recall that the Lie group $U(E_3)$ (resp. $SU(E_3)$) is isomorphic to the Lie group $O(3)$ (resp. $SO(3)$). The Lie groups $U(E_3)$ and $SU(E_3)$ coincide on a sufficiently small neighborhood of the unit element I. Therefore, $U(E_3)$ and $SU(E_3)$ have the same Lie algebra, namely, $su(E_3)$. Moreover, the real 3-dimensional Lie algebra $su(E_3)$ is isomorphic to the Lie algebra $so(3)$ of the Lie group $SO(3)$. Recall that $so(3)$ consists of all the real skew-symmetric (3×3)-matrices.

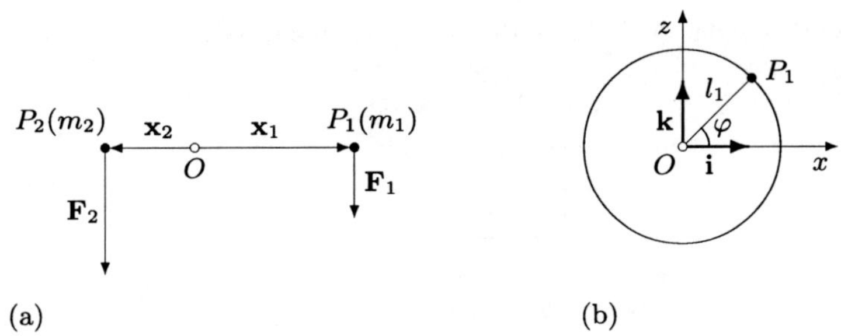

Fig. 6.2. Lever principle

6.4 Constraints in Classical Physics

Our goal is to explain the relation between infinitesimal rotations and constrained problems.

6.4.1 Archimedes' Lever Principle

Consider a balance as depicted in Fig. 6.2(a). Suppose that the points P_1 and P_2 have the masses m_1 and m_2, respectively. Finally, suppose that l_j is the length of the balance beam $\overrightarrow{OP}_j$. In ancient times, Archimedes discovered that equilibrium states of a balance are characterized by the following condition:

$$\boxed{m_1 g \cdot l_1 = m_2 g \cdot l_2.} \tag{6.12}$$

This is called the lever principle. Let us discuss this in terms of classical mechanics created by Newton (1643–1727) and his successors. Condition (6.12) is also called the torque condition, as we will explain below.[3]

Constraints, virtual trajectories, virtual velocities, and virtual accelerations. The two mass points P_1 and P_2 do not move freely, but their motion is governed by the constraints

$$\boxed{\mathbf{x}_1^2 = l_1^2, \qquad \mathbf{x}_2 = -\frac{l_2}{l_1}\mathbf{x}_1.} \tag{6.13}$$

Let $t \mapsto (\mathbf{x}_1(t), \mathbf{x}_2(t))$ be a trajectory which satisfies the constraints (6.13), that is,

$$\boxed{\mathbf{x}_1(t)^2 = l_1^2, \qquad \mathbf{x}_2(t) = -\frac{l_2}{l_1}\mathbf{x}_1(t), \qquad t \in \mathbb{R}.} \tag{6.14}$$

Define

$$\mathbf{v}_j := \dot{\mathbf{x}}_j(t) \qquad \text{and} \qquad \mathbf{a}_j := \ddot{\mathbf{x}}_j(t), \quad j = 1, 2.$$

The trajectories satisfying the constraints are called virtual trajectories. Moreover, $\mathbf{v}_j$ is called a virtual velocity vector, and $\mathbf{a}_j$ is called a virtual acceleration vector.

> *Virtual trajectories concern the possible motions which respect the constraints, but not necessarily the physical forces.*

[3] Torque has the physical dimension 'force times length'.

Differentiating equation (6.14) with respect to time t, we get

- $\mathbf{x}_1(t)\dot{\mathbf{x}}_1(t) = 0,\ l_1\dot{\mathbf{x}}_2(t) = -l_2\dot{\mathbf{x}}_1(t)$,
- $\dot{\mathbf{x}}_1(t)^2 + \mathbf{x}_1(t)\ddot{\mathbf{x}}_1(t) = 0,\ l_1\ddot{\mathbf{x}}_2(t) = -l_2\ddot{\mathbf{x}}_1(t)$

for all times $t \in \mathbb{R}$. This implies the relations

$$\boxed{\mathbf{x}_1(t)\mathbf{v}_1 = 0, \qquad l_1\mathbf{v}_2 = -l_2\mathbf{v}_1} \tag{6.15}$$

for the virtual velocity vectors $\mathbf{v}_1$ and $\mathbf{v}_2$ at the position $\mathbf{x}_1(t)$ and $\mathbf{x}_2(t)$, respectively. Explicitly, the virtual trajectories read as

$$\mathbf{x}_1 = l_1(\cos\varphi(t)\ \mathbf{i} + \sin\varphi(t)\ \mathbf{k}), \quad \mathbf{x}_2 = -l_2(\cos\varphi(t)\ \mathbf{i} + \sin\varphi(t)\ \mathbf{k}). \tag{6.16}$$

Here, $\mathbf{x}_j$ is the position vector $\overrightarrow{OP_j}$ at the origin O, and $\mathbf{i}, \mathbf{k}$ is a right-handed orthonormal basis of the (x, z)-plane (Fig. 6.2(b)). Moreover, we get the following equations

$$\dot{\mathbf{x}}_1(t)^2 + \mathbf{x}_1(t)\mathbf{a}_1 = 0, \quad l_1\mathbf{a}_2 = -l_2\mathbf{a}_1, \qquad t \in \mathbb{R} \tag{6.17}$$

for the virtual accelerations $\mathbf{a}_1, \mathbf{a}_2$ at the positions $\mathbf{x}_1(t), \mathbf{x}_2(t)$, respectively. Note that we do not study the specific molecular forces of the balance beam, but we summarize these forces by taking the constraints into account. Let us first study equilibrium states of the balance.

The relation to torque. The gravitational force of earth exerting on a point of mass m near the surface of earth sis approximately equal to

$$\mathbf{F} = -mg\mathbf{k}$$

where $g = 9.8\ \mathrm{m/s}^2$ (gravitational acceleration on the surface of earth in the SI system). Using the so-called torque

$$\mathbf{T}_j := \mathbf{x}_j \times \mathbf{F}_j, \qquad j = 1, 2$$

of the force $\mathbf{F}_j$ with respect to the point P_j, the lever condition (6.12) for equilibrium states of the balance is equivalent to the vanishing of the total torque:

$$\boxed{\mathbf{T}_1 + \mathbf{T}_2 = 0.}$$

In fact, explicitly $\mathbf{T}_1 + \mathbf{T}_2 = (m_2gl_2 - m_1gl_1)\cos\varphi \cdot \mathbf{i} \times \mathbf{k}$.

The relation to potential energy. Setting $U := mgz$, we get

$$\mathbf{F} = -\,\mathbf{grad}\, U.$$

The function U is called the potential energy of a mass point of mass m and height z in the gravitational field of earth near the surface of earth. The total potential energy of the mass points P_1 and P_2 is given by

$$U(\varphi) = (m_1gl_1 - m_2gl_2)\ \sin\varphi.$$

Then:

- $U'(\varphi) = (m_1gl_1 - m_2gl_2)\ \cos\varphi$, and
- $U''(\varphi) = (m_2gl_2 - m_1gl_1)\ \sin\varphi$.

This implies the following:

Fix the angle $\varphi \in]-\frac{\pi}{2}, \frac{\pi}{2}[$. Then the lever condition (6.12) *is equivalent to the critical potential energy condition: $U'(\varphi) = 0$.*

The problem of stability. In physics, stability of an equilibrium state S means, roughly speaking, that the system returns to the state S after small perturbations.

Stability is fundamental for nature and technology.

For example, only stable bridges are useful. Unstable situations in nature are responsible for ecological catastrophes. In classical mechanics, strict local minima of the potential energy correspond to locally stable equilibrium states. In the present case of the balance, we have to distinguish the following three cases:

(i) Unstable equilibrium state: Fix the angle $\varphi_0 \in]-\frac{\pi}{2}, \frac{\pi}{2}[$. If the lever condition $m_1gl_1 = m_2gl_2$ is satisfied, then the corresponding equilibrium state of the balance satisfies the condition $U'(\varphi_0) = 0$ together with $U''(\varphi_0) = 0$ and $U(\varphi) \equiv 0$ for all $\varphi \in]-\frac{\pi}{2}, \frac{\pi}{2}[$. This reflects the instability of the equilibrium state of the balance, as expected by daily experience.

(ii) Stable equilibrium state: Fix the angle $\varphi_0 = -\frac{\pi}{2}$, and let $m_1l_1 > m_2l_2$. Then the corresponding state of the balance has the property $U'(\varphi_0) = 0$ and

$$U''(\varphi_0) > 0.$$

This is a stable equilibrium state of the balance, as expected by daily life.

(iii) Fix the angle $\varphi_0 = \frac{\pi}{2}$, and let $m_1l_1 > m_2l_2$. Then the corresponding state of the balance has the property $U'(\varphi_0) = 0$ and $U''(\varphi_0) < 0$. This is an unstable equilibrium state of the balance, as expected by daily life.

6.4.2 d'Alembert's Principle of Virtual Power

Let us compute the total work

$$W(t) = \int_{t_0}^{t} \sum_{j=1}^{2} \mathbf{F}_j d\mathbf{x}_j = \int_{t_0}^{t} \sum_{j=1}^{2} \mathbf{F}_j \dot{\mathbf{x}}_j(\tau)\, d\tau$$

along the trajectories $t \mapsto \mathbf{x}_j(t)$ given by (6.16) with the initial condition $\varphi(0) = \varphi_0$. Explicitly,

$$\dot{W}(t) = (m_1gl_1 - m_2gl_2)\, \cos\varphi(t) \cdot \dot{\varphi}(t).$$

Choosing the time $t = 0$ and noting that $\varphi(0) = 0$, we get the following.

Proposition 6.3 *Fix the angle* $\varphi_0 \in]-\frac{\pi}{2}, \frac{\pi}{2}[$. *Then the lever equilibrium condition* (6.12) *is equivalent to* $\frac{dW(0)}{dt} = 0$.

Proof. Observe that we can choose the function $\varphi(t) := \omega t + \varphi_0$ with arbitrary real number ω. Then $\dot{\varphi}(0) = \omega$. □

Proposition 6.3 is called d'Alembert's principle of virtual power. This designation is motivated by the fact that $W(t)$ has the physical dimension of work (energy), and the time derivative $\frac{dW(t)}{dt}$ has the physical dimension of energy per time which is also called power. Note that we compute the virtual work $W(t)$ by using virtual trajectories. Such trajectories obey the constraints, however, they do not necessarily describe physical motions of the mass points governed by the acting force. In other words, virtual trajectories are geometric objects, but not necessarily physical objects.

The principle of virtual power is a general principle in classical mechanics for describing constrained mechanical systems.

The statical principle of virtual power. The equilibrium condition

$$\frac{dW(0)}{dt} = 0$$

can equivalently be written as

$$\boxed{\mathbf{F}_1(P_1)\mathbf{v}_1(P_1) + \mathbf{F}_2(P_2)\mathbf{v}_2(P_2) = 0.} \tag{6.18}$$

More precisely, the pair (P_1, P_2) of points represents an equilibrium state of the balance iff the condition (6.18) is satisfied for all the virtual velocity vectors $\mathbf{v}_j$ at the point P_j, $j = 1, 2$, Note that $\mathbf{F}_j\mathbf{v}_j$ has the physical dimension 'force times velocity' which is also called 'power' (energy per time).

The dynamical principle of virtual power. As a consequence of the general Gaussian principle of least constraint, we will show below that the modified condition

$$\boxed{(\mathbf{F}_1 - m_1\ddot{\mathbf{x}}_1(t))\,\mathbf{v}_1 + (\mathbf{F}_2 - m_2\ddot{\mathbf{x}}_2(t))\,\mathbf{v}_2 = 0,} \tag{6.19}$$

which has to be valid for all virtual velocity vectors $\mathbf{v}_1$ and $\mathbf{v}_2$ at the points $P_1(t)$ and $P(_2(t)$, respectively, describes the dynamics of the balance. d'Alembert postulated that the principle (6.18) remains true if we add the inertial force $-m_j\ddot{\mathbf{x}}_j$ to the force $\mathbf{F}_j$. This way, d'Alembert (1717–1783) reduced dynamics to statics.

6.4.3 d'Alembert's Principle of Virtual Work

In 1743, at the age of 26, d'Alembert published his important *Traité de dynamic* (Treatise on dynamics) which contains the famous 'Alembert principle'.

Folklore

Let us pass from the physical dimension 'power' to 'work' (energy). To this end, we set

$$\delta\mathbf{x}_j := \mathbf{v}_j \cdot \Delta t, \qquad j = 1, 2,$$

where Δt is an arbitrary real number equipped with the physical dimension of time. Multiplying equation (6.19) by time Δt, we get

$$\boxed{(\mathbf{F}_1 - m_1\ddot{\mathbf{x}}_1)\,\delta\mathbf{x}_1 + (\mathbf{F}_2 - m_2\ddot{\mathbf{x}}_2)\,\delta\mathbf{x}_2 = 0.} \tag{6.20}$$

This is called d'Alembert's principle of virtual work. Physicists frequently write δt instead of Δt.

6.4.4 The Gaussian Principle of Least Constraint and Constraining Forces

The Gaussian principle of least constraint is the most general principle of classical mechanics. This principle is a variant of the famous method of least squares, which was published by Gauss (1777–1855) in his 1809 treatise *On the Motion of Celestial Bodies* (in Latin). For a long time, this treatise was the Bible of astronomers. Gauss showed how one can compute the orbit of a celestial body if one has only a few observed positions at hand. This was motivated by the rediscovery of the planetoid Ceres in 1802.

On January 1, 1801, in Palermo (Italy), the astronomer Piazzi observed an unknown celestial body for some weeks, but then the celestial body was lost. Gauss solved the hard problem to compute the orbit of this celestial body on the basis of only minimal information about its positions. Using the method of least squares for an elliptic orbit, Gauss had to solve algebraic equations of eights order. On January 1, 1802, in Gotha (Germany), the astronomer von Zach rediscovered the planetoid Ceres on the basis of Gauss' prediction.

Folklore

We want to study the motion of a balance under the action of the gravitational force of earth. To this end, we will use the Gaussian principle of least constraint. In what follows, we will distinguish between

- holonomic constraints, and
- nonholonomic constraints.

Holonomic constraints depend on the possible trajectories, but not on the possible velocity vectors. In contrast to this, nonholonomic constraints depend on both the possible trajectories and the possible velocity vectors, and they cannot be reduced to holonomic constraints by integration. In terms of geometry, holonomic (resp. nonholonomic) constraints can be handled by passing to a submanifold of the position manifold (resp. tangent bundle of the position manifold).

Holonomic Constraints

Holonomic constraints generate the manifold $\mathcal{M}$ of constrained positions. The virtual velocities are precisely the tangent vectors of $\mathcal{M}$.

Folklore

Generalizing a balance, let us consider the motion

$$\mathbf{x}_j = \mathbf{x}_j(t), \qquad t \in \mathbb{R}, \quad j = 1, 2$$

of two mass points P_1 and P_2 of mass m_1 and m_2, respectively. We are given the following constraints

$$\boxed{f_k(\mathbf{x}_1, \mathbf{x}_2) = 0, \qquad (\mathbf{x}_1, \mathbf{x}_2) \in \mathcal{U}, \; k = 1, \ldots, K} \tag{6.21}$$

where the given functions $f_1, \ldots, f_K : \mathcal{U} \to \mathbb{R}$ are assumed to be smooth, and $\mathcal{U}$ is assumed to be a nonempty, open, arcwise connected subset of $E_3 \times E_3$. Finally, let $1 \leq K < 6$. Note that the constraints (6.21) do not depend on the velocities of the particles. Such constraints are called holonomic. In the special case of a balance, we have $K = 5$. Explicitly,

- $f_j(\mathbf{x}_1, \mathbf{x}_2) := l_2 x_1^j + l_1 x_2^j, \; j = 1, 2, 3,$
- $f_4(\mathbf{x}_1, \mathbf{x}_2) := \mathbf{x}_1^2 - l_1^2,$
- $f_5(\mathbf{x}_1, \mathbf{x}_2) := x_1^3.$

Here, x_m^1, x_m^2, x_m^3 are the Cartesian components of the vector $\mathbf{x}_m$, $m = 1, 2$. These five constraints determine a 1-dimensional submanifold $\mathcal{M}$ of the 6-dimensional linear space $E_3 \times E_3$. This submanifold $\mathcal{M}$ is diffeomorphic to the unit circle, by (6.16).

The linear manifold of virtual accelerations. Differentiating twice the equation

$$f_k(\mathbf{x}_1(t), \mathbf{x}_2(t)) = 0, \qquad k = 1, \ldots, K$$

with respect to time t, and setting $\mathbf{v}_j := \dot{\mathbf{x}}_j(t)$ and $\mathbf{a}_j := \ddot{\mathbf{x}}_j(t)$, $j = 1, 2$, we get

$$\mathbf{v}_1 \cdot \mathbf{grad}_{\mathbf{x}_1} f_k(\mathbf{x}_1, \mathbf{x}_2) + \mathbf{v}_2 \cdot \mathbf{grad}_{\mathbf{x}_2} f_k(\mathbf{x}_1, \mathbf{x}_2) = 0, \quad k = 1, \ldots, K, \tag{6.22}$$

and

$$\mathbf{a}_1 \cdot \mathbf{grad}_{\mathbf{x}_1} f_k(\mathbf{x}_1, \mathbf{x}_2) + \mathbf{a}_2 \cdot \mathbf{grad}_{\mathbf{x}_2} f_k(\mathbf{x}_1, \mathbf{x}_2) = \mathbf{A}_k(\mathbf{x}_1, \mathbf{x}_2, \mathbf{v}_1, \mathbf{v}_2), \tag{6.23}$$

where $k = 1, \ldots, K$, and $(\mathbf{x}_1, \mathbf{x}_2) \in \mathcal{U}$.[4] Here, the system (6.22) (resp. (6.23)) is a linear system for the virtual velocities $\mathbf{v}_1, \mathbf{v}_2$ (resp. virtual accelerations $\mathbf{a}_1, \mathbf{a}_2$). The following regularity assumption is crucial:

We assume that the linearized constraints (6.22) *cannot be reduced to a smaller number of linear constraints.*

More precisely, this means that, for any given point $(\mathbf{x}_1, \mathbf{x}_2, \mathbf{v}_1, \mathbf{v}_2)$ in $\mathcal{U} \times E_3 \times E_3$, the rank r of the linear system (6.22) is maximal (i.e., $r = K$). This means that the virtual velocities form a linear $(6-K)$-dimensional submanifold of the 6-dimensional linear space $E_3 \times E_3$. Thus, by the nonlinear rank theorem on page 1080, this implies that the constraints (6.21) describe a $(6-K)$-dimensional submanifold of $E_3 \times E_3$. In addition, by (6.23), the virtual accelerations form a linear $(6-K)$-dimensional manifold.

The Gaussian minimum problem. Gauss studied the following minimum problem:

$$\boxed{\frac{1}{2m_1}(\mathbf{F}_1(\mathbf{x}_1, \mathbf{x}_2) - m_1\ddot{\mathbf{x}}_1)^2 + \frac{1}{2m_2}(\mathbf{F}_2(\mathbf{x}_1, \mathbf{x}_2) - m_2\ddot{\mathbf{x}}_2)^2 = \min!} \tag{6.24}$$

together with the linearized constraints

$$\ddot{\mathbf{x}}_1 \cdot \mathbf{grad}_{\mathbf{x}_1} f_k(\mathbf{x}_1, \mathbf{x}_2) + \ddot{\mathbf{x}}_2 \cdot \mathbf{grad}_{\mathbf{x}_2} f_k(\mathbf{x}_1, \mathbf{x}_2) = \mathbf{A}_k(\mathbf{x}_1, \mathbf{x}_2, \dot{\mathbf{x}}_1, \dot{\mathbf{x}}_2), \tag{6.25}$$

where $k = 1, \ldots, K$. For given $(\mathbf{x}_1, \mathbf{x}_2, \dot{\mathbf{x}}_1, \dot{\mathbf{x}}_2) \in \mathcal{U} \times E_3 \times E_3$, we are looking for a solution $\ddot{\mathbf{x}}_1, \ddot{\mathbf{x}}_2$ of the minimum problem (6.24) with the linear constraints (6.25).

The Gaussian principle of least constraint selects the physical accelerations among the virtual accelerations determined by the constraints.

It was the philosophy of Gauss that the physical accelerations are determined in such a way that the so-called constraint

$$\frac{1}{2m_1}(\mathbf{F}_1(\mathbf{x}_1, \mathbf{x}_2) - m_1\ddot{\mathbf{x}}_1)^2 + \frac{1}{2m_2}(\mathbf{F}_2(\mathbf{x}_1, \mathbf{x}_2) - m_2\ddot{\mathbf{x}}_2)^2$$

is minimal. That is, the inertial forces $m_1\ddot{\mathbf{x}}_1$ and $m_2\ddot{\mathbf{x}}_2$ fit best the external forces $\mathbf{F}_1$ and $\mathbf{F}_2$ acting on the point P_1 and P_2, respectively. By (6.26) below, this means that the constraining forces $\mathbf{C}_1$ and $\mathbf{C}_2$ are minimal.

Constraining force. For problems in technology (e.g., the construction of rotating machines), it is important to know the strength of the so-called constraining forces which are generated by the constraints. If the constraining forces are too strong, then they will destroy the machine. The Gaussian principle of least constraint allows us to compute the equations of motion including the constraining forces.

---

[4] The expressions $\mathbf{A}_k$, $k = 1, \ldots, K$, are explicitly known, but the proof of Theorem 6.4 on the constraining forces does not need this explicit information.

Theorem 6.4 *Suppose that we know a solution of the Gaussian constrained minimum problem* (6.24), (6.25). *Then there exist uniquely determined real-valued functions* $\lambda_j : \mathcal{U} \to \mathbb{R}$, $j = 1, 2$, *such that*

$$m_j \ddot{\mathbf{x}}_j = \mathbf{F}_j(\mathbf{x}_1, \mathbf{x}_2) + \mathbf{C}_j(\mathbf{x}_1, \mathbf{x}_2), \qquad j = 1, 2 \tag{6.26}$$

with the constraining forces

$$\mathbf{C}_j(\mathbf{x}_1, \mathbf{x}_2) = \sum_{k=1}^{K} \lambda_k(\mathbf{x}_1, \mathbf{x}_2) \cdot \mathbf{grad}_{\mathbf{x}_j} f_k(\mathbf{x}_1, \mathbf{x}_2), \quad j = 1, 2. \tag{6.27}$$

Regarding (6.26) *as a system of differential equations for* $\mathbf{x}_j = \mathbf{x}_j(t), t \in \mathbb{R}$, *we get the corresponding equation of motion.*

Proof. We will use a standard argument from linear algebra. Fix $\mathbf{x}_1$ and $\mathbf{x}_2$. Suppose that $\ddot{\mathbf{x}}_1, \ddot{\mathbf{x}}_2$ is a solution of the minimum problem (6.24), (6.25). Replacing $\ddot{\mathbf{x}}_1, \ddot{\mathbf{x}}_2$ by

$$\ddot{\mathbf{x}}_1 + \tau \mathbf{a}_1, \quad \ddot{\mathbf{x}}_2 + \tau \mathbf{a}_2, \qquad \tau \in \mathbb{R},$$

we obtain that the function

$$\frac{1}{2m_1}(\mathbf{F}_1(\mathbf{x}_1, \mathbf{x}_2) - m_1(\ddot{\mathbf{x}}_1 + \tau \mathbf{a}_1))^2 + \frac{1}{2m_2}(\mathbf{F}_2(\mathbf{x}_1, \mathbf{x}_2) - m_2(\ddot{\mathbf{x}}_2 + \tau \mathbf{a}_2))^2$$

of the real variable τ has a minimum at the point $\tau = 0$ with respect to the constraints

$$(\ddot{\mathbf{x}}_1 + \tau \mathbf{a}_1)\,\mathbf{grad}_{\mathbf{x}_1} f_k(\mathbf{x}_1, \mathbf{x}_2) + (\ddot{\mathbf{x}}_2 + \tau \mathbf{a}_2)\,\mathbf{grad}_{\mathbf{x}_2} f_k(\mathbf{x}_1, \mathbf{x}_2) = \mathbf{A}_k(\mathbf{x}_1, \mathbf{x}_2, \dot{\mathbf{x}}_1, \dot{\mathbf{x}}_2),$$

where $k = 1, \ldots, K$. Differentiation with respect to the variable τ at the point $\tau = 0$ yields

$$\boxed{(\mathbf{F}_1(\mathbf{x}_1, \mathbf{x}_1) - m_1 \ddot{\mathbf{x}}_1)\, \mathbf{a}_1 + (\mathbf{F}_2(\mathbf{x}_1, \mathbf{x}_2) - m_2 \ddot{\mathbf{x}}_2)\, \mathbf{a}_2 = 0} \tag{6.28}$$

for all vectors $\mathbf{a}_1, \mathbf{a}_2$ satisfying the linear system

$$\mathbf{a}_1 \cdot \mathbf{grad}_{\mathbf{x}_1} f_k(\mathbf{x}_1, \mathbf{x}_2) + \mathbf{a}_2 \cdot \mathbf{grad}_{\mathbf{x}_2} f_k(\mathbf{x}_1, \mathbf{x}_2) = 0, \quad k = 1, \ldots, K.$$

Now the claim follows from the same Lagrange multiplier argument as used in Problem 6.1 on page 420. □

The Gaussian principle of virtual acceleration. The preceding proof shows that the physical acceleration vectors $\ddot{\mathbf{x}}_1, \ddot{\mathbf{x}}_2$ at time t satisfy the equation (6.28) for all virtual acceleration vectors $\mathbf{a}_1, \mathbf{a}_2$. This is called the Gaussian principle of virtual acceleration.

Note that the Gaussian principle of virtual acceleration implies the equations (6.26) of motion. Moreover, in the case of holonomic constraints, the Gaussian principle of virtual acceleration is equivalent to the d'Alembert principle of virtual power, as we will show next.

d'Alembert's principle of virtual power. The physical acceleration vectors $\ddot{\mathbf{x}}_1, \ddot{\mathbf{x}}_2$ at any fixed time satisfy the equation

$$\boxed{(\mathbf{F}_1(\mathbf{x}_1, \mathbf{x}_2) - m_1 \ddot{\mathbf{x}}_1)\, \mathbf{v}_1 + (\mathbf{F}_2(\mathbf{x}_1, \mathbf{x}_2) - m_2 \ddot{\mathbf{x}}_2)\, \mathbf{v}_2 = 0} \tag{6.29}$$

for all virtual velocity vectors $\mathbf{v}_1, \mathbf{v}_2$, that is,

$$\mathbf{v}_1 \cdot \mathbf{grad}_{\mathbf{x}_1} f_k(\mathbf{x}_1, \mathbf{x}_2) + \mathbf{v}_2 \cdot \mathbf{grad}_{\mathbf{x}_2} f_k(\mathbf{x}_1, \mathbf{x}_2) = 0, \quad k = 1, \ldots, K. \tag{6.30}$$

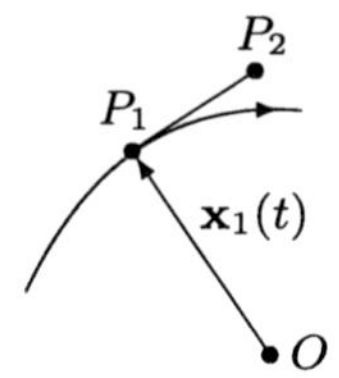

Fig. 6.3. Nonholonomic constraint (ice skating)

To prove this, multiply the equation of motion (6.26) by the virtual velocity $\mathbf{v}_j$ and sum up. Then the terms containing the constraining are cancelled by (6.27) and (6.30).

In terms of mathematics, the equations (6.21) for the constraints characterize a manifold $\mathcal{M}$, and the virtual velocity vectors $\mathbf{v}_1, \mathbf{v}_2$ are precisely the tangent vectors of the manifold $\mathcal{M}$ of constraints.

This shows that d'Alembert's principle of virtual power is closely related to the tangent spaces of the manifold $\mathcal{M}$ generated by the constraints.

If the set of constrained position is not a manifold, then there appears a pathological situation which complicates the approach. In the 19th and 20th century, the development of the theory of manifolds was strongly influenced by the study of complicated problems in celestial and technological mechanics (e.g., the motion of n planets or the motion of spinning tops).

Remark on holonomic constraints. Holonomic constraints are also called integrable constraints. To explain this, consider the constraint

$$a(x,y)dx + b(x,y)dy = 0, \qquad (x,y) \in \mathbb{R}^2. \tag{6.31}$$

This is called an integrable (or holonomic) constraint iff there exists a smooth function $U : \mathbb{R}^2 \to \mathbb{R}$ such that $a(x,y) = U_x(x,y)$ and $b(x,y) = U_y(x,y)$ on $\mathbb{R}^2$. Then the constraint $U(x,y) = \text{const}$ implies (6.31). This point of view was emphasized by Heinrich Hertz, Principles of Mechanics, 1891 (in German).

Nonholonomic Constraints

The Gaussian approach based on the principle of least constraint with respect to virtual accelerations has the advantage that it also applies to nonholonomic constraints, say,

$$f_k(\mathbf{x}_1, \mathbf{x}_2, \dot{\mathbf{x}}_1, \dot{\mathbf{x}}_2) = 0, \quad k = 1, \ldots, K \tag{6.32}$$

which depend on the velocity of the mass points. We assume that $1 \le K < 6$. For example, ice-skating can be modelled by Fig. 6.3 with the constraints

$$(\mathbf{x}_1 - \mathbf{x}_2) \times \dot{\mathbf{x}}_1 = 0, \quad (\mathbf{x}_1 - \mathbf{x}_2)^2 = l_1^2.$$

Here, $\mathbf{x}_j := \overrightarrow{OP_j}, j = 1, 2$. The ice skate corresponds to the directed segment $\overrightarrow{P_1P_2}$.

Set $\mathbf{X} := (\mathbf{x}_1, \mathbf{x}_2)$. Then the Gaussian problem of least constraint reads as

$$\frac{1}{2m_1}\Big(\mathbf{F}_1(\mathbf{X}, \dot{\mathbf{X}}) - m_1\ddot{\mathbf{x}}_1\Big)^2 + \frac{1}{2m_2}\Big(\mathbf{F}_2(\mathbf{X}, \dot{\mathbf{X}}) - m_2\ddot{\mathbf{x}}_2\Big)^2 = \min! \tag{6.33}$$

together with the linearized constraints

$$\sum_{j=1}^{2} \ddot{\mathbf{x}}_j \cdot \mathbf{grad}_{\dot{\mathbf{x}}_j} f_k(\mathbf{X}, \dot{\mathbf{X}}) + \dot{\mathbf{x}}_j \cdot \mathbf{grad}_{\mathbf{x}_j} f_k(\mathbf{X}, \dot{\mathbf{X}}) = 0, \quad k = 1, \ldots, K. \quad (6.34)$$

We fix $\mathbf{X}, \dot{\mathbf{X}}$, and we are looking for $\ddot{\mathbf{x}}_1, \ddot{\mathbf{x}}_2$. We assume that the system (6.34) of linear equations with respect to the variables $\ddot{\mathbf{x}}_1, \ddot{\mathbf{x}}_2$ has the maximal rank $r = K$.[5] As in the holonomic case above, we obtain the equation of motion

$$\boxed{m_j \ddot{\mathbf{x}}_j = \mathbf{F}_j(\mathbf{X}, \dot{\mathbf{X}}) + \mathbf{C}_j(\mathbf{X}, \dot{\mathbf{X}}), \qquad j = 1, 2} \quad (6.35)$$

with the constraining forces

$$\mathbf{C}_j(\mathbf{X}, \dot{\mathbf{X}}) = \sum_{k=1}^{K} \lambda_k(\mathbf{X}, \dot{\mathbf{X}}) \cdot \mathbf{grad}_{\dot{\mathbf{x}}_j} f_k(\mathbf{X}, \dot{\mathbf{X}}), \quad j = 1, 2.$$

Furthermore, the Gaussian principle of virtual acceleration tells us that we have

$$(\mathbf{F}_1(\mathbf{X}, \dot{\mathbf{X}}) - m_1 \ddot{\mathbf{x}}_1)\, \mathbf{a}_1 + (\mathbf{F}_2(\mathbf{X}, \dot{\mathbf{X}}) - m_2 \ddot{\mathbf{x}}_2)\, \mathbf{a}_2 = 0$$

for all virtual accelerations $\mathbf{a}_1, \mathbf{a}_2$ which satisfy the following linear system

$$\sum_{j=1}^{2} \mathbf{a}_j \cdot \mathbf{grad}_{\dot{\mathbf{x}}_j} f_k(\mathbf{X}, \dot{\mathbf{X}}) = 0, \qquad k = 1, \ldots, K.$$

In the most general setting, the Gaussian principle of least constraint is studied in E. Zeidler (1995), Vol. IV, page 45, quoted on page 396 below. This concerns mixed holonomic/nonholonomic time-dependent constraints.

6.4.5 Manifolds and Lagrange's Variational Principle

> The Lagrangian trick is to use a variational principle with respect to local coordinates for the manifold of constrained positions generated by the constraints. This way, the constraints drop out.
>
> Folklore

One degree of freedom. Let us again consider the balance depicted in Fig. 6.2 on page 375. The motion $\varphi = \varphi(t), t \in \mathbb{R}$, of the balance is uniquely determined by the rotation angle φ. From the trajectories

$$\mathbf{x}_1(t) = l_1(\cos\varphi(t)\, \mathbf{i} + \sin\varphi(t)\, \mathbf{j}), \qquad \mathbf{x}_2(t) = -l_2(\cos\varphi(t)\, \mathbf{i} + \sin\varphi(t)\, \mathbf{j}),$$

we get the velocity vector

$$\dot{\mathbf{x}}_1(t) = l_1(-\sin\varphi(t)\, \mathbf{i} + \cos\varphi(t)\, \mathbf{j}) \cdot \dot{\varphi}(t).$$

Hence $\dot{\mathbf{x}}_1(t)^2 = l_1^2(\sin^2\varphi(t) + \cos^2\varphi(t)) \cdot \dot{\varphi}(t)^2$. This way, we get the total kinetic energy

$$E_{\text{kin}} := \tfrac{1}{2} m_1 \dot{\mathbf{x}}_1^2 + \tfrac{1}{2} m_2 \dot{\mathbf{x}}_2^2 = \tfrac{1}{2}(m_1 l_1^2 + m_2 l_2^2)\, \dot{\varphi}^2$$

[5] In other words, the solutions $\ddot{\mathbf{x}}_1, \ddot{\mathbf{x}}_2$ of (6.34) form a $(6 - K)$-dimensional linear submanifold of $E_3 \times E_3$.

and the total potential energy

$$E_{\text{pot}} := m_1 g z_1 + m_2 g z_2 = (m_1 g l_1 - m_2 g l_2)\ \sin\varphi$$

of the balance. Let us assume that $m_1 g l_1 > m_2 g l_2$.

The principle of critical action. Introducing the Lagrangian

$$L(\varphi, \dot{\varphi}) := E_{\text{kin}} - E_{\text{pot}} = \tfrac{1}{2}(m_1 l_1^2 + m_2 l_2^2)\ \dot{\varphi}^2 - (m_1 g l_1 - m_2 g l_2)\ \sin\varphi,$$

the principle of critical action leads to the variational problem

$$\boxed{\int_{t_0}^{t_1} L(\varphi(t), \dot{\varphi}(t))\ d\varphi = \text{critical!}} \tag{6.36}$$

together with the boundary conditions: $\varphi(t_0)$ and $\varphi(t_1)$ are fixed. Note that this variational problem does not contain any constraints. A solution of (6.36) satisfies the Euler–Lagrange equation $\frac{d}{dt} L_{\dot{\varphi}} - L_\varphi = 0$. Explicitly,

$$\ddot{\varphi}(t) + \omega^2 \cos\varphi(t) = 0, \quad t \in \mathbb{R}, \qquad \omega := \left(\frac{m_1 g l_1 - m_2 g l_2}{m_1 l_1^2 + m_2 l_2^2} \right)^{1/2}. \tag{6.37}$$

This is the equation of motion for the balance. Let us add the following initial condition: $\varphi(0) = \varphi_0, \dot{\varphi}(0) = \varphi_1$.

6.4.6 The Method of Perturbation Theory

> Celestial mechanics, quantum mechanics, and quantum field theory are governed by perturbation theory.
>
> Folklore

We want to study the initial-value problem

$$\ddot{\varphi}(t) + \omega^2 \cos\varphi(t) = 0, \quad t \in \mathbb{R}, \quad \varphi(0) = -\frac{\pi}{2}, \ \dot{\varphi}(0) = \varepsilon, \tag{6.38}$$

where ε is a small real number. More precisely, we assume that the dimensionless quantity $\varepsilon/\omega(l_1 + l_2)$ is small compared with 1. Fix the finite time interval $[-t_1, t_1]$. The general theory of ordinary differential equations tells us that the solution of the initial-value problem (6.38) depends analytically on the parameter ε. Explicitly, the solution of (6.38) reads as

$$\varphi(t) = -\frac{\pi}{2} + \varepsilon\varphi_1(t) + \varepsilon^2\varphi_2(t) + \varepsilon^3\varphi_3(t) + \ldots, \qquad t \in [-t_1, t_1].$$

Proposition 6.5 *The unique solution of the initial-value problem* (6.38) *for the balance is given by*

$$\varphi(t) = -\frac{\pi}{2} + \frac{\varepsilon}{\omega}\ \sin\omega t + O(\varepsilon^3), \qquad \varepsilon \to 0, \quad t \in [-t_1, t_1].$$

Proof. Using the Taylor expansion

$$\cos\left(-\frac{\pi}{2} + \psi\right) = \sin\psi = \psi - \frac{\psi^3}{3!} + \frac{\psi^5}{5!} + \ldots,$$

comparison of the coefficients yields the following initial-value problems:

- $\ddot{\varphi}_1(t) + \omega^2 \varphi_1(t) = 0, \ \ \varphi_1(0) = 0, \dot{\varphi}_1(0) = 1,$
- $\ddot{\varphi}_2(t) + \omega^2 \varphi_2(t) = 0, \ \ \varphi_2(0) = \dot{\varphi}_2(0) = 0,$
- $\ddot{\varphi}_3(t) + \omega^2 \varphi_3(t) = \frac{1}{3!}\varphi_1(t)^3, \ \ \varphi_3(0) = \dot{\varphi}_3(0) = 0,$

and so on. This yields $\varphi_1(t) = \frac{1}{\omega}\sin \omega t$ and $\varphi_2(t) \equiv 0$. □

Note that the initial-value problem (6.37) coincides with the equation for the motion of a circular pendulum. The solution can explicitly be written down in terms of elliptic functions. This can be found in Sect. 6.7.2 of Vol. II.

6.4.7 Further Reading on Perturbation Theory and its Applications

Many problems in physics are handled by means of perturbation theory, including quantum field theory. The literature on perturbation theory is vast. Let us summarize some of the most important references.

Classic works:

Ptolemeaus, Almagest, 150 A.D. See M. Adler (Ed.), Ptolemy of Alexandria, Almagest: Great Books of the Western World (60 Volumes), Chicago: Encyclopedia Britannica, 1994.

N. Copernicus, De revolutionibus orbium coelestium libri vi (in Latin) (Six books concerning the revolutions of the heavenly orbs), 1543. See M. Adler (Ed.), Great Books of the Western World (60 Volumes), Chicago: Encyclopedia Britannica, 1994.

J. Kepler, Harmonices Mundi (in Latin) (The Harmonies of the World). See M. Adler (Ed.), Great Books of the Western World (60 Volumes), Chicago: Encyclopedia Britannica, 1994.

J. Kepler, Astronomia nova (in Latin), 1609. English translation: New Astronomy, edited by W. Donahue, 1991.

J. Kepler, Epitome Astronomiae Copernicanae (in Latin) (Epitome of Copernican Astronomy), 1618/21. See M. Adler (Ed.), Great Books of the Western World (60 Volumes), Chicago: Encyclopedia Britannica, 1994.

J. Kepler, Tabulae Rudolphinae (in Latin) (the Rudolphine Tables), 1627. English edition: London, 1675 (standard tool of astronomers for a long time).

G. Galilei, Discorsi e dimonstrazioni matematiche intorno à due nuove scienze (in Italian) (Dialogues and mathematical proofs concerning two new sciences), 1638. English translation: S. Drake, Two New Sciences, 1989.

I. Newton, Philosophiae naturalis principia mathematica, London, 1687. See S. Chandrasekhar, Newton's Principia for the Common Reader, Oxford University Press, 1997.

L. Euler, Mechanica sive motus scientia analytice exposita (in Latin) (Mechanics or the science of the motion exposed in analytic terms), Vols. 1, 2, 1736 (first textbook).
Internet: `http://www.math.dartmouth.edu` (Euler Archive)

L. Euler, Theoria motuum planetarum and cometarum (in Latin) (Theory of the motion of planets and comets), 1744.

J. de Lagrange, Méchanique analitique, Paris, 1788.
New edition: Mécanique analytique, Paris, 1811/1815 (after the orthographic reform of the Paris Academy in about 1800).

English edition: Analytical Mechanics, Kluwer, Dordrecht, 1997.

P. Laplace, Mécanique celeste (Celestial Mechanics), Vols. 1–5, Paris, 1799/1825.

C. Gauss, Theoria motus corporum coelestium in sectionibus conibus solem ambientum (in Latin) (Theory of the motion of celestial bodies on conic sections about the sun), Hamburg, 1809. German edition: Hannover, 1865.

J. Rayleigh, The Theory of Sound, Macmillan, London, 1896. Reprint: Dover, New York 1945.

H. Poincaré (1892), Les méthodes nouvelles de la mécanique céleste (New methods in celestial mechanics), Vols. 1-3, Gauthier-Villars, Paris, 1892/1899. Reprint: Dover 1957.

Modern works:

E. Schrödinger, Quantization as an eigenvalue problem, Part III: Perturbation theory with applications to the Starck effect of the Balmer spectral lines, Ann. Physik **80** (1926), 437–490 (in German) (Schrödinger used Rayleigh's method).

F. Rellich, Perturbation theory for the spectral decomposition, Math. Annalen I–V, **113** (1937), 600–619, 677–685; **116** (1939), 555–570; **117** (1940); 356–382; **118** (1942), 462–484 (in German).

F. Rellich, Perturbation Theory of Eigenvalue Problems, Gordon and Breach, New York. 1969.

V. Arnold, Small divisors and problems of stability of motion in classical and celestial mechanics, Uspekhi Mat. Nauk **18**, 91-196 (in Russian). English translation: Russian Math. Surveys **18** (1963).

V. Arnold, Proof of A.N. Kolmogorov's theorem on the preservation of quasi-periodic motions under small perturbations of the Hamiltonian, Russian Math. Surveys **18** (5)(1963), 9–36.

J. Moser, A rapidly convergent iteration method and nonlinear partial differential equations I, II, Ann. Scuola Norm. Sup. Pisa **20** (1966), 226–315, 449–535.

J. Moser, Convergent series expansions for quasi-periodic motions, Math. Ann. **169** (1967), 136–176.

L. Lyusternik and M. Vishik, Regular degeneration and boundary layer for linear differential equations with small parameters, Uspekhi Mat. Nauk **12**(5), (1957), 3–122. Transl. Ser. 2, Amer. Math. Soc. **20** (1962), 239–364.

O. Oleinik, Mathematical problems in boundary-layer theory, Russian Mathematical Surveys **23** (1968), 1–66.

T. Kato, Perturbation Theory for Linear Operators, Springer, Berlin, 1966.

M. Reed and B. Simon, Methods of Modern Mathematical Physics,
Vol. 4: Perturbation Theory, Academic Press, New York, 1979 (modern standard textbook).

W. Faris, Perturbations and non-normalizable eigenvectors, Helv. Phys. Acta **44** (1971), 930–936.

S. Gustafson and I. Sigal, Mathematical Concepts of Quantum Mechanics, Springer, Berlin, 2003.

L. Takhtajan, Quantum Mechanics for Mathematicians, Amer. Math. Soc., Providence, Rhode Island, 2008.

E. Lieb and R. Seiringer, The Stability of Matter in Quantum Mechanics, Cambridge University Press, 2009.

Methods of perturbation theory in quantum field theory:

F. Dyson, Advanced Quantum Mechanics. Dyson's Cornell Lecture Notes from 1951 (Cornell University, Ithaca, New York). World Scientific, Singapore, 2007.

N. Bogoliubov and J. Mitropolskii, Asymptotic Methods in Nonlinear Oscillation Problems, Gordon and Breach, New York, 1961.

N. Bogoliubov and D. Shirkov, Introduction to Quantum Field Theory, Interscience, New York, 1980.

N. Bogoliubov, Selected Works. Part I: Dynamical Theory; Part II: Quantum and Classical Statistical Mechanics; Part III: Nonlinear Mechanics and Pure Mathematics; Part IV: Quantum Field Theory; Gordon and Breach, London, 1990/95.

G. Scharf, Finite Quantum Electrodynamics: the Causal Approach, Springer, New York, 1995.

J. Feldman et al., QED (Quantum Electrodynamics): A Proof of Renormalizability, Springer, Berlin, 1988.

J. Feldman and E. Trubowitz, Renormalization in classical mechanics and many body quantum field theory, Jerusalem J. d'Analyse Mathématique **52** (1992), 213–247.
Internet: `http://www.math.ubc.ca.~feldman/research.html`

S. Adler, Perturbation theory anomalies, pp. 1–164. In: S. Deser et al. (Eds.), Lectures on Elementary Particles and Quantum Field Theory, Proceedings of the 1970 Brandeis Summer Institute in Theoretical Physics, MIT Press, Cambridge, Massachusetts, 1970.

S. Adler, Adventures in Theoretical Physics, Selected Papers with Commentaries, World Scientific, Singapore, 2006.

R. Brunetti, M. Dütsch, and K. Fredenhagen, Perturbative algebraic quantum field theory and the renormalization groups, Adv. Theor. Math. Phys. **13** (2009), 1–56.

I. Ioffe et al., Quantum Chromodynamics: Perturbative and Non-Perturbative Aspects, Cambridge University Press, 2010.

Dynamical systems:

W. Neutsch and K. Scherer, Celestial Mechanics: An Introduction to Classical and Contemporary Methods, Wissenschaftsverlag, Mannheim (Germany), 1992.

C. Siegel and J. Moser, Lectures on Celestial Mechanics, Springer, Berlin, 1971.

D. Boccaletti and G. Pucacco, Theory of Orbits: Vol 1: Integrable Systems and Non-Perturbative Methods, Vol. 2: Perturbative and Geometrical Methods, Springer, Berlin, 1996.

Y. Hagihara, Celestial Mechanics, Vol. I: Dynamical Principles and Transformation Theory; Vol. II: Perturbation Theory, MIT Press, Cambridge, Massachusetts, 1970; Vol. III: Differential Equations in Celestial Mechanics, Vol. IV: Periodic and Quasi-Periodic Motions (KAM Theory), Vol. V: Topology of the Three-Body Problem. Japan Society for the Promotion of Sciences, Tokyo, 1974–1976.

V. Arnold et al., Singularities of Differentiable Maps, Vols. 1, 2, Birkhäuser, Basel, 1985.

V. Arnold, Catastrophe Theory, Springer, New York, 1986.

V. Arnold et al., Dynamical Systems, Vols. 1–10. Encyclopedia of Mathematics, Springer, Berlin, 1987.

B. Cordani, The Kepler Problem: Group Theoretical Aspects, Regularization and Quantization, with Applications to the Study of Perturbations, Birkhäuser, Basel, 2002.

H. Kielhöfer, Bifurcation Theory: An Introduction with Applications to Partial Differential Equations, Springer, New York, 2004.

Furthermore, we recommend:

A. Nayfeh, Perturbation Methods, Wiley, New York, 1973.

A. Nayfeh and D. Mook, Nonlinear Oscillations, Wiley, New York, 1979.

J. Kevorkian and J. Cole, Perturbation Methods in Applied Mathematics, Springer, New York, 1981.

W. Eckhaus, Asymptotic Analysis of Singular Perturbation, North-Holland, Amsterdam, 1979.

H. Baumgärtel, Analytic Perturbation Theory for Matrices and Operators, Birkhäuser, Boston, 1985.

M. Vainberg and V. Trenogin, Theory of Branching of Solutions of Nonlinear Equations, Noordhoof, Leyden, The Netherlands, 1974.

R. White, Asymptotic Analysis of Differential Equations, World Scientific, Singapore, 2010.

S. Krantz and H. Parks, The Implicit Function Theorem: History, Theory, and Applications, Birkhäuser, Boston, 2002 (including the hard implicit function theorem/Nash–Moser theorem).

S. Alinhac and P. Gérard, Pseudo-Differential Operators and the Nash–Moser Theorem, Amer. Math. Soc. Providence, Rhode Island, 2007.

See also the comprehensive list of references on renormalization theory to be found in Sect. 19.11 of Vol. II.

6.5 Application to the Motion of a Rigid Body

The electron spin to be studied later on is obtained by quantizing the classical angular momentum (see Chap. 7). We want to show that the classical angular momentum governs the motion of rigid bodies (e.g., the spinning earth or spinning tops).

Fix $n = 3, 4, \ldots$, and fix the origin P_0. In what follows, let us consider a model of a rigid body consisting of n moving mass points $P_1, \ldots, P_n$ of the Euclidean manifold $\mathbb{E}^3$. Here, m_j is the mass of the point P_j. We assume the following:

- The distances between all the points are constant for all times $t \in \mathbb{R}$.
- The points are in general position, that is, three different points always span a plane.

Let us describe the motion of the point P_j by the trajectory

$$\mathbf{x}_j = \mathbf{x}_j(t), \quad t \in \mathbb{R}, \quad j = 1, \ldots, n$$

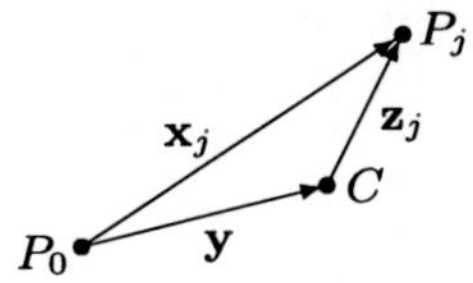

Fig. 6.4. Center C of gravity

where the position vector $\mathbf{x}_j(t) = \overrightarrow{P_0P_j(t)}$ is located at the origin P_0. By our assumption made above, we have the constraints

$$\boxed{(\mathbf{x}_i(t) - \mathbf{x}_j(t))^2 = r_{ij}^2, \quad t \in \mathbb{R}, \qquad i, j = 1, \ldots, n, \, i \neq j.}$$

Here, all the positive distances r_{ij} do not depend on time t.

6.5.1 The Center of Gravity

> The center of gravity moves like a mass point equipped with the total mass under the action of the total force.
>
> Folklore

Let $m := \sum_{j=1}^{n} m_j$ denote the total mass of the rigid body. We define the position vector $\mathbf{y} = \overrightarrow{P_0C}$ by setting

$$\mathbf{y} := \frac{m_1\mathbf{x}_1 + \ldots + m_n\mathbf{x}_n}{m}.$$

The point C is called the center of gravity of the rigid body. Our goal is to separate the motion of the points of the rigid body from the center C of gravity. To this end, we introduce the position vectors $\mathbf{z}_j$ by means of the superposition relation

$$\boxed{\mathbf{x}_j = \mathbf{y} + \mathbf{z}_j, \qquad j = 1, \ldots, n.}$$

The trajectory $\mathbf{z}_j = \mathbf{z}_j(t)$ describes the motion of the point P_j with respect to the center C of gravity (Fig. 6.4).

6.5.2 Moving Orthonormal Frames and Infinitesimal Rotations

> The rigid body is governed by the Lie group $SO(3)$ (rotations) and its Lie algebra $so(3)$ (infinitesimal rotations).
>
> Folklore

In order to describe the motion of the rigid body, it is convenient to choose a right-handed orthonormal system $\mathbf{e}_1, \mathbf{e}_2, \mathbf{e}_3$ located at the center C of gravity such that the points $P_1(t), \ldots, P_n(t)$ of the rigid body are fixed with respect to the basis vectors $\mathbf{e}_1(t), \mathbf{e}_2(t), \mathbf{e}_3(t)$. This means that

$$\mathbf{z}_j(t) = \sum_{k=1}^{3} z_j^k \mathbf{e}_j(t), \qquad t \in \mathbb{R}, \quad j = 1, \ldots, n \tag{6.39}$$

where all the components z_j^1, z_j^2, z_j^3 do not depend on time t. We want to show that

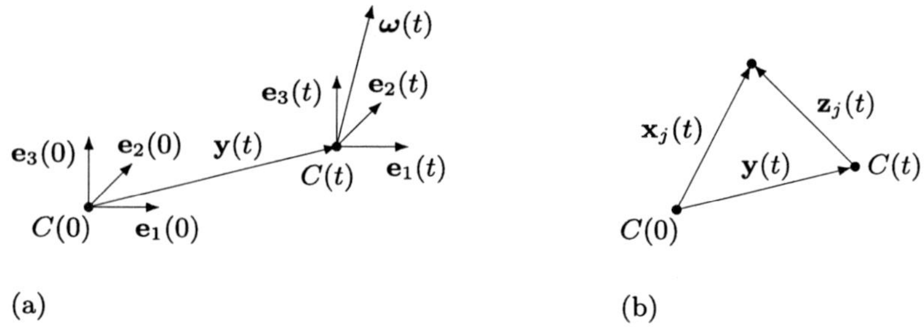

Fig. 6.5. Moving rigid body

$$\boxed{\dot{\mathbf{e}}_j(t) = \boldsymbol{\omega}(t) \times \mathbf{e}_j(t), \quad t \in \mathbb{R}, \quad j = 1, 2, 3.} \tag{6.40}$$

This is an infinitesimal rotation with the angular velocity vector $\boldsymbol{\omega}(t)$ at time t (Fig. 6.5).

Proof. We will use the Lie group $SO(3)$ and its Lie algebra $so(3)$. Since the three vectors $\mathbf{e}_1(t), \mathbf{e}_2(t), \mathbf{e}_3(t)$ form a right-handed orthonormal system, they are obtained from $\mathbf{e}_1(0), \mathbf{e}_2(0), \mathbf{e}_3(0)$ by a rotation (see Prop. 9.1 on page 559). Thus, there exists a matrix $G(t) \in SO(3)$ such that

$$\begin{pmatrix} \mathbf{e}_1(t) \\ \mathbf{e}_2(t) \\ \mathbf{e}_3(t) \end{pmatrix} = G(t) \begin{pmatrix} \mathbf{e}_1(0) \\ \mathbf{e}_2(0) \\ \mathbf{e}_3(0) \end{pmatrix}, \qquad t \in \mathbb{R}.$$

Differentiating this with respect to time t, we get

$$\begin{pmatrix} \dot{\mathbf{e}}_1(t) \\ \dot{\mathbf{e}}_2(t) \\ \dot{\mathbf{e}}_3(t) \end{pmatrix} = \dot{G}(t) \begin{pmatrix} \mathbf{e}_1(0) \\ \mathbf{e}_2(0) \\ \mathbf{e}_3(0) \end{pmatrix} = \dot{G}(t)G(t)^{-1} \begin{pmatrix} \mathbf{e}_1(t) \\ \mathbf{e}_2(t) \\ \mathbf{e}_3(t) \end{pmatrix}.$$

Set $A(t) := \dot{G}(t)G(t)^{-1}$. Since the smooth map $t \mapsto G(t)$ represents a curve in the Lie group $SO(3)$, the derivative $\dot{G}(t)$ is a tangent vector at the point $G(t)$. Thus, $\dot{G}(t)G(t)^{-1}$ is a tangent vector of the manifold $SO(3)$ at the unit element. In other words, $A(t)$ is an element of the Lie algebra $so(3)$ of real skew-symmetric (3×3)-matrices. Consequently,

$$A(t) = \begin{pmatrix} 0 & \omega^3(t) & -\omega^2(t) \\ -\omega^3(t) & 0 & \omega^1(t) \\ \omega^2(t) & -\omega^1(t) & 0 \end{pmatrix}$$

where $\omega^1(t), \omega^2(t), \omega^3(t)$ are fixed real numbers for fixed time t. This yields the claim (6.40). □

It follows from (6.40) that

$$\boxed{\dot{\mathbf{z}}_j(t) = \boldsymbol{\omega}(t) \times \mathbf{z}_j(t), \qquad t \in \mathbb{R}, \quad j = 1, \ldots, n.} \tag{6.41}$$

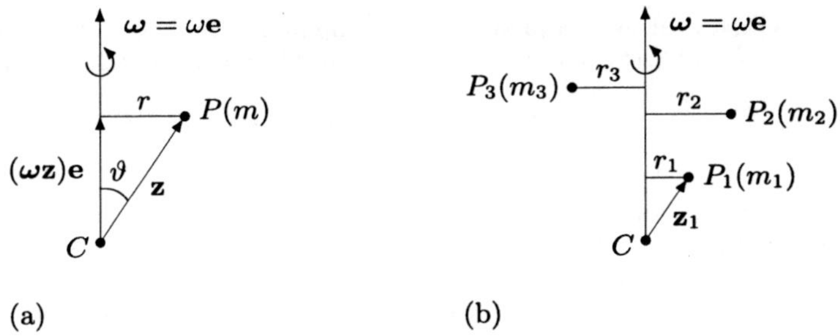

Fig. 6.6. Inertia tensor

6.5.3 Kinetic Energy and the Inertia Tensor

In order to understand the motion of rigid bodies, Euler (1707–1783) studied the principal axes of the symmetric inertia tensor; this tensor generalizes the moment of inertia of a single rotating mass point to a system of rotating mass points.
In the history of physics and mathematics, the inertia tensor (resp. the Gaussian theory of quadratic forms in number theory[6]) motivated the formulation of the theorem of principal axes for finite-dimensional symmetric operators by Cauchy (1789–1857) in 1826 (resp. by Hermite (1822–1901) in 1855). Quantum mechanics is essentially based on an infinite-dimensional version of the theorem of principal axes due to Hilbert (1862–1943) in 1904 (bounded self-adjoint operators) and due to von Neumann (1903–1957) in 1928 (unbounded self-adjoint operators).

Folklore

The kinetic energy and the moment of inertia of a single point rotating about an axis. Consider the situation depicted in Fig. 6.6(a). Let $\mathbf{z} = \overline{CP}$, and let $\mathbf{e}$ be a position unit vector located at the center C of gravity. Choose a real number ω, and set $\boldsymbol{\omega} := \omega\mathbf{e}$. Then the equation

$$\dot{\mathbf{z}}(t) = \boldsymbol{\omega} \times \mathbf{z}(t), \qquad t \in \mathbb{R} \tag{6.42}$$

describes a rotation of the point P about an axis through the point C; the direction of the axis is given by the unit vector $\mathbf{e}$. The constant angular velocity of this rotation is equal to ω. Explicitly, choose a right-handed orthonormal system $\mathbf{i}, \mathbf{j}, \mathbf{e}$. Then

$$\mathbf{z}(t) = (\mathbf{ze})\mathbf{e} + r(\cos\omega t\,\mathbf{i} + \sin\omega t\,\mathbf{j}), \qquad t \in \mathbb{R}$$

satisfies the differential equation (6.42). For the velocity, we get

$$|\dot{\mathbf{z}}(t)| = r\omega.$$

This yields the kinetic energy

$$E_{\text{kin}} = \tfrac{1}{2}m\dot{\mathbf{z}}(t)^2 = \tfrac{1}{2}\theta\omega^2.$$

The quantity

$$\boxed{\theta := mr^2}$$

[6] C. Gauss, Disquisitiones arithmeticae, 1801.

is called the moment of inertia of the rotating point P. Here, m is the mass of the point P, and r is the distance of the point P from the rotation axis. Alternatively,[7]

$$E_{\text{kin}} = \tfrac{1}{2}m(\boldsymbol{\omega} \times \mathbf{z}(t))^2 = \tfrac{1}{2}m(\boldsymbol{\omega}^2\mathbf{z}^2 - (\boldsymbol{\omega}\mathbf{z})^2) = \tfrac{1}{2}\theta\omega^2$$

with $\theta = m(\mathbf{z}^2 - (\mathbf{z}\mathbf{e})^2)$.

The inertia tensor Θ. Consider now the same situation for the rigid body which rotates with the constant angular velocity ω about an axis through the center of gravity (Fig. 6.6(b)); the direction of the axis is given by the unit vector $\mathbf{e}$. Then we have $\dot{\mathbf{z}}_j(t) = \boldsymbol{\omega} \times \mathbf{z}_j(t)$ for all indices. This yields the time-independent kinetic energy of the rigid body:

$$E_{\text{kin}} = \sum_{j=1}^{n} \tfrac{1}{2}m_j\dot{\mathbf{z}}_j^2 = \sum_{j=1}^{n} \tfrac{1}{2}m_j(\mathbf{z}_j^2\boldsymbol{\omega}^2 - (\mathbf{z}_j\boldsymbol{\omega})^2) = \sum_{j=1}^{n} \tfrac{1}{2}m_jr_j^2\omega^2.$$

Define

$$\boxed{\Theta\boldsymbol{\omega} = \sum_{j=1}^{n} \tfrac{1}{2}m_j(\mathbf{z}_j^2 \cdot \boldsymbol{\omega} - (\boldsymbol{\omega}\mathbf{z}_j)\mathbf{z}_j) \qquad \text{for all} \quad \boldsymbol{\omega} \in E_3.}$$

The linear operator $\Theta : E_3 \to E_3$ is called the inertia tensor of the rigid body. For the kinetic energy of the rigid body rotating with constant angular velocity, we get

$$E_{\text{kin}} = \tfrac{1}{2}\boldsymbol{\omega}(\Theta\boldsymbol{\omega}).$$

Proposition 6.6 *The inertia tensor Θ is self-adjoint.*

Proof. If $\boldsymbol{\omega} \in E_3$, then $\boldsymbol{\sigma}(\Theta\boldsymbol{\omega}) = \sum_{j=1}^{n} \frac{1}{2}m_j(\mathbf{z}_j^2 \cdot \boldsymbol{\sigma}\boldsymbol{\omega} - \boldsymbol{\sigma}\mathbf{z}_j \cdot \boldsymbol{\omega}\mathbf{z}_j)$. This expression is symmetric with respect to $\boldsymbol{\sigma}$ and $\boldsymbol{\omega}$. □

The principal moments and the principal axes of inertia. If $\boldsymbol{\omega} \neq 0$, then

$$\boldsymbol{\omega}(\Theta\boldsymbol{\omega}) = \sum_{j=1}^{n} m_jr_j^2\omega^2 > 0,$$

since the points $P_1, \ldots, P_n$ of the rigid body are assumed to be in general position. Consequently, the quadratic form $\boldsymbol{\omega} \mapsto \boldsymbol{\omega}(\Theta\boldsymbol{\omega})$ is positive definite, and the equation

$$\boldsymbol{\omega}(\Theta\boldsymbol{\omega}) = 1, \qquad \boldsymbol{\omega} \in E_3 \tag{6.43}$$

describes an ellipsoid called the ellipsoid of inertia of the rigid body. Let us reformulate the theorem of principal axes for finite-dimensional self-adjoint operators due to Cauchy and Hermite in terms of the rigid body (see Theorem 3.6 on page 202).

Proposition 6.7 *The inertia tensor Θ of the rigid body possesses an orthonormal system of eigenvectors $\mathbf{e}_1, \mathbf{e}_2, \mathbf{e}_3$ in the Euclidean Hilbert space E_3 with positive eigenvalues $\lambda_1, \lambda_2, \lambda_3$, respectively. Thus,*

$$\Theta\mathbf{e}_k = \lambda_k\mathbf{e}_k, \qquad k = 1, 2, 3.$$

[7] Note Lagrange's identity $(\mathbf{a} \times \mathbf{b})(\mathbf{c} \times \mathbf{d}) = \mathbf{ac} - \mathbf{bd}$.

The positive eigenvalues $\lambda_1, \lambda_2, \lambda_3$ are called the principal moments of inertia, and the eigenvectors of the inertia tensor are called principal axes of inertia (e.g., $\mathbf{e}_1, \mathbf{e}_2, \mathbf{e}_3$). Setting $\boldsymbol{\omega} = \omega_1\mathbf{e}_1 + \omega_2\mathbf{e}_2 + \omega_3\mathbf{e}_3$, the ellipsoid of inertia (6.43) looks like[8]

$$\lambda_1\omega_1^2 + \lambda_2\omega_2^2 + \lambda_3\omega_3^2 = 1, \qquad (\omega_1, \omega_2, \omega_3) \in \mathbb{R}^3.$$

This yields the kinetic energy

$$E_{\text{kin}} = \tfrac{1}{2}\boldsymbol{\omega}(\Theta\boldsymbol{\omega}) = \tfrac{1}{2}(\lambda_1\omega_1^2 + \lambda_2\omega_2^2 + \lambda_3\omega_3^2).$$

In order to compute the principal moments of inertia, we choose an orthonormal basis $\mathbf{b}_1, \mathbf{b}_2, \mathbf{b}_3$ of the Euclidean Hilbert space E_3. Define

$$\theta_{kl} := \mathbf{b}_k(\Theta\mathbf{b}_l), \qquad k, l = 1, 2, 3,$$

and consider the real (3×3)-matrix $T = (\theta_{kl})$. Then the principal values $\lambda_1, \lambda_2, \lambda_3$ of inertia are the solutions of Lagrange's secular equation

$$\det(T - \lambda I) = 0, \qquad \lambda \in \mathbb{R}.$$

Moreover, setting $\xi := (\xi^1, \xi^2, \xi^3)$, the solutions ξ of the linear equation

$$\xi(T - \lambda_k I) = \xi, \qquad k = 1, 2, 3, \quad \xi \neq 0$$

yield the principal axes vectors $\sum_{k=1}^3 \xi^k \mathbf{b}_k$ with respect to λ_k.

The total angular momentum with respect to the center of gravity. Define the vector of total angular momentum

$$\mathbf{A} = \sum_{j=1}^n \mathbf{z}_j \times m_j\dot{\mathbf{z}}_j.$$

We claim that $\mathbf{A} = \Theta\boldsymbol{\omega}$.

Proof. This follows from $\dot{\mathbf{z}}_j = \boldsymbol{\omega} \times \mathbf{z}_j$ together with Grassmann's identity

$$\mathbf{z}_j \times (\boldsymbol{\omega} \times \mathbf{z}_j) = \boldsymbol{\omega} \cdot \mathbf{z}_j^2 - (\boldsymbol{\omega}\mathbf{z}_j)\mathbf{z}_j.$$

□

6.5.4 The Equations of Motion – the Existence and Uniqueness Theorem

We assume that the smooth force $\mathbf{F}_j$ and the torque $\mathbf{T}_j := \mathbf{x}_j \times \mathbf{F}_j$ act on the point P_j. This yields

- the total force $\mathbf{F}(\mathbf{x}_1, \ldots, \mathbf{x}_n) := \sum_{j=1}^n \mathbf{F}_j(\mathbf{x}_1, \ldots, \mathbf{x}_n)$, and
- the total torque $\mathbf{T} := \sum_{j=1}^n \mathbf{x}_j \times \mathbf{F}_j(\mathbf{x}_1, \ldots, \mathbf{x}_n)$,

where $\mathbf{x}_j := \underline{P_0P_j}$ is the position vector of the point P_j (Fig. 6.4 on page 389). The equations of motion for the rigid body read as follows:

(M1) Center of gravity: $\ddot{\mathbf{y}} = \mathbf{F}$.
(M2) Time rate of change of the total angular momentum: $\dot{\mathbf{A}} = \mathbf{T}$.
(M3) Angular velocity/angular momentum relation: $\mathbf{A} = \Theta\boldsymbol{\omega}$.

[8] If $\lambda_1 = \lambda_2 = \lambda_3$, then the ellipsoid of inertia is a ball, and we have $\Theta\boldsymbol{\omega} = \lambda_1\boldsymbol{\omega}$ for all $\boldsymbol{\omega} \in E_3$, that is, all the nonzero vectors of E_3 are principal axes of inertia.

(M4) Infinitesimal rotations: $\dot{\mathbf{z}}_j = \boldsymbol{\omega} \times \mathbf{z}_j$, $j = 1, \ldots, n$.
Recall that $\mathbf{x}_j = \mathbf{y} + \mathbf{z}_j$, $j = 1, \ldots, n$.

Initial values. We are given:

- $\mathbf{x}_1(0), \ldots, \mathbf{x}_n(0)$ (positions of the points $P_1, \ldots, P_n$ of the rigid body at time $t = 0$),
- $\dot{\mathbf{y}}(0)$ (velocity vector of the center of gravity at time $t = 0$),
- $\mathbf{e}$ (rotation axis at time $t = 0$; $|\mathbf{e}| = 1$),
- $\omega(0)$ (angular velocity at time $t = 0$).

This yields

- $\mathbf{y}(0) := \frac{1}{m}(\mathbf{x}_1(0) + \ldots + \mathbf{x}_n(0))$ (position of the center of gravity at time $t = 0$),
- $\mathbf{z}_j(0) = \mathbf{x}_j(0) - \mathbf{y}(0)$, $j = 1, \ldots, n$,
- $\boldsymbol{\omega}(0) = \omega(0)\mathbf{e}$ (angular velocity vector at time $t = 0$),
- $\mathbf{A}(0) = (\Theta\boldsymbol{\omega})(0) = \sum_{j=1}^{n} m_j(\mathbf{z}_j(0)^2 \cdot \boldsymbol{\omega}(0) - \boldsymbol{\omega}(0)\mathbf{z}_j(0) \cdot \mathbf{z}_j(0))$ (total angular momentum vector of the rigid body at time $t = 0$).

We are looking for the map

$$t \mapsto (\mathbf{y}(t), \mathbf{A}(t), \boldsymbol{\omega}(t), \mathbf{z}_1(t), \ldots, \mathbf{z}_n(t)). \tag{6.44}$$

Theorem 6.8 *For a sufficiently small time interval, the initial-value problem for the motion of the rigid body has a unique smooth solution.*

Proof. In terms of components, the system (M1)–(M3) consists of $9 + 3n$ equations for $9 + 3n$ unknown functions given by (6.44). Moreover, note that $\boldsymbol{\omega}(t) = \Theta^{-1}\mathbf{A}(t)$. Thus, the system (M1)–(M3) represents a first-order system of ordinary differential equations for the unknown functions

$$\mathbf{y} = \mathbf{y}(t), \quad \mathbf{A} = \mathbf{A}(t), \quad \mathbf{z}_j = \mathbf{z}_j(t), \quad j = 1, \ldots, n.$$

The claim follows now from the standard existence and uniqueness theorem for systems of ordinary differential equations (see the Picard–Lindelöf theorem in Sect. 1.5 of E. Zeidler, Nonlinear Functional Analysis, Vol. I, Springer, New York, 1998). □

Motivation of the equations of motion. Step 1: d'Alembert's principle of virtual power. Differentiating the constraints $(\mathbf{x}_i(t) - \mathbf{x}_j)^2(t)^2 = r_{ij}^2$ with respect to time t and setting $\mathbf{v}_i := \dot{\mathbf{x}}_i(t)$, we get the equations

$$\boxed{(\mathbf{x}_i - \mathbf{x}_j)(\mathbf{v}_i - \mathbf{v}_j) = 0, \qquad i, j = 1, \ldots, n,\ i \neq j} \tag{6.45}$$

for the virtual velocity vectors $\mathbf{v}_1, \ldots, \mathbf{v}_n$. Let $\mathbf{a}, \boldsymbol{\omega} \in E_3$. Choose

$$\mathbf{v}_j := \mathbf{a} + \boldsymbol{\omega} \times \mathbf{x}_j, \qquad j = 1, \ldots, n.$$

Then the vectors $\mathbf{v}_1, \ldots, \mathbf{v}_n$ satisfy the equation (6.45) because of $\mathbf{c}(\boldsymbol{\omega} \times \mathbf{c}) = 0$ for all vectors $\mathbf{c}, \boldsymbol{\omega} \in E_3$. d'Alembert's principle of virtual power tells us that

$$\sum_{j=1}^{n}(\mathbf{F}_j - m_j\ddot{\mathbf{x}}_j)\mathbf{v}_j = 0.$$

Using $\mathbf{b}(\boldsymbol{\omega} \times \mathbf{x}_j) = \boldsymbol{\omega}(\mathbf{x}_j \times \mathbf{b})$, we get

$$\left(\sum_{j=1}^{n}\mathbf{F}_j - m_j\ddot{\mathbf{x}}_j\right)\mathbf{a} + \left(\sum_{j=1}^{n}\mathbf{x}_j \times (\mathbf{F}_j - m_j\ddot{\mathbf{x}}_j)\right)\boldsymbol{\omega} = 0$$

for all $\mathbf{a}, \boldsymbol{\omega} \in E_3$. Hence we get the force balance equation

$$\boxed{\sum_{j=1}^{n} (\mathbf{F}_j - m_j \ddot{\mathbf{x}}_j) = 0,} \tag{6.46}$$

and the torque balance equation

$$\boxed{\sum_{j=1}^{n} \mathbf{x}_j \times (\mathbf{F}_j - m_j \ddot{\mathbf{x}}_j) = 0.} \tag{6.47}$$

Note that these fundamental equations for the motion of the rigid body include the so-called inertia forces depending on the acceleration vectors $\ddot{\mathbf{x}}_j$.

Step 2: Separation from the motion of the center of gravity. We will use the superposition $\mathbf{x}_j = \mathbf{y} + \mathbf{z}_j$ together with the following relations:

- $\frac{d}{dt}(\mathbf{y} \times \dot{\mathbf{y}}) = \dot{\mathbf{y}} \times \dot{\mathbf{y}} + \mathbf{y} \times \ddot{\mathbf{y}} = \mathbf{y} \times \ddot{\mathbf{y}}$;
- $\frac{d}{dt}(\mathbf{z}_j \times \dot{\mathbf{z}}_j) = \mathbf{z}_j \times \ddot{\mathbf{z}}_j$;
- from $\sum_{j=1}^{n} m_j \mathbf{z}_j = \sum_{j=1}^{n} m_j \mathbf{y} - m_j \mathbf{x}_j = 0$ we get $\sum_{j=1}^{n} m_j \dot{\mathbf{z}} = 0$.

It follows from $m\ddot{\mathbf{y}} = \mathbf{F}$ that

$$\frac{d}{dt}(\mathbf{y} \times m\dot{\mathbf{y}}) = \mathbf{y} \times \mathbf{F}.$$

Similarly, equation (6.47) implies

$$\frac{d}{dt} \sum_j \mathbf{x}_j \times m_j \dot{\mathbf{x}}_j = \sum_j \mathbf{x}_j \times m_j \ddot{\mathbf{x}}_j = \sum_j \mathbf{x}_j \times \mathbf{F}_j = \mathbf{y} \times \mathbf{F} + \mathbf{T}. \tag{6.48}$$

Recall that $\mathbf{T} = \sum_j \mathbf{z}_j \times \mathbf{F}_j$ (total torque). Equation (6.48) describes the time rate of change of the total angular momentum with respect to the origin P_0.

We want to reformulate this in terms of the center of gravity. To this end, note that $\sum_j m_j \mathbf{z}_j = \sum_j m_j \dot{\mathbf{z}}_j = 0$. Hence

$$\sum_j \mathbf{x}_j \times m_j \dot{\mathbf{x}}_j = \sum_j (\mathbf{y} + \mathbf{z}_j) \times m_j (\dot{\mathbf{y}} + \dot{\mathbf{z}}_j) = \mathbf{y} \times m\dot{\mathbf{y}} + \sum_j \mathbf{z}_j \times m_j \dot{\mathbf{z}}_j.$$

By (6.48), this implies $\dot{\mathbf{A}} = \frac{d}{dt} \sum_j \mathbf{z}_j \times m_j \dot{\mathbf{z}}_j = \mathbf{T}$. □

6.5.5 Euler's Equation of the Spinning Top

> The approach to problems in physics can be simplified by passing to an appropriate system of reference which fits best the physical situation.
>
> Folklore

Euler (1707–1783) used the most natural system of reference Σ_* for the motion of a rigid body. This is a right-handed Cartesian (x_*, y_*, z_*)-coordinate system with the center of gravity as origin. Moreover, the right-handed orthonormal system $\mathbf{i}_*, \mathbf{j}_*, \mathbf{k}_*$

represents three principal axes of inertia of the rigid body.[9] The equation of motion of the rigid body with respect to the coordinate system Σ_* reads as follows:

$$\begin{aligned} \lambda_1\dot{\omega}^1 + (\lambda_3 - \lambda_2)\omega^2\omega^3 &= T^1, \\ \lambda_2\dot{\omega}^2 + (\lambda_1 - \lambda_3)\omega^3\omega^1 &= T^2, \\ \lambda_1\dot{\omega}^3 + (\lambda_2 - \lambda_1)\omega^1\omega^2 &= T^3. \end{aligned} \tag{6.49}$$

We are given the components T^1, T^2, T^3 of the total torque vector $\mathbf{T}$ observed in the system Σ_*. We are looking for the functions

$$t \mapsto (\omega^1(t), \omega^2(t), \omega^3(t))$$

where $\omega^1(t), \omega^2(t), \omega^3(t)$ are the components of the angular velocity vector $\boldsymbol{\omega}(t)$ observed in the system Σ_* at time t. The positive numbers $\lambda_1, \lambda_2, \lambda_3$ are the principal moments of inertia of the rigid body.

Equation (6.49) is the famous Euler equation for the spinning top. Prototypes of spinning tops are the rotating earth and the gyrocompass. In Einstein's theory of special relativity, inertial system play a key role (see page 905). The moving system $\sum_*$ of reference is the prototype of a non-inertial system. On the rotating earth, the rotation causes two additional forces to the gravitational force, namely, the centrifugal force and the Coriolis force. This is thoroughly studied in E. Zeidler (1995), page 30, quoted below.

Proof. Choose the right-handed orthonormal system of vectors $\mathbf{e}_1, \mathbf{e}_2, \mathbf{e}_3$ such that they represent principal axes of inertia of the rigid body. Decompose

- $\boldsymbol{\omega}(t) = \omega^1(t)\mathbf{e}_1(t) + \omega^2(t)\mathbf{e}_2(t) + \omega^3(t)\mathbf{e}_3(t)$, and
- $\mathbf{T}(t) = T^1(t)\mathbf{e}_1(t) + T^2(t)\mathbf{e}_2(t) + T^3(t)\mathbf{e}_3(t)$.

Note that

$$\mathbf{A}(t) = \Theta\boldsymbol{\omega}(t) = \sum_{j=1}^{3} \lambda_j\omega^j(t)\mathbf{e}_j(t).$$

Hence

$$\dot{\mathbf{A}}(t) = \sum_{j=1}^{3} \lambda_j\big(\dot{\omega}^j(t)\mathbf{e}_j(t) + \omega^j(t)\dot{\mathbf{e}}_j(t)\big).$$

Using the equation $\dot{\mathbf{e}}_j(t) = \boldsymbol{\omega}(t) \times \mathbf{e}_j(t)$, of moving frames, the claim follows from $\dot{\mathbf{A}}(t) = \mathbf{T}$. □

The Euler equation for a continuous rigid body will be considered in Problem 6.3.

Further Reading

E. Zeidler, Nonlinear Functional Analysis and its Applications, Vol. IV: Applications to Mathematical Physics, Springer, New York, 1995 (the first two chapters).

L. Landau and E. Lifshitz, Course of Theoretical Physics, Vol 1: Mechanics, Butterworth-Heinemann, Oxford, 1983.

[9] In particular, if we choose the rotating earth, then the origin is the center of earth, the x_*-axis connects the center of earth with a fixed equatorial point, and the z_*-axis connects the center of earth with the North Pole. The unit vector $\mathbf{k}_*$ points in the direction of the z_*-axis, and so on.

V. Arnold, Mathematical Theory of Classical Mechanics, Springer, Berlin, 1978.

M. Audin, Spinning Tops, Cambridge University Press, 1996.

F. Klein and A. Sommerfeld, The Theory of the Top (classic monograph). English edition: Vol. 1: Birkhäuser, Basel, 2008, Vol. 2: Springer, Berlin, 2010. German edition: Springer, Berlin, 1897.

6.5.6 Equilibrium States and Torque

Assume that the forces $(\mathbf{x}_1, \ldots, \mathbf{x}_n) \mapsto \mathbf{F}_j(\mathbf{x}_1, \ldots, \mathbf{x}_n)$ are smooth, $j = 1, \ldots, n$.

Proposition 6.9 *For the positions $\mathbf{x}_{10}, \ldots, \mathbf{x}_{n0}$, the rigid body is in an equilibrium state iff the following three conditions are satisfied:*
(i) *The total force is zero: $\sum_{j=1}^n \mathbf{F}_j(\mathbf{x}_{10}, \ldots, \mathbf{x}_{n0}) = 0$.*
(ii) *The total torque is zero: $\sum_{j=1}^n \mathbf{x}_{j0} \times \mathbf{F}_j(\mathbf{x}_{10}, \ldots, \mathbf{x}_{n0}) = 0$.*
(iii) *The body is at rest at time $t = 0$: $\dot{\mathbf{x}}_{j0}(0) = 0$, $j = 1, \ldots, n$.*

Proof. (I) Suppose that the rigid body is in an equilibrium state, that is, the constant trajectories

$$\mathbf{x}_j(t) :\equiv \mathbf{x}_{j0}, \quad j = 1, \ldots, n \tag{6.50}$$

are solutions of the equations of motion

$$\ddot{\mathbf{y}} = \sum_j \mathbf{F}_j, \qquad \dot{\mathbf{A}} = \mathbf{T}.$$

It follows from $\dot{\mathbf{y}} = 0$ that (i) is satisfied. Moreover, we have $\dot{\mathbf{z}}_j = \dot{\mathbf{y}} - \dot{\mathbf{x}}_j = 0$ if $j = 1, \ldots, n$. Hence $\mathbf{A} = \sum_j \mathbf{z}_j \times m_j \dot{\mathbf{z}}_j = 0$. This implies $\mathbf{T} = \dot{\mathbf{A}} = 0$ which yields the condition (ii).

(II) Conversely, assume that the properties (i), (ii), and (iii) are satisfied. Then the trajectories (6.50) are a solution of the equations of motion. By Theorem 6.8 on page 394, this solution is unique. □

6.5.7 The Principal Bundle $\mathbb{R}^3 \times SO(3)$ – the Position Space of a Rigid Body

The position space of a rigid body. Let us fix the right-handed Cartesian (x, y, z)-coordinate system at the point P_0 with the right-handed orthonormal basis $\mathbf{i}, \mathbf{j}, \mathbf{k}$ at the point P_0. Parallel transport of $\mathbf{i}, \mathbf{j}, \mathbf{k}$ from the point P_0 to the point C yields the vectors $\mathbf{i}_C, \mathbf{j}_C, \mathbf{k}_C$. The state of a rigid body can be described by the tuple

$$(C; \mathbf{e}_1, \mathbf{e}_2, \mathbf{e}_3) \tag{6.51}$$

where C is the center of gravity of the rigid body, and the unit vectors $\mathbf{e}_1, \mathbf{e}_2, \mathbf{e}_3$ represent a distinguished right-handed orthonormal system of the rigid body at the point C (e.g., three fixed principal inertial axes of the rigid body). Then there exists a matrix $G \in SO(3)$ such that

$$\begin{pmatrix} \mathbf{e}_1 \\ \mathbf{e}_2 \\ \mathbf{e}_3 \end{pmatrix} = G \begin{pmatrix} \mathbf{i}_C \\ \mathbf{j}_C \\ \mathbf{k}_C \end{pmatrix}. \tag{6.52}$$

Thus, the state (6.51) can be described by the coordinates

$$(\mathcal{C}, G) \in \mathbb{R}^3 \times SO(3)$$

where $\mathcal{C} = (x_C, y_C, z_C)$ are the coordinates of the center C of gravity in the (x, y, z)-coordinate system, and G describes the rotation (6.52). The product set

$$\mathbb{R}^3 \times SO(3)$$

is called a (trivial) principal bundle with the linear space $\mathbb{R}^3$ as base manifold, and the Lie group $SO(3)$ as typical fiber.

The motion of the rigid body. A trajectory

$$t \mapsto (\mathcal{C}(t), G(t))$$

on the principal bundle $\mathbb{R}^3 \times SO(3)$ describes the motion of a rigid body.

Gauge transformations. If we replace the right-handed orthonormal frame $\mathbf{e}_1, \mathbf{e}_2, \mathbf{e}_3$ at the point C by the right-handed orthonormal frame $\mathbf{e}_1^+, \mathbf{e}_2^+, \mathbf{e}_3^+$ at the point C, then there exists a matrix $G^+(C) \in SO(3)$ such that

$$\begin{pmatrix} \mathbf{e}_1^+ \\ \mathbf{e}_2^+ \\ \mathbf{e}_3^+ \end{pmatrix} = G^+(C) \begin{pmatrix} \mathbf{e}_1 \\ \mathbf{e}_2 \\ \mathbf{e}_3 \end{pmatrix}.$$

Hence

$$\begin{pmatrix} \mathbf{e}_1^+ \\ \mathbf{e}_2^+ \\ \mathbf{e}_3^+ \end{pmatrix} = G^+(C)G \begin{pmatrix} \mathbf{i}_C \\ \mathbf{j}_C \\ \mathbf{k}_C \end{pmatrix}.$$

Thus the coordinate $(\mathcal{C}, G)$ of the point (6.51) is replaced by the coordinate $(\mathcal{C}, G^+(C)G)$. The transformation

$$G \mapsto G^+(C)G$$

is called a gauge transformation.

6.6 A Glance at Constraints in Quantum Field Theory

In classical and modern physics, gauge theories possess special features. In this section, we only want to sketch some basic ideas. A detailed study will be carried out in Vols. IV and V. The point is that, typically, the local symmetry of the Lagrangian generates constraints for the Euler–Lagrange equations, and the invertibility of the Legendre transformation breaks down.[10]

[10] As an introduction, we refer to H. Rothe and K. Rothe, Classical and Quantum Dynamics of Constrained Hamiltonian Systems, World Scientific Lecture Notes in Physics, World Scientific, Singapore, 2010.

6.6.1 Gauge Transformations and Virtual Degrees of Freedom in Gauge Theory

Potentials in classical mechanics and gauge fixing. As a prototype, consider the gravitational force at the surface of earth,

$$\mathbf{F} = -gm\mathbf{k}.$$

This force acts on a stone of mass m. As usual, we use a right-handed Cartesian (x, y, z)-coordinate system with the right-handed orthonormal basis $\mathbf{i}, \mathbf{j}, \mathbf{k}$. Introducing the function

$$U(x, y, z) := mgz + U_0, \qquad (x, y, z) \in \mathbb{R}^3, \tag{6.53}$$

we get

$$\boxed{\mathbf{F} = -\,\mathbf{grad}\, U.}$$

The function U is called the potential of the force $\mathbf{F}$. Moving the stone of mass m along the curve C from the point (x, y, z) to the point (x_0, y_0, z_0), we gain the energy

$$-\int_C \mathbf{F}d\mathbf{x} = \int_C dU = U(x, y, z) - U(x_0, y_0, z_0) = mg(z - z_0).$$

Therefore, the function U is also called the potential energy of the stone at hight z. If a stone falls down from the height $z > 0$ to the height $z_0 = 0$ of the surface of earth, then the potential energy of the stone is transformed into the heat energy mgz. The following points are crucial:

- The potential U from (6.53) is not uniquely determined; it is only determined up to the constant U_0.
- The choice of the constant U_0 is called a gauge fixing.
- The change of the constant U_0, that is, the transformation

$$mgz + U_0 \mapsto mgz + U_1$$

  is called a gauge transformation.
- The potential energy U does not possess any physical meaning. However, the difference $U(x, y, z) - U(x_0, y_0, z_0)$ possesses a physical meaning, namely, the gain of energy.

The main principle of gauge theory reads as follows:

Only gauge-invariant quantities possess a physical meaning.

For example, the force $\mathbf{F}$ is gauge-invariant. In other words, the force $\mathbf{F}$ does not depend on the choice of the constant U_0. The constant U_0 is called a virtual degree of freedom.

Potentials in Maxwell's electrodynamics and gauge fixing. In Maxwell's theory of electromagnetism, one uses the scalar potential U and the vector potential $\mathbf{A}$ in order to present the electromagnetic field in an inertial system by the relation

$$\mathbf{E} = -\,\mathbf{grad}\, U - \frac{\partial \mathbf{A}}{\partial t}, \qquad \mathbf{B} = \mathbf{curl}\, \mathbf{A}.$$

We will discuss this later on. The transformation

$$U \mapsto U - \frac{\partial \chi}{\partial t}, \qquad \mathbf{A} \mapsto \mathbf{A} + \mathbf{grad}\, \chi,$$

is called a gauge transformation. Because of $\mathbf{curl}\,\mathbf{grad}\, \chi = 0$, the gauge transformation does not change the electromagnetic field $\mathbf{E}, \mathbf{B}$. Physicists use the following two gauge conditions for gauge fixing:

- $\operatorname{div} \mathbf{A} + \frac{1}{c^2} \frac{\partial U}{\partial t} = 0$ (Lorentz gauge), and
- $\operatorname{div} \mathbf{A} = 0$ (Coulomb gauge).

In the Coulomb gauge, the Maxwell equation $\varepsilon_0 \operatorname{div} \mathbf{E} = \varrho$ in a vacuum yields

$$-\varepsilon_0 \operatorname{div} \mathbf{grad}\, U = \varrho.$$

Equivalently, $\varepsilon_0 \Delta U = \varrho$. This is the same Poisson equation as in electrostatics.

6.6.2 Elimination of Unphysical States (Ghosts)

> The introduction of additional unphysical states called ghosts helps to obtain a clear presentation of quantized gauge theories. The key point is that, roughly speaking, the physics observed in experiments does not depend on the ghosts.
>
> Folklore

In classical mechanics, one uses virtual trajectories related to the constraints in order to characterize physical trajectories by means of d'Alemberts principle of critical virtual power. In terms of mathematics, the virtual velocities are elements of the tangent spaces of the constrained position space described by the constraints. Generalizing this classical approach, the quantization of classical gauge field theories leads to virtual states called ghosts. The final goal is to add specific constraints which eliminate the influence of ghosts on physical phenomena observed in experiments.

Gupta–Bleuler ghosts in quantum electrodynamics. In Volume II, we have studied quantum electrodynamics; there we have used the Fourier expansion of classical free fields in order to construct the creation and annihilation operators for photons, electrons, and positrons (method of Fourier quantization). More precisely, we have obtained photons with transversal and longitudinal polarization. The point is that

The Lorenz gauge condition forbids unphysical longitudinal photons.

Roughly speaking, it turns out that photons of transversal and longitudinal polarization form an indefinite Hilbert space X. To overcome this difficulty in Vol. II, we have used the method of Gupta–Bleuler. The basic idea is to construct a linear operator $Q : X \to X$ such that the state $|s\rangle$ in X is a physical state iff

$$Q|s\rangle = 0. \tag{6.54}$$

BRST ghosts. Quantum electrodynamics is a gauge theory with the commutative Lie group $U(1)$ as gauge group. It was discovered by Becchi, Rouet, Stora, and Tyutin in the early 1970s that the method (6.54) can be generalized to gauge theories with noncommutative gauge groups (e.g., the groups $SU(2)$ and $SU(3)$ which are basic for the Standard Model in particle physics). It turns out that there exists a linear operator $Q : X \to X$ with the key property

$$\boxed{Q^2 = 0} \tag{6.55}$$

such that the equation (6.54) characterizes the physical states. The property (6.55) is typical for cohomology (see Sect. 16.8 of Vol. I on the power of cohomology). In terms of mathematics, equation (6.54) tells us that the physical states are precisely the cocycles with respect to the cohomology operator Q. The construction of the operator Q is based on a new type of symmetry called the BRST symmetry (see Sect. 16.7 of Vol. I).

Faddeev–Popov ghosts. The basic idea of this approach is to use the Feynman path integral. Here, one has to integrate over all the physical states. The point is that gauge transformations leave invariant physical states. Therefore, physical states are equivalence classes modulo the gauge group. In other words, physical states are orbits under the action of the gauge group. In 1967, Faddeev and Popov invented a method for computing path integrals on orbit spaces by separating a specific determinant. In addition, they introduced additional fields (so-called ghost fields) in order to enforce the crucial unitarity of the S-matrix (see Sect. 16.6 of Vol. I).

The field-antifield approach due to Batalin–Vilkovisky. This approach emerged in the early 1980s. The basic idea is to extend the given action S to an action S_{ext} which depends on fields and additional anti-fields. The symmetry properties of S_{ext} are governed by a so-called master equation. These symmetry properties of S_{ext} lead to Noether identities. It turns out that the right choice of S_{ext} embodies all the information about the gauge symmetries of the original functional S. This nice procedure can be used for the quantization of gauge theories. As an introduction, we recommend Chap. 12 of H. Rothe and K. Rothe (2010), quoted on page 419. The anti-fields can be regarded as ghosts. Observe the following peculiarity:

It happens frequently, that both the Batalin–Vilkovisky ghosts and the Faddeev–Popov ghosts are fermions described by non-classical Grassmann variables.

This way, the quantization of gauge theories is closely related to Grassmann's 1844 idea of introducing quantities in mathematics which have the key product property

$$a \vee b = -b \vee a.$$

This is a beautiful idea in modern mathematics strongly influenced by the work of physicists. For example, there exists a supersymmetric proof of the Atiyah–Singer index theorem.[11] In superstring theory, supersymmetric models are based on Grassmann's idea.[12] Nowadays it is an open question whether such models are realized in nature. Physicists hope that the LHC (Large Hadron Collider) at CERN (Geneva, Switzerland) will give an answer in the near future.

6.6.3 Degenerate Minimum Problems

Free minimum problem. Fix $\alpha \geq 0$. Consider the minimum problem

$$\tfrac{1}{2}x^2 + \tfrac{1}{2}\alpha^2 y^2 = \min!, \qquad (x, y) \in \mathbb{R}^2. \tag{6.56}$$

[11] See the supersymmetric proof of the Gauss–Bonnet–Chern theorem in Chap. 12 of B. Cycon, R. Froese, W. Kirsch and B. Simon (Eds.), Schrödinger Operators, Springer, New York, 1986.
We also refer to E. Getzler, A short proof of the local Atiyah–Singer index theorem, Topology **25** (1986), 111–117.

[12] K. Becker, M. Becker, and J. Schwarz, String Theory and M-Theory, Cambridge University Press, 2006.
J. Schwarz (Ed.), Superstrings: The First 15 Years of Superstring Theory, Vols. 1, 2, World Scientific, Singapore, 1985.
K. Efetov, Supersymmetry in Disorder and Chaos, Cambridge University Press, 1997.

- Regular case: $\alpha > 0$. Problem (6.56) has the unique solution $(x, y) = (0, 0)$.
- Singular case: $\alpha = 0$. Problem (6.56) has not a unique solution. Precisely all the points $(0, y)$ located on the y-axis are solutions of (6.56).

Constrained minimum problem and the Lagrangian multiplier. Consider now the minimum problem

$$\tfrac{1}{2}x^2 + \tfrac{1}{2}\alpha^2 y^2 = \min!, \qquad (x, y) \in \mathbb{R}^2, \qquad x^2 + y^2 - 1 = 0. \tag{6.57}$$

This means that we add the constraint $x^2 + y^2 - 1 = 0$ to (6.56). Let us define $f(x) := \frac{1}{2}x^2 + \frac{1}{2}\alpha^2 y^2$. Geometrically, we are looking for a minimal value of the function f on the unit circle.

If $0 \le \alpha < 1$, *then the minimum problem* (6.57) *has the solutions* $(x_0, y_0) = (0, \pm 1)$ *with the minimal value* $f(x_0, y_0) = \frac{1}{2}\alpha^2$.

Proof. We will use two different methods which are typical for handling constrained minimum problems.

(I) The method of parametrizing the constraint (reduction to a free minimum problem). Using $x = \cos\varphi, y = \sin\varphi$, and setting

$$g(\varphi) := \tfrac{1}{2}\cos^2\varphi + \tfrac{1}{2}\alpha^2 \sin^2\varphi,$$

we get the free minimum problem

$$g(\varphi) = \min!, \qquad \varphi \in \mathbb{R}.$$

If φ_0 is a solution, then

$$0 = g'(\varphi_0) = (\alpha^2 - 1)\sin\varphi_0 \cos\varphi_0 = \tfrac{1}{2}(\alpha^2 - 1)\sin 2\varphi_0.$$

Hence $\varphi_0 = k\pi,\ k\frac{\pi}{2}$ where $k = 0, \pm 1, \pm 2, \ldots$ Thus, $(x_0, y_0) = (\pm 1, 0), (0, \pm 1)$. Finally, note that $f(\pm 1, 0) = \frac{1}{2}$ and $f(0, \pm 1) = \frac{1}{2}\alpha^2$.

(II) The method of the Lagrangian multiplier. Suppose that (x_0, y_0) is a solution of (6.57). Choose a curve $x = x(t), y = y(t), -t_1 < t < t_1$, with $x(0) = x_0, y(0) = y_0$ such that the curve satisfies the constraint

$$x(t)^2 + y(t)^2 - 1 = 0.$$

Differentiating this with respect to time t at the point $t = 0$, we get the linearized constraint $2x_0\dot{x}(0) + 2y_0\dot{y}(0) = 0$. Moreover, set $\chi(t) := f(x(t), y(t))$. Then the function χ has a minimum at the point $t = 0$. Hence

$$0 = \chi'(0) = f_x(x_0, y_0)\dot{x}(0) + f_y(x_0, y_0)\dot{y}(0).$$

Setting $v = \dot{x}(0)$ and $w := \dot{y}(0)$, we get the linear system

$$\begin{aligned} x_0 v + y_0 w &= 0, \\ x_0 v + \alpha^2 y_0 w &= 0. \end{aligned} \tag{6.58}$$

Note that $x_0^2 + y_0^2 = 1$. Thus, the rank of the coefficient matrix

$$A := \begin{pmatrix} x_0 & y_0 \\ x_0 & \alpha^2 y_0 \end{pmatrix}$$

is at least equal to one. However, we know that every solution of the first equation of (6.58) is also a solution of the second equation. Thus, the rank of the coefficient matrix A is equal to one (see Problem 6.1). Consequently, there exists a real number such that

$$(x_0, y_0) = \lambda(x_0, \alpha^2 y_0).$$

Hence we get the following three equations

$$x_0 = \lambda x_0, \quad \alpha^2 y_0 = \lambda y_0, \quad x_0^2 + y_0^2 = 1$$

for computing the three quantities x_0, y_0, λ. If $x_0 \neq 0$, then $\lambda = 1$. Hence $y_0 = 0$, which implies $x_0 = \pm 1$, and so on. Therefore, the solutions are

- $x_0 = 0, y_0 = \pm 1, \lambda = \alpha^2$, and
- $x_0 = \pm 1, y_0 = 0, \lambda = \frac{1}{2}$.

Thus, we get the two candidates $(x_0, y_0) = (\pm 1, 0), (0, \pm 1)$. Finally, note that $f(\pm 1, 0) = \frac{1}{2}$ and $f(0, \pm 1) = \frac{1}{2}\alpha^2$. □

In addition, the same argument shows that if $\alpha = 1$, then all the points on the unit circle are solutions of (6.57). If $\alpha > 1$, then the minimum problem (6.57) has the solutions $(x_0, y_0) = (\pm 1, 0)$ with the minimal value $f(x_0, y_0) = \frac{1}{2}$.

Variations of real-valued function. As a preparation for the study of local symmetries in the calculus of variations and its relation to degenerate minimum problems, let us consider the smooth function

$$S : \mathbb{R}^2 \to \mathbb{R}.$$

For fixed real numbers h and k, set

$$\chi(\varepsilon) := S(x + \varepsilon h, y + \varepsilon k), \qquad \varepsilon \in \mathbb{R}.$$

By Taylor expansion,

$$\chi(\varepsilon) = \chi(0) + \varepsilon\chi'(0) + \tfrac{1}{2}\varepsilon^2\chi''(0) + \ldots \tag{6.59}$$

This motivates the definitions $\delta x := \varepsilon h, \delta y := \varepsilon k$, and

$$\delta S(x, y; \delta x, \delta y) := \varepsilon\chi'(0), \qquad \delta^n S(x, y; \delta x, \delta y) := \varepsilon^n \chi^{(n)}(0), \qquad n = 2, 3, \ldots$$

Here, $\delta^n S(x, y; \delta, \delta y)$ is called the nth variation of the function S at the point (x, y) in direction $(\delta x, \delta y)$. Explicitly, set

$$S'(x, y) := (S_x(x, y), S_y(x, y)), \qquad S''(x, y) := \begin{pmatrix} S_{xx}(x, y) & S_{xy}(x, y) \\ S_{xy}(x, y) & S_{yy}(x, y) \end{pmatrix}.$$

Then, we obtain the first variation

$$\delta S(x, y; \delta x, \delta y) = S'(x, y) \begin{pmatrix} \delta x \\ \delta y \end{pmatrix} = S_x(x, y)\, \delta x + S_y(x, y)\, \delta y.$$

For the second variation of the function S, we get

$$\delta^2 S(x, y; \delta x, \delta y) = S_{xx}(x, y)\, (\delta x)^2 + 2S_{xy}\, \delta x \delta y + S_{yy}(x, y)\, (\delta y)^2.$$

In terms of the Hessian matrix, this means that

$$\delta^2 S(x, y; \delta x, \delta y) = (\delta x, \delta y)\, S''(x_0, y_0) \begin{pmatrix} \delta x \\ \delta y \end{pmatrix}.$$

Regular minimum problem. *Let $S'(x_0, y_0) = 0$. Suppose that the Hessian matrix $S''(x_0, y_0)$ has only positive eigenvalues. Then the function S has a strict local minimum at the point (x_0, y_0). In particular, if S is a quadratic function, then S has a global minimum at the point (x_0, y_0).*

This result (together with far-reaching generalizations to functionals on infinite-dimensional Banach spaces) can be found in E. Zeidler, Nonlinear Functional Analysis and its Applications, Vol. III: Variational Methods and Optimization, Springer, New York, 1986. The Hessian matrix was studied by Hesse (1811–1874).

Singular minimum problem. *Let $S'(x_0, y_0) = 0$. Suppose that the Hessian matrix $S''(x_0, y_0)$ has the eigenvalues $\lambda_1 > 0$ and $\lambda_2 = 0$, and suppose that (h, k) is an eigenvector of $S''(x_0, y_0)$ corresponding to the degenerate eigenvalue $\lambda_2 = 0$. Then*

$$S(x_0 + \varepsilon h, y_0 + \varepsilon k) = S(x_0, y_0) + O(\varepsilon^3), \quad \varepsilon \to 0.$$

In particular, if S is a quadratic function, then

$$S(x_0 + \varepsilon h, y_0 + \varepsilon k = S(x_0, y_0) \qquad \text{for all} \quad \varepsilon \in \mathbb{R}.$$

Thus, all the points located on the line $\{(x_0 + \varepsilon h, y_0 + \varepsilon k) : \ \varepsilon \in \mathbb{R}\}$ are minimal points of the function S.

Proof. By the Taylor expansion (6.59),

$$\chi(\varepsilon) = \chi(0) + \varepsilon\chi'(0) + \tfrac{1}{2}\varepsilon^2\chi''(0) + O(\varepsilon^3), \qquad \varepsilon \to 0.$$

Note that $\chi'(0) = \chi''(0) = 0$. If the function χ is quadratic, then the term $O(\varepsilon^3)$ drops out. □

In terms of mechanics, regular (resp. singular) minimal points are models for stable (resp. unstable) equilibrium points.

6.6.4 Variation of the Action Functional

> In modern mathematics, the variations $\delta x, \delta y$ and δS are not mystical infinitesimals, but well-defined finite mathematical quantities.
>
> Folklore

The use of the nonzero function δx together with $(\delta x)^2 = 0$ is confusing in the classical calculus papers and in many physics textbooks. We want to discuss how to overcome this confusion in a simple way.

The principle of minimal action. Fix the finite time interval $[t_0, t_1]$. Let us consider the principle of minimal action

$$\boxed{S(x, y) = \min!, \qquad x, y \in C^\infty[t_0, t_1]} \tag{6.60}$$

with the boundary condition: $x(t_0), x(t_1), y(t_0), y(t_1)$ are fixed. We will study the special case where we have the action functional

$$S(x, y) = \int_{t_0}^{t_1} L(x(t), y(t), \dot{x}(t), \dot{y}(t))\, dt, \qquad x, y \in C^\infty[t_0, t_1]$$

with the Lagrangian

$$\boxed{L(x, y, \dot{x}, \dot{y}) := \tfrac{1}{2}\dot{x}^2 + \tfrac{1}{2}\alpha^2\dot{y}^2 + \dot{x}y - \tfrac{1}{2}(x - y)^2.} \tag{6.61}$$

Here, we fix the parameter $\alpha \geq 0$. We will distinguish between the regular case ($\alpha > 0$) and the singular case ($\alpha = 0$). Recall that the space $C^\infty[t_0, t_1]$ consists of all the smooth functions $x : [t_0, t_1] \to \mathbb{R}$.

The Taylor expansion and the language of mathematicians. Choose the functions $h, k \in C^\infty[t_0, t_1]$. Set

$$\chi(\varepsilon) := S(x + \varepsilon h, y + \varepsilon k), \qquad \varepsilon \in \mathbb{R}.$$

Taylor expansion yields

$$\chi(\varepsilon) = \chi(0) + \varepsilon\chi'(0) + \tfrac{1}{2}\varepsilon^2\chi''(0) + O(\varepsilon^3), \quad \varepsilon \to 0.$$

In order to pass to Lagrange's symbol δ, we define

- $\delta x := \varepsilon h$, $\delta y := \varepsilon k$,
- $\delta S := \varepsilon\chi'(0)$, and
- $\delta^n S = \varepsilon^n \chi^{(n)}(0)$, $n = 2, 3, \ldots$

Note that $\delta^n S$ depends on x, y and $\delta x, \delta y$. To simplify notation, we frequently write $\delta^n S$ instead of $\delta^n S(x, y; \delta x, \delta y)$. This yields

$$\boxed{S(x + \delta x, y + \delta y) = S(x, y) + \delta S(x, y; \delta x, \delta y) + O(\varepsilon^2), \quad \varepsilon \to 0} \tag{6.62}$$

and

$$S(x + \delta x, y + \delta y) = S(x, y) + \delta S(x, y; \delta x, \delta y) + \tfrac{1}{2}\delta^2 S(x, y; \delta x, \delta y) + O(\varepsilon^3), \quad \varepsilon \to 0.$$

Here, $\delta S(x, y; \delta x, \delta y)$ (resp. $\delta^2 S(x, y; \delta x, \delta y)$) is called the first (resp. second) variation of the functional S at the point (x, y) in direction $(\delta x, \delta y)$.

Example. Choose the Lagrangian L from (6.61). Fix $\alpha \geq 0$. Let ε be a real number. Replacing x (resp. y) by $x + \varepsilon h$ (resp. $y + \varepsilon k$), we get the function

$$L_\varepsilon := \tfrac{1}{2}(\dot{x} + \varepsilon\dot{h})^2 + \tfrac{1}{2}\alpha^2(\dot{y} + \varepsilon\dot{k})^2 + (\dot{x} + \varepsilon\dot{h})(y + \varepsilon k) - \tfrac{1}{2}(x + \varepsilon h - y - \varepsilon k)^2$$

of the real variable ε. Differentiation with respect to ε at the point $\varepsilon = 0$ yields

$$\frac{dL_\varepsilon}{d\varepsilon}_{|\varepsilon=0} = \dot{x}\dot{h} + \alpha^2\dot{y}\dot{k} + \dot{h}y + \dot{x}k - (x - y)(h - k).$$

Using the Leibniz rule $\frac{d}{dt}(\dot{x}h) = \ddot{x}h + \dot{x}\dot{h}$, we get

$$\frac{dL_\varepsilon}{d\varepsilon}_{|\varepsilon=0} = \frac{d}{dt}(\dot{x}h + \alpha^2\dot{y}k + hy) - h(\ddot{x} + \dot{y} + x - y) - k(\alpha^2\ddot{y} - \dot{x} + y - x).$$

Noting that $\chi'(0) = \int_{t_0}^{t_1} \frac{dL_\varepsilon}{d\varepsilon}_{|\varepsilon=0}\, dt$, we obtain the key formula

$$\begin{aligned}\chi'(0) = -\int_{t_0}^{t_1} &\left[(\ddot{x} + \dot{y} + x - y)h - (\alpha^2\ddot{y} - \dot{x} + y - x)k\right](t)\, dt \\ &+ (\dot{x}h + \alpha^2\dot{y}k + hy)(t_1) - (\dot{x}h + \alpha^2\dot{y}k + hy)(t_0).\end{aligned}$$

Proposition 6.10 *If x, y is a solution of the variational problem* (6.60) *of minimal action, then we have the Euler–Lagrange equations*

$$\ddot{x} + \dot{y} + x - y = 0, \quad \alpha^2\ddot{y} - \dot{x} + y - x = 0 \quad \text{on} \quad [t_0, t_1]. \tag{6.63}$$

Proof. By the general theory, the Euler–Lagrange equations read as

$$\frac{d}{dt}L_{\dot{x}} = L_x, \quad \frac{d}{dt}L_{\dot{y}} = L_y.$$

This yields (6.63). □

For the convenience of the reader, let us repeat the standard argument with respect to the present special case. Let x, y be a solution of (6.60). Choose the functions $h, k \in C^\infty[t_0, t_1]$ with the boundary conditions

$$h(t_0) = h(t_1) = 0, \quad k(t_0) = k(t_1) = 0. \tag{6.64}$$

Then the functions $x + \varepsilon h$ and x (resp. $y + \varepsilon k$ and k) have the same boundary values. Therefore, the real-valued function $\varepsilon \to \chi(\varepsilon)$ has the critical point $\varepsilon = 0$. This implies $\chi'(0) = 0$. Explicitly,

$$-\int_{t_0}^{t_1} \left[(\ddot{x} + \dot{y} + x - y)h + (\alpha^2\ddot{y} - \dot{x} + y - x)k\right](t)\, dt = 0$$

for all functions $h, k \in C^\infty[t_0.t_1]$ with the boundary conditions (6.64). Finally, applying the variational lemma (see Sect. 7.20.2 of Vol. I), we get the claim (6.63). This finishes the proof.

Let us reformulate this in terms of the symbol δ. Since $\delta S = \varepsilon\chi'(0)$, we get the key formula

$$\begin{aligned}\delta S = &-\int_{t_0}^{t_1} \left[(\ddot{x} + \dot{y} + x - y)\,\delta x + (\alpha^2\ddot{y} - \dot{x} + y - x)\,\delta y\right](t)\, dt \\ &+(\dot{x}\,\delta x + \alpha^2\dot{y}\,\delta y + y\,\delta x)(t_1) - (\dot{x}\,\delta x + \alpha^2\dot{y}\,\delta y + y\,\delta x)(t_0).\end{aligned} \tag{6.65}$$

Finally, we want to pass to the elegant formula

$$\boxed{\delta S = \frac{\delta S(x,y)}{\delta x}\,\delta x + \frac{\delta S(x,y)}{\delta y}\,\delta y.} \tag{6.66}$$

To this end, we define

$$\begin{aligned}\frac{\delta S(x,y)}{\delta x}h := &-\int_{t_0}^{t_1} \left[(\ddot{x} + \dot{y} + x - y)h\right](t)\; dt \\ &+(\dot{x}h + yh)(t_1) - (\dot{x}h + yh)(t_0),\end{aligned}$$

and

$$\begin{aligned}\frac{\delta S(x,y)}{\delta y}k := &-\int_{t_0}^{t_1} \left[(\alpha^2\ddot{y} - \dot{x} + y - x)k\right](t)\, dt \\ &+\alpha^2(\dot{y}k)(t_1) - \alpha^2(\dot{y}k)(t_0)\end{aligned}$$

for all functions $h, k \in C^\infty[t_0, t_1]$. The symbol $\frac{\delta S(x,y)}{\delta x}$ stands for the linear map $h \mapsto \frac{\delta S(x,y)}{\delta x}h$. This yields the linear functional

$$\frac{\delta S(x,y)}{\delta x} : C^\infty[t_0, t_1] \to \mathbb{R}$$

which is called the partial functional derivative (with respect to the variable x) of the functional S at the point (x, y). Similarly, $\frac{\delta S(x,y)}{\delta y}$ is a linear functional on the real linear space $C^\infty[t_0, t_1]$.

The language of physicists. In order to pass to the language used by physicists, one has to replace

$$f = O(\varepsilon^2), \qquad \varepsilon \to 0$$

by writing $f = 0$. In other words, all the terms of order $O(\varepsilon^2)$ drop out. For example, instead of $(\delta x)^2 = \varepsilon^2 h^2 = O(\varepsilon^2), \varepsilon \to 0$, we briefly write

$$\boxed{(\delta x)^2 = 0.} \tag{6.67}$$

From (6.62) we get the two variation formulas

$$S(x + \delta x, y + \delta y) = S(x, y) + \delta S(x, y; \delta x, \delta y)$$

and

$$S(x + \delta x, y + \delta y) = S(x, y) + \delta S(x, y; \delta x, \delta y) + \tfrac{1}{2}\delta^2 S(x, y; \delta x, \delta y)$$

used by physicists. Then, Proposition 6.10 can be elegantly formulated in the following way:

If the pair of functions x, y is a solution of the variational problem (6.60) *of minimal action, then*

$$\delta S = 0$$

for all variations $\delta x, \delta y$ which vanish at the boundary points t_0 and t_1.

Local functional derivatives (densities). Our goal is to write

$$\frac{\delta S(x,y)}{\delta x}\,\delta x = \int_{t_0}^{t_1} \frac{\delta S(x,y)}{\delta x(t)}\,\delta x(t)\,dt, \qquad \frac{\delta S(x,y)}{\delta y}\,\delta y = \int_{t_0}^{t_1} \frac{\delta S(x,y)}{\delta y(t)}\,\delta y(t)\,dt.$$

This yields

$$\boxed{\delta S = \int_{t_0}^{t_1} \left(\frac{\delta S(x,y)}{\delta x(t)}\,\delta x(t) + \frac{\delta S(x,y)}{\delta y(t)}\,\delta y(t) \right) dt.} \tag{6.68}$$

(a) Regular density case: If the functions δx and δy vanish at the boundary points t_0 and t_1, then it follows from the key formula (6.65) that

$$\frac{\delta S(x,y)}{\delta x(t)} = -(\ddot{x} + \dot{y} + x - y)(t), \qquad \frac{\delta S(x,y)}{\delta y(t)} = (\alpha^2 \ddot{y} - \dot{x} + y - x)(t), \quad t \in [t_0, t_1].$$

(b) Singular density case: If the boundary values of the functions δx and δy are arbitrary, then we use Dirac's mnemonic density formula

$$\int_{t_0}^{t_1} \delta(t - t_*) f(t)\,dt = f(t_*) \qquad \text{for all} \quad f \in C^\infty[t_0, t_1], \;\; t_* \in [t_0, t_1].$$

In this sense, for all times $t \in [t_0, t_1]$, we get

$$\frac{\delta S(x,y)}{\delta x(t)} = -(\ddot{x} + \dot{y} + x - y)(t) + (\dot{x} + y)(t_1) \cdot \delta(t - t_1) - (\dot{x} + y)(t_0) \cdot \delta(t - t_0),$$

$$\frac{\delta S(x,y)}{\delta y(t)} = -(\alpha^2 \ddot{y} - \dot{x} + y - x)(t) + \alpha^2 \dot{y}(t_1) \cdot \delta(t - t_1) - \alpha^2 \dot{y}(t_0) \cdot \delta(t - t_0).$$

6.6.5 Degenerate Lagrangian and Constraints

Consider the singular case where $\alpha = 0$. From (6.63) we get the Euler–Lagrange equations

$$\ddot{x} + \dot{y} + x - y = 0, \quad \dot{x} + x - y = 0 \quad \text{on} \quad [t_0, t_1]. \tag{6.69}$$

The point is that these two equations are not independent equations. There exists the additional relation

$$\dot{y} = \dot{x}. \tag{6.70}$$

Proof. Differentiating the Euler–Lagrange equation $\dot{x} = y - x$ with respect to time, we get

$$\ddot{x} = \dot{y} - \dot{x}.$$

Using the Euler–Lagrange equation $\ddot{x} = y - x - \dot{y}$, we obtain

$$\dot{y} - \dot{x} = y - x - \dot{y}.$$

This implies $2\dot{y} = 2(y - x)$. Thus, $\dot{y} = y - x = \dot{x}$. □

Proposition 6.11 *The solutions of the Euler–Lagrange equations* (6.69) *are precisely given by the family of linear functions*

$$x = x_0 + vt, \quad y = y_0 + vt, \quad t \in \mathbb{R} \tag{6.71}$$

with the real parameters x_0, y_0 and v where $y_0 - x_0 = v$.

Proof. If x, y is a solution of (6.69), then the relation (6.70) is valid. Hence

$$\ddot{x} = y - x - \dot{x} = 0.$$

Thus, $x = x_0 + vt$ and $y = y_0 + vt$. Moreover, $\dot{x} = y - x = y_0 - x_0 = v$.
Conversely, the family (6.71) is a solution of (6.69). □

6.6.6 Degenerate Legendre Transformation

Consider again the Lagrangian

$$L(x, y, \dot{x}, \dot{y}) := \tfrac{1}{2}\dot{x}^2 + \tfrac{1}{2}\alpha^2\dot{y}^2 + \dot{x}y - \tfrac{1}{2}(x - y)^2 \tag{6.72}$$

together with the Euler–Lagrange equations

$$\ddot{x} + \dot{y} + x - y = 0, \quad \alpha^2\ddot{y} - \dot{x} + y - x = 0 \quad \text{on} \quad \mathbb{R}. \tag{6.73}$$

In Chap. 6 of Vol. II, we have studied the Hamiltonian approach to mechanics. Let us apply this to the special Lagrangian L from (6.72). Define the function

$$\mathcal{H}(x, y, \dot{x}, \dot{y}) := \dot{x}L_{\dot{x}}(x, y, \dot{x}, \dot{y}) + \dot{y}L_{\dot{y}}(x, y, \dot{x}, \dot{y}) - L(x, y, \dot{x}, \dot{y}).$$

Proposition 6.12 *If $t \mapsto (x(t), y(t))$ is a solution of the Euler–Lagrange equations* (6.69), *then $\mathcal{H}$ is a conserved quantity.*

Proof. We have to show that

$$\mathcal{H}(x(t), y(t), \dot{x}(t), \dot{y}(t)) = \text{const} \qquad \text{for all} \quad t \in \mathbb{R}.$$

In fact, by the chain rule,

$$\begin{aligned}\frac{d}{dt}\mathcal{H} &= \ddot{x}L_{\dot{x}} + \dot{x}\frac{d}{dt}L_{\dot{x}} + \ddot{y}L_{\dot{y}} + \dot{y}\frac{d}{dt}L_{\dot{y}} - L_x\dot{x} - L_{\dot{x}}\ddot{x} - L_y\dot{y} - L_{\dot{y}}\ddot{y} \\ &= \dot{x}\left(\frac{d}{dt}L_{\dot{x}} - L_x\right) + \dot{y}\left(\frac{d}{dt}L_{\dot{y}} - L_y\right) \equiv 0\end{aligned}$$

because of the Euler–Lagrange equations. □

Next we want to emphasize the distinction between the regular case and the singular case. The latter is a paradigm for gauge theories.

The regular case. Let $\alpha > 0$. We introduce the generalized momenta p and r given by

$$p := L_{\dot{x}}(x, y, \dot{x}, \dot{y}), \quad r := L_{\dot{y}}(x, y, \dot{x}, \dot{y}).$$

Explicitly,

$$\boxed{p = \dot{x} + y, \quad r = \alpha^2\dot{y}.} \tag{6.74}$$

The bijective map $(x, y, \dot{x}, \dot{y}) \mapsto (x, y, p, r)$ from $\mathbb{R}^4$ onto $\mathbb{R}^4$ is called the Legendre transformation. The inverse Legendre transformation reads as

$$\dot{x} = p - y, \quad \dot{y} = \frac{r}{\alpha^2}.$$

In terms of geometry, we use the following terminology:

- The tuples (x, y) form the position space $M = \mathbb{R}^2$.
- The tuples $(x, y, \dot{x}, \dot{y})$ form the state space which coincides with the tangent bundle TM of the position space M.
- The tuples (x, y, p, r) form the phase space $\mathbb{R}^4$.
- The triplets $(x, y, pdx + rdy)$ form the cotangent bundle T^*M of the position space M.

We have the linear isomorphisms $TM \simeq \mathbb{R}^4$ and $T^*M \simeq \mathbb{R}^4$ (phase space). The Legendre transformation maps diffeomorphically the state space TM onto the phase space T^*M. The function $\mathcal{H}$ passes over to the Hamiltonian

$$H(x, y, p, r) = \frac{(p - y)^2}{2} + \frac{r^2}{2\alpha^2} + \frac{(x - y)^2}{2}.$$

The Euler–Lagrange equations (6.73) are transformed into the so-called canonical equations

$$\dot{p} = -H_x, \quad \dot{r} = -H_y, \quad \dot{x} = H_p, \quad \dot{y} = H_r.$$

Explicitly,

$$\dot{p} = y - x, \quad \dot{r} = p + x - 2y, \quad \dot{x} = p - y, \quad \dot{y} = \frac{r}{\alpha^2}.$$

The singular case. Let $\alpha = 0$. In this case, we have the degenerate Lagrangian

$$L(x, y, \dot{x}, \dot{y}) := \tfrac{1}{2}\dot{x}^2 + \dot{x}y - \tfrac{1}{2}(x - y)^2$$

with the degenerate quadratic form $\frac{1}{2}\dot{x}^2$. The Euler–Lagrange equations read as

$$\ddot{x} + \dot{y} + x - y = 0, \quad \dot{x} - y + x = 0 \quad \text{on } [t_0, t_1]. \tag{6.75}$$

The function $\mathcal{H}$ looks like

$$\mathcal{H}(x, y, \dot{x}) = \dot{x}L_{\dot{x}} - L = \dot{x}(\dot{x} + y) - \tfrac{1}{2}\dot{x}^2 - \dot{x}y + \tfrac{1}{2}(x-y)^2 = \tfrac{1}{2}\dot{x}^2 + \tfrac{1}{2}(x-y)^2.$$

For the generalized momenta, we get

$$p := L_{\dot{x}}(x, y, \dot{x}), \qquad r = L_{\dot{y}}(x, y, \dot{x}).$$

Explicitly,

$$\boxed{p = \dot{x} + y, \qquad r = 0.} \tag{6.76}$$

In this case, the Legendre transformation (6.74) breaks down. In fact, the Legendre transformation (6.74) cannot be inverted, since the generalized momentum r vanishes. The equation $r = 0$ is called a constraint. The vanishing of generalized momenta is typical for gauge theories.

In order to overcome this difficulty, we first construct the Hamiltonian

$$H(x, y, p) = \frac{(p-y)^2}{2} + \frac{(x-y)^2}{2}.$$

Motivated by the constraint $r = 0$, we pass to the extended Hamiltonian

$$\boxed{H^+(x, y, p) = \frac{(p-y)^2}{2} + \frac{(x-y)^2}{2} + \lambda r}$$

where the Lagrangian multiplier λ is a real number. Using the function H^+, we get the following canonical equations:

$$\dot{p} = -H_x^+, \quad \dot{r} = -H_y^+, \quad \dot{x} = H_p^+, \qquad \dot{y} = H_r^+.$$

Explicitly, setting $\lambda = v$ we obtain

$$\dot{p} = y - x, \quad \dot{r} = p + x - 2y, \quad \dot{x} = p - y, \quad \dot{y} = v$$

with the constraint $r = 0$. From this constraint together with the equation of motion $\dot{r} = p + x - 2y$, we get the additional constraint

$$p + x - 2y = 0.$$

Summarizing, we obtain the following modified canonical equations:

$$\boxed{\dot{x} = p - y, \quad \dot{y} = v, \quad \dot{p} = y - x, \quad p + x - 2y = 0.} \tag{6.77}$$

Here, v is an arbitrary real parameter. From the physical point of view, the Lagrangian multiplier $\lambda = v$ is the constant velocity of the trajectory $y = y_0 + vt, t \in \mathbb{R}$.

Proposition 6.13 *The modified canonical equations* (6.77) *are equivalent to the Euler–Lagrange equations* (6.75).

Proof. (I) Let x, y, p be a solution of (6.77). If $\dot{x} = p - y$, then

$$\ddot{x} = \dot{p} - \dot{y} = y - x - \dot{y}.$$

This is the first Euler–Lagrange equation from (6.75). Moreover, it follows from $\dot{x} = p - y$ that $\dot{x} = 2y - x - y = x - y$. This is the second Euler–Lagrange equation.

(II) Conversely, if x, y is a solution of the Euler–Lagrange equations (6.75), then

$$x = x_0 + vt, \qquad y_0 + vt, \quad t \in \mathbb{R}, \quad x_0 - y_0 = v,$$

by Prop. 6.11 on page 408. Setting $p(t) := 2y(t) - x(t)$, we get a solution of (6.77), by explicit computation. □

Dirac's classification of Hamiltonian constraints. The constraint

$$r = 0 \tag{6.78}$$

is called a static (or primary) constraint, whereas the constraint

$$p + x - 2y = 0 \tag{6.79}$$

is called a dynamic (or secondary) constraint. Observe that the static constraint (6.78) is independent of the equations of motion. In contrast to this, the dynamic constraint (6.79) depends on the equations of motion and the static constraint (6.78).

6.6.7 Global and Local Symmetries

> Symmetries of the action functional lead to symmetries of the Euler–Lagrange equations. In particular, invariance of the action functional under time translations is responsible for the conservation of energy.
> Degeneracy of the second variation generates local symmetries also called gauge symmetries. The use of symmetries is basic for modern physics.
>
> Folklore

One has to distinguish between global and local symmetries. A global symmetry corresponds to a Lie group with a finite number of parameters. Local symmetries depend on functions of space and time variables. Let us discuss typical features by considering the Lagrangian

$$L(x, y, \dot{x}, \dot{y}) = \tfrac{1}{2}\dot{x}^2 + \dot{x}y + \tfrac{1}{2}(x - y)^2. \tag{6.80}$$

This Lagrangian is a degenerate quadratic function with respect to $(\dot{x}, \dot{y})$. The Euler–Lagrange equations $\frac{d}{dt}L_{\dot{x}} = L_x$, $\frac{d}{dt}L_{\dot{y}} = L_y$ read as

$$\ddot{x} = x - y - \dot{y}, \qquad \dot{x} = x - y \qquad \text{on } \mathbb{R}.$$

Equivalently,

$$\boxed{\dot{x} = x - y, \qquad \ddot{x} = \dot{x} - \dot{y} \quad \text{on } \mathbb{R}.} \tag{6.81}$$

(i) Global symmetry (conservation of energy): The Euler–Lagrange equations (6.81) are invariant under the transformation

$$\boxed{x^+ = x, \quad y^+ = y, \quad t^+ = t + \delta t} \tag{6.82}$$

where δt is a fixed real number. This time translation sends every solution

$$t \mapsto (x(t), y(t))$$

of (6.81) to the solution $t \mapsto (x(t+\delta t), y(t+\delta t))$ of (6.81). This is a consequence of the fact that the Lagrangian L from (6.80) does not explicitly depend on time t. In addition, this symmetry property of the Lagrangian implies the conservation law

$$\frac{d}{dt}\mathcal{H}(x(t), y(t), \dot{x}(t)) = 0 \qquad \text{for all} \quad t \in \mathbb{R} \tag{6.83}$$

where $\mathcal{H} := \dot{x}L_{\dot{x}} - L$. This follows as in the proof of Prop. 6.12 on page 408. Explicitly,

$$\mathcal{H}(x, y, \dot{x}) = \tfrac{1}{2}\dot{x}^2 + \tfrac{1}{2}(x-y)^2.$$

In physics, this is called the energy to the Lagrangian L.

(ii) Local symmetry (gauge symmetry): Note that the two Euler–Lagrange equations (6.81) are not independent equations. The second equation is obtained from the first equation by differentiation. Consider the transformation

$$\boxed{x^+ = x + \delta x(t), \quad y^+ = y + \delta y(t), \quad t^+ = t} \tag{6.84}$$

where $\delta x : \mathbb{R} \to \mathbb{R}$ is a fixed smooth function, and the function δy is given by the constraint

$$\delta y(t) = \delta x(t) - \frac{d}{dt}(\delta x(t)), \qquad t \in \mathbb{R}.$$

One checks easily that if $t \mapsto (x(t), y(t))$ is a solution of (6.81), then so is

$$t \mapsto (x(t+\delta t), y(t+\delta t)).$$

In other words, the Euler–Lagrange equations are invariant under the symmetry transformation (6.84). This is called a local symmetry transformation, since δx and δy are not constants, but they depend on time t.

Next let us investigate the action functional $S = \int_{t_0}^{t_1} L dt$. Explicitly,

$$S(x,y) := \int_{t_0}^{t_1} \left[\tfrac{1}{2}\dot{x}^2 + \dot{x}y + \tfrac{1}{2}(x-y)^2\right](t)\, dt \qquad \text{for all} \quad x, y \in C^\infty[t_0, t_1].$$

We want to show that the transformations (6.82) and (6.84) describe symmetries of the action functional S.

Global symmetry of the action functional. We will use the following result.

Lemma 6.14 *If the continuous function $f : \mathbb{R} \to \mathbb{R}$ satisfies the condition*

$$\int_{t_0}^{t_1} f(t)dt = 0$$

for all finite intervals $[t_0, t_1]$, then $f(t) = 0$ for all $t \in \mathbb{R}$.

Proof. $f(t_0) = \lim_{t_1 \to t_0} \frac{1}{t_1 - t_0} \int_{t_0}^{t_1} f(t)dt = 0.$ □

Fix $x, y \in C^\infty[t_0, t_1]$. For all real numbers ε, define

$$\chi(\varepsilon) := \int_{t_0-\varepsilon}^{t_1-\varepsilon} L(x(t+\varepsilon), y(t+\varepsilon), \dot{x}(t+\varepsilon), \dot{y}(t+\varepsilon))\, dt.$$

Since the Lagrangian L does not depend explicitly on time t, the substitution rule tells us that the function χ is a constant. Hence

$$\boxed{\chi'(0) = 0.}$$

The calculus rule

$$\frac{d}{d\varepsilon}\int_{g(\varepsilon)}^{f(\varepsilon)} F(t,\varepsilon)\,dt = \int_{g(\varepsilon)}^{f(\varepsilon)} F_\varepsilon(t,\varepsilon)\,dt + F(f(\varepsilon),\varepsilon)\cdot f'(\varepsilon) - F(g(\varepsilon),\varepsilon)\cdot g'(\varepsilon)$$

yields

$$0 = \chi'(0) = \int_{t_0}^{t_1} \left[L_x\dot{x} + L_y\dot{y} + L_{\dot{x}}\ddot{x} + L_{\dot{y}}\ddot{y}\right](P(t))\,dt - L(P(t_1)) + L(P(t_0)).$$

Here, we set $P(t) := (x(t), y(t), \dot{x}(t), \dot{y}(t))$. Using both the Leibniz rule

$$\frac{d}{dt}(\dot{x}L_{\dot{x}}) = \ddot{x}L_{\dot{x}} + \dot{x}\frac{d}{dt}L_{\dot{x}}$$

and the fundamental theorem of calculus,

$$\int_{t_0}^{t_1} \frac{d}{dt}L(P(t))\,dt = L(P(t_1)) - L(P(t_0)),$$

we obtain the integral identity

$$\int_{t_0}^{t_1} \left[\frac{d}{dt}(\dot{x}L_{\dot{x}} + \dot{y}L_{\dot{y}} - L) + \dot{x}\left(L_x - \frac{d}{dt}L_{\dot{x}}\right) + \dot{y}\left(L_y - \frac{d}{dt}L_{\dot{y}}\right)\right](P(t))\,dt = 0.$$

If x, y is a solution of the Euler–Lagrange equations, $\frac{d}{dt}L_{\dot{x}} = L_x$, $\frac{d}{dt}L_{\dot{y}} = L_y$, then

$$\int_{t_0}^{t_1} \left[\frac{d}{dt}(\dot{x}L_{\dot{x}} + \dot{y}L_{\dot{y}} - L)\right](P(t))\,dt = 0.$$

By Lemma 6.14, this implies the desired conservation law

$$\frac{d}{dt}\left(\dot{x}L_{\dot{x}} + \dot{y}L_{\dot{y}} - L\right) = 0 \qquad \text{for all} \quad t \in \mathbb{R}$$

which coincides with (6.83). This is a special case of the general Noether theorem on conservation laws (see Sect. 6. 6.2 of Vol. II). This theorem was obtained by Emmy Noether (1882–1935) in 1918. In Vol. IV we will show that the relativistic invariance of the action functional (i.e., the invariance under the Poincaré group) leads to conservation laws for the energy momentum tensor and the angular momentum tensor.

Local symmetry of the action functional. Fix $x, y, h, k \in C^\infty[t_0, t_1]$. Set $\delta x := \varepsilon h, \delta y := \varepsilon k$.

Proposition 6.15 *If $\delta k = \delta h - \frac{d}{dt}(\delta h)$ and $h(t_1) = h(t_0) = 0$, then*

$$S(x + \delta x, y + \delta y) = S(x, y).$$

In particular, if (x, y) is a minimal point of the action functional S, then so is $(x + \delta x, y + \delta y)$.

Proof. We want to study the function

$$\chi(\varepsilon) := S(x+\varepsilon h, y+\varepsilon k), \qquad \varepsilon \in \mathbb{R}.$$

This means that $\chi(\varepsilon) = \int_{t_0}^{t_1} L_\varepsilon dt$ with the perturbed Lagrangian

$$L_\varepsilon := \tfrac{1}{2}(\dot{x}+\varepsilon\dot{h})^2 + (\dot{x}+\varepsilon\dot{h})(y+\varepsilon k) + \tfrac{1}{2}(x+\varepsilon h - y - \varepsilon k)^2.$$

(I) First variation $\delta S = \varepsilon\chi'(0)$. Differentiation with respect to ε at the point $\varepsilon = 0$ yields

$$\begin{aligned} \frac{dL_\varepsilon}{d\varepsilon}_{|\varepsilon=0} &= \dot{x}\dot{h} + \dot{x}k + y\dot{h} + (x-y)(h-k) \\ &= (\dot{x}+y-x)(\dot{h}-h+k) + \frac{d}{dt}(xh) = \frac{d}{dt}(xh). \end{aligned}$$

Hence

$$\chi'(0) = \int_{t_0}^{t_1} \frac{dL_\varepsilon}{d\varepsilon}_{|\varepsilon=0} dt = x(t_1)h(t_1) - x(t_0)h(t_0) = 0.$$

(II) Second variation $\delta^2 S = \varepsilon^2\chi''(0)$. From the second derivative

$$\frac{d^2L_\varepsilon}{d\varepsilon^2}_{|\varepsilon=0} = \dot{h}^2 + 2\dot{h}k + (k-h)^2 = \dot{h}^2 + 2\dot{h}(h-\dot{h}) + \dot{h}^2 = \frac{d}{dt}(h^2)$$

we get

$$\chi''(0) = \int_{t_0}^{t_1} \frac{d^2L_\varepsilon}{d\varepsilon^2}_{|\varepsilon=0} dt = h(t_1)^2 - h(t_0)^2 = 0.$$

(III) Taylor expansion. The claim follows now from

$$\chi(\varepsilon) = \chi(0) + \varepsilon\chi'(0) + \tfrac{1}{2}\varepsilon^2\chi''(0) = \chi(0).$$

□

The argument of the proof is a special case of Jacobi's method of the accessory quadratic variational problem (see Sect. 6.5.3 of Vol. II). More general results can be found in H. Rothe and K. Rothe (2010), quoted on page 419.

6.6.8 Quantum Symmetries and Anomalies

Symmetry of the action functional and conservation laws. Classical field theories are based on the principle of critical action,

$$S[\psi] = \text{critical!}$$

In 1918, Emmy Noether established the general mathematical principle that

Symmetries of the action functional $\psi \mapsto S[\psi]$ are responsible for conservation laws.

For example, the invariance of S under time-translations leads to the conservation of energy (see Sect. 6.6 of Vol. II). In Einstein's theory of general relativity, the invariance under time-translations can be violated. Therefore, the notion of energy is a nontrivial concept in general relativity. The point is that, roughly speaking, the following hold:

- The energy of the gravitational field interacting with a system of masses and matter fields can be introduced if the space-time manifold is asymptotically flat, that is, it approaches the flat 4-dimensional Minkowski space-time manifold $\mathbb{M}^4$ at spatial infinity.
- The energy of the gravitational field is not localized, that is, a uniquely defined energy density does not exist.
- But it is possible to show that the total energy of the gravitational field is non-negative, and it vanishes only in the absence of matter and gravitational waves. In this special case, the metric of the space-time manifold coincides with the flat metric of the Minkowski space-time manifold $\mathbb{M}^4$. This is the main content of the famous positive energy theorem in general relativity due to Schoen and Yau.

We will study this in Vol. IV.[13] In Einstein's theory of special relativity, the action functional S is invariant under the Poincaré group. This leads to the existence of the energy–momentum tensor and the angular momentum tensor. The general theory will be considered in Vol. IV on quantum mathematics.[14] The prototype is the electromagnetic field. The energy-momentum tensor of the electromagnetic field describes the conservation of energy and momentum including the production of heat (the Joule law) and the light pressure (see Sect. 19.6.3 on page 972).

The generalized Ward identities for the correlation functions of a quantum field. If one wants to quantize a classic field theory, then physicists frequently use the formal, but extremely elegant Feynman functional integral

$$Z(J) = \mathcal{N} \int \mathrm{e}^{\mathrm{i}S[\psi]/\hbar}\mathrm{e}^{\mathrm{i}J[\psi]/\hbar}\, \mathcal{D}\psi \tag{6.85}$$

where $\psi \mapsto S[\psi]$ is the action functional, and $\psi \mapsto J[\psi]$ is the so-called source functional. The explicit form of the normalization factor $\mathcal{N}$ follows from the normalization condition $Z(0) = 1$. We refer to Sect. 13.6 of Vol. I. In particular, we have discussed in Vol. I that the functional Z generalizes the partition function in statistical physics. The key point is that functional differentiation with respect to the source functional J allows us to compute the correlation functions of the quantum field ψ. The correlation functions are also called the Green functions of the quantum field ψ.

The functional integral (6.85) *contains all the information about the quantum field, at least on a formal level.*

One has to distinguish between the following two cases:

(i) Regular case: The functional integral (6.85) is invariant under all the symmetries of the corresponding classical field theory.
(ii) Singular case (anomaly): The functional integral (6.85) is not invariant under all of the symmetries of the classical field theory.

[13] At this point, we refer to L. Faddeev, The energy problem in Einstein's theory of gravitation, Uspekhi Fiz. Nauk **136** (1982), 435–457. This survey article contains Witten's elegant proof of the positive energy theorem by using the Dirac operator.
R. Schoen and S. Yau, On the proof of the positive mass conjecture in general relativity, Commun. Math. Phys. **65** (1979), 45–76; **79** (1981), 231–260.
E. Witten, A new proof of the positive energy theorem, Commun. Math. Phys. **80**, 381–396.

[14] We recommend M. Forger and H. Römer, Currents and the energy-momentum tensor in classical field theory: a fresh look at an old problem, Annals of Physics **309** (2004), 306-389.

In case (i), the symmetries of the functional integral (6.85) induce relations for the correlation functions which play a fundamental role for the renormalization of the quantum field theory. In 1950, Ward discovered such identities in quantum electrodynamics. In general gauge field theory, the relations between the correlation functions are called the Ward–Takahashi identities (commutative gauge group) and Slavnov–Taylor identities (non-commutative gauge group). There is also a close relation to the BRTS-symmetry. We will study this in Vol. IV on quantum mathematics. At this point, we refer to:

J. Jost, Geometry and Physics, Springer, Berlin, 2009,

and

L. Ryder, Quantum Field Theory, Cambridge University Press, 1999.

A. Das, Lectures on Quantum Field Theory, World Scientific, Singapore, 2008.

L. Faddeev and A. Slavnov, Gauge Fields, Benjamin, Reading, Massachusetts, 1982.

M. Böhm, A. Denner, and H. Joos, Gauge Theories of the Strong and Electroweak Interaction, Teubner, Stuttgart, 2001.

T. Kugo, Gauge Field Theory, Springer, Berlin, 1997 (translated from Japanese into German).

O. Piquet and S. Sorella, Algebraic Renormalization, Springer, Berlin, 1995.

Anomalies. Consider the decay

$$\pi^0 \to \gamma + \gamma \tag{6.86}$$

of a pion into two photons. Physicists noticed that there is a substantial discrepancy between the usual computations based on Feynman diagrams and the values measured in experiments.

In the late 1960s, it was discovered that the passage from the classical field theory to the corresponding quantum field theory may destroy classical symmetries.

This shocking phenomenon is described by the sketch word 'anomaly'.[15] In terms of physics, this means that quantum fluctuations of the quantum field destroy some classical symmetries such that classical conservation laws are violated on the quantum level. In particular, the so-called chiral anomaly essentially contributes to the process (6.86). We will study this in Vol. V on the physics of the Standard Model. As an introduction, we recommend:

A. Zee, Quantum Field Theory in a Nutshell, Princeton University Press, 2003 (Chap. IV.7 on the chiral anomaly).

S. Weinberg, Quantum Field Theory, Vol. 2, Chap. 22, Cambridge University Press, 1996.

P. Langacker, The Standard Model and Beyond, CRC Press, Boca Raton, Florida, 2010.

Much additional material can be found in:

---

[15] S. Adler, Axial vector vertex in spinor electrodynamics, Phys. Rev. **177** (1969), 2426–2438.
S. Adler and W. Bardeen, Absence of higher order corrections in the anomalous axial vector divergence equation, Phys. Rev. **182** (1969), 1517–1536.

K. Fujikawa and H. Suzuki, Path Integrals and Quantum Anomalies, Oxford University Press, 2004.

S. Adler, Adventures in Theoretical Physics, Selected Papers with Commentaries, World Scientific, Singapore, 2006.

6.7 Perspectives

6.7.1 Topological Constraints in Maxwell's Theory of Electromagnetism

We will show in Chap. 23 that Maxwell's theory of the electromagnetic field is dominated by constraints. The point is that the structure of the electromagnetic field $\mathbf{E}, \mathbf{B}$ on an open subset $\mathcal{O}$ of the Euclidean manifold $\mathbb{E}^3$ essentially depends on the topological structure of $\mathcal{O}$. In particular, the Betti numbers of $\mathcal{O}$ play a crucial role. It turns out that two cornerstones of modern topology, namely,

- the de Rham cohomology,
- and the Hodge theory about harmonic differential forms on Riemannian and pseudo-Riemannian manifolds,

represent far-reaching generalizations of properties of the electromagnetic field. Since gauge theories generalize the Maxwell theory of electromagnetism, the topology of gauge theories has its roots in the topological properties of the electromagnetic field.

6.7.2 Constraints in Einstein's Theory of General Relativity

We will show in Vol. IV that the initial-value problem for the Einstein equations in general relativity is only solvable if the initial values at time $t = 0$ satisfy appropriate constraints. We refer to:

A. Rendall, Partial Differential Equations in General Relativity, Oxford University Press, 2008.

Y. Choquet–Bruhat, General Relativity and the Einstein Equations, Oxford University Press, Oxford, 2008.

B. Schmidt (Ed.), Einstein's Field Equations and their Physical Implications, Springer, Berlin, 2000.

M. Kriele, Space-Time: Foundations of General Relativity and Differential Geometry, Springer, Berlin, 2000.

J. Baez and J. Muniain, Gauge Fields, Knots and Gravity, World Scientific, Singapore, 1994 (Ashtekar's new variables).

T. Thiemann, Modern Canonical Quantum General Relativity, Cambridge University Press, 2007 (the Ashtekar program).

6.7.3 Hilbert's Algebraic Theory of Relations (Syzygies)

In order to study the algebraic structure of invariants, Hilbert wrote fundamental papers on commutative algebra in about 1890. In a general setting, Hilbert proved that a large class of polynomial rings of invariants (with respect to matrix transformation groups) can be generated by a finite number of basic invariants. This was a revolutionary result which strongly influenced the development of algebraic geometry. In this connection, Hilbert studied relations between invariants. His idea was to pass from relations to 'relations of relations', and so on.[16] Hilbert showed that in important cases this procedure leads to trivial relations after a finite number of steps (finite resolution of the relations). Generalizations of such ideas are also used in gauge theory. We recommend:

M. Henneaux and C. Teitelboim, Quantization of Gauge Systems, Princeton University Press, 1993.

Hilbert did not use constructive methods, but he gave abstract existence proofs. Paul Gordan (1837–1912) said: "This is not mathematics, but theology." Constructive methods can be found in:

D. Cox, J. Little, and D. O'Shea, Using Algebraic Geometry, Springer, New York, 1998 (theory and computer algorithms).

B. Sturmfels, Algorithms in Invariant Theory, Springer, New York, 1993.

In order to compute scattering processes in particle accelerators, one uses automated loop-computations based on the theory of renormalization. Here, the Laporta algorithm is close to the Buchberger algorithm in algebraic geometry (see the hints for further reading on page 978 of Vol. II). Furthermore, we refer to:

D. Hilbert, Theory of Algebraic Invariants, Cambridge University Press, 1993.

D. Hilbert, Hilbert's Invariant Theory Papers, Lie Groups: History, Frontiers and Applications, VIII, Math. Sci. Press, Brooklyn, Massachusetts, 1978.

H. Weyl, The Classical Groups; Their Invariants and Representations, Princeton University Press, 1946.

S. Abhyankar, Algebraic Geometry for Scientists and Engineers, Amer. Math. Soc., Providence, Rhode Island, 1990.

D. Eisenbud, Commutative Algebra with a View to Algebraic Geometry, Springer, New York, 1994.

D. Eisenbud, The Geometry of Syzygies: A Second Course in Commutative Algebra and Algebraic Geometry, Springer, New York, 2005.

S. Lang, Algebra, Chap. 21 (finite free resolutions), Springer, New York, 2002.

6.8 Further Reading

Classic references:

H. Grassmann, The Calculus of Extension, 1844 (in German). Reprint: Chelsea Publ. Company, 1969.

S. Lie and F. Engel, Theory of Transformation Groups (in German), Vols. 1–3, Teubner, Leipzig, 1888. Reprint: Chelsea Publ. Company, 1970.

[16] Relations are also called *syzygies*. The Greek word $\sigma\upsilon\zeta\upsilon\gamma\iota\alpha$ means *yoke*. In ancient astronomy, syzygies described special relations between heavenly bodies, namely, conjunction and opposition. In mathematics, syzygies were introduced by Sylvester (1814–1897) in 1853.

É. Cartan, On certain differential expressions and Pfaff's problem (in French), Annales École Normale Supérieure **16** (1899), 239–332.

E. Noether, Invariant variational problems (in German), Göttinger Nachrichten, Math.-phys. Klasse 1918, 235–257.

F. Fasso and N. Sansonetto, Elemental overview of the non-holonomic Noether theorem, Int. J. of Geometric Methods in Modern Physics, **68** (2009), 1343–1355.

C. Taylor (Ed.), Gauge Theories in the Twentieth Century, World Scientific, Singapore, 2001 (collection of fundamental papers).

Constraints and the quantization of classic theories:

P. Dirac, Lectures on Quantum Mechanics, Belfer Graduate School of Science, Yeshiva University, New York. 1964 (quantization of classical field theories with constraints).

L. Faddeev and A. Slavnov, Gauge Fields, Benjamin, Reading, Massachusetts, 1980 (the Feynman path integral approach, the factorization of gauge orbits, and the Faddeev–Popov ghosts).

M. Henneaux and C. Teitelboim, Quantization of Gauge Systems, Princeton University Press, 1993 (comprehensive monograph: the classical Dirac theory, the BRST-approach and cohomology, the field-anti-field approach, path integrals).

H. Rothe and K. Rothe, Classical and Quantum Dynamics of Constrained Hamiltonian Systems, World Scientific Lecture Notes in Physics, World Scientific, Singapore, 2010 (Dirac's classification of constraints, algorithms for computing local symmetries of Lagrangians, construction of constrained Hamiltonians, BRST-approach, the field-antifield formalism due to Batalin and Vilkovisky).

M. Böhm, A. Denner, and H. Joos, Gauge Theories of the Strong and Electroweak Interaction, Teubner, Stuttgart, 2001 (comprehensive presentation including the Standard Model in particle physics).

O. Piguet and S. Sorella, Algebraic Renormalization, Springer, Berlin, 1995 (BRST-approach and renormalization).

T. Kugo, Eichfeldtheorie (Gauge Field Theory), Springer, Berlin, 1997 (translated from Japanese into German) (comprehensive presentation of gauge theory).

L. Ryder, Quantum Field Theory, Cambridge University Press, 1999 (introduction to the BRST-approach).

A. Das, Lectures on Quantum Field Theory, World Scientific, Singapore, 2008 (introduction to the BRST-approach).

C. Becchi, Lectures on the Renormalization of Gauge Theories. In: Relativity, Groups, and Topology II, pp. 787–821, Les Houches, 1983. Edited by B. DeWitt and R. Stora, Elsevier, Amsterdam, 1984 (BRST-approach).

M. Henneaux, Lectures on the antifield-BRST formalism for gauge theories, Nucl. Phys. B (Proc. Suppl.) **18A**, 47–105.

J. Zinn-Justin, Quantum Field Theory and Critical Phenomena, 4th edition, Clarendon Press, Oxford, 2003 (extensive presentation of about 1000 pages based on the path-integral technique including many specific models).

A. Connes, K. Gawędzki, and J. Zinn-Justin (Eds.), Quantum Symmetries, Les Houches 1995, North-Holland, Amsterdam, 1998 (noncommutative geometry, strings and duality, integrable models, supersymmetry and super algebras, quantum groups, loop spaces, conformal field theory, Poisson algebras).

Problems

6.1 *Lagrange's multiplier rule.* Consider the matrix

$$A = \begin{pmatrix} a_1 & a_2 & a_3 \\ b_1 & b_2 & b_3 \\ c_1 & c_2 & c_3 \end{pmatrix}$$

with real numbers $a_1, a_2, \ldots$ as entries. Suppose that every real solution x, y, z of the system

$$\begin{aligned} a_1 x + a_2 y + a_3 z &= 0, \\ b_1 x + b_2 y + b_3 z &= 0 \end{aligned} \tag{6.87}$$

is also a solution of the equation

$$c_1 x + c_2 y + c_3 z = 0. \tag{6.88}$$

Moreover, suppose that the rank of the coefficient matrix of (6.87) is maximal. Show that then the third row of the matrix A is linearly dependent of the first and second row of A. In other words, there are real numbers λ and μ such that

$$c_j = \lambda a_j + \mu b_j, \qquad j = 1, 2, 3.$$

Solution: The rank of the coefficient matrix of (6.87) is equal to 2. Thus, for the rank r of the matrix A, we have $r = 2$ or $r = 3$.
Case 1: $r = 2$. Then the third row of A is linearly dependent of the first and second row of A, and we are done.
Case 2: $r = 3$. We show that this is impossible. Indeed, if $r = 3$, then the system (6.87), (6.88) has only the trivial solution $x = y = z = 0$. By our assumption, the system (6.87) has a 1-dimensional solution space, and hence the equation (6.88) has at least a 1-dimensional solution space, a contradiction.

6.2 *Infinitesimal motions and virtual velocities of the rigid body.* Fix $n = 3, 4, \ldots$ Let $\mathbf{x}_1, \ldots, \mathbf{x}_n$ be nonzero vectors in the Euclidean space E_3 which have nonzero distances and which span the space E_3. Show that the vectors $\mathbf{v}_1, \ldots, \mathbf{v}_n \in E_3$ satisfy the equations

$$(\mathbf{x}_i - \mathbf{x}_j)(\mathbf{v}_i - \mathbf{v}_j) = 0, \qquad i, j = 1, \ldots, n, \ i \neq j \tag{6.89}$$

iff there exist two vectors $\mathbf{a}, \boldsymbol{\omega} \in E_3$ such that

$$\mathbf{v}_j = \mathbf{a} + \boldsymbol{\omega} \times \mathbf{x}_j, \qquad j = 1, \ldots, n. \tag{6.90}$$

This is the superposition of a translation with an infinitesimal rotation.
Hint: It follows from (6.90) that

$$(\mathbf{x}_i - \mathbf{x}_j)(\mathbf{v}_i - \mathbf{v}_j) = (\mathbf{x}_i - \mathbf{x}_j)\big(\boldsymbol{\omega} \times (\mathbf{x}_i - \mathbf{x}_j)\big) = 0.$$

Conversely, every solution of (6.89) looks like (6.90). To show this, argue by induction with respect to the number of points $n = 3, 4, \ldots$ See Zeidler (1995), page 90, quoted on page 396.

6.3 *Continuous rigid body.* Formulate the Euler equations

$$m\ddot{\mathbf{y}} = \mathbf{F}, \qquad \dot{\mathbf{A}} = \mathbf{T}, \qquad \mathbf{A} = \Theta\boldsymbol{\omega} \tag{6.91}$$

for the motion of a continuous rigid body with continuous mass density μ. Use the Euler equations for a finite number of mass points studied on page 393.

Solution: We have to pass from a finite number of mass points to a continuous mass distribution. To this end, we have to reformulate the following definitions:

- m (total mass),
- $\mathbf{y} = \overrightarrow{P_0C}$ (position vector of the center C of gravity; $P_0 := (0,0,0)$),
- Θ (tensor of inertia), and
- $\mathbf{A}$ (total angular momentum vector).

Consider a right-handed Cartesian (x,y,z)-coordinate system with the right-handed orthonormal basis vectors $\mathbf{i}, \mathbf{j}, \mathbf{k}$. Set $\mathbf{e}_1 := \mathbf{i}, \mathbf{e}_2 := \mathbf{j}, \mathbf{e}_3 := \mathbf{k}$. Divide the continuous body into small pieces of volume $\Delta x \Delta y \Delta z$. Apply the method from Sect. 6.5 on 388 to the discretized rigid body. Finally, carry out the limit $\Delta x \Delta y \Delta z \to 0$. Then we get the following:

- $m = \int_{\mathcal{B}} \mu(x,y,z)\, dxdydz$ (total mass),
- $\mathbf{y} = \frac{1}{m} \int_{\mathcal{B}} \varrho(x,y,z)(x\mathbf{i} + y\mathbf{j} + z\mathbf{k})\, dxdydz$ (position vector $\overrightarrow{P_0C}$ of the center C of gravity),
- $\theta_{33} = \int_{\mathcal{B}} \varrho(x,y,z)\,(x^2 + y^2)\, dxdydz$ (moment of inertia with respect to the z-axis),
- $\theta_{12} = -\int_{\mathcal{B}} \varrho(x,y,z) \cdot xy\, dxdydz$.
- Similarly, we define θ_{22}, θ_{33} (resp. θ_{ij} if $i \neq j$) by using the cyclic permutations $1 \mapsto 2 \mapsto 3 \mapsto 1$ and $x \mapsto y \mapsto z \mapsto x$. This yields the symmetric matrix of the moments of inertia

$$(\theta_{ij}) := \begin{pmatrix} \theta_{11} & \theta_{12} & \theta_{13} \\ \theta_{21} & \theta_{22} & \theta_{23} \\ \theta_{31} & \theta_{32} & \theta_{33} \end{pmatrix}$$

where $\theta_{ij} = \theta_{ji}$ for all indices. The eigenvalues $\lambda_1, \lambda_2, \lambda_3$ of this matrix are called the principal moments of inertia of the rigid body with respect to the (x,y,z)-coordinate system. In particular, if (θ_{ij}) is a diagonal matrix,

$$(\theta_{ij}) = \begin{pmatrix} \theta_{11} & 0 & 0 \\ 0 & \theta_{22} & 0 \\ 0 & 0 & \theta_{33} \end{pmatrix},$$

then we get the principal moments $\lambda_j = \theta_{jj}, j = 1,2,3$. Moreover, the three coordinate axes (i.e., the x-axis, the y-axis, and the z-axis) are principal axes of inertia. We get:

- $\Theta = \sum_{i,j=1}^{3} \theta_{ij} \mathbf{e}_i \otimes \mathbf{e}_j$ (tensor of inertia);
- $\mathbf{A} = \Theta\boldsymbol{\omega}$ (total angular momentum vector),
- $\Theta\left(\sum_{k=1}^{3} \omega_k \mathbf{e}_k\right) := \sum_{i=1}^{3} \left(\sum_{j=1}^{3} \theta_{ij}\omega_j\right) \mathbf{e}_i$.

For computing Θ, we use $(\mathbf{e}_i \otimes \mathbf{e}_j)\mathbf{e}_k := \mathbf{e}_i(\mathbf{e}_j\mathbf{e}_k) = \mathbf{e}_i\delta_{jk}$.

6.4 *The moments of inertia for the ball and the circular cylinder.* Choose a right-handed Cartesian (x, y, z)-system. Consider the following compact subsets of the space $\mathbb{R}^3$:

- $\mathcal{B} := \{(x, y, z) \in \mathbb{R}^3 : x^2 + y^2 + z^2 \le R^2\}$ (ball of radius R), and
- $\mathcal{B} := \{(x, y, z) \in \mathbb{R}^3 : x^2 + y^2 \le R^2, \ -\frac{H}{2} \le z \le \frac{H}{2}\}$ (circular cylinder of radius R and height H).

For constant mass density μ, compute the following quantities: total mass m, center of gravity, principal axes of inertia, and the principal moments of inertia.
Solution: (I) Ball: Using spherical coordinates, we get the total mass

$$m = \int_{r=0}^{R} \int_{\vartheta=-\frac{\pi}{2}}^{\frac{\pi}{2}} \int_{\varphi=-\pi}^{\pi} \mu r^2 \cos\vartheta \, d\vartheta \, d\varphi \, dr = \mu \cdot \frac{4\pi R^3}{3}.$$

Furthermore, we obtain the following:

- $(0, 0, 0)$ (center of gravity),
- $\theta_{11} = \theta_{22} = \theta_{33} = \frac{2}{5} m R^2$ (principal moment of inertia);
- $\theta_{ij} = 0$ if $i \ne j$;
- every axis through the origin is an axis of inertia, by symmetry.

Note the following trick: the sum $3\theta_{11} = \theta_{11} + \theta_{22} + \theta_{33}$ is equal to

$$2\mu \int_{\mathcal{B}} (x^2 + y^2 + z^2) \, dx dy dz = 2\mu \int_{\mathcal{B}} r^4 \cos\vartheta \, d\vartheta \, d\varphi \, dr = \mu \cdot \frac{8\pi R^5}{5}.$$

(II) Circular cylinder: Using cylindrical coordinates, we get the total mass

$$m = \int_{\varphi=-\pi}^{\pi} \int_{\varrho=0}^{R} \int_{z=-H/2}^{H/2} \mu \varrho \, d\varphi \, d\varrho \, dz = \mu \cdot \pi R^2 H,$$

and the moment of inertia with respect to the z-axis:

$$\theta_{33} = \int_{\varphi=-\pi}^{\pi} \int_{\varrho=0}^{R} \int_{z=-H/2}^{H/2} \mu \varrho^2 \cdot \varrho \, d\varphi \, d\varrho \, dz = \tfrac{1}{2} m R^2.$$

Furthermore, we obtain the following:

- $(0, 0, 0)$ (center of gravity),
- $\theta_{11} = \theta_{22} = \frac{1}{4} m R^2 + \frac{1}{12} m H^2, \theta_{33} = \frac{1}{2} m R^2$ (principal moments of inertia);
- $\theta_{ij} = 0$ if $i \ne j$;
- the x-axis, the y-axis, and the z-axis are principal axes of inertia. In addition, by symmetry, every axis lying in the (x, y)-plane and passing through the origin is a principal axis of inertia of the circular cylinder.

6.5 *The three Euler angles.* Euler found a unique parametrization of the group $SO(3)$ by three angles α, β, γ. To explain this, set

$$R_3(\gamma) := \begin{pmatrix} \cos\gamma & -\sin\gamma & 0 \\ \sin\gamma & \cos\gamma & 0 \\ 0 & 0 & 1 \end{pmatrix}, \quad R_1(\beta) = \begin{pmatrix} 1 & 0 & 0 \\ 0 & \cos\beta & -\sin\beta \\ 0 & \sin\beta & \cos\beta \end{pmatrix}.$$

Show that every matrix $B \in SO(3)$ can be uniquely represented by the matrix product

$$B = R_3(\alpha) R_1(\beta) R_3(\gamma), \qquad -\pi < \alpha, \gamma \le \pi, \quad -\frac{\pi}{2} \le \beta \le \frac{\pi}{2}.$$

Conversely, every such product yields a matrix B in $SO(3)$. In other words, the map

$$(\alpha, \beta, \gamma) \mapsto R_3(\alpha) R_1(\beta) R_3(\gamma)$$

is a bijective map $]-\pi, \pi] \times [-\frac{\pi}{2}, \frac{\pi}{2}] \times]-\pi, \pi] \to SO(3)$. This claim tells us that every rotation of the Euclidean manifold $\mathbb{E}^3$ about the origin can be uniquely obtained as

- a rotation about the z-axis with the angle $\gamma \in]-\pi, \pi]$,
- a rotation about the x-axis with the angle $\beta \in [-\frac{\pi}{2}, \frac{\pi}{2}]$, and
- a rotation about the z-axis with the angle $\alpha \in]-\pi, \pi]$.

Hint: See W. Hein, Introduction to the Structure Theory and the Representation Theory of the Classical Groups, p. 57, Springer, Berlin, 1990 (in German).

7. Rotations, Quaternions, the Universal Covering Group, and the Electron Spin

The human eyes are able to observe the rotational symmetry in nature. Mathematics – the cosmic eye of human beings – is able to see the universal covering group $SU(2)$ of the rotational group $SO(3)$ via the quaternions introduced by Hamilton (1805–1865). Nature also sees this universal covering group via the electron spin.
Lie's notion of infinitesimal rotations connects completely different phenomena in physics like the lever principle due to Archimedes of Syracus (287–212 B.C.) together with the spinning top due to Euler (1701)–1783) (the classical angular momentum) and the 1927 approach to the electron spin due to Pauli (1900–1958) (the quantized angular momentum). From the mathematical point of view, one has to study the irreducible representations of the Lie algebra $su(2)$. On the infinitesimal level, we have $so(3) \simeq su(2)$, that is, the Lie algebra $so(3)$ of the rotation group $SO(3)$ is isomorphic to the Lie algebra $su(2)$ of the universal covering group $SU(2)$.
Dirac (1902–1984) discovered in 1928 that the electron spin is a mathematical consequence of combining Einstein's theory of special relativity with quantum mechanics.

Folklore

7.1 Quaternions and the Cayley–Hamilton Rotation Formula

Euler's rotation formula (6.6) can be elegantly written by using Hamilton's quaternions. This was discovered independently by Hamilton and Cayley in 1844, one year after Hamilton's discovery of quaternions. In the language of quaternions, Euler's rotation formula (6.6) reads elegantly as

$$\boxed{\mathbf{x}' = q \cdot \mathbf{x} \cdot q^\dagger, \qquad \mathbf{x} \in E_3.} \tag{7.1}$$

Here, the given quaternion

$$q := \cos\tfrac{\varphi}{2} + \sin\tfrac{\varphi}{2}\,\mathbf{n}$$

contains the information about the rotation angle φ and the rotation axis vector $\mathbf{n}$ of length one. In particular, for the norm of the quaternion q we get

$$|q| = \sqrt{\cos^2\tfrac{\varphi}{2} + \mathbf{n}^2 \sin^2\tfrac{\varphi}{2}} = 1.$$

Hence $q \in U(1, \mathbb{H})$.

E. Zeidler, *Quantum Field Theory III: Gauge Theory*,
DOI 10.1007/978-3-642-22421-8_8,

Proof. Recall the quaternionic product

$$(\mathbf{x}+t)\cdot(\mathbf{y}+s) := \mathbf{x}\times\mathbf{y} - \mathbf{x}\mathbf{y} + ts + t\mathbf{y} + s\mathbf{x}$$

for all vectors $\mathbf{x},\mathbf{y}\in E_3$ and all real numbers t, s. Using the trigonometrical addition theorems

$$\cos\varphi = \cos^2\tfrac{\varphi}{2} - \sin^2\tfrac{\varphi}{2}, \qquad \sin\varphi = 2\sin\tfrac{\varphi}{2}\cos\tfrac{\varphi}{2}, \quad \cos^2\tfrac{\varphi}{2} + \sin^2\tfrac{\varphi}{2} = 1,$$

and the Grassmann expansion formula $(\mathbf{n}\times\mathbf{x})\times\mathbf{n} = \mathbf{n}^2\mathbf{x} - (\mathbf{x}\mathbf{n})\mathbf{n}$ along with $\mathbf{n}^2 = 1$, we get

- $\mathbf{x}\cdot\mathbf{n} = \mathbf{x}\times\mathbf{n} - \mathbf{x}\mathbf{n}$,
- $(\mathbf{n}\times\mathbf{x})\cdot\mathbf{n} = (\mathbf{n}\times\mathbf{x})\times\mathbf{n} - (\mathbf{n}\times\mathbf{x})\mathbf{n} = (\mathbf{n}\times\mathbf{x})\times\mathbf{n} = \mathbf{x} - (\mathbf{x}\mathbf{n})\mathbf{n}$.

Hence $q\cdot\mathbf{x} = (\cos\frac{\varphi}{2} + \mathbf{n}\sin\frac{\varphi}{2})\cdot\mathbf{x} = \mathbf{x}\cos\frac{\varphi}{2} + (\mathbf{n}\times\mathbf{x})\sin\frac{\varphi}{2} - (\mathbf{n}\mathbf{x})\sin\frac{\varphi}{2}$. This implies

$$\begin{aligned} q\cdot\mathbf{x}\cdot q^\dagger &= \left(\mathbf{x}\cos\tfrac{\varphi}{2} + (\mathbf{n}\times\mathbf{x})\sin\tfrac{\varphi}{2} - (\mathbf{n}\mathbf{x})\sin\tfrac{\varphi}{2}\right)\cdot\left(\cos\tfrac{\varphi}{2} - \mathbf{n}\sin\tfrac{\varphi}{2}\right) \\ &= \mathbf{x}(\cos^2\tfrac{\varphi}{2} - \sin^2\tfrac{\varphi}{2}) + 2(\mathbf{n}\times\mathbf{x})\sin\tfrac{\varphi}{2}\cos\tfrac{\varphi}{2} + 2\mathbf{n}(\mathbf{x}\mathbf{n})\sin^2\tfrac{\varphi}{2} \\ &= \mathbf{x}\cos\varphi + (\mathbf{n}\times\mathbf{x})\sin\varphi + \mathbf{n}(\mathbf{x}\mathbf{n})(1-\cos\varphi). \end{aligned}$$

This is the Euler rotation formula. □

7.2 The Universal Covering Group $SU(2)$

The Lie group $U(1,\mathbb{H})$. The Cayley–Hamilton formula (7.1) allows us to parametrize the Lie group $SU(E_3)$ by the quaternions $q\in U(1,\mathbb{H})$. The point is that this is not a one-to-one parametrization. Obviously, the quaternions q and $-q$ generate the same rotation of the Hilbert space E_3. More precisely, set $\chi(q)\mathbf{x} := q\cdot\mathbf{x}\cdot q^\dagger$ for all $\mathbf{x}\in E_3$.

Proposition 7.1 *The map $\chi: U(1,\mathbb{H})\to SU(E_3)$ is a surjective group morphism with the kernel $\chi^{-1}(I) = \{-1,1\}$.*

Proof. Let $q, r\in U(1,\mathbb{H})$. Then

$$\chi(q\cdot r)\mathbf{x} = (q\cdot r)\cdot\mathbf{x}\cdot(q\cdot r)^\dagger = q\cdot(r\cdot\mathbf{x}\cdot r^\dagger)\cdot q^\dagger = \chi(q)(\chi(r)\mathbf{x}).$$

Hence $\chi(q\cdot r) = \chi(q)\chi(r)$. Moreover, if $\chi(q) = I$, then

$$\chi(q)\mathbf{x} = \mathbf{x}\cos\varphi + (\mathbf{n}\times\mathbf{x})\sin\varphi + (1-\cos\varphi)(\mathbf{n}\mathbf{x})\mathbf{n} = \mathbf{x}$$

for all $\mathbf{x}\in E_3$. The identical rotation corresponds to the rotation angle $\varphi = k\cdot 2\pi$, $k = 0, \pm1, \pm2, \ldots$ This implies $q = \cos\frac{\varphi}{2} + \mathbf{n}\sin\frac{\varphi}{2} = \pm1$. □

It follows from Prop. 7.1 that we obtain the group isomorphism

$$\boxed{SU(E_3) \simeq U(1,\mathbb{H})/\{-1,1\}.} \tag{7.2}$$

The Lie group $U(1,\mathbb{H})$ is called the universal covering group of the Lie group E_3. Locally, the groups $SU(E_3)$ and $U(1,\mathbb{H})$ coincide on a sufficiently small neighborhood of the unit element. But globally, $SU(E_3)$ and $U(1,\mathbb{H})$ are different Lie groups.

> *The Lie group $U(1,\mathbb{H})$ is diffeomorphic to a 3-dimensional sphere. Therefore, $U(1,\mathbb{H})$ is arcwise connected and simply connected.*

In contrast to this, it can be shown that the Lie group $SU(E_3)$ is arcwise connected, but not simply connected. In other words, the universal covering group $U(1, \mathbb{H})$ possesses a simpler topological structure than the original Lie group $SU(E_3)$ (group of rotations). On an infinitesimal level, the Lie group $SU(E_3)$ coincides with the Lie group $U(1, \mathbb{H})$. This means that the two Lie groups $SU(E_3)$ and $U(1, \mathbb{H})$ have isomorphic Lie algebras, that is,

$$su(E_3) \simeq u(1, \mathbb{H}).$$

The Lie group $SU(2)$. By page 103, we have the following Lie group isomorphism

$$\boxed{U(1, \mathbb{H}) \simeq SU(2).} \tag{7.3}$$

In addition, we have the Lie group isomorphism

$$SU(E_3) \simeq SO(3).$$

By Prop. 7.1, we get the surjective Lie group morphism

$$\boxed{\chi : SU(2) \to SO(3)}$$

with the kernel $\chi^{-1}(I) = \{-1, 1\}$. This implies the group isomorphism

$$SU(2) \simeq SO(3)/\{-I, I\}.$$

The Lie group $SU(2)$ is called the universal covering group of the Lie group $SO(3)$. The Lie groups $SO(3)$ and $SU(2)$ have the Lie algebras $so(3)$ and $su(2)$, respectively; these Lie algebras are isomorphic,

$$\boxed{so(3) \simeq su(2).} \tag{7.4}$$

Explicitly, the real Lie algebra $so(3)$ consists of all the real skew-adjoint (3×3)-matrices. The three matrices I^1, I^2, I^3 from (7.10) on page 430 form a basis of $so(3)$. Moreover, the real Lie algebra $su(2)$ consists of all the complex skew-adjoint traceless (2×2)-matrices. The three matrices $-\frac{\mathrm{i}}{2}\sigma^1, -\frac{\mathrm{i}}{2}\sigma^2, -\frac{\mathrm{i}}{2}\sigma^3$ from (7.5) on page 428 form a basis of $su(2)$. The Lie algebra isomorphism (7.4) is given by the map

$$I^k \mapsto -\tfrac{\mathrm{i}}{2}\sigma^k, \qquad k = 1, 2, 3.$$

In fact, setting either $A^k := I^k$ or $A^k := \frac{\mathrm{i}}{2}\sigma^k$, $k = 1, 2, 3$, we get the commutation relations.[1]

$$[A^1, A^2]_- = A^3, \quad [A^2, A^3]_- = A^1, \quad [A^3, A^1]_- = A^2.$$

7.3 Irreducible Unitary Representations of the Group $SU(2)$ and the Spin

In nature, we observe quantum states with integer and half-integer spin. Mathematically, this is related to irreducible unitary representations of the universal covering group $SU(2)$ of the rotation group $SO(3)$.

Folklore

[1] Recall that $[A^1, A^2]_- := A^1A^2 - A^2A^1$.

7.3.1 The Spin Quantum Numbers

The Lie group $SU(2)$, the Lie algebra $su(2)$, and the spin operators. Recall that $SU(2)$ denotes the Lie group of all complex unitary (2×2)-matrices G with $\det G = 1$. Let $\mathbb{C}^2$ be the 2-dimensional complex Hilbert space of all the matrices

$$\psi = \begin{pmatrix} \psi^1 \\ \psi^2 \end{pmatrix}, \qquad \psi^1, \psi^2 \in \mathbb{C} \tag{7.5}$$

equipped with the inner product $\langle \varphi | \psi \rangle := \varphi^\dagger \psi = (\varphi^1)^\dagger \psi^1 + (\varphi^2)^\dagger \psi^2$. The matrices $\psi_{1/2} := \begin{pmatrix} 1 \\ 0 \end{pmatrix}$ and $\psi_{-1/2} := \begin{pmatrix} 0 \\ 1 \end{pmatrix}$ form an orthonormal basis of $\mathbb{C}^2$. Using the Pauli matrices

$$\sigma^0 := \begin{pmatrix} 1 & 0 \\ 0 & 1 \end{pmatrix}, \quad \sigma^1 := \begin{pmatrix} 0 & 1 \\ 1 & 0 \end{pmatrix}, \quad \sigma^2 := \begin{pmatrix} 0 & -\mathrm{i} \\ \mathrm{i} & 0 \end{pmatrix}, \quad \sigma^3 := \begin{pmatrix} 1 & 0 \\ 0 & -1 \end{pmatrix},$$

we set $\mathcal{S}^k := \frac{\hbar}{2}\sigma^k, k = 1, 2, 3$. Then we get the commutation rules

$$\boxed{[\mathcal{S}^1, \mathcal{S}^2]_- = \mathrm{i}\hbar\mathcal{S}^3, \quad [\mathcal{S}^2, \mathcal{S}^3]_- = \mathrm{i}\hbar\mathcal{S}^1, \quad [\mathcal{S}^3, \mathcal{S}^1]_- = \mathrm{i}\hbar\mathcal{S}^2,} \tag{7.6}$$

which are fundamental for describing the spin of quantum states in physics. Recall that $[\mathcal{S}^1, \mathcal{S}^2]_- := \mathcal{S}^1\mathcal{S}^2 - \mathcal{S}^2\mathcal{S}^1$, and so on. We have

$$\mathcal{S}^3\psi_m = \hbar m \cdot \psi_m, \quad \left((\mathcal{S}^1)^2 + (\mathcal{S}^2)^2 + (\mathcal{S}^3)^2\right)\psi_m = \frac{3\hbar^2}{4}\psi_m, \qquad m = \tfrac{1}{2}, -\tfrac{1}{2}.$$

In terms of physics, $\psi_{1/2}$ and $\psi_{-1/2}$ describe spin states with the spin $\hbar/2$ and $-\hbar/2$, respectively. The three self-adjoint operators $\mathcal{S}^1, \mathcal{S}^2, \mathcal{S}^3$ were introduced by Pauli in 1927. In terms of mathematics, these three operators do not form a Lie algebra. In order to pass to a Lie algebra, we set $\mathcal{S}^k = \mathrm{i}\hbar\mathcal{I}^k, k = 1, 2, 3$. Hence

$$\mathcal{I}^k := -\frac{\mathrm{i}}{2}\sigma^k, \qquad k = 1, 2, 3.$$

We obtain the following commutation relations

$$\boxed{[\mathcal{I}^1, \mathcal{I}^2]_- = \mathcal{I}^3, \quad [\mathcal{I}^2, \mathcal{I}^3]_- = \mathcal{I}^1, \quad [\mathcal{I}^3, \mathcal{I}^1]_- = \mathcal{I}^2.}$$

The matrices $\mathcal{I}^1, \mathcal{I}^2, \mathcal{I}^3$ are skew-adjoint (i.e., $(\mathcal{I}^k)^\dagger = -\mathcal{I}^k$) and traceless. Precisely, all the matrices

$$\mathcal{A} = \alpha^1\mathcal{I}^1 + \alpha^2\mathcal{I}^2 + \alpha^3\mathcal{I}^3, \qquad \alpha^1, \alpha^2, \alpha^3 \in \mathbb{R}$$

form the Lie algebra $su(2)$. In what follows, it will be convenient to change the coefficients by setting

$$\mathcal{A} = \varphi(n^1\mathcal{I}^1 + n^2\mathcal{I}^2 + n^3\mathcal{I}^3),$$

where φ, n^1, n^2, n^3 are real numbers with $(n^1)^2 + (n^2)^2 + (n^3)^2 = 1$. Furthermore, precisely all the matrices

$$\mathcal{B} = \mathrm{e}^{\mathcal{A}}, \qquad \mathcal{A} \in su(2)$$

form the Lie group $SU(2)$. Explicitly,

$$e^{\mathcal{A}} = e^{-\frac{\varphi}{2}i\cdot\mathbf{n}\boldsymbol{\sigma}} = \sigma^0 \cos\frac{\varphi}{2} - i\cdot\mathbf{n}\boldsymbol{\sigma}\sin\frac{\varphi}{2}, \qquad \varphi \in \mathbb{R}. \tag{7.7}$$

Here, $\mathbf{n}\boldsymbol{\sigma} := \sum_{k=1}^3 n^k\sigma^k$ with $\sum_{k=1}^3 (n^k)^2 = 1$. Moreover,

$$e^{-\frac{\varphi}{2}\cdot\mathbf{n}\boldsymbol{\sigma}} = \sigma^0 \cosh\frac{\varphi}{2} - \mathbf{n}\boldsymbol{\sigma}\sinh\frac{\varphi}{2}, \qquad \varphi \in \mathbb{R}. \tag{7.8}$$

This will be proved in Problem 8.1.

Our goal is to describe more general spin states by generalizing the commutation relations (7.6) *to higher-dimensional complex Hilbert spaces.*

To this end, we will use the representation theory of the group $SU(2)$, and the representation theory of the corresponding Lie algebra $su(2)$.

Representations. Let $\varrho : SU(2) \to U(X)$ be an irreducible unitary continuous representation of the Lie group $SU(2)$ on the complex Hilbert space X. Then the following hold.

(i) Dimension of the Hilbert space: The dimension of X is finite. There exists a number $s = 0, \frac{1}{2}, 1, \frac{3}{2}, \ldots$ such that $\dim X = 2s + 1$.

(ii) Spin operators: There exist self-adjoint operators $S^k : X \to X$, $k = 1, 2, 3$, such that

$$\boxed{[S^1, S^2]_- = i\hbar S^3, \quad [S^2, S^3]_- = i\hbar S^1, \quad [S^3, S^1]_- = i\hbar S^2,} \tag{7.9}$$

and there exists an orthonormal basis $\psi_s, \psi_{s-1}, \ldots, \psi_{-s}$ of the Hilbert space X such that

$$S^3\psi_m = \hbar m \cdot \psi_m, \qquad \left((S^1)^2 + (S^2)^2 + (S^3)^2\right)\psi_m = \hbar^2 s(s+1)\cdot\psi_m$$

for all $m = s, s-1, \ldots, -s$. In terms of physics, ψ_m describes a spin state with the spin $\hbar m$ and the spin quantum number s. The numbers $m = s, s-1, \ldots - s$ are called magnetic quantum numbers.[2] The operator $(S^1)^2 + (S^2)^2 + (S^3)^2$ is called the Casimir operator of the representation ϱ.

(iii) Lie algebra: Set $S^k := i\hbar I^k$. Then the operators $I^k : X \to X$, $k = 1, 2, 3$, are skew-adjoint (i.e., $(I^k)^\dagger = -I^k$), and

$$\boxed{[I^1, I^2]_- = I^3, \quad [I^2, I^3]_- = I^1, \quad [I^3, I^1]_- = I^2.}$$

Define

$$\mu\left(\sum_{k=1}^3 \alpha^k \mathcal{I}^k\right) := \sum_{k=1}^3 \alpha^k I^k, \qquad \alpha^1, \alpha^2, \alpha^3 \in \mathbb{R}.$$

The map μ is a representation of the Lie algebra $su(2)$, and the representation $\varrho : SU(2) \to U(X)$ is given by

$$\varrho\left(e^{\sum_{k=1}^3 \alpha^k \mathcal{I}^k}\right) = e^{\sum_{k=1}^3 \alpha^k I^k}, \qquad \alpha^1, \alpha^2, \alpha^3 \in \mathbb{R}.$$

(iv) Completeness of the spin quantum numbers: For every $s = 0, \frac{1}{2}, 1, \frac{3}{2}, \ldots$, there exists a representation of $SU(2)$ as described above.

---

[2] As we will show in Vol. IV, the number m is responsible for the splitting of the spectrum of the hydrogen atom in a magnetic field (Zeeman effect). In mathematics, the number m is called the weight of the irreducible representation ϱ.

The proof can be found in van der Waerden, Group Theory and Quantum Mechanics, Springer, New York 1974. See also I. Gelfand et al., Representations of the Rotation and Lorentz Groups and Their Applications, Pergamon Press, New York, 1963, and M. Naimark, Linear Representations of the Lorentz Group, Macmillan, New York, 1964. We will study this in greater detail in Vol. IV on quantum mathematics. Let us discuss the special case $s = 1$.

Quantum states with spin number $s = 1$ and the rotation group $SO(3)$. The key formulas are

$$I^1 := \begin{pmatrix} 0 & 0 & 0 \\ 0 & 0 & -1 \\ 0 & 1 & 0 \end{pmatrix}, \quad I^2 := \begin{pmatrix} 0 & 0 & 1 \\ 0 & 0 & 0 \\ -1 & 0 & 0 \end{pmatrix}, \quad I^3 := \begin{pmatrix} 0 & -1 & 0 \\ 1 & 0 & 0 \\ 0 & 0 & 0 \end{pmatrix}, \tag{7.10}$$

and

$$e^{\varphi I^1} = \begin{pmatrix} 1 & 0 & 0 \\ 0 & \cos\varphi & -\sin\varphi \\ 0 & \sin\varphi & \cos\varphi \end{pmatrix}, \quad e^{\varphi I^2} := \begin{pmatrix} \cos\varphi & 0 & \sin\varphi \\ 0 & 1 & 0 \\ -\sin\varphi & 0 & \cos\varphi \end{pmatrix}, \quad e^{\varphi I^3} = \begin{pmatrix} \cos\varphi & -\sin\varphi & 0 \\ \sin\varphi & \cos\varphi & 0 \\ 0 & 0 & 1 \end{pmatrix}$$

for all real numbers φ. For the proof, we refer to Problem 8.2. The real matrices I^1, I^2, I^3 are skew-symmetric (and hence skew-adjoint). Moreover, the real matrices $e^{\varphi I^k}, k = 1, 2, 3$, are orthogonal (and hence unitary). Introduce the self-adjoint spin matrices

$$S^k := \mathrm{i}\hbar I^k, \qquad k = 1, 2, 3.$$

Then we have the commutation relations

$$\boxed{[I^1, I^2]_- = I^3, \quad [I^2, I^3]_- = I^1, \quad [I^3, I^1]_- = I^2,}$$

and

$$\boxed{[S^1, S^2]_- = \mathrm{i}\hbar S^3, \quad [S^2, S^3]_- = \mathrm{i}\hbar S^1, \quad [S^3, S^1]_- = \mathrm{i}\hbar S^2.}$$

All the matrices $\mathcal{A} = \sum_{k=1}^3 \alpha^k I^k$ with real coefficients $\alpha^1, \alpha^2, \alpha^3$ form the real Lie algebra $so(3)$. Furthermore, the matrices $e^{\theta I^k}$, $k = 1, 2, 3$, represent rotations which generate the Lie group $SO(3)$. The matrices I^1, I^2, I^3 represent infinitesimal rotations. The eigenvalues of the matrix I^3 are $\mathrm{i}, -\mathrm{i}, 0$, and hence the eigenvalues of S^3 are $\hbar, -\hbar, 0$.

Let us choose the 3-dimensional complex Hilbert space $X = \mathbb{C}^3$ which consists of all the matrices

$$\psi = \begin{pmatrix} \psi^1 \\ \psi^2 \\ \psi^3 \end{pmatrix}, \qquad \psi^1, \psi^2, \psi^3 \in \mathbb{C}$$

equipped with the inner product $\langle\chi|\psi\rangle := \chi^\dagger\psi = \sum_{k=1}^3 (\chi^k)^\dagger \psi^k$. The matrices

$$\psi_1 = \frac{1}{\sqrt{2}}\begin{pmatrix} 1 \\ \mathrm{i} \\ 0 \end{pmatrix}, \quad \psi_{-1} = \frac{1}{\sqrt{2}}\begin{pmatrix} 1 \\ -\mathrm{i} \\ 0 \end{pmatrix}, \quad \psi_0 = \begin{pmatrix} 0 \\ 0 \\ 1 \end{pmatrix},$$

(a) active rotation of the point P (b) passive rotation of the point P

Fig. 7.1. Rotation

form an orthonormal basis of X. The operators $S^k : X \to X$ are self-adjoint, $k = 1, 2, 3$, and we have

$$S^3\psi_m = m\hbar \cdot \psi_m, \quad \left((S^1)^2 + (S^2)^2 + (S^3)^2\right)\psi_m = 2\hbar^2\psi_m$$

with the weights $m = 1, 0, -1$. Next let us discuss the geometric meaning of the irreducible unitary representation $\varrho : SU(2) \to U(X)$ with $X = \mathbb{C}^3$.

Active rotation. In physics, one distinguishes between active and passive rotations of position vectors (points). Let us discuss this (see Fig. 7.1). Let Σ be a right-handed Cartesian (x, y, z)-coordinate system with the right-handed orthonormal basis $\mathbf{i}, \mathbf{j}, \mathbf{k}$. Suppose that the operator $U_{\mathbf{n}}(\varphi)$ describes the counter-clockwise rotation of the position vector $\mathbf{x} = x\mathbf{i} + y\mathbf{j} + z\mathbf{k}$ at the origin about the unit axis-vector $\mathbf{n}$ with the rotation angle φ (see Fig. 7.1(a) with $\mathbf{n} = \mathbf{k}$). Setting

$$U_{\mathbf{n}}(\varphi)\mathbf{x} := x^+\mathbf{i} + y^+\mathbf{j} + z^+\mathbf{k},$$

we obtain

$$\begin{pmatrix} x^+ \\ y^+ \\ z^+ \end{pmatrix} = \mathrm{e}^{\varphi \sum_{k=1}^3 n^k I^k} \begin{pmatrix} x \\ y \\ z \end{pmatrix} \tag{7.11}$$

where $\mathbf{n} = n^1\mathbf{i} + n^2\mathbf{j} + n^3\mathbf{k}$ with $\mathbf{n}^2 = 1$, and φ is a real number. This is called an active rotation of the position vector $\mathbf{x}$. Noting that

$$\mathrm{e}^{\varphi \sum_{k=1}^3 n^k \mathcal{I}^k} = \sigma^0 \cos\frac{\varphi}{2} - \mathrm{i}\sum_{k=1}^3 n^k\sigma^k \sin\frac{\varphi}{2},$$

the map $\varrho : SU(2) \to SO(3)$ given by

$$\boxed{\varrho\left(\mathrm{e}^{\varphi \sum_{k=1}^3 n^k \mathcal{I}^k}\right) := \mathrm{e}^{\varphi \sum_{k=1}^3 n^k I^k}}$$

is a surjective Lie group morphism which corresponds to an irreducible unitary representation of $SU(2)$.

Changing the observer and passive rotation. Let us change the system of reference Σ to the system Σ' of reference. We assume that Σ' is a right-handed Cartesian (x', y', z')-system with the right-handed orthonormal basis $\mathbf{i}', \mathbf{j}', \mathbf{k}'$. More precisely, we assume that

$$U_{\mathbf{n}}(\varphi)\mathbf{i} = \mathbf{i}', \quad U_{\mathbf{n}}(\varphi)\mathbf{j} = \mathbf{j}', \quad U_{\mathbf{n}}(\varphi)\mathbf{k} = \mathbf{k}'.$$

Then

$$\mathbf{x} = x\mathbf{i} + y\mathbf{j} + z\mathbf{k} = x'\mathbf{i}' + y'\mathbf{j}' + z'\mathbf{k}'.$$

This means that an observer in Σ (resp. Σ') measures the components x, y, z (resp. x', y', z') of the position vector $\mathbf{x}$ corresponding to the point P (see Fig. 7.1(b)). Explicitly, we get

$$\begin{pmatrix} x' \\ y' \\ z' \end{pmatrix} = \mathrm{e}^{-\varphi \sum_{k=1}^{3} n^k I^k} \begin{pmatrix} x \\ y \\ z \end{pmatrix}. \tag{7.12}$$

This is called a passive rotation of the position vector $\mathbf{x}$. Note that the formula (7.11) passes over to (7.12) by changing the sign of the rotation angle φ.

The operator of angular momentum in quantum mechanics. Let us show that the operators of angular momentum and momentum in quantum mechanics are obtained by the action of rotations and translations on physical fields (on an infinitesimal level).[3]

We are given the smooth complex-valued function $\psi : E_3 \to \mathbb{C}$ on the Euclidean space E_3. In terms of physics, this is a physical field. Set

$$\boxed{(\mathcal{U}_{\mathbf{n}}(\varphi)\psi)(\mathbf{x}) := \psi(U_{\mathbf{n}}(\varphi)^{-1}\mathbf{x}), \qquad \mathbf{x} \in E_3.} \tag{7.13}$$

Note that we use the inverse operator $U_{\mathbf{n}}(\varphi)^{-1}$ in order to guarantee that the map $U_{\mathbf{n}}(\varphi) \mapsto \mathcal{U}_{\mathbf{n}}(\varphi)$ is a representation, that is, it respects products. In fact,

$$\big(\mathcal{U}(\mathcal{V}\psi)\big)(\mathbf{x}) = (\mathcal{V}\psi)(U^{-1}\mathbf{x}) = \psi(V^{-1}U^{-1}\mathbf{x}).$$

It follows from $(UV)^{-1} = V^{-1}U^{-1}$ that

$$\big((\mathcal{U}\mathcal{V})\psi\big)(\mathbf{x}) = \psi(V^{-1}U^{-1}\mathbf{x}).$$

Hence $\mathcal{U}(\mathcal{V}\psi) = (\mathcal{U}\mathcal{V})\psi$.

Consider the special case where $\mathbf{n} = \mathbf{k}$, that is, $U_{\mathbf{n}}(\varphi)$ represents a counterclockwise rotation about the z-axis with the angle φ. Define the operator $\mathcal{S}^3$ by setting

$$\boxed{(\mathcal{S}^3\psi)(\mathbf{x}) := \mathrm{i}\hbar \frac{d}{d\varphi}\mathcal{U}_{\mathbf{k}}(\varphi)\psi(\mathbf{x})_{|\varphi=0}.} \tag{7.14}$$

This corresponds to an infinitesimal rotation about the z-axis. Using the system Σ of reference, we write $\psi(x, y, z)$ instead of $\psi(\mathbf{x})$ where $\mathbf{x} = x\mathbf{i} + y\mathbf{j} + z\mathbf{k}$. We claim that

$$\boxed{(\mathcal{S}^3\psi)(x, y, z) = \mathrm{i}\hbar \left(y\frac{\partial\psi(x, y, z)}{\partial x} - x\frac{\partial\psi(x, y, z)}{\partial y} \right).} \tag{7.15}$$

Proof. By (7.12),

$$(\mathcal{U}_{\mathbf{k}}(\varphi)\psi)(x, y, z) = \psi(x', y', z').$$

This is equal to $\psi(x\cos\varphi + y\sin\varphi, y\cos\varphi - x\sin\varphi, z)$. Differentiating this with respect to φ at the point $\varphi = 0$, we get (7.15). □

[3] This is the prototype for more general situations in quantum field theory.

Replacing $U_{\mathbf{k}}(\varphi)$ by $U_{\mathbf{i}}(\varphi)$ (resp. $U_{\mathbf{j}}(\varphi)$), we get the operator $\mathcal{S}^1$ (resp. $\mathcal{S}^2$). Explicitly,

$$(\mathcal{S}^1\psi)(x,y,z) = \mathrm{i}\hbar\left(z\frac{\partial\psi(x,y,z)}{\partial y} - y\frac{\partial\psi(x,y,z)}{\partial z}\right)$$

and

$$(\mathcal{S}^2\psi)(x,y,z) = \mathrm{i}\hbar\left(y\frac{\partial\psi(x,y,z)}{\partial z} - z\frac{\partial\psi(x,y,z)}{\partial x}\right).$$

The operator $\mathbf{S} = \mathcal{S}^1\mathbf{i} + \mathcal{S}^2\mathbf{j} + \mathcal{S}^3\mathbf{k}$ is called the operator of angular momentum in quantum mechanics. Explicitly,

$$\mathbf{S} = \begin{vmatrix} \mathbf{i} & \mathbf{j} & \mathbf{k} \\ x & y & z \\ -\mathrm{i}\hbar\frac{\partial}{\partial x} & -\mathrm{i}\hbar\frac{\partial}{\partial y} & -\mathrm{i}\hbar\frac{\partial}{\partial z} \end{vmatrix} = \mathbf{x} \times \mathbf{P} \tag{7.16}$$

with the momentum operator

$$\mathbf{P} = \mathcal{P}^1\mathbf{i} + \mathcal{P}^2\mathbf{j} + \mathcal{P}^3\mathbf{k} = -\mathrm{i}\hbar\left(\mathbf{i}\frac{\partial}{\partial x} + \mathbf{j}\frac{\partial}{\partial y} + \mathbf{k}\frac{\partial}{\partial z}\right).$$

Note that

$$(\mathcal{P}^1\psi)(\mathbf{x}) = \mathrm{i}\hbar\frac{d}{da}\psi(\mathbf{x} - a\mathbf{i})_{|a=0} = -\mathrm{i}\hbar\frac{\partial\psi(\mathbf{x})}{\partial x}.$$

Comparing this with (7.14), we obtain that the momentum operator $\mathcal{P}^1$ is generated by the translation $\mathbf{x} \mapsto \mathbf{x} + a\mathbf{i}$ (on an infinitesimal level).

Summary of the surjective Lie group morphism $\varrho : SU(2) \to SO(3)$. Consider a right-handed Cartesian (x,y,z)-coordinate system. The formula

$$\boxed{x' = x\,\cos\varphi + y\,\sin\varphi, \quad y' = -x\,\sin\varphi + y\,\cos\varphi, \quad z' = z} \tag{7.17}$$

describes a clockwise rotation of the (x,y,z)-coordinate system about the z-axis with the rotation angle φ (Fig. 7.1(b)). Using the language of matrices, this rotation can be written as

$$\begin{pmatrix} x' \\ y' \\ z' \end{pmatrix} = \mathrm{e}^{-\varphi I^3}\begin{pmatrix} x \\ y \\ z \end{pmatrix} \tag{7.18}$$

where $\mathrm{e}^{-\varphi I^3} = \cos\varphi\; I - \sin\varphi\; I^3$. The corresponding Pauli spinor transformation reads as

$$\begin{pmatrix} \psi^{1'} \\ \psi^{2'} \end{pmatrix} = \mathrm{e}^{-\varphi \mathcal{I}^3}\begin{pmatrix} \psi^1 \\ \psi^2 \end{pmatrix} \tag{7.19}$$

where the components ψ^1, ψ^2 are complex numbers. Here,

$$\mathrm{e}^{-\varphi\mathcal{I}^3} = \cos\frac{\varphi}{2}\,\sigma^0 - \mathrm{i}\sin\frac{\varphi}{2}\sigma^3.$$

The surjective Lie group morphism $\varrho : SU(2) \longrightarrow SO(3)$ specializes to

$$\boxed{\varrho(\mathrm{e}^{-\varphi \mathcal{I}^3}) = \mathrm{e}^{-\varphi I^3}.} \tag{7.20}$$

Note the following peculiarity. For given rotation (7.18), there exist precisely two matrices $A \in SU(2)$ such that $\varrho(A) = \mathrm{e}^{-\varphi I^3}$, namely,

$$A = \pm \mathrm{e}^{-\varphi \mathcal{I}^3}.$$

The remaining formulas for rotations about the x-axis and the y-axis are obtained by the cyclic permutations

$$x \mapsto y \mapsto z \mapsto x \qquad \text{and} \qquad 1 \mapsto 2 \mapsto 3 \mapsto 1.$$

Using the Euler angles α, β, γ, every rotation R of the (x, y, z)-coordinate system can be uniquely represented by the product

$$R = \mathrm{e}^{-\alpha I^3} \mathrm{e}^{-\beta I^1} \mathrm{e}^{-\gamma I^3}$$

where $-\pi \leq \alpha, \gamma < \pi$ and $-\frac{\pi}{2} \leq \gamma \leq \frac{\pi}{2}$ (see page 422). Then

$$\boxed{\varrho(\pm \mathrm{e}^{-\alpha \mathcal{I}^3} \mathrm{e}^{-\beta \mathcal{I}^1} \mathrm{e}^{-\gamma \mathcal{I}^3}) = R.} \tag{7.21}$$

Applications to the Pauli equation on the non-relativistic spinning electron can be found in Sect. 19.1.7 on page 948.

7.3.2 The Addition Theorem for the Spin

Let X_s and Y_r be complex Hilbert spaces of dimension s and r, respectively, where $s, r = 0, \frac{1}{2}, 1, \frac{3}{2}, \ldots$ Suppose that there acts an irreducible unitary representation of the group $SU(2)$ on X_s and Y_r with the spin quantum number s and r, respectively. In terms of physics, the Hilbert space X_s (resp. Y_r) describes quantum states of spin s (resp. r). Consider now the tensor product $X_s \otimes Y_r$. Then the group $SU(2)$ acts on $X_s \otimes Y_r$ as an unitary representation. Since $SU(2)$ is a compact Lie group, the representation on $X_s \otimes Y_r$ is completely reducible.

Theorem 7.2 *There exist linear subspaces Z_j of $X_s \otimes Y_r$ such that we have the direct sum decomposition*

$$X_s \otimes Y_r = Z_{|s-r|} \oplus Z_{|s-r|+1} \oplus Z_{|s-r|+2} \oplus \cdots \oplus Z_{s+r}.$$

All the complex Hilbert spaces Z_j are irreducible under the action of the group $SU(2)$ with the spin quantum number j.

The proof can be found in van der Waerden (1974) quoted on page 430. In terms of physics, this means that if two elementary particles are described by the spin quantum numbers s and r, then the composed particle has one of the spin quantum numbers

$$j = s + r, s + r - 1, s + r - 2, \ldots, |s - r|.$$

This is called the addition theorem for the spin.

Example. Consider two elementary particles with the spin quantum numbers $s = r = \frac{1}{2}$. Then the composed particle has either the spin quantum number $s + r = 1$ or $|s - r| = 0$.

7.3.3 The Model of Homogeneous Polynomials

In representation theory, one uses frequently spaces of polynomials in order to construct irreducible representations of Lie matrix groups. It turns out that the use of special polynomial models yields formulas in representation theory which are valid in a universal way. As a typical example, let us discuss this for the group $SU(2)$.[4]

The Hilbert space $\mathbb{C}_n[x,y]$. Let $n = 0, 1, 2, \ldots$ The symbol $\mathbb{C}_n[x,y]$ denotes the complex linear space of all the complex homogeneous polynomials of degree n with respect to the variables x and y. Explicitly, the elements of $\mathbb{C}_n[x,y]$ are given by

$$p(x,y) := \sum_{k=0}^{n} c_k p_k^n(x,y)$$

where the coefficients $c_0, c_1, \ldots, c_n$ are complex numbers, and

$$p_k^n(x,y) := \frac{x^k y^{n-k}}{\sqrt{k!(n-k)!}}, \qquad k = 0, 1, \ldots, n. \tag{7.22}$$

Introducing the inner product

$$\langle p|p' \rangle := \sum_{k=0}^{n} c_k^\dagger c_k', \qquad p, p' \in \mathbb{C}_n[x,y],$$

the space $\mathbb{C}_n[x,y]$ becomes a complex $(n+1)$-dimensional Hilbert space. For fixed n, the polynomials p_k^n, $k = 0, 1, \ldots, n$ form an orthonormal basis of $\mathbb{C}_n[x,y]$. The normalization factors are introduced in order to get unitary representations of $SU(2)$ in what follows.

Irreducible representation of the Lie group $SU(2)$ on the Hilbert space $\mathbb{C}_n[x,y]$). Consider the transformation

$$\begin{pmatrix} x' \\ y' \end{pmatrix} = A \begin{pmatrix} x \\ y \end{pmatrix}, \qquad A \in SU(2).$$

Naturally enough, setting

$$(\varrho(A)p)(x,y) := p(x',y'),$$

we get the linear transformation $\varrho(A) : \mathbb{C}_n[x,y] \to \mathbb{C}_n[x,y]$.

Theorem 7.3 *The map $A \mapsto \varrho(A)$ is an irreducible unitary representation of the group $SU(2)$ on the $(n+1)$-dimensional complex Hilbert space $\mathbb{C}_n[x,y]$, which has the spin number $s = \frac{n}{2}$.*

In order to get contact to physics, define

$$\psi_m^s(x,y) := p_k^n(x,y) = \frac{x^{s+m} y^{s-m}}{\sqrt{(s+m)!(s-m)!}}, \qquad s = \frac{n}{2}, \ m = k - s.$$

Then, for fixed spin quantum number $s = 0, \frac{1}{2}, 1, \frac{3}{2}, \ldots$, the polynomials ψ_m^s with $m = -s, -s+1, \ldots, s$ form an orthonormal basis of the Hilbert space $\mathbb{C}_{2s}[x,y]$.

[4] The proofs of the following statements including the Clebsch–Gordan coefficients can be found in van der Waerden (1974) quoted on page 430.

Moreover, the irreducible unitary representation ϱ of $SU(2)$ on $\mathbb{C}_{2s}[x,y]$ corresponds to the situation (7.9) described on page 429.[5]

Irreducible representations of the Lie group $SO(3)$. For the spin number $s = 1$, the irreducible unitary representation ϱ of the group $SU(2)$ on the complex 3-dimensional Hilbert space $\mathbb{C}_2[x,y]$ yields a group of linear unitary operators on $\mathbb{C}_2[x,y]$ which is isomorphic to the group $SO(3)$.

For integer spin numbers $s = 0, 1, 2, \ldots$, the irreducible unitary representation ϱ of the group $SU(2)$ on the $(2s+1)$-dimensional complex Hilbert space $C_{2s}[x,y]$ is also an irreducible unitary representation of the group $SO(3)$ on $\mathbb{C}_{2s}[x,y]$.

7.3.4 The Clebsch–Gordan Coefficients

We want to investigate the addition theorem for spin in terms of the polynomial model.

Special case. The Hilbert space $\mathbb{C}_1[x,y] \otimes \mathbb{C}_1[u,v]$ has the four polynomials

$$p_m^{1/2}(x,y)p_{m'}^{1/2}(u,v), \qquad m, m' = \tfrac{1}{2}, -\tfrac{1}{2}$$

as basis elements. The group $SU(2)$ acts on $\mathbb{C}_1[x,y] \otimes \mathbb{C}_1[u,v]$ in a natural way. Our goal is to decompose the representation of $SU(2)$ on $\mathbb{C}_1[x,y] \otimes \mathbb{C}_1[u,v]$ into irreducible components. Moreover, we want to construct explicitly the basis polynomials for the irreducible components. As a special case of the general procedure to be described below, we choose $q_0^0(x,y,u,v) := xv - yu$, and

$$q_1^1(x,y,u,v) := \sqrt{2} \cdot xu, \quad q_0^1(x,y,u,v) := xv + yu, \quad q_{-1}^1(x,y,u,v) := \sqrt{2} \cdot yv.$$

Set $X_0 := \operatorname{span}\{q_0^0\}$ and $X_1 := \operatorname{span}\{q_1^1, q_0^1, q_{-1}^1\}$. Then

$$\mathbb{C}_1[x,y] \otimes \mathbb{C}_1[u,v] = X_0 \oplus X_1.$$

The linear subspaces X_0 and X_1 are invariant under the action of the group $SU(2)$ on $\mathbb{C}_1[x,y] \otimes \mathbb{C}_1[u,v]$. The corresponding representations of $SU(2)$ are irreducible and unitary, and they possess the spin number $s = 0$ and $s = 1$ on X_0 and X_1, respectively.

General case. Fix $s, s' = 0, \frac{1}{2}, 1, \frac{3}{2}, \ldots$ The Hilbert space $\mathbb{C}_{2s}[x,y] \otimes \mathbb{C}_{2s'}[u,v]$ has the basis polynomials

$$p_m^s(x,y)p_{m'}^{s'}(u,v), \qquad m = -s, -s+1, \ldots, s, \;\; m' = -s', -s'+1, \ldots, s'.$$

Let $S = s + s' - \lambda$. Fix $\lambda := 0, 1, \ldots s + s' - |s - s'|$. Consider the polynomial

$$p(\xi, \eta) := (xv - yu)^\lambda (x\xi + y\eta)^{2s-\lambda}(u\xi + v\eta)^{2s'-\lambda}.$$

Let q_M^S be the coefficient of $p(\xi,\eta)$ with respect to the term

$$p_M^S(\xi,\eta) = \frac{\xi^{S+M}\eta^{S-M}}{\sqrt{(S+M)!(S-M)!}}, \qquad S = s + s' - \lambda, M = -S, -S+1, \ldots, S.$$

Define $X_S := \operatorname{span}\{q_M^S\}_{M=-S,-S+1,\ldots,S}$. Then we have the decomposition

$$\boxed{\mathbb{C}_{2s}[x,y] \otimes \mathbb{C}_{2s'}[u,v] = X_{s+s'} \oplus X_{s+s'-1} \oplus \cdots \oplus X_{|s-s'|}.}$$

[5] The state ψ_m in (7.9) coincides with ψ_m^s.

The linear subspaces $X_{s+s'}, \ldots$ are invariant under the action of the group $SU(2)$ on $\mathbb{C}_{2s}[x,y] \otimes \mathbb{C}_{2s'}[u,v]$. The corresponding representations of $SU(2)$ are irreducible and unitary, and they possess the spin number $s + s' - \lambda$ on $X_{s+s'-\lambda}$. Explicitly, we get

$$q_M^S(x,y,u,v) = \lambda!(2s-\lambda)!(2s'-\lambda)! \sum_{m+m'=M} c_{mm'}^S p_m^s(x,y) p_{m'}^{s'}(u,v).$$

The complex numbers $c_{mm'}^S$ are explicitly known; they are called the Clebsch–Gordan coefficients.

7.4 Heisenberg's Isospin

Consider the situation (7.6) on page 428. The equations

$$\mathcal{S}^3 \psi_m = m\hbar\psi_m, \qquad \left((\mathcal{S}^1)^2 + (\mathcal{S}^2)^2 + (\mathcal{S}^3)^2\right)\psi_m = \frac{3\hbar^2}{4}\psi_m$$

with $m = \frac{1}{2}, -\frac{1}{2}$ describe the quantum states $\psi_{1/2}$ and $\psi_{-1/2}$ of an electron with the spin components $\hbar/2$ and $-\hbar/2$, respectively, and the spin number $s = \frac{1}{2}$. In 1932, Heisenberg used this formalism in order to introduce a new quantum number called isospin. In this setting, we replace the spin operators $\mathcal{S}^k = \frac{\hbar}{2}\sigma^k$, $k = 1,2,3$, by the so-called isospin operators $\mathcal{T}^k := \frac{1}{2}\sigma^k$, $k = 1,2,3$. Then we get

$$\mathcal{T}^3 \psi_m = m\psi_m, \quad \left((\mathcal{T}^1)^2 + (\mathcal{T}^2)^2 + (\mathcal{T}^3)^2\right)\psi_m = \frac{3}{4}\psi_m$$

with $m = \frac{1}{2}, -\frac{1}{2}$.[6] Then the quantum state $\psi_{1/2}$ (resp. $\psi_{-1/2}$) describes a proton (resp. neutron).

> *According to Heisenberg, the proton and the neutron are two different quantum states of a so-called nucleon.*

In the Standard Model of particle physics, the proton and the neutron are members of the baryon octet (see Fig 3.3 on page 228).

Problems

7.1 *Proof of* (7.7). Solution: Fix $k = 1,2,3$. Set $f(\varphi) := e^{\varphi i \sigma^k}$. Then

$$f'(\varphi) = \mathrm{i}\,\sigma^k f(\varphi), \quad \varphi \in \mathbb{R}, \quad f(0) = \sigma^k. \tag{7.23}$$

Since $(\sigma^k)^2 = \sigma^0$, the function $g(\varphi) = \cos\varphi + \mathrm{i}\sigma^k \sin\varphi$ is a solution of (7.23). Since the solution of this differential equation is unique, we get $f = g$. The proof of the general case (7.7) proceeds similarly.

7.2 *Proof of the explicit formulas* (7.10) *for* $e^{\varphi I^k}$. Hint: Proceed as in Problem 7.1. Note that $\frac{d}{d\varphi} e^{\varphi I^k} = I^k e^{\varphi I^k}$.

---

6 Concerning the general representations of $SU(2)$ considered on page 428, we obtain the isospin operators $\mathcal{T}^1, \mathcal{T}^2, \mathcal{T}^3$ by setting $\mathcal{T}^k := \hbar^{-1}\mathcal{S}^k$.

8. Changing Observers – A Glance at Invariant Theory Based on the Principle of the Correct Index Picture

It is worth noting that the notation facilitates discovery. This, in a most wonderful way, reduces the mind's labor.

Gottfried Wilhelm Leibniz (1646–1716)

Use only equations which possess the correct index picture!

Golden rule

8.1 A Glance at the History of Invariant Theory

Geometry is the theory of invariants of a transformation group.

Felix Klein (1849–1925)
Erlangen program, 1872

All roads lead to Rome, so I find in my own case at least that all algebraic inquiries, sooner or later, end at the Capitol of Modern Algebra over whose shining portal is inscribed the Theory of Invariants.

James Sylvester, 1884

Invariant theory plays a crucial role in all branches of mathematics and in modern physics. Invariant theory has its roots in celestial mechanics (Lagrange's contributions to the three-body problem),[1] the motion of rigid bodies (Euler's spinning top), Cauchy's theory of elasticity, number theory, projective geometry, and differential geometry. In his fundamental work *Disquisitiones arithmeticae* on number theory from 1801, Gauss (1777–1855) studied invariants of quadratic forms under unimodular linear substitutions with integral coefficients. Later on, more general results on quadratic forms were obtained by Jacobi (1804–1851), Sylvester (1814–1897), and Hermite (1822–1901).

In his 1827 *Disquisitiones generales circa superficies curvas* (general theory of surfaces), Gauss founded the differential geometry of 2-dimensional surfaces by using the invariants of two quadratic forms (the metric and the curvature form). This was the beginning of a fascinating development in differential geometry strongly influenced by Riemann (1826–1866), Ricci-Curbastro (1853–1925), Élie Cartan (1869–1951), Levi-Civita (1873–1941), Hermann Weyl (1885–1955), and Ehresmann (1905–1979). In terms of physics, this development culminated in both Einstein's theory of general relativity (Standard Model in cosmology) and the Standard Model in particle physics.

Projective geometry played a crucial role in the 19th century. In the 1850s, Cayley (1821–1895) gave a complete classification for cubic and biquadratic invariants

[1] Euler (1707–1783), Lagrange (1736–1813), Cauchy (1789–1857).

E. Zeidler, *Quantum Field Theory III: Gauge Theory*,
DOI 10.1007/978-3-642-22421-8_9,

in projective geometry. His work was continued by Sylvester (1814–1897), Salmon (1819–1904), Clebsch (1833–1872), and Gordan (1837–1912).

Invariant theory was revolutionized by Hilbert (1862–1943) in the 1890s. Hilbert solved the main problem of invariant theory by showing that there exists only a finite number of basic invariants for a large class of problems. The predecessors of Hilbert tried to compute explicitly the basic invariants. In contrast of this, Hilbert gave an abstract existence proof of great generality (Hilbert's basis theorem and syzygies). Nowadays computer algebra is used in order to compute invariants. The algorithms are of high complexity. In 1904, Hilbert created functional analysis by studying infinite-dimensional quadratic forms. In 1928, von Neumann (1903–1957) generalized Hilbert's approach to the spectral theory for unbounded self-adjoint operators in Hilbert space. This is the basic mathematical tool in quantum mechanics.

The theory of differential invariants was founded by Lie (1849–1899). Using results due to Killing (1847–1923), Élie Cartan completely classified the representations of the semisimple complex Lie algebras in his 1894 thesis. The representation theory of groups was independently founded by Burnside (1852–1927) and Frobenius (1849–1917) in about 1900. This approach was simplified by Schur (1856–1932). In the 1920s, Weyl (1885–1955) created a general analytic approach to the representation theory of the classic Lie groups. This work was presented in the following fundamental monograph:

> H. Weyl, The Classical Groups: Their Invariants and Representations, Princeton University Press, 1938 (15th edition, 1997).

This book represents a high-light in mathematics.

Integral invariants were used by Poincaré (celestial mechanics), Hilbert (invariant integral in the calculus of variations), and Élie Cartan. In topology, the most important task is the construction of powerful topological invariants. Algebraic topology was founded by Poincaré (1854–1912) in 1895. The development of mathematics in the 20th was strongly influenced by solving problems arising in algebraic topology and by their relations to algebraic geometry and global differential geometry.

8.2 The Basic Philosophy

All of the results investigated in this chapter can be obtained by means of completely elementary computations. One has only to use

- the chain rule in classical calculus, and
- algebraic relations based on permutations.

For historical reasons, there exist apparently different approaches in the mathematical and physical literature. However, it turns out that all the approaches are equivalent to each other. In this chapter, for the convenience of the reader, we would like to give a survey in the spirit of most textbooks in physics; this dates back to Einstein's classic 1915/16 papers on the theory of general theory of relativity. We emphasize the analytical and computational aspects.

> *The advantage is that the calculus presented in this chapter works by its own.*

Therefore, it is a perfect calculus in the sense of Leibniz (1646–1716). The disadvantage is that the geometric and physical intuition behind the analytical approach is not visible; the intuitive meaning will be discussed in later chapters.

> *Our main task is the construction of invariants.*

We will discuss this for the following objects:

- real physical fields (tensor analysis and differential forms),
- symmetries of physical fields and gauge transformations,
- complex physical fields and complex (resp. almost complex) geometry (e.g., spinor analysis).

In an axiomatic way, the notion of the *connection on a vector bundle* (i.e., the notion of a covariant directional derivative) is behind modern differential geometry. For pedagogical reasons, we will first study concrete examples before passing to the axioms later on.

Analytic approach. In analysis and physics, differential operators play a key role. In this chapter, we will investigate linear differential operators of order r given by

$$\boxed{D := T^{i_1 i_2 \ldots i_r}(x)\,\frac{\partial^r}{\partial x^{i_1}\partial x^{i_2}\cdots\partial x^{i_r}}.} \tag{8.1}$$

Here, we sum over $i_1, \ldots, i_r = 1, \ldots, n$, and we set $x := (x^1, \ldots, x^n)$. The smooth coefficient functions $T^{i_1 i_2 \ldots i_r}$ are called a tensorial family. These coefficient functions depend on the choice of the local coordinate system; in terms of physics, they depend on the choice of the observer. In contrast to this, the differential operator D itself is an invariant object which does not depend on the choice of the local coordinates. Introducing the symbol $\partial_i := \frac{\partial}{\partial x^i}$, we get

$$D := T^{i_1 i_2 \ldots i_r}(x)\,\partial_{i_1}\partial_{i_2}\cdots\partial_{i_r}. \tag{8.2}$$

Equivalently, we will also write

$$D = T^{i_1 i_2 \ldots i_r}(x)\,\partial_{i_1}\otimes\partial_{i_2}\otimes\cdots\otimes\partial_{i_r}.$$

Geometric approach and velocity vector fields. In geometry, one considers tangent vectors $\mathbf{v} := v^i\mathbf{b}_i$ at the point P of the manifold M under consideration. Here, $\mathbf{b}_1, \ldots, \mathbf{b}_n$ are basic vectors of the tangent space of M at the point P, and we sum over $i = 1, \ldots, n$. In physics, $\mathbf{v}$ is a velocity vector at the point P. In the present case, the relation between analysis, geometry, and physics is given by the fact that one can use the identification

$$\boxed{v^i\frac{\partial}{\partial x^i} \Leftrightarrow v^i\mathbf{b}_i.}$$

That is, velocity vectors can be identified with linear first-order partial differential operators. From the mnemonic point of view, one prefers the notation (8.1) in modern mathematics, since the well-known classic formulas

$$\boxed{\frac{\partial}{\partial x^{i'}} = \frac{\partial x^i}{\partial x^{i'}}\cdot\frac{\partial}{\partial x^i} \qquad \text{and} \qquad dx^{i'} = \frac{\partial x^{i'}}{\partial x^i}\cdot dx^i}$$

yield automatically the correct transformation laws. In addition, $dx^i(\partial_j) = \delta^i_j$ if $i, j = 1, 2, \ldots, n$. This coincides with

$$\mathbf{b}_{i'} = \frac{\partial x^i}{\partial x^{i'}}\cdot\mathbf{b}_i, \qquad \text{and} \qquad v^{i'} = \frac{\partial x^{i'}}{\partial x^i}\cdot v^i,$$

as well as $dx^i(\mathbf{b}_j) = \delta^i_j$ if $i, j, = 1, 2, \ldots, n$. Then the differential operator D corresponds to the tensor

$$T^{i_1 i_2 \ldots i_r}\, \mathbf{b}_{i_1} \otimes \mathbf{b}_{i_2} \otimes \cdots \otimes \mathbf{b}_{i_r}.$$

In the dual setting, we have the r-linear functional

$$T_{i_1 i_1 \ldots i_r}\, dx^{i_1} \otimes \cdots \otimes dx^{i_r}$$

where we sum over $i_1, \ldots, i_r = 1, \ldots, n$.

The mathematical description of physical processes is based on some fundamental differential equations (e.g., the Maxwell equations in electromagnetism, the Einstein equations in the theory of general relativity, the Dirac equation for the relativistic electron, the Weyl equation for the relativistic massless neutrino, the Standard Model in particle physics generalizing the Dirac equation and the Weyl equation). The change of observers corresponds to a transformation of the fundamental differential equations. The main goal of invariant theory is to find differential expressions which possess nice transformation laws.

> *Thus, it is possible to formulate the differential equations in physics in such a way that they are valid for arbitrary observers (Einstein's principle of general relativity).*

Our goal is to present an approach which realizes this program in a mnemonically very elegant way. From the practical point of view, this culminates in the index principle of mathematical physics (see Sect. 9.3 on page 574).

Changing observers. In what follows $x^1, \ldots, x^n$ are real coordinates measured by an observer O in the system Σ of reference. In terms of physics, $x^1, \ldots, x^n$ are space coordinates or space-time coordinates. We set

$$x := \begin{pmatrix} x^1 \\ \vdots \\ x^n \end{pmatrix}.$$

Another observer O' will measure $x^{1'}, \ldots, x^{n'}$ in the system of reference Σ'. The equation

$$\boxed{x' = x'(x), \qquad x \in \Omega} \tag{8.3}$$

describes the change of the coordinates from the observer O to the observer O'. Explicitly,

$$x^{i'} = x^{i'}(x^1, \ldots, x^n), \quad i = 1, \ldots, n.$$

We assume that the map $x \mapsto x'(x)$ is a diffeomorphism from Ω onto Ω', where Ω and Ω' are nonempty, open, arcwise connected subsets of $\mathbb{R}^n$. We set

$$G(x) := \frac{dx'(x)}{dx} = \begin{pmatrix} \frac{\partial x^{1'}(x)}{\partial x^1} & \cdots & \frac{\partial x^{1'}(x)}{\partial x^n} \\ \vdots & \cdots & \vdots \\ \frac{\partial x^{n'}(x)}{\partial x^1} & \cdots & \frac{\partial x^{n'}(x)}{\partial x^n} \end{pmatrix}.$$

Since the map $x \mapsto G(x)$ is continuous, the integer-valued map $x \mapsto \operatorname{sgn}(\det G(x))$ on the arcwise connected open set Ω is constant. This integer is called the sign of the map $x \mapsto x'(x)$ denoted by $\operatorname{sgn}\det(G)$ (or $\operatorname{sgn}\det\left(\frac{dx'}{dx}\right)$). The map $x \mapsto x'(x)$ is called orientation-preserving iff $\operatorname{sgn}\det(G) = 1$.

Different types of indices. Let $n, N = 1, 2, \ldots$ In order to describe real-valued and complex-valued physical fields, we will use the following types of indices:

- Lower case Latin indices run from 1 to n. For example, $k = 1, \ldots, n$.
- Upper case Latin indices run from 1 to N. For example, $A = 1, \ldots N$.
- Dotted and overlined lower case Latin indices run from 1 to n. For example, $\dot{k} = \dot{1}, \ldots \dot{n}$, and $\bar{k} = \bar{1}, \ldots, \bar{n}$.
- Dotted and overlined upper case Latin indices run from 1 to N. For example, $\dot{A} = \dot{1}, \ldots \dot{N}$, and $\bar{A} = \bar{1}, \ldots, \bar{N}$.

Einstein's summation convention. We sum over equal lower (resp. upper) case Latin indices from 1 to n (resp. from 1 to N). For example,

$$T^k_{kl} = \sum_{k=1}^{n} T^k_{kl}, \qquad \psi^A_{AB} = \sum_{A=1}^{N} \psi^A_{AB}.$$

The same convention remains true for dotted and overlined indices. For example,

- $T^{\dot{k}}_{\dot{k}l} = \sum_{\dot{k}=1}^{n} T^{\dot{k}}_{\dot{k}l}, \qquad \psi^{\dot{A}}_{\dot{A}B} = \sum_{\dot{A}=1}^{N} \psi^{\dot{A}}_{\dot{A}B}$, and
- $T^{\bar{k}}_{\bar{k}l} = \sum_{\bar{k}=1}^{n} T^{\bar{k}}_{\bar{k}l}, \qquad \psi^{\bar{A}}_{\bar{A}B} = \sum_{\bar{A}=1}^{N} \psi^{\bar{A}}_{\bar{A}B}$.

Note that we only sum over indices of the same type: lower case Latin indices, upper case Latin indices, dotted lower case Latin indices, dotted upper case Latin indices, overlined lower case Latin indices, overlined upper case Latin indices. An index is called free iff we do not sum over the index. For example, the expressions T^k_{kj} and $\psi^{\bar{A}\dot{B}j}_{\bar{A}\dot{B}}$ possess the free index j.

Dotted indices play a crucial role in spinor calculus. Overlined indices are critically used in complex geometry (e.g., Kähler manifolds), conformal quantum field theory, and string theory. This will be thoroughly studied in Vol. IV on quantum mathematics. In this volume, we will concentrate on Riemannian and pseudo-Riemannian geometry (tangent bundle of a real manifold) and gauge theory on real manifolds (vector and principal bundles over real manifolds).

Different types of partial derivatives. We set

$$\partial_i := \frac{\partial}{\partial x^i}, \qquad \partial_{i'} := \frac{\partial}{\partial x^{i'}}.$$

Later on, we will also use the complex coordinates

$$z^k := x^k + \mathrm{i}y^k, \qquad \bar{z}^k = x^k - \mathrm{i}y^k, \qquad k = 1, \ldots, n$$

where $x^1, y^1, \ldots, x^n, y^n$ are real coordinates. Following Poincaré, we also set

$$\frac{\partial}{\partial z^k} := \frac{\partial}{\partial x^k} - \mathrm{i}\frac{\partial}{\partial y^k}, \quad \frac{\partial}{\partial \bar{z}^k} := \frac{\partial}{\partial x^k} + \mathrm{i}\frac{\partial}{\partial y^k}.$$

Furthermore, we set $\partial_{z^k} := \frac{\partial}{\partial z^k}$ and $\partial_{\bar{z}^k} := \frac{\partial}{\partial \bar{z}^k}$.

8.3 The Mnemonic Principle of the Correct Index Picture

> If we do scientific work, we must often step down from our high horse of grand principles, and dig in the dirt with our noses. When we achieve our purpose, we cover the tracks of our efforts in order to appear as gods of clear thought.
>
> Albert Einstein (1879–1955)

As we will show below, the classical tensor calculus works by its own if one only uses tensorial families, and all the additive terms of the equations possess the same free lower and upper indices. For example, the equation

$$T^k_{kj} + S^{\dot{A}} T_{\dot{A}j} + R^{\bar{A}\dot{B}l}_{\bar{A}\dot{B}lj} = 0$$

is correctly formulated. The three additive terms possess the same free lower index j and no other free indices. We also say that the remaining indices $k, \dot{A}, \bar{A}, \dot{B}, l$ are killed (by summation). In contrast to this, the expression

$$T^k_{kj} + S^{kj}_k = 0$$

does not have the correct index picture, since the free index j appears as lower and upper index. As we will show below, if the index picture is right, then all the additive terms transform the same way under a change of the system of reference from Σ to Σ'. In what follows, the reader should always check that the equations have the correct index picture (see Problem 8.1).

> *The Einstein summation convention together with the principle of the correct index picture are beautiful mnemonic tools in classical tensor algebra and tensor analysis.*

In what follows, the principle of the correct index picture will be briefly called the index principle. The final version of the index principle (including the inverse index principle) will be formulated in Sect. 8.8.2 on page 493. Experienced mathematicians and physicists use the index principle in order to detect errors of formulas in tensor calculus (see Problem 8.1).

According to Felix Klein's Erlangen program from 1872, differential geometry is the invariant theory of transformation groups (symmetry group of the geometry). We want to show how to construct explicitly such invariants in a systematic way.

8.4 Real-Valued Physical Fields

For the convenience of the reader, we restrict ourselves to the simplest situation in order to explain many interrelationships between apparently different approaches used in mathematics and physics. Our starting point is the study of tensor calculus based on velocity vector fields $\mathbf{v}$. In terms of modern mathematics, velocity vector fields are sections of the tangent bundle of manifolds. A reader who understands the special approach based on velocity vector fields (and, equivalently, based on linear first-order differential operators) will easily understand the general approach

$$\mathbf{v} \Rightarrow \psi$$

where velocity vector fields $\mathbf{v}$ are replaced by general real-valued and complex-valued physical fields ψ. In terms of modern mathematics, the physical fields ψ are sections of vector bundles over manifolds which generalize the tangent bundle of a manifold. Einstein's theory of general relativity for gravitation is related to the $\mathbf{v}$-approach, whereas Maxwell's theory of electromagnetism and the more general Standard Model in particle physics are related to the ψ-approach.

8.4.1 The Chain Rule and the Key Duality Relation

Invariant theory is essentially based on the Leibniz chain rule in differential calculus.

Folklore

Let us consider the diffeomorphism $x' = x(x)$ from Ω to Ω', and the inverse map $x = x(x')$ from Ω' onto Ω. Then we have the so-called duality relation

$$\boxed{\frac{\partial x^i(x')}{\partial x^{i'}} \cdot \frac{\partial x^{i'}(x)}{\partial x^j} = \delta^i_j, \quad i,j = 1,\ldots,n, \quad x' = x'(x).} \tag{8.4}$$

This follows from $\frac{\partial x^i}{\partial x^j} = \delta^i_j$ together with the chain rule. Interchanging x and x', we also get

$$\frac{\partial x^{i'}(x)}{\partial x^i} \cdot \frac{\partial x^i(x')}{\partial x^{j'}} = \delta^{i'}_{j'}, \quad i', j' = 1', \ldots, n', \quad x' = x'(x). \tag{8.5}$$

Let us reformulate the duality relation (8.4) into the language of matrices. To this end, we set

$$G^{i'}_i(x) := \frac{\partial x^{i'}(x)}{\partial x^i}, \qquad G^i_{i'}(x) := \frac{\partial x^i(x')}{\partial x^{i'}}_{|x'=x'(x)}.$$

Then

$$G(x) := \begin{pmatrix} G^{1'}_1(x) & \ldots & g^{1'}_n(x) \\ \vdots & \ldots & \vdots \\ G^{n'}_1(x) & \ldots & G^{n'}_n(x) \end{pmatrix}. \tag{8.6}$$

Furthermore, it follows from (8.5) that

$$(G(x)^{-1})^d = \begin{pmatrix} G^1_{1'}(x) & \ldots & G^n_{1'}(x) \\ \vdots & \ldots & \vdots \\ G^1_{n'}(x) & \ldots & G^n_{n'}(x) \end{pmatrix}, \quad G(x)^{-1} = \begin{pmatrix} G^1_{1'}(x) & \ldots & G^1_{n'}(x) \\ \vdots & \ldots & \vdots \\ G^n_{1'}(x) & \ldots & G^n_{n'}(x) \end{pmatrix}.$$

Recall that the Jacobian of the transformation $x' = x'(x)$ is defined by

$$\frac{\partial(x^{i'}, \ldots, x^{n'})}{\partial(x^1, \ldots, x^n)}(x) := \det G(x), \quad x \in \Omega.$$

The sign of a coordinate transformation. Since the map $x \mapsto G(x)$ is continuous on the arcwise connected set Ω, the integer-valued continuous map $x \mapsto \det G(x)$ is constant on Ω. We define

$$\operatorname{sgn}\left(\frac{dx'}{dx}\right) := \operatorname{sgn}(\det G(x)). \tag{8.7}$$

This is called the sign of the diffeomorphism $x'(.) : \Omega \to \Omega'$. This sign will play a crucial role below if orientation will enter the scene.

The key relations of the classical tensor calculus. Mnemonically, the two key relations read as

$$\boxed{dx^{i'} = \frac{\partial x^{i'}(x)}{\partial x^i}\, dx^i, \qquad \frac{\partial}{\partial x^{i'}} = \frac{\partial x^i(x')}{\partial x^{i'}} \frac{\partial}{\partial x^i}, \qquad i' = 1', \ldots, n'.} \tag{8.8}$$

Equivalently,

$$dx^{i'} = G_i^{i'}(x)\, dx^i, \qquad \partial_{i'} = G_{i'}^i(x)\, \partial_i, \qquad i' = 1', \ldots, n'.$$

Introducing the column matrices

$$dx := \begin{pmatrix} dx^1 \\ \vdots \\ dx^n \end{pmatrix}, \qquad \partial := \begin{pmatrix} \partial_1 \\ \vdots \\ \partial_n \end{pmatrix},$$

the key relations (8.8) read as

$$\boxed{dx' = \frac{dx'}{dx}\, dx = G(x)\, dx, \qquad \partial' = (G(x)^{-1})^d\, \partial, \qquad x \in \Omega.} \tag{8.9}$$

Let us discuss this.

8.4.2 Linear Differential Operators

Sophus Lie (1842–1899) based his approach to group theory and to differential geometry on linear differential operators. We will use linear differential operators as a point of departure for classical tensor calculus. To this end, consider the smooth maps

$$\Theta_O : \Omega_O \to \mathbb{R}, \qquad \Theta_{O'} : \Omega_{O'} \to \mathbb{R}.$$

In terms of physics,

- $\Theta_O(x)$ is the temperature measured by the observer O at the point x;
- $\Theta_{O'}(x')$ is the temperature measured by the observer O' at the point x'.

We postulate that

$$\Theta_{O'}(x') = \Theta_O(x) \qquad \text{for all} \quad x \in \Omega, \;\; x' = x'(x).$$

This means that the two observers O and O' measure the same temperature at the same point described by the local coordinate x and x', respectively. The family $\{\Theta_O\}$ is called a temperature field with respect to the system $\{O\}$ of observers.[2]

- The observer O computes the linear differential operator

$$\boxed{(L_O\Theta_O)(x) := v^i(x)\, \partial_i \Theta_O(x), \qquad x \in \Omega}$$

  where the functions $v^i : \Omega \to \mathbb{R}, i = 1, \ldots, n$, are smooth.
- The observer O' computes

$$(L_{O'}\Theta_{O'})(x') := v^{i'}(x')\, \partial_{i'} \Theta_{O'}(x'), \qquad x' \in \Omega'$$

  where the functions $v^{i'} : \Omega' \to \mathbb{R}, i' = 1', \ldots, n'$, are smooth.

[2] In the sense of tensorial families to be introduced on page 452, we briefly say that Θ is a scalar tensorial family.

We postulate that

$$(L_{O'}\Theta_{O'})(x') = (L_O\Theta_O)(x) \qquad \text{for all} \quad x \in \Omega, \ \ x' = x'(x). \tag{8.10}$$

Our goal is to compute the transformation from the coefficients $v^1(x), \ldots, v^n(x)$ measured by O to the coefficients $v^{1'}(x'), \ldots, v^{n'}(x')$ measured by O'. The chain rule tells us that

$$\partial_{i'}\Theta_{O'}(x') = \frac{\partial x^i(x')}{\partial x^{i'}}\, \partial_i\Theta_O(x). \tag{8.11}$$

Interchanging the observers, we get

$$\partial_i\Theta_O(x) = \frac{\partial x^{i'}(x)}{\partial x^i}\, \partial_{i'}\Theta_{O'}(x'). \tag{8.12}$$

This implies

$$\boxed{v^{i'}(x') = \frac{\partial x^{i'}(x)}{\partial x^i}\, v^i(x), \qquad i' = 1', \ldots, n', \quad x' = x'(x).} \tag{8.13}$$

Proof. By (8.12), $v^i\partial_i\Theta_O = v^i \frac{\partial x^{i'}}{\partial x^i}\, \partial_{i'}\Theta_{O'}$. It follows from (8.10) that

$$v^{i'} \cdot \partial_{i'}\Theta_{O'} = v^i\partial_i\Theta_O = \left(\frac{\partial x^{i'}}{\partial x^i}\, v^i\right) \cdot \partial_{i'}\Theta_{O'}.$$

This is valid for all smooth functions $\Theta_{O'}$. Therefore, we get (8.13). □

8.4.3 Duality and Differentials

Basic definition. We define $dx^1, \ldots, dx^n$ by setting

$$dx^i(v^j(x)\,\partial_j) := v^i(x), \qquad i = 1, \ldots, n. \tag{8.14}$$

This implies the transformation law

$$dx^{i'} = \frac{\partial x^{i'}(x)}{\partial x^i}\, dx^i. \tag{8.15}$$

Proof. Note that

$$dx^{i'}(v^{j'}\partial_{j'}) = v^{i'}, \qquad dx^i(v^j\partial_j) = v^i.$$

Since $v^{j'}\partial_{j'} = v^j\partial_j$, the linear functional dx^i transforms like v^i. Thus, the claim follows from (8.13). □

Note that the transformation laws (8.11) and (8.15) yield the mnemonic key relations (8.8) on page 446.

Velocity vector fields. Consider the partial differential equation

$$\boxed{v^i(x)\,\partial_i\Theta(x) = 0} \tag{8.16}$$

for the scalar tensorial family Θ together with the following system of ordinary differential equations:

$$\frac{dx^i(t)}{dt} = v^i(x(t)), \qquad -t_0 < t < t_0, \quad i = 1, \ldots, n. \tag{8.17}$$

In mathematics, the system (8.17) is called the characteristic system to (8.16). In terms of physics, the solutions $x = x(t)$ of (8.17) describe the motion of fluid particles with the velocity components $\frac{dx^i(t)}{dt}$, $i = 1, \ldots, n$, at time t. The smooth function $x \mapsto v(x)$ represents the velocity vector field of a fluid with the velocity components $v^1(x), \ldots, v^n(x)$ at the point x in the system of reference Σ.

If the temperature function Θ is a smooth solution of (8.16), *then the temperature is constant along the trajectories of the fluid particles.*

This follows from

$$\frac{d\Theta(x(t))}{dt} = \frac{dx^i(t)}{dt}(\partial_i\Theta)(x(t)) = v^i(x(t))(\partial_i\Theta)(x(t)) \equiv 0.$$

Wedge product. We define $dx^i \wedge dx^j$ by setting

$$(dx^i \wedge dx^j)(v, w) := dx^i(v)dx^j(w) - dx^i(w)dx^j(v) = v^i(x)w^j(x) - w^i(x)v^j(x)$$

for all linear first-order differential operators $v = v^i\partial_i$ and $w = w^i\partial_i$ with smooth coefficients. More general, let $p = 1, \ldots, n$. By definition,

$$(dx^{i_1} \wedge \cdots \wedge dx^{i_p})(v_1, \ldots, v_p) := \varepsilon_{i_1 \ldots i_p}\, dx^{i_1}(v_1) \cdots dx^{i_p}(v_p)$$

for all linear first-order differential operators $v_1, \ldots, v_p$ with smooth coefficients.

Terminology. Let Ω be an open nonempty subset of $\mathbb{R}^n$. Then:

- $C^\infty(\Omega, \mathbb{R})$ denotes the real linear space of all the smooth functions $f : \Omega \to \mathbb{R}$. Alternatively, we write $\Lambda^0(\Omega)$ instead of $C^\infty(\Omega, \mathbb{R})$.
- $C_0^\infty(\Omega, \mathbb{R})$ denotes the real linear space of all the smooth functions $f : \Omega \to \mathbb{R}$ which have a compact support, that is, they vanish outside a compact subset of Ω (e.g., a closed ball).
- $\mathsf{Diff}^1(\Omega)$ denotes the real linear space of all the linear first-order differential operators
$$v : C^\infty(\Omega) \to \mathbb{R}$$
with smooth coefficient functions, that is, $v = v^i\partial_i$ and $v^i \in C^\infty(\Omega)$ for all $i = 1, \ldots, n$.[3]
- $\mathsf{Diff}^m(\Omega)$ denotes the real linear space of all the linear mth-order differential operators
$$V : C^\infty(\Omega) \to \mathbb{R}$$
with smooth coefficient functions, that is, $V = v^{i_1 i_2 \ldots i_m}\partial_{i_1}\partial_{i_2}\cdots\partial_{i_m}$ with $v^{i_1 i_2 \ldots i_m} \in C^\infty(\Omega)$ for all $i_1, \ldots, i_m = 1, \ldots, n$. Here, $m = 1, 2, \ldots$
- The differential dx^i introduced above is a linear functional
$$dx^i : \mathsf{Diff}^1(\Omega) \to \mathbb{R}, \qquad i = 1, \ldots, n.$$
- $\Lambda^1(\Omega)$ denotes the real linear space of all the linear functionals
$$v_i dx^i : \mathsf{Diff}^1(\Omega) \to \mathbb{R}$$
with smooth coefficient functions, that is, $v_i \in C^\infty(\Omega)$ if $i = 1, \ldots, n$.

[3] We will show in Sect. 8.8.1 on page 487 that $\mathsf{Diff}^1(\Omega)$ is a real Lie algebra.

- $\Lambda^p(\Omega)$ with $p = 1, 2, 3, \ldots, n$ denotes the real linear space of all the smooth p-forms
$$\omega := \frac{1}{p!}\omega_{i_1 \ldots i_p}\, dx^{i_1} \wedge \cdots \wedge dx^{i_p}$$
where $\omega_{i_1 \ldots i_p} \in C^\infty(\Omega)$ for all $i_1, \ldots, i_p = 1, \ldots, n$, and $\omega_{i_1 \ldots i_p}$ is antisymmetric with respect to all the indices. The elements ω of $\Lambda^2(\Omega)$ are bilinear antisymmetric functionals of the form
$$\omega : \mathsf{Diff}^1(\Omega) \times \mathsf{Diff}^1(\Omega) \to \mathbb{R}.$$
For $p = 1, 2, \ldots$, the elements ω of $\Lambda^p(\Omega)$ are p-linear antisymmetric functionals of the type $\omega : \mathsf{Diff}^1(\Omega) \times \cdots \times \mathsf{Diff}^1(\Omega) \to \mathbb{R}$.

8.4.4 Admissible Systems of Observers

We want to describe the transformations between several observers. The reader who wants to parallel the general setting with a concrete example in Euclidean geometry should compare the following material with Sect. 9.1 on classical vector analysis. Fix $n = 1, 2, \ldots$ The point of departure is the transformation law

$$\boxed{x' = x'(x), \qquad x \in \Omega, \quad \Omega \subseteq \mathbb{R}^n.} \tag{8.18}$$

The compatibility condition for observers. An admissible system $\mathcal{O}$ of observers is characterized by the following situation: There exists a nonempty set of symbols $O, O', O'', \ldots$ called observers.

- For any observer O, there exists a uniquely determined open, arcwise connected, nonempty subset Ω_O of $\mathbb{R}^n$ called the coordinate system of O.
- For any ordered pair (O, O') of observers, there exists a uniquely determined diffeomorphism
$$f_{O',O} : \Omega_O \to \Omega_{O'}$$
called a coordinate transformation (e.g., a transformation $x \mapsto x'$ of space coordinates or space-time coordinates in Einstein's theory of general relativity). In addition, suppose that
$$f_{O,O'} = f^{-1}_{O',O}$$
for all observers O, O'. In particular, $f_{O,O} = \mathrm{id}$.
- For any ordered triplet (O, O', O'') of observers, the following diagram is commutative:

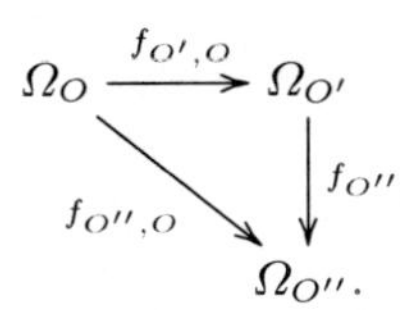

This means that we get the composition rule[4]

$$\boxed{f_{O'',O'} \circ f_{O',O} = f_{O'',O}} \tag{8.19}$$

for the change $O \Rightarrow O' \Rightarrow O''$ of observers. In terms of physics, the observer O measures the coordinate x, and the observer O' measures the coordinate $x' = f_{O',O}(x)$.

[4] Mnemonically, one has to read this equation from right to left.

To simplify notation, we frequently write $x'(x)$ instead of $f_{O',O}(x)$. The relation (8.19) tells us that the measurements carried out by three observers O, O', O'' are compatible with each other. In fact, the following hold:

- If the observer O measures the coordinate x, then the observer O' measures the coordinate $x' = f_{O',O}(x)$, and the observer O'' measures
$$x'' = f_{O'',O}(x). \tag{8.20}$$
- The passage from the observer O' to the observer O'' yields
$$x'' = f_{O'',O'}(x'). \tag{8.21}$$

Naturally enough, the relation (8.19) tells us that the two values (8.20) and (8.21) coincide.

Linearization of the diffeomorphism $f_{O',O}$ at the point x yields

$$\frac{dx'(x)}{dx} = f'_{O',O}(x) := \left(\frac{\partial x^{i'}(x)}{\partial x^i}\right).$$

To simplify notation, we will frequently write $G_{O',O}(x)$ or briefly $G(x)$ instead of $f'_{O',O}(x)$ (see (8.6) above). By the chain rule, linearization of the compatibility relation (8.19) yields

$$\boxed{f'_{O'',O'}(x') \circ f'_{O',O}(x) = f'_{O'',O}(x)} \tag{8.22}$$

where $x' = f_{O',O}(x)$ and $x'' = f_{O'',O}(x)$. Equivalently, we have the matrix product

$$\boxed{G_{O'',O'}(x')G_{O',O}(x) = G_{O'',O}(x)} \tag{8.23}$$

for the change $O \Rightarrow O' \Rightarrow O''$ of observers.

Duality. Using the matrix rules $(AB)^{-1} = B^{-1}A^{-1}$ and $(AB)^d = B^dA^d$, it follows from (8.23) that

$$\boxed{H_{O'',O'}(x')H_{O',O}(x) = H_{O'',O}(x)} \tag{8.24}$$

where $H_{O',O}(x) := (G_{O',O}(x)^{-1})^d$ denotes the contragredient matrix to $G_{O',O}(x)$. Relation (8.24) is called the dual linearized compatibility condition. As we will show below, the use of both (8.23) and (8.24) is crucial for classical tensor calculus. Recall that $GL(n,\mathbb{R})$ denotes the group of all real invertible $(n \times n)$-matrices. Note the following. If we set $\chi(G) := (G^{-1})^d$ for all $G \in GL(n,\mathbb{R})$, then the map

$$\chi : GL(n,\mathbb{R}) \to GL(n,\mathbb{R})$$

is a group isomorphism which is called the dual (or contragredient) representation of the group $GL(n,\mathbb{R})$. As we will show in Sect. 8.4.5, the dual representation is crucial for the construction of invariants (principle of killing indices).

Symmetry. Let $\mathcal{G}$ be a subgroup of the group $GL(n,\mathbb{R})$. The admissible system of observers is called a $\mathcal{G}$-system iff

$$G_{O',O}(x) \in \mathcal{G} \qquad \text{for all} \quad x \in \Omega_O$$

and for all pairs (O, O') of observers.

Examples. (i) Consider the linear transformation law

$$x' = Gx, \qquad x \in \mathbb{R}^n \tag{8.25}$$

for all matrices G in the given subgroup $\mathcal{G}$ of the group $GL(n,\mathbb{R})$ (for example, choose $\mathcal{G} = GL(n,\mathbb{R})$ or $\mathcal{G} = SO(n)$).[5] Intuitively, the group $\mathcal{G}$ describes symmetry properties of the transformations between the observers. For example, in Einstein's theory of special relativity, we have $n = 4$ (four space-time coordinates), and $\mathcal{G}$ equals the orthochronous Lorentz group $SO^{\uparrow}(1,3)$. In this case, the equation (8.25) describes the change of inertial systems where the orientation of time is preserved (see Chap. 18).

The corresponding admissible system of observers can be easily obtained in the following way. We choose all the symbols O_G with $G \in \mathcal{G}$ as observers, that is, the observers are labelled by the group elements. We assign to every pair (O_I, O_G) of observers the linear isomorphism

$$G : \mathbb{R}^n_I \to \mathbb{R}^n_G.$$

More generally, we assign to any pair (O_G, O_H) of observers the linear isomorphism

$$HG^{-1} : \mathbb{R}^n_G \to \mathbb{R}^n_H.$$

Then the following diagram is commutative:

$$\begin{array}{ccc} \mathbb{R}^n_I & \xrightarrow{G} & \mathbb{R}^n_G \\ & {\scriptstyle H}\searrow & \downarrow {\scriptstyle HG^{-1}} \\ & & \mathbb{R}^n_H . \end{array}$$

This corresponds to the three observers O_I, O_G, O_H where

- the composed passage $O_I \to O_G \to O_H$ corresponds to
- the passage $O_I \to O_H$.

In fact, we have the composition rule $HG^{-1} \cdot G = H$. Moreover, the commutativity of the following diagram

$$\begin{array}{ccc} \mathbb{R}^n_J & \xrightarrow{GJ^{-1}} & \mathbb{R}^n_G \\ & {\scriptstyle HJ^{-1}}\searrow & \downarrow {\scriptstyle HG^{-1}} \\ & & \mathbb{R}^n_H \end{array}$$

describes the compatibility of the three observers O_J, O_G, O_H. This means that the composed passage $O_J \to O_G \to O_H$ corresponds to the passage $O_J \to O_H$. In fact,

$$(HG^{-1})(GJ^{-1}) = HJ^{-1}.$$

(ii) Consider the nonlinear transformation law

$$x' = x'(x), \qquad x \in \mathbb{R}^n. \tag{8.26}$$

Here, the map

[5] Note that, in the present special case, the matrix G does not depend on the variable x.

$$f : \mathbb{R}^n \to \mathbb{R}^n$$

is an arbitrary diffeomorphism, and we set $x'(x) := f(x)$. Then

$$dx' = f'(x)dx.$$

Replacing G in (i) by the linearization f', we get an admissible system of observers $O_{f'}$ labelled by f'.

Orientation. Consider an admissible system of observers. We assign to each observer O a number $\iota_O = 1$ or $\iota_O = -1$ such that

$$\boxed{\iota_{O'} = \operatorname{sgn}\left(\frac{dx'}{dx}\right) \cdot \iota_O} \tag{8.27}$$

for all the changes from the observer O to the observer O'. Here, $\operatorname{sgn}(\frac{dx'}{dx})$ is the sign of the change $x \mapsto x'(x)$ of local coordinates (see (8.7) on page 445). The integer ι_O is called the orientation number (or parity number) of the observer O. The function

$$O \mapsto \iota_O$$

is called an orientation function of the given admissible system of observers. Such a function always exists. To show this, fix an observer O_0. Define $\iota_{O_0} := 1$. Moreover, define

$$\iota_O := \operatorname{sgn}\left(\frac{dx}{dx_0}\right)$$

for all observers O. By the chain rule, $\frac{dx'}{dx_0} = \frac{dx'}{dx} \cdot \frac{dx}{dx_0}$. Hence

$$\iota_{O'} := \operatorname{sgn}\left(\frac{dx'}{dx} \cdot \frac{dx}{dx_0}\right).$$

Because of the product rules $\det(AB) = \det A \cdot \det B$ and $\operatorname{sgn}(ab) = \operatorname{sgn} a \cdot \operatorname{sgn} b$, we get

$$\iota_{O'} := \operatorname{sgn}\left(\frac{dx'}{dx}\right) \cdot \operatorname{sgn}\left(\frac{dx}{dx_0}\right) = \operatorname{sgn}\left(\frac{dx'}{dx}\right) \cdot \iota_O.$$

This yields (8.27) which finishes the proof.

The family $\{\iota_O\}$ equipped with the transformation law (8.27) is called a pseudo-invariant. An admissible system of observers is called *oriented* iff all the diffeomorphisms $x \mapsto x'(x)$ (which describe the change of local coordinates of the observers) have a positive sign.

8.4.5 Tensorial Families and the Construction of Invariants via the Basic Trick of Index Killing

Tensorial families. Let $\mathcal{O}$ be an admissible system of observers. By definition, a tensorial family

$$T^{i_1 i_2 \ldots i_r}_{j_1 j_2 \ldots j_s} \tag{8.28}$$

transforms like the product

$$dx^{i_1} dx^{i_2} \cdots dx^{i_r} \partial_{j_1} \partial_{j_2} \cdots \partial_{j_s}. \tag{8.29}$$

More precisely, let $r, s = 1, 2, \ldots$ The indices $i_1, \ldots, i_r, j_1, \ldots, j_s$ run from 1 to n. We assign to each set $\Omega \in \mathcal{O}$ a family of smooth functions

$$T^{i_1 \ldots i_r}_{j_1 \ldots j_s} : \Omega \to \mathbb{R},$$

and the passage from Ω to Ω' corresponds to transformation formulas which are the same as the formulas for (8.29). For example, by the key relations (8.8), we have the transformation formula

$$dx^{i'} \partial_{j'} = G^{i'}_i(x) g^j_{j'}(x) \cdot dx^i \partial_j$$

from Ω to Ω'. Consequently, the tensorial family T^i_j transforms like

$$T^{i'}_{j'}(x') = G^{i'}_i(x) G^j_{j'}(x) \cdot T^i_j(x), \qquad x' = x'(x).$$

The tensorial family (8.28) is called r-fold contravariant and s-fold covariant. We also say that the tensorial family (8.28) is of type (r, s).

Examples of tensorial families. Define $\delta^i_j = \delta_{ij} = \delta^{ij} := 1$ (resp. $= 0$) if $i = j$ (resp. $i \neq j$), and

$$\varepsilon^{i_1 i_2 \ldots i_n} = \varepsilon_{i_1 i_2 \ldots i_n} := \operatorname{sgn} \begin{pmatrix} 1 & 2 \ldots & n \\ i_1 & i_2 \ldots & i_n \end{pmatrix}$$

where $i, j, i_1, \ldots, i_n = 1, \ldots, n$. For example,

- $\varepsilon^{12 \ldots n} = \varepsilon_{12 \ldots n} = 1$,
- $\varepsilon^{2134 \ldots n} = -\varepsilon^{1234 \ldots n} = -1$, and $\varepsilon^{1134 \ldots n} = 0$.

Note that $\varepsilon^{i_1 \ldots i_n}$ changes sign (resp. remains unchanged) under an odd (resp. even) permutation of the indices. Then the following hold:

(i) δ^i_j is a tensorial family for every admissible system $\mathcal{O}$ of observers.

(ii) Let $\mathcal{O}$ be an admissible system of observers as given by (8.25) with respect to the special orthogonal group $\mathcal{G} = SO(n)$. Then

$$\delta^{ij}, \quad \delta_{ij}, \quad \varepsilon_{i_1 \ldots i_n}, \quad \varepsilon^{i_1 \ldots i_n}$$

are tensorial families. Note that these tensorial families have the same position-independent values in every system of reference. Such important tensorial families are called form-invariant.

(iii) Let $\mathcal{O}$ be an admissible system of observers as given by (8.25) with respect to the orthogonal group $\mathcal{G} = O(n)$. Then δ_{ij} and δ^{ij} are tensorial families. However, $\varepsilon_{i_1 \ldots i_n}$ and $\varepsilon^{i_1 \ldots i_n}$ are not tensorial families, but they are pseudo-tensorial families in the sense of the definition given below.

(iv) Let $\mathcal{O}$ be an arbitrary admissible system of observers. Then, in contrast to δ^i_j, the family δ_{ij} is not always a tensorial family.

These examples show that tensorial families depend critically on the choice of the admissible system of observers and on the index picture. Changing the index picture may destroy tensorial families (e.g., the change from δ^i_j to δ_{ij}).

Proof. Ad (i). It follows from the key duality relation (8.5) that

$$\delta^{i'}_{j'} = G^{i'}_i G^j_{j'} \delta^i_j.$$

Ad (ii), (iii). If $G \in O(n)$, then $GG^d = I$. This implies

$$\delta^{i'j'} = G^{i'}_i G^{j'}_j \, \delta^{ij}.$$

Moreover, it follows from $(G^{-1})^d G^{-1} = I$ that

$$\delta_{i'j'} = G^i_{i'} G^j_{j'} \delta_{ij}.$$

Furthermore, we will use the determinant formulas

$$\det G \cdot \varepsilon^{i'_1 i'_2 \ldots i'_n} = G^{i'_1}_{i_1} G^{i'_2}_{i_2} \cdots G^{i'_n}_{i_n} \varepsilon^{i_1 i_2 \ldots i_n}, \tag{8.30}$$

and

$$\det G^{-1} \cdot \varepsilon_{i'_1 i'_2 \ldots i'_n} = G^{i_1}_{i'_1} g^{i_2}_{i'_2} \cdots G^{i_n}_{i'_n} \varepsilon_{i_1 i_2 \ldots i_n} \tag{8.31}$$

which are valid for all complex $(n \times n)$-matrices G. Thus, if $\det G = \pm 1$, then

$$\varepsilon^{i'_1 i'_2 \ldots i'_n} = \operatorname{sgn}(\det G) \cdot G^{i'_1}_{i_1} G^{i'_2}_{i_2} \cdots G^{i'_n}_{i_n} \cdot \varepsilon^{i_1 i_2 \ldots i_n}, \tag{8.32}$$

and

$$\varepsilon_{i'_1 i'_2 \ldots i'_n} = \operatorname{sgn}(\det G) \cdot G^{i_1}_{i'_1} G^{i_2}_{i'_2} \cdots G^{i_n}_{i'_n} \cdot \varepsilon_{i_1 i_2 \ldots i_n}. \tag{8.33}$$

If $G \in SO(n)$, then $\det G = 1$. Hence $\operatorname{sgn}(\det G) = 1$. Thus, the claim (ii) follows from (8.32) and (8.33).

If $G \in O(n)$, then $\operatorname{sgn}(\det G) = \pm 1$, and the transformation laws (8.32) and (8.33) depend on the sign of the determinant of the transformation matrix G. By the definition given below, $\varepsilon^{i_1 i_2 \ldots i_n}$ and $\varepsilon_{i_1 i_2 \ldots i_n}$ are pseudo-tensorial families.

Ad (iv). Let $G = 2I$. Then $\delta_{i'j'} = 4G^i_{i'} G^j_{j'} \delta_{ij}$. This is not a tensorial transformation law. □

A real or complex $(n \times n)$-matrix is called unimodular iff $\det G = 1$. In particular, the matrices $G \in SL(n, \mathbb{C})$ are unimodular.

Tensorial families with two indices. Using the language of matrices, the transformation laws for the tensorial families T^i_j, T_{ij}, T^{ij} read as follows:

(i) $(T^{i'}_{j'}(x')) = G(x)(T^i_j(x))G(x)^{-1}$,
(ii) $(T_{i'j'}(x')) = (G(x)^d)^{-1}(T_{ij}(x))G(x)^{-1}$,
(iii) $(T^{i'j'}(x')) = G(x)(T^{ij}(x))G(x)^d$.

Here, i', i are row indices, and j, j' are column indices. Moreover, $x' = x'(x)$.
Proof. Ad (i). By (8.8), we get

$$\begin{pmatrix} dx^{1'} \\ \vdots \\ dx^{n'} \end{pmatrix} (\partial_{1'}, \ldots \partial_{n'}) = G(x) \begin{pmatrix} dx^1 \\ \vdots \\ dx^n \end{pmatrix} (\partial_1, \ldots, \partial_n) G(x)^{-1}.$$

Hence

$$(dx^{i'} \partial_{j'}) = G(x)(dx^i \partial_j) G(x)^{-1}.$$

Finally, note that T^i_j transforms like $dx^i \partial_j$.

Ad (ii), (iii). Argue similarly. □

Using the trace of matrices, it follows from $\operatorname{tr}(ABC) = \operatorname{tr}(BCA)$ and (i) that

$$T^{i'}_{i'}(x') = \operatorname{tr}(T^{i'}_{j'}(x')) = \operatorname{tr}(T^i_j(x)) = T^i_i(x).$$

Furthermore, for the determinants it follows from $\det(AB) = \det A \cdot \det B$ that

(a) $\det(T^{i'}_{j'}(x')) = \det(T^i_j(x))$,
(b) $\det(T_{i'j'}(x')) = (\det G(x))^{-2} \det(T_{ij}(x))$,
(c) $\det(T^{i'j'}(x')) = (\det G(x))^2 \det(T^{ij}(x))$.

If $G(x) = 1$ for all $x \in \Omega$ (i.e., the coordinate transformation is unimodular), then the determinants $\det(T_{ij}(x))$ and $\det(T^{ij}(x))$ remain unchanged under the passage from the system Σ of reference to Σ'.

Let T_{ij} be a tensorial family with $\det(T_{ij}(x)) \neq 0$ for all $x \in \Omega$. Let T^{ij} denote the entries of the inverse matrix, that is,

$$(T^{ij}(x)) := (T_{ij}(x))^{-1}, \qquad x \in \mathbb{R} \tag{8.34}$$

where i (resp. j) denotes the row (resp. column) index. Then T^{ij} is a tensorial family.

Proof. Passing to the inverse matrices, relation (i) on page 454 implies that

$$(T_{i'j'}(x))^{-1} = G(x)(T_{ij}(x))^{-1}G(x)^d.$$

Now the claim follows from (iii) on page 454. □

The contraction principle and the construction of invariants. The following procedure represents the heart of tensor algebra. Replacing the tensorial family

$$T^{ij}_k$$

by T^i_{ik} is called a contraction. This way, the free indices i, j, k are reduced to the free index k.[6] In the general case, let

$$T^{i_1 \ldots i_r}_{j_1 \ldots j_s}$$

be a tensorial family with respect to the admissible system $\mathcal{O}$ of observers. By definition, a contraction of this tensorial family is obtained by summing over a fixed pair of an upper and a lower index. The position of the indices does not play any role. For example, the contraction of the pair i_1, j_2 yields

$$T^{i_1\, i_2\, i_3 \ldots i_r}_{j_1\, i_1\, j_3 \ldots j_s}.$$

This operation can be repeated. For example, T^{ij}_{ijm} is a contraction of T^{ij}_{klm}. A contraction is called complete iff there are no free indices after contraction. For example, T^i_i and T^{ij}_{ij} are complete contractions.

Theorem 8.1 *The contraction of a tensorial family yields again a tensorial family with respect to the same admissible system of observers. A complete contraction yields an invariant.*

This is called the 'index principle' in classical tensor calculus. Supplemented by the so-called 'inverse index principle', the final formulation of the index principle can be found on page 493.

Proof. (I) As a typical example, consider the tensorial family T^i_j which transforms like

$$v^i \partial_j.$$

Consequently, T^i_i transforms like $v^i \partial_i$. By Sect. 8.4.2,

$$v^{i'}(x')\partial_{i'} = v^i(x)\partial_i, \qquad x' = x'(x).$$

[6] Recall that $T^i_{ik} := \sum_{i=1}^n T^i_{ik}$, by Einstein's summation convention.

Thus, $T_{i'}^{i'}(x') = T_i^i(x)$. This means that T_i^i is an invariant.

Alternatively, this invariance property follows from the transformation law

$$T_{j'}^{i'}(x') = G_i^{i'}(x)G_{j'}^j(x)T_j^i(x)$$

which implies

$$T_{i'}^{i'} = G_i^{i'}(x)G_{i'}^j(x)T_j^i(x) = \delta_i^j T_j^i = T_i^i,$$

by the duality relation (8.4).

(II) As another example, consider the tensorial family T_{jk}^i which transforms like $v^i\partial_j\partial_k$. Thus, T_{ik}^i transforms like $v^i\partial_i\partial_k$. By (I), T_{ik}^i transforms like ∂_k.

(III) The general case proceeds analogously. □

Algebraic properties of tensorial families. Let $\mathcal{O}$ be an admissible system of observers.

- Sum rule: The sum of two tensorial fields of the same type (with respect to $\mathcal{O}$) yields a tensorial family (with respect to $\mathcal{O}$).
- Linear combinations: The real linear combination of two tensorial fields of the same type (with respect to $\mathcal{O}$) yields a tensorial family (with respect to $\mathcal{O}$).
- Product rule: The product of two tensorial fields of arbitrary type (with respect to $\mathcal{O}$) yields a tensorial family (with respect to $\mathcal{O}$).
- Multiplying a tensorial field (with respect to $\mathcal{O}$) by a real number yields a tensorial family (with respect to $\mathcal{O}$).

Forming sums and real linear combinations of tensorial families does not change the type of the tensorial families. Moreover, the type of the product of two tensorial families is indicated by the index picture. For example, if $T_{jk}^i, S_{jk}^i, R_q^p$ are tensorial families on $\mathcal{O}$ and if α, β are real numbers, then

$$\alpha T_{jk}^i + \beta S_{jk}^i, \quad T_{jk}^i R_q^p$$

are also tensorial fields on $\mathcal{O}$. The proof follows immediately from the definition of tensorial families. For example, T_{jk}^i and R_q^p transform like $dx^i\partial_j\partial_k$ and $dx^p\partial_q$, respectively. Therefore, $T_{jk}^i R_q^p$ transforms like $dx^i\partial_j\partial_k \cdot dx^p\partial_q$.

Lifting and lowering of indices. Since the position of the indices of a tensorial family is crucial, one has to distinguish between lower and upper indices. However, in many situations, one has a tensorial family g_{ij} at hand with the property that the matrix $(g_{ij}(x))$ is invertible for all points $x \in \Omega$.[7] As shown above, the entries $g^{ij}(x)$ of the inverse matrix,

$$(g^{ij}(x)) := (g_{ij}(x))^{-1},$$

generate the tensorial family g^{ij} with

$$g^{is}(x)g_{sj}(x) = \delta_j^i \qquad \text{for all} \quad x \in \Omega. \tag{8.35}$$

This can be used in order to construct new tensorial families by the lifting or lowering of indices. For example, consider the tensorial field T^{ijk}. Set

$$T_i^{jk} := g_{is}T^{sjk}. \tag{8.36}$$

[7] For example, this concerns the Euclidean geometry, the symplectic geometry, the Minkowski geometry in Einstein's theory of special relativity, and Einstein's theory of general relativity.

Then T_i^{jk} is a tensorial family, by the product rule. Moreover, it follows from the contraction principle that T_i^{ik} is also a tensorial family. Furthermore, setting

$$S^{ijk} := g^{is} T_s^{jk},$$

we get the tensorial family S^{ijk}. In addition, it follows from (8.35) and (8.36) that $S^{ijk} = T^{ijk}$.[8] Similarly, it is possible to lift or to lower several indices. For example,

$$S_j^i = g^{is} g_{jr} T_s^r.$$

This transforms the tensorial family T_i^j into the tensorial family S_j^i, by lifting the index i, and by lowering the index j.

Symmetry properties of tensorial families. Permutations of upper (resp. lower) indices send tensorial families again to tensorial families. For example, if T_{ij} is a tensorial family, then so is T_{ji}. Consequently, setting

$$S_{ij} := \tfrac{1}{2}(T_{ij} + T_{ji}),$$

we get a new tensorial family S_{ij} which is symmetric, that is, $S_{ij} = S_{ji}$ for all $i, j = 1, \ldots, n$. We write

$$T_{(i,j)} := \tfrac{1}{2}(T_{ij} + T_{ji}),$$

and we call $T_{(i,j)}$ the symmetrization of T_{ij}. Similarly, setting

$$A_{ij} := \tfrac{1}{2}(T_{ij} - T_{ji}),$$

we get a new tensorial family A_{ij} which is antisymmetric (also called skew-symmetric), that is, $A_{ij} = -A_{ji}$ for all $i, j = 1, \ldots, n$.

If $T_{i_1 \ldots i_r}$ is a tensorial family, then we set

$$\boxed{T_{(i_1 \ldots i_r)} := \frac{1}{r!} \sum_\pi T_{\pi(i_1) \ldots \pi(i_r)}}$$

where we sum over all permutations π of the indices $i_1, \ldots, i_r$. The tensorial family $T_{i_1 \ldots i_r}$ is called symmetric iff

$$T_{i_1 \ldots i_r} = T_{(i_1 \ldots i_r)}$$

for all indices. Equivalently, this means that $T_{i_1 \ldots i_r}$ remains unchanged under a permutation of all the indices. The tensorial field $T_{(i_1 \ldots i_r)}$ is called the symmetrization of $T_{i_1 \ldots i_r}$. In particular,

$$T_{(\pi(i_1) \ldots \pi(i_r))} = T_{(i_1 \ldots i_r)}.$$

In order to describe antisymmetry, we set

$$\boxed{T_{[i_1 \ldots i_r]} := \frac{1}{r!} \sum_\pi \operatorname{sgn}(\pi) \cdot T_{\pi(i_1) \ldots \pi(i_r)}.}$$

For example, $T_{[ij]} = \tfrac{1}{2}(T_{ij} - T_{ji})$. The tensorial family $T_{i_1 \ldots i_r}$ is called antisymmetric iff

$$T_{i_1 \ldots i_r} = T_{[i_1 \ldots i_r]}$$

[8] In fact, $T^{ijk} = g^{is} g_{sr} S^{rjk} = \delta_r^i S^{rjk} = S^{ijk}$.

for all indices. Equivalently, this means that $T_{i_1\ldots i_r}$ changes sign under an odd permutation of the indices, and $T_{i_1\ldots i_r}$ remains unchanged under an even permutation of the indices. The tensorial field $T_{[i_1\ldots i_r]}$ is called the antisymmetrization of $T_{i_1\ldots i_r}$. We have

$$T_{[\pi(i_1)\ldots\pi(i_r)]} = \text{sgn}(\pi)\cdot T_{[i_1\ldots i_r]}. \tag{8.37}$$

For example, if $T_{i_1\ldots i_r}$ remains unchanged under a transposition of two indices, then $T_{[i_1\ldots i_r]}$ vanishes. In particular,

$$T_{[iji_3\ldots i_n]} = 0 \qquad \text{if} \qquad T_{iji_3\ldots i_n} = T_{jii_3\ldots i_n}. \tag{8.38}$$

Sometimes, we will also use a similar procedure in order to symmetrize (or to antisymmetrize) a tensorial family with respect to a subset of indices. For example, we set

$$T_{i(jk)} := \tfrac{1}{2}(T_{ijk} + T_{ikj}), \qquad T_{i[jk]} := \tfrac{1}{2}(T_{ijk} - T_{ikj}).$$

The existence theorem for tensorial families. Suppose that we are given an admissible system $\mathcal{O}$ of observers. We want to show that tensorial families of arbitrary type exist with respect to $\mathcal{O}$. Let $r, s = 1, 2, \ldots$ Choose a fixed observer O_0 of $\mathcal{O}$, and choose an arbitrary smooth family of functions

$$T^{i_{01}\ldots i_{0r}}_{j_{01}\ldots j_{0s}} : \Omega_{O_0} \to \mathbb{R} \tag{8.39}$$

where the indices run from 1 to n.

Theorem 8.2 *The family* (8.39) *of functions can be uniquely extended to a tensorial family with respect to the admissible system $\mathcal{O}$ of observers.*

Proof. The basic trick of the proof is the use of the compatibility condition (8.41).

(I) We consider first the special case T^i. Let us proceed in the following three steps:

- From the observer O_0 to the observer O: Define

  $$T^i_O(x) := G^i_{i_0}(x_0)T^{i_0}(x_0), \qquad x_0 \in \Omega_{O_0}, \quad x = x(x_0).$$

  Let $T_O(x)$ denote the column matrix to $(T^1_O(x), \ldots, T^n_O(x))$. Then

  $$T_O(x) = G_{O,O_0}(x_0)T_{O_0}(x_0).$$

  Thus, our definition implies the right transformation law for the change from the observer O_0 to the observer O.
- From the observer O_0 to the observer O': For the change from O_0 to O', we get

  $$T_{O'}(x') = G_{O',O_0}(x_0)T_{O_0}(x_0).$$

- From the observer O to the observer O': It remains to show that for two arbitrary observers O and O', we have the transformation law

  $$T_{O'}(x') = G_{O',O}(x)T_O(x). \tag{8.40}$$

However, this is a consequence of the compatibility condition (8.23). In fact, it follows from

$$\boxed{G_{O',O}(x)G_{O,O_0}(x_0) = G_{O',O_0}(x_0)} \tag{8.41}$$

that

$$G_{O',O}(x)T_O(x) = G_{O',O}(x)G_{O,O_0}(x_0)T_{O_0}(x_0) = G_{O',O_0}(x_0)T_{O_0}(x_0) = T_{O'}(x').$$

(II) We consider next the special case T_i. For the observer O, define

$$(T_i)_O(x) := G_i^{i_0}(x_0)T_{i_0}(x_0), \qquad x \in \Omega_{O_0}.$$

Let $T_O(x)$ denote the column matrix to $((T_1)_O(x), \ldots, (T_n)_O(x))$. Then

$$T_O(x) = H_{O,O_0}(x_0)T_{O_0}(x_0)$$

where $H_{O,O_0}(x_0) := (G_{O,O_0}(x_0)^{-1})^d$. Argue now as in (I) above, by using the dual compatibility relation (8.24).

(III) Finally, we consider the general case. For an arbitrary observer O, we set

$$T^{i_1 \ldots i_r}_{j_1 \ldots j_s}(x) := g^{i_1}_{i_{01}}(x_0) \cdots G^{i_r}_{i_{0r}}(x_0) \cdot G^{j_{01}}_{j_1}(x_0) \cdots G^{j_{0s}}_{j_s}(x_0) \cdot T^{i_{01} \ldots i_{0r}}_{j_{01} \ldots j_{0s}}(x_0)$$

where the transformation functions $G^i_{i_0}(x_0)$ correspond to the transformation matrix $G_{O,O_0}(x_0)$ describing the passage from the observer O_0 to the observer O (see (8.6) on page 445). It follows as in (I) and (II) above that this induces the right transformation law from the observer O to the observer O'. □

Linear independence. Recall that $\partial_i := \frac{\partial}{\partial x^i}$ and $dx^i(\partial_j) = \delta^i_j$.

Proposition 8.3 *$\partial_1, \ldots, \partial_n$ are linearly independent.*

More precisely, this means the following. If v^i is a tensorial family, then it follows from $v^i\partial_i = 0$ on Ω_O for a fixed observer O that $v^i \equiv 0$ for all $i = 1, \ldots, n$.

Proof. Let

$$v^i(x)\, \partial_i \Theta(x) = 0 \qquad \text{for all} \quad x \in \Omega_O$$

and for all smooth functions $\Theta : \Omega_O \to \mathbb{R}$. Fix the point $x_0 \in \Omega_O$, and fix the index $k = 1, \ldots, n$. Choose the function $\Theta(x) := x^k$. Then $\partial_i\Theta(x) = \delta_{ik}$. Hence $v^i(x)\delta_{ik} = 0$. This implies $v^k(x) = 0$. □

Analogously one proves that the family of products $\partial_i\partial_j, i, j = 1, \ldots, n$, is linearly independent. In the general case, fix $r = 1, 2, \ldots$ Then the family of products

$$\partial_{i_1}\partial_{i_2} \cdots \partial_{i_r}, \qquad i_1, \ldots, i_r = 1, \ldots, n$$

is linearly independent. More precisely, if $v^{i_1 \ldots i_r}$ is a tensorial family, then it follows from

$$v^{i_1 i_2 \ldots i_r}(x)\, \partial_{i_1}\partial_{i_2} \ldots \partial_{i_r} = 0 \qquad \text{for all} \quad x \in \Omega_O$$

with respect to a fixed observer O that all the coefficient functions $v^{i_1 i_2 \ldots i_r}$ vanish identically.

Proposition 8.4 *$dx^1, \ldots, dx^n$ are linearly independent.*

More precisely, this means the following. If v_i is a tensorial family, then it follows from $v_i dx^i = 0$ on Ω_O for a fixed observer O that $v_i \equiv 0$ for all $i = 1, \ldots, n$.

Proof. If $v_i dx^i = 0$, then $(v_i(x)\, dx^i)(\partial_j) = 0$. Hence

$$v_i(x)\, dx^i(\partial_j) = v_i(x)\delta^i_j = v_j(x) = 0.$$

□

Similarly, noting that $(dx^i \otimes dx^j)(\partial_k, \partial_l) = dx^i(\partial_k)dx^j(\partial_l) = \delta^i_k \delta^j_l$, we obtain that the family of tensor product $dx^i \otimes dx^j$, $i, j = 1, \ldots, n$, is linearly independent. In the general case, fix $r = 1, 2, \ldots, n$. Then the family of tensor products

$$dx^{i_1} \otimes dx^{i_2} \otimes \cdots \otimes dx^{i_r} \qquad i_1, \ldots, i_r = 1, \ldots, n$$

is linearly independent. More precisely, if $v_{i_1 \ldots i_r}$ is a tensorial family, then it follows from

$$v_{i_1 i_2 \ldots i_r}(x)\, dx^{i_1} \otimes dx^{i_2} \otimes \cdots \otimes dx^{i_r} = 0 \qquad \text{for all} \quad x \in \Omega_O$$

with respect of a fixed observer O that all the coefficient functions $v_{i_1 i_2 \ldots i_r}$ vanish identically.

8.4.6 Orientation, Pseudo-Tensorial Families, and the Levi-Civita Duality

> Orientation plays a fundamental role in physics. In particular, there are physical processes which depend critically on the choice of orientation. Mathematically, this is closely related to the Levi-Civita pseudo-tensorial family.
>
> Folklore

The Levi-Civita tensorial families. Let $\mathcal{O}$ be an admissible system of observers. Recall the matrix notation introduced on page 454. In particular, we will critically use the Jacobian

$$\frac{\partial(x^{1'}, \ldots, x^{n'})}{\partial(x^1, \ldots, x^n)}(x) = \det\left(\frac{dx'(x)}{dx}\right) = \det G(x), \qquad x \in \Omega$$

and its sign in order to describe the change of orientation. Suppose that g_{ij} is a symmetric tensorial family with respect to $\mathcal{O}$. Set

$$g(x) := \det(g_{ij}(x)), \quad x \in \Omega.$$

Assume that $g(x) \neq 0$ for all $x \in \Omega$. For $i_1, \ldots, i_n = 1, \ldots, n$, define

$$\boxed{\mathcal{E}_{i_1 \ldots i_n}(x) := \sqrt{|g(x)|} \cdot \varepsilon_{i_1 \ldots i_n},} \tag{8.42}$$

and

$$\boxed{\mathcal{E}^{i_1 \ldots i_n}(x) := \frac{\operatorname{sgn} g}{\sqrt{|g(x)|}} \cdot \varepsilon^{i_1 \ldots i_n}, \qquad x \in \Omega.} \tag{8.43}$$

We add the sign $\operatorname{sgn} g$ to the definition in order to get convenient lifting and lowering properties of the indices (see (8.52) below). In particular, $\mathcal{E}_{12 \ldots n} = \sqrt{|g|}$, and $\mathcal{E}^{12 \ldots n} = \frac{\operatorname{sgn} g}{\sqrt{|g|}}$. In Einstein's theory of special relativity, we will have $\operatorname{sgn} g = -1$ (see Sect. 18.4.1).

Proposition 8.5 *If all the transformations $x' = x'(x)$ of local coordinates have positive sign, then $\mathcal{E}_{i_1 \ldots i_n}$ and $\mathcal{E}^{i_1 \cdots i_n}$ are tensorial families with respect to $\mathcal{O}$.*

Proof. (I) Our point of departure is the transformation law

$$g_{i'j'}(x') = G^i_{i'}(x)G^j_{j'}(x) \cdot g_{ij}(x) \qquad \text{for all} \quad x \in \Omega$$

from the observer O to the observer O'. By page 454, for the determinants we get[9]

$$g'(x') = (\det G(x))^{-2} g(x). \tag{8.44}$$

This implies

$$\operatorname{sgn} g'(x') = \operatorname{sgn} g(x), \qquad x \in \Omega. \tag{8.45}$$

Since $g(x) \neq 0$ for all $x \in \Omega$, the sign $\operatorname{sgn}(g(x))$ is an invariant which has the value 1 or -1. Furthermore, we get

$$\sqrt{|g'(x')|} = |\det G(x)|^{-1} \cdot \sqrt{|g(x)|}, \qquad x \in \Omega, \quad x' = x'(x). \tag{8.46}$$

By (8.31),

$$(\det G(x))^{-1} \cdot \varepsilon_{i'_1 \ldots i'_n} = G^{i_1}_{i'_1}(x) \ldots G^{i_n}_{i'_n}(x) \cdot \varepsilon_{i_1 \ldots i_n}.$$

Note that $\det G(x) = \operatorname{sgn}(\det(G(x))) \cdot |\det G(x)|$. By (8.46),

$$\sqrt{|g'(x')|} \cdot \varepsilon_{i'_1 \ldots i'_n} = \operatorname{sgn}(\det G(x))\, G^{i_1}_{i'_1}(x) \cdots G^{i_n}_{i'_n}(x) \cdot \sqrt{|g(x)|}\, \varepsilon_{i_1 \ldots i_n}.$$

Equivalently, if $x' = x'(x)$ and $x \in \Omega$, then

$$\boxed{\mathcal{E}_{i'_1 \ldots i'_n}(x') = \operatorname{sgn}(\det G(x)) \cdot G^{i_1}_{i'_1}(x) \cdots G^{i_n}_{i'_n}(x) \cdot \mathcal{E}_{i_1 \ldots i_n}(x).} \tag{8.47}$$

Since we only consider observer transformations with $\operatorname{sgn}\det(G(x)) = 1$, the sign drops out, and we have a tensorial transformation law (8.47) at hand.

(II) Similarly, we get

$$\frac{\varepsilon^{i'_1 \ldots i'_n}}{\sqrt{|g'(x')|}} = \operatorname{sgn}(\det G(x)) \cdot G^{i'_1}_{i_1}(x) \ldots G^{i'_n}_{i_n}(x) \cdot \frac{\varepsilon^{i_1 \ldots i_n}}{\sqrt{|g(x)|}}.$$

Equivalently, if $x' = x'(x)$, then

$$\boxed{\mathcal{E}^{i'_1 \ldots i'_n}(x') = \operatorname{sgn}(\det G(x)) \cdot G^{i'_1}_{i_1}(x) \cdots G^{i'_n}_{i_n}(x) \cdot \mathcal{E}^{i_1 \ldots i_n}(x).} \tag{8.48}$$

□

Mass density. Suppose that we have a family of smooth functions $\varrho_O : O_\Omega \to \mathbb{R}$ together with the transformation law

$$\varrho_{O'}(x') = \left| \frac{\partial(x^1, \ldots, x^n)}{\partial(x^{1'}, \ldots, \partial x^{n'})}(x') \right| \cdot \varrho_O(x), \quad x' = x'(x).$$

This is the transformation law of a mass density ϱ_O. In fact, if U is a compact subset of Ω_O (e.g., a ball), then

[9] Note that the prime of g' refers to the observer O'; the symbol g' does not stand for a derivative.

$$\boxed{\int_{U'} \varrho_{O'}(x')\, dx^{1'} dx^{2'} \cdots dx^{n'} = \int_{O} \varrho_O(x)\, dx^1 dx^2 \cdots dx^n.} \tag{8.49}$$

This follows from the substitution rule for integrals. In terms of physics, the function ϱ_O is the mass density of a fluid measured by the observer O. Moreover, the integral

$$\int_U \varrho_O(x)\, dx^1 dx^2 \cdots dx^n$$

equals the mass contained in the set U. The equation (8.49) tells us that the observers O and O' measure the same mass in the sets U and U', respectively. Here, $U' = x'(U)$ (i.e., the map $x \mapsto x'$ sends the set U to U').

Pseudo-tensorial families. Motivated by (8.47), we modify the notion of a tensorial family in the following way. To begin with, introduce the notation

$$\sigma := \operatorname{sgn}\left(\frac{dx'}{dx}\right),$$

and hence $\sigma = \operatorname{sgn}(\det G(x)) = \operatorname{sgn}\left(\frac{\partial(x^{1'},\ldots,x^{n'})}{\partial(x^1,\ldots,x^n)}(x)\right)$. Recall that $\sigma = \pm 1$ is the sign of the change $x \mapsto x'(x)$ of local coordinates introduced on page 445. If

$$T^{i_1\ldots i_r}_{j_1\ldots j_s}(x') \tag{8.50}$$

is a tensorial family, then $T^{i'_1\ldots i'_r}_{j'_1\ldots j'_s}(x')$ is equal to

$$\alpha \cdot G^{i'_1}_{i_1}(x) \cdots G^{i'_r}_{i_r}(x) G^{j_1}_{j'_1}(x) \cdots G^{j_s}_{j'_s}(x) \cdot T^{i_1\ldots i_r}_{j_1\ldots j_s}(x) \tag{8.51}$$

where $\alpha = 1$. Now suppose that all the transformation laws (8.51) hold by setting

$$\alpha := \sigma.$$

Then, $T^{i_1\ldots i_r}_{j_1\ldots j_s}$ is called a pseudo-tensorial family.

Tensorial density families and pseudo-tensorial density families. Let us consider the following modifications:

(a) $\alpha := \left|\frac{\partial(x^1,\ldots,x^n)}{\partial(x^{1'},\ldots,\partial x^{n'})}(x')\right|^w$ (equivalently, $\alpha = |\det G(x)|^{-w}$),

(b) $\alpha := \sigma \cdot \left|\frac{\partial(x^1,\ldots,x^n)}{\partial(x^{1'},\ldots,\partial x^{n'})}(x')\right|^w$ (equivalently, $\alpha = \sigma \cdot |\det G(x)|^{-w}$).

If the transformation law (8.51) holds by using (a) (resp. (b)), then $T^{i_1\ldots i_r}_{j_1\ldots j_s}$ is called a tensorial density family of weight w (resp. a pseudo-tensorial density family of weight w). For example, it follows from (8.44) that

- the mass density ϱ above is a tensorial density of weight 1,
- $g = \det(g_{ij})$ is a tensorial density of weight $w = 2$,
- $\sqrt{|g|}$ is a tensorial density of weight 1 by (8.46),
- $\frac{1}{\sqrt{|g|}}$ is a tensorial density of weight -1 by (8.46), and
- $\operatorname{sgn} g$ is a tensorial invariant by (8.45).

The Levi-Civita pseudo-tensorial families. The relations (8.47) and (8.48) tell us the following.

Proposition 8.6 *$\mathcal{E}^{i_1 \dots i_n}$ and $\mathcal{E}_{i_1 \dots i_n}$ are pseudo-tensorial families with respect to the admissible system $\mathcal{O}$ of references.*

In addition, lowering and lifting of indices yields

$$\mathcal{E}_{i_1 \dots i_n} = g_{i_1 j_1} \cdots g_{i_n j_n} \mathcal{E}^{j_1 \dots j_n}, \qquad \mathcal{E}^{i_1 \dots i_n} = g^{i_1 j_1} \cdots g^{i_n j_n} \mathcal{E}_{j_1 \dots j_n}. \tag{8.52}$$

This follows from

$$g_{i_1 j_1} \cdots g_{i_n j_n} \mathcal{E}^{j_1 \dots j_n} = \frac{\det(g_{ij}) \cdot \operatorname{sgn} g}{\sqrt{|g|}} \, \varepsilon_{i_1 \dots i_n} = \sqrt{|g|} \cdot \varepsilon_{i_1 \dots i_n}. \tag{8.53}$$

Proposition 8.7 *If $O \mapsto \iota_O$ is an orientation function of the admissible system $\mathcal{O}$ of observers, then $\iota_O \cdot \mathcal{E}^{i_1 \dots i_n}$ and $\iota_O \cdot \mathcal{E}_{i_1 \dots i_n}$ are tensorial families with respect to the admissible system of observers $\mathcal{O}$.*

This follows from the transformation law (8.27) which tells us that ι_O is a pseudo-tensorial invariant. The tensorial family $\iota_O \cdot \mathcal{E}_{i_1 \dots i_n}$ is called the tensorial volume family of the admissible system $\mathcal{O}$ of observers with respect to the choice of the orientation function

$$O \mapsto \iota_O.$$

Note that if the admissible system of observers $\mathcal{O}$ is oriented, then we can choose $\iota_O = 1$ for all observers O. In this case, $\mathcal{E}_{i_1 \dots i_n}$ is a tensorial family.

Levi-Civita duality. Our goal is to use the Levi-Civita pseudo-tensorial families in order to construct a duality between tensorial families and pseudo-tensorial families which contains Hodge duality as a special case. Let $p = 0, 1, \dots, n$. Define

$$(*T)_{i_{p+1} \dots i_n}(x) := \frac{1}{(n-p)!} \, \mathcal{E}_{i_1 \dots i_p i_{p+1} \dots i_n}(x) \, T^{i_1 \dots i_p}(x), \qquad x \in \Omega_O, \tag{8.54}$$

and

$$(*S)^{i_{p+1} \dots i_n}(x) := \frac{1}{(n-p)!} \, \mathcal{E}^{i_1 \dots i_p i_{p+1} \dots i_n}(x) \, S_{i_1 \dots i_p}(x). \tag{8.55}$$

We add the normalization factors in order to fit Hodge duality below. Then the following hold:

- If $T^{i_1 \dots i_p}$ is a tensorial family, then $(*T)_{i_{p+1} \dots i_n}$ is a pseudo-tensorial family, and

  $$\iota_O \cdot (*T)_{i_{p+1} \dots i_n}$$

  is a tensorial family for every orientation function ι.
- If $S_{i_1 \dots i_p}$ is a tensorial family, then $(*T)^{i_{p+1} \dots i_n}$ is a pseudo-tensorial family, and

  $$\iota_O \cdot (*T)^{i_{p+1} \dots i_n}$$

  is a tensorial family for every orientation function ι.

Examples. If $p = n$, then

- $(*T)(x) := \mathcal{E}_{i_1 \dots i_n}(x) \, T^{i_1 \dots i_n}(x)$ and
- $(*S)(x) := \mathcal{E}^{i_1 \dots i_n}(x) \, S_{i_1 \dots i_n}(x)$, $x \in \Omega_O$

are pseudo-invariant functions. Moreover, $\iota_O \cdot *T$ and $\iota_O \cdot *S$ are invariant functions.

8.5 Differential Forms (Exterior Product)

8.5.1 Cartan Families and the Cartan Differential

Modern global analysis is based on the Cartan calculus of differential forms. The Cartan calculus is a special case of the tensor calculus; it is equivalent to the calculus of antisymmetric covariant tensorial families.

Folklore

Cartan families. The elegance of the Cartan calculus of differential forms relies on the nice properties of antisymmetric covariant tensorial families which correspond to the classical theory of determinants. Let $\mathcal{O}$ be an admissible system of observers. Fix $n = 1, 2, \ldots$ By definition, a Cartan family

$$\omega_{i_1 \ldots i_p}$$

with respect to $\mathcal{O}$ is an antisymmetric tensorial family with respect to $\mathcal{O}$. The number $p = 0, 1, \ldots, n$ is called the degree of the Cartan family.[10] All the Cartan families with respect to the admissible system $\mathcal{O}$ of observers form a graded real algebra with differential. This means the following:

(i) Sum: If $\omega_{i_1 \ldots i_p}$ and $\mu_{i_1 \ldots i_p}$ are Cartan families of the same degree and α, β are real numbers, then

$$\alpha\omega_{i_1 \ldots i_p} + \beta\mu_{i_1 \ldots i_p}$$

is also a Cartan family. This remains true if α and β are Cartan families of degree zero.

(ii) Product: If $\omega_{i_1 \ldots i_p}$ and $\mu_{j_1 \ldots j_q}$ are Cartan families, then the antisymmetrization

$$\omega_{[i_1 \ldots i_p}\mu_{j_1 \ldots j_q]}$$

of the product $\omega_{i_1 \ldots i_p}\mu_{j_1 \ldots j_q}$ is again a Cartan family. We define

$$\boxed{\omega_{i_1 \ldots i_p} \wedge \mu_{j_1 \ldots j_q} := \omega_{[i_1 \ldots i_p}\mu_{j_1 \ldots j_q]}.}$$

This is called the wedge product. For Cartan families and real numbers α, β, we have

- the distributive law

$$(\alpha\omega_{i_1 \ldots i_p} + \beta\mu_{i_1 \ldots i_p}) \wedge \nu_{k_1 \ldots k_r} = \alpha\omega_{i_1 \ldots i_p} \wedge \nu_{k_1 \ldots k_r} + \beta\mu_{i_1 \ldots i_p} \wedge \nu_{k_1 \ldots k_r},$$

- the associative law

$$(\omega_{i_1 \ldots i_p} \wedge \mu_{j_1 \ldots j_q}) \wedge \nu_{k_1 \ldots k_r} = \omega_{i_1 \ldots i_p} \wedge (\mu_{j_1 \ldots j_q} \wedge \nu_{k_1 \ldots k_r}), \tag{8.56}$$

- and the supercommutativity law

$$\omega_{i_1 \ldots i_p} \wedge \mu_{j_1 \ldots j_q} = (-1)^{pq}\mu_{j_1 \ldots j_q} \wedge \omega_{i_1 \ldots i_p}. \tag{8.57}$$

This can be proved by using the permutation property (8.37).

(iii) Cartan derivative: If $\omega_{i_1 \ldots i_p}$ is a Cartan family, then

$$\boxed{d_k\omega_{i_1 \ldots i_p} := \partial_{[k}\omega_{i_1 \ldots i_p]}} \tag{8.58}$$

is also a Cartan family.

[10] If $p = 0$, then the Cartan family is a family of invariant functions $U_O : \Omega_O \to \mathbb{R}$ with $U_{O'}(x') = U_O(x)$ for all $x \in \Omega_O$, and $x' = x'(x)$.

The proof of (iii) will be given in Sect. 8.11.2 on page 523. At this point, let us only consider a special case. We want to show that if ω_i is a tensorial family, then

$$\partial_{[k}\omega_{i]} = \tfrac{1}{2}(\partial_k\omega_i - \partial_i\omega_k) \tag{8.59}$$

is again a tensorial family. The trick is that some nasty terms of the transformation law cancel each other because of the antisymmetrization. Explicitly, it follows from the transformation law

$$\omega_{i'} = \frac{\partial x^i}{\partial x^{i'}}\,\omega_i$$

together with the product rule and the chain rule that

$$\partial_{k'}\omega_{i'} = \frac{\partial}{\partial x^{k'}}\left(\frac{\partial x^i}{\partial x^{i'}}\,\omega_i\right) = \frac{\partial^2 x^i}{\partial x^{k'}\partial x^{i'}}\,\omega_i + \frac{\partial x^i}{\partial x^{i'}}\frac{\partial x^k}{\partial x^{k'}}\frac{\partial \omega_i}{\partial x^k}.$$

Analogously,

$$\partial_{i'}\omega_{k'} = \frac{\partial^2 x^i}{\partial x^{i'}\partial x^{k'}}\,\omega_i + \frac{\partial x^i}{\partial x^{k'}}\frac{\partial x^k}{\partial x^{i'}}\frac{\partial \omega_i}{\partial x^k}.$$

Since $\frac{\partial^2 x^i}{\partial x^{k'}\partial x^{i'}} = \frac{\partial^2 x^i}{\partial x^{i'}\partial x^{k'}}$, we get

$$\partial_{k'}\omega_{i'} - \partial_{i'}\omega_{k'} = \frac{\partial x^k}{\partial x^{k'}}\frac{\partial x^i}{\partial x^{i'}}\cdot(\partial_k\omega_i - \partial_i\omega_k).$$

Thus, $\partial_k\omega_i - \partial_i\omega_k$ is a tensorial family. Obviously, this family is antisymmetric. Thus, it is a Cartan family. □

For Cartan families, we have

- the graded Leibniz rule

$$d_i(\omega_{i_1\ldots i_p}\wedge\mu_{j_1\ldots j_q}) = d_i\omega_{i_1\ldots i_p}\wedge\mu_{j_1\ldots j_q} + (-1)^p\omega_{i_1\ldots i_p}\wedge d_i\mu_{j_1\ldots j_q} \tag{8.60}$$

- and the Poincaré cohomology rule

$$\boxed{d_i(d_j\omega_{i_1\ldots i_p}) = 0.} \tag{8.61}$$

Let us prove (8.61). For example, consider the special case where ω_k is a Cartan family of degree 1. Then

$$d_j\omega_k = \partial_{[j}\omega_{k]} = \tfrac{1}{2}(\partial_j\omega_k - \partial_k\omega_j).$$

This implies $\partial_i(d_j\omega_k) = \frac{1}{2}(\partial_i\partial_j\omega_k - \partial_i\partial_k\omega_j)$. Hence

$$d_i(d_j\omega_k) = \tfrac{1}{2}(\partial_{[i}\partial_j\omega_{k]} - \partial_{[i}\partial_k\omega_{j]}).$$

Thus, $d_i(d_j\omega_k) = \partial_{[i}\partial_j\omega_{k]}$. Since we have $\partial_i\partial_j = \partial_j\partial_i$, it follows from (8.38) that $d_i(d_j\omega_k) = \partial_{[i}\partial_j\omega_{k]} = 0$. This is the claim. The general case proceeds analogously. □

Let us mention three further operations for Cartan families:

(a) Contraction with a velocity field: We set

$$v^{i_1}\rfloor\,\omega_{i_1 i_2\ldots i_p} := v^{i_1}\omega_{i_1 i_2\ldots i_p}.$$

If v^i is a tensorial family and $\omega_{i_1\ldots i_p}$ is a Cartan family, then $v^{i_1}\omega_{i_1 i_2\ldots i_p}$ is again a Cartan family, by the index principle.

(b) Hodge duality: If $\omega_{i_1 \ldots i_p}$ is a Cartan family, then

$$\iota_O \cdot \mathcal{E}_{i_1 \ldots i_p i_{p+1} \ldots i_n} \, \omega^{i_1 \ldots i_p}$$

is again a Cartan family which is called the dual Cartan family to $\omega_{i_1 \ldots i_p}$. In contrast to (i)–(iii) on page 464, Hodge duality depends on the choice of a metric tensorial family (g_{ij}), and on the choice of an orientation function ι. In particular, we use g_{ij} in order to lift the indices of $\omega_{i_1 \ldots i_p}$.[11]

Differential forms. We are given the admissible system $\mathcal{O}$ of observers. We want to reformulate the calculus for Cartan families in terms of differential forms. Our goal is to assign to the Cartan family $\omega_{i_1 \ldots i_p}$ the multilinear (i.e., p-linear) real functional

$$\boxed{\omega = \frac{1}{p!} \, \omega_{i_1 i_2 \ldots i_p} \, dx^{i_1} \wedge dx^{i_2} \wedge \cdots \wedge dx^{i_p}.} \tag{8.62}$$

Our strategy is to use invariant expressions which do not depend on the choice of the observer, by means of the principle of the correct index picture.

Let us discuss this. In what follows, we will use the notation introduced on page 448. In particular, we will use the real linear spaces $\Lambda^p(\Omega_O)$ where $p = 0, 1, 2, \ldots n$. If $f \in \Lambda^0(\Omega_O)$, then we define

$$\boxed{df := \partial_j f \cdot dx^i}$$

motivated by the classical differential for smooth functions $f : \Omega_O \to \mathbb{R}$. Note the following:

In modern analysis, differentials df are linear functionals.

In our notation, $df \in \Lambda^1(\Omega_O)$. This way, we obtain the linear operator

$$d : \Lambda^0(\Omega_O) \to \Lambda^1(\Omega_O).$$

Now let us give the basic definition. Fix $p = 1, 2, \ldots$ For every antisymmetric tensorial family $\omega_{i_1 \ldots i_n}$, we define

$$\omega_{x,O} := \frac{1}{p!} \, \omega_{i_1 i_2 \ldots i_p}(x) \, dx^{i_1} \wedge dx^{i_2} \wedge \cdots \wedge dx^{i_p}, \qquad x \in \Omega_O.$$

Proposition 8.8 *The definition of the p-linear real functional ω_O does not depend on the choice of the observer O.*

Proof. (I) First consider the special case where $p = 2$. Then $\omega_O := \frac{1}{2}\omega_{ij} \, dx^i \wedge dx^j$. For the observer O', we define

$$\omega_{x',O'} := \tfrac{1}{2}\omega_{i'j'}(x') \, dx^{i'} \wedge dx^{j'}, \qquad x' = x'(x).$$

The differential operator $v^i(x)\partial_i$ passes over to $v^{i'}(x')\partial_{i'}$. By (8.15),

$$dx^{i'} = \frac{\partial x^{i'}}{\partial x^i} \, dx^i.$$

[11] Explicitly, $\omega^{i_1 \ldots i_p} = g^{i_1 j_1} g^{i_2 j_2} \cdots g^{i_p j_p} \cdot \omega_{j_1 j_2 \ldots j_p}$.

By the distributive law for the wedge product,

$$dx^{i'} \wedge dx^{j'} = \frac{\partial x^{i'}}{\partial x^i}\frac{\partial x^{j'}}{\partial x^j} \cdot dx^i \wedge dx^j .$$

This tells us that $dx^i \wedge dx^j$ transforms like a tensorial family T^{ij}. Consequently, by the index principle, we get

$$\omega_{x',O'} = \tfrac{1}{2}\omega_{i'j'}(x')\, dx^{i'} \wedge dx^{j'} = \tfrac{1}{2}\omega_{ij}(x)\, dx^i \wedge dx^j .$$

(II) In the general case, we proceed analogously. □

Motivated by Prop. 8.8 we briefly write ω instead of ω_O. We call this a differential form of degree p with respect to the admissible system $\mathcal{O}$ of observers (or, briefly, a p-form). Choose an element f_O of $\Lambda^0(\Omega_O)$. We also set

$$f_{O'}(x') := f_O(x) \qquad \text{for all} \quad x \in \Omega_O, \quad x' = x'(x).$$

Then $f_{O'} \in \Lambda^0(\Omega_{O'})$. Naturally enough, the elements of $\Lambda^0(\Omega_O)$ are called differential forms of degree $p = 0$. We briefly write f instead of f_O. Let us add the following definitions:

(i) Sum: If α and β are real numbers, then

$$\alpha\omega_{ij}\, dx^i \wedge dx^j + \beta\mu_{ij}\, dx^i \wedge dx^j := (\alpha\omega_{ij} + \beta\mu_{ij})\, dx^i \wedge dx^j .$$

More general, $\alpha\omega_{i_1\ldots i_p} dx^{i_1} \wedge \cdots \wedge dx^{i_p} + \beta\mu_{i_1\ldots i_p} dx^{i_1} \wedge \cdots \wedge dx^{i_p}$ is equal to

$$(\alpha\omega_{i_1\ldots i_p} + \beta\mu_{i_1\ldots i_p})\, dx^{i_1} \wedge \cdots \wedge dx^{i_p} .$$

This coincides with the linear combination of multilinear functionals.

(ii) Product: Using the wedge product for Cartan families, we define

$$\boxed{\omega_i dx^i \wedge \left(\mu_{jk}\, dx^j \wedge dx^k\right) := \omega_{[i}\mu_{jk]}\, dx^i \wedge dx^j \wedge dx^k .}$$

More generally, the wedge product

$$\omega_{i_1\ldots i_p}\, dx^{i_1} \wedge \cdots \wedge dx^{i_p} \wedge \left(\mu_{j_1\ldots j_q}\, dx^{j_1} \wedge \cdots \wedge dx^{j_q}\right)$$

of differential forms is defined to be

$$\omega_{[i_1\ldots i_p}\mu_{j_1\ldots j_q]} \cdot dx^{i_1} \wedge \cdots \wedge dx^{i_p} \wedge dx^{j_1} \wedge \cdots \wedge dx^{j_q} .$$

This definition coincides with the wedge product for multilinear functionals (see Sect. 2.1.2). For example, since $dx^i \wedge dx^j = -dx^j \wedge dx^i$, we get

$$\omega_i\, dx^i \wedge \mu_j\, dx^j = \tfrac{1}{2}(\omega_i\mu_j - \omega_j\mu_i)\, dx^i \wedge dx^j = \omega_{[i}\mu_{j]}\, dx^i \wedge dx^j .$$

(iii) Cartan differential: We define

$$\boxed{d(\omega_j\, dx^j) := d_i\omega_j \cdot dx^i \wedge dx^j}$$

where $d_i\omega_j := \partial_{[i}\omega_{j]}$. Since $d_i\omega_j$ is a tensorial family, it follows from the index principle that this definition does not depend on the choice of the observer. Since $dx^i \wedge dx^j = -dx^j \wedge dx^i$, we get

$$d(\omega_j\, dx^j) = \tfrac{1}{2}(\partial_i\omega_j - \partial_j\omega_i)\cdot dx^i \wedge dx^j = \partial_i\omega_j \cdot\, dx^i \wedge dx^j.$$

Setting $d\omega_j := \partial_i\omega_j \cdot dx^j$, we obtain

$$d(\omega_j\, dx^j) = d\omega_j \wedge dx^j.$$

More general, let $\omega = \frac{1}{p!}\omega_{i_1\ldots i_p}\, dx^{i_1} \wedge \cdots \wedge dx^{i_p}$. Then we define

$$d\omega := \frac{1}{p!}\, d_i\omega_{i_1\ldots i_p} \cdot dx^i \wedge dx^{i_1} \wedge \cdots \wedge dx^{i_p}$$

where $d_i\omega_{i_1\ldots i_p} = \partial_{[i}\omega_{i_1\ldots i_p]}$ is a tensorial family. Again, by the index principle, the definition of $d\omega$ does not depend on the choice of the observer. Using antisymmetrization, we get

$$d\omega = \frac{1}{p!}\, \partial_i\omega_{i_1\ldots i_p}\, dx^i \wedge dx^{i_1} \wedge \cdots \wedge dx^{i_p}.$$

Equivalently,

$$\boxed{d\omega = \frac{1}{p!}\, d\omega_{i_1\ldots i_p} \wedge dx^{i_1} \wedge \cdots \wedge dx^{i_p}}$$

where $d\omega_{i_1\ldots i_p} := \partial_i\omega_{i_1\ldots i_p}\, dx^i$.

Let $\omega = \frac{1}{2}\omega_{ij}\, dx^i \wedge dx^j$ be a 2-form. For all linear first-order differential operators $v = v^i\partial_i$ and $w = w^i\partial_i$, we get

$$\omega(v,w) = \tfrac{1}{2}\omega_{ij}(dx^i(v)dx^j(w) - dx^i(w)dx^j(v)) = \tfrac{1}{2}\omega_{ij}(v^iw^j - v^jw^i) = \omega_{ij}v^iw^j.$$

For a p-form $\omega = \frac{1}{p!}\,\omega_{i_1\ldots i_p}\, dx^{i_1} \wedge \cdots \wedge dx^{i_p}$ with $p = 1, 2, \ldots$, we obtain

$$\boxed{\omega(v_1, \ldots, v_p) = \omega_{i_1\ldots i_p} v_1^{i_1} v_2^{i_2} \cdots v_p^{i_p}} \tag{8.63}$$

for all linear first-order differential operators $v_k = v_k^i\partial_i$ with $k = 1, \ldots, p$.

The contraction product. If v^i is a tensorial family, then we define

$$v\rfloor\omega := \frac{1}{(p-1)!}\, v^i\omega_{ii_2\ldots i_p}\, dx^{i_2} \wedge \cdots \wedge dx^{i_p}. \tag{8.64}$$

We also write $i_v\omega$ instead of $v\rfloor\omega$. Parallel to (8.63), we get

$$(i_v\omega)(v_2, \ldots, v_p) = \omega(v, v_2, \ldots, v_p) \tag{8.65}$$

for all p-forms with $p = 1, 2, \ldots, n$ and all linear first-order partial differential operators $v = v^i\partial_i$, and $v_K = v_K^i\partial_i$ with $K = 2, \ldots, p$. If $p = 0$, then $i_v\omega := \omega$.

Summary. Let ω, μ, ν be differential forms of degree $p, q, r = 0, 1 \ldots, n$, respectively, and let α, β be real numbers. Then:

- $(\omega \wedge \mu) \wedge \nu = \omega \wedge (\mu \wedge \nu)$ (associative law);
- $\omega \wedge \mu = (-1)^{pq}\mu \wedge \omega$ (supercommutative law);[12]
- If $q = r$, then $\omega \wedge (\alpha\mu + \beta\nu) = \alpha\omega \wedge \mu + \beta\omega \wedge \nu$ (distributive law);

[12] If $p = 0$, then $\omega \wedge \mu = \mu \wedge \omega := \omega\mu$. Note that ω is a function, in this special case.

- $d(\omega \wedge \mu) = d\omega \wedge \mu + (-1)^p \omega \wedge d\mu$ (graded Leibniz rule);[13]
- $d(d\omega) = 0$ (Poincaré's cohomology rule).

In particular, $dx^i \wedge dx^j = -dx^j \wedge dx^i$. This rule is sufficient for computing all the possible products by using both the associative law and the distributive law. For example, $dx^i \wedge dx^i = 0$, and $dx^i \wedge dx^j \wedge dx^i = -dx^i \wedge dx^i \wedge dx^j = 0$.

8.5.2 Hodge Duality, the Hodge Codifferential, and the Laplacian (Hodge's Star Operator)

> In classical mathematical physics, the Laplace equation $\Delta U = 0$ plays a fundamental role. In the 1930s, Hodge (1903–1975) discovered how the classic theory can be generalized to differential forms. Hodge theory generalizes the work of Riemann (1826–1866) about Abelian integrals on Riemann surfaces to harmonic integrals on Riemannian manifolds.[14]
>
> Folklore

Metric tensorial family. By definition, a metric tensorial family is a symmetric tensorial family g_{ij} with $\det(g_{ij}(x)) \neq 0$ for all $x \in \Omega_O$. Recall

$$g(x) := \det(g_{ij}(x)) \qquad \text{for all} \quad x \in \Omega_O.$$

By definition, the Morse index μ of the symmetric matrix $(g_{ij}(x))$ is equal to the number of negative eigenvalues of $(g_{ij}(x))$. Since the set Ω_O is assumed to be arcwise connected and the functions $x \mapsto g_{ij}(x)$ are smooth, the Morse index μ does not depend on the point $x \in \Omega_O$. If we pass from the observer O to the observer O', then the Morse index μ remains unchanged. The tensorial family is called of Riemannian (resp. pseudo-Riemannian) type iff $\mu = 0$ (resp. $0 < \mu < n$). The inverse matrix $(g^{ij}(x)) := (g_{ij}(x))^{-1}$ yields a symmetric tensorial family g^{ij}. The tuple $(n - \mu, \mu)$ is called the signature of g_{ij}. For example, a Riemannian metric tensorial family has the signature $(n, 0)$.

The volume form. Define

$$\boxed{v := \frac{1}{n!}\, \mathcal{E}_{i_1,\ldots,i_n}\, dx^{i_1} \wedge \cdots \wedge dx^{i_n}.}$$

Since $\iota_O \mathcal{E}_{i_1 \ldots i_p}$ is a tensorial family, the differential form $\iota_O \cdot v$ does not depend on the choice of the observer. This implies that

> *The volume form v changes sign if we change the orientation of the observer.*

Explicitly, $v = \sqrt{|g|}\, dx^1 \wedge \cdots \wedge dx^n$.

Hodge duality. The basic idea of Hodge duality is to use the Levi-Civita duality and the volume form v. Fix $p = 1, 2, \ldots$ Let $\omega = \frac{1}{p!}\, \omega_{i_1 \ldots i_p}\, dx^{i_1} \wedge \cdots \wedge dx^{i_p}$ be a p-form. We define

---

[13] In modern mathematics and physics, this is also called the supersymmetric Leibniz rule.

[14] W. Hodge, The Theory and and Applications of Harmonic Integrals, Cambridge University Press, 1941 (second revised edition 1951). The fundamental Hodge existence theorem based on the modern functional analytic approach to elliptic partial differential equations can be found in J. Jost, Riemannian Geometry and Geometric Analysis, Chap. 2, Springer, Berlin, 2008.

$$\boxed{*\omega = \frac{1}{p!(n-p)!}\, \mathcal{E}_{i_1 \ldots i_p \ldots i_n} \omega^{i_1 \ldots i_p}\, dx^{i_{p+1}} \wedge \cdots \wedge dx^{i_n}.} \tag{8.66}$$

This is called the Hodge star operator. If $p = 0$, then the 0-form ω is a function, and we define

$$*\omega := \omega \cdot v.$$

For $p = 0, 1, \ldots$, the definition of the $(n-p)$-form $\iota_O \cdot *\omega$ does not depend on the choice of the observer O, by the index principle. In particular, we have

$$*1 = v, \qquad *v = \operatorname{sgn}(g) \cdot 1$$

where 1 denotes the constant function $f(x) \equiv 1$. More general,

$$* *\omega = (-1)^{p(n-p)} \operatorname{sgn}(g) \cdot \omega. \tag{8.67}$$

The linear operator

$$* : \Lambda^p(\Omega_O) \to \Lambda^{(n-p)}(\Omega_O), \qquad p = 0, 1, \ldots, n$$

is invertible with the inverse operator

$$*^{-1} : \Lambda^{n-p}(\Omega_O) \to \Lambda^p(\Omega_O), \qquad p = 0, 1, \ldots, n.$$

Here, for the $(n-p)$-form ϱ, we get $*^{-1}\varrho = (-1)^{p(n-p)} \operatorname{sgn}(g) * \varrho$.

The Hodge codifferential d^*. Using the Hodge star operator, let us introduce the Hodge codifferential d^* (also called the dual Cartan differential) by setting

$$\boxed{d^*\omega := (-1)^p *^{-1} (d * \omega)}$$

for all p-forms ω with $p = 0, 1, \ldots, n$. Equivalently,

$$d^*\omega = (-1)^{n(p+1)+1} \operatorname{sgn}(g) \cdot *(d * \omega).$$

This corresponds to the commutative diagram

$$\begin{array}{ccc} \Lambda^p(\Omega) & \xrightarrow{d^*} & \Lambda^{p-1}(\Omega) \\ {\scriptstyle *}\downarrow & & \uparrow{\scriptstyle *^{-1}} \\ \Lambda^{n-p}(\Omega) & \xrightarrow{(-1)^p d} & \Lambda^{n-p+1}(\Omega) \end{array}$$

with $p = 0, 1, \ldots, n$. Here, we set $\Lambda^{-1}(\Omega) = \Lambda^{n+1}(\Omega) := \{0\}$. This way, we get the following two chains of maps

$$0 \xrightarrow{d} \Lambda^0(\Omega) \xrightarrow{d} \Lambda^1(\Omega) \xrightarrow{d} \ldots \xrightarrow{d} \Lambda^n(\Omega) \xrightarrow{d} 0,$$

and

$$0 \xleftarrow{d^*} \Lambda^0(\Omega) \xleftarrow{d^*} \Lambda^1(\Omega) \xleftarrow{d^*} \ldots \xleftarrow{d^*} \Lambda^n(\Omega) \xleftarrow{d^*} 0.$$

For the composition of these operators, for all p-forms with $p = 0, 1, \ldots, n$, we have:

- $d(d\omega) = 0$ (Poincaré's cohomology rule),
- $d^*(d^*\omega) = 0$ (Hodge's homology rule).

Mnemonically, we write $dd = 0$ and $d^*d^* = 0$.

The Hodge Laplacian Δ. Let ω be a p-form with $p = 0, 1, \ldots, n$. We define

$$\boxed{\Delta\omega := (d^*d + dd^*)\omega.}$$

This yields the maps

$$\Delta : \Lambda^p(\Omega) \to \Lambda^p(\Omega), \qquad p = 1, 2, \ldots, n.$$

Consider first the special case where $p = 0$. Here, we have $\omega = U$ where U is a real-valued function. Then $\Delta U = d^* d\omega$. This implies

$$\boxed{\Delta U = -\frac{1}{\sqrt{|g|}}\, \partial_i(\sqrt{|g|}\, g^{ij}\partial_j U).} \tag{8.68}$$

In particular, if $g_{ij} = \delta_{ij}$ for a fixed observer, then

$$\boxed{\Delta U = -\delta^{ij}\partial_i\partial_j U.} \tag{8.69}$$

Up to the sign, this is the classical Laplacian. Thus, the Hodge Laplacian $\Delta\omega$ represents an elegantly formulated generalization of the classical Laplacian to differential forms. Hodge theory plays a crucial role in modern differential geometry.[15]

In this monograph, we will always use the modern sign convention (8.69).

In terms of functional analysis, our sign convention ensures the quite natural property that the linear operator

$$\Delta : \mathcal{D}(\mathbb{R}^3) \to \mathcal{D}(\mathbb{R}^3)$$

is symmetric and positive definite on the dense subset $\mathcal{D}(\mathbb{R}^3)$ of the Hilbert space $L_2(\mathbb{R}^3)$.

Proof of (8.68). Note that $\Delta U = d^*dU = -\operatorname{sgn}(g) *d{*}dU$. Compute successively,

- $dU = \partial_j U \cdot dx^j$,
- $*dU = \frac{1}{(n-1)!}\, \varepsilon_{ii_2\ldots i_n}\partial^i U\, dx^{i_2} \wedge \cdots \wedge dx^{i_n}$,
- $d(*dU) = \frac{1}{(n-1)!}\, \partial_k\big(\sqrt{|g|}\, \varepsilon_{ii_2\ldots i_n}\partial^i U\big)\, dx^k \wedge dx^{i_2} \wedge \cdots \wedge dx^{i_n}$,
- $d(*dU) = \partial_k\big(\sqrt{|g|}\, \partial^k U\big)\, dx^1 \wedge \cdots \wedge dx^n = \frac{1}{\sqrt{|g|}}\partial_k\big(\sqrt{|g|}\, \partial^k U\big) \cdot v$,
- $*(d * dU) = \frac{\operatorname{sgn}(g)}{\sqrt{|g|}}\, \partial_k\big(\sqrt{|g|}\, \partial^k U\big)$,
- $\Delta U = -\operatorname{sgn}(g) * (d * dU) = -\frac{1}{\sqrt{|g|}}\, \partial_i\big(\sqrt{|g|}\, \partial^i U\big) = -\frac{1}{\sqrt{|g|}}\, \partial_i\big(\sqrt{|g|}\, g^{ij}\partial_j U\big)$.

□

Consider the p-form $\omega = \frac{1}{p!}\, \omega_{i_1\ldots i_p}\, dx^{i_1} \wedge \cdots \wedge dx^{i_p}$ with $p = 1, 2, \ldots, n$. Suppose that the metric tensorial family g_{ij} is equal to δ_{ij} with respect to the fixed observer O. Then

---

15 See P. Gilkey, Heat Kernel and the Atiyah–Singer theorem, CRC Press, Boca Raton, Florida, J. Jost, Riemannian Geometry and Geometric Analysis, Springer, Berlin, 2008, P. Griffith and J. Harris, Principles of Algebraic Geometry, Wiley, New York, 1979, and C. Voisin, Hodge Theory and Complex Algebraic Theory I, II, Cambridge University Press, 2002.

- $d^*\omega = -\frac{1}{(p-1)!}\,\delta^{ij}\,\partial_j\omega_{i_1\ldots i_p}\,dx^{i_1}\wedge\cdots\wedge dx^{i_p}$,
- $\Delta\omega = -\frac{1}{p!}\,\delta^{ij}\partial_i\partial_j\omega_{i_1\ldots i_p}\,dx^{i_1}\wedge\cdots\wedge dx^{i_p}$.

The proof will be given in Problem 8.8. The explicit expressions for general metric tensorial families g_{ij} in terms of covariant partial derivatives can be found in Sect. 9.3 on page 574, as an application of the index principle in mathematical physics.

Harmonic forms. In classical analysis, the smooth function U is called harmonic on the open set Ω iff $\Delta U = 0$ on Ω. Similarly, the p-form ω is called harmonic iff $\Delta\omega = 0$. In terms of de Rham cohomology, a basic result on harmonic forms tells us that every de Rham cohomology class of a compact Riemannian manifold contains many differential forms as representatives, but precisely one harmonic form. This implies a one-to-one correspondence between de Rham cohomology groups and groups of harmonic forms on compact Riemannian manifolds.

The Maxwell–Hodge–Yang–Mills equation. If ω is a solution of the so-called homogeneous Maxwell–Hodge–Yang–Mills equations

$$\boxed{d\omega = 0, \qquad d^*\omega = 0,} \tag{8.70}$$

then $\Delta\omega = 0$. In the setting of differential forms on compact Riemannian manifolds, we have the stronger result that the Maxwell–Hodge–Yang–Mills equation (8.70) is equivalent to $\Delta\omega = 0$. As we will show later on, the system (8.70) generalizes the Maxwell equations in electrodynamics and the Yang–Mills equations. These equations correspond to the pseudo-Riemannian Minkowski metric tensorial family.

The inner product for differential forms. If $\omega = \frac{1}{p!}\omega_{i_1\ldots i_p}\,dx^{i_1}\wedge\cdots\wedge dx^{i_p}$ and $\mu = \frac{1}{p!}\mu_{i_1\ldots i_p}\,dx^{i_1}\wedge\cdots\wedge dx^{i_p}$ are p-forms, then we define

$$(\omega|\mu) := \frac{1}{p!}\,\omega_{i_1\ldots i_p}\mu^{i_1\ldots i_p}.$$

For example, if $p = n$, then $\omega = \omega_{12\ldots n}dx^1\wedge\cdots\wedge dx^n$, $\mu = \mu_{12\ldots n}\,dx^1\wedge\cdots\wedge dx^n$, and

$$(\omega|\mu) = \omega_{12\ldots n}\mu^{12\ldots n}.$$

The relation to the volume form v is given by

$$\boxed{\omega\wedge *\mu = (\omega|\mu)\cdot v.} \tag{8.71}$$

If the coefficient functions of ω and μ have compact support on the open set Ω_O, then we define[16]

$$\boxed{\langle\omega|\mu\rangle_O := \int_{\Omega_O}(\omega|\mu)\,v.}$$

Observe that $\iota_O\cdot\langle\omega|\mu\rangle_O$ does not depend on the choice of the observer O. Therefore, the inner product $\langle om|\mu\rangle\_O$ changes sign if the observer changes the orientation. In addition,

$$(*\omega| *\mu) = (\omega|\mu). \tag{8.72}$$

The relation

$$\boxed{\langle d\omega|\mu\rangle = \langle\omega|d^*\mu\rangle} \tag{8.73}$$

shows that the codifferential operator d^* is dual to the differential operator d. For p-forms ω, we have the following commutation relations:

[16] The integral will be introduced in Sect. 8.7.

- $*\Delta\omega = \Delta(*\omega)$,
- $d(\Delta\omega) = \Delta(d\omega)$ and $d^*(\Delta\omega) = \Delta(d^*\omega)$.

The contraction product $v\rfloor\omega$**.** Let $\omega = \frac{1}{p!}\,\omega_{i_1\ldots i_p}\,dx^{i_1}\wedge\cdots\wedge dx^{i_p}$ be a p-form, and let $v = v^i\partial_i$ be a a linear first-order differential operator. We define

$$v\rfloor\omega := v^i\omega_{ii_2\ldots i_p}\,dx^{i_2}\wedge\cdots dx^p.$$

This definition does not depend on the choice of the observer, by the index principle. Synonymously, we write $i_v\omega$ instead of $v\rfloor\omega$. We have

$$i_v\omega(v_2,\ldots,v_p) = \omega(v, v_2\ldots v_p)$$

for all linear first-order differential operators $v, v_1,\ldots,v_p$.
Proof. For example, let $\omega = \frac{1}{2}\omega_{ij}\,dx^i\wedge dx^j$. Then

$$(v\rfloor\omega)(w) = v^i\omega_{ij}\,dx^j(w) = v^i\omega_{ij}w^j,$$

and $\omega(v,w) = \frac{1}{2}\omega_{ij}(dx^i\wedge dx^j)(v,w) = \frac{1}{2}\omega_{ij}(v^iw^j - v^jw^i) = \omega_{ij}v^iw^j$. □

8.6 The Kähler–Clifford Calculus and the Dirac Operator (Interior Product)

Both the Cartan exterior differential calculus and Hodge duality have their physical roots in the 1864 Maxwell theory of electromagnetism (see Chap. 18). In about 1960, when studying the 1928 Dirac equation for the relativistic electron, Erich Kähler (1906–2000) discovered the crucial fact that Cartan's exterior differential calculus can be complemented by a dual interior differential calculus.[17] Cartan's exterior differential calculus has its mathematical roots in the work of Grassmann (1809–1877), whereas Kähler's interior differential calculus has its mathematical roots in the work of Hamilton (1805–1865) and Clifford (1845-1879) on quaternions and Clifford algebras, respectively. Indeed, Cartan's exterior differential calculus is based on the Grassmann relation

$$dx^k\wedge dx^l + dx^l\wedge dx^k = 0,$$

and Kähler's interior differential calculus is based on the Clifford relation

$$dx^k\vee dx^l + dx^l\vee dx^k = 2g^{kl}$$

---

[17] E. Kähler, The Dirac equation, Abhandlungen der Deutschen Akademie der Wissenschaften, Berlin, Klasse für Mathematik, Physik und Technik, 1961, No. 1 (in German).
E. Kähler, The interior differential calculus, Rend. Mat. Appl. **21** (5), 425–523 (in German) (see also E. Kähler, Mathematical Works, de Gruyter, Berlin, pp. 499–595). This paper contains a complete representation of the theory together with a long list of all the formulas of the calculus. In this section, we summarize the Kähler calculus. All of the missing proofs can be found in Kähler's paper.
The Kähler calculus is closely related to modern spin geometry (Clifford modules and the connection of spin bundles). See H. Lawson and M. Michelsohn, Spin Geometry, Princeton University Press, 1994, and J. Jost, Riemannian Geometry and Geometric Analysis, Springer, Berlin, 2008.

where we use the definition $dx^k \vee dx^l := dx^k \wedge dx^l + g^{kl}$. The Cartan exterior differential $d\omega$ corresponds to the Kähler interior differential $d_\vee \omega$. The Kähler differential calculus comprehends the Cartan calculus, and the Kähler duality is a modification of Hodge duality. The basic ingredient of the Kähler calculus are the exterior $\wedge$-product, the interior $\vee$-product, the Cartan differential $d\omega$, the Kähler differential $d_\vee \omega$, the Kähler star operator

$$\star\omega := \omega \vee v,$$

and the Kähler codifferential $d^\star \omega := \operatorname{sgn}(g) \star^{-1} d\star\omega$. Here, v is the volume form. Dirac based his approach to the relativistic electron on the square root of the Laplacian for the Minkowski metric (see Chap. 20). The Kähler relation

$$d_\vee(d_\vee \omega) = -\Delta\omega \tag{8.74}$$

is the key to Kähler's theory for the Dirac equation $d_\vee \omega = a \vee \omega$ of the relativistic electron in the electromagnetic field. Equation (8.74) shows that the Dirac–Kähler differential operator $d_\vee$ is more fundamental than the Hodge Laplacian Δ. For example, the desire to find an index theorem for the elliptic Dirac operator was the starting point for the famous Atiyah–Singer theorem for general elliptic operators.[18] Mnemonically, we have the following rules: $dd = 0$ (Poincaré's cohomology rule), and

$$d^\star d^\star = 0, \quad d_\vee = d + d^\star, \quad d_\vee d_\vee = -\Delta.$$

Moreover, $\star\star = (-1)^{n(n-1)/2} \operatorname{sgn}(g)$ and $\star\, d_\vee = \operatorname{sgn}(g) \cdot d_\vee \star$ where n is the dimension of the coordinate space. In gauge theory, this will be complemented by $D = d + \mathcal{A}$ (covariant differential), $DD = \mathcal{F}$ (Cartan's curvature rule), and $DDD = 0$ (Bianchi identity).

Folklore

Let Ω be a fixed nonempty open subset of $\mathbb{R}^n$ where $n = 1, 2, \ldots$ Our goal is to construct both

- the exterior differential algebra $\bigwedge(\Omega)$, and
- the interior differential algebra $\bigvee(\Omega)$.

Suppose that Ω' is another open subset of $\mathbb{R}^m$ with $m = 1, 2, \ldots$, and suppose that we are given a smooth map

$$s : \Omega' \to \Omega,$$

then we get two algebra morphisms

$$s^* : \bigwedge(\Omega) \to \bigwedge(\Omega') \quad \text{and} \quad s^* : \bigvee(\Omega) \to \bigvee(\Omega')$$

called pull-back (see page 475). In particular, if there is an admissible system $\mathcal{O}$ of observers, and if s is a diffeomorphism from $\Omega' := \Omega_{\mathcal{O}'}$ onto $\Omega := \Omega_{\mathcal{O}}$ (with $m = n$), then it will turn out that the transformation laws correspond to tensorial families.

---

[18] M. Atiyah and I. Singer, The index of elliptic operators on compact manifolds, Bull. Amer.Math. Soc. **69** (1963), 422-433. See also J. Roe, Elliptic Operators, Topology, and Asymptotic Methods, Longman, Harlow, United Kingdom, 1988, and P. Gilkey, Invariance Theory, the Heat Equation, and the Atiyah–Singer Index Theorem, CRC Press, Boca Raton, Florida, 1995.

8.6.1 The Exterior Differential Algebra

Let $p = 1, \ldots, n$. We consider all the possible p-forms

$$\omega_p := \frac{1}{p!} \, \omega_{i_1 \ldots i_p} \, dx^{i_1} \wedge \cdots \wedge dx^{i_p}$$

where $\omega_{i_1 \ldots i_p} : \Omega \to \mathbb{R}$ are smooth functions which are antisymmetric with respect to the indices $i_1, \ldots, i_p = 1, \ldots, n$. In addition, if $p = 0$, then $\omega_0 : \Omega \to \mathbb{R}$ is a smooth function. Let $\bigwedge(\Omega)$ denote the set of all the finite sums

$$\omega := \omega_0 + \omega_1 + \omega_2 + \ldots$$

In addition, we set

- $\theta\omega := \omega_0 - \omega_1 + \omega_2 - \omega_3 + \omega_4 - \ldots,$
- $\zeta\omega := (\omega_0 + \omega_1) - (\omega_2 + \omega_3) + (\omega_4 + \omega_5) - \ldots$

For all $\omega, \mu \in \bigwedge(\Omega)$, we have

$$\theta(\omega \wedge \mu) = \theta\omega \wedge \theta\mu, \qquad \zeta(\omega \wedge \mu) = \zeta\mu \wedge \zeta\omega.$$

The real algebra $\bigwedge(\Omega)$. With respect to the exterior product "$\wedge$", the set $\bigwedge(\Omega)$ becomes a real algebra. The map θ (resp. ζ) is an algebra automorphism (resp. algebra anti-automorphism) on $\bigwedge(\Omega)$. In addition we have the linear operator

$$d : \bigwedge(\Omega) \to \bigwedge(\Omega)$$

given by $d\omega := d\omega_0 + d\omega_1 + \ldots$ where $d\omega_0 := \partial_i \omega_0 \cdot dx^i$, and

$$d\omega_p := \frac{1}{p!} \, \partial_{[i} \omega_{i_1 \ldots i_p]} \, dx^i \wedge dx^{i_1} \wedge \cdots \wedge dx^{i_p}, \qquad p = 1, \ldots, n.$$

In addition, we have $d(d\omega) = 0$, and there holds the graded Leibniz rule

$$\boxed{d(\omega \wedge \mu) = d\omega \wedge \mu + \theta\omega \wedge d\mu.}$$

Because of the existence of the operator $\omega \mapsto d\omega$, the algebra $\bigwedge(\Omega)$ is called a 'differential algebra'.

Algebra morphism (pull–back $s^*\omega$). We want to consider the change of coordinates which is crucial for invariant theory. To this end, fix $m, n = 1, 2, \ldots$ Let

$$\boxed{s : \Omega' \to \Omega}$$

be a smooth map defined on the nonempty open subset Ω' of $\mathbb{R}^m$:

$$x^i = s^i(u^1, \ldots, u^m), \qquad i = 1, \ldots, n.$$

To simplify notation, we briefly write $x^i = x^i(u^1, \ldots, u^m)$. Motivated by classical analysis, we define[19]

$$s^*(\omega_i \, dx^i) := \omega_i \cdot \frac{\partial x^i}{\partial u^j} \, du^j.$$

In a natural way, this can be extended to an algebra morphism

---

[19] More precisely, this means that $s^*(\omega_i \, dx^i)(u) = \omega_i(x(u)) \cdot \frac{\partial x^i(u)}{\partial u^j} \, du^j$.

$$\boxed{s^* : \bigwedge(\Omega) \to \bigwedge(\Omega').}$$

Explicitly, if $p = 1, \ldots n$, then we set

$$s^*\omega_p := \frac{1}{p!}\omega_{i_1 \ldots i_p} \cdot \frac{\partial x^{i_1}}{\partial u^{j_1}} \frac{\partial x^{i_2}}{\partial u^{j_2}} \cdots \frac{\partial x^{i_p}}{\partial u^{j_p}}\, du^{j_1} \wedge du^{j_2} \wedge \cdots \wedge du^{j_p}.$$

Here, we sum over $i_1, \ldots i_p = 1, \ldots, n$, and $j_1, \ldots, j_p = 1, \ldots, m$.[20] If $p = 0$, then we define $(s^*\omega_0)(u) := \omega_0(x(u))$.

Mnemonically, one changes the variables according to the classical transformation law for differentials.

For example, if the map s is given by $x = x(u, v), y = y(u, v)$, then

$$s^*(adx \wedge bdy) = a(x_u du + x_v dv) \wedge b(y_u du + y_v dv) = ab(x_u y_v - x_v y_u)\, du \wedge dv.$$

It is crucial that the pull-back transformation respects the Cartan differential, that is,

$$\boxed{s^*(d\omega) = d(s^*\omega).}$$

In particular, if $\chi : \Omega \to \Omega'$ is a diffeomorphism, then choosing $s := \chi^{-1}$, we get

$$s^*\omega = \frac{1}{p!}\, \omega_{i_{1'} \ldots i_{p'}}\, dx^{i_{1'}} \wedge \cdots \wedge dx^{i_{p'}}$$

where

$$\omega_{i_{1'} \ldots i_{p'}} = \frac{\partial x^{i_1}}{\partial x^{i_{1'}}} \frac{\partial x^{i_1}}{\partial x^{i_{1'}}} \cdots \frac{\partial x^{i_p}}{\partial x^{i_{p'}}} \cdot \omega_{i_1 \ldots i_p}.$$

This is the transformation law of a tensorial family.

Differentiation with respect to the basis differential dx^k. Let us start with the antisymmetric normal form

$$\omega_p := \frac{1}{p!}\, \omega_{i_1 \ldots i_p}\, dx^{i_1} \wedge \cdots \wedge dx^{i_p}, \qquad p = 1, \ldots, n.$$

Then we define

$$\delta_k \omega_p := \frac{1}{(p-1)!}\, \omega_{k i_2 \ldots i_p}\, dx^{i_2} \wedge \cdots \wedge dx^{i_p}, \qquad k = 1, \ldots, n. \tag{8.75}$$

Intuitively, we call $\delta_k \omega_p$ the differentiation of the p-form ω_p with respect to the basis differential dx^k. If $p = 0$, then we set $\delta_k \omega_0 := 0$. Note that the choice of the antisymmetric normal form is important in order to define uniquely the derivative $\delta_k \omega_p$. Moreover, we set

$$\delta_k(\omega_0 + \omega_1 + \ldots + \omega_n) := \delta_k \omega_1 + \ldots + \delta_k \omega_n, \qquad k = 1, \ldots, n.$$

Examples. Let $a, b, c : \Omega \to \mathbb{R}^n$ be smooth functions. Then:

- $\delta_k a = 0$ if $k = 1, \ldots, n$.
- $\delta_1(adx^1 + bdx^2 + cdx^3) = a$.
- $\delta_1(adx^1 \wedge dx^2) = adx^2$ and $\delta_1(bdx^3 \wedge dx^1) = \delta_1(-bdx^1 \wedge dx^3) = -bdx^3$.
- $\delta_2(adx^1 \wedge dx^2 \wedge dx^3) = \delta_2(-adx^2 \wedge dx^1 \wedge dx^3) = -adx^1 \wedge dx^3$, and
- $\delta_3(dx^1 \wedge dx^2 \wedge dx^4) = 0$.

[20] Obviously, $s^*\omega_p = 0$ if $p > m$.

Mnemonically, in order to compute δ_k put dx^k at the first position by interchanging the factors (using $dx \wedge dy = -dy \wedge dx$), and then cancel dx^k. To prove this, we need the antisymmetric normal form. In particular, let $n = 3$. Then

$$\omega := adx^1 \wedge dx^2 + bdx^3 \wedge dx^1 = \tfrac{1}{2}\omega_{ij}\, dx^i \wedge dx^j$$

where $\omega_{12} = -\omega_{21} := a, \omega_{31} = -\omega_{13} := b$, and $\omega_{23} = -\omega_{32} := 0$. Moreover, $\omega_{kk} := 0$ if $k = 1, 2, 3$. By (8.75), $\delta_1\omega = \omega_{12}dx^2 + \omega_{13}dx^3 = adx^2 - bdx^3$. □

There are the following rules at hand:

- $\delta_k(dx^k) = 1$, and $\delta_k(dx^l) = 0$ if $k \neq l$.
- $\delta_k(\mu) = 0$ if μ does not contain dx^k.
- $\delta_k(\omega \wedge \mu) = (\delta_k\omega) \wedge \mu + \theta\omega \wedge \delta_k\mu$ (graded Leibniz rule) if $\omega, \mu \in \bigwedge(\Omega)$.
- In particular, $\delta_k(\omega_0\mu) = \omega_0\delta_k\mu$ if ω_0 is a 0-form.
- $\delta_k(dx^k \wedge \mu) = \mu$ if μ does not contain dx^k.

8.6.2 The Interior Differential Algebra

We are given a metric tensorial family g_{kl} of arbitrary signature on the nonempty, open, arcwise connected subset Ω of $\mathbb{R}^n$.[21] Recall that $(g^{kl}) := (g_{kl})^{-1}$, and $g := \det(g_{kl})$. Then $\det(g^{kl}) = g^{-1}$. We will use g_{kl} in order to lift and lower indices in the usual way. In particular, we define

$$\delta^k\omega := g^{ks}\delta_s\omega.$$

For example, choose $n = 2$, and set $x := x^1, y := x^2$. Then:

- $\delta^l(dx \wedge dy) = g^{l1}dy - g^{l2}dx$,
- $\delta^k\delta^l(dx \wedge dy) = g^{k2}g^{l1} - g^{k1}g^{l2}$,
- $\delta_1(dx \wedge dy) = dy$, $\delta_2(dx \wedge dy) = -dx$,
- $\delta_1\delta_2(dx \wedge dy) = -1, \delta_2\delta_1(dx \wedge dy) = 1, \delta_k\delta_k(dx \wedge dy) = 0$ if $k = 1, 2$.

In fact, $\delta^l(dx \wedge dy) = g^{ls}\delta_s(dx \wedge dy)$. This is equal to

$$g^{l1}\delta_1(dx \wedge dy) + g^{l2}\delta_2(dx \wedge dy) = g^{l1}dy - g^{l2}\, dx.$$

Hence $\delta^k\delta^l(dx \wedge dy)$ is equal to $g^{ks}\delta_s(g^{l1}dy - g^{l2}dx) = -g^{k1}g^{l2} + g^{k2}g^{l1}$.

The interior $\vee$-product and the real algebra $\bigvee(\Omega)$. If ω_0 is a 0-form, then we define

$$\omega_0 \vee dx^l = dx^l \vee \omega_0 := \omega_0 dx^l.$$

Now fix $p = 1, \ldots, n$. Let $\omega, \mu \in \bigwedge(\Omega)$. The basic idea is to define an interior product "$\vee$" by setting

$$\boxed{dx^k \vee dx^l := dx^k \wedge dx^l + g^{kl}} \tag{8.76}$$

and

$$\omega \vee dx^l := \omega \wedge dx^l + \theta\delta^l\omega.$$

In particular, for a p-form $\omega_p = \frac{1}{p}\, dx^{i_1} \wedge \cdots \wedge dx^{i_p}$ with antisymmetric coefficients $\omega_{i_1 \ldots i_p}$, we get

[21] This means that g_{kl} transforms like a tensorial family with respect to all the diffeomorphisms $\chi : \Omega_O \to \Omega_{O'}$, where $\Omega_O \equiv \Omega$.

$$\omega_p \vee dx^l = \frac{1}{p!}\,\omega_{k_1\ldots k_p}\,dx^{k_1}\wedge\cdots\wedge dx^{k_p}\wedge dx^l + \frac{(-1)^{p-1}}{(p-1)!}\,g^{lk}\,\omega_{kk_2\ldots k_p}\,dx^{k_2}\wedge\cdots\wedge dx^{k_p}.$$

These are special cases of Kähler's general definition of the $\vee$-product which reads as follows:

$$\omega \vee \mu = \sum_{s=0}^{n}(-1)^{s(s-1)/2}\,\frac{\theta^s}{s!}\,(\delta_{i_1}\delta_{i_2}\cdots\delta_{i_s}\omega)\wedge(\delta^{i_1}\delta^{i_2}\cdots\delta^{i_s}\mu). \tag{8.77}$$

Theorem 8.9 *The set $\bigwedge(\Omega)$ becomes a real algebra with respect to the $\vee$-product. This algebra is generated by the 0-forms and the differentials $dx^1, \ldots, dx^n$.*

This real algebra denoted by $\bigvee(\Omega)$ is called the interior differential algebra over Ω. The designation Clifford product for the $\vee$-product is justified by the Clifford relation

$$\boxed{dx^k \vee dx^l + dx^l \vee dx^k = 2g^{kl}.} \tag{8.78}$$

Examples. Let $n = 2$. Set $x := x^1$, $y := x^2$. Fix the orientation by choosing the volume form $\upsilon := \sqrt{|g|}\,dx \wedge dy$ on the nonempty open subset Ω of $\mathbb{R}^2$. Then:

(i) $(dx \vee dy) \vee (dx \vee dy) = 2g^{12}dx \vee dy - g^{11}g^{22}$,
(ii) $(dx \wedge dy) \vee (dx \wedge dy) = g^{12}g^{12} - g^{11}g^{22} = -g^{-1}$,
(iii) $\upsilon \vee \upsilon = -\operatorname{sgn}(g)$.

Proof. We will not use the general definition of the $\vee$-product. In order to get inside, we will use the relations (8.76) and (8.78) together with the distributive and associative law. The following convenient method can be used in the general case. Ad (i). It follows from the Clifford relation (8.78) that

- $dy \vee dx = -dx \vee dy + 2g^{21}$,
- $dx \vee dx = g^{11}$ and $dy \vee dy = g^{22}$.

Hence

$$\begin{aligned} dx \vee (dy \vee dx) \vee dy &= -dx \vee dx \vee dy \vee dy + 2g^{12}\,dx \vee dy \\ &= -g^{11}g^{22} + 2g^{12}\,dx \vee dy. \end{aligned}$$

Ad (ii). By (8.76), $dx \wedge dy = dx \vee dy - g^{12}$. It follows from (i) that

$$\begin{aligned} &(dx \wedge dy) \vee (dx \wedge dy) = (dx \vee dy - g^{12}) \vee (dx \vee dy - g^{12}) \\ &= dx \vee dy \vee dx \vee dy - 2g^{12}dx \vee dy + g^{12}g^{12} = g^{12}g^{12} - g^{11}g^{22} = -g^{-1}. \end{aligned}$$

Ad (iii). By (ii), $(\sqrt{|g|}\,dx \wedge dy) \vee (\sqrt{|g|}\,dx \wedge dy) = -|g|g^{-1} = -\operatorname{sgn}(g)$. □

The Kähler differential $d_\vee\omega$. For a 0-form ω, we define

$$d_\vee\omega_0 := d\omega_0 = \partial_k\omega_0 \cdot dx^k.$$

Now fix $p = 1, \ldots, n$. Let $\omega_p = \frac{1}{p!}\,\omega_{i_1\ldots i_p}\,dx^{i_1}\wedge\cdots\wedge dx^{i_p}$ be a p-form with antisymmetric coefficients $\omega_{i_1\ldots i_p}$. In what follows, we will use the covariant partial derivative ∇_i for the Levi-Civita connection with respect to g_{kl} to be introduced below. Explicitly, $\nabla_i\omega_{i_1\ldots i_p}$ is given by (8.114) on page 497 together with the Christoffel symbols (8.146) on page 512. The point is that the Cartan differential of the p-form ω_p can be written as[22]

$$d\omega_p := dx^i \wedge \left(\frac{1}{p!} \nabla_i \, \omega_{i_1 \ldots i_p} \, dx^{i_1} \wedge \cdots \wedge dx^{i_p} \right).$$

Replacing the $\wedge$-product by the $\vee$-product, we define the Kähler differential

$$\boxed{d_\vee \omega_p := dx^i \vee \left(\frac{1}{p!} \nabla_i \, \omega_{i_1 \ldots i_p} \, dx^{i_1} \wedge \cdots \wedge dx^{i_p} \right).} \tag{8.79}$$

The definition is invariant under diffeomorphisms. This follows from the index principle by noting that $\nabla_i \omega_{i_1 \ldots i_p}$ is a tensorial family. In other words, if $s : \Omega' \to \Omega$ is a diffeomorphism and if ω is an element of $\bigwedge(\Omega)$, then

$$s^*(d_\vee \omega) = d_\vee(s^* \omega).$$

Here, if $\omega = \omega_0 + \omega_1 + \ldots + \omega_n$, then we set

$$d_\vee \omega := d\omega_0 + d_\vee \omega_1 + \ldots + d_\vee \omega_n.$$

The Kähler codifferential. If ω_0 is a 0-form, set $d^\star \omega_0 := 0$. For $p = 1, \ldots n$, define

$$d^\star \omega_p := \delta^i (\nabla_i \omega_{i_1 \ldots i_p} dx^{i_1} \wedge \cdots dx^{i_p}),$$

and $d^\star \omega := d^\star \omega_1 + \ldots + d^\star \omega_n$.

Theorem 8.10 *$d_\vee \omega = d\omega + d^\star \omega$ if $\omega \in \bigwedge(\Omega)$.*

In addition, we get

- $dd\omega = 0$ and $d^\star d^\star \omega = 0$,
- $d_\vee d_\vee \omega = (dd^\star + d^\star d)\omega = -\Delta \omega$.

8.6.3 Kähler Duality

Fix the volume form

$$\upsilon := \sqrt{|g|} \, dx^1 \wedge \cdots \wedge dx^n$$

on the nonempty, open, arcwise connected subset Ω of $\mathbb{R}^n$. If $s : \Omega' \to \Omega$ is a diffeomorphism, then we have

$$s^* \upsilon = \operatorname{sgn}(s) \cdot \sqrt{|g'|} \, dx^{1'} \wedge \cdots \wedge dx^{n'},$$

by Sect. 8.5.2. Here, $\operatorname{sgn}(s)$ is the sign of the Jacobian of the diffeomorphism s. We call $s^* \upsilon$ the volume form of $\bigvee(\Omega')$. We have

$$\upsilon \vee \upsilon = (-1)^{n(n-1)/2} \operatorname{sgn}(g).$$

For $\omega \in \bigwedge(\Omega)$, we define the Kähler star operator $\omega \mapsto \star\omega$ by setting

$$\boxed{\star\omega := \omega \vee \upsilon.}$$

It follows from $\omega \vee \upsilon \vee \upsilon = \operatorname{sgn}(g) \cdot (-1)^{n(n-1)/2} \omega$ that

---

[22] See Sect. 8.11.2 on page 523.

$$\boxed{\star\star\,\omega = \mathrm{sgn}(g)\cdot(-1)^{n(n-1)/2}\,\omega.} \tag{8.80}$$

This way, we obtain a linear bijective operator

$$\star : \bigvee(\Omega) \to \bigvee(\Omega)$$

with the inverse operator $\star^{-1} = \mathrm{sgn}(g)\cdot(-1)^{n(n-1)/2}\,\star$.[23] In particular,

$$\star 1 = v, \quad \star v = \mathrm{sgn}(g)\cdot(-1)^{n(n-1)}.$$

If $\omega_p = \frac{1}{p!}\omega_{i_1\ldots i_p}\, dx^{i_1}\wedge dx^{i_p}$ is a p-form with antisymmetric coefficients $\omega_{i_1\ldots i_p}$, then $\star\omega_p = \frac{1}{p!}\omega_{i_1\ldots i_p}\delta^{i_1}\delta^{i_2}\cdots\delta^{i_p}v$.

Theorem 8.11 *$d^\star\omega = \mathrm{sgn}(g)\star^{-1} d\star\omega$ if $\omega\in\bigvee(\Omega)$.*

8.6.4 Applications to Fundamental Differential Equations in Physics

The generalized Laplacian. If $\omega\in\bigwedge(\Omega)$, then

$$\boxed{d_\vee(d_\vee\omega) = -\Delta\omega.} \tag{8.81}$$

In particular, if

$$d_\vee\omega = 0,$$

then $\Delta\omega = 0$ by (8.81), that is, ω is a harmonic differential form.

The generalized Maxwell–Yang–Mills system. The system

$$d\omega = 0, \quad d_\vee\omega = 0 \tag{8.82}$$

is equivalent to

$$\boxed{d\omega = 0, \quad d^\star\omega = 0.}$$

In turn, this is equivalent to

$$d\omega = 0, \quad d(\star\omega) = 0.$$

This follows from $d_\vee\omega = d\omega + d^\star\omega$, and $d^\star\omega = \mathrm{sgn}(g)\star^{-1} d\star\omega$. The Maxwell equations will be studied in Sect. 19.8.3 on page 982.

The generalized Dirac equation. We are given the 1-form a. The equation

$$\boxed{d_\vee\omega = a\vee\omega}$$

is a generalization of the Dirac equation for the relativistic electron in an electromagnetic field. Here, the 1-form a corresponds to the 4-potential of the electromagnetic field. If the electromagnetic field vanishes, then the equation

$$\boxed{d_\vee\omega = 0}$$

generalizes the Dirac equation for the free relativistic electron.

Summarizing, Cartan's exterior calculus combined with Kähler's interior calculus allows us to formulate fundamental equations in physics in an extremely elegant way. This concerns

[23] Relation (8.80) shows that the Kähler star operator differs from the Hodge star operator; see (8.67) on page 470.

- the Maxwell equations in electromagnetism,
- the Dirac equation of the relativistic electron in an electromagnetic field, and
- the Yang–Mills equations which emerged implicitly first in classical fluid dynamics (see Sect. 12.2.6 on page 698), and which lead us to the Standard Model in elementary particle physics.

In the special case of rotational symmetry, it is possible to construct special solutions of the equations above which generalize the classic spherical harmonics of Laplace. Explicit formulas can be found in Kähler (1962/2004) quoted in the footnote on page 473.

8.6.5 The Potential Equation and the Importance of the de Rham Cohomology

We want to show how the solution of the generalized Maxwell–Yang–Mills equations can be simplified by solving the so-called potential equation

$$\boxed{dA = \omega \qquad \text{on } \Omega.} \tag{8.83}$$

We are given the differential form ω, and we are looking for the differential form A. In terms of physics, we want to find a potential A to the force ω. This is a crucial problem in gauge theory dating back to the work of Lagrange (1736–1813) and Gauss (1777–1855).

> *The solvability of the potential equation* (8.83) *critically depends on the topology of the set Ω.*

In the history of mathematics, the equation (8.83) was studied by Poincaré (1854–1912), Volterra (1860–1950), de Rham (1903–1990), and Hodge (1903–1975). This culminates in the de Rham cohomology (see Chaps. 22ff). At this point, let us only discuss the key ideas.

If A is a solution of (8.83), then it follows from the Poincaré cohomology rule $d(dA) = 0$ that

$$d\omega = 0 \qquad \text{on } \Omega.$$

This is a necessary condition for the solvability of the potential equation (8.83). There arises the following question:

> *Is the condition $d\omega = 0$ also a sufficient condition for the solvability of the potential equation* (8.83)*?*

De Rham cohomology tells us that the condition $d\omega = 0$ is a sufficient solvability condition for (8.83) if the set Ω is continuously contractible to a point (e.g., the open subset Ω of $\mathbb{R}^n$ is a ball or, more generally, a convex set). Otherwise, one has to add additional integral conditions of the type $\int_C \omega = 0$ which, roughly speaking, depend on the number of holes of the set Ω. In the terminology used in cohomology theory, the equation $dA = \omega$ tells us that A is a coboundary of ω, whereas the equation $d\omega = 0$ tells us that ω is a cocycle. De Rham theory studies the equivalence classes cocycles modulo coboundaries. If the set Ω is contractible to a point, then the de Rham cohomology of Ω is trivial, that is, every cocycle is a coboundary.

Let us consider a typical application. We want to solve the generalized Maxwell–Yang–Mills equations

$$d\omega = 0, \qquad d^\star\omega = 0 \qquad \text{on } \Omega. \tag{8.84}$$

Suppose that the set Ω is continuously contractible to a point. Then every solution ω of (8.84) can be represented as $\omega = dA$.

(i) Gauge condition: If the differential form A satisfies the two conditions
  - $\Delta A = 0$ (harmonic form), and
  - $d^\star A = 0$ (gauge condition),

  then the differential form $\omega = dA$ is a solution of (8.84). In order to prove this, note that $d\omega = ddA = 0$, and

$$d^\star \omega = d^\star dA = d^\star dA + dd^\star A = -\Delta A = 0.$$

(ii) Self-dual solution: If $\omega = dA$ and $\omega = \star\omega$, then ω is a solution of (8.84). To prove this, note that $d\omega = ddA = 0$ and

$$d^\star \omega = \operatorname{sgn}(g) \star^{-1} d \star \omega = \operatorname{sgn}(g) \star^{-1} d\omega = 0.$$

8.6.6 Tensorial Differential Forms

Einstein discovered that the gravitational force is intimately related to the Riemann curvature tensorial family R^k_{ijl} of the 4-dimensional space-time manifold. Élie Cartan combined the Riemann curvature theory with his exterior calculus of differential forms by introducing the connection forms $\omega^k_l := \Gamma^k_{il} dx^i$ and the curvature forms

$$\Omega^k_l := R^k_{ijl}\, dx^i \wedge dx^j, \qquad k, l = 1, \ldots, n.$$

At this point, let us only sketch the relation to Kähler's interior differential calculus. To this end, let us introduce expressions of the form

$$\mathsf{T}^{k_1 \ldots k_r}_{l_1 \ldots l_s} := T^{k_1 \ldots k_r}_{l_1 \ldots l_s; i_1 \ldots i_p}\, dx^{i_1} \wedge \cdots \wedge dx^{i_p}$$

which are called tensorial differential forms. The indices $k_1, l_1, i_1, \ldots$ run from 1 to n. The calculus is designed in such a way that the expression

$$\mathsf{T}^{k_1 \ldots k_r}_{l_1 \ldots l_s}\, dx^{l_1} \otimes \cdots \otimes dx^{l_s} \otimes \partial_{k_1} \otimes \cdots \otimes \partial_{k_r}$$

possesses an invariant meaning under diffeomorphisms $s : \Omega' \to \Omega$.[24] To this end, we assume that the smooth coefficient functions

$$T^{k_1 \ldots k_r}_{l_1 \ldots l_s; i_1 \ldots i_p}$$

represent a tensorial family which is antisymmetric with respect to the indices $i_1, \ldots, i_p$. As in Sect. 8.6.2 on page 479, we will use the covariant partial derivative ∇_i with respect to the Christoffel symbols related to the metric tensorial family g_{kl} (Levi-Civita connection). We define

$$\mathsf{D}_i \mathsf{T}^{k_1 \ldots k_r}_{l_1 \ldots l_s} := \nabla_i T^{k_1 \ldots k_r}_{l_1 \ldots l_s; i_1 \ldots i_p}\, dx^{i_1} \wedge \cdots \wedge dx^{i_p}.$$

This yields the definition of both the exterior differential

$$d\mathsf{T}^{k_1 \ldots k_r}_{l_1 \ldots l_s} := dx^i \wedge \mathsf{D}_i T^{k_1 \ldots k_r}_{l_1 \ldots l_s}$$

and the interior differential

$$d_\vee \mathsf{T}^{k_1 \ldots k_r}_{l_1 \ldots l_s} := dx^i \vee \mathsf{D}_i T^{k_1 \ldots k_r}_{l_1 \ldots l_s}.$$

Setting $d^\star \mathsf{T}^{k_1 \ldots k_r}_{l_1 \ldots l_s} := \delta^i \mathsf{D}_i \mathsf{T}^{k_1 \ldots k_r}_{l_1 \ldots l_s}$, we get the decomposition

$$\boxed{d_\vee \mathsf{T}^{k_1 \ldots k_r}_{l_1 \ldots l_s} = (d + d^\star) \mathsf{T}^{k_1 \ldots k_r}_{l_1 \ldots l_s}.}$$

---

[24] Recall that $dx^i \wedge dx^j = dx^i \otimes dx^j - dx^j \otimes dx^i$.

8.7 Integrals over Differential Forms

Mnemonically, Cartan's differential calculus yields the correct substitution rule for integrals via the Jacobian. The sign of the integral depends on the orientation of the observer.

Folklore

To simplify notation, in what follows we will write Ω, Ω', ι and ι' instead of $\Omega_O, \Omega_{O'}, \iota_O$ and $\iota_{O'}$, respectively. Let Ω', Ω' be nonempty, open, arcwise connected subsets of $\mathbb{R}^n$, $n = 1, 2, \ldots$

Special case. We define

$$\boxed{J = \int_\Omega f(x)\, dx^1 \wedge dx^2 \wedge \cdots \wedge dx^n := \int_\Omega f(x)\, dx^1\, dx^2 \cdots dx^n.}$$

This way, the integral over an n-form is reduced to a classical integral, by definition.[25]

Proposition 8.12 *The product $\iota \cdot J$ does not depend on the choice of the observer.*

Proof. (I) The classical substitution rule for n-dimensional integrals tells us that

$$\int_\Omega f(x)\, dx^1 dx^2 \cdots dx^n = \int_{\Omega'} f(x(x')) \left| \frac{\partial(x^1 \ldots x^n)}{\partial(x^{1'} \ldots x^{n'})}(x') \right| dx^{1'} dx^{2'} \cdots dx^{n'}.$$

This is equal to

$$\sigma \int_{\Omega'} f(x(x')) \frac{\partial(x^1 \ldots x^n)}{\partial(x^{1'} \ldots x^{n'})}(x')\, dx^{1'} dx^{2'} \cdots dx^{n'}$$

where σ is the sign of the Jacobian.

(II) The observer Ω computes the integral

$$J = \int_\Omega f(x)\, dx^1 dx^2 \cdots dx^n.$$

(III) Jacobian: The observer O' uses the transformed differential form

$$dx^1 \wedge \cdots \wedge dx^n = \frac{\partial x^1}{\partial x^{i'_1}} \cdots \frac{\partial x^n}{\partial x^{i'_n}}\, dx^{i'_1} \wedge \cdots \wedge dx^{i'_n}.$$

Hence

$$\boxed{dx^1 \wedge \cdots \wedge dx^n = \frac{\partial(x^1 \ldots x^n)}{\partial(x^{1'} \ldots x^{n'})}(x')\, dx^{1'} \wedge \cdots \wedge dx^{n'}.}$$

Moreover, the observer O' computes the integral

$$J' = \int_{\Omega'} f(x(x'))\, dx^1 \wedge \cdots \wedge dx^n$$

[25] In order to ensure the existence of the integrals to be considered in this section, we assume that the smooth functions $f, \omega_{i_1 \ldots i_n}$ have compact support on Ω. If this is true for one specific observer O, then it is true for all the other observers, since compact sets remain invariant under diffeomorphisms.

by using the transformed differential form. Hence

$$J' = \int_{\Omega'} f(x'(x)) \cdot \frac{\partial(x^1 \ldots x^n)}{\partial(x^{1'} \ldots x^{n'})}(x')\, dx^{1'} \cdots dx^{n'} = \sigma J.$$

Since $\iota' = \sigma \cdot \iota$, we obtain $\iota' \cdot J' = \iota \cdot J$. □

General case. We are given the n-form $\omega = \frac{1}{n!}\omega_{i_1 \ldots i_n}\, dx^{i_1} \wedge \cdots \wedge dx^{i_n}$. Since we have $\omega = \omega_{1 \ldots n}\, dx^1 \wedge \cdots \wedge dx^n$, we define

$$\boxed{\mathcal{J} = \int_\Omega \omega := \int_\Omega \omega_{1 \ldots n}\, dx^1\, dx^2 \cdots dx^n.}$$

Proposition 8.13 *The product $\iota \cdot \mathcal{J}$ does not depend on the choice of the observer.*

Proof. By (8.46), $\sqrt{|g'(x')|} = \frac{\partial(x^1, \ldots, x^n)}{\partial(x^{1'}, \ldots, x^{n'})}(x')\, \sqrt{|g(x)|}$ where $x' = x'(x)$. Thus, the classical substitution rule for integrals yields

$$\int_\Omega f(x)\sqrt{|g(x)|}\, dx^1 \cdots dx^n = \sigma \int_{\Omega'} f(x(x'))\, \sqrt{|g'(x')|}\, dx^{1'} \cdots dx^{n'} \qquad (8.85)$$

for all smooth functions $f : \Omega \to \mathbb{R}$ having compact support. Note that

$$\omega = \frac{1}{n!}\omega_{i_1 \ldots i_n}(x)\, \mathcal{E}^{i_1 \ldots i_n}(x) \cdot \sqrt{|g(x)|}\, dx^1 \wedge \cdots \wedge dx^n.$$

The observers O and O' compute the integrals

- $\iota \cdot \mathcal{J} = \int_\Omega \frac{1}{n!}\, \omega_{i_1 \ldots i_n} \cdot \iota \mathcal{E}^{i_1 \ldots i_n} \cdot \sqrt{|g|}\, dx^1\, dx^2 \cdots dx^n$, and
- $\iota' \cdot \mathcal{J}' = \int_{\Omega'} \frac{1}{n!}\, \omega_{i'_1 \ldots i'_n} \cdot \iota' \mathcal{E}^{i'_1 \ldots i'_n} \cdot \sqrt{|g'|}\, dx^{1'}\, dx^{2'} \cdots dx^{n'}$,

respectively. Set $f := \omega_{i_1 \ldots i_n} \cdot \iota \mathcal{E}^{i_1 \ldots i_n}$. Since $\iota \mathcal{E}^{i_1 \ldots i_n}$ is a tensorial family, the function f is invariant, by the index principle. Thus, it follows from (8.85) that $\iota \cdot \mathcal{J} = \iota' \cdot \mathcal{J}'$. □

The pull-back formula. Let $s : \Omega' \to \Omega$ be an orientation-preserving diffeomorphism. The classical substitution rule for integrals can elegantly be written as

$$\boxed{\int_\Omega \omega = \int_{s^*\Omega} s^*\omega} \qquad (8.86)$$

where ω is an n-form with smooth coefficient functions which have compact support. In addition, for streamlining the formula, we set $s^*\Omega := \Omega'$. The definition of the pull-back $s^*\omega$ of the n-form ω can be found on page 476.

8.8 Derivatives of Tensorial Families

In the 18th and 19th century, physicists and mathematicians investigated physical problems by using curvilinear coordinates (e.g., spherical coordinates in celestial mechanics). Frequently, the computations were clumsy. To simplify the approach, one was looking for modified partial derivatives which possess nice transformation laws. The main tool is the covariant derivative which sends tensorial families again to tensorial families. In the 20th century, mathematicians noticed that it is possible to formulate differentiation processes in a basis-free (i.e., index-free) way. This allows the natural generalization from finite-dimensional linear spaces to infinite-dimensional Hilbert and Banach spaces.

Folklore

We are given the tensorial family

$$T^{i_1 \ldots i_r}_{j_1 \ldots j_s}(x), \qquad x \in \Omega \tag{8.87}$$

of type (r, s). Our main goal is to find derivatives which send this tensorial family again to a tensorial family. To begin with, let us consider the simplest situation.

Linear transformation law. Consider the linear transformation law

$$x^{i'} = G^{i'}_i x^i, \qquad i' = 1', \ldots, n'$$

where the $(n \times n)$-matrix $G = (G^{i'}_i)$ does not depend on the position x.

Proposition 8.14 *If all the admissible coordinate transformations are linear, then*

$$\partial_k T^{i_1 \ldots i_r}_{j_1 \ldots j_s}(x), \qquad x \in \Omega$$

is again a tensorial family.

Proof. For example, consider the tensorial family T^k with the transformation law

$$T^{k'}(x') = \frac{\partial x^{k'}}{\partial x^k} \cdot T^k(x).$$

The point is that $\frac{\partial x^{k'}}{\partial x^k} = C^{k'}_k$ does not depend on the variables $x^1, \ldots, x^n$. Differentiation with respect to the variable $x^{j'}$ yields

$$\frac{\partial T^{k'}(x')}{\partial x^{j'}} = \frac{\partial x^{k'}}{\partial x^k} \cdot \frac{\partial x^j}{\partial x^{j'}} \frac{\partial T^k(x)}{\partial x^j},$$

by the chain rule. Hence

$$\partial_{j'} T^{k'}(x') = G^j_{j'} G^{k'}_k \cdot \partial_j T^k(x).$$

This is a tensorial transformation law. In the general case, the same argument can be applied. □

The trouble with the remainder. The argument above fails if the transformation law reads as

$$T^{k'}(x') = \frac{\partial x^{k'}(x)}{\partial x^k} \cdot T^k(x)$$

where $\frac{\partial x^{k'}(x)}{\partial x^k}$ depends on the position x. As a rule, this is the case if one uses curvilinear coordinates (e.g., spherical coordinates). In this situation, by the chain rule, differentiation with respect to the variable $x^{j'}$ yields

$$\frac{\partial T^{k'}(x')}{\partial x^{j'}} = \frac{\partial x^{k'}}{\partial x^k} \frac{\partial x^j}{\partial x^{j'}} \cdot \frac{\partial T^k(x)}{\partial x^j} + r^{k'}_{j'}(x)$$

with the remainder

$$r_{j'}^{k'}(x) := \frac{\partial^2 x^{k'}(x)}{\partial x^j \partial x^k} \cdot \frac{\partial x^j}{\partial x^{j'}} \cdot T^k(x).$$

In the 19th century, mathematicians introduced modifications of partial derivatives which compensate the remainder terms. The main trick for the compensation of remainders is to use the symmetry property

$$\boxed{\frac{\partial^2 f}{\partial x \partial y} = \frac{\partial^2 f}{\partial y \partial x}}$$

of the second partial derivatives for smooth real-valued functions f. The prototype of such an argument can be found in Problem 8.2. Typically, there are two possibilities:

- the modification of the definition of $\partial_i T^k$ by adding so-called connection terms in order to get the so-called covariant partial derivative

$$\boxed{\nabla_i T^k(x) := \partial_i T^k(x) + \Gamma_{ij}^k(x) T^j(x),} \qquad (8.88)$$

- or the use of linear combinations like $\partial_k T_i - \partial_k T_i$ (see the Lie derivative, the Cartan derivative, and the Weyl derivative below).

From the physical point of view, we will use additional physical quantities in order to modify partial derivatives. We will discuss the following possibilities:

- Lie derivative (additional velocity vector field),
- covariant partial derivative (Weyl connection) (additional gauge potential, e.g., the 4-potential in electromagnetism),
- covariant partial derivative (Levi-Civita connection) (additional metric tensorial family; e.g., the space-time metric in Einstein's theory of general relativity),
- covariant partial derivative in complex geometry (Kähler connection) (additional Kähler potential for getting the metric tensorial family; e.g., in string theory).

In contrast to this, both the Cartan derivative and the Weyl derivative possess a universal meaning; they do not depend on the choice of additional physical objects.

Historical remarks. Concerning (8.88), let us mention that symbols of the type Γ_{ij}^k appeared first in Gauss' *Disquisitiones generales circa superficies curvas* from 1827 (foundation of surface theory). In 1854, Riemann (1826–1866) generalized the Gaussian differential geometry to higher dimensions. Motivated by Riemann's work published in 1866, Christoffel (1829–1900) generalized Riemann's theory to more general Riemannian metric tensors. In particular, in 1869 Christoffel introduced the so-called Christoffel symbols Γ_{ij}^k.

Motivated by physics, in 1918 Weyl (1885–1955) generalized the Christoffel symbols by introducing the notion of an affine connection.[26] In the 1920s, Élie Cartan (1869–1951) established a general theory of connections which was the basis for modern differential geometry in terms of fiber bundles created by Ehresmann (1905–1979) in 1950.[27] In 1928 Dirac formulated the Dirac equation for the relativistic electron.

---

[26] H. Weyl, Pure infinitesimal geometry, Math. Z. **2** (1918), 384–411. In 1918, Weyl published his famous textbook *Space, Time, Matter* on Einstein's 1915 theory of general relativity, Springer, Berlin. Weyl's theory of affine connections was included in the third edition of his textbook which appeared in 1923.

[27] C. Ehresmann, Infinitesimal connections in differentiable fiber spaces, Colloque de Topologie, Bruxelles, 1950, pp. 29–55 (in French).

In this setting, a typical problem is to prove the relativistic invariance of the Dirac equation.

This is a nontrivial task in invariant theory because of the existence of a universal covering group to the Lorentz group. In 1929, van der Waerden (1903–1998) invented his relativistic spinor calculus in order to be able to formulate relativistically invariant partial differential equations in a systematic way.

For describing the Dirac electron in Einstein's general theory of relativity, one needs a covariant differentiation on curved pseudo-Riemannian space-time manifolds which respects the symplectic group $SL(2,\mathbb{C})$ (i.e., the universal covering group of the proper orthochronous Lorentz group $SO^{\uparrow}(1,3)$).[28]

In an Euclidean setting, Élie Cartan (1869–1951) studied the Lie algebra $so(n)$, $n = 3, 4, \ldots$, in the 1920s. In 1935, Brauer (1901–1977) and Weyl (1885–1955) constructed the universal covering group $Spin(n)$ of $SO(n)$ by using Clifford algebras.[29] Both the groups $Spin(n)$ and $SO(n)$ have the same Lie algebra, namely, $so(n)$. Brauer and Weyl showed that Cartan's strange double-valued representations of the group $SO(n)$ can be understood best as representations of the group $Spin(n)$.

8.8.1 The Lie Algebra of Linear Differential Operators and the Lie Derivative

It was the goal of Lie (1842–1899) to introduce linear combinations of partial derivatives which are independent of the choice of the observer. The Lie derivative for arbitrary tensorial families depends on the choice of a tensorial family v^i. In terms of physics, the Lie derivative is closely related to the flow of fluid particles governed by a given velocity vector field $\mathbf{v}$.
Furthermore, in the 19th century, mathematicians discovered that it is crucial to study second partial derivatives and their antisymmetrization. This leads to the Riemann curvature tensor and the concept of Lie algebra. The antisymmetrization generates nice transformation laws. In the 20th century, physicists discovered that the Riemann curvature tensor and its generalizations can be used in order to describe the fundamental forces in nature (Einstein's theory of general relativity and the Standard Model in elementary particle physics). On an infinitesimal level, the fundamental symmetries in nature are described by Lie algebras.

Folklore

The Lie algebra $\mathsf{Diff}^1(\Omega_O)$. Consider an admissible system $\mathcal{O}$ of observers. Let Θ be a scalar tensorial family (e.g., a temperature function measured by different observers). Recall that $\partial_i := \frac{\partial}{\partial x^i}$. Changing the observer, we have the transformation law

---

[28] B. van der Waerden, Spinor analysis, Nachr. Königl. Ges. Wiss. Göttingen 1929, pp. 100–109 (in German).
B. van der Waerden and L. Infeld, The wave equation of the electron in general relativity, Sitzungsber. Preußische Akad. Wiss. Berlin, Math.-Phys. Klasse **9** (1933), pp. 308–401 (in German).

[29] R. Brauer and H. Weyl, Spinors in n dimensions, Amer. J. Math. **57** (1935), 425–449.

$$\Theta_{O'}(x') = \Theta_O(x), \qquad x' = x'(x).$$

Let v^i and w^i be two tensorial families (with respect to $\mathcal{O}$) together with the linear first-order differential operators

$$v := v^i \partial_i \qquad \text{and} \qquad w := w^i \partial_i.$$

Observe that

$$\begin{aligned}(v \cdot w)(\Theta) = v(w(\Theta)) &= v^i \partial_i (w^j \partial_j \Theta) \\ &= v^i w^j \cdot \partial_i \partial_j \Theta + (v^i \partial_i w^j)\, \partial_j \Theta.\end{aligned}$$

Therefore, as a rule, the composition $v \cdot w$ of the two linear first-order differential operators v and w is not a first order differential operator, that is, the real linear space $\mathsf{Diff}^1(\Omega_O)$ is not an algebra with respect to the usual operator product. However, since $\partial_i \partial_j = \partial_j \partial_i$, the remainder $v^i w^j \cdot \partial_i \partial_j \Theta$ can be cancelled if we pass to the Lie bracket

$$\boxed{[v, w](\Theta) := (v \cdot w - w \cdot v)(\Theta).} \tag{8.89}$$

Explicitly,

$$[v, w](\Theta) = (v^i \partial_i w^j - w^i \partial_i v^j)\, \partial_j \Theta.$$

This tells us, that the Lie bracket $[v, w]$ is again a first-order linear operator.

The real linear space $\mathsf{Diff}^1(\Omega_O)$ of linear first-order differential operators on the real linear space $C^\infty(\Omega_O)$ becomes a real Lie algebra equipped with the Lie product $[v, w] := v \cdot w - w \cdot v$.

This simple observation is the key to Lie's approach to differential geometry.

Proposition 8.15 *If v^i and w^i are tensorial families, then so is $v^s \partial_s w^i - w^s \partial_s v^i$.*

Proof. We have to show that

$$v^{i'} \partial_{i'} w^{j'} - w^{i'} \partial_{i'} v^{j'} = \frac{\partial x^{j'}}{\partial x^j} \cdot (v^i \partial_i w^j - w^i \partial_i v^j). \tag{8.90}$$

This transformation law can be established by an explicit computation (see Problem 8.2). However, the claim can be proved without using any computation. To this end, note that the definition (8.89) does no depend on the choice of the observer. Hence

$$(v^{i'} \partial_{i'} w^{j'} - w^{i'} \partial_{i'} v^{j'}) \cdot \partial_{j'} \Theta_{O'}(x') = (v^i \partial_i w^j - w^i \partial_i v^j) \cdot \partial_j \Theta_O(x),$$

where $x' = x'(x)$. Using the transformation law $\partial_j \Theta_O = \frac{\partial x^{j'}}{\partial x^j} \partial_{j'} \Theta_{O'}$, we get

$$(v^{i'} \partial_{i'} w^{j'} - w^{i'} \partial_{i'} v^{j'}) \cdot \partial_{j'} \Theta_{O'}(x') = \left(v^i \partial_i w^j - w^i \partial_i v^j\right) \frac{\partial x^{j'}}{\partial x^j} \cdot \partial_{j'} \Theta_{O'}(x').$$

By Prop. 8.3 on page 459, the operators $\partial_{1'}, \ldots, \partial_{n'}$ are linearly independent. This implies (8.90). □

This elegant argument is the special case of the 'inverse index principle' to be considered in Sect. 8.8.2. In what follows, we suppose that we are given a fixed tensorial family v^i.

The Lie derivative of a temperature field Θ. We define

$$\boxed{(\mathcal{L}_{\mathbf{v}}\Theta)(x) := v^i(x)\,\partial_i\Theta(x), \qquad x \in \Omega_O.}$$

This is called the Lie derivative of the scalar tensorial family Θ at the point x with respect to the velocity vector $\mathbf{v}(x) = v^i(x)\partial_i$ at the point x. Briefly, $\mathcal{L}_{\mathbf{v}}\Theta := v^i\partial_i\Theta$. In mathematics, $\mathcal{L}_{\mathbf{v}}\Theta(x)$ is also called the directional derivative of the function Θ at the point x with respect to the velocity vector field $\mathbf{v}$ at the point x.[30] By (8.89),

$$\mathcal{L}_{\mathbf{v}}\mathcal{L}_{\mathbf{w}}\Theta - \mathcal{L}_{\mathbf{w}}\mathcal{L}_{\mathbf{v}}\Theta = \mathcal{L}_{[\mathbf{v},\mathbf{w}]}\Theta.$$

Consider the tensorial family

$$T^{i_1\ldots i_r}_{j_1\ldots j_s}.$$

Our goal is to define a new tensorial family denoted by

$$\mathcal{L}_{\mathbf{v}}T^{i_1\ldots i_r}_{j_1\ldots j_s}$$

which is called the Lie derivative and which generalizes $\mathcal{L}_{\mathbf{v}}\Theta$. We will proceed step by step.

The Lie derivative of a velocity vector field w. Let w^i be a tensorial family. Motivated by Prop. 8.15, we define

$$\boxed{\mathcal{L}_{\mathbf{v}}w^j := v^i\partial_i w^j - w^i\partial_i v^j, \qquad j = 1, 2, \ldots, n.}$$

This is again a tensorial family.

The Lie derivative of a covector field. Let ω_i be a tensorial family. We want to introduce $\mathcal{L}_{\mathbf{v}}\omega_i$ in such a way that we get the product rule

$$\mathcal{L}_{\mathbf{v}}(w^i\omega_i) = (\mathcal{L}_{\mathbf{v}}w^i)\omega_i + w^i\mathcal{L}_{\mathbf{v}}\omega_i. \tag{8.91}$$

To this end, we define

$$\boxed{\mathcal{L}_{\mathbf{v}}\omega_i := v^s\partial_s\omega_i + \omega_s\partial_i v^s.} \tag{8.92}$$

Proposition 8.16 *If ω_i is a tensorial family, then so is $\mathcal{L}_{\mathbf{v}}\omega_i$.*

Proof. This can be checked by an explicit computation, as in Problem 8.2. However, we want to use an argument which is the special case of a general principle called the 'inverse index principle'. Since $\Theta = w^i\omega_i$ is an invariant, by the index principle, we get

$$\mathcal{L}_{\mathbf{v}}(w^i\omega_i) = v^s\partial_s(w^i\omega_i).$$

It follows from the definition (8.92) that

- $(\mathcal{L}_{\mathbf{v}}w^i)\omega_i = \omega_i v^s\partial_s w^i - \omega_i w^s\partial_s w^i$,
- $w^i\mathcal{L}_{\mathbf{v}}\omega_i = w^i v^s\partial_s\omega_i + w^i\omega_s\partial_i v^s$.

[30] In what follows, the linear first-order differential operator $v = v^i\partial_i$ will be denoted by the vector symbol $\mathbf{v}$ in order to get contact to physical intuition. In Chap. 11 we will study velocity vector fields on the Euclidean manifold. In particular, there exists a one-to-one relation between velocity vectors $\mathbf{v} = v^i\mathbf{b}_i$ of our physical intuition and linear first-order differential operators $\mathbf{v} = v^i\partial_i$.

This implies (8.91). Now to the point. By (8.91), we have

$$w^i \mathcal{L}_{\mathbf{v}} \omega_i = \mathcal{L}_{\mathbf{v}}(w^i \omega_i) - \omega_i \mathcal{L}_{\mathbf{v}} w^i.$$

The right-hand side is a tensorial family. Consequently, $w^i \mathcal{L}_{\mathbf{v}} \omega_i$ is a tensorial family for all tensorial families w^i. Hence

$$w^{i'} \mathcal{L}_{\mathbf{v}} \omega_{i'} = w^i \mathcal{L}_{\mathbf{v}} \omega_i.$$

Using the transformation law $w^i = \frac{\partial x^i}{\partial x^{i'}} w^{i'}$, we get

$$w^{i'} \cdot \mathcal{L}_{\mathbf{v}} \omega_{i'} = w^{i'} \cdot \frac{\partial x^i}{\partial x^{i'}} \mathcal{L}_{\mathbf{v}} \omega_i \qquad \text{on } \Omega_{O'}.$$

Fix a point $x' \in \Omega_{O'}$, and fix the index $k' = 1, \ldots, n'$. By Theorem 8.2 on page 458, there exists a tensorial family $w^{i'}$ with $w^{i'}(x') = \delta^{i'}_{k'}$. Hence

$$\mathcal{L}_{\mathbf{v}} \omega_{k'} = \frac{\partial x^i}{\partial x^{k'}} \mathcal{L}_{\mathbf{v}} \omega_i.$$

This is the desired tensorial transformation law for $\mathcal{L}_{\mathbf{v}} \omega_i$. □

The Lie derivative of a general tensorial family via the Leibniz rule strategy. Consider first a special example. Let S^i, T_j, T^{ij} be tensorial families. We define

$$\boxed{\mathcal{L}_{\mathbf{v}} T^i_j := v^s \partial_s T^i_j + T^i_s \partial_j v^s - T^s_j \partial_s v^i.} \tag{8.93}$$

Note that this formula has the correct index picture. In order to motivate this definition, we first define $\mathcal{L}_{\mathbf{v}}(S^i T_j)$ by using the Leibniz rule

$$\mathcal{L}_{\mathbf{v}}(S^i T_j) := (\mathcal{L}_{\mathbf{v}} S^i) T_j + S^i \mathcal{L}_{\mathbf{v}} T_j.$$

Explicitly, this yields

$$\begin{aligned}\mathcal{L}_{\mathbf{v}}(S^i T_j) &= (v^s \partial_s S^i) T_j - (S^s \partial_s v^i) T_j + S^i v^s \partial_s T_j + S^i T_s \partial_j v^s \\ &= v^s \partial_s (S^i T_j) + S^i T_s \partial_j v^s - S^s T_j \partial_s v^i.\end{aligned}$$

Replacing $S^i T_j$ by T^i_j, we get (8.93). Finally, we postulate that definition (8.93) remains true for all the tensorial families T^i_j which do not have any special product structure. In particular, it follows from (8.93) that

$$\mathcal{L}_{\mathbf{v}} \delta^i_j = \delta^i_s \partial_j v^s - \delta^s_j \partial_s v^i = \partial_j v^i - \partial_j v^i = 0. \tag{8.94}$$

Similarly, we get

- $\mathcal{L}_{\mathbf{v}} T_{ij} = v^s \partial_s T_{ij} + T_{sj} \partial_i v^s + T_{is} \partial_j v^s$;
- $\mathcal{L}_{\mathbf{v}} T^{ij} = v^s \partial_s T^{ij} - T^{sj} \partial_s v^i - T^{is} \partial_s v^j$ (see Problem 8.3).

The general definition reads as follows:

$$\mathcal{L}_{\mathbf{v}} T^{i_1 \ldots i_k}_{j_1 \ldots j_l} := v^s \partial_s T^{i_1 \ldots i_k}_{j_1 \ldots j_l} + \sum_{\sigma=1}^{l} T^{i_1 \ldots i_k}_{j_1 \ldots s \ldots j_l} \partial_{j_\sigma} v^s - \sum_{\sigma=1}^{k} T^{i_1 \ldots s \ldots i_k}_{j_1 \ldots j_l} \partial_s v^{i_\sigma}.$$

Here, we replace the index j_σ (resp. i_σ) of $T^{i_1 \ldots i_k}_{j_1 \ldots j_l}$ by the index s, and we sum over the values $s = 1, \ldots, n$. Mnemonically, for every index of $T^{i_1 \ldots i_k}_{j_1 \ldots j_l}$, the directional derivative

$$v^s \partial_s T^{i_1 \ldots i_k}_{j_1 \ldots j_l}$$

is supplemented by an additional summand which respects the principle of the correct index picture.

Proposition 8.17 *The Lie derivative* $\mathcal{L}_{\mathbf{v}}T^{i_1\ldots i_k}_{j_1\ldots j_l}$ *of a tensorial family* $T^{i_1\ldots i_k}_{j_1\ldots j_l}$ *is again a tensorial family.*

By Prop. 8.16, this is true for $\mathcal{L}_{\mathbf{v}}T^i$ and $\mathcal{L}_{\mathbf{v}}T_j$. In the general case, note that the product construction used above respects the tensorial transformation laws.

The Leibniz rule and the contraction rule. The Lie derivative

$$\mathcal{L}_{\mathbf{v}}\left(T^{i_1\ldots i_r}_{j_1\ldots j_s}\cdot S^{k_1\ldots k_a}_{l_1\ldots l_b}\right)$$

of the product $T^{i_1\ldots i_r}_{j_1\ldots j_s}\cdot S^{k_1\ldots k_a}_{l_1\ldots l_b}$ is equal to

$$\mathcal{L}_{\mathbf{v}}\left(T^{i_1\ldots i_r}_{j_1\ldots j_s}\right)S^{k_1\ldots k_a}_{l_1\ldots l_b}+T^{i_1\ldots i_r}_{j_1\ldots j_s}\mathcal{L}_{\mathbf{v}}\left(S^{k_1\ldots k_a}_{l_1\ldots l_b}\right). \tag{8.95}$$

This Leibniz rule for the Lie derivative follows immediately from our product strategy used above. Note that this Leibniz rule remains valid if one pair or several pairs of indices are contracted. For example,

$$\mathcal{L}_{\mathbf{v}}(T^i\cdot S^k_i)=\mathcal{L}_{\mathbf{v}}T^i\cdot S^k_i+T^i\mathcal{L}_{\mathbf{v}}S^k_i.$$

The Lie derivative of a differential form. Suppose that we are given the p-form $\omega=\frac{1}{p!}\omega_{i_1\ldots i_p}\,dx^{i_1}\wedge\cdots\wedge dx^{i_p}$ where $\omega_{i_1\ldots i_p}$ is a Cartan family. By definition, the Lie derivative $\mathcal{L}_{\mathbf{v}}\omega$ of ω is given by be quite natural formula

$$\boxed{\mathcal{L}_{\mathbf{v}}\omega=\mathcal{L}_{\mathbf{v}}\omega_{i_1\ldots i_p}\cdot dx^{i_1}\wedge\cdots\wedge dx^{i_p}.}$$

By the index principle, this definition does not depend on the choice of the observer. For example,

$$\boxed{\mathcal{L}_{\mathbf{v}}dx^j=\partial_s v^j\cdot dx^s.} \tag{8.96}$$

Proof. Use $\mathcal{L}_{\mathbf{v}}(\omega_i dx^i)=\mathcal{L}_{\mathbf{v}}\omega_i\cdot dx^i=(v^s\partial_s\omega_i+\omega_s\partial_i v^s)\,dx^i$ with $\omega_i:=\delta_{ij}$. □

From the Leibniz rule for the Lie derivative of tensorial families (8.95), we obtain the Leibniz rule for the Lie derivative of differential forms:

$$\boxed{\mathcal{L}_{\mathbf{v}}(\omega\wedge\mu)=(\mathcal{L}_{\mathbf{v}}\omega)\wedge\mu+\omega\wedge\mathcal{L}_{\mathbf{v}}\mu.} \tag{8.97}$$

Using (8.96), this Leibniz rule allows us to compute the Lie derivative $\mathcal{L}_{\mathbf{v}}\omega$ in the following nice way:

$$\mathcal{L}_{\mathbf{v}}(\omega_i\,dx^i)=\mathcal{L}_{\mathbf{v}}\omega_i\cdot dx^i+\omega_i\mathcal{L}_{\mathbf{v}}dx^i=v^s\partial_s\omega_i\cdot dx^i+\omega_i\partial_s v^i\cdot dx^s.$$

Changing the names of the indices, we get

$$\mathcal{L}_{\mathbf{v}}\omega=(v^s\partial_s\omega_i+\omega_s\partial_i v^s)\cdot dx^i.$$

This coincides with $\mathcal{L}_{\mathbf{v}}\omega=\mathcal{L}_{\mathbf{v}}\omega_i\cdot dx^i$.

The Lie–Cartan formula. Let $\omega=\omega_j dx^j$ be a 1-form. Then

$$\boxed{d\omega(\mathbf{v},\mathbf{w})=\mathcal{L}_{\mathbf{v}}(\omega(\mathbf{w}))-\mathcal{L}_{\mathbf{w}}(\omega(\mathbf{v}))-\omega([\mathbf{v},\mathbf{w}])} \tag{8.98}$$

for all first-order differential operators $\mathbf{v}=v^j\partial_j$ and $\mathbf{w}=w^j\partial_j$. Moreover, we use the Lie bracket $[\mathbf{v},\mathbf{w}]:=(v^i\partial_i w^j-w^i\partial_i v^j)\partial_j$.

The Lie–Cartan formula (8.98) tells us that the Cartan differential $d\omega$ can be expressed by the Lie derivative and the Lie bracket.

In turn, the Lie bracket is a special Lie derivative. In fact,

$$[\mathbf{v}, \mathbf{w}] = \mathcal{L}_{\mathbf{v}}\mathbf{w}.$$

This is a special case of the definition (8.100) given below. Summarizing, the Cartan differential $d\omega$ can be expressed by the Lie derivative of velocity vector fields.[31]

Proof. We get the following expressions:

- $d\omega = \partial_i \omega_j \, dx^i \wedge dx^j$;
- $d\omega(\mathbf{v}, \mathbf{w}) = \partial_i\omega_j \cdot (dx^i(\mathbf{v})dx^j(\mathbf{w}) - dx^i(\mathbf{w})dx^j(\mathbf{v})) = \partial_i\omega_j \cdot (v^i w^j - v^j w^i)$;
- $\omega(\mathbf{w}) = \omega_j dx^j(\mathbf{w}) = \omega_j w^j$;
- $\mathcal{L}_{\mathbf{v}}(\omega(\mathbf{w})) = v^i\partial_i(\omega_j w^j) = (v^i\partial_i\omega_j)w^j + (v^i\partial_i w^j)\omega_j$;
- $\mathcal{L}_{\mathbf{w}}(\omega(\mathbf{v})) = w^i\partial_i(\omega_j v^j) = (w^i\partial_i\omega_j)v^j + (w^i\partial_i v^j)\omega_j$;
- $\omega([\mathbf{v}, \mathbf{w}]) = \omega_j(v^i\partial_i w^j - w^i\partial_i v^j)$.

This yields the claim. □

For $p = 2, 3, \ldots, n$, the general Lie–Cartan formula reads as

$$d\omega(\mathbf{v}_0, \mathbf{v}_1, \ldots, \mathbf{v}_p) = \sum_{i=0}^{p}(-1)^i \mathcal{L}_{\mathbf{v}_i}\omega(\mathbf{v}_0, \ldots, \hat{\mathbf{v}}_i, \ldots, \mathbf{v}_p)$$
$$+\sum_{i<j}(-1)^{i+j}\omega([\mathbf{v}_i, \mathbf{v}_j], \mathbf{v}_0, \ldots, \hat{\mathbf{v}}_i, \ldots, \hat{\mathbf{v}}_j, \ldots \mathbf{v}_p).$$

By convention, the terms $\hat{\mathbf{v}}_i$ and $\hat{\mathbf{v}}_j$ have to be cancelled.

Cartan's magic formula. For differential forms of degree $p = 1, 2, \ldots$, we get

$$\boxed{\mathcal{L}_{\mathbf{v}}\omega = i_{\mathbf{v}}(d\omega) + d(i_{\mathbf{v}}\omega).} \tag{8.99}$$

Proof. Consider first the special case where $\omega = \omega_i dx^i$. Then:

- $\mathcal{L}_{\mathbf{v}}\omega = (v^s\partial_s\omega_j + \omega_s\partial_j v^s)\, dx^j$;
- $d\omega = \partial_{[i}\omega_{j]}\, dx^i \wedge dx^j = \frac{1}{2}(\partial_i\omega_j - \partial_j\omega_i)\, dx^i \wedge dx^j$;
- $i_{\mathbf{v}}(d\omega) = v^i(\partial_i\omega_j - \partial_j\omega_i)\, dx^j$;
- $d(i_{\mathbf{v}}\omega) = d(v^i\omega_i) = (\partial_j v^i)\omega_i\, dx^j + v^i\partial_j\omega_i\, dx^j$.

The general case proceeds similarly (see Problem 8.5). An elegant index-free proof can be found in Problem 8.6. □

The Lie derivative of a tensor field. For a tensor field

$$T = T^{i_1 i_2 \ldots i_r}_{j_1 \ldots j_s}\, dx^{j_1} \otimes dx^{j_2} \otimes \cdots \otimes dx^{j_s} \otimes \partial_{i_1} \otimes \partial_{i_2} \otimes \cdots \otimes \partial_{i_r},$$

we define the Lie derivative by setting

$$\mathcal{L}_{\mathbf{v}}T := \mathcal{L}_{\mathbf{v}}T^{i_1 i_2 \ldots i_r}_{j_1 \ldots j_s} \cdot dx^{j_1} \otimes dx^{j_2} \otimes \cdots \otimes dx^{j_s} \otimes \partial_{i_1} \otimes \partial_{i_2} \otimes \cdots \otimes \partial_{i_r}. \tag{8.100}$$

For example,

---

31 This allows us to define the Cartan differential $d\omega$ on infinite-dimensional Banach spaces and Banach manifolds. See R. Abraham, J. Marsden, and T. Ratiu, Manifolds, Tensor Analysis, and Applications, Springer, New York, 1988.

$$\boxed{\mathcal{L}_{\mathbf{v}}(dx^i) = \partial_s v^i \cdot dx^s, \qquad \mathcal{L}_{\mathbf{v}}(\partial_i) = -\partial_i v^s \cdot \partial_s.} \tag{8.101}$$

Proof. Use $(\mathcal{L}_{\mathbf{v}} w^j)\partial_j = (v^s \partial_s w^j - w^s \partial_s v^j)\,\partial_j$ with $\omega^j := \delta^{ij}$. □

As for differential forms above, we get the following Leibniz rule for tensors:

$$\mathcal{L}_{\mathbf{v}}(T \otimes S) = \mathcal{L}_{\mathbf{v}} T \otimes S + T \otimes \mathcal{L}_{\mathbf{v}} S.$$

This rule combined with (8.101) allows us to compute the Lie derivative of arbitrary tensors. For example,[32]

$$\mathcal{L}_{\mathbf{v}}(T^{ij}\partial_i \otimes \partial_j) = \mathcal{L}_{\mathbf{v}}(T^{ij}) \cdot \partial_i \otimes \partial_j + T^{ij}\, \mathcal{L}_{\mathbf{v}}(\partial_i) \otimes \partial_j + T^{ij}\, \partial_i \otimes \mathcal{L}_{\mathbf{v}}(\partial_j).$$

Hence

$$\mathcal{L}_{\mathbf{v}}(T^{ij}\partial_i \otimes \partial_j) = v^s \partial_s T^{ij} \cdot \partial_i \otimes \partial_j - T^{ij}\partial_i v^s \cdot \partial_s \otimes \partial_j - T^{ij}\partial_j v^s \cdot \partial_i \otimes \partial_s.$$

Changing the names of the indices and writing $\partial_i\partial_j$ instead of $\partial_i \otimes \partial_j$, we get

$$\mathcal{L}_{\mathbf{v}}(T^{ij}\partial_i\partial_j) = (v^s \partial_s T^{ij} - T^{sj}\partial_s v^i - T^{is}\partial_s v^j)\,\partial_i\partial_j.$$

This coincides with $\mathcal{L}_{\mathbf{v}} T^{ij} \cdot \partial_i\partial_j$.

Physical interpretation. The approach used above is very nice from the mnemonic point of view, but it does not give any insight. In fact, the Lie derivative can be understood best by using the transport of physical quantities by means of the flow of fluid particles. This will be studied in Chap. 11.

8.8.2 The Inverse Index Principle

> The inverse index principle saves a lot of time by avoiding clumsy computations with a lot of cancellations at the end of the day. A spectacular application concerns the Riemann curvature tensor to be considered in Sect. 8.9.
>
> Folklore

Consider a fixed admissible system $\mathcal{O}$ of observers. Then the following hold:

The family of smooth functions ω_i is a tensorial family iff $\omega_i w^i$ is a tensorial family for all tensorial families w^i.

Proof. (I) Index principle: If ω_i and w^i is a tensorial family, then so is $w^i\omega_i$, by Theorem 8.1 on page 455.

(II) Inverse index principle: Conversely, if ω_i is a family of smooth functions and $w^i\omega_i$ is a tensorial family for all tensorial families, then ω_i is a tensorial family, by the proof of Prop. 8.16 on page 489. The same argument shows that this result remains true in the general case.

Proposition 8.18 *The family $T^{i_1\ldots i_r}_{j_1\ldots j_s}$ of smooth functions is a tensorial family iff the product*

$$T^{i_1\ldots i_r}_{j_1\ldots j_s} \cdot S^{k_1\ldots k_q}_{l_1\ldots l_p}$$

(or a fixed contraction of this product) is a tensorial family for all tensorial families $S^{k_1\ldots k_q}_{l_1\ldots l_p}$.

---

[32] The symbol $\mathcal{L}_{\mathbf{v}}(T^{ij})$ stands for $\mathcal{L}_{\mathbf{v}}\Theta$ with $\Theta := T^{ij}$.

This result is called the (full) index principle. For example, the family T_{ij}^k of smooth functions is a tensorial family iff $T_{ij}^k S_k^i$ is a tensorial family for all tensorial families S_k^i. Let us consider a further application.

The natural basis. Let $T_{j_1 \ldots j_s}^{i_1 \ldots i_r}$ be a family of smooth functions. Then

$$T := T_{j_1 j_2 \ldots j_s}^{i_1 i_2 \ldots i_r} \, dx^{j_1} \otimes dx^{j_2} \otimes \cdots \otimes dx^{j_s} \otimes \partial_{i_1} \otimes \partial_{i_2} \cdots \otimes \partial_{i_r}$$

does not depend on the choice of the observer iff $T_{j_1 j_2 \ldots j_s}^{i_1 i_2 \ldots i_r}$ is a tensorial family.
Proof. For example, consider

$$T := T_i^{jk} dx^i \otimes \partial_j \otimes \partial_k.$$

If $\mathbf{v} = v^k \partial_k$, then

$$T(\mathbf{v}) = T_i^{jk} \, dx^i(\mathbf{v}) \partial_j \partial_k = T_i^{jk} v^i \cdot \partial_j \partial_k.$$

This is a linear second-order differential operator. Thus, the map $\mathbf{v} \mapsto T(\mathbf{v})$ is a linear operator of the form

$$T : \mathsf{Diff}^1(\Omega_O) \to \mathsf{Diff}^2(\Omega_O).$$

The differential operator $\partial_j \partial_k$ transforms like a tensorial family. If T_i^{jk} is a tensorial family, then $T(\mathbf{v}) = T_i^{jk} v^i \partial_j \partial_k$ does not depend on the observer, by the index principle.

Conversely, let T_i^{jk} be a family of smooth functions such that T does not depend on the observer. Then

$$T_i^{jk} v^i \cdot \partial_j \partial_k$$

transforms like a scalar tensorial family. Using the transformation law for ∂_j and ∂_k and the fact that all the differential operators $\partial_j \partial_k$ with $j, k = 1, \ldots, n$ are linearly independent, it follows as in the proof of Prop. 8.16 on page 489 that

$$T_i^{jk} v^i$$

is a tensorial family for all tensorial families v^i. Thus, T_i^{jk} is a tensorial family, by Prop. 8.18. □

8.8.3 The Covariant Derivative (Weyl's Affine Connection)

> The Christoffel symbols Γ_{ij}^k do not form a tensorial family; they form a connection family. In terms of physics, this corresponds to the components of a gauge potential.
>
> Folklore

Choose an admissible system $\mathcal{O}$ of observers. For a scalar tensorial family Θ, we define

$$\nabla_i \Theta := \partial_i \Theta.$$

This is a tensorial family, by Prop. 8.14 on page 485. In terms of physics, this describes the gradient of the temperature field Θ.

Covariant partial derivative of a velocity vector field. We are given the tensorial family v^k. In contrast to $\partial_i \Theta$, the partial derivative

$$\partial_i v^k$$

is not a tensorial family, as a rule. This is a serious defect of the classical partial derivative in mathematics and physics. Our goal is to cure this defect by modifying the partial derivative. To this end, fix the observer O, and define the covariant partial derivative $\nabla_i v^k$ by setting

$$\boxed{\nabla_i v^k(x) := \partial_i v^k(x) + \Gamma^k_{is}(x) v^s(x), \qquad x \in \Omega_O} \tag{8.102}$$

where $i, j = 1, \ldots, n$. That is, we add the term $\Gamma^k_{is}(x) v^s(x)$ to the partial derivative $\partial_i v^k(x)$. We assume that the family of smooth functions $\Gamma^k_{ij} : \Omega_O \to \mathbb{R}$ does not depend on the choice of the tensorial family v^k. The functions Γ^k_{ij} are called the Christoffel symbols (or the connection symbols). For the observer O', we define

$$\nabla_{i'} v^{k'}(x') := \partial_{i'} v^{k'}(x') + \Gamma^{k'}_{i's'}(x') v^{s'}(x'), \qquad x' \in \Omega_{O'} \tag{8.103}$$

where $x' = x'(x)$. The following is basic for gauge theory.

Our goal is to choose the Christoffel symbols Γ^k_{ij} and $\Gamma^{k'}_{i'j'}$ in such a way that $\nabla_i v^k$ is a tensorial family.

Explicitly, this means that we want to obtain the transformation law

$$\nabla_{i'} v^{k'}(x') = G^i_{i'}(x) G^{k'}_k(x) \cdot \nabla_i v^k(x), \qquad x \in \Omega_O, \qquad x' = x'(x). \tag{8.104}$$

The key formula reads as

$$\Gamma^{k'}_{i'j'}(x') = (G^i_{i'} G^j_{j'} G^{k'}_k \cdot \Gamma^k_{ij} - G^i_{i'} G^j_{j'} \partial_i G^{k'}_j)(x), \quad x \in \Omega_O, \quad x' = x'(x). \tag{8.105}$$

The definition of the transformation coefficients $G^{i'}_i(x)$ and $G^i_{i'}(x)$ can be found on page 445. Note that (8.105) is a tensorial transformation law if $\partial_i G^{k'}_j \equiv 0$, that is, the transformation coefficients do not depend on the position x. Otherwise, (8.105) is not a tensorial transformation law, as a rule.

Proposition 8.19 *Suppose that v^k is a tensorial family. If the Christoffel symbols transform like* (8.105), *then $\nabla_i v^k$ is a tensorial family.*

Proof. We have to show that

$$\partial_{i'} v^{k'} + \Gamma^{k'}_{i'j'} v^{j'} = G^i_{i'} G^{k'}_k (\partial_i v^k + \Gamma^k_{ij} v^j). \tag{8.106}$$

In fact, by the chain rule, it follows from $v^{k'}(x') = G^{k'}_k(x) v^k(x)$ that

$$\frac{\partial v^{k'}(x')}{\partial x^{i'}} = \frac{\partial G^{k'}_k(x)}{\partial x^i} \frac{\partial x^i(x')}{\partial x^{i'}} \cdot v^k(x) + G^{k'}_k(x) \frac{\partial v^k(x)}{\partial x^i} \frac{\partial x^i(x')}{\partial x^{i'}}.$$

Hence

$$\partial_{i'} v^{k'} = (\partial_i G^{k'}_k \cdot v^k + G^{k'}_k \partial_i v^k) G^i_{i'}.$$

Moreover, relation (8.105) implies

$$\Gamma^{k'}_{i'j'} v^{j'} = G^i_{i'} G^{k'}_k v^j \cdot \Gamma^k_{ij} - G^i_{i'} v^j \partial_i G^{k'}_j.$$

Adding up, we obtain the claim (8.106). □

The language of matrices. The key transformation formula (8.105) looks clumsy. In order to write down this formula more elegantly, let us introduce the real $(n \times n)$-matrices

$$\boxed{\mathcal{A}_i := (\Gamma^k_{ij}), \qquad i = 1, \ldots, n}$$

where k is the row index, and j is the column index, that is,

$$\mathcal{A}_i := \begin{pmatrix} \Gamma^1_{i1} & \Gamma^1_{i2} & \ldots & \Gamma^1_{in} \\ \vdots & \vdots & \ldots & \vdots \\ \Gamma^n_{i1} & \Gamma^n_{i2} & \ldots & \Gamma^n_{in} \end{pmatrix}.$$

Using the matrix product, the key transformation formula (8.105) is equivalent to the following matrix formula:

$$\mathcal{A}_{i'}(x') = G^i_{i'}(x) \cdot (G\mathcal{A}_i G^{-1} - \partial_i G \cdot G^{-1})(x), \qquad x' = x'(x) \tag{8.107}$$

which is crucial for gauge theory. Introducing the so-called gauge transformation formula

$$\mathcal{A}^+_i(x) := (G\mathcal{A}_i G^{-1} - \partial_i G \cdot G^{-1})(x), \qquad x \in \Omega_O, \quad x' = x'(x), \tag{8.108}$$

the complete transformation formula (8.107) can be written as

$$\boxed{\mathcal{A}_{i'}(x') = G^i_{i'}(x) \cdot \mathcal{A}^+_i(x), \qquad x \in \Omega_O, \quad x' = x'(x).} \tag{8.109}$$

This tells us that the complete transformation $\mathcal{A}_i(x) \mapsto \mathcal{A}_{i'}(x')$ from the original observer O to the observer O' is the superposition of

- the gauge transformation $\mathcal{A}_i(x) \mapsto \mathcal{A}^+_i(x)$ performed by the observer O, and
- the tensorial law $\mathcal{A}^+_i(x) \mapsto \mathcal{A}_{i'}(x')$ which corresponds to the passage from the observer O to the observer O'.

It follows from $G^{-1}G = I$ that $\partial_i G^{-1} \cdot G + G^{-1}\partial_i G = 0$. Consequently, the gauge transformation (8.108) is equivalent to

$$\mathcal{A}^+_i(x) := (G\mathcal{A}_i G^{-1} + G\partial_i G^{-1})(x), \qquad x \in \Omega_O, \quad x' = x'(x), \tag{8.110}$$

If we introduce the differential forms

- $\mathcal{A}(x) := \mathcal{A}_i(x)dx^i$,
- $\mathcal{A}^+(x) := \mathcal{A}^+_i(x)dx^i$,
- $\mathcal{A}'(x') := \mathcal{A}_{i'}dx^{i'}$,

then

$$\boxed{\mathcal{A}'(x') = \mathcal{A}^+(x), \quad x \in O_\Omega, \; x' = x'(x).} \tag{8.111}$$

Weyl's affine connection. Suppose that, for every observer O of the admissible system $\mathcal{O}$ of observers, we have a family

$$\Gamma^k_{ij} : \Omega_O \to \mathbb{R}, \qquad i, j, k = 1, \ldots n$$

of smooth functions together with the transformation law (8.105) which is equivalent to the tensorial transformation law (8.109) for the local coordinates together with the gauge transformation (8.108). Then Γ^k_{ij} is called an affine connection family.[33]

We now proceed as for the Lie derivative in Sect. 8.8.1 on page 489. The proofs of the following results are the same as in Sect. 8.8.1.

Covariant partial derivative of a covector field. In order to enforce the Leibniz rule

$$\nabla_i(w^s\omega_s) = (\nabla_i w^s)\omega_j + w^s\nabla_i\omega_s, \tag{8.112}$$

we define

$$\boxed{\nabla_i\omega_j := \partial_i\omega_j - \Gamma^s_{ij}\omega_s.}$$

Proposition 8.20 *If ω_j is a tensorial family, then so is $\nabla_i\omega_j$.*

Duality between velocity vector fields and covector fields. Set

$$v := \begin{pmatrix} v^1 \\ \vdots \\ v^n \end{pmatrix} \quad \text{and} \quad \omega := \begin{pmatrix} \omega_1 \\ \vdots \\ \omega_n \end{pmatrix}.$$

Using the matrices $\mathcal{A}_1, \ldots, \mathcal{A}_n$, the covariant partial derivatives look like

$$\boxed{\nabla_i v = \partial_i v + \mathcal{A}_i v, \quad i = 1, \ldots, n}$$

and

$$\boxed{\nabla_i \omega = \partial_i\omega - \mathcal{A}_i^d\omega, \quad i = 1, \ldots, n.}$$

This displays the duality between $\nabla_i v^k$ and $\nabla_i\omega_k$.

> *The covariant derivative of a velocity vector field and the covariant derivative of a covector field are dual concepts.*

Covariant partial derivative of a general tensorial family via the Leibniz rule strategy. For a tensorial family T^k_j, we define

$$\boxed{\nabla_i T^k_j := \partial_i T^k_j + \Gamma^k_{is}T^s_j - \Gamma^s_{ij}T^k_s.} \tag{8.113}$$

This is motivated in the following way. If S^k and T_j are tensorial families, then we define

$$\nabla_i(S^kT_j) := (\nabla_i S^k)T_j + S^k(\nabla_i T_j), \tag{8.114}$$

motivated by the Leibniz rule. Hence

$$\nabla_i(S^kT_j) = (\partial_i S^k)T_j + S^k(\partial_i T_j) + \Gamma^k_{is}S^sT_j - \Gamma^s_{ij}S^kT_s.$$

By the Leibniz rule for classical partial derivatives, we get

$$\nabla_i(S^kT_j) = \partial_i(S^kT_j) + \Gamma^k_{is}(S^sT_j) - \Gamma^s_{ij}(S^kT_s).$$

---

[33] This notion was introduced by Weyl (1885–1955) in 1918 (see the footnote on page 486).

Setting $T_j^k := S^k T_j$, we get (8.113). Finally, we postulate that (8.113) remains valid for arbitrary tensorial families T_j^k which do not have any special product structure.

Similarly, we get

$$\boxed{\nabla_i T^{jk} = \partial_i T^{jk} + \Gamma_{is}^j T^{sk} + \Gamma_{is}^k T^{js}.}$$

For the general tensorial family $T^{i_1 \ldots i_k}_{j_1 \ldots j_l}$, the definition reads as

$$\nabla_i T^{i_1 \ldots i_k}_{j_1 \ldots j_l} := \partial_i T^{i_1 \ldots i_k}_{j_1 \ldots j_l} + \sum_{\sigma=1}^{k} \Gamma^{i_\sigma}_{is} T^{i_1 \ldots s \ldots i_k}_{j_1 \ldots j_l} - \sum_{\sigma=1}^{l} \Gamma^{s}_{ij_\sigma} T^{i_1 \ldots i_k}_{j_1 \ldots s \ldots j_l}.$$

Here, we replace the index i_σ (resp. j_σ) of $T^{i_1 \ldots s \ldots i_k}_{j_1 \ldots j_l}$ by the index s, and we sum over $s = 1, \ldots, n$. Mnemonically, note that the index picture is correct.

The Leibniz rule. The covariant partial derivative

$$\nabla_i \left(T^{i_1 \ldots i_r}_{j_1 \ldots j_s} \cdot S^{k_1 \ldots k_a}_{l_1 \ldots l_b} \right)$$

of the product $T^{i_1 \ldots i_r}_{j_1 \ldots j_s} \cdot S^{k_1 \ldots k_a}_{l_1 \ldots l_b}$ of tensorial families is equal to

$$\nabla_i \left(T^{i_1 \ldots i_r}_{j_1 \ldots j_s} \right) \cdot S^{k_1 \ldots k_a}_{l_1 \ldots l_b} + T^{i_1 \ldots i_r}_{j_1 \ldots j_s} \cdot \left(\nabla_i S^{k_1 \ldots k_a}_{l_1 \ldots l_b} \right). \tag{8.115}$$

This Leibniz rule follows immediately from our product strategy used above.

The contraction principle. The Leibniz rule (8.8.3) remains valid if one pair or several pairs of indices are contracted. For example,

$$\nabla_i (T^s S_s^k) = \nabla_i T^s \cdot S_s^k + T^s \cdot \nabla_i S_s^k.$$

Let us discuss this. To this end, we will prove the following general principle:

The operation of contraction can be interchanged with the operation of covariant differentiation.

As a typical simple example, let T_l^k be a tensorial family. Then

$$S_{il}^k := \nabla_i (T_l^k) = \partial_i T_l^k + \Gamma_{il}^s T_s^k - \Gamma_{is}^k T_l^s.$$

Contracting the index k with the index l yields

$$S_{ik}^k = \partial_i (T_k^k) + \Gamma_{ik}^s T_s^k - \Gamma_{is}^k T_k^s = \partial_i (T_k^k).$$

Here, the Christoffel symbols cancel each other.

Otherwise, for the function $\Theta := T_k^k$, we get

$$\nabla_i (T_k^k) = \nabla_i \Theta = \partial_i \Theta = \partial_i (T_k^k) = S_{ik}^k.$$

This shows that the two different ways of computing $\nabla_i T_k^k$ yield the same result. The same argument applies to the contraction of several indices of the tensorial family

$$\nabla_i T^{j_1 \ldots j_r}_{i_1 \ldots i_s}.$$

The torsion tensorial family. By (8.105), as a rule, the Christoffel symbols Γ_{ij}^k do not form any tensorial family. However,

$$\boxed{T^k_{ij} := \Gamma^k_{ij} - \Gamma^k_{ji}} \tag{8.116}$$

is a tensorial family called the torsion family. In fact, $T^{k'}_{i'j'} = G^{k'}_k G^i_{i'} G^j_{j'} \Gamma^k_{ij}$.

Antisymmetrization and torsion. Let ω_j be a tensorial family. Then

$$\nabla_{[i}\omega_{j]} = \partial_{[i}\omega_{j]} - \tfrac{1}{2} T^k_{ij} S_k.$$

This motivates the designation 'torsion'.
Proof. We have $\nabla_i S_j = \partial_i S_j - \Gamma^k_{ij} S_k$ and

$$\nabla_j S_i = \partial_j S_i - \Gamma^k_{ji} S_k.$$

Moreover, recall that $\nabla_{[i} S_{j]} = \frac{1}{2}(\nabla_i S_j - \nabla_j S_i)$. □

Proposition 8.21 *Let $\omega_{i_1 \ldots i_p}$ be an antisymmetric tensorial family. Then*

$$\nabla_{[i}\omega_{i_1 \ldots i_p]} = \partial_{[i}\omega_{i_1 \ldots i_p]} - \tfrac{1}{2} \mathrm{Alt}_{ii_1 \ldots i_p} \left(\sum_{\sigma=1}^{p} T^s_{i i_\sigma} \omega_{i_1 \ldots i_{\sigma-1} s i_{\sigma+1} \ldots i_p} \right).$$

In the special case where the torsion vanishes, that is, $T^k_{ij} \equiv 0$ for all indices, we have

$$\nabla_{[i}\omega_{i_1 \ldots i_p]} = \partial_{[i}\omega_{i_1 \ldots i_p]}. \tag{8.117}$$

This tells us that in the torsion-free case, the alternating covariant partial derivative of an antisymmetric covariant tensorial family coincides with the Cartan derivative.

Here, the symbol $\mathrm{Alt}_{[ii_1 \ldots i_p]}$ means that we antisymmetrize with respect to the $p+1$ indices $i, i_1 \ldots, i_p$. For example,

$$\mathrm{Alt}_{ij}(S_{ijk}) := \tfrac{1}{2}(S_{ijk} - S_{jik}).$$

Similarly, $\mathrm{Sym}_{ij}(S_{ijk}) := \frac{1}{2}(S_{ijk} + S_{jik})$. The proof of Prop. 8.21 will be given in Problem 8.9 on page 554.

Further definitions based on the covariant partial derivative. In what follows, let $T^{i_1 \ldots i_r}_{j_1 \ldots j_s}$ and v^i be tensorial families. Note that all of the following definitions do not depend on the choice of the observer, by the index principle. Set $\mathbf{v} = v^i \partial_i$.

(i) The covariant directional derivative: The tensorial family

$$D_{\mathbf{v}} T^{i_1 \ldots i_r}_{j_1 \ldots j_s} := v^i \nabla_i T^{i_1 \ldots i_r}_{j_1 \ldots j_s}$$

is called the covariant directional derivative family of $T^{i_1 \ldots i_r}_{j_1 \ldots j_s}$. For example,

$$D_{\mathbf{v}} T^k_j = v^i \nabla_i T^k_j.$$

(ii) The covariant time derivative: Let $x^i = x^i(t), i = 1, \ldots, n$, be a smooth curve where $t \in]-t_0, t_0[$. We define

$$\frac{D}{dt} T^{i_1 \ldots i_r}_{j_1 \ldots j_s}(x(t)) = \dot{x}^i(t) \nabla_i T^{i_1 \ldots i_r}_{j_1 \ldots j_s}(x(t)).$$

(iii) The covariant directional derivative of a tensor field: Let

$$T = T^{i_1 \ldots i_r}_{j_1 \ldots j_s} \, dx^{j_1} \otimes \cdots \otimes dx^{j_s} \otimes \partial_{i_1} \otimes \cdots \otimes \partial_{i_r}$$

be a tensor field of type (r, s). Naturally enough, we set

$$\boxed{D_{\mathbf{v}} T := v^j \nabla_j T^{i_1 \ldots i_r}_{j_1 \ldots j_s} \cdot dx^{j_1} \otimes \cdots \otimes dx^{j_s} \otimes \partial_{i_1} \otimes \cdots \otimes \partial_{i_r}.} \tag{8.118}$$

(iv) The covariant differential of a tensor field: The mapping $\mathbf{v} \mapsto D_{\mathbf{v}} T$ is linear. We set

$$(DT)(\mathbf{v}) := D_{\mathbf{v}} T \qquad \text{for all} \quad \mathbf{v} = v^i \partial_i.$$

Explicitly,

$$\boxed{DT = \nabla_j T^{i_1 \ldots i_r}_{j_1 \ldots j_s} \cdot dx^j \otimes dx^{j_1} \otimes \cdots \otimes dx^{j_s} \otimes \partial_{i_1} \otimes \cdots \otimes \partial_{i_r}.}$$

(v) Covariant time derivative of a tensor field along a curve:

$$\frac{DT(x(t))}{dt} := DT(\dot{x}(t)) = \dot{x}^j(t) \nabla_j T^{i_1 \ldots i_r}_{j_1 \ldots j_s} \cdot dx^{j_1} \otimes \cdots \otimes dx^{j_s} \otimes \partial_{i_1} \otimes \cdots \otimes \partial_{i_r}.$$

Summary. Let S, T, U, V be tensorial fields, where U and V have the same type (r, s). Moreover, let Θ a scalar tensorial family. Then, we have the following key rules:

- $D(U + V) = DU + DV$ (sum rule),
- $D(S \otimes T) = DS \otimes T + S \otimes DT$ (Leibniz rule);
- $D(\Theta T) = D\Theta \cdot T + \Theta \cdot DT$, and $D\Theta = d\Theta$ (special Leibniz rule).

This tells us that it is possible to introduce a covariant differential for tensor fields which possesses quite natural properties. This is a far-reaching generalization of the classical Leibniz calculus. In particular, we get[34]

- $D_{\mathbf{v}}(\partial_j) = \Gamma^k_{ij} v^i \cdot \partial_k$, and
- $D(\partial_j) = \Gamma^k_{ij} \, dx^i \otimes \partial_k$.

This explains the meaning of the Christoffel symbols. Dually, we get

- $D_{\mathbf{v}}(dx^k) = -\Gamma^k_{ij} v^i \cdot dx^j$, and
- $D(dx^k) = -\Gamma^k_{ij} \, dx^i \otimes dx^j$.

By (8.118), we get the following:

If T is a tensor field of type (r, s), then $D_{\mathbf{v}} T$ is a tensor field of type (r, s), and DT is a tensor field of type $(r, s+1)$.

Examples. (i) We are given $T = T^k \partial_k$. This is a tensor field of type $(1, 0)$. We want to compute the covariant directional derivative $D_{\mathbf{v}} T$ by using general rules. To begin with, note that the Leibniz rule yields

$$D_{\mathbf{v}}(T^k \partial_k) = D_{\mathbf{v}}(T^k) \cdot \partial_k + T^k \cdot D_{\mathbf{v}}(\partial_k).$$

Mnemonically, we use the symbol $D_{\mathbf{v}}(T^k)$. However, observe that $D_{\mathbf{v}}(T_k)$ does not denote the covariant partial derivative $D_{\mathbf{v}} T^k$ of the tensorial family T^k, but

[34] Note that $\partial_j = T^k \partial_k$ where $T^k := \delta^k_j$ for fixed j. Hence $D_{\mathbf{v}}(\partial_j) = v^i \nabla_i T^k \partial_k$. Explicitly, $v^i \nabla_i T^k = v^i (\partial_i T^k + \Gamma^k_{is} T^s) = v^i \Gamma^k_{is} \delta^s_j = v^i \Gamma^k_{ij}$.

the classical directional derivative, $d_{\mathbf{v}}T^k$, of the function $x \mapsto T^k(x)$. Explicitly, $d_{\mathbf{v}}T^k = v^i\partial_i T^k$. To avoid misunderstandings, we write

$$\boxed{D_{\mathbf{v}}(T^k\partial_k) = d_{\mathbf{v}}T^k \cdot \partial_k + T^k \cdot D_{\mathbf{v}}(\partial_k).}$$

Hence

$$D_{\mathbf{v}}(T^k\partial_k) = (v^i\partial_i T^k)\partial_k + T^k(\Gamma^s_{ik}v^i\partial_s).$$

This implies the final result

$$D_{\mathbf{v}}T = (v^i\partial_i T^k + v^i\Gamma^k_{ij}T^j)\,\partial_k. \tag{8.119}$$

Using the covariant partial derivative ∇_i, we get the tensor field of type $(1,0)$,

$$\boxed{D_{\mathbf{v}}T = v^i\nabla_i T^k \cdot \partial_k,}$$

which coincides with (8.118). From (8.119), we obtain the tensor field of type $(1,1)$,

$$\boxed{DT = \nabla_i T^k\; dx^i \otimes \partial_k.}$$

Alternatively, we get $DT = dT^k \cdot \partial_k + T^k D(\partial_k)$. Hence

$$DT = \partial_i T^k\; dx^i \otimes \partial_k + \Gamma^k_{is}T^s\; dx^i \otimes \partial_k = \nabla_i T^k\; dx^i \otimes dx^k. \tag{8.120}$$

(ii) Let $T := T_k dx^k$. This is a tensor field of type $(0,1)$. We want to show that

$$\boxed{D_{\mathbf{v}}T = (v^i\nabla_i T_k)\; dx^k.} \tag{8.121}$$

This is a tensor field of type $(0,1)$. In fact, by the Leibniz rule,

$$D_{\mathbf{v}}(T_k dx^k) = d_{\mathbf{v}}T_k \cdot dx^k + T_k \cdot D_{\mathbf{v}}(dx^k).$$

Hence

$$D_{\mathbf{v}}(T_k dx^k) = (v^i\partial_i T_k)\cdot dx^k - T_k \cdot (\Gamma^k_{ij}v^i\; dx^j).$$

This implies the final result

$$D_{\mathbf{v}}T = (v^i\partial_i T_k - v^i\Gamma^s_{ik}T_s)\cdot dx^k = v^i\nabla_i T_k \cdot dx^k$$

which coincides with (8.118). In addition , we get

$$\boxed{DT = \nabla_i T_k\; dx^i \otimes dx^k.}$$

This is a tensor field of type (0,2).

(iii) Consider the tensor field $T := T_{kl}dx^k \otimes dx^l$ of type $(0,2)$. By the Leibniz rule,

$$D(dx^k \otimes dx^l) = D(dx^k)\otimes dx^l + dx^k \otimes D(dx^l).$$

Hence

$$D_{\mathbf{v}}T = d_{\mathbf{v}}T_{kl}\cdot dx^k \otimes dx^l + T_{kl}\cdot D_{\mathbf{v}}(dx^k)\otimes dx^l + T_{kl}\cdot dx^k \otimes D_{\mathbf{v}}(dx^l).$$

Noting that $d_{\mathbf{v}}(T_{kl}) = v^i \partial_i T_{kl}$. Hence

$$D_{\mathbf{v}} T = v^i \partial_i T_{kl} \cdot dx^k \otimes dx^l - T_{kl} v^i \Gamma^k_{is} dx^s \otimes dx^l - T_{kl} v^i \Gamma^l_{is} dx^k \otimes dx^s.$$

This implies

$$D_{\mathbf{v}}(T_{kl}) = (v^i \partial_i T_{kl} - v^i \Gamma^s_{ik} T_{sl} - v^i \Gamma^s_{il} T_{ks}) \cdot dx^k \otimes dx^l.$$

Hence $D_{\mathbf{v}}(T_{kl}) = v^i \nabla_i T_{kl} dx^k \otimes dx^l$. In addition, we obtain the tensor field of type $(1, 2)$,

$$\boxed{DT = \nabla_i T_{kl}\, dx^i \otimes dx^k \otimes dx^l.}$$

The existence theorem for affine connections. From the mathematical point of view, the following theorem tells us that there exist infinitely many possibilities to introduce an affine connection family, and hence to get a covariant derivative for tensorial families. In physics, the problem is to find specific covariant derivatives which describe the forces acting in the universe (e.g., the gravitational force in Einstein's theory of general relativity). We are given an admissible system $\mathcal{O}$ of observers. Suppose that a fixed observer O chooses a family of smooth functions $\Gamma^k_{ij} : \Omega_O \to \mathbb{R}$, $i, j, k = 1, \ldots, n$.

Theorem 8.22 *The given family Γ^k_{ij} can be uniquely extended to a connection family of $\mathcal{O}$.*

Proof. Motivated by (8.104), for the observers O' and O'' we define

- $\Gamma^{k'}_{i'j'}(x') := \left(G^i_{i'} G^j_{j'} G^{k'}_k \cdot \Gamma^k_{ij} - G^i_{i'} G^j_{j'} \partial_i G^{k'}_j\right)(x)$, and
- $\Gamma^{k''}_{i''j''}(x'') := \left(G^i_{i''} G^j_{j''} G^{k''}_k \cdot \Gamma^k_{ij} - G^i_{i''} G^j_{j''} \partial_i G^{k''}_j\right)(x)$,

respectively. Here, $x' = x'(x)$ and $x'' = x''(x)$. We have to show that the passage from the observer O' to the observer O'' is given by the following transformation law:

$$\Gamma^{k''}_{i''j''}(x'') = \left(G^{i'}_{i''} G^{j'}_{j''} G^{k''}_{k'} \cdot \Gamma^{k'}_{i'j'} - G^{i'}_{i''} G^{j'}_{j''} \partial_{i'} G^{k''}_{j'}\right)(x') \tag{8.122}$$

where $x'' = x''(x')$. This can be obtained by an explicit computation based on the chain rule for second-order partial derivatives.

In order to get insight, let us use the language of matrices. Set $\mathcal{A}_i := (\Gamma^k_{ij})$. By (8.107), for the observers O' and O'' we define

- $\mathcal{A}_{i'}(x') := G^i_{i'}(x')(G\mathcal{A}_i G^{-1} - \partial_i G \cdot G^{-1})(x)$, and
- $\mathcal{A}_{i''}(x'') = J^i_{i''}(x)(J\mathcal{A}_i J^{-1} - \partial_i J \cdot J^{-1})(x)$,

respectively. The change of observers

$$O \overset{G}{\Longrightarrow} O' \overset{H}{\Longrightarrow} O''$$

corresponds to the product $J = HG$ of the linearized maps. We have to show that the passage from the observer O' to the observer O'' corresponds to the transformation law

$$\mathcal{A}_{i''}(x'') = G^{i'}_{i''}(x')(H\mathcal{A}_{i'} H^{-1} - \partial_{i'} H \cdot H^{-1})(x').$$

In fact, this follows from

$$(HG)^{-1} = G^{-1} H^{-1}, \qquad \partial_i(HG) = \partial_i H \cdot G + H \cdot \partial_i G$$

together with $G^i_{i'} G^{i'}_{i''} = G^i_{i''}$, and $\partial_{i'} = \frac{\partial}{\partial x^{i'}} = \frac{\partial x^i}{\partial x^{i'}}\, \partial_i$. □

8.9 The Riemann–Weyl Curvature Tensor

It was Riemann, who probably more than anyone else, enriched mathematics with new ideas. These ideas display an unusual degree of vitality and impulse to the whole of mathematics as well as to many branches of physics.[35]

Krzysztof Maurin, 1997

From the analytic point of view, the Riemann–Weyl curvature tensorial family R^l_{ijk} relies on the relation

$$\nabla_i \nabla_j v^l - \nabla_j \nabla_i v^l = R^l_{ijk} v^k \tag{8.123}$$

in the torsion-free case. Here, the crucial point is that the left-hand side $\nabla_i \nabla_j v^l - \nabla_j \nabla_i v^l$ contains first-order and second-order partial derivatives of the velocity vector field v^l, whereas the right-hand side $R^l_{ijk} v^k$ is linear in v^k. There appear magic cancellations. In general, in mathematics or physics, nontrivial cancellations always indicate that there is a nice structure behind the clumsy formulas. In the renormalization of quantum fields, for example, physicists observe incredible cancellations which result from hidden or visible quantum symmetries (e.g., Ward identities).

Folklore

Summary. Using the invariant language of operator theory, the two key relations of modern differential geometry read elegantly as follows:

(i) Lie bracket and Riemann curvature:

$$\boxed{D_{\mathbf{u}} D_{\mathbf{v}} \mathbf{w} - D_{\mathbf{v}} D_{\mathbf{u}} \mathbf{w} = D_{[\mathbf{u},\mathbf{v}]} \mathbf{w} + \mathbf{R}(\mathbf{u}, \mathbf{v}, \mathbf{w}).} \tag{8.124}$$

(ii) Lie bracket and torsion:

$$\boxed{D_{\mathbf{u}} \mathbf{v} - D_{\mathbf{v}} \mathbf{u} = [\mathbf{u}, \mathbf{v}] + \mathbf{T}(\mathbf{u}, \mathbf{v}).} \tag{8.125}$$

The main idea is to study the velocity vector fields $\mathbf{u}, \mathbf{v}, \mathbf{w}$ on an infinitesimal level (up to second order) by using directional derivatives. More precisely, the crucial symbol

$$D_{\mathbf{v}} \mathbf{w}(P)$$

denotes the covariant directional derivative of the velocity vector field $\mathbf{w} = w^i \partial_i$ at the point P in direction of the velocity vector $\mathbf{v} = v^i \partial_i$ at the point P. In the special case of the Euclidean manifold $\mathbb{E}^3$ with the metric tensorial family $g_{ij} := \delta_{ij}$, $i, j = 1, 2, 3$, we have

$$D_{\mathbf{v}} \mathbf{w} = \sum_{i=1}^{3} v^i \partial_i \mathbf{w},$$

as well as $\mathbf{R} \equiv 0$ and $\mathbf{T} \equiv 0$ (i.e., curvature $\mathbf{R}$ and torsion $\mathbf{T}$ vanish identically). Consequently, according to (8.124) and (8.125), curvature and torsion measure the deviation from the situation on the Euclidean manifold $\mathbb{E}^3$. Note that (8.124) generalizes the formula (8.123) which can be found in the classic 1915/1916 Einstein papers on general relativity based on the Ricci calculus created by Ricci-Curbastro

[35] K. Maurin, The Riemann Legacy, Kluwer, Dordrecht, 1997.

(1853–1925) in the 1880s. In Einstein's theory of special relativity, one uses the metric tensorial family

$$g_{\alpha\beta} \equiv \eta_{\alpha\beta}, \qquad \alpha, \beta = 0, 1, 2, 3$$

on the 4-dimensional Minkowski manifold $\mathbb{M}^4$.[36] In the theory of special relativity, curvature and torsion vanish identically. In contrast to this, in Einsteins theory of general relativity, the Riemann curvature tensorial family R^l_{ijk} describes the gravitational field which, as a rule, does not vanish (but torsion vanishes identically). In order to describe mathematically the Standard Model in elementary particle physics, one has to modify the curvature formula (8.124) in the following way:

$$\boxed{D_{\mathbf{u}} D_{\mathbf{v}} \psi - D_{\mathbf{v}} D_{\mathbf{u}} \psi = D_{[\mathbf{u},\mathbf{v}]} \psi + \mathbf{F}(\mathbf{u}, \mathbf{v}) \psi.} \tag{8.126}$$

That is, we replace the velocity vector field $\mathbf{w}$ by a general physical field ψ (e.g., the field of a relativistic electron in quantum electrodynamics). The symbol

$$D_{\mathbf{v}} \psi(P)$$

denotes the covariant directional derivative of the physical field ψ at the point P in direction of the velocity vector $\mathbf{v} = v^i \partial_i$ at the point P. For fixed velocity vector fields $\mathbf{u}$ and $\mathbf{v}$, the operator

$$\psi \mapsto F(\mathbf{u}, \mathbf{v}) \psi$$

is called the Riemann curvature operator $\mathbf{F}(\mathbf{u}, \mathbf{v})$. From the physical point of view, as we will show later on, the map $\psi \mapsto F(\mathbf{u}, \mathbf{v})\psi$ describes the parallel transport of the physical field ψ along a small triangle spanned by the vectors $\mathbf{u}$ and $\mathbf{v}$ (see Prop. 13.9 on 827). This is a possibility for measuring forces in nature. The geometric idea of parallel transport lurks behind the following fundamental principle in modern physics:

force = *curvature.*

Internal symmetries of physical fields can be described by gauge transformations. As we will show later on, the symmetry of the physical field ψ is included into the definition of the covariant derivative $D_{\mathbf{v}}\psi$. In modern differential geometry, the directional derivative $D_{\mathbf{v}}\psi$ of ψ is the basic notion.

8.9.1 Second-Order Covariant Partial Derivatives

The two key formulas for temperature fields and velocity vector fields. These formulas read as

$$\boxed{\nabla_i(\nabla_j \Theta) - \nabla_j(\nabla_i \Theta) = -T^s_{ij} \nabla_s \Theta,} \tag{8.127}$$

and

$$\boxed{\nabla_i(\nabla_j v^l) - \nabla_j(\nabla_i v^l) = R^l_{ijk} v^k - T^s_{ij} \nabla_s v^l.} \tag{8.128}$$

Here, we use:

(i) The tensorial family of the torsion:

$$T^k_{ij} := \Gamma^k_{ij} - \Gamma^k_{ji}. \tag{8.129}$$

---

[36] $\eta_{00} := 1, \eta_{11} = \eta_{22} = \eta_{33} := -1,\ \eta_{\alpha\beta} := 0$ if $\alpha \neq \beta$.

(ii) The tensorial family of the Riemann curvature:

$$R^l_{ijk} := \partial_i \Gamma^l_{jk} - \partial_j \Gamma^l_{ik} + \Gamma^l_{is}\Gamma^s_{jk} - \Gamma^l_{js}\Gamma^s_{ik}. \tag{8.130}$$

Proof. Ad (8.127). Note that $\nabla_i \Theta = \partial_i \Theta$. Hence

$$\nabla_i(\nabla_j \Theta) = \partial_i(\partial_j \Theta) - \Gamma^s_{ij}\partial_s \Theta.$$

Since $\partial_i \partial_j \Theta = \partial_j \partial_i \Theta$, we get

$$\nabla_i(\nabla_j \Theta) - \nabla_j(\nabla_i \Theta) = -(\Gamma^s_{ij} - \Gamma^s_{ji})\partial_s \Theta.$$

This is (8.127). In addition, we have shown on page 499 that T^k_{ij} is a tensorial family.

Ad (8.128). Recall that $\nabla_j v^l = \partial_j v^l + \Gamma^l_{jk} v^k$. Setting $T^l_j := \nabla_j v^l$, it follows from

$$\nabla_i T^l_j = \partial_i T^l_j + \Gamma^l_{is} T^s_j - \Gamma^s_{ij} T^l_s$$

that

$$\begin{aligned}\nabla_i(\nabla_j v^l) = \; & \partial_i \partial_j v^l + \partial_i \Gamma^l_{jk} \cdot v^k + \Gamma^l_{jk} \cdot \partial_i v^k \\ & + \Gamma^l_{is}(\partial_j v^s + \Gamma^s_{jk} v^k) - \Gamma^s_{ij}(\partial_s v^l + \Gamma^l_{sk} v^k).\end{aligned}$$

Interchanging the indices i and j, we get

$$\begin{aligned}\nabla_j(\nabla_i v^l) = \; & \partial_j \partial_i v^l + \partial_j \Gamma^l_{ik} \cdot v^k + \Gamma^l_{ik} \cdot \partial_j v^k \\ & + \Gamma^l_{js}(\partial_i v^s + \Gamma^s_{ik} v^k) - \Gamma^s_{ji}(\partial_s v^l + \Gamma^l_{sk} v^k).\end{aligned}$$

Computing $\nabla_i(\nabla_j v^l) - \nabla_j(\nabla_i v^l)$ and noting cancellations, we get the claim (8.128).

It remains to show that R^l_{ijk} is a tensorial family. In fact, it follows from (8.128) that

$$R^l_{ijk} v^k = \nabla_i(\nabla_j v^l) - \nabla_j(\nabla_i v^l) + T^s_{ij} \nabla_s v^l.$$

Since the right-hand side is the sum of tensorial families, $R^l_{ijk} v^k$ is a tensorial family for all tensorial families v^l. By the inverse index principle, R^l_{ijk} is also a tensorial family. □

In the special case where $n = 4$, the Riemann curvature tensorial family R^k_{ijl} has $4^4 = 256$ components. Because of the antisymmetry relation (8.131) below, this reduces to $6 \cdot 4^2 = 96$ components. Using a Riemannian (or pseudo-Riemannian) metric, there appear additional symmetry properties such that we have 20 essential components in Einstein's theory of general relativity (see Sect. 8.10.1). In the special case where $n = 2$, there exists only one essential component of the Riemann curvature tensor which is closely related to the Gaussian curvature K of a 2-dimensional submanifold of the Euclidean manifold $\mathbb{E}^3$ (see Sect. 8.10.1).

The crucial antisymmetry relations. For all the indices $i, j, k, l = 1, \ldots, n$, the following hold:

(i) The Riemann-Weyl antisymmetry relation:

$$R^l_{ijk} = -R^l_{jik}. \tag{8.131}$$

(ii) The Ricci–Weyl antisymmetry relations: Setting $U^l_{ijk} := -T^s_{ik} T^l_{sj}$, we get

$$R^l_{[ijk]} = \nabla_{[i} T^l_{jk]} + U^l_{[ijk]}. \tag{8.132}$$

(iii) The Bianchi–Weyl antisymmetry relations: Setting $V^l_{hijk} := -T^s_{hi} R^l_{sjk}$, we get

$$\nabla_{[h} R^l_{ij]k} = V^l_{[hij]k}. \qquad (8.133)$$

In the special case where the affine connection is torsion-free, we have $T^k_{ij} \equiv 0$, by definition. Hence $U^l_{ijk} \equiv 0$ and $V^l_{hijk} \equiv 0$.[37] The proofs of (i)–(iii) can be given by explicit computation based on the definitions of R^l_{ijk} and T^k_{ij}. We refer to P. Dirac, General Theory of Relativity, Princeton University Press, 1996, p. 23.

> *Historically, the modern development of differential geometry can be understood best by the desire of mathematicians to simplify the formalism in order to get insight.*

This will be discussed in Sect. 8.14 on page 529.

8.9.2 Local Flatness

Trivial affine connection – global flatness. The affine connection is called trivial iff there exists an observer O such that all the corresponding Christoffel symbols vanish identically, that is,

$$\Gamma^k_{ij} \equiv 0 \qquad \text{on } \Omega_O$$

for all indices i, j, k. The prototype of a trivial affine connection is the classical connection on the Euclidean manifold $\mathbb{E}^3$ (Euclidean connection). Using a Cartesian coordinate system, the Christoffel symbols vanish identically. However, with respect to curvilinear coordinates (e.g., cylindrical or spherical coordinates), the Christoffel symbols do not vanish identically, as a rule (see Sect. 9.2).

Proposition 8.23 *If the affine connection is trivial, then both the Riemann–Weyl curvature tensorial family and the torsion family vanish identically for all the observers.*

Proof. Let $\Gamma^k_{il} \equiv 0$ with respect to the fixed observer O. Hence $R^l_{ijk} \equiv 0$ with respect to O. Similarly, $T^k_{ij} = \Gamma^k_{ij} - \Gamma^k_{ji} \equiv 0$. Since R^l_{ijk} is a tensorial family, the observer O' gets

$$R^{l'}_{i'j'k'} = G^i_{i'} G^j_{j'} G^k_{k'} G^{l'}_l R^l_{ijk} \equiv 0.$$

Analogously, $T^{k'}_{i'j'} = G^{k'}_k G^i_{i'} G^j_{j'} T^k_{ij} \equiv 0$. This finishes the proof. The same argument shows the following:

> *If a tensorial family vanishes identically with respect to a fixed observer, then it vanishes identically for all observers.*

This simple fact is responsible for the importance of tensorial families in mathematics and physics. For example, in Einstein's theory of general relativity, a trivial affine connection on the 4-dimensional space-time manifold corresponds to a flat universe without any gravitational forces.

Local flatness. Let us now reverse Prop. 8.23. This is possible in local terms.

---

[37] As we will show later on, Einstein's theory of general relativity is based on a torsion-free affine connection, namely, the canonical Levi-Civita metric connection of a pseudo-Riemannian manifold.

Theorem 8.24 *Consider an affine connection. We are given the point P. Suppose that*

$$R^l_{ijk} \equiv 0 \quad \text{and} \quad T^k_{ij} \equiv 0$$

on a sufficient small neighborhood of the point P, for all indices. Then there exists a local diffeomorphism at the point P such that

$$\Gamma^{k'}_{i'j'} \equiv 0$$

for all indices. That is, the transformed Christoffel symbols vanish identically on a sufficiently small neighborhood of the point P. We say that the affine connection is trivial (or flat) near the point P.

In other words, the Riemann–Weyl curvature tensorial family and the torsion family measure the deviation from the locally trivial situation. For the proof, we refer to E. Zeidler, Nonlinear Functional Analysis and Its Applications, Vol. IV, page 653 and page 684, Springer, New York, 1995. The study of local flatness dates back to Riemann. Several approaches to the fundamental flatness problem can be found in M. Spivak, A Comprehensive Introduction to Differential Geometry, Vol. II, Publish or Perish, Boston, 1979.

8.9.3 The Method of Differential Forms (Cartan's Structural Equations)

An important contribution to the development of modern differential geometry was made by Élie Cartan (1869–1951) in the 1920s. Let us discuss this next.

In mathematics and physics, antisymmetric expressions are frequently used in order to construct invariants. The antisymmetry helps us to cancel nasty terms which are generated by the transformation law. Important examples are:

- Leibniz's determinants and the volume in Euclidean geometry,
- the Euler characteristic in topology,
- the Riemann curvature tensor,
- Poincaré's cohomology rule (de Rham cohomology), and Hodge's homology rule (Hodge homology) in Cartan's exterior differential calculus,
- characteristic classes in the topology of fiber bundles.

In physics,

- the electromagnetic field tensor is antisymmetric,
- the 'gauge field tensor' for the 12 interacting particles (gluons in strong interaction, the photon, and W^+, W^-, Z in weak interaction) in the Standard model of elementary particle physics is antisymmetric, and
- the antisymmetry properties of the Riemann curvature tensor describe antisymmetry properties of the gravitational field in Einstein's theory of general relativity.

In this connection, Grassmann's wedge product $\wedge$ plays a crucial rule. For example, Élie Cartan's exterior differential calculus is based on Grassmann's wedge product. It was the ingenious geometer Élie Cartan who showed that the monstrous index formulas in classical Riemannian geometry can be elegantly formulated in terms of differential forms by employing the antisymmetry properties of both the Riemann curvature tensorial family and the torsion tensorial family. Let $k, l = 1, \ldots, n$. Cartan introduced the following differential forms:

- $\omega^l_k := \Gamma^l_{ik}\, dx^i$ (connection forms),

- $\Omega^l_k := \frac{1}{2} R^l_{ijk}\, dx^i \wedge dx^j$ (curvature forms),
- $\tau^l := \frac{1}{2} T^l_{ij}\, dx^i \wedge dx^j$ (torsion forms).

This is motivated by $R^l_{ijk} = -R^l_{jik}$ and $T^l_{ij} = -T^l_{ji}$. The key relations (8.129) and (8.130) are equivalent to the so-called Cartan structural equations:[38]

$$\begin{aligned} \tau^l &= \omega^l_k \wedge dx^k, \\ \Omega^l_k &= d\omega^l_k + \omega^l_s \wedge \omega^s_k, \qquad k, l = 1, \ldots, n. \end{aligned} \tag{8.134}$$

Applying the Cartan differential and using the Poincaré cohomology rule $dd\omega = 0$ together with the graded Leibniz rule, we get the integrability conditions:

$$\begin{aligned} d\tau^l &= \Omega^l_s \wedge dx^s - \omega^l_s \wedge \tau^s, \\ d\Omega^l_k &= \Omega^l_s \wedge \omega^s_k - \omega^l_s \wedge \Omega^s_k. \end{aligned} \tag{8.135}$$

Proof of (8.135). (I) Noting that $d(dx^k) = 0$, the first equation from (8.134) implies

$$d\tau^l = d\omega^l_k \wedge dx^k + \omega^l_k \wedge d(dx^k) = d\omega^l_k \wedge dx^k.$$

By the second equation from (8.134),

$$d\omega^l_k = \Omega^l_k - \omega^l_s \wedge \omega^s_k. \tag{8.136}$$

Using $\tau^l = \omega^l_k \wedge dx^k$, we get the first equation from (8.135).

(II) Similarly, the second equation from (8.134) yields

$$d\Omega^l_k = d\omega^l_s \wedge \omega^s_k - \omega^l_s \wedge d\omega^s_k.$$

Using (8.136), we get the claim (8.135).

□

The reader should note that the preceding proof is based on cancellations via antisymmetry.

> *Cartan's calculus of differential forms allows us to perform the cancellations in an extremely elegant manner.*

An explicit computation shows that the first (resp. second) equation of (8.135) implies the Ricci–Weyl antisymmetry relations (8.132) (resp. the Bianchi–Weyl antisymmetry relations (8.133)).

The elegant language of matrices. Introducing the matrices

- $\mathcal{A}_i := (\Gamma^l_{ik})$ (connection matrices), and
- $\mathcal{F}_{ij} := (R^l_{ijk})$ (curvature matrices),

the key relation (8.130), $R^l_{ijk} = \partial_i \Gamma^l_{jk} - \partial_j \Gamma^l_{ik} + \Gamma^l_{is}\Gamma^s_{jk} - \Gamma^l_{js}\Gamma^s_{ik}$, can be elegantly written as

$$\mathcal{F}_{ij} = \partial_i \mathcal{A}_j - \partial_j \mathcal{A}_i + \mathcal{A}_i \mathcal{A}_j - \mathcal{A}_j \mathcal{A}_i, \qquad i, j = 1, \ldots, n.$$

Using the Lie bracket $[\mathcal{A}_i, \mathcal{A}_j]_- := \mathcal{A}_i \mathcal{A}_j - \mathcal{A}_j \mathcal{A}_i$, this is equivalent to

$$\boxed{\mathcal{F}_{ij} = \partial_i \mathcal{A}_j - \partial_j \mathcal{A}_i + [\mathcal{A}_i, \mathcal{A}_j]_-, \qquad i, j = 1, \ldots, n.} \tag{8.137}$$

[38] For example, $\omega^l_k \wedge dx^k = \Gamma^l_{ik}\, dx^i \wedge dx^k = \frac{1}{2}(\Gamma^l_{ik} - \Gamma^l_{ki})\, dx^i \wedge dx^k = \frac{1}{2} T^l_{ik}\, dx^i \wedge dx^k$.

This way, we discover the Lie structure behind the Riemann curvature tensorial family.[39] Explicitly,

$$\mathcal{A}_i := \begin{pmatrix} \Gamma^1_{i1} \ \Gamma^1_{i2} \ \dots \ \Gamma^1_{in} \\ \vdots \ \ \vdots \ \ \dots \ \ \vdots \\ \Gamma^n_{i1} \ \Gamma^n_{i2} \ \dots \ \Gamma^n_{in} \end{pmatrix}, \qquad \mathcal{F}_{ij} := \begin{pmatrix} R^1_{ij1} \ R^1_{ij2} \ \dots \ R^1_{ijn} \\ \vdots \ \ \vdots \ \ \dots \ \ \vdots \\ R^n_{ij1} \ R^n_{ij2} \ \dots \ R^n_{ijn} \end{pmatrix}.$$

Matrices with differential forms as entries. Set

- $\mathcal{A} := \mathcal{A}_i \, dx^i$ (local connection form),
- $\mathcal{F} := \frac{1}{2}\mathcal{F}_{ij} \, dx^i \wedge dx^j$ (curvature form),
- τ (local torsion form).

This is motivated by $\mathcal{F}_{ij} = -\mathcal{F}_{ji}$. We get $\mathcal{F} = (\Omega^l_k)$ and $\mathcal{A} = (\omega^l_k)$. Explicitly,

$$\mathcal{A} := \begin{pmatrix} \omega^1_1 \ \dots \ \omega^1_n \\ \vdots \ \ \vdots \ \ \dots \vdots \\ \omega^n_1 \ \dots \ \omega^n_n \end{pmatrix}, \quad \mathcal{F} := \begin{pmatrix} \Omega^1_1 \ \dots \ \Omega^1_n \\ \vdots \ \ \dots \ \ \vdots \\ \Omega^n_1 \ \dots \ \Omega^n_n \end{pmatrix} \quad \tau := \begin{pmatrix} \tau^1 \\ \vdots \\ \tau^n \end{pmatrix}, \quad dx := \begin{pmatrix} dx^1 \\ \vdots \\ dx^n \end{pmatrix}.$$

Using the wedge product of matrices with differential forms as entries, the systems (8.134), (8.135) can be elegantly written in the following way:

(i) Cartan's local structural equations:

$$\begin{aligned} \tau &= \mathcal{A} \wedge dx, \\ \mathcal{F} &= d\mathcal{A} + \mathcal{A} \wedge \mathcal{A}. \end{aligned} \tag{8.138}$$

(ii) The Bianchi integrability conditions:

$$\begin{aligned} d\tau &= \mathcal{F} \wedge dx - \mathcal{A} \wedge \tau, \\ d\mathcal{F} &= \mathcal{F} \wedge \mathcal{A} - \mathcal{A} \wedge \mathcal{F}. \end{aligned} \tag{8.139}$$

Lack of invariance. Note the following. The approach considered in this section depends on the choice of the natural basis ∂_i and dx^i, since Γ^k_{ij} and R^l_{ijk} depend on the choice of this natural basis. However, due to Cartan it is possible to extend this to an invariant formulation. This global setting will be discussed in Chap. 17. Alternatively, there exists a different invariant approach in the spirit of functional analysis (i.e., operator theory):

This approach will be based on the covariant derivative $D_{\mathbf{v}}\mathbf{w}$ of a velocity vector field $\mathbf{w}$ with respect to another velocity vector field $\mathbf{v}$. We will study the properties of the operator $D_{\mathbf{v}}$ in terms of the Lie bracket and the Jacobi identity.

In the next section, we will show that this operator-theoretic setting does not depend on the choice of the observer.

[39] Mnemonically, one should use formula (8.137) in order to recall the definition (8.130) of the Riemann–Weyl curvature tensor.

8.9.4 The Operator Method

> The most important derivative of a physical field is the covariant directional derivative which measures the small change of a physical field in direction of a velocity vector. The covariant directional derivative only depends on the physical interaction, but not on the choice of the observer.
>
> Curvature is closely related to the Lie structure of the covariant directional differential operators.
>
> Folklore

In this section, as a prototype, we consider the special case where the physical fields are velocity vector fields. The prototype of velocity vector fields are the velocity vector fields of fluids (see Chap. 10). The main topic of this volume is to modify the following approach in such a way that it applies to general physical fields.

The covariant directional derivative. We are given an admissible system $\mathcal{O}$ of observers. Let $\mathbf{u} = u^i\partial_i$, $\mathbf{v} = v^i\partial_i$ and $\mathbf{w} = w^i\partial_i$ where u^i, v^i, and w^i be tensorial families. For the observer O, we define

$$\boxed{D_{\mathbf{v}}\mathbf{w}(P) := v^i(x)\nabla_i w^k(x)\cdot\partial_k, \qquad x \in \Omega_O}$$

where x is the local coordinate of the point P. This yields the linear operator

$$D_{\mathbf{v}} : \mathsf{Diff}^1(\Omega_O) \to \mathsf{Diff}^1(\Omega_O).$$

It is crucial that

The definition of $D_{\mathbf{v}}\mathbf{w}$ does not depend on the choice of the observer.

In fact, the observer O' measures the linear operator

$$D_{\mathbf{v}} : \mathsf{Diff}^1(\Omega_{O'}) \to \mathsf{Diff}^1(\Omega_{O'})$$

with $(D_{\mathbf{v}}\mathbf{w})(P) := v^{i'}(x')\nabla_{i'}v^{k'}(x')\cdot\partial_{k'}$. Moreover, by the index principle we have

$$v^i(x)\nabla_i v^k(x)\cdot\partial_k = v^{i'}(x')\nabla_{i'}v^{k'}(x')\cdot\partial_{k'}, \qquad x' = x'(x).$$

Thus, the observer O and the observer O' compute the same value $(D_{\mathbf{v}}\mathbf{w})(P)$ which is called the directional derivative of the velocity vector field $\mathbf{w}$ at the point P in direction of the velocity vector $\mathbf{v}(P)$.

The key formulas. The following three elegant formulas lie at the heart of modern differential geometry.

(i) Curvature (force):

$$\boxed{\mathbf{F}(\mathbf{u},\mathbf{v})\mathbf{w} = D_{\mathbf{u}}D_{\mathbf{v}}\mathbf{w} - D_{\mathbf{v}}D_{\mathbf{u}}\mathbf{w} - D_{[\mathbf{u},\mathbf{v}]}\mathbf{w}.} \tag{8.140}$$

Therefore, the Lie bracket of the covariant directional derivatives satisfies the relation

$$[D_{\mathbf{u}}, D_{\mathbf{v}}]_- = F(\mathbf{u},\mathbf{v}) + D_{[\mathbf{u},\mathbf{v}]}. \tag{8.141}$$

(ii) Jacobi identity:

$$[D_{\mathbf{u}},[D_{\mathbf{v}},D_{\mathbf{w}}]_-]_- + [D_{\mathbf{v}},[D_{\mathbf{w}},D_{\mathbf{u}}]_-]_- + [D_{\mathbf{w}},[D_{\mathbf{u}},D_{\mathbf{v}}]_-]_- = 0. \tag{8.142}$$

(iii) Torsion: $\mathbf{T}(\mathbf{u},\mathbf{v}) = D_{\mathbf{u}}\mathbf{v} - D_{\mathbf{v}}\mathbf{u} - D_{[\mathbf{u},\mathbf{v}]}$.

Here, we use the following expressions:

- $\mathbf{F}(\mathbf{u},\mathbf{v})\mathbf{w} = \mathbf{R}(\mathbf{u},\mathbf{v},\mathbf{w}) := R^l_{ijk}u^i v^j w^k \cdot \partial_l$,
- $\mathbf{T}(\mathbf{u},\mathbf{v}) := T^l_{ij}u^i v^j \cdot \partial_l$.

Proof. Ad (i). Setting $W^l := \partial_i w^l + \Gamma^l_{is}w^s$, we get

- $D_\mathbf{v}\mathbf{w} = v^i(\partial_i w^l + \Gamma^l_{is}w^s)\,\partial_l = W^l\partial_l$,
- $D_\mathbf{u}(D_\mathbf{v}\mathbf{w}) = u^i(\partial_i W^l + \Gamma^l_{is}W^s)\,\partial_l$,
- $D_{[\mathbf{u},\mathbf{v}]}\mathbf{w} = (u^r\partial_r v^i - v^r\partial_r u^i)(\partial_i w^l + \Gamma^l_{is}w^s)\,\partial_l$.

An explicit computation based on the classical Leibniz rule together with (8.130) on page 505 yields (i).

Ad (ii). The Jacobi identity is valid for arbitrary linear operators.

Ad (iii). Argue as in the proof to (i), and use $T^l_{ij} = \Gamma^l_{ij} - \Gamma^l_{ji}$. □

For fixed velocity vectors $\mathbf{u}$ and $\mathbf{v}$, the linear operator

$$\mathbf{F}(\mathbf{u},\mathbf{v}) : \mathsf{Diff}^1(\Omega_O) \to \mathsf{Diff}^1(\Omega_O)$$

is called the Riemann curvature operator with respect to $\mathbf{u}$ and $\mathbf{v}$.[40] Moreover, we have

$$\mathbf{T}(\mathbf{u},\mathbf{v}) \in \mathsf{Diff}^1(\Omega_O).$$

Here, the velocity vector field $\mathbf{T}(\mathbf{u},\mathbf{v})$ is called the torsion velocity vector field with respect to $\mathbf{u}$ and $\mathbf{v}$.

The Bianchi identity. We want to show that the Jacobi identity implies the Bianchi identity. For fixed velocity vectors $\mathbf{u},\mathbf{v},\mathbf{w}$, we define the linear operator

$$\mathbf{D}_\mathbf{u}\mathbf{F}(\mathbf{v},\mathbf{w}) : \mathsf{Diff}^1(\Omega_O) \to \mathsf{Diff}^1(\Omega_O)$$

by setting $\big(\mathbf{D}_\mathbf{u}\mathbf{F}(\mathbf{v},\mathbf{w})\big)\mathbf{z} := D_\mathbf{u}(\mathbf{F}(\mathbf{v},\mathbf{w})\mathbf{z}) - \mathbf{F}(\mathbf{v},\mathbf{w})(D_\mathbf{u}\mathbf{z})$. In other words,

$$\mathbf{D}_\mathbf{u}\mathbf{F}(\mathbf{v},\mathbf{w}) = [D_\mathbf{u}, \mathbf{F}(\mathbf{v},\mathbf{w})]_-.$$

Proposition 8.25 *There holds the Bianchi identity*

$$\mathbf{D}_\mathbf{u}\mathbf{F}(\mathbf{v},\mathbf{w}) + \mathbf{D}_\mathbf{v}\mathbf{F}(\mathbf{w},\mathbf{u}) + \mathbf{D}_\mathbf{w}\mathbf{F}(\mathbf{u},\mathbf{v}) = \mathbf{G}(\mathbf{u},\mathbf{v},\mathbf{w}) \qquad (8.143)$$

where $\mathbf{G}(\mathbf{u},\mathbf{v},\mathbf{w}) := -[D_\mathbf{u}, D_{[\mathbf{v},\mathbf{w}]}]_- - [D_\mathbf{v}, D_{[\mathbf{w},\mathbf{u}]}]_- - [D_\mathbf{w}, D_{[\mathbf{u},\mathbf{v}]}]_-$.

In particular, we have $\mathbf{G}(\mathbf{u},\mathbf{v},\mathbf{w}) = 0$, if the Lie brackets vanish identically, that is, $[\mathbf{u},\mathbf{v}] = [\mathbf{v},\mathbf{w}] = [\mathbf{w},\mathbf{u}] = 0$ (e.g., $\mathbf{u} = \partial_i, \mathbf{v} = \partial_j, \mathbf{w} = \partial_k$).

Proof. Use the Lie bracket (8.141) together with the Jacobi identity (8.142). □

8.10 The Riemann–Christoffel Curvature Tensor

> If a metric tensorial family g_{ij} is available, then it is possible to construct uniquely a connection which generalizes the classic Gauss surface theory. The symmetry properties of g_{ij} imply additional symmetry properties of the Riemann curvature tensorial family.
>
> Folklore

In this section we assume that there is given an admissible system $\mathcal{O}$ of references.

---

[40] The definition of the space $\mathsf{Diff}^1(\Omega_O)$ of linear differential operators can be found on page 448.

8.10.1 The Levi-Civita Metric Connection

Inner product and inner pseudo-product for velocity vectors. Suppose that we are given a metric tensorial family g_{ij} of Riemannian or pseudo-Riemannian type. The tensor field

$$\mathbf{g} := g_{ij}\, dx^i \otimes dx^j$$

is called the metric tensor field. For all velocity vectors $\mathbf{u} = u^i \partial_i$ and $\mathbf{v} = v^i \partial_i$, we set

$$\langle \mathbf{u}|\mathbf{v}\rangle := \mathbf{g}(\mathbf{u}, \mathbf{v}).$$

Explicitly,

$$\langle \mathbf{u}|\mathbf{v}\rangle = u^i g_{ij} v^j.$$

If g_{ij} is of Riemannian type, then the following hold for all points $x \in \Omega_O$:

- $\langle \mathbf{u}(x)|\mathbf{u}(x)\rangle \geq 0$ for all $\mathbf{u}$ (positivity), and
- $\langle \mathbf{u}(x)|\mathbf{u}(x)\rangle = 0$ iff $\mathbf{u}(x) = 0$ (definiteness).

In other words, $\langle \mathbf{u}(x)|\mathbf{v}(x)\rangle$ is an inner product.

If g_{ij} is of pseudo-Riemannian type, then $\langle \mathbf{u}|\mathbf{v}\rangle$ is called a pseudo-inner product. This means that the following hold for all points $x \in \Omega_O$:

- The real map $(\mathbf{u}(x), \mathbf{v}(x)) \mapsto \langle \mathbf{u}(x)|\mathbf{v}(x)\rangle$ is bilinear and symmetric,
- there exist vectors $\mathbf{u}(x)$ and $\mathbf{v}(x)$ with $\langle \mathbf{u}(x)|\mathbf{u}(x)\rangle > 0$ and $\langle \mathbf{v}(x)|\mathbf{v}(x)\rangle < 0$ (indefiniteness), and
- if $\langle \mathbf{u}(x)|\mathbf{v}(x)\rangle = 0$ for all $\mathbf{v}(x)$, then $\mathbf{u}(x) = 0$ (nondegeneracy).

The affine connection family Γ^k_{ij} is called compatible with the metric tensorial family g_{ij} iff

$$\boxed{D\mathbf{g} \equiv 0.} \tag{8.144}$$

It follows from $D\mathbf{g} = \nabla_i g_{jk} dx^i \otimes dx^j \otimes dx^k$ that the compatibility condition (8.144) is equivalent to

$$\nabla_i g_{jk} \equiv 0, \qquad i, j = 1, \ldots, n. \tag{8.145}$$

The existence-and uniqueness theorem. The following connection is frequently used (e.g., in Einstein's theory of general relativity).

Theorem 8.26 *There exists precisely one torsion-free connection family Γ^k_{ij} which is compatible with the given metric tensorial family g_{ij}. Explicitly,*

$$\Gamma^k_{ij} = \tfrac{1}{2}(\partial_i g_{js} + \partial_j g_{is} - \partial_s g_{ij}) g^{sk}. \tag{8.146}$$

This connection is called the Levi-Civita connection.[41]

[41] T. Levi-Civita and G. Ricci, The absolute differential calculus and its applications, Math. Ann. **54** (1901), 125–201 (in French).
T. Levi-Civita, The notion of parallel transport in manifolds and its geometric consequences for the Riemann curvature, Rend. Palermo **42** (1917), 73–205 (in Italian).

Proof. (I) Necessary condition: Suppose that Γ_{ij}^k is torsion-free and compatible with g_{ij}. This means that

$$\Gamma_{ij}^k = \Gamma_{ji}^k, \qquad \text{and} \qquad \nabla_i g_{jk} = \partial_i g_{jk} - \Gamma_{ij}^s g_{sk} - \Gamma_{ik}^s g_{js} = 0. \qquad (8.147)$$

Lowering the index, we get $\Gamma_{ijk} := \Gamma_{ij}^r g_{rk}$. Hence

$$\partial_i g_{jk} = \Gamma_{ijk} + \Gamma_{kij}.$$

Note that $\Gamma_{ijk} = \Gamma_{jik}$. Using cyclic permutation, we obtain

$$\partial_i g_{jk} + \partial_j g_{ki} - \partial_k g_{ij} = 2\Gamma_{ijk}.$$

Lifting the index k, we get (8.146).

(II) Sufficient condition: Conversely, an explicit computation shows that the tensorial transformation law

$$g_{i'j'} = \frac{\partial x^i}{\partial x^{i'}} \frac{\partial x^j}{\partial x^{j'}} \cdot g_{ij}$$

implies that Γ_{ij}^k transforms like a connection family (see (8.105) on page 495). □

The Ricci lemma. The covariant partial derivative ∇_i with respect to the Levi-Civita connection generated by g_{jk} satisfies the two conditions

$$\nabla_i g_{jk} \equiv 0 \quad \text{and} \quad \nabla_i g^{jk} \equiv 0 \quad \text{on } \Omega_O, \quad i,j,k = 1,\ldots,n \qquad (8.148)$$

for all observers O.

Proof. Use (8.145), and note that the relation $g_{kl}g^{lr} = \delta_k^r$ together with the Leibniz rule implies

$$0 = \nabla_i \delta_k^r = (\nabla_i g_{kl})\, g^{lr} + g_{kl}(\nabla_i g^{lr}) = g_{kl} \nabla_i g^{lr}.$$

Hence $\nabla_i g^{sr} = \delta_l^s \nabla_i g^{lr} = g^{sk} g_{kl} \nabla_i g^{lr} = 0$. □

Lifting and lowering of indices. For given Levi-Civita connection, the covariant partial derivative ∇_i can be interchanged with the lifting or lowering of indices. For example, we have

$$\boxed{\nabla_i (g^{js} T_s) = g^{js} \nabla_i T_s.} \qquad (8.149)$$

This is a consequence of the Ricci lemma. In fact, the Leibniz rule combined with (8.148) yields

$$\nabla_i (g^{js} T_s) = (\nabla_i g^{js}) T_s + g^{js} (\nabla_i T_s) = g^{js} \nabla_i T_s.$$

Similarly, we get

$$\nabla_i (g_{js} T^s) = g_{js} \nabla_i T^s. \qquad (8.150)$$

8.10.2 Levi-Civita's Parallel Transport

> The Levi-Civita parallel transport of velocity vectors along the same curve respects the inner product (or the pseudo-inner product).
>
> Folklore

Let Γ_{ij}^k be an affine connection family. By definition, the tensor field

$$T = T^{i_1 \dots i_r}_{j_1 \dots j_s} \, dx^{j_1} \otimes \dots \otimes dx^{j_s} \otimes \partial_{i_1} \otimes \dots \otimes \partial_{i_r}$$

is parallel with respect to the given affine connection Γ_{ij}^k iff

$$\boxed{DT \equiv 0.} \tag{8.151}$$

This is equivalent to

$$\nabla_i T^{i_1 \dots i_r}_{j_1 \dots j_s} \equiv 0 \qquad \text{on } \Omega_O, \quad i = 1, \dots n \tag{8.152}$$

for all indices $i_1, \dots, i_r, j_1, \dots, j_s = 1, \dots, n$. This definition does not depend on the choice of the observer O. In the Euclidean setting, all the functions Γ_{ij}^k vanish identically; in this special case, condition (8.152) means that

$$\partial_i T^{i_1 \dots i_r}_{j_1 \dots j_s} \equiv 0 \qquad \text{on } \Omega_O, \quad i = 1, \dots n.$$

Since the set Ω_O is assumed to be arcwise connected, all of the functions $T^{i_1 \dots i_r}_{j_1 \dots j_s}$ are constant on the domain Ω_O.

Let $C : x^i = x^i(t),\ t \in]-t_0, t_0[,\ i = 1, \dots, n$, be a smooth curve. The tensor field T is parallel along the curve C iff

$$\dot{x}^i(t) \nabla_i T^{i_1 \dots i_r}_{j_1 \dots j_s} = 0, \qquad t \in]-t_0, t_0[$$

for all indices $i_1, \dots, i_r, j_1, \dots, j_s = 1, \dots, n$. We briefly write

$$\frac{DT(x(t))}{dt} = 0, \qquad t \in]-t_0, t_0[.$$

In the Euclidean setting, this means that all of the functions $T^{i_1 \dots i_r}_{j_1 \dots j_s}$ are constant along the curve C.

Proposition 8.27 *Concerning the Levi-Civita connection, the parallel transport of two velocity vector fields along the same curve preserves the inner product (or the pseudo-inner product).*

Proof. Let u^i and v^i be tensorial families. The equations

$$\dot{x}^s(t) \nabla_s u^i_{|x(t)} = 0 \quad \text{and} \quad \dot{x}^s(t) \nabla_s v^i_{|x(t)} = 0, \quad t \in]-t_0, t_0[\tag{8.153}$$

tell us that the tensorial families u^i and v^i are parallel along the curve C. Hence

$$\frac{d}{dt} \langle \mathbf{u}(x(t)) | \mathbf{v}(x(t)) \rangle = 0, \qquad t \in]-t_0, t_0[.$$

In fact, we have

$$\frac{d}{dt} \left(u^i g_{ij} v^j \right)_{|x(t)} = \dot{x}^s(t) \nabla_s \left(u^i g_{ij} v^j \right)_{|x(t)}.$$

This is equal to zero by the covariant Leibniz rule,

$$\nabla_s (u^i g_{ij} v^j) = \nabla_s u^i \cdot g_{ij} v^j + u^i \nabla_s g_{ij} \cdot v^j + u^i g_{ij} \nabla_s v^j,$$

and by noting (8.153) together with the Ricci lemma, $\nabla_s g_{ij} = 0$. □

The intuitive meaning of the parallel transport will be discussed in Sect. 13.4 (parallel transport of the local phase of a physical field) and in Sect. 9.5 (parallel transport of velocity vectors on the surface of earth).

In physics, parallel transport corresponds to the transport of physical information.

8.10.3 Symmetry Properties of the Riemann–Christoffel Curvature Tensor

The Riemann curvature operator $\mathbf{w} \mapsto \mathbf{F}(\mathbf{u}, \mathbf{v})\mathbf{w}$ sends velocity vectors to velocity vectors. Using the inner (or pseudo-inner) product $\langle .|. \rangle$, it is quite natural to define

$$\boxed{\mathcal{R}(\mathbf{u}, \mathbf{v}, \mathbf{w}, \mathbf{z}) := \langle F(\mathbf{u}, \mathbf{v})\mathbf{w}|\mathbf{z}\rangle.}$$

Here, $\mathcal{R}(\mathbf{u}, \mathbf{v}, \mathbf{w}, \mathbf{z})$ is a real number. The 4-linear map $(\mathbf{u}, \mathbf{v}, \mathbf{w}, \mathbf{z}) \mapsto \mathcal{R}(\mathbf{u}, \mathbf{v}, \mathbf{w}, \mathbf{z})$ is called the Riemann–Christoffel tensor field corresponding to the metric tensor field $\mathbf{g}$. Explicitly,

$$\mathcal{R}(\mathbf{u}, \mathbf{v}, \mathbf{w}, \mathbf{z}) := R_{ijkl}u^i v^j w^k z^l.$$

It follows from

$$\mathcal{R}(\mathbf{u}, \mathbf{v}, \mathbf{w}, \mathbf{z}) := \langle F(\mathbf{u}, \mathbf{v})\mathbf{w}|\mathbf{z}\rangle = R^s_{ijk} g_{sl} u^i v^j w^k z^l$$

that

$$\boxed{R_{ijkl} = R^s_{ijk} g_{sl}.}$$

Here, R_{ijkl} is called the Riemann–Christoffel curvature tensorial family corresponding to the metric tensorial family g_{ij}. For all velocity vector fields $\mathbf{u}, \mathbf{v}, \mathbf{w}$, and $\mathbf{z}$, we have the following relations:

- $\mathbf{F}(\mathbf{u}, \mathbf{v})\mathbf{w} + \mathbf{F}(\mathbf{v}, \mathbf{u})\mathbf{w} + \mathbf{F}(\mathbf{w}, \mathbf{v})\mathbf{u} = 0$,
- $\mathbf{F}(\mathbf{u}, \mathbf{v})\mathbf{w} = -\mathbf{F}(\mathbf{v}, \mathbf{u})\mathbf{w}$, and hence $\mathcal{R}(\mathbf{u}, \mathbf{v}, \mathbf{w}, \mathbf{z}) = -\mathcal{R}(\mathbf{v}, \mathbf{u}, \mathbf{w}, \mathbf{z})$,
- $\mathcal{R}(\mathbf{u}, \mathbf{v}, \mathbf{w}, \mathbf{z}) = -\mathcal{R}(\mathbf{u}, \mathbf{v}, \mathbf{z}, \mathbf{w})$,
- $\mathcal{R}(\mathbf{u}, \mathbf{v}, \mathbf{w}, \mathbf{z}) = \mathcal{R}(\mathbf{w}, \mathbf{z}, \mathbf{u}, \mathbf{v})$.

In terms of components, for all the indices $i, j, k, l, s = 1, \ldots, n$, this implies the following:

(i) $R_{[ijk]l} = 0$ (Ricci identity),
(ii) $R_{ijkl} = -R_{jikl}$ (first antisymmetry relation),
(iii) $R_{ijkl} = -R_{ijlk}$ (second antisymmetry relation),
(iv) $R_{ijkl} = R_{klij}$ (symmetry relation),
(v) $\nabla_{[h} R_{ij]kl} = 0$ (Bianchi identity).

The Bianchi identity (v) is equivalent to

$$\partial_{[h} R_{ij]kl} = 0. \tag{8.154}$$

In addition, we have

$$\boxed{R_{ijkl} = \tfrac{1}{2}(\partial_i\partial_k g_{jl} + \partial_j\partial_l g_{ik} - \partial_i\partial_l g_{jk} - \partial_j\partial_k g_{il}).} \tag{8.155}$$

This formula is frequently used for computing the Riemann curvature tensorial family R_{ijkl} in terms of second-order partial derivatives of the metric tensorial family g_{ij}.

The proofs can be carried out by straightforward, but somewhat lengthy computations based on the corresponding definitions. More elegant proofs can be obtained by introducing special local coordinates called normal coordinates at the point P_0. These coordinates have the crucial property that

$$\partial_h g_{ij}(P_0) = 0, \qquad h, i, j = 1, \ldots n,$$

and hence $\Gamma^k_{ij}(P_0) = 0$ for all $i, j, k = 1, \ldots, n$. This simplifies substantially the computations. See J. Jost, Riemannian Geometry and Geometric Analysis, Sect. 3.3, Springer, Berlin, 2008. If v^l and v_k are tensorial families, then we have the formulas

$$\nabla_i \nabla_j v^l - \nabla_j \nabla_i v^l = R^l_{ijk} v^k,$$

and

$$\nabla_i \nabla_j v_k - \nabla_j \nabla_i v_k = -R^s_{ijk} v_s.$$

8.10.4 The Ricci Curvature Tensor and the Einstein Tensor

Our goal is to use the Riemann–Christoffel tensorial family R_{ijkl} in order to construct invariants which simplify the investigation of R_{ijkl}. The key formula reads as

$$\boxed{R = R^i_{ikl} g^{kl}.}$$

By the index principle, this is an invariant which is called the scalar curvature. Let us discuss the relation of the scalar curvature to the Ricci curvature. To this end, we set

- $\mathrm{Ric}(\mathbf{u}, \mathbf{z}) := g^{jk} \mathcal{R}(\mathbf{u}, \partial_j, \partial_k, \mathbf{z})$ (Ricci curvature tensor Ric),
- $R_{il} := \mathrm{Ric}(\partial_i, \partial_l)$ (Ricci curvature tensorial family),
- $R = g^{il} \mathrm{Ric}(\partial_i, \partial_l)$ (Ricci scalar curvature).

Therefore, the Ricci curvature tensor is obtained by averaging the Riemann–Christoffel curvature tensor. In turn, the scalar curvature R is obtained by averaging the Ricci curvature tensor. It follows from $\mathcal{R}(\mathbf{u}, \mathbf{v}, \mathbf{w}, \mathbf{z}) := R_{ijkl} u^i v^j w^k z^l$ that

$$\boxed{\mathrm{Ric}(\mathbf{u}, \mathbf{z}) = R_{il} u^i z^l}$$

with $R_{il} = R_{ijkl} g^{jk}$ and

$$\boxed{R = g^{il} R_{il}.}$$

We have the symmetry property

$$\mathrm{Ric}(\mathbf{u}, \mathbf{z}) = \mathrm{Ric}(\mathbf{z}, \mathbf{u})$$

for all velocity vector fields $\mathbf{u}$ and $\mathbf{z}$. The tensor field

$$\mathbf{G} := \mathrm{Ric} - \tfrac{1}{2} R \mathbf{g}$$

is called the Einstein tensor field which plays a fundamental role in Einstein's theory of general relativity. Explicitly,

$$G_{ij} := R_{ij} - \tfrac{1}{2} R \, g_{ij}.$$

For Riemannian metric tensorial families g_{ij}, Riemann defined

$$\boxed{K_P(\mathbf{u}, \mathbf{v}) := \frac{\mathcal{R}(\mathbf{u}, \mathbf{v}, \mathbf{v}, \mathbf{u})}{\langle \mathbf{u}|\mathbf{u}\rangle \langle \mathbf{v}|\mathbf{v}\rangle - \langle \mathbf{u}|\mathbf{v}\rangle^2}.}$$

This is called the sectional curvature at the point P with respect to the plane spanned by the vectors $\mathbf{u}$ and $\mathbf{v}$.

The special two-dimensional case due to Gauss. If $n = 2$, then the sectional curvature $K_P(\mathbf{u}, \mathbf{v})$ does not depend on the choice of the vectors $\mathbf{u}$ and

v. Then K_P coincides with the Gaussian curvature at the point P. Gauss' famous theorema egregium tells us that

$$K_P = \frac{R_P}{2} = \frac{R_{1221}(P)}{\det(g_{ij}(P))}.$$

This means that the Gaussian curvature K_P can be expressed by the metric tensorial family g_{ij} together with its first-order and second-order partial derivatives. In particular, this means that we can compute the Gaussian curvature of the earth at every point by measuring the distances on earth; it is not necessary to use measurements about the surrounding 3-dimensional space. The point is that because of the symmetry properties of the Riemann–Christoffel tensor R_{ijkl}, there is only one essential component, namely, R_{1221} if $n = 2$. Explicitly,

- $R_{1221} = -R_{2121} = -R_{1212} = R_{2112}$,
- $R_{ij11} = R_{ij22} = R_{11ij} = R_{22ij} = 0$, $i, j = 1, 2$.

Historically, in 1827 Gauss studied the 2-dimensional case (i.e., classical surface theory). In 1854 Riemann generalized this to higher dimensions (Riemannian manifolds).

8.10.5 The Conformal Weyl Curvature Tensor

In the two-dimensional and three-dimensional case, the Ricci curvature tensorial family R_{ij} completely determines the Riemann–Christoffel curvature family R_{ijkl}. Explicitly,

- $R_{ij} = Rg_{ij} = 2Kg_{ij}$, $R_{1221} = \frac{1}{2}R(g_{11}g_{22} - (g_{12})^2)$ if $n = 2$,
- $R_{ijkl} = R_{il}g_{jk} + R_{jk}g_{il} - R_{ik}g_{jl} - R_{jl}g_{ik} + \frac{1}{2}R(g_{ik}g_{jl} - g_{il}g_{jk})$ if $n = 3$.

Let $n = 4, 5, \ldots$ Then the situation changes, as we will show below. To begin with, note that the Riemann–Christoffel curvature tensorial family R_{ijkl} allows the following splitting

$$\boxed{R_{ijkl} = R^*_{ijkl} + W_{ijkl}, \qquad i, j, k, l = 1, \ldots, n}$$

where R^*_{ijkl} is defined by

$$\frac{1}{n-2}\big(R_{il}g_{jk} + R_{jk}g_{il} - R_{ik}g_{jl} - R_{jl}g_{ik}\big) + \frac{R}{(n-1)(n-2)}\,(g_{ik}g_{jl} - g_{il}g_{jk}).$$

Here, R^*_{ijkl} is called the Riemann–Ricci curvature tensorial family, and W_{ijkl} is called the Weyl conformal curvature tensorial family. We have

$$W^k_{ijk.} \equiv 0$$

where $W^k_{ijk.} := g^{kl}W_{ijkl}$. We say that W_{ijkl} is the trace-free contribution to R_{ijkl}.

Geometric meaning of the curvature tensor. Next let us briefly discuss the geometric meaning of both the Riemann–Christoffel curvature tensor and the Weyl conformal curvature tensor. Roughly speaking, these tensorial families are obstructions for introducing special local coordinates which correspond to simple local geometries.

To begin with, let us mention that Gauss proved the following two fundamental theorems for 2-dimensional smooth surfaces: Let M be a 2-dimensional submanifold of the Euclidean manifold $\mathbb{E}^3$ (e.g., the surface of earth). Then the following hold:

- It is always possible to locally introduce new coordinates ξ, η such that

$$ds^2 = \mathrm{e}^{a(\xi,\eta)}(d\xi^2 + d\eta^2).$$

  Here, the positive scaling factor e^a depends on the smooth function a. We say that the submanifold M is conformally flat.
- If the Gaussian curvature vanishes identically, $K \equiv 0$, then it is possible to locally introduce new coordinates ξ, η such that

$$ds^2 = d\xi^2 + d\eta^2.$$

  We say that the submanifold M is locally flat. In this special case, the geometry of the surface is locally Euclidean (e.g., the surface of a cone).

Riemann and Weyl generalized this to higher dimensions. To explain this, let us introduce the following definitions.

- The Riemannian metric tensorial family g_{ij} is called locally flat at the point P iff there exists a diffeomorphism on a sufficiently small neighborhood of the point P such that the transformed tensorial family satisfies the condition

$$g_{i'j'}(x') = \delta_{i'j'}$$

  on a sufficiently small neighborhood of the point P, for all indices. Intuitively, we can construct an observer O' who sees a trivial Euclidean metric near the point P.
- The Riemannian metric tensorial family g_{ij} is called locally conformally flat at the point P iff there exists both a diffeomorphism on a sufficiently small neighborhood of the point P and a smooth function a near P such that

$$g_{i'j'}(x') = \mathrm{e}^{a(x')}\delta_{i'j'}$$

  on a sufficiently small neighborhood of the point P, for all indices. Intuitively, we can construct an observer O' who sees a trivial metric near the point P, up to a positive scaling factor e^a which depends on the position.

Theorem 8.28 *Let g_{ij} be a Riemannian metric tensorial family on a nonempty open arcwise connected subset Ω of $\mathbb{R}^n$. Then the following hold:*

(i) *Let $n \geq 2$. The metric family g_{ij} is locally flat at all the points of Ω iff $R_{ijkl} \equiv 0$ for all indices.*

(ii) *If $n = 2$, then g_{ij} is always locally conformally flat at all the points of Ω (the Gauss theorem).*

(iii) *Let $n \geq 4$. The metric family g_{ij} is locally conformally flat at all the points of Ω iff $W_{ijkl} \equiv 0$ for all indices.*

(iv). *Let $n = 3$. In this exceptional case, we have $W_{ijkl} \equiv 0$ for all indices. The metric family g_{ij} is locally conformally flat at all the points of Ω iff $S_{ijk} \equiv 0$ for all indices.*

Here, we define

- $S_{jk} := R_{jk} - \frac{1}{4}Rg_{jk}$ (Schouten tensorial family),
- $S_{ijk} := \nabla_i S_{jk} - \nabla_j S_{ik}$ (Weyl–Schouten tensorial family).

For given metric tensorial family g_{ij}, the tensorial families $R_{ijkl}, R_{ij}, R, W_{ijkl}$ are distinguished by the theory of differential invariants. A detailed discussion can be found in R. Weitzenböck, Invariantentheorie, Sect. 13ff, Noordhoff, Groningen (in German) (classic approach), and in R. Goodman and N. Wallach, Symmetry, Representations, and Invariants, Sect. 10.3.1, Springer, New York (modern approach).

The metric tensorial family g_{ij} is called Ricci flat iff $R_{ij} \equiv 0$ for all indices. Theorem 8.28 tells us the following:

If the dimension is greater or equal to four, then the Riemann–Christoffel curvature tensorial family R^l_{ijk} cannot always be recovered from the Ricci tensorial family R_{ij}.

8.10.6 The Hodge Codifferential and the Covariant Partial Derivative

Let $p = 1, \ldots n$. Consider the p-form $\omega = \frac{1}{p!}\omega_{i_1\ldots i_p} dx^{i_1} \wedge \cdots \wedge dx^{i_p}$ where $\omega_{i_1\ldots i_p}$ is an antisymmetric tensorial family. Using the covariant partial derivative, we get

$$\begin{aligned} d\omega &= \frac{1}{p!}\,\nabla_i \omega_{i_1\ldots i_p}\, dx^i \wedge dx^{i_1} \wedge \cdots \wedge dx^{i_p} \\ &= \frac{1}{p!}\partial_{[i}\omega_{i_1\ldots i_p]}\, dx^i \wedge dx^{i_1} \wedge \cdots \wedge dx^{i_p} \end{aligned} \tag{8.156}$$

and

$$\boxed{d^*\omega = -\frac{1}{(p-1)!}\,\nabla^i \omega_{i i_2\ldots i_p}\, dx^{i_2} \wedge \cdots \wedge dx^{i_p}.} \tag{8.157}$$

Here, the covariant partial derivative ∇_i refers to the Christoffel symbols (8.146) of the Levi-Civita connection on page 512, and

$$\nabla^i := g^{is}\nabla_s.$$

If $p = 0$, then $d\omega_O = \partial_i\omega_O \cdot dx^i$ and $d^*\omega_O = 0$.

Relation (8.156) will be proven in Sect. 8.11.2 on page 523. For the proof of (8.157), we refer to Problem 8.11 on page 554. Recall that the Hodge Laplacian is given by

$$\Delta\omega = (d^*d + dd^*)\omega.$$

Since $d(d\omega) = 0$ and $d^*(d^*\omega) = 0$, we get

$$\boxed{\Delta\omega = (d + d^*)(d + d^*)\omega.}$$

The operator $d + d^*$ is called the Hodge square root of the Hodge Laplacian Δ.

8.10.7 The Weitzenböck Formula for the Hodge Laplacian

Our goal is to write the Hodge Laplacian $\Delta\omega$ for differential forms ω in a very elegant way by using a biorthogonal frame. We assume that the metric tensorial family g_{ij} is of Riemannian type. Let $\mathbf{e}_1, \ldots, \mathbf{e}_n$ be smooth velocity vector fields on Ω_O such that

$$\langle \mathbf{e}_k | \mathbf{e}_l \rangle = \delta_{kl}, \qquad k, l = 1, \ldots, n.$$

That is, $\mathbf{e}_k \in \mathsf{Diff}^1(\Omega_O)$ if $k = 1, \ldots, n$. Moreover, let $\mu^k : \mathsf{Diff}^1(\Omega_O) \to \mathbb{R}$ be linear functionals such that

$$\mu^k(\mathbf{e}_l) = \delta_l^k, \qquad k, l = 1, \ldots, n.$$

We call $\mathbf{e}_1, \ldots, \mathbf{e}_n, \mu^1, \ldots, \mu^n$ a biorthogonal frame on Ω_O.

Theorem 8.29 *For $p = 1, 2, \ldots, n$, let $\omega = \frac{1}{p!}\omega_{i_1 \ldots i_p}\, dx^{i_1} \wedge \cdots \wedge dx^{i_p}$ be a p-form. Then*

$$\Delta\omega = -D_{\mathbf{e}_k}(D_{\mathbf{e}_k}\omega) - D_{D_{\mathbf{e}_k}\mathbf{e}_k}\omega - \mu^k \wedge i_{\mathbf{F}(\mathbf{e}_k, \mathbf{e}_l)\mathbf{e}_l}\omega.$$

Here, we sum over $k, l = 1, \ldots, n$.

This theorem tells us that there appear specific curvature terms depending on $\mathbf{F}$. The proof can be found in J. Jost, Riemannian Geometry and Geometric Analysis, Sect. 3.6, Springer, Berlin, 2008. The theorem dates back to Weitzenböck (1885–1955).

8.10.8 The One-Dimensional σ-Model and Affine Geodesics

We want to generalize the concept of straight-line in Euclidean geometry. Let Γ_{ij}^k be an affine connection family. The smooth curve

$$C : x^i = x^i(t), \qquad t_0 \le t \le t_1, \quad i = 1, \ldots, n \tag{8.158}$$

on the set Ω_O is called an affine geodesic iff

$$\boxed{\ddot{x}^k(t) + \dot{x}^i(t)\, \Gamma_{ij}^k(x(t))\, \dot{x}^j(t) = 0, \qquad t_0 \le t \le t_1, \quad k = 1, \ldots, n.} \tag{8.159}$$

It follows from the transformation law (8.105) for the Christoffel symbols on page 495 that this definition does not depend on the choice of the observer.

The spray of an affine connection family. Let $\mathbf{v} = v^i \partial_i$ be a smooth velocity vector field. We set

$$V^k(x, \mathbf{v}) := -v^i(x)\, \Gamma_{ij}^k(x)\, v^j(x).$$

The velocity vector field $\mathbf{V} = V^k \partial_k$ is called the spray of the velocity vector field $\mathbf{v}$ generated by the connection family Γ_{ij}^k. Then the equation of motion (8.159) can be written as

$$\boxed{\ddot{x}^k(t) = V^k(x(t), \dot{\mathbf{x}}(t)), \qquad k = 1, \ldots, n.} \tag{8.160}$$

This generalizes the Newtonian equation of motion.

The variational problem. We are given the metric tensorial family g_{ij} of arbitrary signature. Consider the variational problem

$$\boxed{\int_{t_0}^{t_1} \dot{x}^i(t)\, g_{ij}(x(t))\, \dot{x}^j(t)\, dt = \text{critical!}} \tag{8.161}$$

with the boundary condition: $x(t_0)$ and $x(t_1)$ are fixed.

Theorem 8.30 *If the curve $x = x(t)$ is a smooth solution of the variational problem* (8.161), *then it is an affine geodesic* (8.159) *with respect to the Levi-Civita connection* (8.146) *on page* 512.

Proof. Introduce the Lagrangian $L(x, \dot{x}) := \dot{x}^i g_{ij}(x)\, \dot{x}^j$. If $x = x(t)$ is a solution of (8.161), then it satisfies the Euler–Lagrange equation

$$\frac{d}{dt} L_{\dot{x}^i} = L_{x^i}, \qquad i = 1, \ldots, n.$$

Explicitly,

$$\frac{d}{dt}\left(g_{ik}\dot{x}^k + g_{ji}\dot{x}^j\right) = \partial_i g_{jk} \cdot \dot{x}^j \dot{x}^k.$$

Here, g_{ik} stands for $g_{ik}(x(t))$, and $\dot{x}^k$ stands for $\dot{x}^k(t)$. Hence

$$g_{ik}\ddot{x}^k + g_{ji}\ddot{x}^j + \partial_l g_{ik} \cdot \dot{x}^l \dot{x}^k + \partial_l g_{ji} \cdot \dot{x}^l \dot{x}^j = \partial_i g_{jk} \cdot \dot{x}^j \dot{x}^k.$$

Using $g_{ij} = g_{ji}$ and changing the notation of some indices, we obtain

$$2g_{sr}\ddot{x}^r + \left(\partial_i g_{js} + \partial_j g_{is} - \partial_s g_{ij}\right) \dot{x}^i \dot{x}^j = 0.$$

Multiplying this by g^{ks}, it follows from (8.146) that

$$\ddot{x}^k + \Gamma^k_{ij} \dot{x}^i \dot{x}^j = 0.$$

This is the claim. □

The variational problem (8.161) is frequently used in order to compute the Christoffel symbols Γ^k_{ij} of the Levi-Civita connection with respect to the metric tensorial family g_{ij}.

The σ-model. The variational problem (8.161) can be generalized to higher dimensions. In modern physics, this is called the σ-model. We will study this in Vol. IV. In terms of mathematics (resp. physics), this is closely related to minimal surfaces, harmonic maps, and Kähler manifolds (resp. string theory). We refer to:

U. Dierkes, S. Hildebrandt, and F. Sauvigny, Minimal Surfaces, Vol. 1, Springer, Berlin.

U. Dierkes, S. Hildebrandt, and A. Tromba, Minimal Surfaces, Vol 2: Regularity of Minimal Surfaces, Vol. 3: Global Analysis of Minimal Surfaces, Springer, Berlin, 2010.

J. Jost, Riemannian Geometry and Geometric Analysis, Springer, Berlin, 2008.

J. Jost, Harmonic Mappings between Riemannian Manifolds, ANU Press, Canberra, Australia, 1984.

J. Jost, Geometry and Physics, Springer, Berlin, 2009.

S. Ketov, Quantum Non-Linear Sigma Models: From Quantum Field Theory to Supersymmetry, Conformal Field Theory, Black Holes and Strings, Springer, Berlin, 2000.

8.11 The Beauty of Connection-Free Derivatives

The Lie derivative $\mathcal{L}_{\mathbf{v}}T$ of a tensor field T (e.g., the electric field $\mathbf{E}$ or the magnetic field $\mathbf{B}$) generalizes the directional derivative with respect to a velocity vector field $\mathbf{v}$.
The Cartan derivative and the Weyl derivative generalize the curl and the divergence,

$$\mathbf{curl}\,\mathbf{v} \quad \text{and} \quad \operatorname{div}\mathbf{v},$$

of a velocity vector field $\mathbf{v}$, respectively. These derivatives only depend on partial derivatives, but neither on the choice of a metric tensorial family nor on the choice of a connection. In particular, this explains why the Cartan calculus of differential forms can be applied to arbitrary manifolds which are not equipped with any additional structure.

Folklore

The basic strategy of cancelling terms containing Christoffel symbols. Our point of departure is the observation that there holds the identity

$$\boxed{v^s\nabla_s w^i - w^s\nabla_s v^i = v^s\partial_s w^i - w^s\partial_s v^i} \tag{8.162}$$

which will be proved in Sect. 8.11.1. This concerns tensorial families v^i and w^i. Moreover, ∇_i refers to the Levi-Civita connection. Note the following. Observe that it is not obvious that

$$v^s\partial_s w^i - w^s\partial_s v^i$$

is a tensorial family. But it is obvious that

$$v^s\nabla_s w^i - w^s\nabla_s v^i$$

is a tensorial family, by the index principle. Consequently, relation (8.162) tells us immediately that $v^s\partial_s w^i - w^s\partial_s v^i$ is a tensorial family; this coincides with the Lie derivative $\mathcal{L}_{\mathbf{v}}w^i$. Similarly, if ω_j is a tensorial family, then the relation

$$\nabla_i\omega_j - \nabla_j\omega_i = \partial_i\omega_j - \partial_j\omega_i$$

tells us that $\partial_i\omega_j - \partial_j\omega_i$ is a tensorial family.

This simple trick is the key to Cartan's powerful calculus of differential forms.

Our general strategy will be the following:

- Choose a fixed metric tensorial family g_{ij}. Note that such a tensorial family always exists. For example, fix an observer O and choose $g_{ij} := \delta_{ij}$. Moreover, extend this to a tensorial family by Theorem 8.2 on page 458.
- Construct the Levi-Civita connection to g_{ij}.
- Using the corresponding covariant partial derivative ∇_i, assign to a given tensorial family a new tensorial family.
- Show that the new tensorial family does not depend on the choice of g_{ij}, since the terms containing Christoffel symbols cancel each other.

Here, we will use the symmetry relation $\Gamma_{ij}^k = \Gamma_{ji}^k$ (i.e., the Levi-Civita connection is torsion-free). This way, the combination of covariant partial derivatives under consideration only depends on the classical partial derivative ∂_i.

We will use this strategy in order to get the fundamental connection-free derivatives due to Lie, Cartan, and Weyl.

Note that, in contrast to the derivatives due to Lie, Cartan, and Weyl, the Hodge codifferential and the Hodge Laplacian are not connection-free notions; indeed they depend on the specific choice of the Levi-Civita connection.

8.11.1 The Lie Derivative

We are given the tensorial families v^i and w^i. Define

$$T^k := v^s \nabla_s w^k - w^s \nabla_s v^k .$$

By the index principle, T^k is a tensorial family. Explicitly, we get

$$T^k = v^s \partial_s w^k + v^s \Gamma^k_{sr} w^r - w^s \partial_s v^k - w^s \Gamma^k_{sr} v^r .$$

Since the Levi-Civita connection is torsion-free, $\Gamma^k_{ij} = \Gamma^k_{ji}$. Thus, the terms containing Christoffel symbols cancel each other, and we get

$$T^k = v^s \partial_s w^k - w^s \partial_s v^k .$$

This coincides with the Lie derivative $\mathcal{L}_{\mathbf{v}} w^k = T^k$. In contrast to the proof of Prop. 8.15 on 488 based on an explicit checking of the transformation law, the present method yields immediately the fact that $\mathcal{L}_{\mathbf{v}} w^k$ is a tensorial family. Similarly, we set

$$T_k := v^s \nabla_s \omega_k + \omega_s \nabla_k v^s .$$

By the index principle, T_k is a tensorial family. Explicitly,

$$T_k = v^s \partial_s \omega_k - v^s \Gamma^r_{sk} \omega_r + \omega_s \partial_k v^s + \omega_s \Gamma^s_{kr} v^r = v^s \partial_s \omega_k + \omega_s \partial_k v^s .$$

Thus, T_k coincides with the Lie derivative $\mathcal{L}_{\mathbf{v}} \omega_k = T_k$. The same method applies to the Lie derivative for general tensorial families. For example,

$$\mathcal{L}_{\mathbf{v}} T^i_j := v^s \partial_s T^i_j + T^i_s \partial_j v^s - T^s_j \partial_s v^i = v^s \nabla_s T^i_j + T^i_s \nabla_j v^s - T^s_j \nabla_s v^i .$$

8.11.2 The Cartan Derivative

Let $p = 1, \ldots, n$. Recall that a Cartan family $\omega_{i_1 \ldots i_p}$ is an antisymmetric tensorial family. The Cartan derivative is defined by

$$\boxed{d_i \omega_{i_1 \ldots i_p} := \partial_{[i} \omega_{i_1 \ldots i_p]} .} \tag{8.163}$$

Theorem 8.31 *The Cartan derivative $d_i \omega_{i_1 \ldots i_p}$ of a Cartan family $\omega_{i_1 \ldots i_p}$ is again a Cartan family.*

In particular, this theorem tells us that if $\omega_{i_1 \ldots i_p}$ is an antisymmetric tensorial family, then so is $d_i \omega_{i_1 \ldots i_p}$.

Proof. Choose a metric tensorial family g_{ij}. Define $T_{ii_1 \ldots i_p} := \nabla_{[i} \omega_{i_1 \ldots i_p]}$. This is a tensorial family, by the index principle. It remains to show that

$$\nabla_{[i} \omega_{i_1 \ldots i_p]} = \partial_{[i} \omega_{i_1 \ldots i_p]} .$$

This is an immediate consequence of (8.117) on page 499. □

In order to get insight into the structure of the cancellations of the Christoffel symbols, let us explicitly consider two special cases.

(I) Let $p = 1$. Then $\nabla_i \omega_j = \partial_i \omega_j - \Gamma^s_{ij} \omega_s$. Since $\Gamma^s_{ij} = \Gamma^s_{ji}$, we get

$$\nabla_i \omega_j - \nabla_j \omega_i = \partial_i \omega_j - \partial_j \omega_i .$$

(II) Let $p = 2$. Then

- $\nabla_i \omega_{jk} = \partial_i \omega_{jk} - \Gamma^s_{ij} \omega_{sk} - \Gamma^s_{ik} \omega_{js}$,
- $-\nabla_j \omega_{ik} = -\partial_j \omega_{ik} + \Gamma^s_{ji} \omega_{sk} + \Gamma^s_{jk} \omega_{is}$,
- $\nabla_k \omega_{ij} = \partial_k \omega_{ij} - \Gamma^s_{ki} \omega_{sj} - \Gamma^s_{kj} \omega_{is}$.

Since $\Gamma^c_{ab} = \Gamma^c_{ba}$ and $\omega_{ab} = -\omega_{ba}$, we get

$$\nabla_i \omega_{jk} - \nabla_j \omega_{ik} + \nabla_k \omega_{ij} = \partial_i \omega_{jk} - \partial_j \omega_{ik} + \partial_k \omega_{ij}.$$

Again using $\omega_{ab} = -\omega_{ba}$, we obtain $\nabla_{[i} \omega_{jk]} = \partial_{[i} \omega_{jk]}$.

8.11.3 The Weyl Derivative

Weyl family. Let $p = 1, 2, \ldots, n$. By definition, a Weyl family

$$\mathcal{W}^{i_1 \ldots i_p} \tag{8.164}$$

is an antisymmetric tensorial density family of weight 1 (see page 462). The prototype of a Weyl family

$$\mathcal{W}^i = \varrho v^i, \qquad i = 1, \ldots, n$$

is the product of the mass density ϱ (scalar tensorial density family of weight 1) and the tensorial family v^i (e.g., the components of a velocity vector field). The Weyl derivative of the Weyl family (8.164) is defined by setting

$$\boxed{(\delta \mathcal{W})^{i_2 \ldots i_p} := \partial_i \mathcal{W}^{i i_2 \ldots i_p}.} \tag{8.165}$$

In particular, if $p = 1$, then $\delta \mathcal{W} := \partial_i \mathcal{W}^i$.

Theorem 8.32 *The Weyl derivative $\partial_i \mathcal{W}^{i i_2 \ldots i_p}$ of a Weyl family $\mathcal{W}^{i_1 i_2 \ldots i_p}$ is again a Weyl family.*

The proof will be given below after finishing some necessary preparations.

The divergence formula. Let g_{ij} be a metric tensorial family of arbitrary signature. Let us use the Levi-Civita connection. If v^i is a tensorial family, then

$$\nabla_i v^i = \frac{1}{\sqrt{|g|}} \, \partial_i (\sqrt{|g|} \cdot v^i). \tag{8.166}$$

Proof. (I) Identity for the determinant $g(x) = \det(g_{ij}(x))$. Recall that the matrix $(g^{ij}(x))$ is the inverse matrix to $(g_{ij}(x))$. This implies the key formula

$$\partial_k g(x) = g(x) g^{ij}(x) \cdot \partial_k g_{ij}(x). \tag{8.167}$$

The proof (based on the Laplace expansion theorem for determinants) will be given in Problem 8.10 on page 554.

(II) The contraction formula for the Christoffel symbols of the Levi-Civita connection. Suppose that $g > 0$. Then it follows from

$$\Gamma^k_{ij} = \tfrac{1}{2} g^{ks} (\partial_i g_{js} + \partial_j g_{is} - \partial_s g_{ij})$$

that $\Gamma^k_{ik} = \frac{1}{2} g^{ks} \partial_i g_{ks}$, by renaming the indices of the second and third term on the right-hand side. Using (8.167), we get

$$\Gamma^k_{ik} = \frac{\partial_i g}{2g} = \partial_i (\ln \sqrt{g}).$$

(III) The definition of the covariant partial derivative yields

$$\nabla_i v^i = \partial_i v^i + \Gamma^i_{is} v^s = \partial_i v^i + v^s \partial_s (\ln \sqrt{g}) = \partial_i v^i + \frac{1}{\sqrt{g}} v^i \partial_i \sqrt{g}.$$

Using the Leibniz rule, we get the desired formula

$$\nabla_i v^i = \frac{1}{\sqrt{g}}\, \partial_i \big(v^i \sqrt{g} \big).$$

If $g < 0$, then use $|g| = -g$. □

Let $p = 2, \ldots, n$. If $T^{i_1 \ldots i_p}$ is an antisymmetric tensorial family, then

$$\boxed{\nabla_i T^{ii_2 \ldots i_p} = \frac{1}{\sqrt{|g|}}\, \partial_i \big(\sqrt{|g|}\ T^{ii_2 \ldots i_p} \big).} \tag{8.168}$$

Proof. (I) Let $p = 2$. By definition of the covariant partial derivative,

$$\nabla_i T^{ik} = \partial_i T^{ik} + \Gamma^i_{is} T^{sk} + \Gamma^k_{is} T^{is}.$$

The last term vanishes by the symmetry of Γ^i_{is} and the antisymmetry of T^{is} with respect to the indices i and s. Hence

$$\nabla_i T^{ik} = \partial_i T^{ik} + \Gamma^i_{is} T^{sk}.$$

Now use the same argument as in the proof of (8.166) above.

(II) If $p = 3, \ldots, n$, then the proof proceeds analogously to (I). □

Proof of Theorem 8.32. Note that $\sqrt{|g|}$ is a tensorial density of weight one. Thus, if $\mathcal{W}^{i_1 \ldots i_p}$ is a Weyl family (i.e., an antisymmetric tensorial density of weight one), then

$$T^{i_1 i_2 \ldots i_p} := \frac{1}{\sqrt{|g|}} \cdot \mathcal{W}^{i_1 i_2 \ldots i_p}$$

is an antisymmetric tensorial family. Thus, $\nabla_i T^{ii_2 \ldots i_p}$ is a tensorial family. By (8.168),

$$\partial_i \mathcal{W}^{ii_2 \ldots i_p} = \sqrt{|g|} \cdot \nabla_i T^{ii_2 \ldots i_p}.$$

Hence $\partial_i \mathcal{W}^{ii_2 \ldots i_p}$ is an antisymmetric tensorial density of weight one. □

Weyl duality. The preceding proof shows the following. If $T^{i_1 \ldots i_p}$ is an antisymmetric tensorial family, then

$$\mathcal{W}^{i_1 \ldots i_p} := \sqrt{|g|} \cdot T^{i_1 \ldots i_p}$$

is a Weyl family. This duality between antisymmetric tensorial families and Weyl families is called Weyl duality.

Weyl duality plays a crucial role in describing the electromagnetic field in continuous media.

We will show this in Sect. 19.8.

8.12 Global Analysis

Global analysis is invariant theory in action.
Folklore

The term global analysis is used for the analysis and differential topology on manifolds (see the Appendix on page 1069). Observe that all of the preceding investigations apply immediately to manifolds in the following way:

- Manifolds (e.g., the surface of earth) are described by local coordinates (e.g., geographic charts of earth).
- The change of local coordinates corresponds to the change of observers by means of diffeomorphisms.
- Properties of the manifold are described by invariants. Using local coordinates, note that a local-coordinate expression represents an invariant of the manifold if it has the correct index picture in the sense of the calculus developed above.

For example, a p-form of a manifold is described by

$$\omega = \frac{1}{p!}\omega_{i_1 \ldots i_p}\, dx^{i_1} \wedge \cdots \wedge dx^{i_p}$$

where the coefficient functions $\omega_{i_1 \ldots i_p}$ transform like an antisymmetric tensorial family under a change of local coordinates.

Riemannian manifold. A manifold $\mathcal{M}$ can be equipped with an additional structure. For example, the real n-dimensional manifold $\mathcal{M}$ is called a Riemannian manifold iff there exists a system of functions g_{ij} which form a metric tensorial family of Riemann type under a change of local coordinates.[42] This allows us to define invariantly the notion of the length of a curve on the manifold. To this end, we fix a local $(x^1, \ldots, x^n)$-coordinate system, and we assume that the curve is given by the equation $x^i = x^i(t)$, $t_0 \le t \le t_1$, $i = 1, \ldots, n$. Then the length of the curve is defined by the integral

$$\int_{t_0}^{t_1} \sqrt{\dot{x}^i(t) g_{ij}(x(t)) \dot{x}^j(t)} \cdot dt.$$

By the index principle, this integral does not depend on the choice of the local coordinate system.

Orientation. The manifold $\mathcal{M}$ is called oriented iff the change of local coordinate systems is described by diffeomorphisms of positive sign (i.e., the Jacobian has positive sign). In this case, we can also use pseudo-tensorial families because they become tensorial families with respect to the distinguished change of local coordinates. For example, using the pseudo-tensorial family $\mathcal{E}_{i_1 \ldots i_n}$ we can define

$$v := \frac{1}{n!}\, \mathcal{E}_{i_1 \ldots i_n}\, dx^{i_1} \wedge \cdots \wedge dx^{i_n}. \qquad (8.169)$$

If the manifold $\mathcal{M}$ is oriented then $\mathcal{E}_{i_1 \ldots i_n}$ transforms like a tensorial family under diffeomorphisms of positive sign. Therefore, the differential form v possesses an invariant meaning on the manifold, by the index principle. This n-form is called the volume form of the Riemannian manifold $\mathcal{M}$. We get

$$v = \sqrt{|g|}\, dx^1 \wedge \cdots \wedge dx^n.$$

[42] If g_{ij} is a metric family of pseudo-Riemannian type, then we obtain a pseudo-Riemannian manifold.

In contrast to (8.169), this formula is very convenient for doing computations, but it does not display the invariant character of this expression. We refer the reader to Sect. 8.14 where we will discuss the two different philosophies in global analysis, namely, the index-based and the index-free method.

Integration. Let ω be an n-form on the oriented real n-dimensional manifold $\mathcal{M}$. The integral $\int_{\mathcal{M}} \omega$ is obtained by setting

$$\int_{\mathcal{M}} \omega := \sum_{k=1}^{K} \int_{\mathcal{M}} f_k \omega.$$

Here, the smooth, compactly supported functions $f_1, \ldots, f_K : \mathcal{M} \to \mathbb{R}$ with $\sum_{k=1}^{K} f_k = 1$ form a partition of unity on $\mathcal{M}$ (see the Appendix on page 1077). We choose the supports of the smooth functions $f_1, \ldots, f_K$ sufficiently small (i.e., the functions $f_1, \ldots, f_K$ vanish outside sufficiently small compact subsets of $\mathcal{M}$). Then the integrals $\int_{\mathcal{M}} f_k \omega$ can be computed by using local coordinates. The point is that the local integrals do not depend on the choice of the local coordinates, and the global integral $\int_{\mathcal{M}} \omega$ does not depend on the choice of the partition of unity.

8.13 Summary of Notation

> What's in a name? That which we call a rose
> by any other word would smell as sweet.
>
> William Shakespeare (1564–1616)
> *Romeo and Juliet* 2,2

Our goal is to help the reader to memorize the basic formulas in the Riemann curvature theory. Note that different definitions are used for the Riemann curvature tensorial family

$$R_{ijkl}$$

which reduce to different sign conventions, after permuting indices of R_{ijkl} if necessary.[43] Every choice of convention has their own advantages and disadvantages. The convention used in this monograph is based on the following mnemonic philosophy. The point of departure is the relation

$$\boxed{\mathbf{F}(\mathbf{u}, \mathbf{v})\mathbf{w} = \mathbf{R}(\mathbf{u}, \mathbf{v}, \mathbf{w})}$$

for the Riemann curvature operator $\mathbf{F}(\mathbf{u}, \mathbf{v})$, which is the geometric and physical basic object. Using the inner product $\langle . | . \rangle$, we set

$$\mathcal{R}(\mathbf{u}, \mathbf{v}, \mathbf{w}, \mathbf{z}) := \langle \mathbf{R}(\mathbf{u}, \mathbf{v}, \mathbf{w}) | \mathbf{z} \rangle.$$

This motivates

- $\mathcal{R}(\mathbf{u}, \mathbf{v}, \mathbf{w}, \mathbf{z}) = R_{ijkl} u^i v^j w^k z^l$, and hence $R_{ijkl} := \mathcal{R}(\partial_i, \partial_j, \partial_k, \partial_l)$,
- $\mathbf{R}(\mathbf{u}, \mathbf{v}, \mathbf{w}) = R^l_{ijk} u^i v^j w^k \cdot \partial_l$, and hence $R^l_{ijk} = dx^l(\mathbf{R}(\partial_i, \partial_j, \partial_k))$,
- $D_{\mathbf{u}} \mathbf{v} = u^i \Gamma^k_{ij} v^j \cdot \partial_k$, and hence $\Gamma^k_{ij} = dx^k(D_{\partial_i} \partial_j)$.

[43] A table of sign conventions used by about forty selected authors can be found on the front page of the standard textbook by C. Misner, K. Thorne, and A. Wheeler, Gravitation, Freeman, San Francisco, 1973.

Since $R_{ijkl} = R^s_{ijk} g_{sl}$, we also write $R^{\;\;\;\;l}_{ijk.}$ instead of R^l_{ijk} if necessary.

In order to pass to the language of differential forms, we introduce the following matrices

$$\mathcal{A}_i := (\Gamma^l_{ik}) \quad \text{and} \quad \mathcal{F}_{ij} = (R^l_{ijk}).$$

Here, the upper index l is the row index, and the lower index k is the column index of the matrices. Based on this convention, we get the key curvature relation

$$\boxed{\mathcal{F}_{ij} = \partial_i \mathcal{A}_j - \partial_j \mathcal{A}_i + \mathcal{A}_i \mathcal{A}_j - \mathcal{A}_j \mathcal{A}_i}$$

which is equivalent to

$$R^l_{ijk} = \partial_i \Gamma^l_{jk} - \partial_j \Gamma^l_{ik} + \Gamma^l_{is} \Gamma^s_{jk} - \Gamma^l_{js} \Gamma^s_{ik}.$$

Introducing

- $\mathcal{F} := \frac{1}{2} \mathcal{F}_{ij}\, dx^i \wedge dx^j$ (curvature form), and
- $\mathcal{A} := \mathcal{A}_i\, dx^i$ (connection form),

the letter $\mathcal{F}$ stands for 'force'. Moreover, the symbol $\mathcal{A}_i$ resembles the component A_i of the 4-potential in Maxwell's theory of electromagnetism.

Furthermore, the Riemann curvature tensorial family R^l_{ijk} allows the construction of the following crucial invariant

$$\boxed{R := R^i_{ijk} g^{jk}}$$

by lifting and contracting of indices. Here, R is the scalar curvature. In terms of R_{ijkl}, we obtain

$$R = g^{il} R_{ijkl} g^{jk}.$$

Introducing $R_{il} := R_{ijkl} g^{jk}$, we get

$$R = g^{il} R_{il}.$$

Here, R_{il} is called the Ricci curvature tensorial family. Moreover, setting

$$\mathrm{Ric}(\mathbf{u}, \mathbf{v}) := R_{ij} u^i v^j,$$

we get $\mathrm{Ric}(\mathbf{u}, \mathbf{z}) = g^{kl} \mathcal{R}(\mathbf{u}, \partial_k, \partial_l, \mathbf{z})$.

Observe that the transformation

$$R^l_{ijk} \Rightarrow R^l_{kij}, \quad R_{ijkl} \Rightarrow -R_{ijkl}, \quad R_{\alpha\beta} \Rightarrow R_{\alpha\beta}$$

changes our notation to the notation used in the following textbooks:

C. Misner, K. Thorne, and A. Wheeler, Gravitation, Freeman, San Francisco, 1973.

L. Landau and E. Lifshitz, Course of Theoretical Physics, Vol. 2: The Classical Theory of Fields, Butterworth–Heinemann, Oxford, 1982.

J. Jost, Riemannian Geometry and Geometric Analysis, Springer, Berlin, 2008.

For example, our symbol R_{1221} passes over to $-R_{1221}$ which is equal to R_{1212}. Moreover, the transformation

$$R^l_{ijk} \Rightarrow R^l_{kij}, \quad R_{ijkl} \Rightarrow R_{ijkl}, \quad R_{\alpha\beta} \Rightarrow R_{\alpha\beta}$$

changes our notation to the notation used in the textbook by

Y. Choquet-Bruhat, C. DeWitt-Morette, and M. Dillard-Bleick, Analysis, Manifolds, and Physics, Vols. 1, 2, Elsevier, Amsterdam, 1996.

8.14 Two Strategies in Invariant Theory

Invariants play a fundamental role in mathematics and physics. There are two methods of constructing invariants, namely,

- the index method based on the principle of killing indices as described above,
- and the index-free method.

Physicists prefer the index method, whereas mathematicians like the index-free method. In order to be flexible, the reader should master not only one approach, but the two approaches. Note the following:

- If one wants to simplify the computation of physical effects by using special coordinates, then the index method is the right tool. Most textbooks in physics use the index method.
- The index-free method gives insight. Most textbooks in modern mathematics use the index-free method.

In the next chapters, we will study physical fields on the Euclidean manifold (temperature fields, velocity fields, and differential forms). This should help the reader to understand the physical and geometric background of modern physics and modern differential geometry. Later on, we will investigate the relation between gauge theory in physics and the modern theory of fiber bundles.

Note the following. In this chapter, the definitions are formulated by the indexed components, and the index-free expressions follow by the aid of the index principle. In the next chapters, the definitions will be given in an invariant (i.e., index-free) way. Passing to the natural basis $\partial_1, \ldots, \partial_n$ and $dx^1, \ldots, dx^n$, one obtains the expressions considered in the present chapter. The invariant approach has the advantage that, in contrast to the natural basis, one can use an arbitrary basis, and the approach can be generalized to infinite dimensions (e.g., global analysis on infinite-dimensional Hilbert spaces, Banach spaces and Banach manifolds). See R. Abraham, J. Marsden, and T. Ratiu, Manifolds, Tensor Analysis, and Applications, Springer, New York, 1988, and Zeidler (1986), Vol. IV, quoted on page 1089.

In the next section, as a typical example, we want to study the invariant formulation of the notion of a tangent vector of a manifold by derivations. The point is that the notion of a linear differential operator applied to a temperature field on a manifold can be defined in a purely algebraic way. This approach was used by Claude Chevalley (1909–1984) in his fundamental monograph *Theory of Lie Groups,* Princeton University Press, 1946 (15th edition, 1999).

To motivate the following discussion, note that the earth is part of the surrounding universe. Therefore, we can use notions which depend on the universe (e.g., the notion of a tangent plane at a point of the surface of earth). However, in cosmology we want to use intrinsic notions which do not depend on a hypothetical higher-dimensional super universe. Such super universes are discussed in L. Randall, Warped Passages: Unravelling the Mysteries of the Universe's Hidden Dimensions, Ecco, New York, 2005. See also S. Yau and S. Nadis, The Shape of Inner Space: String Theory and the Geometry of the Universe's Hidden Dimensions, Basic Books, New York, 2010.

8.15 Intrinsic Tangent Vectors and Derivations

> Tangent vectors correspond to velocity vectors. On finite-dimensional manifolds, velocity vectors are equivalent to linear differential operators (derivations of temperature fields).
>
> Folklore

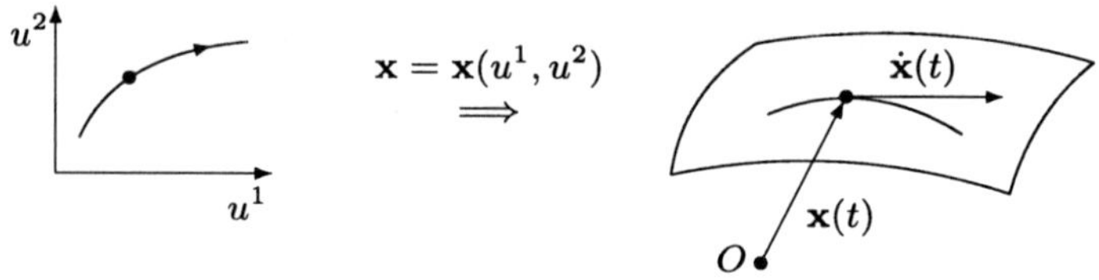

(a) $u^1 = u^1(t),\ u^2 = u^2(t)$ (b) $\mathbf{x}(t) = \mathbf{x}\big(u^1(t), u^2(t)\big)$

Fig. 8.1. Trajectory of a ship

Extrinsic velocity vector on the surface of earth (Fig. 8.1). Intuitively, the velocity vector of a moving ship on the ocean points into the universe. This is called an extrinsic velocity vector. Let us first discuss this notion. Consider the sphere

$$\mathbb{S}^2_R := \{P \in \mathbb{E}^3 : d(O, P) = R\}$$

of positive radius R centered at the origin O. Here, $d(O, P)$ denotes the Euclidean distance between the origin O and the point P. Using the position vector $\mathbf{x} = \overrightarrow{OP}$ pointing from O to P, we can write

$$\mathbb{S}^2_R = \{P \in \mathbb{E}^3 : \mathbf{x}^2 = R^2\}.$$

Intuitively, we regard the sphere $\mathbb{S}^2_R$ as the surface of earth. Let

$$\mathcal{C}: \ P = P(t), \qquad t \in \mathcal{R},$$

be a smooth curve on $\mathbb{S}^2_R$ (e.g., the trajectory of a ship). Here, $\mathcal{R}$ is an open interval of $\mathbb{R}$ which contains the point t_0. Set $P_0 := P(t_0)$.[44] Alternatively, we write

$$\mathcal{C}: \ \mathbf{x} = \mathbf{x}(t), \qquad t \in \mathcal{R}.$$

The time derivative

$$\dot{\mathbf{x}}(t_0) = \lim_{\Delta t \to 0} \frac{\mathbf{x}(t_0 + \Delta t) - \mathbf{x}(t_0)}{\Delta t}$$

is called the velocity vector $\mathbf{v}(P_0)$ (of the ship) at time t_0. Let us describe the motion of the ship on a geographic chart with the real Cartesian coordinates (u^1, u^2). The point (u^1, u^2) of the geographic chart corresponds to the point

$$P = P(u^1, u^2)$$

on the surface of earth. We assume that $P_0 = (u^1_0, u^2_0)$. In terms of position vectors, the point $P(u^1, u^2)$ corresponds to the vector $\mathbf{x}(u^1, u^2)$. The two vectors

$$\mathbf{b}_j(u^1, u^2) := \frac{\partial \mathbf{x}(u^1, u^2)}{\partial u^j}, \qquad j = 1, 2$$

span the tangent space $T_P\mathbb{S}^2_R$ of the sphere $\mathbb{S}^2_R$ at the point $P(u^1, u^2)$ (Fig 8.2). This yields the decomposition

$$\mathbf{v}(P_0) = v^1(P_0)\, \mathbf{b}_1(P_0) + v^2(P_0)\, \mathbf{b}_2(P_0).$$

[44] To simply notation, we choose $t_0 := 0$ and $u^1_0 = u^2_0 := 0$.

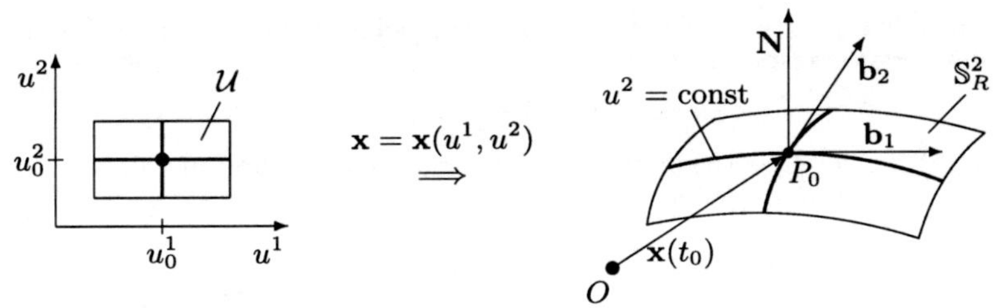

Fig. 8.2. Local parametrization of the sphere $\mathbb{S}_R^2$

The real numbers $v^1(P_0), v^2(P_0)$ are called the coordinates of the velocity vector $\mathbf{v}(P_0)$ (with respect to the local (u^1, u^2)-coordinates). If we pass from the local (u^1, u^2)-coordinates to the local $(u^{1'}, u^{2'})$-coordinates by the local diffeomorphism

$$u^{1'} = u^{1'}(u^1, u^2), \quad u^{2'} = u^{2'}(u^1, u^2),$$

then it follows from the chain rule that[45]

$$\mathbf{b}_{i'}(P_0) = \frac{\partial \mathbf{x}(P_0)}{\partial u^i} \frac{\partial u^i(P_0)}{\partial u^{i'}} = \frac{\partial u^i(P_0)}{\partial u^{i'}} \mathbf{b}_i(P_0).$$

Hence we obtain the transformation law

$$\boxed{v^{i'}(P_0) = \frac{\partial u^{i'}(P_0)}{\partial u^i} v^i(P_0), \qquad i' = 1', 2'} \tag{8.170}$$

for the velocity components. The velocity vector $\mathbf{v}(P_0)$ is called an extrinsic velocity vector, since the definition of $\mathbf{v}(P_0)$ uses the surrounding Euclidean manifold $\mathbb{E}^3$.

Our goal is to define the notion of a tangent vector in an intrinsic way by not using the surrounding Euclidean manifold $\mathbb{E}^3$.

Intrinsic velocity vectors on the surface of earth. We will discuss the two equivalent approaches for the sphere $\mathbb{S}_R^2$:

(G) The geometric approach based on equivalence classes of trajectories.
(A) The algebraic-analytic approach based on linear partial differential operators (derivations).

Ad (G). Let $\mathcal{C} : P = P(t)$ and $\mathcal{C}_* : P = P_*(t)$ with $t \in \mathcal{R}$ be two trajectories on the sphere $\mathbb{S}_R^2$ which pass through the point P_0 at time $t_0 := 0$. We write

$$\mathcal{C} \sim \mathcal{C}_* \qquad \text{at the point } P_0$$

if and only if, in a fixed geographic chart, the two trajectories have the same velocity vector at time $t = 0$, that is,

$$(v^1(0), v^2(0)) = (v_*^1(0), v_*^2(0)).$$

The point is that this equivalence relation does not depend on the choice of the geographic chart because of the transformation law (8.170). The equivalence classes

---

[45] We sum over $i = 1, 2$.

$[\mathcal{C}]_{P_0}$ are called intrinsic tangent vectors at the point P_0 of the sphere. There exists the one-to-one correspondence

$$\mathbf{v}(P_0) \Leftrightarrow [\mathcal{C}]_{P_0}.$$

Ad (A). Let $C^\infty_{P_0}$ be the space of all smooth temperature functions

$$\Theta : \mathcal{O}(P_0) \to \mathbb{R}$$

on some open neighborhood $\mathcal{O}(P_0)$ of the point P_0 on the sphere $\mathbb{S}^2_R$ (the set $\mathcal{O}(P_0)$ depends on the function Θ). We identify two functions of $C^\infty_{P_0}$ iff they coincide on some open neighborhood of P_0. The real linear space $C^\infty_{P_0}$ is called the space of smooth germs at the point P_0. A linear map

$$\mathrm{der} : C^\infty_{P_0} \to \mathbb{R}$$

is called a derivation at the point P_0 iff we have the Leibniz rule

$$\boxed{\mathrm{der}(\Theta\Phi) = \mathrm{der}(\Theta) \cdot \Phi(P_0) + \Theta(P_0) \cdot \mathrm{der}(\Phi)} \tag{8.171}$$

for all $\Theta, \Phi \in C^\infty_{P_0}$. Obviously, this is an invariant definition. We want to determine all the possible derivations at the point P_0. To this end, we choose a local (u^1, u^2)-coordinate system in a neighborhood of the point P_0. Recall that $\partial_i := \frac{\partial}{\partial u^i}$. If v^1, v^2 are real numbers, then

$$\mathrm{der}(\Theta) := v^1\partial_1\Theta(P_0) + v^2\partial_2\Theta(P_0) \tag{8.172}$$

is a derivation at the point P_0; this is an immediate consequence of the classic Leibniz rule $\partial_i(\Theta\Phi) = \partial_i\Theta \cdot \Phi + \Theta \cdot \partial_i\Phi$. We briefly write

$$\boxed{\mathrm{der} = v^i(P_0)\partial_i}$$

where we sum over $i = 1, 2$. This is a linear differential operator which acts on temperature fields on the surface of earth, $\mathbb{S}^2_R$.

Proposition 8.33 *Every derivation at the point P_0 has the form* (8.172).

Proof. Suppose that we are given a derivation, der. We will use the Taylor expansion with the integral form of the remainder. Using the constant temperature field $\Theta \equiv 1$ in a neighborhood of P_0, the Leibniz rule yields

$$\mathrm{der}(1) = 2 \cdot \mathrm{der}(1),$$

and hence $\mathrm{der}(1) = 0$. Moreover, let $\Theta \equiv \alpha$ where $\alpha \in \mathbb{R}$. Then the linearity of the derivation implies that $\mathrm{der}(\Theta) = \alpha \cdot \mathrm{der}(1) = 0$. If $\Theta \in C^\infty_{P_0}$, then

$$\Theta(u) - \Theta(0) = \int_0^1 \frac{d}{d\tau}\Theta(\tau u)\, d\tau = u^i \int_0^1 \partial_i\Theta(\tau u)\, d\tau.$$

This implies the decomposition

$$\Theta(u) = \Theta(0) + u^i R_i(u)$$

with the remainder $R_i \in C^\infty_{P_0}$ where $R_i(0) = \partial_i\Theta_i(0)$, $i = 1, 2$. Hence, by the Leibniz rule,

$$\mathrm{der}(\Theta) = \mathrm{der}(u^i R_i(u)) = \mathrm{der}(u^i) \cdot R_i(0).$$

Setting $v^i := \mathrm{der}(u^i)$, and noting that $R_i(0) = \partial_i\Theta(0)$, we get the claim (8.172). □

Intrinsic velocity vectors on a finite-dimensional manifold. Let $\mathcal{M}$ be a real n-dimensional manifold. The preceding discussion allows an immediate generalization from the sphere $\mathbb{S}^2_R$ to $\mathcal{M}$. In what follows we sum over $i = 1, \ldots, n$.

(G) The geometric approach: Fix the point $P \in \mathcal{M}$. Consider local $(u^1, \ldots, u^n)$-coordinate systems for a neighborhood of the point P. By the definition of a manifold (see Sect. 5.4 of Vol. I), the change of local coordinates is described by a local diffeomorphism,

$$u^{i'} = u^{i'}(u^1, \ldots, u^n), \qquad i' = 1', \ldots, n'.$$

We consider tuples $(P, v^1, \ldots, v^n) \in \mathcal{M} \times \mathbb{R}^n$, and we write

$$(P, v^1, \ldots, v^n) \sim (P, v^{1'}, \ldots, v^{n'})$$

if and only if

$$v^{i'} = \frac{\partial u^{i'}(P)}{\partial u^i} v^i, \qquad i' = 1', \ldots, n'.$$

By definition, the equivalence classes $[(P; v^1, \ldots, v^n)]$ are called tangent vectors of the manifold $\mathcal{M}$ at the point P (or velocity vectors). In particular, if

$$\mathcal{C} : u^i = u^i(t), \qquad i = 1, \ldots, n, \quad t \in \mathcal{R}$$

is a curve on the manifold $\mathcal{M}$ which passes through the point P at time $t = 0$, then setting $v^i := \dot{u}^i(0)$, $i = 1, \ldots, n$, we get the tangent vector (velocity vector)

$$[(P, v^1, \ldots v^n)] \qquad \text{at the point } P.$$

(A) The algebraic-analytic approach: Let C_P^∞ be the space of all smooth (temperature) functions

$$\Theta : \mathcal{O}(P) \to \mathbb{R}$$

on some open neighborhood $\mathcal{O}(P)$ of the point P of $\mathcal{M}$ (the set $\mathcal{O}(P)$ depends on the function Θ). We identify two functions of C_P^∞ iff they coincide on some open neighborhood of the point P. The real linear space C_P^∞ is called the space of smooth germs of the manifold $\mathcal{M}$ at the point P. By definition, a linear map

$$\text{der} : C_P^\infty \to \mathbb{R}$$

is called a derivation at the point P iff we have the Leibniz rule

$$\text{der}(\Theta\Phi) = \text{der}(\Theta) \cdot \Phi(P) + \Theta(P) \cdot \text{der}(\Phi)$$

for all $\Theta, \Phi \in C_P^\infty$. Choose a fixed local $(u^1, \ldots, u^n)$-coordinate system in a neighborhood of the point P. As above for the sphere, one shows that a derivation at the point P has precisely the form

$$\boxed{\text{der} = v^i(P)\partial_i} \tag{8.173}$$

where $v^1(P), \ldots, v^n(P)$ are real numbers, and we sum over $i = 1, \ldots, n$. Obviously, there exists a one-to-one relation

$$v^i(P)\partial_i \Leftrightarrow [(P; v^1(P), \ldots, v^n(P))].$$

That is, tangent vectors at the point P of the manifold $\mathcal{M}$ can be identified with derivations at P.

Remark. On infinite-dimensional Banach manifolds, the two definitions (G) and (A) are not equivalent. In this case, one uses the geometric approach (G) as basic definition for tangent vectors (see Zeidler (1986), Vol. IV, quoted on page 1089).

8.16 Further Reading on Symmetry and Invariants

Symmetry, Sciences, and Human Culture

H. Weyl, Symmetry, Princeton University Press, 1952.

C. Yang, Symmetry and physics, pp. 11–33. In: G. Ekspong (Ed.), The Oskar Klein (1894–1977) Memorial Lectures, Vol. 1, World Scientific, Singapore, 1991.

E. Wigner, Philosophical Reflections and Syntheses, annotated by G. Emch, Springer, New York, 1995.

R. Feynman, The Character of Physical Law, MIT Press, Cambridge, Massachusetts, 1966.

M. Gell-Mann, The Quark and the Jaguar, Freeman, New York, 1994.

P. Cartier, A mad day's work: From Grothendieck to Connes and Kontsevich – The evolution of concepts of space and symmetry, Bull. Amer. Math. Soc. **38**(4) (2001), 389–408.

A. Zee, Fearful Symmetry: The Search for Beauty in Modern Physics, Princeton University Press, 1999.

L. Lederman and C. Hill, Symmetry and the Beautiful Universe, Prometheus Books, New York, 2008.

T. Fujita, Symmetry and Its Breaking in Quantum Field Theory, Nova Science Publisher, New York, 2007.

M. Golubitsky and I. Stewart, The Symmetry Perspective from Equilibrium to Chaos in Phase Space and Physical Space, Birkhäuser, Basel, 2002.

S. Hildebrandt and T. Tromba, The Parsimonious Universe: Shape and Form in the Natural World, Copernicus, New York, 1996.

J. Conway and H. Burgiel, The Symmetry of Things, CRC Press, Boca Raton, Florida, 2008.

B. Greene, The Elegant Universe: Supersymmetric Strings, Hidden Dimensions, and the Quest for the Ultimate Theory, Norton, New York, 1999.

P. Binétruy, Supersymmetry: Theory, Experiment, and Cosmology, Oxford University Press, 2006.

G. Wagnière, On Chirality and the Universal Asymmetry: Reflections on Image and Mirror Image, Wiley-VHC, Zürich, 2007 (including applications to biology).

D. Hofstadter, Gödel, Escher, Bach: An Eternal Golden Braid, Basic Books, New York, 1979.

H. Peitgen and P. Richter, The Beauty of Fractals, Springer, Berlin, 1986.

C. Bovill, Fractal Geometry in Architecture and Design, Birkhäuser, Basel, 1995.

G. Mazolla, The Topos of Music, Birkhäuser, Basel, 2002.

D. Washburn and D. Crowe (Eds.), Symmetry Comes of Age: The Role of Pattern in Culture, University of Washington Press, 2004.

J. Rosen, How Science and Nature are, Springer, Berlin, 2008.

I. Stewart, Why Beauty is Truth: A History of Symmetry, Basic Books, New York, 2008.

Harmonic Analysis

Harmonic analysis is a huge mathematical subject which plays a crucial role in mathematical physics. As surveys, we recommend:

G. Mackey , Harmonic analysis as the exploitation of symmetry – a historical survey, Bull. Amer. Math. Soc. **3** (1980), 543–698.

G. Mackey, The Scope and History of Commutative and Noncommutative Harmonic Analysis, Amer. Math. Soc., Providence, Rhode Island, 1992.

As textbooks, we recommend:

E. Stein and R. Shakarchi, Princeton Lectures in Analysis I: Fourier Analysis, Princeton University Press, 2003.

G. Folland, Harmonic Analysis in Phase Space, Princeton University, 1989 (Heisenberg group, quantization and pseudo-differential operators, Weyl operational calculus, the Stone-von Neumann theorem in quantum mechanics, metaplectic representations).

G. Folland, A Course in Abstract Harmonic Analysis, CRC Press, Boca Raton, Florida, 1995.

W. Rudin, Fourier Analysis on Groups, Interscience Publishers, New York, 1962.

K. Maurin, Generalized Eigenfunction Expansions and Unitary Representations of Topological Groups, Polish Scientific Publishers, Warsaw, 1968.

Furthermore, we recommend:

E. Stein, Harmonic Analysis, Princeton University Press, 1993.

G. Folland, Introduction to Partial Differential Equations, Princeton University Press, 1995.

G. Folland, Fourier Analysis and Its Applications, Amer. Math. Soc., Rhode Island, 2009.

G. Folland, Quantum Field Theory: A Tourist Guide for Mathematicians, Amer. Math. Soc., Providence, Rhode Island, 2008.

V. Varadarajan, Geometry of Quantum Theory, Springer, New York, 2007.

N. Wallach, Symplectic Geometry and Fourier Analysis, Math. Sci. Press, Brookline, Massachusetts, 1977.

M. Naimark, Normed Rings, Noordhoff, Groningen, 1964.

M. Gracia-Bondia, J. Várilly, and H. Figueroa, Elements of Noncommutative Geometry, Birkhäuser, Boston, 2001.

J. Várilly, Lectures on Noncommutative Geometry, European Mathematical Society, 2006.

A. Connes and M. Marcolli, Noncommutative Geometry, Quantum Fields, and Motives, Amer. Math. Soc., Providence, Rhode Island, 2008 (affine group schemes, the Tannakian category in algebraic geometry, and the cosmic renormalization group in quantum field theory, the Riemann *zeta*-Function).

M. Marcolli, Feynman Motives: Renormalization, Algebraic Varieties, and Galois Symmetries, World Scientific, Singapore, 2009.

P. Deligne, Catégories tannakiennes. In: *Grothendieck Festschrift*, Vol. 2, pp. 111–195, Birkhäuser, Basel, 1990 (in French).

Valery Volchkov and Vitaly Volchkov, Harmonic Analysis of Mean Periodic Functions of Symmetric Spaces and the Heisenberg Group, Springer, Berlin, 2009.

The Spirit of Modern Mathematics

As a bridge between classic and modern differential geometry, we recommend:

M. Spivak, A Comprehensive Introduction to Differential Geometry, Vols. 1–5, Publish or Perish, Boston, 1979.

Beautiful relations between algebra, analysis, geometry, number theory, and string theory can be seen by studying elliptic curves. We recommend:

D. Husemoller, Elliptic Curves, Springer, New York, 2002 (elliptic functions, elliptic integrals, theta functions, modular functions, hypergeometric functions, lattices, harmonic analysis on finite fields, Diophantine equations, Galois cohomology, Tate's reduction theory, Fermat's last theorem, Dirichlet's L-series, the Birch and Swinnerton–Dyer conjecture (one of the seven Millenium problems formulated by the Clay Institute in Boston in 2000), complex line bundles and Chern classes, elliptic curves and cryptography).

This book contains many beautiful concrete examples. On the other hand, the study of elliptic curves initiated far-reaching generalizations in the history of mathematics. Therefore, this book serves as an introduction to modern mathematics. In an appendix written by Stefan Theissen, the relations between Calabi–Yau manifolds (higher-dimensional analogues of elliptic curves) and string theory are discussed. As other sources for understanding modern mathematics, we recommend:

K. Maurin, The Riemann Legacy: Riemannian Ideas in Mathematics and Physics of the 20th Century, Kluwer, Dordrecht, 1997.

K. Maurin, Plato's cave parable and the development of modern physics, Rend. Sem. Mat. Univ. Politec. Torino **40** (1982), 1–31.

V. Varadarajan, Euler Through Time: A New Look at Old Themes, Amer. Math. Soc., Providence, Rhode Island, 2006 (e.g., the Langlands program).

À. Lozano-Robledo, Elliptic Curves, Modular forms, and Their L-Functions, Amer. Math. Soc., Providence, Rhode Island, 2011.

S. Chern and F. Hirzebruch (Eds.), Wolf Prize in Mathematics, Vols. 1, 2, World Scientific, Singapore, 2001.

M. Atiyah and D. Iagolnitzer (Eds.), Fields Medallists' Lectures, World Scientific, Singapore, 2003.

M. Monastirsky, Riemann, Topology, and Physics, Birkhäuser, Basel, 1987.

M. Monastirsky, Topology of Gauge Fields and Condensed Matter, Plenum Press, New York, 1993.

M. Monastirsky, Modern Mathematics in the Light of the Fields Medals, Peters, Wellersley, Massachusetts, 1997.

K. Marathe, Topics in Physical Mathematics, Springer, London, 2010.

Introduction to Lie Groups and Lie Algebras

Elementary introduction:

J. Stillwell, Naive Lie Theory, Springer, New York, 2008.

J. Humphreys, Introduction to Lie Algebras and Representation Theory, Springer, Berlin, 1978.

A. Balachandran, S. Jo, and G. Marmo, Group Theory and Hopf Algebras: Lectures for Physicists, World Scientific, Singapore, 2010.

D. Leites (Ed.), Representation Theory, Vol.1: Finite and Compact Groups, Simple Lie Algebras, and an Application, Vol. II: Lie Super Algebras, Abdus Salam School of Mathematical Sciences, Lahore, Pakistan, 2009.

Introduction to the application of Lie group theory in physics:

B. van der Waerden, Group Theory and Quantum Mechanics, Springer, New York, 1974.

S. Sternberg, Group Theory and Physics, Cambridge University Press, 1995.

H. Jones, Groups, Representations, and Physics, Institute of Physics, Bristol, 1998.

Introduction to Lie matrix groups:

B. Hall, Lie Groups, Lie Algebras, and Representations: An Elementary Introduction, Springer, New York, 2003 (emphasizing the Lie groups $SU(2)$ and $SU(3)$ which are basic for the Standard Model in particle physics).

B. Simon, Representations of Finite and Compact Groups, Amer. Math. Soc., Providence, Rhode Island, 1996 (a lot of information on the classical groups).

W. Hein, An Introduction to Structure and Representation of the Classical Groups, Springer, Berlin, 1990 (in German).

Introduction to the general theory of Lie algebras and Lie groups:

A. Kirillov, Jr., An Introduction to Lie Groups and Lie Algebras, Cambridge University Press, 2008.

Introduction to infinite-dimensional groups:

B. Khesin and R. Wendt, The Geometry of Infinite-Dimensional Groups, Springer, Berlin, 2009.

Representations of the rotation group, the Lorentz group, and the Poincaré group:

I. Gelfand, R. Minlos, and Ya. Shapiro, Representations of the Rotation and Lorentz Groups and Their Applications, Pergamon Press, New York, 1963.

M. Naimark, Linear Representations of the Lorentz Group, Macmillan, New York, 1964.

S. Sternberg, Group Theory and Physics, Cambridge University Press, 1995.

A. Barut and R. Rączka, Theory of Group Representations and Applications, World Scientific, Singapore, 1996.

A. Barut (Ed.), Quantum Theory, Groups, Fields, and Particles, Springer, Berlin, 2002.

A. Wightman, Invariance in relativistic quantum mechanics (in French), pp. 159–206. In: Les Houches, Vol. X, Relations de dispersion et particules élémentaires, Wiley, 1960 (survey article).

Y. Ohnuki, Unitary Representations of the Poincaré Group and Relativistic Wave Equations, World Scientific Singapore, 1988.

Y. Kim and M. Noz, Theory and Applications of the Poincaré Group, Reidel, Dordrecht, 1986.

V. Varadarajan, Geometry of Quantum Theory, Springer, New York, 2007.

Comprehensive monograph:

R. Goodman and N. Wallach, Symmetry, Representations, and Invariants, Springer, New York, 2009.

Further References to Lie Groups and Lie Algebras

We recommend:

M. Curtis, Matrix Groups, Springer, New York, 1987.

A. Baker, Matrix Groups: An Introduction to Lie Group Theory, Springer, New York, 2002.

W. Fulton and J. Harris, Representation Theory: A First Course, Springer, Berlin, 1991.

C. Curtis and I. Reiner, Representation Theory of Finite Groups and Associative Algebras, Interscience, New York, 1962.

V. Varadarajan, Lie Groups, Lie Algebras, and Their Representations, Springer, New York, 1984.

J. Serre, Complex Semisimple Lie Algebras, Springer, 1972.

A. Knapp, Representation Theory of Semisimple Groups: An Overview Based on Examples, Princeton University Press, 1986.

A. Knapp, Lie Groups Beyond an Introduction, Birkhäuser, Boston, 2002.

A. Onishnik and E. Vinberg, Lie Groups and Algebraic Groups, Springer, Berlin, 1990.

J. Duistermaat and J. Kolk, Lie Groups, Springer, Berlin, 2000.

R. Carter, Lie Algebras of Finite and Affine Type, Cambridge University Press, 2005.

K. Erdmann and M. Wildon, Introduction to Lie Algebras, Springer, Berlin, 2006.

M. Sepanski, Compact Lie Groups, Springer, Berlin, 2007.

M. Stroppel, Locally Compact Groups, European Mathematical Society, 2006.

S. Lang, $SL(2,\mathbb{R})$, Addison–Wesley, Reading, Massachusetts, 1975.

Bilinear Forms and Lattices

J. Milnor and D. Husemoller, Symmetric Bilinear Forms, Springer, Berlin, 1973.

Theory of Invariants

Classical invariant theory:

R. Weitzenböck, Invariantentheorie, Noordhoff, Groningen, 1923 (in German).

H. Weyl, The Classical Groups: Their Invariants and Representations, Princeton University Press, 1938 (2nd edition with supplement, 1946; 15th printing, 1997).

H. Weyl, Invariants, Duke Math. J. **5** (1939), 489–502 (survey article).

J. Littlewood, The Theory of Group Characters and Matrix Representations of Groups, Amer. Math. Society, Providence, Rhode Island, 2006 (originally published by Clarendon Press, Oxford, 1940).

P. Olver, Applications of Lie Groups to Differential Equations. Springer, New York, 1993.

P. Olver, Classical Invariant Theory, Cambridge University Press, 1999.

H. Kraft and C. Procesi, Classical Invariant Theory: A Primer.
Internet `http://www.math.unibas.ch`

Modern invariant theory: As comprehensive introductions together with many applications, we recommend:

C. Procesi, Lie Groups: An Approach Through Invariants and Representations, Springer, New York, 2007.

R. Goodman and N. Wallach, Symmetry, Representations, and Invariants, Springer, New York, 2009.

Furthermore, we refer to:

I. Macdonald, Symmetric Functions and Hall Polynomials, Oxford University Press, 1995.

B. Sagan, The Symmetric Group: Representations, Combinatorial Algorithms, and Symmetric Functions, Springer, Berlin, 2001.

P. Olver, Equivalence, Invariants, and Symmetry, Cambridge University Press, 1995.

H. Kraft, Geometrical Methods in Invariant Theory, Vieweg, Braunschweig, 1984 (in German).

H. Kraft, P. Slodowy, and T. Springer, Algebraic Transformation Groups and Invariant Theory, Birkhäuser, Basel, 1989.

T. Ceccherini-Silberstein, F. Scarabotti, and F. Tolli, Representation Theory of the Symmetric Groups, Cambridge University Press, 2010.

Applications to quantum field theory (BFFO approach):

B. Fauser, On the Hopf algebraic origin of Wick normal-ordering, J. Phys. A: Math. General **34** (2001), 105–116.

C. Brouder, B. Fauser, A. Frabetti, and R. Oeckl (BFFO), Quantum field theory and Hopf algebra cohomology, J. Phys. A: Math. General **37** (2004), 5895–5927.

B. Fauser, P. Jarvis, R. King, and B. Wybourne, New branchings induced by plethysms, J. Phys. A: Math. General **39** (2006), 2611–2655.

R. Caroll, Fluctuations, Information, Gravity and the Quantum Potential, Kluwer, Dordrecht, 2005.

Special Functions

Many special functions can be understood best by using their symmetry properties:

A. Wawrzyńczyk, Group Representations and Special Functions, Reidel, Dordrecht, 1984.

N. Vilenkin and A. Klimyk, Special Functions and Representations of Lie Groups, Vols. 1–4, Kluwer, Dordrecht, 1991.

Applications to Differential Equations

The solution of differential equations can be simplified by using symmetry properties:

N. Ibragimov, CRC Handbook of Lie Group Analysis of Differential Equations, CRC Press, Boca Raton, Florida, 1993.

L. Ovsiannikov, Group Analysis of Differential Equations, Academic Press, New York, 1982.

H. Stephani, Differential Equations: Their Solution Using Symmetries, Cambridge University Press, 1989.

H. Stephani, Exact Solutions of Einstein's Field Equations, Cambridge University Press, 2003.

P. Olver, Applications of Lie Groups to Differential Equations, Springer, New York, 1993.

V. Fushtshikh and A. Nikitin, Symmetries of the Equations in Quantum Mechanics, Allerton Press, New York, 1994.

F. Finster, N. Kamran, J. Smoller, and S. Yau, Linear waves in the Kerr geometry: a mathematical voyage to black hole physics, Bull. Amer. Math. Soc. **46**(4) (2009), 635–658.

Applications to Geometry

S. Kobayashi, Transformation Groups in Differential Geometry, Springer, Berlin, 1972.

S. Helgason, Differential Geometry, Lie Groups, and Symmetric Spaces, Academic Press, New York, 1978.

J. Wolf, Spaces of Constant Curvature, Publish or Perish, Boston, 1974.

Applications to Physics

As an introduction, we recommend:

S. Sternberg, Group Theory and Physics, Cambridge University Press, 1994.

Furthermore, we recommend the classic monographs:

H. Weyl, The Theory of Groups and Quantum Mechanics, Dover, New York 1931.

B. van der Waerden, Group Theory and Quantum Mechanics, Springer, New York 1974. (German edition: Springer, Berlin, 1932).

G. Ljubarski, The Application of Group Theory in Physics, Pergamon Press, Oxford, 1960.

In addition, we refer to:

L. Fonda and G. Chirardi (1970), Symmetry Principles in Quantum Physics, Marcel Dekker, New York.

M. Hammermesh, Group Theory and its Applications to Physical Problems, Dover, New York, 1989.

J. Fuchs and C. Schweigert, Symmetries, Lie Algebras, and Representations: A Graduate Course for Physicists, Cambridge University Press, 1997.

J. Fuchs, Affine Lie Algebras and Quantum Groups: An Introduction with Applications in Conformal Field Theory, Cambridge University Press, 1992.

H. Jones, Groups, Representations, and Physics, Institute of Physics, Bristol, 1998.

M. Fecko, Differential Geometry and Lie Groups for Physicists, Cambridge University Press, 2006.

W. Neutsch, Coordinates: Theory and Applications, Spektrum, Heidelberg, 1350 pages (in German).

R. Herrman, Lie Groups for Physicists, Benjamin, New York, 1966.

G. Mackey, Induced Representations of Groups and Quantum Mechanics, Benjamin New York, 1968.

G. Mackey, Unitary Group Representations in Physics, Probability, and Number Theory, Benjamin, Reading, Massachusetts. 1978.

M. Mizushima, Quantum Mechanics of Atomic Spectra and Atomic Structure, Benjamin, New York, 1970.

W. Miller, Symmetry Groups and Their Applications, Academic Press, New York, 1972.

R. Gilmore, Lie Groups, Lie Algebras, and Some of Their Applications, Wiley, New York, 1974.

B. Wybourne, The Classical Groups for Physicists, Wiley, New York, 1974.

A. Barut and R. Rączka, Theory of Group Representations and Applications, World Scientific, Singapore, 1986.

W. Greiner and B. Müller, Quantum Mechanics: Symmetries, Springer, New York, 1995.

W. Falter and C. Ludwig, Symmetries in Physics: Group Theory Applied to Physical Problems, Springer, Berlin, 1996.

M. Wagner, Group-Theoretical Methods in Physics, Vieweg, Wiesbaden, 1998 (in German).

U. Mosel, Fields, Symmetries, and Quarks, Springer, Berlin, 1999.

F. Scheck, Quantum Physics, Part II: From Symmetry in Quantum Physics to Electroweak Interactions, Springer, Berlin, 2007.

V. Varadarajan, Geometry of Quantum Theory, Springer, New York, 2007.

B. Khesin and R. Wendt, The Geometry of Infinite-Dimensional Groups, Springer, Berlin, 2009.

Applications to classical mechanics can be found in:

R. Abraham and J. Marsden, Foundations of Mechanics, Addison-Wesley, Reading, Massachusetts, 1978.

J. Marsden and T. Ratiu, Introduction to Mechanics and Symmetry, Springer, New York, 1999.

E. Binz and S. Pods, The Geometry of Heisenberg Groups: With Applications in Signal Theory, Optics, Quantization, and Field Quantization, Amer. Math. Soc., Providence, Rhode Island, 2009.

W. Neutsch, Coordinates (in German), Spektrum, Heidelberg, 1995 (1350 pages).

Concerning the Noether theorem and the energy-momentum tensor, we refer to:

E. Noether, Invariant variational problems, Göttinger Nachrichten, Math.-phys. Klasse 1918, 235–257 (in German).

M. Forger and H. Römer, Currents and the energy-momentum tensor in classical field theory: a fresh look at an old problem, Annals of Physics **309** (2004), 306–389.

Classification of the crystallographic groups:

S. Novikov and A. Fomenko, Basic Elements of Differential Geometry and Topology, Kluwer, Dordrecht, 1987.

Clifford Algebras and Spin Geometry

S. Lang, Algebra, Springer, New York, 2002.

J. Jost, Riemannian Geometry and Geometric Analysis, Springer, Berlin, 2008.

J. Moore, Lectures on Seiberg–Witten Invariants, Springer, Berlin, 1996.

T. Friedrich, Dirac Operators in Riemannian Geometry, Amer. Math. Soc., Providence, Rhode Island, 2000.

M. Atiyah, R. Bott, and A. Shapiro, Clifford modules, Topology **3** (1964), 3–38.

H. Lawson and M. Michelsohn, Spin Geometry, Princeton University Press, 1994.

P. Gilkey, Invariance Theory, the Heat Equation, and the Atiyah–Singer Index Theorem, CRC Press, Boca Raton, Florida, 1995.

P. Gilkey, The spectral geometry of Dirac and Laplace type, pp. 289–326. In: Handbook of Global Analysis. Edited by D. Krupka and D. Saunders, Elsevier, Amsterdam, 2008.

N. Berline, E. Getzler, and M. Vergne, Heat Kernels and Dirac Operators, Springer, New York, 1991.

Applications to quantum field theory:

B. Fauser, A treatise on quantum Clifford algebras, postdoctoral thesis, University of Konstanz (Germany), 2002.
Internet: `http://arxiv.org/math.QA/0202059`

B. Fauser, On an easy transition from operator dynamics to generating functionals by Clifford algebras, J. Math. Phys. **39** (1998), 4928–4947.
Internet: http://arxiv.org/hep-th/9710186

B. Fauser, On the relation of Manin's quantum plane and quantum Clifford algebras, Czechosl. J. Physics **50**(1) (2000), 1221–1228.
Internet: `http://arxiv.org/math.QA/0007137`

B. Fauser, Clifford geometric quantization of inequivalent vacua, Math. Meth. Appl. Sci. **24** (2001), 885–912.
Internet: `http://arxiv.org/hep-th/9719947`

B. Fauser and R. Ablamowicz, Clifford and Grassmann Hopf algebras via the BIGEBRA package for Maple, Computer Physics Communications **170**(2) (2005), 115–130. Internet: `http://arxiv.org/math-ph/0212032`

Riemann Surfaces

The theory of Riemann surfaces combines analysis, algebra, geometry, algebraic geometry, and number theory with each other in a beautiful way. We recommend:

M. Waldschmidt, P. Moussa, J. Luck, and C. Itzykson (Eds.), From Number Theory to Physics, Springer, New York, 1995 (collection of survey articles).

L. Ahlfors, Complex Analysis, McGraw Hill, 1966 (classic textbook).

J. Jost, Compact Riemann Surfaces: An Introduction to Contemporary Mathematics, Springer, Berlin, 1997.

O. Forster, Lectures on Riemann Surfaces, Springer, Berlin, 1981.

R. Narasimhan, Compact Riemann Surfaces. Lectures given at the ETH Zurich, Birkhäuser, Basel, 1997.

M. Farkas and I. Kra, Riemann Surfaces, Springer, New York, 1992.

M. Farkas and I. Kra, Theta Constants, Riemann Surfaces and the Modular Group: An Introduction with Applications to Uniformization Theorems, Partition Identities and Combinatorial Number Theory, Amer. Math. Soc., Providence, Rhode Island, 2001.

Conformal Field Theory and Infinite-Dimensional Lie Algebras

H. Kastrup, On the advancement of conformal transformations and their associated symmetries in geometry and theoretical physics, Ann. Phys. (Berlin) **17** (2008), 631–690.

V. Kac, Infinite-Dimensional Lie Algebras, Cambridge University Press, 1990.

J. Fuchs, Affine Lie Algebras and Quantum Groups: An Introduction with Applications in Conformal Field Theory, Cambridge University Press, 1992.

P. Di Francesco, P. Mathieu, and D. Sénéchal, Conformal Field Theory, Springer, New York, 1997.

Supersymmetry

As an introduction to supersymmetry including supersymmetric Riemann surfaces, we recommend:

J. Jost, Geometry and Physics (functorial approach to supersymmetry), Springer, Berlin, 2009.

The supersymmetric version of the Standard Model in particle physics can be found in:

S. Weinberg, Quantum Field Theory, Vol. III, Cambridge University Press, 2000.

W. Hollik, E. Kraus, M. Roth, C. Rupp, K. Sibold, and D. Stöckinger, Renormalization of the minimal supersymmetric standard model, Nuclear Physics B **639** (2002), 3–65.

D. Bailin and A. Love, Supersymmetric Gauge Field Theory and String Theory, Institute of Physics, Bristol, 1996.

Furthermore, we recommend:

J. Lopuszanski, An Introduction to Symmetry and Supersymmetry in Quantum Field Theory, World Scientific, Singapore, 1991.

M. Chaichian and R. Hagedorn, Symmetries in Quantum Mechanics: From Angular Momentum to Supersymmetry, Institute of Physics, Bristol, 1998.

I. Buchbinder and S. Kuzenko, Ideas and Methods of Supersymmetry and Supergravity or A Walk Through Superspace, Institute of Physics, Bristol, 1995.

A. Khrennikov, Superanalysis, Kluwer, Dordrecht, 1999.

J. Wess and J. Bagger, Supersymmetry and Supergravity, Princeton University Press, 1991.

V. Varadarajan, Supersymmetry for Mathematicians, Courant Lecture Notes, Amer. Math. Soc., Providence, Rhode Island, 2004.

D. Freed, Five Lectures on Supersymmetry, Amer. Math. Soc., Providence, Rhode Island, 1999.

D. Freed, D. Morrison, and I. Singer (Eds.), Quantum Field Theory, Supersymmetry, and Enumerative Geometry, Amer. Math. Soc., Providence, Rhode Island, 2006.

P. Deligne, E. Witten et al. (Eds.), Lectures on Quantum Field Theory: A Course for Mathematicians Given at the Institute for Advanced Study in Princeton, Vols. 1, 2, Amer. Math. Soc., Providence, Rhode Island, 1999.

V. Guillemin and S. Sternberg, Supersymmetry and Equivariant de Rham Theory, Springer, Berlin, 1999.

S. Bellucci, S. Ferrara, and A. Marrani, Supersymmetric Mechanics, Vol. 1: Supersymmetry, Noncommutativity, and Matrix Models, Vol. 2: The Attractor Mechanism and Space Time Singularities, Springer, Berlin, 2006.

P. Binétruy, Supersymmetry: Theory, Experiment, and Cosmology, Oxford University Press, 2006.

P. Srivasta, Supersymmetry, Superfields and Supergravity: An Introduction, Adam Hilger, Bristol, 1985.

V. Cortés (Ed.), Handbook of Pseudo-Riemannian Geometry and Supersymmetry, European Mathematical Society, Zurich, 2010.

Classic Monographs

S. Lie and F. Engel, Theory of Transformation Groups, Vols. 1–3, Teubner, Leipzig, 1888. Reprint: Chelsea Publ. Company, 1970 (foundation of the local theory of Lie groups and Lie algebras) (in German).

C. Chevalley, Theory of Lie Groups, Princeton University Press, 1946 (15th printing, 1999) (foundation of the global theory).

H. Weyl, The Theory of Groups and Quantum Mechanics, Dover, New York, 1931 (German edition: Springer, Berlin, 1929).

H. Weyl, The Classical Groups: Their Invariants and Representations, Princeton University Press, 1938 (2nd edition with supplement, 1946; 15th printing, 1997).

H. Weyl, Symmetry, Princeton University Press, 1952.

E. Wigner, Group Theory and its Applications to the Quantum Mechanics of Atomic Spectra, Academic Press, New York, 1959. (German edition: Springer, Berlin, 1931).

E. Wigner, Symmetries and Reflections, Indiana University Press, Bloomington, 1970.

L. Pontryagin, Topological Groups, Gordon and Breach, 1966 (Russian edition: 1938).

D. Montgomery and L. Zippin, Topological Transformation Groups, Interscience Publishers, New York, 1955.

N. Jacobson, Lie Algebras, Dover, New York, 1962.

V. Bargmann, Representations in Mathematics and Physics, Springer, Berlin, 1970.

F. Warner, Foundations of Differentiable Manifolds and Lie Groups, Scott-Foresman, Glenview, Illinois, 1971.

J. Serre, Linear Representations of Finite Groups, Springer, New York, 1977.

J. Serre, Lie Algebras and Lie Groups, Springer, Berlin, 1992.

G. Hochschild, Basic Theory of Algebraic Groups and Lie Algebras, Springer, New York, 1981.

M. Hammermesh, Group Theory and its Applications to Physical Problems, Dover, New York, 1989.

S. Lang, $SL(2,\mathbb{R})$, Addison–Wesley, Reading, Massachusetts, 1975.

V. Varadarajan, Lie Groups, Lie Algebras, and Their Representations, Springer, New York, 1984.

T. Bröcker and T. tom Dieck, Representation Theory of Compact Lie Groups, Springer, Berlin, 1985.

Locally Compact Groups

E. Wigner, On unitary representations of the inhomogeneous Lorentz group, Ann. of Math. **40** (1939), 149–204.

V. Bargmann, Irreducible representations of the Lorentz group, Ann. of Math. **48** (1947), 568–640.

V. Bargmann, On unitary ray representations of continuous groups, Ann. of Math. **59** (1954), 1–46.

I. Gelfand, R. Minlos, and Ya. Shapiro, Representations of the Rotation and Lorentz Groups and Their Applications, Pergamon Press, New York, 1963.

Y. Ohnuki, Unitary Representations of the Poincaré Group and Relativistic Wave Equations, World Scientific, Singapore, 1987.

A. Knapp, Representation Theory of Semi-Simple Groups, Princeton University Press, 1986.

A. Knapp, Lie Groups, Lie Algebras, and Cohomology, Princeton University Press, 1988.

A. Knapp, Lie Groups Beyond an Introduction, Birkhäuser, Boston, 1996.

M. Stroppel, Locally Compact Groups, European Mathematical Society, Zurich, 2006.

A lot of material can be found in:

N. Bourbaki, Lie Groups and Lie Algebras, Chaps. 1–3, Springer, New York, 1989.

N. Bourbaki, Lie Groups and Lie Algebras, Chaps. 4–6, Springer, New York, 2002.

A. Onishchik et al. (Eds.), Lie Groups and Lie Algebras I–III, Encyclopedia of Mathematical Sciences, Springer, New York, 1993.

Quantum Groups

As an introduction, we recommend:

C. Kassel, M. Rosso, and V. Turaev, Quantum Groups and Knot Invariants, Société Mathématique de France, Paris, 1997.

Furthermore, we recommend:

A. Klimyk and K. Schmüdgen, Quantum Groups and Their Representations, Springer, Berlin, 1997 (many concrete examples).

S. Shnider and S. Sternberg, Quantum Groups. From Coalgebras to Drinfeld Algebras. A Guided Tour, International Press, Boston, 1997.

M. Majid, Foundations of Quantum Group Theory, Cambridge University Press, 1995.

T. Timmermann, An Invitation to Quantum Groups and Duality: From Hopf Algebras to Multiplicative Unitaries and Beyond, European Mathematical Society, 2008 (compact and locally compact quantum groups; approach via operator algebras, generalization of Pontryagin duality).

In addition, we recommend:

Yu. Manin, Topics in Noncommutative Geometry, Princeton University Press, 1991.

S. Woronowicz, Tannaka–Krein duality for compact matrix pseudogroups. Twisted $SU(N)$ groups, Invent. math. **93** (1987), 35–76.

S. Woronowicz, Compact quantum groups. Lectures given at 'Les Houches 1995', pp. 845–884. See the next quotation.

A. Connes, K. Gawędzki, and J. Zinn-Justin (Eds.), Quantum Symmetries, Les Houches, 1995, North-Holland, Amsterdam, 1998.

A. Pressley, Quantum Groups and Lie Theory, Cambridge University Press, 2001.

R. Street, Quantum Groups: A Path to Current Algebra, Cambridge University Press, 2007 (theory of categories).

The theory of quantum groups allows many applications in mathematics and physics:

C. Yang, Some exact results for the many-body problem in one dimension with repulsive delta-function interaction, Phys. Rev. Lett. **19** (1967), 1312–1315.

R. Baxter, Exactly Solved Models in Statistical Mechanics, Academic Press, New York, 1982.

L. Faddeev, Integrable models in 1 + 1-dimensional quantum field theory. Lectures given at 'Les Houches 1982', pp. 561–608. Edited by R. Stora and B. Zuber, North-Holland, Amsterdam, 1984.

L. Faddeev, How the algebraic Bethe ansatz works for integrable models. Lectures given at 'Les Houches 1995', pp. 149–220. Edited by A. Connes et al., North-Holland, Amsterdam, 1998.

V. Turaev, Quantum Invariants of Knots and 3-Manifolds, de Gruyter, Berlin, 1994.

H. de Vega, Integrable Quantum Field Theories and Statistical Models: Yang–Baxter and Kac–Moody Algebras, World Scientific, Singapore, 2000.

Tables

For working with Lie groups and semisimple Lie algebras along with their representations, it is useful to use material summarized in tables. This can be found in:

P. Atkins, M. Child, and C. Philips, Tables for Group Theory, Oxford University Press, 1978.

B. Slansky, Group theory for unified model building, Physics Reports **79**(1) (1981), 1–128.

N. Bourbaki, Lie Groups and Lie Algebras, Chaps. 4–6, Springer, New York, 2002.

R. Gilmore, Lie Groups, Lie Algebras, and Some of Their Applications, Wiley, New York, 1974.

B. Simon, Representations of Finite and Compact Groups, Amer. Math. Soc., Providence, Rhode Island, 1996.

G. Ljubarskij, The Application of Group Theory in Physics, Pergamon Press, Oxford.

L. Frappat, A. Sciarinno, and P. Sorba, Dictionary of Lie Algebras and Super Lie Algebras, Academic Press, New York, 2000.

A. Onishchik, Lectures on Real Semisimple Lie Algebras and Their Representations, European Mathematical Society, 2004.

G. Koster, J. Dimmock, R. Wheeler, and H. Statz, Properties of the Thirty-Two Point Groups, MIT Press, Cambridge, Massachusetts, 1969.

W. Neutsch, Coordinates: Theory and Applications, Spektrum, Heidelberg, 1350 pages (in German).

For realizations of the exceptional Lie algebras and their Lie groups, we refer to Jacobson (1962) quoted on page 545 and to:

N. Jacobson, The Exceptional Lie Algebras, Mimeographed Lecture Notes, Yale University, New Haven, Connecticut, 1957.

J. Adams, Lectures on Exceptional Lie Groups, University Chicago Press, 1996.

Differential Geometry and Gauge Theory

Standard textbooks in modern differential geometry:

S. Kobayashi and K. Nomizu, Foundations of Differential Geometry, Vols. 1, 2, Wiley, New York, 1963.

M. Spivak, A Comprehensive Introduction to Differential Geometry, Vols. 1–5, Publish or Perish, Boston, 1979.

J. Jost, Riemannian Geometry and Geometric Analysis, 5th edition, Springer, Berlin, 2008.

Furthermore, we recommend:

S. Novikov and T. Taimanov, Geometric Structures and Fields, Amer. Math. Soc., Providence, Rhode Island, 2006.

B. Dubrovin, A. Fomenko, and S. Novikov, Modern Geometry: Methods and Applications, Vols. 1–3, Springer, New York, 1992 (including topological methods).

Y. Choquet-Bruhat, C. DeWitt-Morette, and M. Dillard-Bleick, Analysis, Manifolds, and Physics. Vol. 1: Basics; Vol. 2: 92 Applications, Elsevier, Amsterdam, 1996.

T. Frankel, The Geometry of Physics, Cambridge University Press, 2004.

B. Felsager, Geometry, Particles, and Fields, Springer, New York, 1997.

V. Ivancevic and T. Invancevic, Applied Differential Geometry: A Modern Introduction, World Scientific, Singapore, 2007.

E. Bick and F. Steffen (Eds.), Topology and Geometry in Physics, Springer, Berlin, 2005.

Fiber bundles and characteristic classes:

J. Milnor and J. Stasheff, Characteristic Classes, Princeton University Press, 1974.

D. Husemoller, Fibre Bundles, Springer, New York, 1994.

Mathematical approach to gauge theory:

K. Marathe, Topics in Physical Mathematics, Springer, London, 2010.

M. Henneaux and C. Teitelboim, Quantization of Gauge Systems, Princeton University Press, 1993.

G. Naber, Topology, Geometry, and Gauge Fields, Springer, New York, 1997.

Gauge theory, solitons, and the topology of manifolds:

R. Rajaraman, Solitons and Instantons: An Introduction to Solitons and Instantons in Quantum Field Theory, Elsevier, Amsterdam, 1987.

A. Kasmann, Glimpses of Soliton Theory: The Algebra and Geometry of Nonlinear Partial Differential Equations, Amer. Math. Soc., Providence, Rhode Island, 2011.

D. Freed and K. Uhlenbeck, Instantons and Four-Manifolds, Springer, New York, 1984.

Y. Yang, Solitons in Field Theory and Nonlinear Analysis, Springer, New York, 2001.

J. Moore, Lectures on Seiberg–Witten Invariants, Springer, Berlin, 1996.

J. Morgan, The Seiberg–Witten Equations and Applications to the Topology of Four-Manifolds, Princeton University Press, 1996.

S. Donaldson and P. Kronheimer, The Geometry of Four-Manifolds, Oxford University Press, 1990.

P. Kronheimer and T. Mirowka, Monopoles and Three-Manifolds, Cambridge University Press, 2007.

S. Donaldson, Floer Homology Groups, Cambridge University Press, 2002.

M. Atiyah, Collected Works, Vol. V: Gauge Theories, Cambridge University Press, 2004.

C. Yang, Hermann Weyl's contributions to physics, pp. 7–21. In: Hermann Weyl (1885–1955), Springer, Berlin, 1985.

Gauge theory in physics:

C. Taylor (Ed.), Gauge Theories in the Twentieth Century, World Scientific, Singapore, 2001.

M. Monastirsky, Topology of Gauge Fields and Condensed Matter, Plenum Press, New York, 1993.

L. Faddeev and A. Slavnov, Gauge Fields, Benjamin, Reading, Massachusetts, 1980.

A. Das, Lectures on Quantum Field Theory, World Scientific, Singapore, 2008.

M. Böhm, A. Denner, and H. Joos, Gauge Theories of the Strong and Electroweak Interaction, Teubner, Stuttgart, 2001.

T. Kugo, Gauge Field Theory, Springer, Berlin, 1997 (translated from Japanese into German).

Yu. Makeenko, Methods of Contemporary Gauge Theory, Cambridge University Press, 2002.

I. Atchinson and A. Hey, Gauge Theories in Particle Physics, Institute of Physics, Bristol, 1993.

D. Bailin and A. Love, Introduction to Gauge Field Theory, Institute of Physics, Bristol, 1996.

D. Bailin and A. Love, Supersymmetric Gauge Field Theory and String Theory, Institute of Physics, Bristol, 1996.

B. Zwiebach, A First Course in String Theory, Cambridge University Press, 2004.

K. Becker, M. Becker, and J. Schwarz, String Theory and M-Theory, Cambridge University Press, 2006.

S. Hollands, Renormalized Yang–Mills fields in curved spacetime, Rev. Math. Phys. **20**(9) (2008), 1033–1172.
Internet: `http://arxiv.org/0705.3340`

P. Langacker, The Standard Model and Beyond, CRC Press, Boca Raton, Florida, 2010.
Supplementary material: `http://www.sns.ias.edu/ pgl/SMB/`

C. Yang, Selected Papers, 1945–1980, Freeman, New York, 1983.

Concerning the Standard Model in elementary particle physics, see also the references given on page 346.

Number Theory

Yu. Manin and A. Panchishkin, Introduction to Modern Number Theory, Encyclopedia of Mathematical Sciences, Vol. 49, Springer, Berlin, 2005 (survey).

Z. Borevič and I. Šafarevič, Number Theory, Academic Press, New York, 1967.

S. Lang, Algebraic Number Theory, Springer, New York, 1986.

T. Apostol, Introduction to Analytic Number Theory, Springer, New York, 1986.

T. Apostol, Modular Functions and Dirichlet Series in Number Theory, Springer, New York, 1990.

M. Waldschmidt, P. Moussa, J. Luck, and C. Itzykson (Eds.), From Number Theory to Physics, Springer, New York, 1995 (survey articles).

J. Brunier, G. van der Geer, G. Harder, and D. Zagier, The 1-2-3 of Modular Forms. Lectures at a Summer School in Nordfjordeid, Norway, 2008, Springer, Heidelberg, 2009 (survey articles).[46]

K. Kedlaya, p-adic Differential Equations, Cambridge University Press, 2010.

History

The mathematics of the nineteenth century strongly influenced the mathematics of the 20th century and the 21st century. We recommend:

F. Klein, Vorlesungen über die Entwicklung der Mathematik im 19. Jahrhundert, Vols. 1, 2, Springer, Berlin, 1926 (in German). English edition: Development of Mathematics in the 19th Century, with a large appendix by R. Hermann, Math. Sci. Press, New York, 1979.

For the fascinating history of Lie groups, Lie algebras and algebraic groups, we recommend the essays written by Armand Borel (1923–2003) who worked at the Institute for Advanced Study in Princeton:

A. Borel, Essays in the History of Lie Groups and Algebraic Groups, History of Mathematics, Vol. 21, Amer. Math. Soc., Providence, Rhode Island, 2001.

We also refer to the following book review which contains a survey on the history of Lie groups and Lie algebras:

V. Varadarajan, Book review on "Lie Groups: An Approach through Invariants and Representations" by Claudio Procesi, Springer, New York, Bull. Amer. Math. Soc. **45**(4) (2008), 661–674.

Concerning the history of manifolds, we recommend:

[46] In 1842 Sophus Lie was born in Nordfjordeide which is located at the Eidsfjord – a branch of the Nordfjord in Norway. Sophus was the sixth child of a preacher. For a long time, Lie lived in Germany. He was a professor of mathematics at Leipzig University from 1886 until 1898. But all the time he was missing the beauty of his homeland Norway. As a critically ill man, he returned to Norway in 1898 where he died in 1899.

E. Scholz, The concept of manifold, 1850–1950, pp. 25–64. In: I. James (Ed.), History of Topology, Oxford University Press, 1999.

C. Nash, Topology and physics – a historical essay, pp. 359–416. In: I. James (Ed.), History of Topology, Elsevier, Amsterdam, 1999.

E. Scholz (Ed.), Hermann Weyl's 'Space-Time-Matter' and a General Introduction to his Scientific Work, Birkhäuser, Basel, 2001.

J. Lützen (Ed.), The Interaction between Mathematics, Physics, and Philosophy from 1850 to 1940, Kluwer, Dordrecht, 2004.

D. Flament et al. (Eds.), Géométrie au XXième siecle (1930–2000): Histoire et horizons, Hermann, Paris, 2005.

Problems

8.1 *The correct index picture.* Consider the following equations:

- $T^{ij} + S^{ij} = U^{ij},\ \ T^i A_{ij} = T^k_{ir} B^{ri}_{kj},\ \ T^{ij}_{ik} + S^{ij}_{ki} = U^{rj}_{kr},\ \ T^i + S^i = U_i,$
- $\varepsilon_{ijk} v^i w^j \mathbf{e}^k,\ \varepsilon_{ijk} v^i w^j \mathbf{e}_k,\ T^{ij}_{ij} + S^{kr}_{kr} = U^{ab} V_{ba},\ \ A^i_i + B_{kil} C^{kli} = B^j C_j,$
- $B^r A_{rs} = B_k C^{ks},\ g_{ij} v^i v^j,\ g_{ij} v^i v_j.$

Which of these equations do not have the correct index picture (i.e., every additive term has the same free indices)?
Solution: There are precisely five equations which do not have the correct index picture (namely, number four, six, nine, and eleven).

8.2 *Special transformation law (Lie derivative).* Let v^i and w^i be tensorial families. Use an explicit computation in order to show that $v^i \partial_i w^j - w^i \partial_i v^j$ is again a tensorial family.
Solution: We have to show that

$$v^{i'} \partial_{i'} w^{j'} - w^{i'} \partial_{i'} v^{j'} = \frac{\partial x^{j'}}{\partial x^j} \cdot (v^i \partial_i w^j - w^i \partial_i v^j). \tag{8.174}$$

In fact, it follows from $v^{i'} = \frac{\partial x^{i'}}{\partial x^i} \cdot v^i$ and $w^{j'} = \frac{\partial x^{j'}}{\partial x^j} \cdot w^j$ together with the chain rule that

$$v^{i'} \frac{\partial w^{j'}}{\partial x^{i'}} = v^i \frac{\partial x^{i'}}{\partial x^i} \cdot \frac{\partial w^{j'}}{\partial x^{i'}} = v^i \frac{\partial w^{j'}}{\partial x^i} = \frac{\partial^2 x^{j'}}{\partial x^i \partial x^j} v^i w^j + \frac{\partial x^{j'}}{\partial x^j} v^i \frac{\partial w^j}{\partial x^i}.$$

Since $\frac{\partial^2 x^{j'}}{\partial x^i \partial x^j} v^i w^j = \frac{\partial^2 x^{j'}}{\partial x^j \partial x^i}$, we get (8.174).

8.3 *The Lie derivative.* Compute $\mathcal{L}_{\mathbf{v}} T^{ij},\ \mathcal{L}_{\mathbf{v}} T_{jk},$ and $\mathcal{L}_{\mathbf{v}} T^l_{jk}$.
Solution: From $\mathcal{L}_{\mathbf{v}}(S^i T^j) = (\mathcal{L}_{\mathbf{v}} S^i) T^j + S^i \mathcal{L}_{\mathbf{v}} T^j$ we get

$$\begin{aligned}\mathcal{L}_{\mathbf{v}}(S^i T^j) &= (v^s \partial_s S^i) T^j - (S^s \partial_s v^j) T^j + S^i (v^s \partial_s T^j) - S^i T^s \partial_s v^j) \\ &= v^s \partial_s (S^i T^j) - S^s T^j \partial_s v^i - S^i T^s \partial_s v^j.\end{aligned}$$

Hence $\mathcal{L}_{\mathbf{v}} T^{ij} = v^s \partial_s T^{ij} - T^{sj} \partial_s v^i - T^{is} \partial_s v^j$.
Similarly, we get the following expressions:

- $\mathcal{L}_{\mathbf{v}} T_{jk} = v^s \partial_s T_{jk} - T_{sj} \partial_s v^s - T^{js} \partial_k v^s,$

- $\mathcal{L}_{\mathbf{v}} T^l_{jk} = v^s \partial_s T^l jk - T^l_{sk} \partial_j v^s - T^l_{js} \partial_k v^s + T^s_{jk} \partial_s v^l$.

8.4 *The covariant partial derivative.* Compute $\nabla_i T^{jk}$, $\nabla_i T_{jk}$, and $\nabla_i T^l_{jk}$.
Solution: A similar argument as in Problem 8.3 yields

$$\begin{aligned}\nabla_i(S^j T^k) &= (\nabla_i S^j) T^k + S^j (\nabla_i T^k) = (\partial_i S^j + \Gamma^j_{ir} S^r) T^k + S^j (\partial_i T^k + \Gamma^k_{ir} T^r) \\ &= \partial_i (S^j T^k) + \Gamma^j_{ir} S^r T^k + \Gamma^k_{ir} S^j T^r.\end{aligned}$$

Hence $\nabla_i T^{jk} = \partial_i T^{jk} + \Gamma^j_{ir} T^{rk} + \Gamma^k_{ir} T^{jr}$.
Analogously, we get the following:

- $\nabla_i T_{jk} = \partial_i T_{jk} - \Gamma^s_{ij} T_{sk} - \Gamma^s_{ik} T_{js}$,
- $\nabla_i T^l_{jk} = \partial_i T^l_{jk} - \Gamma^s_{ij} T^l_{sk} - \Gamma^s_{ik} T^l_{js} + \Gamma^l_{is} T^s_{jk}$.

8.5 *Cartan's magic formula – the brute force approach.* Use an explicit computation in order to prove

$$\mathcal{L}_{\mathbf{v}} \omega = i_{\mathbf{v}}(d\omega) + d(i_{\mathbf{v}} \omega) \tag{8.175}$$

for the special case where $\omega = \omega_{ij}\, dx^i \wedge dx^j$ with $\omega_{ij} = -\omega_{ji}$.
Solution: Note that

- $\mathcal{L}_{\mathbf{v}} \omega_{ij} = v^s \partial_s \omega_{ij} + \partial_i v^s \cdot \omega_{sj} + \partial_j v^s \cdot \omega_{is}$,
- $\mathcal{L}_{\mathbf{v}} \omega_{ij} = v^s \partial_s \omega_{ij} + \partial_i v^s \cdot \omega_{sj} - \partial_j v^s \cdot \omega_{si}$,
- $\mathcal{L}_{\mathbf{v}} \omega = \mathcal{L}_{\mathbf{v}} \omega_{ij} \cdot dx^i \wedge dx^j = (v^s \partial_s \omega_{ij} + 2\partial_i v^s \cdot \omega_{sj})\, dx^i \wedge dx^j$,
- $i_{\mathbf{v}} \omega = 2 v^s \omega_{sj} dx^j$ (by (8.64)),
- $d(i_{\mathbf{v}} \omega) = (2\partial_i v^s \cdot \omega_{sj} + 2 v^s \partial_i \omega_{sj})\, dx^i \wedge dx^j$

$$= (2\partial_i v^s \cdot \omega_{sj} - v^s \partial_i \omega_{js} + v^s \partial_j \omega_{is})\, dx^i \wedge dx^j,$$

- $d\omega = \partial_{[s} \omega_{ij]}\, dx^s \wedge dx^i \wedge dx^j$,
- $i_{\mathbf{v}}(d\omega) = 3 v^s \partial_{[s} \omega_{ij]} dx^i \wedge dx^j = 3 v^s \partial_{[i} \omega_{js]} dx^i \wedge dx^j$.

Recall that $\partial_{[s} \omega_{ij]}$ denotes the antisymmetrization of $\partial_s \omega_{ij}$. Using antisymmetry, we get the claim (8.175).

8.6 *Cartan's magic formula – the elegant index-free inductive approach.* Use the Leibniz rule (8.97) for the Lie derivative of differential forms on page 491 in order to prove (8.175).
Solution: Let ω be a p-form. Because of (8.97), the proof can be reduced to the special cases where $p = 0$ and $p = 1$, by induction.

- $p = 0$: Use $i_{\mathbf{v}} \omega = 0$.
- $p = 1$: The formula is true for $\omega := dx^k$. In fact, using $dd = 0$ and the commutation relation $d(\mathcal{L}_{\mathbf{v}} \mu) = \mathcal{L}_{\mathbf{v}}(d\mu)$, we get

$$(i_{\mathbf{v}} d + d i_{\mathbf{v}})(d\Theta) = d(i_{\mathbf{v}} d\Theta) = d(\mathcal{L}_{\mathbf{v}} \Theta) = \mathcal{L}_{\mathbf{v}}(d\Theta).$$

Finally, set $\Theta := x^k$.

8.7 *The special case of the Euclidean manifold* $\mathbb{E}^3$. Set $n = 3$ and $g_{ij} := \delta_{ij}$, as well as $x^1 := x, x^2 := y, x^3 := z$. Let U, u, v, w be smooth real-valued functions.[47] Show that:

- $*1 = dx \wedge dy \wedge dz$ and $*(dx \wedge dy \wedge dz) = 1$;
- $*dx = dy \wedge dz$ and $*(dy \wedge dz) = dx$.

---

[47] On the Euclidean manifold $\mathbb{E}^3$, one has not to distinguish between lower and upper indices. For example, $T_k = g_{kl} T^l = \delta_{kl} T^l = T^k$, and so on.

The remaining relations follow by using the cyclic permutation $x \Rightarrow y \Rightarrow z \Rightarrow x$. In particular, $* * \omega = \omega$ for all p-forms ω, $p = 0, 1, 2, 3$. Furthermore, recalling $d^*\omega = (-1)^p * d * \omega$, show that:

- $dU = U_x dx + U_y dy + U_z dz$ and $d^*U = 0$;
- $d(udx + vdy + wdz) = (v_x - u_y) dx \wedge dy + (w_y - v_z)\, dy \wedge dz + (u_z - w_x)\, dz \wedge dx$;
- $d(u\, dy \wedge dz + v\, dz \wedge dx + w\, dx \wedge dy) = (u_x + v_y + w_z)\, dx \wedge dy \wedge dz$;
- $d(U\, dx \wedge dy \wedge dz) = 0$;
- $d^*(udx + vdy + wdz) = -u_x - v_y - w_z$;
- $d^*(u\, dy \wedge dz + v\, dz \wedge dx + w\, dx \wedge dy) = (w_y - v_z)\, dx + (u_z - w_x) dy + (v_x - u_y)\, dz$;
- $d^*(U\, dx \wedge dy \wedge dz) = -U_x\, dy \wedge dz - U_y\, dz \wedge dx - U_z\, dx \wedge dy$.

These operations are closely related to:

- $\mathbf{v} = u\mathbf{i} + v\mathbf{j} + w\mathbf{k}$ and $\operatorname{div} \mathbf{v} = u_x + v_y + w_z$;
- $\mathbf{curl}\, \mathbf{v} = (w_y - v_z)\, \mathbf{i} + (u_z - w_x)\, \mathbf{j} + (v_x - u_y)\, \mathbf{k}$, and $\mathbf{grad}\, U = U_x \mathbf{i} + U_y \mathbf{j} + U_z \mathbf{k}$.

Finally, recalling $\Delta := d^*d + dd^*$, show that:

- $\Delta U = d^*dU = -\operatorname{div} \mathbf{grad}\, U = -U_{xx} - U_{yy} - U_{zz}$;
- $\Delta(udx + vdy + wdz) = \Delta u \cdot dx + \Delta v \cdot dy + \Delta w \cdot dz$;
- $\Delta(u\, dy \wedge dz + v\, dz \wedge dx + w\, dx \wedge dy) = \Delta u \cdot dy \wedge dz + \Delta v \cdot dz \wedge dx + \Delta w \cdot dx \wedge dy$;
- $\Delta(U\, dx \wedge dy \wedge dz) = \Delta U \cdot dx \wedge dy \wedge dz$.

Solution: If $\omega = \frac{1}{3!} \omega_{ijk}\, dx^i \wedge dx^j \wedge dx^k = \omega_{123}\, dx \wedge dy \wedge dz$, then

$$*\omega = \tfrac{1}{3!} \omega_{ijk} \varepsilon^{ijk} = \omega_{123}.$$

Moreover, if $\omega = \frac{1}{2} \omega_{ij}\, dx^i \wedge dx^j$ with $\omega_{ij} = -\omega ji$, then

$$\omega = \omega_{12}\, dx \wedge dy + \omega_{31}\, dz \wedge dy + \omega_{23}\, dy \wedge dz.$$

Hence

$$*\omega = \tfrac{1}{2} \varepsilon_{ijk} \omega^{ij} \cdot dx^k = \omega_{23}\, dx + \omega_{31}\, dy + \omega_{12}\, dz.$$

Moreover, if $\omega = udx + vdy + wdz$, then

$$d\omega = du \wedge dx + dv \wedge dy + dw \wedge dz.$$

Since $du = u_x dx + u_y dy + u_z dz$, we get $du \wedge dx = u_y dy \wedge dx + u_z dz \wedge dx$. In addition, $dy \wedge dx = -dx \wedge dy$.

Concerning ΔU, see the next problem in a more general setting.

8.8 *The Hodge codifferential and the Hodge Laplacian in n-dimensional Cartesian coordinates.* Choose a tensorial family g_{ij} with $g_{ij} := \delta_{ij}$, $i, j = 1, \ldots, n$ for a fixed observer O. Let $\omega = \frac{1}{p!}\, \omega_{i_1 \ldots i_p}\, dx^{i_1} \wedge \cdots \wedge dx^{i_p}$ where $\omega_{i_1 \ldots i_p}$ is an antisymmetric tensorial family. Show that, for the observer O, we get

(i) $d^*\omega = -\frac{1}{(p-1)!}\, \delta^{ij} \partial_j \omega_{i i_2 \ldots i_p}\, dx^{i_2} \wedge \cdots \wedge dx^{i_p}$, and

(ii) $\Delta\omega = -\frac{1}{p!}\, \delta^{ij} \partial_i \partial_j \omega_{i_1 \ldots i_p}\, dx^{i_1} \wedge \cdots \wedge dx^{i_p}$.

Solution: Ad (i). Consider the special case $\omega = \omega_i dx^i$ where $p = 1$. Then:

- $*\omega = \frac{1}{(n-1)!}\, \varepsilon_{i i_2 \ldots i_n} \omega^i \cdot dx^{i_2} \wedge \cdots \wedge dx^{i_n}$;
- $d(*\omega) = \frac{1}{(n-1)!}\, \varepsilon_{i i_2 \ldots i_n}\, \partial_k \omega^i \cdot dx^k \wedge dx^{i_2} \wedge \cdots \wedge dx^{i_n} = \partial_i \omega^i\, dx^1 \wedge dx^2 \wedge \cdots \wedge dx^n$;
- $d^*\omega = -*d(*\omega) = -\partial^i \omega_i$.

Argue similarly if $p = 2, \ldots, n$.

Ad (ii). Again let $\omega = \omega_i dx^i$. Then:

- $d^*\omega = -\partial_i \omega^i$;
- $dd^*\omega = -\partial_k \partial^i \omega_i \, dx^k$;
- $d\omega = \partial_k \omega_i \, dx^k \wedge dx^i = \partial_{[k}\omega_{i]} \, dx^k \wedge dx^i$;
- $d^* d\omega = -2\partial^k \partial_{[k}\omega_{i]} \, dx^i = (-\partial^k \partial_k \omega_i + \partial^k \partial_i \omega_k) \, dx^i$;
- $(dd^* + d^*d)\omega = -\partial^k \partial_k \omega_i \cdot dx^i$.

Argue similarly if $p = 2, \ldots, n$.

8.9 *Covariant partial derivative and Cartan derivative.* Prove Prop. 8.21 on page 499. Solution: By definition of the covariant partial derivative,

$$\nabla_i \omega_{i_1 \ldots i_p} = \partial_i \omega_{i_1 \ldots i_p} - \sum_{\sigma=1}^{p} \Gamma^s_{ii_\sigma} \omega_{i_1 \ldots i_{\sigma-1} s i_{\sigma+1} \ldots i_p}.$$

Antisymmetrization yields

$$\nabla_{[i} \omega_{i_1 \ldots i_p]} = \partial_{[i} \omega_{i_1 \ldots i_p]} - \mathrm{Alt}_{ii_1 \ldots i_p} \sum_{\sigma=1}^{p} \Gamma^s_{ii_\sigma} \omega_{i_1 \ldots i_{\sigma-1} s i_{\sigma+1} \ldots i_p}.$$

Interchanging the indices i and i_σ, the sign changes. Hence

$$\nabla_{[i} \omega_{i_1 \ldots i_p]} = \partial_{[i} \omega_{i_1 \ldots i_p]} + \mathrm{Alt}_{ii_1 \ldots i_p} \sum_{\sigma=1}^{p} \Gamma^s_{i_\sigma i} \omega_{i_1 \ldots i_{\sigma-1} s i_{\sigma+1} \ldots i_p}.$$

Summing up, we get

$$2\nabla_{[i} \omega_{i_1 \ldots i_p]} = 2\partial_{[i} \omega_{i_1 \ldots i_p]} - \mathrm{Alt}_{ii_1 \ldots i_p} \sum_{\sigma=1}^{p} T^s_{ii_\sigma} \omega_{i_1 \ldots i_{\sigma-1} s i_{\sigma+1} \ldots i_p},$$

by using $\Gamma^s_{ii_\sigma} - \Gamma^s_{i_\sigma i} = T^s_{ii_\sigma}$.

8.10 *Proof of the determinant identity* (8.167) *on page* 524. Solution: Consider the special case where $n = 2$. By the Leibniz rule, we have the partial derivative

$$\frac{\partial}{\partial x^k} \begin{vmatrix} g_{11} & g_{12} \\ g_{21} & g_{22} \end{vmatrix} = \begin{vmatrix} \partial_k g_{11} & \partial_k g_{12} \\ g_{21} & g_{22} \end{vmatrix} + \begin{vmatrix} g_{11} & g_{12} \\ \partial_k g_{21} & \partial_k g_{22} \end{vmatrix}.$$

By the Laplace expansion formula, this is equal to

$$\partial_k g_{11} \cdot \mathcal{A}_{11} + \partial_k g_{12} \cdot \mathcal{A}_{12} + \partial_k g_{21} \cdot \mathcal{A}_{21} + \partial_{22} \cdot \mathcal{A}_{22}$$

where $\mathcal{A}_{ij}$ denotes the adjunct of the determinant g to the element g_{ij}. By (1.14) on page 77, $\mathcal{A}_{ij} = gg^{ji}$. Since $g_{ij} = g_{ji}$, we also have $g^{ij} = g^{ji}$. Summarizing,

$$\partial_k g = gg^{ij} \partial_k g_{ij}.$$

This is the claim if $n = 2$. In the general case where $n = 3, 4, \ldots$, the proof proceeds analogously.

8.11 *The Hodge codifferential in terms of the covariant partial derivative.* Prove (8.157) on page 519. Hint: See Choquet–Bruhat et al., Analysis, Manifolds, and Physics, page 317, Vol. 1, Elsevier, Amsterdam, 1996.

8.12 *Summary of important identities for differential forms.* Let ω, μ, ν be differential forms of degree $p, q, r = 0, 1, \ldots$, respectively, and let α, β be real numbers. Prove some of the following formulas:

- $(\omega \wedge \mu) \wedge \nu = \omega \wedge (\mu \wedge \nu)$ (associative law);
- $\omega \wedge \mu = (-1)^{pq} \mu \wedge \omega$ (supercommutative law);[48]
- If $q = r$, then $\omega \wedge (\alpha\mu + \beta\nu) = \alpha\omega \wedge \mu + \beta\omega \wedge \nu$ (distributive law);
- $d(\omega \wedge \mu) = d\omega \wedge \mu + (-1)^p \omega \wedge d\mu$ (graded Leibniz rule);
- $d(d\omega) = 0$ (Poincaré's cohomology rule);
- $d^*(d^*\omega) = 0$ (Hodge's homology rule);
- $\mathcal{L}_{\mathbf{v}}(d\omega) = d(\mathcal{L}_v \omega)$;
- $\mathcal{L}_{\mathbf{v}}\omega = i_{\mathbf{v}}(d\omega) + d(i_{\mathbf{v}}\omega)$ (Cartan's magic formula);
- $d\omega(\mathbf{v}, \mathbf{w}) = \mathcal{L}_{\mathbf{v}}(\omega(\mathbf{w})) - \mathcal{L}_{\mathbf{w}}(\omega(\mathbf{v})) - \omega([\mathbf{v}, \mathbf{w}])$ (special Cartan–Lie formula);
- For $p = 2, 3, \ldots, n$, the general Lie–Cartan formula reads as follows:

$$d\omega(\mathbf{v}_0, \mathbf{v}_1, \ldots, \mathbf{v}_p) = \sum_{i=0}^{p} (-1)^i \mathcal{L}_{\mathbf{v}_i} \omega(\mathbf{v}_0, \ldots, \hat{\mathbf{v}}_i, \ldots, \mathbf{v}_p) + \sum_{i<j} (-1)^{i+j} \omega([\mathbf{v}_i, \mathbf{v}_j], \mathbf{v}_0, \ldots, \hat{\mathbf{v}}_i, \ldots, \hat{\mathbf{v}}_j, \ldots \mathbf{v}_p).$$

By convention, the terms $\hat{\mathbf{v}}_i$ and $\hat{\mathbf{v}}_j$ have to be cancelled. Here, we define $[\mathbf{v}, \mathbf{w}] := \mathcal{L}_{\mathbf{v}}\mathbf{w}$ (Lie bracket). Explicitly, $\mathbf{v} = v^i \partial_i, \mathbf{w} = w^j \partial_j$, and

$$[\mathbf{v}, \mathbf{w}] = v^i \partial_i w^j - w^i \partial_i v^j.$$

- $* * \omega = (-1)^{p(n-p)} \operatorname{sgn}(g) \cdot \omega$ (Hodge star operator);
- $\omega \wedge *\mu = (\omega|\mu) \cdot v$ (where $p = q$, and v is the volume form);
- $(*\omega| * \mu) = (\omega|\mu)$ and $\langle *\omega| * \mu\rangle = \langle\omega|\mu\rangle$ (where $p = q$);
- $d^*\omega = (-1)^p *^{-1} d * \omega = (-1)^{n(p+1)+1} \operatorname{sgn}(g) \; * d * \omega$ (Hodge codifferential);
- $\langle d\omega|\mu\rangle = \langle\omega|d^*\mu\rangle$ and $\langle\Delta\omega|\mu\rangle = \langle\omega|\Delta\mu\rangle$ $(p = q)$;
- $d(f^*\omega) = f^*(d\omega)$ (pull-back; f is a smooth map – see page 476);
- $\int_{f^*\Omega} f^*\omega = \int_\Omega \omega$ (pull-back; f is a diffeomorphism);
- $f^*(\mathcal{L}_{\mathbf{v}}\omega) = \mathcal{L}_{f^*\mathbf{v}}(f^*\omega)$ (f is a diffeomorphism);
- $f^*(i_{\mathbf{v}}\omega) = i_{f^*\mathbf{v}}(f^*\omega)$ (f is a diffeomorphism);
- $i_{\mathbf{v}}(i_{\mathbf{v}}\omega) = 0$ and $i_{\mathbf{v}}(\omega \wedge \mu) = (i_{\mathbf{v}}\omega) \wedge \mu + (-1)^p \omega \wedge i_{\mathbf{v}}\mu$ (contraction product);
- $\mathcal{L}_{f \cdot \mathbf{v}}\omega = f \cdot \mathcal{L}_{\mathbf{v}}\omega + df \wedge i_{\mathbf{v}}\omega$ (f is a smooth map);
- $\mathcal{L}_{\mathbf{v}}(\omega \wedge \mu) = (\mathcal{L}_{\mathbf{v}}\omega) \wedge \mu + \omega \wedge \mathcal{L}_{\mathbf{v}}\mu$ (Leibniz rule);
- $\mathcal{L}_{\mathbf{v}}(d\omega) = d(\mathcal{L}_{\mathbf{v}}\omega)$ and $\mathcal{L}_{\mathbf{v}}(i_{\mathbf{v}}\omega) = i_{\mathbf{v}}(\mathcal{L}_{\mathbf{v}}\omega)$;
- $\mathcal{L}_{[\mathbf{v},\mathbf{w}]}\omega = \mathcal{L}_{\mathbf{v}}(\mathcal{L}_{\mathbf{w}}\omega) - \mathcal{L}_{\mathbf{w}}(\mathcal{L}_{\mathbf{v}})\omega$;
- $f_*([\mathbf{v}, \mathbf{w}]) = [f_*\mathbf{v}, f_*\mathbf{w}]$ (f is a diffeomorphism);[49]
- $(f \circ g)_*\mathbf{v} = f_*(g_*\mathbf{v})$ (f is a diffeomorphism);
- $i_{[\mathbf{v},\mathbf{w}]}\omega = \mathcal{L}_{\mathbf{v}}(i_{\mathbf{w}}\omega) - i_{\mathbf{w}}(\mathcal{L}_{\mathbf{v}}\omega)$;
- $*(\Delta\omega) = \Delta(*\omega)$;
- $d(\Delta\omega) = \Delta(d\omega)$ and $d^*(\Delta\omega) = \Delta(d^*\omega)$.

---

[48] If $p = 0$, then $\omega \wedge \mu = \mu \wedge \omega := \omega\mu$. Note that in this special case, ω is a function.

[49] The definition of $f_*\mathbf{v}$ (push-forward) and $f^*\mathbf{v}$ (pull-back) of a velocity vector field $\mathbf{v}$ can be found on pages 661 and 662, respectively.

Hint: Consider first the simple cases where we have p-differential forms with the special values $p = 1, 2$. Use explicit, completely elementary computations. This will motivate the proofs in the general case. A lot of material can be found in H. Flanders, Differential Forms with Applications to Physical Sciences, Academic Press, New York, 1989, Y. Choquet-Bruhat et al., Analysis, Manifolds, and Physics, Vols. 1, 2, Elsevier, Amsterdam, 1996, and in T. Frankel, The Geometry of Physics, Cambridge University Press, 2004.

8.13 *The universal Kähler interior differential calculus.* Study the very detailed paper by E. Kähler, The interior differential calculus, Rend. Mat. Appl. **21** (5), 425–523 (in German). See also E. Kähler, Mathematical Works, pp. 483–595, de Gruyter, Berlin, 2004.

9. Applications of Invariant Theory to the Rotation Group

Geometry has to be independent of the choice of the observer.
Folklore

9.1 The Method of Orthonormal Frames on the Euclidean Manifold

We want to use the method of orthonormal frames in order to define

- the gradient $\mathbf{grad}\,\Theta$ of a smooth temperature field Θ, and
- both the divergence, $\operatorname{div}\mathbf{v}$, and the curl, $\mathbf{curl}\,\mathbf{v}$, of a smooth velocity vector field $\mathbf{v}$ on the Euclidean manifold $\mathbb{E}^3$.

The physical meaning of $\mathbf{grad}\,\Theta$, $\operatorname{div}\mathbf{v}$, and $\mathbf{curl}\,\mathbf{v}$ will be discussed in Sect. 9.1.4.

Einstein's summation convention. In this chapter, we sum over equal upper and lower indices from 1 to 3. For example, $x^i\mathbf{e}_i = \sum_{i=1}^3 x^i\mathbf{e}_i$.

9.1.1 Hamilton's Quaternionic Analysis

Consider a fixed right-handed Cartesian (x, y, z)-coordinate system of the Euclidean manifold $\mathbb{E}^3$ with the right-handed orthonormal basis $\mathbf{i}, \mathbf{j}, \mathbf{k}$ at the origin P_0. Let $\mathbf{i}_P, \mathbf{j}_P, \mathbf{k}_P$ be a right-handed orthonormal basis of the tangent space $T_P\mathbb{E}^3$ at the point P, which is obtained from the basis vectors at the origin $\mathbf{i}, \mathbf{j}, \mathbf{k}$ by translation (Fig. 9.1). In about 1850, Hamilton (1805–1865) introduced the differential operator

$$\mathcal{D} := \frac{\partial}{\partial t} + \frac{\partial}{\partial x}\mathbf{i}_P + \frac{\partial}{\partial y}\mathbf{j}_P + \frac{\partial}{\partial z}\mathbf{k}_P$$

and applied it to the quaternionic function

$$Q(t, x, y, z) := \Theta(t, x, y, z) + u(t, x, y, z)\mathbf{i}_P + v(t, x, y, z)\mathbf{j}_P + w(t, x, y, z)\mathbf{k}_P.$$

The point P has the Cartesian coordinates (x, y, z). To simplify notation, we replace $\mathbf{i}_P, \mathbf{j}_P, \mathbf{k}_P$ by $\mathbf{i}, \mathbf{j}, \mathbf{k}$, respectively. Furthermore, we set

- $\boldsymbol{\nabla} := \frac{\partial}{\partial x}\mathbf{i} + \frac{\partial}{\partial y}\mathbf{j} + \frac{\partial}{\partial x}\mathbf{k}$ (Hamilton's nabla operator), and
- $\mathbf{v}(P) := u(P)\mathbf{i} + v(P)\mathbf{j} + w(P)\mathbf{k}$.

Finally, since the symbol ∇_i denotes the covariant partial derivative in modern tensor analysis, we replace the vector symbol $\boldsymbol{\nabla}$ by

$$\boldsymbol{\partial} := \frac{\partial}{\partial x}\mathbf{i} + \frac{\partial}{\partial y}\mathbf{j} + \frac{\partial}{\partial z}\mathbf{k}.$$

E. Zeidler, *Quantum Field Theory III: Gauge Theory*,
DOI 10.1007/978-3-642-22421-8_10,

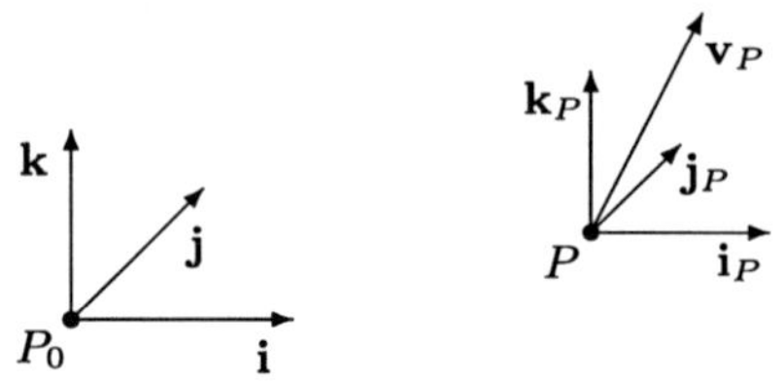

Fig. 9.1. Orthonormal basis of the tangent space $T_P\mathbb{E}^3$

Setting $\partial_t := \frac{\partial}{\partial t}$, we get

$$\mathcal{D} = \partial_t + \boldsymbol{\partial} \qquad \text{and} \qquad Q(t,P) = \Theta(t,P) + \mathbf{v}(t,P).$$

Hamilton investigated the quaternionic product

$$\mathcal{D} \cdot Q = (\partial_t + \boldsymbol{\partial}) \cdot (\Theta + \mathbf{v}) = \partial_t\Theta + \partial_t\mathbf{v} + \boldsymbol{\partial}\Theta - \boldsymbol{\partial}\mathbf{v} + \boldsymbol{\partial} \times \mathbf{v}.$$

This way, we get

- $\boldsymbol{\partial}\Theta = \mathbf{grad}\,\Theta := \frac{\partial\Theta}{\partial x}\mathbf{i} + \frac{\partial\Theta}{\partial y}\mathbf{j} + \frac{\partial\Theta}{\partial z}\mathbf{k}$ (gradient of the temperature field Θ),
- $\boldsymbol{\partial}\mathbf{v} = \operatorname{div}\mathbf{v} := \frac{\partial u}{\partial x} + \frac{\partial v}{\partial y} + \frac{\partial w}{\partial z}$ (divergence of the velocity vector field $\mathbf{v}$),
- $\boldsymbol{\partial} \times \mathbf{v} = \mathbf{curl}\,\mathbf{v}$ (curl of the velocity vector field $\mathbf{v}$). Explicitly,

$$\mathbf{curl}\,\mathbf{v} := \begin{vmatrix} \mathbf{i} & \mathbf{j} & \mathbf{k} \\ \frac{\partial}{\partial x} & \frac{\partial}{\partial y} & \frac{\partial}{\partial z} \\ u & v & w \end{vmatrix}. \tag{9.1}$$

Hence

$$\mathbf{curl}\,\mathbf{v} = \left(\frac{\partial w}{\partial y} - \frac{\partial v}{\partial z}\right)\mathbf{i} + \left(\frac{\partial u}{\partial z} - \frac{\partial w}{\partial x}\right)\mathbf{j} + \left(\frac{\partial v}{\partial x} - \frac{\partial u}{\partial y}\right)\mathbf{k}.$$

- $(\mathbf{v}\boldsymbol{\partial})\Theta = (\mathbf{v}\,\mathbf{grad})\Theta := \left(u\frac{\partial}{\partial x} + v\frac{\partial}{\partial y} + w\frac{\partial}{\partial z}\right)\Theta$. Here, $(\mathbf{v}(P)\,\mathbf{grad})\Theta(P)$ is called the directional derivative of the temperature field Θ at the point P in direction of the velocity vector $\mathbf{v}(P)$ at the point P.
- $(\mathbf{v}\boldsymbol{\partial})\mathbf{E} := (\mathbf{v}\,\mathbf{grad})\mathbf{E} = \left(u\frac{\partial}{\partial x} + v\frac{\partial}{\partial y} + w\frac{\partial}{\partial z}\right)\mathbf{E}$. Here, $(\mathbf{v}(P)\,\mathbf{grad})\mathbf{E}(P)$ is called the directional derivative of the electric field $\mathbf{E}$ at the point P in direction of the velocity vector $\mathbf{v}(P)$ at the point P.
- $\Delta\Theta = -\boldsymbol{\partial}^2\Theta := -\left(\frac{\partial^2}{\partial^2 x} + \frac{\partial^2}{\partial^2 y} + \frac{\partial^2}{\partial^2 z}\right)\Theta$ (Laplacian Δ applied to the temperature field Θ).[1]
- $\Delta\mathbf{E} = -\boldsymbol{\partial}^2 \cdot \mathbf{E} := -\left(\frac{\partial^2}{\partial^2 x} + \frac{\partial^2}{\partial^2 y} + \frac{\partial^2}{\partial^2 z}\right)\mathbf{E}$.

The definitions of

$$\mathbf{grad}\,\Theta, \;\operatorname{div}\mathbf{v}, \;\mathbf{curl}\,\mathbf{v}, \;(\mathbf{v}\,\mathbf{grad})\Theta, \;(\mathbf{v}\,\mathbf{grad})\mathbf{E}, \;\Delta\Theta, \;\Delta\mathbf{E}$$

given above depend on the choice of the right-handed Cartesian (x,y,z)-coordinate system.

[1] Concerning our sign convention for the Laplacian, see page 471.

However, we will show below that the definitions are indeed independent of the choice of the right-handed Cartesian coordinate system.

To this end, we will use the method of orthonormal frames which is the prototype for the use of invariant theory in geometry and analysis. The idea of this method is to define quantities for a fixed right-handed Cartesian coordinate system. Then we show next that the quantity under consideration is independent of the choice of the right-handed Cartesian coordinate system. To this end, we set

$$x^1 := x,\ x^2 := y,\ x^3 := z,\quad \partial_i := \frac{\partial}{\partial x^i} \qquad \text{and} \qquad \mathbf{e}_1 := \mathbf{i},\ \mathbf{e}_2 := \mathbf{j},\ \mathbf{e}_3 := \mathbf{k}.$$

9.1.2 Transformation of Orthonormal Frames

To begin with, let us study the change of orthonormal systems. Let $\mathbf{e}_1, \mathbf{e}_2, \mathbf{e}_3$ be a right-handed orthonormal system in the Euclidean Hilbert space E_3. Furthermore choose three arbitrary vectors $\mathbf{e}_{1'}, \mathbf{e}_{2'}, \mathbf{e}_{3'}$ in E_3 such that

$$\begin{pmatrix} \mathbf{e}_{1'} \\ \mathbf{e}_{2'} \\ \mathbf{e}_{3'} \end{pmatrix} = G \begin{pmatrix} \mathbf{e}_1 \\ \mathbf{e}_2 \\ \mathbf{e}_3 \end{pmatrix} \tag{9.2}$$

where G is an invertible real (3×3)-matrix.

Proposition 9.1 *The transformed vectors $\mathbf{e}_{1'}, \mathbf{e}_{2'}, \mathbf{e}_{3'}$ form a right-handed orthonormal basis in the Euclidean space E_3 iff the transformation matrix G is an element of the Lie group $SO(3)$, that is, $GG^d = I$ and* $\det G = 1$.

Proof. (I) Let $\mathbf{e}_{1'}, \mathbf{e}_{2'}, \mathbf{e}_{3'}$ be a right-handed orthonormal system. Then, we have the orthonormality condition,

$$\mathbf{e}_{i'}\mathbf{e}_{j'} = \delta_{i'j'}, \qquad i, j = 1, 2, 3,$$

and the volume product satisfies the relation $(\mathbf{e}_{1'}\mathbf{e}_{2'}\mathbf{e}_{3'}) = 1$ because of the right-handed orientation. Hence

$$\begin{pmatrix} \mathbf{e}_{1'} \\ \mathbf{e}_{2'} \\ \mathbf{e}_{3'} \end{pmatrix} \begin{pmatrix} \mathbf{e}_{1'}, \mathbf{e}_{2'}, \mathbf{e}_{3'} \end{pmatrix} = \begin{pmatrix} \mathbf{e}_{1'}\mathbf{e}_{1'} & \mathbf{e}_{1'}\mathbf{e}_{2'} & \mathbf{e}_{1'}\mathbf{e}_{3'} \\ \mathbf{e}_{2'}\mathbf{e}_{1'} & \mathbf{e}_{2'}\mathbf{e}_{2'} & \mathbf{e}_{2'}\mathbf{e}_{3'} \\ \mathbf{e}_{3'}\mathbf{e}_{1'} & \mathbf{e}_{3'}\mathbf{e}_{2'} & \mathbf{e}_{3'}\mathbf{e}_{3'}. \end{pmatrix} = \begin{pmatrix} 1 & 0 & 0 \\ 0 & 1 & 0 \\ 0 & 0 & 1 \end{pmatrix} = I.$$

This is equal to

$$G \begin{pmatrix} \mathbf{e}_1 \\ \mathbf{e}_2 \\ \mathbf{e}_3 \end{pmatrix} \begin{pmatrix} \mathbf{e}_1, \mathbf{e}_2, \mathbf{e}_3 \end{pmatrix} G^d = GIG^d = GG^d.$$

Hence $GG^d = I$. Finally, by (9.2), we get

$$1 = (\mathbf{e}_{1'}\mathbf{e}_{2'}\mathbf{e}_{3'}) = \det G \cdot (\mathbf{e}_1\mathbf{e}_2\mathbf{e}_3) = \det G.$$

(II) Conversely, if $GG^d = I$ and $\det G = 1$, then the same argument shows that $\mathbf{e}_{1'}, \mathbf{e}_{2'}, \mathbf{e}_{3'}$ is a right-handed orthonormal system. □

Corollary 9.2 *The vectors* $\mathbf{e}_{1'}, \mathbf{e}_{2'}, \mathbf{e}_{3'}$ *form a left-handed orthonormal basis in the Euclidean space* E_3 *iff the transformation matrix* G *is an element of the Lie group* $O(3)$ *(that is,* $GG^d = I$*) with* $\det G = -1$.

Proof. Note that $(\mathbf{e}_{1'}\mathbf{e}_{2'}\mathbf{e}_{3'}) = -1$ if $\mathbf{e}_{1'}, \mathbf{e}_{2'}, \mathbf{e}_{3'}$ is a left-handed orthonormal basis. □

Set $\mathbf{x} = x^{i'}\mathbf{e}_{i'}$. Here, $x^{1'}, x^{2'}, x^{3'}$ are the coordinates of the position vector $\mathbf{x}$ with respect to the basis $\mathbf{e}_{1'}, \mathbf{e}_{2'}, \mathbf{e}_{3'}$. By (2.84) on page 164, it follows from $x^{i'}\mathbf{e}_{i'} = x^i\mathbf{e}_i$ that

$$\begin{pmatrix} x^{1'} \\ x^{2'} \\ x^{3'} \end{pmatrix} = (G^{-1})^d \begin{pmatrix} x^1 \\ x^2 \\ x^3 \end{pmatrix}. \tag{9.3}$$

If $\mathbf{e}_{1'}, \mathbf{e}_{2'}, \mathbf{e}_{3'}$ is an orthonormal basis, then $(G^{-1})^d = G$. This implies the following specific property of orthonormal frames (without taking orientation into account).

Proposition 9.3 *Under a change of orthonormal frames, the three basis vectors* $\mathbf{e}_1, \mathbf{e}_2, \mathbf{e}_3$ *and the corresponding Cartesian coordinates* x^1, x^2, x^3 *transform themselves in the same way.*

9.1.3 The Coordinate-Dependent Approach ($SO(3)$-Tensor Calculus)

We are now able to prove the main result of Hamilton's vector analysis.

Theorem 9.4 *The definitions of* $\mathbf{grad}\,\Theta$, $\operatorname{div}\mathbf{v}$, $\mathbf{curl}\,\mathbf{v}$, $(\mathbf{v}\,\mathbf{grad})\Theta$, $(\mathbf{v}\,\mathbf{grad})\mathbf{E}$, $\Delta\Theta$, *and* $\Delta\mathbf{E}$ *do not depend on the choice of the right-handed Cartesian coordinate system.*

Proof. The passage from a right-handed Cartesian coordinate system to another right-handed Cartesian coordinate system corresponds to an $SO(3)$-transformation. Therefore, we will use the $SO(3)$-tensor calculus introduced on page 453. In particular, we have the form-invariant tensorial families

$$\delta_{ij}, \quad \delta^{ij}, \quad \delta^i_j, \quad \varepsilon^{ijk}, \quad \varepsilon_{ijk}. \tag{9.4}$$

The basis vectors $\mathbf{e}_i$ transform like a tensorial family. Lifting and lowering of indices can be performed by means of δ^{ij} and δ_{ij}. For example, $\mathbf{e}^i := \delta^{ij}\mathbf{e}_j$. Furthermore, since the transformation formula for the coordinates x^i is given by a matrix which does not depend on the position of the point on the Euclidean manifold $\mathbb{E}^3$, the differential operator ∂_i sends tensorial families again to tensorial families. Note that

- $\mathbf{v} := v^i\mathbf{e}_i$, $\mathbf{E} = E^i\mathbf{e}_i$,
- $\mathbf{grad}\,\Theta = \partial_i\Theta \cdot \mathbf{e}^i$, $\quad \operatorname{div}\mathbf{v} = \partial_i v^i$, $\quad \mathbf{curl}\,\mathbf{v} = \varepsilon^{ijk}\partial_i v_j \cdot \mathbf{e}_k$,
- $(\mathbf{v}\,\mathbf{grad})\Theta = v^i\partial_i\Theta$, $\quad (\mathbf{v}\,\mathbf{grad})\mathbf{E} = (v^i\partial_i)E^j\mathbf{e}_j$,
- $\Delta\Theta = -\delta^{ij}\partial_i\partial_j\Theta = -\partial^j\partial_j\Theta,$, $\quad \Delta\mathbf{E} = -(\delta^{ij}\partial_i\partial_j)E^k\mathbf{e}_k$.

(a) div **v** (b) **curl v**

Fig. 9.2. Measuring velocity vector fields

All the expressions do not have any free indices. Thus, the claim follows immediately from the principle of index killing. □

If we allow the use of both right-handed and left-handed Cartesian coordinate systems, then we have to pass to the $O(3)$-tensor calculus. Let us assign to right-handed (resp. left-handed) coordinate systems the orientation number $\iota = 1$ (resp. $\iota = -1$). Then we have to use the $O(3)$-tensorial families

$$\delta_{ij}, \ \delta^{ij}, \ \delta^i_j, \ \iota \cdot \varepsilon^{ijk}, \ \iota \cdot \varepsilon_{ijk}, \ \mathbf{e}_i, \ x^i.$$

In particular, we write

$$\mathbf{curl}\,\mathbf{v} = \iota \cdot \varepsilon^{ijk}\partial_i v_j \, \mathbf{e}_k.$$

All the other expressions considered above remain unchanged. In addition, for the vector product we get

$$\mathbf{v} \times \mathbf{w} = \iota \cdot \varepsilon_{ijk} v^i w^j \mathbf{e}^k.$$

Examples. Let a be a real number, and let $\mathbf{a}, \boldsymbol{\omega}$ be fixed vectors. Furthermore, let $\mathbf{x} := x\mathbf{i} + y\mathbf{j} + z\mathbf{k}$, as well as $r := |\mathbf{x}| = \sqrt{x^2 + y^2 + z^2}$. Then:

- $\mathbf{grad}(\mathbf{ax}) = \mathbf{a}$,
- $\mathbf{grad}\, U(r) = U'(r)\frac{\mathbf{x}}{r}$,
- $\operatorname{div}\left(\frac{a}{3}\mathbf{x}\right) = a$,
- $\operatorname{div}(U(r)\mathbf{x}) = 3U(r) + rU'(r)$,
- $\mathbf{curl}(\frac{1}{2}\boldsymbol{\omega} \times \mathbf{x}) = \boldsymbol{\omega}$.

9.1.4 The Coordinate-Free Approach

The physical interpretation of the temperature gradient grad Θ. This will be discussed in Sect. 10.1 on page 645. Roughly speaking, the vector $\mathbf{grad}\,\Theta(P)$ points to the direction of the maximal growth of the temperature Θ at the point P, and the length of the vector $\mathbf{grad}\,\Theta(P)$ measures the maximal growth rate of the temperature Θ at the point P.

The physical interpretation of div **v and curl v.** Let **v** be a smooth velocity vector field defined in an open neighborhood of the point P_0 in the Euclidean manifold $\mathbb{E}^3$. So far, we have defined div **v** and **curl v** by using a right-handed Cartesian coordinate system. It follows from tensor analysis that this definition does not depend on the choice of the right-handed Cartesian coordinate system. It is also possible to determine div **v** and **curl v** in an invariant way by the following limits (Fig. 9.2).

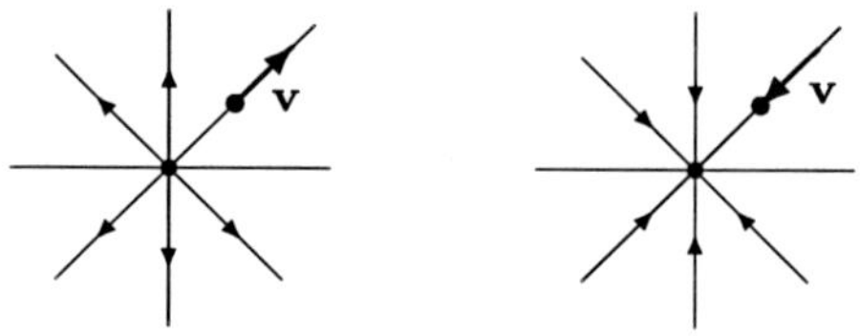

(a) source (b) sink

Fig. 9.3. Special velocity vector fields

Theorem 9.5 *Consider a ball of radius R about the point P_0. Contracting the ball to the point P_0, we get*

$$\operatorname{div}\mathbf{v}(P_0) = \lim_{R\to 0}\frac{3}{4\pi R^3}\int_{\mathbb{S}^2_R(P_0)} \mathbf{v}\mathbf{n}\, dS.$$

Here, $\mathbf{n}$ denotes the outer unit normal vector on the boundary of the ball. Similarly, consider a disk of radius R about the point P_0 which is perpendicular to the unit vector $\mathbf{n}$. Contracting the disk to the point P_0, we get

$$\mathbf{n}\,\mathbf{curl}\,\mathbf{v}(P_0) = \lim_{R\to 0}\frac{1}{\pi R^2}\int_{\mathbb{S}^1_R(P_0)} \mathbf{v}d\mathbf{x}.$$

Proof. By the mean theorem for integrals,

$$\int_{|\mathbf{x}-\mathbf{x}_0|\le R} \operatorname{div}\mathbf{v}\; dxdydz = \frac{4\pi R^3}{3}\operatorname{div}\mathbf{v}(P_1)$$

where P_1 is a suitable point of the ball of radius R about the point P_0. The Gauss–Ostrogradsky integral theorem on page 680 tells us that

$$\frac{3}{4\pi R^3}\int_{|\mathbf{x}-\mathbf{x}_0|\le R} \operatorname{div}\mathbf{v}\; dxdydz = \frac{3}{4\pi R^3}\int_{\mathbb{S}^2_R(P_0)} \mathbf{n}\mathbf{v}\; dS.$$

Letting $R \to 0$, we get $\operatorname{div}\mathbf{v}(P_0)$. Similarly, we obtain $\mathbf{n}\,\mathbf{curl}\,\mathbf{v}(P_0)$ by using the Stokes integral theorem on page 680:

$$\int_{\mathbb{B}^2_R(P_0)} \mathbf{n}\,\mathbf{curl}\,\mathbf{v}\; dS = \int_{\mathbb{S}^1_R(P_0)} \mathbf{v}d\mathbf{x}.$$

□

Examples. Choose the origin, $P_0 := O$. Consider the smooth map $P \mapsto \mathbf{v}_P$. In terms of physics, this is a smooth velocity vector field on the Euclidean manifold $\mathbb{E}^3$. By definition, the streamline $t \mapsto \mathbf{x}(t)$ passing through the point P_0 at time t_0 is given by the solution of the differential equation

$$\dot{\mathbf{x}}(t) = \mathbf{v}(\mathbf{x}(t)), \quad t \in J, \quad \mathbf{x}(t_0) = \mathbf{x}_0, \tag{9.5}$$

where J is an open interval on the real line which contains the point t_0.
Let us consider the prototypes of velocity vector fields.

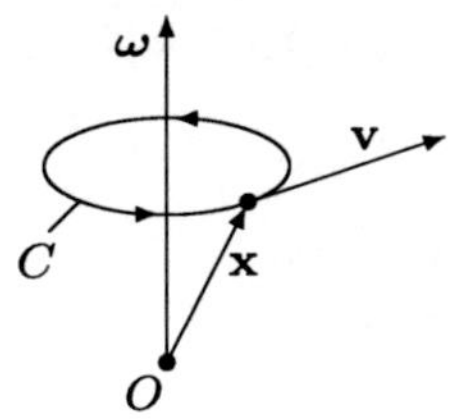

Fig. 9.4. Rotational velocity vector field

- Source at the origin (Fig. 9.3(a)): Choose the velocity vector field $\mathbf{v}(\mathbf{x}) := \frac{a}{3}\mathbf{x}$ with $a > 0$. Then
$$\frac{3}{4\pi R^3}\int_{\mathbb{S}^2_R(O)} \mathbf{vn}\, dS = \frac{3}{4\pi R^3}\cdot\frac{aR}{3}\int_{\mathbb{S}^2_R(O)} dS = a.$$
Letting $R \to 0$, we get $\operatorname{div}\mathbf{v}(O) = a$, by Theorem 9.5. The origin is a source for the streamlines of the velocity vector field, and $\operatorname{div}\mathbf{v}(O)$ measures the strength of this source.
- Sink at the origin (Fig. 9.3(b)): Let $a < 0$. Again we get $\operatorname{div}\mathbf{v}(O) = a$. In this case, the origin is a sink for the streamlines of the velocity vector field.
- Circulation around the z-axis (Fig. 9.4): Let us choose a right-handed Cartesian (x, y, z)-coordinate system with the right-handed orthonormal basis $\mathbf{i}, \mathbf{j}, \mathbf{k}$ at the origin O. Let $\boldsymbol{\omega} := \omega\mathbf{k}$ with $\omega > 0$. Consider the velocity vector field
$$\mathbf{v}(\mathbf{x}) := \tfrac{1}{2}(\boldsymbol{\omega}\times\mathbf{x}).$$
This corresponds to the counter-clockwise rotation of fluid particles about the z-axis with the angular velocity ω. The streamlines are circles parallel to the (x, y)-plane centered at points of the z-axis.

Since the velocity vectors are tangent vectors to the streamlines, we get
$$\frac{1}{\pi R^2}\int_{\mathbb{S}^1_R(O)} \mathbf{v}d\mathbf{x} = \frac{1}{\pi R^2}\cdot\frac{\omega R^2}{2}\int_{\mathbb{S}^1_R(O)} ds = \omega.$$
Letting $R \to 0$, we get $\mathbf{k}\,\mathbf{curl}\,\mathbf{v}(O) = \omega$, by Theorem 9.5. Thus, the z-component of the vector $\mathbf{curl}\,\mathbf{v}(O)$ measures the angular velocity of the fluid particles near the origin.

9.1.5 Hamilton's Nabla Calculus

To begin with, let us summarize the key relations in classical vector calculus. Let $\Theta, \Upsilon : \mathbb{R}^3 \to \mathbb{R}$ and $\mathbf{v}, \mathbf{w} : \mathbb{R}^3 \to E_3$ be smooth temperature functions and smooth velocity vector fields, respectively.

Proposition 9.6 *The following hold:*

(i) $\mathbf{curl}\,\mathbf{grad}\,\Theta = 0$,

(ii) $\operatorname{div}\mathbf{curl}\,\mathbf{v} = 0$,

(iii) $\mathbf{grad}(\Theta + \Upsilon) = \mathbf{grad}\,\Theta + \mathbf{grad}\,\Upsilon$,

(iv) $\mathbf{grad}(\Theta\Upsilon) = (\mathbf{grad}\,\Theta)\Upsilon + \Theta\,\mathbf{grad}\,\Upsilon$,

(v) $\mathbf{grad}(\mathbf{vw}) = (\mathbf{v}\,\mathbf{grad})\mathbf{w} + (\mathbf{w}\,\mathbf{grad})\mathbf{v} + \mathbf{v} \times \mathbf{curl}\,\mathbf{w} + \mathbf{w} \times \mathbf{curl}\,\mathbf{v}$,
(vi) $\mathrm{div}(\mathbf{v} + \mathbf{w}) = \mathrm{div}\,\mathbf{v} + \mathrm{div}\,\mathbf{w}$,
(vii) $\mathrm{div}(\Theta\mathbf{v}) = \mathbf{v}(\mathbf{grad}\,\Theta) + \Theta\,\mathrm{div}\,\mathbf{v}$,
(viii) $\mathrm{div}(\mathbf{v} \times \mathbf{w}) = \mathbf{w}\,\mathbf{curl}\,\mathbf{v} - \mathbf{v}\,\mathbf{curl}\,\mathbf{w}$,
(ix) $\mathbf{curl}(\mathbf{v} + \mathbf{w}) = \mathbf{curl}\,\mathbf{v} + \mathbf{curl}\,\mathbf{w}$,
(x) $\mathbf{curl}(\Theta\mathbf{v}) = (\mathbf{grad}\,\Theta) \times \mathbf{v} + \Theta\,\mathbf{curl}\,\mathbf{v}$,
(xi) $\mathbf{curl}(\mathbf{v} \times \mathbf{w}) = (\mathbf{w}\,\mathbf{grad})\mathbf{v} - (\mathbf{v}\,\mathbf{grad})\mathbf{w} + \mathbf{v}\,\mathrm{div}\,\mathbf{w} - \mathbf{w}\,\mathrm{div}\,\mathbf{v}$,
(xii) $\Delta\Theta = -\,\mathrm{div}\,\mathbf{grad}\,\Theta$,
(xiii) $\Delta\mathbf{v} = \mathbf{curl}\,\mathbf{curl}\,\mathbf{v} - \mathbf{grad}\,\mathrm{div}\,\mathbf{v}$,
(xiv) $2(\mathbf{v}\,\mathbf{grad})\mathbf{w}$ *is equal to*

$$\mathbf{grad}(\mathbf{vw}) + \mathbf{v}\,\mathrm{div}\,\mathbf{w} - \mathbf{w}\,\mathrm{div}\,\mathbf{v} - \mathbf{curl}(\mathbf{v} \times \mathbf{w}) - \mathbf{v} \times \mathbf{curl}\,\mathbf{w} - \mathbf{w} \times \mathbf{curl}\,\mathbf{v}.$$

(xv) $\mathbf{v}(\mathbf{x} + \mathbf{h}) = \mathbf{v}(\mathbf{x}) + (\mathbf{h}\,\mathbf{grad})\mathbf{v}(\mathbf{x}) + o(|\mathbf{h}|), \quad \mathbf{h} \to 0$ *(Taylor expansion).*

The relations (xii)–(xiv) show that $\Delta\Theta$, $\Delta\mathbf{v}$ and $(\mathbf{v}\,\mathbf{grad})\mathbf{w}$ can be reduced to '**grad**', 'div', and '**curl**'. All the relations (i)–(xiv) above can be verified by straightforward computations using a right-handed Cartesian coordinate system. However, the nabla calculus works more effectively. In this connection, we take into account that the nabla operator $\boldsymbol{\partial} = \mathbf{i}\frac{\partial}{\partial x} + \mathbf{j}\frac{\partial}{\partial y} + \mathbf{k}\frac{\partial}{\partial z}$ is both a differential operator and a vector. Therefore, mnemonically, we will proceed as follows:

- Step 1: Apply the Leibniz product rule by decorating the terms with dots.
- Step 2: Use algebraic vector operations in order to move all the dotted (resp. undotted) terms to the right (resp. left) of the nabla operator $\boldsymbol{\partial}$.

Proof. Ad (i), (ii). It follows from $\mathbf{a} \times \Theta\mathbf{a} = 0$ and $\mathbf{a}(\mathbf{a} \times \mathbf{b}) = 0$ that

$$\boldsymbol{\partial} \times \boldsymbol{\partial}\Theta = 0 \quad \text{and} \quad \boldsymbol{\partial}(\boldsymbol{\partial} \times \mathbf{v}) = 0.$$

Hence $\mathbf{curl}\,\mathbf{grad}\,\Theta = 0$ and $\mathrm{div}\,\mathbf{curl}\,\mathbf{v} = 0$.

Ad (iv). By the Leibniz product rule,

$$\boldsymbol{\partial}(\Theta\Upsilon) = \boldsymbol{\partial}(\dot{\Theta}\Upsilon) + \boldsymbol{\partial}(\Theta\dot{\Upsilon}).$$

Moving the undotted quantities to the left of the nabla operator, we get

$$\boldsymbol{\partial}(\Theta\Upsilon) = \Upsilon(\boldsymbol{\partial}\dot{\Theta}) + \Theta(\boldsymbol{\partial}\dot{\Upsilon}).$$

Hence $\mathbf{grad}(\Theta\Upsilon) = \Upsilon\,\mathbf{grad}\,\Theta + \Theta\,\mathbf{grad}\,\Upsilon$.

Ad (xi). By the Leibniz rule,

$$\boldsymbol{\partial} \times (\mathbf{v} \times \mathbf{w}) = \boldsymbol{\partial} \times (\dot{\mathbf{v}} \times \mathbf{w}) + \boldsymbol{\partial} \times (\mathbf{v} \times \dot{\mathbf{w}}).$$

Using the Grassmann expansion formula $\mathbf{a} \times (\mathbf{b} \times \mathbf{c}) = \mathbf{b}(\mathbf{ac}) - \mathbf{c}(\mathbf{ab})$, we get

$$\boldsymbol{\partial} \times (\mathbf{v} \times \mathbf{w}) = \dot{\mathbf{v}}(\boldsymbol{\partial}\mathbf{w}) - \mathbf{w}(\boldsymbol{\partial}\dot{\mathbf{v}}) + \mathbf{v}(\boldsymbol{\partial}\dot{\mathbf{w}}) - \dot{\mathbf{w}}(\boldsymbol{\partial}\mathbf{v}).$$

Finally, moving the undotted terms to the left of the nabla operator $\boldsymbol{\partial}$ by respecting the rules of vector algebra, we get

$$\boldsymbol{\partial} \times (\mathbf{v} \times \mathbf{w}) = (\mathbf{w}\boldsymbol{\partial})\dot{\mathbf{v}} - \mathbf{w}(\boldsymbol{\partial}\dot{\mathbf{v}}) + \mathbf{v}(\boldsymbol{\partial}\dot{\mathbf{w}}) - (\mathbf{v}\boldsymbol{\partial})\dot{\mathbf{w}}.$$

This is the claim (xi).

Ad (v). Use the Grassmann expansion formula $\mathbf{b}(\mathbf{ac}) = \mathbf{a}(\mathbf{bc}) + \mathbf{a} \times (\mathbf{b} \times \mathbf{c})$. The remaining proofs are recommended to the reader as an exercise. □

9.1.6 Rotations and Cauchy's Invariant Functions

Consider a right-handed Cartesian (x, y, z)-coordinate system with the right-handed orthonormal basis $\mathbf{e}_1, \mathbf{e}_2, \mathbf{e}_3$. Let $\mathbf{x}, \mathbf{y}, \mathbf{z} \in E_3$, and let $\mathbf{x} = x^i\mathbf{e}_i$, $\mathbf{y} = y^i\mathbf{e}_i$, and $\mathbf{z} = z^i\mathbf{e}_i$. Then the inner product

$$\mathbf{xy} = \delta_{ij}x^iy^j$$

and the volume product

$$(\mathbf{xyz}) = \varepsilon_{ijk}x^iy^jz^k$$

are invariants under the change of right-handed Cartesian coordinate systems. If we consider the more general case of arbitrary Cartesian (x, y, z)-coordinate systems with an arbitrary orthonormal basis $\mathbf{e}_1, \mathbf{e}_2, \mathbf{e}_3$, then the inner product $\mathbf{xy}$ remains an invariant. However, this is not true anymore for the volume product $(\mathbf{xyz})$ which changes sign under a change of orientation. One of the main results of classic invariant theory tells us that these invariants are the only ones in Euclidean geometry. Let us formulate this in precise terms.

The Cauchy theorem on isotropic functions. The real-valued function $f : E_3 \times \cdots \times E_3 \to \mathbb{R}$ is called isotropic iff

$$\boxed{f(G\mathbf{x}_1, \ldots, G\mathbf{x}_n) = f(\mathbf{x}_1, \ldots, \mathbf{x}_n)} \tag{9.6}$$

for all vectors $\mathbf{x}_1, \ldots, \mathbf{x}_n \in E_3$ and all unitary operators $G \in U(E_3)$. Moreover, the function f is called proper isotropic iff we have the relation (9.6) for all rotations $G \in SU(E_3)$. Note that a function is isotropic iff it is invariant under all rotations and reflections $\mathbf{x} \mapsto -\mathbf{x}$.

Theorem 9.7 (i) *If the function f is isotropic, then it only depends on all the possible inner products*

$$\mathbf{x}_i\mathbf{x}_j, \qquad i, j = 1, \ldots, n. \tag{9.7}$$

(ii) *If the function f is proper isotropic, then it only depends on all the possible inner products* (9.7), *and all the possible volume products* $(\mathbf{x}_i\mathbf{x}_j\mathbf{x}_k), i, j, k = 1, \ldots, n$.

The polynomial ring of invariants. The function $f : E_3 \times \cdots \times E_3 \to \mathbb{R}$ considered above is called a polynomial function iff it is a real polynomial with respect to the Cartesian coordinates of the vectors $\mathbf{x}_1, \ldots, \mathbf{x}_n$. Since the change of Cartesian coordinates is described by linear transformations, this definition does not depend on the choice of the Cartesian coordinate system.

Corollary 9.8 *If the polynomial function f is proper isotropic, then it is a real polynomial of all the possible inner products $\mathbf{x}_i\mathbf{x}_j$, $i, j = 1, \ldots, n$, and all the possible volume products $(\mathbf{x}_i\mathbf{x}_j\mathbf{x}_k), i, j, k = 1, \ldots, n$.*

For the classic proofs of Theorem 9.7 and Corollary 9.8, we refer to the references given in Problem 9.5.

Examples. (a) Every proper isotropic, polynomial function $f : E_3 \to \mathbb{R}$ has the form

$$f(\mathbf{x}) = p(\mathbf{x}^2) \qquad \text{for all} \quad \mathbf{x} \in E_3$$

where p is a polynomial of one variable with real coefficients. Such a function is also isotropic.

(b) Every proper isotropic, polynomial function $f : E_3 \times E_3 \to \mathbb{R}$ has the form

$$f(\mathbf{x},\mathbf{y}) = p(\mathbf{x}^2,\mathbf{y}^2,\mathbf{xy}), \qquad \text{for all} \quad \mathbf{x},\mathbf{y} \in E_3$$

where p is a real polynomial of three variables.[2] Such a function is also isotropic.

(c) Every proper isotropic, polynomial function $f : E_3 \times E_3 \times E_3 \to \mathbb{R}$ has the form

$$f(\mathbf{x},\mathbf{y},\mathbf{z}) = p(\mathbf{x}^2,\mathbf{y}^2,\mathbf{z}^2,\mathbf{xy},\mathbf{xz},\mathbf{yz},(\mathbf{xyz}))$$

for all vectors $\mathbf{x},\mathbf{y},\mathbf{z} \in E_3$. Here, p is a real polynomial of seven variables.

(d) Set $f(\mathbf{x},\mathbf{y},\mathbf{z}) := (\mathbf{xyz})^2$. This polynomial function is isotropic. By Theorem 9.7, we know that f only depends on all the possible inner products of the vectors $\mathbf{x},\mathbf{y},\mathbf{z}$. Explicitly,

$$f(\mathbf{x},\mathbf{y},\mathbf{z}) = \begin{vmatrix} \mathbf{x}^2 & \mathbf{xy} & \mathbf{xz} \\ \mathbf{yx} & \mathbf{y}^2 & \mathbf{yz} \\ \mathbf{zx} & \mathbf{zy} & \mathbf{z}^2 \end{vmatrix}.$$

This is the Gram determinant.

Let $\mathcal{P}(SU(E_3))$ denote the set of all the real polynomials with respect to the variables

$$\mathbf{x}_i\mathbf{x}_j, \quad (\mathbf{x}_i\mathbf{x}_j\mathbf{x}_k), \qquad i,j,k = 1,\ldots,n, \quad n = 1,2,\ldots$$

This set is closed under addition and multiplication, hence it is a commutative ring. The commutative ring $\mathcal{P}(SU(E_3))$ is called the polynomial ring of invariants of the Lie group $SU(E_3)$.

The Rivlin–Ericksen theorem on isotropic, symmetric tensor functions in elasticity theory. Let $L_{\text{sym}}(E_3)$ denote the set of all linear self-adjoint operators

$$A : E_3 \to E_3$$

on the real Hilbert space E_3. The linear operator $T : L_{\text{sym}}(E_3) \to L_{\text{sym}}(E_3)$ is called an isotropic tensor function iff we have

$$R^{-1}T(A)R = T(R^{-1}AR)$$

for all linear operators $A \in L_{\text{sym}}(E_3)$ and all rotations $R \in SU(E_3)$.

Theorem 9.9 *Let T be an isotropic tensor function. Then there exist real functions $a,b,c : \mathbb{R}^3 \to \mathbb{R}$ such that*

$$T(A) = aI + bA + cA^2 \qquad \textit{for all} \quad A \in L_{\text{sym}}(E_3)$$

where $a = a(\operatorname{tr}(A), \operatorname{tr}(A^2), \det A)$ together with analogous expressions for b and c.

Note the following: If $\lambda_1,\lambda_2,\lambda_3$ are the eigenvalues of the operator A, then

$$\operatorname{tr}(A) = \lambda_1 + \lambda_2 + \lambda_3, \quad \operatorname{tr}(A^2) = \lambda_1^2 + \lambda_2^2 + \lambda_3^2, \quad \det(A) = \lambda_1\lambda_2\lambda_3.$$

The proof of Theorem 9.9 together with applications to the formulation of general constitutive laws for elastic material (generalizing the classic Hooke's law) can be found in Zeidler (1986), p. 204, quoted on page 1089.[3]

[2] Note that $(\mathbf{xxy}) = (\mathbf{yyx}) = 0$. Therefore, the volume products disappear.

[3] R. Rivlin and J. Ericksen, Stress-deformation relations for isotropic materials, J. Rat. Mech. Anal. **4** (1955), 681–702.

9.2 Curvilinear Coordinates

Mathematicians and physicists use curvilinear coordinates in order to simplify computations based on symmetry.

Folklore

9.2.1 Local Observers

Let us consider the Euclidean manifold $\mathbb{E}^3$. Fix a right-handed Cartesian (x, y, z)-coordinate system equipped with the right-handed orthonormal basis $\mathbf{i}_P, \mathbf{j}_P, \mathbf{k}_P$ at the point P of $\mathbb{E}^3$. The orthonormal basis at the origin P_0 is denoted by $\mathbf{i}, \mathbf{j}, \mathbf{k}$. As depicted in Fig. 9.1 on page 558, the vectors $\mathbf{i}_P, \mathbf{j}_P, \mathbf{k}_P$ are obtained from $\mathbf{i}, \mathbf{j}, \mathbf{k}$ by parallel transport. The change of coordinates is described by the equation

$$\boxed{x^i = x^i(x, y, z), \qquad i = 1, 2, 3, \quad (x, y, z) \in \mathcal{O}_+.}$$

Suppose that

- the set $\mathcal{O}_+$ (resp. $\mathcal{O}$) is a nonempty, open, arcwise connected subset of $\mathbb{R}^3$, and
- the map $(x, y, z) \mapsto (x^1, x^2, x^3)$ is a diffeomorphism from $\mathcal{O}_+$ onto the subset $\mathcal{O}$ of $\mathbb{R}^3$.

In terms of physics, the observer O_+ (resp. O) measures the coordinates (x, y, z) (resp. (x^1, x^2, x^3)) of the point $P \in \mathbb{E}^3$. We set $x_+^1 := x, x_+^2 := y$, and $x_+^3 := z$.

Typical transformation laws. The following transformation laws are crucial.

(i) Temperature field Θ: The observer O_+ (resp. O) measures the temperature $\Theta(x, y, z)$ (resp. $\Theta(x^1, x^2, x^3)$). By the chain rule,

$$\frac{\partial \Theta(P)}{\partial x^i} = \frac{\partial x_+^i(P)}{\partial x^i} \cdot \frac{\partial \Theta(P)}{\partial x_+^i}.$$

This is the transformation law for the temperature derivatives. The transformation law from the observer O to the observer O' reads as

$$\frac{\partial \Theta(P)}{\partial x^{i'}} = \frac{\partial x^i(P)}{\partial x^{i'}} \cdot \frac{\partial \Theta(P)}{\partial x^i}.$$

This shows that $\partial_i \Theta$ is a tensorial family.

(ii) Velocity components $\dot{x}^i(t)$: Let the parameter t denote time. The observer O_+ (resp. O) measures the curve

$$x = x(t), \quad y = y(t), \quad z = z(t), \qquad i = 1, 2, 3, \quad t \in]-t_0, t_0[$$

(resp. $x^i = x^i(x(t), y(t), z(t))$, $i = 1, 2, 3$). Set $\dot{x}^i(t) := \frac{dx^i(t)}{dt}$. Using the chain rule, differentiation with respect to time t yields

$$\dot{x}^i(t) = \frac{\partial x^i(P(t))}{\partial x_+^i} \cdot \dot{x}_+^i(t), \qquad i = 1, 2, 3.$$

This is the transformation law for the velocity components. The transformation law from the observer O to the observer O' reads as

$$\dot{x}^{i'}(t) = \frac{\partial x^{i'}(P(t))}{\partial x^i} \cdot \dot{x}^i(t), \quad i' = 1', 2', 3'.$$

This shows that $\dot{x}^i(t)$ is a tensorial family.

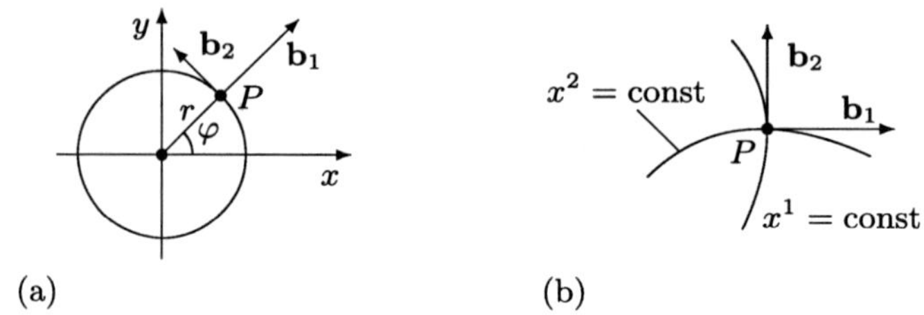

Fig. 9.5. Curvilinear coordinates

The natural frame. Set $\mathbf{b}_1^+(P) := \mathbf{i}_P$, $\mathbf{b}_2^+(P) := \mathbf{j}_P$, $\mathbf{b}_3^+(P) := \mathbf{k}_P$. We define

$$\boxed{\mathbf{b}_i(P) := \frac{\partial x^i(P)}{\partial x_+^i}\,\mathbf{b}_i^+(P), \qquad i = 1,2,3.}$$

The vectors $\mathbf{b}_1(P), \mathbf{b}_2(P), \mathbf{b}_3(P)$ form a basis at the point P. This basis of the tangent space $T_P\mathbb{E}^3$ of the Euclidean manifold $\mathbb{E}^3$ at the point P is called the natural basis of the observer O at the point P. In terms of geometry, the basis vector $\mathbf{b}_1(P)$ is the tangent vector of the curve

$$t \mapsto \big(x(t,x^2,x^3), y(t,x^2,x^3), z(t,x^2,x^3)\big)$$

at the point P. This curve is called the x^1-coordinate line passing through the point P (Fig. 9.5(b)). Similarly, we get the basis vectors $\mathbf{b}_2(P)$ (resp. $\mathbf{b}_3(P)$) as tangent vectors of the x^2-coordinate (resp. x^3-coordinate) line. Passing to another observer O', the chain rule yields the following transformation law from the observer O to the observer O':

$$\mathbf{b}_{i'}(P) = \frac{\partial x^{i'}(P)}{\partial x^i}\,\mathbf{b}_i(P), \qquad i' = 1', 2', 3'.$$

This shows that $\mathbf{b}_i$ is a tensorial family. A vector field on the Euclidean manifold is given by

$$\mathbf{v}(P) = v^i(P)\,\mathbf{b}_i(P).$$

Since $\mathbf{v}(P)$ is an invariant quantity, it follows from the inverse index principle that v^i is a tensorial family.

The natural coframe. Fix the point P, and define

$$dx^i(\mathbf{v}(P)) = v^i(P), \qquad i = 1,2,3.$$

Then, the map $dx^i : T_P\mathbb{E}^3 \to \mathbb{R}$ is a linear functional, and dx^1, dx^2, dx^3 is a basis of the cotangent space $T_P^*\mathbb{E}^3$. The functionals dx^i are transformed like v^i. Thus, dx^i is a tensorial family.

9.2.2 The Metric Tensor

For the observer O, we define

$$\boxed{g_{ij}(P) := \mathbf{b}_i(P)\mathbf{b}_j(P), \qquad i,j = 1,2,3.} \tag{9.8}$$

For the vectors $\mathbf{v}(P) = v^i(P)\mathbf{b}_i(P)$ and $\mathbf{w}(P) = w^i(P)\mathbf{b}_i(P)$, the inner product reads as

$$\mathbf{v}(P)\mathbf{w}(P) = v^i(P)g_{ij}w^j(P).$$

Recall that $g(P) := \det(g_{ij}(P))$. The metric tensor field is given by

$$\boxed{\mathbf{g}_P = g_{ij}(P)\, dx^i \otimes dx^j.}$$

For the observer O_+, we get

$$g^+_{ij}(P) = \mathbf{b}^+_i(P)\mathbf{b}^+_j(P) = \delta_{ij}, \qquad i,j = 1,2,3.$$

Hence

$$\mathbf{g}_P = \delta_{ij} dx^i_+ \otimes dx^j_+ = dx \otimes dx + dy \otimes dy + dz \otimes dz,$$

and $g = \det(\delta_{ij}) = 1$. By the index principle, $\mathbf{g}_P$ does not depend on the choice of the observer.

9.2.3 The Volume Form

Using the metric tensorial family g_{ij}, we are able to introduce the volume form υ of the Euclidean manifold $\mathbb{E}^3$ by setting

$$\boxed{\upsilon(P) := \iota \cdot \mathcal{E}_{ijk}(P)\, dx^i \wedge dx^j \wedge dx^k.}$$

Recall that $\mathcal{E}_{ijk} = \sqrt{g}\ \varepsilon_{ijk}$, and ι denotes the orientation number of the local (x^1, x^2, x^3)-coordinate system. Explicitly,[4]

$$\iota := \operatorname{sgn}\left(\frac{\partial(x^1,x^2,x^3)}{\partial(x,y,z)}(P)\right), \qquad P \in \mathcal{O}_+.$$

Since the set $\mathcal{O}_+$ is arcwise connected, the number ι does not depend on the choice of the point P in $\mathcal{O}_+$. Recall that $\iota \cdot \mathcal{E}_{ijk}$ is a tensorial family (see page 463). Thus, the differential form υ is an invariant, by the index principle. That is, the differential form υ does not depend on the choice of the observer (local coordinates). For the observer O_+, we get

$$\upsilon = dx \wedge dy \wedge dz.$$

9.2.4 Special Coordinates

Let us consider typical examples for curvilinear coordinates, namely, cylindrical coordinates, polar coordinates, and spherical coordinates. We distinguish between

- singular coordinates (i.e., the metric matrix $(g_{ij}(P))$ is not invertible at all the points P, and hence the transformation law is not generated by a diffeomorphism), and
- regular coordinates (i.e., the transformation law is generated by a diffeomorphism).

From the general point of view, one has to use only regular coordinates. However, from the practical point of view, one frequently uses singular coordinates. In a rigorous setting, one has to use regular coordinates combined with a limit process at the singular points of the coordinates (e.g., the North Pole and the South Pole of earth are singular points with respect to spherical coordinates; see (9.12)).

[4] Note that $\frac{\partial(x^1,x^2,x^3)}{\partial(x,y,z)} = \left(\frac{\partial(x,y,z)}{\partial(x^1,x^2,x^3)}\right)^{-1}$. This follows from the fact that the transformation $(x,y,z) \mapsto (x^1,x^2,x^3)$ is a diffeomorphism.

Cylindrical Coordinates

Cylindrical coordinates are used for studying physical systems which are symmetric under rotations about the z-axis.

Singular cylindrical coordinates. The basic transformation law reads as

$$\boxed{x = \varrho\cos\varphi, \quad y = \varrho\sin\varphi, \quad z = z} \tag{9.9}$$

where $-\pi \le \varphi \le \pi$, $\varrho \ge 0$, and $-\infty < z < \infty$ (see Fig. 9.5(a) on page 568). Here, we set $x^1 := \varrho, x^2 := \varphi$, and $x^3 := z$.

Regular cylindrical coordinates. Note that the map

$$(x, y, z) \mapsto (\varrho, \varphi, z) \tag{9.10}$$

is not a diffeomorphism defined on the set $\mathbb{R}^3$. In fact, this map is not bijective, since the point $x = -1, y = z = 0$ has the two angular coordinates $\varphi = \pi$ and $\varphi = -\pi$. To cure this defect, we choose a subset $\mathcal{O}$ of $\mathbb{R}^3$. Explicitly,

$$\mathcal{O} := \mathbb{R}^3 \setminus \{(x, y, z) : x \le 0, y = 0, -\infty < z < \infty\}.$$

This means that we remove a closed half-plane spanned by the negative x-axis and the z-axis. Then the map (9.10) is a diffeomorphism from $\mathcal{O}_+$ onto the set

$$\mathcal{O} := \{(\varrho, \varphi, z) \in \mathbb{R}^3 : \varrho > 0, -\pi < \varphi < \pi, -\infty < z < \infty\}.$$

For the Jacobian, we get

$$\frac{\partial(x, y, z)}{\partial(\varrho, \varphi, z)} = \begin{vmatrix} x_\varrho & x_\varphi & x_z \\ y_\varrho & y_\varphi & y_z \\ z_\varrho & z_\varphi & z_z \end{vmatrix} = \begin{vmatrix} \cos\varphi & -\varrho\sin\varphi & 0 \\ \sin\varphi & \varrho\cos\varphi & 0 \\ 0 & 0 & 1 \end{vmatrix} = \varrho.$$

The sign of the Jacobian equals the orientation number ι. Since $\varrho > 0$, we get

$$\iota = 1$$

for the orientation number of cylindrical coordinates.

The natural frame. Using $\mathbf{x} = x\mathbf{i}_P + y\mathbf{j}_P + z\mathbf{k}_P$, we get

- $\mathbf{b}_1(P) = \mathbf{x}_\varrho = \cos\varphi\,\mathbf{i}_P + \sin\varphi\,\mathbf{j}_P$,
- $\mathbf{b}_2(P) = \mathbf{x}_\varphi = \varrho(-\sin\varphi\,\mathbf{i}_P + \cos\varphi\,\mathbf{j}_P)$,
- $\mathbf{b}_3(P) = \mathbf{x}_z = \mathbf{k}_P$.

The vectors $\mathbf{b}_1(P), \mathbf{b}_2(P), \mathbf{b}_3(P)$ form the natural basis of cylindrical coordinates at the point P (see Fig. 9.5(a) on page 568). The vector $\mathbf{b}_3(P)$ points to the direction of the z-axis. Note the following peculiarity: The basis vectors $\mathbf{b}_1(P), \mathbf{b}_2(P), \mathbf{b}_3(P)$ form an orthogonal system, but they do not form an orthonormal system. For example, the vector $\mathbf{b}_2(P)$ is not a unit vector if $\varrho \neq 1$.

It is not wise, to normalize the natural basis vectors of curvilinear coordinate systems.

In fact, normalization destroys the beauty of the index principle in mathematical physics to be discussed in Sect. 9.3.2 on page 575.

The metric tensor. Setting $g_{ij} := \mathbf{b}_i\mathbf{b}_j$, we obtain

$$\begin{pmatrix} g_{11} & g_{12} & g_{13} \\ g_{21} & g_{22} & g_{23} \\ g_{31} & g_{32} & g_{33} \end{pmatrix} = \begin{pmatrix} 1 & 0 & 0 \\ 0 & \varrho^2 & 0 \\ 0 & 0 & 1 \end{pmatrix}.$$

Thus, $g = \det(g_{ij}) = \varrho^2$. This yields the metric tensor field $\mathbf{g} = g_{ij}dx^i \otimes dx^j$. Hence

$$\mathbf{g} = dx \otimes dx + dy \otimes dy + dz \otimes dz = d\varrho \otimes d\varrho + \varrho^2 d\varphi \otimes d\varphi + dz \otimes dz.$$

Mnemonically,

$$\boxed{ds^2 = dx^2 + dy^2 + dz^2 = d\varrho^2 + \varrho^2 d\varphi^2 + dz^2.}$$

This tells us that the length l of a curve $\varrho = \varrho(t), \varphi = \varphi(t), z = z(t), \quad t_0 \le t \le t_1$, is given by the integral

$$l = \int_{t_0}^{t_1} \frac{ds(t)}{dt}\, dt = \int_{t_0}^{t_1} \sqrt{\dot\varrho(t)^2 + \varrho(t)^2\dot\varphi(t)^2 + \dot z(t)^2}\, dt.$$

The inverse matrix of (g_{ij}) reads as

$$\begin{pmatrix} g^{11} & g^{12} & g^{13} \\ g^{21} & g^{22} & g^{23} \\ g^{31} & g^{32} & g^{33} \end{pmatrix} = \begin{pmatrix} 1 & 0 & 0 \\ 0 & \varrho^{-2} & 0 \\ 0 & 0 & 1 \end{pmatrix}.$$

For the volume form, we get

$$v = dx \wedge dy \wedge dz = \varrho d\varrho \wedge d\varphi \wedge dz.$$

The cylindric set $\mathcal{U} := \{(x, y, z) \in \mathbb{R}^3 : x^2 + y^2 \le R^2, 0 \le z \le h\}$ of radius R and height h has the volume

$$\text{meas}(\mathcal{U}) = \pi R^2 h.$$

Using regular cylindrical coordinates, this is obtained by the limit process

$$\lim_{\varepsilon \to +0} \int_{\mathcal{U}_\varepsilon} v = \lim_{\varepsilon \to +0} \int_{\varrho=0}^{R} \int_{\varphi=-\pi+\varepsilon}^{\varphi=\pi-\varepsilon} \int_{z=0}^{h} \varrho \, d\varrho d\varphi dz = \pi R^2 h.$$

Here, we use the truncated set

$$\mathcal{U}_\varepsilon := \{(\varrho, \varphi, z) \in \mathbb{R}^3 : \varrho > 0, \ -\pi + \varepsilon < \varphi < \pi - \varepsilon, \ 0 \le z \le h\}$$

with respect to cylindrical coordinates. Mnemonically,

$$\boxed{dxdydz = d\varrho \cdot \varrho d\varphi \cdot dz.}$$

Polar Coordinates

Setting $z = 0$, cylindrical coordinates pass over to polar coordinates of the Cartesian (x, y)-plane (see Fig. 9.5(a) on page 568). For example, the metric tensor of the Cartesian (x, y)-plane is given by

$$\mathbf{g} = dx \otimes dx + dy \otimes dy = d\varrho \otimes d\varrho + \varrho^2 d\varphi \otimes d\varphi,$$

and the volume form reads as

$$v = dx \wedge dy = \varrho \, d\varrho \wedge d\varphi.$$

Mnemonically, $ds^2 = dx^2 + dy^2 = d\varrho^2 + \varrho \, d\varphi^2$, and $dxdy = d\varrho \cdot \varrho d\varphi$.

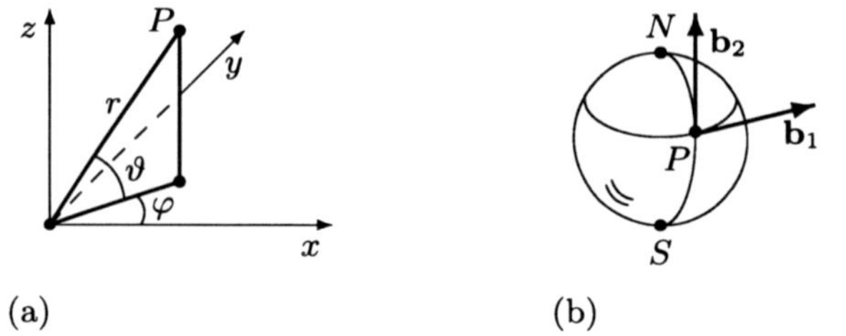

Fig. 9.6. Spherical coordinates

Spherical Coordinates

Spherical coordinates are used for studying physical systems which are symmetric with respect to rotations about the origin.

Singular spherical coordinates. The basic transformation law reads as

$$\boxed{x = r\cos\vartheta\cos\varphi, \qquad y = r\cos\vartheta\sin\varphi, \qquad z = r\sin\vartheta}$$

where $r \geq 0$ and

- $-\pi \leq \varphi \leq \pi$ (geographic length),
- $-\frac{\pi}{2} \leq \vartheta \leq \frac{\pi}{2}$ (geographic latitude).

Our choice of the parameter values is dictated by geography. For fixed radius $r > 0$, we get a sphere (e.g., the surface of earth). Then:

- $\vartheta = 0$ (equator), $\vartheta = \frac{\pi}{2}$ (North Pole), $\vartheta = -\frac{\pi}{2}$ (South Pole).

Moreover, we set $x^1 := \varphi,, x^2 := \vartheta, x^3 := r$ In this singular setting, the North Pole of the earth has the coordinates $r = R$, $\vartheta = \frac{\pi}{2}$ and $-\pi \leq \varphi \leq \pi$. Thus, the map

$$(x, y, z) \mapsto (\varphi, \vartheta, r) \tag{9.11}$$

is not a diffeomorphism defined on the total space $\mathbb{R}^3$.

Regular spherical coordinates. Setting

$$\mathcal{O}_+ = \mathbb{R}^3 \setminus \{(x, y, z) : x \leq 0,\ y = 0,\ -\infty < z < \infty\},$$

the map (9.11) is a diffeomorphism from the truncated space $\mathcal{O}_+$ onto the set

$$\mathcal{O} := \{(\varphi, \vartheta, r) \in \mathbb{R}^3 : -\pi < \varphi < \pi,\ -\frac{\pi}{2} < \vartheta < \frac{\pi}{2},\ r > 0\}.$$

For the Jacobian, we get

$$\frac{\partial(x,y,z)}{\partial(\varphi,\vartheta,r)} = \begin{vmatrix} x_\varphi & x_\vartheta & x_r \\ y_\varphi & y_\vartheta & y_r \\ z_\varphi & z_\vartheta & z_r \end{vmatrix} = \begin{vmatrix} -r\cos\vartheta\ \sin\varphi & -r\sin\vartheta\cos\varphi & \cos\vartheta\cos\varphi \\ r\cos\vartheta\cos\varphi & -r\sin\vartheta\sin\varphi & \cos\vartheta\sin\varphi \\ 0 & r\cos\vartheta & \sin\vartheta \end{vmatrix} = r^2\cos\vartheta.$$

The sign of the Jacobian equals the orientation number ι. Since $r > 0$, we get

$$\iota = 1$$

for the orientation number of spherical coordinates.

Natural frame. Using

$$\mathbf{x} = x\mathbf{i}_P + y\mathbf{j}_P + z\mathbf{k}_P = r\cos\vartheta\cos\varphi\ \mathbf{i}_P + r\cos\vartheta\sin\varphi\ \mathbf{j}_P + r\sin\vartheta\ \mathbf{k}_P,$$

we get

- $\mathbf{b}_1 = \mathbf{x}_\varphi = -r\cos\vartheta\sin\varphi\ \mathbf{i}_P + r\cos\vartheta\cos\varphi\ \mathbf{j}_P$,
- $\mathbf{b}_2 = \mathbf{x}_\vartheta = -r\sin\vartheta\cos\varphi\ \mathbf{i}_P - r\sin\vartheta\sin\varphi\mathbf{j}_P + r\cos\vartheta\ \mathbf{k}_P$.
- $\mathbf{b}_3 = \mathbf{x}_r = \cos\vartheta\cos\varphi\ \mathbf{i}_P + \cos\vartheta\sin\varphi\ \mathbf{j}_P + \sin\vartheta\ \mathbf{k}_P$.

The natural basis vector $\mathbf{b}_1(P)$ at the point is a tangent vector of the latitude circle through the point P (Fig. 9.6(b)). The natural basis vector $\mathbf{b}_2(P)$ at the point P is a tangent vector of the meridian through the point P. Finally, the natural basis vector $\mathbf{b}_3(P)$ at the point P points to the outer radial direction.

Metric tensor. Setting $g_{ij} := \mathbf{b}_i\mathbf{b}_j$, we get

$$\begin{pmatrix} g_{11} & g_{12} & g_{13} \\ g_{21} & g_{22} & g_{23} \\ g_{31} & g_{32} & g_{33} \end{pmatrix} = \begin{pmatrix} r^2\cos^2\vartheta & 0 & 0 \\ 0 & r^2 & 0 \\ 0 & 0 & 1 \end{pmatrix}.$$

This shows that the natural basis vectors $\mathbf{b}_1, \mathbf{b}_j, \mathbf{b}_j$ form an orthogonal system. The metric tensor field $\mathbf{g} = g_{ij}dx^i \otimes dx^j$ reads as

$$\mathbf{g} = dx \otimes dx + dy \otimes dy + dz \otimes dz = r^2\cos^2\vartheta\ d\varphi \otimes d\varphi + r^2 d\vartheta \otimes d\vartheta + dr \otimes dr.$$

Moreover, $g = \det(g_{ij}) = r^4\cos^2\vartheta$. Mnemonically,

$$\boxed{ds^2 = r^2\cos^2\vartheta\ d\varphi^2 + r^2\ d\vartheta^2 + dr^2.}$$

This tells us that the length of a curve $\varphi = \varphi(t), \vartheta = \vartheta(t), r = r(t), t_0 \le t \le t_1$, is given by the integral

$$\int_{t_0}^{t_1} \frac{ds(t)}{dt}\ dt = \int_{t_0}^{t_1} \sqrt{r(t)^2\cos^2\vartheta(t)\cdot\dot\varphi(t)^2 + r(t)^2\dot\vartheta(t)^2 + \dot r(t)^2}\ dt.$$

The inverse matrix of (g_{ij}) reads as

$$\begin{pmatrix} g^{11} & g^{12} & g^{13} \\ g^{21} & g^{22} & g^{23} \\ g^{31} & g^{32} & g^{33} \end{pmatrix} = \begin{pmatrix} (r\cos\vartheta)^{-2} & 0 & 0 \\ 0 & r^{-2} & 0 \\ 0 & 0 & 1 \end{pmatrix}.$$

The volume form. For the volume form, we get $\upsilon = \mathcal{E}_{ijk}\ dx^i \wedge dx^j \wedge dx^k$ with $\mathcal{E}_{ijk} = \sqrt{g}\ \varepsilon_{ijk} = r^2\cos\vartheta\ \varepsilon_{ijk}$. Hence

$$\upsilon = dx \wedge dy \wedge dz = r^2\cos\vartheta\ d\varphi \wedge d\vartheta \wedge dr.$$

For example, the ball $\mathbb{B}^3_R(0)$ of radius $R > 0$ centered at the origin has the volume

$$\text{meas}(\mathbb{B}^3_R(0)) = \frac{4\pi R^3}{3}.$$

Using regular spherical coordinates, this is obtained from the following limit process:

$$\lim_{\varepsilon\to+0}\int_{\mathbb{B}^3_{R,\varepsilon}(0)} \upsilon = \lim_{\varepsilon\to+0}\int_{\varphi=-\pi+\varepsilon}^{\pi-\varepsilon}\int_{\vartheta=-\pi/2}^{\pi/2}\int_{r=0}^{R} r^2\cos^2\vartheta\ d\varphi d\vartheta dr = \frac{4\pi R^3}{3}. \quad (9.12)$$

Mnemonically,

$$\boxed{dxdydz = r^2\cos\vartheta\ d\varphi d\vartheta dr.}$$

Further reading. A lot of material about special coordinates and their various applications in geometry and physics can be found in the monumental monograph by

W. Neutsch, Coordinates: Theory and Applications, Spektrum, Heidelberg, 1350 pages (in German).

9.3 The Index Principle of Mathematical Physics

Replace partial derivatives by covariant partial derivatives, and use only equations which possess the correct index picture.

Golden rule

9.3.1 The Basic Trick

Let us start with the Poisson equation

$$\boxed{-\varepsilon_0(U_{xx}+U_{yy}+U_{zz})=\varrho \qquad \text{on } \mathcal{O}_+} \tag{9.13}$$

with the positive dielectricity constant ε_0. Here, we fix a right-handed Cartesian (x,y,z)-coordinate system on the Euclidean manifold $\mathbb{E}^3$. The functions ϱ and U depend on the variables x,y,z. We are given the function $\varrho:\mathcal{O}_+\to\mathbb{R}$. We are looking for the function $U:\mathcal{O}_+\to\mathbb{R}$. In Maxwell's theory of electrostatics, the electric field $\mathbf{E}$ is given by the equation

$$\mathbf{E}=-\,\mathbf{grad}\,U$$

where U is called the potential of the electric field $\mathbf{E}$. This way, the equation (9.13) passes over to the first Maxwell equation

$$-\varepsilon_0\,\mathrm{div}\,\mathbf{E}=\varrho$$

which tells us the crucial physical fact that the electric charge density ϱ is the source for the electric field $\mathbf{E}$.

Our goal is to transform the given equation (9.13) into arbitrary local coordinates. To this end, we proceed as follows.

Step 1: Use a right-handed (resp. left-handed) Cartesian (x,y,z)-system with the right-handed (resp. left-handed) orthonormal basis $\mathbf{b}_1,\mathbf{b}_2,\mathbf{b}_3$. Write the given equation as an $O(3)$-tensor equation by using the $O(3)$-tensorial families

$$\delta_{ij},\ \delta^{ij},\ \delta^i_j,\ \iota\cdot\varepsilon_{ijk},\ \iota\cdot\varepsilon^{ijk},\ \partial_i,\ \mathbf{b}_i,\ \mathbf{b}^i,\ dx^i$$

and the orientation number ι of the Cartesian coordinate system. In particular, for equation (9.13) we get

$$-\,\varepsilon_0\delta^{ij}\partial_i\partial_j U=\varrho. \tag{9.14}$$

Note that this equation has the correct index picture. Therefore, the equation is valid in every right-handed or left-handed Cartesian coordinate system.

Step 2: Choose a local (curvilinear) (x^1,x^2,x^3)-coordinate system. Write the $O(3)$-tensor equation as a general tensor equation by using the following replacements:

- $\delta_{ij}\Rightarrow g_{ij},\ \delta^{ij}\Rightarrow g^{ij},\ \delta^i_j\Rightarrow\delta^i_j,$
- $\iota\cdot\varepsilon_{ijk}\Rightarrow\iota\cdot\mathcal{E}_{ijk},\ \iota\cdot\varepsilon^{ijk}\Rightarrow\iota\cdot\mathcal{E}^{ijk},$
- $\mathbf{b}_i\Rightarrow\mathbf{b}_i$ (natural basis), $dx^i\Rightarrow dx^i$ (natural cobasis),
- $\mathbf{b}^i\Rightarrow\mathbf{b}^i$ (lifting of indices).[5]

[5] This means $\delta^{is}\mathbf{b}_s\Rightarrow g^{is}\mathbf{b}_s$.

The crucial point is that we replace partial derivatives by covariant partial derivatives, that is,

$$\boxed{\partial_i \Rightarrow \nabla_i, \qquad \partial^i \Rightarrow \nabla^i}$$

where we set $\partial^i := \delta^{is}\partial_s$ and $\nabla^i := g^{is}\nabla_s$ (lifting of indices). This way, the initial $O(3)$-tensor equation passes over to a general tensor equation which is valid in every local (curvilinear) coordinate system. For example, the equation (9.13) passes over to

$$\boxed{-\varepsilon_0 \, g^{ij}\nabla_i\nabla_j U = \varrho.} \tag{9.15}$$

This equation is valid in every local coordinate system. In a right-handed or left-handed Cartesian coordinate system, equation (9.15) coincides with (9.14).

As another example, consider the vector product $\mathbf{v}\times\mathbf{w}$. In a right-handed (resp. left-handed) Cartesian coordinate system, we have

$$\mathbf{v} \times \mathbf{w} = \iota \cdot \varepsilon_{ijk} v^i w^j \mathbf{b}^k$$

where $\mathbf{v} = v^i\mathbf{b}_i$ and $\mathbf{w} = w^i\mathbf{b}_i$. The replacement described above yields

$$\mathbf{v} \times \mathbf{w} = \iota \cdot \mathcal{E}_{ijk} v^i w^j \mathbf{b}^k.$$

This equation is valid in every local coordinate system. Recall that $\mathcal{E}_{ijk} = \sqrt{g}\,\varepsilon_{ijk}$, and $\mathbf{b}^k = g^{ks}\mathbf{b}_s$.

9.3.2 Applications to Vector Analysis

Let us consider the temperature field Θ, the velocity vector field $\mathbf{v} = v^i\mathbf{b}_i$, and the electric field $\mathbf{E} = E^i\mathbf{b}_i$. By page 560, we have the following $O(3)$-tensor equations in Cartesian coordinate systems:

- $\mathbf{grad}\,\Theta = \partial_i\Theta\ \mathbf{b}^i$,
- $\operatorname{div}\mathbf{v} = \partial_i v^i$,
- $\mathbf{curl}\,\mathbf{v} = \iota\cdot\varepsilon^{ijk}\partial_i v_j\ \mathbf{b}_k$,
- $D_\mathbf{v}\Theta = (\mathbf{v}\,\mathbf{grad})\Theta = v^i\partial_i\Theta$,
- $D_\mathbf{v}\mathbf{E} = (v^i\partial_i E^j)\mathbf{b}_j$,
- $D\mathbf{E} = \partial_i E^j\ dx^i \otimes \mathbf{b}_j$,
- $\Delta\mathbf{E} = -(\delta^{ij}\partial_i\partial_j E^k)\mathbf{b}_k$.

Using the replacements described above, we get the following relations which are valid in arbitrary local coordinate systems:

- $\mathbf{grad}\,\Theta = \partial_i\Theta\ \mathbf{b}^i$,[6]
- $\operatorname{div}\mathbf{v} = \nabla_i v^i$,
- $\mathbf{curl}\,\mathbf{v} = \iota\cdot\mathcal{E}^{ijk}\nabla_i v_j\ \mathbf{b}_k$,
- $D_\mathbf{v}\Theta = (\mathbf{v}\,\mathbf{grad})\Theta = v^i\partial_i\Theta$,
- $D_\mathbf{v}\mathbf{E} = (\mathbf{v}\,\mathbf{grad})\mathbf{E} = v^i(\nabla_i E^j)\mathbf{b}_j$,
- $D\mathbf{E} = \nabla_i E^j\ dx^i \otimes \mathbf{b}_j$,
- $\Delta\mathbf{E} = -g^{ij}\nabla_i(\nabla_j E^k)\mathbf{b}_k$,

[6] Note that $\nabla_i\Theta = \partial_i\Theta$ if Θ is a scalar field.

Alternatively, recalling that $g = (g_{ij})$, we obtain

$$\operatorname{div} \mathbf{v} = \nabla_i v^i = \frac{1}{\sqrt{g}}\, \partial_i(\sqrt{g}\, v^i). \tag{9.16}$$

This follows from (8.168) on page 525. Hence

$$\Delta U = -\frac{1}{\sqrt{g}}\, \partial_i(\sqrt{g}\, \partial^i U).$$

In fact, both sides of this equation are tensorial invariants, by (8.168), and they coincide with respect to any Cartesian coordinate system. Thus, they coincide with respect to any local coordinate system. Consequently, the original Poisson equation (9.13) can be written as

$$\boxed{-\frac{\varepsilon_0}{\sqrt{g}}\, \partial_i(\sqrt{g}\, \partial^i U) = \varrho.} \tag{9.17}$$

The point is that this equation is valid in arbitrary local coordinate systems. On the other hand, this invariant equation only contains classical partial derivatives. This underlines the importance of the Weyl derivative in mathematical physics.

In Problem 9.1, we will show how equation (9.17) can be elegantly obtained by using the Dirichlet variational problem about the minimal electrostatic energy. The history of the famous Dirichlet problem is discussed in Vol. I, Sect. 10.4.

9.4 The Euclidean Connection and Gauge Theory

> The goal of geometers is to describe the geometry of the Euclidean manifold (and of more general manifolds) by formulas which are valid in arbitrary local (curvilinear) coordinate systems. Note that simple geometric properties can be hidden by using local coordinates which generate clumsy Christoffel symbols.
>
> Folklore

The geometry of the Euclidean manifold $\mathbb{E}^3$ is trivial, since

- there exists a global parallel transport on $\mathbb{E}^3$, and
- the curvature of $\mathbb{E}^3$ vanishes identically.

Nevertheless, let us formulate this trivial geometry in terms of Cartan's language which can be generalized to the geometry of curved manifolds with respect to a symmetry group in modern differential geometry (realization of Klein's 1872 Erlangen program in differential geometry). This section serves as an intuitive motivation for the general theory which will be considered in Chaps. 13 through 17 (Ariadne's thread in gauge theory).

By definition, the Euclidean connection is the Levi-Civita connection with respect to the metric tensor of the Euclidean manifold $\mathbb{E}^3$. Explicitly, the Christoffel symbols for the observer O with respect to the local coordinates (x^1, x^2, x^3) read as follows:

$$\Gamma^k_{ij} = \tfrac{1}{2} g^{ks} \left(\frac{\partial g_{js}}{\partial x^i} + \frac{\partial g_{is}}{\partial x^j} - \frac{\partial g_{ij}}{\partial x^s} \right), \qquad i, j, k = 1, 2, 3. \tag{9.18}$$

With respect to the observer O_+ (Cartesian (x, y, z)-coordinates), we have $g_{ij} = \delta_{ij}$, and hence the Christoffel symbols vanish identically.

9.4.1 Covariant Partial Derivative

According to Sect. 8.8.3 on page 494, the Christoffel symbols induce the covariant partial derivative. In particular, let $\mathbf{v}(P) = v^j(P)\mathbf{b}_j(P)$ and $\mathbf{w}(P) = w^i(P)\mathbf{b}_i(P)$ be velocity vector fields on the Euclidean manifold $\mathbb{E}^3$. Then we have the tensorial families v^i and w^i at hand. This yields the tensorial family

$$\boxed{\nabla_i v^j = \partial_i v^j + \Gamma^j_{is} v^s, \qquad i, j = 1, 2, 3.}$$

Recall that $\partial_i = \frac{\partial}{\partial x^i}$. In addition, we get the directional derivative[7]

$$\boxed{D_{\mathbf{v}}\mathbf{w} := v^i \nabla_i w^j \mathbf{b}_j.}$$

By the index principle, this definition does not depend on the choice of the observer. For a special observer O_+ using right-handed Cartesian (x, y, z)-coordinates, we get

- $\nabla_i w^j = \partial_i w^j, i, j = 1, 2, 3$, and
- $D_{\mathbf{v}}\mathbf{w} = (\mathbf{v}\,\mathbf{grad})\mathbf{w} = v^i \partial_i w^j \mathbf{b}_j^+$;[8]
- $\partial_i \mathbf{b}_j^+ \equiv 0$ for all indices.

9.4.2 Curves of Least Kinectic Energy (Affine Geodesics)

> In Euclidean geometry, straight lines can be characterized by both the principle of least kinetic energy and the principle of minimal length. This is the paradigm for the fundamental principle of least action in physics. In terms of mathematics, this includes the theory of geodesics on Riemannian and pseudo-Riemannian manifolds.
>
> Folklore

The principle of least kinetic energy. Let $-\infty < t_0 < t_1 < \infty$. Consider the motion $\mathbf{x} = \mathbf{x}(t), t \in [t_0, t_1]$ of a point of mass $m > 0$ on the Euclidean manifold $\mathbb{E}^3$. Here, $\mathbf{x}$ denotes a position vector located at the origin. Let us consider the variational principle

$$\boxed{\int_{t_0}^{t_1} \tfrac{1}{2} m\, \dot{\mathbf{x}}(t)^2 \, dt = \min!} \tag{9.19}$$

together with the boundary conditions: $\mathbf{x}(t_0) = \mathbf{x}_0$ and $\mathbf{x}(t_1) = \mathbf{x}_1$. That is, the initial position $\mathbf{x}(t_0)$ of the mass point at time t_0 and the terminal position $\mathbf{x}(t_1)$ of the mass point at time t_1 are fixed. In terms of physics, we are looking for the trajectories of the motion with minimal kinetic energy. The solutions of (9.19) are called energetic geodesics, affine geodesics, or trajectories of minimal kinetic energy. Intuitively, we expect that the energetic geodesics are straight lines, since no forces are acting. Let us prove this rigorously.

Theorem 9.10 *On the Euclidean manifold $\mathbb{E}^3$, precisely the segments of straight lines are affine geodesics.*

[7] More precisely, we have to write $D_{\mathbf{v}(P)}\mathbf{w}(P) := v^i(P)\nabla_i w^j(P) \cdot \mathbf{b}_j(P)$.

[8] In order to emphasize that $D_{\mathbf{v}}\mathbf{w}$ concerns the Euclidean connection, we frequently replace the symbol $D_{\mathbf{v}}\mathbf{w}$ by $d_{\mathbf{v}}\mathbf{w}$. This convention coincides with the notation used in finite-dimensional and infinite-dimensional Banach spaces.

Proof. (I) Necessary condition. Set $\mathcal{L}(\dot{\mathbf{x}}) := \frac{1}{2}m\dot{\mathbf{x}}^2$. If the trajectory $t \mapsto \mathbf{x}(t)$ is a solution of (9.19), then we have the Euler–Lagrange equation

$$\frac{d}{dt}\mathcal{L}_{\dot{\mathbf{x}}}(\dot{\mathbf{x}}(t)) = 0, \qquad t_0 \le t \le t_1.$$

Hence $\ddot{\mathbf{x}}(t) = 0$, $t_0 \le t \le t$. This implies that $t \mapsto \mathbf{x}(t)$ is the segment of a straight line.

(I) Sufficient condition. Since the original minimum problem (9.19) is of quadratic type, Jacobi's accessory minimum problems based on the second variation is also quadratic. Explicitly, we get

$$\int_{t_0}^{t_1} \tfrac{1}{2}m\dot{\mathbf{h}}(t)^2\, dt = \text{min!} \tag{9.20}$$

together with the boundary conditions: $\mathbf{h}(t_0) = 0$ and $\mathbf{h}(t_1) = 0$. The same argument as in (I) shows that the problem (9.24) has only the trivial solution $\mathbf{h} \equiv 0$. Consequently, every solution of the Euler–Lagrange equation to (9.19) is a solution of (9.19) (see Sect. 6.5.3 of Vol. II). □

Curvilinear coordinates. Using the local coordinates (x^1, x^2, x^3), we get $\mathcal{L}(\dot{x}^1, \dot{x}^2, \dot{x}^3) = g_{ij}(x^1, x^2, x^3)\dot{x}^i\dot{x}^j$. The original minimum problem (9.24) reads as

$$\int_{t_0}^{t_1} \mathcal{L}(\dot{x}^1(t), \dot{x}^2(t), \dot{x}^3(t))\, dt = \text{min!} \tag{9.21}$$

together with the boundary conditions: $x^i(t_0) = x_0^i$ and $x^i(t_1) = x_1^i, i = 1, 2, 3$. By (8.159) on page 520, the solution of (9.19) satisfies the Euler–Lagrange equations

$$\boxed{\ddot{x}(t) + \dot{x}^i(t)\mathcal{A}_i(x(t))\,\dot{x}(t) = 0, \qquad t_0 \le t \le t_1}$$

with the so-called connection matrices

$$\mathcal{A}_i := \begin{pmatrix} \Gamma_{i1}^1 & \Gamma_{i2}^1 & \Gamma_{i3}^1 \\ \Gamma_{i1}^2 & \Gamma_{i2}^2 & \Gamma_{i3}^2 \\ \Gamma_{i1}^3 & \Gamma_{i2}^3 & \Gamma_{i3}^3 \end{pmatrix}, \qquad i = 1, 2, 3.$$

Explicitly,

$$\ddot{x}^k(t) = -\dot{x}^i(t)\Gamma_{ij}^k(x^1(t), x^2(t), x^3(t)) \cdot \dot{x}^j(t), \qquad k = 1, 2, 3 \tag{9.22}$$

which are the equations of motion of the mass point with respect to the local (x^1, x^2, x^3)-coordinates. In terms of physics, the Christoffel symbols describe fictive friction forces generated by the choice of the observer. In terms of geometry, the differential equations (9.22) describe affine geodesics which are segments of straight lines with respect to Cartesian coordinates.

Cylindrical coordinates. The variational problem (9.21) can be used in order to compute effectively the Christoffel symbols. Let us explain this in the special case of cylindrical coordinates. Set $x^1 := \varrho, x^2. = \varphi, x^3. = z$. Then

$$\mathcal{L}(\varrho, \varphi, z) = \tfrac{1}{2}m(\dot{\varrho}^2 + \varrho^2\dot{\varphi}^2 + \dot{z}^2).$$

The Euler–Lagrange equations read as $\frac{d}{dt}\mathcal{L}_{\dot{\varrho}} = \mathcal{L}_\varrho$, $\frac{d}{dt}\mathcal{L}_{\dot{\varphi}} = \mathcal{L}_\varphi$, and $\frac{d}{dt}\mathcal{L}_{\dot{z}} = \mathcal{L}_z$. Explicitly, we get the differential equations

$$\ddot{\varrho} = \varrho \cdot \dot{\varphi}^2, \qquad \ddot{\varphi} = -\frac{2}{\varrho} \cdot \dot{\varrho}\dot{\varphi}, \qquad \ddot{z} = 0$$

for the affine geodesics $t \mapsto (\varrho(t), \varphi(t), z(t))$. This yields

$$\Gamma^1_{22} = -\varrho, \qquad \Gamma^2_{12} = \Gamma^2_{21} = \frac{1}{\varrho}.$$

All the other Christoffel symbols vanish identically. Setting

$$\mathcal{A}_i = \begin{pmatrix} \Gamma^1_{i1} & \Gamma^1_{i2} & \Gamma^1_{i3} \\ \Gamma^2_{i1} & \Gamma^2_{i2} & \Gamma^2_{i3} \\ \Gamma^3_{i1} & \Gamma^3_{i2} & \Gamma^3_{i3} \end{pmatrix}, \qquad i = 1, 2, 3,$$

we obtain the so-called connection matrices

$$\mathcal{A}_1 = \begin{pmatrix} 0 & 0 & 0 \\ 0 & \frac{1}{\varrho} & 0 \\ 0 & 0 & 0 \end{pmatrix}, \quad \mathcal{A}_2 = \begin{pmatrix} 0 & -\varrho & 0 \\ \frac{1}{\varrho} & 0 & 0 \\ 0 & 0 & 0 \end{pmatrix}, \qquad \mathcal{A}_3 = 0.$$

For the local connection form $\mathcal{A}$, we get

$$\mathcal{A} = \mathcal{A}_i dx^i = \begin{pmatrix} 0 & -\varrho d\varphi & 0 \\ \frac{d\varphi}{\varrho} & \frac{d\varrho}{\varrho} & 0 \\ 0 & 0 & 0 \end{pmatrix}. \tag{9.23}$$

Hence

$$\mathcal{F}_{ij} = \partial_i \mathcal{A}_i - \partial_j \mathcal{A}_j + \mathcal{A}_i \mathcal{A}_j - \mathcal{A}_j \mathcal{A}_i = 0, \qquad i, j = 1, 2.$$

This yields the trivial curvature form $\mathcal{F} = \frac{1}{2}\mathcal{F}_{ij} dx^i \wedge dx^j = 0$, as expected. For spherical coordinates, we refer to Problem 9.2.

9.4.3 Curves of Minimal Length

Parallel to the principle of least kinetic energy (9.19), let us study the variational problem

$$\boxed{\int_{t_0}^{t_1} |\dot{\mathbf{x}}(t)| \, dt = \min!} \tag{9.24}$$

together with the boundary conditions: $\mathbf{x}(t_0) = \mathbf{x}_0$ and $\mathbf{x}(t_1) = \mathbf{x}_1$. That is, the initial position $\mathbf{x}(t_0)$ and the terminal position $\mathbf{x}(t_1)$ of the curve $\mathbf{x} = \mathbf{x}(t)$ with the parameter $t \in [t_0, t_1]$ are fixed. The solutions of (9.24) are called curves of minimal length.

Theorem 9.11 *On the Euclidean manifold $\mathbb{E}^3$, precisely the segments of straight lines are curves of minimal length.*

Proof. Choose a Cartesian (x, y, z)-coordinate system. Note that

$$|\dot{\mathbf{x}}(t)| = \sqrt{\dot{x}(t)^2 + \dot{y}(t)^2 + \dot{z}(t)^2}.$$

Suppose that the map $t \mapsto (x(t), y(t), z(t))$ is a solution of (9.24). Then the Euler–Lagrange equation reads as

$$\frac{d}{dt}\left(\frac{1}{|\mathbf{x}(t)|}\frac{d\mathbf{x}(t)}{dt}\right) = 0.$$

Introducing the arc length s as the curve parameter, it follows from $\frac{ds(t)}{dt} = |\dot{\mathbf{x}}(t)|$ that

$$\frac{d^2\mathbf{x}(s)}{ds^2} = 0.$$

Consequently, if there exists a solution of (9.24), then it is the segment of a straight line, and the minimal value of the integral from (9.24) is the distance between the endpoints of the given position vectors $\mathbf{x}_0$ and $\mathbf{x}_1$ at the origin. Since the segment under consideration realizes this distance, it is indeed a solution of (9.24). □

Further reading. We recommend:

J. Jost, Riemannian Geometry and Geometric Analysis, Springer, Berlin, 2008.

W. Klingenberg, Lectures on Closed Geodesics, Springer, Berlin, 1978.

W. Klingenberg, Riemannian Geometry, de Gruyter, Berlin, 1982.

9.4.4 The Gauss Equations of Moving Frames

The integrability conditions for the Gauss equations of moving frames on the Euclidean manifold $\mathbb{E}^3$ yield the vanishing of the Riemann–Christoffel curvature tensor (flatness of the Euclidean connection).

Folklore

Consider a fixed local (x^1, x^2, x^3)-coordinate system. Let (x^1, x^2, x^3) denote the coordinates of the point P.

Theorem 9.12 *The natural basis vectors* $\mathbf{b}_1, \mathbf{b}_2, \mathbf{b}_3$ *satisfy the Gauss equations of moving frames:*

$$\partial_i \mathbf{b}_j(P) = \Gamma_{ij}^l(P)\, \mathbf{b}_l(P), \qquad i, j, = 1, 2, 3. \tag{9.25}$$

This tells us that the Christoffel symbols describe the infinitesimal change of the natural basis vectors.

Proof. Recall that $\mathbf{b}_j = \frac{\partial \mathbf{x}}{\partial x^j}$. Since $\frac{\partial^2 \mathbf{x}}{\partial x^i \partial x^j} = \frac{\partial^2 \mathbf{x}}{\partial x^j \partial x^i}$, we get

$$g_{jr} = \mathbf{b}_j \mathbf{b}_r = \mathbf{b}_r \mathbf{b}_j = g_{rj}.$$

This proves the symmetry of the metric tensorial family g_{ij}. Since the three vectors $\mathbf{b}_1(P), \mathbf{b}_2(P)$, and $\mathbf{b}_3(P)$ form a basis at the point P, there exist real numbers $B_{ij}^k(P)$ such that

$$\partial_i \mathbf{b}_j(P) = B_{ij}^k(P)\, \mathbf{b}_k(P), \qquad i, j - 1, 2, 3.$$

From $\mathbf{b}_k \mathbf{b}_r = g_{kr}$, we get $\partial_i \mathbf{b}_j \cdot \mathbf{b}_r = B_{ij}^k g_{kr}$. Multiplication with g^{kr} yields

$$B_{ij}^k = \delta_m^k B_{ij}^m = g^{kr} g_{rm} B_{ij}^m = g^{kr} \partial_i \mathbf{b}_j \cdot \mathbf{b}_r.$$

Differentiation of $\mathbf{b}_r \mathbf{b}_j = g_{rj}$ gives

$$\partial_i \mathbf{b}_r \cdot \mathbf{b}_j + \mathbf{b}_r \partial_i \mathbf{b}_j = \partial_i g_{rj}.$$

Interchanging the indices and summing the terms yields

$$\partial_i \mathbf{b}_j \cdot \mathbf{b}_r = \tfrac{1}{2}(\partial_i g_{rj} + \partial_j g_{ir} - \partial_r g_{ij}).$$

This implies $B^k_{ij} = \Gamma^k_{ij}$. □

The language of classic tensor analysis. Using the matrices $\mathcal{A}_i := (\Gamma^l_{ik})$ and $\mathcal{F}_{ij} := (R^l_{ijk})$ (the upper index l is the row index, and the lower index k is the column index), we get

$$\mathcal{F}_{ij} = \partial_i \mathcal{A}_j - \partial_j \mathcal{A}_i + \mathcal{A}_i \mathcal{A}_j - \mathcal{A}_j \mathcal{A}_i, \qquad i, j = 1, 2, 3. \tag{9.26}$$

Explicitly, we get both the components of the Riemann–Christoffel tensorial family,

$$R^l_{ijk} = \partial_i \Gamma^l_{jk} - \partial_j \Gamma^l_{ik} + \Gamma^l_{is}\Gamma^s_{jk} - \Gamma^l_{js}\Gamma^s_{ik} \tag{9.27}$$

and the components of the torsion tensor tensorial family, $T^k_{ij} := \Gamma^k_{ij} - \Gamma^k_{ji}$.

Proposition 9.13 *For all indices, we have*
(i) $R^l_{ijk} \equiv 0$ *(vanishing curvature of* $\mathbb{E}^3$*), and*
(ii) $T^k_{ij} \equiv 0$ *(vanishing torsion of* $\mathbb{E}^3$*).*

Proof. If we use Cartesian coordinates on $\mathbb{E}^3$, then the Christoffel symbols Γ^k_{ij} vanish identically. Hence $T^k_{ij} \equiv 0$ and $R^k_{ijl} \equiv 0$. In an arbitrary local coordinate system on $\mathbb{E}^3$, the Christoffel symbols do not vanish identically, as a rule. But, R^k_{ijl} and T^k_{ij} are tensorial families (see Sect. 8.9.1 on page 504). Finally, recall the following: If a tensorial family vanishes identically in a special local coordinate system, then it vanishes identically in all local coordinate systems. □

From the analytic point of view we want to show that:

> *The integrability conditions for the Gauss equations* (9.25) *of moving frames yield* $R^k_{ijl} \equiv 0$.

In mathematics, integrability conditions are always obtained form the commutativity relation $\partial_i\partial_j = \partial_j\partial_i$ for partial derivatives. In particular, it follows from the Gauss equation $\partial_j \mathbf{b}_k = \Gamma^l_{jk}\mathbf{b}_l$ of moving frames that

$$\partial_i\partial_j \mathbf{b}_k = \partial_i \Gamma^l_{jk}\mathbf{b}_l + \Gamma^l_{jk}\partial_i \mathbf{b}_l = (\partial_i \Gamma^l_{jk} + \Gamma^l_{ir}\Gamma^r_{jk})\mathbf{b}_l.$$

Using $\partial_i\partial_j \mathbf{b}_k \equiv \partial_j\partial_i \mathbf{b}_k$, we obtain $R^l_{ijk}\mathbf{b}_l \equiv 0$. Hence $R^l_{ijk} \equiv 0$.

9.4.5 Parallel Transport of a Velocity Vector and Cartan's Propagator Equation

We are given the smooth curve

$$C : P = P(t), \qquad t \in \mathcal{R}.$$

We assign to every point P of the curve C a velocity vector $\mathbf{v}(P)$. We say that the family $\{\mathbf{v}(P)\}_{P\in C}$ is parallel along the curve C iff it is parallel in the usual sense. We set $\mathbf{v}(t) := \mathbf{v}(P(t))$, and we assume that $t \mapsto \mathbf{v}(t)$ is a smooth map from the open interval $\mathcal{R}$ to E_3. Consider a local (x^1, x^2, x^3)-coordinate system which describes the curve in the form

$$C : x^k = x^k(t), \qquad k = 1, 2, 3, \quad t \in \mathcal{R}.$$

With respect to the natural basis $\mathbf{b}_1, \mathbf{b}_2, \mathbf{b}_3$ of the local coordinate system, we get

$$\mathbf{v}(t) = v^j(t)\, \mathbf{b}_j(x^1(t), x^2(t), x^3(t)), \qquad t \in \mathcal{R}. \tag{9.28}$$

In order to simplify the notation, set $v := \begin{pmatrix} v^1 \\ v^2 \\ v^3 \end{pmatrix}$.

Proposition 9.14 *The family* $\{\mathbf{v}(P(t))\}_{t\in\mathcal{R}}$ *of velocity vectors is parallel along the curve* C *if and only if*

$$\dot{v}(t) + \dot{x}(t)^i \mathcal{A}_i(P(t)) \cdot v(t) = 0, \qquad t \in \mathcal{R}. \tag{9.29}$$

Introducing the so-called local connection form $\mathcal{A} := \mathcal{A}_i dx^i$, the equation (9.29) reads as

$$\boxed{\dot{v}(t) + \mathcal{A}_{x(t)}(\dot{x}(t)) \cdot v(t) = 0, \qquad t \in \mathcal{R}.}$$

Explicitly, the differential equation (9.29) of parallel transport for velocity vectors reads as

$$\dot{v}^k(t) + \dot{x}^i(t)\Gamma^k_{ij}(P(t)) \cdot v^j(t) = 0, \qquad k = 1, 2, 3, \quad t \in \mathcal{R}. \tag{9.30}$$

Proof. The family $\{\mathbf{v}(t)\}$ of velocity vectors is parallel along C iff

$$\dot{\mathbf{v}}(t) = 0, \qquad t \in \mathcal{R}.$$

By (9.28), this is equivalent to

$$\dot{v}^j(t)\mathbf{b}_j(P(t)) + v^j(t) \cdot \partial_i \mathbf{b}_j(P(t))\, \dot{x}^i(t) = 0, \qquad j = 1, 2, 3, \quad t \in \mathcal{R}.$$

By the Gauss frame equation (9.25),

$$\{\dot{v}^k(t) + \dot{x}^i(t)\Gamma^k_{ij}(P(t)) \cdot v^j(t)\}\, \mathbf{b}_k(P(t)) = 0, \qquad t \in \mathbb{R}.$$

□

This result shows us that the following hold:

> *A simple geometric fact can lead to a complicated formula if one does not use the appropriate system of local coordinates.*

For example, if a geometric or physical problem has a certain symmetry, then one should use local coordinates which reflect this symmetry. This is, roughly speaking, the philosophy behind the use of Lie groups in geometry and physics by mathematicians and physicists.

Cartan's propagator equation. Recall that $GL(3, \mathbb{R})$ denotes the 9-dimensional Lie group of real invertible (3×3)-matrices. Suppose that the map $t \mapsto G(t)$ is a smooth map from the open time interval $\mathcal{R}$ to the Lie group $GL(3, \mathbb{R})$ which satisfies the so-called Cartan propagator equation:

$$\boxed{\dot{G}(t) + \dot{x}^i(t)\mathcal{A}_i(x(t)) \cdot G(t) = 0, \quad t \in \mathcal{R}, \ G(0) = I} \tag{9.31}$$

which is is a matrix differential equation. Then, for every initial value $v_0 \in \mathbb{R}^3$, the function

$$\boxed{v(t) = G(t)v_0, \qquad t \in \mathcal{R}}$$

is a solution of the differential equation (9.29) of parallel transport.

The function $t \mapsto G(t)$ from the open interval $\mathcal{R}$ to the Lie matrix group $GL(3,\mathbb{R})$ is called the Cartan propagator of the parallel transport of the velocity vector v_0.

Propagators are frequently used in physics. They describe the propagation of physical information (e.g., the Feynman propagator in quantum field theory). In Yang–Mills gauge theory, $G(t)$ corresponds to a so-called local phase factor (see page 821 and page 847 concerning Ariadne's thread in gauge theory).

Proof. Using $\dot{v}(t) = \dot{G}(t)v_0$, equation (9.31) implies (9.29). □

We will show below that it is convenient to write the Cartan propagator equation (9.31) in the equivalent form

$$\boxed{G(t)^{-1}\dot{G}(t) + \dot{x}^i(t)G(t)^{-1}\mathcal{A}_i(x(t))\, G(t) = 0.} \tag{9.32}$$

This simple trick is crucial for Cartan's approach to differential geometry via principal bundles. The point is that we only use notions which possess an intrinsic meaning on Lie groups and Lie algebras:

- We will discuss below that $G^{-1}dG$ is the Maurer–Cartan form of the Lie matrix group $GL(3,\mathbb{R})$,
- and we have
$$\mathrm{ad}(G^{-1})\mathcal{A}_i = G^{-1}\mathcal{A}_iG$$
where $G \mapsto \mathrm{ad}(G)$ is the adjoint representation of the Lie group $Gl(3,\mathbb{R})$ on its Lie algebra $gl(3,\mathbb{R})$.

Using the local connection form $\mathcal{A} = \mathcal{A}_i dx^i$, the Cartan propagator equation (9.32) can be written as

$$G(t)^{-1}\dot{G}(t) + G(t)^{-1}\mathcal{A}_{x(t)}(\dot{x}(t))\, G(t) = 0, \qquad t \in \mathcal{R}. \tag{9.33}$$

In Sect. 9.4.8, we will introduce the global connection form A. This is a differential 1-form (with values in the Lie algebra $gl(3,\mathbb{R})$) on the frame bundle $F\mathbb{E}^3$ over the Euclidean manifold $\mathbb{E}^3$. Equivalently, Cartan's propagator equation (9.33) can be written as

$$\boxed{\mathsf{A}_{Q(t)}(\dot{Q}(t)) = 0, \qquad t \in \mathcal{R}}$$

where $t \mapsto Q(t)$ is a curve on the frame bundle $F\mathbb{E}^3$.

This is the most elegant formulation of Cartan's method of moving frames.

We will show in Sect. 9.4.10 that we can replace the Lie group $GL(3,\mathbb{R})$ by its subgroup $SO(3)$ (by passing to right-handed orthonormal frames). This reflects the fact that Euclidean geometry is invariant under rotations.

Summarizing, there exist two variants of parallel transport which are closely related to each other:

(i) the parallel transport of velocity vectors, and
(ii) the parallel transport of local phase factors.

In the general theory to be considered in Chap. 17, (i) and (ii) correspond to

- the parallel transport on the tangent bundle,
- and the parallel transport on the associated principal bundle,

respectively. In terms of the Standard Model in particle physics, (i) and (ii) correspond to

- the fields of the 12 fundamental particles (electron, 6 quarks, 3 neutrinos, muon, tau),
- and the fields of the 12 interaction particles (photon, 3 vector bosons, 8 gluons),

respectively.

9.4.6 The Dual Cartan Equations of Moving Frames

> Élie Cartan's approach to differential geometry can be viewed as a dual version of the classic approach due to Gauss and Riemann. Gauss used two symmetric tensor fields (quadratic forms) in order to describe metric properties and curvature properties of two-dimensional surfaces. The Riemann–Christoffel curvature tensor in higher dimensions possesses a crucial antisymmetry property. The Cartan calculus of differential forms is based on antisymmetry. Therefore, Cartan was able to relate the Riemann–Christoffel curvature tensor to differential forms. The advantage is that crucial integrability conditions can be elegantly formulated in terms of Poincaré's cohomology rule: $dd\omega = 0$.
>
> Folklore

In what follows, we will use the wedge product for matrices with differential forms as entries (see page 509). Let us specialize the general results from page 509 to the Euclidean manifold $\mathbb{E}^3$.

The Gauss frame equations (9.25) for the natural basis vectors can be written in the form of the following matrix equations:

$$\partial_i(\mathbf{b}_1, \mathbf{b}_2, \mathbf{b}_3) = (\mathbf{b}_1, \mathbf{b}_2, \mathbf{b}_3)\mathcal{A}_i, \qquad i = 1, 2, 3. \tag{9.34}$$

Here, we use the local connection matrices

$$\mathcal{A}_i := \begin{pmatrix} \Gamma^1_{i1} & \Gamma^1_{i2} & \Gamma^1_{i3} \\ \Gamma^2_{i1} & \Gamma^2_{i2} & \Gamma^2_{i3} \\ \Gamma^3_{i1} & \Gamma^3_{i2} & \Gamma^3_{i3} \end{pmatrix}, \qquad i = 1, 2, 3,$$

whose entries are the Christoffel symbols Γ^l_{ik}.

The trick of killing indices. Cartan's approach can be motivated by the method of killing indices. The final goal is to obtain a completely invariant formulation. In the present special case of the Euclidean manifold, the approach is very simple.

(i) Christoffel symbols (local connection form): In order to kill the indices of the Christoffel symbols Γ^l_{ik}, we introduce the differential forms

$$\omega^l_k := \Gamma^l_{ik} dx^i.$$

This yields the matrix

$$\mathcal{A} := \begin{pmatrix} \omega^1_1 & \omega^1_2 & \omega^1_3 \\ \omega^2_1 & \omega^2_2 & \omega^2_3 \\ \omega^3_1 & \omega^3_2 & \omega^3_3 \end{pmatrix} \tag{9.35}$$

which coincides with the local connection form $\mathcal{A} := \mathcal{A}_i dx^i$ introduced on page 582. Note that this form depends on the choice of local coordinates.

(ii) Riemann–Christoffel curvature tensorial family (local curvature form): the components R^l_{ijk} from (9.27) on page 581 are antisymmetric with respect to the indices i and j. Therefore, we introduce the differential forms

$$\Omega^l_k := \tfrac{1}{2} R^l_{ijk}\, dx^i \wedge dx^k,$$

and the matrix

$$\mathcal{F} := \begin{pmatrix} \Omega_1^1 & \Omega_2^1 & \Omega_3^1 \\ \Omega_1^2 & \Omega_2^2 & \Omega_3^2 \\ \Omega_1^3 & \Omega_2^3 & \Omega_3^3 \end{pmatrix}. \tag{9.36}$$

Equivalently, $\mathcal{F} = \frac{1}{2}\mathcal{F}_{ij}\, dx^i \wedge dx^j$ with

$$\mathcal{F}_{ij} = \partial_i \mathcal{A}_j - \partial_j \mathcal{A}_i + \mathcal{A}_i \mathcal{A}_j - \mathcal{A}_j \mathcal{A}_i.$$

The key formula reads as

$$\boxed{\mathcal{F} = d\mathcal{A} + \mathcal{A} \wedge \mathcal{A}.} \tag{9.37}$$

(iii) Torsion tensorial family: The torsion tensorial family $T_{ij}^k := \Gamma_{ij}^k - \Gamma_{ji}^k$ is antisymmetric with respect to the indices i and j. Therefore, we introduce the differential forms $\tau^k := T_{ij}^k\, dx^i \wedge dx^j$ and the matrices

$$\tau := \begin{pmatrix} \tau^1 \\ \tau^2 \\ \tau^3 \end{pmatrix} \qquad \text{and} \qquad dx := \begin{pmatrix} dx^1 \\ dx^2 \\ dx^3 \end{pmatrix}.$$

This yields

$$\boxed{\tau = \mathcal{A} \wedge dx.}$$

Cartan's local frame equations. As a dual variant to the Gauss frame equations, the following hold.

Theorem 9.15 *We have the local Cartan frame equations*

$$d\mathcal{A} + \mathcal{A} \wedge \mathcal{A} = 0, \tag{9.38}$$

$$\mathcal{A} \wedge dx = 0. \tag{9.39}$$

Proof. By Prop. 9.13, $R_{ijk}^l \equiv 0$ and $T_{ij}^k \equiv 0$. These two equations are equivalent to (9.38) and (9.39), respectively, since $\mathcal{F} \equiv 0$ and $\tau \equiv 0$. □

Note that the integrability conditions (Bianchi relations)

$$d\mathcal{F} = \mathcal{F} \wedge \mathcal{A} - \mathcal{A} \wedge \mathcal{F}$$

and $d\tau = \mathcal{F} \wedge dx - \mathcal{A} \wedge \tau$ from (8.138) are satisfied automatically because of $\mathcal{F} \equiv 0$ and $\tau \equiv 0$.

9.4.7 Global Parallel Transport on Lie Groups and the Maurer–Cartan Form

In contrast to the Euclidean manifold $\mathbb{E}^3$, there is no global parallel transport on general manifolds, as a rule. However, Cartan used the crucial fact that there is a global parallel transport on every Lie group called left translation (which generalizes the translations on the Euclidean manifold). The left translations on a Lie group are generated by the tangent vectors of the Lie group at the unit element (i.e., the elements of the Lie algebra). Analytically, on an infinitesimal level, the left translation is governed by the Maurer–Cartan differential form M (mnemonically, $\mathsf{M}_G = G^{-1}dG$).

Folklore

On manifolds (e.g., the Euclidean manifold, spheres, and Lie groups), we will use the following notions synonymously:

- tangent vector, velocity vector, and
- tangent vector field, velocity vector field, vector field.

Let $\mathcal{G}$ be the Lie group $GL(3,\mathbb{R})$ or $SO(3)$.[9] Recall that:

- $GL(3,\mathbb{R})$ consists of all the invertible real (3×3)-matrices, and
- $SO(3)$ consists of all the matrices $R \in GL(3,\mathbb{R})$ with $(R^{-1})^d = R$ and $\det(R) = 1$.

The Lie algebra $\mathcal{LG}$ of the Lie group $\mathcal{G}$ consists of the tangent space $T_{\mathbf{1}}\mathcal{G}$ of the manifold $\mathcal{G}$ at the unit element $\mathbf{1}$ equipped with the Lie bracket $[A,B] := AB - BA$. In particular,

- $\mathcal{L}GL(3,\mathbb{R}) = gl(3,\mathbb{R})$ (all the real (3×3)-matrices), and
- $\mathcal{L}SO(3) = so(3)$ (all the matrices $A \in gl(3,\mathbb{R})$ with $A^d = -A$ and $\mathrm{tr}(A) = 0$).

An introduction to Lie groups and Lie algebras can be found in Chap. 7 of Vol. I.

Left translation on the Lie group $\mathcal{G}$. Fix the point G_0 on the Lie group $\mathcal{G}$. Define the map $L_{G_0} : \mathcal{G} \to \mathcal{G}$ by setting

$$\boxed{L_{G_0}G := G_0 G \qquad \text{for all} \quad G \in \mathcal{G}.}$$

The map $L_{G_0} : \mathcal{G} \to \mathcal{G}$ is a diffeomorphism. For all $G_0, G_1 \in \mathcal{G}$, we have

$$L_{G_1 G_0} = L_{G_1} L_{G_0}, \tag{9.40}$$

and $L_{\mathbf{1}}$ is the identical map id on the group $\mathcal{G}$. The transformation $G \mapsto G_0 G$ is called a left translation of the Lie group $\mathcal{G}$ (generated by the group element $G_0 \in \mathcal{G}$).[10]

The Maurer–Cartan form. Following Sophus Lie (1842–1899), we want to study the left translation on the Lie group $\mathcal{G}$ on an infinitesimal level (i.e., in terms of velocity vectors on $\mathcal{G}$). Fix the point $G_0 \in \mathcal{G}$. Let

$$\mathcal{C} : G = G(t), \qquad -t_0 < t < t_0, \;\; G(0) = G_0$$

be a smooth curve on the Lie group $\mathcal{G}$ which passes through the point G_0 at time $t = 0$. Set $\dot{G}_0 := \frac{d}{dt}G(t)_{|t=0}$. This is the velocity vector of the curve $\mathcal{C}$ at the point G_0. The left translation $G \mapsto G_0^{-1}G$ (which sends the point G_0 to the unit element) yields the translated curve

$$G_0^{-1}\mathcal{C} : H(t) = G_0^{-1}G(t), \qquad -t_0 < t < t_0, \;\; G(0) = G_0.$$

At time $t = 0$, this curve passes through the point $\mathbf{1}$, and it has the velocity vector

$$\dot{H}_0 := \dot{H}(0) = G_0^{-1}\dot{G}_0.$$

We define the Maurer–Cartan form M_{G_0} at the point G_0 by

$$\boxed{\mathsf{M}_{G_0}(\dot{G}_0) := \dot{H}_0.}$$

[9] The following results are valid for arbitrary Lie groups. This will be studied later on. See page 804.

[10] Similarly, setting $R_{G_0}G := GG_0$, we get the right translation $R_{G_0} : \mathcal{G} \to \mathcal{G}$. Here, $R_{G_0 G_1} = R_{G_1} R_{G_0}$, in contrast to (9.40).

We will also write $G_0^{-1}dG$ instead of M_{G_0}. This is motivated by the mnemonic formula $dG(\dot{G}_0) = \dot{G}_0$, and hence

$$G_0^{-1}dG(\dot{G}_0) = G_0^{-1}\dot{G}_0.$$

Proposition 9.16 *The map $\mathsf{M}_{G_0} : T_{G_0}\mathcal{G} \to T_1\mathcal{G}$ is a linear isomorphism from the tangent space of the Lie group $\mathcal{G}$ at the point G_0 onto the tangent space of $\mathcal{G}$ at the unit element* $\mathbf{1}$.

Proof. Choose the element A of the Lie algebra $\mathcal{LG}$. Define

$$G(t) := G_0 e^{tA}, \qquad t \in \mathbb{R}.$$

This is a curve on the group $\mathcal{G}$ which passes through the point G_0 at time $t = 0$, and which has the velocity vector

$$\dot{G}(0) = G_0 A$$

at time $t = 0$. Obviously, $\mathsf{M}_{G_0}(G_0 A) = A$. Thus, the map $G_0 A \mapsto A$ is a bijective map from $T_{G_0}\mathcal{G}$ onto $T_1\mathcal{G}$. In other words, the map $\mathsf{M}_{G_0} : T_{G_0}\mathcal{G} \to T_1\mathcal{G}$ is bijective and linear. □

Since the Lie algebra $\mathcal{LG}$ coincides with the tangent space $T_1\mathcal{G}$ of $\mathcal{G}$ at the unit element $\mathbf{1}$ of $\mathcal{G}$, we get the linear isomorphism

$$\boxed{\mathsf{M}_{G_0} : T_{G_0}\mathcal{G} \to \mathcal{LG}.}$$

We call this a linear operator on the tangent space $T_{G_0}\mathcal{G}$ with values in the Lie algebra $\mathcal{LG}$. The map

$$G_0 \mapsto \mathsf{M}_{G_0}$$

is called a differential 1-form M on the Lie group $\mathcal{G}$ with values in the Lie algebra $\mathcal{LG}$. The differential form M is called the Maurer–Cartan form of the Lie group $\mathcal{G}$.

Left-invariant velocity vector fields on the Lie group $\mathcal{G}$. Let $A \in \mathcal{LG}$. We define

$$V_A(G) := GA \qquad \text{for all} \quad G \in \mathcal{G}.$$

This is a smooth velocity vector field V_A on the Lie group $\mathcal{G}$ (generated by the element A of the Lie algebra $\mathcal{LG}$). Note that

$$V_A(L_{G_0}G) = L_{G_0}V_A(G) \qquad \text{for all} \quad G \in \mathcal{G}.$$

A velocity vector field on $\mathcal{G}$ with this property is called a left-invariant velocity vector field. All possible left-invariant velocity vector fields on $\mathcal{G}$ are obtained by V_A with $A \in \mathcal{LG}$.

9.4.8 Cartan's Global Connection Form on the Frame Bundle of the Euclidean Manifold

One has to distinguish between local and global connection forms in gauge theory. In order to get the global connection form of the Euclidean manifold $\mathbb{E}^3$, one has to pass to the frame bundle $F\mathbb{E}^3$ of $\mathbb{E}^3$ which is isomorphic to the 12-dimensional manifold $\mathbb{E}^3 \times GL(3,\mathbb{R})$. This frame bundle can be reduced to the orthonormal frame bundle which is diffeomorphic to the 6-dimensional manifold $\mathbb{E}^3 \times SO(3)$.

> It happens quite often in mathematics, that a deeper understanding of the mathematical structures is only possible by passing to abstract objects in higher dimensions and by using projections onto lower dimensional objects. In terms of philosophy, this is related to the famous cave parable from Plato's *Politea* (The Republic).[11] In this parable, the prisoner is only able to see shadows on the cave's walls which are generated by objects that exist in the 'realm of ideas'.
>
> Folklore

Fibration of the frame bundle. Recall that the frame bundle $F\mathbb{E}^3$ of the Euclidean manifold $\mathbb{E}^3$ consists of all the possible tuples

$$(P; \mathbf{b}_1, \mathbf{b}_2, \mathbf{b}_3)$$

where P is a point of $\mathbb{E}^3$, and $\mathbf{b}_1, \mathbf{b}_2, \mathbf{b}_3$ is an arbitrary frame at the point P (i.e., $\mathbf{b}_1, \mathbf{b}_2, \mathbf{b}_3$ are linearly independent position vectors located at the point P). Fix the point P. Then the set

$$F_P := \{(P; \mathbf{b}_1, \mathbf{b}_2, \mathbf{b}_3) \in F\mathbb{E}^3\}$$

is called the fiber at the point P. This coincides with the set of all the possible frames at the point P. Setting $\pi(P; \mathbf{b}_1, \mathbf{b}_2, \mathbf{b}_3) := P$, we get the so-called projection map

$$\pi : F\mathbb{E}^3 \to \mathbb{E}^3$$

of the frame bundle $F\mathbb{E}^3$. The map

$$s : \mathbb{E}^3 \to F\mathbb{E}^3$$

is called a section iff $s(P) \in F_P$ for all points $P \in \mathbb{E}^3$. This section assigns to every point of the Euclidean manifold a frame.

Intrinsic characterization of the fibers. The Lie group $GL(3, \mathbb{R})$ acts on the frame bundle $F\mathbb{E}^3$, and the fibers are the orbits of this action. Explicitly, let $G \in GL(3, \mathbb{R})$. Fix the point $P \in \mathbb{E}^3$, and define the operator R_G by means of the following matrix equation:

$$R_G(\mathbf{b}_1, \mathbf{b}_2, \mathbf{b}_3) := (\mathbf{b}_1, \mathbf{b}_2, \mathbf{b}_3)G.$$

The orbit of this action is the set of all the possible frames at the point P. This set coincides with the fiber F_P at the point P. For all elements G, H of the Lie group $GL(3, \mathbb{R})$, we have

$$\boxed{R_{GH} = R_H R_G.}$$

Therefore, we say that the group $GL(3, \mathbb{R})$ acts on the frame bundle manifold $F\mathbb{E}^3$ from the right. In Section 17.2, we will introduce axiomatically the notion of a principal (fiber) bundle. The frame bundle $F\mathbb{E}^3$ of the Euclidean manifold $\mathbb{E}^3$ is the prototype of a principal bundle with the Lie group $GL(3, \mathbb{R})$ as typical fiber.

The Cartesian gauge and the parametrization of the frame bundle $F\mathbb{E}^3$. We fix both a point O (called the origin) of the Euclidean manifold $\mathbb{E}^3$ and a right-handed orthonormal system $\mathbf{i}, \mathbf{j}, \mathbf{k}$ at the point O. For any point P of the

[11] Plato (427–347 B.C.) The correct Greek name is 'Platon'. Plato's Academy in Athens had unparalleled importance for Greek thought. The greatest philosophers, mathematicians, and astronomers worked there. For example, Aristotle (384-322 B.C) studied there as a young man. In 529 A.D., the Academy was closed by the Roman emperor Justitian.

Euclidean manifold $\mathbb{E}^3$, we select a frame $\mathbf{i}_P, \mathbf{j}_P, \mathbf{k}_P$ at P which is parallel to $\mathbf{i}, \mathbf{j}, \mathbf{k}$ (Fig. 9.1 on page 558). This is called a Cartesian gauge. For every frame $\mathbf{b}_1, \mathbf{b}_2, \mathbf{b}_3$ at the point P, there exists a matrix $G \in GL(3, \mathbb{R})$ such that we have the following matrix equation:

$$(\mathbf{b}_1, \mathbf{b}_2, \mathbf{b}_3) = (\mathbf{i}_P, \mathbf{j}_P, \mathbf{k}_P)G.$$

We call (P, G) the bundle coordinate of the point $(P; \mathbf{b}_1, \mathbf{b}_2, \mathbf{b}_3)$ of the frame bundle $F\mathbb{E}^3$.[12] If we use the Cartesian coordinates (x, y, z) with respect to $\mathbf{i}, \mathbf{j}, \mathbf{k}$, then we can describe the bundle coordinate (P, G) by $(x, y, z; G)$.

If we choose another Cartesian gauge based on the right-handed orthonormal system $\mathbf{i}^+, \mathbf{j}^+, \mathbf{k}^+$, then we get the matrix equation

$$(\mathbf{i}, \mathbf{j}, \mathbf{k}) = (\mathbf{i}^+, \mathbf{j}^+, \mathbf{k}^+)G_0.$$

Hence

$$(\mathbf{b}_1, \mathbf{b}_2, \mathbf{b}_3) = (\mathbf{i}_P^+, \mathbf{j}_P^+, \mathbf{k}_P^+)G_0 G.$$

This means that the bundle coordinate (P, G) passes over to (P, G^+) with

$$\boxed{G^+ = G_0 G.}$$

This is called a gauge transformation. Moreover, the bundle coordinate $(x, y, z; G)$ passes over to (x^+, y^+, z^+, G^+) where

$$(x^+, y^+, z^+) = (x, y, z)G.$$

Since the change of the bundle coordinates $(x, y, z; G)$ is described by a diffeomorphism, we get the following.

Proposition 9.17 *The bundle space $F\mathbb{E}^3$ is a 12-dimensional real manifold which is diffeomorphic to $\mathbb{E}^3 \times GL(3, \mathbb{R})$.*

The global connection form A on the frame bundle $F\mathbb{E}^3$. We define

$$\boxed{\mathsf{A}_Q := G^{-1}dG \qquad \text{for all} \quad Q \in F\mathbb{E}^3.}$$

This is to be understood as follows. We choose a fixed Cartesian gauge. Then the point Q has the bundle coordinate (P, G), and we use the Maurer–Cartan form $G^{-1}dG$ on $GL(3, \mathbb{R})$. If

$$C : Q = Q(t), \qquad t \in \mathcal{R},$$

is a curve on $F\mathbb{E}^3$, then

$$\boxed{\mathsf{A}_{Q(t)}(\dot{Q}(t)) = G(t)^{-1}\dot{G}(t), \qquad t \in \mathcal{R}.} \tag{9.41}$$

We have to show that:

> *The global connection form A is an invariant geometric object on the manifold $F\mathbb{E}^3$.*

[12] The symbol $(\mathbf{i}_P, \mathbf{j}_P, \mathbf{k}_P)G$ is to be understood as matrix product. For example, the point $(P; \mathbf{i}_P, \mathbf{j}_P, \mathbf{k}_P)$ has the bundle coordinate (P, I) where I is the (3×3)-unit matrix.

This means that A_Q does not depend on the choice of the Cartesian gauge. In fact, changing the Cartesian gauge means passing from $G(t)$ to $G_0G(t)$ where G_0 does not depend on time t. This implies

$$(G_0G(t))^{-1}\frac{d}{dt}(G_0G(t)) = G(t)^{-1}G_0^{-1}G_0\dot{G}(t) = G(t)^{-1}\frac{d}{dt}G(t).$$

Parallel transport. Let $C : Q = Q(t), t \in \mathcal{R}$, be a smooth curve on the frame bundle $F\mathbb{E}^3$. Explicitly, this is a map

$$t \mapsto (P(t); \mathbf{b}_1(t), \mathbf{b}_2(t), \mathbf{b}_3(t))$$

from the open time interval $\mathcal{R}$ to $F\mathbb{E}^3$. Intuitively, this is a smooth curve $t \mapsto P(t)$ on the Euclidean manifold $\mathbb{E}^3$, and we assign to every curve point $P(t)$ a frame (in a smooth way). By definition, the curve C represents a parallel transport iff all the frames are parallel to each other (in the usual sense). Obviously, this is equivalent to

$$\dot{G}(t) = 0 \qquad \text{for all} \quad t \in \mathcal{R}$$

with respect to the bundle coordinates of any Cartesian gauge. Equivalently, by (9.41), the curve C represents a parallel transport if and only if

$$\boxed{\mathsf{A}_{Q(t)}(\dot{Q}(t)) = 0 \qquad \text{for all} \quad t \in \mathcal{R}.} \tag{9.42}$$

The advantage of this invariant formulation is that, by passing to coordinates, we obtain immediately the differential equation of parallel transport for both arbitrary local coordinates on the Euclidean manifold $\mathbb{E}^3$ and arbitrary gauge fixing of the frames. This will be studied in the next section.

The Cartan structural equation. The global curvature form F of the frame bundle $F\mathbb{E}^3$ is defined by

$$\mathsf{F} := d\mathsf{A} + \mathsf{A} \wedge \mathsf{A}.$$

Theorem 9.18 *The global curvature form of the frame bundle $F\mathbb{E}^3$ of the Euclidean manifold $\mathbb{E}^3$ vanishes identically,* $\mathsf{F} \equiv 0$.

The proof follows from the Cartan structural equation

$$d\mathsf{M} + \mathsf{M} \wedge \mathsf{M} = 0$$

for the Maurer–Cartan form M of a Lie group. This will be proved on page 806.

9.4.9 The Relation to Gauge Theory

> It was the goal of Élie Cartan to obtain an invariant formulation of differential geometry which takes the symmetries into account and which yields quickly the formulas with respect to local coordinates. The point is that the local coordinates of the base manifold and the local coordinates of the frames can be chosen independently. This yields maximal flexibility in differential geometry. Cartan's elegant theory can be viewed as a gauge theory in geometry.
>
> Folklore

General gauge fixing. Let $\mathcal{O}$ be a nonempty open subset of the Euclidean manifold $\mathbb{E}^3$. We are given the smooth section

$$s : \mathcal{O} \to F\mathbb{E}^3$$

where $\mathcal{O}$ is an open subset of the base manifold $\mathbb{E}^3$. Set

$$s(P) = (P; \mathbf{b}_1^+(P), \mathbf{b}_2^+(P), \mathbf{b}_3^+(P)), \qquad P \in \mathcal{O}.$$

Thus, the section s fixes a frame at each point P of $\mathbb{E}^3$.

> *It is our goal to use these distinguished frames in order to introduce local bundle coordinates* (P, G).

In fact, for every point

$$Q = (P; \mathbf{b}_1, \mathbf{b}_2, \mathbf{b}_3)$$

of the frame bundle $F\mathbb{E}^3$, there exists a uniquely determined matrix $G \in GL(3, \mathbb{R})$ such that

$$(\mathbf{b}_1, \mathbf{b}_2, \mathbf{b}_3) = (\mathbf{b}_1^+(P), \mathbf{b}_2^+(P), \mathbf{b}_3^+(P))\, G. \tag{9.43}$$

This way, we assign to the bundle point Q the bundle coordinate

$$(P, G) \in \mathbb{E}^3 \times GL(3, \mathbb{R})$$

with respect to the s-gauge. Moreover, let us introduce a local coordinate system on the open subset $\mathcal{O}$ of the Euclidean manifold $\mathbb{E}^3$. Then we assign to the base point P the coordinate $x = (x^1, x^2, x^3)$, and we assign to the bundle point Q the bundle coordinate

$$(x, G) \in \mathbb{R}^3 \times GL(3, \mathbb{R}).$$

For computing the global connection form A with respect to the local bundle coordinates (x, G), we need the Gauss frame equations

$$\boxed{\partial_i \mathbf{b}_j^+(P) = \gamma_{ij}^l(P)\, \mathbf{b}_l(P), \qquad i, j = 1, 2, 3} \tag{9.44}$$

with respect to the s-gauge. The real numbers $\gamma_{ik}^l(P)$ are called the connection coefficients of the s-gauge.[13] Introducing the matrix $\mathcal{A}_i := (\gamma_{ik}^l)$, the Gauss frame equations pass over to the following matrix equation:

$$\partial_i(\mathbf{b}_1^+(P), \mathbf{b}_2^+(P), \mathbf{b}_3^+(P)) = (\mathbf{b}_1^+(P), \mathbf{b}_2^+(P), \mathbf{b}_3^+(P))\, \mathcal{A}_i, \qquad i = 1, 2, 3.$$

Roughly speaking, the connection coefficients γ_{ik}^l describe the connection between the gauge frames on an infinitesimal level.

Theorem 9.19 (i) *In terms of local bundle coordinates, Cartan's connection form* A *on the frame bundle* $F\mathbb{E}^3$ *looks like*

$$\mathsf{A}_Q = G^{-1}dG + G^{-1}\mathcal{A}(x)G$$

[13] In the special case where the vectors $\mathbf{b}_1^+(P), \mathbf{b}_2^+(P), \mathbf{b}_3^+(P)$ are the natural basis vectors corresponding to a local coordinate system on the Euclidean manifold $\mathbb{E}^3$, we get $\gamma_{ij}^l = \Gamma_{ij}^l$ for all indices. That is, the connection coefficients γ_{ij}^l generalize the classic Christoffel symbols Γ_{ij}^l depending on the metric tensor (see (9.18) on page 576).

where $\mathcal{A}(x) := \mathcal{A}_i(x)dx^i$, *and* $\mathcal{A}_i = (\gamma^l_{ik})$.

(ii) *The curve* $C : Q = Q(t)$, $t \in \mathcal{R}$, *on the frame bundle* $F\mathbb{E}^3$ *represents a parallel transport iff* $\mathsf{A}_{Q(t)}(\dot{Q}(t)) = 0, t \in \mathcal{R}$, *that is,*

$$G(t)^{-1}\dot{G}(t) + G(t)^{-1}\dot{x}^i(t)\mathcal{A}_i(x(t))\, G(t) = 0, \qquad t \in \mathcal{R}.$$

Proof. Ad (i). The idea of the proof is to pass from the s-gauge (9.43) to a Cartesian gauge, since then the differential form A can be computed in a very simple manner, by its definition. To begin with, observe that, with respect to a Cartesian gauge, we get the matrix equation

$$(\mathbf{b}_1^+(P), \mathbf{b}_2^+(P), \mathbf{b}_3^+(P)) = (\mathbf{i}_P, \mathbf{j}_P, \mathbf{k}_P) \cdot G_0(P)$$

where $G_0(P) \in GL(3, \mathbb{R})$. It follows from (9.43) that

$$(\mathbf{b}_1, \mathbf{b}_2, \mathbf{b}_3) = (\mathbf{i}_P, \mathbf{j}_P, \mathbf{k}_P) \cdot G_0(P)G(P).$$

Thus, the bundle point

$$(P; \mathbf{b}_1, \mathbf{b}_2, \mathbf{b}_3)$$

has the bundle coordinate $(P, G_0(P)G(P))$ with respect to the Cartesian gauge. Moreover,

$$\begin{aligned}\partial_i(\mathbf{b}_1^+(P), \mathbf{b}_2^+(P), \mathbf{b}_3^+(P)) &= (\mathbf{i}_P, \mathbf{j}_P, \mathbf{k}_P) \cdot \partial_i G_0(P) \\ &= (\mathbf{b}_1^+(P), \mathbf{b}_2^+(P), \mathbf{b}_3^+(P)) \cdot G_0(P)^{-1}\partial_i G_0(P).\end{aligned}$$

Comparing this with (9.44), we get $G_0^{-1}\partial_i G_0 = \mathcal{A}_i$. By definition of the global connection form A, we have

$$\mathsf{A}_Q = (G_0 G)^{-1} d(G_0 G).$$

Since $d(G_0 G) = dG_0 \cdot G + G_0 dG$ and $(G_0 G)^{-1} = G^{-1}G_0^{-1}$, we obtain

$$\mathsf{A}_Q = G^{-1}(G_0^{-1}dG_0)G + G^{-1}dG = G^{-1}\mathcal{A}_i dx^i G + G^{-1}dG.$$

Ad (ii). This follows from (i). □

The pull-back operation. Consider the curve

$$x = x(t), \qquad t \in \mathbb{R}$$

on the Euclidean manifold $\mathbb{E}^3$. Use the open set $\mathcal{O}$ of the Euclidean manifold (e.g., $\mathcal{O} = \mathbb{E}^3$). Use the smooth section $s : \mathcal{O} \to F\mathbb{E}^3$ for fixing the gauge, and introduce the local bundle coordinates $(x, G) \in \mathcal{O} \times GL(3, \mathbb{R})$. The section s generates the curve

$$Q = s(x(t)) = (x(t), \mathbf{1}), \qquad t \in \mathcal{R}$$

on the frame bundle $F\mathbb{E}^3$. Note that $s(P)$ is a gauge frame which has the bundle coordinate $(P, \mathbf{1})$. Then $\dot{Q}(t) = (\dot{x}(t), 0)$. With respect to local bundle coordinates generated by the s-gauge, the local pull-back $s^*\mathsf{A}$ of the global Cartan form A looks like

$$(s^*\mathsf{A})_{x(t)}(\dot{x}(t)) = \mathsf{A}_{x(t)}(\dot{Q}(t)), \qquad t \in \mathcal{R}.$$

Explicitly,

$$(s^*\mathsf{A})_{x(t)} = \dot{\mathbf{x}}^i(t)\mathcal{A}_i(x(t)), \qquad t \in \mathcal{R}.$$

Hence

$$\boxed{(s^*\mathsf{A})_P = \mathcal{A}(P) = \mathcal{A}_i(P)dx^i, \qquad P \in \mathcal{O}.}$$

This is called the local connection form of the Euclidean manifold $\mathbb{E}^3$ (with respect to the open subset $\mathcal{O}$ of $\mathbb{E}^3$).

9.4.10 The Reduction of the Frame Bundle to the Orthonormal Frame Bundle

The orthonormal frame bundle of the Euclidean manifold takes the rotational symmetry of the Euclidean manifold into account.

Folklore

By definition, the orthonormal frame bundle $F\mathbb{E}^3(SO(3))$ of the Euclidean manifold $\mathbb{E}^3$ consists of all the tuples

$$(P; \mathbf{b}_1, \mathbf{b}_2, \mathbf{b}_3)$$

where P is a point of the Euclidean manifold $\mathbb{E}^3$, and all the frames $\mathbf{b}_1, \mathbf{b}_2, \mathbf{b}_3$ at the point P are right-handed orthonormal systems.

In terms of physics, this means that the observers (at all points P of the Euclidean manifold) only use right-handed orthonormal frames for describing their observations.

As above, the Cartesian gauge yields the matrix equation

$$(\mathbf{b}_1, \mathbf{b}_2, \mathbf{b}_3) = (\mathbf{i}_P, \mathbf{j}_P, \mathbf{k}_P)G$$

where $G \in SO(3)$. Thus, we assign to the bundle point $(P; \mathbf{b}_1, \mathbf{b}_2, \mathbf{b}_3)$ (of the orthonormal frame bundle) the bundle coordinate

$$(P, G) \in \mathbb{E}^3 \times SO(3).$$

A gauge transformation has the form $(P, G) \mapsto (P, G^+)$ with

$$\boxed{G^+ = G_0 G}$$

where G_0 is a fixed element of the Lie group $SO(3)$. In fact, the results obtained above remain valid by using the replacement

$$GL(3, \mathbb{R}) \Rightarrow SO(3),$$

that is, we replace the Lie group $GL(3, \mathbb{R})$ by its subgroup $SO(3)$. For example, the global connection form A of the orthonormal frame bundle $F\mathbb{E}^3(SO(3))$ is given by the Maurer–Cartan form $G^{-1}dG$ of the Lie group $SO(3)$.

The orthonormal frame bundle $F\mathbb{E}^3(SO(3))$ is a principal bundle with the rotation group $SO(3)$ as typical fiber (see the discussion on page 588).

9.5 The Sphere as a Paradigm in Riemannian Geometry and Gauge Theory

We need an analysis which is of geometric nature and describes physical situations as directly as algebra expresses quantities.

Gottfried Wilhelm Leibniz (1646–1716)

We live on the surface of earth – a submanifold of the Euclidean manifold $\mathbb{E}^3$. Approximately, the surface of earth is a sphere. Over the centuries, mathematicians and physicists tried to understand the geometry of the earth.

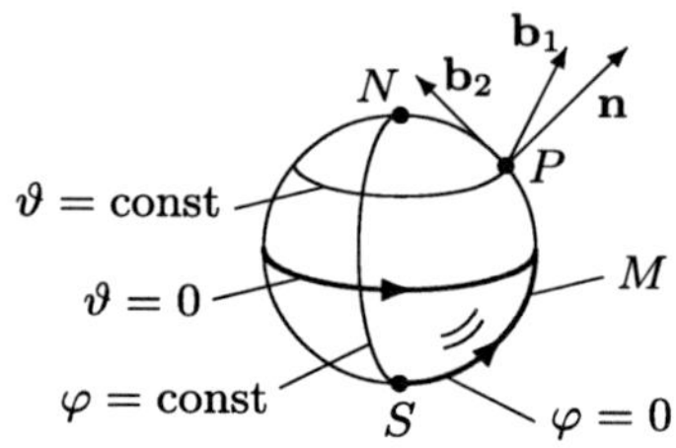

Fig. 9.7. Spherical coordinates

> The differential geometry of the 2-dimensional sphere can be obtained from the surrounding 3-dimensional Euclidean manifold $\mathbb{E}^3$ by using orthogonal projection onto the tangent spaces of the sphere. In terms of physics, this leads to the notion of covariant acceleration measured by an observer on the sphere. In terms of mathematics, this leads to the notion of covariant directional derivative (also called a connection).
> In geometry, one has to distinguish between linear and nonlinear objects. The curvature measures the deviation of a nonlinear object from linearity. In terms of physics, Newton measured forces by the deviation of the trajectories from straight lines. In terms of mathematics, the sphere is the simplest nonlinear geometric object.
> Gauss discovered that the curvature of a 2-dimensional surface in $\mathbb{E}^3$ can be measured intrinsically on the surface without using the surrounding space (theorema egregium).
> There exist far-reaching generalizations in modern mathematics. Here,
> - spheres are replaced by general manifolds of finite or infinite dimensions,
> - and velocity vector fields on spheres are replaced by more general physical fields (sections of vector bundles over manifolds).
>
> In topology, one generalizes this by replacing manifolds (i.e., smooth structures) with topological spaces (i.e., continuous structures) (see Vol. IV).
>
> Folklore

We want to show that the geometry of a sphere $\mathbb{S}_r^2$ of radius r can be understood best by using two bundles, namely,

- the tangent bundle $T\mathbb{S}_r^2$ (vector bundle) and
- the frame bundle $F\mathbb{S}_r^2$ (principal bundle).

Both the bundles are dual to each other. The tangent bundle allows us to study the parallel transport of velocity vectors, whereas the frame bundle allows us to describe the symmetry properties of the sphere. One has to distinguish between

- the extrinsic approach and
- the intrinsic approach.

In the extrinsic approach, we use the surrounding space (universe) in order to describe the geometry of the earth. Typically, we use tangent vectors and normal vectors. In the intrinsic approach, we do not use the surrounding space.

In what follows, we will only use invariant formulas, that is, we will not use local coordinates. However, all the proofs can be obtained by using a special local coordinate system (e.g., spherical coordinates). We recommend the reader to give all the missing proofs explicitly by using spherical coordinates (see page 572 and Fig. 9.7). Indeed, the proofs will be given in the next section for general 2-dimensional surfaces. It is also possible to give the missing proofs in an invariant way by applying

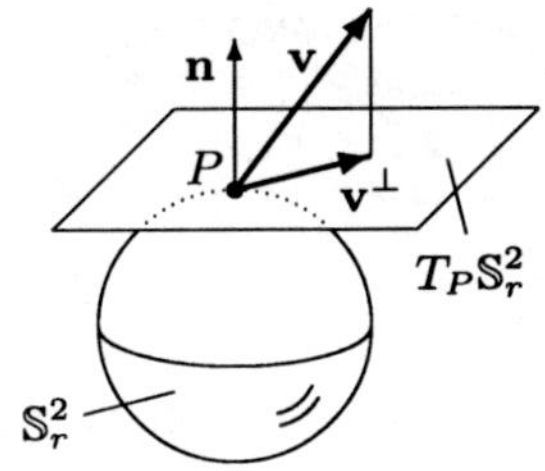

Fig. 9.8. Orthogonal projection

vector analysis to the key formula (9.73) on page 611 for the covariant directional derivative $D_{\mathbf{v}}$.

9.5.1 The Newtonian Equation of Motion and Levi-Civita's Parallel Transport

> The dynamics on the sphere can be understood best by using the notion of covariant acceleration. The Levi-Civita parallel transport of a velocity vector along a curve means that the covariant acceleration vanishes.[14]
>
> Folklore

We will use an approach to the geometry of spheres which is based on physics. Let $\mathcal{R}$ be an open (time) interval on the real line. Consider the motion

$$\mathbf{x} = \mathbf{x}(t), \qquad t \in \mathcal{R}$$

of a point of mass $m > 0$ on the sphere

$$\mathbb{S}_r^2 := \{P \in \mathbb{E}^3 : \ \mathbf{x}^2 = r^2\}.$$

Here, $\mathbf{x} = \overrightarrow{OP}$ is the position vector pointing from the origin O to the point P of the Euclidean manifold $\mathbb{E}^3$. The Newton equation of the mass point in the Euclidean manifold reads as

$$m\ddot{\mathbf{x}}(t) = \mathbf{F}(\mathbf{x}(t)), \qquad t \in \mathcal{R}. \tag{9.45}$$

The basic equation for the motion of the mass point on the sphere $\mathbb{S}_r^2$ reads as

$$m\ddot{\mathbf{x}}(t)^{\perp} = \mathbf{F}(\mathbf{x}(t))^{\perp}, \qquad t \in \mathcal{R}. \tag{9.46}$$

The symbol $\perp$ means that we project orthogonally the classic acceleration vector $\ddot{\mathbf{x}}(t)$ (in the Euclidean manifold $\mathbb{E}^3$) onto the tangent plane $T_{P(t)}\mathbb{S}_r^2$ of the sphere at the point $P(t)$ (which corresponds to the final point of the position vector $\mathbf{x}(t)$; see Fig. 9.8). In contrast to the equation of motion (9.45) in the Euclidean manifold, the equation of motion (9.46) on the sphere takes the constraining forces into account. This additional force guarantees that the mass point does not leave the sphere (earth). Introducing the covariant acceleration vector

[14] T. Levi-Civita, Parallel transport and Riemannian curvature, Rend. Palermo **42**, (1917), 73-205.

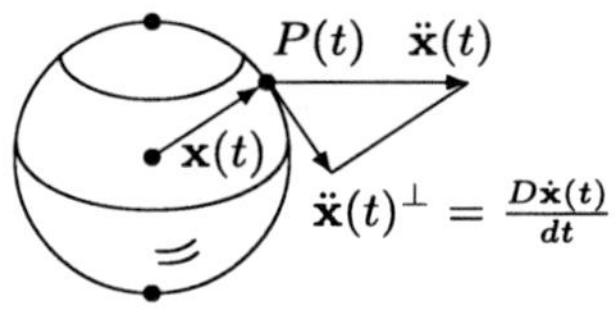

Fig. 9.9. Covariant acceleration vector

$$\frac{D\dot{\mathbf{x}}(t)}{dt} := \dot{\mathbf{x}}(t)^{\perp},$$

the equation (9.46) of motion on the sphere can be written as

$$\boxed{m\frac{D\dot{\mathbf{x}}(t)}{dt} = \mathbf{F}(\mathbf{x}(t))^{\perp}, \qquad t \in \mathcal{R},} \tag{9.47}$$

where $\mathbf{F}^{\perp}$ is the tangential component of the force $\mathbf{F}$ (Fig. 9.9). If the force vanishes, $\mathbf{F} \equiv 0$, then we get

$$\frac{D\dot{\mathbf{x}}(t)}{dt} \equiv 0, \qquad t \in \mathcal{R}. \tag{9.48}$$

On the Euclidean manifold $\mathbb{E}^3$, a trajectory is a straight line iff the acceleration vector vanishes identically. Similarly, a smooth curve

$$C : P = P(t), \qquad t \in \mathcal{R}$$

on the sphere $\mathbb{S}^2_r$ is called a generalized straight line iff the covariant acceleration vector vanishes identically, that is, the equation (9.48) is satisfied. Generalized straight lines are also called affine geodesics (or briefly geodesics).

For example, the motion of a point along the equator with constant angular velocity ω is given by the equation

$$\mathbf{x}(t) := r(\cos\omega t\ \mathbf{i} + \sin\omega t\ \mathbf{j}), \qquad t \in \mathbb{R}.$$

This situation is depicted in Fig. 9.10. We expect that the velocity vectors represent a parallel transport along the equator. In fact, differentiation with respect to times yields

$$\dot{\mathbf{x}}(t) = \omega r(-\sin\omega t\ \mathbf{i} + \cos\omega t\ \mathbf{j}), \quad \ddot{\mathbf{x}}(t) = -\omega^2 r\mathbf{x}(t), \qquad t \in \mathbb{R}.$$

Thus, the tangent component of the acceleration vector vanishes, that is, the trajectory represents a geodesic.

Explicitly, the covariant acceleration vector reads as

$$\frac{D\dot{\mathbf{x}}(t)}{dt} = \ddot{\mathbf{x}}(t) - (\mathbf{n}_{P(t)}\ddot{\mathbf{x}}(t)) \cdot \mathbf{n}_{P(t)} \tag{9.49}$$

where $\dot{\mathbf{x}}(t) = \frac{d\mathbf{x}(t)}{dt}$ (classical velocity vector in $\mathbb{E}^3$), and $\ddot{\mathbf{x}}(t) = \frac{d^2\mathbf{x}(t)}{dt^2}$ (classical acceleration vector in $\mathbb{E}^3$). In addition, $\mathbf{n}_P$ is the outer unit normal vector of the sphere at the point P.

Levi-Civita parallel transport of a velocity vector along a curve on the sphere. We are given the smooth curve $C : \mathbf{x} = \mathbf{x}(t),\ t \in \mathcal{R}$. Let $\mathbf{v}$ be a smooth velocity vector field along the curve C. We set

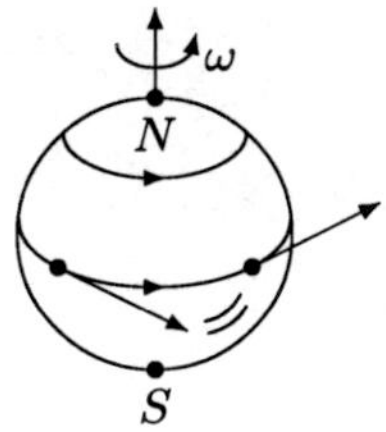

Fig. 9.10. Parallel transport of velocity vectors along the equator

$$\mathbf{v}(t) := \mathbf{v}(P(t)), \qquad t \in \mathcal{R}.$$

By definition, the velocity vector field $t \mapsto \mathbf{v}(t)$ is parallel along the curve C if and only if

$$\boxed{\frac{D\mathbf{v}(t)}{dt} := 0, \qquad t \in \mathcal{R}.} \tag{9.50}$$

This is the crucial Levi-Civita parallel transport on the sphere $\mathbb{S}_r^2$. Explicitly, by (9.49), this means that

$$\frac{D\mathbf{v}(t)}{dt} = \frac{d\mathbf{v}(t)}{dt} - \left(\mathbf{n}_{P(t)}\frac{d\mathbf{v}(t)}{dt}\right)\mathbf{n}_{P(t)}.$$

Covariant acceleration and the geodesic curvature of a curve on the sphere. Consider the trajectory $C : \mathbf{x} = \mathbf{x}(t), t \in \mathcal{R}$, on the sphere $\mathbb{S}_r^2$ where the time t equals the arc length s. We define

$$\kappa_g(s) := \frac{D\dot{\mathbf{x}}(s)}{ds}\big(\mathbf{n}_{P(s)} \times \dot{\mathbf{x}}(s)\big).$$

The real number $\kappa_g(s)$ is called the geodesic curvature of the trajectory at the curve point $P(s)$.[15] Using both the inner product on the tangent space $T_{P(s)}\mathbb{S}_r^2$ at the curve point $P(s)$ and the rotation operator J in the tangent space $T_{P(s)}\mathbb{S}_r^2$ (counter-clockwise rotation about the angle $\frac{\pi}{2}$), then

$$\boxed{\kappa_g(s) = \left\langle \frac{D\dot{\mathbf{x}}(s)}{ds}\middle|\mathsf{J}\dot{\mathbf{x}}(s)\right\rangle.} \tag{9.51}$$

This shows that the geodesic curvature κ_g is defined intrinsically. Since

$$|\kappa_g(s)| = \left|\frac{D\dot{\mathbf{x}}(s)}{ds}\right|,$$

the geodesic curvature κ_g measures the strength of the covariant acceleration.

[15] This is a quite natural generalization of the curvature κ of a curve in the Euclidean plane with the unit normal vector $\mathbf{n}$. Then, $\kappa(s) = \ddot{\mathbf{x}}(s)(\mathbf{n}\times\dot{\mathbf{x}}(s))$. Equivalently, $\kappa(s) = \mathbf{n}(\dot{\mathbf{x}}(s) \times \ddot{\mathbf{x}}(s))$. Since $\dot{\mathbf{x}}(s)$ is a unit vector which is orthogonal to the vector $\mathbf{n}$, we get $|\kappa(s)| = |\ddot{\mathbf{x}}(s)|$.

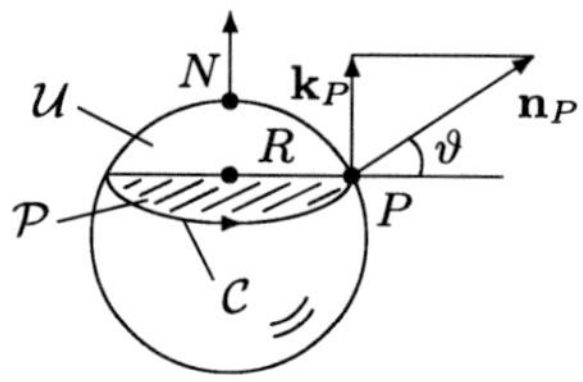

Fig. 9.11. Geodesic curvature of a latitude circle

Geometrically, the geodesic curvature measures the deviation from a geodesic.

In fact, the curve C is a geodesic iff the covariant acceleration vanishes identically, $\kappa_g \equiv 0$. Finally, let us mention that

$$\kappa_g(s) = (\dot{\mathbf{x}}(s) \times \ddot{\mathbf{x}}(s))\, \mathbf{n}_{P(s)}. \tag{9.52}$$

This follows from

$$\kappa_g(s) = \ddot{\mathbf{x}}(s)^{\perp}(\mathbf{n}_{P(s)} \times \dot{\mathbf{x}}(s)) = \ddot{\mathbf{x}}(s)(\mathbf{n}_{P(s)} \times \dot{\mathbf{x}}(s)).$$

Example (latitude circle). Consider the latitude circle of geographic latitude ϑ depicted in Fig. 9.11:

$$\mathbf{x}(s) = R\left(\cos\tfrac{s}{R}\,\mathbf{i} + \cos\tfrac{s}{R}\,\mathbf{j}\right) + r\cos\vartheta\,\mathbf{k}$$

with $s = R\varphi$ and $R = r\cos\vartheta$. We claim that

$$\boxed{\kappa_g(s) = \frac{\sin\vartheta}{R} = \frac{\tan\vartheta}{r}.}$$

Here, $\frac{1}{R}$ is the curvature of the latitude circle regarded as a plane curve. Thus, the equator is the only latitude circle about the north pole with vanishing geodesic curvature.

Proof. It follows from

- $\dot{\mathbf{x}}(s) = (-\sin\frac{s}{R}\,\mathbf{i} + \cos\frac{s}{R})\,\mathbf{j}$, and
- $\ddot{\mathbf{x}}(s) = \frac{1}{R}(-\cos\frac{s}{R}\,\mathbf{i} - \sin\frac{s}{R}\,\mathbf{j})$

that $\dot{\mathbf{x}}(s) \times \ddot{\mathbf{x}}(s) = \frac{1}{R}\mathbf{k}_P$. This implies $\kappa_g(s) = \frac{1}{R}\mathbf{k}_P\mathbf{n}_P = \frac{\sin\vartheta}{R}$. □

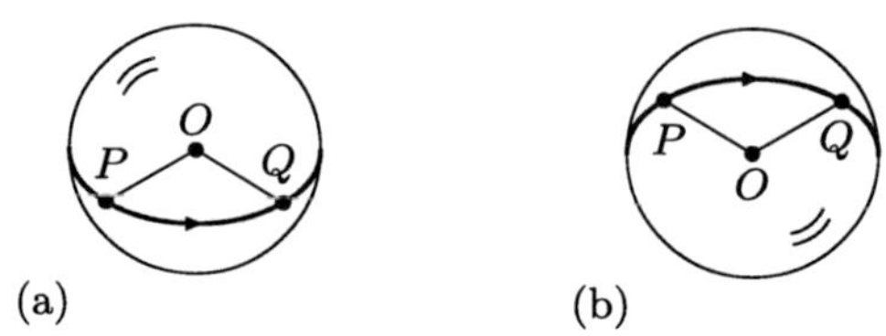

Fig. 9.12. Great circles on the sphere

(a) two focal points (b) no focal point

Fig. 9.13. Focal points on the sphere

9.5.2 Geodesic Triangles and the Gaussian Curvature

Euclidean geometry in the plane is based on the following notions: straight line, triangle, circle. In spherical geometry this is replaced by geodesic line, geodesic triangle, geodesic circle, respectively. Spherical geometry was founded by the astronomer Menelaos of Alexandria in about 100 A.D.

Geodesic lines. Consider two points P and Q on the equator (Fig. 9.12(a)). Intuitively, the equator segment PQ is the curve of minimal length which connects the points P and Q. If we are given two arbitrary points on the sphere $\mathbb{S}_r^2$, then we reduce the situation to the preceding one. To this end, we choose a new equator E_* in such a way that the two points lie on E_*. This new equator can be obtained by choosing a plane through the three points P, Q, O (origin) (Fig. 9.12(b)). The intersection between the plane and the sphere $\mathbb{S}_r^2$ yields E_*. All the circles obtained this way are called great circles. On the earth, for example, all the meridians and the equator are great circles. The latitude circles different from the equator are not great circles. An aircraft always flies along a great circle arc in order to connect P with Q. If the aircraft flies from the North Pole to the South Pole, then it can use any meridian. In this case, the arc of minimal length is not uniquely determined. This is the phenomenon of focal points known in geometric optics (Fig. 9.13).

Geodesic triangles. In the Euclidean plane, consider the triangle depicted in Fig. 9.14(a). For the sum of inner angles α, β, γ, we get

$$\alpha + \beta + \gamma = \pi.$$

In contrast to this situation in Euclidean geometry, we obtain

$$\boxed{\alpha + \beta + \gamma = \pi + K \cdot \text{meas}(\mathcal{U})} \tag{9.53}$$

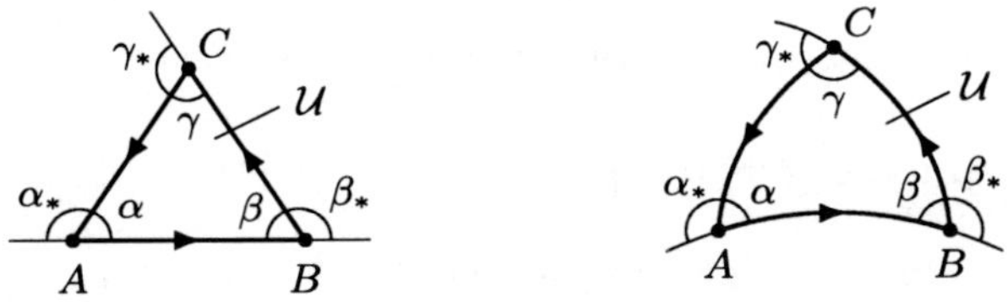

(a) (b)

Fig. 9.14. Geodesic triangles

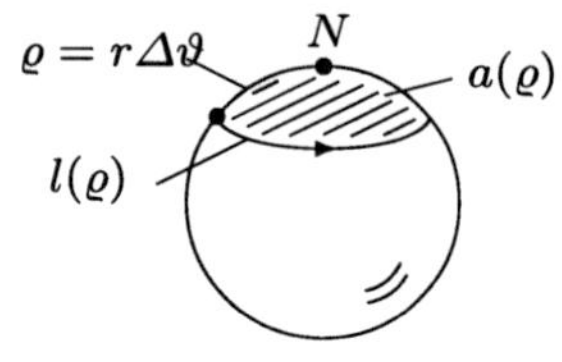

Fig. 9.15. Geodesic circle

for the geodesic triangle on the sphere $\mathbb{S}^2$ depicted in Fig. 9.14(b). Here, we assume that the three sides of the triangle are generalized line segments (i.e., great circle arcs). Moreover,

$$K = \frac{1}{r^2}$$

is the Gaussian curvature of the sphere $\mathbb{S}^2_r$, and meas($\mathcal{U}$) is the measure of the solid triangle $\mathcal{U}$ on the sphere. The relation (9.53) is a special case of the famous Gauss–Bonnet theorem (see page 635).

9.5.3 Geodesic Circles and the Gaussian Curvature

Geodesic circles on a sphere correspond to circles in the Euclidean plane. If the center of the geodesic circle is called the north pole, then a geodesic circle of radius ϱ centered at the north pole N corresponds to a latitude circle of geographic latitude ϑ (Fig. 9.15). Here,

$$\varrho = r(\tfrac{\pi}{2} - \vartheta).$$

This latitude circle has the following properties:

- $l(\varrho) = 2\pi r \sin\frac{\varrho}{r}$ (circumference of the circle),
- $a(\varrho) = 2\pi r^2(1 - \cos\frac{\varrho}{r})$ (surface area of the corresponding disc).

By Taylor expansion, we get

- $l(\varrho) = 2\pi\varrho - \frac{\pi}{3}K\varrho^3 + O(\varrho^4)$, $\varrho \to 0$,
- $a(\varrho) = \pi\varrho^2 - \frac{\pi}{12}K\varrho^4 + O(\varrho^5)$, $\varrho \to 0$.

This shows how the deviations from Euclidean geometry depend on the Gaussian curvature K for small geodesic circles. This implies

$$\boxed{K = \frac{3}{\pi}\lim_{\varrho\to 0}\frac{2\pi\varrho - l(\varrho)}{\varrho^3} = \frac{12}{\pi}\lim_{\varrho\to 0}\frac{\pi\varrho^2 - a(\varrho)}{\varrho^4}.} \tag{9.54}$$

Formula (9.54) remains valid on arbitrary real 2-dimensional Riemannian manifolds.

9.5.4 The Spherical Pendulum

> The dynamics of the spherical pendulum under the influence of the gravitational force on the surface of earth is described by means of elliptic integrals. If the gravitational force is switched off, then the trajectories of the pendulum point are affine geodesics (generalized straight lines).
>
> Folklore

Extrinsic Approach – Lagrange Multiplier and Constraining Force

Consider the motion of a particle of mass m on the sphere $\mathbb{S}_r^2$ under the influence of the vertical gravitational force $\mathbf{F} = -ma\mathbf{k}$ on the surface of the earth. Here, we use $a = 9.81\text{m/s}^2$. The particle has the kinetic energy

$$E_{\text{kin}} := \tfrac{1}{2}m\dot{\mathbf{x}}(t)^2$$

and the potential energy $U = maz$ defined by $\mathbf{F} = -\,\mathbf{grad}\,U$. The principle of critical action reads as

$$\int_{t_0}^{t_1} \left(\tfrac{1}{2}m\dot{\mathbf{x}}(t)^2 - maz(t)\right) dt = \text{critical!} \tag{9.55}$$

with the constraint

$$\mathbf{x}(t)^2 - r^2 = 0, \qquad t_0 \le t \le t_1, \tag{9.56}$$

and the following boundary condition: $\mathbf{x}(t_0)$ and $\mathbf{x}(t_1)$ are fixed on the sphere $\mathbb{S}_r^2$. Introduce the following two functions:

- $L := E_{\text{kin}} - U = \frac{1}{2}m\dot{\mathbf{x}}^2 - maz$ (Lagrangian),
- $\mathcal{L} := L + \frac{1}{2}\lambda(\mathbf{x}^2 - r^2)$ (constrained Lagrangian).

Let $t \mapsto \mathbf{x}(t)$ be a solution of (9.55), (9.56). By the Lagrange multiplier rule, there exists a function $t \mapsto \lambda(t)$ such that

$$\frac{d}{dt}\mathcal{L}_{\dot{\mathbf{x}}} = \mathcal{L}_{\mathbf{x}}$$

along the trajectory $t \mapsto \mathbf{x}(t)$. Since $\mathcal{L}_{\dot{\mathbf{x}}} = m\dot{\mathbf{x}}$, and

$$\mathcal{L}_{\mathbf{x}} = \mathbf{grad}\,\mathcal{L} = -ma\mathbf{k} + \lambda\mathbf{x},$$

this yields the equation of motion

$$\boxed{m\ddot{\mathbf{x}}(t) = -ma\mathbf{k} + \lambda(t)\mathbf{x}(t).} \tag{9.57}$$

In terms of physics, we get the additional force $\lambda(t)\mathbf{x}(t)$. This so-called constraining force keeps the particle on the sphere $\mathbb{S}_r^2$. The Lagrange multiplier λ has to be chosen in such a way that the normal component of the total force vanishes. It follows from $(-ma\mathbf{k} + \lambda\mathbf{x})\mathbf{x} = 0$ that

$$\lambda(t) = \frac{ma}{r^2}\,\mathbf{k}\mathbf{x}(t).$$

Hence $\ddot{\mathbf{x}}(t)^{\perp} = -ma\mathbf{k}^{\perp} = -ma\mathbf{k}_{\text{tang}}(P(t))$. Finally, this yields the equation of motion

$$\frac{D\dot{\mathbf{x}}(t)}{dt} = -ma\mathbf{k}_{\text{tang}}(P(t)).$$

With respect to spherical coordinates, we get

$$-ma(\mathbf{k})_{\text{tang}}(P(t)) = -\frac{ma\cos\vartheta(t)}{r}\,\mathbf{b}_2(P(t)).$$

Note that $\mathbf{k}\mathbf{b}_1 = 0$ (see page 572).

Intrinsic Approach

The extrinsic approach gives insight into the physical structure of the total force. For computing explicitly the motion of the spherical pendulum, it is more convenient to use the following intrinsic approach. To this end, let us use regular spherical coordinates. Then the Lagrangian

$$L = E_{\text{kin}} - U = \tfrac{1}{2}m\dot{\mathbf{x}}^2 - maz$$

passes over to

$$L(\varphi, \vartheta, \dot{\vartheta}) = \tfrac{1}{2}mr^2 \cos^2\vartheta\, \dot{\varphi}^2 + \tfrac{1}{2}mr^2\dot{\vartheta}^2 - amr\sin\vartheta.$$

Now the principle of critical action reads as

$$\int_{t_0}^{t_1} L(\varphi(t), \vartheta(t), \dot{\vartheta}(t))\, dt = \text{critical!} \tag{9.58}$$

together with the following boundary conditions: $\varphi(t_0), \varphi(t_1)$ and $\vartheta(t_0), \vartheta(t_1)$ are fixed. In contrast to the extrinsic approach, there do not appear any constraints.

> *It was discovered by Lagrange (1736–1813) that the use of suitable intrinsic coordinates avoids constraints.*

This was the beginning of the use of n-dimensional manifolds in physics. Here, the dimension n is the number of degrees of freedom. For a gas considered as a mechanical system of molecules, the number n is extremely large ($n \sim 10^{23}$).

The Euler–Lagrange equations. Assume that $\varphi = \varphi(t), \vartheta = \vartheta(t)$ for all times $t \in [t_0, t_1]$ is a solution of the variational problem (9.58). Then the solution satisfies the Euler–Lagrange equations

$$\frac{d}{dt}L_{\dot{\varphi}} = L_{\varphi}, \qquad \frac{d}{dt}L_{\dot{\vartheta}} = L_{\vartheta}.$$

Explicitly, we get

$$\frac{d}{dt}(\dot{\varphi}(t)\cos^2\vartheta(t)) = 0, \tag{9.59}$$

and

$$\ddot{\vartheta}(t) = -\dot{\varphi}(t)^2 \sin\vartheta(t)\cos\vartheta(t) - \frac{am}{r}\cos\vartheta(t). \tag{9.60}$$

Conserved quantities. In order to solve (9.59), (9.60), one uses the conservation of both the energy E and the z-component A_z of angular momentum. Let us discuss this by using spherical coordinates (see page 572). The energy is given by

$$E := E_{\text{kin}} + U = \tfrac{1}{2}mr^2\dot{\varphi}(t)^2\cos^2\vartheta(t) + \tfrac{1}{2}mr^2\dot{\vartheta}(t)^2 + amr\sin\vartheta(t),$$

and the z-component of angular momentum is given by

$$A_z := m\big(\mathbf{x}(t) \times \dot{\mathbf{x}}(t)\big)\mathbf{k} = m\big(x(t)\dot{y}(t) - \dot{x}(t)y(t)\big) = 2mr^2\dot{\varphi}(t)\,\cos^2\vartheta(t).$$

Observe the following symmetry properties of the spherical pendulum:

- The Lagrangian L is invariant under rotations about the z-axis, that is, L does not depend on the angle φ. This yields (9.59) which implies that A_z is a conserved quantity.

- The Lagrangian L is invariant under time translations, that is, it does not depend on time t. By the Noether theorem (see Sect. 6.6.2 of Vol. II), this implies that
$$\dot{\varphi}\,\frac{\partial L}{\partial \dot{\varphi}} + \dot{\vartheta}\,\frac{\partial L}{\partial \dot{\vartheta}} - L$$
is a conserved quantity. An explicit computation shows that this coincides with energy conservation.

We will show in Problem 9.7 that the conservation of E and A_z is sufficient for computing the motion of the spherical pendulum. At this point, let us summarize the final result.

The dynamics of the spherical pendulum. To simplify notation, let us choose a system of units where $r = m = 1$. We are given $\varphi(0), \dot{\varphi}(0), \vartheta(0)$, and $\dot{\vartheta}(0)$ at the initial time $t = 0$. This yields the energy E and the z-component A_z of the angular momentum. Let $z_0 := \sin\vartheta(0)$. Introduce the third-order polynomial

$$P(z) := 2(E - az)(1 - z^2) - \frac{A_z^2}{4}.$$

Proposition 9.20 *The motion $t = t(z), \varphi = \varphi(z)$ of the spherical pendulum is given by the elliptic integrals*

$$t(z) = \int_{z_0}^{z} \frac{d\zeta}{\sqrt{P(\zeta)}}, \qquad \varphi(z) = \varphi(0) + \int_{z_0}^{z} \frac{A_z\, d\zeta}{2(1 - z^2)\sqrt{P(\zeta)}}.$$

This shows that the motion of a spherical pendulum can be rather complex. Next we want to study the motion of a spherical pendulum if the gravitational force vanishes. This leads us to geodesics which play a fundamental role in differential geometry and in the theory of general relativity. Many interesting applications in classical mechanics and celestial mechanics can be found in:

W. Neutsch and K. Scherer, Celestial Mechanics: An Introduction to Classical and Contemporary Methods, Wissenschaftsverlag, Mannheim, 1992.

R. Abraham and J. Marsden, Foundations of Mechanics, Addison-Wesley, Reading, Massachusetts, 1978.

F. Klein and A. Sommerfeld, The Theory of the Top. English edition: Vol. 1: Birkhäuser, Basel, 2008; Vol. 2: Springer, Berlin, 2010. German edition: Teubner, Leipzig, Parts 1–4, 1897.

9.5.5 Geodesics and Gauge Transformations

In order to save fuel, aircrafts fly along geodesics of minimal arc length.

Folklore

The Principle of Critical Action

Consider the principle of critical action

$$\int_{t_0}^{t_1} \tfrac{1}{2} m\dot{\mathbf{x}}(t)^2 dt = \text{critical!} \tag{9.61}$$

with the constraint $\mathbf{x}(t)^2 - r^2 = 0, t_0 \leq t \leq t_1$, and the following boundary condition: $\mathbf{x}(t_0)$ and $\mathbf{x}(t_1)$ are fixed. In terms of physics, this corresponds to the motion of a spherical pendulum of mass $m > 0$ and vanishing gravitational force, that is, $a = 0$. By (9.57), every solution of (9.61) satisfies the equation

$$\boxed{\frac{D\dot{\mathbf{x}}(t)}{dt} = 0, \qquad t_0 \le t \le t_1.} \tag{9.62}$$

Precisely the solutions of this equation are called geodesics. In terms of physics, geodesics are characterized by vanishing covariant acceleration. Therefore, geodesics are also called generalized straight lines.

Proposition 9.21 *For the solutions of equation* (9.62), *the parameter t is proportional to arc length s.*

Proof. By the covariant Leibniz rule (9.75) on page 612, we get

$$\frac{d}{dt}\left\langle \dot{\mathbf{x}}(t) \middle| \dot{\mathbf{x}}(t) \right\rangle = 2\left\langle \frac{D\dot{\mathbf{x}}}{dt} \middle| \dot{\mathbf{x}}(t) \right\rangle = 0.$$

Hence $\frac{ds(t)}{dt} = |\dot{\mathbf{x}}(t)| = \text{const}$ for all times $t \in [t_0, t_1]$. □

Spherical coordinates. With respect to regular spherical coordinates, equation (9.62) reads as

$$\ddot{\varphi} - \dot{\varphi}\,\dot{\vartheta}\,\tan\vartheta = 0, \qquad \ddot{\vartheta} + \dot{\varphi}^2 \sin\vartheta\cos\vartheta = 0. \tag{9.63}$$

Geodesics as subarcs of great circles. A geodesic is uniquely determined by the initial condition $\mathbf{x}(0) = \mathbf{x}_0$ and $\dot{\mathbf{x}}(0) = \dot{\mathbf{x}}_0$. After rotation, if necessary, we may assume that $\varphi(0) = 0$, $\dot{\varphi}(0) = \dot{\varphi}_0$, and $\vartheta(0) = \dot{\vartheta}(0) = 0$. Then the unique solution of (9.63) reads as

$$\varphi(t) = \dot{\varphi}_0 t, \quad \vartheta(t) = 0, \qquad t_0 \le t \le t_1. \tag{9.64}$$

This describes the motion of a particle along the equator with constant angular speed $\dot{\varphi}_0$.

For given points P_0 and P_1 on the sphere, the great circle passing through P_0 and P_1 is defined to be the intersection between the sphere $\mathbb{S}^2_r$ (centered at the origin O) and the plane through the three points P_0, P_1 and O (see Fig. 9.12 on page 598). For example, the equator and the meridians are great circles. It follows from (9.64) that geodesics are parts of great circles (or parts of multiples of great circles). For example, the curve winding twice around the equator is a geodesic. In the next section, we will study geodesics of minimal arc length.

The Christoffel symbols. Set $\xi^1 := \varphi, \xi^2 := \vartheta$. Comparing the Euler–Lagrange equation (9.63) with $\ddot{\xi}^k + \Gamma^k_{ij}\dot{\xi}^i\dot{\xi}^j = 0, k = 1, 2$, we get

$$\Gamma^1_{12} = \Gamma^1_{21} = -\tan\vartheta, \qquad \Gamma^2_{11} = \sin\vartheta\cos\vartheta. \tag{9.65}$$

The remaining Γ^k_{ij} vanish identically. This is a very effective method for computing the Christoffel symbols.

Curves of Minimal Arc Length and Gauge Transformations

We want to study geodesics which are curves of minimal arc length. Using the compact time interval $[t_0, t_1]$, let us introduce the following two functionals:

- $S(C) := \int_{t_0}^{t_1} \frac{1}{2} m\dot{\mathbf{x}}(t)^2 \, dt$ (action functional),[16]
- $L(C) := \int_{t_0}^{t_1} |\dot{\mathbf{x}}(t)| \, dt$ (length functional).

We want to study the minimum problem

$$\boxed{L(C) = \min!, \qquad C \in \mathcal{C}.} \tag{9.66}$$

Here, the symbol $\mathcal{C}$ denotes the space of all smooth curves $C : \mathbf{x} = \mathbf{x}(t), t \in [t_0, t_1]$, on the sphere $\mathbb{S}_r^2$ with fixed end points $\mathbf{x}(t_0)$ and $\mathbf{x}(t_1)$. In addition, the curves C in the space $\mathcal{C}$ are assumed to have a tangent at each curve point, that is, $\dot{\mathbf{x}}(t) \neq 0$ for all $t \in [t_0, t_1]$. Naturally enough, the solutions of (9.66) are called curves of minimal arc length.

Theorem 9.22 *The problem* (9.66) *of minimal arc length is equivalent to the problem*

$$S(C) = \min!, \qquad C \in \mathcal{C} \tag{9.67}$$

of minimal action. In particular, every curve of minimal arc length on the sphere $\mathbb{S}_r^2$ *is a geodesic.*

Proof. (I) (9.66) $\Rightarrow$ (9.67). Let C_0 be a solution of (9.66). The Schwarz inequality tells us that

$$\left(\int_{t_0}^{t_1} |\dot{\mathbf{x}}(t)| \, dt \right)^2 \leq \int_{t_0}^{t_1} |\dot{\mathbf{x}}(t)|^2 \, dt \int_{t_0}^{t_1} dt$$

with equality iff $|\dot{\mathbf{x}}(t)| \equiv \text{const}$. Hence

$$L(C)^2 \leq 2(t_1 - t_0)S(C) \qquad \text{for all} \quad C \in \mathcal{C}. \tag{9.68}$$

The integral $L(C)$ is invariant under changing the parameter t by a diffeomorphism. Since the curves C in $\mathcal{C}$ have the property that $\dot{\mathbf{x}}(t) \neq 0$ for all $t \in [t_1, t_2]$, the curve C can be parametrized by the arc length. Hence $|\dot{\mathbf{x}}(t)| \equiv 1$ if the parameter t equals the arc length. This implies

$$L(C)^2 = 2(t_1 - t_0)S(C) \qquad \text{for all} \quad C \in \mathcal{C}. \tag{9.69}$$

In particular,

$$L(C_0)^2 = 2(t_1 - t_0)S(C_0). \tag{9.70}$$

This implies $2(t_1 - t_0)S(C_0) = L(C_0)^2 \leq L(C)^2 \leq 2(t_1 - t_0)S(C)$ for all $C \in \mathcal{C}$. Thus, the curve C_0 is a solution of the minimum problem (9.67).

(II) (9.67) $\Rightarrow$ (9.66). Let C_0 be a solution of (9.67). By Prop. 9.21, we have $|\dot{\mathbf{x}}_0(t)| \equiv \text{const}$ for all $t \in [t_0, t_1]$. This implies (9.70). It follows from (9.69) that C_0 is a solution of (9.66). □

Gauge transformations. Motivated by physics, the transformation of the curve parameter t is called a gauge transformation. Specializing the parameter t to arc length is called a gauge fixing. We say that the length functional $C \mapsto L(C)$ is invariant under gauge transformations. This is not true for the action functional $C \mapsto S(C)$. However, the solutions of the problem (9.67) of minimal action have the nice property that the Euler–Lagrange equation fixes the gauge, by Prop. 9.21.

[16] To simplify notation, we set $m = 1$

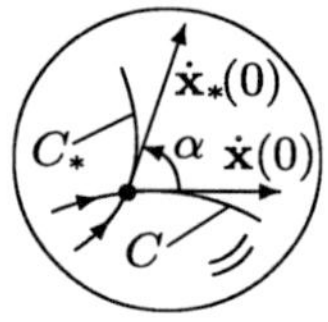

Fig. 9.16. Intersection angle α

This fact (called the gauge fixing trick) is essentially used in the proof of Theorem 9.22 above.

In Volume IV we will show that the gauge fixing trick with respect to conformal transformations is essential for solving the problem for surfaces with minimal area (i.e., minimal surfaces).[17]

> *The theory of both geodesics and minimal surfaces are important examples for simplifying mathematical existence proofs by gauge fixing.*

Gauge invariance (i.e., conformal invariance) plays a crucial role in string theory, too. Again we refer to Volume IV.

9.5.6 The Local Hilbert Space Structure

Metric tensor. Fix the point $P \in \mathbb{S}_r^2$. We define

$$\mathbf{g}_P(\mathbf{u},\mathbf{v}) := \mathbf{u}\mathbf{v} \qquad \text{for all} \quad \mathbf{u},\mathbf{v} \in T_P\mathbb{S}_r^2.$$

Every tangent space of the sphere becomes a real 2-dimensional Hilbert space with respect to the inner product $\mathbf{g}_P$. We also write $\langle \mathbf{u}|\mathbf{v}\rangle_P$ instead of $\mathbf{g}_P(\mathbf{u},\mathbf{v})$. This allows us to define the angle α between two intersecting curves (Fig. 9.16). Explicitly, suppose that the two curves

$$C : \mathbf{x} = \mathbf{x}(t), \qquad C_* : \mathbf{x}_* = \mathbf{x}_*(t), \qquad -t_0 < t < t_0$$

intersect each other at the point P at time $t = 0$. Set $\mathbf{u} := \dot{\mathbf{x}}(0)$ and $\mathbf{v} := \dot{\mathbf{x}}_*(0)$. Then

$$\boxed{\cos\alpha = \frac{\langle \mathbf{u}|\mathbf{v}\rangle}{|\mathbf{u}|\cdot|\mathbf{v}|}.}$$

Arc length. If $C : \mathbf{x} = \mathbf{x}(t)$, $t_0 \le t \le t_1$, is a curve on the sphere, then the integral

$$\boxed{s = \int_{t_0}^{t_1} |\dot{\mathbf{x}}(t)|\, dt}$$

is defined to be the arc length of the curve.

The volume form (symplectic form). For all tangent vectors $\mathbf{u},\mathbf{v} \in T_P\mathbb{S}_r^2$, we define the volume form by setting

$$v_P(\mathbf{u},\mathbf{v}) := (\mathbf{u}\times\mathbf{v})\mathbf{n}$$

[17] In about 1930, this trick was used by Jesse Douglas (1897–1965) in order to solve the classical minimal surface problem. For this seminal contribution to mathematics, Douglas was awarded the first Fields medal in 1936.

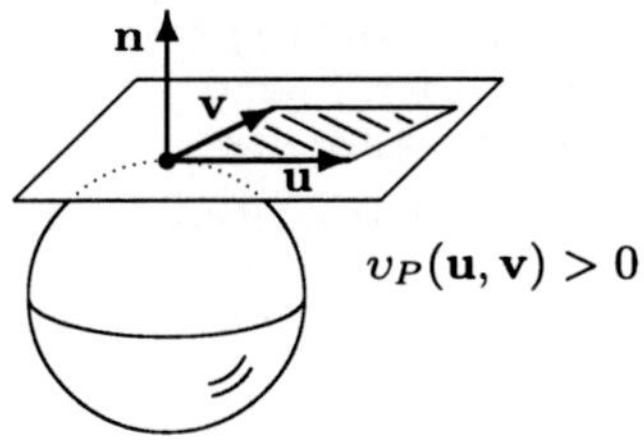

Fig. 9.17. Volume form

where $\mathbf{n}$ is the outer unit normal vector of the sphere (Fig. 9.17). Note that v is antisymmetric. Let $\mathcal{U}$ be an open or closed subset of the sphere $\mathbb{S}_r^2$. The integral

$$\boxed{\operatorname{meas}(\mathcal{U}) = \int_{\mathcal{U}} v}$$

is defined to be the surface area of the set $\mathcal{U}$. For example, in regular spherical coordinates, we get

$$\mathbf{g}_P = r^2 \cos^2 \vartheta \, d\varphi \otimes d\varphi + r^2 d\vartheta \otimes d\vartheta, \qquad v = r^2 \cos \vartheta \, d\varphi \wedge d\vartheta.$$

9.5.7 The Almost Complex Structure

Consider the operator

$$\mathsf{J} : T_P\mathbb{S}_r^2 \to T_P\mathbb{S}_r^2$$

which describes a counter-clockwise rotation of the tangent plane at the point P with the rotation angle $\frac{\pi}{2}$.[18] If $\mathbf{e}$ is a unit tangent vector at the point P, then the two vectors

$$\mathbf{e}, \ \mathsf{J}\mathbf{e}$$

form a right-handed orthonormal basis of the tangent space $T_P\mathbb{S}_r^2$. Every tangent vector $\mathbf{v}$ can be uniquely written as

$$\mathbf{v} = a\mathbf{e} + b \cdot \mathsf{J}\mathbf{e}, \qquad a, b \in \mathbb{R}.$$

Moreover, $\mathsf{J}^2 = -\mathrm{id}$. This resembles the relation $\mathrm{i}^2 = -1$ for the complex number i. In fact, the map

$$a\mathbf{e} + b \cdot \mathsf{J}\mathbf{e} \mapsto a + b\mathrm{i}$$

is a linear bijective operator from the 2-dimensional real tangent space $T_P\mathbb{S}_r^2$ onto the complex plane $\mathbb{C}$ (regarded as a 2-dimensional linear space). In addition,

$$\mathsf{J}(a\mathbf{e} + b \cdot \mathsf{J}\mathbf{e}) = a \cdot \mathsf{J}\mathbf{e} - b\mathbf{e}$$

corresponds to $\mathrm{i}(a + b\mathrm{i}) = a\mathrm{i} - b$. Therefore, we say that the tangent plane $T_P\mathbb{S}_r^2$ is equipped with an almost complex structure by the operator J. For all tangent vectors $\mathbf{u}, \mathbf{v} \in \mathrm{Vect}(\mathbb{S}_r^2)$, we have

$$\boxed{v_P(\mathbf{u}, \mathbf{v}) = -\langle \mathbf{u} | \mathsf{J}\mathbf{v} \rangle_P.} \tag{9.71}$$

This is the fundamental relation between

[18] Explicitly, $\mathsf{J}\mathbf{v} = \mathbf{n}_P \times \mathbf{v}$ for all $\mathbf{v} \in T_P\mathbb{S}_r^2$.

- the Hilbert space structure (inner product),
- the symplectic structure (volume form), and
- the almost complex structure (operator J)

of the tangent spaces of the sphere. The relation (9.71) follows from

$$v_P(\mathbf{u}, \mathbf{v}) = (\mathbf{u} \times \mathbf{v})\mathbf{n} = (\mathbf{uvn}) = (\mathbf{vnu}) = (\mathbf{v} \times \mathbf{n})\mathbf{u} = -\langle \mathbf{u}|\mathsf{J}\mathbf{v}\rangle.$$

9.5.8 The Levi-Civita Connection on the Tangent Bundle and the Riemann Curvature Tensor

> Newton (1643–1727) and Leibniz (1646–1716) introduced the classical derivative of functions. This yields the directional derivative $d_{\mathbf{v}}\mathbf{w}$ of velocity vector fields $\mathbf{w}$ on the Euclidean manifold $\mathbb{E}^3$. In order to describe the curvature of a manifold $\mathcal{M}$, one has to replace the Euclidean directional derivative $d_{\mathbf{v}}\mathbf{w}$ by the covariant directional derivative
>
> $$D_{\mathbf{v}}\mathbf{w}$$
>
> where $\mathbf{v}$ and $\mathbf{w}$ are smooth velocity vector fields on the base manifold $\mathcal{M}$. Thus, the infinitesimal change of velocity vector fields determines the curvature of the underlying manifold $\mathcal{M}$. Both the points of $\mathcal{M}$ (e.g., a sphere) and the velocity vectors on $\mathcal{M}$ are described by a global manifold
>
> $$T\mathcal{M}$$
>
> which is called the tangential bundle of the base manifold $\mathcal{M}$. The generalization of the tangent bundle leads to vector bundles which allow us to replace velocity vector fields $\mathbf{v}$ by general physical fields ψ on $\mathcal{M}$. This generalization is the key to the Standard Model in particle physics.
>
> Folklore

The tangent bundle $T\mathbb{S}_r^2$ of the sphere $\mathbb{S}_r^2$. By definition, the tangent bundle $T\mathbb{S}_r^2$ consists of all the tuples

$$(P, \mathbf{v})$$

where P is a point of the sphere $\mathbb{S}_r^2$, and $\mathbf{v}$ is a velocity vector (synonymously, tangent vector) of the sphere $\mathbb{S}_r^2$ at the point P. The Levi-Civita parallel transport describes a special curve

$$t \mapsto (P(t), \mathbf{v}(t))$$

on the tangent bundle $T\mathbb{S}_r^2$. Fix the point $P \in \mathbb{S}_r^2$. The set

$$F_P := \{(P, \mathbf{v})\}$$

is called the fiber of the tangent bundle $T\mathbb{S}_r^2$. Obviously, $F_P = T_P\mathbb{S}_r^2$. In other words, the fibers are the tangent spaces of the sphere. Set $\pi(P, \mathbf{v}) := P$. The operator

$$\pi : T\mathbb{S}_r^2 \to \mathbb{S}_r^2$$

is called the projection of the bundle space $T\mathbb{S}_r^2 \to \mathbb{S}_r^2$ onto the base manifold $\mathbb{S}_r^2$. The map

$$\boxed{s : \mathbb{S}_r^2 \to T\mathbb{S}_r^2}$$

is called a section of the tangent bundle $T\mathbb{S}_r^2$ iff $s(P) \in F_P$ for all $P \in \mathbb{S}_r^2$. In other words, a section is a tangent vector field on the sphere $\mathbb{S}_r^2$. In order to parametrize the tangent bundle $T\mathbb{S}_r^2$, we choose a local (u^1, u^2)-coordinate system. Then, the point P of the sphere can locally be described by the coordinates (u^1, u^2). Moreover, if $\mathbf{x} = \mathbf{x}(u^1, u^2)$, then the natural basis vectors

$$\mathbf{b}_j(P) := \frac{\partial \mathbf{x}(u^1, u^2)}{\partial u^j}, \qquad j = 1, 2$$

at the point $P(u^1, u^2)$ span the tangent space $T_P\mathbb{S}_r^2$. Every tangent vector $\mathbf{v} \in T_P\mathbb{S}_r^2$ can be uniquely represented as

$$\mathbf{v} = v^1\mathbf{b}_1(P) + v^2\mathbf{b}_2(P). \tag{9.72}$$

This way, we assign to the point $(P, \mathbf{v}) \in T\mathbb{S}_r^2$ the local bundle coordinate

$$(u^1, u^2, v^1, v^2) \in \mathbb{R}^4.$$

Using local bundle coordinates, the tangent bundle $T\mathbb{S}_r^2$ becomes a 4-dimensional real manifold which is locally isomorphic to $\mathcal{O} \times \mathbb{R}^2$ where $\mathcal{O}$ is an open subset of the sphere $\mathbb{S}_r^2$. The symbol

$$\mathrm{Sect}(T\mathbb{S}_r^2)$$

describes the set of all the smooth sections $s : \mathbb{S}_r^2 \to T\mathbb{S}_r^2$. Synonymously, we also use the symbol

$$\mathrm{Vect}(\mathbb{S}_r^2).$$

In terms of geography, we describe the tangent bundle $T\mathbb{S}_r^2$ by

- an atlas of 2-dimensional geographic charts for the earth $\mathbb{S}_r^2$, and
- an atlas of 4-dimensional geographic charts for the tangent bundle $T\mathbb{S}_r^2$ which describes the points of earth and the possible velocity vectors at the points.

The cotangent bundle $T^*\mathbb{S}_r^2$. This bundle consists of all the pairs

$$(P, \omega) \qquad \text{with} \qquad P \in \mathbb{S}_r^2, \;\; \omega \in T_P^*\mathbb{S}_r^2$$

where $\omega : T_P\mathbb{S}_r^2 \to \mathbb{R}$ is a linear functional on the tangent space $T_P\mathbb{S}_r^2$ at the point P. In other words, ω is an element of the cotangent space $T_P^*\mathbb{S}_r^2$ of the sphere at the point P. The elements of $T_P^*\mathbb{S}_r^2$ are called covectors (or differential 1-forms).

Using a local (u^1, u^2)-coordinate system and the decomposition (9.72), we define

$$du^i(\mathbf{v}(P)) = v^i(P), \qquad i = 1, 2, \;\; P \in \mathbb{S}_r^2.$$

In particular, $du^i(\mathbf{b}_j)(P) = \delta_i^j$, $i, j = 1, 2$. Then every element ω of the cotangent space $T_P^*\mathbb{S}_r^2$ of the sphere $\mathbb{S}_r^2$ at the point P can be written as

$$\omega = \omega_1 du^1 + \omega_2 du^2$$

where the real coordinates ω_1, ω_2 are uniquely determined by ω. Thus, a point (P, ω) of the cotangent bundle $T_P^*\mathbb{S}_r^2$ can be locally parametrized by the real coordinates

$$(u^1, u^2; \omega_1, \omega_2).$$

This way, the bundle space $T^*\mathbb{S}_r^2$ becomes a 4-dimensional manifold which is locally diffeomorphic to the product manifold $\mathcal{O} \times \mathbb{R}^2$ where $\mathcal{O}$ is an open subset of $\mathbb{R}^2$. A smooth covector field (or a smooth differential 1-form) on the sphere $\mathbb{S}_r^2$ is a smooth section

$$\boxed{s : \mathbb{S}_r^2 \to T^*\mathbb{S}_r^2}$$

of the cotangent bundle $T^*\mathbb{S}_r^2$. That is, we assign to every point P of the sphere $\mathbb{S}_r^2$ a differential 1-form ω_P at the point P, and the local coordinates of ω_P depend smoothly on the point P of the sphere $\mathbb{S}_r^2$.

The strategy of translating linear algebra to vector bundles (parametrized linear algebra). The tangent space $T_P\mathbb{S}_r^2$ is a real linear space. All the objects A on $T_P\mathbb{S}_r^2$ from linear algebra (e.g., vectors, linear functionals, linear operators, multilinear functionals) become a parametrized family

$$\mathcal{B} := \{A\}_{P \in \mathbb{S}_r^2}.$$

This is also called an abstract bundle. The family of tuples

$$(P, A)$$

with the points $P \in \mathbb{S}_r^2$ and the admissible objects A form the bundle space $\mathcal{B}$. Set $\pi(P, A) := P$. The map

$$\pi : \mathcal{B} \to \mathbb{S}_r^2$$

is called the projection map. The set

$$F_P := \pi^{-1}(P)$$

is called the fiber of the bundle space $\mathcal{B}$ over the base point P. Fibers with distinct base points are disjoint sets. This way, we get the fibration

$$\mathcal{B} = \cup_{P \in \mathbb{S}_r^2} F_P$$

of the bundle space $\mathcal{B}$. The map

$$\boxed{s : \mathbb{S}_r^2 \to \mathcal{B}}$$

is called a section iff $s(P) \in F_P$ for all points $P \in \mathbb{S}_r^2$. Physical fields on the sphere $\mathbb{S}_r^2$ can be described by sections of appropriate bundles (velocity vector fields, electric and magnetic fields, or tensor fields). In order to be able to describe the smoothness of sections (physical fields), it is necessary to introduce local coordinates on the bundle space $\mathcal{B}$.

> *The same way, all the structures appearing in mathematics can be replaced by bundles (families of parametrized structures).*

This idea was very fruitful for the mathematics and physics of the 20th century.

The vector bundle $\mathrm{End}(T\mathbb{S}_r^2)$**.** As an example, consider the space $\mathrm{End}(T_P\mathbb{S}_r^2)$ of all the linear operators

$$A : T_P\mathbb{S}_r^2 \to T_P\mathbb{S}_r^2.$$

The bundle space $\mathrm{End}(T\mathbb{S}_r^2)$ consists of all the tuples

$$(P, A) \qquad \text{with } \; P \in \mathbb{S}_r^2 \; A \in \mathrm{End}(T_P\mathbb{S}_r^2).$$

Using local coordinates and describing the linear operators by matrices, the bundle space $\mathrm{End}(T\mathbb{S}_r^2)$ becomes the structure of a 8-dimensional real manifold.[19]

The tensor bundle $T_n^m(\mathbb{S}_r^2)$**.** The bundle space $\mathcal{B}$ consists of all the tuples

[19] Note that the dimension of the linear space $\mathrm{End}(T_P\mathbb{S}_r^2)$ is equal to four.

$$(P, M) \qquad \text{with } P \in \mathbb{S}_r^2, \;\; M \in \bigotimes\nolimits_n^m (T_P\mathbb{S}_r^2).$$

Here, M is a tensor of type (m, n) on the tangent space $T_P(\mathbb{S}_r^2)$. For example, if $m = 2, n = 1$, then the map

$$M : X \times X^d \times X^d \to \mathbb{R}$$

is trilinear where $X := T_P\mathbb{S}_r^2$. With respect to local coordinates on $\mathbb{S}_r^2$, we get

$$M = T_i^{jk} du^i \otimes \mathbf{b}_j \otimes \mathbf{b}_k.$$

Thus, $M(\mathbf{v}, \omega, \mu) = T_i^{jk} v^i \omega_j \mu_k$.[20] This way, the bundle space $\mathcal{B}$ can locally be described by the local bundle coordinates

$$(u^1, u^2; T_i^{jk})_{i,j,k=1,2}.$$

In particular, $T_0^1(\mathbb{S}_r^2)$ (resp. $T_1^0(\mathbb{S}_r^2)$) coincides with the tangent bundle $T\mathbb{S}_r^2$ (resp. cotangent bundle $T^*\mathbb{S}_r^2$).The smooth sections

$$\boxed{s : \mathbb{S}_r^2 \to T_n^m(\mathbb{S}_r^2)}$$

are smooth tensor fields of type (m, n) on the sphere $\mathbb{S}_r^2$. The set of all these smooth tensor fields is denoted by $\bigotimes_n^m(\mathbb{S}_r^2)$.

Covariant Directional Derivative and Tensor Analysis

The constant Gaussian curvature of a sphere $\mathbb{S}_r^2$ of radius r is given by

$$\boxed{K = \frac{1}{r^2}.}$$

This definition depends on the surrounding Euclidean manifold $\mathbb{E}^3$. We want to construct an intrinsic curvature theory based on a covariant directional derivative $D_\mathbf{v}\mathbf{w}$ (connection on the tangent bundle $T\mathbb{S}_r^2$). Mnemonically, the key formula reads as follows:

$$\boxed{D_\mathbf{v}\mathbf{w} := d_\mathbf{v}\mathbf{w}^\perp \qquad \text{for all} \quad \mathbf{v}, \mathbf{w} \in \text{Vect}(\mathbb{S}_r^2).}$$

The symbol $\perp$ stands for orthogonal projection onto the tangent plane. Explicitly,

$$\boxed{D_\mathbf{v}\mathbf{w} = d_\mathbf{v}\mathbf{w} - (\mathbf{n} d_\mathbf{v}\mathbf{w})\,\mathbf{n}.} \tag{9.73}$$

This mnemonic formula means that

$$\boxed{(D_\mathbf{v}\mathbf{w})_P = (d_\mathbf{v}\mathbf{w})_P - (\mathbf{n}_P (d_\mathbf{v}\mathbf{w})_P)\,\mathbf{n}_P} \tag{9.74}$$

where $\mathbf{n}_P$ is the outer unit normal vector of the sphere $\mathbb{S}_r^2$ at the point $P \in \mathbb{S}_r^2$. It follows from (9.73) that the following hold for all smooth velocity vector fields $\mathbf{v}, \mathbf{w}, \mathbf{z} \in \text{Vect}(\mathbb{S}_r^2)$ and all smooth real-valued functions f, g on $\mathbb{S}_r^2$:

[20] Note that $du^i(\mathbf{v}) = v^i$ and $\mathbf{b}_j(\omega) = \omega_j$.

- $D_{f\mathbf{v}+g\mathbf{w}}\mathbf{z} = fD_{\mathbf{v}}\mathbf{z} + gD_{\mathbf{w}}\mathbf{z}$ (linearity),
- $D_{\mathbf{v}}(\mathbf{w}+\mathbf{z}) = D_{\mathbf{v}}\mathbf{w} + D_{\mathbf{v}}\mathbf{z}$ (linearity),
- $D_{\mathbf{v}}(f\mathbf{w}) = d_{\mathbf{v}}f \cdot \mathbf{w} + fD_{\mathbf{v}}\mathbf{w}$ (special Leibniz rule).

The Leibniz rule and parallel transport. We have

$$\boxed{d_{\mathbf{u}}(\langle \mathbf{v}|\mathbf{w}\rangle_P) = \langle D_{\mathbf{u}}\mathbf{v}_P|\mathbf{w}_P\rangle + \langle \mathbf{v}_P|D_{\mathbf{u}}\mathbf{w}_P\rangle} \tag{9.75}$$

for all velocity vector fields $\mathbf{u}, \mathbf{v}, \mathbf{w} \in \mathrm{Vect}(\mathbb{S}^2_r)$ and all points $P \in \mathbb{S}^2_r$. This follows from the classic Leibniz rule together with orthogonal projection:

$$d_{\mathbf{u}}(\langle \mathbf{v}|\mathbf{w}\rangle_P) = \langle d_{\mathbf{u}}\mathbf{v}_P|\mathbf{w}_P\rangle + \langle \mathbf{v}_P|d_{\mathbf{u}}\mathbf{w}_P\rangle = \langle d_{\mathbf{u}}\mathbf{v}_P^{\perp}|\mathbf{w}_P\rangle + \langle \mathbf{v}_P|d_{\mathbf{u}}\mathbf{w}_P^{\perp}\rangle.$$

In Riemannian geometry, this is also called the Ricci lemma for the metric tensor.

Theorem 9.23 *The Levi-Civita parallel transport of two velocity vectors along the same curve preserves the inner product on the tangent spaces.*

This implies that the length of the velocity vectors and the angle between the two velocity vectors remains unchanged (with respect to the inner product on the tangent spaces). In other words:

> *The Levi-Civita parallel transport respects the local Hilbert space structure of the sphere.*

Extension of the covariant directional derivative. Let us show that it is possible to extend the covariant directional derivative $D_{\mathbf{v}}$ to other physical fields on the sphere $\mathbb{S}^2$. Let $\mathbf{v}, \mathbf{w}, \mathbf{z} \in \mathrm{Vect}(\mathbb{S}^2_r)$. In what follows, we will sum over equal upper and lower indices from 1 to 2.

(i) Covector field (differential 1-forms) ω on $\mathbb{S}^2_r$: We want to ensure the validity of the following Leibniz rule:

$$d_{\mathbf{v}}\big(\omega(\mathbf{w})\big) = (D_{\mathbf{v}}\omega)(\mathbf{w}) + \omega(D_{\mathbf{v}}\mathbf{w}).$$

To this end, we define $(D_{\mathbf{v}}\omega)(\mathbf{w}) := d_{\mathbf{v}}\big(\omega(\mathbf{w})\big) - \omega(D_{\mathbf{v}}\mathbf{w})$.

(ii) Field of smooth 2-differential forms ω on $\mathbb{S}^2_r$: To get the Leibniz rule

$$d_{\mathbf{v}}\big(\omega(\mathbf{w},\mathbf{z})\big) = (D_{\mathbf{v}}\omega)(\mathbf{w},\mathbf{z}) + \omega(D_{\mathbf{v}}\mathbf{w},\mathbf{z}) + \omega(\mathbf{w},D_{\mathbf{v}}\mathbf{z}),$$

we define

$$(D_{\mathbf{v}}\omega)(\mathbf{w},\mathbf{z}) := d_{\mathbf{v}}\big(\omega(\mathbf{w},\mathbf{z})\big) - \omega(D_{\mathbf{v}}\mathbf{w},\mathbf{z}) - \omega(\mathbf{w},D_{\mathbf{v}}\mathbf{z}).$$

(iii) Smooth operator field $\mathrm{End}(T\mathbb{S}^2_r)$ on $\mathbb{S}^2_r$: To get the Leibniz rule

$$D_{\mathbf{v}}(A\mathbf{w}) = (D_{\mathbf{v}}A)\mathbf{w} + A(D_{\mathbf{v}}\mathbf{w}),$$

we define $(D_{\mathbf{v}}A)\mathbf{w} := D_{\mathbf{v}}(A\mathbf{w}) - A(D_{\mathbf{v}}\mathbf{w})$.

(iv) Smooth tensor field of type (m,n) on $\mathbb{S}^2_r$: For example, let $m=2, n=1$. Similarly, as above, we define $D_{\mathbf{v}}M$ in such a way that there holds the following Leibniz rule:

$$\begin{aligned} d_{\mathbf{v}}(M(\mathbf{w},\omega,\mu)) &= (D_{\mathbf{v}}M)(\mathbf{w},\omega,\mu) + M(D_{\mathbf{v}}\mathbf{w},\omega,\mu) \\ &\quad + M(\mathbf{w},D_{\mathbf{v}}\omega,\mu) + M(\mathbf{w},\omega,D_{\mathbf{v}}\mu) \end{aligned}$$

for all smooth differential 1-forms ω and μ, and all smooth velocity vector fields $\mathbf{w}$ on the sphere $\mathbb{S}^2_r$.

Tensor products. The definition (iv) above implies the following Leibniz rule

$$\boxed{D_{\mathbf{v}}(M \otimes N) = (D_{\mathbf{v}}M) \otimes N + M \otimes D_{\mathbf{v}}N}$$

for all smooth tensor fields M and N. If M and N are of the same type (m,n), then we have the following sum rule:

$$D_{\mathbf{v}}(M + N) = D_{\mathbf{v}}M + D_{\mathbf{v}}N.$$

The covariant differential. We are given the smooth velocity vector field $\mathbf{w} \in \mathrm{Vect}(\mathbb{S}^2_r)$. Then the covariant differential $D\mathbf{w}$ is defined by

$$\boxed{(D\mathbf{w})(\mathbf{v}) := D_{\mathbf{v}}\mathbf{w}}$$

for all smooth velocity vector fields $\mathbf{v} \in \mathrm{Vect}(\mathbb{S}^2_r)$.

Local coordinates. Choose local (u^1, u^2)-coordinates. Let us define the Christoffel symbols by setting

$$\Gamma^k_{ij}(P) := du^k(D_{\mathbf{b}_i}\mathbf{b_j}), \qquad i,j = 1,2.$$

Then

$$\boxed{D_{\mathbf{b}_i}\mathbf{b}_j := \Gamma^k_{ij}\mathbf{b}_k, \qquad i,j = 1,2.}$$

Setting $g_{ij} = \mathbf{b}_i\mathbf{b}_j$, it follows as in the proof of Theorem 9.12 on page 580 that

$$\Gamma^k_{ij} = \tfrac{1}{2}g^{ks}(\partial_i g_{js} + \partial_j g_{is} - \partial_s g_{ij}).$$

Recall the covariant partial derivative ∇_i from (8.102) on page 495:

$$\nabla_i w^k = \partial_i w^k + \Gamma^k_{ij}w^j.$$

If $\mathbf{v} = v^i\mathbf{b}_i$ and $\mathbf{w} = w^i\mathbf{b}_i$, then it follows from the Leibniz rule that

$$\boxed{D_{\mathbf{v}}\mathbf{w} = (v^s\nabla_s w^k)\,\mathbf{b}_k.}$$

In fact, we get

$$D_{\mathbf{v}}\mathbf{w} = (d_{\mathbf{v}}w^j)\,\mathbf{b}_j + w^j D_{\mathbf{v}}\mathbf{b}_j = (v^s\partial_s w^k + v^i\Gamma^k_{ij}w^j)\,\mathbf{b}_k = (v^s\nabla_s w^k)\,\mathbf{b}_k.$$

It turns out that

The covariant differential calculus described above coincides with the covariant differential calculus based on ∇_i in Chap. 8.

For example,

$$D_{\mathbf{v}}(T^k_{ij}\,\mathbf{b}_k \otimes dx^i \otimes dx^j) = (v^s\nabla_s T^k_{ij})\,\mathbf{b}_k \otimes dx^i \otimes dx^j.$$

Regular spherical coordinates. Set $u^1 := \varphi, u^2 := \vartheta$. In the special case of spherical coordinates, the sphere $\mathbb{S}^2_r$ is locally represented by the equation

$$\mathbf{x} = r(\cos\vartheta\cos\varphi\,\mathbf{i} + \cos\vartheta\sin\varphi\,\mathbf{j} + \sin\vartheta\,\mathbf{k})$$

where $-\pi < \varphi < \pi$ and $-\frac{\pi}{2} < \vartheta < \frac{\pi}{2}$ (see the discussion about regular and singular spherical coordinates on page 572). This yields the natural basis vectors

- $\mathbf{b}_1(\varphi, \vartheta) = \mathbf{x}_\varphi = r(-\cos\vartheta \sin\varphi\,\mathbf{i} + \cos\vartheta\cos\varphi\,\mathbf{j})$,
- $\mathbf{b}_2(\varphi, \vartheta) = \mathbf{x}_\vartheta = r(-\sin\vartheta\cos\varphi\,\mathbf{i} - \sin\vartheta\sin\varphi\,\mathbf{j} + \cos\vartheta\,\mathbf{k})$ (see Fig. 9.6).

Let us add the outer unit normal vector $\mathbf{n} = (\mathbf{b}_1 \times \mathbf{b}_2)/|\mathbf{b}_1 \times \mathbf{b}_2|$. Explicitly,

$$\mathbf{n} = \cos\varphi\cos\vartheta\,\mathbf{i} + \sin\varphi\cos\vartheta\,\mathbf{j} + \sin\vartheta\,\mathbf{k}.$$

For the metric tensorial family, we get

$$\begin{pmatrix} g_{11} & g_{12} \\ g_{21} & g_{22} \end{pmatrix} = \begin{pmatrix} r^2\cos^2\vartheta & 0 \\ 0 & r^2 \end{pmatrix}.$$

Define

$$\mathcal{A}_i := \begin{pmatrix} \Gamma^1_{i1} & \Gamma^1_{i2} \\ \Gamma^2_{i1} & \Gamma^2_{i2} \end{pmatrix}, \qquad \mathcal{A} = \mathcal{A}_i du^i = \begin{pmatrix} \omega^1_1 & \omega^1_2 \\ \omega^2_1 & \omega^2_2 \end{pmatrix}, \qquad i = 1, 2.$$

Then

$$\mathcal{A}_1 = \begin{pmatrix} 0 & -\tan\vartheta \\ -\frac{1}{2}\sin 2\vartheta & 0 \end{pmatrix}, \quad \mathcal{A}_2 = \begin{pmatrix} -\tan\vartheta & 0 \\ 0 & 0 \end{pmatrix},$$

and

$$\mathcal{A} = \begin{pmatrix} -\tan\vartheta\, d\vartheta & -\tan\vartheta\, d\varphi \\ -\frac{1}{2}\sin 2\vartheta\, d\varphi & 0 \end{pmatrix}.$$

The Lie Algebra Vect($\mathbb{S}^2_r$) of Velocity Vector Fields

The space Vect($\mathbb{S}^2$) of smooth velocity vector fields on the sphere $\mathbb{S}^2_r$ becomes a real Lie algebra equipped with the following Lie product.

$$\boxed{[\mathbf{v}, \mathbf{w}] := D_\mathbf{v}\mathbf{w} - D_\mathbf{w}\mathbf{v}.}$$

In Sect. 11.2, we will show that $[\mathbf{v}, \mathbf{w}] = \mathcal{L}_\mathbf{v}\mathbf{w}$ where the Lie derivative $\mathcal{L}_\mathbf{v}\mathbf{w}$ is defined by means of the flow of fluid particles on the sphere $\mathbb{S}^2_r$ which is generated by the velocity vector field $\mathbf{v}$.

Local coordinates and cancellations. Consider a local (u^1, u^2)-coordinate system on the sphere $\mathbb{S}^2_r$ together with the natural basis $\mathbf{b}_1, \mathbf{b}_2$. Then, the symmetry property $\Gamma^k_{ij} = \Gamma^k_{ji}$ of the Christoffel symbols yields

$$\boxed{[\mathbf{v}, \mathbf{w}] = (v^i\partial_i w^k - w^i\partial_i v^k)\mathbf{b}_k.} \tag{9.76}$$

In fact, let $\mathbf{v} = v^i\mathbf{b}_i$ and $\mathbf{w} = w^i\mathbf{b}_i$. It follows from

$$D_\mathbf{v}\mathbf{w} = (v^i\partial_i w^k + v^i\Gamma^k_{ij}w^j)\,\mathbf{b}_k$$

that

$$[\mathbf{v}, \mathbf{w}] = (v^i\partial_i w^k - w^i\partial_i v^k)\mathbf{b}_k + (\Gamma^k_{ij}v^i w^j - \Gamma^k_{ij}w^i v^j)\mathbf{b}_k.$$

□

In particular, we obtain the special Lie product

$$\boxed{[\mathbf{b}_i, \mathbf{b}_j] = 0, \qquad i, j = 1, 2.} \tag{9.77}$$

The Riemann Curvature Operator $\mathbf{F}(\mathbf{u}, \mathbf{v})$

> The Riemann curvature operator on the sphere is the prototype for describing forces in gauge theory (i.e., Einstein's theory of general relativity on gravitative interaction, Maxwell's theory on electromagnetic interaction, and the Standard Model in particle physics on strong and electroweak interaction).
>
> Folklore

In what follows, let $\mathbf{u}, \mathbf{v}, \mathbf{w}, \mathbf{z} \in \text{Vect}(\mathbb{S}_r^2)$ be smooth velocity vector fields on the sphere $\mathbb{S}_r^2$. In order to get insight, we first only consider invariant formulas by using the langauge of vector analysis. Then we will give the proofs by passing to local coordinates. The starting point in our approach is the crucial analytic definition of the Riemann curvature operator:

$$\boxed{\mathbf{F}(\mathbf{u}, \mathbf{v})\mathbf{w} := D_{\mathbf{u}} D_{\mathbf{v}} \mathbf{w} - D_{\mathbf{v}} D_{\mathbf{u}} \mathbf{w} - D_{[\mathbf{u}, \mathbf{v}]} \mathbf{w}.} \tag{9.78}$$

More precisely, for fixed velocity vector fields $\mathbf{u}, \mathbf{v} \in \text{Vect}(\mathbb{S}_r^2)$, the operator

$$\mathbf{F}_P(\mathbf{u}, \mathbf{v}) : T_P\mathbb{S}_r^2 \to T_P\mathbb{S}_r^2$$

is called the Riemann curvature operator on the tangent space of the sphere at the point P. For fixed velocity vectors $\mathbf{u}, \mathbf{v} \in \text{Vect}(\mathbb{S}_r^2)$, the linear operator

$$\mathbf{F}_P(\mathbf{u}, \mathbf{v}) : T_P\mathbb{S}_r^2 \to T_P\mathbb{S}_r^2$$

is called the Riemann curvature operator at the point P (with respect to the tangent vectors $\mathbf{u}$ and $\mathbf{v}$). Mnemonically, we write

$$\mathbf{F}(\mathbf{u}, \mathbf{v}) := D_{\mathbf{u}} D_{\mathbf{v}} - D_{\mathbf{v}} D_{\mathbf{u}} - D_{[\mathbf{u}, \mathbf{v}]}.$$

Big surprise – incredible cancellations. The analytic definition (9.78) contains derivatives of the velocity vector fields of first and second order. The following theorem shows that all of the derivatives of the velocity vector fields are cancelled.

Theorem 9.24 *For all points $P \in \mathbb{S}_r^2$,*

$$\mathbf{F}_P(\mathbf{u}, \mathbf{v})\mathbf{w} = K \cdot (\langle \mathbf{v}|\mathbf{w}\rangle_P \, \mathbf{u} - \langle \mathbf{u}|\mathbf{w}\rangle_P \, \mathbf{v}). \tag{9.79}$$

Therefore, the Riemann curvature operator is a purely algebraic object; it relates the local Hilbert spacc structure of the sphere to the curvature of the sphere. This is a fundamental property of general Riemannian manifolds. Based on local coordinates, the proof of Theorem 9.24 will be given on page 617. The proof shows that, in local coordinates, $\mathbf{F}_P(\mathbf{u}, \mathbf{v})\mathbf{w}$ only depends on the velocity vectors $\mathbf{u}, \mathbf{v}, \mathbf{w}$ and the first and second partial derivatives of the metric tensorial family g_{ij}. Consequently, $\mathbf{F}_P(\mathbf{u}, \mathbf{v})\mathbf{w}$ is an intrinsic quantity. Thus, Theorem 9.24 tells us that:

> *The Gaussian curvature K of the sphere $\mathbb{S}_r^2$ is an intrinsic geometric quantity.*

This is a special case of the famous theorema egregium of Gauss (see Sect. 9.6.6 on page 632.

Gauge fixing. In order to understand best the geometric meaning of the Riemann curvature operator, we choose a special basis $\mathbf{e}_1, \mathbf{e}_2$ of the tangent space.

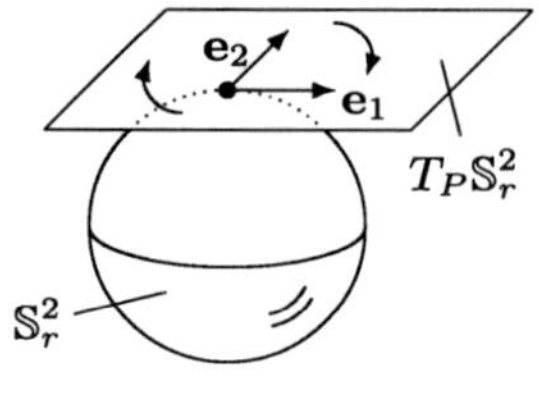

Fig. 9.18. Rotation of an orthonormal frame

More precisely, let $\mathbf{e}_1, \mathbf{e}_2$ be a right-handed orthonormal frame at the point P of the sphere $\mathbb{S}_r^2$ (Fig. 9.18). Explicitly,

$$\langle \mathbf{e}_i | \mathbf{e}_j \rangle_P = \delta_{ij}, \qquad i, j = 1, 2,$$

and $(\mathbf{e}_1 \times \mathbf{e}_2)\mathbf{n} > 0$. By Theorem 9.24, we get

$$\mathbf{F}_P(\mathbf{e}_1, \mathbf{e}_2)\, \mathbf{e}_1 = -K\mathbf{e}_2, \qquad \mathbf{F}_P(\mathbf{e}_1, \mathbf{e}_2)\, \mathbf{e}_2 = K\mathbf{e}_1,$$

and $\mathbf{F}_P(\mathbf{e}_1, \mathbf{e}_2) = -\mathbf{F}_P(\mathbf{e}_2, \mathbf{e}_1)$. Thus, the Riemann curvature operator $\mathbf{F}_P(\mathbf{e}_1, \mathbf{e}_2)$ sends the basis vector $\mathbf{e}_1$ (resp. $\mathbf{e}_2$) to the vector $-K\mathbf{e}_2$ (resp. $K\mathbf{e}_1$). This is a clockwise rotation in the tangent space with the angle $\pi/2$ combined with a stretching by the factor K (Fig. 9.18). By using the linearity with respect to the vectors $\mathbf{u}, \mathbf{v}, \mathbf{w}$, we get

$$\boxed{\mathbf{F}_P(\mathbf{u}, \mathbf{v})\mathbf{w} = -K v_P(\mathbf{u}, \mathbf{v}) \cdot \mathsf{J}\mathbf{w}.} \tag{9.80}$$

Recall that $v_P(\mathbf{u}, \mathbf{v})$ is the volume spanned by the two vectors $\mathbf{u}$ and $\mathbf{v}$, K is the Gaussian curvature, and the vector $\mathsf{J}\mathbf{w}$ is obtained from the vector $\mathbf{w}$ by a counterclockwise rotation with the angle $\pi/2$. Formula (9.80) implies the antisymmetry relation

$$\mathbf{F}_P(\mathbf{u}, \mathbf{v}) = -\mathbf{F}_P(\mathbf{v}, \mathbf{u}).$$

This antisymmetry property is responsible for the fact that Riemannian differential geometry can be based on the language of differential forms (the dual Cartan approach to Riemann's approach). Summarizing:

Metric properties are described by symmetry (metric tensor), and curvature is described by antisymmetry (Riemann's curvature operator).

The combination of the two kinds of symmetry yields the Riemann curvature tensor.

The Riemann Curvature Tensor $\mathcal{R}$

In physics, one measures real numbers. Therefore, we use the Hilbert space structure (i.e., the inner product) of the tangent space $T_P\mathbb{S}_r^2$ in order to define

$$\boxed{\mathcal{R}_P(\mathbf{u}, \mathbf{v}, \mathbf{w}, \mathbf{z}) := \langle \mathbf{F}_P(\mathbf{u}, \mathbf{v})\mathbf{w} | \mathbf{z} \rangle_P.}$$

The 4-linear map

$$\mathcal{R}_P : T_P\mathbb{S}_r^2 \times T_P\mathbb{S}_r^2 \times T_P\mathbb{S}_r^2 \times T_P\mathbb{S}_r^2 \mapsto \mathbb{R}$$

is called the Riemann curvature tensor $\mathcal{R}_P$ at the point P. It follows from Theorem 9.24 that

$$\boxed{\mathcal{R}_P(\mathbf{u}, \mathbf{v}; \mathbf{w}, \mathbf{z}) = K \cdot (\langle \mathbf{v}|\mathbf{w}\rangle_P \langle \mathbf{u}|\mathbf{z}\rangle_P - \langle \mathbf{u}|\mathbf{w}\rangle_P \langle \mathbf{v}|\mathbf{z}\rangle_P).} \tag{9.81}$$

This implies the following symmetries:

- $\mathcal{R}_P(\mathbf{u}, \mathbf{v}; \mathbf{w}, \mathbf{z}) = -\mathcal{R}_P(\mathbf{v}, \mathbf{u}; \mathbf{w}, \mathbf{z}) \quad (\mathbf{u} \Leftrightarrow \mathbf{v})$,
- $\mathcal{R}_P(\mathbf{u}, \mathbf{v}; \mathbf{w}, \mathbf{z}) = -\mathcal{R}_P(\mathbf{u}, \mathbf{v}; \mathbf{z}, \mathbf{w}) \quad (\mathbf{w} \Leftrightarrow \mathbf{z})$,
- $\mathcal{R}_P(\mathbf{u}, \mathbf{v}; \mathbf{w}, \mathbf{z}) = \mathcal{R}_P(\mathbf{w}, \mathbf{z};, \mathbf{u}, \mathbf{v}) \quad ((\mathbf{u}, \mathbf{v}) \Leftrightarrow (\mathbf{w}, \mathbf{z}))$.

From (9.71) and (9.80) we get

$$\boxed{\mathcal{R}_P(\mathbf{u}, \mathbf{v}; \mathbf{w}, \mathbf{z}) = -K\langle \mathbf{u}|\mathsf{J}\mathbf{v}\rangle_P \, \langle \mathbf{w}|\mathsf{J}\mathbf{z}\rangle_P = -K\upsilon_P(\mathbf{u}, \mathbf{v}) \, \upsilon_P(\mathbf{w}, \mathbf{z}).} \tag{9.82}$$

The beautiful formulas (9.81) and (9.82) show how curvature (the Riemann curvature tensor $\mathcal{R}$ and the Gaussian scalar curvature K), Hilbert space structure (inner product), almost complex structure (operator J), and symplectic structure (volume form υ) are related to each other. In particular, we get

$$\boxed{K = \mathcal{R}(\mathbf{e}_1, \mathbf{e}_2; \mathbf{e}_2, \mathbf{e}_1).} \tag{9.83}$$

This is Gauss' theorema egregium for the sphere $\mathbb{S}^2_r$.

9.5.9 The Components of the Riemann Curvature Tensor and Gauge Fixing

> The use of matrices and differential forms simplifies the formulas for the Riemann curvature tensor with respect to local coordinates. The history of differential geometry was strongly influenced by the desire of mathematicians to simplify time-consuming ugly computations by more elegant arguments based on getting more insight into the mathematical structure behind the formulas. Most progress in mathematics is stimulated by the desire of getting insight.
>
> Folklore

Local coordinates. Choose local (u^1, u^2)-coordinates and the corresponding natural basis $\mathbf{b}_1, \mathbf{b}_2$.

Proposition 9.25 (i) $\mathbf{F}(\mathbf{u}, \mathbf{v})\mathbf{w} = (R^s_{ijk} u^i v^j w^k)\mathbf{b}_s$ *where*

$$R^s_{ijk} = \partial_i \Gamma^s_{jk} - \partial_j \Gamma^s_{ik} + \Gamma^s_{ir}\Gamma^r_{jk} - \Gamma^s_{jr}\Gamma^r_{ik}. \tag{9.84}$$

(ii) $\mathcal{R}(\mathbf{u}, \mathbf{v}, \mathbf{w}, \mathbf{z}) = R_{ijkl} u^i v^j w^k z^l$ *where* $R_{ijkl} = R^s_{ijk} g_{sl}$.

Proof. Ad (i). (I) Brute force argument: Recall that

$$\mathbf{F}(\mathbf{u}, \mathbf{v})\mathbf{w} := D_\mathbf{u} D_\mathbf{v} \mathbf{w} - D_\mathbf{v} D_\mathbf{u} \mathbf{w} - D_{[\mathbf{u},\mathbf{v}]}\mathbf{w},$$

and

$$D_\mathbf{v}\mathbf{w} = v^i(\partial_i w^k + \Gamma^k_{ij} w^j) \, \mathbf{b}_k.$$

We recommend the reader to prove the claim (i) by a straightforward computation. There occur incredible cancellations.

(II) Refined argument: Note that $[\mathbf{b}_i, \mathbf{b}_j] = 0$. Hence $D_{[\mathbf{b}_i,\mathbf{b}_j]} = 0$. Furthermore, set $r^s_{ijk} := \mathbf{F}(\mathbf{b}_i, \mathbf{b}_j)\mathbf{b}_k$. By linearity,

$$\mathbf{F}(\mathbf{u}, \mathbf{v})\mathbf{w} = r^s_{ijk} u^i v^j w^k \mathbf{b}_s.$$

Note that $D_{\mathbf{b}_j}\mathbf{b}_k = v^s\mathbf{b}_s$ with $v^s := \Gamma^s_{jk}$. It follows from

$$D_{\mathbf{b}_i}(D_{\mathbf{b}_j}\mathbf{b}_k) = (\partial_i v^s + \Gamma^s_{ir}v^r)\mathbf{b}_s = (\partial_i \Gamma^s_{jk} + \Gamma^s_{ir}\Gamma^r_{jk})\mathbf{b}_s$$

and

$$D_{\mathbf{b}_j}(D_{\mathbf{b}_i}\mathbf{b}_k) = (\partial_j \Gamma^s_{ik} + \Gamma^s_{jr}\Gamma^r_{ik})\mathbf{b}_s$$

that $r^s_{ijk} = R^s_{ijk}$.

Ad (ii). By definition, $\mathcal{R}(\mathbf{u},\mathbf{v},\mathbf{w},\mathbf{z})$ is equal to

$$\langle \mathbf{F}(\mathbf{u},\mathbf{v})\mathbf{w}|\mathbf{z}\rangle = R^s_{ijk}u^i v^j w^k g_{sl} z^l.$$

□

The language of matrices, and the language of differential forms. In order to simplify the clumsy formula (9.84), we set

$$\mathcal{F}_{ij} = (R^l_{ijk}), \qquad \mathcal{F} := \tfrac{1}{2}\mathcal{F}_{ij}\, du^i \wedge du^j, \qquad \Omega^l_k := R^l_{ijk}\, du^i \wedge du^j.$$

Then

$$\mathcal{F}_{ij} = \begin{pmatrix} R^1_{ij1} & R^1_{ij2} \\ R^2_{ij1} & R^2_{ij2} \end{pmatrix}, \qquad \mathcal{F} = \begin{pmatrix} \Omega^1_1 & \Omega^1_2 \\ \Omega^2_1 & \Omega^2_2 \end{pmatrix}.$$

Recalling the definitions $\mathcal{A}_i := (\Gamma^l_{ik})$ and $\mathcal{A} := \mathcal{A}_i du^i$, the clumsy equation (9.84) passes over to Élie Cartan's elegant structural equation

$$\boxed{\mathcal{F} = d\mathcal{A} + \mathcal{A} \wedge \mathcal{A}.}$$

Since $\mathcal{F}$ is a 2-form on a 2-dimensional manifold, the integrability condition (Bianchi identity)

$$d\mathcal{F} = 0$$

is automatically satisfied.

Regular spherical coordinates. In this case, we get

$$\mathcal{F}_{12} = \begin{pmatrix} 0 & 1 \\ -\cos^2\vartheta & 0 \end{pmatrix}, \qquad \mathcal{F}_{21} = -\mathcal{F}_{12}, \qquad \mathcal{F}_{11} = \mathcal{F}_{22} = 0. \tag{9.85}$$

Hence

$$\mathcal{F} = \begin{pmatrix} \Omega^1_1 & \Omega^1_2 \\ \Omega^2_1 & \Omega^2_2 \end{pmatrix} = \begin{pmatrix} 0 & d\varphi \wedge d\vartheta \\ -\cos^2\vartheta\, d\varphi \wedge d\vartheta & 0 \end{pmatrix}.$$

Observe that $R_{1221} = R^s_{122}g_{s1} = R^1_{112}g_{11}$. Similarly,

$$R_{2112} = -R_{1212} = -R_{2121} = R_{1221} = r^2\cos^2\vartheta. \tag{9.86}$$

The remaining components R_{ijkl} vanish identically. Noting that $g = g_{11}g_{22} - g^2_{12}$, and hence $g = r^4\cos^2\vartheta$, we get

$$\boxed{K = \frac{R_{1221}}{g} = \frac{1}{r^2}.} \tag{9.87}$$

This relation for the sphere is a special case of the theorema egregium of Gauss.

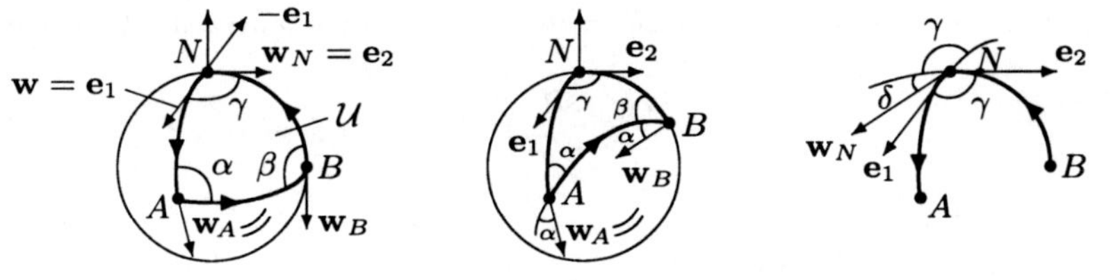

(a) $\alpha = \beta = \gamma = \frac{\pi}{2}$ (b) $\gamma = \frac{\pi}{2}$ (c) $\delta = \alpha + \beta$, $\mathbf{w}_N = \Pi \mathbf{w}$

Fig. 9.19. Parallel transport along a geodesic triangle

9.5.10 Computing the Riemann Curvature Operator via Parallel Transport Along Loops

> It is of fundamental importance for applications of gauge theory to modern physics that the curvature can be measured by the parallel transport of tangent vectors (i.e., velocity vectors) along a loop.
>
> Folklore

We want to show that the Gaussian curvature, the Riemann curvature operator, and the Riemann curvature tensor can be computed by measuring the parallel transport of a velocity vector along a geodesic triangle. This follows from the fact that the parallel transport of velocity vectors preserves both the length of the velocity vectors (measured in tangent spaces) and the angle between two velocity vectors. Since the tangent vectors of a geodesic curve are parallel along the curves, the following holds:

(P) *The parallel transport of a velocity vector along a geodesic curve preserves the length of the velocity vector and the angle between the velocity vector and the geodesic curve.*

Moreover, we will critically use the Gauss-Bonnet theorem (Gauss' theorema elegantissimum).

The prototype. Let $\mathbf{e}_1, \mathbf{e}_2$ be a right-handed orthonormal basis of the tangent space $T_N\mathbb{S}^2_r$ of the sphere $\mathbb{S}^2_r$ at the north pole N. Consider Fig. 9.19(a). We want to investigate the parallel transport

$$\mathbf{w} = \mathbf{e}_1 \Rightarrow \mathbf{w}_A \Rightarrow \mathbf{w}_B \Rightarrow \mathbf{w}_N$$

of the tangent vector $\mathbf{e}_1$ along the geodesic triangle $NABN$. We define the operator Π of parallel transport by setting $\Pi\mathbf{w} := \mathbf{w}_N$. We claim that

$$\Pi \mathbf{e}_1 = \mathbf{e}_2.$$

Proof. In what follows, we will use property (P) above.

- The starting vector $\mathbf{w} = \mathbf{e}_1$ is a tangent vector of the geodesic line NA at the point N. Therefore, $\mathbf{w}_A$ is also a tangent vector of NA at the point A.
- The angle between the vector $\mathbf{w}_A$ and the equator arc AB is equal to $\frac{\pi}{2}$. Thus, $\mathbf{w}_B$ is a tangent vector of the geodesic line BN at the point B.
- Finally, $\mathbf{w}_N$ is a tangent vector of BN at the point N. Hence $\mathbf{w}_N = \mathbf{e}_2$. □

Knowing $\Pi \mathbf{e}_1$, we claim that we can determine the Gaussian curvature by using

$$\boxed{K = \frac{\angle(\mathbf{e}_1, \Pi \mathbf{e}_1)}{\text{meas}(\mathcal{U})}.} \tag{9.88}$$

Here, the symbol $\angle(\mathbf{e}_1, \Pi \mathbf{e}_1)$ is the angle $\theta \in]-\pi, \pi]$ which sends the vector $\mathbf{e}_1$ to the vector $\Pi \mathbf{e}_1$ by counter-clockwise rotation. Moreover, meas$(\mathcal{U})$ is the surface measure of the triangle domain. In the present situation, we have

- $\angle(\mathbf{e}_1, \Pi \mathbf{e}_1)) = \frac{\pi}{2}$,
- $\text{meas}(\mathcal{U}) = \frac{1}{8}\,\text{meas}(\mathbb{S}^2_r) = \frac{\pi}{2} r^2$, and $K = \frac{1}{r^2}$.

If we know the Gaussian curvature K, then we get

- $\mathbf{F}_N(\mathbf{u}, \mathbf{v})\mathbf{w} = -K v_N(\mathbf{u}, \mathbf{v})\, \mathsf{J}\mathbf{w}$ (Riemann curvature operator), and
- $\mathcal{R}_N(\mathbf{u}, \mathbf{v}; \mathbf{w}, \mathbf{z}) = \langle \mathbf{F}_N(\mathbf{u}, \mathbf{v})\mathbf{w} | \mathbf{z} \rangle_N$ (Riemann curvature tensor)

for all tangent vectors $\mathbf{u}, \mathbf{v}, \mathbf{w}, \mathbf{z} \in T_N\mathbb{S}^2_r$ at the north pole. In particular,

$$\mathbf{F}_N(\mathbf{e}_1, \mathbf{e}_2)\, \mathbf{e}_1 = -K\mathbf{e}_2, \;\; \mathbf{F}_N(\mathbf{e}_1, \mathbf{e}_2)\, \mathbf{e}_2 = K\mathbf{e}_1, \;\; \mathcal{R}_N(\mathbf{e}_1, \mathbf{e}_2; \mathbf{e}_2, \mathbf{e}_1) = K.$$

Recall that v is the volume form of the sphere $\mathbb{S}^2_r$, and $\mathsf{J}\mathbf{w}$ is obtained from $\mathbf{w}$ by counter-clockwise rotation with the angle $\frac{\pi}{2}$.

Parallel transport along a general triangle. Consider the triangles depicted in Fig 9.19(b), (c). We claim that the formula (9.88) and the formulas above for the Riemann curvature operator and the Riemann curvature tensor remain valid.

Proof. By Fig. 9.19(b), (c), we have the following angle relations:

- $\angle(\mathbf{w}_A, AB) = \angle(\mathbf{w}_B, AB) = \pi - \alpha$,
- $\angle(BN, \mathbf{w}_B) = \angle(BN, \mathbf{w}_N) = \alpha + \beta$,
- $\angle(\mathbf{w}_N, \mathbf{e}_1) = \pi - \alpha - \beta - \gamma$, $\angle(\mathbf{e}_1, \mathbf{w}_N) = \alpha + \beta + \gamma - \pi$.

By the Gauss–Bonnet theorem, $\alpha + \beta + \gamma - \pi = \int_{\mathcal{U}} Kv = K\,\text{meas}(\mathcal{U})$. Therefore,

$$\frac{\angle(\mathbf{e}_1, \Pi \mathbf{e}_1)}{\text{meas}(\mathcal{U})} = \frac{\alpha + \beta + \gamma - \pi}{\text{meas}(\mathcal{U})} = K.$$

□

9.5.11 The Connection on the Frame Bundle and Parallel Transport

Moving frames. Consider a smooth curve

$$C : P = P(t), \qquad t \in \mathcal{R},$$

on the sphere $\mathbb{S}^2_r$ where $\mathcal{R}$ is an open time interval which contains the time point t_0. By definition, a frame at the point P is an ordered pair $(\mathbf{e}_1, \mathbf{e}_2)$ of linearly independent tangent vectors at the point P. Consider a smooth map

$$t \mapsto (\mathbf{e}_1(t), \mathbf{e}_2(t))$$

which assigns to each time $t \in \mathcal{R}$ a frame $(\mathbf{e}_1(t), \mathbf{e}_2(t))$ at the curve point $P(t)$. In other words, the frames move along the curve C. We assume that the following system of differential equations is satisfied:

$$\boxed{\frac{D\mathbf{e}_i(t)}{dt} = 0, \qquad t \in \mathcal{R}, \quad i = 1, 2.} \tag{9.89}$$

We are given $\mathbf{e}_1(t_0)$ and $\mathbf{e}_2(t_0)$. Geometrically, this means that the tangent vector $\mathbf{e}_i(t)$ at the point $P(t)$ is obtained from the tangent vector $\mathbf{e}_i(t_0)$ at the initial point $P(t_0)$ by parallel transport along the curve C.

Theorem 9.26 *If the initial frame $(\mathbf{e}_1(t_0), \mathbf{e}_2(t_0))$ is a right-handed orthonormal system on the tangent space $T_{P(t_0)}\mathbb{S}^2_r$, then every frame $(\mathbf{e}_1(t), \mathbf{e}_2(t))$ at time $t \in \mathcal{R}$ is a right-handed orthonormal system on every tangent space $T_{P(t)}\mathbb{S}^2_r$.*

Proof. This follows from the fact that parallel transport preserves the length of vectors and the angle between two vectors. Moreover, by continuity, the orientation of a frame is preserved. □

Consequently, if $(\mathbf{e}_1(t_0), \mathbf{e}_2(t_0))$ is a right-handed orthonormal frame, then we have the following matrix equation

$$(\mathbf{e}_1(t), \mathbf{e}_2(t)) = (\mathbf{e}_1(t_0), \mathbf{e}_2(t_0))\, G(t), \qquad t \in \mathcal{R}$$

where $G(t) \in SO(2)$ for all $t \in \mathcal{R}$, and the map $t \mapsto G(t)$ from the time interval $\mathcal{R}$ to the Lie group $SO(2)$ is smooth. Theorem 9.26 tells us that:

> *The global parallel transport on the Euclidean manifold $\mathbb{E}^3$ (Euclidean geometry) is replaced by the Levi-Civita parallel transport on the sphere $\mathbb{S}^2_r$ (locally Euclidean geometry).*

Let v^1, v^2 be real numbers. Consider the velocity vectors

$$\mathbf{v}(t) = v^i \mathbf{e}_i(t), \qquad t \in \mathcal{R}.$$

Then, $\mathbf{v}$ is parallel along the curve C. In fact,

$$\frac{D\mathbf{v}(t)}{dt} = v^i \frac{D\mathbf{e}_i(t)}{dt} = 0.$$

Thus, the frame $(\mathbf{e}_1(t), \mathbf{e}_2(t))$, which moves along the curve C, replaces a global Cartesian coordinate system on the Euclidean manifold. Let us translate the preceding results into the language of fiber bundles.

The frame bundle $F\mathbb{S}^2_r$. By definition, the frame bundle $F\mathbb{S}^2_r$ of the sphere $\mathbb{S}^2_r$ consists of all the tuples

$$(P, \mathbf{e}_1, \mathbf{e}_2)$$

where P is a point of the sphere $\mathbb{S}^2_r$, and $(\mathbf{e}_1, \mathbf{e}_2)$ is a frame at the point P. The parallel transport considered above describes a special curve

$$t \mapsto (P(t), \mathbf{e}_1(t), \mathbf{e}_2(t))$$

on the tangent bundle $T\mathbb{S}^2_r$. Fix the point $P \in \mathbb{S}^2_r$. The set

$$F_P := \{(P, \mathbf{e}_1, \mathbf{e}_2)\}$$

is called the fiber of the frame bundle $F\mathbb{S}^2_r$ over the base point $P \in \mathbb{S}^2_r$. Set $\pi(P, \mathbf{e}_1, \mathbf{e}_2) := P$. The operator

$$\pi : F\mathbb{S}^2_r \to \mathbb{S}^2_r$$

is called the projection of the bundle space $F\mathbb{S}^2_r$ onto the base manifold $\mathbb{S}^2_r$. The map

$$\boxed{s : \mathbb{S}^2_r \to F\mathbb{S}^2_r}$$

is called a section of the frame bundle $F\mathbb{S}^2_r$ iff $s(P) \in F_P$ for all $P \in \mathbb{S}^2_r$. In order to parametrize the frame bundle $F\mathbb{S}^2_r$, we choose a local (u^1, u^2)-coordinate system. Then, the point P of the sphere can locally be described by the coordinates (u^1, u^2). Moreover, if $\mathbf{x} = \mathbf{x}(u^1, u^2)$, then the natural basis vectors

$$\mathbf{b}_j(P) := \frac{\partial \mathbf{x}(u^1, u^2)}{\partial u^j}, \qquad j = 1, 2$$

at the point $P(u^1, u^2)$ are a basis for the tangent space $T_P\mathbb{S}^2_r$. Every frame $(\mathbf{e}_1, \mathbf{e}_2)$ at the point P can be written in the form of the matrix equation

$$(\mathbf{e}_1, \mathbf{e}_2) = (\mathbf{b}_1, \mathbf{b}_2)G$$

where $G \in GL(2, \mathbb{R})$. The tuple

$$(P, G)$$

is called the bundle coordinate of the bundle point $(P, \mathbf{e}_1, \mathbf{e}_2)$. Using bundle coordinates, the frame bundle $F\mathbb{S}^2_r$ becomes a real manifold.

The orthonormal frame bundle. By definition, the orthonormal frame bundle $F^\perp\mathbb{S}^2_r$ of the sphere $\mathbb{S}^2_r$ consists of all the tuples

$$(P, \mathbf{e}_1, \mathbf{e}_2)$$

where P is a point of the sphere $\mathbb{S}^2_r$, and $(\mathbf{e}_1, \mathbf{e}_2)$ is a right-handed orthonormal frame of tangent vectors at the point P. In order to parametrize $F^\perp(\mathbb{S}^2_r)$, we restrict ourselves to choosing orthonormal local coordinates. This means that the natural basis vectors $\mathbf{b}_1(u^1, u^2), \mathbf{b}_2(u^1, u^2)$ form a right-handed orthonormal system. Then the bundle coordinates are of the form (P, G) where $G \in SO(2)$.

Using bundle coordinates, the orthonormal frame bundle $F^\perp\mathbb{S}^2_r$ becomes a three-dimensional real manifold which is locally diffeomorphic to a product bundle.

The gauge construction. Let

$$s : \mathcal{O} \to F\mathbb{S}^2_r$$

be a smooth section where $\mathcal{O}$ is a nonempty open subset of the sphere $\mathbb{S}^2_r$. Then the map s assigns to any point $P \in \mathcal{O}$ a frame $\mathbf{e}_1^+, \mathbf{e}_2^+$. Every point $(P, \mathbf{e}_1, \mathbf{e}_2)$ can be uniquely represented by the matrix equation

$$(\mathbf{e}_1, \mathbf{e}_2) = (\mathbf{e}_1^+, \mathbf{e}_2^+)G$$

where $G \in GL(2, \mathbb{R})$. This way, we assign to the bundle point $(P, \mathbf{e}_1, \mathbf{e}_2)$ the bundle coordinate (P, G). The map s is called a gauge fixing of the frame bundle $F\mathbb{S}^2_r$.

Proposition 9.27 *There does not exist a continuous section*

$$s : \mathbb{S}^2 \to F\mathbb{S}^2.$$

This crucial fact means that it is not possible to choose a global gauge fixing. Physicists noticed this phenomenon in physical gauge theories (Gribov ambiguity). Proposition 9.27 follows from Poincaré's no-go theorem to be considered next.

9.5.12 Poincaré's Topological No-Go Theorem for Velocity Vector Fields on a Sphere

Theorem 9.28 *Every continuous velocity vector field on the sphere* $\mathbb{S}^2_r$ *has a stationary point where the velocity vector vanishes.*

Alternatively, every smooth section $s : \mathbb{S}^2_r \to T\mathbb{S}^2_r$ of the tangent bundle of the sphere has a zero. This means that there exists a point $(P_0, \mathbf{v}(P_0))$ of the tangent bundle of the sphere with $\mathbf{v}(P_0) = 0$.

The classic proof based on the mapping degree can be found in E. Zeidler, Nonlinear Functional Analysis and its Applications. Vol 1: Fixed Point Theory, p. 558, Springer, New York, 1995. The Poincaré no-go theorem allows far reaching generalizations in modern topology in terms of the theory of characteristic classes for vector bundles. We will study this in Vol. IV on quantum mathematics. At this point, we recommend J. Milnor and J. Stasheff, Characteristic Classes, Princeton University Press, 1972.[21]

9.6 Gauss' Theorema Egregium

> I am ill mannered, for I take a lively interest in a mathematical object only where I see a prospect of a clever connection of ideas or of results recommended by elegance or generality.
>
> Carl Friedrich Gauss (1777–1855)

> The greatest mathematicians, such as Archimedes (ca.285–212 B.C.), Newton (1643–1727), and Gauss (1777–1855), always united theory and applications in equal manner.
>
> Felix Klein (1849–1925)

The differential geometry of the sphere considered in the preceding section can be generalized in a straightforward manner to 2-dimensional surfaces. In this section, we will present two different approaches in order to prove the Gauss theorema egregium. The first approach works in the spirit of Gauss and Riemann, the second approach is due to Élie Cartan. The basic tool are differential equations for frames and the corresponding integrability conditions which imply the theorema egregium. Cartan's approach has the advantage that the integrability conditions are an immediate consequence of Poincaré's cohomology rule $dd\omega = 0$ for differential forms.

Summation convention. In this section, we sum over equal upper and lower indices from 1 to 2. In addition, we will use g_{ij} (resp. g^{ij}) introduced below in order to lower (resp. to lift) indices. For example, $h^i_j = g^{is}h_{sj}$ and $h_{ij} = g_{is}h^s_j$.

9.6.1 The Natural Basis and Cobasis

Let us consider the smooth surface

$$\boxed{S: \ \mathbf{x} = \mathbf{x}(u^1, u^2), \qquad (u^1, u^2) \in \mathcal{U}} \tag{9.90}$$

[21] For his seminal contributions to topology, John Milnor (born 1931) was awarded the Fields medal in 1962, the Wolf prize in 1989, and the Abel prize in 2011. We refer to J. Milnor, Collected Works, Vols. 1–5, Amer. Math. Soc., Providence, Rhode Island, 2011.

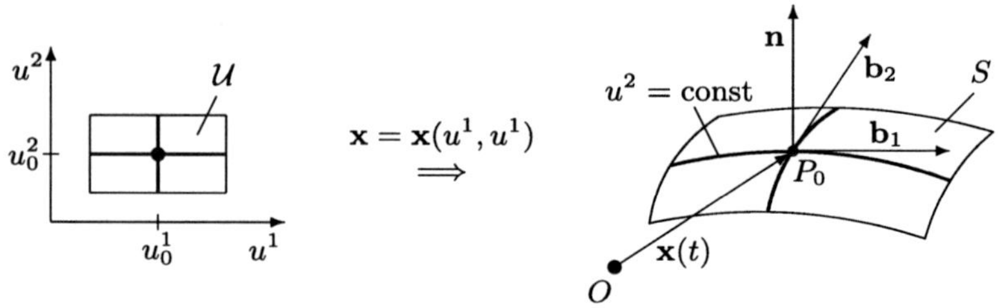

Fig. 9.20. Surface parametrization

where $\mathcal{U}$ is a nonempty, open, arcwise connected subset of $\mathbb{R}^2$. Intuitively, the coordinates (u^1, u^2) describe the surface in a geographic chart (Fig. 9.20). The change of local coordinates

$$u^{i'} = u^{i'}(u^1, u^2), \qquad i' = 1', 2', \tag{9.91}$$

is given by a diffeomorphism $(u^1, u^2) \mapsto (u^{1'}, u^{2'})$ from the open subset $\mathcal{U}$ of $\mathbb{R}^2$ onto the open subset $\mathcal{U}'$ of $\mathbb{R}^2$. We will use the classic tensor calculus from Chap. 8 in order to guarantee that the approach does not depend on the choice of the local coordinates.

The index principle tells us that expressions without free indices are invariants.

Moreover, we will use the language of vector analysis on the Euclidean manifold $\mathbb{E}^3$ in order to guarantee that the results do not depend on the choice of the coordinates in $\mathbb{E}^3$.

Natural basis. We introduce the natural basis vectors

$$\mathbf{b}_i(u^1, u^2) := \frac{\partial \mathbf{x}(u^1, u^2)}{\partial u^i}, \qquad i = 1, 2,$$

and the normal unit vector

$$\mathbf{n}(u^1, u^2) := \frac{\mathbf{b}_1(P) \times \mathbf{b}_2(P)}{|\mathbf{b}_1(P) \times \mathbf{b}_2(P)|}.$$

Here, the point P on the surface S corresponds to the local coordinates (u^1, u^2). Naturally enough, we assume that $\mathbf{b}_1(P) \times \mathbf{b}_2(P) \neq 0$ for all $(u^1, u^2) \in \mathcal{U}$.[22]

Orientation. Our choice of the extrinsic frame $\mathbf{b}_1, \mathbf{b}_2, \mathbf{n}$ as a right-handed system fixes a positive orientation, by definition. The positive orientation is preserved iff the Jacobian of the local coordinate transformation is positive, that is,

$$\frac{\partial(u^{1'} u^{2'})}{\partial(u^1, u^2)} > 0 \qquad \text{for all} \quad (u^1, u^2) \in \mathcal{U}.$$

Natural cobasis. We introduce the natural covector basis du^1, du^2 by setting

$$du^i(\mathbf{v}) := v^i, \qquad i = 1, 2$$

[22] This condition means that the natural basis vectors $\mathbf{b}_1$ and $\mathbf{b}_2$ span a two-dimensional tangent plane. In modern terminology, this condition guaranties that the surface S is a 2-dimensional submanifold of the Euclidean manifold $\mathbb{E}^3$.

if $\mathbf{v} := v^j \mathbf{b}_j$. In terms of physics, this is the i-th velocity component measured by an observer who uses the natural basis. Furthermore, we define

$$\boxed{g_{ij} := \mathbf{b}_i \mathbf{b}_j, \qquad h_{ij} := -\mathbf{b}_i \mathbf{n}_j \quad i,j = 1,2.}$$

Here, we set $\mathbf{n}_j := \frac{\partial \mathbf{n}}{\partial u^j}$.

Proposition 9.29 *The families $\mathbf{b}_i, du^i, g_{ij}$ are tensorial families with respect to a change of local coordinates (u^1, u^2).*

The families $\mathbf{n}_i$ and h_{ij} are pseudo-tensorial families. Moreover, the families g_{ij} and h_{ij} are symmetric. In addition, $h_{ij} = \mathbf{n}\partial_i \mathbf{b}_j$ for all indices.

Proof. Note that by the chain rule,

$$\frac{\partial \mathbf{x}}{\partial u^{i'}} = \frac{\partial u^i}{\partial u^{i'}} \frac{\partial \mathbf{x}}{\partial u^i}.$$

Thus, $\mathbf{b}_i$ is a tensorial family. Similarly, $\mathbf{n}_i$ is a tensorial family if the orientation is preserved under the change of local coordinates. If the orientation is changed, then $\mathbf{n}$ passes over to $-\mathbf{n}$. Thus, $\mathbf{n}_i$ and h_{ij} are pseudo-tensorial families.

Obviously, g_{ij} is symmetric. By the Leibniz rule, it follows from $\mathbf{n}\mathbf{b}_j = 0$ that

$$0 = \frac{\partial}{\partial u^i}(\mathbf{n}\mathbf{b}_j) = \mathbf{n}_i \mathbf{b}_j + \mathbf{n}\frac{\partial^2 \mathbf{x}}{\partial u^i \partial u^j}.$$

Thus, $h_{ij} = \mathbf{n}\frac{\partial^2 \mathbf{x}}{\partial u^i \partial u^j}$. Hence $h_{ij} = h_{ji}$. □

Gauss used the following notation:

$$\begin{pmatrix} E & F \\ F & G \end{pmatrix} = \begin{pmatrix} g_{11} & g_{12} \\ g_{21} & g_{22} \end{pmatrix}, \qquad \begin{pmatrix} L & M \\ M & N \end{pmatrix} = \begin{pmatrix} h_{11} & h_{12} \\ h_{21} & h_{22} \end{pmatrix}. \tag{9.92}$$

Corollary 9.30 *The quantities $g := \det(g_{ij})$ and $h := \det(h_{ij})$ are tensorial density families of weight 2. Therefore, the quotient $\frac{h}{g}$ is an invariant under the change of local coordinates of the surface S.*

We will show on page 629 that the quotient $\frac{h}{g}$ is the Gaussian curvature.

Proof. It follows from the tensorial transformation law $g_{ij} = \frac{\partial u^{i'}}{\partial u^i}\frac{\partial u^{j'}}{\partial u^j} g_{i'j'}$ that

$$\begin{pmatrix} g_{11} & g_{12} \\ g_{21} & g_{22} \end{pmatrix} = G^d \begin{pmatrix} g_{1'1'} & g_{1'2'} \\ g_{2'1'} & g_{2'2'} \end{pmatrix} G \qquad \text{where } G := \begin{pmatrix} \frac{\partial u^{1'}}{\partial u^1} & \frac{\partial u^{1'}}{\partial u^2} \\ \frac{\partial u^{2'}}{\partial u^1} & \frac{\partial u^{2'}}{\partial u^2} \end{pmatrix}.$$

Hence $g = \det(G^d) g' \det(G) = (\det(G))^2 g'$.

Analogously, we get $h = (\det(G))^2 h'$. This implies $\frac{h}{g} = \frac{h'}{g'}$. □

According to Gauss, we introduce the following two symmetric tensors:

- $\mathbf{g} := g_{ij}(u^1, u^2)\, du^i \otimes du^j$, and
- $\mathbf{h} := h_{ij}(u^1, u^2)\, du^i \otimes du^j$.

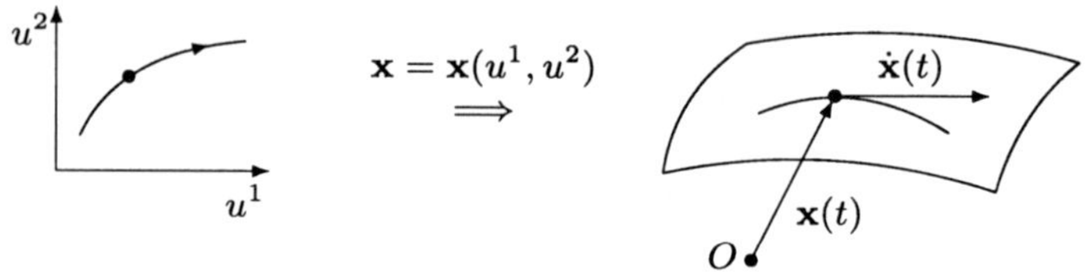

(a) $u^1 = u^1(t)$, $u^2 = u^2(t)$ (b) $\mathbf{x}(t) = \mathbf{x}\big(u^1(t), u^2(t)\big)$

Fig. 9.21. Curve on the surface

Suppose that $\mathbf{v}$ and $\mathbf{w}$ are smooth velocity vector fields on the surface S with $\mathbf{v} = v^i\mathbf{b}_i$ and $\mathbf{w} = w^j\mathbf{b}_j$, then

$$\mathbf{g}(\mathbf{v}, \mathbf{w}) = g_{ij}v^iw^j, \qquad \mathbf{h}(\mathbf{v}, \mathbf{w}) = h_{ij}v^iw^j.$$

As we will show below,

- the so-called first fundamental tensor (or the metric tensor) $\mathbf{g}$ determines the metric properties of the surface S, and
- the so-called second fundamental tensor $\mathbf{h}$ determines the curvature properties of the surface S.

Curve on the surface. We are given the smooth curve

$$\boxed{C: \;\; u^i = u^i(t), \qquad t_0 \le t \le t_1} \tag{9.93}$$

which depends on the time parameter t. This corresponds to the motion

$$\mathbf{x} = \mathbf{x}(t), \qquad t_0 \le t \le t_1$$

on the surface S where $\mathbf{x}(t) := \mathbf{x}(u^1(t), u^2(t))$ (Fig. 9.21). The motion of the unit normal vector along the curve C is given by

$$\mathbf{n} = \mathbf{n}(t), \qquad t_0 \le t \le t.$$

Using the differentials

- $d\mathbf{x} = \frac{\partial \mathbf{x}}{\partial u^i}\, du^i = \mathbf{b}_i du^i$, and
- $d\mathbf{n} = \frac{\partial \mathbf{n}}{\partial u^i}\, du^i = \mathbf{n}_i du^i$,

and using the chain rule, we get the following two formulas for the time derivatives:

- $\dot{\mathbf{x}}(t) = \mathbf{b}_i(u^1(t), u^2(t))\, \dot{u}^i(t)$, and
- $\dot{\mathbf{n}}(t) = \mathbf{n}_i(u^1(t), u^2(t))\, \dot{u}^i(t)$.

Hence

- $\left(\frac{d\mathbf{x}(t)}{dt}\right)^2 = g_{ij}(u^1(t), u^2(t))\, \dot{u}^i(t)\dot{u}^j(t)$, and
- $\frac{d\mathbf{x}(t)}{dt}\frac{d\mathbf{n}(t)}{dt} = -h_{ij}(u^1(t), u^2(t))\, \dot{u}^i(t)\dot{u}^j(t)$.

Mnemonically, we write

- $ds^2 = (d\mathbf{x})^2 = g_{ij}\; du^idu^j$, and
- $-d\mathbf{x}\; d\mathbf{n} = h_{ij}\; du^idu^j$.

The modern language of tensor products yields:

- $\mathbf{g} = d\mathbf{x} \otimes d\mathbf{x} = g_{ij}du^i \otimes du^j$ (first fundamental form of Gauss), and
- $\mathbf{h} = -d\mathbf{x} \otimes d\mathbf{n} = h_{ij}\; du^i \otimes du^j$ (second fundamental form of Gauss).

9.6.2 Intrinsic Metric Properties

The length of a curve on the surface. Consider the curve (9.93) on the surface S. By definition, the length l of this curve is given by

$$\boxed{l := \int_{t_0}^{t_1} |\mathbf{x}(t)|\, dt.}$$

With respect to local (u^1, u^2)-coordinates, we get

$$l = \int_{t_0}^{t_1} \sqrt{g_{ij}(u^1(t), u^2(t))\, \dot{u}^i(t)\dot{u}^j(t)}\ dt.$$

The intersection angle between two curves on the surface. Consider the two curves $\mathbf{x} = \mathbf{x}(t)$ and $\mathbf{x}_* = \mathbf{x}_*(t)$ parametrized by

$$u^i = u^i(t) \qquad \text{and} \qquad u^i_* = u^i_*(t), \qquad -t_1 < t < t_1,$$

respectively. Suppose that the two curves intersect each other at the point P at time $t = 0$. Then the intersection angle α is defined by

$$\boxed{\cos\alpha := \frac{\dot{\mathbf{x}}(0)\dot{\mathbf{x}}_*(0)}{|\dot{\mathbf{x}}(0)| \cdot |\dot{\mathbf{x}}_*(0)|}.}$$

Here, we use the two tangent vectors $\dot{\mathbf{x}}(0)$ and $\dot{\mathbf{x}}_*(0)$ at the intersection point P at time $t = 0$ (see Fig. 9.16 on page 606). With respect to local coordinates, we get

$$\cos\alpha = \frac{g_{ij}(P)\ \dot{u}^i(0)\dot{u}^j_*(0)}{\sqrt{g_{ij}(P)\ \dot{u}^i(0)\dot{u}^j(0)}\ \sqrt{g_{ij}(P)\,\dot{u}^i_*(0)\dot{u}^j_*(0)}}$$

where the intersection point P corresponds to the local coordinates $(u^1(0), u^2(0))$.

Surface measure. On the geographic chart $\mathcal{U}$ (Fig. 9.20 on page 624), consider the rectangle spanned by the points

$$(u_0^1, u_0^2),\ \ (u_0^1 + \Delta u^1, u_0^2),\ \ (u_0^1, u_0^2 + \Delta u^2).$$

This rectangle has the surface area $\Delta u^1 \Delta u^2$. If the positive numbers Δu^1 and Δu^2 are sufficiently small, and if the basis vector $\mathbf{b}_1(u_0^1, u_0^2)$ is orthogonal to the basis vector $\mathbf{b}_2(u_0^1, u_0^2)$, then the corresponding curved rectangle on the surface S has the approximate surface area

$$\Delta S = |\mathbf{b}_1(P)|\ \Delta u^1 \cdot |\mathbf{b}_2(P)|\ \Delta u^2$$

where $P = (u_0^1, u_0^2)$. In the general case, we get

$$\Delta S = |\mathbf{b}_1(P) \times \mathbf{b}_2(P)|\ \Delta u^1 \Delta u^2.$$

Since $|\mathbf{a} \times \mathbf{b}| = \sqrt{\mathbf{a}^2\mathbf{b}^2 - (\mathbf{ab})^2}$, we obtain

$$\Delta S = \sqrt{\det(g)}\ \Delta u^1 \Delta u^2.$$

After this intuitive motivation, let $\mathcal{O}$ be a bounded open subset of the geographic chart $\mathcal{U}$ which corresponds to the set $S_{\mathcal{O}}$ on the surface S. By definition, the surface area of $S_{\mathcal{O}}$ is equal to

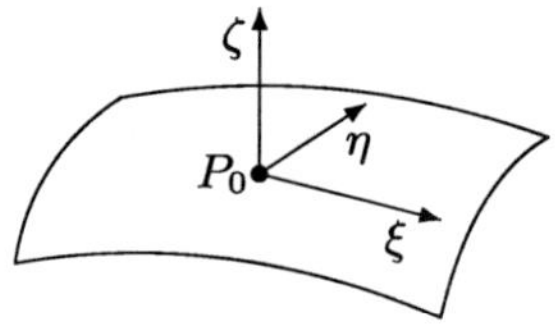

Fig. 9.22. Canonical local Cartesian coordinate system of a surface

$$\boxed{\text{meas}(S_{\mathcal{O}}) := \int_{\mathcal{O}} \sqrt{g(u^1, u^2)}\, du^1 du^2.}$$

Since $\sqrt{g}$ is a tensorial density of weight one, the integral does not depend on the choice of positively oriented local coordinates, by the classical substitution rule. If we change the orientation of the local coordinates, then the integral changes sign.

The volume form. The 2-form

$$v := \iota\sqrt{g}\, du^1 \wedge du^2$$

is called the volume form of the surface S. Here, $\iota = \pm 1$ is the orientation number of the surface S. Note that v does not depend on the choice of local coordinates. We have

$$\text{meas}(S_{\mathcal{O}}) = \int_{S_{\mathcal{O}}} v.$$

9.6.3 The Extrinsic Definition of the Gaussian Curvature

In order to understand the geometric meaning of the Gaussian curvature K, we will start with a geometric definition based on using a distinguished coordinate system together with the Taylor expansion up to second order. Then we will use the tensor calculus in order to get an invariant expression for K. As a prototype, note that a sphere of radius R has the Gaussian curvature $K = \frac{1}{R^2}$, as we will show below.

The canonical local coordinates. Fix the surface point P_0. Choose a right-handed Cartesian (ξ, η, ζ)-coordinate system with the point P_0 as origin (Fig. 9.22). That is, the (ξ, η)-plane coincides with the tangent plane of the surface at the point P_0. By convention, the normal unit vector $\mathbf{n}$ at the point P_0 points in direction of the ζ-axis. In a sufficiently small neighborhood of the point P_0, the equation of the surface is given by

$$\zeta = \zeta(\xi, \eta)$$

with $\zeta(0,0) = 0$. Since the equation $\zeta = 0$ describes the tangent plane at the point P_0, the Taylor expansion tells us that

$$\zeta = \tfrac{1}{2}\zeta_{\xi\xi}(0,0)\, \xi^2 + \zeta_{\xi\eta}(0,0)\, \xi\eta + \tfrac{1}{2}\zeta_{\eta\eta}(0,0)\, \eta^2 + o(\xi^2 + \eta^2), \qquad \xi^2 + \eta^2 \to 0.$$

By the principal axis theorem, there exists a rotation of the (ξ, η)-system such that we get

$$\boxed{\zeta = \tfrac{1}{2}(\lambda_1 \xi^2 + \lambda_2 \eta^2) + o(\xi^2 + \eta^2), \qquad \xi^2 + \eta^2 \to 0.} \qquad (9.94)$$

The numbers λ_1, λ_2 are the eigenvalues of the symmetric matrix

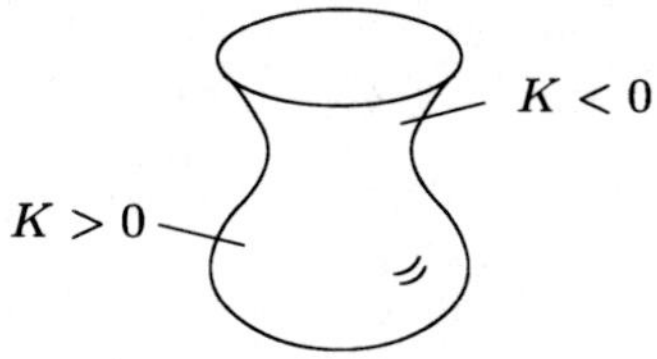

Fig. 9.23. Positive and negative Gaussian curvature

$$\begin{pmatrix} \zeta_{\xi\xi}(0,0) & \zeta_{\xi\eta}(0,0) \\ \zeta_{\xi\eta}(0,0) & \zeta_{\eta\eta}(0,0) \end{pmatrix}.$$

Explicitly, the real numbers λ_1, λ_2 are the zeros of the equation

$$\begin{vmatrix} \zeta_{\xi\xi}(0,0) - \lambda & \zeta_{\xi\eta}(0,0) \\ \zeta_{\xi\eta}(0,0) & \zeta_{\eta\eta}(0,0) - \lambda \end{vmatrix} = 0.$$

We define the Gaussian curvature $K(P_0)$ of the surface S at the point P_0 by setting

$$\boxed{K(P_0) = \lambda_1\lambda_2.}$$

The number $H = \frac{1}{2}(\lambda_1+\lambda_2)$ is called the mean curvature of the surface. If we change the orientation of the surface, then ζ has to be replaced by $-\zeta$. This corresponds to replacing λ_1, λ_2 by $-\lambda_1, -\lambda_2$, respectively. Consequently,

- the Gaussian curvature is invariant under changing the orientation, whereas
- the mean curvature changes sign under a change of orientation.

Example. For a sphere $x^2+y^2+z^2 = R^2$ of radius R, we get the constant Gaussian curvature $K = \frac{1}{R^2}$, and the mean curvature $H = \frac{1}{R}$.

Proof. Without any loss of generality, consider the south pole $x = y = 0, z = -R$. Near the south pole, we obtain the power series expansion

$$z = -\sqrt{R^2\left(1 - \frac{x^2}{R^2} - \frac{y^2}{R^2}\right)} = -R\left(1 - \frac{x^2}{2R^2} - \frac{y^2}{2R^2}\right) + o(x^2+y^2), \quad x^2+y^2 \to 0.$$

Setting $\zeta := z + R, \xi := x, \eta := y$, we get (9.94) with $\lambda_1 = \lambda_2 = \frac{1}{R}$. □

Similarly, an ellipsoid (resp. hyperboloid) is locally described by (9.94) with $\lambda_1 > 0, \lambda_2 > 0$ (resp. $\lambda_1 > 0, \lambda_2 < 0$). Thus, the Gaussian curvature of an ellipsoid (resp. hyperboloid) is positive (resp. negative). For example, an Etruscan vase has points of both positive and negative Gaussian curvature (Fig. 9.23).

A torus (resp. the surface of a cylinder) looks locally like (9.94) with $\lambda_1 > 0$ and $\lambda_2 = 0$. Hence the Gaussian curvature vanishes identically, $K \equiv 0$.

The invariant definition of the Gaussian curvature. The Gaussian curvature at the point P of the surface S is given by

$$\boxed{K(P) = \frac{h(P)}{g(P)}.} \tag{9.95}$$

Using Gauss' notation, this means that

$$K = \frac{h_{11}h_{22} - h_{12}^2}{g_{11}g_{22} - g_{12}^2} = \frac{LN - M^2}{EG - F^2}.$$

Proof. By Corollary 9.30, the quotient $\frac{h}{g}$ does not depend on the choice of the local coordinates. Therefore, we can choose a special local coordinate system. We will use the canonical local coordinate system depicted in Fig. 9.22. After changing the notation, $\xi \Rightarrow x, \eta \Rightarrow y, \zeta \Rightarrow z$, the equation of the surface looks locally like

$$\mathbf{x} = x\mathbf{i} + y\mathbf{j} + z(x,y)\mathbf{k}$$

in a neighborhood of the origin $x = y = z = 0$. Then:

- $\mathbf{x}_x = \mathbf{i} + z_x\mathbf{k}$, $\mathbf{x}_y = \mathbf{j} + z_y\mathbf{k}$,
- $\mathbf{x}_{xx} = z_{xx}\mathbf{k}$, $\mathbf{x}_{xy} = z_{xy}\mathbf{k}$, $\mathbf{x}_{yy} = z_{yy}\mathbf{k}$.

Hence

$$\mathbf{n} = \frac{\mathbf{x}_x \times \mathbf{x}_y}{|\mathbf{x}_x \times \mathbf{x}_y|} = \mu(\mathbf{k} - z_x\mathbf{i} - z_y\mathbf{j})$$

with $\mu := 1/\sqrt{1 + z_x^2 + z_y^2}$. This implies

- $E = \mathbf{x}_x^2 = 1 + z_x^2$, $F = z_x z_y$, $G = 1 + z_y^2$,
- $L = \mathbf{n}\mathbf{x}_{xx} = \mu z_{xx}$, $M = \mu z_{xy}$, $N = \mu z_{yy}$.

Hence

$$\frac{LN - M^2}{EG - F^2} = \frac{z_{xx}z_{yy} - z_{xy}^2}{(1 + z_x^2 + z_y^2)^2}. \tag{9.96}$$

It follows from

$$z = \tfrac{1}{2}(\lambda_1 x^2 + \lambda_2 y^2) + o(x^2 + y^2), \qquad x^2 + y^2 \to 0$$

that $z_x(0,0) = z_y(0,0) = z_{xy}(0,0) = 0$, and $z_{xx}(0,0) = \lambda_1$, $z_{yy}(0,0) = \lambda_2$. Hence the quotient (9.96) is equal to $\lambda_1\lambda_2$ at the origin. This coincides with K. □

Analogously, for the mean curvature H, we get

$$H(P) = \tfrac{1}{2}g^{ij}h_{ij}.$$

Here, g^{ij} are the entries of the inverse matrix to the symmetric matrix (g_{ij}). By (8.34) on page 455, g^{ij} is a tensorial family.

The Gaussian curvature $K(P)$ is a scalar under a change of local coordinates, but the mean curvature $H(P)$ is only a pseudo-scalar.

9.6.4 The Gauss–Weingarten Equations for Moving Frames

We want to show that the partial derivatives of the frame $\mathbf{b}_1, \mathbf{b}_2, \mathbf{n}$ with respect to the local (u^1, u^2)-coordinates satisfy the following partial differential equations:

$$\partial_i \mathbf{b}_j = \Gamma_{ij}^s \mathbf{b}_s + h_{ij}\mathbf{n} \qquad \text{(Gauss, 1827)}, \tag{9.97}$$

$$\partial_i \mathbf{n} = -h_i^s \mathbf{b}_s \qquad \text{(Weingarten, 1861)}. \tag{9.98}$$

Here, we use the Christoffel symbols

$$\Gamma_{ij}^k := \tfrac{1}{2}g^{ks}(\partial_i g_{js} + \partial_j g_{is} - \partial_s g_{ij}).$$

Proof. Ad (9.97). We determine the coefficients α and β in the decomposition

$$\partial_i \mathbf{b}_j = \alpha_{ij}^k \mathbf{b}_k + \beta_{ij}\mathbf{n}. \tag{9.99}$$

Multiplication with $\mathbf{n}$ yields $\beta_{ij} = \mathbf{n}\partial_i\mathbf{b}_j = h_{ij}$.

Furthermore, $g_{js} = \mathbf{b}_j\mathbf{b}_s$ implies that

$$\partial_i g_{js} = \mathbf{b}_j\partial_i\mathbf{b}_s + \mathbf{b}_s\partial_i\mathbf{b}_j.$$

Note that $\partial_i\mathbf{b}_j = \partial_i\partial_j\mathbf{x} = \partial_j\mathbf{b}_i$. Interchanging the indices and summation gives

$$\mathbf{b}_s\partial_i\mathbf{b}_j = \tfrac{1}{2}(\partial_i g_{js} + \partial_j g_{is} - \partial_s g_{ij}).$$

Multiplication of (9.99) with $\mathbf{b}_s$ yields

$$\mathbf{b}_s\partial_i\mathbf{b}_j = \alpha_{ij}^r g_{sr}.$$

Using $g^{ks}g_{sr} = \delta_r^k$, we get $\alpha_{ij}^k = g^{ks}\mathbf{b}_s\partial_i\mathbf{b}_j$. This is the claim.

Ad (9.98). Differentiation of $\mathbf{n}^2 = 1$ with respect to u^i yields $2\mathbf{n}\partial_i\mathbf{n} = 0$. Thus, the vector $\partial_i\mathbf{n}$ is orthogonal to the normal vector $\mathbf{n}$. Hence

$$\partial_i\mathbf{n} = c_i^j\mathbf{b}_j$$

where the numbers c_i^j remain to be determined. It follows from $h_{si} = -\mathbf{b}_s\partial_i\mathbf{n}$ that

$$-h_{si} = c_i^j\mathbf{b}_s\mathbf{b}_j = c_i^j g_{sj}.$$

Using $g^{ms}g_{sj} = \delta_j^m$, we get $c_i^m = -g^{ms}h_{si} = -h_i^m$.

□

9.6.5 The Integrability Conditions and the Riemann Curvature Tensor

Suppose that we are given a smooth surface $S : \mathbf{x} = \mathbf{x}(u^1, u^2), \quad (u^1, u^2) \in \mathbb{R}^2$. Then the Gauss–Weingarten frame equations (9.97) and (9.98) are satisfied. Then the integrability conditions $\partial_r\partial_i\mathbf{b}_j = \partial_i\partial_r\mathbf{b}_j$ and $\partial_r\partial_i\mathbf{n} = \partial_i\partial_r\mathbf{n}$ must be satisfied. This yields

$$\boxed{R_{rij}^k = h_r^k h_{ij} - h_i^k h_{rj},} \tag{9.100}$$

and

$$\boxed{\nabla_r h_{ij} = \nabla_i h_{rj}} \tag{9.101}$$

where all the indices run from 1 to 2. Explicitly,

- $R_{rij}^k := \partial_r\Gamma_{ij}^k - \partial_i\Gamma_{rj}^k + \Gamma_{rs}^k\Gamma_{ij}^s - \Gamma_{is}^k\Gamma_{rj}^s$,
- $\nabla_i h_{jk} := \partial_i h_{jk} - \Gamma_{ij}^s h_{sk} - \Gamma_{ik}^s h_{js}$.

The equations (9.100) are called the theorema egregium of Gauss, and the equations (9.101) are called the Codazzi–Mainardi equations.

Proof. (I) We will use

$$\partial_r\partial_i\mathbf{b}_j = \partial_i\partial_r\mathbf{b}_j. \tag{9.102}$$

By the Leibniz rule, it follows from (9.97) that

$$\partial_r\partial_i\mathbf{b}_j = \partial_r h_{ij}\mathbf{n} + h_{ij}\partial_r\mathbf{n} + \partial_r\Gamma^k_{ij}\mathbf{b}_k + \Gamma^k_{ij}\partial_r\mathbf{b}_k.$$

Again by (9.97),

$$\Gamma^k_{ij}\partial_r\mathbf{b}_k = \Gamma^k_{ij}h_{rk}\mathbf{n} + \Gamma^k_{ij}\Gamma^s_{rk}\mathbf{b}_s.$$

Thus, it follows from (9.102) together with the symmetry properties $\Gamma^k_{ij} = \Gamma^k_{ji}$ and $h_{ij} = h_{ji}$ that

$$\left(R^k_{rij} - h^k_r h_{ij} + h^k_i h_{rj}\right)\mathbf{b_k} + (\nabla_r h_{ij} - \nabla_i h_{rj})\mathbf{n} = 0.$$

This yields (9.100) and (9.101).

(II) We will use $\partial_r\partial_i\mathbf{n} = \partial_i\partial_r\mathbf{n}$. As in (I), this yields

$$\nabla_i h^k_r = \nabla_r h^k_i. \tag{9.103}$$

By the Ricci lemma, the covariant derivative interchanges with lifting and lowering of indices (see page 513). Therefore, the equation (9.103) is equivalent to the equation (9.101). □

Cartan's simplification of the classical approach by using orthonormal frames. The computation above is straightforward, but clumsy. We have carried out this computation in order to show how the Riemann curvature tensor emerges in a quite natural way. Cartan noticed that the classical approach can be substantially simplified if one replaces the natural basis $\mathbf{b}_1, \mathbf{b}_2, \mathbf{n}$ by an arbitrary orthonormal basis $\mathbf{e}_1, \mathbf{e}_2, \mathbf{n}$. This will be studied in Sect. 9.6.10 on page 636.

9.6.6 The Intrinsic Characterization of the Gaussian Curvature (Theorema Egregium)

Define $R_{rijk} = R^s_{rij}g_{ks}$. By (9.100), we get

$$R_{rijk} = h_{kr}h_{ij} - h_{ki}h_{rj}.$$

The Riemann curvature tensor R_{rijk} has 16 components. Since we have the following symmetry relations

$$R_{abcd} = R_{cdab} = -R_{bacd} = -R_{abcd},$$

there is only one essential component, namely, $R_{1221} = h_{11}h_{22} - h^2_{12} = h$. This implies the following fundamental relation.

Theorem 9.31 $K = \frac{R_{1221}}{g}$ *(theorema egregium).*

This theorem tells us that the Gaussian curvature K only depends on the metric tensorial family g_{ij} and its first-order and second-order partial derivatives. In other words:

The Gaussian curvature of a surface is an intrinsic property of the surface.

Set $u = u^1$ and u^2, and let us use the Gauss notation. Explicitly, the Gaussian curvature

$$K = \frac{LN - M^2}{EG - F^2}$$

can be written as

$$K = -\frac{1}{4g^2}\begin{vmatrix} E & F & G \\ E_u & F_u & G_u \\ E_v & F_v & G_v \end{vmatrix} - \frac{1}{2\sqrt{g}}\left[\left(\frac{E_v - F_u}{\sqrt{g}}\right)_v + \left(\frac{G_u - F_v}{\sqrt{g}}\right)_u\right]. \tag{9.104}$$

Thus, K only depends on the functions E, F, G of the first fundamental form which describes the metric properties of the surface. In particular, if we choose local coordinates such that $F(P_0) = M(P_0) = 0$ at the point P_0, then the theorema egregium and the Codazzi–Mainardi equations at the point P_0 read as follows:

$$K = \frac{LN}{EG} = -\tfrac{1}{2}A\left((AE_v)_v + (AG_u)_u\right)$$

$$L_v = \frac{E_v}{2}\left(\frac{L}{E} + \frac{N}{G}\right), \qquad N_u = \frac{G_u}{2}\left(\frac{L}{E} + \frac{N}{G}\right) \tag{9.105}$$

with $A := 1/\sqrt{EG}$.

9.6.7 Differential Invariants and the Existence and Uniqueness Theorem of Classic Surface Theory

We want to answer the following important question:

Which invariants determine a 2-dimensional surface up to translations and rotations?

Roughly speaking, the answer reads as follows: We need the first and second fundamental form of Gauss together with the integrability conditions (9.100) and (9.101). Because of symmetries, there are only three essential integrability conditions, namely, the theorema egregium of Gauss and two Codazzi–Mainardi equations (see (9.105)). The precise formulation reads as follows:

Suppose that we are given six smooth functions

$$g_{ij}, h_{ij} : \mathcal{U} \to \mathbb{R}, \qquad i, j = 1, 2,$$

with $g_{12} = g_{21}$ and $h_{12} = h_{21}$ which satisfy the integrability conditions (9.100) and (9.101) on the nonempty, arcwise connected, simply connected, open subset $\mathcal{U}$ of $\mathbb{R}^2$. Furthermore, assume that the eigenvalues of the symmetric matrix $\det(g_{ij})$ are positive on $\mathcal{U}$. Then the following existence result holds true.

Theorem 9.32 *There exists a smooth 2-dimensional surface*

$$S : \mathbf{x} = \mathbf{x}(u^1, u^2), \qquad (u_1, u_2) \in \mathcal{U}$$

which has the first fundamental form $g_{ij}du^i \otimes du^j$ and the second fundamental form $h_{ij}du^i \otimes du^j$. This surface is unique, up to translations and rotations in the Euclidean manifold $\mathbb{E}^3$.

(a) geodesic triangle (b) non-geodesic triangle

Fig. 9.24. Spherical triangles

This fundamental existence-and uniqueness theorem of classic surface theory was proved by Bonnet in 1867.[23] The idea of the proof is to use the famous Frobenius theorem which says that

- the necessary solvability conditions (i. e., the integrability conditions) for partial differential equations of the type of the Gauss–Weingarten equations (9.97) and (9.98)
- are also *sufficient conditions* for the existence of local solutions (see Sect. 12.11 on page 767).

Since the Gauss-Weingarten frame equations (9.97), (9.98) represent a linear system of partial differential equations with respect to $\mathbf{b}_1, \mathbf{b}_2, \mathbf{n}$, a standard argument allows us to continue the local solution to a global one. Note that we assume that the set $\mathcal{U}$ is simply connected. For the proof, we refer to E. Zeidler, Nonlinear Functional Analysis, Vol. IV, p. 640, Springer, New York, 1995 (reprinted: Beijing 2009), and J. Eschenburg and J. Jost, Differential Geometry and Minimal Surfaces (in German), p. 194, Springer, Berlin.

Finally, let us count the number of unknowns and the number of conditions:

- We have to determine the six functions E, F, G, L, M, N (in the Gauss notation (9.92)), and
- we have to solve the nine Gauss–Weingarten frame equations together with three highly nonlinear constraints given by the integrability conditions (i.e., the Gauss theorema egregium and the two Codazzi–Mainardi equations; see (9.105).

Summarizing, we obtain six conditions for six unknown functions, as naively expected.

9.6.8 Gauss' Theorema Elegantissimum and the Gauss–Bonnet Theorem

A triangle on a 2-dimensional smooth surface is called a geodesic triangle iff the sides of the triangle are geodesic lines (see Fig. 9.24(a) in the special case of a sphere). In 1827, Gauss proved for geodesic triangles on smooth 2-dimensional surfaces that the sum of the angles α, β, γ satisfies the relation

$$\alpha + \beta + \gamma = \pi + \int_{\mathcal{U}} K dS. \tag{9.106}$$

[23] Gauss (1777–1855), Mainardi (1800-1879), Bonnet (1819–1892), Codazzi (1824–1873), Riemann (1826–1869), Frobenius (1849–1917).

Gauss called this the theorema elegantissimum (i.e., the most elegant theorem). For general (geodesic or non-geodesic) triangles on smooth 2-dimensional surfaces, Bonnet proved the following relation in 1848:

$$\boxed{\alpha + \beta + \gamma = \pi + \int_{\mathcal{U}} K dS + \int_{\partial \mathcal{U}} \kappa_g ds.} \tag{9.107}$$

Here, the triangle $\partial\mathcal{U} = ABC$ is counter-clockwise oriented. In the special case of Fig. 9.24(b), the latitude segment AB is not a geodesic. The boundary integral $\int_{\partial\mathcal{U}} \kappa_g(s)\, ds$ takes this into account. The definition of the geodesic curvature κ_g on a general 2-dimensional smooth surface is given as for the sphere (see (9.52) on page 598). Recall that the geodesic curvature generalizes the curvature of a plane curve to curves on surfaces. In particular, we have $\kappa_g \equiv 0$ along a geodesic line. The geodesic curvature is an intrinsic quantity of the surface which measures the deviation of a curve on the surface from a geodesic line.

Relation (9.107) represents the famous Gauss–Bonnet theorem. The proof of the Gauss–Bonnet theorem can be found in J. Stoker, Differential Geometry, Wiley, 1969/1989, p. 195. The Gauss–Bonnet theorem is closely related to the integral theorem of Stokes for differential forms and the homotopy theory for the winding number of plane curves due to Whitney and Hopf in about 1940. This is studied in J. Eschenburg and J. Jost, Differential Geometry and Minimal Surfaces (in German), Springer, Berlin, 2007.

9.6.9 Gauss' Total Curvature and Topological Charges

Let $\mathcal{M}$ be a 2-dimensional, compact, arcwise connected, oriented submanifold (without boundary) of the Euclidean manifold $\mathbb{E}^3$ (e.g., a sphere or a torus). Then the Euler characteristic $\chi(\mathcal{M})$ of $\mathcal{M}$ can be represented by the integral formula

$$\boxed{\chi(\mathcal{M}) = \frac{1}{2\pi} \int_{\mathcal{M}} K v} \tag{9.108}$$

where K (resp. v) is the Gaussian curvature (resp. the volume form) of $\mathcal{M}$. Equivalently, this can be written as

$$\chi(\mathcal{M}) = \int_{\mathcal{M}} \varrho(P)\, dS$$

with the so-called topological charge $\chi(\mathcal{M})$ and the topological charge density $\varrho(P) := K(P)/2\pi$.

For example, if $\mathcal{M}$ is a sphere of radius R, then

$$\chi(\mathcal{M}) = \frac{1}{2\pi} \cdot \frac{1}{R^2} \int_{\mathcal{M}} v = \frac{4\pi R^2}{2\pi R^2} = 2.$$

For the torus, we have $K \equiv 0$. Hence $\chi(\mathcal{M}) = 0$.

The global Gauss–Bonnet theorem (9.108) is the prototype of a topological charge theorem. This theorem allows far-reaching generalization in modern topology which will be investigated in Vol. IV on quantum mathematics.

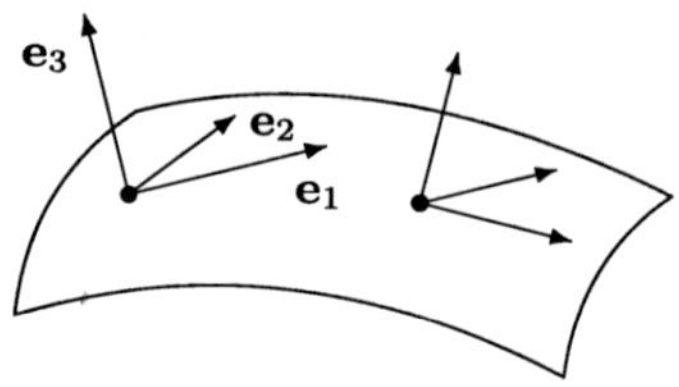

Fig. 9.25. Moving orthonormal frames

9.6.10 Cartan's Method of Moving Orthonormal Frames

> Cartan's method of using orthonormal frames fits best the local Euclidean structure of the surface. This substantially simplifies the theory of surfaces.[24]
>
> Folklore

Moving orthonormal frames. Until now, we have been used the local representation

$$\mathbf{x} = \mathbf{x}(u^1, u^2), \qquad (u^1, u^2) \in \mathcal{U}$$

of the surface along with the natural frame

$$\mathbf{b}_1 = \partial_1 \mathbf{x}, \quad \mathbf{b}_2 = \partial_2 \mathbf{x}, \quad \mathbf{n} := \frac{\mathbf{b}_1 \times \mathbf{b}_2}{|\mathbf{b}_1 \times \mathbf{b}_2|}.$$

It was the idea of Élie Cartan to simplify many considerations in surface theory by replacing the natural frame by a right-handed orthonormal frame

$$\boxed{\mathbf{e}_1(P),\ \mathbf{e}_2(P),\ \mathbf{e}_3(P)}$$

depending smoothly on the point P on suitable open subsets of the surface. Here, the vectors $\mathbf{e}_1, \mathbf{e}_2$ span the tangent plane at the point P (Fig. 9.25). Cartan called this a moving frame (repère mobile in French).[25] In terms of the moving frame, we get the Cartan frame equations

$$\boxed{d\mathbf{x} = \sigma_1 \mathbf{e}_1 + \sigma_2 \mathbf{e}_2, \quad d\mathbf{e}_i = \omega_{i1}\mathbf{e}_1 + \omega_{i2}\mathbf{e}_2 + \omega_{i3}\mathbf{e}_3, \qquad i = 1, 2, 3.} \tag{9.109}$$

Equivalently,

$$\sigma_i = \mathbf{e}_i d\mathbf{x} \qquad \text{and} \qquad \omega_{ij} = \mathbf{e}_j d\mathbf{e}_i, \qquad i, j = 1, 2.$$

It follows from the orthogonality relation $\mathbf{e}_i \mathbf{e}_j = \delta_{ij}$ that

$$d\mathbf{e}_i \cdot \mathbf{e}_j + \mathbf{e}_i d\mathbf{e}_j = 0,$$

[24] See É. Cartan, Riemannian Geometry in an Orthogonal Frame: From lectures delivered by Élie Cartan at Sorbonne (Paris) in 1926–1927. World Scientific, Singapore, 2001.

[25] Physicists call this the method of tetrads, since one has four basis vectors on a 4-dimensional space-time manifold. This tetrad formalism was introduced by Weyl in 1929.

and hence $\omega_{ij} = -\omega_{ji}$ for all $i, j = 1, 2, 3$. The Cartan frame equations replace the Gauss–Weingarten frame equations. To solve the Cartan frame equations (9.109) means that we are looking for a surface together with an orthonormal frame that fits the surface.

Our goal is to compute the integrability conditions for the Cartan frame equations. By the Poincaré cohomology rule, we have $d(d\mathbf{x}) = 0$ and $d(d\mathbf{e}_i) = 0$. Hence

$$\begin{aligned} 0 = dd\mathbf{x} &= d\sigma_1 \cdot \mathbf{e}_1 + \sigma_1 \wedge d\mathbf{e}_1 + d\sigma_2 \cdot \mathbf{e}_2 + \sigma_2 \wedge d\mathbf{e}_2 \\ &= (d\sigma_1 + \sigma_2 \wedge \omega_{21})\,\mathbf{e}_1 + (d\sigma_2 + \sigma_1 \wedge \omega_{12})\,\mathbf{e}_2 + (\sigma_1 \wedge \omega_{13} + \sigma_2 \wedge \omega_{23})\,\mathbf{e}_3, \end{aligned}$$

and

$$0 = dd\mathbf{e}_i = \sum_{j=1}^{3} d\omega_{ij} \cdot \mathbf{e}_j + \omega_{ij} \wedge d\mathbf{e}_j = \sum_{k=1}^{3} (d\omega_{ik} + \sum_{j=1}^{3} \omega_{ij} \wedge \omega_{kj})\,\mathbf{e}_k.$$

This implies the integrability conditions

$$\begin{aligned} &d\sigma_1 = \omega_{12} \wedge \sigma_2, \quad d\sigma_2 = \sigma_1 \wedge \omega_{12}, \quad \sigma_1 \wedge \omega_{13} + \sigma_2 \wedge \omega_{23} = 0, \\ &d\omega_{ij} = \sum_{k=1}^{3} \omega_{ik} \wedge \omega_{kj}, \qquad i, j = 1, 2, 3. \end{aligned} \tag{9.110}$$

The integrability conditions are also called the Cartan structural equations. They replace the Gauss theorema egregium and the Codazzi–Mainardi equations.

The structural equations can be used in order to prove an alternative formulation of Theorem 9.32 on the construction of surfaces. We are given the right-handed orthonormal frame $\mathbf{e}_1, \mathbf{e}_2, \mathbf{e}_3$ at a fixed point P_0 and five differential 1-forms $\sigma_1, \sigma_2, \omega_{12}, \omega_{13}, \omega_{23}$ along with $\omega_{ij} = -\omega_{ji}$ for all i, j which satisfy the integrability conditions (9.110) on a nonempty, open, arcwise connected, simply connected subset $\mathcal{U}$ of $\mathbb{R}^2$. Moreover, suppose that $\sigma_1 \wedge \sigma_2 \neq 0$ on $\mathcal{U}$.

Then there exists a unique surface

$$\mathbf{x} = \mathbf{x}(u^1, u^2), \qquad (u^1, u^2) \in \mathcal{U}$$

which has the following properties: the Cartan frame equations (9.109) *are satisfied, the surface passes through the given point P_0, and the given orthonormal frame $\mathbf{e}_1(P_0), \mathbf{e}_2(P_0), \mathbf{e}_3(P_0)$ is a frame of the surface at the point P_0.*

The proof can be found in I. Agricola and T. Friedrich, Global Analysis: Differential Forms in Analysis, Geometry and Physics, Sect. 5.2, Amer. Math. Soc., Providence, Rhode Island, 2002. The basis idea of the proof is to use the Frobenius theorem which tells us that necessary solvability conditions are also sufficient solvability conditions.

The intrinsic differential form $\omega_{12} = \mathbf{e}_2 d\mathbf{e}_1$. The differential form ω_{12} is completely determined by the differential forms σ_1 and σ_2. In fact, since $\sigma_1 \wedge \sigma_2 \neq 0$, the 2-forms $d\sigma_1$ and $d\sigma_2$ are proportional to $\sigma_1 \wedge \sigma_2$. Hence

$$d\sigma_1 = a\sigma_1 \wedge \sigma_2 \qquad \text{and} \qquad d\sigma_2 = b\sigma_1 \wedge \sigma_2.$$

By the integrability conditions $d\sigma_1 = \omega_{12} \wedge \sigma_2$ and $d\sigma_2 = \sigma_1 \wedge \omega_{12}$, we get

$$\omega_{12} \wedge \sigma_1 = a\sigma_1 \wedge \sigma_2, \qquad \sigma_1 \wedge \omega_{12} = b\sigma_1 \wedge \sigma_2.$$

Hence

$$\omega_{12} = a\sigma_1 + b\sigma_2.$$

Consequently, the differential form ω_{12} is an intrinsic quantity. The same is true for $d\omega_{12}$. Since this is a 2-form, we get $d\omega_{12} = k\sigma_1 \wedge \sigma_2$. We expect that the real-valued function $P \mapsto k(P)$ possesses a geometric meaning. It turns out that k equals $-K$ where K is the Gaussian curvature.

The theorema egregium. Gauss' theorema egregium reads as follows:

$$d\omega_{12} = -K\sigma_1 \wedge \sigma_2. \tag{9.111}$$

Here, $v = \sigma_1 \wedge \sigma_2$ is the volume form of the surface, and $K(P)$ is the Gaussian curvature of the surface at the point P. Equivalently,

$$\boxed{d(\mathbf{e}_1 d\mathbf{e}_2) = K \cdot \mathbf{e}_1 d\mathbf{x} \wedge \mathbf{e}_2 d\mathbf{x}.}$$

Note that the Gaussian curvature K is uniquely determined by this equation. The proof can be found in J. Stoker, Differential Geometry, Wiley, p. 347, New York, 1969/89.

9.7 Parallel Transport in Physics

Perspectives. In Einstein's theory of special relativity, one has to replace the three-dimensional Euclidean manifold $\mathbb{E}^3$ by the four-dimensional Minkowski manifold $\mathbb{M}^4$ (space-time manifold).

- In Chap. 13, we will study the Weyl $U(1)$-gauge theory. Here the curvature of the principal bundle $\mathbb{M}^4 \times U(1)$ describes the electromagnetic field (Maxwell's classic theory of electromagnetism), and mesons in an electromagnetic field.
- In Chap. 14, we will apply the $U(1)$-gauge theory to superconductivity.
- In Chap. 15, we will replace the commutative gauge group $U(1)$ by the noncommutative gauge group $SU(N)$, $N = 2, 3, \ldots$. In the special case where $N = 2$, we get the Yang–Mills theory introduced in 1954 (the local phase factor $G(x, y, z, t)$ is an element of the group $SU(2)$).
- In the special case of the gauge group $SU(3)$, we will obtain the prototype of quantum chromodynamics (colored quarks in strong interaction).
- The axiomatic approach to the curvature theory of vector bundles and principal bundles will be investigated in Chap. 17.

All the generalizations to be considered in Chaps. 13 through 17 (Ariadne's thread in gauge theory) are generalizations of the method of moving frames for the Euclidean manifold. The point is that non-vanishing curvature has far-reaching consequences in geometry and physics.

9.8 Finsler Geometry

In Riemannian geometry, measurements are made with both yardsticks and protractors. These tools are represented by a family of inner products. In Riemann–Finsler geometry (or Finsler geometry for short), one is in principle equipped with only a family of Minkowski norms. So yardsticks are assigned, but protractors are not. With such a limited tool kit, it is natural to wonder, just how much geometry one can uncover and describe?

It now appears that there is a reasonable answer. Finsler geometry encompasses a solid repertoire of rigidity and comparison theorems, most of them founded upon a fruitful analogue of the sectional curvature. There is also a bewildering array of explicit examples, illustrating many phenomena which admit only Finslerian interpretation.[26]

David Bao, Shing-Shen Chern, and Zhongmin Shen, 2000

Fix $n = 1, 2, \ldots$ Let $x = (x^1, \ldots, x^n)$. Consider the curve

$$C : x^i = x^i(t), \qquad t_0 \leq t \leq t_1, \; i = 1, \ldots, n.$$

The Euclidean length of this curve on $\mathbb{R}^n$ is given by the formula

$$s = \int_{t_0}^{t_1} \left(\sum_{i=1}^{n} |\dot{x}^i(t)|^2 \right)^{1/2} dt. \tag{9.112}$$

The length of the curve C in the Finsler geometry with respect to the Lagrangian L reads as

$$\boxed{s := \int_{t_0}^{t_1} L(x^1(t), \ldots, x^n(t), \dot{x}^1(t), \ldots, \dot{x}^n(t))\, dt.}$$

Mnemonically, we write

$$ds = L(x^1, \ldots, x^n, \dot{x}^1, \ldots, \dot{x}^n)\, dt.$$

Here, we assume that the Lagrangian

$$L : \mathbb{R}^n \times \mathbb{R}^n \to \mathbb{R}$$

is continuous, and L is convex and homogenous of degree one with respect to the velocity vector $\dot{x}$. Moreover, we assume that L is symmetric with respect to the velocity components $\dot{x}^1, \ldots, \dot{x}^n$. For example, we may choose

$$L(\dot{x}) := |\dot{x}^1| + \ldots + |\dot{x}^n|. \tag{9.113}$$

Then the corresponding Finsler geometry is not a Riemannian geometry. In particular, the concept of orthogonality in tangent spaces is not available. In contrast to (9.113), the Lagrangian

$$L(\dot{x}) = \left(|\dot{x}^1|^2 + \ldots + |\dot{x}^n|^2 \right)^{1/2}$$

from (9.112) generates a Riemannian geometry.

Example. Consider the space $\mathbb{R}^2$. The equation

$$x^2 + y^2 = 1, \qquad (x, y) \in \mathbb{R}^2$$

describes the usual unit circle. However, the equation

$$|x| + |y| = 1, \qquad (x, y) \in \mathbb{R}^2$$

describes the unit circle in the prototype of a Finsler geometry on $\mathbb{R}^2$; this geometry is generated by the norm $|(x, y)| := |x| + |y|$ on $\mathbb{R}^2$. With respect to this norm, the

[26] D. Bao, S. Chern, and Z. Shen, An Introduction to Riemann–Finsler Geometry, Springer, New York, 2000 (reprinted with permission).

linear space $\mathbb{R}^2$ becomes a Banach space, which is not a Hilbert space. The distance between the two points $P = (x, y)$ and $P' = (x', y')$ is given by

$$\mathrm{dist}(P, P') := |x - x'| + |y - y'|.$$

In this Finsler geometry, the 'unit circle' is the boundary of a square with the vertices $(1,0), (0,1), (-1,0)$, and $(0,-1)$.

Mnemonically, as a rule, the circles in Finsler geometry are not round.

Finsler geometry was introduced by Paul Finsler in his 1918 dissertation supervised by Carthéodory. In fact, Carathéodory was strongly motivated by his geometric approach to the calculus of variations (see Sect. 5.4 of Vol. II on Carathéodory's Royal Road to the calculus of variations). Note that the idea of the Riemann–Finsler geometry had its genesis in Riemann's 1854 habilitation address:

> "Über die Hypothesen, welche der Geometrie zu Grunde liegen" (On the hypotheses which lie at the foundations of geometry).

An English translation of Riemann's seminal lecture together with a commentary by Michael Spivak can be found in M. Spivak (1979), Vol. II, quoted below.

9.9 Further Reading

A general overview on the mathematics of the 20th century can be found in:

> K Maurin, The Riemann Legacy: Riemannian Ideas in Mathematics and Physics of the 20th Century, Kluwer, Dordrecht, 1997.

As a handbook on modern differential geometry and its applications to physics, we recommend:

> V. Ivancevic and T. Ivancevic, Differential Geometry: A Modern Introduction, World Scientific, Singapore, 2009.

The relation between the classical and the modern approach to differential geometry is studied in:

> M. Spivak, Comprehensive Introduction to Differential Geometry, Vols. 1–5, Publish or Perish, Boston, 1979.

Furthermore, we recommend:

> J. Jost, Riemannian Geometry and Geometric Analysis, 5th edition, Springer, Berlin, 2008.

> Y. Choquet-Bruhat, C. DeWitt-Morette, and M. Dillard-Bleick, Analysis, Manifolds, and Physics. Vol. 1: Basics; Vol. 2: 92 Applications, Elsevier, Amsterdam, 1996.

> S. Novikov, and T. Taimanov, Geometric Structures and (Physical) Fields, Amer. Math. Soc., Providence, Rhode Island, 2006.

> T. Frankel, The Geometry of Physics, Cambridge University Press, 2004.

> B. Dubrovin, A. Fomenko, and S. Novikov, Modern Geometry: Methods and Applications, Vols. 1–3, Springer, New York, 1992.

> C. Misner, K. Thorne, and A. Wheeler, Gravitation, Freeman, San Francisco, 1973.

> Ø. Grøn and S. Hervik, Einstein's Theory of General Relativity: with Modern Applications in Cosmology, Springer, New York, 2007.

T. Padmanabhan, Gravitation: Foundations and Frontiers, Cambridge University Press, 2010.

V. Varadarajan, Geometry of Quantum Theory, Springer, New York, 2007.

For classical differential geometry, we refer to:

J. Stoker, Differential Geometry, Wiley, New York, 1969. Reprinted in 1989.

J. Eschenburg and J. Jost, Differential Geometry and Minimal Surfaces (in German), Springer, Berlin, 2007.

The Cartan approach to differential geometry is studied in:

É. Cartan, Riemannian Geometry in an Orthogonal Frame: From lectures delivered by Élie Cartan at the Sorbonne in Paris, 1926–1927, World Scientific, Singapore, 2001.

I Agricola and T. Friedrich, Global Analysis: Differential Forms in Analysis, Geometry and Physics, Amer. Math. Soc., Providence, Rhode Island, 2002.

T. Ivey and J. Landsberg, Cartan for Beginners: Differential Geometry via Moving Frames and Exterior Differential Systems, American Mathematical Society, Providence, Rhode Island, 2003.

R. Sharpe, Differential Geometry: Cartan's Generalization of Klein's Erlangen Program, Springer, New York, 1997.

For Finsler geometry, we refer to:

D. Bao, S. Chern, and Z. Shen, An Introduction to Riemann–Finsler Geometry, Springer, New York, 2000.

Problems

9.1 *The Dirichlet variational problem on minimal electrostatic energy and the transformation of the Poisson equation.* We want to describe a nice general method for transforming partial differential equations into new coordinates. The basic idea reads as follows: In order to save time, one does not transform the partial differential equation itself, but one transforms the corresponding variational problem. This method applies to Euler–Lagrange equations. Let us explain this with a special example. Let $\mathcal{O}$ be a nonempty bounded open subset of $\mathbb{R}^3$ (e.g., a ball). Consider the variational problem

$$\int_{\mathcal{O}} \left(\frac{\varepsilon_0}{2}(U_x^2 + U_y^2 + U_z^2) - \varrho U\right)\, dxdydz = \min! \tag{9.114}$$

with the boundary condition $U = 0$ on $\partial\mathcal{O}$. Observe that the following hold: if U is a solution of (9.114), then the function U satisfies the Euler–Lagrange equation

$$\frac{\partial}{\partial x}\frac{\partial \mathcal{L}}{\partial U_x} + \frac{\partial}{\partial y}\frac{\partial \mathcal{L}}{\partial U_y} + \frac{\partial}{\partial z}\frac{\partial \mathcal{L}}{\partial U_z} = \frac{\partial \mathcal{L}}{\partial U}$$

with the Lagrangian $\mathcal{L} := \frac{1}{2}\varepsilon_0(U_x^2 + U_y^2 + U_z^2) - \varrho U$. This yields

$$-\varepsilon_0(U_{xx} + U_{yy} + U_{zz}) = \varrho.$$

Use arbitrary curvilinear coordinates x^1, x^2, x^3. Transform the integral, and compute the new Lagrangian together with the new Euler–Lagrange equation. This yields the transformed Euler–Lagrange equation. The advantage of this variational approach is that it is easier to transform the Lagrangian than the Euler–Lagrange equation.
Solution: Using the transformation $x^i \to x^{i'}$ with $\mathcal{O} \to \mathcal{O}'$, and changing the notation from $x^{i'}$ (resp. $\mathcal{O}'$) to x^i (resp. $\mathcal{O}$), the transformed variational problem reads as

$$\int_{\mathcal{O}} \left(\tfrac{1}{2}\varepsilon_0 g^{ij}\partial_i U \partial_j U - \varrho U\right)\sqrt{|g|}\, dx^1 dx^2 dx^3 = \text{critical!}, \qquad U = 0 \text{ on } \partial\mathcal{O}.$$

This yields the Euler–Lagrange equation

$$-\varepsilon_0 \partial_i\left(g^{ij}\sqrt{|g|}\partial_j U\right) = \sqrt{|g|}\,\varrho \qquad \text{on } \mathcal{O}.$$

9.2 *Affine geodesics.* Compute the Christoffel symbols of the Euclidean manifold $\mathbb{E}^3$ with respect to spherical coordinates.
Hint: Use the Lagrangian

$$\mathcal{L}(\varphi, \vartheta, r) = \tfrac{1}{2}m(r^2\cos^2\vartheta \cdot \dot{\varphi}^2 + r^2\dot{\vartheta}^2 + \dot{r}^2),$$

and argue as for cylindrical coordinates on page 579.

9.3 *Physical fields.* Compute the examples summarized on page 561.

9.4 *Hamilton's nabla calculus.* Prove Proposition 9.6 on page 563.

9.5 *The Cauchy theorem on isotropic functions.* Study the proofs of the theorems on isotropic functions formulated in Sect. 9.1.6. We refer to:

H. Weyl, The Classical Groups and Their Invariants, pages 31 and 52, Princeton University Press, 1946,

C. Truesdell and W. Noll, The nonlinear field theories in mechanics. In: S. Flügge (Ed.), Handbook of Physics, p. 29, Vol. III/3, Springer, Berlin, 1956.

G. Eisenreich, Lectures on Vector and Tensor Calculus, p. 96, Teubner, Leipzig, 1971 (in German).

9.6 *The Rivlin–Ericksen theorem on isotropic tensor functions.* Study the proof of Theorem 9.9 on page 566 together with applications to Hooke's law and to more general constitutive laws for elastic material. We refer to E. Zeidler, Nonlinear Functional Analysis and its Applications, Vol. IV: Applications to Mathematical Physics, p. 204, Springer, New York, 1995.

9.7 *Proof of Proposition 9.20 on page 603.* Hint: Use the Hamiltonian equations.
Solution: Using

$$p_\varphi := \frac{\partial L}{\partial \dot{\varphi}} = \frac{A_z}{2}, \quad p_\vartheta := \frac{\partial L}{\partial \dot{\vartheta}},$$

we introduce the Hamiltonian

$$H = \dot{\varphi}p_\varphi + \dot{\vartheta}p_\vartheta - L = \frac{p_\varphi^2}{2\cos^2\vartheta} + \frac{p_\vartheta^2}{2} + a\sin\vartheta.$$

Then the Euler–Lagrange equations pass over to the Hamiltonian equations

$$\dot{p}_\varphi = -\frac{\partial H}{\partial \varphi}, \quad \dot{p}_\vartheta = -\frac{\partial H}{\partial \vartheta}, \quad \dot{\varphi} = \frac{\partial H}{\partial p_\varphi}, \quad \dot{\vartheta} = \frac{\partial H}{\partial p_\vartheta}.$$

Since the Hamiltonian H does not depend on time t, it is a conserved quantity. Moreover, since H does not depend on the angle φ, we get $p_\varphi = \text{const}$. Set $z := \sin\vartheta$. Then

$$\frac{dz}{dt} = \cos\vartheta\,\frac{d\vartheta}{dt} = \sqrt{1-z^2}\,\frac{d\vartheta}{dt}.$$

Hence

$$H = \frac{p_\varphi^2}{2(1-z^2)} + \frac{1}{2(1-z^2)}\left(\frac{dz}{dt}\right)^2 + az = E.$$

This implies

$$\left(\frac{dz}{dt}\right)^2 = 2(E-az)(1-z^2) - p_\varphi^2 =: P(z).$$

Hence $\frac{dt}{dz} = \frac{1}{\sqrt{P(z)}}$. Finally,

$$\frac{d\varphi}{dz} = \frac{d\varphi}{dt}\cdot\frac{dt}{dz} = \frac{p_\varphi}{\cos^2\vartheta}\cdot\frac{dt}{dz} = \frac{p_\varphi}{(1-z^2)\sqrt{P(z)}}.$$

10. Temperature Fields on the Euclidean Manifold $\mathbb{E}^3$

In physics, temperature fields are prototypes of scalar fields, whereas velocity vector fields are prototypes of vector fields.

Folklore

10.1 The Directional Derivative

In mathematics and physics, differentiation describes the linearization of analytic objects like physical fields. In this chapter, let us study the directional derivative of a temperature field Θ on the Euclidean manifold $\mathbb{E}^3$. To this end, let

$$\Theta : \mathbb{E}^3 \to \mathbb{R}$$

be a smooth function. In terms of physics, we regard $\Theta(P)$ as the temperature at the point P on $\mathbb{E}^3$. We are given the smooth curve

$$C : P = P(t), \qquad t \in \mathbb{R}$$

on $\mathbb{E}^3$ with $P_0 := P(0)$. In terms of position vectors at the origin, we describe the curve C by the smooth vector function $\mathbf{x} = \mathbf{x}(t), t \in \mathbb{R}$. The derivative

$$\boxed{d_{\mathbf{h}}\Theta(P_0) := \frac{d\Theta(\mathbf{x}(t))}{dt}_{|t=0}}$$

is called the directional derivative of the temperature field Θ along the trajectory C at the point P_0. We will show that this time derivative only depends on the velocity vector $\mathbf{h} = \dot{\mathbf{x}}(0)$ of the trajectory C at time $t = 0$. To this end, consider an arbitrary right-handed Cartesian (x, y, z)-coordinate system with the right-handed orthonormal basis $\mathbf{i}, \mathbf{j}, \mathbf{k}$. Then, we have

$$\Theta = \Theta(x, y, z), \qquad (x, y, z) \in \mathbb{R},$$

and $\mathbf{x}(t) = x(t)\mathbf{i} + y(t)\mathbf{j} + z(t)\mathbf{k}$ together with

$$\mathbf{h} = \dot{\mathbf{x}}(0) = \dot{x}(0)\mathbf{i} + \dot{y}(0)\mathbf{j} + \dot{x}(0)\mathbf{k}.$$

By the chain rule,

$$d_{\mathbf{h}}\Theta(P_0) = \Theta_x(x_0, y_0, z_0)\, \dot{x}(0) + \Theta_y(x_0, y_0, z_0)\, \dot{y}(0) + \Theta_z(x_0, y_0, z_0)\, \dot{z}(0),$$

where Θ_x denotes the partial derivative with respect to x. Synonymously, we write

E. Zeidler, *Quantum Field Theory III: Gauge Theory*,
DOI 10.1007/978-3-642-22421-8_11,

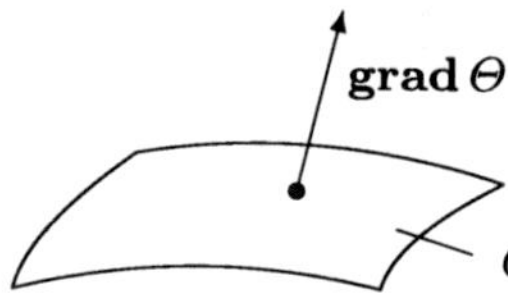

Fig. 10.1. Isothermal surface

$$\boxed{d_{\mathbf{h}}\Theta(P_0) = \Theta'(P_0)\mathbf{h}.} \tag{10.1}$$

Using the gradient of the temperature field,

$$\mathbf{grad}\,\Theta(P_0) := \Theta_x(x_0, y_0, z_0)\,\mathbf{i} + \Theta_y(x_0, y_0, z_0)\,\mathbf{j} + \Theta_z(x_0, y_0, z_0)\,\mathbf{k}, \tag{10.2}$$

we get the following inner product

$$d_{\mathbf{h}}\Theta(P_0) = \mathbf{grad}\,\Theta(P_0) \cdot \mathbf{h}.$$

The map $\mathbf{h} \mapsto d_{\mathbf{h}}\Theta(P_0)$ is a linear functional on the tangent space of $\mathbb{E}^3$ at the point P_0. This linear functional is denoted by

$$d\Theta(P_0) : T_{P_0}\mathbb{E}^3 \to \mathbb{R},$$

and $d\Theta(P_0)$ is called the differential of the temperature field Θ at the point P_0.[1] In other words, we have $d\Theta(P_0) \in T^*_{P_0}\mathbb{E}^3$, that is, $d\Theta(P_0)$ is an element of the cotangent space of $\mathbb{E}^3$ at the point P_0. We write

$$\boxed{d\Theta_{P_0}(\mathbf{h}) = (d_{\mathbf{h}}\Theta)(P_0).}$$

For historical reasons, different notations are used in the literature. Summarizing, we get

$$d_{\mathbf{h}}\Theta(P_0) = d\Theta_{P_0}(\mathbf{h}) = \Theta'(P_0)\mathbf{h} = \mathbf{grad}\,\Theta(P_0) \cdot \mathbf{h}.$$

This motivates the notation

$$\boxed{\Theta'(P_0) = \mathbf{grad}\,\Theta(P_0).}$$

Below we will show that we also have

$$d_{\mathbf{h}}\Theta(P_0) = \mathcal{L}_{\mathbf{h}}\Theta(P_0) = \delta\Theta(P_0; \mathbf{h}) = \frac{\delta\Theta(P_0)}{\delta\mathbf{x}} \cdot \mathbf{h}$$

for the Lie derivative $\mathcal{L}_{\mathbf{h}}\Theta$, the first variation $\delta\Theta$, and the functional derivative $\frac{\delta\Theta}{\delta\mathbf{x}}$.

Isothermal surface. In order to understand the intuitive meaning of the gradient of the temperature field, consider the surface of constant temperature through the point P_0 given by the equation

$$\Theta(P) = \text{const}, \qquad P \in \mathbb{E}^3 \tag{10.3}$$

where the constant is equal to $\Theta(P_0)$. By classical differential geometry, the vector $\mathbf{grad}\,\Theta(P_0)$ is a normal vector of this isothermal surface at the point P_0, and it points to the direction of the strongest increase of the temperature at the point P_0. This follows from (10.1) (Fig. 10.1).

[1] One also writes $d\Theta_{P_0}$ instead of $d\Theta(P_0)$.

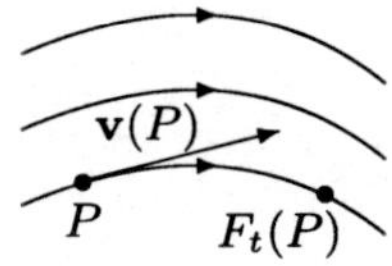

Fig. 10.2. Flow of fluid particles

10.2 The Lie Derivative of a Temperature Field along the Flow of Fluid Particles

> Sophus Lie (1842–1899) based his approach to differential geometry on the physical picture of the flow of fluid particles.
>
> Folklore

The Lie derivative plays a fundamental role in the calculus on manifolds. At this point, let us discuss the physical meaning of the Lie derivative of a smooth temperature field on the Euclidean manifold $\mathbb{E}^3$. In terms of physics, we will study the temperature Θ along the trajectories of fluid particles. We will show that

$$\mathcal{L}_{\mathbf{v}}\Theta(P) = d_{\mathbf{v}(P)}\Theta(P).$$

This tells us that the Lie derivative of the temperature field Θ at the point P coincides with the directional derivative with respect to the velocity vector $\mathbf{v}(P)$ at the point P.

10.2.1 The Flow

> In terms of physics, the prototype of a flow is the parallel flow of fluid particles along straight lines with constant velocity (or the rotation of fluid particles about a fixed axis with constant angular velocity).
> In terms of mathematics, a flow on the Euclidean manifold $\mathbb{E}^3$ is a one-dimensional additive Lie group of diffeomorphisms from $\mathbb{E}^3$ onto $\mathbb{E}^3$.
>
> Folklore

We are given the smooth velocity vector field

$$\mathbf{v} = \mathbf{v}(P), \qquad P \in \mathbb{E}^3$$

on the Euclidean manifold $\mathbb{E}^3$, that is, $\mathbf{v}(P) \in T_P\mathbb{E}^3$ for all points $P \in \mathbb{E}^3$. In terms of physics, we consider the flow of a fluid on the Euclidean manifold $\mathbb{E}^3$ generated by the velocity vector field $P \mapsto \mathbf{v}(P)$ (Fig. 10.2). The trajectory

$$C : P = P(t), \qquad t \in \mathbb{R}$$

of a fluid particle satisfies the ordinary differential equation

$$\boxed{\dot{P}(t) = \mathbf{v}(P(t)), \quad t \in \mathbb{R}, \quad P(0) = P_0.} \qquad (10.4)$$

Intuitively, this equation tells us that the velocity vector $\dot{P}(t)$ of the fluid particle at time t equals the velocity vector $\mathbf{v}(P(t))$ of the velocity vector field $\mathbf{v}$ at the point $P(t)$. The trajectories are also called the streamlines (or field lines) of the velocity vector field (Fig. 10.3). We define

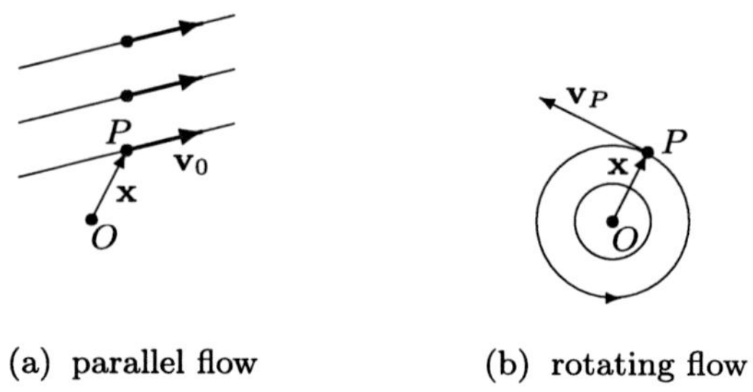

Fig. 10.3. Special flows of fluid particles

$$\boxed{F_t(P_0) := P(t), \qquad t \in \mathbb{R}.} \tag{10.5}$$

That is, for any fixed time t, the flow operator

$$F_t : \mathbb{E}^3 \to \mathbb{E}^3$$

sends the position P_0 of the fluid particle at time $t = 0$ to the position $P(t)$ of the particle at time t. Obviously, $F_0 = \text{id}$ (identity operator) and

$$\boxed{F_{s+t} = F_s F_t \qquad \text{for all} \quad s, t \in \mathbb{R}.} \tag{10.6}$$

This means that $F_{s+t}P_0 = F_s(F_t(P_0)) = F_s(P(t))$ for all points $P_0 \in \mathbb{E}^3$.

Global and local flow. There arises the following difficulty. We have to distinguish between

- local flow, and
- global flow.

A global flow corresponds to the situation where the solution $P = P(t)$ of (10.4) exists for all times $t \in \mathbb{R}$. Otherwise, the flow is called local. In order to avoid technicalities, we assume that we have a global flow at hand.[2] That is, the operator F_t is well-defined for all $t \in \mathbb{R}$, and the global group property (10.6) is valid. In addition, the theory of ordinary differential equations tells us that, for any time $t \in \mathbb{R}$, the flow operator $F_t : \mathbb{E}^3 \to \mathbb{E}^3$ is a diffeomorphism with the inverse map $F_{-t} : \mathbb{E}^3 \to \mathbb{E}^3$.

To simplify the approach in the Euclidean setting, we replace the point P by the position vector $\mathbf{x} = \overrightarrow{OP}$ which points from the origin O to the point P. Then the equation of motion (10.4) for the fluid particles reads as

$$\dot{\mathbf{x}}(t) = \mathbf{v}(\mathbf{x}(t)), \quad t \in \mathbb{R}, \quad \mathbf{x}(0) = \mathbf{x}_0, \tag{10.7}$$

and the flow operator F_t is given by

$$F_t\mathbf{x}_0 := \mathbf{x}(t), \qquad t \in \mathbb{R}.$$

Concerning the differential equation (10.7), note that the vector $\mathbf{v}(\mathbf{x}(t))$ located at the origin is obtained from the velocity vector $\mathbf{v}(P(t))$ at the point $P(t)$ by global Euclidean parallel transport from $P(t)$ to O.

[2] Observe that for introducing the Lie derivative below, we only need a local flow. To simplify terminology, 'global flows' are briefly called 'flows'.

Complete velocity vector fields. The smooth velocity vector field $\mathbf{v}$ on $\mathbb{E}^3$ is called complete iff the trajectories exist for all times, that is, the flow operator F_t is defined for all times $t \in \mathbb{R}$.

Note that the flow may blow up in finite time if the velocity $|\mathbf{v}(P)|$ increases strongly when the point P approaches infinity. On the real line, the simplest example for blowing-up is given by the differential equation

$$\dot{x}(t) = 1 + x(t)^2, \qquad t \in \mathbb{R}, \qquad x(0) = 0$$

with the solution $x(t) = \tan t$. Here, $\lim_{t \to \pi/2-0} x(t) = +\infty$. The solution only exists on the finite open time interval $]-\pi/2, \pi/2[$.

Cartesian coordinate system. Choosing a right-handed Cartesian (x, y, z)-coordinate system with the right-handed orthonormal basis $\mathbf{i}, \mathbf{j}, \mathbf{k}$ at the origin. By global parallel transport, we get the right-handed orthonormal basis $\mathbf{i}_P, \mathbf{j}_P, \mathbf{k}_P$ of the tangent space $T_P\mathbb{E}^3$ at the point P (see Fig. 4.3 on page 323). Then:

- $\mathbf{x} = x\mathbf{i} + y\mathbf{j} + z\mathbf{k}$,
- $\mathbf{v}(P) = u(x, y, z)\mathbf{i}_P + v(x, y, z)\mathbf{j}_P + w(x, y, z)\mathbf{k}_P$,
- $\mathbf{v}(\mathbf{x}) = u(x, y, z)\mathbf{i} + v(x, y, z)\mathbf{j} + w(x, y, z)\mathbf{k}$.

The key differential equation (10.4) reads as

$$\dot{x}(t) = u(x(t), y(t), z(t)), \quad \dot{y}(t) = u(x(t), y(t), z(t)), \quad \dot{z}(t) = w(x(t), y(t), z(t))$$

for all times $t \in \mathbb{R}$ with the initial condition $x(0) = x_0, y(0) = y_0, z(0) = z_0$.

Parallel flow. Suppose that the components u, v, w of the velocity vector $\mathbf{v}(P)$ do not depend on the point P, that is, the functions u, v, w are constant. Then the trajectories are straight lines:

$$x(t) = x_0 + tu, \quad y(t) = y_0 + tv, \quad z(t) = z_0 + tw, \qquad t \in \mathbb{R}.$$

We call this a parallel flow (Fig. 10.3(a)). The trajectories can also be written as

$$\mathbf{x}(t) = \mathbf{x}_0 + \mathbf{v}_0 t, \qquad t \in \mathbb{R}.$$

Then $\dot{\mathbf{x}}(t) = \text{const} = \mathbf{v}_0$.

Rotating fluid particles. Fix $\omega > 0$, and choose $\boldsymbol{\omega} := \omega\mathbf{k}$. The velocity vector field

$$\mathbf{v}(\mathbf{x}) := \boldsymbol{\omega} \times \mathbf{x}$$

describes fluid particles which counter-clockwise rotate about the z-axis with the angular velocity $\omega > 0$. The trajectories are given by

$$x(t) = x_0 \cos\omega t - y_0 \sin\omega t, \quad y(t) = x_0 \sin\omega t + y_0 \cos\omega t, \quad z = z_0, \quad t \in \mathbb{R}.$$

Fig. 10.3(b) shows the rotation of the fluid particles in the (x, y)-plane.

Linear flow and the exponential function. Let $A : E_3 \to E_3$ be a linear operator. Consider the vector field

$$\mathbf{v}(\mathbf{x}) := A\mathbf{x}.$$

Then the linear differential equation (10.7) yields the following trajectories of the fluid particles:

$$\mathbf{x}(t) = \mathrm{e}^{tA}\mathbf{x}_0, \qquad t \in \mathbb{R}.$$

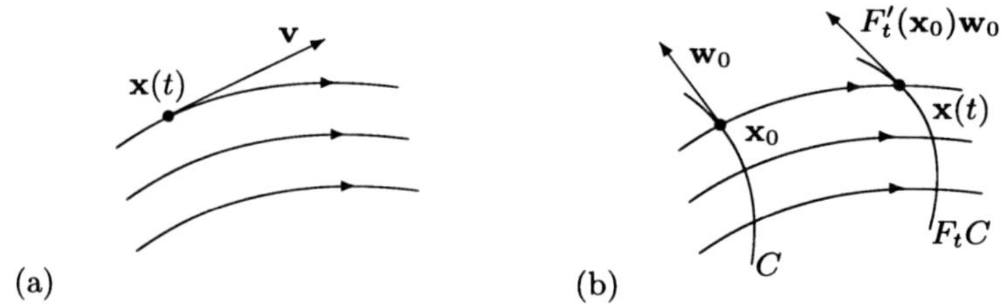

Fig. 10.4. Lie's parallel transport along a flow

10.2.2 The Linearized Flow

The basic idea. Fix time t. We want to linearize the flow operator

$$F_t : \mathbb{E}^3 \to \mathbb{E}^3$$

at the point P_0 of the Euclidean manifold $\mathbb{E}^3$. This yields the linear operator

$$F_t'(P_0) : T_{P_0}\mathbb{E}^3 \to T_{F_t(P_0)}(\mathbb{E}^3).$$

The basic idea reads as follows: The flow operator F_t transports points and hence curves; this implies the transport of the tangent vectors of the curves. Here, the transport of tangent vectors is described by the linearized flow operator $F_t'(P_0)$ (Fig. 10.4). The key formula reads as

$$\boxed{F_t'(P_0)\dot{Q}(0) = \dot{\mathcal{Q}}(0).} \tag{10.8}$$

More precisely, we start with the smooth curve

- $C : Q = Q(\tau), \quad \tau \in \mathbb{R}, \quad Q(0) = P_0.$

Setting $\mathcal{Q}(\tau) := F_tQ(\tau)$, the flow sends the curve C to the smooth curve

- $F_tC : \mathcal{Q} = \mathcal{Q}(\tau), \quad \tau \in \mathbb{R}, \quad \mathcal{Q}(0) = F_t(P_0).$

This implies the transport of the tangent vector $\dot{Q}(0)$ which yields (10.8). Let us write the moved trajectory F_tC at time t by the equation

$$F_tC : \mathbf{x} = \mathbf{x}(t, \tau), \qquad \tau \in \mathbb{R}.$$

The linear differential equation of the linearized flow. In terms of position vectors, we set $\mathbf{w}_0 := \frac{\partial}{\partial\tau}\mathbf{x}_0(0, \tau)_{|\tau=0}$, and we consider the initial-value problem

$$\dot{\mathbf{y}}(t) = \mathbf{v}'(F_t\mathbf{x}_0)\mathbf{y}(t), \quad t \in \mathbb{R}, \quad \mathbf{y}(0) = \mathbf{w}_0. \tag{10.9}$$

This is a linear system of ordinary differential equations with smooth coefficient functions. By the theory of linear differential equations, there exists a unique solution $\mathbf{y} = \mathbf{y}(t)$ for all times $t \in \mathbb{R}$.

Proposition 10.1 *$F_t'(P_0)\mathbf{w}_0 = \mathbf{y}(t)$, $t \in \mathbb{R}$, and*

$$\frac{d}{dt}F_t'(P_0)_{|t=0} = \mathbf{v}'(P_0), \tag{10.10}$$

as well as $\frac{d}{dt}F_t'(P_0)^{-1}_{|t=0} = -\mathbf{v}'(P_0)$.

Proof. Set $\mathbf{x}(t,\tau) := F_t\mathbf{x}(0,\tau)$. Then

$$\mathbf{x}_t(t,\tau) = \mathbf{v}(\mathbf{x}(t,\tau)), \qquad \mathbf{x}(0,\tau) = \mathbf{x}_0(0,\tau). \tag{10.11}$$

Since the initial values $\mathbf{x}_0(0,\tau)$ depend smoothly on the parameter τ, it is admissible to differentiate the equation (10.11) with respect to τ. This yields

$$\frac{\partial}{\partial t}\mathbf{x}_\tau(t,\tau) = \mathbf{v}'(\mathbf{x}(t,\tau))\,\mathbf{x}_\tau(t,\tau).$$

Choosing $\tau = 0$ and setting $\mathbf{y}(t) := \mathbf{x}_\tau(t,0)$, we obtain (10.9). Hence

$$\frac{d}{dt}F_t'(\mathbf{x}_0)\mathbf{w}_{0|t=0} = \dot{\mathbf{y}}(0) = \mathbf{v}'(\mathbf{x}_0)\mathbf{w}_0.$$

This implies (10.10). Finally, noting that $F_0'(P_0) = I$ and differentiating

$$F_t'(P_0)F_t'(P_0)^{-1} = I, \qquad t \in \mathbb{R}$$

with respect to time t at the point $t = 0$, we get

$$\frac{d}{dt}F_t'(P_0)_{|t=0} + \frac{d}{dt}F_t'(P_0)^{-1}_{|t=0} = 0.$$

□

Relation (10.10) will be used in Sect. 11.2.1 in order to compute the Lie derivative of vector fields.

10.2.3 The Lie Derivative

> The Lie derivative studies the behavior of the smooth temperature field Θ along the trajectories of the fluid particles on an infinitesimal time level.
>
> Folklore

We are given the smooth velocity vector field $\mathbf{v}$ on the Euclidean manifold $\mathbb{E}^3$. Let $\{F_t\}_{t\in\mathbb{R}}$ be the flow generated by $\mathbf{v}$. Fix the time $t \in \mathbb{R}$. Set

$$(F_t^*\Theta)(P) := \Theta(F_tP) \qquad \text{for all} \quad P \in \mathbb{E}^3.$$

The temperature field $F_t^*\Theta$ is called the pull-back of the original temperature field Θ with respect to the flow operator F_t. We define the Lie derivative $\mathcal{L}_\mathbf{v}\Theta$ by setting

$$\boxed{(\mathcal{L}_\mathbf{v}\Theta)(P_0) := \frac{d}{dt}F_t^*\Theta(P_0)_{|t=0}.} \tag{10.12}$$

This is a smooth function $\mathcal{L}_\mathbf{v}\Theta : \mathbb{E}^3 \to \mathbb{R}$. The value $(\mathcal{L}_\mathbf{v}\Theta)(P_0)$ is called the Lie derivative of the temperature field Θ at the point P_0 with respect to the velocity vector field $\mathbf{v}$. By Sect. 10.1,

$$\mathcal{L}_\mathbf{v}\Theta(P_0) := d_{\mathbf{v}(P_0)}\Theta(P_0). \tag{10.13}$$

This tells us that, in the special case of a temperature field Θ, the Lie derivative $\mathcal{L}_\mathbf{v}\Theta(P_0)$ coincides with the directional derivative $d_{\mathbf{v}(P_0)}\Theta(P_0)$ with respect to the direction $\mathbf{v}(P_0)$ of the velocity vector field of the fluid at the point P_0. Choosing a right-handed Cartesian (x,y,z)-system with the right-handed orthonormal basis $\mathbf{i},\mathbf{j},\mathbf{k}$ at the origin, then $\mathbf{v}_P = a(P)\mathbf{i}_P + b(P)\mathbf{j}_P + c(P)\mathbf{k}_P$ (see Fig. 4.3 on page 323), and we get

$$(\mathcal{L}_\mathbf{v}\Theta)(P) = a(P)\frac{\partial\Theta(P)}{\partial x} + b(P)\frac{\partial\Theta(P)}{\partial y} + c(P)\frac{\partial\Theta(P)}{\partial z}. \tag{10.14}$$

10.2.4 Conservation Laws

The condition

$$\mathcal{L}_{\mathbf{v}}\Theta(P) = 0 \qquad \text{for all} \quad P \in \mathbb{E}^3$$

is equivalent to the fact that the temperature field Θ is constant along all the trajectories of the fluid particles.

10.3 Higher Variations of a Temperature Field and the Taylor Expansion

Higher variations of the smooth temperature field $\Theta : \mathbb{E}^3 \to \mathbb{R}$ allow us to study approximations of Θ which are more precise than the procedure of linearization. To begin with, consider the smooth function

$$\chi : \mathbb{R} \to \mathbb{R}.$$

For $n = 1, 2, \ldots$, we have the Taylor expansion[3]

$$\chi(t) = \chi(0) + \sum_{k=1}^{n} \frac{t^k}{k!}\chi^{(k)}(0) + \int_0^1 \frac{(1-\tau)^n}{n!}\chi^{(n+1)}(\tau t)\, d\tau \tag{10.15}$$

for all $t \in \mathbb{R}$. This motivates the following definition.

The Taylor expansion of the temperature field. For fixed tangent vector $\mathbf{h} \in T_{P_0}\mathbb{E}^3$ at the point P_0, we set

$$\chi(t) := \Theta(P_0 + t\mathbf{h}), \qquad t \in \mathbb{R}.$$

We use the n-th derivative $\chi^{(n)}$ in order to define

$$\boxed{\delta^n\Theta(P_0 + t\mathbf{h}; \mathbf{h}) := \chi^{(n)}(t), \qquad n = 1, 2, \ldots}$$

Here, $\delta^n\Theta(P_0; \mathbf{h})$ is called the n-th variation of the temperature field Θ at the point P_0 with respect of the direction $\mathbf{h}$. From (10.15) we get

$$\Theta(P_0 + t\mathbf{h}) = \Theta(P_0) + \sum_{k=1}^{n} \frac{t^k}{k!}\,\delta^k\Theta(P_0; \mathbf{h}) + \int_0^1 \frac{(1-\tau)^n}{n!}\,\delta^{n+1}\Theta(P_0 + \tau t\mathbf{h}; \mathbf{h})\, d\tau$$

for all $t \in \mathbb{R}$.

The first variation. By Sect. 10.1,

$$\boxed{\delta\Theta(P_0; \mathbf{h}) = d_{\mathbf{h}}\Theta(P_0).} \tag{10.16}$$

This means that the first variation coincides with the directional derivative. The linear map $\mathbf{h} \mapsto \delta\Theta(P_0; \mathbf{h})$ is denoted by $\Theta'(P_0)$. Thus, for all $\mathbf{h} \in T_{P_0}\mathbb{E}^3$, we get

$$\delta\Theta(P_0; \mathbf{h}) = \Theta'(P_0)\mathbf{h}.$$

The second variation. Choosing a Cartesian (x, y, z)-coordinate system with the orthonormal basis $\mathbf{i}, \mathbf{j}, \mathbf{k}$, we set

[3] Newton (1643–1727), Taylor (1685–1731).

$$\mathbf{h} = \Delta x\, \mathbf{i} + \Delta y\, \mathbf{j} + \Delta z\, \mathbf{k}.$$

By the chain rule, $\delta^2 \Theta(P_0; \mathbf{h})$ is equal to

$$\Theta_{xx}(P_0)(\Delta x)^2 + \Theta_{yy}(P_0)(\Delta y)^2 + \Theta_{zz}(P_0)(\Delta z)^2 \\ +2\Theta_{xy}(P_0)\Delta x \Delta y + 2\Theta_{xz}(P_0)\Delta x \Delta z + 2\Theta_{yz}(P_0)\Delta y \Delta z.$$

The corresponding symmetric bilinear form is denoted by $\Theta''(P_0)$. Hence

$$\boxed{\delta^2 \Theta(P_0; \mathbf{h}) = \Theta''(P_0)(\mathbf{h}, \mathbf{h}).} \tag{10.17}$$

For this, we also briefly write $\Theta''(P_0)\mathbf{h}^2$.

10.4 The Fréchet Derivative

By definition, the temperature field $\Theta : \mathbb{E}^3 \to \mathbb{R}$ has a Fréchet derivative at the point P_0 iff there exists a linear operator $L : T_{P_0}\mathbb{E}^3 \to \mathbb{R}$ such that

$$\Theta(P_0 + \mathbf{h}) = L\mathbf{h} + r(\mathbf{h}) \qquad \text{for all} \quad \mathbf{h} \in T_{P_0}\mathbb{E}^3$$

where the remainder $r(\mathbf{h})$ is of order $o(|\mathbf{h}|) \cdot |\mathbf{h}|$ as $\mathbf{h} \to 0$, that is,

$$\lim_{\mathbf{h} \to 0} \frac{|r(\mathbf{h})|}{|\mathbf{h}|} = 0.$$

The operator L is uniquely determined; it is called the Fréchet derivative of the temperature field Θ at the point P_0.[4] Motivated by classical calculus, we write $\Theta'(P_0)$ instead of L. Moreover, the linear map $\mathbf{h} \mapsto \Theta'(P_0)\mathbf{h}$ is also called the Fréchet differential of Θ at the point P_0. We write

$$d\Theta_{P_0}\mathbf{h} = \Theta'(P_0)\mathbf{h},$$

that is, $d\Theta_{P_0} = \Theta'(P_0) = L$. If the temperature field Θ is smooth, then we have

$$\delta\Theta(P_0; \mathbf{h}) = d\Theta_{P_0}(\mathbf{h}).$$

Frequently, physicists prefer the use of the functional derivative

$$\frac{\delta\Theta(P_0)}{\delta\mathbf{x}} := \Theta'(P_0) = d\Theta_{P_0}.$$

This implies

$$\delta\Theta(P_0; \mathbf{h}) = \frac{\delta\Theta(P_0)}{\delta\mathbf{x}} \cdot \mathbf{h}.$$

The fact that quite different symbols denote the same mathematical object is the result of a long fight in the history of mathematics for establishing a general calculus in finite and infinite dimensions. The differential calculus in finite-dimensional and infinite-dimensional Banach spaces (resp. Banach manifolds), based on the Fréchet derivative, is thoroughly studied in E. Zeidler, Nonlinear Functional Analysis and its Applications, Vol. I (local analysis), Vol. IV (global analysis), Springer, New York, 1993/1997.

---

[4] Fréchet (1878–1973)

10.5 Global Linearization of Smooth Maps and the Tangent Bundle

> The mathematical investigations are simplified when replacing the position space by the state space which is the tangent bundle of the position space. In terms of physics, the state space enriches the position space by adding the possible velocities of particles.
>
> Folklore

Basic idea. We are given the smooth function

$$f : \mathbb{R} \to \mathbb{R}.$$

Setting

$$\boxed{(Tf)(x,v) := (f(x), f'(x)v),} \tag{10.18}$$

we introduce the map

$$Tf : \mathbb{R}^2 \to \mathbb{R}^2$$

which is called the global linearization of f. This notion has the advantage that Leibniz's classical chain rule for the differentiation of composed maps gets a simpler form. This allows the formulation of a global chain rule for general finite-dimensional and infinite-dimensional manifolds. In order to explain the simple basic idea, consider the composed map

$$g \circ f : \mathbb{R} \xrightarrow{f} \mathbb{R} \xrightarrow{g} \mathbb{R}.$$

Then the global linearization $T(g \circ f)$ is obtained by the elegant composition formula

$$T(g \circ f) : \mathbb{R} \xrightarrow{Tf} \mathbb{R} \xrightarrow{Tg} \mathbb{R}.$$

Explicitly, $T(g \circ f) = Tg \circ Tf$. This tells us that:

The global linearization respects the composition of maps.

Let us prove this by using the classical chain rule

$$\frac{d(g \circ f)(x)}{dx} = \frac{dg(f(x))}{dx} = \frac{dg(y)}{dy}\frac{dy}{dx}_{|y=f(x)} = g'(f(x))f'(x).$$

By definition (10.18) of the operator T, we get

$$Th(x,v) = (h(x), h'(x)v).$$

Choosing $h(x) := (g \circ f)(x) = g(f(x))$, we obtain

$$T(g \circ f)(x,v) = (g(f(x)), g'(f(x))f'(x)v).$$

Moreover,

$$(Tg)(f(x), f'(x)v) = (g(f(x)), g'(f(x))f'(x)v).$$

Hence $T(g \circ f) = Tg \circ Tf$.

The tangent bundle of the real line and the global tangent map. In order to get contact with the general theory of manifolds, consider the real line $\mathbb{R}$ as a manifold. Then the tangent bundle of $\mathbb{R}$ reads as

$$T\mathbb{R} := \{(x,v) : \; x \in \mathbb{R}, \; v \in \mathbb{R}\}.$$

In terms of physics, the variable x describes a point on the real line, and the variable v describes the possible particle velocity at the point x of some particle that moves on the real line. Obviously, $T\mathbb{R} = \mathbb{R}^2$. Using this terminology, the global linearization Tf of the map $f : \mathbb{R} \to \mathbb{R}$ reads as

$$\boxed{Tf : T\mathbb{R} \to T\mathbb{R}.}$$

The map Tf is also called the global tangent map of f.

> *Mnemonically, the passage from the original map $f : \mathbb{R} \to \mathbb{R}$ to the tangent map $Tf : T\mathbb{R} \to T\mathbb{R}$ is obtained by replacing the basic manifold $\mathbb{R}$ by its tangent bundle $T\mathbb{R}$.*

This underlines the observation that the notion of the tangent bundle is a quite natural concept. The global chain rule

$$\boxed{T(g \circ f) = Tg \circ Tf}$$

corresponds elegantly to the passage from the diagram

$$g \circ f : \mathbb{R} \xrightarrow{f} \mathbb{R} \xrightarrow{g} \mathbb{R}$$

to the diagram

$$T(g \circ f) : T\mathbb{R} \xrightarrow{Tf} T\mathbb{R} \xrightarrow{Tg} T\mathbb{R}.$$

That is, we only add the symbol T which stands for 'tangent'.

The global tangent map on the Euclidean manifold. We are given the smooth map

$$f : \mathbb{E}^3 \to \mathbb{E}^3.$$

Recall that the tangent bundle of $\mathbb{E}^3$ is given by

$$T\mathbb{E}^3 := \{(P, \mathbf{v}) : \ P \in \mathbb{E}^3, \ \mathbf{v} \in T_P\mathbb{E}^3\}.$$

In terms of physics, P is the position of a particle, and $\mathbf{v}$ is the velocity vector of a particle at the point P. In other words, $\mathbb{E}^3$ is the position space, and $T\mathbb{E}^3$ is the state space for the motion of particles on the Euclidean manifold $\mathbb{E}^3$. Our goal is to construct the global linearization

$$Tf : T\mathbb{E}^3 \to T\mathbb{E}^3 \tag{10.19}$$

which is also called the global tangent map of f. We start with a fixed velocity vector $\mathbf{v}$ at the point P_0. First let us construct the local linearization

$$\boxed{T_{P_0}f : T_{P_0}\mathbb{E}^3 \to T_{f(P_0)}\mathbb{E}^3}$$

which sends the tangent vector $\mathbf{v}$ at the point P_0 to the tangent vector $T_{f(P_0)}\mathbf{v}$ at the image point $f(P_0)$. To this end, we consider a fixed trajectory

$$C : P = P(t), \qquad t \in \mathbb{R}, \qquad P(0) = P_0$$

of a particle which passes through the point P_0 at time $t = 0$ with the velocity vector $\mathbf{v}$. In terms of position vectors located at the origin, the given trajectory reads as

$$C : \mathbf{x} = \mathbf{x}(t), \qquad t \in \mathbb{R}, \qquad \mathbf{x}(0) = \mathbf{x}_0.$$

The map f sends the trajectory C to the trajectory

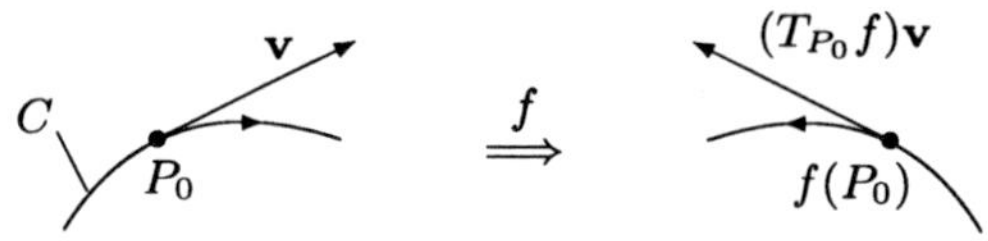

Fig. 10.5. The tangent map $T_{P_0}f$

$$f(C) : P = f(P(t)), \qquad t \in \mathbb{R}.$$

Naturally enough, we consider the velocity vector $\mathbf{w}$ of the transformed trajectory $f(C)$ at the point $f(P_0)$ (Fig. 10.5), and we define

$$\boxed{(T_{P_0}f)\mathbf{v} := \mathbf{w}.}$$

Thus, $T_{P_0}f$ is the map for velocity vectors which is induced by the original map f in a natural manner.[5]

It is crucial that the velocity vector $\mathbf{w}$ *only depends on the given velocity vector* $\mathbf{v}$, *but not on the choice of the trajectory* C.

To show this, consider a right-handed Cartesian (x, y, z)-coordinate system with the orthonormal basis $\mathbf{i}, \mathbf{j}, \mathbf{k}$. With respect to this coordinate system, the map

$$f : \mathbb{E}^3 \to \mathbb{E}^3$$

has the form

$$f(x, y, z) = (X(x, y, z), Y(x, y, z), Z(x, y, z)), \qquad (x, y, z) \in \mathbb{R}^3,$$

that is, the image point $f(P)$ has the Cartesian coordinates X, Y, Z depending on the Cartesian coordinates x, y, z of the original point P. Moreover, we need the velocity vector $\mathbf{v} = u\,\mathbf{i} + v\,\mathbf{j} + w\,\mathbf{k}$. Then we get

$$(T_{P_0}f)\mathbf{v} = U\mathbf{i} + V\mathbf{j} + W\mathbf{k}$$

with

$$\begin{pmatrix} U \\ V \\ W \end{pmatrix} = \begin{pmatrix} X_x(x_0, y_0, z_0) & X_y(x_0, y_0, z_0) & X_z(x_0, y_0, z_0) \\ Y_x(x_0, y_0, z_0) & Y_y(x_0, y_0, z_0) & Y_z(x_0, y_0, z_0) \\ W_x(x_0, y_0, z_0) & W_y(x_0, y_0, z_0) & W_z(x_0, y_0, z_0) \end{pmatrix} \begin{pmatrix} u \\ v \\ w \end{pmatrix}. \qquad (10.20)$$

For the proof, we refer to Problem 10.1. Finally, we define

$$\boxed{T(P_0, \mathbf{v}) := (f(P_0), (T_{P_0}f)\mathbf{v}).}$$

This yields the global tangent map (10.19).

[5] The map T_{P_0} is also called the tangent map of f at the point P_0 (or the linearization of the map f at the point P_0).

10.6 The Global Chain Rule

We are given the smooth maps $f : \mathbb{E}^3 \to \mathbb{E}^3$ and $g : \mathbb{E}^3 \to \mathbb{E}^3$.

Theorem 10.2 *The global linearization of the composed map*

$$g \circ f : \mathbb{E}^3 \xrightarrow{f} \mathbb{E}^3 \xrightarrow{g} \mathbb{E}^3$$

reads as $T(g \circ f) : T\mathbb{E}^3 \xrightarrow{Tf} T\mathbb{E}^3 \xrightarrow{Tg} T\mathbb{E}^3$.

In other words, $T(g \circ f) = Tg \circ Tf$. For the proof, see Problem 10.2 on page 658.

10.7 The Transformation of Temperature Fields

We are given the smooth map

$$\boxed{\tau : \mathbb{E}^3 \to \mathbb{E}^3} \tag{10.21}$$

(e.g., a rotation). We want to use this transformation in order to transplant the given smooth temperature field $\Theta : \mathbb{E}^3 \to \mathbb{R}$. There are the following two possibilities:

- $(\tau^*\Theta)(Q) := \Theta(\tau(Q))$ for all points $Q \in \mathbb{E}^3$ (pull-back),
- $(\tau_*\Theta)(P) := \Theta(\tau^{-1}(P))$ for all points $P \in \mathbb{E}^3$ (push-forward).

Letting $P = \tau(Q)$, we get $(\tau^*\Theta)(Q) = \Theta(P)$ and $(\tau_*\Theta)(P) = \Theta(Q)$.

In order to define the push-forward $\tau_*\Theta$, we have to assume additionally that the transformation (10.21) is a diffeomorphism. Choose a fixed (x, y, z)-Cartesian coordinate system, and assume that $P = \tau(Q)$ corresponds to

$$x = x(\xi, \eta, \zeta), \quad y = y(\xi, \eta, \zeta), \quad z = z(\xi, \eta, \zeta), \qquad (\xi, \eta, \zeta) \in \mathbb{R}^3.$$

That is, the point P (resp. Q) has the Cartesian coordinates (x, y, z) (resp. (ξ, η, ζ)). Then, for all $(\xi, \eta, \zeta) \in \mathbb{R}^3$, we have:

- $(\tau^*\Theta)(\xi, \eta, \zeta) = \Theta(x(\xi, \eta, \zeta), y(\xi, \eta, \zeta), z(\xi, \eta, \zeta))$,
- $(\tau_*\Theta)(x(\xi, \eta, \zeta), y(\xi, \eta, \zeta), z(\xi, \eta, \zeta)) = \Theta(\xi, \eta, \zeta)$.

In terms of commutative diagrams, the following hold.

Pull-back $\tau^*\Theta$ of the temperature field Θ. The transformed temperature field $\tau^*\Theta$ is defined by the commutativity of the following diagram:

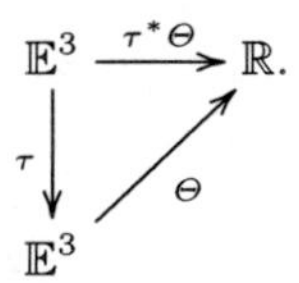

(10.22)

In other words, $\tau^*\Theta = \Theta \circ \tau$.

Push-forward $\tau_*\Theta$ of the temperature field Θ. Suppose that $\tau : \mathbb{E}^3 \to \mathbb{E}^3$ is a diffeomorphism. Then the transformed temperature field $\tau_*\Theta$ is defined by the commutativity of the following diagram:

$$\begin{array}{ccc} \mathbb{E}^3 & \xrightarrow{\tau_*\Theta} & \mathbb{R}. \\ {\scriptstyle \tau^{-1}}\big\downarrow & \nearrow{\scriptstyle \Theta} & \\ \mathbb{E}^3 & & \end{array} \tag{10.23}$$

In other words, $\tau_*\Theta = \Theta \circ \tau^{-1}$. Note that:

The push-forward $\tau_*\Theta$ with respect to the map τ is equal to the pull-back $(\tau^{-1})^*\Theta$ with respect to the inverse map τ^{-1}.

Problems

10.1 *The local linearization $T_{P_0}f$.* Prove (10.20).
Solution: In terms of position vectors at the origin, the original trajectory

$$C : \mathbf{x}(t) = x(t)\,\mathbf{i} + y(t)\,\mathbf{j} + z(t)\,\mathbf{k}, \qquad t \in \mathbb{R}$$

is transformed into the trajectory $f(C)$,

$$\mathbf{X}(t) = X(x(t), y(t), z(t))\,\mathbf{i} + Y(x(t), y(t), z(t))\,\mathbf{j} + Z(x(t), y(t), z(t))\,\mathbf{k}.$$

Then $(T_{P_0}f)\mathbf{v} = \dot{\mathbf{X}}(0)$. This implies (10.20), by using the chain rule.

10.2 *Proof of the global chain rule.* Prove Theorem 10.2. Hint: See E. Zeidler (1986), Vol. IV, p. 603, quoted on page 1089.

11. Velocity Vector Fields on the Euclidean Manifold $\mathbb{E}^3$

Sophus Lie (1842–1899) emphasized the importance of velocity vector fields for the study of manifolds and their symmetries in terms of the linearization procedure. This leads to two fundamental concepts: the Lie derivative and the Lie algebra of velocity vector fields, and the Lie algebra of a Lie group. These concepts are basic for modern mathematics and physics; they connect differential geometry with the theory of dynamical systems.

Folklore

We want to study vector fields

$$\mathbf{w} = \mathbf{w}(P), \quad P \in \mathbb{E}^3$$

on the 3-dimensional Euclidean manifold $\mathbb{E}^3$. For example, this concerns velocity vector fields or force fields like

- Newton's gravitational field $\mathbf{w} = \mathbf{F}_{\text{grav}}$,
- Maxwell's electric field $\mathbf{w} = \mathbf{E}$, or
- Maxwell's magnetic field $\mathbf{w} = \mathbf{B}$.

We will frequently use the intuitive picture of the velocity vector field of a fluid. For such vector fields $\mathbf{w}$ on $\mathbb{E}^3$, one has to distinguish between

- the covariant directional derivative $D_\mathbf{v}\mathbf{w}$, and
- the Lie derivative $\mathcal{L}_\mathbf{v}\mathbf{w} = D_\mathbf{v}\mathbf{w} - D_\mathbf{w}\mathbf{v}$. Here, $\mathbf{v}$ is the velocity field of the flow of fluid particles on $\mathbb{E}^3$.

For example, the Lie derivative describes continuous symmetries on an infinitesimal level. In Chapter 12, dualizing the concept of velocity vector fields, we will introduce covector fields which lie at the heart of Cartan's calculus of differential forms. Note that

- the Lie derivative of velocity vector fields and
- Cartan's calculus of differential forms

can be introduced on arbitrary manifolds. In particular, the Lie derivative $\mathcal{L}_\mathbf{v}\mathbf{w}$ of a velocity vector field $\mathbf{w}$ with respect to the velocity vector field $\mathbf{v}$ allows us to equip the space of velocity vector fields on an arbitrary manifold with the structure of a Lie algebra, by introducing the Lie product

$$[\mathbf{v}, \mathbf{w}] := \mathcal{L}_\mathbf{v}\mathbf{w}$$

in an invariant way. This underlines the importance of Lie algebras for the theory of manifolds.

On special classes of manifolds (e.g., Riemannian and pseudo-Riemannian manifolds), it is possible to introduce the following additional notions:

- the Hodge codifferential $d^*\omega$ which is dual to the Cartan differential $d\omega$ of a differential form ω (see Chap. 12),

E. Zeidler, *Quantum Field Theory III: Gauge Theory*,
DOI 10.1007/978-3-642-22421-8_12,

- the covariant directional derivative $D_{\mathbf{v}}\mathbf{w}$ via parallel transport,
- and the covariant Cartan differential $D\omega$ of differential forms ω on principal fiber bundles or vector bundles (see Chaps. 15 and 17).

Élie Cartan's calculus of differential forms reflects crucial topological (i.e., global) properties of manifolds on the basis of the de Rham cohomology. In terms of physics, de Rham cohomology generalizes important properties of the electromagnetic field (see Chap. 23).

Using Hodge duality, the de Rham cohomology passes over to Hodge homology. In turn, this corresponds to the original Poincaré homology based on Poincaré's triangulation of manifolds and the computing of both the Euler characteristic and the Betti numbers, which represent fundamental topological invariants. The generalization of this approach leads to the modern theory of characteristic classes (e.g., Chern classes).[1]

Covariant directional derivatives $D_{\mathbf{v}}\mathbf{w}$ of velocity vector fields $\mathbf{w}$ (and more general tensor fields) and covariant differentials $D\omega$ of differential forms ω play a fundamental role in gauge theory (see Chap. 15). Observe that one has to distinguish between

- local parallel transport, and
- global parallel transport

of velocity vectors on manifolds. On general manifolds, one can introduce a local parallel transport of velocity vector fields (in the language of modern mathematics, this is a connection on the tangent bundle of the manifold). However, only special manifolds possess a global parallel transport. In particular, the classical parallel transport of velocity vectors on the Euclidean manifold $\mathbb{E}^3$ is global. This yields the Euclidean covariant directional derivative

$$D_{\mathbf{v}}\mathbf{w} = (\mathbf{v}\boldsymbol{\partial})\mathbf{w} \qquad \text{on } \mathbb{E}^3,$$

where $\boldsymbol{\partial}$ is Hamilton's nabla operator.

Perspectives (curvature and torsion). Observe the following special feature.

(i) The curvature operator. Let $\mathbf{u}, \mathbf{v}, \mathbf{w}, \mathbf{z}$ be smooth velocity vector fields. Introduce the so-called curvature operator $\mathbf{w} \mapsto \mathbf{F}(\mathbf{u}, \mathbf{v})\mathbf{w}$ by setting

$$\boxed{\mathbf{F}(\mathbf{u}, \mathbf{v})\mathbf{w} := D_{\mathbf{u}}D_{\mathbf{v}}\mathbf{w} - D_{\mathbf{v}}D_{\mathbf{u}}\mathbf{w} - D_{[\mathbf{u},\mathbf{v}]}\mathbf{w}.}$$

In addition, let us introduce the Riemann curvature tensor $\mathcal{R}$ by setting

$$\mathcal{R}(\mathbf{u}, \mathbf{v}, \mathbf{w}, \mathbf{z}) := \langle \mathbf{F}(\mathbf{u}, \mathbf{v})\mathbf{w} | \, \mathbf{z} \rangle. \tag{11.1}$$

Here, we assume that the tangent spaces of the manifold under consideration are equipped with the inner product $\langle .|. \rangle$. In the special case of the Euclidean manifold $\mathbb{E}^3$, we have

$$\mathbf{F}(\mathbf{u}, \mathbf{v}) \equiv 0 \quad \text{on } \mathbb{E}^3,$$

and $\mathcal{R} \equiv 0$ on $\mathbb{E}^3$. This reflects the flatness of the Euclidean manifold $\mathbb{E}^3$. In the general case, the Riemann curvature operator and the Riemann curvature tensor do not vanish, and they measure the curvature of the manifold under consideration. The prototype of a manifold with nonvanishing curvature is the sphere (see Sect. 9.5).

[1] Lie (1842–1899), Poincaré (1854–1912), Élie Cartan (1869–1951), de Rham (1903–1990), Hodge (1903–1975), Chern (1911–2004).

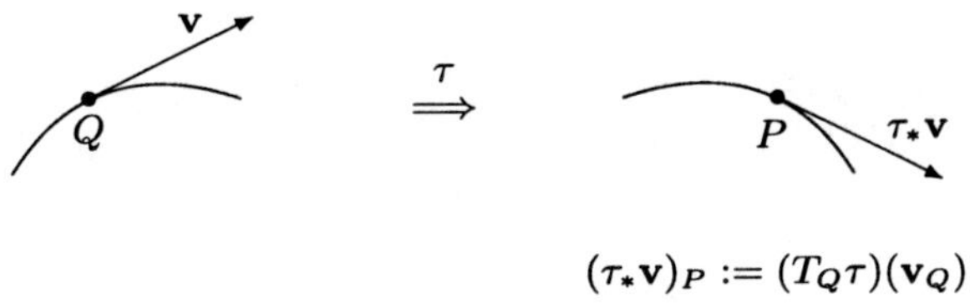

Fig. 11.1. Push-forward $\tau_*\mathbf{v}$ of the velocity vector field $\mathbf{v}$

(ii) The torsion operator. Define the so-called torsion operator $\mathbf{T}$ by setting

$$\boxed{\mathbf{T}(\mathbf{v},\mathbf{w}) := D_{\mathbf{v}}\mathbf{w} - D_{\mathbf{w}}\mathbf{v} - \mathcal{L}_{\mathbf{v}}\mathbf{w}.} \tag{11.2}$$

In the special case of the Euclidean manifold $\mathbb{E}^3$, we get

$$\mathbf{T}(\mathbf{v},\mathbf{w}) \equiv 0 \qquad \text{on } \mathbb{E}^3. \tag{11.3}$$

In the general case, there exist covariant directional derivatives on certain classes of manifolds where the torsion operator $\mathbf{T}$ does not vanish.

The importance of velocity vector fields. In the history of physics, velocity vector fields emerged in the 18th century in order to describe the motion of fluid particles. The modern development of mathematics and physics has been shown that the highly intuitive idea of a velocity vector field is of fundamental importance for all branches of modern physics. In fact, crucial time-dependent processes in nature are described by dynamical systems, and velocity vector fields are nothing other than dynamical systems on an infinitesimal level.

11.1 The Transformation of Velocity Vector Fields

Let us consider the smooth map

$$\boxed{\tau : \mathbb{E}^3 \to \mathbb{E}^3} \tag{11.4}$$

(e.g., a rotation). Assume that the map τ sends the point Q to the point $P = \tau(Q)$.

Push-forward $\tau_*\mathbf{v}$ of the velocity vector field v. We are given the smooth vector field $\mathbf{v}$ on the Euclidean manifold $\mathbb{E}^3$. We want to transplant the velocity vector from the point Q to the point P (Fig. 11.1). To this end. observe the following. The linearization $T_Q\tau$ of the map τ sends

- the velocity vector $\mathbf{v}_Q$ at the point Q
- to the velocity vector $\mathbf{w}_P$ at the point P where $\mathbf{w}_P := (T_Q\tau)(\mathbf{v}_Q)$.

Naturally enough, we define

$$\boxed{(\tau_*\mathbf{v})_P := (T_Q\tau)(\mathbf{v}_Q) \qquad \text{for all} \quad Q \in \mathbb{E}^3.} \tag{11.5}$$

Choose a fixed right-handed (x, y, z)-Cartesian coordinate system with the right-handed orthonormal basis $\mathbf{i}, \mathbf{j}, \mathbf{k}$ at the origin O. Parallel transport of $\mathbf{i}, \mathbf{j}, \mathbf{k}$ to the point P yields the orthonormal basis $\mathbf{i}_P, \mathbf{j}_P, \mathbf{k}_P$ of the tangent space $T_P\mathbb{E}^3$ of the Euclidean plane $\mathbb{E}^3$ at the point P (see Fig. 9.1 on page 558). Assume that the equation $P = \tau(Q)$ corresponds to

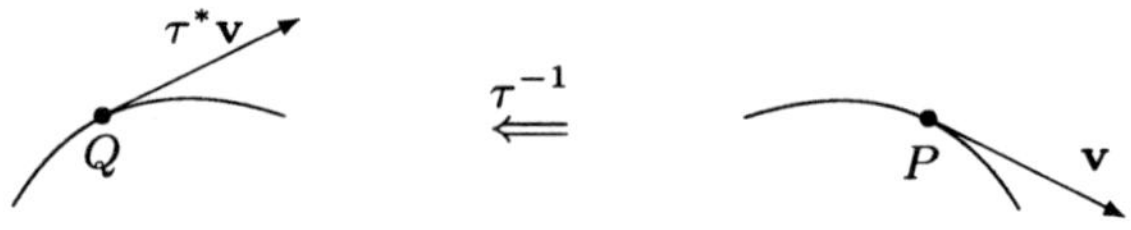

$$(\tau^*\mathbf{v})_Q = (T_P\tau^{-1})(\mathbf{v}_P)$$

Fig. 11.2. Pull-back $\tau^*\mathbf{v}$ of the velocity vector field $\mathbf{v}$

$$x = x(\xi,\eta,\zeta), \quad y = y(\xi,\eta,\zeta), \quad z = z(\xi,\eta,\zeta), \qquad (\xi,\eta,\zeta) \in \mathbb{R}^3.$$

That is, the point P (resp. Q) has the Cartesian coordinates (x, y, z) (resp. (ξ, η, ζ)). Let $\mathbf{v}_Q = a(Q)\mathbf{i}_Q + b(Q)\mathbf{j}_Q + c(Q)\mathbf{k}_Q$. Then, it follows from (10.20) on page 656 that

$$(\tau_*\mathbf{v})_P = A(P)\mathbf{i}_P + B(P)\mathbf{j}_P + C(P)\mathbf{k}_P$$

with

$$\begin{pmatrix} A(P) \\ B(P) \\ C(P) \end{pmatrix} := \begin{pmatrix} x_\xi(Q) & x_\eta(Q) & x_\zeta(Q) \\ y_\xi(Q) & y_\eta(Q) & y_\zeta(Q) \\ z_\xi(Q) & z_\eta(Q) & z_\zeta(Q) \end{pmatrix} \begin{pmatrix} a(Q) \\ b(Q) \\ c(Q) \end{pmatrix}. \tag{11.6}$$

The following diagram is commutative:

$$\begin{array}{ccc} \mathbb{E}^3 & \xrightarrow{\tau_*\mathbf{v}} & T\mathbb{E}^3 \\ {\scriptstyle\tau}\downarrow & & \downarrow{\scriptstyle T\tau} \\ \mathbb{E}^3 & \xrightarrow{\mathbf{v}} & T\mathbb{E}^3. \end{array} \tag{11.7}$$

Here, we regard the velocity vector field $\mathbf{v}$ as a section of the tangent bundle $T\mathbb{E}^3$ (see Sect. 4.5.1 on page 325).

Pull-back $\tau^*\mathbf{v}$ of the velocity vector field $\mathbf{v}$. Let the map τ from (11.4) be a diffeomorphism. The transformed vector field $\tau^*\mathbf{v}$ is defined by

$$\boxed{\tau^*\mathbf{v} := (\tau^{-1})_*\mathbf{v}.} \tag{11.8}$$

Locally, $(\tau^*\mathbf{v})_Q := (T_P\tau^{-1})(\mathbf{v}_P)$ for all points $Q \in \mathbb{E}^3$ (Fig. 11.2). This means that the following diagram is commutative:

$$\begin{array}{ccc} \mathbb{E}^3 & \xrightarrow{\tau^*\mathbf{v}} & T\mathbb{E}^3 \\ {\scriptstyle\tau^{-1}}\downarrow & & \downarrow{\scriptstyle T\tau^{-1}} \\ \mathbb{E}^3 & \xrightarrow{\mathbf{v}} & T\mathbb{E}^3. \end{array} \tag{11.9}$$

In terms of physics, the streamlines of the transformed velocity vector field $\tau_*\mathbf{v}$ (resp. $\tau^*\mathbf{v}$) are obtained from the streamlines of the original velocity vector field $\mathbf{v}$ by applying the map τ (resp. τ^{-1}). The transformed vector fields $\tau_*\mathbf{v}$ and $\tau^*\mathbf{v}$ will be frequently used (e.g., this concerns the definition of the Lie derivative of vector fields, the transformation of differential forms, or the definition of the Lie derivative of differential forms).

11.2 The Lie Derivative of an Electric Field along the Flow of Fluid Particles

> The Lie derivative of a vector field (e.g., an electric field) describes the transport of the velocity vector field along the flow of fluid particles on an infinitesimal level.
>
> Folklore

11.2.1 The Lie Derivative

We are given the smooth velocity vector field $\mathbf{v}$ on the Euclidean manifold $\mathbb{E}^3$. This vector field generates the flow operator $F_t : \mathbb{E}^3 \to \mathbb{E}^3$ which is a diffeomorphism for any time t (see Sect. 10.2.1 on page 647). Let $\mathbf{E}$ be a smooth vector field on $\mathbb{E}^3$ (e.g., an electric field). The Lie derivative $\mathcal{L}_{\mathbf{v}}\mathbf{E}$ of the electric field $\mathbf{E}$ with respect to the velocity vector field $\mathbf{v} \in \mathrm{Vect}(\mathbb{E}^3)$ is defined by the time derivative

$$\boxed{\mathcal{L}_{\mathbf{v}}\mathbf{E} := \frac{d}{dt}F_t^*\mathbf{E}_{|t=0}.} \tag{11.10}$$

Proposition 11.1 *For all points $P \in \mathbb{E}^3$,*

$$(\mathcal{L}_{\mathbf{v}}\mathbf{E})_P = \mathbf{E}'(P)\mathbf{v}(P) - \mathbf{v}'(P)\mathbf{E}(P).$$

Proof. By Sect. 11.1, the pull-back reads as

$$(F_t^*\mathbf{E})_P = F_t'(P)^{-1}\mathbf{E}(F_t(P)), \qquad t \in \mathbb{R}.$$

Using the Leibniz product rule and the chain rule together with (10.10) on page 650, we get

$$(\mathcal{L}_{\mathbf{v}}\mathbf{E})_P = \frac{d}{dt}F_t'(P)^{-1}_{t=0}\mathbf{E}(P) + \frac{d}{dt}\mathbf{E}(F_t(P))_{t=0} = -\mathbf{v}'(P)\mathbf{E}(P) + \mathbf{E}'(P)\mathbf{v}(P).$$

□

Cartesian coordinates. Consider a right-handed Cartesian (x, y, z)-coordinate system with the orthonormal basis $\mathbf{e}_1 := \mathbf{i}_P, \mathbf{e}_2 := \mathbf{j}_P, \mathbf{e}_3 := \mathbf{k}_P$ which is obtained from the right-handed orthonormal basis $\mathbf{i}, \mathbf{j}, \mathbf{k}$ at the origin by parallel transport (Fig. 9.1 on page 558). Set $\mathbf{v} = v^i\mathbf{e}_i$ and $\mathbf{E} = E^i\mathbf{e}_i$. Then[2]

$$(\mathcal{L}_{\mathbf{v}}\mathbf{E})_P = \left(v^i\partial_i E^j - E^i\partial_i v^j\right)_P \mathbf{e}_j. \tag{11.11}$$

11.2.2 Conservation Laws

Let $\mathbf{v}$ be a smooth complete velocity vector field on the Euclidean manifold $\mathbb{E}^3$.

Theorem 11.2 *If $\mathcal{L}_{\mathbf{v}}\mathbf{E} = 0$ on $\mathbb{E}^3$, then*[3]

$$\mathbf{E}_{F_tP} = F_t'(P)\mathbf{E}_P$$

for all times t and all points $P \in \mathbb{E}^3$.

[2] As usual, we set $x^1 := x, x^2 := y, x^3 := z$, and $\partial_i := \partial/\partial x^i$. Moreover, we sum over equal upper and lower indices from 1 to 3.

[3] If the velocity vector field $\mathbf{v}$ is not complete, then only a local form of the statement is valid, that is, the claim is only valid for an appropriate open time interval J with $0 \in J$. Here, J depends on the point P.

Equivalently, we say that the electric field $\mathbf{E}$ is Lie parallel along the trajectories of the fluid particles of the flow.

Proof. Since $F_{s+t} = F_s F_t$, we get $F^*_{s+t} = F^*_t F^*_s$. Hence

$$\frac{d}{ds} F^*_{s+t} \mathbf{E}_{|s=0} = F^*_t \mathcal{L}_\mathbf{v} \mathbf{E} = 0 \qquad \text{for all} \quad t \in \mathbb{R}.$$

Consequently, $F^*_t \mathbf{E}$ does not depend on time t. Hence $F'_t(P)^{-1} \mathbf{E}_{F_t P} = \mathbf{E}_P$ for all $t \in \mathbb{R}$. □

11.2.3 The Lie Algebra of Velocity Vector Fields

> Lie's approach to manifolds and Lie groups is based on the key fact that the set Vect($\mathcal{M}$) of smooth velocity vector fields on the manifold $\mathcal{M}$ is a Lie algebra. The corresponding Lie product $[\mathbf{v}, \mathbf{w}]$ is obtained by the Lie derivative $\mathcal{L}_\mathbf{v} \mathbf{w}$.
>
> Folklore

Let Vect($\mathbb{E}^3$) denote the space of all smooth velocity vector fields on the Euclidean manifold $\mathbb{E}^3$. Set $[\mathbf{v}, \mathbf{w}] := \mathcal{L}_\mathbf{v} \mathbf{w}$ for all $\mathbf{v}, \mathbf{w} \in$ Vect($\mathbb{E}^3$).

Proposition 11.3 *The real linear space* Vect($\mathbb{E}^3$) *becomes a Lie algebra equipped with the Lie product* $[\mathbf{v}, \mathbf{w}]$.

Proof. We have to show that for all $\mathbf{u}, \mathbf{v}, \mathbf{w} \in$ Vect($\mathbb{E}^3$) and all real numbers α, β, the following hold:

- $[\alpha \mathbf{u} + \beta \mathbf{v}, \mathbf{w}] = \alpha[\mathbf{u}, \mathbf{w}] + \beta[\mathbf{v}, \mathbf{w}]$ and $[\mathbf{u}, \mathbf{v}] = -[\mathbf{v}, \mathbf{u}]$,
- $[\mathbf{u}, [\mathbf{v}, \mathbf{w}]] + [\mathbf{v}, [\mathbf{w}, \mathbf{u}]] + [\mathbf{w}, [\mathbf{u}, \mathbf{v}]]$ (Jacobi identity).

This follows by an explicit computation. To this end, use a Cartesian (x^1, x^2, x^3)-coordinate system, as in (11.11). Then, $\mathbf{v} = v^i \mathbf{e}_i$, $\mathbf{w} = w^i \mathbf{e}_i$, and

$$[\mathbf{v}, \mathbf{w}] = (v^i \partial_i w^j - w^i \partial_i) \mathbf{e}_j.$$

□

12. Covector Fields and Cartan's Exterior Differential – the Beauty of Differential Forms

> Covector fields are dual objects to vector fields. The Cartan calculus of differential forms is based on the fact that there exists the Grassmann product $\omega \wedge \mu$ for covector fields ω and μ. Dualizing the Lie derivative for vector fields, we will get the Cartan differential $d\omega$ for differential forms ω. Élie Cartan's calculus for differential forms (also called the exterior differential calculus) is one of the most beautiful tools in mathematics.
>
> Folklore

The calculus of differential forms was introduced by Élie Cartan (1869–1951) in 1899. It was Cartan's goal to study Pfaff systems

$$\sum_{k=1}^{n} a_{jk}(x^1, \ldots, x^n) dx^k = 0, \qquad j = 1, \ldots, m$$

by using a symbolic method.[1] It turns out that:

> *Cartan's calculus is the proper language of generalizing the classical calculus due to Newton (1643–1727) and Leibniz (1646–1716) to real and complex functions with n variables.*

The key idea is to combine the notion of the Leibniz differential df with the alternating product $a \wedge b$ due to Grassmann (1809–1877). Cartan's calculus has its roots in physics. It emerged in the study of point mechanics, elasticity, fluid mechanics, heat conduction, and electromagnetism. It turns out that Cartan's differential calculus is the most important analytic tool in modern differential geometry and differential topology, and hence Cartan's calculus plays a crucial role in modern physics (gauge theory, theory of general relativity, the Standard Model in particle physics). In particular, as we will show in Chap. 19, the language of differential forms shows that Maxwell's theory of electromagnetism fits Einstein's theory of special relativity, whereas the language of classical vector calculus conceals the relativistic invariance of the Maxwell equations.

Convention. If the contrary is not stated explicitly, all the functions are assumed to be smooth in this chapter. The notion of smooth function on an open or closed set is discussed in the Appendix (see page 1070).

Coordinate transformations. We want to use the coordinate transformation

$$x = \tau(u), \qquad u \in \mathcal{U} \tag{12.1}$$

[1] É. Cartan, Sur certaines expressions différentielles et sur le problème de Pfaff, Annales École Normale **16** (1899), 239–332.

E. Zeidler, *Quantum Field Theory III: Gauge Theory*,
DOI 10.1007/978-3-642-22421-8_13,

where $u = (u^1, \ldots, u^n)$ and $x = (x^1, \ldots, x^n)$ are elements of $\mathbb{R}^n$. Let $\mathcal{U}$ and $\mathcal{M}$ be open subsets of $\mathbb{R}^n$ (or let $\mathcal{U}$ and $\mathcal{M}$ be the closure of open subsets of $\mathbb{R}^n$). The map

$$\tau : \mathcal{U} \to \mathcal{M} \tag{12.2}$$

is called a diffeomorphism iff the following hold: τ is bijective and smooth, and the inverse map $\tau^{-1} : \mathcal{M} \to \mathcal{U}$ is also smooth. The diffeomorphism τ is called orientation-preserving iff

$$\det \tau'(P) > 0 \qquad \text{for all} \quad P \in \operatorname{int} \mathcal{U}.$$

Recall that the Jacobian is given by

$$\det \tau'(P) = \frac{\partial(\tau^1, \ldots, \tau^n)}{\partial(u^1, \ldots, u^n)}(u) := \begin{vmatrix} \frac{\partial \tau^1(u)}{\partial u^1} & \ldots & \frac{\partial \tau^1(u)}{\partial u^n} \\ \vdots & \vdots & \vdots \\ \frac{\partial \tau^n(u)}{\partial u^1} & \ldots & \frac{\partial \tau^n(u)}{\partial u^n} \end{vmatrix}.$$

For example, let $n = 1$. If $\mathcal{U}$ and $\mathcal{M}$ are compact intervals, then the map (12) is an orientation-preserving diffeomorphism iff τ is smooth, and $\tau'(u) > 0$ for all $u \in \operatorname{int} \mathcal{U}$.

12.1 Ariadne's Thread

> It is worth noting that notation facilitates discovery. This, in a most wonderful way, reduces the mind's labors.
> We need an analysis which is of geometric nature and describes physical situations as directly as algebra expresses quantities.
>
> Gottfried Wilhelm Leibniz (1646–1716)

For the convenience of the reader, let us start by describing some basic ideas. We want to show how classical integral formulas can be uniformly written in the language of differential forms. The point is that the use of differential forms allows straightforward generalizations to higher dimensions.

12.1.1 One Dimension

Classical formulas. Consider the compact intervals $\mathcal{M} := [a, b]$ and $\mathcal{U} := [\alpha, \beta]$.

(i) The main theorem of calculus: If $f : \mathcal{M} \to \mathbb{R}$ is smooth, then

$$\int_{\mathcal{M}} f'(x)dx = f(b) - f(a). \tag{12.3}$$

(ii) Integration by parts: If $F, G : \mathcal{M} \to \mathbb{R}$ are smooth functions, then

$$\int_{\mathcal{M}} F'(x)G(x)\, dx = -\int_{\mathcal{M}} F(x)G'(x)dx + F(b)G(b) - F(a)G(a). \tag{12.4}$$

This follows from (i) by choosing $f := FG$.

(iii) The substitution rule: If $\tau : \mathcal{U} \to \mathcal{M}$ is an orientation-preserving diffeomorphism, then

$$\int_{\mathcal{M}} f(x)dx = \int_{\mathcal{U}} f(\tau(u)) \cdot \tau'(u)du. \tag{12.5}$$

The formal language of differential forms. To begin with, we will use the Leibniz differential in a formal way. We set

- $\omega := f(x)$ (0-form),
- $\gamma := g(x)dx$ (1-form),

where $f, g : \mathbb{R} \to \mathbb{R}$ are smooth functions.[2] We define the differential by setting

- $d\omega := f'(x)dx$,
- $d\gamma := dg(x) \wedge dx$.

Then $d\gamma = g'(x)dx \wedge dx = 0$. Hence

$$\boxed{d(d\omega) = 0.} \tag{12.6}$$

As we will show later on, this so-called Poincaré cohomology rule is the key to the de Rham theory in differential topology. The main theorem of calculus (12.1.1) can be written as

$$\boxed{\int_{\mathcal{M}} d\omega = \int_{\partial\mathcal{M}} \omega.} \tag{12.7}$$

This is the prototype of the generalized Stokes theorem to be considered in Sect. 12.7 below. Finally, let us investigate the transformation of differential forms under the coordinate transformation

$$x = \tau(u), \qquad u \in \mathcal{U}.$$

Motivated by $dx = \tau'(u)du$, we set

- $(\tau^*\omega)(u) := f(\tau(u))$,
- $(\tau^*\gamma)(u) := g(\tau(u)) \cdot \tau'(u)du$.

The differential form $\tau^*\omega$ (resp. $\tau^*\gamma$) is called the pull-back of ω (resp. γ). Replacing f by g, the substitution rule (12.1.1) can be written as

$$\boxed{\int_{\mathcal{M}} \gamma = \int_{\tau^*\mathcal{M}} \tau^*\gamma.} \tag{12.8}$$

Naturally enough, the set $\tau^*\mathcal{M} := \tau^{-1}(\mathcal{M})$ is called the pull-back of the set $\mathcal{M}$. The differential has the following crucial invariance property

$$\boxed{d(\tau^*\omega) = \tau^*(d\omega),} \tag{12.9}$$

which is responsible for the great flexibility of the Cartan calculus for differential forms. Similarly, $d(\tau^*\gamma) = \tau^*(d\gamma)$.[3]

The rigorous meaning of differential forms. We want to define differential forms on the one-dimensional Euclidean manifold $\mathbb{E}^1$ in an invariant way. We will use a language which can be generalized to arbitrary manifolds later on.

---

[2] In order to indicate the dependence on the point x, we should write $\omega_x := f(x)$ and $\gamma_x := g(x)dx$. To simplify notation, we will omit the index x.

[3] Note that $\tau^*(d\omega) = f'(\tau(u)) \cdot \tau'(u)du$ and $d\gamma = 0$.

(a) (b)

Fig. 12.1. The tangent space $T_P\mathbb{E}^1$

(i) Tangent space $T_P\mathbb{E}^1$: To begin with, let $T_P\mathbb{E}^1$ denote the tangent space of the real line $\mathbb{E}^1$ at the point P (with the coordinate x) (Fig. 12.1). This tangent space consists of all velocity vectors $\mathbf{v}_P$ at the point P. Choosing the unit vector $\mathbf{i}_P$ as basis vector at the point P, we have

$$\mathbf{v}_P := V_P\mathbf{i}_P$$

where V_P is a real number. In terms of coordinates, we write $V(x) := V_P$.

(ii) Cotangent space $T_P^*\mathbb{E}^1$: Define

$$\boxed{dx_P(\mathbf{v}_P) := V_P \qquad \text{for all} \quad \mathbf{v}_P \in T_P\mathbb{E}^1.}$$

This means that

$$dx_P : T_P\mathbb{E}^1 \to \mathbb{R}$$

is a linear functional on the tangent space $T_P\mathbb{E}^1$. In terms of physics, dx_P assigns to the velocity vector $\mathbf{v}_P$ its velocity component V_P with respect to the basis vector $\mathbf{i}_P$. All the linear functionals

$$\gamma_P : T_P\mathbb{E}^1 \to \mathbb{R}$$

are called covectors of the real line $\mathbb{E}^1$ at the point P. In particular, dx_P is a covector at P. All the covectors at P form the cotangent space $T_P^*\mathbb{E}^1$ of $\mathbb{E}^1$ at the point P. We have $T_P^*\mathbb{E}^1 = \{\alpha dx_P : \alpha \in \mathbb{R}\}$.

(iii) Differential forms: Let us use the following terminology.

- $\mathrm{Vect}(\mathbb{E}^1)$ denotes the space of smooth vector fields

$$P \mapsto \mathbf{v}_P$$

on the real line $\mathbb{E}^1$. That is, the function $P \mapsto V_P$ is smooth on $\mathbb{E}^1$.

- $\Lambda^0(\mathbb{E}^1)$ denotes the space of all smooth functions $f : \mathbb{E}^1 \to \mathbb{R}$ on the real line.
- $\Lambda^1(\mathbb{E}^1)$ denotes the space of all smooth covector fields

$$P \mapsto \gamma_P$$

on $\mathbb{E}^1$. That is, $\gamma_P = g(x)dx_P$, and the function $g : \mathbb{R} \to \mathbb{R}$ is smooth. Covector fields are also called differential 1-forms or briefly 1-forms.

- $\Lambda^2(\mathbb{E}^1)$ denotes the set of all maps

$$P \mapsto \nu_P$$

where $\nu_P : T_P\mathbb{E}^1 \times T_P\mathbb{E}^1 \to \mathbb{R}$ is a bilinear antisymmetric functional (for all $P \in \mathbb{E}^1$). Because of $\nu_P = 0$ the space $\Lambda^2(\mathbb{E}^1)$ is trivial.

Note that $\mathrm{Vect}(\mathbb{E}^1)$ and $\Lambda^p(\mathbb{E}^1)$, $p = 0, 1$, are infinite-dimensional real linear spaces.

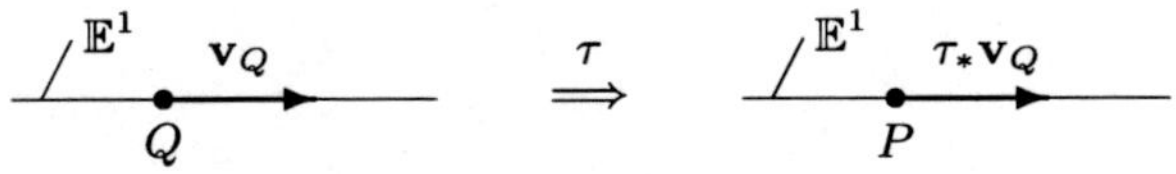

Fig. 12.2. Transformation of the real line

(iv) The Cartan differential operator: We want to introduce the linear operator

$$d : \Lambda^p(\mathbb{E}^1) \to \Lambda^{p+1}(\mathbb{E}^1), \qquad p = 0, 1.$$

In particular, we will get the following:
- If $f \in \Lambda^0(\mathbb{E}^1)$, then $df \in \Lambda^1(\mathbb{E}^1)$, and
- if $\gamma \in \Lambda^1(\mathbb{E}^1)$, then $d\gamma = 0$.

Explicitly, we define

$$\boxed{(df)(\mathbf{v}) := \mathcal{L}_{\mathbf{v}} f \qquad \text{for all} \quad \mathbf{v} \in \mathrm{Vect}(\mathbb{E}^1)}$$

where $\mathcal{L}_{\mathbf{v}} f$ denotes the Lie derivative of the function f with respect to the velocity vector field $\mathbf{v}$. This means that

$$(df)_P(\mathbf{v}_P) := (\mathcal{L}_{\mathbf{v}_P} f)(P)$$

for all points $P \in \mathbb{E}^1$ and all smooth velocity vector fields $\mathbf{v} \in \mathrm{Vect}(\mathbb{E}^1)$. In terms of coordinates, we get

$$\boxed{(df)_P = f'(x) dx_P \qquad \text{for all} \quad P \in \mathbb{E}^1} \tag{12.10}$$

where we assume that the real number x is the coordinate of the point P.

Proof. Note that $f'(x)dx_P(\mathbf{v}_P) = f'(x)V_P$ and $(\mathcal{L}_{\mathbf{v}_P})(P) = V_P f'(x)$. □

(v) Transformation of global coordinates (transplantation of velocity vector fields and covector fields): Let us consider the diffeomorphism

$$\tau : \mathbb{E}^1 \to \mathbb{E}^1$$

of the real line onto itself. We regard this as a coordinate transformation

$$x = \tau(u), \qquad u \in \mathbb{R}.$$

Set $P = \tau(Q)$ (Fig. 12.2). Assume that the point Q (resp. P) has the coordinate u (resp. $x = \tau(u)$). Naturally enough, the transformation τ induces two linear transformations of velocity vector fields and dual covector fields.
- Push-forward of velocity vector fields: $\tau_* : \mathrm{Vect}(\mathbb{E}^1) \to \mathrm{Vect}(\mathbb{E}^1)$. The linearization $T_Q\tau : T_Q\mathbb{E}^1 \to T_P\mathbb{E}^1$ of the map τ at the point Q sends the tangent vector $\mathbf{v}_Q$ at the point Q to the tangent vector $(T_Q\tau)(\mathbf{v}_Q)$ at the point P. We define the transformed velocity vector field $\tau_*\mathbf{v}$ by setting

$$\boxed{(\tau_*\mathbf{v})_P := (T_Q\tau)(\mathbf{v}_Q) \qquad \text{for all} \quad Q \in \mathbb{E}^1.}$$

Mnemonically, we write $\tau_*\mathbf{v} = (T\tau)\mathbf{v}$. In terms of coordinates,

$$\mathbf{v}_Q = V(u)\mathbf{i}_Q,$$

and $(\tau_*\mathbf{v})_P = W(x)\mathbf{i}_P$ with the relation $W(x) = W(\tau(u)) = \tau'(u)V(u)$. In other words,

$$(\tau_*\mathbf{v})(\tau(u)) = \tau'(u)V(u)\mathbf{i}_P \qquad \text{for all} \quad u \in \mathbb{R}.$$

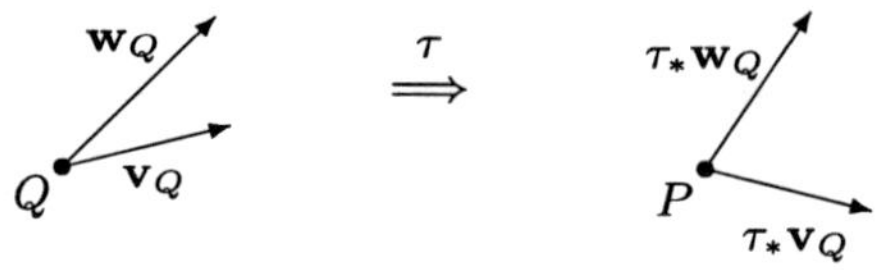

Fig. 12.3. Transformation of the Euclidean plane $\mathbb{E}^2$

- Pull-back of covector fields: $\tau^* : \Lambda^1(\mathbb{E}^1) \to \Lambda^1(\mathbb{E}^1)$. Consider the smooth covector field $\gamma \in \Lambda^1(\mathbb{E}^1)$. Naturally enough, the transformation of velocity vector fields induces a transformation of linear functionals (covectors). Explicitly, we define the transformed covector field $\tau^*\gamma$ by setting[4]

$$\boxed{(\tau^*\gamma)_Q(\mathbf{v}) := \gamma_P((T_Q\tau)\mathbf{v}) \qquad \text{for all} \quad \mathbf{v} \in T_Q\mathbb{E}^1.} \tag{12.11}$$

Mnemonically, we write $(\tau^*\gamma)(\mathbf{v}) = \gamma_\tau(\tau_*\mathbf{v})$. In terms of coordinates, suppose that

$$\gamma = g(x)dx_P.$$

Then

$$(\tau^*\gamma)(u) = g(\tau(u)) \cdot \tau'(u)du_Q. \tag{12.12}$$

Proof. Note that
- $\mathbf{v}_Q = V(u)\mathbf{i}_Q$ and $\mathbf{v}_P = V(\tau(u))\mathbf{i}_P$,
- $(\tau_*\mathbf{v})_P = \tau'(u)V(u)\mathbf{i}_P$,
- $\gamma_P(\mathbf{v}_P) = g(x)dx_P(\mathbf{v}_P) = g(x)V(x)$.

Now the claim (12.12) follows from $\gamma_P((\tau_*\mathbf{v})_P) = g(x)\tau'(u)V(u)$ and

$$g(\tau(u))\tau'(u)du_Q(\mathbf{v}_Q) = g(x)\tau'(u)V(u).$$

□

(vi) Transformation of local coordinates: The same argument applies to local coordinates, that is, the map $\tau : \mathcal{U} \to \mathcal{M}$ is a diffeomorphism where $\mathcal{U}$ and $\mathcal{M}$ are open (or closed) intervals.

Our goal is to generalize this to higher dimensions such that the four key relations (12.6) through (12.9) above remain valid. The crucial trick reads as follows:

Integrals over manifolds (curves, surfaces, regions with boundary like discs or balls) are integrals over differential forms.

12.1.2 Two Dimensions

Consider a right-handed (x, y)-Cartesian coordinate system of the Euclidean plane $\mathbb{E}^2$ with the right-handed orthonormal basis $\mathbf{i}, \mathbf{j}$. Consider the subsets $\mathcal{M}$ and $\mathcal{U}$ of $\mathbb{E}^2$ which are compact, arcwise connected, oriented, 2-dimensional submanifolds of $\mathbb{E}^2$ with coherently oriented boundary $\partial\mathcal{M}$ (Fig. 12.4).

[4] To simplify notation, we write $\mathbf{v}$ instead of $\mathbf{v}_Q$.

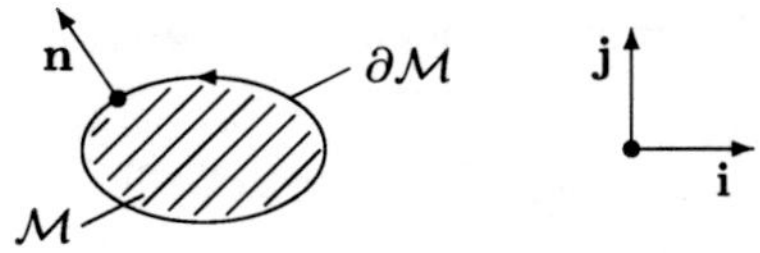

Fig. 12.4. Coherently oriented boundary

(i) The main theorem of calculus: If the functions $U, V : \mathcal{M} \to \mathbb{R}$ are smooth, then

$$\int_{\mathcal{M}} (U_x - V_y)\, dxdy = \int_{\partial\mathcal{M}} U dx + V dy. \tag{12.13}$$

Here, U_x denotes the partial derivative of U with respect to x. Equivalently, we also get

$$\int_{\mathcal{M}} (U_x + V_y)\, dxdy = \int_{\partial\mathcal{M}} \big(U(P)n^1(P) + V(P)n^2(P)\big)\, ds. \tag{12.14}$$

Here, $\mathbf{n} = n^1(P)\ \mathbf{i} + n^2(P)\mathbf{j}$ is the outer normal unit vector at the boundary point P, and the real parameter s denotes the arc length.

(ii) Integration by parts:

$$\int_{\mathcal{M}} U_x V dxdy = -\int_{\mathcal{M}} UV_x dxdy + \int_{\partial\mathcal{M}} U(P)V(P)n^1(P)ds \tag{12.15}$$

and

$$\int_{\mathcal{M}} U_y V dxdy = -\int_{\mathcal{M}} UV_y dxdy + \int_{\partial\mathcal{M}} U(P)V(P)n^2(P)ds. \tag{12.16}$$

This follows from (12.14) by replacing U (resp. V) by the product UV.

(iii) The substitution rule: If $\tau : \mathcal{U} \to \mathcal{M}$ is an orientation-preserving diffeomorphism, then

$$\int_{\mathcal{M}} V(x,y)\, dxdy = \int_{\mathcal{U}} V(\tau(u,v)) \cdot \det \tau'(u,v)\, dudv. \tag{12.17}$$

Equivalently,

$$\int_{\mathcal{M}} V(x,y)\, dxdy = \int_{\mathcal{U}} V(x(u,v), y(u,v)) \cdot \frac{\partial(x,y)}{\partial(u,v)}(u,v)\, dudv. \tag{12.18}$$

For the proof of (i), we refer to Problem 12.1 on page 801. Formula (12.13) is called the Cauchy–Green formula. It was implicitly (resp. explicitly) formulated by Green in 1828 (resp. Cauchy in 1846). Cauchy used this formula for studying the path-independence of the integral $\int_C f'(z)dz$ in the case where f is a holomorphic function on the Gaussian plane (see page 687).

The formal language of differential forms. We set

- f (0-form),
- $\omega := U dx + V dy$ (1-form),
- $\gamma := W\ dx \wedge dy$ (2-form),
- $dx \wedge dy = -dy \wedge dx,\ dx \wedge dx = 0,\ dy \wedge dy = 0$ (Grassmann relations).

Here, $f, U, V, W : \mathbb{R}^2 \to \mathbb{R}$ are smooth functions.

(i) Cartan's exterior differential: We define the Cartan exterior differential by setting
- $df = f_x dx + f_y dy$,
- $d \wedge \omega := dU \wedge dx + dV \wedge dy$,
- $d \wedge \gamma := dW \wedge dx \wedge dy$.

To simplify notation, we replace $d \wedge \omega$ (resp. $d \wedge \gamma$) by $d\omega$ (resp. $d\gamma$). Then $d\omega = (U_x dx + U_y dy) \wedge dx + (V_x dx + V_y dy) \wedge dy$. By the Grassmann relations,

$$d\omega = (V_x - U_y)\, dx \wedge dy.$$

Moreover, by the Grassmann relations, the wedge product of three basis differentials is always equal to zero on the Euclidean plane $\mathbb{E}^2$. For example, $dx \wedge dy \wedge dx = -dx \wedge dx \wedge dy = 0$. This implies

$$d\gamma = (W_x dx + W_y dy) \wedge dx \wedge dy = 0.$$

(ii) Poincaré's cohomology rule: This rule reads as

$$\boxed{d(df) = 0, \qquad d(d\omega) = 0, \qquad d(d\gamma) = 0.} \tag{12.19}$$

Mnemonically, $d \wedge d \wedge \omega = 0$ (or briefly $d \wedge d = 0$). To prove (12.19), note that $f_{xy} = f_{yx}$. This implies

$$\begin{aligned} d(df) = df_x \wedge dx + df_y \wedge dy &= (f_{xx} dx + f_{xy} dy) \wedge dx \\ &+ (f_{yx} dx + f_{yy} dy) \wedge dy = (f_{yx} - f_{xy}) dx \wedge dy = 0. \end{aligned}$$

(iii) Integrals: We define the integral over differential forms in the following way:
- $\int_C \omega = \int_C U dx + V dy$ (classical line integral),
- $\int_{\mathcal{M}} \gamma = \int_{\mathcal{M}} W\, dx \wedge dy := \int_{\mathcal{M}} W(x,y) dx dy$ (classical 2-dimensional integral).

Mnemonically, we replace the wedge product $dx \wedge dy$ by $dxdy$.

(iv) The main theorem of calculus (12.13) (Cauchy–Green theorem): This can elegantly be written as

$$\boxed{\int_{\mathcal{M}} d\omega = \int_{\partial\mathcal{M}} \omega.} \tag{12.20}$$

(v) Coordinate transformations (pull-back): Consider the coordinate transformation

$$x = x(u,v), \quad y = y(u,v), \qquad (u,v) \in \mathcal{U}.$$

We assume that the map $\tau : \mathcal{U} \to \mathcal{M}$ is smooth. Motivated by the Leibniz differential,

$$dx = x_u(u,v) du + x_v(u,v) dv, \qquad dy = y_u(u,v) du + y_v(u,v) dv,$$

we define
- $(\tau^* f)(u,v) := f(x,y)$ where $x = x(u,v)$ and $y = y(u,v)$,
- $(\tau^* \omega)(u,v)$ is equal to

$$U(x,y)(x_u(u,v)du + x_v(u,v)dv) + V(x,y)(y_u(u,v)du + y_v(u,v)dv).$$

Hence

$$\begin{aligned} (\tau^*\omega)(u,v) := \; & \big(U(x,y)x_u(u,v) + V(x,y)y_u(u,v)\big)\, du \\ & + \big(U(x,y)x_v(u,v) + V(x,y)y_v(u,v)\big)\, dv \end{aligned}$$

where $x = x(u,v)$ and $y = y(u,v)$.

- $(\tau^*\gamma)(u,v)$ is equal to

$$W(x,y)(x_u(u,v)du + x_v(u,v)dv) \wedge (y_u(u,v)du + y_v(u,v)dv).$$

This yields

$$\begin{aligned}(\tau^*\gamma)(u,v) :&= W(x,y)(x_u(u,v)y_v(u,v) - x_v(u,v)y_u(u,v))\, du \wedge dv \\ &= W(x,y)\frac{\partial(x,y)}{\partial(u,v)}(u,v)\, du \wedge dv\end{aligned}$$

where $x = x(u,v)$ and $y = y(u,v)$.

The differential form $\tau^*\omega$ (resp. $\tau^*\gamma$) is called the pull-back of ω (resp. γ).

(vi) Substitution rule for integrals: Suppose that the map $\tau : \mathcal{U} \to \mathcal{M}$ is an orientation-preserving diffeomorphism. Then the classical substitution rule for integrals (12.18) can be elegantly written as

$$\boxed{\int_{\mathcal{M}} \gamma = \int_{\tau^*\mathcal{M}} \tau^*\gamma.} \tag{12.21}$$

Naturally enough, the set $\tau^*\mathcal{M} := \tau^{-1}(\mathcal{M})$ is called the pull-back of the original set $\mathcal{M}$. As another example, consider the smooth map $\tau : [a,b] \to \mathbb{R}^2$ on the compact interval $[a,b]$. The equation $(x,y) = \tau(t)$, that is,

$$x = x(t), \quad y = y(t), \qquad a \le t \le b$$

describes a smooth curve $\mathcal{C}$. Motivated by the Leibniz differential

$$dx = \dot{x}(t)\, dt, \qquad dy = \dot{y}(t)\, dt,$$

we define the pull-back

- $(\tau^*\omega)(t) := \big(U(x(t),y(t))\, \dot{x}(t) + U(x(t),y(t))\, \dot{y}(t)\big)\, dt.$

The classical parameter formula for line integrals can be written as

$$\boxed{\int_{\mathcal{C}} \omega = \int_{\tau^*\mathcal{C}} \tau^*\omega.}$$

The reader should note that:

Mnemonically, the Cartan exterior calculus tells us the right substitution rule for the integral.

(vii) Invariance of the exterior differential under pull-back: The differential has the following crucial invariance property

$$\boxed{d(\tau^*\omega) = \tau^*(d\omega),} \tag{12.22}$$

which is responsible for the great flexibility of the Cartan calculus for differential forms. Similarly,

$$d(\tau^* f) = \tau^*(df) \qquad \text{and} \qquad d(\tau^*\gamma) = \tau^*(d\gamma). \tag{12.23}$$

Proof. Let us prove $d(\tau^* f) = \tau^*(df)$. By the chain rule, it follows from

$$(\tau^* f)(u,v) = f(x(u,v), y(u,v))$$

that $d(\tau^* f) = (f_x x_u + f_y y_u)du + (f_x x_v + f_y y_v)dv$. Moreover, $df = f_x dx + f_y dy$. Hence

$$\tau^*(df) = f_x(x_u du + x_v dv) + f_y(y_u du + y_v dv) = d(\tau^* f).$$

This finishes the proof. □

Since $d\gamma = 0$, the claim $d(\tau^*\gamma) = \tau^*(d\gamma) = 0$ is trivial. The proof of (12.22) will be given in Problem 12.2 on page 801. The proofs show that the invariance properties (12.22) and (12.23) are consequences of the chain rule.

(viii) The Hodge star operation: We define

- $*1 := dx \wedge dy$ and $*(dx \wedge dy) := 1$,
- $*dx := dy$ and $*dy := -dx$.

The integral $\int_{\mathcal{M}} *1 = \int_{\mathcal{M}} dxdy$ equals the area measure of the set $\mathcal{M}$. Therefore, $*1$ is called the volume form of the Euclidean plane $\mathbb{E}^2$. By linear extension, we get the following definitions:

- $(*f)(x,y) = f(x,y)dx \wedge dy$ and $*(W(x,y)\, dx \wedge dy) := W(x,y)$,
- $*(U(x,y)dx + V(x,y)dy) := U(x,y)dy - V(x,y)dx$.

(ix) The Hodge codifferential: We define

$$\boxed{d^*\varrho := - * d * \varrho}$$

where ϱ is a p-form with $p = 0, 1, 2$. Explicitly,

- $d^* f := 0$,
- $d^*(Udx + Vdy) = -U_x - V_y$,
- $d^*(Wdx \wedge dy) = W_y dx - W_x dy$.

In fact, $*(Udx + Vdy) = Udy - Vdx$. Hence

$$- * d(Udy - Vdx) = - * \big((U_x + V_y)\, dx \wedge dy\big) = -U_x - V_y.$$

Analogously, we get the other statements.

(x) The Hodge Laplacian Δ: Following Hodge, we set

$$\boxed{\Delta\varrho := (dd^* + d^*d)\varrho}$$

where ϱ is an arbitrary (smooth) p-form with $p = 0, 1, 2$. Explicitly,

- $\Delta f = -f_{xx} - f_{yy}$,
- $\Delta(Udx + Vdy) = \Delta U \cdot dx + \Delta V \cdot dy$,
- $\Delta\big(W\, dx \wedge dy\big) = \Delta W \cdot dx \wedge dy$.

We call Δ the Hodge Laplacian (or briefly the Laplacian) of the Euclidean plane $\mathbb{E}^2$. In fact, $dd^*f = 0$, and $d^*df = d^*(f_x dx + f_y dy) = -f_{xx} - f_{yy}$. This implies $\Delta f = -f_{xx} - f_{yy}$. The proofs of the other statements can be found in Problem 12.5 on page 802. Note that, for a smooth function f, the Hodge Laplacian differs from the classical Laplacian by sign. In this monograph, we will always use the Hodge Laplacian. This fits the terminology used in modern differential geometry.

The rigorous language of differential forms. It is not difficult to give the approach above a sound basis. There exist two different, but equivalent possibilities:

(A) Kähler's algebraic approach, and the

(G) geometric approach based on covectors and the wedge product of antisymmetric multilinear forms.

The algebraic approach. Let us sketch this.

(a) The Grassmann algebra $\Lambda(\mathbb{E}^2)$ and the exterior differential calculus: Let $\Lambda(\mathbb{E}^2)$ denote the set of all symbols of the form

$$f + Udx + Vdy + W\ dx \wedge dy$$

where the coefficients f, U, V, W are smooth real-valued functions on the Euclidean plane $\mathbb{E}^2$. The set $\Lambda(\mathbb{E}^2)$ becomes a real associative algebra by introducing the $\wedge$-product as above. That is, we define the following algebraic relations for the four basis elements $1, dx, dy, dx \wedge dy$:

- $1 \wedge \varrho = \varrho \wedge 1 = \varrho$ if $\varrho = 1, dx, dy, dx \wedge dy$,
- $dy \wedge dx = -dx \wedge dy$,
- $dx \wedge dx = dy \wedge dy = 0$,
- $dx \wedge \mu = dy \wedge \mu = 0$ if $\mu = dx \wedge dy, dy \wedge dx$.

Naturally enough, we extend these definitions by linearity. For example,

$$(Udx + Vdy) \wedge (f + gdx + Wdx \wedge dy) = Ufdx + Vfdy - Vg\ dx \wedge dy.$$

(b) Linear operators: We introduce the Cartan differential d, the Hodge star operator, and the Hodge codifferential d^* by setting

- $d \wedge (f + Udx + Vdy + W\ dx \wedge dy) := df + dU \wedge dx + dV \wedge dy + dW \wedge dx \wedge dy$,
- $*(f + Udx + Vdy + W\ dx \wedge dy) := f\ dx \wedge dy + Udy - Vdx + W$,
- $d^* \varrho := - * d * \varrho$ for all $\varrho \in \Lambda(\mathbb{E}^2)$.

This yields the linear operators $d^*, *, d : \Lambda(\mathbb{E}^2) \to \Lambda(\mathbb{E}^2)$.

(c) Invariance under pull-back: We consider the smooth map

$$\tau : \mathcal{U} \to \mathcal{M}$$

where $\mathcal{U}$ and $\mathcal{M}$ are open subsets of $\mathbb{E}^2$ (or the closure of open subsets of $\mathbb{E}^2$). For all $\varrho \in \Lambda(\mathbb{E}^2)$, we define the pull-back $\tau^* \varrho$ as above. In particular, $\tau^* \varrho \in \Lambda(\mathbb{E}^2)$. For all $\varrho, \mu \in \Lambda(\mathbb{E}^2)$, we get:[5]

- $\tau^*(\varrho \wedge \mu) = \tau^* \varrho \wedge \tau^* \mu$,
- $*(\tau^* \varrho) = \tau^*(*\varrho)$,
- $d(\tau^* \varrho) = \tau^*(d\varrho)$ and $d^*(\tau^* \varrho) = \tau^*(d^* \varrho)$.

These crucial relations show that all the operations introduced above are invariant under the pull-back operation. In particular, they do not depend on the choice of right-handed Cartesian coordinates on the Euclidean plane $\mathbb{E}^2$.

(d) Kähler's interior differential calculus: The Grassmann algebra $\Lambda(\mathbb{E}^2)$ can be equipped with an additional $\vee$-product. This way, it becomes a real associative Clifford algebra. To begin with, we define the $\vee$-product for the four basis elements $1, dx, dy, dx \wedge dy$:

- $1 \vee \varrho = \varrho \vee 1 = \varrho$ if $\varrho = 1, dx, dy, dx \wedge dy$,
- $dx \vee dy := dx \wedge dy$ and $dy \vee dx = -dx \vee dy$,
- $dx \vee dx = dy \vee dy = 1$.[6]

The other definitions of n-fold $\vee$-products with basis elements $1, dx, dy, dx \wedge dy$ as factors can be obtained by using the associative law as a mnemonic tool.[7] For example,

- $dx \vee dx \vee dy := dy$ and $dx \vee dy \vee dx := -dy$,[8]

---

[5] Concerning the Hodge star operator and the Hodge codifferential d^*, we assume that τ is a rotation.

[6] Setting $x^1 := x$ and $x^2 := y$, we get $dx^i \vee dx^j + dx^j \vee dx^i = 2\delta^{ij}, i, j = 1, 2$. This is the so-called Clifford relation.

[7] Of course, at the end of the procedure, one has to show that the general $\vee$-product defined above obeys indeed the associative law. But this is an easy consequence of our construction.

[8] Mnemonically, $(dx \vee dx) \vee dy = 1 \vee dy = dy$ and $dx \vee (dy \vee dx) = -dx \vee (dx \vee dy)$. This is equal to $-(dx \vee dx) \vee dy = -dy$.

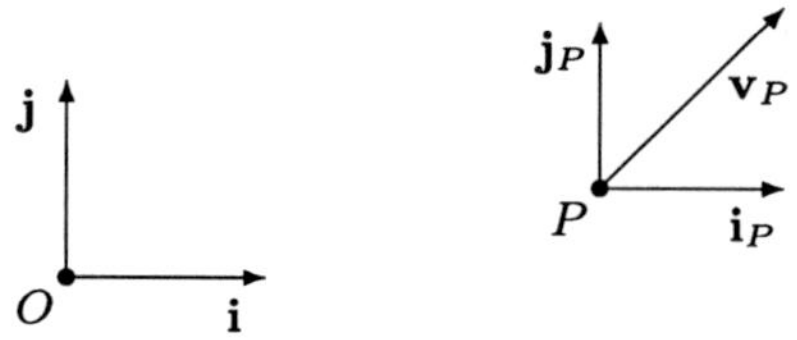

Fig. 12.5. Orthonormal basis of the tangent space $T_P\mathbb{E}^2$

- $dx \vee dy \vee dx \vee dy := -1$.

Replacing the $\wedge$-product by the $\vee$-product, we define the interior Kähler differential by setting

$$d \vee (f + Udx + Vdy + W\ dx \wedge dy) : = df + dU \vee dx + dV \vee dy \\ + dW \vee (dx \wedge dy).$$

With respect to the pull-back for rotations τ, we have the following invariance properties for all $\varrho, \mu \in \Lambda(\mathbb{E}^2)$:

- $\tau^*(\varrho \vee \mu) = \tau^* \varrho \vee \tau^* \mu$,
- $d \vee \tau^* \varrho = \tau^*(d \vee \varrho)$.

For example,

$$\begin{aligned} d \vee (Udx) &= dU \vee dx = (U_x dx + U_y dy) \vee dx \\ &= U_x dx \vee dx + U_y dy \vee dx = U_x - U_y dx \wedge dy. \end{aligned}$$

The notations $d \wedge \varrho$ and $d \vee \varrho$ display the similarities between the exterior and interior differential calculus. To simplify notation, we will replace $d \wedge \varrho$ by $d\varrho$, that is, we briefly write

$$d(f + Udx + Vdy + W\ dx \wedge dy) := df + dU \wedge dx + dV \wedge dy. \quad (12.24)$$

The geometric approach. In Sect. 12.3, we will thoroughly study this. At this point, let us only describe the main ideas. Using the Lie derivative $\mathcal{L}_\mathbf{v}$ with respect to a smooth velocity vector field $\mathbf{v}$ on the Euclidean plane $\mathbb{E}^2$, the key formula reads as

$$\boxed{(d\omega)(\mathbf{v}, \mathbf{w}) := \mathcal{L}_\mathbf{v}\omega(\mathbf{w}) - \mathcal{L}_\mathbf{w}\omega(\mathbf{v}) - \omega([\mathbf{v}, \mathbf{w}])} \quad (12.25)$$

for all smooth velocity vectors fields $\mathbf{v}, \mathbf{w}$ on $\mathbb{E}^2$. Let us discuss this. Recall that the tangent space $T_P\mathbb{E}^2$ of the Euclidean plane $\mathbb{E}^2$ consists of all velocity vectors $\mathbf{v}$ at the point P. All the linear functionals

$$\omega : T_P\mathbb{E}^2 \to \mathbb{R}$$

are called covectors at the point P. They form the cotangent space $T_P^*\mathbb{E}^2$ of $\mathbb{E}^2$ at the point P. If $\omega, \mu \in T_P^*\mathbb{E}^2$, when the wedge product $\omega \wedge \mu$ is defined by

$$(\omega \wedge \mu)(\mathbf{v}, \mathbf{w}) := \omega(\mathbf{v})\mu(\mathbf{w}) - \omega(\mathbf{w})\mu(\mathbf{v}) \qquad \text{for all} \quad \mathbf{v}, \mathbf{w} \in T_P\mathbb{E}^2.$$

This is a real bilinear antisymmetric functional on $T_P\mathbb{E}^2 \times T_P\mathbb{E}^2$.

In order to formulate this in terms of Cartesian coordinates, choose a fixed right-handed Cartesian (x, y)-coordinate system with the right-handed orthonormal

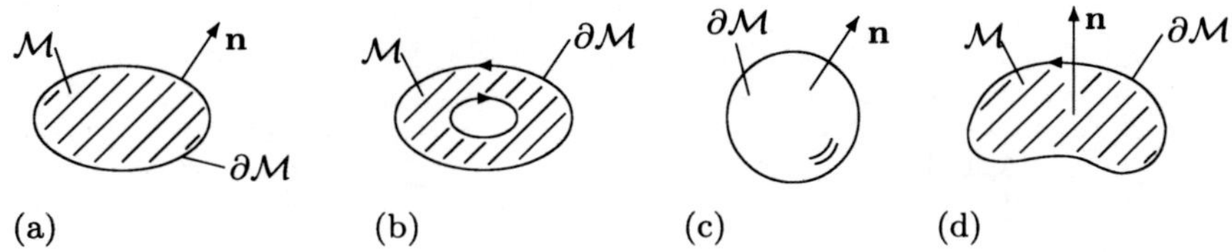

Fig. 12.6. Coherent orientation

basis $\mathbf{i}, \mathbf{j}$ at the origin O (Fig. 12.5). Each velocity vector $\mathbf{v} \in T_P\mathbb{E}^2$ can be uniquely represented as

$$\mathbf{v} = a\mathbf{i}_P + b\mathbf{j}_P$$

with real numbers a and b. We define $dx_P, dy_P \in T_P\mathbb{E}^2$ by setting:

$$dx_P(\mathbf{v}) := a \qquad \text{and} \qquad dy_P(\mathbf{v}) := b \qquad \text{for all} \quad \mathbf{v} \in T_P\mathbb{E}^2.$$

A smooth velocity vector field $\mathbf{v}$ on the Euclidean plane $\mathbb{E}^2$ can be represented as

$$\mathbf{v}_P = a(x,y)\mathbf{i}_P + b(x,y)\mathbf{j}_P.$$

Here, the point P has the Cartesian coordinates (x, y), and the coefficient functions $a, b : \mathbb{R}^2 \to \mathbb{R}$ are smooth. Similarly, a smooth covector field on $\mathbb{E}^2$ can be written as

$$\omega_P = U(x,y)dx_P + V(x,y)dy_P$$

where the functions $U, V : \mathbb{R}^2 \to \mathbb{R}$ are smooth. It follows from (12.25) that

$$\boxed{d\omega_P = (V_x(x,y) - U_y(x,y))\, dx_P \wedge dy_P.} \tag{12.26}$$

This is the same expression as obtained above by the algebraic method. The proof can be found in Problem 12.6 on page 802. For the pull-back $\tau^*\omega$, we refer to Sect. 12.3.4 on page 705.

12.1.3 Three Dimensions

Consider a right-handed (x, y, y)-Cartesian coordinate system of the Euclidean manifold $\mathbb{E}^3$ with the right-handed orthonormal basis $\mathbf{i}, \mathbf{j}, \mathbf{k}$ at the origin O (Fig. 4.3 on page 323). In what follows, all the functions are assumed to be smooth. The domains of integration are depicted in Figs. 12.6 and 12.7.

Let $\mathcal{M}$ and $\mathcal{U}$ be 3-dimensional subsets of $\mathbb{E}^3$. More precisely, suppose that $\mathcal{M}$ (resp. $\mathcal{U}$) is a compact, arcwise connected, oriented, 3-dimensional submanifold of $\mathbb{E}^3$ with coherently oriented boundary $\partial\mathcal{M}$ (resp. $\partial\mathcal{U}$) (e.g., balls; see Fig. 12.6(c)).

(i) The main theorem of calculus: If the functions $U, V, W : \mathcal{M} \to \mathbb{R}$ are smooth, then

$$\boxed{\int_{\mathcal{M}} (U_x + V_y + W_z)\, dxdydz = \int_{\partial\mathcal{M}} (Un^1 + Vn^2 + Wn^3)\, dS.} \tag{12.27}$$

Here, $\mathbf{n} = n^1(P)\mathbf{i} + n^2(P)\mathbf{j} + n^3(P)\mathbf{k}$ is the outer normal unit vector at the boundary point P, and dS denotes the surface differential to be considered below. Recall that U_x denotes the partial derivative of U with respect to x, and so on.

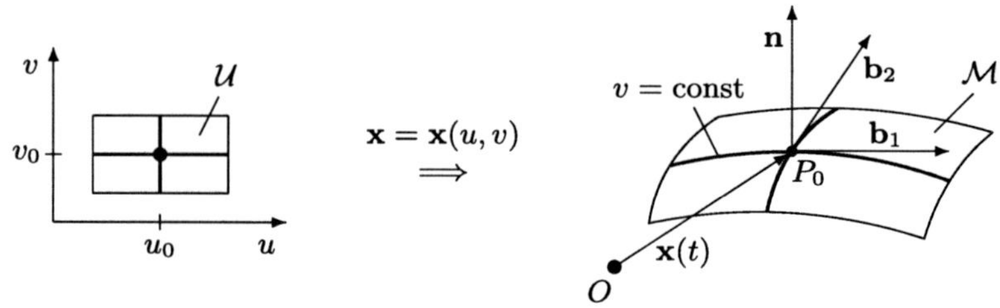

Fig. 12.7. Surface parametrization

(ii) Integration by parts: We have

$$\int_{\mathcal{M}} U_x V \, dxdydz = -\int_{\mathcal{M}} UV_x \, dxdydz + \int_{\partial\mathcal{M}} U(P)V(P)n^1(P) \, dS. \quad (12.28)$$

Similarly,

- $\int_{\mathcal{M}} U_y V \, dxdydz = -\int_{\mathcal{M}} UV_y \, dxdydz + \int_{\partial\mathcal{M}} UVn^2 \, dS,$
- $\int_{\mathcal{M}} U_z V \, dxdydz = -\int_{\mathcal{M}} UV_z \, dxdydz + \int_{\partial\mathcal{M}} UVn^3 \, dS.$

This follows from (12.27) by replacing U (resp. V, W) by the product UV.

(iii) The substitution rule: If $\tau : \mathcal{U} \to \mathcal{M}$ is an orientation-preserving diffeomorphism, then

$$\int_{\mathcal{M}} V(x, y, z) \, dxdydz = \int_{\mathcal{U}} V(\tau(u, v, w)) \cdot \det \tau'(u, v, w) \, dudvdw. \quad (12.29)$$

Equivalently, $\int_{\mathcal{M}} V(x, y, z) \, dxdydz$ is equal to

$$\int_{\mathcal{U}} V(x(u, v, w), y(u, v, w), z(u, v, w)) \cdot \frac{\partial(x, y, z)}{\partial(u, v, w)}(u, v, w) \, dudvdw.$$

(iv) Velocity vector fields: We are given the smooth velocity vector field

$$\mathbf{v}_P = U(x, y, z)\mathbf{i}_P + V(x, y, z)\mathbf{j}_P + W(x, y, z)\mathbf{k}_P$$

on the Euclidean manifold $\mathbb{E}^3$. We want to use the following expressions:

- $\operatorname{div} \mathbf{v} = U_x + V_y + W_z$ (divergence),
- $\mathbf{curl}\, \mathbf{v} = (W_y - V_z)\mathbf{i}_P + (U_z - W_x)\mathbf{j}_P + (V_x - U_y)\mathbf{k}_P$ (curl),
- $\mathbf{grad}\, U = U_x\mathbf{i}_P + U_y\mathbf{j}_P + U_z\mathbf{k}_P$ (gradient),
- $\Delta U = -\operatorname{div} \mathbf{grad}\, U = -U_{xx} - U_{yy} - U_{zz}$ (Laplacian),
- $d_{\mathbf{n}} U = \mathbf{n}\, \mathbf{grad}\, U = n^1 U_x + n^2 U_y + n^3 U_z$ (directional derivative).

(v) Surface integrals: Consider the surface $\mathcal{M}$ described by the equation

$$\mathbf{x} = \mathbf{x}(u, v), \qquad (u, v) \in \mathcal{U}.$$

See Fig. 12.7. The point P_0 has the coordinate (u_0, v_0). The equation

$$\mathbf{x} = \mathbf{x}(u, v_0), \qquad u \in \mathbb{R}$$

describes a curve passing through the point P with the tangent vector $\mathbf{x}_u(u_0, v_0)$ at the point P_0. This curve is called the u-coordinate line through the point P_0 on the surface $\mathcal{M}$ Similarly, the equation

$$\mathbf{x} = \mathbf{x}(u_0, v), \qquad v \in \mathbb{R}$$

describes a curve passing through the point P_0 with the tangent vector $\mathbf{x}_v(u_0, v_0)$ at the point P_0. We call this curve the v-coordinate line through the point P_0. We set

$$\mathbf{b}_1 := \mathbf{x}_u(u_0, v_0), \qquad \mathbf{b}_2 := \mathbf{x}_v(u_0, v_0).$$

Here, $\mathbf{x}_u$ denotes the partial derivative with respect to the real parameter u. We assume that $\mathbf{x}_u(u, v)$ and $\mathbf{x}_v(u, v)$ are linearly independent. In other words,

$$\mathbf{x}_u(u, v) \times \mathbf{x}_v(u, v) \neq 0 \qquad \text{for all} \quad (u, v) \in \mathcal{U}.$$

Then, the two vectors $\mathbf{b}_1, \mathbf{b}_1$ form a basis of the tangent plane at the point P_0, and the vector

$$\mathbf{N} := \mathbf{b}_1 \times \mathbf{b}_2$$

is the normal vector at the point P_0. Then

$$\mathbf{N} = \begin{vmatrix} \mathbf{i}_{P_0} & \mathbf{j}_{P_0} & \mathbf{k}_{P_0} \\ x_u(u_0, v_0) & y_u(u_0, v_0) & z_u(u_0, v_0) \\ x_v(u_0, v_0) & y_v(u_0, v_0) & z_v(u_0, v_0) \end{vmatrix},$$

and we have the following quantities:

- Length of the normal vector $\mathbf{N}$ at the point P_0:

$$|\mathbf{N}| := \sqrt{\left(\frac{\partial(x, y)}{\partial(u, v)}\right)^2 + \left(\frac{\partial(y, z)}{\partial(u, v)}\right)^2 + \left(\frac{\partial(z, x)}{\partial(u, v)}\right)^2}.$$

  Alternatively, by the Lagrange identity (1.30) on page 84,

$$|\mathbf{N}|^2 = (\mathbf{b}_1 \times \mathbf{b}_2)(\mathbf{b}_1 \times \mathbf{b}_2) = \mathbf{b}_1^2\mathbf{b}_2^2 - (\mathbf{b}_1\mathbf{b}_2)^2.$$

- Surface element: $\Delta S = |\mathbf{b}_1 \Delta u \times \mathbf{b}_2 \Delta v| = \Delta u \Delta v |\mathbf{N}|$.
- Surface differential:

$$dS = |\mathbf{N}| \, dudv = \sqrt{\left(\frac{\partial(x, y)}{\partial(u, v)}\right)^2 + \left(\frac{\partial(y, z)}{\partial(u, v)}\right)^2 + \left(\frac{\partial(z, x)}{\partial(u, v)}\right)^2} \; dudv.$$

- Unit normal vector at the point P_0: $\mathbf{n} := \frac{\mathbf{N}}{|\mathbf{N}|}$.
- $\mathbf{n}\Delta S = \mathbf{N}\Delta u \Delta v$.

It follows from $\int_{\mathcal{M}} \mathbf{vn} \, dS = \lim_{\Delta S \to 0} \sum \mathbf{vn} \, \Delta S$ that

$$\int_{\mathcal{M}} \mathbf{vn} \, dS = \int_{\mathcal{U}} \mathbf{v}(\mathbf{x}(u, v))\mathbf{N}(\mathbf{x}(u, v)) \, dudv.$$

Hence

$$\int_{\mathcal{M}} \mathbf{vn} \, dS = \int_{\mathcal{U}} \left(U \frac{\partial(y, z)}{\partial(u, v)} + V \frac{\partial(z, x)}{\partial(u, v)} + W \frac{\partial(x, y)}{\partial(u, v)} \right) dudv. \tag{12.30}$$

We will show below that this integral can be elegantly written as

$$\int_{\mathcal{M}} \omega = \int_{\tau^*\mathcal{M}} \tau^* \omega \tag{12.31}$$

with

- $\omega = U\, dy \wedge dz + V\, dz \wedge dx + W\, dx \wedge dy$,
- $x = x(u, v), y = y(u, v), z = z(u, v)$.

(vi) The divergence theorem: Formula (12.27) can be written as

$$\boxed{\int_{\mathcal{M}} \operatorname{div} \mathbf{v}\; dxdydz = \int_{\partial\mathcal{M}} \mathbf{v}(P)\mathbf{n}(P)\; dS.} \tag{12.32}$$

This is called the divergence theorem (or the Gauss–Ostrogradsky integral theorem). The famous formula (12.27) was implicitly (resp. explicitly) used by Lagrange in 1760 and Gauss in 1813 (resp. Ostrogradsky in 1826).

(vii) The circulation theorem: Let $\mathcal{M}$ be a surface of the Euclidean manifold $\mathbb{E}^3$ with coherently oriented boundary (Fig. 12.6(d) on page 677). More precisely, let $\mathcal{M}$ be a compact, arcwise connected, oriented, 2-dimensional submanifold of $\mathbb{E}^3$ with coherently oriented boundary $\partial\mathcal{M}$. Then

$$\boxed{\int_{\mathcal{M}} \mathbf{n}\,\mathbf{curl}\,\mathbf{v}\; dS = \int_{\partial\mathcal{M}} \mathbf{v}d\mathbf{x}.} \tag{12.33}$$

This is incorrectly called the Stokes integral theorem. In fact, this theorem was discovered by Thomson (the later Lord Kelvin) in 1850 (see the historical discussion on page 782).

(viii) The Green's formula: We are given the 3-dimensional set $\mathcal{M}$ as in (i) on page 677. Then

$$\int_{\mathcal{M}} (U\Delta V - V\Delta U)\; dxdydz = \int_{\partial\mathcal{M}} (Vd_{\mathbf{n}}U - Ud_{\mathbf{n}}V)\; dS. \tag{12.34}$$

Proof. Repeated integration by parts yields

$$\begin{aligned}\int_{\mathcal{M}} UV_{xx}\; dxdydz &= -\int_{\mathcal{M}} U_xV_x\; dxdydz + \int_{\partial\mathcal{M}} UV_xn^1\; dS\\ &= \int_{\mathcal{M}} U_{xx}V\; dxdydz + \int_{\partial\mathcal{M}} (UV_xn^1 - VU_xn^1)\; dS.\end{aligned}$$

□

The formal differential calculus.[9] Let us introduce, the following symbols:

- f (0-form),
- $\mu = Udx + Vdy + Wdz$ (1-form),
- $\omega = U\, dy \wedge dz + V\, dz \wedge dx + W\, dx \wedge dy$ (2-form),
- $\varrho = f\, dx \wedge dy \wedge dz$ (3-form).

(i) Differential: To begin with, we define

$$df := f_xdx + f_ydy + f_zdz.$$

Furthermore, we define

- $d\mu := dU \wedge dx + dV \wedge dy + dW \wedge dz$,
- $d\omega := dU \wedge dy \wedge dz + dV \wedge dz \wedge dx + dW \wedge dx \wedge dy$,
- $d\varrho := df \wedge dx \wedge dy \wedge dz = 0$.

This yields

- $d\mu = (W_y - V_z)\, dy \wedge dz + (U_z - W_x)\, dz \wedge dx + (V_x - U_y)\, dx \wedge dy$,
- $d\omega = (U_x + V_y + W_z)\, dx \wedge dy \wedge dz$.

[9] The rigorous 3-dimensional differential calculus will be studied in Sect. 12.3ff.

(ii) Relation to classical vector calculus: Set $\mathbf{v} = U\mathbf{i} + V\mathbf{j} + W\mathbf{k}$. Then:
- $df = d\mathbf{x}\,\mathbf{grad}\, f$,
- $d\mu$ corresponds to $\mathbf{curl}\,\mathbf{v}$,
- $d\omega$ corresponds to $\operatorname{div}\mathbf{v}$.

(iii) The Poincaré cohomology rule:
- $d(df) = 0$ corresponds to $\mathbf{curl}\,\mathbf{grad}\, f = 0$,
- $d(d\mu) = 0$ corresponds to $\operatorname{div}\mathbf{curl}\,\mathbf{v} = 0$,
- $d(d\varrho) = 0$.

(iv) Coordinate transformation (pull-back): Set $u^1 := u, u^2 := v, u^3 := w$. Fix $r = 1, 2, 3$. Consider the smooth map $\tau : \mathcal{U} \to \mathbb{E}^3$ given by

$$x = x(u^1, \ldots, u^r), \quad y = y(u^1, \ldots, u^r), \quad z = z(u^1, \ldots, u^r)$$

where $\mathcal{U}$ is an open subset of $\mathbb{R}^r$. Set $P = \tau(Q)$. This yields

$$dx = x_{u^1} du^1 + \ldots + x_{u^r} du^r$$

and similar formulas for dy and dz together with the following pull-back transformation formulas:
- $(\tau^* f)(Q) := f(P)$ $(r = 3)$,
- $(\tau^*\mu)_Q := (U(P)x_u(Q) + V(P)y_u(Q) + W(P)z_u(Q))\, du$ $(r = 1)$,
- $(\tau^*\omega)_Q := \left(U(P)\frac{\partial(y,z)}{\partial(u,v)}(Q) + V(P)\frac{\partial(z,x)}{\partial(u,v)} + W(P)\frac{\partial(x,y)}{\partial(u,v)}\right) du \wedge dv$ $(r = 2)$,
- $(\tau^*\varrho)_Q := \varrho(P)\, \frac{\partial(x,y,z)}{\partial(u,v,w)}(Q)\, du \wedge dv \wedge dw$ $(r = 3)$.

For example, let $r = 2$. Then

$$\begin{aligned} dy \wedge dz &= (y_u du + y_v dv) \wedge (z_u du + z_v dv) \\ &= (y_u z_v - y_v z_u)\, du \wedge dv = \frac{\partial(y,z)}{\partial(u,v)}\, du \wedge dv. \end{aligned}$$

12.1.4 Integration over Manifolds

3-dimensional submanifolds. Let $\mathcal{M}$ be the closure of a bounded open subset of the Euclidean manifold $\mathbb{E}^3$ (e.g., a ball), and let $f : \mathbb{E}^3 \to \mathbb{R}$ be a smooth function. We define

$$\boxed{\int_{\mathcal{M}} f(x,y,z)\, dx \wedge dy \wedge dz := \int_{\mathcal{M}} f(x,y,z)\, dxdydz.} \qquad (12.35)$$

In other words, the integral over the 3-form $\varrho = f\, dx \wedge dy \wedge dz$ is defined to be the corresponding classical integral over the function f (i.e., mnemonically, we replace the wedge product $dx \wedge dy \wedge dz$ by the classical volume differential $dxdydz$).

Substitution rule. Let $\mathcal{U}$ be the closure of a bounded open subset of $\mathbb{R}^3$. Let $\tau : \mathcal{U} \to \mathcal{M}$ be an orientation-preserving diffeomorphism given by[10]

$$x = x(u,v,w), \quad y = y(u,v,w), \quad z = z(u,v,w), \qquad (u,v,w) \in \mathcal{U}.$$

Then

[10] This means that the Jacobian is positive: $\frac{\partial(x,y,z)}{\partial(u,v,w)} > 0$ on $\mathcal{U}$.

$$\boxed{\int_{\mathcal{M}} \varrho = \int_{\tau^*\mathcal{M}} \tau^* \varrho} \tag{12.36}$$

where $\tau^*\mathcal{M} := \tau^{-1}(\mathcal{M}) = \mathcal{U}$. This coincides with the classical substitution rule (12.29). Mnemonically, we get

$$f\, dx \wedge dy \wedge dz = f\, \frac{\partial(x,y,z)}{\partial(u,v,w)}\, du \wedge dv \wedge dw.$$

Hence

$$\int_{\mathcal{M}} f\, dx \wedge dy \wedge dz = \int_{\mathcal{U}} f\, \frac{\partial(x,y,z)}{\partial(u,v,w)}\, du \wedge dv \wedge dw.$$

In classical terms, this means that

$$\int_{\mathcal{M}} f\, dxdydz = \int_{\mathcal{U}} f\, \frac{\partial(x,y,z)}{\partial(u,v,w)}\, dudvdw.$$

Thus, the calculus of differential forms fits best the classical substitution rule if we restrict ourselves to parameter transformations with positive Jacobian.

Integrals over parametrized curves and surfaces. Fix $r = 1, 2$. Let us set $u^1 := u, u^2 := v$. We want to integrate smooth r-forms ω over r-dimensional submanifolds $\mathcal{M}$ of the Euclidean manifold $\mathbb{E}^3$. If $r = 1$ (resp. $r = 2$), then this concerns integrals over 1-forms on a curve (resp. 2-forms on a surface). Motivated by (12.36), the key formula reads as

$$\boxed{\int_{\mathcal{M}} \omega := \int_{\mathcal{U}} \tau^* \omega.} \tag{12.37}$$

Here, we assume that there exists an orientation-preserving diffeomorphism

$$\tau : \mathcal{U} \to \mathcal{M} \tag{12.38}$$

where the following hold:

- The domain of integration $\mathcal{M}$ is an r-dimensional submanifold (with boundary) of the Euclidean manifold $\mathbb{E}^3$. In addition, $\mathcal{M}$ is compact, arcwise connected, and oriented.
- The parameter space $\mathcal{U}$ is the closure of an open bounded subset of $\mathbb{R}^r$ (e.g., a compact interval if $r = 1$, or a closed rectangle if $r = 2$).

Explicitly, the definition (12.37) of the integrals reads as follows:

- $r = 1$: $\int_{\mathcal{M}} Udx + Vdy + Wdz := \int_{\mathcal{U}} \left(U\frac{dx}{du} + V\frac{dy}{du} + W\frac{dz}{du}\right)\, du.$
- $r = 2$: $\int_{\mathcal{M}} U\, dy \wedge dz + V\, dz \wedge dx + W\, dx \wedge dy$ is equal to

$$\int_{\mathcal{U}} \left(U\, \frac{\partial(y,z)}{\partial(u,v)} + V\, \frac{\partial(z,x)}{\partial(u,v)} + W\, \frac{\partial(x,y)}{\partial(u,v)} \right)\, dudv.$$

The crucial invariance property of the integrals over differential forms. Note that the integral from (12.37) does not depend on the choice of the parametrization (12.38). If we use another parametrization, then the integral remains unchanged because of the substitution rule for r-dimensional integrals of the classical type $\int_{\mathcal{U}} g(u)\, du$ or $\int_{\mathcal{U}} g(u,v)\, dudv$. Explicitly, we have the classical product rule for the Jacobian:

$$\frac{\partial(x,y)}{\partial(u',v')} = \frac{\partial(x,y)}{\partial(u,v)} \cdot \frac{\partial(u,v)}{\partial(u',v')}.$$

By the substitution rule,

$$\int_{\mathcal{U}} W(P) \frac{\partial(x,y)}{\partial(u,v)}\, dudv = \int_{\mathcal{U}'} W(P') \frac{\partial(x,y)}{\partial(u,v)} \cdot \frac{\partial(u,v)}{\partial(u',v')}\, du'dv' \tag{12.39}$$

$$= \int_{\mathcal{U}'} W(P') \frac{\partial(x,y)}{\partial(u',v')}\, du'dv'. \tag{12.40}$$

Observe that the substitution rule is only valid if the Jacobian is positive, that is,

$$\frac{\partial(u',v')}{\partial(u,v)} > 0 \quad \text{on } \mathcal{U}.$$

The point is that integrals depend on the orientation of the domain of integration.

Therefore, we have to restrict ourselves to parameter transformations which preserve the orientation.

In order to underline the elegance of the Cartan calculus, note that mnemonically we get

$$W(P)\, dx \wedge dy = W(P) \frac{\partial(x,y)}{\partial(u,v)}\, du \wedge dv = W(P') \frac{\partial(x,y)}{\partial(u',v')}\, du' \wedge dv',$$

and hence

$$\int_{\mathcal{M}} W(P)\, dx \wedge dy = \int_{\mathcal{U}} W(P) \frac{\partial(x,y)}{\partial(u,v)}\, du \wedge dv = \int_{\mathcal{U}'} W(P') \frac{\partial(x,y)}{\partial(u',v')}\, du' \wedge dv'$$

which corresponds to the classical relation

$$\int_{\mathcal{U}} W(P) \frac{\partial(x,y)}{\partial(u,v)}\, dudv = \int_{\mathcal{U}'} W(P') \frac{\partial(x,y)}{\partial(u',v')}\, du'dv'$$

called the substitution rule.

The Cartan calculus yields elegantly the correct parameter change for integrals over submanifolds.

The integral over the surface of earth. Let $\mathcal{M}$ be the surface of earth, and let ω be a smooth 2-form on $\mathcal{M}$. We want to compute the integral $\int_{\mathcal{M}} \omega$. To do this, we will take into account that the surface of earth can be locally represented by geographic charts, and integrals over chart coordinates (u,v) are classical integrals of the form $\int_{\mathcal{U}} g\, dudv$. More precisely, let us choose the decomposition

$$\mathcal{M} = \bigcup_{j=1}^{m} \mathcal{M}_j \tag{12.41}$$

of the surface of earth where every subset $\mathcal{M}_j$ of $\mathcal{M}$ has the following properties:[11]

- $\mathcal{M}_j$ is the closure of an open subset $\mathcal{O}_j$ of $\mathcal{M}$.

[11] As an example, consider the case $m = 2$ where $\mathcal{M}_1$ (resp, $\mathcal{M}_2$) is the northern (resp. southern) hemisphere. The intersection $\mathcal{M}_1 \cup \mathcal{M}_2$ is the equator which has the 2-dimensional surface measure zero.

- The set $\mathcal{M}_j$ on the surface of earth can be represented by a geographic chart, that is, there exists a chart map

$$\tau_j : \mathcal{U}_j \to \mathcal{M}_j$$

which is an orientation-preserving diffeomorphism; the set $\mathcal{U}_j$ is the closure of a bounded open subset of $\mathbb{R}^2$.
- The sets $\mathcal{O}_1, \ldots, \mathcal{O}_m$ are pairewise disjoint.

Naturally enough, we define

$$\boxed{\int_{\mathcal{M}} \omega := \sum_{j=1}^{m} \int_{\mathcal{M}_j} \omega = \sum_{j=1}^{m} \int_{\mathcal{U}_j} \tau_j^* \omega.} \tag{12.42}$$

One has to show that this definition does not depend on the choice of the decomposition (12.41) of the surface of earth.

This simple construction can be used in order to define the integral $\int_{\mathcal{M}} \omega$ on general finite-dimensional manifolds $\mathcal{M}$ with boundary.

12.1.5 Integration over Singular Chains

The following generalization of the definition of the integral $\int_{\mathcal{M}} \omega$ plays a key role in modern differential topology (the de Rham cohomology theory for manifolds). Fix $r = 1, 2, 3$. Let

$$\tau : \mathcal{U} \to \mathbb{E}^3$$

be a smooth map where $\mathcal{U}$ is the closure of a bounded open subset in $\mathbb{R}^r$. In topology, the map τ is called a singular r-chain of the Euclidean manifold $\mathbb{E}^3$. Let ω be a smooth r-form on $\mathbb{E}^3$. We define the integral

$$\int_\tau \omega := \int_{\mathcal{U}} \tau^* \omega.$$

If $\sigma, \tau : \mathcal{U} \to \mathbb{E}^3$ are smooth maps and α, β are real numbers, then

$$\int_{\alpha\sigma + \beta\tau} \omega = \alpha \int_\sigma \omega + \beta \int_\tau \omega.$$

The point is that the integral $\int_\tau \omega$ depends on the singular chain τ. But, as a rule, it is not an integral over a subset $\mathcal{M}$ of $\mathbb{E}^3$. Only if τ is a regular chain, then we have

$$\int_\tau \omega = \int_{\mathcal{M}} \omega.$$

with $\mathcal{M} := \tau(\mathcal{U})$. More precisely, we have to assume the following: the image $\tau(\mathcal{U})$ is an r-dimensional, compact, oriented submanifold (with boundary) of $\mathbb{E}^3$, and the map τ is an orientation-preserving diffeomorphism onto $\tau(\mathcal{U})$.

12.2 Applications to Physics

12.2.1 Single-Valued Potentials and Gauge Transformations

Potentials play a key role in gauge theory. Let us discuss some basic ideas. To this end, choose a right-handed Cartesian (x, y)-coordinate system on the Euclidean plane $\mathbb{E}^2$ with the orthonormal basis vectors $\mathbf{i}, \mathbf{j}$ at the origin (Fig. 12.5 on page 676). We are given the smooth force field

$$\mathbf{F}_P = A(x,y)\mathbf{i}_P + B(x,y)\mathbf{j}_P \qquad \text{on } \mathbb{E}^2.$$

Here, the point P has the coordinates (x, y), and the functions $A, B : \mathbb{R}^2 \to \mathbb{R}$ are smooth. Consider a smooth curve $C : x = x(t), y = y(t),\ t_0 \le t \le t_1$, which connects the initial point P_0 with the terminal point P_1. The line integral

$$W = \int_C \mathbf{F}d\mathbf{x} \tag{12.43}$$

equals the work done by the force field $\mathbf{F}$ if it moves a point from the initial position P_0 to the final position P_1 along the curve C. After finishing this motion, we gain the amount W of energy. Explicitly,

$$W = \int_C Adx + Bdy = \int_{t_0}^{t_1} \big(A(x(t), y(t)) \cdot \dot{x}(t) + B(x(t), y(t)) \cdot \dot{y}(t)\big)\, dt.$$

The integral (12.43) is said to be independent of the path of integration iff it only depends on the initial point P_0 and the terminal point P_1, but not on the choice of the smooth curve C itself.

Proposition 12.1 *The integral $\int_C \mathbf{F}d\mathbf{x}$ is independent of the path of integration iff the so-called integrability condition*

$$A_y = B_x \quad \text{on } \mathbb{E}^2 \tag{12.44}$$

is satisfied. This is equivalent to the existence of a smooth function $U : \mathbb{R}^2 \to \mathbb{R}$ such that

$$\mathbf{F} = -\,\mathbf{grad}\, U \qquad \text{on } \mathbb{E}^2. \tag{12.45}$$

The function U is called the potential of the force field $\mathbf{F}$. The potential U is uniquely determined by the force field $\mathbf{F}$ up to an additive constant. The fixing of the additive constant of U is called the gauge fixing of the potential.

Proof. (I) We will show that the existence of the potential U implies the path-independence of the integral $W = \int_C \mathbf{F}d\mathbf{x}$. In fact, it follows from (12.45) that

$$\begin{aligned} W &= -\int_C \mathbf{grad}\, U\, d\mathbf{x} = -\int_{t_0}^{t_1} \mathbf{grad}\, U(\mathbf{x}(t)) \cdot \dot{\mathbf{x}}(t)\, dt \\ &= -\int_{t_0}^{t_1} \frac{d}{dt} U(\mathbf{x}(t))\, dt = U(P_0) - U(P_1). \end{aligned}$$

(II) If $\mathbf{F} = -\,\mathbf{grad}\, U$ and $\mathbf{F} = -\,\mathbf{grad}\, V$, then $(U - V)_x = (U - V)_y = 0$ on $\mathbb{R}^2$. Hence $U - V = \text{const}$.

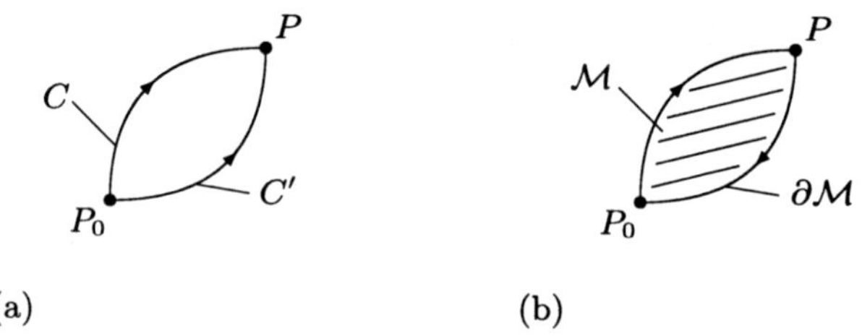

Fig. 12.8. Closed path

(III) Suppose that the integral (12.43) is path-independent. We want to show that this implies the integrability condition (12.44). To this end, we construct the function

$$U(x, y) := -\int_{(0,0)}^{(x,y)} Adx + Bdy. \tag{12.46}$$

Since the integral is path-independent, we may integrate along an arbitrary smooth curve C which connects the origin $(0, 0)$ with the point (x, y) (e.g., a straight line). Hence

$$U(x + \Delta x, y) = U(x, y) - \int_{(x,y)}^{(x+\Delta x,y)} Adx + Bdy.$$

Integrating along a straight line, we get

$$U_x(x, y) = -\lim_{\Delta x \to 0} \frac{1}{\Delta x} \int_x^{x+\Delta x} A(\xi, y)d\xi = -A(x, y).$$

Similarly, $U_y(x, y) = -B(x, y)$. From $U_{xy} = U_{yx}$ we get $A_y = B_x$.

(IV) Assume that the integrability condition (12.44) holds. We want to show that the integral (12.43) is path-independent. To this end, we will use the function

$$U(x, y) := -\int_0^1 \big(A(tx, ty)x + B(tx, ty)y\big)\, dt$$

which corresponds to the integral (12.46) along a straight line connecting the origin $(0, 0)$ with the point (x, y)). Since $A_y = B_x$, we get

$$\begin{aligned} U_x(x, y) &= -\int_0^1 \big(A_x(tx, ty)tx + A(tx, ty) + B_x(tx, ty)ty\big)\, dt \\ &= -\int_0^1 \frac{d}{dt}\big(A(tx, ty)t\big)\, dt = -A(x, y). \end{aligned}$$

Similarly, $U_y(x, y) = -B(x, y)$. This implies (12.45). By (I), the integral $\int_C \mathbf{F}d\mathbf{x}$ is path-independent. □

Heuristic motivation based on the Stokes integral theorem. To get insight, let us motivate Prop. 12.1 by using a heuristic argument. Consider the situation depicted in Fig. 12.8. The Stokes integral theorem tells us that

$$\int_C Adx + By - \int_{C'} Adx + Bdy = \int_{\partial\mathcal{M}} Adx + Bdy = \int_{\mathcal{M}} (B_x - A_y)\, dxdy.$$

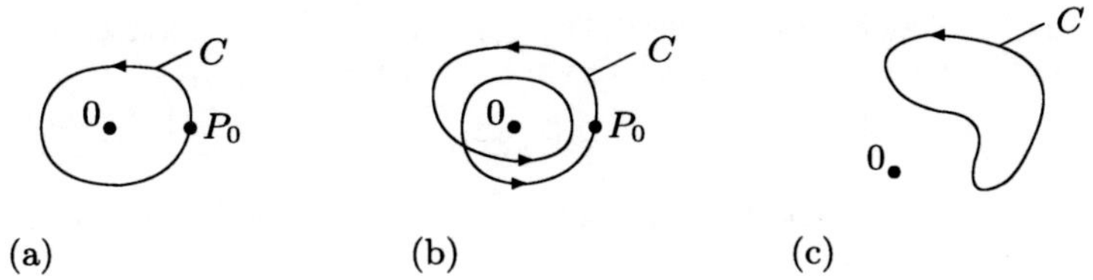

Fig. 12.9. Surrounding a singularity at the origin

If $A_y = B_x$, then $\int_C Adx + By = \int_{C'} Adx + Bdy$, that is, the integral is path-independent. Conversely, if the integral is path-independent, then

$$\frac{1}{\text{meas}(\mathcal{M})} \int_{\mathcal{M}} (B_x - A_y)\, dxdy = 0.$$

Contracting this to the point P, we get $B_x(P) - A_y(P) = 0$. This is the integrability condition.

The importance of the Cauchy–Riemann differential equations. Let $f : \mathbb{C} \to \mathbb{C}$ be a holomorphic function. Then the integral

$$\int_C f(z)dz$$

is path-independent.

Proof. The function $f(z) = u(x,y) + \mathrm{i}v(x,y)$ with smooth functions $u, v : \mathbb{R}^2 \to \mathbb{R}$ satisfies the Cauchy–Riemann differential equations $u_x = v_y, u_y = -v_x$ on $\mathbb{R}^2$. We have

$$\int_C f(z)dz = \int_C (u + \mathrm{i}v)(dx + \mathrm{i}dy) = \int_C udx - vdy + \mathrm{i}\int_C vdx + udy.$$

Because of the Cauchy–Riemann differential equations, the integrability conditions for the two line integrals are satisfied. □

12.2.2 Multi-Valued Potentials and Riemann Surfaces

> As a rule, singularities of a force field destroy the path-independence of the work integral. This leads to multi-valued potentials.
>
> Folklore

Cauchy's residue theorem. Let a be a nonzero complex number. As a prototype, let us consider the line integral

$$\mathcal{W} := \int_C \frac{adz}{z}.$$

We want to show that this integral is path-dependent because of the singularity of the integrand at the point $z = 0$. Motivated by Fig. 12.8(b), we have to find a closed curve $\partial\mathcal{M}$ with $\int_{\partial\mathcal{M}} \frac{adz}{z} \neq 0$. To this end, choose a circle C of radius R about the origin. By Cauchy's residue theorem (see Sect. 4.4 of Vol. I), we get

$$\mathcal{W} = 2\pi\mathrm{i}a.$$

The curves C depicted in Fig. 12.9 (a), (b), (c) yield $\mathcal{W} = 2\pi\mathrm{i}a, 4\pi\mathrm{i}a, 0$, respectively. If the curve C winds m times about the origin, then

$$\mathcal{W} = m \cdot 2\pi\mathrm{i}a.$$

Here, counter-clockwise (resp. clockwise) surrounding of the origin is counted positively (resp. negatively).

The Riemann surface $\mathcal{R}$ of the arg-function. Let $z = x + \mathrm{i}y$ be a point of the Gaussian plane $\mathbb{C}$. We assign to $z = r\mathrm{e}^{\mathrm{i}\varphi}$ the polar coordinates

$$r := |z| \qquad \text{and} \qquad \varphi := \arg(z) \quad \text{with} \quad -\pi < \varphi \leq \pi.$$

Here, $\arg(z)$ is called the principal value of the argument of the complex number z. Note that the function $z \mapsto \arg(z)$ is not continuous. In order to get rid of this defect, let us introduce the Riemann surface

$$\mathcal{R} := \bigcup_{m \in \mathbb{Z}} \mathcal{S}_m.$$

Here, for any integer m, the so-called mth sheet $\mathcal{S}_m$ consists of all the points

$$P := (z, m) \quad \text{with} \quad z \in \mathbb{C} \setminus \{0\}.$$

We define

$$\boxed{r(P) := |z|, \qquad \arg(P) := \arg(z) + 2m\pi.}$$

Following Riemann, we equip the set $\mathcal{R}$ with a topology such that the function

$$\arg : \mathcal{R} \to \mathbb{R} \tag{12.47}$$

is continuous. Intuitively, this topology can be obtained by

- cutting all the sheets $\mathcal{S}_m$ along the negative real axis and
- gluing the upper strand of the sheet $\mathcal{S}_m$ with the lower strand of the sheet $\mathcal{S}_{m+1}$ for all $m = 0, \pm 1, \pm 2, \ldots$ (Fig. 12.10).

Analytically, for example, choose the point $(x_0, 0, m)$ of the mth sheet with $x_0 < 0$. By definition, for sufficiently small $\varepsilon > 0$, the ε-neighborhood $\mathcal{U}_\varepsilon(x_0, 0, m)$ of the point $(x_0, 0, m)$ consists of all the points

- (x, y, m) on the mth sheet $\mathcal{S}_m$ with $|x - x_0| < \varepsilon$, $0 \leq y < \varepsilon$, and all points
- $(x, y, m+1)$ on the $(m+1)$th sheet $\mathcal{S}_{m+1}$ with $|x - x_0| < \varepsilon$, $-\varepsilon < y < 0$.

Similarly, the topological space $\mathcal{R}$ becomes a one-dimensional, arcwise connected, complex manifold. For example, we assign to the points (x, y, m) and $(x, y, m+1)$ of $\mathcal{U}_\varepsilon(x_0, 0, m)$ the complex coordinate $z = x + \mathrm{i}y$.

The Riemann surface $\mathcal{R}$ of the logarithmic function. We define the function $\ln : \mathcal{R} \to \mathbb{C}$ by setting

$$\boxed{\ln P := \ln r(P) + \mathrm{i} \arg(P).} \tag{12.48}$$

That is, $\ln(z, m) := \ln|z| + \mathrm{i}\arg(z) + 2m\pi\mathrm{i}$. It can be shown that the function $\ln : \mathcal{R} \to \mathbb{C}$ is holomorphic. Let C be a smooth curve on $\mathcal{R}$ with the initial point P_0 and the terminal point P. Then

$$\boxed{\int_C \frac{dz}{z} = \ln(P) - \ln(P_0)} \tag{12.49}$$

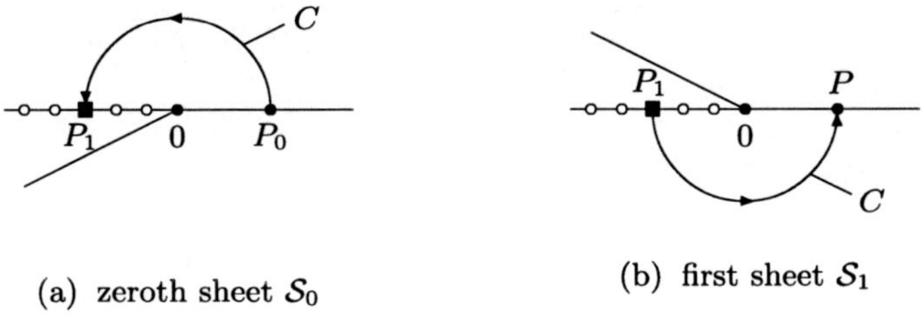

Fig. 12.10. Riemann surface of the map $P \mapsto \ln P$

if the integral is taken with respect to local complex coordinates on $\mathcal{R}$. For example, consider the curve $C = P_0P_1P$ on $\mathcal{R}$ depicted in Fig. 12.10. Then, $\ln(P)$ is equal to $\ln(P_0) + 2\pi\mathrm{i}$. Hence

$$\int_C \frac{dz}{z} = \ln(P) - \ln(P_0) = 2\pi\mathrm{i}.$$

Example of a force field possessing a multi-valued potential. The force field

$$\mathbf{F}_P = -\frac{y}{x^2+y^2}\,\mathbf{i}_P + \frac{x}{x^2+y^2}\,\mathbf{j}_P, \qquad P \in \mathbb{E}^2$$

on the Euclidean plane $\mathbb{E}^2$ has circles about the origin as field lines (Fig. 12.11). Then $|\mathbf{F}_P| = 1/\sqrt{x^2+y^2}$. Let C be a smooth curve on the Riemann surface $\mathcal{R}$ with the initial point P_0 and the terminal point P. Then

$$\boxed{\int_C \mathbf{F}d\mathbf{x} = \arg(P) - \arg(P_0).} \tag{12.50}$$

We say that the function $\arg : \mathcal{R} \to \mathbb{R}$ is a multi-valued potential of the force field $\mathbf{F}$. The surjective map

$$\pi : \mathcal{R} \to \mathbb{E}^2 \setminus \{O\}$$

given by $\pi(x + \mathrm{i}y, m) := (x, y)$ is called a fiber bundle (or a covering space) over the punctured Euclidean plane $\mathbb{E}^2 \setminus \{O\}$.

Proof. It follows from $\Im\left(\frac{dz}{z}\right) = \Im\left(\frac{(x-\mathrm{i}y)(dx+\mathrm{i}dy)}{x^2+y^2}\right)$ that

$$\int_C \mathbf{F}d\mathbf{x} = \int_C \frac{xdy - ydx}{x^2+y^2} = \Im\left(\int_C \frac{dz}{z}\right).$$

This yields the claim (12.50). □

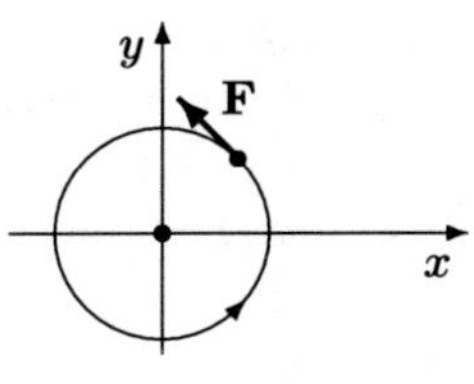

Fig. 12.11. Force field with circular field lines

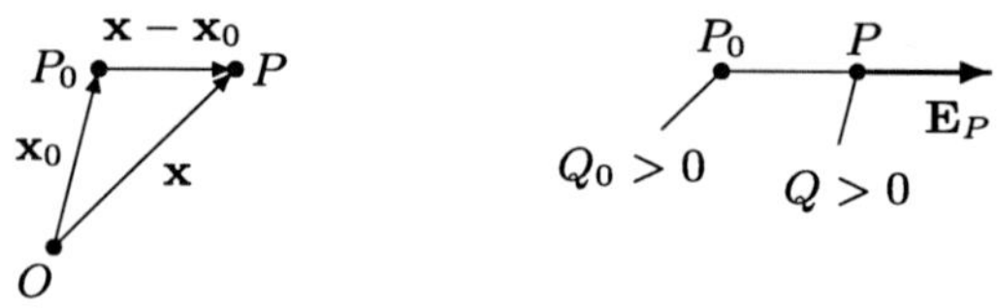

(a) (b) $\mathbf{F}_P = Q\mathbf{E}_P$

Fig. 12.12. Coulomb law

12.2.3 The Electrostatic Coulomb Force and the Dirac Delta Distribution

> Coulomb (1736–1806) noticed that the electrostatic force between two charges is similar to Newton's gravitational force, up to the sign.
>
> Folklore

Let P_0 and P be points of the Euclidean manifold $\mathbb{E}^3$. Consider the two position vectors $\mathbf{x} = \overrightarrow{OP}$ and $\mathbf{x}_0 = \overrightarrow{OP_0}$. Suppose that a particle of electric charge Q_0 is located at the point P_0. This charge induces the electric field vector

$$\mathbf{E}_P = \frac{Q_0(\mathbf{x} - \mathbf{x}_0)}{4\pi\varepsilon_0|\mathbf{x} - \mathbf{x}_0|^3} \tag{12.51}$$

at the point P (Fig. 12.12). Here, ε_0 denotes the electric field constant of a vacuum. Let another particle of charge Q be located at the point P. Then the electric field $\mathbf{E}$ exerts the force

$$\boxed{\mathbf{F}_P = Q\mathbf{E}_P} \tag{12.52}$$

on the particle of charge Q at the point P. If the charges Q_0 and Q have the same (resp. the opposite) sign, then the force $\mathbf{F}$ is repulsive (resp. attractive). This law is due to Coulomb. To simplify notation, we briefly write

$$\mathbf{E}(\mathbf{x}) = \frac{Q_0(\mathbf{x} - \mathbf{x}_0)}{4\pi\varepsilon_0|\mathbf{x} - \mathbf{x}_0|^3}$$

and $\mathbf{F}(\mathbf{x}) = Q\mathbf{E}(\mathbf{x})$. Introducing the so-called electric potential

$$U(\mathbf{x}) := \frac{Q_0}{4\pi\varepsilon_0|\mathbf{x} - \mathbf{x}_0|}, \tag{12.53}$$

we obtain

$$\mathbf{E} = -\,\mathbf{grad}\,U.$$

In fact, $\frac{\partial}{\partial x}(x^2+y^2+z^2)^{-1/2} = -x(x^2+y^2+z^2)^{-3/2}$, by using Cartesian coordinates. If the particle of charge Q moves from the point A to the point B, then we gain the energy

$$\int_A^B \mathbf{F}\,d\mathbf{x} = -\int_A^B Q\,\mathbf{grad}\,U\;d\mathbf{x} = Q(U(A) - U(B)).$$

The potential difference

$$V = U(B) - U(A) = -\int_A^B \mathbf{E}\,d\mathbf{x}$$

is called the voltage between the points B and A.

The Dirac distribution. Using Cartesian coordinates, we get the classical equation

$$\Delta U = 0 \quad \text{on } \mathbb{E}^3 \setminus \{P_0\}.$$

Using the language of distributions due to Laurent Schwartz (1915–2002), we obtain

$$\boxed{\varepsilon_0 \Delta U = Q_0 \delta_{P_0} \quad \text{on } \mathbb{E}^3} \tag{12.54}$$

where δ_{P_0} denotes the Dirac delta distribution at the point P_0. This equation describes the singularity of the electrostatic potential U at the point P_0 in precise terms. Explicitly, this means

$$\varepsilon_0 \int_{\mathbb{R}^3} U \Delta\varphi \, d^3x = Q_0 \delta_{P_0}(\varphi) = Q_0 \varphi(P_0)$$

for all smooth test functions $\varphi : \mathbb{R}^3 \to \mathbb{R}$ with compact support, that is, $\varphi \in \mathcal{D}(\mathbb{R}^3)$. This is thoroughly discussed in Sect. 10.4.8 of Vol. I. In particular, the proof of (12.54) is based on the Green's formula (12.34) on page 680. From (12.54) we get the Maxwell equation

$$\boxed{\varepsilon_0 \operatorname{div} \mathbf{E} = Q_0 \delta_{P_0} \quad \text{on } \mathbb{E}^3.} \tag{12.55}$$

Physicists write

$$\varepsilon_0 \operatorname{div} \mathbf{E}(\mathbf{x}) = Q_0 \delta(\mathbf{x} - \mathbf{x}_0).$$

In terms of physics, the Dirac function $\mathbf{x} \mapsto \delta(\mathbf{x} - \mathbf{x}_0)$ describes the charge density of a point charge Q_0 located at the point P_0. Formally, physicists use the following Dirac formulas:

- $\delta(\mathbf{x} - \mathbf{x}_0) = 0$ if $\mathbf{x} \neq \mathbf{x}_0$, and
- $\int_{\mathbb{R}^3} \delta(\mathbf{x} - \mathbf{x}_0)\varphi(\mathbf{x}) \, d^3x = \varphi(\mathbf{x}_0)$ for all continuous functions $\varphi : \mathbb{R}^3 \to \mathbb{R}$.

Dirac knew very well that there is no classical function which has these properties. But he emphasized that, from the mnemonic point of view, the formulas are extremely useful for passing from continuous structures to discrete ones in physics.

12.2.4 The Magic Green's Function and the Dirac Delta Distribution

> The Green's function is a crucial tool in both classical physics and modern physics (quantum field theory and solid state physics). In terms of physics, the Green's function encodes the propagation of physical information.[12]
>
> The Green's function allows us to describe general physical fields as the superposition of special physical fields generated by sources sharply located in space and time. In electrostatics, for example, the Green's function of a ball describes the electrostatic potential of a normalized point charge inside a metallic sphere (working as metallic conductor). In quantum field theory, the Green's function coincides with the Feynman propagator.
>
> Folklore

[12] Newton's grave in Westminster Abbey (London) is framed by five smaller gravestones with the names of Faraday (1791–1867), Green (1793–1841), Thomson (Lord Kelvin) (1824–1907), Maxwell (1831–1879), and Dirac (1902–1984).

In what follows, let us consider the Euclidean manifold $\mathbb{E}^3$. We will frequently identify the points P and P_0 of $\mathbb{E}^3$ with the position vectors $\mathbf{x} = \overrightarrow{OP}$ and $\mathbf{x}_0 = \overrightarrow{OP_0}$, respectively. For example, we will write $U(\mathbf{x})$ instead of $U(P)$.

The boundary-value problem for the Poisson equation. Let $\mathbb{B}^3$ denote the closed unit ball about the origin O with the boundary $\partial\mathbb{B}^3 = \mathbb{S}^2$. We are given the smooth functions $U_0 : \mathbb{S}^2 \to \mathbb{R}$ and $\varrho : \mathbb{B}^3 \to \mathbb{R}$. We are looking for a smooth function $U : \mathbb{B}^3 \to \mathbb{R}$ such that

$$\boxed{\varepsilon_0 \Delta U = \varrho \text{ on } \mathbb{B}^3, \qquad U = U_0 \text{ on } \mathbb{S}^2.} \tag{12.56}$$

In terms of physics, we are looking for the electrostatic potential U generated by the electric charge density ϱ and the values of the potential on the sphere $\mathbb{S}^2$. For example, if the sphere $\mathbb{S}^2$ is a metallic conductor, then $U_0 \equiv 0$. We want to show how this problem can be solved by using the idea of the Green's function. Let us start with a heuristic argument in the spirit of Green and Dirac.

Heuristic argument based on the Green integral formula and the Dirac delta function. Fix the interior point $P_0 \in \text{int}(\mathbb{B}^3)$. Suppose that the function U is a solution of the original boundary-value problem (12.56). Consider the Green formula

$$\varepsilon_0 \int_{\mathbb{B}^3} (U\Delta G - G\Delta U)\, d^3x = \varepsilon_0 \int_{\partial\mathbb{S}^2} \left(G\frac{\partial U}{\partial r} - U\frac{\partial G}{\partial r} \right) dS. \tag{12.57}$$

Assume that we know a function $\mathbf{x} \mapsto G(\mathbf{x}, \mathbf{x}_0)$ with the following two properties:

- $\varepsilon_0 \Delta_{\mathbf{x}} G(\mathbf{x}, \mathbf{x}_0) = \delta(\mathbf{x} - \mathbf{x}_0)$ on $\mathbb{B}^3$, and
- $G(\mathbf{x}, \mathbf{x}_0) = 0$ on $\mathbb{S}^2$.

In terms of physics, the Green's function G describes the electrostatic potential of a point charge ($Q_0 = 1$) located at the point P_0 inside the metallic unit sphere. Using $U(\mathbf{x}_0) = \int_{\mathbb{B}^3} \delta(\mathbf{x} - \mathbf{x}_0) U(\mathbf{x})\, d^3x$, it follows from (12.57) that

$$\boxed{U(\mathbf{x}_0) = \int_{\mathbb{B}^3} G(\mathbf{x}, \mathbf{x}_0)\varrho(\mathbf{x})\, d^3x - \varepsilon_0 \int_{\mathbb{S}^2} U_0(\mathbf{x}) \frac{\partial G(\mathbf{x}, \mathbf{x}_0)}{\partial |\mathbf{x}|}\, dS_{\mathbf{x}}.} \tag{12.58}$$

The conformal Kelvin transformation and the Green's function of the unit ball. In 1845, Thomson (later Lord Kelvin) used the transformation $P \mapsto P^*$ given by

$$\mathbf{x}^* := \frac{1}{|\mathbf{x}|} \cdot \frac{\mathbf{x}}{|\mathbf{x}|} \tag{12.59}$$

for all points $P \in \mathbb{E}^3 \setminus \{O\}$ in order to solve problems in potential theory for special bodies. With respect to a Cartesian coordinate system, we get

$$x^* = \frac{x}{x^2 + y^2 + z^2}, \qquad y^* = \frac{y}{x^2 + y^2 + z^2}, \qquad z^* = \frac{z}{x^2 + y^2 + z^2}.$$

This transformation is called inversion with respect to the unit sphere (or Kelvin transformation). The Kelvin transformation sends solutions of the Laplace equation $\Delta V = 0$ to solutions of the same equation, and it is conformal (i.e., it preserves the intersection angle of curves by changing the orientation).[13] Again let us fix the point $P_0 \in \text{int}(\mathbb{B}^3)$. Then the Green's function of the unit ball $\mathbb{B}^3$ is given by

[13] The conformal property of this transformation was known in ancient times.

$$G(\mathbf{x}, \mathbf{x}_0) := \frac{Q_0}{4\pi\varepsilon_0|\mathbf{x} - \mathbf{x}_0|} + \frac{Q_0^*}{4\pi\varepsilon_0|\mathbf{x} - \mathbf{x}_0^*|} \tag{12.60}$$

for all points $P \in \mathbb{B}^3 \setminus \{P_0\}$. Here, $Q_0 := 1$ and $Q_0^* := -Q_0|\mathbf{x}_0^*|/|\mathbf{x}_0|$. An elementary geometric argument shows that $G(\mathbf{x}, \mathbf{x}_0) = 0$ for all points $P \in \mathbb{S}^2$ (see Sect. 10.4.8 of Vol. I). Moreover, G is the superposition of two electric charges ($Q_0 = 1$ and Q_0^*) located inside the ball and outside the ball, respectively. Therefore, we have

$$\Delta_{\mathbf{x}} G(\mathbf{x}, \mathbf{x}_0) = 0 \qquad \text{for all} \quad P \in \mathbb{B}^3 \setminus \{P_0\}.$$

Using the language of distributions, it follows from (12.54) that

$$\varepsilon_0 \Delta G = \delta_{P_0} \quad \text{on } \mathbb{B}^3. \tag{12.61}$$

Explicitly, this means that

$$\varepsilon_0 \int_{\mathbb{R}^3} G \Delta\varphi \, d^3x = Q_0 \delta_{P_0}(\varphi) = Q_0 \varphi(P_0)$$

for all smooth test functions $\varphi : \mathbb{R}^3 \to \mathbb{R}$ whose support is contained in the ball $\mathbb{B}^3$.

The rigorous existence-and uniqueness theorem. The following hold.

Proposition 12.2 *The boundary-value problem* (12.56) *has a unique solution which is given by the integral formula* (12.58) *together with the Green's function* (12.60).

This is a special case of the modern functional analytic theory of elliptic partial differential equations. We refer to the detailed discussion in Sect. 10. 4 of Vol. I.

The Faraday cage. For example, if the unit sphere $\mathbb{S}^2$ is a metallic conductor and there is no electric charge in the ball $\mathbb{B}^3$, then $U_0 \equiv 0$ and $\varrho \equiv 0$. By Prop. 12.2, the unique solution U of (12.56) vanishes on the ball $\mathbb{B}^3$. Hence the electrostatic field $\mathbf{E} = -\,\mathbf{grad}\, U$ vanishes on $\mathbb{B}^3$, too. The ball inside the metallic unit sphere $\mathbb{S}^2$ is called the Faraday cage.

The principle of minimal electrostatic energy (Dirichlet principle). The electrostatic energy of a smooth electric field $\mathbf{E}$ on the ball $\mathbb{B}^3$ is given by the integral

$$\frac{\varepsilon_0}{2} \int_{\mathbb{B}^3} \mathbf{E}^2 d^3x = \frac{\varepsilon_0}{2} \int_{\mathbb{B}^3} |\,\mathbf{grad}\, U|^2 d^3x = \frac{\varepsilon_0}{2} \int_{\mathbb{B}^3} (U_x^2 + U_y^2 + U_z^2)\, d^3x.$$

Let $C^\infty(\mathbb{B}^3)$ denote the linear space of all smooth functions $U : \mathbb{B}^3 \to \mathbb{R}$.

Theorem 12.3 *We are given the functions* $\varrho \in C^\infty(\mathbb{B}^3)$ *and* $U_0 \in C^\infty(\mathbb{S}^2)$. *Then the variational problem*

$$\int_{\mathbb{B}^3} \left(\frac{\varepsilon_0}{2} |\,\mathbf{grad}\, U|^2 - \varrho U \right) d^3x = \min!, \quad U \in C^\infty(\mathbb{B}^3), \quad U = U_0 \text{ on } \mathbb{S}^2 \tag{12.62}$$

of minimal electrostatic energy has a unique solution U *which coincides with the unique solution of the boundary-value problem* (12.56).

Historical remarks. Problems (12.56), (12.62) and their generalizations played a crucial role in the history of mathematics. In his lectures, Dirichlet told his students that a solution of the variational problem (12.62) is a solution of the boundary-value problem (12.56), which is frequently called the Dirichlet problem.

But Dirichlet did not prove the existence of a solution of the variational problem. In the 1840s, it seems that Gauss, Thompson (Lord Kelvin) and Dirichlet took the existence of a solution of (12.62) for granted motivated by the physical interpretation of the minimum problem. Riemann attended Dirichlet's lectures in Berlin in the late 1840s. In his seminal 1857 paper on Abelian integrals, Riemann motivated the existence of analytic functions on Riemann surfaces by using a physical argument which he called the 'Dirichlet principle'. In Riemann's model, the sheets of the Riemann surface are thin metallic foils, and the branching points of the Riemann surface correspond to galvanic elements. In fact, Riemann generalized the variational problem (12.62) and its physical interpretation in electrostatics to Riemann surfaces. This way, Riemann created his beautiful theory of algebraic functions and their Abelian integrals on compact Riemann surfaces.

In 1870, Weierstrass published a counter example which showed that there are variational problems which do not possess any solution. This way, the rigorous justification of the Dirichlet principle became a famous open problem. This problem was solved by Hilbert in 1899. He proved Theorem 12.3 above. In 1940, Weyl showed that the Dirichlet problem and related variational problems can be solved by using

- the method of orthogonal projection in infinite-dimensional Hilbert spaces (the generalized Pythagorean theorem)
- together with a regularization argument based on the so-called Weyl lemma (see Sect. 11.3.2 of Vol. I).

A detailed discussion can be found in Sect. 10.4 of Vol. I. Let us only mention that the modern functional analytic proof of Theorem 12.3 above consists of the following two steps:

- Step 1: Generalized solution of a quadratic variational problem on a Sobolev space: We replace the classical variational problem (12.62) by the generalized problem

$$\int_{\mathbb{B}^3} \left(\frac{\varepsilon_0}{2}|\,\mathbf{grad}\,U|^2 - \varrho U\right)\, d^3x = \min!, \qquad U \in W_2^1(\mathbb{B}^3) \tag{12.63}$$

  with the boundary condition $BU = U_0$ on $\mathbb{S}^2$. More precisely, we equip the linear space $C^\infty(\mathbb{B}^3)$ with the inner product

$$\langle U|V\rangle := \int_{\mathbb{B}^3} (UV + U_xV_x + U_yV_y + U_zV_z)\, d^3x.$$

  The Sobolev space $W_2^1(\mathbb{B}^3)$ is the smallest real Hilbert space which contains the pre-Hilbert space $C^\infty(\mathbb{B}^3)$. In addition, the classical boundary operator

$$B : C^\infty(\mathbb{B}^3) \to C^\infty(\mathbb{S}^2),$$

  which assigns to the smooth function $U \in C^\infty(\mathbb{B}^3)$ the corresponding boundary function U_0 on $\mathbb{S}^2$, can be uniquely extended to the linear continuous operator

$$B : W_2^1(\mathbb{B}^3) \to L_2(\mathbb{B}^3).$$

  Using the main theorem on quadratic variational problems in Hilbert space (method of orthogonal projection), we obtain the unique solution U of (12.63).
- Step 2: Regularization: We show that the solution U of the generalized problem (12.63) is indeed a smooth function, and hence it is a solution of the classical problem (12.62).

An introduction to the modern functional analytic theory of the Dirichlet principle can be found in:

E. Zeidler, Applied Functional Analysis, Vol 1: Applications to Mathematical Physics, Springer, New York, 1997.

For the general functional analytic theory of linear partial differential equations, we recommend:

E. Zeidler, Nonlinear Functional Analysis and its Applications, Vol. IIA: Linear Monotone Operators, Springer, New York, 1997.

L. Evans, Partial Differential Equations, Amer. Math. Soc., Providence, Rhode Island, 1998.

J. Jost, Partial Differential Equations, Springer, New York, 2002.

The history of the Dirichlet principle is described in:

S. Hildebrandt, Remarks on the Dirichlet principle (in German). In: H. Weyl, The Concept of a Riemann Surface (in German), Teubner, Leipzig, 1913. New edition with commentaries supervised by R. Remmert, Teubner, Leipzig, 1997, pp. 197–217.

A. Monna, Dirichlet's Principle, a Mathematical Comedy of Errors and its Influence on the Development of Analysis, Oosthoek, Utrecht, 1975.

The classical Dirichlet problem demonstrates that it is possible to justify physical intuition by rigorous mathematics. However, it needs time to find the right mathematical tools.

We hope that the same will happen with quantum field theory.

12.2.5 Conservation of Heat Energy – the Paradigm of Conservation Laws in Physics

> The flow of heat energy investigated by Fourier (1768–1830) is the paradigm for conservation laws in physics (e.g., conservation of heat energy, mass, electric charge, or electromagnetic energy).
>
> Folklore

The basic equation reads as follows:

$$\boxed{\dot{\mathcal{E}} + \operatorname{div} \mathbf{J} = \mathcal{P}.} \tag{12.64}$$

This so-called continuity equation tells us that the gain of heat energy of the domain $\mathcal{M}$ is compensated by both

- the flow of heat energy into the domain $\mathcal{M}$ described by the current density vector $\mathbf{J}$, and
- the heat energy production in the domain $\mathcal{M}$ described by the production function $\mathcal{P}$.

Here, we use the following notation:

- $\mathcal{E}$ (heat energy density);
- $E(t) := \int_{\mathcal{M}} \mathcal{E}(P,t)\, d^3x$ (heat energy located in the domain $\mathcal{M}$ at time t);[14]
- $\mathbf{n}$ (exterior normal unit vector of the boundary surface $\partial\mathcal{M}$),
- $-\mathbf{n}$ (interior normal unit vector of $\partial\mathcal{M}$),
- $\mathbf{J}$ (current density vector of heat energy flow),

[14] We assume that $\mathcal{M}$ is a 3-dimensional compact submanifold $\mathcal{M}$ of the Euclidean manifold $\mathbb{E}^3$ with boundary ∂M (see Fig. 12.6(c) on page 677).

- $-\int_{t_0}^{t_1} dt \int_{\partial\mathcal{M}} \mathbf{J}(P,t)\mathbf{n}(P)\, dS$ (flow of heat energy into the domain $\mathcal{M}$ during the time interval $[t_0, t_1]$),
- $\int_{t_0}^{t_1} dt \int_{\mathcal{M}} \mathcal{P}(P,t)\, dx^3$ (production of heat energy in the domain $\mathcal{M}$ during the time interval $[t_0, t_1]$).

The quantities have the following physical dimensions:

- $\mathcal{E}$: energy/volume,
- $|\mathbf{J}|$: energy/(area· time) = (energy/volume)·velocity,
- $\mathcal{P}$: energy/(volume · time).

By the Gauss–Ostrogradsky integral theorem (12.32), it follows from (12.64) that

$$\frac{d}{dt}\int_{\mathcal{M}} \mathcal{E}(P,t)\, d^3x = \int_{\mathcal{M}} \mathcal{P}(P,t)\, d^3x - \int_{\partial\mathcal{M}} \mathbf{J}(P,t)\mathbf{n}(P)\, dS.$$

This implies

$$\boxed{\dot{E}(t) = \int_{\mathcal{M}} \mathcal{P}(P,t)\, d^3x - \int_{\partial\mathcal{M}} \mathbf{J}(P,t)\mathbf{n}(P)\, dS.} \tag{12.65}$$

Integration over the time interval $[t_0, t_1]$ yields

$$E(t_1) - E(t_0) = \int_{t_0}^{t_1} dt \int_{\mathcal{M}} \mathcal{P}(P,t)\, d^3x - \int_{t_0}^{t_1} dt \int_{\partial\mathcal{M}} \mathbf{J}(P,t)\mathbf{n}(P)\, dS.$$

In particular, if the vector $\mathbf{J}$ is constant on the Euclidean manifold $\mathbb{E}^3$, then the flow of heat energy through a piece of a plane during the time interval $[t_0, t_1]$ is equal to

$$(t_1 - t_0) \cdot |\mathbf{J}| \cdot \Delta S.$$

Here, we assume that the plane is perpendicular to the vector $\mathbf{J}$ whose direction coincides with the flow direction, and ΔS is the surface area of the piece of the plane. This motivates the name 'current density' vector for $\mathbf{J}$.

The heat conduction equation. Let T denote the absolute temperature. Following Fourier, we assume that the material under consideration possesses the following properties:

- $\Delta E = \gamma(T)\mu\Delta V \cdot \Delta T$ (the change ΔV of volume and the change ΔT of temperature cause the change ΔE of heat energy),
- $\mathbf{J} = -\kappa(T)\,\mathbf{grad}\,T$ (Fourier's law of heat conduction),
- μ (mass density), γ (specific heat), κ (heat conductivity).

By (12.64),

$$\gamma(T)\mu\dot{T} - \operatorname{div}(\kappa(T)\,\mathbf{grad}\,T) = \mathcal{P}.$$

Assume that, for a fixed temperature interval $[T_0, T_1]$, the specific heat γ and the heat conductivity κ are constant (i.e., they are independent of the temperature T). Then, we get

$$T_t + \kappa a \Delta T = a\mathcal{P} \tag{12.66}$$

with the positive material constant $a := 1/\gamma\mu$. This is the classical heat conduction equation.

Flow of electrically charged particles. Let $\mathbf{v}$ be the velocity vector field of electrically charged particles, and let ϱ be the electric charge density. Set

$$\mathbf{J} := \varrho\mathbf{v}.$$

Then the continuity equation

$$\boxed{\dot{\varrho} + \operatorname{div} \mathbf{J} = \mathcal{P}} \tag{12.67}$$

describes the conservation of electric charge. The term $\mathcal{P}$ corresponds to the change of electric charges (e.g., chemical reactions produce new particles). Now we have equation (12.65) where $E(t)$ denotes the electric charge located in the domain $\mathcal{M}$ at time t. In a vacuum, we have $\mathcal{P} \equiv 0$.

Flow of fluid particles with mass conservation. The equation (12.67) with $\mathbf{J} := \varrho\mathbf{v}$ and $\mathcal{P} \equiv 0$ describes the flow of fluid particles with mass density ϱ governed by the velocity vector field $\mathbf{v}$. Equation (12.67) implies (12.65) where $E(t)$ denotes the mass located in the domain $\mathcal{M}$ at time t. Since $\mathcal{P} \equiv 0$, the only change of mass in the domain $\mathcal{M}$ is caused by flow of mass through the boundary $\partial\mathcal{M}$. The dynamics of the transport of heat energy, charge, and mass will be studied in Sect. 12.8.4 on page 735 in the setting of the transport theorem.

Conservation of the particle number in diffusion processes. Let ϱ denote the particle density of a diffusion process. Then

$$\mathbf{J} = -D\,\mathbf{grad}\,\varrho$$

where D is called the diffusion constant. This empirical diffusion law was formulated by Fick in the second half of the 19th century. The continuity equation $\dot{\varrho} + \operatorname{div} \mathbf{J} = 0$ yields the classical diffusion equation

$$\dot{\varrho}_t + D\Delta\varrho = 0.$$

The conservation of particle number is described by equation (12.65) with $\mathcal{P} \equiv 0$ where $E(t)$ denotes the number of particles in the domain $\mathcal{M}$ at time t.

Conservation of probability in quantum mechanics. The Schrödinger equation

$$\mathrm{i}\hbar\dot{\psi} = \left(\frac{\mathbf{P}^2}{2m} + U\right)\psi \tag{12.68}$$

with the momentum operator $\mathbf{P} := -\mathrm{i}\hbar\boldsymbol{\partial}$ and the real potential U describes the motion of a quantum particle of mass m on the Euclidean manifold $\mathbb{E}^3$ under the influence of the force field $\mathbf{F} = -\mathbf{grad}\,U$. Let $\psi = \psi(P,t)$ be a smooth solution of (12.68) which satisfies the normalization condition $\int_{\mathbb{E}^3} \varrho(P,0)\,dx^3 = 1$. Here, we set

$$\varrho := \psi\psi^\dagger, \qquad \mathbf{J} := \Re\left(\psi^\dagger\mathbf{V}\psi\right), \quad \mathbf{P} := m\mathbf{V}.$$

Proposition 12.4 $\dot{\varrho}(P,t) + \operatorname{div}\mathbf{J}(P,t) = 0$ *on* $\mathbb{E}^3$ *for all times* $t \in \mathbb{R}$.

The proof will be given in Problem 12.9 on page 802. This allows the following physical interpretation. The integral

$$E(t) := \int_{\mathcal{M}} \varrho(P,t)\,d^3x$$

represents the probability for finding the quantum particle in the compact submanifold $\mathcal{M}$ of $\mathbb{E}^3$. Equation (12.65) with $\mathcal{P} \equiv 0$ tells us that probability is preserved.

Conservation of energy, momentum, and angular momentum (spin) of the classical electromagnetic field. This will be studied in Sect. 19.6.3.

12.2.6 The Classical Predecessors of the Yang–Mills Equations in Gauge Theory (Fluid Dynamics and Electrodynamics)

The Yang–Mills equations read as

$$\boxed{d^*\omega = -f, \qquad d\omega = \Gamma} \tag{12.69}$$

where we are looking for the differential form ω. We will show later on that the Standard Model in particle physics is closely related to this equation. At this point, let us only mention that the classical problem

$$\operatorname{div} \mathbf{v} = f, \qquad \mathbf{curl}\, \mathbf{v} = \mathbf{g} \tag{12.70}$$

is a special case of (12.69) with

- $\mathbf{v} := a\mathbf{i} + b\mathbf{j} + c\mathbf{k}$ and $\omega := adx + bdy + cdz$,
- $\mathbf{g} := A\mathbf{i} + B\mathbf{j} + C\mathbf{k}$, and $\Gamma := Ady \wedge dz + Bdz \wedge dx + Cdx \wedge dy$ (see page 774).

Here, we are given the functions f and $\mathbf{g}$ (i.e., sources and vorticities), and we are looking for the velocity vector field $\mathbf{v}$ (see the main theorem of vector analysis on page 766).

Furthermore, the Maxwell equations in electrodynamics are a special case of (12.69); they read as

$$\boxed{d^*\mathcal{F} = -\mu_0 \mathcal{J}, \qquad d\mathcal{F} = 0,}$$

where the differential 2-form $\mathcal{F}$ on the 4-dimensional space-time manifold describes the electromagnetic field, and the differential 1-form $\mathcal{J}$ corresponds to the electric charges and electric currents (see page 962).

12.2.7 Thermodynamics and the Pfaff Problem

Following Gibbs (1839–1903), the basic equation of a gas in thermodynamical equilibrium reads as follows:

$$\boxed{TdS = dE + PdV.} \tag{12.71}$$

Here, we use the following notation:

- V volume, P pressure, E inner energy,
- T absolute temperature, S entropy.

Equation (12.71) is the prototype of a so-called Pfaff problem first investigated by Pfaff in 1815.[15]

(i) One-dimensional solution manifold: By definition, the smooth functions
 - $V = V(t)$, $P = P(t)$, $E = E(t)$, $T = T(t)$, $S = S(t)$ with $t \in [t_0, t_1]$

 are a solution of (12.71) iff

$$T(t)\dot{S}(t)dt = \dot{E}(t)dt + P(t)\dot{V}(t)dt, \qquad t_0 \le t \le t_1,$$

 that is,

$$T(t)\dot{S}(t) = \dot{E}(t) + P(t)\dot{V}(t), \qquad t_0 \le t \le t_1.$$

[15] Pfaff (1765–1824) was the academic teacher and promoter of Gauss (1777–1855).

In terms of physics, this is a quasi-stationary process for the gas. The process proceeds very slowly such that the gas is in thermodynamical equilibrium at each point of time t. The integral

$$\int_{t_0}^{t_1} (\dot{E}(t) + P(t)\dot{V}(t))\,dt$$

is the amount of heat energy produced during the time interval $[t_0, t_1]$.[16]

(ii) Two-dimensional solution manifold: Let $\mathcal{U}$ be an open subset of $\mathbb{R}^2$. The functions
- $V = V(u,v),\ P = P(u,v),\ E = E(u,v),\ T = T(u,v),\ S = S(u,v)$ with the parameters $(u,v) \in \mathcal{U}$

are a solution of (12.71) iff

$$TS_u du + TS_v dv = E_u du + E_v dv + PV_u du + PV_v dv \qquad \text{on } \mathcal{U},$$

that is,

$$TS_u = E_u + PV_u, \qquad TS_v = E_v + PV_v \quad \text{on } \mathcal{U}.$$

Examples. For an ideal one-atomic gas consisting of N atoms of mass m, we have:
- $P = \frac{NkT}{V}$ (state equation),
- $E = \frac{3}{2}NkT$ (equipartition of energy),
- $S = kN\left[\frac{5}{2} + \ln \frac{V}{N}\left(\frac{2\pi mkT}{h^2}\right)^{3/2}\right]$ (entropy),
- $F = E - TS$ (free energy).

For the photon gas in the universe, we get:
- $E = \int_0^\infty \frac{8\pi hcV\,d\lambda}{e^{hc/kT\lambda}-1} = aT^4V$ (total energy) with $a := 8\pi^5k^4/15c^3h^3$,[17]
- $P = E/3V$ (light pressure),
- $F = -E/3$ (free energy),
- $S = (E-F)/T$ (entropy).

These expressions satisfy the Gibbs equation (12.71); they follow from statistical physics. The photon gas was very hot shortly after the Big Bang. Nowadays it is very cold ($T = 3\text{K}$) by the expansion of the universe.[18] The NASA experiment WMAP (Wilkinson Microwave Anisotropy Probe) measured the anisotropic structure of this background radiation. The knowledge of this huge amount of data allows us to compute the qualitative structure of the universe 370 000 years after the Big Bang. Moreover, the WMAP experiment, too, tells us that the Big Bang happened $13.7\cdot 10^9$ years ago. In 2009, the European Space Agency (ESA) launched the Planck satellite which will provide us with further information on the structure of the early cosmos.

Second law of thermodynamics. We will show in Sect. 12.11.4 that, under reasonable assumptions, the differential form $dE + P(E,V)dV$ can be written as

$$dE + P(E,V)dV = T(E,V)dS.$$

[16] Basic material on equilibrium thermodynamics can be found in Sect. 5.7.4 of Vol. II (first and second law of thermodynamics, thermodynamic potentials and the Legendre transformation, Legendre manifolds).

[17] We use the following notation: c velocity of light in a vacuum, k Boltzmann constant, h Planck quantum of action, λ wavelength of the photon.

[18] Based on statistical physics, this is thoroughly studied in E. Zeidler, Nonlinear Functional Analysis, Vol. IV, Springer, New York, 1997.

This is a special case of the Pfaff normal form problem. The temperature function $T = T(E, V)$ is called an Euler multiplicator. In terms of physics, this result guarantees the existence of entropy $S = S(E, V)$.

12.2.8 Classical Mechanics and Symplectic Geometry

The Hamiltonian approach to classical mechanics is based on symplectic geometry. In this context, Cartan's calculus of differential forms is the natural language. This is studied in Sect. 6.8 of Vol. II.

12.2.9 The Universality of Differential Forms

As a prototype, consider the first-order partial differential equation

$$F(x, y, u, u_x, u_y) = 0. \tag{12.72}$$

Setting $q := u_x, p := u_y$, equation (12.72) is equivalent to the system

$$\boxed{F(x, y, u, q, p) = 0, \qquad du = qdx + pdy}$$

of differential forms of degree $k = 0$ and $k = 1$. Similarly, by introducing new variables, every system of partial differential equations can be written as a system of differential forms.

12.2.10 Cartan's Covariant Differential and the Four Fundamental Interactions in Nature

> The main trick of gauge theory is to replace the Cartan exterior differential $d\omega$ by the covariant differential $D\omega$.
>
> Folklore

As we will discuss later on in great detail, gauge theories are based on the fundamental Cartan structural equation

$$\boxed{\mathsf{F} = D\mathsf{A}} \tag{12.73}$$

together with the corresponding integrability condition

$$D\mathsf{F} = 0. \tag{12.74}$$

In terms of mathematics, the equation (12.73) describes the *curvature* F. In particular, (12.73) generalizes Gauss's theorema egregium and the differential relation between the Riemann curvature tensor and the Christoffel symbols in Riemannian and pseudo-Riemannian geometry. Moreover, (12.73) generalizes Cartan's structural equation for the Maurer–Cartan 1-form of Lie groups, and it governs Cartan's method of moving frames, which is basic for modern differential geometry.

The covariant differential $D\mathsf{A}$ generalizes the Cartan exterior differential $d\mathsf{A}$. Furthermore, the integrability condition (12.74) generalizes the Poincaré cohomology rule $d(d\mathsf{A}) = 0$. In Riemannian and pseudo-Riemannian geometry, (12.74) generalizes the Bianchi identities for the Riemann curvature tensor.

> *Roughly speaking, the covariant differential* $D\mathsf{A}$ *is a 'deformation' of the Cartan exterior differential* $d\mathsf{A}$.

By deformation, we understand perturbation of $d\mathsf{A}$ which is not necessarily small.

In terms of physics, the equation (12.73) describes the *four fundamental forces* F in nature (electromagnetic, strong, weak, and gravitative interaction). For example, in electromagnetism, F corresponds to the electromagnetic field, and A corresponds to the 4-potential of the electromagnetic field (see Chap. 13).

Observe that the Cartan exterior differential $d\omega$ is defined on every finite-dimensional manifold. In contrast to this, the covariant differential $D\omega$ describes an additional structure which can be introduced on principal bundles and vector bundles.

This additional structure is called a connection.

In terms of geometry, a connection represents the local parallel transport of geometric objects (e.g., tangent vectors or tensors); this implies the existence of the covariant directional derivative of fields of geometric objects (e.g., tangent vector fields or tensor fields). In terms of physics, a connection describes the local parallel transport of physical information. This will be studied in Chaps. 15ff.

12.3 Cartan's Algebra of Alternating Differential Forms

The Cartan calculus is governed by the wedge product for antisymmetric multilinear functionals. The Cartan differential $d\omega$ respects the wedge product via the supersymmetric Leibniz rule.[19]

12.3.1 The Geometric Approach

Differentials are linear functionals on the tangent space.
Folklore

In what follows, we will use geometric objects like tangent vectors, cotangent vectors, and k-linear antisymmetric functionals (also called k-forms). In order to describe these quantities in terms of Cartesian coordinates, choose a fixed right-handed Cartesian (x, y, z)-coordinate system with the right-handed orthonormal basis $\mathbf{i}, \mathbf{j}, \mathbf{k}$ at the origin O (Fig. 9.1 on page 558). Since we use geometric objects, the coordinate formulas below do not depend on the choice of the Cartesian coordinate system.

(i) Local objects at the point P of the Euclidean manifold $\mathbb{E}^3$:
- $T_P\mathbb{E}^3$ denotes the tangent space of $\mathbb{E}^3$ at the point P (with the Cartesian coordinates (x, y, z)). This space consists of all velocity vectors $\mathbf{v}_P$ at the point P. The 3-dimensional real linear Hilbert space $T_P\mathbb{E}^3$ has the right-handed orthonormal basis $\mathbf{i}_p, \mathbf{j}_P, \mathbf{k}_P$ which is obtained from $\mathbf{i}, \mathbf{j}, \mathbf{k}$ by parallel transport (Fig. 9.1 on page 558). We have

$$\mathbf{v}_P = a\mathbf{i}_P + b\mathbf{j}_P + c\mathbf{k}_P.$$

The real numbers a, b, c are called the components of the velocity vector $\mathbf{v}_P$ at the point P. Thus, $\dim T_P\mathbb{E}^3 = 3$.

[19] In classical algebra, the supersymmetric Leibniz rule is called the graded Leibniz rule.

- $T_P^*\mathbb{E}^3$ denotes the cotangent space of $\mathbb{E}^3$ at the point P. The elements of the real linear space $T_P^*\mathbb{E}^3$ are linear functionals of the form

$$\omega_P : T_P\mathbb{E}^3 \to \mathbb{R}$$

which are also called covectors (or 1-forms) at the point P. Defining

$$dx_P(\mathbf{v}_P) := a, \quad dy_P(\mathbf{v}_P) := b, \quad dz_P(\mathbf{v}_\mathbf{P}) := c$$

for all velocity vectors $\mathbf{v}_P \in T_P\mathbb{E}^3$ at the point P, we get a basis dx_P, dy_P, dz_P of $T_P^*\mathbb{E}^3$. That is, every element ω_P of $T_P^*\mathbb{E}^3$ can be written as

$$\omega_P = U\ dx_P + V\ dy_P + W\ dz_P.$$

The uniquely determined real numbers U, V, W are called the components of the covector ω_P at the point P. Thus, $\dim T_P^*\mathbb{E}^3 = 3$.
The wedge product $\omega_P \wedge \mu_P$ of two covectors ω_P and μ_P at the point P is defined by

$$(\omega_P \wedge \mu_P)(\mathbf{v}, \mathbf{w}) := \omega_P(\mathbf{v})\mu_P(\mathbf{w}) - \omega_P(\mathbf{w})\mu_P(\mathbf{v})$$

for all vectors $\mathbf{v}, \mathbf{w} \in T_P\mathbb{E}^3$.[20] We have

$$\boxed{\omega_P \wedge \mu_P = -\mu_P \wedge \omega_P}$$

for all $\mu_P, \omega_P \in T_P^*\mathbb{E}^3$. In particular, $dx_P \wedge dy_P = -dy_P \wedge dx_P$, and

$$dy_P \wedge dz_P = -dz_P \wedge dy_P, \qquad dz_P \wedge dx_P = -dx_P \wedge dz_P.$$

- To streamline the terminology, we write $\bigwedge^1(T_P^*\mathbb{E}^3)$ instead of $T_P^*\mathbb{E}^3$, and we set $\bigwedge^0(T_P^*\mathbb{E}^3) := \mathbb{R}$.
- $\bigwedge^2(T_P^*\mathbb{E}^3)$ denotes the real linear space of all bilinear antisymmetric functionals

$$\gamma_P : T_P\mathbb{E}^3 \times T_P\mathbb{E}^3 \to \mathbb{R}$$

which are also called 2-forms at the point P. Every element γ_P of $\bigwedge^2(T_P^*\mathbb{E}^3)$ can be written as

$$\gamma_P = A\ dy_P \wedge dz_P + B\ dz_P \wedge dx_P + C\ dx_P \wedge dy_P.$$

The uniquely determined real numbers A, B, C are called the components of the 2-form γ_P at the point P. Therefore, $\dim \bigwedge^2(T_P^*\mathbb{E}^3) = 3$.
If $\omega_P, \mu_P \in T_P^*\mathbb{E}^3$, then the wedge product $\omega_P \wedge \mu_P$ is contained in the linear space $\bigwedge^2(T_P^*\mathbb{E}^3)$.
- $\bigwedge^3(T_P^*\mathbb{E}^3)$ denotes the real linear space of all 3-linear antisymmetric functionals

$$\varrho_P : T_P\mathbb{E}^3 \times T_P\mathbb{E}^3 \times T_P\mathbb{E}^3 \to \mathbb{R}$$

which are also called 3-forms at the point P. Every element ϱ_P of $\bigwedge^3(T_P^*\mathbb{E}^3)$ can be written as

$$\varrho_P = D dx_P \wedge dy_P \wedge dz_P.$$

The uniquely determined real number D is called the component of the 3-form ϱ_P at the point P. Thus, $\dim \bigwedge^3(T_P^*\mathbb{E}^3) = 1$.

[20] To simplify notation, we write $\mathbf{v}$ instead of $\mathbf{v}_P$.

- $\bigwedge^k(T_P^*\mathbb{E}^3) := \{0\}$ if $k = 4, 5, \ldots$ This definition is motivated by the fact that the k-linear antisymmetric functionals $\mu : T_P\mathbb{E}^3 \times \cdots \times T_P\mathbb{E}^3 \to \mathbb{R}$ are trivial if $k > 3$.

(ii) The wedge product and the local Grassmann algebra $\bigwedge(T_P^*\mathbb{E}^3)$: Parallel to Sect. 2.5.2, we define the direct sum

$$\bigwedge(T_P^*\mathbb{E}^3) := \bigoplus_{k=0}^{\infty} \bigwedge^k(T_P^*\mathbb{E}^3).$$

The elements of $\bigwedge(T_P^*\mathbb{E}^3)$ are given by the sums

$$\alpha_P + \omega_P + \gamma_P + \varrho_P$$

where α_P is a real number. We have $\dim \bigwedge(T_P^*\mathbb{E}^3) = 1 + 3 + 3 + 1 = 8$. With respect to the wedge product $\mu \wedge \nu$, the real linear space $\bigwedge(T_P^*\mathbb{E}^3)$ becomes a real algebra. If $\mu \in \bigwedge^k(T_P^*\mathbb{E}^3)$ and $\nu \in \bigwedge^m(T_P^*\mathbb{E}^3)$ with $k, m = 0, 1, 2, 3$, then

$$\boxed{\mu_P \wedge \nu_P = (-1)^{km}\nu_P \wedge \mu_P.} \tag{12.75}$$

This is called the graded product rule or the supersymmetric product rule (or supercommutativity).

(iii) Global smooth objects on the manifold $\mathbb{E}^3$:

- $\mathrm{Vect}(\mathbb{E}^3)$ denotes the set of all smooth velocity vector fields $P \mapsto \mathbf{v}_P$ on $\mathbb{E}^3$. That is, we have

$$\mathbf{v}_P = a(x, y, y)\mathbf{i}_P + b(x, y, z)\mathbf{j}_P + c(x, y, z)\mathbf{k}_P,$$

and the components $a, b, c : \mathbb{R}^3 \to \mathbb{R}$ are smooth functions.
- $\Lambda^0(\mathbb{E}^3)$ denotes the set of all smooth functions $f : \mathbb{E}^3 \to \mathbb{R}$.
- $\Lambda^1(\mathbb{E}^3)$ denotes the set of all smooth covector fields $P \mapsto \omega_P$ on $\mathbb{E}^3$. That is,

$$\omega_P = U(x, y, z)\, dx_P + V(x, y, z)\, dy_P + W(x, y, z)\, dz_P,$$

and the coefficient functions $U, V, W : \mathbb{R}^3 \to \mathbb{R}$ are smooth.
- $\Lambda^2(\mathbb{E}^3)$ denotes the set of all smooth differential 2-forms $P \mapsto \gamma_P$ on $\mathbb{E}^3$. That is,

$$\gamma_P = A(x, y, z)\, dy_P \wedge dz_P + B(x, y, z)\, dz_P \wedge dx_P + C(x, y, z)\, dx_P \wedge dy_P,$$

and the coefficient functions $A, B, C : \mathbb{R}^3 \to \mathbb{R}$ are smooth.
- $\Lambda^3(\mathbb{E}^3)$ denotes the set of all smooth differential 3-forms $P \mapsto \varrho_P$ on $\mathbb{E}^3$. That is,

$$\varrho_P = D(x, y, z)\, dx_P \wedge dy_P \wedge dz_P,$$

and the coefficient function $D : \mathbb{R}^3 \to \mathbb{R}$ is smooth.
- $\Lambda^k(\mathbb{E}^3)$ denotes the set of all smooth k-forms on the Euclidean manifold $\mathbb{E}^3$. If $k = 4, 5, \ldots$, then the elements of $\Lambda^k(\mathbb{E}^3)$ are the trivial functions $P \mapsto 0$.

Note that $\mathrm{Vect}(\mathbb{E}^3)$ and $\Lambda^k(\mathbb{E}^3)$, $k = 0, 1, 2, 3$ are infinite-dimensional real linear spaces. Let $\mu \in \Lambda^k(\mathbb{E}^3)$ and $\nu \in \Lambda^m(\mathbb{E}^3)$. Naturally enough, the wedge product $\mu \wedge \nu$ is defined by

$$(\mu \wedge \nu)_P := \mu_P \wedge \nu_P \qquad \text{for all} \quad P \in \mathbb{E}^3.$$

Then, we have the supersymmetry product rule

$$\boxed{\mu \wedge \nu = (-1)^{km}\nu \wedge \mu.} \tag{12.76}$$

(iv) The Cartan exterior algebra $\Lambda(\mathbb{E}^3)$ of differential forms on the Euclidean manifold $\mathbb{E}^3$: The elements of the direct sum

$$\Lambda(\mathbb{E}^3) = \bigoplus_{k=0}^{\infty} \Lambda^k(\mathbb{E}^3)$$

are given by the sums

$$f + \omega + \gamma + \varrho$$

where $f, \omega, \gamma, \varrho$ are elements of $\Lambda^k(\mathbb{E}^3)$ with $k = 0, 1, 2, 3$, respectively. The real linear space $\Lambda(\mathbb{E}^3)$ becomes a real infinite-dimensional algebra with respect to the wedge product. This algebra is called the Cartan exterior algebra of the Euclidean manifold $\mathbb{E}^3$. The Cartan algebra possesses the supersymmetric grading (12.76).

12.3.2 The Grassmann Bundle

Modern differential geometry is based on the notion of 'bundle'. In this setting, differential forms on a manifold are sections of the Grassmann bundle of the manifold. Similarly, tensor fields are sections of the tensor bundle of the manifold.

Folklore

The Grassmann bundle $\mathsf{G}(\mathbb{E}^3)$ of the Euclidean manifold $\mathbb{E}^3$ is defined to be the set

$$\mathsf{G}(\mathbb{E}^3) := \{(P, \omega) : \ P \in \mathbb{E}^3, \ \omega \in \bigwedge(T_P^*\mathbb{E}^3)\}.$$

That is, ω is a finite sum of real numbers α and terms of the form

$$\omega^1 \wedge \cdots \wedge \omega^m, \qquad m = 1, 2, 3$$

where $\omega^j \in T_P^*\mathbb{E}^3$ for all indices j. This is a generalization of the cotangent bundle

$$T^*\mathbb{E}^3 = \{(P, \omega) : \ P \in \mathbb{E}^3, \ \omega \in T_P^*\mathbb{E}^3\}$$

of the Euclidean manifold $\mathbb{E}^3$. Setting $\pi(P, \omega) := P$, we get the projection map

$$\boxed{\pi : \mathsf{G}(\mathbb{E}^3) \to \mathbb{E}^3} \qquad (12.77)$$

of the Grassmann bundle. The set $\mathsf{G}(\mathbb{E}^3)$ is called the bundle space.[21] In terms of the coproduct of sets, we have

$$\mathsf{G}(\mathbb{E}^3) = \coprod_{P \in \mathbb{E}^3} \bigwedge(T_P^*\mathbb{E}^3).$$

In other words, the Grassmann bundle space $\mathsf{G}(\mathbb{E}^3)$ is the disjoint union of the Grassmann algebras $\bigwedge(T_P^*\mathbb{E}^3)$ taken over the points P of $\mathbb{E}^3$. If we assign to (P, ω) the three Cartesian coordinates (x, y, z) of the point P and the eight real components of ω, then the Grassmann bundle space $\mathsf{G}(\mathbb{E}^3)$ becomes a real manifold of dimension $\dim \mathsf{G}(\mathbb{E}^3) = 3 + 8 = 11$.

[21] In order to simplify the terminology, sometimes we do not distinguish between the Grassmann bundle space $\mathsf{G}(\mathbb{E}^3)$ and the Grassmann bundle $\pi : \mathsf{G}(\mathbb{E}^3) \to \mathbb{E}^3$, which is a surjective map by definition.

By definition, a smooth section $s : \mathbb{E}^3 \to \mathsf{G}(\mathbb{E}^3)$ of the Grassmann bundle is a smooth map of the form

$$P \mapsto (P, \omega_P).$$

This implies the smooth differential form $P \mapsto \omega_P$. Thus, smooth differential forms $\omega \in \Lambda(\mathbb{E}^3)$ on the Euclidean manifold $\mathbb{E}^3$ coincide with smooth sections of the Grassmann bundle.

12.3.3 The Tensor Bundle

A typical tensor field on the Euclidean manifold $\mathbb{E}^3$ is the metric tensor field $P \mapsto \mathbf{g}_P$ defined by

$$\mathbf{g}_P(\mathbf{u}, \mathbf{v}) := \mathbf{u}\mathbf{v} \qquad \text{for all} \quad \mathbf{u}, \mathbf{v} \in T_P\mathbb{E}^3, \quad P \in \mathbb{E}^3.$$

In a Cartesian (x, y, z)-coordinate system on $\mathbb{E}^3$, we get

$$\mathbf{g}_P = dx \otimes dx + dy \otimes dy + dz \otimes dz.$$

The set

$$\mathsf{T}^0_2(\mathbb{E}^3) := \{(P, \mathbf{T}) : \ P \in \mathbb{E}^3, \ \ \mathbf{T} \in T^*_P\mathbb{E}^3 \otimes T^*_P\mathbb{E}^3\}$$

is called the tensor bundle of type $(0, 2)$ of $\mathbb{E}^3$. The metric tensor field $P \mapsto \mathbf{g}_P$ corresponds to the section

$$s : \mathbb{E}^3 \to \mathsf{T}^0_2(\mathbb{E}^3)$$

with $s(P) := (P, \mathbf{g}_P)$ for all points $P \in \mathbb{E}^3$. In the general case, the tensor bundle $\mathsf{T}(\mathbb{E}^3)$ of the Euclidean manifold $\mathbb{E}^3$ consists of all the pairs

$$(P, \mathbf{T})$$

where $P \in \mathbb{E}^3$, and $\mathbf{T}$ is a finite sum of real numbers and elements of the form

$$\mathbf{v}_1 \otimes \cdots \otimes \mathbf{v}_m \otimes \omega^1 \otimes \cdots \otimes \omega^n \qquad \text{(elements of type } (m, n))$$

where $\mathbf{v}_i \in T_P\mathbb{E}^3$ and $\omega^j \in T^*_P\mathbb{E}^3$ for $i = 1, \ldots, m$ and $j = 1, \ldots, n$. Moreover, $m, n = 0, 1, 2, \ldots$

12.3.4 The Transformation of Covector Fields

> There exists a perfect duality between velocity vector fields $\mathbf{v}$ and covector fields ω.
>
> Folklore

Let us consider the smooth map

$$\boxed{\tau : \mathbb{E}^3 \to \mathbb{E}^3} \tag{12.78}$$

(e.g., a rotation). Assume that the map τ sends the point Q to the point $P = \tau(Q)$.

Pull-back $\tau^*\omega$ of the covector field ω. We are given the smooth covector field ω on the Euclidean manifold $\mathbb{E}^3$. We want to transplant the covector from the point P to the point Q. To this end, we will use the concept of duality for linear functionals (Fig. 12.13). We are given the point $Q \in \mathbb{E}^3$. Define

$$\boxed{(\tau^*\omega)_Q(\mathbf{v}) := \omega_P((T_Q\tau)\mathbf{v}) \qquad \text{for all} \quad \mathbf{v} \in T_Q\mathbb{E}^3.}$$

Mnemonically, we write

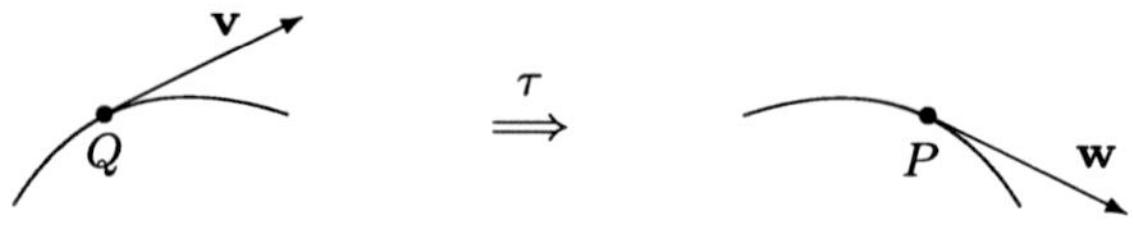

$$(\tau^*\omega)_Q(\mathbf{v}) = \omega_P(\mathbf{w})$$

Fig. 12.13. Pull-back $\tau^*\omega$ of the covector field ω

$$(\tau^*\omega)(\mathbf{v}) = \omega_\tau(\tau_*\mathbf{v}). \tag{12.79}$$

In terms of Cartesian coordinates, the following hold. Choose a fixed right-handed (x, y, z)-Cartesian coordinate system with the right-handed orthonormal basis $\mathbf{i}, \mathbf{j}, \mathbf{k}$ at the origin O. Parallel transport of $\mathbf{i}, \mathbf{j}, \mathbf{k}$ to the point P yields the orthonormal basis $\mathbf{i}_P, \mathbf{j}_P, \mathbf{k}_P$ of the tangent space $T_P\mathbb{E}^3$ of the Euclidean plane $\mathbb{E}^3$ at the point P (Fig. 9.1 on page 558). Assume that the equation $P = \tau(Q)$ corresponds to

$$x = x(\xi, \eta, \zeta), \quad y = y(\xi, \eta, \zeta), \quad z = z(\xi, \eta, \zeta), \qquad (\xi, \eta, \zeta) \in \mathbb{R}^3.$$

That is, the point P (resp. Q) has the Cartesian coordinates (x, y, z) (resp. (ξ, η, ζ)). Let $\omega_P = U(P)dx_P + V(P)dy_P + W(P)dz_P$. Then

$$\begin{aligned} (\tau^*\omega)_Q &= U(P)(x_\xi d\xi + x_\eta d\eta + x_\zeta d\zeta) \\ &+ V(P)(y_\xi d\xi + y_\eta d\eta + y_\zeta d\zeta) \\ &+ W(P)(z_\xi d\xi + z_\eta d\eta + z_\zeta d\zeta). \end{aligned} \tag{12.80}$$

For the proof, we refer to Problem 12.3. Mnemonically, the transformation law (12.80) corresponds to the Leibniz differential rule $dx = x_\xi d\xi + x_\eta d\eta + x_\zeta d\zeta$, and so on.

Proposition 12.5 *If ω and ϱ are differential forms, then $\tau^*(\omega \wedge \varrho) = \tau^*\omega \wedge \tau^*\varrho$.*

This tells us that the pull-back operation respects the wedge product. In algebraic terms, the map $\tau^* : \Lambda(\mathbb{E}^3) \to \Lambda(\mathbb{E}^3)$ is an endomorphism of the Grassmann algebra $\Lambda(\mathbb{E}^3)$. For the proof, we refer to Problem 12.4.

Push-forward $\tau_*\omega$ of the covector field ω. Let the map $\tau : \mathbb{E}^3 \to \mathbb{E}^3$ be a diffeomorphism. The transformed covector field $\tau_*\omega$ is defined by

$$\boxed{\tau_*\omega := (\tau^{-1})^*\omega.}$$

The situation is depicted in Fig. 12.14.

12.4 Cartan's Exterior Differential

The Cartan exterior differential $d\omega$ for covector fields and more general differential forms ω is dual to the Lie derivative $\mathcal{L}_{\mathbf{v}}\mathbf{w} = [\mathbf{v}, \mathbf{w}]$ of velocity vector fields $\mathbf{v}, \mathbf{w}$.

Folklore

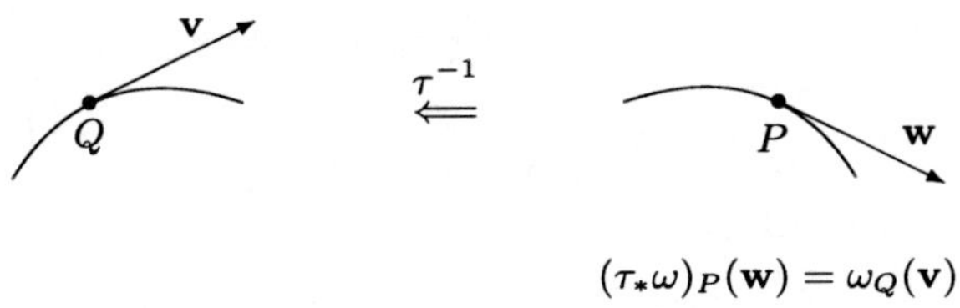

Fig. 12.14. Push-forward $\tau_*\omega$ of the covector field ω

12.4.1 Invariant Definition via the Lie Algebra of Velocity Vector Fields

Let $f : \mathbb{E}^3 \to \mathbb{R}$ be a smooth function on the Euclidean manifold $\mathbb{E}^3$, that is, $f \in \Lambda^0(\mathbb{E}^3)$. For all smooth velocity vector fields $\mathbf{v} \in \mathrm{Vect}(\mathbb{E}^3)$, we define

$$(df)(\mathbf{v}) := \mathcal{L}_{\mathbf{v}} f \tag{12.81}$$

where $\mathcal{L}_{\mathbf{v}} f$ denotes the Lie derivative (i.e., the directional derivative) of the function f. Explicitly, this means that

$$df_P(\mathbf{v}) := (\mathcal{L}_{\mathbf{v}} f)(P)$$

for all points $P \in \mathbb{E}^3$. This is a special case of the following more general definition. Let $\omega \in \Lambda^p(\mathbb{E}^3)$ be a smooth p-form on the Euclidean manifold $\mathbb{E}^3$ with $p = 0, 1, 2, 3$. We define the linear operator $d : \Lambda^p(\mathbb{E}^3) \to \Lambda^{p+1}(\mathbb{E}^3)$ by setting:

$$\begin{aligned}(d\omega)(\mathbf{v}_1, \ldots, \mathbf{v}_{p+1}) &= \sum_{i=1}^{p+1} (-1)^{i+1} \mathcal{L}_{\mathbf{v}_i}\big(\omega(\mathbf{v}_1, \ldots, \hat{\mathbf{v}}_i, \ldots, \mathbf{v}_{p+1})\big) \\ &\quad + \sum_{i<j} (-1)^{i+j} \omega([\mathbf{v}_i, \mathbf{v}_j], \mathbf{v}_1, \ldots, \hat{\mathbf{v}}_i, \ldots, \hat{\mathbf{v}}_j, \ldots, \mathbf{v}_{p+1}).\end{aligned} \tag{12.82}$$

This is the key definition of Cartan's calculus of alternating differential forms. Here, by convention, the terms equipped with a hat (e.g., $\hat{\mathbf{v}}_i$) have to be cancelled. Explicitly, the definition (12.82) means the following for all smooth velocity vector fields $\mathbf{u}, \mathbf{v}, \mathbf{w} \in \mathrm{Vect}(\mathbb{E}^3)$:

- $p = 1$: $(d\omega)(\mathbf{u}, \mathbf{v}) = \mathcal{L}_{\mathbf{u}}\big(\omega(\mathbf{v})\big) - \mathcal{L}_{\mathbf{v}}\big(\omega(\mathbf{u})\big) - \omega([\mathbf{u}, \mathbf{v}])$.
- $p = 2$: $$\begin{aligned}(d\omega)(\mathbf{u}, \mathbf{v}, \mathbf{w}) &= \mathcal{L}_{\mathbf{u}}\big(\omega(\mathbf{v}, \mathbf{w})\big) - \mathcal{L}_{\mathbf{v}}\big(\omega(\mathbf{u}, \mathbf{w})\big) + \mathcal{L}_{\mathbf{w}}\big(\omega(\mathbf{u}, \mathbf{v})\big) \\ &\quad - \omega([\mathbf{u}, \mathbf{v}], \mathbf{w}) + \omega([\mathbf{u}, \mathbf{w}], \mathbf{v}) - \omega([\mathbf{v}, \mathbf{w}], \mathbf{u}).\end{aligned} \tag{12.83}$$
- $p = 3$: $d\omega \equiv 0$.

Cartesian coordinates. In a right-handed Cartesian (x, y, z)-coordinate system, where the point P has the coordinates (x, y, z) and the velocity vector field is given by $\mathbf{v}_P = a(P)\mathbf{i}_P + b(P)\mathbf{j}_P + c(P)\mathbf{k}_P$, we get

$$df_P(\mathbf{v}) = a(x, y, z) f_x(x, y, z) + b(x, y, z) f_y(x, y, z) + c(x, y, z) f_z(x, y, z).$$

Hence

$$df_P = f_x(P)dx_P + f_y(P)dy_P + f_z(P)dz_P. \tag{12.84}$$

To simplify notation, we will briefly write

$$df_P = f_x(P)dx + f_y(P)dy + f_z(P)dz.$$

Set $x^1 := x, x^2 := y, x^3 := z$, and $\partial_j := \partial/\partial x^j$. Then a p-form reads as

$$\omega = \omega_{i_1 \ldots i_p}\, dx^{i_1} \wedge \cdots \wedge dx^{i_p}, \qquad p = 1, 2, 3. \tag{12.85}$$

Here, we sum over equal indices from 1 to 3, and the smooth coefficient functions $\omega_{i_1 \ldots i_p}$ are antisymmetric with respect to the indices $i_1, \ldots, i_p$.[22] Then

$$\boxed{d\omega = \partial_i \omega_{i_1 \ldots i_p}\, dx^i \wedge dx^{i_1} \wedge \cdots \wedge dx^{i_p}.} \tag{12.86}$$

Equivalently,

$$d\omega = d\omega_{i_1 \ldots i_p} \wedge dx^{i_1} \wedge \cdots \wedge dx^{i_p}.$$

This coincides with the expression used in Sect. 12.1.3 on page 680. Mnemonically, one also writes $d \wedge \omega$ instead of $d\omega$.

Proof. Let us consider the case where $p = 1$. Set $\mathbf{e}_1 := \mathbf{i}_P,\ \mathbf{e}_2 := \mathbf{j}_P,\ \mathbf{e}_3 := \mathbf{k}_P$. Let $\omega = \omega_k\, dx^k$. We have to show that $d\omega = d\omega_k \wedge dx^k$. Summing over equal indices from 1 to 3, we get the following:

- $\mathbf{u} = u^i \mathbf{e}_i,\ \mathbf{v} = v^j \mathbf{e}_j$,
- $\omega(\mathbf{v}) = \omega_k\, dx^k(\mathbf{v}) = \omega_k v^k$,
- $\mathcal{L}_{\mathbf{u}}\big(\omega(\mathbf{v})\big) = u^i \partial_i(\omega_k v^k)$,
- $[\mathbf{u}, \mathbf{v}] = \mathcal{L}_{\mathbf{u}} \mathbf{v} = (u^i \partial_i v^k - v^i \partial_i u^k)\, \mathbf{e}_k$,
- $\omega([\mathbf{u}, \mathbf{v}]) = \omega_k (u^i \partial_i v^k - v^i \partial_i u^k)$.

Hence

$$\begin{aligned} d\omega(\mathbf{u}, \mathbf{v}) &= \mathcal{L}_{\mathbf{u}}\big(\omega(\mathbf{v})\big) - \mathcal{L}_{\mathbf{v}}\big(\omega(\mathbf{u})\big) - \omega([\mathbf{u}, \mathbf{v}]) \\ &= u^i \partial_i(\omega_k v^k) - v^i \partial_i(\omega_k u^k) - \omega_k(u^i \partial_i v^k - v^i \partial_i u^k). \end{aligned}$$

By the product rule, $d\omega(\mathbf{u}, \mathbf{v}) = \partial_i \omega_k \cdot (u^i v^k - v^i u^k)$. On the other hand,

$$\begin{aligned} (\partial_i \omega_k\, dx^i \wedge dx^k)(\mathbf{u}, \mathbf{v}) &= \partial_i \omega_k\, \big(dx^i(\mathbf{u})\, dx^k(\mathbf{v}) - dx^i(\mathbf{v})\, dx^k(\mathbf{u})\big) \\ &= \partial_i \omega_k \cdot (u^i v^k - v^i u^k) = d\omega(\mathbf{u}, \mathbf{v}). \end{aligned}$$

Hence $d\omega = d\omega_k \wedge dx^k$. □

The fundamental pull-back invariance of the Cartan differential. Let $\tau : \mathbb{E}^3 \to \mathbb{E}^3$ be a smooth map. Then for all p-forms $\omega \in \Lambda^p(\mathbb{E}^3)$, $p = 0, 1, 2, 3$, the following holds.

Proposition 12.6 *$d(\tau^* \omega) = \tau^*(d\omega)$ on $\mathbb{E}^3$.*

---

[22] For example, if $p = 2$, then it follows from $dx^i \wedge dx^j = -dx^j \wedge dx^i$ and $\omega_{ij} = -\omega_{ji}$ that

$$\omega = \omega_{ij} dx^i \wedge dy^j = 2(\omega_{12} dx^1 \wedge dx^2 + \omega_{23} dx^2 \wedge dx^3 + \omega_{31} dx^3 \wedge dx^1).$$

Proof. This can be proven by using (12.86) together with an explicit computation as performed in Sect. 12.1.2 on page 673. More elegantly, one can use the invariance properties of the Lie derivative. □

In other words, the smooth map $\tau : \mathbb{E}^3 \to \mathbb{E}^3$ generates the following commutative diagram:

$$\begin{array}{ccc} \Lambda(\mathbb{E}^3) & \xrightarrow{\tau^*} & \Lambda(\mathbb{E}^3) \\ {\scriptstyle d}\downarrow & & \downarrow{\scriptstyle d} \\ \Lambda(\mathbb{E}^3) & \xrightarrow{\tau^*} & \Lambda(\mathbb{E}^3). \end{array} \tag{12.87}$$

That is, the morphism $\tau^* : \Lambda(\mathbb{E}^3) \to \Lambda(\mathbb{E}^3)$ of the Grassmann algebra $\Lambda(\mathbb{E}^3)$ commutes with the Cartan differential operator d. Therefore, τ^* is called a differential morphism of the Grassmann algebra $\Lambda(\mathbb{E}^3)$ over the Euclidean manifold $\mathbb{E}^3$.

12.4.2 The Supersymmetric Leibniz Rule

Proposition 12.7 *If $\omega \in \Lambda^p(\mathbb{E}^3)$ and $\varrho \in \Lambda^q(\mathbb{E}^3)$ with $p, q = 0, 1, 2, 3$, then*

$$d(\omega \wedge \varrho) = d\omega \wedge \varrho + (-1)^p \omega \wedge d\varrho \quad \text{on } \mathbb{E}^3. \tag{12.88}$$

This is called the supersymmetric (or graded) Leibniz rule.[23]

Proof This follows from both the classical Leibniz rule $d(fg) = (df)g + fdg$ for smooth functions f, g and the graded anticommutativity of the Grassmann product. For example, let $\omega := fdx$ and $\varrho := gdy$. Since $dg \wedge dx = -dx \wedge dg$, we get

$$\begin{aligned} d(fg\, dx \wedge dy) &= d(fg) \wedge dx \wedge dy = g(df) \wedge dx \wedge dy + f(dg) \wedge dx \wedge dy \\ &= df \wedge dx \wedge g\, dy - f\, dx \wedge dg \wedge dy. \end{aligned}$$

Hence $d(f\, dx \wedge g\, dy) = d(f\, dx) \wedge g\, dy - f\, dx \wedge d(g\, dy)$. □

Example. Let us choose a right-handed Cartesian (x, y, z)-coordinate system on the Euclidean manifold $\mathbb{E}^3$.

(I) If $p = 0$, then $\omega = f$ is a smooth real-valued function on $\mathbb{E}^3$, and

$$d(f\varrho) = df \wedge \varrho + fd\varrho.$$

In the special case where $q = 0$ (i.e., $\varrho = g$ is a real-valued smooth function), we get $d(f\varrho) = df \cdot \varrho + fd\varrho$. In terms of vector analysis, this reads as

$$\boxed{\mathbf{grad}(fg) = g\,\mathbf{grad}\, f + f\,\mathbf{grad}\, g.}$$

(II) Consider the smooth differential 1-forms

$$\omega := adx + bdy + cdz, \qquad \varrho := Adx + Bdy + Cdz,$$

and the corresponding vector fields

$$\mathbf{v} = a\mathbf{i} + b\mathbf{j} + c\mathbf{k}, \qquad \mathbf{w} = A\mathbf{i} + B\mathbf{j} + C\mathbf{k}.$$

We want to show that the supersymmetric Leibniz rule $d(\omega \wedge \varrho) = d\omega \wedge \varrho - \omega \wedge d\varrho$ corresponds to the following formula in vector analysis:

[23] Note that $\omega \wedge \varrho = \omega\varrho$ if $p = 0$ or $q = 0$.

$$\boxed{\operatorname{div}(\mathbf{v} \times \mathbf{w}) = \mathbf{w}\,\mathbf{curl}\,\mathbf{v} - \mathbf{v}\,\mathbf{curl}\,\mathbf{w}.} \tag{12.89}$$

In fact, we have the following relations:[24]

- $\omega \wedge \varrho = (bC - cB)\, dy \wedge dz + (cA - aC)\, dz \wedge dx + (aB - bA)\, dx \wedge dy$,
- $d(\omega \wedge \varrho) = \big[(bC - cB)_x + (cA - aC)_y + (aB - bA)_z\big]\, dx \wedge dy \wedge dz$,
- $d\omega = (c_y - b_z)\, dy \wedge dz + (a_z - c_x)\, dz \wedge dx + (b_x - a_y)\, dx \wedge dy$,
- $d\omega \wedge \varrho = \big[(c_y - b_z)A + (a_z - c_x)B + (b_x - a_y)C\big]\, dx \wedge dy \wedge dz$.

Noting that

$$\mathbf{v} \times \mathbf{w} = \begin{vmatrix} \mathbf{i} & \mathbf{j} & \mathbf{k} \\ a & b & c \\ A & B & C \end{vmatrix} = (bC - cB)\mathbf{i} + (cA - aC)\mathbf{j} + (aB - bA)\mathbf{k},$$

we get the following equivalent expressions:

- $\omega \simeq \mathbf{v}$, $\varrho \simeq \mathbf{w}$, $\omega \wedge \varrho \simeq \mathbf{v} \times \mathbf{w}$,
- $d(\omega \wedge \varrho) \simeq \operatorname{div}(\mathbf{v} \times \mathbf{w})$, $\quad d\omega \wedge \varrho \simeq \mathbf{w}\,\mathbf{curl}\,\mathbf{v}$, $\quad \omega \wedge d\varrho \simeq \mathbf{v}\,\mathbf{curl}\,\mathbf{w}$.

This yields the claim (12.89).

12.4.3 The Poincaré Cohomology Rule

Lemma 12.8 *If* $\omega \in \Lambda^p(\mathbb{E}^3)$, $p = 0, 1, 2, 3$, *then*

$$d(d\omega) = 0 \quad \text{on } \mathbb{E}^3.$$

This is called the Poincaré cohomology rule (or the Poincaré cohomology lemma).

Proof. It follows from (12.86) on page 708 that

$$d(d\omega) = \partial_j \partial_i \omega_{i_1 \ldots i_p}\, dx^j \wedge dx^i \wedge dx^{i_1} \wedge \cdots \wedge dx^{i_p}.$$

This is equal to zero, since $\partial_j \partial_i = \partial_i \partial_j$ and $dx^j \wedge dx^i = -dx^j \wedge dx^i$. □

Example. Using the smooth function $\omega := U$ or the 1-form $\omega = adx + bdy + cdz$ together with the velocity vector field $\mathbf{v} = a\mathbf{i} + b\mathbf{j} + c\mathbf{k}$, we get

$$\mathbf{curl}\,\mathbf{grad}\,U = 0, \qquad \operatorname{div}\mathbf{curl}\,\mathbf{v} = 0.$$

12.4.4 The Axiomatic Approach

The theory of p-forms on the Euclidean manifold $\mathbb{E}^3$ can be generalized straightforward to real n-dimensional manifolds $\mathcal{M}$ with $n = 1, 2, \ldots$ Let $\Lambda^p(\mathcal{M})$ denote the real linear space of all the smooth p-forms on $\mathcal{M}$ with $p = 0, 1, 2, \ldots$ Note that $\Lambda^p(\mathcal{M}) = \{0\}$ if $p > n$. Let

$$\Lambda(\mathcal{M}) := \bigoplus_{p=0}^{\infty} \Lambda^p(\mathcal{M}).$$

[24] Observe that all the formulas are highly symmetric via cyclic permutations.

Theorem 12.9 *There exists precisely one linear operator*

$$d : \Lambda(\mathcal{M}) \to \Lambda(\mathcal{M})$$

which, for all differential forms $\omega \in \Lambda^p(\mathcal{M})$ *and* $\varrho \in \Lambda^q(\mathcal{M})$ *with* $p, q = 0, 1, 2, \ldots$, *has the following properties:*

(i) $d\omega \in \Lambda^{p+1}(\mathcal{M})$ *(grading).*

(ii) $d(d\omega) = 0$ *(Poincaré's cohomology rule).*

(iii) $d(\omega \wedge \varrho) = d\omega \wedge \varrho + (-1)^p \omega \wedge d\varrho$ *(supersymmetric Leibniz rule).*

(iv) *Leibniz differential: If* $f \in \Lambda^0(\mathcal{M})$, *then*

$$(df)_P(\mathbf{v}) = (\mathcal{L}_{\mathbf{v}} f)(P)$$

for all points $P \in \mathcal{M}$ *and all smooth velocity vector fields* $\mathbf{v} \in \mathrm{Vect}(\mathcal{M})$ *on the manifold* $\mathcal{M}$.

(v) *Locality: The operator d is local, that is, if two p-forms* ω *and* μ *coincide on an open subset* $\mathcal{U}$ *of the manifold* $\mathcal{M}$, *then* $d\omega = d\mu$ *on* $\mathcal{U}$.

Proof. Step 1: Special case. Let $\mathcal{M} = \mathbb{E}^3$.

(I) Uniqueness. By (ii), $d(dx^i) = 0$. It follows from the product rule (iii) that

$$d(dx^i \wedge dx^j) = d(dx^i) \wedge dx^j - dx^i \wedge d(dx^j) = 0.$$

Similarly, $d(dx^{i_1} \wedge \cdots \wedge dx^{i_p}) = 0$. Choose $\omega = \omega_{i_1 \ldots i_p} dx^{i_1} \wedge \cdots dx^{i_p}$ as on page 708. By (iii),

$$d\omega = d\omega_{i_1 \ldots i_p} \wedge dx^{i_1} \wedge \cdots \wedge dx^{i_p}.$$

Therefore, by (iv),

$$d\omega = \partial_i \omega_{i_1 \ldots i_p} \, dx^i \wedge dx^{i_1} \wedge \cdots \wedge dx^{i_p}. \tag{12.90}$$

Consequently, the operator d is uniquely determined.

(II) Existence. Our investigations in Sect. 12.4.1ff above show that there exists an invariantly defined operator d which has the properties (i) through (v) above.

Step 2: General case. Let $\mathcal{M}$ be a real n-dimensional manifold. By (v), we can restrict ourselves to local coordinates on $\mathcal{M}$.

(I) Uniqueness. Argue as in Step 1. This way, we get (12.90).

(II) Existence. Define $d\omega$ by (12.90). We have to show that this definition does not depend on the choice of local coordinates. This can be shown by an explicit computation. However, there exists a more elegant approach which we have considered in Sect. 8.11.2 on page 523. □

Proposition 12.10 *Let* $\tau : \mathcal{M} \to \mathcal{N}$ *be a smooth map from the real n-dimensional manifold to the real m-dimensional manifold* $\mathcal{N}$ *with* $n, m = 1, 2, \ldots$ *Then, for all differential forms* ω *on* $\mathcal{N}$, *we have*

$$\tau^*(d\omega) = d(\tau^*\omega).$$

This tells us that the Cartan differential respects smooth mappings (e.g., smooth changes of local coordinates).

Summary. Because of the graded product rule (iii) (supersymmetric Leibniz rule), the Grassmann algebra $\Lambda(\mathcal{M})$ over the manifold $\mathcal{M}$ is called a graded differential algebra. The smooth map

$$\tau : \mathcal{M} \to \mathcal{N}$$

induces the algebra morphism $\tau^* : \Lambda(\mathcal{N}) \to \Lambda(\mathcal{M})$, and the following diagram is commutative:

$$\begin{array}{ccc} \Lambda(\mathcal{M}) & \xleftarrow{\tau^*} & \Lambda(\mathcal{N}) \\ \downarrow d & & \downarrow d \\ \Lambda(\mathcal{M}) & \xleftarrow{\tau^*} & \Lambda(\mathcal{N}). \end{array} \tag{12.91}$$

Therefore, τ^* is called a differential algebra morphism from the Grassmann algebra $\Lambda(\mathcal{N})$ to the Grassmann algebra $\Lambda(\mathcal{M})$.

12.5 The Lie Derivative of Differential Forms

> It was the basic strategy of Sophus Lie (1842–1899) to study operations on manifolds (e.g., transport processes or symmetry transformations) on an infinitesimal level. This leads to the fundamental notion of Lie derivative and Lie algebra.
> Lie discovered that symmetry properties are responsible for the crucial fact that the infinitesimal level knows all about the local level.
> Lie's successors like Élie Cartan (1869–1951), Georges de Rham (1903–1990), and Claude Chevalley (1909–1984) investigated the global level. They found out that one needs additional topological information on the global level.
>
> Folklore

Suppose that we are given the smooth velocity vector field $\mathbf{v}$ on the Euclidean manifold $\mathbb{E}^3$. In terms of physics, this velocity vector $\mathbf{v}$ field describes the transport of fluid particles along the trajectories $P = P(t)$ given by the differential equation

$$\dot{\mathbf{x}}(t) = \mathbf{v}(P(t)), \qquad P(0) = P_0$$

with respect to time t. We set $F_tP_0 := P(t)$. For fixed time t, the flow operator

$$F_t : \mathcal{U}(P_0) \to \mathbb{E}^3$$

is a smooth operator defined on a sufficiently small neighborhood $\mathcal{U}(P_0)$ of the point $P_0 \in \mathbb{E}^3$.

The Lie derivative $\mathcal{L}_{\mathbf{v}}\omega$ describes the flow transport of the p-form ω on an infinitesimal level.

12.5.1 Invariant Definition via the Flow of Fluid Particles

Let $\omega \in \Lambda^p(\mathbb{E}^3)$ be a differential p-form with $p = 0, 1, 2, 3$. The Lie derivative $\mathcal{L}_{\mathbf{v}}\omega$ of ω with respect to the smooth velocity vector field $\mathbf{v} \in \mathrm{Vect}(\mathbb{E}^3)$ is defined by the time derivative

$$\boxed{\mathcal{L}_{\mathbf{v}}\omega := \frac{d}{dt} F_t^*\omega_{|t=0}.} \tag{12.92}$$

If $p = 0$, then $\omega = f$ is a smooth real-valued function on $\mathbb{E}^3$, and $\mathcal{L}_{\mathbf{v}} f$ coincides with the Lie derivative for temperature fields introduced in Chap. 10 above. Let $p \geq 1$. Then

$$\mathcal{L}_{\mathbf{v}}\omega(\mathbf{u}_1, \ldots, \mathbf{u}_p) = \frac{d}{dt} F_t^* \omega(\mathbf{u}_1, \ldots, \mathbf{u}_p)_{|t=0}$$

for all $\mathbf{u}_1, \ldots, \mathbf{u}_p \in \text{Vect}(\mathbb{E}^3)$. In terms of the short time limit $t \to 0$, we obtain

$$\mathcal{L}_{\mathbf{v}}\omega_P(\mathbf{u}_1, \ldots, \mathbf{u}_p) = \lim_{t \to 0} \frac{F_t^* \omega_P(\mathbf{u}_1, \ldots, \mathbf{u}_p) - \omega_P(\mathbf{u}_1, \ldots, \mathbf{u}_p)}{t} \tag{12.93}$$

for all points $P \in \mathbb{E}^3$, and all tangent vectors $\mathbf{u}_1, \ldots, \mathbf{u}_p \in T_P\mathbb{E}^3$.

The product rule. Let $\omega \in \Lambda^p(\mathbb{E}^3)$ and $\varrho \in \Lambda^q(\mathbb{E}^3)$ with $p, q = 0, 1, 2, 3$.

Proposition 12.11 $\mathcal{L}_{\mathbf{v}}(\omega \wedge \varrho) = \mathcal{L}_{\mathbf{v}}\omega \wedge \varrho + \omega \wedge \mathcal{L}_{\mathbf{v}}\varrho$.

Proof. Differentiate $F_t^*(\omega \wedge \varrho) = F_t^*\omega \wedge F_t^*\varrho$ with respect to time t at $t = 0$. □

Cartesian coordinates. Choose a right-handed Cartesian (x, y, z)-coordinate system together with the notation introduced on page 708. In what follows, we will sum over equal indices from 1 to 3. Let $\mathbf{v} = v^i \mathbf{e}_i$. Recall that if $f \in \Lambda^0(\mathbb{E}^3)$, then

$$\mathcal{L}_{\mathbf{v}} f = v^s \partial_s f.$$

Proposition 12.12 $\mathcal{L}_{\mathbf{v}}(dx^i) = \partial_s v^i \cdot dx^s = \mathbf{grad}\, v^i \cdot d\mathbf{x}$ *if* $i = 1, 2, 3$.

Proof. Note that $(F_t^* dx^i)_P(\mathbf{u}) = dx^i_{F_t P}(F_t'(P)\mathbf{u}) = dx^i(F_t'(P)\mathbf{u})$. Using the linearized flow studied on page 650, we get

$$\frac{d}{dt}\left(F_t'(P)\mathbf{u}\right)_{|t=0} = \mathbf{v}'(P)\mathbf{u} = \partial_s v^k u^s \mathbf{e}_k.$$

Therefore, differentiation at time $t = 0$ yields

$$\frac{d}{dt}(F_t^* dx^i)_P(\mathbf{u}) = dx^i(\mathbf{v}'(P)\mathbf{u}) = u^s \partial_s v^i = \partial_s v^i \cdot dx^s(\mathbf{u}).$$

□

Let $\omega \in \Lambda^p(\mathbb{E}^3)$ with $p = 1, 2, 3$. Then

$$\omega = \omega_{i_1 \ldots i_p}\, dx^{i_1} \wedge \cdots \wedge dx^{i_p}.$$

We assume that $\omega_{i_1 \ldots i_p}$ is antisymmetric with respect to the indices $i_1, \ldots, i_p$.

Proposition 12.13 $\mathcal{L}_{\mathbf{v}}\omega = (v^s \partial_s \omega_{i_1 \ldots i_p} + p \cdot \partial_{i_1} v^s \omega_{s i_2 \ldots i_p})\, dx^{i_1} \wedge \cdots \wedge dx^{i_p}$.

Proof. For example, let $p = 2$. By the product rule,

$$\mathcal{L}_{\mathbf{v}}(\omega_{ij}\, dx^i \wedge dx^j) = (\mathcal{L}_{\mathbf{v}}\omega_{ij})\, dx^i \wedge dx^j + \omega_{ij}(\mathcal{L}_{\mathbf{v}} dx^i) \wedge dx^j + \omega_{ij}\, dx^i \wedge (\mathcal{L}_{\mathbf{v}} dx^j).$$

By Prop. 12.12, this is equal to

$$v^s \partial_s \omega_{ij}\, dx^i \wedge dx^j + \omega_{ij} \partial_s v^i dx^s \wedge dx^j + \omega_{ij} \partial_s v^j dx^i \wedge dx^s.$$

Noting that $\omega_{ij} = -\omega_{ji}$ and $dx^i \wedge dx^j = -dx^j \wedge dx^i$, and changing the summation indices, we get the claim. □

For example, if $f \in \Lambda^0(\mathbb{E}^3)$, then

$$\mathcal{L}_{\mathbf{v}}(f\, dx \wedge dy \wedge dz) = (v^s \partial_s f + \partial_s v^s)\, dx \wedge dy \wedge dz. \tag{12.94}$$

Let $\omega \in \Lambda^p(\mathbb{E}^3)$ with $p = 0, 1, 2, 3$, and let $\mathbf{v} \in \text{Vect}(\mathbb{E}^3)$.

Proposition 12.14 $d(\mathcal{L}_{\mathbf{v}}\omega) = \mathcal{L}_{\mathbf{v}}(d\omega)$.

This follows from Prop. 12.13 by a straightforward computation.

12.5.2 The Contraction Product between Velocity Vector Fields and Differential Forms

Let $\mathbf{v} \in \text{Vect}(\mathbb{E}^3)$. We define the linear operator[25]

$$i_{\mathbf{v}} : \Lambda^p(\mathbb{E}^3) \to \Lambda^{p-1}(\mathbb{E}^3), \qquad p = 0, 1, 2, 3$$

by setting

$$(i_{\mathbf{v}}\omega)(\mathbf{u}_1, \ldots \mathbf{u}_{p-1}) := \omega(\mathbf{v}, \mathbf{u}_1, \ldots \mathbf{u}_{p-1}) \qquad \text{for all} \quad \mathbf{u}_1, \ldots, \mathbf{u}_{p-1}.$$

In particular, $i_{\mathbf{v}}\omega := 0$ if $p = 0$, and $i_{\mathbf{v}}\omega := \omega(\mathbf{v})$ if $p = 1$. We also write $\mathbf{v}\rfloor\omega$ instead of $i_{\mathbf{v}}\omega$, and we call this the contraction product of $\mathbf{v}$ with ω.[26]

Cartesian coordinates. In a right-handed Cartesian (x, y, z)-coordinate system, we get:

- $i_{\mathbf{v}}(\omega_i dx^i) = v^i\omega_i$,
- $i_{\mathbf{v}}(\frac{1}{2}\omega_{ij}\, dx^i \wedge dx^j) = v^i\omega_{ij}\, dx^j$,
- $i_{\mathbf{v}}(\frac{1}{3!}\omega_{ijk}\, dx^i \wedge dx^j \wedge dx^k) = \frac{1}{2}v^i\omega_{ijk}\, dx^j \wedge dx^k$.

Here, we assume that ω_{ij} and ω_{ijk} are antisymmetric with respect to the indices i, j, k. The expressions above justify the term 'contraction product'.

Proof. For example, let $\omega = \frac{1}{2}\omega_{ij}\, dx^i \wedge dx^j$. Then

$$i_{\mathbf{v}}(\mathbf{u}) = \tfrac{1}{2}\omega_{ij}dx^i(\mathbf{v})dx^j(\mathbf{u}) - \tfrac{1}{2}\omega_{ij}dx^j(\mathbf{u})dx^i(\mathbf{v}) = \omega_{ij}v^iu^j = \omega_{ij}v^idx^j(\mathbf{u}).$$

□

Let $\omega \in \Lambda^p(\mathbb{E}^3)$ with $p = 0, 1, 2, 3$, and let $\mathbf{v}, \mathbf{w} \in \text{Vect}(\mathbb{E}^3)$. Using Cartesian coordinates, a straightforward computation yields the following.

Proposition 12.15 $\mathcal{L}_{\mathbf{v}}(i_{\mathbf{w}}\omega) - i_{\mathbf{w}}(\mathcal{L}_{\mathbf{v}}\omega) = i_{[\mathbf{v},\mathbf{w}]}\omega$.

In particular, $\mathcal{L}_{\mathbf{v}}(i_{\mathbf{v}}\omega) = i_{\mathbf{v}}(\mathcal{L}_{\mathbf{v}}\omega)$.

12.5.3 Cartan's Magic Formula

Theorem 12.16 *Let* $\mathbf{v} \in \text{Vect}(\mathbb{E}^3)$ *and* $\omega \in \Lambda^p(\mathbb{E}^3)$ *with* $p = 0, 1, 2, 3$. *Then*

$$\mathcal{L}_{\mathbf{v}}\omega = i_{\mathbf{v}}(d\omega) + d(i_{\mathbf{v}}\omega). \tag{12.95}$$

This is a special case of (8.99) on page 492. Applications of this magic formula to conservation laws for the flow of fluid particles will be given in Sect. 12.8 on page 731.

[25] By the usual convention, $\Lambda^r(\mathbb{E}^3) := \{0\}$ if $r = -1, -2, \ldots,$ and $r = 4, 5, \ldots$

[26] This is also called the interior product. However, we will reserve the term 'exterior' and 'interior' product for $\omega \wedge \varrho$ and $\omega \vee \varrho$, respectively (Kähler's interior differential calculus).

12.5.4 The Lie Derivative of the Volume Form

The volume form υ of the Euclidean manifold $\mathbb{E}^3$ is defined by

$$\upsilon_P(\mathbf{u},\mathbf{v},\mathbf{w}) := \mathbf{u}_P(\mathbf{v}_P \times \mathbf{w}_P)$$

for all points $P \in \mathbb{E}^3$ and all smooth velocity vector fields $\mathbf{u},\mathbf{v},\mathbf{w} \in \mathrm{Vect}(\mathbb{E}^3)$. This is the volume of the parallelepiped spanned by the position vectors $\mathbf{u},\mathbf{v},\mathbf{w}$ at the point P. Obviously, $\upsilon \in \Lambda^3(\mathbb{E}^3)$. In a right-handed Cartesian (x,y,z)-coordinate system, we have

$$\upsilon = dx \wedge dy \wedge dz.$$

Let $\varrho : \mathbb{E}^3 \to \mathbb{R}$ be a smooth function. By (12.94), we get the following:

Proposition 12.17 $\mathcal{L}_{\mathbf{v}}(\varrho\upsilon) = \mathrm{div}(\varrho\mathbf{v}) \cdot \upsilon$.

In particular, $\mathcal{L}_{\mathbf{v}}\upsilon = \mathrm{div}\,\mathbf{v} \cdot \upsilon$. This relates the Lie derivative of the volume form to the divergence $\mathrm{div}\,\mathbf{v}$.

12.5.5 The Lie Derivative of the Metric Tensor Field

Let $P \in \mathbb{E}^3$. Define

$$\boxed{g_P(\mathbf{a},\mathbf{b}) := \mathbf{ab} \qquad \text{for all} \quad \mathbf{a},\mathbf{b} \in T_P\mathbb{E}^3.}$$

This is the inner product on the tangent space $T_P\mathbb{E}^3$ which is a real 3-dimensional Hilbert space. The map $P \mapsto g_P$ is called the metric tensor field on the Euclidean manifold $\mathbb{E}^3$. Let $\mathbf{v}$ be a smooth complete velocity vector field on $\mathbb{E}^3$ which generates the flow $\{F_t\}_{t\in\mathbb{R}}$. The pull-back F_t^*g of g is defined by

$$(F_t^*g)_P(\mathbf{a},\mathbf{b}) = g_{F_t(P)}(F_t'(P)\mathbf{a}, F_t'(P)\mathbf{b}). \tag{12.96}$$

This yields the definition of the Lie derivative:

$$(\mathcal{L}_{\mathbf{v}}g)_P(\mathbf{a},\mathbf{b}) := \frac{d}{dt}(F_t^*g)_P(\mathbf{a},\mathbf{b})_{|t=0} \qquad \text{for all} \quad \mathbf{a},\mathbf{b} \in T_P\mathbb{E}^3.$$

Mnemonically, we briefly write $\mathcal{L}_{\mathbf{v}}g := \frac{d}{dt}F_t^*g_{|t=0}$. Let us introduce the linear self-adjoint operator $\mathsf{D}_P : T_P\mathbb{E}^3 \to T_P\mathbb{E}^3$ by setting

$$\mathsf{D}_P := \tfrac{1}{2}(\mathbf{v}'(P) + \mathbf{v}'(P)^d). \tag{12.97}$$

In hydrodynamics, the crucial operator D_P is called the rate-of-deformation operator (or the rate-of-strain tensor) at the point P.

Proposition 12.18 $(\mathcal{L}_{\mathbf{v}}g)_P = 2\mathsf{D}_P$ *for all points* $P \in \mathbb{E}^3$.

Explicitly, this means that

$$(\mathcal{L}_{\mathbf{v}}g)_P(\mathbf{a},\mathbf{b}) = 2\mathbf{a}(\mathsf{D}_P\mathbf{b}) \qquad \text{for all} \quad \mathbf{a},\mathbf{b} \in T_P\mathbb{E}^3. \tag{12.98}$$

Proof. Formula (12.98) follows from

$$\frac{d}{dt}(F_t^* g)_{|t=0} = \frac{d}{dt}(F_t'(P)\mathbf{a} \cdot F_t'(P)\mathbf{b})_{|t=0} = \mathbf{v}'(P)\mathbf{a} \cdot \mathbf{b} + \mathbf{a} \cdot \mathbf{v}'(P)\mathbf{b},$$

by using the Leibniz product rule together with (10.10) on page 650. □

Cartesian coordinate system. Let us choose a right-handed Cartesian (x, y, z)-coordinate system (see page 708). Then

$$g = \delta_{ij} dx^i \otimes dx^j, \qquad \mathbf{v} = v^i \mathbf{e}_i, \quad v_i = v^i, \quad \partial^i = \partial_i = \frac{\partial}{\partial x^i},$$

and

$$\mathcal{L}_\mathbf{v} g = (\partial_i v_j + \partial_j v_i)\, dx^i \otimes dx^j.$$

Moreover, the operator $\mathsf{D}_P = \mathsf{D}^i_j(P)\, \mathbf{e}_i \otimes dx^j$ has the matrix elements

$$\mathsf{D}^i_j(P) = \tfrac{1}{2}(\partial^i v_j(P) + \partial_j v^i(P)), \qquad i, j = 1, 2, 3.$$

Hence

$$\mathrm{tr}(\mathsf{D}_P) = \mathrm{div}\, \mathbf{v}(P).$$

Applications of the Lie derivative $\mathcal{L}_\mathbf{v} g$ to the strain tensor in elasticity, the rate-of-strain tensor in hydrodynamics, and to infinitesimal isometries will be considered below.

12.5.6 The Lie Derivative of Linear Operator Fields

Invariant definition. Let A be a smooth linear operator field on the Euclidean manifold $\mathbb{E}^3$. This means that the operator

$$A_P : T_P\mathbb{E}^3 \to T_P\mathbb{E}^3$$

is linear for all points $P \in \mathbb{E}^3$, and all the matrix elements $A^i_j(P)$ of A_P (with respect to an arbitrary Cartesian coordinate system) depend smoothly on the point P. Let $\mathbf{v} \in \mathrm{Vect}(\mathbb{E}^3)$. We want to define the Lie derivative $\mathcal{L}_\mathbf{v} A$ in such a way that we get the product rule:

$$\mathcal{L}_\mathbf{v}(A\mathbf{w}) = (\mathcal{L}_\mathbf{v} A)\mathbf{w} + A(\mathcal{L}_\mathbf{v}\mathbf{w}).$$

To this end, we define

$$(\mathcal{L}_\mathbf{v} A)\mathbf{w} := \mathcal{L}_\mathbf{v}(A\mathbf{w}) - A(\mathcal{L}_\mathbf{v}\mathbf{w}) \qquad \text{for all} \quad \mathbf{w} \in \mathrm{Vect}(\mathbb{E}^3). \tag{12.99}$$

That is, we reduce the definition of $\mathcal{L}_\mathbf{v} A$ to the known Lie derivative for vector fields. In what follows, we will show that the Lie derivative $\mathcal{L}_\mathbf{v} A$ is a smooth linear operator field on $\mathbb{E}^3$ which is uniquely defined by (12.99).

Cartesian coordinates. Choose a right-handed Cartesian (x, y, z)-coordinate system. Then $A_P = A^i_j(P)\, \mathbf{e}_i \otimes dx^j$ (see page 708).

Proposition 12.19 *There holds*

$$\mathcal{L}_\mathbf{v} A = (\mathcal{L}_\mathbf{v} A^i_j)\, \mathbf{e}_i \otimes dx^j + A^i_j\, \mathcal{L}_\mathbf{v}\mathbf{e}_i \otimes dx^j + A^i_j\, \mathbf{e}_i \otimes \mathcal{L}_\mathbf{v}(dx^j)$$

with $\mathcal{L}_\mathbf{v}\mathbf{e}_i = -\partial_i v^s \mathbf{e}_s$, *and* $\mathcal{L}_\mathbf{v}(dx^j) = (\partial_s v^j)\, dx^s$.

Explicitly, we get

$$\mathcal{L}_{\mathbf{v}}A = (v^s\partial_s A^i_j - \partial_s v^i \cdot A^s_j + \partial_j v^s \cdot A^i_s)\,\mathbf{e}_i \otimes dx^j.$$

Proof. Set $\mathbf{u} := A\mathbf{w}$. Then $u^i = A^i_j w^j$. By definition of the Lie derivative of vector fields, we get

$$\begin{aligned}\mathcal{L}_{\mathbf{v}}(A\mathbf{w}) &= (v^s\partial_s u^i - u^s\partial_s v^i)\,\mathbf{e}_i\\ &= (v^s\partial_s A^i_j \cdot w^j + v^s A^i_j\partial_s w^j - A^s_j\partial_s v^i \cdot w^j)\,\mathbf{e}_i,\end{aligned}$$

and $A(\mathcal{L}_{\mathbf{v}}\mathbf{w}) = A^i_j(v^s\partial_s w^j - w^s\partial_s v^j)\,\mathbf{e}_i$. Hence

$$(\mathcal{L}_{\mathbf{v}}A)\mathbf{w} = \mathcal{L}_{\mathbf{v}}(A\mathbf{w}) - A(\mathcal{L}_{\mathbf{v}}\mathbf{w}) = (v^s\partial_s A^i_j - \partial_s v^i \cdot A^s_j + \partial_j v^s \cdot A^i_s)w^j\,\mathbf{e}_i.$$

□

12.6 Diffeomorphisms and the Mechanics of Continua – the Prototype of an Effective Theory in Physics

> The theory of elasticity and fluid dynamics are prototypes of *effective* theories in physics. This means the following. We do not know the precise molecular interaction forces in elastic bodies and fluids. Therefore, we describe these forces by the stress tensor and constitutive laws which average the interaction. The use of effective theories is very successful in physics and quantum chemistry.
> Many physicists believe that quantum field theory is nothing else than an effective theory at a low energy scale compared to the hot universe after the Big Bang.
> The Standard Model in cosmology is an effective theory where the mass and energy distribution of the universe is described by an ideal fluid.[27]
> Nowadays the universe is expanding in an accelerated manner. This is caused by a negative pressure described by the cosmological constant in the Einstein equations in general relativity. It is a famous open problem to understand the physics behind the cosmological constant.
>
> Folklore

We want to understand the relation between, pressure, negative pressure, the deformation tensor, the Lie derivative of the metric tensor field, and the mathematical

[27] Elasticity, hydrodynamics, thermodynamics, statistical physics, and their applications to Einstein's theory of general relativity and cosmology are thoroughly studied in E. Zeidler, Nonlinear Functional Analysis and its Applications, Vol. IV: Applications to Mathematical Physics, Springer, New York, 1997. Recent developments in cosmology will be discussed in Vol. IV of the present monograph. See also the comprehensive monograph by S. Weinberg, Cosmology, Oxford University Press, 2008.
For effective theories in quantum chemistry (e.g., the density functional theory for large molecules), we recommend the extensive monograph by L. Piela, Ideas of Quantum Chemistry, Elsevier, Amsterdam, 1086 pages, Elsevier, Amsterdam, 2007. Furthermore, we recommend the handbook edited by G. Drake, Handbook of Atomic, Molecular, and Optical Physics, Springer, Berlin, 2005.

structure theory for diffeomorphisms. To this end, let us investigate the local and global behavior of smooth maps[28]

$$F : \mathbb{E}^3 \to \mathbb{E}^3. \tag{12.100}$$

The main idea is to use the Taylor expansion and to study the linear approximation

$$\boxed{F(P) = F(P_0) + F'(P)(\mathbf{x} - \mathbf{x}_0) + o(|\mathbf{x} - \mathbf{x}_0|), \qquad \mathbf{x} \to \mathbf{x}_0.} \tag{12.101}$$

Here, we fix the origin O, and we use the position vectors $\mathbf{x} = \overrightarrow{OP}$ and $\mathbf{x}_0 = \overrightarrow{OP_0}$. The linearization of the map F at the point P_0 is defined by the map

$$F_{\text{lin}}(P) = F(P_0) + F'(P_0)(\mathbf{x} - \mathbf{x}_0) \qquad \text{for all} \quad P \in \mathbb{E}^3. \tag{12.102}$$

Recall that the map (12.100) is called a diffeomorphism iff it is smooth, bijective, and the inverse map is also smooth. Moreover, the map (12.102) is called a linear diffeomorphism iff the linear operator $F'(P_0) : T_{P_0}\mathbb{E}^3 \to T_{P_0}\mathbb{E}^3$ is bijective, that is, $\det(F'(P_0)) \neq 0$. Our first goal is to study the linearization F_{lin}. After that we will pass from the local behavior to the global behavior by using topological methods.

12.6.1 Linear Diffeomorphisms and Deformation Operators

Affine transformations and similarity transformations. The tangent spaces of the Euclidean manifold $\mathbb{E}^3$ are Hilbert spaces which are isomorphic to the Hilbert space E_3. Let us first study affine transformations $T : E_3 \to E_3$, that is, we have

$$\boxed{T(\mathbf{x}) := A\mathbf{x} + \mathbf{a} \qquad \text{for all} \quad \mathbf{x} \in E_3,} \tag{12.103}$$

where $A : E_3 \to E_3$ is a linear bijective operator, and $\mathbf{a}$ is a fixed vector in E_3. Let us introduce the following terminology:

- A is called orientation-preserving iff $\det(A) > 0$.
- A is called volume-preserving iff $\det(A) = 1$.
- A is called a similarity operator iff it is self-adjoint, and all the eigenvalues $\lambda^1, \lambda^2, \lambda^3$ of A are positive.
- D is called a deformation operator on E_3 iff $I + \mathsf{D}$ is a similarity operator.
- The deformation operator D is called volume-preserving on an infinitesimal level iff $\text{tr}(\mathsf{D}) = 0$.

Let us discuss this. To begin with, let $\mathbf{b}, \mathbf{c}, \mathbf{d} \in E_3$ be three linearly independent vectors. They span the oriented volume $v(\mathbf{a}, \mathbf{b}, \mathbf{c}) = \mathbf{a}(\mathbf{b} \times \mathbf{c})$. Then

$$v(A\mathbf{a}, A\mathbf{b}, A\mathbf{c}) = \det(A) \cdot v(\mathbf{a}, \mathbf{b}, \mathbf{c}).$$

If A is a similarity operator, then the principal axis theorem tells us that there exists a right-handed orthonormal vector basis $\mathbf{i}, \mathbf{j}, \mathbf{k}$ such that

$$A(x\mathbf{i} + y\mathbf{j} + z\mathbf{k}) = \lambda^1 x\mathbf{i} + \lambda^2 y\mathbf{j} + \lambda^3 z\mathbf{k}, \qquad x, y, z \in \mathbb{R}.$$

This is a so-called similarity transformation. For example, the x-axis is stretched by the factor $\lambda^1 > 0$. The similarity operator A is called an expansion (resp. compression) operator iff $\lambda^j > 1$ (resp. $\lambda^j < 1$) for $j = 1, 2, 3$.

[28] In order to explain the main ideas in an intuitive setting, we restrict ourselves to the Euclidean manifold. Far-reaching generalizations to finite-dimensional and infinite-dimensional manifolds can be found in E. Zeidler (1997), Vol. IV.

The relative volume change. Naturally enough, the relative volume change $\eta(A)$ of the linear similarity operator A is defined by

$$\eta(A) := \frac{v(A\mathbf{a}, A\mathbf{b}, A\mathbf{c}) - v(\mathbf{a}, \mathbf{b}, \mathbf{c})}{v(\mathbf{a}, \mathbf{b}, \mathbf{c})} = \det(A) - 1 = \lambda^1\lambda^2\lambda^3 - 1.$$

If D is a deformation operator, then

$$\eta(I + \mathsf{D}) = \det(I + \mathsf{D}) - 1 = \operatorname{tr}(\mathsf{D}) + o(||\mathsf{D}||), \qquad ||\mathsf{D}|| \to 0.$$

Explicitly, let $A = I + \mathsf{D}$. Then the deformation operator D has the eigenvalues $\lambda^1 - 1, \lambda^2 - 1, \lambda^3 - 1 > -1$. Hence

$$\operatorname{tr}(\mathsf{D}) = (\lambda^1 - 1) + (\lambda^2 - 1) + (\lambda^3 - 1) = \lambda^1 + \lambda^2 + \lambda^3 - 3,$$

and $||\mathsf{D}||^2 = \operatorname{tr}(\mathsf{D}^2) = (\lambda^1 - 1)^2 + (\lambda^2 - 1)^2 + (\lambda^3 - 1)^2$.

The normal form theorem. The key formula reads as

$$\boxed{A = \operatorname{sgn}\det(A) \cdot R(I + \mathsf{D}).} \tag{12.104}$$

Theorem 12.20 *If the operator $A : E_3 \to E_3$ is linear, then there exist both a rotation operator $R \in U(E_3)$ and a deformation operator D such that the factorization formula* (12.104) *holds. Explicitly, $I + \mathsf{D} = \sqrt{AA^d}$.*

The proof can be found in Zeidler (1997), p. 169, quoted on page 717. In terms of geometry, this theorem tells us that an orientation-preserving affine transformation (12.103) is the superposition of a translation, a rotation, and a similarity transformation. If the affine transformation is not orientation-preserving, then one has to add a reflection $\mathbf{x} \mapsto -\mathbf{x}$.

12.6.2 Local Diffeomorphisms

The map $F : \mathbb{E}^3 \to \mathbb{E}^3$ is called a local diffeomorphism at the point P_0 iff there exist open neighborhoods $U(P_0)$ and $V(F(P_0))$ of the points P_0 and $F(P_0)$, respectively, such that the map $F : U(P_0) \to V(F(P_0))$ is a diffeomorphism.

Proposition 12.21 *The smooth map $F : \mathbb{E}^3 \to \mathbb{E}^3$ is a local diffeomorphism at the point P_0 iff $\det F'(P_0) \neq 0$.*

This generalizes the following elementary theorem: The smooth map

$$F : \mathbb{R} \to \mathbb{R} \tag{12.105}$$

is a local diffeomorphism at the point x_0 iff $F'(x_0) \neq 0$. Next we want to generalize the following two classical theorems:

(i) The smooth map (12.105) is a diffeomorphism iff it is a local diffeomorphism at each point $x \in \mathbb{R}$ and $\lim_{|x| \to \infty} |F(x)| = \infty$ (Fig. 12.15).

(ii) If the smooth map (12.105) is strictly monotone and $\lim_{|x| \to \infty} |F(x)| = \infty$, then we have $F'(x) > 0$ for all $x \in \mathbb{R}$, and hence F is a local diffeomorphism at each point $x \in \mathbb{R}$. By (i), F is a diffeomorphism.

(a) local diffeomorphism (b) global diffeomorphism

Fig. 12.15. Diffeomorphism

12.6.3 Proper Maps and Hadamard's Theorem on Diffeomorphisms

The map $F : X \to Y$ from the topological space X to the topological space Y is called proper iff it is continuous and the preimages of compact sets are again compact. For example, homeomorphisms are proper. Moreover, the smooth map $F : \mathbb{E}^3 \to \mathbb{E}^3$ is proper iff $\lim_{d(P,O)\to\infty} |F(P)| = \infty$.[29] The following theorem is called the global inverse function theorem.

Theorem 12.22 *The smooth map* $F : \mathbb{E}^3 \to \mathbb{E}^3$ *is a diffeomorphism iff* F *is proper and* $\det F'(P) \neq 0$ *for all* $P \in \mathbb{E}^3$.

The proof of the more general Banach–Mazur theorem for finite-dimensional and infinite-dimensional Banach spaces can be found in M. Berger, Nonlinearity and Functional Analysis, Academic Press, 1977 (Sect. 5.1A). The topological proof uses covering spaces.[30]

12.6.4 Monotone Operators and Diffeomorphisms

The operator $A : E_3 \to E_3$ is called strictly monotone iff

$$\langle A(\mathbf{x}) - A(\mathbf{y})|\mathbf{x}-\mathbf{y}\rangle > 0 \qquad \text{for all} \quad \mathbf{x},\mathbf{y} \in E_3, \quad \mathbf{x} \neq \mathbf{y}.$$

Moreover, the operator A is called coercive iff $\lim_{|\mathbf{x}|\to\infty} \frac{|F(\mathbf{x})|}{|\mathbf{x}|} = \infty$.

Theorem 12.23 *The smooth operator* $A : E_3 \to E_3$ *is a diffeomorphism if it is strictly monotone and coercive.*

Let $F : \mathbb{E}^3 \to \mathbb{E}^3$ be a smooth map. In a quite natural way, we assign to F the operator $A : E_3 \to E_3$ by identifying the points P and $F(P)$ with the corresponding position vectors $\overrightarrow{OP}$ and $\overrightarrow{OF(P)}$ at the origin, respectively, and by identifying the tangent space $T_O\mathbb{E}^3$ at the origin with the Hilbert space E_3. By Theorem 12.23, we get the following:

The smooth map $F : \mathbb{E}^3 \to \mathbb{E}^3$ *is a diffeomorphism if the corresponding operator* $A : E_3 \to E_3$ *is strictly monotone and coercive.*

---

[29] Here, $d(P,O)$ denotes the distance between the origin O and the point P.

[30] S. Banach and S. Mazur, On multi-valued continuous mappings, Studia Math. **5** (1934), 174–178 (in German).

The general theory of monotone operators in finite-dimensional and infinite-dimensional Banach spaces was developed in the 1960s and 1970s under the strong influence of Felix Browder. Monotone operators play a fundamental role in solving classes of nonlinear partial differential equations arising in elasticity, plasticity, and hydrodynamics. We refer to E. Zeidler, Nonlinear Functional Analysis and its Applications, Vols. I–IV, Springer, New York, 1986ff. In particular, the proof of Theorem 12.23 on page 557 of Vol. IIB uses the Brouwer fixed-point theorem which is a typical topological result.

12.6.5 Sard's Theorem on the Genericity of Regular Solution Sets

We are given the smooth map $F : \mathbb{E}^3 \to \mathbb{E}^3$. Let us consider the equation

$$F(P) = Q, \qquad P \in \mathbb{E}^3. \tag{12.106}$$

If F is a diffeomorphism, then for every given point $Q \in \mathbb{E}^3$, the equation (12.106) has a unique solution, and this solution depends smoothly on Q. In the general case, equation (12.106) may possess an infinite number of solutions. In 1942, Sard discovered that the situation is much better. He showed that, generically, equation (12.106) has at most a finite number of solutions. The precise formulation will be given in Theorem 12.24 below.[31] To begin with, let us introduce the following terminology:

- Regular image point: The point $Q \in \mathbb{E}^3$ is called a regular image point (or a regular value) of the map F iff the map F is a local diffeomorphism at every solution point P of equation (12.106). In other words, we have $\det F'(P) \neq 0$ for all solutions P of (12.106).
- Singular image point: The point $Q \in \mathbb{E}^3$ is called a singular image point (or a singular value) of the map F iff it is not a regular image point of F.

Then the following hold:

(i) The set S of singular image points of the smooth map $F : \mathbb{E}^3 \to \mathbb{E}^3$ has the Lebesgue measure zero on $\mathbb{E}^3$ (Sard's theorem).
(ii) If, in addition, the map F is proper, then for every point $Q \in \mathbb{E}^3 \setminus S$, the solution set of (12.106) is compact and consists of isolated points, by (i). Therefore, the equation (12.106) has at most a finite number of solutions.

This implies the following result.

Theorem 12.24 *Let $F : \mathbb{E}^3 \to \mathbb{E}^3$ be a smooth and proper map. For every given point $Q_0 \in \mathbb{E}^3$ and every given number $\varepsilon > 0$, there exists a point $Q \in \mathbb{E}^3$ with distance $d(Q, Q_0) < \varepsilon$ such that the equation* (12.106) *has at most a finite number of solutions.*

For a smooth map $F : \mathbb{R} \to \mathbb{R}$, the intuitive meaning of the theorem is depicted in Fig. 12.16. The value y_0 is regular iff the horizontal straight line $y = y_0$ intersects transversally the graph of the function F. It follows geometrically from Fig. 12.16(b) that an arbitrarily small perturbation of the value y_0 changes the singular intersection picture into a regular one.

[31] A.Sard, The measure of critical values of differentiable maps, Bull. Amer. Math. Soc. **48** (1942), 883–890. The proof of Sard's theorem together with many applications in differential topology can be found in V. Guillemin and A. Pollack, Differential Topology, Prentice Hall, Englewood Cliffs, New Jersey, 1974.

(a) regular (b) singular

Fig. 12.16. Regular and singular values y_0

12.6.6 The Strain Tensor and the Stress Tensor in Cauchy's Theory of Elasticity

Ut tensio sic vis.
Robert Hooke, 1678[32]

In elasticity, the strain tensor locally separates the deformations from the rotations. The strain tensor $\mathsf{D} = \frac{1}{2}\mathcal{L}_{\mathbf{u}}g$ is proportional to the Lie derivative of the metric tensor field g with respect to the displacement vector field $\mathbf{u}$. In order to understand the light pressure and the Maxwell stress tensor of the electromagnetic field, one has to understand Cauchy's pressure tensor in elasticity.[33]

Folklore

Let us start with elastostatics where the deformation field $P \mapsto \mathbf{u}_P$ does not depend on time t. Elastodynamics and hydrodynamics will be considered below.

The displacement vector field. Consider the smooth map $\Phi : \mathbb{E}^3 \to \mathbb{E}^3$. Set $Q := \Phi(P)$. Introducing the position vectors $\mathbf{x} := \overrightarrow{OP}$ and $\mathbf{y} = \overrightarrow{OQ}$ at the origin O, and the position vector $\mathbf{u}(P) := \overrightarrow{PQ}$ at the point P (Fig. 12.17(a)), the given transformation $P \mapsto \Phi(P)$ can be written as

$$\boxed{\mathbf{y} = \mathbf{x} + \mathbf{u}(P).}$$

The map $P \mapsto \mathbf{u}(P)$ is called the displacement vector field of the map Φ. Next we want to study the local behavior of the map Φ near the point P_0. By Taylor expansion, we get

$$\mathbf{y} = \mathbf{x}_0 + (I + \mathbf{u}'(P_0))(\mathbf{x} - \mathbf{x}_0) + o(|\mathbf{x} - \mathbf{x}_0|), \qquad \mathbf{x} \to \mathbf{x}_0.$$

Following Cauchy, we use the operator decomposition

$$\mathbf{u}'(P) = \tfrac{1}{2}(\mathbf{u}'(P) + \mathbf{u}'(P)^d) + \tfrac{1}{2}(\mathbf{u}'(P) - \mathbf{u}'(P)^d)$$

into a symmetric and an antisymmetric part.

The strain tensor field. Introducing the notation

- $\mathsf{D}_P := \frac{1}{2}(\mathbf{u}'(P) + \mathbf{u}'(P)^d)$ and

---

[32] The force of a spring is proportional to its relative extension (Hooke's law).

[33] The basic papers in hydrodynamics (resp. elasticity) were written by Euler (1755), Navier (1822), and Stokes (1845) (resp. Cauchy (1827)). See the detailed references quoted on page 792.

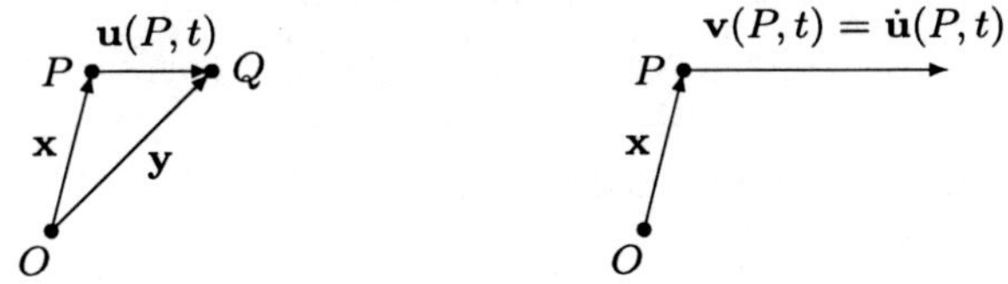

(a) displacement (b) displacement velocity

Fig. 12.17. Deformation

- $\mathsf{R}_P := \frac{1}{2}(\mathbf{u}'(P) - \mathbf{u}'(P)^d)$,

we obtain

$$\mathbf{y} = \mathbf{x}_0 + (I + \mathsf{D}_{P_0} + \mathsf{R}_{P_0})(\mathbf{x} - \mathbf{x}_0) + o(|\mathbf{x} - \mathbf{x}_0|), \qquad \mathbf{x} \to \mathbf{x}_0.$$

In a right-handed Cartesian (x, y, z)-coordinate system, we get

$$\mathsf{D}_{P_0} = D^i_j(P_0)\, \mathbf{e}_i \otimes dx^j, \qquad \mathsf{R}_{P_0} = R^i_j(P_0)\, \mathbf{e}_i \otimes dx^j$$

with the matrix elements

$$D^i_j(P_0) = \tfrac{1}{2}\big(\partial_j u^i(P_0) + \partial^i u_j(P_0)\big), \qquad R^i_j(P_0) = \tfrac{1}{2}\big(\partial_j u^i(P_0) - \partial^i u_j(P_0)\big)$$

where $i, j = 1, 2, 3$. Hence, as $\mathbf{x} \to \mathbf{x}_0$, we obtain the key formula

$$\boxed{\mathbf{y} = \mathbf{x}_0 + (I + \mathsf{D}_{P_0})(\mathbf{x} - \mathbf{x}_0) + \tfrac{1}{2}\,\mathbf{curl}\,\mathbf{u}(P_0) \times (\mathbf{x} - \mathbf{x}_0) + o(|\mathbf{x} - \mathbf{x}_0|).} \tag{12.107}$$

This is the superposition of a translation, a similarity transformation, and an infinitesimal rotation near the point P_0, up to terms of higher order. Moreover,

$$\mathrm{tr}(\mathsf{D}_{P_0}) = \mathrm{div}\,\mathbf{u}(P_0).$$

We assume that the linear self-adjoint operator

$$\mathsf{D}_{P_0} : T_{P_0}\mathbb{E}^3 \to T_{P_0}\mathbb{E}^3$$

is a deformation operator for all points $P_0 \in \mathbb{E}^3$. This is the case if the eigenvalues of D_P are sufficiently small, that is, $||\mathsf{D}_{P_0}|| < 1$ for all $P_0 \in \mathbb{E}^3$. The map $P \mapsto \mathsf{D}_P$ is called the linear strain tensor field of the original map Φ. By (12.98) on page 715, we get the relation between the linear strain tensor and the Lie derivative $\mathcal{L}_{\mathbf{u}} g$ of the metric tensor field g with respect to the displacement vector field $\mathbf{u}$:

$$\boxed{\mathsf{D}_P = \tfrac{1}{2}(\mathcal{L}_{\mathbf{u}} g)_P \qquad \text{for all} \quad P \in \mathbb{E}^3.}$$

Homogeneous and isotropic linear elastic material. The deformation induces an internal stress force. Our goal is to compute this force. To this end, we define

$$\boxed{\mathsf{S} := \mu\, \mathrm{tr}(\mathsf{D}) \cdot I + 2\kappa \mathsf{D}} \tag{12.108}$$

with positive material constants μ and κ. If $\mathcal{M}$ is a compact 3-dimensional submanifold of $\mathbb{E}^3$, then the following elastic force acts on the deformed domain $\Phi(\mathcal{M})$:

$$\boxed{\mathbf{F}_{\mathrm{el}}(\Phi(\mathcal{M})) = \int_{\partial\mathcal{M}} \mathsf{S}\mathbf{n}\, dS} \tag{12.109}$$

where $\mathbf{n}$ denotes the outer unit normal vector. This elastic force is a so-called surface force. The operator $\mathsf{S}_P : T_P\mathbb{E}^3 \to T_P\mathbb{E}^3$ is called the stress tensor at the point P, and the constitutive law (12.108) is called Hooke's law. By the Gauss–Ostrogradsky theorem, we get[34]

$$\mathbf{F}_{\mathrm{el}}(\Phi(\mathcal{M})) = \int_{\mathcal{M}} \operatorname{div} \mathsf{S}\, d^3x.$$

This elastic force is based on molecular interactions. We suppose that there acts additionally the external force

$$\mathbf{F}_{\mathrm{ext}}(\Phi(\mathcal{M})) = \int_{\mathcal{M}} \mathbf{f}_{\mathrm{ext}}\, d^3x$$

on the deformed domain $\Phi(\mathcal{M})$. Here, the map $P \mapsto \mathbf{f}_{\mathrm{ext}}(P)$ is assumed to be a smooth vector field on $\mathbb{E}^3$. Naturally enough, we postulate that

$$\mathbf{F}_{\mathrm{el}}(\Phi(\mathcal{M})) + \mathbf{F}_{\mathrm{ext}}(\Phi(\mathcal{M})) = 0 \qquad \text{(equilibrium of forces)}.$$

Contracting the domain $\mathcal{M}$ to the point P, we get

$$\operatorname{div} \mathsf{S} + \mathbf{f}_{\mathrm{ext}} = 0 \quad \text{on } \mathbb{E}^3.$$

This is the basic equation of linear elastostatics. In terms of the displacement field $\mathbf{u}$, this equation reads as

$$\kappa\Delta\mathbf{u} - (\kappa+\mu)\,\mathbf{grad}\operatorname{div}\mathbf{u} = \mathbf{f}_{\mathrm{ext}} \quad \text{on } \mathbb{E}^3.$$

Consider an elastic body which is a deformation of the 3-dimensional compact submanifold $\mathcal{M}_0$ of $\mathbb{E}^3$. Then the basic boundary-value problem reads as

$$\boxed{\kappa\Delta\mathbf{u} - (\kappa+\mu)\,\mathbf{grad}\operatorname{div}\mathbf{u} = \mathbf{f}_{\mathrm{ext}} \text{ on } \mathcal{M}_0, \quad \mathbf{u} = \mathbf{u}_0 \text{ on } \partial\mathcal{M}_0.} \tag{12.110}$$

We are given the displacement vector field $\mathbf{u}_0$ on the boundary, and we are looking for the displacement vector field $\mathbf{u}$ of the body.

The displacement velocity vector field. Suppose that the displacement vector $\mathbf{u}(P,t)$ at the point $P \in \mathbb{E}^3$ depends on time $t \in \mathbb{R}$. The time derivative of the displacement vector field

$$\mathbf{v}(P,t) := \dot{\mathbf{u}}(P,t), \qquad P \in \mathbb{E}^3, \ t \in \mathbb{R}$$

is called the displacement velocity vector field of the deformation process. The time derivative of the strain tensor field

$$\mathbf{D}(P,t) := \dot{\mathsf{D}}(P,t), \qquad P \in \mathbb{E}^3, \ t \in \mathbb{R}$$

is called the rate-of-strain tensor field of the deformation process. Explicitly,

$$\mathbf{D}(P,t) = \tfrac{1}{2}(\mathbf{v}'(P,t) + \mathbf{v}'(P,t)^d), \qquad P \in \mathbb{E}^3, \ t \in \mathbb{R}.$$

[34] In a right-handed Cartesian coordinate system, we have $\mathsf{S}_P = S^i_j(P)\, \mathbf{e}_i \otimes dx^j$, and $\operatorname{div} \mathsf{S} = \operatorname{div} \mathsf{S}^i \cdot \mathbf{e}_i = \partial^j S^i_j\, \mathbf{e}_i$.

In a right-handed Cartesian (x, y, z)-coordinate system, we get $\mathbf{v} = v^j \mathbf{e}_j$ by introducing the time derivative $v^j := \dot{u}^j$. Moreover,

$$\mathbf{D} = \tfrac{1}{2}(\partial_j v^i + \partial^i v_j)\, \mathbf{e}_i \otimes dx^j.$$

The basic equations in linear elastodynamics. These equations read as

$$\varrho_0 \ddot{\mathbf{u}} + \kappa \Delta \mathbf{u} - (\kappa + \mu)\, \mathbf{grad}\, \mathrm{div}\, \mathbf{u} = \mathbf{f}_{\mathrm{ext}} \quad \text{on } \mathcal{M}_0 \times [0, \infty[.$$

One has to add the following side conditions:

- $\mathbf{u}(P,t) = \mathbf{u}_0(P,t)$ on $\partial\mathcal{M}_0$, $t \geq 0$ (boundary condition),
- $\mathbf{u}(P,0) = \mathbf{u}_1(P)$ and $\dot{\mathbf{u}}(P,0) = \mathbf{v}_1(P)$ on $\mathcal{M}_0$ (initial condition).

We are given $\mathbf{u}_0, \mathbf{u}_1$ and $\mathbf{v}_1$, and we are looking for the displacement vector field $\mathbf{u}$ of the elastic body for all times $t \geq 0$. The positive real number ϱ_0 is the constant mass density of the undeformed elastic body.

Linear and nonlinear theory of elasticity. A critical discussion shows that the approach considered above is only a first-order approximation which is reasonable if the norms $|\mathbf{u}_P|$ and $||\mathbf{u}'_P||$ of the displacement vector field $\mathbf{u}$ and its derivative are small compared with the length scale of the compact elastic body under consideration. The complete nonlinear theory has to refer to the geometry of the deformed elastic body. In particular, one has to use the stress tensor S_* given by

$$\mathbf{F}_{\mathrm{el}}(\Phi(\mathcal{M})) = \int_{\partial\Phi(\mathcal{M})} \mathsf{S}_* \mathbf{n}\, dS.$$

Moreover, one has to use nonlinear constitutive laws. A detailed investigation together with important applications can be found in Zeidler (1997), Vol. IV, quoted on page 717, and in J. Marsden and T. Hughes, Mathematical Foundations of Elasticity, Prentice-Hall, Englewood Cliffs, New Jersey, 1983. For a detailed study of the stress tensor in a general setting, we refer to F. Schuricht, A new mathematical foundation for contact interactions in continuum physics, Archive Rat. Mech. Anal. **184** (2007), 495–551.

12.6.7 The Rate-of-Strain Tensor and the Stress Tensor in the Hydrodynamics of Viscous Fluids

Let us consider a fluid on the Euclidean manifold $\mathbb{E}^3$. The vector $\mathbf{v}(P,t)$ describes the velocity vector of a fluid particle which is located at the point P at time t. We assume that the velocity vector field $(P,t) \mapsto \mathbf{v}(P,t)$ is a smooth map on $\mathbb{E}^3 \times \mathbb{R}$. The point P corresponds to the position vector $\mathbf{x} = \overrightarrow{OP}$ at the origin O. By Taylor expansion near the point P_0, we get

$$\begin{aligned} \mathbf{v}(P) = \mathbf{v}(P_0) &+ \tfrac{1}{2}(\mathbf{v}'(P_0) + \mathbf{v}'(P_0)^d)(\mathbf{x} - \mathbf{x}_0) \\ &+ \tfrac{1}{2}(\mathbf{v}(P_0) - \mathbf{v}'(P_0)^d)(\mathbf{x} - \mathbf{x}_0) + o(|\mathbf{x} - \mathbf{x}_0|), \qquad \mathbf{x} \to \mathbf{x}_0. \end{aligned}$$

Similarly as in (12.107), this implies

$$\mathbf{v}(P) = \mathbf{v}(P_0) + \mathsf{D}_{P_0}(\mathbf{x} - \mathbf{x}_0) + \tfrac{1}{2}\, \mathbf{curl}\, \mathbf{v}(P_0) \times (\mathbf{x} - \mathbf{x}_0) + o(|\mathbf{x} - \mathbf{x}_0|), \quad \mathbf{x} \to \mathbf{x}_0$$

with the rate-of-strain tensor $\mathbf{D}_{P_0} := \frac{1}{2}(\mathbf{v}'(P_0) + \mathbf{v}'(P_0)^d)$ at the point P_0. This yields the linearization

$$\mathbf{v}(P) = \mathbf{v}_{\mathrm{lin}}(P) + o(|\mathbf{x} - \mathbf{x}_0|), \qquad \mathbf{x} \to \mathbf{x}_0$$

with the linearized velocity vector field

$$\boxed{\mathbf{v}_{\text{lin}}(P) := -\,\mathbf{grad}\,U(P) + \boldsymbol{\omega}_0 \times (\mathbf{x} - \mathbf{x}_0).} \tag{12.111}$$

Here, we introduce the rotation vector $\boldsymbol{\omega}_0 := \frac{1}{2}\,\mathbf{curl}\,\mathbf{v}(P_0)$, and the velocity potential

$$U(P) := -\mathbf{v}_0(\mathbf{x} - \mathbf{x}_0) - \tfrac{1}{2}(\mathbf{x} - \mathbf{x}_0)\big(\mathbf{D}_{P_0}(\mathbf{x} - \mathbf{x}_0)\big).$$

The linearized velocity vector field (12.111) is the superposition of a potential flow (e.g., a parallel flow with the velocity vector $\mathbf{v}(P_0)$ if $\mathsf{D}_{P_0} = 0$) and a rotational flow about the axis $\boldsymbol{\omega}_0$ with the angular velocity $\omega_0 = |\boldsymbol{\omega}_0|$. We have

$$\operatorname{div}\mathbf{v}_{\text{lin}}(P) = \Delta U(P), \qquad \mathbf{curl}\,\mathbf{v}_{\text{lin}}(P) = 2\boldsymbol{\omega}_0, \qquad P \in \mathbb{E}^3.$$

Thus, if U is a harmonic function, that is, $\Delta U = 0$ on $\mathbb{E}^3$, then $\operatorname{div}\mathbf{v}_{\text{lin}} = 0$ on $\mathbb{E}^3$.

The stress tensor field. Let $\mathcal{M}$ be a 3-dimensional compact submanifold of the Euclidean manifold $\mathbb{E}^3$. Then the internal force acting on the domain $\mathcal{M}$ is given by

$$\mathbf{F}_{\text{int}}(\mathcal{M}) = -\int_{\partial\mathcal{M}} p\cdot\mathbf{n}\,dS + \int_{\partial\mathcal{M}} \mathsf{S}_{\text{frict}}\mathbf{n}\,dS.$$

Here, the smooth function $p : \mathbb{E}^3 \to [0,\infty[$ is the pressure, and $\mathbf{n}$ is the outer unit normal vector. This internal force is based on molecular interactions. For the stress tensor at the point P, we get

$$\mathsf{S}_P = -p(P)\cdot I + \mathsf{S}_{\text{frict}}(P)$$

where $\mathsf{S}_{\text{frict}}(P)$ is called the stress tensor of inner friction at the point P. The standard constitutive law for viscous fluids due to Navier and Stokes reads as

$$\mathsf{S}_{\text{frict}} = 2\eta(\mathbf{D} - \operatorname{tr}(\mathbf{D})\cdot I) + \gamma\operatorname{tr}(\mathbf{D})\cdot I$$

where η and γ are nonnegative material constants which measure the strength of the viscosity of the fluid.[35] In particular, we have

$$\operatorname{tr}(\mathsf{S}) = \gamma\operatorname{tr}(\mathbf{D}).$$

By the Gauss–Ostrogradsky theorem, we get

$$\begin{aligned}\mathbf{F}_{\text{int}}(\mathcal{M}) &= \int_{\mathcal{M}} (-\,\mathbf{grad}\,p + \operatorname{div}\mathsf{S}_{\text{frict}})\,d^3x\\ &= \int_{\mathcal{M}} (-\,\mathbf{grad}\,p - \eta\Delta\mathbf{v} + (\gamma-\eta)\,\mathbf{grad}\operatorname{div}\mathbf{v})\,d^3x.\end{aligned}$$

We suppose that there acts additionally the external force

$$\mathbf{F}_{\text{ext}}(\mathcal{M}) = \int_{\mathcal{M}} \mathbf{f}_{\text{ext}}\,d^3x$$

on the domain $\mathcal{M}$. The map $P \mapsto \mathbf{f}_{\text{ext}}(P)$ is assumed to be a smooth vector field on $\mathbb{E}^3$. Here, $\mathbf{f}_{\text{ext}}$ is the external force density. There acts the total force

[35] A detailed motivation of this constitutive law and the study of the Navier–Stokes equations in terms of nonlinear functional analysis can be found in Zeidler (1997), Vol. IV quoted on page 717.

$$\mathbf{F}_{\text{int}}(\mathcal{M}) + \mathbf{F}_{\text{ext}}(\mathcal{M})$$

on the domain $\mathcal{M}$. The fluid is called ideal iff the inner friction vanishes identically, that is, $\mathsf{S}_{\text{frict}} = 0$ on $\mathbb{E}^3$. In other words, there is no viscosity ($\eta = \gamma = 0$).

The Navier–Stokes equations. Let $\mathcal{M}_0$ be a 3-dimensional compact submanifold of the Euclidean manifold $\mathbb{E}^3$. The basic equations for a viscous fluid on the domain $\mathcal{M}_0$ read as follows:

(i) Equation of motion:

$$\varrho\dot{\mathbf{v}} + \varrho\,\mathbf{grad}\,\frac{\mathbf{v}^2}{2} - \varrho\mathbf{v} \times \mathbf{curl}\,\mathbf{v} = \mathbf{f}_{\text{ext}} - \mathbf{grad}\,p - \eta\Delta\mathbf{v} + (\gamma - \eta)\,\mathbf{grad}\,\text{div}\,\mathbf{v}.$$

(ii) Mass conservation: $\dot{\varrho} + \text{div}(\varrho\mathbf{v}) = 0$ (continuity equation).
(iii) Density-pressure relation: $\varrho = \varrho(p)$.

We have to add boundary conditions and initial conditions. Here, ϱ denotes the mass density, and the material constants η and γ measure the viscosity of the fluid.

If the mass density ϱ is constant, that is, ϱ does not depend on the pressure p, then the fluid is called incompressible. By the continuity equation (ii), this implies

$$\text{div}\,\mathbf{v} = 0 \quad \text{on } \mathcal{M}_0.$$

The flow is called stationary iff the velocity vector field $\mathbf{v}$, the mass density ϱ, and the pressure p do not depend on time.

Euler's equation of motion for an ideal fluid. We use (i)–(iii) above, but we set $\eta = \gamma = 0$ (no viscosity). Furthermore, we assume that the external force density $\mathbf{f}_{\text{ext}}$ has a potential U, that is, it can be written as

$$\mathbf{f}_{\text{ext}} = -\varrho\,\mathbf{grad}\,U.$$

Then the basic equations for an ideal fluid read as follows:

(i) Equation of motion: $\dot{\mathbf{v}} + \mathbf{grad}\left(\frac{\mathbf{v}^2}{2} + U + \int_{p_0}^{p} \frac{dp}{\varrho(p)}\right) = \mathbf{v} \times \mathbf{curl}\,\mathbf{v}$.
(ii) Mass conservation: $\dot{\varrho} + \text{div}(\varrho\mathbf{v}) = 0$ (continuity equation).
(iii) Density-pressure relation: $\varrho = \varrho(p)$.

We have to add boundary conditions and initial conditions.

Proposition 12.25 *For the stationary irrotational flow of an ideal fluid in an arcwise connected and simply connected open subset of $\mathbb{E}^3$ (e.g., a ball), we have the conservation law of energy, that is, the expression*

$$\tfrac{1}{2}\mathbf{v}^2 + U + \int_{p_0}^{p} \frac{dp}{\varrho(p)} \tag{12.112}$$

does not depend on position and time.

This is called the Bernoulli law. If the fluid is incompressible, then

$$\tfrac{1}{2}\varrho\mathbf{v}^2 + \varrho U + p = \text{const}.$$

Proof. Let L denote the expression (12.112). Since the flow is irrotational, we have $\mathbf{curl}\,\mathbf{v} = 0$. It follows from the equation of motion (i) above that $\mathbf{grad}\,L = 0$. Hence $L = \text{const}$. □

Turbulence and the Millenium Prize Problems. In 2000, the Clay Institute in Boston, Massachusetts (U.S.A), formulated seven famous open problems.

The solution of such a problem will be awarded by 1 million dollars. One of the problems concerns the mathematical theory of turbulence. In physical experiments, turbulence is observed for sufficiently large Reynolds numbers. The millennium problem is to analyze the existence and regularity of the solutions of the Navier–Stokes equations for large Reynolds numbers. We refer to:

> C. Fefferman, Existence and smoothness of the Navier–Stokes equations. In: J. Carlson, A. Jaffe, and A. Wiles (Eds.), The Millennium Prize Problems, Clay Mathematics Institute, Cambridge Massachusetts, 2006, pp. 57–70.

Note that one of the Millennium Prize Problems concerns gauge theory. This can be found in:

> A. Jaffe and E. Witten, Quantum Yang–Mills theory. In: J. Carlson, A. Jaffe, and A. Wiles (Eds.), pp. 129–152.

12.6.8 Vorticity Lines of a Fluid

Set $\boldsymbol{\omega} := \frac{1}{2}\,\mathbf{curl}\,\mathbf{v}$. The vector field $\boldsymbol{\omega}$ measures the vorticity of the velocity vector field $\mathbf{v}$ of the fluid. The differential equation

$$\dot{\mathbf{x}}(t) = \boldsymbol{\omega}(\mathbf{x}(t)), \qquad t \in \mathbb{R}, \qquad \mathbf{x}(0) = \mathbf{x}_0$$

describes the so-called vorticity lines. In other words, the vorticity lines are the streamlines of the vector field $\boldsymbol{\omega}$.

12.6.9 The Lie Derivative of the Covector Field

We are given the smooth velocity vector field $\mathbf{v}$ on the Euclidean manifold $\mathbb{E}^3$. The 1-form

$$\chi := \mathbf{v}d\mathbf{x}$$

is called the covelocity (or covector) field to the velocity vector field $\mathbf{v}$. Explicitly,

$$\chi_P(\mathbf{w}) = (\mathbf{v}\mathbf{w})_P \qquad \text{for all} \quad \mathbf{w} \in \mathrm{Vect}(\mathbb{E}^3), \;\; P \in \mathbb{E}^3.$$

By (12.86) on page 708,

$$d\chi(\mathbf{u}, \mathbf{w}) = (\mathbf{u} \times \mathbf{w})\,\mathbf{curl}\,\mathbf{v} \qquad \text{for all} \quad \mathbf{u}, \mathbf{w} \in \mathrm{Vect}(\mathbb{E}^3).$$

This implies

$$i_{\mathbf{u}} d\chi = (-\mathbf{u} \times \mathbf{curl}\,\mathbf{v})d\mathbf{x}.$$

In fact, $(i_{\mathbf{u}} d\chi)(\mathbf{w}) = d\chi(\mathbf{u}, \mathbf{w}) = -(\mathbf{u} \times \mathbf{curl}\,\mathbf{v})\mathbf{w}$.

Set $\boldsymbol{\omega} := \frac{1}{2}\,\mathbf{curl}\,\mathbf{v}$. The streamlines of the vector field $\boldsymbol{\omega}$ are the vorticity lines of the fluid corresponding to the given velocity vector field $\mathbf{v}$.

Proposition 12.26 (i) $d(\mathcal{L}_{\boldsymbol{\omega}}\chi) = 0$ *on* $\mathbb{E}^3$.

(ii) $d(\mathcal{L}_{\mathbf{v}}\chi) = 0$ *on* $\mathbb{E}^3$ *if the velocity vector field* $\mathbf{v}$ *corresponds to the stationary flow of an ideal fluid.*

We will show in Sect. 12.8.3 on page 735 that (i) and (ii) imply the classical vorticity theorems due to Helmholtz and Thomson, respectively.
Proof. Ad (i). Noting that $d(d\chi) = 0$, it follows from Cartan's magic formula on page 714 that

$$d(\mathcal{L}_\omega \chi) = \mathcal{L}_\omega (d\chi) = i_\omega d(d\chi) + d\big(i_\omega d\chi\big) = d\big(i_\omega d\chi\big).$$

Therefore, it is sufficient to prove that $d\big(i_\omega d\chi\big) = 0$. This follows from

$$i_\omega d\chi = -(\boldsymbol{\omega} \times \mathbf{curl}\,\mathbf{v})\; d\mathbf{x} = -2(\boldsymbol{\omega} \times \boldsymbol{\omega})\; d\mathbf{x} = 0.$$

Ad (ii). Again it is sufficient to show that $d\big(i_\mathbf{v} d\chi\big) = 0$. By the equation of motion (i) above, we get

$$i_\mathbf{v} d\chi = -(\mathbf{v} \times \mathbf{curl}\,\mathbf{v})\; d\mathbf{x} = -\,\mathbf{grad}\, L\; d\mathbf{x} = -dL.$$

Hence $d\big(i_\mathbf{v} d\chi\big) = -d(dL) = 0.$ □

12.7 The Generalized Stokes Theorem (Main Theorem of Calculus)

We want to generalize the main formula of calculus (12.1.1) on page 666. The key formula reads as

$$\boxed{\int_\mathcal{M} d\omega = \int_{\partial\mathcal{M}} \omega.} \tag{12.113}$$

This extremely elegant formula displays a complete symmetry between the boundary operator ∂ and the Cartan differential operator d. As we will show later on, this is the root of the duality between homology and cohomology in differential topology.

> *The generalized Stokes integral theorem reflects a deep relation between mathematics and physics.*

Let us first consider the prototype of the general theorem.

Proposition 12.27 *Let $p = 1, 2, 3$. Equation* (12.113) *holds true if ω is a smooth $(p-1)$-form on the Euclidean manifold $\mathbb{E}^3$, and $\mathcal{M}$ is a p-dimensional compact oriented submanifold of $\mathbb{E}^3$ with coherently oriented boundary.*

In the special case where $p = 3$ and $p = 2$, we get

- the Gauss–Ostrogradsky theorem (see (12.32) on page 680) and
- the Stokes theorem (see (12.33) on page 680), respectively.

If $p = 1$, then (12.113) represents the classical main theorem of calculus due to Leibniz and Newton (see (12.1.1) on page 666).

Concerning an intuitive interpretation of the term 'coherent orientation', we refer to Fig. 12.6 on page 677. The general definition of coherently oriented submanifolds can be found in the Appendix on page 1075.

Generalization to higher dimensions. In order to obtain conservation laws in the theory of special relativity, one needs the validity of formula (12.113) on unbounded submanifolds $\mathcal{M}$ of the 4-dimensional Minkowski manifold $\mathbb{M}^4$. In order to guarantee the existence of the integrals, one uses differential forms ω with compact support, that is, ω vanishes outside a compact subset of $\mathcal{M}$. In the special case where the submanifold $\mathcal{M}$ is compact, every differential form on $\mathcal{M}$ has compact support. The general form of the Stokes integral theorem reads as follows.

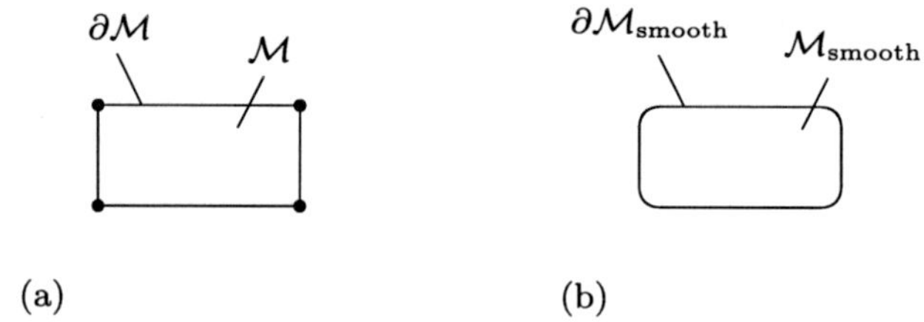

Fig. 12.18. Smoothing of singular boundary parts

Theorem 12.28 *Let* $n = 1, 2, \ldots$ *Equation* (12.113) *holds true if* ω *is a smooth differential* $(n-1)$*-form with compact support on the real* n*-dimensional oriented manifold* $\mathcal{M}$ *with coherently oriented boundary.*

The fairly short standard proof can be found in any modern textbook on calculus. We refer to V. Zorich, Analysis II, Springer, New York, 2003, Sect. 15.3.5.

Admissible and pathological boundary points. Proposition 12.27 cannot be applied to the closed rectangle $\mathcal{M}$ depicted in Fig. 12.18(a). Because of violation of smoothness at the vertices, $\mathcal{M}$ is not a submanifold with boundary of the Euclidean manifold $\mathbb{E}^3$. In this quite natural situation, one can use a smoothing limit. To this end, we first consider the smoothed rectangle $\mathcal{M}_{\text{smooth}}$ from Fig. 12.18(b). Applying Prop. 12.27, we get

$$\int_{\mathcal{M}_{\text{smooth}}} d\omega = \int_{\partial\mathcal{M}_{\text{smooth}}} \omega. \tag{12.114}$$

Then we use the limit $\mathcal{M}_{\text{smooth}} \to \mathcal{M}$; this implies the desired formula:

$$\int_{\mathcal{M}} d\omega = \int_{\partial\mathcal{M}} \omega. \tag{12.115}$$

This limit exists, since the corresponding integrals only differ on sets of small measure. This approach can be generalized to submanifolds of $\mathbb{E}^3$ with 'piecewise smooth

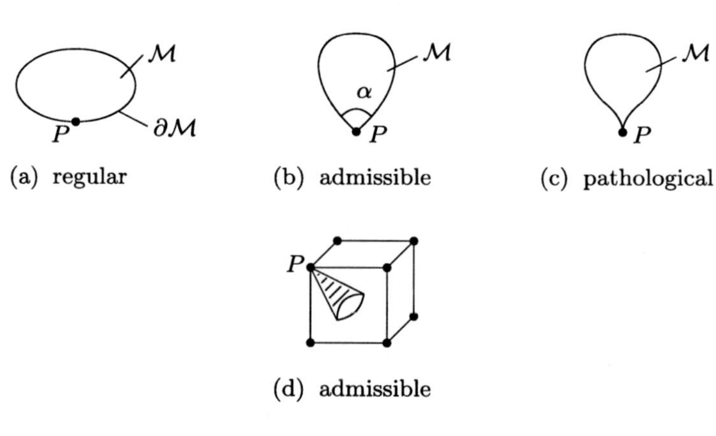

Fig. 12.19. Regular and singular boundary points

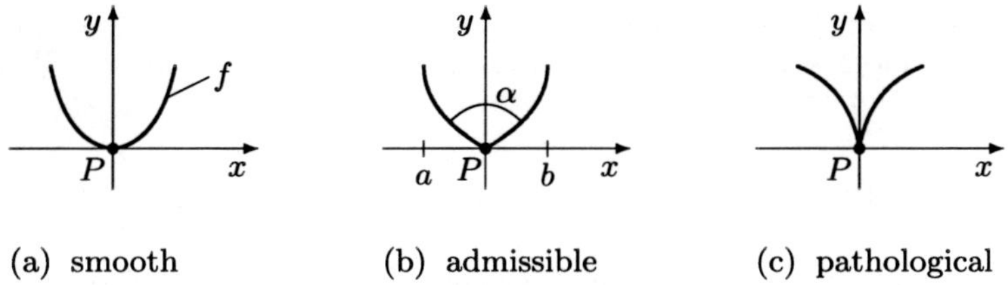

Fig. 12.20. Smooth and non-smooth functions

boundary'. Intuitively, this means that the boundary is smooth up to a finite number of vertices and edges without any zero angles (Fig. 12.19(b), (d)). For example, this concerns triangles, rectangles, polygons, cubes, polyhedra, cones, and finite cylinders. The term 'zero angle' is explained by Fig. 12.19(b) (angle $\alpha > 0$; no zero angle at the vertex P) and Fig. 12.19(c) (zero angle at P). In terms of functions, consider Fig. 12.20. The function f depicted in (b) is not smooth, but it is Lipschitz continuous on the compact interval $[a, b]$.[36] This means that there exists a nonnegative real number L such that

$$|f(x) - f(y)| \leq L|x - y| \qquad \text{for all} \quad x, y \in [a, b].$$

The smallest possible number L is called the Lipschitz constant of f. In Fig. 12.19(c), we depict the function $f(x) := x^{2/3}$ which has a zero angle at the point $(0,0)$. In fact,

$$\lim_{x \to \pm 0} f'(x) = \frac{2}{3} \lim_{x \to \pm 0} \frac{1}{x^{1/3}} = \pm\infty.$$

Thus, as $x \to +0$ and $x \to -0$, the one-sided limits of the tangent lines at the point $(0,0)$ are vertical and parallel.

A general variant of Prop. 12.27 and its generalization to higher dimensions can be found in R. Abraham, J. Marsden, and T. Ratiu, Manifolds, Tensor Analysis, and Applications, Springer, New York, 1988, and in H. Amann and J. Escher, Analysis III, Birkhäuser, Basel, 2001.

12.8 Conservation Laws

> In the 19th century, Thomson (later Lord Kelvin), Helmholtz, Liouville, and Poincaré discovered important conservation laws in fluid dynamics and celestial mechanics based on integral invariants.
> Maxwell applied this to the conservation of electromagnetic energy in the 1870s. The Gibbs approach to statistical mechanics in the 1890s is based on the conservation of the phase-space volume (which is intimately related to the symplectic structure of the Hamiltonian flow on the phase space). In quantum mechanics, the dynamics of the Schrödinger wave function leads to a flow of probability density where probability is preserved like mass is preserved in a fluid.
> The modern approach founded by Élie Cartan in 1922 uses the Lie derivative of differential forms combined with Cartan's magic formula and the generalized Stokes integral theorem.
>
> Folklore

[36] Lipschitz (1832–1903)

Let $\mathbf{v}$ be a smooth complete velocity vector field on the Euclidean manifold $\mathbb{E}^3$, let $\mathbf{u}$ be a smooth vector field (e.g., a displacement vector field) on $\mathbb{E}^3$, and let $\omega \in \Lambda^p(\mathbb{E}^3)$ with $p = 0, 1, 2, 3$.

12.8.1 Infinitesimal Isometries (Metric Killing Vector Fields)

> Infintesimal isometries of the Euclidean manifold $\mathbb{E}^3$ describe the isometries (i.e., translations and rotations) of $\mathbb{E}^3$ on an infinitesimal level near the identity transformation.
>
> Folklore

The smooth velocity vector field $\mathbf{v}$ on $\mathbb{E}^3$ is called an infinitesimal isometry iff

$$\mathcal{L}_{\mathbf{v}} g = 0 \quad \text{on } \mathbb{E}^3$$

where g is the metric tensor field on $\mathbb{E}^3$. Infinitesimal isometries are also called metric Killing vector fields. Fix the point $P \in \mathbb{E}^3$, and fix the tangent vectors $\mathbf{a}, \mathbf{b} \in T_P\mathbb{E}^3$. The flow $\{F_t\}_{t\in\mathbb{R}}$ of a metric Killing vector field has the property that

$$g_P(\mathbf{a}, \mathbf{b}) = g_{F_t(P)}(F_t'(P)\mathbf{a}, F_t'(P)\mathbf{b}) \qquad \text{for all} \quad t \in \mathbb{R}.$$

Explicitly,

$$\mathbf{ab} = (F_t'\mathbf{a})(F_t'\mathbf{b}) \qquad \text{for all} \quad t \in \mathbb{R}.$$

This means that the flow preserves the Hilbert space structure of the tangent spaces along the streamlines.

Proposition 12.29 *The infinitesimal isometries of the Euclidean manifold $\mathbb{E}^3$ form a real Lie subalgebra of* $\mathrm{Vect}(\mathbb{E}^3)$ *denoted by $\mathcal{L}_{\mathrm{isom}}(\mathbb{E}^3)$.*

Proof. Let $\mathbf{v}$ and $\mathbf{w}$ be infinitesimal isometries. If $\mathcal{L}_{\mathbf{v}} g = 0$ and $\mathcal{L}_{\mathbf{w}} g = 0$, then

$$\mathcal{L}_{[\mathbf{v},\mathbf{w}]} g = \mathcal{L}_{\mathbf{v}}(\mathcal{L}_{\mathbf{w}} g) - \mathcal{L}_{\mathbf{w}}(\mathcal{L}_{\mathbf{v}} g) = 0.$$

Thus, the Lie product $[\mathbf{v}, \mathbf{w}]$ is also an infinitesimal isometry. □

Physical interpretation. Fix the origin O of $\mathbb{E}^3$. Choose the position vectors $\mathbf{x}_0 = \overrightarrow{OP_0}$ and $\boldsymbol{\omega}_0$ at the origin. Parallel transport of the position vector $\mathbf{v}_0 + \boldsymbol{\omega}_0 \times \mathbf{x}$ to the point P yields the smooth velocity vector field

$$\boxed{\mathbf{v}_P = \mathbf{v}_0 + \boldsymbol{\omega}_0 \times \mathbf{x}.} \tag{12.116}$$

Using a Cartesian coordinate system, one shows easily that the following hold:

- $\mathcal{L}_{\mathbf{v}} g = 0$ on $\mathbb{E}^3$ (Killing vector field),
- $\operatorname{div} \mathbf{v} = 0$ on $\mathbb{E}^3$ (incompressible fluid), and
- $\mathbf{curl}\, \mathbf{v} = 2\boldsymbol{\omega}_0$ on $\mathbb{E}^3$ (constant vorticity).

The streamlines of the corresponding flow correspond to the superposition of a parallel flow with the velocity vector $\mathbf{v}_0$ and a clockwise rotational flow about the axis $\boldsymbol{\omega}_0$ with angular velocity $\omega_0 := |\boldsymbol{\omega}_0|$. One can show that all the infinitesimal isometries of the Euclidean manifold $\mathbb{E}^3$ are given by (12.116). Since the vectors $\mathbf{v}_0$ and $\boldsymbol{\omega}_0$ have six components,

$$\dim \mathcal{L}_{\mathrm{isom}}(\mathbb{E}^3) = 6.$$

In addition, we have the Lie algebra isomorphism

$$\boxed{\mathcal{L}_{\text{isom}}(\mathbb{E}^3) \simeq o(3) \rtimes \mathcal{L}(\mathbb{R}^3.}$$

Here, $o(3) \rtimes \mathcal{L}(\mathbb{R}^3$ denotes the Lie algebra of the isometry group $O(3) \rtimes \mathbb{R}^3$ of $\mathbb{E}^3$.

Generalization. Isometries are diffeomorphisms which preserve the length. Let $n = 1, 2, \ldots$ The proof of the following fundamental theorem can be found in S. Kobayashi and K. Nomizu, Foundations of Differential Geometry, Vol. 1, Wiley, New York, 1963.

Theorem 12.30 *The isometries of an n-dimensional, arcwise connected, Riemannian manifold $\mathcal{M}$ form a Lie group denoted by* Isom($\mathcal{M}$).

The metric Killing vector fields on $\mathcal{M}$ form a real Lie algebra which is isomorphic to the Lie algebra $\mathcal{L}$(Isom($\mathcal{M}$)) of the isometry group Isom($\mathcal{M}$). *For the dimension of the isometry group, we have the following inequality:*

$$\dim \mathrm{Isom}(\mathcal{M}) \leq \tfrac{1}{2}n(n+1).$$

If the dimension of Isom($\mathcal{M}$) is equal to $m := \frac{1}{2}n(n+1)$, then we say that the manifold $\mathcal{M}$ has an isometry group of maximal dimension. For $n = 1, 2, 3, 4$, we get the maximal dimensions $m = 1, 3, 6, 10$, respectively. The following Riemannian manifolds possess an isometry group of maximal dimension:

- $n = 1$: the real line $\mathbb{E}^1$, the unit circle $\mathbb{S}^1$, and the projective line $\mathbb{P}^1 = \mathbb{S}^1/\{\pm I\}$;
- $n = 2$: the Euclidean plane $\mathbb{E}^2$, the 2-dimensional unit sphere $\mathbb{S}^2$, the projective plane $\mathbb{P}^2 = \mathbb{S}^2/\{\pm I\}$, and the hyperbolic upper half-plane of constant negative curvature, $R = -1$ (see the Poincaré model of non-Euclidean hyperbolic geometry in Sect. 5.10 of Vol. II);
- $n = 3$: the 3-dimensional Euclidean manifold $\mathbb{E}^3$, the 3-dimensional unit sphere $\mathbb{S}^3$, the 3-dimensional projective space $\mathbb{P}^3 = \mathbb{S}^3/\{\pm I\}$, and the 3-dimensional hyperbolic space of constant negative curvature, $R = -1$;
- $n = 4, 5, \ldots$: the n-dimensional Euclidean manifold $\mathbb{E}^n$, the n-dimensional unit sphere $\mathbb{S}^n$, the n-dimensional real projective space $\mathbb{P}^n = \mathbb{S}^n/\{\pm I\}$, and the n-dimensional hyperbolic space of constant negative curvature, $R = -1$. The latter is given by the n-dimensional open unit ball int($\mathbb{B}^n$) equipped with the following metric introduced by Beltrami in 1868:[37]

$$ds^2 = \frac{dx_1^2 + \ldots + dx_n^2 + dt^2}{t^2}$$

where $t := 1 - \sqrt{x_1^2 + \ldots + x_n^2}$.

Integral invariants. We want to show that the conditions

- $\mathcal{L}_{\mathbf{v}}\omega = 0$ and
- $d(\mathcal{L}_{\mathbf{v}}\omega) = 0$

for the Lie derivative $\mathcal{L}_{\mathbf{v}}\omega$ of a differential form ω lead to

- absolute integral invariants $\int_{\mathcal{M}} \omega$ and
- so-called relative integral invariants $\int_{\partial\mathcal{M}} \omega$

with respect to the flow $\{F_t\}_{t\in\mathbb{R}}$ generated by the velocity vector field $\mathbf{v}$, respectively. In the 19th century, Thomson (later Lord Kelvin), Helmholtz, Liouville, and Poincaré discovered such invariants in fluid dynamics and celestial mechanics.[38] Prototypes of an integral invariant are

---

[37] A detailed historical discussion can be found in E. Scholz, History of Manifolds from Riemann to Poincaré, Birkhäuser, Basel, 1980 (in German).

[38] We recommend R. Abraham and J. Marsden, Foundations of Mechanics, Addison-Wesley, Reading, Massachusetts, 1978.

- the volume of an incompressible fluid (Liouville's absolute integral invariant along streamlines) and
- the circulation of an ideal fluid (Thomson's relative integral invariant along streamlines),

and their generalizations to Hamiltonian flow in Hamiltonian mechanics. A general approach to integral invariants was created by Élie Cartan in 1922 (see Cartan (1922) quoted on page 795). In 1918, Emmy Noether proved the fundamental Noether theorem. In terms of physics, this theorem tells us the following. If the action integral is an integral invariant under some flow, then the solutions of the Euler–Lagrange field equations satisfy a specific conservation law (see Sect. 12.8.5). All the crucial continuous conservation laws in physics are obtained this way (e.g., conservation of energy, momentum, and angular momentum). In 1937, Carathéodory discovered his 'royal road' to the calculus of variations based on Huygens' principle in geometrical optics and Hilbert's invariant integral (see Sect. 12.9.9).

12.8.2 Absolute Integral Invariants and Incompressible Fluids

Let ω be a smooth differential r-form on the Euclidean manifold $\mathbb{E}^3$, and let $\mathcal{M}$ be a compact r-dimensional submanifold of $\mathbb{E}^3$ where $r = 1, 2, 3$. Let $\mathbf{v}$ be a smooth complete velocity vector field on $\mathbb{E}^3$ which generates the flow $\{F_t\}_{t\in\mathbb{R}}$.

Theorem 12.31 *Suppose that $\mathcal{L}_\mathbf{v}\omega = 0$ on $\mathbb{E}^3$. Then, the pull-back $(F_t^*\omega)_P$ does not depend on time t for all points $P \in \mathbb{E}^3$. Furthermore, ω is an absolute integral invariant, that is, we have*[39]

$$\int_\mathcal{M} \omega = \int_{F_t(\mathcal{M})} \omega \qquad \textit{for all} \quad t \in \mathbb{R}. \tag{12.117}$$

Proof. Since $F_{s+t} = F_s F_t$, we get $F_{s+t}^* = F_t^* F_s^*$. Hence

$$\frac{d}{ds} F_{s+t}^*\omega_{|s=0} = F_t^* \mathcal{L}_\mathbf{v}\omega = 0.$$

Therefore, $\frac{d}{dt}F_t^*\omega = 0$ for all $t \in \mathbb{R}$. Noting that $F_t^*\omega = \omega$ if $t = 0$, we get

$$\int_\mathcal{M} \omega = \int_\mathcal{M} F_t^*\omega = \int_{F_t(\mathcal{M})} \omega.$$

□

Example. If $\operatorname{div} \mathbf{v} = 0$ on $\mathbb{E}^3$, then the flow $\{F_t\}_{t\in\mathbb{R}}$ preserves the volume of compact 3-dimensional submanifolds $\mathcal{M}$ of $\mathbb{E}^3$. Explicitly, if υ denotes the volume form of $\mathbb{E}^3$, then

$$\int_\mathcal{M} \upsilon = \int_{F_t(\mathcal{M})} \upsilon \qquad \text{for all} \quad t \in \mathbb{R}.$$

Proof. By Prop. 12.17 on page 715, $\mathcal{L}_\mathbf{v}\upsilon = \operatorname{div} \mathbf{v} \cdot \upsilon = 0$. □

[39] If the velocity vector field is not complete, the claims are only valid for appropriate open time intervals J with $0 \in J$. Here, J depends on the point P and the set $\mathcal{M}$.

12.8.3 Relative Integral Invariants and the Vorticity Theorems for Fluids due to Thomson and Helmholtz

Let ω be a smooth differential r-form on the Euclidean manifold $\mathbb{E}^3$, and let $\mathcal{M}$ be a compact, $(r+1)$-dimensional, oriented submanifold of $\mathbb{E}^3$ with coherently oriented boundary $\partial\mathcal{M}$ where $r = 1, 2$. Let $\mathbf{v}$ be a smooth complete velocity vector field on $\mathbb{E}^3$ which generates the flow $\{F_t\}_{t\in\mathbb{R}}$.

Theorem 12.32 *Suppose that $d(\mathcal{L}_{\mathbf{v}}\omega) = 0$ on $\mathbb{E}^3$. Then, $d\omega$ is an absolute integral invariant, and we have*

$$\int_{\partial\mathcal{M}} \omega = \int_{\partial(F_t\mathcal{M})} \omega \qquad \textit{for all} \quad t \in \mathbb{R}. \tag{12.118}$$

Proof. Note that $\mathcal{L}_{\mathbf{v}}(d\omega) = d(\mathcal{L}_{\mathbf{v}}\omega) = 0$. By Theorem 12.31 above, $d\omega$ is an absolute integral invariant, that is, for all $t \in \mathbb{R}$,

$$\int_{\mathcal{M}} d\omega = \int_{F_t(\mathcal{M})} d\omega.$$

The general Stokes theorem yields the claim (12.118). □

The Thomson vorticity theorem. Let $\mathbf{v}$ be a smooth complete velocity vector field on the Euclidean manifold $\mathbb{E}^3$, and let $\mathcal{M}$ be a compact, 2-dimensional, oriented submanifold of $\mathbb{E}^3$ with coherently oriented boundary $\partial\mathcal{M}$. Suppose that the flow $\{F_t\}_{t\in\mathbb{R}}$, generated by the velocity vector field $\mathbf{v}$, is the stationary flow of an ideal fluid. Then the circulation along the oriented boundary curve $\partial\mathcal{M}$ is preserved, that is,

$$\int_{\partial\mathcal{M}} \mathbf{v}d\mathbf{x} = \int_{\partial(F_t\mathcal{M})} \mathbf{v}d\mathbf{x} \qquad \text{for all} \quad t \in \mathbb{R}.$$

Proof. Use Theorem 12.32 above and note that $d(\mathcal{L}_{\mathbf{v}}(\mathbf{v}d\mathbf{x})) = 0$, by Prop. 12.26 on page 728. □

The Helmholtz vorticity theorem. Let $\mathbf{v}$ be a smooth velocity vector field on $\mathbb{E}^3$. Set $\boldsymbol{\omega} := \frac{1}{2}\,\mathbf{curl}\,\mathbf{v}$. Suppose that $\boldsymbol{\omega}$ is a complete vector field on $\mathbb{E}^3$, and let $\{\mathcal{F}_t\}_{t\in\mathbb{R}}$ be the flow generated by $\boldsymbol{\omega}$. The streamlines of $\{\mathcal{F}_t\}_{t\in\mathbb{R}}$ are the vorticity lines of the velocity vector field $\mathbf{v}$. Let $\mathcal{M}$ be a compact 2-dimensional oriented submanifold of $\mathbb{E}^3$ with coherently oriented boundary $\partial\mathcal{M}$. Then the circulation of the velocity vector field $\mathbf{v}$ along the oriented curve $\partial\mathcal{M}$ satisfies the relation

$$\int_{\partial\mathcal{M}} \mathbf{v}d\mathbf{x} = \int_{\partial(\mathcal{F}_t\mathcal{M})} \mathbf{v}d\mathbf{x} \qquad \text{for all} \quad t \in \mathbb{R}.$$

Proof. Use Theorem 12.32 above and note that $d(\mathcal{L}_{\boldsymbol{\omega}}(\mathbf{v}d\mathbf{x})) = 0$, by Prop. 12.26 on page 728. □

12.8.4 The Transport Theorem

We want to study the time derivative of the following integral

$$\boxed{M(t) := \int_{F_t(\mathcal{C})} \varrho(P,t)\, \upsilon, \qquad t \in \mathbb{R}} \tag{12.119}$$

where $\mathcal{C}$ is a compact subset of the Euclidean manifold $\mathbb{E}^3$, and υ is the volume form on $\mathbb{E}^3$.[40] We are given the smooth density function $\varrho : \mathbb{E}^3 \times \mathbb{R} \to \mathbb{R}$ and the smooth complete velocity vector field $\mathbf{v}$ on $\mathbb{E}^3$ which generates the flow $\{F_t\}_{t\in\mathbb{R}}$. In terms of physics, we regard $\varrho(P,t)$ as mass density (or electric charge density) at the point P at time t. The integral $M(0) = \int_{\mathcal{C}} \varrho(P,0)\ \upsilon$ is the mass of the domain $\mathcal{C}$ at time $t = 0$. The flow transports the domain $\mathcal{C}$ at time $t = 0$ to the domain $F_t(\mathcal{C})$ at time t, and $M(t)$ is the mass of the domain $F_t(\mathcal{C})$ at time t.

Theorem 12.33 *For all times $t \in \mathbb{R}$, we have the time derivative*

$$\dot{M}(t) = \int_{F_t(\mathcal{C})} (\dot{\varrho} + \operatorname{div} \varrho\mathbf{v})(P,t)\ \upsilon. \qquad (12.120)$$

Proof. Note that $M(t) = \int_{\mathcal{C}} F_t^*(\varrho(P,t)\ \upsilon)$. Fix time $t = 0$. The product rule yields

$$\frac{d}{dt}\big(F_t^*\varrho(P,t)\ \upsilon\big)_{|t=0} = \frac{d}{dt}\big(F_t^*\varrho(P,0)\ \upsilon\big)_{|t=0} + \dot{\varrho}(P,0)\ \upsilon.$$

By Prop. 12.17 on page 715, this is equal to

$$(\mathcal{L}_{\mathbf{v}}\varrho\upsilon)(P,0) + \dot{\varrho}(P,0)\upsilon = (\operatorname{div} \varrho\mathbf{v} + \dot{\varrho})(P,0)\ \upsilon.$$

Now consider an arbitrary time t. The same argument as in the proof of Theorem 12.31 on page 734 yields

$$\frac{d}{dt}\big(F_t^*\varrho(P,t)\ \upsilon\big) = F_t^*(\operatorname{div} \varrho\mathbf{v} + \dot{\varrho})(P,t)\ \upsilon.$$

Integrating this over $F_t(\mathcal{C})$, we obtain $\dot{M}(t)$. □

Mass conservation and the continuity equation. Theorem 12.33 is called the transport theorem. This implies the following crucial result: Suppose that there exists a smooth function $\mathcal{P}_{\text{ext}} : \mathbb{E}^3 \times \mathbb{R} \to \mathbb{R}$ such that

$$\dot{M}(t) = \int_{F_t(\mathcal{C})} \mathcal{P}_{\text{ext}}(P,t)\ \upsilon$$

for all times $t \in \mathbb{R}$ and all compact subsets $\mathcal{C}$ of $\mathbb{E}^3$. Then we have the continuity equation

$$\dot{\varrho}(P,t) + (\operatorname{div} \varrho\mathbf{v})(P,t) = \mathcal{P}_{\text{ext}}(P,t) \qquad \text{for all} \quad P \in \mathbb{E}^3, \quad t \in \mathbb{R}. \quad (12.121)$$

In particular, if $\dot{M}(t) = 0$ for all times and all compact subsets $\mathcal{C}$ of $\mathbb{E}^3$, then there holds the continuity equation (12.121) with $\mathcal{P}_{\text{ext}} \equiv 0$. In terms of physics, the function $\mathcal{P}_{\text{ext}}$ describes the external mass production (e.g., the change of the number of specific particles caused by chemical reactions).
Proof. For fixed time t and balls $\mathcal{C}$, use Theorem 12.33, and contract the sets $F_t(\mathcal{C})$ to the point P. □

[40] In a Cartesian (x,y,z)-coordinate system, $M(t) = \int_{F_t(\mathcal{C})} \varrho(P,t)\ dx \wedge dy \wedge dz$.

12.8.5 The Noether Theorem – Symmetry Implies Conservation Laws in the Calculus of Variations

Roughly speaking, the fundamental Noether principle tells us that:

> *If the variational integral of a variational problem is invariant under a one-parameter family of transformations of the independent and dependent variables, then the solutions of the variational problem satisfy a conservation law.*[41]

This principle is fundamental for relativistic quantum field theories. The relativistic invariance of the action integral implies the invariance under time translations, space translations, space rotations, and special Lorentz transformations. This yields conservation of energy, momentum, and angular momentum, which will be studied later on. At this point, we want to illustrate the basic idea in terms of the flow of fluid particles.

> *We will show that the Noether theorem is a straightforward consequence of the transport theorem.*

In terms of mathematics, Noether's conservation law represents a constraint for the Euler–Lagrange equations of motion.

The variational problem. We want to investigate the following variational problem:

$$\boxed{\int_{\mathcal{M}} L(\mathbf{x}, \mathbf{q}(\mathbf{x}), \mathbf{q}'(\mathbf{x}))\ v = \text{critical!}, \qquad \mathbf{q} = \mathbf{q}_0 \ \text{ on } \partial\mathcal{M}.} \tag{12.122}$$

We are given the vector field $\mathbf{q}_0$ on the boundary $\partial\mathcal{M}$ of the 3-dimensional submanifold $\mathcal{M}$ of the Euclidean manifold $\mathbb{E}^3$. Here, v denotes the volume form on $\mathbb{E}^3$. Moreover, we frequently replace the point P by the position vector $\mathbf{x} = \overrightarrow{OP}$ where O denotes the origin. In what follows, all the functions are supposed to be smooth. In particular, we assume that the Lagrangian L is a smooth real-valued function of all its arguments. If the smooth vector field $\mathbf{q} = \mathbf{q}(\mathbf{x})$ on $\mathcal{M}$ is a solution of (12.122), then it satisfies the Euler–Lagrange equation

$$\boxed{\operatorname{div} L_{\mathbf{q}'}(\mathcal{Q}) = L_{\mathbf{q}}(\mathcal{Q}) \qquad \text{on } \operatorname{int}(\mathcal{M})} \tag{12.123}$$

with $\mathcal{Q} := (\mathbf{x}, \mathbf{q}(\mathbf{x}), \mathbf{q}'(\mathbf{x}))$. Conversely, if the smooth solution of (12.123) satisfies the boundary condition $\mathbf{q} = \mathbf{q}_0$ on $\partial\mathcal{M}$, then it is a solution of the variational problem (12.122).

Cartesian coordinate system. In a Cartesian (x, y, z)-coordinate system on the Euclidean manifold $\mathbb{E}^3$, we set

$$P := (x, y, z), \quad \mathbf{x} := x\mathbf{i} + y\mathbf{j} + z\mathbf{k}, \quad \mathbf{q}(P) := u(P)\mathbf{i}_P + v(P)\mathbf{j}_P + w(P)\mathbf{k}_P,$$

and $\partial u(P) := (u_x(P), u_y(P), u_z(P))$ (see (Fig. 4.3 on page 323). Then the variational problem (12.122) reads as

$$\int_{\mathcal{M}} L(P, u(P), v(P), w(P), \partial u(P), \partial v(P), \partial w(P))\, dx \wedge dy \wedge dz = \text{critical!}$$

[41] E. Noether, Invariant variational problems, Göttinger Nachrichten, Math.-phys. Klasse 1918, 235–257 (in German).

together with the boundary condition $u = u_0, v = v_0, w = w_0$ on $\partial\mathcal{M}$. The Euler–Lagrange equation (12.123) corresponds to the system

$$\begin{aligned}\frac{\partial}{\partial x}L_{u_x}(\mathcal{Q}) + \frac{\partial}{\partial y}L_{u_y}(\mathcal{Q}) + \frac{\partial}{\partial z}L_{u_z}(\mathcal{Q}) &= L_u(\mathcal{Q}),\\ \frac{\partial}{\partial x}L_{v_x}(\mathcal{Q}) + \frac{\partial}{\partial y}L_{v_y}(\mathcal{Q}) + \frac{\partial}{\partial z}L_{v_z}(\mathcal{Q}) &= L_v(\mathcal{Q}),\\ \frac{\partial}{\partial x}L_{w_x}(\mathcal{Q}) + \frac{\partial}{\partial y}L_{w_y}(\mathcal{Q}) + \frac{\partial}{\partial z}L_{w_z}(\mathcal{Q}) &= L_w(\mathcal{Q})\end{aligned} \tag{12.124}$$

with $\mathcal{Q} := (P, u(P), v(P), w(P), \partial u(P), \partial v(P), \partial w(P))$.

To write this concisely, we set $x^1 := x, x^2 := y, x^3 := z$, $\partial_j := \partial/\partial x^j$, and $q^1 := u, q^2 := v, q^3 := w$, $\mathbf{e}_1 := \mathbf{i}_P, \mathbf{e}_2 := \mathbf{j}_P, \mathbf{e}_3 := \mathbf{k}_P$, as well as $\mathbf{q}(P) = q^i(P)\mathbf{e}_i$. Then the Euler–Lagrange equation (12.124) can be written as[42]

$$\partial_s L_{\partial_s q^i}(\mathcal{Q}) = L_{q^i}(\mathcal{Q}), \quad i = 1, 2, 3.$$

We have $\operatorname{div} L_{\mathbf{q}'}(\mathcal{Q}) = \partial_s L_{\partial_s q^i}(\mathcal{Q})\ \mathbf{e}_i$ and $L_{\mathbf{q}} = L_{q^i}\mathbf{e}_i$. These expressions do not depend on the choice of the Cartesian coordinate system.

The prototype of the Noether theorem. Suppose that the Lagrangian L does not depend on the variable $\mathbf{q}$. Then $L_{\mathbf{q}} \equiv 0$. It follows from the Euler–Lagrange equation (12.123) that

$$\operatorname{div} L_{\mathbf{q}'}\big(\mathbf{x}, \mathbf{q}(\mathbf{x}), \mathbf{q}'(\mathbf{x})\big) = 0 \qquad \text{on } \operatorname{int}(\mathcal{M}). \tag{12.125}$$

It follows from (12.124) with $L_u \equiv 0, L_v \equiv 0$, and $L_w \equiv 0$ that equation (12.125) represents the three conservation laws

$$\operatorname{div} L_{\partial u}(\mathcal{Q}) = 0, \quad \operatorname{div} L_{\partial v}(\mathcal{Q}) = 0, \quad \operatorname{div} L_{\partial w}(\mathcal{Q}) = 0 \qquad \text{on } \operatorname{int}(\mathcal{M}).$$

Equivalently, $\partial_s L_{\partial_s q^i}(\mathcal{Q}) = 0$ on $\operatorname{int}(\mathcal{M})$ if $i = 1, 2, 3$. As we will show below, these three conservation laws follow from the invariance of the variational integral $\int_{\mathcal{M}} L(\mathbf{x}, \mathbf{q}'(\mathbf{x}))\ v$ under the translations $\mathbf{q} \mapsto \mathbf{q}+\mathbf{a}$ with three translation parameters given by the three components of the translation vector $\mathbf{a}$.

The elegant correspondence principle in calculus. The symbol $L_{\mathbf{q}}$ denotes the Fréchet derivative with respect to the variable $\mathbf{q}$ on the Euclidean Hilbert space E_3. There exists a general calculus on Banach spaces which is designed in such a way that the following mnemonic correspondence principle holds. To explain this, consider the Euler–Lagrange equation (12.123). If we regard $\mathbf{x}$ and $\mathbf{q}$ as real variables, then (12.123) coincides with the Euler–Lagrange equation derived in Sect. 6.5.2 of Vol. II. The extremely useful correspondence principle tells us the following:

The formulas in classical calculus remain valid when passing to the calculus in finite-dimensional or infinite-dimensional Banach spaces.

The calculus in Banach spaces is thoroughly studied in Zeidler (1986), Vol. I, quoted on page 1089. The reader, who does not know this calculus, should use both the correspondence principle as a useful mnemonic tool and Cartesian coordinates for rigorous justification.

Transformation of the variational integral under the flow of fluid particles. We are given the smooth complete velocity vector field $\mathbf{v}$ on the Euclidean manifold $\mathbb{E}^3$. This velocity vector field generates a flow $\{F_t\}_{t\in\mathbb{R}}$ of fluid particles. In terms of mathematics, we have the family of diffeomorphisms

[42] We sum over equal indices from 1 to 3.

$$F_t : \mathbb{E}^3 \to \mathbb{E}^3, \qquad t \in \mathbb{R},$$

with the group property $F_0 = I$ (identity map) and $F_{t+s} = F_t F_s$ for all $t, s \in \mathbb{R}$. This is called a one-parameter group of diffeomorphisms. The velocity vector field $\mathbf{v}$ is called the infinitesimal flow. The trajectory $t \mapsto \mathbf{x}(t)$ of a fluid particle is given by $\mathbf{x}(t) := F_t(\mathbf{x})$ for all $t \in \mathbb{R}$.

- Transformation: Let us pass from the original variables $\mathbf{x}, \mathbf{q}$ on E_3 to the new variables $\mathbf{X}, \mathbf{Q}$ on E_3 given by the smooth transformation

$$\boxed{\mathbf{X} = \mathbf{X}(\mathbf{x}; t), \qquad \mathbf{Q} = \mathbf{Q}(\mathbf{x}, \mathbf{q}; t), \qquad t \in \mathbb{R}.} \tag{12.126}$$

  These transformation formulas depend on the real parameter t regarded as time. We assume that the transformation is the identity map at the initial time $t = 0$, that is, $\mathbf{X}(\mathbf{x}; 0) = \mathbf{x}$ and $\mathbf{Q}(\mathbf{x}, \mathbf{q}; 0) = \mathbf{q}$.
- Naturally enough, we choose the transformation $\mathbf{X}(\mathbf{x}, t) := F_t(\mathbf{x})$. This yields the time derivative $\mathbf{X}_t(\mathbf{x}, 0) = \mathbf{v}(\mathbf{x})$. By Taylor expansion,

$$\mathbf{X}(\mathbf{x}, t) = \mathbf{x} + \mathbf{X}_t(\mathbf{x}, 0)t + o(t), \quad \mathbf{Q}(\mathbf{x}, \mathbf{q}, t) = \mathbf{q} + \mathbf{Q}_t(\mathbf{x}, \mathbf{q}, 0)t + o(t), \quad t \to 0.$$

- Linearized transformation: Motivated by the Taylor expansion, the transformation

$$\boxed{\mathbf{X} = \mathbf{x} + \mathbf{v}(\mathbf{x})t, \qquad \mathbf{Q} = \mathbf{q} + \mathbf{Q}_t(\mathbf{x}, \mathbf{q}, 0)t} \tag{12.127}$$

  is called the linearization of the original transformation (12.126) at time $t = 0$ (or the infinitesimal transformation to (12.126)).
- The language of physicists: Setting $\delta\mathbf{x} := \mathbf{v}(\mathbf{x})t$ and $\delta\mathbf{q} := \mathbf{Q}_t(\mathbf{x}, \mathbf{q}, 0)t$, the infinitesimal transformation reads as

$$\mathbf{X} = \mathbf{x} + \delta\mathbf{x}, \qquad \mathbf{Q} = \mathbf{q} + \delta\mathbf{q}.$$

- Perturbed curve: For fixed time t, the curve $\mathbf{x} \mapsto \mathbf{q}(\mathbf{x})$ is transformed into the curve $\mathbf{X} \mapsto \mathbf{q}(\mathbf{X}, t)$. Explicitly,

$$\mathbf{q}(\mathbf{X}, t) := \mathbf{Q}(\mathbf{x}, \mathbf{q}(\mathbf{x}), t), \qquad \mathbf{X} = F_t(\mathbf{x}).$$

  In particular, $\mathbf{q}(\mathbf{x}, 0) = \mathbf{q}(\mathbf{x})$. This means that, naturally enough, we get the unperturbed curve at time $t = 0$.
- The linearization of the perturbed curve is defined by

$$\boxed{\delta\mathbf{q}(\mathbf{x}) := \mathbf{Q}_t(\mathbf{x}, \mathbf{q}(\mathbf{x}), 0) \cdot t.} \tag{12.128}$$

  This is the crucial quantity which will appear in Noether's conservation law (12.134) on page 741. By Taylor expansion,

$$\mathbf{q}(F_t\mathbf{x}, t) = \mathbf{Q}(\mathbf{x}, \mathbf{q}(\mathbf{x}), t) = \mathbf{q}(\mathbf{x}) + \delta\mathbf{q}(\mathbf{x}) + o(t), \qquad t \to 0.$$

- The language of physicists: Mnemonically, physicists write

$$\boxed{\delta\mathbf{x} = \mathbf{v}(\mathbf{x})\,\delta t, \qquad \delta\mathbf{q}(\mathbf{x}) = \mathbf{Q}_t(\mathbf{x}, \mathbf{q}, 0)\,\delta t.}$$

  Here, the symbol t is replaced by δt. For historical reasons, following Leibniz and his successors, the symbol δt is frequently called an 'infinitesimally small' quantity.

The reader should note that δt is not a mystical infinitesimally small quantity, but a well-defined mathematical quantity, namely, a real number. Similarly, $\delta \mathbf{x}$ and $\delta \mathbf{q}(\mathbf{x})$ are well-defined mathematical quantities, namely, vectors.[43]

- Transformation of the variational integral: By definition, the given variational integral $\int_{\mathcal{U}} L(\mathbf{x}, \mathbf{q}(\mathbf{x}), \mathbf{q}'(\mathbf{x}))\, dx \wedge dy \wedge dz$ is transformed into[44]

$$\int_{F_t(\mathcal{U})} L\big(\mathbf{X}, \mathbf{q}(\mathbf{X}, t), \mathbf{q}_{\mathbf{X}}(\mathbf{X}, t)\big)\, dX \wedge dY \wedge dZ. \tag{12.129}$$

Now to the point. We assume that, for all open subsets $\mathcal{U}$ of $\mathbb{E}^3$, the variational integral is invariant, that is, the transformed integral (12.129) does not depend on time t. Let $\mathbf{x} \mapsto \mathbf{q}(\mathbf{x})$ be a smooth vector field on $\mathbb{E}^3$. Introducing the density function

$$\varrho(\mathbf{X}, t) := L(\mathbf{X}, \mathbf{q}(\mathbf{X}, t), \mathbf{q}_{\mathbf{X}}(\mathbf{X}, t)),$$

the invariance of the variational integral means that

$$\int_{F_t(\mathcal{U})} \varrho(\mathbf{X}, t)\, dX \wedge dY \wedge dZ = \text{const} \qquad \text{for all} \quad t \in \mathbb{R}.$$

This is equivalent to the condition

$$\int_{\mathcal{U}} \mathbf{X}^* \left(\varrho(\mathbf{X}, t)\, dX \wedge dY \wedge dZ\right) = \text{const} \qquad \text{for all} \quad t \in \mathbb{R}.$$

Contracting the open set $\mathcal{U}$ to some point, this is equivalent to the local condition

$$\varrho(\mathbf{X}(\mathbf{x}, t), t)\, \frac{\partial(X, Y, Z)}{\partial(x, y, z)}(\mathbf{x}, t) = \varrho(\mathbf{x}, 0) \qquad \text{for all} \quad t \in \mathbb{R}, \;\; (x, y, z) \in \mathbb{R}^3.$$

The Noether constraint via the transport theorem. By the transport theorem on page 735, we get the continuity equation

$$\varrho_t(\mathbf{X}, t) + \operatorname{div}\left(\varrho(\mathbf{X}, t)\mathbf{v}(\mathbf{X})\right) = 0. \tag{12.130}$$

We only need this equation at time $t = 0$.

Proposition 12.34 *It follows from* (12.130) *at time* $t = 0$ *that*

$$L_{\mathbf{q}}(\mathcal{Q})\mathbf{u}(\mathbf{x}) + L_{\mathbf{q}'}(\mathcal{Q}) \cdot \mathbf{u}'(\mathbf{x}) + \operatorname{div}\big(L(\mathcal{Q})\mathbf{v}(\mathbf{x})\big) = 0 \qquad \text{on } \mathbb{E}^3 \tag{12.131}$$

where we set $\mathcal{Q} := \big(\mathbf{x}, \mathbf{q}(\mathbf{x}), \mathbf{q}'(\mathbf{x})\big)$, *and* $\delta \mathbf{q}(\mathbf{x}) := \mathbf{Q}_t(\mathbf{x}, \mathbf{q}(\mathbf{x}), 0)$, *as well as*

$$\mathbf{u}(\mathbf{x}) := \delta \mathbf{q}(\mathbf{x}) - \mathbf{q}'(\mathbf{x})\mathbf{v}(\mathbf{x}).$$

In a Cartesian coordinate system, equation (12.131) reads as

$$L_{q^i}(\mathcal{Q})u^i(\mathbf{x}) + L_{\partial_s q^i}(\mathcal{Q})\partial_s u^i + \partial_s(L(\mathcal{Q})v^s(\mathbf{x})) = 0 \tag{12.132}$$

with $u^i := \delta q^i - (\partial_s q^i)v^s$.

---

[43] In Sect. 4.6 of Vol. II, we investigate the rigorous justification of infinitesimally small quantities in terms of non-standard analysis.

[44] The symbol $\mathbf{q}_{\mathbf{X}}$ denotes the partial derivative with respect to the variable $\mathbf{X}$.

Proof. Noting that $\mathbf{X} = \mathbf{x}$ at time $t = 0$, it follows from (12.130) that

$$\varrho_t(\mathbf{x}, 0) + \operatorname{div}\big(\varrho(\mathbf{x}, 0)\mathbf{v}(\mathbf{x})\big) = 0.$$

It remains to compute $\varrho_t(\mathbf{x}, 0)$.

(I) It follows from $\varrho(\mathbf{X}, t) = L(\mathbf{X}, \mathbf{q}(\mathbf{X}, t), \mathbf{q_X}(\mathbf{X}, t))$ that

$$\varrho_t(\mathbf{X}, t) = L_\mathbf{q}(.)\mathbf{q}_t(\mathbf{X}, t) + L_{\mathbf{q}'}(.) \cdot \mathbf{q}_{\mathbf{X}t}(\mathbf{X}, t).$$

By the chain rule, the definition $\mathbf{q}(\mathbf{X}, t) := \mathbf{Q}(\mathbf{x}, \mathbf{q}(\mathbf{x}), t)$ yields

$$\begin{aligned}\mathbf{q}_t(\mathbf{X}, t) &= \frac{\partial}{\partial \mathbf{x}}\mathbf{Q}(\mathbf{x}, \mathbf{q}(\mathbf{x}), t)\frac{\partial \mathbf{x}(\mathbf{X}, t)}{\partial t} + \mathbf{Q}_t(.) \\ &= \big(\mathbf{Q_x}(.) + \mathbf{Q_q}(.)\mathbf{q}'(\mathbf{x})\big)\frac{\partial \mathbf{x}(\mathbf{X}, t)}{\partial t} + \mathbf{Q}_t(.).\end{aligned}$$

Since $\mathbf{X} = F_t(\mathbf{x})$, we obtain $\mathbf{x} = F_{-t}(\mathbf{X})$. Hence $\frac{\partial \mathbf{x}(\mathbf{X},0)}{\partial t} = -\mathbf{v}(\mathbf{X}) = -\mathbf{v}(\mathbf{x})$. Since we have the identity map $\mathbf{Q}(\mathbf{x}, \mathbf{q}, 0) = \mathbf{q}$ at time $t = 0$, we obtain

$$\mathbf{Q_q}(\mathbf{x}, \mathbf{q}, 0) = I, \qquad \mathbf{Q_x}(\mathbf{x}, \mathbf{q}, 0) = 0.$$

Recall the definition $\delta\mathbf{q}(\mathbf{x}) := \mathbf{Q}_t(\mathbf{x}, \mathbf{q}(\mathbf{x}), 0)t$. Fix $t = 1$. Thus, at the point $t = 0$, we get

$$\mathbf{q}_t(\mathbf{x}, 0) = \delta\mathbf{q}(\mathbf{x}) - \mathbf{q}'(\mathbf{x})\mathbf{v}(\mathbf{x}).$$

(II) Note that $\frac{\partial^2 \mathbf{q}}{\partial t \partial \mathbf{X}} = \frac{\partial^2 \mathbf{q}}{\partial \mathbf{X} \partial t}$. Thus, at the point $t = 0$, it follows from (I) that

$$\mathbf{q}_{\mathbf{X}t}(\mathbf{x}, 0) = \frac{d}{d\mathbf{x}}\big(\delta\mathbf{q}(\mathbf{x}) - \mathbf{q}'(\mathbf{x})\mathbf{v}(\mathbf{x})\big).$$

□

The conservation law (Noether theorem). Finally, we will combine the Noether constraint (12.131) with the Euler–Lagrange equation. Let us use the notation introduced on page 739. We are given the smooth complete vector field $\mathbf{x} \mapsto \mathbf{v}(\mathbf{x})$ on the Euclidean manifold $\mathbb{E}^3$ and the transformation (12.126). This transformation is related to the flow $\{F_t\}_{t\in\mathbb{R}}$ generated by the vector field $\mathbf{v}$.

Theorem 12.35 *Suppose that the variational integral $\int_\mathcal{U} L(\mathcal{Q})\, v$ is invariant under the transformation* (12.126) *for all open subsets $\mathcal{U}$ of the Euclidean manifold $\mathbb{E}^3$. Then every smooth solution $\mathbf{x} \mapsto \mathbf{q}(\mathbf{x})$ of the Euler–Lagrange equation*

$$\operatorname{div} L_{\mathbf{q}'}(\mathcal{Q}) = L_\mathbf{q}(\mathcal{Q}) \qquad \text{on } \operatorname{int}(\mathcal{M}) \tag{12.133}$$

satisfies the conservation law $\operatorname{div}\mathbf{J}(\mathbf{x}) = 0$ *on* $\operatorname{int}(\mathcal{M})$ *with the current density vector*

$$\mathbf{J}(\mathbf{x}) := L(\mathcal{Q})\mathbf{v}(\mathbf{x}) + L_{\mathbf{q}'}(\mathcal{Q})\mathbf{u}(\mathbf{x}) \tag{12.134}$$

where $\mathbf{u}(\mathbf{x}) := \delta\mathbf{q}(\mathbf{x}) - \mathbf{q}'(\mathbf{x})\mathbf{v}(\mathbf{x})$, *and* $\delta\mathbf{q}(\mathbf{x}) := \mathbf{Q}_t(\mathbf{x}, \mathbf{q}, 0)$.

In a Cartesian coordinate system, the conservation law reads as

$$\partial_s\big(L(\mathcal{Q})v^s(\mathbf{x}) + L_{\partial_s q^i}(\mathcal{Q})u^i(\mathbf{x})\big) = 0 \qquad \text{on } \operatorname{int}(\mathcal{M}) \tag{12.135}$$

with $u^i := \delta q^i - (\partial_s q^i)v^s$, $i = 1, 2, 3$.

Proof. By the Euler–Lagrange equation, $L_{q^i} = \partial_s L_{\partial^s q^i}$. The Noether constraint (12.132) yields

$$u^i\partial_s L_{\partial_s q^i} + L_{\partial_s q^i}\partial_s u^i + \partial_s(Lv^s) = 0.$$

Using the product rule, we get the claim (12.135). □

In the following examples, let $\mathbf{q} = \mathbf{q}(\mathbf{x})$ denote an arbitrary solution of the Euler–Lagrange equation (12.133) above.

Example 1 (translation of $\mathbf{q}$). If the Lagrangian $L = L(\mathbf{x}, \mathbf{q}')$ does not depend on the variable $\mathbf{q}$, then we get the conservation law

$$\operatorname{div} L_{\mathbf{q}'}(\mathcal{Q})\,\mathbf{a} = 0 \qquad \text{on } \mathcal{M}$$

with $\mathcal{Q} := (\mathbf{x}, \mathbf{q}'(\mathbf{x}))$ for all vectors $\mathbf{a} \in E_3$. Choosing $\mathbf{a} = \mathbf{i}, \mathbf{j}, \mathbf{k}$, we get the following three conservation laws:

$$\operatorname{div} L_{\mathbf{q}'}(\mathcal{Q})\,\mathbf{i} = 0, \quad \operatorname{div} L_{\mathbf{q}'}(\mathcal{Q})\,\mathbf{j} = 0, \quad \operatorname{div} L_{\mathbf{q}'}(\mathcal{Q})\,\mathbf{k} = 0 \qquad \text{on int } (\mathcal{M}).$$

This coincides with (12.125) on page 738.

Proof. Choose the trivial vector field $\mathbf{v}(\mathbf{x}) \equiv 0$ which generates the trivial flow $F_t(\mathbf{x}) = \mathbf{x}$ for all $t \in \mathbb{R}$. Fix the vector $\mathbf{a}$. Define the transformation

$$\mathbf{X}(\mathbf{x}, t) := \mathbf{x}, \qquad \mathbf{Q}(\mathbf{x}, \mathbf{q}, t) := \mathbf{q} + t\mathbf{a}, \quad t \in \mathbb{R}.$$

The transformed function reads as $\mathbf{q}(\mathbf{x}, t) := \mathbf{q}(\mathbf{x}) + t\mathbf{a}$ with the perturbation parameter t. It follows from

$$\int_{\mathcal{U}} L(\mathbf{x}, \mathbf{q}_{\mathbf{x}}(\mathbf{x}, t))\, dx \wedge dy \wedge dz = \int_{\mathcal{U}} L(\mathbf{x}, \mathbf{q}'(\mathbf{x}))\, dx \wedge dy \wedge dz \qquad \text{for all} \quad t \in \mathbb{R}$$

that the variational integral is invariant under the transformation $(\mathbf{x}, \mathbf{q}) \mapsto (\mathbf{X}, \mathbf{Q})$. From $\delta\mathbf{q}(\mathbf{x}) := \mathbf{Q}_t(\mathbf{x}, \mathbf{q}, 0) = \mathbf{a}$ and the Noether theorem (12.134), we get the claim. □

Example 2 (translation of $\mathbf{x}$). If the Lagrangian $L = L(\mathbf{q}, \mathbf{q}')$ does not depend on the position vector $\mathbf{x}$, then we get the conservation law

$$\operatorname{div}\big(L(\mathcal{Q})\mathbf{v}_0 - L_{\mathbf{q}'}(\mathcal{Q})\mathbf{q}'(\mathbf{x})\mathbf{v}_0\big) = 0 \qquad \text{on int}(\mathcal{M}) \tag{12.136}$$

for all vectors $\mathbf{v}_0 \in E_3$. Here, $\mathcal{Q} := (\mathbf{q}(\mathbf{x}), \mathbf{q}'(\mathbf{x}))$. Choosing $\mathbf{v}_0 = \mathbf{i}, \mathbf{j}, \mathbf{k}$, we obtain the following three conservation laws:

$$\operatorname{div}\big(L(\mathcal{Q})\mathbf{i} - L_{\mathbf{q}'}(\mathcal{Q})\mathbf{q}'(\mathbf{x})\mathbf{i}\big) = 0, \quad \operatorname{div}\big(L(\mathcal{Q})\mathbf{j} - L_{\mathbf{q}'}(\mathcal{Q})\mathbf{q}'(\mathbf{x})\mathbf{j}\big) = 0,$$
$$\operatorname{div}\big(L(\mathcal{Q})\mathbf{k} - L_{\mathbf{q}'}(\mathcal{Q})\mathbf{q}'(\mathbf{x})\mathbf{k}\big) = 0 \qquad \text{on int}(\mathcal{M}).$$

Proof. Choose the constant vector field $\mathbf{v}(\mathbf{x}) := \mathbf{v}_0$. This generates the flow $F_t(\mathbf{x}) := \mathbf{x} + t\mathbf{v}_0$. Define the transformation

$$\mathbf{X}(\mathbf{x}, t) := \mathbf{x} + t\mathbf{v}_0, \qquad \mathbf{Q}(\mathbf{x}, \mathbf{q}, t) := \mathbf{q}, \qquad t \in \mathbb{R}.$$

This yields the transformed curve $\mathbf{q}(\mathbf{X}, t) := \mathbf{Q}(\mathbf{x}, \mathbf{q}(\mathbf{x}), t) = \mathbf{q}(\mathbf{x})$. The translation invariance of the integral tells us that

$$\int_{F_t(\mathcal{U})} L\big(\mathbf{q}(\mathbf{X}, t), q_{\mathbf{X}}(\mathbf{X}, t)\big)\, dX \wedge dY \wedge dZ = \int_{\mathcal{U}} L\big(\mathbf{q}(\mathbf{x}), \mathbf{q}'(\mathbf{x})\big)\, dx \wedge dy \wedge dz.$$

This is the invariance of the variational integral with respect to the transformation $(\mathbf{x}, \mathbf{q}) \mapsto (\mathbf{X}, \mathbf{Q})$. Since $\delta\mathbf{q}(\mathbf{x}) := \mathbf{Q}_t(\mathbf{x}, \mathbf{q}, 0) = 0$, the Noether theorem (12.134) yields the claim. □

Example 3 (induced flow of vector fields). We are given the smooth complete vector field $\mathbf{x} \mapsto \mathbf{v}(\mathbf{x})$ on the Euclidean manifold $\mathbb{E}^3$ which generates the flow $\{F_t\}_{t\in\mathbb{R}}$. Consider the transformation

$$\mathbf{X}(\mathbf{x},t) := F_t(\mathbf{x}), \qquad \mathbf{Q}(\mathbf{x},\mathbf{q},t) := F_t'(\mathbf{x})\mathbf{q}, \quad t \in \mathbb{R}. \tag{12.137}$$

By (10.10) on page 650, this corresponds to the linearized transformation

$$\mathbf{X}(\mathbf{x},t) = \mathbf{x} + \mathbf{v}(\mathbf{x}) \cdot t, \qquad \mathbf{Q}(\mathbf{x},\mathbf{q},t) = \mathbf{q} + \mathbf{v}'(\mathbf{x})\mathbf{q} \cdot t, \quad t \in \mathbb{R}.$$

If the variational integral (12.122) on page 737 is invariant under the transformation (12.137), then we get the conservation law[45]

$$\operatorname{div}\big(L(\mathcal{Q})\mathbf{v}(\mathbf{x}) + L_{\mathbf{q}'}(\mathcal{Q})\,[\mathbf{q},\mathbf{v}](\mathbf{x})\big) = 0 \qquad \text{on } \operatorname{int}(\mathcal{M}). \tag{12.138}$$

Proof. Use $\delta\mathbf{q}(\mathbf{x}) = \mathbf{v}'(\mathbf{x})\mathbf{q}(\mathbf{x})$ and the Noether theorem (12.134). □

Example 4 (rotational invariance). As a special case of Example 3, let us consider a rotation about the origin. Fix the nonzero vector $\boldsymbol{\omega}$. We are given the velocity vector field

$$\mathbf{v}(\mathbf{x}) := \boldsymbol{\omega} \times \mathbf{x}$$

which generates the flow $F_t(\mathbf{x}_0) := \mathbf{x}(t)$ with $\dot{\mathbf{x}}(t) = \boldsymbol{\omega} \times \mathbf{x}(t)$, $t \in \mathbb{R}$, and $\mathbf{x}(0) = \mathbf{x}_0$. Let us use the transformation (12.137) which represents a clockwise rotation of points and vector fields about the axis $\boldsymbol{\omega}$ through the origin with the angular velocity $\omega = |\boldsymbol{\omega}|$. Since

$$\mathbf{v}'(\mathbf{x})\mathbf{h} = \frac{d}{d\sigma}\mathbf{v}(\mathbf{x} + \sigma\mathbf{h})_{|\sigma=0} = \boldsymbol{\omega} \times \mathbf{h},$$

the linearized transformation reads as

$$\mathbf{X}(\mathbf{x},t) := \mathbf{x} + (\boldsymbol{\omega} \times \mathbf{x})\, t, \qquad \mathbf{Q}(\mathbf{x},\mathbf{q},t) = \mathbf{q} + (\boldsymbol{\omega} \times \mathbf{q})\, t.$$

Thus, $\delta\mathbf{x} := (\boldsymbol{\omega} \times \mathbf{x})\, \delta t$ and $\delta\mathbf{q} := (\boldsymbol{\omega} \times \mathbf{q})\, \delta t$. Then the conservation law (12.138) passes over to

$$\operatorname{div}\Big(L(\mathcal{Q})(\boldsymbol{\omega} \times \mathbf{x}) + L_{\mathbf{q}'}(\mathcal{Q})\big(\boldsymbol{\omega} \times \mathbf{q}(\mathbf{x}) - \mathbf{q}'(\mathbf{x})(\boldsymbol{\omega} \times \mathbf{x})\big)\Big) = 0 \qquad \text{on } \operatorname{int}(\mathcal{M})$$

for all vectors $\boldsymbol{\omega} \in E_3$. Choosing $\boldsymbol{\omega} = \mathbf{i}, \mathbf{j}, \mathbf{k}$, we get three conservation laws. For example, the variational integral

$$\int_{\mathcal{M}} \left((\operatorname{div}\mathbf{q})^2 - \mathbf{q}^2\right) dx \wedge dy \wedge dz$$

is invariant under rotations.

The linearized version of the Noether theorem. Sometimes it is useful to apply the following corollary of the Noether theorem. For the proof of the corollary, note that we only need a local variant of the arguments used above. In particular, it is sufficient to have a local flow near time $t = 0$ at hand. Set $\mathcal{Q} := \big(\mathbf{x}, \mathbf{q}(\mathbf{x}), \mathbf{q}'(\mathbf{x})\big)$, and $\mathbf{u}(\mathbf{x}) := \delta\mathbf{q}(\mathbf{x}) - \mathbf{q}'(\mathbf{x})\mathbf{v}(\mathbf{x})$.

[45] Recall the Lie bracket $[\mathbf{q},\mathbf{v}](\mathbf{x}) = \mathbf{v}'(\mathbf{x})\mathbf{q}(\mathbf{x}) - \mathbf{q}'(\mathbf{x})\mathbf{v}(\mathbf{x})$.

Corollary 12.36 *Suppose that we are given two smooth vector fields* $\mathbf{x} \mapsto \mathbf{v}(\mathbf{x})$ *and* $\mathbf{x} \mapsto \delta\mathbf{q}(\mathbf{x})$ *on the open set* $\mathcal{O}$ *of the Euclidean manifold* $\mathbb{E}^3$ *which satisfy the Noether constraint*

$$L_{\mathbf{q}}(\mathcal{Q})\mathbf{u}(\mathbf{x}) + L_{\mathbf{q}'}(\mathcal{Q}) \cdot \mathbf{u}'(\mathbf{x}) + \operatorname{div}\big(L(\mathcal{Q})\mathbf{v}(\mathbf{x})\big) = 0 \qquad \text{on } \mathcal{O} \tag{12.139}$$

for all smooth maps $\mathbf{x} \mapsto \mathbf{q}(\mathbf{x})$ *on* $\mathcal{O}$. *Then we get the conservation law*

$$\operatorname{div}\big(L(\mathcal{Q})\mathbf{v}(\mathbf{x}) + L_{\mathbf{q}'}(\mathcal{Q})\mathbf{u}(\mathbf{x})\big) = 0 \qquad \text{on } \mathcal{O} \tag{12.140}$$

for all solutions $\mathbf{x} \mapsto \mathbf{q}(\mathbf{x})$ *of the Euler–Lagrange equation* (12.133) *on* $\mathcal{O}$.

In a Cartesian coordinate system on $\mathbb{E}^3$, the formulation of (12.139) (resp. (12.140)) can be found in (12.132) (resp. (12.135)).

Example. If the Lagrangian $L = L(\mathbf{x}, \mathbf{q}')$ does not depend on the variable $\mathbf{q}$, then we get the conservation law

$$\operatorname{div} L_{\mathbf{q}'}(\mathbf{x}, \mathbf{q}'(\mathbf{x}))\, \mathbf{a} = 0 \qquad \text{on } \mathcal{O}$$

for all vectors $\mathbf{a} \in E_3$ and all solutions $\mathbf{q} = \mathbf{q}(\mathbf{x})$ of the Euler–Lagrange equation (12.133) on $\mathcal{O}$. Choosing the special vectors $\mathbf{a} = \mathbf{i}, \mathbf{j}, \mathbf{k}$, this yields three conservation laws as in Example 1 on page 742.

Proof. Fix the vector $\mathbf{a}$. Set $\mathbf{v}(\mathbf{x}) :\equiv 0$ and $\delta\mathbf{q}(\mathbf{x}) := \mathbf{a}$. Then $\mathbf{u}(\mathbf{x}) = \mathbf{a}$. Hence $\mathbf{u}'(\mathbf{x}) \equiv 0$. Obviously, the Noether constraint (12.139) is satisfied, and the claim follows from (12.140). □

Lie's strategy of linearization and the fundamental role of infinitesimal transformations in physics. Lie discovered that the local theory of symmetry is completely governed by studying the corresponding linearizations. Corollary 12.36 establishes this for the Noether symmetry principle. This is the reason why the restriction to infinitesimal transformations is successful in the physics literature.

12.9 The Hamiltonian Flow on the Euclidean Manifold – a Paradigm of Hamiltonian Mechanics

> The most elegant formulation of classical mechanics is based on Cartan's calculus of exterior differential forms starting with the Poincaré–Cartan 1-form $pdq - Hdt$.
>
> Folklore

The classical problem. We want to study the motion

$$\boxed{q = q(t), \qquad t \in \mathbb{R}} \tag{12.141}$$

of a point of mass $m > 0$ on the real line. Here, q and t denote position and time, respectively. We want to study this problem in the setting of Hamiltonian mechanics based on the key Hamiltonian function

$$H = H(q, p, \tau).$$

In terms of physics, H represents the energy function depending on position q, momentum p, and time τ. As starting point, we choose the Hamiltonian equation of motion

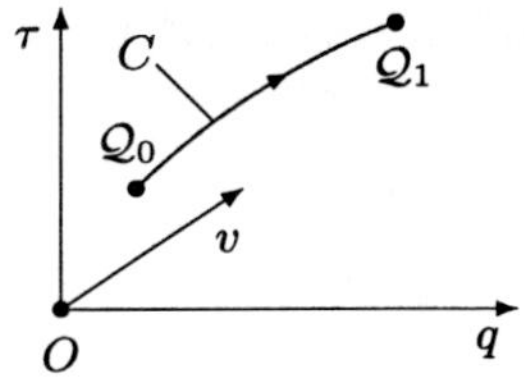

(a) Extended state space

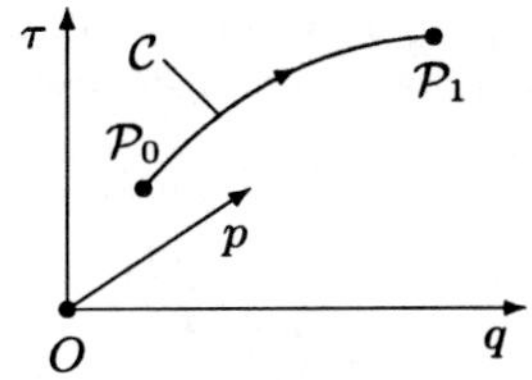

(b) Extended phase space

Fig. 12.21. Legendre transformation

$$\boxed{\dot{q}(t) = H_p(P(t)), \quad \dot{p}(t) = -H_q(P(t)), \quad \dot{\tau}(t) = 1, \qquad t \in \mathbb{R}} \tag{12.142}$$

with the initial condition $q(t_0) = q_0, p(t_0) = p_0, \tau(t_0) = t_0$. Here, we introduce the point $P := (q, p, \tau)$ with the real coordinates

- q (position), p (momentum), v (velocity), t, τ (time).

In addition, we set

- (q, v) (point of the state space $\mathbb{E}^2$),
- $Q := (q, v, \tau)$ (point of the extended state space $\mathbb{E}^3$),
- (q, p) (point of the phase space $\mathbb{E}^2$),
- (q, p, τ) (point of the extended phase space $\mathbb{E}^3$).

The Hamiltonian function $P \mapsto H(P)$ is assumed to be a smooth function

$$H : \mathbb{E}^3 \to \mathbb{R}$$

on the extended phase space. Here, as depicted in Fig. 12.21(b), we fix a right-handed Cartesian (q, p, τ)-coordinate system on $\mathbb{E}^3$ with the right-handed orthonormal basis $\mathbf{i}, \mathbf{j}, \mathbf{k}$ at the origin O. By parallel transport, this yields the basis $\mathbf{i}_P, \mathbf{j}_P, \mathbf{k}_P$ of the tangent space $T_P\mathbb{E}^3$ at the point P (see Fig 4.3 on page 323). In this setting, the Hamiltonian equation of motion (12.142) describes a curve in the extended phase space $\mathbb{E}^3$. The gauge condition $\dot{\tau}(t) = 1$ together with $\tau(t_0) = t_0$ implies $\tau(t) = t$ for all $t \in \mathbb{R}$.

For example, the Hamiltonian function

$$H(q, p) := \frac{p^2}{2m} + \frac{\omega^2 q^2}{2}, \qquad q, p \in \mathbb{R}$$

describes the so-called harmonic oscillator on the real line. The Hamiltonian equation of motion (12.142) reads as

$$\dot{q}(t) = \frac{p(t)}{m}, \quad \dot{p}(t) = -\omega^2 q(t), \quad \dot{\tau}(t) = 1, \qquad t \in \mathbb{R}$$

with $q(0) = q_0, p(0) = p_0, \tau(0) = 0$. The unique solution (12.142) is given by

$$q(t) = q_0 \sin \omega t, \qquad p(t) = m\dot{q}(t), \qquad \tau(t) = t \qquad \text{for all} \quad t \in \mathbb{R}.$$

This is an oscillation on the real line with angular frequency $\omega > 0$. For given initial momentum p_0, we get the initial velocity $\dot{q}(0) = p_0/m$.

The language of geometry in higher dimensions. It was discovered in the history of classical mechanics that it is necessary to pass to higher dimensions in order to get insight. To discuss this, let us start with the trajectory (12.141) on the real line. Introducing the velocity $v(t) := \dot{q}(t)$, we get the curve

$$C : q = q(t), \quad v = v(t), \quad \tau = t, \qquad t \in \mathbb{R} \tag{12.143}$$

in the 3-dimensional extended state space (Fig. 12.21(a)). Introducing the momentum $p(t) := m\dot{q}(t)$, the motion (12.143) passes over to the curve

$$\mathcal{C} : q = q(t), \quad p = p(t), \quad \tau = \tau(t), \qquad t \in \mathbb{R}$$

in the 3-dimensional extended phase space (Fig. 12.21(b)). Our goal is to simplify the integration of the Hamiltonian equation of motion (12.142) by using

- a variational problem for the curve $\mathcal{C}$ (principle of critical action), and
- a geometric transformation called canonical transformation (Lie's contact geometry with respect to the Poincaré–Cartan contact 2-form).

The Legendre transformation. There exist two approaches to classical mechanics, namely,

- the Hamiltonian approach on the extended phase space, and
- the Lagrangian approach on the extended state space.

Historically, the Hamiltonian approach due to Hamilton (1805–1865) was derived from the Lagrangian approach due to Lagrange (1736–1813). In what follows, for pedagogical reasons, we will reverse the order. We will first concentrate on the Hamiltonian approach based on the extended phase space. In Sect. 12.9.8, we will use the Legendre transformation in order to pass from the Hamiltonian approach to the Lagrangian approach and vice versa. In fact, the two approaches are equivalent. In Vol. II, concerning quantum mechanics we studied in detail

- the Hamiltonian approach due to Heisenberg, Schrödinger, Dirac, and von Neumann (operator theory and the Schrödinger equation), and
- the Lagrangian approach due to Dirac and Feynman (the Feynman path integral).

12.9.1 Hamilton's Principle of Critical Action

We will work in the extended phase space $\mathbb{E}^3$ introduced above. We are given the points $P_0 = (q_0, p_0, t_0)$ and $P_1 = (q_1, p_1, t_1)$ of $\mathbb{E}^3$. Here, $-\infty < t_0 < t_1 < \infty$. The curve

$$\mathcal{C} : \ P = P(t), \qquad t \in \mathbb{R}$$

is called admissible iff it is smooth and $P(t_0) = P_0, P(t_1) = P_1$. Explicitly, the curve $\mathcal{C}$ is given by

$$\mathcal{C} : q = q(t), \quad p = p(t), \quad \tau(t) = t, \qquad t \in \mathbb{R}. \tag{12.144}$$

The symbol A denotes the set of all admissible curves. The variational problem

$$\boxed{\int_{\mathcal{C}} pdq - Hd\tau = \text{critical!}, \qquad \mathcal{C} \in \mathsf{A}} \tag{12.145}$$

is called Hamilton's principle of critical action. Note that $pdq - Hd\tau$ has the physical dimension of energy $\times$ time $=$ action.

The extremals. We are looking for admissible curves $\mathcal{C}$ such that the line integral $\int_{\mathcal{C}} pdq - Hd\tau$ becomes critical.

Theorem 12.37 *The admissible curve* $\mathcal{C}$ *is a solution of* (12.145) *iff it corresponds to a solution of the Hamiltonian equation of motion* (12.142).

Proof. Using the parametrization (12.144), the variational problem (12.145) reads as

$$\int_{t_0}^{t_1} (p(t)\dot{q}(t) - H(q(t), p(t), t))\, dt = \text{critical!} \tag{12.146}$$

with the boundary conditions $q(t_j) = q_j$ and $p(t_j) = p_j$ if $j = 0, 1$. By the basic result of the calculus of variations, the solutions of (12.146) are characterized by the Euler–Lagrange equations

$$\frac{d}{dt}\mathsf{L}_{\dot{q}} = \mathsf{L}_q, \quad \frac{d}{dt}\mathsf{L}_{\dot{p}} = \mathsf{L}_p$$

with $\mathsf{L}(q, \dot{q}, p, t) := p\dot{q} - H(q, p, t)$. Hence $\frac{d}{dt}p = -H_q$ and $0 = \dot{q} - H_p$. This is (12.142). □

The key formulas for the extremals (Hamiltonian flow). In the next sections, we will investigate the extremals of the principle of critical action, that is, the solutions of the Hamiltonian equation of motion (12.142). At this point, let us summarize the key formulas which will be used below. In order to get an intuitive physical picture, let us write (12.142) in the succinct form

$$\dot{\mathbf{x}}(t) = \mathbf{v}(P(t)), \qquad t \in \mathbb{R}.$$

This describes the motion of fluid particles on the extended phase space. The velocity vector field $\mathbf{v}$ is called the Hamiltonian vector field, and the corresponding flow is called the Hamiltonian flow. Introducing the Poincaré–Cartan 1-form

$$\chi := pdq - Hd\tau,$$

which is the integrand of the action integral in (12.145), we get the equation

$$\boxed{i_{\mathbf{v}}(d\chi) = 0,}$$

as we will show below. In order to solve the equation of motion (12.142), we will use canonical transformations F based on the formula

$$F^*\chi = \chi + dS.$$

This is equivalent to $F^*(d\chi) = d\chi$ (contact transformation with respect to the Poincaré–Cartan 2-form $d\chi$). The solution procedure for the Hamiltonian equation of motion (12.142) will be based on the solution of the Hamilton–Jacobi partial differential equation

$$\boxed{S_t(q, t) + H(q, S_q(q, t), t) = 0} \tag{12.147}$$

by a family $S = S(q, t; \mathcal{Q})$ of solutions which depends on the real parameter $\mathcal{Q}$. The basic idea is to use the function S in order to generate a canonical transformation which transforms the Hamiltonian equation of motion into a system which can be easily solved.

This is the prototype of a close relationship between first-order partial differential equations (e.g., (12.147)) and systems of ordinary differential equations (e.g., (12.142)). In terms of geometric optics, this corresponds to the relation between light rays and wave fronts of light (see Chap. 5 of Vol. II).

12.9.2 Basic Formulas

Let us introduce the following quantities:

- $\mathbf{x} := q\mathbf{i} + p\mathbf{j} + \tau\mathbf{k}$ (position vector $\mathbf{x} = \overrightarrow{OP}$ at the origin O with $P = (q, p, \tau)$),
- $\mathbf{v}_P = H_p(P)\mathbf{i}_P - H_q(P)\mathbf{j}_P + \mathbf{k}_P$ (Hamiltonian velocity vector field $\mathbf{v}$ on $\mathbb{E}^3$ at the point P),
- $V_P := H_p(P)\frac{\partial}{\partial q} - H_q(P)\frac{\partial}{\partial p} + \frac{\partial}{\partial \tau}$ (Hamiltonian derivation at the point P – linear partial differential operator of first order on $C^\infty(\mathbb{E}^3, \mathbb{R})$),[46]
- $\{\Theta, H\} := \Theta_q H_p - \Theta_p H_q$ (Poisson bracket),
- $\sigma := dq \wedge dp$ (symplectic 2-form on $\mathbb{E}^3$),
- $\chi := pdq - Hd\tau$ (Poincaré–Cartan 1-form).

The Hamiltonian flow. The differential equation

$$\boxed{\dot{\mathbf{x}}(t) = \mathbf{v}(P(t)), \quad t \in \mathbb{R}, \quad \mathbf{x}(0) = \mathbf{x}_0} \tag{12.148}$$

is equivalent to the Hamiltonian equation of motion (12.142). We assume that the Hamiltonian velocity vector field $\mathbf{v}$ is complete. The corresponding flow is defined by

$$F_t\mathbf{x}_0 := \mathbf{x}(t), \qquad t \in \mathbb{R}$$

where $\mathbf{x} = \mathbf{x}(t)$ is the unique solution of (12.148). In order to study conservation laws with respect to the Hamiltonian flow $\{F_t\}_{t\in\mathbb{R}}$, one has to investigate the Lie derivative $\mathcal{L}_\mathbf{v}\omega$ of differential forms. This will be done next.

The Lie derivative. Let $\Theta : \mathbb{E}^3 \to \mathbb{R}$ be a smooth temperature field on $\mathbb{E}^3$. There hold the following formulas on $\mathbb{E}^3$:

(i) $\mathcal{L}_\mathbf{v}\Theta = V(\Theta) = \{\Theta, H\} + \Theta_\tau$,
(ii) $\mathcal{L}_\mathbf{v}H = H_\tau$,
(iii) $i_\mathbf{v}(dq) = H_p$, $i_\mathbf{v}(dp) = -H_q$, $i_\mathbf{v}(d\tau) = 1$,
(iv) $\mathcal{L}_\mathbf{v}(dq) = dH_p$ and $\mathcal{L}_\mathbf{v}(dp) = -dH_q$,
(v) $i_\mathbf{v}(dH) = H_\tau$,
(vi) $d\sigma = 0$,
(vii) $i_\mathbf{v}\sigma = dH - H_\tau d\tau$,
(viii) $\mathcal{L}_\mathbf{v}\sigma = -dH_\tau \wedge d\tau$,
(ix) $d\chi = -\sigma - dH \wedge d\tau$,
(x) $i_\mathbf{v}(d\chi) = 0$,
(xi) $d(\mathcal{L}_\mathbf{v}\chi) = \mathcal{L}_\mathbf{v}(d\chi) = 0$.

---

[46] In modern differential geometry, one identifies the Hamiltonian velocity vector field $\mathbf{v}$ on $\mathbb{E}^3$ with the Hamiltonian derivation V on the space $C^\infty(\mathbb{E}^3, \mathbb{R})$ of smooth functions $\Theta : \mathbb{E}^3 \to \mathbb{R}$. Mnemonically, this is very convenient (see Sect. 8.15 on page 529).
Moreover, observe the following. In modern mathematical physics, Hamiltonian mechanics is formulated in terms of the cotangent bundle of the position space. In the present situation, the position space is the real line $\mathbb{E}^1$, and the cotangent bundle

$$T^*\mathbb{E}^1 = \mathbb{E}^1 \times (\mathbb{E}^1)^d$$

is called the phase space. In order to get an intuitive interpretation, we identify $T^*\mathbb{E}^1$ with the Euclidean (q,p)-plane $\mathbb{E}^2$, and the extended phase space $T^*\mathbb{E}^1 \times \mathbb{R}$ is identified with the Euclidean (q, p, τ)-manifold $\mathbb{E}^3$. All the identifications are based on linear isomorphisms between the relevant linear spaces.

Proof. We will use the following formulas: Let α (resp. β) be an r-form (resp. s-form) on $\mathbb{E}^3$ with $r, s = 0, 1, 2, 3$. Then:

- $d(\alpha \wedge \beta) = d\alpha \wedge \beta + (-1)^r \alpha \wedge d\beta$ (antiderivation),
- $i_{\mathbf{v}}(\alpha \wedge \beta) = i_{\mathbf{v}}\alpha \wedge \beta + (-1)^r \alpha \wedge i_{\mathbf{v}}\beta$ (antiderivation),[47]
- $\mathcal{L}_{\mathbf{v}}(\alpha \wedge \beta) = \mathcal{L}_{\mathbf{v}}\alpha \wedge \beta + \alpha \wedge \mathcal{L}_{\mathbf{v}}\beta$ (derivation),
- $d(d\alpha) = 0$ (Poincaré's cohomology rule),
- $\mathcal{L}_{\mathbf{v}}\alpha = i_{\mathbf{v}}(d\alpha) + d(i_{\mathbf{v}}\alpha)$ (Cartan's magic formula),
- $\mathcal{L}_{\mathbf{v}}(d\beta) = d(i_{\mathbf{v}}d\beta)$, since $d(d\beta) = 0$.

Ad (i), (ii). See (10.14) on page 651.
Ad (iii). $i_{\mathbf{v}}(dq) = dq(\mathbf{v}) = H_p$.
Ad (iv). $\mathcal{L}_{\mathbf{v}}(dq) = d(i_{\mathbf{v}}dq) = dH_p$.
Ad (v). $i_{\mathbf{v}}(dH) = i_{\mathbf{v}}(H_q dq) + i_{\mathbf{v}}(H_p dp) + i_{\mathbf{v}}(H_\tau d\tau)$. This is equal to

$$H_q i_{\mathbf{v}}(dq) + H_p i_{\mathbf{v}}(dp) + H_\tau i_{\mathbf{v}}(d\tau) = H_q H_p - H_p H_q + H_\tau i_{\mathbf{v}}(d\tau).$$

Ad (vi). $d(dq \wedge dp) = d(dq) \wedge dp - dq \wedge d(dp) = 0$.
Ad (vii). $i_{\mathbf{v}}\sigma = i_{\mathbf{v}}(dq) \wedge dp - dq \wedge i_{\mathbf{v}}(dp) = H_p dp + H_q dq = dH - H_\tau d\tau$.
Ad (viii). $\mathcal{L}_{\mathbf{v}}\sigma = i_{\mathbf{v}}(d\sigma) + d(i_{\mathbf{v}}\sigma) = d(i_{\mathbf{v}}\sigma) = d(dH - H_\tau d\tau) = -dH_\tau \wedge d\tau$.
Ad (ix). $d\chi = d(pdq - Hd\tau) = dp \wedge dq - dH \wedge d\tau$.
Ad (x). $i_{\mathbf{v}}(d\chi) = -i_{\mathbf{v}}(\sigma) - i_{\mathbf{v}}(dH) \wedge d\tau + dH \wedge i_{\mathbf{v}}(d\tau)$. This is equal to $-dH + H_\tau d\tau - H_\tau d\tau + dH = 0$.
Ad (xi). $\mathcal{L}_{\mathbf{v}}(d\chi) = d(i_{\mathbf{v}}(d\chi)) = 0$. □

12.9.3 The Poincaré–Cartan Integral Invariant

Let $\mathbf{v}$ be a smooth complete Hamiltonian velocity vector field on the Euclidean manifold $\mathbb{E}^3$ which generates the Hamiltonian flow $\{F_t\}_{t\in\mathbb{R}}$. Let $\mathcal{M}$ be a compact 2-dimensional oriented submanifold of $\mathbb{E}^3$ with coherently oriented boundary $\partial\mathcal{M}$.

Theorem 12.38 *For all times $t \in \mathbb{R}$,*

$$\int_{\partial\mathcal{M}} pdq - Hd\tau = \int_{\partial(F_t\mathcal{M})} pdq - Hd\tau.$$

This is called the Poincaré–Cartan integral invariant.

Proof. Use Theorem 12.32 on page 735 together with $d\mathcal{L}_{\mathbf{v}}(pdq - Hd\tau) = 0$, by (xi) on page 748. □

12.9.4 Energy Conservation and the Liouville Integral Invariant

Let us consider the special case where the Hamiltonian function $H = H(q,p)$ does not depend on time. That is, we want to study the Hamiltonian equation of motion

$$\dot{q}(t) = H_p(q(t), p(t)), \qquad \dot{p}(t) = -H_q(q(t), p(t)), \quad t \in \mathbb{R} \tag{12.149}$$

with the initial condition $q(t_0) = q_0, p(t_0) = p_0$. Set $Q := (q,p)$. The velocity vector field

$$\mathbf{w}_Q := H_p(Q)\mathbf{i}_Q - H_q(Q)\mathbf{j}_Q$$

is called the Hamiltonian vector field on the Euclidean (q,p)-plane $\mathbb{E}^2$ corresponding to H. We assume that $\mathbf{w}$ is complete. Then it generates the flow $\{\mathcal{F}_t\}_{t\in\mathbb{R}}$ which is called the Hamiltonian flow on $\mathbb{E}^2$. Set $\sigma := dq \wedge dp$. Then:

---

[47] Recall that, by definition, $i_{\mathbf{v}}\alpha = 0$ if $r = 0$.

(a) $d\sigma = 0$ on $\mathbb{E}^2$;
(b) $\mathcal{L}_{\mathbf{w}}\sigma = 0$ on $\mathbb{E}^2$;
(c) $\mathcal{L}_{\mathbf{w}}H = \{H, H\} = 0$.

For the proof, use the same argument as on page 748.

Energy conservation. It follows from (c) that the Hamiltonian function is a conserved quantity under the Hamiltonian flow, that is, for the solutions of (12.149), we have

$$H(q(t), p(t)) = H(q_0, p_0) \qquad \text{for all} \quad t \in \mathbb{R}.$$

In physics, H represents the energy.

Conservation of the phase volume. Suppose that $\mathcal{C}$ is a compact subset of the (q,p)-plane $\mathbb{E}^2$. It follows from (b) together with Theorem 12.31 on page 734 that

$$\int_{\mathcal{F}_t(\mathcal{C})} dq \wedge dp = \int_{\mathcal{C}} dq \wedge dp \qquad \text{for all} \quad t \in \mathbb{R}.$$

In classical terms, $\int_{\mathcal{F}_t(\mathcal{C})} dqdp = \int_{\mathcal{C}} dqdp$, that is, the phase volume is preserved. This is Liouville's theorem.

12.9.5 Jacobi's Canonical Transformations, Lie's Contact Geometry, and Symplectic Geometry

> Canonical transformations preserve the structure of the Hamiltonian equations of motion. This can be used in order to get simpler Hamiltonian equations which can be solved explicitly. In the 19th century, mathematicians and physicists constructed more and more canonical transformations in order to attack the two basic problems in celestial mechanics: the explicit solution of the n-body problem ($n \geq 3$), and the stability of our planetary system.[48]
> Time-independent Hamiltonian functions (i.e., energy functions) H are related to symplectic geometry on even-dimensional symplectic phase spaces $\mathcal{M}$ (e.g., $\mathcal{M} = \mathbb{E}^2$). In contrast to this situation, time-dependent Hamiltonian functions are related to the contact geometry on the odd-dimensional product manifold $\mathcal{M} \times \mathbb{R}$ where the space $\mathbb{R}$ describes time (e.g., $\mathbb{E}^2 \times \mathbb{R} = \mathbb{E}^3$).
> Sophus Lie (1842–1899) is the father of symplectic and contact geometry.
> Folklore

Jacobi's crucial trick. Consider again the Hamiltonian equation of motion

$$\dot{q}(t) = H_p(q(t), p(t), \tau(t)), \;\; \dot{p}(t) = -H_q(q(t), p(t), \tau(t)), \;\; \dot{\tau}(t) = 1 \qquad (12.150)$$

for all $t \in \mathbb{R}$. It was Jacobi's idea to use a transformation

$$\mathcal{Q} = \mathcal{Q}(q, p, \tau), \quad \mathcal{P} = \mathcal{P}(q, p, \tau), \quad \mathcal{T} = \tau \qquad (12.151)$$

in order to obtain the transformed equation of motion

$$\dot{\mathcal{Q}}(t) = \mathcal{H}_p(\mathcal{Q}(t), \mathcal{P}(t), \mathcal{T}(t)), \;\; \dot{\mathcal{P}}(t) = -H_{\mathcal{Q}}(\mathcal{Q}(t), \mathcal{P}(t), \mathcal{T}(t)), \;\; \dot{\mathcal{T}}(t) = 1 \; (12.152)$$

[48] See R. Abraham and J. Marsden, Foundations of Mechanics, Addison-Wesley, Reading, Massachusetts, 1978, and D. Boccaletti and G. Pucacco, Theory of Orbits: Vol 1: Integrable Systems and Non-Perturbative Methods, Vol. 2: Perturbative and Geometrical Methods, Springer, Berlin, 1996.

for all $t \in \mathbb{R}$ with the transformed Hamiltonian function

$$\mathcal{H}(\mathcal{Q}, \mathcal{P}, t) := H(q, p, t).$$

Such transformations were called canonical transformations by Jacobi.[49] In particular, suppose that we find such a transformation with $\mathcal{H} \equiv 0$. Then, $\mathcal{Q} = \text{const}$ and $\mathcal{P} = \text{const}$ by (12.152). Reversing the transformation (12.151), we get a solution of (12.150). Jacobi found out that such a transformation exists if we know a solution $S = S(q, t; \mathcal{Q})$ of the Hamilton–Jacobi partial differential equation

$$S_t(q, t; \mathcal{Q}) + H(q, S_q(q, t; \mathcal{Q}), t) = 0$$

which depends on the parameter $\mathcal{Q}$. Let us discuss this.

The rank of a 2-form. Fix $m = 2, 3, \ldots$ Let ω be a 2-form on the m-dimensional manifold $\mathcal{M}$. Consider the skew-symmetric $(m \times m)$-matrix

$$(\omega_P(\mathbf{b}_i, \mathbf{b}_j)), \qquad i, j = 1, \ldots, m \tag{12.153}$$

where $\mathbf{b}_1, \ldots, \mathbf{b}_m$ is a basis of the tangent space $T_P\mathcal{M}$. By definition, the 2-form ω has the rank r iff the matrix (12.153) has the rank r for all points $P \in \mathcal{M}$. This definition does not depend on the choice of the basis vectors $\mathbf{b}_1, \ldots, \mathbf{b}_m$. If m is odd, then the determinant of the matrix (12.153) vanishes. Thus, we have $r \le m - 1$. If m is even, then $r \le m$. We have maximal rank, that is, $r = m$ iff ω_P is non-degenerate for all $P \in \mathcal{M}$.[50]

The symplectic 2-form σ on the Euclidean phase space $\mathbb{E}^2$. The Euclidean (q, p)-plane $\mathbb{E}^2$ of the Euclidean (q, p, τ)-manifold $\mathbb{E}^3$ is called the phase space for the motion on the real line (Fig. 12.21(b) on page 745). The trajectory $q = q(t), t \in \mathbb{R}$, of a particle of mass $m > 0$ moving on the real line corresponds to the curve $q = q(t), p = p(t) = m\dot{q}(t)$ on $\mathbb{E}^2$. Set $\sigma := dq \wedge dp$.

The 2-form σ is symplectic on $\mathbb{E}^2$, that is, $d\sigma = 0$, and the rank of σ is maximal on $\mathbb{E}^2$ $(r = 2)$.

Proof. Set $\mathbf{e}_1 := \mathbf{i}_P, \mathbf{e}_2 := \mathbf{j}_P$. Then

$$\sigma_P(\mathbf{e}_i, \mathbf{e}_j) = dq(\mathbf{e}_i)dp(\mathbf{e}_j) - dq(\mathbf{e}_j)dp(\mathbf{e}_i), \qquad i, j = 1, 2.$$

Hence

$$\begin{pmatrix} \sigma_P(\mathbf{e}_1, \mathbf{e}_1) & \sigma_P(\mathbf{e}_1, \mathbf{e}_2) \\ \sigma_P(\mathbf{e}_2, \mathbf{e}_1) & \sigma_P(\mathbf{e}_2, \mathbf{e}_2) \end{pmatrix} = \begin{pmatrix} 0 & 1 \\ -1 & 0 \end{pmatrix}.$$

□

The Euclidean (q, p)-plane becomes a symplectic manifold equipped with the symplectic form σ.

The map $F : \mathbb{E}^2 \to \mathbb{E}^2$ is called a symplectic isomorphism iff it is a diffeomorphism and $F^*\sigma = \sigma$. Explicitly, the map f is given by $Q = Q(q, p), P = P(q, p)$ with $dQ \wedge dP = dq \wedge dp$.

Proposition 12.39 *The diffeomorphism $F : \mathbb{E}^2 \to \mathbb{E}^2$ is a symplectic isomorphism iff there exists a smooth function $S : \mathbb{E}^2 \to \mathbb{E}^2$ with $F^*(pdq) = pdq - dS$ on $\mathbb{E}^2$.*

---

[49] See C. Jacobi, Lectures on Analytical Mechanics (including Celestial Mechanics). Edited by the German Mathematical Society (DMV), Vieweg, Braunschweig 1996 (in German).

[50] This means that $\omega_P(\mathbf{v}, \mathbf{w}) = 0$ for fixed $\mathbf{v} \in T_P\mathcal{M}$ and all $\mathbf{w} \in T_P\mathcal{M}$ implies $\mathbf{v} = 0$.

The function $S = S(q,p)$ is called the generating function of the symplectic isomorphism F.
Proof. If $F^*(pdq) = pdq - dS$, then

$$F^*\sigma = F^*(dq \wedge dp) = -F^*d(pdq) = -dF^*(pdq) = -d(pdq) = \sigma.$$

Thus, F is a symplectic isomorphism.

Conversely, if $F^*\sigma = \sigma$, then $d(F^*(pdq) - pdq) = 0$ on $\mathbb{E}^2$. It follows from the Poincaré–Volterra theorem (Theorem 12.46 on page 762) that there exists a smooth function $S : \mathbb{E}^2 \to \mathbb{E}^2$ such that $F^*(pdq) - pdq = -dS$. □

The Poincaré–Cartan contact 2-form $d\chi$ on the extended phase space $\mathbb{E}^3$. Our goal is to study the Hamiltonian equations of motion (12.150) for time-dependent Hamiltonian functions $H = H(q,p,t)$. To this end, we need the extended phase space $\mathbb{E}^2 \times \mathbb{R}$. This Euclidean (q,p,τ)-space $\mathbb{E}^3$ is depicted in Fig. 12.21(b) on page 745. Differentiating the Poincaré–Cartan 1-form $\chi := pdq - Hd\tau$, we get the differential $d\chi = dp \wedge dq - dH \wedge d\tau = -\sigma - dH \wedge d\tau$. The dimension of a symplectic manifold is always even. Therefore, the 3-dimensional extended phase space $\mathbb{E}^3$ cannot be equipped with a symplectic structure. However, it can be equipped with a contact structure.

The 2-form $d\chi$ is a contact form on $\mathbb{E}^3$, that is, $d(d\chi) = 0$, and the rank of $d\chi$ is maximal ($r = 2$).

Proof. Note that

$$\begin{pmatrix} d\chi_P(\mathbf{e}_1,\mathbf{e}_1) & d\chi_P(\mathbf{e}_1,\mathbf{e}_2) & d\chi_P(\mathbf{e}_1,\mathbf{e}_3) \\ d\chi_P(\mathbf{e}_2,\mathbf{e}_1) & d\chi_P(\mathbf{e}_2,\mathbf{e}_2) & d\chi_P(\mathbf{e}_2,\mathbf{e}_3) \\ d\chi_P(\mathbf{e}_3,\mathbf{e}_1) & d\chi_P(\mathbf{e}_3,\mathbf{e}_2) & d\chi_P(\mathbf{e}_3,\mathbf{e}_3) \end{pmatrix} = \begin{pmatrix} 0 & -1 & -H_q(P) \\ 1 & 0 & -H_p(P) \\ H_q(P) & H_p(P) & 0 \end{pmatrix}.$$

The determinant of this skew-symmetric matrix vanishes. Thus, the rank of this matrix is equal to 2. The extended (q,p,τ)-phase space $\mathbb{E}^3$ becomes a contact manifold equipped with the Poincaré–Cartan contact 2-form $d\chi$.

Canonical transformation. The map $F : \mathbb{E}^3 \to \mathbb{E}^3$ is called a canonical transformation iff it is a contact isomorphism, that is, F is a diffeomorphism and

$$F^*(d\chi) = d\chi.$$

Explicitly, the map F has the form

$$\mathcal{Q} = \mathcal{Q}(q,p,\tau), \quad \mathcal{P} = \mathcal{P}(q,p,\tau), \quad \mathcal{T} = \mathcal{T}(q,p,\tau) \tag{12.154}$$

with

$$-d\mathcal{Q} \wedge d\mathcal{P} - d\mathcal{H}(\mathcal{Q},\mathcal{P},\mathcal{T}) \wedge d\mathcal{T} = -dq \wedge dp - dH(q,p,\tau) \wedge d\tau$$

where $\mathcal{H}(\mathcal{Q},\mathcal{P},\mathcal{T}) = H(q,p,\tau)$.

Proposition 12.40 *The diffeomorphism $F : \mathbb{E}^3 \to \mathbb{E}^3$ is a contact isomorphism iff there exists a smooth function $S : \mathbb{E}^3 \to \mathbb{E}^3$ with $F^*\chi = \chi - dS$.*

The function $S = S(q,p,\tau)$ is called the generating function of the canonical transformation. The proof follows by using the same argument as in the proof of Prop. 12.39 above.

12.9.6 Hilbert's Invariant Integral

Fix the points P_0 and P_1 in the extended phase space, and recall the definition of the set A of admissible curves $\mathcal{C}$ (see page 746).

Proposition 12.41 *If $F : \mathbb{E}^3 \to \mathbb{E}^3$ is a canonical transformation, then there exists a real constant such that*

$$\int_{\mathcal{C}} F^*\chi = \int_{\mathcal{C}} \chi + \text{const} \qquad \textit{for all} \quad \mathcal{C} \in \mathsf{A}.$$

The integral $\int_{\mathcal{C}} \chi = \int_{\mathcal{C}} pdq - H(q,p,\tau)d\tau$ is called Hilbert's invariant integral. The claim follows from $F^*\chi = \chi - dS$, and hence

$$\int_{\mathcal{C}} F^*\chi = \int_{\mathcal{C}} \chi - S(P_1) + S(P_0).$$

Theorem 12.42 *Canonical transformations send solutions of the Hamiltonian equation of motion to solutions of the transformed Hamiltonian equation of motion.*

Proof. The variational problem of critical action $\int_{\mathcal{C}} \chi =$ critical!, $\mathcal{C} \in \mathsf{A}$ and the transformed problem

$$\int_{\mathcal{C}} F^*\chi = \text{critical!}, \qquad \mathcal{C} \in \mathsf{A}$$

possesses the same solutions, since the integrals only differ by a constant according to Prop. 12.41. Then the claim follows from Theorem 12.37 on page 747. □

12.9.7 Jacobi's Integration Method

Suppose that the smooth function $S : \mathbb{R}^3 \to \mathbb{R}$ is a solution of the Hamilton–Jacobi partial differential equation

$$S_t(q,t;\mathcal{Q}) + H(q, S_q(q,t;\mathcal{Q}), t) = 0, \qquad (q,t,\mathcal{Q}) \in \mathbb{R}^3$$

where $\mathcal{Q}$ plays the role of an additional real parameter. Suppose that the equation

$$p = S_q(q,t;\mathcal{Q}), \qquad \mathcal{P} = -S_{\mathcal{Q}}(q,t;\mathcal{P}) \tag{12.155}$$

can be solved globally such that map $(q,p,t) \mapsto (\mathcal{Q},\mathcal{P},t)$ is a diffeomorphism F from $\mathbb{R}^3$ onto $\mathbb{R}^3$.

Theorem 12.43 *If we fix the real parameters $\mathcal{Q}$ and $\mathcal{P}$, then* (12.155) *yields the solution*

$$q = q(t;\mathcal{Q},\mathcal{P}), \quad p = p(t;\mathcal{Q},\mathcal{P}), \qquad t \in \mathbb{R}$$

of the Hamiltonian equation of motion (12.142) *on page 745.*

Proof. The idea of the proof is to show that F is a canonical transformation with trivial transformed Hamiltonian $\mathcal{H} \equiv 0$. It follows from

$$pdq - Hdt - dS = (p - S_q)dq - (H + S_t)dt - S_{\mathcal{Q}}d\mathcal{Q} = \mathcal{P}d\mathcal{Q}$$

that $F^*(pdq - Hdt) = F^*(\mathcal{P}d\mathcal{Q}) + F^*dS = \mathcal{P}d\mathcal{Q} + d(F^*dS)$. Consequently, the solution of the transformed variational problem

$$\int_{\mathcal{C}} F^*(pdq - Hdt) = \text{critical!}, \qquad \mathcal{C} \in \mathsf{A} \tag{12.156}$$

is equivalent to $\int_{\mathcal{C}} \mathcal{P}d\mathcal{Q} = \text{critical!}$, $\mathcal{C} \in \mathsf{A}$. The solutions satisfy the Hamiltonian equations of motion

$$\dot{\mathcal{Q}}(t) = 0, \qquad \dot{\mathcal{P}}(t) = 0, \qquad t \in \mathbb{R}$$

with vanishing Hamiltonian. The solution reads as $\mathcal{Q}(t) = \text{const}$, $\mathcal{P}(t) = \text{const}$. Using the inverse transformation F^{-1}, we get the claim. □

If the equation (12.155) can be solved only locally, then we obtain a local solution of the Hamiltonian equations of motion.

12.9.8 Legendre Transformation

The Lagrangian equation of motion. Our goal is to construct a diffeomorphism $F : \mathbb{E}^3 \to \mathbb{E}^3$ from the extended state space to the extended phase space (Fig.12.21 on page 745). Explicitly, $(q, p, \tau) = F(q, v, \tau)$. To this end, suppose that the equation

$$\boxed{v = H_p(q, p, \tau)} \tag{12.157}$$

describes a diffeomorphism $(q, p, \tau) \mapsto (q, v, \tau)$ from the extended phase space to the extended state space. The inverse map $(q, v, \tau) \mapsto (q, p, \tau)$ is then the desired diffeomorphism F. Setting

$$\boxed{L := vp - H,}$$

we obtain the Lagrangian $L : \mathbb{E}^3 \to \mathbb{R}$ on the extended state space. Explicitly,

$$L(q, v, \tau) := vp(q, v, \tau) - H(q, p(q, v, \tau), \tau).$$

Suppose that

$$\mathcal{C} : q = q(t), \;\; p = p(t), \;\; \tau = t, \qquad t_0 \le t \le t_1$$

is a smooth curve on the extended phase space. The inverse Legendre transformation sends this to the curve

$$C : q = q(t), \;\; v = v(t), \;\; \tau = t, \qquad t_0 \le t \le t_1$$

on the extended state space.

Proposition 12.44 *If the curve $\mathcal{C}$ satisfies the Hamiltonian equation of motion* (12.142) *on page 745, then the transformed curve C satisfies both the condition $v(t) = \dot{q}(t)$ and the Lagrangian equation of motion*

$$\frac{d}{dt} L_v(q(t), \dot{q}(t), t) = L_q(q(t), \dot{q}(t), t), \qquad t_0 \le t \le t_1. \tag{12.158}$$

Moreover, $\int_C Ldt = \int_{\mathcal{C}} pdq - Hdt$, that is, the action integrals coincide.

Proof. From $L = vp - H$ we get

$$dL = (dv)p + vdp - H_q dq - H_p dp - H_\tau d\tau = pdv - H_q dq - H_\tau d\tau.$$

Hence $L_v = p, L_q = -H_q, L_\tau = -H_\tau$. Explicitly,

$$L_v(q, v, \tau) = p, \quad L_q(q, v, \tau) = -H_q(q, p, \tau), \quad L_\tau(q, v, \tau) = -H_\tau(q, p, \tau).$$

It follows from the Hamiltonian equations of motion $\dot{q} = H_p, \dot{p} = -H_q$ and $v = H_p$ that $v = \dot{q}$ and $\frac{d}{dt}L_v = -H_q = L_q$. This is the Lagrangian equation of motion. In addition, we get

$$\int_C L d\tau = \int_{t_0}^{t_1} L(q(t), \dot{q}(t), t)\, dt = \int_{t_0}^{t_1} (p(t)\dot{q}(t) - H(q(t), p(t), t))\, dt = \int_C pdq - Hdt.$$

□

The Lagrangian principle of critical action. Fix the real numbers t_0, t_1, q_0, and q_1. The Lagrangian principle of critical action reads as

$$\boxed{\int_{t_0}^{t_1} L(q(t), \dot{q}(t), t)\, dt = \text{critical!}, \qquad q \in \mathsf{A}_L.} \tag{12.159}$$

Here, the admissible set A_L consists of all the smooth functions $q : [t_0, t_1] \to \mathbb{R}$ which satisfy the boundary condition $q(t_0) = q_0$, $q_(t_1) = q_1$. A standard result of the calculus of variations, tells us the following:

> *The Lagrangian principle of critical action* (12.159) *is equivalent to the Lagrangian equation of motion* (12.158).

Since the Hamiltonian (resp. Lagrangian) equation of motion is equivalent to the Hamiltonian (resp. Lagrangian) principle of critical action, it follows from Prop. 12.44 that:

> *The Lagrangian principle of critical action* (12.159) *is equivalent to the Hamiltonian principle of critical action* (12.145) *on page 746.*

Local approach. If equation (12.157) only generates a local diffeomorphism (i.e., a local Legendre transformation), then the results obtained above remain valid in a local setting.

12.9.9 Carathéodory's Royal Road to the Calculus of Variations

> In the Hamiltonian setting, Carathéodory's 'royal road' to the calculus of variations does not start from a variational problem, but from a covelocity (or momentum) function $w = w(t, q)$ which satisfies Carathéodory's fundamental equation
>
> $$w^*(d\chi) = 0$$
>
> where $\chi := pdq - H(t, q, p)dt$ is the Poincaré–Cartan 1-form. The Lagrangian setting of this approach is then obtained by Legendre transformation.
>
> This approach was strongly influenced by the analogy between light rays and geodesics in Riemannian geometry, the duality between light rays and wave fronts via the Huygens principle, Lie's contact transformations, and Hilbert's invariant integral.[51]
>
> Folklore

[51] This integral was known to Beltrami in 1864. Hilbert rediscovered it in about 1900.

> Carathéodory realized that his fundamental equation can be used as a key to the calculus of variations. Therefore Boerner has called Carathéodory's approach *a royal road to the calculus of variations.*[52]
> In Carathéodory's treatise[53] this road is somewhat hidden because Carathéodory had discovered it while reading the galley proofs of his book, and only in the last minute he managed to include it into the book.
> Undoubtedly Carathéodory's royal road is nowadays the quickest and most elegant approach to *sufficient* conditions for minimum problems in the calculus of variations (see Sect. 5.4 of Vol. II). In addition, it can easily be carried over to multiple integrals. Thus it may be surprising to learn that Carathéodory was led to his approach by Johann Bernoulli's paper from 1718 (where Bernoulli showed that the minimum for the problem of quickest descent is indeed attained by the cycloid).[54]
>
> Mariano Giaquinta and Stefan Hildebrandt, 1996

Terminology. We will use the following notation:

- q (position), t (time), p (momentum or generalized momentum),
- $H = H(t, q, p)$ (Hamiltonian function),
- $\chi := pdq - H(t, q, p)\, dt$ (Poincaré–Cartan 1-form),
- $d\chi$ (Poincaré–Cartan contact 2-form),
- $w = w(t, q)$ (covelocity field function),
- $S = S(t, q)$ (action function),
- $v(t, q) := H_p(t, q, w(t, q))$ (velocity field function).

The quantities have the following physical dimension: q (length), t (time), H (energy), w, p (covelocity =momentum=energy/velocity), S (energy $\times$ time = action), v (velocity).

Carathéodory's fundamental equation and its consequences. We are given the smooth Hamiltonian function $H : \mathbb{R}^3 \to \mathbb{R}$. Suppose that the smooth covelocity function $w : \mathbb{R}^2 \to \mathbb{R}$ is a solution of Carathéodory's equation $d\chi = 0$ on $\mathbb{R}^3$, that is,

$$\boxed{w^*(d\chi) = 0 \qquad \text{on } \mathbb{R}^2.} \tag{12.160}$$

Theorem 12.45 (i) *The Hamilton–Jacobi partial differential equation: There exists a smooth function $S : \mathbb{R}^2 \to \mathbb{R}$ such that*

$$dS = w^*\chi \quad \text{on } \mathbb{R}^2. \tag{12.161}$$

If the pair (w, S) of smooth functions forms a complete configuration, that is, it satisfies (12.161), *then the action function S is a solution of the following Hamilton–Jacobi differential equation:*

$$S_t(t, q) + H(t, q, S_q(t, q)) = 0, \qquad (t, q) \in \mathbb{R}^2.$$

(ii) *The Hamiltonian equation of motion: Let $q = q(t), t_0 \le t \le t_1$ be a smooth solution of the ordinary differential equation*

[52] H. Boerner, Carathéodory's approach to the calculus of variations, Jahresber. Deutsche Mathem.-Verein. **56** (1953), 31–58 (in German).

[53] C. Carathéodory, Calculus of Variations and Partial Differential Equations of First Order. German edition: Teubner, Leipzig, 1937. English edition: Chelsea, New York, 1982.

[54] M. Giaquinta and S. Hildebrandt, Calculus of Variations I, Springer, Berlin, 1996 (reprinted with permission).

$$\dot{q}(t) = H_p(t, q(t), w(t, q(t))), \qquad t_0 \le t \le t_1.$$

Set $p(t) := w(t, q(t))$. *Then the function* $t \mapsto (q(t), p(t))$ *satisfies the Hamiltonian equation of motion:*

$$\dot{q}(t) = H_p(t, q(t), p(t)), \quad \dot{p}(t) = -H_q(t, q(t), p(t)), \qquad t_0 \le t \le t_1.$$

In addition, if the pair (w, S) *satisfies* (12.161), *then the value of the action integral is given by*

$$\int_{t_0}^{t_1} \big(p(t)\dot{q}(t) - H(t, q(t), p(t))\big)\, dt = S(t_1, q(t_1)) - S(t_0, q(t_0)). \qquad (12.162)$$

This explains the intuitive meaning of the action function S.

Proof. Ad (i). Since $d(w^*\chi) = w^*(d\chi)$, it follows from (12.160) that

$$d(w^*\chi) = 0 \qquad \text{on } \mathbb{R}^2.$$

By the Poincaré–Volterra theorem on page 762, there exists a smooth function $S : \mathbb{R}^2 \to \mathbb{R}^2$ such that $dS = w^*\chi$. Hence

$$S_t(t, q)dt + S_q(t, q)dq = w(t, q)dq - H(t, q, w(t, q)dt.$$

This is equivalent to

$$\boxed{S_t(q, t) = -H(t, q, w(t, q)), \qquad S_q(t, q) = w(t, q),} \qquad (12.163)$$

which immediately yields the Hamilton–Jacobi equation. The integrability condition $S_{qt} = S_{tq}$ implies the key equation

$$\boxed{w_t(q, t) = -H_q(t, q, w(t, q)) - H_p(t, q, w(t, q))\, w_q(t, q).} \qquad (12.164)$$

Ad (ii). It follows from $p(t) = w(t, q(t))$ that

$$\dot{p}(t) = w_t(t, q(t)) + w_q(t, q(t))\, \dot{q}(t).$$

Setting $P(t) := (t, q(t), p(t))$, equation (12.164) tells us that

$$\dot{p}(t) = w_q(t, q(t))H_p(P(t)) - H_q(P(t)) - H_p(P(t))\, w_q(t, q(t)) = -H_p(P(t)).$$

This is the claim. Finally, setting $S(t) := S(t, q(t))$, we get

$$\dot{S}(t) = S_q(t, q(t))\, \dot{q}(t) + S_t(t, q(t)) = p(t)\dot{q}(t) - H(t, q(t), p(t)).$$

This yields the integral relation (12.162). □

It follows from $dS = w^*\chi$ that the line integral

$$\int_C w^*\chi = \int_C w(t, q)dq - H(t, q, w(q, t))$$

does not depend on the path C, but only on the initial point and the terminal point of the curve C. This integral is called Hilbert's invariant integral.

Local approach. If the covelocity field function w is only a local solution of Carathéodory's fundamental equation (12.160), then all the results formulated above remain valid in a local setting.

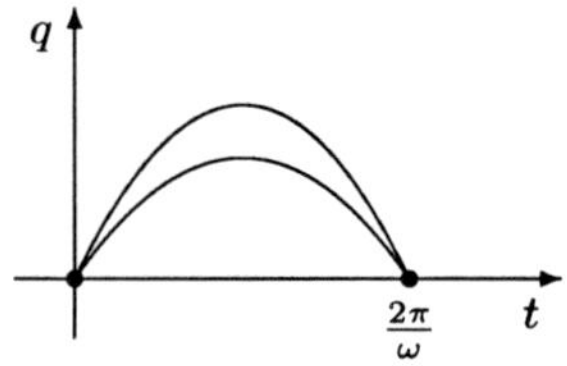

Fig. 12.22. Family of harmonic oscillators

The Legendre transformation. Set $v := H_p(t, q, p)$, and consider the inverse Legendre transformation $(t, q, p) \mapsto (t, q, v)$ as described on page 754. Then the solutions $t \mapsto (q(t), p(t))$ of the Hamiltonian equation of motion pass over to the solutions $t \mapsto q(t)$ of the Lagrangian equation of motion with the Lagrangian

$$L = vp - H.$$

In particular, the covelocity field function $w = w(t, q)$ passes over to the velocity field function

$$v(t, q) = H_p(t, q, w(t, q)).$$

From $\dot{q}(t) = H_p(t, q(t), w(t, q(t))$, we get the differential equation

$$\dot{q}(t) = v(t, q(t)), \qquad t \in \mathbb{R}.$$

This tells us that the value $v(t, q)$ describes the velocity of the particle at the point q at time t.

The harmonic oscillator. In order to explain the intuitive meaning of Carathéodory's approach, let us consider the harmonic oscillator with the Lagrangian

$$L = \frac{mv^2}{2} - \frac{\omega^2 q^2}{2}.$$

Consider the family of solutions

$$q(t) = q_0 \sin \omega t, \qquad 0 < t < \frac{\pi}{\omega}$$

of the Lagrangian equation of motion $\frac{d}{dt} L_v = L_q$ (Fig. 12.22). Fix the point (q_1, t_1) with $0 < t_1 < \frac{\pi}{\omega}$. The unique trajectory passing through the point q_1 at time t_1 is given by

$$q(t) = \frac{q_1}{\sin \omega t_1} \cdot \sin \omega t, \qquad 0 < t < \frac{\pi}{\omega}.$$

We set

- $v(t_1, q_1) := \dot{q}(t_1)$ (velocity function), and
- $w(t_1, q_1) := p(t_1) = m\dot{q}(t_1)$ (covelocity or momentum function).

Explicitly, we get

$$\boxed{v(q, t) = \omega q \cdot \cot \omega t, \qquad w(t, q) = m\omega q \cdot \cot \omega t, \quad 0 < t < \frac{\pi}{\omega}, \ q \in \mathbb{R}.}$$

Focal points. Note that the functions v and w have singularities at $t = 0$ and $t = \frac{\pi}{\omega}$. These singularities correspond to the focal points $(0, 0)$ and $(\frac{\pi}{\omega}, 0)$ depicted in Fig. 12.22.

12.9.10 Geometrical Optics

Hamilton (1805–1865) created Hamiltonian mechanics by using geometrical optics as a paradigm. We will reverse the historical order, that is, we will show that geometrical optics is a special case of Hamiltonian mechanics by changing dramatically the physical interpretation of the mathematical quantities. For example, the time variable τ becomes a space variable x, and the action function S becomes the eikonal function which has the physical dimension of time. We will use the following terminology:

- $\tau = x, q = y$ (space variables of a Cartesian (x, y)-coordinate system),
- $y = y(x)$ (equation of a light ray curve),
- $S = S(x, y)$ (eikonal function),
- $w = w(x, y)$ (coslope function), p (coslope),
- $v = v(x, y)$ (slope function),
- $n(x, y)$ (refraction index; $n \equiv 1$ in a vacuum), $c/n(x, y)$ (velocity of light in the optical substance at the point (x, y)),
- $L(x, y, y') = \frac{n(x,y)}{c}\sqrt{1 + y'^2}$ (Lagrangian of Fermat's principle of critical time).

The quantities have the following physical dimensions: x, y (length), S (time), v (dimensionless), L, H, w, p (time/length).

We start with Fermat's principle of critical time for light rays:

$$\int_{x_0}^{x_1} \frac{n(x,y)}{c}\sqrt{1 + y'(x)^2}\, dx = \text{critical!}, \quad y(x_0) = y_0,\ y(x_1) = y_1. \tag{12.165}$$

We are given the points (x_0, y_0) and $x_1.y_1)$ in the Euclidean plane equipped with an Cartesian (x, y)-coordinate system, and the smooth refraction function

$$n : \mathbb{R}^2 \to]0, \infty[$$

with $\inf_{(x,y)\in\mathbb{R}^2} n(x, y) > 0$. We are looking for a smooth light ray

$$y = y(x), \quad x_0 \le x \le x_1,$$

which passes from the point (x_0, y_0) to the point (x_1, y_1). The integral on the left side of (12.165) is the time needed by the light ray. Let us introduce the Lagrangian

$$L(x, y, v) := \frac{n(x,y)}{c}\sqrt{1 + v^2}.$$

The Legendre transformation $(x, y, v) \mapsto (x, y, p)$ is generated by $p = L_v(x, y, v)$. We also introduce the Hamiltonian function $H(x, y, p) := pv(x, y, p) - L(x, y, v(x, y, p))$. This yields

$$p = \frac{n(x,y)v}{c\sqrt{1 + v^2}}, \qquad H(x, y, p) = -\sqrt{\frac{n(x,y)^2}{c^2} - p^2}.$$

Carathéodory's fundamental equation. In this special case, equation (12.163) reads as

$$S_x(x, y) = \frac{n(x,y)}{c\sqrt{1 + v(x,y)^2}}, \qquad S_y(x, y) = v(x, y)S_x(x, y). \tag{12.166}$$

Here, $w(x, y) = \frac{n(x,y)v(x,y)}{c\sqrt{1+v(x,y)^2}}$. Theorem 12.45 on page 756 tells us the following:

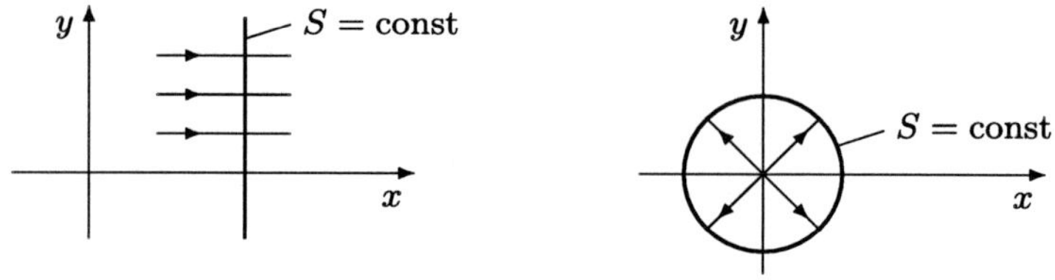

(a) plane wavefront (b) circular wavefront

Fig. 12.23. Light rays and wavefronts

Let v, S be a solution of (12.166) *with $S(x_0, y_0) = 0$, and let $y = y(x)$ be a solution of the differential equation*

$$y'(x) = v(x, y(x)), \quad x_0 \leq x \leq x_1 \qquad y(x_0) = y_0. \tag{12.167}$$

Set $y_1 := y(x_1)$. Then $y = y(x)$ is a solution of the variational problem (12.165). *The value $S(x_1, y_1)$ is the time needed by the light ray to pass from the point (x_0, y_0) to the point (x_1, y_1). The wave fronts starting from the point (x_0, y_0) are given by*

$$S(x, y) = \text{const}.$$

Let us consider two simple examples. Suppose that $n(x, y) = \text{const}$ (i.e., the velocity of the light rays is equal to c/n).

Linear wave front. We are given $v(x, y) \equiv 0$. Equation (12.166) has the solution $S(x, y) = \frac{n}{c}(x - x_0)$. Then equation (12.167) yields the light rays $y(x) = \text{const}$ and the wave fronts $x = \text{const}$ (Fig. 12.23(a)).

Circular wave front. The functions $v(x, y) := \frac{y}{x}$ and

$$S(x, y) := \frac{n}{c}\sqrt{x^2 + y^2}$$

solve the equation (12.166). The differential equation (12.167) yields the light rays $y(x) = \text{const} \cdot x$ and the wave fronts $\frac{n^2}{c^2}(x^2 + y^2) = \text{const}$ (Fig. 12.23(b)). Many applications can be found in:

C. Carathéodory, Geometrical Optics, Springer, Berlin, 1937 (in German).

M. Born and E. Wolf, Principles of Optics, 7th edition, Cambridge University Press, 1999.

12.10 The Main Theorems in Classical Gauge Theory (Existence of Potentials)

We want to study the equation

$$\boxed{d\omega = \mu \qquad \text{on } \mathcal{M}} \tag{12.168}$$

which generalizes the classical potential equation $\mathbf{F} = -\,\mathbf{grad}\, U$ on the Euclidean manifold $\mathbb{E}^3$. We are given the smooth r-form μ on the real n-dimensional manifold $\mathcal{M}$ with $r \geq 1$. We are looking for the smooth $(r-1)$-form on $\mathcal{M}$. If the equation (12.168) has a solution ω, then

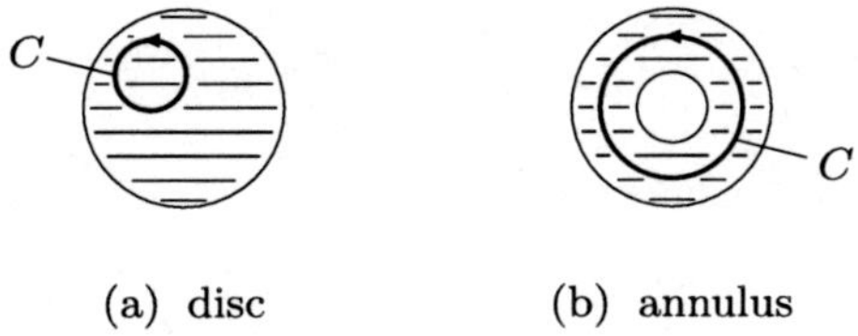

Fig. 12.24. Connectivity

(i) $d\mu = 0$ (local solvability condition), and
(ii) $\int_C \mu = 0$ for all r-dimensional compact submanifolds C (without boundary) of the manifold $\mathcal{M}$ (global solvability condition).

In fact, it follows from the Poincaré cohomology rule $d(d\omega) = 0$ that $d\mu = 0$. Moreover, the generalized Stokes theorem tells us that

$$\int_C \mu = \int_C d\omega = \int_{\partial C} \omega = 0,$$

since the boundary ∂C is empty. We are looking for sufficient solvability conditions of (12.168). Roughly speaking, the necessary solvability conditions (i), (ii) are also sufficient solvability conditions. The point is that special properties of μ and $\mathcal{M}$ simplify the approach. Let us mention some typical simplifications:

- Suppose that C is the boundary of a submanifold $\mathcal{N}$ of $\mathcal{M}$, that is, $C = \partial\mathcal{N}$. Then the local solvability condition $d\mu = 0$ implies

$$\int_C \mu = \int_{\partial\mathcal{N}} \mu = \int_{\mathcal{N}} d\mu = 0.$$

  In this special case, the global solvability condition $\int_C \mu = 0$ is always satisfied.
- If μ is an n-form, then the local solvability condition $d\mu = 0$ is always satisfied.
- If the manifold $\mathcal{M}$ is continuously contractible to a point, then the global solvability condition drops out.

Furthermore, both the number of essential solutions and the number of global solvability conditions of (12.168) depend on the so-called Betti numbers which are topological invariants of the manifold $\mathcal{M}$. In particular, we have to distinguish between contractible and non-contractible manifolds $\mathcal{M}$. For example, the disc (resp. the annulus) is contractible (resp. not contractible) to a point (Fig. 12.24). To begin with, let us study two prototypes.

First prototype. Consider the open interval $\mathcal{M} :=]a, b[$ on the real line. Fix $x_0 \in \mathcal{M}$. We are given the smooth 1-form $\mu = f(x)dx$. Then the general solution of the equation $d\omega = \mu$ on $\mathcal{M}$ (i.e., $\omega'(x) = f(x)$) is given by

$$\omega(x) = \int_{x_0}^{x} f(\xi)d\xi + c \qquad \text{for all} \quad x \in \mathcal{M}$$

where c is an arbitrary real number. This is the main theorem of calculus. Note that there are no solvability conditions for μ. The situation changes if we pass to the unit circle.

Second prototype. Consider the unit circle $\mathcal{M} := \mathbb{S}^1$. If $\omega : \mathbb{S}^1 \to \mathbb{R}$ is a smooth function, then there exists the Fourier expansion

$$\omega(\varphi) = a_0 + \sum_{k=1}^{\infty} a_k \cos k\varphi + b_k \sin k\varphi$$

where a_0, a_k, b_k are real numbers. Hence

$$d\omega(\varphi) = \left(\sum_{k=1}^{\infty} -ka_k \sin k\varphi + kb_k \cos k\varphi \right) v$$

where v is the volume form on $\mathbb{S}^1$. Explicitly,

$$\int_{\mathcal{S}} v = \beta - \alpha$$

if $\mathcal{S} := \{e^{i\varphi} : -\pi \leq \alpha \leq \varphi \leq \beta \leq \pi\}$. Mnemonically, we write $d\varphi$ instead of v. Suppose that we are given the smooth 1-form

$$\mu = \left(A_0 + \sum_{k=1}^{\infty} A_k \cos k\varphi + B_k \sin k\varphi \right) d\varphi$$

on $\mathbb{S}^1$ with real coefficients A_0, A_k, B_k. Then the equation

$$d\omega = \mu \qquad \text{on } \mathbb{S}^1$$

has a solution ω iff

$$\boxed{\int_{\mathbb{S}^1} \mu = 0,} \tag{12.169}$$

that is, $A_0 = 0$. The general solution reads as

$$\omega = a_0 + \sum_{k=1}^{\infty} \frac{A_k}{k} \sin k\varphi - \frac{B_k}{k} \cos k\varphi$$

where a_0 is an arbitrary real number.

The global solvability condition (12.169) is crucial. It is caused by the nontrivial topology of the unit circle, in contrast to the trivial topology of the open interval $]a, b[$. We will show next that:

Global solvability conditions always depend on the nontrivial topology of the manifold $\mathcal{M}$.

For example, as we will discuss below, the appearance of precisely one global solvability condition (12.169) for the unit circle $\mathbb{S}^1$ depends on the fact that the first Betti number of $\mathbb{S}^1$ is equal to one, $\beta_1 = 1$.

12.10.1 Contractible Manifolds (the Poincaré–Volterra Theorem)

Fix $n = 1, 2, \ldots$ Let $r = 1, \ldots, n$. We are given the real n-dimensional manifold $\mathcal{M}$.

Theorem 12.46 *Suppose that $\mathcal{M}$ is continuously contractible to a point. Then, for given smooth r-form μ on $\mathcal{M}$, the equation $d\omega = \mu$ on $\mathcal{M}$ has a solution iff $d\mu = 0$ on $\mathcal{M}$.*

The proof can be found in Frankel (2004), Sect. 5.4, quoted on page 775. See also Amann and Escher (1998), Vol. 3, Sect. XI.3, quoted on page 776.

Examples. (i) Let $\mathcal{M}$ be an open disc. Let $U : \mathcal{M} \to \mathbb{R}$ be a smooth function, and let $\mu = Adx + Bdy$ be a smooth 1-form on $\mathcal{M}$. Then the equation

$$dU = \mu \qquad \text{on } \mathcal{M}$$

has a solution iff $d\mu = 0$. Since $d\mu = (B_x - A_y)\, dx \wedge dy$, we obtain that the system

$$U_x(x,y) = A(x,y), \quad U_y(x,y) = B(x,y) \qquad \text{on } \mathcal{M}$$

has a solution iff $B_x - A_y = 0$, that is, the integrability condition $A_y = B_x$ is satisfied on $\mathcal{M}$.

(ii) Potential U of a force field $\mathbf{F}$: Assume that we are given the smooth vector field $\mathbf{F} \in \mathrm{Vect}(\mathbb{E}^3)$. Then the equation

$$\mathbf{F} = -\,\mathbf{grad}\, U \qquad \text{on } \mathbb{E}^3$$

has a smooth solution $U : \mathbb{E}^3 \to \mathbb{R}$ iff $\mathbf{curl}\,\mathbf{F} = 0$ on $\mathbb{E}^3$.
To prove this, choose $\omega := -U$ and

- $\mu = Adx + Bdy + Cdz$, $\mathbf{F} = A\mathbf{i} + B\mathbf{j} + C\mathbf{k}$.

Furthermore, note that

- $d\omega = -(U_x dx + U_y dy + U_z dz)$, $\mathbf{grad}\, U = U_x\mathbf{i} + U_y\mathbf{j} + U_z\mathbf{k}$,
- $d\mu = (C_y - B_z)\, dy \wedge dz + (A_z - C_x)\, dz \wedge dx + (B_x - A_y)\, dx \wedge dy$,
- $\mathbf{curl}\,\mathbf{F} = (C_y - B_z)\mathbf{i} + (A_z - C_x)\mathbf{j} + (B_x - A_y)\mathbf{k}$.

(iii) Vector potential $\mathbf{A}$ of a magnetic field $\mathbf{B}$: We are given the smooth vector field $\mathbf{B} \in \mathrm{Vect}(\mathbb{E}^3)$. Then the equation

$$\mathbf{B} = \mathbf{curl}\,\mathbf{A} \qquad \text{on } \mathbb{E}^3$$

has a smooth solution $\mathbf{A} \in \mathrm{Vect}(\mathbb{E}^3)$ iff $\mathrm{div}\,\mathbf{B} = 0$ on $\mathbb{E}^3$.
To prove this, choose

- $\mu = A\, dy \wedge dz + B\, dz \wedge dx + C\, dx \wedge dy$, $\mathbf{B} = A\mathbf{i} + B\mathbf{j} + C\mathbf{k}$,
- $\omega = adx + bdy + cdz$, $\mathbf{A} = a\mathbf{i} + b\mathbf{j} + c\mathbf{k}$,
- $d\mu = (A_x + B_y + C_z)\, dx \wedge dy \wedge dz$, $\mathrm{div}\,\mathbf{B} = A_x + B_y + C_z$,
- $d\omega = (c_y - b_z)\, dy \wedge dz + (a_z - c_x)\, dz \wedge dx + (b_x - a_y)\, dx \wedge dy$,
- $\mathbf{curl}\,\mathbf{A} = (c_y - b_z)\mathbf{i} + (a_z - c_x)\mathbf{j} + (b_x - a_y)\mathbf{k}$.

(iv) Electrostatic field: We are given the smooth electric charge density function $\varrho : \mathbb{E}^3 \to \mathbb{R}$. Let $\varepsilon_0 > 0$ be the electric constant of a vacuum. Then there exists a smooth electrostatic vector field $\mathbf{E} \in \mathrm{Vect}(\mathbb{E}^3)$ such that

$$\varepsilon_0\, \mathrm{div}\,\mathbf{E} = \varrho \qquad \text{on } \mathbb{E}^3.$$

To prove this, choose $\mu = \varrho\; dx \wedge dy \wedge dz$ and $\omega = \varepsilon_0(A\; dy \wedge dz + B\; dz \wedge dx + C\; dx \wedge dy)$.
The following result is called the local Poincaré theorem.

Corollary 12.47 *Let P be a point on the real n-dimensional manifold $\mathcal{M}$. For given smooth r-form μ on $\mathcal{M}$, there exists an open neighborhood $\mathcal{U}$ of P such that the equation $d\omega = \mu$ on $\mathcal{U}$ has a solution iff $d\mu = 0$ on $\mathcal{U}$.*

Note that it is always possible to choose an open neighborhood $\mathcal{U}$ of P which is continuously contractible to the point P. Thus, the corollary is an immediate consequence of Theorem 12.46.

12.10.2 Non-Contractible Manifolds and Betti Numbers (De Rham's Theorem on Periods)

Before discussing the meaning of Betti numbers, let us formulate the main result which is a special case of the de Rham theory in differential topology. Fix the dimension $n = 1, 2, \ldots$, and let $r = 1, \ldots, n$. Let $\mathcal{M}$ be a real n-dimensional arcwise connected manifold with or without boundary. Assume that the Betti numbers $\beta_1, \ldots, \beta_n$ of $\mathcal{M}$ are finite (examples will be considered below). We are given the smooth r-form μ on $\mathcal{M}$. We are looking for a smooth $(r-1)$-form ω on $\mathcal{M}$ such that

$$d\omega = \mu \qquad \text{on } \mathcal{M}. \tag{12.170}$$

If $\beta_r > 0$, then suppose that we know

- smooth r-forms $\mu_1, \ldots, \mu_{\beta_r}$ on the manifold $\mathcal{M}$ with $d\mu_1 = \ldots = d\mu_{\beta_r} = 0$, and
- r-dimensional submanifolds (without boundary) $C_1, \ldots, C_{\beta_r}$ of $\mathcal{M}$ such that $\int_{C_j} \mu_k = \delta_{jk}$ if $j, k = 1, \ldots, \beta_r$.

It is a nontrivial result in differential topology that such quantities always exist.

Theorem 12.48 *If $\beta_r = 0$, then the equation* (12.170) *has a solution ω iff $d\mu = 0$ on the manifold $\mathcal{M}$.*

If $\beta_r > 0$, then the equation (12.170) *has a solution ω iff $d\mu = 0$ on $\mathcal{M}$ and the periods of μ vanish, that is, $\int_{C_j} \mu = 0$ if $j = 1, \ldots, \beta_r$.*

The general solution of (12.170) *is the sum of a special solution of* (12.170) *and the general solution ω of the homogenous equation*

$$d\omega = 0 \qquad \text{on } \mathcal{M}. \tag{12.171}$$

Thus it remains to solve (12.171). We want to show that the number of nontrivial solutions of (12.171) is equal to the Betti number β_{r-1} if $r \geq 2$. Let us start with the following definition. If $r \geq 2$ (resp. $r = 1$), then the solution ω of (12.171) is called trivial iff $\omega = d\nu$ where ν is a smooth $(r-2)$-form on $\mathcal{M}$ (resp. $\omega = \text{const}$).

Corollary 12.49 *If either $r = 1$ or $r \geq 2$ and $\beta_{r-1} = 0$, then the equation* (12.171) *has only trivial solutions ω.*

If $r \geq 2$ and $\beta_{r-1} > 0$, then the general solution of (12.171) *reads as*

$$\omega = \alpha_1\omega_1 + \ldots + \alpha_{\beta_{r-1}}\omega_{\beta_{r-1}} + d\nu$$

where $\omega_1, \ldots, \omega_{\beta_{r-1}}$ are smooth $(r-1)$ forms on $\mathcal{M}$ with $d\omega_1 = \ldots = d\omega_{\beta_{r-1}} = 0$, $\alpha_1, \ldots, \alpha_{\beta_{r-1}}$ are real numbers, and ν is a smooth $(r-2)$-form on $\mathcal{M}$. In addition, the solution ω is trivial iff $\alpha_1 = \ldots = \alpha_{\beta_{r-1}} = 0$.

The Betti numbers of the manifold $\mathcal{M}$. The Betti numbers of a topological space will be studied later on. At this point let us only mention the following properties:

(i) The Betti numbers of a point are $\beta_0 = 1, \beta_k = 0$ if $k = 1, 2, \ldots$
(ii) The Betti numbers of an n-dimensional sphere are $\beta_0 = \beta_n = 1$ and $\beta_k = 0$ if $k \neq 0, n$.
(iii) If the real finite-dimensional manifold $\mathcal{M}$ with or without boundary can be continuously deformed into a point (resp. an n-dimensional sphere), then it has the same Betti numbers as a point (resp. an n-dimensional sphere).

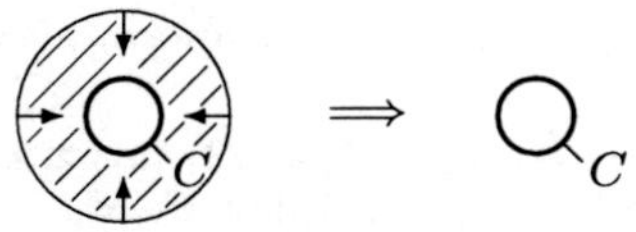

$\beta_0 = \beta_1 = 1$ $\beta_0 = \beta_1 = 1$

Fig. 12.25. Deformation (homotopy)

Let us consider some examples:

- The Euclidean plane $\mathbb{E}^2$, a disc, the Euclidean manifold $\mathbb{E}^3$, and a ball can be continuously deformed into a point. Therefore, they have the Betti numbers of a point, that is, $\beta_0 = 1$ and $\beta_k = 0$ if $k = 1, 2, \ldots$
- An annulus and the punctured Euclidean plane $\mathbb{E}^2 \setminus \{0\}$ can be continuously deformed into a circle (Fig. 12.25). Therefore, they have the same Betti numbers as a circle (i.e., a 1-dimensional sphere), that is, $\beta_0 = \beta_1 = 1$ and $\beta_k = 0$ if $k = 2, 3, \ldots$
- The punctured Euclidean manifold $\mathbb{E}^3 \setminus \{0\}$ can be continuously deformed into a 2-dimensional sphere. Therefore, it has the Betti numbers $\beta_0 = \beta_2 = 1$ and $\beta_k = 0$ if $k \neq 0, 2$.
- Consider the set $\mathcal{M} := \mathbb{E}^3 \setminus \mathcal{Z}$ where $\mathcal{Z}$ denotes the z-axis. This set can be continuously deformed into the punctured Euclidean plane $\mathbb{E}^2 \setminus \{0\}$, and hence $\mathcal{M}$ can be continuously deformed into a circle. Therefore, $\mathcal{M}$ has the Betti numbers $\beta_0 = \beta_1 = 1$ and $\beta_k = 0$ if $k = 2, 3, \ldots$

Potential on the annulus. Let $\mathcal{M}$ be an annulus, and let C be a circle inside the annulus (see Fig. 12.24 on page 761). We are given the smooth functions $A, B : \mathcal{M} \to \mathbb{R}$. Set $\mu := Adx + Bdy$. Then the equation

$$dU = \mu \qquad \text{on } \mathcal{M} \tag{12.172}$$

has a smooth solution $U : \mathcal{M} \to \mathbb{R}$ iff $d\mu = 0$ on $\mathcal{M}$ and $\int_C \mu = 0$ (i.e., the period of μ vanishes). Equivalently, the system

$$U_x = A, \;\; U_y = B \qquad \text{on } \mathcal{M}$$

has a smooth solution U iff $A_y = B_x$ on $\mathcal{M}$ and $\int_C Adx + Bdy = 0$.

Proof. If the equation (12.172) has a solution U, then we obtain $d\mu = d(dU) = 0$ and $\int_C \mu = \int_C dU = \int_{\partial C} U = 0$, since the boundary ∂C is empty.

Conversely, note that the annulus has the first Betti number $\beta_1 = 1$. Therefore, we get precisely one global solvability condition. To see that C is the right submanifold, observe that there exists a 1-form μ_1 such that

$$\int_C \mu_1 = 1.$$

In fact, using appropriate polar coordinates φ, r, we have $\int_C \frac{1}{2\pi} d\varphi = 1$. Now the claim follows from Theorem 12.48. □

Potentials on the punctured Euclidean manifold. Fix $r = 1, 2, 3$. We want to solve the equation

$$d\omega = \mu \qquad \text{on } \mathbb{E}^3 \setminus \{0\}. \tag{12.173}$$

We are given the smooth r-form μ on $\mathbb{E}^3 \setminus \{0\}$, and we are looking for a smooth $(r-1)$-form ω on $\mathbb{E}^3 \setminus \{0\}$. The Betti numbers of $\mathbb{E}^3 \setminus \{0\}$ are $\beta_1 = \beta_3 = 0$ and $\beta_2 = 1$.

Therefore, a global solvability condition only appears if $r = 2$. More precisely, the following hold:

- If $r = 1$ or $r = 3$, then the equation (12.173) has a solution ω iff $d\mu = 0$. This local solvability condition is always satisfied if $r = 3$.
- If $r = 2$, then the equation (12.173) has a solution ω iff $d\mu = 0$ and $\int_{\mathbb{S}^2} \mu = 0$.

In the language of vector analysis, this reads as follows. Suppose that we are given the smooth function $\varrho : \mathbb{E}^3 \setminus \{0\} \to \mathbb{R}$ and the smooth vector fields $\mathbf{F}$ and $\mathbf{B}$ on $\mathbb{E}^3 \setminus \{0\}$. Then the following hold:

- $r = 1$: The equation $\mathbf{F} = -\,\mathbf{grad}\, U$ on $\mathbb{E}^3 \setminus \{0\}$ has a smooth solution U iff $\mathbf{curl}\,\mathbf{F} = 0$ on $\mathbb{E}^3 \setminus \{0\}$.
- $r = 2$: The equation $\mathbf{B} = \mathbf{curl}\,\mathbf{A}$ on $\mathbb{E}^3 \setminus \{0\}$ has a smooth solution $\mathbf{A}$ iff $\operatorname{div} \mathbf{B} = 0$ on $\mathbb{E}^3 \setminus \{0\}$ and
$$\int_{\mathbb{S}^2} \mathbf{Bn} \cdot dS = 0$$
where $\mathbf{n}$ denotes the outer unit normal vector of the unit sphere $\mathbb{S}^2$.
- $r = 3$: The equation $\varepsilon_0 \operatorname{div} \mathbf{E} = \varrho$ on $\mathbb{E}^3 \setminus \{0\}$ has always a smooth solution $\mathbf{E}$ on $\mathbb{E}^3 \setminus \{0\}$.

The homogeneous potential equation on the Euclidean manifold. The Betti numbers of the Euclidean manifold $\mathbb{E}^3$ are $\beta_j = 0$ if $j \geq 1$ Thus, if the k-form ω $(k \geq 0)$ is a solution of the equation

$$d\omega = 0 \qquad \text{on } \mathbb{E}^3,$$

then it is trivial. Explicitly, this means the following:

- The general smooth solution of $\mathbf{grad}\, U = 0$ on $\mathbb{E}^3$ is $U = \text{const}$.
- The general smooth solution of $\mathbf{curl}\,\mathbf{v} = 0$ on $\mathbb{E}^3$ is $\mathbf{v} = \mathbf{grad}\, V$ where $V : \mathbb{E}^3 \to \mathbb{R}$ is an arbitrary smooth function. Here, V is called the potential of the velocity vector field $\mathbf{v}$.
- The general smooth solution of $\operatorname{div} \mathbf{v} = 0$ on $\mathbb{E}^3$ is $\mathbf{v} = \mathbf{curl}\,\mathbf{w}$ where $\mathbf{w}$ is an arbitrary smooth vector field on $\mathbb{E}^3$. Here, $\mathbf{w}$ is called the vector potential of the velocity vector field $\mathbf{v}$.

The same results remain true if we replace the Euclidean manifold $\mathbb{E}^3$ by an open subset of $\mathbb{E}^3$ which is continuously contractible to a point (e.g., a ball).

12.10.3 The Main Theorem for Velocity Vector Fields

We want to show that a velocity vector field $\mathbf{v}$ is given by its sources and circulations together with boundary conditions. To this end, choose a 3-dimensional, compact, arcwise connected submanifold $\mathcal{M}$ of the Euclidean manifold $\mathbb{E}^3$ with the boundary $\partial\mathcal{M}$. Intuitively, this is the closure of a bounded, open, arcwise connected subset of $\mathbb{R}^3$ with regular behavior at the boundary (e.g., a ball). We assume that the submanifold $\mathcal{M}$ and the boundary $\partial\mathcal{M}$ are coherently oriented (see Fig. 12.6(c) on page 677). Consider the system

$$\operatorname{div} \mathbf{v} = f, \qquad \mathbf{curl}\,\mathbf{v} = \mathbf{g} \quad \text{on } \mathcal{M} \tag{12.174}$$

together with the boundary condition

$$\mathbf{nv} = b \qquad \text{on } \partial\mathcal{M}. \tag{12.175}$$

We are given the smooth functions $f : \mathcal{M} \to \mathbb{R}$ and $\mathbf{g} : \mathcal{M} \to E_3$, and the smooth boundary function $b : \partial\mathcal{M} \to \mathbb{R}$ with $\int_{\mathcal{M}} f\, d^3x = \int_{\partial\mathcal{M}} b\, dS$. We are looking for the smooth velocity vector field $\mathbf{v} : \mathcal{M} \to E_3$.

Theorem 12.50 *The boundary-value problem* (12.174), (12.175) *has a unique smooth solution* $\mathbf{v}$.

The proof of this classical result based on potential theory can be found in the monograph by A. Tikhonov and A. Samarski, The Equations of Mathematical Physics, Macmillan, New York, 1963.

12.11 Systems of Differential Forms

12.11.1 Integrability Condition

Prototype. Fix a point P_0 of the Euclidean manifold $\mathbb{E}^3$. Choose a right-handed Cartesian (x, y, z)-coordinate system on $\mathbb{E}^3$. Let us consider the system

$$z_x(x,y) = A(x,y,z(x,y)), \quad z_y(x,y) = B(x,y,z(x,y)) \tag{12.176}$$

locally at the point $P_0 = (x_0, y_0, z_0)$ with $z(x_0, y_0) = z_0$. This means that we are given the smooth real-valued functions A, B on some open neighborhood of the point P_0. We are looking for a smooth real-valued function $(x, y) \mapsto z(x, y)$ on some open neighborhood of the point (x_0, y_0). In terms of geometry, we are looking for a smooth surface $z = z(x, y)$ which passes through the point P_0 and which has prescribed tangent planes given by the functions A and B.[55] This is the prototype of the following fundamental problem in differential geometry:

Construct a submanifold which has prescribed tangent spaces.

The point is that, naturally enough, this problem is not always solvable. One needs additional solvability conditions which are called integrability conditions. Let us discuss this. Set $Q := (x, y)$, and $Q_0 := (x_0, y_0)$.

If $z = z(Q)$ is a solution of (12.176), then it follows from the key condition

$$\boxed{z_{xy}(Q) = z_{yx}(Q)}$$

that

$$A_y(Q, z(Q)) + A_z(Q, z(Q))z_y(Q) = B_x(Q, z(Q)) + B_z(Q, z(Q))z_x(Q).$$

This implies

$$A_y(Q, z(Q)) + A_z(Q, z(Q))B(Q) = B_x(Q, z(Q)) + B_z(Q, z(Q))A(Q)$$

locally at Q_0. This is a necessary solvability condition. Roughly speaking, this is also a sufficient solvability condition.

Proposition 12.51 *Suppose that the so-called integrability condition*

$$A_y(x,y,z) + A_z(x,y,z)B(x,y) = B_x(x,y,z) + B_z(x,y,z)A(x,y) \tag{12.177}$$

is satisfied on some open neighborhood of the point P_0 in the Euclidean manifold $\mathbb{E}^3$. Then the problem (12.176) *has a local solution which is locally unique at the point P_0.*

[55] Note that the tangent plane of the surface $z = z(x, y)$ at the point (x_0, y_0, z_0) is given by the equation

$$z = z_0 + A(x_0, y_0, z_0)(x - x_0) + B(x_0, y_0, z_0)(y - y_0).$$

The proof can be found in Zeidler (1986), Sect. 4.12 of Vol. I, quoted on page 1089. The classical idea of the proof is to construct the surface $z = z(x, y)$ by using curves which are called characteristics. It is a general strategy in the theory of manifolds to construct submanifolds by putting lower dimensional manifolds (characteristic manifolds) together. This strategy dates back to Cauchy (1789–1857).

The language of differential forms. Integrability conditions can be elegantly formulated in terms of differential forms. To explain the basic idea, let us set

$$\omega := dz - A(x, y, z)dx - B(x, y, z)dy.$$

The original problem (12.176) can be written as

$$\boxed{z^*\omega = 0} \qquad (12.178)$$

and the integrability condition (12.177) is equivalent to

$$\boxed{d\omega \wedge \omega = 0.}$$

In fact, since

$$\begin{aligned} z^*\omega &= z_x(x,y)dx + z_y(x,y)dy - A(x,y,z(x,y))dx - B(x,y,z(x,y))dy \\ &= (z_x(x,y) - A(x,y,z(x,y)))\, dx + (z_y(x,y) - B(x,y,z(x,y)))\, dy, \end{aligned}$$

the equation $z^*\omega = 0$ is equivalent to $z_x = A, z_y = B$ which is (12.176). Moreover,

$$\begin{aligned} d\omega &= -(A_x dx + A_y dy + A_z dz) \wedge dx - (B_x dx + B_y dy + B_z dz) \wedge dy \\ &= (A_y - B_x)\, dx \wedge dy - A_z dz \wedge dx + B_z\, dy \wedge dz. \end{aligned}$$

Hence

$$d\omega \wedge \omega = (A_y + A_z B - B_x - B_z A)\, dx \wedge dy \wedge dz = 0$$

is equivalent to $A_y + A_z B = B_x + B_z A$ which is (12.177).

The global Frobenius theorem in the linear case, and the Maurer–Cartan equation. Fix $n = 1, 2, \ldots$, and $N = 1, 2, \ldots$ Let us consider the linear system of partial differential equations

$$\boxed{\partial_j \psi(x) = \psi(x) M_j(x), \quad j = 1, \ldots, n \quad x \in \mathcal{U}, \quad \psi(x_0) = \psi_0.} \qquad (12.179)$$

Here, $\mathcal{U}$ is an open, arcwise connected, simply connected subset of $\mathbb{R}^n$ (e.g., $\mathcal{U} = \mathbb{R}^n$, or $\mathcal{U}$ is an open ball). Furthermore, $x = (x^1, \ldots, x^n)$ with $x \in \mathbb{R}^n$, and $\partial_j := \frac{\partial}{\partial x^j}$. We are given the smooth matrix functions

$$M_j : \mathcal{U} \to gl(N, \mathbb{R}), \qquad j = 1, 2, \ldots$$

We are looking for the smooth matrix function

$$x \mapsto (\psi^1(x), \ldots, \psi^N(x))$$

from $\mathcal{U}$ to $\mathbb{R}^N$ where $\psi(x_0) \in \mathbb{R}^N$ is given. If ψ is a smooth solution of (12.179), then it follows from the integrability condition $\partial_i\partial_j\psi = \partial_j\partial_i\psi$ that[56]

$$\partial_i M_j - \partial_j M_i + M_i M_j - M_j M_i = 0 \qquad \text{on } \mathcal{U}, \quad i, j = 1, \ldots, N. \quad (12.180)$$

[56] Note that $\partial_i\partial_j\psi = \partial_i\psi \cdot M_j + \psi\, \partial_i M_j = \psi(M_i M_j + \partial_i M_j)$.

This is called the Maurer–Cartan equation. Setting $M := M_j dx^j$, we obtain the Cartan differential $dM = \partial_i M_j \, dx^i \wedge dx^j = \frac{1}{2}(\partial_i M_j - \partial_j M_i) \, dx^i \wedge dx^j$. Thus, the Maurer–Cartan equation (12.180) is equivalent to

$$\boxed{dM + M \wedge M = 0.}$$

This equation appears again and again in gauge theory.

Theorem 12.52 *The original problem* (12.179) *has a smooth solution iff the Maurer–Cartan equation* (12.180) *is satisfied. The solution is unique.*

For example, this crucial theorem implies the existence theorem for 2-dimensional surfaces on page 633. For the proof of Theorem 12.52, we refer to Problem 12.23 on page 809.

12.11.2 The Frobenius Theorem for Pfaff Systems

Fix $n = 2, 3, \ldots$ and $r = 2, \ldots, n-1$. Let $\mathcal{M}$ be a real n-dimensional manifold (e.g., $\mathbb{E}^3$). Consider the Pfaff system

$$\omega_1 = 0, \ldots, \omega_r = 0 \tag{12.181}$$

where $\omega_1, \ldots, \omega_r$ are smooth 1-forms on $\mathcal{M}$ which are linearly independent, that is,

$$(\omega_1 \wedge \omega_2 \wedge \cdots \wedge \omega_r)(P) \neq 0 \qquad \text{for all} \quad P \in \mathcal{M}.$$

Let $\mathcal{U}$ be an open subset of $\mathbb{R}^m$. The smooth function $f : \mathcal{U} \to \mathcal{M}$ is a solution of (12.181) iff the pull-backs satisfy the equations

$$f^*\omega_1 = 0, \ldots, f^*\omega_r = 0 \qquad \text{on } \mathcal{U}.$$

This generalizes (12.176), (12.178). Similarly, the submanifold $\mathcal{N}$ of $\mathcal{M}$ is a solution of (12.181) iff

$$\chi^*\omega_1 = 0, \ldots, \chi^*\omega_r = 0 \qquad \text{on } \mathcal{N}$$

where the injective map $\chi : \mathcal{N} \to \mathcal{M}$ is given by $\chi(P) = P$ for all $P \in \mathcal{N}$. The Pfaff system (12.181) is called completely integrable on the manifold $\mathcal{M}$ iff, for every point $P_0 \in \mathcal{M}$, there exists an $(n-r)$-dimensional submanifold $\mathcal{N}$ of $\mathcal{M}$ which is a solution of (12.181) with $P_0 \in \mathcal{N}$, and this solution is locally unique at the point P_0.

Theorem 12.53 *The Pfaff system* (12.181) *is locally, completely integrable on the manifold* $\mathcal{M}$ *iff*

$$d\omega_j \wedge \omega_1 \wedge \cdots \wedge \omega_r = 0 \qquad \text{on } \mathcal{M}, \quad j = 1, \ldots, r.$$

The proofs of Theorems 12.53 and 12.54 can be found in Choquet-Bruhat et al. (1996), Sect. IV.C.6 of Vol. 1, quoted on page 775.

Local coordinates on the manifold $\mathcal{M}$. If the Pfaff system (12.181) is locally, completely integrable, then for every point $P_0 \in \mathcal{M}$, there exists a local coordinate system $u_1, \ldots, u_n$ on the manifold $\mathcal{M}$ at P_0 such that $\omega_j = du^j$ for $j = 1, \ldots, r$. Then, the Pfaff system (12.181) locally passes over to the trivial system

$$du^1 = 0, \ldots, du^r = 0,$$

and the local solution submanifold $\mathcal{N}$ of (12.181) at the point P_0 is given by the equation $u^1 = \ldots = u^r = 0$. Here, $u^{r+1}, \ldots, u^n$ are arbitrary, sufficiently small real numbers.

12.11.3 The Dual Frobenius Theorem

Completely integrable distributions of linear spaces on a manifold. Again let $\mathcal{M}$ be a real n-dimensional manifold with $n = 2, 3, \ldots$ Fix $r = 1, \ldots, n-1$. Suppose that, for every point $P_0 \in \mathcal{M}$, we are given an $(n-r)$-dimensional linear subspace L_{P_0} of the tangent space $T_{P_0}\mathcal{M}$. This is called a distribution of linear spaces on $\mathcal{M}$. This distribution is called completely integrable iff, for every point P_0 on $\mathcal{M}$, there exists an $(n-r)$-dimensional submanifold $\mathcal{N}$ of $\mathcal{M}$ such that $P_0 \in \mathcal{N}$ and

$$T_P\mathcal{N} = L_P \qquad \text{for all} \quad P \in \mathcal{N},$$

and $\mathcal{N}$ is locally unique. This means that the submanifold $\mathcal{N}$ passes through the point P_0, and the tangent spaces of $\mathcal{N}$ coincide with the linear spaces of the distribution.

Lie brackets of velocity vector fields and involutive distributions. By definition, the distribution is in involution iff, for every point $P_0 \in \mathcal{M}$, there exists an open neighborhood $\mathcal{U}$ of the point P_0 on the manifold $\mathcal{M}$ such that the velocity vector fields on $\mathcal{M}$ possess the following local property: If $\mathbf{v}$ and $\mathbf{w}$ are smooth velocity vector fields on $\mathcal{U}$ with

$$\mathbf{v}_P, \mathbf{w}_P \in L_P \qquad \text{for all} \quad P \in \mathcal{U},$$

then the Lie bracket satisfies $[\mathbf{v}, \mathbf{w}]_P \in L_P$ for all $P \in \mathcal{U}$.

Theorem 12.54 *The distribution $\{L_P\}_{P \in \mathcal{M}}$ of linear spaces on the manifold $\mathcal{M}$ is completely integrable iff it is in involution.*

12.11.4 The Pfaff Normal Form and the Second Law of Thermodynamics

Special case. We are given the smooth differential 1-form

$$dE + P(E, V)dV$$

on the open set $\mathcal{O} = \{(E, V) \in \mathbb{R}^2 : E > 0, V > 0\}$. We are looking for smooth functions $T = T(E, V)$ and $S = S(E, V)$ such that

$$\boxed{dE + PdV = TdS.} \tag{12.182}$$

In terms of mathematics, this means that we want to reduce $dE + PdV$ to a simpler expression. In terms of physics, we want to prove the existence of temperature T and entropy S. Choose a point $(E_0, V_0) \in \mathcal{O}$, and suppose that

$$P_E(E_0, V_0) > 0.$$

In terms of physics, this means that the pressure P increases if the inner energy E of a gas increases.

Proposition 12.55 *There exist an open neighborhood $\mathcal{U}$ of (E_0, V_0) and smooth functions $S, T : \mathcal{U} \to \mathbb{R}$ such that the equation (12.182) is satisfied on $\mathcal{U}$.*

Since the rank of the following matrix

$$\begin{pmatrix} 1 & 0 & -P_E(E_0, V_0) \\ P(E_0, V_0) & P_E(E_0, V_0) & 0 \end{pmatrix}$$

is equal to 2, the proposition is a special case of the following general result.

General case. Fix $n = 2, 3, \ldots$ Set $u = (u^1, \ldots, u^n)$ where $u \in \mathbb{R}^n$. We are given the 1-form

$$a_1(u)du^1 + \ldots + a_n(u)du^n$$

where $a_i : \mathcal{O} \to \mathbb{R}$, $i = 1, \ldots, n$, are smooth functions on the open subset $\mathcal{O}$ of $\mathbb{R}^n$. We are looking for smooth functions $F_j = F_j(u)$ and $G_k = G_k(u)$ such that we have either

$$a_1 du^1 + \ldots + a_n du^n = F_1 dG_1 + \ldots + F_m dG_m \tag{12.183}$$

or

$$a_1 du^1 + \ldots + a_n du^n = F_1 dG_1 + \ldots + F_m G_m + dG_{m+1}. \tag{12.184}$$

Fix a point $u_0 \in \mathcal{O}$. Set $a_{ij} := \frac{\partial a_i}{\partial u^j} - \frac{\partial a_j}{\partial u^i}$. Suppose that the matrix

$$\begin{pmatrix} a_1(u_0) & a_{11}(u_0) & \ldots & a_{1n}(u_0) \\ \vdots & \vdots & & \vdots \\ a_n(u_0) & a_{n1}(u_0) & \ldots & a_{nn}(u_0) \end{pmatrix}$$

has the rank r. Then there exist an open neighborhood $\mathcal{U}$ of the point u_0 in $\mathcal{O}$ and smooth functions $F_i, G_k : \mathcal{U} \to \mathbb{R}$ such that the following hold.

Theorem 12.56 *If $r = 2m$ is even (resp. $r = 2m + 1$ is odd), then equation* (12.183) resp. (12.184) *is satisfied on $\mathcal{U}$.*

The proof can be found in C. Carathéodory, Calculus of Variations and Partial Differential Equations of First Order, Chelsea, New York, 1982, Sect. 141.

12.12 Hodge Duality

Hodge duality is based on the Hodge star operator $\omega \mapsto *\omega$ for differential forms on oriented manifolds. This will allow us to generalize the classical Laplacian to differential forms.
Hodge duality implies the Poincaré duality for the homology and cohomology groups of manifolds.

Folklore

General Hodge duality is based on oriented Riemannian (or pseudo-Riemannian) manifolds. In this section, we equip the Euclidean manifold $\mathbb{E}^3$ with a right-handed orientation.

12.12.1 The Hodge Codifferential

Let $\omega \in \Lambda^p(\mathbb{E}^3)$ with $p = 0, 1, 2, 3$. We define

$$\boxed{d^*\omega := (-1)^p * d(*\omega).}$$

This is called the Hodge codifferential $d^*\omega$ of the differential p-form ω (or the dual Cartan differential).

The Hodge $*$-operator. The invariant definition of the Hodge $*$-operator (star operator) on the oriented real 3-dimensional Hilbert space E_3 can be found in Sect. 2.7 on page 138. Since the tangent space $T_P\mathbb{E}^3$ of the Euclidean manifold $\mathbb{E}^3$ at the point P is an oriented real Hilbert space which is isomorphic to E_3, the definitions from E_3 can be immediately translated to the tangent space $T_P\mathbb{E}^3$ and its dual space, and hence to differential forms on $\mathbb{E}^3$. The Hodge $*$-operator on the Euclidean manifold $\mathbb{E}^3$ has the following properties:[57]

- $**\omega = \omega$ for all $\omega \in \Lambda^p(\mathbb{E}^3)$ with $p = 0, 1, 2, 3$ (duality).
- The linear operator

$$* : \Lambda^p(\mathbb{E}^3) \to \Lambda^{3-p}(\mathbb{E}^3), \qquad p = 0, 1, 2, 3 \tag{12.185}$$

  is bijective. The inverse operator

$$*^{-1} : \Lambda^{3-p}(\mathbb{E}^3) \to \Lambda^p(\mathbb{E}^3), \qquad p = 0, 1, 2, 3$$

  coincides with the $*$-operator, that is, $*^{-1}\omega = *\omega$ for all $\omega \in \Lambda^{3-p}(\mathbb{E}^3)$ with $p = 0, 1, 2, 3$.

Cartesian coordinates. Choose a right-handed Cartesian (x, y, z)-coordinate system on $\mathbb{E}^3$. Let us use the following notation:

- $\omega := adx + bdy + cdz$,
- $\varrho := A\, dy \wedge dz + B\, dz \wedge dx + C\, dx \wedge dy$,
- $\upsilon := dx \wedge dy \wedge dz$ (volume form),
- $\Theta : \mathbb{E}^3 \to \mathbb{R}$ (smooth function; e.g., a temperature field).

For the Hodge $*$-operator, we get:

- $*\upsilon = 1$ and $*1 = \upsilon$,
- $*\Theta = \Theta\upsilon$ and $*(\Theta\upsilon) = \Theta$,
- $*(adx + bdy + cdz) = a\, dy \wedge dz + b\, dz \wedge dx + c\, dx \wedge dy$,
- $*(a\, dy \wedge dz + b\, dz \wedge dx + c\, dx \wedge dy) = adx + bdy + cdz$.

Proposition 12.57 (i) $d\Theta = \Theta_x dx + \Theta_y dy + \Theta_z dz$, *and* $d^*\Theta = 0$,
(ii) $d\omega = (c_y - b_z)\, dy \wedge dz + (a_z - c_x)\, dz \wedge dx + (b_x - a_y)\, dx \wedge dy$,
(iii) $d^*\omega = -a_x - b_y - c_z$,
(iv) $d\varrho = (A_x + B_y + C_z)\, dx \wedge dy \wedge dz$,
(v) $d^*\varrho = (C_y - B_z)\, dx + (A_z - C_x)\, dy + (B_x - A_y)\, dz$,
(vi) $d(\Theta\upsilon) = 0$, *and* $d^*(\Theta\upsilon) = - * d\Theta$.

Proof. For example, choose $\omega = adx + bdy + cdz$. It follows from

$$*\omega = a\, dy \wedge dz + b\, dz \wedge dx + c\, dx \wedge dy$$

that $d(*\omega) = (a_x + b_y + c_z)\, dx \wedge dy \wedge dz$. Hence $d^*\omega = - * d(*\omega) = -a_x - b_y - c_z$. □

[57] Later on, we will generalize this operator to oriented n-dimensional Riemannian and pseudo-Riemannian manifolds. Note that the properties of the $*$-operator depend on the dimension n.

12.12.2 The Hodge Homology Rule

The linear operators[58]

$$d : \Lambda^p(\mathbb{E}^3) \to \Lambda^{p+1}(\mathbb{E}^3), \qquad p = 0, \pm 1, \pm 2, \ldots$$

satisfy the Poincaré cohomology rule $d(d\omega) = 0$ for all p-forms ω with integer p. We briefly write $d^2 = 0$. Similarly, the linear operators

$$d^* : \Lambda^p(\mathbb{E}^3) \to \Lambda^{p-1}(\mathbb{E}^3), \qquad p = 0, \pm 1, \pm 2, \ldots$$

satisfy the so-called Hodge homology rule $d^*(d^*\omega) = 0$ for all p-forms ω with integer p. We briefly write $d^{*2} = 0$. In addition, for every $p = 0, 1, 2, 3$, we have the following commutative diagram:

$$\begin{array}{ccc} \Lambda^p(\mathbb{E}^3) & \xrightarrow{\;d^*\;} & \Lambda^{p-1}(\mathbb{E}^3) \\ {\scriptstyle *}\downarrow & & \downarrow{\scriptstyle *} \\ \Lambda^{3-p}(\mathbb{E}^3) & \xrightarrow{(-1)^p d} & \Lambda^{3-p+1}(\mathbb{E}^3). \end{array} \tag{12.186}$$

12.12.3 The Relation between the Cartan–Hodge Calculus and Classical Vector Analysis via Riesz Duality

> On the Euclidean manifold $\mathbb{E}^3$, Hamilton's nabla calculus and the Cartan–Hodge calculus are equivalent. However, in contrast to the Hamilton nabla calculus, the Cartan–Hodge calculus can be immediately generalized to n-dimensional manifolds.
>
> Folklore

Let us introduce the two linear isomorphisms

- $\aleph : \mathrm{Vect}(\mathbb{E}^3) \to \Lambda^1(\mathbb{E}^3)$, and
- $*\aleph : \mathrm{Vect}(\mathbb{E}^3) \to \Lambda^2(\mathbb{E}^3)$

by setting for all smooth velocity vector fields $\mathbf{u}, \mathbf{v}, \mathbf{w} \in \mathrm{Vect}(\mathbb{E}^3)$:

- $(\aleph\mathbf{v})(\mathbf{u}) := \mathbf{vu}$.
- $(*\aleph)\mathbf{v} := *(\aleph\mathbf{v})$.

Explicitly,

- $*(\aleph\mathbf{v})(\mathbf{u}, \mathbf{w}) = \mathbf{v}(\mathbf{u} \times \mathbf{w})$.

More precisely, $\aleph\mathbf{v}$ is a 1-form on $\mathbb{E}^3$ with $(\aleph\mathbf{v})_P = \mathbf{v}_P\mathbf{u}_P$ for all $P \in \mathbb{E}^3$. Furthermore, $(*\aleph)\mathbf{v}$ is a 2-form. The operator $\aleph$ is called the Riesz duality operator on $\mathrm{Vect}(\mathbb{E}^3)$.

> *The Riesz duality operator $\aleph$ (resp. the operator $*\aleph$) sends velocity vector fields to 1-forms (resp. 2-forms).*

Using these linear bijective operators, velocity vector fields on the Euclidean manifold $\mathbb{E}^3$ can be identified with differential 1-forms (resp. 2-forms).

---

[58] As usual, we define $\Lambda^p(\mathbb{E}^3) := \{0\}$ if either $p = 4, 5, \ldots$ or $p = -1, -2, \ldots$

Proposition 12.58 *Let Θ be a smooth temperature field on the Euclidean manifold $\mathbb{E}^3$, and let $\mathbf{v}$ be a smooth velocity vector field on $\mathbb{E}^3$. Then:*
(i) $\mathbf{grad}\,\Theta = \aleph^{-1} d\Theta$,
(ii) $\mathbf{curl}\,\mathbf{v} = (*\aleph)^{-1} d(\aleph\mathbf{v})$,
(iii) $\operatorname{div}\mathbf{v} = -d^*(\aleph\mathbf{v})$.

Cartesian coordinates. All of the invariant formulas above can be easily verified by using Cartesian coordinates. To this end, choose a right-handed Cartesian (x, y, z)-coordinate system. Then:

- $\mathbf{x} = x\mathbf{i} + y\mathbf{j} + z\mathbf{k}$, and $d\mathbf{x} := dx\,\mathbf{i} + dy\,\mathbf{j} + dz\,\mathbf{k}$,
- $\mathbf{v} = a\mathbf{i} + b\mathbf{j} + c\mathbf{k}$, and $\aleph\mathbf{v} = \mathbf{v}d\mathbf{x} = a\,dx + b\,dy + c\,dz$,
- $\aleph\mathbf{i} = dx$, $\aleph\mathbf{j} = dy$, $\aleph\mathbf{k} = dz$ (Riesz duality operator),
- $(*\aleph)\mathbf{v} = *(\aleph\mathbf{v}) = a\,dy\wedge dz + b\,dz\wedge dx + c\,dx\wedge dy$ (Hodge $*$-operator),
- $\boldsymbol{\partial} = \mathbf{i}\frac{\partial}{\partial x} + \mathbf{j}\frac{\partial}{\partial y} + \mathbf{k}\frac{\partial}{\partial z}$ (Hamilton's nabla operator),
- $\upsilon = dx\wedge dy\wedge dz$ (volume form),
- $(i_{\mathbf{v}}\upsilon)(\mathbf{u},\mathbf{w}) = \upsilon(\mathbf{v},\mathbf{u},\mathbf{w}) = \mathbf{v}(\mathbf{u}\times\mathbf{w})$.

This implies:

- $\mathbf{grad}\,\Theta = \aleph^{-1}d\Theta = \aleph^{-1}(\Theta_x dx + \Theta_y dy + \Theta_z dz) = \Theta_x\mathbf{i} + \Theta_y\mathbf{j} + \Theta_z\mathbf{k} = \boldsymbol{\partial}\Theta$,
- $\mathbf{curl}\,\mathbf{v} = (*\aleph)^{-1}d(\mathbf{v}d\mathbf{x}) = \boldsymbol{\partial}\times\mathbf{v}$,
- $\operatorname{div}\mathbf{v} = -d^*(\mathbf{v}d\mathbf{x}) = \boldsymbol{\partial}\mathbf{v}$.

Conversely, using the volume form υ, we get:

- $d\Theta = \aleph(\mathbf{grad}\,\Theta)$, $d^*\Theta = 0$,
- $d\Theta(\mathbf{v}) = \mathcal{L}_{\mathbf{v}}\Theta = \mathbf{v}\,\mathbf{grad}\,\Theta$,
- $d(\aleph\mathbf{v}) = d(\mathbf{v}d\mathbf{x}) = i_{\mathbf{curl}\,\mathbf{v}}\upsilon$,
- $d^*(\aleph\mathbf{v}) = -\operatorname{div}\mathbf{v}$,
- $d(i_{\mathbf{v}}\upsilon) = \mathcal{L}_{\mathbf{v}}\upsilon = \operatorname{div}\mathbf{v}\cdot\upsilon$,
- $d^*(i_{\mathbf{v}}\upsilon) = *i_{\mathbf{curl}\,\mathbf{v}}\upsilon$,
- $d(\aleph\mathbf{v})(\mathbf{u},\mathbf{w}) = (\mathbf{u}\times\mathbf{w})\,\mathbf{curl}\,\mathbf{v}$.

12.12.4 The Classical Prototype of the Yang–Mills Equation in Gauge Theory

Let us introduce the following notation:

- $\mathbf{v} := a\mathbf{i} + b\mathbf{j} + c\mathbf{k}$ and $\omega := \aleph\mathbf{v} = adx + bdy + cdz$,
- $\mathbf{g} = A\mathbf{i} + B\mathbf{j} + C\mathbf{k}$ and $\Gamma := *\aleph\mathbf{g} = A\,dy\wedge dz + B\,dz\wedge dx + C\,dx\wedge dy$.

The equation

$$\operatorname{div}\mathbf{v} = f, \qquad \mathbf{curl}\,\mathbf{v} = \mathbf{g} \tag{12.187}$$

plays an important role in classical mathematical physics. It describes the determination of a velocity field $\mathbf{v}$ by its sources and circulations given by the functions f $\mathbf{g}$, respectively. Using the language of differential forms, equation (12.187) can be written as

$$\boxed{d^*\omega = -f, \qquad d\omega = \Gamma.} \tag{12.188}$$

This is the classical prototype of the Yang–Mills equation (see Theorem 12.50 on page 767). Equivalently, this equation reads as

$$d^*(\aleph\mathbf{v}) = -f, \qquad d(\aleph\mathbf{v}) = i_{\mathbf{g}}\upsilon$$

where υ denotes the volume form on the Euclidean manifold $\mathbb{E}^3$.

12.12.5 The Hodge–Laplace Operator and Harmonic Forms

We want to generalize the classical Laplacian to differential forms ω by setting

$$\boxed{\Delta\omega := (d^*d + dd^*)\omega.}$$

Assume that $\omega \in \Lambda^p(\mathbb{E}^3)$ with $p = 1, 2, 3$, and let $\Theta : \mathbb{E}^3 \to \mathbb{R}$ be a smooth function.

Cartesian coordinates. In what follows, we will sum over equal upper and lower indices from 1 to 3. Note that $\partial_s = \partial^s := \partial/\partial x^s$, and $x^1 := x, x^2 := y, x^3 := z$. Choose a right-handed Cartesian (x, y, z)-coordinate system. For differential forms

$$\omega = \frac{1}{p!}\, w_{i_1 \ldots i_p}\, dx^{i_1} \wedge \cdots \wedge dx^{i_p}, \qquad p = 1, 2, 3$$

with antisymmetric smooth coefficient functions $\omega_{i_1 \ldots i_p}$, the following hold:

Proposition 12.59 (i) $\Delta\Theta = -\Theta_{xx} - \Theta_{yy} - \Theta_{zz}$,
(ii) $d\omega = \frac{1}{p!}\, \partial_i \omega_{i_1 \ldots i_p}\, dx^i \wedge dx^{i_1} \wedge \cdots \wedge dx^{i_p}$,
(iii) $d^*\omega = -\frac{1}{(p-1)!}\, \partial^s \omega_{s i_2 \ldots i_p}\, dx^{i_2} \wedge \ldots \wedge dx^{i_p}$,
(iv) $\Delta\omega = \Delta\omega_{i_1 \ldots i_p} \cdot dx^{i_1} \wedge \cdots \wedge dx^{i_p}$.

The differential form ω is called harmonic iff $\Delta\omega \equiv 0$. This means that all the coefficient functions $\omega_{i_1 \ldots i_p}$ are harmonic. In order to discuss the relation to the classical vector calculus, let us consider the smooth velocity vector field

- $\mathbf{v} = a\mathbf{i} + b\mathbf{j} + c\mathbf{k}$

together with the corresponding differential forms

- $\omega = \aleph\mathbf{v} = adx + bdy + cdz$, and
- $*\aleph\mathbf{v} = a \cdot dy \wedge dz + b \cdot dz \wedge dx + c \cdot dx \wedge dy$.

Then:

- $\Delta\omega = \Delta a \cdot dx + \Delta b \cdot dy + \Delta c \cdot dz$. This is equivalent to

$$\Delta\mathbf{v} = \mathbf{curl}\,\mathbf{curl}\,\mathbf{v} - \mathbf{grad}\,\mathrm{div}\,\mathbf{v}. \qquad (12.189)$$

- $\Delta(*\aleph\mathbf{v}) = \Delta a \cdot dy \wedge dz + \Delta b \cdot dz \wedge dx + \Delta c \cdot dx \wedge dy$. This is also equivalent to the classical relation (12.189).
- $\Delta(\Theta v) = \Delta\Theta \cdot v$.
- $d(d\omega) = 0$ and $d^*(d^* * \aleph\mathbf{v}) = 0$ are equivalent to $\mathrm{div}\,\mathbf{curl}\,\mathbf{v} = 0$.
- $d(d\Theta) = 0$ is equivalent to $\mathbf{curl}\,\mathbf{grad}\,\Theta = 0$.

12.13 Further Reading

As an introduction, we recommend the following modern textbooks:

V. Zorich, Analysis I, II, Springer, New York, 2003.

P. Bamberg and S. Sternberg, A Course in Mathematics for Students of Physics, Cambridge University Press, 1999.

H. Flanders, Differential Forms with Applications to Physical Sciences, Academic Press, New York, 1989.

Y. Choquet-Bruhat, C. DeWitt-Morette, and M. Dillard-Bleick, Analysis, Manifolds, and Physics. Vols. 1, 2, Elsevier, Amsterdam, 1996.

T. Frankel, The Geometry of Physics, Cambridge University Press, 2004.

Furthermore, we recommend:

H. Amann and J. Escher, Analysis, Vols. 1–3, Birkhäuser, Basel, 1998.

K. Maurin, Analysis II: Integration, Distributions, Holomorphic Functions, Tensor and Harmonic Analysis, PWN Polish Scientific Publishers, Warsaw/Reidel, Dordrecht, 1980.

I. Agricola and T. Friedrich, Global Analysis: Differential Forms in Analysis, Geometry and Physics, Amer. Math. Soc., Providence, Rhode Island, 2002.

B. Dubrovin, A. Fomenko, and S. Novikov, Modern Geometry: Methods and Applications, Vols. 1-3, Springer, New York, 1992.

S. Novikov and T. Taimanov, Geometric Structures and Fields, Amer. Math. Soc., Providence, Rhode Island, 2006.

Applications to physics:

R. Abraham and J. Marsden, Foundations of Mechanics, Addison-Wesley, Reading, Massachusetts, 1978.

R. Abraham, J. Marsden, and T. Ratiu, Manifolds, Tensor Analysis, and Applications, Springer, New York, 1988.

J. Marsden, Applications of Global Analysis in Mathematical Physics, Publish or Perish, Boston, 1974.

J. Marsden, Lectures on Geometric Methods in Mathematical Physics, SIAM, Philadelphia, 1981.

J. Marsden and T. Ratiu, Introduction to Mechanics and Symmetry, Springer, New York, 1999.

A. Fomenko, Integrability and Non-Integrability in Geometry and Mechanics, Kluwer, Dordrecht, 1988.

G. Schwarz, Hodge Decomposition–a Method of Solving Boundary Value Problems, Springer, Berlin, 1995.

Applications to differential topology:

V. Guillemin and A. Pollack, Differential Topology, Prentice Hall, Englewood Cliffs, New Jersey, 1974.

R. Bott and L. Tu, Differential Forms in Algebraic Topology, Springer, New York, 1982.

I. Madsen and J. Tornehave, From Calculus to Cohomology: de Rham Cohomology and Characteristic Classes, Cambridge University Press, 1997.

J. Moore, Lectures on Seiberg–Witten Invariants, Springer, Berlin, 1996.

S. Lefschetz, Applications of Algebraic Topology: Graphs and Networks, the Picard–Lefschetz Theory, and Feynman Algorithms, Springer, New York, 1975.

Applications to supersymmetry:

V. Guillemin and S. Sternberg, Supersymmetry and Equivariant De Rham Theory, Springer, Berlin, 1999.

12.14 Historical Remarks

> Over the centuries, mathematics makes progress by solving hard problems. The calculus of differential forms has its roots in physics (classical mechanics, fluid dynamics, electromagnetism, and thermodynamics). In modern mathematics, the Cartan calculus combines algebra, analysis, geometry, and topology with each other. In modern physics, the Cartan calculus is used in order to describe the fundamental forces in nature in terms of gauge theory (Einstein's theory of general relativity and the Standard Model in particle physics).
>
> Folklore

Biographical data. For the convenience of the reader, let us first summarize the biographical data of mathematicians and physicists whose work will be mentioned below.

Hooke (1635–1703), Newton (1643–1727), Leibniz (1646–1716), Euler (1707–1783), Clairaut (1713–1765), Lagrange (1736–1813), Monge (1746–1818), Laplace (1749–1827).

Legendre (1752–1833), Pfaff (1765–1825), Ampére (1775–1836), Gauss (1777–1855), Poisson (1781–1840), Cauchy (1789–1857), Green (1793–1843).

Ostrogradsky (1801–1862), Jacobi (1804–1851), Grassmann (1809–1877), Stokes (1819–1903), Helmholtz (1821–1894), Betti (1823–1892), Thomson (Lord Kelvin since 1893) (1824–1907), Riemann (1826–1866), Maxwell (1831–1879), Lie (1842–1899), Clifford (1845-1879), Klein (1849–1925), Frobenius (1849–1917).

Kovalevskaya (1850–1891), Poincaré (1854–1912), Goursat (1858–1936), Volterra (1860–1940), Hilbert (1862–1943), Élie Cartan (1869–1951), Carathéodory (1873–1950), Emmy Noether (1882–1935), Weyl (1885–1955), Künneth (1892–1975).

Dirac (1902–1984), de Rham (1903–1990), Hodge (1903–1975), Ehresmann (1905–1979), Kähler (1906–2000), Sard (1909–1980), Chern (1911–2004), Yang (born 1922), Singer (born 1924), Hirzebruch (born 1927), Atiyah (born 1929), Gromov (born 1943), Yau (born 1949), Witten (born 1951), Donaldson (born 1957).

In what follows, the quotations (e.g., Gauss (1813)) refer to the references to be found on page 791 below. To avoid lengthy formulas, we will use the language of vector calculus. However, note that this language was only introduced in the second half of the 19th century. Newton used a geometric language, and Lagrange worked with Cartesian coordinates.

The Gravitational Force of a Body

Consider two positive masses m_0 and m (e.g., sun and earth) located at the points P_0 and P, respectively. Let us introduce the origin O and the position vectors $\mathbf{x}_0 := \overrightarrow{OP_0}$ and $\mathbf{x} := \overrightarrow{OP}$ (Fig. 12.26). Newton's gravitational law tells us that the attracting force between two masses m_0 and m acting at the point P is given by

$$\boxed{\mathbf{F}_P = \frac{Gmm_0}{|\mathbf{x}_0 - \mathbf{x}|^2} \cdot \frac{\mathbf{x}_0 - \mathbf{x}}{|\mathbf{x}_0 - \mathbf{x}|}} \tag{12.190}$$

(a) (b) $\mathbf{F}_P = \frac{mm_0(\mathbf{x}_0-\mathbf{x})}{|\mathbf{x}_0-\mathbf{x}|^3}$

Fig. 12.26. The gravitational law

with the gravitational constant G.[59] To simplify notation, we also briefly write

$$\mathbf{F}(\mathbf{x}) = \frac{Gmm_0(\mathbf{x}_0 - \mathbf{x})}{|\mathbf{x}_0 - \mathbf{x}|^3}.$$

In particular, Newton obtained the result that the gravitational force inside a spherical shell vanishes. It was the goal of Lagrange to replace the geometric arguments due to Newton and his successors by analytic methods. Lagrange realized this program in his famous *Méchanique analitique* first published in Paris in 1788.[60] This is one of the most influential works in mathematical physics.

Consider a body of mass density ϱ_0 which occupies the set $\mathcal{M}$. We assume that $\mathcal{M}$ is the closure of a 3-dimensional bounded open set $\mathcal{M}$(e.g., a ball or an ellipsoid). In 1775, Lagrange derived the formula

$$\boxed{\mathbf{F}_{\text{body}}(P) = \int_{\mathcal{M}} \frac{Gm\varrho(P_0)}{|\mathbf{x}_0 - \mathbf{x}|^2} \cdot \frac{\mathbf{x}_0 - \mathbf{x}}{|\mathbf{x}_0 - \mathbf{x}|}\, d^3x_0.} \tag{12.191}$$

This is the gravitational force which the body (e.g., the earth) exerts on a mass m located at the point P (e.g., a stone). In 1777, Lagrange noticed that the components of the gravitational force (12.190) can be represented by the partial derivatives of a single function U. Explicitly, we have

$$\mathbf{F}(P) = \frac{Gmm_0(\mathbf{x}_0 - \mathbf{x})}{|\mathbf{x} - \mathbf{x}_0|^3} = -\,\mathbf{grad}\, U(P) \tag{12.192}$$

with

$$\boxed{U(P) := -\frac{Gmm_0}{|\mathbf{x} - \mathbf{x}_0|}.}$$

The function U was called *potential* by Gauss in 1813. This is one of the most important notions in physics. We will show in Chap. 15 that

[59] Recall that the symbol $|\mathbf{x}_0 - \mathbf{x}|$ denotes the length of the vector $\mathbf{x}_0 - \mathbf{x}$. The gravitational force acting at the point P_0 is equal to

$$\mathbf{F}_{P_0} = \frac{Gmm_0(\mathbf{x} - \mathbf{x}_0)}{|\mathbf{x} - \mathbf{x}_0|^3},$$

which is parallel to $-\mathbf{F}_P$. This is Newton's principle 'actio = reactio'.

[60] After an orthographic reform in France organized by the Paris Academy, the title of the second edition changed into *Mécanique analytique* in 1811.

Gauge theory represents a generalization of classical potential theory to modern physics.

In terms of physics, $U(P)$ is the potential energy of the mass m at the point P with respect to the gravitational field generated by the mass m_0 at the point P_0. By convention, this potential energy of the gravitational field is negative, and it vanishes at infinity. Note that we gain the energy

$$E = U(P_1) - U(P_2) = \int_{P_1}^{P_2} \mathbf{F}\, d\mathbf{x}$$

if the point of mass m moves from the position P_1 to the position P_2 in the gravitational field generated by the mass m_0. This energy does not depend on the trajectory of the moving body. For example, if a stone of mass m is falling from the point P_1 to the point P_2 on earth, then, approximately, we gain the energy $E = gmh$, where h is the initial height of the stone, and g is the gravitational acceleration on earth. The potential corresponding to the force (12.191) reads as

$$\boxed{U_{\text{body}}(P) = -\int_{\mathcal{M}} \frac{Gm\varrho_0(P_0)d^3x_0}{|\mathbf{x} - \mathbf{x}_0|}.} \tag{12.193}$$

In 1785, Laplace showed that the potential satisfies the Laplace equation

$$\Delta U_{\text{body}}(P) = 0 \qquad \text{for all} \quad P \notin \text{cl}(\mathcal{M}) \tag{12.194}$$

where he used spherical coordinates. In 1787, Laplace published the equation

$$\frac{\partial^2 U_{\text{body}}(P)}{\partial x^2} + \frac{\partial^2 U_{\text{body}}(P)}{\partial y^2} + \frac{\partial^2 U_{\text{body}}(P)}{\partial z^2} = 0, \quad P \notin \mathcal{M} \tag{12.195}$$

with respect to Cartesian coordinates. The symbol Δ was introduced by Murphy in 1833. In this monograph, we use the notation

$$\Delta U := -U_{xx} - U_{yy} - U_{zz}.$$

This sign convention corresponds to the definition $\Delta U := (dd^* + d^*d)U$ in modern differential geometry (the Hodge-Laplace operator). In 1813, Poisson obtained that

$$\Delta U_{\text{body}}(P) = -4\pi G m \varrho_0(P) \qquad \text{for all} \quad P \in \text{int}(\mathcal{M}). \tag{12.196}$$

This is called the Poisson equation. In the history of mathematical physics, trouble was caused by the fact that, for a continuous density function $\varrho_0 : \text{cl}(\mathcal{M}) \to \mathbb{R}$, the Poisson equation (12.196) is not always valid in the classical sense, but only in the sense of generalized functions. However, if the mass density ϱ_0 is continuously differentiable on $\text{cl}(\mathcal{M})$, then the Poisson equation (12.196) holds in the classical sense.

In the second edition of his *Mécanique analytique* from 1811, Lagrange introduced general surface integrals, and he used this in order to perform integration by parts for special volume integrals.

The Shape of the Earth

The precise computation of the gravitational force generated by a homogeneous ellipsoid is one of the most difficult problems in astronomy.[61]

Carl Friedrich Gauss, 1813

[61] C. Gauss (1813), On a new method for studying the gravitational force of homogeneous elliptical spheroids, Göttingen (24 pages in Latin).

In his famous *Mathematical Principles of Natural Philosophy*, published in 1687, Newton predicted theoretically the flattening of earth. He found that the equatorial semi-axis would be $\frac{1}{230}$ longer than the polar semi-axis (true value about $\frac{1}{300}$). In the 18th century, the Paris Academy sent expeditions to Lapland and Peru in order to determine experimentally the flattening of earth. In order to compute the gravitational force of earth, mathematicians and physicists tried to compute the integral (12.191) in the case where the body $\mathcal{M}$ is an ellipsoid. Let us assume that the mass density ϱ_0 of the body is constant.

Legendre polynomials and moments of the mass distribution of the body. Let $\mathbf{x} = \overrightarrow{OP}$ with $r := |\mathbf{x}|, r_0 := |\mathbf{x}_0|$, and $\langle \mathbf{x}|\mathbf{x}_0\rangle = rr_0 \cos\vartheta_0$. Then

$$|\mathbf{x} - \mathbf{x}_0| = \sqrt{r^2 - 2rr_0\cos\vartheta_0 + r_0^2}.$$

In 1782, Legendre introduced the following formula

$$\frac{1}{|\mathbf{x} - \mathbf{x}_0|} = \frac{1}{r}\sum_{k=0}^{\infty}\left(\frac{r_0}{r}\right)^k P_k(\cos\vartheta_0). \tag{12.197}$$

This series is convergent if $\frac{r_0}{r} < 1$. The polynomials

$$P_k(z) := \frac{1}{2^k k!}\frac{d^k(z^2-1)^k}{dz^k}, \qquad z \in \mathbb{C}, \ k = 0, 1, 2, \ldots$$

are called the Legendre polynomials. Explicitly, $P_0(z) := 1$, $P_1(z) := z$, and

$$P_2(z) := \tfrac{1}{2}(3z^2 - 1), \quad P_3(z) := \tfrac{1}{2}(5z^3 - 3z), \quad P_4(z) := \tfrac{1}{8}(35z^4 - 30z^2 + 3).$$

Hence

$$\boxed{U_{\text{body}}(P) = -mG\left(\frac{M_0}{r} + \frac{M_1}{r^2} + \frac{M_2}{r^3} + \ldots\right)} \tag{12.198}$$

with the total mass $M_0 := \int_{\mathcal{M}} \varrho_0 \, d^3x_0$ and the moments

$$M_k := \int_{\mathcal{M}} \varrho_0 r_0^k P_k(\cos\vartheta_0)\, dx_0dy_0dz_0, \qquad k = 1, 2, \ldots$$

of the body.[62] In particular, M_2 (resp. M_4) is called the dipole (resp. quadrupole) moment of the body (e.g., earth). If the body is contained in a ball of radius R about the origin, then the formula (12.198) is valid for all points P whose distance from the body is large enough (i.e., $r > R$). In the special case where the body is symmetric under the reflection $\mathbf{x} \mapsto -\mathbf{x}$, then $M_1 = M_3 = M_5 = \ldots = 0$.

In the 1780s, Laplace used this method in order to compute the gravitational potential of general ellipsoids. He devoted a chapter to this subject in his famous *Celestial Mechanics*, vol. II, published in 1799.

Gauss' reduction of the gravitational force to a surface integral. In 1813, Gauss showed that

German translation in: A. Wangerin (Ed.), On the Attraction of Homogeneous Ellipsoids: basic papers by Laplace (1782), Ivory (1809), Gauss (1813), Chasles (1838), and Dirichlet (1839), Ostwalds Klassiker, Vol. 19, Engelmann, Leipzig, 1890.

[62] We set $\cos\vartheta_0 = \frac{z_0}{r_0}$.

$$\mathbf{F}_{\rm body}(P) = \int_{\partial\mathcal{M}} U(P, P_0)\mathbf{n}_{P_0} dS_{P_0} \tag{12.199}$$

with $U(P, P_0) := -\frac{Gm\varrho_0}{|\mathbf{x}-\mathbf{x}_0|}$. Here, $\mathbf{n}_{P_0}$ is the outer normal unit vector at the boundary point P_0 of the surface $\partial\mathcal{M}$ of the ellipsoid. Gauss used a geometric argument. With the aid of Ostrogradsky's divergence theorem (12.200) below, the proof of (12.199) reads as follows:

$$\begin{aligned}\mathbf{F}_{\rm body}(P) &= -\int_{\mathcal{M}} \mathbf{grad}_P\, U(P, P_0)\, d^3x_0 \\ &= \int_{\mathcal{M}} \mathbf{grad}_{P_0}\, U(P, P_0)\, d^3x_0 = \int_{\partial\mathcal{M}} U(P, P_0)\mathbf{n}_{P_0}\, dS_{P_0}.\end{aligned}$$

Integral Theorems

The three integral theorems of Gauss–Ostrogradsky, Green, and Stokes emerged slowly in the 19th century with increasing generality. For a detailed study of the complicated history of these integral theorems, we refer to V. Katz, The history of Stokes' theorem, Mathematics Magazine **52** (1979), 146–156.

Ostrogradsky's divergence formula. The divergence formula in three dimensions,

$$\boxed{\int_{\mathcal{M}} \left(\frac{\partial U}{\partial x} + \frac{\partial V}{\partial y} + \frac{\partial W}{\partial z}\right) dxdydz = \int_{\partial\mathcal{M}} (Un^1 + Vn^2 + Wn^3)\, dS,} \tag{12.200}$$

was first published by Ostrogradsky (1826) (see the references quoted on page 792). Here, n^1, n^2, n^3 are the Cartesian components of the outer normal unit vector $\mathbf{n}$. This formula was generalized to n dimensions by Ostrogradsky (1836).

The Green integral formula. George Green was autodidact in mathematics. In 1828, he published a famous essay on electricity and magnetism where he used the following formula for the Laplacian in three dimensions:

$$\int_{\mathcal{M}} (U\Delta V - V\Delta U)dxdydz = \int_{\partial\mathcal{M}} \left(V\frac{\partial U}{\partial n} - U\frac{\partial V}{\partial n}\right) dS.$$

Here, $\frac{\partial U}{\partial n} = \mathbf{n}\,\mathbf{grad}\,U$. This formula is called the Green formula for the Laplacian. Using integration by parts, it is possible to obtain such formulas for general linear differential operators. Green used this formula in order to develop his method of the Green's functions which plays a fundamental role in both the classical theory of partial differential equations and modern physics (quantum field theory and solid state physics). See Dyson (1993) on page 791.

The Cauchy integral formula. In order to study line integrals, Cauchy (1845) used the formula

$$\int_{\partial\mathcal{M}} Udx + Vdy = \int_{\mathcal{M}} \left(\frac{\partial V}{\partial x} - \frac{\partial U}{\partial y}\right) dxdy. \tag{12.201}$$

Here, $\mathcal{M}$ is a compact 2-dimensional submanifold of the Euclidean plane with coherently oriented boundary $\partial\mathcal{M}$. In particular, Cauchy used the fact that

$$\int_{\partial\mathcal{M}} Udx + Vdy = 0$$

if the integrability condition $U_y = V_x$ is satisfied on $\mathcal{M}$.

The Thomson integral formula. In a letter from July 4th, 1850, Thomson (later Lord Kelvin) communicated two theorems to Stokes who answered: "The theorems which you communicated are very elegant and new to me." In modern terminology, one of the theorems reads as follows:

$$\int_{\mathcal{M}} \mathbf{curl}\,\mathbf{H}\, dS = \int_{\partial\mathcal{M}} \mathbf{H} d\mathbf{x}. \tag{12.202}$$

Here, $\mathcal{M}$ is a compact 2-dimensional submanifold of the 3-dimensional Euclidean manifold with coherently oriented boundary $\partial\mathcal{M}$. Thomson obtained this theorem by studying the properties of magnetic fields.

The theorem (12.202) first appeared in print in 1854. George Stokes had for several years been setting the Smith's Prize Examination at Cambridge University (England). In February 1854, Stokes formulated (12.202) as an examination question. Therefore, formula (12.202) was called the Stokes theorem.

The Generalized Stokes Theorem

In a fundamental paper, Volterra (1889) proved the following two basic results:

(i) The generalization of the Stokes integral theorem to n dimensions, and
(ii) necessary and sufficient conditions for solving the following system of differential equations on $\mathbb{R}^n$ with $n = 2, 3, \ldots$:

$$\partial_{[j_1} V_{j_2 \ldots j_n]} = F_{j_1 \ldots j_n}, \qquad j_1, \ldots, j_n = 1, \ldots, n. \tag{12.203}$$

This system represents a generalization of the potential equation $\mathbf{grad}\, V = \mathbf{F}$. We are given the smooth functions $F_{j_1\ldots j_n}$ on $\mathbb{R}^n$ which are antisymmetric with respect to all the indices. We are looking for smooth solutions $V_{j_2\ldots j_n}$ which satisfy equation (12.203), and which are antisymmetric with respect to all the indices. We set $\partial_j := \frac{\partial}{\partial x^j}$. Moreover, the symbol $\partial_{[j_1} V_{j_2\ldots j_n]}$ indicates the antisymmetrization with respect to the indices $j_1, \ldots, j_n$.[63]

Volterra proved that the system (12.203) has a solution on $\mathbb{R}^n$ iff the following integrability conditions are satisfied:

$$\partial_{[j_0} F_{j_1 \ldots j_n]} = 0 \qquad \text{on } \mathbb{R}^n, \quad j_0, \ldots, j_n = 1, \ldots, n. \tag{12.204}$$

Volterra did not use the language of differential forms which was introduced by Élie Cartan ten years later in 1899. In modern terminology, Volterra proved that

$$\boxed{\int_{\mathcal{M}} d\omega = \int_{\partial\mathcal{M}} \omega} \tag{12.205}$$

is valid on a compact submanifold $\mathcal{M}$ of $\mathbb{R}^n$ with coherently oriented boundary $\partial\mathcal{M}$. The system (12.203) corresponds to

$$\boxed{d\omega = \mu \qquad \text{on } \mathcal{M}} \tag{12.206}$$

where $\mathcal{M} := \mathbb{R}^n$. Here, we set $\mu := F_{j_1\ldots j_n} dx^{j_1} \wedge \cdots \wedge dx^{j_n}$, and[64]

[63] For example, $\partial_{[j} V_{k]} = \frac{1}{2!}(\partial_j V_k - \partial_k V_j)$.
[64] We sum over equal upper and lower indices from 1 to n.

$$\omega := V_{j_2 \ldots j_n} dx^{j_2} \wedge \cdots \wedge dx^{j_n}.$$

If there exists a solution ω of (12.206), then it follows from

$$\boxed{d(d\omega) = 0} \tag{12.207}$$

that $d\mu = 0$. This is the integrability condition (12.204) above. Equation (12.207) is frequently called the Poincaré lemma; it was published by Poincaré (1887) two years before Volterra's paper, but in a different context. Poincaré studied integral invariants (see page 784 below). It follows from the generalized Stokes theorem (12.205) that

$$\int_{\mathcal{C}} d\omega = 0 \tag{12.208}$$

if the boundary $\partial\mathcal{C}$ of $\mathcal{C}$ is empty. This relation motivated Poincaré to study cycles. By definition, a smooth cycle $\mathcal{C}$ is a manifold or a submanifold which has no boundary (e.g., a sphere or a circle). The following observation is crucial. If ω is a solution of the equation (12.206) and $\mathcal{C}$ is a cycle in the manifold $\mathcal{M}$, then

$$\int_{\mathcal{C}} \mu = 0. \tag{12.209}$$

In contrast to the local integrability condition $d\mu = 0$, this is a global integrability condition. At this point, topology enters the story. Poincaré was amazed by the relation between analysis and topology. The final approach was formulated by de Rham in the 1930s.

The Euler Multiplicator, the Pfaff Problem, and Cartan's Exterior Calculus of Differential Forms

Clairaut's total differential. In 1739, Clairaut studied the ordinary differential equation

$$\frac{dy}{dx} = -\frac{U(x,y)}{V(x,y)}. \tag{12.210}$$

Using the formal Leibniz notation, he wrote this as

$$U(x,y)dx + V(x,y)dy = 0. \tag{12.211}$$

Motivated by this reformulation, Clairaut proved that the equation

$$dW = Udx + Vdy$$

has a solution W iff the integrability condition $U_y = V_x$ is satisfied. In this case, the equation $W(x,y) = 0$ yields a solution $y = y(x)$ of (12.210).

The Euler multiplicator. Euler considered the case where the integrability condition $U_y = V_x$ is violated. He showed that it is always possible to find functions W and M of the variables x, y such that

$$M \cdot dW = Udx + Vdy.$$

Then it follows from

$$W(x, y(x)) = 0$$

that $M(x,y(x))\, dW(x,y(x)) = 0$. Hence $U(x,y(x))dx + V(x,y(x))dy = 0$. This implies that the function $y = y(x)$ is a solution of the original differential equation (12.210). The function M is called the Euler multiplicator.

The Pfaff problem. In 1815, Paff posed the following problem: Solve the system

$$f_{j1}(x)\, dx^1 + \ldots + f_{jn}(x)\, dx^n = 0, \qquad j = 1, \ldots, N \tag{12.212}$$

where $x = (x^1, \ldots, x^n)$. In addition, he posed the following reduction problem: Represent the differential form $\omega = f_1 dg_1 + \ldots + f_k dg_k$ as

$$\omega = F_1 dG_1 + \ldots + F_m dG_m,$$

where the number m of functions $F_1, G_1 \ldots, F_m, G_m$ is minimal. In 1877, Frobenius wrote a basic paper on the Pfaff problem. He proved necessary and sufficient conditions for solving (12.212), and he solved the reduction problem. This can be found in Sect. 12.11 on page 767.

Élie Cartan's differential forms. In order to reformulate the Frobenius results on the Pfaff problem by means of a symbolic method, Cartan (1899) introduced the so-called exterior differential calculus. To this end, he combined the Leibniz differential with Grassmann's antisymmetric (exterior) product. In 1909, Carathéodory used the Pfaff problem in order to give phenomenological thermodynamics a sound mathematical basis.[65] In 1917, Goursat studied a generalization of the Pfaff problem in terms of differential forms of higher degree. In this paper, he formulated the generalized Stokes theorem (12.205) in terms of differential forms for the first time.

It seems that the generalized Stokes theorem appeared in a textbook for the first time in 1959.[66] The theory of differential forms shows that progress in mathematics may proceed on a large time scale.

Poincaré's Generalization of Cauchy's Residue Theorem and Riemann's Periods (Topological Charges)

Cauchy's residue theorem. In what follows, we will use Cauchy's residue theorem (see Theorem 4.2 of Vol. I). The curves C, C_1, C_2 depicted in Fig. 12.27 are assumed to be smooth closed curves which are homeomorphic to the unit circle.

(i) Gaussian plane (Fig. 12.27(a)): If the function $f : \mathbb{C} \to \mathbb{C}$ is holomorphic, then

$$\int_C f(z)dz = 0.$$

(ii) Punctured Gaussian plane $\mathbb{C} \setminus \{z_1\}$ (Fig. 12.27(b)): Setting $f(z) := \frac{1}{z-z_1}$ for all $z \in \mathbb{C} \setminus \{z_1\}$, we get

$$\int_{C_1} f(z)dz = 2\pi\mathrm{i}.$$

[65] This is thoroughly investigated in Frankel (2004), Sect. 6.3, quoted on page 775.
[66] H. Nickerson, D. Spencer, and N. Steenrod, Advanced Calculus, van Nostrand, Princeton, 1959.

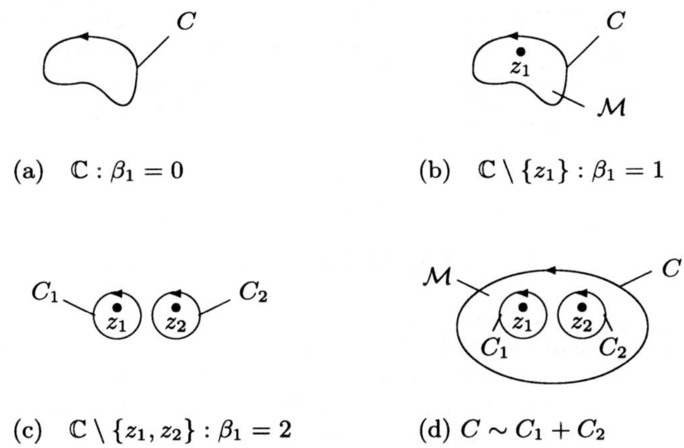

Fig. 12.27. The first Betti number β_1

(iii) Double-punctured Gaussian plane $\mathbb{C} \setminus \{z_1, z_2\}$ (Fig. 12.27(c), (d)): Fix the complex numbers a_1, a_2. Setting

$$f(z) := \frac{a_1}{z - z_1} + \frac{a_2}{z - z_2} \qquad \text{for all} \quad z \in \mathbb{C} \setminus \{z_1, z_2\},$$

we get $\int_{C_j} f(z)dz = 2\pi \mathrm{i} a_j$ with $j = 1, 2$, and

$$\int_C f(z)dz = \int_{C_1} f(z)dz + \int_{C_2} f(z)dz = 2\pi\mathrm{i}(a_1 + a_2). \tag{12.213}$$

The integrals $\int_C f(z)dz$, $\int_{C_j} f(z)dz$, $j = 1, 2$, are called periods (synonymously, integral invariants or topological charges). Observe that the integrals remain unchanged under appropriate deformations of the curves C, C_j.

Riemann's periods. Riemann generalized this by considering line integrals on Riemann surfaces. This way, in his famous paper from 1857, Riemann created a general theory for Abelian integrals

$$\int_C \omega \tag{12.214}$$

where ω is an Abelian differential which looks like $\omega = f(z)dz$ in local coordinates. More precisely, we have $f(z) = R(z, w(z))$ where R is a rational function of the complex variables z and w. In addition, $w(z)$ is given by the algebraic equation

$$P(z, w(z)) = 0 \tag{12.215}$$

where P is a polynomial of the complex variables z and w (e.g., $P(z, w) = w^2 - z^3$). In terms of geometry, equation (12.215) describes an algebraic curve. For example, elliptic integrals are special Abelian integrals related to so-called elliptic curves. Elliptic curves can be globally parametrized by elliptic functions. This generalizes the global parametrization of a circle by trigonometric functions.

Riemann's periods are obtained by integrals of the form (12.214) in the case where the closed curve C cannot be continuously contracted to a point. In the

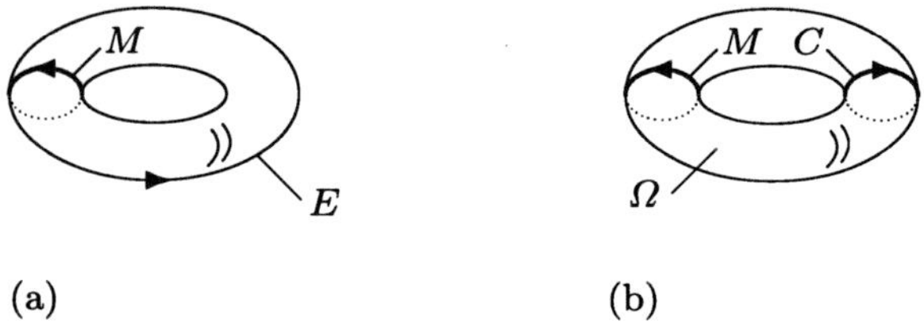

Fig. 12.28. Torus

special case of elliptic integrals, the Riemann surface is a torus, and the two periods of an elliptic integral, $\int_C \omega$, are obtained by integrating along the equator E and some meridian M of the torus (see Fig. 12.28). The periods of the integral $\int_C \omega$ are precisely the periods of the elliptic function which globally parametrizes the integral.

Riemann passed over to Riemann surfaces in order to handle the difficulty that, for fixed z, equation (12.215) has several solutions (i.e., the function $z \mapsto w(z)$ is multi-valued).

The Jacobian variety of a Riemann surface. We want to study the values of line integrals on the compact Riemann surface $\mathcal{R}$ in terms of the periods. If $\mathcal{R}$ has the genus g (e.g., $g = 1$ for the torus), then there exist precisely $2g$ basic 1-cycles $C_1, \ldots C_{2g}$. In addition, there exist g differential forms $\omega_1, \ldots, \omega_g$ on $\mathcal{R}$ such that

$$\int_{C_j} \omega_k = \delta_{jk}, \qquad j,k = 1,\ldots,g.$$

Moreover, the matrix (π_{jk}) with the entries

$$\pi_{jk} := \int_{C_{g+j}} \omega_k, \qquad j,k = 1,\ldots,g$$

is symmetric, and $\Im(\pi_{jk}) > 0$ for all $j,k = 1,\ldots,g$. Fix the point P_0 on the Riemann surface $\mathcal{R}$. For all points P in $\mathcal{R}$, define

$$\varphi(P) := \left(\int_{P_0}^{P} \omega_1, \ldots, \int_{P_0}^{P} \omega_g \right).$$

The line integrals from the point P_0 to the point P are independent of the path of integration. However, because of the periods, there exists a lattice Λ in $\mathbb{C}^g$ such that $\varphi(P)$ is uniquely defined modulo the lattice (i.e., modulo integer-valued linear combinations of the periods). More precisely, there exists a holomorphic map

$$\varphi : \mathcal{R} \to \mathrm{Jac}(\mathcal{R})$$

where the quotient space $\mathrm{Jac}(\mathcal{R}) := \mathbb{C}^g/\Lambda$ is called the Jacobian variety of the Riemann surface $\mathcal{R}$. The Jacobian variety is a compact, commutative, g-dimensional, complex Lie group.

The following remarks are crucial for understanding the creation of algebraic topology by Poincaré in the 1890s.

Poincaré's homology and Betti numbers. According to Poincaré, we want to consider the integrals above from a point of view which can be generalized to higher dimensions. To this end, consider the manifold $\mathcal{N}$ (e.g., the Gaussian plane $\mathbb{C}$, the punctured Gaussian planes $\mathbb{C}\setminus\{z_1\}$ and $\mathbb{C}\setminus\{z_1, z_2\}$, the Euclidean manifold $\mathbb{E}^3$ or an n-dimensional sphere). An m-dimensional submanifold C of $\mathcal{N}$ (without boundary) is called a regular m-cycle. We write

$$\boxed{C \sim 0}$$

iff C is the boundary of an $(m+1)$-dimensional submanifold $\mathcal{M}$ of $\mathcal{N}$ (i.e., $C = \partial\mathcal{M}$). In this special case, we say that C is homologous to zero (or a trivial cycle) in the manifold $\mathcal{N}$. For example, the circle (resp. the m-dimensional sphere) is a regular 1-cycle (resp. m-cycle). The following proposition motivates the terminology.

Proposition 12.60 *Let $f : \mathcal{O} \to \mathbb{C}$ be a holomorphic function on the open subset $\mathcal{O}$ of the Gaussian plane $\mathbb{C}$. Set $\omega := f(z)dz$. Suppose that C is a regular 1-cycle which lies in the set $\mathcal{O}$, and $C \sim 0$. Then*

$$\int_C \omega = 0.$$

Proof. By the Cauchy–Riemann differential equation, $d\omega = 0$ on $\mathcal{O}$. By Fig. 12.27(a) on page 785, the Stokes theorem tells us that

$$\int_C \omega = \int_{\partial\mathcal{M}} \omega = \int_{\mathcal{M}} d\omega = 0.$$

□

For example, let us apply this to Fig 12.27 (d) on page 785. It follows from the boundary decomposition $\partial\mathcal{M} = C - C_1 - C_2$ that

$$0 = \int_{\mathcal{M}} d\omega = \int_{\partial\mathcal{M}} \omega = \int_C \omega - \int_{C_1} \omega - \int_{C_2} \omega.$$

This yields (12.213) above.

By definition, the first Betti number β_1 counts the number of essential 1-cycles. Intuitively, this means that the integral over an arbitrary 1-cycle C can be written as a linear combination (with integer coefficients) over the integrals along the essential 1-cycles $C_1, \ldots, C_{\beta_1}$, that is,

$$\int_C \omega = m_1 \int_{C_1} \omega + \ldots + m_{\beta_1} \int_{C_{\beta_1}} \omega$$

where $m_1, \ldots, m_{\beta_1}$ are integers. For the manifolds $\mathbb{C}, \mathbb{C}\setminus\{z_1\}$, and $\mathbb{C}\setminus\{z_1, z_2\}$, we get the first Betti number $\beta_1 = 0, 1$, and $\beta_1 = 2$, respectively (see Fig. 12.27 on page 785).

Poincaré's periods. In 1887, Poincaré started to extend Riemann's theory of periods to functions of n complex variables. This way, in the late 1880s and in the 1890s, Poincaré discovered implicitly all the relations (12.205) through (12.209) on page 782 without using explicitly the language of differential forms of degree $p > 1$; this language was introduced later by Cartan in 1899. In addition, the problem of periods of n-dimensional integrals was one of the sources for creating algebraic topology by Poincaré in 1895. In this setting, regular n-cycles are n-dimensional submanifolds which are homeomorphic to n-dimensional spheres.[67]

[67] Poincaré studied at the École Polytechnique in Paris. From 1881 until 1912, he worked as professor of mathematics at the Sorbonne in Paris.

The De Rham Theory

In his thesis from 1931, de Rham gave Poincaré's ideas the final form in the setting of differential topology.

Further Developments

Let us briefly sketch some further important developments.

Cartan's method of moving frames. In his *Erlangen program* from 1872, Felix Klein pointed out that[68]

Geometry is the invariant theory of transformation groups.

It was the goal of the great geometer Élie Cartan to generalize Klein's Erlangen program to differential geometry by combining differential calculus with the theory of Lie groups. To this end, Cartan invented his method of moving frames which we will study in this volume.[69] In this setting, differential forms play the crucial role. The final theory was created by Ehresmann (1950) using the language of fiber bundles. The first monograph about this new approach to differential geometry was written by Kobayashi and Nomizu (1963). This is the basis of gauge theory in modern physics.

Kähler manifolds. In 1933, using the language of differential forms, Kähler discovered that there exists a class of n-dimensional complex manifolds which has similar nice properties as Riemann surfaces. The basic idea can be found in Sect. 5.10.1 of Vol. II (the relation between geometric optics, Poincaré's hyperbolic non-Euclidean geometry on the upper half-plane, and Kähler geometry). The point is that the geometry and the topology of a Kähler manifold are governed by a differential form ω with

$$d\omega = 0.$$

This is called the Kähler form. In a very natural way, Kähler geometry combines the complex Hilbert space geometry of the tangent spaces, parallel transport of tangent vectors, real Riemannian geometry, and symplectic geometry. These geometries are fundamental for physics. Moreover, from the practical point of view, the computation of the Riemann curvature tensor can be dramatically simplified by differentiating one special scalar function U called the Kähler potential:

$$\omega = \mathrm{i}\partial\overline{\partial}U.$$

Calabi–Yau manifolds, which are special Kähler manifolds, play a crucial role in string theory. We refer to Jost (2008) (introduction), Voisin (2002), Ballmann (2006), Moroianu (2007), and Becker (2006) (string theory) quoted on page 799. Furthermore, we refer to the survey article:

P. Bourguignon, The unabated vitality of Kählerian geometry.
In: E. Kähler, Mathematical Works, pp. 737–766, de Gruyter, Berlin, 2004.

---

[68] Klein studied at the Universities of Bonn and Göttingen. He worked as professor of mathematics at the following Universities in Germany: Erlangen (1872–1875), Technical University Munich (1875–1880), Leipzig (1880–1886), and Göttingen (1886–1925).

[69] Cartan studied at the École Normale Superieure in Paris. He worked as professor of mathematics at the following Universities in France: Montpellier (1894–1896), Lyon (1896–1903), Nancy (1903–1909), and Sorbonne in Paris (1909–1940).

For example, Calabi conjectured in 1955 the following: A Kähler manifold $\mathcal{M}$ is an Einstein manifold, that is, the Ricci curvature of $\mathcal{M}$ vanishes,

$$\mathrm{Ric}(g) = 0 \qquad \text{on } \mathcal{M},$$

iff the first Chern class of the manifold $\mathcal{M}$ vanishes (i.e., $c_1 = 0$). It was proven by Yau in 1977 that this conjecture is true. To this end, Yau thoroughly studied the complex Monge–Ampère differential equation.[70]

Systems of differential forms and Kähler's differential ideals. In 1934, Kähler wrote a seminal monograph where he studied general systems of differential forms,

$$\boxed{\omega_j = 0 \quad \text{on } \mathcal{M}, \ j = 1, \ldots, m} \tag{12.216}$$

in a holomorphic setting. That is, $\mathcal{M}$ is an n-dimensional complex manifold (with biholomorphic transformations of local coordinates), and the coefficient functions of the differential forms $\omega_1, \ldots, \omega_m$ of finite degree are holomorphic (i.e., the coefficients are power series expansions with respect to local coordinates). Combining elegantly algebra with analysis, Kähler proved a general existence theorem for (12.216) which generalizes the classical Frobenius theorem.

The basic idea reads as follows. First we have to add the integrability conditions $d\omega_j = 0$, that is, we replace (12.216) by the extended system

$$\boxed{\omega_j = 0, \quad d\omega_j = 0, \qquad j = 1, \ldots, m.} \tag{12.217}$$

Next the solution manifold is constructed inductively with respect to increasing dimension. More precisely, in each step, by a biholomorphic coordinate transformation, one has to solve a well-posed initial-value problem with the aid of the Cauchy–Kovalevskaya theorem. If the initial problem is ill-posed, then this corresponds to the generalization of a well-known singular situation in the classical theory of partial differential equations (Cauchy's characteristics).

In algebraic geometry, systems of polynomial equations are described by ideals in polynomial rings. Kähler generalized this strategy by introducing so-called differential ideals. By definition, a set of differential forms is called a differential ideal iff it is invariant under the following operations: linear combinations, wedge products, and passing from ω to $d\omega$. In this setting, problem (12.217) corresponds to the annihilation of a differential ideal generated by $\omega_1, \ldots, \omega_m$.

In 1945, Élie Cartan wrote a monograph on systems of differential forms and their applications to differential geometry. Cartan used the notion of differential ideals in order to supplement the original system (12.216) by the system

$$\theta_1 = 0, \ldots, \theta_r = 0 \quad \text{on } \mathcal{M} \tag{12.218}$$

for differential forms θ_j of first degree. Cartan showed how to use the adjoint system (12.218) (which is always integrable) in order to solve the original system (12.216). This approach generalizes Cauchy's method for solving first-order partial differential equations with the aid of characteristic curves.

The theory described above is called the Cartan–Kähler theory for systems of differential forms. The basic ideas can be found in E. Zeidler, Oxford Users'

[70] For his crucial contributions to geometric analysis (combining nonlinear partial differential equations with geometry), Shing-Tung Yau (born 1949) was awarded the Fields medal in 1983.

Guide to Mathematics, Oxford University Press, 2004, Sect. 1.13 (the Cartan–Kähler theorem). For a detailed study, we refer to Bryant et al. (1991), Sharpe (1997), Ivey and Landsberg (2004) quoted on page 799.

The Dirac equation and Kähler's interior differential calculus. Motivated by the Dirac equation, Kähler (1962) invented his interior differential calculus in order to formulate the Dirac equation in terms of an invariant differential calculus. The idea is the following. First generate the exterior differential algebra by the relation

$$dx^i \wedge dx^j = -dx^j \wedge dx^i,$$

where $x^1, \ldots, x^n$ are local coordinates of the manifold $\mathcal{M}$ under consideration. Suppose now that there exists a Riemannian (or pseudo-Riemannian) tensor (g_{ij}) on $\mathcal{M}$. Then we introduce the interior product $\vee$ by setting

$$dx^i \vee dx^j := dx^i \wedge dx^j + g^{ij}, \qquad i, j = 1, \ldots, n.$$

As usual, (g^{ij}) denotes the inverse matrix to the matrix (g_{ij}). This implies the Clifford relation

$$dx^i \vee dx^j + dx^j \vee dx^i = 2g^{ij}, \qquad i, j = 1, \ldots, n.$$

The point is that this approach possesses an invariant meaning, that is, it does not depend on the choice of the local coordinates.

Nowadays the Dirac equation and its generalizations (like the fundamental Seiberg–Witten equation) are formulated in terms of the universal modern approach to differential geometry. The idea is to construct a covariant differentiation which respects a given symmetry group by using the connection of a principal fiber bundle and to transplant this to the associated vector bundle. In particular, the Dirac equation is described by spin bundles and the corresponding spin geometry. We refer to Kobayashi and Nomizu (1963) on page 797, Jost (2008) (the Seiberg–Witten equation and its relation to the Ginzburg–Landau equation) and Moore (1996) on page 800.

Index theory. Important progress in topology is related to the construction of new classes of topological invariants as the index of differential operators on manifolds (e.g., the Riemann–Roch–Hirzebruch theorem, the Atiyah–Singer index theorem, the Yang–Mills equation and the Donaldson theory of 4-manifolds, the Seiberg–Witten equation and 4-manifolds). We refer to Hirzebruch (1956) on page 797, Shanahan (1978) and Gilkey (1998) on page 800, and Donaldson (1990), (1996), (2002) on page 798.

Differential forms on the manifold $\mathcal{M}$ as a section of the Grassmann bundle on $\mathcal{M}$. In 1899, Cartan introduced differential forms in a symbolic sense. In his 1934 monograph, Kähler gave a rigorous algebraic definition. He introduced the algebra of the following expression:

$$\omega = a_0(x) + a_i(x)dx^i + a_{ij}(x)[dx^i, dx^j] + \ldots$$

where the coefficients $a_0, a_i, a_{ij}, \ldots$ are smooth functions, and we have the relations $[dx^i, dx^j] = -[dx^j, dx^i]$, and so on. Here, we sum over equal upper and lower indices from 1 to n. Moreover, $x = (x^1, \ldots, x^n)$. Nowadays we write

$$\omega = a_0(x) + a_i(x)dx^i + a_{ij}(x)dx^i \wedge dx^j + \ldots \qquad (12.219)$$

Furthermore, Kähler introduced the transformation rules for differential forms (i.e., the pull-back). Moreover, he defined the differential $d\omega$, and he proved that this operation is invariant under coordinate transformations.

In the 1930s and 1940s, the theory of fiber bundles emerged slowly. From the modern point of view, a differential form is a section

$$\sigma : \mathcal{M} \to G(\mathcal{M})$$

of the Grassmann bundle $G(\mathcal{M})$ over the manifold $\mathcal{M}$. As we will show later on, this is a localized variant of the section $\sigma : \mathbb{E}^3 \to G(\mathbb{E}^3)$ discussed on page 704.

Differential forms represent the analytic standard tool of modern geometry, differential topology, and physics.

A panorama of references can be found on page 798ff.

References

Introduction to the history of mathematics:

M. Kline, Mathematical Thought from Ancient to Modern Times, Vols. 1–3, Oxford University Press, 1990.

F. Klein, Development of Mathematics in the 19th Century, Springer, Berlin, 1926 (in German). English edition with a large appendix by R. Hermann, Math. Sci. Press, New York, 1979.

B. van der Waerden, A History of Algebra: From al-Khwarimzi to Emmy Noether, Springer, New York, 1984.

B. van der Waerden, The foundation of algebraic geometry from Severi to André Weil, Archive for History of Exact Sciences **7** (1971), 171–180.

J. Dieudonné, History of Algebraic Geometry, 400 B.C.–1985 A.D., Chapman, New York, 1985.

J. Dieudonnè, A History of Algebraic and Differential Topology, 1900–1960, Birkhäuser, Boston, 1989.

F. Dyson, George Green and Physics, Phys. World **6**, August 1993, 33–38.

Articles on the history of differential forms:[71]

M. Bacharach, On the History of Potential Theory, Vandenhoeck & Ruprecht, Göttingen, 1883 (in German).

J. Cross, Integral theorems in Cambridge mathematical physics, 1830–1855, pp. 122–148. In: P. Harmann, Wranglers and Physicists, Manchester University Press, 1985.

V. Katz, The history of Stokes' theorem, Mathematics Magazine **52** (1979), 146–156.

V. Katz, The history of differential forms from Clairaut to Poincaré, Historia Mathematica **8** (1981), 161–188.

V. Katz, Differential forms – Cartan to de Rham, Archive for History of Exact Sciences **33** (1985), 321–336.

V. Katz, Differential forms, Chap. 5. In: I. James, (Ed.), History of Topology, Elsevier, Amsterdam, 1999.

J. Crowe, A History of Vector Analysis, University of Notre Dame Press, Notre Dame, Indiana, 1967.

[71] I would like to thank Ivor Grattan–Guinness (London) and Erhard Scholz (Wuppertal) for drawing my attention to the papers by Bacharach, Cross, and Katz.

The following papers describe a beautiful long-term development in mathematics together with many interactions between mathematics and physics:

R. Hooke (1678), De potentia restitutiva (On the elastic force) (in Latin), London.

I. Newton (1687), Philosophiae naturalis principia mathematica (in Latin), London. See S. Chandrasekhar, Newton's Principia for the Common Reader, Oxford University Press, 1997.

A. Clairaut (1739), General investigations on integral calculus (in French), Histoire de l'Académie Royale des Science avec les Mémoires de Mathématique et Physique, 425–436.

A. Clairaut (1740), On the integration of first-order ordinary differential equations (in French), Histoire de l'Académie Royale des Science avec les Mémoires de Mathématique et Physique, 293–323.

L. Euler (1755), On the general principles of the motion of fluids (in French), Hist. de l'Acad. Berlin.

J. de Lagrange (1788), Analytical Mechanics (in French), Paris. New edition: 1813/1815. English edition, Kluwer, Dordrecht, 1997.

P. Laplace (1799), Celestial Mechanics, 5 vols. (in French), Paris, 1799–1825. English edition: Chelsea 1966.

C. Gauss (1813), On a new method for studying the gravitational force of homogeneous elliptical spheroids (in Latin), Göttingen, 24 pages.
German translation in: A. Wangerin (Ed.), On the Attraction of Homogeneous Ellipsoids: basic papers by Laplace (1782), Ivory (1809), Gauss (1813), Chasles (1838), and Dirichlet (1839), Ostwalds Klassiker, Vol. 19, Engelmann, Leipzig, 1890 (in German).

C. Navier (1822), On the law of the motion of fluids (in French), Mém. Acad. Sciences, Paris.

M. Ostrogradsky (1826), Proof of a theorem in integral calculus (in French). Paper presented to the Paris Academy in 1826. Russian translation: Istoriko-Matematicheskie Issledovanie, Vol. XVI (1965), 49–96.

C. Gauss (1827), General theory of surfaces (in Latin). In: C. Gauss, Collected Works, Vol. 5, pp. 217–256; 341–347, Göttingen, Vols. 1–12, 1863–1929. See also P. Dombrowski, 150 years after Gauss' 'Disquisitiones generales circa superficies curvas', Astérisque **62**, 1979.

A. Cauchy (1827), On pressure and tension in a solid body (in French), Exercises de mathématique, Vol. 2, 42–56.

A. Cauchy (1828), On the fundamental equations in elasticity (in French), Exercises de mathématique, Vol. 3, 160–187.

G. Green (1828), An Essay on the Application of Mathematical Analysis to the Theories of Electricity and Magnetism, Nottingham, 1828. In: G. Green, Mathematical Papers, pp. 3–115. Reprint, Hermann, Paris, 1903.

M. Ostrogradsky (1836), On the calculus of variations and on multiple integrals (in French), Crelles Journal **15**, 332–354.

C. Gauss (1839), General theorems on attracting and repelling forces which are proportional to the inverse square of distance (in German). In: C. Gauss, Collected Works, Vol. 5, pp. 195–242, Göttingen, 1863.

R. Hamilton (1843), On a new species of imaginary quantities connected with the theory of quaternions, Proc. Royal Irish Academy. See R. Hamilton, Mathematical Papers, Vol. 3, pp. 117–155, Cambridge University Press, 1963.

H. Grassmann (1844), The Calculus of Extension (in German). Reprint: Chelsea 1969.

A. Cayley (1845), On the theory of linear transformations, Cambridge Math. J. **4**, 193–209.

G. Stokes (1845), On the theories of the internal friction of fluids in motion, Cambridge Transactions.

A. Cauchy (1846), On integrals taken over a curve (in French), Comptes Rendus, Paris **23**, 251–255.

C. Jacobi (1846), Lectures on Analytical Dynamics (including Celestial Mechanics). Edited by The German Mathematical Society (DMV), Vieweg, Braunschweig 1996 (in German). English edition: Jacobi Lectures on Dynamics, Hindustan Books Agency, India, 2009.

B. Riemann (1851), Foundations of a general theory of functions of one complex variable (in German), Ph.D. thesis, Göttingen. In: B. Riemann, Collected Mathematical Works with commentaries. Springer, New York/Teubner, Leipzig, 1990.

B. Riemann (1854), On the hypotheses which lie at the foundation of geometry (in German), Göttinger Abhandlungen **13**, 272–287. English translation: see M. Spivak, A Comprehensive Introduction to Differential Geometry, Vol. 2, Publish or Perish, Boston 1979.

G. Stokes (1854), Smith's Prize Examination Paper, February 1854, Cambridge University Calendar. In: G. Stokes, Mathematical and Physical Papers, Vol. V, p. 320, Cambridge, 1905.

B. Riemann (1857), Theory of Abelian functions (in German), Crelles Journal für die reine und angewandte Mathematik **54**, 115–155.

C. Jordan (1870), Treatise on Substitutions and Algebraic Equations (in French), Gauthier-Villars, Paris.

F. Klein (1872), The Erlangen Program (in German). See H. Wußing (Ed.), Ostwalds Klassiker, Vol. **253**, Teubner, Leipzig, 1974. English edition: A Comparative review of recent researches in geometry, Bull. of the New York Math. Soc., July 1893.

J. Maxwell (1873), A Treatise on Electricity and Magnetism, London, 1873, Vols. 1, 2. Reprint: Dover, New York, 1954.

S. Lie (1874), On transformation groups (in German), Nachrichten Königl. Gesellschaft der Wissenschaften, Göttingen, 529–542.

F. Frobenius (1877), On the Pfaff problem (in German), Crelles Journal für die reine und angewandte Mathematik **82**, 230–315.

A. Cayley (1878), The theory of groups, Amer. J. of Math. **1**, 50–52.

W. Clifford (1878), Applications of Grassmann's extensive algebra, Amer. J. of Math. **1**, 350–358.

G. Darboux (1887), Lectures on the General Theory of Surfaces, Vols. 1–4 (in French), Gauthier-Villars, Paris. Reprint: Chelsea, New York, 1972.

H. Poincaré (1887), On the residues of double integrals (in French), Acta Mathematica **9**, 321–380.

S. Lie and F. Engel (1888), Theory of Transformation Groups (in German), Vols. 1–3, Teubner, Leipzig. Reprint: Chelsea Publ. Company, 1970.

V. Volterra (1889), On complex variables in multidimensional spaces (in Italian), Rendiconti della Accad. dei Lincei ser. IV, Vol. V, 158–165.

H. Poincaré (1892), New Methods in Celestial Mechanics, Vols. 1–3 (in French), Gauthier-Villars, Paris, 1892–1899. Reprint: Dover 1957.

É. Cartan (1894), On the structure of finite and continuous transformation groups (in French), Thèse, Nony, Paris.

H. Poincaré (1895), Analysis situs (topology) (in French), J. Math. École Polytechnique **1**, 1–121.

É. Delassus (1896), Extension of Cauchy's theorem to systems of partial differential equations (in French), Ann. Sci. École Norm. Sup. **13** (3), 421–467.

É. Cartan (1899), On certain differential expressions and on the Pfaff problem (in French), Annales École Normale Sup. **16**, 239–332.

D. Hilbert (1899), Foundations of Geometry (in German), Teubner, Leipzig, 12th edition, 1977.

D. Hilbert (1900), Mathematical Problems. Lecture delivered before the Second International Congress of Mathematicians at Paris 1900 (in German). English translation: Bull. Amer. Math. Soc. **8** (1902), 437–479.

D. Hilbert (1900), On the Dirichlet principle (in German), Jahresber. der Deutschen Mathematiker-Vereinigung **8**, 184–188.

É. Cartan (1901), On the solution of systems of total differential equations (in French), Ann. Sci. École Normale Sup. **18** (3), 241–311.

A. Forsythe (1901), A Theory of Differential Equations, Vols. 1–6, Cambridge.

A. Einstein (1905), On the electrodynamics of moving bodies (in German), Annalen der Physik **17**, 891–921. English translation in S. Hawking (Ed.), The Essential Einstein, pp. 4–31, Penguin Books, London, 2008.

D. Hilbert (1906), On the calculus of variations (in German), Math. Ann. **62**, 151–186.

C. Carathéodory (1909), On the foundations of thermodynamics (in German), Math. Ann. **67**, 355–386.

C. Riquier (1910), Systems of Partial Differential Equations (in French), Paris.

H. Weyl (1913), The Concept of a Riemann Surface (in German), Teubner, Leipzig, 1913. New edition with commentaries supervised by R. Remmert, Teubner, Leipzig, 1997. English edition: Addison Wesley, Reading, Massachusetts, 1955.

A. Einstein (1915), On general relativity: the field equations of gravitation (in German), Sitzungsber. Preuss. Akademie Wiss. Berlin, March 11, 1915, and December 2, 1915.

D. Hilbert (1915), The foundations of physics (in German), Nachr. Akad. Wiss. Göttingen, Math.-phys. Kl. 1915, 395–407; 1917, 53–76.

A. Einstein (1916), The foundation of the general theory of relativity (in German), Annalen der Physik **49**, 769–822. English translation in S. Hawking (Ed.), The Essential Einstein, pp. 46–98, Penguin Books, London, 2008.

É. Goursat (1917), On systems of total differential equations and on a generalization of the Pfaff problem (in French), Ann. Fac. Sci. Toulouse **7**(3), 1–58.

E. Noether (1918), Invariant variational problems (in German), Göttinger Nachrichten, Math.-phys. Klasse 1918, 235–257.

H. Weyl (1918), Space-Time-Matter (in German), Springer, Berlin, 1918, 8th edition, 1993. English edition: Dover, New York, 1953.

L. Lichtenstein (1921), New developments in potential theory (in German), Enzyklopädie der Mathematischen Wissenschaften, Vol. II, 3.2, Teubner, Leipzig.

É. Cartan (1922), Lectures on Integral Invariants (in French), Hermann, Paris, 1922.

É. Goursat (1922), Lectures on the Pfaff Problem (in French), Paris, 1922.

C. Carathéodory (1925), The method of geodesically equidistant surfaces (in German), Acta Math. **47**, 199–233.

É. Cartan (1926), Riemannian Geometry in an Orthogonal Frame. From lectures delivered by Élie Cartan at the Sorbonne (Paris) in 1926–1927. World Scientific Singapore, 2001.

F. Peter and H. Weyl (1927), On the completeness of the irreducible representations of compact continuous groups (in German), Math. Ann. **97**, 737–755.

É. Cartan (1928), Geometry of Riemannian Spaces (in French), Gauthier–Villars, Paris, 1928, 3rd printing 1988. English edition: Math. Scientific Press, Brookline, Massachusetts, 1983.

P. Dirac (1928), The quantum theory of the electron, Proc. Royal Soc. London **A117**, 610-624; **A118**, 351–361.

É. Cartan and A. Einstein (1929), Letters on absolute parallelism 1929–1932, Princeton University Press, 1979.

B. van der Waerden (1929), Spinor analysis (in German), Nachr. der königlichen Gesellschaft Göttingen, pp. 100–109.

H. Weyl (1929), Electron and gravitation (in German), Z. Phys. **56**, 330–352. See also H. Weyl, Electron and gravitation, Proc. National Acad. USA **15** (1929), 323–334.

H. Weyl (1929), The Theory of Groups and Quantum Mechanics (in German), Springer, Berlin. English edition: Dover, New York 1931.

G. de Rham (1931), On the topology of n-dimensional manifolds (in French), J. Math. Pures Appl. **10**(9), 115–200.

B. van der Waerden (1932), Group Theory and Quantum Mechanics (in German), Springer, Berlin. English edition: Springer, New York 1974 (spinor calculus and the Dirac equation in relativistic quantum mechanics).

L. Infeld and B. van der Waerden (1933), The wave equation of the electron in general relativity (in German), Sitzungsber. Preußische Akad. Wiss. Berlin, Math.-Phys. Klasse **9**, pp. 308–401.

E. Kähler (1933), On a remarkable Hermitean metric (in German), Abh. Seminar Univ. Hamburg **9**, 173–186.

E. Kähler (1934), Introduction to the Theory of Systems of Differential Forms (in German), Teubner, Leipzig, 1934.

R. Bauer and H. Weyl (1935), Spinors in n dimensions, Amer. J. Mathem. **57**, 425–449.

C. Carathéodory (1935), Calculus of Variations and Partial Differential Equations of First Order (in German), Teubner, 1935. English edition: Chelsea, New York, 1988. New extended German edition: Teubner, 1993. Edited with commentaries by R. Klötzler on new developments including control theory.

C. Carathéodory (1936), Beginning of research in the calculus of variations. Lecture presented at the tercentenary celebration of Harvard University in 1936. Reprinted in C. Carathéodory, Collected Works, Vol. 2, pp. 93–107, Beck, München, 1937.

C. Carathéodory (1937), Geometrical Optics (in German), Springer, Berlin.

G. de Rham (1936), Relations between topology and the theory of multiple integrals (in French), L'Enseignement math. **4**, 213–228.

É. Cartan (1938), Theory of Spinors (in French), Hermann, Paris. English edition: Hermann, Paris, 1966.

H. Weyl (1938), The Classical Groups: Their Invariants and Representations, Princeton University Press, 8th edition 1973.

E. Wigner (1939), On unitary representations of the inhomogeneous Lorentz group, Ann. of Math. **40**, 149–204.

H. Weyl (1940), The method of orthogonal projection in potential theory, Duke Math. J. **7**, 414–444.

W. Hodge (1941), The Theory and Applications of Harmonic Integrals, Cambridge University Press (second revised edition 1952).

H. Weyl (1943), On Hodge's theory of harmonic integrals, Ann. of Math. **44**, 1–6.

É. Cartan (1945), Systems of Exterior Differential Formes and Their Application in Geometry (in French), Hermann, Paris.

S. Chern (1945), A simple intrinsic proof of the Gauss–Bonnet formula for closed Riemannian manifolds, Ann. of Math. **45**, 747–752.

S. Chern (1946), Characteristic classes of Hermitean manifolds, Ann. of Math. **47**, 85–121.

C. Chevalley (1946), Theory of Lie Groups, Princeton University Press, 15th edition 1999.

L. Eisenhart (1946), Riemannian Geometry, Princeton University Press.

V. Bargmann (1947), Irreducible representations of the Lorentz group, Ann. of Math. **48**, 568–640.

C. Ehresmann (1950), Infinitesimal connections in a differentiable fiber space (in French), Colloque de Topologie, Bruxelles, pp. 29–55.

N. Steenrod (1951), The Topology of Fiber Bundles, Princeton University Press.

S. Eilenberg and N. Steenrod (1952), Foundations of Algebraic Topology, Princeton University Press.

V. Bargmann (1954), On unitary ray representations of continuous groups, Ann. of Math. **59**, 1–46.

F. Hirzebruch (1954), Arithmetic genera and the theorem of Riemann–Roch for algebraic varieties, Proc. Nat. Acad. Sci. USA, 110–114.

C. Yang and R. Mills (1954), Conservation of isotopic spin and isotopic spin invariance, Phys. Rev. **96**, 191–195.

G. de Rham (1955), Differentiable Manifolds (in French), Hermann, Paris, 1955.

F. Hirzebruch (1956), Topological Methods in Algebraic Geometry (in German), Springer, Berlin. English enlarged edition: Springer, New York, 1966.

A. Weil (1957), Kähler Manifolds (in French), Hermann, Paris.

E. Kähler (1961), The Dirac equation (in German), Abh. Deutsche Akademie Wiss. Berlin, Klasse für Mathematik, Physik und Technik, 1961, No. 1. See also E. Kähler, Mathematical Works, pp. 449–482, de Gruyter, Berlin, 2004.

E. Kähler (1962), The interior differential calculus (in German), Rend. Mat. Appl. **21** (5) (1962), 425–523. See also E. Kähler, Mathematical Works, pp. 483–595, de Gruyter, Berlin, 2004.

S. Kobayashi and K. Nomizu (1963), Foundations of Differential Geometry, Vols. 1, 2, Wiley, New York.

M. Atiyah and I. Singer (1963), The index of elliptic operators on compact manifolds, Bull. Amer. Math. Soc. **69**, 422–433.

M. Atiyah, R. Bott, and A. Shapiro (1964), Clifford modules, Topology **3**, 3–38.

J. Leray (1965), Differential and integral calculus on finite-dimensional complex manifolds (in French), Bull. Soc. Math. France **87** (59), 81–190 (the Cauchy residue calculus in higher dimensions).[72]

S. Yau (1978), On the Ricci-curvature of a complex Kähler manifold and the complex Monge–Ampère equation, Commun. Pure Appl. Math. **31**, 339–411.

S. Yau (1978), The Kähler–Einstein metric on open manifolds. In: The first Chern class and the solution of the Calabi conjecture, Astérisque **58** (1978), pp. 163–167.

E. Witten (1982), Supersymmetry and Morse theory, J. Diff. Geom. **17**, 661–692.

A. Connes (1985), Noncommutative differential geometry, Publ. Math. IHES **62**, 257–360.

M. Gromov (1985), Pseudo-holomorphic curves in symplectic manifolds, Invent. Math. **82**, 307–347.

E. Witten (1988), Holomorphic curves on manifolds of $SU(3)$-holonomy. In: S. Yau (Ed.), String Theory, World Scientific, Singapore, pp. 145–149.

A. Floer (1989), Witten's complex and infinite-dimensional Morse theory, J. Diff. Geometry **30** (1989), 207–221.

S. Donaldson and P. Kronheimer (1990), The Geometry of Four-Manifolds, Oxford University Press.

A. Connes and J. Lott (1990), Particle models and noncommutative geometry, Nucl. Phys. B (Proc. Suppl.), **18**, 29–47.

[72] See also B. Shabat, Introduction to Complex Analysis, Vol. 2, Amer. Math. Soc., Providence, Rhode Island, 1992, and
R. Hwa and V. Teplitz, Homology and Feynman Diagrams, Benjamin, Reading, Massachusetts, 1966.

A. Connes (1994), Noncommutative Geometry, Academic Press, New York.
N. Seiberg and E. Witten (1994), Electric-magnetic duality, monopole condensation, and confinement in $N = 2$ supersymmetric Yang–Mills theory, Nuclear Phys. **B426**, 19–52.

N. Seiberg and E. Witten (1994), Monopoles, duality and chiral symmetry breaking in $N = 2$ supersymmetric QCD, Nucl. Physics **B431**, 485–550.

E. Witten (1994), Monopoles and four-manifolds, Math. Research Letters **1**, 769–796.

E. Witten (1999), Witten's Lectures on Three-Dimensional Topological Quantum Field Theory. Edited by Sen Hu, World Scientific, Singapore, 1999.

S. Donaldson (1996), The Seiberg–Witten equations and 4-manifold topology, Bull. Amer. Math. Soc. **33**, 45–70.

S. Donaldson (2002), Floer Homology Groups, Cambridge University Press.

C. Taubes (2000), Seiberg–Witten and Gromov Invariants for Symplectic 4-Manifolds, International Press, Boston.

C. Taubes (2007), The Seiberg–Witten equations and the Weinstein conjecture I, II: Geom. Topol. **11** (2007), 2117–2002, **13** (2009), 1337–1417.

P. Kronheimer and T. Mirowka (2007), Monopoles and Three-Manifolds, Cambridge University Press.

B. Chow, P. Lu, and L. Ni (2006), Hamilton's Ricci Flow, Amer. Math. Soc., Providence, Rhode Island.

H. Cao and X. Zhu (2006), A complete proof of the Poincaré and geometrization conjectures – Application of the Hamilton–Perelman Theory of Ricci flow, Asian J. Math. **10**(2).

H. Cao, S. Yau, and X. Zhu (2006), Structure of Three-dimensional Space: The Poincaré and Geometrization Conjectures, International Press, Boston.

J. Morgan and G. Tian (2007), Ricci Flow and the Poincaré Conjecture, Amer. Math. Soc., Providence, Rhode Island/Clay Mathematics Institute, Cambridge, Massachusetts.

J. Morgan and F. Fong, Ricci Flow and Geometrization of 3-Manifolds, Amer. Math. Soc., Rhode Island, 2010 (survey).

A. Connes and M. Marcolli (2008), Noncommutative Geometry, Quantum Fields, and Motives, Amer. Math. Soc., Providence, Rhode Island.

M. Marcolli (2009), Feynman Motives: Renormalization, Algebraic Varieties, and Galois Symmetries, World Scientific, Singapore.

A Panorama of Modern Approaches

General survey:

V. Varadarajan, Euler through Time: A New Look at Old Themes, Amer. Math. Soc., Providence, Rhode Island, 2006.

K. Maurin, The Riemann Legacy: Riemannian Ideas in Mathematics and Physics of the 20th Century, Kluwer, Dordrecht, 1997.

Potential theory:

L. Helms, Potential Theory, Springer, New York, 2009.

Introduction to modern differential geometry:

T. Ivey and J. Landsberg, Cartan for Beginners: Differential Geometry via Moving Frames and Exterior Differential Systems, Amer. Math. Soc., Providence, Rhode Island, 2003.

R. Bryant et al., Exterior Differential Systems, Springer, New York, 1991.

R. Sharpe, Differential Geometry: Cartan's Generalization of Klein's Erlangen Program, Springer, New York, 1997.

J. Jost, Riemannian Geometry and Geometric Analysis, 5th edition, Springer, Berlin, 2008.

V. Ivancevic and T. Invancevic, Differential Geometry: A Modern Introduction (with applications to physics), World Scientific, Singapore, 2007.

Introduction to Riemann surfaces (elliptic and algebraic functions, Jacobi's and Riemann's theta functions, homology and genus, cohomology and characteristic classes, Abelian differentials/Abelian integrals and the Riemann–Roch theorem, divisors and line bundles (fiber bundles), sheaf cohomology, periods and the Jacobian variety, conformal geometry, Riemann's moduli spaces and Teichmüller spaces, elliptic and algebraic curves, uniformization, birational geometry):

M. Waldschmidt et al. (Eds.), From Number Theory to Physics, Springer, New York, 1995 (e.g., detailed survey articles on compact Riemannian surfaces, elliptic curves, modular forms, Jacobian varieties, and Abelian surfaces).

J. Jost, Compact Riemann Surfaces: An Introduction to Contemporary Mathematics, Springer, Berlin, 1997.

O. Forster, Lectures on Riemann Surfaces, Springer, Berlin, 1981.

M. Farkas and I. Kra, Riemann Surfaces, Springer, New York, 1992.

R. Narasimhan, Compact Riemann Surfaces, Birkhäuser, Basel, 1996.

Complex manifolds:

R. Wells, Differential Analysis on Complex Manifolds, Springer, New York, 2008.

P. Griffith and H. Harris, Principles of Algebraic Geometry, Wiley, New York, 1978.

E. Arbarello, M. Cornalba, P. Griffith, and J. Harris, Geometry of Algebraic Curves, Springer, New York, 1985.

W. Ebeling, Functions of Several Complex Variables and Their Singularities, Amer. Math. Soc., Providence Rhode Island, 2007.

Introduction to Kähler manifolds:

J. Jost, Riemannian Geometry and Geometric Analysis, Springer, Berlin, 2008 (introduction).

C. Voisin, Hodge Theory and Complex Algebraic Theory. Vol. I: Compact Kähler Manifolds, Vol. II: Algebraic Varieties, Cambridge University Press, 2002.

W. Ballmann, Lectures on Kähler Manifolds, European Mathematical Society, 2006.

A. Moroianu, Lectures on Kähler Geometry, Cambridge University Press, 2007.

K. Becker, M. Becker, and J. Schwarz, String Theory and M-Theory, Cambridge University Press, 2006.

P. Fre and P. Soriani, The $N = 2$ Wonderland: From Calabi–Yau Manifolds to Topological Field Theories, World Scientific, Singapore, 1995.

Mirror symmetry:

S. Yau, Essays on Mirror Manifolds, International Press, Hong Kong 1993.

C. Voisin, Mirror Symmetry, Soc. Math. de France, Paris, 1996 (in French).

Characteristic classes:

Y. Choquet-Bruhat, C. DeWitt-Morette, and M. Dillard-Bleick, Analysis, Manifolds, and Physics. Vol. 1, Elsevier, Amsterdam, 1996 (introduction).

J. Milnor and J. Stasheff, Characteristic Classes, Princeton University Press, 1994.

D. Husemoller, Fiber Bundles, Springer, New York, 1994.

R. Bott and L. Tu, Differential Forms in Algebraic Topology, Springer, New York, 1982.

I. Madsen and J. Tornehave, From Calculus to Cohomology: de Rham Cohomology and Characteristic Classes, Cambridge University Press, 1997.

Introduction to the Atiyah–Singer index theorem:

P. Shanahan, The Atiyah–Singer Index Theorem, Springer, Berlin, 1978.

P. Gilkey, Invariance Theory, the Heat Equation, and the Atiyah–Singer Index Theorem, CRC Press, Boca Raton, Florida, 1995.

Spin geometry and introduction to the Seiberg–Witten theory:

J. Jost, Riemannian Geometry and Geometric Analysis, 5th edition, Springer, Berlin, 2008.

J. Moore, Lectures on Seiberg–Witten Invariants, Springer, Berlin, 1996.

H. Lawson and M. Michelsohn, Spin Geometry, Princeton University Press, 1994.

Pseudo-holomorphic curves:

D. McDuff and D. Salamon, J-holomorphic Curves and Quantum Cohomology, Amer. Math. Soc., Providence, Rhode Island, 1994.

D. McDuff and D. Salamon, Introduction to Symplectic Topology, Clarendon Press, Oxford, 1998.

Collected works: The following volumes contain fundamental papers which strongly influenced the development of mathematics in the 20th century:

É. Cartan, Oeuvres complètes, Vols. 1–3, Gauthier-Villars, Paris, 1952.

H. Weyl, Collected Works, Vols. 1–4, Springer, New York, 1968.

M. Atiyah, Collected Works, Vols. 1–6, Cambridge University Press, 2004, Vol. 1: no subtitle Vol. 2: K-Theory, Vol. 3: Index Theory 1, Vol. 4: Index Theory 2, Vol. 5: Gauge Theories, Vol. 6 no subtitle.

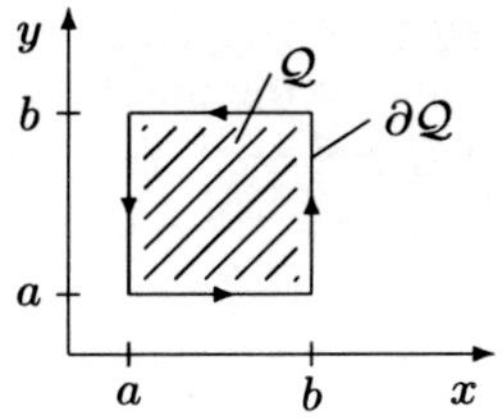

(a) square

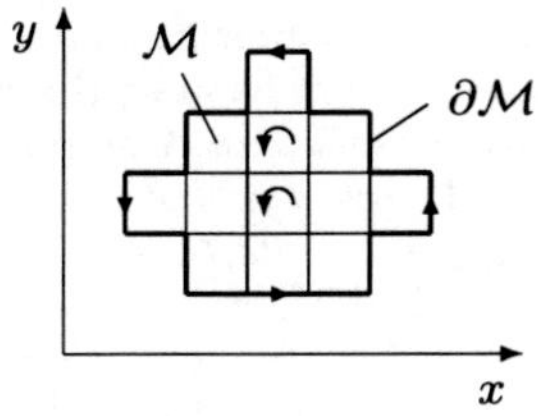

(b) union of squares

Fig. 12.29. Special regions

Problems

12.1 *The main theorem of calculus in two dimensions.*
(a) Prove (12.13) on page 671 for the square $\mathcal{Q} := \{(x, y) \in \mathbb{R}^2 : a \le x, y \le b\}$.
(b) Prove (12.13) for the region $\mathcal{M}$ depicted in Fig. 12.29.
(c) Study the full proof for (12.13) in V. Zorich, Analysis, Vol. II, Sect. 13.3.1, Springer, New York, 2003.

Solution. Ad (a). Integration by parts for one-dimensional integrals yields

$$\int_{\mathcal{Q}} U_x dx dy = \int_a^b \left(\int_a^b U_x(x, y) dx \right) dy$$
$$= \int_a^b (U(b, y) - U(a, y))\, dy = \int_{\partial\mathcal{Q}} U dy.$$

Ad (b). The region $\mathcal{M}$ is the sum of squares $\mathcal{Q}_1, \mathcal{Q}_2, \ldots$ By (a), we get

$$\int_{\mathcal{M}} U_x dx dy = \sum_j \int_{\mathcal{Q}_j} U_x dx dy = \sum_j \int_{\partial\mathcal{Q}_j} U dy.$$

This is equal to $\int_{\partial\mathcal{M}} U dy$, since the other contributions cancel each other by opposite orientation of the boundaries of adjoining squares.

12.2 *Pull-back.* Prove (12.22) on page 673.

Solution: We have
- $\omega = U dx + V dy$ and $d\omega = (V_x - U_y)\, dx \wedge dy$,
- $\tau^*\omega = (U x_u + V y_u) du + (U x_v + V y_v) dv$,
- $d(\tau^*\omega) = \big((U x_v + V y_v)_u - (U x_u + V y_u)_v\big)\, du \wedge dv$.

By the product rule, $(U x_v)_u = U_u x_v + U x_{vu}$, and so on. Noting that $x_{uv} = x_{vu}$, we get

$$d(\tau^*\omega) = (U_u x_v + V_u y_v - U_v x_u - V_v y_u)\, du \wedge dv.$$

By the chain rule, $U_u = U_x x_u + U_y y_u$. This yields

$$d(\tau^*\omega) = (V_x - U_y)(x_u y_v - x_v y_u)\, du \wedge dv = \tau^*(d\omega).$$

12.3 *Pull-back of covector fields.* Prove (12.80) on page 706.

Solution: Use (11.6) on page 662, and argue as in Problem 12.2.

12.4 *Pull-back of differential forms.* Prove Prop. 12.5 on page 706.
Hint: Use the definition of the wedge product and of the pull-back.

12.5 *The Hodge Laplacian.* Compute $\Delta(Udx + Vdy)$ and $\Delta(W\, dx \wedge dy)$.
Solution: (I) Set $\omega := Udx + Vdy$. Then:
- $d\omega = (V_x - U_y)\, dx \wedge dy$,
- $d^*\omega = -U_x - V_y$,
- $dd^*\omega = -(U_{xx} + V_{yx})dx - (U_{xy} + V_{yy})dy$,
- $d^*d\omega = (V_x - U_y)_y dx - (V_x - U_y)_x dy = (V_{xy} - U_{yy})dx + (U_{yx} - V_{xx})dy$.
Hence $\Delta\omega = (dd^* + d^*d)\omega = -(U_{xx} + U_{yy})dx - (V_{xx} + V_{yy})dy$.

(II) Set $\gamma := W\, dx \wedge dy$. Then $d\gamma = 0$. Moreover, we have $d^*\gamma = W_y dx - W_x dy$, and

$$\Delta\gamma = dd^*\gamma = -(W_{yy} + W_{xx})\, dx \wedge dy.$$

12.6 *Invariant definition of the exterior Cartan differential via the Lie derivative and the Lie algebra of velocity vector fields.* Let $\omega_P = U(x,y)dx_P + V(x,y)dy_P$. Use the invariant definition (12.25) in order to show that

$$d\omega_P = (V_x(x,y) - U_y(x,y))\, dx_P \wedge dy_P. \tag{12.220}$$

Solution: Set $\mathbf{v} = a\mathbf{i}_P + b\mathbf{j}_P$, and $\mathbf{w} = A\mathbf{i}_P + B\mathbf{j}_P$. Then:
- $\omega(\mathbf{v}) = Ua + Vb$,
- $\mathcal{L}_{\mathbf{w}}\omega(\mathbf{v}) = \left(A\frac{\partial}{\partial x} + B\frac{\partial}{\partial y}\right)(Ua + Vb) = AU_x a + AUa_x + BV_y b + BVb_y$,
- $\mathcal{L}_{\mathbf{v}}\omega(\mathbf{w}) = aU_x A + aUA_x + bV_y B + bVB_y$,
- $[\mathbf{v}, \mathbf{w}] = \mathcal{L}_{\mathbf{v}}\mathbf{w} - \mathcal{L}_{\mathbf{w}}\mathbf{v} = \left(a\frac{\partial}{\partial x} + b\frac{\partial}{\partial y}\right)\mathbf{w} - \left(A\frac{\partial}{\partial x} + B\frac{\partial}{\partial y}\right)\mathbf{v}$. Hence

$$[\mathbf{v}, \mathbf{w}] = (aA_x + bA_y - Aa_x - Ba_y)\mathbf{i}_P + (aB_x + bB_y - Ab_x - Bb_y)\mathbf{j}_P,$$

- $\omega([\mathbf{v}, \mathbf{w}]) = U(aA_x + bA_y - Aa_x - Ba_y) + V(aB_x + bB_y - Ab_x - Bb_y)$.

By (12.25), $d\omega(\mathbf{v}, \mathbf{w}) = \mathcal{L}_{\mathbf{v}}\omega(\mathbf{w}) - \mathcal{L}_{\mathbf{w}}\omega(\mathbf{v}) - \omega([\mathbf{v}, \mathbf{w}])$. After cancelling many terms, we get

$$d\omega(\mathbf{v}, \mathbf{w}) = (V_x - U_y)(aB - Ab) = (V_x - U_y)\,(dx_P \wedge dy_P)(\mathbf{v}, \mathbf{w}).$$

This is the claim (12.220).

12.7 *Differential forms and the classical vector calculus.* Prove all the formulas summarized in Sect. 12.12.3 on page 773.

12.8 *Harmonic forms.* Prove all the formulas summarized in Sect. 12.12.5 on page 775.

12.9 *Conservation of probability in quantum mechanics.* Prove Proposition 12.4 on page 697.
Solution: To simplify the computation, set $\hbar := 1$ and $2m := 1$. By the Schrödinger equation,

$$\mathrm{i}\dot{\varrho} = \mathrm{i}\dot{\psi}\psi^\dagger + \mathrm{i}\psi\dot{\psi}^\dagger = \psi^\dagger(-\boldsymbol{\partial}^2 + U)\psi - \psi(-\boldsymbol{\partial}^2 + U)\psi^\dagger.$$

Moreover, $\mathrm{i}\,\mathrm{div}\,\mathbf{J} = \mathrm{i}\boldsymbol{\partial}\mathbf{J} = \boldsymbol{\partial}(\psi^\dagger\boldsymbol{\partial}\psi - \psi\boldsymbol{\partial}\psi^\dagger)$. Hence

$$\mathrm{i}\,\mathrm{div}\,\mathbf{J} = \boldsymbol{\partial}\psi^\dagger \cdot \boldsymbol{\partial}\psi + \psi^\dagger\boldsymbol{\partial}^2\psi - \boldsymbol{\partial}\psi \cdot \boldsymbol{\partial}\psi^\dagger - \psi\boldsymbol{\partial}^2\psi^\dagger = -\mathrm{i}\dot{\varrho}.$$

12.10 *Matrix-valued differential forms.* Let

$$A = (a^i_j), \quad B = (b^i_j)$$

be $(n \times n)$-matrices whose entries a^i_j (resp. b^i_j) are differential p-forms (resp. q-forms).[73] Quite naturally, we define both the Cartan differential $dA = (da^i_j)$ and the wedge product

$$A \wedge B = (c^i_j), \qquad \text{where} \quad c^i_j := a^i_s \wedge b^s_j.$$

This corresponds to the formula for the usual matrix product by replacing $a^i_s b^s_j$ with the corresponding wedge product $a^i_s \wedge b^s_j$. Show that

$$A \wedge B = (-1)^{pq} B \wedge A,$$

$$d(A \wedge B) = dA \wedge B + (-1)^p A \wedge dB. \tag{12.221}$$

Hint: Use the corresponding relations for the entries.

12.11 *Lie algebra-valued differential forms.* Let $\mathcal{L}$ be a real n-dimensional algebra with the basis elements $B_1, ..., B_n$. By a differential p-form with values in $\mathcal{L}$, we understand

$$A = a^i B_i$$

where a^i_j are usual differential p-forms. Here, we sum over $i = 1, ..., n$. If, in addition, B is a differential q-form with values in $\mathcal{L}$, then we define the Lie product

$$[A, B] := (a^i \wedge b^j)[B_i, B_j],$$

and the Cartan differential $dA = da^i B_i$. Show that this definition is independent of the choice of the basis $B_1, ..., B_n$ and that the following hold true.
(i) $d[A, B] = [dA, B] + (-1)^p [A, dB]$.
(ii) $[A, B] = -(-1)^{pq}[B, A]$.
(iii) For the Lie algebra $\mathcal{L} = gl(n, \mathbb{R})$ of real $(n \times n)$-matrices,

$$[A, B] = A \wedge B - (-1)^{pq} B \wedge A.$$

In particular, if A is a 1-form, then $[A, A] = 2A \wedge A$.

Motivated by (iii), for a differential 1-form A on a real finite-dimensional manifold M with values in a real finite-dimensional Lie algebra $\mathcal{L}$, we define

$$A \wedge A := \tfrac{1}{2}[A, A].$$

Explicitly, $(A \wedge A)_P(\mathbf{v}, \mathbf{w}) = [A_P(\mathbf{v}), A_P(\mathbf{w})]$ for all velocity vectors $\mathbf{v}, \mathbf{w}$ at the tangent space $T_P M$.

Solution: Ad (i), (ii). Apply the relations for usual differential forms.

Ad (iii). Choose the basis matrices $B^i_j = (\beta^k_l)$ where $\beta^i_j := 1$ and $\beta^k_l = 0$ otherwise. Then

$$A \wedge B = (a^i_s \wedge b^r_j) B^s_i B^j_r.$$

On the other hand, $[A, B] = (a^i_s \wedge b^r_j)(B^s_i B^j_r - B^j_r B^s_i)$. This is equal to

$$(a^i_s \wedge b^r_j) B^s_i B^j_r - (-1)^{pq} (b^r_j \wedge a^i_s) B^j_r B^s_i.$$

12.12 *Group commutator and Lie product.* In the following problems, let $\mathcal{G}$ be a regular Lie matrix group. Consider two smooth curves $G = G(t)$ and $H = H(t)$ on $\mathcal{G}$ with $G(0) = H(0) = I$. Show that

$$G(t)H(t)G(t)^{-1}H(t)^{-1} = I + t^2[\dot{G}(0), \dot{H}(0)] + o(t^2), \quad t \to 0.$$

---

[73] In what follows, we will use the Einstein convention, that is, we sum over equal upper and lower Latin indices from 1 to n.

Solution: The Neumann series tells us that if $||C|| < 1$ then

$$(I+C)^{-1} = I - C + C^2 + ...$$

Hence

$$G(t) = I + tA + t^2B + ..., \qquad G(t)^{-1} = I - tA + t^2(A^2 - B) + ...$$

Use a similar expression for $H(t)$ and multiply the expressions with each other.

12.13 *Global parallel transport on a Lie group.* The theory of Lie groups is governed by the fact that there exist two global parallel transports called left translation and right translation. For each group element $\mathbf{G} \in \mathcal{G}$, define

$$\boxed{L_{\mathbf{G}}G := \mathbf{G}G \qquad \text{for all} \quad G \in \mathcal{G}.}$$

This way, we get the left translation $L_{\mathbf{G}} : \mathcal{G} \to \mathcal{G}$ along with the linearization

$$L'_{\mathbf{G}}(H) : T_H\mathcal{G} \to T_{\mathbf{G}H}\mathcal{G}$$

at the point $H \in \mathcal{G}$. Show the following.
(i) $L'_{\mathbf{G}}(H)V = \mathbf{G}V$ for all tangent vectors $V \in T_H\mathcal{G}$.
(ii) The operator $L'_{\mathbf{G}}(H)$ is a linear isomorphism.
(iii) Each tangent vector W at the point $G \in \mathcal{G}$ can be uniquely represented as

$$W = GV \qquad \text{where} \quad V \in \mathcal{LG}.$$

Solution: Ad (i). Consider a curve $G = G(t)$ on the group $\mathcal{G}$ with

$$G(0) = H \qquad \text{and} \qquad \dot{G}(0) = V.$$

Differentiating $L_{\mathbf{G}}G(t) = \mathbf{G}G(t)$ at time $t = 0$, $L'_{\mathbf{G}}(H)V = \mathbf{G}V$.
Ad (ii), (iii). The inverse operator to the operator $V \mapsto \mathbf{G}V$ is the operator $W \mapsto \mathbf{G}^{-1}W$. □
The inverse operator $L'_{\mathbf{G}}(I)^{-1} : T_{\mathbf{G}}\mathcal{G} \to \mathcal{LG}$ is called the Maurer–Cartan operator $\mathsf{M}_{\mathbf{G}}$ at the point $\mathbf{G}$. Explicitly,

$$\boxed{\mathsf{M}_{\mathbf{G}}(W) = \mathbf{G}^{-1}W \qquad \text{for all} \quad W \in T_{\mathbf{G}}\mathcal{G}.}$$

Alternatively, $\mathsf{M}_{\mathbf{G}}(\mathbf{G}V) = V$ for all $V \in \mathcal{LG}$.
Similarly, we define the right translation $R_{\mathbf{G}} : \mathcal{G} \to \mathcal{G}$ by setting

$$R_{\mathbf{G}}G := G\mathbf{G} \qquad \text{for all} \quad G \in \mathcal{G}$$

and fixed $\mathbf{G} \in \mathcal{G}$. All the results above remain then valid iff we replace left multiplication by right multiplication. In particular, each tangent vector W at the point $G \in \mathcal{G}$ can be uniquely represented as

$$W = VG \qquad \text{where} \quad V \in \mathcal{LG}.$$

12.14 *The directional derivative of a left-invariant temperature field on a Lie group.* Let $T : \mathcal{G} \to \mathbb{R}$ be a smooth function on the Lie group $\mathcal{G}$ which is left invariant, that is, the function T is invariant under left translations,

$$T(\mathbf{G}G) = T(G) \qquad \text{for all} \quad G, \mathbf{G} \in \mathcal{G}.$$

Show that, for all points $H \in \mathcal{G}$ and all tangent vectors $V \in T_H\mathcal{G}$,

$$\boxed{\mathcal{L}_V T(H) = 0.}$$

This result remains true if we replace left translations by right translations. The claim generalizes the fact that, for a smooth function $f : \mathbb{R} \to \mathbb{R}$, it follows from $f(x+a) = f(a)$ for all $x, a \in \mathbb{R}$ that $f'(x) \equiv 0$.
Solution: By definition of the directional derivative,

$$\mathcal{L}_V T(G) = \dot{w}(0)$$

where $w(t) := T(G(t))$. Here, the smooth curve $G = G(t)$ on the Lie group $\mathcal{G}$ has the property

$$G(0) = H, \quad \dot{G}(0) = V.$$

The point is that the directional derivative is independent of the choice of this curve. Therefore, we can use the special curve $G(t) := \mathbf{G}(t)V$ where $\mathbf{G}(0) = I$. Then $w(t) = \text{const}$, and hence $\dot{w}(0) = 0$.

12.15 *The Lie algebra of left-invariant velocity vector fields on a Lie group.* The velocity vector field $\mathbf{V} = \mathbf{V}(G)$ on the Lie group $\mathcal{G}$ is called left invariant iff

$$\boxed{\mathbf{V}(\mathbf{G}G) = \mathbf{G}\mathbf{V}(G) \qquad \text{for all} \quad G, \mathbf{G} \in \mathcal{G}.}$$

Each left-invariant velocity vector field has the form

$$\mathbf{V}(G) = GV \qquad \text{for all} \quad G \in \mathcal{G}$$

where V is a fixed element of the Lie algebra $\mathcal{LG}$. Thus, there exists a one-to-one map between the elements V of the Lie algebra $\mathcal{LG}$ and the left-invariant vector fields $\mathbf{V}(G) = G\mathbf{V}$ on the Lie group $\mathcal{G}$. For fixed $V \in \mathcal{LG}$ and fixed $G_0 \in \mathcal{G}$, consider the differential equation

$$\boxed{\dot{G}(t) = G(t)V, \quad t \in \mathbb{R}, \quad G(0) = G_0}$$

generated by the left-invariant velocity field GV. This differential equation has the unique solution

$$\boxed{G(t) = G_0 \mathrm{e}^{tV} \qquad \text{for all} \quad t \in \mathbb{R}.}$$

In terms of physics, the curve $G = G(t)$ is the trajectory of a fluid particle which has the velocity vector GV at the point G. Define

$$F_t G_0 := G(t), \qquad t \in \mathbb{R},$$

and show the following:
(i) The linearization $F_t'(G_0) : T_{G_0} \to T_{F_t G_0}$ is given by $F_t'(G_0)W = W\mathrm{e}^{tV}$.
(ii) Let $\mathbf{V}(G) := GV$ and $\mathbf{W}(G) := GW$ be left-invariant velocity vector fields for all $G \in \mathcal{G}$ and fixed $V, W \in \mathcal{LG}$. The Lie derivative reads then as

$$\boxed{\mathcal{L}_{\mathbf{V}}\mathbf{W}(G) = G(VW - WV) \qquad \text{for all} \quad G \in \mathcal{G}.}$$

Therefore, the left-invariant velocity vector fields on $\mathcal{G}$ form a Lie algebra which is isomorphic to $\mathcal{LG}$. The isomorphism is given by the map $V \mapsto GV$.
Solution: Ad (i). Consider a curve $G = G(\tau)$ with $G(0) = G_0$ and $\dot{G}(0) = W$. Differentiating the equation

$$F_t G(\tau) = G(\tau)e^{tV}$$

at time $\tau = 0$, we get the claim.
Ad (ii). By definition, the Lie derivative reads as

$$\mathcal{L}_{\mathbf{V}}\mathbf{W}(G) := \frac{dF_t^*(G)}{dt}_{|t=0}$$

where

$$F_t^*\mathbf{W}(G) := F'_{-t}\mathbf{W}(F_tG)$$

is called the pull-back of the velocity vector field $\mathbf{W}$ with respect to the flow F_t generated by the velocity vector field $\mathbf{V}$. Observe that the operator F'_{-t} transports the velocity vector $\mathbf{W}$ at the point F_tG back to the point G. Explicitly,

$$F_t^*\mathbf{W}(G) = (F_tG)We^{-tV} = Ge^{tV}We^{-tV}.$$

Differentiating this at time $t = 0$, we get $G(VW - WV)$.
12.16 *The Maurer–Cartan form* M. Set $\mathsf{M}_{\mathbf{G}} := L'_{\mathbf{G}}$. Explicitly,

$$\mathsf{M}_{\mathbf{G}}(W) = \mathbf{G}^{-1}W \qquad \text{for all} \quad W \in T_H\mathcal{G},\ \mathbf{G}\mathcal{G}.$$

Show that
(i) $L_{\mathbf{G}}^*\mathsf{M} = \mathsf{M}$, and
(ii) $R_{\mathbf{G}}^* = \mathbf{G}^{-1}\mathsf{M}\mathbf{G}$.
Solution: Ad (i). We have to show that

$$\mathsf{M}_{\mathbf{G}H}(\mathbf{G}H) = \mathsf{M}_H(W) \qquad \text{for all} \quad W \in T_H\mathcal{G},\ H, \mathbf{G} \in \mathcal{G}.$$

In fact, $\mathsf{M}_{\mathbf{G}H}(\mathbf{G}H) = (\mathbf{G}H)^{-1}(\mathbf{G}W) = H^{-1}W$.
Ad (ii). We have to show that

$$\mathsf{M}_{H\mathbf{G}}(H\mathbf{G}) = \mathbf{G}^{-1}\mathsf{M}_H(W)\mathbf{G} \qquad \text{for all} \quad W \in T_H\mathcal{G},\ H, \mathbf{G} \in \mathcal{G}.$$

In fact, $\mathsf{M}_{H\mathbf{G}}(H\mathbf{G}) = (H\mathbf{G})^{-1}(W\mathbf{G}) = \mathbf{G}^{-1}H^{-1}W\mathbf{G}$.
12.17 *The trivial curvature of a Lie group.* Show that

$$\boxed{d\mathsf{M} + \mathsf{M} \wedge \mathsf{M} = 0.}$$

In terms of gauge theory, we set

$$\mathsf{F} := d\mathsf{M} + \mathsf{M} \wedge \mathsf{M}.$$

Obviously, $\mathsf{F} = 0$. The Maurer–Cartan form M also called the gauge potential (or the connection form) to the trivial curvature form F of the Lie group $\mathcal{G}$.
Solution: Let us choose two left-invariant velocity vector fields $\mathbf{V}(G) := GV$ and $\mathbf{W}(G) := GW$ on $\mathcal{G}$ for fixed $V, W \in \mathcal{LG}$. By Cartan's magic formula

$$d\mathsf{M}_G(\mathbf{V}, \mathbf{W}) = \mathcal{L}_{\mathbf{V}}\mathsf{M}_G(\mathbf{W}) - \mathcal{L}_{\mathbf{W}}\mathsf{M}_G(\mathbf{V}) - \mathsf{M}_G([\mathbf{V}, \mathbf{W}]).$$

Here, $\mathcal{L}_{\mathbf{V}}\mathsf{M}_G(\mathbf{W})$ denotes the directional derivative of the function

$$G \mapsto \mathsf{M}_G(\mathbf{W})$$

at the point G in direction of the vector $\mathbf{V}(G)$. Observe that this function is left invariant,[74] and hence the corresponding directional derivative vanishes, by Problem 12.14. Therefore,

[74] In fact, $\mathsf{M}_{\mathbf{G}G}(\mathbf{W}(\mathbf{G}G)) = W$ for all $\mathbf{G}, G \in \mathcal{G}$.

$$d\mathsf{M}_G(\mathbf{V}, \mathbf{W}) = -\mathsf{M}_G([\mathbf{V}, \mathbf{W}]).$$

This implies $d\mathsf{M}_G(\mathbf{V}, \mathbf{W}) = -[V, W]$. By Problem 12.11,

$$(\mathsf{M} \wedge \mathsf{M})_G(\mathbf{V}, \mathbf{W}) = [\mathsf{M}_G(\mathbf{V}), \mathsf{M}_G(\mathbf{W})] = [V, W].$$

Hence $d\mathsf{M} = -\mathsf{M} \wedge \mathsf{M}$.
The same result can be obtained by applying a mnemonic approach to the Maurer–Cartan form which is frequently used by physicists. To this end, we write

$$\mathsf{M}_G := G^{-1}dG, \qquad (12.222)$$

and we add the rule $dG(\dot{G}(t)) := \dot{G}(t)$ along with $d(dG) = 0$, and

$$d(G^{-1}) = -G^{-1}dG \cdot G^{-1}.$$

The latter identity can be obtained formally from $GG^{-1} = I$, and hence

$$dG \cdot G^{-1} + Gd(G^{-1}) = 0.$$

The relation $d\mathsf{M} = -\mathsf{M} \wedge \mathsf{M}$ follows then from

$$d\mathsf{M} = d(G^{-1}G) = d(G^{-1}) \wedge dG.$$

This is equal to $-G^{-1}dG \cdot G^{-1} \wedge dG = -G^{-1}dG \wedge G^{-1}dG = -\mathsf{M} \wedge \mathsf{M}$.

12.18 *Duality on Lie groups and the Maurer–Cartan structural equations.* For given differential 1-form ω on the Lie group $\mathcal{G}$, define

$$\mu_H(W) := \omega_{\mathbf{G}H}(\mathbf{G}W)$$

for all tangent vectors $W \in T_H\mathcal{G}$ and all $H, \mathbf{G} \in \mathcal{G}$. The differential 1-form

$$L^*_{\mathbf{G}}\omega := \mu$$

is called the pull-back of ω with respect to the left translation $L_{\mathbf{G}}$. Naturally enough, the differential form ω is called left invariant iff

$$L^*_{\mathbf{G}}\omega = \omega \qquad \text{for all} \quad \mathbf{G} \in \mathcal{G}.$$

Explicitly, $\omega_H(W) = \omega_{\mathbf{G}H}(\mathbf{G}W)$ for all $W \in T_H\mathcal{G}$ and all $H, \mathbf{G} \in \mathcal{G}$. Let $V_1, ..., V_n$ be a basis of the Lie algebra $\mathcal{LG}$. For fixed index $i = 1, ..., n$ and arbitrary real numbers a^j, set

$$V^i(a^jV_j) := a^i.$$

The linear functionals $V^1, .., V^n : \mathcal{LG} \to \mathbb{R}$ form a basis of the dual space to $\mathcal{LG}$ which is called the cobasis to $V_1, ..., V_n$. Letting

$$\theta^i_G(GV) := V^i(V) \qquad \text{for all} \quad V \in \mathcal{LG},\ G \in \mathcal{G},$$

we get left-invariant differential forms $\theta^1, ..., \theta^n$ on the Lie group $\mathcal{G}$ which form a basis for all left-invariant tangent vector fields on $\mathcal{G}$. Show that there hold the following Cartan structural equations

$$\boxed{d\theta^k + \tfrac{1}{2}c^k_{ij}\theta^i \wedge \theta^j = 0, \qquad k = 1, .., n}$$

where the real numbers c_{ij}^k defined by

$$[V_i, V_j] = c_{ij}^k V_k, \qquad i, j = 1, ..., n$$

are called the structure constants of the Lie algebra $\mathcal{LG}$ with respect to the basis $V_1, ..., V_n$.
Solution: Using Cartan's magic formula, it follows as in Problem 12.17 that

$$d\theta^k(\mathbf{V}, \mathbf{W}) = -\theta^k([\mathbf{V}, \mathbf{W}]).$$

Hence $d\theta^k(\mathbf{V}_r, \mathbf{V}_s) = -V^k([V_r, V_s]) = -c_{rs}^k$. On the other hand, noting that $c_{ij}^k = -c_{ji}^k$, the 2-form

$$\tfrac{1}{2}c_{ij}^k(\theta^i \wedge \theta^j)(\mathbf{V}_r, \mathbf{V}_s)$$

is equal to $c_{ij}^k \theta^i(\mathbf{V}_r)\theta^j(\mathbf{V}_s) = c_{ij}^k V^i(V_r)V^j(V_s) = c_{rs}^k$.

12.19 *The action of a Lie group on itself and the internal symmetry of a Lie group.* For fixed group element $\mathbf{G}$ of $\mathcal{G}$, define $\mathbf{Ad}(\mathbf{G}) := L_{\mathbf{G}}(R_{\mathbf{G}})^{-1}$. Explicitly,

$$\mathbf{Ad}(\mathbf{G}) := \mathbf{G}G\mathbf{G}^{-1} \qquad \text{for all} \quad G \in \mathcal{G}.$$

Since $(\mathbf{GH})G(\mathbf{GH})^{-1} = \mathbf{G}(\mathbf{H}G\mathbf{H}^{-1})\mathbf{G}^{-1}$, we get

$$\mathbf{Ad}(\mathbf{GH}) = \mathbf{Ad}(\mathbf{G})\mathbf{Ad}(\mathbf{H}) \qquad \text{for all} \quad \mathbf{G}, \mathbf{H} \in \mathcal{G}.$$

Show the following.
(i) The map $\mathbf{Ad}(\mathbf{G}) : \mathcal{G} \to \mathcal{G}$ is a group automorphism.
(ii) The set of all automorphisms $\chi : \mathcal{G} \to \mathcal{G}$ forms a group called the automorphism group $Aut(\mathcal{G})$ of $\mathcal{G}$.
(iii) The set of all maps $\mathbf{Ad}(\mathbf{G})$ with $\mathbf{G} \in \mathcal{G}$ forms a subgroup of $Aut(\mathcal{G})$ called the symmetry group $Sym(\mathcal{G})$ of $\mathcal{G}$.
(iv) The map $\mathbf{Ad} : \mathcal{G} \to Sym(\mathcal{G})$ given by

$$\mathbf{G} \mapsto \mathbf{Ad}(\mathbf{G})$$

is a group epimorphism called the adjoint representation of the group on itself.
(v) For group elements G, H of $\mathcal{G}$, we write

$$G \sim H \qquad \text{iff} \qquad G = \mathbf{G}H \qquad \text{for some} \quad \mathbf{G} \in \mathcal{G}.$$

This is an equivalence relation. The equivalence class $[G]$ coincides with the orbit of G under the adjoint representation; $[G]$ is also called the conjugacy class of G.

12.20 *The action of a Lie group on its Lie algebra and the external symmetry of a Lie algebra.* For fixed $\mathbf{G} \in \mathcal{G}$, define $\mathrm{Ad}(\mathbf{G}) := L'_{\mathbf{G}}(R'_{\mathbf{G}})^{-1}$. Explicitly,

$$\mathrm{Ad}(\mathbf{G}) = \mathbf{G}V\mathbf{G}^{-1} \qquad \text{for all} \quad V \in \mathcal{LG}.$$

Show the following.
(i) The map $\mathrm{Ad}(\mathbf{G}) : \mathcal{LG} \to \mathcal{LG}$ is a linear operator with

$$\mathrm{Ad}(\mathbf{GH}) = \mathrm{Ad}(\mathbf{G})\mathrm{Ad}(\mathbf{H}) \qquad \text{for all} \quad \mathbf{H}, \mathbf{G} \in \mathcal{LG}.$$

(ii) The map $\mathrm{Ad}(\mathbf{G})$ respects the Lie product. Explicitly,

$$\mathrm{Ad}(\mathbf{G})([V, W]) = [\mathrm{Ad}(\mathbf{G})V, \mathrm{Ad}(\mathbf{G})W] \qquad \text{for all} \quad V, W \in \mathcal{LG}.$$

Recall that $[V, W] := VW - WV$.

The map $\mathbf{G} \mapsto \mathrm{Ad}(\mathbf{G})$ is called the adjoint representation of the Lie group $\mathcal{G}$ on its Lie algebra $\mathcal{LG}$.

12.21 *The action of a Lie algebra on itself and the internal symmetry of a Lie algebra.* For fixed $V \in \mathcal{LG}$, define

$$\mathrm{ad}(V)W := [V, W] \qquad \text{for all} \quad W \in \mathcal{LG}.$$

Show that $\mathrm{ad}(V) : \mathcal{LG} \to \mathcal{LG}$ is a linear operator with

$$\mathrm{ad}([V, W]) = [\mathrm{ad}(V), \mathrm{ad}(W)] \qquad \text{for all} \quad V, W \in \mathcal{LG}.$$

Here, $[\mathrm{ad}(V), \mathrm{ad}(W)] := \mathrm{ad}(V)\mathrm{ad}(W) - \mathrm{ad}(W)\mathrm{ad}(V)$. The map $V \mapsto \mathrm{ad}(V)$ is called the adjoint representation of the Lie algebra $\mathcal{LG}$ on itself.

Solution: By the Jacobi identity,

$$[[V, W], Z] + [W, Z], V] + [[Z, V], W] = 0$$

for all $V, W, Z \in \mathcal{LG}$, the expression

$$\mathrm{ad}(V)\mathrm{ad}(W)Z - \mathrm{ad}(W)\mathrm{ad}](V)Z = [V, [W, Z]] - [W, [V, Z]]$$

is equal to $\mathrm{ad}([V, W])Z = [[V, W], Z]$.

12.22 *The Killing form.* Let $\mathcal{L}$ be a finite-dimensional (real or complex) Lie algebra. Using the linear operators $\mathrm{ad}(V), \mathrm{ad}(W) : \mathcal{L} \to \mathcal{L}$, the Killing form is defined by

$$K(V, W) := \mathrm{tr}(\mathrm{ad}(V)\mathrm{ad}(W)) \qquad \text{for all} \quad V, W \in \mathcal{L}.$$

Let $V_1, ..., V_n$ be a basis of $\mathcal{L}$ along with the structural constants c_{ij}^k given by

$$[V_i, V_j] = c_{ij}^k V_k, \qquad i, j = 1, ..., n.$$

Show that

(i) If $V = v^i V_i$ and $W = w^j V_j$, then

$$K(V, W) = k_{ij} v^i w^j$$

where $k_{ij} := c_{ir}^s c_{js}^r$.

(ii) For the group $SU(2)$, choose the basis vectors $V_j = \mathrm{i}\sigma^j$ where $\sigma^j, j = 1, 2, 3$, are the Pauli matrices. Show that $k_{ij} = -2\delta_{ij}$ for $i, j = 1, 2, 3$.

Solution: Ad (i). Let $V = v^i V_i$ and $Z = z^j V_j$. It follows from

$$\mathrm{ad}(V)Z = [V, Z] = (v^i z^j c_{ij}^k) V_k$$

that $\mathrm{ad}(V)Z = (\mathrm{ad}(V)_j^k z^j) V_k$ where $\mathrm{ad}(V)_j^k := v^i c_{ij}^k$. Thus, for the trace,

$$\mathrm{tr}(\mathrm{ad}(V)\mathrm{ad}(W)) = \mathrm{ad}(V)_r^s \cdot \mathrm{ad}(W)_s^r = v^i c_{ir}^s w^j c_{js}^r.$$

Ad (ii). We have the commutation rules

$$[B_1, B_2] = -B_3, \quad [B_2, B_3] = -B_1, \quad [B_3, B_1] = -B_2.$$

Hence $c_{12}^3 = -1$, etc. □

The Killing form $K : \mathcal{L} \to \mathbb{K}$ of a Lie algebra over $\mathbb{K} = \mathbb{R}, \mathbb{C}$ is bilinear and symmetric; it encodes important structural properties of the Lie algebra.

12.23 *Proof of the linear Frobenius theorem.* Prove Theorem 12.52 on page 769. Hint: Use the Frobenius theorem on page 768 in order to prove a local existence result. Continue this local solution to a global solution by using the fact that the set $\mathcal{U}$ is simply connected. Therefore, the continuation is path-independent. The local existence result can also be obtained by an induction argument (reduction to ordinary differential equations). See J. Stoker, Differential Geometry, Wiley 1969/89, Appendix B, p. 392. See also J. Eschenburg and J. Jost, Differential Geometry and Minimal Surfaces (in German), Appendix, p. 233.

13. The Commutative Weyl $U(1)$-Gauge Theory and the Electromagnetic Field

In modern physics, interactions are a consequence of the principle of local symmetry invariance. This leads to the Standard Model in particle physics.

Folklore

13.1 Basic Ideas

When we do scientific work, we must often step down from our high horse of grand principles, and dig in the dirt with our noses. When we achieve our purpose, we cover the tracks of our efforts in order to appear as gods of clear thought.

Albert Einstein (1879–1955)

In what follows, we will consider the following two transformations:

(i) transformation of the space and time coordinates, and
(ii) gauge transformations of the physical field (local symmetry transformations).

Our final goal is to establish a mathematical formalism which is invariant under both transformations.

This invariant mathematical formalism will be based on the language of fiber bundles and Cartan's language of differential forms.

One has to distinguish between

- commutative gauge groups (e.g, the commutative group $U(1)$), and
- noncommutative gauge groups (e.g., the groups $SU(2), SU(3)$).[1]

The Standard Model in particle physics is based on the gauge group $\mathcal{G}$ given by the direct product

$$\mathcal{G} := U(1) \times SU(2) \times SU(3).$$

The use of noncommutative gauge groups is crucial for the Standard Model. From the analytical point of view, we will use two differential forms, namely,

- the connection 1-form and
- the curvature 2-form.

From the geometrical point of view, the basic notion is the notion of parallel transport. In terms of physics, parallel transport corresponds to the transport of physical information.

The crucial point is that, in the nontrivial case, the parallel transport depends on the path of transport.

[1] Commutative and noncommutative groups are also called Abelian and non-Abelian groups, respectively.

E. Zeidler, *Quantum Field Theory III: Gauge Theory*,
DOI 10.1007/978-3-642-22421-8_14,

This means that the parallel transport along a loop is not trivial, that is, the final state differs from the initial state. This fact can be used in order to measure curvature. In terms of physics, we measure the strength of interaction. In this chapter, we will use the commutative gauge group $U(1)$, and we will show that the corresponding gauge theory is intimately related to the electromagnetic field. The corresponding noncommutative $SU(N)$-gauge theory and gauge theories with general gauge groups will be studied in Chap. 15. The formulas of the commutative $U(1)$-gauge theory are simpler than the formulas for noncommutative gauge theories, since the Lie brackets vanish identically in the $U(1)$-case.

Let us sketch the basic ideas. To this end, let ψ be a physical field defined on the space-time manifold $\mathbb{M}^4$ (Minkowski manifold). The key transformation is given by

$$\boxed{\psi^+(P) = G(P)\psi(P), \qquad P \in \mathbb{M}^4.}$$

This is called a gauge transformation. We have $G(P) \in \mathcal{G}$, that is, the local phase factor $G(P)$ is an element of the so-called gauge group $\mathcal{G}$ which is assumed to be a Lie group, e.g., $\mathcal{G} = U(1), U(N), SU(N), GL(N, \mathbb{C})$ or $\mathcal{G} = U(1) \times SU(2) \times SU(3)$. In terms of physics, our goal is the following:

(V) We want to develop a differential calculus which allows us to formulate gauge-invariant partial differential equations on $\mathbb{M}^4$ (or on more general curved space-time manifolds). This will be based on the notion of covariant directional derivative

$$D_v\psi$$

which has the following two crucial properties:

- Relativistic invariance: $D_v\psi$ is invariant under a change of the inertial system in Einstein's theory of special relativity. In terms of physics, this means that if ψ is a physical field, then $D_v\psi$ describes a new physical field which does not depend on the choice of the observer in an inertial system. In terms of mathematics, if ψ is a section of a vector bundle over the base manifold $\mathbb{M}^4$, then so is $D_v\psi$.
- Compatibility with gauge transformation: $D_v\psi$ transforms like the physical field ψ under the gauge transformation, that is, if $\psi^+(P) = G(P)\psi(P)$, then

$$\boxed{D_v^+\psi^+(P) = G(P)D_v\psi(P), \qquad P \in \mathbb{M}^4.} \tag{13.1}$$

Equivalently, we have the compatibility condition

$$D_v^+\psi^+(x) = (D_v\psi(x))^+.$$

(P) We want to construct a gauge-invariant equation of motion for the phase factor along curves on the space-time manifold. In terms of mathematics, this is called a parallel transport. We postulate that

- the parallel transport does not depend on the choice of the observer (relativistic invariance), and
- the parallel transport is compatible with gauge transformations.

Force and curvature. In terms of modern physics, forces are measured by parallel transport of physical fields along a small loop on the space-time manifold. In terms of geometry, this measures the curvature. Gauge theory realizes the fundamental principle:

$$\boxed{\text{force} = \text{curvature}.} \tag{13.2}$$

The language of bundles. It turns out that the language of bundles is the proper tool in gauge theory. For example, the principle (13.2) refers to the curvature of principal bundles and vector bundles. We will proceed in the following two steps:

Step 1: We use product bundles. Concerning (V), the prototype is given by the product vector bundle

$$\mathbb{M}^4 \times \mathbb{C}^N$$

where the Lie group $\mathcal{G}$ consists of invertible complex $(N \times N)$-matrices. Concerning (P), the prototype is given by the product principal bundle

$$\mathbb{M}^4 \times \mathcal{G}.$$

Step 2: General bundles are obtained by gluing together product bundles. To this end, we will use a cocycle with values in the Lie group $\mathcal{G}$ (or with values in a Lie group $\mathcal{H}$ obtained by a surjective group morphism $r : \mathcal{G} \to \mathcal{H}$).

The bundles from Step 1 are sufficient for formulating the Standard Model in particle physics (see Vol. IV). If one wants to describe gauge theories on curved space-times (e.g., theory of general relativity, string theory, quantum gravity), then one has to use Step 2.

In terms of physics, different observers use different product bundles for describing their measurements.

Cocycles describe the change of observers.

To explain this, let A, B, C be three observers. The typical cocycle condition (see (16.1) on page 871) guarantees that the transformations

$$A \Rightarrow B \qquad \text{and} \qquad B \Rightarrow C$$

from A to B and from B to C are compatible with the transformation

$$A \Rightarrow C.$$

The properties of cocycles depend on both the topology of the space-time manifold and the structure of the gauge group $\mathcal{G}$. The deviation of vector bundles from product bundles (i.e., the twist of the bundles) is measured by so-called characteristic classes (Chern classes, Euler classes, Pontryagin classes, Stiefel–Whitney classes, Thom classes).

Characteristic classes of vector bundles play a fundamental role in modern topology and geometry.

For example, the Chern classes of the tangent bundle of a Riemannian manifold allow us to generalize the Gauss–Bonnet theorem to higher dimensions.

Relation to classical geometry. All the notions discussed above have their origin in classical geometry. We refer to 9.5 on page 593, where we discuss how the theory sketched above is rooted in the geometry of the 2-dimensional sphere (e.g., the surface of earth). In this special case, we use

- the parallel transport of velocity vectors on the tangent bundle (the Levi-Civita connection on the tangent bundle), and
- the parallel transport of frames (the Levi-Civita connection on the frame bundle).

The frame bundle of the sphere is a principal bundle with the gauge group $SO(2)$, and the tangent bundle of the sphere is called the associated vector bundle to the frame bundle.

Summarizing, in the following four chapters we want to show how gauge theory is intimately related to both physical intuition and geometrical intuition.

Historical remarks can be found on page 891. Both the crucial homotopy classification of vector bundles (universal bundles) and the characteristic classes will be thoroughly studied in Vol. IV on quantum mathematics.[2]

In the following chapter, we need elementary properties of the electromagnetic field (e.g., the four-potential of the electromagnetic field and the Minkowski manifold $\mathbb{M}^4$). The reader who is not familiar with this should first glance at Chap. 18 (Einstein's theory of special relativity) and Chap. 19 (the Maxwell equations in electromagnetism).

13.2 The Fundamental Principle of Local Symmetry Invariance in Modern Physics

We want to motivate the local phase factor of a physical field. In quantum mechanics, the local phase factor was introduced by Vladimir Fock (1898–1974) in 1926 in connection with a crucial invariance property of the Klein–Fock–Gorden equation. Nowadays this invariance property is called gauge invariance. In 1935, Yukawa used the Klein–Fock–Gordon equation in order to predict the existence of mesons. In 1949, Hideki Yukawa (1907–1981) was awarded the Noble prize in physics for the prediction of the meson.[3] The modern theory of strong interaction corresponds to quantum chromodynamics in the Standard Model of particle physics.

- In Sect. 13.2.1, we will study the free meson together with the Yukawa potential.
- In Sect. 13.2.2, we will pass to a principle of critical action which is invariant under local symmetry transformations. This leads to adding the potential of an interaction force. It turns out that this potential is the four-potential of the electromagnetic field.

  The principle of the invariance of the action integral under local symmetries leads to the interaction forces.

  This is an universal principle in modern physics which also leads to the Standard Model in elementary particle physics (see Vol. IV).

13.2.1 The Free Meson

> A note in proofreading: When this note was already in press, an elegant work by Oskar Klein *Quantum theory and five-dimensional theory of relativity*, Z. Phys. **37** (1926), 895–906 (in German) reached Leningrad (nowadays St. Petersburg). Its author had obtained results that are in principle identical to those of the current work.[4]
>
> Vladimir Fock, 1926

---

[2] At this point, we recommend D. Husemoller, Fibre Bundles, Springer, New York, 1994.

[3] H. Yukawa, On the interaction of elementary particles, Proc. Phys.-Math. Soc. Japan **17** (1935), 48–57.

[4] V. Fock, On the invariant form of the wave and motion equations for a charged point mass, Z. Phys. **39** (1926), 839–841 (in German). English translation in: Physics Uspekhi **53**(8) (2010), 839–841. Fock (1898–1974) used a 5-dimensional metric, as Oskar Klein did.

The same equation was obtained by Walter Gordon (1893–1939) in 1926. See W. Gordon, The Compton effect according to Schrödinger's theory (in German), Z. Phys. **40** (1926), 117–133. I would like to thank Harald Fritzsch (Munich) for drawing my attention to Fock's paper.

The Klein–Fock–Gordon equation. In Einstein's theory of special relativity, we have the key equation

$$\boxed{E^2 = m_0^2c^4 + c^2\mathbf{p}^2}$$

for the energy E of a particle in an inertial system. Here, m_0 and $\mathbf{p}$ denote the rest mass and the momentum vector of the particle, respectively, and c is the velocity of light in a vacuum. Using Schrödinger's substitution trick

$$E \mapsto \mathrm{i}\hbar\frac{\partial}{\partial t}, \qquad \mathbf{p} \mapsto -\mathrm{i}\hbar\partial,$$

we get the Klein–Fock–Gordon equation

$$\boxed{\Box\psi + \frac{m_0^2c^2}{\hbar^2}\,\psi = 0} \tag{13.3}$$

where $\Box\psi = \frac{1}{c^2}\,\psi_{tt} + \Delta\psi$. In 1935, Yukawa used the Klein–Fock–Gordon equation in order to formulate a phenomenological theory for strong interaction. Introducing the Compton wave length

$$\lambda := \frac{h}{m_0c},$$

and $\lambda\!\!\!^{-} := \lambda/2\pi$, the Klein–Fock–Gordon equation has the stationary solution

$$\boxed{\psi_Y(\mathbf{x}) = \frac{g}{4\pi r}\,\mathrm{e}^{-r/\lambda\!\!\!^{-}}}$$

where $r := |\mathbf{x}|$, and g is a positive constant called the coupling constant of strong interaction. The function ψ_Y is called the Yukawa potential. Letting $m_0 \to 0$, we get the Coulomb potential

$$\psi_C(\mathbf{x}) = \frac{g}{4\pi r}, \qquad g = \frac{Q}{\varepsilon_0}$$

of a particle having the electric charge Q. Let $\mathbf{x} = \overrightarrow{OP}$ be the position vector pointing from the origin O to the point P. The repulsive Coulomb force between two particles of electric charge Q located at the origin and at the point P is equal to

$$\mathbf{F}(\mathbf{x}) = -Q\,\mathbf{grad}\,\psi_C(\mathbf{x}) = \frac{Q^2}{4\pi\varepsilon_0 r^2}\cdot\frac{\mathbf{x}}{|\mathbf{x}|}\,.$$

From the Yukawa potential, we get the attractive Yukawa force

$$\mathbf{F}_Y(\mathbf{x}) = g\,\mathbf{grad}\,\psi_Y = -\frac{g^2}{4\pi}\left(\frac{\mathrm{e}^{-r/\lambda\!\!\!^{-}}}{r^2} + \frac{\mathrm{e}^{-r/\lambda\!\!\!^{-}}}{\lambda\!\!\!^{-}\,r}\right)\frac{\mathbf{x}}{r}$$

between two nucleons located at the origin and at the point P. The characteristic length scale $\lambda\!\!\!^{-}$ is called the range of the Yukawa force. Yukawa formulated the following hypotheses:

One has to distinguish between the German mathematician Felix Klein (1849–1925) (the friend of Sophus Lie (1842–1899) and promoter of David Hilbert (1862–1943)) and the Swedish physicist Oskar Klein (1894–1977). See G. Ekspong (1991) quoted on page 896.

(M1) The range of the strong force between two nucleons equals the radius of the nucleus, $R = 10^{-15}$m. Hence

$$\lambda\!\!\!^{-} = \frac{\hbar}{m_0 c} = R.$$

(M2) Like the electromagnetic interaction is based on the exchange of a massless particle called photon, the strong force is based on the exchange of a massive particle of mass m_0. Yukawa called this hypothetical particle meson. From (M1) we get the meson mass

$$m_0 = \frac{\hbar}{cR}.$$

Explicitly, $m_0 = 100\,\text{MeV}/c^2 = 1/10$ proton mass.

(M3) The meson is an unstable particle. Its mean life time Δt follows from the energy-time uncertainty relation

$$\Delta E\,\Delta t = \frac{\hbar}{2},$$

where $\Delta E = m_0 c^2$. Hence $\Delta t = R/2c$. Explicitly, $\Delta t = 10^{-23}$s.

In 1947, the so-called π-meson of mass $140\text{MeV}/c^2$ was found. Indeed, there are now known to exist three such spinless particles, the π^+, π^-, and π^0 of approximately the same mass, and of electric charge $e, -e, 0$, respectively. Scattering experiments allow us to determine the Yukawa coupling constant,

$$\frac{g^2}{4\pi\hbar c} = 15.$$

For the electromagnetic interaction of two electrons,

$$\frac{e^2}{4\pi\varepsilon_0\hbar c} = \frac{1}{137}.$$

A comparison of these two characteristic dimensionless numbers shows that the strong interaction is much stronger than the electromagnetic interaction.

Consider the variational principle

$$\boxed{S := \int_\Omega \mathcal{L}(\psi, \partial\psi)\, d^4x = \text{critical!}} \tag{13.4}$$

in an inertial system where Ω is a nonempty bounded open subset of $\mathbb{R}^4$, and

$$d^4x := dx^0 dx^1 dx^2 dx^3, \qquad x^0 := ct.$$

Moreover, the smooth field $\psi : \text{cl}(\Omega) \to \mathbb{C}$ is fixed on the boundary. Explicitly,

$$\mathcal{L} = \tfrac{1}{2}\left(\partial_\mu\psi\partial^\mu\psi^\dagger - V(|\psi|^2)\right) = \tfrac{1}{2}\left(\tfrac{1}{c^2}|\psi_t|^2 - |\partial\psi|^2 - V(|\psi|^2)\right).$$

Here, $\partial_\mu := \frac{\partial}{\partial x^\mu}$, and $\partial^0 := \partial_0$, $\partial^j := -\partial_j, j = 1, 2, 3$. We sum over $\mu = 0, 1, 2, 3$.

Proposition 13.1 *Each solution of the variational problem* (13.4) *satisfies the nonlinear Klein–Fock–Gordon equation*

$$\Box\psi + V'(|\psi|^2)\psi = 0.$$

In particular, if we choose the so-called mass term

$$\boxed{V(|\psi|^2) = \frac{m_0^2 c^2}{\hbar^2}\,|\psi|^2,}$$

then we obtain the Klein–Fock–Gordon equation with $V'(|\psi|^2)\psi = \frac{m_0^2c^2}{\hbar^2}\,\psi$.

Proof. To simplify notation, let $c = \hbar = 1$. Set

$$\psi = \varphi + \chi\mathrm{i}$$

where φ, χ are real-valued functions. Then

$$\mathcal{L} = \frac{1}{2}\big(\partial_\mu\varphi\partial^\mu\varphi + \partial_\mu\chi\partial^\mu\chi - V(\varphi^2 + \chi^2)\big).$$

Replacing φ by $\varphi + \sigma h$ with $h \in C_0^\infty(\Omega)$ and differentiating $S(\sigma)$ at $\sigma = 0$, we get

$$0 = S'(0) = \int_\Omega \big(\partial_\mu\varphi\partial^\mu h - V'(\varphi^2 + \chi^2)\varphi h\big)d^4x.$$

Integrating by parts,

$$0 = \int_\Omega \big(\partial^\mu\partial_\mu\varphi + V'(\varphi^2 + \chi^2)\varphi\big)h d^4x$$

for all $h \in C_0^\infty(\Omega)$. By the Variational Lemma,

$$\partial^\mu\partial_\mu\varphi + V'(\varphi^2 + \chi^2)\varphi = 0.$$

The same equation is obtained for χ. □

Energy. Define the energy of the nonlinear Klein–Fock–Gordon equation by

$$\boxed{\mathcal{E}(t) := \int_{\mathbb{R}^3} \frac{1}{2}\left(\frac{|\psi_t|^2}{c^2} + |\boldsymbol{\partial}\psi|^2 + V(|\psi|^2)\right) d^3x} \tag{13.5}$$

where we assume that the field $\psi = \psi(\mathbf{x}, t)$ vanishes outside a sufficiently large ball for each time t.

Proposition 13.2 *If ψ is a solution of the nonlinear Klein–Fock–Gordon equation, then the energy is conserved.*

Proof. Let $c = \hbar = 1$. Noting that $|\psi|^2 = \psi^\dagger\psi$ and using integration by parts,

$$\dot{\mathcal{E}}(t) = \int_{\mathbb{R}^3} \big(\psi_{tt} + \Delta\psi + V'(|\psi|^2)\psi\big)\,\psi_t^\dagger d^3x + c.c.$$

where *c.c.* stands for the complex conjugate value of the integral. By the Klein–Fock–Gordon equation, $\dot{\mathcal{E}}(t) \equiv 0$. □

13.2.2 Local Symmetry and the Charged Meson in an Electromagnetic Field

The replacement trick for the Schrödinger equation. The non-relativistic Schrödinger equation for a free quantum particle with (rest) mass m_0 reads as

$$\mathrm{i}\hbar\frac{\partial\psi}{\partial t}=\frac{\mathbf{p}^2}{2m_0}\psi$$

with the differential operator $\mathbf{p}=-\mathrm{i}\hbar\boldsymbol{\partial}$. In order to pass to the Schrödinger equation for a quantum particle of electric charge Q in the electric field $\mathbf{E},\mathbf{B}$, physicists use the replacement

$$\boxed{\mathrm{i}\hbar\frac{\partial}{\partial t}\Rightarrow\mathrm{i}\hbar\frac{\partial}{\partial t}-QU,\qquad \mathbf{p}\Rightarrow\mathbf{p}-Q\mathbf{A}} \tag{13.6}$$

together with

$$\mathbf{E}=-\,\mathbf{grad}\,U-\mathbf{A}_t,\qquad \mathbf{B}=\mathbf{curl}\,\mathbf{A}$$

where $U,\mathbf{A}$ is the 4-potential of the electromagnetic field. The replacement trick (13.6) is motivated by the Hamiltonian approach in classical electrodynamics (see page 979). In the classical case, one replaces the momentum vector $\mathbf{p}$ by the vector $\mathbf{P}-Q\mathbf{A}$ where $\mathbf{P}$ is the canonical momentum vector. In quantum mechanics, canonical quantization means that $\mathbf{P}$ is replaced by the differential operator $-\mathrm{i}\hbar\boldsymbol{\partial}$. This yields the Schrödinger equation

$$\left(\mathrm{i}\hbar\frac{\partial}{\partial t}-QU\right)\psi=\frac{(\mathbf{p}-Q\mathbf{A})^2}{2m_0}\psi \tag{13.7}$$

for a charged quantum particle in an electromagnetic field. In 1926, Schrödinger used this equation in order to compute the discrete and continuous spectrum of the hydrogen atom (see Vol. IV).

The replacement trick for the Klein–Fock–Gordon equation. Similarly, we apply the replacement trick (13.6) to the free Klein–Fock–Gordon equation (13.3). This yields the Klein–Fock– Gordon equation for a spinless relativistic quantum particle of rest mass m_0 and electric charge Q in an electromagnetic field:

$$\left(\frac{1}{c^2}\left(\mathrm{i}\hbar\frac{\partial}{\partial t}-QU\right)^2-(\mathbf{p}-Q\mathbf{A})^2-m_0^2c^2\right)\psi=0. \tag{13.8}$$

This equation is valid in every inertial system.

Fock's discovery of gauge invariance in quantum mechanics. In 1926, Fock discovered that the equation (13.8) is invariant under the following gauge transformations:

$$\psi^+(\mathbf{x},t)=\mathrm{e}^{\mathrm{i}a(\mathbf{x},t)}\psi(\mathbf{x},t) \tag{13.9}$$

and

$$U^+(\mathbf{x},t)=U(\mathbf{x},t)-\frac{\hbar}{Q}a_t(\mathbf{x},t),\quad \mathbf{A}^+(\mathbf{x},t)=\mathbf{A}(\mathbf{x},t)+\frac{\hbar}{Q}\boldsymbol{\partial}a(\mathbf{x},t). \tag{13.10}$$

Here, the so-called local phase function a is real-valued. Note that this gauge transformation does not change the electromagnetic field, that is, $\mathbf{E}^+=\mathbf{E}$ and $\mathbf{B}^+=\mathbf{B}$.

This gauge invariance can be proved by an explicit computation. However, in order to get insight, we will show how Fock's gauge invariance is related to the setting of a $U(1)$-gauge theory.

The covariant partial derivative. In terms of mathematics, the replacement trick (13.6) of physicists is nothing other than the passage from the partial derivative ∂_α to the covariant partial derivative

$$\boxed{D_\alpha := \partial_\alpha + \mathcal{A}_\alpha, \qquad \alpha = 0,1,2,3} \tag{13.11}$$

with respect to some connection on the vector bundle $\mathbb{M}^4 \times U(1)$ (the product between the Minkowski space-time manifold $\mathbb{M}^4$ and the gauge group $U(1)$). We set

$$\mathcal{A}_\alpha := \frac{\mathrm{i}Q}{\hbar} A_\alpha, \qquad \alpha = 0,1,2,3 \tag{13.12}$$

where $A_0 = U/c$ and $\mathbf{A} = A^1\mathbf{i} + A^2\mathbf{j} + A^3\mathbf{k}$ together with $A_j = -A^j, j = 1,2,3$. Moreover, let $D^0 := D_0$, and $D^j := -D_j, j = 1,2,3$. Then the Klein–Fock–Gordon equation (13.8) reads as

$$\boxed{D_\alpha D^\alpha \psi + \frac{m_0^2 c^2}{\hbar^2} \psi = 0.} \tag{13.13}$$

Here, we sum over $\alpha = 0,1,2,3$. This shows that the Klein–Fock–Gordon equation is valid in every inertial system, by the general index principle applied to the theory of special relativity (see page 443). Now to the point. The gauge transformation (13.10) can be written as

$$\mathcal{A}_\alpha^+ = \mathcal{A}_\alpha - \mathrm{i}\partial_\alpha a, \qquad \alpha = 0,1,2,3.$$

Introducing the transformed covariant partial derivative

$$D_\alpha^+ := \partial_\alpha + \mathcal{A}_\alpha^+, \qquad \alpha = 0,1,2,3,$$

we get the following.[5]

Proposition 13.3 *If the complex-valued function ψ is a solution of the Klein–Fock–Gordon equation* (13.13)*, then ψ^+ is a solution of the transformed equation*

$$D_\alpha^+ D^{+\alpha} \psi^+ + \frac{m_0^2 c^2}{\hbar^2} \psi^+ = 0.$$

Proof. By (13.4) on page 822 below, we have the key relation of the covariant partial derivative in gauge theory:

$$D_\alpha^+ \psi^+ = \mathrm{e}^{\mathrm{i}a} D_\alpha \psi.$$

Then $D_\alpha^+(D^{+\alpha}\psi^+) = D_\alpha^+ \mathrm{e}^{\mathrm{i}a} D^\alpha \psi = \mathrm{e}^{\mathrm{i}a} D_\alpha D^\alpha \psi$. □

The fundamental principle of local symmetry in modern physics. This principle has the crucial property that

It generates physical fields which are responsible for the interaction.

[5] We set $D^{+0} := D_0^+$ and $D^{+j} := -D_j^+, j = 1,2,3$.

The idea is to start with the principle of critical action for the free particle. Then we demand that the action integral is invariant under local symmetry. This can be realized by replacing the classical partial derivative ∂_α by the covariant partial derivative $D_\alpha = \partial_\alpha + \mathcal{A}_\alpha$. Then, $\mathcal{A}_\alpha$ is an additional physical field which is responsible for the interaction of the free particles (e.g., the electromagnetic interaction in the present case). To illustrate this with a concrete example, consider the variational principle

$$\int_\Omega \mathcal{L}\, d^4x = \text{critical!} \tag{13.14}$$

with the boundary condition: $\psi =$ fixed on $\partial\Omega$. The Lagrangian

$$\mathcal{L} := \tfrac{1}{2}\left(\partial_\alpha\psi\partial^\alpha\psi^\dagger - \frac{m_0^2c^2}{\hbar^2}\,\psi\psi^\dagger\right)$$

yields the Euler–Lagrange equation

$$\partial_\alpha\partial^\alpha\psi + \frac{m_0^2c^2}{\hbar^2}\psi = 0.$$

This Klein–Fock–Gordon equation describes a free meson. The Lagrangian $\mathcal{L}$ is not always invariant under the local symmetry transformation

$$\boxed{\psi^+(\mathbf{x},t) := \mathrm{e}^{ia(\mathbf{x},t)}\,\psi(\mathbf{x},t)} \tag{13.15}$$

where ψ is a complex-valued smooth function, and a is a real-valued smooth function. Thus, the local phase factor $\mathrm{e}^{\mathrm{i}a(\mathbf{x},t)}$ is an element of the gauge group $U(1)$. However, setting $D_\alpha := \partial + \mathcal{A}_\alpha$, the modified Lagrangian

$$\mathcal{L} := \tfrac{1}{2}\left(D_\alpha\psi(D^\alpha\psi)^\dagger + \frac{m_0^2c^2}{\hbar^2}\psi\psi^\dagger\right) \tag{13.16}$$

is invariant under the local symmetry transformation (13.15). In fact, it follows from

$$D_\alpha^+\psi^+(D^{+\alpha}\psi)^\dagger = \mathrm{e}^{\mathrm{i}a}D_\alpha\psi(\mathrm{e}^{\mathrm{i}a}D^\alpha\psi)^\dagger = D_\alpha\psi(D^\alpha\psi)^\dagger$$

that $\mathcal{L}^+ = \mathcal{L}$. Suppose that all the values of $\mathcal{A}_\alpha$ lie in the Lie algebra $u(1)$, that is, $\mathcal{A}_\alpha^\dagger = -\mathcal{A}_\alpha$. Then the Euler–Lagrange equation corresponding to the Lagrangian from (13.16) reads as

$$D_\alpha D^\alpha\psi + \frac{m_0^2c^2}{\hbar^2}\psi = 0.$$

Setting $\mathcal{A}_\alpha := \frac{\mathrm{i}Q}{\hbar}A_\alpha, \alpha = 0,1,2,3$, we obtain the Klein–Fock–Gordon equation for a charged meson in an electromagnetic field.

13.3 The Vector Bundle $\mathbb{M}^4 \times \mathbb{C}$, Covariant Directional Derivative, and Curvature

The Minkowski manifold $\mathbb{M}^4$. In this chapter, smooth maps

$$\psi : \mathbb{M}^4 \to \mathbb{C}$$

are called physical fields. We will use inertial systems described by the coordinates

$$x = (x^0, x^1, x^2, x^3)$$

where x^1, x^2, x^3 are Cartesian coordinates, and $x^0 := ct$. Here, t denotes the time, and c denotes the velocity of light in a vacuum. The transformation between two inertial systems is given by

$$x' = \Lambda x + a$$

where $\Lambda \in O(1,3)$ (Lorentz transformation), and $a \in \mathbb{R}^4$. Invariants under these space-time transformations are invariants of the Minkowski manifold $\mathbb{M}^4$. Finally, we set

$$\partial_\alpha := \frac{\partial}{\partial x^\alpha}, \qquad \alpha = 0, 1, 2, 3.$$

We will sum over equal upper and lower Greek indices from 0 to 3.

The vector bundle $\mathbb{M}^4 \times \mathbb{C}$. By definition, the product set

$$\mathbb{M}^4 \times \mathbb{C} := \{(P, \psi) : \ P \in \mathbb{M}^4, \quad \psi \in \mathbb{C}\}$$

is called a vector product bundle over the base manifold $\mathbb{M}^4$ with the fiber

$$F_P := \{(P, \psi) : \psi \in \mathbb{C}\}$$

over the point $P \in \mathbb{M}^4$. There exists a one-to-one correspondence between the fiber F_P and the complex linear space $\mathbb{C}$. More precisely, the surjective map

$$\pi : \mathbb{M}^4 \times \mathbb{C} \to \mathbb{M}^4$$

given by $\pi(P, \psi) := P$ is called a vector bundle with the fiber $F_P = \pi^{-1}(P)$, the bundle space $\mathbb{M}^4 \times \mathbb{C}$, and the base space $\mathbb{M}^4$. For the sake of brevity, we speak of the vector bundle $\mathbb{M}^4 \times \mathbb{C}$. The map

$$s : \mathbb{M}^4 \to \mathbb{M}^4 \times \mathbb{C}$$

is called a section iff $s(P) \in F_P$ for all $P \in \mathbb{M}^4$. Hence

$$s(P) = (P, \psi(P))$$

where ψ is a map from $\mathbb{M}^4$ to $\mathbb{C}$. Thus, physical fields ψ and smooth sections s of the vector bundle $\mathbb{M}^4 \times \mathbb{C}$ can be identified with each other.

The local phase factor and gauge transformations. The transformation

$$\boxed{\psi^+(P) := G_0(P)\psi(P), \qquad P \in \mathbb{M}^4} \tag{13.17}$$

is called a gauge transformation iff $G_0(P) \in U(1)$ for all $P \in \mathbb{M}^4$, and the map

$$G_0 : \mathbb{M}^4 \to U(1)$$

is smooth. Setting

$$T(P, \psi) := (P, G_0(P)\psi),$$

we get the map

$$T : \mathbb{M}^4 \times \mathbb{C} \to \mathbb{M}^4 \times \mathbb{C}.$$

In mathematics, this is called a transition map from the product bundle $\mathbb{M}^4 \times \mathbb{C}$ onto itself. One also says that the map T describes a change of the bundle coordinates, that is, the bundle coordinate (P, ψ) is replaced by the new bundle coordinate $(P, G_0(P)\psi)$.

The covariant directional derivative. Let v be a smooth vector field on $\mathbb{M}^4$. Our goal is to introduce the directional derivative

$$D_v\psi$$

of the physical field ψ which possesses an invariant meaning on the space-time manifold $\mathbb{M}^4$, and which transforms like the physical field ψ under gauge transformations.

To begin with, fix an inertial system. Let us introduce the covariant partial derivatives

$$\boxed{D_\alpha\psi(x) := (\partial_\alpha + \mathcal{A}_\alpha(x))\psi(x), \qquad x \in \mathbb{R}^4, \ \ \alpha = 0,1,2,3.}$$

Here, we assume that the functions $\mathcal{A}_\alpha : \mathbb{R}^4 \to u(1)$, $\alpha = 0,1,2,3$, are smooth (i.e., $\mathcal{A}_\alpha(x)$ is a purely imaginary number). We add the following transformation laws:

- Under a change of inertial systems, $\mathcal{A}_\alpha(x)$ transforms like ∂_α.
- Under the gauge transformation (13.17), we have

  $$D^+_\alpha\psi^+(x) := (\partial_\alpha + \mathcal{A}^+_\alpha(x))\psi^+(x), \qquad x \in \mathbb{R}^4, \ \ \alpha = 0,1,2,3$$

  where

  $$\boxed{\mathcal{A}^+_\alpha(x) := G_0(x)\mathcal{A}_\alpha(x)G_0(x)^{-1} - \partial_\alpha G_0(x) \cdot G_0(x)^{-1}.} \tag{13.18}$$

  In the present commutative case, $G_0(x)\mathcal{A}_\alpha(x)G_0(x)^{-1} = \mathcal{A}_\alpha(x)$. However, formula (13.18) remains valid in the noncommutative case to be considered in Chapter 15.

Proposition 13.4 *The covariant partial derivative $D_\alpha\psi$ transforms like the physical field ψ itself.*

Proof. By the Leibniz rule,

$$\begin{aligned} D^+_\alpha\psi^+ &= D^+_\alpha(G_0\psi) = \partial_\alpha(G_0\psi) + (G_0\mathcal{A}_\alpha G_0^{-1} - \partial_\alpha G_0 \cdot G_0^{-1}) \cdot G_0\psi \\ &= G_0(\partial_\alpha\psi + \mathcal{A}_\alpha\psi) = G_0 D_\alpha\psi. \end{aligned}$$

□

Now we introduce the covariant directional derivative of a physical field ψ by setting

$$\boxed{D_v\psi := v^\alpha D_\alpha\psi \qquad \text{on } \mathbb{M}^4.}$$

By the index principle, this definition does not depend on the choice of the inertial system. Moreover, it follows from Proposition 13.4 that the gauge transformation $\psi^+(x) = G_0(x)\psi(x)$ implies

$$(D^+_v\psi^+)(x) = G_0(x)D_v\psi(x), \qquad x \in \mathbb{R}^4.$$

The curvature form. For a physical field ψ, the Leibniz rule yields

$$\begin{aligned} D_\alpha D_\beta\psi = (\partial_\alpha + \mathcal{A}_\alpha)(\partial_\beta\psi + \mathcal{A}_\beta\psi) &= \partial_\alpha\partial_\beta\psi + \mathcal{A}_\alpha\partial_\beta\psi \\ &+\partial_\alpha\mathcal{A}_\beta \cdot \psi + \mathcal{A}_\beta \cdot \partial_\alpha\psi + \mathcal{A}_\alpha\mathcal{A}_\beta\psi. \end{aligned} \tag{13.19}$$

Since $\mathcal{A}_\alpha\mathcal{A}_\beta = \mathcal{A}_\beta\mathcal{A}_\alpha$, we get

$$(D_\alpha D_\beta - D_\beta D_\alpha)\psi = \mathcal{F}_{\alpha\beta}\psi$$

where

$$\mathcal{F}_{\alpha\beta} := \partial_\alpha \mathcal{A}_\beta - \partial_\beta \mathcal{A}_\alpha. \tag{13.20}$$

The differential 2-form

$$\boxed{\mathcal{F} := \tfrac{1}{2}\mathcal{F}_{\alpha\beta}\, dx^\alpha \wedge dx^\beta}$$

possesses an invariant meaning on $\mathbb{M}^4$. It is called the curvature form on $\mathbb{M}^4$. The differential form

$$\boxed{\mathcal{A} = \mathcal{A}_\beta dx^\beta}$$

is called the connection 1-form on $\mathbb{M}^4$.

Theorem 13.5 *There hold both the Cartan structural equation*

$$\mathcal{F} = d\mathcal{A} \qquad \text{on } \mathbb{M}^4 \tag{13.21}$$

and the Bianchi equation

$$d\mathcal{F} = 0 \qquad \text{on } \mathbb{M}^4. \tag{13.22}$$

Choose an inertial system. As the following proof shows, Cartan's structural equation corresponds to (13.20), and the Bianchi equation is equivalent to

$$\partial_{[\gamma}\mathcal{F}_{\alpha\beta]} = 0, \qquad \alpha, \beta, \gamma = 0, 1, 2, 3.$$

Proof. The structural equation follows from

$$d\mathcal{A} = d\mathcal{A}_\beta \wedge dx^\beta = \partial_\alpha \mathcal{A}_\beta\, dx^\alpha \wedge dx^\beta = \tfrac{1}{2}(\partial_\alpha \mathcal{A}_\beta - \partial_\beta \mathcal{A}_\alpha)\, dx^\alpha \wedge dx^\beta.$$

Furthermore, by the Poincaré cohomology rule, $d\mathcal{F} = 0$. This yields the Bianchi equation. Explicitly,

$$d\mathcal{F} = \tfrac{1}{2} d\mathcal{F}_{\alpha\beta}\, dx^\alpha \wedge dx^\beta = \tfrac{1}{2}\partial_{[\gamma}\mathcal{F}_{\alpha\beta]}\, dx^\gamma \wedge dx^\alpha \wedge dx^\beta = 0.$$

□

Proposition 13.6 $\mathcal{F}^+ = \mathcal{F}$.

Proof. It follows from

$$D_\alpha^+(D_\beta^+\psi^+) = D_\alpha^+\big(G_0(D_\beta\psi)\big) = G_0 D_\alpha D_\beta \psi$$

that

$$\mathcal{F}_{\alpha\beta}^+\psi^+ = D_\alpha^+(D_\beta^+\psi^+) - D_\beta^+(D_\alpha^+\psi^+) = G_0\mathcal{F}_{\alpha\beta}\psi.$$

Hence $\mathcal{F}_{\alpha\beta}^+\psi^+ = G_0\mathcal{F}_{\alpha\beta}G_0^{-1}\psi^+$ for all physical fields ψ. This implies

$$\mathcal{F}^+ = G_0\mathcal{F}G_0^{-1}.$$

Commutativity yields $\mathcal{F}^+ = \mathcal{F}$. □

The electromagnetic field. Define

$$A_\alpha := -\frac{\mathrm{i}\hbar}{Q}\mathcal{A}_\alpha, \qquad F_{\alpha\beta} := -\frac{\mathrm{i}\hbar}{Q}\mathcal{F}_{\alpha\beta}.$$

Motivated by the Klein–Fock–Gordon equation (see (13.12), we have chosen the notation in such a way that the replacement $\partial_\alpha \Rightarrow \partial_\alpha + \mathcal{A}_\alpha$ implies the replacement

$$\mathrm{i}\hbar\partial_\alpha \Rightarrow \mathrm{i}\hbar\partial_\alpha - QA_\alpha.$$

Then

$$F_{\alpha\beta} = \partial_\alpha A_\beta - \partial_\beta A_\alpha, \tag{13.23}$$

$$\partial_{[\gamma} F_{\alpha\beta]} = 0 \tag{13.24}$$

for all indices $\alpha, \beta, \gamma = 0, 1, 2, 3$. Since $\mathcal{A}_\alpha(x) \in u(1)$, that is, $\mathcal{A}_\alpha(x)$ is a purely imaginary number, we get

$$A_\alpha(x) \in \mathbb{R} \quad \text{and} \quad F_{\alpha\beta}(x) \in \mathbb{R}$$

for all $x \in \mathbb{R}^4$ and all indices. As we will show in Sect. 19.3 on page 960, the equation (13.23) describes the relation between the electromagnetic field $F_{\alpha\beta}$ and its 4-potential A_α. Setting

$$F = \tfrac{1}{2} F_{\alpha\beta}\, dx^\alpha \wedge dx^\beta, \qquad A = A_\beta\, dx^\beta,$$

the Cartan structural equation and the Bianchi equation pass over to

$$\boxed{F = dA, \qquad dF = 0 \qquad \text{on } \mathbb{M}^4.}$$

The electromagnetic field corresponding to F is a solution of the Maxwell equations in a vacuum if there holds the additional equation

$$\boxed{d * F = 0 \qquad \text{on } \mathbb{M}^4.}$$

Here, we use the Hodge $*$-operator on $\mathbb{M}^4$ (see page 962).

Parallelism of a physical field along a curve. Let

$$C : P = P(\sigma), \qquad \sigma \in \mathcal{R}$$

be a curve on the Minkowski manifold $\mathbb{M}^4$. With respect to an inertial system, the curve reads as $x = x(\sigma), \sigma \in \mathcal{R}$. Here, $\mathcal{R}$ is an open interval on the real line (e.g., $\mathcal{R} = \mathbb{R}$). By definition, the physical field ψ is parallel along the curve C iff

$$\boxed{D_{\dot{P}(\sigma)}\psi(P(\sigma)) = 0, \qquad \sigma \in \mathcal{R}.} \tag{13.25}$$

This definition does not depend on the choice of the inertial system. In an inertial system, we get

$$D_{\dot{x}(\sigma)}\psi(x(\sigma)) = 0, \qquad \sigma \in \mathcal{R}.$$

Explicitly,

$$\dot{x}^\alpha(\sigma)\partial_\alpha\psi(x(\sigma)) + \dot{x}^\alpha(\sigma)\mathcal{A}_\alpha(x(\sigma)) \cdot \psi(x(\sigma)) = 0, \qquad \sigma \in \mathcal{R}.$$

By the chain rule, this is equivalent to the differential equation

$$\frac{d\psi(\sigma)}{d\sigma} + \dot{x}^\alpha(\sigma)\mathcal{A}_\alpha(x(\sigma)) \cdot \psi(\sigma) = 0, \qquad \sigma \in \mathcal{R}. \tag{13.26}$$

Here we set $\psi(\sigma) := \psi(x(\sigma))$.

Proposition 13.7 *The notion of parallelism of a physical field along a curve is gauge invariant.*

Proof. It follows from

$$D^+_{\dot{x}(\sigma)}\psi^+(x(\sigma)) = G_0(x(\sigma))D_{\dot{x}(\sigma)}\psi(x(\sigma))$$

that $D_{\dot{x}(\sigma)}\psi(x(\sigma)) = 0$ implies $D^+_{\dot{x}(\sigma)}\psi^+(x(\sigma)) = 0$. □

The covariant differential of a physical field. For a physical field ψ, we define

$$(D\psi)_P(v) := D_{v(P)}\psi(P)$$

for all vector fields v on the Minkowski manifold $\mathbb{M}^4$. Explicitly, choosing an inertial system, we get

$$D\psi(x) = D_\alpha\psi(x)\, dx^\alpha, \qquad x \in \mathbb{R}^4.$$

13.4 The Principal Bundle $\mathbb{M}^4 \times U(1)$ and the Parallel Transport of the Local Phase Factor

The principal bundle $\mathbb{M}^4 \times U(1)$ describes the transport of the local phase factor.

Folklore

The transport equation for the local phase factor. This fundamental equation reads as

$$\boxed{\dot{G}(\sigma) = -\mathcal{A}_\alpha(x(\sigma))\dot{x}^\alpha(\sigma) \cdot G(\sigma), \quad \sigma \in \mathcal{R}.} \tag{13.27}$$

We are given the smooth curve $C : P = P(\sigma), \sigma \in \mathcal{R}$ on $\mathbb{M}^4$. We are looking for a smooth function

$$G : \mathcal{R} \to U(1).$$

From the physical point of view, the differential equation (13.27) describes the transport $\sigma \mapsto G(\sigma)$ of a local phase factor along the curve C.

The principal bundle $\mathbb{M}^4 \times U(1)$. Naturally enough, we regard the solution

$$\sigma \mapsto (x(\sigma), G(\sigma))$$

of the transport equation (13.27) as the trajectory of a dynamical system on the principal product bundle

$$\mathbb{M}^4 \times U(1) := \{(P, G) : \; P \in \mathbb{M}^4, \; G \in U(1)\}.$$

Here, the fiber $F_P = \{(P, G) : \; G \in U(1)\}$ is in one-to-one correspondence to the Lie group $U(1)$. This property of the fiber is typical for a principal bundle. Let $G_0 \in U(1)$. The map

$$(P, G) \mapsto (P, G^+(P))$$

with $G^+(P) = G_0(P)G$ is called a gauge transformation of $\mathbb{M}^4 \times U(1)$. This can be regarded as a change of the bundle coordinates from (P, G) to $(P, G^+(P))$.

Proposition 13.8 *The parallel transport of the local phase factor is gauge invariant.*

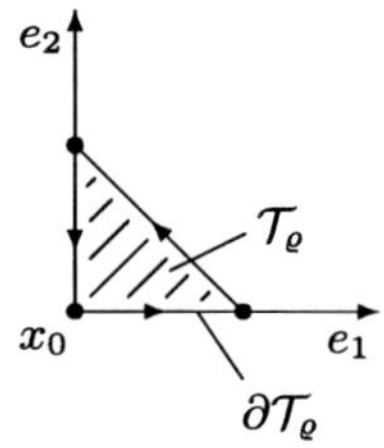

Fig. 13.1. Parallel transport along a loop

Proof. Set $G^+(\sigma) := G_0(x(\sigma))G(\sigma)$. We have to show that

$$\dot{G}^+(\sigma) = -\mathcal{A}_\alpha^+(x(\sigma))\dot{x}^\alpha(\sigma) \cdot G^+(\sigma). \tag{13.28}$$

In fact, by the Leibniz rule, we get

$$\dot{G}^+(\sigma) = \partial_\alpha G_0\big(x(\sigma)\big)\dot{x}^\alpha(\sigma) \cdot G(\sigma) + G_0\big(x(\sigma)\big)\dot{G}(\sigma).$$

By (13.27),

$$\dot{G}^+(\sigma) = \partial_\alpha G_0 \cdot \dot{x}^\alpha G_0^{-1}G^+ - G_0\mathcal{A}_\alpha\dot{x}^\alpha G_0^{-1}G^+.$$

Using the transformation law (13.18) from $\mathcal{A}_\alpha$ to $\mathcal{A}_\alpha^+$, we get the claim (13.28). □

The connection form on the principal bundle $\mathbb{M}^4 \times U(1)$. We define

$$\mathsf{A}_{(x,G)} := \mathcal{A}_\alpha(x)dx^\alpha + G^{-1}dG.$$

This differential 1-form on the principal product bundle $\mathbb{M}^4 \times U(1)$ is called the connection form of $\mathbb{M}^4 \times U(1)$. Let

$$C: \quad x = x(\sigma), \;\; G = G(\sigma), \qquad \sigma \in \mathcal{R} \tag{13.29}$$

be a smooth curve on $\mathbb{M}^4 \times U(1)$. Then

$$\mathsf{A}_{(x(\sigma),G(\sigma))}(\dot{x}(\sigma), \dot{G}(\sigma)) = \mathcal{A}_\alpha(x(\sigma))\dot{x}^\alpha(\sigma) + G(\sigma)^{-1}\dot{G}(\sigma).$$

Thus, the curve C is a solution of the equation (13.27) of parallel transport iff

$$\boxed{\mathsf{A} = 0 \qquad \text{along } C.}$$

Explicitly, $\mathsf{A}_{(x(\sigma),G(\sigma))}(\dot{x}(\sigma), \dot{G}(\sigma)) = 0$ for all $\sigma \in \mathcal{R}$. The 1-form A is called the connection 1-form on the principal bundle $\mathbb{M}^4 \times U(1)$, since A governs the parallel transport on $\mathbb{M}^4 \times U(1)$ which connects the fibers with each other.

The curvature form on $\mathbb{M}^4 \times U(1)$. We set

$$\boxed{\mathsf{F} := d\mathsf{A}.}$$

The differential 2-form F on $\mathbb{M}^4 \times U(1)$ is called the curvature 2-form on the principal bundle $\mathbb{M}^4 \times U(1)$.

13.5 Parallel Transport of Physical Fields – the Propagator Approach

The parallel transport of the local phase factor on the principal bundle $\mathbb{M}^4 \times U(1)$ induces the parallel transport on the associated vector bundle $\mathbb{M}^4 \times \mathbb{C}$. The curvature can be measured by the parallel transport of a physical field along a small loop. The covariant directional derivative of a physical field can be regarded as an infinitesimal parallel transport.

Folklore

Parallel transport for physical fields and the propagator. Choose the curve $C : P = P(\sigma)$, $\sigma \in \mathcal{R}$ on the base manifold $\mathbb{M}^4$. Consider a solution $\sigma \mapsto (x(\sigma), G(\sigma))$ of the transport equation (13.27) with $G(\sigma_0) := 1$ and fixed parameter $\sigma_0 \in \mathcal{R}$. For given $\psi_0 \in \mathbb{C}$, we define

$$\psi(\sigma) := G(\sigma)\psi_0, \qquad \sigma \in \mathcal{R}.$$

Then $\psi(\sigma_0) = \psi_0$. Using the transport equation (13.27), we get the differential equation

$$\dot{\psi}(\sigma) + \mathcal{A}_\alpha(x(\sigma))\dot{x}^\alpha(\sigma) \cdot \psi(\sigma) = 0, \qquad \sigma \in \mathcal{R} \tag{13.30}$$

which is identical with (13.26). We say that the curve

$$\sigma \mapsto (x(\sigma), \psi(\sigma))$$

on $\mathbb{M}^4 \times \mathbb{C}$ describes a parallel transport on the vector bundle $\mathbb{M}^4 \times \mathbb{C}$ which connects the point $(x(\sigma_0), \psi(\sigma_0))$ with the point $(x(\sigma), \psi(\sigma))$. We define the propagator

$$\Pi(\sigma, \sigma_0)(x(\sigma_0), \psi(\sigma_0)) := (x(\sigma), \psi(\sigma)), \qquad \sigma \in \mathcal{R}.$$

To simplify notation, we also briefly write $\Pi(\sigma, \sigma_0)\psi(\sigma_0) := \psi(\sigma)$.

Parallel transport along loops and curvature. We want to show that the curvature form $\mathcal{F}$ can be computed by using the parallel transport along sufficiently small loops. Fix the point $x_0 \in \mathbb{R}^4$. Consider the triangle $\mathcal{T}_\varrho$ depicted in Fig. 13.1. This triangle is contained in a 2-dimensional plane. This plane is located in $\mathbb{R}^4$; it passes through the point x_0, and it is spanned by the unit vectors e_1 and e_2. Explicitly,

$$\mathcal{T}_\varrho := \{x_0 + \xi e_1 + \eta e_2 : 0 \le \xi, \eta \le \varrho, \ \ \xi + \eta \le \varrho\}, \qquad \varrho > 0.$$

We assume that the boundary $\partial \mathcal{T}_\varrho$ of the triangle is positively oriented. Moreover, let $\text{meas}(\mathcal{T}_\varrho) = \frac{1}{2}\varrho^2$ denote the surface area of the triangle $\mathcal{T}_\varrho$.

For given value $\psi_0 \in \mathbb{C}$ of the physical field at the point x_0, let us transport ψ_0 along the positively oriented loop $\partial \mathcal{T}_\varrho$. After surrounding counter-clockwise the triangle once, we get the value $\Pi_{\partial \mathcal{T}_\varrho}\psi_0$ at the final point x_0.

Proposition 13.9 *The curvature component $\mathcal{F}_{12}(x_0)$ is given by the limit*

$$\mathcal{F}_{12}(x_0)\psi_0 = \lim_{\varrho \to 0} \frac{\psi_0 - \Pi_{\partial \mathcal{T}_\varrho}\psi_0}{\text{meas}(\mathcal{T}_\varrho)}.$$

Analogous expressions are obtained for $\mathcal{F}_{\alpha\beta}$ with $\alpha < \beta$.

Proof. Set $\mathcal{A} := \mathcal{A}_\beta dx^\beta$. Since $d\mathcal{A} = \mathcal{F}$, it follows from the Stokes integral theorem that

$$\int_{\partial\mathcal{T}_\varrho} \mathcal{A} = \int_{\mathcal{T}_\varrho} d\mathcal{A} = \int_{\mathcal{T}_\varrho} \mathcal{F} = \int_{\mathcal{T}_\varrho} \mathcal{F}_{12}\, dx^1 \wedge dx^2.$$

By the mean value theorem for integrals,

$$\int_{\partial\mathcal{T}_\varrho} \mathcal{A} = \tfrac{1}{2}\varrho^2 \cdot \mathcal{F}_{12}(x_0) + o(\varrho^2), \quad \varrho \to 0.$$

Integrating the differential equation (13.30), we get

$$\Pi_{\partial\mathcal{T}_\varrho}\psi_0 = \psi_0 - \tfrac{1}{2}\varrho^2\mathcal{F}_{12}(x_0)\psi_0 + o(\varrho^2), \qquad \varrho \to 0.$$

This implies the claim. □

Infinitesimal parallel transport and the covariant directional derivative of a physical field. Let ψ be a physical field on $\mathbb{M}^4$. In other words, ψ is a smooth section of the vector bundle $\mathbb{M}^4 \times \mathbb{C}$. Let $C : P = P(\sigma), \sigma \in P$, be a smooth curve on $\mathbb{M}^4$. Set $\psi(\sigma) := \psi(P(\sigma))$.

Proposition 13.10 *There holds*

$$D_{\dot{P}(0)}\psi(0) = \lim_{\sigma\to 0} \frac{\Pi(0,\sigma)\psi(\sigma) - \psi(0)}{\sigma}.$$

Intuitively, this means that we transport the value $\psi(\sigma)$ of the physical field ψ at the point $P(\sigma)$ along the curve C to the point $P(0)$. This yields $\Pi(0,\sigma)\psi(\sigma)$. Then we compare this value with the value $\psi(0)$ of the physical field ψ at the point $P(0)$ by computing the difference quotient

$$\frac{\Pi(0,\sigma)\psi(\sigma) - \psi(0)}{\sigma}.$$

Finally, let the curve parameter σ go to zero.

Proof. Since the parallel transport does not depend on the choice of the inertial system, we can use a fixed inertial system. We have to show that

$$D_{\dot{x}(0)}\psi(0) = \lim_{\sigma\to 0} \frac{\Pi(0,\sigma)\psi(\sigma) - \psi(0)}{\sigma}.$$

By the construction of the propagator, it follows from the transport equation (13.30) that

$$\Pi(0,\sigma)\psi(\sigma) = \psi(\sigma) + \sigma \cdot \dot{x}^\alpha(\sigma)\mathcal{A}_\alpha(x(\sigma))\psi(\sigma) + o(\sigma), \qquad \sigma \to 0.$$

Using the Taylor expansion, $\psi(\sigma) = \psi(0) + \sigma\dot{\psi}(0) + o(1), \sigma \to 0$, we get

$$\lim_{\sigma\to 0} \frac{\Pi(0,\sigma)\psi(\sigma) - \psi(0)}{\sigma} = \dot{\psi}(0) + \dot{x}^\alpha(0)\mathcal{A}_\alpha(x(0))\psi(0).$$

This is the claim. □

13.6 The Wilson Loop and Holonomy

Let $C : P = P(\sigma)$, $\sigma \in [\sigma_0, \sigma_1]$, be a loop on the base manifold $\mathbb{M}^4$ with

$$P(\sigma_0) = P(\sigma_1).$$

For given point (P, ψ_0) on the vector bundle $\mathbb{M}^4 \times \mathbb{C}$, we define

$$W_C(P(\sigma_0), \psi_0) := (P(\sigma_0), \Pi(\sigma_1, \sigma_0)\psi_0).$$

The operator

$$W_C : F_{P(\sigma_0)} \to F_{P(\sigma_0)}$$

on the fiber $F_{P(\sigma_0)}$ is called the Wilson loop corresponding to the loop C. The map

$$C \to W_C$$

is called the holonomy map (with respect to the fiber $F_{P(\sigma_0)}$). Intuitively, the Wilson loop W_C measures the parallel transport along the loop C. Using Lagrange's variation-of-parameter formula (see Sect. 7.17.1 of Vol. I), we get

$$\Pi(\sigma_1, \sigma_0)\psi_0 := \mathrm{e}^{-\int_{\sigma_0}^{\sigma_1} \dot{x}^\alpha(\sigma)\mathcal{A}_\alpha(x(\sigma))\,d\sigma}\,\psi_0.$$

Hence

$$W_C(P(\sigma_0), \psi_0) = (P(\sigma_0), \mathrm{e}^{-\int_C \mathcal{A}}\,\psi_0).$$

Problems

13.1 *The gauge invariance of the Schrödinger equation.* Show that the Schrödinger equation (13.7) is invariant under the gauge transformations (13.9), (13.10) on page 818.

13.2 *Differential for the inverse matrix.* Fix $N = 1, 2, \ldots$ Suppose that the matrix functions $G, H : \mathbb{R}^4 \to GL(N, \mathbb{C})$ are smooth. Show that

$$\partial_\alpha G(x)^{-1} = -G(x)^{-1}\partial_\alpha G(x) \cdot G(x)^{-1}, \qquad x \in \mathbb{R}^4.$$

Hence

$$dG(x)^{-1} = -G(x)^{-1} dG(x) \cdot G(x)^{-1}, \qquad x \in \mathbb{R}^4.$$

Solution: By the Leibniz rule,

$$\partial_\alpha(GH) = (\partial_\alpha G)H + G\partial_\alpha H.$$

Thus, it follows from $\partial_\alpha(G^{-1}G) = \partial_\alpha I = 0$ that

$$\partial_\alpha G^{-1} \cdot G + G^{-1}\partial_\alpha G = 0.$$

Finally, note that $dH = \partial_\alpha H\, dx^\alpha$.

13.3 *Construction of a connection.* We are given the smooth functions

$$G_0, G_1 : \mathbb{R}^4 \to U(1).$$

Set $G_2(x) := G_1(x)G_0(x)^{-1}$. Then it follows from

$$\psi^+(x) = G_0(x)\psi(x) \quad \text{and} \quad \psi^{++}(x) = G_1(x)\psi(P) \tag{13.31}$$

that

$$\psi^{++}(x) = G_2(x)\psi^+(x). \tag{13.32}$$

Let $\alpha = 0, 1, 2, 3$. Suppose that

$$\mathcal{A}^+_\alpha(x) := \mathcal{A}_\alpha(x) - \partial_\alpha G_0(x) \cdot G_0(x)^{-1},$$

and

$$\mathcal{A}^{++}_\alpha(x) := \mathcal{A}_\alpha(x) - \partial_\alpha G_1(x) \cdot G_1(x)^{-1}.$$

Show that

$$\mathcal{A}^{++}_\alpha(x) := \mathcal{A}^+_\alpha(x) - \partial_\alpha G_2(x) \cdot G_2(x)^{-1}. \tag{13.33}$$

Hint: Use Problem 13.2.
Remark. The transformation property (13.33) allows us to construct the family of connection matrices together with compatible transformation laws. We proceed as follows. Fix an inertial system. We are given the smooth functions

$$\mathcal{A}_\alpha : \mathbb{R}^4 \to u(1), \qquad \alpha = 0, 1, 2, 3.$$

Consider the gauge transformation (13.31), and construct both $\mathcal{A}^+$ and $\mathcal{A}^{++}$, as above. Then, the relation (13.33) fits the gauge transformation (13.32). Under a change of the inertial system, the matrices $\mathcal{A}_\alpha$ are transformed as ∂_α, $\alpha = 0, 1, 2, 3$.

14. Symmetry Breaking

We want to study the typical behavior of physical fields near a ground state (also called vacuum). It happens frequently that the ground state of a many-particle system is not unique. In this case, the system can oscillate near different ground states which, as a rule, corresponds to different physical behavior. Therefore, the choice of the ground state plays a crucial role. Historically, Pauli criticized the formulation of gauge field theories by Yang and Mills in 1954; Pauli emphasized that the corresponding interacting gauge particles are massless, in contrast to physical experiments. This defect of gauge theories could be cured in the 1960s by using the so-called Higgs mechanism which equips the gauge bosons with mass. This way, the $W^{\pm}$-bosons and the Z^0-boson obtain their mass in the Standard Model of particle physics. Physicists speak of symmetry breaking (or loss of symmetry) for the following reason.

- The original theory possesses a family of ground states which can be transformed into each other by using the symmetry group $\mathcal{G}$ of the theory.
- In nature, physical systems oscillate frequently near a distinguished ground state. These realistic states are not anymore symmetric under the original symmetry group $\mathcal{G}$. In this sense, the symmetry group $\mathcal{G}$ is broken.

14.1 The Prototype in Mechanics

Consider the equation of motion

$$m\ddot{q} = K(q) \tag{14.1}$$

for the trajectory $q = q(t)$ of a particle on the real line. Here, the force has the form $K(q) = -U'(q)$ with the Ginzburg–Landau potential

$$\boxed{U(q) := (q^2 - a^2)^2}$$

for fixed $a > 0$. The energy is given by

$$E = \frac{m}{2}\dot{q}^2 + U(q).$$

Each solution $q = q(t)$ of equation (14.1) has constant energy. In fact,

$$\dot{E}(t) = \big(m\ddot{q}(t) + U'(q(t))\big)\dot{q}(t) = 0$$

for all times t. Taylor expansion at the point q_0 yields

$$K = -U'(q_0) - U''(q_0)(q - q_0) + o(q - q_0), \quad q \to q_0.$$

E. Zeidler, *Quantum Field Theory III: Gauge Theory*,
DOI 10.1007/978-3-642-22421-8_15,

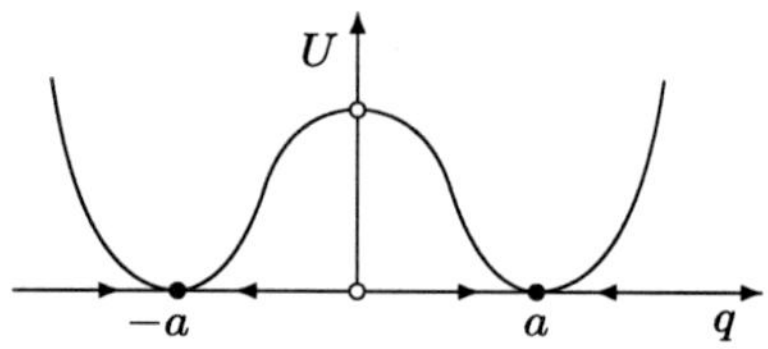

Fig. 14.1. The Ginzburg–Landau potential

Note that $U'(q) = 4(q^2 - a^2)q$ and $U''(q) = 12q^2 - 4a^2$. The critical equation

$$\boxed{U'(q_0) = 0}$$

has the solutions $q_0 = \pm a, 0$. The force vanishes precisely at these equilibrium points. Since $U''(\pm a) > 0$ and $U''(0) < 0$,

- the force $K = -U''(a)(q-a) + \ldots$ is attracting near the equilibrium point $q = a$,
- the force $K = -U''(-a)(q + a) + \ldots$ is attracting near the equilibrium point $q = -a$, and
- the force $K = -U''(0)q$ is repelling near the equilibrium point $q = 0$ (Fig. 14.1).

Consequently, the Ginzburg–Landau potential describes the motion of a particle that has the two stable equilibrium points $q = \pm a$ and one unstable equilibrium point $q = 0$. A passage from $q = a$ to $q = -a$ models a phase transition. To explain this, consider a particle which oscillates near the equilibrium point $q = a$. If the oscillations become so large that the particle passes the point $q = 0$, then the particle is attracted by the equilibrium point $q = -a$, and it starts oscillating near $q = -a$. Generally, essential changes of the qualitative behavior of physical systems are called phase transitions by physicists. The variational problem corresponding to the differential equation (14.1) reads as

$$\int_{t_0}^{t_1} \left(\frac{m}{2}\dot{q}^2 - U(q)\right)dt = \text{critical!}$$

with fixed values $q(t_0)$ and $q(t_1)$.

14.2 The Goldstone-Particle Mechanism

The original variational problem. Let us study a special nonlinear Klein–Fock–Gordon equation based on the Lagrangian

$$\mathcal{L} := \frac{1}{2}\left(\partial_\mu \psi \partial^\mu \psi^\dagger - V(|\psi|^2)\right)$$

with the Ginzburg–Landau potential

$$V(|\psi|^2) := b\left(|\psi|^2 - a^2\right)^2$$

where a and b are positive constants. Consider the corresponding variational problem

$$\boxed{\int_\Omega \mathcal{L}(\psi, \partial\psi)\, d^4x = \text{critical!}} \tag{14.2}$$

in an inertial system where Ω is a nonempty bounded open subset of $\mathbb{R}^4$. Moreover, the smooth field $\psi : \text{cl}(\Omega) \to \mathbb{C}$ is fixed on the boundary. By Prop. 13.1 on page 816, each solution of the variational problem (14.2) satisfies the nonlinear Klein–Fock–Gordon equation

$$\Box\psi + 2b\big(|\psi|^2 - a^2\big)\psi = 0. \tag{14.3}$$

The ground states. The states $\psi_0(\mathbf{x}, t) = \text{const}$ with $|\psi_0| = a$ play a special role. Since $|\psi_0|^2 - a^2 = 0$, the states ψ_0 are solutions of the Klein–Fock–Gordon equation (14.3). Moreover, since $V(|\psi_0|^2) = 0$, it follows from (13.5) on page 817 that the total energy of the states ψ_0 is equal to zero, and hence it is minimal. Thus, the states ψ_0 are ground states.

The global symmetry of the Lagrangian. The Lagrangian $\mathcal{L}$ is invariant under the global gauge transformation

$$\psi_+(\mathbf{x}, t) = \mathrm{e}^{\mathrm{i}\alpha}\psi(\mathbf{x}, t)$$

where α is a fixed real number. The point is that, by a global gauge transformation, we can transform each ground state ψ_0 into the special ground state $\psi_{\text{spec}} := a$.

The modified Lagrangian. Let us study a small perturbation of the special ground state $\psi_{\text{spec}} = a$. To this end, we make the ansatz

$$\boxed{\psi = a + \varphi + \mathrm{i}\chi.}$$

Here, we assume that the small fields φ and χ are real. Substituting this into the Lagrangian, we get[1]

$$\mathcal{L} = \tfrac{1}{2}\partial_\mu\varphi\partial^\mu\varphi + \tfrac{1}{2}\partial_\mu\chi\partial^\mu\chi - \tfrac{1}{2}b(2a\varphi + \varphi^2 + \chi^2)^2.$$

Introducing the mass parameter m_0,

$$\frac{m_0^2c^2}{\hbar^2} := 4a^2b,$$

we write

$$\mathcal{L} = \mathcal{L}_{\text{free}} + \mathcal{L}_{\text{int}}.$$

Here, the free Lagrangian,

$$\mathcal{L}_{\text{free}} = \frac{1}{2}\left(\partial_\mu\varphi\partial^\mu\varphi - \frac{m_0^2c^2}{\hbar^2}\varphi^2\right) + \tfrac{1}{2}\partial_\mu\chi\partial^\mu\chi,$$

contains all the quadratic terms of the Lagrangian $\mathcal{L}$. The remaining higher-order terms form the Lagrangian $\mathcal{L}_{\text{int}}$ of interaction. For the free Lagrangian $\mathcal{L}_{\text{free}}$, the Euler–Lagrange equations read as

$$\boxed{\Box\varphi + \frac{m_0^2c^2}{\hbar^2}\varphi = 0, \qquad \Box\chi = 0.}$$

This allows us to give the following physical interpretation. We have obtained two interacting real fields φ and χ, whereas φ corresponds to a particle of rest mass m_0, and χ corresponds to a massless particle. Since χ is a scalar field, the particle is spinless. Therefore, physicists summarize this procedure by saying that

[1] Note that $|\psi|^2 = (a + \varphi)^2 + \chi^2 = a^2 + 2a\varphi + \varphi^2 + \chi^2$.

Global symmetry breaking produces a massless Goldstone boson χ.

We will now show that local symmetry breaking generates both a massive Higgs boson and a massive gauge boson.

14.3 The Higgs-Particle Mechanism

Consider the modified Lagrangian

$$\mathcal{L} = \frac{1}{2}\left(D_\mu\psi(D^\mu\psi)^\dagger - V(|\psi|^2) - \frac{1}{4}F_{\mu\nu}F^{\mu\nu}\right)$$

with the covariant partial derivative

$$D_\mu := \partial_\mu + \frac{\mathrm{i}Q}{\hbar}A_\mu$$

where A_μ is the four-potential of the electromagnetic field,

$$F_{\mu\nu} = \partial_\mu A_\nu - \partial_\nu A_\mu.$$

Again we assume that

$$V(|\psi|^2) := b\big(|\psi|^2 - a^2\big)^2$$

where a and b are positive real numbers. The Lagrangian is invariant under the gauge transformation

$$\psi_+(\mathbf{x},t) = \psi(\mathbf{x},t)\mathrm{e}^{\mathrm{i}a(\mathbf{x},t)}, \quad A_\mu^+ = A_\mu - \frac{\hbar}{Q}\,\partial_\mu a.$$

The point is that:

We can assume that the field ψ is real-valued, after a gauge transformation.

Let us make the ansatz

$$\psi(\mathbf{x},t) = a + \varphi(\mathbf{x},t)$$

where the field φ is real-valued. For the Lagrangian, we get

$$\mathcal{L} = \frac{1}{2}\left(\partial_\mu + \frac{\mathrm{i}Q}{\hbar}A_\mu\right)(a+\varphi)\left(\partial^\mu - \frac{\mathrm{i}Q}{\hbar}A^\mu\right)(a+\varphi) \\ - \frac{1}{2}\,b\big(2a\varphi + \varphi^2\big)^2 - \frac{1}{8}F_{\mu\nu}F^{\mu\nu}.$$

This yields

$$\mathcal{L} = \mathcal{L}_{\text{free}} + \mathcal{L}_{\text{int}}$$

where the free Lagrangian contains the quadratic terms,

$$\mathcal{L}_{\text{free}} = \frac{1}{2}\left(\partial_\mu\varphi\partial^\mu\varphi - \frac{m_0^2c^2}{\hbar^2}\varphi^2 - \frac{1}{4}F_{\mu\nu}F^{\mu\nu} + \frac{1}{2}\,\frac{M_0^2c^2}{\hbar^2}\,A_\mu A^\mu\right).$$

Here, we introduce the masses m_0 and M_0,

$$\frac{m_0^2c^2}{\hbar^2} := 4ba^2, \quad \frac{M_0^2c^2}{\hbar^2} := \frac{Q^2}{\hbar^2}\,.$$

Recall that $F_{\mu\nu} = \partial_\mu A_\nu - \partial_\nu A_\mu$. The Lagrangian of interaction contains terms of higher order. The Euler–Lagrange equations for the free Lagrangian read as

$$\boxed{\Box\varphi + \frac{m_0^2 c^2}{\hbar^2}\varphi = 0}$$

and

$$\boxed{\Box A_\mu + \frac{M_0^2 c^2}{\hbar^2} A_\mu = \partial_\mu(\partial^\nu A_\nu), \qquad \mu = 0, 1, 2, 3.}$$

The latter equation is called the Proca equation. Explicitly,

$$\partial_\nu \partial^\nu A_\mu + \frac{M_0^2 c^2}{\hbar^2} A_\mu = \partial_\mu(\partial^\nu A_\nu).$$

If A_μ is a solution of this equation, then applying ∂^μ, we get

$$\partial_\nu \partial^\nu (\partial^\mu A_\mu) + \frac{M_0 c^2}{\hbar^2} \partial^\mu A_\mu = \partial^\mu \partial_\mu(\partial^\nu A_\nu).$$

Since $M_0 \neq 0$, $\partial^\mu A_\mu = 0$. Hence

$$\Box A_\mu + \frac{M_0^2 c^2}{\hbar^2} A_\mu = 0.$$

Summarizing, we have obtained

- a real field φ which corresponds to a so-called Higgs boson with mass m_0, and
- the gauge field A_μ corresponds to a particle with mass M_0 called massive gauge boson.

14.4 Dimensional Reduction and the Kaluza–Klein Approach

We want to show how variational problems in five dimensions generate physical fields in four-dimensional space-time manifolds. This idea was first used by Kaluza and Klein in the early 1920s. Consider a flat 4-dimensional space-time manifold with Cartesian coordinates x^1, x^2, x^3, and the time coordinate $x^0 = ct$. Let us add an additional coordinate x^4. As a simple example, let us discuss the following variational problem[2]

$$\boxed{\int_{\Omega\times]0,1[} \left(-\frac{1}{4} G_{ij} G^{ij} + V(B, \partial B)\right) d^5x = \text{critical!}}$$

where

$$G_{ij} := \partial_i B_j - \partial_j B_i, \qquad i, j = 0, 1, 2, 3, 4,$$

and the fields B_i are fixed on the boundary of the product set $\Omega\times]0,1[$. Here, Ω is a nonempty bounded open subset of $\mathbb{R}^4$. Furthermore, $d^5x := dx^0 dx^1 dx^2 dx^3 dx^4$. Now to the point.

---

[2] We sum over equal upper and lower Latin (resp. Greek indices) from 0 to 4 (resp. 0 to 3).

Assume that the gauge potential B does not depend on the additional variable x^4.

Let us define the so-called Higgs field

$$\boxed{\varphi(x) := B_4(x)}$$

where $x = (x^0, x^1, x^2, x^3)$. Then

$$G_{44} = 0, \quad G_{\mu 4} = -G_{4\mu} = \partial_\mu \varphi.$$

Moreover, letting $A_\mu := B_\mu$ and $F_{\mu\nu} := G_{\mu\nu}$ for $\mu, \nu = 0, 1, 2, 3$, we get

$$G_{ij}G^{ij} = F_{\mu\nu}F^{\mu\nu} + 2\partial_\mu\varphi\partial^\mu\varphi.$$

This way, we obtain the reduced variational problem

$$\boxed{\int_\Omega \left(-\frac{1}{4}F_{\mu\nu}F^{\mu\nu} - \frac{1}{2}\partial_\mu\varphi\partial^\mu\varphi + V(\varphi, \partial\varphi, A, \partial A) \right) d^4x = \text{critical!}}$$

This problem corresponds to the variational problem for a gauge field $A_0, \ldots, A_3$ and a Higgs field φ. The term V is responsible for interactions between the Higgs field and the gauge field.

14.5 Superconductivity and the Ginzburg–Landau Equation

> As you know, very many metals become superconducting below a certain temperature first discovered by Onnes in 1911.[3] The critical temperature is different for different metals. When you reduce the temperature sufficiently, the metals conduct electricity without any resistance... It took a very long time to understand what was going on inside a superconductor. It turns out that due to the interactions in the lattice, there is a small net attraction between electrons. The result is that the electrons form together, if I may speak very qualitatively and crudely, bound pairs (called Cooper pairs). Now you know that a single electron is a Fermi particle (fermion). But a bound pair acts as a Bose particle (boson)... This fundamental point in the theory of superconductivity was first explained by Bardeen, Cooper, and Schrieffer in 1957.[4]
>
> Richard Feynman, 1963

[3] The following physicists were awarded the Nobel prize in physics for their contributions to superconductivity: Heike Kammerlingh Onnes 1913, Lev Landau 1962, John Bardeen, Leon Cooper, and Robert Schrieffer 1972 (quantum field theory of Cooper pairs), Georg Bednorz and Alexander Müller 1987 (experimental discovery of high-temperature superconductivity in ceramic materials), Ginzburg 2003. The phenomenological Ginzburg–Landau theory for superconductivity is a typical nonlinear gauge field theory. Typical features of the Ginzburg–Landau model are also used in the Standard Model in particle physics (the Higgs mechanism).

[4] R. Feynman, R. Leighton, and M. Sands, The Feynman Lectures in Physics, Addison-Wesley, Reading, Massachusetts (reprinted with permission).

> The existing phenomenological theory of superconductivity due to Fritz and Heinz London from the year 1935 is unsatisfactory since it does not allow us to determine the surface tension at the boundary between the normal and the superconducting phases and does not allow for the possibility to describe correctly the destruction of superconductivity by a magnetic field or current. In the present paper a theory is constructed which is free of these faults. We find equations for the effective wave function ψ of the 'superconducting electrons' which we introduced and for the vector potential $\mathbf{A}$. We have solved these equations for the one-dimensional case. The essence of the matter lies in the fact that the function ψ is in no way a true wave function of the electrons in the metal, but is a certain average quantity.[5]
>
> Vitaly Ginzburg and Lev Landau, 1950

> A superconductor is simply a material in which electromagnetic gauge invariance is spontaneously broken. Detailed dynamical theories are needed to explain why and at what temperature this symmetry breaking occurs, but they are not needed to derive the most striking aspects of superconductivity: exclusion of magnetic fields, flux quantization, zero resistivity, and alternating electric currents at a gap between superconductors held at different voltages. As we will see here, these consequences of broken gauge invariance can be worked out in a manner somewhat like our treatment of soft pions, solely on the basis of general properties of the Goldstone mode.[6]
>
> Steven Weinberg, 1996

Experimentally, superconductivity was discovered by Kamerlingh Onnes in 1908. This phenomenon consists in the vanishing of electric resistance at low temperatures. We want to show that superconductivity can be understood qualitatively in terms of gauge theory. The first phenomenological theory of superconductivity was formulated by Fritz and Heinz London in 1935. They supposed that, in a superconductor, the electric current density vector $\mathbf{J}$ is related to the electric field $\mathbf{E}$ by the constitutive equation law

$$\boxed{\frac{\partial \mathbf{J}}{\partial t} = \gamma \mathbf{E}}$$

where the positive constant γ only depends on temperature. Recall that, in a normal metallic conductor, Ohm's law tells us that

$$\mathbf{J} = \sigma \mathbf{E}$$

where the positive constant σ is called electrical conductivity. In a normal conductor, a constant electric field $\mathbf{E}$ generates a constant electric current. In contrast to this classical behavior, a constant electric field generates an increasing electric current in a superconductor. The London model was substantially improved by Ginzburg and Landau in 1950. They included thermodynamical effects on a phenomenological level. The final microscopic theory was formulated by Bardeen, Cooper, and Schrieffer in 1957 (Nobel prize in physics in 1972). They assumed that

[5] V. Ginzburg and L. Landau, On the theory of superconductivity, J. Experimental and Theoretical Physics **20**, 1064–1082 (in Russian). English translation in L. Landau, Collected Papers, pp. 546–568, Pergamon Press, Oxford (reprinted with permission).

[6] S. Weinberg, The Quantum Theory of Fields, Vol.II, Cambridge University Press, 1996 (reprinted with permission).

pairs of electrons called Cooper pairs are responsible for superconductivity. First of all, in contrast to electrons, Cooper pairs are bosons (with zero spin). Therefore, the number of Cooper pairs in a physical state is not restricted by Pauli's exclusion principle, telling us that two fermions (e.g., electrons) can never be in the same physical state. The size of a Cooper pair is a few hundred nanometers, far larger than the size of an atom. The Cooper pairs move with the same drift speed. From the physical point of view, the formation of Cooper pairs causes a gap in the energy distribution of the free electrons. This energy gap prevents Cooper pairs from moving to higher energy levels. Consequently, Cooper pairs cannot transform electric energy into heat energy by scattering processes, as in normal conductors.

In a phenomenological theory, the Cooper pairs are described by a complex-valued function

$$\psi(\mathbf{x},t) = \sqrt{\varrho(\mathbf{x},t)}\,\mathrm{e}^{\mathrm{i}\alpha(\mathbf{x},t)}$$

with the real phase function α; the real function $\varrho = \psi^\dagger\psi$ describes the density of Cooper pairs. We will motivate below that it is reasonable to assume that the electric current density vector $\mathbf{J}$ is given by

$$\boxed{\mathbf{J} = \frac{\hbar}{m}\left(\mathbf{grad}\,\alpha - \frac{Q}{\hbar}\mathbf{A}\right)\varrho} \tag{14.4}$$

where $\mathbf{A}$ is the vector potential of the magnetic field,

$$\mathbf{B} = \mathbf{curl}\,\mathbf{A}.$$

The effective charge Q is the charge of the Cooper pairs, $Q = -2e$, and m is the effective mass of the Cooper pairs to be determined by experiment.

The Ginzburg–Landau equation. We assume that the function ψ satisfies the Schrödinger–Maxwell equation[7]

$$\boxed{\mathrm{i}\hbar\,\psi_t = \frac{(\mathbf{P} - Q\mathbf{A})^2}{2m}\psi + QU\psi + V(|\psi|^2)} \tag{14.5}$$

where $\mathbf{P} := -\mathrm{i}\hbar\boldsymbol{\partial}$. Moreover, $(U, \mathbf{A})$ is the four-potential of the electromagnetic field,

$$\mathbf{B} = \mathbf{curl}\,\mathbf{A}, \quad \mathbf{E} = -\,\mathbf{grad}\,U - \mathbf{A}_t,$$

and V is the Ginzburg–Landau potential (i.e., the free energy of the Cooper pairs). Explicitly, we make the ansatz

$$V(|\psi|^2) = \mathrm{const} - a|\psi|^2 + b|\psi|^4,$$

and we assume that for temperatures T below the critical temperature T_{crit},

$$a(T) > 0, \quad b(T) > 0.$$

For $T > T_{\mathrm{crit}}$, suppose that

$$a(T) = 0, \quad b(T) > 0.$$

This has the following important consequence.

(i) If $T > T_{\mathrm{crit}}$, then the free energy V has a minimum at $|\psi| = 0$. Thus, there are no Cooper pairs in thermodynamical equilibrium.

[7] This equation is also called the Ginzburg–Landau equation.

(ii) If $T < T_{\rm crit}$, then the free energy V has a minimum at

$$|\psi|^2 = \frac{a(T)}{b(T)}$$

which corresponds to the density of Cooper pairs in thermodynamical equilibrium at temperature T.

By an order parameter, physicists understand a quantity η which vanishes in the disordered phase and is positive in the ordered phase. In this terminology, the density $|\psi|^2$ of Cooper pairs is an order parameter for the phase transition to superconductivity.[8]

The continuity equation. We want to show the following.

If ψ is a solution of the Schrödinger–Maxwell equation (14.5), *then we have the continuity equation*

$$\varrho_t + \operatorname{div} \mathbf{J} = 0$$

where

$$\mathbf{J} := \frac{\psi^\dagger}{2m}(\mathbf{P} - Q\mathbf{A})\psi + \frac{\psi}{2m}\left((\mathbf{P} - Q\mathbf{A})\psi\right)^\dagger. \tag{14.6}$$

The proof follows from

$$\varrho_t = \psi_t\psi^\dagger + \psi\psi_t^\dagger$$

by using (14.5) for computing ψ_t and $\psi_t^\dagger$. By definition, $\mathbf{J}$ is the electric current density vector of Cooper pairs. Inserting $\psi(\mathbf{x}, t) = \sqrt{\varrho(\mathbf{x}, t)}\, \mathrm{e}^{\mathrm{i}\alpha(\mathbf{x},t)}$ into (14.6), we obtain the key equation (14.4). We now want to use the expression (14.4) in order to understand typical properties of superconductors.

The London equation. Consider a simply connected region of a superconductor (e.g., a ball). We assume that all of the Cooper pairs are in the same state. This means that the phase function α of Cooper pairs is constant. By (14.4),

$$\mathbf{J} = -\frac{Q}{m}\mathbf{A}\varrho.$$

Suppose that there is no electric potential, $U \equiv 0$, and the density ϱ of Cooper pairs is constant. Then $\mathbf{E} = -\mathbf{A}_t$. Hence

$$\mathbf{J}_t = -\frac{Q}{m}\mathbf{A}_t\varrho = \frac{Q\varrho}{m}\mathbf{E}.$$

This is London's material law for superconductors mentioned above.

The Meissner effect. It was discovered experimentally by Meissner and Ochsenfeld in 1933 that, in a superconductor, nonvanishing magnetic fields only exist in a thin boundary layer. To understand this phenomenon, consider a stationary magnetic field $\mathbf{B} = \mathbf{curl}\,\mathbf{A}$ which satisfies the gauge condition $\operatorname{div} \mathbf{A} = 0$. It follows from the wave equation

$$\Box\mathbf{A} = \mu_0\mathbf{J}$$

along with the London equation $\mathbf{J} = -\frac{Q}{m}\mathbf{A}\varrho$ that

$$\boxed{\Delta\mathbf{A} = -\frac{\mathbf{A}}{\delta} \qquad \text{on} \quad \Omega}$$

[8] Ginzburg and Landau introduced $|\psi|^2$ by physical intuition, without knowing the physical interpretation as density of Cooper pairs.

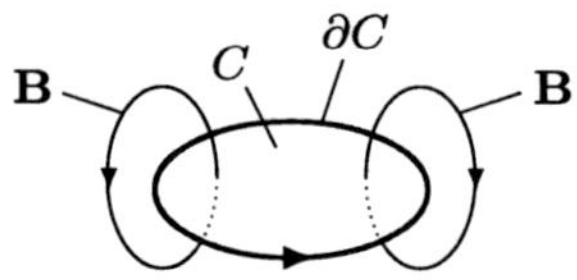

Fig. 14.2. Flux quantization

where $\delta := m/\mu_0 Q\varrho$. Consider a box $\Omega := \{(x, y, z) : 0 \leq x,\ y,\ z \leq 1\}$ and assume that the density ϱ of Cooper pairs is constant. We then get the special solution

$$\mathbf{A}(\mathbf{x}) = a\big(\mathrm{e}^{-x/\delta} + \mathrm{e}^{-(1-x)/\delta}\big)\mathbf{i} + b\big(\mathrm{e}^{-y/\delta} + \mathrm{e}^{-(1-y)/\delta}\big)\mathbf{j} + c\big(\mathrm{e}^{-z/\delta} + \mathrm{e}^{-(1-z)/\delta}\big)\mathbf{k}$$

where a, b, c are real constants. Now to the point. It turns out that δ is a small quantity. Thus, the field $\mathbf{A}$ vanishes approximately outside a δ-neighborhood of the boundary. This is the Meissner effect.

Flux quantization and Cooper pairs. Consider the situation pictured in Fig. 14.2. Assume that there is no electric current in the closed ring ∂C of a superconductor, $\mathbf{J} \equiv 0$. For the magnetic flux, we then get

$$\boxed{\int_C \mathbf{Bn}\, dS = \frac{2\pi n\hbar}{Q}, \qquad n = 0, \pm 1, \pm 2, \ldots} \tag{14.7}$$

In fact, it follows from (14.4) and $\mathbf{J} \equiv 0$ that

$$\mathbf{A} = \frac{\hbar}{Q}\,\mathbf{grad}\,\alpha.$$

By the Stokes integral theorem,

$$\int_C \mathbf{Bn}\, dS = \int_{\partial C} \mathbf{A}\, d\mathbf{x} = \frac{\hbar \Delta\alpha}{Q}.$$

If we move the phase function α along the closed curve ∂C, then the function ψ does not change. Hence

$$\Delta\alpha = 2\pi n, \quad n = 0, \pm 1, \pm 2, \ldots$$

The experiments show that $Q = -2e$ where $-e$ is the electric charge of an electron. This establishes experimentally the existence of Cooper pairs consisting of two electrons.

14.6 The Idea of Effective Theories in Physics

Typically, physical theories describe phenomena by concentrating on essential features. To this end, physicists introduce effective quantities which summarize subtle interactions in nature. The Ginzburg–Landau theory serves as a typical example for an effective theory. In this setting, we do not consider the microscopic processes which are responsible for superconduction, but we introduce the function ψ which carries a sufficient amount of physical information.

The strategy of effective theories has been very successful in physics in order to govern more and more complicated processes in nature. The task of physicists is to single out the right effective quantities by physical intuition, and to refine the effective theories on the basis of improved experiments.

There exists the dream of a final theory in physics which allows the deduction of all physical results from the fundamental interactions. The present experience of physicists indicates that it seems to be hopeless to realize such an ambitious program. For example, there is no hope to understand the properties of large molecules on the basis of a final theory for the fundamental interactions in nature. In quantum chemistry, physicists and chemists use the so-called density functional method for large molecules. This is an effective theory for describing the solutions of the Schrödinger equation for large molecules. As an introduction, we recommend the monograph by H. Eschrig, The Fundamentals of Density Functional Theory, Teubner, Leipzig, 2003. In 1998, Walter Kohn was awarded the Nobel prize in chemistry for the development of the density functional method.

15. The Noncommutative Yang–Mills $SU(N)$-Gauge Theory

Fix $N = 1, 2, \ldots$ Let $\mathcal{G}$ be a closed subgroup of the Lie matrix group $GL(N, \mathbb{C})$ of invertible complex $(N \times N)$-matrices. Then, $\mathcal{G}$ is a Lie group. As a prototype, the reader should have the special case in mind where $N = 2$ and

$$\mathcal{G} = SU(2).$$

This gauge group was used by Yang and Mills in 1954. Recall that the Lie group $SU(2)$ consists of all the unitary (2×2)-matrices U with $\det(U) = 1$. The corresponding Lie algebra $su(2)$ consists of all the complex (2×2)-matrices A with $A^\dagger = -A$ and $\mathrm{tr}(A) = 0$. Further examples for the Lie group $\mathcal{G}$ are the Lie groups $U(N), SU(N), GL(N, \mathbb{C}), SL(N, \mathbb{C})$, and $SO(3)$ with $N = 3$. We will show how the $U(1)$-gauge theory from Chap. 13 has to be modified in the case of a noncommutative gauge group $\mathcal{G}$ (e.g., $\mathcal{G} = SU(N)$ with $N = 2, 3, \ldots$).

In order to help the reader to understand the simple intuitive ideas behind the theory of vector bundles and principal bundles, in this chapter and the following two chapters, we will discuss the interrelationship between different approaches to the curvature theory of bundles.

15.1 The Vector Bundle $\mathbb{M}^4 \times \mathbb{C}^N$, Covariant Directional Derivative, and Curvature

We choose the Minkowski manifold $\mathbb{M}^4$ as in Sect. 13.3, and we set

$$\psi = \begin{pmatrix} \psi^1 \\ \vdots \\ \psi^N \end{pmatrix}$$

where $\psi^1, \ldots, \psi^N$ are complex numbers. We write $\psi \in \mathbb{C}^N$. In what follows, precisely the smooth maps $\psi : \mathbb{M}^4 \to \mathbb{C}^N$ are called physical fields.

The vector bundle $\mathbb{M}^4 \times \mathbb{C}^N$. By definition, the product set

$$\mathbb{M}^4 \times \mathbb{C}^N := \{(P, \psi) : \; P \in \mathbb{M}^4, \;\; \psi \in \mathbb{C}^N\}$$

is called a vector product bundle over the base manifold $\mathbb{M}^4$ with the fiber

$$F_P := \{(P, \psi) : \psi \in \mathbb{C}^N\}$$

over the point $P \in \mathbb{M}^4$. There exists a one-to-one correspondence between the fiber F_P and the complex linear space $\mathbb{C}^N$. More precisely, the surjective map

E. Zeidler, *Quantum Field Theory III: Gauge Theory*,
DOI 10.1007/978-3-642-22421-8_16,

$$\pi : \mathbb{M}^4 \times \mathbb{C}^N \to \mathbb{M}^4$$

given by $\pi(P, \psi) := P$ is called a vector bundle with the fiber $F_P = \pi^{-1}(P)$, the bundle space $\mathbb{M}^4 \times \mathbb{C}^N$, and the base space $\mathbb{M}^4$. For the sake of brevity, we speak of the vector bundle $\mathbb{M}^4 \times \mathbb{C}^N$. The map

$$s : \mathbb{M}^4 \to \mathbb{M}^4 \times \mathbb{C}^N$$

is called a section iff $s(P) \in F_P$ for all $P \in \mathbb{M}^4$. Hence

$$s(P) = (P, \psi(P))$$

where ψ is a map from $\mathbb{M}^4$ to $\mathbb{C}^N$. Thus, physical fields ψ and smooth sections s of the vector bundle $\mathbb{M}^4 \times \mathbb{C}^N$ can be identified with each other.

The local phase factor and gauge transformations. The transformation

$$\psi^+(P) := G_0(P)\psi(P), \qquad P \in \mathbb{M}^4 \tag{15.1}$$

is called a gauge transformation iff $G_0(P) \in \mathcal{G}$ and the map $G_0 : \mathbb{M}^4 \to \mathcal{G}$ is smooth. Setting

$$T(P, \psi) := (P, G_0(P)\psi),$$

we get the map

$$T : \mathbb{M}^4 \times \mathbb{C}^N \to \mathbb{M}^4 \times \mathbb{C}^N.$$

This is called a transition map from the product bundle $\mathbb{M}^4 \times \mathbb{C}^N$ onto itself. One also says that the map T describes a change of the bundle coordinates, that is, the bundle coordinate (P, ψ) is replaced by the new bundle coordinate $(P, G_0\psi)$.

The covariant directional derivative. Let v be a vector field on $\mathbb{M}^4$. Our goal is to introduce the directional derivative

$$D_v\psi$$

of the physical field ψ which possesses an invariant meaning on the space-time manifold $\mathbb{M}^4$, and which transforms like the physical field ψ under gauge transformations.

To begin with, fix an inertial system. Let us introduce the covariant partial derivatives

$$D_\alpha\psi(x) := (\partial_\alpha + \mathcal{A}_\alpha(x))\psi(x), \qquad x \in \mathbb{R}^4, \ \ \alpha = 0, 1, 2, 3.$$

Here, we assume that the matrix functions $\mathcal{A}_\alpha : \mathbb{R}^4 \to \mathcal{LG}$, $\alpha = 0, 1, 2, 3$, are smooth (i.e., $\mathcal{A}_\alpha(x)$ is an element of the Lie algebra $\mathcal{LG}$). We add the following transformation laws:

- Under a change of inertial systems, $\mathcal{A}_\alpha$ transforms like ∂_α.
- Under the gauge transformation (15.1), we have

$$D_\alpha^+\psi^+(x) := (\partial_\alpha + \mathcal{A}_\alpha^+(x))\psi^+(x), \qquad x \in \mathbb{R}^4, \ \ \alpha = 0, 1, 2, 3$$

  where

$$\mathcal{A}_\alpha^+(x) := G_0(x)\mathcal{A}_\alpha(x)G_0(x)^{-1} - \partial_\alpha G_0(x) \cdot G_0(x)^{-1}. \tag{15.2}$$

Note that $\mathcal{A}_\alpha^+(x)$ is an element of the Lie algebra $\mathcal{LG}$. Set

$$\mathcal{A} = \mathcal{A}_\alpha \, dx^\alpha.$$

This differential form is called the connection 1-form on the base manifold $\mathbb{M}^4$. From (15.2) we get the transformation law

$$\mathcal{A}^+(x) = G_0(x)\mathcal{A}(x)G_0(x)^{-1} - dG_0 \cdot G_0(x)^{-1}.$$

Proposition 15.1 *The covariant partial derivative $D_\alpha\psi$ transforms like the physical field ψ itself.*

The proof proceeds as on page 822. Now we introduce the covariant directional derivative of a physical field ψ by setting

$$\boxed{D_v\psi := v^\alpha D_\alpha\psi \qquad \text{on } \mathbb{M}^4.}$$

By the index principle, this definition does not depend on the choice of the inertial system. Moreover, it follows from Proposition 15.1 that the gauge transformation $\psi^+(x) = G_0(x)\psi(x)$ implies

$$(D_v^+\psi^+)(x) = G_0(x)D_v\psi(x), \qquad x \in \mathbb{R}^4.$$

The curvature form. For a physical field ψ, the Leibniz rule yields

$$\begin{aligned} D_\alpha D_\beta\psi = (\partial_\alpha + \mathcal{A}_\alpha)(\partial_\beta\psi + \mathcal{A}_\beta\psi) &= \partial_\alpha\partial_\beta\psi + \mathcal{A}_\alpha\partial_\beta\psi \\ &+\partial_\alpha\mathcal{A}_\beta \cdot \psi + \mathcal{A}_\beta \cdot \partial_\alpha\psi + \mathcal{A}_\alpha\mathcal{A}_\beta\psi. \end{aligned} \tag{15.3}$$

This implies

$$(D_\alpha D_\beta - D_\beta D_\alpha)\psi = \mathcal{F}_{\alpha\beta}\psi$$

where

$$\mathcal{F}_{\alpha\beta} := \partial_\alpha\mathcal{A}_\beta - \partial_\beta\mathcal{A}_\alpha + [\mathcal{A}_\alpha, \mathcal{A}_\beta]_- \tag{15.4}$$

with the Lie bracket $[\mathcal{A}_\alpha, \mathcal{A}_\beta]_- = \mathcal{A}_\alpha\mathcal{A}_\beta - \mathcal{A}_\beta\mathcal{A}_\alpha$. The point is that $\mathcal{F}_{\alpha\beta}$ does not depend on the second partial derivatives of $\mathcal{A}_\alpha$, $\alpha = 0, 1, 2, 3$. Moreover, $\mathcal{F}_{\alpha\beta}$ is not a differential operator acting on ψ, but it is a matrix multiplication operator. The differential 2-form

$$\boxed{\mathcal{F} := \tfrac{1}{2}\mathcal{F}_{\alpha\beta}\, dx^\alpha \wedge dx^\beta}$$

possesses an invariant meaning on $\mathbb{M}^4$. It is called the curvature 2-form on the base manifold $\mathbb{M}^4$.

Theorem 15.2 *There hold the Cartan structural equation*

$$\mathcal{F} = d\mathcal{A} + \mathcal{A} \wedge \mathcal{A} \qquad \text{on } \mathbb{M}^4 \tag{15.5}$$

and the Bianchi equation (integrability condition)

$$d\mathcal{F} = \mathcal{F} \wedge \mathcal{A} - \mathcal{A} \wedge \mathcal{F} \qquad \text{on } \mathbb{M}^4. \tag{15.6}$$

Here, $\mathcal{A}$ and $\mathcal{F}$ are matrices with differential forms as entries. The wedge product of such matrices is the usual matrix product where the classic product of the entries is replaced by the wedge product of the entries. For example,

$$\begin{pmatrix} \omega_{11} & \omega_{12} \\ \omega_{21} & \omega_{22} \end{pmatrix} \wedge \begin{pmatrix} \mu_{11} & \mu_{12} \\ \mu_{21} & \mu_{22} \end{pmatrix} = \begin{pmatrix} \omega_{11} \wedge \mu_{11} + \omega_{12} \wedge \mu_{21} & \omega_{11} \wedge \mu_{12} + \omega_{12} \wedge \mu_{22} \\ \omega_{21} \wedge \mu_{11} + \omega_{22} \wedge \mu_{21} & \omega_{21} \wedge \mu_{12} + \omega_{22} \wedge \mu_{22} \end{pmatrix}.$$

Proof. Ad (15.5). We get

$$d\mathcal{A} = d\mathcal{A}_\beta \wedge dx^\beta = \partial_\alpha\mathcal{A}_\beta\, dx^\alpha \wedge dx^\beta = \tfrac{1}{2}(\partial_\alpha\mathcal{A}_\beta - \partial_\beta\mathcal{A}_\alpha)\, dx^\alpha \wedge dx^\beta,$$

and $\mathcal{A}_\alpha \wedge \mathcal{A}_\beta = \mathcal{A}_\alpha\mathcal{A}_\beta\, dx^\alpha \wedge dx^\beta = \frac{1}{2}(\mathcal{A}_\alpha\mathcal{A}_\beta - \mathcal{A}_\beta\mathcal{A}_\alpha)\, dx^\alpha \wedge dx^\beta$.

Ad (15.6). From Cartan's structural equation (15.5) we get

$$d\mathcal{F} = d\mathcal{A} \wedge \mathcal{A} - \mathcal{A} \wedge d\mathcal{A}.$$

Since $d\mathcal{A} = \mathcal{F} - \mathcal{A} \wedge \mathcal{A}$, we obtain the claim (15.6). □

Proposition 15.3 *For all vector fields v and w on $\mathbb{M}^4$ and all physical fields ψ on $\mathbb{M}^4$, we have*

$$\mathcal{F}(v,w) = d\mathcal{A}(v,w) + [\mathcal{A}(v), \mathcal{A}(w)]_- \tag{15.7}$$

and

$$\mathcal{F}(v,w)\psi := (D_v D_w - D_w D_v - D_{[v,w]})\psi. \tag{15.8}$$

Proof. Ad (15.7). Use the structural equation (15.5) and

$$(\mathcal{A} \wedge \mathcal{A})(v,w) = \mathcal{A}(v)\mathcal{A}(w) - \mathcal{A}(w)\mathcal{A}(v).$$

Ad (15.8). Use $D_v(D_w\psi) = (v^\alpha \partial_\alpha + v^\alpha \mathcal{A}_\alpha)(w^\beta \partial_\beta \psi + w^\beta \mathcal{A}_\beta \psi)$ and

$$D_{[v,w]}\psi = (v^\mu \partial_\mu w^\gamma - w^\mu \partial_\mu v^\gamma)(\partial_\gamma \psi + \mathcal{A}_\gamma \psi).$$

□

Gauge transformation. In contrast to the $U(1)$-case, the curvature form $\mathcal{F}$ is not always invariant under gauge transformations. But there exists a simple transformation law.

Proposition 15.4 $\mathcal{F}^+(P) = G_0(P)\mathcal{F}(P)G_0(P)^{-1}$.

This follows as in the proof of Prop. 13.6 on page 823. In particular, we have

$$\mathrm{tr}(\mathcal{F}^+(P)) = \mathrm{tr}(\mathcal{F}(P)).$$

This invariance property is crucial for Chern classes.

Parallelism of a physical field along a curve. Let

$$C : P = P(\sigma), \qquad \sigma \in \mathcal{R}$$

be a curve on the Minkowski manifold $\mathbb{M}^4$. With respect to an inertial system, the curve reads as $x = x(\sigma), \sigma \in \mathcal{R}$. Here, $\mathcal{R}$ is an open interval on the real line (e.g., $\mathcal{R} = \mathbb{R}$). By definition, the physical field ψ is parallel along the curve C iff

$$\boxed{D_{\dot{P}(\sigma)}\psi(P(\sigma)) = 0, \qquad \sigma \in \mathcal{R}.} \tag{15.9}$$

This definition does not depend on the choice of the inertial system. In an inertial system, we get

$$D_{\dot{x}(\sigma)}\psi(x(\sigma)) = 0, \qquad \sigma \in \mathcal{R}.$$

Explicitly,

$$\dot{x}^\alpha(\sigma)\partial_\alpha\psi(x(\sigma)) + \dot{x}^\alpha(\sigma)\mathcal{A}_\alpha(x(\sigma))\, \psi(x(\sigma)) = 0, \qquad \sigma \in \mathcal{R}.$$

By the chain rule, this is equivalent to the differential equation

$$\frac{d\psi(\sigma)}{d\sigma} + \dot{x}^\alpha(\sigma)\mathcal{A}_\alpha(x(\sigma))\, \psi(\sigma) = 0, \qquad \sigma \in \mathcal{R}. \tag{15.10}$$

Here we set $\psi(\sigma) := \psi(x(\sigma))$.

Proposition 15.5 *The notion of parallelism of a physical field along a curve is gauge invariant.*

Proof. It follows from

$$D^+_{\dot{x}(\sigma)}\psi^+(x(\sigma)) = G_0(x(\sigma))D_{\dot{x}(\sigma)}\psi(x(\sigma))$$

that $D_{\dot{x}(\sigma)}\psi(x(\sigma)) = 0$ implies $D^+_{\dot{x}(\sigma)}\psi^+(x(\sigma)) = 0$. $\square$

The covariant differential of a physical field. For a physical field ψ, we define

$$(D\psi)_P(v) := D_{v(P)}\psi(P)$$

for all vector fields v on the Minkowski manifold $\mathbb{M}^4$. Explicitly, choosing an inertial system, we get

$$D\psi(x) = D_\alpha\psi(x)\, dx^\alpha, \qquad x \in \mathbb{R}^4.$$

Components. Note that $\mathcal{A}_\alpha(x)$ and $\mathcal{F}_{\alpha\beta}(x)$ are $(N \times N)$-matrices. For the entries, we write

$$\mathcal{A}_\alpha(x) = (\Gamma^K_{\alpha L}(x)) \qquad \text{and} \qquad \mathcal{F}_{\alpha\beta}(x) = (R^K_{\alpha\beta L}(x))$$

where $K, L = 1, 2, \ldots, N$. Here, K (resp. L) is the index of the rows (resp. columns).

15.2 The Principal Bundle $\mathbb{M}^4 \times \mathcal{G}$ and the Parallel Transport of the Local Phase Factor

The transport equation for the local phase factor. This fundamental equation reads as

$$\dot{G}(\sigma) = -\mathcal{A}_\alpha\big(x(\sigma)\big)\dot{x}^\alpha(\sigma) \cdot G(\sigma), \quad \sigma \in \mathcal{R}. \tag{15.11}$$

We are given the curve $C : P = P(\sigma), \sigma \in \mathcal{R}$ on the manifold $\mathbb{M}^4$. We are looking for a smooth function

$$G : \mathcal{R} \to \mathcal{G}.$$

From the physical point of view, the equation (15.11) describes the transport

$$\sigma \mapsto G(\sigma)$$

of a local phase factor along the curve C. Note the following. Since $\mathcal{A}_\alpha(x)$ is an element of the Lie algebra $\mathcal{LG}$, the matrix $\mathcal{A}_\alpha(x(\sigma))G(\sigma)$ is a tangent vector of the Lie group $\mathcal{G}$ at the point $G(\sigma)$. Consequently, the solution $\sigma \mapsto G(\sigma)$ of the matrix differential equation (15.11) is a curve in the Lie group $\mathcal{G}$. Furthermore,

$$G(\sigma)^{-1}\mathcal{A}_\alpha(x(\sigma))G(\sigma) \in \mathcal{LG} \quad \text{and} \quad G(\sigma)^{-1}\dot{G}(\sigma) \in \mathcal{LG}.$$

Equivalently, the transport equation (15.11) can be written as

$$\boxed{G(\sigma)^{-1}\dot{G}(\sigma) + G(\sigma)^{-1}\mathcal{A}_\alpha\big(x(\sigma)\big)\dot{x}^\alpha(\sigma)\, G(\sigma) = 0, \quad \sigma \in \mathcal{R}.} \tag{15.12}$$

The equations (15.11) and (15.12) do not depend on the choice of the inertial system.

Gauge transformation. Let $G_0 : \mathbb{M}^4 \to \mathcal{G}$ be a smooth map. We choose an inertial system, and we set[1]

[1] To simplify notation, we will synonymously use the symbols G_+ and G^+.

$$G^+(x) := G_0(x)G(x), \qquad x \in \mathbb{R}^4. \tag{15.13}$$

With respect to the curve $C : x = x(\sigma), \sigma \in \mathcal{R}$, we define

$$G^+(\sigma) := G_0(x(\sigma))G(\sigma), \qquad \sigma \in \mathcal{R}.$$

The following simple identity is the key to Cartan's approach to curvature.

Theorem 15.6 *For all parameters $\sigma \in \mathcal{R}$, we get the crucial formula*

$$\begin{aligned} &G_+(\sigma)^{-1}\dot{G}_+(\sigma) + G_+(\sigma)^{-1}\dot{x}^\alpha(\sigma)\mathcal{A}^+_\alpha(x(\sigma))\, G_+(\sigma) \\ &= G(\sigma)^{-1}\dot{G}(\sigma) + G(\sigma)^{-1}\dot{x}^\alpha(\sigma)\mathcal{A}_\alpha(x(\sigma))\, G(\sigma). \end{aligned} \tag{15.14}$$

Proof. By the chain rule, $\dot{G}_0(\sigma) = \dot{x}^\alpha(\sigma)\partial_\alpha G_0(x(\sigma))$. It follows from (15.2) that

$$\dot{x}^\alpha(\sigma)\mathcal{A}^+_\alpha(x(\sigma)) = G_0(\sigma)\mathcal{A}_\alpha(x(\sigma))\, G_0(\sigma)^{-1} - \dot{G}_0(\sigma)G_0(\sigma)^{-1}.$$

Note that $G_+^{-1} = (G_0G)^{-1} = G^{-1}G_0^{-1}$ and $\dot{G}_+ = \dot{G}_0G + G_0\dot{G}$. For all parameters $\sigma \in \mathcal{R}$, this implies

$$\begin{aligned} G_+^{-1}\dot{G}_+ + G_+^{-1}\dot{x}^\alpha\mathcal{A}^+_\alpha\, G_+ &= G_+^{-1}(\dot{G}_0G + G_0\dot{G} - \dot{G}_0G) + G^{-1}\dot{x}^\alpha\mathcal{A}_\alpha G \\ &= G^{-1}\dot{G} + G^{-1}\dot{x}^\alpha\mathcal{A}_\alpha\, G. \end{aligned}$$

□

This theorem implies that if $\sigma \mapsto (x(\sigma), G(\sigma))$ is a solution of the transport equation (15.12), then the transformed curve $\sigma \mapsto (x(\sigma), G_+(\sigma))$ is a solution of the transformed equation

$$G_+(\sigma)^{-1}\dot{G}_+(\sigma) + G_+(\sigma)^{-1}\dot{x}^\alpha(\sigma)\mathcal{A}^+_\alpha(x(\sigma))\, G_+(\sigma) = 0.$$

The language of differential forms. Differential forms are invariant objects on manifolds.

> *Therefore, it is our next goal to describe the transport of the local phase factor in terms of differential forms on the manifold $\mathbb{M}^4 \times \mathcal{G}$.*

To this end, we will use the concept of the connection form A on a principal bundle.

The principal bundle $\mathcal{P}$. In what follows, we set

$$\mathcal{P} = \mathbb{M}^4 \times \mathcal{G} := \{(P, G) : \ P \in \mathbb{M}^4, \ G \in \mathcal{G}\}.$$

The solution $\sigma \mapsto (x(\sigma), G(\sigma))$ of the transport equation (15.12) is a trajectory on the principal product bundle $\mathcal{P}$. Here, all the fibers $F_P = \{(P, G) : \ G \in \mathcal{G}\}$ are in one-to-one correspondence to the Lie group $\mathcal{G}$. This property of the fiber is typical for a principal bundle. Setting $\pi(P, G) := P$, we get the surjective map

$$\boxed{\pi : \mathcal{P} \to \mathbb{M}^4}$$

with $\pi^{-1}(P) = F_P$ for all $P \in \mathbb{M}^4$. Choose the smooth map $G_0 : \mathbb{M}^4 \to \mathcal{G}$. Then the map

$$(P, G) \mapsto (P, G^+(P))$$

with $G^+(P) = G_0(P)G$ is called a gauge transformation of $\mathcal{P}$. This can also be regarded as a change of the bundle coordinates from (P, G) to $(P, G^+(P))$.

The action of the gauge group $\mathcal{G}$ on the manifold $\mathcal{P}$. For every element G_1 of the gauge group $\mathcal{G}$, we set

$$\boxed{R_{G_1}(P,G) := (P, GG_1).}$$

The map $R_{G_1} : \mathcal{P} \to \mathcal{P}$ is an action of the Lie group $\mathcal{G}$ on the principal bundle $\mathcal{P}$ from the right, that is, we have

$$R_{G_2 G_1} = R_{G_1} R_{G_2} \qquad \text{for all} \quad G_1, G_2 \in \mathcal{G}.$$

In addition, this action of $\mathcal{G}$ on $\mathcal{P}$ preserves the fibers. Finally, the action is compatible with gauge transformations. In fact, we have

$$R_{G_1}(G_0(P)G) = G_0(P)GG_1 = G_0(P)R_{G_1}(G).$$

The connection form on $\mathcal{P}$. For $(P,G) \in \mathcal{P}$, we define

$$\boxed{\mathsf{A}_{(P,G)} := G^{-1}\mathcal{A}(P)G + G^{-1}dG.} \tag{15.15}$$

The differential 1-form A on the manifold $\mathcal{P}$ with values in the Lie algebra $\mathcal{LG}$ is called the connection 1-form on the principal bundle $\mathcal{P}$. With respect to an inertial system, we get

$$\mathsf{A}_{(x,G)} := G^{-1}\mathcal{A}_\alpha(x)G\, dx^\alpha + G^{-1}dG.$$

Here, $G^{-1}dG$ is the Maurer–Cartan form on the Lie group $\mathcal{G}$. Explicitly, let

$$\sigma \mapsto (x(\sigma), G(\sigma))$$

be a curve on $\mathbb{M}^4 \times \mathcal{G}$ with the curve parameter $\sigma \in \mathcal{R}$. Then

$$(G(\sigma)^{-1}dG)(\dot{G}(\sigma)) = G(\sigma)^{-1}\dot{G}(\sigma),$$

and hence $\mathsf{A}_{(x(\sigma),G(\sigma))}(\dot{x}(\sigma), \dot{G}(\sigma))$ is equal to

$$G(\sigma)^{-1}\dot{x}^\alpha(\sigma)\mathcal{A}_\alpha(x(\sigma))G(\sigma) + G(\sigma)^{-1}\dot{G}(\sigma), \qquad \sigma \in \mathcal{R}.$$

Parallel transport. By definition, the curve $\sigma \mapsto (x(\sigma), G(\sigma))$ represents a parallel transport on the principal bundle $\mathcal{P}$ iff

$$\boxed{\mathsf{A}_{(x(\sigma),G(\sigma))}(\dot{x}(\sigma), \dot{G}(\sigma)) = 0, \qquad \sigma \in \mathcal{R}.} \tag{15.16}$$

This is identical with the transport equation (15.12). Concerning a gauge transformation, we get

$$\mathsf{A}^+_{(x(\sigma),G^+(\sigma))}(\dot{x}(\sigma), \dot{G}^+(\sigma)) = \mathsf{A}_{(x(\sigma),G(\sigma))}(\dot{x}(\sigma), \dot{G}(\sigma)) \tag{15.17}$$

for all $\sigma \in \mathcal{R}$. This is precisely the key formula (15.14). This tells us that

Parallel transport is gauge invariant.

Furthermore, it follows from (15.17) that

$$\mathsf{A}^+_{(x,G^+)}(\dot{x}, \dot{G}^+) = \mathsf{A}_{(x,G)}(\dot{x}, \dot{G}) \tag{15.18}$$

for all points (x,G) and all tangent vectors $(\dot{x}, \dot{G})$ of $\mathcal{P}$ at the point (x,G). Here

$$(x, G^+) = (x, G_0(x)G), \tag{15.19}$$

and the tangent vector $(\dot{x}, \dot{G}^+)$ is obtained from the tangent vector $(\dot{x}, \dot{G})$ by linearization, that is,

$$(\dot{x}, \dot{G}^+) = (\dot{x}, G_0'(x)\dot{G}).$$

Alternatively, one can choose a curve $\sigma \mapsto (x(\sigma), G(\sigma))$ with $x(0) = x$, $G(0) = G$, as well as $\dot{x}(0) = \dot{x}$, $\dot{G}(0) = \dot{G}$. Then

$$\dot{G}^+ = \frac{d}{d\sigma} G_0(x(\sigma))G(\sigma)_{|\sigma=0}.$$

The curvature form. Motivated by (15.7), we define

$$\boxed{\mathsf{F} := d\mathsf{A} + \mathsf{A} \wedge \mathsf{A}.} \tag{15.20}$$

By convention, this expression means that

$$\mathsf{F}_Q(V, W) := d\mathsf{A}_Q(V, W) + [\mathsf{A}_Q(V), \mathsf{A}_Q(W)]_-$$

for all points $Q \in \mathcal{P}$ and all tangent vectors V, W of the manifold $\mathcal{P}$ at the point Q. Here, $[.,.]_-$ denotes the Lie bracket on $\mathcal{LG}$. Obviously,

$$\mathsf{F}_Q(V, W) = -\mathsf{F}_Q(W, V)$$

for all tangent vectors V, W at the point Q. Thus, F is a differential 2-form on the manifold $\mathcal{P}$ with values in the Lie algebra $\mathcal{LG}$. This is called the curvature form on the manifold $\mathcal{P}$. The proof of the following proposition will be given in Problem 15.3.

Proposition 15.7 *For all points (x, G) of the bundle manifold $\mathcal{P}$, we get*

$$\mathsf{F}_{(x,G)} = G^{-1}\mathcal{F}_x G.$$

Vertical and horizontal tangent vectors on $\mathcal{P}$. The following notions play a crucial role in the general theory of principal bundles to be considered in Sect. 17.2. Consider the tangent vector $(\dot{x}, \dot{G})$ of the manifold $\mathcal{P}$ at the point (x, G) Then:

- $(\dot{x}, \dot{G})$ is called vertical iff $\dot{x} = 0$;
- $(\dot{x}, \dot{G})$ is called horizontal iff $\mathsf{A}_{(x,G)}(\dot{x}, \dot{G}) = 0$.

Let $\mathsf{A}_{(x,G)}(\dot{x}, \dot{G}) = A$. Then, A is an element of the Lie algebra $\mathcal{LG}$, and we get the unique decomposition

$$(\dot{x}, \dot{G}) = (0, GA) + (\dot{x}, \dot{G} - GA)$$

where $(0, GA)$ is a vertical tangent vector and $(\dot{x}, \dot{G} - GA)$ is a horizontal tangent vector. In fact,

$$\mathsf{A}_{(x,G)}(0, GA) = G^{-1}dG(GA) = G^{-1}GA = A.$$

Hence $\mathsf{A}_{(x,G)}(\dot{x}, \dot{G} - GA) = A - A = 0$. Define

$$\text{hor}(\dot{x}, \dot{G}) := (\dot{x}, \dot{G} - GA).$$

Then, we get the linear operator

$$\text{hor} : T_{(x,G)}\mathcal{P} \to T_{(x,G)}\mathcal{P}_{\text{hor}}$$

which assigns to each tangent vector of the manifold $\mathcal{P}$ at the point (x, G) a uniquely determined horizontal tangent vector.

Fundamental velocity vector field on $\mathcal{P}$. Let $A \in \mathcal{LG}$. Set

$$V_A(x, G) := (0, GA), \qquad (x, G) \in \mathcal{P}.$$

Then, V_A is a vertical tangent vector field on $\mathcal{P}$ which is called the fundamental vector field on $\mathcal{P}$ generated by the element A of the Lie algebra $\mathcal{LG}$. Obviously,

$$\mathsf{A}_{(x,G)}(V_A) = A.$$

Symmetries of the connection form A on $\mathcal{P}$. For all points (P, G) of $\mathcal{P}$, all elements H of the Lie group $\mathcal{G}$, and all elements A of the Lie algebra $\mathcal{LG}$, the following hold:

(C1) $\mathsf{A}_{(x,G)}(V_A) = A$ (fundamental vector field V_A);
(C2) $R_H^* \mathsf{A} = H^{-1}\mathsf{A}H$.

Proof. It remains to prove (C2). In fact, the pull-back $(R_H^*\mathsf{A})_{(x,G)}(\dot{x}, \dot{G})$ is equal to

$$\begin{aligned}\mathsf{A}_{(x,GH)}(\dot{x}, \dot{G}H) &= (GH)^{-1}(\dot{G}H) + (GH)^{-1}\mathcal{A}(x)(GH) \\ &= H^{-1}(G^{-1}\dot{G} + G^{-1}\mathcal{A}(x)G)H = H^{-1}\mathsf{A}_{(x,G)}(\dot{x}, \dot{G})H.\end{aligned}$$

□

The covariant Cartan differential $D\omega$. Let ω be a differential p-form on the manifold $\mathcal{P}$ with values in the Lie algebra $\mathcal{LG}$. For all tangent vectors $V_1, \ldots, V_p$ of the manifold $\mathcal{P}$ at the point (x, G), we define

$$\boxed{D\omega_{(x,G)}(V_1, \ldots, V_p) := d\omega_{(x,G)}(\mathrm{hor}(V_1), \ldots, \mathrm{hor}(V_p)).}$$

Theorem 15.8 *For the curvature form F, we have the elegant Cartan structural equation*

$$\mathsf{F} = D\mathsf{A} \quad \text{on } \mathcal{P} \tag{15.21}$$

together with the integrability condition (Bianchi relation)

$$D\mathsf{F} = 0 \quad \text{on } \mathcal{P}. \tag{15.22}$$

Proof. Ad (15.21). Let V, W, Z be tangent vectors of $\mathcal{P}$ at the point (x, G). Then

$$(\mathsf{A} \wedge \mathsf{A})(V, W) = \mathsf{A}(V)\mathsf{A}(W) - \mathsf{A}(W)\mathsf{A}(V).$$

Since $\mathsf{A}(\mathrm{hor}(V)) = 0$, we get $(\mathsf{A} \wedge \mathsf{A})(\mathrm{hor}V, \mathrm{hor}(W)) = 0$. By Prop. 15.7,

$$\mathsf{F}_{(x,G)}(\mathrm{hor}(V), \mathrm{hor}(W)) = \mathsf{F}_{(x,G)}(V, W).$$

Finally, it follows from $d\mathsf{A} = \mathsf{F} - \mathsf{A} \wedge \mathsf{A}$ that

$$(D\mathsf{A})(V, W) = d\mathsf{A}(\mathrm{hor}(V), \mathrm{hor}(W)) = \mathsf{F}(V, W).$$

Ad (15.22). It follows from $\mathsf{F} = d\mathsf{A} + \mathsf{A} \wedge \mathsf{A}$ that

$$d\mathsf{F} = d\mathsf{A} \wedge \mathsf{A} - \mathsf{A} \wedge d\mathsf{A}.$$

Hence $d\mathsf{F} = (\mathsf{F} - \mathsf{A} \wedge \mathsf{A}) \wedge \mathsf{A} - \mathsf{A} \wedge (\mathsf{F} - \mathsf{A} \wedge \mathsf{A})$. Thus,

$$d\mathsf{F} = \mathsf{F} \wedge \mathsf{A} - \mathsf{A} \wedge \mathsf{F}.$$

This implies

$$(D\mathsf{F})(V, W, Z) = d\mathsf{F}(\mathrm{hor}(V), \mathrm{hor}(W), \mathrm{hor}(Z)) = 0.$$

□

The local pull-back construction in gauge theory. Consider an open subset $\mathcal{O}$ of the base manifold $\mathbb{M}^4$. Let us consider a smooth section

$$s : \mathcal{O} \to \mathbb{M}^4 \times \mathcal{G}$$

given by $s(P) := (P, \mathbf{1})$ where $\mathbf{1}$ is the unit element of the Lie group $\mathcal{G}$. Then

$$\mathcal{A} = s^*\mathsf{A} \qquad \text{and} \qquad \mathcal{F} = s^*\mathsf{F} \qquad \text{on } \mathcal{O}.$$

This allows us to reconstruct the connection form $\mathcal{A}$ and the curvature form $\mathcal{F}$ on $\mathcal{O}$. In the present case, it is possible to choose $\mathcal{O} = \mathbb{M}^4$. However, in the general bundle case, the bundle looks locally like a product bundle, but globally it is not a product bundle. Therefore, the reconstruction described above is only a local procedure in the general case.

Proof. The pull-back $(s^*\mathsf{A})_x(\dot{x})$ is equal to $\mathsf{A}_{(x,\mathbf{1})}(\dot{x}, 0) = \mathcal{A}(x)$. Moreover,

$$s^*\mathsf{F} = d(s^*\mathsf{A}) + s^*\mathsf{A} \wedge s^*\mathsf{A} = d\mathcal{A} + \mathcal{A} \wedge \mathcal{A} = \mathcal{F}.$$

□

For a general smooth section $s : \mathcal{O} \to \mathbb{M}^4 \times \mathcal{G}$, it follows from Prop. 15.7 that

$$(s^*\mathsf{F})_P = G(P)^{-1}\mathcal{F}_P G(P)$$

where $s(P) = (P, G(P))$. The section s is called a local gauge fixing. Furthermore,

$$(s^*\mathsf{A})_x(\dot{x}) = G(x)^{-1}\dot{x}^\alpha \mathcal{A}_\alpha(x) G(x) + G(x)^{-1}G'(x)\dot{x}$$

where $s(x) = (x, G(x))$.

15.3 Parallel Transport of Physical Fields – the Propagator Approach

Parallel transport for physical fields and the propagator. Choose the curve $C : P = P(\sigma)$, $\sigma \in \mathcal{R}$ on the base manifold $\mathbb{M}^4$. Consider a solution $\sigma \mapsto (x(\sigma), G(\sigma))$ of the transport equation (15.11) with $G(\sigma_0) := \mathbf{1}$ and fixed parameter $\sigma_0 \in \mathcal{R}$. For given $\psi_0 \in \mathbb{C}^N$, we define

$$\psi(\sigma) := G(\sigma)\psi_0, \qquad \sigma \in \mathcal{R}.$$

Then $\psi(\sigma_0) = \psi_0$. Using the transport equation (15.11), we get the differential equation

$$\dot{\psi}(\sigma) + \mathcal{A}_\alpha(x(\sigma))\dot{x}^\alpha(\sigma)\, \psi(\sigma) = 0, \qquad \sigma \in \mathcal{R} \tag{15.23}$$

which is identical with (15.10). We say that the curve

$$\sigma \mapsto (x(\sigma), \psi(\sigma))$$

on $\mathbb{M}^4 \times \mathbb{C}^N$ describes a parallel transport on the vector bundle $\mathbb{M}^4 \times \mathbb{C}^N$ which connects the point $(x(\sigma_0), \psi(\sigma_0))$ with the point $(x(\sigma), \psi(\sigma))$. We define the propagator

$$\Pi(\sigma, \sigma_0)(x(\sigma_0), \psi(\sigma_0)) := (x(\sigma), \psi(\sigma)), \qquad \sigma \in \mathcal{R}.$$

To simplify notation, we also briefly write $\Pi(\sigma, \sigma_0)\psi(\sigma_0) := \psi(\sigma)$.

Parallel transport along loops and curvature. We want to show that the curvature form $\mathcal{F}$ can be computed by using the parallel transport along sufficiently small loops. Fix the point $x_0 \in \mathbb{R}^4$. Consider the triangle $\mathcal{T}_\varrho$ depicted in Fig. 13.1 on page 826. This triangle is contained in a 2-dimensional plane in $\mathbb{R}^4$ passing through the point x_0 and spanned by the unit vectors e_1 and e_2. Explicitly,

$$\mathcal{T}_\varrho := \{x_0 + \xi e_1 + \eta e_2 : 0 \le \xi, \eta \le \varrho, \;\; \xi + \eta \le \varrho\}, \qquad \varrho > 0.$$

We assume that the boundary $\partial\mathcal{T}_\varrho$ of the triangle is positively oriented. Moreover, let $\mathrm{meas}(\mathcal{T}_\varrho) = \frac{1}{2}\varrho^2$ denote the surface area of the triangle $\mathcal{T}_\varrho$.

For given value $\psi_0 \in \mathbb{C}^N$ of the physical field at the point x_0, let us transport ψ_0 along the positively oriented loop $\partial\mathcal{T}_\varrho$. After surrounding counter-clockwise the triangle once, we get the value $\Pi_{\partial\mathcal{T}_\varrho}\psi_0$ at the final point x_0.

Proposition 15.9 *The curvature component $\mathcal{F}_{12}(x_0)$ is given by the limit*

$$\mathcal{F}_{12}(x_0)\psi_0 = \lim_{\varrho \to 0} \frac{\psi_0 - \Pi_{\partial\mathcal{T}_\varrho}\psi_0}{\mathrm{meas}(\mathcal{T}_\varrho)}.$$

Analogous expressions are obtained for $\mathcal{F}_{\alpha\beta}$ with $\alpha < \beta$.

Proof. We will modify the proof given on page 827. Set $x_0 := 0$. The differential equation for the counter-clockwise parallel transport along the boundary $\partial\mathcal{T}_\varrho$ of the triangle $\mathcal{T}_\varrho$ reads as

$$\dot{\psi}(\sigma) = -\dot{x}^\alpha(\sigma)\mathcal{A}_\alpha(P(\sigma))\psi(\sigma), \quad 0 \le \sigma \le \sigma_1, \quad \psi(0) = \psi_0. \tag{15.24}$$

Hence

$$\Pi_{\partial\mathcal{T}_\varrho}\psi_0 - \psi_0 = \int_0^{\sigma_1} \dot{\psi}(\sigma)d\sigma = -\int_{\partial\mathcal{T}_\varrho} dx^\alpha \mathcal{A}_\alpha(P)\psi(P).$$

The basic trick of the proof is to extend the values of ψ on the boundary $\partial\mathcal{T}_\varrho$ to the triangle $\mathcal{T}_\varrho$ in a smooth way. Set

$$A := dx^\alpha \mathcal{A}_\alpha \psi.$$

By the Stokes integral theorem, we have

$$\int_{\partial\mathcal{T}_\varrho} A = \int_{\mathcal{T}_\varrho} dA.$$

Hence

$$\int_{\partial\mathcal{T}_\varrho} A = \int_{\mathcal{T}_\varrho} \partial_\alpha(\mathcal{A}_\beta\psi)\, dx^\alpha \wedge dx^\beta = \int_{\mathcal{T}_\varrho} (\partial_1(\mathcal{A}_2\psi) - \partial_2(\mathcal{A}_1\psi))\, dx^1 dx^2.$$

Set

$$\mathcal{B} := \lim_{\varrho \to 0} \frac{\int_{\mathcal{T}_\varrho} (\partial_1(\mathcal{A}_2\psi) - \partial_2(\mathcal{A}_1\psi))\, dx^1 dx^2}{\text{meas}(\mathcal{T}_\varrho)}.$$

Noting that $\partial_\alpha(\mathcal{A}_\beta\psi) = \partial_\alpha\mathcal{A}_\beta \cdot \psi + \mathcal{A}_\beta\partial_\alpha\psi$, we get

$$\mathcal{B} = \partial_1\mathcal{A}_2(0)\psi_0 - \partial_2\mathcal{A}_1(0)\psi_0 + \mathcal{A}_2(0)\partial_1\psi(0) - \mathcal{A}_1(0)\partial_2\psi(0).$$

By (15.24), $\partial_1\psi(0) = -\mathcal{A}_1(0)\psi_0$ and $\partial_2\psi(0) = -\mathcal{A}_2(0)\psi_0$. Hence

$$\mathcal{B} = \{\partial_1\mathcal{A}_2(0) - \partial_2\mathcal{A}_1(0) + \mathcal{A}_1(0)\mathcal{A}_2(0) - \mathcal{A}_2(0)\mathcal{A}_1(0)\}\psi_0 = \mathcal{F}_{12}(0)\psi_0.$$

□

Infinitesimal parallel transport and the covariant directional derivative of a physical field. Let ψ be a physical field on $\mathbb{M}^4$. In other words, ψ is a smooth section of the vector bundle $\mathbb{M}^4 \times \mathbb{C}^N$. Let $C : P = P(\sigma), \sigma \in P$, be a smooth curve on $\mathbb{M}^4$. Set $\psi(\sigma) := \psi(P(\sigma))$. An analogous argument as in the proof of Prop. 13.10 on page 828 yields the following.

Proposition 15.10 *There holds*

$$D_{\dot{P}(0)}\psi(0) = \lim_{\sigma \to 0} \frac{\Pi(0,\sigma)\psi(\sigma) - \psi(0)}{\sigma}.$$

15.4 The Principle of Critical Action and the Yang–Mills Equations

The following variational problem (15.25) with $N = 1$ (concerning the gauge group $U(1)$ with the Lie algebra $u(1)$) describes classical electrodynamics, as we will show in Sect. 19.7.1.

In the Standard Model in particle physics, the following variational principle with $N = 3$ (concerning the gauge group $SU(3)$ with the Lie algebra $su(3)$) describes the gluons in quantum chromodynamics. In Volume IV, we will study the complete Standard Model in particle physics. Roughly speaking, the basic variational problem (principle of least action) is of the type (15.25) for the 12 interaction particles (8 gluons, photon, 3 vector bosons). In order to describe the 12 fundamental particles (6 quarks, electron, myon, tau, 3 neutrinos), one has to add 12 fields ψ to the Lagrangian. Finally, one has to add a field φ to the Lagrangian which describes the Higgs particle. The Higgs field φ generates the terms for the masses of the 3 vector bosons in electroweak interaction. In what follows, we will sum over equal lower and upper Greek indices from 0 to 3.

We consider the variational problem

$$\boxed{\int_{\mathcal{O}} (\, \langle \mathcal{F}_{\mu\nu}|\mathcal{F}^{\mu\nu}\rangle + 4\langle \mathcal{A}_\nu|\mathcal{J}^\nu\rangle \,)\; dx^4 = \text{critical!}} \tag{15.25}$$

together with

$$\boxed{\mathcal{F}_{\mu\nu} = \partial_\mu\mathcal{A}_\nu - \partial_\nu\mathcal{A}_\mu + [\mathcal{A}_\mu, \mathcal{A}_\nu]_-, \qquad \mu,\nu = 0,1,2,3,} \tag{15.26}$$

and the boundary condition

$$\boxed{\mathcal{A}_\mu = \text{fixed} \quad \text{on } \partial\mathcal{O}, \quad \mu = 0,1,2,3.} \tag{15.27}$$

Recall that $[A, B]_- := AB - BA$. Here, $\mathcal{O}$ is a nonempty, bounded, open subset of the Minkowski manifold $\mathbb{M}^4$. Fix $N = 1, 2, \ldots$ We are given the smooth functions

$$\mathcal{J}_\mu : \mathrm{cl}(\mathcal{O}) \to u(N).$$

We are looking for a smooth solution

$$\mathcal{A}_\mu : \mathrm{cl}(\mathcal{O}) \to u(N), \qquad \mu = 0, 1, 2, 3.$$

Concerning the notation, we use the symbol

$$\langle A|B\rangle := -\operatorname{tr}(AB), \qquad A, B \in u(N).$$

This is an inner product on the Lie algebra $u(N)$.[2] This way, $u(N)$ becomes a real Hilbert space (see Problem 15.4). Moreover, we equip the Minkowski manifold $\mathbb{M}^4$ with the metric tensor

$$g = \eta_{\alpha\beta}\, dx^\alpha \otimes dx^\beta$$

where we use an arbitrary inertial system (see (18.30) on page 924). Explicitly, $1 = \eta_{00} = -\eta_{11} = -\eta_{22} = -\eta_{33}$, and $\eta_{\alpha\beta} = 0$ if $\alpha \neq \beta$. Furthermore, $\eta^{\alpha\beta} = \eta_{\alpha\beta}$. We use the metric tensorial families $\eta_{\alpha\beta}$ and $\eta^{\alpha\beta}$ in order to lower and to lift indices. For example,

$$\mathcal{F}^{\mu\nu} = \eta^{\mu\alpha}\eta^{\nu\beta}\mathcal{F}_{\alpha\beta}.$$

Hence

$$\tfrac{1}{2}\langle \mathcal{F}_{\mu\nu}|\mathcal{F}^{\mu\nu}\rangle = \sum_{j=1}^{3}\langle \mathcal{F}_{0j}|\mathcal{F}_{0j}\rangle - \sum_{1\le i<j\le 3}\langle \mathcal{F}_{ij}|\mathcal{F}_{ij}\rangle.$$

Theorem 15.11 *Every smooth solution $\mathcal{A}_0, \mathcal{A}_1, \mathcal{A}_2, \mathcal{A}_3$ of the variational problem* (15.25) *through* (15.27) *satisfies both the Yang–Mills equations*

$$D_\mu \mathcal{F}^{\mu\nu} = \mathcal{J}^\nu, \qquad \nu = 0, 1, 2, 3, \tag{15.28}$$

and the Bianchi equations

$$D_{[\lambda}\mathcal{F}_{\mu\nu]} = 0, \qquad \lambda, \mu, \nu = 0, 1, 2, 3. \tag{15.29}$$

The theorem remains true if we replace the real Lie algebra $u(N)$ by its Lie subalgebra $su(N)$, $N = 2, 3, \ldots$ The Bianchi equations are equivalent to

$$D_\lambda \mathcal{F}_{\mu\nu} + D_\mu \mathcal{F}_{\nu\lambda} + D_\nu \mathcal{F}_{\lambda\mu} = 0, \qquad \lambda, \mu, \nu = 0, 1, 2, 3,$$

since $\mathcal{F}_{\mu\nu} = -\mathcal{F}_{\nu\mu}$ for all indices. Here, we set

$$D_\lambda \mathcal{F}_{\mu\nu} := \partial_\lambda \mathcal{F}_{\mu\nu} + [\mathcal{A}_\lambda, \mathcal{F}_{\mu\nu}]_-.$$

Proof. (I) Symmetry of the trace. For all $A, B, C \in u(N)$, we have

$$\langle\, A|\ [B, C]_-\rangle = \langle\, [C, A]_-|B\rangle. \tag{15.30}$$

In fact, the trace $\operatorname{tr}(ABC)$ is invariant under a cyclic permutation of the factors. Hence

$$\operatorname{tr}(ABC) = \operatorname{tr}(CAB).$$

---

[2] At the same time, the inner product $\langle .|.\rangle$ is proportional to the negative Killing form of the real Lie algebra $u(N)$.

This implies (15.30).

(II) Generalized variational lemma. Let $C : \mathcal{O} \to u(N)$ be a smooth map. Suppose that

$$\int_{\mathcal{O}} \langle C|B\rangle \, dx^4 = 0$$

for all smooth functions $B : \mathcal{O} \to u(N)$ which vanish outside a compact subset of $\mathcal{O}$. Then $C \equiv 0$.

To prove this, let $B_1, \ldots, B_r$ be a basis of the real Lie algebra $u(N)$. Choose $B := \chi B_j$ where the smooth function $\chi : \mathcal{O} \to \mathbb{R}$ vanishes outside a compact subset of $\mathcal{O}$. Then

$$\int_{\mathcal{O}} \langle C|B_j\rangle \, \chi dx^4 = 0$$

for all $\chi \in C_0^\infty(\mathcal{O})$. By the classic variational lemma, this implies $\langle C|B_j\rangle \equiv 0$ for all indices $j = 1, \ldots, r$. Since $\langle .|.\rangle$ is an inner product, we get $C \equiv 0$.

(III) First variation. Choose smooth functions $\delta\mathcal{A}_\mu : \mathcal{O} \to u(N)$ which vanish outside a compact subset of $\mathcal{O}$. Replace $\mathcal{A}_\mu$ by $\mathcal{A}_\mu + \tau\delta\mathcal{A}_u, \tau \in \mathbb{R}$. Set

$$\sigma(\tau) := \int_{\mathcal{O}} \left(\langle \mathcal{F}^+_{\mu\nu}|\mathcal{F}^{\mu\nu}_+\rangle + 4\langle \mathcal{A}_\nu|\mathcal{J}^\nu\rangle \right) dx^4, \qquad \tau \in \mathbb{R}$$

where

$$\mathcal{F}^+_{\mu\nu} := \partial_\mu(\mathcal{A}_\nu + \tau\delta\mathcal{A}_\nu) - \partial_\nu(\mathcal{A}_\mu + \tau\delta\mathcal{A}_\mu) + [\mathcal{A}_\mu + \tau\delta\mathcal{A}_\mu, \mathcal{A}_\nu + \tau\delta\mathcal{A}_\nu]_-.$$

Differentiating this with respect to the real parameter τ at the point $\tau = 0$, we get

$$\delta\mathcal{F}_{\mu\nu} := \partial_\mu\delta\mathcal{A}_\nu - \partial_\nu\delta\mathcal{A}_\mu + [\delta\mathcal{A}_\mu, \mathcal{A}_\nu]_- - [\delta\mathcal{A}_\nu, \mathcal{A}_\mu].$$

Suppose that $\mathcal{A}_0, \mathcal{A}_1, \mathcal{A}_2, \mathcal{A}_3$ is a solution of the variational problem (15.25), then $\dot{\sigma}(0) = 0$. This implies

$$\int_{\mathcal{O}} (\langle \delta\mathcal{F}_{\mu\nu}|\mathcal{F}^{\mu\nu}\rangle + \langle \mathcal{F}_{\mu\nu}|\, \delta\mathcal{F}^{\mu\nu}\rangle + 4\langle \delta\mathcal{A}_\nu|\mathcal{J}^\nu\rangle) \; dx^4 = 0.$$

Hence

$$\int_{\mathcal{O}} (2\langle \delta\mathcal{F}_{\mu\nu}|\mathcal{F}^{\mu\nu}\rangle + 4\langle \delta\mathcal{A}_\nu|\mathcal{J}^\nu\rangle) \; dx^4 = 0.$$

Using integration by parts combined with the trace formula (15.30), we get

$$\int_{\mathcal{O}} (-\langle \, \delta\mathcal{A}_\nu|\partial_\mu F^{\mu\nu}\rangle - \langle \, \delta\mathcal{A}_\nu| \; [\mathcal{A}_\mu, \mathcal{F}^{\mu\nu}] \, \rangle + \langle \, \delta\mathcal{A}_\nu|\mathcal{J}^\nu\rangle) \; dx^4 = 0.$$

By the generalized variation lemma (II), we get

$$-\partial_\mu\mathcal{F}^{\mu\nu} - [\mathcal{A}_\mu, \mathcal{F}^{\mu\nu}] + \mathcal{J}^\nu = 0.$$

This yields the Yang–Mills equations (15.28).

(IV) The Bianchi equations (15.29) follow from (15.26) by an explicit computation. See Problem 15.2.

□

Equivalent formulation. We set

$$\mathcal{A} := \mathcal{A}_\alpha dx^\alpha, \qquad \mathcal{F} := \tfrac{1}{2}\mathcal{F}_{\mu\nu} \, dx^\mu \wedge dx^\nu.$$

Moreover,

- $*\mathcal{F} = \frac{1}{2}(*\mathcal{F})_{\mu\nu}\, dx^\mu \wedge dx^\nu$ where $(*\mathcal{F})_{\mu\nu} = \frac{1}{2}\varepsilon_{\alpha\beta\mu\nu}\mathcal{F}^{\alpha\beta}$ (Hodge star operator),
- $D\mathcal{F} = \frac{1}{6}D_{[\lambda}\mathcal{F}_{\mu\nu]}\, dx^\lambda \wedge dx^\mu \wedge dx^\nu$,
- $D(*\mathcal{F}) = \frac{1}{6}D_{[\lambda}(*\mathcal{F})_{\mu\nu]}\, dx^\lambda \wedge dx^\mu \wedge dx^\nu$,
- $D^*\mathcal{F} = *^{-1}D(*\mathcal{F})$,
- $|\mathcal{F}|^2 := \langle \mathcal{F}_{\mu\nu}|\mathcal{F}^{\mu\nu}\rangle$,
- $\mathcal{J} := \mathcal{J}_\nu dx^\nu$,
- $\langle \mathcal{A}|\mathcal{J}\rangle := \langle \mathcal{A}_\mu \mathcal{J}^\mu \rangle$.

The following theorem is equivalent to Theorem 15.11. We consider the variational problem

$$\boxed{\int_{\mathcal{O}} (|\mathcal{F}|^2 + 4\langle \mathcal{A}|\mathcal{J}\rangle\ )\ dx^4 = \text{critical!}} \tag{15.31}$$

together with

$$\boxed{\mathcal{F} = d\mathcal{A} + \mathcal{A} \wedge \mathcal{A},} \tag{15.32}$$

and the boundary condition

$$\boxed{\mathcal{A} = \text{fixed} \quad \text{on } \partial\mathcal{O}.} \tag{15.33}$$

Fix $N = 1, 2, \ldots$ We are given the smooth differential 1-form $\mathcal{J}$ on the closure of the subset of $\mathcal{O}$ with values in the Lie algebra $u(N)$. We are looking for a smooth differential 1-form $\mathcal{A}$ on the closure of $\mathcal{O}$ with values in the Lie algebra $u(N)$.

Theorem 15.12 *Every smooth solution $\mathcal{A}$ of the variational problem* (15.31) *through* (15.33) *satisfies both the Yang–Mills equation*

$$-\ D^*\mathcal{F} = \mathcal{J}, \tag{15.34}$$

and the Bianchi equation

$$D\mathcal{F} = 0. \tag{15.35}$$

This is the most elegant formulation of the Yang–Mills equation. This formulation shows clearly that the Yang–Mills equation is based on Hodge duality. The Yang–Mills equation (15.34) is also equivalent to[3]

$$-D(*\mathcal{F}) = *\mathcal{J}.$$

The theorem remains true if we replace the real Lie algebra $u(N)$ by its Lie subalgebra $su(N), N = 2, 3, \ldots$

The variational problem (15.11) above leads quite naturally to the definition of $\mathcal{D}_\lambda \mathcal{F}_{\mu\nu}$. Setting $\mathcal{F} := \frac{1}{2}\mathcal{F}_{\mu\nu}\, dx^\mu \wedge dx^\nu$, we define

$$\boxed{D\mathcal{F} := D_\lambda \mathcal{F}_{\mu\nu} dx^\lambda \wedge dx^\mu \wedge dx^\nu} \tag{15.36}$$

with $D\mathcal{F}_{\mu\nu} := D_\lambda \mathcal{F}_{\mu\nu}\, dx^\lambda$. In the next two sections, we will systematically study covariant differentials. In particular, the definition (15.36) will appear as a special case in the general setting. We will present two equivalent strategies:

- the universal extension strategy via the Leibniz rule (invariant approach),
- the index principle in bundle theory – tensorial families of differential forms (observer-oriented approach).

[3] Note that $-*D^*\mathcal{F} = -**^{-1}D(*\mathcal{F}) = -D(*\mathcal{F}) = *\mathcal{J}$.

15.5 The Universal Extension Strategy via the Leibniz Rule

> Using the Leibniz rule as a universal guiding principle, it is possible to extend the covariant directional derivative $D_v\psi$ for physical fields $\psi : \mathbb{M}^4 \to \mathbb{C}^N$ to the covariant directional derivative $D_v\Psi$ for more general physical fields Ψ (e.g., $\Psi : \mathbb{M}^4 \to gl(N, \mathbb{C})$). In addition, setting
>
> $$(D\Psi)(v) := D_v\Psi$$
>
> for all smooth velocity vector fields v on the base manifold $\mathbb{M}^4$, we get the covariant differential $D\Psi$.
>
> Folklore

Let v be a smooth velocity vector field on the Minkowski manifold $\mathbb{M}^4$. Let $C^\infty(\mathbb{M}^4, \mathbb{C}^N)$ be the space of all smooth maps[4]

$$\psi : \mathbb{M}^4 \to \mathbb{C}^N.$$

Moreover, the symbol $C^\infty(\mathbb{M}^4, gl(N, \mathbb{C}))$ denotes the space of all smooth maps

$$\Psi : \mathbb{M}^4 \to gl(N, \mathbb{C})$$

with values in the Lie algebra $gl(N, \mathbb{C})$. Finally, recall that $\mathrm{Vect}(\mathbb{M}^4)$ denotes the space of all the smooth vector fields on the Minkowski manifold $\mathbb{M}^4$.

Dual field to $\psi \in C^\infty(\mathbb{M}^4, \mathbb{C}^N)$. Let

$$\omega : \mathbb{M}^4 \to (\mathbb{C}^N)^d$$

be a smooth map where $(\mathbb{C}^N)^d$ is the dual space to $\mathbb{C}^N$. For any smooth map $\psi \in C^\infty(\mathbb{M}^4, \mathbb{C}^N)$, we get the smooth map

$$P \mapsto \omega_P(\psi(P))$$

from $\mathbb{M}^4$ to $\mathbb{C}$. We want to define the covariant directional derivative $D_v\omega$. In order to ensure the Leibniz rule

$$\boxed{d_v(\omega(\psi)) = (D_v\omega)(\psi) + \omega(D_v\psi) \qquad \text{for all} \quad \psi \in C^\infty(\mathbb{M}^4, \mathbb{C}^N),}$$

we define $D_v\omega$ by setting

$$(D_v\omega)(\psi) := d_v(\omega(\psi)) - \omega(D_v\psi).$$

Differential p-form ω on $\mathbb{M}^4$ with values in $C^\infty(\mathbb{M}^4, \mathbb{C}^N)$.

(i) Let $p = 1$. Then, $\omega(w) \in C^\infty(\mathbb{M}^4, \mathbb{C}^N)$ for all $w \in \mathrm{Vect}(\mathbb{M}^4)$. In order to ensure the Leibniz rule

$$D_v(\omega(w)) = (D_v\omega)(w) + \omega(D_v w) \qquad \text{for all} \quad w \in \mathrm{Vect}(\mathbb{M}^4),$$

we define $D_v\omega$ by setting

$$(D_v\omega)(w) := D_v(\omega(w)) - \omega(D_v w).$$

[4] The space $C^\infty(\mathbb{M}^4, \mathbb{C}^N)$ coincides with the space $\mathrm{Sect}(\mathbb{M}^4 \times \mathbb{C}^N)$ of all the smooth sections of the vector bundle $\mathbb{M}^4 \times \mathbb{C}^N$.

(ii) Let $p = 2$. Then $\omega(w, z) \in C^\infty(\mathbb{M}^4, \mathbb{C}^N)$ for all $w, z \in \mathrm{Vect}(\mathbb{M}^4)$. In order to ensure the Leibniz rule

$$D_v(\omega(w,z)) = (D_v\omega)(w,z) + \omega(D_v w, z) + \omega(w, D_v z) \qquad \text{for all} \quad w, z \in \mathrm{Vect}(\mathbb{M}^4),$$

we define $D_v\omega$ by setting

$$(D_v\omega)(w,z) := D_v(\omega(w,z)) - \omega(D_v w, z) - \omega(w, D_v z).$$

In the case where $p > 2$, we proceed similarly.

Smooth maps on $\mathbb{M}^4$ with values in the Lie algebra $gl(N, \mathbb{C})$. Suppose that $\Psi : \mathbb{M}^4 \to gl(N, \mathbb{C})$ is a smooth map which sends the point $x \in \mathbb{M}^4$ to the complex $(N \times N)$-matrix $\Psi(x)$. Naturally enough, define $\Psi\psi$ by setting

$$(\Psi\psi)(x) := \Psi(x)\psi(x), \qquad x \in \mathbb{M}^4$$

for all physical fields $\psi \in C^\infty(\mathbb{M}^4, \mathbb{C}^N)$. Then

$$\Psi\psi \in C^\infty(\mathbb{M}^4, \mathbb{C}^N) \qquad \text{for all} \quad \psi \in C^\infty(\mathbb{M}^4, \mathbb{C}^N).$$

Therefore, $D_v(\Psi\psi)$ is well-defined. In order to ensure the Leibniz rule

$$D_v(\Psi\psi) = (D_v\Psi)(\psi) + \Psi(D_v\psi) \qquad \text{for all} \quad \psi \in C^\infty(\mathbb{M}^4),$$

we define $D_v\Psi$ by setting

$$\boxed{(D_v\Psi)(\psi) := D_v(\Psi\psi) - \Psi(D_v\psi).} \tag{15.37}$$

Differential p-form on $\mathbb{M}^4$ with values in $C^\infty(\mathbb{M}^4, gl(N, \mathbb{C}))$. For example, let $p = 2$. Then,

$$\omega(w,z) \in C^\infty(\mathbb{M}^4, gl(N, \mathbb{C})) \qquad \text{for all} \quad w, z \in \mathrm{Vect}(\mathbb{M}^4).$$

Therefore, $D_v(\omega(w,z))$ is well-defined by (15.37). In order to ensure the Leibniz rule

$$D_v(\omega(w,z)) = (D_v\omega)(w,z) + \omega(D_v w, z) + \omega(w, D_v z), \quad w, z \in \mathrm{Vect}(\mathbb{M}^4),$$

we define $D_v\omega$ by setting

$$\boxed{(D_v\omega)(w,z) := D_v(\omega(w,z)) - \omega(D_v w, z) - \omega(w, D_v z).} \tag{15.38}$$

The local expressions with respect to bundle coordinates will be considered in the next section.

15.6 Tensor Calculus on Vector Bundles

In Sect. 8.14, we have discussed the distinction between

- the index-free (i.e., invariant) approach and
- the index approach

in differential geometry. In fact, it is useful for the reader to master the two approaches and to understand the close interrelations between them. In this section, we will generalize the index approach from Chap. 8 to bundles. For didactic reasons, as a prototype, we will consider the product bundle

$$\mathbb{M}^4 \times \mathbb{C}^N. \tag{15.39}$$

Note that the following calculus can be translated straightforward to general vector bundles.

In the bundle case (15.39), *we will use two types of indices. Greek indices run from 0 to 3, and Latin indices run from 1 to N.*

We will sum over equal upper and lower Greek (resp. Latin) indices from 0 to 3 (resp. 1 to N). Let $\mathcal{O}$ be an open subset of the Minkowski manifold $\mathbb{M}^4$. We assign to the subset

$$\mathcal{O} \times \mathbb{C}^N$$

of the vector bundle $\mathbb{M}^4 \times \mathbb{C}^N$ the bundle chart

$$\mathcal{U} \times \mathbb{C}^N \quad \text{with } \mathcal{U} \subseteq \mathbb{R}^4.$$

More precisely, we assign to the point $P \in \mathcal{O} \times \mathbb{C}^N$ the point

$$(x, \psi) \in \mathbb{R}^4 \times \mathbb{C}^N$$

as bundle coordinate. Here, x is a local coordinate of the point P on the Minkowski manifold $\mathbb{M}^4$.

15.6.1 Tensor Algebra

Fix $N = 1, 2, \ldots$. We set

$$\psi := \begin{pmatrix} \psi^1 \\ \vdots \\ \psi^N \end{pmatrix} \qquad \text{and} \qquad \varphi := (\varphi_1, \ldots \varphi_N).$$

The two fundamental gauge transformation laws (contravariant and covariant). Let $\mathcal{G}$ be a closed subgroup of the matrix group $GL(N, \mathbb{C})$ (e.g., we choose $\mathcal{G} = U(N)$). We are given the smooth map

$$G : \mathcal{U} \to \mathcal{G}.$$

We will use the smooth maps

$$\psi : \mathcal{U} \to \mathbb{C}^N \qquad \text{and} \qquad \varphi : \mathcal{U} \to (\mathbb{C}^N)^d.$$

Here, the symbol $(\mathbb{C}^N)^d$ refers to the fact that φ is a row matrix (in contrast to the column matrix ψ). We define the so-called contravariant gauge transformation law

$$\boxed{\psi^+(x) = G(x)\psi(x), \qquad x \in \mathcal{U},} \tag{15.40}$$

and the dual covariant transformation law

$$\boxed{\varphi^+(x) = \varphi(x)G^{-1}(x), \quad x \in \mathcal{U}.}$$

Here, we use matrix products. In terms of components, this is equivalent to

$$\psi^{i^+}(x) = G_i^{i^+}(x)\psi^i(x), \qquad i^+ = 1^+, \ldots, N^+,$$

and

$$\varphi_{i^+}(x) = G_{i^+}^i(x)\varphi_i(x), \qquad i^+ = 1^+, \ldots, N^+.$$

Theorem 15.13 *For all $x \in \mathcal{U}$, we have the two key matrix relations*[5]

$$\varphi^+(x)\psi^+(x) = \varphi(x)\psi(x)$$

and $\psi^+(x)\varphi^+(x) = G(x) \cdot \big(\psi(x)\varphi(x)\big) \cdot G^{-1}(x)$.

In terms of matrix components, this means that

- $\psi^{i^+}(x)\varphi_{i^+}(x) = \psi^i(x)\varphi_i(x)$, and
- $(\psi^{i^+}(x)\varphi_{j^+}(x)) = G(x) \cdot \big(\psi^i(x)\varphi_j(x)\big) \cdot G(x)^{-1}$.

Equivalently,

$$\psi^{i^+}(x)\varphi_{j^+}(x) = G_i^{i^+}(x)G_{j^+}^j(x) \cdot \psi^i(x)\varphi_j(x), \qquad i^+, j^+ = 1^+, \ldots, N^+.$$

Tensorial families. The family of complex-valued functions

$$T^{\alpha_1 \ldots \alpha_m i_1 \ldots i_k}_{\beta_1 \ldots \beta_n j_1 \ldots j_l} \tag{15.41}$$

is called a tensorial family iff it transforms like the product

$$\dot{x}^{\alpha_1} \cdots \dot{x}^{\alpha_m} \cdot \partial_{\beta_1} \cdots \partial_{\beta_n} \cdot \psi^{i_1} \cdots \psi^{i_k} \cdot \varphi_{j_1} \cdots \varphi_{j_l}$$

where

- $\dot{x}^\alpha := \frac{dx^\alpha}{dt}$ (time derivative), and
- $\partial_\beta := \frac{\partial}{\partial x^\beta}$ (partial derivative).

Similarly, as in Sect. 8.3, we have the index principle concerning Greek (resp. Latin) indices. For example, if $T^i_{\alpha j}$, S^j_i, and v^α are tensorial families, then the tensorial family

$$\chi(x) := T^i_{\alpha j}(x)S^j_i(x)v^\alpha(x), \qquad x \in \mathcal{U}$$

has no free indices, and hence it is a function from $\mathcal{U}$ to $\mathbb{C}$ which is invariant under both gauge transformations and transformations of the local space-time coordinates on the base manifold $\mathbb{M}^4$.

Invariant approach. As in classical vector algebra, we write

$$\boxed{\psi = \psi^i \mathbf{e}_i, \qquad \varphi = \varphi_j \mathbf{e}^j}$$

with the row matrices

$$\mathbf{e}^1 := (1, 0, 0, \ldots, 0), \;\; \mathbf{e}^2 := (0, 1, 0, \ldots, 0), \ldots, \mathbf{e}^N := (0, 0, \ldots, 0, 1)$$

and the column matrices

[5] Note that $\varphi^+\psi^+ = \varphi G^{-1}G\psi = \varphi\psi$.

$$\mathbf{e}_1 := \begin{pmatrix} 1 \\ 0 \\ \vdots \\ 0 \end{pmatrix} = (1, 0, \ldots, 0)^d,$$

and

$$\mathbf{e}_2 := (0, 1, 0, \ldots, 0)^d, \ldots, \mathbf{e}_N := (0, 0, \ldots, 0, 1)^d.$$

Here, $\mathbf{e}_1, \ldots, \mathbf{e}_N$ is a basis of the complex linear space $\mathbb{C}^N$, and $\mathbf{e}^1, \ldots, \mathbf{e}^N$ is a basis of the dual space $(\mathbb{C}^N)^d$. Under gauge transformations, we transform $\mathbf{e}_i$ and $\mathbf{e}^i$ as tensorial families. Explicitly,

$$\mathbf{e}_{i^+} = G^i_{i^+}(x)\mathbf{e}_i, \quad \mathbf{e}^{i^+} = G^{i^+}_i(x)\mathbf{e}^i,$$

if $x \in \mathcal{U}$ and $i^+ = 1^+, \ldots N^+$. We replace ψ and φ by

$$\boldsymbol{\psi} = \psi^i \mathbf{e}_i \qquad \text{and} \qquad \boldsymbol{\varphi} = \varphi_i \mathbf{e}^i,$$

respectively. By the index principle,

$$\boldsymbol{\psi}(x) = \psi^i(x)\mathbf{e}_i = \psi^{i^+}(x)\mathbf{e}_{i^+}, \qquad x \in \mathcal{U}.$$

Thus, $\boldsymbol{\psi}$ possesses an invariant meaning on $\mathcal{O} \times \mathbb{C}^N$. Similarly,

$$\boldsymbol{\varphi}(x) = \varphi_i(x)\mathbf{e}^i = \varphi_{i^+}(x)\mathbf{e}^{i^+}, \qquad x \in \mathcal{U}.$$

In the general case, for the tensorial family (15.41), we define the tensor

$$\begin{aligned}\mathbf{T} : = T^{\alpha_1 \ldots \alpha_m i_1 \ldots i_k}_{\beta_1 \ldots \beta_n j_1 \ldots j_l} \, dx^{\beta_1} \otimes \cdots \otimes dx^{\beta_n} \otimes \partial_{\alpha_1} \otimes \cdots \otimes \partial_{\alpha_m} \\ \otimes \mathbf{e}^{j_1} \otimes \cdots \otimes \mathbf{e}^{j_l} \otimes \mathbf{e}_{i_1} \otimes \cdots \otimes \mathbf{e}_{i_k}\end{aligned}$$

in an invariant way. This means that $\mathbf{T}$ does not depend on the choice of the bundle coordinates, by the index principle.

In particular, a tensorial family of differential forms looks like

$$\omega^{i_1 \ldots i_k}_{j_1 \ldots j_l} := \frac{1}{n!} T^{i_1 \ldots i_k}_{\beta_1 \ldots \beta_n j_1 \ldots j_l} \, dx^{\beta_1} \wedge \cdots \wedge dx^{\beta_n}.$$

Here, we assume that the tensorial family $T^{i_1 \ldots i_k}_{\beta_1 \ldots \beta_n j_1 \ldots j_l}$ is antisymmetric with respect to the Greek space-time indices $\beta_1, \ldots, \beta_n$. The corresponding invariant tensor is given by

$$\boxed{\mathbf{T} := \omega^{i_1 \ldots i_k}_{j_1 \ldots j_l} \otimes \mathbf{e}_{i_1} \otimes \cdots \otimes \mathbf{e}_{i_k} \otimes \mathbf{e}^{j_1} \otimes \cdots \otimes \mathbf{e}^{j_l}.} \tag{15.42}$$

This is also called a differential form of tensor type. Recall that

$$dx^\alpha \wedge dx^\beta = dx^\alpha \otimes dx^\beta - dx^\beta \otimes dx^\alpha,$$

and so on.

15.6.2 Connection and Christoffel Symbols

Suppose that we are given the complex $(N \times N)$-matrices

$$\mathcal{A}_\alpha(x) \in \mathcal{LG}, \qquad \alpha = 0, 1, 2, 3, \quad x \in \mathcal{U}.$$

For example, if $\mathcal{G} = GL(N, \mathbb{C})$, then $\mathcal{LG} = gl(N, \mathbb{C})$. Suppose that

$$x \mapsto \mathcal{A}_\alpha(x), \qquad \alpha = 0, 1, 2, 3$$

is a smooth map from $\mathcal{U}$ to $\mathcal{LG}$. We set $\mathcal{A}_\alpha(x) = (\Gamma^i_{\alpha j}(x))$, and

$$\mathcal{A}(x) := \mathcal{A}_\alpha(x)\, dx^\alpha, \qquad x \in \mathcal{U}.$$

The point is that we assign to the gauge transformation

$$\boxed{\psi^+(x) = G(x)\psi(x), \qquad x \in \mathcal{U}}$$

of the physical field ψ the gauge transformation

$$\boxed{\mathcal{A}^+(x) = G(x)\mathcal{A}(x)G(x)^{-1} - dG(x) \cdot G(x)^{-1}} \qquad (15.43)$$

of the connection form $\mathcal{A}$ where $dG(x) = \partial_\alpha G(x)\, dx^\alpha$. Furthermore, we assume that:

The matrix $\mathcal{A}_\alpha$ transforms like the partial derivative ∂_α under a change of the local space-time coordinates on the base manifold $\mathbb{M}^4$.

Thus, the differential 1-form $\mathcal{A}$ is an invariant under a change of local space-time coordinates, by the index principle.

The covariant differential $D\psi$. We are given $\psi = \psi^i \mathbf{e}_i$ where ψ^i is a tensorial family. We first consider the column matrix $\psi = (\psi^1, \ldots, \psi^n)^d$, and we define $D\psi$ by setting

$$(D_\alpha\psi)(x) := (\partial_\alpha\psi)(x) + \mathcal{A}_\alpha(x)\psi(x), \qquad x \in \mathcal{U},$$

and

$$(D\psi)(x) := (D_\alpha\psi)(x)\, dx^\alpha.$$

In terms of components, we get

$$D_\alpha\psi^i := \partial_\alpha\psi^i + \Gamma^i_{\alpha s}\psi^s,$$

and $D\psi^i := D_\alpha\psi^i\, dx^\alpha$. By Prop. 15.1 on page 845, we get

$$D^+_\alpha\psi^+(x) = G(x)D_\alpha\psi(x), \qquad x \in \mathcal{U}, \;\; \alpha = 0, 1, 2, 3. \qquad (15.44)$$

This shows that $D_\alpha\psi^i$ is a tensorial family. Finally, we define the covariant differential $D\psi$ by setting

$$\boxed{(D\psi)(P) := (D\psi^i)(x) \otimes \mathbf{e}_i, \qquad P \in \mathcal{O}.} \qquad (15.45)$$

This definition does not depend on the choice of the bundle coordinates.

15.6.3 Covariant Differential for Differential Forms of Tensor Type

> If we use tensorial families of differential forms on the base manifold, then the differentiation processes with respect to space-time time variables (Greek indices) and gauge variables (Latin indices) fit together.
>
> Folklore

Consider the following differential form of tensor type

$$\mathbf{T} = \tfrac{1}{n!}\, T^{i_1\ldots i_k}_{\beta_1\ldots\beta_n j_1\cdots j_l}\, dx^{\beta_1}\wedge\cdots\wedge dx^{\beta_n}\otimes \mathbf{e}_{i_1}\otimes\cdots\otimes\mathbf{e}_{i_k}\otimes\mathbf{e}^{j_1}\otimes\cdots\otimes\mathbf{e}^{j_l}.$$

Here, $T^{i_1\ldots i_k}_{\beta_1\ldots\beta_n j_1\cdots j_l}$ is a tensorial family which is antisymmetric with respect to the Greek space-time indices $\beta_1,\ldots,\beta_n$. We define the covariant differential $D\mathbf{T}$ by setting

$$D\mathbf{T} := \tfrac{1}{n!}\left(D_\alpha T^{i_1\ldots i_k}_{\beta_1\ldots\beta_n j_1\cdots j_l}\, dx^\alpha\right)\wedge dx^{\beta_1}\wedge\cdots\wedge dx^{\beta_n}$$
$$\otimes\mathbf{e}_{i_1}\otimes\cdots\otimes\mathbf{e}_{i_k}\otimes\mathbf{e}^{j_1}\otimes\cdots\otimes\mathbf{e}^{j_l}$$

where we introduce[6]

$$D_\alpha T^{i_1\ldots i_k}_{\beta_1\ldots\beta_n j_1\cdots j_l} := \partial_\alpha T^{i_1\ldots i_k}_{\beta_1\ldots\beta_n j_1\cdots j_l} + \sum_{\sigma=1}^{k}\Gamma^{i_\sigma}_{\alpha s}\cdot T^{i_1\ldots s\ldots i_k}_{\beta_1\ldots\beta_n j_1\cdots j_l}$$
$$-\sum_{\sigma=1}^{l}\Gamma^{s}_{\alpha j_\sigma}\cdot T^{i_1\ldots i_k}_{\beta_1\ldots\beta_n j_1\ldots s\ldots j_l}.$$

Here, we replace the Latin index i_σ (resp. j_σ) of $T^{i_1\ldots i_k}_{\ldots j_1\cdots j_l}$ by the index s, and we sum over $s = 1,\ldots,N$. Mnemonically, note that the index picture is correct. The formulas remain valid if the Greek space-time indices drop out. In addition, we define

$$DT^{i_1\ldots i_k}_{\beta_1\ldots\beta_n j_1\cdots j_l} := D_\alpha T^{i_1\ldots i_k}_{\beta_1\ldots\beta_n j_1\cdots j_l}\, dx^\alpha.$$

Theorem 15.14 *The definition of $D\mathbf{T}$ does not depend on the choice of the bundle coordinates.*

The simple proof (based on the Leibniz rule) will be given below.

The Leibniz rule. Let us consider the following special cases of the general definition above:

(D1) $D_\alpha\psi^i := \partial_\alpha\psi^i + \Gamma^i_{\alpha s}\psi^s$,

(D2) $D_\alpha\varphi_j := \partial_\alpha\varphi_j - \Gamma^s_{\alpha j}\varphi_s$,

(D3) $D_\alpha(\psi^i\varphi_j) := \partial_\alpha(\psi^i\varphi_j) + \Gamma^i_{\alpha s}\psi^s\varphi_j - \Gamma^s_{\alpha j}\psi^i\varphi_s$,

(D4) $D_\alpha\Psi^i_j := \partial_\alpha\Psi^i_j + \Gamma^i_{\alpha s}\Psi^s_j - \Gamma^s_{\alpha j}\Psi^i_s$.

Proposition 15.15 *If ψ^i, φ_j, and Ψ^i_j are tensorial families, then the families* (D1) *through* (D4) *are also tensorial families.*

[6] This definition is an obvious modification of the definition given on page 498 for the covariant partial derivative ∇_α.

Proof. Ad (D1). See (15.44) above.

Ad (D2). It follows from the definition of $D_\alpha \varphi_j$ that

$$\partial_\alpha(\psi^i \varphi_i) = (D_\alpha \psi^i)\varphi_i + \psi^i(D_\alpha \varphi_i)$$

for all tensorial families ψ^i. Since $\psi^i \varphi_i$ is a scalar function, the family $\partial_\alpha(\psi^i \varphi_i)$ is a tensorial family. Moreover, we know by (D1) that $(D_\alpha \psi^i)\varphi_i$ is a tensorial family. Thus, $\psi^i(D_\alpha \varphi_i)$ is a tensorial family for all tensorial families ψ^i. By the inverse index principle from Sect. 8.8.2 on page 493, we obtain that $D_\alpha \varphi_i$ is a tensorial family.

Ad (D3). The definition of $D_\alpha(\psi^i \varphi_j)$ yields the Leibniz rule

$$D_\alpha(\psi^i \varphi_j) = (D_\alpha \psi^i)\varphi_j + \psi^i(D_\alpha \varphi_j).$$

Since the right-hand side is a tensorial family by the statement concerning (D1) and (D2), the left-hand side $D_\alpha(\psi^i \varphi_j)$ is also a tensorial family.

Ad (D4). Note that Ψ^i_j transforms like $\psi^i \varphi_j$. □

The covariant differential $D\boldsymbol{\varphi}$. Using the language of matrices, the definitions (D1) and (D2) above can be written as

- $D_\alpha \psi = \partial_\alpha \psi + \mathcal{A}_\alpha \psi$, and
- $D_\alpha \varphi = \partial_\varphi - \varphi \mathcal{A}_\alpha$,

respectively. For the covariant differential, this implies

- $D\psi = D_\alpha \psi \, dx^\alpha = d\psi + \mathcal{A}\psi$, and
- $D\varphi = D_\alpha \varphi \, dx^\alpha = d\varphi - \varphi \mathcal{A}$,

respectively. Setting $\boldsymbol{\varphi} := \varphi_j \mathbf{e}^j$, we get

$$\boxed{D\boldsymbol{\varphi} := D_\alpha \varphi_j \, dx^\alpha \otimes \mathbf{e}^j.} \tag{15.46}$$

By the index principle, this definition does not depend on the choice of the bundle coordinates.

The covariant differential $D\boldsymbol{\Psi}$. Set $\boldsymbol{\Psi} := \Psi^i_j \, \mathbf{e}_i \otimes \mathbf{e}^j$. Then

$$\boxed{D\boldsymbol{\Psi} := D_\alpha \Psi^i_j \, dx^\alpha \otimes \mathbf{e}_i \otimes \mathbf{e}^j.} \tag{15.47}$$

By the index principle, this definition does not depend on the choice of the bundle coordinates.

Differential forms of tensor type. Suppose that

$$\psi^i_{\beta_1 \ldots \beta_n}, \quad \varphi_{\beta_1, \ldots, \beta_n j} \qquad \text{and} \qquad \Psi^i_{\beta_1 \ldots \beta_n j}$$

are tensorial families which are antisymmetric with respect to the Greek space-time indices $\beta_1, \ldots, \beta_n$. Similarly, as in (D1) through (D4) above, we define

- $D_\alpha \psi^i_{\beta_1 \ldots \beta_n} := \partial_\alpha \psi^i_{\beta_1 \ldots \beta_n} + \Gamma^i_{\alpha s} \psi^s_{\beta_1 \ldots \beta_n}$,
- $D_\alpha \varphi_{\beta_1 \ldots \beta_n j} := \partial_\alpha \varphi_{\beta_1 \ldots \beta_n j} - \Gamma^s_{\alpha j} \varphi_{\beta_1 \ldots \beta_n s}$,
- $D_\alpha \Psi^i_{\beta_1 \ldots \beta_n j} := \partial_\alpha \Psi^i_{\beta_1 \ldots \beta_n j} + \Gamma^i_{\alpha s} \Psi^s_{\beta_1 \ldots \beta_n j} - \Gamma^s_{\alpha j} \Psi^i_{\beta_1 \ldots \beta_n s}$.

Proposition 15.16 *Antisymmetrization with respect to the Greek indices yields the tensorial families*

$$D_{[\alpha} \psi^i_{\beta_1 \ldots \beta_n]}, \quad D_{[\alpha} \varphi_{\beta_1 \ldots \beta_n] j}, \quad D_{[\alpha} \Psi^i_{\beta_1 \ldots \beta_n] j}.$$

Proof. The basic idea of the proof is to use the Cartan derivative for antisymmetric covariant tensor families (see page 523) with respect to the Greek indices. For example, consider

$$D_\alpha \psi^i_{\beta_1 \ldots \beta_n} = \partial_\alpha \psi^i_{\beta_1 \ldots \beta_n} + \Gamma^i_{\alpha s} \psi^s_{\beta_1 \ldots \beta_n}.$$

By Prop. 15.15, we know that $D_\alpha \psi^i_{\beta_1 \ldots \beta_n}$ is a tensorial family with respect to the Latin index. Moreover, the Cartan derivative

$$\partial_{[\alpha} \psi^i_{\beta_1 \ldots \beta_n]}$$

is a tensorial family with respect to the Greek indices. Finally, since $\mathcal{A}_\alpha$ transforms like ∂_α, as postulated on page 863, the family

$$\Gamma^i_{\alpha s} \psi^s_{\beta_1 \ldots \beta_n}$$

is a tensorial family with respect to the Greek indices, and the same property has the antisymmetrization with respect to Greek indices. This yields the claim for $D_{[\alpha} \psi^i_{\beta_1 \ldots \beta_n]}$. The other claims are proved analogously. □

Example. Let $\Psi^i_{\mu\nu j}$ be a tensorial family which is antisymmetric with respect to the Greek indices μ and ν. Define

$$\boldsymbol{\Psi} := \tfrac{1}{2} \Psi^i_{\mu\nu j}\, dx^\mu \wedge dx^\nu \otimes \mathbf{e}_i \otimes \mathbf{e}^j.$$

Then

$$D\boldsymbol{\Psi} := \tfrac{1}{2} \left(D_{[\lambda} \Psi^i_{\mu\nu]j} dx^\lambda \wedge dx^\mu \wedge dx^\nu \right) \otimes \mathbf{e}_i \otimes \mathbf{e}^j.$$

By Prop. 15.16, this definition does not depend on the choice of bundle coordinates. By antisymmetry,

$$\boxed{D\boldsymbol{\Psi} := \tfrac{1}{2} \left(D_\lambda \Psi^i_{\mu\nu j}\, dx^\lambda \wedge dx^\mu \wedge dx^\nu \right) \otimes \mathbf{e}_i \otimes \mathbf{e}^j.}$$

Explicitly, introducing the matrix $\mathcal{T}_{\mu\nu} := (\Psi^i_{\mu\nu j})$, we have

$$D_\lambda \Psi^i_{\mu\nu j} = \partial_\lambda \Psi^i_{\mu\nu} + \Gamma^i_{\lambda s} \Psi^s_{\mu\nu j} - \Gamma^s_{\lambda j} \Psi^i_{\mu\nu s},$$

and hence

$$D_\lambda \mathcal{T}_{\mu\nu} = \partial_\lambda \mathcal{T}_{\mu\nu} + \mathcal{A}_\lambda \mathcal{T}_{\mu\nu} - \mathcal{T}_{\mu\nu} \mathcal{A}_\lambda. \tag{15.48}$$

Setting $\mathcal{T} := \frac{1}{2} \mathcal{T}_{\mu\nu}\, dx^\mu \wedge dx^\nu$ and $\mathcal{A} := \mathcal{A}_\lambda dx^\lambda$, we get

$$D\mathcal{T} := \tfrac{1}{2} D_\lambda \mathcal{T}_{\mu\nu}\, dx^\lambda \wedge dx^\mu \wedge dx^\nu.$$

This implies

$$\boxed{D\mathcal{T} = d\mathcal{T} + \mathcal{A} \wedge \mathcal{T} - \mathcal{T} \wedge \mathcal{A}.}$$

Remark. Similar arguments show that Theorem 15.14 above is always true.

15.6.4 Application to the Riemann Curvature Operator

Invariant approach. Our goal is to use the tensor calculus introduced above in order to get the Riemann curvature operator in an extremely elegant, invariant way. Let

$$\psi : \mathbb{M}^4 \to \mathbb{M}^4 \times \mathbb{C}^N$$

be a section of the vector bundle $\pi : \mathbb{M}^4 \times \mathbb{C}^N \to \mathbb{M}^4$. With respect to local bundle coordinates, we have

$$\psi = \psi^i \mathbf{e}_i$$

where ψ^i is a tensorial family. Note that ψ does not depend on the choice of local bundle coordinates, by the index principle. Moreover, let

$$v, w \in \mathrm{Vect}(\mathbb{M}^4)$$

be smooth velocity vector fields on the base manifold $\mathbb{M}^4$. Here, $v = v^\alpha \partial_\alpha$ and $w = w^\alpha \partial_\alpha$, with respect to local coordinates on $\mathbb{M}^4$.

Theorem 15.17 *We have the invariant curvature relation*

$$D(D\psi) = \mathbf{F}\psi, \tag{15.49}$$

and the Bianchi relation

$$D\mathbf{F} = 0. \tag{15.50}$$

Mnemonically, we write

$$\boxed{DD = \mathbf{F}, \qquad DDD = 0.}$$

Comparing this with the Poincaré cohomology rule, $dd = 0$, we see that the Riemann curvature operator $\mathbf{F}$ measures the deviation of the covariant differential from the trivial relation $DD = 0$ which corresponds to flatness.

Local bundle coordinates. The proof of Theorem 15.17 will be given in Problem 15.6. This proof shows that the theorem is equivalent to the approach studied above in Sect. 15.1. Explicitly, using local bundle coordinates, we get

- $\mathbf{F}\psi = \frac{1}{2} R^i_{\mu\nu j}\psi^j \, dx^\mu \wedge dx^\nu \otimes \mathbf{e}_i$,
- $\mathbf{F} = \frac{1}{2} R^i_{\mu\nu j} \, dx^\mu \wedge dx^\nu \otimes \mathbf{e}_i \otimes \mathbf{e}^j$, and
- $\mathbf{F}(v, w)\psi = (R^i_{\mu\nu j} v^\mu w^\nu \psi^j)\, \mathbf{e}_i$.

The point is that $\mathbf{F}\psi$ does not contain any partial derivatives of ψ^i; they cancel each other. The coefficient functions $R^i_{\mu\nu j}$ follow from the matrix equation

$$\boxed{\mathcal{F}_{\mu\nu} = \partial_\mu \mathcal{A}_\nu - \partial_\nu \mathcal{A}_\mu + \mathcal{A}_\mu \mathcal{A}_\nu - \mathcal{A}_\nu \mathcal{A}_\mu, \qquad \mu, \nu = 0, 1, 2, 3}$$

together with the $(N \times N)$-matrices $\mathcal{A}_\mu = (\Gamma^i_{\mu j})$ and $\mathcal{F}_{\mu\nu} = (R^i_{\mu\nu j})$. Here, the index i (resp. j) counts the rows (resp. columns). The Bianchi identity (15.50) reads as

$$\boxed{D_{[\lambda}\mathcal{F}_{\mu\nu]} = 0, \qquad \lambda, \mu, \nu = 0, 1, 2, 3}$$

where $D_\lambda \mathcal{F}_{\mu\nu} = \partial_\lambda \mathcal{F}_{\mu\nu} + \mathcal{A}_\lambda \mathcal{F}_{\mu\nu} - \mathcal{F}_{\mu\nu}\mathcal{A}_\lambda$.

The language of local tensorial differential forms. Set

$$\omega^i_j := \Gamma^i_{\mu j}\, dx^\mu, \qquad \Omega^i_j := \tfrac{1}{2} R^i_{\mu\nu j}\, dx^\mu \wedge dx^\nu, \quad i,j = 1,\ldots,N.$$

Then

$$\mathcal{A} := \mathcal{A}_\mu\, dx^\mu = (\omega^i_j), \qquad \mathcal{F} := \tfrac{1}{2}\mathcal{F}_{\mu\nu}\, dx^\mu \wedge dx^\nu = (\Omega^i_j).$$

The curvature relation (15.49) reads as

$$\boxed{\mathcal{F} = d\mathcal{A} + \mathcal{A} \wedge \mathcal{A},}$$

and the Bianchi relation (15.50) reads as

$$\boxed{d\mathcal{F} + \mathcal{A} \wedge \mathcal{F} - \mathcal{F} \wedge \mathcal{A} = 0.}$$

Further reading. The Kähler–Clifford calculus (introduced in Sect. 8.6 based on the exterior Cartan product and the interior Clifford product) can be easily extended to vector bundles (tensorial differential forms). This can be found in E. Kähler, The interior differential calculus (in German), Collected Works, de Gruyter, Berlin, 2003, pp. 497–595.

Problems

15.1 *Construction of a connection.* Fix $N = 1, 2, \ldots$, and fix an inertial system on $\mathbb{M}^4$. We are given the smooth matrix functions

$$G_0, G_1 : \mathbb{R}^4 \to \mathcal{G}$$

where $\mathcal{G}$ is a closed subgroup of $GL(N, \mathbb{C})$ (e.g., $\mathcal{G} = U(N)$). Moreover, set $G_2(x) := G_1(x)G_0(x)^{-1}$. Then it follows from

$$\psi^+(x) = G_0(x)\psi(x) \quad \text{and} \quad \psi^{++}(x) = G_1(x)\psi(x) \tag{15.51}$$

that

$$\psi^{++}(x) = G_2(x)\psi^+(x). \tag{15.52}$$

Moreover, let $\mathcal{A}_\alpha(x)$ be a complex $(N \times N)$-matrix which is an element of the Lie algebra $\mathcal{LG}$ of the Lie group (e.g., if $\mathcal{G} = U(n)$, then $\mathcal{LG} = u(n)$). Suppose that

$$\mathcal{A}^+_\alpha(x) := G_0(x)\mathcal{A}_\alpha(x)G_0(x)^{-1} - \partial_\alpha G_0(x) \cdot G_0(x)^{-1},$$

and

$$\mathcal{A}^{++}_\alpha(x) := G_1(x)\mathcal{A}_\alpha(x)G_1(x)^{-1} - \partial_\alpha G_1(x) \cdot G_1(x)^{-1}.$$

Show that

$$\mathcal{A}^{++}_\alpha(x) := G_2(x)\mathcal{A}^+_\alpha(x)G_2(x)^{-1} - \partial_\alpha G_2(x) \cdot G_2(x)^{-1}. \tag{15.53}$$

Hint: Use Problem 13.3 together with $(GH)^{-1} = H^{-1}G^{-1}$.
Remark. The transformation property (15.53) allows us to construct the connection matrices with the correct transformation law under both gauge transformations and transformations of inertial systems. We proceed as follows. We choose a fixed inertial system. We are given the smooth matrix functions

$$\mathcal{A}_\alpha : \mathbb{R}^4 \to \mathcal{LG}, \qquad \alpha = 0, 1, 2, 3.$$

Consider the gauge transformations (15.51), and construct both $\mathcal{A}^+$ and $\mathcal{A}^{++}$, as above. Then, the relation (15.53) fits the gauge transformation (15.52). Under a change of the inertial system, the matrices $\mathcal{A}_\alpha$ are transformed like ∂_α, $\alpha = 0, 1, 2, 3$.

15.2 *Proof of the Bianchi identity.* Let $\mathcal{L}$ be a Lie subalgebra of the real Lie algebra $gl(N, \mathbb{C}), N = 1, 2, \ldots$ Set

$$\mathcal{F}_{\mu\nu} := \partial_\mu \mathcal{A}_\nu - \partial_\nu \mathcal{A}_\mu + [\mathcal{A}_\mu, \mathcal{A}_\nu]_-, \qquad \mu, \nu = 0.1, 2, 3.$$

Define

$$D_\varrho \mathcal{F}_{\mu\nu} := \partial_\varrho \mathcal{F}_{\mu\nu} + [\mathcal{A}_\varrho, \mathcal{F}_{\mu\nu}]_-, \quad \varrho, \mu, \nu = 0, 1, 2, 3.$$

Show that there holds the Bianchi identity:

$$D_\varrho \mathcal{F}_{\mu\nu} + D_\mu \mathcal{F}_{\nu\varrho} + D_\nu \mathcal{F}_{\varrho\mu} = 0, \quad \varrho, \mu, \nu = 0, 1, 2, 3. \tag{15.54}$$

Equivalently, $D_{[\varrho} \mathcal{F}_{\mu\nu]} = 0$ (antisymmetrization).
Solution: We will write $[A, B]$ instead of $[A, B]_-$. The main trick is to use the Jacobi identity:

$$[\mathcal{A}_\varrho, [\mathcal{A}_\mu, \mathcal{A}_\nu]] + [\mathcal{A}_\mu, [\mathcal{A}_\nu, \mathcal{A}_\varrho]] + [\mathcal{A}_\nu, [\mathcal{A}_\varrho, \mathcal{A}_\mu]] = 0.$$

By the Leibniz rule,

$$D_\varrho \mathcal{F}_{\mu\nu} = \partial_\varrho \partial_\mu \mathcal{A}_\nu - \partial_\varrho \partial_\nu \mathcal{A}_\mu + [\partial_\varrho \mathcal{A}_\mu, \mathcal{A}_\nu] + [\mathcal{A}_\mu, \partial_\varrho \mathcal{A}_\nu] + [\mathcal{A}_\varrho, [\mathcal{A}_\mu, \mathcal{A}_\nu]].$$

This implies the claim (15.54), by using the Jacobi identity.

15.3 *Proof of Proposition 15.7 on page 850.* Use a similar argument as in the proof of Problem 12.17 about the Maurer–Cartan form $\mathsf{M} := G^{-1} dG$ (see page 806).
Solution: Using the Leibniz rule, it follows from $\mathsf{A} = G^{-1} \mathcal{A} G + \mathsf{M}$ that

$$d\mathsf{A} = dG^{-1} \wedge \mathcal{A} \cdot G + G^{-1} d\mathcal{A} \cdot G - G^{-1} \mathcal{A} \wedge dG + d\mathsf{M}.$$

By Problem 12.17, $dG^{-1} = -G^{-1} dG \cdot G^{-1}$ and $d\mathsf{M} = -\mathsf{M} \wedge \mathsf{M}$. Hence

$$dG^{-1} \wedge \mathcal{A} \cdot G + G^{-1} \mathcal{A} \wedge dG = -\mathsf{M} \wedge G^{-1} \mathcal{A}\, G - G^{-1} \mathcal{A} \cdot G \wedge \mathsf{M} = 0.$$

This implies $d\mathsf{A} = G^{-1} d\mathcal{A} \cdot G - \mathsf{M} \wedge \mathsf{M}$. Furthermore,

$$\mathsf{A} \wedge \mathsf{A} = (G^{-1} \mathcal{A} G + \mathsf{M}) \wedge (G^{-1} \mathcal{A} G + \mathsf{M}) = G^{-1} (\mathcal{A} \wedge \mathcal{A}) G + \mathsf{M} \wedge \mathsf{M}.$$

Finally, $\mathsf{F} = d\mathsf{A} + \mathsf{A} \wedge \mathsf{A} = G^{-1} (d\mathcal{A} + \mathcal{A} \wedge \mathcal{A}) G = G^{-1} \mathcal{F} G.$

15.4 *The Hilbert space structure of the real Lie algebra $u(N)$, $N = 1, 2, \ldots$* Set

$$\langle A|B \rangle := -\mathrm{tr}(AB) \qquad \text{for all} \quad A, B \in u(N).$$

Show that this is an inner product on $u(N)$. This way, the real Lie algebra $u(N)$ becomes a real Hilbert space.
Solution: The complex $(N \times N)$-matrix A is an element of $u(N)$ iff $A^\dagger = -A$. Equivalently, $A = \mathrm{i}A'$ where A' is self-adjoint. By the principal axis theorem, there exists a unitary matrix G such that $A' = G^{-1} \Lambda G$ where Λ is a diagonal matrix with real entries. Noting that $\mathrm{tr}(ABC) = \mathrm{tr}(CAB)$ and $G^{-1} G = I$, we get

$$\langle A|A \rangle = -\langle \mathrm{i}G\Lambda G^{-1} | \mathrm{i}G\Lambda G^{-1} \rangle = \mathrm{tr}(\Lambda^2) \geq 0.$$

15.5 *The Hodge star operator.* Let $\mathcal{F} = \frac{1}{2} \mathcal{F}_{\mu\nu}\, dx^\mu \wedge dx^\nu$ where $\mathcal{F}_{\mu\nu} \in u(N)$ for $\mu, \nu = 0, 1, 2, 3$. Compute $*\mathcal{F}$ where the Hodge $*$-operator refers to the metric tensor $g = \eta_{\alpha\beta}\, dx^\alpha \otimes dx^\beta$ on the Minkowski manifold $\mathbb{M}^4$ (see Sect. 18.4.1).
Solution: Since $\det(\eta_{\alpha\beta}) = -1$, it follows from the definition (8.66) on page 470 that

$$*\mathcal{F} = \tfrac{1}{4} \varepsilon_{\mu\nu\alpha\beta} \mathcal{F}^{\mu\nu}\, dx^\alpha \wedge dx^\beta.$$

15.6 *The Riemann curvature operator.* Prove Theorem 15.17 on page 867.
Solution: Ad (15.49). It follows from $\boldsymbol{\psi} = \psi^i \mathbf{e}_i$ and

$$D\boldsymbol{\psi} = D_\nu \psi^i \, dx^\nu \otimes \mathbf{e}_i$$

with $D_\nu \psi^i = \partial_\nu \psi^i + \Gamma^i_{\nu s} \psi^s$ that

$$\begin{aligned} D(D\boldsymbol{\psi}) &= D_\mu D_\nu \psi^i \, dx^\mu \wedge dx^\nu \otimes \mathbf{e}_i \\ &= \tfrac{1}{2}(D_\mu D_\nu - D_\nu D_\mu)\psi^i \, dx^\mu \wedge dx^\nu \otimes \mathbf{e}_i . \end{aligned}$$

As in (15.3) on page 845, we get

$$(D_\mu D_\nu - D_\nu D_\mu)\psi = (\partial_\mu \mathcal{A}_\nu - \partial_\nu \mathcal{A}_\mu + \mathcal{A}_\mu \mathcal{A}_\nu - \mathcal{A}_\nu \mathcal{A}_\mu)\psi.$$

Explicitly,

$$(D_\mu D_\nu - D_\nu D_\mu)\psi^i = (\partial_\mu \Gamma^i_{\nu s} - \partial_\nu \Gamma^i_{\mu s} + \Gamma^i_{\mu r}\Gamma^r_{\nu s} - \Gamma^i_{\nu r}\Gamma^r_{\mu s})\psi^s .$$

Ad (15.50). We have

$$\begin{aligned} D\mathbf{F} &= \tfrac{1}{2} D_\lambda R^i_{\mu\nu j} \, dx^\lambda \wedge dx^\mu \wedge dx^\nu \otimes \mathbf{e}_i \otimes \mathbf{e}^j \\ &= \tfrac{1}{2} D_{[\lambda} R^i_{\mu\nu]j} \, dx^\lambda \wedge dx^\mu \wedge dx^\nu \otimes \mathbf{e}_i \otimes \mathbf{e}^j . \end{aligned}$$

It remains to show that

$$D_{[\lambda} R^i_{\mu\nu]j} = 0.$$

In fact, set $\mathcal{F}_{\mu\nu} := (R^i_{\mu\nu j})$. By (15.48) on page 866,

$$D_\lambda \mathcal{F}_{\mu\nu} = \partial_\lambda \mathcal{F}_{\mu\nu} + \mathcal{A}_\lambda \mathcal{F}_{\mu\nu} - \mathcal{F}_{\mu\nu}\mathcal{A}_\lambda .$$

Finally, it follows from Problem 15.2 that $D_{[\lambda}\mathcal{F}_{\mu\nu]} = 0$.

16. Cocycles and Observers

There exist two approaches to the theory of bundles, namely,

(i) the observer approach based on cocycles, and
(ii) the axiomatic geometric approach.

In (i), roughly speaking, we will use a cocycle in order to glue together product bundles. This will be considered in the present chapter. The geometric approach will be considered in the next chapter; this is the most elegant approach based on a few geometric axioms. For didactic reasons, we start with (i). The idea is to use bundle charts (product bundles) and to describe the change of bundle coordinates by a cocycle. In fact, the two approaches (i) and (ii) are equivalent to each other. In (ii), the cocycle corresponds to the transition maps between the bundle charts.

16.1 Cocycles

> Cocycles describe the exchange of physical information between three observers. This exchange has to satisfy quite natural compatibility conditions which characterize cocycles. The notion of cocycle is of fundamental importance for modern geometry and topology.
>
> Folklore

Let $\mathcal{M}$ be a finite-dimensional real manifold (e.g., $\mathcal{M} = \mathbb{M}^4$). Furthermore, we assume that

$$\mathcal{M} = \cup_{j=1}^{\mathcal{J}} \mathcal{O}_j$$

where $\mathcal{O}_1, \mathcal{O}_1, \ldots, \mathcal{O}_{\mathcal{J}}$ are open subsets of $\mathcal{M}$. Let $\mathcal{G}$ be a closed subgroup of the Lie group $GL(N, \mathbb{C})$. By definition, the family of maps[1]

$$G_{kj} : \mathcal{O}_k \cap \mathcal{O}_j \to \mathcal{G}, \qquad j, k = 1, \ldots, \mathcal{J}$$

is called a cocycle (with values in the Lie group $\mathcal{G}$) iff the following conditions are satisfied for all indices $j, k, l = 1, \ldots, \mathcal{J}$:

- $G_{jj}(P) = \mathbf{1}$ for all $P \in \mathcal{O}_j$,
- $G_{jk}(P) = G_{kj}(P)^{-1}$ for all $P \in \mathcal{O}_k \cap \mathcal{O}_j$, and

$$G_{lk}(P)G_{kj}(P) = G_{lj}(P) \qquad \text{for all} \quad P \in \mathcal{O}_l \cap \mathcal{O}_k \cap \mathcal{O}_j. \tag{16.1}$$

[1] We only consider such index pairs j, k where the intersection $\mathcal{O}_k \cap \mathcal{O}_j$ is not empty.

E. Zeidler, *Quantum Field Theory III: Gauge Theory*,
DOI 10.1007/978-3-642-22421-8_17,

16.2 Physical Fields via the Cocycle Strategy

Observers and physical fields. Fix $N = 1, 2, \ldots$ By definition, a physical field

$$\psi = \{\psi_j\}$$

on the manifold $\mathcal{M}$ (e.g., $\mathcal{M} = \mathbb{M}^4$) with values in $\mathbb{C}^N$ (and with respect to the local observers $O_1, \ldots, O_J$) is a family of smooth maps

$$\psi_j : \mathcal{O}_j \to \mathbb{C}^N, \qquad j = 1, \ldots, J$$

together with the transformation laws

$$\boxed{\psi_k(P) = G_{kj}(P)\psi_j(P), \qquad P \in \mathcal{O}_k \cap \mathcal{O}_j.} \tag{16.2}$$

Here, $\{G_{kj}\}$ is a cocycle with values in the Lie group $\mathcal{G}$ of complex $(N \times N)$-matrices. In terms of physics, the observer O_j (resp. O_k) uses the physical field ψ_j (resp. ψ_k). If the point P is an element of both the sets $\mathcal{O}_j$ and $\mathcal{O}_k$, then we need a transformation law which is given by (16.2). This transformation law corresponds to the gauge transformation $\psi^+(P) = G_0(P)\psi(P)$ from (15.1) on page 844.

The cocycle strategy. This strategy reads as follows:

- We use product bundles as in Sects. 15.1 through 15.3.
- We study gauge transformations for product bundles as in Sects. 15.1 through 15.3.
- We replace the gauge transformations by the corresponding transformations (16.2) with respect to the cocycle $\{G_{kj}\}$.

This way, it is possible to generalize straightforward the operations introduced in Sects. 15.1 through 15.3.

Example (covariant directional derivative of physical fields). Let v be a smooth velocity vector field on $\mathcal{M}$. The covariant directional derivative $D_v\psi$ of $\psi = \{\psi_j\}$ is given by the family $\{D_v\psi_j\}$ of maps

$$D_v\psi_j : \mathcal{O}_j \to \mathbb{C}^N, \qquad j = 1, \ldots, J$$

with the transformation laws

$$D_v\psi_k(P) = G_{kj}(P)D_v\psi_j(P), \qquad P \in \mathcal{O}_k \cap \mathcal{O}_j.$$

The map $D_v\psi_j$ is constructed as in Sect. 15.1. Our definition is motivated by the gauge transformation $D_v^+\psi^+(P) = G_0(P)D_v\psi(P)$.

The bundle manifold $\mathcal{V}$. We write

$$(P, \psi_k) \sim (P, \psi_j)$$

iff $P \in \mathcal{O}_k \cap \mathcal{O}_j$ and $\psi_k = G_{kj}(P)\psi_j$. This is an equivalence relation. By definition, all the equivalence classes $[(P, \psi_j)]$ form the bundle space

$$\mathcal{V} := \{[(P, \psi_j)]\}.$$

By definition, if $P \in \mathcal{O}_j$, then the point $[(P, \psi_j)]$ of $\mathcal{V}$ has the bundle coordinate

$$(P, \psi_j) \in \mathcal{O}_j \times \mathbb{C}^N.$$

Setting $\pi([(P, \psi_j)]) := P$, we get the surjective map

$$\pi : \mathcal{V} \to \mathcal{M}. \tag{16.3}$$

Since $\mathcal{O}_j \times \mathbb{C}^N$ is a real manifold, the bundle space $\mathcal{V}$ becomes the structure of a real manifold, and the map π is smooth. The map (16.3) relates the approach above to the notion of vector bundle to be considered in the next chapter.

The approach works analogously if we have a family $\{\mathcal{O}_j\}_{j\in\mathcal{J}}$ with general index set $\mathcal{J}$. The relation between different observers, cocycles, and bundles is discussed in great detail in Sect. 4.4.2 of Vol. II.

16.3 Local Phase Factors via the Cocycle Strategy

Let us now replace the typical fiber $\mathbb{C}^N$ by the Lie group $\mathcal{G}$. In particular, the transformation law (16.2) is replaced by the transformation law

$$\boxed{G_k(P) = G_{kj}(P)G_j(P), \qquad P \in \mathcal{O}_k \cap \mathcal{O}_j.} \tag{16.4}$$

This is motivated by the gauge transformation $G^+(P) = G_0(P)G(P)$ from (15.13) on page 848.

By definition, a phase factor field $G = \{G_j\}$ on the manifold $\mathcal{M}$ (e.g., $\mathcal{M} = \mathbb{M}^4$) with values in the Lie group $\mathcal{G}$ (and with respect to the local observers $O_1, \ldots, O_{\mathcal{J}}$) is a family of smooth maps

$$G_j : \mathcal{O}_j \to \mathbb{C}^N, \qquad j = 1, \ldots, J$$

together with the transformation laws (16.4).

The bundle manifold $\mathcal{P}$. We write

$$(P, G_k) \sim (P, G_j)$$

iff $P \in \mathcal{O}_j \cap \mathcal{O}_k$ and $G_k = G_{kj}(P)G_j$. This is an equivalence relation. By definition, all the equivalence classes $[P, G_j]$ form the bundle space

$$\mathcal{P} := \{[(P, G_j)]\}.$$

By definition, if $P \in \mathcal{O}_j$, then the point $[(P, G_j)]$ has the bundle coordinate

$$(P, G_j) \in \mathcal{O}_j \times \mathcal{G}.$$

Setting $\pi([(P, \psi_j)]) := P$, we get the surjective map

$$\pi : \mathcal{P} \to \mathcal{M}. \tag{16.5}$$

Since $\mathcal{O}_j \times \mathcal{G}$ is a real manifold, the bundle space $\mathcal{P}$ becomes the structure of a real manifold, and the map π is smooth. The map (16.5) relates the approach above to the notion of principal bundle to be considered in the next chapter. Since the transformation laws (16.2) and (16.4) are based on the same cocycle $\{G_{kj}\}$, we say that

The vector bundle $\pi : \mathcal{V} \to \mathcal{M}$ is associated to the principal bundle $\pi : \mathcal{P} \to \mathcal{M}$ and vice versa.

17. The Axiomatic Geometric Approach to Bundles

Before you axiomatize, there must be mathematical substance.
Hermann Weyl (1885–1955)

Our strategy is to define the notion of vector bundles and principal bundles in an invariant way by only using geometric properties of manifolds. In order to prove further geometric properties of these manifolds (e.g., curvature or parallel transport), we use the fact that, by definition, these properties do not depend on the choice of local bundle coordinates. Therefore, we can pass to special bundle coordinates. This is the situation of product bundles considered in Sects. 15.1 through 15.3. This way, the general results are immediate consequences of our special results about product bundles.

In this chapter, we tacitly assume that all the objects are smooth, that is, they are described by smooth functions with respect to local coordinates.

17.1 Connection on a Vector Bundle

From the geometric point of view, a smooth vector bundle is a real manifold $\mathcal{V}$ which possesses a fibration. The fibers are linearly isomorphic to a real linear space X (e.g., $X = \mathbb{R}^n$) called the typical fiber.[1] The fibers are parametrized by the base manifold $\mathcal{M}$. The map

$$\pi : \mathcal{V} \to \mathcal{M}$$

assigns to the points of a fiber the corresponding parameter of the fiber. Locally, the fibration of $\mathcal{V}$ is trivial. That is, the bundle manifold $\mathcal{V}$ looks locally like the product $\mathcal{O} \times X$ where $\mathcal{O}$ is an open subset of the base manifold $\mathcal{M}$. Roughly speaking, a vector bundle is a regularly parametrized family of linear spaces. The theory of vector bundles is nothing other than parametrized linear algebra.
A connection of $\mathcal{V}$ connects the fibers with each other by parallel transport. On the infinitesimal level, such a parallel transport is given by a directional derivative $D_v\psi$ for sections $\psi : \mathcal{M} \to \mathcal{V}$.

Folklore

Intuitively, a very simple situation is depicted in Fig. 17.1. Here, we have the bundle manifold $\mathcal{V} = \mathbb{R}^2$. The fibers F_x are the straight lines parallel to the y-axis. The base manifold is the x-axis. The map $\pi : \mathcal{V} \to \mathcal{M}$ is given by the projection $\pi(x, y) := x$. A section $s : \mathcal{M} \to \mathcal{V}$ is given by the map $s(x) = (x, \psi(x))$. A parallel transport

[1] In the case where $X = \mathbb{C}^N$, the set $\mathbb{C}^N$ is considered as a real linear space (of dimension $2N$).

E. Zeidler, *Quantum Field Theory III: Gauge Theory*,
DOI 10.1007/978-3-642-22421-8_18,

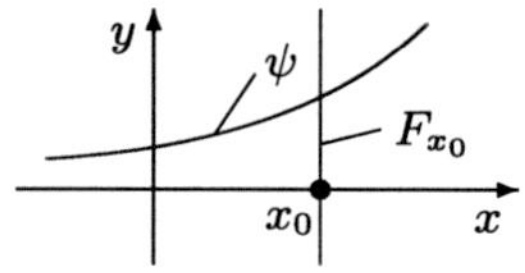

Fig. 17.1. Section ψ of a vector bundle

sends a point (x_1, y_1) of the fiber F_{x_1} to the point (x_2, y_2) of the fiber F_{x_2} along the curve

$$x \mapsto (x, y(x)), \qquad x \in [x_1, x_2]$$

with $y(x_1) = y_1$ and $y(x_2) = y_2$.

Definition of a smooth vector bundle. The smooth surjective map

$$\boxed{\pi : \mathcal{V} \to \mathcal{M}} \tag{17.1}$$

is called a real smooth vector bundle of rank n iff the following conditions are satisfied:

(V1) Bundle manifold: The so-called bundle manifold $\mathcal{V}$ and the so-called base manifold $\mathcal{M}$ are real finite-dimensional manifolds.

(V2) Linear fibration: For every point P of the base manifold $\mathcal{M}$, the so-called fiber

$$F_P := \pi^{-1}(P)$$

over the base point P is a real n-dimensional linear space.

(V3) Local triviality (bundle coordinates): The bundle manifold $\mathcal{V}$ is locally parametrized by so-called bundle coordinates (P, ψ_j) which lie in $\mathcal{O}_j \times \mathbb{R}^n$. Explicitly, there exists a family of nonempty open subsets $\mathcal{O}_j$, $j \in \mathcal{J}$, of the base manifold $\mathcal{M}$ which cover $\mathcal{M}$. Moreover, there exist diffeomorphisms[2]

$$\beta_j : \pi^{-1}(\mathcal{O}_j) \to \mathcal{O}_j \times \mathbb{R}^n, \qquad j \in \mathcal{J},$$

which respect the linear fiber structure. That is, writing

$$\beta_j(Q) = (P, \psi_j), \qquad Q \in F_P, \qquad P \in \mathcal{O}_j, \psi_j \in \mathbb{R}^n,$$

the operator $Q \mapsto \psi_j$ is a linear isomorphism from the fiber F_P onto $\mathbb{R}^n$ for all the base points $P \in \mathcal{O}_j$.

The map β_j is called a bundle chart map, and $\beta_j(Q) = (P, \psi_j)$ is called the local bundle coordinate of the point $Q \in \mathcal{V}$. Note that this local bundle coordinate of the point Q depends on the choice of the open subset $\mathcal{O}_j$ of the base manifold $\mathcal{M}$.

Sections (physical fields). The map

$$\boxed{\psi : \mathcal{M} \to \mathcal{V}} \tag{17.2}$$

is called a cross-section (or, briefly, a section) iff $\psi(P) \in F_P$ for all $P \in \mathcal{M}$. In terms of physics, sections are physical fields. The prototype of a section is depicted in Fig. 17.1. The symbol Sect($\mathcal{V}$) denotes the space of all the smooth sections of the form (17.2).

---

[2] Note that $\pi^{-1}(\mathcal{O}_j) = \cup_{P \in \mathcal{O}_j} F_P$.

The cocycle of transition maps. As a rule, the bundle coordinates of a point Q are not uniquely determined. The change of the bundle coordinates is described by so-called transition maps

$$\psi_k = G_{kj}(P)\psi_j \tag{17.3}$$

where $\psi_j, \psi_k \in \mathbb{R}^n$ and $G_{kj}(P) \in GL(n, \mathbb{R})$. More precisely, the transition map T_{kj} is defined by the commutativity of the following diagram:

$$\begin{array}{ccc} \pi^{-1}(\mathcal{O}_j \cap \mathcal{O}_k) & \xrightarrow{\beta_j} & (\mathcal{O}_j \cap \mathcal{O}_k) \times \mathbb{R}^n \\ & \searrow^{\beta_k} & \downarrow T_{kj} \\ & & (\mathcal{O}_j \cap \mathcal{O}_k) \times \mathbb{R}^n . \end{array}$$

Explicitly, $T_{kj} := \beta_k \circ \beta_j^{-1}$, and $T_{kj}(P, \psi_j) = (P, \psi_k)$ with (17.3). The maps

$$G_{kj} : \mathcal{O}_k \cap \mathcal{O}_j \to GL(n, \mathbb{R})$$

form a cocycle. If there exists a closed subgroup $\mathcal{G}$ of $GL(n, \mathbb{R})$ such that always $G_{kj}(P) \in \mathcal{G}$, then we say that $\mathcal{G}$ is the symmetry group of the vector bundle $\pi : \mathcal{V} \to \mathcal{M}$.

Definition of a connection. For a general vector bundle of rank n, all the fibers are linearly isomorphic to $\mathbb{R}^n$. But one has not a canonical linear isomorphism between different fibers at hand. In order to get such a canonical linear isomorphism between the fibers, we need an additional structure called a connection. More precisely, let v be a smooth velocity vector field on the base manifold $\mathcal{M}$. By definition, a directional derivative is a linear map

$$D_v : \text{Sect}(\mathcal{V}) \to \text{Sect}(\mathcal{V})$$

on the real linear space $\text{Sect}(\mathcal{V})$ of smooth sections (physical fields) of the vector bundle $\pi : \mathcal{V} \to \mathcal{M}$ which satisfies the linearity condition

- $D_{fv+gw}\psi = fD_v\psi + gD_w\psi$

and the Leibniz rule

- $D_v(f\psi) = d_v f \cdot \psi + fD_v\psi$.

We assume that this is satisfied for both all the smooth sections $\psi \in \text{Sect}(\mathcal{V})$ and all the smooth maps $f, g : \mathcal{M} \to \mathbb{R}$ with compact support. By (17.4) below, $(D_v\psi)(P)$ only depends on the tangent vector $v(P)$. Recall that the classical directional derivative $d_v f$ coincides with the Lie derivative $\mathcal{L}_{v(P)}\psi(P))$.

Curvature. The following result is crucial for the theory of vector bundles in modern mathematics.

Theorem 17.1 *There exists a differential 2-form $\mathcal{F}$ on the base manifold $\mathcal{M}$ such that*

$$(D_v D_w - D_w D_v - D_{[v,w]})\psi = \mathcal{F}(v, w)\psi$$

for all smooth velocity vector fields v, w on the base manifold $\mathcal{M}$ and for all smooth sections $\psi \in \text{Sect}(\mathcal{V})$. In addition, the map $\psi \mapsto \mathcal{F}(v, w)\psi$ is a linear operator on the real linear space $\text{Sect}(\mathcal{V})$ *of smooth sections of the vector bundle $\pi : \mathcal{V} \mapsto \mathcal{M}$.*

The proof can be reduced to the case of a product bundle considered in (15.8) on page 846. To this end, one uses local bundle coordinates, and one shows that

$$D_v\psi = v^\alpha \partial_\alpha \psi + v^\alpha \mathcal{A}_\alpha \psi. \tag{17.4}$$

This will be proved in Problem 17.1.

Parallel transport. Let

$$C : P = P(\sigma), \qquad \sigma \in \mathcal{R},$$

be a smooth curve on the base manifold $\mathcal{M}$ where $\mathcal{R}$ is an open interval of $\mathbb{R}$. Moreover, let

$$\psi : \mathcal{M} \to \mathcal{V}$$

be a smooth section of the vector bundle $\mathcal{V}$. We set $\psi(\sigma) := \psi(P(\sigma))$. By definition, the section ψ is parallel along the curve C iff

$$\boxed{(D_{\dot{P}(\sigma)}\psi)(P(\sigma)) = 0 \qquad \text{for all} \quad \sigma \in \mathcal{R}.} \tag{17.5}$$

Using bundle coordinates (x, ψ_j), this means that

$$\dot{x}^\alpha(\sigma)(\partial_\alpha \psi_j)(x(\sigma)) + \dot{x}^\alpha(\sigma)\mathcal{A}_\alpha(x(\sigma)) \cdot \psi_j(x) = 0, \qquad \sigma \in \mathcal{R}.$$

By the chain rule, this is equivalent to

$$\frac{D\psi_j(\sigma)}{d\sigma} = 0, \qquad \sigma \in \mathcal{R}, \tag{17.6}$$

where we define

$$\frac{D\psi_j(\sigma)}{d\sigma} := \frac{d\psi_j(\sigma)}{d\sigma} + \dot{x}^\alpha(\sigma)\mathcal{A}_\alpha(x(\sigma)) \cdot \psi_j(\sigma). \tag{17.7}$$

If we regard σ as time parameter, then

$$\frac{D\psi(\sigma)}{d\sigma} := (D_{\dot{P}(\sigma)}\psi)(P(\sigma))$$

is called the covariant time derivative of the map $\sigma \mapsto \psi(P(\sigma))$.

Now let us consider a slightly more general situation. Let

$$\psi : \mathcal{R} \to \mathcal{V} \tag{17.8}$$

be a smooth curve such that $\pi(\psi(\sigma)) = P(\sigma)$ for all $\sigma \in \mathcal{R}$. In order to define the covariant time derivative

$$\frac{D\psi(\sigma)}{d\sigma}, \qquad \sigma \in \mathcal{R},$$

we use local bundle coordinates, and we define

$$\frac{D\psi_j(\sigma)}{d\sigma} := \frac{d\psi_j(\sigma)}{d\sigma} + \dot{x}^\alpha(\sigma)\mathcal{A}_\alpha(x(\sigma)) \cdot \psi_j(\sigma).$$

Proposition 17.2 *This definition does not depend on the choice of the bundle coordinates.*

For the proof, we refer to Problem 17.3. Mnemonically, we write

$$\frac{D\psi(\sigma)}{d\sigma} = D_{\dot{P}(\sigma)}\psi(\sigma).$$

Let $\sigma_1, \sigma_2 \in \mathcal{R}$. We say that the curve (17.8) represents a parallel transport

- from the point $\psi(\sigma_1)$ of the fiber $F_{P(\sigma_1)}$
- to the point $\psi(\sigma_2)$ of the fiber $F_{P(\sigma_2)}$

along the curve C iff the covariant 'time' derivative vanishes identically, that is,

$$\frac{D\psi(\sigma)}{d\sigma} = 0, \qquad \sigma_1 \leq \sigma \leq \sigma_2.$$

17.2 Connection on a Principal Bundle

From the geometric point of view, a principal bundle is a manifold $\mathcal{P}$ which possesses a Lie group $\mathcal{G}$ as symmetry group. This group acts freely on $\mathcal{P}$ from the right. This induces orbits on $\mathcal{P}$. Every orbit is in one-to-one correspondence to the group $\mathcal{G}$. The orbits are the fibers of a fibration of the bundle manifold $\mathcal{P}$. The fibers are parametrized by the base manifold $\mathcal{M}$. The map

$$\pi : \mathcal{P} \to \mathcal{M}$$

assigns to the points of a fiber F_P the corresponding parameter $P \in \mathcal{M}$ of the fiber. We postulate that, locally, the fibration of $\mathcal{P}$ is trivial. That is, the bundle manifold $\mathcal{P}$ looks locally like the product $\mathcal{O} \times \mathcal{G}$ where $\mathcal{O}$ is an open subset of the base manifold $\mathcal{M}$. Roughly speaking, a principal fiber bundle is a regularly parametrized orbit space generated by the symmetry group $\mathcal{G}$.

A connection on $\mathcal{G}$ connects the fibers with each other by parallel transport. Such a parallel transport from the point $Q_1 \in F_{P_1}$ to the point $Q_2 \in F_{P_2}$ is given by a curve on the bundle manifold $\mathcal{P}$,

$$\mathcal{C} : Q = Q(\sigma), \qquad \sigma_1 \leq \sigma \leq \sigma_2,$$

with $Q(\sigma_1) = Q_1$ and $Q(\sigma_2) = Q_2$. This curve $\mathcal{C}$ is characterized by the fact that its tangent vectors are so-called horizontal tangent vectors on the bundle manifold $\mathcal{P}$. The crucial connection differential 1-form A on $\mathcal{P}$ allows us to define horizontal tangent vectors. In fact, A assigns to every tangent vector of $\mathcal{P}$ a uniquely determined horizontal tangent vector by projection. This projection from tangent vectors to horizontal tangent vectors can be used to assign to every Cartan differential

$$d\omega$$

of a differential form ω on the bundle manifold $\mathcal{P}$ a covariant Cartan differential

$$D\omega.$$

This yields elegantly the curvature differential 2-form $\mathsf{F} = D\mathsf{A}$ on $\mathcal{P}$ (Cartan's structural equation) together with the integrability condition $D\mathsf{F} = 0$ on $\mathcal{P}$ (Bianchi identity).

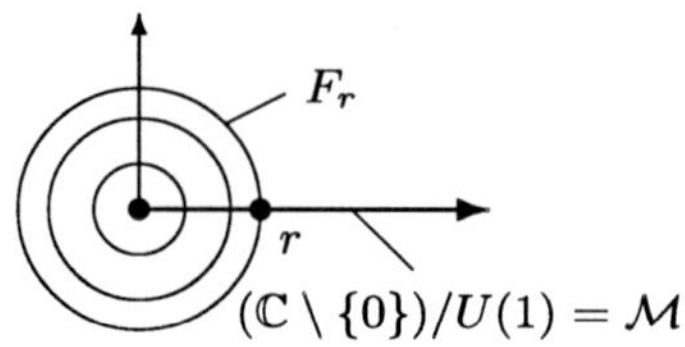

Fig. 17.2. Action of the rotation group $U(1)$ on the Gaussian plane

In terms of mathematics, Cartan's structural equation generalizes Gauss' theorema egregium. In terms of physics, the curvature 2-form F generalizes the electromagnetic field, and the connection 1-form A generalizes the four-potential of the electromagnetic field.

Folklore

Intuitively, a very simple situation is depicted in Fig. 17.2. Here, we use the bundle manifold $\mathcal{P} := \mathbb{C} \setminus \{0\}$ (pointed Gaussian plane). The action of the group $U(1)$ corresponds to rotations about the origin $z = 0$. The orbits F_r (fibers) are circles about the origin parametrized by the radius $r \in \mathcal{M}$ where $\mathcal{M} =]0, \infty[$ (positive x-axis). The map

$$\pi : \mathcal{P} \to \mathcal{M}$$

assigns to every point of the fiber F_r the radius r of the circle. A parallel transport sends a point Q_1 of the fiber F_{r_1} to a point Q_2 of the fiber F_{r_2} along the curve

$$r \mapsto z(r), \qquad r \in [r_1, r_2]$$

with $z(r) \in F_r$ for all r, and $z(r_1) = Q_1, z(r_2) = Q_2$.

Definition of a principal bundle. Let $\mathcal{G}$ be a closed subgroup of the Lie group $GL(n, \mathbb{R})$. The smooth surjective map

$$\boxed{\pi : \mathcal{P} \to \mathcal{M}} \tag{17.9}$$

is called a principal bundle with the symmetry group $\mathcal{G}$ iff the following conditions are satisfied:

(P1) Symmetry of the bundle manifold: The so-called bundle manifold $\mathcal{P}$ and the so-called base manifold $\mathcal{M}$ are real finite-dimensional manifolds. There exists a closed subgroup $\mathcal{G}$ of $GL(n, \mathbb{R})$ which acts freely on the bundle manifold $\mathcal{P}$ from the right. Explicitly, this means that for every group element $G \in \mathcal{G}$, there exists a diffeomorphism $R_G : \mathcal{P} \to \mathcal{P}$ such that

$$R_{GH}(Q) = R_H(R_G(Q)) \qquad \text{for all} \quad Q \in \mathcal{P},\ G, H \in \mathcal{G}.$$

Moreover, $R_1(Q) = Q$ for all $Q \in \mathcal{P}$, and if R_G has a fixed point, then $G = \mathbf{1}$.

(P2) Regular fibration by orbits: The set $\{R_G Q : G \in \mathcal{G}\}$ is called the orbit through the point $Q \in \mathcal{P}$. For every point $P \in \mathcal{M}$, the preimage $F_P := \pi^{-1}(P)$ is an orbit. In other words, the orbits (fibers) of the bundle manifold $\mathcal{P}$ are parametrized by the points P of the manifold $\mathcal{M}$.

(P3) Local triviality (bundle coordinates): The bundle manifold $\mathcal{P}$ is locally parametrized by so-called bundle coordinates (P, G) which lie in $\mathcal{O}_j \times \mathcal{G}$. More precisely, there exists a family of nonempty open subsets $\mathcal{O}_j, j \in \mathcal{J}$, of the base manifold $\mathcal{M}$ which cover $\mathcal{M}$. In addition, there exist diffeomorphisms

$$\beta_j : \pi^{-1}(\mathcal{O}_j) \to \mathcal{O}_j \times \mathcal{G}, \qquad j \in \mathcal{J}$$

of the form $\beta_j(Q) = (\pi(Q), G)$ which respect the group action, that is, $\beta_j(R_H(Q)) = (\pi(Q), GH)$ for all $H \in \mathcal{G}$.

Remark. In physics, we frequently encounter vector bundles with the typical fiber $\mathbb{C}^N$, and the symmetry group $\mathcal{G}$ is a closed subgroup of $GL(N, \mathbb{C})$. In this case, we identify $\mathbb{C}^N$ with $\mathbb{R}^{2N}$, and we use the fact that every closed subgroup of $GL(N, \mathbb{C})$ is isomorphic to a closed subgroup of $GL(2N, \mathbb{R})$. For example, $\mathbb{C}$ will be identified with $\mathbb{R}^2$, and $U(1)$ is isomorphic to the rotation group $SO(2)$.

Sections (phase factor fields). The map $s : \mathcal{M} \to \mathcal{P}$ is called a cross-section (or, briefly, a section) iff $s(P) \in F_P$ for all $P \in \mathcal{M}$.

The cocycle of transition maps. The change of the bundle coordinates from (P, G_j) to (P, G_k) is described by so-called transition maps

$$G_k = G_{kj}(P) G_j \tag{17.10}$$

where $G_j, G_k \in \mathcal{G}$ and $G_{kj}(P) \in \mathcal{G}$. More precisely, the transition map T_{kj} is defined by the commutativity of the following diagram:

$$\begin{array}{ccc} \pi^{-1}(\mathcal{O}_j \cap \mathcal{O}_k) & \xrightarrow{\beta_j} & (\mathcal{O}_j \cap \mathcal{O}_k) \times \mathcal{G} \\ & \searrow^{\beta_k} & \downarrow T_{kj} \\ & & (\mathcal{O}_j \cap \mathcal{O}_k) \times \mathcal{G}. \end{array}$$

Explicitly, $T_{kj} := \beta_k \circ \beta_j^{-1}$, and $T_{kj}(P, G_j) = (P, G_k)$ with (17.10). The maps

$$G_{kj} : \mathcal{O}_k \cap \mathcal{O}_j \to \mathcal{G}$$

form a cocycle.

A vector bundle and a principal bundle are called associated to each other iff they have the same cocycle of transition maps.

Fundamental velocity vector fields on $\mathcal{P}$. On the infinitesimal level, the action of the symmetry group of $\mathcal{G}$ on the bundle manifold $\mathcal{P}$ generates special velocity vector fields V_A on $\mathcal{P}$ which are called fundamental velocity vector fields; they are labelled by the elements A of the Lie algebra $\mathcal{LG}$. Explicitly, let $A \in \mathcal{LG}$. The smooth curve $\sigma \mapsto e^{\sigma A}$ on the Lie group $\mathcal{G}$ generates the smooth curve

$$\sigma \mapsto R_{e^{\sigma A}}(Q)$$

on the bundle manifold $\mathcal{P}$; this curve passes through the point Q if $\sigma = 0$. This yields the fundamental velocity vector field

$$V_A(Q) := \frac{d}{d\sigma} R_{e^{\sigma A}}(Q)_{|\sigma=0}.$$

The connection 1-form A on the bundle manifold $\mathcal{P}$. The differential 1-form A on $\mathcal{P}$ with values in the Lie algebra $\mathcal{LG}$ is called a connection form iff the following hold for all $G \in \mathcal{G}$ and all $A \in \mathcal{LG}$:

(C1) $\mathsf{A}(V_A) = A$ (fundamental velocity vector field V_A),
(C2) $R_G^* \mathsf{A} = G^{-1} \mathsf{A} G$ (symmetry).

Explicitly, this means the following. We have $\mathsf{A}_Q(V) \in \mathcal{LG}$ for all points $Q \in \mathcal{P}$ and for all tangent vectors $V \in T_Q\mathcal{P}$. In addition, we have the linear map $V \mapsto \mathsf{A}_Q(V)$ from the tangent space $T_Q\mathcal{P}$ to the Lie algebra $\mathcal{LG}$, and we have the smooth map

$$Q \mapsto \mathsf{A}_Q(V)$$

from $\mathcal{P}$ to $\mathcal{LG}$ for all smooth velocity vector fields V on $\mathcal{P}$. Condition (C1) means that $\mathsf{A}_Q(V_A(Q)) = A$ for all $Q \in \mathcal{P}$ and all $A \in \mathcal{LG}$. Condition (C2) tells us that

$$\mathsf{A}_{R_G(Q)}(R'_G(V)) = G^{-1}\mathsf{A}_Q(V)G$$

for all points $Q \in \mathcal{P}$ and all tangent vectors $V \in T_Q\mathcal{P}$.

Parallel transport on the bundle manifold $\mathcal{P}$. By definition, the curve $\mathcal{C}: Q = Q(\sigma), \sigma \in \mathcal{R}$, on the bundle manifold $\mathcal{P}$ represents a parallel transport iff

$$\boxed{\mathsf{A}_{Q(\sigma)}(\dot{Q}(\sigma)) = 0 \qquad \text{for all} \quad \sigma \in \mathcal{R}.}$$

Horizontal tangent vectors on $\mathcal{P}$. Motivated by the definition of parallel transport, the tangent vector $\dot{Q} \in T_Q\mathcal{P}$ of the bundle manifold $\mathcal{P}$ at the point Q is called horizontal iff

$$\mathsf{A}_Q(\dot{Q}) = 0.$$

The horizontal tangent vectors of $\mathcal{P}$ at the point Q form a linear subspace $T_Q\mathcal{P}_{\text{hor}}$ of the tangent space $T_Q\mathcal{P}$. In particular, the curve $\mathcal{C}$ represents a parallel transport on $\mathcal{P}$ iff all the tangent vectors of the curve are horizontal.

The following construction is basic for the geometry of principal bundles. Let $V \in T_Q\mathcal{P}$ be a tangent vector of the bundle manifold $\mathcal{P}$ at the point Q. Compute $A := \mathsf{A}_Q(V)$. Using the fundamental velocity vector field V_A, define

$$\text{ver}(V) := V_A(Q), \qquad \text{hor}(V) := V - V_A(Q).$$

Here, ver(V) (resp. hor(V)) is called the trivial vertical (resp. nontrivial horizontal) part of the tangent vector V at the point Q. To motivate this terminology, observe that it follows from $\mathsf{A}_Q(V - V_A(Q)) = A - A = 0$ that $V - V_A(Q)$ is a horizontal tangent vector. Thus, we have the key decomposition

$$\boxed{V = \text{hor}(V) + \text{ver}(V) \qquad \text{for all} \quad V \in T_Q\mathcal{P}.} \tag{17.11}$$

The tangent vector V is horizontal iff ver$(V) = 0$. The tangent vector V is called vertical iff hor$(V) = 0$. By (17.11), we get the direct sum

$$T_Q\mathcal{P} = T_Q\mathcal{P}_{\text{hor}} \oplus T_Q\mathcal{P}_{\text{ver}}$$

where the linear space $T_Q\mathcal{P}_{\text{hor}}$ (resp. $T_Q\mathcal{P}_{\text{ver}}$) consists of all the horizontal (resp. vertical) tangent vectors of $\mathcal{P}$ at the point Q.

The covariant Cartan differential $D\omega$ on $\mathcal{P}$. Let ω be a differential p-form on the bundle manifold $\mathcal{P}$. Then the classical Cartan differential $d\omega$ on the manifold $\mathcal{P}$ is well defined. We use $d\omega$ in order to define

$$\boxed{D_Q\omega(V_1, \ldots, V_p) := d\omega_Q(\text{hor}(V_1), \ldots, \text{hor}(V_p))}$$

for all velocity vectors $V_1, \ldots, V_p \in T_Q\mathcal{P}$. The same definition applies to differential forms ω with values in the Lie algebra $\mathcal{LG}$ (e.g., $\omega = \mathsf{A}$).

The curvature 2-form F on the bundle manifold $\mathcal{P}$. We will use the covariant Cartan differential in order to define

$$\boxed{\mathsf{F} := D\mathsf{A} \qquad \text{on } \mathcal{P}.} \tag{17.12}$$

Theorem 17.3 *For all smooth velocity vector fields V, W on the bundle manifold $\mathcal{P}$, we have the Cartan structural equation*

$$\mathsf{F}_Q(V,W) = d\mathsf{A}_Q(V,W) + [\mathsf{A}_Q(V), \mathsf{A}_Q(W)]_-, \qquad Q \in \mathcal{P}$$

together with the integrability condition (Bianchi identity)

$$D\mathsf{F} = 0 \qquad \text{on } \mathcal{P}.$$

Gauge fixing and localization of the curvature form. Choose both an open subset $\mathcal{O}_j$ of the base manifold $\mathcal{M}$ and a section

$$s : \mathcal{O}_j \to \mathcal{P}.$$

Define

$$\mathcal{F} := s^*\mathsf{F} \qquad \text{and} \qquad \mathcal{A} := s^*\mathsf{A}$$

by using the pull-back operation for differential forms. Then we get the local Cartan structural equation

$$\mathcal{F}_P(v,w) = d\mathcal{A}_P(v,w) + [\mathcal{A}_P(v), \mathcal{A}_P(w)]_-, \qquad P \in \mathcal{O}_j$$

for all smooth velocity vector fields v, w on $\mathcal{O}_j$. Equivalently,

$$\mathcal{F} = d\mathcal{A} + \mathcal{A} \wedge \mathcal{A}.$$

The integrability condition (Bianchi identity) reads as

$$d\mathcal{F} = \mathcal{F} \wedge \mathcal{A} - \mathcal{A} \wedge \mathcal{F}.$$

This is the local variant (on the base manifold $\mathcal{M}$) of Cartan's elegant global theory (on the bundle manifold $\mathcal{P}$). Note that $\mathcal{F}$ and $\mathcal{A}$ depend on the choice of both the open subset $\mathcal{O}_j$ of the base manifold $\mathcal{M}$ and the section s with $s(P) = (P, G(P))$. In terms of physics, this corresponds to the choice of both a local observer O_j on the space-time manifold $\mathcal{M}$ and a gauge fixing by means of s.

Using bundle charts, this local theory corresponds to the product bundle theory considered in Sects. 15.1 through 15.3.

Global gauge transformation on a principal bundle. Recall that a typical property of a principal bundle $\pi : \mathcal{P} \to \mathcal{M}$ is the fact that the Lie group $\mathcal{G}$ acts on the bundle space $\mathcal{P}$ from the right. By definition, the diffeomorphism

$$f : \mathcal{P} \to \mathcal{P}$$

is called a global gauge transformation iff it respects the action of the group $\mathcal{G}$ on the bundle space $\mathcal{P}$ (from the right). Explicitly, this means that

- $\pi \circ f = \pi$ (preservation of the fibers of $\mathcal{P}$), and
- $f(R_G(P)) = R_G(f(P))$ for all $P \in \mathcal{P}$ and all $G \in \mathcal{G}$ (the map f commutes with the action of $\mathcal{G}$ on $\mathcal{P}$).

17.3 The Philosophy of Parallel Transport

Parallel transport plays a crucial role in gauge theory in order to describe the transport of physical information in modern physics. In what follows, we will discuss the following concepts:

- associated vector bundle to a principal bundle,
- horizontal tangent vectors on a principal bundle, and
- lifting of curves.

In Sect. 17.2, the axioms for a connection on a principal bundle were based on the differential 1-form A with values in the Lie algebra $\mathcal{LG}$ of the gauge group $\mathcal{G}$. This is an analytic approach. In terms of physics, the connection differential 1-form A generalizes the four-potential in electromagnetism.

In this section, we want to discuss a geometric approach. The basic idea is to formulate axioms which describe the parallel transport on an infinitesimal level:

> *We regard the parallel transport as a dynamical system on the principal bundle $\mathcal{P}$. In general, dynamical systems on a manifold are described by differential equations based on velocity vector fields. In the case of parallel transport on $\mathcal{P}$, we use so-called lifted horizontal vector fields.*

The goal is to characterize axiomatically the horizontal tangent vectors of $\mathcal{P}$. The passage from the geometric axioms below to the analytic approach considered in Sect. 17.2 is quite simple. One only has to show how to construct the connection form A. The concept of an associated vector bundle allows us quite naturally to transplant

- the trajectory of parallel transport on the principal bundle to
- the trajectory of parallel transport on the associated vector bundle.

17.3.1 Vector Bundles Associated to a Principal Bundle

> If the symmetry group $\mathcal{G}$ of a principal bundle $\pi : \mathcal{P} \to \mathcal{M}$ acts on a linear space F, then it is possible to construct a vector bundle
>
> $$\pi : \mathcal{V} \to \mathcal{M} \tag{17.13}$$
>
> with typical fiber F such that $\mathcal{G}$ (or a representation of $\mathcal{G}$) is the symmetry group of $\mathcal{V}$. In terms of physics, it is possible to construct a gauge theory with gauge group $\mathcal{G}$ for the sections
>
> $$s : \mathcal{M} \to \mathcal{V}$$
>
> of the vector bundle $\mathcal{V}$. These sections are physical fields on the base manifold $\mathcal{M}$ with values in the linear space F (for a local observer). The vector bundle (17.13) is called the vector bundle (with typical fiber F) associated to the principal bundle $\mathcal{P}$.
>
> Folklore

The cocycle strategy (observer strategy). Vector bundles and principal bundles can be described by a cocycle on the base manifold $\mathcal{M}$ with values in the symmetry Lie group $\mathcal{G}$. The cocycle represents the transition maps between the bundle coordinates (see Chap. 16). If the Lie group $\mathcal{G}$ is a closed subgroup of the matrix group $GL(N, \mathbb{C})$, then it is possible to assign to the local transition map

$$\boxed{(P, G) \mapsto (P, G_0(P)G)} \tag{17.14}$$

of the principal bundle $\pi : \mathcal{P} \to \mathcal{M}$ the so-called associated local transition map

$$(P, \psi) \mapsto (P, G_0(P)\psi)$$

where $\psi \in \mathbb{C}^N$. More generally, if we have a surjective group morphism

$$r : \mathcal{G} \to \mathcal{H}$$

where $\mathcal{H}$ is a closed subgroup of $GL(K, \mathbb{C})$, then it is possible to assign to the transition map (17.14) the associated transition map

$$\boxed{(P, \Psi) \mapsto (P, r(G_0(P))\, \Psi)}$$

where $\Psi \in \mathbb{C}^K$. This way, it is possible to assign to the given principal bundle

$$\pi : \mathcal{P} \to \mathcal{M} \qquad (17.15)$$

with the symmetry group $\mathcal{G}$ the so-called associated vector bundle

$$\boxed{\pi : \mathcal{V} \to \mathcal{M}} \qquad (17.16)$$

with the typical fiber $F = \mathbb{C}^K$ and the symmetry group $r(\mathcal{G}) = \mathcal{H}$. From the physical point of view, the principal bundle (17.15) with the symmetry group $\mathcal{G}$ allows us to construct physical fields Ψ with values in the fiber $F = \mathbb{C}^K$ (and with the symmetry group $r(\mathcal{G})$). The physical fields look locally like

$$\Psi : \mathcal{O} \to \mathbb{C}^K$$

where $\mathcal{O}$ is an open neighborhood of the point P_0 on the base manifold $\mathcal{M}$. Globally, the physical fields are sections

$$\Psi : \mathcal{M} \to \mathcal{V}$$

of the associated vector bundle (17.16).

The vector bundle $\mathcal{V}$ associated to the principal bundle $\mathcal{P}$ depends on a representation r of the symmetry group $\mathcal{G}$ on the linear space $\mathbb{C}^K$.

For example, the electron in an electromagnetic field is described by a physical field

$$\Psi : \mathbb{M}^4 \to \mathbb{C}^4$$

which satisfies the Dirac equation. Equivalently, this is a section

$$\Psi : \mathbb{M}^4 \to \mathbb{M}^4 \times \mathbb{C}^4$$

of the vector bundle $\mathbb{M}^4 \times \mathbb{C}^4$. This situation corresponds to

- the principal bundle $\pi : \mathbb{M}^4 \times U(1) \to \mathbb{M}^4$ with the symmetry group $U(1)$ (photon), and
- the associated vector bundle $\boldsymbol{\pi} : \mathbb{M}^4 \times \mathbb{C}^4 \to \mathbb{M}^4$ (electron).

The transition maps (gauge transformations) for the associated vector bundle describe the change

$$\Psi(ct, x, y, z) \mapsto \mathrm{e}^{\mathrm{i}a(ct,x,y,z,t)}\Psi(ct, x, y, z)$$

by multiplying the field Ψ with a local phase factor.

Transformation of curves. The notion of associated vector bundle is important in order to transform curves on the principal bundle $\mathcal{P}$ into curves on the vector bundle $\mathcal{V}$. Mnemonically, we will use the key relation

$$\boxed{(P(t), G(t)) \Rightarrow (P(t), G(t)\psi_0).}$$

(i) Principal bundle $\pi : \mathcal{P} \to \mathcal{M}$: Suppose that we are given the curve

$$t \mapsto Q(t)$$

on the principal bundle $\mathcal{P}$. Using a bundle chart, we get the curve

$$t \mapsto (P(t), G(t)). \tag{17.17}$$

Set $G_0(t) := G_0(P(t))$. Changing the bundle chart, we get

$$t \mapsto (P(t), G_0(t)G(t)). \tag{17.18}$$

(ii) Associated vector bundle: $\pi : \mathcal{V} \to \mathcal{M}$: Choose the point $(P(0), G(0)\psi_0)$ on the bundle space $\mathcal{V}$. The curve (17.17) passes over to the curve

$$t \mapsto (P(t), G(t)\psi_0)$$

on the corresponding bundle chart of the vector bundle $\mathcal{V}$. This passage from the principal bundle $\mathcal{P}$ to the vector bundle $\mathcal{V}$ respects the change of bundle charts. In fact, the curve (17.18) corresponds to

$$t \mapsto (P(t), G_0(t)G(t)\psi_0).$$

This way, we obtain a curve

$$t \mapsto V(t)$$

on the associated vector bundle $\mathcal{V}$. This general procedure for curves can be used in order to transform the trajectories of parallel transport on $\mathcal{P}$ into the trajectories of parallel transport on $\mathcal{V}$.

The invariant strategy for constructing the associated vector bundle. The construction does not use any cocycles, but it is based on the following commutative diagram:

$$\begin{array}{ccc} \mathcal{P} \times F & \xleftarrow{\text{proj}} & \mathcal{V} = \mathcal{P} \times_{\mathcal{G}} F \\ {\scriptstyle \text{proj}_1}\downarrow & & \downarrow{\scriptstyle \pi} \\ \mathcal{P} & \xrightarrow{\pi} & \mathcal{M}. \end{array} \tag{17.19}$$

We are given the principal bundle $\pi : \mathcal{P} \to \mathcal{M}$ with the symmetry group $\mathcal{G}$ which is a Lie group. Furthermore, we are given the finite-dimensional linear space F. Suppose that the group $\mathcal{G}$ acts on the linear space F from the left. In other words, there exists a smooth representation r of $\mathcal{G}$ on F.[3] Now consider the product set

$$\mathcal{P} \times F.$$

The elements of $\mathcal{P} \times F$ are denoted by $(Q, \mathbf{f})$ where $Q \in \mathcal{P}$ and $\mathbf{f} \in F$. The group $\mathcal{G}$ acts on the product set $\mathcal{P} \times F$ from the right by setting

[3] Explicitly, there exists a smooth group morphism $r : \mathcal{G} \to GL(F)$. That is, for every group element $G \in \mathcal{G}$, there exists a linear operator

$$r(G) : F \to F$$

such that $r(GG') = r(G)r(G')$ for all $G, G' \in \mathcal{G}$. In addition, the map $\mathcal{G} \mapsto GL(F)$ is smooth. If $\mathbf{f} \in F$, then it is very suggestive to write $G\mathbf{f}$ instead of $r(G)\mathbf{f}$.

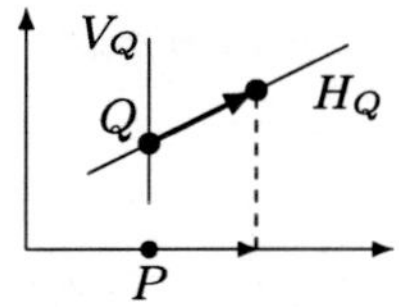

Fig. 17.3. Vertical space V_Q and horizontal space H_Q

$$(Q, \mathbf{f})G := (Q, G^{-1}\mathbf{f}), \qquad G \in \mathcal{G}.$$

The corresponding orbit space $\mathcal{P} \times_{\mathcal{G}} F$ is obtained in the following way. We write

$$(Q, \mathbf{f}_1) \sim (Q, \mathbf{f}_2)$$

iff there exists a group element $G \in \mathcal{G}$ such that $(Q, \mathbf{f}_2) = (Q, \mathbf{f}_1)G$. This is an equivalence relation. The equivalence classes $[(Q, \mathbf{f})]$ are called orbits, and the space of all the orbits is denoted by $\mathcal{P} \times_{\mathcal{G}} F$. Observe that there are two natural projections, namely,

- $\text{proj}_1(Q, \mathbf{f}) := Q$, and
- $\text{proj}([(Q, \mathbf{f})]) := (Q, \mathbf{f})$ (canonical projection).

Set $\mathcal{V} := \mathcal{P} \times_{\mathcal{G}} F$. Constructing the operator $\boldsymbol{\pi}$ by the commutative diagram (17.19), we obtain the vector bundle

$$\boldsymbol{\pi} : \mathcal{V} \to \mathcal{M} \tag{17.20}$$

with typical fiber F. The vector bundle (17.20) is called the associated vector bundle to the principal bundle $\pi : \mathcal{P} \to \mathcal{M}$.

17.3.2 Horizontal Vector Fields on a Principal Bundle

Let $\pi : \mathcal{P} \to \mathcal{M}$ be a principal bundle. The idea is to study the decomposition of the tangent space $T_Q\mathcal{P}$ of the bundle manifold $\mathcal{P}$ at the point $Q \in \mathcal{P}$ (Fig. 17.3):

$$\boxed{T_Q\mathcal{P} = V_Q \oplus H_Q.} \tag{17.21}$$

For any tangent vector $V \in T_Q\mathcal{P}$, we have the decomposition

$$V = \text{ver}(V) + \text{hor}(V), \qquad \text{ver}(V) \in V_Q, \;\; \text{hor}(V) \in H_Q.$$

By definition, the linear subspace V_Q of $T_Q\mathcal{P}$ is the tangent space of the fiber F_P at the point Q (space of vertical tangent vectors at the point Q). Thus, the linear space V_Q is given canonically by the fiber structure of $\mathcal{P}$. According to linear algebra, there exists always a direct sum decomposition of the form (17.21). However, the point is that the complementary linear subspace H_Q is not uniquely determined. We need an additional structure (called connection) on the principal bundle $\mathcal{P}$ in order to construct uniquely the space H_Q of horizontal tangent vectors.

Axioms for horizontal vector fields. We postulate that, for every point $Q \in \mathcal{P}$, there exists a linear subspace H_Q of the tangent space $T_Q\mathcal{P}$ of the bundle manifold $\mathcal{P}$ at the point Q such that the following hold:

(H1) Direct sum: We have the direct sum decomposition (17.21).

(H2) The (restricted) linearized projection map

$$\pi' : H_Q \to T_P\mathcal{M}$$

is a linear isomorphism if $P = \pi(Q)$.

(H3) Symmetry: The action of the symmetry group $\mathcal{G}$ on the bundle manifold $\mathcal{P}$ from the right is respected. That is, for all group elements $G_0 \in \mathcal{G}$, we have

$$H_{GG_0} = R'_{G_0} H_G$$

where R'_{G_0} is the linearization of the map $R_{G_0} : \mathcal{P} \to \mathcal{P}$.[4]

(H4) Smoothness: If V is a smooth velocity vector field on $\mathcal{P}$, then $\text{ver}(V)$ and $\text{hor}(V)$ are also smooth velocity vector fields on $\mathcal{P}$.

Remark. The axiom (H1) is a consequence of the axiom (H2), by using the fiber structure of $\mathcal{P}$. In fact, let $V \in T_Q\mathcal{P}$. By (H2), there exists precisely one tangent vector $W \in H_Q$ such that $\pi'(W) = \pi'(V)$. Hence $\pi'(V - W) = 0$, that is, $V - W \in V_Q$. Finally, $V = (V - W) + W$.

17.3.3 The Lifting of Curves in Fiber Bundles

> In the history of topology, the lifting of curves in fiber bundles played an important role.[5]
>
> Folklore

To explain the intuitive idea, consider the tangent bundle $T\mathbb{S}^2_r$ of a sphere $\mathbb{S}^2_r$. The parallel transport of a velocity vector $\mathbf{v}$ along a curve

$$C : P = P(t), \qquad t \in \mathcal{R}$$

on the sphere $\mathbb{S}^2_r$ is a curve

$$t \mapsto (P(t), \mathbf{v}(t)) \tag{17.22}$$

on the bundle manifold $T\mathbb{S}^2_r$. We say that the curve (17.22) is obtained from the curve C on the base manifold $\mathbb{S}^2_r$ by lifting. Similarly, replacing the tangent bundle $T\mathbb{S}^2_r$ by the frame bundle $F\mathbb{S}^2_r$, the parallel transport

$$t \mapsto (P(t), \mathbf{e}_1(t), \mathbf{e}_2(t))$$

of a frame along the curve C is called a lift of the curve C to the bundle space $F\mathbb{S}^2_r$.

Definition. Consider a principal bundle $\pi : \mathcal{P} \to \mathcal{M}$. The curve

$$Q = Q(t), \qquad t \in \mathcal{R}$$

on the bundle manifold $\mathcal{P}$ is called a lift of the curve

$$C : P = P(t), \qquad t \in \mathcal{R}$$

on the base manifold iff $P(t) = \pi(Q(t))$ for all times $t \in \mathcal{R}$.

Lifting. We are given the principal bundle

$$\pi : \mathcal{P} \to \mathcal{M}$$

[4] Recall that $R_{G_0}Q := QG_0$ for all $Q \in \mathcal{P}$.

[5] See M. Zisman, Fibre Bundles, Fibre Maps, pp. 605–629. In: I. James (Ed.), History of Topology, Elsevier, Amsterdam, 1999.

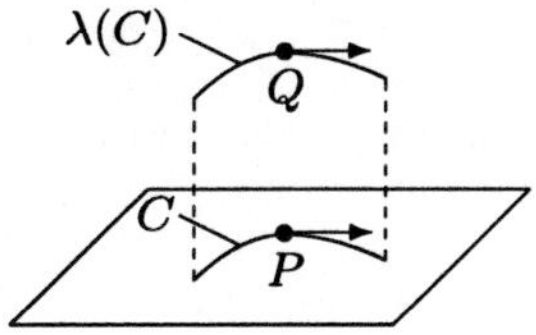

Fig. 17.4. Lifting of the curve C

which satisfies the axioms (H1) through (H4) above. Then the map

$$\pi' : H_P \to T_P\mathcal{M}$$

with $P = \pi(Q)$ is a linear isomorphism. The inverse map

$$\lambda : T_P\mathcal{M} \to H_Q$$

is called a lifting.

Parallel transport. We are given the curve

$$C : P = P(t), \qquad t \in \mathcal{R}.$$

We lift the tangent vector $\dot{P}(t)$ of the curve C at the point $P(t) \in \mathcal{M}$ to the tangent vector

$$\lambda_{P(t)}(\dot{P}(t))$$

of $\mathcal{P}$ at the point $Q(t)$ (Fig. 17.4). The solution $\lambda(C) : Q = Q(t), t \in \mathcal{P}$, of the differential equation

$$\boxed{\dot{Q}(t) = \lambda_{P(t)}(\dot{P}(t)), \qquad t \in \mathcal{R}, \quad Q(0) = Q_0}$$

is called a parallel transport passing through the bundle point Q_0 along the curve C on the base manifold C (Fig. 17.4).

Connection 1-form A. For any tangent vector $V \in T_Q\mathcal{P}$ of the bundle manifold $\mathcal{P}$ at the point Q, we define

$$\boxed{\mathsf{A}_Q(V) := \mathrm{ver}(V).}$$

Local bundle coordinates. Let us choose local bundle coordinates for the principal bundle $\mathcal{P}$, and let us sketch the approach in this local setting. Then the bundle manifold $\mathcal{P}$ looks locally like the product set

$$\mathcal{O} \times \mathcal{G}$$

where $\mathcal{O}$ is an open subset of $\mathbb{R}^n$. The point $Q \in \mathcal{P}$ is described by the bundle coordinate

$$(x, G) \qquad \text{where } x \in \mathcal{O} \text{ and } G \in \mathcal{G}$$

with the projection $\pi(x, G) = x$. A curve $t \mapsto Q(t)$ on the bundle manifold $\mathcal{P}$, which passes through the point Q_0 at time $t = 0$, is locally described by the map

$$t \mapsto (x(t), G(t))$$

where $Q(0) = Q_0$, and $x(0) = x_0$, $G(0) = G_0$. Thus, a tangent vector of the bundle manifold $\mathcal{P}$ at the point Q_0 looks locally like

$$(\dot{x}, \dot{G}) \text{ where } \dot{x} \in \mathbb{R}^n \text{ and } \dot{G} \in \mathcal{LG}.$$

The space H_Q of horizontal tangent vectors of the bundle manifold $\mathcal{P}$ at the point Q consists of all the points $(\dot{x}, \dot{G}) \in \mathbb{R}^n \times \mathcal{LG}$ with

$$\dot{x}^i G^{-1} \mathcal{A}_i(x) G + G^{-1} \dot{G} = 0.$$

Here, we sum over $i = 1, \ldots, n$, and the functions $x \mapsto \mathcal{A}_i(x)$, $i = 1, \ldots, n$ are smooth functions from $\mathcal{O}$ to the Lie algebra $\mathcal{LG}$. Finally, in local bundle coordinates, the connection differential 1-form $\mathsf{A}_{(x,G)}$ looks like

$$\boxed{dx^i G^{-1} \mathcal{A}_i(x) G + \mathsf{M}_G.}$$

Here, $\mathsf{M}_G(\dot{G}) = G^{-1} dG(\dot{G}) = G^{-1}\dot{G}$ is the Maurer–Cartan form of the Lie group $\mathcal{G}$. Let $G_0 \in \mathcal{G}$. Then the pull-back of the Maurer–Cartan form looks like

$$(R^*_{G_0}\mathsf{M})_G(\dot{G}) = \mathsf{M}_{GG_0}(\dot{G}G_0) = (GG_0)^{-1}\dot{G}G_0 = G_0^{-1}G^{-1}\dot{G}G_0 = G_0^{-1}\mathsf{M}_G(\dot{G})G_0.$$

Hence

$$\mathsf{A}_{(x,GG_0)}(\dot{x}, \dot{G}G_0) = G_0^{-1}\mathsf{A}_{(x,GG_0)}(\dot{x}, \dot{G})G_0.$$

This implies the symmetry property of the connection form A:

$$\boxed{R^*_{G_0}\mathsf{A} = G_0^{-1}\mathsf{A}G_0.}$$

Further Reading

The standard textbook on the axiomatic approach to modern differential geometry is the monograph by

> S. Kobayashi and K. Nomizu, Foundations of Differential Geometry, Vols. 1, 2, Wiley, New York, 1963.

In addition, we recommend

> Y. Choquet-Bruhat, C. DeWitt-Morette, and M. Dillard-Bleick, Analysis, Manifolds, and Physics. Vol. 1: Basics; Vol. 2: 92 Applications, Elsevier, Amsterdam, 1996,
>
> J. Jost, Riemannian Geometry and Geometric Analysis, Springer, Berlin, 2008,

and

> M. Crampin and F. Pirani, Applicable Differential Geometry, Cambridge University Press, 1987,
>
> S. Novikov and T. Taimanov, Geometric Structures and Fields, Amer. Math. Soc., Providence, Rhode Island, 2006,
>
> V. Ivancevic and T. Invancevic, Differential Geometry: A Modern Introduction, World Scientific, Singapore, 2007.

The standard textbook on fiber bundles is the monograph by

> D. Husemoller, Fiber Bundles, Springer, New York, 1994.

17.4 A Glance at the History of Gauge Theory

> Wider expanses and greater depths are now exposed to the searching eye of knowledge, regions of which we had not even a presentiment. It has brought us much nearer to grasping the plan that underlies all physical happening.[6]
>
> Hermann Weyl, 1918

17.4.1 From Weyl's Gauge Theory in Gravity to the Standard Model in Particle Physics

Let us first summarize some important papers:

H. Weyl (1918), Raum, Zeit, Materie, Springer, Berlin, 1918; 8th German edition, 1993 (English edition: Space-Time-Matter, Dover, New York, 1990).

H. Weyl (1918), Gravitation and electricity (in German), Sitzungsbericht der Königlich-Preussischen Akademie zu Berlin, pp. 465–480 (the idea of a real-valued gauge theory by changing the length scale).

H. Weyl (1919), A new extension of the theory of general relativity (in German), Z. Phys. **59**, 101–133.

T. Kaluza (1921), On the problem of the unification of physics (in German), Berliner Berichte, 1921, pp. 966–972.

O. Klein (1926), Quantum theory and five-dimensional theory of relativity, Z. Physik **37**, 895–906 (in German). English translation in: G. Ekspong (1991), The Oskar Klein (1894–1977) Memorial Lectures, pp. 67–80, World Scientific, Singapore.

O. Klein (1926), The atomicity of electricity as a quantum theory law, Nature, 516–518. Reprinted in: G. Ekspong (1991), pp. 81–83.

V. Fock (1926), On the invariant form of the wave and motion equations for a charged point mass, Z. Phys. **39**, 839–841.

W. Gordon (1926), The Compton effect according to Schrödinger's theory (in German), Z. Phys. **40**, 117–133.

F. London (1927), Quantum-mechanical interpretation of Weyl's theory (in German), Z. Phys. **42**, 375–389.

H. Weyl (1929), Electron and gravitation (in German), Z. Phys. **56** (1929), 330–352. See also: Gravitation and the electron, Proc. Nat. Acad. Sci. USA **15** (1929), 323–334.

O. Klein (1938), On the theory of charged fields, pp. 895–906 (in French). In: New Theories in Physics, Conference organized in Warsaw, 1938. English translation in: G. Ekspong (1991), pp. 85–102 (see O. Klein (1926) above).

C. Ehresmann (1950), Infinitesimal connections in differentiable fiber spaces, Colloque de Topologie, Bruxelles, pp. 29–55 (in French).

W. Pauli (1953), Meson–nucleon interaction and differential geometry. In: W. Pauli (1999), Wissenschaftlicher Briefwechsel (Scientific correspondence), Vol. IV, Part II, letters 1614 and 1682, Springer, Berlin.

[6] From the Preface to the first edition of H. Weyl, Space, Time, Matter (in German), Springer, Berlin, 1918.

C. Yang and R. Mills (1954), Conservation of isotopic spin and isotopic spin invariance, Phys. Rev. **96**, 191–195.

J. Goldstone (1960), Field theories with "superconductor" solutions, Nuovo Cimento **19**, 154–164.

S. Kobayashi and K. Nomizu (1963), Foundations of Differential Geometry, Vols. 1, 2, Wiley, New York.

G. Guralnik, C. Hagen, and T. Kibble (1964), Global conservation laws and massless particles, Phys. Rev. Letters, **13**, 585–587.

P. Higgs (1964), Broken symmetry and the masses of gauge bosons, Phys. Rev. Lett. **13**, 508–509.

L. Faddeev and V. Popov (1967), Feynman diagrams for the Yang–Mills field, Phys. Lett. **25B**, 29–30.

G. 't Hooft, Renormalization of massless Yang-Mills fields, Nuclear Phys. B **33**, 173–199.

G. t'Hooft, Renormalizable Lagrangians for massive Yang–Mills fields, Nucl. Phys. **B35**(1), 167–188.

G. 't Hooft and M. Veltman (1972), Regularization and renormalization of gauge fields, Nucl. Phys. **B44**, 189–213.

T. Wu and C. Yang (1975), Concept of non-integrable phase factors and global formulation of gauge fields, Phys. Rev. **D12**, 3845–3857.

C. Yang (1986), Selected Papers, 1945–1980 with Commentary, Freeman, San Francisco.

Nobel Lectures in 1979:

S. Glashow, Towards a unified theory – threads in a tapestry, Rev. Mod. Phys. **52**(3) (1980), 539–543.

S. Salam, Gauge unification of fundamental forces, Rev. Mod. Physics **52**(3) (1980), 525–538.

S. Weinberg, Conceptual foundations of the unified theory of weak and electromagnetic interaction, Rev. Mod. Phys. **52** (1980), 515–523.

Collection of important papers:

G. 't Hooft, Under the Spell of the Gauge Principle, World Scientific, Singapore, 1994.

O'Raifeartaigh, The Dawning of Gauge Theory, Princeton University Press, 1997.

C. Taylor (Ed.), Gauge Theories in the Twentieth Century, World Scientific, Singapore, 2001.

Remarks. Let us briefly comment this list of references above. In 1918, Weyl used scaling transformations in order to generalize Einstein's theory of general relativity to a theory of both gravitation and electromagnetism. In terms of physics, this original approach was not successful. In 1926, Erwin Schrödinger published his famous non-relativistic wave equation. In the same year, Oskar Klein, Vladimir Fock, and Walter Gordon published independently the same relativistic wave equation (13.8) called the Klein–Fock–Gordon equation (or the Klein–Gordon equation). In fact, Schrödinger was the first physicist who tried to use this equation in order to understand the spectrum of the hydrogen atom. But he obtained the wrong spectrum compared with the experimental values. Therefore, he passed over to the non-relativistic Schrödinger equation. Nowadays we know that the electron of the relativistic hydrogen has the spin $\frac{1}{2}\hbar$; therefore, we have to use the Dirac equation introduced by Dirac in 1928. The Klein–Gordon–Fock equation describes spin-less particles like the mesons π^+, π^-, π^-. In 1926, Fock discovered for the first time that the Klein–Fock–Gordon equation is invariant under the complex-valued gauge transformation

$$\psi^+(\mathbf{x}, t) = \mathrm{e}^{\mathrm{i}a(\mathbf{x},t)}\psi(\mathbf{x}, t)$$

and the gauge transformation

$$U^+(\mathbf{x}, t) = U(\mathbf{x}, t) - \frac{\hbar}{Q}a_t(\mathbf{x}, t), \quad \mathbf{A}^+(\mathbf{x}, t) = \mathbf{A}(\mathbf{x}, t) + \frac{\hbar}{Q}\boldsymbol{\partial} a(\mathbf{x}, t).$$

In 1927, London showed that Weyl's real gauge theory from 1918 can be reformulated as a complex-valued gauge theory which is useful for the new quantum mechanics. In 1929, Weyl wrote a seminal paper entitled *Electron and gravitation* (in German). In this paper, the $U(1)$-gauge theory is founded, and Weyl emphasized the importance of the gauge principle for the relativistic electron in an electromagnetic field. He writes:

> From the arbitrariness of the gauge factor of the wave function ψ appears the necessity of introducing the electromagnetic 4-potential.

This underlines the key point of gauge theories:

> *The principle of local gauge symmetry in physics determines the interaction.*

In 1938, Oskar Klein formulated a gauge theory in order to describe mesons in the setting of a 5-dimensional Kaluza–Klein theory. This paper contains the broken symmetry of a $SU(2)$-gauge group. In 1953, Pauli wrote a letter to Pais in Princeton. In this letter, Pauli uses a 6-dimensional approach in order to model the meson-nucleon interaction. By symmetry breaking, Pauli arrives at a $SU(2)$-gauge theory. The first general $SU(2)$-gauge theory was published by Yang and Mills in 1954. They generalized Weyl's gauge theory from the commutative gauge group $U(1)$ to the noncommutative gauge group $SU(2)$. From the physical point of view, Pauli criticized the Yang–Mills approach. The point is that the messenger particles of a $SU(2)$-gauge theory are massless like the photon in electrodynamics. But in the 1950s, except for the photon, all known messenger particles were massive. This mass problem was theoretically solved in the 1960s by inventing the Higgs mechanism in gauge theory based on the idea of spontaneous symmetry breaking.

Electroweak interaction. In the 1960s, Glashow, Salam, and Weinberg unified the electromagnetic interaction with the weak interaction in the framework of the theory of the electroweak interaction. This is a gauge theory with the gauge group

$$SU(2) \times U(1).$$

This theory predicts the existence of four messenger particles, namely, the massless photon and the three massive weak bosons W^+ (positive electric charge), W^- (negative electric charge), Z^0 (no electric charge).[7] In contrast to the classic Fermi theory for the weak interaction from the 1930s, the electroweak theory allows the scattering process

$$e^- + \nu_\mu \to e^- + \nu_\mu \tag{17.23}$$

between electrons and muon neutrinos. The Feynman diagram of this scattering process contains the neutral weak boson Z^0. Physicists call this a weak current process.

[7] Weak bosons are also called vector bosons

In the 1970s, the weak current process (17.23) was experimentally established with more and more higher precision. Glashow, Salam, and Weinberg were awarded the Nobel prize in physics in 1979. The three massive weak bosons W^+, W^- and Z^0 were experimentally established at the CERN collider in 1983. The rest masses of the proton p, and the weak bosons $W^\pm, Z^0$ are equal to

$$1\text{GeV}/c^2, \quad 80\text{GeV}/c^2, \quad 90\text{GeV}/c^2,$$

respectively. In the mathematical setting of the electroweak theory, one adds to the Lagrangian an additional scalar field of a spin-less massive particle (called the Higgs boson) which generates the mass terms for the three weak bosons $W^\pm, Z^0$. In near future, physicists hope to establish experimentally the existence of the missing Higgs boson at the LHC (Large Hadron Collider)/Geneva, Switzerland.

Strong interaction and quantum chromodynamics. The decisive impact for the creation of quantum chromodynamics came from Gell-Mann and Fritzsch in the early 1970s. Quantum chromodynamics is a gauge theory with the gauge symmetry group

$$SU(3)$$

which refers to the color of quarks (i.e., the electric charge is replaced by the so-called color charge). This theory predicts the existence of 8 massless messenger particles called gluons which are responsible for the strong interaction with respect to the quark colors.

Physicists postulate that it is impossible to observe colored quarks and colored gluons as isolated objects (quark confinement).

We are only able to observe bound states of colored quarks (baryons and mesons). These bound states are colorless (also called white). A proton consists of three quarks. Mesons consist of quark-antiquark pairs (see Sect. 3.14). In 1973/74 Gross and Wilczek, as well as Politzer applied the method of the renormalization group to quantum chromodynamics in order to show that the coupling constant of quantum chromodynamics goes to zero if the energy goes to infinity. This means that quarks behave as free particles at very high energies. This is called the asymptotic freedom of quantum chromodynamics. Gross, Politzer, and Wilczek were awarded the Nobel prize in physics in 2004.

The Standard Model in particle physics. The combination of the electroweak theory with quantum chromodynamics yields the Standard Model in particle physics with the gauge group

$$\boxed{SU(3) \times SU(2) \times U(1).}$$

This is a noncommutative Lie group, in contrast to the commutative Lie group $U(1)$ in electrodynamics. The Standard Model contains

- 12 fundamental particles (6 quarks with three colors and 6 leptons), and
- 12 messenger particles (the massless photon, the three massive weak bosons, W^+, W^-, Z^0, and the eight massless gluons which possess three colors).

The messenger particles are responsible for the interaction. Therefore, they are also called the interaction particles. The leptons are the electron e^-, the muon, μ, the tau lepton τ, and the three corresponding neutrinos (the electron neutrino ν_e, the muon neutrino ν_μ, and the tau neutrino ν_τ). In 1930, the existence of the electron neutrino was postulated by Pauli in order to explain the observed energy loss in beta decay (a form of radioactivity). The electron neutrino was discovered experimentally by Cowan and Reines in 1956. The neutrinos possess a small mass. We have to add the antiparticles.

- The 12 fundamental particles are fermions (with spin $\frac{1}{2}\hbar$), whereas
- the 12 messenger particles are bosons (with spin $\hbar$).

Thus, the Pauli exclusion principle is valid for the fundamental particles, but not for the messenger particles. For the formulation of the Standard Model in particle physics, the following principles play a crucial role:

- relativistic invariance with respect to space and time (Einstein's principle of special relativity and the Poincaré group),
- local symmetry (gauge symmetry),
- violation of spatial reflection symmetry (parity violation) and time reflection symmetry in weak interaction,
- generation of particle masses by spontaneous symmetry breaking,
- universal CPT-symmetry (combined charge conjugation, space reflection, and time reflection),
- renormalizability.

The quantization of gauge theories was formulated by Faddeev and Popov in 1967 by using a factorization of Feynman functional integrals (with respect to the orbits of the gauge symmetry group). In 1971, t'Hooft proved that the electroweak gauge theory is renormalizable, in contrast to the classical Fermi theory from the 1930s. In 1999, t'Hooft and Veltman were awarded the Nobel prize in physics for elucidating the quantum structure of electroweak interactions by means of renormalization. For the final formulation of the Standard Model in particle physics, many fundamental experimental results played a crucial role. Let us mention the following Nobel prizes in physics which mainly refer to experimental discoveries:

- Hess, 1936 (discovery of cosmic radiation in 1912), Anderson, 1936 (discovery of the positron in cosmic rays in 1932),
- Fermi, 1938 (beta decay in weak interaction),
- Stern, 1943 (discovery of the magnetic moment of the proton in 1932),
- Pauli, 1945 (exclusion principle),
- Yukawa, 1949 (theory of nuclear forces and prediction of mesons in 1935),
- Powell, 1950 (discovery of the π-meson in 1947),
- Kusch, 1955 (precision determination of the anomalous magnetic moment of the electron),
- Yang and Lee, 1957 (prediction of the violation of the parity symmetry in weak interaction),[8]
- Segrè and Chamberlain, 1959 (discovery of the antiproton in 1955),
- Wigner, 1963 (fundamental symmetry principles),
- Feynman, Schwinger, Tomonga, 1965 (quantum electrodynamics),
- Gell–Mann 1969 (classification of elementary particles and their interactions by using unitary symmetry/the eightfold way),
- Richter and Ting, 1976 (discovery of the heavy J/ψ particle in 1974),
- Penzias and Wilson, 1978 (discovery of the cosmic microwave background radiation in 1964),
- Cronin and Fitch, 1980 (discovery of CP-violation[9] in the decay of neutral K-mesons in 1964),
- Rubbia and van der Meer, 1984 (discovery of the weak bosons W^+, W^-, Z^0 in 1983),

[8] Parity violation was experimentally established by Mrs. Wu in 1957.

[9] The symbol 'CP' stands for charge conjugation/parity. Since the CPT-symmetry is assumed to be a universal symmetry, the CP-violation implies the violation of time reflection symmetry T.

- Lederman, Schwartz, and Steinberger, 1988 (discovery of the muon neutrino in 1961),
- Friedman, Kendall, and Taylor, 1990 (detection of quarks by deep inelastic scattering of electrons on protons in 1968),
- Reines, 1995 (discovery of the electron neutrino in 1956), Perl, 1995 (discovery of the tau lepton in 1975),
- Mather and Smoot, 2006 (anisotropy of the cosmic microwave background radiation),
- Nambu, Kobayashi, and Maskawa, 2008 (broken symmetry).

The Standard Model will be studied in Vol. IV on quantum mathematics.

Further Reading

We recommend:

H. Fritzsch, Quarks: The Stuff of Matter, Penguin, London, 1992. Revised German edition: Piper, Munich, 2008 (a beautiful popular history of modern elementary particle physics).

M. Veltman, Facts and Mysteries in Elementary Particle Physics, World Scientific, Singapore, 2003.

C. Sutton, Spaceship Neutrino, Cambridge University Press, 1992.

L. Brown, M. Dresden, L. Hoddeson, and M. Riordan (Eds.), The Rise of the Standard Model, Cambridge University Press, 1995.

L. O'Raifeartaigh and N. Straumann, Gauge theory: Historical origins and some modern developments, Rev. Mod. Phys. **72** (2000), 1–23.

J. Jackson and L. Okun, Historical roots of gauge invariance, Rev. Mod. Phys. **73**(2001), 663–680.

L. Okun, V. A. Fock and gauge symmetry, Physics Uspekhi **53**(8) (2010), 835–837.

G. Guralnik, The history of the Guralnik, Hagen, and Kibble development of the theory of spontaneous symmetry breaking and gauge particles, Intern. J. Modern Phys. **A24** (2009), 2601–2627.

G. Ekspong (Ed.), The Oskar Klein (1894–1977) Memorial Lectures, World Scientific, Singapore, 1991.

O. Klein, From my life of physics. In: G. Ekspong (Ed.) (1991), pp. 103–117.

C. Yang, Symmetry and Physics. In: G. Ekspong (Ed.) (1991), pp. 11–33.

17.4.2 From Gauss' Theorema Egregium to Modern Differential Geometry

In 1827, Gauss founded the differential geometry of two-dimensional curved surfaces by publishing his *General Investigations of Curved Surfaces* (in Latin). The main result was the theorema egregium which says that the Gaussian curvature is an intrinsic property of the surface. We refer to:

P. Dombrowski, 150 years after Gauss' 'Disquisitiones generales circa superficies curvas', Astérisque **62**, 1979.

The theorema egregium paved the way to fascinating developments in physics in the 20th century. In 1854, Riemann generalized Gauss' theory to higher dimensions. Roughly speaking, the Riemann curvature tensor at the point P of a Riemannian manifold $\mathcal{M}$ collects the information on the Gaussian curvature $K(P)$ of all the two-dimensional submanifolds of $\mathcal{M}$ which pass through the point P. In 1915, Einstein used the Riemann curvature tensor in order to formulate his theory of general relativity. The point is that the curvature of the four-dimensional space-time manifold describes gravitation. Two years later, the Levi-Civita connection was introduced by

T. Levi-Civita, The notion of parallel transport in Riemannian geometry, Rend. Palermo **42** (1917), 73–205 (in Italian).

Weyl wanted to generalize Einstein's theory of general relativity by including electromagnetism. To this end, he generalized the Levi-Civita parallel transport by a more general parallel transport based on generalized Christoffel symbols which do not depend on a metric tensor:

H. Weyl, Gravitation and Electricity, Sitzungsberichte Preuss. Akademie der Wiss. **65** (1918), pp. 465–480 (in German).[10]

In modern terminology, this is an affine connection. In the 1920s, Élie Cartan developed the theory of connections for a class of symmetry groups based on his method of moving frames. Cartan's classical method can be found in:

É. Cartan, Riemannian Geometry in an Orthogonal Frame: From Lectures Delivered by Élie Cartan at Sorbonne (Paris) in 1926–1927. World Scientific, Singapore, 2001.

In 1927, under the influence of Cartan, Friedrichs used a special connection in order to study the passage from Einstein's theory of general relativity to the Newtonian gravitational theory based on the limit $c \to \infty$ (i.e., the speed of light goes to infinity):

K. Friedrichs, An invariant formulation of Newton's gravitational law and the passage from Einstein's theory of general relativity to Newton's classical theory, Math. Ann. **98** (1927), 566–575 (in German).[11]

The further development was strongly influenced by Dirac's fundamental paper on the relativistic electron:

P. Dirac, The quantum theory of the electron, Proc. Royal Soc. London **A117** (1928), 610–624; **A118** (1928), 351–361.

The formulation of Maxwell's theory of electromagnetism as a $U(1)$-gauge theory (based on a $U(1)$-connection) and the relation to the Dirac equation for the relativistic electron was emphasized by Weyl in 1929:

H. Weyl, Electron and gravitation, Z. Phys. **56** (1929), 330–352 (in German). See also H. Weyl, Electron and gravitation, Proc. National Acad. U.S.A. **15** (1929), 323–334.

---

[10] See also the fifth German edition of H. Weyl, Raum, Zeit, Materie, Springer, Berlin, 1918, which appeared in 1922. (English edition: Space, Time, Matter, Dover, New York, 1953.)

[11] We also refer to A. Rendall, The Newtonian limit for asymptotically flat solutions of the Vlasov–Einstein system (in plasma physics), Commun. Math. Phys. **163** (1974), 89–112.

Motivated by Dirac's paper, van der Waerden invented a generalization of the classical tensor calculus called spinor analysis which shows automatically that both the Dirac equations and the Dirac–Einstein equations are relativistically invariant:

B. van der Waerden, Spinor analysis, Nachr. der königlichen Gesellschaft Göttingen **1929**, pp. 100–109 (in German).

B. van der Waerden and L. Infeld, The wave equation of the electron in general relativity, Sitzungsber. Preußische Akad. Wiss. Berlin, Math.-Phys. Klasse **9** (1933), pp. 308–401 (in German).

In classical terms, one has to find Christoffel symbols which generate a covariant directional derivative that respects the relativistic symmetry group, that is, the covariant directional derivative respects the Lorentz group (more precisely, it respects the universal covering group $SL(2,\mathbb{C})$ of the proper Lorentz group). In modern terminology, this is a connection with $SL(2,\mathbb{C})$-symmetry. We will study this in Vol. IV on quantum mathematics.[12] The work of Cartan and van der Waerden on spinors motivated Bauer and Weyl to generalize spinors to higher dimensions:

H. Weyl and H. Bauer, Spinors in n dimensions, Amer. J. Math. **57** (1935), 425–449.

Based on Clifford algebras, Bauer and Weyl constructed the universal covering group $Spin(n)$ to the rotation group $SO(n)$ in n dimensions. If $n \geq 4$, the Lie group $Spin(n)$ is not a matrix group. Therefore, in order to develop a connection with $Spin(n)$-symmetry, one needs a generalization of Cartan's method of moving frames. In order to be able to apply Cartan's method of moving frames, one needs a reduction of the $SO(n)$-frame bundle to the structure group $Spin(n)$. This is only possible if both the first and second Stiefel–Whitney class of the base manifold vanishes. This shows why characteristic classes from topology play a crucial role in modern differential geometry. The theory of characteristic classes (Chern classes) was developed by Chern in order to generalize the Gauss–Bonnet theorem for two-dimensional surfaces (Gauss' theorema elegantissimum) to higher-dimensional Riemannian manifolds in an intrinsic way:

S. Chern, A simple intrinsic proof of the Gauss–Bonnet formula for closed Riemannian manifolds, Ann. of Math. **45** (1945), 747–752.

S. Chern, Characteristic classes of Hermitean manifolds, Ann. of Math. **47** (1946), 85–121.

The final general mathematical theory of connections with arbitrary symmetry groups was developed by

C. Ehresmann, Infinitesimal connections in differentiable fiber spaces, Colloque de Topologie, Bruxelles (Belgium), 1950, pp. 29–55 (in French).

This theory can be found in the standard textbooks by:

S. Kobayashi and K. Nomizu, Foundations of Differential Geometry, Vols. 1, 2, Wiley, New York, 1963.

Y. Choquet-Bruhat, C. DeWitt-Morette, and M. Dillard-Bleick, Analysis, Manifolds, and Physics. Vol. 1: Basics; Vol 2: 92 Applications, Elsevier, Amsterdam, 1996.

T. Frankel, The Geometry of Physics, Cambridge University Press, 2004.

[12] As a young man, van der Waerden assisted Weitzenböck in preparing his classic textbook *Invariant Theory*, Groningen, the Netherlands, 1923. Therefore, he was well prepared for applying methods from invariant theory to spinor analysis.

J. Jost, Riemannian Geometry and Geometric Analysis, Springer, Berlin, 2008 (including Clifford algebras, the groups $Spin(n)$ and $Pin(n)$, spin geometry, the Yang–Mills equations, and the Seiberg–Witten equations as generalized Ginzburg–Landau equations).

D. Husemoller, Fibre Bundles, Springer, New York, 1994.

Riemann studied the structure of the set of all compact Riemann surfaces (i.e., the moduli space of compact Riemann surfaces). This moduli space is investigated by means of Teichmüller spaces in:

J. Jost, Compact Riemann Surfaces: An Introduction to Contemporary Mathematics, Springer, Berlin, 1997.

Similarly, one can study the topological structure of the moduli space of all connections on a base manifold (e.g., the sphere). This deep relationship between topology and gauge theory is summarized in:

K. Marathe and G. Martucci, The Mathematical Foundations of Gauge Theories, North-Holland, Amsterdam, 1992.

K. Marathe, Topics in Physical Mathematics, Springer, London, 2010.

G. Naber, Topology, Geometry, and Gauge Fields, Springer, New York, 1997.

Fundamental papers can be found in:

M. Atiyah, Collected Works, Vol. 5: Gauge Theories, Cambridge University Press, 2004.

The relations between the Atiyah–Singer index theorem and the Dirac equation are discussed in:

M. Atiyah, The Dirac equation and geometry, pp. 108–124. In: P. Goddard (Ed.), Paul Dirac – the Man and his Work, Cambridge University Press, 1999.

In 1954, Yang and Mills generalized Weyl's $U(1)$-gauge theory for the electromagnetic field to the gauge group $SU(2)$:

C. Yang and R. Mills, Conservation of isotopic spin and isotopic spin invariance, Phys. Rev. **96** (1954), 191–195.

This opened the door for a fascinating development in theoretical physics culminating in the creation of the Standard Model in particle physics in the 1960s and 1970s. From the mathematical point of view, the Standard Model in particle physics is based on a 'connection' with the symmetry group $U(1) \times SU(2) \times SU(3)$. This internal symmetry of elementary particles was discovered on the basis of a huge amount of experimental data obtained in particle accelerators.

The discovery of the $U(1) \times SU(2) \times SU(3)$ symmetry in nature is one of the greatest achievements of the physics of the twentieth century.

When creating Yang–Mills theory, the authors did not know the equivalent mathematical theory of connections founded by Ehresmann. This relationship was only discovered twenty years later:

T. Wu and C. Yang, Concept of non-integrable phase factors and global formulation of gauge fields, Phys. Rev. **D12** (1975), 3845–3857.

More details can be found in the Prologue on page 5.

Felix Klein's Erlangen program. In 1871, Felix Klein formulated his Erlangen program:

Geometry is the invariant theory of transformation groups.

Modern differential geometry can be regarded as a far-reaching completion of this program for general classes of symmetry groups.

17.4.3 The Work of Hermann Weyl

In his work, Hermann Weyl (1885–1955) emphasized the relations between mathematics, physics, philosophy, and aesthetics. During an interview in 2010, Sir Michael Atiyah was asked the following question:[13] Which mathematician do you most admire/respect, and why? Sir Michael answered:

> Among past mathematicians Hermann Weyl is the mathematician I admire most. The breadth of his interests and and the elegance of his style have been my model.

A detailed study of Weyl's work can be found in:

> E. Scholz (Ed.), Hermann Weyl's "Space-Time-Matter" and a General Introduction to his Scientific Work, Birkhäuser, Basel, 2001.

In particular, we recommend the article by:

> R. Coleman and H. Korté, Hermann Weyl: Mathematician, physicist, philosopher, pp. 159–398. In: E. Scholz (Ed.) (2001).

We also recommend:

> C. Chevalley and A. Weil (1957), Hermann Weyl (1885–1955), L'Enseignement mathémathique, tome III, fasc. 3 (1957) (in French) (obituary).
>
> Hermann Weyl (1885–1955), Springer, Berlin, 1985.
>
> C. Yang, Hermann Weyl's contributions to physics, pp. 7–21. In: Hermann Weyl (1885–1955), Springer, Berlin, 1985.
>
> J. Ehlers, Hermann Weyl's contributions to the general theory of relativity, pp. 83–105. In: W. Deppert and K. Hübner, (Eds.), Exact Sciences and Their Philosophical Foundations (International Hermann–Weyl Symposium), Peter Lang, Frankfurt/Main, 1988.
>
> F. Dyson, Birds and frogs in mathematics and physics, Einstein lecture 2008, Notices Amer. Math. Soc. **56** (2) (2009), 212–223.

Weyl's collected works comprehend four volumes:

> H. Weyl, Gesammelte Werke (Collected Works), Vols. 1–4. Edited by K. Chandrasekharan, Springer, New York, 1968.

The following list of selected publications by Hermann Weyl emphasizes papers which are closely related to physics:

> H. Weyl (1908), Singular integral equations, Mathem. Annalen **66** (Ph.D. dissertation supervised by Hilbert in Göttingen).
>
> H. Weyl (1910), On ordinary differential equations with singularities (in German), Math. Ann. **68**, 220–269.[14]
>
> H. Weyl (1911), On the asymptotic distribution of eigenvalues, Göttinger Nachrichten, pp. 110–117.

---

[13] International Center of Mathematics (CIM), Portugal, Bulletin **29**, January 2011, pp. 23–27.

[14] Weyl based the theory of singular differential equations on his theory of singular integral equations. We will show in Vol. IV that the discrete and continuous energy spectrum of the non-relativistic and relativistic hydrogen atom is a special case of Weyl's theory which was the predecessor of John von Neumann's spectral theory for unbounded self-adjoint operators in Hilbert space. See J. Dieudonné, History of Functional Analysis, North-Holland, Amsterdam, 1981.

H. Weyl (1912), On the spectrum of the black body radiation, J. Reine und Angew. Mathematik **41**, 163–181.[15]

H. Weyl (1912), Henri Poincaré (1854–1912), Mathematisch-naturwissenschaftliche Blätter 1912, pp. 161–163 (obituary).

H. Weyl (1913), Die Idee der Riemannschen Fläche (in German), Teubner-Verlag, Leipzig. New edition with commentaries supervised by R. Remmert, Teubner, Leipzig, 1997. English edition: The Concept of a Riemann Surface, Addison Wesley, Reading, Massachusetts, 1955.

H. Weyl (1915), On the asymptotic law for the frequencies of the eigensolutions of an arbitrary elastic body (in German), Rend. Circolo Mat. Palermo **39**, 1–50.

H. Weyl (1916), On the equipartition of numbers mod 1 (in German), Mathem. Annalen **77**, 313–352.

H. Weyl (1918), Das Kontinuum, Veit, Leipzig. Englisch translation: The Continuum, Dover, New York, 1994.

H. Weyl (1918), Gravitation and electricity (in German), Sitzungsbericht der Königlich-Preussischen Akademie zu Berlin, pp. 465–480 (the idea of a real-valued gauge theory by changing the length scale).

H. Weyl (1918), Raum, Zeit, Materie (in German), Springer, Berlin.

H. Weyl (1919), A new extension of the theory of general relativity (in German), Z. Phys. **59**, 101–133.

H. Weyl (1921), Raum, Zeit, Materie, 4th essentially revised edition, Springer, Berlin. (English translation: Space, Time Matter, Methuen, London, 1922). 7th German edition with notes by J. Ehlers, Springer, Berlin, 1988.

H. Weyl (1922), Das Raumproblem (The space problem) (in German), Jahresbericht DMV (Deutsche Mathematiker-Vereinigung) **31**, pp. 328–344.

H. Weyl (1923), Repartición de corriente en una red conductora (Distribution of an electric current in a network), Revista Matematica Hispano-Americana **5**, pp. 153–164 (in Spanish). In: H. Weyl, Collected Works, (1968), Vol. II, pp. 368–389, Springer, 1968. English translation: George Washington University Logistics Research Project (1951).

H. Weyl (1925), Representation theory for continuous semisimple groups by linear transformations I, II, III, Math. Zeitschrift **23**, pp. 271–309; **24**, pp. 328–376, 377–395, 789–791 (in German).

H. Weyl and F. Peter (1927), On the completeness of the irreducible representations of compact continuous groups, Math.Ann. **97**, 737–755 (in German).

H. Weyl (1929), Die gruppentheoretische Methode in der Quantenmechanik, Springer, Berlin. (English translation: The Theory of Groups and Quantum Mechanics, Dover, New York 1931.)

H. Weyl (1929), Elektron und Gravitation (in German), Z. Phys. **56** (1929), 330–352. See also H. Weyl, Gravitation and the electron, Proc. Nat. Acad. Sci. USA **15** (1929), 323–334.

[15] This paper showed that Planck's method of computing the eigensolutions of the Laplacian in a cubic box does not depend on the shape of the box if the volume goes to infinity.

H. Weyl (1935), Emmy Noether (1882–1935), Scripta mathematica, pp. 201–220 (obituary).

H. Weyl and R. Brauer (1935), Spinors in n dimensions, Amer. J. Math. **57**, pp. 425–449.

H. Weyl (1938), The Classical Groups: Their Invariants and Representations, 2nd edition with supplement, 1946, 15th printing, 1997, Princeton University Press.

H. Weyl (1940), The method of orthogonal projection in potential theory, Duke Math. J. **7**, pp. 414–444.

H. Weyl (1943), On Hodge's theory of harmonic integrals, Ann. of Math. **44**, pp. 1–6.

H. Weyl (1944), David Hilbert and his mathematical work, Bull. Amer. Math. Soc. **50**, pp. 612–654 (obituary).

H.Weyl (1949), Philosophy of Mathematics and Natural Sciences, Princeton University Press, (3rd edition, 2009).

H. Weyl (1952), Symmetry, Princeton University Press, 1952/1983.

Problems

17.1 *The covariant directional derivative $D_v\psi$ with respect to local bundle coordinates.* Prove that the connection axioms imply the local formula (17.4) for the covariant directional derivative $D_v\psi$.
Solution: In what follows, we will sum over $\alpha = 1, \ldots, m$ and $j, k, l = 1, \ldots, n$. Let us describe the bundle chart $\mathcal{O} \times \mathbb{R}^n$ by local coordinates living in $\mathbb{R}^m \times \mathbb{R}^n$. Let $x = (x^1, \ldots, x^m)$. Choose the standard basis $e_1, \ldots, e_n$ of $\mathbb{R}^n$,

$$e_1 := \begin{pmatrix} 1 \\ 0 \\ \vdots \\ 0 \end{pmatrix}, \ldots, \quad e_n := \begin{pmatrix} 0 \\ 0 \\ \vdots \\ 1 \end{pmatrix}.$$

This generates the constant maps $s_j : \mathcal{O} \to \mathbb{R}^n$ with $s_j(P) := e_j$ for all $P \in \mathcal{O}$. To simplify notation, we will write e_j instead of s_j. Use the basis representation $v = v^\alpha \partial_\alpha$ and $\psi(x) = \psi^j(x)e_j$. Then, by both linearity and the Leibniz rule for $D_v\psi$, we get

$$D_v\psi = v^\alpha D_{\partial_\alpha}(\psi^l e_l) = v^\alpha(\partial_\alpha \psi^l)e_l + v^\alpha(D_{\partial_\alpha} e_l).$$

Since $e_1, \ldots, e_n$ is a basis of $\mathbb{R}^n$, there exist real numbers $\Gamma^k_{\alpha l}(x)$ such that

$$D_{\partial_\alpha} e_l = \Gamma^k_{\alpha l}(x) e_k, \qquad \alpha = 1, \ldots, m.$$

Hence

$$D_v\psi = (v^\alpha \partial_\alpha \psi^l)e_l + \Gamma^k_{\alpha l}(x)\psi^l \cdot e_k = v^\alpha(\partial_\alpha + \mathcal{A}_\alpha)\psi.$$

17.2 *Transformation law for the connection matrices.* Let $\psi^+(x) = G(x)\psi(x)$ be a change of bundle coordinates. Determine the transformation law for the matrices $\mathcal{A}_\alpha$.
Solution: Since $D_v\psi$ transforms like ψ, we get

$$D_v^+ \psi^+ = G D_v \psi.$$

Hence

$$v^\alpha(\partial_\alpha + \mathcal{A}_\alpha^+)(G\psi) = G v^\alpha(\partial_\alpha + \mathcal{A}_\alpha)\psi.$$

This is true for all velocity vector fields v. Thus,

$$(\partial_\alpha + \mathcal{A}_\alpha^+)(G\psi) = G(\partial_\alpha + \mathcal{A}_\alpha)\psi.$$

Hence $\partial_\alpha G \cdot \psi + \mathcal{A}_\alpha^+ G\psi = G\mathcal{A}_\alpha \psi$ for all ψ. This yields

$$\mathcal{A}_\alpha^+ = G\mathcal{A}_\alpha G^{-1} - \partial_\alpha G \cdot G^{-1}.$$

17.3 *Covariant time derivative.* Prove Prop. 17.2 on page 878.
Hint: Use Problem 17.2 together with the chain rule.

17.4 *Proof of Theorem 17.12.* Hint: There are two possibilities to prove this theorem.
(i) The local approach. Use local bundle charts and reduce the proof to product bundles.
(ii) The global approach. Use the formula (12.82) (definition of $d\omega$ in terms of the Lie derivative). See Y. Choquet et al., Analysis, Manifolds, and Physics. Vol. 1, p. 373, Elsevier, Amsterdam, 1996.

18. Inertial Systems and Einstein's Principle of Special Relativity

It is known that Maxwell's electrodynamics, when applied to moving bodies, leads to asymmetries which do not appear to be inherent in the phenomena. Take, for example, the electrodynamic interaction between a magnet and a conductor. The observable phenomenon here depends only on the relative motion of the conductor and the magnet, whereas the customary view draws a sharp distinction between the two cases in which either the one or the other of these bodies is in motion ...
The unsuccessful attempts to discover any motion of the earth relatively to the "light medium" (ether), suggest that the phenomena of electrodynamics as well as of mechanics possess no properties corresponding to the idea of absolute rest.[1]

Albert Einstein, 1905

Henceforth space by itself and time by itself are doomed to fade away into mere shadows and only a kind of union of the two will preserve an independent reality.[2]

Hermann Minkowski, 1908

The Einstein convention. In what follows, we will sum over equal upper and lower Greek (resp. Latin) indices from 0 to 3 (resp. 1 to 3). For example,

$$a^\mu b_\mu = a^0 b_0 + a^1 b_1 + a^2 b_2 + a^3 b_3 = a^0 b_0 + a^j b_j .$$

Classical mechanics and the Galilean transformation. Consider the situation depicted in Fig. 18.1. Let Σ and Σ' be two right-handed Cartesian coordinate systems which coincide at time $t = 0$. Here, (x, y, z) and (x', y', z') denote the coordinates of Σ and Σ', respectively. Newton assumed that there exists an absolute system of reference. Suppose that this absolute system corresponds to Σ. Furthermore, we assume that the system Σ' moves with the velocity V parallel to the system Σ in direction of the x-axis of Σ. In classical mechanics, a system of reference is called an inertial system iff a force-free mass point rests or it moves with

[1] This is the beginning of one of the most influential papers in physics: A. Einstein, On the electrodynamics of moving bodies (in German), Ann. Phys. **17** (1905), 891–921. In this paper, the young Einstein (1879–1955) founded the theory of special relativity. See also S. Hawking (Ed.), The Essential Einstein: his Greatest Works, pp. 4–31, Penguin Books, London, 2009.

[2] H. Minkowski, Space and time (in German), Lecture to the 80th Assembly of Natural Scientists and Physicians, Cologne 1908, Phys. Z. **10** (1909), 104–111. English translation: The principle of relativity, Aberdeen Univ. Press, Aberdeen, 1923. Minkowski (1864–1909) gave this lecture one year before his early death.

E. Zeidler, *Quantum Field Theory III: Gauge Theory*,

DOI 10.1007/978-3-642-22421-8_19,

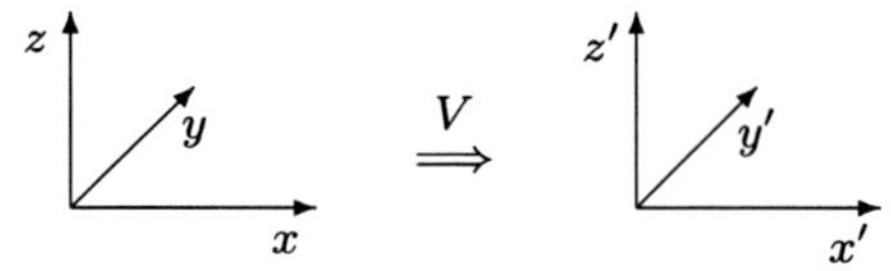

Fig. 18.1. Lorentz boost

constant speed along a straight line. Newton assumed that the absolute system of reference is an inertial system. In classical mechanics, the transformation law from the system Σ to Σ' is given by the Galilean transformation

$$x' = x - Vt, \quad y' = y, \quad z' = z, \quad t' = t. \tag{18.1}$$

Moreover, it is assumed that $m = m'$, that is, mass is a conserved quantity. The motion

$$x = x(t), \quad y = y(t), \quad z = z(t), \qquad t \in \mathbb{R}$$

of a point of mass m in the original system of reference Σ is transformed into the motion

$$x' = x'(t'), \quad y' = y'(t'), \quad z' = z'(t'), \qquad t' = t$$

in the new system of reference Σ'. For the transformed velocities, we get

$$\frac{dx'(t)}{dt} = \frac{dx(t)}{dt} - V, \quad \frac{dy'(t)}{dt} = \frac{dy(t)}{dt}, \quad \frac{dz'(t)}{dt} = \frac{dz(t)}{dt}, \qquad t \in \mathbb{R}.$$

This is called the Galilean addition theorem for velocities. For the acceleration, we get

$$\frac{d^2x'(t)}{dt^2} = \frac{d^2x(t)}{dt^2}, \quad \frac{d^2y'(t)}{dt^2} = \frac{d^2y(t)}{dt^2}, \quad \frac{d^2z'(t)}{dt^2} = \frac{d^2z(t)}{dt^2}, \qquad t \in \mathbb{R}.$$

Because of the invariance of mass, $m' = m$, we obtain

$$m\frac{d^2x'(t)}{dt^2} = m\frac{d^2x(t)}{dt^2}, \quad m\frac{d^2y'(t)}{dt^2} = m\frac{d^2y(t)}{dt^2}, \quad m\frac{d^2z'(t)}{dt^2} = m\frac{d^2z(t)}{dt^2}, \qquad t \in \mathbb{R}.$$

This tells us that the forces remain invariant, and Σ' is an inertial system.

Einstein's goal in 1905. Einstein wanted to obtain the transformation law for the electric field vector **E** and the magnetic field vector **B** when passing from the system of reference Σ to the moving system Σ'. His main results read as follows:

(i) There is no absolute system of reference.
(ii) Lorentz transformation: If Σ and Σ' are inertial systems, then the Galilean transformation law (18.1) has to be replaced by the Lorentz transformation

$$x' = \frac{x - Vt}{\sqrt{1 - V^2/c^2}}, \quad y' = y, \quad z' = z, \quad t' = \frac{t - xV/c^2}{\sqrt{1 - V^2/c^2}}. \tag{18.2}$$

Here, c is the velocity of light in a vacuum. Let us set
- $x^0 := ct,\ x^1 := x,\ x^2 := y,\ x^3 := z$, and
- $x_0 := x^0, x_1 := -x^1, x_2 := -x^2, x_3 := -x^3$.

Then

$$\boxed{x^{0'} = \frac{x^0 - x^1 V/c}{\sqrt{1 - V^2/c^2}}, \quad x^{1'} = \frac{x^1 - x^0 V/c}{\sqrt{1 - V^2/c^2}}, \quad x^{2'} = x^2, \quad x^{3'} = x^3.}$$

(iii) Electromagnetic field tensor: Einstein discovered that one has to replace the pair of field vectors $\mathbf{E} = E^1\mathbf{i} + E^2\mathbf{j} + E^3\mathbf{k}$ and $\mathbf{B} = B^1\mathbf{i} + B^2\mathbf{j} + B^3\mathbf{k}$ by the antisymmetric matrix

$$\begin{pmatrix} F_{00} & F_{01} & F_{02} & F_{03} \\ F_{10} & F_{11} & F_{12} & F_{13} \\ F_{20} & F_{21} & F_{22} & F_{23} \\ F_{30} & F_{31} & F_{32} & F_{33} \end{pmatrix} := \begin{pmatrix} 0 & E^1/c & E^2/c & E^3/c \\ -E^1/c & 0 & -B^3 & B^2 \\ -E^2/c & B^3 & 0 & -B^1 \\ -E^3/c & -B^2 & B^1 & 0 \end{pmatrix}.$$

According to Einstein, this is an antisymmetric tensor. This means that $F_{\alpha\beta}$ transforms like the product $x_\alpha x_\beta$ under Lorentz transformations. The corresponding transformation laws for $\mathbf{E}, \mathbf{B}$ can be found in Sect. 19.4 on page 967. Equivalently, the differential form

$$\boxed{\mathcal{F} = \tfrac{1}{2} F_{\alpha\beta}\, dx^\alpha \wedge dx^\beta}$$

possesses an invariant meaning for all inertial systems.

In addition, it is convenient to introduce the symmetric matrix

$$\begin{pmatrix} \eta_{00} & \eta_{01} & \eta_{02} & \eta_{03} \\ \eta_{10} & \eta_{11} & \eta_{12} & \eta_{13} \\ \eta_{20} & \eta_{21} & \eta_{22} & \eta_{23} \\ \eta_{30} & \eta_{31} & \eta_{32} & \eta_{33} \end{pmatrix} = \begin{pmatrix} 1 & 0 & 0 & 0 \\ 0 & -1 & 0 & 0 \\ 0 & 0 & -1 & 0 \\ 0 & 0 & 0 & -1 \end{pmatrix}. \tag{18.3}$$

As we will show later on, $\eta_{\alpha\beta}$ is a symmetric, numerically invariant tensorial family with respect to Lorentz transformations. That is, if we transform $\eta_{\alpha\beta}$ like $x_\alpha x_\beta$, then the values of $\eta_{\alpha\beta}$ remain unchanged. The tensor

$$g = \eta_{\alpha\beta}\, dx^\alpha \otimes dx^\beta$$

is the metric tensor of the Minkowski manifold $\mathbb{M}^4$ to be introduced below. We also set

$$\eta^{\alpha\beta} := \eta_{\alpha\beta}, \qquad \alpha, \beta = 0, 1, 2, 3.$$

Then, $\eta^{\alpha\beta}$ is also a tensorial family with respect to Lorentz transformations. The tensorial families $\eta_{\alpha\beta}$ and $\eta^{\alpha\beta}$ can be used for the lowering or the lifting of indices with respect to Lorentz transformations, that is, we set

$$x_\alpha = \eta_{\alpha\beta} x^\beta \quad \text{and} \quad x^\alpha = \eta^{\alpha\beta} x_\beta.$$

The deformation of classical mechanics. Note the following crucial fact.

- If the velocity V is small compared with the velocity of light c, that is,

$$V/c \ll 1,$$

then the Lorentz transformation (18.2) is approximated by the Galilean transformation (18.1). More precisely, if $V/c \to 0$, then the Lorentz transformation converges to the Galilean transformation. In this sense, the theory of special relativity is a deformation of classical mechanics. For example, if a car moves with the speed V of 100 km/hour, then $V/c = 10^{-8}$. For an aircraft with the speed of $V = 1000$ km/hour, we get $V/c = 10^{-7}$. Therefore, the theory of special relativity does not play any role in daily life.[3]

- Similarly, quantum mechanics is a deformation of classical mechanics with respect to the small parameter

$$h/S \ll 1$$

where h is Planck's quantum of action, and S is the typical action of processes in daily life. In the SI system, we obtain approximately $S = 1\cdot$Js$= 1\cdot$ kg m^2/s, and we have precisely $h = 6.625 \cdot 10^{-34}$ Js. Hence $h/S = 6.625 \cdot 10^{-34}$.

18.1 The Principle of Special Relativity

Inertial systems. In physics, the notion of an inertial system plays a crucial role:

An (x, y, z)-Cartesian coordinate system with time t is called an inertial system iff every mass particle rests or moves along a straight line with constant velocity provided no force is acting.

The prototype of an inertial system is a spaceship which moves at a far distance from both our solar system and other stars and which flies without rocket propulsion.

The invariance principle in classical mechanics. Newton (1643–1717) postulated the existence of a distinguished inertial system which he called the absolute system of reference. In particular, the time t measured in the absolute system of reference plays the role of an absolute world time. Newton's equations of motion refer to the (hypothetical) absolute inertial system. In classical mechanics, a system of reference is an inertial system iff it moves relatively to the absolute inertial system with a constant velocity vector. The classical Galilean principle of relativity reads as follows: All inertial systems are physically equivalent with respect to processes in classical mechanics, that is, physical processes concerning classical mechanics are the same in all inertial systems when the initial and boundary conditions are the same.

Maxwell (1831–1875) assumed that his equations for the electromagnetic field (first formulated in 1864) are valid in a distinguished inertial system which he called the ether system.[4] In 1881 Michelson performed an experiment in order to measure the relative motion of earth to the ether system. To this end, he measured the speed of light at different time points of the year. By the classical Galilean additional theorem for velocities, Michelson expected different values $V - c$ and $V + c$ of the speed of light after six months (Fig. 18.2). But he observed the constancy of the speed of light during the year. In 1887 Michelson and Morley made refined measurements, but the result was the same. Therefore, Einstein postulated in 1905:

In every inertial system light travels with the same constant velocity c in every direction in a vacuum, that is, the speed of light is a universal constant in nature.

---

[3] This is not completely true. To be precise, the GPS (Global Positioning System) uses both Einstein's theory of special relativity and Einstein's theory of general relativity concerning the gravitational field of the earth.

[4] J. Maxwell, A Treatise on Electricity and Magnetism, London, 1873. Reprinted by Dover, New York, 1954.

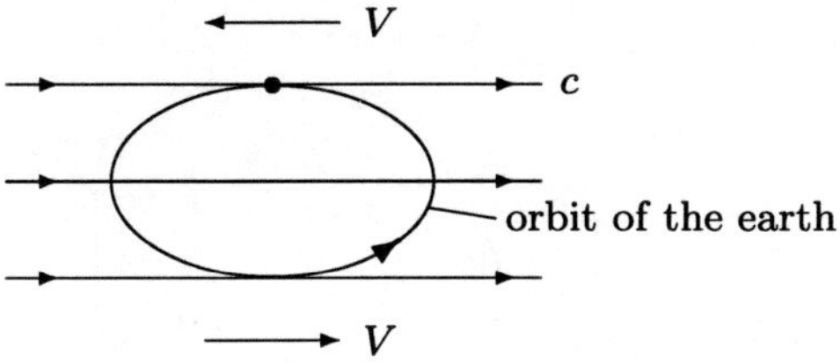

Fig. 18.2. The Michelson experiment

Einstein's general invariance principle. More general, in 1905 Einstein formulated the following fundamental principle in physics:

All inertial systems are physically equivalent, that is, all the physical processes are the same in all inertial systems when initial and boundary conditions are the same.

As we will show below, this principle changed completely our philosophy about space and time. In particular, time depends on the choice of the inertial system.

18.1.1 The Lorentz Boost

Consider two parallel right-handed inertial systems Σ and Σ' as depicted in Fig. 18.1 on page 906. We assume that

- the system Σ' with the coordinates x', y', z', t' moves with the relative velocity $V > 0$ with respect to
- the system Σ with the coordinates x, y, z, t.

We postulate that the following quite natural conditions are satisfied:

(H1) The transformation from Σ to Σ' is a linear invertible transformation of the form $x' = \alpha x + \beta t, \;\; t' = \gamma x + \delta t, \;\; y' = y, \;\; z' = z$.
(H2) The equation $x = ct$ of a light ray in the system Σ corresponds to the equation $x' = ct'$ in Σ' (constancy of the speed of light).
(H3) The origin $x' = 0, t' \in \mathbb{R}$ in Σ' corresponds to $x = Vt, t \in \mathbb{R}$ in Σ.
(H4) $x > 0$ and $t = 0$ implies $x' > 0$.
(H5) The transformation $\Sigma' \Rightarrow \Sigma$ is obtained from $\Sigma \Rightarrow \Sigma'$ by replacing V by $-V$.

We will show in Problem 18.1 on page 933 that these conditions uniquely determine the so-called Lorentz boost (in direction of the x-axis):

$$\boxed{x' = \frac{x - Vt}{\sqrt{1 - V^2/c^2}}, \quad y' = y, \quad z' = z, \quad t' = \frac{t - Vx/c^2}{\sqrt{1 - V^2/c^2}}.} \tag{18.4}$$

Here, c is the velocity of light in a vacuum, and we assume that $V < c$. Introducing the new notation $x^0 := ct, x^1 := x, x^2 := y, x^3 := z$, we get the following matrix transformation

$$\begin{pmatrix} x^{0'} \\ x^{1'} \\ x^{2'} \\ x^{3'} \end{pmatrix} = \begin{pmatrix} \cosh\chi & -\sinh\chi & 0 & 0 \\ -\sinh\chi & \cosh\chi & 0 & 0 \\ 0 & 0 & 1 & 0 \\ 0 & 0 & 0 & 1 \end{pmatrix} \begin{pmatrix} x^0 \\ x^1 \\ x^2 \\ x^3 \end{pmatrix} \tag{18.5}$$

where $\cosh\chi := \frac{1}{\sqrt{1-\gamma^2}}$, $\sinh\chi := \frac{\gamma}{\sqrt{1-\gamma^2}}$ with $\gamma := \frac{V}{c}$. We set

$$L_1(\chi) := \begin{pmatrix} \cosh\chi & -\sinh\chi & 0 & 0 \\ -\sinh\chi & \cosh\chi & 0 & 0 \\ 0 & 0 & 1 & 0 \\ 0 & 0 & 0 & 1 \end{pmatrix} \tag{18.6}$$

and call this the Lorentz boost matrix in direction of the x-axis.

If the relative velocity V is small compared with the velocity of light c in a vacuum, that is, $\gamma = V/c \ll 1$, then the Lorentz boost (18.4) reads as

$$x^{1'} = x^1 - Vt + O(\gamma^2), \quad t' = t + O(\gamma^2), \quad \gamma \to 0.$$

Neglecting terms of order $O(\gamma^2)$, we get

$$x' = x - Vt, \quad t' = t.$$

This is the Galilean transformation in classical mechanics.

18.1.2 The Transformation of Velocities

Suppose that the motion of a massive point particle is given by the equation

$$\mathbf{x} = \mathbf{x}(t) \quad \text{in the inertial system } \Sigma,$$

and by the equation

$$\mathbf{x}' = \mathbf{x}'(t') \quad \text{in the inertial system } \Sigma'.$$

Thus one observes the velocity vectors

$$\mathbf{v} = \frac{d\mathbf{x}(t)}{dt} \quad \text{and} \quad \mathbf{v}' = \frac{d\mathbf{x}'(t')}{dt'}$$

in Σ and Σ', respectively (see Fig. 18.1 on page 906). Set

$$\mathbf{v} = v_1\mathbf{i} + v_2\mathbf{j} + v_3\mathbf{k},$$

and $\mathbf{v}' = v_1'\mathbf{i}' + v_2'\mathbf{j}' + v_3'\mathbf{k}'$. We assume that $|\mathbf{v}| < c$. Then, the Lorentz boost (18.4) implies the following transformation law for the velocity components:

$$v_1' = \frac{v_1 - V}{1 - v_1V/c^2}, \quad v_2' = \frac{\sqrt{1 - V^2/c^2}\, v_2}{1 - v_1V/c^2}, \quad v_3' = \frac{\sqrt{1 - V^2/c^2}\, v_3}{1 - v_1V/c^2}. \tag{18.7}$$

If the velocities are small compared with the velocity of light (i.e., $|v_1V/c^2| \ll 1$ and $V/c \ll 1$), then the first approximation yields

$$v_1' = v_1 - V, \quad v_2' = v_2, \quad v_3' = v_3.$$

This is the classical Galilean addition theorem of velocities used in daily life.
Proof. From $y'(t') = y(t), z'(t') = z(t)$, and

$$x' = \frac{x - Vt}{\sqrt{1 - V^2/c^2}}, \qquad t' = \frac{t - Vx/c^2}{\sqrt{1 - V^2/c^2}}$$

it follows by differentiation that

$$\frac{dx'}{dt'} = \frac{dx'}{dt} : \frac{dt'}{dt} = \frac{\dot{x}(t) - V}{1 - V\dot{x}(t)/c^2}.$$

Similarly, we compute $\frac{dy'}{dt'}$ and $\frac{dz'}{dt'}$. □

Equation (18.7) implies

$$c^2 - |\mathbf{v}'|^2 = \frac{(c^2 - |\mathbf{v}|^2)(1 - V^2/c^2)}{(1 - v_1 V/c^2)^2}.$$

Hence, because of $0 \leq V < c$ we have the following results:

(i) From $|\mathbf{v}| < c$ there follows $|\mathbf{v}'| < c$, that is, subvelocity of light remains subvelocity of light.
(ii) From $|\mathbf{v}| = c$ there follows $|\mathbf{v}'| = c$, that is, velocity of light remains velocity of light.
(iii) For super-velocities of light with $|v_1| > c$ in an inertial system Σ, there always exists an inertial system Σ' as depicted in Fig. 18.1 with the relative velocity $0 < V < c$ for which the transformation law (18.7) becomes singular, that is, $|\mathbf{v}'|$ becomes infinitely large. Therefore, we exclude super-velocities of light.

18.1.3 Time Dilatation

Suppose that, in the inertial system Σ, two events take place at the same position (x, y, z) and at different times t_0 and $t_1 = t_0 + \Delta t$. An observer P' in the inertial system Σ' measures the time difference

$$\Delta t' = \frac{\Delta t}{\sqrt{1 - V^2/c^2}}.$$

In fact, P' observes the two events

$$x_0' = \frac{x - Vt_0}{\sqrt{1 - V^2/c^2}}, \quad y_0' = y, \quad z_0' = z, \quad t_0' = \frac{t_0 - Vx/c^2}{\sqrt{1 - V^2/c^2}}$$

and

$$x_1' = \frac{x - Vt_1}{\sqrt{1 - V^2/c^2}}, \quad y_1' = y, \quad z_1' = z, \quad t_1' = \frac{t_1 - Vx/c^2}{\sqrt{1 - V^2/c^2}}.$$

Finally, note that $\Delta t' = t_1' - t_0'$.

18.1.4 Length Contraction

Consider a rod of length l which is at rest on the x-axis of the inertial system Σ. An observer P' in the inertial system Σ' measures the length

$$l' = l\,\sqrt{1 - V^2/c^2}.$$

To show this, note that the endpoints $(x_0, y_0 = 0, z_0 = 0)$ and $(x_1 = x_0 + l, y_0, z_0)$ of the rod in Σ move in Σ' according to the equations

$$x_0' = \frac{x_0 - Vt}{\sqrt{1 - V^2/c^2}}, \quad y_0 = z_0 = 0, \quad t_0' = \frac{t - Vx_0/c^2}{\sqrt{1 - V^2/c^2}}$$

and

$$x_1' = \frac{x_1 - Vt}{\sqrt{1 - V^2/c^2}}, \quad y_1 = z_1 = 0, \quad t_1' = \frac{t - Vx_1/c^2}{\sqrt{1 - V^2/c^2}}.$$

Hence

$$x_0 = \frac{x_0' + Vt_0'}{\sqrt{1 - V^2/c^2}}, \qquad x_1 = \frac{x_1' + Vt_1'}{\sqrt{1 - V^2/c^2}}. \tag{18.8}$$

It is important to note that the observer P' measures the length $l' = x_1' - x_0'$ of the rod not at the same t-time, but at the same t'-time. Letting $t_1' = t_0'$ in (18.8), we obtain $l = x_1 - x_0 = l'/\sqrt{1 - V^2/c^2}$. This is the claim.

18.1.5 The Synchronization of Clocks

Consider two clocks C and D in the inertial system Σ at the points, say, $(x_0, 0, 0)$ and $(x_1, 0, 0)$, respectively. In order to synchronize the clocks, we send a light signal from C to D which returns immediately back to C. Suppose that the light signal departs the point $(x_0, 0, 0)$ at time t_0, and it returns to the point $(x_0, 0, 0)$ at time t_1. The clock D is synchronized with the clock C iff the time t of arrival of the light signal at the point $(x_1, 0, 0)$ is equal to

$$t = t_0 + \frac{t_1 - t_0}{2} = t_0 + \frac{c}{2(x_1 - x_0)}.$$

Such a synchronization is always possible in the inertial system Σ.

Now let us pass to the inertial system Σ' (see Fig. 18.1 on page 906). The point is that two events

$$(x_0, y_0 = 0, z_0 = 0, t) \qquad \text{and} \qquad (x_1, 0, 0, t) \tag{18.9}$$

which happen at different points in Σ at the same time $t_0 = t_1 = t$ are transformed into

$$x_0' = \frac{x_0 - Vt}{\sqrt{1 - V^2/c^2}}, \quad y_0' = z_0' = 0, \quad t_0' = \frac{t - Vx_0/c^2}{\sqrt{1 - V^2/c^2}}$$

and

$$x_1' = \frac{x_1 - Vt}{\sqrt{1 - V^2/c^2}}, \quad y_1' = z_1' = 0, \quad t_1' = \frac{t - Vx_1/c^2}{\sqrt{1 - V^2/c^2}}.$$

Obviously,

$$t_0' \neq t_1',$$

that is, the two simultaneous events (18.9) in Σ happen in the inertial system Σ' at different times. Therefore, the notion of simultaneousness depends on the observer.

18.1.6 General Change of Inertial Systems in Terms of Physics

Let Σ and Σ' be two inertial systems with right-handed Cartesian coordinates x, y, z and x', y', z', respectively. Moreover, we will use the time t and t' in Σ and Σ', respectively. Heuristically, we expect that a physicist in Σ observes that the origin of Σ' moves with constant velocity V along a straight line, that is, this motion is described by the equation

$$x = u(t - t_0) + x_0, \quad y = v(t - t_0) + y_0, \quad z = w(t - t_0) + z_0,$$

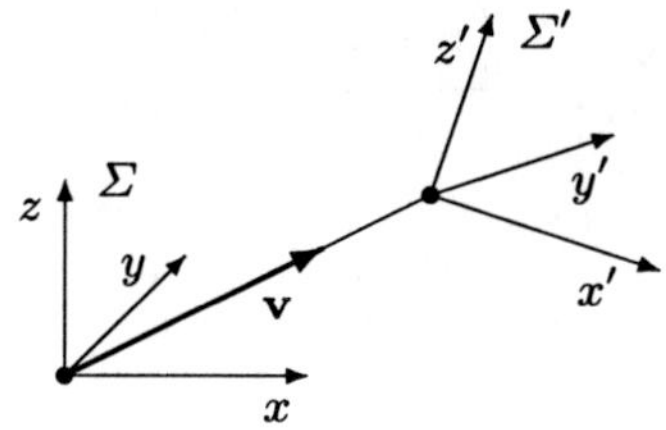

Fig. 18.3. Inertial systems Σ and Σ'

with $V = \sqrt{u^2+v^2+w^2}$ (Fig. 18.3). After a suitable rotation of Σ and Σ' and after a translation of the space and time coordinates, we get the situation of the Lorentz boost depicted in Fig. 18.1 on page 906. This yields

$$\begin{pmatrix} ct' \\ x' \\ y' \\ z' \end{pmatrix} = RL_1(\chi)S \begin{pmatrix} c(t-t_0) \\ x-x_0 \\ y-y_0 \\ z-z_0 \end{pmatrix} \tag{18.10}$$

where R and S are rotations of spatial coordinates, and $L_1(\chi)$ is the Lorentz boost matrix (18.6). As we will show below, the homogeneous transformations of the form (18.10) with $t_0 = x_0 = y_0 = z_0 = 0$ are precisely the transformations of the proper orthochronous Lorentz group $SO^{\uparrow}(1,3)$ (see Prop. 18.2). If we pass to left-handed Cartesian coordinates or if we reverse the time direction, then we have to add space reflections and time reflections, respectively. In this general case, the inhomogeneous (resp. homogeneous) transformations (18.5) will form the Poincaré group $P(1,3)$ (resp. the Lorentz group $O(1,3)$). In the next sections, it is our goal to replace these heuristic arguments by the rigorous approach due to Minkowski. Before doing this, let us reformulate the Lorentz transformation in terms of a coordinate-free approach.

Invariant formulation of the transformation of position and time. Suppose that an observer in the inertial system Σ (resp. Σ') measures the position vector $\mathbf{x}$ at the origin O and the time t (resp. the position vector $\mathbf{x}'$ at the origin O' and the time t'). Then the transformation law reads as

$$\mathbf{x}' = \mathbf{x} - \mathbf{V}t + \left(\frac{1}{\sqrt{1-\mathbf{V}^2/c^2}} - 1\right)((\mathbf{n}\mathbf{x})\mathbf{n} - \mathbf{V}t) + \mathbf{x}'_0,$$

$$t' = \frac{t - \mathbf{V}\mathbf{x}/c^2}{\sqrt{1-\mathbf{V}^2/c^2}} + t'_0 \tag{18.11}$$

where $\mathbf{n} := \mathbf{V}/|\mathbf{V}|$. In particular, $\mathbf{x} = 0, t = 0$ corresponds to $\mathbf{x}'_0, t'_0$. In the special case where $\mathbf{V} = V\mathbf{i}$, this coincides with the Lorentz boost (18.4). If the velocity $|\mathbf{V}|$ is small compared with the velocity of light c, then we get the classical Galilean transformation

$$\mathbf{x}' = \mathbf{x} - \mathbf{V}t, \quad t' = t,$$

as a first approximation.

The addition theorem for velocities. If an observer in the inertial system Σ (resp. Σ') measures the velocity vector $\mathbf{v}$ (resp. $\mathbf{v}'$) of a massive particle or of a light ray (i.e., $|\mathbf{v}| \le c$), then

$$\mathbf{v}' = \frac{\mathbf{v} - \mathbf{V}}{1 - \mathbf{V}\mathbf{v}/c^2} - \frac{\left(1 - \sqrt{1 - \mathbf{V}^2/c^2}\right)(\mathbf{v} - (\mathbf{v}\mathbf{n})\mathbf{n})}{1 - \mathbf{V}\mathbf{v}/c^2}.$$

This corresponds to (18.7). If the velocities $|\mathbf{v}|$ and $|\mathbf{V}|$ are small compared with the velocity of light c, then we get the classical addition theorem $\mathbf{v}' = \mathbf{v} - \mathbf{V}$ for velocity vectors, as a first approximation.

18.2 Matrix Groups

18.2.1 The Group $O(1,1)$

The following simple matrix relations are the key to Minkowski's approach. Let $\alpha, \beta, \gamma, \delta$ be real numbers. Let $\chi \in \mathbb{R}$. We set

$$\mathcal{G} := \begin{pmatrix} \alpha & \beta \\ \gamma & \delta \end{pmatrix}, \quad \eta := \begin{pmatrix} 1 & 0 \\ 0 & -1 \end{pmatrix}, \quad L(\chi) := \begin{pmatrix} \cosh\chi & -\sinh\chi \\ -\sinh\chi & \cosh\chi \end{pmatrix}, \quad I := \begin{pmatrix} 1 & 0 \\ 0 & 1 \end{pmatrix}.$$

Proposition 18.1 *The transformation* $\begin{pmatrix} ct' \\ x' \end{pmatrix} = \mathcal{G}\begin{pmatrix} ct \\ x \end{pmatrix}$ *satisfies the Minkowski condition*

$$c^2t'^2 - x'^2 = c^2t^2 - x^2 \qquad \text{for all} \quad x, t \in \mathbb{R} \tag{18.12}$$

iff $\mathcal{G} = RL(\chi)$ *where* $L(\chi)$ *is a Lorentz boost for some real parameter* χ, *and* R *equals one of the following matrices:* $R = I$ *(unit matrix),* $R = \eta$ *(space reflection),* $R = -\eta$ *(time reflection),* $R = -I$ *(space-time reflection).*

All these real (2×2)*-matrices* $\mathcal{G}$ *form a group denoted by* $O(1,1)$.

Proof. (I) Equivalent formulation. The Minkowski condition (18.12) is equivalent to

$$(ct', x')\,\eta \begin{pmatrix} ct' \\ x' \end{pmatrix} = (ct, x)\,\mathcal{G}^d\eta\mathcal{G}\begin{pmatrix} ct \\ x \end{pmatrix} = (ct, x)\,\eta\begin{pmatrix} ct \\ x \end{pmatrix}$$

for all $t, x \in \mathbb{R}$. In turn, this is equivalent to the matrix equation

$$\mathcal{G}^d\eta\,\mathcal{G} = \eta. \tag{18.13}$$

(II) Necessary condition. Suppose that condition (18.13) is satisfied. Then $\det\mathcal{G}^d \det\eta \det\mathcal{G} = \det\eta$. Hence $(\det\mathcal{G})^2 = 1$. Thus, $\det\mathcal{G} = \pm 1$. Moreover,

$$\begin{pmatrix} \alpha^2 - \gamma^2 & \alpha\beta - \gamma\delta \\ \alpha\beta - \gamma\delta & \beta^2 - \delta^2 \end{pmatrix} = \begin{pmatrix} 1 & 0 \\ 0 & -1 \end{pmatrix}.$$

Hence $\alpha \neq 0$ and $\delta \neq 0$. First suppose that $\alpha > 0$.It follows from $\alpha^2 - \gamma^2 = 1$ that there exists a real number χ such that

$$\alpha = \cosh\chi, \qquad \gamma = -\sinh\chi.$$

Moreover, $\frac{\gamma}{\alpha} = \frac{\beta}{\delta}$ implies

$$\beta = \lambda\gamma, \qquad \delta = \lambda\alpha, \quad \lambda \in \mathbb{R}.$$

Since $1 = \delta^2 - \beta^2 = \lambda^2(\alpha^2 - \gamma^2)$, we get $\lambda^2 = 1$. Hence $\lambda = \pm 1$.

If we assume that $\alpha < 0$, the same argument yields $\alpha = -\cosh\chi, \gamma = \sinh\chi$.

(III) Sufficient condition. Conversely, $\mathcal{G} = RL$ yields (18.13).

(IV) Group property. If the matrices $\mathcal{G}$ and $\mathcal{G}'$ satisfy the Minkowski condition (18.13), then the product $\mathcal{G}\mathcal{G}'$ also satisfies the condition (18.13). In fact,

$$(\mathcal{G}\mathcal{G}')^d\eta(\mathcal{G}\mathcal{G}') = \mathcal{G}'^d(\mathcal{G}^d\eta\mathcal{G})\mathcal{G}' = \mathcal{G}'^d\eta\mathcal{G}' = \eta.$$

□

Let us add the following definitions of the subgroups $SO(1,1)$ and $SO^\uparrow(1,1)$ of the group $O(1,1)$:

- $\mathcal{G} \in SO(1,1)$ iff $\mathcal{G} \in O(1,1)$ and $\det\mathcal{G} = 1$. Obviously, $\mathcal{G} \in SO(1,1)$ iff $\mathcal{G} = \pm L(\chi)$ for some real parameter χ.
- $\mathcal{G} \in SO^\uparrow(1,1)$ iff $\mathcal{G} \in SO(1,1)$ and $\alpha > 0$ (i.e., the transformation matrix $\mathcal{G}$ does not change the direction of time). Obviously, $\mathcal{G} \in SO^\uparrow(1,1)$ iff $\mathcal{G} = L(\chi)$ for some $\chi \in \mathbb{R}$.

The Lie groups $O(1,1)$ and $SO(1,1)$ are not connected, since these groups contain reflections. However, the Lie group $SO^\uparrow(1,1)$ is connected, since the matrices $L(\chi)$ depend continuously on the parameter χ. More precisely, $SO^\uparrow(1,1)$ is the component of the unit element I of the Lie group $O(1,1)$.

The Lie algebra $so(1,1)$**.** For small parameters χ, we get

$$L(\chi) = I + \chi A + o(\chi), \quad \chi \to 0, \quad \text{where} \quad A := \begin{pmatrix} 0 & 1 \\ 1 & 0 \end{pmatrix}.$$

The matrices χA with the real parameter χ form the real 1-dimensional Lie algebra $so(1,1)$. The matrix A is a basis of $so(1,1)$, and we have the trivial Lie product $[A,A] := 0$. Conversely,

$$L(\chi) = \mathrm{e}^{\chi A} \qquad \text{for all} \quad \chi \in \mathbb{R}.$$

The group $P(1,1)$**.** Let $\mathcal{G} \in O(1,1)$, and let $a \in \mathbb{R}^2$, that is, $a := \begin{pmatrix} a^0 \\ a^1 \end{pmatrix}$. All the transformations $(ct, x) \mapsto (ct', x')$ of the type

$$\begin{pmatrix} ct' \\ x' \end{pmatrix} = \mathcal{G}\begin{pmatrix} ct \\ x \end{pmatrix} + a, \quad x, t \in \mathbb{R} \tag{18.14}$$

form a group of linear transformations from $\mathbb{R}^2$ to $\mathbb{R}^2$ (with respect to the composition of maps). This group is denoted by $P(1,1)$. We have $P(1,1) = O(1,1) \rtimes \mathbb{R}^2$ (semidirect product). If we write the transformation (18.14) as

$$\mathbf{x}' = \mathcal{G}\mathbf{x} + a,$$

then the composition with the map $\mathbf{y}' = \mathcal{G}'\mathbf{x}' + a'$ yields

$$\mathbf{y}' = \mathcal{G}'(\mathcal{G}\mathbf{x} + a) + a' = \mathcal{G}'\mathcal{G}\mathbf{x} + (\mathcal{G}'a + a').$$

Alternatively, the Poincaré transformation (18.14) can be written as

$$\begin{pmatrix} ct' \\ x \\ 1 \end{pmatrix} = \begin{pmatrix} \mathcal{G} & a \\ 0 & 1 \end{pmatrix} \begin{pmatrix} ct \\ x \\ 1 \end{pmatrix}.$$

The product formula

$$\begin{pmatrix} \mathcal{G}' & a' \\ 0 & 1 \end{pmatrix} \begin{pmatrix} \mathcal{G} & a \\ 0 & 1 \end{pmatrix} = \begin{pmatrix} \mathcal{G}'\mathcal{G} & \mathcal{G}'a + a' \\ 0 & 1 \end{pmatrix}$$

corresponds to the composition of maps of the type (18.14). Therefore, the group $P(1,1)$ is isomorphic to the group of all the matrices

$$\begin{pmatrix} \mathcal{G} & a \\ 0 & 1 \end{pmatrix} \quad \text{where} \quad \mathcal{G} \in O(1,1),\ a \in \mathbb{R}^2.$$

Setting $\mathcal{G} = L(\chi)$, linearization yields the matrices

$$\begin{pmatrix} L(\chi) & a \\ 0 & 1 \end{pmatrix} = I + \mathcal{A}(\chi, a) + o(\chi), \quad \chi \to 0, \quad \text{where} \quad \mathcal{A}(\chi, a) := \begin{pmatrix} A(\chi) & a \\ 0 & 0 \end{pmatrix}.$$

All the matrices $\mathcal{A}(\chi, a)$ form the real 3-dimensional Lie algebra $p(1,1)$ of the Lie group $P(1,1)$. This Lie algebra has the basis

$$A := \begin{pmatrix} 0 & 1 & 0 \\ 1 & 0 & 0 \\ 0 & 0 & 0 \end{pmatrix}, \quad B := \begin{pmatrix} 0 & 0 & 1 \\ 0 & 0 & 0 \\ 0 & 0 & 0 \end{pmatrix}, \quad C := \begin{pmatrix} 0 & 0 & 0 \\ 0 & 0 & 1 \\ 0 & 0 & 0 \end{pmatrix}$$

with the Lie products

$$[A, B] = C, \quad [B, C] = 0, \quad [C, A] = 0$$

where $[A, B] := AB - BA$, and so on.

18.2.2 The Lorentz Group $O(1, 3)$

Fix $c > 0$. Let $\mathcal{G}$ be a real (4×4)-matrix. We write $\mathcal{G} \in O(1,3)$ iff the linear transformation

$$\begin{pmatrix} ct' \\ x' \\ y' \\ z' \end{pmatrix} = \mathcal{G} \begin{pmatrix} ct \\ x \\ y \\ z \end{pmatrix} \tag{18.15}$$

satisfies the Minkowski condition

$$c^2 t'^2 - x'^2 - y'^2 - z'^2 = c^2 t^2 - x^2 - y^2 - z^2$$

for all $x, y, z, t \in \mathbb{R}$. Setting

$$\eta := (\eta_{\alpha\beta}) = \begin{pmatrix} 1 & 0 & 0 & 0 \\ 0 & -1 & 0 & 0 \\ 0 & 0 & -1 & 0 \\ 0 & 0 & 0 & -1 \end{pmatrix}, \tag{18.16}$$

we have $\mathcal{G} \in O(1,3)$ iff

$$\mathcal{G}^d \eta \mathcal{G} = \eta. \tag{18.17}$$

As in Sect. 18.17, it follows from (18.16) that $\det \mathcal{G} = \pm 1$. Hence the elements of $O(1,3)$ are invertible matrices. Therefore,

- $O(1,3)$ is a group called the Lorentz matrix group;
- the elements $\mathcal{G}$ of $O(1,3)$ with $\det \mathcal{G} = 1$ form a subgroup of $O(1,3)$ called the proper matrix Lorentz group, $SO(1,3)$.

The Lorentz boost matrix $L_1(\chi)$ from (18.6) is an element of $SO(1,3)$ for every real parameter χ. Furthermore, the matrices

$$R_- := \begin{pmatrix} 1 & 0 & 0 & 0 \\ 0 & -1 & 0 & 0 \\ 0 & 0 & -1 & 0 \\ 0 & 0 & 0 & -1 \end{pmatrix}, \qquad T_- := \begin{pmatrix} -1 & 0 & 0 & 0 \\ 0 & 1 & 0 & 0 \\ 0 & 0 & 1 & 0 \\ 0 & 0 & 0 & 1 \end{pmatrix} \tag{18.18}$$

are elements of $O(1,3)$. By (18.15), the matrices R_- and T_- correspond to the transformations

- $t' = t, x' = -x, y' = -y, z' = -z$ (spatial reflection), and
- $t' = -t, x' = x, y' = y, z' = z$ (time reflection), respectively.

Proposition 18.2 *The real (4×4)-matrix $\mathcal{G}$ is an element of the Lorentz matrix group $O(1,3)$ iff it can be represented as the following matrix product*

$$\mathcal{G} = MRL_1(\chi)S \tag{18.19}$$

for some real parameter χ. Here, the (4×4)-matrices R and S correspond to rotations of the spatial variables, and $M = I$ (identity matrix), $M = R_-$ (spatial reflection), or $M = T_-$ (time reflection).

All the matrices (18.19) with $M = I$ form a subgroup $SO^{\uparrow}(1,3)$ of the Lie group $O(1,3)$ which is called the proper orthochronous Lorentz matrix group. This is the component of the unit element in $O(1,3)$ and $SO(1,3)$. We have $\mathcal{G} \in SO^{\uparrow}(1,3)$ iff the transformation (18.15) does not change the direction of time. As we will show, this is equivalent to

$$\frac{\partial t'(t,x,y,z)}{\partial t} \geq 1.$$

The matrix $\mathcal{G}$ from (18.19) is an element of the proper Lorentz matrix group $SO(1,3)$ iff either $M = I$ or $M = -I$.

The proofs of all the statements will be given in Problem 18.4 on page 934.

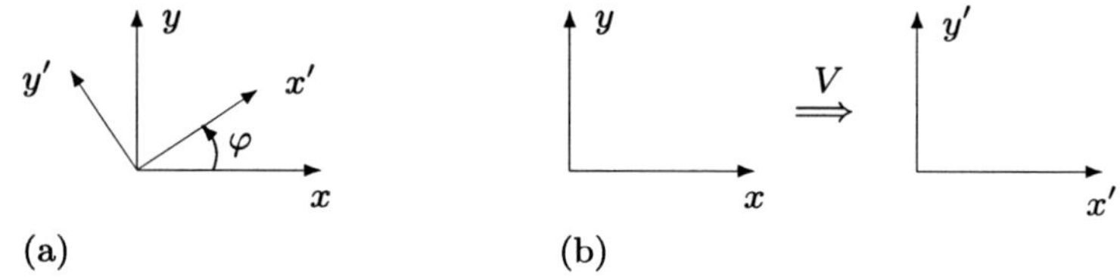

Fig. 18.4. Rotation and Lorentz boost

18.3 Infinitesimal Transformations

18.3.1 The Lie Algebra $o(1,3)$ of the Lorentz Group $O(1,3)$

The Lie group $SO^{\uparrow}(1,3)$ of proper orthochronous Lorentz transformations. The group $SO^{\uparrow}(1,3)$ contains the following three rotations:

- $x' = x\cos\varphi + y\sin\varphi,\ y' = -x\sin\varphi + y\cos\varphi,\ z' = z,\ t' = t$ (counter-clockwise rotation of Σ about the z-axis with the rotation angle φ; see Fig. 18.4(a));
- $y' = y\cos\varphi + z\sin\varphi,\ z' = -y\sin\varphi + z\cos\varphi,\ x' = x,\ t' = t$ (counter-clockwise rotation of Σ about the x-axis with the rotation angle φ);
- $z' = z\cos\varphi + x\sin\varphi,\ x' = -z\sin\varphi + x\cos\varphi,\ y' = y,\ t' = t$ (counter-clockwise rotation of Σ about the y-axis with the rotation angle φ).

Furthermore, the group $SO^{\uparrow}(1,3)$ contains the following three Lorentz boosts:

- $ct' = ct\cosh\chi - x\sinh\chi, x' = -ct\sinh\chi + x\cosh\chi,\ y' = y, z' = z$ (translation of Σ' in direction of the x-axis with the velocity V; see Fig. 18.4(b));
- $ct' = ct\cosh\chi - y\sinh\chi, y' = -ct\sinh\chi + y\cosh\chi, z' = z, x' = x$ (translation of Σ' in direction of the y-axis with the velocity V);
- $ct' = ct\cosh\chi - z\sinh\chi, z' = -ct\sinh\chi + z\cosh, x' = x, y' = y$ (translation of Σ' in direction of the z-axis with velocity V).

Here, we set $\tanh\chi = V/c$. Then, $\cosh\chi = \frac{1}{\sqrt{1-V^2/c^2}}$ and $\sinh\chi = \frac{V/c}{\sqrt{1-V^2/c^2}}$. This corresponds to the transformation

$$\begin{pmatrix} ct' \\ x' \\ y' \\ z' \end{pmatrix} = \mathcal{G} \begin{pmatrix} ct \\ x \\ y \\ z \end{pmatrix}$$

from the inertial system Σ to the inertial system Σ' with the following transformation matrices:

$$R_1(\varphi) := \begin{pmatrix} 1 & 0 & 0 & 0 \\ 0 & 1 & 0 & 0 \\ 0 & 0 & \cos\varphi & \sin\varphi \\ 0 & 0 & -\sin\varphi & \cos\varphi \end{pmatrix}, \quad R_2(\varphi) := \begin{pmatrix} 1 & 0 & 0 & 0 \\ 0 & \cos\varphi & 0 & -\sin\varphi \\ 0 & 0 & 1 & 0 \\ 0 & \sin\varphi & 0 & \cos\varphi \end{pmatrix},$$

$$R_3(\varphi) := \begin{pmatrix} 1 & 0 & 0 & 0 \\ 0 & \cos\varphi & \sin\varphi & 0 \\ 0 & -\sin\varphi & \cos\varphi & 0 \\ 0 & 0 & 0 & 1 \end{pmatrix},$$

and

$$L_1(\chi) := \begin{pmatrix} \cosh\chi & -\sinh\chi & 0 & 0 \\ -\sinh\chi & \cosh\chi & 0 & 0 \\ 0 & 0 & 1 & 0 \\ 0 & 0 & 0 & 1 \end{pmatrix}, \quad L_2(\chi) := \begin{pmatrix} \cosh\chi & 0 & -\sinh\chi & 0 \\ 0 & 1 & 0 & 0 \\ -\sinh\chi & 0 & \cosh\chi & 0 \\ 0 & 0 & 0 & 1 \end{pmatrix},$$

$$L_3(\chi) := \begin{pmatrix} \cosh\chi & 0 & 0 & -\sinh\chi \\ 0 & 1 & 0 & 0 \\ 0 & 0 & 1 & 0 \\ -\sinh\chi & 0 & 0 & \cosh\chi \end{pmatrix},$$

For example, Fig. 18.4(a) and (b) corresponds to $\mathcal{G} = R_3(\varphi)$ and $\mathcal{G} = L_1(\chi)$, respectively. The elements of the Lie group $SO^{\uparrow}(1,3)$ are finite products of the matrices $R_j(\varphi), L_k(\varphi)$ with $j,k = 1,2,3$.

Infinitesimal Lorentz transformations and the Lie algebra $o(1,3)$. We set $A_j := -R_j'(0)$ and $B_j := -L_j'(0)$, $j = 1,2,3$. Hence

$$A_1 := \begin{pmatrix} 0 & 0 & 0 & 0 \\ 0 & 0 & 0 & 0 \\ 0 & 0 & 0 & -1 \\ 0 & 0 & 1 & 0 \end{pmatrix}, \quad A_2 := \begin{pmatrix} 0 & 0 & 0 & 0 \\ 0 & 0 & 0 & 1 \\ 0 & 0 & 0 & 0 \\ 0 & -1 & 0 & 0 \end{pmatrix},$$

$$A_3 := \begin{pmatrix} 0 & 0 & 0 & 0 \\ 0 & 0 & -1 & 0 \\ 0 & 1 & 0 & 0 \\ 0 & 0 & 0 & 0 \end{pmatrix},$$

and

$$B_1 := \begin{pmatrix} 0 & 1 & 0 & 0 \\ 1 & 0 & 0 & 0 \\ 0 & 0 & 1 & 0 \\ 0 & 0 & 0 & 0 \end{pmatrix}, \quad B_2 := \begin{pmatrix} 0 & 0 & 1 & 0 \\ 0 & 0 & 0 & 0 \\ 1 & 0 & 0 & 0 \\ 0 & 0 & 0 & 0 \end{pmatrix},$$

$$B_3 := \begin{pmatrix} 0 & 0 & 0 & 1 \\ 0 & 0 & 0 & 0 \\ 0 & 0 & 0 & 0 \\ 1 & 0 & 0 & 0 \end{pmatrix}.$$

We have the following commutation relations

$$[A_1, A_2]_- = A_3, \quad [B_1, B_2]_- = -A_3, \tag{18.20}$$

and

$$[A_1, B_1]_- = 0, \quad [A_1, B_2]_- = B_3, \quad [A_1, B_3]_- = -B_2 \tag{18.21}$$

together with all the commutation relations which are obtained from this by using the cyclic permutations $1 \to 2 \to 3 \to 1$.

The six matrices $A_1, A_2, A_3, B_1, B_2, B_3$ are the basis of a real 6-dimensional Lie algebra which is the Lie algebra to the Lie group $SO^{\uparrow}(1,3)$.

Since the construction of the Lie algebra of a Lie group only uses the component of the unit element, the Lie algebra $so^{\uparrow}(1,3)$ of $SO^{\uparrow}(1,3)$ coincides with the Lie algebra $o(1,3)$ of the Lorentz group $O(1,3)$. This Lie algebra is called the real Lorentz Lie algebra. An elementary computation shows that we have the differential equations

$$R_j'(\varphi) = -A_j R_j(\varphi), \quad R_j(0) = I, \quad L_j'(\chi) = -B_j L_j(\chi), \quad L_j(0) = I, \quad j = 1,2,3.$$

These equations have the unique solution

$$R_j(\varphi) = \mathrm{e}^{-\varphi A_j}, \quad \varphi \in \mathbb{R} \qquad L_j(\chi) = \mathrm{e}^{-\chi B_j}, \quad \chi \in \mathbb{R}.$$

The real Lie algebra $sl(2,\mathbb{C})$. The real Lie algebra $sl(2,\mathbb{C})$ consists of all the complex traceless (2×2)-matrices equipped with the Lie product

$$[C,D]_- := CD - DC.$$

If $A \in sl(2,\mathbb{C})$, then we set

$$a := \tfrac{1}{2}(A - A^\dagger), \qquad b := \tfrac{1}{2}(A + A^\dagger).$$

That is, the traceless matrix a (resp. b) is skew-adjoint (resp. self-adjoint). The Pauli matrices

$$\sigma^1 = \begin{pmatrix} 0 & 1 \\ 1 & 0 \end{pmatrix}, \quad \sigma^2 = \begin{pmatrix} 0 & -\mathrm{i} \\ \mathrm{i} & 0 \end{pmatrix} \quad \sigma^3 = \begin{pmatrix} 1 & 0 \\ 0 & -1 \end{pmatrix} \tag{18.22}$$

form a basis of the real linear space of complex traceless self-adjoint (2×2)-matrices. Furthermore, the matrices $\mathrm{i}\sigma^1, \mathrm{i}\sigma^2, \mathrm{i}\sigma^3$ form a basis of the real linear space of complex skew-adjoint (2×2)-matrices. Thus, the six matrices

$$a_j := -\tfrac{1}{2}\mathrm{i}\sigma^j, \qquad b_j := \tfrac{1}{2}\sigma^j, \qquad j = 1,2,3$$

form a basis of the real 6-dimensional linear space $sl(2,\mathbb{C})$. In addition, the matrices $a_1, a_2, a_3, b_1, b_2, b_3$ satisfy the same commutation relations as the matrices $A_1, A_2, A_3, B_1, B_2, B_3$. For example,

$$[a_1, a_2]_- = a_3, \quad [b_1, b_2]_- = -b_3,$$

and $[a_1, b_1]_- = 0$, $[a_1, b_2]_- = b_3$, $[a_1, b_3]_- = -b_2$. This tells us that the map $a_j \mapsto A_j, b_j \mapsto B_j$, $j = 1,2,3$, yields the real Lie algebra isomorphism

$$so^{\uparrow}(1,3) \simeq sl(2,\mathbb{C}). \tag{18.23}$$

The complex Lie algebra $sl_{\mathbb{C}}(2,\mathbb{C})$. The complex Lie algebra $sl_{\mathbb{C}}(2,\mathbb{C})$ consists of all the complex traceless (2×2)-matrices. Every matrix $a \in sl(2,C)$ is a real linear combination of the matrices $\mathrm{i}\sigma^j, \sigma^j$, $j = 1,2,3$. Consequently, the matrix can be represented as a complex linear combination of the matrices $\mathrm{i}\sigma^j$, $j = 1,2,3$. Hence the matrices a_1, a_2, a_3 form a basis of $sl_{\mathbb{C}}(2,\mathbb{C})$ with the Lie products,

$$[a_1, a_2]_- = a_3, \quad [a_2, a_3]_- = a_1, \quad [a_3, a_1]_- = a_1. \tag{18.24}$$

18.3.2 The Lie Algebra $p(1,3)$ of the Poincaré Group $P(1,3)$

The Poincaré group $P(1,3)$ is the basic symmetry group of relativistic quantum field theory.

Folklore

The semidirect product $P(1,3) = O(1,3) \rtimes \mathbb{R}^4$. We want to study the Poincaré transformation $x \mapsto x'$ from $\mathbb{R}^4$ to $\mathbb{R}^4$ given by

$$\boxed{x' = \mathcal{G}x + a, \qquad x \in \mathbb{R}^4.} \tag{18.25}$$

This means that we first perform a Lorentz transformation $x \mapsto \mathcal{G}x$ and then a space-time translation $y \mapsto y + a$. Here,

$$x = \begin{pmatrix} x^0 \\ x^1 \\ x^2 \\ x^3 \end{pmatrix}, \quad x' = \begin{pmatrix} x^{0'} \\ x^{1'} \\ x^{2'} \\ x^{3'} \end{pmatrix}, \quad a = \begin{pmatrix} a^0 \\ a^1 \\ a^2 \\ a^3 \end{pmatrix},$$

and the real (4×4)-matrix $\mathcal{G}$ is an element of the Lorentz matrix group $O(1,3)$. The composition of two Poincaré transformations yields

$$x'' = \mathcal{G}'x' + a' = \mathcal{G}'(\mathcal{G}x + a) + a' = \mathcal{G}'\mathcal{G}x + (\mathcal{G}'a + a').$$

If we denote the Poincaré transformation (18.25) by the symbol $(\mathcal{G}, a)$, then we get the product

$$(\mathcal{G}', a')(\mathcal{G}, a) = (\mathcal{G}'\mathcal{G}, \mathcal{G}'a + a'). \tag{18.26}$$

For example, $(I,0)(\mathcal{G},a) = (I,0)$. Thus, $(I,0)$ is the unit element, and

$$(\mathcal{G}^{-1}, -a)(\mathcal{G}, a) = (I, 0)$$

tells us that $(\mathcal{G},a)^{-1} = (\mathcal{G}^{-1}, -a)$. This corresponds to the so-called semidirect product $O(1,3) \rtimes \mathbb{R}^4$.

Proposition 18.3 *The Poincaré group $P(1,3)$ is isomorphic to a subgroup of the matrix group $GL(5,\mathbb{R})$ which consists of all the matrices*

$$\begin{pmatrix} \mathcal{G} & a \\ 0 & 1 \end{pmatrix}, \qquad \mathcal{G} \in O(1,3), \; a \in \mathbb{R}^4. \tag{18.27}$$

Proof. We write the Poincaré transformation (18.25) as

$$\begin{pmatrix} x' \\ 1 \end{pmatrix} = \begin{pmatrix} \mathcal{G} & a \\ 0 & 1 \end{pmatrix} \begin{pmatrix} x \\ 1 \end{pmatrix}. \tag{18.28}$$

The matrix product yields

$$\begin{pmatrix} \mathcal{G}' & a' \\ 0 & 1 \end{pmatrix} \begin{pmatrix} \mathcal{G} & a \\ 0 & 1 \end{pmatrix} = \begin{pmatrix} \mathcal{G}'\mathcal{G} & \mathcal{G}'a + a' \\ 0 & 1 \end{pmatrix}$$

which corresponds to (18.26). □

The Poincaré group is the symmetry group of relativistic quantum field theory.

Important subgroups of the Poincaré group. Let us consider special cases of the matrices (18.27).

(i) Lorentz group: The group $\mathcal{O}(1,3)$ consists of all the real (5×5)-matrices (18.27) with $\mathcal{G} \in O(1,3)$ and $a = 0$. We have the group isomorphism $\mathcal{O}(1,3) \simeq O(1,3)$.

(ii) The proper Lorentz group: The group $\mathcal{SO}(1,3)$ consists of all the matrices (18.27) with $\mathcal{G} \in SO(1,3)$ and $a = 0$, that is, $\mathcal{G} \in O(1,3)$ and $\det \mathcal{G} = 1$. We have the group isomorphism $\mathcal{SO}(1,3) \simeq SO(1,3)$.

(iii) The proper orthochronous Lorentz group: The group $\mathcal{SO}^{\uparrow}(1,3)$ consists of all the matrices (18.27) with $\mathcal{G} \in SO^{\uparrow}(1,3)$ and $a = 0$, that is, $\mathcal{SO}^{\uparrow}(1,3)$ is the component of the unit element in $\mathcal{SO}(1,3)$. We have the group isomorphism $\mathcal{SO}^{\uparrow}(1,3) \simeq SO^{\uparrow}(1,3)$.

(iv) The group of space-time translations: The group $\mathcal{T}(\mathbb{R}^4)$ consists of all the matrices (18.27) with $\mathcal{G} = I$ and

$$a = \overset{0}{a} e_0 + \overset{1}{a} e_1 + \overset{2}{a} e_2 + \overset{3}{a} e_3, \qquad \overset{0}{a}, \overset{1}{a}, \overset{2}{a}, \overset{3}{a} \in \mathbb{R}.$$

Here, we set

$$e_0 := \begin{pmatrix} 1 \\ 0 \\ 0 \\ 0 \end{pmatrix}, \quad e_1 := \begin{pmatrix} 0 \\ 1 \\ 0 \\ 0 \end{pmatrix}, \quad e_2 := \begin{pmatrix} 0 \\ 0 \\ 1 \\ 0 \end{pmatrix}, \quad e_3 := \begin{pmatrix} 0 \\ 0 \\ 0 \\ 1 \end{pmatrix}.$$

We have the additive group isomorphism $\mathcal{T}(\mathbb{R}^4) \simeq \mathbb{R}^4$.

(v) The group of space translations: The group $\mathcal{T}(\mathbb{R}^3)$ consists of all the elements of $\mathcal{T}(\mathbb{R}^4)$ with $\overset{0}{a} = 0$. We have the additive group isomorphism $\mathcal{T}(\mathbb{R}^3) \simeq \mathbb{R}^3$.

(vi) The group of time translations: The group $\mathcal{T}(\mathbb{R})$ consists of all the elements of $\mathcal{T}(\mathbb{R}^4)$ with $\overset{1}{a} = \overset{2}{a} = \overset{3}{a} = 0$. We have the additive group isomorphism $\mathcal{T}(\mathbb{R}) \simeq \mathbb{R}$.

(vii) The space reflection group (or the parity group): The group $\mathcal{R}(\mathbb{R}^3)$ consists of all the matrices (18.27) with $\mathcal{G} = I, R_-$ and $a = 0$.[5] We have the multiplicative group isomorphism $\mathcal{R}(\mathbb{R}^3) \simeq \{1, -1\}$.

(viii) The time reflection group: The group $\mathcal{R}(\mathbb{R})$ consists of all the matrices (18.27) with $\mathcal{G} = I, T_-$ and $a = 0$. We have the multiplicative group isomorphism $\mathcal{R}(\mathbb{R}) \simeq \{1, -1\}$.

The Poincaré algebra $p(1,3)$. Each transformation of the Poincaré group $P(1,3)$ can be represented as a product of finitely many matrices of the form

$$\begin{pmatrix} \mathrm{e}^{\varphi_j A_j} & 0 \\ 0 & 1 \end{pmatrix}, \quad \begin{pmatrix} \mathrm{e}^{\chi_j B_j} & 0 \\ 0 & 1 \end{pmatrix}, \quad \begin{pmatrix} I & \overset{\mu}{a} e_\mu \\ 0 & 1 \end{pmatrix}, \quad \begin{pmatrix} R_- & 0 \\ 0 & 1 \end{pmatrix}, \quad \begin{pmatrix} T_- & 0 \\ 0 & 1 \end{pmatrix}$$

with real parameters φ_j, χ_j, $\overset{\mu}{a}$, and the indices $j = 1,2,3$, $\mu = 0,1,2,3$. Linearization at the unit element yields the following matrices:

$$\mathcal{A}_j := \begin{pmatrix} A_j & 0 \\ 0 & 0 \end{pmatrix}, \quad \mathcal{B}_j := \begin{pmatrix} B_j & 0 \\ 0 & 0 \end{pmatrix}, \quad \mathcal{C}_\mu := \begin{pmatrix} 0 & e_\mu \\ 0 & 0 \end{pmatrix}, \quad j = 1,2,3, \quad \mu = 0,1,2,3.$$

[5] The definition of the reflection matrices R_- and T_- can be found in (18.18) on page 917.

Explicitly, the elements $\mathcal{A}$ of the Poincaré group $P(1,3)$ in a sufficiently small neighborhood of the unit element $\mathcal{I} = \begin{pmatrix} I & 0 \\ 0 & 1 \end{pmatrix}$ can be represented as

$$\mathcal{A} = \mathcal{I} + \varphi^j \mathcal{A}_j + \chi^j \mathcal{B}_j + a^\mu \mathcal{C}_\mu + \ldots$$

Here, the dots stand for terms of higher order with respect to the real parameters $\varphi^j, \chi^j a^\mu$. Let us study the commutation relations (Lie products). First of all we obtain the same commutation relations as for the real Lorentz algebra $o(1,3)$, namely,

$$\begin{array}{lll} [\mathcal{A}_1, \mathcal{A}_2]_- = \mathcal{A}_3, & [\mathcal{B}_1, \mathcal{B}_2]_- = -\mathcal{A}_3, & \\ [\mathcal{A}_1, \mathcal{B}_1]_- = 0, & [\mathcal{A}_1, \mathcal{B}_2]_- = \mathcal{B}_3, & [\mathcal{A}_1, \mathcal{B}_3]_- = -\mathcal{B}_2. \end{array}$$

Furthermore, $[\mathcal{C}_\mu, \mathcal{C}_\nu]_- = 0$ for all $\mu, \nu = 1,2,3,4$, and

$$\begin{array}{lll} [\mathcal{A}_1, \mathcal{C}_1]_- = [\mathcal{A}_1, \mathcal{C}_0]_- = 0, & [\mathcal{A}_1, \mathcal{C}_2]_- = \mathcal{C}_3, & [\mathcal{A}_1, \mathcal{C}_3]_- = -\mathcal{C}_2, \\ [\mathcal{B}_1, \mathcal{C}_2]_- = [\mathcal{B}_1, \mathcal{C}_3]_- = 0, & [\mathcal{B}_1, \mathcal{C}_1]_- = \mathcal{C}_0, & [\mathcal{B}_1, \mathcal{C}_0]_- = \mathcal{C}_1. \end{array}$$

The remaining commutation rules are obtained by using the cyclic permutations $1 \mapsto 2 \mapsto 3 \mapsto 1$. For example,

$$\begin{aligned} [\mathcal{B}_1, \mathcal{C}_1]_- &= \begin{pmatrix} B_1 & 0 \\ 0 & 0 \end{pmatrix} \begin{pmatrix} 0 & e_1 \\ 0 & 0 \end{pmatrix} - \begin{pmatrix} 0 & e_1 \\ 0 & 0 \end{pmatrix} \begin{pmatrix} B_1 & 0 \\ 0 & 0 \end{pmatrix} \\ &= \begin{pmatrix} 0 & B_1 e_1 \\ 0 & 0 \end{pmatrix} = \begin{pmatrix} 0 & e_0 \\ 0 & 0 \end{pmatrix} = \mathcal{C}_0. \end{aligned}$$

Summarizing, we get the following:

The real linear hull of the real (5×5)*-matrices*

$$\mathcal{A}_1, \mathcal{A}_2, \mathcal{A}_3; \quad \mathcal{B}_1, \mathcal{B}_2, \mathcal{B}_3; \quad \mathcal{C}_0, \mathcal{C}_1, \mathcal{C}_2, \mathcal{C}_3 \tag{18.29}$$

forms a real 10-dimensional Lie algebra called the Poincaré algebra $p(1,3)$.

The matrices (18.29) correspond to infinitesimal rotations, infinitesimal Lorentz boosts, and infinitesimal space-time translations. In terms of semidirect products, $p(1,3) = o(1,3) \rtimes \mathbb{R}^4$.

The complexified Poincaré algebra $\mathbb{C} \otimes p(1,3)$**.** This 10-dimensional complex Lie algebra consists of all the complex linear combinations of the symbols

$$1 \otimes \mathcal{A}_j, \quad 1 \otimes \mathcal{B}_j, \quad 1 \otimes \mathcal{C}_\mu, \quad j = 1,2,3, \quad \mu = 0,1,2,3.$$

18.4 The Minkowski Space M_4

18.4.1 Pseudo-Orthonormal Systems and Inertial Systems

One passes from the Euclidean space E_3 to the Minkowski space M_4 by replacing orthonormal basis systems by pseudo-orthonormal basis systems.

Folklore

Indefinite Hilbert space and its signature. By definition, the crucial Minkowski space M_4 is a real 4-dimensional linear space equipped with the structure of an indefinite Hilbert space of signature $(1,3)$. Explicitly, this means that there exists a bilinear map

$$g : M_4 \times M_4 \to \mathbb{R}$$

with the property that the space M_4 has a basis $\mathbf{e}_0, \mathbf{e}_1, \mathbf{e}_2, \mathbf{e}_3$ such that

$$g(\mathbf{e}_\alpha, \mathbf{e}_\beta) = \eta_{\alpha\beta}, \qquad \alpha, \beta = 0, 1, 2, 3. \tag{18.30}$$

Equivalently,

$$g = \eta_{\alpha\beta}\, dx^\alpha \otimes dx^\beta.$$

Here, we use the Minkowski matrix

$$\eta = (\eta_{\alpha\beta}) = \begin{pmatrix} 1 & 0 & 0 & 0 \\ 0 & -1 & 0 & 0 \\ 0 & 0 & -1 & 0 \\ 0 & 0 & 0 & -1 \end{pmatrix}, \qquad \alpha, \beta = 0, 1, 2, 3. \tag{18.31}$$

By definition, a basis $\mathbf{e}_0, \mathbf{e}_1, \mathbf{e}_2, \mathbf{e}_3$ of M_4 is called a pseudo-orthonormal basis iff the condition (18.30) is satisfied. The symbol $\eta_{\mu\nu}$ is called the Minkowski symbol.[6] We also introduce the inverse matrix

$$\eta^{-1} = (\eta^{\alpha\beta}),$$

that is, $\eta^{\alpha\beta} = \eta_{\alpha\beta}$ for all $\alpha, \beta = 0, 1, 2, 3$. The distinction between upper and lower indices is crucial for constructing relativistic invariants.

In order to equip the Minkowski space with an orientation, we fix a specific pseudo-orthonormal basis as a positive basis. The physical meaning will be discussed below. The map g is called the metric tensor of the Minkowski space.

The Lorentz group $L(M_4)$. By definition, the set $L(M_4)$ consists of all the linear operators $G : M_4 \to M_4$ with

$$\boxed{g(Gx, Gy) = g(x, y) \qquad \text{for all} \quad x, y \in M_4.}$$

This is a group called the Lorentz group. We have the Lie group isomorphism

$$L(M_4) \simeq O(1,3) \tag{18.32}$$

between the Lorentz group $L(M_4)$ (acting on the Minkowski space M_4) and the Lorentz matrix group $O(1,3)$. The Lie group isomorphism (18.32), $G \mapsto \mathcal{G}$, assigns to any Lorentz transformation $G : M_4 \to M_4$ the matrix $\mathcal{G}$ with respect to a fixed pseudo-orthonormal basis; that is,

$$G\mathbf{e}_\alpha = \mathcal{G}^\beta_\alpha \mathbf{e}_\beta, \quad \alpha = 0, 1, 2, 3.$$

The Poincaré group $P(M_4)$ consists of all the transformations $x \mapsto y$ given by

$$y = Gx + x_0, \qquad \text{for all} \quad x \in M_4$$

---

[6] Our sign convention of η corresponds to the convention used by Einstein. This convention is motivated by the fact that the proper time of a massive particle is positive. Most physicists use this convention in elementary particle physics. From the mathematical point of view, one would prefer the replacement $\eta \Rightarrow -\eta$. Then, g is an extension of the Euclidean metric tensor.

where $G \in L(M_4)$ and $x_0 \in M_4$. This means that

$$P(M_4) := L(M_4) \rtimes T(M_4).$$

Thus, the Poincaré group is the semidirect product of the Lorentz group $L(M_4)$ with the translation group $T(M_4)$ given by the translations $x \mapsto x + x_0$ of the Minkowski space M_4.

> *The Poincaré group $P(M_4)$ generalizes the Euclidean group of motions on the Euclidean space E_3 to the Minkowski space M_4.*

The Poincaré group $P(M_4)$ is isomorphic to the group $O(1,3) \rtimes \mathbb{R}^4$ given by the matrix transformations $x \mapsto y$ from $\mathbb{R}^4$ to $\mathbb{R}^4$ where

$$y = \mathcal{G}x + x_0$$

and $\mathcal{G} \in O(1,3), x_0 \in \mathbb{R}^4$. The Poincaré group is a 10-dimensional Lie group.

In particular, the transformation matrix $\mathcal{G}$ from (18.5) corresponding to the special Lorentz transformation is an element of the group $SO^{\uparrow}(1,3)$. The matrix $\mathcal{G}$ is an element of the group $O(1,3)$ iff it is the finite product of the following matrices:

- special Lorentz transformation (18.5),
- rotation of the spatial coordinates x, y, z,
- space reflection $x' = -x, y' = -y, z' = -z$,
- time reflection $t' = -t$.

Moreover, the matrix $\mathcal{G}$ is an element of the group $SO^{\uparrow}(1,3)$ iff the space reflections and the time reflections drop out.

Inertial system and pseudo-orthonormal basis. In terms of mathematics, precisely every pseudo-orthonormal basis $\mathbf{e}_0, \mathbf{e}_1, \mathbf{e}_2, \mathbf{e}_3$ of the Minkowski space M_4 represents an inertial system. In this case, every point $x \in M_4$ can be uniquely represented as

$$x = x^{\alpha}\mathbf{e}_{\alpha} = x^0\mathbf{e}_0 + x^1\mathbf{e}_1 + x^2\mathbf{e}_2 + x^3\mathbf{e}_3.$$

Here, $x^1 = x, x^2 = y, x^3 = z$ are (right-handed or left-handed) Cartesian coordinates of an inertial system Σ. Moreover, we have

$$x^0 = ct$$

where t is the time observed in Σ, and c is the velocity of light in a vacuum measured in Σ.

Transformation of inertial systems in Einstein's theory of special relativity. Let $\mathbf{e}_0, \mathbf{e}_1, \mathbf{e}_2, \mathbf{e}_3$ and $\mathbf{e}_{0'}, \mathbf{e}_{1'}, \mathbf{e}_{2'}, \mathbf{e}_{3'}$ be two pseudo-orthonormal basis systems of the Minkowski space M_4. The points x of M_4 are called events. They are characterized by space and time coordinates. The transformation of the coordinates is given by

$$x = x^{\alpha}\mathbf{e}_{\alpha} = x^{\alpha'}\mathbf{e}_{\alpha'}.$$

Using transformation matrices, we get

$$\begin{pmatrix} x^{0'} \\ x^{1'} \\ x^{2'} \\ x^{3'} \end{pmatrix} = \mathcal{G} \begin{pmatrix} x^0 \\ x^1 \\ x^2 \\ x^3 \end{pmatrix}, \qquad \begin{pmatrix} \mathbf{e}_{0'} \\ \mathbf{e}_{1'} \\ \mathbf{e}_{2'} \\ \mathbf{e}_{3'} \end{pmatrix} = \mathcal{H} \begin{pmatrix} \mathbf{e}_0 \\ \mathbf{e}_1 \\ \mathbf{e}_2 \\ \mathbf{e}_3 \end{pmatrix}, \qquad \mathcal{H} = (\mathcal{G}^d)^{-1}. \qquad (18.33)$$

Proposition 18.4 *If the system $\mathbf{e}_0, \mathbf{e}_1, \mathbf{e}_2, \mathbf{e}_3$ is a pseudo-orthonormal basis of M_4, then $\mathbf{e}_{0'}, \mathbf{e}_{1'}, \mathbf{e}_{2'}, \mathbf{e}_{3'}$ is a pseudo-orthonormal basis of M_4 iff $\mathcal{G} \in O(1,3)$.*

Proof. For a general basis, we have $\mathbf{e}_{\alpha'} = \mathcal{H}^{\alpha}_{\alpha'}\mathbf{e}_{\alpha}$. Hence

$$x^{\alpha'} = \mathcal{G}^{\alpha'}_{\alpha}\, x^{\alpha}$$

together with $\mathcal{H}^{\alpha}_{\alpha'}\mathcal{G}^{\alpha'}_{\beta} = \delta^{\alpha}_{\beta}$. This implies

$$g(\mathbf{e}_{\alpha'}, \mathbf{e}_{\beta'}) = \mathcal{H}^{\alpha}_{\alpha'}\mathcal{H}^{\beta}_{\beta'}\; g(\mathbf{e}_{\alpha}, \mathbf{e}_{\beta}).$$

The basis $\mathbf{e}_{0'}, \mathbf{e}_{1'}, \mathbf{e}_{2'}, \mathbf{e}_{3'}$ is a pseudo-orthonormal system iff $\eta_{\alpha'\beta'} = \mathcal{H}^{\alpha}_{\alpha'}\mathcal{H}^{\beta}_{\beta'}\eta_{\alpha\beta}$. This is equivalent to the matrix equation

$$\eta = \mathcal{H}\eta\mathcal{H}^{d}.$$

In turn, this is equivalent to $\mathcal{G}^{d}\eta\mathcal{G} = \eta$, that is, $\mathcal{G} \in O(1,3)$. This is equivalent to $\mathcal{H} \in O(1,3)$. □

18.4.2 Orientation

Let us fix the pseudo-orthonormal basis $\mathbf{e}_0, \mathbf{e}_1, \mathbf{e}_2, \mathbf{e}_3$ of M_4. By definition, the pseudo-orthonormal basis $\mathbf{e}_{0'}, \mathbf{e}_{1'}, \mathbf{e}_{2'}, \mathbf{e}_{3'}$ is called positively oriented iff $\mathcal{G} \in O(1,3)$ and $\det \mathcal{G} = 1$, that is,

$$\mathcal{G} \in SO(1,3).$$

This is equivalent to $\mathcal{H} \in SO(1,3)$. In terms of physics, the reference frame $\mathbf{e}_0, \mathbf{e}_1, \mathbf{e}_2, \mathbf{e}_3$ corresponds to a right-handed Cartesian (x,y,z)-system, and the time is positively oriented in the usual sense, that is, time t flows from past to future. The transformation

$$x \mapsto -x, \qquad y \mapsto -y, \qquad z \mapsto -z, \qquad t \mapsto -t$$

combines the space reflection with a time reflection. This transformation does not change the orientation of the coordinate system on $\mathbb{M}^4$, since the determinant of the transformation matrix $\mathrm{diag}\,(-1,-1,-1,-1)$ is equal to one.

We say that the inertial system is strictly positively oriented iff the Cartesian (x,y,z)-coordinate system is positively oriented (i.e., it is right handed) and the time is positively oriented. The pseudo-orthonormal basis $\mathbf{e}_{0'}, \mathbf{e}_{1'}, \mathbf{e}_{2'}, \mathbf{e}_{3'}$ is strictly positively oriented iff $\mathcal{G} \in SO^{\uparrow}(1,3)$.

18.4.3 Proper Time and the Twin Paradox

Proper time. Following Minkowski, Einstein's principle of special relativity reads as follows in terms of mathematics:

In the theory of special relativity, physical quantities are geometric invariants of the Minkowski space M_4.

Since the symmetry group of the Minkowski space is the Poincaré group, physical quantities are invariants of the Poincaré group. As an example, let us consider the proper time. Let

$$x = x(\sigma), \qquad \sigma_0 \le \sigma \le \sigma_1 \tag{18.34}$$

be a smooth curve on the Minkowski space M_4 such that $g(\dot{x}(\sigma), \dot{x}(\sigma)) > 0$ for all $\sigma \in [\sigma_0, \sigma_1]$. The arc length of the curve is defined by

$$\boxed{s := \int_{\sigma_0}^{\sigma_1} \sqrt{g(\dot{x}(\sigma), \dot{x}(\sigma))}\, d\sigma.} \tag{18.35}$$

In addition, the so-called proper time τ is defined by

$$\tau := \frac{s}{c}.$$

To discuss the physical meaning of τ, choose a fixed inertial system described by the pseudo-orthonormal basis $\mathbf{e}_0, \mathbf{e}_1, \mathbf{e}_2, \mathbf{e}_3$. Then $x(\sigma) = x^\alpha(\sigma)\mathbf{e}_\alpha$. Hence

$$g(\dot{x}(\sigma), \dot{x}(\sigma)) = \dot{x}^\alpha(\sigma)\dot{x}^\beta(\sigma) \cdot g(\mathbf{e}_\alpha, \mathbf{e}_\beta) = \dot{x}^\alpha(\sigma)\dot{x}^\beta(\sigma) \cdot \eta_{\alpha\beta}.$$

This yields

$$\tau = \frac{1}{c}\int_{\sigma_0}^{\sigma_1} \sqrt{\dot{x}^\alpha(\sigma)\dot{x}^\beta(\sigma)\, \eta_{\alpha\beta}}\, d\sigma.$$

If we choose the parameter $\sigma = t$, then

$$\tau = \int_{t_0}^{t_1} \sqrt{1 - \frac{\dot{x}^2(t) + \dot{y}(t)^2 + \dot{z}(t)^2}{c^2}}\, dt. \tag{18.36}$$

Physicists postulate that the proper time τ is the time shown by an atomic clock which moves along the trajectory $x = x(t), y = y(t), z = z(t)$, $t_0 \le t \le t_1$, in an inertial system.

The twin paradox. Suppose that at time $t = t_0$ and at the origin O of an inertial system the twins T_1 and T_2 are born. Shortly thereafter, T_2 is brought to a spaceship and begins a journey through the universe while T_1 remains at O. After several years, T_2 returns to T_1 at time $t = t_1$. Both are surprised that T_2 is much younger than T_1. This fact can be easily explained if one assumes that the biological clock of the twins shows the proper time. For T_2, it follows from (18.36) that $\tau_2 < t_1 - t_0$, whereas for T_1 we get $\tau_1 = t_1 - t_0$, since $\dot{x}(t) = \dot{y}(t) = \dot{z}(t) \equiv 0$.

18.4.4 The Free Relativistic Particle and the Energy-Mass Equivalence

We want to motivate the fundamental Einstein relation

$$\boxed{E^2 = m_0^2c^4 + c^2\mathbf{p}^2} \tag{18.37}$$

for the energy E of a free particle with positive rest mass m_0 moving in an inertial system with momentum vector $\mathbf{p}$. If the particle rests, then $\mathbf{p} = 0$. Hence

$$\boxed{E = m_0c^2.} \tag{18.38}$$

This is Einstein's famous formula from 1905, stating the *equivalence between mass and energy.* The energy production of all stars is based upon (18.38). For example, during the synthesis of helium from hydrogen in the sun, mass is transformed into energy. Formula (18.38) is a triumph for the mental ability of man; in a frightening way it also allows the self-destruction of mankind by atomic bombs.

In order to obtain (18.37), we use the variational principle of critical arc length

$$-\, m_0c \int_{\sigma_0}^{\sigma_1} \sqrt{g(\dot{x}(\sigma), \dot{x}(\sigma))}\, d\sigma = \text{critical!}, \quad x(\sigma_0) = a, \quad x(\sigma_1) = b. \tag{18.39}$$

We are looking for a smooth curve $x = x(\sigma)$, $\sigma_0 \le \sigma \le \sigma_1$, with fixed end points a and b. In addition, we demand that $g(\dot{x}(\sigma), \dot{x}(\sigma)) > 0$ along the curve. This means that the speed of the particle is less than the speed of light. We add the factor $-m_0 c$ in order to get below the right approximation of the classical principle of critical action. In a fixed inertial system, the variational principle reads as

$$\int_{t_0}^{t_1} L(\dot{\mathbf{x}}(t))\, dt = \text{critical!}, \quad \mathbf{x}(t_0) = \mathbf{a}, \ \mathbf{x}(t_1) = \mathbf{b} \tag{18.40}$$

with the Lagrangian

$$L(\dot{\mathbf{x}}) := -m_0 c^2 \sqrt{1 - \frac{\dot{\mathbf{x}}^2}{c^2}}.$$

This follows as in (18.36) by setting $\mathbf{x}(t) = x(t)\mathbf{e}_1 + y(t)\mathbf{e}_2 + z(t)\mathbf{e}_3$.

The Lagrangian approach. Every solution of (18.40) satisfies the Euler–Lagrange equation of motion

$$\frac{d}{dt} L_{\dot{\mathbf{x}}} = L_{\mathbf{x}}.$$

Noting that $L_{\mathbf{x}} = 0$, we get

$$\frac{d}{dt}\big(m(t)\dot{\mathbf{x}}(t)\big) = 0, \qquad t_0 \le t \le t_1$$

with the mass function

$$m(t) = \frac{m_0}{\sqrt{1 - \dot{\mathbf{x}}(t)^2/c^2}}. \tag{18.41}$$

This tells us that the mass $m(t)$ of the free particle depends on its velocity, in contrast to classical mechanics.

The Hamiltonian approach. Introducing the momentum vector

$$\mathbf{p} := L_{\dot{\mathbf{x}}} = m\dot{\mathbf{x}}$$

and the Hamiltonian

$$H = \dot{\mathbf{x}}\mathbf{p} - L = mc^2 = \sqrt{m_0^2 c^4 + c^2 \mathbf{p}^2},$$

we get the Hamiltonian equations of motion

$$\dot{\mathbf{p}}(t) = -H_{\dot{\mathbf{x}}}(\mathbf{p}(t)) \equiv 0, \qquad \dot{\mathbf{x}}(t) = H_{\mathbf{p}}(\mathbf{p}(t)),$$

and the conservation law

$$H(\mathbf{p}(t)) = \text{const}, \qquad t_0 \le t \le t_1.$$

Note that, in the setting of the Hamiltonian approach to mechanics, the Hamiltonian function H always represents the energy of the particle. This way, we get the claim (18.37).

The classical approximation. If the velocity of the particle is small compared with the velocity of light, that is, $|\dot{\mathbf{x}}(t)/c| \ll 1$, then[7]

$$L = -m_0 c^2 \left(1 - \frac{\dot{\mathbf{x}}(t)^2}{2c^2} + \ldots\right) = -m_0 c^2 + \tfrac{1}{2} m_0 \dot{\mathbf{x}}(t)^2 + \ldots$$

This is approximately the classical Lagrangian of a free particle, up to an additive constant, $-m_0 c^2$ (negative rest energy), which does not influence the solutions of the variational problem (18.40).

[7] Note that $\sqrt{1-\alpha} = 1 - \frac{1}{2}\alpha + O(\alpha^2)$, $\alpha \to 0$.

18.4.5 The Photon

If the rest mass m_0 of the particle is equal to zero, then the energy equation (18.37) passes over to the limit $E^2 = c^2\mathbf{p}^2$. Hence

$$E = c|\mathbf{p}|.$$

In 1905, Einstein postulated that light in a vacuum consists of particles with the energy

$$E = h\nu$$

where ν is the frequency of the light, and h is the Planck quantum of action. Noting that $c = \lambda\nu$, we get

$$|\mathbf{p}| = \frac{h}{\lambda}$$

for the momentum of the light particle where λ is the wave length of the light. Einstein used this light particle hypothesis in order to derive the Planck radiation law via statistical physics and to explain the photoelectric effect (see Sect. 1.1 of Vol. I). In 1921 Einstein was awarded the Nobel prize in physics for his services to Theoretical Physics, and especially for his discovery of the law of the photoelectric effect. The name photon was coined by the chemical physicist Lewis in 1926.

18.5 The Minkowski Manifold $\mathbb{M}^4$

In Euclidean geometry, one uses both the Euclidean space E_3 (real 3-dimensional Hilbert space) and the Euclidean manifold $\mathbb{E}_3$ (real 3-dimensional manifold). The two notions are closely related to each other. Similarly, in Einstein's theory of special relativity we use the Minkowski space M_4 (real 4-dimensional indefinite Hilbert space) and the Minkowski manifold $\mathbb{M}_4$ (real 4-dimensional manifold).

Let us discuss how to introduce the Minkowski manifold $\mathbb{M}_4$. We are given the Minkowski space M_4. Choose a fixed pseudo-orthonormal basis $\mathbf{e}_0^+, \mathbf{e}_1^+, \mathbf{e}_2^+, \mathbf{e}_3^+$. For all elements x of M_4, we have the unique decomposition

$$x = x^\alpha \mathbf{e}_\alpha^+.$$

By definition, the points P of $\mathbb{M}_4$ are precisely the tuples

$$(x^0, x^1, x^2, x^3)$$

of the real numbers x^0, x^1, x^2, x^3.

- A curve on the Minkowski manifold $\mathbb{M}_4$ passing through the point P_0 is given by the smooth map
$$\sigma \mapsto (x^0(\sigma), x^1(\sigma), x^2(\sigma), x^3(\sigma))$$
defined on the open interval $]-\sigma_1, \sigma_1[$ with $\sigma_1 > 0$, and $x^\alpha(0) = x_0^\alpha$, $\alpha = 0, 1, 2, 3$.
- The tangent space $T_{P_0}\mathbb{M}_4$ of the Minkowski manifold $\mathbb{M}_4$ at the point P_0 consists of all the possible velocity vectors
$$v = (\dot{x}^0(0), \dot{x}^1(0), \dot{x}^2(0), \dot{x}^3(0)).$$
The tangent space $T_{P_0}\mathbb{M}_4$ is a real 4-dimensional space which can be equipped with the structure of an indefinite Hilbert space of signature $(1, 3)$. Explicitly, we define $g_{P_0}(v, w) := v^\alpha w^\beta\ \eta_{\alpha\beta}$ for all $v, w \in T_{P_0}\mathbb{M}_4$.

Physical interpretation. Every point P of the Minkowski manifold represents an event characterized by the local coordinates (x^0, x^1, x^2, x^3) measured in the fixed inertial system $\mathbf{e}_0^+, \mathbf{e}_1^+, \mathbf{e}_2^+, \mathbf{e}_3^+$. Now fix the point P_0. Choose a pseudo-orthonormal basis $\mathbf{e}_0, \mathbf{e}_1, \mathbf{e}_2, \mathbf{e}_3$ of the tangent space $T_{P_0}\mathbb{M}_4$. This represents an inertial system which assigns to the event P_0 the local coordinate $(0, 0, 0, 0)$.

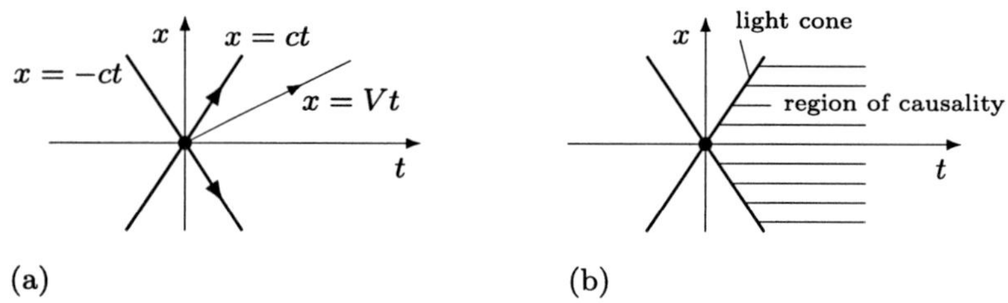

Fig. 18.5. Causality

18.5.1 Causality and the Maximal Signal Velocity

Choose a strictly positively oriented inertial system with the coordinates x, y, z, t. Fix the real number V such that $-c < V < c$. The equation

$$x = Vt, \quad y = z = 0$$

describes the motion of a particle with velocity V (Fig. 18.5). Moreover, the equation

$$x = ct$$

or $x = -ct$ describes the motion of a photon with the speed of light c in a vacuum. In 1905, Einstein postulated the following for inertial systems:

The speed of massive particles is always less than the speed of light. Physical signals can travel at most with the velocity of light c.

In science fiction, there appear frequently so-called tachyons. Such particles travel faster than light. So far there is no experimental evidence for the existence of such particles.[8]

Classification of pairs of events. For two points x and y of the Minkowski space M_4, we define the following:

- The pair x, y of events is time-like iff $g(x-y, x-y) > 0$.
- The pair x, y of events is space-like iff $g(x-y, x-y) < 0$.
- The pair x, y of events is light-like iff $g(x-y, x-y) = 0$.

In terms of a fixed inertial system, we get

$$g(x-y, x-y) = (x^0 - y^0)^2 - (x^1 - y^1)^2 - (x^2 - y^2)^2 - (x^3 - y^3)^2$$

where x^0, x^1, x^2, x^3 and y^0, y^1, y^2, y^3 are the coordinates of x and y, respectively.

Examples. (i) Time-like events: If

$$x^0 = ct_1, \;\; y^0 = ct_2; \quad x^1 = y^1, x^2 = y^2, x^3 = y^3, \qquad t_1 \neq t_2,$$

then the events x and y take place at the same position, but at different moments. Thus,

$$g(x-y, x-y) = c^2(t_1 - t_2)^2 > 0,$$

that is, the pair x, y of events is time-like.

(ii) Space-like events: If

[8] For example, see R. Hughes and G. Stephenson, Against tachyonic neutrinos, Phys. Lett. **B244** (1990), 95–100.

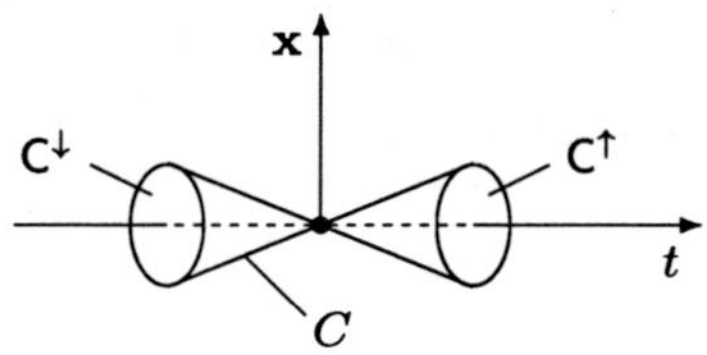

Fig. 18.6. Light cone

$$x^0 = y^0 = ct, \qquad x^1 \neq y^1,$$

then the events x and y take place at the same time, but at different positions. Thus,

$$g(x-y, x-y) = -(x^1-y^1)^2 - (x^2-y^2)^2 - (x^3-y^3)^2 < 0,$$

that is, the pair x, y of events is space-like.

(iii) Light-like events: Fix the velocity V. If

$$x^0 = x^1 = x^2 = x^3 = 0, \qquad y^0 = ct, \;\; y^1 = Vt, \;\; y^2 = y^3 = 0, \quad t > 0,$$

then the events x and y correspond to the beginning and the end of the motion of a particle with the velocity V. Then

$$g(x-y, x-y) = (c^2 - V^2)t^2.$$

Thus, the following hold:

- The pair x, y of events is light-like iff $|V| = c$ (light ray).
- The pair x, y is time-like iff $|V| < c$ (massive particle).
- The pair x, y of events is space-like iff $|V| > c$ (tachyon).

Let us introduce the following sets (Fig. 18.6):

- $C := \{(ct, x^1, x^2, x^3) \in \mathbb{R}^4 : c^2t^2 - (x^1)^2 - (x^2)^2 - (x^3)^2 - (x^3)^2 = 0\}$ (light cone);
- $C^\uparrow := \{(ct, x^1, x^2, x^3) \in C : \; t \geq 0\}$ (forward light cone);
- $\mathsf{C}^\uparrow := \{(ct, x^1, x^2, x^3) \in \mathbb{R}^4 : c^2t^2 - (x^1)^2 - (x^2)^2 - (x^3)^2 \geq 0, \;\; t \geq 0\}$ (causality cone).

Obviously, the forward light cone $C^\uparrow$ is the boundary of the causality cone $\mathsf{C}^\uparrow$.

Einstein's principle of causality. We postulate the following:

If the pair x, y of events is space-like, then there exists no causal connection between x and y.

Intuitively, this is motivated by the fact that the maximal velocity of physical signals is the velocity c of light.

18.5.2 Hodge Duality

Let us choose a strictly positively oriented inertial system with the right-handed Cartesian (x, y, z)-coordinate system. Let $x^0 := ct$. With respect to the metric tensorial family $\eta_{\alpha\beta}$, the Hodge star operator (see page 470) reads as follows:

(i) volume form: $\upsilon := dx^0 \wedge dx \wedge dy \wedge dz$, and $\;*1 = \upsilon, \quad *\upsilon = -1.$
(ii) 1-forms: $*dx^0 = dx \wedge dy \wedge dz$, and

$$*dx = dx^0 \wedge dy \wedge dz, \quad *dy = dx^0 \wedge dz \wedge dx, \quad *dz = dx^0 \wedge dx \wedge dy.$$

(iii) 2-forms: $*(dx \wedge dy) = dx^0 \wedge dz$, and

$$*(dy \wedge dz) = dx^0 \wedge dx, \quad *(dz \wedge dx) = dx^0 \wedge dy,$$

as well as $*(dx^0 \wedge dz) = -dx \wedge dy$, and

$$*\,(dx^0 \wedge dx) = -dy \wedge dz, \quad *(dx^0 \wedge dy) = -dz \wedge dx. \tag{18.42}$$

(iv) 3-forms: $*(dx \wedge dy \wedge dz) = dx^0$, $*(dx^0 \wedge dy \wedge dz) = dx$, and

$$*(dx^0 \wedge dz \wedge dx) = dy, \qquad *(dx^0 \wedge dx \wedge dy) = dz.$$

These formulas are invariant under the cyclic permutation $dx \mapsto dy \mapsto dz \mapsto dx$ together with $dx^0 \mapsto dx^0$.

For all the differential k-forms ω, $k = 0, 1, 2, 3$, on the Minkowski manifold $\mathbb{M}_4$, we define the Hodge codifferential

$$\boxed{d^*\omega := (-1)^k *^{-1} d * \omega.}$$

Then, we have the following relations:

- $**\omega = -(-1)^k\omega$ (duality relation), and hence
- $*^{-1}\omega = -(-1)^k * \omega$,
- $d^*\omega = (-1)^k * d * \omega$,
- $dd\omega = 0$ (Poincaré's cohomology rule),
- $d^*d^*\omega = 0$ Hodge's homology rule).

Note that $d^*d^*\omega = (-1)^{k-1}(-1)^k *^{-1} d(**^{-1})d*\omega = -*^{-1}dd(*\omega) = 0$. Applications to the Maxwell equations will be considered in Sect. 19.3 on page 960.

18.5.3 Arbitrary Local Coordinates

Let $\mathcal{O}$ and $\mathcal{O}'$ be nonempty open subsets of $\mathbb{R}^4$, and let

$$f : \mathcal{O} \to \mathcal{O}'$$

be a diffeomorphism (i.e., f is bijective and both f and f^{-1} are smooth). Let $x^0 := ct, x^1 := x, x^2 := y, x^3 := z$ be the coordinates of a strictly positively oriented inertial system. The transformation

$$x^{\alpha'} = f^{\alpha'}(x^0, x^1, x^2, x^3), \qquad \alpha' = 0', 1', 2', 3'$$

introduces new local coordinates on the Minkowski manifold $\mathbb{M}^4$. This transformation is orientation-preserving iff the functional determinant $\det(f')$ is positive on $\mathcal{O}$, that is,

$$\frac{\partial(x^{0'}, x^{1'}, x^{2'}, x^{3'})}{\partial(x^0, x^1, x^2, x^3)}(x) > 0 \qquad \text{for all} \quad x \in \mathcal{O}.$$

We say that the transformation preserves the time orientation iff

$$\frac{\partial x^{0'}}{\partial x^0}(x) > 0 \qquad \text{for all} \quad x \in \mathcal{O}.$$

According to Theorem 8.2 on page 458, we construct the metric tensorial family $g_{\alpha\beta}$ by setting

$$\boxed{g_{\alpha'\beta'} := \frac{\partial x^\alpha}{\partial x^{\alpha'}} \frac{\partial x^\beta}{\partial x^{\beta'}} \eta_{\alpha\beta}, \qquad \alpha', \beta' = 0', 1', 2', 3'.}$$

On inertial systems, we have $g_{\alpha\beta} = \eta_{\alpha\beta}$ for all indices. Using the metric tensorial family $g_{\alpha\beta}$, we can apply the index principle from Sect. 8.3 on page 443 in order to formulate relativistically invariant equations.

Problems

18.1 *The Lorentz boost.* Prove the boost relation (18.4) on page 909.
Solution: Using

$$x' = \alpha x + \beta t, \quad t' = \gamma x + \delta t$$

and noting that $x = ct$ implies $x' = ct'$, we get the key relation

$$c\alpha + \beta = c^2\gamma + c\delta.$$

By (H3), $x = Vt$ implies $x' = 0$ for all t'. Hence

$$\alpha V + \beta = 0.$$

The inverse transformation reads as

$$x = \frac{\delta x' - \beta t'}{\alpha\delta - \beta\gamma}, \qquad t = \frac{\alpha t' - \gamma x'}{\alpha\delta - \beta\gamma}.$$

By (H5), $x = 0, t \in \mathbb{R}$ yields $x' = -Vt', t' \in \mathbb{R}$. Hence $\delta V + \beta = 0$. This implies

$$\beta = -\alpha V, \quad \delta = \alpha, \quad \gamma = -\alpha V/c^2, \quad \alpha\delta - \beta\gamma = \alpha^2(1 - V^2/c^2).$$

By (H4), if $x > 0, t = 0$, then $x' > 0$. Thus, $\alpha > 0$. Summarizing,

$$x' = \alpha(x - Vt), \qquad x = \frac{\alpha(x' + Vt')}{\alpha\delta - \beta\gamma}.$$

Finally, in order to determine the free parameter α, we use (H5), saying that $\alpha\delta - \beta\gamma = 1$. Hence $\alpha = 1/\sqrt{1 - V^2/c^2}$.

18.2 *The structure of the Lorentz matrices.* Note that every real (4×4)-matrix can be written as

$$\mathcal{G} = \begin{pmatrix} \alpha & a^d \\ b & G \end{pmatrix}$$

where $\alpha \in \mathbb{R}, a, b \in \mathbb{R}^3$, and G is a real (3×3)-matrix. Show that $\mathcal{G} \in O(1,3)$ iff

$$\alpha^2 = 1 + b^d b, \quad GG^d = I + aa^d, \qquad G^d b = \alpha a. \tag{18.43}$$

Solution: Recall that $\mathcal{G} \in O(1,3)$ iff $\mathcal{G}^d \eta \mathcal{G} = \eta$. This is equivalent to

$$\begin{pmatrix} \alpha & b^d \\ a & G^d \end{pmatrix} \begin{pmatrix} 1 & 0 \\ 0 & -I \end{pmatrix} \begin{pmatrix} \alpha & a^d \\ b & G \end{pmatrix} = \begin{pmatrix} \alpha^2 - b^d b & \alpha a^d - b^d G \\ \alpha a - G^d b & aa^d - GG^d \end{pmatrix} = \begin{pmatrix} 1 & 0 \\ 0 & -I \end{pmatrix}.$$

18.3 *Characterization of Euclidean rotations.* Fix $e_0 := (1,0,0,0,0)^d$. Suppose that $\mathcal{G} \in O(1,3)$. Show that

$$\mathcal{G} = \begin{pmatrix} 1 & 0 \\ 0 & G \end{pmatrix}, \quad G \in SO(3) \quad \text{iff} \quad \mathcal{G}e_0 = e_0.$$

Solution: If $\mathcal{G}e_0 = e_0$, then $b = 0$. By (18.43), $a = 0$ and $G^d G = I$.

18.4 *Proof of Proposition 18.2.* Solution: Let $\mathcal{G} \in O(1,3)$. We already know that $\mathcal{G}^d \eta \mathcal{G} = \eta$ implies $\det \mathcal{G} = \pm 1$. By (18.43), $\alpha^2 \geq 1$.

(I) We consider first the special case where $\alpha \geq 1$, and $\det \mathcal{G} = 1$. By (18.43),

$$\alpha^2 - |b|^2 = 1.$$

Thus, there exists a real parameter χ such that $\alpha = \cosh \chi$ and $|b| = \sinh \chi$. Consequently, there exists a rotation R_0 such that

$$R_0^{-1} b = \begin{pmatrix} -\sinh \chi \\ 0 \\ 0 \end{pmatrix}.$$

Define

$$S := L_1(-\chi) \begin{pmatrix} 1 & 0 \\ 0 & R_0^{-1} \end{pmatrix} \mathcal{G} = \begin{pmatrix} \cosh \chi & \sinh \chi & 0 & 0 \\ \sinh \chi & \cosh \chi & 0 & 0 \\ 0 & 0 & 1 & 0 \\ 0 & 0 & 0 & 1 \end{pmatrix} \begin{pmatrix} \cosh \chi & a^d \\ R_0^{-1} b & R_0^{-1} G \end{pmatrix}.$$

Hence $Se_0 = e_0$. By Problem 18.3, there exists a matrix $S_0 \in SO(3)$ such that

$$S = \begin{pmatrix} 1 & 0 \\ 0 & S_0 \end{pmatrix}.$$

Noting that $L_1(-\chi)^{-1} = L_1(\chi)$, we get

$$\mathcal{G} = \begin{pmatrix} 1 & 0 \\ 0 & R_0 \end{pmatrix} L_1(\chi) \begin{pmatrix} 1 & 0 \\ 0 & S_0 \end{pmatrix}.$$

(II) The remaining cases (where $\alpha \geq 1, \det \mathcal{G} = -1$, or $\alpha \leq -1, \det \mathcal{G} = \pm 1$) can be reduced to (I) by passing from the matrix $\mathcal{G}$ to the matrix $M^{-1}\mathcal{G}$. Here, M is a reflection.

19. The Relativistic Invariance of the Maxwell Equations

> Maxwell's work is the most profound and the most fruitful work that physics has experienced since the time of Newton in the 17th century.
>
> Albert Einstein, 1931
>
> (On the occasion of the 100th anniversary of Maxwell's birth)

Consider a fixed inertial system. We have to distinguish between

- the Maxwell equations in a vacuum, and
- the Maxwell equations in materials.

The basic quantities are

- the electric vector field $\mathbf{E}$,
- the magnetic vector field $\mathbf{B}$,
- the electric charge density ϱ and
- the electric current density vector $\mathbf{J}$.

Resting electric charges produce electric fields, whereas moving electric charges (electric currents) produce magnetic fields. The Maxwell equations in a vacuum describe the interaction between electric charges, electric currents and both electric and magnetic fields. It is important that there exist electric and magnetic waves in the absence of electric charges and electric currents. In particular, visible light and radio waves are electromagnetic waves based on the interaction between electric and magnetic fields in a vacuum. The quantization of such electromagnetic waves leads to the concept of the photon predicted by Einstein in 1905. Einstein was awarded the 1921 Nobel prize in physics for his services to Theoretical Physics, and especially for his discovery of the law of the photoelectric effect based on the light quantum (photon) hypothesis. The interaction between electrons, positrons, and photons is studied in quantum electrodynamics (see Vol. II). Quantum electrodynamics was independently created by Feynman, Schwinger, and Tomonaga in about 1946 (Nobel prize in physics in 1965). If a particle with electric charge Q and the rest mass m_0 moves with the velocity vector $\mathbf{v}$, then the electromagnetic field $\mathbf{E}, \mathbf{B}$ exerts the Lorentz force

$$\boxed{\mathbf{F}(P) = Q\mathbf{E}(P) + Q\mathbf{v}(P) \times \mathbf{B}(P)} \tag{19.1}$$

on the particle at the point P. The relativistic equation of motion of the particle (e.g., in a particle accelerator) reads as

$$\frac{d}{dt}\left(\frac{m_0\dot{\mathbf{x}}(t)}{\sqrt{1-\dot{\mathbf{x}}^2(t)/c^2}}\right) = Q\mathbf{E}(\mathbf{x}(t),t) + Q\dot{\mathbf{x}}(t) \times \mathbf{B}(\mathbf{x}(t),t), \qquad t \in \mathbb{R} \tag{19.2}$$

where c is the velocity of light in a vacuum. We will use SI physical units (Système International) (see the Appendix of Vol. I). In this system, the basic physical quantities have the following physical dimensions:

E. Zeidler, *Quantum Field Theory III: Gauge Theory*,
DOI 10.1007/978-3-642-22421-8_20,

- electric current strength: $[\mathsf{J}]=$ A (ampere),
- electric charge: $[Q] =$ C= As (coulomb),
- electric current density: $[\mathbf{J}] = \mathrm{A/m^2 \cdot s}$,
- electric field strength: $[\mathbf{E}]:=$ N/C (force per electric charge),
- magnetic field strength: $[\mathbf{B}] = [\mathbf{F}]/[Q][\mathbf{v}]=$ N/Am (force per magnetic charge).[1]

In a thin metallic wire with the small radius r_0 of the cross section, we have the electric current strength

$$\mathsf{J} = |\mathbf{J}| \cdot \pi r_0^2 \qquad \text{(electric charge per time).} \tag{19.3}$$

Dividing the force $\mathbf{F}$ by volume, we get the force density

$$\mathbf{f} = \varrho \mathbf{E} + \mathbf{J} \times \mathbf{B} \tag{19.4}$$

with $\mathbf{J} := \varrho \mathbf{v}$. This chapter is organized as follows:

- In Sect. 19.2, we will start with the Maxwell equations in a vacuum. In terms of modern mathematics, the Maxwell equations in a vacuum are based on the Hodge theory for differential forms. This formulation is independent of the system of reference.
- In technology, materials consist of molecules which generate additional electric and magnetic fields. In Sect 19.8, the Maxwell equations for materials will be based on Weyl duality. Here, the electric polarization $\mathbf{P}_{\mathrm{el}}$ and the magnetic polarization (magnetization) $\mathbf{M}$ of the materials are described by Weyl fields (antisymmetric contravariant tensor densities of weight one).

In this connection, we will use the following physical quantities:

- electric dipole moment (electric charge times length): Cm=Asm;
- magnetic dipole moment (magnetic charge times length): Am· m = Am^2;
- polarization (electric dipole moment per volume): C/m^2=As/m^2;
- magnetization (magnetic dipole moment per volume): A/m.

For the polarization $\mathbf{P}_{\mathrm{el}}$ and magnetization $\mathbf{M}$ of materials in external electric and magnetic fields, $\mathbf{E}$ and $\mathbf{B}$, we have the following constitutive laws:

$$\mathbf{P}_{\mathrm{el}} = \mathbf{P}_{\mathrm{el}}(\mathbf{E}, \mathbf{B}), \qquad \mathbf{M} = \mathbf{M}(\mathbf{E}, \mathbf{B}).$$

Details will be studied in Sect. 19.8.

In quantum electrodynamics, the polarization of the vacuum plays a fundamental role.

This will be thoroughly studied in Vol. IV (renormalization of the electric charge of the electron and the anomalous magnetic moment of the electron).

19.1 Historical Background

Oerstedt's and Faraday's brilliant discoveries have opened up a new world of scientific research, whose enchanted gardens will fill us with admiration; these rich fields can only be conquered under the art of measurement.

Carl Friedrich Gauss, 1836[2]

The legendary father of the sciences, the great Greek philosopher Thales of Miletus (600 B.C.) knew that there are black stones[3] which attract iron. Such stones

[1] Further material can be found in Sect. 19.9 on page 983.

[2] Gauss remarked this during a popular lecture given in 1836. The author took this quotation from the beautiful book *Electrodynamics from Ampère to Einstein* by Olivier Darrigol, Oxford University Press, New York, 2000.

[3] Magnetite is a very common iron oxide mineral, Fe_3O_4.

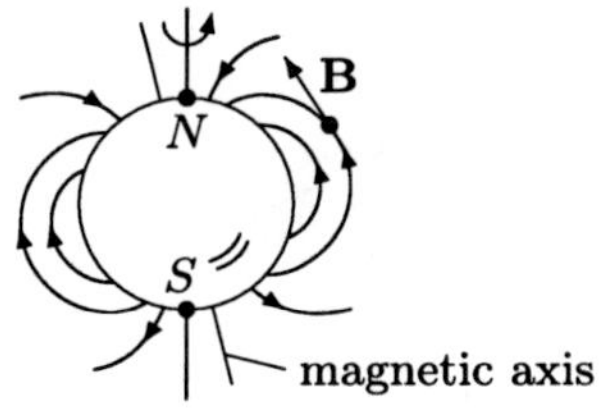

Fig. 19.1. The magnetic field of earth

were called magnets. The Greek word $\mu\alpha\acute{\gamma}\nu\eta\tau\eta\zeta$ (magnetez) for 'magnet' is related to the Greek word $\mu\alpha\gamma\varepsilon\iota\alpha$ (mageia) for 'magic'. In ancient Greece it was also known that amber possesses electrostatic properties. The Greek word for amber is $\acute{\eta}\lambda\varepsilon\kappa\tau\rho$ (electro). After 1200 A.D. magnetic compasses were used for navigation. Four hundred years later, in 1600 William Gilbert (1544–1603) wrote the book *On the Magnet, Magnetic Bodies, and the Great Magnet of the Earth* (see Fig. 19.1). Nowadays it is assumed that birds use the magnetic field of the earth for orienting themselves during bird migration.

19.1.1 The Coulomb Force and the Gauss Law

> In a hydrogen atom, the electrostatic attracting Coulomb force between the proton and the electron is 10^{39} larger than the corresponding gravitational force.
>
> Folklore

In 1790, Coulomb formulated the Coulomb law. Consider two particles with electric charges Q_0 and Q located at the points P_0 and P, respectively. Then the particle at the point P_0 exerts the electrostatic force $\mathbf{F}$ on the particle at the point P (Fig. 19.2). Explicitly,

$$\boxed{\mathbf{F}(\mathbf{x}) = \frac{QQ_0}{4\pi\varepsilon_0|\mathbf{x}-\mathbf{x}_0|^2}\cdot\frac{\mathbf{x}-\mathbf{x}_0}{|\mathbf{x}-\mathbf{x}_0|}, \qquad \mathbf{x}\neq\mathbf{x}_0.} \tag{19.5}$$

This force is repelling (resp. attracting) if $QQ_0 > 0$ (same sign) (resp. $QQ_0 < 0$ (different sign)). The Coulomb force resembles Newton's gravitational force, but the

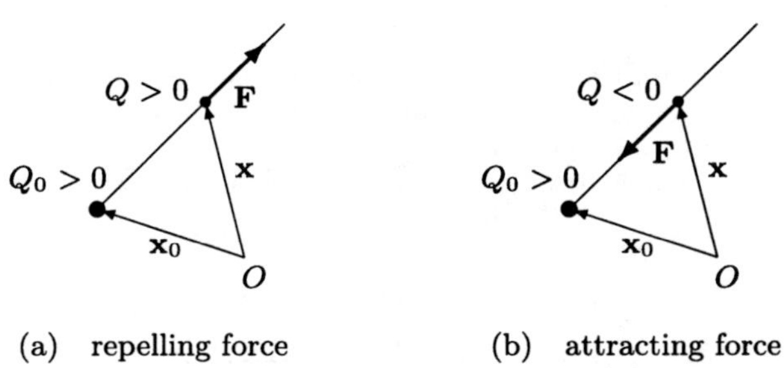

Fig. 19.2. The Coulomb force

(a) charge $Q_0 > 0$ at the point P_0 (b) charge $Q_0 < 0$ at the point P_0

Fig. 19.3. Electrostatic Coulomb field

two forces possess different sign. The Newton force between two positive masses is attracting, whereas the Coulomb force between two positive charges is repelling.

The electric field E of the Coulomb monopole. Define

$$\mathbf{E}(\mathbf{x}) := \frac{\mathbf{F}(\mathbf{x})}{Q} \qquad \text{for all} \quad \mathbf{x} \neq \mathbf{x}_0. \tag{19.6}$$

This is called the electric field at the point P (Fig. 19.3). Explicitly,

$$\mathbf{E}(\mathbf{x}) := \frac{Q_0}{4\pi\varepsilon_0|\mathbf{x}-\mathbf{x}_0|^2} \cdot \frac{\mathbf{x}-\mathbf{x}_0}{|\mathbf{x}-\mathbf{x}_0|}. \tag{19.7}$$

In other words, the charged particle at the point P_0 generates the electric field $\mathbf{E}(P)$ at the point P.[4] The electric field $\mathbf{E}(\mathbf{x})$ can be determined by measuring the force

$$\boxed{\mathbf{F}(\mathbf{x}) = Q\mathbf{E}(\mathbf{x})}$$

which is exerted on a particle located at the point P with electric charge Q. By Newton's law 'actio = reactio' (in Latin), the charged particle located at the point P exerts the force

$$\mathbf{F}_{\text{reactio}}(\mathbf{x}_0) = \frac{QQ_0}{4\pi\varepsilon_0|\mathbf{x}_0-\mathbf{x}|^2} \cdot \frac{\mathbf{x}_0-\mathbf{x}}{|\mathbf{x}-\mathbf{x}_0|}, \qquad \mathbf{x} \neq \mathbf{x}_0$$

on the particle at the point P_0. The two forces $\mathbf{F}(\mathbf{x})$ and $\mathbf{F}_{\text{reactio}}(\mathbf{x}_0)$ have the same strength, but opposite directions. By definition, if the force between the two positive charges Q and $Q_0 = Q$ is equal to 1 N (newton), then the charge Q is equal to 1 C (coulomb) in the SI system. Explicitly,

$$\varepsilon_0 = 8.854 \text{ C}^2/\text{Nm}^2.$$

Faraday's electric field lines. By definition, an electric field line is a curve such that the electrical field vector is tangential to the curve at all the curve points (see Fig. 19.3 and Fig. 19.8 on page 946).

The electric Gauss law. Let $\mathbb{B}^3_R(P_0)$ be a ball of radius R centered at the point P_0 with the boundary $\mathbb{S}^2_R(P_0)$. Then

$$\boxed{\varepsilon_0 \int_{\mathbb{S}^2_R(P_0)} \mathbf{E}(\mathbf{x})\mathbf{n}\, dS = Q_0} \tag{19.8}$$

[4] It is convenient to write $\mathbf{E}(\mathbf{x})$ instead of $\mathbf{E}(P)$ by using the position vector $\mathbf{x}$ pointing from the origin O to the point P.

where $\mathbf{n}$ denotes the outer normal unit vector of the sphere $\mathbb{S}^2_R(P_0)$. The integral relation (19.8) for the electric Coulomb field $\mathbf{E}$ from (19.7) is called the Gauss law.
Proof. Noting that $\mathbf{n} = (\mathbf{x} - \mathbf{x}_0)/|\mathbf{x} - \mathbf{x}_0|$, we get

$$\varepsilon_0 \int_{\mathbb{S}^2_R(P_0)} \mathbf{En}\, dS = \frac{Q_0}{4\pi R^2} \int_{\mathbb{S}^2_R(P_0)} dS = Q_0.$$

□

An elementary computation shows that the Coulomb field $\mathbf{E}$ from (19.7) satisfies the two Maxwell equations

$$\operatorname{div} \mathbf{E} = 0 \quad \text{and} \quad \mathbf{curl}\, \mathbf{E} = 0 \qquad \text{on} \quad \mathbb{E}^3 \setminus \{P_0\}. \tag{19.9}$$

The proof will be given in Problem 19.1

In 1839, Gauss wrote his paper *General theorems on attracting and repelling forces which are proportional to the inverse of the square of the distance* (in German). In this paper, Gauss introduced the term *potential.*[5] This paper founded potential theory as a special field in mathematics.

The Dirac delta distribution. The Gauss law (19.8) describes the singularity of the electric Coulomb field at the point P_0 in global terms. In order to get the corresponding local formulation, we have to use the theory of generalized functions introduced by Laurent Schwartz in about 1945. Using the Dirac delta distribution δ_{P_0} with support at the point P_0, the Coulomb field $\mathbf{E}$ satisfies the following equations in the sense of generalized functions:

$$\varepsilon_0 \operatorname{div} \mathbf{E} = Q_0 \cdot \delta_{P_0} \quad \text{and} \quad \mathbf{curl}\, \mathbf{E} = 0 \qquad \text{on} \ \mathbb{E}^3.$$

Mnemonically, physicists write

$$\varepsilon_0 \operatorname{div} \mathbf{E}(\mathbf{x}) = Q_0 \cdot \delta(\mathbf{x} - \mathbf{x}_0) \quad \text{and} \quad \mathbf{curl}\, \mathbf{E}(\mathbf{x}) = 0 \qquad \text{on} \ \mathbb{E}^3.$$

The proof will be given in Problem 19.2.

The topological invariance of the Gauss integral. Let $\mathcal{M}$ be a compact, 3-dimensional, positively oriented submanifold of $\mathbb{E}^3$ with coherently oriented boundary $\partial\mathcal{M}$; let P_0 be an interior point of $\mathcal{M}$. Then

$$\boxed{\varepsilon_0 \int_{\partial\mathcal{M}} \mathbf{En}\, dS = Q_0.} \tag{19.10}$$

This means that the domain of integration of the Gauss integral (19.10) can be deformed without changing the value of the integral.
Proof. Set $\mathcal{N} := \mathcal{M} \setminus \mathbb{B}^3_R(P_0)$. It follows from $\operatorname{div} \mathbf{E} = 0$ on $\mathbb{E}^3 \setminus \{0\}$ and the Gauss–Ostrogradski integral theorem that

$$0 = \int_{\mathcal{N}} \operatorname{div} \mathbf{E}\, d^3x = \int_{\partial\mathcal{N}} \mathbf{En}\, dS = \int_{\partial\mathcal{M}} \mathbf{En}\, dS - \int_{\mathbb{S}^2_R(P_0)} \mathbf{En}\, dS.$$

Now use (19.8). Here, we assume that the radius R is sufficiently small. □

The electrostatic potential and the voltage. We want to translate the idea of potential energy in Newton's mechanics to electrostatics. Define

$$U(\mathbf{x}) := \frac{Q_0}{|\mathbf{x} - \mathbf{x}_0|}, \qquad \mathbf{x} \neq \mathbf{x}_0.$$

[5] See C. F. Gauß, Collected Works with commentaries, Vol. 5, pp. 195–242, Göttingen 1863/1933.

Then the Coulomb field can be written as

$$\mathbf{E} = -\,\mathbf{grad}\, U.$$

For the Coulomb force, we get $\mathbf{F} = -\,\mathbf{grad}(QU)$. Thus, the Coulomb force has the potential energy QU. The function U is called the electrostatic potential of the electric Coulomb field. The path-independent integral

$$W = \int_{P_0}^{P} \mathbf{F}(\mathbf{x})d\mathbf{x}$$

equals the work done by the electric Coulomb field $\mathbf{E}$ by moving a particle of electric charge Q from the point P_0 to the point P. We get

$$W = -\int_{P_0}^{P} \mathbf{grad}(QU)\; d\mathbf{x} = Q \cdot (U(P_0) - U(P)).$$

The quantity

$$V := U(P) - U(P_0) \tag{19.11}$$

is called the voltage of the oriented segment (P_0P). An electron with the charge $-e$ has the potential energy $W = eV$. In the SI system, if the voltage V is equal to one volt, then the electron has the potential energy

$$W = \text{eV} = 1.602 \cdot 10^{-19}\text{J} \tag{19.12}$$

where J=Nm (joule). The physical unit eV is called electron volt. This unit is used in accelerator physics. The rest mass of an electron is approximately equal to

$$0.5 \text{ MeV} = 500\,000 \text{ eV}.$$

The rest mass of a proton is approximately equal to 1 GeV $= 10^9$ eV. The vector bosons W^+, W^-, Z have a rest energy of approximately 100 GeV. The LHC (Large Hadron Collider) at CERN (Geneva, Switzerland) will reach particle energies of

$$14\,\text{TeV} = 14\,000 \text{ GeV}.$$

The symbols MeV (resp. GeV, TeV) stand for mega (resp. giga, tera) electron volt.[6]

Gauge transformation. Let U_0 be a real number. The transformation

$$U_+(P) = U(P) + U_0$$

from U to U_+ is called a gauge transformation of the electrostatic potential U. By (19.11),

$$V = U(P) - U(P_0) = U_+(P) - U_+(P_0).$$

This means that the voltage is gauge invariant, whereas the electrostatic potential is not gauge invariant. Physical quantities have to be gauge invariant. Consequently, the voltage has a physical meaning, whereas the electrostatic potential has no immediate physical meaning. In the SI system, voltage is measured in volt (see (19.22)).

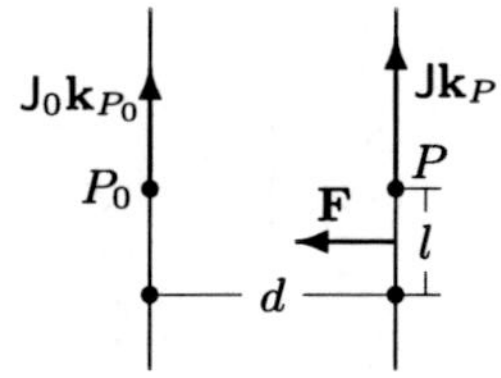

Fig. 19.4. The Ampère force **F**

19.1.2 The Ampère Force and the Ampère Law

In 1819, the Danish physicist Oerstedt noticed that an electric current is able to move a magnetic needle. In 1820, Ampère, Biot, and Savart began research on this phenomenon in Paris. The strongest results were obtained by Ampère. In 1826, Ampére published his treatise entitled *Memoir on the Mathematical Theory of Electrodynamic Phenomena, Uniquely Deduced from Experience* (in French). Interestingly enough, Ampère did not get any school education. At the age of 13, Ampère began to read the volumes of the French Encyclopedia edited by the philosopher Diderot (1713–1784) and the mathematician d'Alembert (1717–1783).[7] The contemporaries called Ampère a universal genius.

The Ampère force law. Consider the situation depicted in Fig. 19.4. In 1820, Ampère discovered that two electric currents of strength J_0 and J attract each other if there are charges of the same sign which flow the same direction. More precisely, a segment of the wire C_0 of length l attracts a segment of the wire C of length l by the force strength

$$|\mathbf{F}| = \frac{\mu_0 l}{2\pi d} \cdot \mathsf{J}_0 \mathsf{J}. \tag{19.13}$$

Here, the symbol d denotes the distance between the two wires. The force vector $\mathbf{F}$ is orthogonal to the wire C, and it lies in the plane spanned by the two wires. This law can be used in order to fix the physical unit of current strength, the ampere, in the SI system. By definition,

$$\mu_0 := 4\pi \cdot 10^{-7} \mathrm{N/A}^2.$$

This means the following. If the electric current of one ampere flows through the two wires in the same direction, then each of the two wires attracts the other one by the force 1 N (newton).

The Ampère law for the magnetic field. Consider a fixed right-handed Cartesian (x, y, z)-coordinate system with the right-handed orthonormal basis vectors $\mathbf{i}, \mathbf{j}, \mathbf{k}$ at the origin, and consider the parallel frame $\mathbf{i}_P, \mathbf{j}_P, \mathbf{k}_P$ at the point P (Fig. 4.3 on page 323). Let us study the situation depicted in Fig. 19.5. An electric current of the strength J flows along the z-axis. The current generates the magnetic field

$$\mathbf{B}(P) = B(r)\, \mathbf{e}_\varphi(P) \tag{19.14}$$

[6] The Greek words μεγάλος (megalos) and γιγαντας (gigantas) mean 'large' and 'giant', respectively. The Greek word τέρας (teras) means 'monster'.

[7] Encyclopedia, or a Descriptive Dictionary of the Sciences, Arts, and Trades, Paris, 1751–1772 (in French).

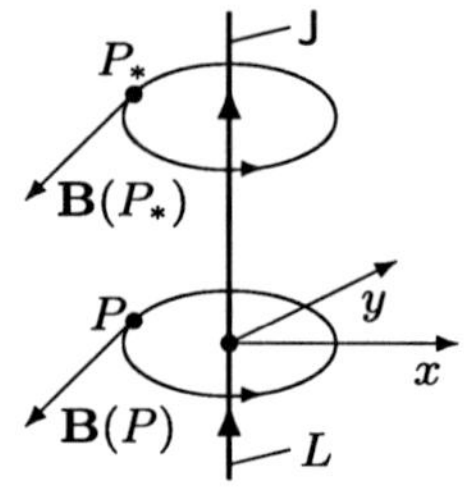

Fig. 19.5. Ampère's law for magnetic fields

where the point P has the cylindrical coordinates r, φ, z with $r := \sqrt{x^2 + y^2}$ and

$$\mathbf{e}_r := \cos\varphi\,\mathbf{i}_P + \sin\varphi\,\mathbf{j}_P, \qquad \mathbf{e}_\varphi(P) := -\sin\varphi\,\mathbf{i}_P + \cos\varphi\,\mathbf{j}_P.$$

By cylindrical symmetry, the magnetic field lines are concentric circles about the z-axis. The strength of the magnetic field is given by the Ampère law

$$\int_{\mathbb{S}^1_r} \mathbf{B} d\mathbf{x} = \mu_0 \mathsf{J}. \tag{19.15}$$

This yields $\int_{-\pi}^{\pi} B(r) r \, d\varphi = \mu_0 \mathsf{J}$. Hence

$$B(r) = \frac{\mu_0 \mathsf{J}}{2\pi r}, \qquad r > 0. \tag{19.16}$$

The Dirac distribution on the (x, y)-plane. The magnetic field $\mathbf{B}$ from (19.14) satisfies the two Maxwell equations

$$\mathbf{curl}\,\mathbf{B} = 0 \quad \text{and} \quad \operatorname{div}\mathbf{B} = 0 \qquad \text{on } \mathbb{E}^3 \setminus L \tag{19.17}$$

where L denotes the z-axis. The singularity corresponding to the electric current can be described by using the language of generalized functions. Then

$$\mathbf{curl}\,\mathbf{B} = \mu_0 \mathsf{J} \cdot \delta_{(0,0)}\,\mathbf{k} \quad \text{and} \quad \operatorname{div}\mathbf{B} = 0 \qquad \text{on } \mathbb{E}^2.$$

Here, $\delta_{(0,0)}$ denotes the Dirac delta distribution on the (x, y)-plane with support at the origin $(0, 0)$. The proof will be given in Problem 19.5. Mnemonically, physicists write

$$\mathbf{curl}\,\mathbf{B}(\mathbf{x}) = \mu_0 \mathsf{J} \cdot \delta(x)\delta(y) \quad \text{and} \quad \operatorname{div}\mathbf{B}(\mathbf{x}) = 0 \qquad \text{on } \mathbb{E}^2.$$

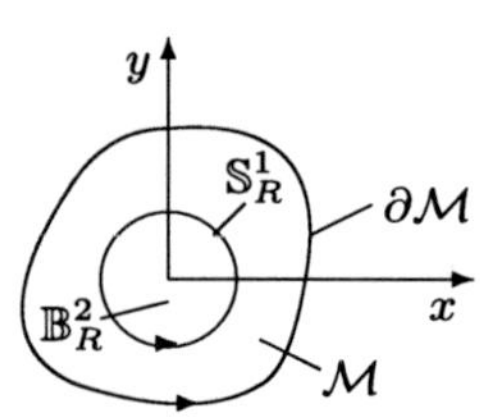

Fig. 19.6. Topological invariance of the Ampère integral

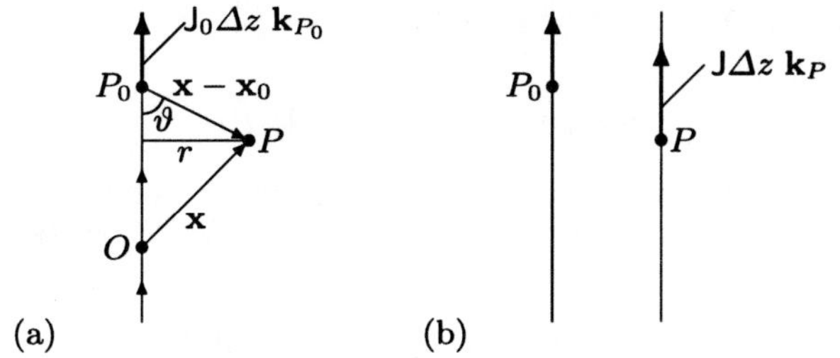

Fig. 19.7. The Biot-Savart law for electric currents in thin wires

Observe that the magnetic field $\mathbf{B}$ only depends on the real variables x and y.

The topological invariance of the Ampère integral. Let $\mathcal{M}$ be a compact, 2-dimensional, positively oriented submanifold of the Euclidean (x, y)-plane $\mathbb{E}^2$ with the coherently oriented boundary $\partial\mathcal{M}$; let the origin $(0, 0)$ be an interior point of $\mathcal{M}$ (Fig. 19.6). Then

$$\boxed{\int_{\partial\mathcal{M}} \mathbf{B} d\mathbf{x} = \mu_0 \mathsf{J}.} \tag{19.18}$$

This means that the domain of integration of the Ampère integral (19.15) can be deformed without changing the value of the integral.

Proof. Set $\mathcal{N} := \mathcal{M} \setminus \mathbb{B}^2_R$ (Fig. 19.6). It follows from $\mathbf{curl}\,\mathbf{B} = 0$ on $\mathbb{E}^2 \setminus \{0\}$ and from the Stokes integral theorem that

$$0 = \int_{\mathcal{N}} \mathbf{curl}\,\mathbf{B}\; dx dy = \int_{\partial\mathcal{N}} \mathbf{B} d\mathbf{x} = \int_{\partial\mathcal{M}} \mathbf{B} d\mathbf{x} - \int_{\mathbb{S}^1_R} \mathbf{B} d\mathbf{x}.$$

Here, we assume that the radius R is sufficiently small. □

Note the following crucial fact:

> *From the topological point of view, the electric Coulomb field* $\mathbf{E}$ *(resp. the magnetic Ampère field* $\mathbf{B}$*) is closely related to the topology (i.e., the Betti numbers) of the pointed Euclidean manifold* $\mathbb{E}^3 \setminus \{P_0\}$ *(resp. the manifold* $\mathbb{E}^3 \setminus L$*).*

This will be discussed in Sect. 23.4 in terms of the de Rham cohomology. As we will show in Sect. 19.2.1, this is the topological key to the Maxwell equations.

The Biot–Savart law for thin metallic wires. In order to understand Oerstedt's experiment, in 1820 Biot and Savart formulated a magnetic analogue of the Coulomb law (Fig. 19.7). They replaced the electric charge Q_0 of Coulomb's law by the directed electric current element

$$\mathsf{J}_0 \Delta z\, \mathbf{k}_{P_0} \tag{19.19}$$

at the point P_0 with the electric current strength $\mathsf{J}_0 > 0$, the length Δz, and the unit vector $\mathbf{k}_{P_0}$ which indicates the direction of the flow of positive electric charges. The Biot-Savart law tells us that the directed electric current element (19.19) generates the magnetic field

$$\Delta\mathbf{B}(P) := \frac{\mu_0 \mathsf{J}_0 \Delta z\, \mathbf{k}_{P_0} \times (\mathbf{x} - \mathbf{x}_0)}{4\pi|\mathbf{x} - \mathbf{x}_0|^3} \tag{19.20}$$

at the point P. Moreover, the directed current element (19.19) at the point P_0 exerts the force

$$\Delta\mathbf{F}(P) = \mathsf{J}\Delta z\,\mathbf{k}_P \times \Delta\mathbf{B}(P) \tag{19.21}$$

on the directed electric current element $\mathsf{J}\Delta z\ \mathbf{k}_P$ at the point P. As a typical application, let us show that the Biot–Savart law (19.20) yields the magnetic field (19.14) together with (19.16). In fact, since $|\mathbf{x} - \mathbf{x}_0| = (r^2 + z^2)^{1/2}$, we get

$$\Delta\mathbf{B}(\mathbf{x}) = \frac{\mu_0 \mathsf{J}_0 \Delta z \cdot \sin\vartheta}{4\pi|\mathbf{x} - \mathbf{x}_0|^2}\,\mathbf{j}_P = \frac{\mu_0 \mathsf{J}_0 r \Delta z}{4\pi(r^2 + z^2)^{3/2}}\,\mathbf{j}_P.$$

By superposition,

$$\mathbf{B}(\mathbf{x}) = \int_{-\infty}^{\infty} \Delta\mathbf{B}(\mathbf{x})\,dz = \frac{\mu_0 \mathsf{J}_0 r}{4\pi}\int_{-\infty}^{\infty} \frac{dz}{(r^2 + z^2)^{3/2}}\,\mathbf{j}_P = \frac{\mu_0 \mathsf{J}_0}{2\pi r}\,\mathbf{j}_P.$$

This coincides with the magnetic field (19.16).

19.1.3 Joule's Heat Energy Law

Heat energy produced in electric circuits. In 1840, Joule discovered that an electric current, which flows through a thin metallic wire, produces the heat energy

$$E_{\text{heat}} = \mathsf{J}V(t_1 - t_0)$$

in the oriented segment (P_0P) of the wire during the time interval $[t_0, t_1]$. This is called the Joule law. Here, J denotes the electric current strength (charge per time), and V denotes the positive voltage between the points P and P_0 of the wire. For the physical dimension in the SI system, we get

$$[E_{\text{heat}}] = \text{AVs} = \text{Ws} = \text{J} = \text{Nm} \tag{19.22}$$

where the symbols J, V, A, W, and N stand for joule (energy), volt (voltage), ampere (current strength), watt (power=enery per time), and newton (force), respectively.

Ohm's law. In 1827, Ohm (1789–1854) experimentally discovered the law

$$V = R\mathsf{J} \tag{19.23}$$

in thin metallic wires. Here, R is the so-called resistance of the wire measured in ohm, $\mathcal{O} = \text{V/A}$, in the SI system. Hence

$$E_{\text{heat}} = R\mathsf{J}^2(t_1 - t_0).$$

From the physical point of view, electrons flow through the wire. There is friction which transforms the mechanical energy of the electrons into heat energy.

19.1.4 Faraday's Induction Law

The physicist and chemist Michael Faraday (1791–1867) was one of the greatest scientists in the nineteenth century. He discovered fundamental laws in electricity and electrochemistry; he also invented the first electric motor and dynamo. Faraday paved the way for Maxwell's theory of electromagnetism. The following four treatises are cornerstones in the history of electromagnetism:

A. Ampère, Memoir on the Mathematical Theory of Electrodynamic Phenomena, Uniquely Deduced from Experience, Paris, 1826 (in French).[8]

M. Faraday, Experimental Researches in Electricity, Vols. 1–3, London, 1839–1855 (reissued in 1965).

M. Faraday, Experimental Researches in Chemistry and Physics, London, 1859 (reissued 1991).

J. Maxwell, A Treatise on Electricity and Magnetism, London, 1973. Reprinted by Dover, Vols. 1, 2, New York, 1954.

Note that both Ampère and Faraday were self-taught persons. Faraday started work as a bookbinder-apprentice when he was 13 years old. At the age of twenty, Faraday attended a lecture given by Sir Humphrey Davy on electricity. This changed his life. Later on Faraday became an assistant of Davy at the Royal Institution.[9]

Maxwell formulated Faraday's electromagnetic induction law in the following local form:

$$\mathbf{curl}\,\mathbf{E} = -\frac{\partial \mathbf{B}}{\partial t}. \tag{19.24}$$

Intuitively, a time-varying magnetic field generates closed electric field lines. Maxwell postulated that also the converse is true. He formulated the law

$$\mathbf{curl}\,\mathbf{B} = \frac{1}{c^2}\frac{\partial \mathbf{E}}{\partial t} \tag{19.25}$$

in a vacuum; this law holds true if there are no electric currents. As we will show in Sect. 19.5 on page 969, the two laws (19.24) and (19.25) govern the propagation of light. Intuitively, light consists of time-varying electric and magnetic fields which interact with each other. The strength of interaction depends on the rate of changing electric and magnetic fields. Here, $c = 1/\sqrt{\varepsilon_0\mu_0}$ is the velocity of light in a vacuum. By adding the Ampère law, the complete Maxwell equation reads as

$$\mathbf{curl}\,\mathbf{B} = \mu_0\mathbf{J} + \frac{1}{c^2}\frac{\partial \mathbf{E}}{\partial t}.$$

19.1.5 Electric Dipoles

For describing the polarization of substances, electric dipoles are crucial. By definition, the electric field

$$\boxed{\mathbf{E}(\mathbf{x}) := \frac{3(\mathbf{p}_{\mathrm{el}}\mathbf{x})\mathbf{x} - \mathbf{x}^2 \cdot \mathbf{p}_{\mathrm{el}}}{4\pi\varepsilon_0|\mathbf{x}|^5}} \tag{19.26}$$

is called the electric field of a dipole located at the origin with the electric dipole moment vector $\mathbf{p}_{\mathrm{el}}$. An explicit computation shows that

$$\mathbf{E} = -\,\mathbf{grad}\,U, \qquad U(\mathbf{x}) = \frac{\mathbf{p}_{\mathrm{el}}\mathbf{x}}{4\pi\varepsilon_0|\mathbf{x}|^3}, \qquad \mathbf{x} \neq 0. \tag{19.27}$$

The function U is called the electrostatic potential of the dipole. Moreover, we get

---

[8] The term 'electrodynamics' was coined by Ampère.

[9] The whole story can be found in the book by D. Bodanis, $E = mc^2$: A Biography of the World's Most Famous Equation, Berkeley Publishing Group, New York, 2000.

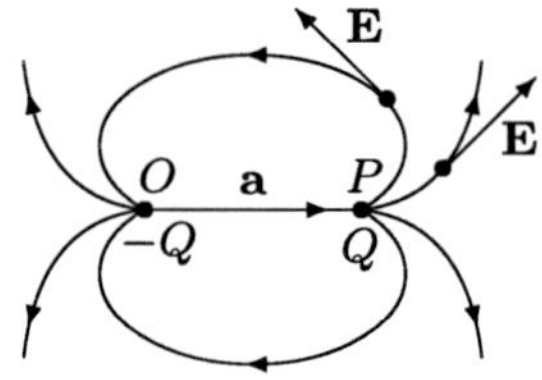

Fig. 19.8. Electric field of an electric dipole

$$\operatorname{div} \mathbf{E} = 0 \qquad \text{and} \qquad \mathbf{curl}\, \mathbf{E} = 0 \qquad \text{on } \mathbb{E}^3 \setminus \{0\}$$

together with the vanishing Gauss integral

$$\int_{\partial\mathcal{M}} \mathbf{E}\mathbf{n}\, dS = 0.$$

Here, we assume that $\mathcal{M}$ is a compact, 3-dimensional, positively oriented submanifold of the Euclidean manifold $\mathbb{E}^3$ with coherently oriented boundary $\partial\mathcal{M}$; let the origin be an interior point of $\mathcal{M}$ (e.g., $\mathcal{M}$ is a ball centered at the origin). Using the language of generalized functions, the singularity of the dipole at the origin can be described by the equation

$$\varepsilon_0 \operatorname{div} \mathbf{E} = -(\mathbf{p}_{\mathrm{el}}\, \mathbf{grad})\, \delta_O, \qquad \mathbf{curl}\, \mathbf{E} = 0 \qquad \text{on } \mathbb{E}^3.$$

Mnemonically, physicists write

$$\varepsilon_0 \operatorname{div} \mathbf{E}(\mathbf{x}) = -(\mathbf{p}_{\mathrm{el}}\, \mathbf{grad})\, \delta(\mathbf{x}), \qquad \mathbf{curl}\, \mathbf{E} = 0 \qquad \text{on } \mathbb{E}^3.$$

Let us motivate this. Consider the situation depicted in Fig. 19.8. The positive charge Q is located at the point P, and the negative charge $-Q$ is located at the origin O. The vector $\mathbf{p}_{\mathrm{el}} = Q\mathbf{a}$ is called the dipole moment vector. By superposition, the electrostatic potential is given by

$$\mathcal{U}(\mathbf{x}) := U(\mathbf{x}) - U(\mathbf{x} - \mathbf{p}_{\mathrm{el}}/Q) = Q\left(\frac{1}{4\pi\varepsilon_0|\mathbf{x}|} - \frac{1}{4\pi\varepsilon_0|\mathbf{x} - \mathbf{p}_{\mathrm{el}}/Q|}\right).$$

Fix the vector $\mathbf{p}_{\mathrm{el}}$. For large charges Q, we get

$$\mathcal{U}(\mathbf{x}) = -(\mathbf{p}_{\mathrm{el}}\, \mathbf{grad})\left(\frac{1}{4\pi\varepsilon_0|\mathbf{x}|}\right) = \frac{\mathbf{p}_{\mathrm{el}}\mathbf{x}}{4\pi\varepsilon_0|\mathbf{x}|^3},$$

up to terms of order $o\left(\frac{1}{Q}\right)$ as $Q \to \infty$. Letting $Q \to \infty$, we obtain the electrostatic dipole potential from (19.27).

Summarizing, the dipole potential U from (19.27) *is the negative directional derivative of the Coulomb potential with electric charge $Q_0 = 1$.*

A more general motivation goes like this. Let ϱ be a smooth electric charge density which vanishes outside a sufficiently large ball of radius R centered at the origin. Then the corresponding electrostatic potential reads as

$$U(\mathbf{x}) = \int_{\mathbb{B}^3_R} \frac{\varrho(\mathbf{x}_0)}{4\pi\varepsilon_0|\mathbf{x} - \mathbf{x}_0|}\, d^3x_0.$$

Then, for large distances, that is, $|\mathbf{x}|/R \gg 1$, the Taylor expansion yields

$$U(\mathbf{x}) = \frac{Q_0}{4\pi\varepsilon_0|\mathbf{x}|} + \frac{\mathbf{p}_{\mathrm{el}}\mathbf{x}}{4\pi\varepsilon_0|\mathbf{x}|^3} + o\left(\frac{1}{|\mathbf{x}|^2}\right), \qquad |\mathbf{x}| \to \infty$$

where

$$Q_0 = \int_{\mathbb{B}^3_R} \varrho(\mathbf{x}_0)\, d^3x_0 \qquad \text{and} \qquad \mathbf{p}_{\mathrm{el}} = \int_{\mathbb{B}^3_R} \varrho(\mathbf{x}_0)\mathbf{x}_0\, d^3x_0.$$

This tells us that far away from the ball $\mathbb{B}^3_R$, the electrostatic potential (generated by the charge density ϱ) looks like the superposition of the Coulomb potential with the electric charge Q_0 at the origin and the dipole potential of a dipole at the origin with dipole moment vector $\mathbf{p}_{\mathrm{el}}$. Sometimes it is necessary to consider higher approximations. This leads to quadrupole moments, and so on.

19.1.6 Magnetic Dipoles

Consider a circular electric current of current strength J_0 as depicted in Fig. 19.9. By the Biot–Savart law, the current generates the magnetic field

$$\mathbf{B}(\mathbf{x}) = \frac{\mu_0 \mathsf{J}_0}{4\pi} \int_{\mathbb{S}^1_R} \frac{d\mathbf{x}_0 \times (\mathbf{x} - \mathbf{x}_0)}{|\mathbf{x} - \mathbf{x}_0|^3}. \tag{19.28}$$

Far away from the electric current, that is, for $|\mathbf{x}|/R \gg 1$, the magnetic field looks like

$$\boxed{\mathbf{B}(\mathbf{x}) = \mu_0 \frac{3(\mathbf{m}\mathbf{x})\mathbf{x} - \mathbf{x}^2 \cdot \mathbf{m}}{4\pi|\mathbf{x}|^5}.} \tag{19.29}$$

This is a so-called magnetic dipole field with the magnetic moment vector $\mathbf{m}$. Explicitly,

$$\mathbf{m} = \pi R^2 \mathsf{J}_0\, \mathbf{k}.$$

Let us summarize properties of the magnetic dipole field (19.29):

- The Maxwell equations: We have

$$\operatorname{div} \mathbf{B} = 0 \qquad \text{and} \qquad \mathbf{curl}\, \mathbf{B} = 0 \qquad \text{on } \mathbb{E}^3 \setminus \{0\}.$$

  In the language of generalized functions, we get

$$\operatorname{div} \mathbf{B} = -\mu_0(\mathbf{m}\,\mathbf{grad})\, \delta_O \qquad \text{and} \qquad \mathbf{curl}\, \mathbf{B} = 0 \qquad \text{on } \mathbb{E}^3.$$

  Mnemonically, physicists write

$$\operatorname{div} \mathbf{B}(\mathbf{x}) = -\mu_0(\mathbf{m}\,\mathbf{grad})\, \delta(\mathbf{x}) \qquad \text{and} \qquad \mathbf{curl}\, \mathbf{B}(\mathbf{x}) = 0 \qquad \text{on } \mathbb{E}^3.$$

- The magnetic Gauss law: Let $\mathcal{M}$ be a compact, 3-dimensional, positively oriented submanifold of the Euclidean manifold $\mathbb{E}^3$ with coherently oriented boundary $\partial\mathcal{M}$. If the origin is an interior point of $\mathcal{M}$, then

$$\int_{\partial\mathcal{M}} \mathbf{B}\, dS = 0.$$

- The vector potential $\mathbf{A}$: Setting $\mathbf{A} := \frac{\mu_0(\mathbf{m}\times\mathbf{x})}{4\pi|\mathbf{x}|^3}$, we get $\mathbf{B} = \mathbf{curl}\, \mathbf{A}$.

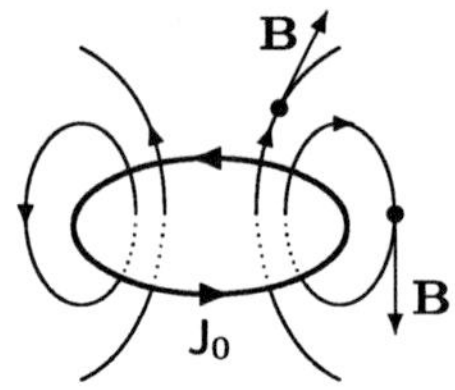

Fig. 19.9. Magnetic dipole generated by an electric circular current

19.1.7 The Electron Spin

> The spin of elementary particles plays a fundamental role in modern technology and medicine (e.g., MRI/magnetic resonance imaging).
>
> Folklore

As a particle, the electron was identified by Joseph John Thomson (1856–1940) in 1897 by performing cathode ray experiments. In 1906, Thomson was awarded the Nobel prize in physics in recognition of his great merits of his theoretical and experimental investigations on the conduction of electricity. In 1909, the electric charge of a single electron was first measured by Robert Millikan (1868–1953) by means of his famous oil-drop experiment. In 1923, Millikan was awarded the Nobel prize in physics for his study of the elementary electronic charge and the photoelectric effect.

In 1921, Stern (1888–1969) and Gerlach (1986–1979) experimentally investigated the splitting of a beam of silver atoms caused by an inhomogeneous magnetic field.[10] In 1925, Goudsmit (1902–1978) and Uhlenbeck (1900–1988) explained the Stern–Gerlach experiment by suggesting that the electron possesses an intrinsic angular momentum called spin.

Pauli's non-relativistic equation of the spinning electron. In 1927, Pauli formulated a modification of the non-relativistic 1926 Schrödinger equation which includes the electron spin. Set

$$\psi(x,y,z,t) := \begin{pmatrix} \psi^1(x,y,z,t) \\ \psi^2(x,y,z,t) \end{pmatrix}, \qquad \psi_+ := \begin{pmatrix} \psi^1 \\ 0 \end{pmatrix}, \qquad \psi_- := \begin{pmatrix} 0 \\ \psi^2 \end{pmatrix}.$$

We choose a right-handed Cartesian (x,y,z)-coordinate system with the right-handed orthonormal basis vectors $\mathbf{i},\mathbf{j},\mathbf{k}$. We assume that this system of reference is an inertial system. Pauli's equation for the spinning electron reads as follows:

$$\boxed{\mathrm{i}\hbar \frac{\partial \psi}{\partial t} = H\psi} \tag{19.30}$$

with the energy operator (Hamiltonian)

$$H := \frac{\mathbf{P}^2}{2m_e} - eU - \mathbf{mB}$$

and the following operators:

[10] In 1943, Otto Stern was awarded the Nobel prize in physics for his contribution to the molecular ray method and his 1933 discovery of the magnetic moment of the proton.

- $\mathbf{P} := -\mathrm{i}\hbar\boldsymbol{\partial}$ (momentum),
- $\mathbf{L} := \mathbf{x} \times \mathbf{P}$ (orbital angular momentum),
- $\mathbf{S} := \mathcal{S}^1\mathbf{i} + \mathcal{S}^2\mathbf{j} + \mathcal{S}^3\mathbf{k}$ (spin) with $\mathcal{S}^k := \frac{\hbar}{2}\sigma^k$, $k = 1, 2, 3$ (spin components),
- $\mathbf{L} + \mathbf{S}$ (total angular momentum),
- $\mathbf{m} = \mathbf{m_L} + \mathbf{m_S}$ (total magnetic moment of the electron),
- $\mathbf{m_L} := -\frac{e}{2m_e}\mathbf{L}$ (magnetic moment of the electron generated by the orbital angular momentum $\mathbf{L}$),
- $\mathbf{m_S} := -\frac{e}{m_e}\mathbf{S}$ (magnetic moment of the electron corresponding to the intrinsic spin of the electron).

We assume that the electric field $\mathbf{E} = -\,\mathbf{grad}\,U$ and the magnetic field $\mathbf{B}$ act on the electron of mass m_e and electric charge $-e$. Recall the Pauli matrices

$$\sigma^0 := \begin{pmatrix} 1 & 0 \\ 0 & 1 \end{pmatrix}, \quad \sigma^1 := \begin{pmatrix} 0 & 1 \\ 1 & 0 \end{pmatrix}, \quad \sigma^2 := \begin{pmatrix} 0 & -\mathrm{i} \\ \mathrm{i} & 0 \end{pmatrix}, \quad \sigma^3 := \begin{pmatrix} 1 & 0 \\ 0 & -1 \end{pmatrix}. \tag{19.31}$$

Finally, we introduce the inner product

$$\langle\psi|\psi_*\rangle := \int_{\mathbb{R}^3} \psi(x, y, z, t)^\dagger \psi_*(x, y, z, t)\, dxdydz.$$

Expectation values. Let ψ be a solution of the Pauli equation (19.30) with the normalization condition $\langle\psi|\psi\rangle = 1$. The mean energy $\bar{E}$, the mean orbital angular momentum vector $\bar{\mathbf{L}}$, and the mean spin vector $\bar{\mathbf{S}}$ of the state ψ are given by

$$\bar{E} := \langle\psi|H\psi\rangle, \quad \bar{\mathbf{L}} := \langle\psi|\mathbf{L}\psi\rangle, \quad \bar{\mathbf{S}} := \langle\psi|\mathbf{S}\psi\rangle.$$

It follows from

$$\mathcal{S}^3\psi_+ = \frac{\hbar}{2}\psi_+, \qquad \mathcal{S}^3\psi_- = -\frac{\hbar}{2}\psi_-$$

that the state ψ_+ (resp. (ψ_-) has the sharp spin component $\frac{\hbar}{2}$ (resp. $-\frac{\hbar}{2}$) in direction of the z-axis.

The commutation relations. The components

$$L^1 = \mathrm{i}\hbar\left(z\frac{\partial}{\partial y} - y\frac{\partial}{\partial z}\right), \quad L^2 = \mathrm{i}\hbar\left(x\frac{\partial}{\partial z} - z\frac{\partial}{\partial x}\right), \quad L^3 = \mathrm{i}\hbar\left(y\frac{\partial}{\partial x} - x\frac{\partial}{\partial y}\right)$$

of the orbital angular momentum operator $\mathbf{L} = L^1\mathbf{i} + L^2\mathbf{j} + L^3\mathbf{k}$ satisfy the following commutation relations:[11]

$$\boxed{[L^1, L^2]_- = \mathrm{i}\hbar L^3, \quad [L^2, L^3]_- = \mathrm{i}\hbar L^2, \quad [L^3, L^1]_- = \mathrm{i}\hbar L^3.}$$

In 1927, Pauli postulated that the components of the spin operator satisfy the same commutation relations as L^1, L^2, L^3, that is,

$$\boxed{[\mathcal{S}^1, \mathcal{S}^2]_- = \mathrm{i}\hbar\mathcal{S}^3, \quad [\mathcal{S}^2, \mathcal{S}^3]_- = \mathrm{i}\hbar\mathcal{S}^2, \quad [\mathcal{S}^3, \mathcal{S}^1]_- = \mathrm{i}\hbar\mathcal{S}^3.} \tag{19.32}$$

To get this, Pauli made the ansatz $\mathcal{S}^k := \frac{\hbar}{2}\sigma^k$, and he introduced the matrices (19.31). In fact, this yields (19.32).

[11] Recall that $[A, B]_- := AB - BA$.

Transformation of the Pauli equation under rotations of the Cartesian coordinate system. For physical reasons, it is quite natural to postulate that the Pauli equation (19.30) is invariant under rotations of the Cartesian coordinate system. We want to show that it is possible to fulfill this condition by using the universal covering group $SU(2)$ of the rotation group $SO(3)$. More precisely, we will use the surjective group morphism

$$\varrho : SU(2) \to SO(3)$$

considered in (7.21) on page 434 with the kernel $\ker(\varrho) = \{I, -I\}$. Set

$$X := \begin{pmatrix} x \\ y \\ z \end{pmatrix}.$$

Let $A \in SU(2)$. This implies the rotation

$$X' = \varrho(A)X$$

of the Cartesian coordinate system. Finally, we define the key transformation law

$$\boxed{\psi'(X') = A\psi(X).} \tag{19.33}$$

Proposition 19.1 *The Pauli transformation law* (19.33) *leaves the Pauli equation* (19.30) *invariant. In addition, the expectation value* $\bar{\mathbf{S}}$ *transforms like a vector.*

The proof will be given in Problem 19.8. The proof is based on the following key fact:

> *The adjoint representation of the Lie group* $SU(2)$ *on the Lie algebra* $su(2)$ *is equivalent to the representation of the Lie group* $SO(3)$ *on* $\mathbb{R}^3$ *given by* $X' = \varrho(A)X$.

Dirac's relativistic equation for the electron. In 1928, Dirac formulated the relativistic equation for the electron, and he showed that the electron spin is a relativistic effect (see Sect. 20.3 on page 999). The Pauli spin equation (19.30) can be obtained as a non-relativistic approximation of the Dirac equation.

Pauli's exclusion principle. In 1924, Pauli formulated the fundamental principle that two electrons of an atom can never be in the same quantum state. This allows us to justify the periodic table of elements in chemistry. The point is that only a certain number of electrons can occupy the same energy level.

Pauli's spin-statistics principle. This principle is closely related to the exclusion principle. Recall that an elementary particle is either

- a fermion (half-integer spin $k\hbar$ where $2k$ is a positive integer) or
- a boson (integer spin $n\hbar$ where n is a nonnegative integer).

For example, protons, neutrons, electrons, and quarks are fermions, whereas photons and gluons are bosons. In 1940, Pauli formulated the spin-statistics principle. This principle says that two fermions of the same type can never be in the same quantum state, and hence

- fermions of the same type (e.g., the electrons in a neutron star) are governed by the Fermi–Dirac statistics, whereas
- bosons of the same type (e.g., the photons in the universe) are governed by the Bose–Einstein statistics.

In 1945, Wolfgang Pauli (1900–1958) was awarded the Nobel prize in physics for the discovery of the exclusion principle also called the Pauli principle.

Further Reading

W. Pauli, On the quantum mechanics of the magnetic electron, Z. Phys. **43** (1927), 603–623 (in German).

P. Dirac, The quantum theory of the electron, Proc. Royal Soc. London **A117** (1928), 610–624; **A118**, 351–361.

W. Pauli, The connection between spin and statistics, Phys. Rev. **58** (1940), 716–722; Progr. Theor. Phys. **5** (1950), 526–543.

W. Pauli, Exclusion principle, Lorentz group and reflection of space-time and charge, pp. 30–51. In: W. Pauli (Ed.), Niels Bohr and and the Development of Physics, Pergamon Press, New York, 1955.

Furthermore, we refer to:

B. van der Waerden, Group Theory and Quantum Mechanics, Springer, New York 1974 (spectra of atoms and molecules).

M. Mizushima, Quantum Mechanics of Atomic Spectra and Atomic Structure, Benjamin, New York, 1970.

G. Drake (Ed.), Springer Handbook of Atomic, Molecular, and Optical Physics, Springer, Berlin, 2005.

R. Streater and A. Whightman, PCT, Spin, Statistics, and All That, Benjamin, New York, 1968.

N. Bogoliubov, A. Logunov, A. Orsak, and I. Todorov, General Principles of Quantum Field Theory, Kluwer, Dordrecht, 1990.

I. Duck and E. Sudarshan, Pauli and the Spin-Statistics Theorem, World Scientific, Singapore, 1997.

D. Buchholz and H. Epstein, Spin and statistics of quantum topological charges, Fizika, **17** (1985), 329–343.

R. Verch, A spin-statistics theorem for quantum fields on manifolds in a generally covariant framework, Commun. Math. Phys. **223** (2001), 261–288.

Poincaré Seminar 2007: The Spin. Edited by B. Duplantier, J. Raimond, and V. Rivasseau, Birkhäuser, Basel, 2009.

S. Bass, The Spin Structure of the Proton, World Scientific, Singapore, 2008.

We will come back to this topic in later volumes.

19.1.8 The Dirac Magnetic Monopole

In classical electromagnetism, there are no magnetic charges (monopoles), but only magnetic dipoles. Physicists expect that there exist high-energy magnetic monopoles in the universe which were produced at the time of the Big-Bang ($13.7 \cdot 10^9$ years ago). The existence of magnetic monopoles was predicted by Dirac in 1931. The magnetic field of a magnetic monopole is given by

$$\mathbf{B}(\mathbf{x}) := \frac{\mu_0 q}{4\pi|\mathbf{x}-\mathbf{x}_0|^2} \cdot \frac{\mathbf{x}-\mathbf{x}_0}{|\mathbf{x}-\mathbf{x}_0|}. \tag{19.34}$$

In the SI system, the magnetic charge q has the physical dimension Am (ampere meter).

19.1.9 Vacuum Polarization in Quantum Electrodynamics

Based on Feynman diagrams, the method of perturbation theory combined with the method of renormalization yields the following modification of the Coulomb field $\mathbf{E} = -\operatorname{\mathbf{grad}} U$ with the modified electrostatic potential

$$U(\mathbf{x}) = \frac{Q_0 s(\mathbf{x})}{4\pi\varepsilon_0|\mathbf{x} - \mathbf{x}_0|}.$$

Here, we use the function

$$s(\mathbf{x}) := 1 + \frac{2\alpha}{3\pi}\int_1^\infty \mathrm{e}^{-2m_0c\,|\mathbf{x}-\mathbf{x}_0|\zeta/\hbar}\left(1 + \frac{1}{2\zeta^2}\right)\frac{\sqrt{\zeta^2-1}}{\zeta^2}\,d\zeta.$$

This law was first obtained by Uehling and Serber in 1935. Here, m_0 and Q_0 are the rest mass and the electric charge of the particle, respectively; they generate the Coulomb field $\mathbf{E}(\mathbf{x})$ at the point P (Fig. 19.3). Intuitively, the particle is surrounded by a cloud of virtual electron-positron pairs. This additional electric dipole density is called vacuum polarization. Effectively, this vacuum polarization is described by the function $\mathbf{x} \mapsto s(\mathbf{x})$. If the distance $|\mathbf{x} - \mathbf{x}_0|$ is large compared with the reduced Compton length $\lambda\!\!\!^{-}_C := \hbar/m_0c$ of the particle, then we approximately get

$$s(\mathbf{x}) = 1 + \frac{\alpha}{4\sqrt{\pi}}\,\frac{\mathrm{e}^{-2|\mathbf{x}-\mathbf{x}_0|/\lambda\!\!\!^{-}_C}}{(|\mathbf{x}-\mathbf{x}_0|/\lambda\!\!\!^{-}_C)^{3/2}}, \qquad |\mathbf{x}-\mathbf{x}_0| \gg \lambda\!\!\!^{-}_C.$$

Here, we use the fundamental fine structure constant

$$\alpha = \frac{e^2}{4\pi\varepsilon_0\hbar c} = \frac{1}{137.04}.$$

Recent high-precision laser experiments seem to indicate that the fine structure constant varies slowly in time.

Vacuum polarization is also responsible for the anomalous magnetic moment of the electron

$$|\mathbf{m}| = \frac{e\hbar}{2m_e}\left(1 + \frac{\alpha}{2\pi} - 0.328\,\frac{\alpha^2}{\pi^2}\right) \tag{19.35}$$

where m_e is the rest mass of the electron. The value $\frac{e\hbar}{2m_e}\left(1 + \frac{\alpha}{2\pi}\right)$ was first obtained by Schwinger in 1949. We will study this in greater detail in Vol. IV (see also L. Landau and E. Lifshitz, Course of Theoretical Physics, Vol 4: Quantum Electrodynamics, Butterworth–Heinemann, Oxford, 1982).

Historical remarks. Let us sketch the fascinating history of electromagnetism by the following summary:

- Gilbert (1544–1603) (magnetic field of the earth), Newton (1643–1727),
- Coulomb (1736–1806), (Coulomb's law in 1790), Galvani (1737–1798), Volta (1745–1845),
- Oerstedt (1777–1851) (he observed in 1819 the interaction of an electric current with a magnetic needle), Biot (1774–1862) and Savart (1791–1841), (Biot–Savart law in 1820), Ampère (1775–1836) (the Ampère force in 1820, the Ampère flux law in 1823),
- Laplace (1749–1827) (Laplace equation), Gauss (1777–1855) (potential theory), Weber (1795–1878) (magnetic Gauss–Weber experiments), Poisson (1781–1840) (Poisson equation), Hamilton (1788–1856) (quaternions and nabla calculus), Thomson (1824–1907) (later Lord Kelvin), Stokes (1819-1903), Carl Neumann (1832–1925) (potential theory),

- Ohm (1789–1854) (electric resistance of metallic wires; Ohm's law in 1827), Joule (1818–1889) (heat production in electric wires; Joule's law in 1840),
- Faraday (1791–1867) (formulated the idea of the field lines of the electromagnetic field in 1810, discovery of electromagnetic induction in 1831),
- Maxwell (1831–1879) (formulation of the Maxwell equations for the electromagnetic field in 1864), Heinrich Hertz (1857–1894) (experimental existence proof for electromagnetic waves in 1888),
- Heaviside (1850–1925) (telegraphy and electric transmission, prediction of the ionosphere of the earth), Tesla (1856–1943) (construction of alternating-current dynamos, transformers, and motors),
- Ludvig Lorenz (1829–1891) (Lorenz gauge condition), Hendrik Lorentz (1853–1928) (Lorentz transformation), Poincaré (1854–1912) (Poincaré group),
- Einstein (1879–1955) (theory of special relativity in 1905), Dirac (1902–1984) (Dirac equation of the relativistic electron in 1928),
- Élie Cartan (1869–1951), de Rham (1903–1990), Hodge (1903–1975) (differential forms and differential topology),
- Laurent Schwartz (1915–2002) (creation of the theory of generalized functions in the late 1940s) (For this achievement, Schwartz was awarded the Fields medal in 1950),
- Pauli (1900–1958) (exclusion principle for electrons in 1924, shell structure of atoms and molecules, the spin–statistics principle for bosons and fermions in 1940) (Nobel prize in physics for the exclusion principle in 1945),
- Weyl (1885–1955) (introduction of the idea of gauge theory in 1918, as a generalization of Einstein's theory of general relativity),
- Fock (1898–1974) (discovery of the local phase factor for the Klein–Fock–Gordon equation in 1926), London (1954–1954) (local phase factor in 1927), and Weyl (commutative $U(1)$-gauge theory in quantum mechanics in 1929),
- Oskar Klein (1894–1977) (noncommutative pre-gauge theory in 1938),
- Pauli (noncommutative gauge theory formulated in a letter to Pais in 1953),
- Yang (born 1922) and Mills (1927–1999) (noncommutative $SU(2)$-gauge theory in 1954; Yang–Mills theory),
- Lamb (1913–2008) (experimental discovery of the Lamb shift together with Retherford in 1947 – hyperfine structure of the spectrum of the hydrogen atom); Kusch (1911–1993) (precision measurement of the anomalous magnetic moment of the electron) (Lamb and Kusch were awarded the Nobel prize in physics in 1955);
- Bethe (1906–2005) (theory of the Lamb shift via renormalization in 1947) (In 1967, Bethe was awarded the Nobel prize in physics for his contributions to the theory of nuclear reactions, especially his discoveries concerning the energy production in the sun and in stars),
- Schwinger computed the anomalous magnetic moment of the electron in 1949 on the basis of his approach to quantum field theory,
- Feynman (1918–1988), Schwinger (1918–1994), and Tomonaga (1906–1979) (quantum electrodynamics; Nobel prize in physics in 1965),
- Glashow (born 1932), Salam (1926–1996), and Weinberg (born 1933) (electroweak interaction; Nobel prize in physics in 1979),
- 't Hooft (born 1946) and Veltman (born 1931) (renormalization of the electroweak interaction in 1971; Nobel prize in physics in 1999).

In 1864, Maxwell unified the electric interaction with the magnetic interaction. Approximately hundred years later, Glashow, Salam, and Weinberg unified the electromagnetic interaction with the weak interaction (e.g., radioactive decay) in particle physics.

Modern high technology is strongly influenced by the discovery of new properties of the electromagnetic field. This is reflected by the following (incomplete) list of Nobel prizes in physics:

- 1956: Shockley, Bardeen, and Houser (discovery of the transistor effect and semiconductors – the basic tool for constructing computers),
- 1958: Cherenkov, Frank, and Tamm (discovery and theoretical interpretation of the Cherenkov effect),
- 1964: Basov, Prokhorov, and Townes (quantum electronics, maser–laser principle),
- 1972: Bardeen, Cooper, and Schrieffer (BCS-theory of superconductivity),
- 1978: Penzias and Wilson (discovery of cosmic microwave background radiation),
- 1981: Bloembergen and Schawlow (contributions to laser spectroscopy), Siegbahn (high-resolution electron spectroscopy),
- 1985: von Klitzing (discovery of the quantized Hall effect, quantization of electric resistance),
- 1986: Ruska (design of the first electron microscope in 1933); Binnig and Rohrer (design of the scanning tunneling microscope),
- 1987: Bednorz and A. Müller (discovery of superconductivity in ceramic metals – high-temperature superconductivity); Chu, Cohen-Tannoudji, Phillips, (development of methods to cool and trap atoms with laser light),
- 1989: Ramsey (atomic clocks); Dehmelt and Paul (development of the ion trap technique),
- 1998: Laughlin, Störmer, and Tsui (discovery of a new form of quantum fluid with fractionally charged excitations),
- 2000: Alferov and Kroemer (semiconductor heterostructures used in high-speed electronics and opto-electronics); Kilby (invention of the integrated circuit),
- 2001: Cornell, Ketterle, and Wieman (Bose–Einstein condensation in dilute gases),
- 2004: Glauber (quantum theory of optical coherence–laser beams); Hall and Hänsch (development of laser-based precision spectroscopy),
- 2006: Mather and Smoot (discovery of the anisotropy of the cosmic microwave background radiation),
- 2007: Fert and Grünberg (discovery of giant magnetic resistance; the electron spin is used to store and transport information),
- 2009: Kao (transmission of light in fibers); Boyle and Smith (invention of an imaging semiconductor circuit).

Let us also mention the following two Nobel prizes in medicine:

- 1962: Crick, Watson, and Wilkins (discoveries concerning the molecular structure of nuclear acids and its significance for information transfer in living material),
- 1979: Cormack and Hounsfield (development of computer assisted tomography).

In mathematics, the physical ideas coming from electrostatic phenomena strongly influenced the modern theory of Riemann surfaces, the calculus of variations, and the theory of elliptic partial differential equations (see the historical remarks on the Dirichlet principle in Sect. 10.4 of Vol. I).

19.2 The Maxwell Equations in a Vacuum

Maxwell based his theory of electromagnetism on a precise mathematical formulation of Faraday's idea of electric and magnetic field lines.

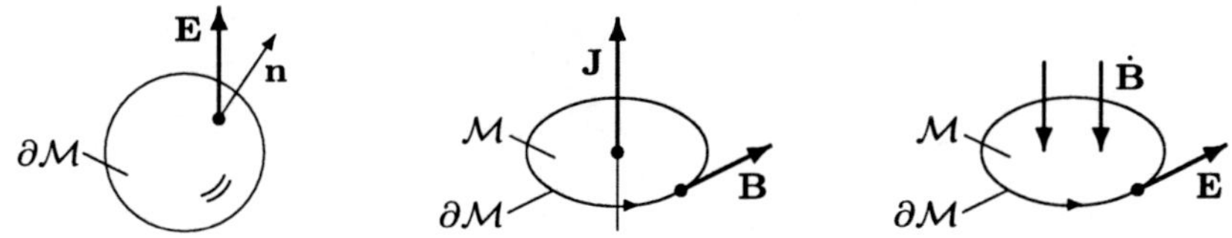

(a) Gauss' law (b) Ampère's law (c) Faraday's law

Fig. 19.10. The Maxwell equations in a vacuum

19.2.1 The Global Maxwell Equations Based on Electric and Magnetic Flux

In an arbitrary inertial system, the four global Maxwell equations read as follows:[12]

(i) The electric Gauss law (Fig. 19.10(a)):

$$\varepsilon_0 \int_{\partial\mathcal{M}} \mathbf{En} \cdot dS = Q. \tag{19.36}$$

This means that electric charges located in the compact, positively oriented, 3-dimensional manifold $\mathcal{M}$ (e.g., a ball) generate an electric field. The flow of the electric field $\mathbf{E}$ through the coherently oriented boundary $\partial\mathcal{M}$ (e.g., a sphere) measures the electric charges located in the manifold $\mathcal{M}$.

(ii) The magnetic Gauss law:

$$\int_{\partial\mathcal{M}} \mathbf{Bn} \cdot dS = 0. \tag{19.37}$$

This tells us that there are no magnetic monopole charges in classical electrodynamics. However, physicists conjecture that there exist high-energy magnetic monopoles in the universe as a relict of the Big-Bang.

(iii) The Ampère–Maxwell law (Fig. 19.10(b)):

$$\int_{\partial\mathcal{M}} \mathbf{B}d\mathbf{x} = \mu_0 \int_{\mathcal{M}} (\mathbf{J} + \mathbf{J}_{\text{polar}})\mathbf{n} \cdot dS \cdot \tag{19.38}$$

This means that the flow of electric charges through the compact, oriented, two-dimensional manifold $\mathcal{M}$ with coherently oriented boundary $\partial\mathcal{M}$ generates a magnetic field $\mathbf{B}$. Maxwell introduced the so-called electric polarization current[13]

$$\mathbf{J}_{\text{polar}} := \varepsilon_0 \dot{\mathbf{E}}$$

in order to obtain electromagnetic waves as solutions of the Maxwell equations. Maxwell conjectured that these waves describe the propagation of light. In 1888, Heinrich Hertz experimentally established the existence of electromagnetic waves in nature. In this case, the Maxwell law reads as

---

[12] Positive orientation refers to a fixed right-handed Cartesian (x, y, z)-coordinate system. Coherent orientation of the boundary $\partial\mathcal{M}$ is depicted in Figure 12.6 on page 677.

[13] This is also called the electric displacement current.

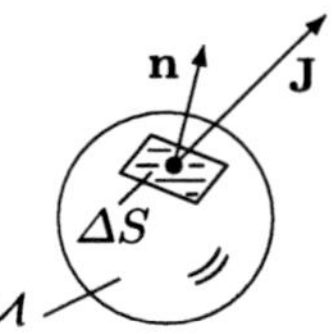

Fig. 19.11. Flow of electric charge

$$\int_{\partial \mathcal{M}} \mathbf{B} d\mathbf{x} = \frac{1}{c^2} \frac{d}{dt} \int_{\mathcal{M}} \mathbf{E}\mathbf{n} \cdot dS.$$

(iv) The Faraday induction law (Fig. 19.10(c)):

$$\int_{\partial \mathcal{M}} \mathbf{E} d\mathbf{x} = -\frac{d}{dt} \int_{\mathcal{M}} \mathbf{B}\mathbf{n} \cdot dS. \tag{19.39}$$

This means that the time-varying magnetic flux trough the compact, oriented, 2-dimensional manifold $\mathcal{M}$ induces an electric field on the coherently oriented boundary $\partial \mathcal{M}$.

Conservation laws. Electric charge, energy, momentum, and angular momentum of the Maxwell $(\mathbf{E}, \mathbf{B}, \varrho, \mathbf{J})$-system in a vacuum are conserved quantities. Explicitly, the following hold:

(a) Conservation of electric charge (Fig. 19.11):

$$\frac{d}{dt} \int_{\mathcal{M}} \varrho \, d^3x = -\int_{\partial \mathcal{M}} \mathbf{J}\mathbf{n} \, dS. \tag{19.40}$$

(b) Conservation of energy:

$$\frac{d}{dt} \int_{\mathcal{M}} \eta \, d^3x = -\int_{\partial \mathcal{M}} \mathbf{S}\mathbf{n} \, dS - \int_{\mathcal{M}} \mathbf{E}\mathbf{J} \, d^3x. \tag{19.41}$$

(c) Conservation of momentum (balance of forces):

$$\frac{d}{dt} \int_{\partial \mathcal{M}} \tfrac{1}{c^2} \mathbf{S} \, d^3x = -\int_{\mathcal{M}} \mathbf{f} \, d^3x + \int_{\partial \mathcal{M}} \mathsf{T}\mathbf{n} \, dS. \tag{19.42}$$

(d) Conservation of angular momentum (balance of torque):

$$\frac{d}{dt} \int_{\mathcal{M}} \left(\mathbf{x} \times \tfrac{1}{c^2} \mathbf{S}\right) d^3x = -\int_{\mathcal{M}} (\mathbf{x} \times \mathbf{f}) \, d^3x + \int_{\partial \mathcal{M}} (\mathbf{x} \times \mathsf{T}\mathbf{n}) \, dS. \tag{19.43}$$

Here, we use the following expressions:

- $\mathbf{D} := \varepsilon_0 \mathbf{E}$, $\mathbf{H} := \frac{1}{\mu_0} \mathbf{B}$,
- $\eta := \frac{1}{2}(\mathbf{ED} + \mathbf{BH})$ (energy density);
- $\mathbf{S} := \mathbf{E} \times \mathbf{H}$ (energy current density vector);
- $\frac{1}{c^2} \mathbf{S} := \mathbf{D} \times \mathbf{B}$ (momentum density vector);
- $\mathbf{JE}$ (rate of Joule's energy density);
- $\mathbf{f} := \varrho \mathbf{E} + \mathbf{J} \times \mathbf{B}$ (Lorentz force density vector);
- $\mathsf{T} := \mathbf{D} \otimes \mathbf{E} + \mathbf{B} \otimes \mathbf{H} - \eta I$ (Maxwell's stress tensor);
- $\mathsf{T}\mathbf{n} := (\mathbf{En})\mathbf{D} + (\mathbf{Hn})\mathbf{B}$.

19.2.2 The Local Maxwell Equations Formulated in Maxwell's Language of Vector Calculus

Fix an inertial system. The Maxwell equations for the electric vector field **E** and the magnetic vector field **B** in a vacuum read as follows:

(i) The electric Gauss law (Fig. 19.10(a)):

$$\varepsilon_0 \operatorname{div} \mathbf{E} = \varrho. \tag{19.44}$$

(ii) The magnetic Gauss law:

$$\operatorname{div} \mathbf{B} = 0. \tag{19.45}$$

(iii) The Ampère–Maxwell law (Fig. 19.10(b)):

$$\mathbf{curl}\,\mathbf{B} = \mu_0(\mathbf{J} + \mathbf{J}_{\text{polar}}). \tag{19.46}$$

Here, $\mathbf{J}_{\text{polar}} := \varepsilon_0 \dot{\mathbf{E}}$.

(iv) The Faraday induction law (Fig. 19.10(c)):

$$\mathbf{curl}\,\mathbf{E} = -\dot{\mathbf{B}} \tag{19.47}$$

Here, ϱ is the electric charge density function, and **J** is the electric current density vector field. All the functions depend on space and time. Moreover, ε_0 (resp. μ_0) is the electric (resp. magnetic) field constant of a vacuum.

These local Maxwell equations follow from the global Maxwell equations by using the integral theorems of Gauss–Ostrogradski and Stokes. For example, set $Q = \int_{\mathcal{M}} \varrho \, d^3x$. Then

$$\varepsilon_0 \int_{\partial\mathcal{M}} \mathbf{E}\mathbf{n} \, dS = \varepsilon_0 \int_{\mathcal{M}} \operatorname{div} \mathbf{E} \, d^3x = \int_{\mathcal{M}} \varrho \, d^3x.$$

Choose the set $\mathcal{M}$ as a ball of radius r centered at the point P. Contracting the ball to the point P by letting $r \to 0$, we get $\varepsilon_0 \operatorname{div} \mathbf{E}(P) = \varrho(P)$.

Moreover, choose a right-handed Cartesian (x, y, z)-system with the right-handed orthonormal vector basis $\mathbf{i}, \mathbf{j}, \mathbf{k}$. In the (x, y)-plane, let $\mathcal{M}$ be a disc of radius r centered at the point P. The Ampère–Maxwell law tells us that

$$\int_{\mathcal{M}} \mathbf{k}\,\mathbf{curl}\,\mathbf{B} \, dxdy = \int_{\partial\mathcal{M}} \mathbf{B}d\mathbf{x} = \int_{\mathcal{M}} \mu_0 \mathbf{k}(\mathbf{J} + \mathbf{J}_{\text{polar}}) \, dxdy.$$

Contracting the disc to the point P, we get

$$\mathbf{k}\,\mathbf{curl}\,\mathbf{B}(P) = \mu_0 \mathbf{k}(\mathbf{J} + \mathbf{J}_{\text{polar}})(P).$$

Since the (x, y, z)-system can be chosen arbitrarily, the vector **k** is an arbitrary unit vector. Hence $\mathbf{curl}\,\mathbf{B}(P) = \mu_0(\mathbf{J} + \mathbf{J}_{\text{polar}})(P)$.

Conversely, integrating the local Maxwell equations, we get the global Maxwell equations (19.36). Thus, for smooth functions, the local and global variants of the Maxwell equations are equivalent to each other. For non-smooth functions, one has to use the global Maxwell equations (or the formulation of the local Maxwell equations in terms of generalized functions; see Sect. 23.5.1).

Conservation laws. Similarly, the global conservation laws on page 956 are equivalent to the following local conservation laws:

(a) Conservation of electric charge:

$$\frac{\partial \varrho}{\partial t} + \operatorname{div} \mathbf{J} = 0. \tag{19.48}$$

(b) Conservation of energy:

$$\frac{\partial \eta}{\partial t} + \operatorname{div} \mathbf{S} = -\mathbf{J}\mathbf{E}. \tag{19.49}$$

(c) Conservation of momentum:

$$\frac{1}{c^2}\frac{\partial \mathbf{S}}{\partial t} + \mathbf{f} = \operatorname{div} \mathsf{T}. \tag{19.50}$$

(d) Conservation of angular momentum:

$$\frac{1}{c^2}\frac{\partial}{\partial t}(\mathbf{x} \times \mathbf{S}) + \mathbf{x} \times \mathbf{f} = \mathbf{x} \times \operatorname{div} \mathsf{T}. \tag{19.51}$$

The conservation laws are immediate consequences of the Maxwell equation (i)–(iv). For example, the Gauss law combined with the Ampère–Maxwell law implies

$$\dot{\varrho} = \varepsilon_0 \operatorname{div} \dot{\mathbf{E}} = \operatorname{div}\left(\frac{1}{\mu_0}\,\mathbf{curl}\,\mathbf{B} - \mathbf{J}\right) = -\operatorname{div} \mathbf{J}.$$

The other conservation laws can be most elegantly obtained by using tensor analysis (see Sect. 19.6.3).

19.2.3 Discrete Symmetries and *CPT*

Let us introduce the following three operators C, P, T.

(i) Charge conjugation $Q \mapsto -Q$:

$$\begin{aligned} (C\varrho)(\mathbf{x},t) &:= -\varrho(\mathbf{x},t), \qquad & (C\mathbf{J})(\mathbf{x},t) &:= -\mathbf{J}(\mathbf{x},t),\\ (C\mathbf{E})(\mathbf{x},t) &:= -\mathbf{E}(\mathbf{x},t), \qquad & (C\mathbf{B})(\mathbf{x},t) &:= -\mathbf{B}(\mathbf{x},t). \end{aligned}$$

(ii) Parity transformation $\mathbf{x} \mapsto -\mathbf{x}$:

$$\begin{aligned} (P\varrho)(\mathbf{x},t) &:= \varrho(-\mathbf{x},t), \qquad & (P\mathbf{J})(\mathbf{x},t) &:= -\mathbf{J}(-\mathbf{x},t),\\ (P\mathbf{E})(\mathbf{x},t) &:= -\mathbf{E}(-\mathbf{x},t), \qquad & (P\mathbf{B})(\mathbf{x},t) &:= \mathbf{B}(-\mathbf{x},t). \end{aligned}$$

(iii) Time reversal $t \mapsto -t$:

$$\begin{aligned} (T\varrho)(\mathbf{x},t) &:= \varrho(\mathbf{x},-t), \qquad & (T\mathbf{J})(\mathbf{x},-t) &:= -\mathbf{J}(\mathbf{x},-t),\\ (T\mathbf{E})(\mathbf{x},t) &:= \mathbf{E}(\mathbf{x},-t), \qquad & (T\mathbf{B})(\mathbf{x},t) &:= -\mathbf{B}(\mathbf{x},-t). \end{aligned}$$

A simple computation shows that

The Maxwell equations (19.44)–(19.47) *are invariant under the transformations C, P, and T.*

For the transformation T, this means the following: If $\mathbf{E}, \mathbf{B}, \varrho, \mathbf{J}$ are solutions of the Maxwell equations (19.44)–(19.47), then so are

$$T\mathbf{E},\ T\mathbf{B},\ T\varrho,\ T\mathbf{J}.$$

The same is true, if we replace T by either C or P. The transformation P is motivated by the electric Coulomb field:

$$\mathbf{E}(\mathbf{x}) = \frac{Q_0\mathbf{x}}{4\pi\varepsilon_0|\mathbf{x}|^3}.$$

The transformation law T is motivated by the Coulomb field and by the fact that the time reversal $t \mapsto -t$ changes the sign of the velocity vector, $\mathbf{v} \mapsto -\mathbf{v}$, and hence $\varrho\mathbf{v} \mapsto -\varrho\mathbf{v}$, which implies $\mathbf{J} \mapsto -\mathbf{J}$. For the combined transformation CPT, we get:

$$(CPT\varrho)(\mathbf{x}, t) := -\varrho(-\mathbf{x}, -t), \qquad (CPT\mathbf{J})(-\mathbf{x}, -t) := -\mathbf{J}(-\mathbf{x}, -t),$$

and

$$(CPT\mathbf{E})(\mathbf{x}, t) := \mathbf{E}(-\mathbf{x}, -t), \qquad (CPT\mathbf{B})(\mathbf{x}, t) := \mathbf{B}(-\mathbf{x}, -t).$$

Obviously, the Maxwell equations are invariant under the transformation CPT.

We will show in Sect. 19.3.3 that the electromagnetic field can be represented in the form

$$\boxed{\mathbf{E} = -\,\mathbf{grad}\,U - \dot{\mathbf{A}}, \qquad \mathbf{B} = \mathbf{curl}\,\mathbf{A}.}$$

Here, U and $\mathbf{A}$ are called the scalar and the vector potential of the electromagnetic field, respectively. Let us define the operators C, P, T in the following way:

(i) Charge conjugation $Q \mapsto -Q$:

$$(CU)(\mathbf{x}, t) := -U(\mathbf{x}, t), \qquad (C\mathbf{A})(\mathbf{x}, t) := -\mathbf{A}(\mathbf{x}, t).$$

(ii) Parity transformation $\mathbf{x} \mapsto -\mathbf{x}$:

$$(PU)(\mathbf{x}, t) := U(-\mathbf{x}, t), \qquad (P\mathbf{A})(\mathbf{x}, t) := -\mathbf{A}(-\mathbf{x}, t).$$

(iii) Time reversal $t \mapsto -t$:

$$(TU)(\mathbf{x}, t) := U(\mathbf{x}, -t), \qquad (T\mathbf{A})(\mathbf{x}, t) := -\mathbf{A}(\mathbf{x}, -t).$$

This yields the CPT transformation

$$(CPTU)(\mathbf{x}, t) := -U(-\mathbf{x}, -t), \qquad (CPT\mathbf{A})(\mathbf{x}, t) := -\mathbf{A}(-\mathbf{x}, -t).$$

These operations imply the corresponding operations for the electromagnetic field $\mathbf{E}, \mathbf{B}$ introduced above. The CPT transformation describes a fundamental symmetry of relativistic quantum field theories, as we will show in Vol. IV.

In 1905, Einstein published the following fundamental paper:

A. Einstein, Zur Elektrodynamik bewegter Körper (On the electrodynamics of moving bodies), Ann. Phys. **17**, 891–921.

The English translation can be found in S. Hawking, The Essential Einstein: His Greatest Works. Edited, with commentary, by Stephen Hawking, Penguin Books, London, 2008. It was Einstein's goal to find a relativistically invariant formulation of the Maxwell equations which obeys the principle of special relativity. That is, the Maxwell equations have the same form in all inertial systems, and the transformation laws between different inertial systems are known for the electric field, the electric charge densities, and the electric current densities. In the next section, we will study this.

19.3 Invariant Formulation of the Maxwell Equations in a Vacuum

In what follows, we will use the Einstein convention introduced on page 905.

19.3.1 Einstein's Language of Tensor Calculus

We will use the tensor calculus introduced in Chap. 8. The Maxwell equations on the Minkowski manifold $\mathbb{M}^4$ read as follows:

$$\boxed{\nabla_\alpha F^{\alpha\beta} = \mu_0 \mathcal{J}^\beta, \quad \nabla_{[\gamma} F_{\lambda\mu]} = 0, \qquad \beta, \gamma, \lambda, \mu = 0, 1, 2, 3.} \tag{19.52}$$

This is a tensor equation on $\mathbb{M}^4$ with respect to the metric tensorial family $g_{\alpha\beta}$ (which coincides with $\eta_{\alpha\beta}$ on inertial systems). We assume that $F^{\alpha\beta}$ and $\mathcal{J}^\beta$ are tensorial families on $\mathbb{M}^4$. Lowering indices, we get

$$F_{\alpha\beta} = g_{\alpha\lambda} g_{\beta\mu} F^{\lambda\mu}, \qquad \alpha, \beta = 0, 1, 2, 3,$$

and $\mathcal{J}_\alpha = g_{\alpha\beta}\mathcal{J}^\beta$, $\alpha = 0, 1, 2, 3$. Since the equations from (19.52) possess the correct index picture, they are valid for all local coordinate systems on $\mathbb{M}^4$.

In order to get Maxwell's classical notation, we define

$$E^k := \sigma F^{k0}, \quad B^k := \sigma \mathcal{E}^{0k\lambda\mu} F_{\lambda\mu}, \quad J^k := \sigma \mathcal{J}^k, \quad \varrho c := \sigma \mathcal{J}^0 \tag{19.53}$$

where $k = 1, 2, 3$. Here, $\sigma = \pm 1$ denotes the time orientation of the local coordinate system on $\mathbb{M}^4$. Note that $\mathcal{E}^{\alpha\beta\mu\lambda} F_{\lambda\mu}$ is a pseudo-tensorial family (see page 460). This is responsible for the crucial fact that the electric field and the magnetic field are differently transformed by changing the orientation (i.e., either the spatial orientation or the time orientation).

Inertial systems. Let us choose a strictly positively oriented inertial system with the right-handed Cartesian (x, y, z)-system and positive time orientation in the usual sense (i.e., $\sigma = 1$). Let $\mathbf{i}, \mathbf{j}, \mathbf{k}$ be the corresponding right-handed orthonormal basis Set $\mathbf{e}_1 := \mathbf{i}, \mathbf{e}_2 := \mathbf{j}, \mathbf{e}_3 := \mathbf{k}$. In this case, we have $g_{\alpha\beta} = \eta_{\alpha\beta}$, and

$$\mathcal{E}^{\alpha\beta\lambda\mu} = -\operatorname{sgn}\begin{pmatrix} 0 & 1 & 2 & 3 \\ \alpha & \beta & \lambda & \mu \end{pmatrix}$$

for all indices. Then, we get the electromagnetic field

$$\mathbf{E} = E^1\mathbf{i} + E^2\mathbf{j} + E^3\mathbf{k} = E^k\mathbf{e}_k, \qquad \mathbf{B} = B^k\mathbf{e}_k,$$

the electric charge density ϱ, and the electric current density vector $\mathbf{J} = J^k\mathbf{e}_k$. In addition, we set $J^0 := \varrho c$. Then

$$\begin{pmatrix} F_{00} & F_{01} & F_{02} & F_{03} \\ F_{10} & F_{11} & F_{12} & F_{13} \\ F_{20} & F_{21} & F_{22} & F_{23} \\ F_{30} & F_{31} & F_{32} & F_{33} \end{pmatrix} := \begin{pmatrix} 0 & E^1/c & E^2/c & E^3/c \\ -E^1/c & 0 & -B^3 & B^2 \\ -E^2/c & B^3 & 0 & -B^1 \\ -E^3/c & -B^2 & B^1 & 0 \end{pmatrix}, \tag{19.54}$$

as well as $J^\beta = \mathcal{J}^\beta$, $\beta = 0, 1, 2, 3$, $\varrho c = \mathcal{J}_0 = \mathcal{J}^0$, $J^k = -\mathcal{J}_k, k = 1, 2, 3$, and

$$\begin{pmatrix} F^{00} & F^{01} & F^{02} & F^{03} \\ F^{10} & F^{11} & F^{12} & F^{13} \\ F^{20} & F^{21} & F^{22} & F^{23} \\ F^{30} & F^{31} & F^{32} & F^{33} \end{pmatrix} := \begin{pmatrix} 0 & -E^1/c & -E^2/c & -E^3/c \\ E^1/c & 0 & -B^3 & B^2 \\ E^2/c & B^3 & 0 & -B^1 \\ E^3/c & -B^2 & B^1 & 0 \end{pmatrix}. \qquad (19.55)$$

Theorem 19.2 *The Maxwell equations* (19.52) *are equivalent to Maxwell's formulation in terms of vector calculus.*

Proof. In the inertial system, we have $1 = \eta_{00} = -\eta_{11} = -\eta_{22} = -\eta_{33}$, and $\eta_{\alpha\beta} = 0$ otherwise. Consequently, $\nabla_\alpha = \partial_\alpha$ for all α. Note that $F_{\alpha\beta} = -F_{\beta\alpha}$. Thus, the equations (19.52) pass over to

- $\partial_\alpha F^{\alpha\beta} = \mu_0 J^\beta$, $\beta = 0, 1, 2, 3$ (the Maxwell source equations), and
- $\partial_\gamma F_{\lambda\mu} + \partial_\lambda F_{\mu\gamma} + \partial_\mu F_{\gamma\lambda} = 0$, $\gamma, \lambda, \mu = 0, 1, 2, 3$ (the Maxwell–Bianchi equations).

Here, $\partial_0 := \partial/\partial x^0$, and $\partial_k := \partial/\partial x^k, k = 1, 2, 3$. This yields

- $\partial_0 F^{01} + \partial_2 F^{21} + \partial_3 F^{31} = \mu_0 J^1$,
- $\partial_0 F_{12} + \partial_1 F_{20} + \partial_2 F_{01} = 0$,
- $\partial_1 F_{23} + \partial_2 F_{31} + \partial_3 F_{12} = 0$.

The other equations are obtained by the cyclic permutations $1 \mapsto 2 \mapsto 3 \mapsto 1$, and by $0 \mapsto 0$. Using $\varepsilon_0 \mu_0 c^2 = 1$, we get the desired classical Maxwell equations:

- $\varepsilon_0 \operatorname{div} \mathbf{E} = \varrho$ and $\mathbf{curl}\, \mathbf{B} = \mu_0(\mathbf{J} + \varepsilon_0 \dot{\mathbf{E}})$ (the Maxwell source equations),
- $\operatorname{div} \mathbf{B} = 0$ and $\mathbf{curl}\, \mathbf{E} = -\dot{\mathbf{B}}$ (the Maxwell–Bianchi equations).

□

Changing orientation. Let Σ denote the inertial system considered above.

(i) Space reflection: We want to pass from Σ to a left-handed coordinate system Σ' by using the following coordinate transformation:

$$x' = -x, \quad y' = -y, \quad z' = -z, \quad t' = t.$$

The new basis vectors read as $\mathbf{e}_{k'} := -\mathbf{e}_k, k = 1, 2, 3$. The vectors $\mathbf{e}_{1'}, \mathbf{e}_{2'}, \mathbf{e}_{3'}$ form a left-handed orthonormal basis. Setting $P := (x, y, z, t)$ and $P' := (x', y', z', t')$, from (19.53) we get the following transformation formulas:

$$E^{k'}(P') = -E^k(P), \quad B^{k'}(P') = B^k(P), \quad J^{k'}(P') = -J^k(P), \quad \varrho'(P') = \varrho(P)$$

where $k = 1, 2, 3$. An observer in Σ' measures the following vectors:

- $E^{k'}(P')\mathbf{e}_{k'} = -E^k(P)\mathbf{e}_{k'}$ (electric field vector),
- $B^{k'}(P')\mathbf{e}_{k'} = B^k(P)\mathbf{e}_{k'}$ (magnetic field vector),
- $J^{k'}(P')\mathbf{e}_{k'} = -J^k(P)\mathbf{e}_{k'}$ (electric current density vector).

(ii) Time reflection: We pass from Σ to the inertial system Σ' by using the following coordinate transformation:

$$x' = x, \quad y' = y, \quad z' = z, \quad t' = -t.$$

Moreover, $\mathbf{e}_{k'} = \mathbf{e}_k, k = 1, 2, 3$. Setting $P := (x, y, z, t)$ and $P' := (x', y', z', t')$, we get the following transformation formulas:

$$E^{k'}(P') = E^k(P), \quad B^{k'}(P') = -B^k(P), \quad J^{k'}(P') = -J^k(P), \quad \varrho'(P') = \varrho(P)$$

where $k = 1, 2, 3$. An observer in Σ' measures the following vectors:

- $E^{k'}(P')\mathbf{e}_{k'} = E^k(P)\mathbf{e}_k$ (electric field vector),
- $B^{k'}(P')\mathbf{e}_{k'} = -B^k(P)\mathbf{e}_k$ (magnetic field vector),
- $J^{k'}(P')\mathbf{e}_{k'} = -J^k(P)\mathbf{e}_k$ (electric current density vector).

Charge conjugation. This transformation does not change the coordinates of the inertial system Σ, but the sign of the electric charges is changed. We define

$$CF_{\alpha\beta} := -F_{\alpha\beta}, \qquad C\mathcal{J}^\beta = -\mathcal{J}^\beta, \qquad \lambda, \beta = 0, 1, 2, 3.$$

Obviously, the Maxwell equations (19.52) are invariant under this transformation. By (19.53), we get

$$CE^k = -E^k, \quad CB^k := -B^k, \quad CJ^k := -J^k, \quad C\varrho := -\varrho$$

where $k = 1, 2, 3$.

19.3.2 The Language of Differential Forms and Hodge Duality

The Maxwell equations on the Minkowski manifold read as follows:

$$\boxed{-d^* F = \mu_0 \mathcal{J}, \qquad dF = 0 \qquad \text{on } \mathbb{M}^4.} \tag{19.56}$$

We are given the 1-form $\mathcal{J}$ on $\mathbb{M}^4$, and we are looking for the 2-form F. Since differential forms and their Hodge duals possess an invariant meaning on positively oriented pseudo-Riemannian manifolds, this formulation of the Maxwell equations is valid for all local coordinate systems on the Minkowski manifold $\mathbb{M}^4$.

Equivalent formulation. The Maxwell equations (19.56) are equivalent to the system

$$\boxed{-d * F = \mu_0 * \mathcal{J}, \qquad dF = 0 \qquad \text{on } \mathbb{M}^4.} \tag{19.57}$$

In fact, by Sect. 18.5.2 on page 931,

$$\mu_0 * \mathcal{J} = - * *^{-1} d * F = -(d * F).$$

Inertial system. Choose the strictly positively oriented inertial system as in Sect. 19.3.1 on page 960. Set

$$F := \tfrac{1}{2} F_{\alpha\beta} dx^\alpha \wedge dx^\beta, \qquad \mathcal{J} = \mathcal{J}_\alpha dx^\alpha.$$

Explicitly, we get

$$\begin{aligned} F = &-\tfrac{1}{c}(E^1 dx + E^2 dy + E^3 dz) \wedge dx^0 \\ &-B^1 dy \wedge dz - B^2 dz \wedge dx - B^3 dx \wedge dy, \end{aligned} \tag{19.58}$$

and

$$\mathcal{J} = \varrho c dx^0 - J^1 dx - J^2 dy - J^3 dz. \tag{19.59}$$

By Hodge duality (see Sect. 18.5.2 on page 931), we get

$$\begin{aligned} * F = &-\tfrac{1}{c}(E^1 dy \wedge dz + E^2 dz \wedge dx + E^3 dx \wedge dy) \\ &+(B^1 dx + B^2 dy + B^3 dz) \wedge dx^0. \end{aligned} \tag{19.60}$$

Thus, the transformation $F \Rightarrow *F$ corresponds to the transformations

$$\mathbf{E} \Rightarrow -c\mathbf{B}, \qquad \text{and} \qquad c\mathbf{B} \Rightarrow \mathbf{E}. \tag{19.61}$$

Moreover,

$$*\mathcal{J} = \varrho c dx \wedge dy \wedge dz - J^1 dx^0 \wedge dy \wedge dz - J^2 dx^0 \wedge dz \wedge dx - J^3 dx^0 \wedge dx \wedge dy.$$

Theorem 19.3 *The Maxwell equations* (19.57) *are equivalent to Maxwell's formulation in terms of vector calculus.*

Proof. We get

$$\begin{aligned}-dF = {} & \tfrac{1}{c}(dE^1 \wedge dx + dE^2 \wedge dy + dE^3 \wedge dz) \wedge dx^0 \\ & + dB^1 \wedge dy \wedge dz + dB^2 \wedge dz \wedge dx + dB^3 \wedge dx \wedge dy.\end{aligned}$$

Hence

$$\begin{aligned}-dF = {} & \tfrac{1}{c}\big((E_x^2 - E_y^1 + B_t^3)\, dx \wedge dy + (E_y^3 - E_z^2 + B_t^1)\, dy \wedge dz\big) \wedge dx^0 \\ & + \tfrac{1}{c}(E_z^1 - E_x^3 + B_t^2)\, dz \wedge dx \wedge dx^0 + (B_x^1 + B_y^2 + B_z^3)\, dx \wedge dy \wedge dz.\end{aligned}$$

Using the Hodge duality transformation (19.61), we obtain

$$\begin{aligned}-d * F = {} & -\big((B_x^2 - B_y^1 - \tfrac{1}{c^2}E_t^3)\, dx \wedge dy + (B_y^3 - B_z^2 - \tfrac{1}{c^2}E_t^1)\, dy \wedge dz\big) \wedge dx^0 \\ & -(B_z^1 - B_x^3 - \tfrac{1}{c^2}E_t^2)\, dz \wedge dx \wedge dx^0 + \frac{1}{c}(E_x^1 + E_y^2 + E_z^3)\, dx \wedge dy \wedge dz.\end{aligned}$$

Therefore, the equation $dF = 0$ yields

$$\mathbf{curl}\,\mathbf{E} + \mathbf{B}_t = 0, \qquad \operatorname{div}\mathbf{B} = 0.$$

Furthermore, $-d * F = \mu_0 * \mathcal{J}$ yields

$$\operatorname{div}\mathbf{E} = \mu_0 c^2 \varrho, \qquad \mathbf{curl}\,\mathbf{B} - \tfrac{1}{c^2}\mathbf{E}_t = \mu_0 \mathbf{J}.$$

Finally, note that $\varepsilon_0\mu_0 c^2 = 1$. □

Hodge duality. *If F is a solution of the homogeneous Maxwell equations*

$$d * F = 0, \qquad dF = 0 \qquad \text{on } \mathbb{M}^4,$$

then so is $*F$.

In fact, this follows from $*(*F) = -F$. In order to formulate this in terms of physics, note that the Hodge transformation $F \Rightarrow *F$ corresponds to the transformation (19.61). Obviously, the homogeneous Maxwell equations

$$\operatorname{div}\mathbf{E} = 0, \quad \operatorname{div}\mathbf{B} = 0, \quad \mathbf{curl}\,\mathbf{E} = -\dot{\mathbf{B}}, \quad \mathbf{curl}\,\mathbf{B} = \tfrac{1}{c^2}\dot{\mathbf{E}}$$

are left invariant under the Hodge transformation (19.61).

19.3.3 De Rham Cohomology and the Four-Potential of the Electromagnetic Field

> The Minkowski manifold $\mathbb{M}^4$ is continuously contractible to a point. Thus, from the topological point of view, the Minkowski manifold is trivial. From the analytical point view, this has the nice consequence that a smooth electromagnetic field F has always a 4-potential A with $F = dA$. As already Maxwell noticed in his 1873 treatise on electricity and magnetism, the existence of a 4-potential considerably simplifies the construction of solutions of the Maxwell equations.
>
> Folklore

The topological triviality of the Minkowski manifold $\mathbb{M}^4$. Consider the key equation

$$\boxed{dA = F \qquad \text{on } \mathbb{M}^4.} \tag{19.62}$$

We are given the smooth differential 2-form F. We are looking for the smooth differential 1-form A. If the equation (19.62) has a solution, then it follows from Poincaré's cohomology rule that $dF = d(dA) = 0$.

Proposition 19.4 *The equation* (19.62) *has a solution A iff $dF = 0$ on $\mathbb{M}^4$. If A_{special} is a special solution of* (19.62)*, then the general solution of* (19.62) *reads as*

$$A = A_{\mathrm{special}} + d\chi \tag{19.63}$$

where $\chi : \mathbb{M}^4 \to \mathbb{R}$ is an arbitrary smooth function.

This is a special case of the fact that $H^2(\mathbb{M}^4) = \{0\}$, that is, the second de Rham cohomology group of $\mathbb{M}^4$ is trivial (see Sect. 23.4). More precisely, we have

$$H^0(\mathbb{M}^4) = \mathbb{R}, \quad H^k(\mathbb{M}^4) = \{0\}, \qquad k = 1, 2, \ldots$$

Observe the following. Since $dd\chi = 0$, we get $F = dA = dA_{\mathrm{special}}$. The transformation

$$A_{\mathrm{special}} \mapsto A_{\mathrm{special}} + d\chi \tag{19.64}$$

is called a gauge transformation. This gauge transformation changes the 4-potential, but the electromagnetic field F remains unchanged.

Solution of the Maxwell equations via 4-potential. We are given the smooth differential 1-form $\mathcal{J}$ on $\mathbb{M}^4$. Suppose that the smooth differential 2-form F is a solution of the Maxwell equations

$$dF = 0, \qquad -d^*F = \mu_0 \mathcal{J} \qquad \text{on } \mathbb{M}^4. \tag{19.65}$$

It follows from $dF = 0$ that there exists a differential 1-form A such that $F = dA$. Since $-d^*F = \mu_0\mathcal{J}$, we get

$$-d^*dA = \mu_0\mathcal{J}.$$

Suppose that the 4-potential A of the electromagnetic field F satisfies the so-called Lorenz gauge condition

$$\boxed{d^*A = 0 \qquad \text{on } \mathbb{M}^4.} \tag{19.66}$$

Then the 4-potential A satisfies the equation

$$\boxed{-(d^*d + dd^*)A = \mu_0 \mathcal{J}.} \qquad (19.67)$$

In terms of Hodge theory, the operator $d^*d + dd^*$ is the Laplacian of the Minkowski manifold $\mathbb{M}^4$. In an inertial system, the key equation (19.67) represents an inhomogeneous wave equation (see (19.71) below). The argument above can be easily reversed.

Theorem 19.5 *If the smooth differential 1-form A satisfies the wave equation* (19.67) *together with the Lorenz gauge condition* (19.66)*, then $F = dA$ is a solution of the Maxwell equations* (19.65).

The differential forms F and A are invariant, that is, they do not depend on the choice of the local coordinate system on the Minkowski manifold $\mathbb{M}^4$. This means that $F_{\alpha\beta}$ and A_α are tensorial families on $\mathbb{M}^4$. Lifting the indices, we get

$$A^\alpha := g^{\alpha\beta} A_\beta, \qquad \alpha = 0, 1, 2, 3.$$

In order to obtain Maxwell's classical notation below, we define

$$\mathsf{A}^\alpha := \sigma A^\alpha, \qquad \alpha = 0, 1, 2, 3$$

where $\sigma = \pm 1$ is the time orientation of the local coordinate system.

Charge conservation. Again we are given the smooth differential 1-form $\mathcal{J}$. If there exists a solution of the Maxwell equations (19.65), then

$$\boxed{d^*\mathcal{J} = 0 \qquad \text{on } \mathbb{M}^4.} \qquad (19.68)$$

In fact, $\mu_0 d^*\mathcal{J} = -d^*d^*F = 0$ (see Sect. 18.5.2 on page 931). We will show below that the equation (19.68) describes the conservation of electric charge.

Inertial systems. Choose a strictly positively oriented inertial system Σ as in Sect. 19.3.1 on page 960. Set

$$\mathbf{A} = \mathsf{A}^1\mathbf{i} + \mathsf{A}^2\mathbf{j} + \mathsf{A}^3\mathbf{k}, \qquad \mathsf{A}^0 := \tfrac{1}{c}U,$$

and $A_\alpha := \eta_{\alpha\beta}A^\beta$, $\alpha = 0, 1, 2, 3$. Explicitly, $A_0 = A^0 = \frac{1}{c}U$, and $A_k = -A^k = -\mathsf{A}^k$ where $k = 1, 2, 3$. Define

$$A = A_\beta dx^\beta.$$

Then, $dA = \partial_\alpha A_\beta dx^\alpha \wedge dx^\beta = \frac{1}{2}(\partial_\alpha A_\beta - \partial_\beta A_\alpha)\, dx^\alpha \wedge dx^\beta$. It follows from

$$F = \tfrac{1}{2} F_{\alpha\beta}\, dx^\alpha \wedge dx^\beta$$

that

$$\boxed{F_{\alpha\beta} = \partial_\alpha A_\beta - \partial_\beta A_\alpha, \qquad \alpha, \beta = 0, 1, 2, 3.} \qquad (19.69)$$

By (19.54) on page 960, this means that

$$\mathbf{E} = -\,\mathbf{grad}\, U - \dot{\mathbf{A}}, \qquad \mathbf{B} = \mathbf{curl}\,\mathbf{A}.$$

Proposition 19.6 *The Lorenz gauge condition $d^*A = 0$ is equivalent to the equation $\partial_\alpha A^\alpha = 0$.*

Proof. It follows from

$$*A = A_0\, dx \wedge dy \wedge dz + A_1\, dx^0 \wedge dy \wedge dz + A_2\, dx^0 \wedge dz \wedge dx + A_3\, dx^0 \wedge dx \wedge dy$$

that $d(*A) = \partial_\alpha A^\alpha\, dx^0 \wedge dx \wedge dy \wedge dz$. Note that $d^*A = *^{-1}d * A$. Thus, $d^*A = 0$ is equivalent to $d * A = 0$. □

Proposition 19.7 *The equation* $-(d^*d + dd^*)A = \mu_0 \mathcal{J}$ *together with* $d^*A = 0$ *reads as*

$$\partial_\alpha \partial^\alpha A^\beta = \mu_0 \mathcal{J}^\beta, \qquad \beta = 0, 1, 2, 3 \tag{19.70}$$

together with $\partial_\alpha A^\alpha = 0$.

Proof. We have to show that $-d^*dA = \mu_0 \mathcal{J}$. This corresponds to the Maxwell equation

$$\partial_\alpha F^{\alpha\beta} = \mu_0 \mathcal{J}^\beta, \quad \beta = 0, 1, 2, 3.$$

By (19.69), we have

$$\partial_\alpha(\partial^\alpha A^\beta - \partial^\beta A^\alpha) = \mu_0 \mathcal{J}^\beta \quad \text{and} \quad \partial_\alpha A^\alpha = 0, \qquad \beta = 0, 1, 2, 3.$$

This yields (19.70). □

In the language of vector calculus, the Lorenz gauge condition $\partial_\alpha A^\alpha = 0$ reads as

$$\boxed{\frac{1}{c^2}\frac{\partial U}{\partial t} + \operatorname{div} \mathbf{A} = 0.}$$

Furthermore, the equation (19.70) reads as

$$\boxed{\varepsilon_0 \Box U = \varrho, \qquad \Box \mathbf{A} = \mu_0 \mathbf{J}} \tag{19.71}$$

with the wave operator $\Box := \frac{1}{c^2}\frac{\partial^2}{\partial t^2} - \frac{\partial^2}{\partial x^2} - \frac{\partial^2}{\partial y^2} - \frac{\partial^2}{\partial z^2}$. The gauge transformation (19.64) reads as

$$U \mapsto U + \frac{\partial \chi}{\partial t}, \qquad \mathbf{A} \mapsto \mathbf{A} - \mathbf{grad}\, \chi.$$

Proposition 19.8 *The equation* $d^*\mathcal{J} = 0$ *means that* $\partial_\alpha \mathcal{J}^\alpha = 0$.

In the language of vector calculus, this reads as

$$\boxed{\dot{\varrho} + \operatorname{div} \mathbf{J} = 0.}$$

This equation describes the conservation of the electric charge.

Changing orientation. Let Σ denote the positively oriented inertial system considered above. Recall that this corresponds to a right-handed Cartesian (x, y, z)-coordinate system, and the time t is positively oriented.

(i) Space reflection: We want to pass from Σ to a left-handed coordinate system Σ' by using the following coordinate transformation:

$$x' = -x, \quad y' = -y, \quad z' = -z, \quad t' = t.$$

The new basis vectors read as $\mathbf{e}_{k'} := -\mathbf{e}_k, k = 1, 2, 3$. Setting $P := (x, y, z, t)$ and $P' := (x', y', z', t')$, we get the following transformation formulas:

$$\mathsf{A}^{0'}(P') = \mathsf{A}^0(P), \qquad \mathsf{A}^{k'}(P') = -\mathsf{A}^k(P), \qquad k = 1, 2, 3.$$

An observer in Σ' measures the scalar potential $U'(P') = U(P)$ and the vector potential $\mathsf{A}^{k'}(P')\mathbf{e}_{k'} = -\mathsf{A}^k(P)\mathbf{e}_{k'}$.

(ii) Time reflection: We pass from Σ to the inertial system Σ' by using the following coordinate transformation:

$$x' = x, \quad y' = y, \quad z' = z, \quad t' = -t.$$

Moreover, $\mathbf{e}_{k'} = \mathbf{e}_k, k = 1, 2, 3$. Setting $P := (x, y, z, t)$ and $P' : s = (x', y', z', t')$, we get the following transformation formulas:

$$\mathsf{A}^{0'}(P') = \mathsf{A}^0(P), \quad \mathsf{A}^{k'}(P') = -\mathsf{A}^k(P), \qquad k = 1, 2, 3.$$

An observer in Σ' measures the scalar potential $U'(P') = U(P)$ and the vector potential $\mathsf{A}^{k'}(P')\mathbf{e}_{k'} = -\mathsf{A}^k(P)\mathbf{e}_{k'}$.

Charge conjugation. This transformation does not change the coordinates of Σ, but the sign of the electric charges is changed. We define

$$CA_\alpha := -A_\alpha, \qquad \alpha = 0, 1, 2, 3.$$

This implies $CU = -U$ and $C\mathsf{A}^k = -\mathsf{A}^k$, $k = 1, 2, 3$. Obviously, the Maxwell equations (19.52) are invariant under this transformation.

Arbitrary local coordinates on the Minkowski manifold $\mathbb{M}^4$. Then we have to replace the partial derivative ∂_α by the covariant partial derivative ∇_α with respect to the metric tensorial family $g_{\alpha\beta}$. Summarizing, we get the following:

- $F_{\alpha\beta} = \nabla_\alpha A_\beta - \nabla_\beta A_\alpha$ (relation between the electromagnetic field and the four-potential),
- $\nabla_\alpha A^\alpha = 0$ (Lorenz gauge condition),
- $\nabla_\alpha \nabla^\alpha A^\beta = \mu_0 \mathcal{J}^\beta$ (inhomogeneous wave equation for the four-potential),
- $\nabla_\beta \mathcal{J}^\beta = 0$ (conservation of electric charge).

19.3.4 The Language of Fiber Bundles

We have shown in Sects. 13.3 and 13.4 that the key relation (19.69) between the electromagnetic field $F_{\alpha\beta}$ and the 4-potential A_α is a consequence of Cartan's structural equation

$$\mathsf{F} = d\mathsf{A}$$

which describes the relation between the curvature 2-form F and the connection 1-form A of the principal bundle $\mathbb{M}^4 \times U(1)$.

19.4 The Transformation Law for the Electromagnetic Field

Choose a strictly positively oriented inertial system Σ (resp. Σ') with the right-handed Cartesian coordinates x, y, z (resp. $x'y', z'$), and the positively oriented time t (resp. t'). As usually, we set $x^1 := x, x^2 := y, x^3 := z, x^0 := ct$ where c is the velocity of light in a vacuum. Furthermore, let $\mathbf{i}, \mathbf{j}, \mathbf{k}$ (resp. $\mathbf{i}', \mathbf{j}', \mathbf{k}'$) be the corresponding orthonormal basis of Σ (resp. Σ'). Suppose that the inertial system

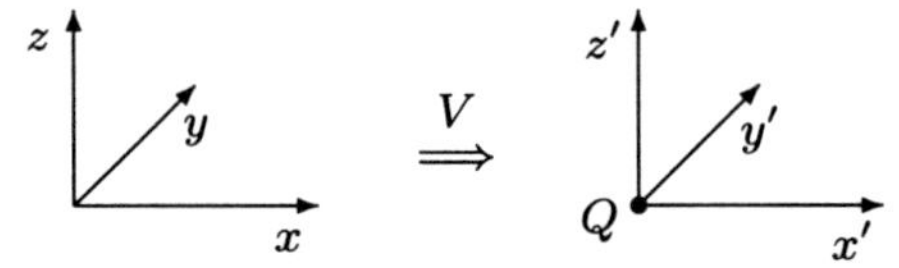

Fig. 19.12. Moving electric charge

Σ and Σ' coincide at time $t = t' = 0$, and the origin of Σ' moves with the velocity V along the positive x-axis of Σ (Fig. 19.12). Then we have the Lorentz transformation

$$x^{0'} = \frac{x^0 - Vx^1/c}{\sqrt{1 - V^2/c^2}}, \quad x^{1'} = \frac{x^1 - Vx^0/c}{\sqrt{1 - V^2/c^2}}, \quad x^{2'} = x^2, \quad x^{3'} = x^3.$$

An observer in Σ measures the electromagnetic field

$$\mathbf{E} = E^1\mathbf{i} + E^2\mathbf{j} + E^3\mathbf{k}, \qquad \mathbf{B} = B^1\mathbf{i} + B^2\mathbf{j} + B^3\mathbf{k},$$

the charge density ϱ, and the current density vector $\mathbf{J} = J^1\mathbf{i} + J^2\mathbf{j} + J^3\mathbf{k}$. Similarly, an observer in Σ' measures

$$\mathbf{E}' = E^{1'}\mathbf{i}' + E^{2'}\mathbf{j}' + E^{3'}\mathbf{k}', \qquad \mathbf{B}' = B^{1'}\mathbf{i}' + B^{2'}\mathbf{j}' + B^{3'}\mathbf{k}',$$

as well as ϱ', and $\mathbf{J}' = J^{1'}\mathbf{i}' + J^{2'}\mathbf{j}' + J^{3'}\mathbf{k}'$. The tensorial families J^α and $F^{\alpha\beta}$ transform like x^α and $x^\alpha x^\beta$, respectively. Explicitly, using (19.55) on page 961, we get

$$\varrho' = \frac{\varrho - J^1V/c^2}{\sqrt{1 - V^2/c^2}}, \quad J^{1'} = \frac{J^1 - \varrho V}{\sqrt{1 - V^2/c^2}}, \quad J^{2'} = J^2, \quad J^{3'} = J^3, \tag{19.72}$$

as well as

$$E^{1'} = E^1, \quad E^{2'} = \frac{E^2 - B^3V}{\sqrt{1 - V^2/c^2}}, \quad E^{3'} = \frac{E^3 + B^2V}{\sqrt{1 - V^2/c^2}} \tag{19.73}$$

and

$$B^{1'} = B^1, \quad B^{2'} = \frac{B^2 + E^3V/c^2}{\sqrt{1 - V^2/c^2}}, \quad B^{3'} = \frac{B^3 - E^2V/c^2}{\sqrt{1 - V^2/c^2}}. \tag{19.74}$$

The inverse formulas are obtained by replacing V by $-V$.

Example. Consider a particle of electric charge Q which rests at the origin of the inertial system Σ' for all times.

- The observer in Σ' measures the electric Coulomb field $\mathbf{E}'(x', y', z')$ and no magnetic field, $\mathbf{B}' \equiv 0$.
- The observer in Σ observes a charge moving with the constant velocity V along the x-axis. He measures the electromagnetic field $\mathbf{E}(x, y, z, t), \mathbf{B}(x, y, z, t)$ with the following components:

$$E^1 = E^{1'}, \quad E^2 = \frac{E^{2'}}{\sqrt{1 - V^2/c^2}}, \quad E^3 = \frac{E^{3'}}{\sqrt{1 - V^2/c^2}},$$

$$B^1 = 0, \quad B^2 = -\frac{E^{3'}V/c^2}{\sqrt{1 - V^2/c^2}}, \quad B^3 = \frac{E^{2'}V/c^2}{\sqrt{1 - V^2/c^2}}.$$

19.5 Electromagnetic Waves

Light rays are the streamlines of the electromagnetic energy flow.
Folklore

Consider a strictly positively oriented inertial system as on page 960. Let $\mathbf{n}$ and $\mathbf{e}$ be a pair of orthogonal unit vectors (i.e., $\mathbf{ne} = 0$). Let $f : \mathbb{M}^4 \to \mathbb{R}$ be a smooth function.

Theorem 19.9 *The electromagnetic field*

$$\mathbf{E}(\mathbf{x}, t) := f(\mathbf{nx} - ct)\mathbf{e}, \qquad \mathbf{B}(\mathbf{x}, t) := \tfrac{1}{c}(\mathbf{n} \times \mathbf{E}(\mathbf{x}, t)) \tag{19.75}$$

is a solution of the Maxwell equations with $\varrho \equiv 0$ *and* $\mathbf{J} \equiv 0$ *(no electric charges and no electric currents).*

Using Hamilton's nabla calculus, the proof will be given in Problem 19.9.

Wave fronts. Let a be a fixed real number. The equation

$$\mathbf{nx} - ct = a$$

describes a plane with the normal vector $\mathbf{n}$ which propagates in direction of $\mathbf{n}$ with the velocity c. These planes are called wave fronts of the electromagnetic wave (19.75). Both the electric field and the magnetic field of (19.75) are orthogonal to the direction $\mathbf{n}$ of propagation, and they are constant on the wave fronts.

Light rays. The flow of energy is described by the energy current density vector $\mathbf{S} = \frac{1}{\mu_0}(\mathbf{E} \times \mathbf{B})$. Explicitly,

$$\mathbf{S}(\mathbf{x}, t) = \frac{f(\mathbf{nx} - ct)^2}{\mu_0 c}\,\mathbf{n}.$$

This means that the energy flows in direction of the vector $\mathbf{n}$. The trajectories of the energy flow are straight lines which are orthogonal to the moving wave fronts. These trajectories are called light rays.

19.6 Invariants of the Electromagnetic Field

We want to show that simple arguments from invariant theory yield crucial properties of the electromagnetic field.

Invariants under rotations. By Cauchy's theorem on isotropic functions (see Theorem 9.7 page 565), a real valued function $f : E_3 \times E_3 \to$ with

$$f(\mathbf{E}, \mathbf{B}) = f(R\mathbf{E}, R\mathbf{B})$$

for all rotations R depends only on the inner products

$$\mathbf{E}^2, \quad \mathbf{B}^2, \quad \mathbf{EB}.$$

In addition, the space reflection $\mathbf{x} \mapsto -\mathbf{x}$ induces the transformations

$$\mathbf{E} \mapsto -\mathbf{E}, \quad \mathbf{B} \mapsto \mathbf{B}, \quad \mathbf{J} \mapsto -\mathbf{J}, \quad \mathbf{A} \mapsto -\mathbf{A},$$

and the time reflection $t \mapsto -t$ yields

$$\mathbf{E} \mapsto \mathbf{E}, \quad \mathbf{B} \mapsto -\mathbf{B}, \quad \mathbf{J} \mapsto -\mathbf{J}, \quad \mathbf{A} \mapsto -\mathbf{A}.$$

Consequently, the two functions

- $\eta := \frac{1}{2\mu_0}\left(\frac{1}{c^2}\mathbf{E}^2 + \mathbf{B}^2\right)$ and
- $\mathcal{L} := \frac{1}{2}\left(\mathbf{B}^2 - \frac{1}{c^2}\mathbf{E}^2\right) + \mu_0 U\varrho - \mu_0 \mathbf{AJ}$

are invariant under rotations, space reflections, and time reflections. We will motivate below that η (resp. $\mathcal{L}$) represents the energy density (resp. the Lagrangian) of the electromagnetic field. The function

$$g(\mathbf{E}, \mathbf{B}) := \mathbf{EB}$$

is invariant under rotations, but g is not invariant under space (resp. time) reflections.

Relativistic invariants. On the Minkowski manifold $\mathbb{M}^4$, the tensorial family $F^{\alpha\beta}$ allows us to construct the invariant

$$F_{\alpha\beta}F^{\alpha\beta},$$

by the index principle. In an inertial system, this invariant is equal to $2(\mathbf{B}^2 - \frac{1}{c^2}\mathbf{E}^2)$ (see (19.54)). Thus, the function $\mathcal{L}$ above can be written as

$$\mathcal{L} = \tfrac{1}{4}F_{\alpha\beta}F^{\alpha\beta} + \mu_0 A_\beta \mathcal{J}^\beta.$$

We will show on page 976 that the corresponding principle of critical action yields the Maxwell equations.

19.6.1 The Motion of a Charged Particle and the Lorentz Force

Consider a particle with rest mass m_0 and electric charge Q. The equation of motion for the particle reads as

$$\boxed{\frac{dp^\alpha(\tau)}{d\tau} = QF^{\alpha\beta}(x(\tau))\,\mathcal{J}_\beta(x(\tau)), \qquad \alpha = 0,1,2,3, \quad \tau \in \mathbb{R}} \tag{19.76}$$

where $p^\alpha := m_0 \frac{dx^\alpha(\tau)}{d\tau}$, $\alpha = 0,1,2,3$ (4-momentum). The trajectory of the particle is given by the curve

$$x^\alpha = x^\alpha(\tau), \qquad \alpha = 0,1,2,3, \quad \tau \in \mathbb{R}$$

on the Minkowski manifold $\mathbb{M}^4$. In addition, we assume that

- the particle moves with subvelocity of light, that is,

$$g_{\alpha\beta}(x(\tau))\, p^\alpha(\tau)p^\beta(\tau) > 0, \qquad \tau \in \mathbb{R},$$

- and the parameter τ is the proper time of the particle, that is, $s = \tau/c$ represents the arc length of the curve on $\mathbb{M}^4$.

Motivation. The equation of motion (19.76) possesses an invariant meaning on the Minkowski manifold $\mathbb{M}^4$, that is, it is valid for arbitrary local coordinates. In order to motivate this equation, it remains to check a simple special case. To this end, let Σ be a strictly positively oriented inertial system as used on page 960. Suppose that the magnetic field vanishes, $\mathbf{B} \equiv 0$. Then, the proper τ is given by

$$\tau(t_1) = \int_0^{t_1} \sqrt{1 - \dot{\mathbf{x}}^2(t)/c^2}\, dt,$$

and the equation (19.76) of motion reads as

$$\frac{d}{dt}\big(m(t)\dot{\mathbf{x}}(t)\big) = Q\mathbf{E}(\mathbf{x}(t)), \qquad t \in \mathbb{R} \tag{19.77}$$

with Einstein's relativistic mass

$$m(t) := \frac{m_0}{\sqrt{1 - \dot{\mathbf{x}}^2(t)/c^2}}.$$

This is a quite natural result which motivates (19.76).

For a general electromagnetic field $\mathbf{E}, \mathbf{B}$, the equation (19.76) of motion reads as

$$\boxed{\frac{d}{dt}\big(m(t)\dot{\mathbf{x}}(t)\big) = Q\mathbf{E}\big(\mathbf{x}(t)\big) + Q\dot{\mathbf{x}}(t) \times \mathbf{B}\big(\mathbf{x}(t)\big), \qquad t \in \mathbb{R}} \tag{19.78}$$

and

$$\boxed{\frac{d}{dt}E_{\text{mech}}(t) = Q\dot{\mathbf{x}}(t)\mathbf{E}(\mathbf{x}(t)), \qquad t \in \mathbb{R}.} \tag{19.79}$$

Here, we use Einstein's mechanical energy

$$E_{\text{mech}} := m(t)c^2.$$

Equation (19.79) describes the change of mechanical energy.

Summarizing, the equation of motion (19.76) *is the simplest relativistically invariant equation which generalizes the quite natural equation* (19.77).

This way, we obtain the Lorentz force

$$\boxed{\mathbf{F} = Q\mathbf{E} + Q\dot{\mathbf{x}} \times \mathbf{B}.}$$

19.6.2 The Energy Density and the Energy-Momentum Tensor

We want to show that the notion of energy density η of an electromagnetic field is not a relativistic invariant. But the energy-momentum tensor $T = T^{\alpha\beta}\,\partial_\alpha \otimes \partial_\beta$ is a relativistic invariant, and we have $\eta = T^{00}$.

Energy density. Consider a strictly positively oriented inertial system as used on page 960. For the energy density η of the electromagnetic field $\mathbf{E}, \mathbf{B}$, we make the ansatz

$$\eta = \tfrac{a}{2}\left(\tfrac{1}{c^2}\mathbf{E}^2 + \mathbf{B}^2\right)$$

where a is an unknown positive universal constant. This ansatz is the simplest expression which has the following properties:

- η is never negative,
- η is invariant under rotations, space reflections, and time reflections,
- η is invariant under the Hodge duality transformation: $\mathbf{E} \Rightarrow -c\mathbf{B}$ and $\mathbf{B} \Rightarrow \frac{1}{c}\mathbf{E}$.

It follows from the Maxwell equations

$$\mathbf{curl}\,\mathbf{E} = -\dot{\mathbf{B}}, \qquad \mathbf{curl}\,\mathbf{B} = \mu_0\mathbf{J} + \tfrac{1}{c^2}\dot{\mathbf{E}}$$

that

$$\dot{\eta} + \operatorname{div}(a\mu_0 \mathbf{S}) + a\mu_0 \mathbf{JE} = 0 \tag{19.80}$$

where $\mathbf{S} := \frac{1}{\mu_0}(\mathbf{E} \times \mathbf{B})$. In fact,

$$\begin{aligned} \dot{\eta} = \tfrac{a}{c^2}\mathbf{E}\dot{\mathbf{E}} + a\mathbf{B}\dot{\mathbf{B}} &= a\mathbf{E}(\mathbf{curl}\,\mathbf{B} - \mu_0\mathbf{J}) - a\mathbf{B}\,\mathbf{curl}\,\mathbf{E} \\ &= -a\mu_0\mathbf{JE} - a\operatorname{div}(\mathbf{E}\times\mathbf{B}). \end{aligned}$$

Motivated by (19.79) and (19.80), we choose $a := 1/\mu_0$. This way, we get the energy density

$$\boxed{\eta = \tfrac{1}{2\mu_0}\left(\tfrac{1}{c^2}\mathbf{E}^2 + \mathbf{B}^2\right)}$$

of the electromagnetic field $\mathbf{E}, \mathbf{B}$.

The energy–momentum tensor. Let us now pass over to relativistic invariance. To this end, we write

$$\eta = \tfrac{1}{2\mu_0}\left(\mathbf{B}^2 - \tfrac{1}{c^2}\mathbf{E}^2 + \tfrac{2}{c^2}\mathbf{E}^2\right) = \tfrac{1}{4\mu_0}(F^{\lambda\mu}F_{\lambda\mu} + 4F^{0k}F_{0k}). \tag{19.81}$$

Motivated by (19.81), we define the tensorial family

$$\boxed{T^{\alpha\beta} := \tfrac{1}{4\mu_0}(g^{\alpha\beta}F^{\lambda\mu}F_{\lambda\mu} + 4F^{\alpha\lambda}F_{\lambda\mu}g^{\mu\beta}).} \tag{19.82}$$

Then

$$\boxed{\eta = T^{00}.} \tag{19.83}$$

The tensor

$$T := T^{\alpha\beta}\,\partial_\alpha \otimes \partial_\beta$$

is called the energy-momentum tensor of the electromagnetic field

$$F = \tfrac{1}{2}F_{\alpha\beta}\,dx^\alpha \wedge dx^\beta.$$

Proposition 19.10 *The energy-momentum tensor T is symmetric and trace free.*

This means that $T^{\alpha\beta} = T^{\beta\alpha}$, and hence $T_{\alpha\beta} = T_{\beta\alpha}$ for all indices. Moreover, the trace T^α_α vanishes.

Proof. (I) We show that $T^{\alpha\beta} = T^{\beta\alpha}$. Note that $F^{\alpha\lambda}F_{\lambda\mu}g^{\mu\beta} = g^{\alpha\varrho}g^{\beta\mu}g^{\lambda\sigma}F_{\varrho\sigma}F_{\lambda\mu}$. This remains unchanged under the index transformation $\alpha \Leftrightarrow \beta$, $\varrho \Leftrightarrow \mu$, and $\sigma \Leftrightarrow \lambda$. Moreover, $g^{\alpha\beta} = g^{\beta\alpha}$.

(II) We show that $T^\alpha_\alpha = 0$. This follows from $\delta^\alpha_\alpha = 4$, and

$$4\mu_0 T^\alpha_\beta = \delta^\alpha_\beta\, F^{\lambda\mu}F_{\lambda\mu} + 4F^{\alpha\lambda}F_{\lambda\beta} = \delta^\alpha_\beta\, F^{\lambda\mu}F_{\lambda\mu} - 4F^{\alpha\lambda}F_{\beta\lambda}.$$

□

19.6.3 Conservation Laws

Perpetual motion machines have fascinated many people over hundreds of years. The first perpetual motion device was suggested in the 13th century by the French architect Villard de Honnecourt.
In 1842, Robert Mayer (1814–1878) formulated the universal law of energy conservation in nature, and he determined experimentally the relation between mechanical energy and heat energy. The first law of thermodynamics

(energy conservation) and the second law of thermodynamics tell us that it is impossible to construct perpetual motion machines.
In 1918, Emmy Noether (1882–1935) proved in terms of mathematics that conservation laws are based on symmetries. Conservation laws and symmetries are fundamental concepts of the modern philosophy of sciences.

Folklore

Consider an arbitrary inertial system. Define

- $T^{\alpha\beta} := \frac{1}{4\mu_0}(g^{\alpha\beta}F^{\lambda\mu}F_{\lambda\mu} + 4F^{\alpha\lambda}F_{\lambda\mu}g^{\mu\beta})$,
- $S^{\alpha\beta\gamma} := x^\alpha T^{\beta\gamma} - x^\beta T^{\alpha\gamma}$,
- $f^\alpha := F^{\alpha\beta}\mathcal{J}_\beta$,
- $s^{\alpha\beta} := x^\alpha f^\beta - x^\beta f^\alpha$.

According to Theorem 8.2 on page 458, these expressions can be extended to tensorial families on the Minkowski manifold $\mathbb{M}^4$ with respect to arbitrary local coordinate systems. Using these tensorial families, we introduce the following tensors on $\mathbb{M}^4$:

- $T = T^{\alpha\beta}\,\partial_\alpha \otimes \partial_\beta$ (energy–momentum tensor),
- $S = S^{\alpha\beta\gamma}\,\partial_\alpha \otimes \partial_\beta \otimes \partial_\gamma$ (angular momentum tensor),
- $f^\alpha\partial_\alpha$ (Lorentz force 4-vector),
- $s^{\alpha\beta}\,\partial_\alpha \otimes \partial_\beta$ (Lorentz torque tensor),
- $J = \mathcal{J}^\beta\partial_\beta$ (electric current density 4-vector).

Then we have the following conservation laws:

(i) Conservation of electric charge: $\operatorname{div} J = 0$, that is,

$$\nabla_\beta \mathcal{J}^\beta = 0, \qquad \beta = 0,1,2,3. \tag{19.84}$$

(ii) Conservation of energy and momentum: $\operatorname{div} T + f = 0$, that is,

$$\nabla_\beta T^{\alpha\beta} + f^\alpha = 0, \qquad \alpha = 0,1,2,3. \tag{19.85}$$

(iii) Conservation of angular momentum: $\operatorname{div}(x \wedge T) + x \wedge f = 0$, that is,

$$\nabla_\gamma S^{\alpha\beta\gamma} + s^{\alpha\beta} = 0, \qquad \alpha,\beta = 0,1,2,3. \tag{19.86}$$

These equations are valid for arbitrary local coordinates on the Minkowski manifold $\mathbb{M}^4$. In an inertial system, we get $\nabla_\alpha = \partial_\alpha$, $\alpha = 0,1,2,3$.

Theorem 19.11 *If the tensorial families $F^{\alpha\beta}$ and $\mathcal{J}^\beta$ satisfy the Maxwell equations*

$$\nabla_\alpha F^{\alpha\beta} = \mu_0\mathcal{J}^\beta, \qquad \nabla_{[\gamma}F_{\lambda\mu]} = 0, \qquad \beta,\gamma,\lambda,\mu = 0,1,2,3,$$

then the conservation laws (i)–(iii) *are valid.*

Proof. Since the equations (i)–(iii) are valid in arbitrary local coordinate systems, let us verify them by choosing an inertial system. Then $g_{\alpha\beta} = \eta_{\alpha\beta}$, and $\nabla_\alpha = \partial_\alpha$.

Ad (i). Since $F^{\alpha\beta} = -F^{\beta\alpha}$ and $\partial_\alpha\partial_\beta = \partial_\beta\partial_\alpha$, we get $\mu_0\partial_\beta\mathcal{J}^\beta = \partial_\beta\partial_\alpha F^{\alpha\beta} = 0$.

Ad (ii). We will use the same argument as in classical mechanics. To recall the classical idea, consider the equation of motion

$$m\frac{d}{dt}\dot{x} = F \tag{19.87}$$

on the real line. Hence $\dot{x}m\frac{d}{dt}\dot{x} = \dot{x}F$. The elegant trick is to use the Leibniz rule in order to get the identity

$$\frac{d}{dt}(\tfrac{1}{2}\dot{x}^2) = \tfrac{1}{2}\ddot{x}\dot{x} + \tfrac{1}{2}\dot{x}\ddot{x} = \dot{x}\ddot{x}. \tag{19.88}$$

This yields

$$\frac{d}{dt}(\tfrac{1}{2}m\dot{x}^2) = \dot{x}F$$

which describes the conservation of energy.[14]

The idea of our proof is to replace the equation of motion (19.87) by the Maxwell equations

$$\partial_\alpha F^{\alpha\beta} = \mu_0 \mathcal{J}^\beta. \tag{19.89}$$

The generalization of the key identity (19.88) will be based on the Maxwell–Bianchi equations

$$\partial_\kappa F_{\lambda\mu} + \partial_\lambda F_{\mu\kappa} + \partial_\mu F_{\kappa\lambda} = 0, \qquad F_{\lambda\mu} = -F_{\mu\lambda}. \tag{19.90}$$

(I) Equation of motion. From (19.89), we get

$$\eta^{\varrho\beta} F_{\sigma\varrho}\partial_\alpha F^{\alpha\sigma} = \mu_0 \mathcal{J}^\sigma F_{\sigma\varrho}\eta^{\varrho\beta}.$$

Since $\mathcal{J}^\sigma F_{\sigma\varrho}\eta^{\varrho\beta} = \mathcal{J}_\sigma F^{\sigma\beta} = -\mu_0 f^\beta$, we obtain

$$\eta^{\varrho\beta} F_{\sigma\varrho}\partial_\alpha F^{\alpha\sigma} = -\mu_0 f^\beta.$$

(II) The key identity. We replace the identity (19.88) by

$$\mu_0\partial_\alpha T^{\alpha\beta} = \eta^{\varrho\beta}F_{\sigma\varrho}\partial_\alpha F^{\alpha\sigma} \tag{19.91}$$

where $\mu_0 T^{\alpha\beta} := \frac{1}{4}\eta^{\alpha\beta}F^{\varrho\sigma}F_{\varrho\sigma} + F^{\alpha\sigma}F_{\sigma\varrho}\eta^{\varrho\beta}$. This implies the claim $\mu_0\partial_\alpha T^{\alpha\beta} = -f^\beta$.

(III) Proof of (19.91). By the Leibniz rule, we get

$$\mu_0\partial_\alpha T^{\alpha\beta} = \tfrac{1}{4}\eta^{\alpha\beta}\partial_\alpha(F^{\varrho\sigma}F_{\varrho\sigma}) + (\partial_\alpha F^{\alpha\sigma})F_{\sigma\varrho}\eta^{\varrho\beta} + F^{\alpha\sigma}\partial_\alpha(F_{\sigma\varrho}\eta^{\varrho\beta}).$$

Setting $\mathcal{R} := \frac{1}{4}\eta^{\alpha\beta}\partial_\alpha(F^{\varrho\sigma}F_{\varrho\sigma}) + F^{\alpha\sigma}\partial_\alpha(F_{\sigma\varrho}\eta^{\varrho\beta})$, we get

$$\mu_0\partial_\alpha T^{\alpha\beta} = \mathcal{R} + (\partial_\alpha F^{\alpha\sigma})F_{\sigma\varrho}\eta^{\varrho\beta}.$$

It remains to show that $\mathcal{R} = 0$. To this end, note that

$$\mathcal{R} = \tfrac{1}{2}\eta^{\alpha\beta}(\partial_\alpha F^{\varrho\sigma})F_{\varrho\sigma} + F_{\alpha\sigma}\partial^\alpha F^{\sigma\varrho}\eta_{\varrho\beta} = \tfrac{1}{2}(\partial^\beta F^{\varrho\sigma})F_{\varrho\sigma} + F_{\sigma\alpha}\partial^\alpha F^{\beta\sigma},$$

after using the Leibniz rule and lifting indices. Again lifting indices, the Maxwell–Bianchi equations (19.90) read as

$$\partial^\kappa F\lambda\mu + \partial^\lambda F^{\mu\kappa} + \partial^\mu F^{\kappa\lambda} = 0, \qquad F^{\lambda\mu} = -F^{\mu\lambda}.$$

Hence

$$\mathcal{R} = -\tfrac{1}{2}(\partial^\varrho F^{\sigma\beta} + \partial^\sigma F^{\beta\varrho})F_{\varrho\sigma} + F_{\sigma\alpha}\partial^\alpha F^{\beta\sigma}.$$

---

[14] In the special case where $F(x) = -U'(x)$, we get $\frac{d}{dt}(\frac{1}{2}m\dot{x}^2(t) + U(x(t))) = 0$.

Finally, changing the summation indices and using $F^{\alpha\beta} = -F^{\beta\alpha}$, we obtain

$$\mathcal{R} = -\tfrac{1}{2}F_{\varrho\sigma}\partial^\varrho F^{\sigma\beta} - \tfrac{1}{2}F_{\varrho\sigma}\partial^\varrho F^{\sigma\beta} + F_{\varrho\sigma}\partial^\varrho F^{\sigma\beta} = 0.$$

Ad (iii). Recall that $T^{\alpha\beta} = T^{\beta\alpha}$. By the Leibniz rule,

$$\begin{aligned}\partial_\gamma S^{\alpha\beta\gamma} &= \partial_\gamma(x^\alpha T^{\beta\gamma} - x^\beta T^{\alpha\gamma}) = \delta^\alpha_\gamma T^{\beta\gamma} + x^\alpha\partial_\gamma T^{\beta\gamma} - \delta^\beta_\gamma T^{\alpha\gamma} - x^\beta\partial_\gamma T^{\alpha\gamma} \\ &= T^{\beta\alpha} - \mu_0 x^\alpha f^\beta - T^{\alpha\beta} + \mu_0 x^\beta f^\alpha = -\mu_0(x^\alpha f^\beta - x^\beta f^\alpha) = -\mu_0 s^{\alpha\beta}.\end{aligned}$$

□

Inertial systems. Consider a strictly positively oriented inertial system as used on page 960. Maxwell tried to understand the electromagnetic field by using a model from elasticity theory. In this context, he introduced the so-called Maxwell stress tensor

$$\mathsf{T} = \mathbf{D}\otimes\mathbf{E} + \mathbf{B}\otimes\mathbf{H} - \eta I.$$

Set

- $\mathbf{E} = E^k\mathbf{e}_k$, $\mathbf{D} = D^k\mathbf{e}_k$, $\mathbf{D} = \varepsilon_0\mathbf{E}$,
- $\mathbf{B} = B^k\mathbf{e}_k$, $\mathbf{H} = H^k\mathbf{e}_k$, $\mathbf{B} = \mu_0\mathbf{H}$,
- $\eta = \frac{1}{2}(\mathbf{DE} + \mathbf{BH})$ (energy density of the electromagnetic field),
- $\mathbf{S} = \mathbf{E}\times\mathbf{H}$ (energy current density vector),
- $\boldsymbol{\pi} := \mathbf{D}\times\mathbf{E}$ (momentum density vector), $\boldsymbol{\pi} = \frac{1}{c^2}\mathbf{S}$,
- $\mathbf{f} = \varrho\mathbf{E} + \mathbf{J}\times\mathbf{B}$ (Lorentz force density vector),
- $cf^0 = \mathbf{JE}$ (rate of the Joule energy density),
- $\mathbf{S} = S^k\mathbf{e}_k, \boldsymbol{\pi} = \pi^k\mathbf{e}_k$,
- $\tau^{ij} := D^iE^j + B^iH^j$.

Then $\mathsf{T} = \left(\tau^{ij} - \eta\delta^{ij}\right)\mathbf{e}_i\otimes\mathbf{e}_j$, and

$$\begin{pmatrix} T^{00} & T^{01} & T^{02} & T^{03} \\ T^{10} & T^{11} & T^{12} & T^{13} \\ T^{20} & T^{21} & T^{22} & T^{23} \\ T^{30} & T^{31} & T^{32} & T^{33}\end{pmatrix} = \begin{pmatrix} \eta & S^1/c & S^2/c & S^3/c \\ c\pi^1 & \eta-\tau^{11} & -\tau^{12} & -\tau^{13} \\ c\pi^2 & -\tau^{12} & \eta-\tau^{22} & -\tau^{23} \\ c\pi^3 & -\tau^{13} & -\tau^{23} & \eta-\tau^{33}\end{pmatrix}.$$

Note that $T^{\alpha\beta} = T^{\beta\alpha}$, $\alpha,\beta = 0,1,2,3$ (symmetry of the energy-momentum tensor). Then we get the following conservation laws:

(i) Charge conservation: $\dot{\varrho} + \operatorname{div}\mathbf{J} = 0$.
(ii) Energy conservation: $\dot{\eta} + \operatorname{div}\mathbf{S} + \mathbf{JE} = 0$.
  Momentum conservation: $\dot{\boldsymbol{\pi}} + \mathbf{f} = \operatorname{div}\mathsf{T}$.
(iii) Angular momentum conservation: $\mathbf{x}\times\dot{\boldsymbol{\pi}} + \mathbf{x}\times\mathbf{f} = \mathbf{x}\times\operatorname{div}\mathsf{T}$.

If electrons flow through a metallic wire, then heat is produced by the collisions of the electrons with the molecules of the metal. This heat production is described by the term **JE**. More precisely, the integral

$$\int_{t_0}^{t_1}\left(\int_{\mathcal{U}}\mathbf{J}(\mathbf{x},t)\mathbf{E}(\mathbf{x},t)\,dxdydz\right)dt$$

is equal to the heat energy produced in the region $\mathcal{U}$ during the time interval $[t_0, t_1]$.

The Noether theorem and the energy-momentum tensor. It turns out that the conservation laws for energy, momentum, and angular momentum are consequences of the invariance of the Maxwell equations under the Poincaré group. This follows from the Noether theorem and the fact that the variational integral

(19.92) of the principle of critical action is invariant under the Poincaré group. The Poincaré group has 10 parameters. Consequently, we get 10 conservation laws.[15] This is the special case of a general result in mathematical physics which can be applied to all relativistically invariant field theories. We will study this in Vol. IV on quantum mathematics.

Linear material. In Sect. 19.8, we will investigate linear materials. In an inertial system, all the formulas given above for a vacuum remain valid for linear material provided we use the following replacements:

$$\varepsilon_0 \Rightarrow \varepsilon, \qquad \mu_0 \Rightarrow \mu \qquad c \Rightarrow c_*$$

where c_* is the velocity of light in the material. In particular, we get $\mathbf{D} = \varepsilon\mathbf{E}$ and $\mathbf{B} = \mu\mathbf{H}$.

19.7 The Principle of Critical Action

> The principle of critical action does not depend on the electromagnetic field itself, but on the four-potential. However, the variational problem is invariant under gauge transformations. This is a typical feature of gauge theories.
>
> Folklore

19.7.1 The Electromagnetic Field

Consider an inertial system. The principle of critical action for the electromagnetic field reads as

$$\int_{\mathcal{O}} \left(\tfrac{1}{4}F_{\alpha\beta}F^{\alpha\beta} + \mu_0 A_\beta \mathcal{J}^\beta\right) d^4x = \text{critical!} \qquad (19.92)$$

with the boundary condition: A_α = fixed on $\partial\mathcal{O}$, $\alpha = 0,1,2,3$. Here, we assume that $\mathcal{O}$ is a nonempty bounded open subset of $\mathbb{R}^4$ with a smooth boundary $\partial\mathcal{O}$ and the closure $\mathrm{cl}(\mathcal{O}) = \mathcal{O} \cup \partial\mathcal{O}$ (e.g., $\mathcal{O}$ is a 4-dimensional ball). In addition, we set

$$F_{\alpha\beta} := \partial_\alpha A_\beta - \partial_\beta A_\alpha, \qquad \alpha,\beta = 0,1,2,3.$$

The Euler–Lagrange equations. Naturally enough, we are given the smooth functions $\mathcal{J}^\alpha : \mathrm{cl}(\mathcal{O}) \to \mathbb{R}$, $\alpha = 0,1,2,3,4$.

Theorem 19.12 *If the functions A_0, A_1, A_2, A_3 are a smooth solution of the variational problem* (19.92)*, then these functions satisfy the Maxwell equations*

$$\partial_\alpha F^{\alpha\beta} = \mu_0 \mathcal{J}^\beta, \qquad \partial_{[\gamma}F_{\lambda\mu]} = 0, \qquad \beta,\gamma,\lambda,\mu = 0,1,2,3.$$

---

[15] In fact, the equations (19.85) and (19.86) represent 10 equations. However, it follows from (19.85) with $\alpha = 0, \beta = 1,2,3$ that

$$t(\dot{\boldsymbol{\pi}} + \mathbf{f} - \operatorname{div}\mathsf{T}) - \mathbf{x}(\dot{\eta} + \operatorname{div}\mathbf{S} + \mathbf{JE}) = 0,$$

and this conservation law is a consequence of the other conservation laws.

The proof will be given in Problem 19.10. Using the language of vector calculus, the principle (19.92) of critical action reads as

$$\int_{\mathcal{O}} \left(\tfrac{1}{2}(\mathbf{B}^2 - \tfrac{1}{c^2}\mathbf{E}^2) - \mu_0 \mathbf{A}\mathbf{J} + \mu_0 U \varrho\right) dtdxdydz = \text{critical!} \qquad (19.93)$$

with the boundary condition: $\mathbf{A}, U$ = fixed on $\partial\mathcal{O}$. In addition, we set

$$\mathbf{E} = -\,\mathbf{grad}\, U - \dot{\mathbf{A}}, \qquad \mathbf{B} = \mathbf{curl}\,\mathbf{A}.$$

We are given the electric charge density ϱ and the electric current density vector $\mathbf{J}$. We are looking for both the scalar potential U and the vector potential $\mathbf{A}$.

Gauge invariance. The variational integral (19.92) is independent of the choice of the inertial system, by the index principle. In addition, the variational integral (19.92) is also invariant under the gauge transformation

$$A_\alpha \Rightarrow A_\alpha + \partial_\alpha \chi, \qquad \alpha = 0, 1, 2, 3.$$

In fact, since $\partial_\beta \mathcal{J}^\beta = 0$, integration by parts yields

$$\int_{\mathcal{O}} (A_\beta + \partial_\beta \chi)\mathcal{J}^\beta \, d^4x = \int_{\mathcal{O}} (A_\beta \mathcal{J}^\beta - \partial_\beta \mathcal{J}^\beta \cdot \chi)\, d^4x = \int_{\mathcal{O}} A_\beta \mathcal{J}^\beta \, d^4x.$$

19.7.2 Motion of Charged Particles and Gauge Transformations

Consider a particle of rest mass m_0 and electric charge Q. The motion of the particle under the influence of an electromagnetic field is described by the following variational problem

$$\int_{\sigma_0}^{\sigma_1} \left(-m_0 c \frac{ds(\sigma)}{d\sigma} - Q A_\alpha(x(\sigma)) \, \frac{dx^\alpha(\sigma)}{d\sigma} \right) d\sigma = \text{critical!} \qquad (19.94)$$

on the Minkowski manifold $\mathbb{M}^4$. We have to add the following boundary condition: $x^\alpha(\sigma_0)$ and $x^\alpha(\sigma_1)$ are fixed if $\alpha = 0, 1, 2, 3$. We are looking for the trajectory

$$x^\alpha = x^\alpha(\sigma), \qquad \alpha = 0, 1, 2, 3$$

on $\mathbb{M}^4$. In addition, we assume that the particle moves with subvelocity of light, that is,

$$g_{\alpha\beta}(x(\tau)) \, \frac{dx^\alpha(\sigma)}{d\sigma} \frac{dx^\beta(\sigma)}{d\sigma} > 0, \qquad \sigma \in [\sigma_0, \sigma_1].$$

The parameter s is the arc length of the curve on $\mathbb{M}^4$, and the parameter interval $[\sigma_0, \sigma_1]$ is compact.

The variational problem (19.94) does not depend on the choice of the local coordinate system on the Minkowski manifold $\mathbb{M}^4$.

Proposition 19.13 *The principle of critical action* (19.94) *is invariant under the gauge transformation*

$$A_\alpha \Rightarrow A_\alpha + \partial_\alpha \chi, \qquad \alpha = 0, 1, 2, 3.$$

Proof. Observe that

$$\int_{\sigma_0}^{\sigma_1} (\mathcal{A}_\alpha(x(\sigma)) + \partial_\alpha \chi(x(\sigma)))\, \frac{dx^\alpha(\sigma)}{d\sigma}\, d\sigma = \int_{\sigma_0}^{\sigma_1} \mathcal{A}_\alpha(x(\sigma))\, \frac{dx^\alpha(\sigma)}{d\sigma}\, d\sigma + \text{const},$$

since $\partial_\alpha \chi(x(\sigma))\, \frac{dx^\alpha(\sigma)}{d\sigma} = \frac{d\chi(x(\sigma))}{d\sigma}$. Thus the action integral changes by a constant. However, this does not change the possible solutions of the problem. □

The Lagrangian approach in an inertial system. Let us consider a strictly positively oriented inertial system Σ as used on page 960. Let us choose the parameter $\sigma = t$. Then the principle of critical action (19.94) reads as

$$\int_{t_0}^{t_1} L(\mathbf{x}(t), \dot{\mathbf{x}}(t), t)\, dt = \text{critical!} \tag{19.95}$$

with the Lagrangian

$$L(\mathbf{x}, \dot{\mathbf{x}}, t) := -m_0 c^2 \sqrt{1 - \tfrac{\dot{\mathbf{x}}^2}{c^2}} + Q\dot{\mathbf{x}}\mathbf{A}(\mathbf{x}, t) - QU(\mathbf{x}, t).$$

We have to add the following boundary conditions: $\mathbf{x}(t_0)$ and $\mathbf{x}(t_1)$ are fixed. We are given the 4-potential $U, \mathbf{A}$ of the electromagnetic field $\mathbf{E}, \mathbf{B}$, that is,

$$\mathbf{E} = -\,\mathbf{grad}\, U - \dot{\mathbf{A}}, \qquad \mathbf{B} = \mathbf{curl}\, \mathbf{A}.$$

We are looking for the curve $\mathbf{x} = \mathbf{x}(t)$, $t \in [t_0, t_1]$.

Theorem 19.14 *Every solution of the variational problem* (19.95) *satisfies the differential equation*

$$\frac{d}{dt}\big(m(t)\dot{\mathbf{x}}(t)\big) = Q\mathbf{E}\big(\mathbf{x}(t)\big) + Q\dot{\mathbf{x}}(t) \times \mathbf{B}\big(\mathbf{x}(t)\big) \tag{19.96}$$

with the relativistic mass $m(t) := \frac{m_0}{\sqrt{1 - \dot{\mathbf{x}}^2(t)/c^2}}$. *This differential equation coincides with the equation of motion* (19.78).

Proof. An explicit computation shows that the Euler–Lagrange equation

$$\frac{d}{dt} L_{\dot{\mathbf{x}}} = L_{\mathbf{x}}$$

coincides with (19.96) (see Problem 19.11). □

The Hamiltonian approach in an inertial system. Let us use the Hamiltonian approach introduced in Sect. 6.8 of Vol. II. The idea is to pass from the variables $\mathbf{x}, \dot{\mathbf{x}}$ to the new variables $\mathbf{x}, \mathbf{P}$ by setting

$$\mathbf{P} = L_{\dot{\mathbf{x}}}(\mathbf{x}, \dot{\mathbf{x}}, t).$$

Furthermore, we introduce the Hamiltonian

$$\mathcal{H} = \dot{\mathbf{x}} L_{\dot{\mathbf{x}}} - L.$$

Using the momentum vector $\mathbf{p} = m\dot{\mathbf{x}}$ with the relativistic mass $m = \frac{m_0}{\sqrt{1 - \dot{\mathbf{x}}^2/c^2}}$, we get the canonical (or generalized) momentum vector

$$\mathbf{P} = \mathbf{p} + Q\mathbf{A},$$

and the Hamiltonian

$$\mathcal{H}(\mathbf{x},\mathbf{P},t)=\sqrt{m_0^2c^4+c^2(\mathbf{P}-Q\mathbf{A}(\mathbf{x},t))^2}+QU(\mathbf{x},t).$$

The solutions of the Euler–Lagrange equation (19.96) pass over to the solutions of the canonical equations

$$\boxed{\dot{\mathbf{P}}=-\mathcal{H}_{\mathbf{x}},\quad \dot{\mathbf{x}}=\mathcal{H}_{\mathbf{P}}.} \tag{19.97}$$

We are looking for the trajectory $\mathbf{x}=\mathbf{x}(t),\mathbf{P}=\mathbf{P}(t),t\in[t_0,t_1]$.

Gauge transformation. Let us use the transformation

$$U^+(\mathbf{x},t)=U(\mathbf{x},t)+\chi_t(\mathbf{x},t),\quad \mathbf{A}_+(\mathbf{x},t)=\mathbf{A}(\mathbf{x},t)-\mathbf{grad}\,\chi(\mathbf{x},t)$$

together with the transformed Hamiltonian

$$\mathcal{H}^+(\mathbf{x},\mathbf{P}^+,t)=\sqrt{m_0^2c^4+c^2(\mathbf{P}^+-Q\mathbf{A}^+(\mathbf{x},t))^2}+QU^+(\mathbf{x},t).$$

Suppose that all the functions are smooth.

Proposition 19.15 *If the trajectory* $\mathbf{x}=\mathbf{x}(t),\mathbf{P}=\mathbf{P}(t),t\in[t_0,t_1]$ *is a solution of the canonical equation* (19.97)*, then the transformed trajectory*

$$\mathbf{x}=\mathbf{x}(t),\quad \mathbf{P}^+=\mathbf{P}^+(t),\qquad t\in[t_0,t_1]$$

is a solution of the new canonical equation

$$\dot{\mathbf{P}}^+=-\mathcal{H}^+_{\mathbf{x}},\quad \dot{\mathbf{x}}=\mathcal{H}^+_{\mathbf{P}^+}. \tag{19.98}$$

Proof. This can be proven by a direct calculation. To get insight, we define the so-called generating function

$$\mathcal{S}(\mathbf{x},\mathbf{P}^+,t):=\mathbf{x}\mathbf{P}^++Q\chi(\mathbf{x},t),$$

and we set

$$\mathbf{P}=\mathcal{S}_{\mathbf{x}},\quad \mathbf{x}^+=\mathcal{S}_{\mathbf{P}^+},\quad \mathcal{H}^+=\mathcal{H}+\mathcal{S}_t.$$

Then, $\mathbf{P}=\mathbf{P}^++Q\,\mathbf{grad}\,\chi$, $\mathbf{x}^+=\mathbf{x}$, and $\mathcal{H}^+=\mathcal{H}+Q\chi_t$. The general theory tells us that this is a canonical transformation which leads to the new canonical equation (19.98). □

Summarizing, the gauge transformation considered above is nothing other than a special canonical transformation in the sense of classical mechanics.

19.8 Weyl Duality and the Maxwell Equations in Materials

Let us consider electric and magnetic properties of moving material. Typically, such a moving body consists of molecules, and the electric and magnetic properties of the molecules have to be taken into account by formulating the so-called constitutive laws. We will proceed in the following two steps:

Step 1: We consider the body in its rest system Σ_{rest}. We assume that this is an inertial system, and we formulate the Maxwell equations in Σ_{rest} including the constitutive law.

Step 2: Using Weyl densities, we formulate the Maxwell equations in such a way that they are valid in each inertial system.

19.8.1 The Maxwell Equations in the Rest System

The Maxwell equations in the rest inertial system Σ_{rest} read as follows:
(i) The Maxwell source equations:

$$\operatorname{div} \mathbf{D} = \varrho, \qquad \mathbf{curl}\, \mathbf{H} = \mathbf{J} + \dot{\mathbf{D}}. \tag{19.99}$$

(ii) The Maxwell–Bianchi equations:

$$\operatorname{div} \mathbf{B} = 0, \qquad \mathbf{curl}\, \mathbf{E} = -\dot{\mathbf{B}}. \tag{19.100}$$

(iii) The constitutive law: $(\mathbf{D}, \mathbf{H}, \mathbf{J}) = \Lambda(\mathbf{E}, \mathbf{B}, T)$. Here, T denotes the temperature.

In addition, we introduce the polarization vector $\mathbf{P}_{\text{el}}$ and the magnetization vector $\mathbf{M}$ by setting

- $\mathbf{P}_{\text{el}} := \mathbf{D} - \varepsilon_0 \mathbf{E}$, and
- $\mathbf{M} := \mu_0^{-1}\mathbf{B} - \mathbf{H}$.

Here, $\mathbf{P}_{\text{el}}$ (resp. $\mathbf{M}$) describes the electric (resp. magnetic) dipole moment density which is generated by the action of the electric field $\mathbf{E}$ (resp. the magnetic field $\mathbf{B}$) on the molecules of the material. Equivalently, equations (i), (ii) can be written as

$$\begin{aligned} \varepsilon_0 \operatorname{div} \mathbf{E} &= \varrho + \varrho_{\text{el,micro}}, & \mathbf{curl}\, \mathbf{H} &= \mathbf{J} + \mathbf{J}_{\text{micro}} + \varepsilon_0 \dot{\mathbf{E}}, \\ \operatorname{div} \mathbf{H} &= \varrho_{\text{magn,micro}}, & \mathbf{curl}\, \mathbf{E} &= -\mu_0 \dot{\mathbf{H}} - \mu_0 \dot{\mathbf{M}}. \end{aligned} \tag{19.101}$$

Here, we set:

- $\varrho_{\text{el,micro}} := -\operatorname{div} \mathbf{P}_{\text{el}}$ (microscopic effective electric charge density),
- $\varrho_{\text{magn,micro}} := -\operatorname{div} \mathbf{M}$ (microscopic effective magnetic charge density), and
- $\mathbf{J}_{\text{micro}} := \dot{\mathbf{P}}_{\text{el}}$ (microscopic electric current density vector).

In a vacuum, we have $\mathbf{P}_{\text{el}} \equiv 0$ and $\mathbf{M} \equiv 0$. Hence $\varrho_{\text{el,micro}} \equiv 0$, $\varrho_{\text{magn,micro}} \equiv 0$, and $\mathbf{J}_{\text{micro}} \equiv 0$. In materials, we get additional microscopic charges which correspond to electric and magnetic properties of the molecules generated by both the electric field $\mathbf{E}$ and the magnetic field $\mathbf{B}$.

Note that the electric field $\mathbf{E}$ and the magnetic field $\mathbf{B}$ are the fundamental fields. The derived electric field $\mathbf{D}$ and the derived magnetic field $\mathbf{H}$ are generated by additional electric and magnetic dipole moments of the molecules of the material.

19.8.2 Typical Examples of Constitutive Laws

The following constitutive laws are frequently encountered:

(a) Linear electric material: $\mathbf{D} = \varepsilon \mathbf{E}$.
(b) Linear magnetic material: $\mathbf{H} = \lambda \mathbf{B}$, $\lambda = 1/\mu$.
(c) Ohm's law: $\mathbf{J} = \nu \mathbf{E}$.
(d) Ferromagnetic material: $\mathbf{B} = \mu_0 \mathbf{H} + \mu_0 \mathbf{M}(\mathbf{H}, T)$ (multi-valued constitutive law; hysteresis).

Let us discuss this.

Ad (a). We set $\varepsilon = \varepsilon_0(1 + \chi_{\text{el}})$. The constant χ_{el} is called the electric susceptibility. In a vacuum, we have $\chi_{\text{el}} = 0$. Otherwise $\chi_{\text{el}} > 0$. Moreover, ε is called the dielectric constant. It follows from

$$\mathbf{D} = \varepsilon_0 \mathbf{E} + \mathbf{P}_{\text{el}} = \varepsilon_0 \mathbf{E} + \varepsilon_0 \chi_{\text{el}} \mathbf{E}$$

that the polarization is given by $\mathbf{P}_{\text{el}} = \varepsilon_0 \chi_{\text{el}} \mathbf{E}$.

Ad (b). We set $\mu = \mu_0(1 + \chi_{\text{magn}})$. We have to distinguish between the following cases:

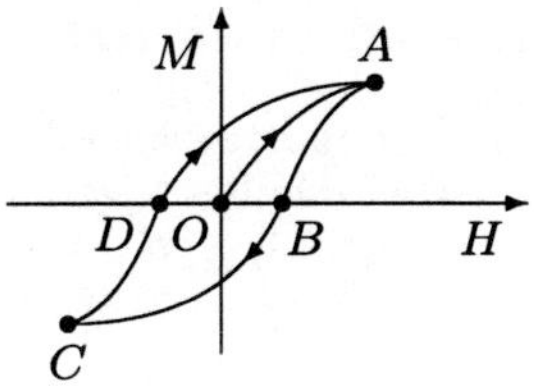

Fig. 19.13. Hysteresis

- $\mu = \mu_0$ (in a vacuum),
- $0 < \mu < \mu_0$ (diamagnetic material),
- $\mu > \mu_0$ (paramagnetic material).

Here, μ (resp. χ_{magn}) is called the magnetic permeability (resp. magnetic susceptibility). It follows from $\mathbf{B} = \mu_0(\mathbf{H} + \mathbf{M})$ that $\mathbf{M} = \chi_{\text{magn}}\mathbf{H}$. For linear material, the Maxwell equations read as:

$$\varepsilon \operatorname{div} \mathbf{E} = \varrho, \qquad \mathbf{curl}\, \mathbf{B} = \mu \mathbf{J} + \tfrac{1}{c_*^2} \dot{\mathbf{E}},$$
$$\operatorname{div} \mathbf{B} = 0, \qquad \mathbf{curl}\, \mathbf{E} = -\dot{\mathbf{B}}. \tag{19.102}$$

Here, we set $c_*^2 := 1/\varepsilon\mu$, and c_* is the velocity of light in the material.

The equations (19.102) are obtained from the Maxwell equations in a vacuum by replacing the 'bare' constants ε_0, μ_0 by the 'effective constants' ε, μ, respectively. This is the prototype of renormalization in classical physics.

Ad (c). The Ohm law governs the motion of electrons in metallic conductors.

Ad (d). The phenomenon of ferromagnetism was discovered in ancient times. For example, iron possesses the ferromagnetic property.[16] Above the so-called Curie temperature of 1017 K, iron looses the ferromagnetic property. Ferromagnetism depends on the spin of the electrons. We recommend:

A. Aharoni, Introduction to the Theory of Ferromagnetism, Oxford University Press, 2000.

B. McCoy and Tai-Tsu Wu, The Two-Dimensional Ising Model, Harvard University Press, Cambridge, Massachusetts, 1997.

Hysteresis and multi-valued constitutive laws. Figure 19.13 displays a so-called multi-valued constitutive law between the derived magnetic field strength H and the magnetization M of ferromagnetic material. If H is increased from zero to some positive value, then M increases from zero to some positive value along the curve from O to A. If H is diminished, then M decreases along a different curve ABC. If H is increased again, then M increases along the different curve CDA. This phenomenon is called hysteresis. Multi-valued constitutive laws can be described by the modern theory of variational inequalities. This can be found in E. Zeidler, Nonlinear Functional Analysis and its Applications, Vol. IV, Springer, New York, 1995 (e.g., applications to plasticity). We also recommend:

E. Brokate and J. Sprekels, Hysteresis and Phase Transitions, Springer, Berlin, 1996.

[16] The Latin word 'ferrum' means iron.

19.8.3 The Maxwell Equations in an Arbitrary Inertial System

In what follows, we will use the terminology introduced in Sect. 8.11 on page 522 concerning Cartan families and Weyl families together with their Cartan and Weyl derivatives, respectively. Choose a strictly positively oriented inertial system. Let us introduce the tensorial family of the electromagnetic field:

$$(F_{\alpha\beta}) := \begin{pmatrix} 0 & E^1/c & E^2/c & E^3/c \\ -E^1/c & 0 & -B^3 & B^2 \\ -E^2/c & B^3 & 0 & -B^1 \\ -E^3/c & -B^2 & B^1 & 0 \end{pmatrix}, \tag{19.103}$$

together with the lifted tensorial family $F^{\alpha\beta}$ with respect to $\eta_{\alpha\beta}$:

$$(F^{\alpha\beta}) := \begin{pmatrix} 0 & -E^1/c & -E^2/c & -E^3/c \\ E^1/c & 0 & -B^3 & B^2 \\ E^2/c & B^3 & 0 & -B^1 \\ E^3/c & -B^2 & B^1 & 0 \end{pmatrix}. \tag{19.104}$$

Furthermore, we introduce the Weyl family:

$$(W^{\alpha\beta}) := \begin{pmatrix} 0 & -D^1c & -D^2c & -D^3c \\ D^1c & 0 & -H^3 & H^2 \\ D^2c & H^3 & 0 & -H^1 \\ D^3c & -H^2 & H^1 & 0 \end{pmatrix}. \tag{19.105}$$

More precisely, we assume that:

- $F_{\alpha\beta}$ is a smooth Cartan family (i.e., an antisymmetric tensorial family), and $\mathcal{J}^\alpha$ is a smooth tensorial family on the Minkowski manifold $\mathbb{M}^4$. Set

$$F := \tfrac{1}{2}F_{\beta\gamma}\, dx^\beta \wedge dx^\gamma.$$

- $W^{\alpha\beta}$ is a smooth Weyl family (i.e., an antisymmetric tensorial density family of weight one) on $\mathbb{M}^4$.

Using Weyl duality, we set $w^\beta := \sqrt{|g|}\mathcal{J}^\beta$ for all $\beta = 0, 1, 2, 3$. Then the Maxwell equations read as follows:

(i) The Maxwell source equation: $-\delta W = w$.
(ii) The Maxwell–Bianchi equation: $dF = 0$.
(iii) Constitutive law: $W = \Lambda(F, T)$ in the rest inertial system Σ_{rest}.[17]

Here, $d_\mu F_{\beta\gamma} = \partial_{[\mu}F_{\beta\gamma]}$ is the Cartan derivative, and $(\delta W)^\beta = \partial_\alpha W^{\alpha\beta}$ is the Weyl derivative. These derivatives are defined in an invariant way, that is, the choice of the local coordinates on $\mathbb{M}^4$ does not matter. Explicitly, the Maxwell equations (i) and (ii) read as:

$$-\partial_\alpha W^{\alpha\beta} = \sqrt{|g|}\, J^\beta, \qquad \partial_{[\mu}F_{\beta\gamma]} = 0, \quad \beta, \gamma, \mu = 0, 1, 2, 3.$$

This is equivalent to

$$\boxed{-\partial_\alpha W^{\alpha\beta} = \sqrt{|g|}\, J^\beta, \qquad \partial_\mu F_{\beta\gamma} + \partial_\beta F_{\gamma\mu} + \partial_\gamma F_{\mu\beta} = 0} \tag{19.106}$$

[17] Explicitly, $W^{\alpha\beta} = \Lambda^{\alpha\beta}(F^{01}, F^{0,2}, \ldots, F^{33})$ for all $\alpha, \beta = 0, 1, 2, 3$.

for all indices $\beta, \gamma, \mu = 0, 1, 2, 3$. These Maxwell equations do not depend on the choice of the local coordinate system on $\mathbb{M}^4$. More precisely, we compute $W^{\alpha\beta}$ in the distinguished rest inertial system Σ_{rest} by the aid of $F_{\alpha\beta}$. In an arbitrary local coordinate system Σ', we then obtain $W^{\alpha'\beta'}$ from $W^{\alpha\beta}$ by using the transformation law of a Weyl family. Recall that $g_{\alpha\beta}$ is the metric tensorial family on $\mathbb{M}^4$. For an inertial system, we have $g_{\alpha\beta} = \eta_{\alpha\beta}$ for all indices.

Examples. The Maxwell equations in a vacuum are given by (19.106) with $W^{\alpha\beta} = F^{\alpha\beta}/\mu_0$ and $c^2 = 1/\varepsilon_0\mu_0$. Using the replacement

$$\varepsilon_0 \Rightarrow \varepsilon, \qquad \mu_0 \Rightarrow \mu,$$

we get the Maxwell equations for linear material. Because of the constitutive law for linear material, we have $\mu W^{\alpha\beta} = F^{\alpha\beta}$ in the special inertial system Σ_{rest} with $\sqrt{|g|} = 1$. In an arbitrary local coordinate system on $\mathbb{M}^4$, we get

$$\mu W^{\alpha\beta} = \sqrt{|g|}\; F^{\alpha\beta},$$

since $\sqrt{|g|}\; F^{\alpha\beta}$ transforms like an antisymmetric tensorial density family of weight one. Thus, the Maxwell equations can be elegantly written as

$$\boxed{-\partial_\alpha(\sqrt{|g|}\; F^{\alpha\beta}) = \mu\sqrt{|g|}\; \mathcal{J}^\beta, \qquad \partial_\mu F_{\beta\gamma} + \partial_\beta F_{\gamma\mu} + \partial_\gamma F_{\mu\beta} = 0} \qquad (19.107)$$

for all indices $\beta, \gamma, \mu = 0, 1, 2, 3$. These equations are valid for all local coordinate systems of $\mathbb{M}^4$. In a vacuum, we have to replace μ by μ_0.

19.9 Physical Units

19.9.1 The SI System

The basic units can be found in Table 19.1. For the derived units in electromagnetism, we refer to Table 19.2 on page 984.

Table 19.1. Basic units

length	m	meter
time	s	second
mass	kg	kilogram
temperature	K	Kelvin
electric current strength	A	ampere
amount of substance	mol	1 mol $= 6.026 \cdot 10^{23}$ pieces
luminous intensity	cd	candela

Table 19.2. Derived units

velocity	– –	– –	m/s $\left(c = \frac{1}{\sqrt{\varepsilon_0 \mu_0}} = 2.99 \cdot 10^8 \text{m/s}\right)$
acceleration (velocity per time)	–	–	m/s^2
force	N	newton	$\text{N} = \text{kg} \cdot \text{m/s}^2 = \text{J/m}$
energy (work)	J	joule	$\text{J} = \text{Nm} = \text{Ws} = \text{VAs}$ $= \text{CV} = \text{kg} \cdot \text{m}^2/\text{s}^2$
	eV	electron volt	$1\ \text{eV} = 1.6 \cdot 10^{-19}\text{J}$
action (energy times time)	–	–	Js $(\text{h} = 6.625 \cdot 10^{-34}\text{Js})$
power (energy per time)	W	watt	$\text{W} = \text{J/s} = \text{Nm/s} = \text{VA}$
torque of a force (force times length)	–	–	$\text{Nm} = \text{J}$ (energy)
momentum of a particle (mass times velocity)	–	–	$\text{Ns} = \text{kg} \cdot \text{m/s}$
angular momentum of a particle (spin) (momentum times length)			$\text{Ns} \cdot \text{m} = \text{Js}$ (action)
voltage	V	volt	$\text{V} = \text{W/A} = \text{J/As} = \text{J/C}$
electric charge	C e	coulomb electric charge of the proton	$\text{C} = \text{As}$ $e = 1.602 \cdot 10^{-19}\text{C}$
magnetic charge	D	dirac	$\text{D} = \text{Am}$
electric field strength (force per electric charge)	–	–	$[\mathbf{E}] = \text{N/C} = \text{V/m}$ $= \text{T} \cdot \text{m/s}$
magnetic field strength (force per magnetic charge)	T	tesla	$[\mathbf{B}] = \text{N/D} = \text{N/Am}$ $= \text{Vs/m}^2 = \text{T}$
electric dipole moment (electric charge times length)	–	–	$[\mathbf{p}_{\text{el}}] = \text{Cm} = \text{J}/[\mathbf{E}] = \text{Jm/V}$
magnetic dipole moment (magnetic charge times length)	–	–	$[\mathbf{m}] = \text{Dm} = \text{J}/[\mathbf{B}] = \text{J/T}$

continued on next page

Table 19.2. Derived units (continued)

electric polarization (electric dipole moment per volume)	–	–	$[\mathbf{P}_{\mathrm{el}}] = [\mathbf{p}_{\mathrm{el}}]/\mathrm{m}^3 = \mathrm{C}/\mathrm{m}^2 = \mathrm{As}/\mathrm{m}^2$
magnetic polarization (magnetic dipole moment per volume–magnetization)	–	–	$[\mathbf{M}] = [\mathbf{m}]/\mathrm{m}^3 = \mathrm{D}/\mathrm{m}^2 = \mathrm{A}/\mathrm{m}$
derived electric field	–	–	$[\mathbf{D}] = [\mathbf{P}_{\mathrm{el}}]$
derived magnetic field	–	–	$[\mathbf{H}] = [\mathbf{M}]$
electric charge density	–	–	$\mathrm{C}/\mathrm{m}^3 = \mathrm{As}/\mathrm{m}^3$
electric current density (electric charge per area and time)	–	–	$[\mathbf{J}] = \mathrm{C}/\mathrm{m}^2\mathrm{s} = \mathrm{A}/\mathrm{m}^2$
electric current strength (electric charge per time)	–	–	$[\mathrm{J}] = \mathrm{A}/\mathrm{Cs}$
magnetic flow (magnetic field strength times area)	Wb	weber	$\mathrm{Wb} = \mathrm{Tm}^2 = \mathrm{Vs}$
electric resistance	Ω	ohm	$\Omega = \mathrm{V}/\mathrm{A}$
inductance	H	henry	$\mathrm{H} = \mathrm{Wb}/\mathrm{A}$
capacity (electric charge per voltage)	F	farad	$\mathrm{F} = \mathrm{C}/\mathrm{V}$
light pressure (force per area)	Pa	pascal	$\mathrm{Pa} = \mathrm{N}/\mathrm{m}^2$
electromagnetic energy density (energy per volume)	–	–	$[\mathbf{E}][\mathbf{D}] = [\mathbf{B}][\mathbf{H}] = \mathrm{J}/\mathrm{m}^3$
electromagnetic energy current density (energy per area and time)	–	–	$[\mathbf{E}][\mathbf{H}] = \mathrm{J}/\mathrm{m}^2\mathrm{s}$
electric field constant in a vacuum	–	–	$\varepsilon_0 = 8.854 \cdot 10^{-12}\mathrm{A}^2\mathrm{s}^2/\mathrm{Nm}^2$
magnetic field constant in a vacuum	–	–	$\mu_0 = 4\pi \cdot 10^{-7}\mathrm{N}/\mathrm{A}^2$ $\varepsilon_0\mu_0 = \frac{1}{c^2}$

19.9.2 The Universal Approach

If one wants to formulate the Maxwell equations in a universal way, then one has to use the three constants μ_0, ε_0, and κ with

$$\varepsilon_0 \mu_0 \kappa^2 = \frac{1}{c^2}$$

where c is the velocity of light in a vacuum. Then the Maxwell equations in a vacuum read as follows:

$$\varepsilon_0 \operatorname{div} \mathbf{E} = \varrho, \qquad \mathbf{curl}\, \mathbf{B} = \kappa\mu_0(\mathbf{J} + \varepsilon_0 \dot{\mathbf{E}}),$$
$$\mathbf{curl}\, \mathbf{E} = -\kappa\dot{\mathbf{B}}, \qquad \operatorname{div} \mathbf{B} = 0.$$

By specializing the constants $\varepsilon_0, \mu_0, \kappa$, we get the following systems of units:
(i) SI system: $\kappa := 1$; $\mu_0 := 4\pi \cdot 10^{-7}\ \mathrm{N/A^2}$, $\varepsilon_0 := \frac{1}{\mu_0 c^2}$,
(ii) Gauss system: $\kappa := \frac{1}{c}$; $\varepsilon_0 := \frac{1}{4\pi}, \mu_0 := 4\pi$,
(iii) Heaviside system: $\kappa := \frac{1}{c}$; $\varepsilon_0 = \mu_0 := 1$.
The Coulomb law (19.5) on page 937 reads as

$$\mathbf{F}(\mathbf{x}) = \frac{QQ_0}{4\pi\varepsilon_0|\mathbf{x} - \mathbf{x}_0|^2} \cdot \frac{\mathbf{x} - \mathbf{x}_0}{|\mathbf{x} - \mathbf{x}_0|},$$

and the Ampère law (19.13) on page 941 reads as

$$|\mathbf{F}| = \frac{\kappa^2 \mu_0 l}{2\pi d} \cdot \mathsf{J}_0 \mathsf{J}.$$

19.10 Further Reading

A lot of classic material can be found in:

J. Maxwell, A Treatise on Electricity and Magnetism, London, 1873. Reprinted by Dover, Vols. 1, 2, New York, 1954.

J. Jackson, Classical Electrodynamics, Wiley, New York, 1995.

Furthermore, we recommend:

R. Feynman, R. Leighton, and M. Sands, The Feynman Lectures in Physics, Vol. II, Parts 1, 2: Electromagnetism and Matter, Addison-Wesley, Reading, Massachusetts, 1963.

P. Tipler, Physics for Scientists and Engineers, Freeman, New York, 1999 (general physics; 1360 pages).

J. Marsden and A. Tromba, Vector Calculus, Freeman, New York, 1996.

G. Scharf, From Electrostatics to Optics, Springer, Berlin, 1994.

G. Scharf, Finite Quantum Electrodynamics: the Causal Approach, Springer, Berlin, 1995.

A. Sommerfeld, Lectures on Theoretical Physics, Vol. 3: Electrodynamics, Academic Press, New York, 1949.

L. Landau and E. Lifshitz, Course of Theoretical Physics, Vol 2: The Classical Theory of Fields, Vol. 4: Quantum Electrodynamics, Vol. 8: Electrodynamics of Continuous Media, Butterworth–Heinemann, Oxford, 1982.

F. Dyson, Feynman's proof of the Maxwell equations, Amer. J. Phys. **58**(3) (1990), 209–211.

H. Römer and M. Forger, Elementary Field Theory (in German), Wiley-VCH, Weinheim, 1993.

H. Goenner, The Special Theory of Relativity and the Classical Theory of Fields (in German), Spektrum, Heidelberg, 2004.

Special Applications:

J. Schwinger, Classical Electrodynamics, Perseus Books, Reading, Massachusetts, 1998 (many important applications; e.g., waveguides, synchrotron radiation, scattering, diffraction, and reflection of electromagnetic waves, antennas).

M. Born and E. Wolf, Principles of Optics, 7th edition, Cambridge University Press, 1999.

F. Natterer, The Mathematics of Computerized Tomography, Wiley and Teubner/Stuttgart, 1986.

A. Friedman, Variational Principles and Free-Boundary Problems, Wiley, New York, 1988.

M. Brokate and J. Sprekels, Hysteresis and Phase Transitions, Springer, Berlin, 1996.

H. Spohn, Dynamics of Charged Particles and Their Radiation Field, Cambridge University Press, 2005.

History:

O. Darrigol, Electrodynamics from Ampère to Einstein, Oxford University Press, 2003.

C. Everitt, James Clerk Maxwell, Physicist and Natural Philosopher, Scribner, New York, 1975.

C. Domb, James Clerk Maxwell (1831–1879): 100 years later, Nature **182** (1979), 235–239.

R. Jost, Michael Faraday – 150 years after the discovery of electromagnetic induction, pp. 117–130. In: R. Jost, The Fairy Tale about the Ivory Tower: Essays and Lectures (in German and partly in English), Springer, Berlin, 1995.

D. Bodanis, $E = mc^2$: A Biography of the World's Most Famous Equation, Walker, New York, 2000.

S. Schweber, QED (Quantum Electrodynamics) and the Men Who Made It: Dyson, Feynman, Schwinger, and Tomonaga. Princeton University Press, 1994 (history of quantum electrodynamics).

Problems

19.1 *Explicit computation.* In Sect. 19.1, we mention a number of formulas which can be obtained by explicit computation by means of classical derivatives. Carry out these computations by using Hamilton's nabla calculus. In particular, prove the validity of the Maxwell equations (19.9) and (19.17) of the Coulomb field **E** and the Ampère field **B**, respectively.
Solution: Let $f(\mathbf{x}) := \frac{1}{|\mathbf{x}|^3}$. For all $\mathbf{x} \neq 0$, we get

- $\mathbf{grad}\, f = \mathbf{grad}\,(x^2 + y^2 + z^2)^{-3/2} = -\frac{3\mathbf{x}}{|\mathbf{x}|^5}$;

- $\operatorname{div}(f\mathbf{x}) = f \operatorname{div}\mathbf{x} + \mathbf{x}\cdot\mathbf{grad}\, f = 0$;
- $\mathbf{curl}(f\mathbf{x} = f\,\mathbf{curl}\,\mathbf{x} - \mathbf{x}\times\mathbf{grad}\, f = 0$.

Note that $\operatorname{div}\mathbf{x} = 3$, $\mathbf{curl}\,\mathbf{x} = 0$, and $\mathbf{x}\times\mathbf{x} = 0$.

Additional material can be found in E. Zeidler (Ed.), Oxford Users' Guide to Mathematics, Oxford University Press, 2004. The relations for generalized functions will be considered in the following problems. The main trick is to use integration by parts (see Sect. 10.4.2 of Vol. I).

19.2 *The Maxwell equations in the language of generalized functions.* In what follows, we will use the space $\mathcal{D}'(\mathbb{R}^3)$ of generalized functions introduced in Sect. 11.3.2 of Vol. I. Suppose that the electric charge density ϱ and the components of the vector fields $\mathbf{E}, \mathbf{B}$, and $\mathbf{J}$ are integrable over each ball of the Euclidean manifold $\mathbb{E}^3$. Show that the equations

$$\varepsilon_0 \operatorname{div}\mathbf{E} = \varrho \qquad \text{and} \qquad \mathbf{curl}\,\mathbf{B} = \mu_0\mathbf{J} \tag{19.108}$$

on the space $\mathcal{D}'(\mathbb{R}^3)$ of generalized functions mean that

$$-\varepsilon_0 \int_{\mathbb{E}^3} \mathbf{E}\cdot\mathbf{grad}\,\varphi\; d^3x = \int_{\mathbb{E}^3} \varrho\varphi\; d^3x$$

and

$$\int_{\mathbb{E}^3} \mathbf{B}\,\times\mathbf{grad}\,\varphi\; d^3x = \mu_0 \int_{\mathbb{E}^3} \mathbf{J}\varphi\; d^3x$$

for all test functions $\varphi \in \mathcal{D}(\mathbb{R}^3)$ (i.e., φ is smooth and vanishes outside a sufficiently large ball).
Hint: Multiply equation (19.108) by the function φ, and use integration by parts.

19.3 *The Maxwell equations for the Coulomb field in the language of generalized functions.* Show that the Coulomb field

$$\mathbf{E}(\mathbf{x}) = \frac{Q_0}{4\pi\varepsilon_0|\mathbf{x}|^2}\cdot\frac{\mathbf{x}}{|\mathbf{x}|} \tag{19.109}$$

corresponds to a generalized function, that is, the components of $\mathbf{E}$ are elements of $\mathcal{D}'(\mathbb{E}^3)$. Recall that such a generalized function has partial derivatives of all orders. Explicitly, show that there hold the equations

$$\varepsilon_0 \operatorname{div}\mathbf{E} = Q_0\delta_{P_O} \qquad \text{and} \qquad \mathbf{curl}\,\mathbf{E} = 0$$

on the space $\mathcal{D}'(\mathbb{R}^3)$ of generalized functions.
Solution: The integral

$$\mathbf{E}(\varphi) := \int_{\mathbb{E}^3} \mathbf{E}(\mathbf{x})\varphi(\mathbf{x})\; d^3x$$

exists for all test functions $\varphi \in \mathcal{D}(\mathbb{R}^3)$. To show this use spherical coordinates and observe that the singularity $\frac{1}{r^2}$ of $\mathbf{E}$ at the origin is compensated by $r^2\cos\vartheta d\varphi\; d\vartheta\; dr$. Let P_0 be the origin. We have to show that

$$-\,\varepsilon_0 \int_{\mathbb{E}^3} \mathbf{E}\cdot\mathbf{grad}\,\varphi\; d^3x = Q_0\varphi(0) \tag{19.110}$$

and

$$\int_{\mathbb{E}^3} (\mathbf{E} \times \mathbf{grad}\, \varphi)\, d^3x = 0 \tag{19.111}$$

for all test functions $\varphi \in \mathcal{D}(\mathbb{R}^3)$.
Ad (19.110). (I) Key relation. Using spherical coordinates and the mean-value theorem for integrals, we get

$$\varepsilon_0 \int_{|\mathbf{x}|=r} \varphi \cdot (\mathbf{En})\, dS = \frac{Q_0}{4\pi} \int_{|\mathbf{x}|=r} \varphi \cdot \frac{dS}{r^2} = Q_0 \varphi(\mathbf{x}_0)$$

for some $\mathbf{x}_0$ with $|\mathbf{x}_0| = r$. Consequently,

$$\lim_{r \to 0} \varepsilon_0 \int_{|\mathbf{x}|=r} \varphi \cdot (\mathbf{En})\, dS = Q_0 \varphi(0).$$

(II) Integration by parts. Let $\mathcal{M} := \{(x,y,z) \in \mathbb{R}^3 : r^2 < x^2 + y^2 + z^2 < R^2\}$ with respect to a right-handed Cartesian (x,y,z)-coordinate system. By the nabla calculus, we get

$$\operatorname{div}(\varphi \mathbf{E}) = \varphi \operatorname{div} \mathbf{E} + \mathbf{E}\, \mathbf{grad}\, \varphi = \mathbf{E}\, \mathbf{grad}\, \varphi \qquad \text{on } \mathcal{M},$$

since $\operatorname{div} \mathbf{E} = 0$ on $\mathcal{M}$. The Gauss–Ostrogradski theorem yields

$$-\varepsilon_0 \int_{\mathcal{M}} \mathbf{E}\, \mathbf{grad}\, \varphi\, d^3x = -\varepsilon_0 \int_{|\mathbf{x}|=R} \varphi \cdot (\mathbf{En})\, dS + \varepsilon_0 \int_{|\mathbf{x}|=r} \varphi \cdot (\mathbf{En})\, dS.$$

Note that the test function φ vanishes outside a sufficiently large ball. Thus, letting $R \to \infty$ and $r \to 0$, we get the claim (19.110).
Ad (19.111). Using cartesian components, integration by parts yields

$$\int_{\mathcal{M}} (\mathbf{E} \times \mathbf{grad}\, \varphi)\, d^3x = \int_{\mathcal{M}} (\mathbf{curl}\, \mathbf{E} \cdot \varphi)\, d^3x + \int_{|\mathbf{x}|=r} (\mathbf{E} \times \mathbf{n})\, dS - \int_{|\mathbf{x}|=R} (\mathbf{E} \times \mathbf{n})\, dS.$$

Note that $\mathbf{E} \times \mathbf{n} = 0$ and $\mathbf{curl}\, \mathbf{E} = 0$ on $\mathcal{M}$. Letting $R \to 0$ and $r \to 0$, we get the claim (19.111).

19.4 *The Maxwell equations for the electric dipole in the language of generalized functions.* Fix the vector $\mathbf{p}_{\text{el}}$. Consider the Coulomb field $\mathbf{E}$ from (19.109). Define the directional derivative

$$\mathbf{E}_{\text{dipole}} := -(\mathbf{p}_{\text{el}}\, \mathbf{grad})\mathbf{E},$$

in the sense of generalized functions on $\mathbb{E}^3$. Show that $\mathbf{E}_{\text{dipole}}$ coincides with the classical expression given in (19.26) on page 945.
Hint: Observe that the Coulomb field $\mathbf{E}$ is smooth on $\mathbb{E}^3 \setminus \{0\}$. Thus, the classical partial derivatives on $\mathbb{E}^3 \setminus \{0\}$ coincide with the corresponding derivatives in the sense of distributions.

19.5 *The Ampère magnetic field in the language of generalized functions.* Show that the Ampère field

$$\mathbf{B}(\varphi, r) = B(r)\, \mathbf{e}_\varphi, \qquad B(r) = \frac{\mu_0 \mathsf{J}}{2\pi r}$$

from (19.14) is a generalized function on the (x,y)-plane. Moreover, we have the Maxwell equations

$$\mathbf{curl}\,\mathbf{B} = \mu_0 \mathsf{J}\delta_{(0,0)}\mathbf{k} \qquad \text{and} \qquad \operatorname{div}\mathbf{B} = 0 \qquad \text{on } \mathbb{E}^2.$$

Solution: We have to show that

$$\int_{\mathbb{E}^2} \mathbf{B} \times \mathbf{grad}\,\chi\; dxdy = \mu_0 \mathsf{J}\chi(0,0), \tag{19.112}$$

and

$$\int_{\mathbb{E}^2} \mathbf{B}\,\mathbf{grad}\,\chi\; dxdy = 0 \tag{19.113}$$

for all test functions $\chi \in \mathcal{D}(\mathbb{R}^2)$.
Ad (19.112). (I) Key relation. On the (x,y)-plane, we get

$$\int_{|\mathbf{x}|=r} \chi \mathbf{B} d\mathbf{x} = \int_{-\pi}^{\pi} \chi B(r)\cdot r d\varphi = \frac{\mu_0 \mathsf{J}}{2\pi}\int_{-\pi}^{\pi} \chi\, d\varphi = \mu_0 \mathsf{J}\chi(\mathbf{x}_0)$$

for some $\mathbf{x}_0$ with $|\mathbf{x}_0| = r$. Hence $\lim_{r\to 0}\int_{|\mathbf{x}|=r}\chi\mathbf{B}d\mathbf{x} = \mu_0\mathsf{J}\chi(0,0)$.
(II) Integration by parts. Choose $\mathcal{M} := \{(x,y)\in\mathbb{R}^2,\ r^2 < x^2+y^2 < R^2\}$. By the nabla calculus,

$$\mathbf{curl}(\chi\mathbf{B}) = \chi\,\mathbf{curl}\,\mathbf{B} - \mathbf{B}\times\mathbf{grad}\,\chi = -\mathbf{B}\times\mathbf{grad}\,\chi \quad \text{on}\mathcal{M},$$

since $\mathbf{curl}\,\mathbf{B} = 0$ on $\mathcal{M}$. The Stokes integral theorem yields

$$\int_{\mathcal{M}} \mathbf{B}\times\mathbf{grad}\,\chi\; dxdy = \int_{|\mathbf{x}|=r} \mathbf{B}d\mathbf{x} - \int_{|\mathbf{x}|=R} \mathbf{B}d\mathbf{x}.$$

Letting $R \to 0$ and $r \to 0$, we obtain the claim (19.112).
Ad (19.113). By the nabla calculus,

$$\operatorname{div}(\chi\mathbf{B}) = \mathbf{B}\,\mathbf{grad}\,\chi + \chi\operatorname{div}\mathbf{B} = \mathbf{B}\,\mathbf{grad}\,\chi \qquad \text{on } \mathcal{M},$$

since $\operatorname{div}\mathbf{B} = 0$ on $\mathcal{M}$. The Gauss–Ostrogradski integral theorem yields

$$\int_{\mathcal{M}} \mathbf{B}\,\mathbf{grad}\,\chi\; dxdy = \int_{|\mathbf{x}|=R} \chi\,\mathbf{Bn}\; ds - \int_{|\mathbf{x}|=r} \chi\,\mathbf{Bn}\; ds.$$

Since the magnetic field $\mathbf{B}$ is orthogonal to the outer normal unit vector $\mathbf{n}$ of the circle of radius r about the origin, we have $\mathbf{Bn} = 0$. Letting $R \to \infty$ and $r \to \infty$, we get the claim (19.113).

19.6 *Special transformation law.* Consider the transformation

$$x' = \cos\varphi\cdot x + \sin\varphi\cdot y, \quad y' = -\sin\varphi\cdot x + \cos\varphi\cdot y, \quad z' = z \tag{19.114}$$

which describes a clockwise rotation of the right-handed Cartesian (x,y,z)-coordinate system about the z-axis with the rotation angle φ. Consider the matrix $\mathrm{e}^{-\varphi\mathcal{I}^3} = \cos\frac{\varphi}{2}\,\sigma^0 - \mathrm{i}\sin\frac{\varphi}{2}\,\sigma^3$. Define

$$\sigma^{k'} := A^{-1}\sigma^k A, \qquad k = 1,2,3.$$

Show that

$$\sigma^{1'} = \cos\varphi\cdot\sigma^1 + \sin\varphi\cdot\sigma^2, \quad \sigma^{2'} = -\sin\varphi\cdot\sigma^1 + \cos\varphi\cdot\sigma^2, \quad \sigma^{3'} = \sigma^3.$$

Note that this is the same transformation law as (19.114). Show that the cyclic permutations $x \mapsto y \mapsto z \mapsto x$ and $1 \mapsto 2 \mapsto 3 \mapsto 1$ yield formulas which are also valid.
Hint: Use $\sigma^1\sigma^2 = -\sigma^2\sigma^1 = \mathrm{i}\sigma^3$ together with the formulas obtained by cyclic permutation of the indices. Moreover, observe that $\sin\varphi = 2\sin\frac{\varphi}{2}\cos\frac{\varphi}{2}$, and $\cos\varphi = \cos^2\frac{\varphi}{2} - \sin^2\frac{\varphi}{2}$.

19.7 *The adjoint representation of the Lie group $SU(2)$ on the Lie algebra $su(2)$.* Recall that $\mathrm{i}\sigma^1, \mathrm{i}\sigma^2, \mathrm{i}\sigma^3$ is a basis of $su(2)$. For all matrices $A \in SU(2)$, define

$$\chi_A\left(\sum_{k=1}^{3} \alpha_k \cdot \mathrm{i}\sigma^k\right) := A^{-1}\left(\sum_{k=1}^{3} \alpha_k \cdot \mathrm{i}\sigma^k\right) A, \qquad \alpha_1, \alpha_2, \alpha_3 \in \mathbb{R}.$$

Show that $\chi_A : su(2) \to su(2)$ is a linear operator, and show that the map $A \mapsto \chi_A$ is a representation of $SU(2)$ on $su(2)$. In addition, show that this representation is equivalent to the natural representation of $SO(3)$ on $\mathbb{R}^3$.
Hint: Set $\mathrm{i}\sigma^{k'} := A^{-1}(\mathrm{i}\sigma^k)A, k = 1, 2, 3$. By Problem 19.6, the transformation formulas from $\mathrm{i}\sigma^1, \mathrm{i}\sigma^2, \mathrm{i}\sigma^3$ to $\mathrm{i}\sigma^{1'}, \mathrm{i}\sigma^{2'}, \mathrm{i}\sigma^{3'}$ are the same as the transformation formulas from x, y, z to $x'y', z'$ under the rotation $\varrho(A)$ which corresponds to the group epimorphism $\varrho : SU(2) \to SO(3)$ (see (7.21) on page 434).

19.8 *Transformation law for the Pauli equation.* Prove Prop. 19.33 on page 950.
Hint: For example, let us consider the transformation $\psi' = A\psi$ with the matrix $A := \mathrm{e}^{-\varphi\mathcal{I}^3}$. This corresponds to the rotation (19.114) about the z-axis. Set $x^1 := x, x^2 := y, x^3 := z$. For the components of vectors, the inner product is an invariant under rotations. Hence

$$\sum_{k=1}^{3} x^k B^k = \sum_{k=1}^{3} x^{k'} B^{k'}.$$

By Problem 19.6,

$$\mathbf{mB} = \sum_{k=1}^{3} B^k\sigma^k = \sum_{k=1}^{3} B^{k'}\sigma^{k'}.$$

Set $(\mathbf{mB})' := B^{1'}\sigma^1 + B^{2'}\sigma^2 + B^{3'}\sigma^3$. Then $\mathbf{mB} = A^{-1}(\mathbf{mB})'A$.
(I) Transformation of the Pauli equation (19.30). Let

$$\mathrm{i}\hbar\psi_t = \left(\frac{\mathbf{P}^2}{2m_e} - eU - \mathbf{mB}\right)\psi. \tag{19.115}$$

We have to show that

$$\mathrm{i}\hbar\psi'_t = \left(\frac{\mathbf{P}^2}{2m_e} - eU - (\mathbf{mB})'\right)\psi'. \tag{19.116}$$

In fact, it follows from (19.115) that

$$\mathrm{i}\hbar\psi_t = A^{-1}\left(\frac{\mathbf{P}^2}{2m_e} - eU - (\mathbf{mB}'\right) A\psi.$$

This implies (19.116).
(II) Expectation values. Let $X = (x, y, z)$. Choose $k = 1, 2, 3$. Let us consider the expectation value

$$\bar{\mathcal{S}}^k = \frac{\hbar}{2}\int_{\mathbb{R}^3} \psi^\dagger(X,t)\, \sigma^k\, \psi(X,t)\, d^3x$$

of the spin component operator $\mathcal{S}^k$ with respect to the state ψ with $\langle\psi|\psi\rangle = 1$. The transformed state $\psi' = A\psi$ yields the transformed expectation value

$$\bar{\mathcal{S}}^{k'} = \frac{\hbar}{2}\int_{\mathbb{R}^3} (A\psi)^\dagger(X',t)\, \sigma^k \cdot (A\psi)(X',t)\, d^3x'.$$

Since the matrix A is unitary, we get $\langle\psi'|\psi'\rangle = 1$, and $A^\dagger = A^{-1}$. Hence

$$\bar{\mathcal{S}}^{k'} = \frac{\hbar}{2}\int_{\mathbb{R}^3} \psi^\dagger(X',t)\, (A^{-1}\sigma^k A)\, \psi(X',t)\, d^3x'.$$

Noting that $A^{-1}\sigma^k A = \sigma^{k'}$ and using the invariance of the integral under rotations, we get

$$\bar{\mathcal{S}}^{k'} = \frac{\hbar}{2}\int_{\mathbb{R}^3} \psi^\dagger(X,t)\, \sigma^{k'}\, \psi(X,t)\, d^3x.$$

By Problem 19.8, the transformation law from $\mathcal{S}^1, \mathcal{S}^2, \mathcal{S}^3$ to $\mathcal{S}^{1'}, \mathcal{S}^{2'}, \mathcal{S}^{3'}$ is the same as the transformation law from x, y, z to x', y', z'.

19.9 *Electromagnetic waves and the nabla calculus.* Prove Theorem 19.9 on page 969.

Solution: Let $\mathbf{E}(\mathbf{x},t) := f(\mathbf{nx} - ct)\mathbf{e}$ and $\mathbf{B}(\mathbf{x},t) := \frac{1}{c}(\mathbf{n} \times \mathbf{E}(\mathbf{x},t))$. Then:

- $\operatorname{div}\mathbf{E} = \boldsymbol{\partial}(f\mathbf{e}) = \boldsymbol{\partial}f \cdot \mathbf{e} = f'\mathbf{en} = 0$.
- $\operatorname{\mathbf{curl}}\mathbf{E} = \boldsymbol{\partial} \times f\mathbf{e} = \boldsymbol{\partial}f \times \mathbf{e} = f'\mathbf{n} \times \mathbf{e}$.
- $c\operatorname{div}\mathbf{B} = \boldsymbol{\partial}(\mathbf{n} \times \mathbf{E}) = -\mathbf{n}(\boldsymbol{\partial} \times \mathbf{E}) = -f'\mathbf{n}(\mathbf{n} \times \mathbf{e}) = 0$.
- $c\operatorname{\mathbf{curl}}\mathbf{B} = \boldsymbol{\partial} \times (\mathbf{n} \times \mathbf{E}) = \mathbf{n}(\boldsymbol{\partial}\mathbf{E}) - (\mathbf{n}\boldsymbol{\partial})\mathbf{E} = -(\mathbf{n}\boldsymbol{\partial})\mathbf{E} = -(\mathbf{n}\boldsymbol{\partial})f\mathbf{e}$ $= -f'\mathbf{n}^2 \cdot \mathbf{e} = -f'\mathbf{e}$.
- $\dot{\mathbf{E}} = -cf'\mathbf{e}$.
- $\dot{\mathbf{B}} = \frac{1}{c}(\mathbf{n} \times \dot{\mathbf{E}}) = -f'(\mathbf{n} \times \mathbf{e})$.

Hence $\operatorname{\mathbf{curl}}\mathbf{E} = -\dot{\mathbf{B}}$ and $c^2 \operatorname{\mathbf{curl}}\mathbf{B} = \dot{\mathbf{E}}$.

19.10 *The Euler–Lagrange equations of the electromagnetic field.* Prove Theorem 19.12 on page 976.

Solution: Choose the test functions $h_\alpha \in C_0^\infty(\mathcal{O},\mathbb{R})$, $\alpha = 0,1,2,3$ (i.e., $h_\alpha : \mathcal{O} \to \mathbb{R}$ is smooth and has compact support (see Vol. I)). Define

$$\chi(\sigma) := \int_{\mathcal{O}} \left(\tfrac{1}{4}\mathcal{F}_{\alpha\beta}\mathcal{F}^{\alpha\beta} + \mu_0 A_\alpha \mathcal{J}^\alpha\right) d^4x, \qquad \sigma \in \mathbb{R}$$

where $\mathcal{F}_{\alpha\beta} := \partial_\alpha(A_\beta + \sigma h_\beta) - \partial_\beta(A_\alpha + \sigma h_\alpha)$. If A_0, A_1, A_2, A_3 is a smooth solution of (19.92), then $\chi'(0) = 0$. Hence

$$\chi'(0) = \int_{\mathcal{O}} \left(\tfrac{1}{2}(\partial_\alpha h_\beta - \partial_\beta h_\alpha)F^{\alpha\beta} + \mu_0 h_\beta \mathcal{J}^\beta\right) d^4x = 0.$$

Since $F^{\alpha\beta} = -F^{\beta\alpha}$, we get $\int_{\mathcal{O}}(\partial_\alpha h_\beta \cdot F^{\alpha\beta} + \mu_0 h_\beta \mathcal{J}^\beta)\, d^4x = 0$. Integration by parts yields

$$\int_{\mathcal{O}} (-\partial_\alpha F^{\alpha\beta} + \mu_0 \mathcal{J}^\beta) h_\beta \, d^4x = 0$$

for all $h_\beta \in C_0^\infty(\mathcal{O},\mathbb{R})$). Hence

$$-\partial_\alpha F^{\alpha\beta} + \mu_0 \mathcal{J}^\beta = 0, \qquad \beta = 0,1,2,3.$$

In addition, it follows from $F_{\beta\gamma} = \partial_\beta A_\gamma - \partial_\gamma A_\beta$ that

$$\partial_\alpha F_{\beta\gamma} + \partial_\beta F_{\gamma\alpha} + \partial_\gamma F_{\alpha\beta} = 0, \qquad \alpha, \beta, \gamma = 0, 1, 2, 3.$$

19.11 *The principle of critical action for a charged particle.* Prove Theorem 19.14 on page 978.
Solution: Use the Euler–Lagrange equation $\frac{d}{dt} L_{\dot{\mathbf{x}}} = L_{\mathbf{x}}$, and observe that

- $L_{\dot{\mathbf{x}}} = \frac{m_0 \dot{\mathbf{x}}}{\sqrt{1-\dot{\mathbf{x}}^2/c^2}} + Q\mathbf{A} = \mathbf{p} + Q\mathbf{A}$;
- $L_{\mathbf{x}} = Q\,\mathbf{grad}_{\mathbf{x}}(\dot{\mathbf{x}}\mathbf{A}) - Q\,\mathbf{grad}\,U = Q\dot{\mathbf{x}} \times \mathbf{curl}\,\mathbf{A} + Q(\dot{\mathbf{x}}\,\mathbf{grad}_{\mathbf{x}})\mathbf{A} - Q\,\mathbf{grad}\,U$;
- $\frac{d}{dt}\mathbf{A}(\mathbf{x}(t), t)) = (\dot{\mathbf{x}}(t)\,\mathbf{grad}_{\mathbf{x}})\mathbf{A}(\mathbf{x}(t), t) + \mathbf{A}_t(\mathbf{x}(t), t)$.

20. The Relativistic Invariance of the Dirac Equation and the Electron Spin

Dirac discovered in 1928 that the electron spin is a relativistic effect. Combining Einstein's theory of special relativity with quantum mechanics, Dirac replaced Schrödinger's non-relativistic equation for the electron by the relativistic equation for the electron. The transformation of the Dirac wave function under rotations yields the operator for the electron spin. All the fundamental particles of the Standard Model in particle physics are described by the Dirac equation with additional interaction terms related to the messenger particles.

Folklore

In this chapter, we restrict ourselves on discussing the basic ideas. A detailed investigation of the Dirac equation together with the Seiberg–Witten equation and the relations to spinor calculus, Clifford algebras, and spin geometry can be found in Vol. IV on quantum mathematics.

Convention. We will use positively oriented inertial systems with the right-handed Cartesian coordinates x, y, z and the time t. We set $x^1 := x, x^2 := y$ and $x^3 := z, x^0 := ct$. Moreover, $\partial_\mu := \frac{\partial}{\partial x^\mu}$. We sum over equal upper and lower Greek indices from 0 to 3.

20.1 The Dirac Equation

The Dirac equation for the free relativistic electron with positive rest mass m_0 in a positively oriented inertial system reads as follows:

$$\boxed{(i\hbar\gamma^\mu\partial_\mu - m_0c)\,\psi = 0.} \tag{20.1}$$

We are looking for the Dirac wave function $(x, y, z, ct) \mapsto \psi(x, y, z, ct)$ from $\mathbb{R}^4$ to $\mathbb{C}^4$. Here,

$$\psi = \begin{pmatrix} \psi^1 \\ \psi^2 \\ \psi^3 \\ \psi^4 \end{pmatrix}.$$

In addition, we introduce the complex-valued (4×4)-Dirac–Pauli matrices

$$\gamma^0 := \begin{pmatrix} \sigma^0 & 0 \\ 0 & -\sigma^0 \end{pmatrix}, \quad \gamma^j := \begin{pmatrix} 0 & \sigma^j \\ -\sigma^j & 0 \end{pmatrix}, \qquad j = 1, 2, 3, \tag{20.2}$$

E. Zeidler, *Quantum Field Theory III: Gauge Theory*,
DOI 10.1007/978-3-642-22421-8_21,

where $\sigma^0, \sigma^1, \sigma^2, \sigma^3$ are the Pauli matrices. Explicitly,

$$\sigma^0 = \begin{pmatrix} 1 & 0 \\ 0 & 1 \end{pmatrix}, \quad \sigma^1 = \begin{pmatrix} 0 & 1 \\ 1 & 0 \end{pmatrix}, \quad \sigma^2 = \begin{pmatrix} 0 & -\mathrm{i} \\ \mathrm{i} & 0 \end{pmatrix}, \quad \sigma^3 = \begin{pmatrix} 1 & 0 \\ 0 & -1 \end{pmatrix}.$$

In addition, we set $\overline{\psi} := \psi^\dagger \gamma^0$ (Dirac's adjoint wave function).

The Dirac–Pauli matrices satisfy the following Clifford relations:

$$\boxed{\gamma^\mu \gamma^\nu + \gamma^\nu \gamma^\mu = 2g^{\mu\nu}\mathbf{1}, \qquad \mu, \nu = 0, 1, 2, 3.} \tag{20.3}$$

Recall that $g^{00} = 1, g^{jj} := -1, j = 1, 2, 3$, and $g^{\mu\nu} = 0$ if $\mu \neq \nu$. The symbol $\mathbf{1}$ denotes the (4×4)-unit matrix. Explicitly,

$$(\gamma^0)^2 = \mathbf{1}, \ (\gamma^1)^2 = (\gamma^2)^2 = (\gamma^3)^2 = -\mathbf{1}, \ \gamma^\mu \gamma^\nu = -\gamma^\nu \gamma^\mu, \quad \mu \neq \nu.$$

The Clifford relations play the key role in the theory of the Dirac equation (see Vol. IV).

Motivation of the Dirac equation (Dirac's square root trick). To begin with, consider the relativistic energy relation

$$\boxed{E^2 = m_0^2 c^4 + c^2 \mathbf{p}^2} \tag{20.4}$$

for a relativistic free particle with the rest mass m_0 and the momentum vector $\mathbf{p}$. As for the Schrödinger equation, we use the following quantization rule:

$$E \Rightarrow \mathrm{i}\hbar c \partial_0, \qquad p^j \Rightarrow -\mathrm{i}\hbar \partial_j, \qquad j = 1, 2, 3. \tag{20.5}$$

Note that the contravariant energy-momentum 4-vector p^μ is given by $p^0 = E/c$ and the momentum vector $\mathbf{p} = p^1\mathbf{i} + p^2\mathbf{j} + p^3\mathbf{k}$. Lowering the index, we get the covariant energy-momentum 4-vector $p_\alpha := g_{\alpha\beta} p^\beta$. Explicitly,

$$p_0 = p^0 = \frac{E}{c}, \quad p_j = -p^j, \quad j = 1, 2, 3.$$

Using this, the quantization rule (20.5) reads as

$$p_\alpha \Rightarrow \mathrm{i}\hbar \partial_\alpha, \qquad \alpha = 0, 1, 2, 3. \tag{20.6}$$

If we apply this quantization rule to (20.4), then we obtain the Klein–Gordon–Fock equation (see Sect. 13.2.1). This is a second-order partial differential equation. This differential equation does not yield the spectrum of the hydrogen atom observed in experiment. Therefore, Dirac tried to find a first-order partial differential equation. To this end, we first write the energy relation (20.4) in the form

$$\frac{E^2}{c^2} - p_1^2 - p_2^2 - p_3^2 = m_0^2 c^2.$$

This is equivalent to

$$p_\mu g^{\mu\nu} p_\nu = m_0 c^2.$$

Now to the point. We want to determine the quantities $\gamma^0, \gamma^1, \gamma^2, \gamma^3$ in such a way that there holds the square root relation:

$$\boxed{(\gamma^\mu p_\mu)(\gamma^\nu p_\nu) = p_\mu g^{\mu\nu} p_\nu \mathbf{1}.}$$

Since $p_\mu \gamma^\mu \gamma^\nu p_\nu = \frac{1}{2} p_\mu (\gamma^\mu \gamma^\nu + \gamma^\nu \gamma^\mu) p_\nu$, we have to realize the Clifford relations

$$\boxed{(\gamma^\mu \gamma^\nu + \gamma^\nu \gamma^\mu) = 2g^{\mu\nu}\mathbf{1}, \qquad \mu, \nu = 0, 1, 2, 3.}$$

Using the Pauli spin matrices introduced by Pauli in 1927, Dirac constructed the matrices (20.2) which satisfy the Clifford relations.

Proposition 20.1 *Choose the Dirac–Pauli matrices* (20.2). *If* p_0, p_1, p_2, p_3 *are real numbers with* $\gamma^\mu p_\mu = m_0 c\mathbf{1}$, *then* $(\gamma^\mu p_\mu)^2 = p_\mu g^{\mu\nu} p_\nu \mathbf{1} = m_0^2 c^2 \mathbf{1}$.

Using the quantization rule $p_\mu \Rightarrow \mathrm{i}\hbar\partial_\mu$, Dirac arrived at his famous first-order partial diffenertial equation $\mathrm{i}\hbar\gamma^\mu\partial_\mu\psi = m_0 c\psi$.

Roughly speaking, Dirac obtained his electron equation by computing the square root of the Klein–Fock–Gordon operator.

Nowadays there exists a well-developed calculus of pseudo-differential operators which allows us to construct broad classes of functions of differential operators. We refer to G. Hsiao and W. Wendland, Boundary Integral Equations, Chap. 6, Springer, New York, 2008, and L. Hörmander, The Analysis of Linear Partial Differential Operators, Vol. 3: Pseudodifferential Operators, Springer, New York, 1983.

Historical remark. The Dirac equation is the result of a long fascinating development in mathematics and physics. Hamilton discovered quaternions in 1843. Thirty five years later, Clifford generalized the algebra of quaternions to Clifford algebras. Finally, in 1928, Dirac used the idea of Clifford algebra in order to formulate his equation for the relativistic electron. In the 20th century, the Dirac equation together with the idea of local symmetry (gauge theory) was the decisive tool for the formulation of the Standard Model in particle physics. In terms of mathematics, the Dirac equation played a crucial role in the discovery of the Atiyah–Singer index theorem in the early 1960s and for the formulation of spin geometry. In terms of physics, the Seiberg–Witten equation was formulated in the context of the quark confinement. In terms of mathematics, the Seiberg–Witten equation plays a key role in the theory of 4-dimensional manifolds (see Vol. IV).

20.2 Changing the Inertial System

Dirac proved in 1928 that the Dirac equation is valid in all inertial systems.

Folklore

The key problem is to find the right transformation law for the Dirac wave function under a change of the inertial system. This is a nontrivial task because of topological peculiarities. The crucial fact is that there exists a surjective group morphism

$$\varrho : SL(2,\mathbb{C}) \to SO^\uparrow(1,3)$$

from the symplectic group $SL(2,\mathbb{C})$ onto the component $SO^\uparrow(1,3)$ of the unit element of the Lorentz group $O(1,3)$.[1] Recall that $SL(2,\mathbb{C})$ consists of all complex (2×2)-matrices A with $\det(A) = 1$. The relativistic invariance of the Dirac equation follows then by using the representation theory of the group $SL(2,\mathbb{C})$ (spinor calculus). The group $SO^\uparrow(1,3)$ is not simply connected, but the group $SL(2,\mathbb{C})$ is

[1] The Lie group $SO^\uparrow(1,3)$ is also called the orthochronous special Lorentz group. The symplectic group $SL(2,\mathbb{C})$ is also denoted by $Sp(2,\mathbb{C})$.

simply connected; the Lie group $SL(2,\mathbb{C})$ is the universal covering group of the Lie group $SO^{\uparrow}(1,3)$, and we have the Lie group isomorphism

$$\boxed{SL(2,\mathbb{C})/\{\mathbf{1},-\mathbf{1}\} \simeq SO^{\uparrow}(1,3).} \tag{20.7}$$

This will be investigated in Vol. IV. There we will also study space reflections P, time reflections T, and charge conjugation C. At this point, we will summarize some important transformation laws concerning Lorentz boosts and rotations.

Special Lorentz transformations. Consider two positively oriented inertial systems Σ and Σ' which coincide at time $t = 0$. Suppose that Σ' moves along the x-axis of Σ with the positive velocity V (see Fig.18.1 on page 906). The passage from Σ to Σ' is described by the formula

$$x' = x\ \cosh\chi - ct\ \sinh\chi,\ \ y' = y,\ \ z' = z,\ \ ct' = ct\ \cosh\chi - x\sinh\chi$$

with $\cosh\chi := \frac{1}{\sqrt{1-V^2/c^2}}$ and $\sinh\chi := \frac{V}{c\sqrt{1-V^2/c^2}}$. The corresponding transformation formula for the Dirac wave function reads as

$$\boxed{\psi'(x',y',z',ct') = \mathrm{e}^{-\frac{\chi}{2}\gamma^0\gamma^1}\psi(x,y,z,ct).} \tag{20.8}$$

Here, we have

$$\mathrm{e}^{-\frac{\chi}{2}\gamma^0\gamma^1} = \cosh\frac{\chi}{2}\cdot\mathbf{1} - \sinh\frac{\chi}{2}\cdot\gamma^0\gamma^1. \tag{20.9}$$

The proof of this formula can be found in Problem 20.1. If we replace the x-axis by the y-axis (resp. z-axis), then we have to use the cyclic permutations:

$$x \Rightarrow y \Rightarrow z \Rightarrow x \quad \text{and} \quad \gamma^1 \Rightarrow \gamma^2 \Rightarrow \gamma^3 \Rightarrow \gamma^1. \tag{20.10}$$

Rotations. Suppose that the inertial system Σ' is obtained from the inertial system Σ by a counter-clockwise rotation of Σ about the z-axis with rotation angle φ. Then the transformation formulas read as follows:

$$x' = x\cos\varphi + y\sin\varphi,\ \ y' = -x\sin\varphi + y\cos\varphi,\ \ z' = z,\ \ t' = t, \tag{20.11}$$

and

$$\boxed{\psi'(x',y',z',ct') = \mathrm{e}^{-\frac{\varphi}{2}\gamma^1\gamma^2}\psi(x,y,z,ct).} \tag{20.12}$$

Here,

$$\mathrm{e}^{-\frac{\varphi}{2}\gamma^1\gamma^2} = \cos\frac{\varphi}{2}\cdot\mathbf{1} - \sin\frac{\varphi}{2}\cdot\gamma^1\gamma^2. \tag{20.13}$$

If we replace the z-axis by the x-axis (resp. y-axis), then we have to use the cyclic permutations (20.10). The proof of the following proposition will be given in Problem 20.3.

Proposition 20.2 *If the wave function ψ satisfies the Dirac equation* (20.1) *in Σ, then the transformed wave functions ψ' from* (20.8) *and* (20.12) *satisfy the Dirac equation in Σ'.*

Topological peculiarity. Consider the transformation formula (20.12).

- For the rotation angle $\varphi = 2\pi$, we get $x' = x, y' = y, z = z', t' = t$, but
$$\psi'(x', y', z', ct') = -\psi(x, y, z, ct).$$
- For the rotation angle $\varphi = 4\pi$, we get $x' = x, y' = y, z = z', t' = t$, and
$$\psi'(x', y', z', ct') = \psi(x, y, z, ct).$$

Thus, the transformed wave function ψ' is only given up to sign. In Vol. IV, we will show that this is a consequence of the universal-covering-group isomorphism (20.7).

Mnemonically, the Dirac wave function ψ sees the universal covering group $SL(2, \mathbb{C})$ of the proper orthochronous Lorentz group $SO^{\uparrow}(1, 3)$.

20.3 The Electron Spin

Infinitesimal rotation. Angular momentum and spin in physics are always related to infinitesimal rotations of the physical quantities.

Proposition 20.3 *For small rotation angle φ, we get*
$$\psi'(x, y, z, ct) = \psi(x, y, z, ct) + \tfrac{i\varphi}{\hbar}(L_z + S_z)\psi(x, y, z, ct) + o(\varphi), \quad \varphi \to 0$$
with the differential operator
$$L_z := -i\hbar\left(x\frac{\partial}{\partial y} - y\frac{\partial}{\partial x}\right)$$
and the matrix operator
$$S_z := \frac{\hbar}{2}\begin{pmatrix} \sigma^3 & 0 \\ 0 & \sigma^3 \end{pmatrix} = \frac{\hbar}{2}\begin{pmatrix} 1 & 0 & 0 & 0 \\ 0 & -1 & 0 & 0 \\ 0 & 0 & 1 & 0 \\ 0 & 0 & 0 & -1 \end{pmatrix}.$$

Proof. Replacing the rotation angle φ by $-\varphi$, we get
$$x = x' - \varphi y' + o(\varphi), \quad y = y' + \varphi x' + o(\varphi), \quad \varphi \to 0, \quad z' = z, \ t' = t,$$
and
$$\psi'(x', y', z', ct') = \left(\mathbf{1} - \tfrac{\varphi}{2}\gamma^1\gamma^2\right)\psi(x, y, z, ct) + o(\varphi), \quad \varphi \to 0.$$
Note that $\gamma^1\gamma^2 = -\mathrm{i}\begin{pmatrix} \sigma^3 & 0 \\ 0 & \sigma^3 \end{pmatrix}$. By the Taylor expansion,
$$\begin{aligned}\psi(x, y, z, ct) &= \psi(x' - \varphi y' + o(\varphi), y' + \varphi x' + o(\varphi), z', ct') \\ &= \varphi\left(-y'\frac{\partial\psi}{\partial x} + x'\frac{\partial\psi}{\partial y}\right)(x', y'z', ct') + o(\varphi), \quad \varphi \to 0.\end{aligned}$$
Replacing $x' \Rightarrow x, y' \Rightarrow y$, and $z' \Rightarrow z$, we get the claim. □

The spin operator. Note that the operator of angular momentum in Schrödinger's quantum mechanics is given by

$$\mathbf{L} = \mathbf{x} \times \mathbf{p} = L_x\mathbf{i} + L_y\mathbf{j} + L_z\mathbf{k}$$

with the momentum operator $\mathbf{p} = -i\hbar\boldsymbol{\partial}$. Proposition 20.3 shows that, in Dirac's relativistic quantum mechanics, we have to replace the operator L_z by the operator sum $L_z + S_z$ (total angular momentum). By cyclic permutation, we define the spin operator

$$\mathbf{S} = S_x\mathbf{i} + S_y\mathbf{j} + S_z\mathbf{k}$$

with

$$S_x := \frac{\hbar}{2}\begin{pmatrix} \sigma^1 & 0 \\ 0 & \sigma^1 \end{pmatrix}, \quad S_y := \frac{\hbar}{2}\begin{pmatrix} \sigma^2 & 0 \\ 0 & \sigma^2 \end{pmatrix}, \quad S_z := \frac{\hbar}{2}\begin{pmatrix} \sigma^3 & 0 \\ 0 & \sigma^3 \end{pmatrix}.$$

Spin states. Let us introduce the following states of the relativistic electron:

$$\psi_+^{(1)} := \begin{pmatrix} 1 \\ 0 \\ 0 \\ 0 \end{pmatrix}, \quad \psi_-^{(1)} := \begin{pmatrix} 0 \\ 1 \\ 0 \\ 0 \end{pmatrix}, \quad \psi_+^{(2)} := \begin{pmatrix} 0 \\ 0 \\ 1 \\ 0 \end{pmatrix}, \quad \psi_-^{(2)} := \begin{pmatrix} 0 \\ 0 \\ 0 \\ 1 \end{pmatrix}.$$

Then

$$S_z\psi_+^{(j)} = \frac{\hbar}{2}\psi_+^{(j)}, \qquad S_z\psi_-^{(j)} = -\frac{\hbar}{2}\psi_-^{(j)}, \quad j = 1, 2.$$

Thus, $\psi_+^{(1)}, \psi_+^{(2)}$ are eigenstates of the electron with the spin vector

$$\mathbf{S} = \tfrac{\hbar}{2}\mathbf{k},$$

and $\psi_-^{(1)}, \psi_-^{(2)}$ are eigenstates of the electron with the spin vector $\mathbf{S} = -\frac{\hbar}{2}\mathbf{k}$.

Problems

20.1 *The exponential function.* Prove (20.9) and (20.13).
Solution: It follows from $\gamma^0\gamma^1\gamma^0\gamma^1 = -\gamma^0\gamma^0\gamma^1\gamma^1 = \mathbf{1}$ that

$$e^{-\frac{\chi}{2}\gamma^0\gamma^1} = \mathbf{1} - \tfrac{\chi}{2}\gamma^0\gamma^1 + \tfrac{1}{2}\left(\tfrac{\chi}{2}\right)^2 - \ldots = \cosh\tfrac{\chi}{2} \cdot \mathbf{1} - \sinh\tfrac{\chi}{2} \cdot \gamma^0\gamma^1.$$

20.2 *Transformation law for the Dirac equation.* Consider the linear coordinate transformation

$$\boxed{x'^\nu = \Lambda^\nu_\mu x^\mu, \qquad \nu = 0, 1, 2, 3} \tag{20.14}$$

from the inertial system Σ to the inertial system Σ'. Set $\partial'_\mu := \frac{\partial}{\partial x'^\mu}$, and set $X := (x, y, z, ct)$. Consider the transformation

$$\psi'(X') = T\psi(X)$$

where T is an invertible complex (4×4)-matrix which has the following property:

$$\boxed{T^{-1}\gamma^\nu T = \Lambda^\nu_\mu \gamma^\mu, \qquad \nu = 0,1,2,3.} \tag{20.15}$$

Show that if ψ is a solution of the Dirac equation

$$\mathrm{i}\hbar\gamma^\mu \partial_\mu \psi(X) = m_0 c\, \psi(X) \tag{20.16}$$

in the inertial system Σ, then ψ' is a solution of the Dirac equation in Σ', that is,

$$\mathrm{i}\hbar\gamma^\mu \partial'_\mu \psi'(X') = m_0 c\, \psi'(X'). \tag{20.17}$$

Note that the key transformation law (20.15) parallels the transformation law (20.14) for the coordinates.
Solution: By the chain rule $\partial_\mu = \frac{\partial x'^\nu}{\partial x^\mu}\partial'_\nu = \Lambda^\nu_\mu \partial'_\nu$. It follows from (20.16) that

$$\mathrm{i}\hbar\gamma^\mu \Lambda^\nu_\mu \partial'_\nu T^{-1}\psi'(X') = m_0 c \cdot T^{-1}\psi'(X').$$

Hence

$$\mathrm{i}\hbar(T\gamma^\mu \Lambda^\mu_\nu T^{-1})\, \partial'_\nu \psi'(X') = m_0 c\, \psi'(X').$$

Since $T\gamma^\mu \Lambda^\nu_\mu T^{-1} = \gamma^\nu$, we get (20.17).

20.3 *Proof of Proposition 20.2.* Use Problem 20.2 in order to prove Prop. 20.2.
Solution: Let us prove the claim for the transformation (20.8). Note that

$$ct' = ct\cosh\chi - x\sinh\chi, \quad x' = x\cosh\chi - ct\sinh\chi, \quad y' = y, \quad z' = z.$$

Choose $T := \mathrm{e}^{-\frac{\chi}{2}\gamma^0\gamma^1} = \cosh\frac{\chi}{2}\cdot\mathbf{1} - \sinh\frac{\chi}{2}\cdot\gamma^0\gamma^1$. We have to show that:

- $T^{-1}\gamma^0 T = \gamma^0\cosh\chi - \gamma^1\sinh\chi$,
- $T^{-1}\gamma^1 T = \gamma^1\cosh\chi - \gamma^0\sinh\chi$,
- $T^{-1}\gamma^2 T = \gamma^2$ and $T^{-1}\gamma^3 T = \gamma^3$.

In fact, these equations follow from the Clifford relations. For example, since $\gamma^0\gamma^0\gamma^1 = \gamma^1$, we get

$$\gamma^0 T = \gamma^0\cosh\tfrac{\chi}{2} - \gamma^1\sinh\tfrac{\chi}{2},$$

and

$$\begin{aligned} T^{-1}\gamma^0 T &= (\mathbf{1}\cosh\tfrac{\chi}{2} + \gamma^0\gamma^1\sinh\tfrac{\chi}{2})(\gamma^0\cosh\tfrac{\chi}{2} - \gamma^1\sinh\tfrac{\chi}{2}) \\ &= \gamma^0(\cosh^2\tfrac{\chi}{2} + \sinh^2\tfrac{\chi}{2}) - 2\gamma^1\sinh\tfrac{\chi}{2}\cosh\tfrac{\chi}{2} = \gamma^0\cosh\chi - \gamma^1\sinh\chi. \end{aligned}$$

Similarly, proceed for the claim concerning the transformation (20.12).

20.4 *The real matrix algebra* $\mathrm{Mat}(2,\mathbb{C})_\mathbb{R}$. Let $\sigma^0,\sigma^1,\sigma^2,\sigma^3$ be the four Pauli matrices. Let $\mathrm{Mat}(2,\mathbb{C})_\mathbb{R}$ denote the real algebra of all the complex (2×2)-matrices. Show that the eight matrices

$$\sigma^0,\sigma^1,\sigma^2,\sigma^3,\mathrm{i}\sigma^0,\mathrm{i}\sigma^1,\mathrm{i}\sigma^2,\mathrm{i}\sigma^3$$

form a basis of $\mathrm{Mat}(2,\mathbb{C})_\mathbb{R}$. Furthermore, show that the four Pauli matrices

$$\sigma^0,\sigma^1,\sigma^2,\sigma^3$$

form a basis of the real linear space X which consists of all the complex self-adjoint (2×2)-matrices.

20.5 *The real Clifford algebra* $\mathrm{Mat}_{\mathbb{R}}(4,\mathbb{C})$. Let $\gamma^0,\gamma^1,\gamma^2,\gamma^3$ be the Dirac–Pauli matrices. Prove that the sixteen matrices

$$\gamma^0,\ \gamma^1,\ \gamma^2,\ \gamma^3,\quad \gamma^0\gamma^0,\ \gamma^0\gamma^1,\ \gamma^0\gamma^2, \gamma^0\gamma^3,\quad \gamma^1\gamma^2,\ \gamma^1\gamma^3,\ \gamma^2\gamma^3,$$
$$\gamma^0\gamma^1\gamma^2,\ \gamma^0\gamma^1\gamma^3,\ \gamma^0\gamma^2\gamma^3,\quad \gamma^1\gamma^2\gamma^3,\qquad \gamma^0\gamma^1\gamma^2\gamma^3$$

are linearly independent (with respect to real linear combinations). Consequently, the real matrix algebra $\mathrm{Mat}(4,\mathbb{C})_{\mathbb{R}}$ of all the complex (4×4)-matrices coincides with the real Clifford algebra generated by the four Dirac–Pauli matrices.

20.6 *Reduction of products.* Use the Clifford relations in order to reduce the following products

$$\gamma^3\gamma^0\gamma^3\gamma^0\gamma^1,\qquad \gamma^3\gamma^2\gamma^1\gamma^0$$

to the basis products from Problem 20.5, up to sign.
Solution: We get $\gamma^3\gamma^0\gamma^3\gamma^0\gamma^1 = -\gamma^0\gamma^3\gamma^3\gamma^1 = \gamma^0\gamma^1$. Similarly,

$$\gamma^3\gamma^2\gamma^1\gamma^0 = -\gamma^2\gamma^3\gamma^1\gamma^0 = -\gamma^1\gamma^2\gamma^3\gamma^0 = \gamma^0\gamma^1\gamma^2\gamma^3.$$

21. The Language of Exact Sequences

> Exact sequences play a crucial role in modern algebra, geometry, number theory, and topology (homological algebra).
>
> Folklore

We will show that exact sequences are a basic tool for the following topics:

- systems of linear equations (linear operator equations),
- homology and cohomology in topology,
- the general potential equation in gauge theory; here, Betti numbers are counting the linearly independent constraints for the existence of physical fields on manifolds: Betti numbers are also counting the gauge degrees of freedom of the potentials (modulo cohomology).

21.1 Applications to Linear Algebra

As a prototype, let us reformulate the theory of linear operators in the language of exact sequences. We choose $\mathbb{K} = \mathbb{R}$ (field of real numbers) or $\mathbb{K} = \mathbb{C}$ (field of complex numbers). Let X, Y, Z be linear spaces over $\mathbb{K}$, and let A, B, C linear operators. We write

$$X \simeq Y \tag{21.1}$$

iff there exists a linear isomorphism $A : X \to Y$. We also say that the linear space X is linearly equivalent to the linear space Y. If the dimensions of X and Y are finite, then we have $X \simeq Y$ iff $\dim(X) = \dim(Y)$.

For a linear operator $A : X \to Y$, we set

$$\ker(A) := \{x \in X : Ax = 0\} \quad \text{and} \quad \operatorname{im}(A) := \{Ax : \ x \in X\}.$$

This is called the kernel $\ker(A)$ and the image $\operatorname{im}(A)$ of the linear operator A, respectively. The sequence of linear operators

$$X \xrightarrow{A} Y \xrightarrow{B} Z \tag{21.2}$$

is called exact iff

$$\boxed{\operatorname{im}(A) = \ker(B).}$$

In other words, the linear equation

$$Ax = y, \qquad x \in X$$

has a solution x iff $By = 0$. To simplify notation,

E. Zeidler, *Quantum Field Theory III: Gauge Theory*,
DOI 10.1007/978-3-642-22421-8_22,

- the symbol $X \to 0$ denotes the trivial operator $A : X \to \{0\}$ with $Ax := 0$ for all $x \in X$;
- the symbol $0 \to X$ denotes the trivial operator $B : \{0\} \to X$ with $B(0) = 0$.

Moreover, we frequently write $\ker A$ (resp. im A) instead of $\ker(A)$ (resp. $\operatorname{im}(A)$).

Proposition 21.1 *Let $A : X \to Y$ be a linear operator. Then:*

(i) *The sequence $X \xrightarrow{A} Y \to 0$ is exact iff A is surjective.*

(ii) *The sequence $0 \to X \xrightarrow{A} Y$ is exact iff A is injective.*

(iii) *The sequence $0 \to X \xrightarrow{A} Y \to 0$ is exact iff A is bijective.*

Proof. Ad (i). The sequence $X \xrightarrow{A} Y \to 0$ is exact iff $\operatorname{im} A = Y$.

Ad (ii). The sequence $0 \to X \xrightarrow{A} Y$ is exact iff $\ker A = \{0\}$.

Ad (iii). The linear operator $A : X \to Y$ is bijective iff it is both surjective and injective. □

Linear quotient spaces and short exact sequences. Consider the surjective linear operator

$$B : Y \to Z.$$

Let $y, y' \in Y$. We write $y \sim y'$ iff $B(y - y') = 0$. This is an equivalence relation. The equivalence classes $[y]$ form the linear quotient space $Y/\ker(B)$ (see Sect. 4.1.4 of Vol. II). The map $[y] \mapsto By$ yields the linear isomorphism

$$\boxed{Y/\ker B \simeq Z.}$$

Introducing the trivial injective map $i : \ker B \to Y$ by setting $i(x) := x$ for all $x \in \ker B$, we get the exact sequence

$$0 \to \ker B \xrightarrow{i} Y \xrightarrow{B} Z \to 0.$$

Conversely, if the sequence

$$0 \to X \xrightarrow{A} Y \xrightarrow{B} Z \to 0 \tag{21.3}$$

is exact, then the map $B : Y \to Z$ is surjective, and the map $A : X \to Y$ is injective with $\operatorname{im}(A) = \ker B$. Thus, the map $A : X \to \ker B$ is a linear isomorphism. Identifying linear isomorphic linear spaces with each other, we get the linear isomorphism

$$\boxed{Y/X \simeq Z.}$$

By definition, a short exact sequence is an exact sequence of the form (21.3).

Direct sums and split exact sequences. Consider the direct sum

$$Z = X \oplus Y$$

where X and Y are linear subspaces of the linear space Z over $\mathbb{K}$. Recall that, for all $z \in Z$, we have the unique decomposition

$$z = x + y, \qquad x \in X,\ y \in Y.$$

Define

- $i(x) := x$ for all $x \in X$ (canonical injection operator on X),
- $j(y) := y$ for all $y \in Y$ (canonical injection operator on Y),

- $\pi(z) := y$ for all $z \in Z$ (projection operator onto Y).

Then the sequence

$$0 \longrightarrow X \xrightarrow{i} X \oplus Y \xrightarrow{\pi} Y \longrightarrow 0 \tag{21.4}$$

is exact. Furthermore, we have

$$0 \longrightarrow X \xrightarrow{i} X \oplus Y \underset{j}{\overset{\pi}{\rightleftarrows}} Y \longrightarrow 0 \tag{21.5}$$

where $j : Y \to X \oplus Y$ is the trivial injective map $j(y) := 0 + y = y$. Motivated by this situation, we say that the sequence

$$0 \longrightarrow X \xrightarrow{i} Z \xrightarrow{\pi} Y \longrightarrow 0 \tag{21.6}$$

is a split exact sequence iff it is an exact sequence, and the map π has a right inverse, that is, there exists an injective map $j : Y \to Z$ such that $\pi \circ j = \mathrm{id}_Y$. We also write

$$0 \longrightarrow X \xrightarrow{i} Z \underset{j}{\overset{\pi}{\rightleftarrows}} Y \longrightarrow 0. \tag{21.7}$$

Proposition 21.2 *The sequence* (21.6) *is a split exact sequence iff* $Z \simeq X \oplus Y$.

Proof. Suppose that (21.6) is a split exact sequence. There exists a linear subspace $\ker(\pi)^\perp$ of Z such that

$$Z = \ker(\pi) \oplus \ker(\pi)^\perp. \tag{21.8}$$

The map π is surjective. Hence $\mathrm{im}(\pi) = Y$, and $Y \simeq Z/\ker(\pi)$. Since the map i is injective, we get $\mathrm{im}(i) \simeq X$. By exactness, $\mathrm{im}(i) \simeq \ker(\pi)$. Finally, (21.8) yields the claim. □

21.2 The Fredholm Alternative

Linear operator equations. Let $A : X \to Y$ be a linear operator. Then the following hold:

- Choose a linear subspace $(\ker A)^\perp$ of X and a linear subspace $(\mathrm{im}\ A)^\perp$ of Y such that
$$X = \ker A \oplus (\ker A)^\perp, \qquad Y = \mathrm{im}\ A \oplus (\mathrm{im}\ A)^\perp.$$
Set $A_*x := Ax$ for all $x \in (\ker A)^\perp$. Then the linear operator
$$\boxed{A_* : (\ker A)^\perp \to \mathrm{im}\ A}$$
is bijective. This means that all the solutions of the linear equation
$$Ax = y, \qquad x \in X$$
are given by
$$x = A_*^{-1}y + x_0$$
where x_0 is an arbitrary solution of the homogeneous equation $Ax_0 = 0$. This tells us that the general solution of a linear operator equation (e.g., a system of linear equations) is obtained as the sum of a special solution plus the general solution of the homogeneous equation.

- We have the linear equivalences

$$X/\ker A \simeq (\ker A)^{\perp}, \qquad Y/\operatorname{im} \mathrm{A} \simeq (\operatorname{im} \mathrm{A})^{\perp}.$$

We define

- coker $A := Y/\operatorname{im} A$ (cokernel of A), and
- coim $A := X/\ker A$ (coimage of A).

Then, A is injective (resp. surjective) iff $\ker A = \{0\}$ (resp. coker $A = \{0\}$).

In the language of exact sequences, this means that we have the following two exact sequences

$$0 \longrightarrow \ker A \xrightarrow{j} X \xrightarrow{A} \operatorname{im} A \longrightarrow 0 \tag{21.9}$$

and

$$0 \longrightarrow \ker A \xrightarrow{j} X \xrightarrow{A} Y \xrightarrow{\pi} \operatorname{coker} A \longrightarrow 0. \tag{21.10}$$

Here, we set $j(x) := x$ for all $x \in \ker A$ and $\pi(y) := [y]$ for all $y \in Y$. Moreover, $[y] \in Y/\operatorname{im} A$.

Duality and the Fredholm alternative. We are given the linear operator $A : X \to Y$. Let

$$A^d : Y^d \to X^d$$

be the corresponding dual linear operator given by $(A^d f)(x) = f(Ax)$ for all $x \in X$ and all $f \in Y^d$. Recall that Y^d denotes the dual space to Y. Suppose that the dimension of X and Y is finite. The Fredholm alternative tells us that the linear operator equation

$$Ax = y, \qquad x \in X$$

has a solution iff $f(y) = 0$ for all linear functionals $f : Y \to \mathbb{K}$ with $A^d f = 0$. We have the linear equivalences

$$\operatorname{coker}(A) \simeq \ker(A^d), \qquad \operatorname{coim}(A) \simeq \operatorname{im}(A^d).$$

This justifies the designation 'cokernel' and 'coimage'. The dimension of im(A) is called the rank of the operator A, that is, $\dim(\operatorname{im}(A)) = \operatorname{rank}(A)$. The rank theorem tells us that

$$\operatorname{rank}(A) = \operatorname{rank}(A^d).$$

Dual exact sequence. If the sequence

$$X \xrightarrow{A} Y \xrightarrow{B} Z$$

is exact, then the dual sequence

$$Z^d \xrightarrow{B^d} Y^d \xrightarrow{A^d} X^d \tag{21.11}$$

is also exact. For the proof, we refer to Problem 21.2.

21.3 The Deviation from Exact Sequences and Cohomology

Consider the sequence

$$X \xrightarrow{A} Y \xrightarrow{B} Z \tag{21.12}$$

with the additional property $BA = 0$. This condition tells us that $Ax = 0$ implies $BAx = 0$. This yields

$$\text{im } A \subseteq \ker B.$$

Therefore, it makes sense to define the linear quotient space

$$\boxed{H := \ker B / \text{ im } A.}$$

This is called the cohomology group of the sequence (21.12). To simplify notation, we write $H = 0$ iff $H = \{0\}$.

Proposition 21.3 *The sequence* (21.12) *is exact iff its cohomology group is trivial, that is,* $H = 0$.

Consequently, the cohomology group H measures the deviation of the sequence (21.12) from an exact sequence.

Cohomology is one of the most important notions in modern mathematics and physics.

Long exact sequences. Let X_k be a linear space over $\mathbb{K}$. The sequence of linear operators

$$\ldots \xrightarrow{d_{k-1}} X_k \xrightarrow{d_k} X_{k+1} \xrightarrow{d_{k+1}} X_{k+2} \xrightarrow{d_{k+2}} \ldots \tag{21.13}$$

is called exact iff

$$\ker\ d_k = \text{im } d_{k-1}$$

for all indices k. If $d_k d_{k-1} = 0$ for all indices k, then the linear quotient space

$$\boxed{H^k := \ker d_k / \text{ im } d_{k-1}}$$

is called the kth cohomology group of the sequence (21.13). Applications will be considered in Chap. 22 on electric circuits and in Chap. 23 on the relations between de Rham cohomology for differential forms and the electromagnetic field.

21.4 Perspectives

The definition of exact sequences can be immediately translated to additive groups, modules over a ring, rings, Lie algebras, and vector bundles (i.e., families $\{X_\gamma\}_{\gamma \in \Gamma}$ of linear spaces X_γ). A slight modification allows us also to define exact sequences for groups. To explain this, let

$$X \xrightarrow{A} Y \xrightarrow{B} Z \tag{21.14}$$

be a sequence of groups X, Y, Z and group morphisms A, B, C. Naturally enough, the kernel of the group morphism $A : X \to Y$ is defined by

$$\ker(A) := \{G \in X : \ A(G) = \mathbf{1}\}$$

where $\mathbf{1}$ is the unit element of the group Y. Then the sequence (21.14) is called exact iff $\text{im}(A) = \ker(B)$.

Proposition 21.4 *Let X and Y be groups, and let $A : X \to Y$ be a group morphism. Then: .*

(i) *The sequence $X \xrightarrow{A} Y \to \mathbf{1}$ is exact iff A is surjective.*

(ii) *The sequence $\mathbf{1} \to X \xrightarrow{A} Y$ is exact iff A is injective.*

(iii) *The sequence $\mathbf{1} \to X \xrightarrow{A} Y \to \mathbf{1}$ is exact iff A is bijective.*

We will show in Vol. IV that semidirect products of groups (or Lie algebras) can be introduced by using split exact sequences of the form (21.7) on page 1005. Note that, in the case of groups, we have to replace the zero element '0' by the unit element '**1**'. Moreover, we will show in Vol. IV that exact sequences can be used for computing topological invariants via homology groups, cohomology groups, homotopy groups, and K-theory. A special case will be considered in Sect. 23.6.1 on page 1053 (the Mayer–Vietoris sequence in de Rham cohomology).

Problems

21.1 *Special exact sequence.* Let X, Y, Z be real, finite-dimensional, linear spaces with $\dim(Y) > \dim(X) > 0$. Show that if the sequence

$$0 \xrightarrow{\alpha} X \xrightarrow{\beta} Y \xrightarrow{\gamma} X \xrightarrow{\delta} Z \to 0$$

is exact, then $\dim(Z) < \dim(X)$.
Solution: Since the linear operator $\delta : X \to Z$ is surjective, $\dim(X) \geq \dim(Z)$. Suppose that $\dim(Z) = \dim(X)$. Then the operator δ is bijective. Hence $\ker(\delta) = \{0\}$. This implies

$$\text{im}(\gamma) = \ker(\delta) = \{0\}.$$

Thus, $\gamma = 0$. Hence $\text{im}(\beta) = \ker(\gamma) = Y$. But, $\dim \text{im}(\beta) \leq \dim(X) < \dim(Y)$. This is a contradiction.

21.2 *Exactness of the duality functor.* Prove the exactness of the sequence (21.11). Hint: See E. Zeidler, Applied Functional Analysis, Vol. 2: Main Principles and Their Applications, Sect. 3.11, AMS 109, Springer, New York, 1995 (reprinted in China, 2009).

22. Electrical Circuits as a Paradigm in Homology and Cohomology

The study of electrical networks rests upon preliminary theory of graphs. In the literature this theory has always been dealt with special *ad hoc* methods. My purpose here is to show that actually this theory is nothing else than the first chapter of classical algebraic topology and may be very advantageously treated as such by the well known methods of that science.[1]

Solomon Lefschetz (1884–1972)

We want to show that:

(i) Electric currents J are 1-cycles: $\partial \mathsf{J} = 0$.
(ii) Voltages V are 1-coboundaries: $V = -dU$ (U is the electrostatic potential).
(iii) There exists a duality relation between electric currents and voltages: $\langle V|\mathsf{J}\rangle = 0$ (orthogonality).
(iv) If the electrical circuit is connected, then we get $\beta_0 = 1$ for the zeroth Betti number. In the general case, β_0 is equal to the number of connectivity components of the electrical circuit.
(v) If the electrical circuit has s_0 nodes and s_1 connections, then the Euler characteristic is given by $\chi = s_0 - s_1$.
(vi) This yields the first Betti number $\beta_1 = \beta_0 - \chi$.
(vii) The space of electric currents is a linear space of dimension β_1.

Modern computers are based on huge electrical circuits. In this section, we would like to study the theory of electrical circuits as a paradigm for important generalizations in modern physics and mathematics.

22.1 Basic Equations

The electrical circuit depicted in Fig. 22.1(a) is governed by the following two ordinary differential equations:

$$\boxed{R\mathsf{J}(t) - \frac{Q(t)}{C} + L\dot{\mathsf{J}}(t) = F(t), \quad \dot{Q}(t) = -\mathsf{J}(t), \quad t > 0} \qquad (22.1)$$

along with the initial condition

$$\mathsf{J}(+0) = \mathsf{J}_0, \qquad Q(+0) = Q_0.$$

[1] S. Lefschetz, Applications of Algebraic Topology: Graphs and Networks, the Picard–Lefschetz Theory, and Feynman Algorithms, Springer, New York, 1975. Lefschetz (born in Moscow) was professor of mathematics at Princeton University from 1925 until 1953.

E. Zeidler, *Quantum Field Theory III: Gauge Theory*,

DOI 10.1007/978-3-642-22421-8_23,

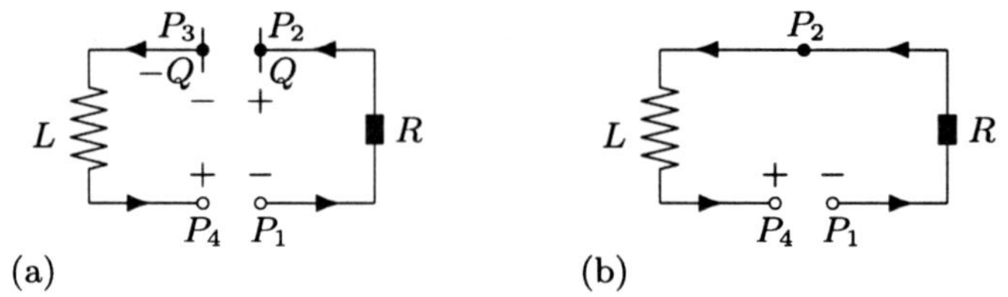

Fig. 22.1. Electrical circuits

This means that $\mathsf{J}(t) \to \mathsf{J}_0$ and $Q(t) \to Q_0$ as $t \to +0$. This electric circuit consists of a resistor P_1P_2, a capacitor P_2P_3, a coil P_3P_4, and an electric current source P_4P_1 (e.g., an electrical socket or a battery). The symbols possess the following meaning:

- t (time),
- J (absolute value of the electric current at the resistor of physical dimension A (ampere)),
- Q (positive electric charge of physical dimension C (coulomb) at the positive capacitor plate),
- V (voltage at the capacitor of physical dimension V (volt)), $Q = CV$,
- F (electromotive force of the electric current source of physical dimension V (volt)),
- the positive constants R, C, and L are the resistance of the resistor (e.g., a light bulb), the capacitance of the capacitor, and the inductance of the coil, respectively. The physical dimensions are V/A, C/V, Vs/A, respectively.

Multiplying the basic equation (22.1) by J, we get

$$F(t)\mathsf{J}(t) = R\mathsf{J}(t)^2 + \frac{d}{dt}\left(\frac{Q(t)^2}{2C} + \frac{L\mathsf{J}(t)^2}{2}\right) \tag{22.2}$$

for all times t. This equation describes the conservation of energy. To show this, let us introduce the following energy functions:

- $E_F(t_0, t) := \int_{t_0}^t F\mathsf{J}\, dt$ (electric energy transferred from the electric current source to the circuit during the time interval $[t_0, t]$);
- $E_R(t_0, t) := \int_{t_0}^t R\mathsf{J}^2 dt$ (heat energy produced at the resistor during the time interval $[t_0, t]$);[2]
- $E_C(t_0, t) := (Q(t)^2 - Q(t_0)^2)/2C$ (electric energy transferred to the electric field of the capacitor during the time interval $[t_0, t]$);
- $E_L(t_0, t) := \frac{1}{2}(L\mathsf{J}(t)^2 - L\mathsf{J}(t_0)^2)$ (electric energy transferred to the electric field – induced by the electric current flowing through the coil – during the time interval $[t_0, t]$).

It follows from equation (22.2) that

$$\frac{\partial}{\partial t}E_F(t_0, t) = \frac{\partial}{\partial t}\big(E_R(t_0, t) + E_C(t_0, t) + E_L(t_0, t)\big), \quad t \geq t_0.$$

This implies the conservation of energy:

[2] Observe that it does not make any sense of speaking about heat energy at a fixed time. Therefore, we have to use a heat energy function depending on time intervals.

$$\boxed{E_F(t_0,t) = E_R(t_0,t) + E_C(t_0,t) + E_L(t_0,t), \qquad t \geq t_0.}$$

Summarizing, the electric current source transfers electric energy to the electric circuit. On the other hand, the electric circuit looses electric energy by spending energy to the environment, namely,

- heat energy at the resistor,
- electric field energy at the capacitor, and
- electric field energy at the coil.

If the electric current source is a battery, as in a car, then chemical energy is transformed into electric energy. The universal validity of energy conservation was postulated and experimentally established by Robert Mayer (1814–1878) in 1842.

Energy conservation is the most fundamental law in nature.

For many-particle systems, the direction of time-depending processes is governed by entropy. Explicitly, the entropy of a closed system never decreases in time.

Physical motivation. Let us motivate the basic differential equations (22.1). Electromagnetic phenomena are described by an electric field $\mathbf{E}$ and a magnetic field $\mathbf{B}$ depending on position and time. The energy of the electromagnetic field in a region Ω at time t reads as

$$\int_\Omega \left(\frac{\varepsilon \mathbf{E}(\mathbf{x},t)^2}{2} + \frac{\mathbf{B}(\mathbf{x},t)^2}{2\mu} \right) d^3x$$

where ε and μ are the constants of the electric and magnetic field in the homogeneous material under consideration, respectively. This energy expression summarizes the atomic structure of the material. The electromagnetic field is governed by the Maxwell equations. In the present case, we do not need the full power of the Maxwell equations.

It is typical for metals that there exist freely moving electrons because of the specific total potential of the metal molecules. The flow of an electric current in a metallic wire corresponds to the flow of free electrons of negative electric charge.[3] The behavior of an electric circuit can be compared with the flow of petroleum in a pipeline.

Pressure and pressure differences in pipelines correspond to the electric potential U and the voltages V in electric circuits, respectively.

The petroleum flow and the electron flow are driven by the pressure difference and the voltage, respectively. The electron flow can be described by the electric current density vector

$$\mathbf{J} = \varrho \mathbf{v}$$

where $\mathbf{v}$ and ϱ denote the velocity vector and the electric charge density of the free electrons, respectively. Let S be a fixed cross section of the wire with unit normal vector $\mathbf{n}$ in direction of the electron flow. Then the so-called electric current is defined to be

$$\mathsf{J} := \int_S \mathbf{J}\mathbf{n}\, dS.$$

[3] In 1897, the electron was experimentally discovered by Sir Joseph John Thomson (1856–1940); he performed cathode ray experiments at the Cavendish Laboratory (Cambridge, England). In 1906, Thomson was awarded the Nobel prize in physics. The history of this discovery can be found in S. Weinberg, The Discovery of Subatomic Particles, Scientific American Library, New York, 1983.

If $\mathcal{Q}(t_0, t)$ denotes the electric charge flowing through the cross section $\mathcal{S}$ during the time interval $[t_0, t]$, then

$$\boxed{Q(t_0, t) = \int_{t_0}^{t} \mathsf{J}(t)dt.} \tag{22.3}$$

Physical experiments show that the strength of the electron current in a metallic wire depends on the strength of the electric field vector $\mathbf{E}$. Ohm's law tells us that

$$\boxed{\mathbf{J} = \sigma\mathbf{E}.}$$

The positive material constant σ is called the electric conductivity of the wire. As a special case of Maxwell's theory for electromagnetism, the electric field of an electric circuit possesses a potential U given by

$$\boxed{\mathbf{E} = -\,\mathbf{grad}\, U.}$$

The line integral

$$V(AB) := \int_A^B \mathbf{E}d\mathbf{x}$$

along the electric circuit is called the voltage between the points A and B. Since $\mathbf{E} = -\,\mathbf{grad}\, U$,

$$\boxed{V(AB) = U(A) - U(B).}$$

In order to understand the physical meaning of the voltage $V(AB)$, consider the motion of a particle of electric charge Q. If the particle moves from the point A to the point B, then the line integral

$$QV(AB) := \int_A^B Q\mathbf{E}d\mathbf{x}$$

is equal to the work done by the electric field $\mathbf{E}$. Therefore, the voltage V has the physical dimension of "energy divided by electric charge." By (22.3), the electric current strength has the physical dimension of "electric charge divided by time." In the SI system, one uses the units of joule J, coulomb C, ampere A, volt V, and watt W for energy, electric charge, electric current, voltage, and power, respectively. There hold the following relations:

$$\boxed{\mathrm{J} = \mathrm{kg} \cdot \mathrm{m}^2/\mathrm{s}^2 = \mathrm{Ws} = \mathrm{VAs}, \quad \mathrm{C} = \mathrm{As}, \quad \mathrm{W} = \mathrm{AV}.}$$

Let us now consider the specific situation depicted in Fig. 22.1(a).

(i) *Electric current*: The strength of the electron current flowing from the point P_1 to the point P_2 is denoted by $\mathsf{J}(\mathsf{P}_1\mathsf{P}_2)$. Since electrons carry a negative charge, $\mathsf{J}(P_1P_2) < 0$. The integral

$$Q(t, t_0) := \int_{t_0}^{t} \mathsf{J}(P_1P_2)(t)dt \tag{22.4}$$

is equal to the amount of electric charge which flows from the point P_1 to the point P_2 during the time-interval $[t_0, t]$. Conservation of electric charge tells us *Kirchhoff's electric current rule*:

$$\boxed{\mathsf{J}(P_1P_2) = \mathsf{J}(P_3P_4).} \tag{22.5}$$

This means that the same electric current flows through both the resistor and the coil.

(ii) *Voltage*: Mathematically, the electric potential U is determined by the electric field only up to an arbitrary constant which can be fixed by a gauge condition. In a physical experiment, we cannot measure the electric potential $U = U(P)$ at a single point, but only the potential difference

$$\boxed{V(P_1P_2) := U(P_1) - U(P_2)}$$

which is called the voltage between the points P_1 and P_2. From the trivial identity

$$U(P_1) - U(P_2) + U(P_2) - U(P_3) + U(P_3) - U(P_4) + U(P_4) - U(P_1) = 0$$

we get the following *Kirchhoff voltage rule*:

$$\boxed{V(P_1P_2) + V(P_2P_3) + V(P_3P_4) + V(P_4P_1) = 0.} \tag{22.6}$$

The voltage $F := V(P_4P_1)$ is called the electromotive force of the electric current source.

(iii) *Resistor*: Ohm's law tells us that

$$\boxed{V(P_1P_2) = R(P_1P_2)\mathsf{J}(P_1, P_2).} \tag{22.7}$$

The material constant $R(P_1P_2) > 0$ is called the resistance between the nodes P_1 and P_2. Resistance has the physical dimension V/A. To simplify notation, set $R := R(P_1P_2)$. The magnitude of the resistance R depends on the friction of the freely moving electrons in the metal. The friction transforms the kinetic energy of the free electrons into heat energy. The integral

$$E_R(t_0, t) = \int_{t_0}^{t} R\mathsf{J}(P_1P_2)(t)^2 dt$$

is equal to the amount of heat energy produced during the time interval $[t_0, t]$. For the rate of heat energy production, we get

$$\frac{\partial}{\partial t} E_R(t_0, t) = R\mathsf{J}(P_1P_2)(t)^2.$$

(iv) *Capacitor:* A plate capacitor P_2P_3 consists of two parallel plates. Suppose that the plates located at the points P_2 and P_3 carry the positive and negative electric charge $Q(t)$ and $-Q(t)$ at time t, respectively. The capacitor law tells us that

$$\boxed{Q = CV(P_2P_3).}$$

The positive constant C is called capacitance. The positive and negative charges of the capacitor generate an electric field between the two plates given by

$$E = \frac{Q}{\varepsilon S} = \frac{V(P_2P_3)}{d}.$$

Here, S is the surface area of a single plate, d is the distance between the two plates, and ε is the electric field constant of the substance located between the two plates. The electric field is directed from the positively charged plate to the negatively charged plate; it has the electric energy $E_C(t) := Q(t)^2/2C$ at time t. Negatively charged electrons arrive at the capacitor plate P_2. Therefore, the positive electric charge $Q(t)$ on the plate P_2 is decreasing, and hence the negative electric charge $-Q(t)$ on the plate P_3 is increasing; this means that electrons flow from the point P_3 through the coil to the point P_4.

(vi) *Coil*: Faraday's law of self-induction tells us that

$$\boxed{V(P_3P_4)(t) = L\dot{\mathsf{J}}(P_3P_4)(t).} \tag{22.8}$$

The electric current $\mathsf{J}(P_3P_4)$ flowing through the coil generates a magnetic field $\mathbf{B}$. If the electric current changes in time, then so does the magnetic field. This induces an electric field in the coil; $V(P_3P_4)$ is the voltage of the induced electric field. This way the coil produces an electric field which has the energy

$$E_L(t) := \tfrac{1}{2}L\dot{\mathsf{J}}(t)^2.$$

(vi) *Sign convention:* Recall that the electron current $\mathsf{J}(P_1P_2)$ is negative. Physicists and engineers prefer working with positive currents. Therefore, we introduce the positive current

$$\mathsf{J} := -\mathsf{J}(P_1P_2).$$

From Kirchhoff's voltage relation (22.6) along with Ohm's law and Faraday's self-induction law we get

$$-R\mathsf{J} + V(P_2P_3) - L\frac{d\mathsf{J}}{dt} + F = 0.$$

Finally, the capacitor law yields $CV(P_2P_3) = Q$. This implies the desired basic equation (22.1).

Initial-value problem. Consider first the special situation depicted in Fig. 22.1(b). Setting $Q = 0$, we get the differential equation

$$\boxed{R\mathsf{J}(t) + L\dot{\mathsf{J}}(t) = F(t), \qquad t > 0} \tag{22.9}$$

along with the initial-condition $\mathsf{J}(+0) = \mathsf{J}_0$. We are given the continuous electromotive force $F : [0,\infty[\to \mathbb{R}$ of the electric current source and the electric current J_0 at the initial time $t = 0$. The given positive constants R and L represent the resistance of the resistor and the inductance of the coil, respectively. We are looking for an electric current $\mathsf{J} : [0,\infty[\to \mathbb{R}$ such that the differential equation (22.9) is satisfied along with the initial condition $\mathsf{J}(t) \to \mathsf{J}_0$ as $t \to +0$.[4] Let us introduce the retarded propagator (also called the Green function)

$$P^+(t,\tau) := \frac{1}{L}\theta(t-\tau)\mathrm{e}^{-R(t-\tau)/L}, \qquad t,\tau \in \mathbb{R}.$$

Recall that the Heaviside function is defined to be $\theta(t) := 1$ if $t \geq 0$, and $\theta(t) = 0$ if $t < 0$.

Proposition 22.1 *The initial-value problem* (22.9) *has the unique solution*

$$\mathsf{J}(t) = \int_0^t P^+(t,\tau)F(\tau)d\tau + \mathsf{J}(+0)P^+(t,+0), \qquad t \geq 0.$$

[4] More precisely, we assume that the function J is continuous on the closed interval $[0,\infty[$ and differentiable on the open interval $]0,\infty[$.

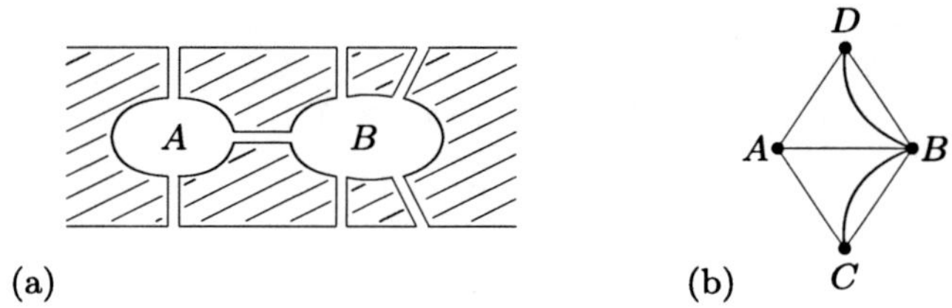

Fig. 22.2. Euler's bridge problem

Consider now the situation depicted in Fig. 22.1(a) along with the differential equation

$$RJ(t) - \frac{Q(t)}{C} + L\dot{J}(t) = F(t), \;\; \dot{Q}(t) = -J(t), \quad t > 0.$$

We are given the electromotive force F, the electric current $J(+0)$, and the electric charge $Q(+0)$ of the capacitor at time $t = 0$. We restrict ourselves to the important special case where $F(t) = \text{const}$ and $J(+0) = Q(+0) = 0$. Moreover, set

$$\alpha := \frac{R}{2L}, \qquad \mu := \frac{1}{LC} - \frac{R^2}{4L^2}$$

and assume that $\mu > 0$. The solution reads then as

$$J(t) = \frac{F}{L\omega} \cdot \mathrm{e}^{-\alpha t} \sin \omega t, \qquad t \geq 0.$$

This is a damped oscillation of the electric current called Thomson oscillation with angular frequency $\omega := \sqrt{\mu}$. For the electric charge of the capacitor,

$$Q(t) = -\int_0^t J(\tau) d\tau, \qquad t \geq 0.$$

The resistor is responsible for the damping of the oscillations. If the resistor drops out, $R = 0$, then we get the harmonic oscillations

$$J(t) = \frac{F}{L\omega} \cdot \sin \omega t, \qquad Q(t) = \frac{F}{L\omega^2} \cdot (\cos \omega t - 1), \qquad t \geq 0.$$

22.2 Euler's Bridge Problem and the Kirchhoff Rules

The bridge problem. Figure 22.2(a) shows a situation located in Königsberg in the eighteenth century (nowadays Kaliningrad). There is a river together with two islands A and B and seven bridges. The problem is to continuously walk over all the bridges without recrossing one. In order to solve the problem, Euler depicted the situation schematically as indicated in Fig. 22.2(b). By definition, the index of a node is equal to the number of connections that hit the node. Euler proved that the desired walk over the bridges is possible iff there exist not more than two nodes of odd index. In Fig. 22.2(b), there are four nodes of odd index. Thus, the Königsberg bridge problem is not solvable.

The Kirchhoff rules. Generalizing Kirchhoff's electric current rule (22.5) and Kirchoff's voltage rule (22.6) in an obvious way, we obtain the following two Kirchhoff rules:

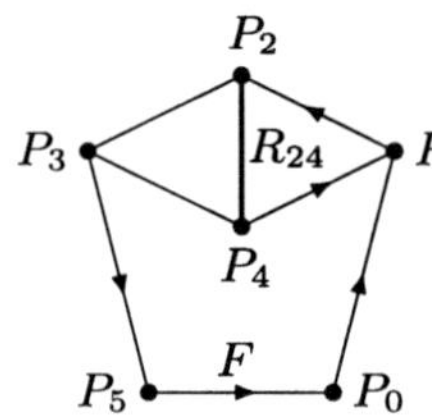

Fig. 22.3. The Wheatstone bridge

(i) Kirchhoff's electric current rule: For each node, the sum of the electric currents, which flow directly to the node, is equal to zero.
(ii) Kirchhoff's voltage rule: For each loop, the sum of voltages is equal to zero.

These rules were formulated by Kirchhoff in 1847.

The Wheatstone bridge problem and constraints. As an example, consider the circuit depicted in Fig. 22.3. To simplify notation, for the electric current, the voltage, and the resistance, we set

$$\mathsf{J}_{ij} := \mathsf{J}(P_iP_j), \quad V_{ij} := U(P_i) - U(P_j), \quad R_{ij} := R(P_iP_j).$$

By Ohm's law, $\mathsf{J}_{ij} = R_{ij}V_{ij}$. Furthermore, the voltage V_{50} of the battery is called the electromotive force F. Note that $\mathsf{J}_{ji} = -\mathsf{J}_{ij}$ and $V_{ji} = -V_{ij}$.

Theorem 22.2 *We are given the electromotive force F (battery voltage) between the nodes P_0 and P_5 and the resistances $R_{ij} > 0$ along the remaining direct connections. Then:*

(i) *There exist a unique electric current (J_{ij}) and a unique voltage (V_{ij}) such that $V_{50} = F$.*

(ii) *For given value U_0 of the electrostatic potential at the node P_0, there exists a unique electrostatic potential (U_i) such that $V_{ij} = U_i - U_j$ for all i, j with $i \neq j$.*

As a corollary, for $F \neq 0$, we obtain that

$$\boxed{\mathsf{J}_{24} = 0 \qquad \text{iff} \qquad R_{14} = R_{34}R_{12}/R_{23}.} \tag{22.10}$$

This relation is used by physicists in order to measure the unknown resistance R_{14}. To this end, the experimentalist changes the two resistances R_{12} and R_{23} as long as $\mathsf{J}_{24} = 0$, that is, no electric current flows over the 'bridge' P_2P_4.

Proof. Ad (i). Using the Kirchhoff rules, we will get a linear system for J_{ij} and V_{ij}. By Ohm's law, V_{ij} can be eliminated. The point is that there exist linear relations between the equations which reflect elementary geometric properties of the graph in Fig. 22.3. In fact, we will make critically use of the two relations

$$c_4 = c_1 + c_2, \qquad c_5 = c_3 - c_4$$

for the cycles (loops)

$$c_1 := P_1P_2P_4P_1 \quad c_2 := P_2P_3P_4P_2, \quad c_3 := P_0P_1P_2P_3P_5P_0$$

as well as $c_4 := P_1P_2P_3P_4P_1$ and $c_5 := P_0P_1P_4P_3P_5P_0$. Thus, the cycles c_1, c_2, c_3 form a basis for all the cycles. By the Kirchhoff current rule,

$$\begin{aligned} \mathsf{J}_{01} + \mathsf{J}_{21} + \mathsf{J}_{41} &= 0 \quad \text{at} \quad P_1, \\ \mathsf{J}_{12} + \mathsf{J}_{32} + \mathsf{J}_{42} &= 0 \quad \text{at} \quad P_2, \\ \mathsf{J}_{23} + \mathsf{J}_{43} + \mathsf{J}_{53} &= 0 \quad \text{at} \quad P_3, \end{aligned} \tag{22.11}$$

as well as

$$\begin{aligned} \mathsf{J}_{14} + \mathsf{J}_{24} + \mathsf{J}_{34} &= 0 \quad \text{at} \quad P_4, \\ \mathsf{J}_{05} + \mathsf{J}_{35} &= 0 \quad \text{at} \quad P_5, \\ \mathsf{J}_{10} + \mathsf{J}_{50} &= 0 \quad \text{at} \quad P_0. \end{aligned} \tag{22.12}$$

In addition, $\mathsf{J}_{ij} = -\mathsf{J}_{ji}$. By the Kirchhoff voltage rule,

$$\begin{aligned} V_{12} + V_{24} + V_{41} &= 0 \quad \text{along} \quad c_1, \\ V_{23} + V_{34} + V_{42} &= 0 \quad \text{along} \quad c_2, \\ V_{01} + V_{12} + V_{23} + V_{35} + V_{50} &= 0 \quad \text{along} \quad c_3, \end{aligned} \tag{22.13}$$

as well as

$$\begin{aligned} V_{12} + V_{23} + V_{34} + V_{41} &= 0 \quad \text{along} \quad c_4, \\ V_{01} + V_{14} + V_{43} + V_{35} + V_{50} &= 0 \quad \text{along} \quad c_5. \end{aligned} \tag{22.14}$$

In addition, $V_{ij} = -V_{ji}$. By Ohm's law, $V_{ij} = R_{ij}\mathsf{J}_{ij}$. We have to show that for given electromotive force (battery voltage) $F = V_{50}$, this system has a unique solution J_{ij}. This is an overdetermined system of 11 equations for 7 unknown electric currents $J_{01}, J_{12}, \ldots$ Let us reduce this to a system of 6 equations for 6 unknowns. In fact, the summation of the three equations from (22.11) yields the first equation of (22.12). Moreover, summing (22.13) according to $c_4 = c_1 + c_2$ and $c_5 = c_3 - c_4$, it follows that (22.14) is a simple consequence of (22.13). This way, we get the final linear system of the form

$$(\mathsf{J}_{01}, \mathsf{J}_{21}, \mathsf{J}_{41}, \mathsf{J}_{32}, \mathsf{J}_{42}, \mathsf{J}_{43})A = (0, 0, 0, 0, 0, F)$$

where A is a real (6×6)-matrix. An explicit computation shows that the determinant of the matrix A does not vanish. Thus, the system has a unique solution. In particular, we get that J_{24} is proportional to $F(R_{14}R_{23} - R_{34}R_{12})$.

Ad (ii). Starting at the node P_0, we construct

$$U(P_1) := U_0 + V_{10}.$$

For a fixed node, say P_2, we choose a path, say $P_0P_1P_2$, which connects P_2 with P_1 (Fig. 22.3). Then, we define

$$U(P_2) := U_0 + V_{10} + V_{21}.$$

If we choose another path, say $P_0P_1P_4P_2$, then we define

$$U(P_2) := U_0 + V_{10} + V_{41} + V_{24}.$$

However, because of Kirchhoff's voltage rule, $V_{12} + V_{24} + V_{41} = 0$. Noting that $V_{12} = -V_{21}$, this yields

$$V_{10} + V_{21} = V_{10} + V_{41} + V_{24}.$$

Thus, the two definitions of $U(P_2)$ coincide. This argument can be used in order to show that $U(P_j)$ can be uniquely constructed in a path-independent way for all nodes P_j. □

This is a brute-force proof which cannot be extended to more complicated electrical circuits. It is our goal to give an elegant proof for general electrical circuits based on algebraic topology.

22.3 Weyl's Theorem on Electrical Circuits

Consider a connected electrical circuit which has one battery and all the other connections are equipped with resistors. In 1923, Weyl proved the following result.[5]

Theorem 22.3 *The Wheatstone Theorem 22.2 remains valid for all connected electrical circuits which possess at least one loop.*

Basic ideas of the proof. The proof of the theorem will be given in the next section. The basic ideas are the following:

(a) The Kirchhoff rules yield an overdetermined linear system of equations.
(b) This system can be reduced to an inhomogeneous linear system of β_1 equations for β_1 unknowns. Explicitly,

$$\boxed{\beta_1 = 1 - s_0 + s_1}$$

where s_0 and s_1 denotes the number of nodes and connections, respectively. The number $\chi := s_0 - s_1$ is called the Euler characteristic of the electric circuit.
(c) The homogeneous linear system with vanishing battery voltage, $F = 0$, has only the trivial solution.
(d) Suppose that the condition (c) holds true for a finite system of linear equations where the number of equations and the number of unknowns coincide (as in (b)). Then the inhomogeneous system has always a unique solution. This is a standard result of linear algebra.

Let us use this example in order to illustrate the different methods of thinking used by physicists and mathematicians.

- *The thinking of physicists*: From the physical point of view, claim (c) is obvious. Indeed, if the battery voltage vanishes, then nothing happens. This is a beautiful heuristic argument motivated by observing nature. But, this is *not* a rigorous mathematical argument.
- *The thinking of mathematicians*: A rigorous mathematical argument can be based on the duality relation

$$\boxed{\sum \mathsf{J}_{ij} V_{ij} = 0} \qquad (22.15)$$

where we sum over all connections. Suppose that the battery voltage vanishes. Then it follows from Ohm's law $V_{ij} = R_{ij} \mathsf{J}_{ij}$ that

$$\sum{}' R_{ij} V_{ij}^2 = 0,$$

where we sum over all connections except for the battery. Hence $V_{ij} = 0$ for all voltages. This implies $\mathsf{J}_{ij} = 0$ for all electric currents. Therefore, the homogeneous linear system has only the trivial solution.

[5] H. Weyl, Repartición de corriente en una red conductora (Distribution of an electric current in a network), Revista Matematica Hispano-Americana 1923, 153–164 (in Spanish). In: H. Weyl, Collected Works, Vol. II, pp. 368–389. English translation: George Washington University Logistics Research Project, 1951.

It remains to prove the claim (b) and the duality relation (22.15). This will be done in the next section by using the elegant method of homology and cohomology from algebraic topology.

The fundamental role played by the basic 1-cycles of an electrical circuit. Loops are also called 1-cycles. For the Wheatstone bridge in Fig. 22.3, we get the Euler characteristic

$$\chi = 6 - 8,$$

and hence we get the first Betti number, $\beta_1 = 3$. This is equal to the number of basic 1-cycles. Explicitly, we can choose c_1, c_2, c_3 as basic 1-cycles. In the preceding 'brute force' proof for the Wheatstone bridge, we used 6 equations for 6 unknowns. Our general proof will show that, in fact, we only need 3 linear equations for 3 unknowns. Explicitly,

- the three equations correspond to the Kirchhoff voltage rules for the 1-cycles (loops) c_1, c_2, c_3, and
- the three unknowns are the electric current strengths in the loops c_1, c_2, c_3.

Generally, an electrical circuit is governed by its basic 1-cycles. It was Poincaré's ingenious idea to base algebraic topology (homology) in arbitrary dimensions on the notion of cycles.

Fredholm operators of index zero. Statement (d) above allows a far-reaching generalization to infinite-dimensional problems in functional analysis. Consider a linear Fredholm operator $A : X \to X$ of index zero on the Banach space X. For given $y \in X$, the equation

$$Ax = y, \qquad x \in X$$

has a unique solution if $Ax = 0$ implies $x = 0$. In other words, uniqueness implies existence. This is one of the main theorems of linear functional analysis. The proof can be found in Zeidler (1995b), Chap. 5 (quoted on page 1089) along with applications to the Navier–Stokes equations for inviscid fluids.

22.4 Homology and Cohomology in Electrical Circuits

Mathematics is the art of avoiding computations.
Folklore

We want to show that

- the electric current J is a 1-cycle,
- the electrostatic potential U is a 0-cocycle, and
- the voltage V is a 1-cocycle.

Moreover, the basic equations of an electric circuit read as

$$\boxed{\partial \mathsf{J} = 0, \qquad V = -dU} \tag{22.16}$$

along with Ohm's law $V = R\mathsf{J}$. These two equations imply the Kirchhoff rules for electric current and voltage. The equation $V = -dU$ represents the simplest case of Cartan's potential equation. We will also show that there holds the discrete Poincaré cohomology rule

$$\partial(\partial \mathsf{J}) = 0, \qquad d(dU) = 0.$$

Therefore, it follows from (22.16) that

$$\boxed{\partial \mathsf{J} = 0, \qquad d(R\mathsf{J}) = 0.}$$

This is called the discrete Yang–Mills equation. Summarizing, we get the following:

- Homology describes the geometry of the electric circuit; in particular, the first Betti number β_1 is equal to the number of essential loops (also called 1-cycles).
- Cohomology describes the physics of the circuit (i.e., cohomology describes the voltage and hence the electric currents, by Ohm's law).
- There exists a crucial duality relation between homology and cohomology which reflects the influence of the geometry of an electrical circuit on its physics (based on the duality relation (22.17) below).

Let us discuss this. Generalizing the Wheatstone bridge (Fig. 22.3), by definition, an electric circuit $\mathcal{C}$ consists of a set of nodes $P_0, ..., P_N$ and a set of connections P_iP_j for some $i, j = 0, ..., N$ with $i \neq j$. Each node P_i is connected at least to one other node. The nodes and the connections are also called 0-simplices and 1-simplices, respectively. In addition, we use the convention $P_iP_j = -P_jP_i$.

The boundary operator. By a 0-chain, we understand a linear combination

$$c = \alpha_0 P_0 + ... + \alpha_N P_N, \qquad \alpha_0, ..., \alpha_N \in \mathbb{R}.$$

For the boundary, $\partial c := 0$. By a 1-chain, we understand a finite linear combination

$$\mathsf{J} = \mathsf{J}_{01} \cdot (P_0P_1) + \mathsf{J}_{12} \cdot (P_1P_2) + ..., \qquad \mathsf{J}_{01}, \mathsf{J}_{12}, ... \in \mathbb{R}$$

where we sum over all connections. We use the convention $\mathsf{J}_{ij} = -\mathsf{J}_{ji}$. For the boundary, we have $\partial(P_iP_j) := P_j - P_i$, and

$$\partial \mathsf{J} := \mathsf{J}_{01}(P_1 - P_0) + \mathsf{J}_{12}(P_2 - P_1) + ...$$

The 1-chain J is called a 1-cycle iff $\partial \mathsf{J} = 0$. Moreover, the 0-chain c is called a 0-boundary iff $c = \partial \mathsf{J}$ for some 1-chain J.

The electrical current is a 1-cycle.

In fact, rearranging $\partial \mathsf{J}$ as a linear combination of nodes

$$\partial \mathsf{J} = (\mathsf{J}_{10} + \mathsf{J}_{20} + ... + \mathsf{J}_{N0})P_0 + ...,$$

we see that $\partial \mathsf{J} = 0$ is equivalent to the Kirchhoff rule for electrical currents.

The coboundary operator. By a 0-cocycle, we understand a real function U which assigns to each node P_i a real number $U(P_i)$. This way, we get a linear functional on the linear space of all 0-chains. Explicitly,

$$U(\alpha_0 P_0 + \alpha_1 P_1 + ...) := \alpha_0 U(P_0) + \alpha_1 U(P_1) + ...$$

This yields the following:

The electrostatic potential is a 0-cochain.

By a 1-cochain, we understand a function V, which assigns to each connection P_iP_j a real number V_{ij}. This way, we get a linear functional on the space of all 1-chains.

The voltage is a 1-cochain.

The coboundary dU of a 0-cochain U is a 1-cochain defined by

$$dU(\mathsf{J}) = U(\partial \mathsf{J}) \qquad \text{for all 1-chains} \quad \mathsf{J}. \tag{22.17}$$

Explicitly, $dU(P_iP_j) = U(P_j - P_i) = U(P_j) - U(P_i)$ which equals $-V_{ij}$. This tells us that:

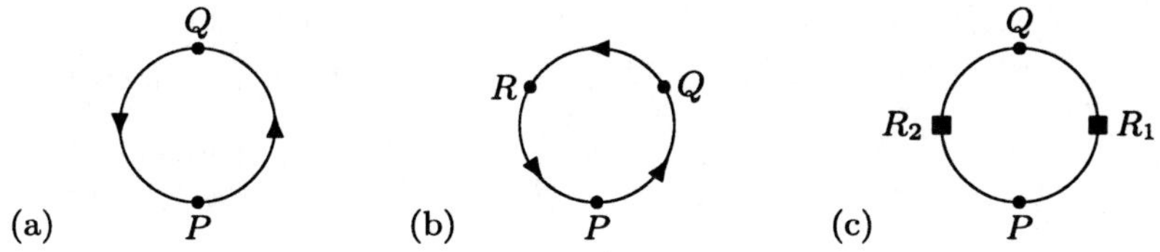

Fig. 22.4. Homology of a circle

For the voltage and the electrostatic potential, we have $V = -dU$. For each electric current J and each voltage V,

$$\boxed{V(\mathsf{J}) = \sum_{ij} \mathsf{J}_{ij} V_{ij} = 0} \tag{22.18}$$

where we sum over all connections. This follows from $\partial \mathsf{J} = 0$ and

$$V(\mathsf{J}) = -dU(\mathsf{J}) = -U(\partial \mathsf{J}) = 0.$$

Equation (22.18) implies the desired relation (22.15) above.

To get contact with the general homology and cohomology theory, note that $\partial c = 0$ for al 0-chains c. Therefore, all the 0-chains are 0-cycles. In addition, by definition, all the 2-chains C vanish with $\partial C := 0$. Therefore, all the 1-boundaries vanish. Similarly, if W is a 1-cochain, then $dW(C) = W(\partial C) = 0$ for all 2-chains C. Thus, all the 1-cochains are 1-cocycles. Finally, by definition, all the 0-coboundaries vanish.

Homology groups. By definition, the real linear space of 1-cycles forms the first homology group $H_1(\mathcal{C})$ of the electrical circuit. The dimension of this space is called the first Betti number β_1 of the electrical circuit.

The zeroth homology group $H_0(\mathcal{C})$ is obtained from the real linear space of all 0-chains by putting 0-boundaries equal to zero. In other words,

$H_0(\mathcal{C}) :=$ 0-chains modulo 0-boundaries.

The dimension of the real linear space $H_0(\mathcal{C})$ is called the zeroth Betti number β_0 of the circuit. Summarizing, up to linear isomorphisms, we have

$$H_k(\mathcal{C}) = \mathbb{R}^{\beta_k}, \qquad k = 0, 1.$$

Application to circular circuits. Consider the circuit pictured in Fig. 22.4. For the homology groups, up to homomorphism,

$$\boxed{H_0 = H_1 = \mathbb{R}.}$$

Thus, for the Betti numbers, $\beta_0 = \beta_1 = 1$. These homology groups do not depend on the choice of the triangulation.

Proof. In fact, the 1-chain $\mathsf{J} = \mathsf{J}_1 PQ + \mathsf{J}_2 QP$ satisfies the condition $\partial \mathsf{J} = 0$ iff

$$\mathsf{J}_1(Q - P) + \mathsf{J}_2(P - Q) = (\mathsf{J}_1 - \mathsf{J}_2)(P - Q) = 0.$$

Hence $\mathsf{J} = \mathsf{J}_1(PQ + QP)$ with $\mathsf{J}_1 \in \mathbb{R}$. Consequently, the space of 1-cycles is a one-dimensional real linear space. Hence $\beta_1 = 1$. Each 0-chain can be represented as

$$c = \alpha P + \beta Q = (\alpha + \beta)P + \beta(Q - P) = (\alpha + \beta)P + \beta\partial(PQ).$$

Thus $c = \gamma P$ modulo 0-boundaries where γ is an arbitrary real number. Hence $\beta_0 = 1$.

To show that the homology groups do not depend on the chosen triangulation, look at the triangulation depicted in Fig. 22.4(b). This yields the 1-cycles

$$\mathsf{J}_1(PQ + QR + RP), \qquad \mathsf{J}_1 \in \mathbb{R}.$$

Thus, again $\beta_1 = 1$. □

Theorem 22.4 *The zeroth Betti number β_0 of an electrical circuit is equal to the number of components of the circuit.*

Proof. If we distinguish one node Q_j in each component of the m components of the electric circuit, then an arbitrary 0-chain is equal to

$$\alpha_1 Q_1 + ... + \alpha_m Q_m$$

modulo 0-boundaries. This follows as for circular electrical circuits. □

Cohomology groups. By definition, the real linear space of 0-cocycles forms the zeroth cohomology group $H^0(\mathcal{C})$ of the electrical circuit $\mathcal{C}$. The dimension of this space is called the zeroth (dual) Betti number β^0 of the electrical circuit.

The first cohomology group $H^1(\mathcal{C})$ of $\mathcal{C}$ is obtained from the real linear space of all 1-cochains by putting 1-coboundaries equal to zero. In other words,

$$H^1(\mathcal{C}) := \text{1-cochains modulo 1-coboundaries.}$$

The dimension of the real linear space $H^1(\mathcal{C})$ is called the first (dual) Betti number β^1 of the circuit. Summarizing, up to linear isomorphisms,

$$H^k(\mathcal{C}) = \mathbb{R}^{\beta^k}, \qquad k = 0, 1.$$

We will show in Sect. 22.5 that $\beta_k = \beta^k$ for $k = 0, 1$ (de Rham duality).

Application to circular circuits. Consider again Fig. 22.4. We want to show that, for the cohomology groups,

$$H^0 = H^1 = \mathbb{R}^1.$$

Thus, for the dual Betti numbers, $\beta^0 = \beta^1 = 1$. These cohomology groups do not depend on the choice of the triangulation.

Proof. Ad H^0: Let α and β be real numbers. Explicitly, the cochains read as follows:

- 0-cochains: $U(\alpha P + \beta Q) = \alpha U(P) + \beta U(Q)$,
- 1-cochains: $V(\alpha PQ + \beta QP) = \alpha V(PQ) + \beta V(QP)$.

The values $U(P), U(Q)$ are fixed, but otherwise arbitrary real numbers. Therefore, a 0-cochain can be characterized by the tuple $(U(P), U(Q))$ of real numbers. Similarly, a 1-cochain can be characterized by the tuple $(V(PQ), V(QP))$ of real numbers. Furthermore, the 0-cocycles read as follows:

$$U(\alpha P + \beta Q) = (\alpha + \beta)U(P).$$

In fact, it follows from

$$\begin{aligned}(dU)(\alpha PQ + \beta QP) &= U(\partial(\alpha PQ + \beta QP)) \\ &= U((\alpha - \beta)(Q - P)) = (\alpha - \beta)(U(Q) - U(P)) = 0\end{aligned}$$

for all $\alpha, \beta \in \mathbb{R}$ that $U(P) = U(Q)$. Thus, the space of 0-cocycles is one-dimensional. Hence $\beta^0 = 1$.

Ad H^1: The 1-coboundaries look like

$$V(\alpha PQ + \beta QP) = \alpha V(PQ) + \beta V(QP), \qquad V(PQ) = -V(QP).$$

In fact, it follows from $V = dU$ that

$$V(\alpha PQ + \beta QP) = (\alpha - \beta)(U(P) - U(Q)).$$

Hence $V(PQ) = U(P) - U(Q) = -V(QP)$. Thus, the 1-coboundaries correspond to tuples $(V(PQ), -V(PQ))$. For each 1-cochain, we have the decomposition

$$(V(PQ), V(QP)) = (V(PQ), -V(PQ)) + (0, V(QP) + V(PQ)).$$

Putting the 1-coboundaries to zero means that we set $(V(PQ), -V(PQ)) = 0$. Hence the 1-cocycles modulo 1-coboundaries are given by $(0, V(QP) + V(PQ))$ for arbitrary real numbers $V(PQ)$ and $V(QP)$. This is a one-dimensional real linear space. Hence $\beta^1 = 1$. □

The structure of electrical currents. Let the 1-cycles $c_1, \ldots, c_{\beta_1}$ be a basis of $H_1(\mathcal{C})$. Then each electric current of the electrical circuit $\mathcal{C}$ has the form

$$\mathsf{J} = \mathsf{J}_1 c_1 + \ldots + \mathsf{J}_{\beta_1} c_{\beta_1}, \qquad \mathsf{J}_1, \ldots, \mathsf{J}_{\beta_1} \in \mathbb{R}. \tag{22.19}$$

The structure of voltage. For given 1-cochain V of the electrical circuit $\mathcal{C}$, the equation

$$\boxed{V = -dU} \tag{22.20}$$

has a 0-cochain U of $\mathcal{C}$ as solution iff

$$V(c_j) = 0, \qquad j = 1, \ldots, c_{\beta_1}. \tag{22.21}$$

The solution U is uniquely determined by prescribing the value of U at a fixed node.

Equation (22.20) is called Cartan's discrete potential equation. In terms of physics, this means that each voltage can be characterized by the fact that the Kirchhoff voltage rule holds true along the basic 1-cycles $c_1, \ldots, c_{\beta_1}$.

Proof. If U is a solution of $V = -dU$, then $V(\mathsf{J}) = -dU(\mathsf{J}) = -U(\partial \mathsf{J}) = 0$ for each 1-cycle J, since $\partial \mathsf{J} = 0$. Conversely, if $V(c_j) = 0$ for all j, then $V(\mathsf{J}) = 0$ for all 1-cycles J. The electrostatic potential can now be constructed as in the proof of Prop. 22.2 for the Wheatstone bridge, by using a discrete line integral. This line integral is path-independent, since V vanishes along each loop. □

Proof of the Main Theorem 22.3. By (22.19), the electric current strengths J_{ij} depend linearly on β_1 real parameters $\mathsf{J}_1, \ldots, \mathsf{J}_{\beta_1}$. According to (22.21), we get β_1 equations for the voltage components V_{ij}.

First set the battery voltage equal to zero. Using Ohm's law, $V_{ij} = R_{ij}\mathsf{J}_{ij}$ for the remaining voltages, we get β_1 linear homogeneous equations for the unknowns $\mathsf{J}_1, \ldots, \mathsf{J}_{\beta_1}$. Prescribing the battery voltage, we get an inhomogeneous linear system $\mathcal{S}$. Finally, use the argument from (22.15) above to show that the homogeneous system $\mathcal{S}$ has only the trivial solution. Since the number of equations β_1 coincides with the number of unknowns, the inhomogeneous equation has a unique solution. This unique solution corresponds to both the unique voltage V and the unique electric current J. Using (22.20), we get the unique solution of the equation

$$\partial \mathsf{J} = 0, \quad V = -dU, \quad U(P_0) = U_0.$$

□

Fig. 22.5. Refinement of triangulation

22.5 Euler Characteristic and Betti Numbers

The Euler characteristic of an electrical circuit is defined to be

$$\boxed{\chi := s_0 - s_1}$$

where s_0 and s_1 is the number of nodes and connections of the circuit, respectively. The crucial property of the Euler characteristic is the fact that it does not depend on the triangulation. This follows from Fig. 22.5. In fact, if we add one node, than the number of connections increases by one. Hence the difference $s_0 - s_1$ remains unchanged.

Theorem 22.5 *For the Euler characteristic,* $\chi = \beta_0 - \beta_1$.

Proof. Let C_k, Z_k, and B_k denote the space of k-chains, k-cycles, and k-boundaries, respectively. [6] Consider the sequence

$$\boxed{0 \xrightarrow{\partial_2} C_1 \xrightarrow{\partial_1} C_0 \xrightarrow{\partial_0} 0.} \tag{22.22}$$

Here, in order to avoid misunderstandings, we equip the boundary operator ∂ with an index k which indicates that the boundary operator $\partial : C_k \to C_{k-1}$ acts on the space C_k. Then, Z_k (resp. B_{k-1}) is the kernel or null space (resp. the image) of the boundary operator ∂_k, that is,

$$Z_k = \ker(\partial_k), \qquad B_k = \operatorname{im}(\partial_{k+1}).$$

Moreover, in terms of factor spaces,

$$H_k = Z_k/B_k, \qquad k = 0, 1.$$

By a standard result of linear algebra on the dimension of factor spaces, we get the first key relation,

$$\boxed{\beta_k = \dim H_k = \dim Z_k - \dim B_k.}$$

Moreover, the fundamental epimorphism theorem for linear operators applied to ∂_k tells us that

$$C_k/\ker(\partial_k) = \operatorname{im}(\partial_k),$$

in the sense of a linear isomorphism. Hence $C_k/Z_k = B_{k-1}$. This implies the second key relation

$$\boxed{\dim C_k - \dim Z_k = \dim B_{k-1}, \qquad k = 0, 1.}$$

Noting that $B_1 = 0$ and $B_{-1} = 0$, and noting that

- $\dim C_0 = s_0$ (number of nodes), and
- $\dim C_1 = s_1$ (number of connections),

[6] We set $C_2 := C_{-1} := 0$. Traditionally, the symbol Z_k is motivated by the German word 'Zyklus' for cycle.

we get

$$\begin{aligned}\beta_0 - \beta_1 &= (\dim Z_0 - \dim B_0) - (\dim Z_1 - \dim B_1)\\ &= (\dim C_0 - \dim B_0) - (\dim C_1 - \dim B_0) = s_0 - s_1.\end{aligned}$$

This completes the proof. □

We will show later on that the Betti numbers are topological invariants. This tells us that:

Homeomorphic electric circuits possess the same Betti numbers, and hence the same Euler characteristic.

Theorem 22.6 *The homology groups of an electric circuit $\mathcal{C}$ are linearly isomorphic to the corresponding cohomology groups, that is,*

$$H_k(\mathcal{C}) = H^k(\mathcal{C}) = \mathbb{R}^{\beta_k}, \qquad k = 0, 1.$$

In terms of Betti numbers, $\beta_k = \beta^k$ for $k = 0, 1$ (de Rham duality).

Proof. Replace the linear space C_k by its dual space C_k^d, and the linear operator $\partial_k : C_k \to C_{k-1}$ by its dual operator $\partial_k^d : C_{k-1}^d \to C_k^d$. Setting $d_k := \partial_k^d$, the sequence (22.22) passes over to the dual sequence

$$\boxed{0 \overset{d_2}{\longleftarrow} C_1^d \overset{d_1}{\longleftarrow} C_0^d \overset{d_0}{\longleftarrow} 0.} \qquad (22.23)$$

The operator $d_k : C_{k-1}^d \to C_k^d$ corresponds to the coboundary operator d introduced above. Then, Z^{k-1} (resp. B_k) is the kernel or null space (resp. the image) of the coboundary operator d_k, that is,

$$Z^k = \ker(d_{k+1}), \quad B^k = \operatorname{im}(d_k).$$

For the kth cohomology group, we get

$$H^k = Z^k/B^k, \qquad k = 0, 1.$$

We are now going to use the Fredholm alternative for linear operators. Explicitly, let $A : X \to Y$ be a linear operator where X and Y are real, finite-dimensional, linear spaces. Then

$$\dim \operatorname{im} A = \dim Y - \dim \ker A^d.$$

Similarly, for the dual operator $A^d : Y^d \to X^d$,

$$\dim \operatorname{im} A^d = \dim X^d - \dim \ker A.$$

Applying this to the boundary operator $\partial_{k+1} : C_{k+1} \to C_k$ and its dual operator $d_{k+1} : C_k^d \to C_{k+1}^d$, we get

$$\dim B_k = \dim C_k - \dim Z^k, \quad \dim B^k = \dim C_k^d - \dim Z_k.$$

A finite-dimensional linear space has always the same dimension as its dual space. Hence $\dim C_k = \dim C_k^d$. This implies

$$\dim Z_k - \dim B_k = \dim Z^k - \dim B^k,$$

telling us that $\dim H_k = \dim H^k$. Finally, note that two real, finite-dimensional, linear spaces are linearly isomorphic iff they possess the same dimension. □

22.6 The Discrete de Rham Theory

The discrete Poincaré rule. Since $\partial P_i = 0$, we get $\partial c = 0$ for each 0-chain c. For a 1-chain J, the boundary $\partial\mathsf{J}$ is a 0-chain. Hence

$$\boxed{\partial(\partial\mathsf{J}) = 0 \qquad \text{for all 1-chains} \quad \mathsf{J}.}$$

Dually, $(ddV)(\mathsf{J}) = dV(\partial\mathsf{J}) = V(\partial\partial\mathsf{J}) = 0$ for all 1-chains J. Hence

$$\boxed{d(dV) = 0 \qquad \text{for all 1-cochains} \quad V.}$$

The discrete Stokes integral theorem. Introduce the discrete integral

$$\int_{\mathsf{J}} V := V(\mathsf{J}).$$

Explicitly, $V(\mathsf{J}) := \sum \mathsf{J}_{ij} V_{ij}$ where we sum over all connections P_iP_j and P_jP_i. Recall that $V_{ij} = -V_{ji}$ and $\mathsf{J}_{ij} = -\mathsf{J}_{ji}$ for all connections. The defining equation for the coboundary operator, $dU(\mathsf{J}) := U(\partial\mathsf{J})$, can be written as

$$\boxed{\int_{\mathsf{J}} dU = \int_{\partial\mathsf{J}} U}$$

for all 1-chains J and all 0-cochains U. This identity is called the discrete Stokes integral theorem.

The discrete de Rham theorem. The following theorem is the reformulation of results proved in the preceding sections.

Theorem 22.7 *Let V be a 1-cochain. Then the potential equation*

$$V = -dU$$

has a solution U iff $\int_{\mathsf{J}} V = 0$ for all 1-cycles J. This is equivalent to the condition

$$\int_{c_j} V = 0 \qquad j = 1, \ldots, \beta_1$$

for all the basic 1-cycles $c_1, \ldots, c_{\beta_1}$.

It was shown in the 20th century that the discrete theory considered in the present chapter allows far-reaching generalizations to topological spaces (general homology and cohomology theory) and to manifolds (de Rham cohomology).

Further reading. We recommend:

P. Bamberg and S. Sternberg, A Course in Mathematics for Students of Physics, Vol. 2, Cambridge University Press, 1999.

23. The Electromagnetic Field and the de Rham Cohomology

De Rham cohomology reformulates and generalizes the fundamental theorem of calculus due to Newton and Leibniz to differential forms on manifolds. In terms of physics, this describes the existence of potentials. The key role is played by Poincaré's cohomology rule and the generalized Stokes integral theorem.

Folklore

23.1 The De Rham Cohomology Groups

23.1.1 Elementary Examples

Gauge theory is based on potentials. The de Rham cohomology of a manifold M relates the existence of potentials to the topology of the manifold M. To begin with, let us explain this for the real line and the unit circle $\mathbb{S}^1$.

Potential on the real line. Consider the differential equation

$$\boxed{F = U' \qquad \text{on } \mathbb{R}.} \tag{23.1}$$

In terms of physics, we are given the force F, and we are looking for the potential U. Let $\Lambda^0(\mathbb{R})$ denote the set of all smooth functions $F : \mathbb{R} \to \mathbb{R}$. The fundamental theorem of calculus tells us the following.

Proposition 23.1 *For given function $F \in \Lambda^0(\mathbb{R})$, the general solution of the differential equation* (23.1) *is given by*

$$U(x) = U_0 + \int_0^x F(\xi)d\xi \qquad \text{for all} \quad x \in \mathbb{R}$$

where U_0 is an arbitrary real number which describes the gauge freedom of the potential U.

The integral $\int_0^x F(\xi)d\xi$ is the work done by the force field F if it moves a particle from the point $x_0 = 0$ to the point x on the real line $\mathbb{R}$. We want to translate this into the language of de Rham cohomology groups. Let $\Lambda^k(\mathbb{R})$, $k = 1, 2, \ldots$, denote the real linear space of all differential k-forms on the real line $\mathbb{R}$. We have $\omega \in \Lambda^1(\mathbb{R})$ iff

$$\omega = f(x)dx, \qquad f \in \Lambda^0(\mathbb{R}).$$

For $k = 2, 3, \ldots$, nontrivial differential k-forms do not exist. Therefore, we set $\Lambda^k(\mathbb{R}) := \{0\}$ if $k = 2, 3, \ldots$ A crucial role is played by the differential operator d.

E. Zeidler, *Quantum Field Theory III: Gauge Theory*,
DOI 10.1007/978-3-642-22421-8_24,

- If $U \in \Lambda^0(\mathbb{R})$, then $dU = U'(x)dx$.
- If $\omega \in \Lambda^1(\mathbb{R})$, then $d\omega = 0$. In fact, $d(fdx) = f'dx \wedge dx = 0$, since $dx \wedge dx = 0$.

Equivalently, the equation (23.1) can be written as

$$\boxed{Fdx = dU \qquad \text{on } \mathbb{R}.}$$

The following linear sequence is basic:

$$0 \xrightarrow{d_{-1}} \Lambda^0(\mathbb{R}) \xrightarrow{d_0} \Lambda^1(\mathbb{R}) \xrightarrow{d_1} 0 \xrightarrow{d_2} 0 \xrightarrow{d_3} \ldots \tag{23.2}$$

Here, we set $d_k\omega := d\omega$ if $\omega \in \Lambda^k(\mathbb{R})$.[1] Obviously, $d_k d_{k-1}\mu = 0$ for all $k = 0, 1, 2, \ldots$ and all $\mu \in \Lambda^{k-1}(\mathbb{R})$. By definition, the linear quotient space

$$H^k(\mathbb{R}) := \ker(d_k)/\mathrm{im}(d_{k-1})$$

is called the kth de Rham cohomology group of the real line.[2]

Proposition 23.2 *$H^0(\mathbb{R}) = \mathbb{R}$ and $H^k(\mathbb{R}) = 0$ if $k = 1, 2 \ldots$*

Proof. By Prop. 23.1, the equation $dU = 0$ on $\mathbb{R}$ has the solution $U \equiv U_0$. Hence

$$\ker(d_0) = \mathbb{R}.$$

Trivially, $\mathrm{im}(d_{-1}) = 0$. This implies $H^0(\mathbb{R}) = \ker(d_0)/\mathrm{im}(d_{-1}) = \mathbb{R}$. Again, by Prop. 23.1, the equation $Fdx = dU$ has always a solution. Hence

$$\mathrm{im}(d_0) = \Lambda^1(\mathbb{R}).$$

Trivially, $\ker(d_1) = \Lambda^1(\mathbb{R})$. This implies $H^1(\mathbb{R}) = \ker(d_1)/\mathrm{im}(d_0) = 0$. □

Potential on an open interval. Replacing the real line $\mathbb{R}$ by the open interval $]a,b[$, the same argument as used above yields

$$H^0(\,]a,b[\,) = \mathbb{R} \qquad \text{and} \qquad H^k(\,]a,b[\,) = 0, \quad k = 1, 2, \ldots$$

Potential on two disjoint open intervals. Consider $M :=]0,1[\, \cup\,]2,3[$. Then the equation

$$F = U' \qquad \text{on } M$$

has the two solutions

- $U(x) = U_0 + \int_{1/2}^x F(x)dx, \quad x \in]0,1[$, and
- $U(x) = U_1 + \int_{3/2}^x F(x)dx, \quad x \in]2,3[$

with the real numbers U_0 and U_1. The same argument as for the real line yields

$$H^0(M) = \mathbb{R}^2 \qquad \text{and} \qquad H^k(M) = 0, \qquad k = 1, 2, \ldots$$

Periodic potentials. Let $C^\infty_{2\pi}(\mathbb{R})$ denote the set of all 2π-periodic smooth functions $F : \mathbb{R} \to \mathbb{R}$.

---

[1] Pedantically, we use the different symbols d_1 and d_2 in order to indicate that the operator d acts on different spaces.

[2] To simplify notation, we will write $H^k(\mathbb{R}) = 0$ instead of $H^k(\mathbb{R}) = \{0\}$.

Proposition 23.3 *For given periodic force* $F \in C^\infty_{2\pi}(\mathbb{R})$, *the equation*

$$F(\varphi) = U'(\varphi) \qquad \text{on } \mathbb{R} \tag{23.3}$$

has a solution $U \in C^\infty_{2\pi}(\mathbb{R})$ *iff the constraint* $\int_{-\pi}^{\pi} F(\varphi)d\varphi = 0$ *is satisfied. In this case, the general solution of* (23.3) *reads as*

$$U(\varphi) = U_0 + \int_{-\pi}^{\varphi} F(\psi)d\psi, \qquad \varphi \in \mathbb{R} \tag{23.4}$$

where U_0 *is an arbitrary real number which describes the gauge freedom of the potential* U.

Proof. If the function U is a solution of (23.3), then

$$\int_{-\pi}^{\pi} F(\varphi)d\varphi = U(\pi) - U(-\pi) = 0,$$

because of the periodicity of the potential U. Conversely, it follows from (23.4) that $U(\pi) = U(-\pi)$. Thus, the function U has the period 2π. □

In order to get insight, let us use the Fourier series

$$F(\varphi) = a_0 + \sum_{k=1}^{\infty} a_k \cos k\varphi + b_k \sin k\varphi, \qquad \varphi \in \mathbb{R}. \tag{23.5}$$

Proposition 23.4 *The equation* (23.3) *has a solution iff* $a_0 = 0$. *The general solution reads as*

$$U(\varphi) = U_0 + \sum_{k=1}^{\infty} \frac{a_k}{k} \sin k\varphi - \frac{b_k}{k} \cos k\varphi, \qquad \varphi \in \mathbb{R}$$

where U_0 *is an arbitrary real number which describes the gauge freedom of the potential* U.

Potentials on the unit circle. Intuitively, the real line and the unit circle $\mathbb{S}^1$ possess a different qualitative geometric structure. We want to show that the de Rham cohomology is able to see the difference. Let $\Lambda^k(\mathbb{S}^1)$ denote the real linear space of all the differential k-forms on the unit circle $\mathbb{S}^1$:

- $U \in \Lambda^0(\mathbb{S}^1)$ iff $U \in C^\infty_{2\pi}(\mathbb{R})$.
- $\omega \in \Lambda^1(\mathbb{S}^1)$ iff $\omega = F(\varphi)d\varphi$ with $F \in C^\infty_{2\pi}(\mathbb{R})$.

Proposition 23.5 *For given* $\omega = F(\varphi)d\varphi$, *the equation*

$$\omega = dU \qquad \text{on } \mathbb{S}^1$$

has a solution iff $\int_{\mathbb{S}^1} \omega = 0$. *The general solution reads as*

$$U = U_0 + \int_{-\pi}^{\varphi} F(\psi)d\psi.$$

This is a reformulation of Prop. 23.3. In order to compute the de Rham cohomology groups of the unit circle $\mathbb{S}^1$, let us start with the following sequence of linear operators

$$0 \xrightarrow{d_{-1}} \Lambda^0(\mathbb{S}^1) \xrightarrow{d_0} \Lambda^1(\mathbb{S}^1) \xrightarrow{d_1} 0 \xrightarrow{d_2} 0 \xrightarrow{d_3} \ldots \tag{23.6}$$

Here, we set $d_k\omega := d\omega$.

- If $U \in \Lambda^0(\mathbb{S}^1)$, then $dU = U'(\varphi)d\varphi$, and
- if $\omega \in \Lambda^1(\mathbb{S}^1)$, then $d\omega = 0$.

Obviously, $d_k d_{k-1}\mu = 0$ for all $k = 0, 1, \ldots,$ and all $\mu \in \Lambda^{k-1}(\mathbb{S}^1)$.

Proposition 23.6 *$H^0(\mathbb{S}^1) = H^1(\mathbb{S}^1) = \mathbb{R}$ and $H^k(\mathbb{S}^1) = 0$ if $k = 2, 3, \ldots$*

Proof. Let $U \in \Lambda^0(\mathbb{S}^1)$. If $dU = 0$ on $\mathbb{S}^1$, then $U \equiv U_0$. As for the real line $\mathbb{R}$, we get $H^0(\mathbb{S}^1) = \mathbb{R}$.

Let $\omega \in \Lambda^1(\mathbb{S}^1)$. Then $\omega = F(\varphi)d\varphi$ with $F \in C^\infty_{2\pi}(\mathbb{R})$. By Prop. 23.3, the equation $F - a_0 = U'$ has a solution $U \in \Lambda^0(\mathbb{S}^1)$. Hence

$$\omega = dU + a_0 d\varphi.$$

Since $dU \in \text{im}(d_0)$, we get the decomposition

$$\Lambda^1(\mathbb{S}^1) = \text{im}(d_0) \oplus \text{span}(a_0 d\varphi) \simeq \text{im}(d_0) \oplus \mathbb{R}.$$

Finally, since $\ker(d_1) = \Lambda^1(\mathbb{S}^1)$, we obtain

$$H^1(\mathbb{S}^1) = \ker(d_1)/\text{im}(d_0) = \Lambda^1(\mathbb{S}^1)/\text{im}(d_0) = \mathbb{R}.$$

□

It is our goal to generalize the preceding simple examples to the general case. To this end, we need the language of differential forms which is the most elegant tool for generalizing the calculus for functions of one real variable to functions of several real variables.

23.1.2 Advanced Examples

Fix $n = 1, 2, \ldots$ Let M be an n-dimensional real manifold. Then we have the sequence of linear operators

$$0 \xrightarrow{d_{-1}} \Lambda^0(M) \xrightarrow{d_0} \ldots \xrightarrow{d_{k-1}} \Lambda^k(M) \xrightarrow{d_k} \Lambda^{k+1}(M) \xrightarrow{d_{k+1}} \ldots \tag{23.7}$$

Here, $\Lambda^k(M)$ denotes the real linear space of all the smooth differential k-forms ω on the manifold M, and we set $d_k\omega := d\omega$ (Cartan differential). By the Poincaré cohomology rule, we have

$$d_{k+1}(d_k\omega) = d(d\omega) = 0 \qquad \text{for all} \quad \omega \in \Lambda^k(M).$$

The kth de Rham cohomology group of the manifold M is defined by

$$\boxed{H^k(M) := \ker(d_k)/\ \text{im}\ (d_{k-1}).} \tag{23.8}$$

This is a real linear space. Furthermore, we set

$$\boxed{\beta_k(M) := \dim H^k(M).} \tag{23.9}$$

This is called the kth Betti number of the manifold M. The number

$$\boxed{\chi(M) := \beta_0 - \beta_1 + \beta_2 - \ldots}$$

is called the Euler characteristic of the manifold M. If $\dim M = n$, then $\beta_k(M) = 0$ if $k > n$. By definition, the Poincaré polynomial is given by

$$p_M(x) := \beta_0 + \beta_1 x + \beta_2 x^2 + \ldots$$

In particular, $\chi(M) = p_M(-1)$. Let us consider the following examples.

(a) The real line $\mathbb{R}$: $H^0(\mathbb{R}) = \mathbb{R}$, and $H^k(\mathbb{S}^1) = 0$ if $k = 2, 3, \ldots$, $\beta_0 = 1, \beta_k = 0$ if $k = 1, 2, \ldots$,

$$p_{\mathbb{R}} = 1, \qquad \chi_{\mathbb{R}} = 1.$$

(b) The union M of m pairwise disjoint open intervals on the real line, $m = 1, 2, \ldots$: $H^0(M) = \mathbb{R}^m$ and $H^1(M) = 0$ if $k = 1, 2, \ldots$, $\beta_0 = m$ and $\beta_k = 0$ otherwise,

$$p_M = m, \qquad \chi_M = m.$$

(c) The unit circle $\mathbb{S}^1$: $H^0(\mathbb{S}^1) = H^1(\mathbb{S}^1) = \mathbb{R}$, and $H^k(\mathbb{S}^1) = 0$ if $k = 2, 3, \ldots$, $\beta_0 = \beta_1 = 1$ and $\beta_k = 0$ if $k = 2, 3, \ldots$,

$$p_{\mathbb{S}^1} = 1 + x, \qquad \chi_{\mathbb{S}^1} = 0.$$

(d) The unit sphere $\mathbb{S}^2$: $H^0(\mathbb{S}^2) = H^2(\mathbb{S}^2) = \mathbb{R}$, $H^1(\mathbb{S}^1) = 0$, and $H^k(\mathbb{S}^2) = 0$ if $k = 3, 4, \ldots$, $\beta_0 = \beta_2 = 1, \beta_1 = 0$, and $\beta_k = 0$ if $k = 3, 4, \ldots$,

$$p_{\mathbb{S}^2} = 1 + x^2, \qquad \chi_{\mathbb{S}^2} = 2.$$

(e) Cylinder $C = \mathbb{S}^1 \times]0, 1[$: $H^k(C) = H^k(\mathbb{S}^1)$ if $k = 0, 1, 2, \ldots$,

$$p_C = 1 + x, \qquad \chi_C = 0.$$

(f) If M is a real 2-dimensional compact manifold of genus $g = 0, 1, 2, \ldots$, then $H^0(M) = H^2(M) = \mathbb{R}$, $H^1(M) = \mathbb{R}^g$, and $H^k(M) = 0$ if $k = 3, 4, \ldots$ Moreover,

$$\beta_0 = \beta_2 = 1, \quad \beta_1 = 2g, \quad p_M = 1 + 2gx + x^2, \quad \chi_M = 2 - 2g.$$

The definition of the genus can be found in Sect. 5.12 of Vol. II. In particular, the sphere (resp. the torus) has the genus $g = 0$ (resp. $g = 1$).

(g) If M is a nonempty open convex set of $\mathbb{R}^n$, $n = 1, 2, \ldots$, then $H^k(M) = H^k(\mathbb{R})$ for all $k = 0, 1, 2, \ldots$,

$$p_M = 1, \qquad \chi_M = 1.$$

(h) If M is the union of m pairwise disjoint, nonempty, convex sets of $\mathbb{R}^n$ with $n = 1, 2, \ldots$, then $H^0(M) = m$ and $H^k(M) = 0$ if $k = 1, 2, \ldots$. Hence

$$p_M = m, \qquad \chi_M = m.$$

Let us sketch the proofs.

Ad (a), (b), (g), (h). This is a special case of the Poincaré theorem (i) below.

Ad (c). See Prop. 23.6 on page 1030.

Ad (d). Use a similar argument as for the unit circle in the proof of Prop. 23.6. Replace the Fourier series by a series with respect to spherical harmonics. The full proof will be given in Vol. IV.

Ad (e). The projection map $(x, y, z) \mapsto (x, y)$ sends the points

$$\{(x, y, z) \in \mathbb{R}^3 : x^2 + y^2 = 1,\ 0 < z < 1\}$$

of the cylinder C to the points of the unit circle $\mathbb{S}^1$. Thus, the manifolds C and $\mathbb{S}^1$ are homotopically equivalent. By the deformation invariance of the de Rham topology formulated below, the cylinder C has the same de Rham cohomology groups as the unit circle.

Ad (f). See Vol. IV. □

Geometric motivation. In order to get insight, let us summarize the following crucial results in topology. Recall that a topological space (e.g., a manifold) is contractible iff there exist both a point $p \in X$ and a continuous map

$$H : X \times [0,1] \to X$$

such that $X(x,0) = x$ for all $x \in X$, and $H(x,1) = p$ for all $x \in X$. For example, a ball is contractible, whereas a circle or a sphere are not contractible.

Theorem 23.7 *Let M be a real finite-dimensional (nonempty) manifold. Then:*

(i) *If the manifold M is contractible, then $H^0(M) = \mathbb{R}$ and $H^k(M) = 0$ if* $k = 1, 2, \ldots$ (Poincaré's theorem).

(ii) *If M is the union of m pairwise disjoint, real, finite-dimensional, contractible manifolds, then $H^0(M) = \mathbb{R}^m$ and $H^k(M) = 0$ if $k = 1, 2, \ldots$*

(iii) *If M is arcwise connected, then we obtain $H^0(M) = \mathbb{R}$ and $\beta_0 = 1$.*

(iv) *The Betti number β_0 is equal to the number of connected components of the manifold M.*

(v) *If M is simply connected, then $H^1(M) = 0$.*

(vi) *If M is a real finite-dimensional compact manifold, then the Betti numbers of M are nonnegative integers, and the Euler characteristic of M is an integer.*

(vii) *If M and N are real finite-dimensional compact manifolds, then we have the following elegant Künneth product formula for the Poincaré polynomials:*[3]

$$p_{M \times N} = p_M \cdot p_N.$$

(viii) *If M is a real, n-dimensional, compact, arcwise connected, oriented manifold $\mathcal{M}$, then*

$$\beta_k(M) = \beta_{n-k}(M), \qquad k = 0, 1, \ldots n.$$

This is called the Poincaré duality.

As an example, consider the cylinder $C = \mathbb{S}^1 \times]0,1[$. Since the unit interval $]0,1[$ is continuously contractible to a point, we get $p_{\,]0,1[} = 1$. By the Künneth product formula,

$$p_C = p_{\,\mathbb{S}^1} \cdot p_{\,]0,1[} = p_{\,\mathbb{S}^1} = 1 + x.$$

Moreover, the 2-dimensional torus $\mathbb{T}^2$ is the Cartesian product of two unit circles (see Fig. 5.27 of Vol. II). It follows from $\mathbb{T}^2 = \mathbb{S}^1 \times \mathbb{S}^1$ that

$$p_{\,\mathbb{T}^2} = p_{\,\mathbb{S}^1} \cdot p_{\,\mathbb{S}^1} = (1+x)^2 = 1 + 2x + x^2.$$

Hence $\beta^0(\mathbb{T}^2) = \beta^2(\mathbb{T}^2) = 1$, and $\beta_1(\mathbb{T}^2) = 2$. This coincides with (f) above by choosing the genus $g = 1$.

23.1.3 Topological Invariance of the de Rham Cohomology Groups

The topological invariance of the de Rham cohomology groups. Recall that two topological spaces X and Y are called topologically equivalent iff there exists a homeomorphism $f : X \to Y$ (i.e., the continuous map f is bijective, and the inverse map is also continuous).

[3] H. Künneth, On the Betti numbers of product manifolds, Math. Ann. **90** (1923), 65–85 (in German); Künneth (1892–1975).

Theorem 23.8 *Two topologically equivalent finite-dimensional real manifolds have the same de Rham cohomology groups.*

This is a deep result in topology. The proof can be found in Bott and Tu (1982) and Lück (2006), quoted on page 1061. In particular, this tells us that the Betti numbers and the Euler characteristic introduced above are topological invariants. For example, an ellipse E is homeomorphic to the unit circle $\mathbb{S}^1$. Therefore,

$$H^k(E) = H^k(\mathbb{S}^1), \quad k = 1, 2, \ldots$$

Hence $\beta_0(E) = \beta_1(E) = 1$ and $\beta_k(E) = 0$ if $k = 2, 3, \ldots$

23.1.4 Homotopical Invariance of the de Rham Cohomology Groups

Homotopies describe deformations. Let X and Y be topological spaces. Recall that two continuous maps

$$f, g : X \to Y$$

are called homotopic iff there exists a continuous map $H : X \times [0,1] \to Y$ such that

$$H(x,0) = f(x) \qquad \text{and} \qquad H(x,1) = g(x) \qquad \text{for all} \quad x \in X.$$

That is, for every fixed time $t \in [0,1]$ we get a map $x \mapsto H(x,t)$ from X to Y which coincides with the map f at time $t = 0$ and the map g at time $t = 1$. The continuity of the map H ensures that the map f is continuously deformed into the map g during the time interval $[0,1]$. If f and g are homotopic, then we write

$$f \simeq g.$$

The topological spaces X and Y are called homotopically equivalent iff there exist continuous maps

$$f : X \to Y \qquad \text{and} \qquad h : Y \to X$$

such that $h \circ f \simeq \mathrm{id}_X$ and $f \circ h \simeq \mathrm{id}_Y$. Roughly speaking, this means that the composed map $h \circ f$ (resp. $f \circ h$) can be continuously deformed into the identity map on X (resp. Y). As a typical example, let us consider contractible spaces. Let p be a point in the topological space X. The space X is contractible to the point p iff it is homotopically equivalent to the trivial topological space $\{p\}$ consisting precisely of the point p. More material on homotopy can be found in Sect. 4.4.4 of Vol. II.

Theorem 23.9 *Two homotopically equivalent real finite-dimensional manifolds have the same de Rham cohomology groups.*

Cycles and Poincaré's idea of homology. Homology theory is based on the boundary operator ∂. Let X be a topological space (e.g., $\mathbb{R}^n$ or a sphere), and let U, V, W be closed subsets of X.

- The set U is called a cycle iff it has no boundary.
- The set U is called a boundary iff there exists a closed set V such that $U = \partial V$.
- A cycle U is called homologically trivial iff it is a boundary.

The goal of homology theory is to describe the qualitative properties of topological spaces by considering homologically nontrivial cycles. In fact, there arise technical problems. Therefore, Poincaré passed to a special class of topological spaces which allow triangulations. He used triangulations in order to introduce the Betti numbers.

> *Roughly speaking, for an oriented topological space, the kth Betti number (in the sense of Poincaré) equals the number of homologically nontrivial k-dimensional cycles.*

In the case of non-oriented spaces, there arise also so-called torsion numbers. Let us discuss the intuitive background by considering some examples.[4]

- The equator of the 2-dimensional unit sphere $\mathbb{S}^2$ has no boundary. Thus, it is a cycle. But it is a homologically trivial 1-cycle, since the equator is the boundary of the northern hemisphere. Every reasonable closed curve on the sphere $\mathbb{S}^2$ is also a boundary. Therefore, we assign intuitively the Betti number $\beta_1 = 0$ to $\mathbb{S}^2$.
- The sphere $\mathbb{S}^2$ itself has no boundary, and it is not a boundary. Thus, it is a homologically nontrivial cycle. We assign intuitively the Betti number $\beta_2 = 1$ to the sphere $\mathbb{S}^2$.
- The outer (or the inner) equator of a 2-dimensional torus $\mathbb{T}^2$ has no boundary, and it is not the boundary of a closed subset of the torus. Thus, in contrast to the sphere $\mathbb{S}^2$, the outer (or inner) equator of a torus is a homologically nontrivial 1-cycle. A more precise formulation of homology theory (to be considered in Vol. IV) yields the fact that the torus $\mathbb{T}^2$ has two 1-cycles up to homological equivalence, namely, the outer equator and a fixed meridian. This implies the Betti number $\beta_1 = 2$.

For the following considerations, we restrict ourselves to so-called regular k-cycles. Let M be a real finite-dimensional compact oriented manifold. By definition, a regular k-cycle is a finite sum

$$C = \alpha_1 S_1 + \alpha_2 S_2 + \ldots$$

where $S_1, S_2, \ldots$ are oriented submanifolds of the manifold M, and $\alpha_1, \alpha_2, \ldots$ are real numbers. The manifold M and the submanifold $S_1, S_2, \ldots$ have no boundary. Therefore, we define

$$\partial C = \alpha_1 \partial S_1 + \alpha_2 \partial S_2 + \ldots := 0.$$

If ω is a differential k-form, then we define the integral

$$\int_C \omega := \alpha_1 \int_{S_1} \omega + \alpha_2 \int_{S_2} \omega + \ldots \tag{23.10}$$

If ω is a differential k-form on M, and S is a k-dimensional submanifold with boundary on M, then the Stokes integral theorem tells us that

$$\boxed{\int_S d\omega = \int_{\partial S} \omega.} \tag{23.11}$$

In particular, it follows from (23.11) that the following hold:

- If $\partial S = \emptyset$, then $\int_S d\omega = 0$.
- If $d\omega = 0$, then $\int_{\partial S} \omega = 0$.

This implies the following result which will be critically used below.

Proposition 23.10 *If C is a regular k-cycle, and μ is a differential k-form with $\mu = d\omega$, then $\int_C \mu = 0$.*

---

[4] The rigorous approach will be considered in Vol. IV.

Proof. Use $\int_{S_j} \mu = \int_{S_j} d\omega = 0$, and apply this to (23.10). □

The Stokes integral theorem (23.11) reflects a fundamental duality between the boundary operator, ∂, and Cartan's differential operator, d. This culminates in the solution theory for the potential equation $d\omega = \mu$ on M to be considered below. Before studying this, we need some preparations.

Cocycles and the de Rham cohomology. Let M be a real finite-dimensional manifold. To begin with, let us introduce the dual concepts to cycles and boundaries.

- The differential k-form ω is called a k-cocycle iff $d\omega = 0$.
- The differential k-form ω is called a coboundary iff there holds $\omega = d\nu$ for some differential $(k-1)$-form ν.

We are interested in cocycles which are not coboundaries. To this end, for two k-cycles ω and σ on M, we write

$$\omega \sim \sigma$$

iff the difference $\omega - \sigma$ is a k-coboundary, that is,

$$\omega - \sigma = d\nu$$

for some differential $(k-1)$-form ν. This is an equivalence relation. For the equivalence classes $[\omega]$ and $[\tau]$, we define

$$\alpha[\omega] + \beta[\tau] = [\alpha\omega + \beta\tau], \qquad \alpha, \beta \in \mathbb{R}.$$

This way, the equivalence classes $[\omega]$ become a linear space which coincides with the kth de Rham cohomology group $H^k(M)$ introduced in (23.8) on page 1030. The equivalence classes $[\omega]$ are called cohohomology classes.

Fundamental system of regular k-cycles. Let M be a real finite-dimensional compact oriented manifold (e.g., a sphere or a torus). Let β_k be the kth Betti number of M, in the sense of (23.9). By definition, the system

$$C_1, \ldots, C_{\beta_k}$$

of regular k-cycles is a fundamental system iff there exist differential k-forms $\omega_1, \omega_2, \ldots \omega_{\beta_k}$ such that we have the following orthogonal relations

$$\boxed{\int_{C_r} \omega_s = \delta_{rs}, \qquad r, s = 1, \ldots, \beta_k.}$$

A deep result in differential topology proved by René Thom (1923–2002) in 1953 tells us that such a fundamental system always exists. If ω is a differential k-form, then the real numbers

$$p_r := \int_{C_r} \omega, \qquad r = 1, 2, \ldots, \beta_k$$

are called periods of ω. Riemann studied such periods for differential forms on Riemann surfaces. This plays a crucial role for understanding the theory of elliptic and Abelian integrals. We say that the differential form ω has no periods iff $p_r = 0$ for all $r = 1, 2, \ldots, \beta_k$.

23.2 The Fundamental Potential Equation in Gauge Theory and the Analytic Meaning of the Betti Numbers

> The topology of a manifold restricts the possible structure of physical fields and their potentials on the manifold. The constraints for the physical fields and the gauge degrees of freedom of the potentials are measured by the Betti numbers of the manifold. One has to distinguish between local and global integrability conditions (constraints) for the physical field.
>
> Folklore

As a generalization of the classic potential equation

$$\mathbf{F} = -\,\mathbf{grad}\,U$$

in mechanics, let us investigate the generalized potential equation

$$\boxed{\mu = d\omega \quad \text{on } M.} \tag{23.12}$$

We want to show that the de Rham cohomology is nothing else than the optimal tool for solving this equation. Let M be a real n-dimensional manifold, $n = 1, 2, \ldots$ Fix $k = 0, 1, 2, \ldots n$. We are given the differential $(k+1)$-form μ, and we are looking for a differential k-form ω.

(i) Homogeneous equation: Let $\mu = 0$. The solutions ω of the homogeneous equation $d\omega = 0$ on M are precisely the k-cocycles on the manifold M. The general solution of this equation looks like

$$\omega = \alpha_1\omega_1 + \alpha\omega_2 + \ldots + \alpha_{\beta_k}\omega_{\beta_k} + d\nu, \qquad \alpha_1, \ldots, \alpha_{\beta_k} \in \mathbb{R}$$

where ν is an arbitrary differential $(k-1)$-form.[5] Moreover, the cohomology classes $[\omega_1], \ldots, [\omega_{\beta_k}]$ form a basis of the kth de Rham cohomology group $H^k(M)$.

In other words, the Betti number β_k tells us the dimension (modulo cohomology) of the solution space of the homogeneous equation $d\omega = 0$ on M.

(ii) The superposition principle for the inhomogeneous equation: If the equation (23.12) has a solution ω_{special}, then the general solution reads as

$$\omega = \omega_{\text{special}} + \omega_{\text{hom}}$$

where ω_{hom} is an arbitrary solution of the corresponding homogeneous problem.

(iii) Necessary solvability condition: If the equation (23.12) has a solution ω, then $d\mu = 0$.[6]

(iv) Poincaré's sufficient solvability condition: If the manifold can be continuously contracted to a point, and if $d\mu = 0$, then the equation (23.12) has a solution.

In order to get a necessary and sufficient solvability condition for (23.12), we assume:

(H) The real manifold M is finite-dimensional, compact, and oriented (e.g., a sphere or a torus).

We choose a fundamental system $C_1, C_2, \ldots, C_{\beta_{k+1}}$ of $(k+1)$-cycles of the manifold M, and we consider the de Rham constraints

$$\int_{C_r} \mu = 0, \qquad r = 1, 2, \ldots, \beta_{k+1}. \tag{23.13}$$

[5] If $k = 0$, then $\nu = 0$.

[6] This means that μ is both a $(k+1)$-coboundary and a $(k+1)$-cocycle. The claim follows immediately from $d\mu = d(d\omega) = 0$ (Poincaré's cohomology rule).

By convention, these constraints drop out if the manifold M can be continuously contracted to a point. Condition (23.13) tells us that the periods of the differential form μ vanish. If the equation $d\omega = \mu$ has a solution ω, then the condition (23.13) is satisfied by the Stokes integral theorem (see Prop. 23.10). The crucial point is that the condition (23.13) together with $d\mu = 0$ is also a sufficient solvability condition.

Theorem 23.11 (de Rham) *The equation* (23.12) *has a solution iff both the local integrability condition* $d\mu = 0$ *on* M *and the global integrability condition* (23.13) *are satisfied.*

This fundamental theorem due to de Rham tells us the following in terms of physics: If the equation $\mu = d\omega$ on M has a solution, then the physical field μ has the potential ω. Let $\beta_0, \beta_1, \ldots$ be the Betti numbers of the manifold M. Suppose that μ is a differential $(k+1)$-form.

> *There are* $\beta_{k+1} + 1$ *constraints (or integrability conditions) for the physical field* μ *to possess a potential. In addition, the potential* ω *has* β_k *degrees of gauge freedom modulo cohomology.*

For the proof, we refer to F. Warner, Foundations of Differentiable Manifolds and Lie Groups, Scott-Foresman, Glenview, Illinois, 1971, and R. Thom, Some global properties of differentiable manifolds (in French), Comm. Math. Helv. **28**, 17–86. English translation in: S. Novikov and I. Taimanov (Eds.), Topological Library, Vol. 1: Cobordisms and their Applications, pp. 131–209. See also the following two classic monographs:

> W. Hodge, The Theory and Applications of Harmonic Integrals, Cambridge University Press, 1941 (second revised edition 1951).
>
> G. de Rham, Variétés différentiables: Formes, courants, formes harmoniques, Hermann, Paris, 1955 (in French).

De Rham and Hodge related algebraic topology to analysis, differential topology, and physics. Elements of this approach can be traced back to the work of Poincaré.

Examples. Consider again the equation

$$\boxed{d\omega = \mu, \qquad \omega \in \Lambda^k(M), \;\; k = 0, 1, \ldots, n, \; n = \dim(M).} \tag{23.14}$$

We are given $\mu \in \Lambda^{k+1}(M)$. If $k = n$, then the equation reduces to the homogeneous equation $d\omega = 0$ on M because of $\Lambda^{n+1}(M) = \{0\}$.

(i) Unit circle: Let $M = \mathbb{S}^1$. Note that $\beta_0 = \beta_1 = 1$ (Betti numbers).
- $k = 0$: The equation (23.14) has a solution iff $d\mu = 0$ and $\int_{\mathbb{S}^1} \mu = 0$. The general solution reads as
$$\omega = \omega_{\text{special}} + \alpha, \qquad \alpha \in \mathbb{R}.$$
- $k = 1$, $\mu = 0$: The solution space of the equation (23.14) has the dimension $\beta_1 = 1$ modulo cohomology. The general solution of (23.14) is given by
$$\omega = \alpha v + dU, \qquad \alpha \in \mathbb{R}$$
where v is the normalized volume form on $\mathbb{S}^1$ (i.e., $\int_{\mathbb{S}^1} v = 1$), and $U : \mathbb{S}^1 \to \mathbb{R}$ is an arbitrary smooth function. The proof of this result can be found on page 1029.

(ii) Unit sphere: Let $M = \mathbb{S}^2$. Note that $\beta_0 = \beta_2 = 1, \beta_1 = 0$.

- $k = 0$: The equation (23.14) has a solution iff $d\mu = 0$. If ω_{special} is a special solution of (23.14), then the general solution of (23.14) reads as

$$\omega = \omega_{\text{special}} + \alpha, \qquad \alpha \in \mathbb{R}.$$

Thus the solution space of (23.14) has the dimension $\beta_0 = 1$.
- $k = 1$: The equation (23.14) has a solution iff $d\mu = 0$ and $\int_{\mathbb{S}^2} \mu = 0$. If ω_{special} is a special solution of (23.14), then the general solution of (23.14) is given by

$$\omega = \omega_{\text{special}} + dU$$

where $U : \mathbb{S}^2 \to \mathbb{R}$ is an arbitrary smooth function. The solution space has the dimension $\beta_1 = 0$ modulo cohomology.
- $k = 2, \mu = 0$: The general solution of (23.14) is given by

$$\omega = \alpha v + d\nu, \qquad \alpha \in \mathbb{R}$$

where v is the normalized volume form (i.e., $\int_{\mathbb{S}^2} v = 1$), and ν is an arbitrary differential 1-form on $\mathbb{S}^2$.

(iii) Torus: Let $M = \mathbb{T}^2$. Note that $\beta_0 = \beta_2 = 1, \beta_1 = 2$. Choose the outer equator C_1 and a meridian C_2 of the torus.

- $k = 0$: The equation (23.14) has a solution iff $d\mu = 0$ and $\int_{C_j} \mu = 0, j = 1, 2$. If ω_{special} is a special solution of (23.14), then the general solution of (23.14) reads as

$$\omega = \omega_{\text{special}} + \alpha, \qquad \alpha \in \mathbb{R}.$$

Thus the solution space of (23.14) has the dimension $\beta_0 = 1$.
- $k = 1$: The equation (23.14) has a solution iff $d\mu = 0$ and $\int_{\mathbb{T}^2} \mu = 0$. If ω_{special} is a special solution of (23.14), then the general solution of (23.14) is given by

$$\omega = \omega_{\text{special}} + \alpha_1\omega_1 + \alpha_2\omega_2 + dU, \qquad \alpha_1, \alpha_2 \in \mathbb{R}$$

where $U : \mathbb{T}^2 \to \mathbb{R}$ is an arbitrary smooth function. Moreover, ω_1 and ω_2 are differential 1-forms on $\mathbb{T}^2$ with $\int_{C_r} \omega_s = \delta_{rs}$ if $r, s = 1, 2$. The solution space of (23.14) has the dimension $\beta_1 = 2$ modulo cohomology.
- $k = 2, \mu = 0$: The general solution of (23.14) is given by

$$\omega = \alpha v + d\sigma, \qquad \alpha \in \mathbb{R}$$

where v is the normalized volume form (i.e., $\int_{\mathbb{T}^2} v = 1$). Moreover, σ is an arbitrary differential 1-form on $\mathbb{T}^2$.

23.3 Hodge Theory (Representing Cohomology Classes by Harmonic Forms)

Hodge theory was created by Hodge (1903–1975) in the 1930s. It was the goal of Hodge to investigate algebraic manifolds by generalizing Riemann's theory for Abelian integrals on Riemann surfaces.

Let M be a real n-dimensional compact Riemannian manifold, $n = 1, 2, \ldots$ (e.g., an n-dimensional sphere). Such a manifold is always oriented.

Theorem 23.12 *If the equation $d\omega = \mu$ on M has a solution ω, then it has precisely one solution with the property $\Delta\omega = 0$, (i.e., ω is a harmonic form).*

Equivalently, the following hold.[7]

> *Every de Rham cohomology class of the real n-dimensional compact Riemannian manifold M contains precisely one harmonic form.*

Consequently, there exists a linear isomorphism

$$\mathcal{H}^k(M) \simeq H^k(M), \qquad k = 0, 1, 2, \ldots$$

between the real linear space $\mathcal{H}^k(M)$ of harmonic k-forms on the manifold M and the de Rham cohomology group $H^k(M)$. This allows us to reduce the de Rham cohomology to solving the Laplace equation

$$\Delta\omega = 0 \text{ on } M, \qquad \omega \in \Lambda^k(M).$$

23.4 The Topology of the Electromagnetic Field and Potentials

> The de Rham cohomology is a far-reaching generalization of Maxwell's theory for the electromagnetic field. If the open set $\mathcal{O}$ is contractible to a point, then the existence of an electric (resp. magnetic) potential is based on the local constraint $\mathbf{curl}\,\mathbf{E} = 0$ (resp. $\operatorname{div}\mathbf{B} = 0$) for the electric field $\mathbf{E}$ (resp. the magnetic field $\mathbf{B}$). The number of global constraints for the existence of electric (resp. magnetic) potentials depends on the first Betti number β_1 (resp. the second Betti number β_2) of the open set $\mathcal{O}$. The Betti number β_1 (resp. β_2) measures the number of essential 1-cycles (resp. 2-cycles) which are not boundaries. The Betti numbers are homotopical invariants, and hence the number of linearly independent constraints is also a homotopical invariant.
>
> Folklore

Fix a strictly positively oriented inertial system. The Maxwell equations for the electric vector field $\mathbf{E}$ and the magnetic vector field $\mathbf{B}$ in a vacuum read as

$$\varepsilon_0 \operatorname{div}\mathbf{E} = \varrho, \qquad \mathbf{curl}\,\mathbf{B} = \mu_0\mathbf{J} + \frac{1}{c^2}\frac{\partial\mathbf{E}}{\partial t},$$
$$\mathbf{curl}\,\mathbf{E} = -\frac{\partial\mathbf{B}}{\partial t}, \qquad \operatorname{div}\mathbf{B} = 0. \tag{23.15}$$

Here, ϱ is the electric charge density, and $\mathbf{J}$ is the electric current density vector. All the functions depend on space and time.[8] Moreover, ε_0 (resp. μ_0) is the electric (resp. magnetic) field constant of a vacuum. This is related to the velocity of light

[7] For the proof, we refer to Jost (2008), Chap. 2, quoted on page 1061. The existence proof is reduced to a variational principle of minimal energy. One uses the same functional-analytic Hilbert space method as applied to the classic Dirichlet problem in Sect. 10.4 of Vol. II (quadratic variational problem on a Sobolev space).

[8] We use the SI system of physical units (see the Appendix of Vol. I).

c in a vacuum by $\mu_0\varepsilon_0 = 1/c^2$. In the special stationary case, all the functions do not depend on time t. This yields the stationary Maxwell equations:

$$\begin{aligned} \varepsilon_0 \operatorname{div} \mathbf{E} &= \varrho, \qquad & \mathbf{curl}\, \mathbf{B} &= \mu_0 \mathbf{J}, \\ \mathbf{curl}\, \mathbf{E} &= 0, \qquad & \operatorname{div} \mathbf{B} &= 0 \qquad \text{on } \mathcal{O} \end{aligned} \tag{23.16}$$

where $\mathcal{O}$ is an open subset of the Euclidean manifold $\mathbb{E}^3$. As we will see below, the equations $\mathbf{curl}\,\mathbf{E} = 0$ and $\operatorname{div}\mathbf{B} = 0$ are constraints for the electrodynamic field which allow us to introduce potentials.

Let us fix a right-handed Cartesian (x, y, z)-coordinate system with the origin O and the right-handed orthonormal basis $\mathbf{i}, \mathbf{j}, \mathbf{k}$ (Fig. 23.1 on page 1043). We will use the position vector $\mathbf{x} = \overrightarrow{OP}$ pointing from the origin O to the point P. It will be convenient to write $\mathbf{E}(\mathbf{x})$ instead of $\mathbf{E}(P)$.

Stationary electric and magnetic potential. The stationary Maxwell equations (23.16) motivate the study of the following four equations:

$$\mathbf{curl}\,\mathbf{E} = 0 \qquad \text{on } \mathcal{O}, \tag{23.17}$$

$$\operatorname{div}\mathbf{B} = 0 \qquad \text{on } \mathcal{O}, \tag{23.18}$$

$$\mathbf{E} = -\,\mathbf{grad}\, U \qquad \text{on } \mathcal{O}, \tag{23.19}$$

$$\mathbf{B} = \mathbf{curl}\,\mathbf{A} \qquad \text{on } \mathcal{O}. \tag{23.20}$$

We want to show that the solutions of these equations depend critically on the topology (i.e., the Betti numbers) of the open subset $\mathcal{O}$ of the Euclidean manifold $\mathbb{E}^3$. In what follows, all the functions are assumed to be smooth on $\mathcal{O}$. Because of the identities $\mathbf{curl}\,\mathbf{grad}\, U = 0$ and $\operatorname{div}\mathbf{curl}\,\mathbf{v} = 0$ on $\mathcal{O}$, there exist trivial solutions and trivial constraints. Explicitly,

- the field $\mathbf{grad}\, U$ (resp. $\mathbf{curl}\,\mathbf{A}$) is a trivial solution of the homogeneous equation (23.17) (resp. (23.18)).
- If the inhomogeneous equation (23.19) (resp. (23.20)) has a solution, then we have $\mathbf{curl}\,\mathbf{E} = 0$ (resp. $\operatorname{div}\mathbf{B} = 0$) on $\mathcal{O}$.

Nontrivial topology generates additional nontrivial solutions of the homogeneous equation and additional global constraints which are an immediate consequence of the classical integral theorems due to Gauss and Stokes. Note that

The electric (resp. magnetic) field sees the Betti number β_1 (resp. β_2).

In other words, the electric (resp. magnetic) field sees the nontrivial 1-cycles (resp. 2-cycles) of the set $\mathcal{O}$. The reason for that is the crucial fact that the electric field corresponds to a differential 1-form, whereas the magnetic field corresponds to a differential 2-form. Summarizing, we obtain the following:

(i) $\mathcal{O} = \mathbb{E}^3$ ($\beta_0 = 1, \beta_1 = \beta_2 = 0$): The general solution of the homogeneous equation (23.17) is given by

$$\mathbf{E} = -\,\mathbf{grad}\, U$$

where $U : \mathbb{E}^3 \to \mathbb{R}$ is an arbitrary smooth function. The general solution of the homogeneous equation (23.18) reads as

$$\mathbf{B} = \mathbf{curl}\,\mathbf{A}$$

where $\mathbf{A} : \mathbb{E}^3 \to E_3$ is an arbitrary smooth vector field.

For given smooth vector field $\mathbf{E}$, the inhomogeneous equation (23.19) has a solution iff $\mathbf{curl}\,\mathbf{E} = 0$ on $\mathbb{E}^3$. The general solution reads as

$$U(\mathbf{x}) = U_0 - \int_{\mathbf{0}}^{\mathbf{x}} \mathbf{E}\, d\mathbf{x} \tag{23.21}$$

where U_0 is an arbitrary real number which describes the gauge freedom. Obviously, $U(0) = U_0$. Note that the integral is independent of the path of integration. The function U is called an electrostatic potential. In physics, the potential difference

$$V = U(\mathbf{x}) - U(\mathbf{x}_0)$$

is called the voltage between the points P and P_0 with respect to the orientation $\overrightarrow{P_0P}$ from the point P_0 to the point P. In contrast to the electrostatic potential, the voltage is gauge invariant, and it possesses a physical meaning (see Sect. 23.5.2).

For given smooth vector field $\mathbf{B}$, the inhomogeneous equation (23.20) has a solution iff $\operatorname{div}\mathbf{B} = 0$ on $\mathbb{E}^3$. The general solution reads as

$$\mathbf{A}(\mathbf{x}) = \int_0^1 (\mathbf{B}(t\mathbf{x}) \times \mathbf{x})\, dt + \mathbf{grad}\, \chi(\mathbf{x})$$

where χ is an arbitrary smooth function $\chi : \mathbb{E}^3 \to \mathbb{R}$ which describes the gauge freedom.

(ii) $\mathcal{O} = \mathbb{E}^3 \setminus \{0\}$ ($\beta_0 = \beta_2 = 1, \beta_1 = 0$): The general solution of the homogeneous equation (23.17) is given by

$$\mathbf{E} = -\,\mathbf{grad}\, V$$

where $V : \mathbb{E}^3 \setminus \{0\} \to \mathbb{R}$ is an arbitrary smooth function. The general solution of the homogeneous equation (23.18) reads as

$$\mathbf{B} = \alpha \mathbf{B}_{\text{monopole}} + \mathbf{curl}\, \mathbf{A}, \qquad \alpha \in \mathbb{R}$$

where $\mathbf{A} : \mathcal{O} \to E_3$ is an arbitrary smooth vector field. Furthermore, $\mathbf{B}_{\text{monopole}}$ is the magnetic field of a magnetic monopole of magnetic charge equal to one (see (19.34) on page 951).

For given smooth vector field $\mathbf{E}$, the inhomogeneous equation (23.19) has a solution iff $\mathbf{curl}\,\mathbf{E} = 0$ on $\mathcal{O}$. The general solution reads as[9]

$$V(\mathbf{x}) = V_{\text{special}}(\mathbf{x}) + V_0$$

where V_0 is an arbitrary real number which describes the gauge freedom. For given smooth vector field $\mathbf{B}$, the inhomogeneous equation (23.20) has a solution iff we have $\operatorname{div}\mathbf{B} = 0$ on $\mathcal{O}$. The general solution reads as

$$\mathbf{A} = \mathbf{A}_{\text{special}} + \mathbf{grad}\, \chi$$

where $\chi : \mathbb{E}^3 \setminus \{0\} \to \mathbb{R}$ is an arbitrary smooth function which describes the gauge freedom.

(iii) $\mathcal{O} = \mathbb{E}^3 \setminus L$ (L is a straight-line, for example, a wire; $\beta_0 = \beta_1 = 1, \beta_2 = 0$). The general solution of the homogeneous equation (23.17) is given by

$$\mathbf{E} = \alpha \mathbf{E}_1 - \mathbf{grad}\, V, \qquad \alpha \in \mathbb{R} \tag{23.22}$$

where $V : \mathbb{E}^3 \setminus L \to \mathbb{R}$ is an arbitrary smooth function. Moreover,

$$\mathbf{E}_1(x, y, z) := \frac{x\mathbf{j} - y\mathbf{i}}{x^2 + y^2}, \qquad (x, y, z) \in \mathbb{R}^3.$$

[9] Explicitly, $V_{\text{special}}(\mathbf{x}) = -\int_{\mathbf{x}_0}^{\mathbf{x}} \mathbf{E} d\mathbf{x}$ where we choose $\mathbf{x}_0 \neq 0$. Note that the integral is independent of the path of integration.

Here, $\mathbf{curl}\,\mathbf{E}_1 = 0$ on $\mathcal{O}$, and $\int_{\mathbb{S}^1} \mathbf{E}_1(\mathbf{x})d\mathbf{x} = 2\pi$. The general solution of the homogeneous equation (23.18) reads as

$$\mathbf{B} = \mathbf{curl}\,\mathbf{A}$$

where $\mathbf{A} : \mathcal{O} \to E_3$ is an arbitrary smooth vector field.

The inhomogeneous equation (23.19) has a solution iff

$$\mathbf{curl}\,\mathbf{E} = 0 \text{ on } \mathcal{O} \quad \text{and} \quad \int_{\mathbb{S}^1} \mathbf{E}d\mathbf{x} = 0$$

where $\mathbb{S}^1$ is a unit circle which surrounds the wire L. The general solution reads as

$$U(\mathbf{x}) = U_{\text{special}}(\mathbf{x}) + U_0$$

where U_0 is an arbitrary real number which describes the gauge freedom. For given smooth vector field $\mathbf{B}$, the inhomogeneous equation (23.20) has a solution iff we have $\operatorname{div}\mathbf{B} = 0$ on $\mathcal{O}$. The general solution reads as

$$\mathbf{A} = \mathbf{A}_{\text{special}} + \alpha\mathbf{E}_1 - \mathbf{grad}\,V, \qquad \alpha \in \mathbb{R}$$

where we use (23.22).

The crucial Betti numbers follow from using a deformation argument. In fact, the set $\mathcal{O}$ from (i) is contractible to a point. The set $\mathcal{O}$ from (ii) is contractible to the sphere $\mathbb{S}^2$, and hence the Betti numbers of $\mathcal{O}$ are the same as for the sphere $\mathbb{S}^2$. Finally, the set $\mathcal{O}$ from (iii) can be orthogonally projected onto a pointed plane $\mathbb{E}^2 \setminus \{0\}$ which is perpendicular to the straight-line L. Moreover, the pointed plane can be continuously contracted to the unit circle. Hence the Betti numbers of the set $\mathcal{O}$ coincide with the Betti numbers of the unit circle $\mathbb{S}^1$.

Sketch of the proof. Let us choose a right-handed Cartesian (x, y, z)-coordinate system with the right-handed orthonormal system $\mathbf{i}, \mathbf{j}, \mathbf{k}$. Set

$$\mathbf{E} = E^1\mathbf{i} + E^2\mathbf{j} + E^3\mathbf{k}, \quad \mathbf{B} = B^1\mathbf{i} + B^2\mathbf{j} + B^3\mathbf{k}, \quad \mathbf{A} = A^1\mathbf{i} + A^2\mathbf{j} + A^3\mathbf{k}.$$

The corresponding differential forms read as

$$\omega_{\mathbf{E}} := E^1dx + E^2dy + E^3dz, \quad \omega_{\mathbf{B}} = B^1dy \wedge dz + B^2dz \wedge dx + B^3dx \wedge dy,$$

and $\omega_{\mathbf{A}} := A^1dx + A^2dy + A^3dz$. The corresponding Cartan differential is given by:

- $dV = V_xdx + V_ydy + V_zdz = (\mathbf{grad}\,V)^1dx + (\mathbf{grad}\,V)^2dy + (\mathbf{grad}\,V)^3dz$,
- $d\omega_{\mathbf{E}} = dE^1 \wedge dx + dE^2 \wedge dy + dE^3 \wedge dz$
  $= (E^3_y - E^2_z)\, dy \wedge dz + (E^1_z - E^3_x)\, dz \wedge dx + (E^2_x - E^1_y)\, dx \wedge dy$
  $= (\mathbf{curl}\,\mathbf{E})^1\, dy \wedge dz + (\mathbf{curl}\,\mathbf{E})^2\, dz \wedge dx + (\mathbf{curl}\,\mathbf{E})^3\, dx \wedge dy$,
- $d\omega_{\mathbf{B}} = dB^1 \wedge dy \wedge dz + dB^2 \wedge dz \wedge dx + dB^3 \wedge dx \wedge dy$
  $= (B^1_x + B^2_y + B^3_z)\, dx \wedge dy \wedge dz = \operatorname{div}\mathbf{B} \cdot dx \wedge dy \wedge dz$.

Therefore, the Poincaré cohomology rule $d(d\omega) = 0$ comprehends the following special cases:

- $d(dV) = 0$ is equivalent to $\mathbf{curl}\,\mathbf{grad}\,V = 0$, and
- $d(d\omega_{\mathbf{E}}) = 0$ is equivalent to $\operatorname{div}\mathbf{curl}\,\mathbf{E} = 0$.
- $d(d\omega_{\mathbf{B}}) = 0$ is always satisfied.

The general Stokes integral theorem $\int_M d\omega = \int_{\partial M} \omega$ for differential forms comprehends the following two classical integral theorems due to Gauss–Ostrogradski and Stokes:

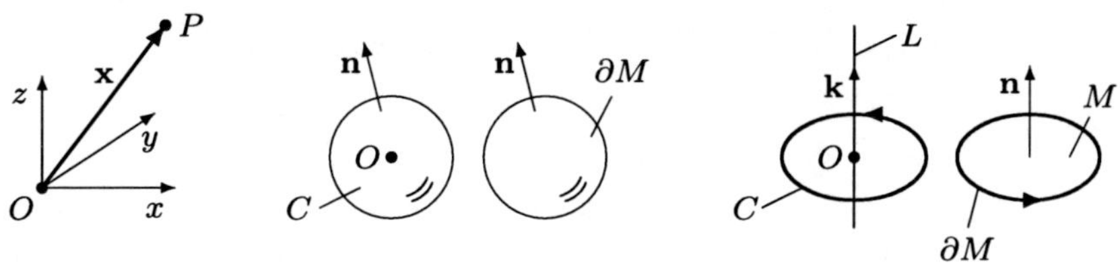

Fig. 23.1. Essential and trivial cycles in electromagnetism

- $\int_{\partial M} \mathbf{Bn} \cdot dS = \int_M \operatorname{div} \mathbf{B}\, dxdydz$ (Fig. 23.1(b)), and
- $\int_{\partial M} \mathbf{E}d\mathbf{x} = \int_M (\mathbf{curl}\,\mathbf{E})\mathbf{n} \cdot dS$ (Fig. 23.1(c)).

Here, $\mathbf{n}$ is the (outer) unit normal vector. The point is that the following hold:

- Fig. 23.1(b): If $\operatorname{div} \mathbf{E} = 0$ on $\mathbb{E}^3 \setminus \{0\}$, then

$$\int_{\partial M} \mathbf{En} = \int_M \operatorname{div} \mathbf{E}\, dxdydz = 0.$$

In terms of Poincaré's homology, let us say that the 2-cycle ∂M depicted in Fig 23.1(b) is trivial (i.e., the 2-cycle ∂M is the boundary of a 3-dimensional submanifold M of the manifold $\mathbb{E}^3 \setminus \{0\}$. Note that ∂M does not surround the origin). But if

$$\mathbf{E}_{\text{special}}(\mathbf{x}) := \frac{Q\mathbf{x}}{4\pi\varepsilon_0|\mathbf{x}|^3} \qquad \text{(Coulomb's law)},$$

then

$$\boxed{\varepsilon_0 \int_C \mathbf{E}_{\text{special}}\mathbf{n} \cdot dS = Q.} \tag{23.23}$$

In terms of Poincaré's homology, let us say that the closed surface $C = \mathbb{S}^2$ depicted in Fig. 23.1(b) is an essential 2-cycle (i.e., C is not the boundary of a 2-dimensional submanifold of the manifold $\mathbb{E}^3 \setminus \{0\}$). In terms of physics, the vector field $\mathbf{E}_{\text{special}}$ is the electric field generated by an electric charge Q at the origin (the famous Coulomb law). We have

$$\mathbf{curl}\,\mathbf{E}_{\text{special}} = 0 \qquad \text{and} \qquad \operatorname{div} \mathbf{E}_{\text{special}} = 0 \quad \text{on } \mathbb{E}^3 \setminus \{0\}.$$

- Fig. 23.1(c): If $\mathbf{curl}\,\mathbf{B} = 0$ on $\mathbb{E}^3 \setminus L$, then

$$\int_{\partial M} \mathbf{B}d\mathbf{x} = \int_M (\mathbf{curl}\,\mathbf{B})\mathbf{n} \cdot dS = 0.$$

We say that the 1-cycle ∂M depicted in Fig 23.1(c) is trivial (i.e., it is the boundary of a 2-dimensional submanifold M of the manifold $\mathbb{E}^3 \setminus L$). Note that ∂M lies outside the wire L. But if

$$\mathbf{B}_{\text{special}}(\mathbf{x}) = \frac{\mu_0 \mathsf{J}}{2\pi|\mathbf{x}|} \cdot \frac{\mathbf{k} \times \mathbf{x}}{|\mathbf{k} \times \mathbf{x}|} \qquad \text{(Ampère's law)}, \tag{23.24}$$

then

$$\boxed{\int_C \mathbf{B}_{\text{special}}(\mathbf{x})\, d\mathbf{x} = \mu_0 \mathsf{J}.} \tag{23.25}$$

In terms of physics, the vector field $\mathbf{B}_{\text{special}}$ is the magnetic field generated by an electric current which flows in the wire L with the current strength J along the z-axis (the famous Ampère law). We say that $C = \mathbb{S}^1$ is an essential 1-cycle (i.e., C is not the boundary of a 2-dimensional submanifold of $\mathbb{E}^3 \setminus L$). We have

$$\operatorname{div} \mathbf{B}_{\text{special}} = 0 \qquad \text{and} \qquad \mathbf{curl}\, \mathbf{B}_{\text{special}} = \mu_0 \mathsf{J}\, \mathbf{k} \quad \text{on } \mathbb{E}^3 \setminus L.$$

Note that

Essential cycles describe the sources of the electric and magnetic field.

In fact, the essential 2-cycle C from Fig. 23.1(b) describes the electric point charge Q via the Coulomb law (23.23) for the electrostatic field $\mathbf{E}_{\text{special}}$. Moreover, the essential 1-cycle C from Fig. 23.1(c) describes the strength J of the electric current via the Ampère law (23.25) for the magnetostatic field $\mathbf{B}_{\text{special}}$.

Ad (i). Let $\mathcal{O} = \mathbb{E}^3$. The claims are consequences of the Poincaré theorem (i) on page 1032. But we want to use a completely elementary approach based on explicit formulas.

(I) If $\mathbf{E} = -\,\mathbf{grad}\, V$, then $\mathbf{curl}\, \mathbf{E} = 0$. Conversely, suppose that $\mathbf{curl}\, \mathbf{E} = 0$ on $\mathcal{O}$. Alternatively to (23.21), define the function

$$V(\mathbf{x}) := -\int_0^1 \mathbf{E}(t\mathbf{x})\mathbf{x} \cdot dt.$$

Then, an elementary computation shows that[10]

$$\mathbf{grad}\, V(\mathbf{x}) = -\int_0^1 \mathbf{grad}(\mathbf{E}(t\mathbf{x})\mathbf{x})\, dt = -\int_0^1 \frac{d}{dt}(t\mathbf{E}(t\mathbf{x}))\, dt = -\mathbf{E}(\mathbf{x}).$$

In addition, if $\mathbf{E} = -\,\mathbf{grad}\, V$ and $\mathbf{E} = -\,\mathbf{grad}\, W$, then $\mathbf{grad}(V - W) = 0$, and hence $V - W = \text{const}$.

(II) If $\mathbf{B} = \mathbf{curl}\, \mathbf{A}$ on $\mathcal{O}$, then $\operatorname{div} \mathbf{B} = \operatorname{div} \mathbf{curl}\, \mathbf{A} = 0$. Conversely, suppose that $\operatorname{div} \mathbf{B} = 0$ on $\mathcal{O}$. Define

$$\mathbf{A}(\mathbf{x}) := \int_0^1 (\mathbf{B}(t\mathbf{x}) \times \mathbf{x})\, dt.$$

Then an elementary computation shows that[11]

$$\mathbf{curl}\, \mathbf{A}(\mathbf{x}) = \int_0^1 \mathbf{curl}(\mathbf{B}(t\mathbf{x}) \times \mathbf{x}) dt = \int_0^1 \frac{d}{dt}(t^2 \mathbf{B}(t\mathbf{x}))\, dt = \mathbf{B}(\mathbf{x}).$$

In what follows we will use the following basic principle:

---

[10] To get insight, use the identity

$$\mathbf{grad}(\mathbf{v}\mathbf{w}) = (\mathbf{v}\,\mathbf{grad})\mathbf{w} + (\mathbf{w}\,\mathbf{grad})\mathbf{v} + \mathbf{v} \times \mathbf{curl}\,\mathbf{w} + \mathbf{w} \times \mathbf{curl}\,\mathbf{v}$$

from Hamilton's nabla calculus (see Sect. 9.1.5 on page 563), and note that both $\mathbf{curl}\, \mathbf{E} = 0$ and $\mathbf{curl}\, \mathbf{x} = 0$.

[11] Use $\mathbf{curl}(\mathbf{v} \times \mathbf{w}) = (\mathbf{w}\,\mathbf{grad})\mathbf{v} - (\mathbf{v}\,\mathbf{grad})\mathbf{w} + \mathbf{v} \operatorname{div} \mathbf{w} - \mathbf{w} \operatorname{div} \mathbf{v}$.

The necessary conditions for the existence of the potentials of stationary electric and magnetic fields are also sufficient conditions, by de Rham cohomology.

Ad (ii). Let $\mathcal{O} = \mathbb{E}^3 \setminus \{0\}$ (Fig. 23.1(b)): The equation $\mathbf{curl}\,\mathbf{E} = 0$ on $\mathcal{O}$ has the trivial solution $\mathbf{E} = -\,\mathbf{grad}\,V$. Since $\mathbf{E}$ corresponds to a 1-form and $\beta_1(\mathcal{O}) = 0$, the space of solutions has the dimension zero modulo cohomology.

Ad (iii). Let $\mathcal{O} = \mathbb{E}^3 \setminus L$ (Fig. 23.1(c)). Since $\beta_1 = 1$, the dimension of the solution space of equation (23.17) is equal to one modulo cohomology. Moreover, since $\beta_2 = 0$, the dimension of the solution space of (23.18) is equal to zero modulo cohomology.

23.5 The Analysis of the Electromagnetic Field

23.5.1 The Main Theorem of Electrostatics, the Dirichlet Principle, and Generalized Functions

> Electrostatics is the physical background for the modern mathematical approach to solving the Laplace equation and the Poisson equation, and for the theory of compact Riemann surfaces. This strongly influenced the development of the modern theory of elliptic partial differential equations which describe stationary processes in nature.
> The theory of generalized functions, initiated by Dirac in about 1930 and founded as a mathematical theory by Laurent Schwartz in about 1945, allows us to describe the electric fields generated by smooth and highly non-smooth electric charge densities ϱ in an elegant uniform way.
>
> Folklore

The Maxwell equations of electrostatics read as

$$\boxed{\varepsilon_0 \operatorname{div} \mathbf{E} = \varrho \qquad \text{and} \qquad \mathbf{curl}\,\mathbf{E} = 0 \qquad \text{on } \mathbb{E}^3.} \tag{23.26}$$

The Coulomb law and the volume potential. To begin with, consider the classic key formula

$$U_{\text{special}}(\mathbf{x}) := \int_{\mathbb{E}^3} \frac{\varrho(\mathbf{x}_0)}{4\pi\varepsilon_0|\mathbf{x} - \mathbf{x}_0|}\, d^3x_0$$

together with the Poisson equation

$$\boxed{\varepsilon_0 \Delta U = \varrho \qquad \text{on } \mathbb{E}^3,} \tag{23.27}$$

and the formula

$$\mathbf{E} = -\,\mathbf{grad}\, U$$

for the electric field $\mathbf{E}$. Consider the position vector $\mathbf{x}_j = \overrightarrow{OP_j}$. Intuitively, the volume potential U_{special} is the volume limit $\Delta x \Delta y \Delta z \to 0$ of the finite sum

$$\sum_j \frac{Q_j}{4\pi\varepsilon_0|\mathbf{x} - \mathbf{x}_j|}$$

with the electric charge $Q_j := \varrho(\mathbf{x}_j)\Delta x \Delta y \Delta z$. This is the superposition of electrostatic Coulomb potentials with the electric charge Q_j at the point P_j. The mathematical problem is to give all the expressions used by physicists a rigorous mathematical foundation.

Theorem 23.13 *Suppose that the electric charge density $\varrho : \mathbb{E}^3 \to \mathbb{R}$ is a smooth function, and it vanishes outside a sufficiently large ball. Then the volume potential $U_{\rm special}$ is the unique smooth solution of the Poisson equation* (23.27) *with the following asymptotic behavior at infinity:*

$$U(\mathbf{x}) = O\left(\frac{1}{|\mathbf{x}|}\right), \qquad |\mathbf{x}| \to \infty.$$

The proof of this classic result can be found in the standard textbook by R. Courant and D. Hilbert, Methods of Mathematical Physics, Wiley, 1989. In addition, using the Taylor expansion, we get the asymptotic formula

$$U_{\rm special}(\mathbf{x}) = \frac{Q_0}{4\pi\varepsilon_0|\mathbf{x}|} + \frac{\mathbf{p}\mathbf{x}}{4\pi\varepsilon_0|\mathbf{x}|^3} + o\left(\frac{1}{|\mathbf{x}|^2}\right), \qquad |\mathbf{x}| \to \infty$$

with the effective charge Q_0 and the effective dipole moment vector $\mathbf{p}$. Explicitly,

$$Q_0 := \int_{\mathbb{E}^3} \varrho(\mathbf{x}_0)\, d^3x_0 \qquad \text{and} \qquad \mathbf{p} := \int_{\mathbb{E}^3} \varrho(\mathbf{x}_0)\mathbf{x}_0\, d^3x_0.$$

This tells us that if the charge density ϱ vanishes outside a ball of radius R centered at the origin, and if we are far away from the ball (i.e., $|\mathbf{x}|/R \gg 1$), then the electric potential looks like the superposition of the electric fields of a monopole and a dipole. Multiplying both sides of the Poisson equation (23.27) with the function φ and using integration by parts, we get the integral relation

$$\varepsilon_0 \int_{\mathbb{E}^3} U(\mathbf{x}_0)\Delta\varphi(\mathbf{x}_0)\, dx_0^3 = \int_{\mathbb{E}^3} \varrho(\mathbf{x}_0)\varphi(\mathbf{x}_0)\, dx_0^3 \tag{23.28}$$

for all test functions $\varphi \in \mathcal{S}(\mathbb{R}^3)$.[12] The equation is called the generalized (or averaged) form of the Poisson equation (23.27). As we will discuss below, it is necessary to pass to a more general variant of the Poisson equation. Let the electric charge ϱ be a tempered distribution, that is, $\varrho \in \mathcal{S}'(\mathbb{R}^3)$. The tempered distribution $U \in \mathcal{S}'(\mathbb{R}^3)$ is a solution of the Poisson equation (23.27) iff this equation holds on the space $\mathcal{S}'(\mathbb{R}^3)$. In particular, the derivatives are to be understood in the sense of tempered distributions. Explicitly, this means that

$$\varepsilon_0 U(\Delta\varphi) = \varrho(\varphi) \qquad \text{for all} \quad \varphi \in \mathcal{S}(\mathbb{R}^3).$$

The classic trouble with continuous and discontinuous charge densities ϱ. Suppose that the electric charge density ϱ is continuous and it vanishes outside a sufficiently large ball. Surprisingly enough, the electrostatic potential $U_{\rm special}$ has first-order partial derivatives, but second-order partial derivatives do not always exist. This means that the Maxwell equations (23.26) are not always satisfied, in the classical sense. However, using tempered distributions, we get an elegant general theory.

[12] The definition of both the space $\mathcal{S}(\mathbb{R}^3)$ (rapidly decreasing smooth functions) and the dual space $\mathcal{S}'(\mathbb{R}^3)$ (tempered distributions) can be found in Sect. 1.3.3 of Vol. I. Physicists and mathematicians use frequently the space $\mathcal{S}'(\mathbb{R}^3)$ instead of $\mathcal{D}'(\mathbb{R}^3)$, since there exists a perfect theory of the Fourier transform for tempered distributions.

Theorem 23.14 *Suppose that the electric charge density $\varrho : \mathbb{E}^3 \to \mathbb{R}$ is integrable over each ball of $\mathbb{E}^3$. Then the general solution U of the Poisson equation (23.27) on the space $\mathcal{S}'(\mathbb{E}^3)$ of tempered distributions reads as*

$$U = U_{\text{special}} + P$$

where P is a polynomial with $\Delta P = 0$.

The proof can be found in H. Triebel, Higher Analysis, Sect. III.14, Barth, Leipzig, 1989.

The formal Dirac delta function. Consider a point-like particle located at the point P_1 which has the electric charge Q_1. In the classic sense, this particle has no classical charge density. In about 1930, Dirac introduced the generalized charge density

$$\varrho(\mathbf{x}) := Q_1\delta(\mathbf{x} - \mathbf{x}_1).$$

Mnemonically, $\delta(\mathbf{x} - \mathbf{x}_1) = 0$ if $\mathbf{x} \neq \mathbf{x}_1$. Moreover, motivated by the physical meaning of the charge density, physicists use the following integral formula

$$\int_{\mathbb{E}^3} \frac{Q_1\delta(\mathbf{x}_0 - \mathbf{x}_1)}{4\pi\varepsilon_0|\mathbf{x} - \mathbf{x}_0|}\, d^3x_0 = \frac{Q_1}{4\pi\varepsilon_0|\mathbf{x} - \mathbf{x}_1|}$$

in order to get the Coulomb potential. Motivated by this procedure, physicists write

$$\varepsilon_0\Delta\left(\frac{Q_1}{4\pi\varepsilon_0|\mathbf{x} - \mathbf{x}_1|}\right) = Q_1\delta(\mathbf{x} - \mathbf{x}_1)$$

for all the position vectors $\mathbf{x}$ whose initial point is located at the origin O.

The rigorous Dirac delta distribution. Rigorously, for all test functions $\varphi \in \mathcal{S}(\mathbb{R}^3)$, we set

$$\delta_{P_1}(\varphi) := \varphi(\mathbf{x}_1) \qquad \text{for all} \quad \varphi \in \mathcal{S}(\mathbb{R}^3),$$

and

$$U(\varphi) := \int_{\mathbb{E}^3} \frac{Q_1\varphi(\mathbf{x})}{4\pi\varepsilon_0|\mathbf{x} - \mathbf{x}_1|}\, d^3x.$$

Then, U and δ_{P_1} are tempered distributions (i.e., they are elements of $\mathcal{S}'(\mathbb{R}^3)$) which satisfy the Poisson equation

$$\varepsilon_0\Delta U = Q_1\delta_{P_1}.$$

Explicitly, this means that

$$\varepsilon_0 U(\Delta\varphi) = Q_1\varphi(\mathbf{x}_1) \qquad \text{for all} \quad \varphi \in \mathcal{S}(\mathbb{R}^3).$$

For the proof, we refer to Prop. 10.22 of Vol. 1.

In what follows, we restrict ourselves to the smooth situation. The proofs of the following two theorems are highly nontrivial. In a first step, one proves the existence of generalized functions in Sobolev spaces. Then one shows that the generalized solutions are smooth if the data are smooth. We refer to Zeidler (1986), Vol. IIA (quoted on page 1089) and to J. Jost, Partial Differential Equations, Springer, New York, 2002. The Dirichlet problem has a long and famous history which is discussed in Sect. 10.4 of Vol. I.

The Dirichlet principle of minimal electrostatic energy (the first boundary-value problem). Let $\mathcal{M}$ be a compact submanifold of the Euclidean (x, y)-plane with boundary $\partial\mathcal{M}$ (Fig. 23.2(a)). Consider the variational problem

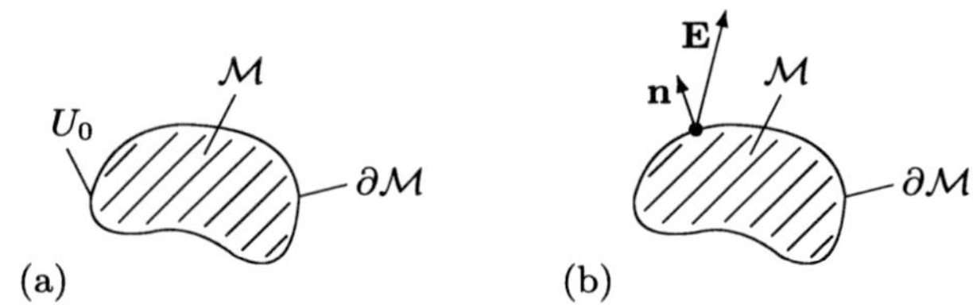

Fig. 23.2. Plane boundary-value problems

$$\boxed{\int_{\mathcal{M}} \left(\tfrac{1}{2}\varepsilon_0(\mathbf{grad}\, U)^2 - \varrho U\right) dxdy = \min!, \qquad U = U_0 \;\text{ on } \partial\mathcal{M}.} \tag{23.29}$$

The corresponding Euler–Lagrange equation reads as

$$\Delta U = \varrho \quad \text{on } \mathcal{M}, \qquad U = U_0 \quad \text{on } \partial\mathcal{M}. \tag{23.30}$$

We are given the smooth charge density $\varrho : \mathcal{M} \to \mathbb{R}$ and the smooth boundary-value function $U_0 : \partial\mathcal{M} \to \mathbb{R}$ of the electrostatic potential U. We are looking for a smooth electrostatic potential $U : \mathcal{M} \to \mathbb{R}$. In terms of physics, we are looking for an electrostatic field $\mathbf{E} = -\,\mathbf{grad}\, U$ of minimal energy.

Theorem 23.15 *The Dirichlet problem* (23.29) *has a unique solution. This is also the unique solution of* (23.30).

The second boundary-value problem. Replace (23.30) by the following boundary-value problem:

$$\boxed{\varepsilon_0 \,\mathrm{div}\, \mathbf{E} = \varrho \quad \text{on } \mathcal{M}, \qquad \mathbf{En} = E_0 \quad \text{on } \partial\mathcal{M}.} \tag{23.31}$$

We are given the smooth charge density $\varrho : \mathcal{M} \to \mathbb{R}$ and the smooth boundary function $E_0 : \partial\mathcal{M} \to \mathbb{R}$ (Fig. 23.2(b)).

Theorem 23.16 *The boundary-value problem* (23.31) *for the electrostatic field has a unique smooth solution* $\mathbf{E} : \mathcal{M} \to E_3$.

This theorem tells us that the electrostatic field $\mathbf{E}$ is uniquely determined by its normal component $\mathbf{En}$ at the boundary and by the charge density. In terms of the electrostatic potential, the problem (23.31) is equivalent to

$$\varepsilon_0 \Delta U = \varrho \quad \text{on } \mathcal{M}, \qquad \frac{\partial U}{\partial n} = -E_0 \quad \text{on } \partial\mathcal{M}.$$

23.5.2 The Coulomb Gauge and the Main Theorem of Magnetostatics

The vector potential. The Maxwell equations in magnetostatics read as

$$\boxed{\mathrm{div}\, \mathbf{B} = 0 \qquad \text{and} \qquad \mathbf{curl}\, \mathbf{B} = \mu_0 \mathbf{J}_0 \qquad \text{on } \mathbb{E}^3.} \tag{23.32}$$

We are given the smooth current density vector field $\mathbf{J}_0$, and we are looking for the smooth magnetic field $\mathbf{B}$. In addition, we postulate that

$$\operatorname{div} \mathbf{J}_0 = 0 \qquad \text{on} \quad \mathbb{E}^3.$$

In fact, if the system (23.32) has a solution $\mathbf{B}$, then it follows from $\operatorname{div} \mathbf{curl}\, \mathbf{B} = 0$ that $\operatorname{div} \mathbf{J}_0 = 0$. The general solution of (23.32) has the form

$$\mathbf{B} = \mathbf{curl}\, \mathbf{A}$$

where the vector potential $\mathbf{A}$ is an arbitrary smooth vector field on $\mathbb{E}^3$ (see Sect. 23.4). We add the so-called Coulomb gauge condition

$$\boxed{\operatorname{div} \mathbf{A} = 0 \qquad \text{on} \quad \mathbb{E}^3.} \tag{23.33}$$

If the function $\chi : \mathbb{E}^3 \to \mathbb{R}$ is smooth, then the transformation

$$\mathbf{A}_+ := \mathbf{A} + \mathbf{grad}\, \chi$$

is called a gauge transformation. If we add the condition $\operatorname{div} \mathbf{grad}\, \chi = 0$ on $\mathbb{E}^3$, then we have $\operatorname{div} \mathbf{A}_+ = 0$, that is, the gauge transformation respects the Coulomb gauge condition (23.33).

Theorem 23.17 *We are given the smooth current density vector field* $\mathbf{J}_0$ *which satisfies the condition* $\operatorname{div} \mathbf{J}_0 = 0$ *on* $\mathbb{E}^3$ *(conservation of electric charge). In addition, we assume that* $\mathbf{J}_0$ *vanishes outside a sufficiently large ball. Then the smooth function*

$$\mathbf{A}(\mathbf{x}) := \frac{\mu_0}{4\pi} \int_{\mathbb{E}^3} \frac{\mathbf{J}_0(\mathbf{x}_0)}{|\mathbf{x} - \mathbf{x}_0|}\, d^3x_0 \tag{23.34}$$

satisfies the gauge condition (23.33). *Moreover, the magnetic field* $\mathbf{B} = \mathbf{curl}\, \mathbf{A}$ *is a solution of the Maxwell equations* (23.32) *of magnetostatics. Explicitly,*

$$\mathbf{B}(\mathbf{x}) = \frac{\mu_0}{4\pi} \int_{\mathbb{E}^3} \frac{\mathbf{J}_0(\mathbf{x}_0) \times (\mathbf{x} - \mathbf{x}_0)}{|\mathbf{x} - \mathbf{x}_0|^3}\, d^3x_0. \tag{23.35}$$

The proof will be given in Problem 23.1 on page 1066. The vector potential $\mathbf{A}$ from (23.34) has the asymptotic form

$$\mathbf{A}(\mathbf{x}) = \frac{\mu_0(\mathbf{m} \times \mathbf{x})}{4\pi|\mathbf{x}|^3} + o\left(\frac{1}{|\mathbf{x}|^2}\right), \qquad |\mathbf{x}| \to \infty \tag{23.36}$$

with the magnetic dipole moment

$$\boxed{\mathbf{m} = \tfrac{1}{2} \int_{\mathbb{E}^3} (\mathbf{x}_0 \times \mathbf{J}_0(\mathbf{x}_0))\, d^3x_0.} \tag{23.37}$$

The corresponding magnetic field reads as

$$\boxed{\mathbf{B}(\mathbf{x}) = \mu_0 \frac{3(\mathbf{m}\mathbf{x})\mathbf{x} - \mathbf{x}^2 \cdot \mathbf{x}}{|\mathbf{x}|^5},} \tag{23.38}$$

up to terms of order $o\left(\frac{1}{|\mathbf{x}|^3}\right)$ as $|\mathbf{x}| \to \infty$. Recall that the formula (23.38) represents the magnetic field of a magnetic dipole with the magnetic moment $\mathbf{m}$.

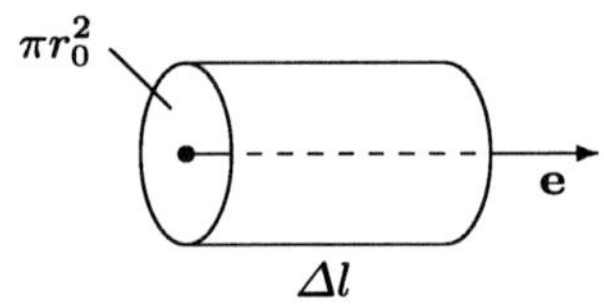

Fig. 23.3. Electric current element

Magnetic force. Suppose that we are given an electric current with the current density vector $\mathbf{J}$. Then the given magnetic field $\mathbf{B}$ exerts the total force $\mathbf{F}$ on the current. Explicitly,

$$\boxed{\mathbf{F} = \int_{\mathbb{E}^3} (\mathbf{J}(\mathbf{x}) \times \mathbf{B}(\mathbf{x}))\, d^3x.} \tag{23.39}$$

Small electric current elements and the Biot–Savart law. Consider the piece of a thin metallic wire as depicted in Fig. 23.3. Then the current density vector $\mathbf{J}$ is related to the current strength J by the formula

$$\mathbf{J}\pi r_0^2 \Delta l\ \mathbf{e} = \mathsf{J}\Delta l\ \mathbf{e}.$$

Here, $\mathbf{e}$ is a unit vector. The key formulas (23.38) and (23.39) can be regarded as the superposition of the physical effects caused by small current elements. Explicitly, we get the following local Biot–Savart law:

- The current element $\mu_0 \mathsf{J}_0 \Delta l_0\ \mathbf{e}_{P_0}$ generates the magnetic field

$$\Delta\mathbf{B}(\mathbf{x}) = \frac{\mu_0 \mathsf{J}_0 \Delta l_0\ \mathbf{e}_{P_0} \times (\mathbf{x} - \mathbf{x}_0)}{4\pi|\mathbf{x} - \mathbf{x}_0|^3}.$$

- The magnetic field $\mathbf{B}$ exerts the force

$$\Delta\mathbf{F}(\mathbf{x}) = \mathsf{J}\Delta l\ \mathbf{e}_P \times \Delta\mathbf{B}(\mathbf{x})$$

  on the current element $\mathsf{J}\Delta l\ \mathbf{e}_P$. Here, $\mathbf{e}_{P_0}$ and $\mathbf{e}_P$ are unit vectors.

Thin metallic wires. Let us use the Biot–Savart law together with the superposition principle. Then we get the following. In Fig. 23.4, the thin metallic wire C_0 with the constant electric current strength J_0 generates the magnetic field

$$\mathbf{B}(\mathbf{x}) = \int_{C_0} \frac{\mu_0 \mathsf{J}_0 d\mathbf{x}_0 \times (\mathbf{x} - \mathbf{x}_0)}{4\pi|\mathbf{x} - \mathbf{x}_0|^3} + o\left(\frac{1}{|\mathbf{x}|^3}\right), \qquad |\mathbf{x}| \to \infty.$$

Let C be a second thin metallic wire with the constant electric current strength J. Then the wire C_0 exerts the force

$$\mathbf{F} = \int_C \mathsf{J} d\mathbf{x} \times \mathbf{B}(\mathbf{x})$$

on the wire C.

The magnetic field generated by a circular electric current. Consider a circle of radius R about the origin in the Cartesian (x, y)-plane. Suppose that an electric current of current strength J_0 flows counter-clockwise along the circle (see Fig. 19.9 on page 948). This current generates the magnetic field

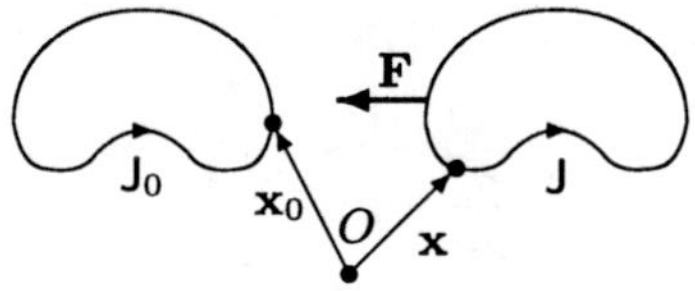

Fig. 23.4. Attracting Ampère force **F**

$$\mathbf{B}(\mathbf{x}) = \frac{\mu_0 \mathsf{J}_0}{4\pi} \int_{\mathbb{S}^1_R} \frac{d\mathbf{x}_0 \times (\mathbf{x} - \mathbf{x}_0)}{|\mathbf{x}|^3}. \tag{23.40}$$

Far away from the electric current (i.e., $|\mathbf{x}|/R \gg 1$), the magnetic field **B** looks like the magnetic field of a magnetic dipole (23.39) with the magnetic moment

$$\mathbf{m} = \mathsf{J}_0 \cdot \pi R^2. \tag{23.41}$$

Rotating Charge

Consider a point-like particle of the electric charge Q. Suppose that the particle moves with constant angular velocity on a circle of radius R about the origin O. This rotating charge generates a magnetic field **B**. Explicitly,

$$\mathbf{B}(\mathbf{x}) = \mu_0 \frac{3(\mathbf{m}\mathbf{x})\mathbf{x} - \mathbf{x}^2 \cdot \mathbf{x}}{4\pi|\mathbf{x}|^5} + o\left(\frac{1}{|\mathbf{x}|^3}\right), \qquad |\mathbf{x}| \to \infty \tag{23.42}$$

with the magnetic moment

$$\boxed{\mathbf{m} = \frac{Q}{2m_{\text{particle}}} \mathbf{L}.} \tag{23.43}$$

Here, **L** is the angular momentum vector of the rotating particle, and m_{particle} is the relativistic mass of the rotating particle. This means that, far away from the particle trajectory (i.e., $|\mathbf{x}|/R \gg 1$), the magnetic field of the rotating particle is the field of a magnetic dipole with magnetic moment (23.43).

Let us motivate this. Choose a right-handed Cartesian (x, y, z)-coordinate system with the right-handed orthonormal basis $\mathbf{i}, \mathbf{j}, \mathbf{k}$. The particle moves counterclockwise in the (x, y)-plane. The trajectory of the rotating particle reads as

$$\mathbf{x}(t) = R(\cos\omega t\, \mathbf{i} + \sin\omega t\, \mathbf{j}).$$

This yields the angular momentum vector

$$\mathbf{L} = \mathbf{x}(t) \times m_{\text{particle}}\, \dot{\mathbf{x}}(t) = m_{\text{particle}}\, \omega R^2\, \mathbf{k}.$$

The particle needs the time $T = 2\pi/\omega$ for surrounding the circle once. The basic idea, used by physicists, is to approximate the rotating particle (with electric charge Q) by an electric current of the strength

$$\mathsf{J} = \frac{Q}{T} = \frac{Q\omega}{2\pi}.$$

By (23.41), we get

$$\mathbf{m} = \mathsf{J}\pi R^2\, \mathbf{k} = \frac{Q\omega R^2}{2}\, \mathbf{k} = \frac{Q}{2m_{\text{particle}}}\, \mathbf{L}.$$

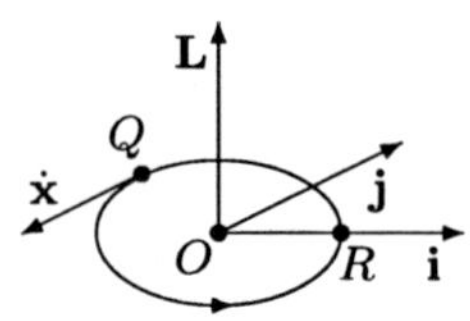

Fig. 23.5. Rotating electric charge

23.5.3 The Main Theorem of Electrodynamics

Choose a strictly positively oriented inertial system. We are given the following quantities:

(H1) the smooth charge density function $\varrho : \mathbb{E}^3 \times \mathbb{R} \to \mathbb{R}$, and
(H2) the smooth electric current density vector field $\mathbf{J} : \mathbb{E}^3 \times \mathbb{R} \to E_3$ such that

$$\varrho_t(\mathbf{x}, t) + \operatorname{div} \mathbf{J}(\mathbf{x}, t) = 0 \qquad \text{on } \mathbb{E}^3 \times \mathbb{R}; \tag{23.44}$$

(H3) there exists a ball $\mathcal{B}_0$ such that ϱ and $\mathbf{J}$ vanish outside the ball $\mathcal{B}_0$ for all times $t \in \mathbb{R}$.

Condition (23.44) describes the conservation of the electric charge on the Euclidean manifold $\mathbb{E}^3$ for all times $t \in \mathbb{R}$.

Special electromagnetic field. The Maxwell equations in a vacuum

$$\frac{\partial \mathbf{B}}{\partial t} = -\,\mathbf{curl}\,\mathbf{E}, \qquad \frac{\partial \mathbf{E}}{\partial t} = c^2\,\mathbf{curl}\,\mathbf{B} - \mu_0 c^2 \mathbf{J}, \tag{23.45}$$

$$\varepsilon_0 \operatorname{div} \mathbf{E} = \varrho, \qquad \operatorname{div} \mathbf{B} = 0 \tag{23.46}$$

possess the special smooth solution

$$\mathbf{E}_1 := -\,\mathbf{grad}\,U - \frac{\partial \mathbf{A}}{\partial t}, \qquad \mathbf{B}_1 := \mathbf{curl}\,\mathbf{A} \qquad \text{on } \mathbb{E}^3 \times \mathbb{R}$$

with the 4-potential

$$U(\mathbf{x}, t) := \frac{1}{4\pi\varepsilon_0} \int_{\mathbb{E}^3} \frac{\varrho(\mathbf{y}, t - |\mathbf{y} - \mathbf{x}|/c)}{|\mathbf{y} - \mathbf{x}|}\, d^3y,$$

$$\mathbf{A}(\mathbf{x}, t) := \frac{\mu_0}{4\pi} \int_{\mathbb{E}^3} \frac{\mathbf{J}(\mathbf{y}, t - |\mathbf{y} - \mathbf{x}|/c)}{|\mathbf{y} - \mathbf{x}|}\, d^3y.$$

This 4-potential satisfies the Lorenz gauge condition $U_t + c^2 \operatorname{div} \mathbf{A} = 0$ on $\mathbb{E}^3 \times \mathbb{R}$.

The initial-value problem for the electromagnetic field. We are looking for a smooth solution $\mathbf{E}, \mathbf{B} : \mathbb{E}^3 \times]0, \infty[\to E_3$ of the Maxwell equations (23.45), (23.46) which satisfies the initial condition

$$\lim_{t \to +0} \mathbf{E}(\mathbf{x}, t) = \mathbf{E}_0(\mathbf{x}), \qquad \lim_{t \to +0} \mathbf{B}(\mathbf{x}, t) = \mathbf{B}_0(\mathbf{x}) \tag{23.47}$$

on the Euclidean manifold $\mathbb{E}^3$ at the initial time $t = 0$. We assume (H1)–(H3). In addition, we are given the smooth electromagnetic field $\mathbf{E}_0, \mathbf{B}_0 : \mathbb{E}^3 \to E_3$ with the constraints

$$\varepsilon_0 \operatorname{div} \mathbf{E}_0(\mathbf{x}) = \varrho(\mathbf{x}), \qquad \operatorname{div} \mathbf{B}_0(\mathbf{x}) = 0 \quad \text{on } \mathbb{E}^3.$$

We assume that $\mathbf{E}_0$ and $\mathbf{B}_0$ vanish outside the sufficiently large ball $\mathcal{B}_0$.

Theorem 23.18 *The initial-value problem for the Maxwell equations in a vacuum possesses a unique smooth solution on* $\mathbb{E}^3 \times]0, \infty[$ *such that the limits* (23.47) *exist. For all times* $t > 0$ *and for all points in* $\mathbb{E}^3$*, the solution is given by the following integral formulas:*

$$\mathbf{E}(\mathbf{x},t) = \mathbf{E}_1(\mathbf{x},t) + \frac{1}{4\pi t}\int_{|\mathbf{y}-\mathbf{x}|=ct} \mathbf{curl}(\mathbf{B}_0(\mathbf{y}) - \mathbf{B}_1(\mathbf{y},0))\, dS$$
$$+\frac{\partial}{\partial t}\left(\frac{1}{4\pi c^2 t}\int_{|\mathbf{y}-\mathbf{x}|=ct} (\mathbf{E}_0(\mathbf{y}) - \mathbf{E}_1(\mathbf{y},0))\, dS\right)$$

and

$$\mathbf{B}(\mathbf{x},t) = \mathbf{B}_1(\mathbf{x},t) - \frac{1}{4\pi c^2 t}\int_{|\mathbf{y}-\mathbf{x}|=ct} \mathbf{curl}\big(\mathbf{E}_0(\mathbf{y}) - \mathbf{E}_1(\mathbf{y},0)\big)\, dS$$
$$+\frac{\partial}{\partial t}\left(\frac{1}{4\pi c^2 t}\int_{|\mathbf{y}-\mathbf{x}|=ct} (\mathbf{B}_0(\mathbf{y}) - \mathbf{B}_1(\mathbf{y},0))\, dS\right).$$

Observe that the integrals only depend on the values of $\mathbf{E}_0, \mathbf{E}_1, \mathbf{B}_0, \mathbf{B}_1$ on the sphere of radius ct about the center P (which corresponds to $\mathbf{x}$). This reflects the fact that the electromagnetic interaction propagates with the speed c of light in a vacuum.

Furthermore, note the following peculiarity. The Maxwell equations consist of the dynamical equations (23.45) and the constraints (23.46). We assume that the constraints are satisfied at the initial time $t = 0$. Then the constraints are valid for all times $t \geq 0$. The proof can be found in H. Triebel, Higher Analysis, Barth, Leipzig, 1989 (translated from German into English).

23.6 Important Tools

23.6.1 The Exact Mayer–Vietoris Sequence and the Computation of the de Rham Cohomology Groups

> There exist relations between the de Rham cohomology groups of the open subsets
>
> $$U, \;V, \;U \cup V, \;U \cap V$$
>
> of a manifold. These relations can be described in terms of an exact sequence. This allows us to compute the cohomology groups of a manifold by using a covering by open contractible sets.
>
> Folklore

Let $\mathcal{M}$ be a real finite-dimensional manifold. Suppose that we have the decomposition

$$\mathcal{M} = U \cup V$$

where U and V are nonempty open subsets of $\mathcal{M}$, and the intersection $U \cap V$ is not empty. Then, for all $k = 1, 2, \ldots$, the following sequence is exact:

$$0 \to H^0(\mathcal{M}) \to H^0(U) \oplus H^0(V) \to H^0(U \cap V) \to \ldots$$
$$\to H^k(\mathcal{M}) \to H^k(U) \oplus H^k(V) \to H^k(U \cap V) \to H^{k+1}(\mathcal{M}) \to \ldots$$

If the intersection $U \cap V$ is empty, then

$$H^p(\mathcal{M}) = H^p(U) \oplus H^p(V), \qquad p = 0, 1, 2, \ldots$$

Note that $H^k(U) = H^k(V) = 0$, $k = 1, 2, \ldots$ if U and V are contractible.

The proof can be found in I. Madsen and J. Tornehave, From Calculus to Cohomology: de Rham Cohomology and Characteristic Classes, Cambridge University Press, 1997, p. 35.

Example. For the unit sphere $\mathbb{S}^2$, we want to show that

$$H^0(\mathbb{S}^2) = H^2(\mathbb{S}^2) = \mathbb{R}, \qquad H^1(\mathbb{S}^2) = 0.$$

Proof. Since $\mathbb{S}^2$ is arcwise connected, we get $H^0(\mathbb{S}^2) = \mathbb{R}$. We will use

$$H^0(\mathbb{S}^1) = H^1(\mathbb{S}^1) = \mathbb{R}.$$

Choose the sets

$$U := \mathbb{S}^2 \setminus \{N\} \qquad \text{and} \qquad V := \mathbb{S}^2 \setminus \{S\}$$

where N (resp. S) is the north (resp. south) pole. Thus, $\mathbb{S}^2 = U \cup V$. We get the exact sequence

$$\begin{aligned} 0 &\to H^0(\mathbb{S}^2) \to H^0(U) \oplus H^0(V) \to H^0(U \cap V) \to H^1(\mathbb{S}^2) \\ &\to H^1(U) \oplus H^1(V) \to H^1(U \cap V) \to H^2(\mathbb{S}^2) \to H^2(U) \oplus H^2(V) \to \ldots \end{aligned}$$

Since the intersection $U \cap V$ is homotopically equivalent to the equator, we get

$$H^p(U \cap V) = H^p(\mathbb{S}^1), \qquad p = 0, 1, 2, \ldots$$

Moreover, the sets U and V are contractible. Thus,

$$H^0(U) = H^0(V) = \mathbb{R}, \qquad H^p(U) = H^p(V) = 0, \qquad p = 1, 2, \ldots$$

This implies the exact sequences

$$0 \to H^0(\mathbb{S}^2) \to \mathbb{R}^2 \to H^0(\mathbb{S}^1) \to H^1(\mathbb{S}^2) \to 0, \tag{23.48}$$

and

$$0 \to H^1(\mathbb{S}^2) \to H^2(\mathbb{S}^2) \to 0. \tag{23.49}$$

From (23.49) we get the isomorphism $H^2(\mathbb{S}^2) = H^1(\mathbb{S}^1)$. Hence $H^2(\mathbb{S}^2) = \mathbb{R}$. From (23.48) we obtain the exact sequence

$$0 \xrightarrow{\alpha} \mathbb{R} \xrightarrow{\beta} \mathbb{R}^2 \xrightarrow{\gamma} \mathbb{R} \xrightarrow{\delta} H^1(\mathbb{S}^2) \to 0.$$

This implies $H^1(\mathbb{S}^2) = 0$ (see the proof given in Problem 23.1 on page 1008). □

23.6.2 The de Rham Cohomology Algebra

Let $\mathcal{M}$ be a real n-dimensional manifold. Set

$$H(\mathcal{M}) := H^0(\mathcal{M}) \oplus H^1(\mathcal{M}) \oplus \ldots \oplus H^n(\mathcal{M}).$$

Then, $H(\mathcal{M})$ becomes the structure of a real finite-dimensional algebra. This is called the de Rham cohomology algebra of the manifold $\mathcal{M}$. The product is generated by the wedge product $\omega \wedge \mu$ of differential forms. More precisely, the following hold: For two differential forms ω and ω', we write

$$\omega \sim \omega' \quad \text{iff} \quad \omega - \omega' = d\mu$$

for some differential form μ. We say that ω is cohomologous to ω'. This is an equivalence relation. The equivalence classes $[\omega]$ are called the de Rham cohomology classes. If ω is a p-form, then $[\omega]$ is an element of $H^p(\mathcal{M})$. The equivalence relation respects both the linear combinations of differential forms of the same degree and the wedge product. Therefore, the definitions

- $[\alpha\omega + \beta\nu] := \alpha[\omega] + \beta[\nu], \ \alpha, \beta \in \mathbb{R}$, and
- $[\omega][\varrho] := [\omega \wedge \varrho]$

do not depend on the choice of the representatives. This way, we get the linear combinations and the products of the cohomology algebra $H(\mathcal{M})$. If $\omega \sim \omega'$, then $\int_{\mathcal{M}} \omega - \omega' = \int_{\mathcal{M}} d\mu = \int_{\partial\mathcal{M}} \mu = 0$, by the generalized Stokes theorem. Note that the boundary $\partial\mathcal{M}$ of the manifold $\mathcal{M}$ is empty. Hence

$$\int_{\mathcal{M}} \omega = \int_{\mathcal{M}} \omega'.$$

This tells us that the integral $\int_{\mathcal{M}} \omega$ only depends on the cohomology class $[\omega]$.

23.7 The Beauty of Partial Differential Equations in Physics, Analysis, and Topology

Many important modern developments in topology are based on partial differential equations which play a crucial role in physics. This concerns:

- the potential equation $d\omega = \mu$: de Rham cohomology,
- the Laplace equation $\Delta\omega = 0$: Hodge theory,
- Hamilton's canonical equations in mechanics: symplectic geometry,
- the Cauchy–Riemann differential equation (Riemann surfaces and Kähler manifolds, symplectic geometry, string theory, conformal quantum field theory),
- the generalized Cauchy-Riemann equations for generalized analytic functions (Gromov's pseudoholomorphic curves in complex geometry),
- the Maxwell equation: Hodge theory,
- the Yang–Mills equation: Donaldson's theory for 4-manifolds,
- the Seiberg–Witten equation (nonlinear elliptic Dirac equation): alternative approach to Donaldson's theory for 4-manifolds,
- the Dirac equation: Atiyah–Singer index theorem, spin geometry,
- the heat equation: Atiyah–Singer index theorem, Ricci flow, solution of the Poincaré conjecture for the 3-dimensional sphere $\mathbb{S}^3$,
- the equation for geodesics: Morse theory, Floer homology,
- the minimal surface equation: Morse theory, string theory,
- the Einstein equations in general relativity: Calabi–Yau manifolds in string theory.

23.8 A Glance at Topological Quantum Field Theory (Statistics for Mathematical Structures)

> Witten's approach to the Jones polynomials of knots is one of the beautiful examples for the flow of ideas from physics to mathematics. This approach is based on a model in gauge theory.
>
> Folklore

The goal of topological quantum field theory (TQFT) is to use models from quantum field theory in order to construct and to compute explicitly topological invariants. The basic idea is to use the partition function[13]

$$Z(W) := \int_\Omega \mathrm{e}^{\mathrm{i}S(\omega)} W(\omega)\, \mathcal{D}\omega \tag{23.50}$$

and to compute the expectation value

$$\boxed{\overline{W} := \frac{Z(W)}{Z(1)}} \tag{23.51}$$

which is the desired topological invariant. We use the following terminology:

- The set Ω is called the state space; the elements ω of Ω are called physical states.[14]
- The function $S : \Omega \to \mathbb{R}$ is called the action functional; $S(\omega)$ is called the action of the physical state ω.
- The function $W : \Omega \to \mathbb{R}$ is called a physical observable (i.e., the real number $W(\omega)$ can be measured in a physical experiment).
- The integral Z is called the partition function (or the Feynman functional integral) of the model.
- This integral has to be computed by using the method of the stationary phase. In order to get the first approximation, one determines the critical points $\omega_1, \omega_2, \ldots$ of the action functional S, that is, the solutions of the equation

  $$S'(\omega) = 0, \qquad \omega \in \Omega,$$

  and one defines

  $$\boxed{Z_0(W) := \sum_j \int_\Omega \mathrm{e}^{\mathrm{i}S_{0j}(\omega)} W(\omega)\, \mathcal{D}\omega} \tag{23.52}$$

  where S_{0j} represents a 'quadratic' approximation of S near the critical state ω_j.[15]

---

[13] We use physical units where $\hbar = 1$.

[14] As a rule, the state space consists of equivalence classes, that is, Ω is a so-called moduli space of mathematical structures. Typical examples are (i) Riemann's moduli space of compact Riemann surfaces modulo conformal equivalence, and (ii) the moduli space of connections on a principal fiber bundle modulo global gauge transformations.

[15] For a classical function $\omega \mapsto S(\omega)$ on the real line, we have $S(\omega) = S_{0j}(\omega) + o(\omega^2)$ as $\omega \to 0$ with $S_{0j}(\omega) := S(\omega_j) + \frac{1}{2}S''(\omega_j)(\omega - \omega_j)^2$. This is a paradigm for the infinite-dimensional situation.

Formula (23.52) means that one replaces the full Feynman functional integral by functional integrals of 'Gaussian type'. Methods for computing infinite-dimensional Gaussian integrals can be found in Sect. 7.9 of Vol. II (zeta function regularization and the Wick trick). The experience of physicists shows that, as a rule, one can replace Z by Z_0 in the expectation-value formula (23.51). This means that the essential topological information is already contained in the approximation

$$\overline{W}_0 = \frac{Z_0(W)}{Z_0(1)}. \tag{23.53}$$

It is shown in the paper by

J. Duistermaat and G. Heckmann, On the variation in the cohomology in the symplectic form of the reduced phase space, Invent. Math. **69** (1982), 259–268; **72** (1983), 153

that there exist models in classical dynamics where $Z = Z_0$, that is, the full integral coincides with the first approximation (in the sense of the stationary-phase method).

The Jones polynomial of a knot K and gauge theory. Witten discovered that the Jones polynomial $V = V(x)$ of a knot K can be obtained by the expectation value

$$\boxed{V(\mathrm{e}^{2\pi\mathrm{i}/\kappa+2}) = \overline{W}_0, \qquad \kappa = 1, 2, \ldots} \tag{23.54}$$

Roughly speaking, the parameter κ is the coupling constant of the Chern–Simons action, and the physical observable W is the Wilson loop functional generated by the knot K.

Let us sketch the physical intuition behind this discovery. We will choose the product bundle

$$\mathbb{E}^3 \times SU(2)$$

with the Euclidean manifold $\mathbb{E}^3 \simeq \mathbb{R}^3$ as base manifold and the Lie group $SU(2)$ as fiber. Later on, we will compactify the base manifold (i.e., we will replace $\mathbb{E}^3$ by the 3-dimensional Riemann unit sphere $\mathbb{S}^3$).

(i) The knot K: By definition, a knot is a continuous embedding

$$\chi : \mathbb{S}^1 \to \mathbb{E}^3 \tag{23.55}$$

of the unit circle $\mathbb{S}^1$ into the 3-dimensional Euclidean manifold $\mathbb{E}^3$. This means that the map $\chi : \mathbb{S}^1 \to \chi(\mathbb{S}^1)$ is a homeomorphism. We have to distinguish between the map χ and the image set $K := \chi(\mathbb{S}^1)$. The latter is called the geometric knot K. Parameterizing the unit circle by the angle variable $\varphi \in [-\pi, \pi]$, we get the function $\chi = \chi(\varphi)$ with $\chi(-\pi) = \chi(\pi)$. Setting $P(t) := \chi(t - \pi)$, the equation

$$\boxed{K: \quad P = P(t), \qquad 0 \le t \le 2\pi, \quad P(0) = P_0} \tag{23.56}$$

describes the motion along the geometric knot K (Fig. 23.6).

(ii) Physical field on the Euclidean manifold and gauge transformation: Let us consider a physical field

$$\psi : \mathbb{E}^3 \to \mathbb{C}^2$$

on the Euclidean manifold $\mathbb{E}^3$. By definition, a gauge transformation of the physical field ψ reads as

$$\psi_+(P) := \mathbf{G}(P)\psi(P) \qquad \text{for all} \quad P \in \mathbb{E}^3 \tag{23.57}$$

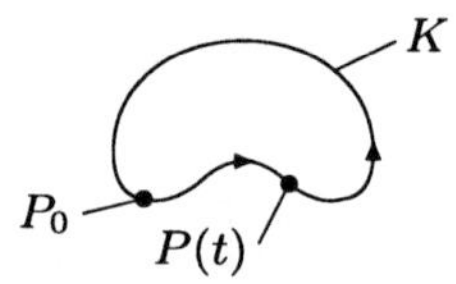

Fig. 23.6. Motion along a knot

where $\mathbf{G}(P) \in SU(2)$ for all $P \in SU(2)$.

(iii) Parallel transport of the physical field: Set $\psi(P(0)) = \psi_0$. The equation

$$\psi(P(t)) = G(t)\psi_0, \qquad 0 \le t \le t_0$$

describes the parallel transport $t \mapsto \psi(P(t))$ of the physical field along the curve (23.56) in the Euclidean manifold $\mathbb{E}^3$. Here, the function $t \mapsto G(t)$ is given by the solution of the differential equation

$$\dot{G}(t) = -\mathcal{A}_{P(t)}(\dot{P}(t)) \cdot G(t), \qquad 0 \le t \le t_0, \quad G(0) = G_0. \tag{23.58}$$

This is called the phase equation.[16]

(iv) The Wilson loop functional: We consider the motion (23.56) along the geometric knot (loop) from the initial point $P(0) = P_0$ to the final point

$$P(2\pi) = P(0)$$

during the time interval $[0, 2\pi]$ (Fig. 23.6). Parallel transport of the point (P_0, G_0) along the loop (23.56) yields the bundle point

$$(P_0, G_K G_0) \qquad \text{for all} \quad G_0 \in SU(2).$$

Finally, choose a representation $\varrho : SU(2) \to L(X, X)$ of the group $SU(2)$ on the finite-dimensional real linear space X. Then we define

$$W := \operatorname{tr} \varrho(G_K).$$

(v) Compactification: To simplify the mathematical situation, we will compactify the Euclidean manifold $\mathbb{E}^3$. In complex function theory, one compactifies the Gaussian plane $\mathbb{C}$ by the Riemann sphere $\mathbb{S}^2$ based on stereographic projection (Fig. 0.1 on page 15). Similarly, we replace the non-compact 3-dimensional Euclidean manifold $\mathbb{E}^3$ by the compact 3-dimensional unit sphere $\mathbb{S}^3$. This implies that we replace the knot map (23.55) by the continuous embedding

$$\boxed{\chi : \mathbb{S}^1 \to \mathbb{S}^3.}$$

(vi) The Chern–Simons action: As an essential ingredient of the model, we choose the Chern–Simons action

$$S := -\frac{\kappa}{4\pi} \int_{\mathbb{S}^3} \operatorname{tr}(\mathcal{A} \wedge d\mathcal{A} + \tfrac{2}{3}(\mathcal{A} \wedge \mathcal{A} \wedge \mathcal{A}))$$

[16] Explicitly $\dot{G}(t) = -\dot{x}^j(t)\mathcal{A}_j(x(t)) \cdot G(t)$ where $\mathcal{A}_j(x)$ lies in the Lie algebra $su(2)$ for all $x \in \mathbb{R}^3$ and all $j = 1, 2, 3$. This implies that $G(t)$ lies in the Lie group $SU(2)$ for all $t \in [0, t_0]$. Here, we sum over $j = 1, 2, 3$.

on the sphere $\mathbb{S}^3$. Here, the gauge potential $\mathcal{A}$ is a differential 1-form on the sphere $\mathbb{S}^3$ with values in the Lie algebra $su(2)$.

(vii) Gauge invariance: Every smooth map $\mathbf{G} : \mathbb{S}^3 \to SU(2)$ induces both the gauge transformation (23.57) of the physical field and the following gauge transformation of the gauge potential:

$$\mathcal{A}^+ := \mathbf{G}^{-1}\mathcal{A}\mathbf{G} - d\mathbf{G} \cdot \mathbf{G}^{-1}.$$

This yields the equivalence relation $\mathcal{A}^+ \sim \mathcal{A}$. By definition, the corresponding equivalence classes $[\mathcal{A}]$ form the elements of the state space Ω appearing in the key partition-function formula (23.50). This makes sense since the Wilson loop functional W is gauge invariant. The action S is not gauge invariant, but the exponential expression $e^{iS(\mathcal{A})}$ is gauge invariant, since the exponential function has the period $2\pi i$, and we choose the coupling constant κ as a positive integer.

In the language of mathematics, the following hold:

- *The state space Ω is the moduli space of the connections on the trivial principal fiber bundle $\mathbb{S}^3 \times SU(2)$ over the 3-sphere $\mathbb{S}^3$ and the Lie group $SU(2)$ as typical fiber.*
- *Using the trace of matrices, the Wilson loop functional is generated by a linear representation of the holonomy group acting on the typical fiber $SU(2)$.*

We refer to the following paper:

E. Witten, Quantum field theory and the Jones polynomials, Commun. Math. Phys. **212** (1989), 359–399.

Furthermore, we recommend:

E. Witten, Topological quantum field theory, Commun. Math. Phys. **117** (1988), 353–386.

E. Witten, Topological sigma models, Commun. Math. Phys. **118** (1988), 411–449.

E. Witten, Global gravitational anomalies, Commun. Math. Phys. **121** (1989), 297–309.

The axiomatic approach in mathematics. Such an approach is outlined in

M. Atiyah, The Geometry and Physics of Knots, Cambridge University Press, 1990.

The axioms are formulated in terms of a functor Z which assigns

(i) a finite-dimensional complex Hilbert space $Z(M)$ (space of quantum states) to each compact oriented d-dimensional manifold M, and

(ii) an element $Z(N)$ of $Z(M)$ for each compact oriented $(d+1)$-dimensional manifold N with boundary $\partial N = M$. Here, $Z(N)$ corresponds to the partition function (Feynman functional integral).

The axioms concern multiplicativity, associativity, involution, and non-triviality (i.e., $Z(\emptyset) := \mathbb{C}$). For the situation of the Jones polynomials, one chooses

$$N := \mathbb{S}^3.$$

Then the boundary $M := \partial\mathbb{S}^3$ is empty. Hence we get the one-dimensional complex Hilbert space $Z(M) = \mathbb{C}$, and $Z(N)$ is a complex number. The values of the Jones polynomials for special arguments correspond to $Z(N)$ depending on the choice of the coupling constant. In this setting, the theory is formulated in the spirit of cobordism theory. See also

M. Atiyah, The impact of Thom's cobordism theory, Bull. Amer. Math. Soc. 41(3) (2004), 337–340.

The discussion shows that the axiomatic approach can be viewed as a quantum theory without any dynamics, that is, the Hamiltonian is trivial, $H = 0$.

The partition function (23.50) represents some statistics over mathematical structures. In the history of mathematics, one observes the following steps:

Step: 1 Quantitative measurements: In ancient times, mathematics used concrete numbers in order to describe quantities that can be measured (e.g., the area of a piece of land).

Step 2: From numbers to letters as symbols for numbers: In the late sixteenth century under the influence of Viète (Vieta) (1550–1603), mathematicians began to use letters as symbols for numbers. That is, the quantity which each symbol represented was left indefinite, while the quality of the object it represented in computations was fixed.

Step 3: Mathematical structures: In the 20th century, there emerged the theory of mathematical structures (e.g., groups, topological spaces, manifolds) and the combination of mathematical structures with each other (e.g., Lie groups are obtained by combining the notion of group with the notion of manifold). In this case, even the quality of the symbols used is left indeterminate, leading to a genuine theory of the operations. For example, the first textbook on modern algebra was written by van der Waerden (1903–1998) in 1930:

B. van der Waerden, Moderne Algebra (in German), Vols. 1, 2, Springer, Berlin, 1930; 8th edition in 1993. English edition: Frederyck Ungar, New York 1975.

This book was based on lectures given by Emmy Noether (1882–1935) and Emil Artin (1898–1962) in Göttingen in the 1920s.

Step 4: Categories and functors: The theory of categories created in the 1940s by Eilenberg (1913–1998) and MacLane (1909–2005) represents a super theory for mathematical structures; functors allow the passage from one mathematical structure to another one. The modern standard textbook on algebra written by Lang (1903–1998) uses extensively the language of categories and functors:

S. Lang, Algebra, Addison-Wesley, Reading, Massachusetts, 1993; 3rd edition, Springer, New York, 2002.

Step 5: Statistics of mathematical structures: Under the influence of modern physics, one studies the statistics over all the objects of a given mathematical structure. For example, there exists the possibility that shortly after the Big Bang the structure of the space-time was random. One possibility to handle this situation is to use a partition function, that is, a Feynman functional integral over all possible pseudo-Riemannian manifolds modulo a specific equivalence relation.

The example of Jones polynomials considered above shows that the strategy of carrying out a statistics over mathematical structures can be very useful for answering purely mathematical questions.

As an introduction to the flow of ideas from modern physics to modern mathematics, we recommend:

K. Marathe, A chapter in physical mathematics: theory of knots in the sciences, pp. 873–888. In: B. Enquist and W. Schmid (Eds.), Mathematics Unlimited—2001 and Beyond, Springer, New York, 2001.

K. Marathe, Topics in Physical Mathematics, Springer, London, 2010.

23.9 Further Reading

Classic calculus and differential forms:

V. Zorich, Mathematical Analysis, Vol. II, Springer, Berlin, 2003.

I. Agricola and T. Friedrich, Global Analysis: Differential Forms in Analysis, Geometry and Physics, Amer. Math. Soc., Providence, Rhode Island, 2002.

Introduction:

C. Guillemin and V. Pollack, Differential Topology, Prentice Hall, Englewood Cliffs, New Jersey, 1974 (the best elementary introduction to differential topology) (lectures given at the Massachusetts Institute of Technology, Cambridge, Massachusetts).

J. Milnor, Topology from the Differential Point of View, University Press of Virginia, Charlottesville, Virginia, 1965.

J. Milnor, Morse Theory, Princeton University Press, 1963.

J. Milnor and J. Stasheff, Characteristic Classes, Princeton University Press, 1974.

Furthermore, we recommend:

S. Matveev, Lectures on Algebraic Topology, European Mathematical Society, Zurich, 2006 (the best elementary introduction to homology and homotopy based on exact sequences and emphasizing geometric intuition).

I. Madsen and J. Tornehave, From Calculus to Cohomology: de Rham Cohomology and Characteristic Classes, Cambridge University Press, 1997.

J. Jost, Riemannian Geometry and Geometric Analysis, Springer, Berlin, 1995 (de Rham cohomology, Hodge theory, Morse theory, and characteristic classes).

J. Jost, Riemannian Geometry and Geometric Analysis, 5th edition, Springer, Berlin, 2008 (including an elementary introduction to Floer homology and Witten's Morse theory, spin geometry and the Seiberg–Witten equation).

J. Jost, Compact Riemann Surfaces: An Introduction to Contemporary Mathematics, 3rd edition, Springer, Berlin, 2006 (including the topology of Riemann surfaces).

C. Kinsey, Topology of Surfaces, Springer, New York, 1993 (emphasizing geometric intuition).

G. Naber, Topological Methods in Euclidean Spaces, Cambridge University Press, 1980 (emphasizing geometric intuition).

R. Bott and L. Tu, Differential Forms in Algebraic Topology, Springer, New York, 1982.

W. Lück, Algebraic Topology (in German), Vieweg, Wiesbaden (exact sequences, singular homology and singular cohomology of general topological spaces, homological algebra, de Rham cohomology of manifolds).

M. Atiyah, Algebraic topology and elliptic operators, Commun. Pure Appl. Math. **20** (1967), 237–249.

F. Hirzebruch, New Topological Methods in Algebraic Geometry, third enlarged edition, Springer, New York, 1966 (English edition) (first German edition, 1956).

V. Turaev, Quantum Invariants of Knots and 3-Manifolds, de Gruyter, Berlin, 1994.

The heat equation and topology:

P. Gilkey, Invariance Theory, the Heat Equation, and the Atiyah–Singer Index Theorem, CRC Press, Boca Raton, Florida, 1995.

Topology in physics:

T. Frankel, The Geometry of Physics, Cambridge University Press, 2004 (including de Rham cohomology, periods of integrals over cycles, and characteristic classes).

K. Marathe, Topics in Physical Mathematics, Springer, London, 2010 (vector bundles, K-theory, topology of gauge fields, knots, 3-manifolds and 4-manifolds).

M. Monastirsky, Topology of Gauge Fields and Condensed Matter, Plenum Press, New York, 1993.

R. Hwa and V. Teplitz, Homology and Feynman Diagrams, Benjamin, Reading, Massachusetts, 1966.

J. Naber, Space-Time and Singularities, Cambridge University Press, 1988.

G. Naber, Topology, Geometry, and Gauge Fields, Springer, New York, 1997.

B. Felsager, Geometry, Particles, and Fields, Springer, New York, 1997.

A. Schwarz, Quantum Field Theory and Topology, Springer, Berlin, 1993.

A. Schwarz, Topology for Physicists, Springer, Berlin, 1993.

C. Nash and S. Sen, Topology and Geometry for Physicists, Academic Press, London, 1983.

C. Nash, Differential Topology and Quantum Field Theory, Academic Press, New York, 1991.

J. Baez and J. Muniain, Gauge Fields, Knots, and Gravity, World Scientific, Singapore, 1994.

P. Bandyopadhyay, Geometry, Topology, and Quantization, Kluwer, Dordrecht, 1996.

E. Bick and F. Steffen (Eds.), Topology and Geometry in Physics, Springer, Berlin, 2005.

M. Atiyah, Geometry of Yang–Mills Fields, Lezioni Fermiani, Academia Nazionale dei Lincei, Scuola Normale Superiore, Pisa, Italia, 1979.

M. Atiyah and N. Hitchin, The Geometry and Dynamics of Magnetic Monopoles, Princeton University Press, 1988.

A. Jaffe and C. Taubes, Vortices and Monopoles: Structure of Static Gauge Theories, Birkhäuser, Boston, 1980.

R. Rajaraman, Solitons and Instantons: An Introduction to Solitons and Instantons in Quantum Field Theory, Elsevier, Amsterdam, 1987.

Y. Yang, Solitons in Field Theory and Nonlinear Analysis, Springer, New York, 2001.

M. Manton and P. Sutcliffe, Topological Solitons, Cambridge University Press, 2004.

N. Hitchin, G. Segal, and R. Ward, Integrable Systems, Twistors, Loop Groups, and Riemann Surfaces, Oxford University Press, 1999.

J. Kock and J. Vainsencher, An Invitation to Quantum Cohomology: Kontsevich's Formula for Plane Curves, Birkhäuser, Basel, 2006.

D. Freed and K. Uhlenbeck, Instantons and Four-Manifolds, Springer, New York, 1984.

S. Donaldson and P. Kronheimer, The Geometry of Four-Manifolds, Oxford University Press, 1990.

J. Moore, Lectures on Seiberg–Witten Invariants, Springer, Berlin, 1996 (elementary introduction).

J. Morgan, The Seiberg–Witten Equations and Applications to the Topology of Four-Manifolds, Princeton University Press, 1996.

S. Donaldson, The Seiberg–Witten equations and 4-manifold topology, Bull. Amer. Math. Soc. **33** (1996), 45–70.

M. Schwarz, Morse Homology, Birkhäuser, Basel, 1993.

S. Donaldson, Floer Homology Groups, Cambridge University Press, 2002.

M. Atiyah, The Geometry and Physics of Knots, Cambridge University Press, 1990 (bordism theory and topological quantum field theory).

E. Witten, Witten's Lectures on Three-Dimensional Topological Quantum Field Theory. Edited by Sen Hu, World Scientific, Singapore 1999.

E. Witten, Physical law and the quest for mathematical understanding, Bull. Amer. Math. Soc. **40** (2003), 21–30.

The Poincaré conjecture and the Ricci flow:

H. Cao, S. Yau, and X. Zhu, Structure of Three-dimensional Space: The Poincaré and Geometrization Conjectures, International Press, Boston, 2006..

J. Morgan and G. Tian, Ricci Flow and the Poincaré Conjecture, Amer. Math. Soc., Providence, Rhode Island/Clay Mathematics Institute, Cambridge, Massachusetts, 2007.

J. Morgan and F. Fong, Ricci Flow and Geometrization of 3-Manifolds, Amer. Math. Soc., Rhode Island, 2010.

Exact sequences, group extensions, and homological algebra:

S. Lang, Algebra, 3rd edition, Springer, New York, 2002.

H. Cartan and S. Eilenberg, Homological Algebra, Princeton University Press, 1956 (classic monograph).

S. Gelfand and Yu. Manin, Homological Algebra, Springer, New York, 1996 (including derived functors).

Exact sequences in sheaf cohomology and complex analysis:

K. Maurin, Methods of Hilbert Spaces, Polish Scientific Publishers, Warsaw, 1972.

O. Forster, Lectures on Riemann Surfaces, Springer, Berlin, 1981.

G. Bredon, Sheaf Theory, Springer, New York, 1998.

Yu. Manin, Gauge Fields and Complex Geometry, Springer, Berlin, 1997.

Yu. Manin, Frobenius manifolds, quantum cohomology and moduli spaces, Amer. Math. Soc., Providence, Rhode Island, 1999.

Exact sequences in topology:

W. Massey, Algebraic Topology: An Introduction, Springer, New York, 1967/1987 (7th edition) (emphasizing geometric intuition).

W. Massey, Singular Homology Theory, Springer, New York, 1980.

G. Bredon, Topology and Geometry, Springer, New York, 1993.

M. Atiyah, K-Theory, Benjamin, New York, 1967.

A. Hatcher, Algebraic Topology, Cambridge University Press, 2002.

A. Hatcher, Vector Bundles and K-Theory (draft), 2010.

Internet: `http://www.math.cornell.edu/~hatcher`

A. Hatcher, Spectral Sequences in Algebraic Topology (draft), 2010. Internet: `http://www.math.cornell.edu/~hatcher`

E. Spanier, Algebraic Topology, Springer, New York, 1989.

F. Warner, Foundations of Differentiable Manifolds and Lie Groups, Scott-Foresman, Glenview, Illinois, 1971.

M. Kreck, Differential Algebraic Topology: From Stratifolds to Exotic Spheres, Amer. Math. Soc., Providence, Rhode Island, 2009.

Exact sequences in the theory of fiber bundles:

D. Husemoller, Fibre Bundles, Springer, New York, 1994.

Exact sequences in algebraic geometry:

I. Shafarevich, Basic Algebraic Geometry, Vols. 1, 2, Springer, Berlin, 1994.

P. Griffith and J. Harris, Principles of Algebraic Geometry, Wiley, New York, 1978.

A. Wallace, Homology Theory of Algebraic Varieties, Pergamon Press, New York, 1958.

R. Friedman, Algebraic Surfaces and Holomorphic Vector Bundles, Springer, New York, 1998.

D. Husemoller, Elliptic Curves, Springer, New York, 2004.

W. Ebeling, Functions of Several Complex Variables, and Their Singularities, Amer. Math. Soc., Providence, Rhode Island, 2007.

Hodge theory:

K. Maurin, Analysis, Vol. 2, Reidel, Dordrecht, 1980.

W. Hodge, The Theory and Applications of Harmonic Integrals, Cambridge University Press, 1941 (second revised edition 1951).

G. de Rham, Differentiable Manifolds: Forms, Currents, Harmonic Forms, Hermann, Paris, 1955 (in French).

C. Voisin, Hodge Theory and Complex Algebraic Theory, Vols. I, II, Cambridge University Press, 2002.

G. Schwarz, Hodge Decomposition– a Method of Solving Boundary Value Problems, Springer, Berlin, 1995.

Exact sequences in number theory:

Yu. Manin and A. Panchishkin, Introduction to Modern Number Theory, Encyclopedia of Mathematical Sciences, Vol. 49, Springer, Berlin, 2005.

C. Bergbauer and D. Kreimer, Hopf algebras in renormalization theory: locality and Dyson–Schwinger equations from Hochschild cohomology, pp.

133–165. In: L. Nyssen (Ed.), Physics and Number Theory, European Mathematical Society, Zurich, 2006.

Exact sequences in non-commutative geometry:

A. Connes and M. Marcolli, Noncommutative Geometry, Quantum Fields, and Motives, Amer. Math. Soc., Providence, Rhode Island, 2008.

M. Marcolli, Feynman Motives: Renormalization, Algebraic Varieties, and Galois Symmetries, World Scientific, Singapore, 2009 (Tate motives) (lectures given at the California Institute of Technology, Pasadena, California).

J. Várilly, Lectures on Noncommutative Geometry, European Mathematical Society, 2006.

M. Gracia-Bondia, J. Várilly, and H. Figueroa, Elements of Noncommutative Geometry, Birkhäuser, Boston, 2001.

Exact sequences and the Whitehead cohomology of Lie algebras:

V. Varadarajan, Lie Groups, Lie Algebras, and Their Representations, Springer, New York, 1984 (Chap. 3: applications to the fundamental theorems of Weyl and Levi-Malcev).

N. Bourbaki, Lie Groups and Lie Algebras, Vols. 1, 2, Springer, New York, 1989/2002.

History

A comprehsive history of topology can be found in:

I. James (Ed.), History of Topology, Elsevier, Amsterdam, 1999 (1050 pages).

J. Dieudonnè, A History of Algebraic and Differential Topology, 1900–1960, Birkhäuser, Boston, 1989.

We recommend:

C. Nash, Topology and physics – a historical essay. In: I. James (Ed.), pp. 359–416 (see above).

K. Maurin, The Riemann Legacy: Riemannian Ideas in Mathematics and Physics of the 20th Century, Kluwer, Dordrecht, 1997.

M. Monastirsky, Riemann, Topology, and Physics, Birkhäuser, Basel, 1987.

Furthermore, we recommend:

E. Scholz, The concept of manifold, 1850–1950. In: I. James (Ed.) (1999), pp. 25–64.

K. Sarkaria, The topological work of Henri Poincaré, 1895–1912. In: I. James (Ed.) (1999), pp. 123–167.

S. Lefschetz, The early development of algebraic topology, 1895–1932 (homology theory and Morse theory). In: I. James (Ed.) (1999), pp. 531–560.

A. Durfee, Singularities (the work of Milnor, Brieskorn, Hirzebruch). In: James (Ed.) (1999), pp. 417–434.

W. Massey, A history of cohomology, (1935–1999). In: I. James (Ed.), pp. 579–604.

S. MacLane, Group extensions for 45 years (homological algebra), Math. Intelligencer **10**(2) (1988), 29–35.

V. Varadarajan, Euler through Time: A New Look at Old Themes, Amer. Math. Soc., Providence, Rhode Island, 2006.

S. Novikov and I. Taimanov (Eds.), Topological Library, Vol. 1: Cobordisms and their Applications, Vol. 2: Characteristic Classes and Smooth Structures, World Scientific, Singapore, 2007/09 (collection of fundamental papers).

M. Atiyah, Mathematics in the 20th Century, Bull. London Math. Soc. **34** (2002), 1–15.

M. Atiyah and D. Iagolnitzer (Eds.), Fields Medallists' Lectures, World Scientific, Singapore, 2003.

M. Atiyah, The Work of Edward Witten, Proc. Intern. Congr. Math. Kyoto 1990, Math. Soc. Japan 1991. In: M. Atiyah, M. and D. Iagolnitzer (Eds.) (2003), pp. 514–518.

M. Atiyah, The Geometry and Physics of Knots, Cambridge University Press, 1990 (cobordism theory and axiomatic topological quantum field theory).

M. Atiyah, The Dirac equation and geometry, pp. 108–124. In. P. Goddard (Ed.), Paul Dirac – the Man and his Work, Cambridge University Press, Cambridge, United Kingdom, 1999.

M. Atiyah, The impact of Thom's cobordism theory, Bull. Amer. Math. Soc. 41 (3) (2004), 337–340.

H. Kastrup, On the advancement of conformal transformations and their associated symmetries in geometry and theoretical physics, Annalen der Physik **17** (2008), 631–690.

F. Dyson, Birds and frogs in mathematics and physics, Einstein lecture 2008, Notices Amer. Math. Soc. **56** (2) (2009), 212–223.

I. Kobzarev and Yu. Manin, Elementary Particles: Mathematics, Physics, and Philosophy, Kluwer, Dordrecht, 1989.

Yu. Manin, Matematika kak metafora (in Russian), Izdatelstvo MTNMO, Moscow, 2008. English edition: Mathematics as a Metaphor, Amer. Math. Soc., Providence Rhode Island, 2007.

Yu. Manin, Strings, Math. Intelligencer **11**(2) (1989), 59–65.

S. Yau and S. Nadis, The Shape of Inner Space: String Theory and the Geometry of the Universe's Hidden Dimensions, Basic Books, New York, 2010.

É. Charpentier, É. Ghys, and A. Lesne, The Scientific Legacy of Henri Poincaré, Amer. Math. Soc., Providence, Rhode Island, 2010.

Problems

23.1 *Proof of Theorem 23.17 on page 1049.* Solution: We first show that $\operatorname{div}\mathbf{A} = 0$. In fact, we get

$$\operatorname{div}\mathbf{A} = \frac{\mu_0}{4\pi}\int_{\mathbb{E}^3} \mathbf{J}_0(\mathbf{x}_0)\,\mathbf{grad}_{\mathbf{x}}\,\frac{1}{|\mathbf{x}-\mathbf{x}_0|}\,dx_0^3.$$

Since $\mathbf{grad}_{\mathbf{x}} \frac{1}{|\mathbf{x}-\mathbf{x}_0|} = -\mathbf{grad}_{\mathbf{x}_0} \frac{1}{|\mathbf{x}-\mathbf{x}_0|}$, integration by parts yields

$$\operatorname{div} \mathbf{A} = \frac{\mu_0}{4\pi} \int_{\mathbb{E}^3} \operatorname{div}_{\mathbf{x}_0} \mathbf{J}_0(\mathbf{x}_0) \cdot \frac{1}{|\mathbf{x}-\mathbf{x}_0|}\, dx_0^3 = 0,$$

because of $\operatorname{div}_{\mathbf{x}_0} \mathbf{J}(\mathbf{x}_0) = 0$. Set $\mathbf{B} := \mathbf{curl}\,\mathbf{A}$. Then $\operatorname{div} \mathbf{B} = 0$ and

$$\mathbf{curl}\,\mathbf{B} = \mathbf{curl}\,\mathbf{curl}\,\mathbf{A} = \Delta\mathbf{A} - \mathbf{grad}\operatorname{div}\mathbf{A} = \Delta\mathbf{A} = \mu_0\mathbf{J},$$

by (23.27) on page 1045.

23.2 *Asymptotic expansion.* Prove (23.36) and (23.38) on page 1049. Hint: Use the Taylor expansion of $\frac{1}{|\mathbf{x}-\mathbf{x}_0|}$.

23.3 *Circular magnetic current.* Prove that the circular electric current of radius R and current strength J_0 generates a magnetic field (23.40). As $|\mathbf{x}| \to \infty$, this field behaves like a magnetic dipole field with magnetic moment $|\mathbf{m}| = \mathsf{J}_0 \cdot \pi R^2$. Hint: Use the Biot–Savart law and the Taylor expansion of $\frac{1}{|\mathbf{x}-\mathbf{x}_0|^3}$.

Appendix

A.1 Manifolds and Diffeomorphisms

Special topological spaces. Topological spaces are sets where the notion of 'open' subset is defined. The precise definition of a topological space X can be found in Sect. 5.5 of Vol. I. In particular, every subset of a topological space is also a topological space (with respect to the natural induced topology). An open neighborhood $U(P)$ of the point P is an open subset of X which contains the point P. A subset $N(P)$ of X is called a neighborhood of the point P iff it contains an open neighborhood $U(P)$ of P. The real line $\mathbb{R}$ is the prototype of a separated topological space. Moreover, the following hold:

- The open interval $]0,1[$ is an open subset of the topological space $\mathbb{R}$.
- The closed unit interval $[0,1]$ is a compact subset of $\mathbb{R}$. A subset of $\mathbb{R}$ is compact iff it is bounded and closed.
- The intervals $]0,1], [0,1[,]0,1[$ are relatively compact subsets of $\mathbb{R}$, but they are not compact subsets of $\mathbb{R}$.
- The real line $\mathbb{R}$ is a locally compact and a paracompact topological space, but $\mathbb{R}$ is not a compact topological space.
- A non-empty subset of $\mathbb{R}$ is arcwise connected iff it is an interval.

The general definitions read as follows: By an open covering of the topological space X, we understand a family $\{U_\alpha\}$ of open subsets U_α of X such that every point P of X is contained in some set U_α.

- The subset C of X is called compact iff every open covering of C contains a finite subcover, that is, finitely many sets of the open covering already cover C.
- The subset S of X is called relatively compact iff the closure of S is a compact subset of X.
- X is called locally compact iff any point $P \in X$ has a compact neighborhood.
- X is called paracompact iff every open covering of X has a locally finite refinement which covers X.[1]
- The topological space X is called separated (or Hausdorff) iff for any two different points P and Q in X there exist open neighborhoods $U(P)$ and $V(Q)$ which are disjoint.

---

[1] This means that there exists an open covering $\{V_\beta\}$ of X such that every set V_β is contained in some set U_α, and every point $P \in X$ has an open neighborhood $W(P)$ such that only finitely many sets V_β intersect $W(P)$.

E. Zeidler, *Quantum Field Theory III: Gauge Theory*, DOI 10.1007/978-3-642-22421-8,

- X is arcwise connected iff, for any pair of points $P, Q \in X$, there exists a continuous map $c : [0,1] \to X$ such that $c(0) = P$ and $c(1) = Q$. Intuitively, this is a continuous curve which connects the point P with the point Q.

As we will show below, any separated paracompact topological space X possesses a partition of unity which allows the globalization of physical fields. Compact topological spaces and metric spaces are paracompact. In particular, every subset of $\mathbb{R}^n$, $n = 1, 2, \ldots$, and every subset of a Hilbert space is paracompact.

Moreover, a separated topological space X is paracompact if it possesses a countable basis $\{V_j\}$ of open sets. This means that every open subset U of X can be represented as the union of some open sets V_j.

Much more material on topological spaces, metric spaces, uniform spaces, Hilbert spaces, Banach spaces, and locally convex spaces can be found in the Appendix to E. Zeidler, Nonlinear Functional Analysis and its Applications, Vol. I, third edition, Springer, New York, 1998. Reprinted: Beijing (China) 2009.

The Zariski topology in algebraic geometry. Most topological spaces appearing in analysis are separated. However, there is a crucial topology used in algebraic geometry which is not separated. This is the Zariski topology.[2] Fix $n = 1, 2, \ldots$ By definition, a subset V of $\mathbb{C}^n$ is called an affine algebraic variety iff there exist complex polynomials $p_j : \mathbb{C}^n \to \mathbb{C}$, $j = 1, \ldots, m$ such that V is precisely the solution set of the system

$$p_j(z_1, \ldots, z_n) = 0, \quad j = 1, \ldots, m, \qquad (z_1, \ldots, z_n) \in \mathbb{C}.$$

By definition, precisely the affine algebraic varieties are the Zariski-closed subsets of $\mathbb{C}^n$. Moreover, a subset of $\mathbb{C}^n$ is called Zariski-open iff it is the complement of an affine algebraic variety. The Zariski-open sets form the open sets of the Zariski topology of $\mathbb{C}^n$. Every Zariski-open subset of $\mathbb{C}^n$ is also an open set of $\mathbb{C}^n$ with respect to the classical topology. However, the converse is not true.

Two non-empty Zariski open subsets of $\mathbb{C}^n$ possess a non-empty intersection.

Therefore, the Zariski topology on $\mathbb{C}^n$ is not separated, in contrast to the classical topology on $\mathbb{C}^n$.

Smooth functions. Let us fix the terminology by using the method of extensions.

(i) $\mathbb{R}$ (real line): Let $-\infty < a < b < \infty$. The function $f :]a,b[\to \mathbb{R}$ is smooth on the open interval $]a,b[$ iff it is continuous and has continuous derivatives of arbitrary order on $]a,b[$ (e.g., the function $f(x) := \sin x$ is smooth on all open intervals).
The function $f : [a,b] \to \mathbb{R}$ is smooth on the closed interval $[a,b]$ iff the following hold: f is smooth on the open interval $]a,b[$, f is continuous on $[a,b]$, and all the continuous derivatives of f on $]a,b[$ can be uniquely extended to continuous functions on $[a,b]$.
For example, the function $f(x) := |x|$ is smooth on $[0,1]$. We have $f'(x) = 1$ for all $x \in]0,1[$. The classical derivative does not exist at the point $x = 0$. But, the derivative f' on $]0,1[$ can be uniquely extended to the closed interval $[0,1]$. Therefore, we write $f'(x) = 1$ for all $x \in [0,1]$.[3]

(ii) $\mathbb{R}^n$, $n = 1, 2, \ldots$: Let the set $\mathcal{C}$ be the closure of the open subset $\mathcal{O}$ of $\mathbb{R}^n$ (e.g, $\mathcal{O}$ is an open ball, and $\mathcal{C}$ is the corresponding closed ball). The function

[2] For his fundamental contributions to modern algebraic topology, Oscar Zariski (1899–1986) was awarded the Wolf prize in 1981.

[3] Set $f(x) := \frac{1}{x}$. The function f is smooth on $]0,1[$, but not smooth on $[0,1]$.

(a) no boundary

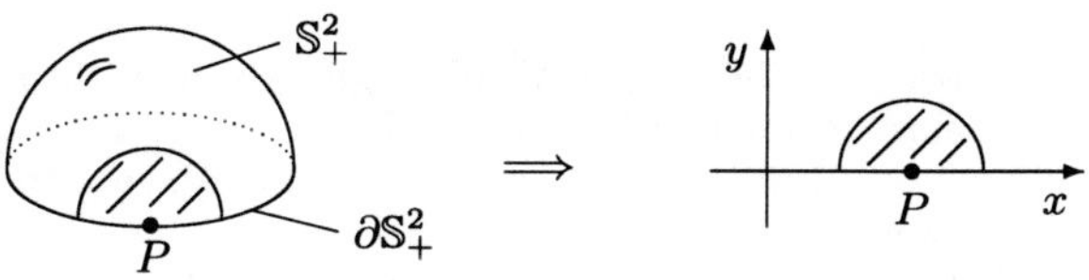

(b) boundary point P

Fig. A.1. The surface of earth

$f : \mathcal{O} \to \mathbb{R}$ is smooth on the open set $\mathcal{O}$ iff it is continuous and has continuous partial derivatives of all orders on $\mathcal{O}$.
The function $f : \mathcal{C} \to \mathbb{R}$ is smooth on the closed set $\mathcal{C}$ iff the following hold: f is smooth on the open set $\mathcal{O}$, f is continuous on $\mathcal{C}$, and all the continuous derivatives of f of arbitrary order on $\mathcal{O}$ can be uniquely extended to continuous functions on $\mathcal{C}$.

The same definition applies to functions $f : \mathcal{C} \to \mathbb{C}$ where $\mathcal{C}$ is the closure of the open subset $\mathcal{O}$ of $\mathbb{C}^n$. Note the following peculiarity. If the function $f : \mathcal{O} \to \mathbb{C}$ is smooth, then it is holomorphic (i.e., for each point $P \in \mathcal{O}$, there exists an open neighborhood $\mathcal{V}$ of P such that f can be represented as an absolutely convergent power series on $\mathcal{V}$). Such a nice result is not valid for real-valued smooth functions.

A.1.1 Manifolds without Boundary

We have to distinguish between

- manifolds with boundary, and
- manifolds without boundary.

In terms of physics, the prototype of a manifold is the surface of earth. This manifold has no boundary. In geography, the surface of earth is described by an atlas of geographic charts. In a chart, an open neighborhood of a point P of the surface of earth looks like the open neighborhood of a point in the Euclidean plane (Fig. A.1(a)).

In contrast to this, the northern hemisphere including the equator is a manifold with boundary where the equator is the boundary. In an appropriate geographic chart, an open neighborhood of the boundary point P looks like an open neighborhood of a boundary point of the closed upper half-plane. By definition, this is the intersection between an open neighborhood in the Euclidean plane and the closed upper half-plane (Fig. A.1(b)). Finally, the northern hemisphere excluding the equator is a manifold without boundary.

These concepts can be straightforward generalized to higher dimensions.

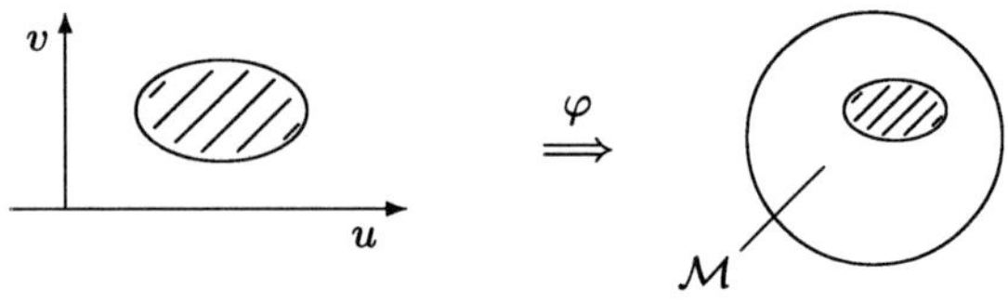

Fig. A.2. Manifold

Smooth manifold. The definition of manifolds given in Sect. 5.4 of Vol. II concerns a so-called smooth manifold, that is, the change of the local coordinates is described by diffeomorphisms between open subsets of the geographic charts.

Topological manifold. By definition, a real n-dimensional topological manifold is a separated paracompact topological space X which locally looks like the Euclidean space $\mathbb{R}^n$. That is, every point P of X has an open neighborhood $U(P)$ which is homeomorphic to an open subset $\mathcal{U}$ of $\mathbb{R}^n$, that is, there exists a homeomorphism

$$\varphi : U(P) \to \mathcal{U}.$$

The map φ is called a geographic chart, and the open subset $\mathcal{U}$ of $\mathbb{R}^n$ is called the Euclidean chart space. If $Q \in U(P)$, then $\varphi(Q) = (x^1, \ldots, x^n)$, and the real numbers $x^1, \ldots, x^n$ are called the local coordinates of the point Q. The change of local coordinates is described by a homeomorphism between open subsets of $\mathbb{R}^n$. The topological manifold is called oriented iff these chart-change homeomorphisms have the topological index one. For diffeomorphisms, the topological index is equal to one iff the determinant of the linearization is equal to one. In the more general case of a homeomorphisms, the local mapping degree has to be equal to one (see Vol. IV). Intuitively, this means that the homeomorphisms can locally be approximated by diffeomorphisms which have the topological index equal to one.

A.1.2 Manifolds with Boundary

Before defining manifolds with boundaries, the reader should notice the following conventions which we will use in order to streamline the terminology and to avoid pathologies.

Terminology and conventions. A manifold $\mathcal{M}$ with boundary $\partial\mathcal{M}$ includes the case that the boundary is empty.

- Smooth manifolds without boundary are simply called manifolds. For example, an open disc is a 2-dimensional real manifold (Fig. A.2). In contrast to this, a closed disc is a manifold with boundary; the boundary is a circle (Fig. A.3).
- We have to distinguish between manifolds (i.e., smooth manifolds) and topological manifolds. Manifolds are always topological manifolds, but the converse is not true.
- Every manifold is a topological space M. A subset U of M is called open iff every point $P \in U$ is contained in some subset $\mathcal{O}$ of U such that the chart image of $\mathcal{O}$ is open in some geographic chart. We always assume that manifolds with boundary are separated. Furthermore, since we assume below that a manifold possesses only an at most countable set of Euclidean charts, by definition, manifolds with boundary are always paracompact topological spaces.
- By abuse of language, a closed manifold is a compact manifold without boundary. For example, the circle $\mathbb{S}^1$, the 2-dimensional sphere $\mathbb{S}^2$ and the n-dimensional sphere $\mathbb{S}^n$, $n = 3, 4, \ldots$ are closed manifolds.

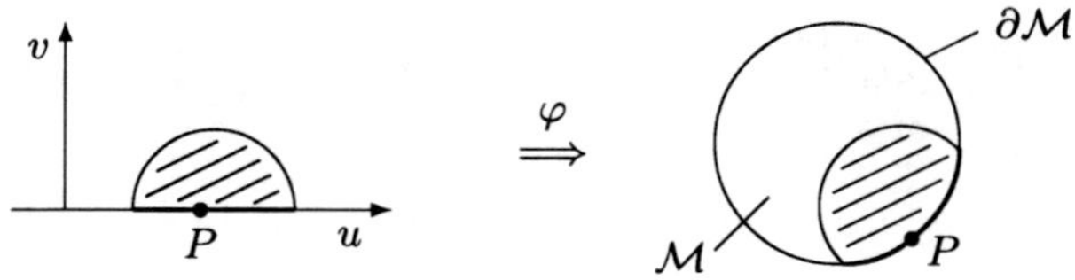

Fig. A.3. Manifold with boundary

- Since the definition of a manifold includes the restriction to an atlas with an at most countable set of geographic charts, manifold are always paracompact topological spaces. This useful convention always guarantees the existence of a partition of unity (see page 1077).
- If we do not state explicitly the contrary, manifolds are always finite-dimensional manifolds.[4]

Open neighborhood of a point in the closed upper half-plane. The following notion is crucial for the definition of manifolds with boundary to be given below. The set

$$\mathbb{R}^2_{\geq} := \{(u,v) \in \mathbb{R}^2 : \ v \geq 0\}$$

is called the closed upper half-plane (Fig. A.3). The set

$$\partial\mathbb{R}^2_{\geq} := \{(u,v) \in \mathbb{R}^2 : \ v = 0\}$$

is called the boundary of the upper half-plane $\mathbb{R}^2_{\geq}$. Moreover, the set

$$\mathrm{int}(\mathbb{R}^2_{\geq}) := \{(u,v) \in \mathbb{R}^2 : \ v > 0\}$$

is called the interior of the upper half-plane $\mathbb{R}^2_{\geq}$. Let $P \in \mathbb{R}^2_{\geq}$. By definition, an open neighborhood $\mathcal{U}(P)$ of the point P is the intersection

$$\mathcal{U}(P) := \mathbb{R}^2_{\geq} \cap \mathcal{O}$$

of the upper half-plane $\mathbb{R}^2_{\geq}$ with an open set $\mathcal{O}$ in $\mathbb{R}^2$ where $P \in \mathcal{O}$. Observe that if P is a boundary point of the upper half-plane, then the set $\mathcal{U}(P)$ contains boundary points of the upper half-plane.[5] For example,

- the disc $\{(u,v) \in \mathbb{R}^2 : (u-1)^2 + (v-2)^2 < r^2\}$ with $0 < r < 2$ is an open neighborhood of the point $(1,2)$ in $\mathbb{R}^2_{\geq}$,
- the semidisc $\mathcal{U} := \{(u,v) \in \mathbb{R}^2 : \ (u-1)^2 + v^2 < r^2, \ v \geq 0\}$ with $r > 0$ is an open neighborhood of the boundary point $(1,0)$ in $\mathbb{R}^2_{\geq}$, and
- the set $\mathcal{V} := \{(u,v) \in \mathbb{R}^2 : \ (u-1)^2 + v^2 < r^2, \ v > 0\}$ is called the interior of the semidisc $\mathcal{U}$.

The function

$$g : \mathcal{U}(P) \to \mathbb{R}$$

is called smooth iff it is smooth in the sense of the general definition given on page 1070 via extension. Explicitly, g is smooth iff the following hold: g is continuous

[4] The theory of infinite-dimensional manifolds and its applications is studied in E. Zeidler, Nonlinear Functional Analysis, Vol. 4, Springer, New York, 1995. Reprinted: Beijing (China), 2009.

[5] In terms of topology, $\mathcal{U}(P)$ is an open subset of $\mathbb{R}^2_{\geq}$ with respect to the topology on the upper half-plane $\mathbb{R}^2_{\geq}$ induced by the topology on the plane $\mathbb{R}^2$.

on $\mathcal{U}(P)$, g is smooth on the interior of $\mathcal{U}(P)$, and all the partial derivatives of g (of arbitrary order defined on the interior of $\mathcal{U}(P)$) can be extended to continuous functions on $\mathcal{U}(P)$. For example, set

$$g(u,v) := u + |v| \qquad \text{for all} \quad u,v \in \mathbb{R}^2_{\geq}.$$

Then the function g is smooth on $\mathbb{R}^2_{\geq}$, but it is not smooth on $\mathbb{R}^2$. Let $\mathcal{U}$ and $\mathcal{V}$ be open neighborhoods of the upper half-space $\mathbb{R}^2_{\geq}$. A map

$$f : \mathcal{U} \to \mathcal{V}$$

is called smooth iff its components are smooth. Moreover, f is called a diffeomorphism iff it is a bijective smooth map, and the inverse map $f^{-1} : \mathcal{V} \to \mathcal{U}$ is also smooth.

This can immediately be generalized to m-dimensional half-spaces

$$\mathbb{R}^m_{\geq} := \{(u^1, \ldots, u^m) \in \mathbb{R}^m : \ u^m \geq 0\}, \qquad m = 1, 2, \ldots$$

Let $P \in \mathbb{R}^m_{\geq}$. By definition, an open neighborhood $\mathcal{U}(P)$ of the point P is the intersection

$$\mathcal{U}(P) := \mathbb{R}^m_{\geq} \cap \mathcal{O}$$

of the upper half-space $\mathbb{R}^m_{\geq}$ with an open set $\mathcal{O}$ in $\mathbb{R}^m$ where $P \in \mathcal{O}$.

Basic definition of an m-dimensional manifold with boundary. Choose $m = 1, 2, \ldots$ Roughly speaking, a real m-dimensional manifold with boundary looks locally like an open neighborhood of an m-dimensional half-space with respect to local coordinates, and the change of local coordinates is given by diffeomorphisms. Boundary points correspond to boundary points in local coordinates (i.e., boundary points of m-dimensional half spaces). Explicitly, the precise definition reads similarly as the definition of real manifolds (without boundary) given in Sect. 5.4 of Vol. I.

We only replace open neighborhoods of the Euclidean space $\mathbb{R}^m$ by open neighborhoods of the half-space $\mathbb{R}^m_{\geq}$.

(M1) Chart maps: A real m-dimensional manifold with boundary is a set $\mathcal{M}$ together with a (finite or countable) family of bijective maps

$$\varphi_A : M_A \to U_A.$$

Here, M_A is a subset of $\mathcal{M}$, and U_A is the open neighborhood of some point of the upper half-space $\mathbb{R}^m_{\geq}$. We call $u_A = \varphi_A(P)$ the local coordinate of the point $P \in \mathcal{M}$, and φ_A is called a chart map. (Naturally enough, we assume that every point P of $\mathcal{M}$ lies in some set M_A.)

(M2) Changing local coordinates: If the point P lies both in the sets M_A and M_B, then the two local coordinates

$$u_A = \varphi_A(P) \quad \text{and} \quad u_B = \varphi_B(P)$$

are assigned to the point P. In this case, we assume that the transition map

$$u_B = \varphi_{BA}(u_A)$$

is a diffeomorphism on its natural domain of definition. Explicitly, we have $\varphi_{BA} := \varphi_B \circ \varphi_A^{-1}$ on $\varphi_A(M_A \cap M_B)$ (see Fig. 5.10 in Sect. 5.4 of Vol. I).

The family $\{\varphi_A\}$ of chart maps is called an atlas of the manifold $\mathcal{M}$. This concept generalizes geographic atlases of the earth.

The set M can be equipped with the structure of a topological space in a natural way, by using local coordinates . A subset U of M is called open iff any point $P \in U$ is contained in a set $V(P)$ whose chart image is open on the closed upper-half space. We always assume that the topological space M is separated.

Manifold morphism. Let $\mathcal{M}$ and $\mathcal{N}$ be manifolds with boundary. By a manifold morphism, we understand a smooth map

$$f : \mathcal{M} \to \mathcal{N}, \tag{A.1}$$

that is, the map is smooth with respect to local coordinates. This definition does not depend on the choice of local coordinates because of (M2) above. Furthermore, the map (A.1) is called a manifold isomorphism iff it is a bijective manifold morphism, and the inverse map

$$f^{-1} : \mathcal{N} \to \mathcal{M}$$

is also a manifold morphism. Manifold isomorphisms are also called diffeomorphisms.

Orientation. The manifold $\mathcal{M}$ is called oriented iff all the transition maps φ_{BA} from (M2) above have a positive Jacobian, that is, $\det \varphi'_{BA}(u) > 0$ on the natural domain of definition $\varphi_A(M_A \cap M_B)$. The explicit form of the Jacobian can be found on page 666.

By definition, two oriented manifolds $\mathcal{M}$ and $\mathcal{N}$ with boundary have the same orientation iff there exists a diffeomorphism

$$f : \mathcal{M} \to \mathcal{N}$$

such that the Jacobian of f is positive with respect to local coordinates. A manifold with boundary is called orientable iff it is diffeomorphic to an oriented manifold.

Coherent orientation of the boundary. If the real m-dimensional manifold $\mathcal{M}$ with boundary is oriented, then the boundary $\partial\mathcal{M}$ is an orientable manifold. In fact, let P be a boundary point $\mathcal{M}$. Then, a neighborhood of the point P on $\mathcal{M}$ can be described by the local coordinates

$$u^1, u^2, \ldots, u^{m-1}, u^m \in \mathbb{R} \tag{A.2}$$

with $|u^j| < \varepsilon$ if $j = 1, \ldots, m-1$, and $0 \le u^m < \varepsilon$ (where ε is a sufficiently small positive number). Furthermore, a neighborhood of P on $\partial\mathcal{M}$ is given by the local coordinates

$$u^1, u^2, \ldots, u^{m-1} \in \mathbb{R}, \text{ and } u^m = 0$$

with $|u^j| < \varepsilon$ if $j = 1, \ldots, m-1$. This yields the oriented manifold $\partial\mathcal{M}$. A detailed proof can be found in Zeidler (1986) (page 585 of Vol. IV) quoted on page 1089.

In order to formulate the generalized Stokes integral theorem, one needs a so-called coherent orientation of the boundary $\partial\mathcal{M}$ which is depicted in Fig. 12.6 on page 677. In order to obtain this, we have to change the notation above. That is, we have to pass from the local coordinates (A.2) to the local coordinates

$$v^1, v^2, \ldots, v^{m-1}, v^m \in \mathbb{R}$$

where $v^1 := u^1, \ldots, v^{m-1} := u^{m-1}$ and $v^m := -u^m$.

Equivalent atlases. Two atlases $\mathcal{M}$ are called equivalent iff their union is again an atlas. Intuitively, two geographic atlases of the earth are always equivalent. In fact, putting the two atlases together, we obtain a larger atlas of the earth.

A.1.3 Submanifolds

We have to distinguish between

- submanifolds and
- submanifolds with boundary.

A submanifold is a manifold (i.e., it has no boundary), whereas a submanifold with boundary is a manifold with boundary. Typical examples are:

- an open ball in the 3-dimensional Euclidean manifold $\mathbb{E}^3$ is a 3-dimensional submanifold of $\mathbb{E}^3$;
- a 2-dimensional sphere in $\mathbb{E}^3$ is a 2-dimensional submanifold of $\mathbb{E}^3$;
- a closed ball in $\mathbb{E}^3$ is a 3-dimensional submanifold with boundary of $\mathbb{E}^3$.

Roughly speaking, we have the following situation:

- A real m-dimensional manifold $\mathcal{M}$ looks locally like an open neighborhood of a point in $\mathbb{R}^m$ (with respect to local coordinates), and
- a real r-dimensional submanifold $\mathcal{N}$ of $\mathcal{M}$ is a subset of $\mathcal{M}$ which looks locally like the intersection
$$\mathcal{U} \cap L$$
where $\mathcal{U}$ is the open neighborhood of a point in $\mathbb{R}^m$, and L is an r-dimensional plane in $\mathbb{R}^m$.

Basic definition. Let $0 < r \le m$. We are given the real m-dimensional manifold $\mathcal{M}$. A subset $\mathcal{N}$ of $\mathcal{M}$ is called a real r-dimensional submanifold of $\mathcal{M}$ iff it has the following properties:

- $\mathcal{N}$ is a real r-dimensional manifold;
- $\mathcal{M}$ and $\mathcal{N}$ have equivalent atlases such that, for every point $P \in \mathcal{N}$, there exist local coordinates $u^1, \ldots, u^m$ in $\mathcal{M}$ such that a neighborhood $\mathcal{U}$ of the point P in $\mathcal{M}$ is described by
$$u^1, \ldots, u^r, u^{r+1}, \ldots, u^m \in \mathbb{R}$$
with $|u^j| < \varepsilon$ if $j = 1, \ldots, m$, and the intersection $\mathcal{U} \cap \mathcal{N}$ is described by
$$u^1, \ldots, u^r \in \mathbb{R}, \quad \text{and} \quad u^{r+1} = \ldots = u^m = 0$$
with $|u^j| < \varepsilon$ if $j = 1, \ldots, r$.[6]

Submanifold with boundary. Consider a closed ball $\mathcal{B}$ which is a subset of the Euclidean manifold $\mathbb{E}^3$. If P is an interior point of $\mathcal{B}$, then there exist local coordinates
$$u^1, u^2, u^3 \in \mathbb{R}$$
of the Euclidean manifold $\mathbb{E}^3$ with $|u^1|, |u^2|, |u^3| < \varepsilon$ which are also local coordinates of the ball $\mathcal{B}$ (near the point P). If P is a boundary point of $\mathcal{B}$, then there exist local coordinates u^1, u^2, u^3 of $\mathbb{E}^3$ (near the point P) such that
$$u^1, u^2, u^3 \in \mathbb{R} \quad \text{with} \quad |u^1|, |u^2| < \varepsilon \quad \text{and} \quad 0 \le u^3 < \varepsilon$$
describe locally the closed ball near the point P. Moreover, the boundary $\partial\mathcal{B}$ of the ball is described by
$$u^1, u^2 \in \mathbb{R} \text{ and } u^3 = 0$$
with $|u^1|, |u^2| < \varepsilon$ (near the point P), and u^1, u^2 are local coordinates of the boundary $\partial\mathcal{B}$ (near the point P).

This highly intuitive argument can be easily extended to a precise definition of r-dimensional submanifolds with boundary.

---

[6] By definition, a 0-dimensional submanifold of $\mathcal{M}$ is a point of $\mathcal{M}$ or a set of isolated points of $\mathcal{M}$.

A.1.4 Partition of Unity and the Globalization of Physical Fields

Compact topological space. Let X be a separated compact topological space. Suppose that we are given a finite open covering $\{U_j\}$ of X, $j = 1, \ldots, J$. Then there exists a system $\{f_j\}$ of continuous functions $f_j : X \to \mathbb{R}$, $j = 1, \ldots, J$, such that the following properties are satisfied for all indices j:

- $0 \le f_j(P) \le 1$ for all $P \in X$, and f_j vanishes outside a closed subset of U_j;
- $\sum_{j=1}^{J} f_j(P) = 1$ for all $P \in X$.

Manifolds. Let X be a real n-dimensional compact manifold, and let $\{U_j\}$ be given as above. Then the functions f_j can be chosen in such a way that they are smooth.

Paracompact topological space. In this case, a slight modification is valid. Let X be a separated paracompact topological space. Suppose that we are given an at most countable, locally finite open covering $\{U_j\}$ of X.[7] Then there exists an at most countable system $\{f_i\}$ of continuous functions $f_i : X \to \mathbb{R}$ such that the following properties are satisfied for all indices i:

- $0 \le f_i(P) \le 1$ for all $P \in X$, and f_i vanishes outside a closed subset of $U_{j(i)}$ for some index $j(i)$ depending on i.
- $\sum_i f_i(P) = 1$ for all $P \in X$.

If X is a manifold, then the functions f_i are smooth.

The main theorem on Riemannian manifolds. As a typical application, let us prove that every real n-dimensional manifold M can be equipped with the structure of a Riemannian manifold.

Proof. (I) Assume that M is compact. Using local coordinates on M, the tangent bundle TM looks locally like the product $U_j \times \mathbb{R}^n$ where U_j is an open subset of M. If we equip $\mathbb{R}^n$ with the structure of a real Hilbert space, then we get the local metric tensor

$$g^j(P) := g^j_{kl} dx^k \otimes dx^l, \qquad P \in U_j.$$

Here, we sum over $k, l = 1, \ldots, n$. Choosing a finite open covering $\{U_j\}$ of M by sufficiently small open sets (with respect to local coordinates), we define

$$g(P) := \sum_j g^j(P) f_j(P) \qquad \text{for all} \quad P \in M.$$

Here, $\{f_j\}$ is a partition of unity corresponding to $\{U_j\}$. Then, g is the desired Riemannian metric tensor on M.

(II) In the general case, the manifold M is paracompact. This implies that every open covering of M possesses a locally finite refinement. Now we argue as in (I) by using a partition of unity with respect to the locally finite refinement. More precisely, we set

$$g(P) := \sum_i g^{j(i)} f_i(P) \qquad \text{for all} \quad P \in M$$

where the compact support of f_i is contained in $U_{j(i)}$ for some index $j(i)$ depending on i. □

[7] Such a covering always exists on a paracompact space.

A.2 The Solution of Nonlinear Equations

A.2.1 Linearization and the Rank Theorem

The rank theorem is fundamental for the theory of finite-dimensional manifolds.

Folklore

A general strategy in mathematics is to transform equations into normal forms in order to display the structure of the solution set. As a typical example, we want to study linear and nonlinear systems of equations.

Linear System

The normal form. Fix $m, n = 1, 2, \ldots$ with $m \leq n$. Consider the linear system

$$Ax = b, \qquad x \in \mathbb{R}^n \tag{A.3}$$

where $A : \mathbb{R}^n \to \mathbb{R}^m$ is a real $(m \times n)$-matrix. Explicitly,

$$\boxed{\sum_{k=1}^{n} A^j_k x^k = b^j, \quad j = 1, \ldots, m, \quad x \in \mathbb{R}^n.}$$

Our goal is to transform this linear system into the simple normal form

$$\boxed{u^1 = 0, \ldots, u^r = 0, \qquad u \in \mathbb{R}^n,} \tag{A.4}$$

by using the transformation

$$u = B(x - x_0), \qquad x \in \mathbb{R}^n \tag{A.5}$$

where $B : \mathbb{R}^n \to \mathbb{R}^n$ is a linear bijective map. Here, r is a fixed number with $r = 1, \ldots, m$. To this end, we set

$$A := \begin{pmatrix} A^1_1 & A^1_2 & \ldots A^1_n \\ \vdots & \vdots & \ldots & \vdots \\ A^m_1 & A^m_2 & \ldots A^m_n \end{pmatrix}, \qquad x := \begin{pmatrix} x^1 \\ \vdots \\ x^n \end{pmatrix}, \qquad b := \begin{pmatrix} b^1 \\ \vdots \\ b^m \end{pmatrix}.$$

Here, A^j_k, x^k, b^j are fixed real numbers. We are looking for $x \in \mathbb{R}^n$.

Theorem A.1 *Assume that the system* (A.3) *has the solution* $x_0 \in \mathbb{R}^n$, *and assume that the rank of the matrix* A *is equal to* r *with* $1 \leq r \leq m$. *Then there exists a linear bijective map* $B : \mathbb{R}^n \to \mathbb{R}^n$ *such that the transformation* (A.5) *sends the original linear system* (A.3) *to the normal form* (A.4).

Consequently, the solution set of the equation (A.3) is given by

$$x = x_0 + B^{-1}u$$

where $u^1 = 0, \ldots, u^r = 0$ and $u^{r+1}, \ldots, u^n \in \mathbb{R}$. In other words, the solution set of (A.3) is a real linear $(n - r)$-dimensional manifold.

The favourable case. If the rank of the matrix A is maximal, that is, $r = m$, then the linear operator $A : \mathbb{R}^n \to \mathbb{R}^m$ is surjective. In particular, for every given $b \in \mathbb{R}^m$, the system (A.3) has a solution, and the solution set is a linear manifold of dimension $n - m$.

Trivial case. If the rank of the matrix A is equal to $r = 0$, then $A = 0$. In this trivial case, the equation (A.3) has a solution iff $b = 0$. Then the solution set of (A.3) consists of all the points $x \in \mathbb{R}^n$.

Example (equation of a plane). Consider the linear equation

$$\alpha x + \beta y + \gamma z = b, \qquad x \in \mathbb{R}^3. \tag{A.6}$$

We are given the real numbers α, β, γ. We are looking for $(x, y, z) \in \mathbb{R}^3$. Suppose that $\alpha^2 + \beta^2 + \gamma^2 \neq 0$, that is, the coefficient matrix $A = (\alpha, \beta, \gamma)$ has the maximal rank $r = 1$. Then the solution set $\mathcal{P}$ of (A.6) is a 2-dimensional linear manifold in $\mathbb{R}^3$, that is, $\mathcal{P}$ is a plane in $\mathbb{R}^3$. Let x_0 be a point on the plane $\mathcal{P}$. Then, there exists a linear bijective map $B : \mathbb{R}^3 \to \mathbb{R}^3$ such that the transformation $u = B(x - x_0)$ sends the equation (A.6) to the normal form

$$u^1 = 0, \qquad (u^1, u^2, u^3) \in \mathbb{R}^3.$$

Then, the plane $\mathcal{P}$ can be described by the parametrization

$$x = x_0 + B^{-1}u, \qquad u = (0, u^2, u^3), \;\; u^2, u^3 \in \mathbb{R}.$$

Nonlinear System

The linearization principle. The rank theorem for linear systems can be generalized to nonlinear systems, if we restrict ourselves to the local behavior of the equations. That is, linear bijective transformation maps are replaced by local diffeomorphisms. Intuitively, this corresponds to the use of appropriate curvilinear coordinates. The key condition is the local constancy of the rank of the linearized coefficient matrix.

Fix $m, n = 1, 2, \ldots$ with $m \leq n$. In what follows, the symbol $\mathcal{V}(x_0)$ (resp. $\mathcal{U}(0)$) denotes a sufficiently small open neighborhood of the point x_0 (resp. $u = 0$) in $\mathbb{R}^n$. Let us consider the nonlinear system

$$f(x) = 0, \quad x \in \mathcal{V}(x_0)$$

with $f(x_0) = 0$ by using the linearized system $f'(x_0)(x - x_0) = 0, x \in \mathbb{R}^n$. In terms of components, we investigate the real nonlinear system

$$\boxed{f_j(x) = 0, \qquad j = 1, \ldots, m, \quad x \in \mathcal{V}(x_0)} \tag{A.7}$$

where $x = (x^1, \ldots, x^n)$ is a point of $\mathbb{R}^n$. We assume that we know a fixed solution x_0 of the system (A.7), and we are looking for solutions x of (A.7) near the point x_0. Our goal is to transform locally the nonlinear system (A.7) into the linear system

$$\boxed{u^1 = 0, \ldots, u^r = 0, \qquad u \in \mathcal{U}(0),} \tag{A.8}$$

by using a local diffeomorphism $u = \varphi(x)$ from $\mathcal{V}(x_0)$ onto $\mathcal{U}(0)$. Here, r is a fixed number with $r = 1, \ldots, m$. To this end, we consider the linearized system

$$\sum_{k=1}^{n} \frac{\partial f_j(x_0)}{\partial x^k}(x^k - x_0^k) = 0, \qquad j = 1, \ldots, m, \quad x \in \mathbb{R}^n,$$

and the corresponding matrix $f'(x) : \mathbb{R}^n \to \mathbb{R}^m$ given by

$$\begin{pmatrix} \frac{\partial f_1(x)}{\partial x^1} & \frac{\partial f_1(x)}{\partial x^2} & \cdots & \frac{\partial f_1(x)}{\partial x^n} \\ \vdots & \vdots & \cdots & \vdots \\ \frac{\partial f_m(x)}{\partial x^1} & \frac{\partial f_m(x)}{\partial x^2} & \cdots & \frac{\partial f_m(x)}{\partial x^n} \end{pmatrix}. \tag{A.9}$$

We assume the following:

(H1) The real-valued functions $f_1, \ldots, f_m$ are smooth on some open neighborhood of the point x_0 in $\mathbb{R}^n$.

(H2) Rank condition: The rank r of $f'(x)$ is locally constant near the point x_0. This means that the rank of the matrix (A.9) is equal to r on a sufficiently small open neighborhood of x_0 in $\mathbb{R}^n$.

Note that $0 \leq r \leq m$.

Theorem A.2 *Locally, the solution set of* (A.7) *is an* $(n-r)$*-dimensional submanifold of* $\mathbb{R}^n$ *which passes through the point* x_0.

Explicitly, this means the following. Let $r \geq 1$. There exists a diffeomorphism

$$\varphi : \mathcal{V}(x_0) \to \mathcal{U}(0)$$

such that the transformation $u = \varphi(x)$ with $\varphi(x_0) = 0$ sends the original nonlinear system (A.7) to the linear system (A.8). Here, $\mathcal{V}(x_0)$ (resp. $\mathcal{U}(0)$) is a sufficiently small neighborhood of the point x_0 (resp. $u = 0$) in $\mathbb{R}^n$. The solutions of (A.7) are given by the back transformation

$$x = \varphi^{-1}(0, \ldots, 0; u^{r+1}, \ldots, u^n), \qquad u \in \mathcal{U}(0).$$

This is a local parametrization of the solution set of (A.7).[8] The proof of Theorem A.2 (together with generalizations to infinite-dimensional Banach spaces) can be found in Zeidler (1986) (page 550 of Vol. IV) quoted on page 1089.

The implicit function theorem (special rank condition). If the matrix $f'(x_0)$ has maximal rank at the point x_0, that is, $r = m$, then the assumption (H2) is satisfied automatically. In this special case, Theorem A.2 is called the implicit function theorem.

The trivial case $r = 0$. Suppose that condition (H2) is satisfied in the special case where $r = 0$. Then the functions $f_1, \ldots, f_m$ are constant on a sufficiently small open neighborhood $\mathcal{V}(x_0)$ of the point x_0. Obviously, if $f_1(x_0) = 0, \ldots, f_m(x_0) = 0$, then the solution set of (A.7) is equal to $\mathcal{V}(x_0)$. In other words, locally, the solution set of (A.7) is an n-dimensional submanifold of $\mathbb{R}^n$ which contains the point x_0.

Example ($n = 3, m = 1, r = 0, 1$). We want to show that Theorem A.2 generalizes the classical method for constructing regular surfaces in the 3-dimensional Euclidean manifold. To this end, consider the equation

$$f(x, y, z) = 0, \qquad (x, y, z) \in \mathcal{V}(x_0, y_0, z_0) \tag{A.10}$$

where the real-valued function f is smooth on the sufficiently small open neighborhood $\mathcal{V}(x_0, y_0, z_0)$ of the point (x_0, y_0, z_0) in $\mathbb{R}^3$. We assume that $f(x_0, y_0, z_0) = 0$. The Taylor expansion

[8] In terms of manifolds, the map φ is called a chart map of $\mathbb{R}^n$, and $u = \varphi(x)$ is called the local coordinate of the point $x \in \mathbb{R}^n$.

$$f(x,y,z) = f_x(x_0,y_0,z_0)(x-x_0) + f_y(x_0,y_0,z_0)(y-y_0) + f_z(x_0,y_0,z_0)(z-z_0) + \ldots$$

of the function f at the point (x_0, y_0, z_0) yields the linearized system

$$f_x(x_0,y_0,z_0)(x-x_0) + f_y(x_0,y_0,z_0)(y-y_0) + f_z(x_0,y_0,z_0)(z-z_0) = 0$$

where $(x,y,z) \in \mathbb{R}^3$. The corresponding coefficient matrix reads as

$$(f_x(x_0,y_0,z_0), f_y(x_0,y_0,z_0), f_z(x_0,y_0,z_0)). \tag{A.11}$$

We assume that the matrix

$$(f_x(x,y,z), f_y(x,y,z), f_z(x,y,z))$$

has the constant rank r on a sufficiently small open neighborhood of the point (x_0, y_0, z_0) in $\mathbb{R}^3$.

Case 1: $r = 1$. Then, locally, the solution manifold of (A.10) is a 2-dimensional submanifold of $\mathbb{R}^3$ which passes through the point (x_0, y_0, z_0). In other words, there exists a diffeomorphism $\varphi : \mathcal{V}(x_0) \to \mathcal{U}(0)$ such that the transformation $u = \varphi(x)$ with $\varphi(x_0) = 0$ sends the equation (A.10) to the linear system

$$u^1 = 0, \qquad u \in \mathcal{U}(0).$$

Here, $\mathcal{V}(x_0)$ (resp. $\mathcal{U}(0)$) is a sufficiently small neighborhood of the point x_0 (resp. $u = 0$) in $\mathbb{R}^3$. The solutions of (A.10) are given by the back transformation

$$x = \varphi^{-1}(0, u^2, u^3), \qquad u \in \mathcal{U}(0).$$

This is a local parametrization of the solution set of (A.10)

Case 2: $r = 0$. Then, we have $f_x(x,y,z) = f_y(x,y,z) = f_z(x,y,z) = 0$ on a sufficiently small open neighborhood of (x_0, y_0, z_0). Thus, f is constant on a sufficiently small open neighborhood of (x_0, y_0, z_0). By assumption, $f(x_0, y_0, z_0) = 0$. Consequently, the solution set of the equation (A.10) contains an open neighborhood of the point (x_0, y_0, z_0) in $\mathbb{R}^3$.

The implicit function theorem (special rank condition). Let us mention the following special feature. If the rank of the matrix

$$(f_x(x_0,y_0,z_0), f_y(x_0,y_0,z), f_z(x_0,y_0,z_0) \tag{A.12}$$

is equal to one, then the rank of the matrix (A.11) is equal to one on a sufficiently small open neighborhood of the point (x_0, y, z_0). To see this, assume that, say,

$$f_z(x_0,y_0,z_0) \neq 0. \tag{A.13}$$

Then we get $f_z(x,y,z) \neq 0$ on a sufficiently small open neighborhood of the point (x_0, y_0, z_0) in $\mathbb{R}^3$, by continuity. More precisely, condition (A.13) implies that the solution set of the equation (A.10) reads as

$$(x, y, z(x,y)) \qquad \text{for all} \quad (x,y) \in \mathcal{W}(x_0, y_0)$$

where $\mathcal{W}(x_0, y_0)$ is a sufficiently small open neighborhood of the point (x_0, y_0) in $\mathbb{R}^2$. In terms of geometry, the equation (A.10) describes the local smooth surface

$$z = z(x,y), \qquad (x,y) \in \mathcal{W}(x_0,y_0)$$

with $z_0 = z(x_0, y_0)$.

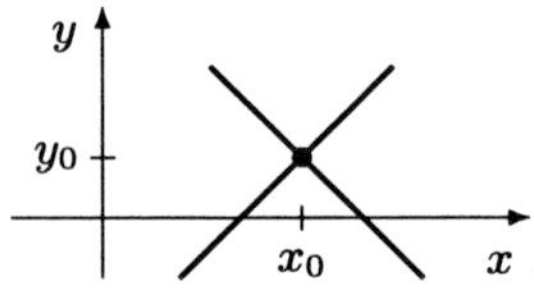

Fig. A.4. Bifurcation

Physical interpretation. Let $\Theta : \mathbb{E}^3 \to \mathbb{R}$ be a smooth temperature field on the 3-dimensional Euclidean manifold $\mathbb{E}^3$. Let us choose a right-handed Cartesian (x, y, z)-coordinate system. Suppose that

$$\mathbf{grad}\,\Theta(x_0, y_0, z_0) \neq 0,$$

that is, the temperature gradient does not vanish at the point (x_0, y_0, z_0). This corresponds to the rank condition (A.12). Then, for given fixed temperature Θ_0, the isothermal surface

$$\Theta(x, y, z) = \Theta_0, \qquad (x, y, z) \in \mathbb{R}^3$$

represents a smooth 2-dimensional submanifold $\mathcal{M}$ of the 3-dimensional Euclidean manifold $\mathbb{E}^3$ in a sufficiently small open ball centered at the point (x_0, y_0, z_0), and the surface $\mathcal{M}$ has the normal vector $\mathbf{grad}\,\Theta(x_0, y_0, z_0)$ at the point (x_0, y_0, z_0). If the z-component of the gradient $\mathbf{grad}\,\Theta(x_0, y_0, z_0)$ does not vanish, that is, $\Theta_z(x_0, y_0, z_0) \neq 0$, then the isothermal surface through the point (x_0, y_0, z_0) is locally given by the equation

$$z = z(x, y), \qquad (x, y) \in \mathcal{W}(x_0, y_0)$$

with $z(x_0, y_0) = z_0$.

A.2.2 Violation of the Rank Condition and Bifurcation

The equation

$$\boxed{(x - x_0)^2 - (y - y_0)^2 = 0, \qquad (x, y) \in \mathbb{R}^2} \tag{A.14}$$

has the two straight lines

$$y = (x - x_0) + y_0, \qquad y = -(x - x_0) + y_0, \qquad x \in \mathbb{R}^2$$

as solution set. By Fig. A.4, this solution set is not a submanifold of $\mathbb{R}^2$ in a neighborhood of the point (x_0, y_0). This follows from the obvious fact that there is no tangent line at the point (x_0, y_0). Naturally enough, the point (x_0, y_0) is called a bifurcation point of the equation (A.14).

The linearization of equation (A.14) at the point (x_1, y_1) reads as

$$2(x_1 - x_0)(x - x_0) + 2(y_1 - y_0)(y - y_0) = 0, \qquad (x, y) \in \mathbb{R}^2$$

with the coefficient matrix

$$(2(x_1 - x_0), 2(y_1 - y_0)).$$

This matrix has the rank $r = 1$ if $(x_1, y_1) \neq (x_0, y_0)$ (resp. $r = 0$ if we have $(x_1, y_1) = (x_0, y_0)$). Thus, the rank jumps at the point (x_0, y_0). This means that the rank condition for the equation (A.14) is violated at the point (x_0, y_0).

Complex linear and nonlinear systems. The results above remain valid if we replace the real variables $x^1, \ldots, x^n$ by complex variables (resp. the real-valued functions $f_1, \ldots, f_m$ by complex-valued functions). That is, we replace $\mathbb{R}^n$ and $\mathbb{R}^m$ by $\mathbb{C}^n$ and $\mathbb{C}^m$, respectively.

A.3 Lie Matrix Groups

An elementary introduction to Lie groups can be found in Chapter 7 of Volume I. Let us summarize additional advanced material. The proofs can be found in:

B. Hall, Lie Groups, Lie Algebras, and Representations: An Elementary Introduction, Springer, New York, 2003.

W. Hein, Introduction to Structure and Representation of the Classical Groups, Springer, Berlin, 1990 (in German).

A. Kirillov, An Introduction to Lie Groups and Lie Algebras, Cambridge University Press, 2008.

The exponential function. Let $n = 1, 2, \ldots$ Recall that $GL(n, \mathbb{C})$ denotes the group of all complex $(n \times n)$-matrices. This is a Lie group (see Section 7.8 of Volume I). The set $gl(n, \mathbb{C})$ consists of all complex $(n \times n)$-matrices. This is a real Lie algebra with respect to the Lie product

$$[A, B]_- = AB - BA \qquad \text{for all} \quad A, B \in gl(n, \mathbb{C}).$$

Fix $t_0 > 0$. Let $G = G(t), -t_0 < t < t_0$, be a smooth curve on $GL(n, \mathbb{C})$ with $G(0) = I$. Then the derivative $\dot{G}(0)$ is an element of the tangent space $T_I GL(n, \mathbb{C})$ of the real manifold $GL(n, \mathbb{C})$ at the unit element I. In addition,

$$T_I GL(n, \mathbb{C}) = gl(n, \mathbb{C}).$$

We call $gl(n, \mathbb{C})$ the Lie algebra of the Lie group $GL(n, \mathbb{C})$. If $A \in gl(n, \mathbb{C})$, then $e^A \in GL(n, \mathbb{C})$. The map

$$A \mapsto e^A \tag{A.15}$$

sends the Lie algebra $gl(n, \mathbb{C})$ to a neighborhood of the unit element of the group $GL(n, \mathbb{C})$. More precisely, the map (A.15) is a local diffeomorphism which maps a sufficiently small open neighborhood of the zero element in $gl(n, \mathbb{C})$ onto a sufficiently small open neighborhood of the unit element in $GL(n, \mathbb{C})$. This tells us that:

A sufficiently small open neighborhood of the unit element I in the Lie group $GL(n, \mathbb{C})$ can be parametrized by a sufficiently small open neighborhood of the Lie algebra $gl(n, \mathbb{C})$.

If $A, B \in gl(n, \mathbb{C})$, then

$$e^{A+B} = \lim_{\Delta t \to +0} \left(e^{\Delta t\, A} e^{\Delta t\, B} \right)^{1/\Delta t} \tag{A.16}$$

and

$$e^{[A,B]_-} = \lim_{\Delta t \to 0} \left(e^{\Delta t A} e^{\Delta t\, B} e^{-\Delta t\, A} e^{-\Delta t\, B} \right)^{1/(\Delta t)^2}. \tag{A.17}$$

Moreover, if $A, B \in gl(n, \mathbb{C})$ and $G \in GL(n, \mathbb{C})$, then we have

- $A = \frac{d}{dt}e^{tA}$ at $t = 0$,
- $GAG^{-1} = \frac{d}{dt}(Ge^{tA}G^{-1})$ at $t = 0$,
- $[A, B]_- = \frac{\partial^2}{\partial t \partial s}(e^{sA}e^{tB}e^{-sA}e^{-tB})$ at $t = 0, s = 0$.

The proof can be found in Hein (1990) and Hall (2003) quoted above.

The beauty of Lie matrix groups. By definition, a Lie matrix group is a closed subgroup of $GL(n, \mathbb{C})$ for some $n = 1, 2, \ldots$ Let $\mathcal{G}$ be a Lie matrix group. By definition, the set $\mathcal{L}$ consists precisely of all the matrices $A \in gl(n, C)$ which have the property that

$$e^{tA} \in \mathcal{G} \qquad \text{for all} \quad t \in \mathbb{R}.$$

Then the following hold:

(i) If $A, B \in \mathcal{L}$ and $\alpha \in \mathbb{R}$, then $\alpha A, A + B, [A, B]_- \in \mathcal{L}$. Thus, $\mathcal{L}$ is a real Lie algebra which is a Lie subalgebra of $gl(n, \mathbb{C})$.

(ii) The restriction of the map (A.15) to $\mathcal{L}$ is a local diffeomorphism from an open neighborhood of the zero element in $\mathcal{L}$ onto an open neighborhood of the unit element in $\mathcal{G}$. Thus, $\mathcal{G}$ is a submanifold of $GL(n, \mathbb{C})$. More precisely, $\mathcal{G}$ is a Lie subgroup of the Lie group $GL(n, \mathbb{C})$, and $\mathcal{L}$ is the Lie algebra of the Lie group $\mathcal{G}$, that is, $\mathcal{LG} = \mathcal{L}$.

The scope of Lie matrix groups. A theorem of Élie Cartan tells us that the Lie subgroups of a Lie group are always closed. Therefore, the statement (ii) shows that we have the following nice result:

A subgroup $\mathcal{G}$ of $GL(n, \mathbb{C})$ is a Lie subgroup iff it is a Lie matrix group.

From the pedagogical point of view, the theory of Lie matrix groups is simpler than the general theory of Lie groups, since one can use matrix calculus. All the classical Lie groups are Lie matrix groups. As an introduction to Lie matrix groups, we recommend Hall (2003) and Hein (1990) quoted above.

Every Lie matrix group is a Lie group, but the converse is not always true.

There are the following counterexamples:

- The universal covering group of the Lie matrix group $SL(2, \mathbb{R})$ is a Lie group, but not a Lie matrix group. The proof can be found in Hall (2003), p. 317, quoted above.
- The quotient group of a Lie matrix group is a Lie group, but not always a Lie matrix (see the Birkhoff–Heisenberg quotient group on page 110).

For the theory of general Lie groups, we refer to Kirillov (2008) quoted above.

The first main theorem on the morphisms of Lie matrix groups. Let $\mathcal{G}$ and $\mathcal{H}$ be Lie matrix groups. Suppose that

$$\varrho : \mathcal{G} \to \mathcal{H}$$

is a smooth group morphism. Let $A \in \mathcal{LG}$. Define $\varrho_*(A) := \frac{d}{dt}\varrho\left(e^{tA}\right)_{|t=0}$. Then the map

$$\varrho_* : \mathcal{LG} \to \mathcal{LH}$$

is a Lie algebra morphism.

The proof uses (A.16) and (A.17). See Hein (1990), p. 45, quoted above. An alternative proof can be found in Hall (2003), p. 125, quoted above.

The second main theorem on the morphisms of Lie matrix groups. Let $\mathcal{G}$ and $\mathcal{H}$ be Lie matrix groups. Consider the Lie algebra morphism

$$\lambda : \mathcal{LG} \to \mathcal{LH}.$$

Suppose that the Lie group $\mathcal{G}$ is arcwise connected and simply connected. Then there exists a unique smooth group morphism $\varrho : \mathcal{G} \to \mathcal{H}$ such that $\varrho_* = \lambda$.

The proof uses the Baker–Campbell–Hausdorff theorem (see Sect. 8.4 of Vol. I). The sophisticated proof can be found in B. Hall (2003) quoted above, p. 76.

A.4 The Main Theorem on the Global Structure of Lie Groups

We are given the real finite-dimensional Lie algebra $\mathcal{L}$. We want to know all the Lie groups whose Lie algebra is isomorphic to $\mathcal{L}$. The following hold:

(i) Existence and uniqueness. There exists an arcwise connected, simply connected Lie group $\mathcal{U}$ whose Lie algebra $\mathcal{LU}$ is isomorphic to $\mathcal{L}$. The Lie group $\mathcal{U}$ is unique, up to Lie group isomorphism.

(ii) If the arcwise connected Lie group $\mathcal{G}$ has a Lie algebra $\mathcal{LG}$ which is isomorphic to $\mathcal{L}$, then there exists a surjective Lie group morphism

$$\varrho : \mathcal{U} \to \mathcal{G}.$$

The kernel $\ker(\varrho)$ (i.e., the preimage $\varrho^{-1}(1)$ of the unit element in $\mathcal{G}$) is a discrete (i.e., an at most countable) normal subgroup of $\mathcal{U}$ contained in the center of $\mathcal{U}$ such that we have the Lie group isomorphism

$$\mathcal{G} \simeq \mathcal{U}/\ker(\varrho).$$

(iii)) Conversely, if $\mathcal{N}$ is a discrete normal subgroup of $\mathcal{U}$ contained in the center of $\mathcal{U}$, then the quotient group $\mathcal{U}/\mathcal{N}$ is an arcwise connected Lie group whose Lie algebra is isomorphic to $\mathcal{L}$.

This theorem tells us that every arcwise connected Lie group $\mathcal{G}$ has a unique universal covering group $\mathcal{U}$, and the group $\mathcal{U}$ knows all the arcwise connected Lie algebras which have the same Lie algebra as $\mathcal{G}$. The sophisticated proof can be found in L. Pontryagin, Topological Groups, Gordon and Breach, 1966, Chap. 10. For a sketch of the proof, see Kirillov (2008), Sect. 3.8, quoted above.

Epilogue

By explanation the scientist means nothing else than a reduction to very few and simple basic rules, which cannot be reduced any further, but which allow a complete deduction of the phenomena.

Gauss in *Electromagnetism and Magnetometer*

I thank you, highly honored Sir, in the name of mankind, for presenting us with a picture of the highest intellectual power and force together with an inspiring and never ending warmth of feeling.

Alexander von Humboldt in a letter to Gauss, 1853

We do not claim for mathematics the prerogative of a Queen of Science; there are other fields which are of the same or even higher importance in education. But mathematics sets the standard of objective truth for all intellectual endeavors. Science and technology bear witness to its practical usefulness. Besides language and music it is one of the primary manifestations of free creative power of the human mind, and it is the universal organ for world-understanding through theoretical construction. Mathematics must therefore remain an essential element of the knowledge and abilities which we have to teach, of the culture we have to transmit, to the next generation. Only he who knows what mathematics is, and what its function in our present civilization, can give sound advice for the improvement of our mathematical teaching.[1]

Hermann Weyl

Knowledge in all physical sciences – astronomy, physics, chemistry – is based on observation. But observation can only ascertain what is. How can we predict what will be? To that end observation must be combined with mathematics.

Hermann Weyl (from a Radio Talk 1947)

Symmetry, Lie groups and gauge invariance are now recognized, through theoretical and experimental developments, to play essential role in determining the basic forces of the physical universe. I have called this the principle that *symmetry dictates interaction*.[2]

Cheng Ning Yang, 1985

---

[1] From the introduction to the *Gesammelte Abhandlungen* (Collected Works) by H. Weyl, Vols. 1–4, Springer, Berlin, 1968 (reprinted with permission).

[2] C. Yang, Hermann Weyl's Contributions to Physics. In: Hermann Weyl (1885–1955), pp. 7–21. Springer, Berlin, 1985.

E. Zeidler, *Quantum Field Theory III: Gauge Theory*,
DOI 10.1007/978-3-642-22421-8,

References

A lot of material including proofs quoted in this volume can be found in the following books:

Zeidler, E., (1986), Nonlinear Functional Analysis and its Applications. Vol. I: Fixed-Point Theorems, 3rd edn. 1998; Vol. IIA: Linear Monotone Operators, 2nd edn. 1997; Vol. IIB: Nonlinear Monotone Operators; Vol. III: Variational Methods and Optimization; Vol. IV: Applications to Mathematical Physics, 2nd edn. 1995, Springer, New York. Reprinted: Beijing (China), 2009.

Zeidler, E. (1995a), Applied Functional Analysis, Vol 1: Applications to Mathematical Physics, 2nd edn. 1997, Applied Mathematical Sciences, AMS 108, Springer, New York. Reprinted: Beijing (China), 2009.

Zeidler, E. (1995b), Applied Functional Analysis, Vol. 2: Main Principles and Their Applications, Applied Mathematical Sciences, AMS 109, Springer, New York. Reprinted: Beijing (China), 2009.

Zeidler, E. (Ed.) (2004), Oxford Users' Guide to Mathematics, Oxford University Press, New York (translated from German into English).

Zeidler, E. (Ed.) (2003), Teubner-Taschenbuch der Mathematik (The Teubner Handbook in Mathematics) (in German).
Vol. 1: 3rd edition, Teubner, Wiesbaden, 2003 (English edition: see Zeidler (2004));
Vol. 2: 8th edition. Edited by G. Grosche, V. Ziegler, D. Ziegler, and E. Zeidler, Teubner, Wiesbaden, 2003.

The complete list of all the papers and books quoted in this volume can be found on the author's homepage:

`http://www.mis.mpg.de/zeidler/qft.html`

E. Zeidler, *Quantum Field Theory III: Gauge Theory*,
DOI 10.1007/978-3-642-22421-8,

List of Symbols

[1] The Greek letters $\alpha, \beta, \gamma, \varGamma, \delta, \varDelta, \varepsilon, \epsilon, \zeta, \eta, \theta, \vartheta, \varTheta, \iota, \kappa, \lambda, \varLambda, \mu, \nu, \xi, \varXi, o, \square$ (omicron), $\pi, \varPi, \varrho, \sigma, \varsigma, \varSigma, \tau, \upsilon, \varUpsilon$ (upsilon), $\varphi, \phi, \varPhi, \chi, \psi, \varPsi, \omega, \varOmega$, the Hebrew letter $\aleph$ (aleph), the Phoenician symbol ∇ (nabla), and some special symbols (e.g., tensor products $\otimes$ and integrals $\int$) can be found at the end of this list. See also the list of symbols in Volumes I and II.

[2] The notation $\dot{x}$ dates back to Newton (1643–1727) who used the 'dot' for indicating the first time derivative. The symbols $\frac{dx}{dt}$, $\frac{\partial f}{\partial t}$ (partial derivative) and δf (variation) were used by Leibniz (1646–1716), Clairaut (1713–1765), and Lagrange (1736–1813), respectively.

E. Zeidler, *Quantum Field Theory III: Gauge Theory*,
DOI 10.1007/978-3-642-22421-8,

[3] Our sign convention for the classical Laplacian coincides with the sign convention for the Hodge Laplacian used in modern differential geometry.

[4] Observe that the notation is not uniform in the literature. Some of the authors use the symbol e for the negative charge of the electron.

Index

E. Zeidler, *Quantum Field Theory III: Gauge Theory*,
DOI 10.1007/978-3-642-22421-8,

Lightning Source UK Ltd.
Milton Keynes UK
UKOW031836081111

181640UK00011B/2/P